INTERNATIONAL MINE VENTILATION CONGRESS 2024

12–16 AUGUST 2024
SYDNEY, AUSTRALIA

The Australasian Institute of Mining and Metallurgy
Publication Series No 4/2024

AusIMM

Published by:
The Australasian Institute of Mining and Metallurgy
Ground Floor, 204 Lygon Street, Carlton Victoria 3053, Australia

ISBN 978-1-922395-34-4

CORE COMMITTEE

Dr Bharath Belle *MAusIMM(CP)*, UNSW, University of Queensland

Duncan Chalmers *MAusIMM*, UNSW

Dr Rick Brake *FAusIMM(CP)*, Mine Ventilation Australia

Prof Ismet Canbulat *FAusIMM*, UNSW

Lou Lander

John Rowland, Dallas Mining Services Pty Ltd

Dr Guangyao Si *MAusIMM*, UNSW

Michael Shearer *MAusIMM*, MVSA

Dr Craig Stewart *MAusIMM*, MVSA

AUSTRALIAN ADVISORY COMMITTEE

Dr Bharath Belle *MAusIMM(CP)*, IMVC2024 Chair

Duncan Chalmers *MAusIMM*, IMVC2024 Deputy Chair, UNSW

Prof Ismet Canbulat *FAusIMM*, UNSW

Mr Livio Salvestro, Howden, A Chart Industries Company

Mr Miguel Alcalde, Zitron

Mr Adrian Uria, Zitron

Mr John Grieves, ACARP

Ms Kerri Melkersson, Resources Safety & Health Queensland

Mr David Gordon, NSW Resources Regulator

Mr Shane Hansen, Kestrel Coal Resources

Mr Paul O'Grady, Glencore

Mr Steven Smyth, Mining and Energy Union

Mr Jacques le Roux, Resources Safety & Health Queensland

Mr Andrew Clough, Health and Safety Commissioner Queensland

Dr Rao Balusu, CSIRO

Dr Gavin Lind, Minerals Resource Council

Mr Keith McLean, Gordon Brothers

Prof Naj Aziz, University of Wollongong

Prof Alex Remennikov, University of Wollongong

Prof Ting Ren, University of Wollongong

Prof Mehmet Kizil, University of Queensland

INTERNATIONAL ADVISORY COMMITTEE

Mr Adrian Uria, *Australia*

Prof Anna Luiza M Ayres da Silva, *Brazil*

Dr Cláudio Lúcio Lopes Pinto, *Brazil*

Dr Cheryl Allen, *Canada*

Prof Juan Pablo Hurtado Cruz, *Chile*

Mr Daniel Sepulveda, *Chile*

Prof Weimin Cheng, *China*

Prof Yuanping Cheng, *China*

TECHNICAL PROGRAM AND EDITORIAL COMMITTEE

Dr Bharath Belle, Hon Editor

Dr Guangyao Si, UNSW, Deputy Editor; and Dr Hao Wu, Deputy Editor

Dr Rick Brake

Dr Gareth Kennedy

Dr Rao Balusu

Dr Jerry Tien

Mr Duncan Chalmers

TOUR SUBCOMMITTEE

Godlieb Combrinck

Livio Salvestro *MAusIMM*

Marle Hooman

Michael Shearer

Murray Jamieson

Dr Craig Stewart *MAusIMM*

Jeff Norris

Claudia Vejrazka *MAusIMM*

Jakes Raubenheimer *MAusIMM*

WORKSHOP SUBCOMMITTEE

Dr Craig Stewart *MAusIMM*, Subcommittee Lead

Mehmet Kizil *MAusIMM*

Nikky LaBranche *FAusIMM(CP)*

Saiied Aminossadati

Jeff Norris

AUSIMM

Julie Allen
Head of Events

Fiona Geoghegan *AAusIMM*
Senior Manager, Events

REVIEWERS

We would like to thank the following people, as well as a few others, for their contribution towards enhancing the quality of the papers included in this volume:

Kayode Ajayi

Cheryl Allen

Sampurna Arya

Prof Naj Aziz

Davood Bahrami

Dr Rao Balusu

Prof/Dr Bharath Belle

Dr Sekhar Bhattacharyya

Christian Boucher

John Bowling

Michael Brady

Dr Rick Brake

Dr Jurgen Brune

David Carey

Duncan Chalmers

Dr Ping Chang

Prof David Cliff

Andrew Derrington

Heather Dougherty

Dr Joey Duan

Thomas Dubaniewicz

Myriam Francoeur

Vasu Gangrade

Arush Habibi

Dr Adrian Halim

Stephen Hardcastle

Marcia Harris

Johan Holtzhausen

Marle Hooman

Runzhe Hu

Hua Jiang

Dr Yonggang Jin

Dr Gerald Joy

Dr Aditya Juganda

Dr Gareth Kennedy

Dr Manoj Khanal

A/Prof Mehmet Kizil

Charles Kocsis

Dr Ashish Ranjan Kumar

Bob Leeming

Prof Shimin Liu

Ian Loomis

Kevin Lownie

Anu Martikainen

Keith McClean

Chris McGuire

Paul Meisburger

Florian Michelin

Prof D P Mishra

Dr Niroj Kumar Mohalik

Moe Momayez

Ramakrishna Morla

Dr Pierre Mousset-Jones

Sean Muller

Paul O'Grady

Durga Charan Panigrahi

Luca Pantano

Dr Partha Sarathi Paul

Dr Eranda Perera

Brian Prosser

Ming Qiao

Dr Qingdong Qu

Dr Aleksandar Rai

Vaibhav Raj

Gavin Ratner

Prof Alex Remennikov

Prof Ting Ren

Pedram Roghanchi

John Rowland

Sergei Sabanov

Glenn Savage

Steven Schafrik

Steven Schatzel

Daniel Sepulvada

Dr Guangyao Si

Fatemeh Soleimani

Srivatsan Sridharan

Dr Craig Stewart

Dr Daniel Stinnette

Craig Stuart

Krishna Tanguturi

Dr P Thakur

Prof Jerry Tien

Jack Trackemas

Prof Purushotham Tukkaraja

Leon Van Den Berg

Claudia Vejrazka

Jon Volkwein

Keith Wallace

Yuehan Wang

Michael Webber

Paul Wild

Darryl Witow

Dr Hsin-Wei Wu

A/Prof Guang Xu

Dave Yantek

Liming Yuan

Hongbin Zhang

Yi Zheng

Lihong Zhou

Dongfeng Zhu

History of the International Mine Ventilation Congress (IMVC) Series

1975 Johannesburg, South Africa

1979 Reno, USA

1984 Harrogate, UK

1988 Brisbane, Australia

1992 Johannesburg, South Africa

1997 Pittsburgh, USA

2001 Cracow, Poland

2005 Brisbane, Australia

2009 New Delhi, India

2014 Sun City, South Africa

2018 Xi'an, China

2024 Sydney, Australia (delayed due to COVID)

WELCOME FROM PREVIOUS HOST – XI'AN UNIVERSITY OF SCIENCE AND TECHNOLOGY

We are honoured to be invited to deliver this greeting at the 12th International Mine Ventilation Congress (IMVC) in Sydney, Australia. The 1st IMVC was initiated by the industry and academic organisations of the world's major mining countries in 1975. It is held every four years and is the international academic conference with the longest history and with the highest academic influence in the field mining. It has been an important platform for the exchange of new technologies, concepts, products and all innovative achievements of ventilation, safety and occupational health in the mining industries.

The 11th IMVC was held for the first time in Xi'an, China, the world's largest coal mining country, from September 14 to 20, 2018. More than 500 delegates in the field of mining from over 39 countries and regions attended the congress, focusing on the theme of mine ventilation technology and management under economic, environmental, health and safety challenges. The participating experts shared their knowledge and scientific research progress in ventilation, mine environment, health and safety, energy efficiency and management in recent years, and jointly published excellent articles combined with Springer. It has made great contributions to the scientific and technological progress of the world's energy and resources.

As the host of the 11th IMVC, we were very honoured to take over the baton from Mr Frank von Glehn, Chairman of the 10th Executive Committee, and to now transfer the hosting right of the 12th Congress to the University of New South Wales (UNSW), the Mine Ventilation Association of Australia, and the Australasian Institute of Mining and Metallurgy (the AusIMM). Recognised as the best engineering University in Australia, UNSW's Mining Engineering programmers possess a high reputation in the mining industry. The Mine Ventilation Association of Australia and the Australasian Institute of Mining and Metallurgy have been committed to providing a world-class technical exchange platform for the global mining and mine ventilation fields, showcasing the latest technologies, innovations and industry best practices, and actively promoting the development of mine ventilation technology.

Due to the global pandemic, the 12th IMVC was postponed to August 2024, and we would like to express our heartfelt thanks to the ventilation engineers and experts in the field of mine safety from all over the world, the organisers of the conference for providing a rare and broad communication platform for mine ventilation safety experts and practitioners from all over the world, and the members of the conference team for their efforts for the smooth convening of this conference, and wish experts and scholars from all over the world the opportunity to exchange and learn during the congress. We wish the 12th IMVC2024 to be a smooth and successful event.

Jun Deng and Shugang Li

President of Xi'an University of Science and Technology

FOREWORD AND EDITORIAL

VENTILATION ENGINEERING – THE HEARTBEAT OF MINING

Worldwide, mining engineering has taken significant strides since the beginning of the last century. Considering the resource boom and the maximum number of mines opening and expanding in recent years, the theme of this Congress 'Ventilation Engineering – the Heartbeat of Mining', is particularly appropriate. Congress like this provide an excellent opportunity for the exchange of knowledge among our colleagues.

Congratulations to you all in the mining and ventilation engineering discipline for this unique specialist mining engineering discipline of technical experts. The IMVC 2024 core committee has come up with a unique logo 'windmill'. The windmill is one of the Australian Outback's most recognised and respected icons. It's simple yet persistent action turns an otherwise inhospitable environment into a safe and productive one. Similarly, the ventilation engineering systems in mines create safe working environments, which is necessary to produce the globally essential minerals for human health, happiness, safety and security of nations.

I believe that the re-gathering of this unique specialist mine ventilation engineering talent to Australia for the 3rd time in its history provides a fitting platform to what is an important occasion for the people in the mining fraternity. I hope that this opportunity for interaction will assist mining operations across the world in the design of a safe and healthy work environment for the future and re-visit the challenges in the current operations.

Since its inception over many centuries, the mining industry with all commodities represented, has made an enormous contribution to humankind, global infrastructure, economy, and to each nation's safety and security. No doubt, mine ventilation has played a crucial role in the frontiers of mining and engineering design for mines. Global mining operations have achieved significant milestones and expanding mining frontiers, demonstrating to the world that what was once considered impossible in mine ventilation engineering is now possible and has become a routine, thanks to the support of other core disciplines. The mining industry worldwide has entered a new era of mining imperative such as Zero Serious Harm, which requires close interaction with all disciplines to address every potential safety, health and environment concern of our mines. In order to achieve this, we must operate our mines without an implicit belief that mines are dangerous. This can be only justified if the evidence supports this notion, which is not the case currently.

This leads to the question of history of the IMVC. The South African mining industry had made memorable and pioneering contributions to improve the health and safety of global mine workers, one being the International Conference on Pneumoconiosis held in Johannesburg in February 1959 and the second being the first International Mine Ventilation Congress (IMVC) in 1975, along with a globally unique mine ventilation engineering society to promote the art and science of mine ventilation engineering in 1945.

What is the context and connection between South Africa and Australia? The discovery of a gold-bearing conglomerate on the Witwatersrand of South Africa in February 1886 and diamonds caused a sensation worldwide, resulting in mines that required ventilation. One can argue that it can be linked to an Australian, Mr. George Harrison, an insignificant gold digger, who had taken a temporary job of building a house for a widow on the farm, Langlaagte, near Johannesburg or its African name Egoli and had unmistakably found the gold. Therefore, in some sense, this Australian IMVC originates in South Africa, which held the 1st IMVC at the University of the Witwatersrand with 661 local and overseas delegates in attendance on 15th Sept. 1975. The 1st IMVC was inaugurated by the South African Minister of Mines and attended by world experts, including Mr Henry Doyle, doyen of Industrial Hygienists in the USA. The IMVC was to be the forerunner of the North American, South American and Australian Mine Ventilation Symposiums, commanding professional respect within the mining community.

So what is the origin of the mine ventilation engineering professional? The history of current specialist mine ventilation engineers can be traced back to erstwhile role of 'dust inspectors' created in 1901 after the Boer War in South Africa, who were given the responsibility of protecting workers from the hazards of the mining environment through adequate ventilation, at the request of employers, workers and the government. This evolved specialist expertise and knowledge in an area that is now fully recognised as integral to the mining profession and its standing globally as the mine ventilation engineer.

Over the years, many speakers at mine ventilation meetings predicted future challenges facing mine ventilation engineering. At the Third US Mine Ventilation Symposium in 1987, Dr. Howard Hartman previewed his vision and stated that the 'Elimination of catastrophic disasters in mines' would be the 'crowning accomplishments-to-be in mine ventilation.' This task often falls heavily on us as mine ventilation engineers where we are faced with providing advice or making decisions that can have profound consequences.

Returning to the present, the mine ventilation challenges we are now facing require innovative solutions, technology and Zero Serious Harm Vision. Where do you lean to for guidance and solutions or advice in situation like these? Over the last two decades, major explosion events and re-identification of black lung and silicosis in Australia have highlighted the deficiencies in managing the catastrophic risks in the ventilation engineering profession. These explosion events have shown that the drift from 'compliance' based legislation to 'risk' based solutions without critical checks and balances is insufficient. The situation demands risk ownership and accountability by ventilation engineers, mine managers, vigilant regulators, and consultants who may be tempted to propose deficient solutions with 'disclaimers.' This is not about a lack of 'trust' or integrity but about the ability to provide assurance to frontline workers and our wider society that our operations are safe. For the mine ventilation engineering[1] profession – *There is no tertium quid*.

Health and safety at work is the core of workers, their families and communities. Mine explosions, fires (including spontaneous combustion) and respirable dust are catastrophic risks with operational challenges and controls managed by the mine ventilation engineering profession. Major mine unsafe events create an undesirable, unique and ambiguous setting for operators and regulators, with the continued presence of constraints in terms of time, complex natural and mining environments, resources, and intuitive decision-making expertise. These require making step changes from compliance-based designs to expert risk-based routes through design intervention and driving out or controlling 'the Bads' when sabotaging the harm. As expert engineers, it is hoped that we all assure the face worker and community by sharing distinctive control of 'Bads' and promoting 'Goods' and cast out any hurdles in front towards Zero Serious Harm.

Winning the bid to host this 12th IMVC some six years ago in China was a recognition of the capability of Australian mine ventilation engineering professionals, academics, researchers, suppliers, and regulatory bodies. We would not have succeeded in this endeavour without the stellar support from Prof Canbulat of the UNSW and Mr Steve Durkin, Ms Melissa Holdsworth and Ms Julie Allen of the AusIMM, the CSIRO, Universities, the mining industry, union members, our advisory committee, and our equipment suppliers and sponsors. The selection of technical papers for this congress has been a mammoth task. There were a total of 182 abstract submissions, with a final paper submission of 90 papers from 17 countries. I would like to thank the Core Committee members, and most importantly, my deputy Editors, Dr Guangyao Si and Dr Hsin-Wei Wu, for their immense contribution to a successful Technical Programme. The technical committee and global voluntary technical peer reviewers are a vital part of the quality assurance program for the papers being presented. We also thank the superlative work by the sub-committee chairs and members organising the field trips, the technical workshops and the social programme.

On a personal note, I also reflect back some 30 years to when I was a young mining engineering student at the Pennsylvania State University, where I had the opportunity to contribute to the IMVC held in Pittsburgh in 1997. It is therefore a particular source of pleasure and gratitude for me to now have the honour to contribute again to this wonderful congress series; this time as the Congress Chair along with Mr Duncan Chalmers as Deputy Chair, whose support has been invaluable to me in my role. I sincerely hope that the energy derived from this Congress will enable the global

---

[1] Belle, B, 2009. Ventilation engineering – There is no *tertium quid,* Mine Ventilation Society of South Africa Presidential Address, Johannesburg, South Africa.

ventilation engineering profession to regain its lost lustre amongst global mining companies. It is hoped that the IMVC 2024 proceedings presented herein will be a valuable reference material.

I am very confident that this Congress will provide renewed impetus and resolve for us all, together, to contribute to develop and apply new solutions to the existing and emerging issues and opportunities that face our industry and our profession. In order to achieve the various challenges presented by all of us here and at our operations globally, we are required to commit to:

1. Providing superior mine ventilation engineering designs that consider every aspect of safety, health and environment towards Zero Serious Harm.

2. Providing exceptional mine designs that give an edge to improve safety with appropriate use of technology, automation, revision of historic design parameters, global leading practices and innovation.

3. Delivering enhanced fail-safe mine designs and systems that were built through close interaction between every core discipline of mining engineering.

4. Providing outstanding technical and engineering assurance to provide a health and safety work environment and rectification of sub-standard situations in the current operating mines.

5. Adopting a non-negotiable winning formula for the success of our profession – Zero Serious Harm Ventilation Engineering Challenge.

6. Providing appropriate use of technology, revising design parameters, and providing innovative ventilation engineering solutions which admirably serve to protect the worker and the community.

7. Mentoring and sharing your wealth of experience with new entrants to the profession locally and globally. As the IMVC Chair, I have taken the initiative to expand the current IMVC Committee to include the other mining nations that were not previously represented in this unique engineering profession.

With the above self-commitments in mind, I hope you will enjoy the next three days and use the occasion to get to know people from other specialist areas, cultures and geographies. Several suggestions, practices and recommendations made during the Congress should be scrutinised with further discussions. We should debate their intentions and the logic behind innovative operational practices to benefit all and improve mine health and safety. It is important that you take away new ideas from this Congress and have the determination to make a difference when you return to your place of work. I hope this international Congress will mark the rejuvenation of Ventilation Engineering as a core stand-alone discipline in every corner of the world.

Best wishes for a successful congress.

Dr Bharath Belle MAusIMM(CP)

Chair and Honorary Editor, IMVC2024

Duncan Chalmers MAusIMM

IMVC2024 Deputy Chair, UNSW

SPONSORS

Congress Dinner Sponsor

Keynote Speaker Sponsor

Coffee Cart Sponsor

Exhibition Lounge Sponsor

Supporting Partner

Note Pads and Pens Sponsor

CONTENTS

Mine fires and emergency response planning

Mine gases

Numerical modelling and integration with planning and remote monitoring

Occupational health (mine dusts, gases, radon, etc)

Spontaneous combustion

Ventilation air methane (VAM) and GHG management

Ventilation economics and optimisation

Ventilation monitoring and control

Ventilation planning for coal and metalliferous mines (case studies)

Diesel emissions control and measurement

Numerical study of the diesel exhaust distributions and ventilation optimisation in an underground development face

P Chang[1], J Alonso Del Rio[2], R Morla[3] and W Wu[4]

1. Lecturer, Curtin University, Kalgoorlie WA 6430. Email: ping.chang@curtin.edu.au
2. Graduate Mining Engineer, Northern Star Resources Limited, Kalgoorlie WA 6430. Email: juliandalonsor@gmail.com
3. Senior Ventilation Engineer, Genesis Minerals Limited, Cloisters Square WA 6850. Email: ramsiit99@gmail.com
4. PhD candidate, Curtin University, Kalgoorlie WA 6430. Email: weifan.wu@postgrad.curtin.edu.au

ABSTRACT

Diesel exhaust is a carcinogenic emission from diesel engines. Due to extensive use of diesel-powered equipment and limited ventilation supplied to the confined areas of underground mines, underground miners are exposed to relatively high concentrations of diesel exhaust. Currently, mine ventilation is still the primary method to control diesel exhaust in the underground mining industry. Most of the states in Australia have outlined specific guidelines to manage diesel exhaust in mines. However, recent research indicates that the legislated volumetric flow rates of 0.05–0.06 m^3/s/kW, while effective in controlling diesel gas emissions, fall short in effectively managing diesel particulate matter. The computational fluid dynamics simulations were conducted to evaluate the effectiveness of both the legislated and alternative flow in controlling the diesel exhaust in an underground development face. Various scenarios were analysed to identify the airflow velocity required to maintain pollutant levels below the permissible limit. Results indicate that the legislated flow rate successfully controls CO, NO, and NO_2 but falls short in managing diesel particulate matter (DPM) below the limit. The study concludes that a volumetric flow rate almost twice as high as the legislated rate is necessary to control DPM effectively. The findings of this study provide vital information to the industry to design effective strategies as well as to government bodies or health agencies when revising and optimising the current legislation for diesel exhaust control in the future.

INTRODUCTION

Since the 1960s, diesel-powered equipment has been widely used in mining industries because of its high performance, durability, and efficiency compared to gasoline engines. However, the downside of the wide implementation of diesel engines is potentially high exposure to diesel particulate matter (DPM). Diesel particulate matter is generated by incomplete combustion of diesel fuel. Since 2012, diesel exhaust has been classified as a Group 1 carcinogen by the International Agency of Research on Cancer (IARC, 2013). It was reported that underground miners who work in confined areas could be exposed to 100 times higher concentrations of DPM than those who work in normal environments (Noll *et al*, 2006). Both long-term and short-term exposures to DPM may result in adverse health effects. Long-term exposure to high concentrations of DPM might result in lung and blader cancer, cardiovascular disease, and cognitive problem. Short-term exposure may contribute to acute health effects such as irritation, asthma, cough, and light-headedness (Chang and Xu, 2017; Lucking *et al*, 2011; US EPA, 2002).

In order to reduce DPM health hazards, personal exposure levels to DPM should be maintained below standards. In Australia, the Australian Institute of Occupational Hygienists (AIOH) has recommended an 8-hr TWA exposure standard for elemental carbon (EC) of 0.1 mg/m^3. Therefore, it is necessary to take effective approaches to reduce the DPM concentration in underground mines. Mine ventilation is one of the primary methods to control DPM in underground mines. To ensure adequate airflow to dilute diesel pollutants in underground mines, the Australian mining industry uses a minimum ventilation rate of 0.05m^3/s/kW in Western Australia and 0.06 m^3/s/kW in Eastern states (based on rated power for the engine) (DMIRS, 2022; New South Wales Department of Primary Industries (NSW DPI), 2008). However, it is worth noting that this volume was initially set-up for diluting diesel exhaust gases, such as NO_x (NSW DPI, 2008). Whether this volume is sufficient to maintain the DPM concentration below the limit is still unclear. To address this issue, computational fluid dynamics (CFD) can be used to investigate the effectiveness of current ventilation requirements on controlling DPM concentrations in underground mines.

CFD simulation is widely used in mining engineering to investigate the behaviours of DPM in underground mines. Zheng *et al* (2011, 2017, 2015a, 2015b, 2015c), Thiruvengadam, Zheng and Tien (2016a) and Thiruvengadam *et al* (2016b) have conducted a series of studies on DPM in underground mines by using CFD methods. A good agreement was achieved between the simulation results and the field test. Several authors simulated DPM distributions in underground mines for different scenarios (Chang and Xu, 2019; Chang, Xu, and Huang, 2020; Chang *et al*, 2020, 2019a, 2019b). It is worth noting that in one of those studies, the current ventilation rate for a loading activity already met the minimum airflow requirement (based on 0.06 $m^3/s/kW$), however, the DPM concentrations were above the limit (0.1 mg/m^3) in most of the areas in the heading (Chang *et al*, 2020). In addition, other researchers also used CFD modelling to evaluate the effects of ventilation system performance on diesel exhaust dilution (Kurnia, Sasmito, and Mujumdar, 2014; Kurnia *et al*, 2014), to study the DPM dispersion and diffusion in a coalmine heading (Duan *et al*, 2021), and to evaluate the impact of air velocities on DPM dispersion (Morla *et al*, 2019). All of the mentioned studies above have demonstrated that CFD is an effective and efficient tool for simulating dispersion of DPM plume in underground mines.

The main objective of this study is to investigate whether the current ventilation requirements are sufficient to dilute the DPM concentration under the limit. The study was carried out based on an on-site experiment in heading in an underground gold mine in Western Australia. The CFD simulation method was used to study the concentration distributions of DPM and toxic gases, such as NO, NO_2, and CO, under different ventilation rates. The optimum ventilation rate for the DPM dilution was recommended based on the simulation results.

MODEL DESCRIPTION

Physical model geometry

An underground heading in a Western Australian gold mine is selected for the physical model. The 3D view for this heading is illustrated in Figure 1. The length of the tunnel is 65.5 m. The opening is 6.13 m wide and 6.50 m high. A 26.4 m cuddy is located 20.1 m from the heading inlet. The diameter of the ventilation duct is 1.4 m. A loader is parked 30 m from the outlet. The model of the loader used in this study is a Caterpillar R1700G powered by a Caterpillar C11 engine rated at 263 kW, which is assumed to work at its maximum capacity.

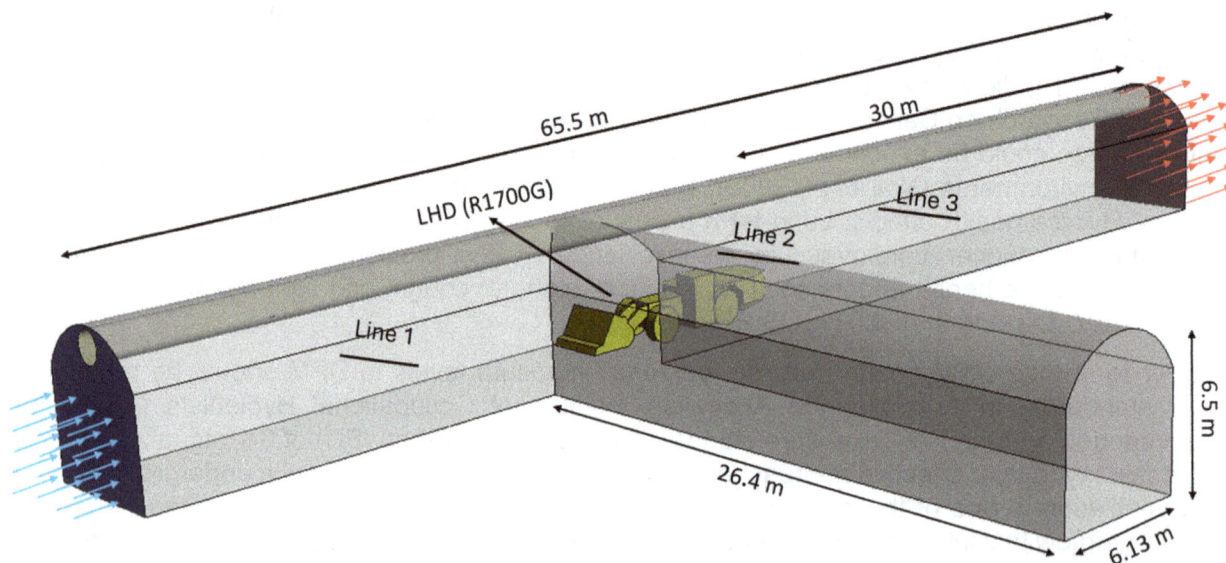

FIG 1 – 3D view of the CFD model geometry.

Boundary conditions

The standard k-ε model is used as the turbulent model to simulate the airflow region. The airflow is first simulated to achieve a steady state by using the SIMPLE algorithm. Then, the toxic gases and DPM are simulated using the species transport model. The boundary conditions and initial conditions are summarised in Table 1.

TABLE 1

Boundary and initial conditions.

Boundary	Parameter	Value
Inlet	Air Velocity	0.51 m/s
Outlet	Static Pressure	0 Pa
	Airflow rate	2.0 m³/s
	CO	282 ppm
Exhaust pipe	NO	349 ppm
	NO₂	19 ppm
	DPM (measured as EC)	2.6 mg/m³
Loader and walls	No slip wall function	N/A

Governing equations

The simulations were conducted with assumption that air is incompressible. Heat transfer was neglected. Navier-Stokes equations are used to describe the movement of airflow. The continuity and momentum equations are given below (Blazek, 2015):

$$\nabla \cdot \vec{u} = 0 \tag{1}$$

$$\nabla \cdot (\vec{u}\vec{u}) = -\frac{1}{\rho}\nabla p + (v + v_t)\nabla^2 \vec{u} + \vec{g} \tag{2}$$

where:

$\vec{u}$	is the air velocity vector in m/s
ρ	is the air density, which is 1.2 kg/m³ in this study
p	is the pressure in Pa
μ	is the air dynamic viscosity in Pa/s
v, v_t	are the air kinematic viscosity and turbulent viscosity, respectively, in m²/s
$\vec{g}$	is the gravity acceleration in m²/s

The species transport model equation is given by:

$$\nabla \cdot (\vec{u}C) = \left(D + \frac{v_t}{S_{ct}}\right)\nabla^2 C \tag{3}$$

where:

D	is the diffusion coefficient (m²/s)
S_{ct}	is the turbulent Schmidt number
C	is a concentration of the respective species, such as DPM or gas

Mesh independence study

The results of CFD simulations rely highly on the mesh quality. Conducting a mesh independence study is important because it guarantees that the results are both accurate and achieved with minimal computational resources. Thus, three mesh sizes, coarse, medium, and fine mesh, are generated for the mesh independence study. The total number of cells for three mesh sizes is 0.48 million, 1.61 million, and 4.17 million, respectively. The medium mesh for the domain is given as a representative in Figure 2.

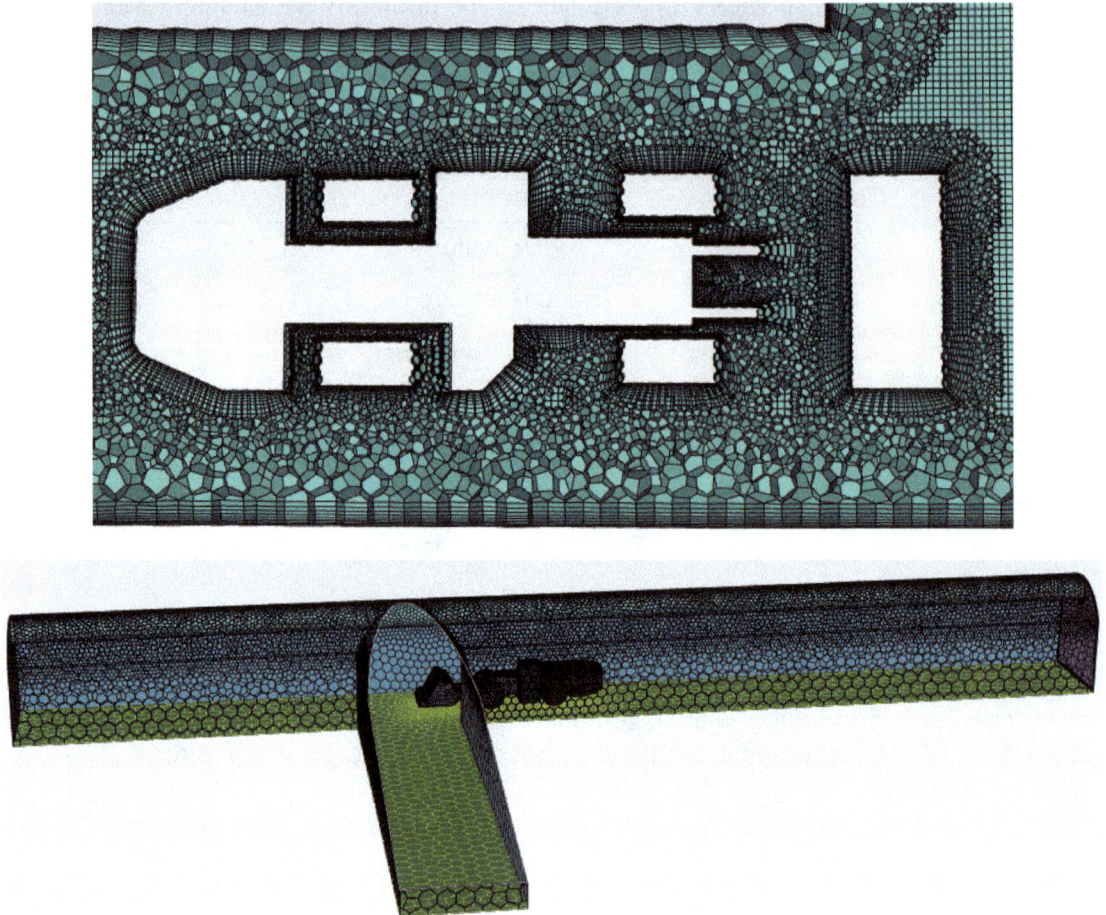

FIG 2 – Medium mesh used in this study.

Velocity profiles for three monitor lines at different locations are compared. The distance between the three lines and the tunnel outlet is 45 m (upstream of the load-haul-dump (LHD) vehicle), 30 m (LHD vehicle), and 20 m (downstream of the LHD vehicle), respectively, as shown in Figure 1. The horizontal lines are 3.25 m (height of the LHD) above the floor. Figure 3 gives the air velocity profiles at three monitor lines for different meshes. As can be seen, the velocity profiles of medium mesh are close to that of fine mesh, which means that mesh independence has been achieved. Thus, the medium mesh is used in this study for the following simulations.

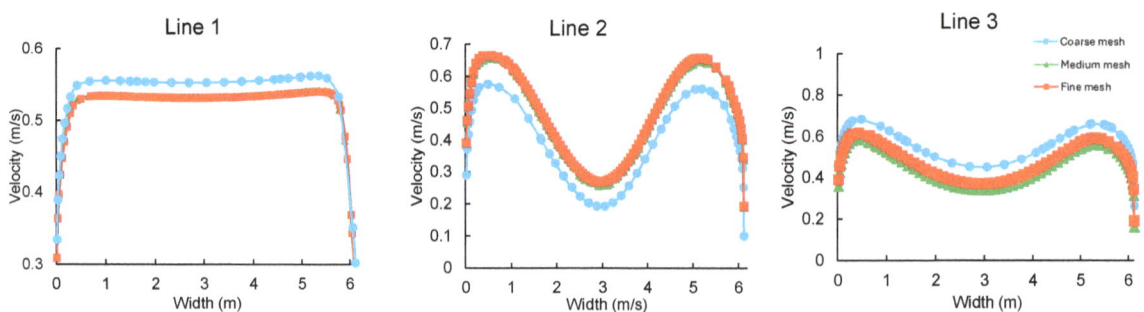

FIG 3 – Velocity profiles at the line monitors.

Model validation

Model validation is an important part of the CFD simulation to ensure the accuracy of simulation results. Both the airflow and DPM concentration simulation results were validated with the on-site test data. In this study, the velocity at nine points on a cross-section (Figure 4) located 5 m behind the loader was measured. The comparison between the CFD results and measured data can be found in Table 2. As can be seen, the average error is 6.7 per cent, and the largest difference is 14 per cent at P8. Considering the complicated underground environment as well as the operator's reading error when measuring the data, such an error is acceptable.

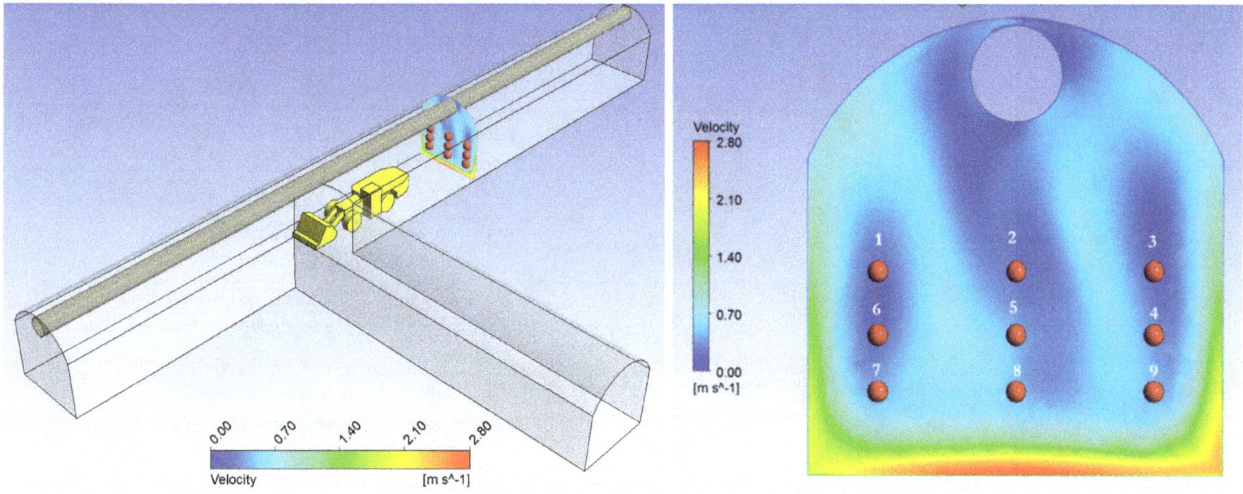

FIG 4 – Velocity points (red dots) on the cross-section.

TABLE 2

Comparison between simulation results and on-site data.

Point number	P1	P2	P3	P4	P5	P6	P7	P8	P9
On-site data (m/s)	0.28	0.26	0.2	0.3	0.5	0.4	0.36	0.49	0.42
Simulation results (m/s)	0.25	0.26	0.185	0.281	0.5	0.38	0.38	0.56	0.46
Error (%)	11%	0%	8%	6%	0%	5%	6%	14%	10%

To validate the accuracy of the species transport model, the on-site DPM data was collected and compared with the CFD simulations. The data was measured at 5 m, 10 m, 15 m, 20 m and 25 m behind the loader. The comparison is shown in Table 3. The average difference is 4.24 per cent, and the largest difference is 6.9 per cent, which are acceptable.

TABLE 3

EC data comparison.

Distance from LHD	5 m	10 m	15 m	20 m	25 m
On-site data ($\mu g/m^3$)	154.955	154.733	152.346	145.354	136.533
Simulation results ($\mu g/m^3$)	161.987	148.040	141.830	137.606	136.252
Error (%)	4.5%	4.3%	6.9%	5.3%	0.2%

RESULT AND DISCUSSION

The air flow quantity in the tunnel was 18.56 m^3/s exceeding the minimum requirements based on the regulation. Figure 5 presents the distributions of the selected diesel exhaust constituents with concentrations higher than the corresponding PEL limits for the case when the current ventilation rate was simulated. the concentrations of three of the studied diesel exhaust gases, CO, NO, and NO_2, were higher than the limits only in the vicinity of the exhaust discharge. However, it is noted that the EC concentrations in most of the downstream areas of the loader are higher than the limit (0.1 mg/m^3). Apparently, the current regulation is not sufficient to dilute the DPM concentration under the limit.

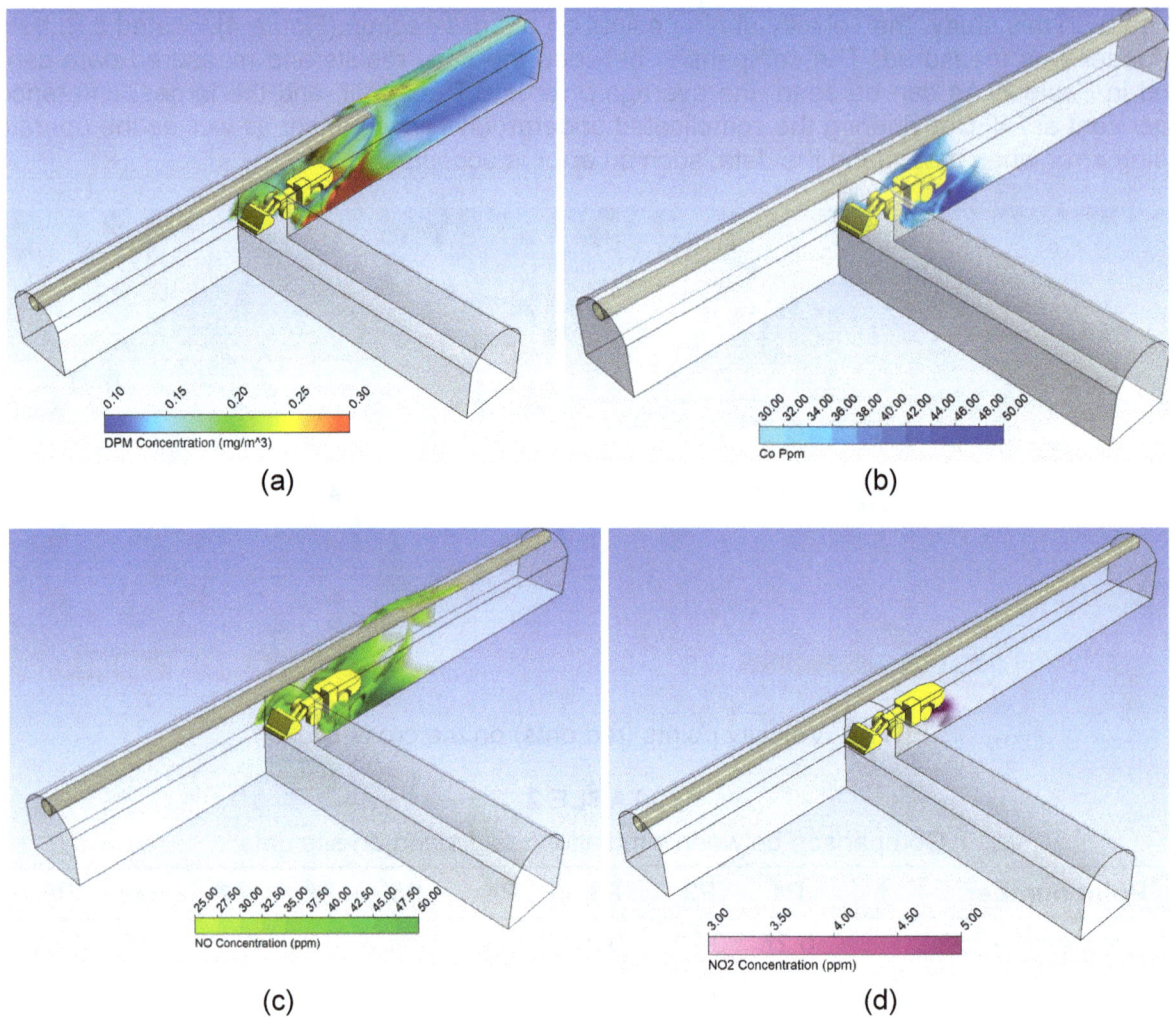

FIG 5 – Concentrations of selected diesel exhaust constituents for current ventilation rate (18.56 m^3/s): (a) DPM distribution (measured as EC) (> 0.1 mg/m^3); (b) CO distribution (> 30 ppm); (c) NO distribution (> 25 ppm); (d) NO_2 distribution (> 3 ppm).

However, as mentioned before, the minimum ventilation regulation of 0.05 to 0.06 $m^3/s/KW$ of the diesel engine power was initially developed to control the toxic gases from the diesel exhaust. To further investigate whether the required ventilation rates are sufficient to dilute the criteria gases, a few more simulations were conducted. The results are shown in Figure 6. The CO and NO_2 were distributed similarly as in case of the actual ventilation rate. However, when the 0.05 $m^3/s/kW$ ventilation rate was applied, the NO concentrations exceed the limit (25 ppm) in certain locations downstream. When the rate was increased to 0.06 $m^3/s/kW$, the NO concentration remained below the limit except in the area next to the LHD. The NO concentration distribution is similar to the distribution under the actual ventilation rate. Only the area at the downstream near the LHD is higher than the limit. Based on the simulation results, it can be concluded that the current ventilation regulation is sufficient to control most of the diesel exhaust gases under the limit. However, this regulation could not effectively control the DPM at a safe level.

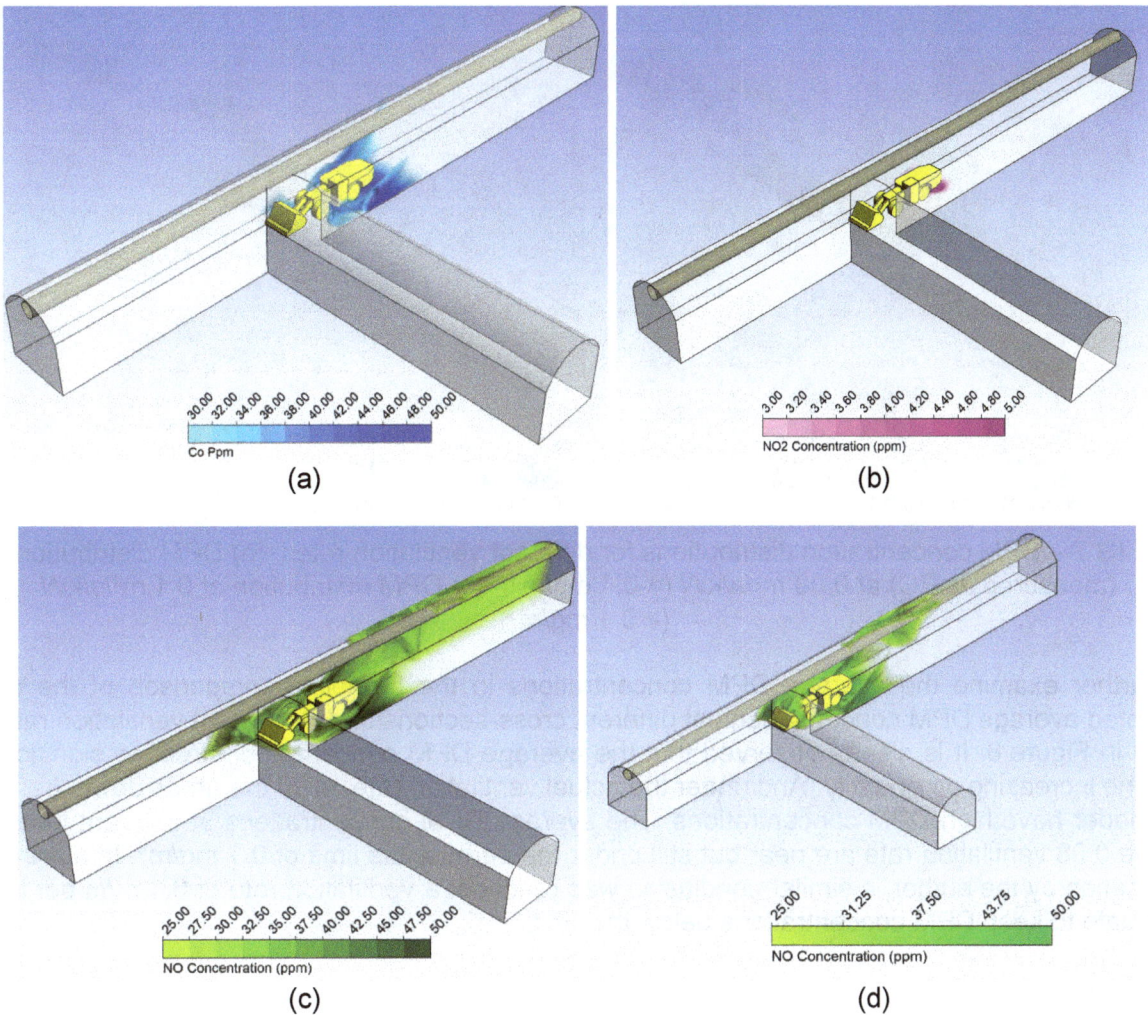

FIG 6 – Diesel exhaust gases concentration distribution under the current ventilation regulations: (a) CO distribution at 0.05 m³/s/kw (>30 ppm); (b) NO₂ distribution at 0.05 m³/s/kw (>25 ppm); (c) NO distribution at 0.05 m³/s/kw (>25 ppm); (d) NO distribution at 0.06 m³/s/kw (>25 ppm).

It is worth noting that diesel exhaust was officially classified as a carcinogenic in 2012, while most of the ventilation regulations were promulgated before that (Halim, 2017). In response to the IARC report, the government in Western Australia published a report named 'Management of diesel emissions in Western Australian mining operations' (DMP, 2013). The report indicated that six to eight times the currently required air volume was necessary to sufficiently dilute the (DPM) below the specified limit. A previous study also suggested that underground mines' ventilation rate should generally range from 0.06 to 0.1 m³/s/kW to dilute the diesel exhaust (Hedges, Djukic and Irving, 2007). To investigate the requirement for ventilation rate to dilute the DPM under the personal exposure limit, two more simulations were conducted with the ventilation rate set as 0.08 m³/s/kW and 0.1 m³/s/kW. The results are presented in Figure 7. With increase in air quantity, the DPM concentration decreased. The DPM concentration downstream of the LHD was near the EC PEL limit (0.1 mg/m³) when the ventilation rate is 0.08 m³/s/kW. When the ventilation rate was 0.1 m³/s/kW, the DPM concentrations were further reduced.

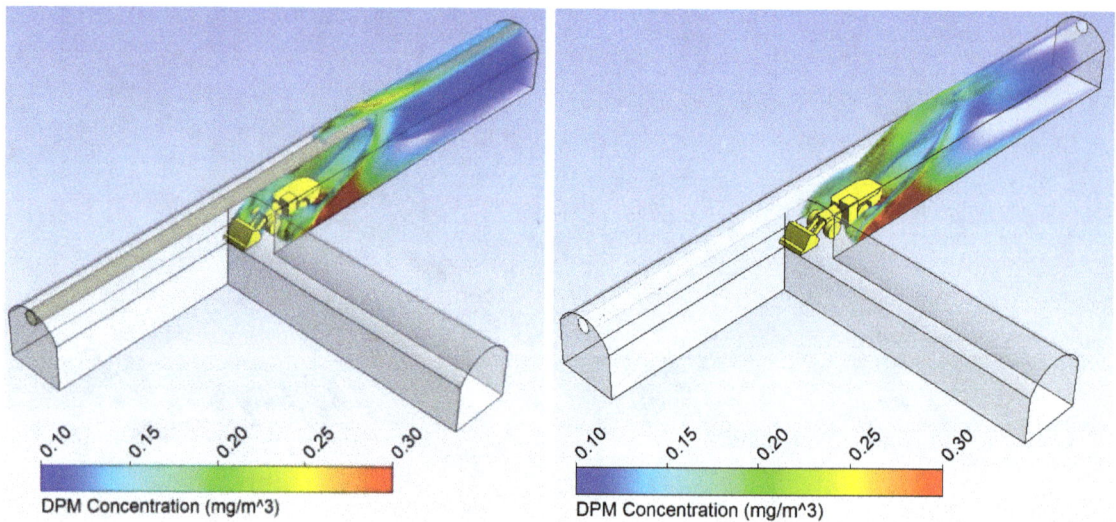

FIG 7 – DPM concentration distributions for different ventilation rates: (a) DPM distribution (measured as EC) at 0.08 m³/s/kW (> 0.1 mg/m³); (b) DPM distribution at 0.1 m³/s/kW (> 0.1 mg/m³).

To further examine the average DPM concentrations in the tunnel, a comparison of the area-weighted average DPM concentrations at different cross-sections under different ventilation rates is given in Figure 8. It is clearly observed that the average DPM concentrations reduce significantly with the increasing air quantity. And under the actual ventilation rate, all of the areas downstream of the loader have high DPM concentrations. The average DPM concentrations at different locations for the 0.08 ventilation rate are near but still under the permissible limit of 0.1 mg/m³. In a previous publication by the author, a similar conclusion was reached: a ventilation rate of 0.1 m³/s per kW is adequate to keep DPM concentrations below the limit.

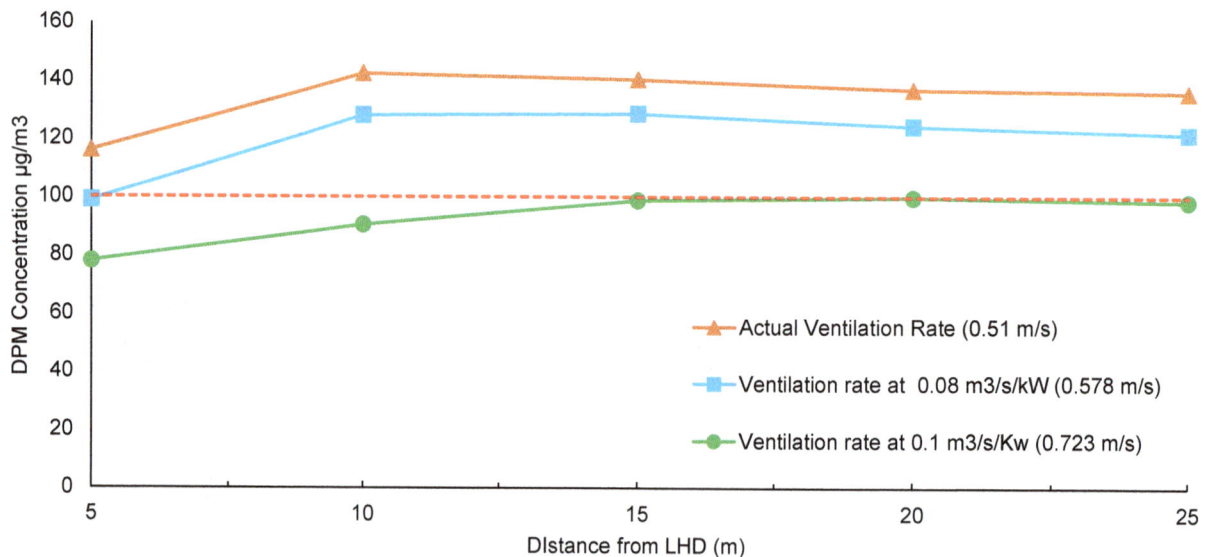

FIG 8 – Area weighted average DPM concentrations at different cross-sections under different ventilation rates.

CONCLUSIONS

This CFD simulation study thoroughly investigated the effectiveness of current ventilation requirements on concentrations of DPM in underground mines. The concentrations of DPM and diesel exhaust gases, namely CO, NO, and NO_2, in an underground development heading were evaluated for various ventilation conditions.

The results showed that the currently required ventilation rates (0.05 and 0.06 m³/s/kW) are sufficient to effectively control CO, NO, and NO_2 below their respective PEL limits. However, those are insufficient to dilute DPM concentrations to the recommended limit. DPM concentrations downstream of the loader exceeded the limit of 0.1 mg/m^3 under the current ventilation rate, indicating the inadequacy of the existing regulations in mitigating DPM exposure risks. Additional simulations were also conducted to evaluate higher ventilation rates, revealing that a 0.1 m^3/s per kW ventilation rate could effectively control DPM concentrations below the 0.1 mg/m^3 limit.

Overall, this study emphasises the critical role of ventilation optimisation in controlling DPM concentrations in underground mining environments and provides valuable insights for future regulatory revisions and ventilation system design considerations.

REFERENCES

Blazek, J, 2015. *Computational fluid dynamics: principles and applications* (Butterworth-Heinemann).

Chang, P and Xu, G, 2017. A review of the health effects and exposure-responsible relationship of diesel particulate matter for underground mines, *International Journal of Mining Science and Technology*, 27(5):831–838.

Chang, P and Xu, G, 2019. Review of Diesel Particulate Matter Control Methods in Underground Mines, Paper presented at the Proceedings of the 11th International Mine Ventilation Congress.

Chang, P, Xu, G and Huang, J, 2020. Numerical study on DPM dispersion and distribution in an underground development face based on dynamic mesh, *International Journal of Mining Science and Technology*, 30(4):471–475.

Chang, P, Xu, G, Mullins, B, Abishek, S and Sharifzadeh, M, 2020. Numerical investigation of diesel particulate matter dispersion in an underground development face during key mining activities, *Advanced Powder Technology*, 31(9):3882–3896. https://doi.org/10.1016/j.apt.2020.07.031

Chang, P, Xu, G, Zhou, F, Mullins, B and Abishek, S, 2019a. Comparison of underground mine DPM simulation using discrete phase and continuous phase models, *Process Safety and Environmental Protection*, 127:45–55.

Chang, P, Xu, G, Zhou, F, Mullins, B, Abishek, S and Chalmers, D, 2019b. Minimizing DPM pollution in an underground mine by optimizing auxiliary ventilation systems using CFD, *Tunnelling and Underground Space Technology*, 87:112–121.

Department of Mines and Petroleum (DMP), 2013. *Management of diesel emissions in Western Australia mining operations - Guideline*, Resources Safety, Department of Mines and Petroleum, Government of Western Australia, 37 p. Available from: <https://www.dmp.wa.gov.au/Safety/MSH_G_DieselEmissions.pdf>

Department of Mines, Industry Regulation and Safety (DMIRS), 2022. *Work Health and Safety (Mines) Regulations 2022*, Government of Western Australia. Available from: <https://www.legislation.wa.gov.au/legislation/statutes.nsf/RedirectURL?OpenAgent&query=mrdoc_47453.pdf>

Duan, J, Zhou, G, Yang, Y, Jing, B and Hu, S, 2021. CFD numerical simulation on diffusion and distribution of diesel exhaust particulates in coal mine heading face, *Advanced Powder Technology*, 32(10):3660–3671.

Halim, A, 2017. Ventilation requirements for diesel equipment in underground mines–Are we using the correct values, paper presented at the 16th North American Mine Ventilation Symposium, Golden, Colorado.

Hedges, K, Djukic, F and Irving, G, 2007. Diesel Particulate Matter in Underground Mines–Controlling the Risk (an update), paper presented at the Queensland Mining Industry Health and Safety Conference.

International Agency for Research on Cancer (IARC), 2013. Diesel and gasoline engine exhausts and some nitroarenes, IARC Monographs on the Evaluation of Carcinogenic Risks to Humans, No. 105. Available from: <https://publications.iarc.fr/Book-And-Report-Series/Iarc-Monographs-On-The-Identification-Of-Carcinogenic-Hazards-To-Humans/Diesel-And-Gasoline-Engine-Exhausts-And-Some-Nitroarenes-2013>

Kurnia, J C, Sasmito, A P and Mujumdar, A S, 2014. Simulation of a novel intermittent ventilation system for underground mines, *Tunnelling and Underground Space Technology*, 42:206–215.

Kurnia, J C, Sasmito, A P, Wong, W Y and Mujumdar, A S, 2014. Prediction and innovative control strategies for oxygen and hazardous gases from diesel emission in underground mines, *Science of the Total Environment*, 481:317–334.

Lucking, A J, Lundbäck, M, Barath, S L, Mills, N L, Sidhu, M K, Langrish, J P, Boon, N A, Pourazar, J, Badimon, J J, Gerlofs-Nijland, M E, Cassee, F R, Boman, C, Donaldson, K, Sandstrom, T, Newby, D E and Blomberg, A, 2011. Particle traps prevent adverse vascular and prothrombotic effects of diesel engine exhaust inhalation in men, *Circulation*, 123(16):1721–1728.

Morla, R, Karekal, S, Godbole, A, Bhattacharjee, R M, Nasina, B and Inumula, S, 2019. Effect of ventilation air velocities on diesel particulate matter dispersion in underground coal mines, *International Journal of Mining and Geo-Engineering*, 53(2):117–121.

New South Wales Department of Primary Industries (NSW DPI), 2008. Guideline for the management of diesel engine pollutants in underground environments, MDG 29, Mine Safety Operations Division, New South Wales Department of Primary Industries. Available from: <https://www.resourcesregulator.nsw.gov.au/sites/default/files/documents/mdg-29.pdf>

Noll, J, Mischler, S, Schnakenberg, G, Bugarski, A, Mutmansky, J and Ramani, R, 2006. Measuring Diesel Particulate Matter in Underground Mines Using Sub Micron Elemental Carbon as a Surrogate, paper presented at the Proceedings for the 11th US North American Mine Ventilation Symposium.

Thiruvengadam, M, Zheng, Y and Tien, J C, 2016a. DPM simulation in an underground entry: Comparison between particle and species models, *International Journal of Mining Science and Technology,* 26(3):487–494.

Thiruvengadam, M, Zheng, Y, Lan, H and Tien, J C, 2016b. A diesel particulate matter dispersion study inside a single dead end entry using dynamic mesh model, *International Journal of Mining and Mineral Engineering,* 7(3):210–223.

US EPA, 2002. *Health assessment document for diesel engine exhaust (Final 2002),* US Environmental Protection Agency, National Center for Environmental Assessment, EPA/600/8-90/057F. Available from: <https://cfpub.epa.gov/ncea/risk/recordisplay.cfm?deid=29060>

Zheng, Y, Lan, H, Thiruvengadam, M and Tien, J, 2011. DPM dissipation experiment at MST's experimental mine and comparison with CFD simulation, *Journal of Coal Science and Engineering (China),* 17(3):285–289. https://doi.org/10.1007/s12404-011-0311-1

Zheng, Y, Lan, H, Thiruvengadam, M, Tien, J C and Li, Y, 2017. Effect of single dead end entry inclination on DPM plume dispersion, *International Journal of Mining Science and Technology,* 27(3):401–406.

Zheng, Y, Thiruvengadam, M, Lan, H and Tien, C J, 2015a. Effect of auxiliary ventilations on diesel particulate matter dispersion inside a dead-end entry, *International Journal of Mining Science and Technology,* 25(6):927–932.

Zheng, Y, Thiruvengadam, M, Lan, H and Tien, C, J, 2015b. Simulation of DPM distribution in a long single entry with buoyancy effect, *International Journal of Mining Science and Technology,* 25(1):47–52. https://doi.org/http://dx.doi.org/10.1016/j.ijmst.2014.11.004

Zheng, Y, Thiruvengadam, M, Lan, H and Tien, J C, 2015c. Design of push–pull system to control diesel particular matter inside a dead-end entry, *International Journal of Coal Science and Technology,* 2(3):237–244.

Analysis of data received from the black carbon portable real-time monitor tested in an underground mine for monitoring DPM

S Sabanov[1], R Korshunova[2], A R Qureshi[3], G Kurmangazy[4] and Z Dauitbay[5]

1. Associate Professor, School of Mining and Geosciences, Nazarbayev University, Astana 010000, Kazakhstan. Email: sergei.sabanov@nu.edu.kz
2. Graduate student, School of Mining and Geosciences, Nazarbayev University, Astana 010000, Kazakhstan. Email: ruslana.korshunova@nu.edu.kz
3. PhD Student, School of Mining and Geosciences, Nazarbayev University, Astana 010000, Kazakhstan. Email: abdullah.qureshi@nu.edu.kz
4. Undergraduate student, School of Mining and Geosciences, Nazarbayev University, Astana 010000, Kazakhstan. Email: gulim.kurmangazy@nu.edu.kz
5. Undergraduate student, School of Mining and Geosciences, Nazarbayev University, Astana 010000, Kazakhstan. Email: zhaudir.dauitbay@nu.edu.kz

ABSTRACT

Diesel particulate matter (DPM) needs to be continuously monitored to provide better control for mine ventilation. Exceeding DPM range in underground mines can be caused adverse miners health effects. DPM appears mainly in the submicron diapason, and generally can be interpreted by two components: elemental carbon (EC) and organic carbon (OC). This study presents results of continuous measurements of black carbon (BC) as a surrogate of DPM produced in an underground metalliferous mine, where diesel equipment is extensively utilised. Aim of this study is to analyse data received from the portable 'MicroAeth® MA200' BC monitor that can be used for continuous monitoring of DPM. Measurements were produced at a distance of 10 m from the dumping point where diesel-powered face haulage loader Caterpillar R1700 (engine model Cat@C11 ACERT, 241 kW, Tier 3/Stage IIIA Equivalent Engine) worked in the active mine face. In addition, a $PM_{0.3}$ and particle diameter have been sampled with help of the 'Naneous Partector 2' instrument to derive a conversion factor with BC. As a result, BC average concentration was 2098 $\mu g/m^3$ and $PM_{0.3}$ was 1018 $\mu g/m^3$. Performance of the mobile real-time BC instrument as a portable tool for monitoring DPM has been assessed. Results of these field measurements will be used for computational fluid dynamics (CFD) modelling of DPM concentration distribution with different auxiliary fan sets considering load-haul-dump (LHD) loader and ventilation duct position in the active mine face zone to optimise mine ventilation system. This will help to improve DPM monitoring approaches for protecting miners' health.

INTRODUCTION

The calculation of airflow amounts in mine ventilation is conducted with the objective of ensuring the maintenance of a safe and healthy environment for miners. The determination of necessary volumes of ventilation air is frequently reliant on the dilution of DPM (Bugarski *et al*, 2011). The emissions and exposure levels of DPM exhibit significant variability, contingent upon mine schedules and activities (Janisko and Noll, 2010). DPM appears mainly in the submicron diapason, and generally can be interpreted by two components: elemental carbon (EC) and organic carbon (OC). For DPM measuring, certain commercially available aerosol monitors or particle counters are commonly employed, utilising the real-time light scattering principle (Arnott *et al*, 2008). In order to ensure compliance, the majority of mining companies employ the widely recognised NIOSH 5040 standard as the most precise approach for assessing the long-term effects of DPM on workers (Schauer, 2003). Volkwein *et al* (2017) highlight the significant advantage of this method because it reduces the likelihood of measurement results being influenced by mineral sources or other flammable substances. Additionally, it allows for the differentiation of carbon content into OC and EC components to calculate a total carbon (TC) value (Birch and Noll, 2004). However, despite its advantages, this approach is hindered by some drawbacks, such as the time-consuming nature and the imprecision in capturing particles (Koponen *et al*, 2023). The NIOSH 5040 approach is inherently incapable of detecting DPM transients during the measuring process (Khan *et al*, 2021). According to Bond *et al* (2013), the environmental community usually uses black carbon as a metric, while Noll *et al* (2006) report that the mining sector utilises elemental carbon as their measurement. These two concepts exhibit a high degree of similarity, but the only clear difference appears to be the sophistication of the analysis. Simultaneously, there are a number of studies that demonstrate the relationship of black carbon (BC) to total carbon, elemental carbon and DPM. For example, Fruin, Winer and Rodes (2004) examined the impact of BC in automobiles as a measure of DPM effects in California. They found that the exposure to BC was 4 µg/m³, which is equivalent to DPM concentrations ranging from 7–23 µg/m³. In comparison with this, Burtscher, Künzel and Hüglin (1998) demonstrated that the proportion of BC to the overall quantity of DPM collected from the tailpipe varied between 45 per cent and 100 per cent. In another study (Barrett *et al*, 2019) that evaluated various DPM field monitors for underground mining applications, the AE33 Aethalometer (AE33) measurements revealed that the ratio of BC to EC is approximately 2.0 at low concentrations and around 1.5 at higher concentrations. Similar results were obtained in a study by Volkwein *et al* (2017) on the measurement of DPM in a large underground metal mining mine. The average levels of BC were below 53 µg/m³. Volkwein *et al* (2017) in another study on the use of an environmental 'black carbon' particle sensor to measure DPM in three underground mines also stated, that the proportion of $BC_{(Aathalometer)}$ to $EC_{(NIOSH5040)}$ was approximately 1.5 (ie BC~15.EC). According to a recent study conducted by Koponen *et al* (2023) on the use of aethalometers to measure the concentration of DPM in underground mines, the average ratio of black carbon to elemental carbon was found to be 2.1±0.7 (for AE33 instrument) and 2.4±1.7 (for MA200 instrument).

This study presents results of continuous measurements of black carbon (BC) as a surrogate of DPM produced in an underground metalliferous mine, where diesel equipment is extensively utilised. The aim of this study is to analyse data received from the portable 'MicroAeth® MA200' BC monitor that can be used for continuous monitoring DPM in conditions of underground polymetallic mines using diesel mobile equipment.

MATERIAL AND METHODS

The study has been produced in a polymetallic mine in Kazakhstan, where the diesel exhaust samples were collected. This mine uses a force-exhaust ventilation system with two main surface fans and numerous auxiliary fans situated in the underground working faces. Measurements were produced at a distance about 10 m from the temporary dumpsite (Figure 1). The diesel-powered LHD loader Caterpillar R1700 (engine model Cat@C11 ACERT, 241 kW, Tier 3/Stage IIIA Equivalent Engine) working in the active mine face, which has a 72 m length, 4.3 m in height and 4.0 m in wide (Figure 1). Average air velocity in the drift was around 0.9 m/s. The loader engine exploits the high-quality diesel fuel DT-L-K2 (standard 'GOST305-82'), which in the first approximation corresponds to the Euro 3 standard and differs only in the cetane number. The auxiliary fans were operating during the active ore haulage process to pull fresh air through the two ventilation ducts. Figure 1 shows the positioning of LHD loader during its operational process. During measurements, the LHD diesel loader was working in its operational cycle: load from the face, haul and maneuverers, and dump to the temporary dumpsite. The main source of diesel fumes was only an exhaust pipe of the LHD loader. The sampling was conducted by the Naneos Partector 2 and the portable 'MicroAeth® MA200' BC monitors.

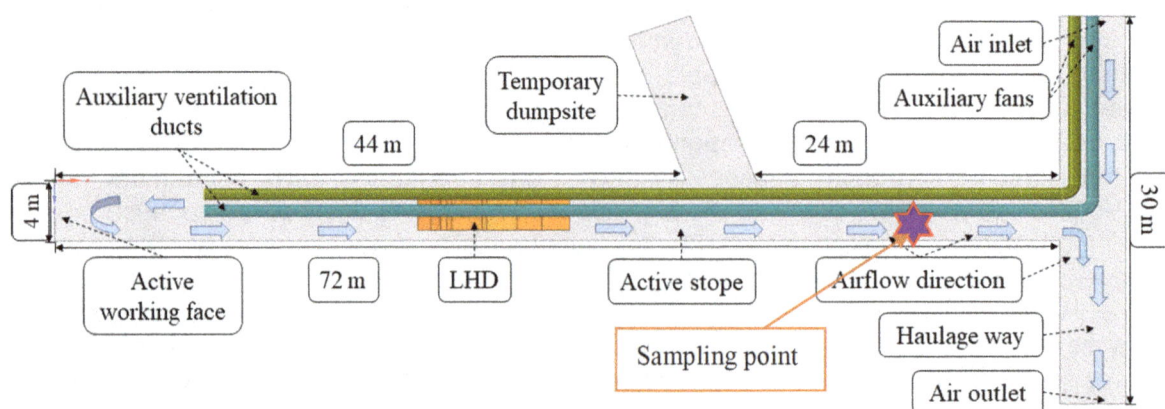

FIG 1 – Schematic layout of the sampling site.

The MicroAeth® MA200 is a self-contained instrument that integrates multiple components, including a pump, flow control mechanism, data storage system, and battery. Furthermore, the device possesses supplementary functionalities such as quantifying relative humidity and temperature. The instrument measures the mass concentration of light absorbing carbonaceous particles in a sampled aerosol. The instrument have five analytical channels each operating at a different wavelength (880 nm, 625 nm, 528 nm, 470 nm, 375 nm). Measurement at 880 nm is interpreted as concentration of BC. Measurement at 375 nm is interpreted as Ultraviolet Particulate Matter (UVPM) indicative of organic sources such as woodsmoke, tobacco, and biomass burning (AethLabs, 3085, San Francisco).

The 'Naneos Partector 2' (Naneos Particle Solutions gmbh, Switzerland) multimetric particle detector uses dual noncontact detection stages to measure the lung deposition surface area (LDSA), particle number concentration (PNC), and average particle diameter. Additionally, it calculates the particle surface area concentration (SA) and the particle mass <0.3 μm ($PM_{0.3}$). The manufacturer provides the following measurements and accuracy ranges for each variable of Partector 2: LDSA = 0–12 000 ±30 per cent $\mu m^2/cm^3$, SA = 0–50 000 ± 30 per cent $\mu m^2/cm^3$, d = 10–300 ± 30 per cent nm, PNC = 0–106 ± 30 per cent pt/cm^3, and $PM_{0.3}$ = 0–1000 ± 50 per cent $\mu g/m^3$. The displayed LDSA value is only accurate in the size range of 20–400 nm; however, the instrument can be used to measure micron-sized particles too. The 20–150 nm uses a fixed deposition voltage, and 10–300 nm uses an adaptive deposition voltage. The particle size range of LDSA is from 10 nm to 10 μm, and for size distribution, it has eight channels between 10 and 300 nm. The noise floor is about 0.5 $\mu m^2/cm^3$ in particle-free air for LDSA (https://www.naneos.ch/partector2.html).

RESULTS AND DISCUSSIONS

Measurements of BC and $PM_{0.3}$ produced during 61 mins and shown in Figure 2. As a result, BC average concentration was 2098 μg/m³ and for $PM_{0.3}$ average concentration was 1018 μg/m³. High peaks are mostly associated with air velocity decreases from 0.9 to 0.7 m/s. Descriptive statistics presented in Table 1.

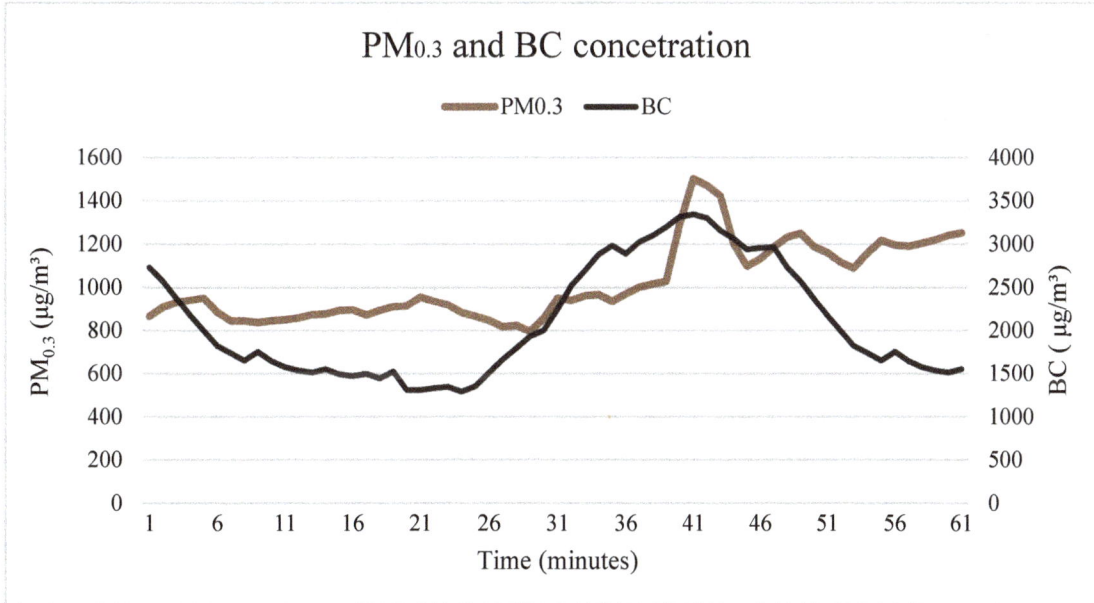

FIG 2 – $PM_{0.3}$ and BC concentrations during the sampling time.

TABLE 1

Descriptive statistics.

Name	PM$_{0.3}$	BC
Mean	1018	2098
Standard Error	23	84
Median	948	1819
Standard Deviation	177	656
Sample Variance	31 266	430 436
Kurtosis	-0.04	-1.22
Skewness	0.90	0.54
Range	707	2052
Minimum	793	1289
Maximum	1500	3341
Sum	62 077	127 961
Count	61	61
Confidence Level (95.0%)	45	168

BC has higher standard deviation and range of concentration numbers comparing to PM$_{0.3}$ (Table 1). A high Sample Variance of BC indicates that the data points are very spread out from the mean, and from one another.

Figures 3 presents a comparison of two different real time instruments and demonstrated a conversion factor between BC and PM$_{0.3}$. The relationship between PM$_{0.3}$ and BC concentrations does not show a great R^2 factor which is 0.23. Based on this it can be problematic for precise predictions uding one instrument separately for such measurements. However, Partector 2 demonstrated their reliability to be used for measurements of gaseous aerosols (Sabanov *et al* 2024). And ration of averaged BC to PM$_{0.3}$ is received about 2, which is pretty equal to ratio of BC to EC stated by Barrett *et al* (2019) and Koponen *et al* (2023).

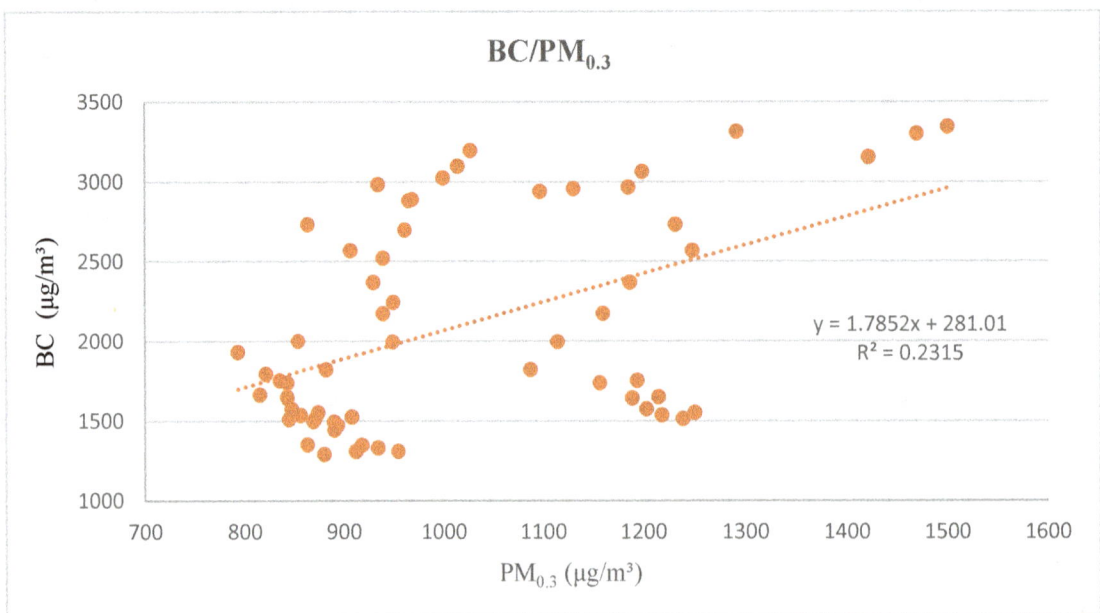

FIG 3 – PM$_{0.3}$ and BC relationship.

Compared to other studies of BC values, this study's results highlight a very high BC concentrations. This require to conduct further test works and additional validations to prove the reliability of the instruments to be used under such conditions.

CONCLUSIONS

This study analysed data received from the portable 'MicroAeth® MA200' BC monitor for its reliability to continuous monitoring of DPM. Measurements were produced at a distance about 10 m from the dumping point with the operational diesel-powered LHD. In addition, a $PM_{0.3}$ and particle diameter have been sampled with help of the 'Naneous Partector 2' instrument to derive a conversion factor with BC. As a result, BC average concentration was 2098 µg/m^3 and $PM_{0.3}$ was 1018 µg/m^3. A comparison of two different real time instruments has been produced to demonstrate a conversion factor between BC and $PM_{0.3}$. The relationship between $PM_{0.3}$ and BC concentrations does not show a great R^2 factor, which was only 0.23. Results of these field measurements can be used for computational fluid dynamics (CFD) modelling of DPM concentration distribution with different auxiliary fan sets considering LHD loader and ventilation duct position in the active mine face zone to optimise mine ventilation system. This will help to improve DPM monitoring approaches for protecting miners' health.

ACKNOWLEDGEMENT

This research was funded by Nazarbayev University Grant Programs: Research Grant #20122022FD4128, and Collaborative Research Project # 091019CRP2104.

REFERENCES

Arnott, W P, Arnold, I J, Mousset-Jones, P, Kins, K and Shaff, S, 2008. Real-time measurements of diesel E C and TC in a Nevada gold mine with photoacoustic and Dusttrak instruments: Comparison with NIOSH 5040 filter results, In 12th US/North American Mine Ventilation Symposium, Reno, Nevada.

Barrett, C, Sarver, E, Cauda, E, Noll, J, Vanderslice, S and Volkwein, J, 2019. Comparison of Several DPM Field Monitors for Use in Underground Mining Applications, *Aerosol and Air Quality Research*, 19:2367–2380. https://doi.org/10.4209/aaqr.2019.06.0319

Birch, M E and Noll, J D, 2004. Submicrometer elemental carbon as a selective measure of diesel particulate matter in coal mines, *Journal of Environmental Monitoring*, 6(10):799–806.

Bond, T C, Doherty, S J, Fahey, D W, Forster, P M, Berntsen, T, DeAngelo, B J, Flanner, M G, Ghan, S, Kärcher, B, Koch, D, Kinne, S, Kondo, Y, Quinn, P K, Sarofim, M C, Schultz, M G, Schulz, M, Venkataraman, C, Zhang, H, Zhang, S, Bellouin, N, Guttikunda, S K, Hopke, P K, Jacobson, M Z, Kaiser, J W, Klimont, Z, Lohmann, U, Schwarz, J P, Shindell, D, Storelvmo, T, Warren, S G and Zender, C S, 2013. Bounding the role of black carbon in the climate system: A scientific assessment, *Journal of Geophysical Research: Atmospheres*, 118(11):5380–5552. https://doi.org/10.1002/jgrd.50171

Bugarski, A D, Janisko, S, Cauda, E, Noll, J D and Mischler, S, 2011. Diesel aerosols and gases in underground mines: Guide to exposure assessment and control. National Institute for Occupational Safety and Health (NIOSH).

Burtscher, H, Künzel, S and Hüglin, C, 1998. Characterization of particles in combustion engine exhaust, *Journal of Aerosol Science*, 29(4):389–396. https://doi.org/10.1016/S0021-8502(97)10001-5

Fruin, S A, Winer, A M and Rodes, C E, 2004. Black carbon concentrations in California vehicles and estimation of in-vehicle diesel exhaust particulate matter exposures, *Atmospheric Environment*, 38(25):4123–4133. https://doi.org/10.1016/j.atmosenv.2004.04.026

Janisko, S and Noll, J D, 2010. Field Evaluation of Diesel Particulate Matter Using Portable Elemental Carbon Monitors. in *Proceedings of the 13th US/North American Mine Ventilation Symposium*, Canada.

Khan, M U, Homan, K O, Saki, S A, Emad, M Z and Raza, M A, 2021. Real-time diesel particulate matter monitoring in underground mines: evolution and applications, *International Journal of Mining, Reclamation and Environment*, 35(4):291–305. https://doi.org/10.1080/17480930.2020.1818937

Koponen, H, Lukkarinen, K, Leppänen, M, Kilpeläinen, L, Väätäinen, S, Jussheikki, P, Karjalainen, A, Ruokolainen, J, Yli-Pirilä, P, Ihalainen, M, Hyttinen, M, Pertti Pasanen, O and Sippula, O, 2023. Applicability of aethalometers for monitoring diesel particulate matter concentrations and exposure in underground mines, *Journal of Aerosol Science*, 106330. https://doi.org/10.1016/j.jaerosci.2023.106330

Noll, J D, Mischler, S E, Schnakenberg Jr, G H and Bugarski, A D, 2006. Measuring Diesel Particulate Matter in Underground Mines Using Submicron Elemental Carbon as a Surrogate, in *Proceedings of the 11th US/North American Mine Ventilation Symposium*, pp 105–110 (Taylor and Francis Group: London).

Sabanov, S, Qureshi, A, Korshunova, R and Kurmangazy, G, 2024. Analysis of Experimental Measurements of Particulate Matter (PM) and Lung Deposition Surface Area (LDSA) in Operational Faces of an Oil Shale Underground Mine, *Atmosphere*, 15(2):200. https://doi.org/10.3390/atmos15020200

Schauer, J J, 2003. Evaluation of elemental carbon as a marker for diesel particulate matter, *Journal of Exposure Science and Environmental Epidemiology*, 13(6):443–453. https://doi.org/10.1038/sj.jea.7500298

Volkwein, J, Barrett, C, Sarver, E and Hansen, A D A, 2017. Application of an Environmental 'Black Carbon' Particulate Sensor for Continuous Measurement of DPM in Three Underground Mines, in *Proceedings Australian Mine Vent Conference 2017*, pp 143–150 (The Australasian Institute of Mining and Metallurgy: Melbourne).

Gas management

More effective pre-drainage of low-permeability coals using indirect hydraulic fracturing (IHF) in co-application with surface-to-inseam (SIS) wells

R L Johnson Jr[1] and M Ramsay[2]

1. General Manager – Technical Services, Novus Fuels, Brisbane Qld 4069.
   Email: ray.johnson@novusfuels.com
2. Drilling Superintendent – Gas, Anglo American, Moranbah Qld 4744.
   Email: matt.ramsay@angloamerican.com

ABSTRACT

Historically, surface-to-inseam (SIS) wells have been effective in pre-drainage underground coalmines; however, as mines encounter lower permeability coals, the ability to utilise SIS wells alone to achieve just-in-time drainage becomes unrealistic and results in earlier implementation of underground-inseam (UIS) drainage wells. To date, there have been limited applications of horizontal wellbores employing multiple hydraulic fractures in mining applications primarily because of the steel equipment required to be placed within potentially mineable coal seams to achieve multi-stage treatments. Johnson Jr, Cheong and Farley (2020) and Johnson Jr et al (2021) detailed single- and multi-stage indirect hydraulic fracturing treatments (IHF) for coal seam gas (CSG) wells and mining applications using a steel packer and frac sleeve string assembly placed in the floor below the seam.

This paper summarises past treatments and details results from a recent application of IHF, a 23-stage treatment was placed in an undrained portion of the mine in advance of pre-drainage in the Moranbah Goonyella Middle (GM) seam. Pressures during this deployment differed from prior applications placed in previously drained areas. However, like prior applications, the monitoring of micro seismicity, pressure observation wells, and pre-positioned SIS boreholes indicated the treatments successfully interacted with a large area thereby creating an extensive stimulated reservoir volume (SRV) for pre-drainage.

Effective IHF design requires a good understanding of the pressure, stress and rock mechanical properties of the coal, the adjoining interval of fracture initiation as well as boundary conditions. This allows excellent control of individual placements based on geological data and pressure management, minimises unintentional fracture propagation and potential roof destabilisation, and optimises placements based on geological data. The authors will discuss the initial reasoning for the application, observed responses based on deployment (ie vertical versus horizontal well), post-IHF SIS production improvements of prior treatments, anecdotal improvements with subsequent UIS deployment and go-forward strategies for further deployments.

BACKGROUND

In the Moranbah North Mine typically SIS wells have been drilled on 60 m spacing and adequately drained sections of the mine to place gate roads and subsequently pre-drain panels to safe levels using UIS wells. In a recent study, empirically it was found that SIS wells had outperformed UIS in the same area based on a gas per metreage rate (see Table 1). However, as the UIS is installed after SIS, the degree of depletion should be normalised to adequately reflect the productivity as a function of drawdown and surface area. Nevertheless, as in other mine drainage cases to Moranbah North Mine (MNM) SIS and later UIS have effectively predrained areas before mining.

TABLE 1

Comparison of SIS to UIS gas predrainage on a normalised length and area basis.

Year	Normalised length (km) SIS wells	Normalised area (m²) SIS wells	Normalised length (km) UIS/SIS wells	Normalised area (m²) UIS/SIS
2020	1.0	1.0	3.2	5.2
2021	1.0	1.0	1.3	2.2
2022	1.0	1.0	2.8	4.5
2023	1.0	1.0	11.3	18.2
Normalised total rate	109%		100%	

In 2019, a study of SIS wells in a lower permeability area of the mine planned for development indicated that inter-well drainage and inadequate recovery would result. This provided the impetus to review alternative strategies for pre-drainage including branching, successful at the adjoining Arrow Energy Moranbah Gas Project (MGP) (Johnson Jr and Mazumder, 2014) and prior fracture stimulations in an adjoining area. In MGP, fracture stimulations had been unsuccessful and had indicated little correlation in job size or total proppant to productivity. Further, coiled tubing, pinpoint fracturing was attempted in an inseam well and had difficulties and was not successful. Thus, an alternative for low permeability coals was needed and other alternatives were considered.

In 2019, Strike Energy reported the successful implementation of an IHF well at the Jaws 1ST CSG well (Branajaya, Archer and Farley, 2019). Olsen *et al* (2007) proposed IHF stimulation to improve hydraulic fracturing in coal seams by initiating fractures in lower-stress clastic rock adjacent to coal seams and allowing a more simplified fracture to propagate in the coals. The well was sidetrack-drilled well placed below the target seam following drilling difficulties in the inseam lateral. The Jaws 1ST well was successfully stimulated in seven stages using six indirect fractures and one direct hydraulic fracture in the heel of the well and utilised several diagnostics including surface deformation tiltmeters, microseismic monitoring, and liquid chemical tracers in each stage.

A full evaluation incorporating a planar, fully 3D frac model (Barree, 1983) was performed to assess the treatment dimensions relative to the diagnostic data (Johnson Jr, Cheong and Farley, 2020). The well was successful in improving production and only one of the six IHF stages was unsuccessful and showed little to no liquid tracer responses post-frac. This stage was believed to have suffered a disconnection from the IHF lateral resulting from significant overflushing. The general observation from this case was that the IHF process could be successful in strike-slip stress regimes, and application would be highly dependent on the wellbore placement as well as the stress and rock mechanical properties of the coal and the adjoining clastic interval of fracture initiation.

Based on the need to implement a timely solution for MNM to achieve target gas contents and favourable potential modelling of IHF well in the pre-existing panel of SIS wells, a trial was planned to use two vertical wells to be mined back followed by the initial IHF placed under the next planned gate road.

INITIAL IMPLEMENTATIONS

Initial vertical well IHF planning, execution and post-treatment evaluation

The first implementation of IHF in this Bowen Basin was on a vertical well SO774 in a pre-drained area of the MNM in the GM seam. The vertical well trials were made to ensure that hydraulic fracturing did not destabilise the roof, despite significant evidence from other mining areas indicating the compatibility of hydraulic fracturing with mining operations (Jeffrey *et al*, 1992, 1998; Mills, Jeffrey and Gale, 2007).

The vertical well was drilled and cased with glass-reinforced epoxy and cemented back across the coal with a frac sleeve placed 2 m below the coal. A diagnostic fracture injection test (DFIT) was performed on the well to calibrate the stress profile (see Figure 1) and the frac effectively placed the desired amount of proppant in formation. The range of stress values in the undrained floor and strike-slip regime were similar to those observed following DFIT testing in an offsetting area (Pallikathekathil *et al*, 2013). However, localised mining was believed to be the result of lowered stress in the 'floor' or area below the seam (and likely some in the 'roof' area) unlike published profiles, which were in undrained areas east of the MNM.

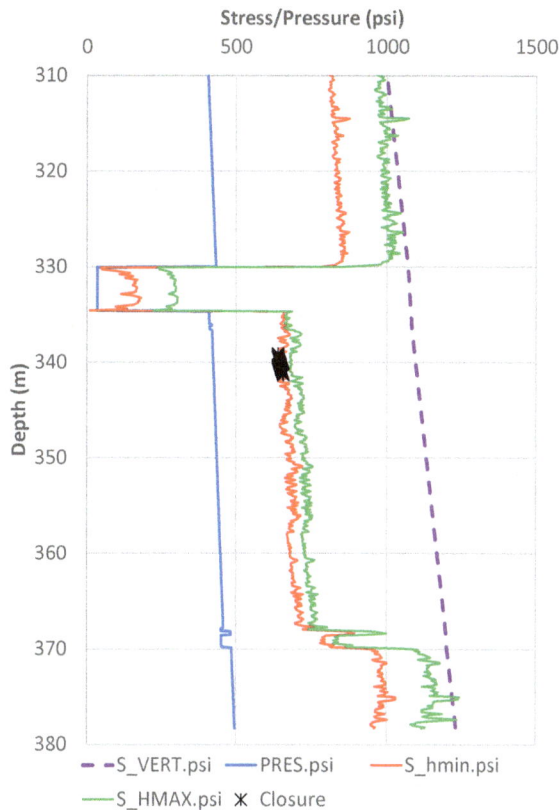

FIG 1 – Development of log-derived 1D stress profile for S0774 based on DFIT and current reservoir pressure (Johnson Jr *et al*, 2021).

The fracture zone surrounding the treatment borehole SO774 was eventually exposed during the longwall retreat and identification of fractures by Geotechnical Engineer inspections. Unmistakable evidence of intersections along the coalface (see black dots, Figure 2) indicated a dominant fracture and orthogonal fractures, likely as the result of activating the natural cleat network.

The trend of the fracture remained generally aligned in the general maximum horizontal stress direction (σ_{HMax}), generally NNE to NE, despite significant depressurisation and stress relaxation in the coal creating near isotropic horizontal stress values at this location. This is believed to be the result of permeability anisotropy of the dominant natural fracturing in the area also striking the NNE to NE direction. The width of the fracture region also closely resembled the width of microseismic responses observed during a prior hydraulic fracturing treatment in another area.

So, the first vertical well trial established confidence in the IHF process and the extent of the propped fracture that would be created. The model was able to history-match the observed pressures (see Figure 3) with some confidence using the DFIT parameters; however, the resulting dimensions were longer and more planar than the mine back data (see Figure 4).

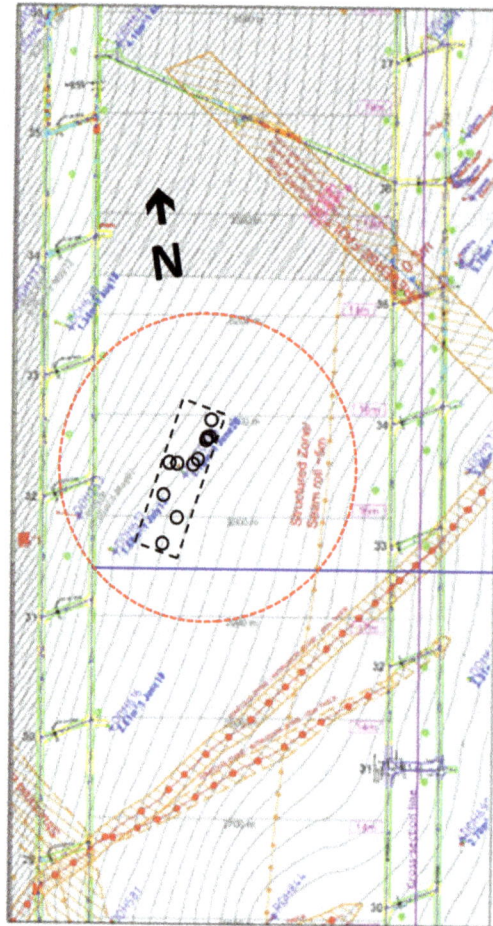

FIG 2 – Orientation and longwall interceptions (black dots) of the hydraulic fracturing in S0774 (Johnson Jr *et al*, 2021).

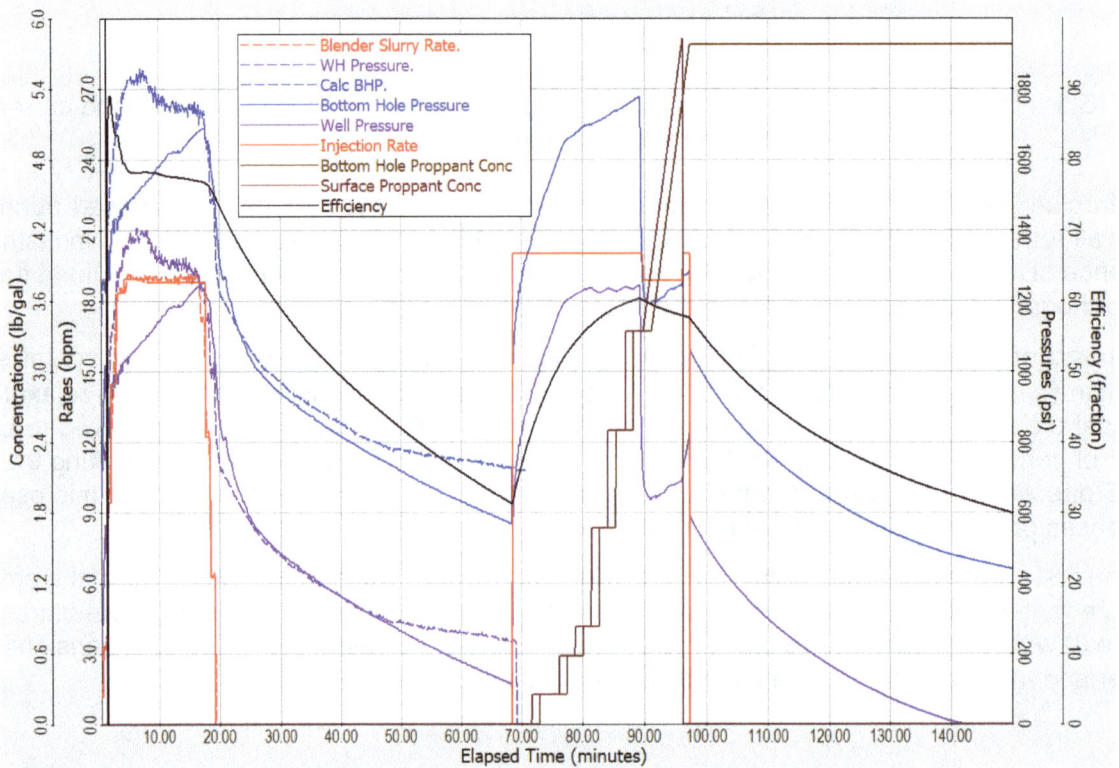

FIG 3 – Planar 3D hydraulic frac model history-match of observed pressures, S0774 (Johnson Jr *et al*, 2021).

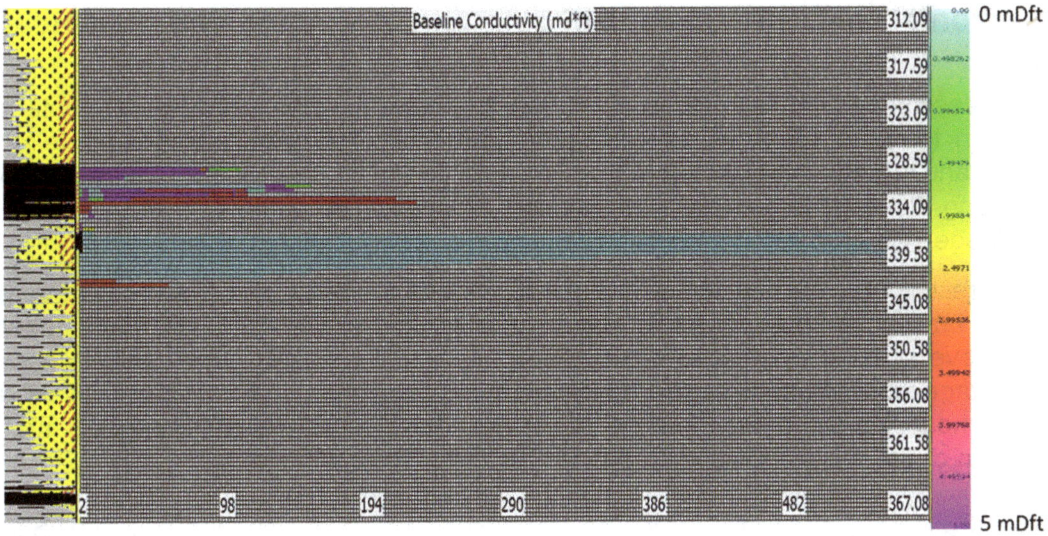

FIG 4 – Planar 3D hydraulic frac model predicted dimensions and conductivity based on history-match, S0774 (Johnson Jr *et al*, 2021).

With this success, a second vertical well SO782 was drilled and stimulated successfully with the IHF process (see Figure 5). The treatment zone in SO782 was also successfully intersected by development mining operations with exposure of fractures aligned to the general σ_{HMax} (NNE to NE) direction. In contrast to the intersection of the previous SO774 treatment zone, the longwall retreat exposure in SO782 was the subject of a several-day period. The second IHF trial fracture zone has remained exposed by mining development activities since its intersection in October 2020 and open roadway intersections have remained structurally stable, belaying geotechnical concerns.

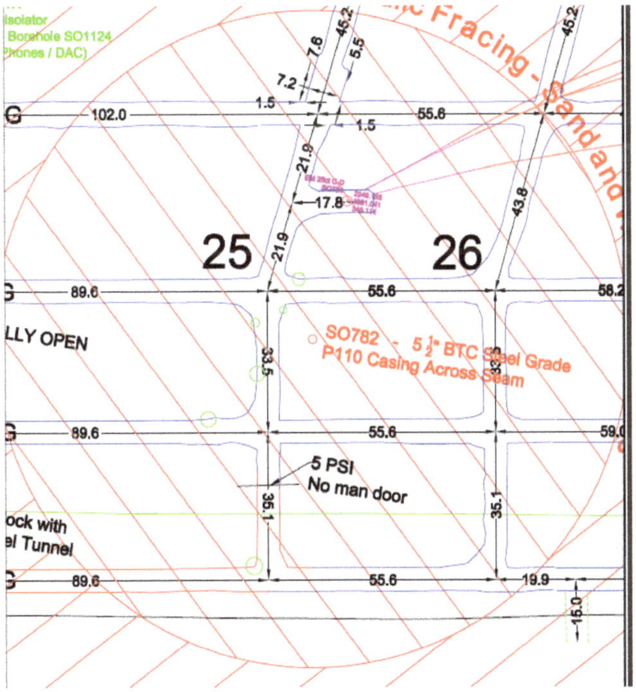

FIG 5 – Frac impact zone S0782 vertical well IHF second trial (Johnson Jr *et al*, 2021).

Additionally, evidence of fracture zone interaction with the coal seam underlying the GM coal seam (GML or GM Lower, see Figure 5) was identified in SO782 by geotechnical and gas monitoring compliance inspections. This data provided unmistakable evidence of intersections with lower seams, indicating potential destressing of lower seams and resulting in their release of gas content, another beneficial outcome of IHF stimulation in the floor.

Initial horizontal well (SG 608-01) IHF planning, execution and post-treatment evaluation

After two successful vertical well IHF applications, the next application was a horizontal well IHF implementation, the SG806 well, which was planned and executed below a planned gate road in the longwall mine and targeting the GM seam from a lateral underlying the seam. This application would be amid existing SIS wells to ascertain the potential co-application of IHF and SIS to enhance mine gas pre-drainage.

Based on the results from the first vertical well trials, the lateral was targeted just below the GM seam based on a 2–4 m standoff (see Figure 6). Based on analytical, probabilistic, reservoir modelling and fracturing, sleeve spacing was set at 40 m with omissions in areas of geotechnical concern. The resulting 23-stage job was based largely on the pumping schedule used in the S0774 well. The pre-frac reservoir modelling determined that the overall planned fracture dimensions would intersect the existing pattern of SIS wells and assist pre-drainage (see Figures 7 and 8).

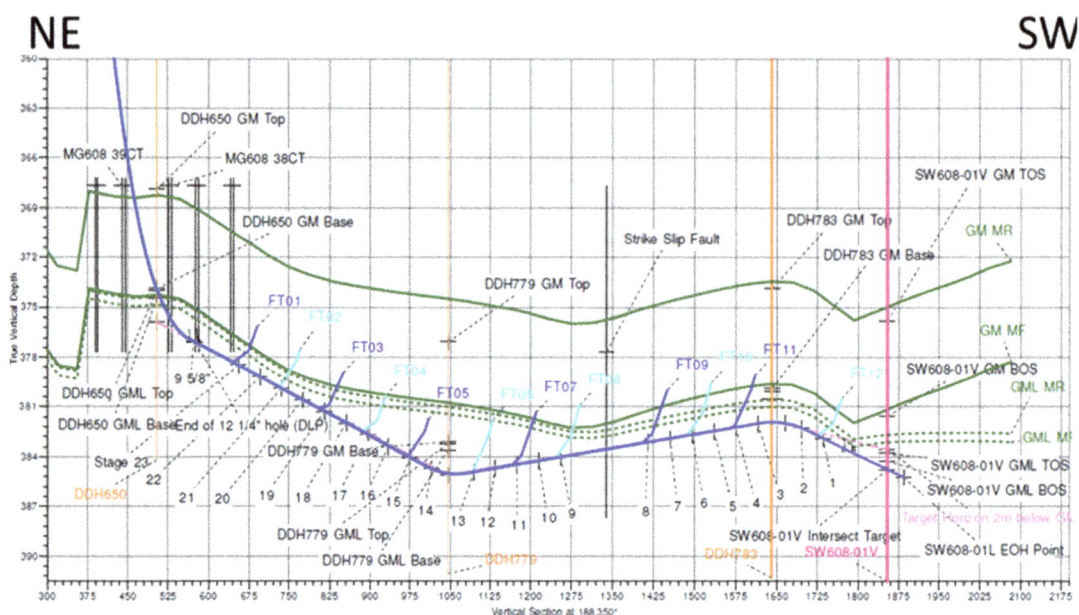

FIG 6 – SG608-01 well trajectory (blue line), GM Seam (dark green line) and intercept vertical well (magenta line) (Johnson Jr *et al*, 2021).

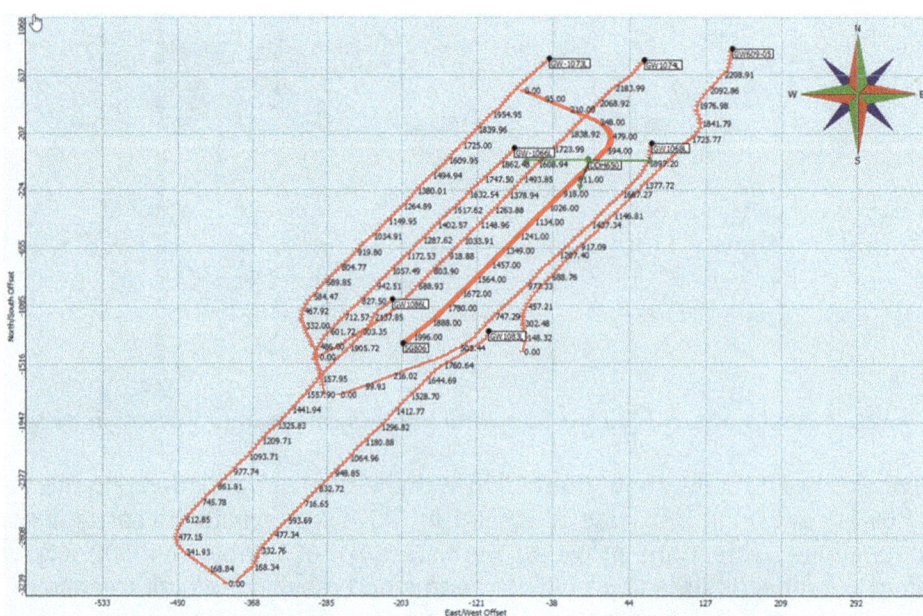

FIG 7 – SG806 well trajectory (darker solid red line) and offsetting SIS wells [red hashed dashed lines) in map view (Johnson Jr *et al*, 2021).

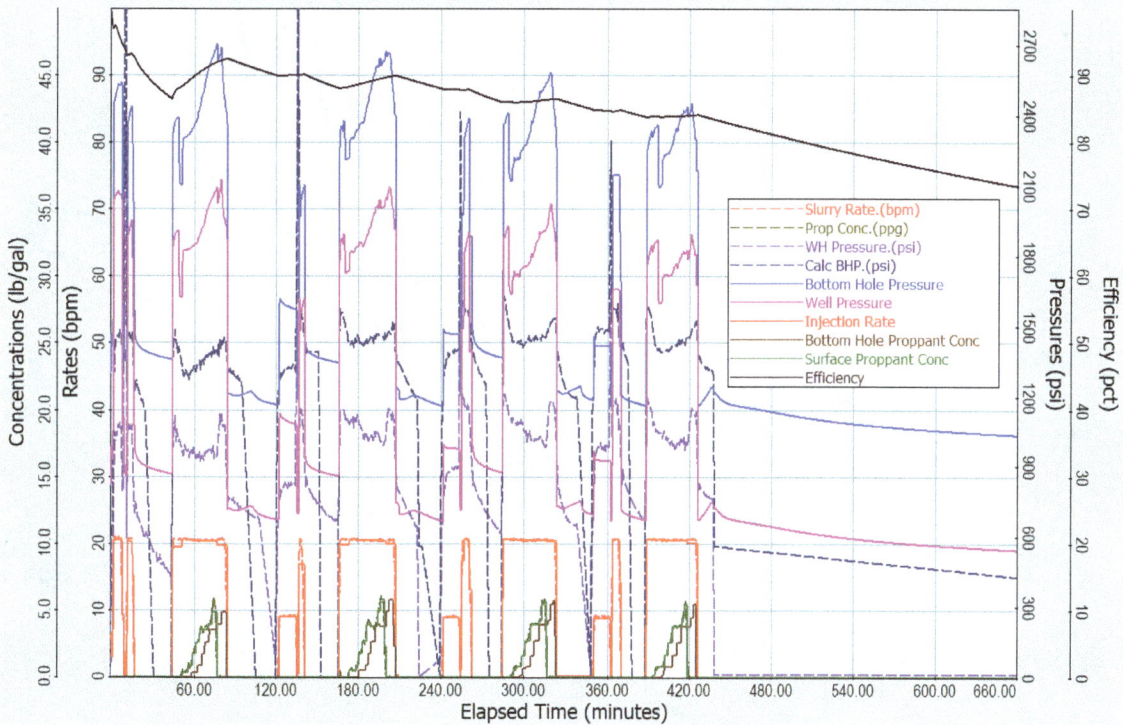

FIG 8 – Modelled versus observed pressures during SG608-01 day 2, stages 4–7 (Johnson Jr *et al*, 2021).

SG 608-01 hydraulic fracture pressure history matching and results

Before the fracturing treatment, a DFIT was performed at the first sleeve location to calibrate the stress model, adjust the leak off, and make any pre-job modifications necessary. As this treatment was being performed on a mine site, limited access precluded coiled tubing from being readily accessible in the case of a screen out. As such, a thorough pre-job QC process was implemented, a decision protocol for each frac execution in the event of abnormal pressure responses was developed, and a pre-frac mini-frac was implemented for each treatment. This allowed us to review each job, adjust any volumes based on observed fracture efficiency and pressures, and assure that proppant concentrations of proppant >2.0 lb/gal sand added could be effectively placed.

Using this stress profile, 23 stages were successfully placed over seven days. The modelled pressures correlated well on some stages with the observed pressures on a stage-by-stage basis and varied positively and negatively on others; an example plot of the day 2 treatments can be seen in Figure 8. This variation could be attributed to several reasons including effects between the clastic and coal interface, the fact that some intervals were connected to the coal through well branches (used for steering and depth control to the seam) and variability in the underlying clastics not effectively captured by a single reference well log used to populate the planar 3D model grid. For example, in Figure 8, the mini-frac shows lower observed pressures than the model. Stage 4 was adjoining a drilling branch into the coal (see Figure 6) and the fracture may have immediately closed upon shut-in. Thereafter, the model showed reasonably good conformance at the end of the job. Notwithstanding, in almost all cases the model predicted surface and calculated BHTP values exceeded the observed pressures even with large reductions in frictional components.

All stages were modelled, and the fracture lengths were determined based on a cut-off of >0.5 lb/ft² or >5 mDft conductivity. To evaluate the post-fracture results microseismic data was collected through the job from a vertical well, located SE of the toe. These data indicated a low distribution of events in the GM Seam and a predominant distribution of events considerably higher and lower than in the GM Seam. The resulting fracture dimensions, locations of offsetting SIS wells, and microseismic responses can be seen in Figure 9. The post-frac dimensions met or exceeded the pre-frac modelled dimensions (see Figure 10).

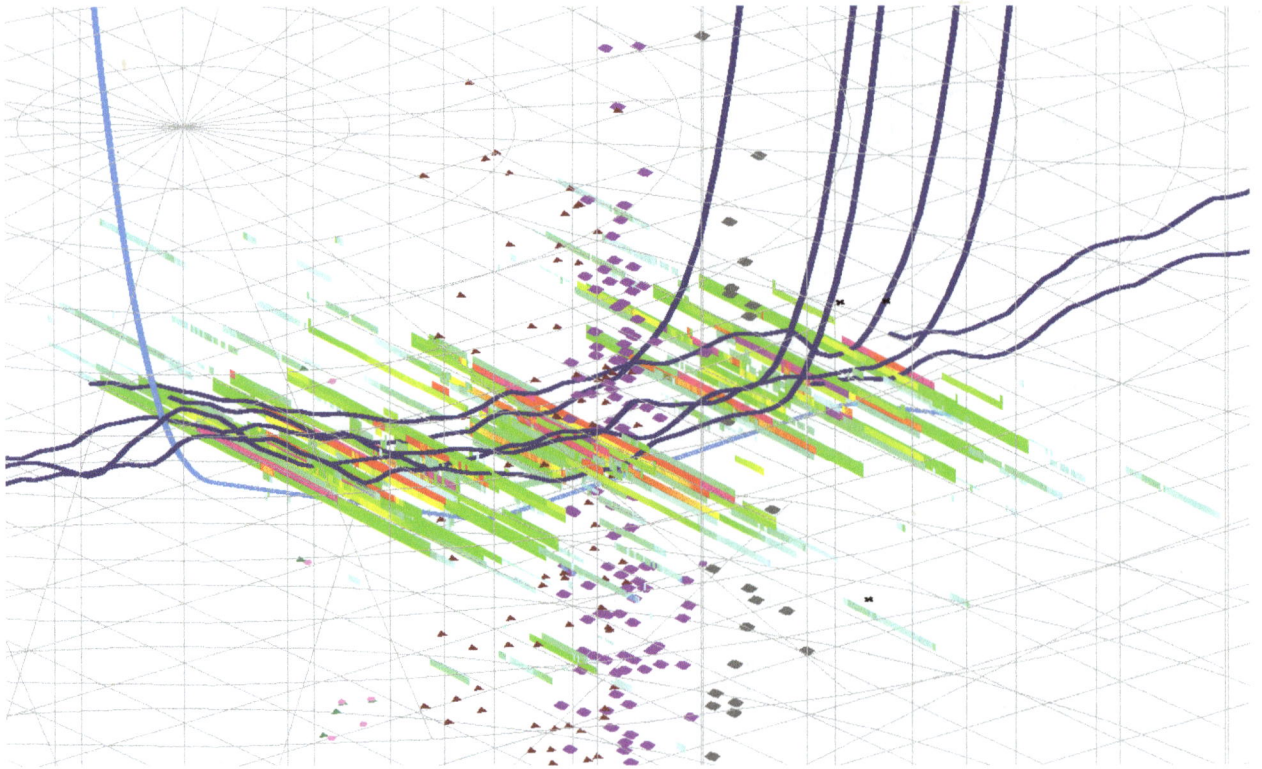

FIG 9 – SG608-01 well and offsetting SIS wells trajectory with hydraulic fracture conductivity and microseismic events (Johnson Jr *et al*, 2021).

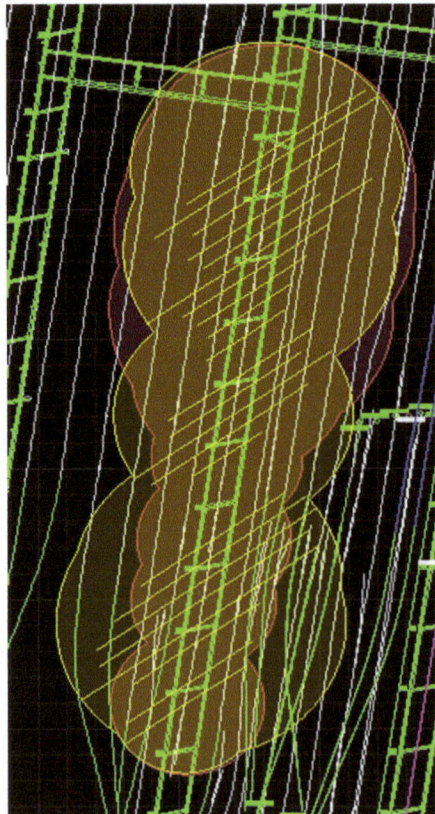

FIG 10 – SG608-01 well trajectory, post-frac modelled hydraulic fracture lengths and SRV (yellow lines and yellow shape), SIS wells (white lines), planned mine gate road (green lines), and post-frac verified SRV based on offsetting well pressure events (red shape) (Johnson Jr *et al*, 2021).

Low incidences of microseismic events in the coal and anomalous events outside the coal have been noted by diagnostics in the WCM as well as examples in North America (Barree, personal communication, 2009; Flottman *et al*, 2013; Johnson Jr *et al*, 2010; Zimmer, 2010). Johnson Jr *et al* (2021) hypothesised that most of these events were unrelated to the fracturing treatment or shear events related to the high degree of depressurisation and destressing of the lower and upper clastics around the GM Seam because of the adjoining mineworks. In the SG609 case, less noise and 'shadowing' above the coal was observed, likely as a result of less depletion and destressing.

608 panel post-frac reservoir modelling and results

Throughout the treatment, pressure, fluid, and sand influx were noted in varying SIS and pressure observation wells during various stages; therefore, interconnections to the SIS wells were successful. The IHF wellbore was only recently turned on to production and is still cleaning up fluids.

The offsetting SIS wells had prior production, and some are awaiting cleanouts at the time of writing. However, some SIS wells are in production and are believed to have shown an improvement. As an example, an offsetting SIS well production is illustrated in Figure 11 and shows:

- the actual normalised gas rate (red solid line, normalised to maximum rate)

- the post-frac normalised forecasted gas rate (dashed red line) based on an analytical model history match done before the frac operations

- the model estimated pre-frac BHFP (dashed grey line) and the modelled post-frac BHFP (black line).

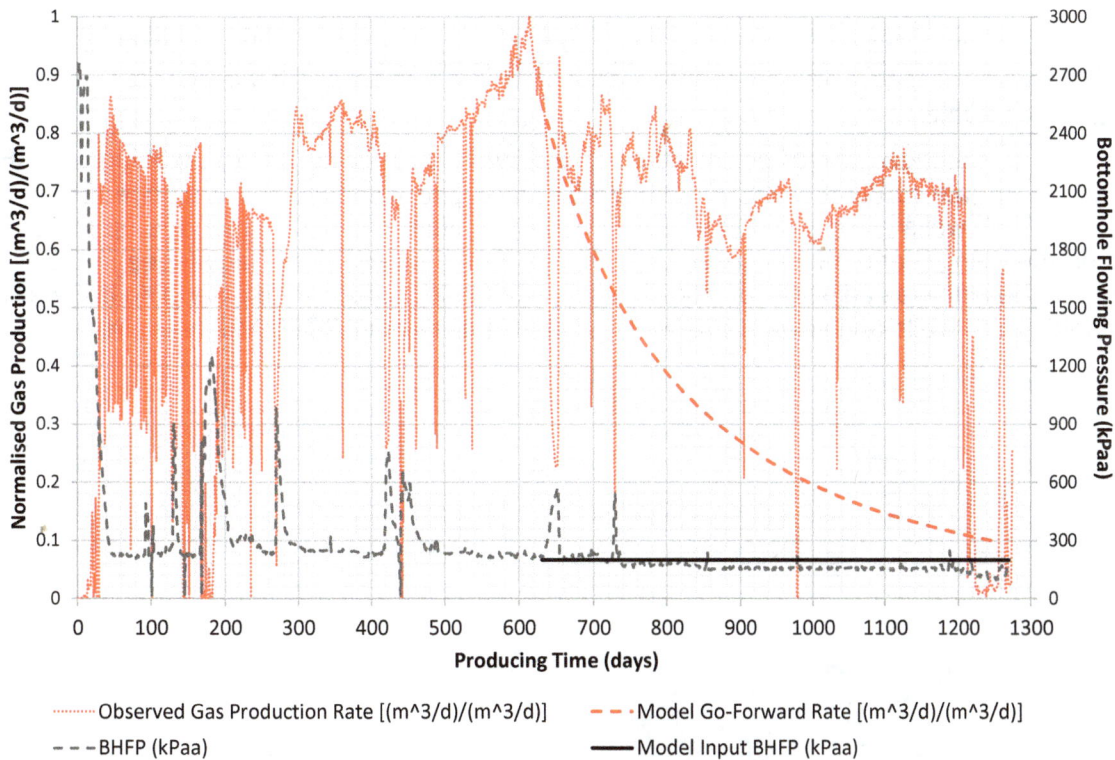

FIG 11 – Production history from SIS well offsetting SG608-01 Well, then model-predicted actual versus forecast gas production.

The pre-frac model history match is anchored on the observed gas rate data, honouring the pre-frac BHFP; the modelled rates post-frac show the expected production, and the forecast does not match observed gas rates post-frac and would not be achievable in the model. Thus, the frac is believed to have improved the SIS well lateral length, increased permeability, decreased skin, or increased drainage area.

Other wells indicated anecdotal decreases in decline, but operational difficulties and inability to work over offsetting SIS wells post-frac meant that UIS drilling was required to achieve the desired drainage targets. Finally, anecdotal evidence from UIS wells in the panel adjoining the initial IHF well indicate that the panel may have been stimulated based on an increase of both UIS and the IHF well productivity relative to offsetting and unaffected SIS production wells (see Table 2). Further, the 1st year production of the IHF well approaches the same volume as the worst-performing UIS with similar permeability and pressure data.

TABLE 2

Normalised Comparison of Well/UIS patter production on adjoining panel to initial IHF well.

Well/UIS pattern	Production start date	Days on production	Normalised average flow rate per day [(m³/day) / m³/day)]	Normalised cumulative volume (m³/m³)
SIS 1	26/03/2019	1427	12%	30%
SIS 2	5/04/2019	1417	40%	98%
UIS 1	27/12/2020	339	74%	100%
IHF Well[1]	21/12/2021[2]	781	51%	30%
UIS 2	18/08/2021	550	89%	84%
UIS 3	18/11/2021	460	100%	79%

1 – Available data only available for the first 339 days of production post frac and dewatering; 2 – Fracture stimulation was performed 18/2/2021 before installation of UIS patterns 2 and 3 and shortly after pattern 1.

SG609-02 HYDRAULIC FRACTURE PRESSURE MODELLING, TREATMENT PRESSURE HISTORY MATCHING AND OFFSET WELL PRESSURE OBSERVATIONS

Progressing a drainage plan of pre-positioning SIS wells in ongoing panels, placing an IHF well under planned gate roads and implementing UIS drilling to complete the drainage. Thus, Anglo planned two IHF wells along the 609 panel. These wells were planned to manage up to 30 stages based on their length and the successful spacing of 50m between prior frac stages in SG 608-01. Eventually, the plan was adjusted to only stimulate SG609-02, a northward extending IHF well along two offsetting SIS wells and several SIS wells south with toes near SG609-02 (see Figure 12).

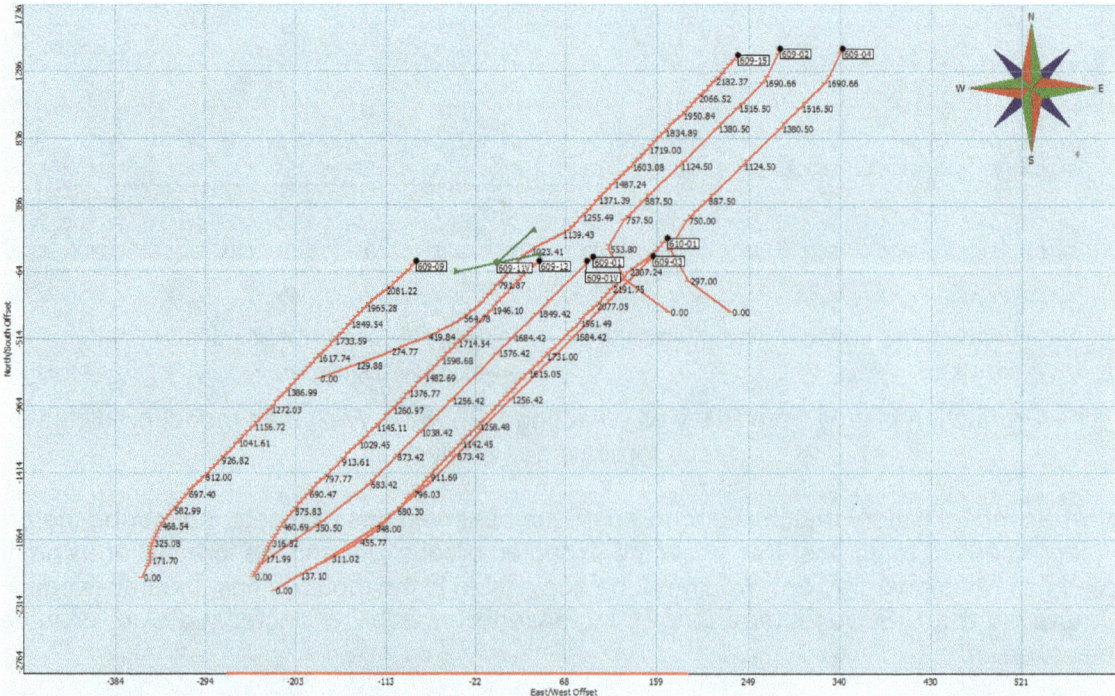

FIG 12 – SG609-02 well trajectory (top centre wellbore) and two directly offsetting SIS wells and offsetting heel wells in map view.

Hydraulic fracture model construction and well trajectory

Rock mechanical properties and geologic data were derived from offsetting vertical well log data (see green arrows, Figure 12), sonic velocity relationships from previous 1D stress profiling (previous and offsetting CSG 1D stress profiling (Johnson Jr *et al*, 2021; Pallikathekathil *et al*, 2013)), and physical mapping of the offsetting well trajectories within the coal seam. Tectonic strains were used from prior 1D stress profiles. A DFIT was performed from the toe and first sleeve. The toe sleeve communicated with the vertical well; however, a clear analysable closure pressure was observed in Stage 1. However, for this case, the 1D stress profile required adjustment to account for the lack of depletion in the 609 panel (see Figure 13).

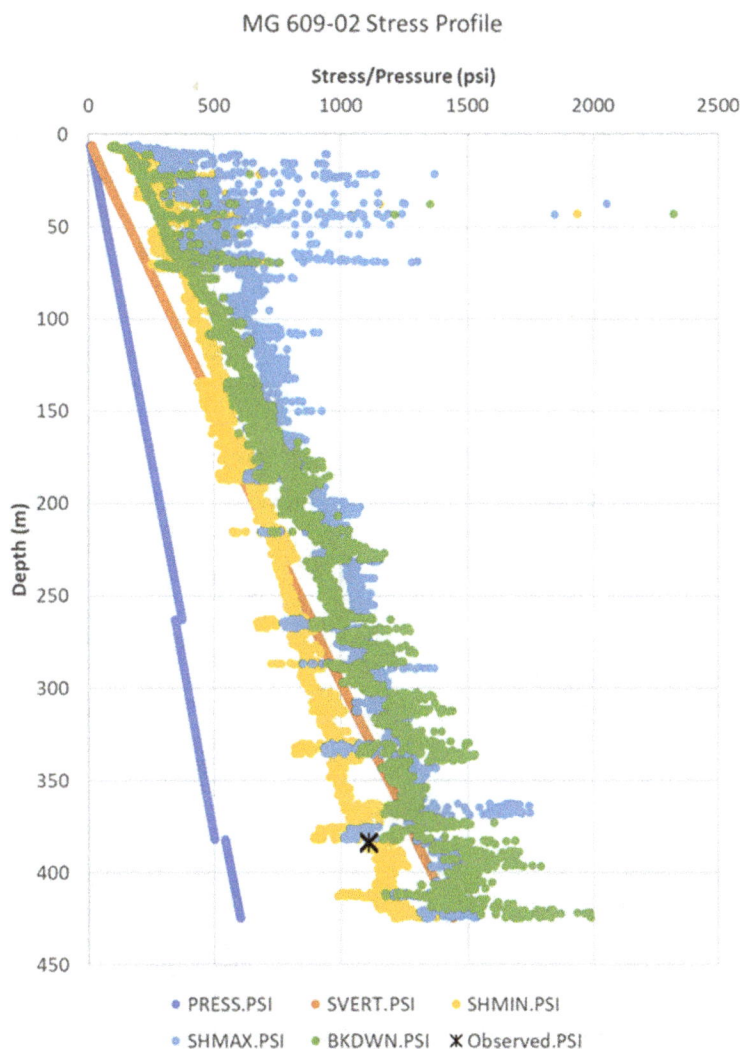

FIG 13 – Development of log-derived 1D stress profile for 609-02 based on observed DFIT closure (black asterisk) and estimated reservoir pressure with minimal depletion (dark blue trace).

As with the previous case, 609-02 was drilled 2–3 m below the GM seam, intercepted a vertical well and had 12 vertical floor touches to aid drilling and fracture interconnection (see Figure 14). It should be noted that the greatest separation between the IHF well and the seam was in the middle of the hole; this may be the cause of the greatest deviation between observed and modelled post-frac pressures, described in the next section.

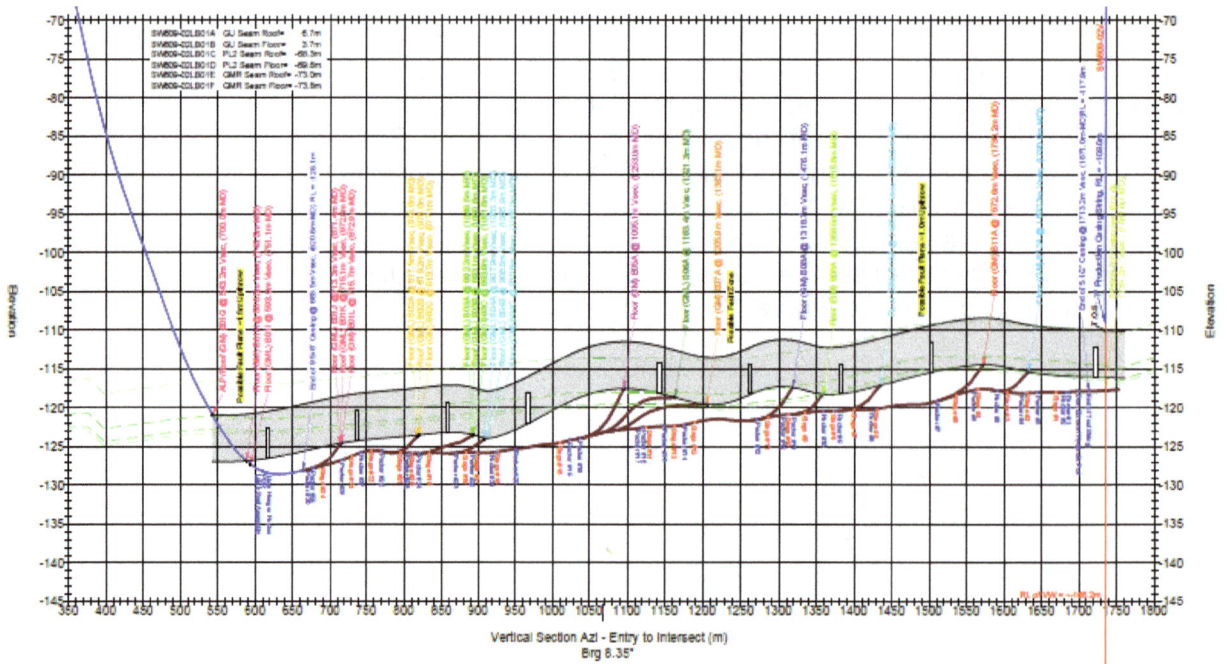

FIG 14 – SG609-02 well trajectory (blue line), GM Seam (grey region) and intercept vertical well (vertical blue line near the toe) (Johnson Jr *et al*, 2021).

Treatment pressure history-matching and propped dimensions

One fluid injection and 23 proppant-laden stages were pumped in the SG 609-20 well. The treatment pressures were generally higher than those observed in stimulating the previous IHF lateral well, likely as a result of lower pre-frac depletion in this well. Days 1–4 or the first 16 stages closely resembled the predicted pressures. Figure 15 illustrates Day 4 or stages 10–16 where observed surface and model-predicted pressures closely matched with values juxtaposing higher and lower positions between successive treatments.

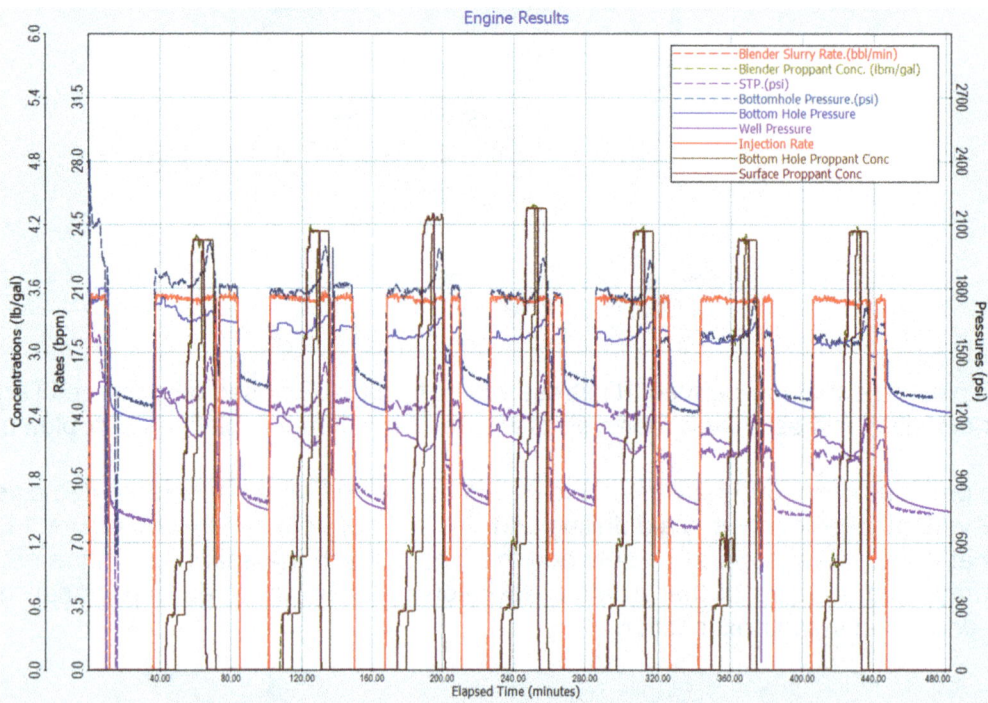

FIG 15 – Treatment pressure history-match of observed surface and model-predicted pressures (dashed and solid magenta lines, respectively, surface and downhole proppant (dark brown and light brown/green dashed lines), the injection rate (red lines), and model-predicted and surface calculated bottom hole treating pressures (BHTP, solid and dashed dark blue lines, respectively).

Days 1, 2, 4 and 5 all produced good transverse fractures achieving the targeted 4 lb/gal proppant concentration resulting in a downhole concentration >0.5 lb/sqft (see Figure 16). This target value was based on prior lab testing of varying concentrations in Bituminous coal to adequate fracture conductivity (Fraser and Johnson Jr, 2018). The model predicted for Day 3 (stages 5–9) some longitudinal components were created in addition to a few transverse fractures. This region corresponded to the region of greatest separation between the seam and the IHF well and was consistent with the yellow traces of microseismic events beginning in stage 5 that are closely adjacent to the IHF wellbore (see Figures 17 and 18).

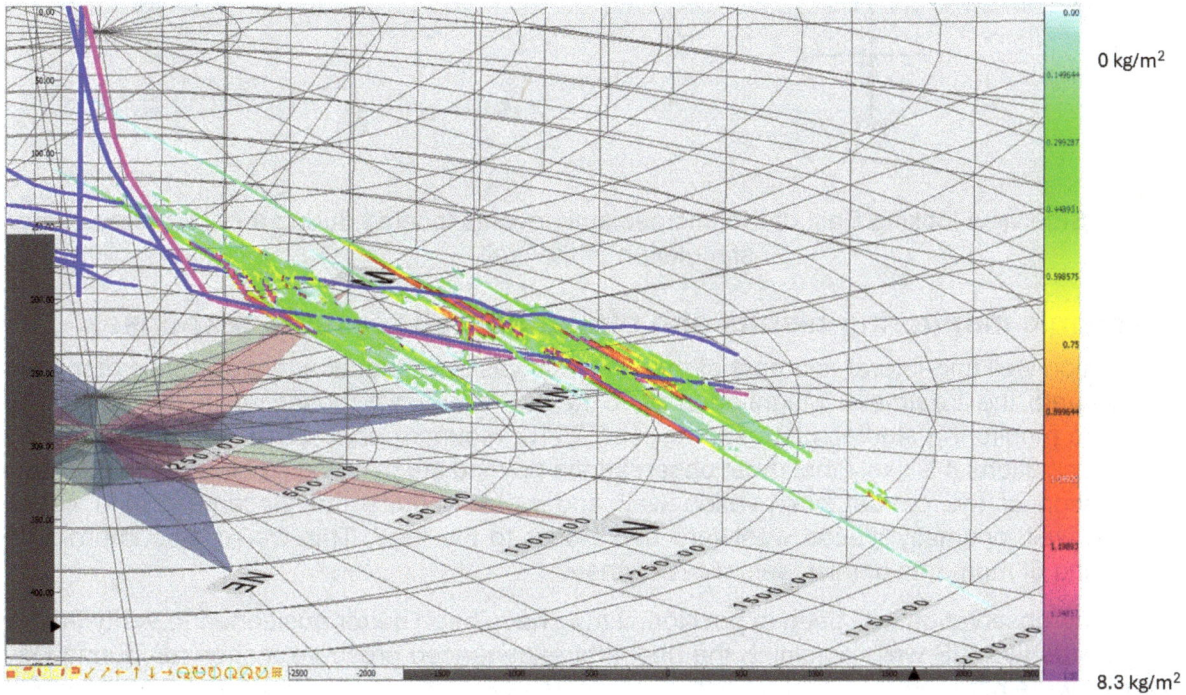

FIG 16 – Graphical representation and distribution of propped fractures around IHF well (magenta) and adjoining SIS wells (blue).

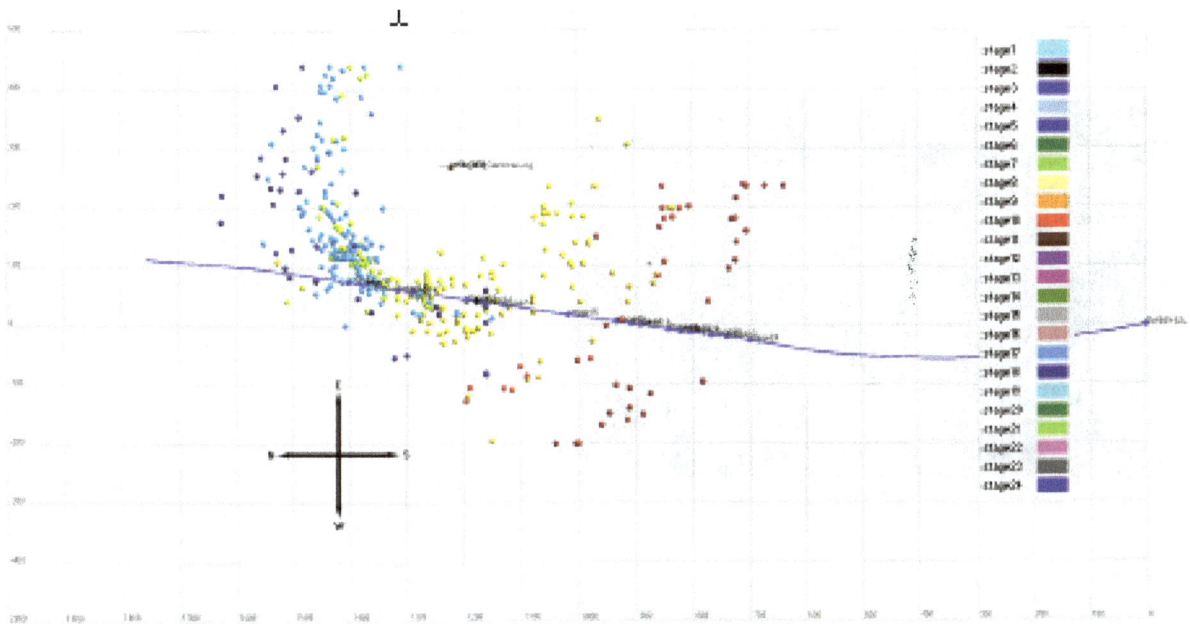

FIG 17 – Map view looking East of the microseismic events located during the 24 stages of the stimulation of the IHF well.

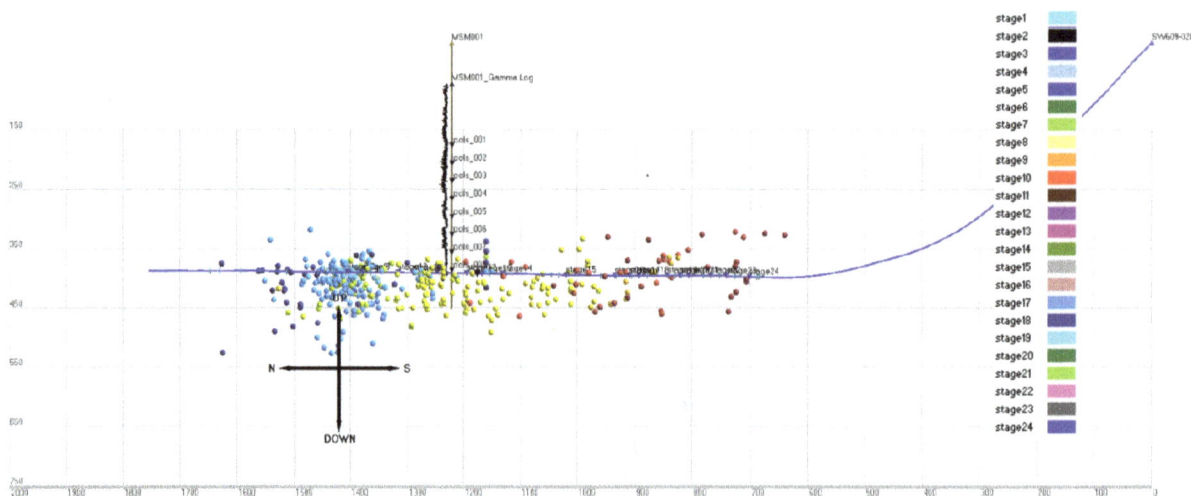

FIG 18 – Side view looking East of the microseismic events located during the 24 stages of the stimulation of the well.

Microseismic monitoring results and offsetting well pressure responses

An observation well mid-lateral and east of the IHF well was selected as a microseismic monitoring well to evaluate the treatment. Magnitudes of -3 to -5 are typically observed in published CSG microseismic monitoring studies (Johnson Jr *et al*, 2010; Flottman *et al*, 2013). Based on pre-well signal determinations it is likely that most observations beyond stage 16 could be outside the radius of investigation of the current observation well and would require the placement of a second observation well, optimally placed westward and closer to the heel. This would also improve the understanding of moments of microseismic responses.

Had a second observation well been in position, it may have been better positioned to verify fracture growth towards two SIS wells adjoining the heel that experienced pressure responses during Days 2 and 4 that were not predicted by the fracture model (see Figures 19 and 20). Conjugate fracturing and fracture complexity are not uncommon in coal frac treatments (Jeffrey *et al*, 1992, 1998; Jeffrey, Settari and Smith, 1995; Johnson Jr *et al*, 2010; Mills and Jeffrey, 2006; Mills, Jeffrey and Gale, 2007; Scott *et al*, 2010). Further, small-width fractures may extend to great lengths but may not remain open without the implementation of small micro-proppants (ie <50 mM particle sizes) (Keshavarz *et al*, 2016, 2014b).

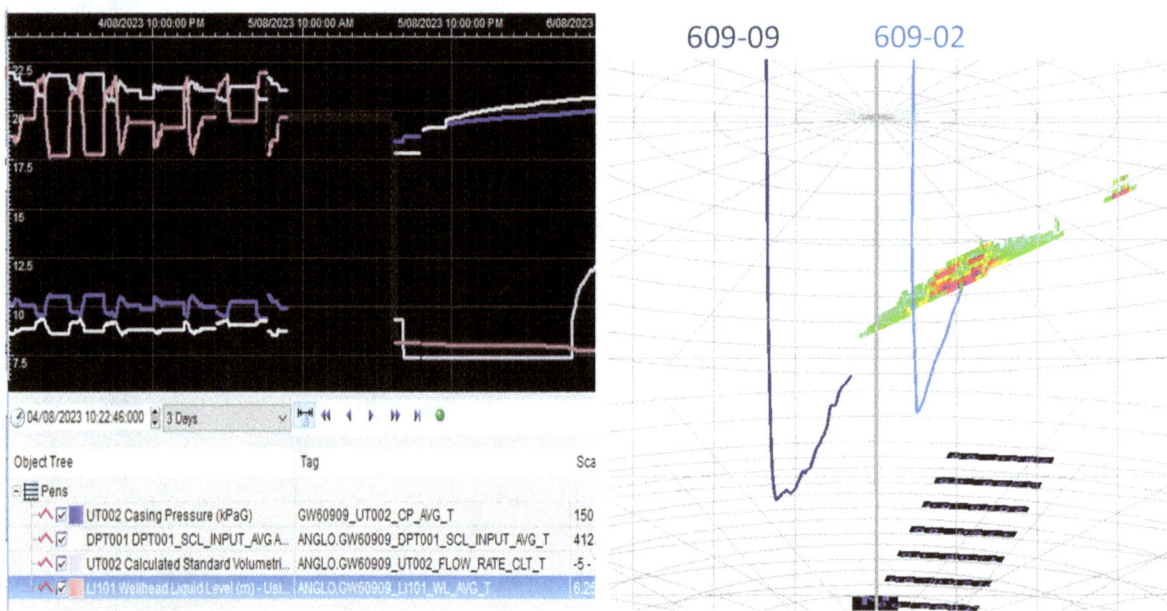

FIG 19 – Pressure responses during day 2, stages 2–4 (left above) in an offsetting SIS well 609-09 (right above).

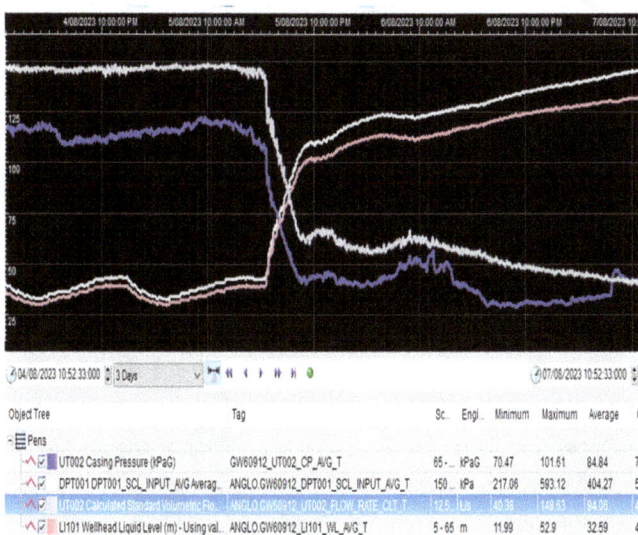

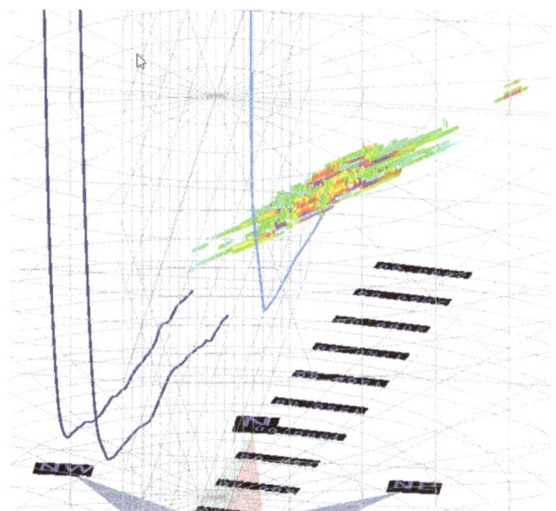

FIG 20 – Pressure responses during day 4, stages 5–9 (left above) in an offsetting SIS well 609-12 (right above).

As aforementioned, two SIS wells adjoining the heel experienced pressure responses during Days 2 and 4 that were not predicted by the fracture model (see Figures 19 and 20, respectively). On Day 2, 609-09 experienced a pressure response early in the day. Proppant profiles did not indicate that the 'frac hit' was likely to be proppant-laden (see Figure 19). Unfortunately, the two adjoining wells to the IHF well did not have pressure monitoring; however, there was a strong indication that the wells were interconnected by propped fractures by model dimensions (see Figure 16) and microseismic events (see Figure 17). Similarly, 609-12 experienced a pressure response early on Day 3 (see Figure 20). Recall that this stage had indications of longitudinal components that may not be adequately mapped as being propped on Figure 20 but may be closely aligned with the heel-oriented 609-12 lateral. This illustrates the value of pre-configuring the adjoining SIS wells with pressure monitoring. Further, offsetting fibre-optic configured laterals may give greater confidence than microseismic wells on fracture propagation by measuring strains created by the fractures, akin to surface deformation tiltmeter measurements (Bourne *et al*, 2021).

CONCLUSIONS

Whilst the process is relatively new and the cost to benefits are still being studied, there are positive signs in the production, treatment observations and diagnostic data that a more effective SRV is being established that encompasses the IHF and adjoining SIS wells.

Whilst only three treatments have been pumped in Australia, all treatments have been initiated from strike-slip regimes into normally stressed coals and all stages effectively placed, except when over displacement has occurred. In over 50 stages pumped between the three wells to date, there has been little evidence of impending screen-out conditions. All IHF treatments have some near-wellbore tortuosity based on the process and indicate some proppant drag effects at >2 lb/gal concentration. The authors believe the process has several potential applications and improvements including:

- Better drainage in areas where direct placement of SIS wells is difficult or unstable.

- Overlying implementation where seam floors are unstable or if tuffaceous siltstones are present.

- Stimulation of floor or rider seams to improve drainage and reduce goaf influx post-mining.

- Potential use of micro-proppants to extend the effective SRV based on laboratory and modelling studies that indicate three-fold improvements are likely on an individual fracture basis (Keshavarz *et al*, 2014a, 2014b) and improved performance in a pattern of SIS wells (Ramanandraibe *et al*, 2022, 2023).

- Optimisation of SIS/IHF well placements to more rapidly drain low permeability areas by increasing hydraulic fracture interconnection and complexity. This could be achieved by adjoining IHF week placements and simultaneous treatments akin to the 'zipper frac' process used in North American shale gas treatments (King and Leonard, 2011; Sierra and Mayerhofer, 2014).

ACKNOWLEDGEMENTS

The authors would like to recognise Novus Fuels and Anglo American for support and permission to publish this work. The authors also want to recognise the efforts of the Anglo field support staff, Will Corbett and the Condor frac crews that make it happen in the field.

REFERENCES

Barree, R D, 1983. A Practical Numerical Simulator for Three-Dimensional Fracture Propagation in Heterogeneous Media, paper presented at the SPE Reservoir Simulation Symposium, SPE 12273.

Bourne, S, Hindriks, K, Savitski, A A, Ugueto, G A and Wojtaszek, M, 2021. Inference of Induced Fracture Geometries Using Fiber-Optic Distributed Strain Sensing in Hydraulic Fracture Test Site 2, paper presented at the SPE/AAPG/SEG Unconventional Resources Technology Conference, paper URTeC 5472. https://doi.org/10.15530/urtec-2021-5472

Branajaya, R, Archer, P and Farley, A, 2019. Pushing the boundaries – deployment of innovative drilling, completion and production technology to advance a deep coal seam play, Extended Abstract AJ18115, *the APPEA Journal*, 59(2):770–775. https://doi.org/10.1071/AJ18115

Flottman, T, Brooke-Barnett, S, Trubshaw, R L, Naidu, S K, Paul, P K, Kirk-Burnnand, E, Paul, P, Busetti, S and Hennings, P, 2013. Influence of in Situ Stresses on Fracture Stimulation in the Surat Basin, Southeast Queensland, paper presented at the SPE Unconventional Resources Conference and Exhibition-Asia Pacific, paper SPE-167064-MS.

Fraser, S A and Johnson Jr, R L, 2018. Impact of Laboratory Testing Variability in Fracture Conductivity for Stimulation Effectiveness in Permian Deep Coal Source Rocks, Cooper Basin, South Australia, paper presented at the SPE Asia Pacific Oil and Gas Conference and Exhibition, paper SPE-191883-MS. https://doi.org/10.2118/191883-MS

Jeffrey, R G, Byrnes, R P, Lynch, P J and Ling, D J, 1992. An Analysis of Hydraulic Fracture and Mineback Data for a Treatment in the German Creek Coal Seam, paper presented at the SPE Rocky Mountain Regional Meeting, SPE 24362.

Jeffrey, R G, Settari, A and Smith, N P, 1995. A Comparison of Hydraulic Fracture Field Experiments, Including Mineback Geometry Data, with Numerical Fracture Model Simulations, paper presented at the SPE Annual Technical Conference and Exhibition, SPE 30508.

Jeffrey, R G, Vlahovic, W, Doyle, R P and Wood, J H, 1998. Propped Fracture Geometry of Three Hydraulic Fractures in Sydney Basin Coal Seams, paper presented at the SPE Asia Pacific Oil and Gas Conference and Exhibition, SPE 50061.

Johnson Jr, R L and Mazumder, S, 2014. Key Factors Differentiate the Success Rate of Coalbed Methane Pilots Outside of North America - Some Australian Experiences, presented at the International Petroleum Technology Conference, paper IPTC-18108-MS.

Johnson Jr, R L, Cheong, S and Farley, A, 2020. Characterizing the Application of Horizontal Wells and Indirect Hydraulic Fracturing for Improved Coal Seam Gas Drainage, *The APPEA Journal*.

Johnson Jr, R L, Ramanandraibe, H M, Ribeiro, A, Ramsay, M, Kaa, T and Corbett, W, 2021. Applications of Indirect Hydraulic Fracturing to Improve Coal Seam Gas Drainage for the Surat and Bowen Basins, Australia, paper presented at the SPE/AAPG/SEG Asia Pacific Unconventional Resources Technology Conference, paper AP-URTEC-2021–208375. https://doi.org/10.15530/AP-URTEC-2021-208375

Johnson Jr, R L, Scott, M, Jeffrey, R G, Chen, Z Y, Bennett, L, Vandenborn, C and Tcherkashnev, S, 2010. Evaluating Hydraulic Fracture Effectiveness in a Coal Seam Gas Reservoir from Surface Tiltmeter and Microseismic Monitoring, paper presented at the SPE Annual Technical Conference and Exhibition, SPE 133063.

Keshavarz, A, Badalyan, A, Carageorgos, T, Johnson Jr, R L and Bedrikovetsky, P, 2014a. Stimulation of Unconventional Naturally Fractured Reservoirs by Graded Proppant Injection: Experimental Study and Mathematical Model, paper presented at the SPE/EAGE European Unconventional Conference and Exhibition, SPE 167757.

Keshavarz, A, Johnson, R, Carageorgos, T, Bedrikovetsky, P and Badalyan, A, 2016. Improving the Conductivity of Natural Fracture Systems in Conjunction with Hydraulic Fracturing in Stress Sensitive Reservoirs, presented at the SPE Asia Pacific Oil & Gas Conference and Exhibition, paper SPE-182306-MS.

Keshavarz, A, Yang, Y, Badalyan, A, Johnson Jr, R L and Bedrikovetsky, P, 2014b. Laboratory-based mathematical modelling of graded proppant injection in CBM reservoirs, *International Journal of Coal Geology*, 136:1–16. https://doi.org/10.1016/j.coal.2014.10.005

King, G E and Leonard, D, 2011. Utilizing Fluid and Proppant Tracer Results to Analyze Multi-Fractured Well Flow Back in Shales: A Framework for Optimizing Fracture Design and Application, paper presented at the SPE Hydraulic Fracturing Technology Conference, Paper SPE-140105-MS. https://doi.org/10.2118/140105-MS

Mills, K and Jeffrey, R G, 2006. Hydraulic Fracturing in Coal Mining, ACARP Report C10010.

Mills, K, Jeffrey, R G and Gale, W, 2007. Implications of Production Scale Hydraulic Fracturing by Coal Bed Methane Drainage Operations on the Subsequent Safe and Efficient Mineability of Coal Seams, ACARP Report C14011.

Olsen, T N, Bratton, T R, Tanner, K V, Donald, A and Koepsell, R, 2007. Application of Indirect Fracturing for efficient Stimulation of Coalbed Methane, presented at the Rocky Mountain Oil & Gas Technology Symposium, paper SPE-107985-MS (Society of Petroleum Engineers).

Pallikathekathil, Z J, Puspitasari, R, Altaf, I, Alboub, M, Mazumder, S, Sur, S, Scott, M and Gan, T, 2013. Calibrated Mechanical Earth Models Answer Questions on Hydraulic Fracture Containment and Wellbore Stability in Some of the CSG Wells in the Bowen Basin, presented at the SPE Unconventional Resources Conference and Exhibition-Asia Pacific, paper SPE-167069-MS.

Ramanandraibe, H M, Johnson, R L, Sedaghat, M and Leonardi, C R, 2023. Co-Application of Indirect Hydraulic Fracturing and Micro-Proppants with Existing Surface-to-Inseam Wells to Improve Pre-Drainage of Low Permeability Coals in Mining Areas, paper presented at the ADIPEC, Paper SPE-216613-MS. https://doi.org/10.2118/216613-MS

Ramanandraibe, H M, Sedaghat, M, Johnson, R and Santiago, V, 2022. Co-application of micro-proppant with horizontal well, multi-stage hydraulic fracturing treatments to improve productivity in the Permian coal measures, Bowen Basin, Australia, *APPEA Journal*, 62:14. https://doi.org/10.1071/AJ21048

Scott, M, Johnson Jr, R L, Datey, A, Vandenborn, C and Woodroof, R A, 2010. Evaluating Hydraulic Fracture Geometry from Sonic Anisotropy and Radioactive Tracer Logs, paper presented at the SPE Asia Pacific Oil and Gas Conference and Exhibition, SPE 133059.

Sierra, L and Mayerhofer, M J, 2014. Evaluating the Benefits of Zipper Fracs in Unconventional Reservoirs, paper presented at the SPE Unconventional Resources Conference, Paper SPE-168977-MS.

Zimmer, U, 2010. Microseismic Mapping of Hydraulic Treatments in Coalbed-Methane (CBM) Formations: Challenges and Solutions, paper, paper presented at the SPE Asia Pacific Oil and Gas Conference and Exhibition, SPE-132958-MS.

Effect of ultrasonic modification of nanopores in coal on the competitive adsorption of CH_4 by CO_2

L Wang[1,2,3], W Yang[4], Z Li[5], C Tian[6], Z Song[7] and Y Zhao[8]

1. Key Laboratory of Theory and Technology on Coal and Rock Dynamic Disaster Prevention and Control, National Mine Safety Administration, China University of Mining and Technology, Xuzhou 221116, China.
2. Key Laboratory of Coal Methane and Fire Control, Ministry of Education, China University of Mining and Technology, Xuzhou 221116, China.
3. School of Safety Engineering, China University of Mining and Technology, Xuzhou 221116, China. Email: wangliang@cumt.edu.cn
4. PhD candidate, China University of Geosciences, Beijing 100083, China. Email: ywei07@163.com
5. PhD candidate, University of Wollongong, Wollongong NSW 2522. Email: zhongbei@uow.edu.au
6. PhD candidate, China University of Mining and Technology, Xuzhou 221116, China. Email: TB22120017A51@cumt.edu.cn
7. PhD candidate, China University of Mining and Technology, Xuzhou 221116, China. Email:TS22120080A31@cumt.edu.cn
8. PhD candidate, China University of Mining and Technology, Xuzhou 221116, China. Email: TS22120110A31LD@cumt.edu.cn

ABSTRACT

The nanometer-sized pores within coal are the primary sites for CH_4 adsorption and competitive adsorption with CO_2. Reasonable modification of the nanopore structure to enhance CH_4 desorption, diffusion rates, and CO_2 competitive adsorption effects can significantly enhance coalbed methane (CBM) production. However, ultrasonic synchronous modification multiple features of nanopores leads to complex and variable gas adsorption behaviours in coal. To reveal the effect of ultrasonic modification of coal nanopores on gas adsorption, pore measurement experiments and molecular simulation studies were conducted. The results show that the volume ratio of diffusion pores to adsorption pores (V_2/V_1) decreased significantly after ultrasonic excitation. With a decrease in the proportion of the volume of diffusion pores, the proportion of CH_4 migration from the pore walls of the adsorption pores continuously increases. The proportion of CH_4 migration from the pore walls of the diffusion pores to the pore space of the diffusion pores continuously decreases.

INTRODUCTION

China's resource endowment of 'lacking gas, scarce oil, and abundant coal' determines that coal will remain the primary energy source for a considerable period in the future (Wang *et al*, 2023). Pre-mining gas extraction eliminates the risk of coalbed CH_4 disasters and ensures safe coal mining. On the other hand, gas, primarily composed of CH_4, is considered a clean and efficient energy source (Karacan *et al*, 2011) and a significant component of greenhouse gas emissions (Kholod *et al*, 2020). Therefore, gas control has become crucial for the safe, clean development of coal resources, ecological and environmental protection, and achieving 'carbon neutrality'. However, the low permeability and weak diffusion characteristics of high-stress coal seams result in extremely poor CH_4 flowability, making the extraction of coalbed CH_4 extremely challenging (Liu *et al*, 2021). In-depth research on efficient extraction theories and technologies is essential to ensure the green, safe, and efficient mining of coal resources in China.

Currently, ultrasonic technology is a promising emerging auxiliary extraction method; however, the significant attenuation of ultrasonic waves in coal seams (Shi, 2018) and the characteristic of increasing permeability after exciting coal bodies restrict its development. Yang *et al* (2024) proposed an array arrangement scheme for ultrasonic stimulation devices to address these limitations and reduce the disadvantage of poor permeability enhancement owing to significant attenuation. Yang *et al* also envisioned the combined assistance of ultrasonics and CO_2 in extraction using the displacement effect of CO_2 to promote CH_4 desorption and diffusion, providing a source of

CH_4 for permeation, thereby compensating for the shortcomings of ultrasonic measures in assisting extraction. Moreover, the injection of CO_2 into the coal seams has a carbon sequestration effect. Coal seams are suitable media for CO_2 sequestration, with coal exhibiting a higher affinity for CO_2 than CH_4 (Mukherjee and Misra, 2018; Pajdak *et al*, 2019). CO_2 was adsorbed onto the coal matrix and not emitted (Ma *et al*, 2022). Therefore, researching ultrasonic-CO_2-ECBM technology holds significant economic and environmental implications.

This study used a self-designed ultrasonic stimulation test platform to treat coal samples for different durations. Subsequently, using low-temperature N_2 adsorption experiments, we determined the typical adsorption and diffusion pore diameters of the coal samples at different treatment times. Based on the pore size distribution curve, we calculated the ratio of the adsorption pore volume to the diffusion pore volume and constructed a dual-pore molecular structure model for the composite experimental parameters. Considering the actual CH_4 pressure and formation temperature at the sampling location, we used molecular simulation software to analyse the impact of CH_4 adsorption and CO_2 injection on the adsorption behaviour. We also explored the differences in the gas occurrence patterns in coal structures. These research results play a crucial role in revealing the impact of ultrasonic stimulation on the microstructural modification of coal reservoirs in CO_2-ECBM. This information will be beneficial for the development of a more effective combination of ultrasonic and CO_2 displacement technologies. Additionally, the results are essential for uncovering the adsorption evolution characteristics of gases in dual-pore structures, thereby providing a convenient approach for investigating the adsorption and diffusion mechanisms of CH_4.

EXPERIMENTAL AND SIMULATION METHODS

Experimental process

The experimental coal samples were collected from the GuXian Coal Mine in Shanxi Province, China. The coal sample is anthracite with high preparation degree. After collection, coal samples were sealed to prevent contamination. Upon arrival at the laboratory, the samples were processed into cylindrical shapes of 50 mm × 50 mm dimensions. The flow of the experiment is illustrated in Figure 1.

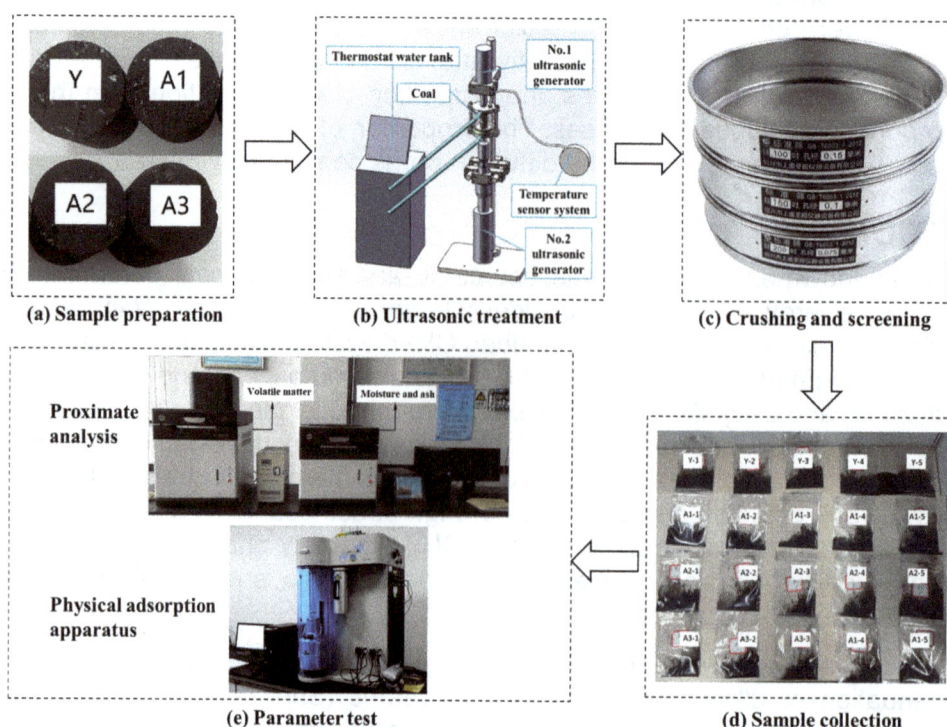

FIG 1 – Pore size distribution.

Simulation method

This study investigates the adsorption behaviour of CH_4 and CO_2 on slit-type coal nanopores using Grand Canonical Monte Carlo (GCMC) and Molecular Dynamics (MD) methods. In the GCMC simulations, the chemical potential, volume, and temperature were maintained constant. Simulations were conducted using the Forcite and Absorption modules. The relationship between the gas adsorption pressure and fugacity follows the Peng-Robinson state equation (Zhang *et al*, 2015), and the converted relationship is shown in Figure 2. The force field used in this simulation was Compass (Jia, Song and Jia, 2024), and the electrostatic and van der Waals forces during the simulation were calculated using Ewald and Atom-Based summation methods, respectively (Jia *et al*, 2023).

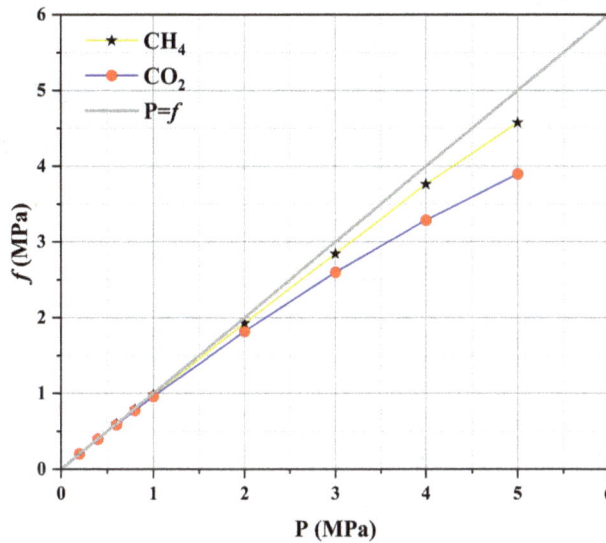

FIG 2 – Transition curve between fugacity and pressure.

MODEL CONSTRUCTION

Based on the pressure-dependent N_2 adsorption data measured at 77 K, the pore size distribution characteristics of the four coal samples were calculated and plotted using the QSDFT model (Figure 3). From Figure 3, it can be observed that the peak pore size in the adsorption pore region for all four coal samples was 1.299 nm, whereas that in the diffusion pore region was 4.678 nm. The volume of adsorption pores fluctuated to some extent with increasing duration of ultrasonic excitation. In contrast, the volume of diffusion pores showed a trend of significant initial decrease, followed by a relative increase.

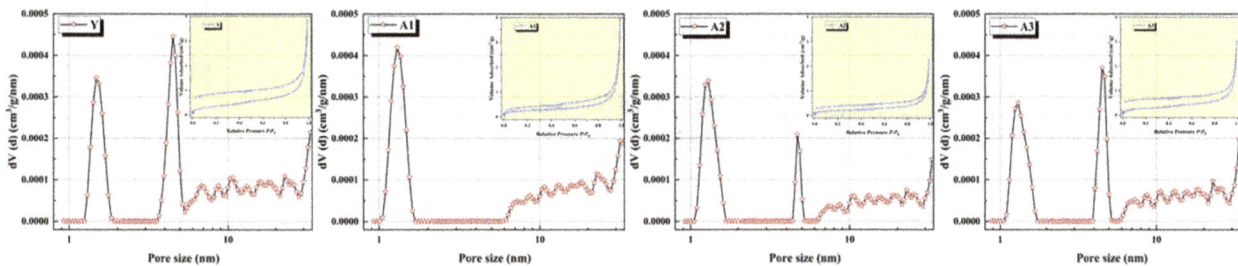

FIG 3 – Pore size distribution.

In this study, the pore sizes corresponding to the peak positions of these two pore regions were considered typical pore sizes. Combined with the proportions of peak areas, a pore model for the four coal samples was constructed. Typical pore sizes and volumes are presented in Table 1. The structural parameters of the original coal samples were 4 nm × 1.299 nm × 4 nm. Using these parameters as benchmarks, different models for each sample were constructed based on the actual total pore volume ratio and the volume ratios of adsorption and diffusion pore as shown in Table 2. The parameters of the pore models for each sample are presented in table, and the constructed pore models are shown in Figure 4. Purple region in the model were filled with helium atoms to prevent the adsorption of gas molecules in these areas during subsequent adsorption calculations. Helium atoms were chosen because they are inert gases with stable chemical properties, weak interactions with coal, CH_4, and CO_2, and minimal influence on the calculation results.

TABLE 1
Typical pore diameter and pore volume parameters.

Samples	Y	A1	A2	A3
Adsorption pore size (nm)	1.299	1.299	1.299	1.299
Adsorption pore volume (cm³/g)	1.096×10^{-4}	1.340×10^{-4}	1.071×10^{-4}	1.006×10^{-4}
Diffusion pore size (nm)	4.678	\	4.678	4.678
Diffusion pore volume (cm³/g)	3.546×10^{-4}	\	0.804×10^{-4}	2.339×10^{-4}
Total pore volume (cm³/g)	4.642×10^{-4}	1.340×10^{-4}	1.875×10^{-4}	3.345×10^{-4}

TABLE 2
Slit pore model parameters.

Samples	x-1	y-1	z-1	x-2	y-2	z-2
Y	3.4 nm	1.299 nm	4 nm	2.9 nm	4.678 nm	4 nm
A1	4 nm	1.299 nm	4 nm	\	\	\
A2	3.2 nm	1.299 nm	4 nm	0.7 nm	4.678 nm	4 nm
A3	2.9 nm	1.299 nm	4 nm	1.9 nm	4.678 nm	4 nm

Note: x-1, y-1 and z-1 are the size parameters of the adsorption pore, x-2, y-2 and z-2 are the size parameters of the diffusion pore.

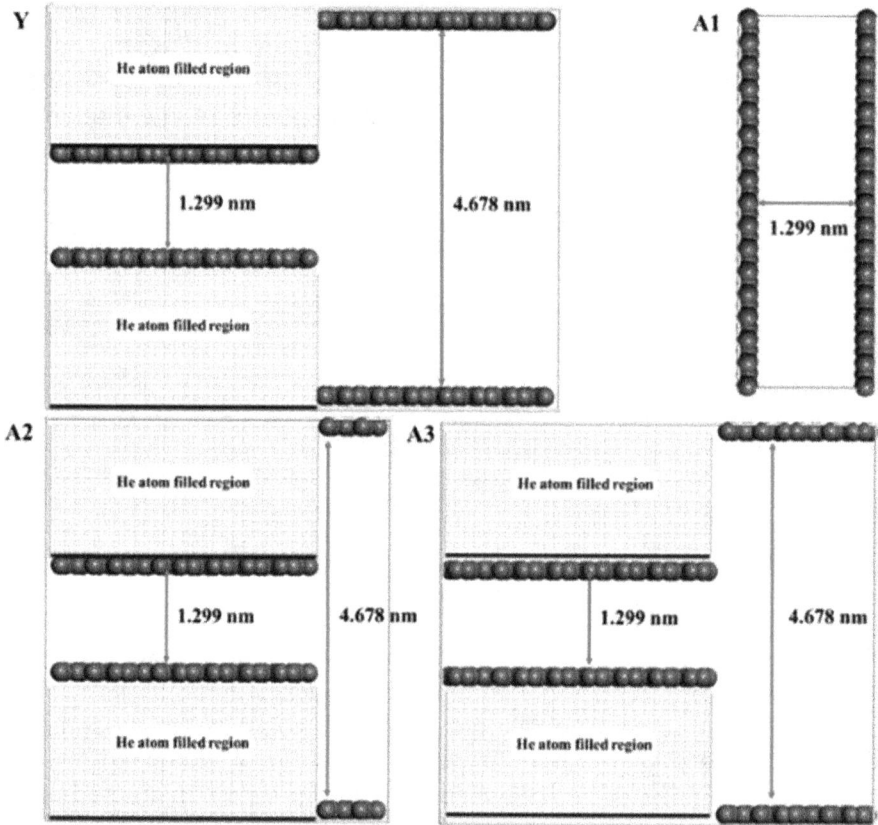

FIG 4 – Composite slit hole model of coal.

The modelling approach for typical composite pore sizes can largely reflect the characteristics of the coal pore structure, highlighting the main differences in the coal pore structure after ultrasonic excitation. This is a reliable and effective modelling method for studying the impact of pore structure evolution on gas adsorption behaviour from a molecular perspective.

RESULTS AND DISCUSSION

Effect of gas content on gas-solid interaction

When the system contains three-phase substances, the formula for calculating the interaction energy between any two components is (You *et al*, 2018):

$$E_{\text{int } A\&B} = \frac{\left(E_{total} - E_A - E_B + E_C + E_{A+B} - E_{A+C} - E_{B+C}\right)}{2} \tag{3}$$

where:

$E_{\text{int } A\&B}$ is the interaction energy between A and B

E_{total} is the total energy of the system

E_A, E_B and E_C denote the energies of the A, B and C, respectively

E_{A+B}, E_{A+C} and E_{B+C} are the total energies of A and B, A and C, and B and C, respectively

CH₄ adsorption behaviour in coal before and after ultrasonic excitation

Figure 5 compares the competitive adsorption effects after the ultrasonic treatment of the coal bodies. Two notable phenomena are illustrated in Figure 5. First, under the same gas conditions, the interaction energy between the coal and CH₄ was minimal in the A1 coal sample. This is because the A1 coal sample only contains adsorption pores (1.299 nm pores), and adsorption pores exhibit a strong gas adsorption capability on the pore walls or inside the pores (Cheng and Hu, 2023).

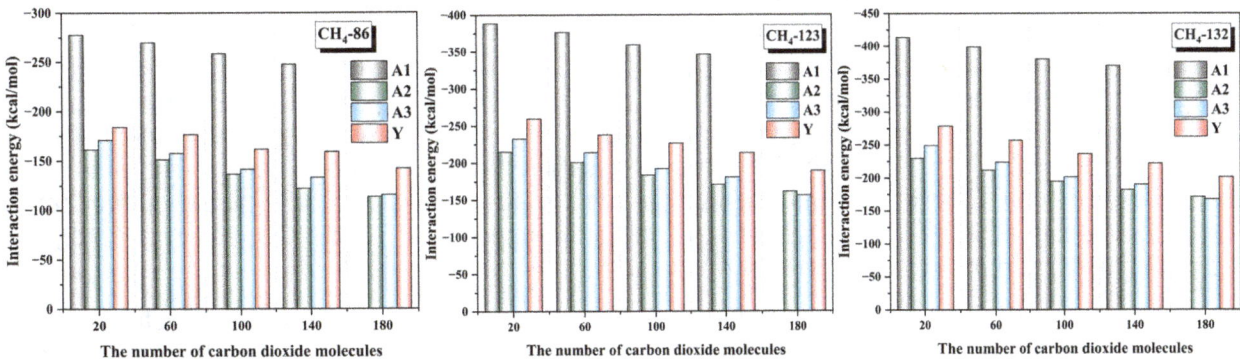

FIG 5 – Interaction energy difference between CH₄ and coal.

Second, in coal samples with both adsorption and diffusion pores (4.678 nm) under identical gas conditions, the rule A2 > A3 > Y exists for the interaction energy between CH₄ and coal. In other words, the binding strength between coal and CH₄ followed the pattern A2 < A3 < Y. This pattern suggests that the CH₄ in the A2 coal sample may be more prone to desorption after competitive adsorption. Among the three coal samples (A2, A3, and Y), there was little difference in the adsorption pore surface area and volume. In contrast, there was a significant difference in the diffusion pore surface area. Therefore, the diffusion pore surface area was the critical parameter determining the difference in the binding strength between coal and CH₄. The size of the diffusion pore surface area followed the pattern A2 < A3 < Y (Table 2), resulting in the observed pattern in Int-E (Coal-CH₄) after the competitive adsorption of the same amount of CO₂ on CH₄.

CO_2 adsorption behaviour in coal before and after ultrasonic excitation

The variation trend of the interaction energy between coal and CO_2 in the different coal pore structure models is consistent with the aforementioned variation trend in the interaction energy between coal and CH_4 (Figure 6). Under the same gas conditions, there was a pattern of A1 < Y < A3 < A2 in the interaction energies between the four coal structures and CO_2. The reason for the consistently minimal Int-E(Coal-CO_2) in A1 is that, despite its limited surface area, it only contains adsorption pores. The CO_2 molecules experienced a force field due to the overlapping of adjacent pore walls, leading to a much stronger adsorption force on the CO_2 molecules compared to the CO_2 molecules on the surface structures of diffusion pores. This results in a far smaller interaction energy with CO_2 than other structures containing diffusion pores. The appearance of the Y < A3 < A2 pattern resulted from the influence of the diffusion pore surface area, which is consistent with the analysis of CH_4.

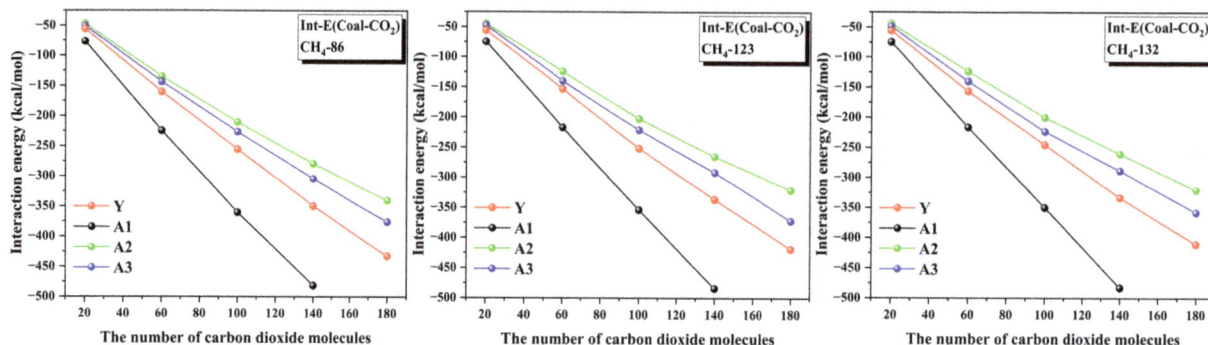

FIG 6 – Interaction energy difference between CO_2 and coal.

From Figure 6, it can also be observed that the difference in the interaction energy between coal and CO_2 among the different coal structures becomes more apparent with an increase in the CO_2 injection amount. The difference influenced this phenomenon in the coal diffusion pore surface area. When the CO_2 injection amount was small, owing to the minimal differences in the dominant adsorption pore surface area and pore volume, all four coal structures could provide sufficient adsorption sites, resulting in little difference in the interaction energy between coal and CO_2. However, owing to the rule Y > A3 > A2 in the diffusion pore surface area of the coal samples, as the CO_2 injection amount increased, the consumption rate of the dominant adsorption points on the surfaces followed the sequence A2 > A3 > Y. When CO_2 starts to adsorb in the diffusion pore space of A2, the CO_2 in A3 and Y can still be adsorbed onto the surface of the diffusion pores. Similarly, when CO_2 begins to adsorb in the diffusion pore space of A3, the CO_2 in Y can still be adsorbed on the surface of the diffusion pores. Gas in diffusion pores mainly exists as single-molecule layer adsorption on the wall surface, exhibiting stronger stability than gas in pore spaces. Therefore, with an increase in the CO_2 injection amount, changes in the adsorption area bring about differences in the gas molecule adsorption stability, causing the differences in Int-E(Coal-CO_2) among Y, A2, and A3 to become more prominent.

In addition, this is visually apparent from Figure 6. There is a lack of data on the CO_2 injection quantity of 180 in the A1 structure. This is because of the limited pore space in A1; when the CO_2 quantity reaches 180, it surpasses its maximum adsorption gas volume, resulting in simulation failure. Thus, although in competitive adsorption, the binding strength between coal and CO_2 molecules is highest in A1, the inability to inject a larger amount of CO_2 may lead to an unsatisfactory displacement effect.

Distribution and evolution of competitively adsorbed gases

Adsorption density

CH_4 adsorption density

As shown in Figure 7, the CH_4 molecules were symmetrically distributed along the slit pore walls, forming a distinct adsorption layer. It was observed that under the same gas conditions, as the ultrasonic excitation time increased, there was a pattern of initially increasing and then decreasing

CH₄ density in the coal adsorption pore walls. In contrast, at the diffusion pore walls, the CH₄ density initially decreases and then increases with increasing ultrasonic excitation time.

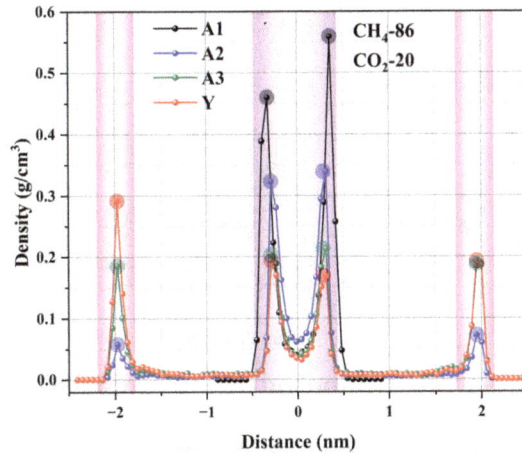

FIG 7 – CH₄ density distribution during competitive adsorption.

Figure 8 shows only the CH₄ adsorption density proportions in the four adsorption regions of each structure, with a CH₄ quantity of 86 as the CO_2 injection varied. The CH₄ adsorption density in the adsorption pore wall and interior regions followed the pattern A1 > A2 > A3 > Y in each coal model, whereas that in the diffusion pore wall and interior regions followed the pattern Y > A3 > A2.

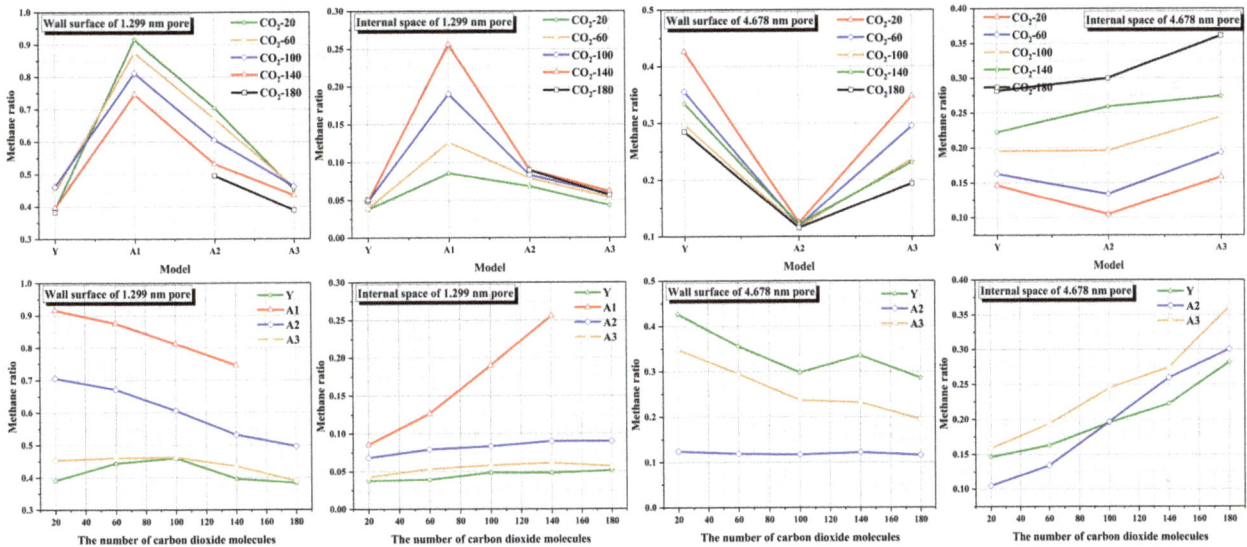

FIG 8 – The proportion of CH₄ density in each region under different CO_2 injection amount.

CO_2 adsorption density

Figure 9 illustrates the distribution of the CO_2 adsorption density during competitive adsorption processes. CO_2 was primarily adsorbed on the pore walls. An adsorption density peak also formed inside the adsorption pores when a large amount of CO_2 was injected. In various coal structures, when the adsorption quantities of CH₄ and CO_2 are equal, a regularity exists in the coal pore structure model, in which the CO_2 density peak on the adsorption pore walls follows the pattern Y < A3 < A2 < A1, whereas the CO_2 density peak on the diffusion pore walls follows the pattern Y > A3 > A1.

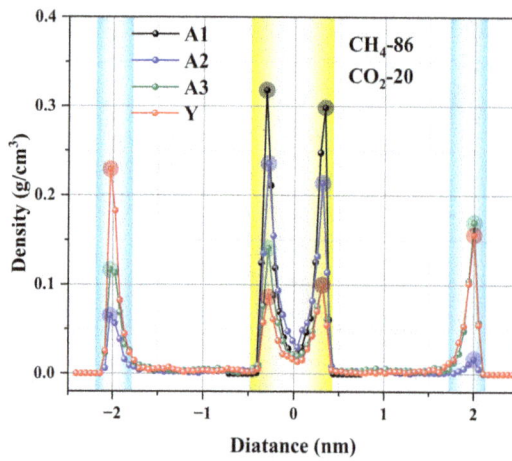

FIG 9 – CO_2 density distribution during competitive adsorption.

Figure 10 shows the CO_2 adsorption density distribution with respect to the variation in the CO_2 injection for each of the four adsorption regions in the different structures, with CH_4 adsorption at 86. It can be visually observed from the graph that the evolution pattern of adsorption density in each adsorption region during the competitive adsorption process of CH_4 and CO_2 is not influenced by the quantity of CH_4. The quantity of CH_4 only affected the relative densities of the four regions. The adsorption pattern of CO_2 differs from that of CH_4. The adsorption behaviour and evolution pattern of CO_2 in different regions and the increase in its injection quantity was consistent with the abovementioned CH_4 adsorption pattern. There were only minor differences in some details, such as a significant increase in the adsorption percentage of internal CO_2 in the adsorption pores in structures A2 and A3 with an increase in the CO_2 injection quantity. In contrast, the increase in the CH_4 adsorption percentage under the same conditions was very weak.

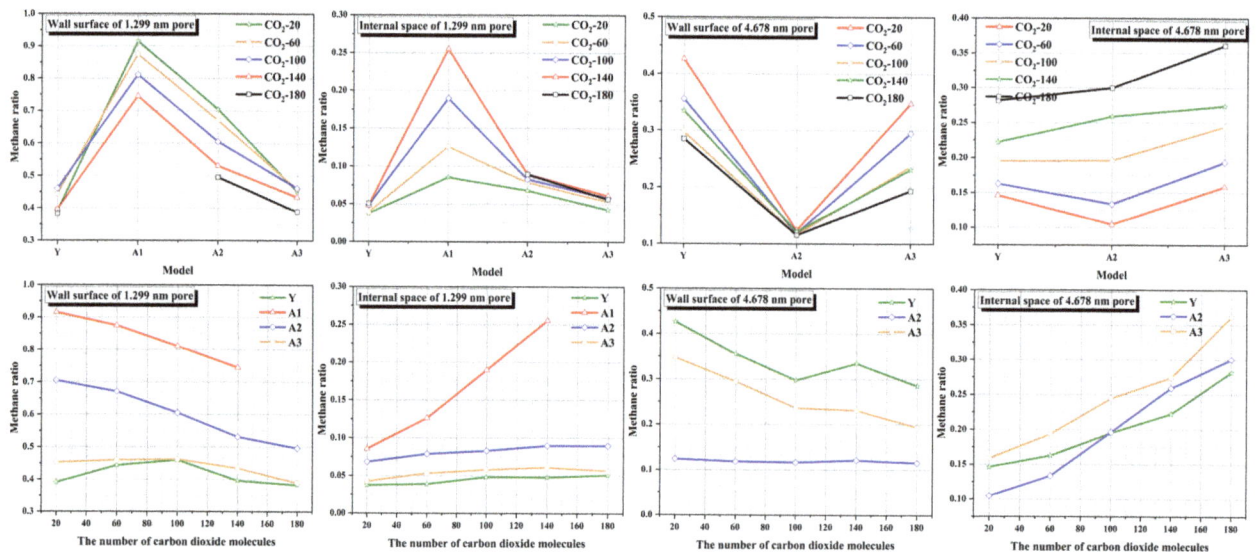

FIG 10 – The proportion of CO_2 density in each region under different CO_2 injection amount.

CONCLUSIONS

- In coal pore structures containing adsorption and diffusion pores, the larger the pore wall area, the stronger the gas adsorption strength. When injecting CO_2 into structures with only adsorption pores, a large amount of CH_4 migrates from the pore wall to the pore space. However, when diffusion pores are present, as the proportion of diffusion pore volume increases, the proportion of CH_4 migration from the pore wall of the adsorption pores continuously decreases, while the proportion of CH_4 migration from the pore wall of the diffusion pores to the pore space of the diffusion pores continuously increases.

- The early-stage decrease in the proportion of diffusion pores due to ultrasonic treatment enhances the adsorption capacity of coal for CO_2 and CH_4, while in the later stage of ultrasonic treatment, as the proportion of diffusion pores increases, the adsorption capacity of coal for CO_2 and CH_4 continuously decreases.

ACKNOWLEDGEMENTS

This work was supported by the National Natural Science Foundation of China (Nos. 52174216 and 51974300) and the Fundamental Research Funds for the Central Universities (No. 2021YCPY0206).

REFERENCES

Cheng, Y P and Hu, B, 2023. A new pore classification method based on the methane occurrence and migration characteristic in coal, *Journal of China Coal Society*, 48(01):212–225. https://doi.org/10.13225/j.cnki.jccs.Z022.1889

Jia, J, Song, H and Jia, P, 2024. Selective adsorption mechanism of CO2/CH4/N2 multi-component gas mixtures by N/S atoms and functional groups in coal. *Process Safety and Environmental Protection*, 182:210–221. https://doi.org/10.1016/j.psep.2023.11.079

Jia, J, Wang, D, Li, B, Wu, Y and Zhao, D, 2023. Molecular simulation study on the effect of coal metamorphism on the competitive adsorption of CO2/CH4 in binary system, *Fuel*, 335. https://doi.org/10.1016/j.fuel.2022.127046

Karacan, C Ö, Ruiz, F A, Cotè, M and Phipps, S, 2011. Coal mine methane: A review of capture and utilization practices with benefits to mining safety and to greenhouse gas reduction, *International Journal of Coal Geology*, 86(2–3):121–156. https://doi.org/10.1016/j.coal.2011.02.009

Kholod, N, Evans, M, Pilcher, R C, Roshchanka, V, Ruiz, F, Coté, M and Collings, R, 2020. Global methane emissions from coal mining to continue growing even with declining coal production, *Journal of Cleaner Production*, 256(A). https://doi.org/10.1016/j.jclepro.2020.120489

Liu, P, Fan, J, Jiang, D and Li, J, 2021. Evaluation of underground coal gas drainage performance: Mine site measurements and parametric sensitivity analysis, *Process Safety and Environmental Protection*, 148:711–723. https://doi.org/10.1016/j.psep.2021.01.054

Ma, R, Yao, Y, Wang, M, Dai, X and Li, A, 2022. CH4 and CO2 Adsorption Characteristics of Low-Rank Coals Containing Water: An Experimental and Comparative Study, *Natural Resources Research*, 31(2):993–1009. https://doi.org/10.1007/s11053-022-10026-x

Mukherjee, M and Misra, S, 2018. A review of experimental research on Enhanced Coal Bed Methane (ECBM) recovery via CO2 sequestration, *Earth-Science Reviews*, 179:392–410. https://doi.org/10.1016/j.earscirev.2018.02.018

Pajdak, A, Kudasik, M, Skoczylas, N, Wierzbicki, M and Teixeira Palla Braga, L, 2019. Studies on the competitive sorption of CO2 and CH4 on hard coal, *Intnl J Greenhouse Gas Control*, 90. https://doi.org/10.1016/j.ijggc.2019.102789

Shi, Q M, 2018. Response and Mechanism of Coal Physical Properties under Ultrasonic Load, PhD thesis, China University of Mining and Technology, China.

Wang, L, Sun, Y, Zheng, S, Shu, L and Zhang, X, 2023. How efficient coal mine methane control can benefit carbon-neutral target: Evidence from China, *Journal of Cleaner Production*, 424. https://doi.org/10.1016/j.jclepro.2023.138895

Yang, W, Wang, L, Yang, K, Fu, S, Tian, C and Pan, R, 2024. Molecular insights on influence of CO2 on CH4 adsorption and diffusion behaviour in coal under ultrasonic excitation, *Fuel*, 355. https://doi.org/10.1016/j.fuel.2023.129519

You, X, He, M, Zhang, W, Wei, H, Lyu, X, He, Q and Li, L, 2018. Molecular dynamics simulations of nonylphenol ethoxylate on the Hatcher model of subbituminous coal surface, *Powder Technology*, 332:323–330. https://doi.org/10.1016/j.powtec.2018.04.004

Zhang, J, Liu, K, Clennell, M B, Dewhurst, D N, Pan, Z, Pervukhina, M and Han, T, 2015. Molecular simulation studies of hydrocarbon and carbon dioxide adsorption on coal, *Petroleum Science*, 12(4):692–704. https://doi.org/10.1007/s12182-015-0052-7

Multiphysics coupling model for gas dynamic evolution in longwall goaf

X Wu[1], G Si[2], Y Jing[3] and P Mostaghimi[4]

1. MPhil Candidate, School of Minerals and Energy Resources Engineering, University of New South Wales, Sydney NSW 2052. Email: xuebin.wu1@student.unsw.edu.au
2. Associate Professor, School of Minerals and Energy Resources Engineering, University of New South Wales, Sydney NSW 2052. Email: g.si@unsw.edu.au
3. Associate Professor, School of Minerals and Energy Resources Engineering, University of New South Wales, Sydney NSW 2052. Email: yu.jing@unsw.edu.au
4. Professor, School of Civil and Environmental Engineering, University of New South Wales, Sydney NSW 2052. Email: peyman@unsw.edu.au

ABSTRACT

Coal spontaneous combustion (CSC) remains a significant hazard in coalmine safety, posing serious challenges to mine operations and environmental protection. Identifying potential CSC risk areas in advance is crucial for mine safety. Traditional monitoring methods often rely on placing monitoring equipment at pre-determined locations, which can be suboptimal. The specific area of spontaneous combustion is usually identified only after the residual coal begins to self-heat and produce gases, by which time optimal preventative measures may be less effective. Additionally, these monitoring results can be influenced by environmental factors such as diesel exhaust emissions from mining equipment, potentially leading to missed early signs of combustion or delayed responses. This research addresses these issues by integrating a coal-oxygen compound reaction mechanism into a goaf flow field model using COMSOL Multiphysics to simulate the spatial distribution of CSC areas. While the model offers a promising approach for identifying high-risk areas and informing the deployment of gas and temperature monitoring equipment, further validation is required to confirm its reliability. The results of this research aim to provide mine managers with a valuable tool for the timely identification and prevention of CSC accidents.

INTRODUCTION

Each year, coal spontaneous combustion (CSC) consumes tens of thousands of tons of coal resources globally, resulting in direct economic losses exceeding 280 million dollars (Wang *et al*, 2023). Additionally, the carbon dioxide emissions from these coal fires account for approximately 0.10 to 0.22 per cent of the global carbon emissions from fossil fuels (Onifade, 2022). Coal oxidation is an inevitable natural process that cannot be completely prevented; however, early warning signs of CSC can be identified to mitigate its impact. Indicator gases are the most common method for monitoring CSC. CO is frequently used as the primary indicator gas due to its detectability, while H_2 can be utilised as a supplementary indicator gas (Cliff, Brady and Watkinson, 2015; Wang, 2020). The tube bundle monitoring system with gas chromatographic spontaneous combustion analysis is the most used detection system for indicator gases of CSC (Cheng *et al*, 2021; Liang *et al*, 2019; Xue and Cui, 2004). It effectively monitors the concentrations of gases such as O_2, C_2H_2, CO, and H_2 present in underground mines. Traditional tube bundle systems require the installation of multiple monitoring points, and the regular maintenance of the tube bundle system necessitates considerable manpower (Zipf *et al*, 2013). In this research, we developed a numerical model coupling the temperature field and gas field to simulate the dynamic evolution of CO and H_2 within the self-heating goaf. This model can predict high-risk spontaneous combustion zones and the migration patterns of indicator gases within the goaf, significantly aiding in the optimal placement of tube bundle systems

MODELLING OF COAL OXIDATIONS

Mathematical model of the gas flow in the longwall goaf

The goaf is a porous medium where the internal flow transitions from turbulent to laminar and finally to creeping flow, rendering traditional Darcy's law inapplicable. The Brinkman equation, which combines Darcy's law and the Navier-Stokes equations, can effectively describe the prominent effects of different factors under complex fluid flow conditions. Therefore, this study employs the Brinkman equation to characterise the airflow within the goaf.

The equation of mass conservation:

$$\frac{\partial}{\partial t}(\varepsilon\rho) + \boldsymbol{\nabla} \cdot (\rho\boldsymbol{u}) = Q_m \tag{1}$$

where:

ε is the porosity

ρ is the density of gas (kg/m^3)

$\boldsymbol{u}$ is the velocity (m/s)

Q_m is the gas mass in the goaf $(kg/(m^3 \cdot s))$

The equation of momentum conservation:

$$\boldsymbol{\nabla} \cdot \boldsymbol{K} = \frac{\rho}{\varepsilon}\left(\frac{\partial\boldsymbol{u}}{\partial t} + (\boldsymbol{u} \cdot \boldsymbol{\nabla})\frac{\boldsymbol{u}}{\varepsilon}\right) + \boldsymbol{\nabla}p + +(\kappa^{-1}\mu + \beta_F\rho_g|\boldsymbol{u}| + \frac{Q_m}{\varepsilon^2})\boldsymbol{u} \tag{2}$$

where:

$$\boldsymbol{K} = [\frac{1}{\varepsilon_p}\{\mu(\boldsymbol{\nabla u} + (\boldsymbol{\nabla u})^{\mathrm{T}}) - \frac{2}{3}\mu(\boldsymbol{\nabla} \cdot \boldsymbol{u})\mathbf{I}\}]$$

μ is the dynamic viscosity of the mixed gas in the goaf $(Pa \cdot s)$

β_F is the Forchheimer coefficient (kg/m^4)

$$\beta_F = \frac{\varepsilon\rho_g C_f}{\sqrt{k}} \tag{3}$$

where:

C_f is the dimensionless coefficient of friction, $C_f = \frac{1.75}{\sqrt{150\varepsilon^3}}$.

In this research, to simulate the changes more accurately in gas composition caused by coal self-heating within the goaf, the mass transfer equation is employed to describe the consumption of oxygen and the generation of other gas products, as shown in Equation 4:

$$\varepsilon\frac{\partial C_i}{\partial t} + \boldsymbol{\nabla} \cdot (\boldsymbol{u}C_i) = \boldsymbol{\nabla} \cdot (D\boldsymbol{\nabla}C_i) + r_i \tag{4}$$

where:

C_i is the concentration of gas component i (mol/m^3)

D represents the diffusion coefficient of the component $\left(\frac{m^2}{s}\right)$

$\boldsymbol{u}$ indicates the velocity field

r_i is defined as the source term of the gas component in (mol/m^3)

Porosity and permeability distribution

Permeability is a critical determinant of the fluid state within the goaf, directly influencing the airflow intensity. The permeability distribution typically exhibits an 'O'-shaped pattern. This is characterised by a gradual increase as one moves laterally from the centre of the working face, and a gradual decrease as one moves towards the deeper parts of the goaf from the working face, as shown in FIG **1Error! Reference source not found.**. The permeability can be calculated by the Kozeny-Carmon Equation 5.

$$\kappa = \frac{d_p{}^2}{150} \cdot \frac{\varepsilon^3}{(1-\varepsilon)^2} \tag{5}$$

where:

d_p is the particle size (m) of coal in the goaf.

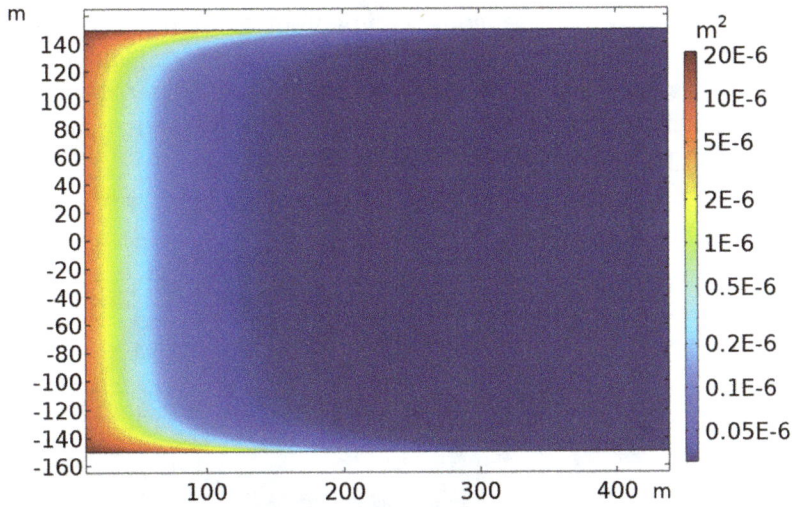

FIG 1 – Assumed permeability distribution in the goaf for the Multiphysics coupled CSC modelling.

Energy conservation in the process of CSC

In the CSC process, the chemical reaction between coal and oxygen is the main source for the heat production. The rate of oxygen consumption can be represented by the Arrhenius form, as shown in Equation 6.

$$r = A * C_{o_2} \exp\left(-E/RT\right) \tag{6}$$

where:

A is the pre-exponential factor (s^{-1})

C_{o_2} is the O_2 concentration (mol/m^3)

E is the activation energy (J/mol)

R is the gas constant $(J/(mol \cdot K))$

T is the coal temperature in goaf (K)

The local thermal equilibrium equation is used to describe the energy transformation of coal self-heating. The energy conservation is expressed as Equations 7 and 8:

$$(\rho c_P)_{eff} \frac{\partial T}{\partial t} + \rho_f c_{p,f} \boldsymbol{u} \cdot \boldsymbol{\nabla} T + \boldsymbol{\nabla} \cdot \left(-k_{eff} \boldsymbol{\nabla} T\right) = Q \tag{7}$$

$$(\rho c_P)_{eff} = \varepsilon \rho_f c_{p.f} + (1 - \varepsilon)\rho_s c_{P,s} \tag{8}$$

where:

s and f represent solid and fluid, respectively

c_p is the specific heat capacity $(J/(kg \cdot K))$

Q is the reaction heat (W/m^3)

k_{eff} is the effective thermal conductivity $W/(m \cdot K))$, as represented by Equation 9:

$$k_{eff} = \varepsilon k_f + (1 - \varepsilon)k_s \tag{9}$$

where:

k_f and k_s are the coefficient of thermal conductivity for gas and coal, respectively

Goaf model set-ups

In this research, a numerical model has been developed to simulate the transport of CSC indicator gases in the goaf. The main parameters of the longwall model are as follows: a strike length of 450 m, a dip length of 300 m, and a height of 25 m, with coal seam having a thickness of 4 m, as shown in FIG 2. This model utilises conventional U-type ventilation, maintaining a 20 m³/s ventilation flow rate and defined the return airway as a free exit. A reduced ventilation flow rate is employed due to the exclusion of gas sources' impact in this experiment. This research is conducted using COMSOL Multiphysics software, selected for its robust performance in solving complex fluid dynamics problems, making it well-suited for this research.

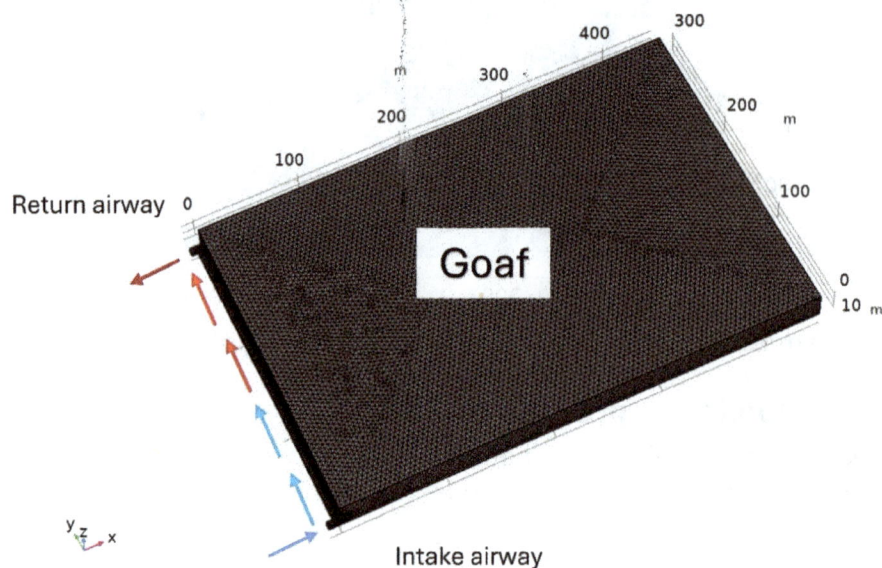

FIG 2 – Geometry model of longwall mining with the U-type ventilation.

CO, due to its regular pattern of temperature change, is widely used as the primary gas for monitoring CSC. Hydrogen (H₂) is an auxiliary indicator for coal self-heating and spontaneous combustion, detected only at high temperatures, indicating the deep oxidation stage with rapid temperature rise. Therefore, this paper focuses on the research and analysis of CO as the main gas product and H₂ as the secondary gas product from coal oxidation in longwall goafs. This dual-gas monitoring approach not only aids in more accurately understanding the dynamic process of CSC but also provides a more reliable basis for its prevention and control. The simulation parameters used in the model are listed in Table 1. The main parameters are obtained from the published literature (Xia et al, 2016).

TABLE 1

Main parameters used in the simulation.

Parameters	Symbols	Values
Reaction heat	Q	$300 \ kJ/mol$
Pre-exponential factor	A	$5 \times 10^6 \ s^{-1}$
Coal size	d_p	0.15 m
Thermal conductivity coefficient of coal-rock	k_s	$1.7 \ W/(m \cdot K)$
Coal density	ρ_s	1570 kg/m^3
Specific heat capacity of coal	$c_{P,s}$	$1360 \ J/(kg \cdot K))$
Wind speed	Q_{in}	$1200 \ m^3/min$
Gas diffusion coefficient	D	$2 \times 10^5 m^2/s$
Dynamic viscosity	μ	$1.87 \times 10^{-5} \ Pa \cdot s$
CO generation rate	r_{CO}	$6.5625 \times 10^4 \times e^{9.141/T} \ mol/m^3/s$
H₂ generation rate	r_{H2}	$5.17 \times 10^{-11} \times e^{0.071T} \ mol/m^3/s$

SIMULATION RESULTS

FIG **3** illustrates the concentration distribution characteristics of O_2 and CO at different heights within the goaf. At the 2 m high above the seam floor, oxygen is predominantly concentrated on the intake side, with higher concentrations observed near the working face. As the depth increases, the oxygen concentration gradually decreases. Within the range of 100 to 200 m, there is a significant gradient in the oxygen concentration. In the same region, the CO concentration reaches its peak, indicating an intensive coal oxidation reaction. At the 8 m height, the distribution gradients of O_2 and CO are more gradual, with the concentrations of both gases being more dispersed, especially with a marked decrease in CO concentration.

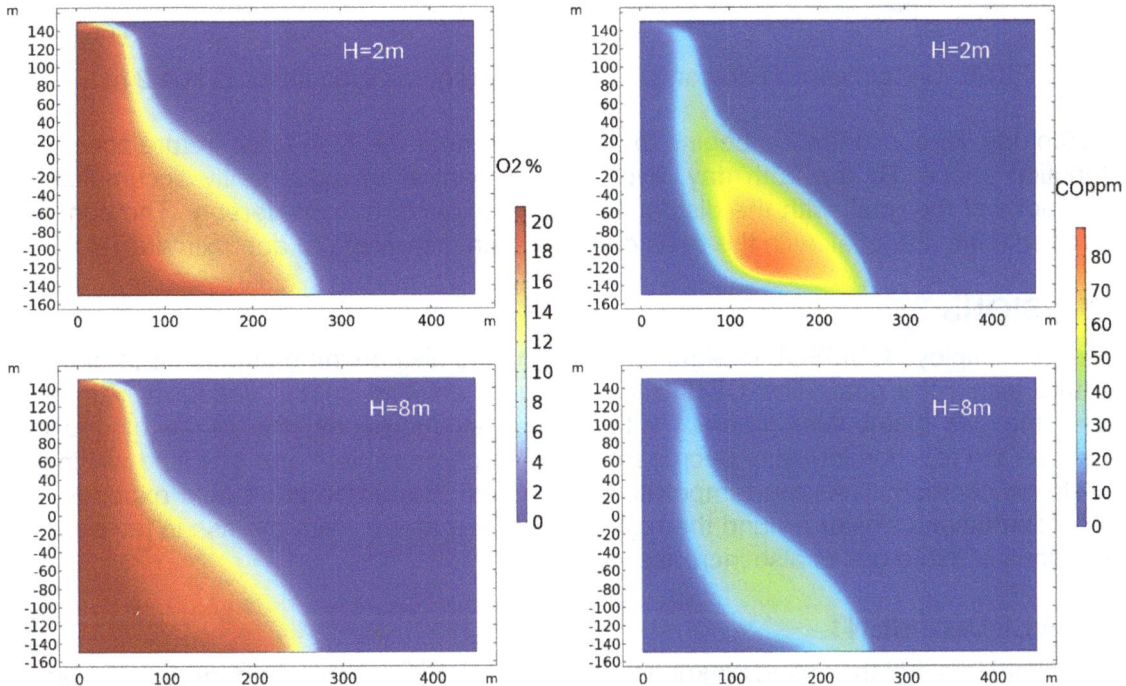

FIG 3 – O_2 (left) and CO (right) concentration contours at different horizontal planes.

Figures 4 and 5 present the temperature distribution within the goaf and the H_2 concentration distribution over different heating time. Based on the oxygen concentration distribution, the goaf is divided into three zones: the cooling zone, the oxidation zone, and the suffocation zone. The cooling zone, located near the working face, exhibits lower temperatures; the oxidation zone, characterised by intense coal oxidation reaction, exhibits the highest temperatures; and the suffocation zone, situated deeper within the goaf, shows gradually decreasing temperatures due to limited oxygen supply, which hinders the oxidation process.

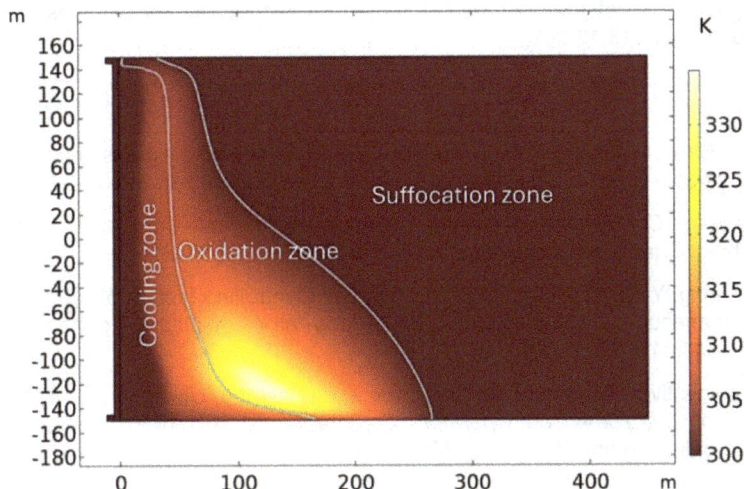

FIG 4 – Temperature distribution in the goaf at the 25th day.

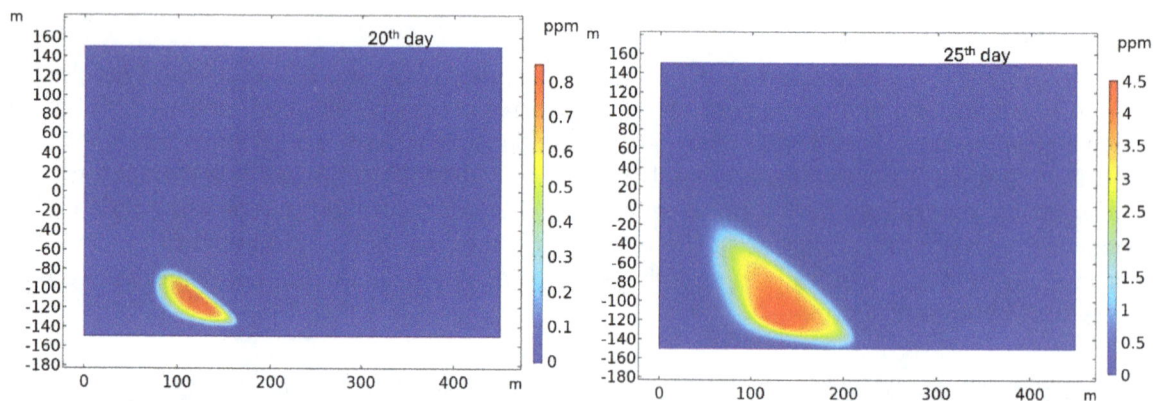

FIG 5 – H_2 concentration contours after different reaction times at H = 2 m.

On the 20th day, H_2 concentration begins to accumulate near the high-temperature zone, mirroring its distribution shape. By the 25th day, hydrogen concentration significantly accumulates in the deeper regions of the goaf, with the distribution range expanding considerably. The concentration extends deeper into the goaf, with the peak concentration increasing to approximately 4.5 ppm.

CONCLUSIONS

This research employs COMSOL to simulate the dynamic distribution characteristics of gases and temperature within the goaf. The CSC within the goaf exhibits spatial variability. As the height increases, the gas distribution gradient within the goaf decreases, the range of gas diffusion expands, and the reaction intensity decreases. The temperature field and gas field within the goaf show a strong correlation. As the temperature increases, the temperature field expands, and CO and H_2 concentrations cluster around the high-temperature areas. With the rise in temperature, the concentrations of these gases also increase.

ACKNOWLEDGEMENT

The authors declare that they have no known competing financial interests or personal relationships that could have appeared to influence the work reported in this paper.

REFERENCES

Cheng, J, Luo, W, Zhao, Z, Jiang, D, Zhao, J, Shi, L, Liu, Y, Zhao, G, Wang, Z, Song, W, Gao, K, Liu, G, Zheng, W, Lu, P and Wang, Y, 2021. Controlling Coal Spontaneous Combustion Fire in Longwall Gob Using Comprehensive Methods—a Case Study, *Mining, Metallurgy and Exploration*, 38:1801–1816. https://doi.org/10.1007/s42461-021-00427-6

Cliff, D, Brady, D and Watkinson, M, 2015. ACARP Project C18013: *The Green Book – Spontaneous Combustion in Australian Coal Mines*, Australian Coal Association Research Program ACARP, Brisbane.

Liang, Y, Zhang, J, Wang, L, Luo, H and Ren, T, 2019. Forecasting spontaneous combustion of coal in underground coal mines by index gases: A review, *Journal of Loss Prevention in the Process Industries*, 57:208–222. https://doi.org/10.1016/j.jlp.2018.12.003

Onifade, M, 2022. Countermeasures against coal spontaneous combustion: a review, *International Journal of Coal Preparation and Utilization*, 42:2953–2975. https://doi.org/10.1080/19392699.2021.1920933

Wang, T, Wang, H, Fang, X, Wang, G, Chen, Y, Xu, Z and Qi, Q, 2023. Research progress and visualization of underground coal fire detection methods, *Environ Sci Pollut Res*, 30:74671–74690. https://doi.org/10.1007/s11356-023-27678-8

Wang, X, 2020. *Spontaneous Combustion of Coal: Characteristics, Evaluation and Risk Assessment* (Springer International Publishing: Cham). https://doi.org/10.1007/978-3-030-33691-2

Xia, T, Zhou, F, Wang, X, Zhang, Y, Li, Y, Kang, J and Liu, J, 2016. Controlling factors of symbiotic disaster between coal gas and spontaneous combustion in longwall mining gobs, *Fuel*, 182:886–896. https://doi.org/10.1016/j.fuel.2016.05.090

Xue, S and Cui, H, 2004. Innovative Techniques for Detection and Control of Underground Spontaneous Combustion of Coal, in *Proceedings of Coal 2004: Coal Operators' Conference*, pp 161–167 (University of Wollongong; and The Australasian Institute of Mining and Metallurgy: Melbourne).

Zipf, R K, Marchewka, W, Mohamed, K, Addis, J and Karnack, F, 2013. Tube bundle system, *Min Eng*, 65:57–63.

Numerical simulation of dynamic evolution of hydration film expansion process on coal surface

M Yan[1], H L Jia[2], S G Li[2], Y B Wang[2], H X Zhang[2] and H F Lin[2]

1. Key Laboratory of Western Mine Exploitation and Hazard Prevention, Ministry of Education, Xi'an University of Science and Technology, Xian, Shaanxi Province, China. Email: minyan1230@xust.edu.cn
2. Key Laboratory of Western Mine Exploitation and Hazard Prevention, Ministry of Education, Xi'an University of Science and Technology, Xian, Shaanxi Province, China.

ABSTRACT

When the coal seam is fractured by hydraulic fracturing, the shape and position of the hydration film on the coal surface directly affect the wettability of the coal surface, and then affect the desorption and diffusion law of gas in the coal body. Therefore, the formation mechanism of coal seam gas desorption hysteresis and water lock effect is not only affected by the structural parameters of coal micro-interface, but also closely related to the morphology of hydration film on the surface of coal. In this study, COMSOL Multiphysics®, version 6.2 (by Stockholm, COMSOL AB) was used to dynamically simulate the diffusion process of hydration film on coal surface, and the level set method was used to track the free surface of water droplets. The effects of physical and chemical parameters and liquid characteristics of coal on the diffusion law of hydration film on coal surface were analysed. The results show that the liquid property parameters and physical and chemical parameters of coal have an effect on the expansion and evolution of the hydration film on the surface of coal, and the contact angle and fractal dimension have the most significant influence on it. The maximum spreading coefficient of droplets with contact angles of 45° and 90° on the surface of coal increased by 13.913 per cent and 15.624 per cent compared with that with contact angle of 135°. The maximum spreading coefficient of droplets with fractal dimension of 1.25 and 1.5 increased by 34.793 per cent and 34.181 per cent compared with the fractal dimension of 1.8. The interaction between contact angle and fractal dimension has a significant effect on D_{max}/D_0. The D_{max}/D_0 in the two-factor level range is between 3 and 6 and has a certain time effect.

INTRODUCTION

In recent years, coal seam water injection and other measures with coal wetting as the necessary link are important technical means to realise the safe simultaneous mining of coal and gas in coalmines. The effect of water injection wetting will greatly affect the realisation of the comprehensive prevention and control function of coal seam water injection composite disaster (Wang, Jin and Sun, 2004; Yang et al, 2023). The hydration film covered on the surface of the coal body will occupy the pore fissure channel on the surface of the coal body, which will cause the water lock effect to varying degrees (Liao, Xu and Hu, 2002). The typical phenomenon of water lock effect is not only affected by the micro-mesoscopic structural parameters of coal interface, but also closely related to the shape and position of hydration film spreading at coal interface (Yang et al, 2023).

In order to study the effect of water lock effect on coal interface, many scholars have carried out preliminary numerical model calculation and experimental research from the perspective of droplet impact on solid wall. Josserand and Thoroddsen (2016) summarised the phenomenon of droplet impact on the wall surface, and considered that the droplet spreading process is not only related to the droplet inertia, viscosity and surface tension, but also related to the interaction of surrounding gas. The bubbles wrapped in the liquid film promote the splash of the droplet. Aboud and Kietzig (2015) tested the impact of droplets on six different inclined surfaces, and found that the spreading morphology asymmetry of droplets on smooth surfaces is much larger than that on rough surfaces. Chen, Wang and Shen (2013) obtained the oscillation expression of droplet spreading radius and the influence of surface tension, viscosity coefficient and other parameters on droplet spreading by analysing the force state of droplet. Ezzatneshan and Khosroabadi (2021) used the lattice Boltzmann method to simulate the impact process of droplets under different equilibrium contact angles. It was found that this method can accurately capture the morphological changes during the droplet impact process and reveal the influence of the contact angle of the coal surface on the droplet spreading.

In order to further explore the morphological evolution law of hydration film spreading at coal interface, many scholars have carried out in-depth exploration according to the research direction of droplet impact on solid wall surface. Chen (2024) studied the dynamic behaviour of cleaning droplets impacting, spreading and infiltrating on random rough surfaces. They believed that compared with smooth surfaces, due to the effect of wall roughness, the spreading edge of droplets will produce many broken small droplets. Zhang *et al* (2023) used numerical simulation method to simulate the contact process between coal surface and liquid under different roughness by establishing a three-dimensional model of rough coal surface. It was found that the increase of roughness will lead to the change of wettability of coal surface, which will affect the spreading behaviour of liquid on the surface. Guan *et al* (2023) used the Volume of Fluid (VOF) algorithm to capture the water droplet interface to ensure the accuracy of the interface change, and analysed the movement law of the water droplet impact process and the change of the internal velocity field in detail. Wang *et al* (2024) used a three-dimensional computational simulation method to analyse the interaction between droplets and random rough surfaces by numerical simulation, considering the dynamic behaviour of droplets and the morphology of surfaces. Yanling *et al* (2023) studied the effects of surface wettability and rough structure on the static and dynamic wetting behaviour of droplets by molecular dynamics method. It is concluded that the structure of rough surface has an inhibitory effect on the diffusion of droplets compared with smooth surface, and it is found that the apparent contact angle of droplets on rough surfaces with different structures is greater than the intrinsic contact angle of droplets on smooth surfaces.

There are many factors affecting the dynamic spreading interface effect of hydration film on coal surface. At present, most of them focus on the influence of single factors such as contact angle and surface tension, but in fact, there are interactions among various factors. In view of this, the author uses the response surface method to design the optimisation scheme of hydration film spreading and diffusion parameters on the surface of coal, establishes the theoretical model of fluid-solid coupling of coal body, and uses COMSOL software to establish a two-dimensional hydration film spreading and diffusion model on the surface of coal body. The influence of single factor and interaction factors on the interface effect of hydration film spreading and diffusion under different interface roughness conditions is studied, and the influencing parameters are optimised. The dynamic spreading prediction model of droplets impacting on the surface of coal body is established, and the optimal spreading and diffusion wetting parameters of droplets are determined.

THEORETICAL MODEL OF FLUID-SOLID COUPLING OF COAL BODY

Fundamental assumption

According to the characteristics of coal seam gas occurrence, the following assumptions are proposed:

- Coal is homogeneous and isotropic.

- The flow is laminar incompressible.

- The coal is composed of spherical solid particles with the same diameter, which are evenly distributed in the porous area.

- The properties of liquid phase, gas phase and solid phase are constant.

Coal fluid flow control equation

The fluid flow of the droplets during the collision, whether outside the coal body or inside the microscopic pores of the coal body, is controlled by the momentum conservation equation and the continuity equation.

Assuming that the porosity is a constant, the mass and momentum conservation equations in the average coal interface region on a local small volume can be expressed as (Tang, 2023):

$$\nabla \cdot u = 0 \tag{1}$$

$$\rho \frac{\partial u_p}{\partial t} + \rho u_p \cdot \nabla u_p = -\nabla p_p + \nabla \cdot \left(\mu \left(\nabla u_p \right) + \left(\nabla u_p \right)^T \right) + \varepsilon \rho g + \varepsilon F_{SV} + B \tag{2}$$

where:

u_p is the Darcy velocity vector, also known as the macroscopic average velocity vector ($u_p = \varepsilon u$) in porous media

p_p is the Darcy pressure, also known as the dimensionless macroscopic average pressure ($p_p = \varepsilon p$) inside porous media

B is the total resistance per unit volume due to the presence of solid particles, representing the sum of pressure resistance and viscous friction

The momentum equation contains the gravity g as well as the surface tension component represented by F_{SV}.

Level set equation

The convection transport equation of the level set function is:

$$\frac{\partial \phi}{\partial t} + u \cdot \left(\nabla \phi \right) = \gamma \nabla \left\{ \xi \nabla \phi - \phi \left(1 - \phi \right) \frac{\nabla \phi}{|\nabla \phi|} \right\} \tag{3}$$

where:

γ is the parameter reinitialised in the level set equation

ξ is the parameter controlling the thickness of the interface

NUMERICAL SIMULATION

Grid independence verification

Grid independence verification is needed before numerical calculation to avoid the influence of grid parameters on the calculation results. In this paper, COMSOL software is used to establish a structured grid model. Figure 1 is a schematic diagram of the grid division of the calculation model under two-phase conditions. The numerical calculation adopts the laminar-level set model, and the mesh refinement is performed at the junction of the fluid domain and the solid domain. In this paper, three sets of grids are divided and calculated. The number of elements is 200 000, 100 000 and 60 000 respectively. The least mesh model is selected as the benchmark to verify the grid independence without affecting the calculation results. The fine mesh is refined along the axial and tangential directions on the basis of coarse mesh division. Figure 2 shows the variation of the maximum spreading coefficient of the hydration film at the coal interface in the three sets of grid simulation calculations. The variation trend of the maximum spreading coefficient of the three sets of grids is almost the same, with only slight differences in the local area. Considering the calculation resources, the time required for calculation and the calculation accuracy, a grid with a number of 100 000 units is finally selected as the general grid for the follow-up work.

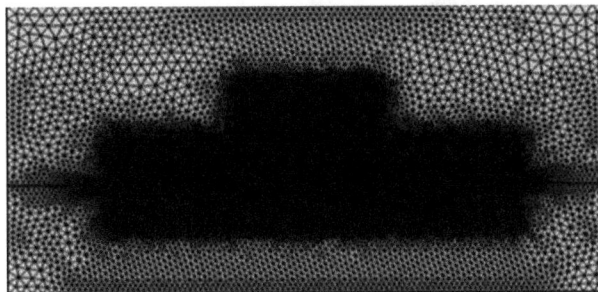

FIG 1 – The grid model used in the calculation.

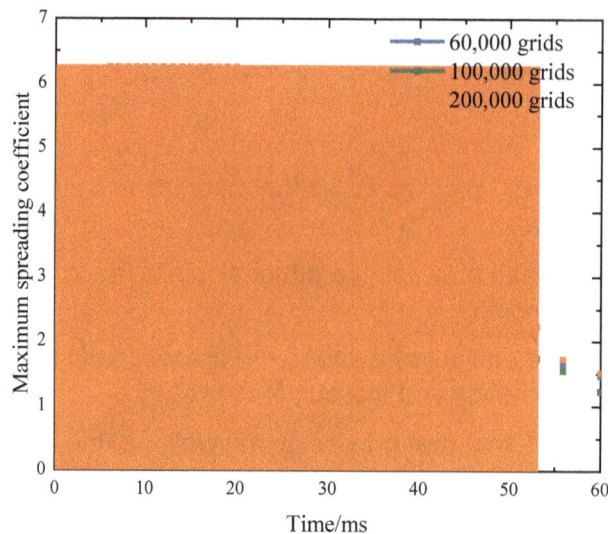

FIG 2 – Three sets of grid maximum spreading coefficient.

Simulation program

Geometric model and definite solution conditions

The two-phase flow model of the following droplets impacting the surface of the coal body is shown in Figure 3. Air is the main phase, droplets are the secondary phase, and the surface of the coal body is a fixed wall. The droplets impact the coal body at a certain initial velocity. Due to the existence of curved boundary in the geometric model of the hydration film spreading coal medium, the free triangular mesh is used to mesh the model. The model is divided into three regions as a whole. The lower part of the model is the coal medium domain and the upper part is the hydration film domain and the air domain. The overall calculation domain is 15 mm × 8 mm and the height of the coal area is 3 mm. In this study, it is necessary to consider the spreading process of the hydration film on the surface of the coal body. Since the droplet will not reach the surrounding air domain during the movement process, the air domain is calibrated as 'fluid dynamics-conventional', the maximum unit size is 0.36 mm, and the maximum unit growth rate is 1.15. The unit size of hydration film and coal medium domain is calibrated to the maximum unit size of 0.03 mm, and the maximum unit growth rate is 1.05. The initial boundary of the hydration film and the three-phase interface are encrypted, and the mesh size is 0.06 mm. The whole model consists of 100909 grids and the average element mass is 0.9059.

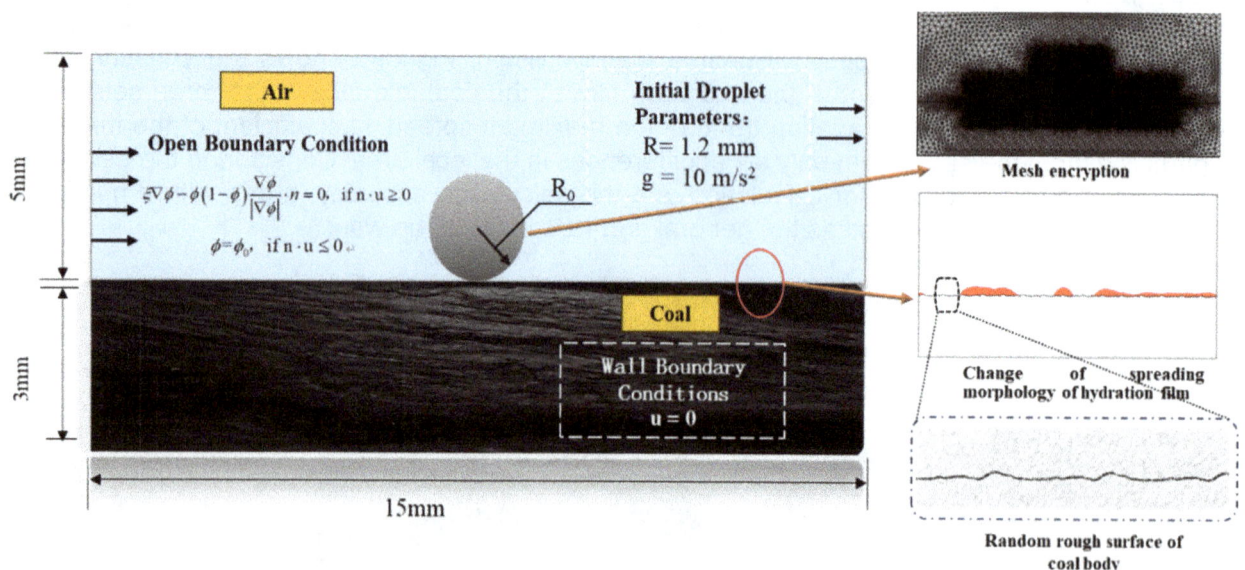

FIG 3 – Initial conditions, boundary conditions and meshing of geometric model.

Response surface method simulation scheme

The response surface analysis method has the advantages of less test times and high prediction accuracy. In the optimisation design, not only the relationship between the response target and the design variables can be obtained, but also the optimal combination of the design variables can be obtained (Gunst, 2008). The Box-Behnken Design response surface analysis method is used to carry out the scheme design. The research shows that (Jing, 2023), the six factors of contact angle, surface tension, dynamic viscosity, particle diameter, fractal dimension and impact velocity have a significant effect on the dynamic spreading interface of hydration film on the surface of coal body. Therefore, the design of six-factor three-level response surface scheme is carried out, which is coded by $(0, \pm 1)$ (Table 1) and a total of 52 groups of optimisation schemes (Table 2).

TABLE 1
Code and level of design factors.

Influencing factors	Variable	Level -1	Level 0	Level +1
contact angle (°)	X_1	45	90	135
surface tension (mN/m)	X_2	72.09	53.073	44.493
dynamic viscosity (mPa·s)	X_3	1.267	1.259	1.254
particle diameter (mm)	X_4	0.05	0.1	0.25
fractal dimension (1)	X_5	1.8	1.25	1.5
stroke speed (m/s)	X_6	1.5	1	2

TABLE 2
Test plan.

Scheme Serial number	X_1	X_2	X_3	X_4	X_5	X_6	D_{max}/D_0	Scheme Serial number	X_1	X_2	X_3	X_4	X_5	X_6	D_{max}/D_0
1	-1	-1	0	-1	0	0	6.1575	27	-1	0	0	1	-1	0	6.25
2	1	-1	0	-1	0	0	5.40542	28	1	0	0	1	-1	0	3.989
3	-1	1	0	-1	0	0	6.1574	29	-1	0	0	-1	1	0	6.21379
4	1	1	0	-1	0	0	5.33303	30	1	0	0	-1	1	0	6.21379
5	-1	-1	0	1	0	0	6.16218	31	-1	0	0	1	1	0	6.19395
6	1	-1	0	1	0	0	4.40895	32	1	0	0	1	1	0	4.97104
7	-1	1	0	1	0	0	6.21379	33	0	-1	0	0	-1	-1	6.16308
8	1	1	0	1	0	0	4.39307	34	0	1	0	0	-1	-1	5.7084
9	0	-1	-1	0	-1	0	4.49325	35	0	-1	0	0	1	-1	6.25
10	0	1	-1	0	-1	0	4.62193	36	0	1	0	0	1	-1	6.25
11	0	-1	1	0	-1	0	4.88938	37	0	-1	0	0	-1	1	6.25
12	0	1	1	0	-1	0	6.18875	38	0	1	0	0	-1	1	6.25
13	0	-1	-1	0	1	0	6.25	39	-1	0	-1	0	0	-1	6.25
14	0	1	-1	0	1	0	6.21383	40	1	0	-1	0	0	-1	6.1489
15	0	-1	1	0	1	0	6.25	41	-1	0	1	0	0	-1	6.25
16	0	1	1	0	1	0	6.20173	42	1	0	1	0	0	-1	6.1526
17	0	0	-1	-1	0	-1	6.22588	43	-1	0	-1	0	0	1	6.2017
18	0	0	1	-1	0	-1	6.2379	44	-1	0	1	0	0	1	6.25

Scheme Serial number	Influencing factor						Maximum spreading coefficient D_{max}/D_0	Scheme Serial number	Influencing factor						Maximum spreading coefficient D_{max}/D_0
	X_1	X_2	X_3	X_4	X_5	X_6			X_1	X_2	X_3	X_4	X_5	X_6	
19	0	0	-1	1	0	-1	6.21383	45	0	0	0	0	0	0	6.10275
20	0	0	1	1	0	-1	6.25	46	0	0	0	0	0	0	6.10275
21	0	0	-1	-1	0	1	6.25	47	0	0	0	0	0	0	6.10275
22	0	0	1	-1	0	1	6.2379	48	0	0	0	0	0	0	6.10275
23	0	0	-1	1	0	1	6.20175	49	0	0	0	0	0	0	6.10275
24	0	0	1	1	0	1	6.25	50	0	0	0	0	0	0	6.10275
25	-1	0	0	-1	-1	0	6.25	51	0	-1	0	-1	0	0	6.25
26	1	0	0	-1	-1	0	3.5974	52	0	1	-1	0	0	0	6.20175

Note: Only five significant digits are retained after the decimal point of the maximum spreading coefficient D_{max}/D_0.

Dynamic spreading and wetting characteristics of coal-water interface

At present, the main index to determine the wetting parameter is contact Angle θ, which is mainly used to characterise the wetting degree of the two-phase interface. For the field process parameters such as the coverage area of the coal body water injection, there is no clear selection method at present, and the superposition effect of the two-phase interface between the coal body and the liquid drop is ignored. The judgment of coal bed gas adsorption and desorption is not accurate enough.

In view of this, the author uses relevant program software to encode the droplet spreading contour, and introduces the maximum spreading coefficient $β_m$ to describe the dynamic spreading and wetting characteristics of the droplet impacting the coal surface. The maximum spreading coefficient of the droplet is the ratio of the maximum contact diameter between the droplet and the coal surface to the initial diameter of the droplet. The calculation formula is as follows:

$$β_m = D_{max}/D_0 \tag{4}$$

where:

$β_m$ is the droplet spreading coefficient

D_{max} is the maximum diameter of the droplet in contact with the coal surface

D_0 is the initial diameter of the droplet

D_{max}/D_0 not only considers the influence of the contact Angle on the gas adsorption and desorption effect of coal seam, but also considers the superposition effect of the coal body and the liquid droplet, which can determine whether the interface effect of the spreading and diffusion of the hydration film on the surface of the coal body is significant. D_{max}/D_0 is close to 6.25, indicating that the interfacial hydration film is completely covered on the surface of coal body, and the wetting effect is ideal. D_{max}/D_0 is close to 0, which indicates that the coverage rate of the hydration film on the coal surface is low and the wetting effect is poor. Taking parameters with coding value 0 among the influencing factors as an example, the dynamic evolution process of coal surface hydration film spreading morphology was analysed, as shown in Figure 4. When t = 7 ms, the hydration film is completely covered on the coal surface, and D_{max}/D_0 = 6.25, the wetting effect is the best. When t = 53 ms, the hydration film is broken, and only part of the droplets stay on the surface of the coal body. At this time, D_{max}/D_0 = 1.02284, the result is close to 0, and the wetting effect is not obvious.

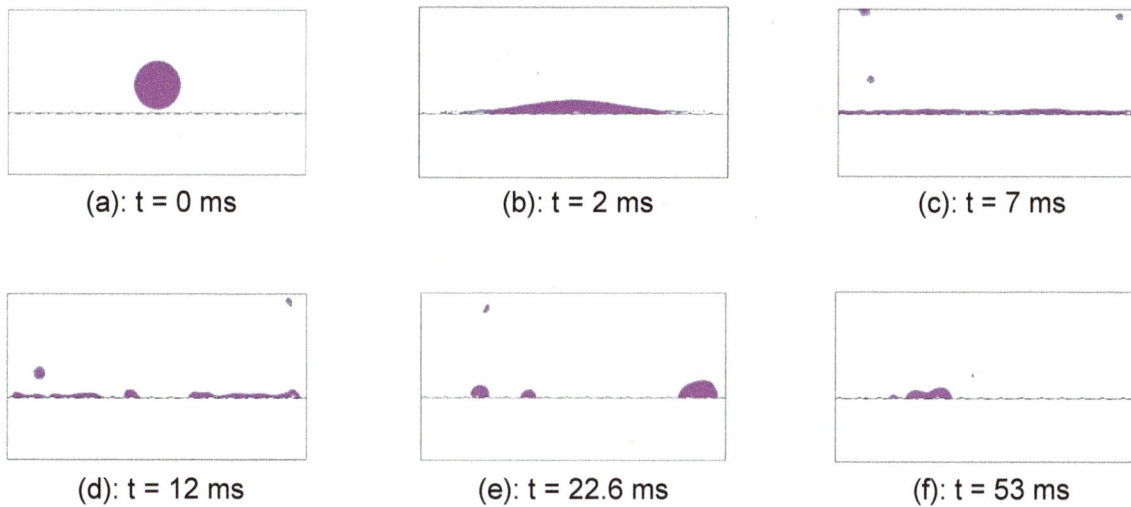

(a): t = 0 ms (b): t = 2 ms (c): t = 7 ms (d): t = 12 ms (e): t = 22.6 ms (f): t = 53 ms

FIG 4 – Evolution process of hydration film spreading morphology on coal surface.

Response model and its applicability analysis

The simulation time of droplet collision with hydration film spreading on coal surface is set to 60 ms. The influence of single factor/interaction factor on the dynamic evolution process of hydration film spreading is analysed in detail, and the optimisation design of wetting characteristic parameters of coal-water two-phase interface is carried out. The test results are shown in Table 2. Multiple regression fitting was performed on 50 sets of test results. By comparing the data of significance test and correlation test of each model, the D_{max}/D_0 response model was determined to be a five-element second-order polynomial. The response model and variance analysis are shown in Table 3. The regression fitting equation of D_{max}/D_0 response model is:

$$\beta_m = 6.13037 - 0.5813X_1 + 0.064259X_2 + 0.069492X_3 - 0.114486X_4 + 0.404461X_5 - 0.024726X_6 - 0.017472X_1X_2 + 0.012398X_1X_3 - 0.181963X_1X_4 + 0.451011X_1X_5 - 0.255585X_1X_6 + 0.117755X_2X_3 + 0.0031357X_2X_4 - 0.055542X_2X_5 + 0.083979X_2X_6 + 0.010562X_3X_4 - 0.277443X_3X_5 + 0.018028X_3X_6 - 0.2228949X_4X_5 - 0.004526X_4X_6 + 0.044833X_5X_6 - 0.433298X_1^2 - 0.057953X_2^2 - 0.201286X_3^2 - 0.076639X_4^2 - 0.211322X_5^2 + 0.393809X_6^2$$

TABLE 3

Analysis of variance of regression model.

Source of variance	Quadratic sum	Degree of freedom	Mean square	F value	P value	Significance
Model	19.37	27	0.7179	4.27	0.0003	significant
A: Contact angle	6.43	1	6.43	38.28	<0.0001	**
B: surface tension	0.0924	1	0.0924	0.5499	0.4655	——
C: viscosity	0.1040	1	0.1040	0.6194	0.4390	——
D: particle diameter	0.3242	1	0.3242	1.93	0.1775	——
E: fractal dimension	3.1	1	3.1	18.46	0.0002	**
F: stroke speed	0.0088	1	0.0088	0.0526	0.8205	——
AB	0.0024	1	0.0024	0.0145	0.9050	——
AC	0.0008	1	0.0008	0.0050	0.9441	——
AD	0.5298	1	0.5298	3.15	0.0884	——
AE	1.63	1	1.63	9.69	0.0047	**
AF	0.2929	1	0.2929	1.74	0.1991	——

Source of variance	Quadratic sum	Degree of freedom	Mean square	F value	P value	Significance
Model	19.37	27	0.7179	4.27	0.0003	significant
BC	0.1219	1	0.1219	0.7256	0.4027	——
BD	0.0086	1	0.0086	0.0514	0.8226	——
BE	0.0402	1	0.0402	0.2392	0.6292	——
BF	0.0387	1	0.0387	0.2305	0.6355	——
CD	0.0009	1	0.0009	0.0053	0.9425	——
CE	0.6158	1	0.6158	3.67	0.0675	——
CF	0.0042	1	0.0042	0.0252	0.8753	——
DE	0.4193	1	0.4193	2.50	0.1272	——
DF	0.0002	1	0.0002	0.0010	0.9753	——
EF	0.0089	1	0.0089	0.0533	0.8195	——
A^2	1.58	1	1.58	9.39	0.0053	*
B^2	0.0343	1	0.0343	0.2045	0.6552	——
C^2	0.3592	1	0.3592	2.14	0.1566	——
D^2	0.0508	1	0.0508	0.3026	0.5874	——
E^2	0.4681	1	0.4681	2.79	0.1080	——
F^2	1.16	1	1.16	6.91	0.0147	*
DE	0.4193	1	0.4193	2.50	0.1272	——
DF	0.0002	1	0.0002	0.0010	0.9753	——
EF	0.0089	1	0.0089	0.0533	0.8195	——
A^2	1.58	1	1.58	9.39	0.0053	*
B^2	0.0343	1	0.0343	0.2045	0.6552	——
C^2	0.3592	1	0.3592	2.14	0.1566	——
D^2	0.0508	1	0.0508	0.3026	0.5874	——
E^2	0.4681	1	0.4681	2.79	0.1080	——
F^2	1.16	1	1.16	6.91	0.0147	*
Residual error	4.03	24	0.1680			
Misfit term	4.03	19	0.2333		0.2644	not significant
Pure error	0	5	0			
Total	23.41	51				
R_2	0.8286					
R^2_{adj}	0.6786					

Note: * and * * are significant at P < 0.05 and P < 0.01 levels, respectively.

Through the variance and error analysis of the selected response model, the significant influence probability of the model is less than 0.0001, and the regression effect of the model is more significant. Among them, R_2 can reflect the difference between the response value and the true value. The greater the R_2, the better the correlation of the model. The predicted $R_2 = 0.6340$ is consistent with the actual $R_2 = 0.8278$, that is, the difference is less than 0.2 (see Table 3); through the residual analysis, the normal probability distribution diagram of the residual and the distribution diagram of the predicted value and the actual value of the test (Figure 5; it can be seen from Figure 6) that the scatter points of each model are located near y = x, indicating that the model has a good fitting degree. The model can be used to analyse the interaction between factors and optimise the wetting parameters of hydration film. In summary, the maximum spreading coefficient in the model has a good fit to the contact angle and fractal dimension.

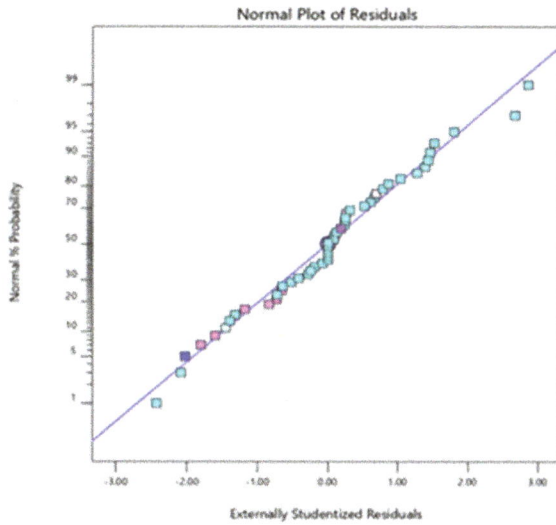

FIG 5 – Normal probability distribution of residuals.

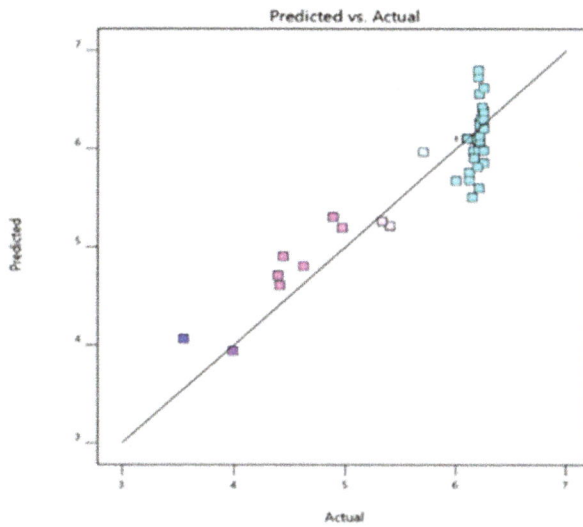

FIG 6 – Response target D_{max}/D_0 test value.

ANALYSIS OF INFLUENCING FACTORS OF SPREADING AND DIFFUSION OF HYDRATION FILM ON COAL SURFACE

The influence of single factor on the spreading and diffusion of hydration film

The factors affecting the dynamic evolution process of hydration film spreading and diffusion on the surface of coal are divided into fluid factors and coal factors, which together determine the wetting effect of coal interface. Among the six factors, X_1, X_2, X_3 and X_6 are fluid factors, X_4 and X_5 are coal factors. By comparing the F values (F test results) of the six factors in Table 3, it can be seen that the sensitivity of D_{max}/D_0 to each factor is the strongest in X_1 and X_5 and X_1 has the most significant effect on D_{max}/D_0. In order to intuitively analyse the influence of single factor on D_{max}/D_0, the parameters of other factors are controlled to be constant, and the horizontal value of each variable is the horizontal axis and the response value is the vertical axis to fit the curve of D_{max}/D_0, as shown in Figures 7 and 8.

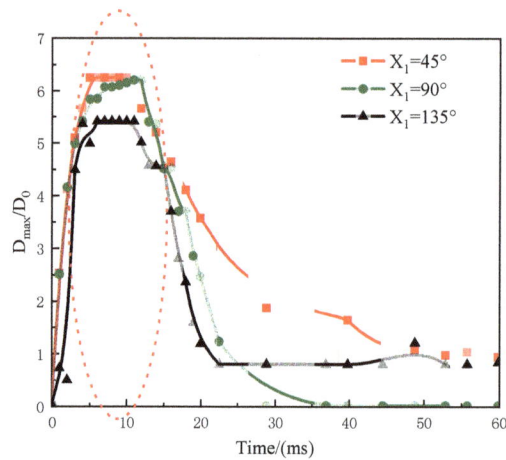

FIG 7 – Effect of contact angle on D_{max}/D_0.

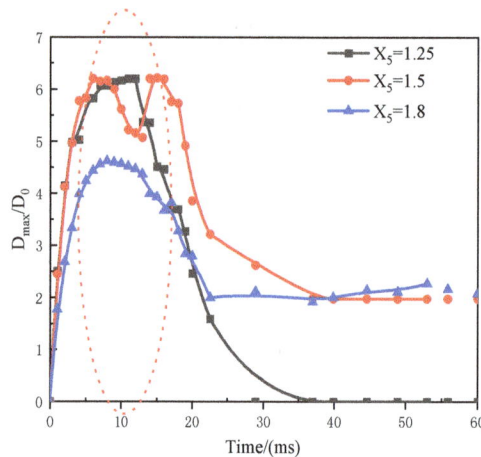

FIG 8 – Effect of fractal dimension on D_{max}/D_0.

By setting the parameters $\theta = 45°$, $90°$ and $135°$, the influence of droplets on the spreading process of coal surfaces with different wettability was studied. Figure 7 shows that the maximum spreading coefficient of hydrophilic coal on the surface of coal and the time to reach the maximum spreading coefficient are greater than those of hydrophobic coal. The maximum spreading coefficient of droplets with contact angles of 45° and 90° on the surface of coal increases by 13.913 per cent and 15.624 per cent compared with the contact angle of 135°.

Similarly, The influence of droplets on the spreading process of different rough coal surfaces was studied by setting the fractal dimension parameters as 1.25, 1.5 and 1.8. Figure 8 shows that the larger the fractal dimension is, the lower the maximum spreading coefficient of hydration film is. The maximum spreading coefficient of droplets with fractal dimension of 1.25 and 1.5 increases by 34.793 per cent and 34.181 per cent compared with that with fractal dimension of 1.8.

The influence of interaction factors on the spreading and diffusion of hydration film

D_{max}/D_0 was not only affected by single factor, but also by the interaction between factors. Through single factor analysis and variance analysis, it was found that X_2, X_3, X_4 and X_6 had little effect on D_{max}/D_0, resulting in the interaction between these four factors was not obvious. Therefore, the interaction between fluid factors and coal factors focuses on the analysis of X_1 and X_5, as shown in Figure 9.

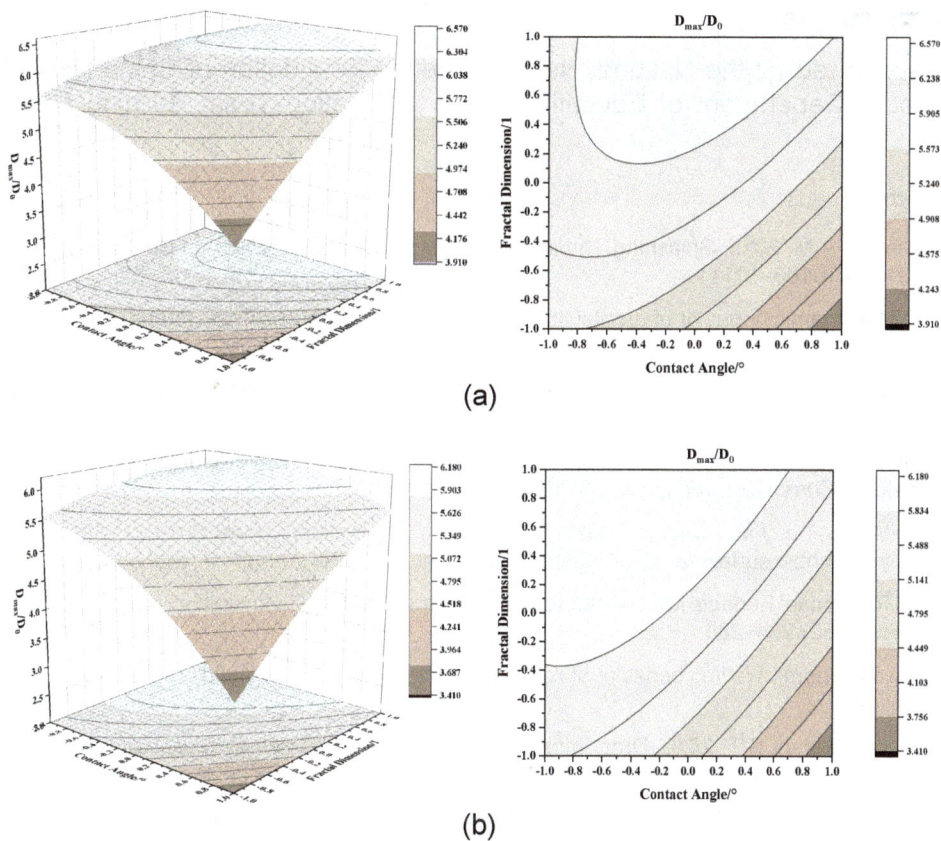

FIG 9 – Response surface and contour of the maximum spreading coefficient under the interaction of X_1 and X_5: (a) when $X_1 = 45°$, $X_5 = 1.25$; (b) when $X_1 = 90°$, $X_5 = 1.8$.

When the contact angle (X_1) and the fractal dimension (X_5) are 45° and 1.25 respectively, the interaction contour is semi-elliptical. It can be seen from Figure 9 that the response surface of X_1 and X_5 is steep and the response contour is distorted. The D_{max}/D_0 in the two-factor level range is between 3 and 6, indicating that X_1 and X_5 have a very significant effect on D_{max}/D_0, and have a significant effect on the value of D_{max}/D_0 and a certain time effect. In addition, the response contour line and contour line of Dm_{ax}/D_0 are gathered in the lower right corner, indicating that X_1 and X_5 are under the interaction, and the influence of X_1 on D_{max}/D_0 is still significant, which is consistent with the results of single factor analysis.

CONCLUSIONS

- Using COMSOL numerical simulation software to construct the fractal dimension of random rough surface, combined with the other five influencing factors, the evolution law of the free water layer spreading morphology at the coal interface is obtained.

- The maximum spreading coefficient of hydration film on coal surface is related to coal factors and fluid factors. Among them, the contact angle and surface fractal dimension have the strongest sensitivity to D_{max}/D_0. The maximum spreading coefficient of hydrophilic coal surface is 14.77 per cent higher than that of hydrophobic coal surface on average. The larger the fractal dimension of coal surface, the higher the complexity of the system, and the smaller the maximum spreading coefficient of hydration film.

- The interaction between contact angle and fractal dimension has a significant effect on D_{max}/D_0, and has a great influence on its value. The D_{max}/D_0 in the range of two factors is between 3 and 6. The interaction of the other factors is not obvious, which has little effect on its value. The interaction between various factors has a certain time effect on D_{max}/D_0.

ACKNOWLEDGEMENTS

This study is supported by the National Natural Science Foundation of China (No. 5227428) and Shaanxi Provincial Department of Education Youth Innovation Team Building Research Project (No. 21JP073).

REFERENCES

Aboud, D G and Kietzig, A M, 2015. Splashing Threshold of Oblique Droplet Impacts on Surfaces of Various Wettability, *Langmuir*, 31(36):10100–10111.

Chen, F H J W, 2024. Dynamic behavior of droplet impact on a random rough surface, *Journal of Civil Aviation University of China*, 42(1):53–58.

Chen, S, Wang, H and Shen, S, 2013. Droplet oscillation model and comparison with numerical simulation, *Acta Physica Sinica*, 62(20):1–6.

Ezzatneshan, E and Khosroabadi, A, 2021. Droplet spreading dynamics on hydrophobic textured surfaces: A lattice Boltzmann study, *Computers and fluids*, 231105063.

Guan, Y, Wang, M, Wu, S, Fu, J and Chen, X, 2023. The spreading and sliding characteristics of droplet impingement on an inclined hydrophobic surface at low Weber numbers, *International Journal of Heat and Fluid Flow*, 100109113.

Gunst, R F, 2008. Response Surface Methodology: Process and Product Optimization Using Designed Experiments, *Technometrics*, 3(50):284–286.

Jing, Y, 2023. Study on dynamic wetting behavior of coal dust surface impacted by liquid droplet, China University of Mining and Technology.

Josserand, C and Thoroddsen, S T, 2016. Drop Impact on a Solid Surface, *Annual Review of Fluid Mechanics*, 48:365–391. https://doi.org/10.1146/annurev-fluid-122414-034401

Liao, R, Xu, Y and Hu, X, 2002. Water lock effect on low permeability reservoir damage and methods for restraining and relieving, *Natural Gas Industry*, (06):87–89;3–2.

Tang, J, 2023. Numerical simulation of dynamics and heat transfer characteristics of droplet impact on porous media, Zhengzhou University of Light Industry.

Wang, F, Guo, F, Tang, M, Zhang, X, Zhang, Z, Li, S and Yang, B, 2024. 3-D computational study of a single droplet impacting the random rough surface: Hydromechanical solidification, *International Journal of Heat and Mass Transfer*, 224125311.

Wang, Q, Jin, L and Sun, J, 2004. Analysis of coal seam water injection process and mechanism of coal body wetting, *Journal of Safety and Environment*, (01):70–73.

Yang, W, Luo, L, Wang, Y and Bai, H, 2023. Chemical regulation of coal microstructure and gas displacement by water injection, *Journal of China Coal*, 48(08):3091–3101.

Yanling, C, Liang, G, Wanchen, S, Ningning, C and Yuying, Y, 2023. Molecular dynamics simulations of wetting behaviors of droplets on surfaces with different rough structures, *International Journal of Multiphase Flow*, 169.

Zhang, J, Xu, B, Wei, J, Zhang, P, Cai, M and Zhang, K, 2023. Numerical simulation of coal surface contact angle based on roughness, *Coal Science and Technology*, 51(4):96–104.

Heat and humidity (refrigeration and air cooling)

Applications and limitations of spot cooler systems in mines

M Beukes[1], R Hattingh[2] and R Funnell[3]

1. Senior Ventilation and Refrigeration Consultant, BBE Consulting Canada, Sudbury Ontario P3E 5S1, Canada. Email: mornebeukes@bbegroup.ca
2. Specialist Refrigeration Engineer, BBE Consulting South Africa, Johannesburg Gauteng 2060, South Africa. Email: rhattingh@bbe.co.za
3. Principal Engineer, BBE Consulting South Africa, Johannesburg Gauteng 2060, South Africa. Email: rfunnell@bbe.co.za

ABSTRACT

Effective management of heat in mining operations is critical in maintaining worker safety, productivity, and operational efficiency. Considering the global trend towards deeper, hotter mines, increased reliance on refrigeration systems to combat excessive heat loads and preserve acceptable working conditions must be expected.

Given the lower positional efficiency of surface air cooling systems, and the high cost and long lead time required to sink additional shafts in existing mines, underground spot cooling, with its apparent simplicity, low initial cost, and rapid implementation timeline, becomes an attractive solution for management of heat stress within localised 'problem areas'. However, it is essential to recognise the practical limitations that may impact its viability and long-term success.

This paper explores the applications and limitations of spot-cooling strategies, with consideration of various spot cooling approaches as well as discussing the various types and strategies of spot cooling as they relate to CAPEX and OPEX, existing infrastructure, heat rejection, complexity of installation, operation, maintenance, efficacy, and operational robustness.

A spot cooler, as defined in this paper, refers to an air cooler and refrigeration system, strategically placed near the area that requires cooling. The typical configuration of a spot cooler system includes three main components: refrigeration unit, air cooler, and heat rejection. These refrigeration units are commonly available as standard factory-packaged systems, and they come in either skid-mounted or containerised formats. The capacities of these spot cooler modules typically range from 200 to 600 kW_R duty, allowing for efficient and targeted cooling in specific underground areas.

INTRODUCTION

Underground mining operations are often complex, necessitating careful design and implementation strategies to ensure safe working environments and efficient operation. Ventilation and cooling play a pivotal role in enabling mining by ensuring suitable workplace conditions are maintained. Ventilation is the primary method for heat removal for underground mines, as cited by McPherson (2009). This process involves the supply of fresh air into the mine to replace the contaminated air generated by underground mining activities. Ventilation in underground mines plays a dual role. Firstly, it is used to dilute contaminated air, which may contain pollutants such as diesel fumes, dust, radon, and other contaminants. Secondly, ventilation is crucial for creating suitable working conditions by providing air at temperatures conducive to work.

In deeper mining operations, maintaining acceptable temperatures in active work zones becomes increasingly challenging, and simply relying on adding fresh air may not be sufficient. Mechanical cooling, which involves using refrigeration equipment to cool the air, becomes a necessary solution. This ensures that the air reaching the work zones remains at temperatures conducive to the well-being and productivity of the underground workers. The reasons for mechanical or artificial cooling are described in some detail by Brake (2001).

Common mine-wide cooling strategies

Ventilation and cooling strategies in mining operations are closely interconnected, with effective cooling relying on the delivery of sufficient ventilation to the areas where the cooling is needed.

Several strategies exist to provide cooling to mining operations where reliance on ambient intake air alone is insufficient. These strategies include:

- **Surface bulk air cooling**: In this approach, ambient intake air is cooled before being drawn into the mine, ensuring that the air entering the mine is at a lower temperature.

- **Underground bulk air cooling:** Fresh air is cooled underground at strategically determined locations, at main intakes.

- **Underground secondary and tertiary air cooling:** In cases where primary bulk air cooling is insufficient, supplementary secondary and tertiary air cooling can be implemented in a distributed network closer to production zones. The chilled water is commonly supplied from a central refrigeration plant, either on surface or underground.

  - **Underground spot cooling**: A spot cooling system refers to an air cooler and refrigeration system, strategically placed near the mining area that requires cooling. Spot cooling should not be confused with a mine-wide cooling strategy, rather it should be considered as a subset of underground secondary and tertiary cooling, focusing on strategically cooling air in specific locations near production zones.

In general, the optimum cooling strategy is strongly related to the depth of the mining operation. Figure 1 shows the general order of precedence for the selection of an appropriate mine cooling strategy, as outlined by Bluhm, Wilson and Biffi (1998) and further developed by Bluhm, von Glehn and Smit (2003). The standard approach is to prioritise surface air cooling initially, aiming to lower the intake air temperature before it enters the mine. Following this, supplemental underground air cooling is employed, either centrally or through distributed air coolers strategically placed throughout the mine. However, it is important to note that there can be exceptions to this standard approach. The determination of the most cost-effective and efficient solution requires careful analysis, considering the specific conditions and requirements of each mining operation.

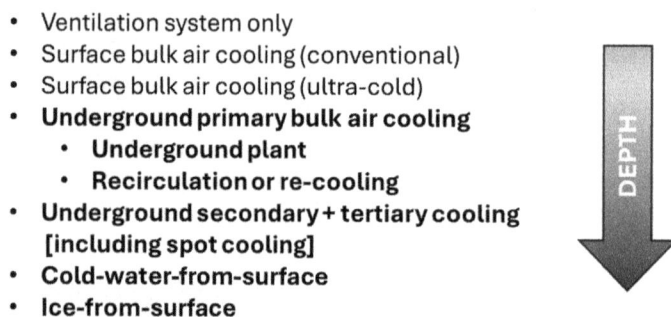

- Ventilation system only
- Surface bulk air cooling (conventional)
- Surface bulk air cooling (ultra-cold)
- **Underground primary bulk air cooling**
  - **Underground plant**
  - **Recirculation or re-cooling**
- **Underground secondary + tertiary cooling [including spot cooling]**
- **Cold-water-from-surface**
- **Ice-from-surface**

FIG 1 – Generic theoretical phases in mine cooling by Bluhm, Wilson and Biffi (1998).

The choice of placement of the refrigeration plant is a crucial consideration to ensure a viable, practical, thermally efficient, and cost-effective cooling solution. For a surface bulk air cooler system, the refrigeration plant is generally located on surface close to the bulk air cooler, which results in a compact arrangement. However, for cooling strategies involving underground air coolers, there are various options for the placement of the refrigeration plant, which can be either on surface or underground. Possible refrigeration strategies include:

- **Underground central refrigeration plants:** Underground refrigeration plants have the advantage of being located closer to the underground air coolers, thereby incurring fewer thermal losses, less pumping power and eliminating the need for high-pressure pumps and shaft piping. The capacity of underground refrigeration plants is generally limited by the quantity of available return airflow, which is used for heat rejection. Centralised underground refrigeration plants can be strategically placed to offer a robust and effective long-term cooling solution. However, these facilities require large excavations and often have high maintenance costs compared to surface plant.

- **Underground localised refrigeration plants (spot cooling):** For mines facing localised hotspots, or requiring temporary cooling, the deployment of smaller localised refrigeration plants in proximity to the hot areas can be a viable solution. However, challenges may arise with installing refrigeration equipment close to active mining operations, primarily due to harsh environmental conditions that may necessitate high maintenance. The design and placement of spot cooling systems need special consideration and must be seamlessly integrated into the mining cycle and activities to ensure effective and reliable performance.

- **Surface refrigeration plants (cold-water-from-surface):** Surface refrigeration plants present a compelling advantage with lower capital cost and superior coefficient of performance compared to their underground counterparts. However, these benefits are offset by the substantial CAPEX and OPEX costs associated with distributing chilled water to underground air coolers and pumping the return water back to surface, including the cost of providing high-pressure pumps and shaft piping. Furthermore, there are notable thermal losses in the chilled water distribution system, requiring additional cooling capacity and associated costs. Sending chilled water to depths beyond about 1000 m incurs a substantial increase in costs, prompting the introduction of energy recovery devices, which are crucial for reducing the static pressure of the water, aligning it with the practical limits of piping. Therefore, this option is generally only considered if there is inadequate heat rejection or another constraints that limit the use of underground plant.

- **Surface ice plants (ice-from-surface):** A surface ice plant achieves economic efficiency by drastically reducing the quantity of required return water that needs to be pumped back to surface, reaching only 15 per cent of the original water volume. This substantial reduction in the demand for return water enhances the cost-effectiveness of the surface ice plant compared to other cooling methods in the context of deep mining operations. For deep mining operations (say > 3000 m), the effectiveness of a cold-water-from-surface system is diminished, and high pumping costs make this option prohibitively expensive. In this case, a surface ice plant emerges as a more cost-effective alternative.

This paper focuses on the underground spot cooling approach, which, with its apparent simplicity, low initial cost, and rapid implementation timeline, becomes an attractive solution for management of heat stress within localised 'problem areas' in mining operations. However, it is essential to recognise the practical limitations that may impact its viability and long-term success. The paper aims to provide insights into various underground spot cooling systems available to mine owners, evaluating their effectiveness, mine-wide implications, and associated ownership costs and considerations such as maintenance, replacement, logistics, and longevity. The objective is to assist ventilation practitioners in choosing and arranging spot coolers to deliver efficient cooling for identified 'hot spots'.

DEFINING SPOT COOLERS AND THEIR PRINCIPLE OF OPERATION

A spot cooler, as defined in this paper, refers to an air cooler and refrigeration system, strategically placed near the mining area that requires cooling. The capacities of spot cooler modules typically range from 200 to 600 kW_R duty. In instances where higher cooling capacities are required, multiple spot coolers are often deployed together in a makeshift bulk air cooling arrangement. The spot cooler systems may utilise either water or refrigerant as the cooling fluid in the air cooler, referred to as a chilled-water-spot-cooler or a refrigerant-spot-cooler, respectively. A description of these two spot cooler types is given below.

Chilled-water-spot-cooler

An example of a chilled-water-spot-cooler is a water chiller supplying chilled water to one or more air coolers modules in a closed circuit. Heat rejection from the water chiller is generally via a condenser water circuit to a condenser spray chamber (direct contact heat exchanger) – with heat rejected to the return air.

Figure 2 shows a typical chilled-water-spot-cooler system which comprises the below major components:

- Water chiller (comprising compressor, evaporator heat exchanger and condenser heat exchanger).

- Air cooler unit(s).
- Chilled water circuit (including piping and pumps).
- Heat rejection facility.
- Condenser water circuit (including piping and pumps).

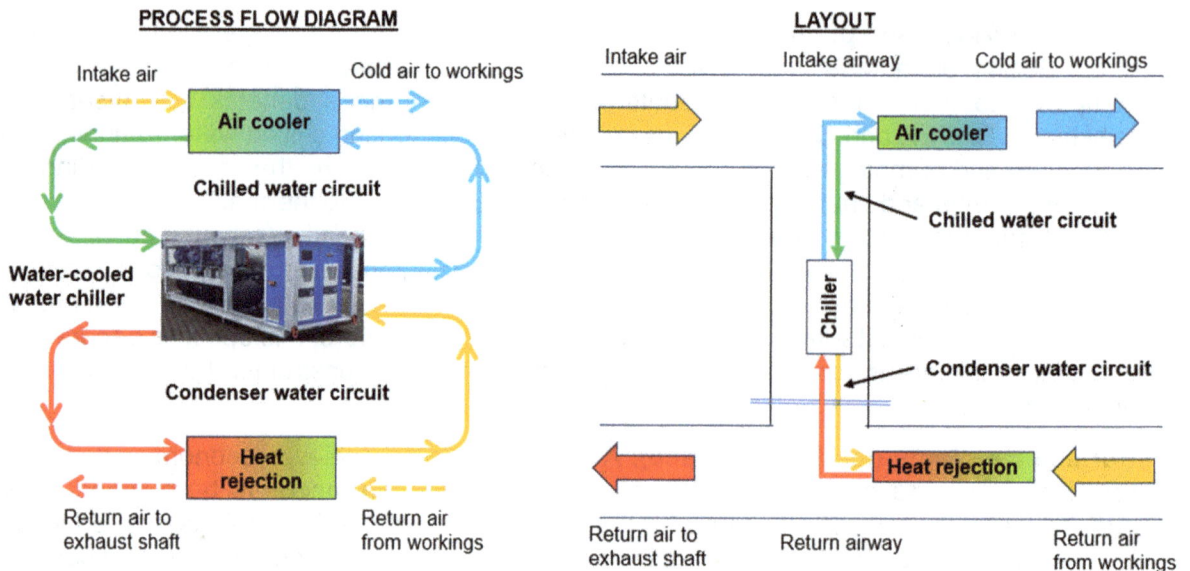

FIG 2 – Chilled-water-spot-cooler system.

A water chiller produces chilled water in the evaporator heat exchanger. The chilled water is distributed in piping to one or more air coolers located in the intake air stream. There are two common types of air coolers used in these applications, namely:

- **Indirect-contact air cooler (coils)**: In this configuration, chilled water circulates through a water-to-air cooling coil, in a closed-circuit arrangement. The coil is often mounted on a railcar chassis and equipped with an auxiliary fan, commonly referred to as a cooling car. One notable advantage of using coils is the consistent cleanliness of the chilled water circuit. Evaporator construction employs standard materials, demanding minimal maintenance. However, these coils are vulnerable to airside fouling and potential damage to the fins, necessitating regular maintenance and refurbishment.

- **Direct-contact air cooler (packed fill type)**: This set-up involves chilled water passing through a water-to-air packed fill, in an open-circuit arrangement. The fill housing is generally mounted on a railcar chassis and equipped with an auxiliary fan. Direct-contact air coolers are generally less prone to airside fouling compared to coils. However, the chilled water circuit is susceptible to fouling and necessitates regular maintenance.

A water chiller (water-cooled type) produces warm water in the condenser heat exchanger, which is distributed in piping to a heat rejection facility, which may comprise a single large, centralised facility or smaller distributed facilities. Common heat rejection sinks in mines, includes:

- **Heat rejection to return air (direct contact type)**: This approach involves circulating warm water through a condenser spray chamber, functioning as a direct contact heat exchanger. This solution is appealing due to its robustness, high performance, and relatively low maintenance requirements.

- **Heat rejection to return air (indirect contact type)**: This method involves circulating warm water through an indirect contact water-to-air coil. The coil, or cooling car, is mounted on a railcar chassis and equipped with an auxiliary fan. Placed in the return airway, this configuration has the potential to expose the heat exchanger fins to rapid fouling from dust, diesel particulates and blast fumes, leading to high maintenance and performance degradation. Therefore, this solution is best suited for applications with relatively clean return air, such as an exploration drill section in a mine.

- **Heat rejection to mine service water**: This method utilises mine service water directly in the condenser heat exchanger for heat rejection. The feasibility of this approach is contingent on the quality and reliability of the mine service water supply.

Refrigerant-spot-cooler

A refrigerant-spot-cooler eliminates the secondary chilled water circuit and associated heat exchanger by directly passing the refrigerant to the air cooler. Commercially available refrigerant-spot-cooler units comprise a complete package, including a refrigerant-to-air coil with an integrated auxiliary fan and refrigeration equipment (comprising compressor, expansion valve, water-cooled condenser, piping, and controls) mounted in a single housing on a railcar chassis. These packaged units, offer simplicity in deployment and installation, allowing for a more rapid set-up. The water-cooled condenser in this configuration operates similarly to the condenser circuit in chilled-water-spot-cooler systems, where warm water is distributed through piping to a heat rejection facility.

Figure 3 shows a typical refrigerant-spot-cooler system which comprises the below major components:

- Air cooler with on-board refrigeration system (comprising compressor, evaporator heat exchanger and condenser heat exchanger).
- Heat rejection facility.
- Condenser water circuit (including piping and pumps).

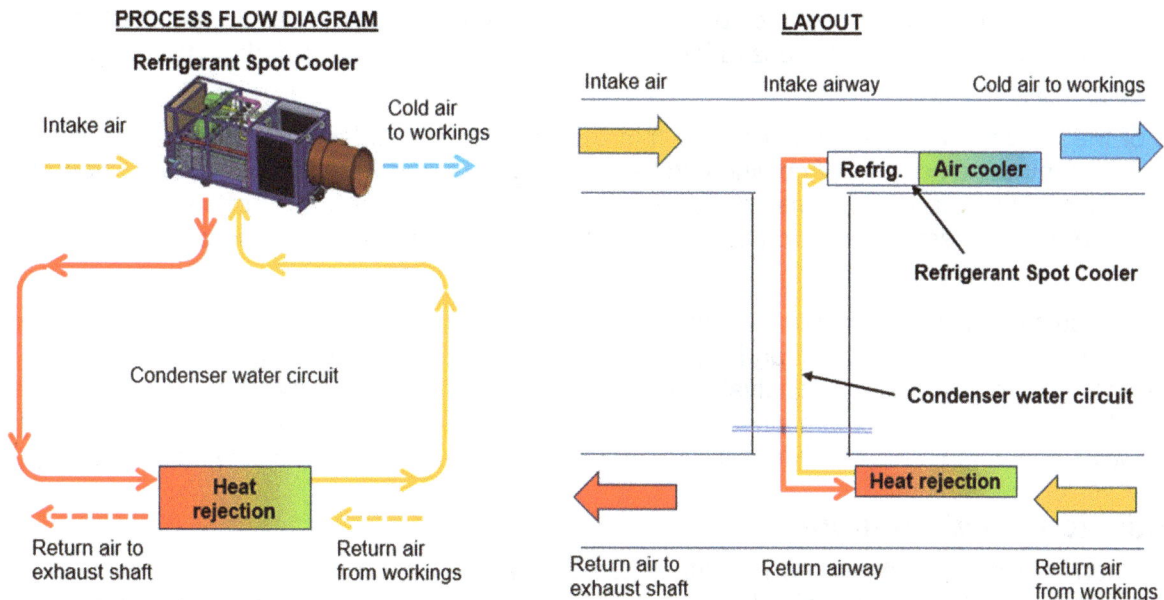

FIG 3 – Refrigerant-spot-cooler system.

Selection considerations for spot-cooler type

From a selection standpoint, two main distinctions exist between chilled-water-spot-cooler and refrigerant-spot-cooler systems:

- **Cooling capacity per unit**: Chilled-water-spot-cooler modules have cooling capacities up to approximately 600 kW$_R$. In comparison, refrigerant-spot-cooler modules are typically smaller capacity units, usually not exceeding 250 kW$_R$. To meet higher cooling requirements, multiple units must be installed. Managing an inventory of multiple refrigerant-spot-cooler units adds complexity in that more units will need to be inspected, cleaned, maintained, and placed on a rotational cycle (which will include removing them from the mine for refurbishing purposes).

- **Packaged units with integrated components**: Chilled-water-spot-coolers have separate components for the air cooler and refrigeration chiller. In contrast, refrigerant-spot-cooler units integrate these components into a single packaged unit. The coils, typically made from copper with aluminium fins, require careful handling during cleaning to prevent fin damage and maintain coil performance. Additionally, refurbishing refrigerant-spot-cooler units involves replacing the entire unit since the coil is integrated, resulting in significant costs.

DESIGN CONSIDERATIONS

Why localised air-cooling strategy?

Localised spot cooling presents significant advantages when targeting specific underground 'hot spots' that exceed the required workplace temperature limit. These compact installations offer minimal distribution losses. In contrast, central bulk air-cooling systems, whether on the surface or underground, cool a larger volume of air rather than a specific location, providing cooling on a macro-scale. Consequently, increasing the cooling duty from a central bulk air cooler will inevitably lead to over-cooling some areas of the mine, resulting in both capital and operating cost implications. Capital costs rise with the need for a larger capacity system, while operating costs increase due to the additional power required to generate the extra cooling duty.

It is common for spot coolers, or even air-cooling cars, to be located close to production zones since these areas are often at the extremities of the mine where the arriving intake air has limited cooling capacity and must be supplemented with refrigeration.

Furthermore, having passed through many airways, in which various mining activities take place, the arriving air quality is often poor. Dust and vehicle fumes affect the performance of the air coolers located in these areas, due to fouling on the outside of the finned coils and/or contamination of the chilled water circuit (for direct-contact systems). These harsh conditions can impact the longevity of the spot cooler equipment, necessitating regular maintenance and refurbishment/replacement, typically every 18 months.

One of the advantages of a spot cooler is that the air coolers are less prone to water distribution issues. In contrast, a network of cooling cars, supplied with chilled water from a central underground refrigeration plant, suffer water distribution issues as these systems are: (i) often not balanced, (ii) have more cooling cars connected than water available, (iii) share high pressure water with mining activities.

Water reticulation systems for central refrigeration plants often become unbalanced because mine cooling systems are not static and must be expanded as mining progresses outward. This necessitates ongoing network analysis to deliver the correct water flows to the air coolers. In contrast, a spot cooler, for example used to cool a development section, may simply be moved closer to the workings to maintain conditions in the heading, and does not require flow balancing.

Ventilation considerations

Ventilation and cooling strategies in mining operations are closely interconnected, with effective cooling relying on the delivery of sufficient ventilation to the areas where the cooling is needed. In cases where mines experience poor ventilation or engage in inadequate ventilation practices, relying solely on spot coolers might be a temporary and incomplete solution. To effectively tackle heat-related challenges, a comprehensive examination of the ventilation system and mining practices is essential to identify and address the underlying issues contributing to the localised heat problem. It is crucial to note that using spot coolers to alleviate heat problems, without thoroughly examining the ventilation system, may not address the root cause of the heat issue.

Heat rejection considerations

The most crucial design aspect of any underground spot cooling system is the availability of suitable heat rejection in close proximity to the area requiring spot cooling. Heat rejection should be to a location or medium where it will not add an additional heat burden on the intake air system.

Generally, heat rejection is to the return air, however there are also applications where the mine service water is used. Figure 4 shows the typical configuration for these two types of heat rejection configurations, namely return air and mine service water.

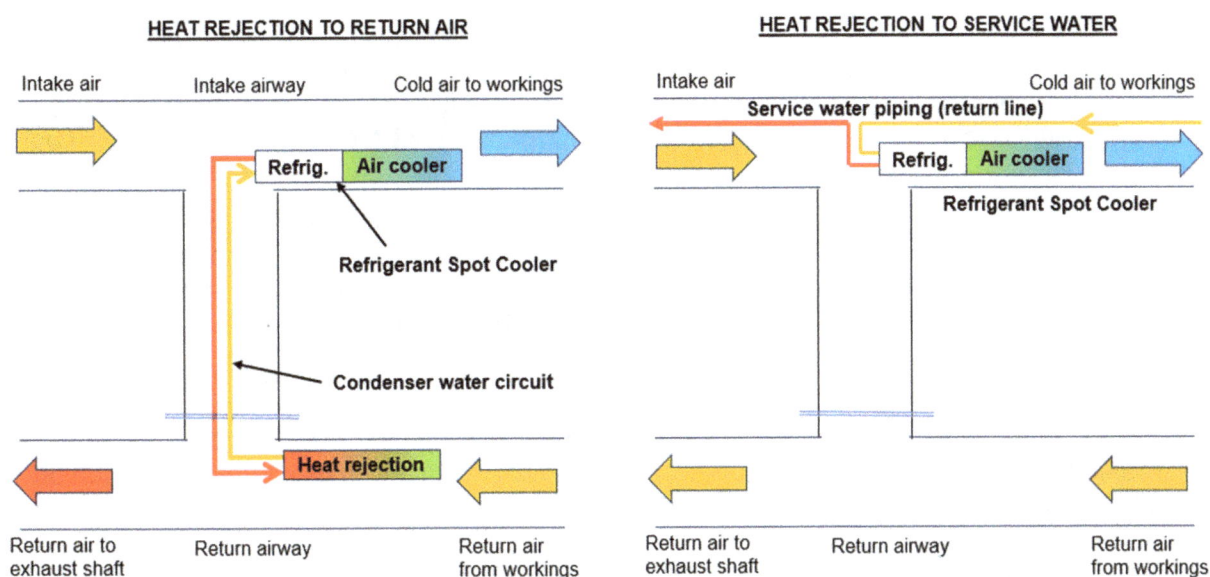

FIG 4 – Heat rejection configurations for spot cooling systems (return air and service water).

The required heat rejection capacity is equal to the evaporator duty plus the absorbed power of the compressor. Consequently, the heat rejection facility plays a critical role in the feasibility of the spot cooling system. Any capacity limitation in the heat rejection facility will lead to insufficient cooling capacity and/or an increase in compressor power, resulting in higher running costs.

Heat rejection into return air

Heat can be rejected into the mine's return air, provided it is not upstream of areas where work activities, such as maintenance, occur. Although heat rejection may be possible in these airways without pushing the air temperatures above the allowable limit, it is unlikely as heat rejection often leads to unacceptably hot and humid conditions in the return air.

Consideration should be given to the amount of available air and its temperature. High air temperatures in the return airways would require operating the chiller at a high condensing temperature, directly impacting capacity, performance, and operating costs. The practical temperature limit for the condenser water supplied to the heat rejection facility is about 45–50°C. Therefore, the upper limit for the return air wet-bulb temperature leaving the heat rejection facility is in the range of 40–45°C, defining the heat rejection capacity limit.

The heat rejection equipment can be either a condenser spray chamber (direct-contact heat exchanger) or a water-to-air coil (indirect-contact heat exchanger). Condenser spray chambers are generally more robust, perform better, and require less maintenance than water-to-air coils. Coils have lower capital costs and are easier to install but require more maintenance.

The return air quality should also be considered to avoid excessive maintenance or damage to the heat rejection equipment, or any other equipment installed in airways downstream of the condenser spray chamber.

Careful ventilation planning and energy balances are essential to properly consider the influence of heat sources in the return air, such as heat rejection facilities.

The proximity of the nearest return airway to the spot cooler location significantly influences the viability of spot cooling. The mine design and layout often drive this proximity.

Rejection into return service water

Heat rejection into the mine's return service water network is an alternative to using return air. The viability of this option depends on the quality and consistency of the service water supply. A stable return service water supply is crucial for the continuous operation of the spot cooler system, as an erratic supply could lead to nuisance tripping of equipment. The general availability of return service water largely depends on the mining method; mechanised mines use relatively low quantities, while conventional narrow reef mines utilise larger quantities. The instantaneous availability of return water will depend on the mining cycle and the pumping strategy employed by the mine, influenced by factors such as dam sizes and power constraints. Consequently, the service water availability may not always align with the mining cycle. These factors can impact the viability of the spot cooling system.

It's essential to consider only service water returning to the surface as a suitable sink for heat rejection. The utilisation of heat rejection into the service water should not impose an additional burden on any central underground refrigeration plant or surface refrigeration plant. Moreover, the impact on any surface precooling towers must be thoroughly understood.

Some spot cooler suppliers promote the use of mining water (run-off water) as a potential means for heat rejection. However, this can be a risky strategy if the path and impact of the warm return water is not clearly understood, as it may lead to unwanted heat leaking back into the intake air. Any warm return service water pipes located in intake airways should be suitably insulated and any open drains should be covered.

Make-up water considerations

The accessibility of suitable make-up water at spot cooling locations frequently poses a limitation to the practicality of rejecting heat into the return air. Water is crucial to replace evaporation losses and blow-down losses in a condenser spray chamber.

Maintaining good-quality water is essential for ensuring the reliable and effective operation of chillers. Poor water quality will lead to performance issues, increased maintenance demands for the cooling system, and may shorten the economic life of the equipment.

Maintenance considerations

In many projects, maintenance considerations are sometimes underestimated, often taking a backseat to capital and power costs. However, overlooking maintenance aspects can elevate the risks of equipment failure, unplanned downtime, and unforeseen repairs, leading to substantial financial and operational implications. Maintenance encompasses more than just implementing inspection, cleaning, or repair programs; it also hinges on the accessibility of air-cooling equipment and the efficiency of maintenance processes.

The success of maintenance programs relies on factors like the ease of conducting tasks. Proactive maintenance practices, including preventive maintenance and predictive maintenance, play a crucial role in minimising risks and ensuring the reliability and longevity of assets.

It is reasonable to assume that maintaining a surface refrigeration plant is comparatively simpler than maintaining an underground refrigeration plant, primarily due to accessibility for inspections and repairs. Similarly, managing smaller, dispersed spot cooling plants in remote locations poses an even greater challenge. The time needed to reach these plants and the complexity of the repair process are exacerbated by the absence of supporting facilities and the harsh environmental conditions characterised by heat, humidity, and dust in which the equipment operates.

From a maintenance perspective, there are several arguments in favour of using fewer spot cooler units (which favours use of larger capacity water-cooled-spot-coolers). Managing maintenance for fewer units results in a simpler maintenance schedule. With fewer units, maintenance tasks such as inspections, repairs, and replacements can be completed more efficiently, leading to increased availability.

In some cases introducing spare units that can be readily substituted for the operational unit undergoing maintenance or refurbishment may be more practical; however, this strategy increases the capital costs.

Mining method considerations

In a mining environment, areas requiring spot cooling may be situated far from the primary air return, necessitating the transportation of waste heat through significant portions of the auxiliary ventilation network. The viability of transporting waste heat through the auxiliary ventilation network, as well as the ultimate rejection of waste heat out the mine through the primary ventilation network, depends on the mining method and ventilation layout in use. Some examples are given below.

- The **room and pillar method** involves a network of rooms or 'stopes' created within the coal seam or ore deposit. The method involves leaving pillars of untouched material to support the roof of the mine, while the rooms are mined out. This creates a grid-like pattern of rooms and pillars, resembling the layout of a checkerboard.

  - **Independently ventilated room and pillar method**: This mining method involves discrete mining panels, each with its own independent ventilation system consisting of intake and return airways. This layout is well-suited for spot cooling applications. However, a key consideration is the potential impact of heat rejection on temperature conditions in the return airway, particularly if these airways are utilised for ore transport.

  - **Room and pillar method with re-use of ventilation**: This mining method, with re-use of ventilation, is commonly used in narrow reef mining operations where all infrastructure is situated in the reef plane. In this method, airflow is minimised by re-using air over multiple levels. This approach is generally unsuitable for spot cooling due to the absence of dedicated return airways on levels. Consequently, the waste heat cannot be easily expelled from working areas requiring refrigeration. Therefore, this mining method may be better suited to a bulk air cooling strategy.

- **Open stoping** or **shrinkage stoping** may be suitable for spot cooling. In these methods, large underground voids are created during the extraction of ore, and these voids may be closer to return airways or designed in a way that allows for more efficient cooling strategies, making spot cooling more viable. The suitability of spot cooling in these methods depends on factors such as the length and configuration of stopes, their proximity to return ventilation infrastructure, and the overall underground layout.

- The ventilation layouts commonly found in **cave mining methods** are generally well-suited to spot cooling. In cave mining, the generation of pollutants, contaminants, and heat is concentrated within relatively small areas. This concentration enables effective heat rejection near the working areas. Return airways in cave mining are typically situated near the working zones, facilitating substantial heat rejection. However, the decision between spot cooling and bulk-cooling depends on various factors, including the extent of temperature issues in different working areas. If temperatures exceed acceptable levels across multiple workings, opting for a bulk-cooling approach might be more economically advantageous. The choice between these strategies involves considerations of cost, efficiency, and the specific conditions within the mine.

Additionally, the choice of return airway type can influence the feasibility of spot cooling. Dedicated return raises are generally preferred over return ramps because the introduction of heat and humidity may affect personnel and equipment operating on the ramp. In older mines, the presence of timber supports can hinder heat rejection, as hot and humid conditions may accelerate wear on the supports and complicate maintenance activities.

Spot cooling can prove to be an effective solution, especially during development phases when heat loads and air flow quantities are lower compared to full production. The flexibility and ease of installing the cooling system become crucial during these initial stages. Typically, the spot cooling system will be installed in fresh through-ventilation, close to the start of the ducted development section with the air cooler in proximity to the intake ventilation duct.

Selection guide

Table 1 offers a concise selection guide to aid in assessing the feasibility of spot cooling within a given mining environment. The table presents key design considerations as questions for ventilation practitioners, accompanied by a range of potential arguments for contemplation, aiming to avoid any potential fatal flaws. Since mining operations vary significantly in design, layout, infrastructure, engineering practices, ventilation approaches, and cost considerations, among other factors, it is crucial for practitioners to conduct a thorough assessment of the specific mine and carefully evaluate all relevant elements to ascertain the suitability of spot cooling as an effective cooling method.

TABLE 1

Selection guide for spot cooling systems.

No.	Checks	Viable case for spot cooling	Obstacles for spot cooling
1	Are there isolated hot spot(s) in the mine?	• Yes, viable case for spot coolers if there are isolated hot spots.	• No, large portions of the mine needs cooling. Rather consider bulk air cooling.
2	Is the ventilation system optimised in the hot areas?	• No extra airflow available. • Increasing the airflow would exceed the design criteria (eg max velocity)	
3	Is the supply air at a temperature suitable to provide cooling?	• Air to be used for cooling is already at, or close to, the design temperature limit (ie there is limited or no air-cooling potential in the supply air)	
4	Have all alternative ventilation strategies been explored?	• Additional fresh air cannot be introduced to 'sweeten' intake air. • Reuse of air from other zones is not possible.	
5	Are the intake and return air streams relatively close to one another?	• Yes, intake and return paths are close, which will allow the straightforward installation of both air cooling and heat rejection facilities. (Tightly integrating the air cooler, chiller, and heat rejection facilities leads to lower capital costs and improved ease of maintenance).	• No, the return air stream is not in proximity therefore the condenser water must be piped to the nearest viable return air stream. Extended pipe runs will result in additional costs. • Extended pipe runs of warm condenser pipes, if not insulated, will introduce heat into the intake air stream, diminishing the effectiveness of the cooling strategy.
6	Is the nearby return airstream suitable for heat rejection?	• Sufficient air quantity is available for the required heat rejection. There is no infrastructure immediately downstream that will be affected by the high temperature and humidity conditions (access, corrosion, suitable operating conditions). Return air off the condenser spray chamber mixes with other return air suitably improving the conditions before entering equipped facilities. The return air stream is suitably far from an active work area.	• Insufficient air quantity is available. Downstream locations where sufficient air is available is far away. Infrastructure such as pipes, supports, booster fans, and pumps are located downstream of the condenser spray chamber (producing conditions that will affect the operation of the equipment, longevity and/or the mines ability to maintain this equipment. The return air is close to an active work area, subject to large quantities of dust and high blast fume concentrations.
7	Has redundancy and maintenance of spot coolers been considered? Spot coolers require regular maintenance. This often involves the removal of the air coolers from the mine.	• The mine makes use of cooling cars and has an inventory capable of supporting the additional needs for the spot cooler. • The mine has an effective air cooler cleaning program to ensure the performance of the air cooler is maintained. • The mine has suitable means to transport cooling cars to and from the shaft. • The mine can set aside shaft time for transport of cooling cars to surface.	• If cooling cars are not standard equipment in the mine, additional units will need to be acquired to facilitate continuous cooling during maintenance periods when the coils are offline. • A cooling car cleaning strategy will need to be developed and staff allocated. • The cooling car design must suit the mine infrastructure. Rail cars can only be used on mines equipped for this means of transport. Alternative means of transport will need to be factored in the design for trackless operations. • Personnel and resources will need to be assigned for cooling cars to be routinely removed from the mine (decoupled, transported on level, lifted out the mine, taken to a repairer).

COST CONSIDERATIONS

Capital cost constraints often drive interest in spot cooling solutions. While spot coolers can serve as a temporary measure to address localised heat issues, there is a temptation to extend their use as a longer-term and more widespread strategy. These small installations may show immediate improvements in localised areas, but the system's longevity and its impact on the overall ventilation system need careful consideration, especially as problems may compound over time. Table 1 highlights various design factors influencing system costs, making it essential to assess and compare all cost aspects, including capital, power, maintenance, and replacement/sustaining costs.

Capital costs (CAPEX) and sustaining costs (SUSEX)

The feasibility of a project often hinges on its capital costs and the timely availability of sufficient funds. Key drivers of capital costs include the cooling capacity needed, which influences the size and layout of the spot cooler system, as well as the proximity of power and make-up water facilities to the intended plant location. Other factors contributing to capital costs involve mining expenses, such as utilising existing tunnels, slyping requirements, or new excavation, along with the spacing between spot cooler components, determining piping and pumping needs.

While capital cost is a crucial factor in assessing a spot cooling solution, it should not be the sole consideration. Spot cooler systems typically include components with a relatively short useful life, requiring frequent replacement. Therefore, a thorough cost analysis should be conducted for the anticipated lifespan of the equipment. In scenarios where cooling is needed for a short duration during peak periods, the intended lifespan may be brief. However, for equipment expected to deliver continuous cooling over an extended period, recurring replacement costs could result in ongoing, unsustainable expenses.

When contemplating short-term spot cooling systems, it's crucial to be mindful of expenses that can be minimised or altogether avoided. Firstly, efforts should be made to minimise mining costs, and leveraging existing excavations is optimal for the cost-effectiveness of spot coolers. Attention should be directed to the physical size of the spot cooling equipment and their space requirements for maintenance. Secondly, the use of permanent civil structures should be minimised, and preference should be given to skid mounting using mobile frames. Thirdly, the utilisation of permanent pipe systems should be minimised, and the use of plastic pipes and hoses should be maximised where permitted. Additionally, efforts should be made to avoid long pipe runs as they can significantly increase costs and undermine the intended simple, low-cost, and rapidly deployable solution. Lastly, the use of non-standard mine fans should be avoided; the design should prioritise the use of standard mine fans to ensure cost-effectiveness.

When evaluating sustaining costs, several crucial factors must be considered. Refurbishment costs carry significant weight, especially for water-to-air heat exchangers, which are prone to fouling and demand intensive interventions for sustained efficiency. The frequency of refurbishment becomes a key consideration, with cooling cars and refrigerant spot cooling systems typically requiring major overhauls and removal from the mine approximately every 18 months. The replacement intervals are also pivotal, given the short lifespan of the equipment. Mining-related costs associated with refurbishment or replacement should be considered where feasible. However, quantifying these costs is challenging as they encompass factors such as shaft time, staffing requirements, and other logistical interventions (both underground and on the surface, including transport between the shaft, mine stores, and vendor). The inclusion of spare holding and its associated costs is essential, with spare cooling cars and refrigerant spot coolers, for instance, being necessary to ensure continuous cooling when these units are removed for refurbishment.

Operating costs (OPEX)

Power cost is undoubtedly a crucial factor when assessing the feasibility of a spot cooling system in comparison to a central cooling system. Generally, spot cooling systems tend to exhibit lower power consumption due to their ability to provide targeted cooling with high positional efficiency. While power cost is significant, it is not the sole determining factor in this comparison.

Maintenance poses a substantial consideration in evaluating spot cooling systems. Although direct costs may not initially seem substantial, challenges arise with systems that are inherently difficult to maintain. Spot coolers are often remote, positioned at the far reaches of the mine, and may experience infrequent maintenance interventions. This can result in reliability issues, as poorly maintained equipment is more prone to trips and component failures. The consequent downtime from non-operational cooling equipment far exceeds the direct maintenance costs, significantly impacting mining operations. When assessing spot cooling systems, factors such as accessibility and travel time to the spot cooler location, the availability of tools, replacement units, and spares, as well as the existence of maintenance facilities at the site, must be carefully considered. Additionally, environmental conditions at the spot cooler site, including dust, fumes, and temperatures, play a crucial role in the ease with which maintenance can be carried out.

In conclusion, operating costs for spot cooling systems in mining demands a strategic balance between immediate cost considerations and the long-term sustainability of operations. While minimising day-to-day expenses is crucial, it is equally essential to anticipate and plan for maintenance, refurbishment, and replacement costs inherent in spot cooling solutions. The accessibility and reliability of maintenance procedures, coupled with the environmental challenges of underground mining, play pivotal roles in the overall OPEX equation. Therefore, a comprehensive and forward-looking approach to OPEX, considering both routine maintenance and unforeseen disruptions, ensures the continued effectiveness and efficiency of spot cooling systems in the dynamic and demanding realm of mining operations.

CASE STUDY

Brief introduction

A case study, outlined here, is provided by the authors. The case study aims to provide valuable insights into the planning considerations for implementing a spot cooler system. Focused on a deepening expansion project within a mechanised gold mine, the mine's shallower sections currently do not demand cooling. However, projections for the deeper section, extending to a depth of 1200 m below the surface, indicate a significant requirement for approximately 1800 kW_R of air cooling duty. In response to this specific and localised cooling need, the proposed solution involves the strategic deployment of a spot cooler system. This choice is underpinned by the concentrated cooling demand in particular areas and the advantageous presence of a return raise system in the ramp, facilitating effective heat rejection.

This case study provides a comprehensive overview, encompassing the process flow and energy balance employed for equipment sizing and power requirement estimation. Additionally, the proposed layout of the spot cooling system within the underground mine is delineated, shedding light on the mining requirements. The study also presents the estimated cost and lead time required for implementing the project.

Process flow and energy balance

The proposed refrigeration system is designed to incorporate three chillers, each with a cooling duty of 608 kW_R. These chillers are configured with condenser and evaporator heat exchangers arranged in parallel. Each chiller is estimated to have an electrical absorbed power of approximately 164 kW_E, with a chiller COP (Coefficient of Performance) of 3.7. The chiller evaporators are designed to provide 43 L/s of chilled water at 11.0°C, which will be distributed to closed-circuit cooling coils arranged in two air cooler units. Each cooling coil unit will be connected to a dedicated fan unit. The two air coolers are tasked with cooling a 60 kg/s airflow, reducing air temperatures from 31.1°Cwb/32.6°Cdb to 24.4°Cwb/25.3°Cdb. The return water from the coils is expected to be around 21.1°C. Four chilled water pumps will be provided to facilitate chilled water circulation, including one standby unit. The number of active chilled water pumps will align with the number of operational chiller modules, ensuring an efficient and reliable cooling system.

The heat rejection process for the three chillers, each with a thermal duty of 781 kW_R, is based on the use of a direct-contact type condenser spray chamber. Collectively, this results in an overall heat rejection duty of 2402 kW_R. Within the condenser spray chamber, 81 L/s of condenser water will be cooled from 42.0°C to 35.1°C. The air in the condenser spray chamber, with an airflow of 69.2 kg/s, will experience an increase in temperature from 31.0°Cwb/33.2°Cdb to 34.8°C saturated. The circulation of condenser water will be facilitated by one condenser water pump, accompanied by a standby pump unit, ensuring the continuous flow of water from the basin through the condenser heat exchangers of the chillers and back to the first spray stage of the condenser spray chamber. In the second spray stage, a single re-spray pump will be employed to re-spray the water.

At the underground site, a supply of fresh make-up water is required to compensate for evaporation losses from the condenser spray chamber. To ensure the water quality in the condenser water circuit is maintained, a small blow-down water flow is implemented, preventing the accumulation of dissolved solids.

Table 2 shows the energy balance analysis for the case study and Table 3 shows the typical equipment specifications. The energy balance is 1824 kW for the chilled water circuit and 2402 kW for the condenser water circuit. The total absorbed power for the cooling system is projected to be 667 kW, delivering a total cooling duty of 1800 kW. Consequently, the system's Coefficient of Performance (COP) is estimated to be approximately 2.7.

TABLE 2
Energy balance for case study.

Chilled water circuit energy balance	Duty, kW	Condenser water circuit energy balance	Duty, kW
Heat added to water circuit:	**1824**	**Heat added to water circuit:**	**2402**
Air cooling duty (excl. fan power)	1800	Chiller no.1 condenser	781
Chilled water pump power	24	Chiller no.2 condenser	781
		Chiller no.3 condenser	781
		Condenser pump power	34
		Respray pump power	25
Heat removed from water circuit:	**1824**	**Heat removed from water circuit:**	**2402**
Chiller no.1 evaporator	608	Condenser spray chamber	2402
Chiller no.2 evaporator	608		
Chiller no.3 evaporator	608		

TABLE 3

Equipment specifications for the case study.

Equipment item	Value
Air coolers:	
Air cooling duty (excluding fan power)	1800 kW
Available intake airflow	50 m³/s (60 kg/s)
Type	Cooling coils (closed-circuit type)
Number of modules	2 off
Module capacity (excluding fan power)	900 kW$_R$
Arriving air temperature	31.1°Cwb/32.6°Cdb
Discharge from air cooler (after fan)	24.4°Cwb/25.3°Cdb
Chilled water flow (total)	43 L/s (11.0°C to 21.1°C)
Chillers:	
Total cooling duty	1824 kW$_R$
Type	Water-cooled-spot-cooler
Number of modules	3 off
Module capacity	608 kW$_R$
Electrical absorbed power (per chiller module)	164 kW$_E$
COP	3.7
Heat rejection:	
Required heat rejection duty	2400 kW
Type	Condenser spray chamber (direct contact)
Available return airflow	60 m³/s (69.2 kg/s)
Arriving return air temperature	31.0°Cwb/33.2°Cdb
Discharge from condenser spray chamber	34.8°C saturated
Condenser water flow (total)	81 L/s (42.0°C to 35.1°C)

Proposed site layout

Figure 5 shows the proposed location for the underground spot cooler plant, which is to be developed in virgin ground off the ramp. This section of the ramp is in its initial stages of development, presenting an opportunity to optimise the site layout, and achieve a compact layout The envisioned layout includes a new plant room and air cooler facility developed in the intake raise system and a new condenser spray chamber developed in the return raise system. A new electrical room is proposed to be developed off the ramp.

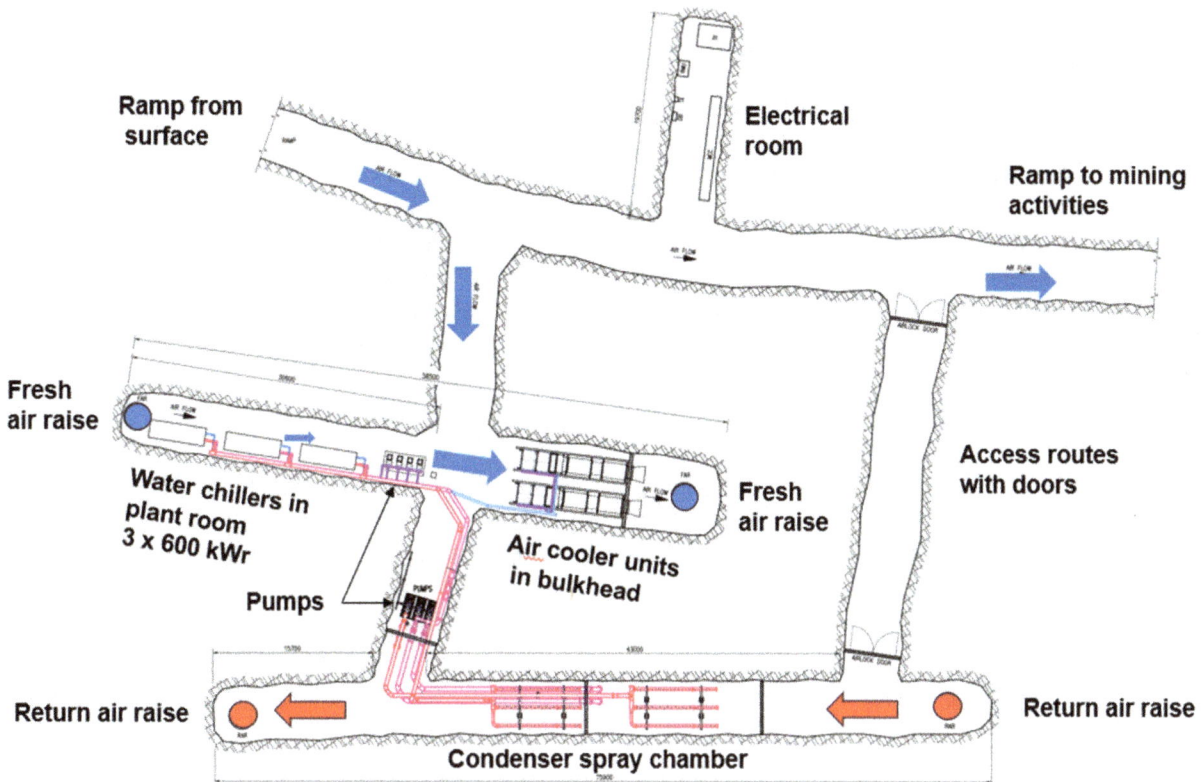

FIG 5 – Typical layout of an 1800 kW$_R$ spot cooler system for the case study.

Cost and schedule

The budget cost estimate for the turnkey supply of the spot cooler system is about USD5.8 million, as provided by a vendor. It's important to note that this estimate does not encompass the provision of high-voltage power supply to the underground site and the supply of make-up water. For this project, the supply of the chiller units has been identified as a critical path item. The planned implementation schedule spans 64 weeks, including the commissioning phase, as shown in Figure 6. The project is strategically planned for implementation in the year 2026.

TASK	DURATION	TIMELINE (WEEKS)
		4	8	12	16	20	24	28	32	36	40	44	48	52	56	60	64
Procurement process for the detailed design scope	6 weeks	■															
Technical specifications for long-lead items	3 weeks	■															
Procurement process for long lead items	5 weeks	■															
Fabrication and supply of spot cooler (chiller unit)	30 weeks		■	■	■	■	■	■	■								
Shipping and clearance	6 weeks											■					
Delivery to mine site	2 weeks												■				
Installation of spot cooler and piping connections	4 weeks												■				
Final electrical installation works	4 weeks													■			
Commissioning, training and handover	4 weeks																■
Total lead time	**64 weeks**																

Note 1 - Only the critical path is shown
Note 2 - For this example, the critical path is constrained by the supply of the chiller unit
Note 3 - The above critical path assumes that suitable excavations are available or are provided in parallel with the chiller supply.

FIG 6 – Critical path schedule for the case study.

CONCLUSIONS

Spot coolers present a viable solution for addressing short-term localised heat challenges, commonly known as hot spots, within mining operations. However, their applicability is not universal, and several key factors must be considered when contemplating their use. Special attention should be given to the availability, quantity, and quality of return air in proximity to the hot spot to ensure the spot cooling design is compact, cost-effective, and efficient. Emphasis is placed on the short-term use of spot coolers due to their accessibility, maintenance-intensive nature, high wear rate of components, and construction, which prioritises speed of delivery over longevity. In summary, spot cooling is a flexible air-cooling solution in mining operations where traditional cooling methods fall short, provided logistical and operational considerations are carefully addressed.

The obstacles to implementing spot cooling in mining environments are multifaceted. Challenges include the potential need to transport condenser water to distant return air streams, which may incur significant costs with long pipe runs. Furthermore, if condenser water pipes are placed in intake airways without insulation, heat may be introduced into the intake air stream, negating part of the cooling effect. Additionally, logistical considerations related to equipment installation, replacement, and maintenance strategies adds complexity to the long-term viability of these installations. These obstacles underscore the importance of meticulous planning and the implementation of effective mitigation strategies when contemplating spot cooling solutions as part of comprehensive cooling strategies in mining operations.

REFERENCES

Bluhm, S J, Wilson, R B and Biffi, M, 1998. Optimised cooling systems for mining at extreme depth, Proceedings CIM/CMM/MIGA Montreal 98, CIM.

Bluhm, S, von Glehn, F and Smit, H, 2003. Important basics of mine ventilation and cooling planning, in *Journal Mine Ventilation Society South Africa*, 57(1):15.

Brake, D J, 2001. The application of refrigeration in mechanised mines, in *Proceedings of the Australasian Institute of Mining and Metallurgy,* 306(1):1–10.

McPherson, M J, 2009. Subsurface ventilation engineering, mine ventilation services (Springer Science and Business).

Examining use of absorption chillers to produce mine air cooling from power generation plant waste heat in remote sites

K Boyd[1], M Brown[2], C McGuire[3] and G Cooper[4]

1. Ventilation EIT, Hatch Ltd., Mississauga ON Canada. Email: boydk@hatch.com
2. Process Engineer, Hatch Ltd., Mississauga ON Canada. Email: melissa.brown@hatch.com
3. Senior Ventilation Engineer, Hatch Ltd., Mississauga ON Canada.
   Email: chris.mcguire@hatch.com
4. Senior Mechanical Engineer, Hatch Ltd., St. John's Newfoundland Canada.
   Email: george.cooper@hatch.com

ABSTRACT

For many Australian underground mining operations, the remote nature of the mine has historically led to a reliance on diesel-powered generation stations for site electrical requirements. In addition, the increasing demand for critical minerals is causing mines to expand and increase production at deeper extents. This, combined with hot ambient surface conditions, often requires the implementation of a mine air cooling system to achieve acceptable underground working conditions. Conventional vapour-compression refrigeration systems for mine air cooling can represent a large electrical power user, straining site resources. With major mining companies targeting 'net-zero' operations by 2050, conventional vapour-compression systems may not be favourable due to the contribution of greenhouse gas emissions from additional diesel consumption to power the refrigeration machines. Large quantities of waste heat are often available from diesel generators, presenting an opportunity to implement a form of waste heat recovery to improve the site energy balance, resulting in reduced power consumption. One such potential technology is absorption chillers, which use a thermo-chemical process driven by an external heat source to drive the refrigeration cycle. This paper compares the use of absorption chillers with traditional vapour-compression machines for a hypothetical Australian mine air cooling application, highlighting the potential reduction in electrical demand, operating cost, diesel consumption, and site greenhouse gas emissions (CO_2-equivalent). Consideration for both diesel and battery electric mobile equipment fleets are examined as part of the case study comparison.

INTRODUCTION

The world is recognising the need to lower greenhouse gas (GHG) emissions through the global energy transition from fossil-fuel based systems to more sustainable, long-term energy strategies. This will likely benefit the mining sector, as many critical minerals (eg cobalt, lithium, nickel, manganese, and copper) are integral to the production of clean energy technologies. Regulatory pressure and pressure from various other stakeholders with respect to reducing emissions have led the largest mining companies including BHP, Rio Tinto and Vale to target 'net zero' operations by 2050, with varying interim carbon reduction goals of 30–50 per cent by 2025–2035. Mines will require creative solutions to expand production to deeper extents to extract critical minerals while simultaneously reducing their GHG emissions.

The primary sources of GHG emissions in the mining industry are power generation and mobile equipment. Some companies are already transitioning from traditional diesel mobile equipment, which typically account for up to 30 percent of on-site Scope 1 GHG emissions, to battery electric vehicles (BEVs) and many others are investigating the switch. However, for many remote operations, including those in Australia, reliance is on diesel-fired generators for site electrical requirements, which degrades the GHG benefits of BEVs. Typically, power generation accounts for 30 to 35 per cent of these operations' CO_2-equivalent emissions. Although changes to sustainable fuels for power generation or switching to renewable energy would help address this issue, these would require significant capital expenditure and technological advancements to meet operational requirements. This long-term opportunity requires further investigation, however there are many short-term decarbonisation actions that can be made in the present to reduce GHG emissions. Short-term actions should focus on readily available technologies with a reasonable payback period. In mining applications, this can include looking for ways to reduce the traditionally high electrical consumption from the ventilation and cooling systems through alternative technologies.

This paper compares the impact of fleet type (diesel or BEV) and chiller technology (vapour-compression or absorption) on a cost and environmental basis for a hypothetical Australian mining application.

BACKGROUND AND LITERATURE REVIEW

Recovery of waste heat for use in mine air heating is currently done in cold climates, where recovered heat can be used to create a hot water/glycol or brine stream to be used in the mine air heating process. Installation of an indirect contact air-liquid heat exchanger is used to pre-heat air and supply significant portions of seasonal mine air heating demands at mines in Canada and Sweden, perhaps others (Jones *et al*, 2022). However, the use of waste heat for underground mines is not as prevalent in hot climates, as there are limited use cases for hot liquid streams when bulk heating is not required.

Absorption chillers are an established technology which are available from many of the same manufacturers as conventional vapour-compression chillers (ie type commonly used for mine air cooling). The operating principle of absorption chillers is identical to that of conventional chillers, where a refrigerant fluid is made to undergo a reversible phase change process which is endothermic when evaporating, taking energy from the process fluid to be chilled, and exothermic when condensing, releasing energy to a heat sink (typically air or a separate cooling water circuit). Conventional chillers use an electrically powered mechanical compressor to drive the refrigeration cycle. Where absorption chillers differ is that the refrigerant is typically pure water, and the refrigeration cycle is achieved using a thermo-chemical process driven by an external heat source. The reaction can be facilitated through the use of a relatively low-quality waste heat source, such as low-pressure steam or water in excess of approximately 70°C in single-effect absorption chillers. A general schematic of a single-effect, indirect fired absorption chiller is shown in Figure 1; this represents the absorption chiller type considered for the case study analysis. Alternative absorption chiller types include double-effect, indirect fired, which operate with higher quality waste heat (eg medium-pressure steam) and utilise a high-pressure second generator to reduce the amount of heat required per kilowatt of refrigeration produced as the driving source of energy (ASHRAE, 2022).

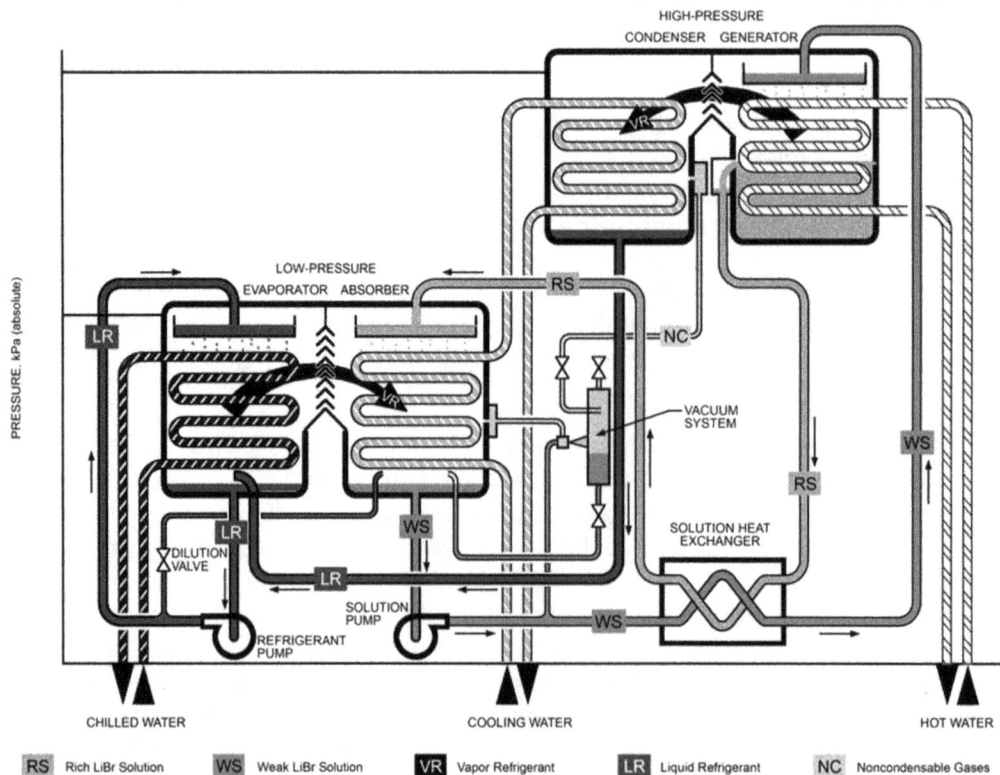

FIG 1 – Single-effect, indirect fired absorption chiller schematic (ASHRAE, 2022).

The absorption chiller technology is promising for use in remote mining applications which are often served by local diesel generators to furnish the site's power needs. Waste heat recovery from diesel generators is a proven technology often applied to combined heat and power (CHP) cogeneration facilities as well as combined heat and power and cooling trigeneration facilities and would be within the typical optional offerings from major equipment suppliers. Waste heat recovery from diesel engines captures heat from the engine oil cooler, charge air coolers, engine jacket coolers and engine exhaust and can be used to produce both hot water and steam for cogeneration and trigeneration applications. This paper explores the potential to include waste heat recovery and absorption chillers to minimise cost and GHG emissions associated with the underground air-cooling demand in remote mine sites.

The authors have reviewed the available literature and identified two previous published papers which consider the use of absorption chillers for use in bulk air cooling of underground mine ventilation air in the hard rock mining context. The first of these, a paper on the Gwalia Gold mine, located in Western Australia, air cooling system, includes a 4.5 MW_R absorption chiller installation supplementing an existing conventional ammonia chiller and was presented at the 2015 AusIMM Mine Ventilation Conference. The paper indicates that an absorption chiller system was selected 'based on the lowest net present value of the operating and capital expenditure coupled with its low maintenance requirements' (Broodryk et al, 2015), but does not include supporting values to quantify the estimated difference. This paper seeks to provide some quantitative comparison of an absorption chiller installation against conventional vapour-compression systems. It is also notable that the Gwalia installation recovers waste heat from the exhaust gases of the gas-fired electrical generators on-site directly as the heat source for the absorption chillers. Through discussion with selected OEM's, the authors found that this direct exhaust gas approach is likely less favourable due to reduced reliability of the chillers and increased risk to the generators due to fluctuating back pressure. To address these risks, this paper is considering a case with hot water as the waste heat source, produced at the generator's jacket cooler and an exhaust recovery shell-and-tube heat exchanger fitted to the exhaust pipe, which is consistent with OEM standard offerings.

The other published work considers a trade-off study where, in the absorption chilling case, new generators are purchased outright to provide the driving heat source for the chillers, and only a small portion of the generator power output is credited as operating cost savings for the refrigeration system. The complete capital and operating costs associated with the generators are included, dominating the analysis and resulting in an unfavourable net present cost (NPC) for the absorption chiller case (del Castillo, 2010).

Other known installations in mining applications are at coalmine sites in Poland and China, where waste heat is recovered from gas generators associated with mine drainage activities (Nadaraju, 2021). Further details on these systems are not described here as the focus of the case study analysis is on technology application in the hard rock mining context.

The key insight under investigation in this paper, which the authors feel has not been explored in the existing literature, is that the capital cost required for the driving heat source, ie the power generation equipment for the mine site, would usually be a sunk cost which is independent of the mine's cooling needs. The analysis that follows seeks to explore the implementation of absorption chillers with existing (or planned) generators, quantifying the incremental increase to site power consumption – and by extension the incremental increase in diesel consumption for power generation – upon commissioning of a mine air cooling plant. A comparison of absorption chillers and conventional vapour-compression chillers is presented on the basis of NPC and GHG emissions to examine the business case for the implementation of such technologies on remote mine sites.

METHODOLOGY

Overview

The analysis presented in this paper is based on a hypothetical underground mining operation which is progressing towards an extended exploration program with definition drilling being conducted from excavations approximately 850 metres below surface. The case study site assumes similar climatic and geotechnical characteristics as would be expected for South Australian mines, to provide a reference for thermal modelling of the underground workplace temperature and humidity conditions. Ventilation demand was identified based on a mobile equipment fleet required to develop, construct and maintain the mine access and egress declines, critical infrastructure, and exploration platform. It is assumed that the mine does not have access to the power grid and is supported by on-site diesel power generation.

A total of four cases have been developed, in which all combinations of the following key parameters are analysed:

- Chiller Technology: Absorption Chiller or Conventional Vapour-Compression Chiller.
- Mobile Equipment Fleet Composition: Diesel or Battery-Electric Vehicle (BEV).

All cases are assessed on an NPC basis, accounting for both the initial capital cost and an annual operating cost complete with power, water and carbon tax impacts. The power cost is inclusive of electrical consumption for primary fans and all refrigeration infrastructure, in addition to the power cost for BEV charging.

Schematics of the examined cases are shown in Figure 2. Additional information on the configurations is outlined in the following sections.

Absorption Chillers (Diesel and BEV Cases)

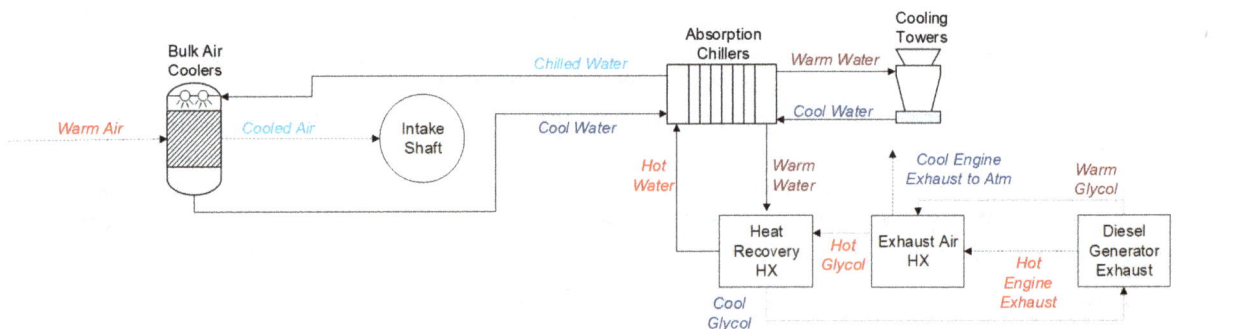

Conventional Chillers (Diesel and BEV Cases)

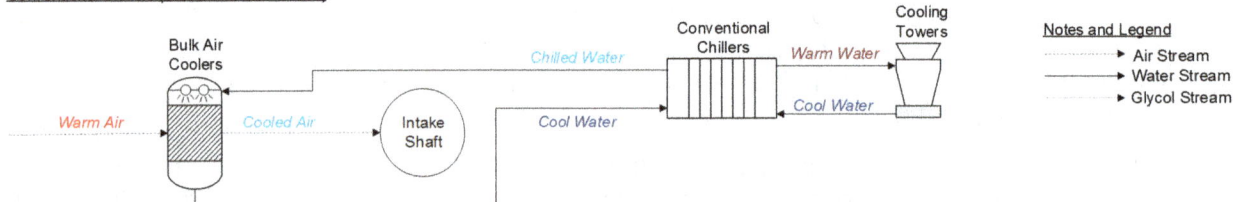

FIG 2 – Cooling schematics.

Design basis and assumptions

The total ventilation rate was obtained from a recent, representative project site for an entirely diesel-powered mobile equipment fleet. The ventilation rate for the BEV fleet was reduced by 20 per cent as a conservative assessment of potential ventilation savings.

Ventilation network modelling (including preliminary thermodynamic modelling) have been developed in VentSim DESIGN™ version 5.4 (by Howden) to identify fan power consumption and to validate the cooling required to meet workplace temperature conditions.

Key parameters applied to the case study analysis are summarised in Table 1.

TABLE 1

Summary of assumed design parameters.

Parameter	Units	Value
Diesel-fired generators		
Generator station size (total)	MW$_e$	12.6
No. of generators operating	No.	7
No. of purchased generators (including standby)	No.	9
Heat recovery per generator	MW	1.39
Ventilation requirements		
Surface air flow (diesel)	m³/s (kg/s dry)	213 (240)
Surface air flow (BEV)	m³/s (kg/s dry)	178 (200)
Surface design ambient temperatures	°C wb/db	20.4/32.5

The following assumptions apply to the analysis and have been applied to all four cases unless specifically noted otherwise:

- Site power generation is by diesel-fired generators, which will be fitted with an exhaust shell-and-tube heat exchanger and associated heat recovery plate-and-frame heat exchanger for the absorption chiller case.

- Absorption chillers will be driven by waste heat provided in the form of hot water from the heat recovery plate-and-frame heat exchanger. It is assumed that the hot water from the heat recovery heat exchanger will be at 90°C.

- Absorption chillers are assumed to have a coefficient of performance (ie ratio of the cooling capacity to the amount of heat from the driving heat source) of 0.8.

- Direct-contact cross-flow packed towers are used for bulk air cooling, supplied with chilled water from the chillers, for all cases.

- Conventional chillers assume R-134a refrigerant. Consideration for low-GWP refrigerants can be made during detailed, site-specific engineering studies in the context of phase down or phase out of hydrofluorocarbon refrigerants depending on the jurisdiction of interest. In addition, ammonia was not considered as a refrigerant for conventional chillers due to toxicity concerns and associated layout impacts. Ultra-low chilled water temperatures afforded with ammonia chillers were also not required in these case study scenarios.

- Conventional chillers are assumed to be water-cooled, to allow as direct a comparison with the absorption chiller case as possible. A trade-off on water-cooled versus air-cooled chillers can and should be made during detailed, site-specific engineering studies:

  - For reference, Gwalia Gold mine investigated alternative heat rejection systems (eg air-to-air heat exchange, mine pit cooling ponds etc) to minimise water requirements for their absorption chillers but found them to be 'costly' (Broodryk *et al*, 2015). Ultimately, they implemented new systems (ie additional boreholes) to provide the required water (Broodryk *et al*, 2015).

- Conventional chillers are assumed to have a coefficient of performance (COP) of 6.0. Detailed studies could look to optimise the cooling circuits by trading off higher COPs with higher associated piping and pumping costs, or alternative chiller piping configurations (eg series-counterflow).

- It is assumed that this mine for the case study will have access to sufficient water supply to support the use of open cooling towers for heat rejection. Water supply is assumed to be via truck, with costs based on a reference project in South Australia:
  - Note that additional water supply required for blowdown to manage dissolved solids has not been considered at this time. This would represent an addition to the stated water consumption rates, with exact quantities depending on site-specific water quality.
- A compact cooling plant layout has been assumed with minimal process impacts from pump heat and thermal losses. An exception to this is the absorption chiller hot water circuit, where the available hot water temperature was reduced by 2°C between the heat recovery heat exchanger and the chiller.

Calculation methodology

Site-wide power and diesel demand were obtained from a recent, representative project and adapted for use within this study. A detailed diesel versus BEV mobile equipment trade-off study previously completed for the representative project site provided the fuel consumption and charging power requirements for both fleet compositions. Similarly, the baseline generator make, model and quantity for site-wide power needs were utilised and the quantity was adjusted as necessary for each of the four studied cases to account for their specific generation capacity requirements including maintaining an excess capacity to handle short-duration demand spikes, and installed spares for redundancy in the event of an outage. The base, site-wide power consumption exclusive of ventilation and refrigeration loads or BEVs is approximately 7.8 MW. Installed base generation capacity for the site is 12.6 MW, which includes 2.8 MW of installed spare capacity and 2 MW available for power station auxiliaries, ventilation and cooling loads and applicable excess capacity.

For absorption chiller cases, each diesel generator was assumed to be fitted with heat recovery infrastructure available from the OEM. The available flow of hot water from the heat recovery heat exchanger was estimated based on the available heat rejection, as provided by the OEM, and the thermal performance of the CHP from a recent, reference project. Based on this calculation, a driving source for absorption chilling of 77 kg/s of 90°C water was assumed with seven generators operating.

A maximum achievable cooling duty of 7.0 MW_R was calculated for the absorption chiller cases based on the waste heat availability and absorption chiller capabilities. This translates to a cooled air outlet temperature of ~10°C for the diesel scenario. This was validated with available ventilation network models to confirm acceptable underground working conditions. The same cooling parameters (7.0 MW_R for a 10°C cooled air temperature) were applied to the diesel conventional chiller case. For the BEV cases, the same cooled air temperature of 10°C was targeted, requiring a cooling duty of 5.9 MW_R at the BEV air flow rate. This is expected to be a conservative estimate of the required cooling duty for the application, given the large heat load reduction expected with BEVs in place of diesel mobile equipment.

Power and diesel consumption are estimated on a quarterly basis. Consumption for a typical year (Year 6 of the project) was selected, as it represents a time with the maximum fleet size and utilisation and the corresponding ventilation and cooling loads are also at their maximum. The entire year is considered to account for seasonal variation in cooling load in determination of annual average emissions and consumptions. Annual cooling loads and associated water requirements are calculated based on average monthly weather data. As mentioned previously, water consumption only accounts for evaporation across the cooling towers, as this is typically the largest contributor to total makeup water requirements. No consideration is given for blowdown as part of water quality management, as this is dependent on site-specific makeup water quality. In addition, for conservatism in water requirements and simplicity in the case study analysis, no credit is given for supplementing make-up water requirements using condensed moisture from the bulk air coolers.

The NPC comparison assumes the site sustains operations at the maximum ventilation and cooling load for five years.

RESULTS AND DISCUSSION

Cooling system sizing

Cooling systems were sized for each of the four cases based on the design basis, assumptions and methodology outlined previously. Key parameters are summarised in Table 2.

TABLE 2
Cooling system comparison.

Parameter	Absorption chiller		Conventional chiller	
	Diesel	BEV	Diesel	BEV
Cooled air flow (dry kg/s)	240	200	240	200
Cooled air temperature (°C)	10.0	10.0	10.0	10.0
Waste heat from generators (MW_T)	8.8	7.4	8.8	7.4
Cooling duty (MW_R)	7.0	5.9	7.0	5.9
Heat rejection duty (MW_T)	16.1	13.4	8.2	6.9
Cooling tower evaporation rate – design (kg/s)	6.6	5.6	3.4	2.8

Key differentiators between the cases include:

- Cooling duties are lower for the BEV cases due to the same cooled air temperature, but lower air flow rate compared to the diesel cases. The diesel fleet requires higher underground ventilation air flows to account for diesel fleet exhaust emissions underground.

- Higher heat rejection requirements to the condenser water circuit for the absorption chiller scenarios by a factor of approximately 2. This ultimately ends up as additional load on the cooling towers and results in higher water evaporation rates (ie higher make-up water requirements).

Cost comparison

Comparative capital costs for the cooling systems are compared in Table 3. Costs assumed constant, or with insignificant differences between cases, are excluded from the analysis. Items included in the developed capital costs are:

- Additional diesel-fired generators, if required.

- Heat recovery systems for diesel-fired generators for the absorption chiller scenarios:
  - Inclusive of exhaust shell-and-tube heat exchanger and heat recovery plate-and-frame heat exchanger.

- Chillers (either absorption or conventional, depending on the case).

- Bulk air coolers complete with fans.

- Conventional cooling towers.

- Chilled water pumps and piping.

- Cooling tower water pumps and piping.

- Hot water pumps and piping (for absorption chiller cases).

- Electrical substation.

- Allowances for additional disciplines including:

- o Civil/structural/architectural.
- o Electrical cabling.
- o Automation and instrumentation.

TABLE 3

Summary of capital costs for all cases (AUD).

Item		Absorption chiller		Conventional chiller	
		Diesel	**BEV**	**Diesel**	**BEV**
Mechanical	Additional generators	-	-	3 080 000	3 080 000
	Heat recovery system	2 040 000	2 040 000	-	-
	Cooling plant	7 290 000	6 250 000	5 510 000	4 740 000
Piping		870 000	780 000	480 000	420 000
Civil/structural/architectural		2 480 000	2 210 000	2 200 000	2 000 000
Electrical		1 400 000	1 240 000	830 000	710 000
Instrumentation		750 000	660 000	440 000	380 000
Total direct cost		14 830 000	13 180 000	12 540 000	11 330 000

As can be seen from Table 3, a key advantage of the absorption chiller system is that the electrical load is sufficiently small such that it can be accommodated within the operating margin of the existing generation system on-site, not requiring additional generators on-site to support. However, retrofit of the existing generators on-site to include heat recovery systems will erode the savings to some degree.

The total direct cost is ~16–18 per cent higher for the absorption chiller system compared to the conventional system for an equivalent fleet composition, owing to the higher capital cost for the absorption chillers themselves, increased heat rejection requirements (ie more cooling towers, and larger cooling tower water pumps and piping) and the addition of hot water pumps and piping.

Comparative operating costs for each option are presented in Table 4. Similar to the capital cost assessment, costs assumed constant or with relatively insignificant differences between cases (eg maintenance and water treatment) are excluded from the analysis. Operating costs quantified for the study include:

- Diesel fuel consumption for the mobile equipment fleet.
- Power generation cost, quantified as diesel fuel consumption at the generator station, for the following:
  - o Power station auxiliaries.
  - o Primary mine ventilation fans.
  - o Mine air-cooling plant.
  - o BEV mobile equipment charging.
- Make-up water to account for evaporation in the cooling towers.

TABLE 4

Operating cost summary for all cases.

Item	Absorption chiller		Conventional chiller	
	Diesel	BEV	Diesel	BEV
Diesel consumption for mobile equipment (L/day)	1855	0	1855	0
Diesel consumption for power generation (L/day)	9239	10 237	11 461	12 115
Subtotal diesel consumption (L/day)	11 094	10 237	13 316	12 115
Average water consumption (ML/month)	5.0	4.2	2.7	2.2
Annual diesel cost (AUD/a)	4 860 000	4 480 000	5 830 000	5 310 000
Annual water cost (AUD/a)	300 000	250 000	160 000	130 000
Total annual operating cost	5 160 000	4 730 000	5 990 000	5 440 000

It is important to note that the sources considered within Table 4 account for less than 20 per cent of the overall site power generation requirements in this case study. For comparison, the base diesel consumption exclusive of refrigeration and mobile equipment is estimated at approximately 48 000 L/day.

Total cost of ownership of the ventilation and cooling system, presented as the NPC of each option over five years of operation assuming an 8 per cent discount rate, is shown in Table 5.

TABLE 5

NPC comparison for all cases.

Item	Absorption chiller		Conventional chiller	
	Diesel	BEV	Diesel	BEV
Total direct cost (M AUD)	14.83	13.18	12.54	11.33
Annual operating cost (M AUD/a)	5.16	4.73	5.99	5.44
Net present cost (M AUD)	37.58	34.09	39.31	35.62
Savings compared to conv – diesel (M AUD)	*1.73*	*5.22*	*-*	*3.69*

The results from Table 5 suggest that the optimal configuration to maximise cost reductions on the mine ventilation and mine air-cooling systems is the use of absorption chillers with BEVs. It is important to note that this comparison does not consider the capital cost of the mobile equipment fleet. A trade-off on diesel versus BEVs can be made during detailed, site-specific engineering studies to determine whether implementation is applicable. Alternatively, if a project is not suitable for the implementation of BEVs, or the BEV equipment costs are sufficient to erode the NPC improvements in the above table, the use of absorption chillers with diesel mobile equipment presents a cost-positive alternative to conventional vapor-compression chillers.

The annual operating cost in this analysis only considers water to support evaporation across the cooling towers, as this typically represents the largest portion of total makeup water requirements. In reality, additional water would be required to support blowdown and maintain acceptable water quality within the cooling circuit, with exact requirements being site-specific, dependent on the makeup water quality. In addition, makeup water requirements could be supplemented through condensation in the bulk air cooling circuit, although this was not accounted for in this study. Overall, based on the assumptions used in this analysis, final makeup water requirements are not expected to fundamentally change the case study conclusions, as annual operating costs are dominated by diesel fuel consumption. However, water consumption, availability and cost should be assessed as part of detailed, site-specific engineering studies.

Note that any costs or pricing attached to CO_2 equivalent emissions are excluded from the above operating costs and NPC values.

Environmental considerations

The study also sought to quantify the GHG emissions associated with the ventilation, cooling, and mobile equipment fleet. Depending on local or internal carbon pricing, whether enforceable by the regulatory authority in question or implemented by the mine owner to meet corporate strategic targets, GHG emissions reduction are expected to become an increasingly important factor in engineering trade-off considerations. In the case of a remote site with diesel-fired generators, reduction in site electrical load has a considerable impact on Scope 1 emissions. This study has considered the following sources of emissions for comparison between cases:

- Diesel generators:
  - Emissions associated with power consumption for the generator station auxiliaries.
  - Emissions associated with power consumption for the primary ventilation fans.
  - Emissions associated with power consumption for the refrigeration system
  - Emissions associated with charging load for the BEVs.
- Mobile equipment:
  - Emissions associated with diesel consumption in the mobile equipment fleet.

All additional diesel fuel consumption for power generation to support operations not listed above (ie for camps, surface facilities, underground infrastructure other than ventilation such as dewatering pumps) is excluded from the scope of this assessment, as modifications to the ventilation and cooling strategy will not have a direct impact on these base loads.

The annual CO_2-equivalent Scope 1 emissions for each option are compared in Table 6.

TABLE 6
CO_2-equivalent scope 1 emissions for all cases.

Item	Absorption chiller		Conventional chiller	
	Diesel	BEV	Diesel	BEV
CO_2e emissions from mobile equipment (t/a)	2000	0	2000	0
CO_2e emissions from power generation (t/a)	9060	10 030	11 230	11 870
Total CO_2e emissions (t/a)	11 060	10 030	13 230	11 870

The results in Table 6 demonstrate a key environmental advantage to the absorption chiller technology, which enables a ~16 per cent reduction in GHG emissions when compared to conventional vapor-compression chillers for a given fleet type. Switching from a diesel mobile equipment to BEVs also shows an advantage, with a ~10 per cent reduction in total CO_2-equivalent production for a given chiller type. Further, absorption chillers with a diesel fleet have lower GHG emissions than conventional chillers with BEVs.

CONCLUSIONS

It has been shown that the use of absorption chillers with BEVs presents a potential opportunity for improved project economics and reduced environmental impact associated with mine ventilation and cooling systems in remote mining applications. Multiple configurations were compared with varying fleet type (diesel versus BEV) and chiller technology (conventional vapour-compression versus absorption), with the mine ventilation and cooling system showing economic and social improvements when altering from the traditional configuration of diesel mobile equipment with conventional vapor compression chillers. The overall conclusion should be assessed on a site-specific basis to confirm the findings as well as the preferred approach for each individual site, including considerations for waste heat availability and quality, as well as water limitations.

This study demonstrates the potential for absorption chillers, either alone or in combination with BEVs, as a positive payback short-term decarbonisation action that sites can implement to improve the project economics, reduce GHG emissions, or both.

REFERENCES

American Society of Heating, Refrigerating and Air-Conditioning Engineers (ASHRAE), 2022. Absorption Equipment, *2022 ASHRAE Handbook – Refrigeration (SI),* Chapter 18, pp 18.1–18.17 (ASHRAE).

Broodryk, A, deVries, J, Kyselica, P and McLean, K, 2015. The Design, Installation and Commissioning of an Absorption Refrigeration System at Gwalia Gold Mine in Western Australia, in *The Australian Mine Ventilation Conference* (ed: B Belle), pp 193–199 (The Australasian Institute of Mining and Metallurgy: Melbourne).

del Castillo, D O, 2010. Demystifying the Use of Cogeneration in Mine Cooling Applications, in 13th United States/North American Mine Ventilation Symposium (MIRARCO).

Jones, T H, Jonsson, L, Bergström, J, Martikainen, A and Engberg, H, 2022. The LKAB transformation: approaches for greater depths and a changing world, in *Proceedings of the Fifth International Conference on Block and Sublevel Caving*, pp 433–442 (Australian Centre for Geomechanics: Perth).

Nadaraju, F J, 2021. Assessment of a Novel Concept for Co-Generation of Heat and Power, PhD thesis (unpublished), The University of Newcastle, Australia.

Precise evaluation of fracture porosity of coal body – key factor in assessing free gas outburst expansion energy

M Cheng[1], Y Cheng[2] and C Wang[3]

1. PhD candidate, School of Safety Engineering, China University of Mining and Technology, Xuzhou 221116, PR China. Email: cm160701112@163.com
2. Professor, School of Safety Engineering, China University of Mining and Technology, Xuzhou 221116, PR China. Email: ypcheng@cumt.edu.cn
3. Associate Professor, Artificial Intelligence Research Institute, Xuzhou 221116, PR China. Email: chenghao@cumt.edu.cn

ABSTRACT

Coal is a complex, porous material, and recent studies have categorised its pores into four distinct types: inaccessible pores (<0.38 nm), filling pores (0.38–1.5 nm), diffusion pores (1.5–100 nm) and seepage pores (>100 nm). The seepage pores, also known as fracture spaces, are crucial as they act as the primary storage for free gas within coal bodies. The fracture porosity significantly influences the free gas expansion energy, a key determinant in the occurrence of coal and gas outbursts. Therefore, precise evaluating the fracture porosity of coal bodies is vital for accurately assessing free gas expansion energy and predicting outbursts. While existing methods are adequate for measuring seepage pores in coal particles, there is a notable gap in effective techniques for assessing fracture porosity in coal bodies, particularly under varied stress conditions. To overcome this challenge, this study designed a series of experimental methods and constructed corresponding computational models. This study employed raw coal sample and remade coal samples as experimental subjects to simulate the conditions of real coal seams. The volume, total porosity, and permeability of the coal samples were measured under different hydrostatic stresses (5, 10, 20, 30, 40 and 50 MPa) and a helium gas pressure of 1 MPa. Ultimately, the results of fracture porosity under various stresses were successfully obtained. The research revealed that as stress increased, the fracture porosity in remade coal samples declined from 25.96 per cent to 6.52 per cent and in the raw coal sample, it dropped from 1.2 per cent to 0 per cent. By incorporating the latest gas expansion energy models, it was determined that the free gas expansion energy in remade coal samples decreased from 913 kJ/t to 35 kJ/t and from 42 kJ/t to 0 kJ/t in the raw coal sample. Additionally, the study delves into the impact and mechanism of free gas expansion energy on coal and gas outbursts.

INTRODUCTION

Coalbed methane (CBM), primarily composed of CH_4, is recognised as a highly efficient and clean energy source, while also posing a significant hazard during the coal mining process (Moore, 2012; Zhao *et al*, 2016; Jin *et al*, 2018; Cheng and Pan, 2020). The extraction and utilisation of CH_4 not only contribute to the healthy development of the energy sector and support the achievement of carbon neutrality goals but also play a crucial role in preventing coal and gas outbursts and reducing environmental pollution (Retallack, 2001; Aguado and Nicieza, 2007; Karacan *et al*, 2011; Lei *et al*, 2021). The complex multiscale porous structure within coal acts as both a reservoir for CH_4 and a pathway for its migration, decisively influencing the forms of CH_4 storage and its movement (Chalmers, Bustin and Power, 2012; Clarkson *et al*, 2013; Zhu *et al*, 2018; Zhang *et al*, 2024). Therefore, studying the pore structure within coal is of paramount importance.

With advancements in pore characterisation technologies, Hu *et al* (2020) and Hu, Cheng and Pan (2023) have deepened their understanding of methane (CH_4) storage and migration within coal seams. They have identified four distinct types of pore structures in coal: inaccessible pores (<0.38 nm), filling pores (0.38–1.50 nm), diffusion pores (1.50–100 nm) and seepage pores (>100 nm). Among these four types of pore structures, seepage pores with a diameter greater than 100 nm provide a space for the storage of free gas within the coal matrix, directly influencing the volume of free gas involved in outbursts (Sengupta *et al*, 2011; Pan and Connell, 2012). The seepage pores within the coal include both intra-matrix pores and inter-matrix fractures (Ju *et al*,

2017; Zhao, Wang and Liu, 2019). Given that the porosity of intra-matrix pores constitutes a minimal proportion of the total porosity within the coal matrix, this study uniformly categorises the pores within the coal with diameters greater than 100 nm as fractures. The volume fraction of these fractures to the total volume of the coal is referred to as the fracture porosity (Zhou *et al*, 2016; Liu *et al*, 2021).

Scholars commonly utilise the mercury intrusion method to measure the fracture space on coal particles (Liu *et al*, 2015; Zhang *et al*, 2018). Although many mercury intrusion porosimeters claim a lower limit of pore diameter measurement between 6–7 nm, the method is likely to cause shrinkage and deformation of the coal matrix at high pressures and may even lead to the collapse and destruction of the coal's pore structure. Consequently, the mercury intrusion method inherently has drawbacks that can lead to significantly overestimated measurement results (Yue *et al*, 2019; Zheng *et al*, 2019). Furthermore, when measuring the fracture space of raw coal sample or remade coal samples, treating the coal as a whole, the existing measurement methods also exhibit severe flaws. Due to limitations in classifying types of coal pores and a lack of understanding of their structure, scholars like Zhang and others have adopted the helium method to measure the porosity of coal (Zhang *et al*, 2015; Liu, Y *et al*, 2019; Zhang *et al*, 2020; Li *et al*, 2022; Xu *et al*, 2022). However, these scholars were unaware of the abundant presence of filling pores in coal with diameters below 1.5 nm. CH_4 in coal is primarily adsorbed in microporous fillings within these pores, which are not spaces for the storage of free gas. Thus, the helium method measures the total pore volume of the coal, not the internal fracture space. The porosity measured by this method for remade coal ranges between 10–40 per cent, significantly higher than the true value (Wang *et al*, 2020; Lei *et al*, 2021; Yi *et al*, 2021). This testing method is even less reasonable for raw coal, which is extremely dense and has fewer internal fractures. Under high-stress conditions, the fractures close, making it impossible to measure the diffusion and filling pores within the coal matrix (Zhu *et al*, 2013; Zhou *et al*, 2020). Therefore, the measurement results not only deviate significantly from the actual fracture space but also show a marked difference from the actual total pore volume.

Given the limited understanding of coal's pore classifications and micro-porous structures, alongside the constraints of existing experimental measurement techniques, past studies have struggled to precisely assess the fracture space and porosity within coal. This knowledge gap extends to the specific relationship between stress levels and the coal's fracture porosity, a critical factor that remains largely undefined. As the investigation into the dynamics of coal and gas outbursts advances, a growing body of laboratory research is now focusing on the creation of remade coal samples from compressed coal particles (Espinoza *et al*, 2016; Geng *et al*, 2017; Niu *et al*, 2017). These models aim to replicate tectonic coal, acknowledged as an essential condition for outburst occurrences. The ability to accurately determine the fracture space and porosity is crucial for a comprehensive analysis of coal's seepage, diffusion, and energy characteristics, applicable to both remade and raw coal sample (An *et al*, 2019; Cao *et al*, 2021). Consequently, the development of a reliable experimental technique and a robust calculation model for the precise evaluation of coal fracture porosity under varying stress scenarios is of paramount practical significance. This approach not only enhances the accuracy and coherence of the research but also strengthens its relevance to real-world applications in coal mining and safety management.

To overcome this challenge, this study first reviews the scholars' achievements in researching the seepage characteristics of coal. Then, based on theoretical model analysis, it identifies the key parameters that play a decisive role in measuring the fracture porosity of coal and proposes experimental methods to measure these parameters. Subsequently, the fracture porosity of coal under different stress conditions is calculated based on the experimental results. This paper also explores the impact and mechanisms of coal fracture porosity on the initiation of outbursts.

THEORY

Coal is a complex fractured porous dual medium. Scholars, aiming to quantitatively analyse the role of stress in the evolution of permeability, have simplified the physical structure of coal when establishing permeability models. The cubic model is currently the most common approach (Cho, Jeong and Sung, 2013; Liu, G *et al*, 2019). This model indicates that the permeability of the coal is directly proportional to the cube of the fracture porosity. If stress is applied to the coal, causing the fractures to contract, then the relationship between the coal's permeability and fracture porosity before and after the application of stress is as follows:

$$\frac{k_0}{k_1} = \left(\frac{\phi_0}{\phi_1}\right)^3 \tag{1}$$

Where k_0 and k_1 are the permeability before and after the application of stress; ϕ_0 and ϕ_1 are the fracture porosity before and after the application of stress.

Transforming the fracture porosity into a relationship between fracture space and coal volume, Equation 1 can be rewritten into Equation 2:

$$\frac{V_{\phi 0}}{V_0} \cdot \frac{V_1}{V_{\phi 1}} = \left(\frac{k_0}{k_1}\right)^{\frac{1}{3}} \tag{2}$$

where $V_{\phi 0}$ and $V_{\phi 1}$ are the fracture space before and after the application of stress; V_0 and V_1 are the volume of coal body before and after the application of stress. Although the permeability of the coal under different stress conditions can be measured using transient methods, Equation 2 only allows us to understand the relationship of changes in fracture porosity without providing a definitive value for the fracture porosity itself.

To further obtain an accurate value for the fracture porosity, an assumption was proposed: the inaccessible pores, filling pores, and diffusion pores within the coal matrix remain unchanged by stress, with only the fracture spaces being affected by stress. When stress is applied to the coal, the majority of the volumetric contraction originates from the closure of fracture spaces, with a minor portion resulting from the elastic deformation of the coal matrix. Consequently, Equation 2 can be further optimised as follows:

$$\frac{V_{\phi 0}}{V_0} \cdot \frac{V_1}{V_{\phi 0} - \Delta V} = \left(\frac{k_0}{k_1}\right)^{\frac{1}{3}} \tag{3}$$

In Equation 3, ΔV is the fracture difference before and after the application of stress.

Based on the assumption that only the fracture spaces within the coal matrix change due to external stress, we can measure the total pore volume of the coal before and after stress changes using the helium method. The difference in total pore volume, represents the change in fracture space, ΔV. Furthermore, the volume of the coal under different stress conditions, V_0 and V_1, can be obtained using stress-strain sensors installed on the coal body. The coal's permeability, k_0 and k_1, under varying stress conditions, can be determined through permeability experiments. By incorporating the results of key parameters into Equation 3, accurate results for the initial stress condition's fracture space and porosity can be obtained and then the results of fracture space and porosity under different stresses can be calculated.

EXPERIMENTS

Experimental design

Upon establishing the key parameters critical for accurately fracture spaces and porosity, this study designed an experimental methodology to quantify these parameters. Standard coal samples, including three remade coal samples and one raw coal sample, were employed to simulate coal seams, thus serving as the experimental medium for fracture space and porosity analysis. The research applied a compressive stress of 100 kN to compact coal powders. According to Equation 3, the volume of coal body (V_c) permeability (k) and total pore volume under different stresses are pivotal in determining fracture space and porosity. To ensure the accuracy of the volume of coal body, all experiments within this study were carried out under controlled hydrostatic stress conditions.

Coal sample preparation

Figure 1 presents the schematic representation of the experimental set-up employed in this study. Fresh coal blocks, designated as XT-4, XT-9 and P8 were sourced from the No. 4 and No. 9 coal seams of the Xintian Coal Mine and Pingdingshan Coal Mine, respectively. As depicted in Figure 1, the initial step involved extracting a 50 mm × 100 mm standard-sized raw coal sample, XT-9 Raw, from the XT-9 coal blocks. Subsequently, the residual coal specimens underwent the processes of crushing and grinding, yielding coal particles segregated into two distinct size ranges: 0.25–0.5 mm and 0.1–0.2 mm. Coal particles in the 0.2–0.25 mm range were earmarked for proximate analysis, the results of which are tabulated in Table 1. Proximate analysis was conducted in strict compliance with the Chinese national standard GB/T212–2008 (Yan *et al*, 2019). The fixed carbon (FCad) percentages for XT-4, XT-9 and P8 were recorded at 70.63 per cent, 87.61 per cent and 70.73 per cent, respectively, categorising all as high-rank coals. Adherence to the Chinese national standard GB/T23561.12–2010 revealed Protodyakonov coefficients of 2.63 for XT-4, 2.74 for XT-9 and 0.33 P8, signifying a hardness level from Xintian Coal Mine and a weak level from Pingdingshan Coal Mine considerably surpassing that of ordinary coal specimens (Wang *et al*, 2018).

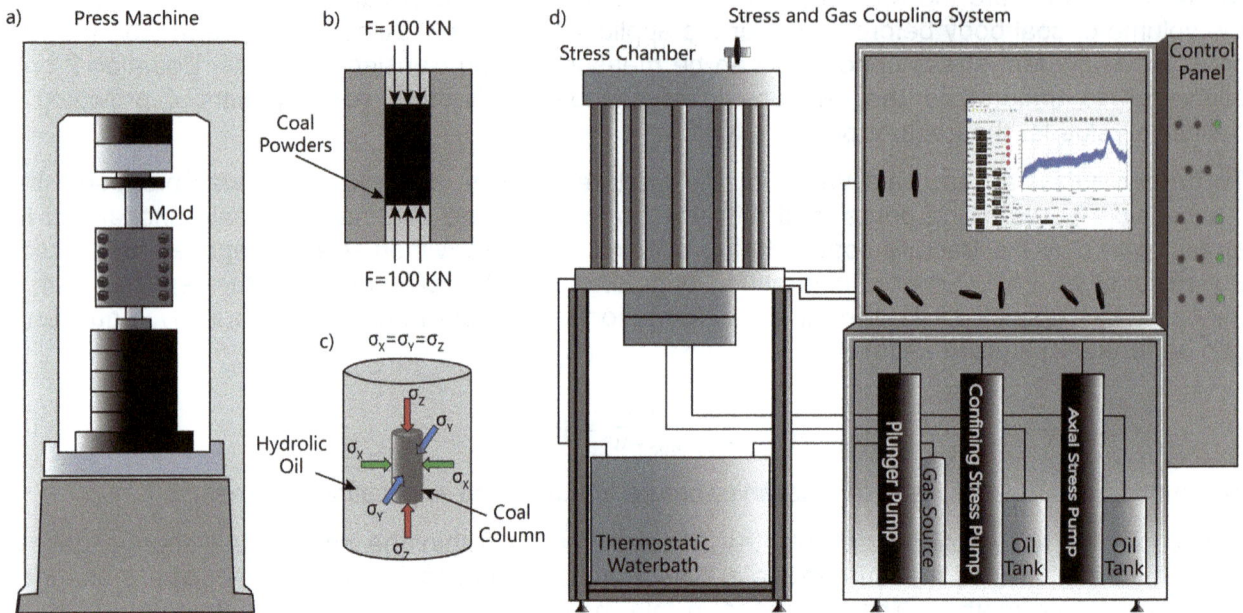

FIG 1 – Schematic diagram of the experimental design.

TABLE 1

Basic properties of coal samples.

Coal sample	Mad (%)	Aad (%)	Vad (%)	FCad (%)	f
XT-4	0.78	24.47	4.12	70.63	2.63
XT-9	0.86	7.92	3.61	87.61	2.74
P8	1.23	8.95	21.35	70.73	0.33

In the procedures illustrated by Figure 1a and 1b, coal particles from XT-4, XT-9 and P8, within the 0.1–0.2 mm size range, were compacted into remade coal samples XT-4 Re, XT-9 Re and P8 Re, using a bidirectional compression technique. Before reconstituting, 15 per cent water was added to the coal particles and mixed thoroughly until they could be formed into clumps. During the reconstituting process, to achieve a higher initial fracture porosity, we used a stress of 100 kN for 30 minutes.

Post-compression, these samples underwent a 24-hr drying phase in a controlled oven environment. Subsequently, key physical parameters such as mass, length, diameter and apparent density of these three coal samples were measured, with the detailed results presented in Table 2.

TABLE 2
Basic properties of coal samples.

Coal Sample	XT-4 Re	XT-9 Re	P8 Re	XT-9 Raw
Length (mm)	95	95.7	105.1	100.90
Diameter (mm)	50	50	50	50.30
Volume (cm^3)	186.44	187.81	206.26	200.80
Mass (g)	228.07	185.29	246.87	282.51
Density (g/cm^3)	1.22	0.99	1.20	1.41

Stress-strain, total pore volume and permeability tests

As depicted in Figure 1c and 1d, the three coal samples underwent an array of tests: stress-strain, total pore volume and permeability, within a stress and gas coupling experimental apparatus. These tests were executed under varying stresses, specifically at 5, 10, 20, 30, 40 and 50 MPa.

Figure 2 illustrates the set-up of the stress chamber, filled with hydraulic oil, employed to uniformly apply hydrostatic stress ($\sigma_x = \sigma_y = \sigma_z$) to the coal samples. First, the coal samples were placed within this chamber, each securely enveloped in a sealed tube, with both ends tightly wrapped with insulation tape, ensuring an airtight seal against hydraulic oil infiltration. Then, the coal samples were equipped with axial and radial deformation sensors, which facilitated real-time monitoring of the samples' deformations under different stress conditions.

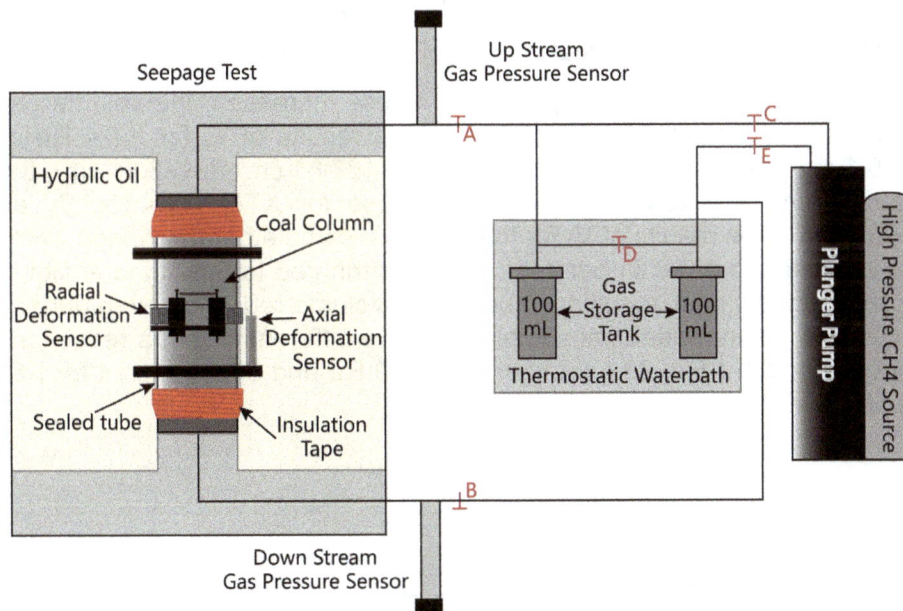

FIG 2 – Schematic diagram of stress-strain, total pore volume and permeability tests.

Furthermore, Figure 2 illustrates the schematic diagrams for total pore volume and permeability tests. The crucial aspect of these tests is to achieve a state of gas equilibrium within the coal samples. However, the time required for different coal samples to reach this state varies significantly. The raw sample takes much longer to reach gas equilibrium compared to the remade coal samples. Given that this study involves measuring a large number of total pore volume and permeability data points, the process is extremely time-consuming. Therefore, considering practical circumstances, a uniform gas equilibrium time of 24 hrs is set.

For total pore volume analysis, each coal sample was subjected to six sets of tests under diverse stresses and 1 MPa helium pressure. Post a 24-hr equilibrium phase, a protocol involving the sequential closure of valve A and B and opening of valve D, followed by pressuring the airway to 1 MPa. Then close valve C and E and reopen valve A and B to start the tests. After 24 hrs, each test concluded with the calculation of total pore volume.

The permeability of each coal sample was assessed using a transient method. This involved conducting six sets of tests under varied stresses and a consistent 1 MPa helium pressure. Following a 24-hr equilibrium period, the process entailed closing valve A and D, raising the upstream pressure to 1.1 MPa, then closing valve C and E and ultimately reopening valve A (Wang *et al*, 2021). The experiment concluded with the recording of pressure changes, forming the basis for permeability calculations.

RESULTS AND DISCUSSION

Results of volume, total pore volume and permeability for coal samples

Determining the volume of coal samples (V_c) under different stresses directly affects the accuracy of fracture porosity calculations. In this study, the volume of coal samples is calculated by measuring the axial and radial deformations under different stresses.

The axial compression was set as a positive value, while the radial compression was set as a negative value. Axial deformation and radial deformation represent a reduction in the length and diameter of coal samples, respectively. Since the experiments were conducted under hydrostatic stress, the lateral surfaces were evenly stressed. It can be assumed that the diameter at every point along the axis is equal to that at the location where radial deformation was measured.

Based on the results of axial and radial deformations, the volume of coal samples can be calculated using the formula for the volume of a cylinder. Then, the ratio of V_c to the initial volume can be calculated using Equation 4:

$$v = \frac{V_{coal}}{V_0} \tag{4}$$

where v is the volume ratio; V_0 is the initial volume of coal samples.

The results of V_c and v are shown in Figure 3. As stress increases, the volume changes of four coal samples exhibit completely different patterns: The results of V_c for XT-4 Re decrease from 186.4 cm³ to 155.1 cm³; those for XT-9 Re reduce from 187.81 cm³ to 128.41 cm³; those for P8 Re reduce from 206.26 cm³ to 246.87 cm³; whereas those for XT-9 Raw slightly decrease from 198.0 cm³ to 195.7 cm³. The results of V_c for the raw coal sample hardly change, with the results of v decreasing to a maximum of 98.81 per cent. This is attributed to the high density, hardness and strength of the selected raw coal, resulting in negligible volume reduction even under a stress of up to 50 MPa. In contrast, the remade coal samples behave differently, with a reduction in the results of v to 83.2 per cent for XT-4 Re, 68.4 per cent for XT-9 Re and 85.1 per cent for P8 Re when the stress reaches 50 MPa.

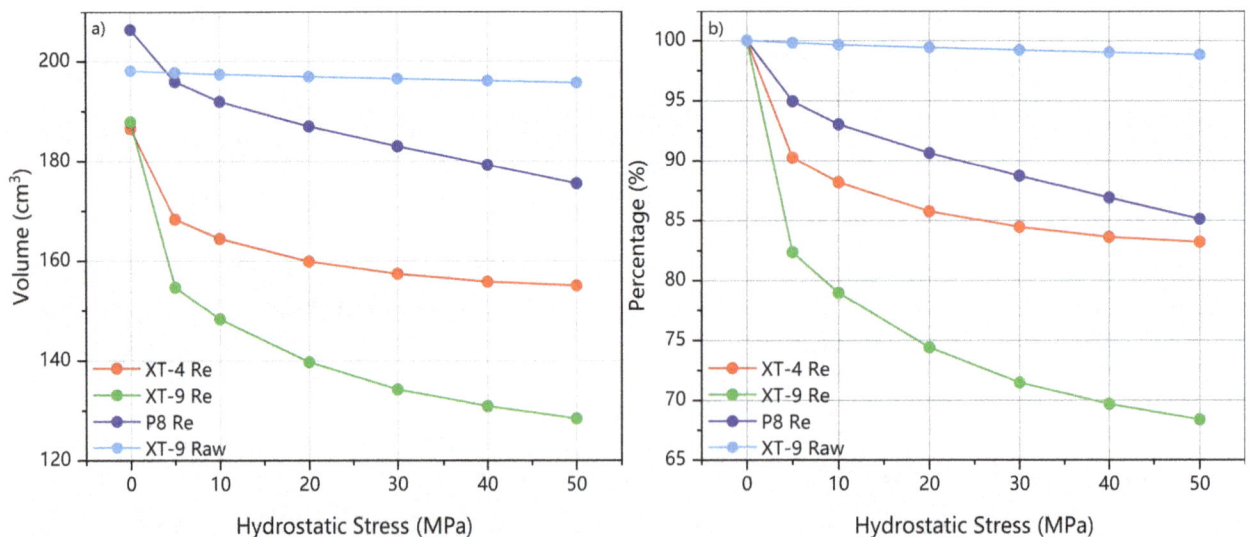

FIG 3 – Volume of coal samples and volume ratio under different stresses.

The results of total porosity (Φ_t) are shown in Figure 4a. As stress increases, the results of total porosity of four coal samples exhibit similar patterns: the results of total porosity for XT-4 Re decrease from 22.3 per cent to 17.1 per cent; those for XT-9 Re reduce from 39.6 per cent to 29.1 per cent; those for P8 Re reduce from 26.8 per cent to 20.1 per cent; whereas those for XT-9 Raw decrease from 5.3 per cent to 3.13 per cent. Among the remade coal samples, XT-9 Re exhibits the highest total porosity, almost twice that of XT-4 Re. Although XT-9 Raw has the lowest total porosity, it experiences the greatest reduction ratio in porosity, with the total porosity at 50 MPa being 58.7 per cent of its value at 5 MPa.

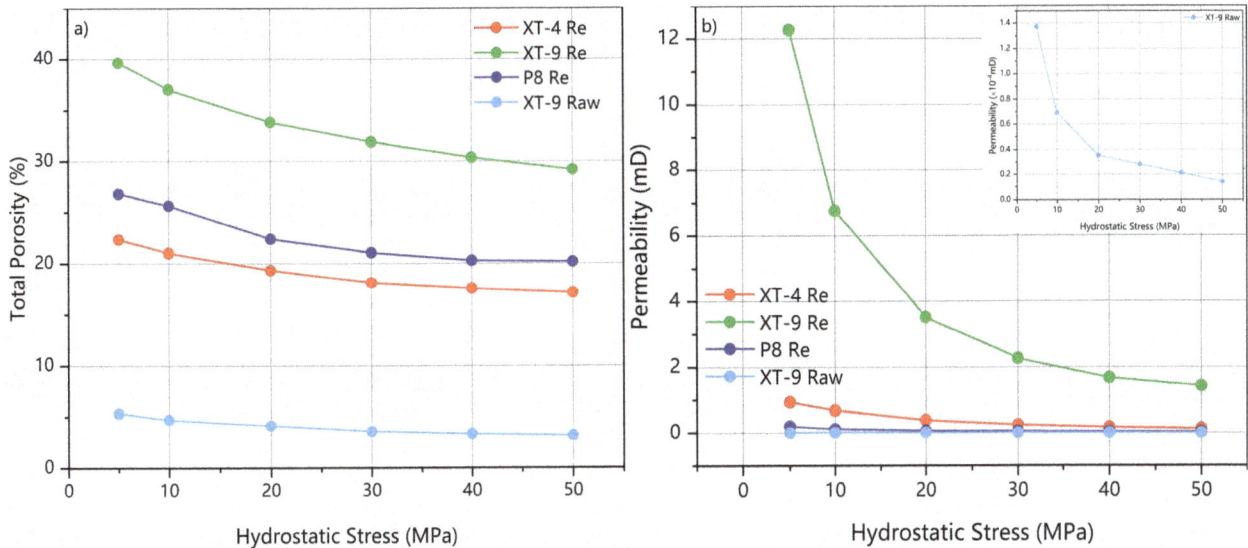

FIG 4 – Total porosity of coal samples under different stresses.

The permeability (k) results of coal samples under different stresses are shown in Figure 4b. Due to the extremely low permeability of XT-9 Raw, its results are presented in an enlarged view. The permeability trends of all four coal samples are generally consistent, each decreasing with increasing stress. However, there are significant differences in the permeability results among the four coal samples. As the stress increases, the results of permeability for XT-4 Re decrease from 0.94 mD to 0.13 mD; those for XT-9 Re drop from 12.25 mD to 1.43 mD; those for P8 Re drop from 0.19 mD to 0.02 mD; whereas those for XT-9 Raw decrease from 1.37×10^{-4} mD to 0.14×10^{-4} mD.

Results of fracture porosity for remade coal samples

Utilising the experimental results and applying Equation 3, it is possible to initially determine the fracture porosity of three variants of remade coal samples at a stress level of 5 MPa. Subsequent calculations allow for determining their fracture porosity across additional stress levels. Figure 5 illustrates these findings. The fracture porosity trend (Φ) across the three remade coal samples exhibits remarkable consistency: the results of Φ for XT-4 Re reduce from 12.6 per cent to 6.5 per cent; for XT-9 Re, the results decline from 26.0 per cent to 12.7 per cent; and for P8 Re, the results decrease from 15.5 per cent to 7.6 per cent. With escalating stress, the fracture porosity of these remade coal samples at 50 MPa significantly reduces to approximately half of their initial values at 5 MPa, highlighting the substantial impact of stress on the fracture porosity.

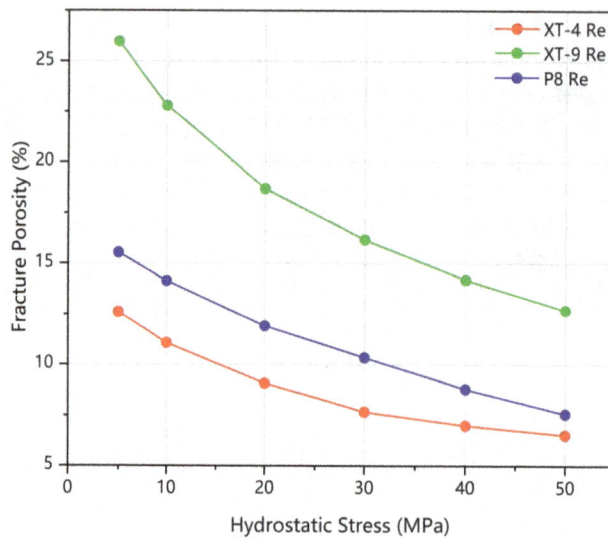

FIG 5 – Fracture porosity of coal samples under different stresses.

By sequentially comparing the ratio of fracture porosity to the cube root of permeability for three types of remade coal samples as the applied stress increases from 5 MPa to 10 MPa, to 20 MPa, to 30 MPa, to 40 MPa and to 50 MPa, we can validate the accuracy of the fracture porosity results measured by this method. The comparison results are shown in Figure 6. The trends in both ratios are almost identical, with minimal differences in the outcomes. This is especially true for the comparison as stress increases from 5 MPa to 50 MPa, where the results of the two ratios for the three types of remade coal samples are nearly the same. This consistency demonstrates the accuracy and effectiveness of the experimental method and calculation model for precisely evaluating coal fracture porosity proposed in this study.

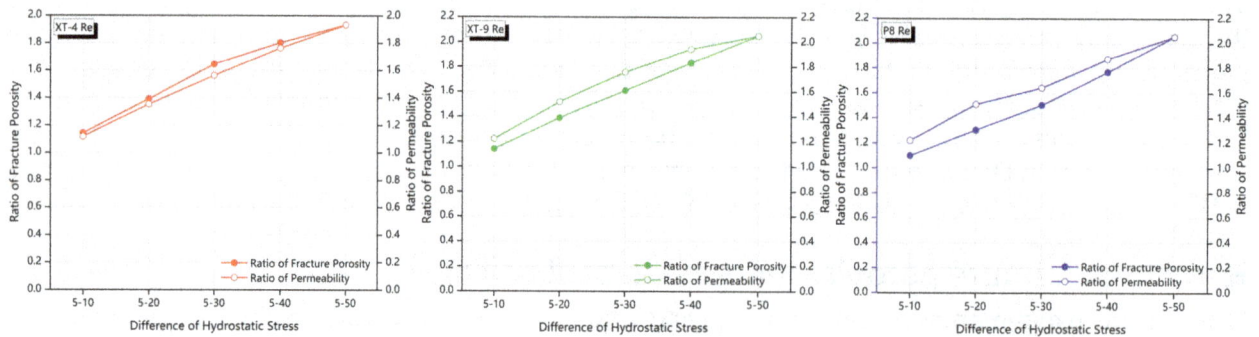

FIG 6 – Validation of the accuracy of evaluating method for fracture porosity.

Results of fracture porosity for raw coal sample

For raw coal sample, the evaluating method for remade coal samples becomes inapplicable due to the coal's inherent density and the scarcity of internal fractures. Under high-stress conditions, fractures close, resulting in the inability to measure all diffusion and filling pores within the coal matrix within 24 hrs. According to the total porosity results, as stress increases, the total porosity of XT-9 Raw decreases from 5.33 per cent to 3.13 per cent, a reduction of 41.3 per cent, which significantly exceeds that observed in the three types of remade coal samples. This substantial decrease is attributed to the dense nature and limited fractures in raw coal. Since the results of measuring total porosity using helium not only diverge significantly from the actual fracture data but also show a pronounced discrepancy from the total pore volume, relying on this data can lead to severe measurement inaccuracies.

However, for raw coal, approaching from the perspective of volume deformation and permeability changes can simplify the process of obtaining its fracture porosity under different stress levels. Based on the results of coal volume deformation and permeability, it is observed that when the stress increases to 20 MPa, the permeability decreases to $3.5×10^{-5}$ mD. At this point, it can be concluded that all fractures in the coal have closed, and the volume deformation beyond 20 MPa originates solely from the elastic deformation of the coal matrix. In other words, beyond 20 MPa, the fracture porosity of the coal is approximately 0 per cent. The fracture porosity at stress levels below 20 MPa can be approximated by the ratio of volume changes, with a fracture porosity of 0.37 per cent at 5 MPa and 0.28 per cent at 10 MPa.

Results of free gas expansion energy for all coal samples

The energy released by gas within a very short time in coal and gas outbursts as it decreases from the pressure inside coal seams to atmospheric pressure is commonly known as gas expansion energy. This energy can be further categorised based on the gas source involved in outbursts into free gas expansion energy and desorbed gas expansion energy. The free gas stored within the fractures of the coal bodies determines the magnitude of free gas expansion energy. The role of free gas expansion energy is pivotal in determining the likelihood of an outburst. Building on the foundation of previous research, Wang and Cheng (2023) proposed Equation 5 for accurately assessing the free gas expansion energy:

$$W_{g,f} = \frac{P_0}{n-1} \emptyset V_c \left(\frac{P_g}{P_0}\right)^{\frac{1}{n}} \left[\left(\frac{P_g}{P_0}\right)^{\frac{n-1}{n}} - 1 \right] \tag{5}$$

For the free gas expansion energy, the core parameters are V_c, Φ and P_g. Based on the results, the V_c and Φ are known. Many studies on gas expansion energy assume that the free gas in coal bodies participating in outbursts is instantly released at the initial moment. In other words, the free gas pressure involved in the outburst trigger process instantly drops from the initial gas pressure to atmospheric pressure. Therefore, Therefore, we assume scenarios where P_g is 2 MPa, 1 MPa and 0.5 MPa, respectively, to calculate the free gas expansion energy for all coal samples. The results of free gas expansion energy are shown in Figure 7.

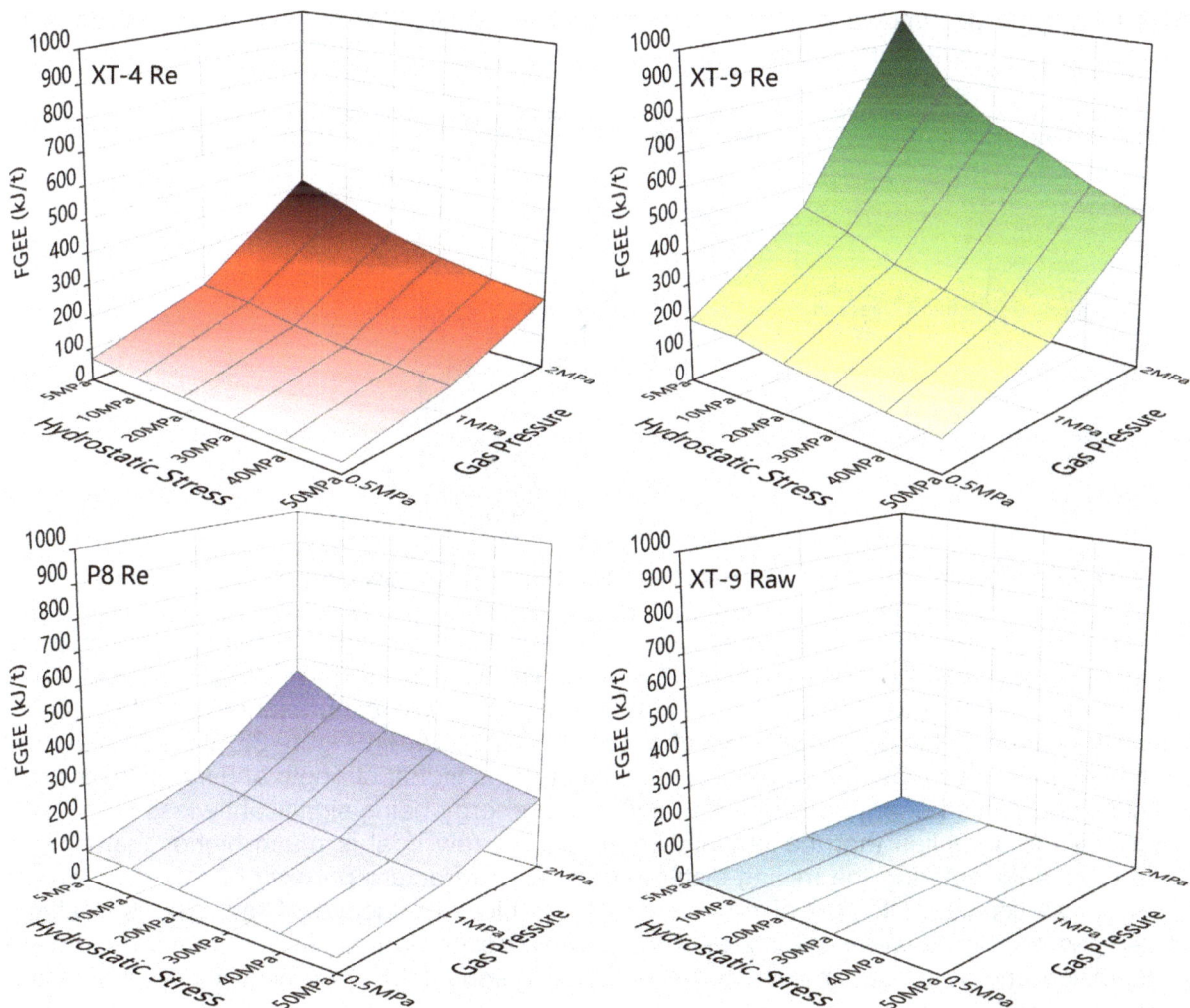

FIG 7 – Free gas expansion energy of coal samples under different stresses and gas pressures.

The results of free gas expansion energy demonstrate a consistent trend across three coal samples. With escalating stress and diminishing gas pressure, energy values for XT-4 Re witnessed a decline from 423 kJ/t to 35 kJ/t. Similarly, XT-9 Re exhibited a decrease from 975 kJ/t to 100 kJ/t, P8 Re exhibited a decrease from 455 kJ/t to 46 kJ/t, while XT-9 Raw showed a reduction from 9 kJ/t to 0 kJ/t. In conditions of steady stress, a downward trajectory is observed in energy values concurrent with a decrease in gas pressure. Specifically, energy values at 0.5 MPa constituted only 6.0 per cent to 20.8 per cent of those recorded at 2 MPa. Inversely, under fixed gas pressure, an increase in stress also resulted in diminished energy values. Energy values at 50 MPa accounted for 42.9 per cent to 52.6 per cent of those at 5 MPa. Collectively, these findings indicate that both a decrease in gas pressure and an increase in stress contribute to a reduction in free gas expansion energy. However, the free gas expansion energy demonstrates a heightened sensitivity to the reduction in gas pressure.

Effects and mechanism of free gas expansion energy on coal and gas outbursts

Figure 8a illustrates the releasing model of free gas expansion energy (FGEE) during the outburst process. During the geological preparation stage, coal bodies are affected by tectonism, forming tectonic coals with a high risk of outbursts. This process is accompanied by increased stress and gas content in coal seams, leading to the accumulation of both coal deformation energy (CDE) and free gas expansion energy. The tectonism significantly weakens the strength of coal bodies and greatly reduces the outburst threshold, providing the necessary conditions for triggering outbursts. Once entering the mining-disturbed stage, the coal seam is subjected to uneven stress loading in the vertical direction, leading to the pre-release of coal deformation energy and the forward movement of coal seams. During this process, coal bodies are re-fractured, further lowering the outburst threshold. The gas within coal bodies rapidly desorbs and fills the newly created fractures, leading to a rapid accumulation of free gas expansion energy. When this energy reaches the outburst threshold, an outburst is triggered.

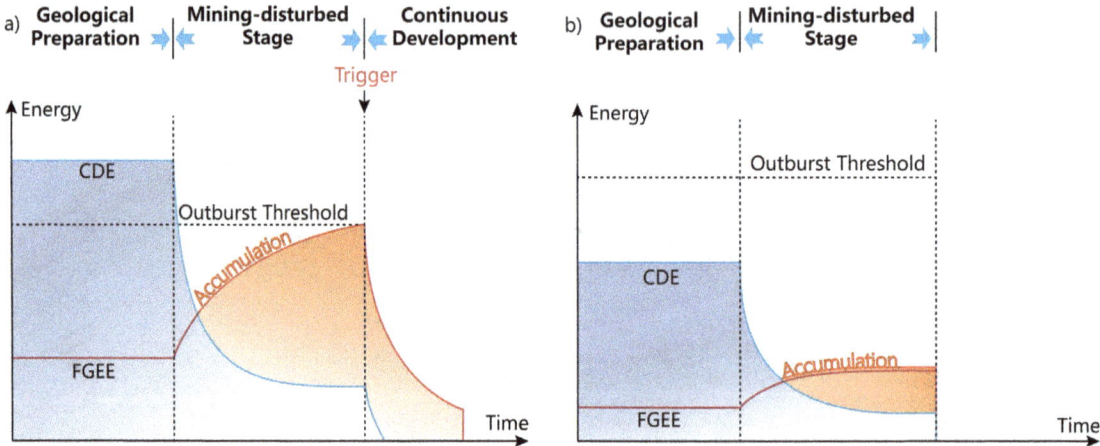

FIG 8 – Releasing model of free gas expansion energy for raw coal and tectonic coal.

Upon reviewing the results of free gas expansion energy for three types of remade coal samples and one raw coal sample, it is evident that the free gas expansion energy in raw coal is significantly lower than that in tectonic coal. This also explains why the presence of tectonic coal is a necessary condition for the occurrence of outbursts. As shown in Figure 8b, because raw coal has not been subjected to tectonism, its coal body maintains higher strength and integrity, resulting in the coal deformation energy and free gas expansion energy being significantly less than that of remade coal. This implies that the outburst threshold for raw coal is much higher than that for remade coal. After entering the mining-disturbed stage, the fracture porosity of raw coal remains relatively small compared to that of remade coal. The blocky structure of raw coal is unable to replenish gas into newly created fractures through rapid desorption, leading to an inability for free gas expansion energy to accumulate to the outburst threshold which is significantly higher than remade coal.

CONCLUSIONS

The fracture porosity within coal body significantly influences the free gas expansion energy which is a key determinant in the occurrence of coal and gas outbursts. Precise measurement of fracture porosity is vital for accurately assessing free gas expansion energy and predicting the probability of coal and gas outbursts. In this study, the key parameters that decisively influence fracture porosity were identified by reviewing theoretical models. Around these key parameters, experimental methods were designed and fracture porosity was calculated based on the experimental results. The main conclusions are summarised as follows:

1. As stress increases, the fracture porosity of XT-4 Re reduces from 12.6 per cent to 6.5 per cent; that of XT-9 Re declines from 26.0 per cent to 12.7 per cent; that of P8 Re decreases from 15.5 per cent to 7.6 per cent; whereas, that of XT-9 Raw declines from 0.37 per cent to 0 per cent.

2. With escalating stress and diminishing gas pressure, the free gas expansion energy for XT-4 Re witnesses a decline from 423 kJ/t to 35 kJ/t. Similarly, XT-9 Re exhibits a decrease from 975 kJ/t to 100 kJ/t, P8 Re exhibits a decrease from 455 kJ/t to 46 kJ/t, while XT-9 Raw shows a reduction from 9 kJ/t to 0 kJ/t. The free gas expansion energy demonstrates a heightened sensitivity to the reduction in gas pressure.

3. The free gas expansion energy determines whether an outburst can occur. The tectonic coal has a lower outburst threshold, and the free gas expansion energy can rapidly accumulate beyond this threshold, ultimately triggering an outburst. In contrast, the outburst threshold for raw coal is higher, and the free gas expansion energy is less likely to exceed this threshold, making outbursts almost impossible.

ACKNOWLEDGEMENTS

The authors gratefully acknowledge financial support from the National Natural Science Foundation of China (No. 52034008, 52374240). Special thanks to Weinan Chen for her support to this work.

REFERENCES

Aguado, M B D and Nicieza, C G, 2007. Control and prevention of gas outbursts in coal mines, Riosa-Olloniego coalfield, Spain, *International Journal of Coal Geology*, 69 (4):253–266.

An, F H, Yuan, Y, Chen, X J, Li, Z Q and Li, L Y, 2019. Expansion energy of coal gas for the initiation of coal and gas outbursts, *Fuel*, 235:551–557.

Cao, J, Hu, Q T, Gao, Y N, Li, M H and Sun, D L, 2021. Gas Expansion Energy Model and Numerical Simulation of Outburst Coal Seam under Multifield Coupling, *Geofluids 2021*.

Chalmers, G R, Bustin, R M and Power, I M, 2012. Characterization of gas shale pore systems by porosimetry, pycnometry, surface area and field emission scanning electron microscopy/transmission electron microscopy image analyses: Examples from the Barnett, Woodford, Haynesville, Marcellus and Doig units, *AAPG Bulletin*, 96 (6):1099–1119.

Cheng, Y and Pan, Z, 2020. Reservoir properties of Chinese tectonic coal: A review, *Fuel*, 260:116350.

Cho, H, Jeong, N and Sung, H J, 2013. Permeability of microscale fibrous porous media using the lattice Boltzmann method, *International Journal of Heat and Fluid Flow*, 44:435–443.

Clarkson, C R, Solano, N, Bustin, R M, Bustin, A M M, Chalmers, G R L, He, L, Melnichenko, Y B, Radlinski, A P and Blach, T P, 2013. Pore structure characterization of North American shale gas reservoirs using USANS/SANS gas adsorption and mercury intrusion, *Fuel*, 103:606–616.

Espinoza, D N, Vandamme, M, Dangla, P, Pereira, J M and Vidal-Gilbert, S, 2016. Adsorptive-mechanical properties of reconstituted granular coal: Experimental characterization and poromechanical modeling, *International Journal of Coal Geology*, 162:158–168.

Geng, Y G, Tang, D Z, Xu, H, Tao, S, Tang, S L, Ma, L and Zhu, X G, 2017. Experimental study on permeability stress sensitivity of reconstituted granular coal with different lithotypes, *Fuel*, 202:12–22.

Hu, B, Cheng, Y P and Pan, Z J, 2023. Classification methods of pore structures in coal: A review and new insight, *Gas Science and Engineering*, vol 110.

Hu, B, Cheng, Y, He, X, Wang, Z, Jiang, Z, Wang, C, Li, W and Wang, L, 2020. New insights into the Ch4 adsorption capacity of coal based on microscopic pore properties, *Fuel*, vol 262.

Jin, K, Cheng, Y, Ren, T, Zhao, W, Tu, Q, Dong, J, Wang, Z and Hu, B, 2018. Experimental investigation on the formation and transport mechanism of outburst coal-gas flow: Implications for the role of gas desorption in the development stage of outburst, *International Journal of Coal Geology*, 194:45–58.

Ju, Y, Zhang, Q G, Zheng, J T, Wang, J G, Chang, C and Gao, F, 2017. Experimental study on CH_4 permeability and its dependence on interior fracture networks of fractured coal under different excavation stress paths, *Fuel*, 202:483–493.

Karacan, C, Ruiz, F A, Cotè, M and Phipps, S, 2011. Coal mine methane: A review of capture and utilization practices with benefits to mining safety and to greenhouse gas reduction, *International Journal of Coal Geology*, 86 (2–3):121–156.

Lei, Y, Cheng, Y, Ren, T, Tu, Q, Li, Y and Shu, L, 2021. Experimental Investigation on the Mechanism of Coal and Gas Outburst: Novel Insights on the Formation and Development of Coal Spallation, *Rock Mechanics and Rock Engineering*, 54 (11):5807–5825.

Li, L, Zhang, S F, Li, Z Q, Chen, X J, Wang, L and Feng, S L, 2022. An Experimental and Numerical Study of Abrupt Changes in Coal Permeability with Gas Flowing through Fracture-Pore Structure, *Energies*, 15 (21).

Liu, C J, Wang, G X, Sang, S X, Gilani, W and Rudolph, V, 2015. Fractal analysis in pore structure of coal under conditions of CO_2 sequestration process, *Fuel*, 139:125–132.

Liu, G N, Liu, L, Liu, J S, Ye, D Y and Gao, F, 2019. A fractal approach to fully-couple coal deformation and gas flow, *Fuel*, 240:219–236.

Liu, X F, Nie, B S, Guo, K Y, Zhang, C P, Wang, Z P and Wang, L K, 2021. Permeability enhancement and porosity change of coal by liquid carbon dioxide phase change fracturing, *Engineering Geology*, vol 287.

Liu, Y B, Yin, G Z, Zhang, D M, Li, M H, Deng, B Z, Liu, C, Zhao, H G and Yin, S Y, 2019. Directional permeability evolution in intact and fractured coal subjected to true-triaxial stresses under dry and water-saturated conditions, *International Journal of Rock Mechanics and Mining Sciences*, 119:22–34.

Moore, T A, 2012. Coalbed methane: A review, *International Journal of Coal Geology*, 101:36–81.

Niu, Q H, Cao, L W, Sang, S X, Zhou, X Z, Wang, Z Z and Wu, Z Y, 2017. The adsorption-swelling and permeability characteristics of natural and reconstituted anthracite coals, *Energy*, 141:2206–2217.

Pan, Z J and Connell, L D, 2012. Modelling permeability for coal reservoirs: A review of analytical models and testing data, *International Journal of Coal Geology*, 92:1–44.

Retallack, G J, 2001. A 300-million-year record of atmospheric carbon dioxide from fossil plant cuticles, *Nature*, 411 (6835):287–290.

Sengupta, R, Bhattacharya, M, Bandyopadhyay, S and Bhowmick, A K, 2011. A review on the mechanical and electrical properties of graphite and modified graphite reinforced polymer composites, *Progress In Polymer Science*, 36 (5):638–670.

Wang, C and Cheng, Y, 2023. Role of coal deformation energy in coal and gas outburst: A review, *Fuel*, 332:126019.

Wang, C, Cheng, Y, Yi, M, Jiang, J and Wang, D, 2021. Threshold pressure gradient for helium seepage in coal and its application to equivalent seepage channel characterization, *Journal of Natural Gas Science and Engineering*, 96:104231.

Wang, C, Cheng, Y, Yi, M, Lei, Y and He, X, 2020. Powder Mass of Coal After Impact Crushing: A New Fractal-Theory-Based Index to Evaluate Rock Firmness, *Rock Mechanics and Rock Engineering*, 53 (9):4251–4270.

Wang, Z, Cheng, Y, Zhang, K, Hao, C, Wang, L, Li, W and Hu, B, 2018. Characteristics of microscopic pore structure and fractal dimension of bituminous coal by cyclic gas adsorption/desorption: An experimental study, *Fuel*, 232:495–505.

Xu, X L, Xu, L Q, Yue, C Q and Liu, G N, 2022. A New Fractal Permeability Model Considering Tortuosity of Rock Fractures, *Processes*, 10 (2).

Yan, F, Xu, J, Lin, B, Peng, S, Zou, Q and Zhang, X, 2019. Effect of moisture content on structural evolution characteristics of bituminous coal subjected to high-voltage electrical pulses, *Fuel*, 241:571–578.

Yi, M, Wang, L, Liu, Q, Hao, C, Wang, Z and Chu, P, 2021. Characteristics of Seepage and Diffusion in Gas Drainage and Its Application for Enhancing the gas utilization rate, *Transport in Porous Media*, 137 (2):417–431.

Yue, J W, Wang, Z F, Chen, J S, Zheng, M H, Wang, Q and Lou, X F, 2019. Investigation of pore structure characteristics and adsorption characteristics of coals with different destruction types, *Adsorption Science and Technology*, 37 (7–8):623–648.

Zhang, B B, Jing, L, Shen, J P and Zhang, S Y, 2024. Pore Structure and Methane Adsorption Characteristics of Primary Structural and Tectonic Coals, *Chemistry and Technology of Fuels and Oils – Innovative Technologies of Oil and Gas*, 59:1279–1289.

Zhang, B, Zhu, J, He, F and Jiang, Y D, 2018. Compressibility and fractal dimension analysis in the bituminous coal specimens, *AIP Advances*, 8 (7).

Zhang, Z T, Xie, H P, Zhang, R, Gao, M Z, Ai, T and Zha, E S, 2020. Size and spatial fractal distributions of coal fracture networks under different mining-induced stress conditions, *International Journal of Rock Mechanics and Mining Sciences*, vol 132.

Zhang, Z T, Zhang, R, Xie, H P and Gao, M Z, 2015. The relationships among stress, effective porosity and permeability of coal considering the distribution of natural fractures: theoretical and experimental analyses, *Environmental Earth Sciences,* 73 (10):5997–6007.

Zhao, H F, Wang, X H and Liu, Z Y, 2019. Experimental investigation of hydraulic sand fracturing on fracture propagation under the influence of coal macrolithotypes in Hancheng block, China, *Journal of Petroleum Science and Engineering*, 175:60–71.

Zhao, W, Cheng, Y, Jiang, H, Jin, K, Wang, H and Wang, L, 2016. Role of the rapid gas desorption of coal powders in the development stage of outbursts, *Journal of Natural Gas Science and Engineering,* 28:491–501.

Zheng, S J, Yao, Y B, Zhang, S S, Liu, Y and Yang, J H, 2019. Insights into Multifractal Characterization of Coals by Mercury Intrusion Porosimetry, *Energies*, 12 (24).

Zhou, M S, Liang, X, Ai, L and Liu, D R, 2016. Analysis of Fracture Porosity of Coalbed with *Nmr* Experiments and Resistivity Logging Data, in *Proceedings of the 2016 International Conference on Energy, Power and Electrical Engineering*, 56:228–231.

Zhou, W, Gao, K, Xue, S, Han, Y C, Shu, C M and Huang, P, 2020. Experimental study of the effects of gas adsorption on the mechanical properties of coal, *Fuel*, vol 281.

Zhu, J, Zhang, B, Zhang, Y, Tang, J and Jiang, Y D, 2018. Coal pore characteristics in different coal mine dynamic disasters, *Arabian Journal of Geosciences*, 11 (17).

Zhu, W C, Wei, C H, Liu, J, Xu, T and Elsworth, D, 2013. Impact of Gas Adsorption Induced Coal Matrix Damage on the Evolution of Coal Permeability, *Rock Mechanics and Rock Engineering,* 46 (6):1353–1366.

Effects of water inflows on mine ventilation system

J Connot[1], P Tukkaraja[2] and S Jayaraman Sridharan[3]

1. Underground Operations Engineer, Sanford Underground Research Facility, Lead, South Dakota 57754, USA. Email: jconnot@sanfordlab.org
2. Associate Professor, South Dakota School of Mines and Technology, Rapid City, South Dakota 57701, USA. Email: pt@sdsmt.edu
3. Research Scientist, South Dakota School of Mines and Technology, Rapid City, South Dakota 57701, USA. Email: srivatsan.jayaramansridharan@sdsmt.edu

ABSTRACT

Underground mine ventilation systems can face significant challenges due to adverse alterations in the subterranean environment. Identifying the precise cause of ventilation changes can be particularly challenging in inaccessible areas that are still connected to the ventilation system. However, strategically installing ventilation monitoring equipment at key exhaust locations makes it possible to directly pinpoint the root cause of ventilation fluctuations and understand their impact on the overall ventilation system.

This paper delves into the specific instances of water inflows affecting an exhaust shaft at the Sanford Underground Research Facility, formerly known as the Homestake Gold Mine in Lead, SD. The case study presented here meticulously examines the various effects of water inflows on the ventilation system. Through the utilisation of real-time sensor data and analysis with VentSim, this study provides valuable insights into the alterations experienced by the mine ventilation system.

INTRODUCTION

Mine ventilation involves controlling air quantity, quality, and distribution to maintain safe and efficient conditions underground. Other processes specifically help to accomplish quality control (eg gas and dust control) or temperature and humidity control (eg air cooling and dehumidification, heating). Ventilation is the only auxiliary operation that can accomplish all three control functions. As stated by Hartman and Mutmansky (2002), 'when the control of the atmospheric environment is complete-that is, when there is the simultaneous control of the quality, quantity, and temperature-humidity of the air in a designated space – then we are employing total air conditioning'. Underground mine ventilation is key to performing an efficient and cost-effective underground mineral extraction process. A well-designed and efficient ventilation system can improve aspects such as dust control, mechanical equipment efficiency, worker productivity, and most importantly, personnel safety. Ventilation system design is a complex process that can create a significant impact to each mining process, while there are also many aspects that can adversely affect the ventilation system. These aspects can range from the outside temperature, barometric pressure, underground geologic conditions, heat and humidity, mine gases, airborne dust, equipment operation, inaccessible workings, and even water inflows. Underground mines situated in areas with substantial groundwater aquifers or experiencing high precipitation may require dewatering pump capacities exceeding 100 000 L/min to prevent inundation of the operational regions (Lottermoser and Lottermoser, 2003).

Water can influence the ventilation system in a number of ways. Water accumulation in upcast shafts, also known as water blanketing, can significantly increase airflow resistance. This phenomenon can potentially lead to ventilation issues and reduced airflow through the mine (Kolesov et al, 2023; Semin and Zaitsev, 2020). With flowing water, the drag force between the water surface and the air in the headspace can be significant. While this phenomenon is less observed in mine ventilation scenarios, it is an important factor in controlling airflow in sewer systems. Wastewater drag, along with natural factors like wind speed and barometric pressure gradient, are major forces responsible for airflow in sanitary sewer channels (Edwini-Bonsu and Steffler, 2004; Nielsen, Hvitved-Jacobsen and Vollertsen, 2012). Previous researchers have developed dynamic ventilation models for gravity sewer networks to predict airflow due to water drag and pressure differences (Wang et al, 2012). Edwini-Bonsu and Steffler (2004) investigated the influence of wastewater drag and natural factors (differential wind speed and barometric pressure gradients) on airflow within sanitary sewer channels using computational fluid dynamics (CFD) models. Plunging water can trap air causing downstream pressure and odour issues in drainage systems (Ma, Zhu and Rajaratnam, 2016). Various mechanisms contribute to this entrainment, including water flow patterns, drop breakup, and even downstream conditions (Ramezani, Karney and Malekpour, 2016). Studying this phenomenon is challenging due to these interacting factors and the limitations of scaled models (Yongfei, Wang and Zhang, 2018).

The current study was conducted at the Sanford Underground Research Facility (SURF) which is in the former Homestake Goldmine in Lead, SD. It was home to the late John Marks, a renowned ventilation expert. SURF is operated by the South Dakota Science and Technology Authority (SDSTA) and is a deep-level (1500 m) underground research lab that is ventilated mechanically through exhausting fans. Ventilation is critical to maintaining a safe and healthy work environment and ensuring quality results for research projects. SURF's ventilation system is made up of two primary intake shafts; Yates and Ross, and two primary exhaust paths; Oro Hondo and #5 Shaft. SURF currently operates two primary fan installations, located at the top of the Oro Hondo Raise system and the No. 5 Shaft, respectively as shown in Figure 1. The primary airflow enters the Ross and Yates shafts and exits the levels that are open to ventilation. The American Davidson at the Oro Honda is the primary fan that exhausts air from the underground. It is a type 1400-S1BAB92-EV single-width centrifugal with backward-curved airfoil blades. The rotating wheel is 3.5 m in diameter and weighs 5760 kg. The fan is powered by a 2.2 MW (3000 HP), 720 revolutions per minute (rev/min) synchronous motor that is directly connected to the fan and is limited by a 1.3 MW (1750 HP) variable frequency drive (VFD). The VFD provides soft-start capabilities and controls the fan speed from 50 percent to 85 percent capacity of the synchronous motor speed. The fan's normal operating point in the past was 390 rev/min, producing 145 m³/s at 1.88 kPa for fan total pressure, with an inlet air density of 0.98 kg/m³. Currently, the fan operates at 600 rev/min, exhausting 233 m³/s at 3.6 kPa for fan total pressure, with an inlet air density of 1.01 kg/m³ in its current underground configuration as of 2 November 2023. The Spendrup 150 fan is the secondary main exhaust fan located at #5 Shaft. The Spendrup 150 is a series 125-07-1800-A vane axial fan that exhausts 20.5 m³/s at 1.79 kPa with an inlet air density of 0.98 kg/m³. Currently, there is a blockage within #5 Shaft located 208 m below the shaft collar that restricts the airflow through the 5.8 m diameter shaft the fan can exhaust from the underground.

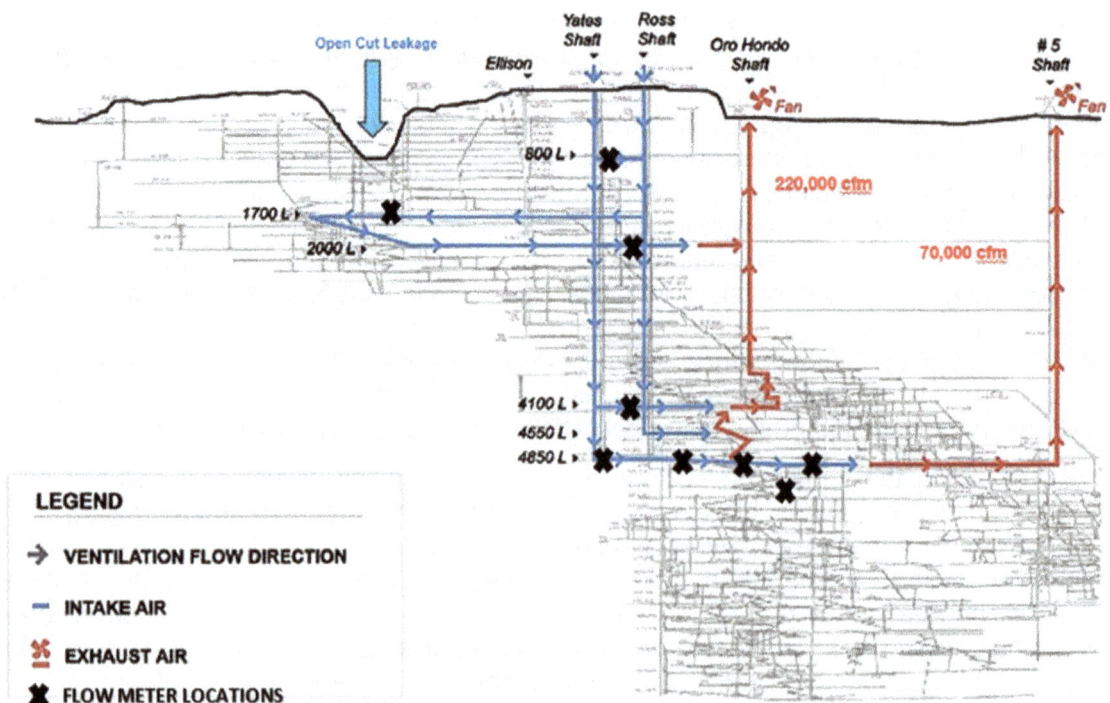

FIG 1 – SURF ventilation network (Artz, Tukkaraja and Pietzyk, 2015).

For this case study, we will look at the ventilation provided to the 1700L which travels across the level from the Yates and down the ramp system before it is exhausted into the Oro Hondo Shaft and #5 shaft on the 2000L. A schematic representation of this circuit is shown in Figure 2. Currently, there is an automated louvered regulator that was purchased from Maestro Digital Mine and provides the ability to control the airflow remotely for the Caterpillar Minestar Lab on the 1700L as well as monitor the intake airflow and the exhaust airflow to the Oro Hondo. When the Maestro regulator is set to its normal position of 30 percent open, SURF ventilation system exhausts approximately 19 m³/s from the 2000L with 12 m³/s exhausting to the Oro Hondo and 7 m³/s exhausts to #5 Shaft.

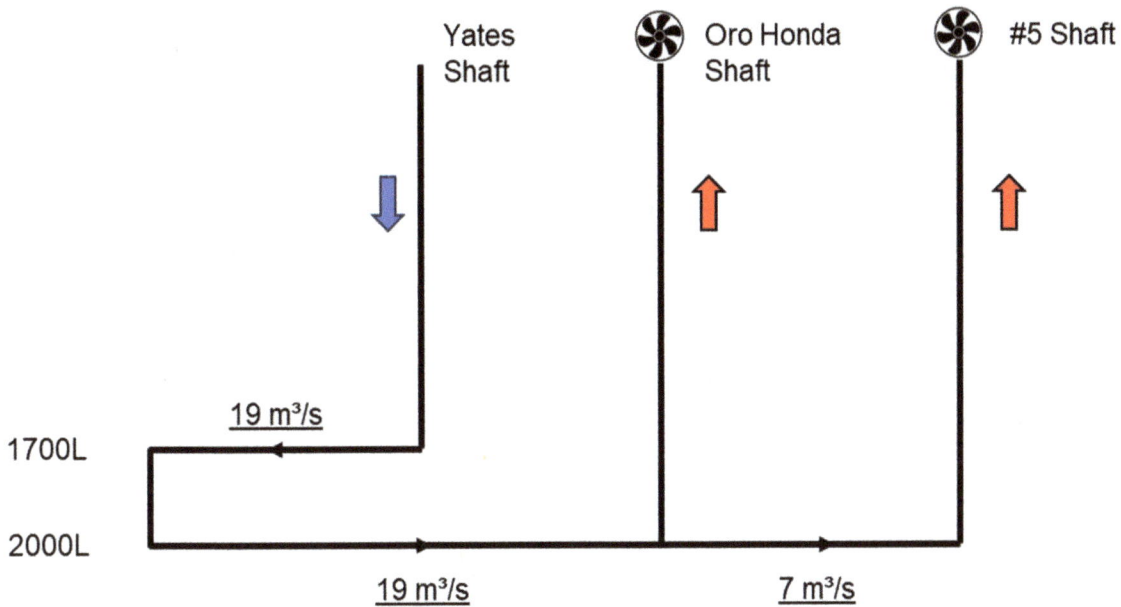

FIG 2 – Schematic representation of 1700L ventilation.

The 2000L is also an important level for our water inflow system. This is the point where the majority of inflows that are captured from the workings above the 1850L are directed to a 16 inch diameter pipe. The water inflows are sent down a 1 m diameter raise from the 1850L to the 2000L where they are typically sent to the 2450L sump to be dewatered up the Ross shaft to the wastewater treatment plant to be processed. In high inflow events, typically during heavy rainfall, or during hot days in the early spring that cause large quick snowmelt, the 2450L sump gets overwhelmed from the influx of water to the underground. Instead of overflowing the sump, SURF installed a concrete sump wall to direct water back to the deep pool utilising #5 shaft through an 18 inch pipe. SURF transformed the drift leading up to #5 shaft into a sump by installing a two-foot-thick concrete wall that is approximately six feet tall in a 2.75 m high by 2.5 m wide drift. The 18 inch pipe will then discharge into the sump as shown in Figure 3.

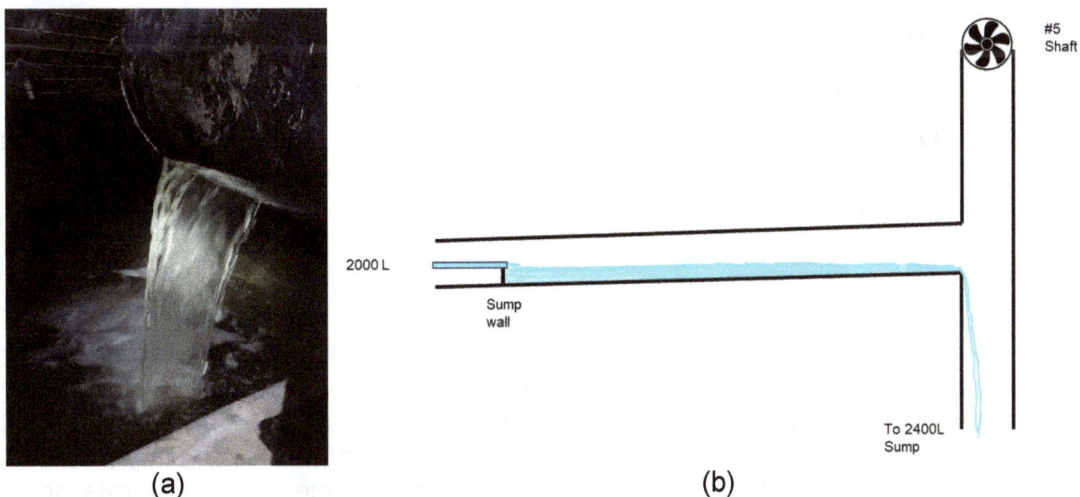

(a)   (b)

FIG 3 – (a) 18 inch discharge pipe; (b) sump wall and water discharge into #5 Shaft.

This concrete wall was needed to push the water to #5 shaft by the water elevation. There is a 0.7 m elevation gain from the Ellison to #5 shaft across a 2.2 km linear foot distance from shaft station to shaft station as the Homestake Mining Company drove all drifts from any shaft to flow towards the Ellison shaft, where their main water inflow collection system was located. Reference Figure 4 that depicts the 2000L map of important features.

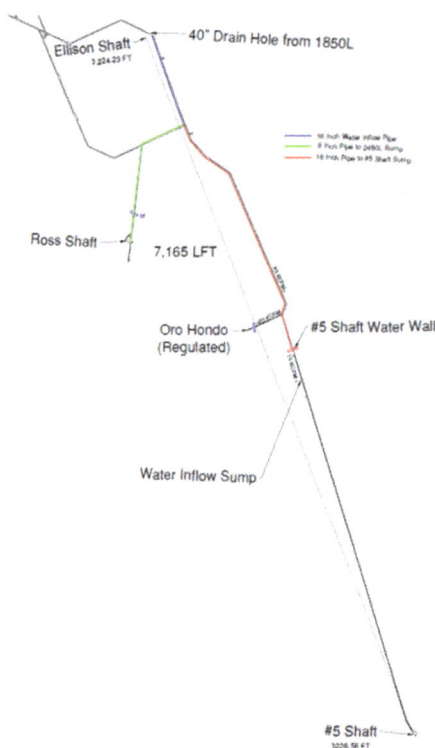

FIG 4 – 2000L map of important features.

During the high-water inflow events, the water is directed to the #5 shaft water inflow sump. Due to the elevation change, the water level in the sump will rise slightly before naturally overflowing into #5 shaft, which free-falls to the deep pool. During water discharge to #5 shaft sump, the ventilation system observed unusual conditions, which was unknown why they were happening. These conditions consisted of the following:

- #5 Shaft Exhaust fan going into stall.
- The 5000L pump room ventilation greatly decreased.
- #4 winze drift on the 4850L would reverse flow direction.
- The 1700L Caterpillar Lab would increase airflow volume.

Once the Maestro Digital Mine airflow sensors were installed at the 2000L, we were able to capture the data and see the trend line of the airflow characteristics to make a correlation between water discharge in gallons per minute (GPM) to level airflow increase in cubic feet per minute (CFM).

METHODS

To make a correlation between the water inflow rate and the airflow rate increase that was observed on the 2000L exhaust, we needed to collect the discharge rate during high water inflow events that was flowing down #5 Shaft. Currently, SURF has water flow metres on the 16 inch pipe coming from the 1850L and the 8 inch pipe that is directed to the Ross shaft and down to the 2450L sump, but no flow metre on the 18 inch pipe to #5 Shaft. The water flow rate to #5 Shaft can simply be calculated by taking the flow rate of the 16 inch pipe and subtracting it by the 8 inch flow rate of the pipe (Equation 1):

$$16' \text{ Pipe flow rate} - 8' \text{ Pipe flow rate} = \#5 \text{ Shaft flow rate} \tag{1}$$

Understanding the discharge rate down #5 shaft gave us the ability to know the exact water flow rate that was free-falling down the shaft back to the deep pool. The next step we needed to take was to be able to understand the airflow characteristics and how they change during the high inflow events where water is redirected to #5 Shaft. This was completed by installing the Maestro Digital Mine Vigilante Air Quality Station (AQS) monitors at the exhaust junction between the Oro Hondo and #5 shaft as shown in Figure 5. At the time of installation, the Maestro Vigilante's purpose was to operate the mine ventilation regulator and control the airflow supplied to the CAT Lab on the 1700L. This allowed SURF to direct more airflow to the 4850L during the LBNF Neutrino Cavern Excavation, where most of the air was needed. Since the airflow monitoring station was installed prior to understanding the water inflow anomaly, we had only placed airflow sensors on the intake drift leading up to the junction of the Oro Hondo and #5 Shaft to capture the total intake airflow. The other sensor was placed towards the drift of the Oro Hondo to determine the primary exhaust path for the ventilation circuit. The airflow to #5 Shaft could then be calculated by balancing the intersection using Kirchhoff's first law, which states the quantity of air leaving a junction must equal the quantity of air entering a junction (Hartman *et al*, 1997). The Maestro sensor placement and airflow balance during normal operating conditions can be observed in Figure 6.

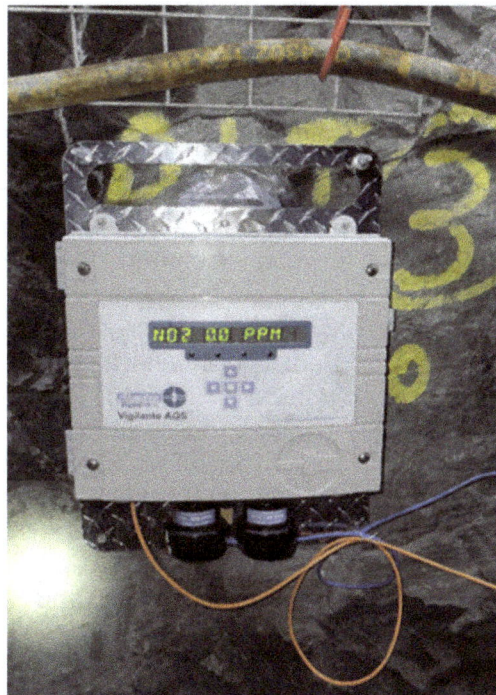

FIG 5 – Maestro AQS monitor.

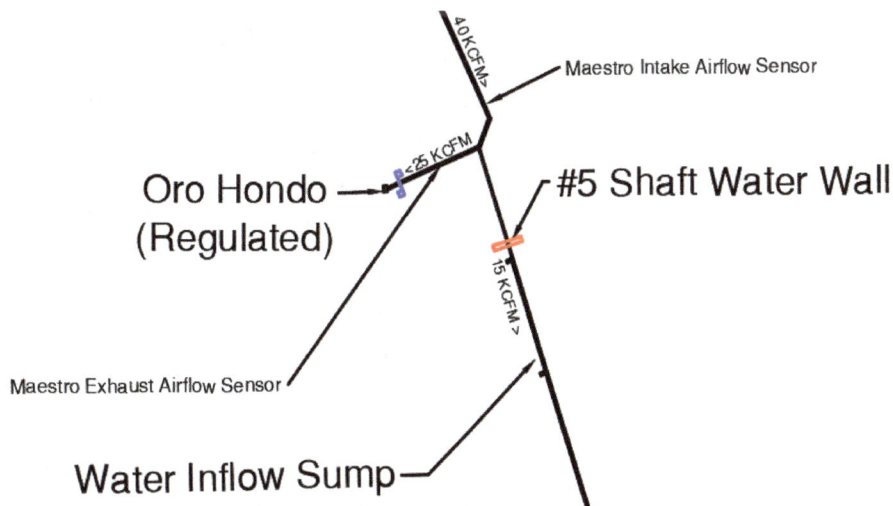

FIG 6 – 2000L normal airflow conditions and maestro sensor placement.

Now the systems were in place to collect the necessary data to make the correlation between water discharge rate and airflow volume increase, we had to wait to capture it during a high-water inflow event where water was directed to #5 shaft back to the deep pool. Over the course of the late spring and throughout the summer of 2023, we were able to capture three separate data sets. They occurred between the following dates below:

- 23 June 2023 through 25 June 2023.

- 27 June 2023 through 2 July 2023.

- 11 July 2023 through 13 July 2023.

The following section will discuss the findings and display the results of the data collected during these three water inflow events and how they affected the ventilation system.

RESULTS AND DISCUSSION

The results from the three data sets were analysed and were further broken down to compare multiple high-water inflows during each data set. The Maestro trending of airflow volume increase can be seen during the same time water was directed to #5 Shaft. The data was extracted from both the South Dakota Science and Technology Authority's water flow metres from the 2000L as well as from the Maestro ventilation monitors and inserted into a central database using Microsoft excel. This provided the ability to compare the two separate data sets in one graph for each of the three events.

After analysing the data sets, it is obvious a correlation can be made that water falling down #5 shaft directly affects airflow increase on the level. As more water cascades into the shaft, the airflow is delayed approximately 3 hrs after the initial water discharge. This can be seen in all events and in event two can be seen multiple times. Reference the plots in Figures 7, 8 and 9. In event one (Figure 7) from 23 June 2023 to 25 June 2023, due to a large rain event, the peak of water discharge of approximately 26.5 L/s occurring at 7:11 am shown with a red line, the airflow increases, which occurs around 10:18 am increasing the flow from 4.7 m^3/s up to 17 m^3/s. As the water flow tapers the airflow begins to taper consistently 3 hrs behind the water flow. During the time of the event, the ventilation system was affected through many aspects. Excavation work was being conducted at this time by Thyssen Mining on the 4850L for the Fermilab Deep Underground Neutrino Experiment. The excavated material was being skipped through the Ross shaft using an orepass on the 4850L and loaded on the 5000L. The orepass system and 5000L are all tied to the #5 shaft exhaust system through #6 winze down to through the workings of the 5300L and 5600L, which are inaccessible, and into #5 shaft. At the time of the water inflow event where the water was being discharged down #5 shaft the 5000L decreased in flow. It was unknown what was causing the flow disruption at the time, but it almost acted like something had blocked our exhaust path. On the 4850L, the exhaust drift to #4 winze and #5 shaft reversed. This was not a favourable condition as the Cap and Powder magazines for the excavation was located in this drift and needed to be exhausted away from the workers for safety reasons. Having the flow reverse in this location if there was a fire, would send the toxic fumes to the work area. The last effect on the ventilation system was on the #5 shaft exhaust fan. As the water pulled air down the shaft, it caused the exhaust fan to go into stall mode.

After a short time later, the lead area received another large rain event during 27 June 2023 and 2 July 2023. Again, the underground ventilation system was in havoc and it was unknown for reasons why. During the second water inflow event the fan was in stall, the 5000L airflow ceased, and the 4850L cap and powder magazine drift reversed. During this event, the water flow was directed back to the Ross shaft dewatering sump located on the 2450L, which is the normal path for the water inflow system. During these high inflow events, the water can be directed to #5 shaft. The sharp drop-offs seen in the blue line that represents water discharge to #5 shaft is mirrored with the airflow through #5 shaft. Once again, each peak water discharge can be replicated in the peak airflow approximately 3 hrs behind the peak water discharge.

During the last rain event displayed in Figure 9, the airflow and water correlate once again. Like the other two events caused the same issues to the ventilation system that was observed before. The airflow seems to mimic the water discharge rate approximately 3 hrs after the peak discharge.

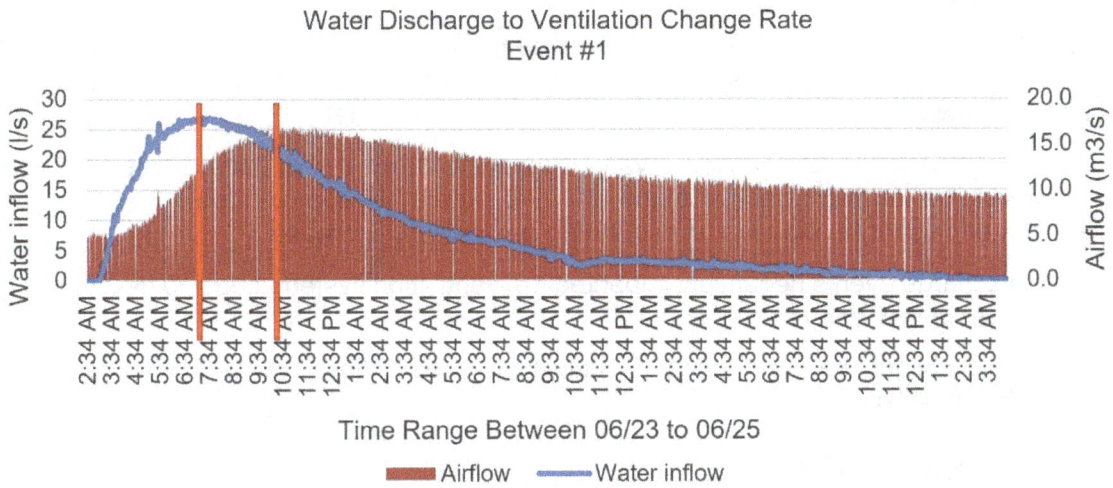

FIG 7 – Event one water discharge correlation to airflow increase.

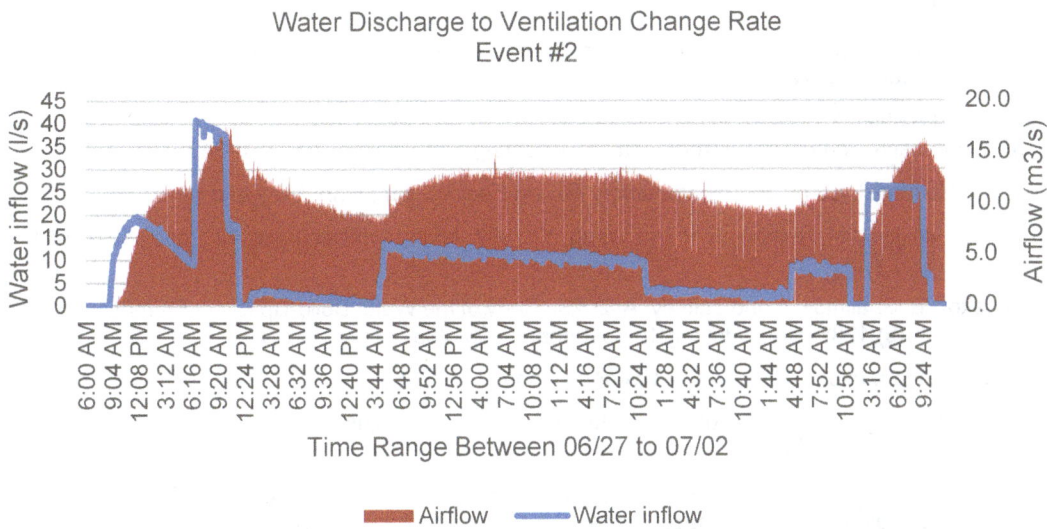

FIG 8 – Event two water discharge correlation to airflow increase.

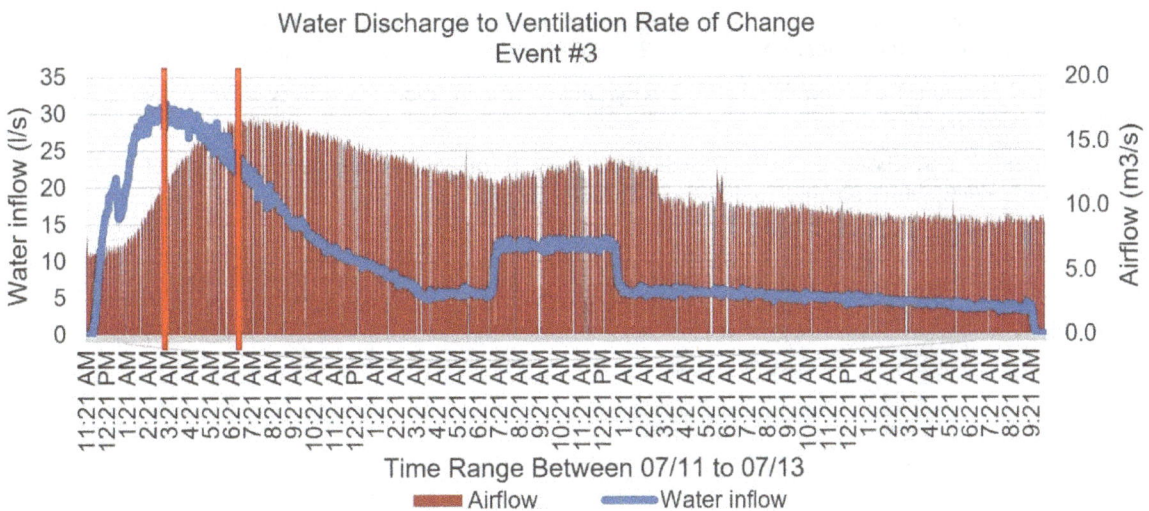

FIG 9 – Event three water discharge correlation to airflow increase.

SUMMARY AND CONCLUSION

This study investigated the correlation between water discharge rates and airflow volume within a mine ventilation system, specifically water discharged down an exhaust shaft. Data was collected during three high-water inflow events in 2023. The data showed that:

- A clear correlation exists between water cascading down the exhaust shaft and increased airflow volume. In this particular study, there is a delay of approximately 3 hrs between the peak water discharge and the corresponding peak airflow increase.

- Water discharge events negatively impacted the ventilation system, causing decreased airflow in the lower levels, and fan stalling.

These findings highlight the importance of managing water inflow events to maintain safe and efficient ventilation within the mine.

ACKNOWLEDGEMENTS

The authors extend their thanks to the Sanford Underground Research Laboratory for granting permission to conduct this study and Maestro Digital Mine for their assistance with data acquisition.

REFERENCES

Artz, T, Tukkaraja, P and Pietzyk, B, 2015. Ventilation System Design for the Sanford Underground Research Facility, Applications of Computers and Operations Research in the Mineral Industry.

Edwini-Bonsu, S and Steffler, P M, 2004. Air flow in sanitary sewer conduits due to wastewater drag: a computational fluid dynamics approach, *Journal of Environmental Engineering and Science*, 3(5):331–342.

Hartman, H L and Mutmansky, J M, 2002. *Introductory Mining Engineering* (John Wiley and Sons).

Hartman, H L, Mutmansky, J M, Ramani, R V and Wang, Y, 1997. *Mine Ventilation and Air Conditioning* (John Wiley and Sons).

Kolesov, E, Kazakov, B, Shalimov, A and Zaitsev, A, 2023. Study of the Water Build-Up Effect Formation in Upcast Shafts, *Mathematics*, 11(6):1288. https://doi.org/10.3390/math11061288

Lottermoser, B and Lottermoser, B, 2003. *Mine Water* (Springer).

Ma, Y, Zhu, D Z and Rajaratnam, N, 2016. Air entrainment in a tall plunging flow dropshaft, *Journal of Hydraulic Engineering*, 142(10):04016038.

Nielsen, A H, Hvitved-Jacobsen, T and Vollertsen, J, 2012. Effect of Sewer Headspace Air-Flow on Hydrogen Sulfide Removal by Corroding Concrete Surfaces, *Water Environment Research*, 84(3):265–273.

Ramezani, L, Karney, B W and Malekpour, A, 2016. Encouraging Effective Air Management in Water Pipelines: A Critical Review, *Journal of Water Resources Planning and Management*. https://doi.org/10.1061/(asce)wr.1943-5452.0000695

Semin, M and Zaitsev, A, 2020. On a possible mechanism for the water build-up formation in mine ventilation shafts, *Thermal Science and Engineering Progress*, 20:100760. https://doi.org/10.1016/j.tsep.2020.100760

Wang, Y, Nobi, N, Nguyen, T and Vorreiter, L, 2012. A dynamic ventilation model for gravity sewer networks, *Water Science and Technology*, 65(1):60–68.

Yongfei, Q, Wang, Y and Zhang, J, 2018. Three-Dimensional Turbulence Numerical Simulation of Flow in a Stepped Dropshaft, *Water*. https://doi.org/10.3390/w11010030

Rock strata heat in time-dependent underground heat modelling

M D Griffith[1]

1. Software Team Leader, Howden Ventsim, South Brisbane Qld 4101.
   Email: martin.griffith@howden.com

ABSTRACT

The paper presents a method to solve for the rock strata radial heat conduction across all the airways of a mine ventilation model. The method allows the investigation of a range of time-varying heat phenomena in underground mines. This contrasts with the practice of modelling the heat flow from rock strata using the Gibson function (1976), which gives the strata heat flow as a function of the temperature and velocity of the airstream, the age of the airway, and a range of rock thermal properties. The major drawback of the Gibson function is that it requires a fixed temperature and velocity over the life of the tunnel. Therefore, using it in modelling any situation requiring changes in these parameters (daily and seasonal atmospheric variation, fires, installation or shutdown of a cooling plant or fan) is problematic and achieved by introducing an often case-specific calibration to match to observed real-world results. Another problem exists in using the Gibson function on airways in isolation when they are interconnected. As heat flow to one airway reduces over time as the rock cools, changing airstream temperatures will affect the heat flow occurring in downstream airways, making the result of the function – in all but the smallest of mine ventilation networks – tend towards underestimating the historical temperature with respect to the actual temperature used in the function. The proposed method in this paper fully simulates the rock wall temperature radial distribution, allowing the heat history of the airway to be included in the heat flow calculated at any moment. From this, the paper provides a quantification of the error in the Gibson function for interconnected airways in a standard sized mine ventilation network. The paper also presents how this new method improves the accuracy of modelling of dynamic heat phenomena (such as seasonal atmospheric variation and fires), removing the need for calibration factors. Also demonstrated is how this enables transient modelling of heat over the mine life and ultimately better prediction of mine environment conditions and requirements for ventilation and cooling.

INTRODUCTION

A major factor in the heat and moisture in an underground mine is the geothermal heat that enters the airstream from the walls of the airways. While the underlying equations for this physical process are well understood, modelling it in an underground mine is difficult due to the three-dimensional layout of the airflow network. The heat transfer from the rock wall to the airstream can be modelled one-dimensionally per airway but modelling it over thousands of connected airways is more difficult. This paper seeks to present the current practice for handling geothermal heat in underground heat network modelling and explore the possibility of increasing the complexity of the modelling techniques and obtaining a more accurate and spatially and temporally resolved simulation.

When an airway is mined, the rock is at the virgin rock temperature (VRT). As soon as it is exposed to the air it begins to cool, with the rock closer to the air cooling first, and the cooling extending deeper into the rock over time. It cools by conduction of the heat from deep in the rock to the rock surface and then convective heat transfer from the rock surface to the airstream. If we simplify the airway to a cylinder the heat flux per square metre of exposed rock surface can be represented by:

$$q = k\left(\frac{\partial \theta}{\partial r}\right)_s = h(\theta_s - \theta_d) \left[\frac{W}{m^2}\right]$$

where:

K	is the thermal conductivity of the rock [W/(m°C)]
θ	is the temperature [°C]
r	is the radial coordinate [m]
h	is the heat transfer coefficient for heat transfer from the rock surface to the air [W/ (m²°C)]

θ_s is the temperature of the rock at its surface, or at its interface with the air

θ_d is the dry bulb temperature of the air (the forcing temperature), with the subscript $_s$ representing a value at the rock surface to air interface

But as heat is transferred between the air and the rock there is change over time of the rock wall temperature. The problem is simplified by assuming the heat conduction in the rock varies only in the radial direction, which allows the simplified radial heat conduction equation:

$$\alpha\left(\frac{\partial^2\theta}{\partial r^2} + \frac{1}{r}\frac{\partial\theta}{\partial r}\right) = \frac{\partial\theta}{\partial t}\left[\frac{°C}{s}\right]$$

where,

α is the rock thermal diffusivity [m²/s]

t is time [s] (Danko *et al*, 2012; McPherson, 1993)

Note, this equation is expressed differently in Danko *et al* (2012), but it simplifies to the same equation. With these two equations we can model the rock temperature variation into the wall. Normally we have a good idea of rock thermal conductivity and diffusivity and we know at $t = 0$ the rock is all at the VRT. The convective heat transfer coefficient, h, can be calculated as a function of the rock surface roughness and the air velocity across the rock surface.

One can solve the equations above numerically to obtain the heat flux at any moment for a given airway age, VRT, rock type, and air velocity and temperature. But it can be computationally expensive, so various short cuts and simplifications are used. The best known is the Gibson function (McPherson, 1993), which was presented as a quick way to obtain the heat flux for a given airway as a function of its age, rock thermal properties, air velocity and temperature. It allows the accurate calculation of heat flux with a handful of steps and is far less intensive than numerically solving all the heat variation into the rock. This is very useful for a network heat simulation which needs to have the heat flux from thousands or tens of thousands of airways. But the function has several simplifications, with the one of most interest being the assumption of a constant air temperature and speed.

In an underground mine environment, the air temperature and speed will not be constant. Temperature will vary with night and day and passing seasons (depending on proximity to the intakes). Also, an underground mine is a dynamic, constantly changing environment; sections of the mine might be opened then closed, the amount of vehicle heat sources might change, or refrigeration or heating units might be installed or replaced. None of these changes in the forcing air temperature of the rock heat transfer can be considered in the Gibson function.

To take daily and seasonal variation first; these effects are largely periodic, therefore their effect can be averaged out and the Gibson function should be able to calculate a good answer regardless. But it cannot tell us anything about those daily and seasonal variations, if they are of interest to us. For example the effect of the heat capacitance cannot be included, which will cause a phase lag in the temperature deeper underground with respect to the outside forcing temperature. Step changes in forcing temperature – from, for example, a refrigeration unit starting up or shutting down – cannot be modelled with the Gibson function assumption of constant temperature.

Another issue is the translation of this one airway method to a complex network heat simulation, where thousands of airways interact. Firstly, an airway downstream will have its forcing air temperature affected by the rock wall heat transfer of airways upstream, meaning in a hot mine, the Gibson function will generally underestimate the dry bulb temperature if used on multiple airways in a network.

The aim of this paper is to investigate these inaccuracies of the Gibson function and then to discuss the possibility of incorporating time dependent rock wall heat transfer into a network heat simulation. This will be computationally expensive, possibly prohibitively so. But doing so would not only allow the investigation of effects of changing forcing temperatures, but also allow better calibration of models with measured survey data. Presently, a heat simulation using the Gibson function simulates an average state, rather than the heat state at any given time; but this average state never exists.

Finding a way to simulate to a given moment, rather than to an average state would be useful for calibrating models with site temperature measurements.

METHOD

To investigate the effect of having the true air temperature history in an airway included in the rock strata heat flux, a numerical solver was written to model the radial heat distribution over time in a single airway. The single airway solver discretises the radial heat equation over the radial distance into the rock wall and then models the rock temperature deep in the wall as it changes over time in response to the air temperature and velocity.

Effort was put into making the solver run as quickly as possible, because it is anticipated that the solver would be run in parallel across hundreds or thousands of airways in a network model. The solver uses central difference finite differences to calculate the 1st and 2nd spatial derivatives in Equation 2, while the solution is stepped forward in time using the Runge-Kutta method, which proved the fastest method for a given accuracy. The radial temperature gradient reaches deeper into the wall over time; once the radial temperature gradient at the deepest point exceeds 0.01°C/m, the domain is extended.

To verify the solver works, we can run it for a constant temperature over the life of the airway and we should arrive near the same result as the Gibson function. For the calibration case shown below, the initial spatial discretisation at the rock wall is 0.01 m, and then increases into the rock wall as the spatial temperature gradients diminish. The time step is 150 secs. Increasing the spatial or temporal resolutions beyond these values produced only negligible further change. For this test, we will use the same airway parameters as used in Danko *et al* (2012). These are:

- Airway: width 3.5 m, height 5.7 m, perimeter, 18.4 m, area, 19.95 m².

- Average air temperature 28°C dry bulb, wet bulb 15.9°C, pressure, 87.15 kPa.

- Rock: VRT 28°C, specific heat 844 J/kg/K, thermal conductivity 3.15 W/m/K, density 2309 kg/m³.

A forcing air temperature of 18°C is chosen. The development of the rock heat over time is shown in Figure 1, while the heat flux over time can be seen in the solid purple line in Figure 2a.

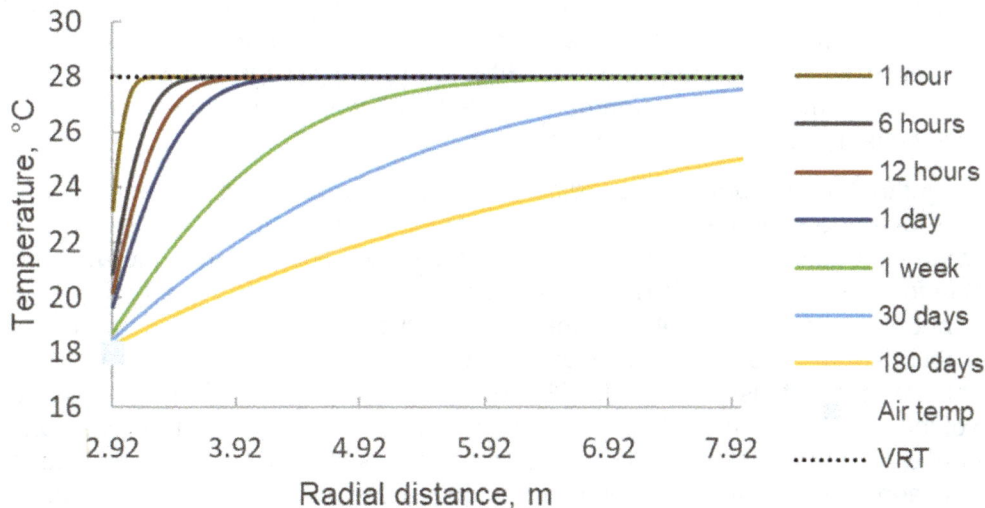

FIG 1 – Variation of rock temperature over time for an airway of hydraulic diameter 2.92 m, of VRT 28°C, exposed to constant air temperature of 18°C, with air flow quantity 50 m³/s.

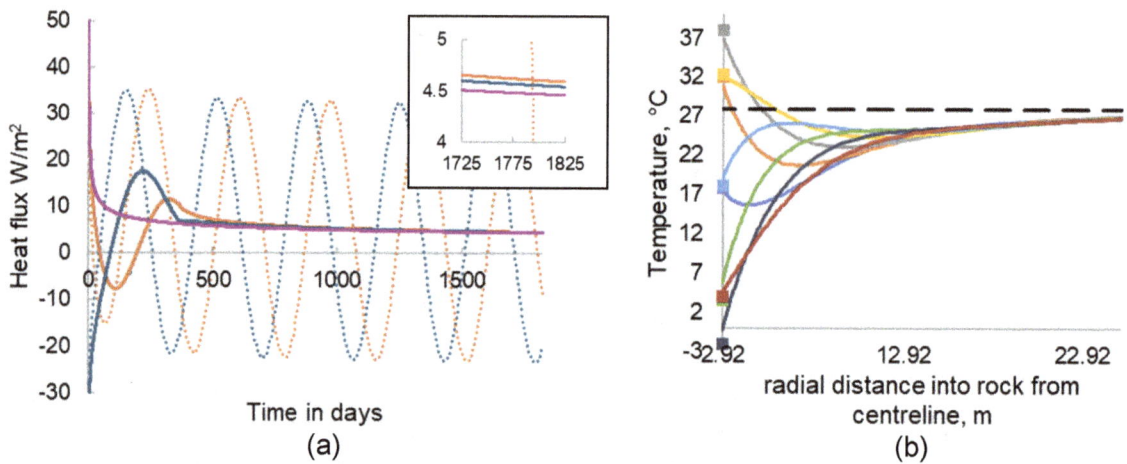

FIG 2 – For a single airway, in (a) the time dependent (dotted) and mean (solid) heat flux for three cases and in (b) the temperature inside the rock wall at eight times over one year for one particular case. More detail in text.

We can verify this solution by calculating the rock wall surface temperature and the heat flux from the Gibson function. The Gibson function for the same inputs returns a heat flux per unit of rock wall area of 7.32 W/m^2, which is within 2 per cent of the heat flux after 180 days for our solver, of 7.45 W/m^2; a similar accuracy is obtained extending out to five years. This is the stated accuracy of the Gibson function, so we can have confidence in the accuracy of the single airway solver presented here.

SINGLE AIRWAY

The advantage that the single airway solver has over the Gibson function is that we can now model variations over time in the input temperature.

Figure 2a plots in purple the heat flux per unit area over five years for the case with constant forcing temperature of 18°C; in dotted orange it shows the heat flux over five years for the same case as in Figure 1, but now with an annual sinusoidal temperature variation of 20°C added to the 18°C average temperature (giving maximum 38°C, minimum -2°C); the running one year average of this heat flux is plotted in the solid orange line; in the dotted and solid blue lines is shown the same case but with the annual variation shifted by three months. Effectively, the orange shows an airway mined in spring, blue an airway mined in the summer. Figure 2b shows the internal rock wall temperature variation at eight evenly spaced times over one year.

It is often thought that the linearity of the radial heat equation means that the annual variation when considering the average state can be ignored, but this is not quite true. On the graph we can see that the average heat flux for the constant forcing temperature is different to the averaged heat flux for the variable forcing. It is very different for the first year, and still different after five years (see the inset graph), but only slightly so; (note: for the Gibson function, this is possibly considered already in its stated accuracy of 2 per cent; the original reference for the function was unavailable to the author and this issue is not discussed in McPherson (1993). Another variable that affects the result is what time of the year the airway is first mined, with Figure 2 showing heat flux for a summer airway and a spring airway. These differences are due to the average cooling over time; the temperature difference between the rock and the air is what drives the heat transfer; what air temperature differences the rock wall is exposed to when it is early in its life are more critical than later in its life, because the rock is on average hotter when the airway is younger. So an airway opened in the summer will hit the colder part of the year earlier in its life than an airway opened in the spring, so it will lose more heat earlier, resulting in less heat flux later; the result of this can be seen in the higher heat flux for the spring airway after five years (see inset graph of Figure 2).

The Gibson function and time averaged result is significantly different from the running average heat flux for both cases until about one year of airway life. Of course the extent of this error would depend on the extent of the annual variation in temperature, and on the airway's exposure to this variation, which would be less the deeper in the mine the airway is.

MULTIPLE AIRWAY SOLVER

Another issue with the use of the Gibson function in network heat simulations is the interconnectedness of the many airways in a standard underground mine. An airway placed downstream, deep in the mine will be affected by the cooling/heating of airways upstream. This effect is not possible to include in a network heat simulation using the Gibson function. So in a hot mine, the Gibson function will tend towards underestimating the air temperature in downstream airways.

The multiple airway solver uses as a base the Ventsim DESIGN Heat Simulation tool, but with the Gibson-based rock strata heat transfer function replaced on every airway with the single airway solver described in the previous section. This will have a number of effects:

- such a solver will account for the effects of heating/cooling airways upstream on airways downstream

- it will be able to model the phase lag in underground temperature peaks with respect to atmospheric temperature peaks (due to the heat capacitance effect of the underground rock walls)

- it will be able to handle transient changes in forcing effects (such as seasonal changes, diurnal changes, and heating or refrigeration changes)

- and finally it will require far more computation.

For even a small mine, running the single airway solver over a few hundred airways for any useful period of time will require considerably more computer memory and operations than the existing standard heat simulations. This will be discussed more later.

To validate the multiple airway solver, we can refer to the work of Danko *et al* (2012). They simulated the varying temperature down a 6 km tunnel over a five year period. The tunnel uses the same parameters as described in the Method section of this paper. The temperature variation is 20°C around an average temperature of 28°C, so – in contrast to the results of Figure 2 which oscillated around 18°C – varying between 8°C and 48°C. Figure 3 plots the temperature variation 6 km into the tunnel over a five year period using the multiple airway solver.

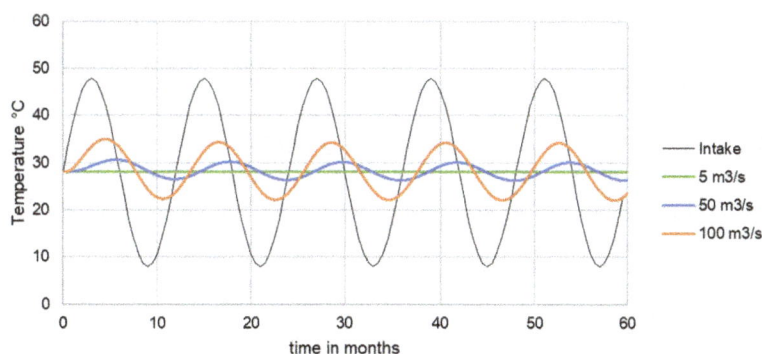

FIG 3 – The temperature variation at the 6 km mark for the test case at three different airflow rates, calculated from the multiple airway solver.

In the graph, the peaks of the temperature variation for the 50 m³/s and 100 m³/s cases lag the peaks of the intake temperature. There is also a reduction in the amplitude of the variation with respect to the intake temperature. This phase lag and amplitude reduction are the effects of the heat capacitance of the rock wall upstream of the point shown in the graph. We can measure the phase lag and amplitude decay as a function of the distance down the tunnel, allowing a direct comparison with figures 10 and 11 of Danko *et al* (2012).

Figure 4 plots the variation of amplitude decay and phase lag along the tunnel for the three cases shown in Figure 3. These plots compare well qualitatively to figures 10 and 11 of Danko *et al* (2012); but there is some difference in values. The amplitude decay in the three cases after 6 km is 0.0006, 0.100 and 0.309 for the 5 m³/s, 50 m³/s and 100 m³/s cases respectively; while for the same values Danko *et al* (2012) report approximately 0, 0.07 and 0.23. For phase lag after 6 km, Figure 4 shows

8.98 months, 2.78 months and 1.563 months for the 5 m³/s, 50 m³/s and 100 m³/s cases respectively; while for the same values Danko *et al* (2012) report approximately 8.2 months, 2.1 months and 1.1 months.

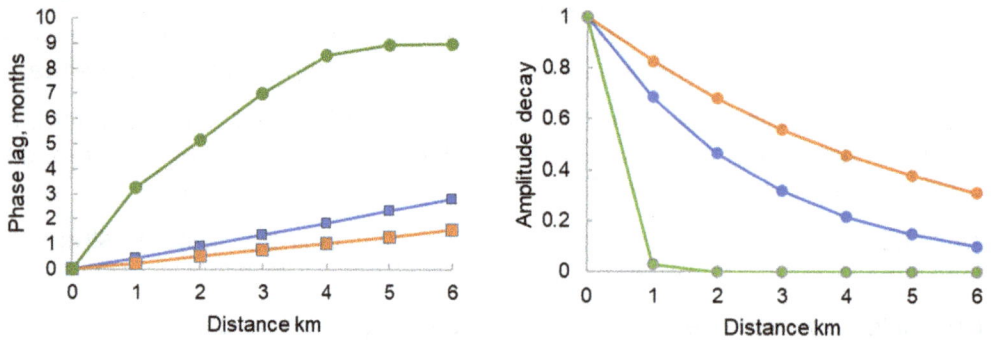

FIG 4 – (a) the variation in phase lag along the tunnel for the three airflow rates tested; (b) the variation of amplitude decay with respect to the intake amplitude for the same cases (green 5 m³/s, blue 50 m³/s, orange 100 m³/s).

The match up in terms of change in phase lag and amplitude decay with distance is encouraging. The discrepancies in absolute values could be due to several factors. Firstly, Danko *et al* (2012) makes no mention of friction factor in their case, which should have an impact on the heat transfer coefficient for the rock wall to air interface (McPherson, 1993). In the current method an Atkinson friction factor of 0.012 kg/m³ was used in the formulation of the heat transfer coefficient. A quick test with the current method shows doubling the Atkinson friction factor causes approximately 5 per cent and 2.5 per cent increases at 6 km in the phase lag and amplitude decay, respectively. secondly, the rock thermal diffusivity used in Danko *et al* (2012) is not consistent with the rock thermal conductivity, specific heat capacity and density presented, so there is some uncertainty. These potential differences in problem formulation between the two methods would likely create quantitative differences in result, without changing overall trends, which is what is seen in Figure 4.

Further cases are run to gauge the effect of more realistic conditions. In the 6 km tunnel shown above, all 6 km of the tunnel are modelled as present at time = 0. In reality, this 6 km tunnel would not entirely appear at one instant but be mined over time. It would also possibly be increasing in-depth, so an increase in VRT along the tunnel is likely.

Figure 5 plots the temperature variation at 6 km, along with the amplitude decay and phase lag for four cases where the VRT has been scaled up as if the tunnel is going deeper underground in the presence of a geothermal gradient.

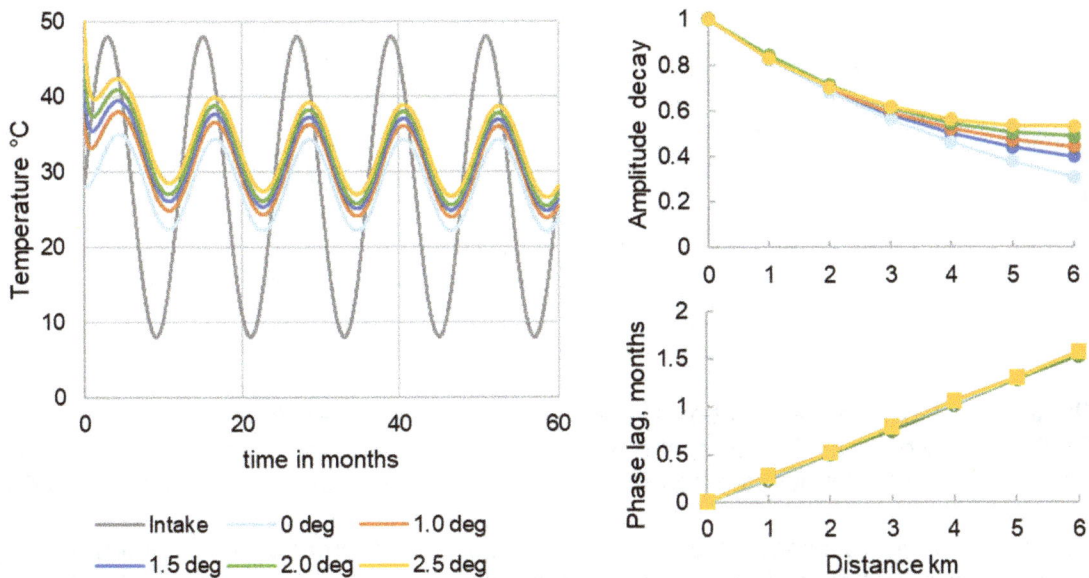

FIG 5 – Results for a flow of 100 m³/s, varying the geothermal gradient along the tunnel. (a) shows the temperature variation at the 6 km mark, while (b) and (c) show the corresponding amplitude decay and phase lag variation.

Each airway was changed from VRT = 28°C to a value corresponding with its distance from the intake and the geothermal gradient and an assumed decline gradient of 15 per cent. Note, the elevation of the airways was not changed, so this case does not include any effect from changes in air density resulting from changes in-depth along the decline. The expected higher temperatures can be seen in the temperature variation at 6 km. The phase lag is not affected by the higher rock temperatures, while the amplitude of the heat variations is sustained further downstream. These results are consistent with there being simply more heat energy available in the rock mass, thus sustaining temperatures. It seems the phase lag remains independent of the amount of heat energy available, instead having a dependence on the speed at which heat energy in the air is carried down the tunnel, as shown in Figure 4.

A further complication comes from the tunnel not instantly appearing at a moment in time, but rather being constructed over a set amount of time. More tests were run whereby a mining date was assigned to each airway, so that airways only took part in the simulation once the running time of the simulation had passed the date mined. Adding this factor had no effect on the phase lag or the amplitude decay. The only effect was to delay the start of the cooling process on each airway, but since in this case downstream airways have no effect on upstream airways, there was no interesting change in results beyond this.

The final case shown here features the same tunnel but with a cooling power introduced at the intake at the two year mark, and then removed at the four year mark. Such an example demonstrates how such a simulation tool could be used to measure transient effects of large changes in temperature. In this example, a geothermal gradient of 2.0°C is applied at a supposed 15 per cent decline gradient (VRT varying from 28°C to 46°C). The cooling unit, when applied, is modelled by setting a maximum intake wet bulb temperature of 20°C, and if adjusted, also setting the dry bulb temperature to 20°C. Figure 6 plots the resultant temperature variation at the 6 km mark.

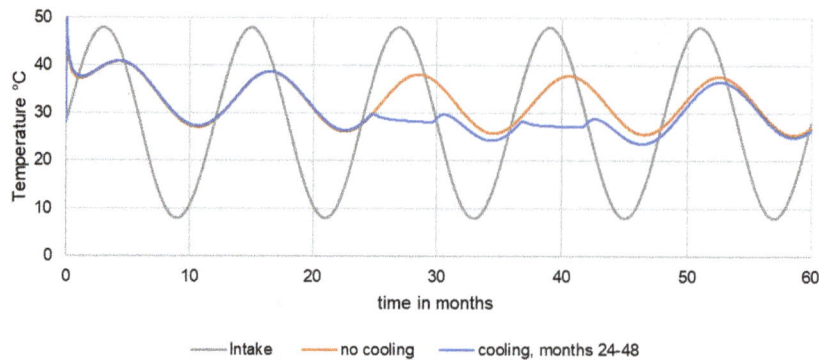

FIG 6 – The temperature variation at the 6 km mark for a case with geothermal gradient 2.0°C/100 m, flow quantity 100 m³/s, and with a cooling unit limiting the intake temperature to a maximum of 20°C wet bulb temperature, but only operating between months 24 and 48.

The effect of the added cooling can be seen in the third and fourth years, and then the lingering effect from the rock heat capacitance seen in the small difference in the fifth year. Contrast such a result with a steady state simulation which uses the Gibson function with an airway age set at five years. In such a case, a constant forcing temperature must be chosen. Firstly, without the cooling power, if the average forcing temperature of 28°C is chosen, this results in a temperature at the 6 km mark of 31.2°C. If the oft-used method of picking a forcing temperature at the 95 per cent percentile of maximum temperatures is used, the temperature at the 6 km mark returned is 42.5°C. If the cooling unit is added, the temperature returned at the 6 km mark is 31.2°C.

These values returned from the steady state heat simulation are useful in determining the general effect and the relative difference with other simulation results, but do not provide nearly as much detail and resolution of information as is available with a simulation performed with the multiple airway solver. With the multiple airway solver, it is no longer reporting time-averaged values of temperature, but rather temperatures at specific times. Time-specific measurements of temperature are difficult to correlate with a simulated time-averaged temperature from a method using the Gibson function but could be matched with a time-resolved temperature from the multiple airway solver, such as the temperatures shown in Figure 6.

Further to calibrating model heat data, such a solver can be used to predict more accurately temperatures deeper underground which will have significant lag or amplitude decay from intake temperature variations. This currently is not accounted for in heat simulations using the Gibson function. Also not accounted for are abrupt changes in forcing temperatures resulting from operational changes in the mine. These abrupt changes could result from: the installation of a cooling or heating plant; a switch from vehicle haulage to a hoisting shaft; the opening up then shutting down of development levels; or a fire.

In underground fire ventilation modelling, simplifications are made in the modelling to account for the absorption by the rock mass of the massive amount of heat coming from the fire. This is an important component of the fire modelling task. The Ventsim VentFIRE method (Brake, 2013) makes an assumption about the thickness of rock wall that heats up in response to the fire, a thickness which is calibrated for the time frame of a fire model, usually about 1 hr. Other fire simulation methods simply take off a fixed factor of the fire heat to account for tunnel heat absorption. Another example of such *ad hoc* methods is in the flywheel solver of Griffith and Stewart (2019) which is calibrated to account for observed phase lag and amplitude decays for seasonal temperature variations. All these methods are calibrated to specific time scales and, to greater or lesser extent, rock thermal properties and are inflexible to varying degrees to changes in total simulation time and rock parameters. The multiple airway solver would perform this task more accurately and with greater flexibility, at all-time scales and across different rock types.

There are however significant drawbacks to using the multiple airway rock heat solver. The amount of computational memory and computing time required is orders of magnitude greater than what mine ventilation modellers are accustomed to. Most of the results in this study were obtained on an airway network consisting of 60 airways. Running the solver for five years on this network using the author's computer required around 10 mins of computation time; typical mine models consist of 500 to 20 000 airways, often more, and the computation time will scale roughly proportionally with the number of airways. However, there remains large scope to improve the speed of the solver; work on speeding up the time-stepping algorithm of the single airway solver was stopped with the intention of obtaining the results presented, so there is still work to do on optimising. There is also the possibility in the future of using analytical solutions of the radial heat equation, rather than the numerical one currently used, that should be faster (Hefni *et al*, 2022). One more thing to consider is the possibility of the engineer adjusting workflows to longer computation times; standard simulation times vary widely across engineering disciplines, with mine ventilation sitting at the extreme shorter end.

Finally, this study has avoided discussing moisture transfer from the rock to the air as it would constitute a complete further study unto itself, and all the results reported feature entirely dry heat transfer at the rock-air interface. There are ad hoc methods for handling moisture (such as in McPherson, 1993) which are not strictly correct and usually modelling with moisture requires after the fact adjustment of wetness fractions or moisture flows to account for the extra heat flow resulting from the presence of the moisture and the latent to sensible heat flow ratio. Incorporating moisture into the solver presented in this study without resorting to generalised moisture fractions or breakups between sensible and latent heat would be a significant challenge and an interesting avenue of further investigation.

CONCLUSIONS

The study has presented a 1D radial heat solver and its calibration against existing methods. The single airway solver has been incorporated into an existing heat network solver to produce time-resolved heat simulations of heat flows along a tunnel. The results have shown agreement with another study in the literature. Further results presented showed the effects of changes to rock parameters and temperature forcing, demonstrating the potential use and high versatility of such a solver to the mine modeller. Finally, the various challenges lying ahead for using such a solver on a standard mine model were discussed.

REFERENCES

Brake, D J, 2013. Fire Modelling in Underground Mines using Ventsim Visual VentFIRE Software, in *Proceedings Australian Mine Ventilation Conference*, pp 265–276 (The Australasian Institute of Mining and Metallurgy: Melbourne).

Danko, G, Bahrami, D, Asante, W K, Rostami, P and Grymko, R, 2012. Temperature Variations In Underground Tunnels in *Proceedings of the 14th United State/North American Mine Ventilation Symposium*, pp 365–373 (Calizaya & Nelson).

Gibson, K L, 1976. The computer simulation of climatic conditions in underground mines, PhD thesis, University of Nottingham.

Griffith, M D and Stewart, C, 2019. Predicting Annual Underground Thermal Flywheel Effects, in *Proceedings Australian Mine Ventilation Conference 2019*, pp 284–291 (The Australasian Institute of Mining and Metallurgy: Melbourne).

Hefni, M A, Xu, M, Zueter, A F, Hassani, F, Eltaher, M A, Ahmed, H M, Saleem, H A, Ahmed, H A M, Hassan, G S A, Ahmed, K I, Moustafa, E B, Ghandourah, E and Sasmito, A P, 2022. A 3D Space-Marching Analytical Model for Geothermal Borehole Systems with Multiple Heat Exchangers, *Applied Thermal Engineering*, 216:119027.

McPherson, M J, 1993. *Subsurface Ventilation and Environmental Engineering* (Chapman & Hall).

A practical method to determine airflow rate in areas affected by underground thermal waters

D Gutiérrez[1]

1. Senior Ventilation Engineer, Glencore, Brisbane City, Qld 4000.
   Email: daniel.gutierrez@glencore.com.au

ABSTRACT

The thermal influence exerted by high-temperature underground water sources needs specialised considerations in determining the precise fresh airflow required to mitigate resultant elevated air temperatures. This analysis excludes the effects of other heat sources, whether of machinery or non-machinery origin.

This study endeavours to construct a thermal profile of the mine within areas directly impacted by thermal water presence. This profile entails the systematic measurement of air temperatures at pertinent locations. Subsequently, these collected data serve as input parameters for the VentSim®, version 5 (by The Howden Group) (predominant ventilation software used in all Peruvian mines), facilitating the derivation of thermal loads. These values are then incorporated into thermodynamic equations governing parameters such as saturation vapor pressure, saturation moisture content, latent heat of evaporation, and sigma heat, thereby enabling the determination of optimal fresh airflow rates necessary for underground temperature reduction.

This method enables the ventilation engineer to compute the required airflow to mitigate elevated air temperatures induced by the existence of thermal waters. It entails the measurement of those air temperatures to derive the associated heat load, rather than relying on direct temperature measurements of the thermal waters.

The presented streamlined approach has been successfully implemented across some of the Volcan mines (Peru) affected by thermal infiltrations, facilitating the development of ventilation systems capable of effectively addressing all heat sources attributed to thermal water.

INTRODUCTION

In Peruvian mining operations affected by the presence of high-temperature thermal waters, the calculation of the requisite volume of fresh air to mitigate resultant elevated air temperatures typically adheres to the directives delineated in Appendix 38, subparagraph c, of the Peruvian Mining Regulations (2017). However, as will be elaborated subsequently, the method prescribed therein proves inadequate. Notably, one of its deficiencies lies in its application only to air dry bulb temperatures ranging between 24 and 29°C, despite empirical evidence indicating that thermal waters yield air dry and wet bulb temperatures surpassing that range.

This paper presents a straightforward and practical approach for determining the requisite fresh airflow based on air wet bulb temperatures stemming from high-temperature thermal waters. While a similar methodology could potentially be employed to evaluate the impact of other subterranean heat sources, such analysis is beyond the scope of this paper.

Theoretical framework

Peruvian mining operations adhere to the methodology outlined in the Peruvian Mining Regulations (2017) to determine the requisite airflow for cooling environments affected by subterranean thermal waters. However, a limitation of this approach is its lack of grounding in thermodynamic principles.

This study employs formulas incorporating a range of thermodynamic parameters, including saturation vapor pressure, saturation moisture content, latent heat of evaporation, and sigma heat. These parameters are elaborated upon as follows:

- **Heat sources**: Those encompass auto-compression, geothermal gradient, strata heat, broken rock, cement, mechanised equipment, fissure water, oxidation, and explosives, though this study focuses solely on thermal water as the main heat source.

- **Wet bulb temperature**, (t_w): It is, according to McPherson (1993a):

   A most important parameter in hot climatic conditions for two separate but inter-related purposes. The first lies in its vital importance in evaluating the ability of the air to remove heat from personnel. The second is the use of the wet bulb temperature in quantifying the humidity of air.

- **Dry bulb temperature**, (t_d): It is the record of the ordinary air temperature.

- **Difference between dry bulb and wet bulb temperatures**, (t_d–t_w): If the evaporation capacity of the air is large, the difference between both bulbs will also be large.

- **Sensible heat**, (q_{sin}): In the absence of water, heat is transmitted to the air causing an increase in temperature that is recorded by a dry bulb thermometer.

- **Latent heat**, (q_L): Energy in the air-vapor mixture produced by the water evaporation process. This increase in heat content is not recorded by a dry bulb thermometer.

- **Total heat**, (q): Sum of sensible and latent heat.

- **Saturation vapor pressure of air**, (e_{sw}): Pressure experienced in a system that no longer accepts any more water vapor, that is, in a saturated system. It only depends on the temperature of the water and not on the presence of other gases.

- **Air saturation moisture content**, (X_s): It is the mass of water vapor associated with one kilogram of dry air.

- **Latent heat of evaporation**, (L_w): Energy required to evaporate 1 kilogram of water.

- **Sigma Heat**, (S): Kilojoules of heat associated with each kilogram of dry air. Its value depends solely on the wet bulb temperature of the air at a certain barometric pressure.

Background

Equation 1 is a tool employed by Peruvian ventilation engineers to calculate the airflow rate that could reduce elevated air temperatures within mine environments. That equation is used for evaluating the impact of subterranean thermal waters on the atmospheric conditions within the mine. That mathematical expression is formally documented within the framework outlined by the Peruvian Mining Regulation (2017).

$$Q_{Te} = V_m \times A \times N \ (m^3 /min) \tag{1}$$

where:

Q_{Te}	Airflow necessary to reduce air temperature (m³/min)
V_m	Minimum air velocity (m/min)
A	Average cross-sectional area of the level (m²)
N	Number of levels exhibiting air dry bulb temperatures surpassing 23°C

In addition, the regulatory stipulations dictate that in instances where air dry bulb temperatures range from 24°C to 29°C, the minimum air velocity, V_m, must be 30 m/min (0.5 m/s).

PROBLEM TO BE ADDRESSED

According to the Peruvian Mining Regulation (2017), when dry air temperatures fall within the range of 24 to 29°C, an air velocity of 30 m/min (0.5 m/s) is mandated for calculating the required airflow volume to mitigate temperatures. However, in instances where the air temperature in a designated area of the mine, impacted by thermal water seepage, surpasses 29°C, what air velocity should be employed?

Although the Peruvian regulation incorporates the document 'Guide 2: Thermal Stress Management', which stipulates the maximum Wet Bulb Globe Temperature (WBGT) permissible within a particular workplace, it is important to clarify that this paper does not address maximum temperature limits. Instead, its objective is to calculate the airflow necessary to reduce air temperatures induced by thermal water sources before operations in the affected area are halted due to high-temperature conditions.

GOAL

This paper explains a method to quantify the requisite airflow to mitigate the elevated air temperatures induced by thermal water sources. Thus, the objective of this paper is delineated as follows:

- Presenting a method for determining the requisite volume of fresh airflow indispensable for mitigating heightened air temperatures resultant from the influence of thermal waters across various operational mining faces within a mine.

CALCULATION OF AIRFLOW REQUIRED DUE TO HIGH AIR TEMPERATURES IN WORK OPERATIONS AS PER PERUVIAN MINING REGULATIONS

In most Peruvian mining operations, specific areas exhibit elevated temperatures exceeding 29°C (dry bulb temperature) due to the influence of geothermal waters. Under such circumstances, the application of the 30 m/min air velocity prescribed by Equation 1 becomes impractical. Determining the appropriate air velocity for air temperatures surpassing 29°C becomes uncertain and necessitates further analysis, as addressed in this paper.

Equation 1 leads the user to think that a mining level is linear and can only contain a single temperature (FIG **1**).

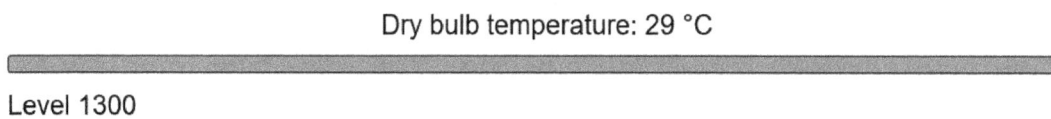

Dry bulb temperature: 29 °C

Level 1300

FIG 1 – Plan view of a simple non-branched level.

In contrast, reality more closely resembles what is depicted in FIG **2**, where within a level, it can be encountered multiple working faces as well as various active stopes, each with different air temperature values depending on the heat sources affecting them.

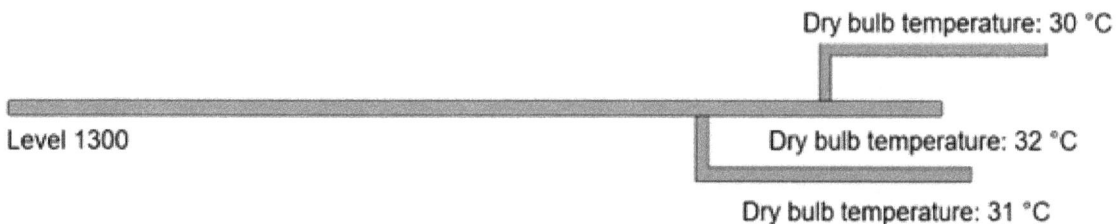

Dry bulb temperature: 30 °C

Level 1300

Dry bulb temperature: 32 °C

Dry bulb temperature: 31 °C

FIG 2 – Plan view of a level with three working faces.

Applying Equation 1 to the conditions depicted in FIG **2** yields airflow rates insufficient for temperature reduction as the equation, in accordance with the Peruvian Mining Regulation (2017), is calibrated for temperatures up to 29°C. Consequently, no provisions are made for temperatures exceeding this threshold.

Hence, assessments should focus on the number of thermal sources emanating from geothermal waters per level, rather than the number of levels, to provide a more accurate representation of the ventilation requirements. Moreover, the use of wet bulb temperatures, as opposed to dry bulb temperatures, is imperative. This adjustment is necessary because calculating the total heat content (Sigma heat) in the air mass to be cooled relies solely on wet bulb temperatures for accurate determination of airflow requirements.

The primary objective of this study is to determine the volume of air required to mitigate high air temperatures caused by the presence of thermal waters. Reference is made to Peruvian Mining Regulation (2017), which stipulates a specific procedure for calculating airflow rate, dividing it into the following steps:

1. Determination of the required airflow based on the number of workers.

2. Determination of the necessary airflow based on wood consumption.

3. Determination of the required airflow based on temperatures in the work environment.

4. Calculation of the necessary airflow for diesel engine equipment.

5. Calculation of the required airflow due to leaks.

This study focuses particularly on point 3, specifically when the influence of thermal waters reaches significant magnitudes, as is the case in the mines analysed by the author.

PROPOSED METHOD

Prior to implementing the method, the establishment of a calibrated mine ventilation model is imperative. A calibrated model is defined as one that shows a percentage difference of ±10 per cent—an acceptable threshold as proposed by McPherson (1993b)—between field measurements and simulated values.

For a thorough comprehension of the proposed method, its procedural steps will be delineated into the subsequent domains: delineation of heat sources through mapping (fieldwork), measurement of air temperatures in areas identified with thermal anomalies (fieldwork), determination of heat loads (desk-based work), and computation of the requisite airflow to establish favourable thermal conditions (desk-based work).

Mapping of underground heat sources related to thermal waters

- Select an area characterised by thermal water-related heat anomalies.

- Within the schematic diagrams of the surveyed area, delineate thermal sources, specifying their dimensions in terms of length and width.

Measuring air wet bulb temperatures

Upon identification of the area experiencing heat issues stemming from thermal water, auxiliary ventilation systems within said area are identified, and air wet bulb temperatures are gauged at strategic spots:

- Determine extant auxiliary ventilation systems within the designated area.

- Ascertain the fresh air intake and return air path of the auxiliary ventilation system.

- Conduct measurements for the average air wet bulb temperature at:

    o fresh airflow intake

    o return airflow path

    o working face

    o system outlet.

Determining heat loads

To ascertain the requisite airflow for environmental cooling within the designated area, an evaluation of heat loads is imperative. This necessitates the utilisation of temperature readings alongside a calibrated VentSim® model. Key steps in this process include:

- Identification of heat-monitored zones within the calibrated model of the mine.

- Adjustment of sensible and latent heat values based on the measured air temperatures, ensuring alignment of the model's wet bulb temperature with field measurements.

- Iterative replication of the preceding procedure across all additional areas of concern.

Calculation of airflow needed to reduce air temperature

The next step involves the utilisation of thermodynamic parameters delineated earlier in this paper. Accurate determination of Sigma heat assumes paramount significance in facilitating the requisite volume of fresh airflow essential to cool the designated environment:

- Total heat loads (q), comprising latent heat and sensible heat as determined in antecedent procedures, are assimilated into the corresponding psychrometric equations. These equations enable the computation of crucial parameters such as saturation vapor pressure (e_{sw}), moisture content at saturation (X_s), latent heat of evaporation (L_w), and heat Sigma (S).

- In conjunction with the heat loads, air density (ρ) is incorporated into the equations. Additionally, wet bulb temperatures (t_w) at the inlet and outlet of various auxiliary systems identified within the zones, along with mean barometric pressure (P), are considered.

- The values encompassing air density, total heat loads, and differences between Sigma heats (S) collectively dictate the requisite airflow rate (Q) imperative for temperature reduction.

RESULTS

In the Peruvian mine site selected to apply the approach of this study, specifically at ramp 025 (5.0 m × 5.0 m), an example of thermal water filtration occurred, resulting in an elevation of the air's wet bulb temperature to 32.9°C. This elevation posed a significant hazard to workers' safety, necessitating the cessation of development activities within the ramp due to the heightened risk of heat stress.

To address this issue in a systematic manner, the ventilation department opted to implement the methodology outlined in this study to ascertain the requisite airflow for mitigating the elevated air temperature and thereby enhancing working conditions for personnel. Utilising the measured air's wet bulb temperature as a reference in the calibrated model of the mine to represent the thermal water infiltration, a total heat load of 131 kW was obtained.

Despite the water being actively pumped to a nearby sump, its ongoing interaction with the surrounding environment facilitated the heating of the atmospheric air, thereby resulting in the attainment of the previously mentioned wet bulb temperature of 32.9°C.

The 131 kW heat load, in conjunction with parameters such as air density, barometric pressure, and Sigma heat values representing the initial and target conditions of the ventilation auxiliary system, form the fundamental components of Equation 2 (McPherson, 1993c):

$$Q = q / [\rho \times (S_1 - S_2)] \times 60 \ (m^3 /min) \quad (2)$$

where:

Q	Airflow required to reduce initial air wet bulb temperature to a specified target wet bulb temperature
q	Total heat flux into the air from thermal source
ρ	mean density of the air
S_1	Sigma heat at the inlet of the auxiliary ventilation system affected by the specified thermal source
S_2	Sigma heat related to the target air wet bulb temperature

The factor $q = Q\rho \times (S_1 - S_2)$ is known as the Heat Removal Capacity (HRC) of the given airflow Q.

The data for the case of the ramp 025 are the following:

- Air wet bulb temperature at the entrance to the ramp system: 22.9°C.

- Target air wet bulb temperature: 27.0°C.

- Air density: 0.77 kg/m³.

- Average barometric pressure: 67.28 kPa.

- Total heat load: 131 kW (value obtained using the airflow-heat calibrated VentSim® model).

Solving Equation 2 yielded a required airflow (Q) of 10 m³/s to reduce the air's wet bulb temperature to the desired 27°C in the ramp just due to the thermal water present there. The air velocity within the ramp corresponded to that airflow was 0.4 m/s (24 m/min). This indicates that the employed methodology facilitated the precise determination of the requisite velocity for cooling purposes, thereby avoiding the reliance on a fixed value, irrespective of it is excessive or not.

To facilitate this temperature adjustment, the implementation of a raise bore (RB 164) was scheduled. This raise bore would introduce fresh airflow into the system, thereby achieving the desired environmental conditions. The determination of the optimal diameter of the raise bore was made possible by considering factors such as airflow requirements for diluting diesel equipment emissions and the necessary air volume for personnel working within the area. The precise knowledge of the airflow required to reduce temperatures, in this case due to thermal waters, is also crucial for selecting the appropriate capacity auxiliary fan as well as the diameter and arrangement of the ventilation ducts.

The airflow calculation, incorporating the influence of thermal waters alongside additional factors such as human presence and diesel-powered equipment, reaffirmed the necessity of retaining the existing 30 m³/s fan for deployment at the base of RB 164.

Figures 3 and 4 portray the problematic situation and the solution adopted in the case of ramp 025.

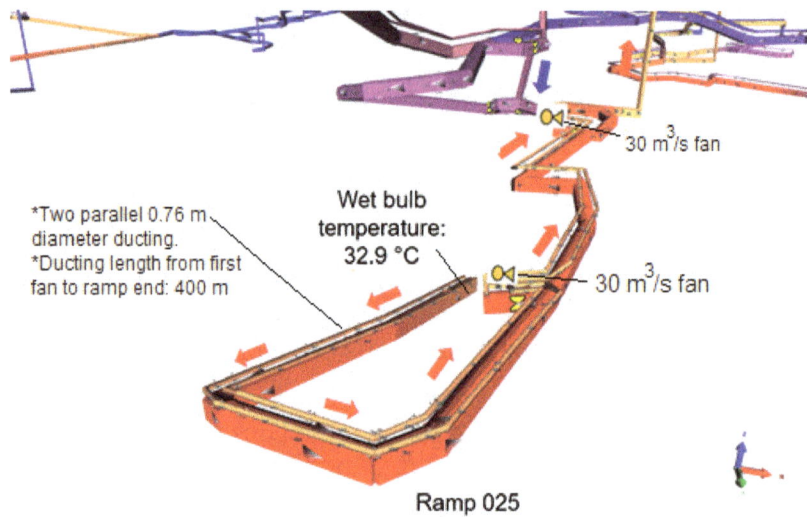

FIG 3 – Problematic situation in ramp 025, with 32.9°C air wet bulb temperature at its face. VentSim® model, version 5.

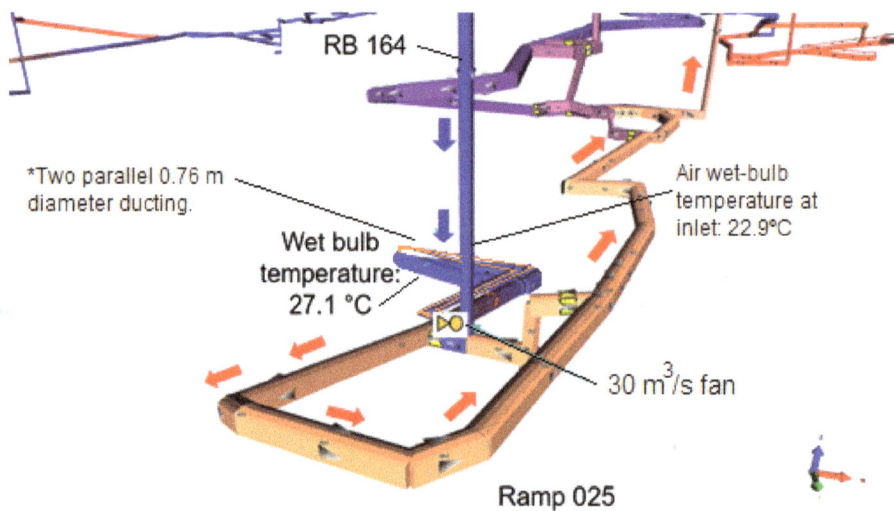

FIG 4 – Solution to the ventilation problem in ramp 025, with air wet bulb temperature of 27.1°C. VentSim® model, version 5.

CONCLUSIONS

A meticulous observation of the thermal water sources, precisely pinpointing their locations on corresponding level schematics alongside the associated air wet bulb temperatures, constitutes a good method for calculating the requisite fresh airflow needed for temperature reduction due solely to those sources. The airflow thus obtained is then included to the air volume determined for people and diesel equipment.

Upon calculating the requisite volume of air necessary to mitigate the temperatures induced by the influence of thermal waters, one can ascertain the total volume of fresh air required to be directed to the ramp face. This enables the selection of the appropriate capacity for the auxiliary fan, as well as the diameter and quantity of flexible ventilation ducts.

With the representation of thermal waters (either pooled or flowing) depicted in the plans of the various levels of the mine, the ventilation engineer can calculate the air flow associated with reducing the elevated air temperatures influenced by these sources. This enables the determination of whether an auxiliary ventilation system can resolve the issue or if, due to the excessive water volume, it is necessary to temporarily close the working face until the water has been completely pumped out.

In the Peruvian Volcan mines, where this methodology has been implemented, ventilation plans delineating the occurrences of thermal water sources are updated monthly due to the frequency of their emergence.

ACKNOWLEDGEMENTS

The author extends acknowledgement to Glencore and Volcan mining companies for granting permission for the publication of this study and for providing the opportunity to conduct the research therein.

REFERENCES

McPherson, M J, 1993a. Theory of the wet bulb thermometer, in *Subsurface Ventilation and Environmental Engineering*, p 480 (Mine Ventilation Services: California).

McPherson, M J, 1993b. Establishment of the basic network, in *Subsurface Ventilation and Environmental Engineering*, p 282 (Mine Ventilation Services: California)

McPherson, M J, 1993c. Airflow requirements and velocity limits, in *Subsurface Ventilation and Environmental Engineering*, p 288 (Mine Ventilation Services: California)

Peruvian Mining Regulations, 2017. D.S. N° 024-2016-EM modified by the D.S. N° 023-2017-EM, Appendix 38, subparagraph c. Available from: <https://www.gob.pe/institucion/minem/informes-publicaciones/4339000-reglamento-de-seguridad-y-salud-ocupacional-en-mineria-ed-2020>

Design of a large 'off the shelf' air cooled mine refrigeration plant

A Hatt[1] and F Nadaraju[2]

1. MAusIMM, Principal Underground Mining Engineer, Newmont Corporation, Melbourne Vic 3004. Email: alex.hatt@newmont.com
2. MAusIMM, PhD, Principal Mechanical Engineer, Resources and Industrial, AECOM, Newcastle West NSW 2310. Email: francis.nadaraju@aecom.com

ABSTRACT

Mine refrigeration is a significant capital and operating cost for underground mines at depth or in adverse climates, traditionally consuming significant quantities of electrical power and water. Cost pressures due to inflation, skilled labour scarcity, and power generation have changed the balance to be struck between capital efficient cooling plant designs and operational cost efficiency. Layered over this is a change in technology and performance for chilled water plants with respect to both power and water efficiency predominantly driven by the rise of green building certification schemes and Environmental, Social and Governance (ESG) requirements.

This paper presents the decision process and design for a large, air-cooled mine refrigeration system using 'off the shelf' chiller units from the commercial heating, ventilation and air-conditioning (HVAC) industry. A series of trade-off studies were completed to inform plant selection including supply of water and power, impacts of electrical and water efficiency, positional efficiency of bulk air coolers, capital costs, and net present cost (NPC) assessment.

INTRODUCTION

Mine ventilation and refrigeration represent significant capital and operating costs for underground mines due to the infrastructure installed as well as the consumption of water and electrical power. Historical benchmark capital costs for mine refrigeration plants are on the order of AUD1–2 million (Loudon, 2021) however depending on the particulars of the system design, size, and site-specific requirements this may vary considerably. Similarly, the operating costs of the mine ventilation system can make up 40–50 per cent of the mine electrical consumption (De Souza, 2018) with the costs to operate a mine cooling system increasing this proportion.

Recent escalation in costs of raw materials, labour, and power have changed the balance between capital and operating costs in developing the 'best' mine cooling plant. For example, the Producer Price Index for Metals and Metal Products (US Bureau of Labor Statistics, 2024) rose 45 per cent between January 2018 and January 2023 after a relatively stable preceding decade and has remained elevated. This has put prior benchmark values 'under stress' with cost estimates deviating from prior typical ranges. This became apparent during a recent technical study for an underground mine ventilation and cooling system where cost estimates came in well above the expected range. The study initially set out to have a 'premium' efficiency plant using Ammonia refrigerant (R717) and air-cooled condensers. The plant configuration arose from multiple site-specific requirements including:

- a lack of the required volume of make-up water for condenser cooling,
- exceptionally poor water quality resulting in prohibitive treatment costs,
- minimising the size of high voltage electrical infrastructure supply to the study location.

While the plant design resulted in a high-performance chiller plant given the constraints, it also came at premium costs. By the end of the study phase occurring in the inflationary period above, the cost of the plant had increased to well past the 'traditional' benchmarks for mine cooling plants and was subsequently identified as a major target for value engineering.

PROJECT CONTEXT

The project under consideration is in a hot arid desert climate, with 95th percentile annual wet bulb temperatures of approximately 24°C and peak ambient dry bulb temperatures of up to 50°C

(Australian Bureau of Meteorology, 2020). The various cases considered during the pre-feasibility study (PFS) stage of the project indicated the requirement for a mine cooling plant in the order of 20 MW$_R$ paired with an 1150 kg/s ventilation system to service working areas at depths from 0.4 to 1.2 km below surface. Both the cooling plant and ventilation system would be a significant consumer of electrical power as well as a significant capital expenditure (CAPEX) cost for the project.

Hydrogeological investigations in the study area indicated that the quality of local groundwater resources was unsuitable for use for use in a mine cooling system without treatment due to total dissolved solids (TDS) in excess of 30 000 ppm as well as elevated chlorides, sulfates, iron, and manganese (Rockwater, 2021). Treatment would require the implementation of a reverse-osmosis plant, precipitation of iron and manganese, and present challenges for the disposal of brines.

COOLING SYSTEM PERFORMANCE

Minimum chiller performance requirements

Over the past 20 years minimum efficiency requirements for water/liquid chilling plant have become more stringent, largely driven by the implementation of 'green' building certification schemes including the Leadership in Energy and Environmental Design (LEED) code in the United States (US Green Building Council, 2024) and the European Ecodesign Directive (European Commission, 2009), with their corresponding standards. These schemes and others came into effect starting in the late 1990s, with minimum performance requirements for chilled water systems increasing over time. An example of the scale of increases in the minimum performance requirements for full load Coefficient of Performance (COP) and Integrated Part Load Value (IPLV) as set out in ASHRAE 90.1 (ASHRAE, 2022; 2004) over the period are shown in Table 1.

TABLE 1

Changes in ASHRAE 90.1 chiller minimum performance requirements.

	Year	Air cooled (with condenser) >527 kW$_R$	Water cooled centrifugal >2110 kW$_R$	Water cooled positive displacement >2110 kW$_R$
Full Load COP[1]	2004	2.80	6.10	5.50
	2022	2.96	6.28	6.28
	Change	6%	3%	14%
Integrated Part Load Value COP (IPLV)[2]	2004	3.05	6.40	6.15
	2022	4.71	9.26	9.26
	Change	54%	45%	51%

1 – 2022 full load COP references Path A (full load optimised) values; 2 – IPLV values for 2022 use Path B (part load optimised) values.

While the full load performance requirements for of chillers have not improved significantly over the period, part load performance has increased substantially. Part load performance is critical to applications that do not require running the chiller system at continuous full load regardless of ambient conditions. The reality for a mine cooling application is that the day/night and seasonal variations in ambient temperature as well as 'shoulder season' cooling requirements result in a system operating away from peak for much of its life.

The ratio of minimum IPLV performance of water cooled to air cooled chillers has effectively remained the same, with water cooled chillers having a minimum COP of about two times air cooled units, however the absolute difference between the two has decreased. This still shows that water cooled chillers retain a performance advantage in a relative sense. Comparing the electrical consumption of a centrifugal water cooled to an air-cooled chiller using the 2004 standards for a theoretical 10 MW$_R$ load gives a difference in electrical power consumption of 1.7 MW$_E$. Making the same comparison for the 2022 standards gives a difference of 1.0 MW$_E$. The water-cooled chiller is still twice as energy efficient as air cooled units, but the electrical consumption has been reduced by the increased efficiency requirements across the board.

It is also important to note that the above table does not include electrical loads for the condensing side of a water-cooled plant (cooling tower fans, pumps, water treatment systems etc) that are already included in the COP of a 'packaged' air cooled unit. These additional loads further decrease the 'whole of plant' COP/IPLV of a water-cooled system, closing the gap in power consumption. While these loads can be minimised in plant design the table above does not present an 'apples to apples' comparison.

Current state of the market

Based on review of packaged Variable Speed Drive (VSD) R134a air cooled chiller units, full load COP at nominal ambient conditions (~30°C DB) ranges between 3.5 and 4.0, however part load performance increases depending on the load of the compressor and ambient conditions. As most of the operating time is well below full load, the chiller system can be optimised to achieve the higher time-weighted COP. Use of ammonia as the refrigerant should yield an improvement of ~10 per cent on the COP due to the higher specific heat of the gas and subsequent reduction in compressor requirement. Based on this, the suggested range of time-weighted plant COPs for evaluation in air cooled systems is 4.0–5.0.

VSD water-cooled chiller plants can achieve plant COPs in the order of 7–9, considering benefits of part load performance of chillers, variable condenser flow rates, and well implemented plant control strategies. These values have been used as the range applicable to water-cooled systems with either R134a or Ammonia as the refrigerant in this analysis.

Cooling demand analysis

VentSim modelling from the PFS indicated a requirement to achieve an intake air sigma heat of 55.2 kJ/kg for the design reject temperature of 29°C WB at the lowest level of the mine. The difference between this requirement and the ambient sigma heat (based on the ambient weather conditions) allows for calculation of the required cooling load at any point in time.

$$Mine\ Cooling\ Demand\ (kW) = \left(Sigma\ Heat_{Ambient} - Sigma\ Heat_{Required}\right) \times MassFlow$$

Required cooling on an hourly basis was calculated using historic weather data from an adjacent Australian Bureau of Meteorology weather station for a typical operating season. Information for the months of November through April (inclusive) was extracted to generate unit cooling bin data and the corresponding whole of system demand for the six summer months that the plant would be operational. This is shown in Table 2.

TABLE 2

Cooling demand estimate.

Min	Max	Per cent	Calendar hours	Avg unit cooling (kJ/kg)	Demand (MW_R)	TW avg	MCDB[1]	MW_rhr	Energy dissipated (MJ)	Water evaporated (m³)
0	1	54.1	2368	0.0	0.0	16.4	30.8	88	317 672	141
1	2	3.5	155	1.5	1.7	19.7	33.5	267	959 764	425
2	3	3.5	154	2.5	2.9	20.0	33.0	441	1 589 140	704
3	4	3.2	141	3.5	4.0	20.3	32.8	570	2 052 628	910
4	5	3.7	160	4.5	5.2	20.6	32.9	828	2 979 655	1321
5	6	3.4	150	5.5	6.3	20.9	32.1	948	3 413 054	1513
6	7	3.1	138	6.5	7.5	21.2	31.9	1031	3 710 915	1645
7	8	3.4	147	7.5	8.6	21.5	32.2	1271	4 576 353	2029
8	9	3.2	139	8.5	9.8	21.7	31.8	1357	4 886 062	2166
9	10	3.1	134	9.5	10.9	22.0	31.2	1466	5 277 521	2339
10	11	2.9	127	10.5	12.1	22.3	31.1	1537	5 534 660	2453
11	12	2.8	121	11.5	13.2	22.6	30.9	1603	5 772 008	2559
12	13	2.4	106	12.5	14.4	22.8	30.9	1522	5 477 406	2428
13	14	2.1	94	13.5	15.5	23.1	30.8	1458	5 249 588	2327
14	15	1.6	72	14.5	16.7	23.3	30.8	1201	4 324 946	1917
15	16	1.3	57	15.5	17.8	23.6	31.1	1014	3 649 015	1617
16	17	0.9	38	16.5	19.0	23.8	30.9	725	2 611 500	1158
17	18	0.6	27	17.5	20.1	24.1	30.8	533	1 917 927	850
18	19	0.4	16	18.5	21.2	24.3	31.0	336	1 210 187	536
19	999	0.8	35	29.1	33.5	26.6	30.7	1168	4 205 143	1864
							TOTAL	19 365	69 715 143	30 902

1 – Mean Coincident Dry Bulb.

The estimation of cooling demand through the plant operating season using the reference plant and ventilation system (20 MW$_R$, 1150 kg/s) indicates that:

- There is no requirement for cooling for over half of the calendar time during the operating season.
  - This is largely due to the day/night variation in ambient temperatures.
- The plant would operate at full load <1 per cent of the time.
  - This aligns to the weightings used to derive the IPLV, where full load operation is weighted as 1 per cent of the operating time.
- The plant would operate >50 per cent load for only ~15 per cent of the time.
  - This emphasises the need for a chilled water plant and operating strategy that is designed to take advantage of part load conditions.

Using this data, total annual cooling consumption (kW$_R$hr) was calculated for the 6 month period of plant operation.

$$Cooling\ Consumption\ (kW_R hr) = Cooling\ Demand\ (kW_R) \times Calendar\ Hours\ (hr)$$

A range of time-weighed COPs were then used to calculate the electrical consumption (kW$_E$hr) of the plant through the operating season. This is analogous to the IPLV performance but is site specific (sometimes referred to as the Non-standard Part Load Value or NPLV). The COPs used in the analysis are whole of plant seasonal weighted averages to account for the variable nature of the plant performance due to changing ambient conditions and loads, as well as accounting for power consumed by ancillary loads (pumps, fans etc) which are not included in the COP for compressor units.

Economic analysis

Economic parameters applied for PV calculations are shown in Table 3. For the purposes of this paper economics are presented in Megawatt-hours of electrical power (MW$_E$hr). Five cycles of concentration was used under the assumption that make-up water would be of potable quality post treatment.

TABLE 3

Economic evaluation parameters.

LOM	10	years
Discount rate	4.5%	
Cycles of concentration	5	
Mass flow	1150	kg/s dry air
Plant size	20	MW$_R$

Using these inputs, an annual electrical power and water consumption was calculated for each COP, shown in Table 4.

TABLE 4

Incremental PV analysis by increasing COP.

Cooling consumption (MW$_R$hr)	Time weighted COP	Annual power consumption (MW$_E$hr)	Incremental savings (MW$_E$hr)	Discounted power savings (MW$_E$hr)	Evaporative loss (air chilling) (m^3)	Evaporative loss (mech work) (m^3)	Blowdown (m^3)	Total water loss (ML)
19 400	4.0	4850	-	-	30 900	7739	9660	48.3
19 400	4.5	4311	539	4265	30 900	6879	9445	47.2
19 400	5.0	3880	970	7675	30 900	6191	9273	46.4
19 400	5.5	3527	1323	10 469	30 900	5628	9132	45.7
19 400	6.0	3233	1617	12 795	30 900	5159	9015	45.1
19 400	6.5	2985	1865	14 757	30 900	4763	8916	44.6
19 400	7.0	2771	2079	16 451	30 900	4422	8830	44.2
19 400	7.5	2587	2263	17 906	30 900	4128	8757	43.8
19 400	8.0	2425	2425	19 188	30 900	3870	8692	43.5
19 400	8.5	2282	2568	20 320	30 900	3641	8635	43.2
19 400	9.0	2156	2694	21 317	30 900	3440	8585	42.9

Using a base COP of 4.0 representing the low end of the packaged air-cooled chiller performance spectrum, incremental analysis of the power consumption indicates that:

- Increasing to COP from 4.0 to 9.0 results in an annual power savings of ~2700 MW$_E$hr.
  - This represents the maximum likely efficiency increase by moving from air cooled to water cooled condensing.
- Using a water-cooled condensing arrangement to achieve a plant COP of 9.0 requires 43 ML of good quality water over six months. This combined with the power maximum power savings identifies the maximum price to acquire water that would be value add.
  - Using 43 ML/6 months consumption to save 2700 MW$_E$hr results in a water price of ~62 MW$_E$hr/1 ML of usable water.
- The 'Net Present Value' of the 2700 MW$_E$hr annual power savings is 21 300 MW$_E$hr.
  - This is the maximum incremental capital spend to establish a water-cooled plant with a COP of 9.0 and secure the required volume/quality of water and supporting infrastructure (eg bores, pipelines, treatment, brine disposal).
  - This assumes that all other operating costs (outside of electrical costs) are the same between the plant options. This is not the case as additional operational costs for water treatment and brine disposal are required.

These numbers provide upper limits to determine if water cooled condensing is value add to the project, specifically:

- If the operating cost of water to supply the cooling plant is >62 MW$_E$hr/ML, then water condensing is not a value accretive option.
- If the incremental capital cost to develop a water supply and construct a water-cooled condenser circuit exceeds the equivalent of 21 300 MW$_E$hr of power, then water condensing is not a value accretive option.

Water supply options

Pipeline

An initial cost estimate for a water pipeline from an existing bore field to the project site had been made by the study team across a range of flow rates. Looking at the flow rates applicable to the supply of water to the cooling plant the capital cost ranged between 83 000 to 250 000 MW$_E$hr. Compared to the 21 300 MW$_E$hr threshold and accounting for the level of accuracy of the estimate this indicated that a pipeline for the sole purpose of supplying water to a cooling plant was not value accretive, even when the operating costs were not considered.

Water carting

Information from early site water supply work (Newcrest Mining Limited, 2020, personal communication) indicated a wet hire rate of 2.1 MW$_E$hr/hr for a 28 m^3 water cart. With an approximate 2 hr round trip, the transportation cost of water by carting would be ~150 MW$_E$hr/ML, well in excess of the 62 MW$_E$hr/ML identified above. This indicated that trucking water to supply the plant was also not value accretive even when the cost of the water itself was excluded.

Local bore field and water treatment

Due to the poor local water quality noted above, use of local water sources requires treatment for use in an evaporative condensing arrangement. Initial feedback from reverse osmosis vendors indicated a recovery of 40–60 per cent based on the expected water quality. Make-up water requirements of ~4 L/s (steady state) and ~13 L/s (peak) would therefore require abstraction rates of 6–10 L/s raw water (steady state) or 22–33 L/s (peak). Site hydrogeological investigations confirmed that local water resources would not be able to support the required flows, however costing of the option was completed in the event that a viable source of water of poor quality could be located.

Initial capital costs for a containerised 1 ML/d reverse osmosis plant and Fe/Mg treatment (sized to the peak water requirements for evaporative condensing) were 12 500 MW$_E$hr including a 40 per cent allowance for installation, civil works, ponds, tanks, piping, and pumping. Given that this was approximately 50 per cent the value of the power savings from a premium efficiency water cooled plant and excluded the cost of establishing bores, pipelines, nor the ongoing operating costs of water treatment and effluent disposal the option was discarded by the study team as highly unlikely to be value accretive.

Outcomes

The results from assessment indicated that a water-cooled condensing circuit would not be a practical and economically viable option for the project. While these trade-offs were focused on the costs of water against power savings, additional confirmatory work was required to close out the issue of plant costs related to the choice of refrigerant at the required level of confidence.

MARKET ENGAGEMENT PROCESS

As a result of the trade-offs completed for water supply, three hypotheses were developed for testing the refrigeration equipment market and the costs associated with the prior design of a 'boutique' plant:

- The improved COP of R717 does not offset the additional CAPEX.

- The capital cost of an integrated air-cooled R134a plant is similar to a water cooled R134a plant.

  - This would make the economic selection of plant based mainly on the cost of water versus cost of power, which had already been established.

- The capital cost of a packaged 'off the shelf' plant is the same or less than a 'conventionally' constructed plant.

  - 'Off the shelf' equipment common to the larger commercial building and non-mining heavy industry market should be more cost competitive than mining specific vendors and options.

  - This would enable a modular approach to construction and phasing in the plant, with an associated decrease in site costs.

Vendor engagement

An engineering service provider was engaged to complete basic process engineering, develop datasheets for major components of the system (chillers, water treatment, cooling towers, dry condensers etc), complete market engagement, and develop costs for different plant configurations. Seven plant configurations were costed for major components (chillers, condensers, bulk air coolers) excluding the balance of plant to determine any cases to be removed before detailed costing as shown below in Table 5 (AECOM Australia, 2023).

TABLE 5

Plant configurations and capital costs.

#	Refrigerant	Chiller	Condenser	MW$_E$hr/ kW$_R$	Notional plant cost (MW$_E$hr)
1	R134a	Modular	Cooling tower	6.7	130 000
2	R717	Modular	Cooling tower	9.7	190 000
3	R134a	Conventional	Cooling tower	7.4	150 000
4	R717	Conventional	Cooling tower	9.5	190 000
5	R134a	Modular	Integrated air cooled	6.8	140 000
6	R717	Modular	Integrated air cooled	9.8	200 000
7	R717	Conventional	External dry condensers	15	290 000

Based on the capital costs of major components in the table above:

- The nominal 10 per cent increase in COP for a R717 air cooled plant does not justify the premium pricing of the plant.
  - Annual electrical savings are in the order of 550 MW$_E$hr to move from a COP of 4.0 to 4.5, with diminishing savings as COP increases.
  - This is substantially less than the 40 000 MW$_E$hr capital cost difference between the closest R717 and R134a options over a ten year period as contemplated in the project.
  - Like for like, R717 plants were 30–50 per cent more expensive than R134a per kW$_R$.
- The cost per kW$_R$ for water and air cooled R134a plants were within 10 per cent of each other.
  - While this comparison of major componentry excludes the cost of pumps and piping for a water-cooled plant, at the level of detail of the estimate this was not considered material.
  - This supports the hypothesis that the cost of plant is 'similar' for R134a plants and that the main value driver is the power savings and cost of water.
- Construction method did not affect the capital costs of components for either R134a or R717 per kW$_R$ (within the accuracy of the cost estimate).
  - This does not account for savings in the construction process arising from the reduced site-based scope for modular plants.
- The cost of the previous base case plant (conventional R717 with external dry condensers) was 50 per cent more expensive than the next highest cost option per kW$_R$, and more than 100 per cent more expensive than the lowest cost option per kW$_R$.
  - This is likely driven by the large amounts of stainless steel required for the condenser circuits for the capacity of plant contemplated.
  - This cost excludes the costs of specialised trades and requirements for management of R717 plants in a mining setting, which would only serve to increase the gap.

POSITIONAL EFFICIENCY AND DIVERSITY

Another challenge encountered during the design of the ventilation system was restrictions in the number of ventilation connections to the surface due to poor ground conditions and the associated costs. Initial estimates put the cost of establishing ventilation shafts through the cover sequence at 5–10× the cost of unlined conventional raise bored airway. This ultimately led to a design with a single primary ventilation intake shaft as shown in Figure 1.

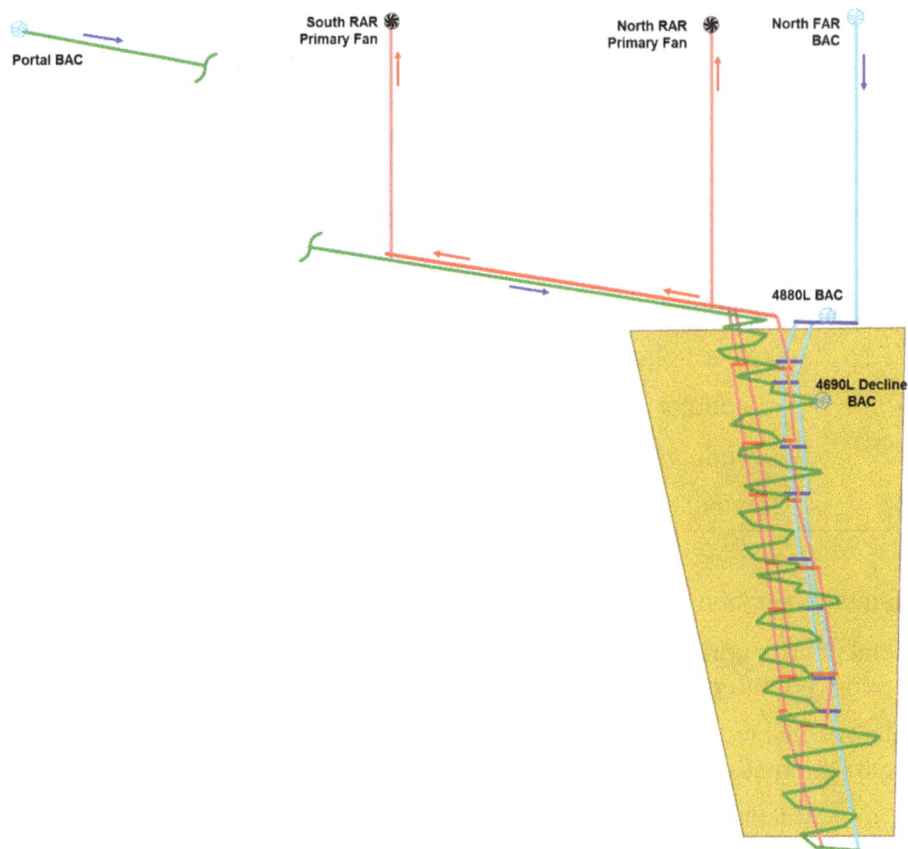

FIG 1 – Primary ventilation design schematic.

A single intake shaft for a chilled air application across a large vertical range (~1 km) of working areas presents an issue of over-chilling the air and the associated capital and operating costs. Air downcast through the North FAR needed to meet the requirements of three different cooling intensities:

- The upper production levels, where parallel single pass ventilation is employed.

- The upper decline with high levels of diesel activity from trucking and series ventilation.

- The 'deep' production levels and decline, where increased cooling is required to offset the effects of auto-compression in addition to the loads that the shallower areas contend with from diesel equipment and the ambient environment.

Each of these require different unit cooling (kJ/kg) applied to meet the target reject temperature. Initial VentSim models using surface only BAC's gave impractical air-off temperatures from BAC's or resulted in very low reject temperatures from the upper levels of the mine, indicating poor positional efficiency by restricting BAC locations to the collar and portal. This drove the development of a mixed BAC strategy where surface BACs are used to chill large volumes of air and meet the requirements of the shallow portions of the mine, and underground BACs are used to intensely chill smaller volumes of air for deep areas or high heat loads.

The savings resulting from the improved positional efficiency were calculated after the fact and are shown below in Table 6. Sizing of the 'indicative surface only cooling load' assumes that the same cumulative unit cooling would be applied to the full flow of the mine (1150 kg/s). The implementation of a mixed BAC strategy nominally decreased the cooling plant size by 50 per cent for the incremental cost of a chilled water loop and underground excavations to locate BACs.

TABLE 6

Positional efficiency estimates.

	4880L	4690L decline
Total mass flow (kg/s dry)	400	200
Cumulative unit cooling applied (kJ/kg)	20	30
Indicative surface only cooling load (MW_R)	23	35
'Mixed' cooling load (MW_R)	8	6

GO FORWARD PLANT DESIGN

Based on the outcomes of the trade-off studies and market engagement conducted, the project selected modular (~1.5 MW_R) integrated air-cooled R134a chillers as the go-forward plant option for the next stage of study. In addition to the economics of the decision, additional study and execution risks were reduced including:

- Reliance on identifying and permitting bore fields if other water sources were located near the project site,

- Elimination of a significant consumer of water, in line with water stewardship targets,

- Increased commercial optionality with >five vendors servicing the Asia-pacific market with chillers of the selected type, ensuring a competitive process in the future.

CONCLUSIONS

Initial cost estimates for an air-cooled R717 plant under study resulted in a review of the current state of chiller performance, benefits of R717 over R134a, and benefits of water cooled and air-cooled condensing. The key finding of the review was that improvements in air cooled chiller performance (driven by minimum efficiency standards) has narrowed operating cost difference between water- and air-cooled chillers.

A cooling demand estimate was made using site specific weather data and a target level of cooling determined from VentSim modelling. Different levels of cooling plant performance were evaluated against the cooling demand estimate for use in assessing the incremental economics of increased plant performance provided by different refrigerants and the use of water-cooled condensing. When compared to the costs of establishing water supply, the potential power savings, for this study, were not value accretive.

Based on these findings an out to market process was completed to test assumptions related to the capital cost of plant for water- and air-cooled configurations. The process identified that:

- The incremental performance (+10 per cent COP) of a R717 plant did not justify the additional capital cost,

- Modular 'off the shelf' R134a chiller packages (both air and water cooled) were the same or lower cost than a 'conventionally' constructed plant.

The points above in conjunction with assessments of positional efficiency resulted in an air-cooled, modular, R134a chilled water plant incorporating both surface and underground BACs to achieve the correct level of cooling for each of the working areas within the proposed mine.

It is worth noting that there are some specific conditions in relation to the project studied that enable the selection of air cooled 'off the shelf' chillers:

- The anticipated mine life (ten years) is within the typical service life of air-cooled chiller packages and as such, does not trigger any sustaining capital to replace or rebuild units that have reached the end of their useful life.

- The area surrounding the chilling plant is expected to have effectively no surface disturbance or influence from dust generating activity that would impact the performance or reliability of the condenser units.

- The project team made a wilful decision to treat each packaged chiller as a 'unit' in terms of electrical and mechanical specifications. No allowance was made to modify the chiller to meet site specific requirements or improve maintainability given the expected project life.

- R134a is in the process of being phased down across many jurisdictions, being replaced with lower Global Warming Potential refrigerants. The anticipated duration and timing of the project is not expected to be impacted by this change, however the modular nature of the cooling plant allows for incremental changes to the chiller units (eg different refrigerant gases in each packaged unit) should the situation change.

The authors contend that even where water is readily available for mine cooling, the use of modern integrated air-cooled units should be given serious consideration. While the efficiency of air-cooled plants compared to water cooled is still lower, the difference in absolute terms has reduced significantly and the question of 'best beneficial use' of water resources requires careful consideration. Global experience over the last decade of extended droughts has demonstrated the importance of importance of groundwater resources to surrounding communities and industry.

The authors also propose a 'rule of thumb' of 60 MW_Ehr/ML of usable water as the approximate tipping point where the cost of water supply results in no value-add from water cooled chillers. This is based on improving the overall COP of the plant from four to nine in line with the notional maximum improvement gained from water cooling over air cooling.

ACKNOWLEDGEMENTS

The authors acknowledge the assistance and effort by the various contributors to this process:

- The Mining Studies and Projects Teams for their willingness to explore alternatives to conventional cooling plant arrangements for mine sites

- The Surface Infrastructure Study Team for their patience with a constantly moving target of plant layouts, water, and power demands.

Lastly, the authors thank Newmont Corporation for permission to publish this paper.

REFERENCES

AECOM Australia, 2023. Secondary Ventilation and Refrigeration Task Report, Document No. 702-2520-ME-REP-1000, Internal Report.

ASHRAE, 2004. Standard 90.1–2004—Energy Standard for Buildings Except Low-Rise Residential Buildings. Available from: <https://store.accuristech.com/ashrae/standards/ashrae-90-1-2004-i-p?product_id=1199725>

ASHRAE, 2022. Standard 90.1–2022—Energy Standard for Sites and Buildings Except Low-Rise Residential Buildings. Available from: <https://www.ashrae.org/technical-resources/bookstore/standard-90-1>

Australian Bureau of Meteorology, 2020. Climate Data Online – Long Term Temperature Data 1 January 1996 to 27 March 2020. Available from: <https://reg.bom.gov.au/climate/data-services/>

De Souza, E, 2018. Cost-saving strategies in mine ventilation, *CIM Journal*, 9(2).

European Commission, 2009. Ecodesign and Energy Labelling – Framework Directives. Available from: <https://single-market-economy.ec.europa.eu/single-market/european-standards/harmonised-standards/ecodesign_en>

Loudon, A, 2021. Webinar: Refrigeration Plant Equipment Options for Hot Mines [online] (The Australasian Institute of Mining and Metallurgy: Melbourne). Available from: <https://www.ausimm.com/videos/community-event/webinar-refrigeration-plant-equipment-options-for-hot-mines/>

Rockwater, 2021. Bore Completion Report. Report No. 6.2/21/02, Internal Report.

US Bureau of Labor Statistics, 2024. Producer Price Index by Commodity: Metals and Metal Products [WPU10], FRED – Federal Reserve Economic Data, Federal Reserve Bank of St Louis. Available from: <https://fred.stlouisfed.org/series/WPU10>.

US Green Building Council, 2024. LEED Rating System [online]. Available from <https://www.usgbc.org/leed>

Understanding heat loads in hot volcanic underground mines

M Hooman[1] and A Kramers[2]

1. Principal Engineer, BBE Consulting Australasia, Brisbane Qld 4000.
   Email: mhooman@bbegroup.com.au
2. Senior Engineer, BBE Consulting Australasia, Brisbane Qld 4000.
   Email: akramers@bbegroup.com.au

ABSTRACT

Volcanic massive sulfide deposits can hold significant base metal deposits. Underground extraction methods of these orebodies include block caving, sub-level caving, sub-level open stoping or hybrid variations. These orebodies can present unique challenges for ventilation and heat management due to fractured rock sublayers with potentially large quantities of hot fissure water and steam, high rock temperatures with large variations both vertically and laterally, and presence of toxic gases such as hydrogen sulfide, sulfur dioxide and other contaminants.

The paper is based on learnings from a recent study for a greenfield project located on the Pacific Rim, with a planned production of 30 Mtpa in a geothermal environment involving hot rock, water and steam at temperatures above 90°C. The fundamental steps that can be used to understand heat loads in hot volcanic underground mines are discussed in detail, and an in-depth look is taken at the unique characteristics of the ore behaviour using an unsteady heat flow model and a simpler Є-NTU-method.

INTRODUCTION

Background

Volcanic arcs are formed by large bodies of magma rising through faults and fissures from the crust of the earth to form near-surface mineral deposits (Herrington, 2019). These deposits hold precious and base metals such as diamonds, gold, silver, copper, iron and molybdenum, and are often surrounded by hot springs.

These deposits are typically near-vertical massive orebodies and are extracted using mining methods such as block caving, sub-level caving, sub-level open stoping or hybrid variations (Hooman *et al*, 2022). Typically, underground mines globally produce less than 10 million tons per annum (Mt/a) (Mining.com, 2021) however, massive mining techniques allow production rates well above 10 Mtpa.

Massive orebodies can present unique challenges for ventilation and heat management, particularly when they are characterised by fractured rock sublayers with high rock temperatures and potentially large quantities of hot fissure water and steam. There may be large variations in the virgin rock temperature (VRT) both vertically and laterally, with the presence of toxic gases such as hydrogen sulfide, sulfur dioxide, and other contaminants present additional challenges.

The paper is aimed at understanding the mining techniques, surrounding rock temperatures, the associated oreflow system, and water inflows and how they can be applied to the underground environment to accurately predict the heat load. Once the heat loads are understood, they can be analysed and applied to specific work areas. Thereafter, aerodynamic and thermodynamic energy balances can be applied to design the ventilation and cooling systems to ensure the health and safety of personnel are maintained despite the hostile conditions.

Scope of work

The authors have been involved in several volcanic-type projects. Unfortunately, specific details of the project cannot be shared due to confidentiality arrangements. However, several general features of great interest can be highlighted. The authors endeavour to use general problem-solving techniques rather than specifics.

Of interest are the following features are discussed in this paper:

- very hot virgin rock temperatures (above 90°C)

- very high production rates at a block cave mine (30 Mt/a)

- high heat flow from broken oreflow

- heat flow from the ingress of hot steam and gases

- use of specific risk management tools in case of water and steam inflows

- possibility of using broad-spectrum refrigeration techniques.

HEAT LOAD DESIGN CRITERIA

Introduction

When designing the ventilation and refrigeration requirements for massive underground mines, ventilation design criteria need to be developed to cover various issues and conditions. These conditions are dictated by the presence of diesel engine emissions, airborne dust, gases and blasts, by minimum airway velocity requirements (personnel exposure) and by heat energy removal rates in production and development drives. Ventilation modelling is used to verify the effect on the air quality and air temperatures (wet-bulb and dry-bulb) in production drifts and other critical areas.

Once the ventilation design criteria are set, external heat sources like the materials handling system, dewatering pump systems, workshops, magazines, and other service areas liberating heat need to be defined.

Thermal design criteria

Many of the planned massive mines or those in operation are at depths of 1000 m and more. Furthermore, they are located in tropical and/or arid regions with humid and high surface air temperatures (Table 1). The depth of the operation will impact the increase in air temperature due to auto-compression and surrounding virgin rock temperatures. The orebodies typically have high geothermal gradients ranging between 0.01°C/m and 0.03°C/m (Hoinkes, Hauzenberger and Schmid, 2005).

Ventilation criteria are set to operate shafts and airways at acceptable velocities. Ventilation-related criteria are not stipulated herein as the focus of this paper is around thermal heat parameters. Generally, the ventilation criteria dictate the ventilation air quantity required, and the thermal criteria dictate the mechanical refrigeration needs when ventilation air only cannot adequately cool the underground environment.

For this study, the thermal design criteria are listed in Table 1.

TABLE 1

Thermal design criteria.

Ambient surface condition		
Wet-bulb (wb) temperature	°Cwb	24.7
Mean coincident dry-bulb (db) temperature	°Cdb	26.9
Barometric pressure	kPa	93.0
Air density	kg/m³	1.07
Thermal criteria		
Maximum mixed return (personnel working outside equipment)	°Cwb	28.5
Maximum (personnel inside equipment cabin)	°Cwb	30.0
Maximum (automated equipment doing work, no personnel)	°Cwb	32.0
Inactive areas	°Cwb	32.0
Maximum dry-bulb temperature	°Cdb	37.0
Geothermal rock properties and gradient		
Virgin rock temperature	°C	28.4 + 2.7°C/100 m
Density	kg/m³	2643
Thermal conductivity	W/m°C	3.75
Heat capacity	J/kg°C	890
Thermal diffusivity	m²/s	1.59×10^{-6}

Heat load factors

The mine design, production rate and ramp-up schedule impact the heat release from freshly exposed broken rock and strata rock. Furthermore, large diesel mobile equipment is used during the early development years to construct the mine before battery-electric vehicles can be considered for permanent production phases.

Infrastructure such as conveyor belts, workshops, crushers, ore transfer points etc, require careful planning to ensure that these areas are ventilated with fresh air that returns directly into return airways without being used in series in other areas. Conveyor airways are often used to return air to surface. All the heat from secondary ventilation fans will report as heat in the airstream.

Heat will be released from hot rock as it passes through the material handling system. The heat release depends on the residence time within the ventilation circuit. Not all the heat is released in the underground mine since the rock will reach surface in a still hot condition.

Hot geothermal rock will introduce significant heat to the broken and strata rock; a typical isotherm for this mine is shown in Figure 1. A typical outline of the production level is shown in purple. Groundwater seepage from the surrounding rock will enter the mine at the virgin rock temperature. The orange shown in Figure 1 shows where hot steam and/or gas will be expected.

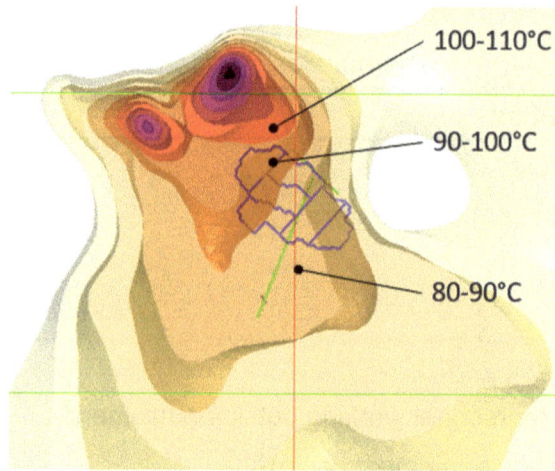

FIG 1 – Geothermal rock temperature isotherms.

Heat indices

There are several heat stress indices used in the mining industry to ensure that worker health and safety are maintained. The four main ones are:

1. Effective temperature, incorporating wet-bulb and dry-bulb temperatures and velocity (Australia, Europe and South Africa).

2. Wet-bulb globe temperature (North America, International Standards Organisation [ISO]).

3. Air cooling power, incorporating wet-bulb temperatures and velocity (South Africa and Australia).

4. Thermal work limits in Australia (Brake and Graham, 2002).

The International Organisation for Standardisation (ISO, 1989; ISO, 2004) has developed standards for heat stress management. These are internationally recognised and describe a rational approach. In this approach, the risk categories (RC) are defined as in Figure 2.

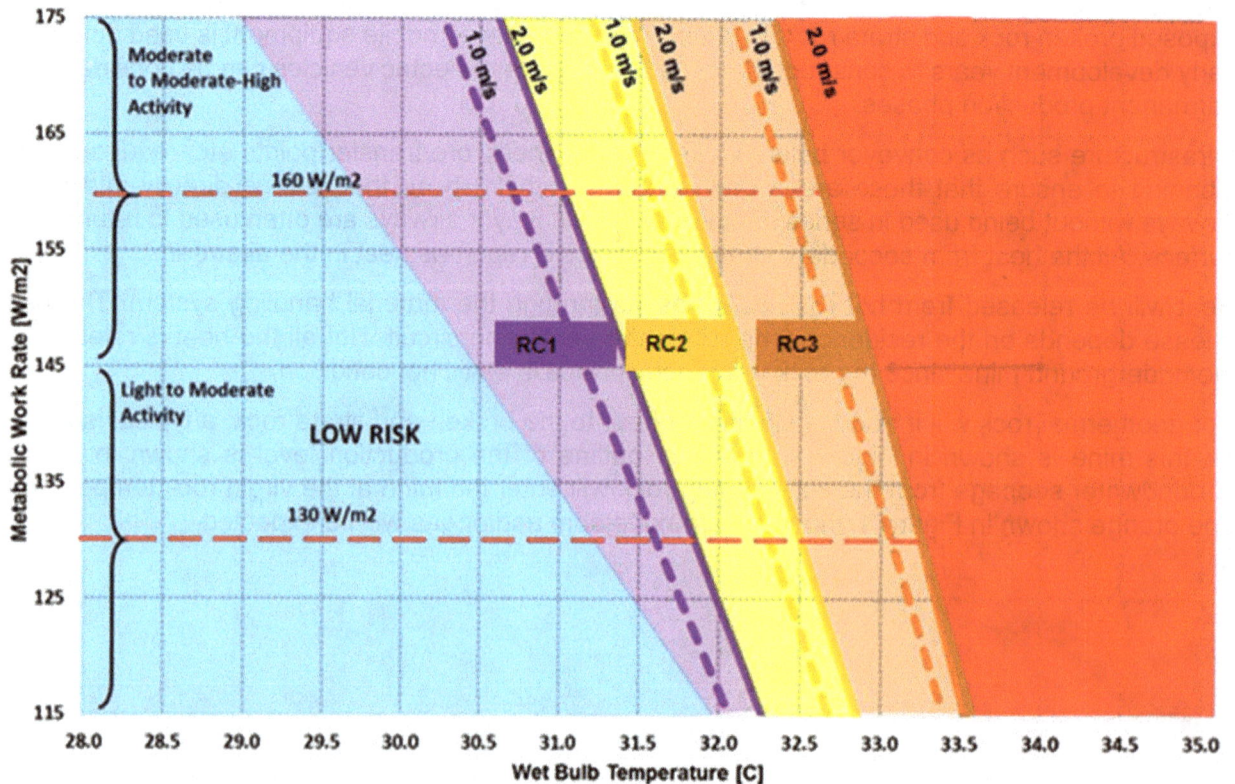

FIG 2 – Relationship between work rate, risk and temperature limits.

The ventilation criteria stipulated above, and the heat loads obtained underground are studied in the section below to determine the impact of these items on the underground environment.

ANALYSING HEAT LOADS

Understanding the heat loads is a critical step towards defining a definitive ventilation and cooling strategy. For the mine defined in the Design Criteria above, the following heat load analysis methodologies were applied.

Heat analysis methodologies – temperature dependent heat

Heat from hot rock

The temperature of rock (VRT) generally increases with increasing mining depths. In most situations, the VRT increases linearly however, projects adjacent to geothermal sources may have VRT that varies both laterally and with depth. In these cases, additional geothermal VRT contour information must be attained from drilling programmes and input to the model manually.

Heat from the shaft and tunnel sidewalls varies depending on the age of the tunnel and the historical ambient temperatures and air velocities it has experienced since first being exposed. The heat transfer is unsteady (varying both space and time). Algorithms have been developed that incorporate semi-infinite solid approximations and, together with empirical data, have been calibrated to simulate rock heat.

The mine design specifies the shaft and tunnel airway lengths and dimensions and sometimes tunnel age, and these can be conveniently imported into ventilation network analysis software programs to estimate the rock heat for any snapshot.

Heat from hot oreflow

The evaluation of the quantity of heat transferred by the ore to the ventilation system can be complex. It depends on many factors, including the type of ore, rock temperature and ambient temperature, residency time underground, degree of exposure to ventilation, air velocity, type of material handling systems, mechanical and friction heat, ore moisture and frequency of application (in dust suppression water sprays) to name a few.

The hot ore also affects the design of the material handling system. For example, standard mine hard rock conveyor belting is rated to a maximum of 65°C. It thus often falls on the ventilation engineer to estimate the ore temperatures likely to be encountered at different points in the material handling system to assist the conveyor engineer in the belt design.

The strategy for cooling ore must be carefully investigated. The surface temperature of ore may approach ambient temperatures quickly, but depending on the period of ore exposure underground, the core temperature of the heap burden pile on the belt may still be quite hot. While the ore is enclosed and 'insulated' against the environment by the belt, its temperature will slowly equalise, becoming hotter in contact with the belt. This effect could reduce any perceived advantages for cooling ore prior to the conveyor, albeit it being acknowledged that there will be some mass transfer by diffusion from the heap (thus providing some internal cooling).

Consider, for example, a 290 mm ore burden heap on a conveyor belt travelling out of the mine. The ore configuration can be likened to a simplified plate model experiencing unsteady cooling. In this study, the initial temperature of the ore was set at 90°C and the mean ambient wet and dry-bulb conditions at 26.0°C/31.5°C. The ore and ventilation are homotropal (flowing in the same direction) with 3 m/s relative velocity difference. The temperature profile through the ore heap over 5 hrs of exposure time is shown in Figure 3.

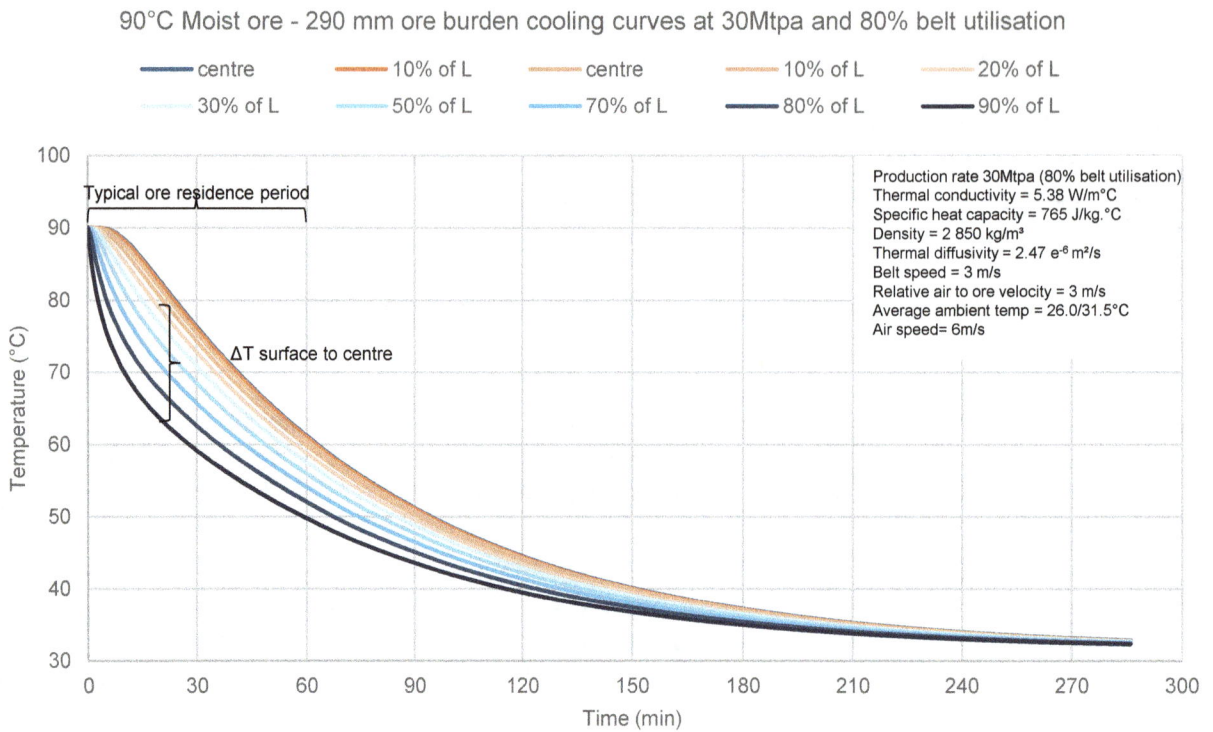

90°C Moist ore - 290 mm ore burden cooling curves at 30Mtpa and 80% belt utilisation

Legend: centre — 10% of L — centre — 10% of L — 20% of L — 30% of L — 50% of L — 70% of L — 80% of L — 90% of L

Production rate 30Mtpa (80% belt utilisation)
Thermal conductivity = 5.38 W/m°C
Specific heat capacity = 765 J/kg.°C
Density = 2 850 kg/m³
Thermal diffusivity = 2.47 e⁻⁶ m²/s
Belt speed = 3 m/s
Relative air to ore velocity = 3 m/s
Average ambient temp = 26.0/31.5°C
Air speed= 6m/s

FIG 3 – Ore pile cooling profile.

Figure 3 shows that there may be a considerable difference in temperature between the surface of the ore heap and the core in the early cooling stages. For the model adopted here, the average temperature of the ore burden pile at any one time was estimated as the average across the ten curves.

However, a simpler calculation approach has been proposed by Van den Berg, Moreby and Kok (2015) that uses an Є-NTU model, as described by Leinhard and Leinhard (2008). The effectiveness (Є) expression was adjusted for homotropal flow to compare with the above unsteady flow with results presented in Figure 4.

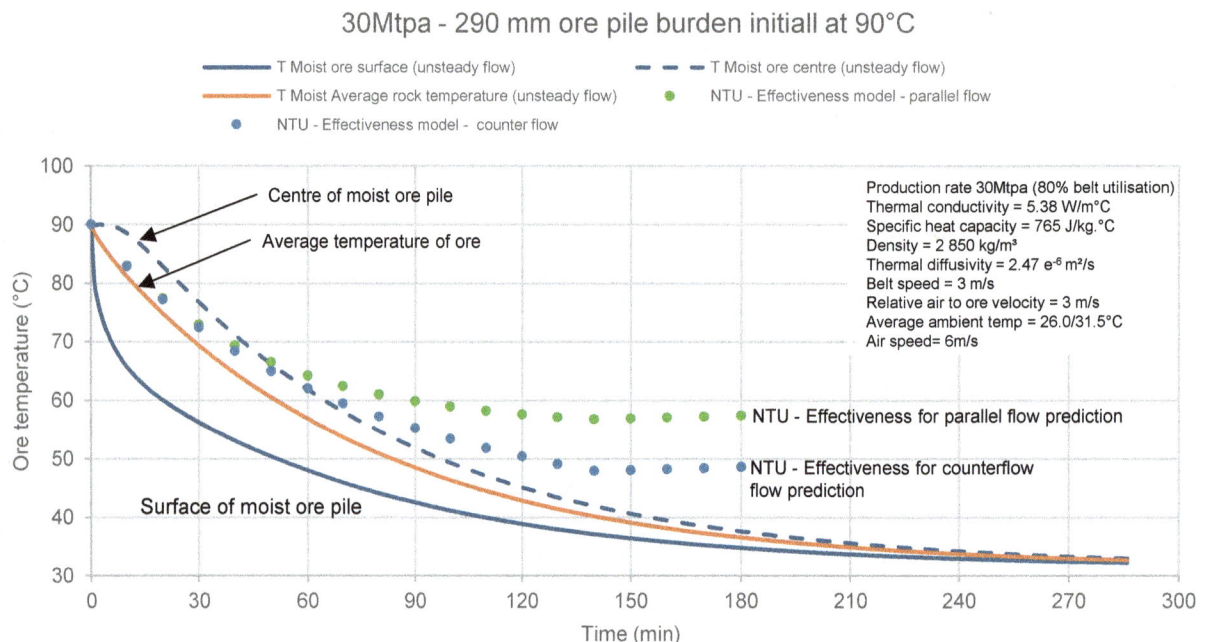

30Mtpa - 290 mm ore pile burden initiall at 90°C

Legend: T Moist ore surface (unsteady flow) — T Moist ore centre (unsteady flow) — T Moist Average rock temperature (unsteady flow) — NTU - Effectiveness model - parallel flow — NTU - Effectiveness model - counter flow

Production rate 30Mtpa (80% belt utilisation)
Thermal conductivity = 5.38 W/m°C
Specific heat capacity = 765 J/kg.°C
Density = 2 850 kg/m³
Thermal diffusivity = 2.47 e⁻⁶ m²/s
Belt speed = 3 m/s
Relative air to ore velocity = 3 m/s
Average ambient temp = 26.0/31.5°C
Air speed= 6m/s

FIG 4 – Comparison of unsteady transient cooling profiles versus NTU effectiveness predictions.

The NTU-effectiveness method compares well with the unsteady model for the first 40 min. The NTU-effectiveness deviates from the unsteady model at longer residency as the NTU-effectiveness method tends to the log mean temperature difference of the rock and air stream- while the unsteady cooling model assumes the air remains at the same average temperature (ie cooling is being supplied to remove heat maintaining a steady air temperature). This confirms the NTU- Effectiveness methodology can be used in a quasi-static manner over short residency periods to predict heat loads in mine material handling systems.

The cumulative heat loss is shown in Figure 5. The cumulative heat loss from the ore (dry, moist pf wet) can be read off the graph at the ore residence time on the conveyor. For this mine, moist ore on a conveyor belt of 7 km from a depth of 1 km below surface with roller and idler frictional heat and motor drive heat, is combined and as distributed sensible and latent heat (in kW/m) over the conveyor belt system.

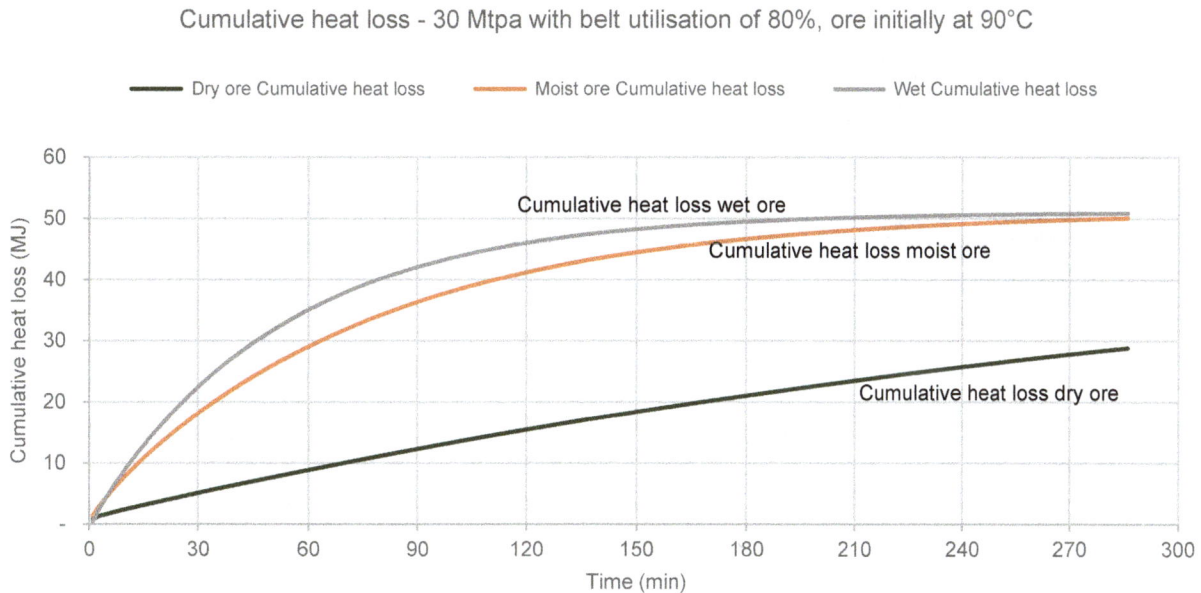

Cumulative heat loss - 30 Mtpa with belt utilisation of 80%, ore initially at 90°C

FIG 5 – Cumulative heat loss from ore at an initial temperature of 90°C.

Heat from hot water sources

Heat from hot fissure water can also be significant depending on the inflow, temperature, mine design and engineering controls in place to remove water from intake airways. Dewatering test programs during concept and feasibility studies will remove some uncertainty, and the mine will be designed to handle a certain maximum inflow. Engineering controls include the following:

- surface and underground dewatering wells and pumps
- sumps with dewatering pumps
- enclosed drains and drainage boreholes
- drainage levels and dams
- separation of hot water from intake airstreams.

Heat from hot fissure water can then be estimated at each life-of-mine snapshot with reasonable assumptions made for the drains, pumps and piping exposed areas in intake airways. This heat must include both sensible and latent components. Figure 6 shows the hot water temperatures at the virgin rock temperature isotherms for specific zones of a typical block cave mine.

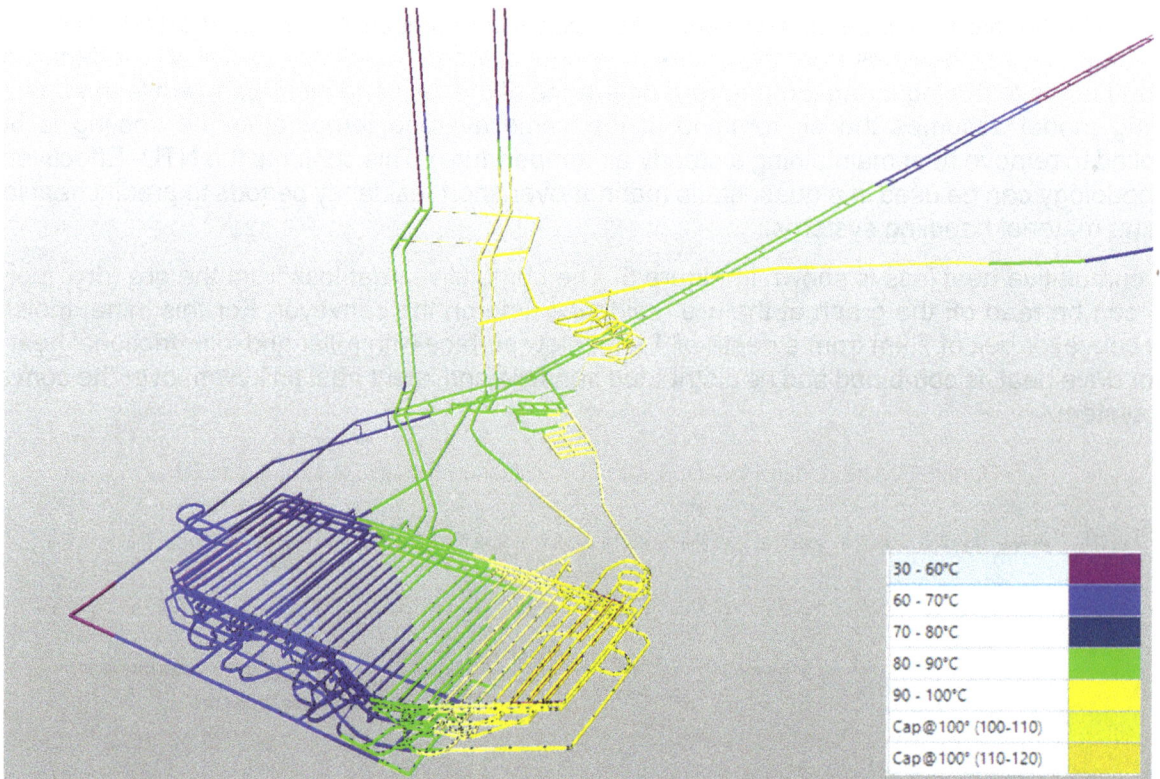

FIG 6 – Hot water temperature zones.

When engineering controls like dewatering pump systems and cool water injection strategies are applied to the underground exposed zones, the water temperature of the strata rock and hot water will reduce, as shown in Figure 7.

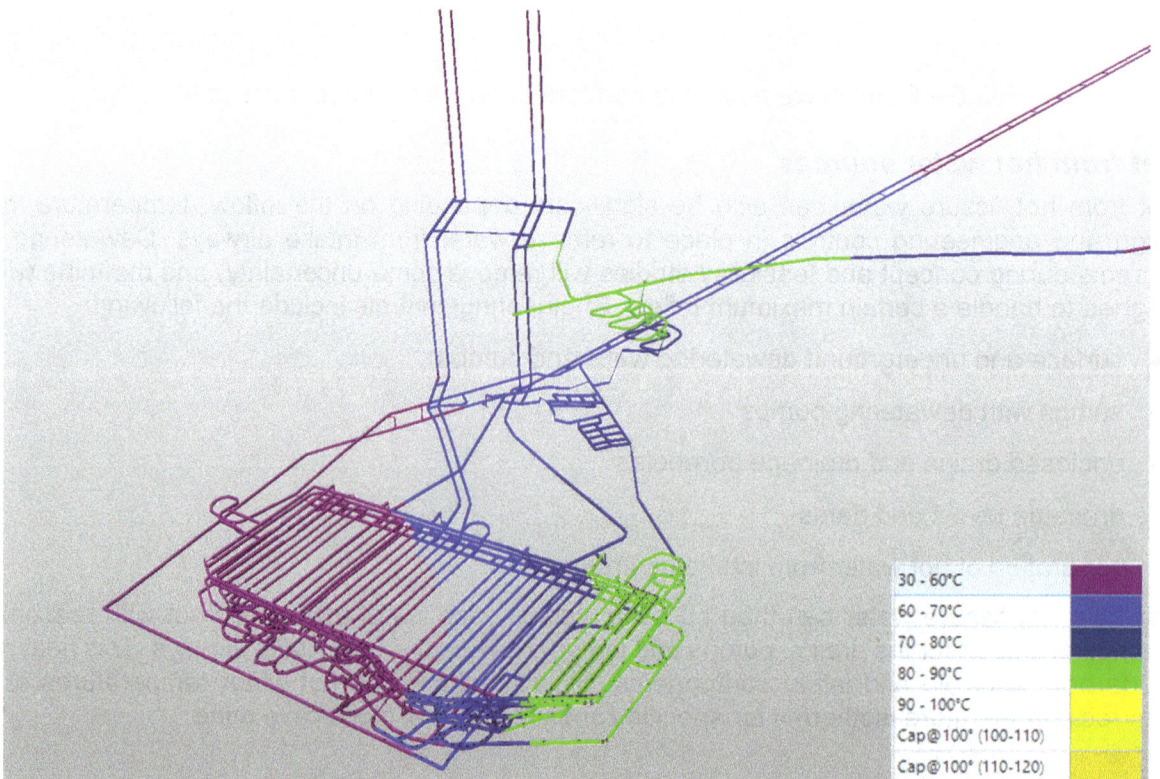

FIG 7 – Hot water temperature zones (-20°C cool water injection).

Heat analysis methodologies – temperature independent heat

Diesel engines have a thermal efficiency of 30 per cent at full load motor rated power. This portion of the heat from a diesel engine is converted to mechanical energy, is not considered useful power, and is dissipated as heat.

The heat from electrical machinery will impact potential and/or mechanical energy. Electrically driven machinery like fans, hoists, lights, motor drives, winches, compressed air, rockdrills etc do not do any useful work and the mechanical power of the units at a specific utilisation and availability factor is added as heat to the underground environment.

A total of 124 MW of heat loads will be present at the Mine when 30 Mtpa of ore is extraction from the block cave at steady state (Year 17), see Figure 8.

30Mtpa block cave production heat distribution

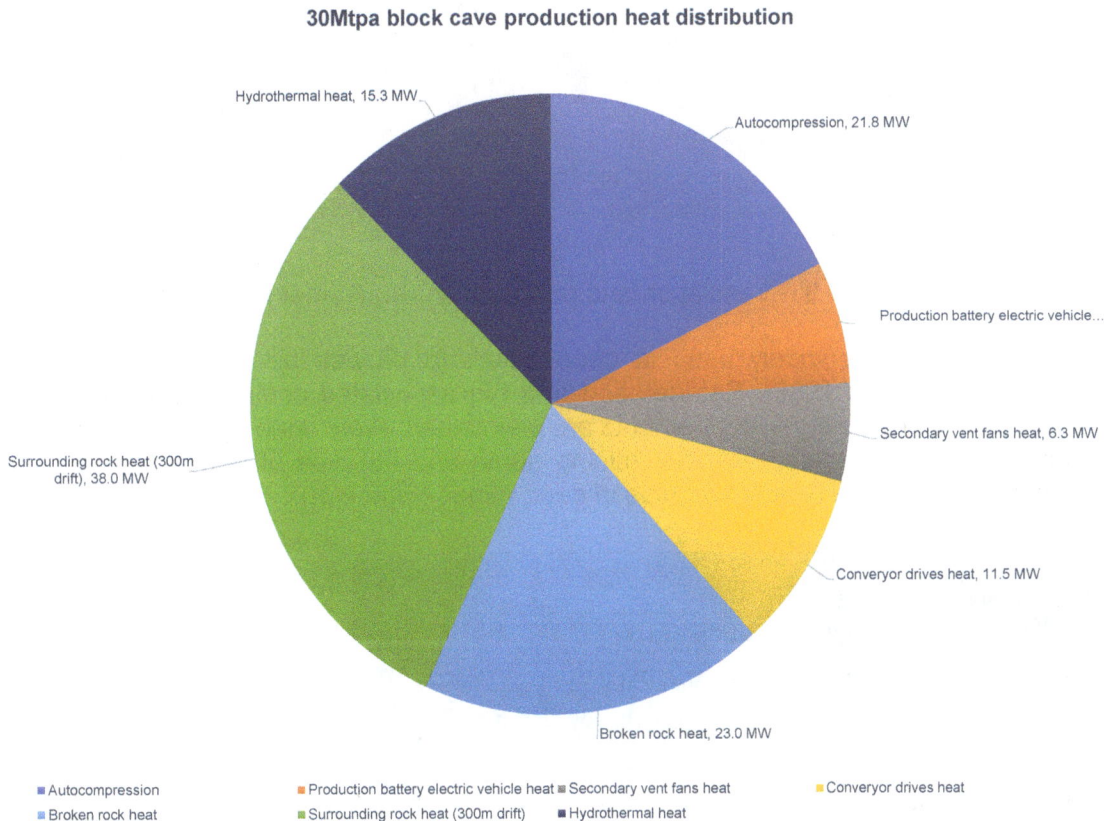

FIG 8 – Heat load distribution.

HEAT MANAGEMENT CONSIDERATIONS

During mine planning of green (new) or brown (existing) field projects, environmental design criteria are set to ensure legal compliance and, thus, a healthy and safe working environment. When ventilation thermal design criteria are exceeded, and ventilation air alone cannot dissipate the mine heat load, refrigeration and associated cooling systems need to be introduced.

For this mine, the ventilation and cooling requirements were determined after an extensive analysis to understand the mine's heat load. The calculated results are introduced to advanced ventilation software simulations tools, that is used to verify the ventilation distribution and mechanical refrigeration requirements. The results of this typical Mine with a 30 Mtpa production rate with hot ore, and hot water/steam, are show in Figure 9. Note that due to the very high rock and water/steam temperatures and high production rates, excessive ventilation and refrigeration infrastructure will be required. This Mine will typically require 2700 m³/s and 100 MWc of air cooling for life-of-mine. Note that air cooling is stated and that the refrigeration demand will be 20 per cent to 40 per cent more than the air cooling needs due to thermal losses between the refrigeration plant and the air cooling installations.

Airflow Profile

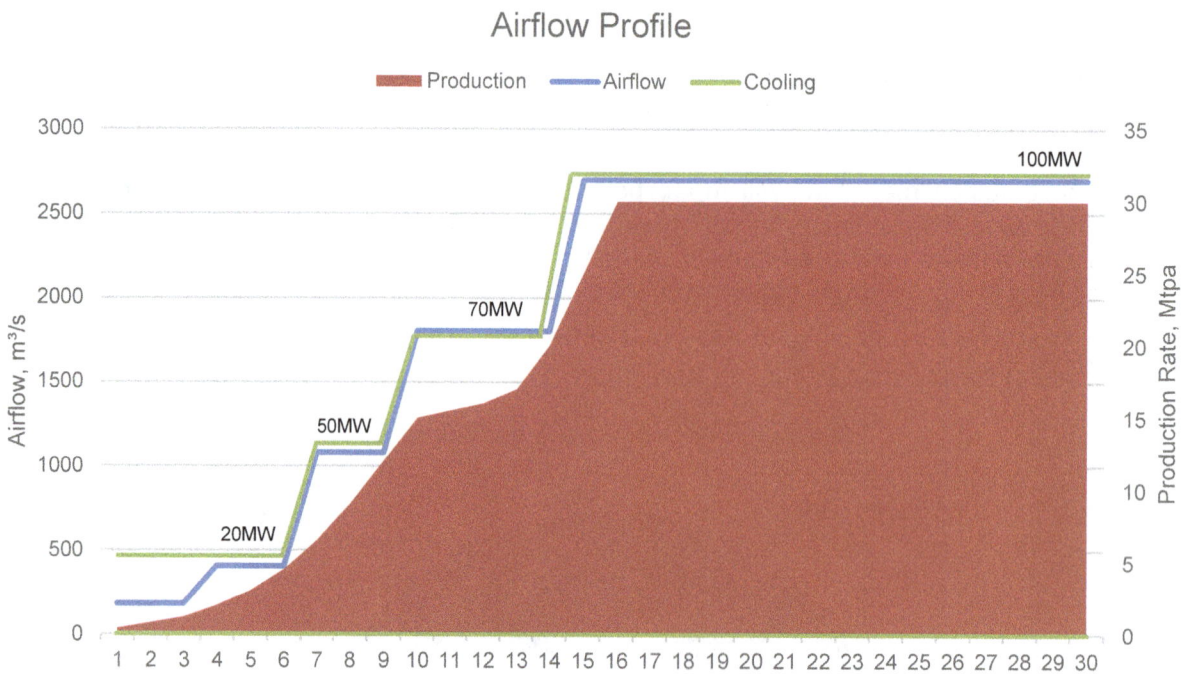

FIG 9 – Ventilation and refrigeration requirements.

Refrigeration systems are widely used in deep mines to provide air cooling for underground operations (Burrows *et al*, 1989). Refrigeration and the associated cooling installations at depth become essential to ensure legal requirements are maintained. Work done by Bluhm and Von Glehn (2010) indicates the relationship between mining depth and the type of refrigeration and cooling system required that can be used as a guide in deep mines (Figure 10).

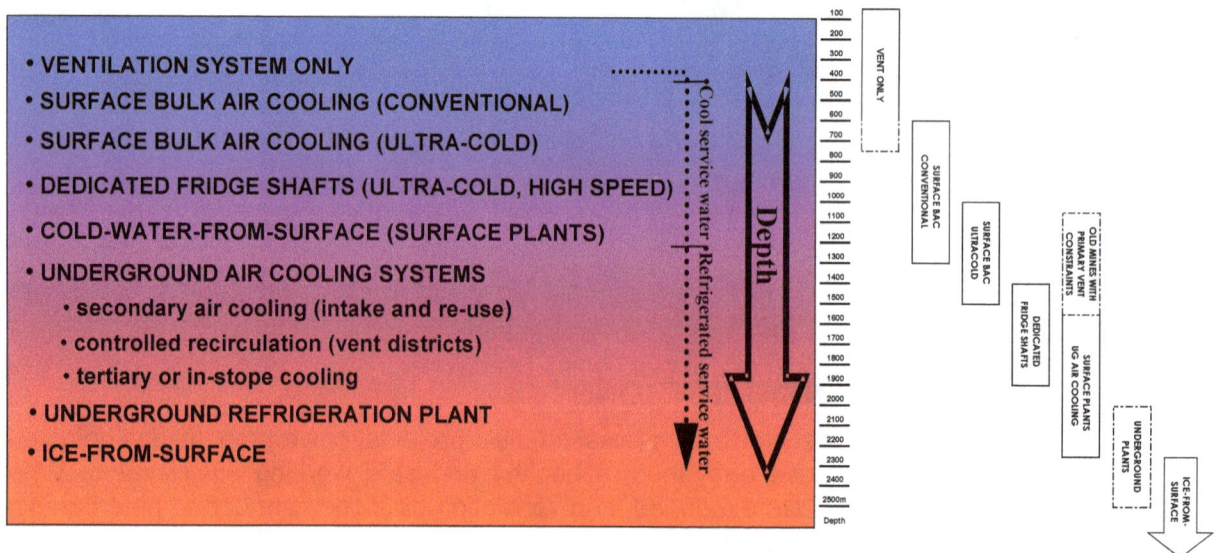

FIG 10 – Depth versus refrigeration system.

Trade-off studies are generally adopted to determine which of the following options are best for the specific site:

- surface refrigeration and surface air cooling
- surface refrigeration and underground air cooling
- underground refrigeration and underground air cooling
- ice from surface.

For each of these options, various types of refrigerants are available in the market that continuously improves in-line with the global warming potential hazards. Furthermore, there are various types of air-cooling heat exchanger options that include spray chambers, packed and unpacked towers, and indirect-contact coil banks. For this mine, a combination of these refrigeration options is required due to the magnitude of the heat load from the rock and water and their application in specific zones underground. Figure 11 shows a typical layout of a block cave mine refrigeration system using surface and underground refrigeration and air cooling systems.

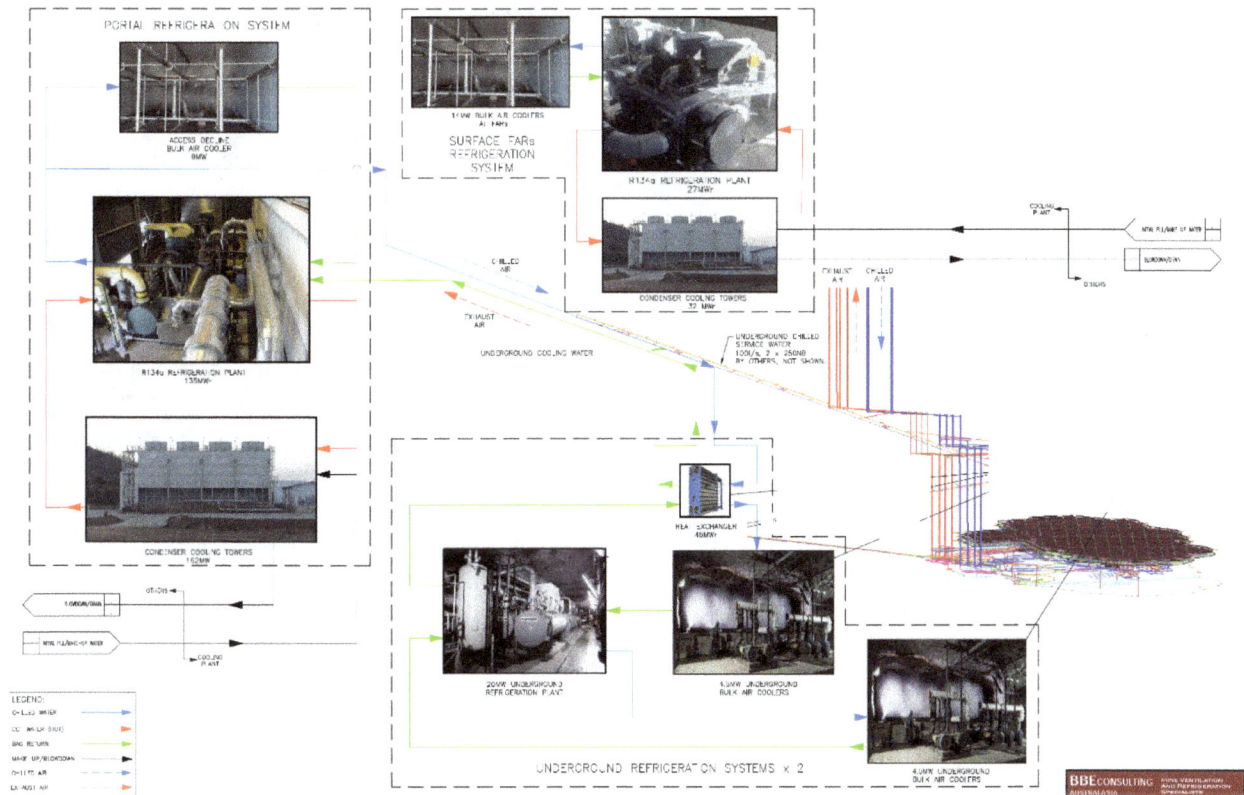

FIG 11 – Typical mine refrigeration system.

Administrative controls are the last resort to manage the heat and includes work-rest cycles, acclimatisation, rest areas, cooling, vests, aircon cabins, and the extent of the application need to be determined for each site.

CONCLUSIONS

The paper provided a summary of the learnings from a recent study for a greenfield mining project located on the Pacific Rim, with a planned production of 30 Mtpa in a geothermal environment involving hot rock, water and steam at temperatures above 90°C. The ventilation design criteria are set at the onset of the project, from which point the mine design is used to determine critical snapshots to consider over life-of-mine.

This study is unique as it investigated the characteristics of the ore behaviour using an unsteady heat model and Є-NTU-method model. The correlation was very close for short residency periods, and it provides insights into the impact the heat from ore will have on the environment and on material handling systems. The heat load analysis steps summarised in this paper provides the ventilation design engineer confidence when planning the ventilation and cooling requirements of the mine over time.

ACKNOWLEDGEMENTS

We would like to thank the Ventilation Principal and Engineering Manager at the Mine for allowing us to publish this paper; in due course, more details of the site will be made available.

REFERENCES

Bluhm, S J and von Glehn, F, 2010. Refrigeration and Cooling Concepts for Ultra-Deep Operations, BBE Report 0301.

Brake, D J and Graham, P B, 2002. Limiting Metabolic Rate (Thermal Work Limit) as an Index of Thermal Stress, *Applied Occupational and Environmental Hygiene*, 17(3):176–186.

Burrows, J, Hemp, R, Holding, W and Stroh, R M, 1989. *Environmental Engineering in South African Mines,* Mine Ventilation Society of South Africa.

Herrington, R, 2019. Mining volcanoes: diamonds, copper and hot water [online], Gresham College. Available from: <https://www.gresham.ac.uk/watch-now/mining-volcanoes> [Accessed 26 January 2024].

Hoinkes, G, Hauzenberger, C A and Schmid, R, 2005. Metamorphic rocks/Classification, nomenclature and formation, *Reference Module in Earth Systems and Environmental Sciences, Encyclopedia of Geology*, pp 386–402. https://doi.org/10.1016/B0-12-369396-9/00478-0

Hooman, M, Van den Berg, L, Möhle, H and Paiken, L, 2022. Ventilation requirements of cave mine, in *Caving* 2022 (ed: Y Potvin), pp 343–354 (Australian Centre of Geomechanics, Perth).

International Standards Organisation, 1989. ISO7243 Hot Environments – Estimation of the heat stress on a working man based on the WBGT-index (wet-bulb globe temperature), August 1989.

International Standards Organisation, 2004. ISO7933 Ergonomics of the thermal environment – Analytical determination and interpretation of heat stress using calculation of the predicted heat strain, August 2004.

Leinhard IV, J H and Leinhard V, J H, 2008. *A Heat Transfer Textbook*, third edition, pp 99–136; 267–338 (Phlogiston Press: Cambridge).

Mining.com, 2021. Ranked: World's 10 biggest underground mines by tonnes of ore milled [online]. Available from: <https://www.mining.com/featured-article/ranked-worlds-10-biggest-underground-mines-by-tonnes-of-ore-milled/> [Accessed: 24 January 2024].

Van den Berg, L, Moreby, R and Kok, J, 2015. Heat Load Estimation from Conveyor Systems in Mass Mines, The Australian Mine Ventilation Conference.

A cooling study for an Australian underground longwall mine

H Kerr[1], M Hooman[2] and K Van Zyl[3]

1. Senior Ventilation Engineer, BBE Consulting Australasia, Joondalup WA 6027.
   Email: hkerr@bbegroup.com.au
2. Principal Engineer, BBE Consulting Australasia, Joondalup WA 6027.
   Email: mhooman@bbegroup.com.au
3. Principal Engineer, BBE Consulting Africa, Johannesburg 2191, South Africa.
   Email: kvanzyl@bbeco.za

ABSTRACT

Longwall mining is the principal extraction method used for underground coalmines in Australia. Although most coalmines are considered shallow, with a depth of less than 600 m, some mines experience extensive heat loads, such as underground coalmines in the Central Queensland region and in deeper mines in New South Wales. This paper describes a cooling study completed for an unspecified underground longwall mine in Australia. Potential cooling strategies considered include a larger ventilation shaft with bulk cooling, a series of small ventilation boreholes with smaller coolers, and chilled water distribution to local underground air coolers. Chilled service water is generally used for dust suppression sprays in the face zone and was not considered part of this work. This paper compares the effectiveness of the different cooling strategies using a calibrated ventilation and heat model using ventilation software. It then compares the capital and operating costs and the maintainability and longevity of the equipment. To date, modelling has shown that the proximity of the air cooler to the heat-affected area is a key factor, however, other factors, such as the surface ambient temperature, mining depth, mine layout and mining equipment, can also impact the optimum cooling strategy.

INTRODUCTION

Background

Longwall (LW) mining was developed in the 17th century, when miners were used to undercut the coal along the width of the coalface (Bauer, 2015). From the 19th century, mechanisation started when conveyor belts were introduced in gate roads and along the mains, increasing the productivity substantially (Vinay, Bhattacharjee and Ghosh, 2022). Today, longwall mining panelpanel typically extend 2 km to 7 km and have a face width of 200 m to 400 m. Although these mines are considered shallow, with a depth of cover of less than 600 m, some mines experience extensive heat loads.

Scope of work

The mining panels of this 'Mine' are between 6 km to 7 km long and at a 600 m depth of cover, which is longer and deeper than most coalmines, leading to higher heat loads. A typical coalmine layout is shown in Figure 1, which shows a LW shearer at mining panel LW103 and two development panels with a Continuous Miner (CM) each at mining panel LW104 and LW105. The face zone is supported by a continuous line of automatic hydraulic roof shields. Coal is loaded onto Armoured Face Conveyor (AFC) and then transferred to feeder breaker that are located in the belt airways in a homotropal arrangement. LW100 and LW101 have been mined out and the goafs are sealed.

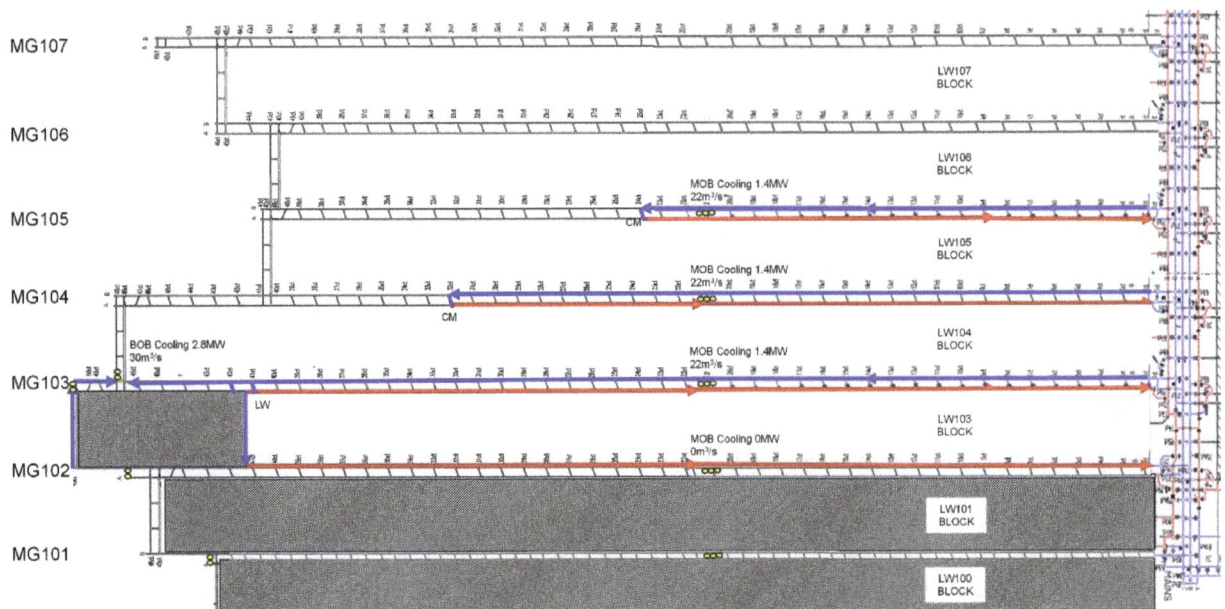

FIG 1 – Typical coalmine with seven LW panels.

The main heat loads for the Mine include the longwall (shearer, AFC and automatic hydraulic roof shields), the development face equipment (continuous miner, shuttle cars and feeder breaker), rock strata, diesel fleet, and main conveyor belts. The face zones are generally the most critical due to the high concentration of heat and limited airflow quantity that needs to be maintained within acceptable thermal limits to ensure worker health and safety.

For this Mine, the heat management strategy is to provide cooling via middle-of-panel (MOP) and back-of-panel (BOP) air coolers. This paper considers the following work:

- Refrigeration alternatives:
  - central chiller plants and chilled water piping to the cooling sites
  - smaller rental chiller plants with chilled water at the cooling sites.
- Gate road ventilation capacity alternatives:
  - single cluster of boreholes servicing longwall panels
  - single blind bored shaft servicing multiple panels.

Cooling strategies were developed for various heat load parameters studied at an underground longwall mine in the Central Queensland region that can be applied to warm and deep coalmines globally.

THERMAL DESIGN CRITERIA

Heat has become an issue in underground coalmines in the Central Queensland region and in deeper mines in New South Wales, Australia, and elsewhere. As a ventilation practitioner, it is important to understand the heat loads and heat indices to manage personnel heat stress exposure to stay within legislated requirements. A summary of the main aspects are discussed below.

Heat indices

According to McPherson (1993), heat stress indices can be classified into three types, which have been developed to ensure underground worker health and safety:

1. Single measurements, eg wet-bulb temperature (wbt), Kata thermometer and others.

2. Empirical methods include effective temperature (Basic and Normal), corrected effective temperature, kata thermometer, wet-bulb temperature, wet-bulb globe temperature and others.

3. Rational indices are based on the rational heat balance equation, eg air cooling power, thermal work limit and others.

In this study, the Queensland heat indices (normal effective temperature) will be applied.

Legislation

Coalmines in Queensland operate under the Coal Mining Safety and Health Regulation 2017, which nominates a wbt of 27.0°C as the threshold for where heat stress provisions must be enacted in the site's health and safety management system. No work is allowed above a Normal Effective Temperature (NET) of 29.4°C other than in rescue or emergency situations or to conduct work to reduce the effective temperature. The legislation requires compliance with Recognised Standard 18: Management of Heat in Underground Mines, which details the hazards of working in heat, potential controls and management system elements, including the development of a heat Trigger Action Response Plan (TARP).

Typical TARPs in use in Queensland operations shown in Figure 2 divide the legislated boundaries into:

- Normal – Less than 27°C wbt and 27°C NET
- Level 1 – between 27°C and 28°C NET requiring job rotation, acclimatisation, fitness for work and hydration
- Level 2 – 28.0–29.4°C NET requiring controls as per level 1 and scheduled work breaks
- Level 3 – Greater than 29.4°C NET requiring work to stop.

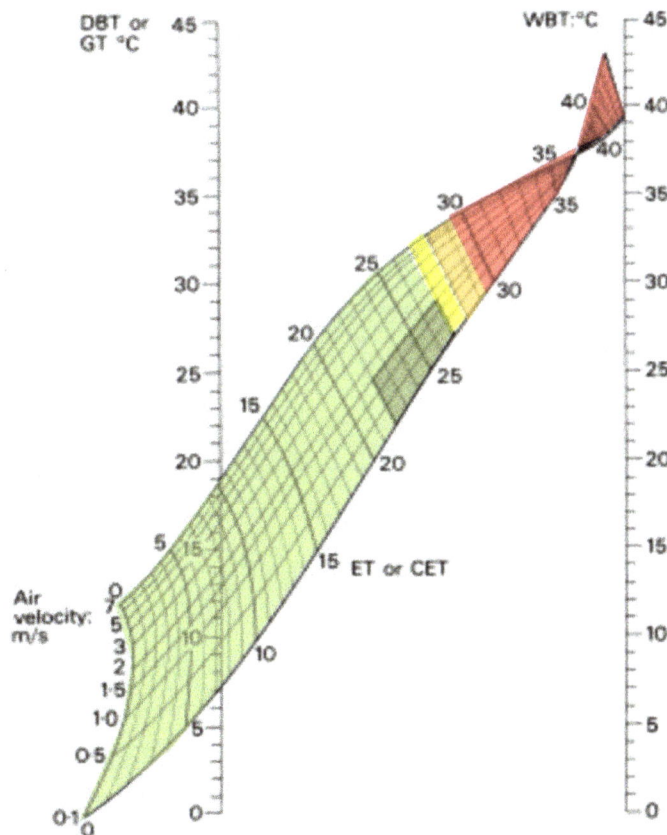

FIG 2 – Normal effective temperature nomogram with TARP levels.

To determine required refrigeration temperatures, design temperatures are set at 28°C NET, which is the Level 2 trigger threshold in this case, but it can be lower for some operations. Given the expected inaccuracy of the heat model (±2°C wbt), this allows for some variability without reaching a stop work condition. This is consistent with design wbt of 30.0°C in the longwall tailgate and 27.0°C in the development panel where the face velocities are lower.

SOURCES OF HEAT

There are a number of heat sources present in an underground longwall mine:

- Warm inlet ambient air from surface (Mitchell, 2003).
- Auto-compression.
- Strata rock heat as a result of high geothermal gradients typically in the order of 1.6°C/100m to 6.0°C/100m (Wang, Liu and Yand, 2020; Karwasiecka, 1996).
- Heat from development equipment.
- Heat from longwall mining face equipment.
- Conveyor heat in the form of friction and drive end motors (both main conveyors and panel conveyors).
- Diesel equipment during panel moves.
- General support diesel equipment.
- Other sources of heat, such as secondary fans, raw water and groundwater.

Strata heat load

The heat transfer from the coal around the roadways is often the mine's largest and most variable heat source through the mine, with Virgin Rock Temperatures (VRT) increasing with depth. Figure 3 has been generated based on borehole temperature logs for the case study mine. The variability of heat transfer relates to the relationship between the coal/air temperature difference. The greater the difference, the greater the heat transfer to the air from the coal. As such, during seasonally colder months or when outbye equipment is not operating, the heat transfer is greater from the coal.

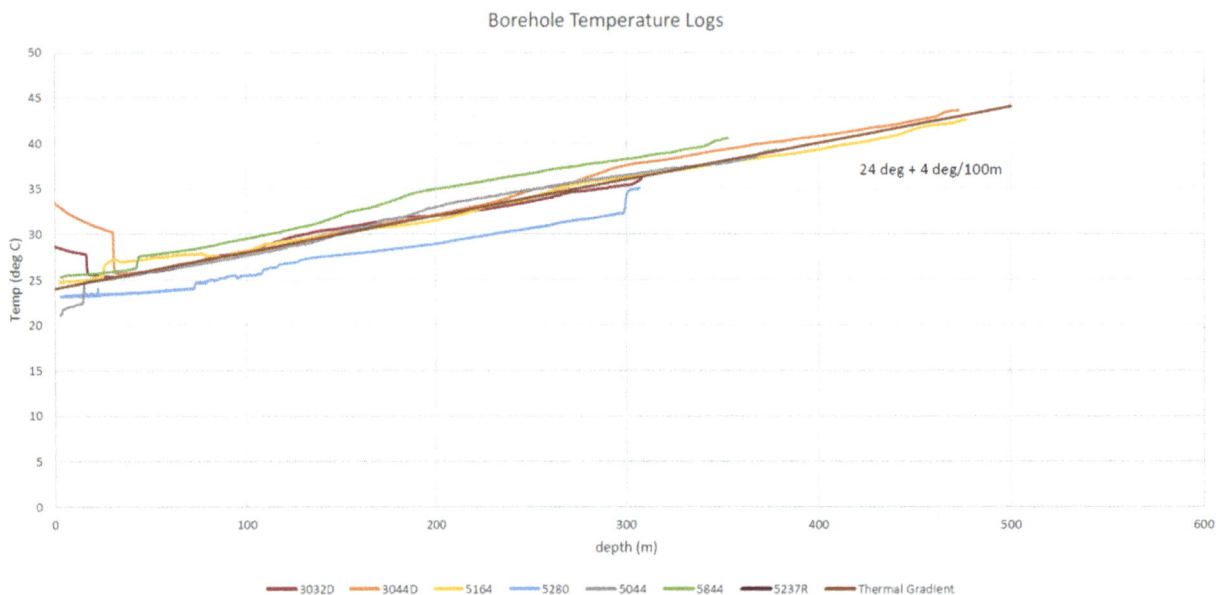

FIG 3 – Virgin rock temperature trend.

Based on the geothermal gradient above, the expected face temperature varies between longwalls over the life-of-mine as the depth of cover varies. The VRT of the Mine has been plotted against available data from Belle and Biffi (2018), which correlates very well with other Central Queensland mines like Oaky Creek, Grasstree and Grovenor (Figure 4).

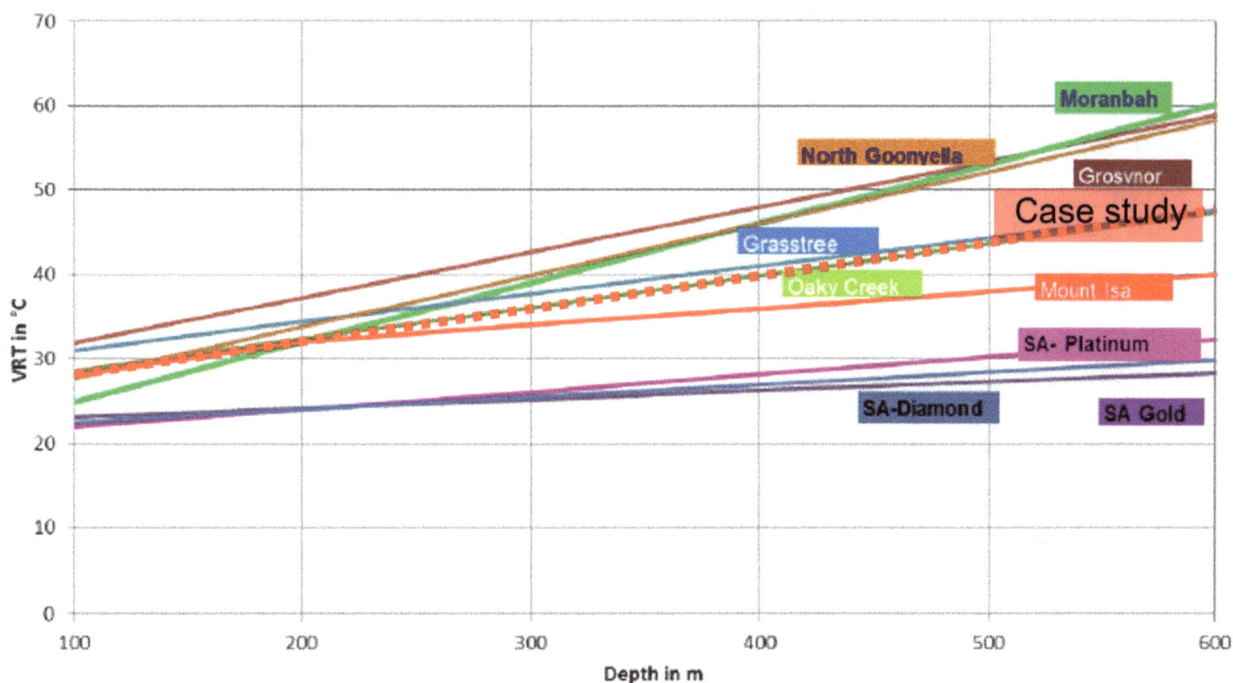

FIG 4 – Gate road thermal gradient in Central Qld mines (Belle and Biffi, 2018).

Development gate road heat load

The heat transfer also changes with the length of a development gateroadway as seen in Figure 5 (intake only, excluding conveyor and return heat). Ventilation survey data is used to show increasing heat load from the start to the end of the panel (which also includes face equipment and conveyor heat sources).

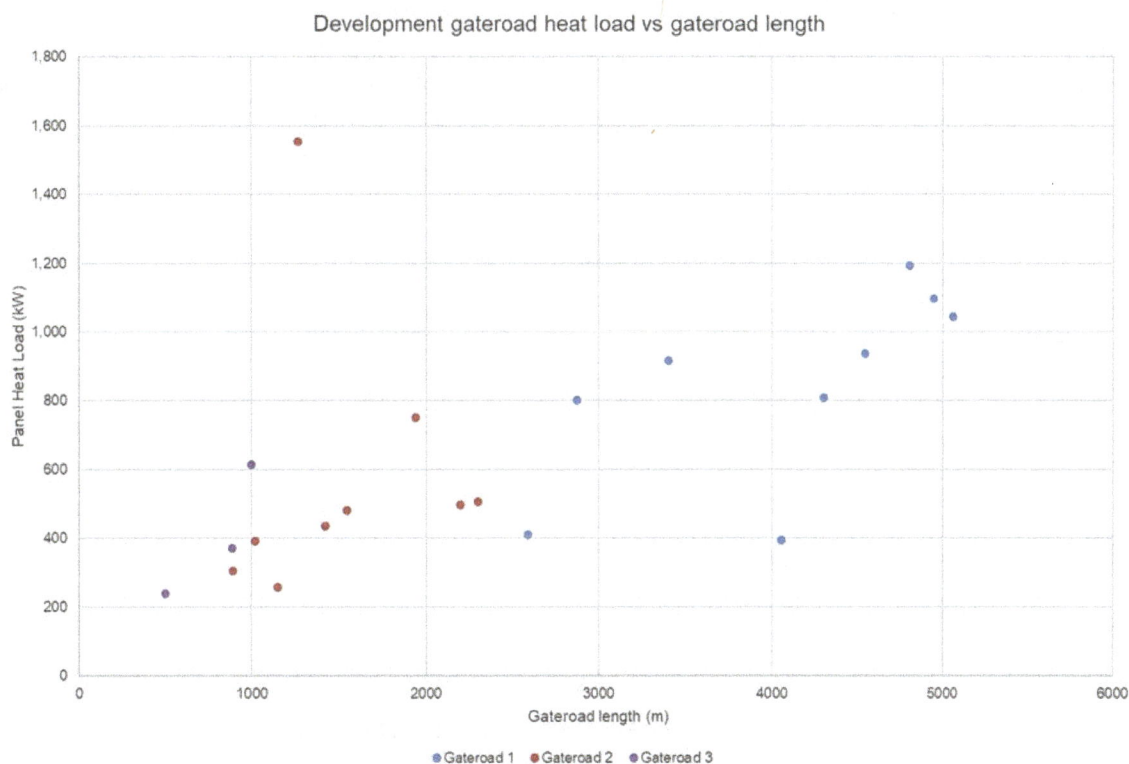

FIG 5 – Development gate road heat load against length.

This same data plotted against intake temperature shows that the heat relationship with the air/coal temperature difference still applies (Figure 6).

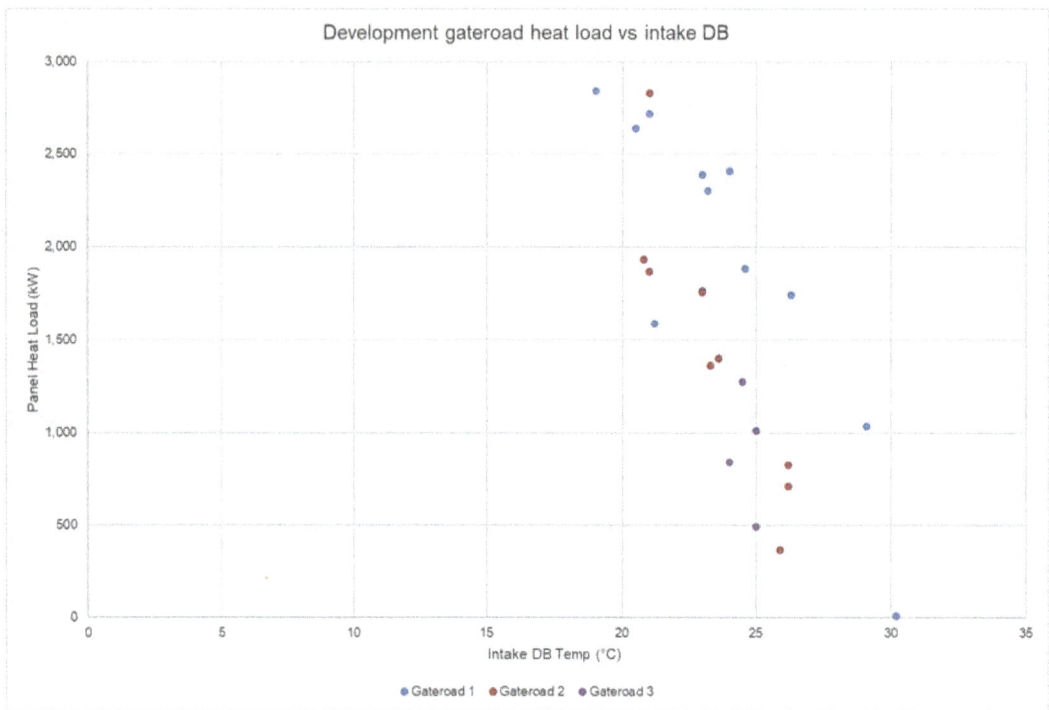

FIG 6 – Development gate road heat load against intake dry-bulb temperature.

Longwall heat load

The measured LW heat load (including strata, goaf heat, broken coal, antitropal conveyor roadway and longwall face equipment heat) measured during ventilation surveys varies between 2000 kW and 5000 kW. This variance does not correlate with face position or remaining panel length but is likely aligned to the production consistency prior to the survey and whether the face equipment and conveyor were operating at the time of the survey.

The rated equipment power for the LW is 7800 kW (at LW start-up). The percentage of rated equipment power to measured heat load at a similar mine is 40 per cent (van den Berg and Olsen, 2017), which is consistent with the range of measured heat at this mine (Figure 7).

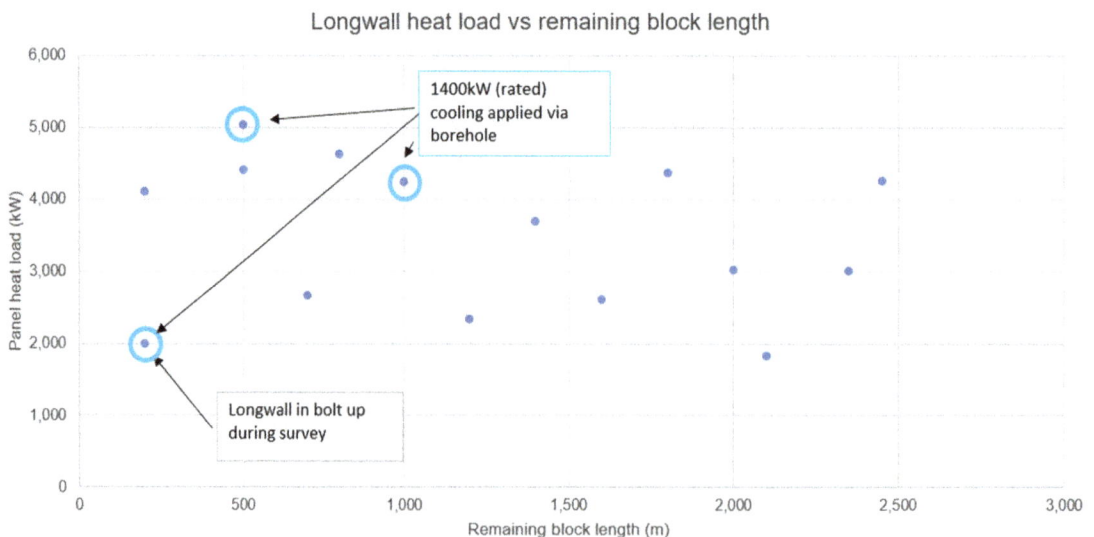

FIG 7 – Longwall heat load with retreating panel.

Development equipment heat load

The development equipment installed power 830 kW for CM and 710 kW for the belt conveyor. Measured gate road heat from ventilation survey data, including the panel conveyor, diesel equipment, face equipment and strata heat, ranged up to 3000 kW, increasing with gate road length.

Using the average monthly diesel fuel consumption for the underground equipment and assuming that all of the energy from the fuel is converted to heat spread evenly across 80 per cent of the month (allowing for shift change and non-production windows, fan outages etc), results in a heat load of 2408 kW. Spread across the 10 km of travel road, this equates to 241 W/m.

Conveyor heat load

Installed conveyor power varied with capacity and lift ranging from 1420 kW for shorter trunk conveyors to 5600 kW for the drift conveyor drives, which are located on surface. It is expected that the heat load for the conveyors will vary over the life-of-mine as shown in Figure 8.

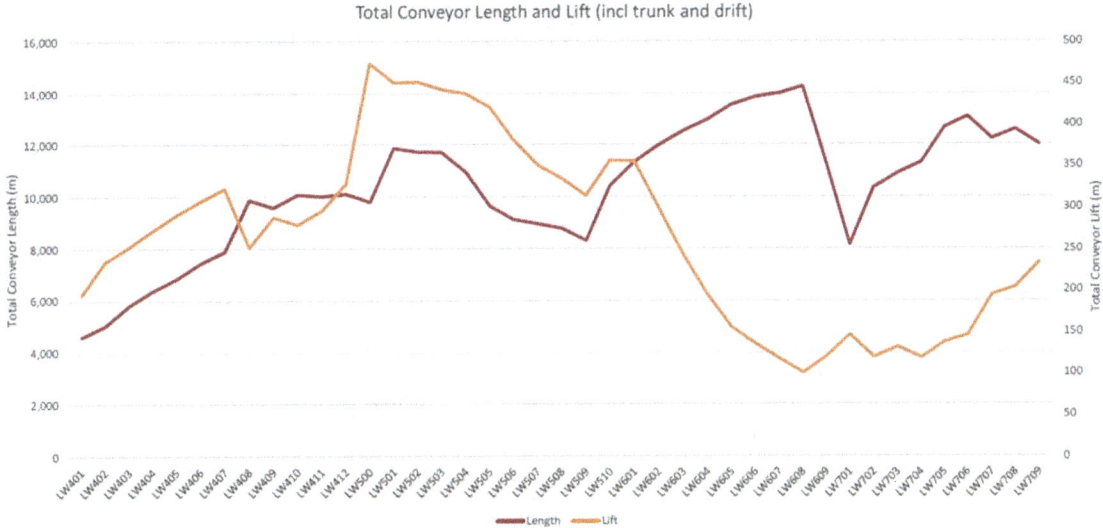

FIG 8 – Conveyor lift and length over the life-of-mine.

MODELLING

Ventilation modelling is a simulation tool that gives users a visual representation of the mine and provides the confidence that calculated results are a close representation of current and future life-of-mine planning. The location of the mine, the depth of cover, the panel lengths, strata rock, and equipment used underground will have a significant impact on the heat load of the mine.

Once the heat loads have been accounted for, the corresponding air cooling units can be added to determine their optimal location. The next section provides insight into the air cooling options. Below is a visual comparison of the placement of the air coolers.

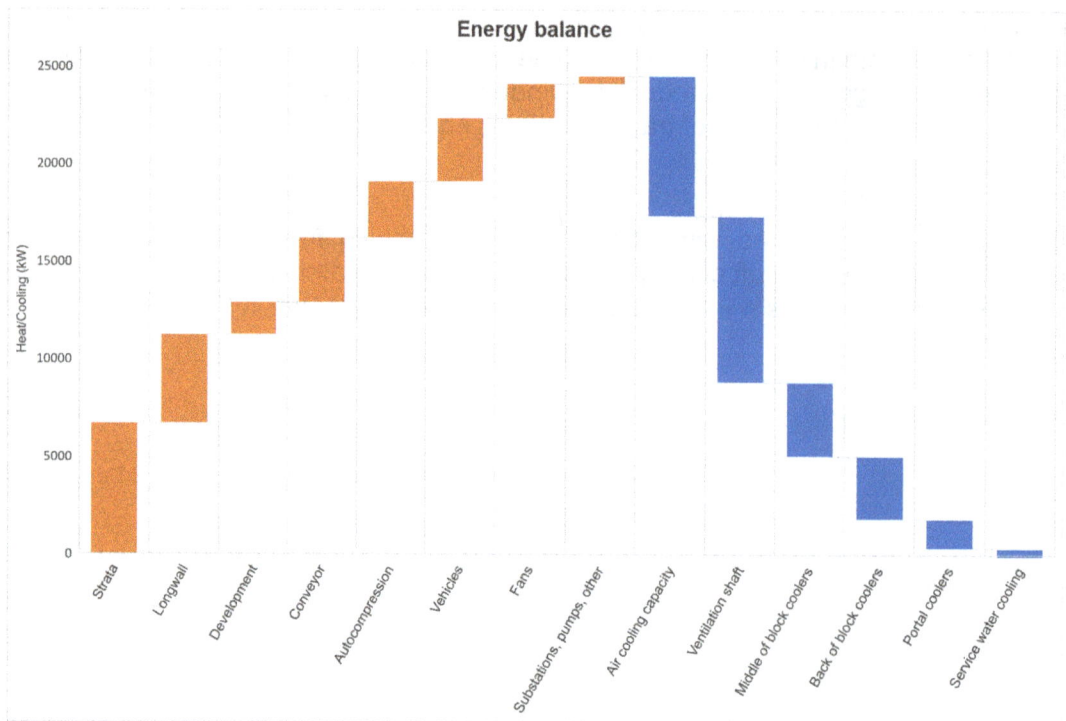

FIG 9 – Heat load and heat sink energy balance.

Modelling has shown that the proximity of the cooling to the affected area is a key factor in determining its net impact at the area of concern. Figure 10 shows positional efficiency in practice with varying face and return temperatures depending on the distance of a 1000 kW cooler from the production face in a gate road. This positional efficiency of applied cooling is due to the loss of chilled air to leakage and non-face air splits (van den Berg and Olsen, 2017). It is also due to strata heat load transfer, which increases at lower intake temperatures due to the greater differential between the strata and air temperature. The introduction of the cooler in the ventilation model increased the strata heat transfer by 300 kW. This is also evident when modelling the mine's heat gain over the year at different ambient temperatures (Figure 11).

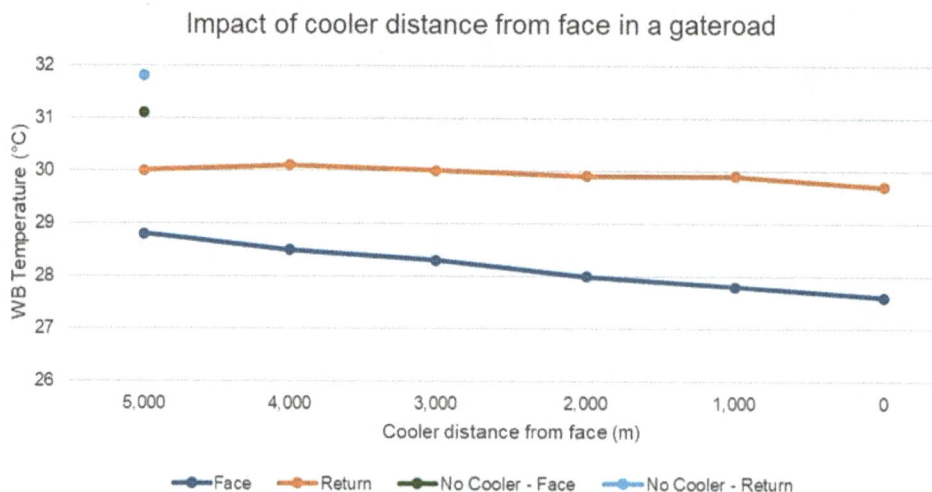

FIG 10 – Cooler distance impact in a gate road.

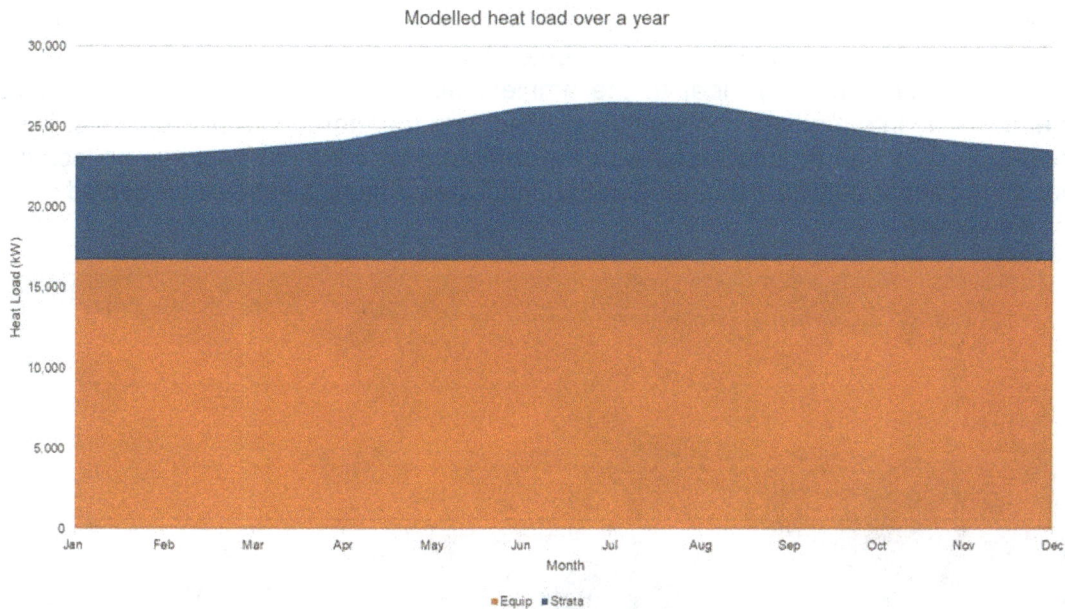

Modelled heat load over a year

FIG 11 – Impact of ambient temperatures on strata heat load.

The optimum location from a cooling efficiency perspective would be to maintain the cooler location close to the area of concern, which is normally the head gate road. The longwall faces retreat with production, and as such, it is not economically feasible to always maintain the cooler location (whether by borehole, shaft or inseam air cooler) close to the face. In practice, installation of a chilled air borehole or shaft is most practical at the gate road intake such that it is available for use for the entire panel life and relocation, therefore a new hole will not be required. The efficiency losses are significantly reduced compared to bulk air cooling at a main intake shaft.

COOLING STRATEGIES

In Central Queensland, more than five coalmines at production rates above 3 Mtpa and a depth of cover over 200 m have introduced cooling ranging between 3 $MW_{(BAC)}$ and 21 $MW_{(BAC)}$. For this Mine, 17 $MW_{(BAC)}$ of mechanical air cooling will be required (excluding thermal losses to the refrigeration chillers). Air-cooled (rental and permanent) and water-cooled permanent refrigeration plant options are available, which will be considered briefly below. Different methods of introducing air cooling into the underground workings were investigated which were via air coolers at the shaft or at the borehole. A comparison was completed on the effectiveness of the placement of the refrigeration installations, the capital and operating costs, the maintainability and longevity of the equipment.

Water-cooled versus air-cooled refrigeration machines

Generally, mine refrigeration systems make use of water-cooled refrigeration machines that reject heat by means of wet Condenser Cooling Towers (CCT). This is the most common method of heat rejection adopted globally in the mining industry due to the lower operating costs associated with water-cooled refrigeration systems. Figure 12 shows a water-cooled refrigeration plant with CCTs (right), refrigeration machines in the shed and Bulk Air Coolers (BAC), (left).

FIG 12 – Water-cooled refrigeration plant (permanent).

An alternative to the use of water-cooled refrigeration machines is air-cooled refrigeration machines that can be procured as permanent capital or rented temporarily. Air-cooled refrigeration machines reject the condenser refrigerant heat to the ambient air via an indirect-contact condenser heat exchanger coils. No intermediate cooling fluid is used, and therefore, no make-up water is required. The finned-tube coils can be integrated with the refrigeration machine package or located in coil banks positioned separately from the refrigeration machines. Figure 13 shows air-cooled chillers with BACs for a coalmine.

FIG 13 – Air-cooled refrigeration plant with BAC coils and air cooled chillers in the background.

A capital and operating total owning cost for 25 years have been completed for this study as shown in Figure 14. The figure shows that rental air-cooled refrigeration plants become more expensive than permanent air-cooled refrigeration plants from 3 years of use. For coalmines, water and air-cooled refrigeration plants are very similar in total owning cost.

Rental vs Permanent Refrigeration System

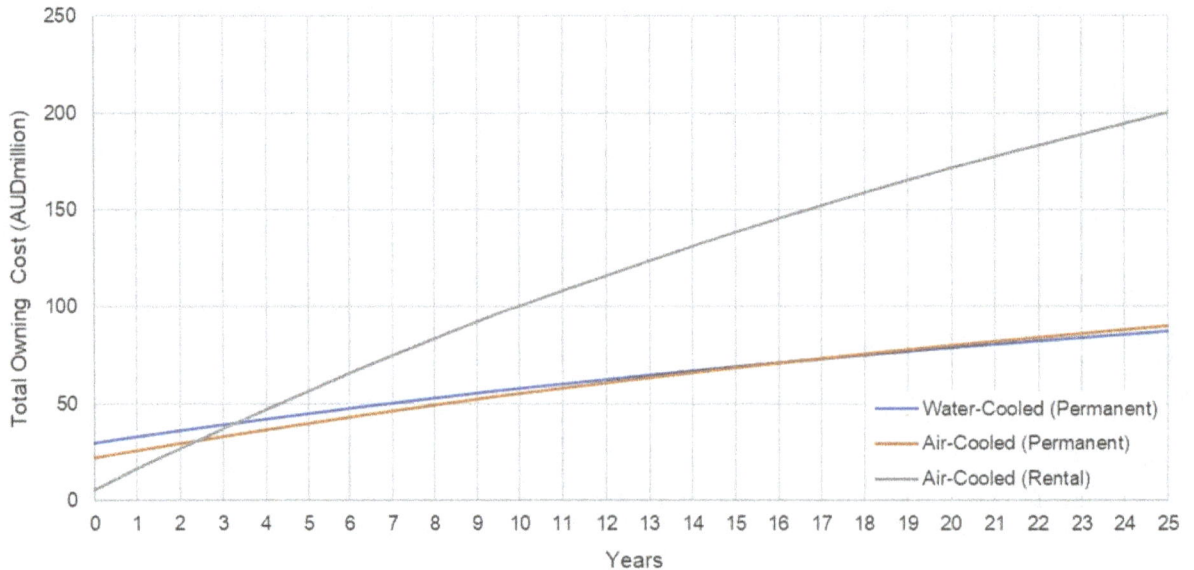

FIG 14 – Air-cooled versus water-cooled break-even analysis.

Air cooling options

The refrigeration plants discussed above provide chilled water to air coolers that are placed near the longwall face to provide the best cooling efficiency throughout the longwall retreating panel. From the modelling completed, utilising borehole clusters located in the gate roads entries to provide chilled air underground maximises positional efficiency for the cooling. Ventilation modelling and site experience have seen that the cooling is less effective at reducing longwall face temperatures when located more than 3 km from the working face. Therefore, strategies to provide optimal air cooling at

MOPpanel and BOPpanel locations were investigated in this study. BAC at the mains will be present but was excluded from the paper.

Air cooling opportunities investigated for the mining panels were:

- surface air coolers via a cluster of boreholes (0.6 mØ)
- surface air coolers via blindbore shafts (3.0 mØ)
- underground air coolers via boreholes (0.6 mØ).

Borehole sizes depend on the site-specific largest hole size possible due to rig constraints (0.6 mØ), whereas blindbore shafts were targeting a minimum feasible diameter given the equipment (3.0 mØ). Boreholes are typically lined to below the tertiary sediments and then unlined to the seam intersection. Blind boreholes will be steel-cased throughout their entire length. Raiseboring is typically not a practical shaft development method due to the depth of tertiary sediments.

For each of the options, a central purchased refrigeration system with three refrigeration chiller modules will be installed, with two running and one on standby. One chiller will serve the back of panel air coolers and one the mid-panel air coolers. For the business case, the refrigeration plant was located outside of the subsidence zone to ensure long-term use of the installation. It would be possible to locate the plant centrally to a mining panel above a chain pillar, designed to be subsided or advanced ahead of the longwall subsidence zone.

Surface air coolers via a cluster of boreholes

Chilled water from the refrigeration plant will be supplied to surface air coolers that will supply cooled air to a cluster of boreholes (4 off) serving two to three longwalls (Figure 15).

Surface Air Cooling System

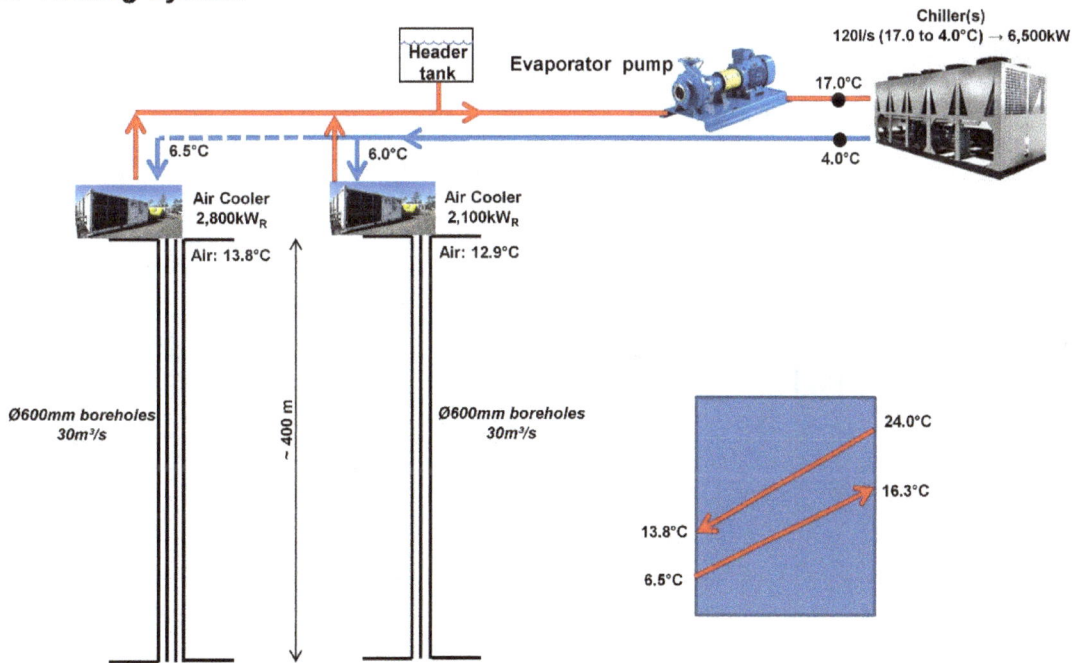

FIG 15 – Surface air coolers via a cluster of boreholes.

Surface air coolers via blindbore shafts

Chilled water from the refrigeration plant will be supplied to surface air coolers that will supply cooled air to a blindbore shaft serving two to three longwalls (Figure 16).

Surface Air Cooling System

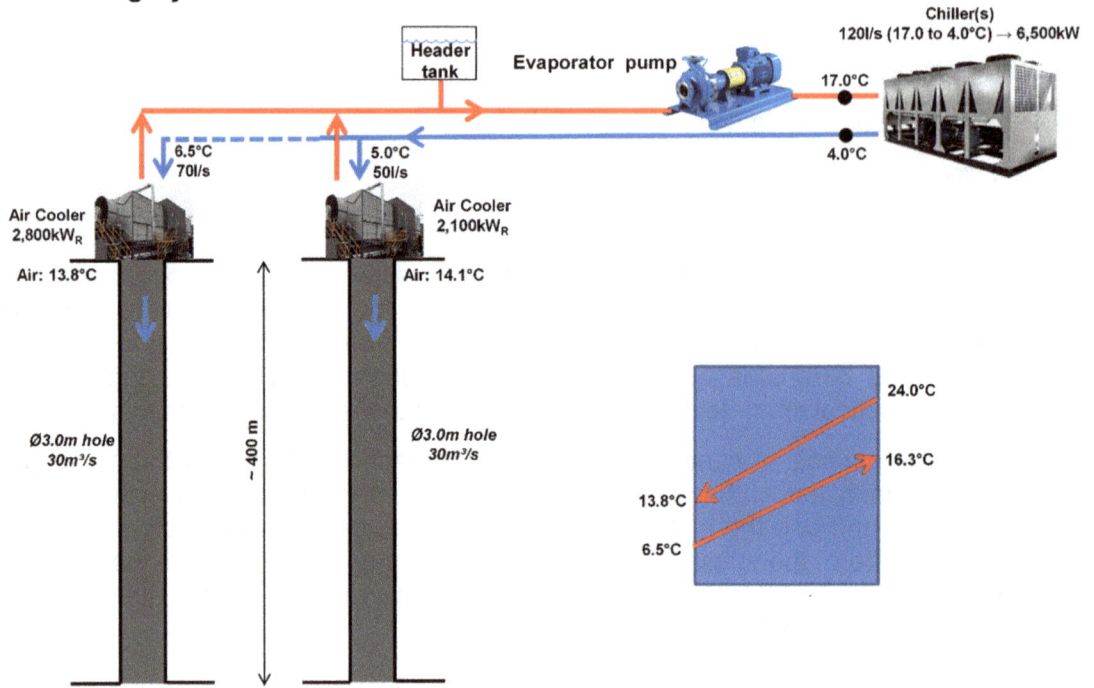

FIG 16 – Surface air cooling system via blindbores.

Underground air coolers via boreholes

Chilled water from the refrigeration plant will be supplied to underground air coolers via chilled water pipes cased in blind boreholes providing chilled water to underground air coolers serving two to three longwalls (Figure 17).

Underground Air-cooling System

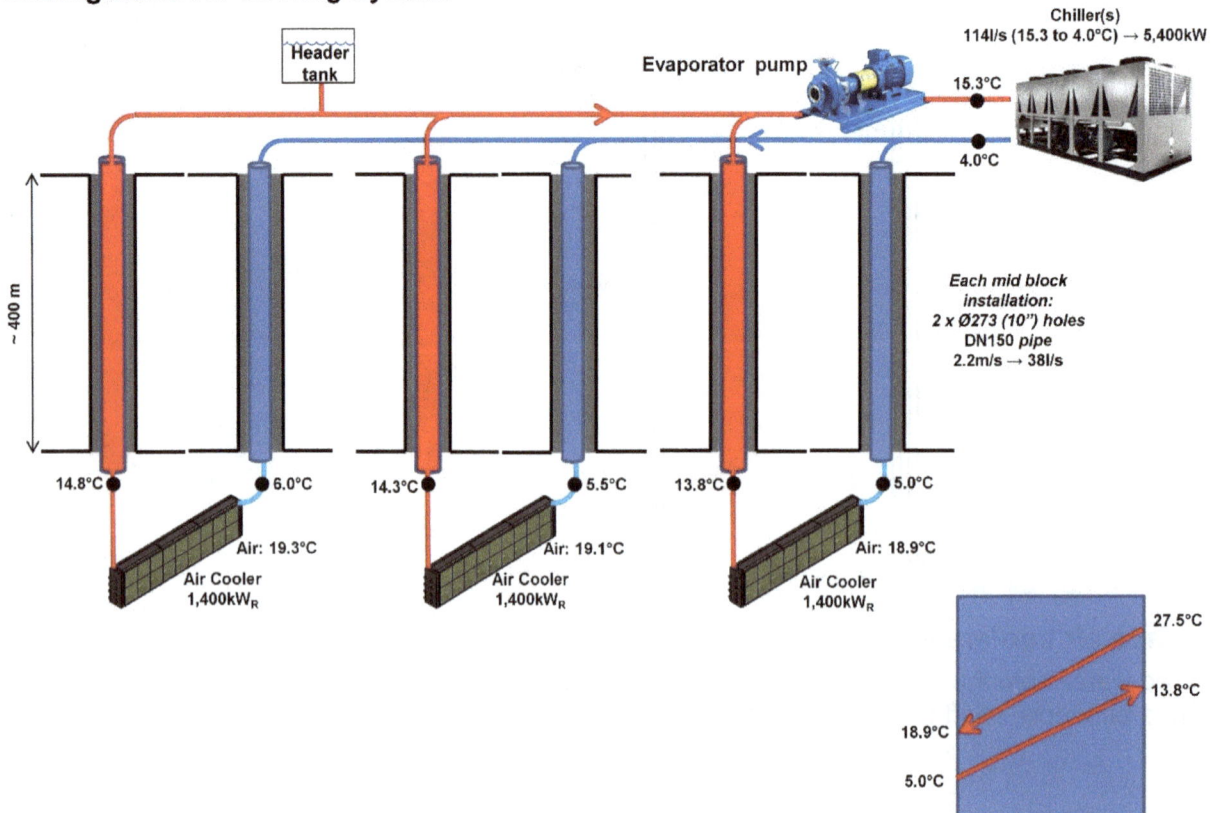

FIG 17 – Underground air cooling system.

It is noted that underground air coolers are not commonly installed in coalmines due to dust accumulation and recirculation that can occur, as well as the requirement for electrical explosion protection measures when installed in a face area. It does, however, have improved positional efficiency compared to the surface air cooling installations. A cost comparison was completed using a decentralised surface refrigeration system and surface mid panel and rear of panel air cooling at clusters of boreholes, against centralised surface refrigeration systems using either surface or underground air cooling (Figure 18).

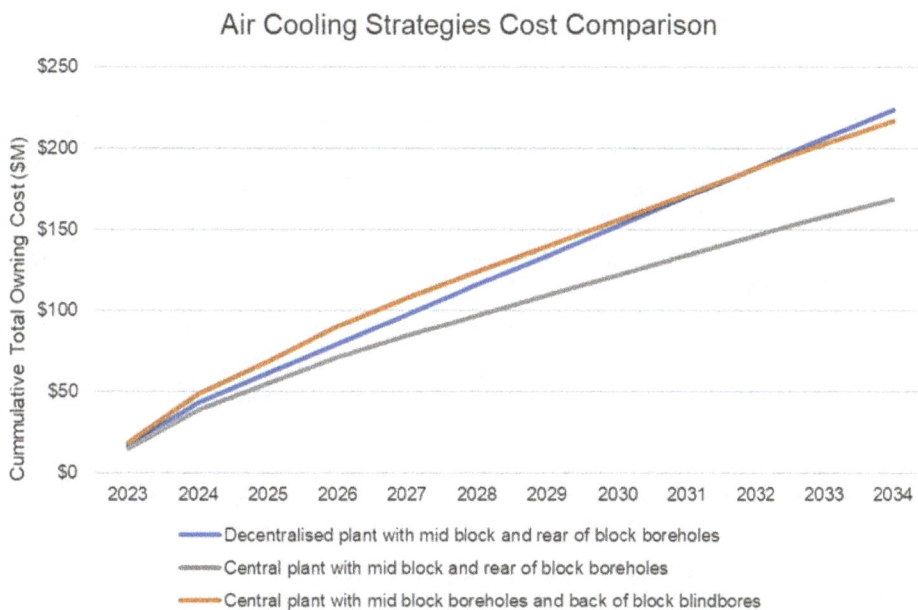

FIG 18 – Air cooling options cost comparison.

Carbon efficiency

A key consideration in the selection of refrigeration equipment and implementation strategy is the carbon efficiency of the proposed system. Whilst it is possible to purchase Australian Carbon Credit Units (ACCU) to offset the carbon emissions or use green electricity, it is preferable to first improve the carbon efficiency of the system. The available options for improving carbon efficiency in a refrigeration system at the Mine include:

- Utilising methane gas currently produced and flared on-site to power gas engines, which power the refrigeration system, or alternatively use the gas for adsorption cooling.

- Purchase or rent efficient coolers. The higher the COP (coefficient of performance), the better the energy (and carbon) efficiency.

- Connect the systems to mains power rather than using diesel-powered generators, which have a lower energy (and carbon) efficiency

- Optimise the positional efficiency of the unit such that the minimum amount of refrigeration is required.

CONCLUSIONS

Positional efficiency of cooling is realised not only due to the loss of chilled air to mine leakage but also due to mining and ventilation equipment and the increase in strata heat transfer due to the greater temperature differential of chilled air. In practice, there is an optimum location for refrigeration, balancing the requirement for continuous advancement of the air coolers or boreholes or shafts versus efficiency trade-offs. Centralised air-cooled chillers with air coolers on a cluster of boreholes have proven to be the best option.

The centralised refrigeration option with the cluster of air coolers at boreholes will have a total owning cost saving of 25 per cent compared to the decentralised plant option as the operating cost of continuous movement of the infrastructure becomes excessive. The centralised refrigeration option with the air cooling system using blindbores provided a total owning cost saving of 3 per cent compared to the decentralised plant option. The latter option provides further benefits in that it can be used for a secondary escape and can introduce more air in the workings if required. Both air cooling strategies reduce the cost of using decentralised refrigeration systems.

ACKNOWLEDGEMENTS

We would like to thank the Project Manager, Ventilation Principle and Engineering Manager at the Mine for allowing us to publish this paper.

REFERENCES

Bauer, R A, 2015. Brief History of Longwall Coal Mining in Illinois, Illinois State Geological Survey [online]. Available from: <https://library.isgs.illinois.edu/Pubs/pdfs/manuscripts/BauerRA-3-2015.pdf> [Accessed: 3 February 2024].

Belle, B and Biffi, M, 2018. Cooling pathways for deep Australian longwall coal mines of the future, *International Journal of Mining Science and Technology*, 28:865–875.

Karwasiecka, M, 1996. Geothermal Prospects of the Upper Silecian Coal Basin (in Polish), *Technika Poszukwan Geologicmych*, 3–4:28–32.

McPherson, M J, 1993. *Subsurface Ventilation and Environmental Engineering* (Chapman Hall: London).

Mitchell, P, 2003. Controlling and Reducing Heat on Longwall Faces, in *Proceedings of the* 2003 *Coal Operators' Conference*, pp 234–245 (The Australasian Institute of Mining and Metallurgy: Melbourne). Available from: <https://ro.uow.edu.au/coal/185>

Van den Berg, L and Olsen, M, 2017. Heat Load Assessment and Mine Cooling Strategies for a Longwall Coal Mine, in *Proceedings of the 2017 Australian Mine Ventilation Conference*, pp 231–236 (The Australasian Institute of Mining and Metallurgy: Melbourne).

Vinay, L S, Bhattacharjee, R M and Ghosh, N, 2022. Underground Coal Mining Methods and Their Impact on Safety, *Natural Hazards – New Insights*, IntechOpen. https://doi.org/10.5772/intechopen.109083

Wang, L T, Liu, Q M and Yang, S, 2020. Research on the Distribution Law and Influencing Factors of Ground Temperature in Xutuan Coal Mine, *Journal of Geoscience and Environment Protection*, 8:88–101. https://doi.org/10.4236/gep.2020.810006

Initial suitability evaluation of some non-conventional cooling methods for the Malmberget Mine

A L Martikainen[1], F K R Klose[2] and A R J van Vuuren[3]

1. Ventilation Specialist, LKAB, Malmberget 98381, Sweden. Email: anu.martikainen@lkab.com
2. Research Engineer, LKAB, Malmberget 98381, Sweden. Email: frederic.klose@lkab.com
3. Principal Engineer, BBE Consulting Australasia, Joondalup WA 6027.
   Email: rjansevanvuuren@bbegroup.com.au

ABSTRACT

In 2022, a scoping study to expand Malmberget Mine was completed. As part of this study, approximate future cooling requirements were defined based on available data and some assumptions to initiate discussion on cooling methods. Conventional cooling methods require large investments and may involve the use of environmentally hazardous chemicals. Non-conventional methods, mostly developed in Canada, were regarded worth studying due to Malmberget Mine's location inside the Arctic circle in Northern Sweden. In addition, LKAB's strong sustainability goals, prioritising environmental safety and strategy aiming for zero carbon dioxide emissions inspired the study.

The evaluated non-conventional cooling methods include ice stopes, snow storage, modular thermal transfer unit (MTTU), heat exchangers, mine water cooling, lake cooling, absorption chilling and propylene glycol storage. The evaluation consisted of an in-depth literature survey and preliminary calculations to analyse the suitability of the methods for Malmberget. The calculations included for example maximum available cooling capacities based on restrictions found in Malmberget and estimated area requirements for potential footprints.

It was found that implementation of the required air cooling estimated for Malmberget would be challenging if relying solely on non-conventional methods. Due to this, combinations of non-conventional technologies as well as combining conventional cooling with non-conventional technologies will be considered in the future. For example, conventional cooling systems could be implemented for the initial cooling requirements whilst non-conventional approaches can be considered for future additional heat loads and when sufficient renewable energy sources become available. Alternatively, as a top-down mining method is expected to be used, one or more of these non-conventional methods could be chosen to provide cooling in the initial stages of the mining expansion, even though they might not provide sufficient cooling capacity as mining expands.

INTRODUCTION

Malmberget Mine is a sublevel caving operation owned by Luossavaara-Kiirunavaara Aktiebolag (LKAB), a Swedish state-owned iron ore mining company, in northern Sweden. The mine has an annual iron ore production of approximately 15 Mtpa, with a capacity of up to 18 Mtpa. The current strategy foresees an expansion which will increase the mine's depth considerably, from the current main level of 1250 down to level 1900 or even deeper. Production capacity increase potential and utilising different mining methods are also being investigated. Currently, mostly hydroelectric power is used at the mine site.

Malmberget Mine is large and complex, with several orebodies and 14 current mining areas. It's divided into deeper Eastern and shallower Western Fields. The industrial area is limited and has significant elevation differences. Currently, the mine has six fresh air stations and 11 exhausts, with a total ventilation capacity of 2300 m³/s. These issues pose challenges for potential heating and cooling solutions. The mine surface including ventilation station locations is shown in Figure 1.

FIG 1 – Overview of the mining area division into the Western and Eastern Fields and the locations of the six intakes (white markers) and 11 exhausts (black markers).

As part of LKAB's transformation process and expansion projects, a scoping study, including ventilation considerations, was completed for the Malmberget Mine in 2022. During the study, cooling requirements were studied based on collected thermal parameter and weather data as well as some assumptions on, for example, machinery, mining method and ventilation capacity. Thermal parameter data indicated a virgin rock temperature of about 15°C on the current main level 1250 and an expected VRT of approximately 26°C at 1900 (Klose, Jones and Martikainen, 2022). A sensitivity evaluation allowed for defining an expected range for cooling requirements, depending on the assumptions. Non-conventional cooling methods, often utilising ice, snow, or water for cooling were considered for studying due to all LKAB mines being located inside the Arctic circle and providing similar weather conditions to those found in Canada where most of these methods have been developed, tested, or used.

CURRENT STATUS AND FUTURE EXPECTATIONS

There is no cooling required for the current production areas in Malmberget Mine. A local cooling system is however located on the main level 1250. Air from a local ventilation shaft is passed through a heat exchanger unit through which mine water with a constant temperature is circulated. A conventional refrigeration unit provides additional cooling to the mine water used. The total cooling capacity of the system is approximately 300 kW. The cooled air is distributed to the office, restaurant and workshop areas via rigid overhead ducts as shown in Figure 2.

However, the cooling that the system can provide is limited. Demand for cooled air at 1250 is increasing due to an ongoing expansion of the office area. During the latest summers, the personnel in the workshop and restaurant areas have reported uncomfortably humid conditions and/or warm temperatures during the summer.

FIG 2 – Local area cooling on level 1250 for workshops, offices and restaurant (left, HEX containers circled), cooled air distribution in workshop (right, circled).

Mining below 1250 is estimated to start in the early 2030s. Mining method, layouts and sequencing are undecided; however, some preliminary options have been evaluated. Based on these, and a prior study on thermal parameters, calculations and preliminary simulations were performed. It was concluded that a significant amount of cooling will be required for the expansion, with estimates ranging from about 30 MWR up to about 100 MWR depending on the options implemented. Based on mining down to level 1900 and implementation of the most likely assumptions, the cooling requirement was estimated at about 60 MWR. This estimate was used as the basis for the study, but requires re-evaluation at the pre-feasibility study stage, when more accurate information is available for planning and the number of assumptions is limited. However, if planning for mining down to level 1900 or below continues, with increased production capacity and thus, higher ventilation requirements, significant cooling requirements can be expected.

EVALUATED NON-CONVENTIONAL COOLING METHODS

Ice stopes

Two ice stope descriptions were found from literature. The ice stope system implemented at the now closed Stobie Mine in Canada, the principal example of this concept being successfully applied, relied on two mined-out stopes located 122 m below surface with a capacity of approximately 170 000 m^3 each. During the winter season, 184 000 m^3 of ice would be produced from unfiltered mine water through spray freezing within the stopes themselves. Overall, this produced an approximate cooling capacity of 7.2 MWR during summer (Allen, Morgan and Rantanen, 2012). Another example of a similar concept are cold stopes found at the Kidd Mine in Canada, capable of supplying 6 MWR, or 8.5 MWR after extension. In this example, ice is created in both the open pit and its connecting airways to the mine and is used to cool air in summer (Hortin and Howes, 2005). Besides the usage of stopes, the seasonal use of old mine workings as a heat exchange area can also reduce cooling demands, as shown for Red Lake Mine by Wallace Jr *et al* (2006).

A local example of the ice stope heat exchange principle can be found in the historical Nautanen Mine, less than 8 km from LKAB Malmberget. The site was visited on 24 July 2021. Ice was found to have naturally accumulated to a thickness of around 10 cm or more near the mine entrance. It can be assumed that the thickness increases over winter as indicated in the bottom right section of Figure 3. Although no readings were taken, temperatures felt noticeably cooler than the approximately 16°C recorded at the time of the visit by the local weather station 'Gällivare A', which is operated by Sveriges Meteorologiska och Hydrologiska Institut (SMHI, Swedish Meteorological and Hydrological Institute).

FIG 3 – Observations at Nautanen Mine.

Another site visit was performed in August and measurements were taken at select locations. Ice was observed, starting a few metres in the entrance, with melted water flowing out of the entrance. The outside temperature during the visit was again approximately 16°C. The measurements were taken at the entrance, entrance area and inside the mine as shown in Figure 3. The recorded dry bulb temperatures and relative humidity values are shown in Table 1. The air velocity in the tunnel both inside and within the entrance area was about 1 m/s.

TABLE 1

Measurement results from Nautanen Mine.

Measurement location	Temperature (°C)	Relative humidity (%)
Entrance	8.0	65
Entrance area	3.0	87
Inside	3.2	90

Previous references (Halim, 2020; Halim, Bolsöy and Klemo, 2020) consider ice stopes unsuitable for the Nordic mines, including Malmberget Mine. However, even if Malmberget Mine doesn't have old stopes available due to the sublevel caving mining method and the continuous growth of the subsidence zone causes instability of the existing open pit, the ice stope method was recognised to have minor potential for further study, inspired by observations made in Nautanen Mine. Two potential cases were identified, namely utilising a purposefully excavated ice stope or utilising existing decommissioned hoisting shafts.

A concept of a purposefully excavated thermal storage, using ice as the storage medium and functioning in much the same way as Stobie's ice stope system has been proposed by Trapani (2019). Here, the suggested ice stope thermal storage utilises a similar concept to Stobie for mine air heating, while the cooling process is modified to include a bulk air cooler, fed by chilled water from the melted ice. To compare the performance of the proposed ice stope thermal storage against the traditional, natural gas heater and chiller refrigeration system, a simulation of ice stope thermal storage was developed. The simulations were run for 8 MW of cooling. To ensure that the mine does not run out of ice, it was stated that monitoring of the ice build-up in winter will be required and if sufficient ice is not available for summer cooling, more ice has to be made with a snow machine or stored from the snow precipitation at surface, which would be possible in Malmberget. Based on the cost calculations presented by Trapani (2019), the comparison was favourable towards this concept instead of traditional cooling. The costs and space requirements for a larger scale system as required for Malmberget will need to be investigated. The stability of such large stope sizes in Malmberget is also questionable, thus the potential further study of this option was decided to be postponed.

The availability of the decommissioned hoisting shafts was evaluated based on the information from mine personnel and Deswik files. The locations of the shafts were also compared with the future subsidence models, giving indication of the length of time these shafts would be usable for potential cooling applications. It was found that the total volume of the currently available unused hoisting shafts within the area of Malmberget Mine adds up to more than 30 000 m³. Some of these shafts are however within areas closed several decades ago and are currently inaccessible. The total volume of the accessible shafts within reasonable distance from the current or potential future ventilation system is approximately 16 000 m³. This volume is significantly less than the volume available for ice creation in the Stobie ice stope system. For any later application, some of the currently used hoisting facilities expected to be replaced in the future were checked for available volume. These added up to about 40 000 m³. If compared to the references, it can be concluded that the currently available volume would most likely provide less than 1 MW of cooling, which won't come even close to the estimated future cooling requirements. Even if the currently used hoisting facilities were included in the system at a later stage, these ice shafts could only be expected to provide a few MWs of cooling in total.

Snow storage

Snow storage is not commonly used for mine cooling, but several other industrial applications exist around the world. The advantages of this cooling method for northern Sweden are the ample availability of the cooling media, being all natural, environmentally friendly and free.

The biggest disadvantages are the large seasonal variance in availability and therefore, the large volumes required and resulting storage areas, as well as the large equipment fleets required to move snow to the storage areas. Additional requirements such as insulation to keep the snow from melting prior to its use and issues caused by the insulation materials such as wood chips or tarpaulins may result in waste residue which in turn may require treatment of the melted water. Equipment, insulation, and water treatment requirements may lead to long payback times even if the cooling media is freely available.

Some examples found in literature include:

- Oslo airport: 2000 m²/MWR (Moe, 2018).
- Sapporo airport: 8000 m²/MWR (Nordell, 2021).
- Canada quarry pit: 1370 m²/MWR (Morofsky, 1981).
- Sundsvall hospital: 5200 m²/MWR (Nordell and Skogsberg, 2007).

Based on the reference information, it can be estimated that for surface storage on a mine site, a minimum of 2000 m²/MWR can be expected to be required.

Evaluation of the industrial area of the Malmberget Mine revealed several challenges for a snow storage application. Firstly, Malmberget Mine is in an area with large elevation differences and steep hills, with ventilation station elevation differences of up to 200 m. Secondly, the available industrial area is very limited and the most accessible parts of it is already in use. Thirdly, the preliminary areas reserved for the transformation project facilities are significant and use up many of the remaining suitable areas. Inaccessible subsidence areas, the pits, take up most of the remaining industrial area. Based on area maps and plans, it was concluded that large enough areas for surface storage of snow for cooling aren't available within reasonable proximity to current and planned ventilation infrastructure.

Modular thermal transfer unit

The Modular Thermal Transfer Unit (MTTU) builds upon the idea of ice stopes and was developed in Canada, based on Stobie ice stope experiences and further research (Li *et al*, 2017; Allen, Morgan and Rantanen, 2012). The idea is to utilise ice buildup and melting for both heating and cooling in a more controlled setting, inside a container unit. The logic behind the MTTU is to heat the cold air in winter by spraying water into the air and generating ice and to cool the hot air in summer by using the ice stored in the unit. A prototype was operated using mine return water and tested under representative weather conditions over a two-year period.

The biggest advantage of this method is the modular concept, which allows for easy adjustment of the system capacity. Unfortunately, several disadvantages were discovered during the research project. It was found that the performance of the unit varied more than expected with the variation of ambient temperature. In addition, maintenance, moisture, and abrasion issues were found during the testing period and cooling was less effective than anticipated, providing only a 4–5°C cooling effect of the ambient air. Due to these issues, development work was never completed, and further development would be required to take the technology from the prototype stage to a fully operational heating and cooling facility.

The prototype was able to provide 1 MW of heating and 12 kWR cooling with three standard 40' containers, covering an area of about 30 m^2. For cooling, this would mean 84 container sets of three for each MWR, which would result in a requirement of approximately 2500 m^2/MWR cooling. This is comparable to the snow storage footprint while also requiring investment in other items including containers, insulation, spray freezing equipment and an extensive control system. It can also be expected to require a high level of maintenance and have an unknown life expectancy, as long-term testing or use was not performed. The shortage of space at Malmberget that would prevent the use of a snow storage system will also render the MTTU system unsuitable.

Heat exchangers

Heat exchangers of different types are utilised by several mining companies around the world for heating. Using heat exchangers for cooling would be very practical and inexpensive in cases where a heat exchanger system is already installed for heating purposes, increasing the utilisation rate of an existing system. For Malmberget, two heat exchanger studies have been completed previously (Linder, 2014; Pöyry, 2017), but only for heating purposes. No installations have been completed despite the positive results, indicating 4-to-10-year payback times depending on the evaluated shaft system.

At Kiruna Mine, some testing of heat exchangers has been completed and a full-scale air-to-air plate heat exchanger was recently commissioned at the mine's largest ventilation station. The purpose of this heat exchanger is primarily to heat the downcast air with the warm exhaust air during the winter period. At the time of procurement, the possibility was also investigated to cool the intake air with the cooler exhaust air during summer. For an estimated exhaust air temperature of 6°C, the cooling effect for ambient air at maximum average summer conditions was estimated at approximately 2.8 MW.

For comparison purposes, Malmberget exhaust shaft temperature measurement results were checked from previous reports. Spot measurements, taken at the top of the shafts from March to June 2014 were found to range between 3.5°C and 10.5°C (depending on the shaft), with an average value of 6.7°C. As heat exchangers are currently being studied in Malmberget for future heating purposes, these numbers were considered promising enough to continue the cooling study as well. In 2023, continuous temperature measurement instruments were installed to record temperature data at several existing exhaust shafts for the purpose of providing more accurate data for future estimates, especially for the coldest and the hottest months when heating and cooling requirements are the highest.

Mine water cooling

As a part of this study and based on the positive experience on the local cooling plant at 1250, possibilities for large-scale cooling by mine water was evaluated. Information on water management, basins and pumping stations was collected. Daily water flows, minimum and maximum values for pumping stations and water flow fluctuations were studied for seven pumping stations and the related water basins. Of these stations, the two located in the Vitåfors shaft area have the highest capacity and flows and therefore could provide the highest cooling capacity. Also, based on the temperature measurements near the basins, the Vitåfors stations were found to be the most likely options to provide reasonable cooling capacities due to the low recorded water temperatures of around 7°C to 10°C. The daily water flow measurements varied considerably, but averaged over a month, the values ranged from about 14 000 m³/d to over 28 000 m³/d for the largest pump station during the summer half of the year, when cooling is expected to be required.

The large volume flows and low water temperature indicates potential, but unfortunately no ventilation infrastructure is located in this area or planned for this area for the expansion currently. Discussions on future use of the Vitåfors hoisting shafts as well as conveyor belt tunnels in the area were had, but the current assumption is that most likely, those won't be available for ventilation use in future but will remain part of the ore handling system. It would be challenging and inefficient to route the cooling media over long distances to the areas where cooling of significant intake airflows will be required. Some of the other concerns for using mine water cooling in Malmberget are water quality issues, underground stability, installation space issues and radon exposure.

The question concerning the use of drilling water, with basin located in the Tingvalskulle area, which is much closer to the planned future ventilation infrastructure, was posed. No capacity, flow, or temperature data was available, but a site visit was performed and more information was collected. The drilling water basin was found to be an old haulage level section including drawpoints, which was converted into a water collection basin by constructing a dam. Calculations showed that the basin has a hypothetical maximum capacity of around 35 000 m³. Currently, temperature data is being collected to enable further evaluation of mine water cooling potential at this location.

From literature, no large scale mine water cooling systems were found. However, smaller installations have been implemented in China. A heating and cooling system using cold mine inflow water was suggested and built in the Jinqu Gold Mine in China, focusing on tertiary cooling with a capacity of about 300 kW (Nie *et al*, 2018). The practical results showed that the temperature on the working face of the target area was maintained at 28°C instead of the previously measured 32.7°C. At the Jisan Coal Mine in China, heat pumps are used in two circuits to cool intake air (Feng *et al*, 2018). A reduction of 2°C from the average air temperature at the working face was reportedly achieved for the month of June at a depth of 700 m below surface.

Lake cooling

As Malmberget is not located near major lakes, piping for considerable distances would result in low cooling system efficiency. Also, the use of lake water for mining activity in large quantities can be expected to result in environmental permit challenges.

Even if viewed favourably, the large quantities of water needed to be transported to and from the lakes over large distances and uneven terrain to multiple shafts can be expected to be problematic. In the example given by Kuyuk *et al* (2019), 550 L/s is required to reach cooling power of about 15 MW for an airflow of 472 m³/s, which is about a fifth of the current airflow capacity of Malmberget Mine. Due to these issues, no further study was conducted for lake cooling.

Absorption chilling

Warm wastewater is expected to be available in the near future in Malmberget from the so called Hybrit process, which is based on an ongoing development project to produce sponge iron using Hydrogen. Preliminary calculations completed for an absorption refrigeration plant utilising Hybrit low temperature wastewater of an estimated 70°C show that potential exists for this cooling method.

Based on the preliminary data received, such as the maximum water flow, the estimated maximum plant capacity would be around 18 MWR. However, losses of approximately 2 MW can be expected. With this solution, water temperature of down to 6°C could be reached for cooling.

Unfortunately, as can be seen from the calculation results, this method will only cover part of the expected cooling requirements. However, high temperature water, also potentially available from the Hybrit process, but in smaller quantities could also be investigated. Using 90°C water can be expected to result in higher system efficiencies and will increase the plant capacity with a decrease in footprint. Due to the relatively reasonable expected capacity and footprint, the evaluation for using absorption chilling in Malmberget continues.

Propylene glycol storage

A brief look into propylene glycol coolth storage showed potential based on the numbers included in the work by Fox (2021) in which the idea was first introduced. Also, the technology is similar to circulation of glycol mixture through heating batteries, which is one of the currently used heating methods in LKAB's mines. Small-scale testing of the system on-site could be possible by reverting the current heating system to provide cooling during the summer. However, the connections to shafts could be challenging and the first calculations indicated a larger footprint than anticipated, which resulted in postponing further work and defining the findings so far inconclusive.

CONCLUSIONS

Out of the evaluated methods, heat exchangers, mine water cooling and absorption chilling showed the most promise for future applications. Further work is required to define the potential of the propylene glycol storage. Snow storage, ice stopes, lake cooling and modular thermal transfer unit were found to be unsuitable for Malmberget based on this high-level evaluation. Due to the preliminary nature of this study, work will continue on these sustainable cooling options to confirm the preliminary results and to move the most promising options to a concept level.

The findings suggest that none of these methods will be able to provide all cooling required for the expected mine life. This will be confirmed in the pre-feasibility stage, with more accurate cooling requirement estimates and more detailed cooling system capacity calculations. The combination of different technologies may provide an opportunity to reach the required cooling capacity, eg using heat exchangers on the surface could be combined with underground mine water cooling.

As Malmberget Mine is large and complex with some current and future ventilation stations located far from the main industrial area, it may be relevant to consider combining conventional and non-conventional cooling. For example, absorption chilling may only be available near the waste heat source in the main process and personnel building zone, which is a few kilometres away from at least two main ventilation stations.

At this time, it's difficult to say when cooling will be required and how much the requirement will be, as the potential mining methods, layouts and sequencing are in very preliminary evaluation stages. If a top-down method is chosen, the cooling requirements will increase over time with the mine getting deeper. In this case, sustainable cooling methods could provide the initial cooling capacity, supported by the later addition of conventional cooling or *vice versa*. The feasibility of the cooling method combinations was defined to warrant further study.

ACKNOWLEDGEMENTS

LKAB is gratefully acknowledged for providing means to conduct these activities and for permitting this paper to be published. We extend our sincere gratitude to our colleagues Robert Hansson, Anders Magnusson, Stina Klemo, and Joakim Jonasson for providing information on previous studies, testing, and existing systems in Malmberget and Kiruna Mines. Thank you to all those people outside of our organisation, with whom we've had good conversations or from whom we have received useful material, clarifications, and additional information. This includes personnel from the BBE offices in Canada, South Africa and Australia.

REFERENCES

Allen, C, Morgan, J and Rantanen, E, 2012. Modular Thermal Transfer Unit (MTTU) - Portable Ice Stope, in Proceedings of the 14th US/North American Mine Symposium.

Feng, X-P, Jia, Z, Liang, H, Wang, Z, Wang, B, Jiang, X, Cao, H and Sun, X, 2018. A full air cooling and heating system based on mine water source, *Applied Thermal Engineering*, 145:610–617. https://doi.org/10.1016/j.applthermaleng.2018.09.047

Fox, J E, 2021. Evaluation of Existing Subsurface Cooling Systems Worldwide and Development of Efficient Cooling Methods and Systems Based on Renewable Energy Sources, Doctoral Thesis, University of Nevada, Reno, United States, 241 p.

Halim, A, 2020. Ventilation and air conditioning challenges in deep Swedish mines, final report, 37 p, Available from: <http://www.diva-portal.se/smash/get/diva2:1503899/FULLTEXT03.pdf>

Halim, A, Bolsöy, T and Klemo, S, 2020. An overview of the Nordic mine ventilation system, *CIM Journal*, 11(2):111–119.

Hortin, K and Howes, M J, 2005. Surface Cooling at Kidd Creek Mine, in *Proceedings of the Eighth International Mine Ventilation Congress*, pp 55–63 (The Australasian Institute of Mining and Metallurgy: Melbourne).

Klose, F K R, Jones, T H and Martikainen, A M, 2022. Geothermal gradient determination for ventilation and air conditioning modelling at Malmberget mine, in *Proceedings of the Australian Mine Ventilation Conference*, pp 66–80 (The Australasian Institute of Mining and Metallurgy: Melbourne).

Kuyuk, A F, Ghoreishi-Madiseh, S A, Sasmito, A P and Hassani, F P, 2019. Designing a Large-Scale Lake Cooling System for an Ultra-Deep Mine: Canadian Case Study, *Energies,* 12(5):811. https://doi.org/10.3390/en12050811

Li, G, Butler, K, Hardcastle, S, Morgan, J and Allen, C, 2017. Testing and Evaluation of a Modular Thermal Transfer Unit for Underground Mine Air Conditioning, in *Proceedings of the 16th North American Mine Ventilation Symposium*, pp 4-1–4-8.

Linder, K, 2014. Återvinning av gruvventilation Malmberget (in Swedish), Internal report, LKAB, Sweden.

Moe, J M, 2018. Using stored snow as cooling at Oslo Airport, Norway, in *Proceedings of the Institution of Civil Engineers – Civil Engineering*, 171 (5):11–16. https://doi.org/10.1680/jcien.17.00041

Morofsky, E, 1981. Project Snowbowl, Public Works of Canada (PWC), Contract EN, 280-0-3650.

Nie, X, Wei, X, Li, X and Lu, C, 2018. Heat Treatment and Ventilation Optimization in a Deep Mine, *Advances in Civil Engineering*, 2018:1529490. https://doi.org/10.1155/2018/1529490

Nordell, B and Skogsberg, K, 2007. The Sundsvall Snow Storage - Six Years of Operation, in *Thermal Energy Storage for Sustainable Energy Consumption* (NATO Science Series), pp 349–366 (Springer Netherlands: Dordrecht).

Nordell, B, 2021. Using ice and snow in thermal energy storage systems, in *Advances in Thermal Energy Storage Systems*, pp 207–220 (Elsevier).

Pöyry, 2017. Förprojektering: Återvinning Evakueringsvärme Gruvventilation, LKAB Malmberget, Förbättrad Ventilation MUJ (in Swedish), Consultant's Final report, 11 p.

Trapani, K, 2019. Techno-economic of an ice stope thermal storage for mine heating and cooling in sub-arctic climates, in *Proceedings of the 17th North American Mine Ventilation Symposium*, 7 p.

Wallace Jr, K G, Tessier, M, Pahkala, M and Sletmoen, L, 2006. Optimization of the Red Lake Mine ventilation system, in *Proceedings of the 11th US/North American Mine Ventilation Symposium* (eds: J M Mutmansky and R V Ramani), pp 61–66.

Onaping depth project – underground refrigeration plant update

C McGuire, K Boyd, T Mehedi and E Pilkington

1. Senior Ventilation Engineer, Hatch Ltd., Mississauga ON, Canada.
   Email: chris.mcguire@hatch.com
2. Ventilation EIT, Hatch Ltd., Mississauga ON, Canada. Email: boydk@hatch.com
3. Ventilation EIT, Hatch Ltd., Sudbury ON, Canada. Email: tanveer.mehedi@hatch.com
4. Mechanical Infrastructure, Mining Projects, Sudbury Integrated Nickel Operations A Glencore
   Company, Sudbury ON, Canada. Email: evan.pilkington@glencore.ca

ABSTRACT

The Onaping Depth Project is a nickel/copper underground mine project currently in execution by Glencore's Sudbury Integrated Nickel Operations. The Onaping Depth orebodies will be accessed via the existing Craig Mine workings after completion of sinking a winze from 1200 to 2635 m below surface. Due to the final depth of the workplace, a 16 MW(R) underground refrigeration plant will be operated at a depth of 1915 m below surface to provide conditioned air to the workplace throughout the project development and life-of-mine production phases. Detailed engineering of the plant was completed by Hatch in 2020, with construction finished in 2022 and commissioning completed in early 2023. This paper will provide an overview of the design, construction, and commissioning of the underground refrigeration plant, including lessons learned in the contracting and construction methodologies. Initial operating performance at low cooling demand will also be presented.

INTRODUCTION

Glencore's Sudbury Integrated Nickel Operations (Sudbury INO) are in the progress of executing the Onaping Depth Project, an ultra-deep nickel/copper mining complex in Sudbury, Ontario, Canada. Hatch and Glencore have been partners on engineering and development of the project since 2015, commencing with pre-feasibility and feasibility studies and continuing into detailed engineering, project and construction management through the execution phase of the project. Hatch completed the initial trade-off studies for the refrigeration plant, the process design and discipline definition of the plant in the feasibility study and continued through execution with the detailed design of the plant.

Mine air cooling has been included in project plans throughout the study phases due to the depth of the orebody (currently 2745 m below surface at the deepest level). Virgin rock temperatures are in excess of 60°C in the deepest levels. The mechanised mining mobile equipment fleet will be entirely battery-electric, allowing for an approximately 35–40 per cent reduction in ventilation volume compared to mining the deposit with a traditional diesel fleet (Wisniewski, 2017).

Due to the orebody depth, coupled with the use of a battery-electric mobile equipment fleet, auto compression heat dominates the underground heat loads, representing 45 per cent of heat added in the fresh air pathway. Through a series of studies and trade-offs, an underground refrigeration plant was identified as the preferred technology to ensure safe and productive working conditions in the deep zone. This location maximises the positional efficiency of the cooling plant and allows for further optimisation of the ventilation quantity, reducing sizes required for many excavations.

Construction and commissioning of the plant has been completed. Major mechanical equipment was purchased in 2017 and delivered to site in 2019–2020. Construction commenced at the 1915L (nominal metres below surface) complex in early 2021 and commissioning of the refrigeration plant was substantially completed in Q1 2023. The plant has been designed with the challenges of underground refrigeration in mind. A unique layout was developed with a centrally-located plant room between the two open spray chambers, minimising excavation requirements and reducing large bore pipe lengths, all while respecting the geotechnical constraints of designing large excavations at a depth of 1915 m below surface.

PROJECT BACKGROUND

The Onaping Depth deposit has been known since the 1990s, when exploration drilling from the existing Craig-Onaping mine complex identified the main orebody. The 18.4 Mt deposit resides in the vicinity of 2.6 km below surface and will be mined at a target production rate of 1.2 million tonnes per annum with a design life of 20 years. The refrigeration plant will operate for the complete life-of-mine in addition to two to three years of the project period required to access the orebody.

Multiple attempts to develop a mine design which would be economical with conventional diesel mechanised mining methods were unsuccessful. Mechanical refrigeration is necessary due to drill-indicated virgin rock temperatures in excess of 50°C (over 60°C when extrapolated to ultimate orebody depth) and auto-compression across more than 2700 m of vertical depth. These factors resulted in excessive workplace temperature conditions in the absence of cooling intervention. In addition, capital and operating costs associated with the volume of air required to support a diesel fleet were found to be prohibitive.

In 2016, the Onaping Depth Project carried out a feasibility study on the basis of an entirely battery-electric fleet for the deep workings. This allowed for a significant reduction in the ventilation volume (approximately 40 per cent; Wisniewski, 2017), and therefore both the ventilation and refrigeration power consumption (approximately 45 per cent; Wisniewski, 2017). In addition, major capital cost savings were realised from a reduction in excavation sizes and development quantities in the primary airways, in addition to the reduction in ventilation and refrigeration infrastructure cost. The project was given approval to commence execution in late 2016, with the first development blasts fired in January 2017.

Lateral development for mine access construction within the first five years of the project timeline was executed with diesel equipment. All of this development was accessible from the existing ramp system in Craig Mine and is supported by existing diesel distribution and fuelling infrastructure. The lower workings of the mine, known as 'the Depth', will only be accessible via a new internal winze excavated from 1150L to 2635L. All mobile equipment in use in the Depth for development or production will be battery-electric vehicles (BEVs), with the exception of some temporary contractor equipment to support early off-shaft development during construction and commissioning of BEV equipment. Temporary workshops have been established near-shaft at each of the three deep shaft stations to allow for battery-electric fleet mobilisation.

Changeover of the winze from a sinking arrangement to its permanent configuration is underway and expected to be completed in Q2 2024, at which time lateral development will re-commence on the project critical path with BEVs only.

REFRIGERATION PLANT DESIGN

During early engineering phases, Hatch, working with Glencore, undertook detailed assessments and trade-offs to determine the optimal cooling plant location. Overall, locating the cooling plant underground at 1915L was deemed the most favourable from a technical and economic standpoint when compared with a surface refrigeration installation. This is largely attributed to the improved positional efficiency of an underground plant location, as a colder temperature can be delivered to the mining zone due to the reduction in auto compression heat loads experienced downstream of the plant (approximately 800 m of elevation change from the 1915L cooling plant versus 2700 m for a surface plant).

The work identified that powerful synergies arise due to the combination of underground cooling with the reduction in primary ventilation volume attainable when regulatory requirements for diesel exhaust dilution do not exist. The reduced primary ventilation volume, when coupled with the lower air temperature delivered to the workplace, results in reductions in both capital and operating costs when compared to surface plants, and is explored in detail by Brown *et al* (2018).

Key process conditions forming the basis of the Onaping Depth underground cooling plant design are summarised in Table 1.

TABLE 1

Summary of air side process parameters for refrigeration plant design.

Parameter	Units	Value
Air Flow (Volumetric)	m³/s	229
Mass Flow (Dry Air)	kg/s	308
Inlet Air Temperature	°C wb/db	27.6/30.5
Chilled Air Temperature	°C wb/db	12.0/12.0
Barometric Pressure	kPa	120.5
Nominal Air Cooling Duty	kW$_T$	14 600
Air Temperature Available for Heat Rejection	°C wb/db	23.1/26.1
Maximum Air Flow Available for Heat Rejection	m³/s	203

Note that these parameters are specific to the life-of-mine (maximum) design condition for air flow and refrigeration duty; the plant will be operated at a series of interim conditions with reduced air flow and a target chilled air temperature of 10°C during the development and construction phases of the project.

The cooling plant consists of the following main components:

- Bulk air cooler (BAC) for air-cooling.

- Refrigeration plant room (Plant Room) housing the refrigeration machines to provide chilled water to the BAC.

- Condenser spray chamber (CSC) for system heat rejection.

- Water pumps and water treatment for the recirculating water circuits between the refrigeration machines and the spray chambers.

Water treatment systems consist of side-stream filters and chemical dosing equipment including corrosion inhibitor, biocide to mitigate legionella risk, and alkali for pH control, with identical chemical dosing systems installed on both circuits. The CSC circuit includes a side stream filter due to exposure of the spray chamber to dusty mine exhaust. The BAC currently does not include side stream filtration as it is located in the fresh air stream, however layout provision for an identical, future filter has been included.

Both the BAC and CSC consist of two-stage, horizontal spray chambers with sloped basin floors to minimise slimes build-up and facilitate cleaning. Nominal dimensions of each spray chamber are 8.0 m wide, with a minimum height of 7.0 m inside shotcrete ground support at either end of the chambers and a maximum height of 8.0 m high at the low point in the sloped floor.

The Plant Room is 9.0 m wide × 8.2 m high and contains three R-134a single-stage centrifugal compressor refrigeration machines (Johnson Controls Model YK), arranged in a series-counterflow configuration. This configuration was selected by Hatch in the feasibility study, allowing the centrifugal chillers to achieve the high lift required of the underground installation (ie mine exhaust air is both finite and typically hotter than ambient air for heat rejection), while not requiring two-stage compressors on the chillers. Two-stage compressors are more expensive and increase the equipment size substantially, driving excavation requirements out of practical limits at this depth. Therefore, single stage compressors were preferred and the piping was optimised to achieve this. The design philosophy from feasibility has been retained through detailed engineering and execution.

A summary of the water side process conditions is provided in Table 2.

TABLE 2

Summary of water side process parameters.

Parameter	Units	Value
Refrigeration Machine Duty (Installed Capacity)	kW(R)	16 200
Refrigeration Machine Duty (Operating Duty)	kW(R)	15 400
BAC Water Flow	L/s	400
BAC Water Inlet Temperature	°C	7.5
BAC Return Water Temperature	°C	16.5
CSC Water Flow	L/s	550
CSC Water Inlet Temperature	°C	44.1
CSC Return Water Temperature	°C	35.8
Heat Rejection Duty	kW(R)	19 600

A 3D rendering of the various system components on 1915L is shown in Figure 1. The plant layout was designed with the following key design aspects in mind:

- Plant room centrally located to minimise pump and piping requirements between refrigeration machines and spray chambers.

- Chillers installed in series counterflow arrangement due to high lift and desire to avoid two-stage centrifugal compressors.

- Sloped spray chamber basin floors to minimise excavation quantity and ease fines clean-out.

- Equipment access and maintenance requirements including overhead crane.

- Geomechanical considerations for drift size and spacing.

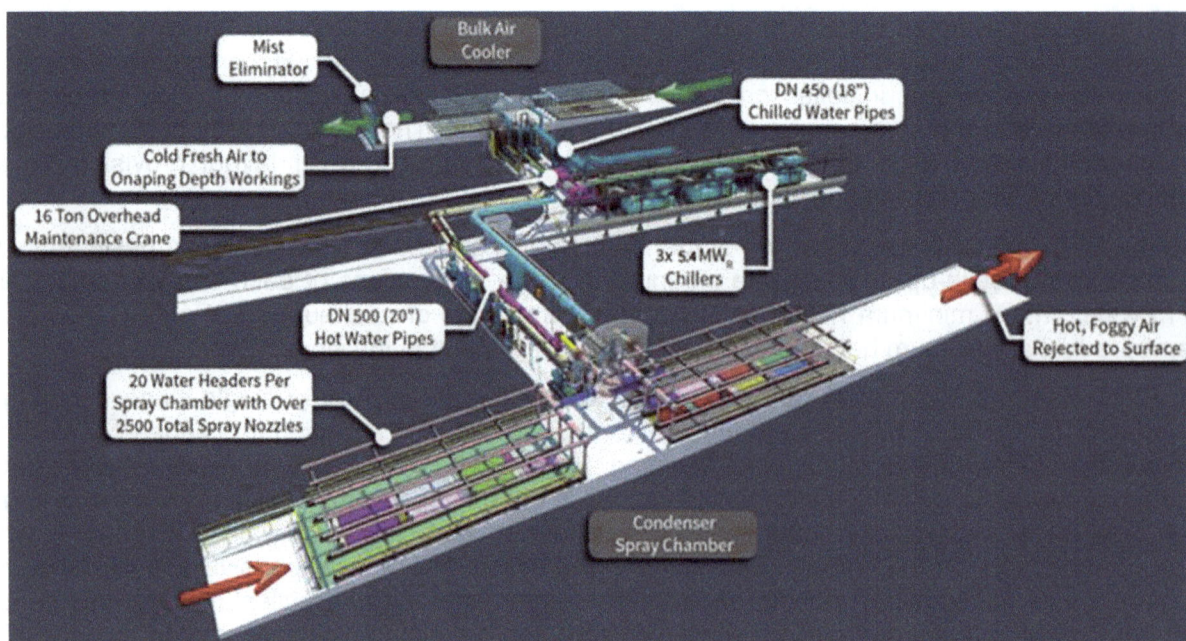

FIG 1 – 3D model overview of Onaping depth 1915L refrigeration plant.

The sloped floor of the BAC can be seen in Figure 2; the CSC is a similar design. In addition to minimising excavation volumes, the sloped concrete basins will facilitate wash-down of collected dust residues during plant outages as needed. Stainless steel work platforms are provided above the water level to allow for inspection and maintenance of the spray pipes without having to drain the full system volume from the basin. The depressed centre section visible is local to the pump suction (see section view in Figure 3). The spray chamber floor is excavated approximately two metres above the adjacent pump room floor in order to provide a suitable minimum suction head of 1.5 m at spray chamber basin low water level.

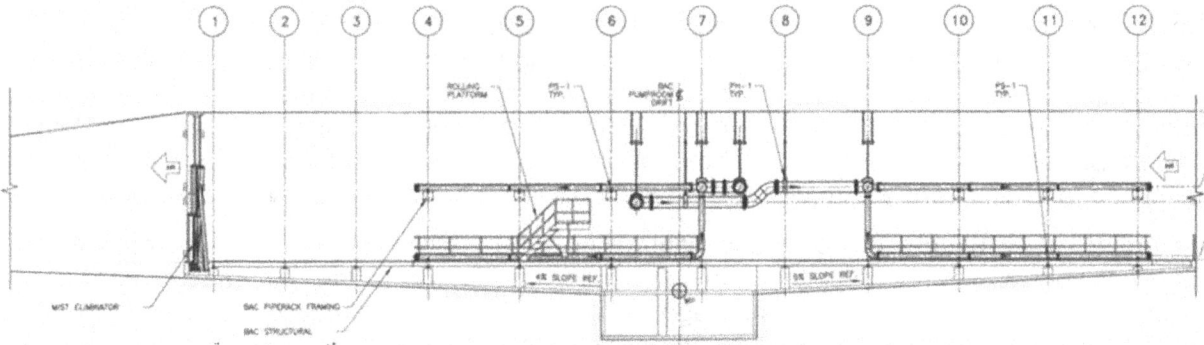

FIG 2 – Long section view of BAC spray chamber.

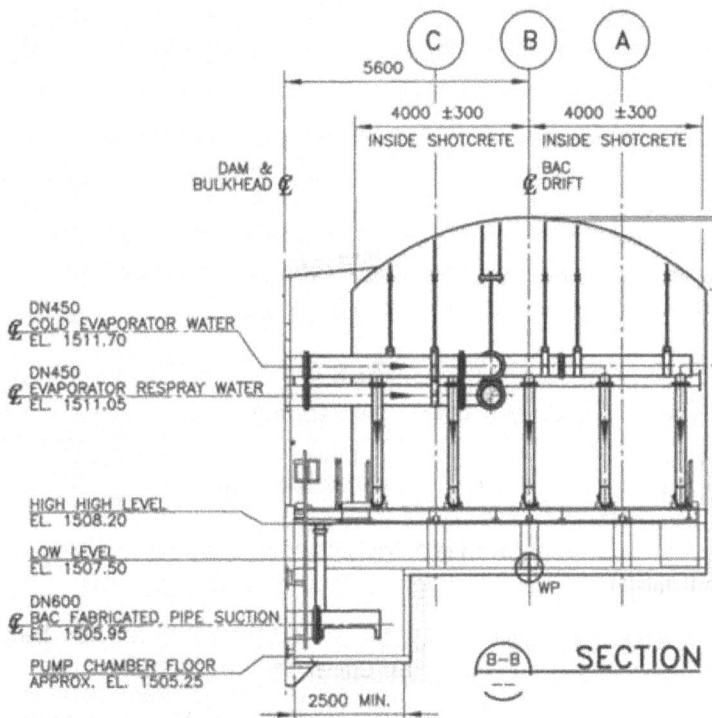

FIG 3 – Section view of sump and pump suction.

Design of the plant prioritised robust, reliable, and safe performance in an underground mining environment. For example, the refrigeration machines were specified with high heat exchanger fouling factors reflective of conditions in underground mining operations. Heat exchanger tubes were specified to be thicker than standard and constructed of more robust copper-nickel alloy material for long-term protection against both the corrosion and the erosion expected in an underground environment with the expected water quality. The chiller units were also specified to use a Class A1 refrigerant with low toxicity and no flame propagation to minimise risk to workers in the event of a leak in the confined underground environment. In addition, the airflow in the plant room is ventilated directly to exhaust, reducing personnel risk from refrigerant leaks, should they occur. Finally, the water treatment system, as described above with chemical treatment and side stream filtration, was configured to improve the operability of the plant and minimise personnel risk to biological contaminants such as legionella (Legionnaire's Disease).

The refrigeration plant was standardised to operate at 4160 V, with motors for all chillers, pumps, and the nearby primary ventilation fans and dewatering pumps all operating at this specification.

Power distribution to the level is accomplished by a pair of 25 kV shaft feeders from surface to the main substation near the entrance to the Plant Room, at which point it is stepped down to 4160 V and 600 V for ancillary loads.

CONTRACTING STRATEGY

Construction of underground refrigeration plants extends beyond the typical skill set of most underground mining trade workforces. In particular, fabrication and construction of custom, large bore pipes (DN450 and DN500) and work with refrigeration equipment are unusual in the underground environment. However, some of the construction activities such as construction of concrete bulkheads, drilling of rock anchors for pipe supports and structural steel tie-backs etc are well-suited to mining contractors with experience in drilling and underground construction.

The project team sought to develop a contracting model which would allow for both the high degree of specialisation associated with the cooling systems and the expertise in underground construction. For the initial tender package, a lump sum contract model was selected with an approved bidders list consisting of underground mining contractors. Contract language was included which ensured that dedicated piping fabricators and tradespeople would be engaged from more traditional 'surface contractors'. The package structure also intended to put the onus on the successful contractor to receive personnel and materials at the shaft station(s) underground and self-provide transport to the refrigeration plant workplace – approximately 3.5 km down ramp from the lowest shaft station.

Comprehensive tender evaluations revealed that the contract arrangement resulted in a detrimental outcome primarily due to subcontracting markups included in all bid prices. In response, an alternate strategy was proposed in which the total scope was divided into multiple subcontracts, each consisting of a single specialised trade. Bidders lists were crafted such that some contracts were only bid to mining contractors, while others were only bid to specialist surface contractors. The division of scope is listed in Table 3.

TABLE 3

Construction contract summary.

Contract	Scope included	Contractor type
Messenger Wire (Catenary) Installation	Installation of electrical messenger/catenary system including grounding and tensioning	Mining
Concrete	Slab on grade, equipment foundations, formed basins, dam walls, shotcrete ventilation bulkheads	Mining or Surface Civil
Structural/Mechanical/ Piping	Structural steel, mechanical equipment (chillers, pumps, auxiliary ventilation), and refrigeration plant piping	Surface Mechanical
Electrical and Instrumentation	Substation construction, cable pulls, instrumentation installation, communications	Surface Electrical

In addition to these four construction contracts, an Underground Logistics contractor was brought onto site who took responsibility for delivery of all construction materials and personnel from the shaft station to the contract workplaces. Thus, the single lump-sum contract became five specialised contracts which were all tendered competitively. Although an increase in project management and contracts administration oversight was required, the reduction in construction cost was approximately 27 per cent.

A downside of the approach of using multiple contractors is the difficulty of managing handover of completed workplaces to subsequent contractors. Delays in early contracts (eg concrete) introduce the risk of delaying mobilisation of subsequent contractors, which can produce financial penalties and/or loss of key personnel. Careful construction management is required which can respond to ongoing disputes, re-prioritise work to allow for handover of isolated workplaces, and generally facilitate discussion and co-operation between multiple contractors sharing the same workplaces.

CONSTRUCTION

As of February 2024, the refrigeration plant had been constructed and in operation for approximately nine months, with commissioning substantially complete in May 2023. Photos below show a construction milestone at the completion of the chiller installation prior to pipe installation, with the overhead bridge crane used for installation and maintenance visible above (Figure 4), and a view of the complete plant room following piping tie-in and insulation (Figure 5).

FIG 4 – Refrigeration plant room following completion of chiller installations.

FIG 5 – View of complete plant room including insulated evaporator water piping.

The BAC and CSC follow similar design and construction methodologies to each other, with lower spray headers supported from the platform structure and upper spray headers supported on beams spanning the width of the spray chamber. This is visible in Figure 6.

FIG 6 – View of CSC pipes, pipe supports and access platform following construction.

PERFORMANCE ASSESSMENT

The cooling plant has been in operation for approximately nine months as of February 2024, spanning one summer season. The mine ventilation volume is at a significantly reduced condition in comparison to the final design conditions, as shown in Table 4.

TABLE 4

Comparison of BAC inlet conditions versus Life-of-Mine.

Case	Current Operation (Summer 2023)	Interim Project Operation (Design Maximum Turndown)	Life-of-Mine Design Case
Volumetric Air Flow (m³/s)	91	94	229
Mass Flow of Dry Air (kg/s)	125	128	308
Wet Bulb Temperature (°C)	23.6	26.3	27.6
Target Chilled Air Temperature (°C)	7.5	10.0	12.0

The current turndown case has a smaller refrigeration duty than the lowest case considered in the initial design, due to a change in lateral development methodology which required work concurrent with shaft sinking. Due to this low refrigeration load, only one of the three chillers is required to be run to satisfy the cooling demand. The Onaping Depth chillers are fixed speed machines, and chiller duty is modulated by adjustment of the pre-rotation vanes (PRVs) at the compressor inlet.

Closing the PRVs too much was found to create compressor stall cycles which are not favourable for long-term operation of the plant. Therefore, the air temperature set point has been reduced to 7.5°C (versus the expected interim design condition of 10°C) to increase the cooling demand and allow the chiller's PRVs to operate in a more stable range, safely away from the stall zone.

Table 5 shows a summary of measured operating conditions on a typical day in July 2023, near the peak loading observed on the plant to date. Of note is that the BAC approach temperature – defined as the differential between the entering chilled water temperature and the leaving air temperature – is very low at approximately 0.5°C. This is not typical of the final design, and is expected to increase as the air velocity through the spray chambers increases with increasing mine ventilation capacity over time. This tight approach temperature allows for the air temperature set point of 7.5°C to be achieved without excessively low chilled water temperatures, which would approach low-pressure conditions in the evaporator refrigerant circuit.

TABLE 5

Typical cooling plant performance in July 2023.

Parameter	Units	BAC	CSC
Mass Flow Dry Air	kg/s	125	103
Inlet Air Flow (Volumetric)	m³/s	91	73
Inlet Air Temperature	°C wb/db	23.6/27.3	18.4/18.6
Outlet Air Temperature	°C wb/db	7.6/7.6	36.7/36.7
Water Flow	kg/s	403	574
Inlet Water Temperature	°C	7.0	35.3
Return Water Temperature	°C	10.1	32.0
Nominal Spray Chamber Duty	kW$_T$	5230	7950

One notable observation from the performance conditions shown in Table 5 is that the CSC inlet air conditions are quite cold in this temporary ventilation condition. This is due to the need to bypass some chilled air from the BAC directly to the CSC in order to maintain stable operation of the cooling plant. Chilled air required for shaft sinking and early off-shaft development is limited to approximately 30 m³/s by the ventilation duct passing through the sinking Galloway and winze equipping stages. The cooling plant is not able to turn down to achieve the low refrigeration duty required if only the required air volume is cooled; this limitation occurs both due to the primary ventilation fans' turndown limit and the chillers' modulation limit with PRVs as mentioned above. The current primary ventilation volume is similar to that of the designed maximum turndown case (Table 4), but only approximately 35 per cent of the chilled air volume is in use in the active workplaces.

Following completion of the shaft sinking phase and removal of the sinking stages, the complete primary ventilation volume can be delivered to the deep workplaces, allowing the plant to operate as per the designed minimum turndown condition. Subsequent increases in refrigeration duty are tied to ventilation raise breakthroughs and construction milestones over the course of the next two to three years. The chilled air target temperature of 10.0°C will remain constant with each of the step changes in ventilation volume, with the increased refrigeration duty tied to the increased mass of air to be cooled at each step. Only after holing of the final return air raise from the main orebody (target in 2027) will the target air temperature be adjusted up to the final design value of 12.0°C.

CONCLUSIONS

For the Onaping Depth Project, replacement of diesel equipment with BEV's in the mining plan allowed reduced airflow and heat loads in the mine.

An underground cooling plant was selected to provide bulk air cooling in close proximity to the deep mining operations. This was shown to be more economical than a surface cooling plant for the specific circumstances of the Onaping Depth mine design.

Equipment selection included single-stage, water cooled, centrifugal mechanical chillers with synthetic refrigerant and shell and tube heat exchangers. Chillers were arranged in series counterflow water circuit arrangement to minimise lift required on each machine and balance loading.

Both bulk air cooling and condenser heat rejection were designed to use direct contact open spray chambers adjacent to the refrigeration plant room.

Contracting strategy for the underground construction works was found to be sub-economic using a single lump sum contract, and cost savings were achieved by dividing work to specialist contractors and intensively coordinating to minimise work clashes between these packages.

Despite needing to operate at a higher turndown than envisioned, the plant operation has been successful to date, meeting the project needs. The overall approach, process design and equipment selection from the feasibility study was carried through the detailed design phase and has resulted in a flexible, robust plant to date.

ACKNOWLEDGEMENTS

Prepared with permission from Sudbury Integrated Nickel Operations: A Glencore Company.

A portion of this paper was presented at the SME 19th North American Mine Ventilation Symposium held in Rapid City, SD, USA in 2023 (Durieux et al, 2023).

REFERENCES

Brown, M, Guse, T, Arsenault, S and Rogers, B, 2018. Effects of an Electric Fleet on Mine Ventilation and Refrigeration with a Comparison to an Equivalent Diesel Mine, paper presented at the Roomvent and Ventilation Conference, Espoo, Finland.

Wisniewski, S, 2017. Onaping Depth Project, Presented at CIM Sudbury General Meeting, Sudbury, Canada.

Durieux, D, McGuire, C, Mehedi, T and Witow, D, 2023. Onaping Depth Project Underground Refrigeration Plant Update. Paper presented at the 19th North American Mine Ventilation Symposium, Rapid City, USA.

Assessment of heat stress and dehydration in underground coalmines using machine learning (ML) techniques to improve occupational health and miners productivity

V Sakinala[1], P S Paul[2], S Anand[3] and R M Bhattacharjee[4]

1. Research Scholar, Department of Mining Engineering, Indian Institute of Technology (ISM), Dhanbad, Jharkhand, India. Email: vikram.19dr0130@me.iitism.ac.in
2. Associate Professor, Department of Mining Engineering, Indian Institute of Technology (ISM), Dhanbad, Jharkhand, India. Email: drpspaul@iitism.ac.in
3. Management Trainee, Coal India Limited, India. Email: isourabhanand@gmail.com
4. Professor, Department of Mining Engineering, Indian Institute of Technology (ISM), Dhanbad, Jharkhand, India. Email: rmbhattacharjee@iitism.ac.in

ABSTRACT

Existing small-capacity mines in India are gradually being converted into large-capacity, deep and extensive mines due to the country's swiftly increasing coal demand. As a result, workers are compelled to operate under challenging conditions, such as heat stress, which has a substantial influence on their overall well-being, safety and productivity. Prediction of heat stress and dehydration levels in underground workplaces at an early stage is crucial in order to formulate an effective strategy for mitigating the aforementioned hazards. Therefore, the present research was intended to build heat stress and dehydration prediction models to mitigate the occupational health hazards in coalmines. The machine learning methods like artificial-neural-network (ANN) and random forest (RF) models were used to predict the heat stress. The RF model showed the highest $R^2=0.861$ in comparison to the ANN model with $R^2=0.779$. Moreover, it was observed that the air velocity at the operating face of the mine was sluggish. It was also observed that 23.91 per cent of the miners were exposed to the high level of methane as per Indian regulations. The average wet bulb globe temperature (WBGT) at the working face of the mine was 31.85°C. The urine analysis results found that the average value of urine specific gravity was 1.021, which states the workers were significantly dehydrated. Finally, predictive equations were developed for assessing heat stress and dehydration. These equations provide valuable tools for the management of mines studied, enabling them to promptly anticipate and minimise the impact of heat stress on employees by forecasting WBGT and dehydration levels of the mine studied. Ultimately, the use of these equations is expected to enhance the efficiency and occupational health of mine workers.

INTRODUCTION

The world population is growing at a faster rate than at any other moment in history, while mineral utilisation is increasing more rapidly than population as more consumers join the mineral market and the global quality of living rises (Kesler *et al*, 2007). Since minerals are valuable natural resources which serve as key raw materials for core industries, the expansion of the mining sector is critical for a country's overall economic development. Every single one of these businesses requires some sort of fossil fuel to function. Coal is currently among the most important fossil fuels utilised in the globe for power production, steel mills and other applications. According to Gui *et al* (2020) and the Public Information Bureau India (2022), coal constitutes a significant proportion of the energy consumption in the two largest coal-producing nations, China and India, accounting for 60 per cent and 55 per cent, respectively. According to an International Energy Agency (IEA) analysis, fossil fuel consumption in India is expected to hit an all-time peak of 1160 Mt in 2023 (IEA, 2022).

Coal demand remains obstinate, and it is expected to hit an all-time high in 2024, increasing global emissions. In the words of the coal ministry of India, the aim for all Indian coal output for the fiscal year (FY) 2023–2024 is 1017 million tonnes (Mt), with a forecast of 1.31 billion tonnes (Bt) for FY 2025 and 1.5 Bt by FY 2030 (Public Information Bureau India, 2022). According to the IEA, the use of energy by industries accounts for about 40 per cent of the world's total final consumption at the present time and continues to be dominated by the use of fossil fuels, especially coal. The energy sector in India is almost entirely reliant on coal. The electricity sector accounts for 67 per cent of total consumption, followed by the steel and iron sector with 13 per cent, and the cement industry, which accounts for 4 per cent of all consumption (Public Information Bureau India, 2022). To satisfy the ever-increasing demand for coal, it will be essential to conduct coal mining operations below the surface of the earth. This is because the vast majority of the coal resource is located quite a distance below the surface of the earth. Underground mining poses significant risks for those who work in the industry. With the current emergence of coal output, the employment environment in underground mining has been related to a significant variety of health risk factors, including physically demanding workload, noise, vibration syndrome, exposure to radiation, spontaneous coal heating, high degree of heat, humidity, poisonous gases, dust and diesel exhaust (Vikram, Paul and Chandrakar, 2020). As a result, poor environmental factors have an influence on miners' wellness, safety and production.

According to the World Health Organization (WHO, 2018), heat that might persist for an extended period can have a substantial effect on society as a whole including an increase in heat-associated health disorders (HAHD). Working in the underground coalmines, the working capacity of underground miners can be reduced due to health degradation caused by repeated and protracted exposure to hot, humid and respirable dust-dominated at mining faces. This may alter the functional capabilities of a miner. As underground coalmines get deeper, more extensive, and excavated with higher-capacity equipment, heat addition due to multiple factors such as geothermal gradient, ventilation and heat expulsion from machinery increases. As a result, heat stress is common in underground coal deposits (Mishra et al, 2022). The measurement of stress caused by heat in underground mining settings is critical since heat disease is becoming more common as mine depth increases. Researchers have suggested many heat stress indicators to measure the thermal stress state in coal underground mines, but still they are not acknowledged worldwide.

The International Standard Organization (ISO) 9886:2004 provides measures of body temperature at the core (Figure 1), heart rate, temperature of the skin and loss of body weight via perspiration for assessing body heat strain. A heat stress indices are one number which incorporates the impacts of the six fundamental variables in any individual's thermal surroundings (such as surrounding temperature, relative humidity (RH), radiation temperature, the velocity of air, metabolic rate, and clothing properties), so that its value differs based on how much thermal stress a person is experiencing (Parsons, 2003). Heat stress is characterised as the aggregate of metabolic heat, which is generated within the body, and ambient heat, which is acquired from the surroundings, subtracted by the amount of heat dissipated from the human body to the environment. Heat stress is caused by a combination of environmental factors, the rate of metabolism and work clothing (Foster et al, 2020). A working individual generates heat inside the body, notably via muscular effort, contributing to heat stress in an elevated temperature environment (Bridger, 2003).

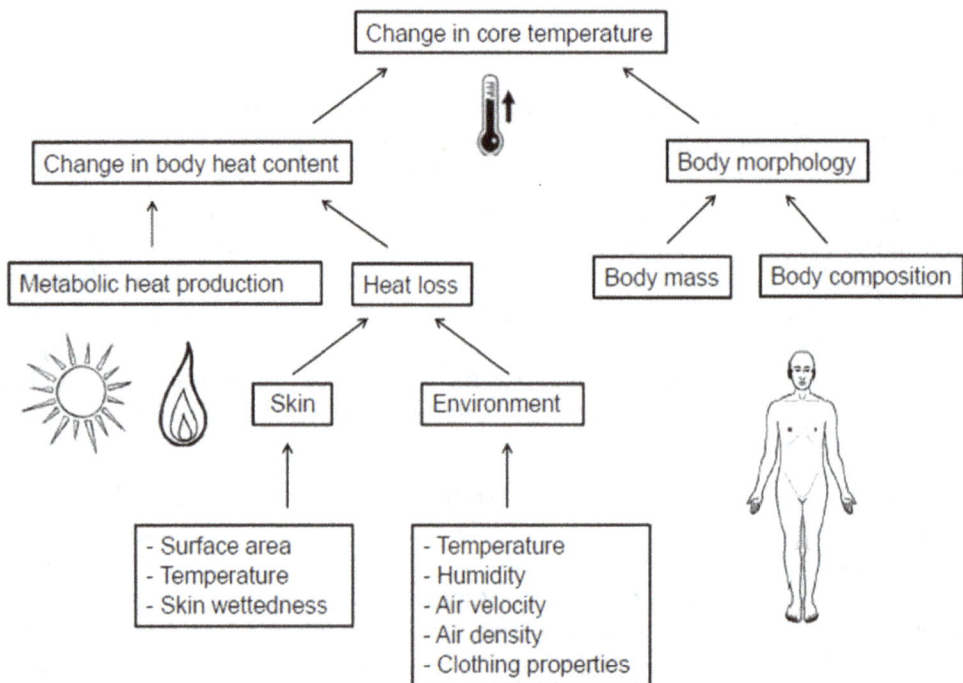

FIG 1 – Biophysical variables that influence core temperature (Bridger, 2003).

In an underground coalmine, collecting all of these metrics will become very difficult and time-consuming. Despite the fact that there are several heat stress indices, there has not yet been a clear technique or approach for selecting the appropriate index for a specific subterranean environment (Roghanchi and Kocsis, 2018). Moreover, an index of one kind cannot function as a 'universal index' on its own. A universal indicator would include a range of comfort levels depending on metabolic rates, however, there are numerous heat stress indices that selecting the proper one for a specific sector or work environment may be difficult (Roghanchi and Kocsis, 2018). As a result, wet bulb globe temperature (WBGT) measurement is by far the most extensively used and recognised empirical metric for controlling occupational heat stress (ACGIH, 2012). Furthermore, hot and humid weather causes thirst because perspiration rises, reducing the quantity of fluids in miners' bodies. With low hydration in the body due to improper water intake and high sweating it can have a negative impact on workers' health and safety. Daily heat stress and the dehydration may be linked to recurrent acute kidney damage, between crop cultivators worldwide (Roncal-Jimenez *et al* 2015). The dehydration is measured using the urine specific gravity (USG).

Heat stress events are becoming increasingly prevalent in underground mines nowadays. Mine management may take the necessary precautions depending on the extent of heat stress in an underground coal workplace if they are aware of it in advance. Since there are multiple ecological and geotechnical characteristics that might impact an underground mine's heat stress, a little shift in these parameters can result in a significant change in the thermal stress level. As a result, the building of a heat stress prediction method capable of anticipating heat stress for a specific underground mining site is required. Therefore, this study is intended to develop the heat stress prediction models using machine learning (ML) techniques and predictive equations of heat stress and dehydration.

METHODOLOGY

First and foremost, a broad literature analysis was undertaken in this study to investigate the effect of WBGT. A few limitations were found based on the available literature analysis which was discussed in section 1. According to the research review, there is a significant need to undertake a study estimating the effects of thermal stress and the subsequent loss of productivity in mine operators working in hazardous subterranean coalfield conditions. After identifying the research gap and establishing the study objectives a comprehensive methodology was developed to achieve the research goals which is shown in Figure 2.

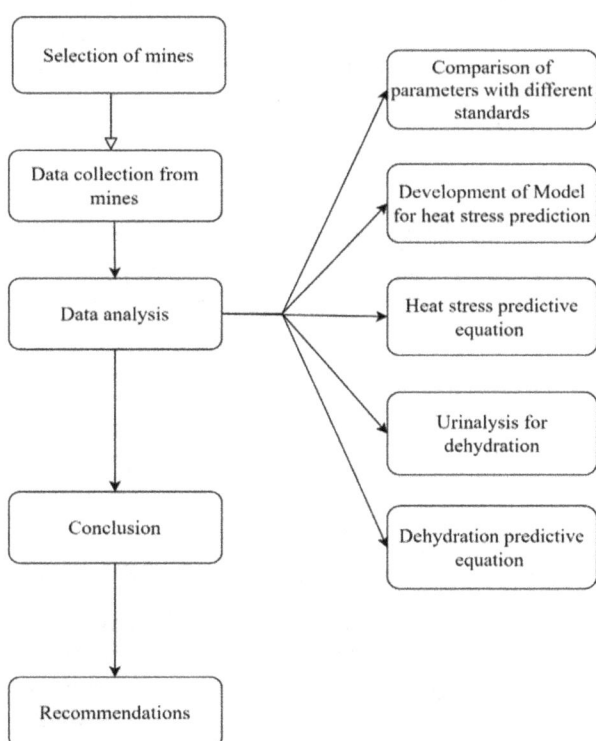

FIG 2 – Research methodology flow chart.

Firstly, pre-set criteria were developed. As per the criteria, the mines should have mechanised underground mining districts with high production. Its working face should be at far from the mine entrance to figure out the ventilation situation. It should be a deep mine to see the effect of geothermal properties, radiation etc, as an effect of heat stress. Therefore, for the field investigation, a deep, vast and highly mechanised underground coalmine of Coal India Ltd., was chosen. In the mine, geotechnical along with environmental data were collected. Parameters that were collected during the field studies include velocity of air (VOA) (m/s), methane percentage CH_4 (%), wet bulb temperature (WBT) (°C), dry bulb temperature (DBT) (°C), RH (%), Depth (m), length of chainage (LOC) (m) and WBGT were considered for model development.

DATA ANALYSES

The chosen mine was operating at 500–700 m depth, and the primary mechanical ventilator installed in said mines has a capacity of drawing 165 m^3 per second of air at a static pressure of 1618 Pascals. Poor VOA was observed at the pre-set points and on comparing the field data to that of standard laid by Coalmines Regulation India, (CMR; Directorate General of Mines Safety (DGMS), 2017). It has been observed that 28.26 per cent of the workers of the mines at 4.5 m from face and 76.08 per cent at 7.5 m from face had to work at VOA below the standard for operation. The mine is operated at a great depth with high gassiness in the coal seam. The study found that the VOA remained relatively low in removing the noxious gases. On comparing the field data to that of standard laid by DGMS (2017) ie 0.75 per cent, it has been observed that 23.91 per cent of the miners were exposed to CH_4. Similarly, the recommended lower limit for CH_4 in indoor environment by American Conference of Governmental Industrial Hygienists (ACGIH) and Occupational Safety and Health Administration (OSHA) is 0.1 per cent, which on comparison finds 100 per cent of the workers under high CH_4 per cent. The average WBT was 31.39°C with the standard deviation of 1.86. The average DBT was 32.77°C with the range of 30.1–35.8°C. While the average RH, depth and LOC were 90.32 per cent, 582.61 m and 150.65 m respectively.

With the deepening of mines along with the use of high mechanisation, the WBGT at the working face of the mine was on the higher side. The lowest and highest WBGT was found to be 28.3°C and 34.6°C with an average value of 31.78°C. When compared with the standards of ISO, 4.3 per cent, 71.73 per cent and 23.91 per cent of the workers were performing their work at above 28°C, 30°C and 33°C respectively. Similarly, 6.52 per cent and 93.47 per cent workers were working above 28°C and 30°C according to National Institute for Occupational Safety and Health (NIOSH) with 36.95 per cent and 63.04 per cent workers were working at WBGT of 28–31°C and above 31°C according to Japan Sports Association (JASA). Further form the standards from Nag (1996), it was found that 54.34 per cent, 23.91 per cent and 21.73 per cent of the workers were working above WBGT of 28°C, 31.5°C and 33.5°C. Various standards for evaluation of heat stress is shown in the Table 1.

TABLE 1

Standards for evaluation of heat stress by international organisations or governments.

ACGIH	JASA	ISO	NIOSH/OSHA	Nag	Workload
25	<21	23	-	28	Very heavy
-	21–25	25	25	31.5	Heavy
26.9	25–28	28	26	33.5	Moderate
30	28–31	30	28	35	Light
	>=31	33	30	37	Resting

Adapted from the American Industrial Hygiene Association (AIHA) (2003), JASA (2013) and Nag (1996).

It's a known fact that our body contains almost 60–70 per cent water and among that brain alone has 75–80 per cent of water. So, being hydrated becomes very much important to facilitate the transfer of signals. On comparing the urine analysis of the sample collected from the mines it was found that 8.96 per cent of the workers were found to be well hydrated, 45.65 per cent of the workers were at minimal dehydration, 21.73 per cent of the workers were at significant dehydration and

23.91 per cent of the workers were at sufficient dehydration level. The pathology report of the samples collected from the coalmines for urine analysis found that the average value of USG was 1.021 and according to the standard, a human body is significantly dehydrated at this value. It reflects limited water intake during the mining activity in hot and humid circumstances. National Athletic Trainers Association (NATA) has prescribed standard for USG for adults which is given in Table 2 and the parameters characteristics of the parameters considered for the study is shown in Table 3.

TABLE 2

Standards for USG.

Condition	USG
Well hydrated	USG<1.010
Minimal dehydration	1.010=<USG=<1.020
Significant dehydration	1.020<USG<1.030
Serious dehydration	USG>=1.030

Source: Casa *et al* (2000).

TABLE 3

Parameters characteristics considered for the study.

Parameter	Range	Mean	Standard deviation
VOA	0.3–0.8	0.56	0.21
CH_4	0.22–0.84	0.54	0.21
WBT	28.3–35.1	31.39	1.86
DBT	30.1–35.8	32.77	1.56
RH	85.6–96.5	90.32	2.59
DEPTH	500–700	582.61	61.41
LOC	80–120	150.65	44.59
WBGT	28.3–34.6	31.78	1.46
USG	1.01–1.03	1.02	0.0064

Random Forest (RF) for heat stress model

RF is a widely accepted tree-based machine-learning algorithm (Ebrahimy *et al*, 2020). In recent years, the application of RF in solving complex relationship between variables has become popular due to its accurate prediction ability (Brito *et al*, 2022). A different kind of sampling technique, called bootstrap sampling, is used in RF algorithm which increases the diversity of sample selection process. Numerous decision trees are created from this bootstrap data set and combined to get a much more accurate and stable prediction. RF does not depend on single decision tree; instead, it takes prediction from an individual tree and predicts the final outcome depending on the majority of votes. Accuracy of prediction by the RF model improves with the increase in number of decision trees in the model. Four stages comprise the design of the RF algorithm: the input stage (Dataset), the concealed stage (decision trees), averaging or majority voting stage and the output stage is shown in the Figure 3.

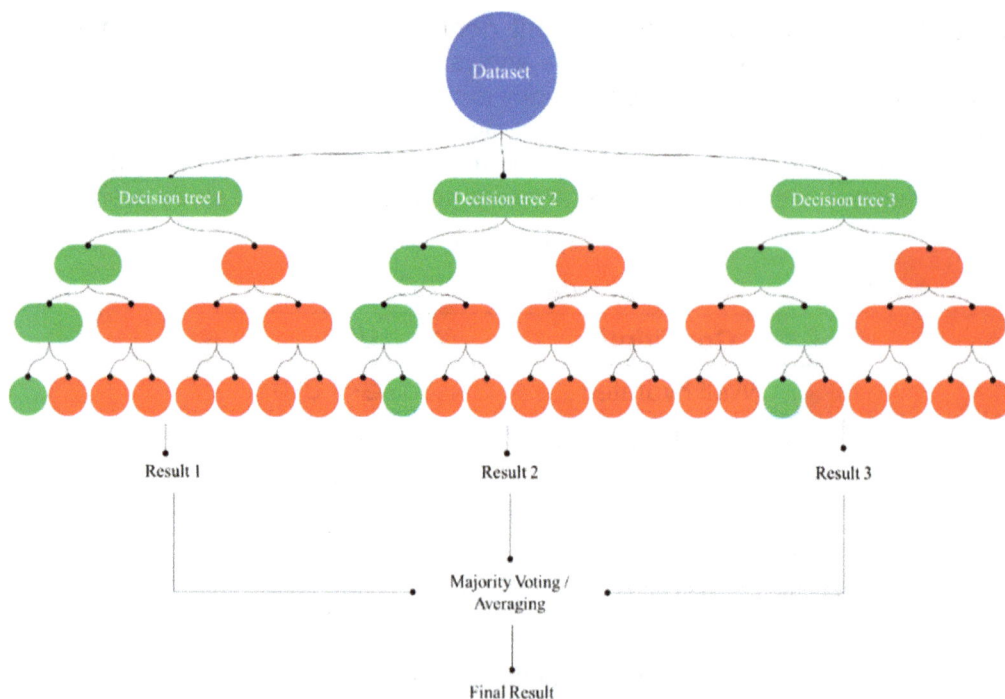

FIG 3 – Schematic diagram showing the process of RF algorithm.

Artificial Neural Network (ANN) for heat stress model

ANN is an artificial intelligence tool used for complex problem solving. It was invented long back based on the mathematical model of the human nervous system, since then it has been used in various areas, mainly in engineering field (Kim, Lee and Yo, 2021). ANN has been utilised in solving complex problems in the mining industry, some of them are prediction of subsidence (Ambrozic and Turk, 2003), prediction of blast-induced vibration (Paneiro *et al*, 2020) and prediction of stability of longwall gate roads (Mahdevari, 2019). Therefore, this study uses ANN for the first time to predict the heat stress in Indian underground mine and its schematic diagram is shown in Figure 4.

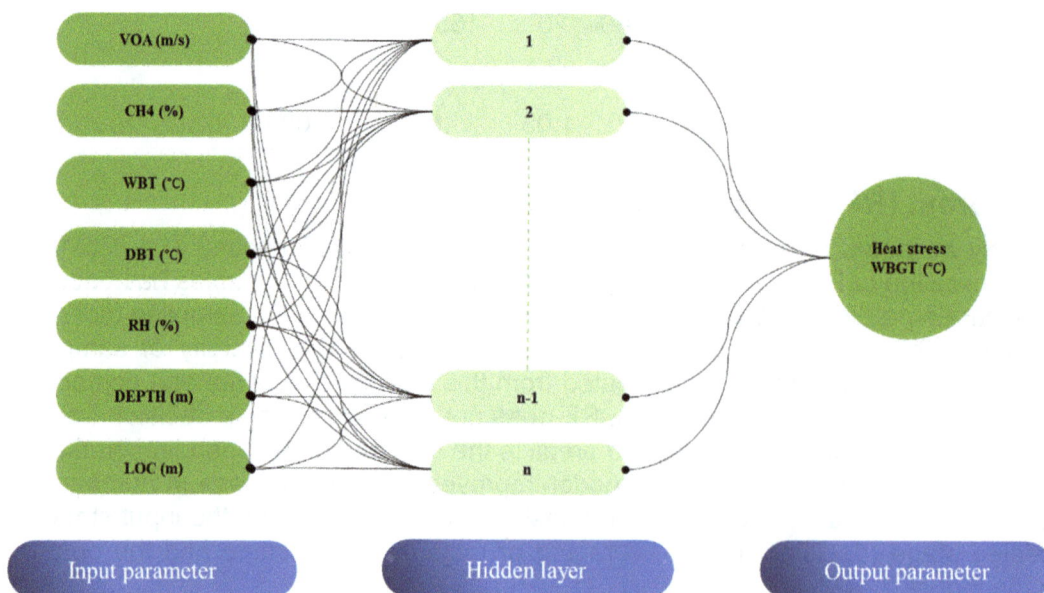

FIG 4 – Schematic diagram showing the process of ANN algorithm.

Predictive ML model analysis

In the course of this study, ML algorithms have been implemented to evaluate the impact of several elements to the experience of thermal stress. The RF and ANN techniques were subjected to a train-test technique in order to determine which of the two models is the superior option for doing data analysis on the given data set. In this research, Python platform was used to develop the RF and ANN model. 70 per cent of the input data set was used for training and each 30 per cent of the remaining data was used for testing. Models were constructed by utilising the train data and then those models were run using the testing data. In order to choose the best model, Coefficient of determination (R^2), root mean squared error (RMSE) and mean absolute error (MAE) catalogues were constructed using the test data set. The ideal model will have a high R^2 value while also having low RMSE and MAE values (Moriasi *et al*, 2007). The results of the ML techniques are shown in the Table 4.

TABLE 4

Comparison of various ML techniques.

ML techniques	Training			Testing		
	R^2	RMSE	MAE	R^2	RMSE	MAE
RF	0.981	0.016	0.012	0.861	0.066	0.049
ANN	0.961	0.024	0.021	0.779	0.083	0.064

The R^2, RMSE and MAE for RF was found to be 0.981, 0.016, 0.012 respectively for training data sets. Similarly, R^2, RMSE, MAE for ANN was found to be 0.961, 0.024, 0.021 respectively for training data sets. Further for testing data sets of RF and ANN, the values of R^2, RMSE, MAE were 0.861, 0.066, 0.049 and 0.779, 0.083, 0.064 respectively. Hence the RF model predicted WBGT accurately than the ANN model.

DEVELOPMENT OF PREDICTIVE EQUATIONS

The predictive equation development provides a basic idea about the heat stress risk exposure of mine workers. Moreover, the utilisation of the predictive equation obviates the need for intricate methodologies in assessing heat stress exposure and pathology lab analysis for dehydration estimation. The developed predictive equation developed for the prediction of heat stress is given by Equation (1). The R^2 value of the developed equation is 0.96 which ratifies the usage of the equation.

$$WBGT = 16.88 - 7.79(VOA) + 1.58(CH4) + 0.14(DBT) + 0.18\,(WBT) + 0.15(RH) +$$
$$0.007(DEPTH) - 0.03(LOC) \tag{1}$$

The developed predictive equation developed for the prediction of dehydration is given by Equation (2). R^2 value between USG and WBGT was found to be equal to 0.862. The predictive equation can be used to check the status of hydration among the mine workers working in underground coalmines where they experience high heat temperatures. Figure 5 depicts the variation in USG of mine labourer collected samples according to WBGT.

$$USG = 0.8906 + 0.0041(WBGT) \tag{2}$$

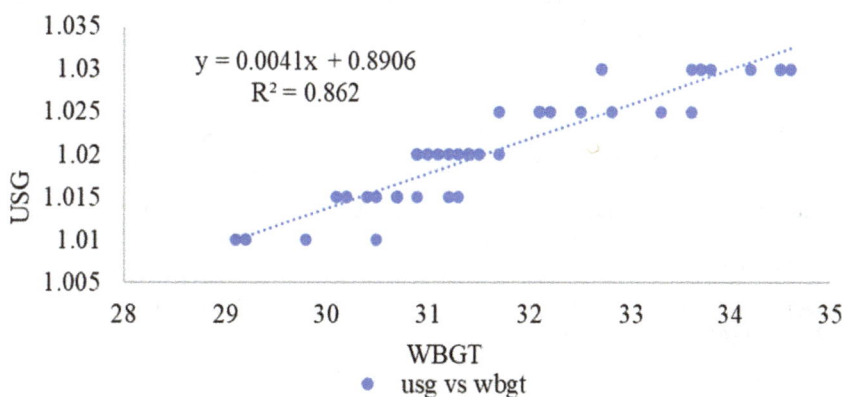

FIG 5 – Variation of USG versus WBGT.

CONCLUSION

The heat stress state of a workplace is determined by various environmental elements, which fluctuate rapidly. As a result, developing an adequate heat stress prediction model that can forecast the heat stress of a given underground mining site is a challenging task. This study assessed underground mine condition by comparing field data to standards established by several organisations. In the field observation, it is found that the extent of the mine working area grows in tandem with the depth of the mines. As the depth of the panel and galleries increases, it becomes vital that the ventilation systems perform suitably. This is due to the machines utilised for development galleries were running in blind directions, the airflow rate measured at the mine final measuring station (near the operating face) was slow due to air short-circuiting caused by non-maintenance of ventilation. With such a little amount of circulation, there is less ability to replace hot and humid air with new fresh air, therefore an adequate working environment cannot be created in mines. As a result, the heat stress (WBGT) at the working face surpasses the standards established by various occupational health and safety agencies. This indicates that as mine depth increases, the proper performance of ventilation systems becomes more crucial. As a result of inadequate ventilation, the thermal stress level at the working faces becomes severe and surpasses the ACGIH recommended threshold value of 28°C.

It was also revealed that mine employees had a tremendous workload and were well acclimatised to the thermally stressed mining environment. To mitigate the negative impact of heat stress, workers should allocate one-quarter of every hour (ie 2 hrs out of an 8-hr shift) to work, with the remaining time designated for rest, according to ACGIH's (2012) suggested threshold value limit, while working at a heavy workload in an underground mine environment with WBGT ranging from 28.3°C to 34.6°C. Employees are susceptible to HAHD as a result of higher WBGT in mines, resulting in a quarter loss in productivity, higher production costs and a drop in overall profit.

The mining sector stands to benefit from improvements in production and efficiency, as well as a decrease in operating costs, if autonomous technologies such as machine learning and artificial intelligence are used. The application of machine learning to anticipate WBGT will help mine management identify areas with particularly high WBGT and make the appropriate technological modifications to provide better thermal stress or WBGT conditions for personnel. This will contribute to higher overall production while also improving employee occupational health and safety. Mine management may improve productivity up to four times and create four times the profit with the forecasting WBGT technique without jeopardising worker health and safety.

According to the pathology report of the samples obtained from the coalmines for urine analysis, the average value of USG was 1.021 and a human body is extremely dehydrated at this value, according to the standards of NATA (Casa *et al*, 2000). It illustrates the restricted water intake during mining operations in hot and humid conditions. Water replenishment is crucial since it is the most vital component in the human organism, accounting for around 60–80 per cent of lean body mass (Rowntree, 1992). Excessive dehydration affects cardiovascular function, thermoregulation, muscle function, fluid volume status and exercise performance (Roncal-Jimenez *et al*, 2015). Moreover, the developed predictive equations of heat stress and dehydration helps the mine management to take appropriate measurements to proactively minimise the effect of heat stress and increase the functional capacity of mine workers.

REFERENCES

Ambrozic, T and Turk, G, 2003. Prediction of subsidence due to underground mining by artificial neural networks, Computational Geosciences, 29:627–637. https://doi.org/10.1016/S0098-3004(03)00044-X

American Conference of Governmental Industrial Hygienists (ACGIH), 2012. T for Diacetyl, in *Documentation of the Threshold Limit Values for Chemical Substances*, 7th edition (American Conference of Governmental Industrial Hygienists (ACGIH): Cincinnati).

American Industrial Hygiene Association (AIHA), 2003. *The Occupational Environment: Its Evaluation, Control and Management*, 2nd edn (AIHA Press: Fairfax).

Bridger, R S, 2003. *Introduction to Ergonomics*, 2nd edn (Taylor and Francis: London). https://doi.org/10.1201/b12640

Brito, M P, Chen, Z, Wise, J and Mortimore, S, 2022. Quantifying the impact of environment factors on the risk of medical responders' stress-related absenteeism, *Risk Analysis*, 42:1834–1851. https://doi.org/10.1111/risa.13909

Casa, D J, Armstrong, L E, Hillman, S K, Montain, S J, Reiff, R V, Rich, B S, Roberts, W O and Stone, J A, 2000. National Athletic Trainers' Association (NATA) position statement: fluid replacement for athletes, *Journal of Athletic Training*, 35(2):212–224. https://www.ncbi.nlm.nih.gov/pmc/articles/PMC1323420/

Directorate General of Mines Safety (DGMS), 2017. The Gazette of India – Extraordinary, Ministry of Labour and Employment Notification, part II, section 3(i), Coalmines Regulation 2017, Government of India – Ministry of Labour and Employment. Available from: <http://www.dgms.net/Coal%20Mines%20Regulation%202017.pdf> [Accessed: 12 February 2024].

Ebrahimy, H, Feizizadeh, B, Salmani, S and Azadi, H, 2020. A comparative study of land subsidence susceptibility mapping of Tasuj plane, Iran, using boosted regression tree, random forest and classification and regression tree methods, *Environmental Earth Sciences*, 79:1–12. https://doi.org/10.1007/s12665-020-08953-0

Foster, J, Hodder, S G, Lloyd, A B and Havenith, G, 2020. Individual responses to heat stress: implications for hyperthermia and physical work capacity, *Front in Phys*, 11:541483. https://doi.org/10.3389/fphys.2020.541483

Gui, C et al, 2020. Gas–solid two-phase flow in an underground mine with an optimized air-curtain system: A numerical study, *Process Safety and Environmental Protection*, (140):137–150. https://doi.org/10.1016/j.psep.2020.04.028

International Energy Agency (IEA), 2022. Global coal consumption, 2020–2023, in Coal Market Update – July 2022. Available from: <https://www.iea.org/data-and-statistics/charts/global-coal-consumption-2020-2023> [Accessed: 12 February 2024].

International Organisation for Standardization (ISO), 2004. ISO 9886:2004. Ergonomics, Evaluation of thermal strain by physiological measurements, International Organisation for Standardization.

Japan Sports Association (JASA), 2013. A Guidebook for the Prevention of Heat Disorder During Sports Activities. Available from: <https://www.wbgt.env.go.jp/en/wbgt.php> [Accessed: 12 February 2024].

Kesler, S E, 2007. Mineral supply and demand into the 21st century, in *Proceedings for a Workshop on Deposit Modelling, Mineral Resource Assessment and their Role in Sustainable Development*, US Geological Survey Circular, 1294:55–62. Available from: <https://pubs.usgs.gov/circ/2007/1294/circ1294.pdf#page=62> [Accessed: 12 February 2024].

Kim, S, Lee, J and Yo, T, 2021. Road surface conditions forecasting in rainy weather using artificial neural networks, *Safety Science*, 140:105302. https://doi.org/10.1016/j.ssci.2021.105302

Mahdevari, S, 2019. Coalmine methane: Control, utilization and abatement, in *Advances in Productive, Safe and Responsible Coal Mining*, pp 179–198. https://doi.org/10.1016/B978-0-08-101288-8.00010-9 (Woodhead Publishing).

Mishra, D P, Roy, S, Bhattacharjee, R M and Agrawal, H, 2022. Genetic Programming for Prediction of Heat Stress Hazard in Underground Coal mine Environment, *Natural Hazards*, preprint, https://doi.org/10.21203/rs.3.rs-1244556/v1

Moriasi, D N, et al, 2007. Model evaluation guidelines for systematic quantification of accuracy in watershed simulations, *Transactions of the ASABE*, 50(3):885–900. https://doi.org/10.13031/2013.23153

Nag, P K, 1996. Criteria for recommended standards for human exposure to environmental heat, A Report, Ministry of Environment and Forests, Government of India, National Institute of Occupational Health, Ahmedabad, India.

Paneiro, G, Durão, F O, Costa e Silva, M and Bernardo, P A, 2020. Neural network approach based on a bilevel optimization for the prediction of underground blast-induced ground vibration amplitudes, *Neural Computing and Applications*, 32:5975–5987. https://doi.org/10.1007/s00521-019-04083-2

Parsons, K, 2003. *Human Thermal Environments*, 2nd edn (Taylor and Francis: London). https://doi.org/10.1201/b16750

Public Information Bureau India, 2022. Domestic Demand of Coal, Ministry of Coal, India. Available from: <https://pib.gov.in/PressReleaseIframePage.aspx?PRID=1847895> [Accessed: 12 February 2024].

Roghanchi, P and Kocsis, K C, 2018. Challenges in selecting an appropriate heat stress index to protect workers in hot and humid underground mines, *Safety & Health at Work*, 9(1):10–16. https://doi.org/10.1016/j.shaw.2017.04.002

Roncal-Jimenez, C, et al, 2015. Mechanisms by which dehydration may lead to chronic kidney disease, *Annals of Nutrition and Metabolism*, 66(suppl:3):10–13. https://doi.org/10.1159/000381239

Rowntree, L G, 1922. The water balance of the body, *Physiological Reviews*, 2(1):116–169. https://doi.org/10.1152/physrev.1922.2.1.116

Vikram, S, Paul, P S and Chandrakar, S, 2022. Assessment of Work Postures and Physical Workload of Machine Operators in Underground Coalmines, *Journal of the Institution of Engineers (India): Series D*, https://doi.org/10.1007/s40033-022-00389-z

World Health Organisation (WHO), 2018. Heat and health. Available from: <https://www.who.int/news-room/fact-sheets/detail/climate-change-heat-and-health> [Accessed: 12 February 2024].

Fog control and prevention in underground mine travelways using refrigeration

C M Stewart[1] and A Loudon[2]

1. Principal Engineer, Minware, Cleveland Qld 4163. Email: craig@minware.com.au
2. Mine Cooling Specialist, Howden, Sydney NSW 2153. Email: alan.loudon@howden.com.au

ABSTRACT

Fog formation in underground mines can cause poor visibility in travel ways, accelerate corrosion of ground support and steel infrastructure, and damage electrical and electronic devices continually exposed to moisture. Fog in mines is usually formed from the adiabatic decompression and cooling of humid air as it ascends, causing water vapour to condense into suspended fine water droplets.

Warm climate mines typically exhaust through dedicated shafts largely negating the hazardous potential of fog, however, cooler climate mines often exhaust warm humid air through travelways to minimise heating requirements and prevent freezing in main travelways during winter periods. Hazards resulting from poor visibility due to fog are typically addressed through reduced traffic speed, guide lights and traffic control procedures, which only partially reduce the risks, and result in reduced productivity and increased operating costs.

Fog can normally only be reduced or eliminated by heating or dehumidifying air, or by the physical removal of water droplets from air. Dehumidification is a common industrial process used in HVAC systems to control air quality and comfort but has yet to be routinely used at an industrial scale in underground mines. This paper describes a process of using refrigeration technology to create a dehumidification process to control large volumes of fog. The process is modelled in Ventsim™ version 6.0 (by Howden Ventsim, Howden) to demonstrate the theoretical effectiveness, efficiency, and cost of a refrigerative dehumidification system compared to an equivalent electric heat system and discusses practical ways to implement the technology in typical underground mines.

INTRODUCTION

Fog is the presence of fine suspended water droplets in the air and is part of a wider psychrometric process related to the behaviour of mixed air and water vapour under differing temperatures and pressures. The definition of fog is internationally recognised by the Federal Coordinator for Meteorology (2005) as the presence of fine water droplets that reduce visibility to less than 1000 m. Assuming fine droplets, this equates to an initial water concentration for light fog of between 50–100 mg/m^3 of air, increasing to over 1000 mg/m^3 for thick fog. This is a simplification, as visibility through fog is also associated with the water droplet size, with finer droplets reducing visibility for the same mass density. If saturated air approaches its dew point temperature, excess water condensates to form suspended liquid water droplets (often assisted by existing nuclei of diesel particulates or dust).

Fog is recognised as a working hazard in mines, particularly in cold climate regions where the exhaust air is often directed through main travelways to reduce heating requirements and eliminate icing hazards (Figure 1). Poor visibility causes hazards for traffic interactions, productivity can suffer due to reduced travel speeds, moisture accelerates the corrosion of steel and ground support, and water can ingress into electrical infrastructure.

There is little research on reducing fog in underground mines. Traditional methods of controlling fog hazards include tolerating the problem through traffic control, procedures, speed restrictions and guide lights, or sometimes in exceptional circumstances closing sections of the mine where fog has become unmanageable.

FIG 1 – Access portal fog emissions traffic control.

FOG REMOVAL METHODS

Fog can be removed via many methods, several of which are described below:

- Agitating air with increased flow and turbulence.
- Mixing with dryer air sources.
- Physical removal of liquid moisture droplets from the air.
- Removal of water vapour (latent heat) from the air.
- Heating the air to increase the water vapour carrying capacity.

Agitation

Schimmelpfennig (1982) researched numerous methods of fog reduction including heating the mine air, chemically drying moist air, refrigeration of air, use of centrifugal fan scrubbers, chilling intake air and increasing airflow using additional fans. Schimmelpfennig suggested the most plausible solution was to increase air velocity to promote air mass mixing and evaporation. Centrifugal fan scrubbers and mist eliminators provided an effective mechanical means of removing water droplets. Coupled with the added fan power and heat generated in the process, the study found the fog could effectively be removed, but at a relatively high capital and operating cost.

Water droplet removal

Martikainen (2007) discusses the analysis of fog in three sub-artic mines in Finland; Pyhäsalmi Mine, Orivesi Mine, and Louhi Mine including the causes and potential for removal or reduction of the fog. Mechanical methods using filters or screens to capture droplets were trialled but found to have only limited success in local fog removal. An aluminium net and a mist eliminator and fibrous filter fabric combination provided the best results but still failed to completely eliminate the fog.

The effectiveness of both agitation and water droplet removal in a mine ramp application is questionable as the methods must capture all airflow and remove no latent heat. In an upcasting ramp system, the fog quickly re-forms once the air further reduces in density and cools.

Heating

Direct electric heating is occasionally used in foggy mines, and Martikainen (2006) notes heating with fans and electric heaters reduces fog by raising the temperature and increasing the moisture-carrying capacity of air while evaporating any existing moisture droplets. Temporary fog-reducing benefits can be seen practically in circumstances such as the operation of large diesel trucks, large fans or waste heat from infrastructure such as compressor stations or transformer stations.

Heating of air however does not remove moisture, and some processes such as diesel equipment can add more moisture through exhaust gases to the air. Heating also tends to be energy inefficient as any rise in temperature potentially causes more rapid heat loss to surrounding rock strata. Ultimately, the fog reforms once the air approaches the dew point temperature again.

Water vapour removal

Any fog elimination process that relies only on liquid water droplet removal, or heats the air without removing moisture, risks having only short-term benefits in an upcast ramp system. Efficient, long-lasting fog elimination requires the removal of water vapour (latent heat) to ensure the psychrometric properties of the air are altered to allow re-cooling without immediate fog formation.

Latent heat removal reduces the specific moisture content (water mass) in the air rather than only reducing relative humidity (% of water vapour carrying capacity). The 'drying' of the air allows any remaining water droplets (fog) to be re-evaporated into the air if not removed. If enough latent heat moisture is removed, it provides a buffer to allow further moisture addition or cooling without fog re-formation. There are two main methods of latent heat removal: desiccant (or absorption) dehumidification, and condensate (or refrigerative) dehumidification.

The first commercial dehumidifier was developed by William Carrier in 1902 (Teitelbaum, Miller and Meggers, 2023) using refrigeration technology and his work led to the understanding of dew-point behaviour and the development of the psychrometric chart in 1911 (Simha, 2012). Carrier (among others before) realised the importance of moisture control in industry and his work sparked a technological revolution in the science of conditioning air and the modernisation of HVAC processes (heating, ventilation, and air conditioning).

Desiccant dehumidification

Desiccant dehumidification requires a moisture-absorbing hydrophilic material such as silica gel to remove water vapour from the air (Figure 2). Moist air passed through the dry (activated) gel is dried, passing moisture to the gel. The saturated gel can then be heated to remove absorbed water, thereby reactivating it for reuse in the process. The method is commonly used in industrial HVAC and dryers to reduce humidity and has the advantage of simplicity and the ability to work in cold temperatures (less than 10°C). The disadvantages are the high relative power cost and the logistics of removing moisture from the desiccant without it re-entering the airflow. If this process was used underground, the humid air from the desiccant regeneration would need to be rejected into an exhaust system to prevent it from re-entering the mine access and causing more fog when cooled.

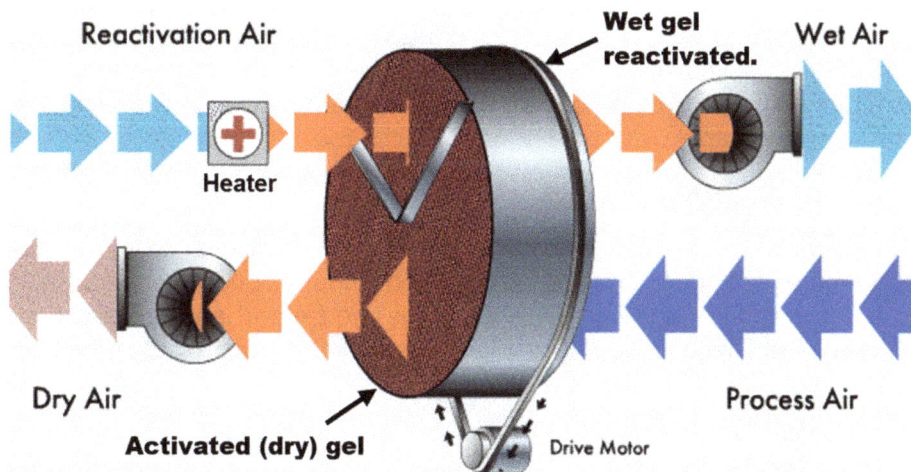

FIG 2 – Example of rotor desiccant dehumidification (Munters, 2024).

Refrigerative dehumidification

Refrigeration dehumidification uses a refrigeration process to chill air to below the dew point temperature, forcing condensate water to be extracted and disposed of in a liquid state as shown in Figure 3. The combination of chilled air, condenser heat rejection and mechanical heat of the refrigeration process is re-added as a mix of warmer dryer air to the more humid surroundings, reducing both relative humidity and specific moisture.

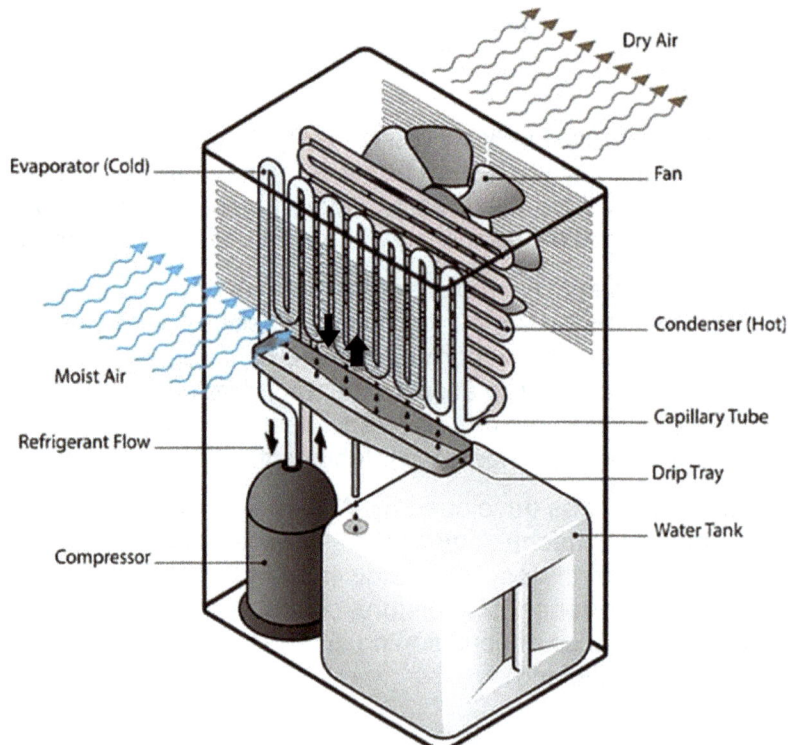

FIG 3 – Example of refrigerative dehumidification (Alorair, 2024).

The method is also commonly used in HVAC processes and has the advantage of higher energy efficiency, but with the disadvantages of greater mechanical complexity, and poorer efficiency in cooler temperatures (under 10°C) due to the limits imposed by evaporator plate icing and reduced heat transfer efficiency.

Mine fog removal with refrigeration

Refrigeration has been successfully used for many decades for cooling in hot underground mines and applied in processes such as bulk air cooling, spot cooling, chilled water and ice manufacture. Refrigerative dehumidification likewise has a long history of successful use in industrial HVAC, business, and residential applications. The application of dehumidification strategies in underground environments to reduce fog is therefore not considered theoretical or speculative as this paper simply researches and models new applications of proven processes.

Mine refrigerative dehumidification

The dehumidification process proposed for mine fog removal is similar to mine refrigerative spot cooling with a few minor changes. The plant is located underground in or near the foggy region, treats a portion of the passing foggy airflow and allows the treated and untreated streams to mix downstream in the normal flow direction. The process should not be disruptive to traffic and can be scaled to fit available infrastructure or excavations accordingly.

A spot-cooling refrigeration plant typically contains two main components: a compressor chiller circuit and a cooling circuit. The compressor pumps refrigerant gas through condenser heat exchanger coils at high pressure which condenses the gas to a liquid, releasing heat. The heat is removed from the coils with air or water. The compressed liquid refrigerant is then piped and evaporated through cooling coils, absorbing heat and chilling any air or water passing around the coils. The chilled air or water is used to cool a portion of the mine air, while the heat rejected from the compressor condenser circuit is sent to the air exhaust or a hot water reject circuit.

A refrigerative dehumidification process uses the same equipment but applies several configuration changes:

- The refrigerant condenser heat exchanger returns heat to the main air circuit instead of discarding it to exhaust. The heat increases the air temperature, decreasing relative humidity and helping reduce or eliminate fog.

- The chilled air produced from the evaporator cooling coils (minus any condensate water that is captured and removed) is used to assist the condenser heat exchanger in removing heat, increasing the efficiency of the refrigerant condensation process.

- Only a portion of the passing air is treated. The 'treated air' that has passed through the cooling circuit and the heat reject circuit is both dryer (because of condensate removed) and warmer and re-enters the remaining 'untreated' passing foggy air, immediately evaporating any remaining water droplets and clearing the air of all fog.

Psychrometric process

The process is efficient and effective for fog as the air is already fully saturated, and most of the energy consumed on the chiller side is used to remove latent heat (water vapour), not reduce ambient air temperature. The latent heat removal (from the evaporator coils) and the equivalent addition of sensible (dry) heat (from the condenser coils), as well as any additional mechanical heat produced in the process, creates a dryer air mixture than by direct heating alone with the equivalent amount of power used as a heater.

An illustrative example of the dehumidification process on a psychrometric chart is shown in Figure 4. In this example, humid air at 150 m³/s @ 20°C, 95 per cent humidity (POINT 0) has a 33 per cent portion of the main airflow (50 m³/s) directed through refrigeration (650 kW using 175 kW power POINT 1) and then reheated (825 kW POINT 2) before finally remixed with the untreated main airflow (POINT 3). The combined air is well inside the fog-forming region and considerably lower in both relative and absolute humidity than the starting condition. Any fog remaining in the untreated air stream is evaporated and quickly cleared.

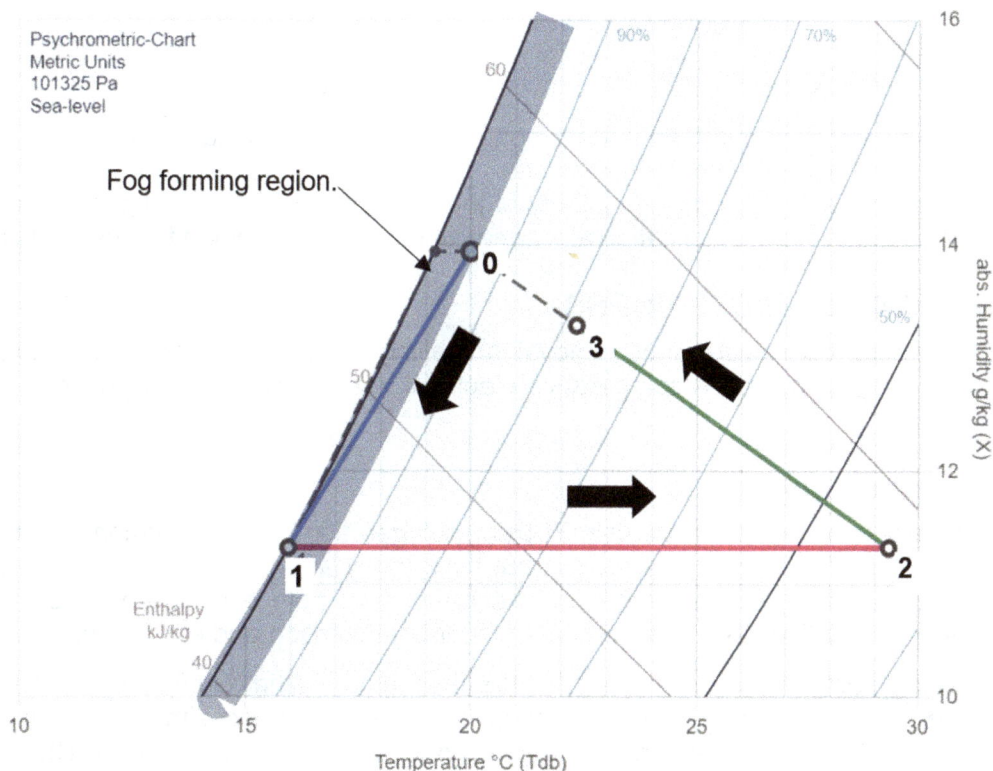

FIG 4 – Psychrometric chart showing dehumidification process (www.psych-chart.com).

Design considerations

The refrigeration design on the condenser side uses an air-cooled heat exchanger as opposed to a water spray-cooled system that would add moisture to the air. The evaporator side could potentially use either a fan-forced coil-finned evaporator or a chilled water system with cooling towers or spray chambers, as both systems will condense and remove water vapour from chilled moist air. The selection of the cooling technology may depend on air and water quality and other factors such as cost, space, efficiency, maintenance and cleaning.

Given the location is likely to be in or near a travel ramp, the size of the apparatus may need to be modular and compact to allow easy placement along the roadway or within a side tunnel without the need to mine extensive chambers. Systems may need to be repeated along the ramp at strategic intervals, particularly if made smaller to fit into tight locations.

An example system adjacent to a main ramp is shown in Figure 5, where a portion of the ramp airflow is drawn in by fans through the intake cooling coils, reheated from the chiller compressor reject, and then injected back onto the ramp to mix with the remaining foggy air. Cold water condensate is generated which would typically be discarded to the drainage system.

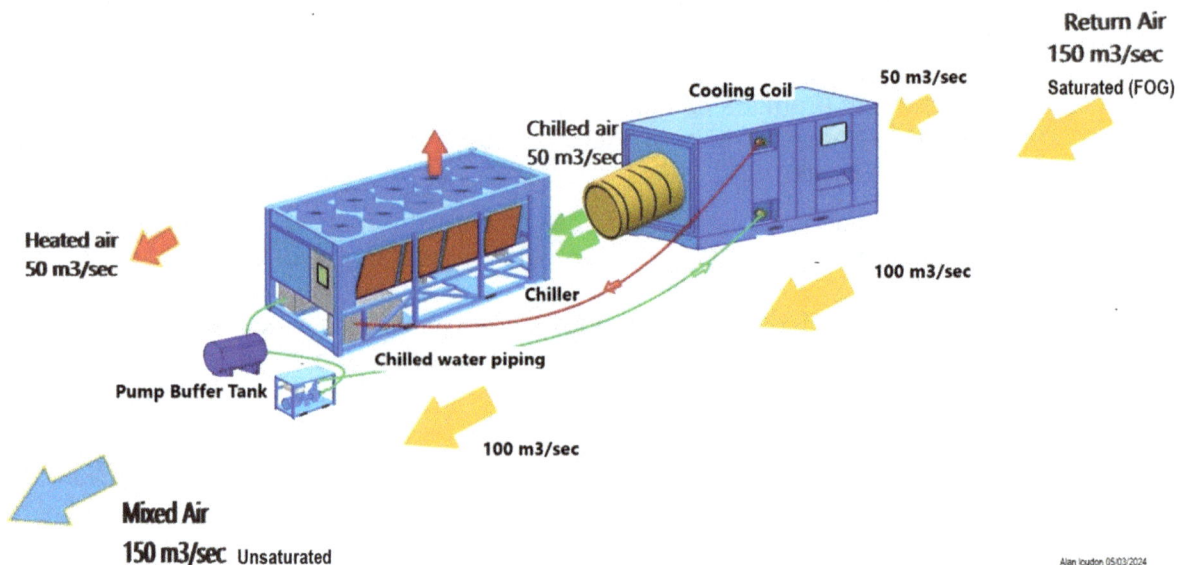

FIG 5 – Example layout of an underground refrigeration dehumidification system.

Another consideration is whether the modularity of the system could be reconfigured and relocated for refrigeration instead of dehumidification at warmer times of the year when fog may not be present.

EFFICIENCY AND EFFECTIVENESS

To determine the potential viability of fog removal, modelling is used to analyse and compare a mine dehumidification system to a base comparative electrical heating system that provides similar fog removal potential.

Modelling methodology

A series of models were created in Ventsim™ using refrigeration plant performance data supplied by Howden Australia. The modelling aims to treat thick foggy air so that the atmosphere remains clear for more than one kilometre further up the ramp, with both a refrigeration system and a conventional electric heating system to compare power consumption and effectiveness.

Ventsim's thermodynamic simulation capability includes fog modelling (Griffith, 2021). The software calculates water condensate formed from cooling saturated air during simulation and transports liquid condensate water within the airflow mass up to a maximum concentration limit. The transported condensate is allowed to re-evaporate if heat is added or it is mixed with dryer air. If condensate exceeds 1200 mg/m³, coarse droplets are assumed to 'fall out' limiting further water accumulation. Modelling by the author at numerous cold climate mines has validated the fog modelling results, provided humidity is correctly calculated using appropriate wetness fractions, diesel equipment inputs and heat assumptions.

The use of a full thermodynamic network modelling approach instead of simple psychrometric analysis has several advantages as it considers other external factors such as downstream heat loss to the surrounding rock strata, additional moisture evaporated from roadways and wet surfaces, and water vapour added from diesel equipment exhaust and activities.

In addition, not only the effectiveness of fog removal can be examined, but also the distance the air can remain fog-free in an upcasting ramp airflow environment. This allows a more realistic comparison of fog reduction effectiveness, given the differing temperature outputs of a direct heating process versus a dehumidification process.

To achieve this goal, a hypothetical cold climate mine is modelled for both technologies:

- Diesel equipment and warm humid air is upcast through a ramp at 150 m³/s.

- Different ambient saturated ramp air temperatures (10°C, 15°C and 20°C) with moderately heavy fog eventually forming (proposed at 400 mg/m³ in density at the test site) are tested to compare process efficiencies and effectiveness at different air temperatures.

- Airflow of 50 m³/s (one-third of the total) will be diverted and treated by dehumidification and remixed downstream with the remaining air on the ramp.

- Howden Australia has provided technical specifications for the performance of a typical refrigeration spot cooling unit.

- The heater system applies sensible heat only and is equalised by iteratively adjusting the output until giving the same fog-free distance as the dehumidification system (Figure 6).

- The resultant power consumption and budget capital costs between each equivalent system will be financially compared.

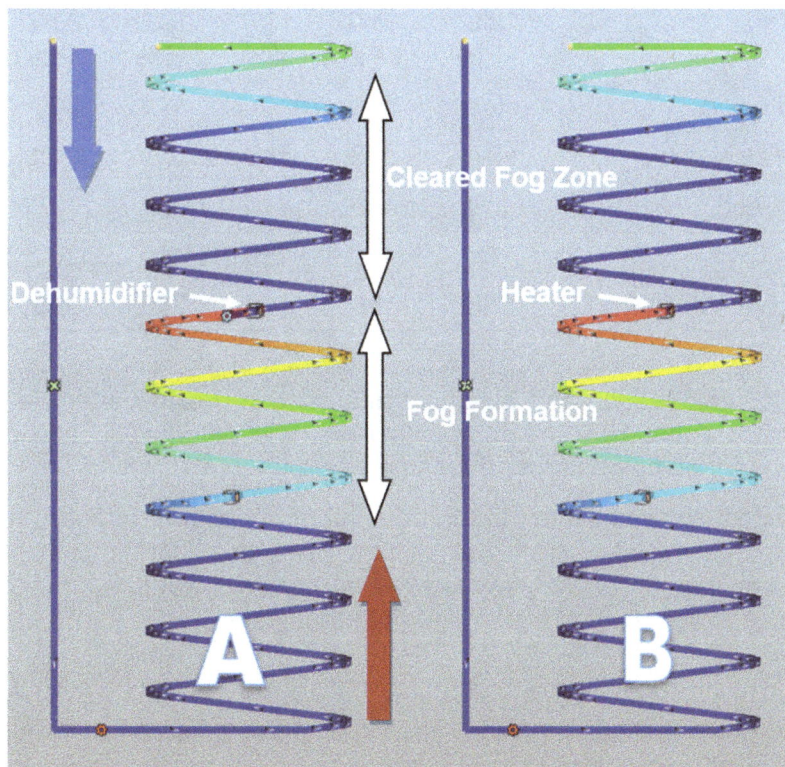

FIG 6 – Ramp cross-section view showing fog modelling of refrigerative dehumidification (a) equalised with direct heaters (b) in Ventsim™ (colour showing fog density, dark blue = clear of fog, red = thick fog).

The modelling arrangement shown in (Figure 6) demonstrates the modelling output of a refrigerative dehumidifier versus an electric heater sized to give the same fog removal distance and performance.

The colours show the relative fog concentration, with dark blue at 0 mg/m³ (clear) and thick fog in red at approximately 400 mg/m³.

RESULTS

The results in Table 1 summarise the fog reformation distance at which the fog reaches 50 mg/m³ (considered a light fog with reasonable visibility) and the distance at which the fog returns to the pre-treatment concentration of 400 mg/m³, together with output temperature and consumed power data.

TABLE 1

Modelling results for refrigerative dehumidification (Refg) versus electric heaters (Heater).

Item	Unit	CASE A		CASE B		CASE C	
Ramp temperature (100% RH)	°C	10.0		15.0		20.0	
Initial fog concentration	mg/m³	400		400		400	
Ramp clear distance (<50 mg/m³)	m	1000		1100		1100	
Ramp clear distance (<400 mg/m³)	m	2400		2300		2200	
Heater kW to Refrig Power kW ratio	ratio	3.87		3.81		3.83	
		Refg	**Heater**	**Refg**	**Heater**	**Refg**	**Heater**
Discharge conditions	wb	11.7	15.7	16.8	21.4	21.6	25.3
	db	18.1	23.3	26.9	33.1	33.1	38.2
	RH	44%	45%	33%	35%	34%	36%
Ramp mixed conditions	wb	10.8	12.1	15.6	17.4	20.5	21.9
	db	13.3	13.8	19.0	20.3	24.4	25.2
	RH	84%	83%	77%	76%	77%	76%
Moisture removed	mL/sec	110	0	179	0	204	0
Process input electrical power	kW	212	840	295	1125	300	1150
Refrigeration produced	kWr	496		717		729	
Estimated CAPEX	$k	450	150	450	200	450	200
Annual power cost (0.10 c/kWh)	$k	186	718	258	985	263	1007

Both refrigeration dehumidification and electrical heating can provide effective ramp fog-clearing results over a significant length of the ramp, however, the dehumidification process uses only 26 per cent of the energy of the heater for the same result.

The refrigeration system showed similar fog removal performance for temperatures between 10°C and 20°C. While a decrease in plant performance was noted for colder temperatures (10°C), the fog clearing distance remained similar likely because of the reduced water vapour carrying capacity of colder air and reduced evaporation of water along the roadways.

DISCUSSION

A refrigerative fog removal system is more efficient, using only a quarter of the power of equivalent electric heaters. While still arguably high in energy and capital costs, the payback period is less than six months compared to using electric heaters. The justification for installing a fog removal system will rely on the mine requirement to control visibility safety hazards, protect infrastructure and electricals from moisture and gain productivity improvements from improved traffic access and speed.

The practical challenges with using a refrigeration system for fog removal will be to design an efficient, compact, low-maintenance system that limits fouling or frequent cleaning requirements. In exhaust conditions with fumes and dust, coil fin heat exchangers will likely require regular cleaning, although a chilled water spray chamber or cooling tower could be used instead if space was available.

No attempt was made by Howden Australia to improve the efficiency of the plant for fog removal applications or maintenance and further inroads could likely be made here with detailed engineering design and optimisation. In addition, mechanical aids such as mesh droplet removal or filtering before the intake may provide a more efficient process and also help remove products that would otherwise foul refrigeration coils.

While only one plant size and airflow application were modelled in this research, a system can scale accordingly so dehumidification could be sized and distributed at strategic locations at the mine.

CONCLUSIONS

Refrigerative dehumidification for the removal of fog in underground mines appears to be an effective application of proven processes. The dehumidification process consumes only one-quarter of the energy of a comparable direct electrical heating system and provides a more effective and longer-lasting fog removal effect than some other methods such as mechanical filters, screens or mesh.

In summary, the use of refrigeration technology to remove fog in underground mines is likely to be more effective and more economically viable compared to existing non-refrigerative methods that have been previously researched, however, practical field test work would be required to confirm performance assumptions and ensure practical issues such as maintenance and cleaning can be reliably achieved.

ACKNOWLEDGEMENTS

Thank you to Howden Australia for supplying the resources and expertise to assist with plant performance concepts and calculations.

REFERENCES

Alorair, 2024. LGR Dehumidifier, Alorair. Available from: <https://www.alorair.com/what-is-the-lgr-dehumidifier.>

Federal Coordinator for Meteorology, 2005. Federal Meteorological Handbook, Washington DC, USA, Office of the Federal Coordinator for Meteorology.

Griffith, M D, 2021. Modelling the effect of fog on air density. Available from: <ttps://ventsim.invisionzone.com/topic/1322-new-feature-modelling-the-effect-of-fog-on-air-density/>

Martikainen, A, 2006. Alternative fog removal methods in mine ramps, Proceedings of the 11th US/North American Mine Ventilation Symposium, The Pennsylvania State University.

Martikainen, A, 2007. Fog removal in the declines of underground mines in sub-arctic regions, PhD Thesis, Helsinki University of Technology, Helsinki University of Technology.

Munters, 2024. Industrial Desiccant Dehumidification Rotor Technology. Available from: <https://www.munters.com/en/solutions/dehumidification/>

Schimmelpfennig, M A, 1982. Fogging in underground mine atmospheres, University of Missouri-Rolla.

Simha, R, 2012. Willis H Carrier: Father of air conditioning, *Resonance*, 17:117–138.

Teitelbaum, E, Miller, C and Meggers, F, 2023. Highway to the Comfort Zone: History of the Psychrometric Chart, *Buildings*, 13:797.

Evaluating the influence of heat stress indices on refrigeration requirements during mine shaft excavations

K Tom[1]

1. Senior Engineer, Howden, St-Bruno QC J3V 6B5, Canada. Email: kevin.tom@howden.com

ABSTRACT

Deep mines often require cooling to comply with local heat stress regulations. The impact of heat-stress indices on ventilation and cooling requirements is poorly understood. This study examines the refrigeration capacity required to develop a deep mineshaft as a function of direct (the psychrometric wet bulb temperature, WB), empirical (wet bulb globe temperature, WBGT), and rational (thermal work limit, TWL) heat stress indices. The TWL algorithm yields skin wettedness values <1 (fully wet) with wet bulb depressions >10°C and air velocities >1 m/s. Consequently, this work amended the definition for heat transfer rate of sweat evaporation from the skin (E_{sk}) for the non-fully wetted skin condition. Considering the interaction between clothing insulation values and evaporation rate, the study adapts wet clothing insulation factor as a function of skin wettedness. Thermodynamic simulations predicted a chiller capacity, sized from a wet bulb or sigma-heat (σ) balance, insufficient for continuous work per WBGT and TWL protocols at working depth. Active engineering controls like increased airflow quantity, evaporative, and secondary cooling support refrigeration equipment constraints, such as minimum collar WB and ventilation design criteria. Additional research is necessary to develop empirical methods that could facilitate the prediction of cooling requirements for mine sites that cover a range of heat stress indices.

INTRODUCTION

Background

As the demand for essential metals and minerals grows, mining operations will venture to greater depths to access ore. Deep underground mines present workers with inherent challenges such as autocompression and strata heat, leading to instances of elevated heat stress due to rising temperatures and humidity. NIOSH (2016) defines heat stress as the combined heat load from environmental and physiological factors, resulting in a net increase in body heat storage. To manage these challenges, mines utilise heat stress indices, a single value that helps quantify the level of heat stress in hot environments, to meet local heat regulations. However, the absence of universal consensus on a standardised index stems from various constraints (Epstein and Moran, 2006; NIOSH, 2016):

- Be accurate and feasible.

- Account for environmental and physiological cursors of heat stress:
  - Environmental cursors include dry bulb temperature (DB), mean radiant temperature (MRT), air velocity, and WB.
  - Physiological cursors include metabolic rate and clothing ensemble.

- Calculations or measurements that are easy to perform.

- Threshold limits that function under a wide range of metabolic and environmental conditions.

- A range that is physiologically or psychologically weighted to inform of an increased risk to health and safety.

Since the early 20th century, over 150 heat stress indices have been developed to assess thermal comfort with varying levels of industry acceptance (Roghanchi *et al*, 2016). Previous research (Graveling, Morris, and Graves, 1988; Brake, 2002; Brake and Bates, 2002; NIOSH, 2016) categorised them into three main groups based on derivation method and complexity: direct, empirical, and rational or physiological indices. Direct indices are the simplest for workplace application as they monitor environmental factors to gauge current thermal comfort levels. Empirical indices are derived from field measurements on heat strain, combining environmental and physiological parameters. Rational indices are the most complex because they attempt to solve the human heat balance shown by Equation 1:

$$S = (M-W) \pm (C + R + K) - E \qquad (1)$$

where:

S	is the net heat storage
M	is the metabolic rate
W	is the mechanical work done on the human body
C	is the convective heat transfer between body and environment
R	is the radiative heat transfer between body and environment
K	is the heat conductance exchange
E	is the evaporation rate

All terms have units in W/m^2.

Roghanchi *et al* (2016) suggested 1.5 m/s air velocity over the skin for optimal thermal comfort. They recommended using rational or empirical indices during design, including a simple index with air velocity for thermal management. Several studies highlighted issues with some empirical and rational indices: the need for continuous physiological monitoring, inaccurate estimates of metabolic rates, time-weighted exposure estimates, or globe temperature measurement (Brake, 2002; Epstein and Moran, 2006; NIOSH, 2016). Modern heat stress monitors simplify globe temperature recording for MRT estimation (NIOSH, 2016). Brake (2002) introduced the rational thermal work limit (TWL) index to address these issues but lacks a predictive method for clothing insulation with fully wetted skin, a crucial assumption. Additionally, TWL may yield skin wettedness values below 1 under specific conditions of high wet bulb depression (DB–WB) and air velocities over 1 m/s.

Motivation

Deep mines that wish to minimise disruptions to production, and maintain compliance with occupational legislations on heat stress, typically rely on engineering controls like increased ventilation and mechanical refrigeration to protect its workers. Beyond a certain depth, artificial cooling becomes the only feasible option. Implementing additional work/rest schedules may reduce exposure times but could impact production for externally paced operations when heat stress exceeds limits. This study explores how TWL and WBGT indices influence ventilation and cooling design compared to a basic heat stress index like the WB. The study adapts a simple regression model of wet clothing insulation factor as a function of skin wettedness to partially address TWL's limitation.

METHOD

Wet bulb temperature

The WB is commonly used to gauge heat strain and is favoured for mine cooling design and regular underground climate monitoring due to its sensitivity and ease of measurement (Webber *et al*, 2003). It is best suited for humid underground environments with sufficient air movement and minimal radiant heat (NIOSH, 2016). Limiting WB varies slightly by region with air velocity used as an additional safety net. Belle and Biffi (2018) correlated productivity loss as a function of WB based on operational data from a gold mine and is given by Equation 2. Although this method demonstrates potential in prescribing a value for the limiting WB, it is unclear if the trend can be extended universally due to differences in population acclimatisation and underground climatic conditions:

$$\text{Productivity Loss (\%)} = 0.0383 \, WB^3 - 1.8787 \, WB^2 + 23.067 \, WB \qquad (2)$$

A distinguishing feature of the WB is that it quantifies the energy of the dry air. By extension, it can predict the air's heat absorption capacity before requiring refrigeration through the cooling power surplus (CPS) described in Equation 3:

$$CPS = \dot{m}(\sigma_{reject} - \sigma_{in}) \tag{3}$$

where:

$\dot{m}$ is the mass flow of dry-air (kg/s)

σ is the sigma heat (kJ/kg) with subscripts denoting the reject and inlet WB, respectively.

Sigma heat is calculated from Equation 4 (McPherson, 1993):

$$\sigma = 1.005WB + \omega(2502.5 - 2.386WB) \tag{4}$$

where ω is the absolute humidity (kg_v/kg_a) with the subscripts, 'v' and 'a', signifying water vapor and dry-air, respectively.

Some mines, particularly those located in South Africa and Australia, may impose additional constraints like a maximum DB (ie 37°C) or an air velocity (V) of at least 0.25 m/s in conjunction to the limiting WB within a risk management plan that objectively categorises an environment as being hot (Webber *et al*, 2003).

Wet bulb globe temperature

Due to the ease of calculation and exposure limits developed from both environmental and physiological predictors of heat stress, the WBGT is widely preferred as the index of choice by many occupational hygiene organisations (Golbabaei *et al*, 2021). In underground mines with air temperatures measured away from radiant heat sources, the index can be expressed by Equation 5:

$$WBGT = 0.7 \times WB_n + 0.3 \times DB \tag{5}$$

where WB_n = natural wet bulb (°C), which can be obtained from Equation 6 (Carter *et al*, 2020):

$$WB_n = DB - C (DB - WB) \tag{6}$$

where $C = 0.85$ for $V < 0.03$ m/s; $C = 0.96 + 0.069 \log_{10} V$ for 0.03 m/s $\leq V \leq 3$ m/s; $C = 1$ for $V > 3$ m/s.

Despite the correlation mentioned above, Brake (2002) showed that WB_n lacked sensitivity to wind speed, highlighting a weakness of the WBGT. The author suggested WBGT's validity in temperate climates but noted its impracticality in the tropics due to overly restrictive exposure limits for continuous work at moderate metabolic rates. Claassen and Kok (2007) investigated WBGT combinations between 24°C to 32°C at 30 per cent and 70 per cent relative humidity (RH), finding that the index underestimated heat strain above 30°C at higher RH. However, their study used a metabolic rate that did not adhere to the recommended protocol on work-rest periods for the index. Ramsey (1978) addressed some shortcomings by introducing adjustment factors for wind speeds above 1.5 m/s, gender, acclimatisation, clothing, body shape, and age. This study adopts these recommendations, with specified limiting values detailed in Table 1.

TABLE 1

OHSA-recommended WBGT exposure limits permitting continuous work for acclimatised workers.

Metabolic rate* (W/m²)	Exposure limit (°C)	
	V < 1.5 m/s	V ≥ 1.5 m/s
Light (<130)	30	32.2
Moderate (130–194)	27.8	30.6
Heavy (>194)	26.1	28.9

*1.8 m² body surface area.

According to Brake and Bates (2001), the metabolic rate is a variable that can be used to compare different classes of heat stress indices. Following the authors' approach, the metabolic rates for continuous work in Table 1 can be expressed more precisely by Equations 7 and 8:

$$M = e^{(55.764-WBGT)/5} \text{ (high velocity)} \tag{7}$$

$$M = e^{(53.764-WBGT)/5} \text{ (low velocity)} \tag{8}$$

NIOSH's recommended exposure limits provide trends for non-continuous work. Additionally, it is important to recognise that exposure limits on the WBGT scale may vary slightly among industrial hygiene organisations due to differences in the cut-off range for metabolic rates (NIOSH, 2016).

Thermal work limit

The rational TWL heat stress index predicts the maximum work rate permissible under a set of environmental and physiological parameters and is most suitable in tropical climates for self-pacing, acclimatised individuals (Brake, 2002). The algorithm simplifies Equation 1 with the storage, mechanical work and conductance terms set to 0 at steady state: $(M - B) \pm (C + R) - E = 0$, where B is an added heat-loss term for respiration. Its main advantages over other indices include:

- Metabolic rate not required as an input. The algorithm is applicable over a wide range of work rates: 60 W/m² (light) to 380 W/m² (heavy).

- Sensitive to radiant heat.

- Sensitive to wind speed between 0.2 m/s to 4 m/s.

- Clothing insulation as an input.

- Sweat rate correlated to weighted sum of skin and deep body core temperature.

A critical assumption is that skin wettedness (w), expressed by Equation 9 as a ratio of actual to maximum evaporation rate of sweat from the skin, is maintained at 1 (ie fully wetted):

$$w = E/E_{max} \tag{9}$$

Consequently, evaporation is always maximised, resulting in less conservative limiting metabolic rates.

On the contrary, the current study demonstrates that non-fully wetted skin occurs under specific environmental conditions (ie wet bulb depressions >10°C and air speed >1 m/s). The actual evaporation rate can be computed from the work by Brake and Bates (2002) with an improved regression function shown in Table 2:

TABLE 2
Estimation of actual evaporation rate of sweat.

Case	E (W/m²)
$\lambda S_r E_{max}^{-1} < 0.46$	λS_r
$0.46 \leq \lambda S_r E_{max}^{-1} \leq 1.7$	$\lambda S_r e^{(0.2704-0.5104\lambda S_r/E_{max})}$
$\lambda S_r E_{max}^{-1} > 1.7$	E_{max}

where λ is latent heat of vaporisation evaluated at 30°C (2430 kJ/kgK) and S_r is the sweat rate (kg/m²/h).

E_{max} can be determined from Equation 10 by setting w to 1:

$$E_{sk} = w\, F_{pcl}\, f_{cl}\, h_e\, (p_{g,sk} - p_v) \tag{10}$$

where F_{pcl} is the permeation efficiency, f_{cl} is the clothing surface area factor, h_e is the evaporative heat transfer coefficient of a nude person (W/m²/°C), $p_{g,sk}$ is the saturated water vapor pressure evaluated at the mean skin temperature (kPa), and p_v is the water vapor pressure in ambient air (kPa).

While clothing thermal insulation decreases when wetted (Brake, 2002), the standard TWL formulation does not readily predict this. Wang et al (2015) studied the relationship between applied water (100 g to 700 g) and the decrease in clothing thermal insulation, termed the apparent wet thermal insulation factor, on various clothing ensembles. This paper adopts this approach, shown by

Equation 11, by equating the maximum moisture content to fully wetted skin and normalising the scale accordingly. Additionally, the author suggests validating the correlation between applied absolute moisture on clothing and skin wettedness, as Wang *et al* (2015) ensured no water dripping during the experiments, while sweat was observed to drip before reaching fully wetted skin (Brake and Bates, 2002).

$$\text{Apparent Wet Thermal Insulation Factor} = 19.277w^3 - 59.78w^2 + 64.75w \qquad (11)$$

Thus, the nominal dry clothing insulation value is specified and the modified TWL algorithm subsequently iterates for both a mean skin temperature and apparent wet thermal insulation factor (both F_{pcl} and f_{cl} depend on clothing thermal insulation) when solving the heat balance equation.

CASE STUDY

Model description

Currently, a non-hard rock operation is sinking multiple shafts using a road header-like machine. Due to uncertainty about the level of heat stress exposure at maximum depth and a lack of agreement on the heat stress index, the mine sought external guidance to update the outdated heat management program. As part of the due diligence, a simplified ventilation model (see Figure 1) was developed to understand the influence of the major heat sources on the cutting machine's service and work decks.

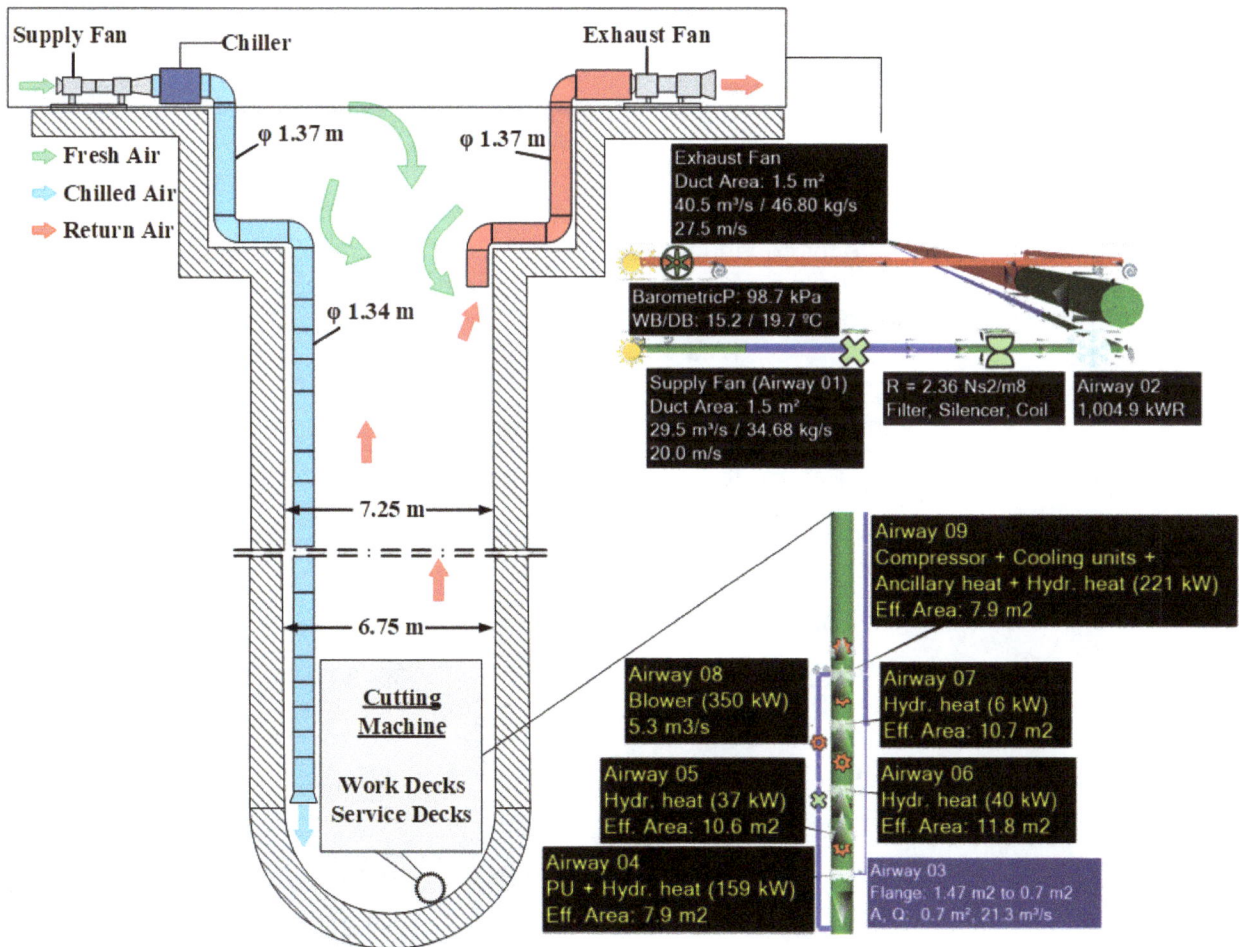

FIG 1 – Shaft sinking operation schematic with inset of ventilation model (amended from Tom, 2023).

The push-pull ventilation system comprises a dual-stage intake auxiliary fan with nominal 500 kW motor, and a single-stage exhaust fan with a 200 kW motor. Fresh air is delivered to the shaft bottom through a single duct column that contracts from 1.37 m to 1.34 m diameter. Downstream of the fan assembly and inline with the duct lies a 2 MWr bulk air cooler (BAC) that chills the air to a specified WB at the shaft collar. The cool air ventilates several service and work decks situated on the cutting machine. The inset of the cutting area highlights the heat sources along the Airways. Notably, air exiting 'Airway 03' splits two ways: a small percentage travels through another duct connected to a muck filtration system ('Airway 08'), while the balance flows upward to the upper decks. The two air streams subsequently mix at 'Airway 09' before exiting the shaft via the exhaust fan.

Design criteria

Design parameters of the ventilation model and input conditions of the TWL algorithm are provided in Table 3. The rationale behind the most impactful values is summarised below:

- A reject WB of 28°C provides a productivity rate of 87 per cent, which includes breaks and time allotted to reach working depth in the shaft.

- Shaft wetness is assumed to be most wetted at the cutter location due to dust mitigation, concrete curation, and heat dissipation at the cutting head. Away from the equipment, the shaft progressively becomes drier.

- Duct friction factor, leakage porosity, and heat transfer parameters are both best practices and informed by the duct supplier.

- The most accepted value for deep body core temperature before excess heat strain sets in is 38.2°C (Brake, 2002). However, the standard TWL algorithm permits the user to adjust the target value with a hardcoded maximum set to 39.5°C.

- Clothing insulation assumed from standard cotton overalls with under garments, and standard safety work boots. This value is marginally greater than the 0.55 specified in previous works by Brake (2002).

- In the absence of globe temperature data, MRT is assumed equal to DB.

- The limiting metabolic rate for continuous work for TWL is 220 W/m². Objectively, this equates to heavy work with a standard body surface area of 1.8 m². It is equivalent to 28.8°C WBGT (high velocity) in the present work.

TABLE 3

Ventilation design and heat stress parameters.

Parameter	Units	Value
Ambient WB/DB	°C	15.1/19.7
Minimum velocity	m/s	0.5
Collar WB, reject WB	°C	6; 28
Shaft wetness	%	30–80
Duct K	kg/m³	0.0013
Duct leakage porosity	mm²/m²	25
Duct thickness	mm	3
Duct thermal conductivity	W/(m°C)	0.48
Deep body core temp	°C	38.2
Dry clothing insulation	Clo	0.6
MRT	°C	= DB
M	W/m²	220
Limiting TWL WB/DB	°C/°C	32/44

RESULTS AND DISCUSSION

Baseline

The baseline scenario involves the energy balance method according to the CPS. Table 4 describes the results in detail. The decrease in air velocity corresponds to an increase in area as the fresh air exits the duct. There is an approximate 20 per cent mass flow leakage from surface (Airway 1) to work deck entry (Airway 3), which proportionally reduces the CPS. It is observed that the CPS decreases asymptotically as the air sequentially absorbs more heat with each work and service deck (Airways 4 to 7, and 9). Consequently, the predicted return WB profile is mostly favourable with a chiller capacity set to 1 MWr.

TABLE 4
Baseline Airway properties.

#	ρ (kg/m³)	Q (m³/s)	V (m/s)	(WB/DB)$_{in}$ (°C/°C)	(WB/DB)$_{out}$ (°C/°C)	CPS (kW)	WBGT$_{in}$ (°C)	WBGT$_{out}$ (°C)	M$_{TWL}$ (W/m²)
1	1.19	29.4	20.0	15.1/19.7	18.1/26.5	1600	16.5	20.6	N/A
2	1.21	28.2	19.2	17.9/26.5	6.0/6.0	1272	20.5	6.0	N/A
3	1.3	21.2	32.1	21.6/41.1	21.5/41.1	621	27.5	27.4	309
4	1.28	16.2	2.1	21.6/41.2	24.0/47.6	470	27.7	31.4	285
5	1.27	16.4	1.6	24.0/47.5	24.5/48.1	309	31.4	32.0	235*
6	1.27	16.7	1.4	24.5/48.0	25.0/48.0	277	32.0	32.4	222*
7	1.27	16.7	1.6	25.0/48.1	25.1/47.0	238	32.3	32.1	216*
9	1.21	23.2	2.9	27.8/57.9	29.7/64.6	24	37.0	40.4	118*

* Presented for completion; however, DB exceeds design limits in TWL algorithm.

Conversely, the DB typically rises faster than the WB due to the predominance of sensible heat. As a result, the DB on most working levels causes WBGT to be non-compliant. The TWL generally allows for continuous work and is perceived as less stringent in assessing heat stress compared to the WBGT. However, the TWL imposes constraints on both WB and DB to prevent incorrect thermal stress evaluations. For example, work duration in Airway 5 and above should be reduced when DB exceeds 44°C, despite optimistic estimates of the limiting metabolic rate.

Apart from strata, the heat sources in Figure 2 were either defined or calculated as sensible heat. Hydration heat was reclassified as a latent source to explore the impact of increased ω on WBGT and TWL scales. This analysis is complemented by scenarios enhancing the CPS through increased refrigeration, airflow, and reduced equipment heat. Duct insulation was deemed impractical due to space limitations in the shaft; alternatively, duct heat transfer can be reduced by increasing flow rate, constrained by the maximum fan motor size and the performance limits of the BAC to maintain a 6°C WB at the shaft collar without exceeding the 2 MWr surface cooling capacity.

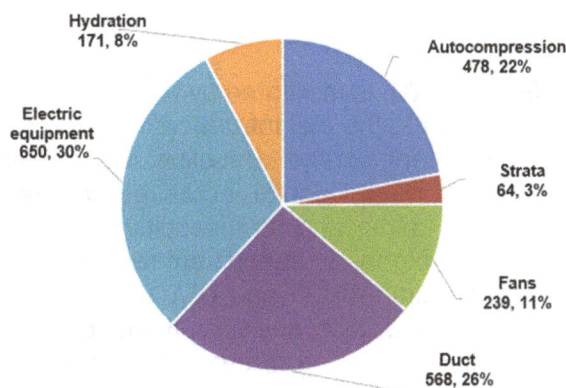

FIG 2 – Baseline heat load distribution (reproduced from Tom, 2023).

Four scenarios have been designed to further investigate WBGT and TWL compliance:

1. Enhanced flow: increase supply fan flow to 38 m³/s.

2. Enhanced flow + chiller: identical to scenario 1, with a remote cooling coil installed mid-shaft. The coil refrigerates the air down to 6°C WB.

3. Enhanced flow + equipment cooling: identical to scenario 1, with separate water cooler that reduces electric equipment heat from 650 kW to 460 kW.

4. Enhanced flow + evaporative cooling: identical to scenario 1, with a moisture load (in ml/s) applied at the outlet of Airway 3.

Energy balance

Energy balance analysis using σ is a straightforward method for sizing refrigeration systems. Figure 3 illustrates the energy balance for all scenarios, with the stacked column representing heat sources and CPS assessed upstream of the supply fan, derated by approximately 20 per cent to account for mass flow leakage. Refrigeration quantity is determined to achieve the WB at the collar and mid-shaft (in the enhanced flow scenario with chiller). For evaporative cooling, a moisture load equivalent to the mid-shaft cooler (430 kWr) is applied before the first work deck. Consequently, the two cases with additional cooling have nearly identical refrigeration capacities. As expected in the baseline, there is minimal overlap between CPS and refrigeration. However, in other cases, overlap arises from pushing ventilation and cooling equipment to their limits. Evaporative cooling exhibits the largest overlap, benefiting from better positional efficiency compared to the mid-shaft chiller option. Additionally, it generates the least duct heat, corresponding to the lowest DB. Though unrelated to heat stress, a significant concern with evaporative cooling is fog formation due to air decompression, necessitating separate treatment. Interestingly, strata heat is highest with the mid-shaft chiller option, likely due to greater thermal and moisture transfer potentials.

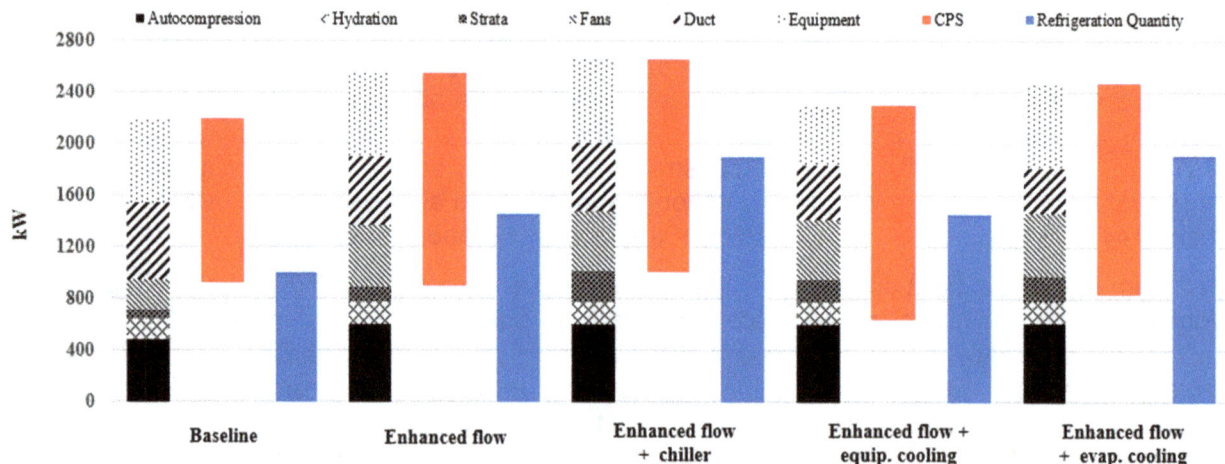

FIG 3 – Energy balance chart.

Limiting metabolic rate

The comparison between WBGT and TWL is facilitated by the limiting metabolic rate, as depicted in Figure 4. In addition to the five scenarios, the plot introduces dry and wet conditions, with the latter treating hydration heat as a fully latent source. Baseline data reveals an unacceptable work environment due to very high radiant heat. Despite increased flow rates, there is only a modest improvement on the WBGT scale, underscoring its insensitivity to air velocity. Conversely, TWL demonstrates significant benefits from higher air speeds but remains non-compliant due to high DB. Additional cooling and humidity render both WBGT and TWL fully compliant, with the latter achieving a greater safety margin. Enhanced evaporation due to humidity plays a favourable role and helps mitigate uncertainty in thermodynamic modelling. Furthermore, equipment capacities can be fine-tuned during the design stage using the limiting metabolic rate (ie 350 kWr instead of 430 kWr mid-shaft chiller). Interestingly, productivity for WBGT dry cases, calculated as a ratio of residual work capacities assuming a resting M of 60 W/m² (ie 110–60/(220–60)), aligns closely with interpolating between continuous and 25 per cent work rate trends using NIOSH data. These considerations can inform the business case for implementing targeted cooling systems and balancing worker comfort and productivity.

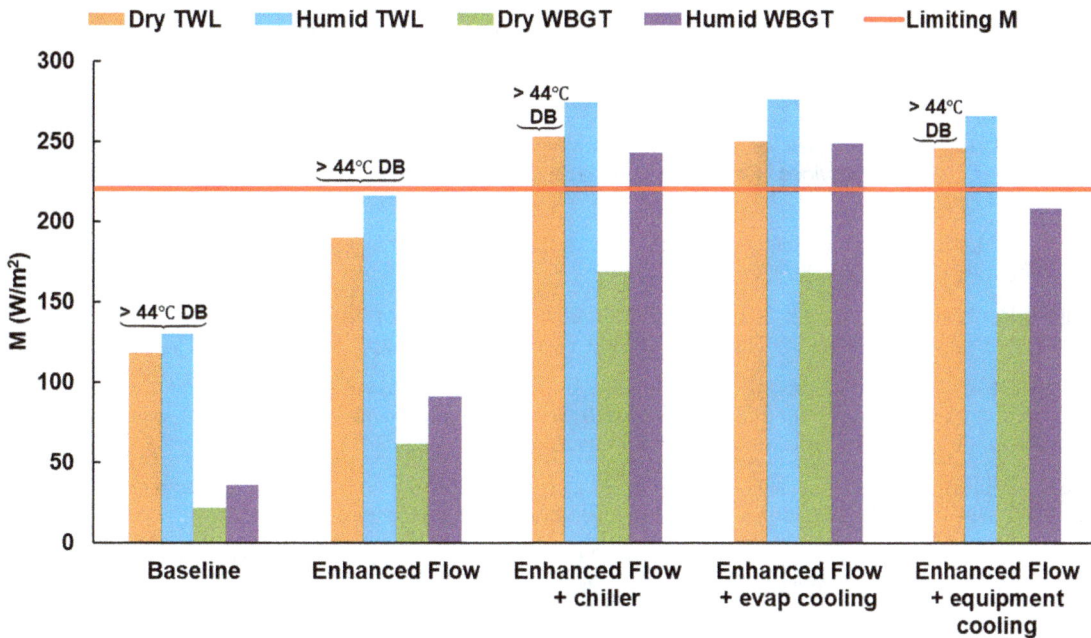

FIG 4 – Limiting metabolic rates in Airway 9.

CONCLUSIONS

Ventilation and refrigeration equipment sizing during shaft sinking at maximum depth was investigated using three different heat stress indices: WB, WBGT, and TWL. While WB provided a theoretical evaluation through the CPS, it failed to address the high level of heat stress at significant wet bulb depressions with large MRT. Applying the apparent wet thermal insulation factor to the standard TWL allowed for estimating clothing insulation as a function of skin wettedness, thereby making the index less conservative. The study's limiting metabolic rate showed that WBGT was insensitive to increased velocity, and both TWL and WBGT derived maximum benefit from increased humidity under the evaporative cooling option. While WBGT proved more conservative than TWL, it was simpler to apply, with only marginal differences in continuous work compliance noted at higher humidity levels. The study demonstrated that equipment capacities could be adjusted iteratively using the limiting metabolic rate. Developing empirical methods to correlate the limiting metabolic rate with equipment capacity changes could improve the design process. Additionally, a separate study to confirm the relationship between skin wettedness and the apparent wet thermal insulation factor using typical mining coveralls is strongly recommended.

REFERENCES

Belle, B and Biffi, M, 2018. Cooling pathways for deep Australian longwall coal mines of the future, *International Journal of Mining Science and Technology,* 28(6):865–875.

Brake, D J, 2002. The deep body core temperatures, physical fatigue and fluid status of thermally stressed workers and the development of thermal work limit as an index of heat stress, Doctoral thesis, Curtin University of Technology. http://hdl.handle.net/20.500.11937/621

Brake, D J and Bates, G, 2001. A valid method for comparing rational and empirical heat stress indices, *Annals of Occupational Hygiene,* 46(2):165–174.

Brake, D J and Bates, G, 2002. Limiting metabolic rate (thermal work limit) as an index of thermal stress, *Applied Occupational and Environmental Hygiene,* 17(3):176–186.

Carter, A W, Zaitchik, B F, Gohlke, J M, Wang, S and Richardson, M B, 2020. Methods for estimating wet bulb globe temperature from remote and low-cost data: A comparative study in Central Alabama, *GeoHealth,* 4:1–16.

Claassen, N and Kok, R, 2007. The accuracy of the WBGT heat stress index at low and high humidity levels, *Occupational Health Southern Africa,* 13(2):12–18.

Epstein, Y and Moran, D S, 2006. Thermal comfort and the heat stress indices, *Industrial Health,* 44:388–398.

Golbabaei, F, Assour, A A, Keyvani, S, Kolahdouzi, M, Mohammadiyan, M and Ramandi, F F, 2021. Limitations of WBGT index for application in industries: A systematic review, *International Journal of Occupational Hygiene,* 13(4):365–388.

Graveling, R A, Morris, L A and Graves, R J, 1988. Working in hot conditions in mining: a literature review (Technical Report TM/88/13), Institute of Occupational Medicine, Edinburgh.

McPherson, M J, 1993. *Subsurface Ventilation and Environmental Engineering* (Springer: New York).

NIOSH, 2016. Criteria for a recommended standard: Occupational exposure to heat and hot environments, US Department of Health and Human Services, Centers for Disease Control and Prevention, National Institute for Occupational Safety and Health.

Ramsey, J D, 1978. Abbreviated guidelines for heat stress exposure, Technical Publication 78–18, Texas Tech University. https://doi.org/10.1080/0002889778507794

Roghanchi, P, Sunkpal, M, Carpenter, K and Kocsis, C, 2016. Application of heat stress indices in underground mines, poster session, presented at the Northern Nevada Section of SME, https://doi.org/10.13140/RG.2.2.18728.16643

Tom, K, 2023. Case study: Refrigeration requirements during mineshaft excavation as a function of heat stress index, in *Underground Ventilation: Proceedings of the 19th North American Mine Ventilation Symposium (NAMVS)* (ed: P Tukkaraja), pp 200–207 (Taylor and Francis). https://doi.org/10.1201/9781003429241

Wang, F, Shi, W, Lu, Y, Song, G, Rossi, R M and Anaheim, S, 2015. Effects of moisture content and clothing fit on clothing apparent 'wet' thermal insulation: A thermal manikin study, *Textile Research Journal*, 7 p.

Webber, R C W, Franz, R M, Marx, W M and Schutte, P C, 2003. A review of local and international heat stress indices, standards and limits with reference to ultra-deep mining, *The Journal of The South African Institute of Mining and Metallurgy*, pp 313–324.

Leakage and goaf and cave management

Characteristics of self-extinction in coalmine dead-end roadway fires with different sealing configurations and sealing ratios

J X Wang[1], B Wu[2], C Li[3] and L S Huang[4]

1. PhD candidate, China University of Mining and Technology, Beijing 100083, China. Email: wjxin0622@126.com
2. Professor, China University of Mining and Technology, Beijing 100083, China. Email: wbelcy@vip.sina.com
3. PhD candidate, China University of Mining and Technology, Beijing 100083, China. Email: cumtb_lic@163.com
4. PhD candidate, China University of Mining and Technology, Beijing 100083, China. Email: anquanhls@163.com

ABSTRACT

The isolated fire extinguishing method is a safe and effective measure to extinguish coalmine dead-end roadway fires. This paper experimentally investigated the combustion states and temperature distribution of dead-end roadway fires under varying sealing conditions using a small-scale model, considering different sealing configurations and sealing ratios as well as the fire size. Results show that under unsealed conditions, the flame exhibits a tilting behaviour and gradually detaches from the oil pan, resulting in a ghosting flame phenomenon. When the dead-end roadway is sealed, dead-end roadway fires always undergo a transition from fuel-controlled combustion to ventilation-controlled combustion, and after a period, the fire source will self-extinguish. The fire self-extinction time is related to sealing conditions. Specifically, larger fire sizes or higher sealing ratios lead to shorter self-extinction times. Notably, downward sealing accelerates extinguishment compared to upward sealing, with minimal differences observed in self-extinction times when the sealing ratio reaches 90 per cent. Moreover, downward sealing induces higher CO levels compared to upward sealing, with a peak difference of approximately 250 ppm. Besides, implementing sealing strategies effectively suppresses fire source combustion and reduces overall smoke temperatures within the dead-end roadway. These findings underscore the efficacy of sealing strategies in inhibiting fire development and enhancing safety in such environments. The insights gleaned from this study provide valuable experimental data that can enhance understanding of dead-end roadway fire developments and help decision-making processes for firefighters engaged in rescue operations.

INTRODUCTION

Safe production is the primary goal of coal mining in every country (Wu *et al*, 2023). As one of the main mine disasters, mine fire has always been a key concern of governments (Fan *et al*, 2017; Wang *et al*, 2021). As a special and widely existing roadway structure in coalmines, the dead-end roadway (ie blind roadway) has one closed end, and another end is usually connected with a main roadway (Yao *et al*, 2023). Due to the large amount of coal and electrical equipment, dead-end roadway fires are very likely to occur close to the heading face. At the initial stage, the dead-end roadway fire can be directly extinguished. However, when the fire is not extinguished immediately, a large amount of smoke will accumulate in the dead-end roadway, and there is a threat to disaster relief personnel by directly extinguishing the fire. At this moment, to control the fire quickly, sealing strategies are often used to cut-off the oxygen supply to extinguish the fire. The main sealing configurations are firewalls and fire shutters in coalmines, which have different placement positions on the roadway cross-section. The firewall is placed upward from the floor, and the fire shutter is placed downward from the ceiling. Different sealing configurations have different inhibitory effects on fire. In this study, to distinguish the two sealing configurations, the definitions of upward sealing and downward sealing are proposed. In addition to the sealing configurations, the sealing distance and sealing ratio are the key factors to inhibit fire development. The closer the sealing position to the fire source, the better the fire extinguishing effect, but also more dangerous. Therefore, the only portal of the dead-end roadway is the common sealing position. In actual rescue, high-temperature toxic gas and collapse of the sealing wall caused by high pressure (You *et al*, 2020) may cause life danger to firefighters, so it is difficult to completely seal the roadway. In fact, when the sealing ratio reaches a certain value, the effect of suffocation extinguishing is basically the same as that of complete sealing. Therefore, it is extremely important to consider combining the sealing configuration and sealing ratio to achieve efficient and safe sealing operation.

Since the single-hole tunnel structure and the dead-end roadway structure are similar to some extent, the research on tunnel fire can play a positive role in dead-end roadway fire control. Using a 1/9 model-scale tunnel, Chen *et al* (2016) conducted an experimental study on the effects of asymmetric and symmetric sealing on fire behaviour and temperature distribution. It was found that when the tunnel portals were asymmetrically sealed, the smoke temperature was higher at the side with less sealing ratio. When the tunnel portals were symmetrically sealed, the temperature beneath the ceiling varied versus the sealing ratio. Additionally, related to the fuel area, there was a critical sealing ratio, at which the ceiling temperature would be at its highest. Yao *et al* (2019a; 2019b) studied the influence of ventilation conditions on fire combustion state by both medium-scale and model-scale tunnels, and the characteristics of ventilation-controlled fire and fire self-extinction were analysed, which provided a theoretical basis for fire rescue. Huang *et al* (2018) numerically observed the impacts of sealing ratio on fire behaviour. It was found that the ceiling temperature increased with the increasing sealing ratio when the heat release rate was small. Additionally, a prediction model for the longitudinal temperature distribution and the maximum ceiling temperature was proposed. Liu *et al* (2019) experimentally studied the temperature distribution in an enclosed utility tunnel fire and put forward a modified ceiling maximum temperature prediction model. Wang *et al* (2019) studied the self-extinction of tunnel fire by using a 1/20 model-scale tunnel. It was found that the smoke layer would descend to the floor at a certain distance away from the fire source in the long tunnel with a large fire size. Due to the entrainment of fresh air and smoke, the backflow of air toward the fire seat was highly impaired. When the oxygen content is reduced to the flammable limit, the fire source will self-extinguish. Han *et al* (2020) experimentally and numerically investigated the smoke movement in an inclined tunnel fire under a restricted environment with one end closed. A prediction model of temperature distribution was proposed.

In recent years, a small number of scholars have initiated research on the sealing of fire areas in mine fires (Fan *et al*, 2017; Guo *et al*, 2022; Wang *et al*, 2023a, 2023b; Yao *et al*, 2023), but this area remains underexplored. In practical mine fire extinguishing applications, the isolation fire extinguishing method is commonly employed. However, the literature review reveals a significant gap in research focusing on fire sealing and extinguishing specifically in coalmine dead-end roadway fires. Given the substantial structural and ventilation mode distinctions between dead-end roadways and other narrow passages in roadways, investigating dead-end roadways holds significant importance for guiding fire rescue efforts in these environments. With this in mind, several small-scale experiments were conducted to examine the combustion state of dead-end roadway fires under various sealing configurations and ratios. Parameters such as flame shape, gas volume fraction, and roadway temperature were meticulously measured and recorded. The outcomes of this study are anticipated to provide valuable guidance for dead-end roadway fire control strategies.

MATERIAL AND METHODS

Figure 1 illustrates the detailed structure of the model roadway. The dead-end roadway measures 0.32 m in width and 0.24 m in height, connecting to the main roadway of the same dimensions at the entrance. The dead-end roadway spans a total length of 6 m, while the main roadway extends for 8 m. Its front side is constructed with 5 mm thick fireproof glass, while the remaining sidewalls consist of 5 mm thick steel plates. Temperature measurements were conducted using Type K thermocouples with a range of 0–1000°C and an error margin of approximately 1.5°C. Evenly distributed were 27 thermocouple trees within the dead-end roadway, spaced 0.2 m apart, with varying intervals from the ceiling ranging from 0.01 to 0.23 m. Notably, the thermocouple tree positioned above the fire source only featured three upper thermocouples to prevent interference with the oil pan. Detailed thermocouple arrangements are depicted in Figure 2. Additionally, three gas analysers (#1, #2, #3) were positioned at distances of 0.4 m, 3.5 m, and 5.5 m from the heading face to measure gas concentrations, with respective intervals from the ceiling.

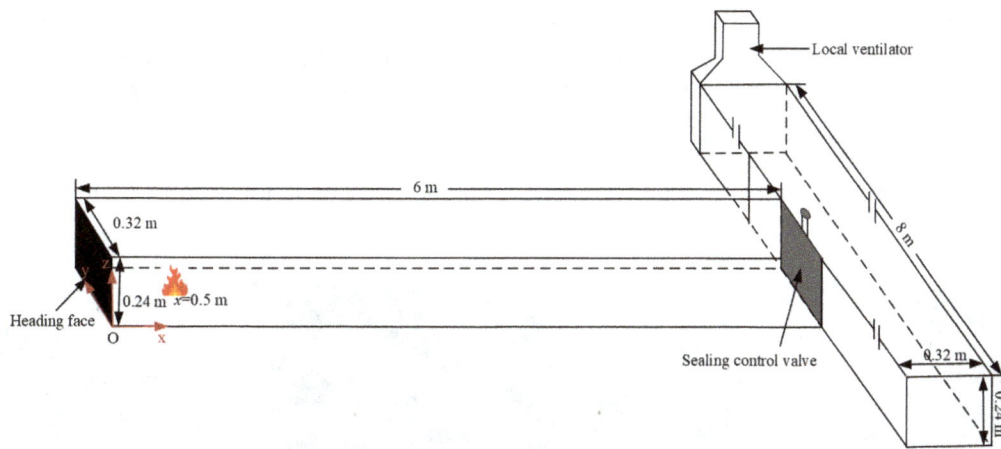

FIG 1 – Dead-end roadway structure.

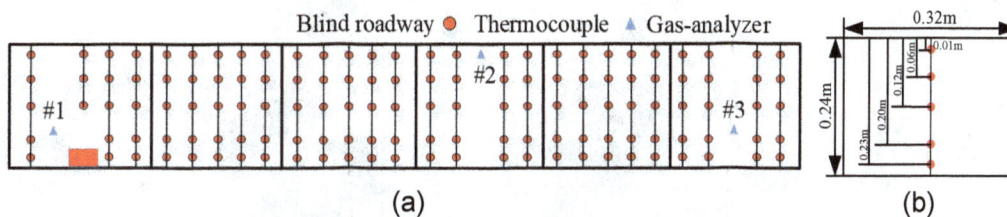

FIG 2 – Measurement point arrangement and Liquid-level stabilisation device: (a) Measurements positions; (b) Thermocouple tree; (c) Liquid-level stabilisation device.

The combustible materials within the coalmine roadway are predominantly solid, including wood, belts, and coal. Additionally, there are liquid combustibles like hydraulic fluid. The solid combustibles found in coalmine roadways are often mixtures with variable properties. Sometimes, the combustion of solid fuels exhibits smoldering, a phenomenon markedly different from liquid fuels in terms of thermal feedback mechanism, combustion mode, and flame shape. Thus, researchers opt for liquid fuels as fire sources in studies related to mine roadway fires. This choice facilitates better observation of flame characteristics, simplifies combustion rate control, and enhances experiment reproducibility and accuracy. To facilitate calculation and control of the combustion rate of the fire source (Yao *et al*, 2019a, 2019b), n-heptane oil pans were utilised in experiments, designed in sizes of 4 cm × 4 cm, 6 cm × 6 cm, and 8 cm × 8 cm, respectively. Employing the communicator principle, the oil pan's bottom was connected to a level stabiliser via a U-shaped hose to maintain a constant fuel level and eliminate fuel supply influence on fire combustion. To monitor the mass loss rate, a liquid-level stabilisation device was positioned on an electronic balance with a precision of 0.1 g, see Figure 2c.

The connection position between the main roadway and the dead-end roadway was a sealing control valve, which could change the sealing configuration and sealing ratio. The sealing configurations included downward sealing and upward sealing. There was sufficient air to support combustion and there was no discernible impact on the development of a fire when the sealing ratio at the only portal of the dead-end roadway was less than 50 per cent (Chen *et al*, 2016). Therefore, as shown in Figure 3, the sealing ratios in this study were set at 0 per cent, 50 per cent, 60 per cent, 70 per cent, 80 per cent, 90 per cent and 100 per cent. There were 36 tests performed, and every step of the experiment was captured by a camera. The experimental conditions are summarised in Table 1. The local ventilation device was not set since the local ventilation device in the dead-end roadway needed to be stopped when the sealing was implemented.

FIG 3 – Flame shapes (sealing ratio is 0).

TABLE 1

Fire scenarios.

Pool size (cm × cm)	Sealing configuration	Sealing ratio (%)
4 cm×4 cm/ 6 cm×6 cm/ 8 cm×8 cm	Downward sealing/ Upward sealing	0%, 50%, 60%, 70%, 80%, 90%, 100%

RESULTS AND DISCUSSION

Flame shape

When the sealing ratio is 0, the obvious ghosting flame was observed during the experiment. Taking the fire source with a size of 6 cm × 6 cm as an example, Figure 3 illustrates the evolution of flame shape over time with a sealing ratio of 0. At 900 sec, the flame notably tilts towards the heading face, with the degree of inclination progressively increasing as combustion proceeds. This tilting phenomenon primarily arises due to smoke accumulation at the heading face, leading to temperature and pressure imbalances on either side of the fire source (Yao *et al*, 2017), which intensify throughout the combustion process. From 1140 sec, the flame shape changes from a relatively steady flame to a fluctuated flame (You *et al*, 2020), with the degree of fluctuation steadily escalating, as shown in Figure 3a–3e. In the time range of 1320–1440 sec, part of the fire flame gradually begins to burn at the edge of the oil pan and swings constantly, as shown in Figure 3f. Subsequently, the flame completely detaches from the oil pan. At about 1500 sec, the flame returns to the edge of the oil pan to burn, as shown in Figure 3f–3h. After 1500 sec, the flame movement repeats the above phenomenon, albeit with less apparent periodicity. When the sealing ratio is not zero, the fire source will extinguish under most working conditions, with no significant change in flame position observed, taking Figure 4 as an example.

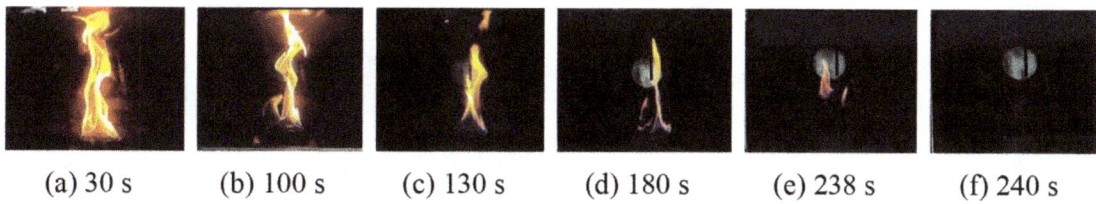

(a) 30 s   (b) 100 s   (c) 130 s   (d) 180 s   (e) 238 s   (f) 240 s

FIG 4 – Flame shapes (6 cm × 6 cm, upward sealing, sealing ratio is 70 per cent).

Self-extinction

The effects of sealing configuration, sealing ratio, and heat release rate on the fire behaviour are shown in Figure 5. The fire self-extinction criterion was determined by the disappearance of luminous flame (You *et al*, 2020). Overall, except for tests lacking sealing and a 4 cm × 4 cm fire source with a 50 per cent sealing ratio, all fires self-extinguished. This is attributed to smoke accumulation between the heading face and the fire source, leading to a decrease in oxygen volume fraction, thereby impeding sustained combustion (Yao *et al*, 2023). Figure 6 illustrates the variation in self-extinction time with sealing ratios across different sealing configurations. It is found that:

- Self-extinction time decreases with increasing sealing ratio, as larger sealing ratios restrict the mass exchange between outside air and internal smoke to a greater extent.

- When the dead-end roadway is downward sealed, self-extinction time is shorter than in upward sealing conditions, suggesting a better inhibitory effect of fire combustion with downward sealing. The reason is that fire smoke, driven by thermal buoyancy, tends to move along the ceiling, and downward sealing directly impedes smoke's outward diffusion. When the sealing ratio reaches 90 per cent, the discrepancy in self-extinction time between the two sealing configurations becomes negligible.

- At a given sealing ratio, larger fire sizes result in faster self-extinguishment, consistent with findings by Wang *et al* (2019).

- Smaller fire sizes exhibit a greater reduction in self-extinction time with increasing sealing ratio.

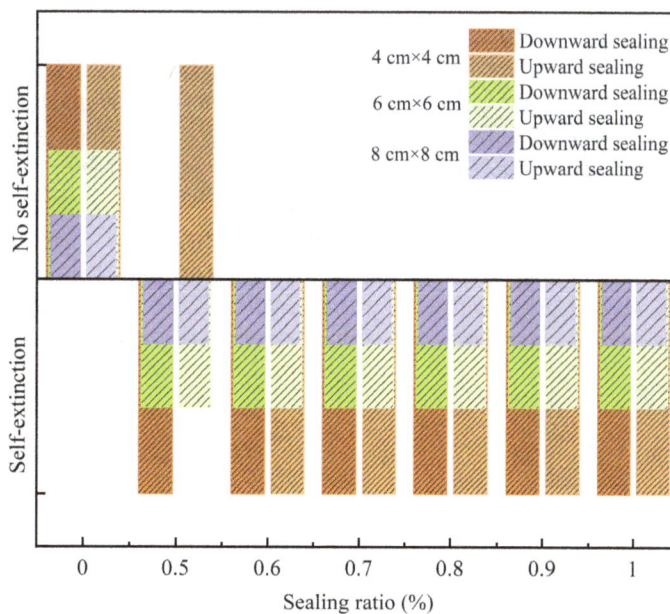

FIG 5 – Fire behaviour in all scenarios.

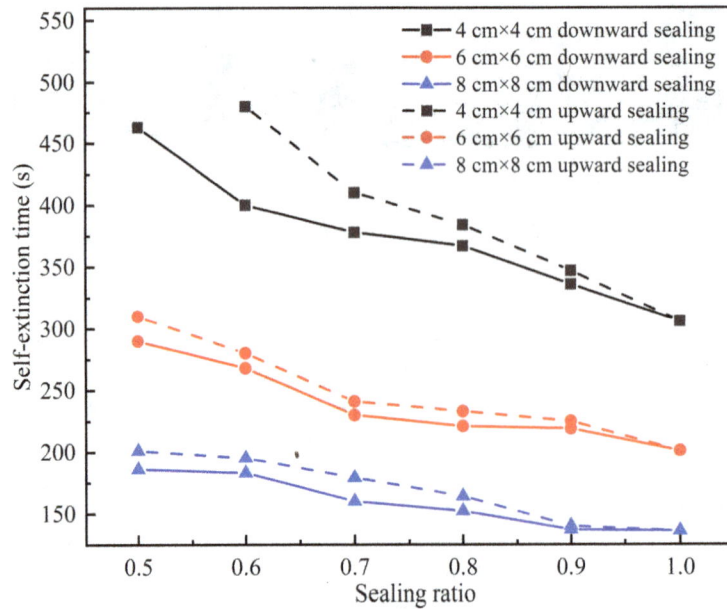

FIG 6 – Self-extinction time.

Gas concentrations

The concentration of gases such as O_2, CO, and CO_2 within the smoke plays a pivotal role in determining the combustion state of the fire source. Figure 7a portrays the fluctuation patterns of O_2 levels at the fire source (#1 point) across diverse operating conditions with a sealing ratio of 0, with the emergence of ghosting flame marked. Ghosting flames emerge when O_2 levels at the fire source surpass a critical threshold. As the oxygen content decreases, the concentration of combustible vapors in the space continues to rise. Areas distant from the oil pool exhibit higher oxygen content compared to those in proximity, facilitating flame spread to regions meeting requisite oxygen and fuel concentration criteria. When the oxygen concentration in a region is inadequate, flame propagation recurs to other viable combustion areas (Li *et al*, 2010), resulting in an unstable ghosting flame accompanied by oscillating oxygen volume fraction. Using the 8 cm × 8 cm oil pan as an example, Figures 7b–7c and Figure 8 illustrate the change in CO content (#2 point) in smoke under various sealing conditions. Observations indicate that when the roadway is sealed, insufficient oxygen content precludes sustained combustion, resulting in rapid self-extinguishment. After extinguishment, gradual gas exchange between the interior and exterior of the dead-end roadway prompts a reversion of gas content to its initial state, with the rate of reversion accelerating as sealing ratios diminish. Furthermore, downward sealing induces higher CO levels compared to upward sealing, as downward sealing precipitates an earlier transition of combustion into ventilation control. Under upward sealing, peak CO production ranges between 200 and 500 ppm across all operational conditions. However, under downward sealing, this range extends from 450 to 650 ppm.

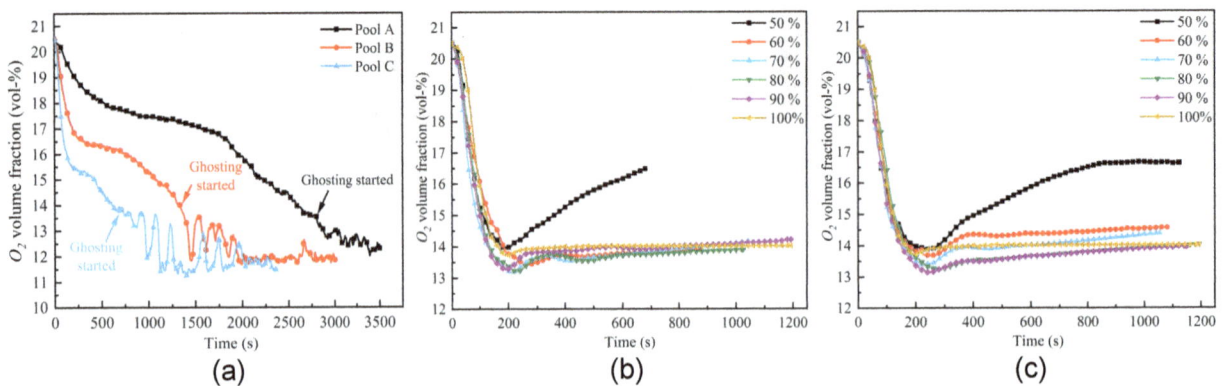

FIG 7 – Curves of O_2 volume fraction over time: (a) no sealing; (b) downward sealing; (c) upward sealing.

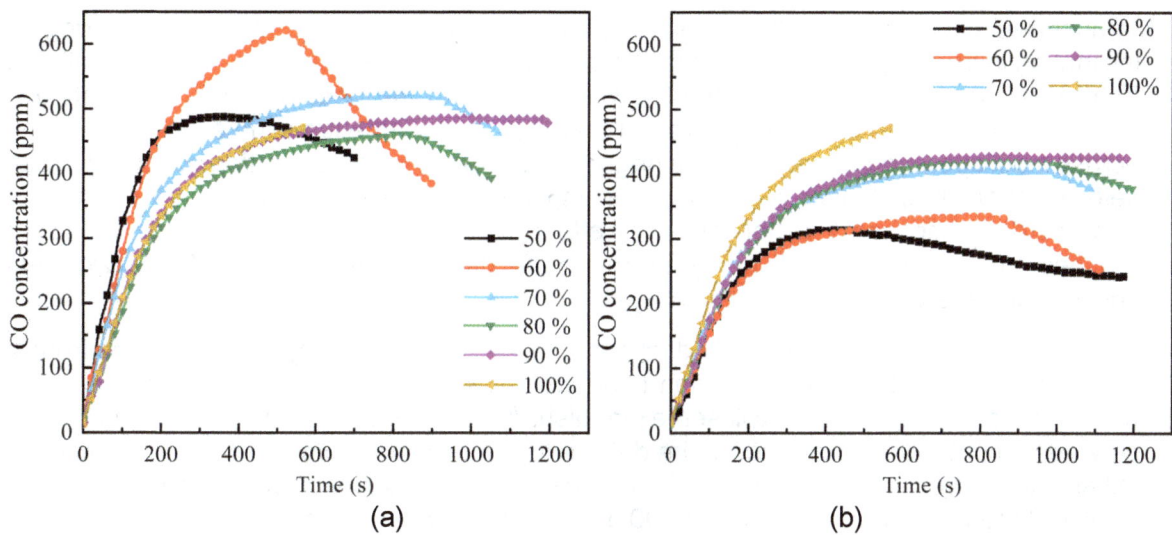

FIG 8 – Curves of CO content over time: (a) downward sealing; (b) upward sealing.

Temperature distribution

Figure 9 illustrates the longitudinal temperature distribution trend beneath the ceiling in the dead-end roadway under various working conditions, using the 8 cm × 8 cm oil pan as an example. Overall, the gas temperature decreases exponentially with increasing distance from the fire source, aligning with previous studies (Li *et al*, 2021; Liu *et al*, 2019; Tang *et al*, 2017; Yao *et al*, 2023). However, some differences arise due to the unique structure of the dead-end roadway. The gas temperature at the measuring point closest to the heading face tends to be slightly higher than that at adjacent points. This discrepancy results from the fact that the smoke will sink and entrain the air below after hitting the heading face and gradually accumulate. Similarly, under downward sealing conditions, the presence of a sealing wall leads to a slightly higher gas temperature at the measuring point closest to it compared to adjacent points. Furthermore, implementing sealing at the entrance of the dead-end roadway effectively suppresses fire source combustion, significantly lowering the roadway's gas temperature compared to unsealed conditions. Besides, in contrast to upward sealing, downward sealing results in higher gas temperatures in the dead-end roadway.

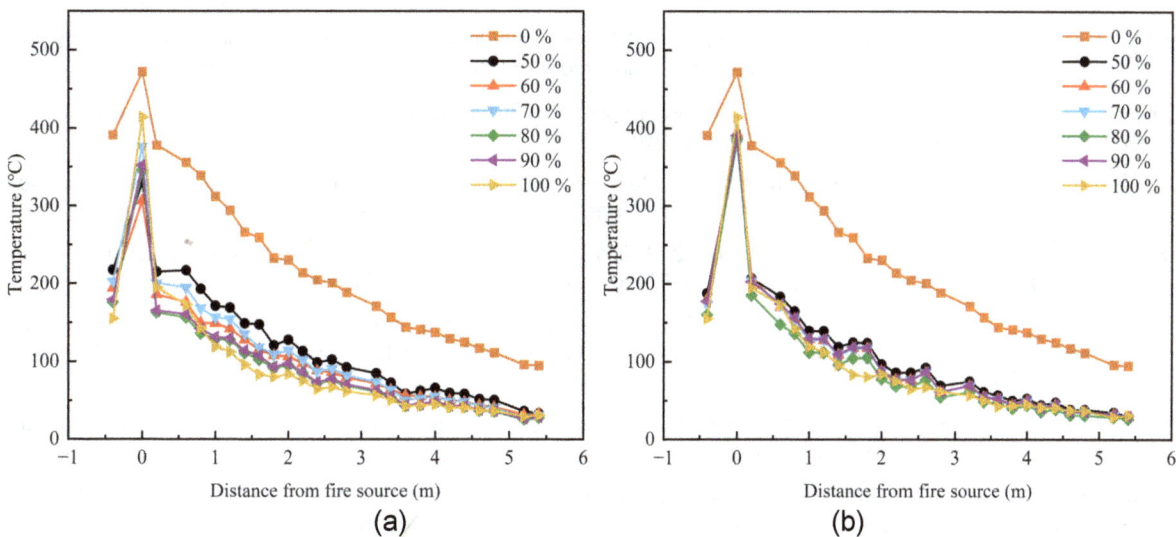

FIG 9 – Longitudinal temperature distribution: (a) downward sealing; (b) upward sealing.

CONCLUSIONS

In this research, a series of n-heptane fire experiments were conducted within a small-scale coalmine dead-end roadway to examine the impact of various sealing conditions on combustion characteristics. The key findings are outlined as follows:

- In unsealed conditions, the flame tends to tilt towards the heading face, with the degree of inclination increasing gradually throughout combustion. After some time, the flame will detach from the oil pan and move between the fire source and the heading face, manifesting as a ghosting flame. When a sealing ratio other than 0 is applied, the fire source tends to extinguish across most operating conditions, with negligible alteration in flame position observed. Notably, self-extinction time diminishes with increasing sealing ratios and pool sizes. Besides, the downward sealing has a better inhibitory effect on fire combustion. When the sealing ratio reaches 90 per cent, the disparity in the self-extinction time between the two sealing configurations is very small.

- As oxygen volume fraction steadily declines, the dead-end roadway fire transitions from fuel-controlled combustion to ventilation-controlled combustion. Below the limiting oxygen concentration, the fire source will self-extinguish. After extinguishment, gradual gas exchange between the interior and exterior of the dead-end roadway prompts gas content to revert to its initial state. Moreover, downward sealing yields elevated CO levels relative to upward sealing. Peak CO production ranges between 200 and 500 ppm under upward sealing conditions, while under downward sealing, this range expands to 450 to 650 ppm.

- Implementing sealing can significantly inhibit fire source combustion and reduce the overall smoke temperature within a dead-end roadway. A downward sealing results in higher gas temperature in the dead-end roadway compared to an upward sealing.

ACKNOWLEDGEMENTS

This work was supported by the National Natural Science Foundation of China [52374252].

REFERENCES

Chen, C K, Zhu, C X, Liu, X Y and Yu, N H, 2016. Experimental investigation on the effect of asymmetrical sealing on tunnel fire behavior, *International Journal of Heat and Mass Transfer*, 92:55–65.

Fan, C G, Li, X Y, Mu, Y, Guo, F Y and Ji, J, 2017. Smoke movement characteristics under stack effect in a mine laneway fire, *Applied Thermal Engineering*, 110:70–79.

Guo, R Z, Ma, L, Fan, J, Wang, W F, Wei, G M and Ren, L F, 2022. Temperature Distribution of Coal Mine Tunnel Fire During Dynamic Sealing Process: A Numerical Study, *Combustion Science and Technology*, pp 1–18.

Han, J Q, Liu, F, Wang, F, Weng, M C and Wang, J, 2020. Study on the smoke movement and downstream temperature distribution in a sloping tunnel with one closed portal, *International Journal of Thermal Sciences*, 149.

Huang, Y B, Li, Y F, Dong, B Y, Li, J M and Liang, Q, 2018. Numerical investigation on the maximum ceiling temperature and longitudinal decay in a sealing tunnel fire, *Tunnelling and Underground Space Technology*, 72:120–130.

Li, C H, Lu, S X, Yuan, M and Zhou, Y, 2010. Studies on ghosting fire from pool fire in closed compartments, *Journal of university of science and technology of China*, 10:751–761.

Li, L J, Zhu, D Q, Gao, Z H, Xu, P and Zhang, W C, 2021. A study on longitudinal distribution of temperature rise and carbon monoxide concentration in tunnel fires with one opening portal, *Case Studies in Thermal Engineering*, 28.

Liu, H N, Zhu, G Q, Pan, R L, Yu, M M and Liang, Z H, 2019. Experimental investigation of fire temperature distribution and ceiling temperature prediction in closed utility tunnel, *Case Studies in Thermal Engineering*, 14.

Tang, F, Li, L J, Chen, W K, Tao, C F and Zhan, Z, 2017. Studies on ceiling maximum thermal smoke temperature and longitudinal decay in a tunnel fire with different transverse gas burner locations, *Applied Thermal Engineering*, 110:1674–1681.

Wang, G Q, Shi, G Q, Wang, Y M and Shen, H Y, 2021. Numerical study on the evolution of methane explosion regions in the process of coal mine fire zone sealing, *Fuel*, 289.

Wang, J X, Fu, Y X, Qu, B L, Chang, C, Wen, X Y, Ma, X H, Wang, G Y and Wu, B, 2023a. Experimental study of fire behaviors influenced by sealing time in coal mine blind roadway fires, *Thermal Science and Engineering Progress*, 45.

Wang, J X, Qu, B L, Li, C, Huang, L S, Fu, Y X, Chang, C andWu, B, 2023b. Numerical study on gas temperature and smoke control in blind roadway fires, *Case Studies in Thermal Engineering*, 50.

Wang, K H, Chen, J M, Wang, Z K, Gao, D L, Wang, G Y and Lin, P, 2019. An experimental study on self-extinction of tunnel fire under natural ventilation condition, *Tunnelling and Underground Space Technology*, 84:177–188.

Wu, B, Wang, J X, Qu, B L, Qi, P Y and Meng, Y, 2023. Development, effectiveness and deficiency of China's Coal Mine Safety Supervision System, *Resources Policy*, 82.

Yao, Y Z, Cheng, X D, Zhang, S G, Zhu, K, Zhang, H P and Shi, L, 2017. Maximum smoke temperature beneath the ceiling in an enclosed channel with different fire locations, *Applied Thermal Engineering*, 111:30–38.

Yao, Y Z, Li, Y Z, Ingason, H and Cheng, X D, 2019a. The characteristics of under-ventilated pool fires in both model and medium-scale tunnels, *Tunnelling and Underground Space Technology*, 87:27–40.

Yao, Y Z, Li, Y Z, Lönnermark, A, Ingason, H and Cheng, X D, 2019b. Study of tunnel fires during construction using a model scale tunnel, *Tunnelling and Underground Space Technology*, 89:50–67.

Yao, Y Z, Wang, J X, Jiang, L, Wu, B and Qu, B L, 2023. Numerical study on fire behavior and temperature distribution in a blind roadway with different sealing situations, *Environmental Science and Pollution Research*, 30:36967–36978.

You, S H, Wang, K H, Shi, J K, Chen, Z N, Gao, D, Chen, J M and Lin, P, 2020. Self-extinction of methanol fire in tunnel with different configuration of blocks, *Tunnelling and Underground Space Technology*, 98.

Mine explosions

Research on the pressure relief effect of gas explosion in vertical shaft explosion proof cover

J L Gao[1], J Z Ren[2] and X B Zhang[3]

1. Professor, College of Safety Science and Engineering, Henan Polytechnic University, Jiaozuo, Henan 454003, China. Email: gao@hpu.edu.cn
2. Lecture, School of Safety Engineering and Emergency Management, Shijiazhuang Tiedao University, Shijiazhuang, Hebei 050043, China. Email: renjingzhang@126.com
3. Associate Professor, College of Safety Science and Engineering, Henan Polytechnic University, Jiaozuo, Henan 454003, China. Email: zhxb@hpu.edu.cn

ABSTRACT

It has been studied that the impact load of the explosion shock wave on the explosion-proof cover and fan blades when a gas explosion occurs in a certain underground excavation face through numerical simulation. Firstly, a small-scale air shaft fan explosion-proof cover explosion propagation experimental system was established, and gas explosion shock wave propagation experiments were carried out. By comparing the experimental results with numerical simulation results, mathematical models for gas explosion shock wave propagation was determined. It has been studied that the effect of explosion-proof cover counterweight on the lifting process of explosion-proof cover and the load acting on the explosion-proof covers and fan blades; and the pressure relief effects under different conditions have been obtained. The results indicate that as the explosion intensity increases, the impact load acting on the fan blades and explosion-proof covers increases. Due to the high load exerted by the explosion shock wave on the explosion-proof cover, the lifting acceleration of the explosion-proof cover is significant, therefore the influence of counterweigh on the impact load exerted to the fan blades and explosion-proof cover is very small. The greater the explosion intensity, the worse the pressure relief effect of the explosion-proof cover, and the passive opening of the explosion-proof cover cannot effectively protect the ventilation fan. The advance opening of explosion-proof cover has a better pressure relief effect than passive opening.

INTRODUCTION

The main ventilation fan on the ground of a mine is the 'lung' of the mine, responsible for pumping fresh air to the underground and discharging toxic and harmful gases. Especially in the event of disasters such as outbursts, explosions, and fires, the continuous, safe, and reliable operation of the main ventilation fan in the mine is of great significance for emergency rescue and ensuring the safety of all underground personnel. According to the 'Coal Mine Safety Regulations' in China, 'explosion-proof cover should be installed at the outlet of the main ventilation fan' (China Coalmine Safety Regulations, 2016). Canada's Occupational Health and Safety Regulations for Coalmines (SOR/90–97) stipulate that 'the main ventilation room should be equipped with pressure relief cover or other pressure relief devices that facilitate the opening of explosive shock waves' (Government of Canada, 2015), while regulations in South Africa and New South Wales, Australia, state that effective measures should be taken to prevent the main ventilation fan from being damaged by explosions (Department of Minerals and Energy, 1996; NSW Government, 2006). The purpose of installing an explosion-proof cover (explosion-proof cover) on the shaft equipped with main ventilation fan in the mine is to release pressure in the event of an explosion underground, protect the fan from being damaged by the explosion shock wave, and enable the ventilation fan to ventilate normally.

However, when an explosion disaster occurs underground, sometimes the explosion-proof cover is not opened in a timely manner or cannot be opened quickly, resulting in damage to the ventilation fan. Sometimes the explosion-proof cover is severely deformed or thrown out, and cannot be reset in a timely manner, resulting in a short circuit of air flow and affecting normal ventilation. There are also phenomena where explosion-proof covers were thrown out under the action of explosion shock waves, and the blades of mine ventilation fans were still severely deformed and damaged. Since 1999, there have been 13 explosion accidents in major coalmines in China, where explosion-proof covers were opened, deformed, damaged, and thrown out under the action of explosion shock waves (Song and Sun, 2015). Although the explosion-proof covers were opened (damaged) under the impact of the explosion shock wave in these two accidents, the main ventilation fan blades were still damaged and unable to ventilate normally. Therefore, It is necessary to study the pressure relief effect of shaft explosion proof cover.

Joonwon and Youngsik (2018) conducted full-scale on-site tests to study the structural dynamic response characteristics of explosion-proof cover under single and multiple explosion loads. They predicted and analysed the distribution of explosion-proof cover under dynamic explosion loads, as well as their bending deformation characteristics and boundary tearing effects under explosion shock waves. Choi, Lee and Yoo (2016) conducted experimental research to investigate the characteristics of explosion-proof cover under the impact of detonation waves, and found that under the action of the explosion shock wave, the cover will bend to form an elliptical opening. An (2013) used finite element analysis software to analyse the stress and strain of self resetting explosion-proof cover under explosive loads, and found that under the action of explosion shock waves, the maximum deformation of explosion-proof cover occurs in the middle part of the cover panel. Wu, Yan and Zhuang (2020) studied the component failure mode and anti-explosion performance of explosion-proof cover under explosive shock waves, and numerically simulated the changes in stress load, impact pressure, and propagation speed over time during the opening process of explosion-proof cover. Song (2018) studied the dynamic response characteristics and laws of the new type of explosion-proof cover invented by him in the process of explosion impact, and formed two new design schemes for coalmine explosion-proof cover: 'guided buffer explosion-proof cover' and 'double cover explosion-proof cover'.

The above-mentioned scholars have conducted extensive research on the propagation of gas explosion shock waves, but there has been no research on the dynamic evolution of impact loads and pressure relief effects of blast cover and fan blades during the dynamic opening process of blast cover under the action of shock waves. The authors studied the dynamic response characteristics of the vertical shaft explosion-proof cover and fan blades under the action of explosion shock waves when different intensities of gas explosions occurred in the excavation face of Yangchangwan Mine II020611 return air channel. The evolution law of the impact load of the explosion-proof cover and fan blades, and the pressure relief effect were analysed. The concept of actively opening the explosion-proof cover in advance was proposed, It is of great significance to enhance the disaster resistance of the main ventilation fans in mines and reduce the loss of life and property caused by explosion disasters.

SELECTION AND EXPERIMENTAL VERIFICATION OF NUMERICAL MODELS FOR GAS EXPLOSIONS

There are multiple mathematical models to describe the propagation of gas explosions, and turbulence models mainly include direct numerical simulation (DNS), large eddy simulation (LES), and Reynolds averaged N-S equations (RANS) models. In terms of combustion models, there are mainly laminar finite rate models (Finite Rate), laminar finite rate/eddy dissipation models (Finite Rate/Eddy Dissipation), eddy dissipation models (EDM), and eddy dissipation conceptual models (EDC). Through comparative analysis, this paper uses LES and EDM to describe the propagation of gas explosions.

A small-scale air shaft fan explosion-proof cover explosion propagation experimental system was designed and constructed based on a 6 m diameter mine prototype scale return air shaft, using a 1:15 similarity ratio of shaft diameter. The experimental system consists of four main parts: explosion propagation pipeline system, gas distribution system, data acquisition system, ignition device, etc.

A geometric model was established according to the experimental size, and the selected theoretical models were used to simulate the gas explosion shock propagation process when the gas filling length was 1m and the methane concentration was 8.0 per cent and 9.5 per cent, respectively. The shape of the overpressure change curves of each monitoring point in numerical simulation is consistent with the experimental measurement curves, and the trend of the increase and attenuation of shock wave overpressure is consistent. The relative error of the maximum overpressure values obtained from numerical simulation and experiments at measuring points arrange between

3.60 per cent~7.18 per cent, which verifies the reliability of the combustion models used in the LES turbulence model and EDM.

DYNAMIC RESPONSE CHARACTERISTICS OF VERTICAL SHAFT EXPLOSION-PROOF COVER AND FAN BLADES UNDER SHOCK WAVE ACTION

The return air route of the II020611 return air tunnel excavation working face in Yangchangwan Mine is shown in Figure 1. The working face is 518 m away from the bottom of the return air shaft, the depth of the return air shaft is 556 m, and the diameter of the shaft is 6 m. The installed main fan model is ANN-2660/1440, with a total of 16 blades and an installation angle of 45°. The vertical distance between the fan blades and the outer wall of the shaft is 16.44 m. The dimensions of the main ventilation fan wind tunnel explosion-proof cover are shown in Figure 2. Fluent software was used to numerically simulate the propagation process of gas explosion shock overpressure waves.

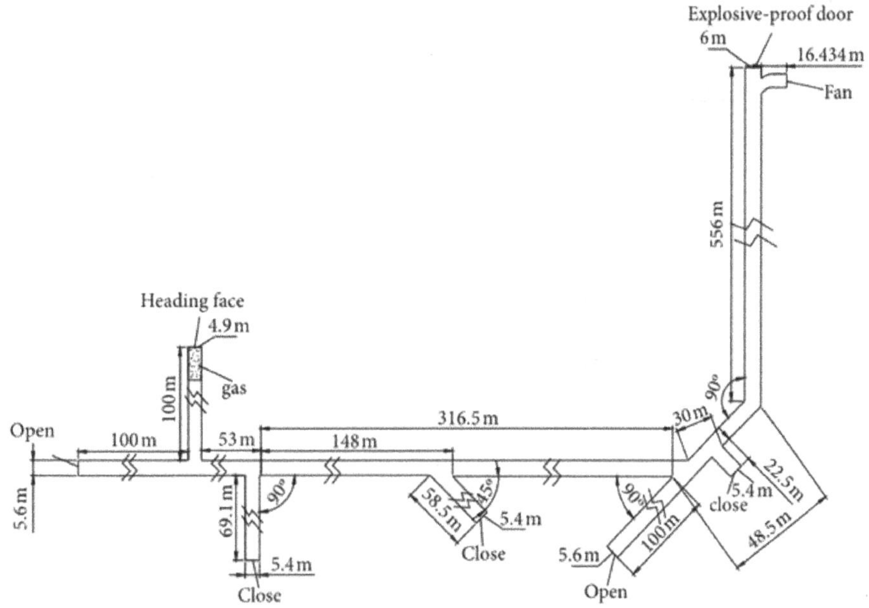

FIG 1 – Geometric model of the heading face to the explosion-proof cover and the wind shaft.

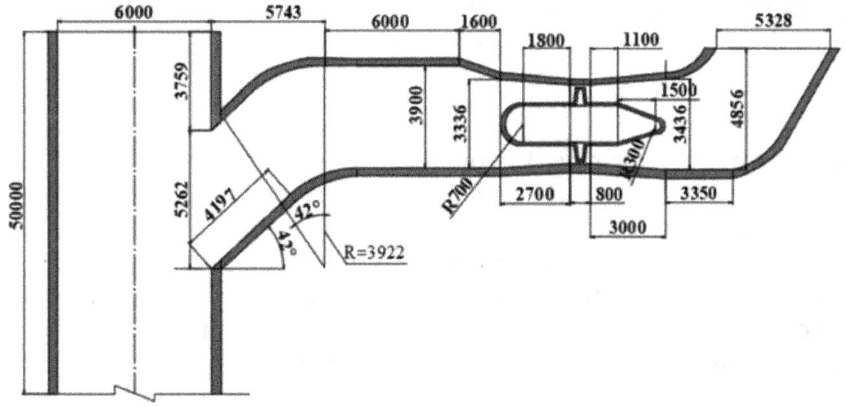

FIG 2 – Geometric model of the main ventilators.

The gas accumulation concentration in the excavation face is 9.5 per cent, and the gas accumulation volume is 155.82 m³, 233.73 m³, and 311.64 m³, respectively. The equivalent TNT explosion equivalent m_{TNT} is 15.66 kg, 23.49 kg, and 31.33 kg, respectively. The Numerical calculations were conducted on the dynamic impact process of explosion shock waves on the explosion-proof cover with a mass of 2700 kg, and fan during the passive opening process of the explosion-proof cover under five counterweight conditions(700 kg, 1700 kg, 3200 kg, 3700 kg, and 4700 kg, respectively). The counterweight is recorded as m_{cw}.

Impact load on explosion-proof cover

The impact load on the explosion-proof cover during the non-opening and opening process of the explosion-proof cover is shown in Figure 3 under the action of three explosion shock waves with different explosion equivalents and five explosion-proof cover counterweight masses.

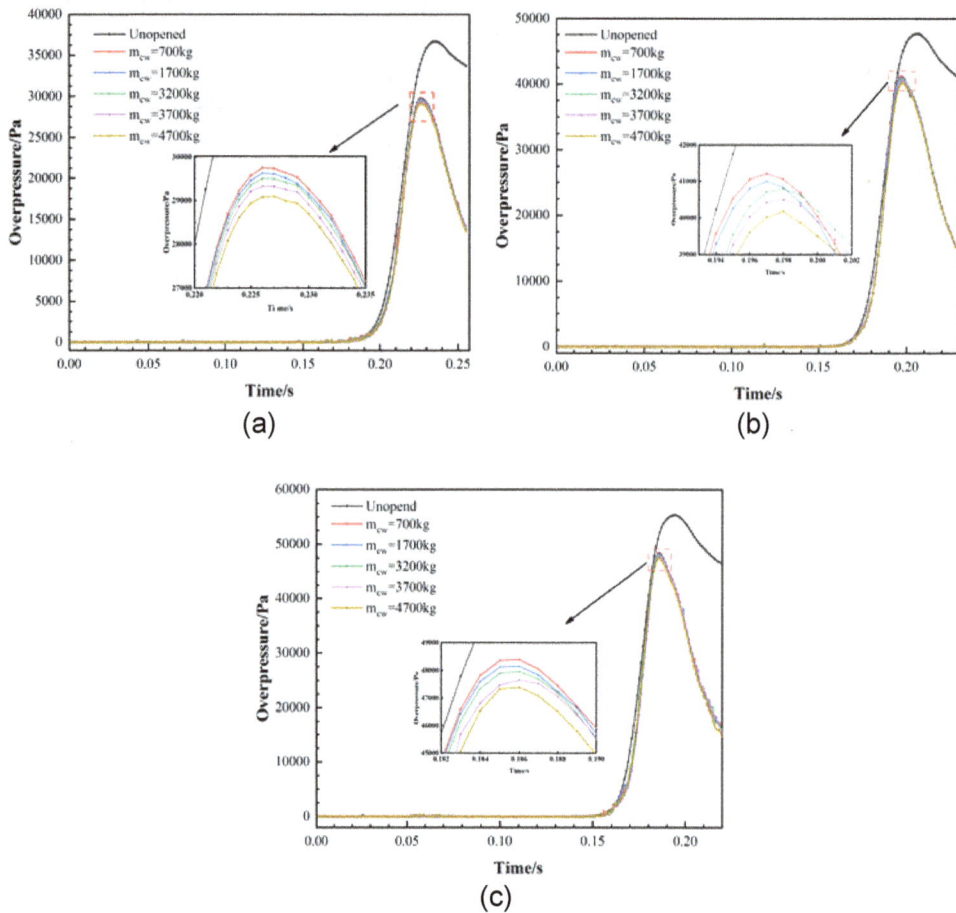

FIG 3 – Overpressure on explosion-proof cover under different conditions: (a) m_{TNT} = 15.66 kg; (b) m_{TNT} = 23.49 kg; (c) m_{TNT} = 31.33 kg.

It can be seen that:

- In the early stage of the explosion, the overpressure of the shock wave is relatively small, and the impact load on the explosion-proof cover slowly increases. As the gas explosion shock wave continues to propagate, the high-pressure area of the shock wave reaches the explosion-proof cover, and the impact load on the explosion-proof cover begins to rapidly increase until it reaches its peak. After the propagation of the high-pressure area of the explosion shock wave, the overpressure of the shock wave begins to decay, and the impact load acting on the explosion-proof cover also begins to decay.

- After the explosion-proof cover is opened, a pressure relief outlet area is formed near the wellhead, and the high-pressure explosive airflow is discharged from the wellhead, releasing the pressure. Therefore, the peak impact load acting on the explosion-proof cover under the condition of opening is smaller than when the explosion-proof cover is not opened.

- As the explosion intensity increases, the peak impact load on the explosion-proof cover increases.

- As the weight of the counterweight increases, the peak impact load on the explosion-proof cover slightly decreases.

Impact load on fan blades

The impact load on the main ventilation fan blades during the opening process of the explosion-proof cover is shown in Figure 4 under the action of explosion shock waves with three explosion equivalents and five explosion-proof cover counterweight masses.

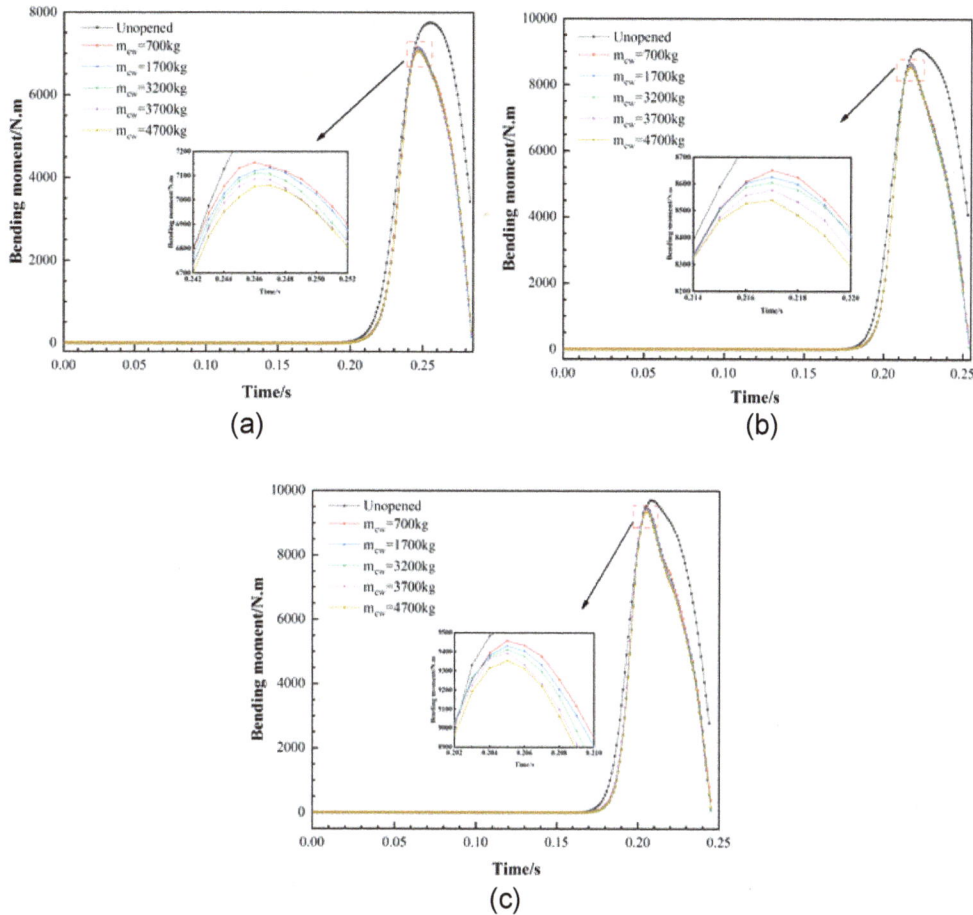

(a)

(b)

(c)

FIG 4 – Bending moment on blade root under different conditions: (a) m_{TNT} = 15.66 kg; (b) m_{TNT} = 23.49 kg; (c) m_{TNT} = 31.33 kg.

Figure 4 shows that:

- In the initial stage when the overpressure of the explosion shock wave propagates along the wind tunnel and duct to the fan blades, the overpressure of the shock wave is relatively small and rises slowly, causing the impact load on the blades to slowly increase. As the high-pressure zone of the shock wave transmitted through the wind tunnel reaches the blades, the impact load on the blades begins to increase rapidly and reaches its peak. As the high-pressure zone of the explosion shock wave dissipates, the impact load acting on the blades also begins to decay.

- During the opening process of the explosion-proof cover, the peak impact load on the fan blades under the condition of opening the explosion-proof cover is smaller than when the explosion-proof cover is not opened.

- As the weight of the counterweight increases, the peak bending moment at the root of the fan blade slightly decreases. As the explosion intensity increases, the peak bending moment at the root of the fan blade increases. When the counterweight mass increased from 700 kg to 4700 kg, the peak bending moment at the blade root decreased by 1.30 per cent,

1.27 per cent, and 1.10 per cent only, corresponding to gas explosion equivalents were 15.66 kg, 23.49 kg, and 31.33 kg, respectively.

Pressure relief effect during the opening process of explosion-proof cover

The pressure relief percentage on the fan blades under different operating conditions is shown in Figure 5. It can be seen that: under the action of gas explosion shock wave, the explosion-proof cover opens, and the shock wave load acting on the fan blades is reduced. The pressure relief effect decreases with the increase of gas explosion equivalent. As the counterweight of the explosion-proof cover increases, the pressure relief effect slightly increases. Overall, under the passive opening condition of the explosion-proof cover under the action of the explosion shock wave, the pressure relief of the impact load on the ventilation fan blades is reduced to a certain extent, but the pressure relief effect is not significant. The explosion-proof cover of the vertical shaft does not effectively protect the main ventilation fan.

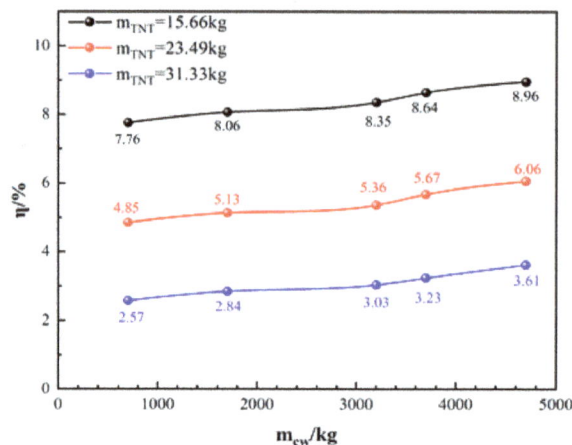

FIG 5 – Pressure relief effect on fan blades under conditions of passive opening of explosion-proof cover.

Pressure relief effect under the condition of early active opening of explosion-proof cover

It can be considered to use the explosion-proof cover to actively open and release pressure in advance. Active early opening means that before the explosion shock wave propagates to the explosion-proof cover, the explosion-proof cover opens actively to provide sufficient pressure relief area. After the explosion shock wave propagates to the wellhead, it releases high-pressure airflow, thereby reducing the explosion shock load at the fan and improving its survival ability.

Time variation curve of blade root bending moment at different heights when the explosion-proof cover is actively opened in advance are shown in Figure 6. The pressure relief effect at the fan after the explosion-proof cover is opened under different opening conditions is shown in Figure 7.

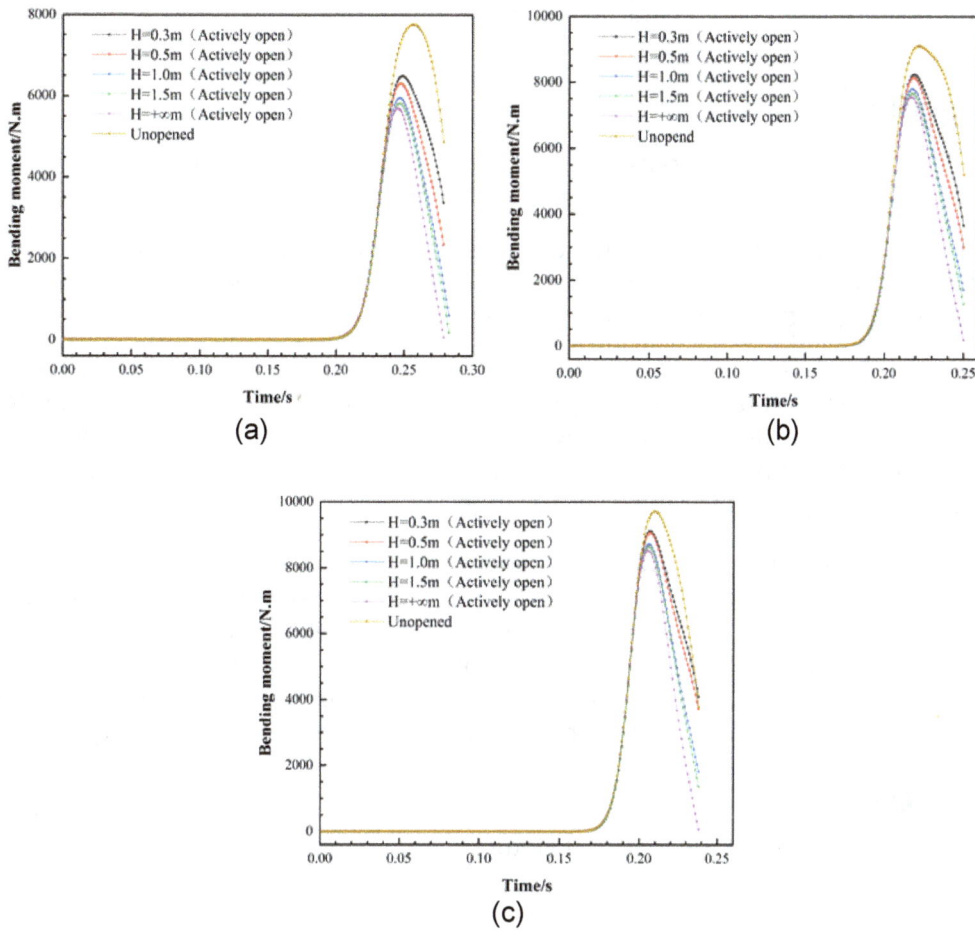

FIG 6 – Blade root bending moment variation curve: (a) m_{TNT} = 15.66 kg; (b) m_{TNT} = 23.49 kg; (c) m_{TNT} = 31.33 kg.

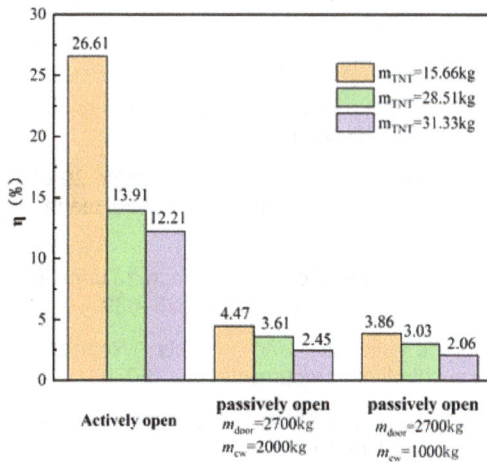

FIG 7 – Pressure relief effect on fan blades under conditions of active opening of explosion-proof cover.

As shown in Figure 7, the pressure relief effect under active opening is much better than that under passive opening. As the gas explosion equivalent increases, the pressure relief effect of the explosion-proof cover after opening gradually deteriorates.

When the explosion-proof cover is actively opened in advance, the impact load borne by the fan blades is reduced, and the probability of the fan being damaged will be greatly reduced. Compared with the pressure relief effect of passive opening of explosion-proof cover under the action of explosion shock waves, actively opening explosion-proof cover in advance can to some extent increase the pressure relief percentage at the fan and improve the survival ability of the fan under disaster conditions.

CONCLUSIONS

- As the explosion intensity increases, the peak impact load of the explosion shock wave on the explosion-proof cover and the peak bending moment at the root of the fan blade increase.

- As the weight of the counterweight increases, the impact load acting on the explosion-proof cover and fan blades during the lifting process of the explosion-proof cover has slightly decreased, but the difference is not significant.

- As the intensity of gas explosion increases, the effect of pressure relief at the explosion-proof cover and blade decreases; the weight of the counterweight has little effect on the pressure relief effect at the explosion-proof cover and fan. When the explosion intensity is high, the passive opening of the explosion-proof cover under the action of the explosion shock wave cannot effectively protect the fan from damage.

- Before the explosion shock wave arrives, the explosion-proof cover actively opens in advance, greatly reducing the impact load borne by the fan blades and reducing the probability of damage to the fan. Compared with the pressure relief effect of passive opening of explosion-proof cover under the action of explosion shock waves, actively opening explosion-proof cover at a certain height in advance will significantly increase the pressure relief percentage at the fan and enhance the fan's disaster resistance ability.

ACKNOWLEDGEMENTS

Thank you for the funding from the National Natural Science Foundation of China. Project approval number: 52274187.

REFERENCES

An, C H, 2013. Finite element simulation analysis of mine self-reclosing blast covers, *Coal Mine Machinery*, 34(08):104–106.

China State Administration of Coal Mine Safety, 2016. Coal mine safety regulations, Beijing: Coal Industry Press.

Choi, Y, Lee, J and Yoo, Y H, 2016. A Study on the behavior of blast proof cover under blast load, *International Journal of Precision Engineering and Manufacturing*, 17(1):119–124.

Government of Canada, 2015. Coal Mining Occupational Health and Safety Regulations (SOR/90–97). Available from: <https://laws-lois.justice.gc.ca/eng/regulations/sor-90-97/index.html>

Department of Minerals And Energy, 1996. Mine Health and Safety Act No 29 of 1996 and Regulations. Available from: <https://mhsc.org.za/sites/default/files/public/publications/Mine%20Health%20and%20Safety%20Act%2029%20of%201996%20and%20Regulations%20Final%20Booklet.pdf>

Joonwon, L and Youngsik, C, 2018. Effects of a Near-Field Explosion in a Tunnel Behind a Blast Proof Cover, International *Journal of Precision Engineering and Manufacturing*, 19(4):625–630.

NSW Government, 2006. Coal Mine Health and Safety Regulation 2006 (NSW). Available from: <https://legislation.nsw.gov.au/view/html/inforce/current/sl-2006-0783>

Song, W B and Sun, Y N, 2015. Analysis on current situation of explosion cover for vertical air shaft in coal mine, *Journal of Safety Science and Technology*, 11(06):108–114.

Song, W B, 2018. Research on the theory and technology of safety protection of explosion-proof covers in coal mine vertical air shafts [D], Jiaozuo, Henan Polytechnic University.

Wu, J, Yan, Q S and Zhuang, T S, 2020. Structural design and damage assessment of a chamber for internal blast with explosion vent, *Mechanics of Advanced Materials and Structures*, 27(24):2052–2058.

Simulation study of gas explosion propagation law in a coal mining face with different ventilation modes

J Liu[1,2,3], Y Zhang[1], S Chen[1], Z Nie[1] and D Yang[1]

1. Henan Polytechnic University, School of Safety Science and Engineering, Jiaozuo Henan 454003, China. Email: liujiajia@hpu.edu.cn
2. Collaborative Innovation Center of Coal Work Safety and Clean High Efficiency Utilization, Jiaozuo Henan 454003, China.
3. Henan Provincial Key Laboratory of Gas Geology and Gas Control Provincial and Ministry of State Key Laboratory Breeding Base, Jiaozuo Henan 454003, China.

ABSTRACT

In ventilation working face, different ventilation way is one of the important factors affecting the propagation of gas explosion shock wave. In order to study the gas explosion shock wave propagation laws in coal mining faces with different ventilation methods, using Fluent simulation software, combined with pipeline gas explosion experiments and the actual situation of gas accumulation area explosion in the corner of 415 coalface in Chenjiashan coalmine, Shaanxi Province, and on the basis of building a three-dimensional mathematical and physical model, the gas explosion simulations under Y-type and W-type ventilation methods were carried out respectively. The results are described as follows:

- The attenuation trend of overpressure peak value of inlet roadway #1 and working face with Y-type ventilation conforms to the power function form. Intake airway #2 and return airway have the same peak overpressure attenuation trend, but the peak value of overpressure in return airway is generally 10.6 per cent higher than that in intake airway #2, and its attenuation rate is a process from low to high and then to low.

- The peak attenuation trend of overpressure in intake airway #1 and #2 with W-type ventilation conforms to the form of power function, and the peak attenuation trend of overpressure in return airway conforms to the form of exponential function. The peak value of overpressure decreases with the increase of the distance from the explosion source, and its attenuation rate decreases gradually.

The research results have important theoretical significance for improving the prevention of gas explosion in coal mining face with different ventilation modes.

INTRODUCTION

Coalmines are an extremely important link in China's energy structure and has always occupied a very important position. When gas explosion accidents occurs underground, high temperature, shock waves and harmful gases are often generated, resulting in serious casualties, damage to equipment and ventilation facilities and roadway collapse. At present, gas explosion accidents in coalmines in China are one of the most serious disasters (Xue, 2019). Therefore, it is very important to study the propagation law of gas explosion shock wave. At present, many domestic and foreign scholars have done a lot of research on the propagation law of gas explosion shock wave. For example, Wang and Du (2021) conducted gas explosion tests under different bifurcation angles and initial overpressure conditions respectively, and systematically studied the propagation rules of shock waves generated by gas explosion in different ventilation pipe networks. Wang *et al* (2015) and Wang, Zhao and Addai, 2017) studied the distribution and propagation of gas explosion shock waves in the pipeline model. Qiu (2018) used a combination of theoretical analysis, experiment and numerical simulation to study the propagation characteristics and attenuation laws of gas explosion shock waves in straight, bend, bifurcated, and abrupt cross-section pipelines, A-type networks, and parallel networks, and on this basis, the process of gas explosion disaster simulation system is developed, and a coalmine in our country as the study of gas explosion accident cases to verify the application effect of the simulation system. Zhang *et al* (2011) simulated the ignition and propagation characteristics of gas explosion by introducing basic chemical reactions. Ye *et al* (2009) used the fluid-solid coupling algorithm of ANSYS/LS-DYNA program and established a roadway gas explosion physical model, numerically simulated the gas explosion process in the roadway space, and studied the gas explosion shock wave in the roadway space propagation characteristics.

Kravtsov *et al* (2015) studied the propagation law and dynamic response characteristics of methane gas explosion inside the pipeline, and concluded that due to the variability of the medium, changes in the explosion process depend on turbulence factors. Ibrahirn and Masri (2001) discussed the influence of different types of obstacles on shock wave propagation, and proved that obstacles can aggravate the propagation of shock wave. Masri *et al* (2000) qualitatively analysed the attenuation law of shock waves in straight roadway, and summarised the propagation characteristics of shock waves in straight roadway. Shao *et al* (2015) found through experiments that under any vacuum condition, the vacuum chamber can effectively suppress explosive flame and overpressure. Jia, Xu and Jing (2015) analysed the influence of overpressure and bifurcation angle before bifurcation of one-way bifurcation pipeline on propagation of gas explosion shock wave in the pipeline. Wu, Zhang and Xu (2005) started from the three-dimensional NS equation and used the TVD format to numerically simulate the process of pressure waves generated by the flame in the process of gas explosion. Ye *et al* (2009) conducted theoretical and experimental studies on the propagation law of flame and shock wave in curves. Gieras *et al* (2006) found in their study that initial temperature has an important influence on the basic parameters of explosion. Yan *et al* (2011) carried out numerical simulation on the variation rule of precursor wave front and flame wave front by using Fluent software for fluid dynamics analysis, and conducted numerical simulation analysis on the overpressure transfer rule in different types of roadway gas explosion.

At present, domestic and foreign scholars have studied the propagation law of gas explosion shock wave in coal mining face mainly for the U-type ventilation mode, but there are few studies on the propagation law of gas explosion shock wave in Y and W ventilation modes, and there are still gaps. Therefore, in this paper, combined with the case of the 11.28 gas explosion accident in Chenjiashan Coalmine in Shanxi Province, fluid dynamics simulation software Fluent was used to build a three-dimensional mathematical physical model, set reasonable boundaries and initial conditions, and study the propagation law of the gas explosion shock wave in coal mining face with Y-type and W-type ventilation mode, and further study and master the difference of propagation law of gas explosion shock wave in coal mining face under different ventilation modes. The research results have important guiding significance for improving the level of mine safety production and reducing the damage caused by gas explosion.

DETERMINE THE MATHEMATICAL AND PHYSICAL MODEL

Reasonable mathematical and physical model is very important for the reliability of experimental results, and reasonable boundary conditions and initial conditions and other key parameters should be determined before numerical simulation. The specific mathematical and physical model are determined as follows:

Establish the physical model

The basic assumptions

Gas explosion is a very complex fluid elastic-plastic process, accompanied by chemical reaction, turbulence changes and other phenomena, so it is difficult to study the propagation law of mine gas explosion considering all the conditions. In order to ensure the reliability of gas explosion numerical simulation as far as possible, some basic assumptions must be made (Liu *et al*, 2020):

- Both premixed gas and combustion products satisfy the ideal gas state equation.
- The specific heat capacity of the mixed gas follows the mixing rule, and the specific heat capacity of each component is a function of temperature.
- The wall of the physical model is a rigid adiabatic wall, which does not produce relative displacement.
- The gas explosion reaction is one-way and irreversible.

- In the physical model, the gas filling area is a normal uniform mixture of gas and air, and it is an ideal state.

Y-type ventilation

The Y-type ventilation method means that there is an intake airway at the upper and lower ends of the coal mining face and a return air roadway on each side of the goaf. Research of Y-type ventilation type gas explosion shock wave propagation law, three into or return air roadway length of 40 m, tilted face length of 175 m, model the overall grid size of 0.5 m, gas filling area and the ignition area partial encryption technology refined mesh grid by grid, a gas filling area encryption, encryption two ignition area. The three-dimensional physical model of Y-type ventilation is shown in Figure 1; intake airway #1, intake airway #2, and return airway. The gas accumulation area of Y-type ventilation is located at the corner of working face and return airway, after one encryption, the grid number of the gas accumulation area is 6768, and the grid number of the ignition area is 226.

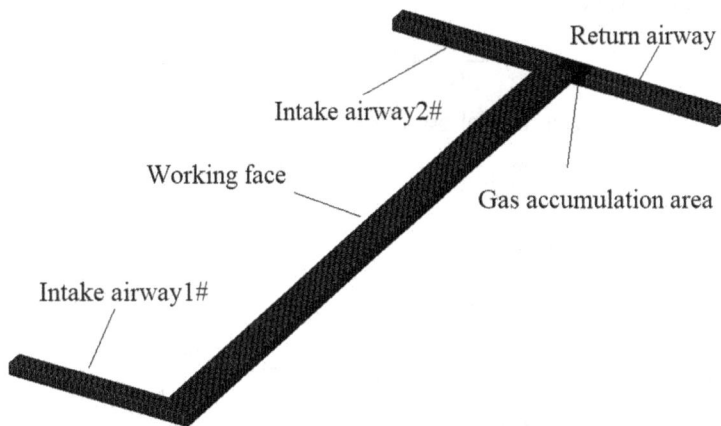

FIG 1 – Three-dimensional physical model of Y-type ventilation.

W-type ventilation

In the W-type ventilation mode, there are three roadways, and two mining faces share one return airway, which is the same as the Y-type ventilation mode. The length of the roadway is 40 m, the inclined length of the working face is 175 m, and the overall mesh size of the model is 0.5 m. The three-dimensional physical model of W-type ventilation is shown in Figure 2. Intake airway #1 and intake airway #2 enter air simultaneously, and return airway returns air. The W-type ventilation mode is different from the Y-type ventilation mode, and the gas accumulation area is located in the middle of the working face. After one encryption, the grid number of the gas accumulation area is 8978, and the grid number of the ignition area is 244.

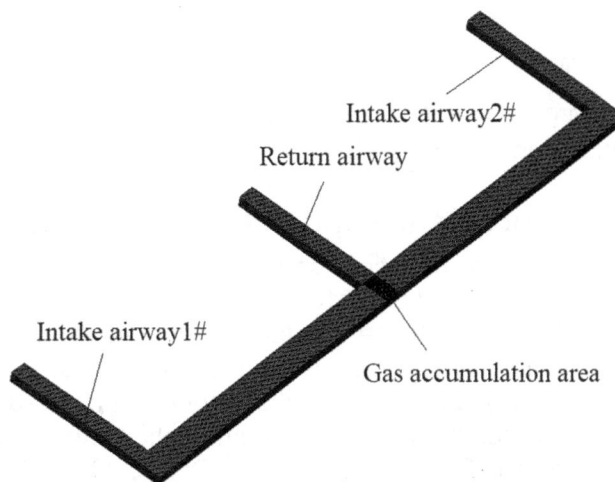

FIG 2 – Three-dimensional physical model of W-type ventilation.

Boundary conditions

Air inlet and return air outlet are set as pressure outlet, and other are standard wall; According to the analysis and identification report of Chen Jiashan Coalmine gas explosion accident, the volume fraction of CH_4 and O_2 in the gas accumulation area are 0.095 and 0.212 respectively. The burned area is a spherical area with a radius of 0.5 m in the centre of the gas accumulation area, in which the volume fraction of CO_2 is 0.1456, the volume fraction of H_2O is 0.11925, the standard atmospheric pressure is 101 325 Pa, and the temperature is 1600 K. The SIMPLE algorithm was used for iterative solution, and the iteration step was 0.0005 sec.

Establish the mathematical I model

The basic equation of aerodynamics is also studied based on the law of conservation of mass and energy etc. Therefore, the basic governing equation used in numerical simulation of gas explosion in this paper is as follows (Wu, Ye and Liu, 2017):

The continuity equation is as follows:

$$\frac{\partial p}{\partial t} + \frac{\partial (pu)}{\partial x} + \frac{\partial (pv)}{\partial y} + \frac{\partial (pw)}{\partial z} = 0 \tag{1}$$

The energy conservation equation is as follows:

$$\frac{\partial e}{\partial t} + u\frac{\partial u}{\partial x} + v\frac{\partial u}{\partial y} + w\frac{\partial e}{\partial z} = 0 \tag{2}$$

The momentum equation is as follows:

$$\left\{ \begin{array}{l} \frac{\partial u}{\partial t} + u\frac{\partial u}{\partial x} + v\frac{\partial u}{\partial y} + w\frac{\partial u}{\partial z} = -\frac{1}{p}\frac{\partial p}{\partial x} \\ \frac{\partial v}{\partial t} + u\frac{\partial v}{\partial x} + v\frac{\partial v}{\partial y} + w\frac{\partial v}{\partial z} = -\frac{1}{p}\frac{\partial p}{\partial y} \\ \frac{\partial w}{\partial t} + u\frac{\partial w}{\partial x} + v\frac{\partial w}{\partial y} + w\frac{\partial w}{\partial z} = -\frac{1}{p}\frac{\partial p}{\partial z} \end{array} \right\} \tag{3}$$

The gas state equation is as follows:

$$p = p(\rho, T) = Prt \tag{4}$$

where:

P	is pressure
T	is time
x, y, z	are rectangular coordinate system parameter
u, v, w	are the velocities in three coordinate directions respectively
ρ	is fluid density
T	is temperature
R	is the gas constant
e	is the specific energy, $e = p/(\gamma - 1) + \rho(u^2 + v^2 + w^2)/2$ (γ is the gas index and a constant).

Reliability verification of turbulence and combustion model

Gas explosion will lead to turbulent combustion, and the flow is unsteady turbulent flow. According to reference (Wen *et al*, 2013, 2016; Li *et al*, 2015), Large Eddy Simulation can better simulate the required results. Therefore, LES turbulence model is adopted in this paper to describe the characteristics of turbulent flow field in the combustion process, and standard wall surface function is adopted for wall surface. The eddy current dissipation model, whose reaction rate is controlled by turbulent mixing, is used for non-premixed flames. In combination with literature (Tian, 2020; Chen, Huang and Wang, 2020; Hoste *et al*, 2019; Halouane and Dehbi, 2017; Hassan *et al*, 2010), the ED eddy dissipation model which applicable to LES turbulence model is adopted for combustion model, and the Simple algorithm is used for iterative solution with an iterative time step of 0.0005 sec. The reliability of the turbulence and combustion model is verified below.

Establish the physical model

According to the previous pipeline gas explosion experiment, the Geometry and Mesh modules in Ansys Workbench are used to construct the full-size three-dimensional physical model and divide the grid. The cross-section of the physical model is shown in Figure 3.

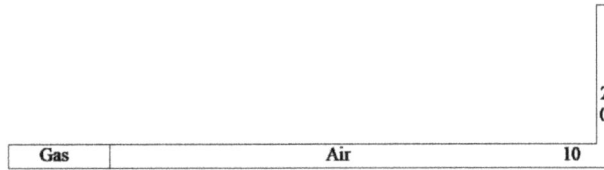

FIG 3 – Section diagram of physical model.

The width and height of the pipe in the physical model are 80 mm, the length is 19.2 m, and the length of the vertical pipe is 5 m. Experimental monitoring point 1 is 19 m away from the leftmost end and located on the middle line of the horizontal pipeline; experimental monitoring point 2 is located on the middle line of the vertical pipeline and 0.5 m away from the middle line of the horizontal pipeline.

Initial and boundary conditions of gas explosion

The initial conditions and boundary conditions of gas explosion numerical simulation are consistent with those of previous pipeline gas explosion experiments, as follows:

- The initial conditions of the burned zone are: T = 1600 K, H_2O volume fraction is 0.118, CO_2 volume fraction is 0.145, and initial pressure is 101 325 Pa. The initial conditions of unburned zone are as follows: T = 300 K, CH_4 volume fraction is 0.053, O_2 volume fraction is 0.21, both H_2O and CO_2 volume fractions are 0. The initial conditions of the air zone are as follows: the volume fraction of CH_4, H_2O and CO_2 are all 0, and the volume fraction of O_2 is 0.233.

- Boundary conditions: the pipe boundary is set as an adiabatic wall, the temperature is 300 K, and the outlet above the vertical pipe is set as a pressure outlet.

- Calculation model: fluid flow is unsteady turbulent flow, LES turbulence equation is used for turbulence model, standard wall function is used for wall surface, ED vortex dissipation model suitable for turbulent combustion is used for chemical reaction of methane combustion, and SIMPLE algorithm is used for iterative solution.

Comparison of simulation results

According to experimental conditions, 4 m, 5.5 m and 7 m gas were filled in the simulation of pipeline gas explosion. The simulation and experimental comparison of measuring points 1 and 2 under different gas filling lengths are shown in Table 1 and Figure 4.

TABLE 1

Comparison between simulation and experiment at different gas filling lengths.

Filling length /m	Point 1 (Simulation) MPa	Point 1 (Experiment) MPa	Average MPa	Measurement point 1 error	Point 2 (simulation) MPa	Point 2 (Experiment) MPa	Mean MPa	Error of point 2
		0.3032				0.2415		
4	0.3619	0.3481	0.349	3.7%	0.2733	0.2761	0.265	3.1%
		0.3943				0.2765		
		0.4118				0.2867		
5.5	0.5073	0.4385	0.462	9.8%	0.3325	0.2956	0.296	3.7%
		0.5343				0.3048		
		0.5736				0.3460		
7	0.7073	0.6344	0.635	11.3%	0.3835	0.3722	0.364	5.1%
		0.6978				0.3737		

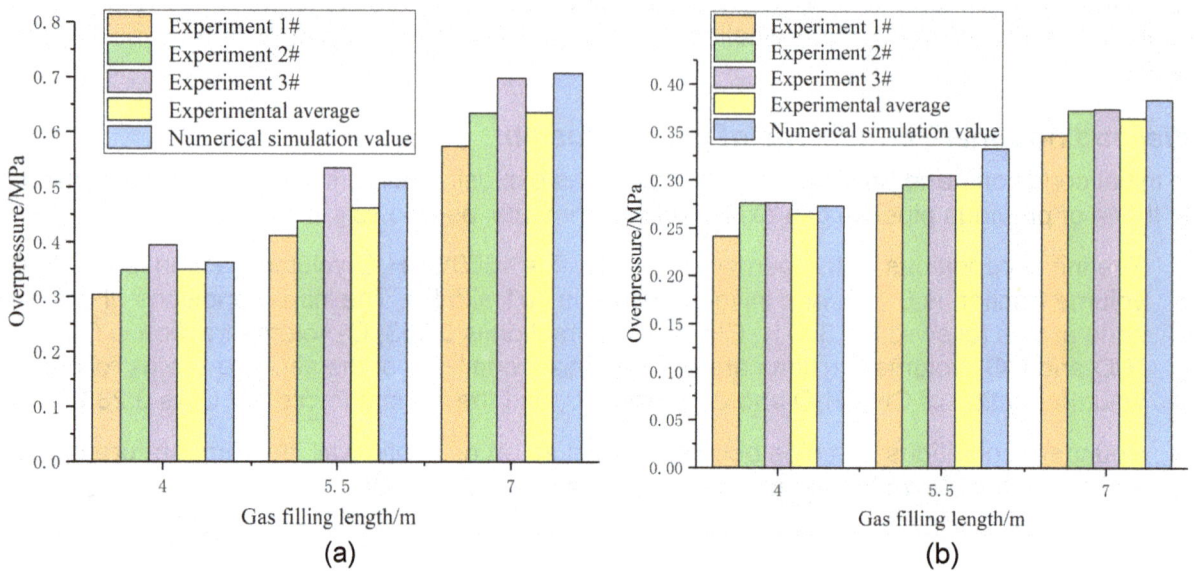

FIG 4 – Comparison between simulation and experiment at different gas filling lengths: (a) measuring point 1; (b) measuring point 2.

By analysing Table 1 and Figure 4, it can be seen that when the gas filling length is 4 m, 5.5 m and 7 m respectively, the numerical simulation results are larger than the experimental results by comparing the overpressure peaks at each measuring point, with the maximum error of 11.3 per cent and the minimum error of 3.1 per cent, both less than 15 per cent. Considering the idealisation of numerical simulation and the allowable error range of engineering, the reliability of LES turbulence model and ED vortex dissipation calculation model suitable for turbulent combustion in gas explosion numerical simulation is verified. Taking into account the idealisation of numerical simulation and the allowable error range of engineering, the reliability of choosing the LES turbulence model and the ED vortex dissipation calculation model that is suitable for turbulent combustion in the numerical simulation of gas explosion is verified.

ANALYSIS OF SIMULATION RESULTS

Y-type ventilation

High temperature ignition point was used to detonate the gas in the gas accumulation area of Y-ventilation mode. The cloud diagram of overpressure distribution of gas explosion shock wave at different times is shown in Figure 5.

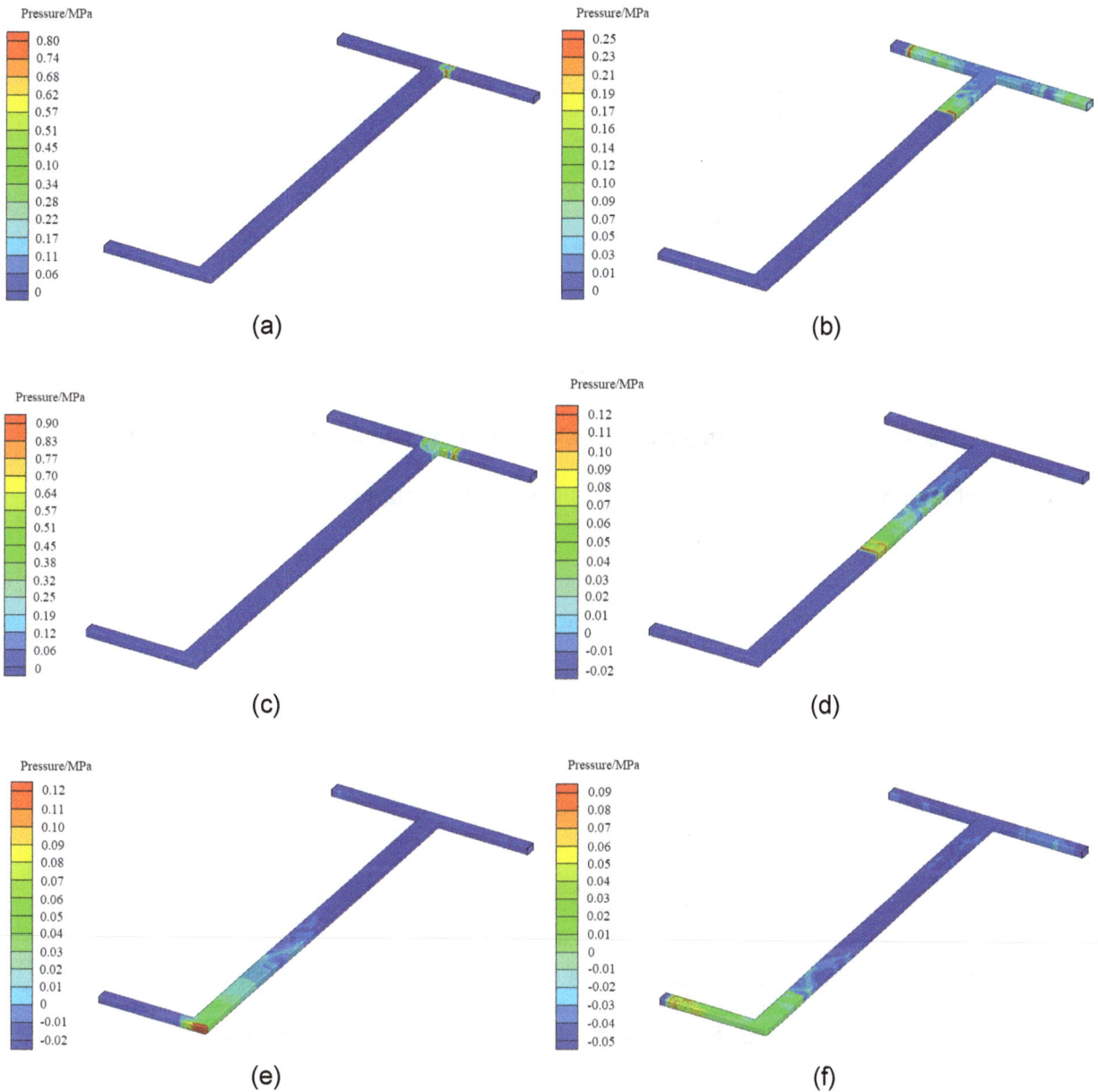

FIG 5 – Cloud diagram of gas explosion shock wave overpressure distribution at different times in Y-type ventilation mode: (a) t = 0.006 sec; (b) t = 0.012 sec; (c) t = 0.06 sec; (d) t = 0.17 sec; (e) t = 0.37 sec; (f) t = 0.46 sec.

After the gas in the gas accumulation area is detonated, the overpressure change of each measuring point is the process of oscillation from small to large to small, so there is a peak overpressure in each position. With the propagation of explosive shock wave, the variation of overpressure peak value in intake airway #1 with distance is shown in Figure 6.

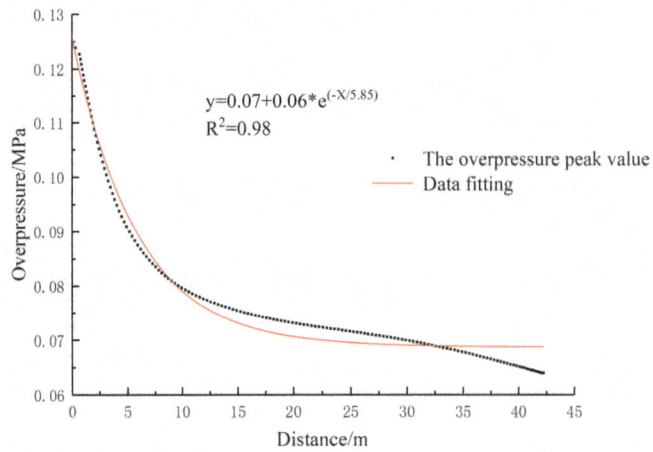

FIG 6 – Variation of overpressure peak value with distance in intake airway #1.

The peak value of overpressure in working face varies with distance as shown in Figure 7.

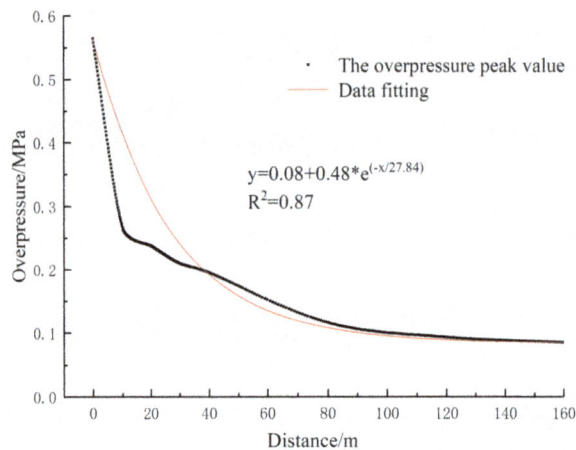

FIG 7 – Variation rule of overpressure peak value with distance in working face.

The variation of overpressure peak value in intake airway #2 and return airway with distance is shown in Figure 8.

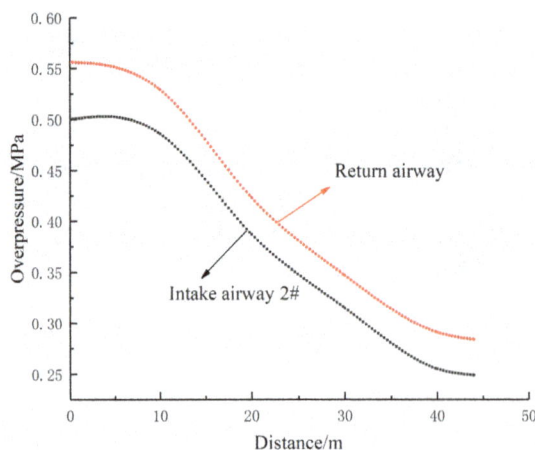

FIG 8 – Variation of peak overpressure with distance in intake airway #2 and return airway.

Analysis of Figure 5 shows that when the gas is detonated, the impact overpressure of the gas explosion near the explosion source is the largest. As the shock wave propagates along the intake airway #1 and the working face, the shock wave overpressure gradually decreases. It can be clearly seen from the cloud image of shock wave propagation that obvious reflection superposition of shock wave occurs at the junction between intake airway #1 and working face, resulting in a temporary increase of shock wave overpressure near the reflection superposition area.

The analysis of Figures 6 and 8 shows that the attenuation of shock wave overpressure in working face and intake airway #1 is a process from intense attenuation to steady attenuation, and its attenuation law is a power function relationship. Through data fitting, the variation law of overpressure peak in working face with distance is $y = 0.08 + 0.48 \times e^{(-x/27.84)}$, and the correlation is 0.87. The variation law of overpressure peak value in intake airway #1 with distance is $y = 0.07 + 0.06 \times e^{(-x/5.85)}$, and the correlation is 0.98. The attenuation trend of shock wave overpressure in intake airway #2 and return airway is completely different from that in intake airway #1 and working face, and the peak value of shock wave overpressure in return airway is obviously larger than that in intake airway #2. Through calculation, the peak value of overpressure in all places in return airway is 10.6 per cent higher than that in intake airway #2 on average. Combined with the shock wave cloud diagram at t = 0.012 sec in Figure 5, part of the explosion shock wave in the gas accumulation area spreads along the working face, resulting in the diversion of the shock wave. Therefore, the peak value of overpressure of the shock wave at each position in the return airway is significantly larger than that in the intake airway #2.

Statistical table of attenuation rate per 10m of peak overpressure in intake airway #1, intake airway #2 and return airway and variation curve of attenuation rate are shown in Table 2 and Figure 9 respectively.

TABLE 2

Statistical table of attenuation rate of peak overpressure per 10 m in intake airway #1, intake airway #2 and return airway.

The name of roadway	0 m	10 m	20 m	30 m	40 m
Intake airway #1	0.197	0.189	0.128	0.043	0.042
Intake airway #2	0.028	0.204	0.186	0.169	0.149
Return airway	0.049	0.200	0.180	0.156	0.142

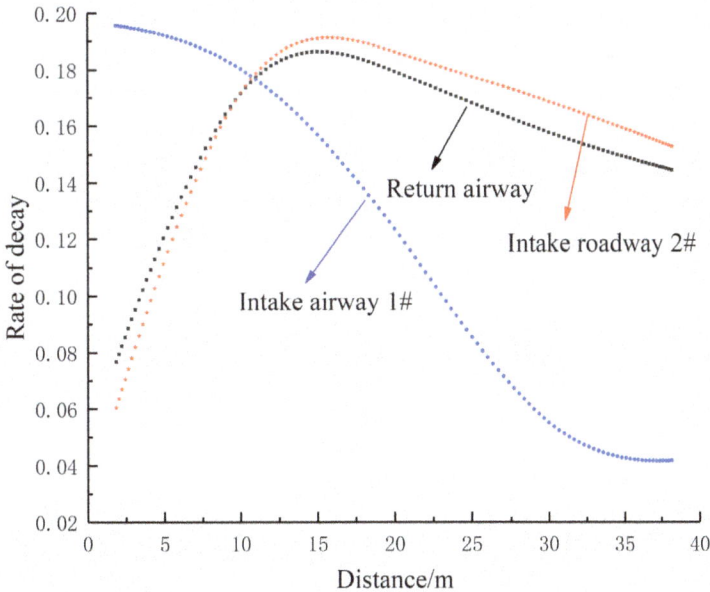

FIG 9 – Variation rule of attenuation rate of peak overpressure per 10m in intake airway #1, intake airway #2 and return airway.

The statistical table of attenuation rate of overpressure peak at every 20m in the working face and the variation curve of attenuation rate are shown in Table 3 and Figure 10 respectively.

TABLE 3

Statistical table of attenuation rate of overpressure peak at every 20 m in working face.

	0 m	20 m	40 m	60 m	80 m	100 m	120 m	140 m	160 m
Working face	0.530	0.345	0.218	0.186	0.158	0.135	0.126	0.118	0.113

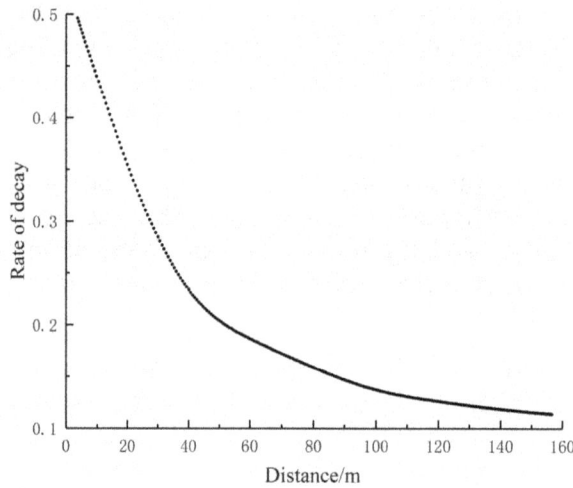

FIG 10 – Variation curve of attenuation rate of overpressure peak at every 20 m in working face.

As can be seen from Figures 9 and 10, the attenuation rate curves of peak overpressure in intake airway #2 and return airway first rise and then decrease, and reach the maximum at 15 m, that is, the peak overpressure attenuates slowly at the beginning, and the attenuation accelerates after the shock wave propagates for a certain distance, and then the attenuation rate decreases again. Because the air intake airway #1 is far from the explosion source, the overpressure in the roadway is relatively low, similar to the back end of the working face. The attenuation rate of the peak overpressure of the intake airway #1 gradually decreases, and it starts to decay steadily when it approaches the fixed value of 0.04 at 40 m of the roadway; The peak value of overpressure in the working face attenuates rapidly from the beginning, and with the propagation of shock wave, its attenuation rate gradually decreases, and the further away from the explosion source, the more gentle the curve of overpressure peak attenuation rate changes, gradually approaching 0.1.

W-type ventilation

The gas accumulated in the middle of the working face was detonated through the high-temperature ignition point, and the cloud diagram of the overpressure distribution of the gas explosion shock wave at different times is shown in Figure 11.

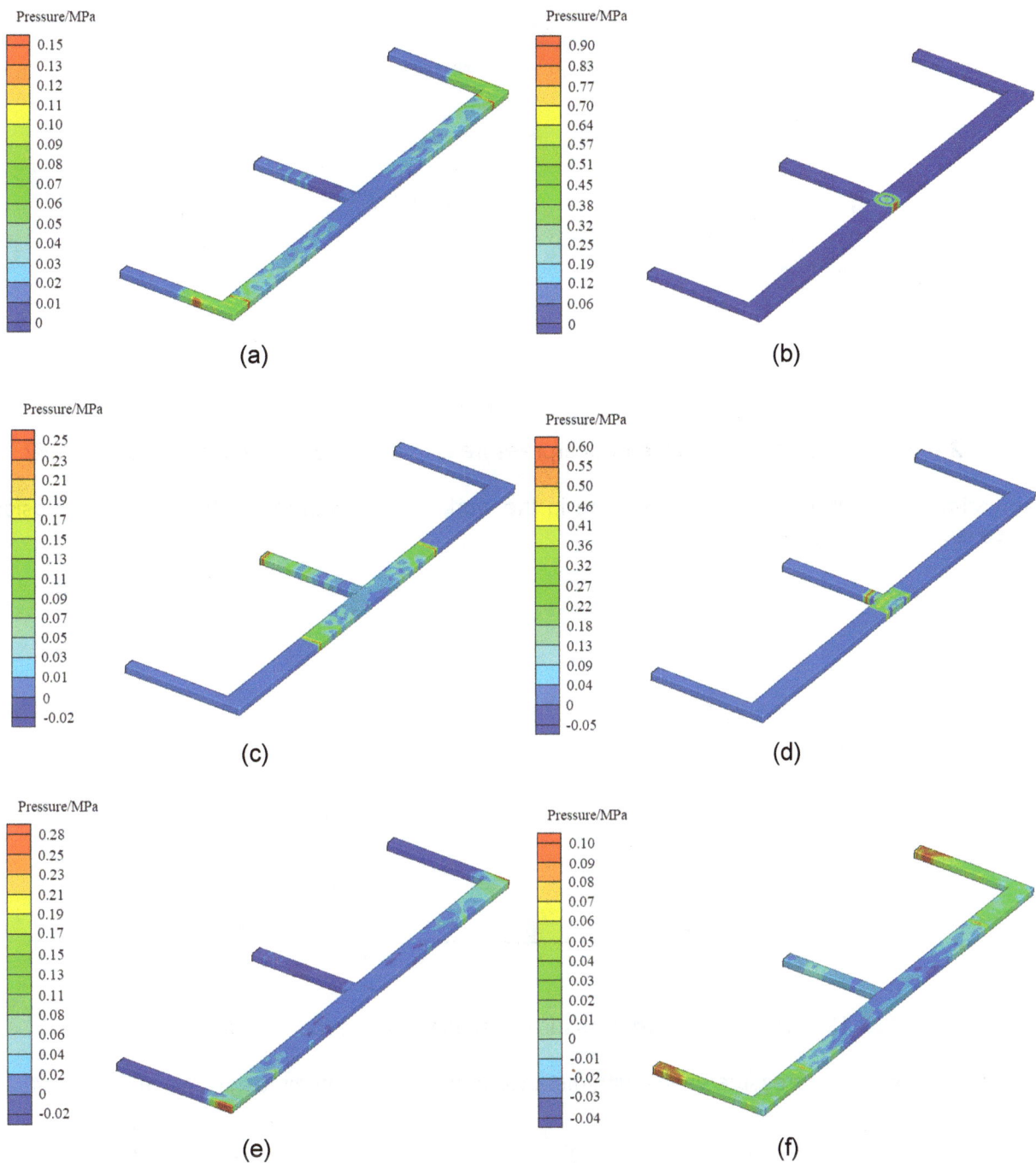

FIG 11 – Cloud diagram of shock wave overpressure distribution at different time of gas explosion: (a) t = 0.006 sec; (b) t = 0.012 sec; (c) t = 0.06 sec; (d) t = 0.17 sec; (e) t = 0.2 sec; (f) t = 0.26 sec.

Due to the symmetry of W-type ventilation mode, when gas explosion occurs in the accumulation area in the middle of the working face, combined with the relevant assumptions of numerical simulation, the overpressure variation law of intake airway #1 and #2 is the same in theory. With the propagation of explosive shock wave, the variation of overpressure peak value in intake airway #1 and #2 with distance is shown in Figure 12.

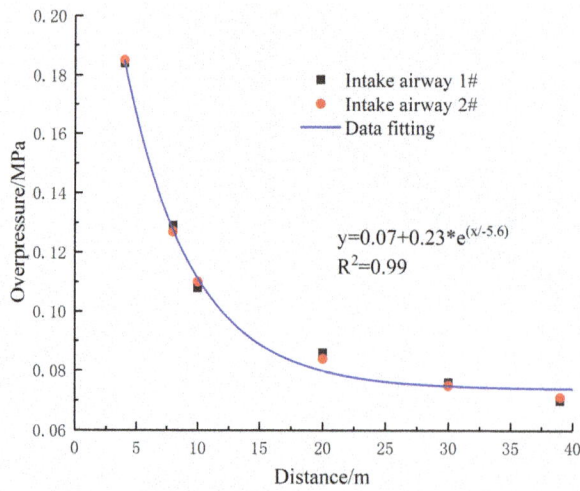

FIG 12 – Variation of the peak value of overpressure with distance in intake airway #1 and #2.

The variation rule of the peak overpressure in the working face with the distance from explosion source is shown in Figure 13.

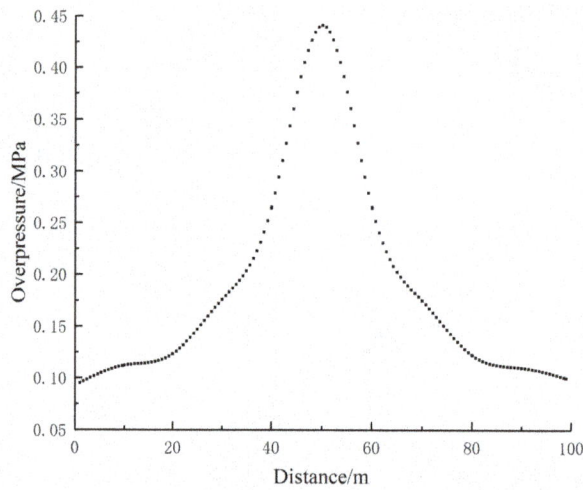

FIG 13 – Variation of overpressure peak in working face with distance.

The peak value of overpressure varies with distance in the return airway, as shown in Figure 14.

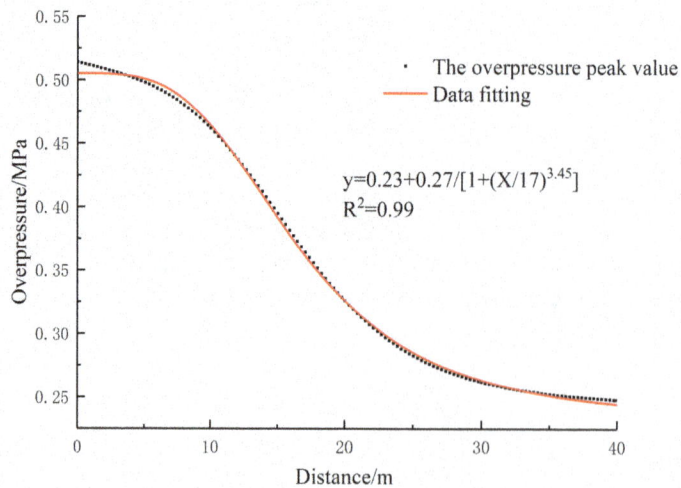

FIG 14 – Variation of peak overpressure with distance in return airway.

Analysing Figures 12 and 14 we can see: in intake airway #1 and #2, the peak value of overpressure of shock wave has the same variation pattern, and its attenuation is relatively rapid at the beginning. With the increase of propagation distance, the attenuation of the peak value of overpressure gradually slows down. Through data fitting, the variation law of overpressure peak value in intake airway #1 and #2 with distance is $y = 0.07 + 0.23 \times e^{(x/-5.6)}$, and the correlation is 0.99. In the working face, after the gas explosion occurs in the gas accumulation area, the blast wave propagates from the explosion source to both sides, and the attenuation law of the overpressure of the shock wave on both sides is the same. In the return airway, the peak value of shock wave overpressure is relatively high as it is close to the explosion source, and its attenuation trend is a process from gentle to severe and then to gentle. Its attenuation law is $y = 0.23 + 0.27 / [1 + (x/17)^{3.45}]$, and the correlation is 0.99. Statistical table of attenuation rates per 10 m of peak overpressure in intake airway #1 and #2 and variation curves of attenuation rates are shown in Table 4 and Figure 15 respectively.

TABLE 4

Statistical table of attenuation rate of peak overpressure per 10 m in intake airway #1 and #2.

The name of roadway	0 m	10 m	20 m	30 m	40 m
Intake airway #1	0.124	0.163	0.214	0.116	0.102
Intake airway #2	0.132	0.158	0.211	0.112	0.104

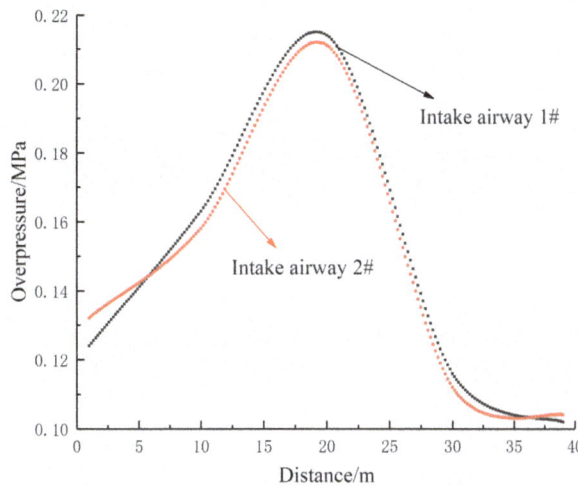

FIG 15 – Variation rule of attenuation rate of peak overpressure per 10 m in intake airway #1 and #2.

The statistical table of attenuation rate of overpressure peak in working face every 20 m and the variation law of attenuation rate are shown in Table 5 and Figure 16 respectively.

TABLE 5

Statistical table of attenuation rate of overpressure peak at every 20 m in working face.

	80 m	60m	40 m	20 m	0 m	20 m	40 m	60 m	80 m
Working face	0.112	0.128	0.303	0.370	0.401	0.365	0.298	0.124	0.110

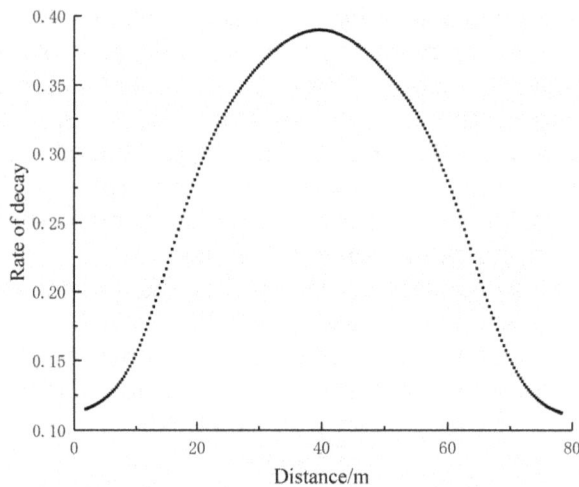

FIG 16 – Variation curve of attenuation rate of overpressure peak at every 20 m in working face.

The statistical table of attenuation rate per 10 m of peak overpressure in the return airway and the variation curve of attenuation rate are shown in Table 6 and Figure 17 respectively.

TABLE 6

Statistical table of attenuation rate of peak overpressure per 10 m in return airway.

The name of roadway	0 m	10 m	20 m	30 m	40 m
Return airway	0.101	0.294	0.254	0.103	0.053

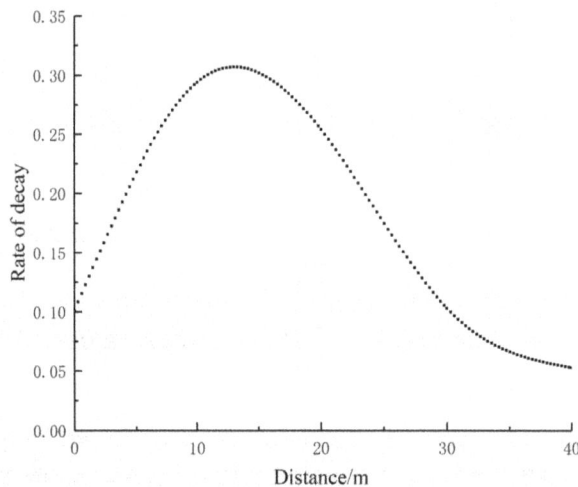

FIG 17 – Variation curve of attenuation rate of peak overpressure per 10 m in return airway.

Analysing Figures 15 and 17, it can be seen when the gas exploded in the working face of W-type ventilation, the peak overpressure attenuation rate in the intake airway #1 and #2 shows a law from low to high to low, reaching the maximum at 20 m and beginning to steadily decay at 40 m with an attenuation rate of approximately 0.1. The peak attenuation rate of overpressure in working face gradually decreases to close to 0.1 with the propagation distance of shock wave. The return airway with W-type ventilation is close to the explosion source, and the overpressure in the roadway is large after the explosion. The attenuation rate of the peak overpressure in the return airway of the W-type ventilation mode first increases and then decreases, and the attenuation rate is the largest at 15 m, and it decays steadily at a rate of about 0.05 from 40 m.

CONCLUSION

- Under the condition of Y-type ventilation mode: in the working face and intake airway #1, the shock wave overpressure is a process of attenuation from violent to steady attenuation, and their attenuation laws are the same and present a power function relationship. The change of pressure peak with distance is $y = 0.08 + 0.48 \times e^{(-x/27.84)}$. The change of the peak overpressure with distance in the intake airway #1 is $y = 0.07 + 0.06 \times e^{(-x/5.85)}$. The attenuation trend of shock wave overpressure in intake airway #2 and return airway is completely different from that in intake airway a and working face, and the peak value of shock wave overpressure in all parts of return airway is 10.6 per cent larger than that in intake airway #2 on average.

- Under the condition of Y-type ventilation mode: the peak attenuation rate curve of overpressure in the intake airway #2 and the return airway first rises and then drops, and the maximum is at 15 m; The overpressure in the air intake airway #1 is relatively low, similar to the back end of the working face. The attenuation rate of the peak overpressure of the intake airway #1 gradually decreases, and it starts to decay steadily when it approaches the fixed value of 0.04 at 40 m of the roadway; The overpressure peak in the working face, with the propagation of shock wave, its attenuation rate gradually decreases, and the farther away from the explosion source, the more gentle the overpressure peak attenuation rate curve, gradually approaching 0.1.

- Under the condition of W-type ventilation mode: in intake airway #1 and #2, the peak value of overpressure of shock wave has the same variation law, and its attenuation is relatively rapid at the beginning. With the increase of propagation distance, the peak value of overpressure attenuation gradually slows down. Through data fitting, the variation law of overpressure peak value in inlet roadway #1 and #2 with distance is $y = 0.07 + 0.23 \times e^{(x/-5.6)}$. In the working face, the shock wave propagates from the detonation source to both sides, and the attenuation law of overpressure of the shock wave on both sides is the same. In the return airway, the peak value of shock wave overpressure is relatively high as it is close to the explosion source, and its attenuation trend is a process from gentle to severe and then to gentle and its attenuation law is $y = 0.23 + 0.27 / [1 + (x/17)^{3.45}]$.

- Under the condition of W-type ventilation mode: the peak attenuation rate of overpressure in intake airway #1 and #2 shows a law from low to high and then to low, reaching the maximum at 20 m and beginning to steadily decay at about 0.1 at 40 m. The peak attenuation rate of overpressure in working face decreases gradually with the propagation distance of shock wave. In the return airway, the attenuation rate of peak overpressure increases at first and then decreases, and the attenuation rate reaches the maximum at 15 m, and starts to decay steadily at about 0.05 from 40 m.

REFERENCES

Chen, Y L, Huang, Q H and Wang, P Y, 2020. Further development of the modified Eddy Dissipation Model, *Journal of Physics: Conference Series*, 1509(1).

Gieras, M, Klemens, R, Rarata, G and Wolanski, P, 2006. Determination of explosion parameters of methane-air mixtures in the chamber of 40dm3 at normal and elevated temperature, *Journal of Loss Prevention in the Process Industries*, 19(2):263–270.

Halouane, Y and Dehbi, A, 2017. CFD simulations of premixed hydrogen combustion using the Eddy Dissipation and the Turbulent Flame Closure models, *International Journal of Hydrogen Energy*, 42(34):21990–22004.

Hassan, I K, Khalid, M S, Hossam, S A, Mohsin, M S and Mazlan, A W, 2010. Implementation of the eddy dissipation model of turbulent non-premixed combustion in OpenFOAM, *International Communications in Heat and Mass Transfer*, 38(3).

Hoste, J J O E, Fossati, M, Taylor, I J and Gollan, R J, 2019. Characterisation of the eddy dissipation model for the analysis of hydrogen-fuelled scramjets, *The Aeronautical Journal*, 123(1262).

Ibrahirn, S S and Masri, A R, 2001.The effects of obstructions on overpressure resulting from premixed flame deflagration, *Journal of Loss Prevention in the Process Industries*, 14(3):213–221.

Jia, Z W, Xu, S M and Jing, G X, 2015. Experimental study on propagation law of gas explosion shock wave in unidirectional bifurcation pipe, *China Safety Science Journal*, 25(12):51–55.

Kravtsov, A N, Zdebski, J, Svoboda, P and Pospichal, V, 2015. Numerical analysis of explosion to deflagration process due to methane gas explosion in underground structures, in *Proceedings of the 2015 International Conference on Military Technologies (ICMT)*, IEEE, pp 1–9.

Li, Z F, Yu, M G, Ji, W T and Wen, X P, 2015. Numerical simulation of turbulent flame induced by obstacles in gas Explosion, *Journal of Henan Polytechnic University (Natural Science Edition)*, 34(2):167–170.

Liu, J J, Chen, S Q, Ren, J Z and Hu, J M, 2020. Research on the influence of gas accumulation amount and Explosion Distance on fan and explosion Door, *China Work Safety Science and Technology*, 16(9):57–63.

Masri, A R, Ibrahim, S S, Nehzat, N and Green, A R, 2000. Experimental study of premixed flame propagation over various solid obstructions, *Experimental Thermal and Fluid Science*, 21(1):109–116.

Qiu, J W, 2018. Simulation study on propagation characteristics and disaster process of gas explosion shock wave in pipe network, academic dissertation, Anhui Polytechnic University.

Shao, H, Jiang, S G, Zhang, X, Wu, Z Y, Wang, K and Zhang, W Q, 2015. Influence of vacuum degree on the effect of gas explosion suppression by vacuum chamber, *Journal of Loss Prevention in the Process Industries*, 38:214–223.

Tian, Z D, 2020. Simulation of HiFIRE-2 Scramjet Based on Vortex Dissipation Model, *Electronic testing*, (7):60–61.

Wang, C L and Du, J, 2021. Experimental study on propagation law of gas explosion shock wave in different ventilation pipe network, *Mine Construction Technology*, 42(4):32–38; 26.

Wang, C, Zhao, Y and Addai, E K, 2017. Investigation on propagation mechanism of large scale mine gas explosions, *Journal of Loss Prevention in the Process Industries*, 49:342–347.

Wang, K, Jiang, S G, Ma, X P, Wu, Z Y and Zhang, W Q, 2015. Study of the destruction of ventilation systems in coal mines due to gas explosions, *Powder Technology*, 286:401–411.

Wen, X P, Xie, M Z, Yu, M G and Liu, Z C, 2013. Dynamic characteristics of gas deflagration in small-scale confined space, *Combustion Science and Technology*, 19(4):347–351.

Wen, X P, Yu, M G, Deng, H X, Chen, J J, Wang, F H and Liu, Z C, 2016. Large Eddy Simulation of gas turbulence deflagration in small scale confined Space, *Journal of Chemical Industry and Technology*, 67(5):1837–1843.

Wu, B, Zhang, L C and Xu, J D, 2005. Numerical simulation of shock wave induced by moving flame for gas explosion, *Journal of China University of Mining and Technology*, 34(4):423–426.

Wu, T R, Ye, Q and Liu, W, 2017. Study on seismic absorption and energy absorption performance of borehole on coal and rock roadway Wall, *China Safety Science and Technology*, 1304):81–86.

Xue, S S, 2019. Study on cause analysis and control of coal mine gas accident based on probabilistic reasoning, academic dissertation, China University of Mining and Technology.

Yan, A H, Nie, B S, Dai, L C, Zhang, Q Q, Liu, X N, Yang, H, Liu, Z and Hu, T Z, 2011. Numerical Simulation on the Gas Explosion Propagation Related to Roadway, *Procedia Engineering*, 26:1563–1570.

Ye, Q, Lin, B Q, Jia, Z Z and Zhu, C J, 2009. Propagation law and analysis of gas explosion in bend duct, *Procedia Earth and Planetary Science*, 1(1):316–321.

Zhang, L C, Zhang, Y L, Xu, J D, Yang, G Y, Hou, S Q and Yang, L, 2011. Numerical Simulation of Shock Wave Structure in Gas Explosion, *Procedia Engineering*, 26:1322–1329.

Goaf explosive gas zones under the impact of intensive goaf gas drainage from vertical boreholes

Y Wang[1], G Si[2], R Hu[3], B Belle[4] and J Oh[5]

1. PhD Candidate and Research Assistant, School of Minerals and Energy Resources Engineering, University of New South Wales, Sydney NSW 2052.
   Email: yuehan.wang@student.unsw.edu.au
2. Associate Professor, School of Minerals and Energy Resources Engineering, University of New South Wales, Sydney NSW 2052. Email: g.si@unsw.edu.au
3. PhD Student, School of Minerals and Energy Resources Engineering, University of New South Wales, Sydney NSW 2052. Email: runzhe.hu@student.unsw.edu.au
4. Simtars, Resources Safety and Health Queensland, Redbank Qld 4301.
   Email: director@61drawings.com
5. Associate Professor, School of Minerals and Energy Resources Engineering, University of New South Wales, Sydney NSW 2052. Email: joung.oh@unsw.edu.au

ABSTRACT

In Australian underground coalmines, goaf gas drainage is extensively employed to mitigate gas emissions and reduce their impact. As mining operations reach greater depths and produce higher levels of gas emissions, narrower spacing between adjacent vertical boreholes and increased suction pressure at the borehole's upper surface are adopted. Consequently, this proactive goaf gas drainage design facilitates enhanced extraction of mine gas. However, there is a concern that an increased amount of ventilation air might be drawn into the deep goaf, potentially resulting in the accumulation of a gas explosive zone composed of methane-air mixtures. Utilising extensive goaf gas drainage data collected from various Australian coalmines, these data sets have been subjected to detailed analysis in prior back analysis studies. These findings are subsequently harnessed to validate a goaf Computational Fluid Dynamics (CFD) model, enabling engineers to visualise the otherwise inaccessible goaf environment. Therefore, the outcomes of the CFD modelling can be integrated with Coward's triangle to demarcate potential gas explosion zones within the goaf. Furthermore, this study conducts an examination of factors influencing gas explosions in the goaf, offering insights to guide strategies for mitigating associated risks. This includes optimising goaf gas drainage efficiency to curtail gas emissions while prioritising mining safety. Thus, this study plays a key role in optimising the efficiency of goaf gas drainage, aiming to minimise gas emissions into the atmosphere while upholding the priority of mining safety.

INTRODUCTION

Longwall mining is a highly productive underground coal extraction method involving the removal of coal in long panels or faces. This globally used technique is efficient but poses challenges related to safety, environmental impacts, and economic viability. A major concern is methane release, the primary gas emitted from coal seams, which permeates the goaf from the surrounding strata (Brodny and Tutak, 2021; Tutak and Brodny, 2017). Longwall gas emissions are rising due to high retreat rates, high gas content in coal seams, and the increased size of longwall panels (Balusu et al, 2005; Tanguturi and Balusu, 2014). When methane mixes with leaked air from the working face, it can form a flammable and explosive methane-air mixture at concentrations between 5–15 per cent (Coward and Jones, 1952; Karacan et al, 2011). If this mixture contacts an ignition source, it poses a significant risk of severe injuries or fatalities among workers, and substantial damage to infrastructure and equipment. In recent years, multiple mine fires and explosions have indicated the presence of an Explosive Gas Zone (EGZ) in the longwall goaf (Brune, 2013). Additionally, methane is 82.5 times more potent than carbon dioxide over 20 years in the atmosphere and 29.8 times more potent over 100 years (Forster et al, 2021). Therefore, mitigating gas emissions from fossil fuels is crucial for addressing climate change, particularly in the short-term. As methane is both a potent greenhouse gas and explosive, its removal is essential for safety and environmental reasons.

Various techniques have been developed to mitigate gas emissions from longwall mining, including horizontal, vertical, and cross-measure boreholes (Karacan *et al*, 2011). In Australian underground coalmines, vertical boreholes are widely used to manage goaf gas drainage when overlying strata collapse into the goaf. This method involves drilling vertical boreholes from the surface to a specific depth to extract gas from the goaf and fractured strata, typically 30 to 70 m from the tailgate edge and 5 to 10 m from the working seam. Vertical boreholes not only recover high-purity gas from multiple coal seams as an energy source but also prevent longwall production delays caused by gas exceedance at the tailgate. However, with deeper mining operations and increased gas emissions, mines have adopted more aggressive goaf drainage designs, with narrower borehole spacing and stronger suction pressure. While this approach enhances the boreholes' gas capture capacity, it can alter goaf pressure distribution and increase ventilation air leakage (Saki, 2016; Saki *et al*, 2017), raising the risk of gas explosions within the goaf. To optimise goaf gas drainage efficiency and ensure underground goaf safety, understanding the inaccessible goaf gas atmosphere and the position and size of the EGZ in the longwall goaf with operating boreholes is necessary.

This paper delineates the 3D distribution of the EGZ in the goaf, considering the influence of intensive gas drainage from surface vertical boreholes. Building on previous research, a CFD model was established for a case study mine with extensive use of vertical boreholes for gas drainage. The CFD model was calibrated using goaf gas profiles from extensive field data collected from Australian underground longwall panels. Moreover, this paper effectively leverages field data to analyse scientific issues, noting that previous studies did not explore the goaf atmosphere using a large amount of drainage production data (Si and Belle, 2019). It employs the simulation results of CH_4 and O_2 concentrations from the CFD model and the Coward triangle (Coward and Jones, 1952) to accurately identify where gas explosion risks may occur in the goaf. Additionally, this paper evaluates the impact of various goaf gas drainage conditions on the size and shape of the EGZ, which is crucial for predicting potential risks and guiding engineers in implementing effective control strategies.

GOAF CFD MODEL DEVELOPMENT

This study employed FLUENT, version 2022R2 (by ANSYS, Inc), advanced CFD software, to model gas distribution in the longwall active goaf influenced by multiple simultaneously operating vertical boreholes. Based on the geometric conditions and U-type ventilation system of the case study longwall panel, a simplified goaf geometry model was established, as shown in FIG **1**. The maingate entry serves as the sole ventilation air inlet, with leaked air exiting through the tailgate entry after circulating the goaf perimeter. Moreover, the model includes eight active vertical boreholes to extract coal mined gas, reflecting the maximum number operating simultaneously in the case study to manage tailgate return gas levels. Each borehole, with a 0.25 m diameter and located 10 m above the coal seam, is labelled BH1 to BH8 by proximity to the working face. BH1 is 50 m from the working face, with the remaining boreholes spaced 50 m apart. After establishing the geometry model, an appropriate mesh is selected, and the entire volume is divided into multiple computational grids. As shown in FIG **2**, the geometry model was divided into approximately 12 million hexahedral elements using the cut-cell method, employing multiple size controls. The final model achieved an excellent quality mesh metric, with more than 98 per cent of elements falling within the range of 0.95 to 1 for orthogonal quality and 0 to 0.25 for skewness. These grids are then employed in FLUENT to solve for the conservation of mass, momentum, and energy using the finite volume method. Detailed dimensions and mesh settings for each section of the model are provided in TABLE 1.

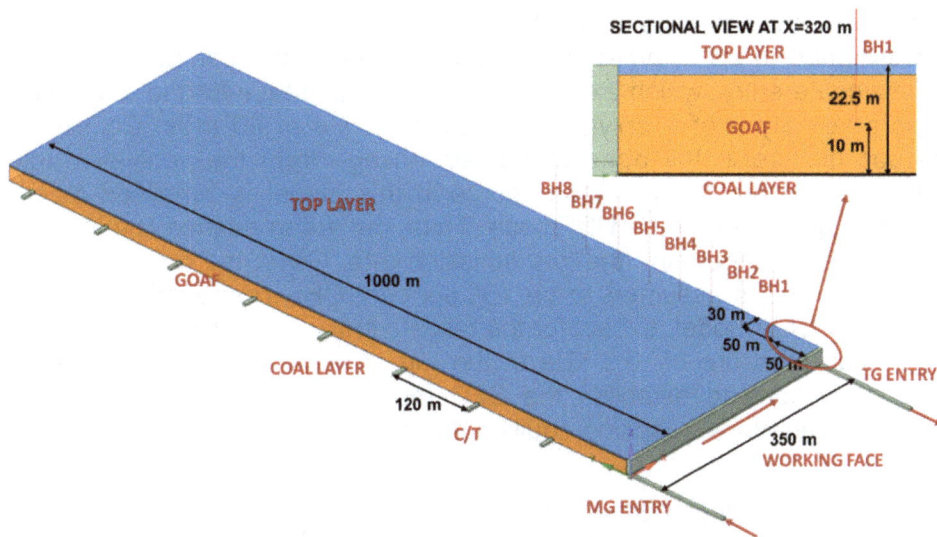

FIG 1 – The simplified goaf geometry model.

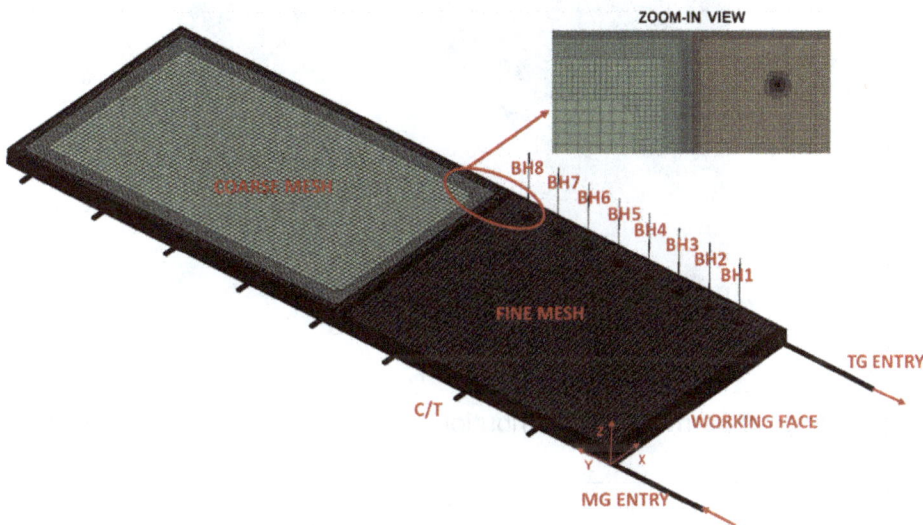

FIG 2 – The goaf mesh model.

TABLE 1

Geometry settings and mesh settings of the goaf CFD model.

Geometry settings (width (m) × length (m) × height (m))	Mesh settings (fine mesh (m) and coarse mesh (m))
Top Layer (350 × 1000 × 2, blue zone)	1.5 and 5
Goaf Layer (350 × 1000 × 20, orange zone)	1.5 and 5
Bottom Coal Layer (350 × 1000 × 0.5, black zone)	1.5
Working Face (350 × 5.4 × 2.8, green zone)	5
Maingate/Tailgate Entry (5.4 × 100 × 2.8, green zone)	5
Vertical Boreholes (Φ 0.25, BH1 to BH8)	0.05

Within the FLUENT solver, fluid parameters and physical properties can be configured. In the specified working face, treated as a free-flow domain in this CFD model, airflow is assumed to be fully turbulent, utilising the Standard k-ε turbulence model and standard wall function (ANSYS Inc, 2020). Conversely, airflow in the goaf area is described as laminar within the porous media domain, with permeability implemented via a User-Defined Function (UDF). FIG **3** depicts the goaf permeability contour at the borehole completion depth, ranging from 1e-4 to 1e-9 m². Additionally, permeability is assumed to be symmetrically distributed along the central lines of X=175 m and Y=500 m. The maingate entry, which exclusively provides air intake for the longwall working face, is configured as a velocity inlet supplying fresh air at a rate of 60 m³/s. Considering the molar fraction of the gas mixture in the atmosphere, air leakage from the ventilation inlet comprises 20.93 per cent O_2. Moreover, the methane source in this model is simplified as a velocity inlet across the top layer, governed by the UDF code. Thus, the green dashed line in FIG **4** represents the fitting curve utilised to determine the gas emission rate. Given the gas reservoir conditions of Mine A, the gas emission is assumed to be 100 per cent CH_4 to streamline the CFD modelling simulation. Within this CFD model, except for the tailgate entry, gas mixture in the goaf is extracted by applying a suction pressure of -6.5 kPa at the top of eight vertical boreholes. Following the aforementioned instructions to establish the CFD model, the FLUENT solver converges after approximately 20 000 iterations, reaching equilibrium with residuals below 1e-3.

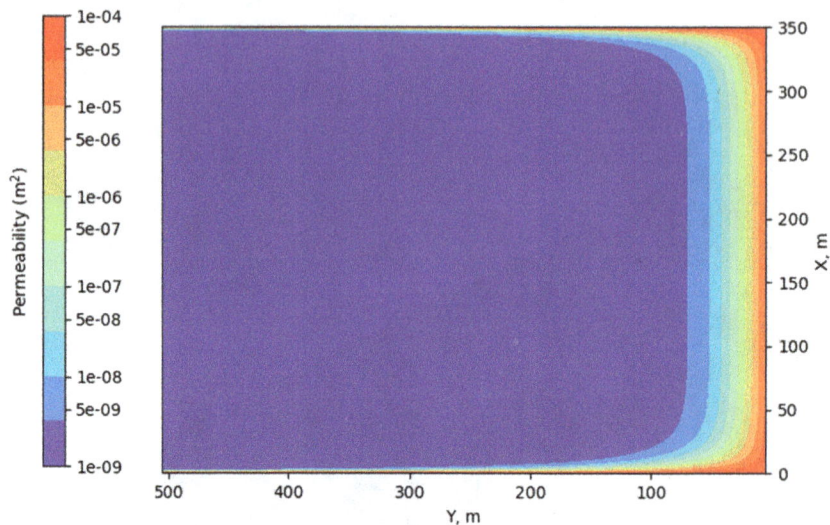

FIG 3 – The goaf permeability distribution at 10 m above the coal seam level.

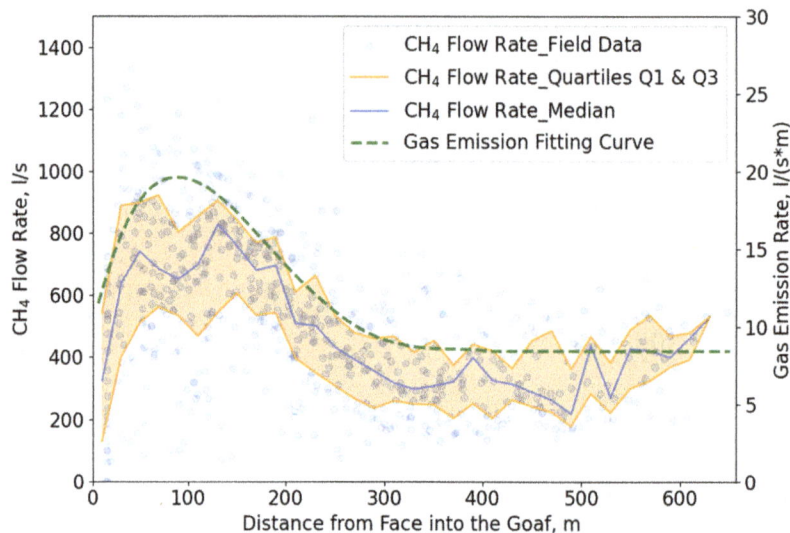

FIG 4 – Methane gas emission trends at the top layer.

GOAF GAS ATMOSPHERE ANALYSIS

Based on the above information, this model enables the simulation and visualisation of gas distributions within the inaccessible goaf area, heavily influenced by intensive goaf gas drainage via vertical boreholes. FIG **5** illustrates the 3D contours of O_2 and CH_4 distributions in the goaf. In FIG **5**a, an O_2-rich zone (exceeding 5 per cent) is evident behind the working face and along the tailgate side goaf, influenced by constant suction pressure applied at the borehole tops. The zoomed-in view of the O_2 contour at 30 m from the tailgate side, reveals that O_2 has risen to the drainage level 10 m above the seam, indicating a high O_2 concentration over 16 per cent. Further inspection of the O_2 ingress zone on the tailgate side goaf at various elevations, it becomes apparent that high-purity O_2 diminishes as elevation increases, due to gas emissions primarily consisting of 100 per cent CH_4 penetrating into the goaf from the top layer. In contrast, CH_4 concentration near both the working face and tailgate side goaf notably decreases, attributed to the ventilation air sweeping effect from the working face, as depicted in FIG 5b. With continuous gas emission from the top layer, the CH_4 concentration within the goaf varies at different depths, gradually increasing with goaf height. Meanwhile, ventilation air dilutes the gas as it moves away from the roof, resulting in a low CH_4 zone (below 20 per cent) in the lower goaf, which extends into the deeper regions of the goaf. Consequently, the diluted gas, consisting of varying components at different depths, flows through slotted casing and into vertical boreholes situated at various locations.

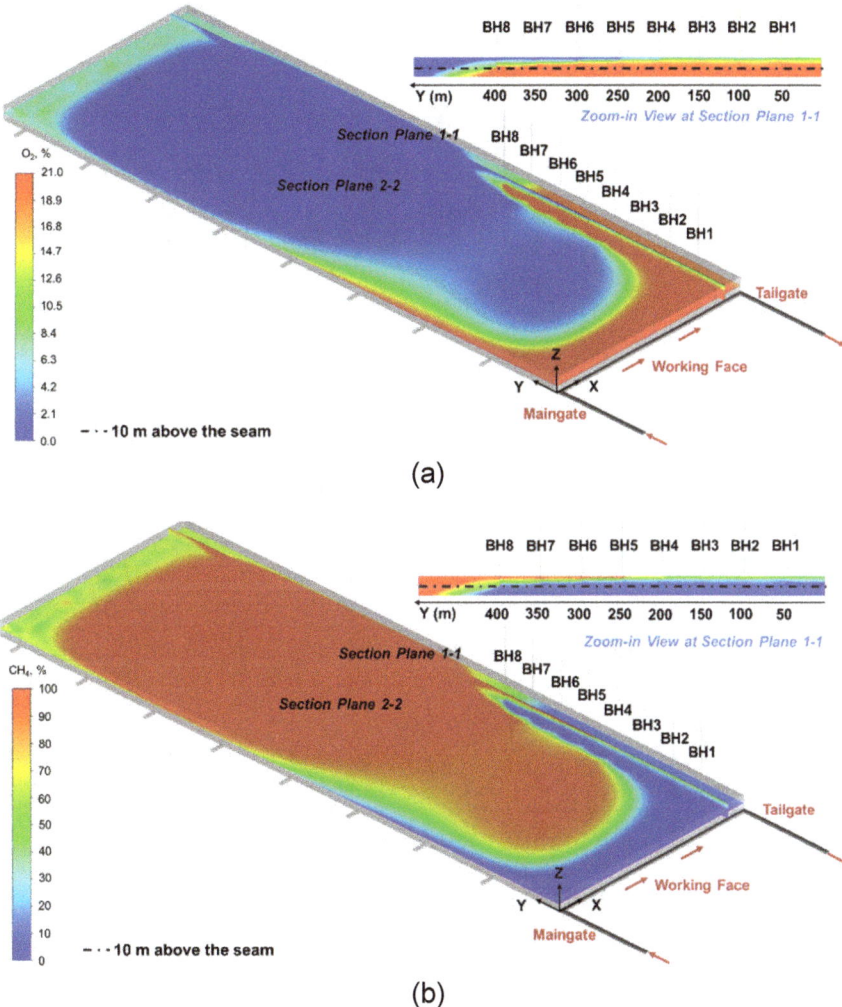

(a)

(b)

FIG 5 – (a) O_2 and (b) CH_4 concentrations: three-dimensional gas distribution contours in the longwall goaf.

The accuracy of the model can be validated by comparing gas concentrations from eight boreholes in the CFD model with field data obtained from tailgate side boreholes. In FIG **6**, O_2 and CH_4 concentrations extracted from field production data are presented. FIG **6**a illustrates daily O_2 concentrations measured at various boreholes located at different distances from the working face, represented by red dots, while FIG **6**b depicts CH_4 concentrations with blue dots. Additionally, O_2 concentration in eight drained boreholes from the CFD simulation are indicated as triangles in FIG **6**. The O_2 concentration decrease from ~12 per cent in BH1 to ~3 per cent in BH8 as one moves into the deep goaf. The O_2 content in these boreholes aligns consistently in magnitude and trend with field data. However, O_2 concentrations exhibit significant fluctuations within 200 m close to the face, with some field data points indicating very low (less than 5 per cent) or high (more than 12 per cent) O_2 levels. Apart from field measurement errors, these variations may be influenced by complex factors related to the goaf's natural characteristics, including permeability distribution and gas emission rate. Further detailed analysis and explanation are provided in Wang *et al* (2023, 2024). Furthermore, the CH_4 purity found in eight drained boreholes obtained from the CFD modelling increases from 40 per cent to 80 per cent, consistent with the rebounding trend observed in the field data. However, CH_4 levels in drained boreholes from the CFD model are higher than those indicated by field drainage data. This discrepancy arises because gas emissions in this model were based on continuous velocity input, while the operational gas emission zone in the field may have limited desorption capacity.

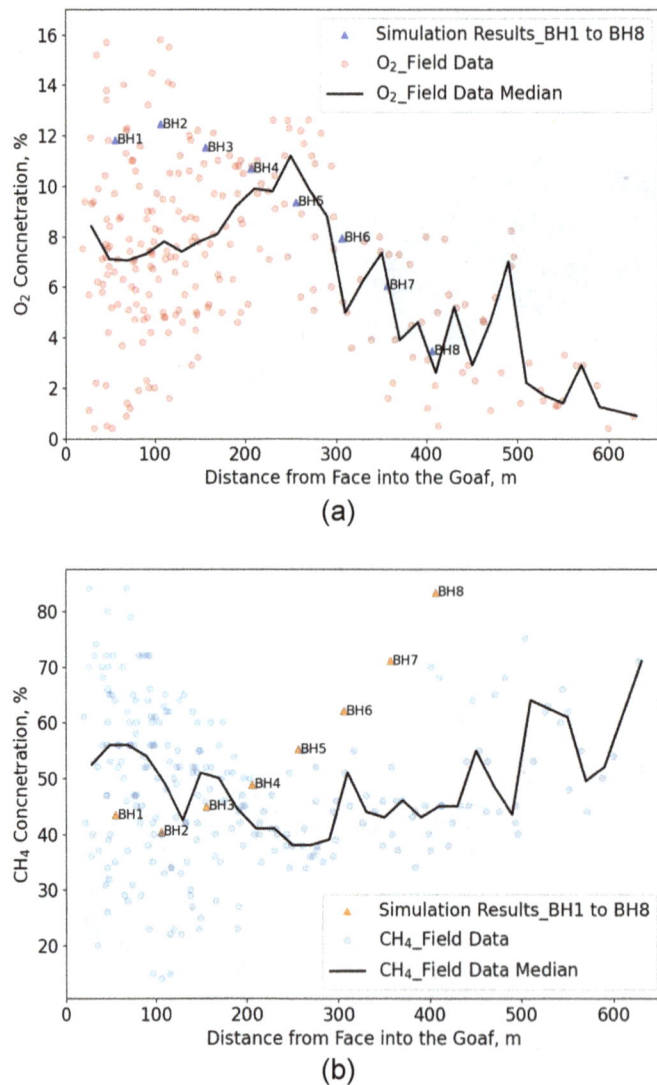

FIG 6 – (a) O_2 and (b) CH_4 concentrations: CFD modelling results versus field goaf gas drainage data in the case study mine.

GAS EXPLOSIVE ZONE ASSESSMENT

According to Coward's triangle, the goaf can be classified into four zones based on O_2 and CH_4 concentrations: (1) the red zone, indicating an explosive methane-air mixture; (2) the green zone, indicating a non-explosive mixture; (3) the yellow zone, signifying a fuel-rich mixture that becomes explosive with added oxygen; and (4) the cyan zone, indicating a fuel-lean mixture that could become explosive with added methane. FIG **7** illustrates potential gas explosion risks at various horizontal and vertical cut-planes from simulation results. Starting at the coal seam level, it becomes clear that the O_2-rich zone following the working face is non-explosive. As depth from the seam increases, the non-explosive zone gradually decreases. Moreover, with distance from the face extending into the deep goaf, O_2 content decreases while CH_4 content increases. Consequently, the fuel-lean zone closely follows the non-explosive zone, indicating an escalating risk of gas explosions with additional gas. The explosive zone at the coal seam level is drawn away from the working face by tailgate side vertical boreholes, which is different from the explosive zone near the working face in the goaf where no drilling is performed (Brune and Saki, 2017). As the distance to the face increases, the leaked air cannot reach the deeper goaf, which exhibits characteristics of a fuel-rich zone. Furthermore, as shown in FIG 7b, the explosive zone at a horizontal cut-plane 10 m above the coal seam, located not only 100 m after the face but also at the tailgate side. This accumulation is due to suction pressure on top of the boreholes, drawing O_2 from the working face upward into the higher goaf and subsequently extracted by these vertical boreholes. With increasing depth of the horizontal plan view, the size of the explosive zone gradually decreases. This phenomenon is also observed in vertical cut-plane contours in FIG 7d, where the explosive zone accumulates at the borehole completion depth on the tailgate side goaf. Moreover, CH_4 concentration on the tailgate side goaf increases as it approaches the methane source from the top layer in this CFD model, characterising these areas as fuel-rich zones with higher gas concentrations. In contrast, the explosive zone rapidly diminishes with increased distance from the coal seam level on the maingate side goaf, due to the absence of goaf gas drainage.

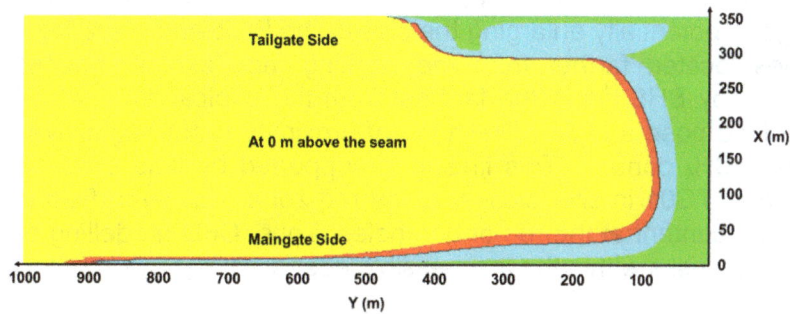

(a)

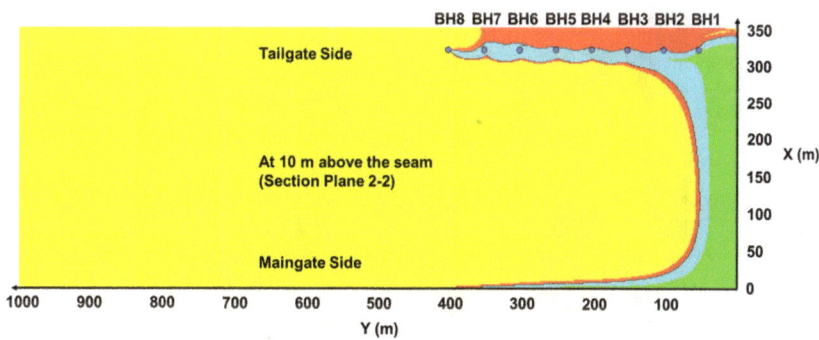

(b)

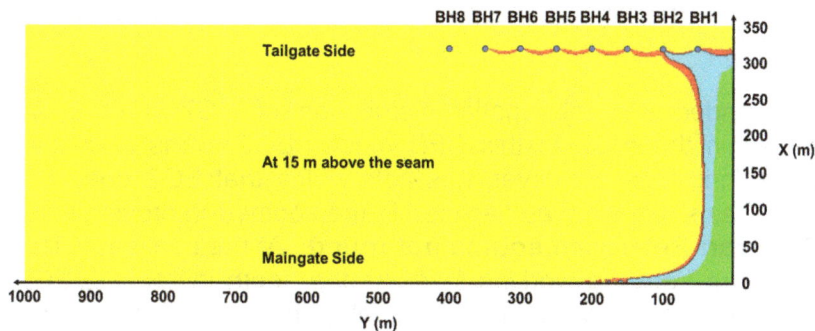

(c)

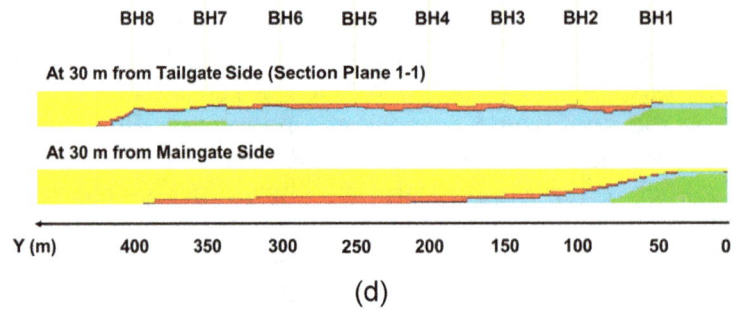

(d)

FIG 7 – Gas explosion zones at various horizontal (a) 0 m, (b) 10 m, (c) 15 m above the coal seam level and (d) vertical cut-planes in the goaf.

Furthermore, simulation results of drained gas composition from goaf boreholes can be used for comparing with field goaf gas production data, as shown in FIG **8**. This figure illustrates gas content within drained boreholes, represented by black triangles for simulation results and blue dots for field measurements. All simulation results and field data points fall within the fuel-rich zone. Triangles BH1 to BH8 are positioned along the boundary line due to coal oxidation omission in this CFD model. With coal oxidation, O_2 consumption in the goaf increases, leading to more leaked air entering the deep goaf, potentially enlarging the size of the EGZ and elevating gas explosion risks. Moreover, boreholes located farther from the working face exhibit reduced potential for gas explosions, evidenced by BH8 being the farthest from the explosive zone in FIG **8**. Conversely, boreholes near the face pose a higher likelihood of proximity to the explosive zone due to higher O_2 content and lower CH_4 content. This finding is supported by field production data, especially data from monitoring at TG01 in-site, closest to the red zone. However, field production data only offers insights into gas content at the top of boreholes. Thus, CFD modelling proves invaluable for assessing gas explosion risks within the inaccessible goaf area.

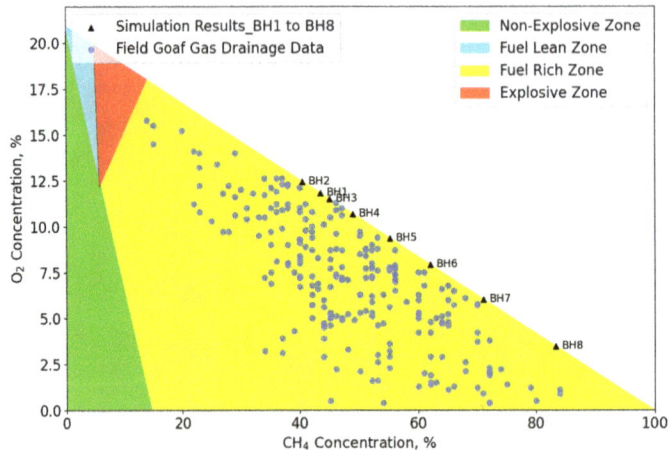

FIG 8 – Gas concentration in drained boreholes: simulation results versus field goaf gas drainage data.

To comprehensively understand the spatial distribution of EGZ, FIG **9** illustrates its three-dimensional shape within the goaf. In this CFD model, EGZ forms a concentrated 'U-shaped' pattern behind the working face. However, it is noteworthy that EGZ does not exhibit a uniform distribution along the Z-axis direction, as depicted in the zoomed-in cross-sectional view at the top right corner of FIG **9Error! Reference source not found.**. At the coal seam level, EGZ retreats to a distance of over ~80 m from the working face. As the depth increases, EGZ expands upward towards the upper section of the working face. Notably, the tailgate side demonstrates a higher EGZ presence compared to the maingate side at the upper goaf, attributed to goaf gas drainage operations. This observation contrasts with findings in the absence of goaf gas drainage, and the influence of operating goaf boreholes will be extensively explored in the next section. Additionally, the size of EGZ within the goaf can be accurately computed using FLUENT volume integrals. This quantitative method proves valuable for assessing gas explosion risks under various goaf gas drainage scenarios.

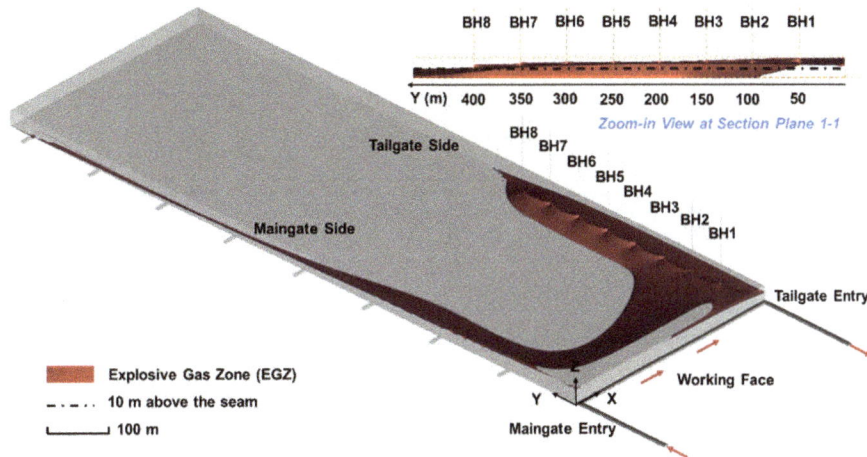

FIG 9 – The three-dimensional distribution of the EGZ in the goaf.

THE IMPACT OF GOAF GAS DRAINAGES

Influence of active boreholes number

During goaf gas drainage operations, various scenarios with different active boreholes have been observed. The simulation results for the scenario with the maximum number of operating boreholes, eight in this case, have been comprehensively examined in previous sections. This section quantifies the impact of varying numbers of active boreholes on the potential for gas explosions in the goaf. Specifically, simulations are conducted with 0, 2 (BH1 and BH2), 4 (BH1 to BH4), and 6 (BH1 to BH6) boreholes, spaced 50 m apart. FIG 10 illustrates the gas explosion zones at the target seam in the goaf under different active boreholes based on Coward's triangle. Without vertical borehole drainage, the non-explosive zone immediately adjacent to the working face (FIG 10a) is characterised by high O_2 content and low CH_4 content. As the distance from the working face increases, the O_2 concentration gradually decreases. Following a brief fuel lean zone, the region transitions into the red-coloured explosive zone, which is consistent with Brune and Saki (2017). Furthermore, increasing the number of operational boreholes gradually draws the explosive zone deeper into the goaf and farther from the working face, encompassing both the goaf tailgate and maingate sides. In FIG 10b, the explosive zone at 30 m from the tailgate side in the vertical direction under the influence of different active boreholes is compared. It is evident that simultaneous extraction from multiple boreholes, especially with six or eight active boreholes, concentrates the explosive zone near the completion depth of these boreholes. The suction applied to the top of the borehole changes the goaf pressure, causing more O_2 to flow into the deeper goaf. Consequently, a higher number of active boreholes increases the potential for gas explosion risks in the goaf.

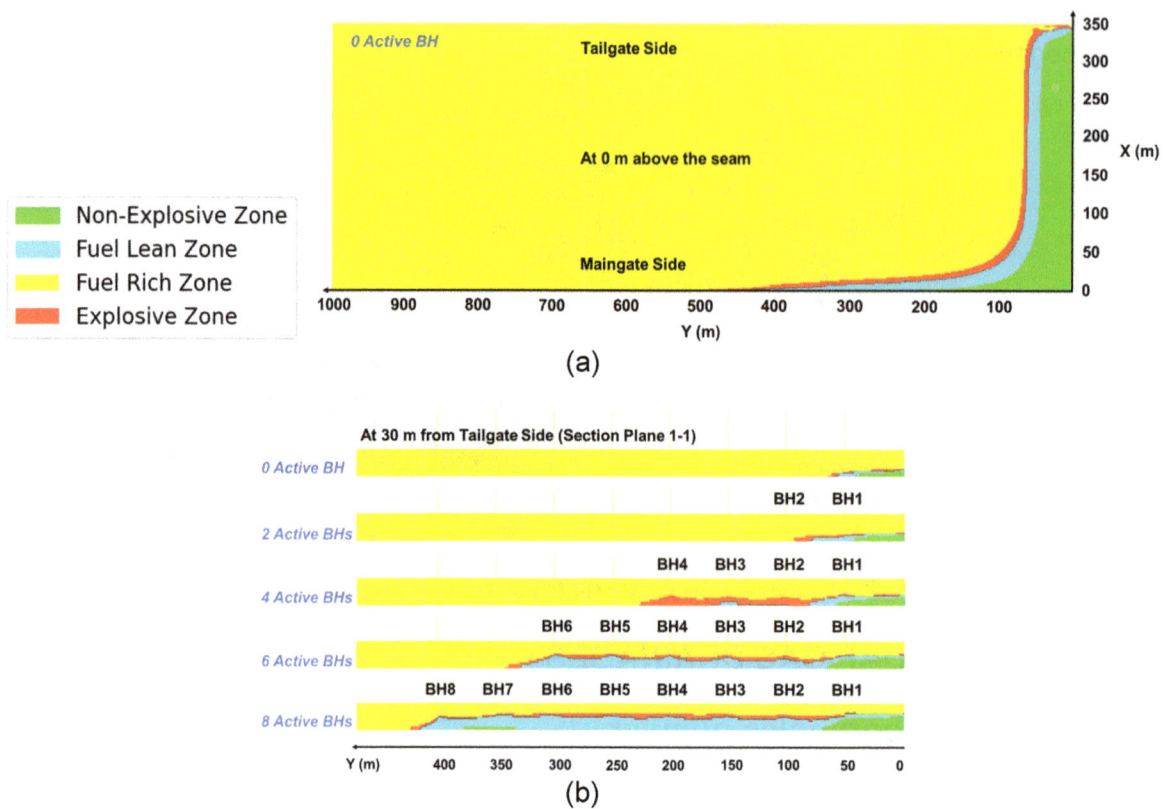

FIG 10 – The explosive zone at the target seam level under various numbers of active boreholes:
(a) 0 active borehole; (b) cross-sectional view at 30 m from tailgate side.

To assess the influence of multiple active boreholes on goaf explosion risks, the size of the EGZ served as an indicator. FIG **11** presents a blue bar graph illustrating the potential EGZ volume within the goaf as a percentage of the total goaf volume. For instance, the value of 1.943 per cent shown in FIG **11** for the case with eight active boreholes was calculated by dividing its corresponding EGZ volume (135 987.61 m³) by the total goaf volume (7 000 000 m³) and then multiplying by 100 per cent. The quantification of gas explosion risks reveals a positive correlation between the number of active boreholes and the EGZ size. Moreover, the min-max normalisation was applied to scale the EGZ size under different scenarios to a fixed range, typically between 0 and 1. As shown in Equation 1, min-max normalisation is calculated by subtracting the minimum value from a specific data point and then dividing by the difference between the maximum and minimum values. The orange triangles in FIG **11**, depict the min-max normalised values for scenarios with 0, 2, 4, 6, and 8 active boreholes, respectively. Additionally, an orange dashed trend line was introduced to enhance the understanding of the relationship between the number of active boreholes and gas explosion risks in the goaf. Therefore, employing a methodology that integrates min-max normalisation and the EGZ percentage provides valuable insights into both the direction and magnitude of correlations. This method not only facilitates the understanding of the relative scales of features related to EGZ size but also serves as a predictive tool for estimating EGZs under various goal gas drainage scenarios based on observed trends.

$$Min - Max\ Normalization = \frac{X - Min(X)}{Max(X) - Min(X)} \tag{1}$$

where:

X   is the original value of the feature

Min(X)   is the minimum value of the feature in the data set

Max(X)   is the maximum value of the feature in the data set

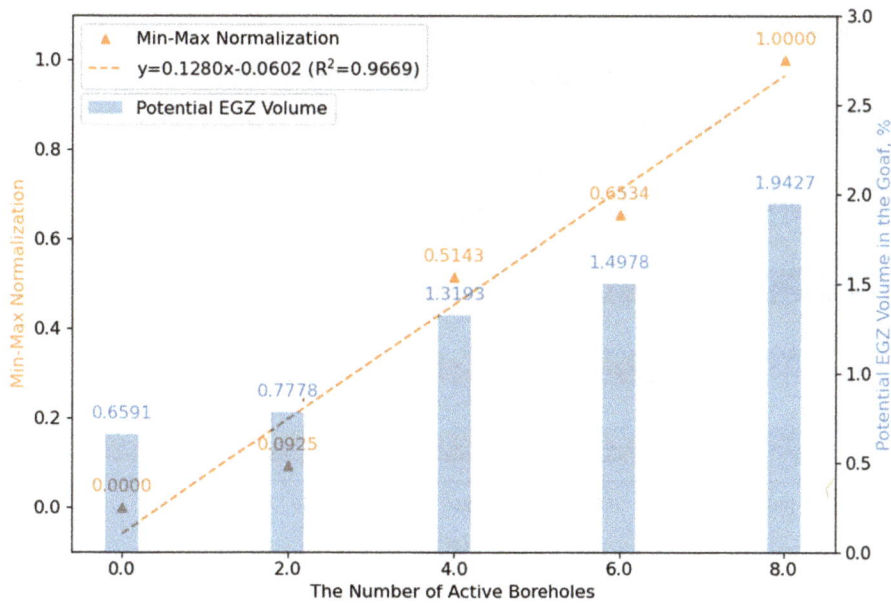

FIG 11 – The min-max normalisation and the potential EGZ volume percentage in the goaf under various numbers of active boreholes.

Influence of boreholes completion depths

The goaf gas drainage method involves drilling a series of vertical boreholes from the surface to a specific depth above the working seam, typically between 5 m and 10 m. Previous studies have focused on the impact of borehole location on drainage efficiency, mainly in horizontal planes at the same depths. However, the effect of vertical borehole completion depths on the goaf atmosphere and gas explosion risks is not well understood. In this CFD model, the bottoms of eight vertical boreholes (BH1-BH8) are positioned 10 m above the seam in the original case. This study examines different borehole completion depths on EGZs in the goaf by simulating CFD models with boreholes completed at depths of 5 m, 7.5 m, 12.5 m, and 15 m above the seam. All other parameters and set-ups are consistent with the 10 m scenario. FIG **12** shows the O_2 contour and EGZs at 30 m from the tailgate side for various borehole depths (5 m, 7.5 m, 10 m, 12.5 m, and 15 m above the working seam). The O_2-rich zone extends higher when the borehole bottom is further above the seam, resulting in a larger EGZ at a higher position. When the borehole completion depth is far from the seam, it can draw O_2 to higher positions, thus increasing gas explosion risks in the goaf. To accurately assess the impact of borehole completion depth on the EGZ in the goaf, FIG **13** depicts the min-max normalisation and EGZ percentage for different completion depths. The calculation method is the same as in the previous section. The figure shows a nonlinear, parabolic relationship between borehole completion depth and EGZ size, described by a quadratic fitting equation. The EGZ volume percentage in the goaf remains around 2 per cent across different depths. Compared to the number of active boreholes, the depth of borehole completion has a lesser impact on the size of the EGZ. This observation can be leveraged for future parameter studies aimed at optimising borehole design, improving drainage efficiency, and ensuring that the risk of gas explosions and increased air leakage resulting from intensive goaf gas drainage remains within a controllable range. Therefore, this paper provides guidance for engineers to strike a balance between drainage efficiency and mine safety. For instance, boreholes situated near the gas emission source bed are likely to achieve higher capture efficiency with minimal impact on EGZ expansion.

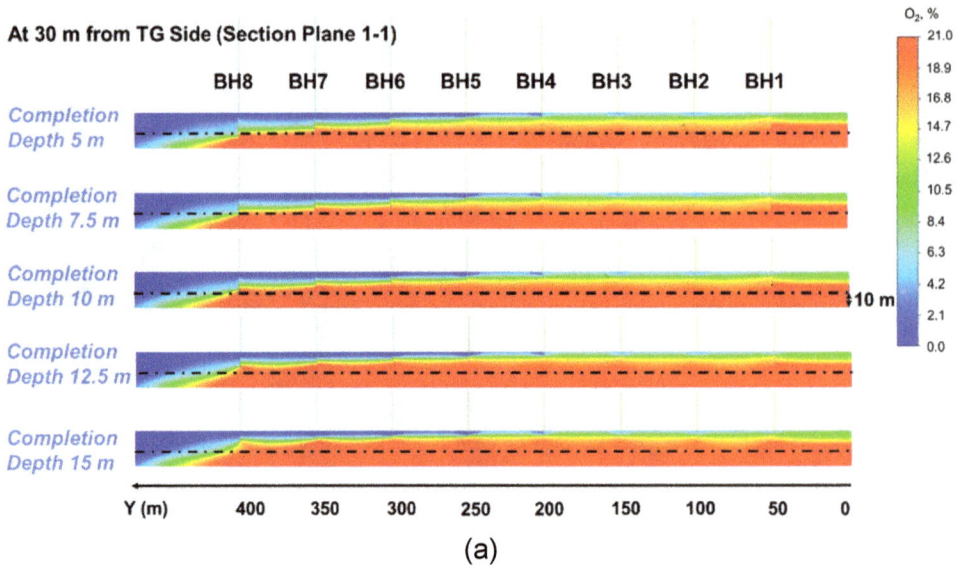

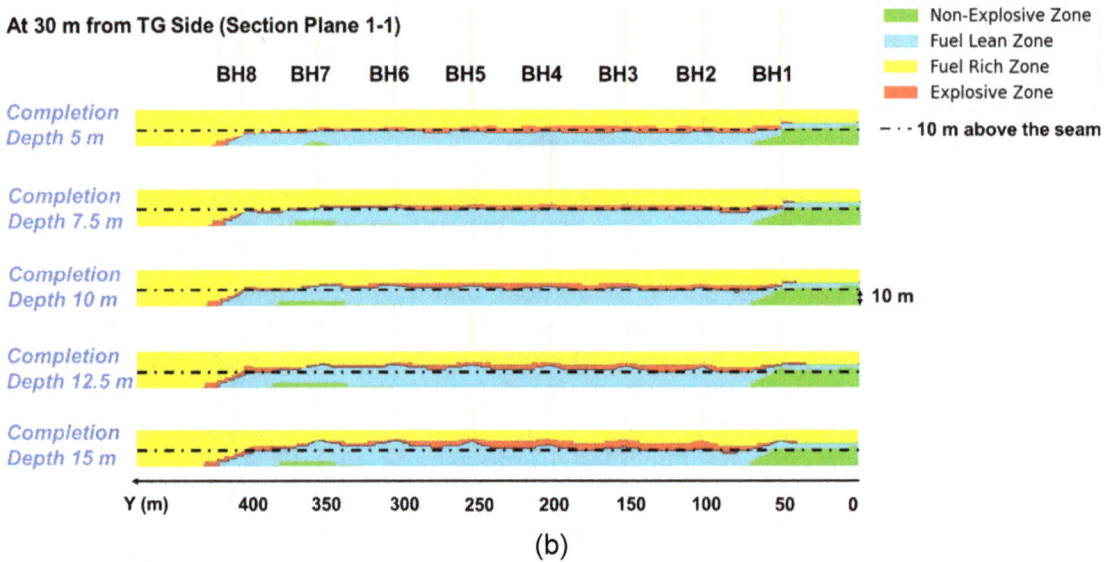

FIG 12 – The (a) O_2 contour and (b) gas explosion zones at 30 m from the tailgate side edge under various completion depths.

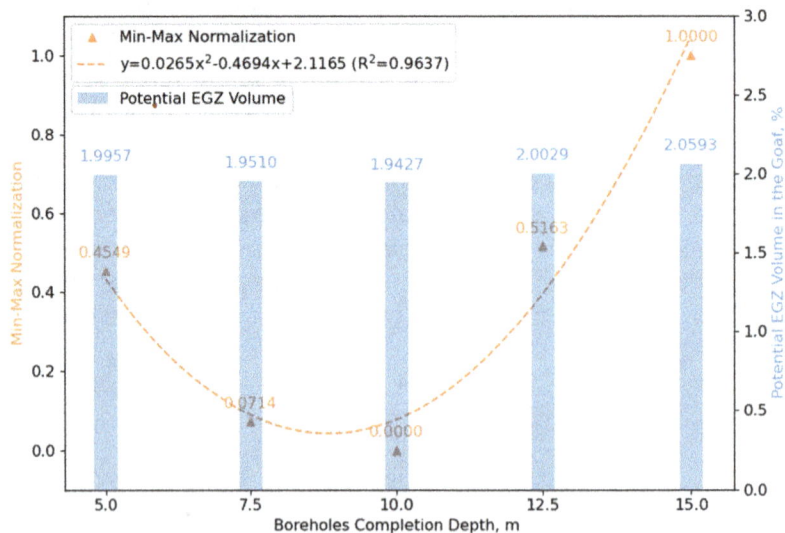

FIG 13 – The min-max normalisation and the potential EGZ volume percentage in the goaf under various completion depths.

Influence of boreholes suction pressures

In the *Goaf CFD Model Development* section, a suction pressure of 6.5 kPa was initially applied to the tops of eight vertical boreholes, based on field goaf gas drainage production data. It is worth noting that this suction pressure changes dynamically during each production period, with pressures fluctuating between 2 kPa and 11 kPa. This section evaluates how varying suction pressures affect the goaf atmosphere and the risk of gas explosions. Besides, the same suction pressure was consistently applied to the borehole top surface, and the bottoms of the eight vertical boreholes are positioned 10 m above the seam, in alignment with the original case. FIG **14** illustrates the EGZ at vertical cut-planes, positioned 30 m from both the tailgate and maingate sides, respectively. Under higher suction pressure, the EGZ defined by the red zone was drawn to a higher position, farther away from the coal seam level. This is because the intensive vertical boreholes with high suction pressure would draw the O_2-rich zone to a higher location, which can be validated by a larger non-explosive zone (green zone) at the bottom of the goaf. Moreover, the EGZ zone on the maingate side is increased with increasing suction pressure and extends deeper into the goaf, as shown in FIG **14**b. To precisely assess the impact of borehole suction pressure on gas explosion risk, FIG **15** provide the calculated min-max normalisation and EGZ percentage for various suction pressure scenarios, similar to the previous evaluation of goaf gas drainage factors. The EGZ volume percentage exhibits a positive correlation with suction pressure, rising from 2 kPa to 8 kPa. However, when the suction pressure exceeds 8 kPa, the EGZ volume percentage decreases slightly, stabilising at approximately 2 per cent. Notably, suction pressure exhibits a more significant influence on the EGZ within the goaf compared to borehole completion depth.

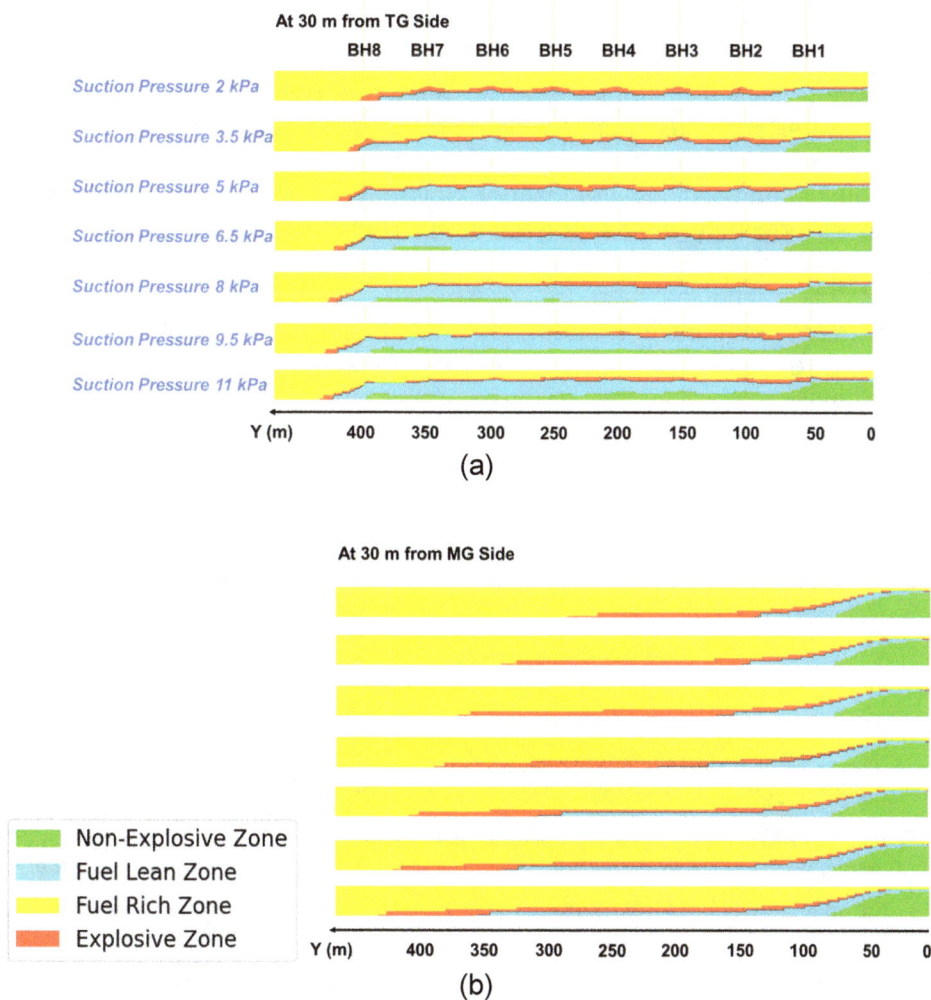

FIG 14 – Gas explosion zones at (a) 30 m from the tailgate side edge and (b) 30 m from the maingate side edge under various suction pressures.

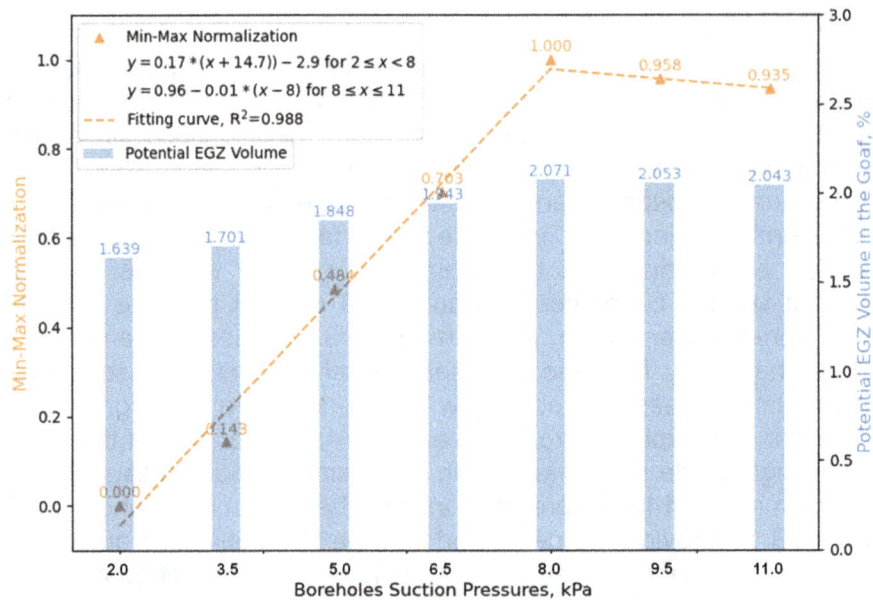

FIG 15 – The min-max normalisation and the potential EGZ volume percentage in the goaf under various suction pressures.

CONCLUSIONS

In conclusion, understanding the impact of goaf gas drainage on the underground goaf atmosphere is crucial for mining safety and productivity. This study utilised back analysis results and CFD modelling simulation results to pinpoint specific areas at risk of gas explosions in the goaf, considering parameters such as active borehole number, suction pressure, and borehole completion depth. The EGZ demonstrates a clear 'U-shaped' distribution, located 100 m from the working face with multiple operational vertical boreholes. It can also expand and deepen with more boreholes and closer proximity to the target seam. These findings inform decision-making for miners and engineers, balancing improved gas drainage efficiency with risks of ventilation air leakage. Moreover, integrating real-time goaf gas monitoring systems is essential for effective risk management. However, variations in EGZ and potential hazards from coal oxidation require further consideration. Future research could delve into the effectiveness of strategies such as injecting inert gases to mitigate explosive gas mixtures, enhancing operational safety while reducing greenhouse gas emissions.

ACKNOWLEDGEMENTS

The authors would like to thank Australian Coal Association Research Program (ACARP) C29017 and C34011 for supporting this work. The authors would also like to thank Australian Research Council Linkage Program (LP200301404) for sponsoring this research.

REFERENCES

ANSYS Inc, 2020. ANSYS Fluent Theory Guide (Pennsylvania: ANSYS Inc, Canonsburg).

Balusu, R, Tuffs, N, Peace, R and Xue, S, 2005. Longwall Goaf Gas Drainage and Control Strategies for Highly Gassy Mines, in *Proceedings of the 8th International Mine Ventilation Congress*, pp 201–209.

Brodny, J and Tutak, M, 2021. Applying computational fluid dynamics in research on ventilation safety during underground hard coal mining: A systematic literature review, *Process Safety and Environmental Protection*.

Brune, J F, 2013. The methane-air explosion hazard within coal mine gobs, *Transactions of the Society for Mining, Metallurgy and Exploration*, 334:376–390.

Brune, J F and Saki, S A, 2017. Prevention of gob ignitions and explosions in longwall mining using dynamic seals, *Int J Min Sci Technol*, 27:999–1003.

Coward, H F and Jones, G W, 1952. Limits of flammability of gases and vapors, Bulletin 503, US Bureau of Mines.

Forster, P T, Storelvmo, K, Armour, W and Collins, J L, 2021. The Earth's Energy Budget, Climate Feedbacks and Climate Sensitivity, in *Climate Change 2021 – The Physical Science Basis*, pp 923–1054 (Cambridge University Press).

Karacan, C Ö, Ruiz, F A, Cotè, M and Phipps, S, 2011. Coal mine methane: A review of capture and utilization practices with benefits to mining safety and to greenhouse gas reduction, *Int J Coal Geol*, 86:2–3:121–156.

Saki, S A, 2016. Gob Ventilation Borehole Design and Performance Optimization for Longwall Coal Mining Using Computational Fluid Dynamics, Doctoral dissertation, Colorado School of Mines.

Saki, S A, Brune, J F, Bogin, G E, Grubb, J W, Emad, M Z and Gilmore, R C, 2017. CFD study of the effect of face ventilation on CH4 in returns and explosive gas zones in progressively sealed longwall gobs, *J South Afr Inst Min Metall*, 117:257–262.

Si, G and Belle, B, 2019. Performance analysis of vertical goaf gas drainage holes using gas indicators in Australian coal mines, *Int J Coal Geol*, 216:103301.

Tanguturi, K and Balusu, R, 2014. CFD Modeling of Methane Gas Distribution and Control Strategies in a Gassy Coal Mine, *The Journal of Computational Multiphase Flows*, 6:65–77.

Tutak, M and Brodny, J, 2017. Analysis of Influence of Goaf Sealing from Tailgate on the Methane Concentration at the Outlet from the Longwall, in IOP Conference Series: Earth and Environmental Science (Institute of Physics Publishing).

Wang, Y, Si, G, Belle, B, Webb, D, Zhao, L and Oh, J, 2024. Impact of goaf gas drainage from surface vertical boreholes on goaf explosive gas zones, *Int J Coal Geol*, 284:104461.

Wang, Y, Si, G, Oh, J and Belle, B, 2023. CFD modelling of longwall goaf atmosphere under vertical boreholes gas drainage, *Int J Coal Geol*, 280:104400.

Mine fans

Power down your fans while powering up your production

L Andrews[1]

1. AAusIMM, Senior Mining Engineer, Mining Plus, Perth WA 6018.
   Email: lisa.andrews@miningplus.com.au

ABSTRACT

Trucks and loaders are often used as the default materials handling options for underground mines as they are a well-known 'safe' technology. With mines looking towards future technologies, so much time and energy is being put into electrifying trucks and loaders but could there be a better way? When completing a materials handling trade-off study including ventilation, that considers all the options available, results show there is potential to save money, reduce costs and increase productivity.

When performing a trade-off, the ventilation costs and reduced CO_2 emissions should be considered to find the most cost-effective solution. Mining Plus has completed multiple materials handling trade-offs to compare the capital and operating costs for truck and loader, Railveyor, conveyor and shafts.

While initially the truck and loader option may seem cheaper and more adaptable, incorporating primary ventilation costs often demonstrates that Railveyor, and occasionally conveyors, can be more cost-effective due to reduced operating costs and primary infrastructure associated with primary fans. Secondary ventilation savings and carbon modelling results further reinforce these cost advantages.

By thoroughly exploring materials handling trade-offs and examining the long-term benefits of alternative materials handling technologies like Railveyor, mining operations can potentially unlock substantial cost savings, enhance environmental sustainability and make significant progress towards achieving net-zero emissions objectives.

INTRODUCTION

Underground mines traditionally rely on trucks and loaders for materials handling as they are well known, flexible and reliable. Mines are now looking towards future technologies largely focusing on electrifying trucks and boggers without considering investing in other existing technologies. The Railveyor system is often overlooked as a materials handling technology due to trade-offs not considering the additional savings from the reduction in ventilation requirements. Materials handling trade-off studies should consider primary ventilation costs at a minimum and the potential for greater savings with secondary ventilation modelling when using any electrified materials handling system.

The materials handling trade-off study for the Gavião Project, owned by Almina and located in Portugal, is discussed as part of this paper to highlight the influence that ventilation has on the capital and operating costs of a mine. This study included diesel truck haulage, Railveyor and traditional belt conveyor. Due to the shallow depth of the mine and high capital costs, shaft haulage and locomotive methods of transport were excluded from this study. Battery-electric trucks and trolley-assist options were also not considered due to lack of availability and current restrictions of the technology.

RAILVEYOR

The Railveyor is a conveyor rail hybrid system that can be utilised as part of the main material handling system in underground mines and on surface, (Railveyor, 2021). It is not restricted by belt lengths thus does not require expensive transfer stations and additional excavations. The system is remotely controlled and it has a series of two wheeled railcars, Figure 1, making up a train. The train is driven by fixed driving stations located along the track.

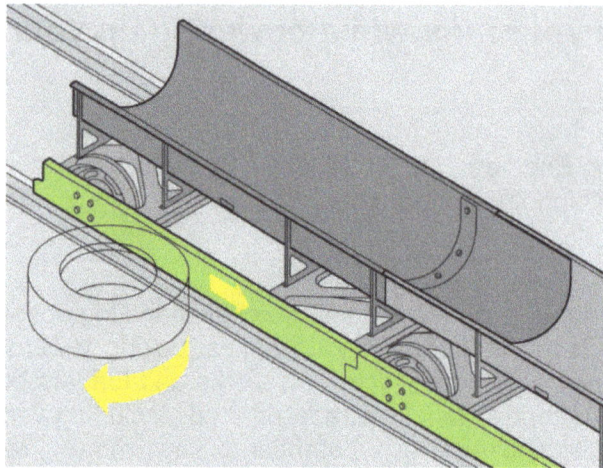

FIG 1 – Individual rail car, which is part of a Railveyor train (source: https://www.railveyor.com/).

The Railveyor does not require ballast or other material to keep the rails in place and can be relocated easily compared to a traditional rail or conveyor system. To increase and optimise the materials handling, a Railveyor can have multiple train systems running on the same track with speeds up to 8 m/s and climb grades of up to 1:5. It is more flexible than a conveyor and, as a result, can be designed in a variety of ways. Crushers are not required as the Railveyor can handle a maximum rock size of 700 mm to 1000 mm depending on the railcar size used. There are multiple discharge options and it can be designed to discharge into multiple locations, Figure 2 shows an example of a discharge loop underground (Railveyor, 2021).

FIG 2 – A Railveyor underground discharge loop (source: https://www.railveyor.com/).

MATERIALS HANDLING COSTS

For the Gavião trade-off study a 'base case' design was created to calculate the costs for all materials handling options. Figure 3 shows the size and orientation of the orebody the mine was designed for. A planned production profile was supplied by Almina assuming a maximum haulage limit of 1.4 Mt of ore per annum achieved in Year 5. All costs were obtained from a database of similar-sized projects and costs were converted to Euros. It was assumed that the Railveyor and conveyor would operate on a 1:7 decline to ensure that it was a fair comparison; however, if one of these options were chosen, it would be optimised for that mine.

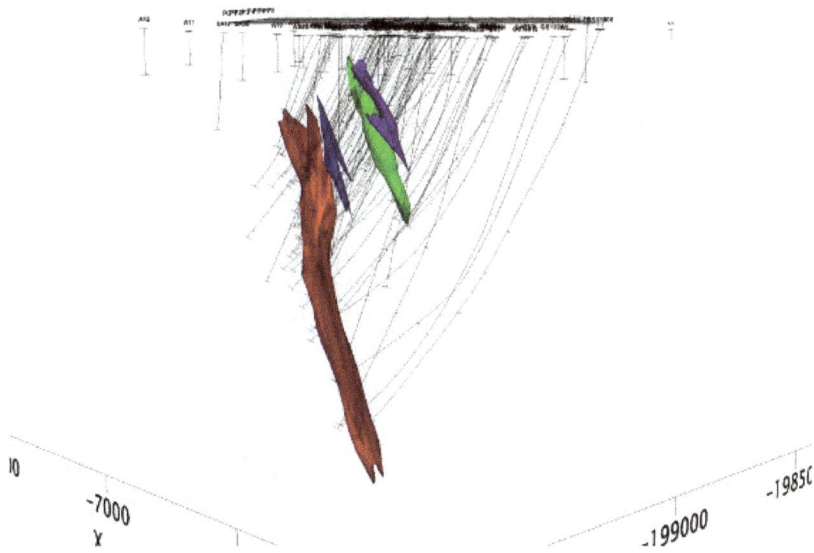

FIG 3 – 3D modelling of the Gavião mineral deposit (source: https://edm.pt/en/mineira/gaviao/).

Sandvik LH517i loaders were used for all scenarios, with an average tramming distance of 350 m. These were paired with Sandvik TH551i trucks requiring three passes to load them (Sandvik, 2024). When running on a single decline additional trucks were factored in to account for delays at the portal and when passing other vehicles in the decline. Trucking rates were calculated based on an average round trip haul distance of 5.2 km and an 80 per cent fill factor. The capital costs for the trucks and loaders were added separately based on a five year life for each piece of equipment.

Conveyor capital costs were based on a 900 mm wide belt, with an ST3150 belt tension rating and an 850 kW power demand. The costs included a gyratory crusher, a 1500 t crushed ore bin, a picking conveyor and a 400 t out loading bin on the surface for direct loading into trucks or an overland conveyor.

Railveyor costs included five 30" rubber trough trains with 107 cars per train, there are 160 drive stations required and both waste and ore surface discharge points were designed. The Railveyor and conveyor costs were added in three instalments as the bottom of each orepass was mined and prepared for haulage from that area.

The design included vertical orepasses to allow material to be tipped by loaders directly on each mining level, reducing vehicle interactions. There are a total of six rises and three dedicated materials handling levels. The orepasses have a grizzly for each tipping point that will be maintained using a mobile rock breaker. To load trucks there will be a pneumatic loading chute whilst to load Railveyor and conveyors there will be an orepass feeding system.

It was assumed that waste would be trucked to the surface using diesel trucks for all methods, although it could be batch hauled using the Railveyor or conveyor. A discount rate of 8 per cent was applied and a summary of the cumulative discounted cash flow costs for each option can be seen in Figure 4.

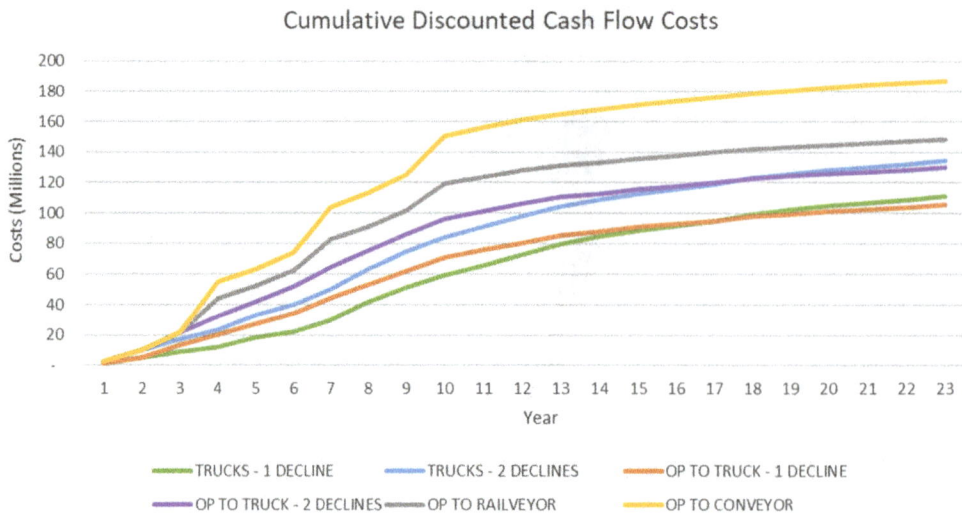

FIG 4 – Pre-tax discounted cash flow cumulative costs.

The results from this showed that having orepasses and trucking resulted in the cheapest cumulative discounted cash flow (DCF) costs over a 23-year mine life, with the single decline option being the cheapest. Table 1 summarises the results; note calculations are only based on the materials handling development and systems.

TABLE 1

Discounted cash flow costs for materials handling methods.

	Total DCF cumulative cost (millions €)	DCF cost per tonne of ore €/t
Trucks 1 decline	€111	€5.01
Trucks 2 declines	€134	€6.05
OP to truck – 1 decline	**€105**	**€4.75**
Op to truck – 2 declines	€130	€5.86
Railveyor	€148	€6.67
Conveyor	€187	€8.43

This project requires two declines to create an exhaust system as a rise is unable to be used early on in the project due to the location of the portals. The single decline options were removed from the trade-off, resulting in trucking with orepasses being the cheapest option based on the DCF cumulative cost. If a trade-off study is concluded here the indirect costs associated with ventilation are not fully realised and thus alternative material handling options are not further considered.

As part of the initial trade-off study ventilation was included as an indirect cost for a 300 kW primary fan running at 75 per cent for all scenarios. The fan rating and cost weren't adjusted based on the requirement for each materials handling method and thus failed to demonstrate the extent that ventilation has on costs.

VENTILATION REQUIREMENTS

Calculating an accurate estimate of airflow requirements is fundamental to the ventilation planning process, however also one of the issues most open to disagreement (Brake, 2020). For the purpose of this trade-off study, primary ventilation requirements were calculated based on the Work Health and Safety (Mines) Regulations 2022 (Department of Mines and Petroleum (DMP), 2022), ensuring that '*each diesel unit has a ventilation volume rate of not less than 0.05 m³ per second per kilowatt of the maximum rated engine output specified by the manufacturer on engine kilowatt rating multiplied by 0.05'* and the assumption that all equipment may need to operate underground simultaneously.

This method oversimplifies the complexities of calculating the ventilation requirements as it does not account for potential dust and heat problems that may occur no matter which materials handling method is chosen. This should be assessed once a ventilation model is created and flows modelled throughout the mine. Calculations were also performed using utilisation and availability of each piece of equipment with similar results obtained; the 'everything all at once' equipment option is discussed in more detail.

For the purposes of this trade-off secondary ventilation costs were not explicitly considered in the final calculations, however some basic cost calculations are discussed in this paper. Using any electrified method will significantly reduce the fan operating costs (McGuire *et al*, 2022). Removing diesel trucks from entering headings means that a much smaller fan can be used and thus costs are reduced. The calculations could easily be performed once a schedule and cost model have been created or could be estimated at a high level during initial trade-off.

Equipment requirements were calculated based on a supplied production profile, with modifications made for each materials handling method. Ancillary equipment numbers were also supplied and modified as appropriate. All options assumed a maximum of two loaders and two trucks would be used for waste haulage and ancillary projects, such as road base transport, as required.

The truck calculations were based on the average cycle time of being loaded with a loader versus using a loading chute from an orepass. Orepass loading is faster thus requiring less trucks over the project life. The calculated primary ventilation requirements over the life-of-mine can be seen in Figure 5.

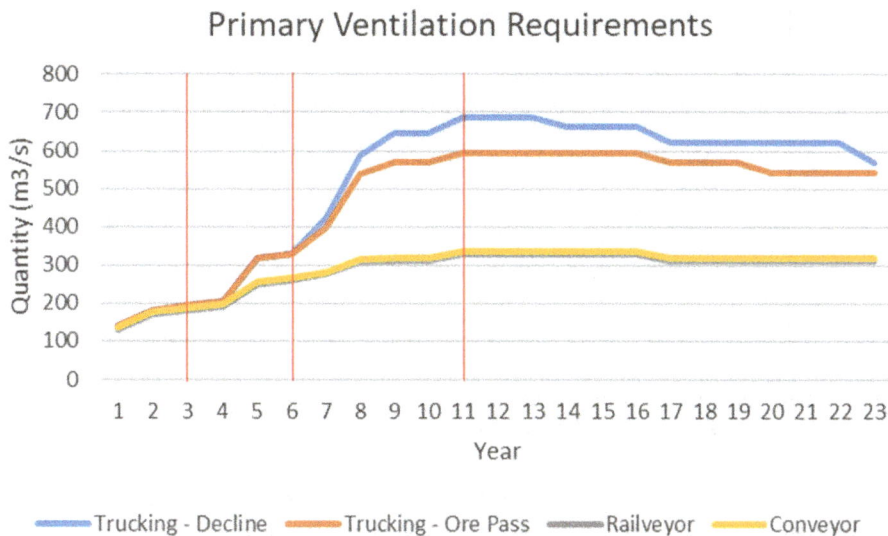

FIG 5 – Primary ventilation requirements for the Gavião project life-of-mine.

VENTILATION MODELLING

The mine design was imported into Ventsim® Design Version 6 and the scenarios modelled for years 3, 6 and 11. The Railveyor and Conveyor scenario is shown in Figure 6 with one rise required from year 6 onwards. The decline and return air drive were designed to be 6 m × 6 m in size to ensure velocity limits weren't exceeded whilst having sufficient airflow. The rises were optimised to have a 5.5 m diameter and the return air raises were assumed to be mined with a longhole at 4.5 m × 4.5 m in size. Clemcorp CC2650/1.55 1250 kW Reaction Bladed Axial Flow fans were located either in the return air portal or in the exhaust rise. Only a single fan is required to be used when exhausting from the portal due to velocity restrictions in the decline and return air drives. These fans would be able to be relocated to be used on top of the return air rise in conjunction with a second fan of the same size in parallel.

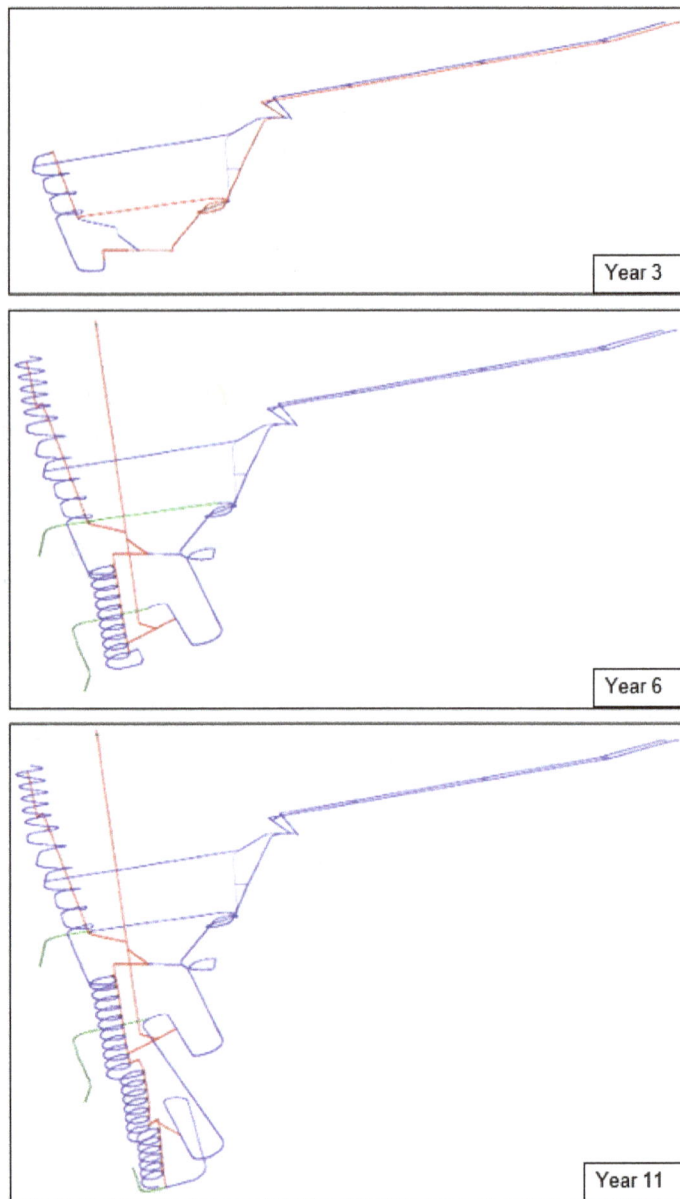

FIG 6 – Ventsim model of the Gavião mine design for Railveyor and conveyor materials handling system, with fresh air in blue and exhaust air in red.

For the trucking scenarios two rises were required from year 6 onwards, shown in Figure 7, to exhaust the additional diesel fumes created when using trucks. This option allows the second decline to remain in exhaust air and the trucks are able to have dedicated routes to ease traffic congestion. For the purposes of this study the materials handling decline design was not modified for each materials handling method however would be optimised at further levels of study.

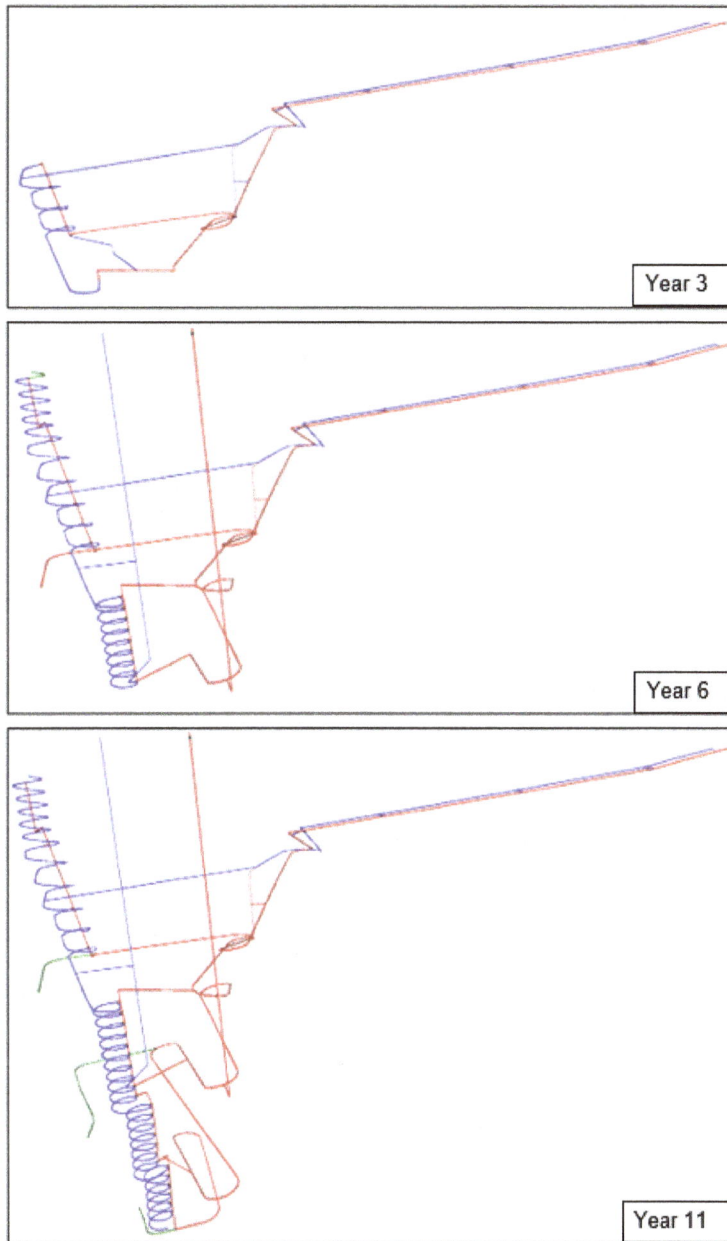

FIG 7 – Ventsim model of the Gavião mine design for the diesel truck materials handling systems with fresh air in blue and exhaust air in red.

The blade angles for the primary fans were adjusted for each scenario to ensure the minimum flow requirement was achieved and could flow-through the mine design. The capital costs for the primary fans were based on a quote from Clemcorp and the operating costs were calculated by Ventsim using a power cost of US$0.110/kWh.

Due to the surface rock type and ground conditions having poor strength down to approximately 70 m below the surface, rises will be costly to install and preference was given to not having any or only having them if essential. The Railveyor and conveyor option required one and the trucking options required two.

The primary fan capital was added as a staged process with the total cost included at the periods that the fan would be installed. The operating costs were added as a yearly cost based on the stages shown above. These costs were added to the trade-off study to show the effect they had on the different materials handling methods, Figure 8. The results show that over the 24 year mine life, with an 8 per cent discounted cash flow (DCF) rate, Railveyor is significantly cheaper due to not requiring a second rise to the surface and the reduced ongoing ventilation costs.

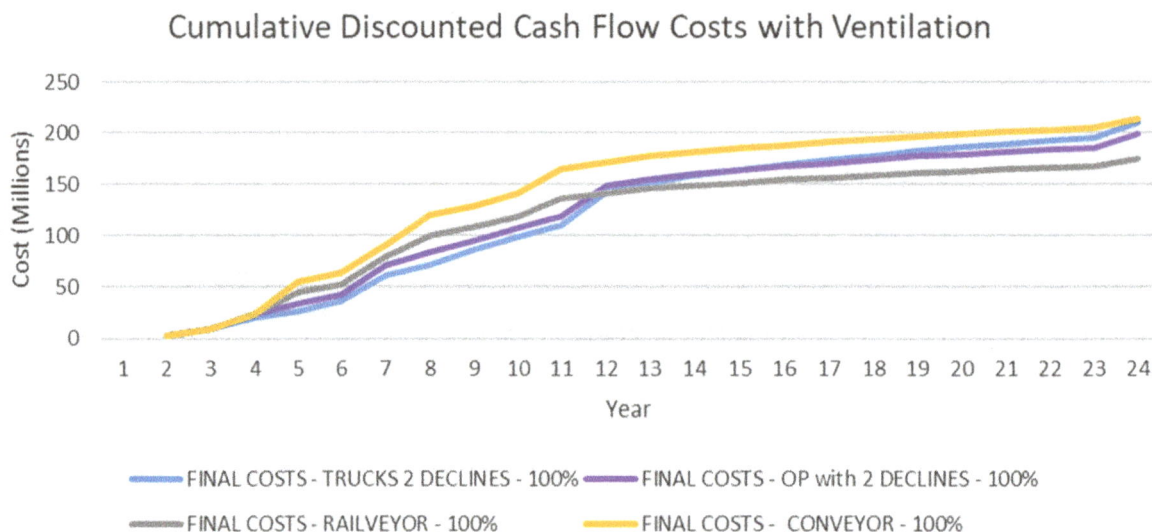

Cumulative Discounted Cash Flow Costs with Ventilation

Legend:
— FINAL COSTS - TRUCKS 2 DECLINES - 100% — FINAL COSTS - OP with 2 DECLINES - 100%
— FINAL COSTS - RAILVEYOR - 100% — FINAL COSTS - CONVEYOR - 100%

FIG 8 – Cumulative discounted cash flow costs including ventilation for materials handling at the Gavião Project showing that Railveyor is cheaper once ventilation costs are included.

This additional work to include ventilation costs shows that over the life-of-mine the Railveyor option is approximately €70 million cheaper than trucking with orepasses. The conveyor option also looks more attractive with an overall cost similar to the 'orepass with twin decline' option.

The additional ventilation costs calculated using Ventsim show that the cost per tonne of ore can increase by up to 56 per cent depending on the materials handling method chosen, summarised in Table 2.

TABLE 2

Costs for materials handling methods including primary ventilation.

	Total DCF cumulative cost including ventilation (Millions €)	DCF cost per tonne of ore excluding ventilation costs (€/t)	DCF cost per tonne of ore including ventilation (€/t)	Percentage increase
Loader to truck	€210	€6.05	€9.45	56%
Ore pass to truck	€198	€5.86	€8.92	52%
Railveyor	**€175**	€6.67	**€7.88**	18%
Conveyor	€213	€8.43	€9.62	14%

The results from adding primary ventilation to the trade-off shows that Railveyor is approximately 21 per cent cheaper than using trucking methods and should be considered at further levels of study.

SECONDARY VENTILATION

Materials handling trade-off studies, such as this one, at a feasibility level generally use a set price for secondary ventilation based on costs from similar sized mines. This means that the cost savings and potential problems with secondary ventilation are easily overlooked when choosing which materials handling methods to move forward with. Without utilising a full mine schedule and design, basic calculations could be performed using the estimated equipment numbers.

During peak production the Gavião project will have four active mine production areas. Ventilating production headings requiring both trucks and boggers, requires a minimum of 42 m³/s with a total cost of approximately €1.3 million per annum. When using orepasses the bogging levels will require 16 m³/s and the truck levels 26 m³/s. Whilst double the amount of fans are required the operating cost are much lower at approximately €639 k per annum. The Railveyor and conveyor systems only require ventilation of the bogger for production areas which would cost approximately €156 k per annum.

These yearly costs are overly simplified and fail to take into account the cost of buying and maintaining the fans or the possibility that fans may need to be run at higher quantities due to less airflow in the primary system. However, this basic calculation demonstrates that secondary ventilation can also have a large impact on the overall running costs when choosing a materials handling system. Relying on both diesel trucks and boggers could result in running costs ten times higher than a system only utilising boggers.

ADDITIONAL TRADE-OFFS

The reduction in costs can also be seen in other materials handling trade-off studies based on different orebody shapes and depths. Mines B and C are de-identified for the purposes of discussing in this paper. Mines B and C were calculated with loader and truck capital costs set at a monthly rate that would be paid to a contractor versus buying the equipment outright. This increased the trucking and bogging rates per tonne however removed the large upfront capital expense of replacing them every five years.

Mine B has a projected annual production rate of 2.75 Mt and extends to approximately 350 m below the surface. Before considering primary ventilation costs for that project, the trade-off showed that the trucking option had the lowest cumulative cost and thus appears as the best financial option to move forward with. For this project there was only additional operating cost for primary ventilation as the planned infrastructure and fans were sufficient for all materials handling methods when operating on different fan curves. These operating costs were added to the trade-off resulting in Railveyor and conveyor having a lower cumulative cost from approximately five years into the mine life, shown in Figure 9.

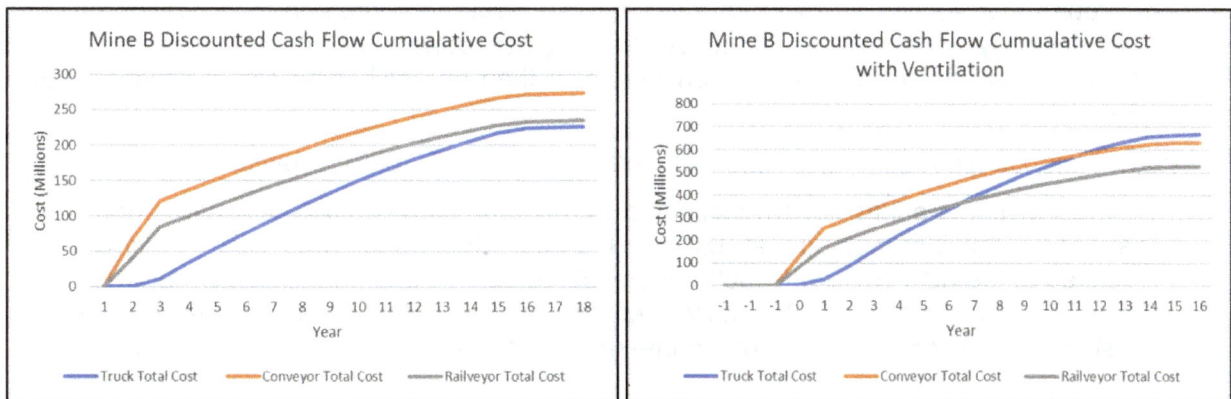

FIG 9 – Mine B cumulative costs before and after including primary ventilation.

Mine C has a projected annual production rate of 3.25 Mt and extends to approximately 500 m below the surface. In this study the Railveyor option had the best overall cumulative cost, however when using an 8 per cent discounted cash flow rate trucking had a lower cumulative cost. Once primary fan operating costs were added the Railveyor has a much lower cumulative discounted cash flow cost, Figure 10.

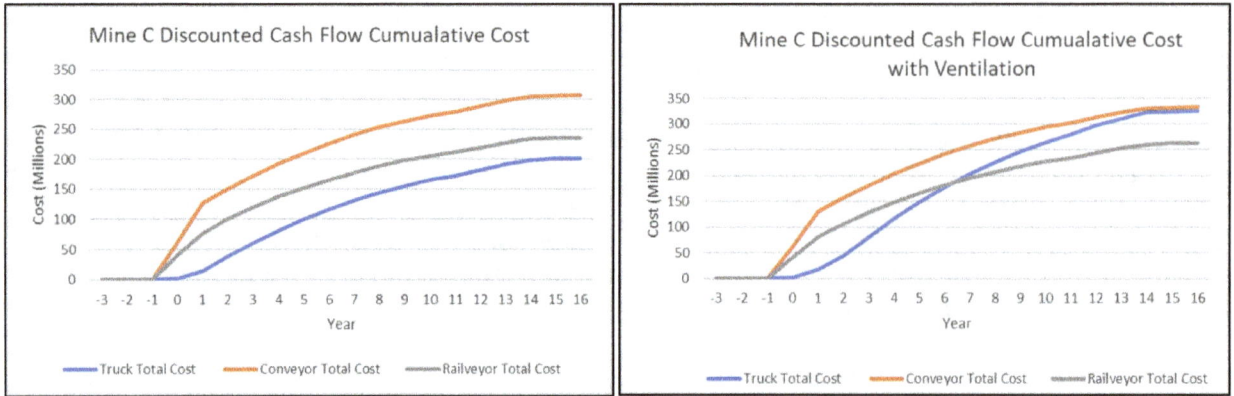

FIG 10 – Mine C cumulative costs before and after including primary ventilation.

INCREASING PRODUCTION

Ventilation infrastructure is a large proportion of cost to most projects, it is usually designed to meet the required production value with little to no allowance for increases (McPherson, 2018). Increasing production is problematic if the ventilation system is running at maximum in mines utilising truck and loaders. The Railveyor system requires minimum ventilation, so if the system needs to be expanded or modified the ventilation system is unlikely to need major upgrades to accommodate the changes.

ENVIRONMENTAL, SOCIAL AND GOVERNANCE CONSIDERATIONS

The removal of diesel equipment which is replaced by electric driven equipment is a known way to reduce the environmental, social and governance (ESG) liabilities of mining projects. The Railveyor system is electrically powered on demand (the drive stations only operate when the 'train' is passing through that section of track) so it is very low powered compared to conveyor systems. With more electricity generation being converted to green power options, the mine using more electric solutions can derive a better carbon reduction than one still using diesel powered trucks and loaders.

The study '*Materials Transport: Racing to Net Zero*' by Bieg *et al* (2022), showed that:

> Diesel-electric haul trucks are estimated to have the highest GHG emissions of the specified hauling methods at 0.096 kilograms CO_2e/tonne-km. This is directly related to the high emissions factor of diesel fuel and limited haul capacity per truck. The hauling method with the lowest emissions is the Railveyor system, with an estimated 0.012 kilograms CO_2e /tonne-km (Figure 11).

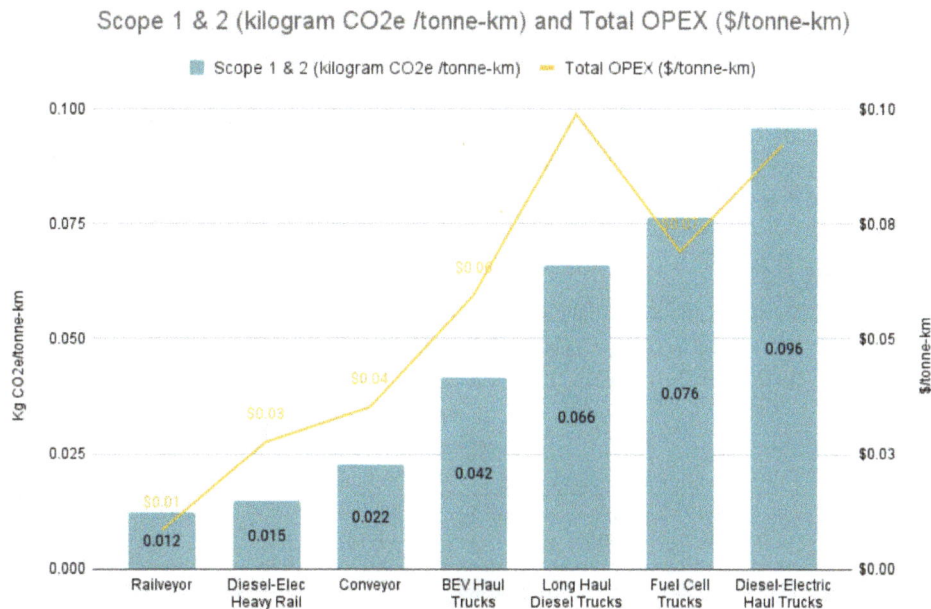

FIG 11 – Scope 1 and 2 and total OPEX comparison, Source: Bieg *et al* (2022).

To reduce the ESG and ventilation costs further battery electric loaders could be used with the chosen materials handling method. These loaders are generally constrained to a level at a time so the battery option becomes more efficient and feasible than a loader requiring tramming along high gradient drives. The ESG costs were not calculated in the studies presented here but further savings could be also achieved if the costs took into account the carbon credits from a more electrified mine.

OTHER EMERGING TECHNOLOGIES

In addition to Railveyor, there are alternative existing and emerging materials handling solutions that should be considered. All of these options should be assessed including ventilation and ESG costs to ensure that the most cost-effective option is selected.

The Riino™ Inc, is a zero-emission monorail currently at the concept stage that plans to address many of the problems faced when using trucks and loaders. They are focusing on decarbonisation, increasing productivity at lower costs, reducing energy consumption, environmental risks and the mining footprint whilst improving operational safety.

Battery electric trucks and loaders are constantly being improved and optimised with a strong belief from the industry that they will be the solution to reducing diesel emissions. They currently perform well in flat areas with long tramming distances, however, are yet to excel in mines with long trams up steep gradients. As of 2024 the availability to get these machines is also a roadblock some mines are facing with large lead times.

Using trolley-assist trucks, either diesel or battery electric, is another technology being considered underground to reduce the diesel emissions. These come with their own set of challenges to install and maintain, however could prove to be a viable solution. Heavy rail is also appropriate to consider in flat orebodies and shafts for deeper orebodies.

Although these studies focused mainly on Railveyor, trade-off studies should consider all potential options based on the orebody characteristics. Including ventilation costs and considering ESG liabilities could change the outcome of any method considered.

CONCLUSION

High-level materials handling trade-off studies often fail to capture the full spectrum of costs associated with each method. When ventilation requirements are included into the costs, the Railveyor generally has a lower discounted cash flow cost due to the reduced ventilation requirements. In some studies, Railveyor was also the cheapest option prior to adding primary ventilation costs however they were not discussed as part of this paper.

By excluding alternative options early into a study, considerable value of the project could be missed due to the traditional truck and loader options appearing to be the better financial option when in actual fact over the life-of-mine the cumulative cost is much higher. Careful consideration should be given to the strengths of different materials handlings methods and whether the modelling is truly allowing these strengths to be accounted for whether through direct costs or harder to quantify benefits.

ACKNOWLEDGEMENTS

The author would like to acknowledge Almina mining company for facilitating access to their data and granting approval for sharing and publishing findings. Furthermore, appreciation to Railveyor for supplying essential data, quotes and information necessary for the completion of this study.

REFERENCES

Bieg, A, Samson, M, Bell, Z, Schuyler, J and Zinsser, A, 2022. Materials Transport: Racing To Net Zero: Comparing Haulage Options for Reduced Greenhouse Gas Emissions and Costs, report prepared by Warm Springs Consulting LLC, 63 p. Available from: <https://www.railveyor.com/featured/wsc/>

Brake, D, 2022. *Mine Ventilation – A Practitioner's Manual*, 76–04 ed, November 2022.

Department of Mines and Petroleum (DMP), 2022. Work Health and Safety (Mines) Regulations 2022, updated March 2022, Government of Western Australia. Available from: <https://www.legislation.wa.gov.au/legislation/statutes.nsf/law_s53266.html>

McGuire, C, Witow, D, Mayhew, M and Bowness, K, 2022. Comparison of heat, noise and ore handling capacity of battery-electric versus diesel LHD, in *Proceedings of the Australian Mine Ventilation Conference*, pp 348–359 (The Australasian Institute of Mining and Metallurgy: Melbourne).

McPherson, M, 2018, *Subsurface Ventilation Engineering Book*, Mine Ventilation Services Incorporated.

Railveyor, 2021. Technology for Material Handling – Railveyor [online]. Available from: <https://www.railveyor.com/what-is-railveyor/technology/>

Sandvik, 2024. Underground Mining Equipment and Surface Equipment [online]. Available from: <https://www.rocktechnology.sandvik/en/products/equipment/>

Power savings through secondary fan management at Granny Smith Mine

A Broodryk[1]

1. Superintendent, Gold Fields – Granny Smith Mine, Laverton WA 3028.
   Email: andre.broodryk@goldfields.com

ABSTRACT

Ventilation accounts for a large portion of energy use in the total power demand of underground mines, with secondary fans being identified as a big part of that. It has also become increasingly important to examine new technologies that enhance energy efficiencies as well as new strategies to reduce power use, not only to reduce costs, but also to reduce CO_2 emissions to align with company environmental, social and governance (ESG) targets. The traditional approach to secondary ventilation involves running fans at a constant speed, often at full capacity. This method is neither energy-efficient nor adaptable to changing mine conditions. By adopting variable speed control, mines can tailor fan operations to meet specific ventilation needs, optimising energy use and reducing costs. Recently Granny Smith Mine has been exploring different new technologies and approaches to reduce secondary fan power use. These include variable speed secondary fans and controls to only run fans needed based on weekly production plans. This case study describes the outcomes and power savings achieved throughout 2023.

INTRODUCTION

The industrial utilisation of energy is extensively documented and understood as a pivotal element, not solely in business operational expenses but also in its globally significant impact on the environment. Gold Fields Ltd has integrated global sustainability standards and reporting guidelines to ensure effective management of our climate-related and other environmental risks, which include, amongst various others, emissions reduction targets through research, innovation, and technology development, as well as energy efficiency initiatives. In December 2021 Gold Fields Ltd published a comprehensive set of 2030 targets for its most material environmental, social and governance (ESG) priorities (Gold Fields Ltd, 2022). Adaptation actions that our Australian mines and projects implemented in 2022 included the investigation of ventilation on demand (VOD) and energy efficient technologies.

Ventilation and refrigeration requirements increase as mines move deeper and ambient temperature increases. Legislation also dictates minimum standards for underground ventilation in terms of diesel equipment dilution rates and subsequent air volume requirements for specific tasks. The concept of VOD aims to channel airflow exclusively to the active working zones within the mine while reducing or isolating airflow to inactive sections. This system can range from a basic set-up that activates ventilation in a zone irrespective of work activity to a relatively complex system that regulates airflow volumes in specific areas based on the ongoing activities within those zones.

The orebody layout and subsequent ventilation system design at Granny Smith Mine largely hinder the implementation of adjustments in primary air flows within specific zones. However, the ventilation system relies on the utilisation of multiple secondary fans, both in series and in parallel, to facilitate the ventilation of the necessary ore drives in its horizontal orebodies. In total, the underground mine currently uses 59 secondary fans with a combined rated power of approximately 8 MW to ventilate development and production areas. For this reason, the mine had to look at technologies and processes that would enable a more efficient use of secondary fans.

Two separate controls were recently implemented to achieve power savings through the management of secondary fans. Phase one is an administrative control that was established where working areas are identified during the weekly planning process. These areas, which do not require ventilation, have their respective fans listed on a no-start list to prevent them from energising during the start-up sequence of the supervisory control and data acquisition (SCADA) site control system. Phase 2 is an engineering control which involves the use of variable voltage, variable frequency drives (VVVF) (also called variable speed) to adjust the secondary fan speed as needed.

PRODUCTION SCHEDULE FAN MANAGEMENT

The first phase of the secondary fan management program that was implemented in June 2022 and involves fan operating requirements driven by the weekly production schedule. Switching secondary fans off when not required will always be the preferred option for power savings, since the recorded total power draw of installed twin stage fans amount to between 50 000 and 56 000 kWh per month each. This project was developed and implemented as a low-cost initiative to identify and remotely manage secondary fans in underground areas that were not required to run for production purposes during each shift. At the beginning of the mine production planning week, information about which fans are not required to run is provided to the mine control team. They then identify and tag the relevant fans on the SCADA site control system (Figure 1). Consequently, the identified fans do not start-up at the commencement of the shift following the underground blast clearing sequence.

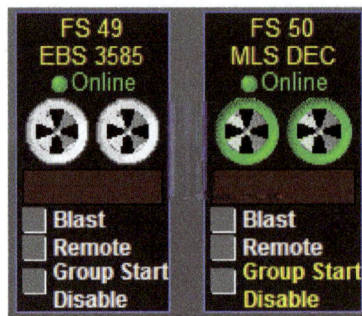

FIG 1 – Fans disabled from start-up.

Each week, an average of approximately 25 secondary fans of various sizes and locations are identified and then excluded from start-up and operation via PITRAM. Prior to commencing the project, the majority of secondary fan starters were not recording the operating run times of the fans. In order to determine the power savings achieved, all fans had to have their run time monitoring repaired and was only fully completed two months before the project commenced. These months were therefor the only baseline available to compare the fan power draws. The results were a 30 per cent per month energy reduction in measured secondary fan power. This level of savings has not only had a positive effect on energy consumption, but it has also provided a decrease in mine operational costs and carbon emission as well. Figure 2 provides the power cost and CO_2 savings on a monthly basis throughout 2023 for Granny Smith Mine.

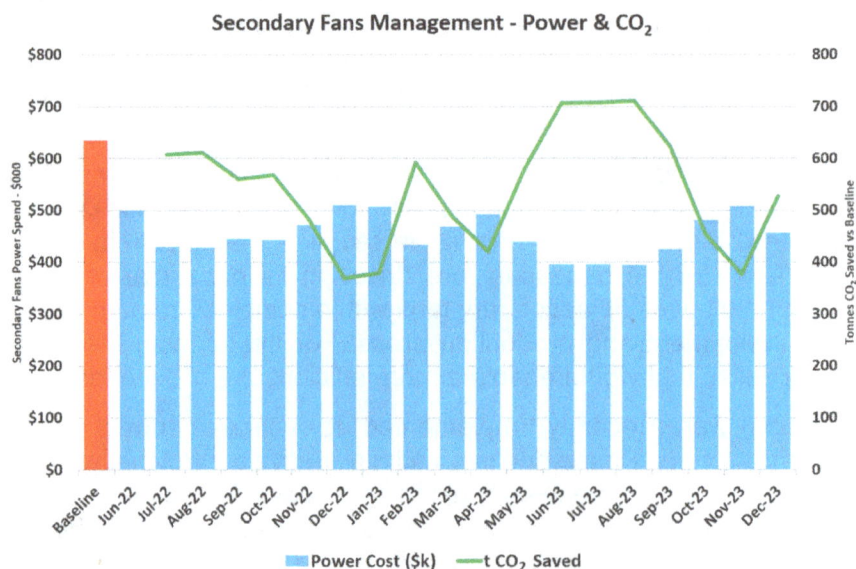

FIG 2 – Secondary fan power and CO_2.

VARIABLE SPEED FAN

For installed fans to accommodate various demand changes, flow is controlled by different methods, such as inlet vanes and speed control. Inlet vanes were more commonly used with centrifugal fans, but this method has recently made its way into the secondary fan space by some manufacturers. Variable speed fans operate by utilising a variable frequency drive (VFD), also referred to as an 'adjustable-frequency drive', 'variable-voltage, variable-frequency (VVVF) drive', 'AC Drive' or 'Inverter drive', among other terms, to regulate AC motor speed and torque. This regulation is achieved by altering the motor's input frequency and voltage. Over the past four decades, advancements in power electronics technology have resulted in reduced VFD costs and sizes, as well as enhanced performance due to progress in semiconductor switching, drive topologies, simulation and control techniques, and control hardware and software. Variable Speed Drives adjust the motor's speed by modifying the voltage and frequency of the power supplied to it. Maintaining the proper power factor and preventing excessive motor heating necessitates the maintenance of the nameplate volts/hertz ratio. This constitutes the primary function of the VFD.

Traditionally, secondary fans in underground mining have been operated at a constant speed, often running at maximum capacity. This approach, while ensuring a baseline level of ventilation, is highly inefficient and leads to substantial energy waste. Constant-speed operation fails to adapt to changing underground conditions, resulting in excessive power consumption, and increased operational costs.

The adoption of variable speed control for secondary fans allows mines to precisely adjust fan speeds in response to changing ventilation requirements. During periods of lower demand, such as when there are no heavy diesel equipment operating within the ventilated drive or when only inspections occur in the drive, these fans can operate at reduced speeds. This flexibility can lead to substantial energy savings.

Actual power savings achieved follows the fan affinity laws. The fan affinity laws are fundamental principles that describe how changes in the operating conditions of a fan, such as speed, affect its performance. These laws are crucial for understanding the relationship between fan speed, airflow, pressure, and power consumption.

Fan affinity laws

The fan affinity laws describe how changes in the speed of a fan affect its performance parameters, including airflow (flow), pressure, and power consumption. There are three primary affinity laws related to speed change.

Flow rate law

The flow rate law states that the airflow or flow rate through a fan is directly proportional to the fan speed. If the speed of the fan changes, the airflow will change in the same proportion (Figure 3).

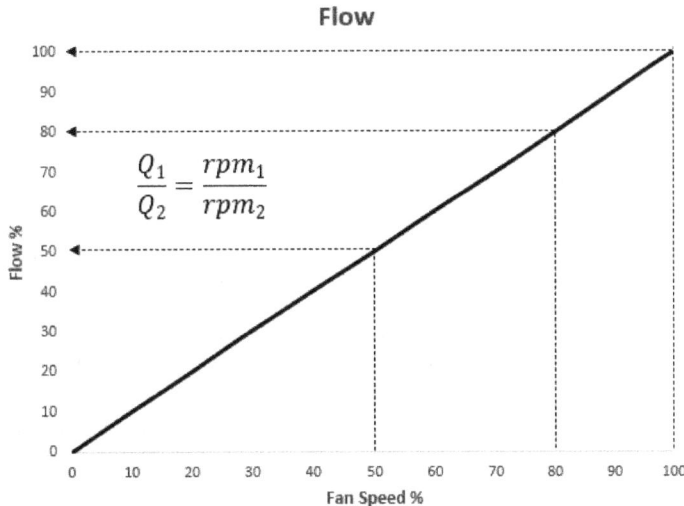

$$\frac{Q_1}{Q_2} = \frac{rpm_1}{rpm_2}$$

FIG 3 – Flow rate relation to fan speed.

Pressure law

The pressure law describes the relationship between fan speed and the pressure generated by the fan. It states that the pressure produced by a fan is proportional to the square of the fan speed. When the fan speed changes, the pressure changes by the square of the speed change (Figure 4).

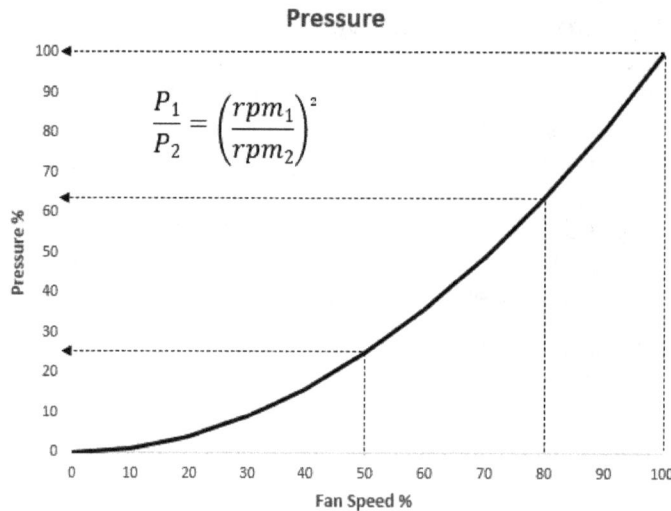

$$\frac{P_1}{P_2} = \left(\frac{rpm_1}{rpm_2}\right)^2$$

FIG 4 – Fan pressure relation to fan speed.

Power law

The power law relates changes in fan speed to the power consumed by the fan. It states that the power consumed by a fan is proportional to the cube of the fan speed. When the fan speed changes, the power consumption changes by the cube of the speed change (Figure 5).

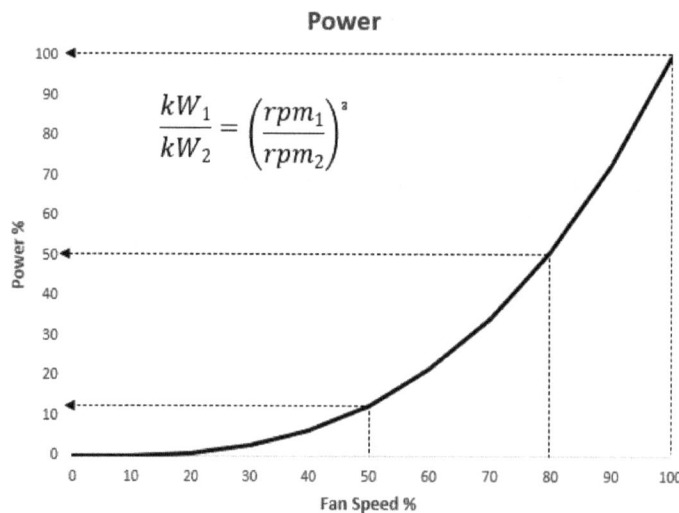

$$\frac{kW_1}{kW_2} = \left(\frac{rpm_1}{rpm_2}\right)^3$$

FIG 5 – Fan power relation to fan speed.

Another benefit of operating fans at reduced speed is the maintenance cost reduction: Constant-speed operation can subject fan components to unnecessary wear and tear, leading to frequent maintenance and replacement expenses. Variable speed control reduces mechanical stress on these components, extending their lifespan and significantly reducing maintenance costs.

Heat produced by secondary fans is also reduced when running at lower speeds. Implementing variable speed control to optimise fan operation, can also help minimise unnecessary heat production while maintaining adequate ventilation. The full capacity of heat reduction from the fan and addition by the VFD drive has not yet been assessed at the time of compiling this paper but would be the same as the electrical power saving achieved.

In the author's experience, the use of variable speed fans underground has historically been isolated to primary fans due to the heat created by variable speed drives and the requirement to keep drives in clean environments. Advancements in technology have altered the capabilities of this equipment, rendering it more adaptable for a broader spectrum of applications.

In 2022 Granny Smith Mine was invited by TLT-Turbo GmbH in conjunction with SMEC Power and Technology to view new technology for secondary fans and 1000 v variable speed, liquid cooled starters being made available to the market. A viewing and inspection of the equipment in operation at their workshop was arranged with an agreement to a six month 'fan and starter' trial initiated soon after.

The VSD Drive topology used in the trial was 3-level switching which was developed specifically to reduce motor winding stress, decreased electrical noise, semiconductor losses, and minimise drive-induced problems associated with long motor cables and premature motor bearing failures. The reduction of motor winding stress is a result of the 3-level switching multi-step voltage waveforms which are one-half of a traditional 2-level switching VVVF drive. The reduced voltage steps, and lower line-to-line voltage decrease the non-uniform voltage distribution among motor winding turns which reduces stress on insulation material. The reduced voltage steps also significantly reduce motor shaft voltages and bearing currents which allows larger motors to be used with standard bearings.

Figure 6 shows a comparison between 3-level versus 2-level voltage wave forms, with the 3-level producing a significantly improved sinusoidal curve (wave form) and one-half voltage steps (Yaskawa Electric America, 2005).

THREE - LEVEL SWITCHING
OUTPUT WAVEFORM

TRADITIONAL TWO - LEVEL SWITCHING
OUTPUT WAVEFORM

FIG 6 – Three-level switching versus 2-level switching waveforms.

Underground installation

The underground installation position was carefully selected to assess the capabilities of the fan and starter in an area known for its elevated temperatures and considerable traffic and activity within the decline. The chosen area was a decline drive under development utilising independent firing during the shift. The rationale for selecting this location lies in the potential for achieving power savings in the mine's most active area, thereby promising even greater benefits elsewhere. The location allowed evaluation of the reliability of the VVVF fan starter in an unfavourable location.

The installation of the secondary fan required no additional requirements to the existing site fan hanging procedure. The fan starter position was located on the intake side of the fan, off the main decline in a connection drive. The location was ventilated by only an 11 kW fan, which also provided air to a substation opposite the starter (Figure 7).

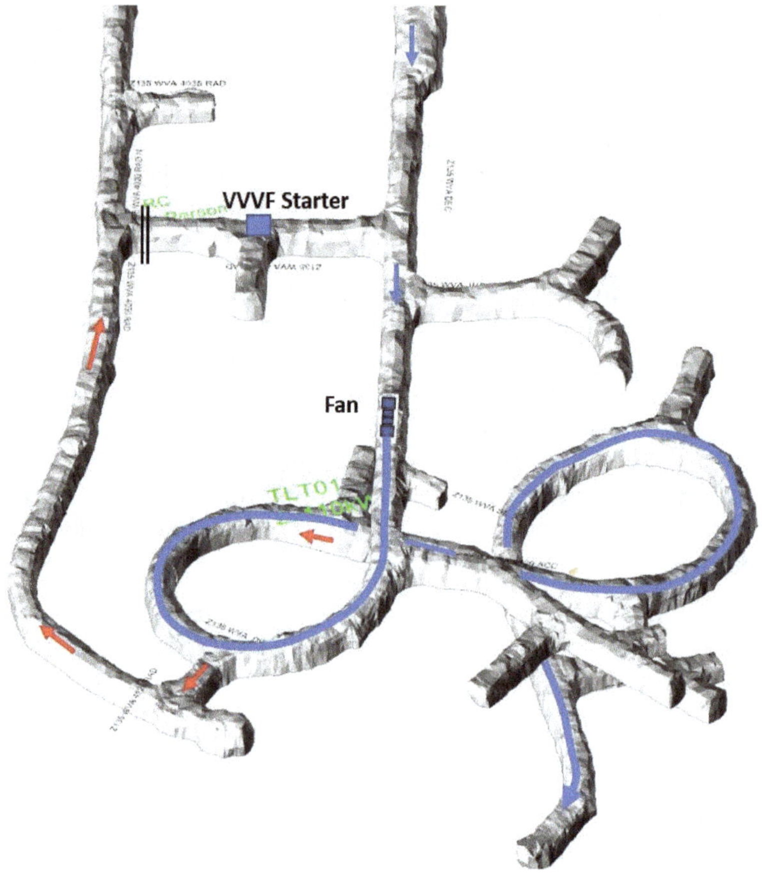

FIG 7 – Initial fan and starter position layout.

During operation of the fan starter, the internal temperature ranged between 45 and 55°C at various speed settings during the trial, while the internal alarm/trip setting is set to 75°C. This is another attribute of the 3-level switching where switching losses (losses are proportional to heat produced) are up to 44 per cent less than traditional 2-level switching drives. Three-level switching has reduced switch losses as a result of their lower switching parameter voltage values.

Operation and results

Both the fan and starter were connected to the underground long-term evolution (LTE) network, with real time monitoring through the site SCADA network. All aspects of the starter operating conditions were monitored, and remote speed adjustment was made available to PITRAM operators through Citect (Figure 8).

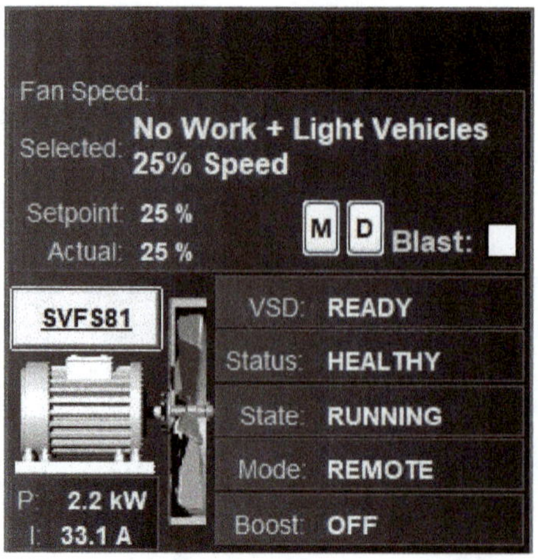

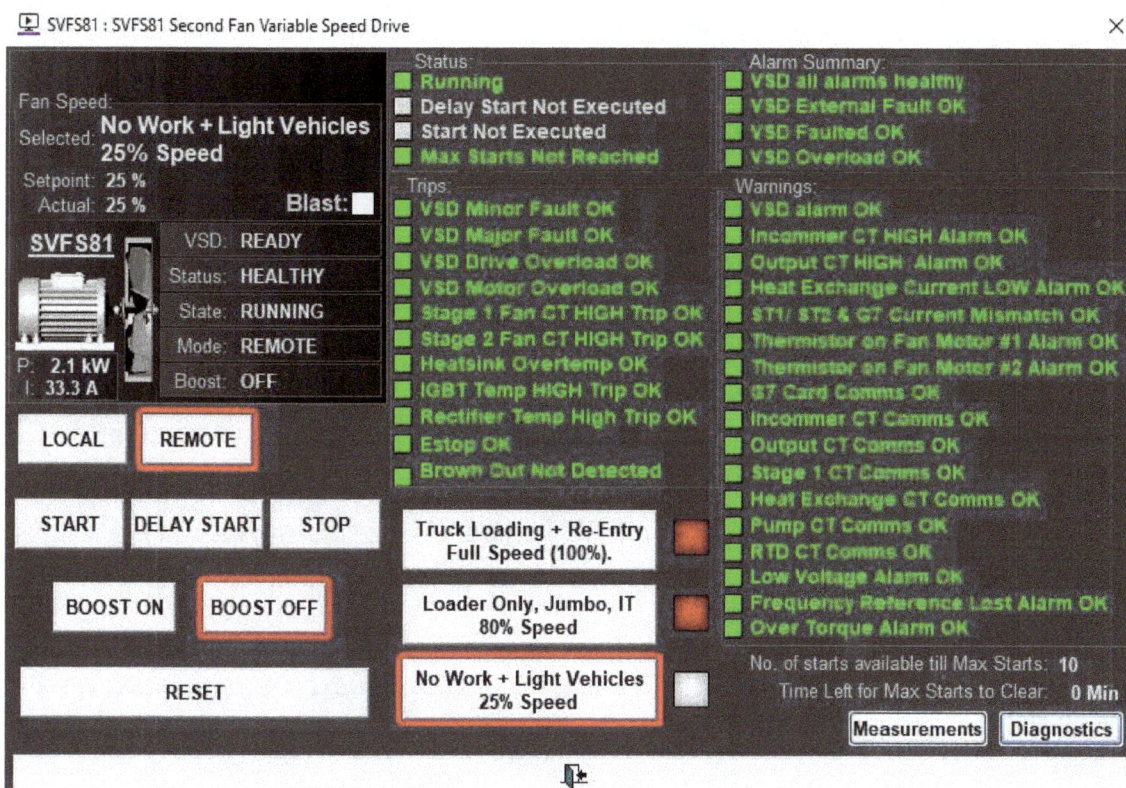

FIG 8 – Variable speed starter remote access.

Although the fan can operate at any required speed through the starter, it was set to operate at three different speeds, 25 per cent, 80 per cent and 100 per cent.

This allowed for three scenarios as shown in Table 1.

TABLE 1

Fan speed settings used.

Fan speed	Fan Flow	Fan Pressure	Fan power draw	Scenario
100%	100%	100%	100%	Loading of trucks, Re-entries
80%	80%	36%	51%	Jumbo drilling, IT's etc
25%	25%	6%	2%	LV's, inspections, vent bag extension

The initial plan ideally aimed to have the fan operate autonomously, relying on vehicle movement tracked through the site's vehicle tracking system. However, this integration was unfortunately unavailable at the time, necessitating manual adjustment of the fan speed by PITRAM operators. This adjustment was based on the work being conducted in the drive at that time and relied on operators notifying PITRAM of their movements in and out of the drive. This resulted in some instances where the fan was running at full speed when not required. For the first two and a half months the fan was operating in the decline development. This was during mid-summer in January 2023. During the start of the trial the vent bag was short, leakage was low and the air discharge volume reduction at 80 per cent fan speed was barely noticeable. As the vent bag length increased in the following months it became evident that a 20 per cent reduction in flow, and 36 per cent reduction in fan pressure became more noticeable at the face during the 80 per cent fan speed setting. This resulted in the fan operating at 100 per cent speed for most of February. Power reduction in January and February of 2023 were 21 per cent and 6 per cent respectively. The power saving was calculated based on the duration of time the fan was operating at the set reduced speeds and power draw calculated as per Table 1 (The fan speed setting is recorded at one-min intervals).

In mid-March 2023, the fan was relocated further up the decline to provide ventilation for an Incline development. This particular drive was shorter and not included in the independent firing zone. The shorter vent bag, combined with the decreased production rate, facilitated significantly greater power savings in the last three months. The power reductions achieved from April to June 2023 were 49 per cent, 53 per cent, and 53 per cent, respectively. Throughout the 196 days of data collection, the fan operated at the speeds described in Table 2.

TABLE 2

Fan speed setting and saving per month.

Fan speed	January	February	March	April	May	June
100%	74%	90%	65%	40%	39%	37%
80%	24%	9%	14%	21%	14%	17%
25%	2%	1%	21%	39%	47%	46%
Total power saving	13%	6%	27%	49%	53%	53%

There were several days when the fan did not operate at all, and these instances were excluded from the calculations. The cumulative reduction in electrical power draw over the data collection period amounted to 35 per cent. The actual power draw of the fan is dependent upon the resistance it encounters; however, compared to a full-time power draw of, say 190 kW for a twin-stage 110 kW fan, the resulting power saving can amount to 48 MW per fan per month, equating to 292 t of CO_2 per annum.

Figure 9 shows the total power saving achieved compared to when a fixed speed fan was to be installed in the same position.

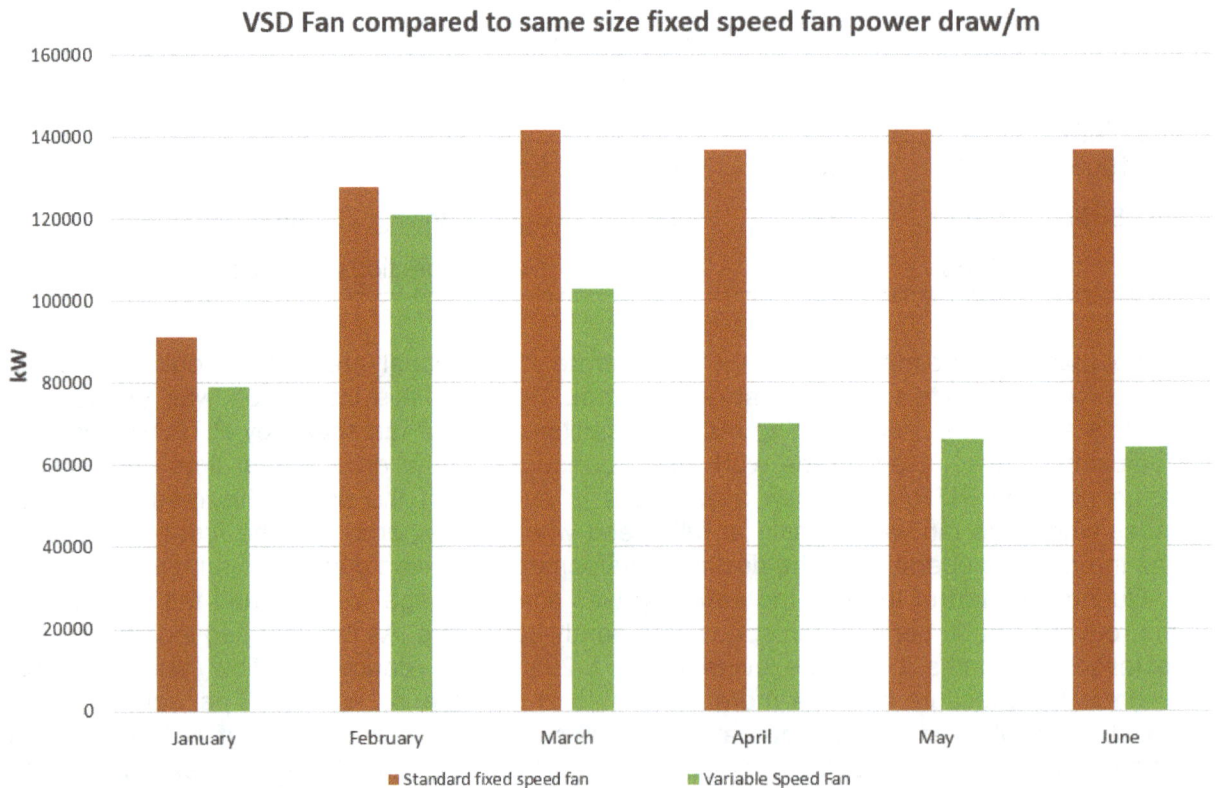

FIG 9 – Power saving of variable speed fan compared to a fixed speed fan of the same size during trial period.

The initial assessment comparing maintenance costs to those of normal direct online (DOL) fan starters has indicated that only fortnightly dust filter checks and cleaning are required, in addition to the existing periodic maintenance checks.

CONCLUSION

The implementation of secondary fan management controls at Granny Smith Mine has been successful in reducing the power use on-site, with the next phase of fan speed automation having the ability to reduce power even further. The magnitude of power savings through variable speed secondary fans depends on each fan's installation location, but if carefully selected, can amount to large power savings. Through the use of variable speed fans, as well as the management of operating fans based on the production schedule, it has thus far shown at Granny Smith Mine that power savings of 30 to 35 per cent are very achievable.

ACKNOWLEDGEMENTS

- Martin Law, Warren Ellis, SMEC Power and Technologies

- Kenny Kramara, Divergent Engineering

- Paul Michetti, TLT Turbo GmbH.

- Data was collected and reproduced with permission from Gold Fields' Granny Smith mine.

REFERENCES

Gold Fields Ltd, 2022. Climate Change Report 2022. Available from: <https://www.goldfields.com/pdf/investors/integrated-annual-reports/2022/ccr-2022-report.pdf> [Accessed: December 2023].

Yaskawa Electric America, 2005. Motor Bearing Current Phenomenon and 3-Level Inverter Technology. Available from: <www.yaskawa.com>

Primary fan selection for a 43 megawatt ventilation system – PT Freeport Indonesia's Kucing Liar and Grasberg Block Cave

A Habibi[1], R Sani[2] and Z Diaz[3]

1. Technical Expert – PT Freeport Indonesia, Papua, Indonesia. Email: ahabibi@fmi.com
2. Ventilation Manager – PT Freeport Indonesia, Papua, Indonesia.
3. Ventilation Superintendent GBC-KL – PT Freeport Indonesia, Papua, Indonesia.

ABSTRACT

Kucing Liar (KL) is the latest block cave in PT Freeport Indonesia (PTFI) operations. The KL mine ventilation system has been designed to operate in parallel to the Grasberg Block Cave (GBC). A detailed ventilation study was conducted to identify the ventilation budget requirements, shortfalls, and opportunities for this complex system. The KL mine is designed to utilise diesel load and haul units whereas the GBC mine is equipped with an electric haulage train system. The optimisation study consisted of identifying the required number of primary main intake and exhaust drifts and raises, the number of main fans and their respective configuration as well as fixed facilities (FF) ventilation design.

The modelling results showed the combined total requirement of 5400 m^3/s for GBC-KL block caves. The current Life-of-mine (LOM) also requires a total of three (7 m × 7 m) exhaust drifts in line with five main 6 m exhaust raises. The requirement of two (7 m × 7 m) intake drift and four main 6 m intake raises to meet the intake requirements. The FF intake was designed to be delivered via two separate intake raises sized at a 4 m diameter. Three primary fans (6.15 MW each) were added to the existing 25 MW GBC system to supply the airflow. This manuscript provides the detailed modelling, fan selection process and challenges for KL-GBC parallel block cave.

INTRODUCTION

Currently, there are three underground mines in the PT Freeport Indonesia (PTFI) district. The Grasberg Block Cave (GBC) mine produces 130–160 kt/d, the Deep Mill Level Zone (DMLZ) with a production rate of 80 kt/d and the Big Gossan (BG) open-stoping mine at 7 kt/d. The DMLZ and BG mines run on a separate ventilation network. Kucing Liar (KL) mine is currently under development with a nameplate rate of 90 kt/d. Figure 1 depicts the PTFI mining district. The KL orebody is positioned at a lower level than the GBC, which required nine vertical 6 m raises to primary fan levels. Currently, five parallel primary five MW Howden mixed flow fans provide 3500 m^3/s of fresh air to the GBC cave. The fresh air passes through four parallel primary intake drifts – Grasberg Ventilation Drifts (GVD1-4) and reports back to the main fans through five parallel exhaust drifts (GVD 5-9). All GVDs are mined at 6.8 m × 9.0 m profile. The distance of fans to the GBC footprint is 2.6 km.

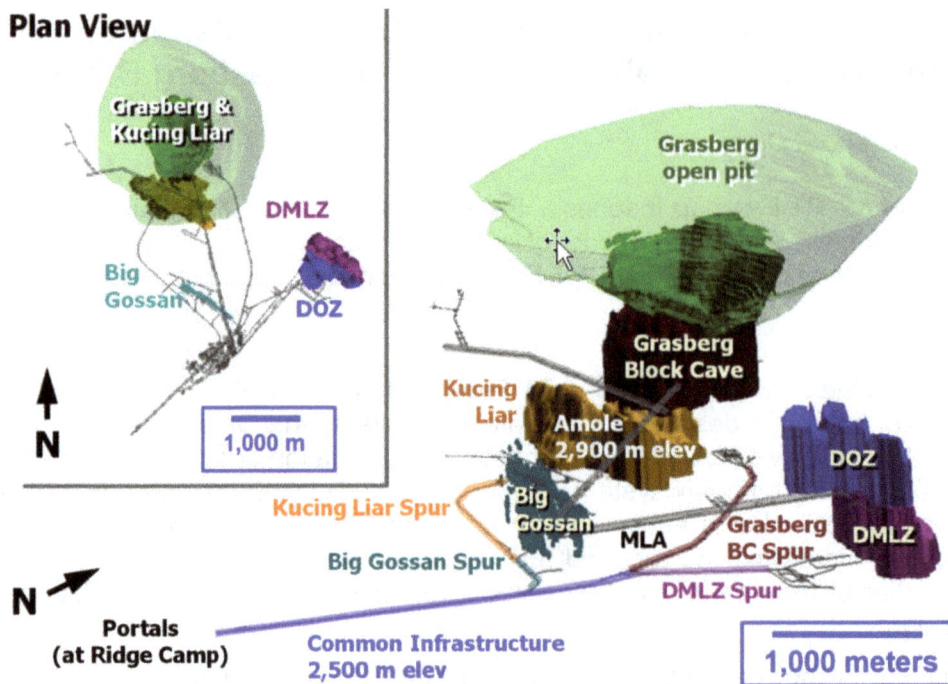

FIG 1 – PT Freeport Indonesia mining district.

Various fan configurations were investigated for KL primary fans. This included scenarios of additional exhaust fans and pull-push configuration using intake booster fans. In addition, utilisation of vane axial fans (airfoil blades) was studied. The axial fans (often adjustable in pitch) generate total pressure rise over the impeller stage by accelerating the air in the direction of flow (axial) providing both velocity and static pressure rise. A diffuser discharge opening partially converts the velocity pressure to more static pressure. The duty of axial fans can be adjusted by changing the blade angles or using different blade profiles or solidity, or variable speed to generate different pressure and flow characteristics.

VENTILATION DESIGN

Budget and design assumptions

The KL mine shares primary exhaust and intake ventilation with the GBC mine through the combined GVD exhaust and intake system. The additional KL production and development combined with continued GBC production will require additional primary airflow (and therefore more fans) through the GVD exhaust system to satisfy both mines. The current ventilation capacity of this system is inadequate for the combined KL and GBC development and production and will need to be progressively expanded.

The portion of airflow received by each mine will change as production and development profiles change in each mine. Over the next ten years, the GBC portion of airflow will decrease as the KL portion increases. This raises challenges in how to effectively split the airflow between each mine without significantly raising primary airflow resistance.

Based on current schedule forecasts, combined GVD ventilation capacity is required to increase from 3500 m^3/s to a peak of 5400 m^3/s by around 2030 due to scheduled production increases and development requirements from the combined mines (Habibi *et al*, 2023).

A comparison of the proposed KL airflow requirements per production tonne with GBC indicates similar airflow productivity (Figure 2). By 2033, the KL mine will be operating with the same airflow (per production tonne) that GBC is operating at in 2024. Given that KL haulage is by diesel truck, this suggests no excess contingency or airflow reduction opportunities exist in the budget estimates unless additional primary fan infrastructure is added.

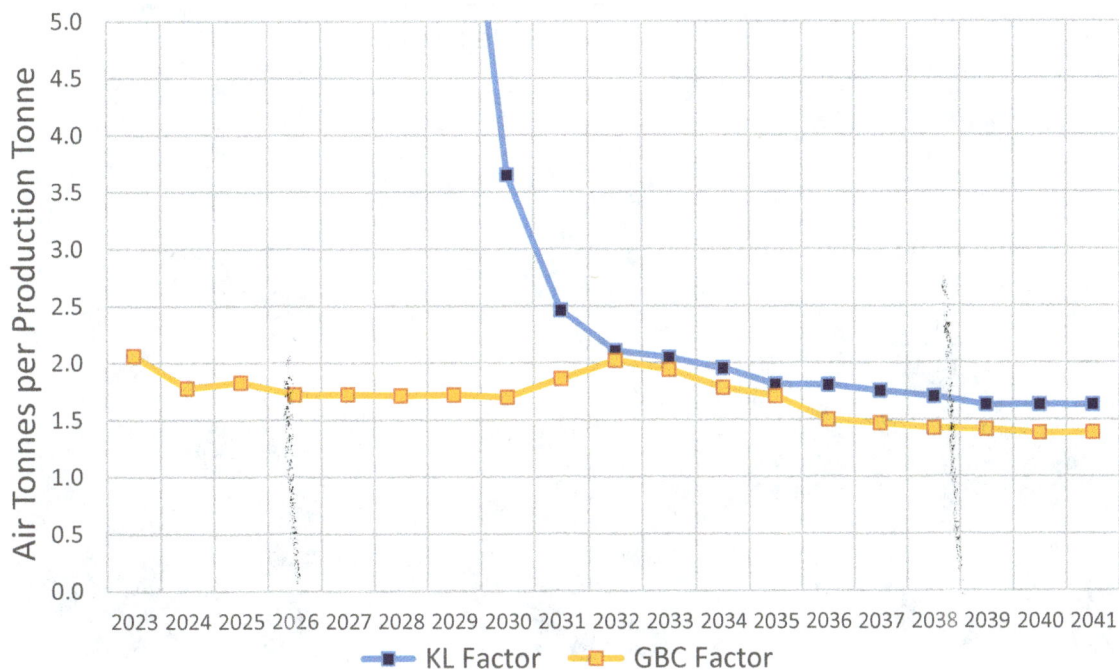

FIG 2 – GBC and KL Mine airflow factor.

Based on the recent LOM baseline estimate, the current GBC mine (which will be operating in parallel with the KL block cave) consumes approximately 25 m³/s/ktpd. Other block cave mines typically have ratios from 17 to 40 m³/s/ktpd (Brannon *et al*, 2020). While the ratio currently sits at the lower end of the scale, a Ventilation On Demand (VOD) system has been put in place to improve the efficiency of air distribution. KL mine airflow requirements (per tonne) are similar to GBC.

Currently, KL is designed for 90 ktpd of KL operation. The current cave sequence is designed for Production Blocks (PBs) close to KL main ventilation raises. As the mine develops to the western section (deeper in the design) additional pressure is required to overcome the linear resistance.

Scenarios and simulation results

Several scenarios were considered to optimise the final design of the KL and GBC complex system. The ventilation model for 2033 was used as a peak requirement design as it is the maximum budget airflow year for KL. Several options to increase and balance airflow between KL and GBC were modelled. Assuming the base case of seven shared (between GBC and KL) primary exhaust fans, the options considered to meet increased airflow requirements were:

- Scenario A – Large regulators in the GBC circuit to increase GBC resistance and direct more airflow to the KL mine.

- Scenario B – Booster fans in the KL circuit to balance pressure with GVD exhaust, decrease pressure on primary fans and increase and regulate airflow.

- Scenario C – Fan #8 and additional exhaust drift.

Scenario A – Regulators to restrict GBC airflow

Utilisation of regulators to increase the resistance and restrict the airflow to GBC will increase the airflow to KL as it increases primary ventilation pressure to the KL circuit. Four regulators were modelled within the GVD exhaust drifts just upstream (downhill) from the connection with the KL exhaust raises as shown in Figure 3. The regulators were closed sufficiently to direct the required 2500 m³/s to KL. This had several major implications. All primary GVD fan pressures were increased by over 1000 Pa, bringing the GVD fans closer to stall and making parallel operation potentially more unstable.

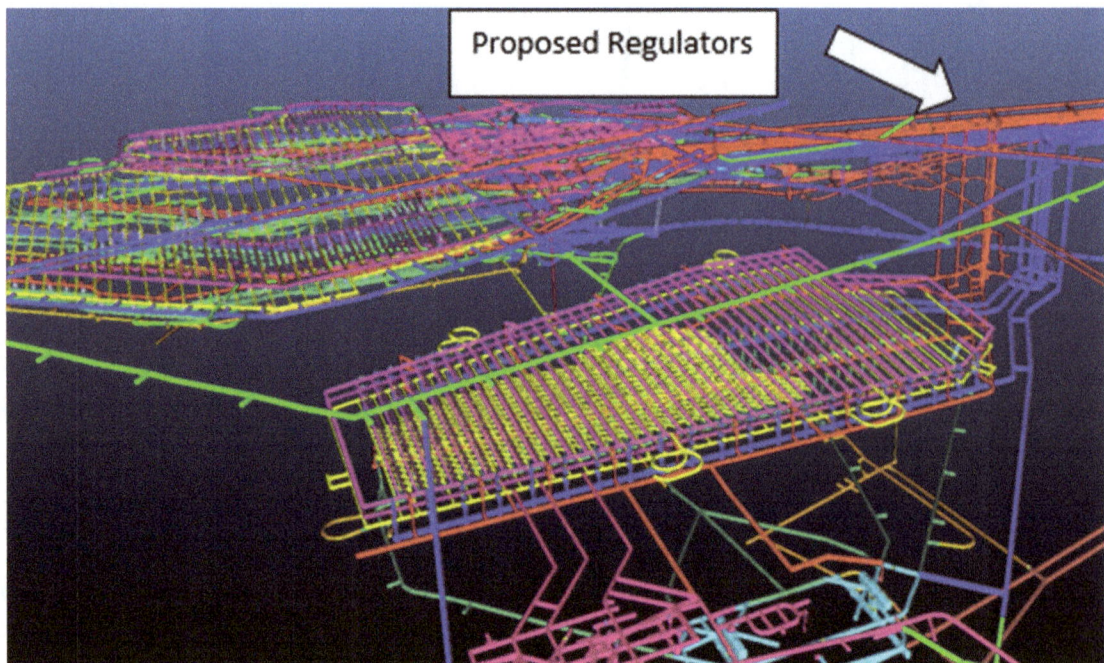

FIG 3 – Location of large regulators.

The required resistance became so high, that GBC airflows were reduced to 2300 m³/s before airflow to KL mine became sufficient, well below the required 2700 m³/s for that year. The energy consumed by the pressure loss across each regulator is equivalent to nearly 1 MW in power each – ie the sum effect of all regulators was to reduce available primary fan power by 4 MW.

Scenario B – Booster fan in KL intake and exhaust drifts

Unlike regulators, the addition of booster fans does not add resistance to a ventilation circuit and therefore overall airflow is not lost. In fact, primary ventilation fan pressure can be reduced, increasing airflow. To overcome the higher resistance and airflow requirement for the KL circuit, booster fans for the circuit were considered:

- Placing the booster fans in the intake at the base of the KL intake raises.
- Placing the booster fans in the exhaust circuit at the base of the KL exhaust raises.

The booster fans were assumed to have a 1 MW capacity each (allowing some contingency) over the four exhaust drifts or the four intake drifts at a fan efficiency of 70 per cent. The fans will only require a low pressure of less than 1500 Pa.

While the exhaust booster option is viable, the intake drift option is preferred because:

- The current geometry and design of the intake drifts (long and straight) assist with airflow entry, losses, and resultant efficiency through the proposed booster fans.
- Placement of the fans in a fresh air circuit simplifies fan construction, maintenance, and design (providing drainage water can be contained and access maintained).
- A push-pull system is sometimes used in caving ventilation to reduce harmful gas emissions from the cave such as radon.
- Finally, the more neutral pressure decreases the leakage of entry doors to KL from the BG drift access and the AB tunnel access.

On the downside, the eventual merging of the KL cave void to the GBC cave void in future years may result in KL pressure tending to push and permeate the air through the cave mass to the GBC mine. Modelling of the two alternatives for booster fans shows that in both cases, the airflow to KL can be raised to 2450 m³/s (close to the target 2500 m³/s) while still maintaining the GBC ventilation supply close to the target. Figure 4 shows the pressure gradient of the booster fan scenario.

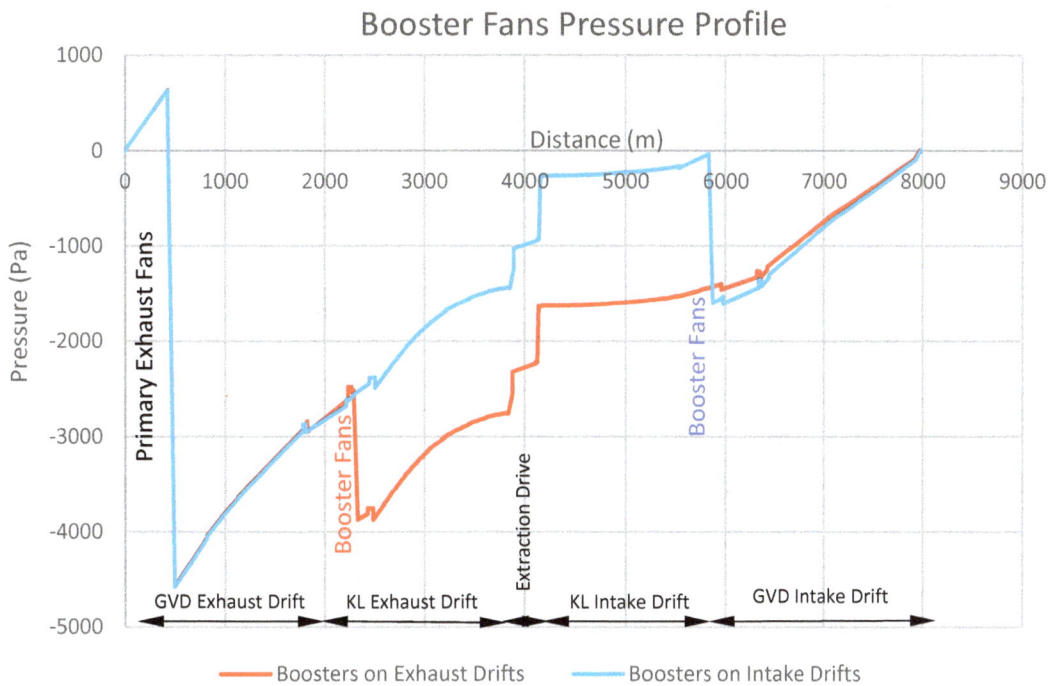

FIG 4 – Booster fan scenario.

Scenario C – Additional primary fan #8 and exhaust drift

The current LOM ventilation plan has a total of seven primary fans installed in the GVD exhaust drifts, five are currently installed in existing exhaust drifts with another two exhaust drifts with primary fans planned. As the total modelled airflow for these seven fans does not meet the airflow requirements for both KL and GBC simultaneously during the peak airflow requirements (a period from 2028 to 2033), an additional primary fan and exhaust drift was modelled.

The increasing KL airflow demand and decreasing GBC airflow demand require the GBC-allocated fans to be transferred to KL circuit duty over time. By 2033, a total of four fans (5, 6, 7 and 8) will need to be allocated to KL circuits, and the respective exhaust drives in the upper GVD and KL circuit; isolated with doors and bulkheads from the lower GVD drifts allocated to the GBC circuit.

Modelling of an eighth exhaust drift and fan achieves the budget requirement airflows for both KL and GBC with an additional 300 m³/s surplus to spare. In the event the surplus is not required, the fan speeds can be reduced for operating cost savings, oversetting the additional capital cost over time.

SWOT analysis – simulation results

A SWOT analysis was completed, and Scenario C was chosen as the LOM design. The main driving factors for this decision were operational efficiencies, flexibility, low OPEX (operating expenses), and flow distribution and contingencies. Figure 5 shows a high-level summary of the SWOT (strengths, weaknesses, opportunities and threats) analysis.

	Scenario C	Scenario A	Scenario B
	FAN8	**Only 7 Fans + GBC Regulators**	**7 Fans + 4 Booster Fans**
Installation Cost	New KL/GVD Exhaust Drift Required	Large, complex and Expensive GBC regulators required	Chambers and Bypass for Booster fans required
Operating Cost	Lowest Operating cost.	Lowest Operating Cost (lowest airflow through current infrastructure)	Highest Operating Cost (increased airflow through existing infrastructure)
Capital Cost	Higher Capital Cost	Lowest Capital Cost	Mid Range Capital Cost
Flow Distribution and Target	Simple Flow Distribution Changes will meet and (if required) exceed target. Contingency in case of Motor Failure.	Difficult Flow Distribution changes between GBC and KL with regulator settings – KL and/or GBC will not achieve requirements	Flexibility for Flow Distribution Changes.
Other Weaknesses	Increases Intake airflow velocity to more hazardous levels. Limited useful life - not required beyond 2034 on current schedules.	Flow shortfall will increase fumes, dust, clearance times and reduce operational diesel machine capacity > reduced production.	Adds complexity to **drainage, GBC&KL Cave Interaction. Blocking INT drift if fan is down.**
Other Strengths	Proven design philosophy that provides some *additional capacity* above targets if required. Beyond 2034 will continue to deliver operational cost savings for LOM due to lower resistance.	Nil	Simplifies distribution of air between GBC and KL – can be removed in future years beyond peak demand period.

FIG 5 – High-level summary of SWOT analysis for GBC/KL LOM design.

The study recommends an eighth fan and GVD exhaust drift option. The financial outcome of eight exhaust drifts and fans is slightly better seven exhaust drift booster fan option. The design also has the capability of producing more flow than budgeted, allowing some contingency, and accommodating additional development or production should it be required. The design is a proven conventional exhausting design that does not deviate from existing successful design philosophies.

Fan #7 will be commissioned by mid-2025 to support development activities. Fan #6 will be commissioned in 1st Qtr 2026. However, as the development rate increases and production ramp up, an additional fan (Fan #8) is required to supply air to ensure we can provide enough ventilation support to safely sustain and expand development works. Figure 6 shows the budget requirements and airflow availability.

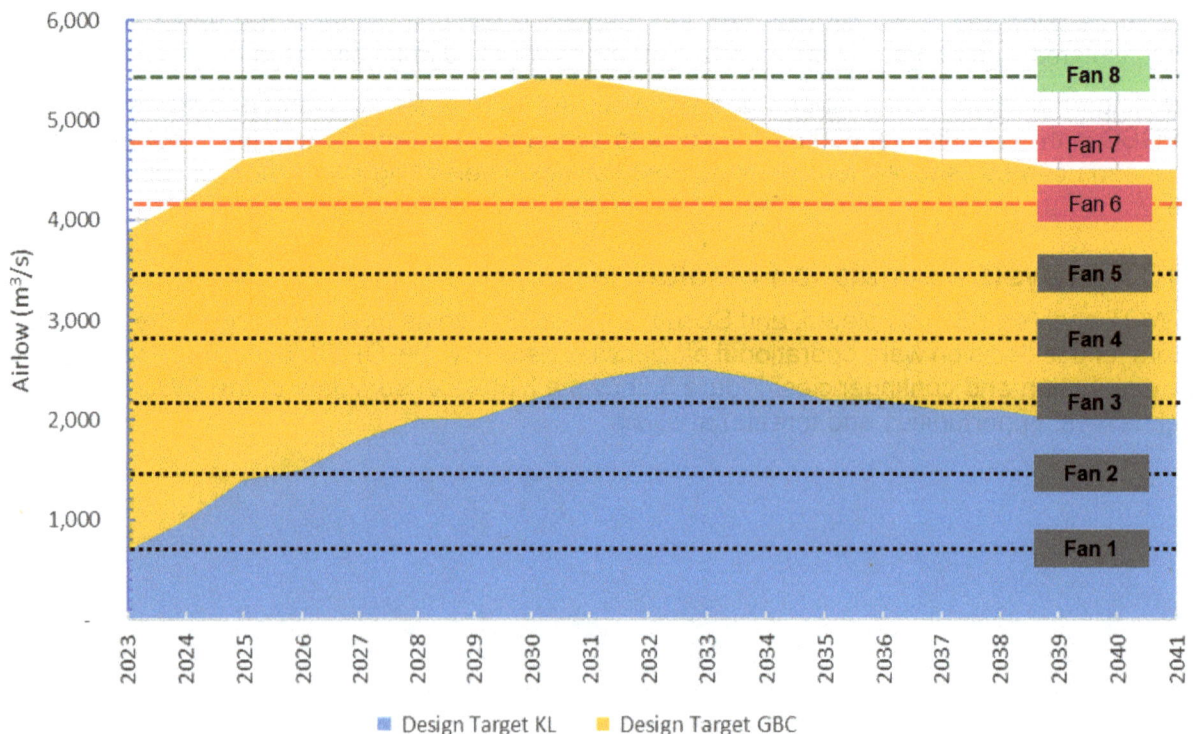

FIG 6 – GBC and KL combined ventilation budget and fan allocation.

FAN SELECTION AND DESIGN

The fan selection process was initiated by designing the LOM operating points. The operating points were determined by the required flow using the simulation results in the LOM model. An extensive tender and engineering review was performed by the PTFI selection committee for the selection of these fans, to satisfy procurement, maintenance, parts, support, and service requirements. The selection committee for this project consisted of representatives from the following departments: Ventilation, Maintenance, Mechanical, Construction services, electrical, Logistics, Contract group, Supply Chain and Corporate office. Operating points contingency was considered to improve pressure contingency for the final KL design, in particular the eighth fan case which requires higher pressure due to KL resistance.

Five operating points were considered for all three additional fans. This is mainly important since all fans are required to operate along with five existing 5 MW mixed-flow fans. Additional difficulties were to consider fan start-up flexibility (for example in case of a blackout) due to negative pressure induced from parallel fans.

Mixed flow (axial impulse) versus axial fans (axial overpressure)

PTFI currently operates five mixed-flow fans (5 MW each) at the GBC mine. The packages received from the tender process included both mixed flow and vane axial fans. The selection committee reviewed all received documents. A comparison of mixed flow and axial was necessary to understand the operational and strategic impact of the selected fan.

Structural differences

The current mixed-flow fans consist of a few main components. Radial Vane Control (RVCs) are located at the inlet section. The RVC is a feature that be adjusted remotely to control the fan operating point. The air in the Anti Stall Chamber (ASC) is discharged through the diffuser. These components along with the motor and fan pedestal also are the main contributors to 'fan losses'. The ventilation model must be adjusted and calibrated against these losses. The fan losses are generally supplied by the fan manufacturer which can also be verified by field measurements. Figure 7 shows the main components of a mixed-flow fan.

FIG 7 – Mixed-flow fan configuration.

The structural components of the tendered vane axial fan are shown in Figure 8. The motor sits in front of the fan inlet. The air enters through the inlet cone and directly into the fan impeller. The discharge section is directly located downstream to the impeller housing. The role of RVC in the axial fan has been replaced by a hydraulically driven variable pitch blade which automatically controls the blade pitch to control the required airflow target. A monorail hoisting system has been designed into the chamber for lifting. A mobile crane will be used as a secondary means of heavy components such as the impeller and motor.

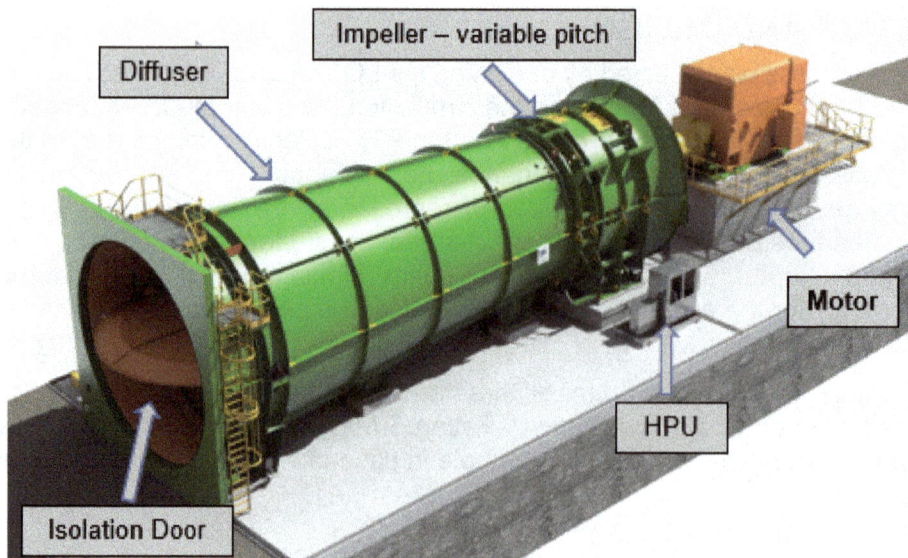

FIG 8 – Vane axial fan components.

Operating principles

Mixed-flow fans use adjustable inlet radial vane deflectors to vector the airflow onto the rotating fixed weld on blades on the impeller to generate both radial and axial flow (mixed). The velocity pressure is then converted to static pressure by using internal vanes to direct and widen the flow-through the diffuser cross-section. While some flexibility in mixed-flow fan duty is available through the RVC adjustment, the system is not as flexible to wider duty changes due to losses in efficiency at lower RVC settings and fixed impeller blade limitations.

Generally, the hub to tip ratio and diffuser sections for mixed-flow fans are larger to generate the required velocity pressure and subsequently static pressure rise. This results in smaller blade opening and hence higher velocity pressure. The above results in higher efficiencies of axial flow compared to mixed-flow fans.

The interior structure of diffuser also varies between two designs. The mixed flow fans equipped with straight or widening tail cone. Inside of the cone usually can be accessible either from the fan discharge (if the fan is not in operation) or from and access tube in the side of the casing. Figures 9 and 10 show the configuration of mixed-flow and axial tail cone. The axial fan designs generally possess a tail cone.

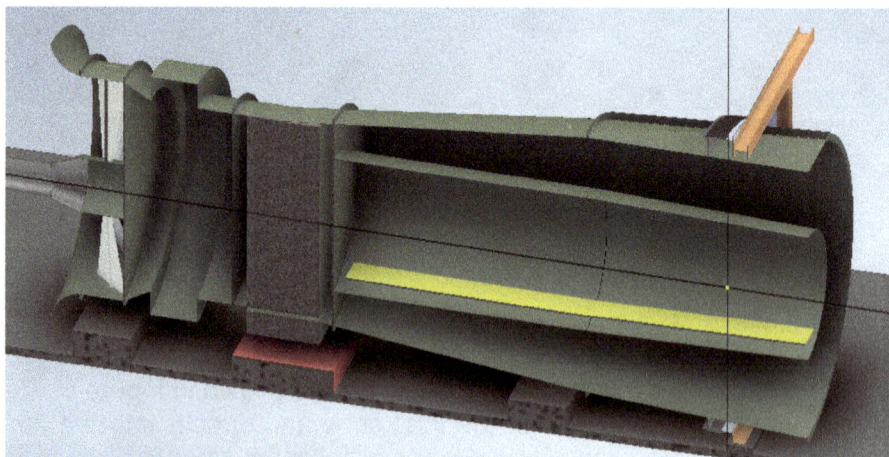

FIG 9 – Mixed flow tail cone configuration.

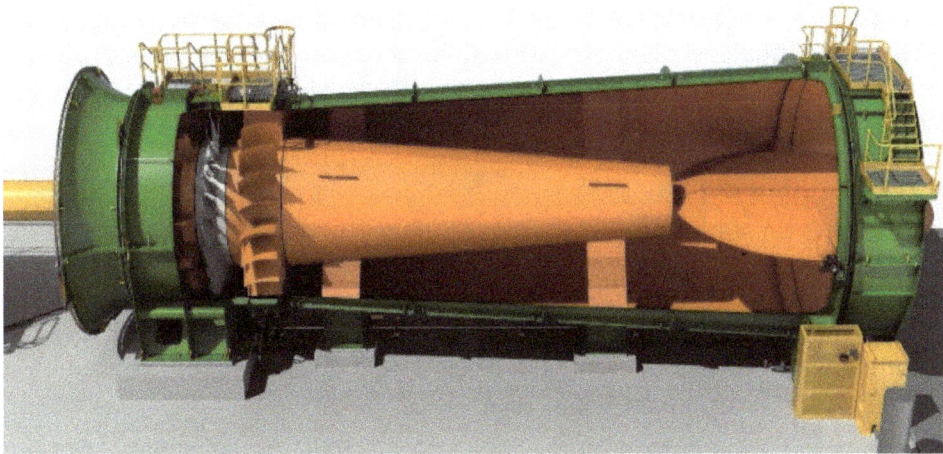

FIG 10 – Axial fan tail cone configuration.

Operational robustness

Mixed-flow fans are considered more rugged, having thicker more robust blades and are therefore more resilient to airborne particles and debris causing damage and blade wear (Loomis and Duckworth, 2012). Mixed-flow fans are often used in challenging applications such as dirty mine exhaust air and powerplants. RVCs on mixed flow fans need to be designed to match the mine pressure and environment which often contain coarse particles and corrosive liquids particularly if installed in exhaust drifts.

Due to more streamlined blades, axial fans are more prone to blade wear although this can be reduced with wear-resistant blade materials or coatings as required. While many examples are used in exhaust applications, scheduled replacement of blades in abrasive environments is normally a consideration. If the blades are manually or hydraulically adjustable in pitch, special consideration must also be taken on wear and corrosion in the moving mechanisms. PTFI has chosen cast Iron blade for KL application instead of aluminium.

Fan stall stabiliser systems

Flow separation and recirculation around the blades (rotating stall) can take place if a fan reaches a limiting pressure and enters the stall region of the duty curve. When a full span of blades enters stall conditions pressure loss occurs and airflow reduces, accompanied by fan vibration and instability.

Design improvements can be made to lift the stall line (saddle) and this is a critical consideration in parallel configurations as the back pressure from other fans may instigate stall or start-up issues on other fans. For mixed flow fans, the ASC unit (as shown in Figure 7) assists with the stability of the fan in parallel operation by preventing the stall from propagating to the full span of the blade by moving it from the tip of the leading edge back into the impeller. The ASC lifts the stall line while leading high-velocity pressure air back into the impeller.

Generally, axial flow fans have a smoother stall saddle for parallel operations. The stall condition for an axial fan is similar to a mixed-flow fan and occurs if as part span and eventually full span of the blade, where no flow can be produced. Similar to ASC, the stall ring reduces the propagation of part span into a deep stall by forcing air around the ring and preventing the rotating stall swirl to propagate. The ring results in breaking the stall condition by pushing the flow into the impeller. Although the axial flow stall ring does not lift the stall saddle compared to ASC, it has less impact on flow disturbance, fan stability, fan losses and therefore efficiency.

Fan efficiency

Fan efficiency is the ratio of power derived from the actual fan total pressure and flow produced versus this input shaft power to the fan impellor, and directly influences the operating cost of the fan over the LOM design. The measurements can be entered into the mine ventilation model for calibration. Fan losses occur due to velocity changes across the inlet and discharge cones, turbulence shock losses around the motor, RVC, stall prevention system, diffuser section, interaction with outlet cone and eventually discharge section. Unexpected losses can also occur if the fan duty operates outside of the peak efficiency zone due to changes in mine resistance. Due to higher efficiency, the energy consumption of axial flow fans is generally lower compared to the mixed flow fans for the same duty.

Start-up stability

In parallel operations, if on start-up the system resistance curve crosses one of the fans operating points in the unstable stall region, the fans will remain in the stall region and will not operate or enter a pendulum flow fluctuation between fans. The starting fan will stall and may even cause the other fan to stall. The KL mine will have eight fans running in parallel conditions and it is critical to have the ability to turn any or all of the fans on after a blackout scenario, or during preventative maintenance.

Mixed-flow fan duties can be adjusted on the fly using the RVC to assist with start-up and reduce stall instability, however, most commonly used axial fans rely on variable speed control which is ineffective by itself in avoiding stall instability on start-up. Figure 11 shows the variable pitch blade set-up. To facilitate a stable start-up two dynamic options can be utilised:

1. Variable frequency drive (VFD): If the mine resistance curve stay the same, VFDs can control the flow while maintaining similar efficiency, assuming the stable blade pitch has already been designed and set. The main advantage of VFD during start-up is lowering starting amperage. This option requires a medium voltage drive which would need to be installed in close vicinity of the fan chamber, this option will require significant capital expenditure. Additionally, the VFD station will require a very clean environment free of dust and diesel particulates.

2. Variable blade pitch angle: consisting of a hydraulic power pack and hydraulic blade servos. On start-up, the fan blades are set to fully feather (no load) to build-up to full rotational speed while preventing stall at low angle of attack an airfoil blade is stable with no stall saddle existing. The outlet isolation door would then open and the blade pitch changed to reach the target flow without entering the unstable stall region.

FIG 11 – Variable pitch blade system for vane axial fan.

Operational flexibility and maintainability

Mixed-flow blades are welded on, and RVCs are used to adjust the flow for various conditions. Mixed flow RVCs require periodic maintenance or upgrades to manage higher pressure. Axial fan duty flexibility can be achieved with variable pitch blades. If a mixed-flow fan blade change is required, the entire impeller assembly is required to be changed out whereas for axial fans the blades can be changed individually. To improve the stability and manage the dust accumulation on the blades, a fully automatic blade washing system has been designed to rinse down the dust particles. The specification of washing frequency will need to be optimised according to field conditions, dust and water droplets in the air.

Motor

The GBC and KL systems will have a combined eight fan motors running in parallel. The existing five GBC 5 MW motors have IP56 protection class, Class F insulation and Class B temperature rating (withstand temperature rise of 115°C). The return air main exhaust drifts, contains high humidity along with corrosive environment. The environment is not particularly dusty since the primary fans are more than 2.5 km from the working areas. As a result of recent motor failures, PTFI has increased new motor requirements to IP66 protection class with Class H insulation. This was considerably challenging as at 6.15 MW motors with this rating are not manufactured by mainstream motor suppliers and would require a custom build.

Chamber optimisation and CFD analysis

Computational Fluid Dynamic (CFD) analysis was conducted on chamber optimisation. Fan chamber development requires significant capex. The simulation results were used to optimise the pressure profile across the chamber and fan components. A mass flow rate of 685 kg/s at a density of 0.83 kg/m^3 was used in a steady-state simulation. The simulation results show a transient flow separation chamber inlet would improve the flow stability. The current chamber design acts similarly to a diffuser and a separated flow tends to move upwards causing turbulence. Although the improved design also has shock losses and turbulence, flow separation (and hence fan stability) is improved due to a narrower flow pattern. A modified improved design is used to develop future fan chambers. Figure 12 shows the modified improved chamber design, in diffusion view.

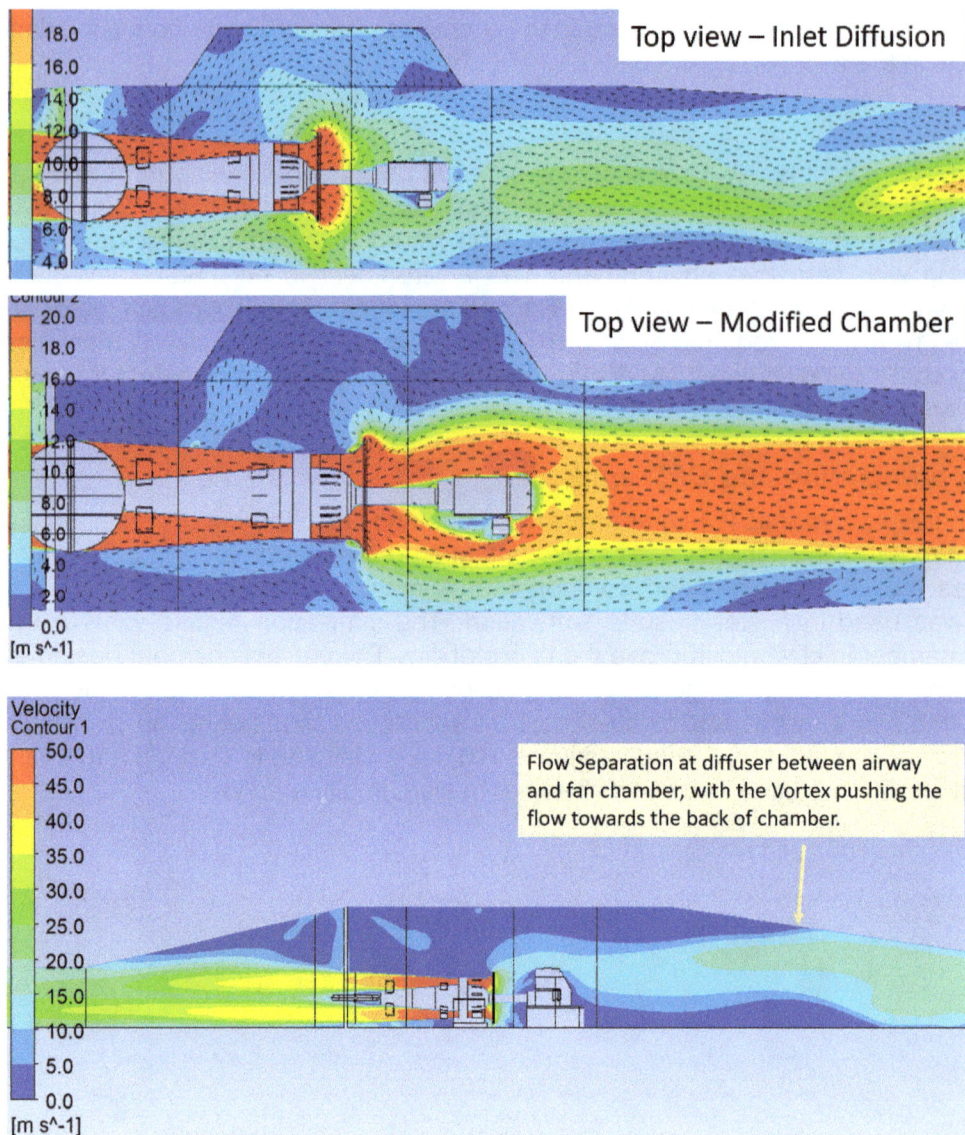

Top view – Inlet Diffusion

Top view – Modified Chamber

Flow Separation at diffuser between airway and fan chamber, with the Vortex pushing the flow towards the back of chamber.

FIG 12 – Primary fan chamber optimisation.

CONCLUSION

The PTFI selection team selected a 6.15 MW axial fan equipped with a variable pitch blade, no stall stabiliser ring and an automated blade washing system. This decision was reached after a detailed LOM ventilation study was carried out to identify the required airflow budget and primary fan infrastructure. Three additional primary fans and exhaust drifts were required for the future design. The PTFI LOM ventilation model was calibrated with updated friction (K) factors obtained from a pressure-quantity survey. Fan losses were also added to the model using results from a collaborative fan survey with the manufacturer. The required LOM future operating duties were tendered and the PTFI selection committee reviewed the bids which included both mixed-flow and axial-flow options.

Although all tender options met the requirements, several criteria were examined to determine the most suitable for the GBC-KL complex. The structural differences from an installation point of view are similar with the main difference between the mixed-flow and axial fan being the configuration of discharge air regulation and louvre regulators versus an isolation door. Although the operational principles for mixed flow and axial are vastly different, the pressure losses across the fan (through the components such as the anti-stall chamber and tail cone) were considered as performance metrics.

The existing mixed-flow fans' operational robustness on-site has been well-regarded to date and blade wear and susceptibility to airborne particles for a new axial fan was a major consideration.

Automatic blade washing systems were specified to improve fan operation against particle accumulation. The frequency of washing will be optimised according to actual field conditions.

The effect of stall stabiliser systems on fan efficiency and fan operating points was examined. The start-up flexibility during blackouts and preventative maintenance is of high value for the PTFI team. Both mixed-flow and axial options provide the flexibility to control flow while the fans reach their respective operative points, however, the variable pitch was preferred compared to the RVC during start-ups as the blade pitch can be fully feathered at start-up and can automatically adjust to ensure the fan duty remains clear of the unstable area in fan curve. The pitch adjustment will also be used to accommodate underground resistance changes and smooth operation with the existing mixed-flow fans, further adding to operational flexibility. Motor requirements of class H/B and IP66 insulation were also a major factor in the selection process favouring the variable pitch axial fan.

The operating and capital expenditures were also examined for LOM and while mixed-flow fans required less capex, the operating costs and overall LOM costs were significantly lower for axial fans. In addition, the fan chamber optimisation was achieved using CFD analysis which resulted in a smaller chamber with a better flow stream, which will further improve fan stability and operation.

ACKNOWLEDGEMENTS

The authors would like to acknowledge assistance of PTFI maintenance, operations, and central services departments. The assistance of Howden Australia group for field measurements and technical discussion is greatly appreciated. The authors also thank TLT-Turbo for their technical assistance in design work and CFD analysis.

REFERENCES

Brannon, C, Brard, D, Pascoe, N and Priantna, A, 2020. Development and production update for the Grasberg Block Cave mine – PT Freeport Indonesia, in *Proceedings of the 8th International Conference and Exhibition on Mass Mining*, virtual conference, pp 747–760.

Habibi, A, Setiawan, I, Prasojo, R and Stewart, C, 2023. Kucing Liar Mine LOM Preliminary Ventilation Design- PT Freeport Indonesia, Proceedings of the 19th US/North American Mine ventilation Symposium 2023 (South Dakota School of Mines).

Loomis, I and Duckworth, I, 2012. Planning and selection of main fans for the Grasberg block cave mine, Proceedings of the 14th US/North American Mine ventilation Symposium 2012 (University of Utah: Salt Lake City).

Computational modelling of heat hazard and methane dispersion in a roadway development with auxiliary ventilation

A Mishra[1], D P Mishra[2], S Rao[3] and U S Shukla[4]

1. Assistant Manager, Nigahi Project, Northern Coalfield Ltd., India.
   Email: aishwaryam.rs.min16@itbhu.ac.in
2. Professor and Head, Department of Mining Engineering, IIT(ISM) Dhanbad, 826004, India.
   Email: dpmishra@iitism.ac.in
3. Senior Manager, Moonidih Colliery, Bharat Coking Coal Ltd., 828129, India; Research Scholar, Department of Mining Engineering, IIT(ISM) Dhanbad 826004, India.
   Email: sunnyrao.coalndia@gmail.com
4. Manager, Dudhichua Project, Northern Coalfield Ltd., India; Research Scholar, Department of Mining Engineering, IIT(ISM) Dhanbad, 826004, India. Email: us.shukla30@gmail.com

ABSTRACT

Excavation of blind heading in strata with high heat load has a severe impact on miners' health, while exposure to methane in such a restricted setting poses a challenge to worker safety against methane explosion events. A computational model has been developed considering the methane dispersion and convective airflow in the roadway with constant wall temperature. The simulation results were validated against the ventilation survey data. The validated model investigated the impact of various auxiliary ventilation schemes, cross-section design, and strata heat load and methane emission rate on the underground mine environment on the longitudinal length of heat transfer and methane dispersion in the roadway. As a result, a novel auxiliary ventilation design has been implemented in the case study mine to address the heat stress and methane hazard issues in the subsurface environment. Thus, with the aim to improve the working condition at the working face, the novel ventilation system reduced the heat load and methane risk, and increased the air velocity in the working zone, thereby minimising the risk and improving the working condition while excavating the blind heading.

INTRODUCTION

The underground mining is a hazardous job. It involves the risks to workers health and life owing to their continuous exposure to particulate matter, heat, humidity and flammable gas mixture. Deep seated coal and shale deposits being the seat to methane and heat, become the main cause of concern during underground mining operations as a large amount of methane is released at the working face area. The severity of methane-related disaster can be judged by the immense loss of man and material due to methane explosion accidents (Brodny and Tutak, 2018; Brune, 2013). Numerous methane-related incidents and accidents with fatalities have attracted significant attention all over the world, highlighting the importance of an effective mine ventilation to ensure a safe and productive environment in an underground mine whilst keeping the energy usage and operating cost at minimum.

Generally, forcing auxiliary ventilation through ducts is used in underground mines to tackle the methane and heat in the blind headings. Given the operational and safety issues, exhaust ventilation and overlapping auxiliary ventilation have also been used in the mines. Computational fluid dynamics had been used by researchers to understand the methane and dust dispersion in the underground tunnels and blind headings. Further, innovative ventilation techniques such as overlapping ventilation system, use of scrubber near the heading, using intermittent ventilation system etc, have been designed to improve the underground environments. Such techniques were meant to eliminate almost all the high-risk zones in the computational domain, thereby improving the underground working condition (Kurnia, Sasmito and Mujumdar, 2014). However, the systems introduced focused only on a single issue like methane or dust or heat. In this study, the authors considered a deep underground mine working having high temperature and methane emission from face. Under such circumstances, researchers have proposed different variants of auxiliary ventilation practices to ventilate the blind heading (refer Table 1).

TABLE 1

Description of different variants of tunnel ventilation.

S No	Variant name	Configuration features
1	Single duct	Conventional, single duct at top corner of tunnel
2	Twin duct	Two ducts at top corner of either side of wall
3	Twin duct staggered	Two ducts, one is at top corner of a wall, Other at the middle of tunnel near another wall. Both ducts have different set back distances from the working face

The dimension of the road header drivage is generally considered as 4.8 m × 2.8 m. However, once the tunnels are driven using robust machines like bolter miner, the width and height of the heading increases to execute the mining process. However, the ventilation aspect gets compromised considering the conventional method of ventilation. The authors of the paper worked in an underground coalmine where the headings were driven by bolter miner. The ventilation of such headings using conventional design were sluggish. This resulted in poor sweeping effect of methane emitted from the walls of the gallery. Additionally, the temperature of the gallery remained high. So, the authors came forward with twin duct ventilation system. In this system, the authors introduced the twin ducts to ventilate the heading. Additionally, different ventilation designs using twin ducts have been simulated to come up with new configuration. It is found that using such a combination of ducts, the working environment improved and the safety enhanced manyfold.

STUDY AREA

The study area is an underground coalmine in the Jharia Coalfield, India. The coal seams dip south-westerly with an average dip angle varying between 7° and 12°. Amongst all the coal seams of Jharia coalfield, XV and XVI seams are most vulnerable to gaseous explosion (Chandra, 1992). The tunnels considered in this study are driven through these coal seams. The blind headings were driven by using road header as well as bolter miner. The size of the heading excavated using road header has the dimension of 4.8 m × 2.8 m, while the headings excavated using the bolter miner has the dimension of 5.5 m × 3.5 m. The ducts used were of diameter 1 m, and around 550 m³/min of air was injected through them to the face. Figure 1 depicts the orthogonal view of the heading being excavated and different ventilation configuration that can be adopted to drive the gallery successfully.

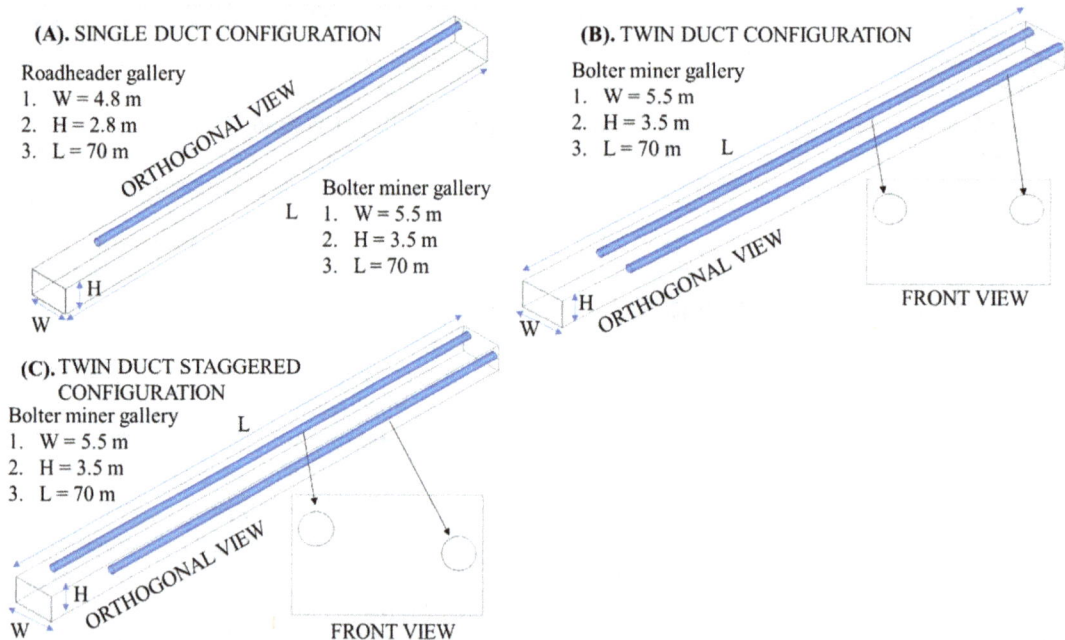

FIG 1 – The representative diagram of different ventilation configurations.

METHODOLOGY

A simulation-based analysis was conducted in this study for a better understanding of the physical process of heat and methane transport in the blind heading, and to evaluate the efficacy of ventilation control design. A CFD simulation study was thus carried out in order to understand the methane and heat hazard in tunnel drivage using conventional single duct ventilation system, thereby analysing the impact of twin duct ventilation system. The aim is to prevent the hazard from becoming a disaster in short-term as well as long-term, thereby improving the working environment at face. The methodology adopted for the computational study is depicted in Figure 2.

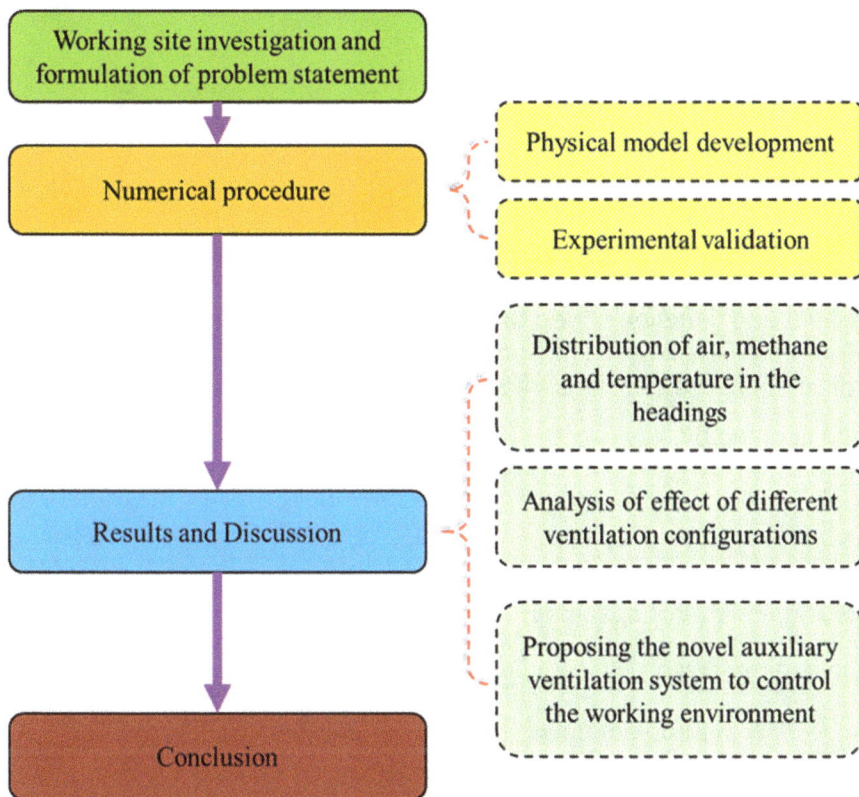

FIG 2 – The flow chart of methodology adopted.

PHYSICAL MODEL

The computational model considered in this study consists of a 70 m long straight heading with cross-sectional areas of 4.8 m × 2.8 m and 5.5 m × 3.5 m. The methane emitted from the closed end of the heading at a rate of 0.0128 kg/m³.s. The tunnel environment was controlled by ventilating the heading using single/double auxiliary ventilation system. The field study was done to estimate the methane concentration in the blind heading. The boundary condition for velocity inlet was adopted at the duct inlet to inject 550 m³/min of air in the blind heading through ducts of 1m diameter. The tunnel outlet was considered as pressure outlet. The cross-sectional wall of tunnel was maintained as methane inlet. The walls of the tunnel were maintained at a constant temperature considering the phenomenon of geothermal gradient. The walls of tunnel were assumed to be smooth, adiabatic, and assigned no-slip boundary conditions. During the iterations, the convergence was set at 1×10^{-4} as a standard of accuracy mentioned in the Ansys Fluent Theory Guide (ANSYS, Inc., 2013).

NUMERICAL METHODOLOGY

The governing equations for conservation of mass, conservation of momentum, transport, gas state, and turbulence considered in this study are given in Equations 1–6 (Versteeg and Malalasekera, 2007; Tutak and Brodny, 2015).

Conservation of mass: The equation represents the mass flux and change in density over time. It is given by:

$$\frac{\partial \rho}{\partial t} + \frac{\partial}{\partial x_i}(\rho u_i) = 0 \tag{1}$$

where:

ρ is the density

u is the velocity

Conservation of momentum equation is given by:

$$\frac{\partial}{\partial x_j}(\rho u_i u_j) = -\frac{\partial p}{\partial x_i} + \frac{\partial \tau_{ij}}{\partial x_j} + \rho g_i + S_i \tag{2}$$

where:

ρ is the density

p   is the pressure

τ_{ij} is the stress tensor

g is the gravity.

The pressure gradient and diffusion added to gravity and other forces give the local acceleration and advection in the fluid flow. The flow of fluid in porous media needs to increase the momentum source term composed of viscous loss term and internal loss term (S_i).

Transport equation is given by:

$$\frac{\partial}{\partial x_i}(\rho \vec{\vartheta} Y_i) = -\frac{\partial}{\partial x_i}\vec{J_i} \tag{3}$$

where:

$\vec{J_i}$ is the diffusion flux of species i

$\vec{J_i} = -\rho D_{i,m}\nabla Y_i$ in laminar flows

$\vec{J_i} = -\left(\rho D_{i,m} + \frac{\mu_t}{Sc_t}\right)\nabla Y_i$ in turbulent flows

Energy Equation is given by:

$$\frac{\partial}{\partial t}(\rho E) + \frac{\partial}{\partial x_i}[u_i(\rho E + p)] = \frac{\partial}{\partial x_i}\left(k_{eff}\frac{\partial T}{\partial x_j} + u_i(\tau_{ij})_{eff}\right) + S_h \tag{4}$$

Turbulence kinetic energy, k equation:

$$\frac{\partial}{\partial t}(\rho k) + \frac{\partial}{\partial x_i}(\rho k u_i) = \frac{\partial}{\partial x_i}\left(\alpha_k \mu_{eff}\frac{\partial k}{\partial x_j}\right) + G_k - \rho\varepsilon \tag{5}$$

Turbulent kinetic energy dissipation, ε equation:

$$\frac{\partial}{\partial t}(\rho\varepsilon) + \frac{\partial}{\partial x_i}(\rho\varepsilon u_i) = \frac{\partial}{\partial x_i}\left(\frac{\alpha_\varepsilon}{\mu_{eff}}\frac{\partial \varepsilon}{\partial x_j}\right) + C_{1\varepsilon}\frac{\varepsilon}{k}G_k - (C_{2\varepsilon})\rho\frac{\varepsilon^2}{k} \tag{6}$$

where:

ρ   is the density, kg/m^3

x   is the space coordinate, m

t   is the time coordinate, s

u   is the velocity, m/s

i and j   are coordinate directions

p   is the static pressure, Pa

τ_{ij} is the viscous stress tensor, kg/m.s^2

k_{eff} is the effective thermal conductivity, W/m.K

E	is the energy of system of gases
T	is the temperature, K
S_h	is the source term of chemical reaction energy, W/m³
μ_{eff}	is the effective turbulent viscosity coefficient, Pa.s
G_k	is the shear force caused by the fluctuation of turbulent kinetic energy rate, kg/m.s³
k	is the turbulent kinetic energy, m²/s³
ε	is the turbulent dissipation rate, m²/s³
$C_{1\varepsilon}, C_{2\varepsilon}, \alpha_k, \alpha_\varepsilon$	are the constants of the model and have default values (Launder and Spalding, 1974)

RESULTS AND DISCUSSION

Selection of turbulence model and model validation

Methane dispersion in tunnel was widely modelled by the researchers using the k-epsilon turbulence model (Mishra, Kumar and Panigrahi, 2016). The overall flow behaviour for design purposes is well reflected by the k-epsilon model, which has been widely used in engineering application, and is found to be suitable whilst maintaining low and fast computation (Kumar *et al*, 2017). We, therefore, proceed further with this turbulence model. The best variant of the k-epsilon turbulence model must be used for the better prediction of methane dispersion. In this work, three versions of k-epsilon turbulence model, namely Standard, RNG and Realisable, were compared for the results of methane dispersion. The measurements were done to measure the methane concentration along the centre of the tunnel running towards the outlet. The average methane concentration inside the tunnel was found to be around 0.5 per cent. The graphical comparison of experimental data with different turbulence models is shown in Figure 3. Up to a distance of 10 m from face, the methane concentration obtained from standard k-epsilon model deviated from field data between 4 per cent and 10 per cent, while for RNG and Realisable k-epsilon models, the error ranges between 10 per cent and 68 per cent, and 25 per cent and 68 per cent, respectively. However, for the distance from 10 m till the outlet, the simulation result is comparable to the field data for the methane concentration. Thus, it may be observed that the results of standard k-epsilon model are in good agreement with the experimental data and has been used by the researchers for further analysis.

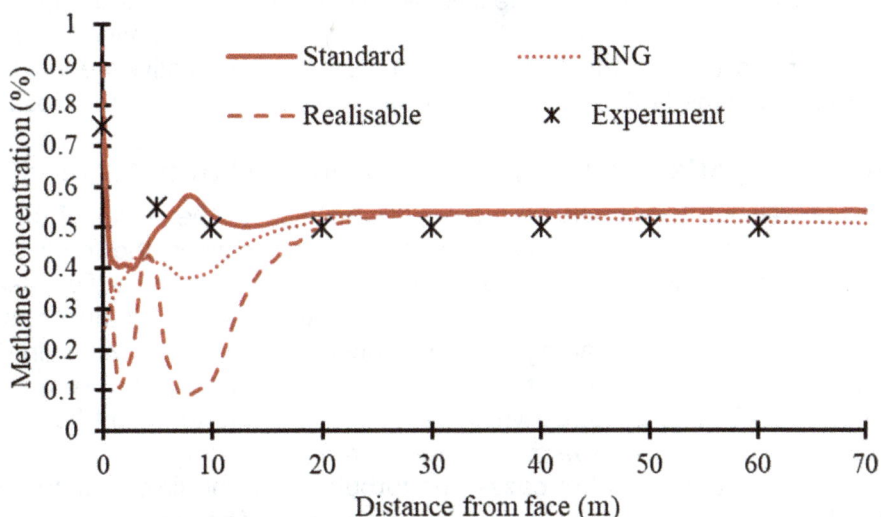

FIG 3 – Validation of the model with field data.

The conventional forced ventilation method has been opted in the gallery of cross-section 4.8 m × 2.8 m. The forced ventilation caused the localised dispersion of methane emitted from the face of the tunnel due to air velocity and its reduction due to sudden increase in the cross-sectional area upon coming out of the duct. However, as we move towards the outlet, the methane disperses equally across the whole cross-section. This is due to reduction in turbulence in the tunnel that is mainly governed by the low Reynolds number of the flow. The different characteristics of the tunnel flow is illustrated in the Figure 4. To understand the thermal condition in an underground tunnel, the tunnel wall is maintained at a constant temperature. The Figure 4c shows that the air temperature in tunnel increases while moving towards the outlet. However, the wall heat transfer coefficient (Figure 4d) reduces while moving towards the outlet. It implies that major exchange of heat takes place near the face, but temperature remains low due to the cooling effect of air from duct outlet. Since, man and machine deployment is at the face, they should be protected from the exposure to methane and heat hazard.

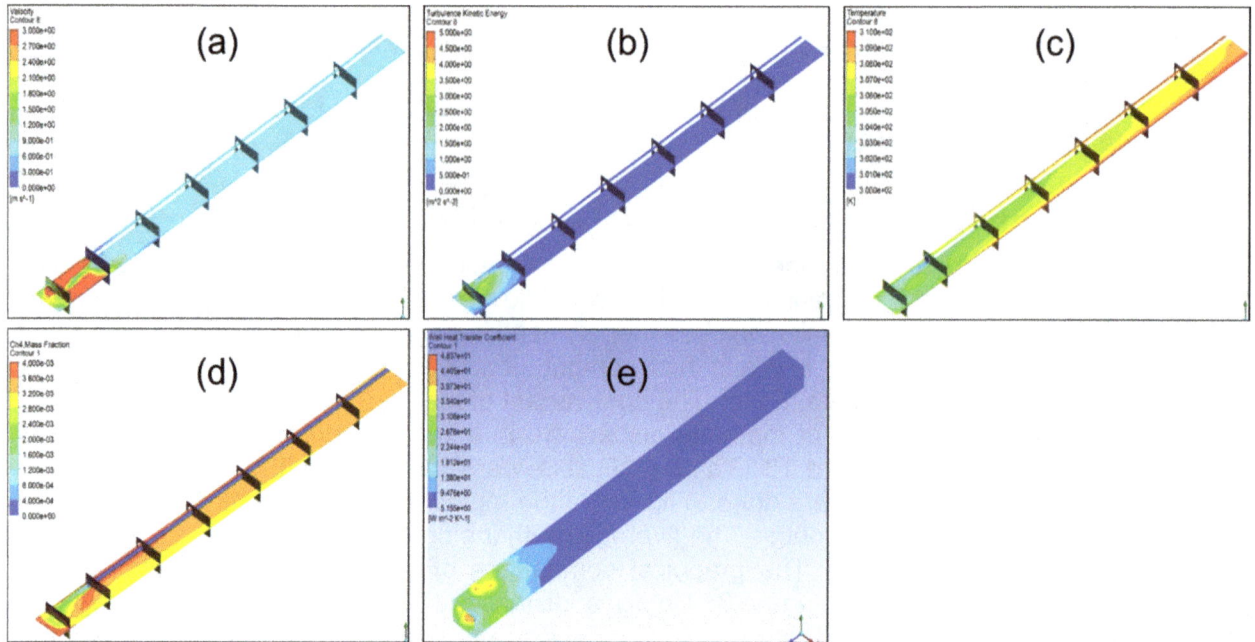

FIG 4 – The contours showing the variations of (a) air velocity, (b) turbulent kinetic energy, (c) temperature, (d) methane concentration, and (e) wall heat transfer coefficient in the tunnel.

For further analysis, a tunnel with larger cross-section (5.5 m × 3.5 m) was considered. The flow behaviour and methane dispersion in the tunnel with different ventilation systems installed were investigated. In the following section, the effectiveness of different variants of forcing ventilation to manage methane dispersion as well as local temperature in underground mine has been evaluated. Thus, this study attempts to evaluate the effectiveness of an innovative ventilation design with respect to the conventional method to drive an underground gallery with larger cross-section.

Effect of ventilation pattern on temperature and heat transfer coefficient

The heat-transfer coefficient decreased from the heading face to the outlet of the heading in all cases, which could be attributed to the intense heat transfer between the fresh cold airflow that emerged from the forcing air duct and the heading face. However, the rock-airflow heat transfer was gradually alleviated while moving towards outlet owing to the uniform distributions of the airflow velocities close to the outlet of the heading. Figure 5 shows that the case with twin duct ventilation system of unequal length has the largest heat-transfer region, and the maximum heat transfer coefficient is 60 W/m²·K. This is because when the airflow exhibits turbulent flow on the heading, turbulence gets properly developed towards the face from duct opening near the surrounding strata (Xin *et al*, 2021). In contrast, in the other cases, the turbulence in the flow was not fully developed. Thus, such cases have shown similar heat-transfer coefficients. Therefore, for twin duct staggered ventilation configuration, measures should be adopted to reduce the heat transfer into the tunnel environment. Consequent to the high heat transfer coefficient in the twin duct ventilation system, the increase in air temperature in such headings is least for heat transfer from heading wall to the air flowing in the heading (see Figure 6). This in turn enhances the quality of the work environment in the heading.

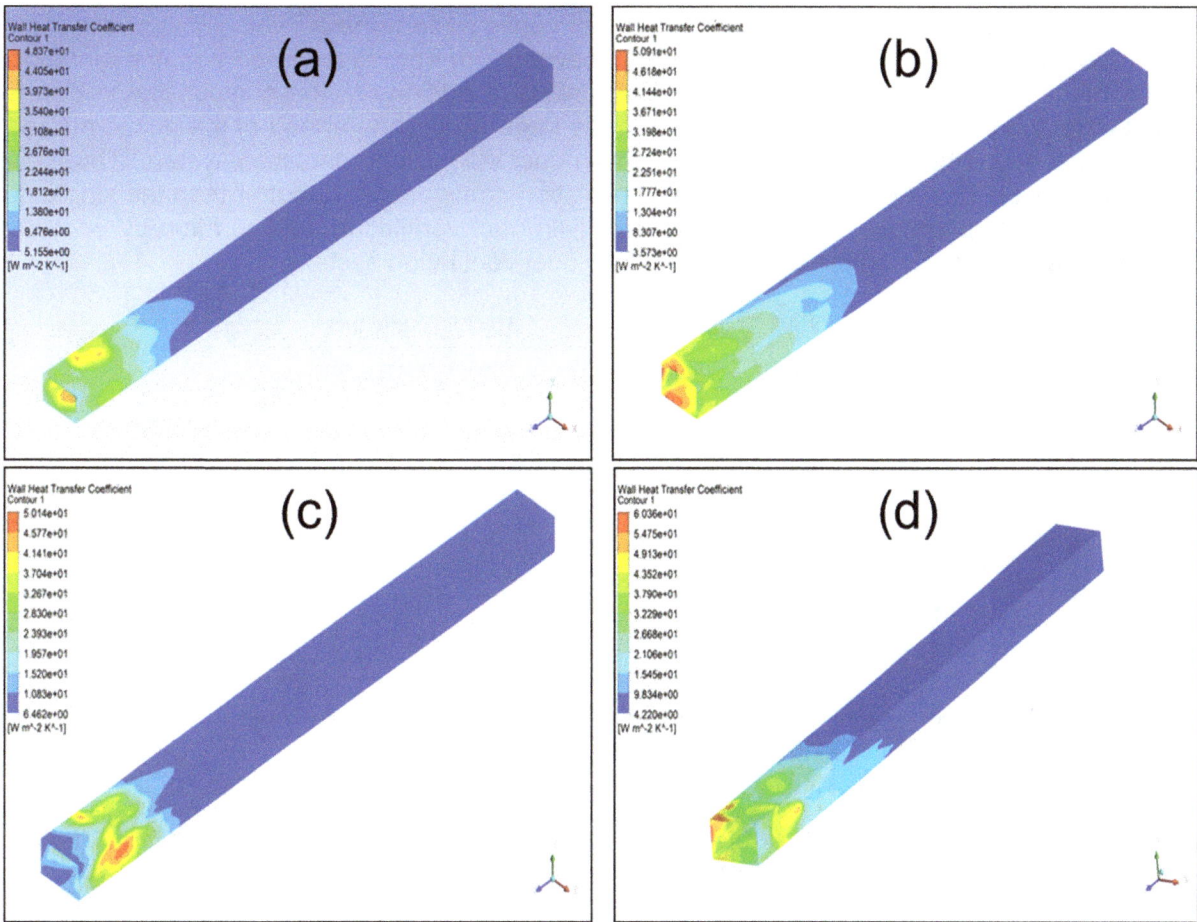

FIG 5 – Contour representation of wall heat transfer coefficient for (a) single duct ventilation system for tunnel of cross-section 4.8 m × 2.8 m, (b) single duct ventilation system for tunnel of cross-section 5.5 m × 3.5 m, (c) twin duct ventilation system for tunnel of cross-section 5.5 m × 3.5 m, and (d) twin duct staggered ventilation system for tunnel of cross-section 5.5 m × 3.5 m.

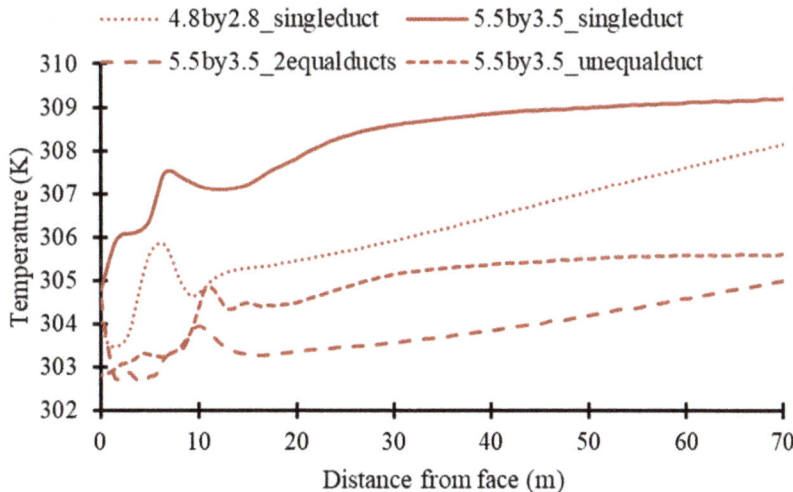

FIG 6 – The variation of air temperature inside tunnel with distance from face for different ventilation configurations.

Effect of ventilation pattern on air velocity in tunnel

The airflow behaviour in different designs of forcing auxiliary ventilation is depicted in Figures 7 and 8. Figure 7 show that for different designs, the air velocity maintains a uniform profile from 20 m behind the face. Within the zone up to 20 m from the face, the air velocity shows non-uniform profile, indicating turbulence in the region. Using conventional single duct and overlapped twin duct staggered configuration, the velocity in the range of 1–4 m/s is found up to a distance of 3 m from the face. However, with the use of overlapped duct configuration, the air velocity near the face is found to be around 1 m/s. However, in the region beyond 3 m from face up to 12 m, the air velocity is in the range of 1–4.5 m/s with twin duct configuration. Additionally, the contour plot in Figure 8 suggests that the overlapped duct configuration has large blind zone ahead of the ducts, making it a seat of methane. This is not found in case of twin duct staggered ventilation system. Thus, even though the conventional twin duct overlapped ventilation configuration is better than the single duct system, staggering improves the effectiveness of twin duct ventilation configuration by ventilating the blind zone created during the conventional twin duct ventilation system.

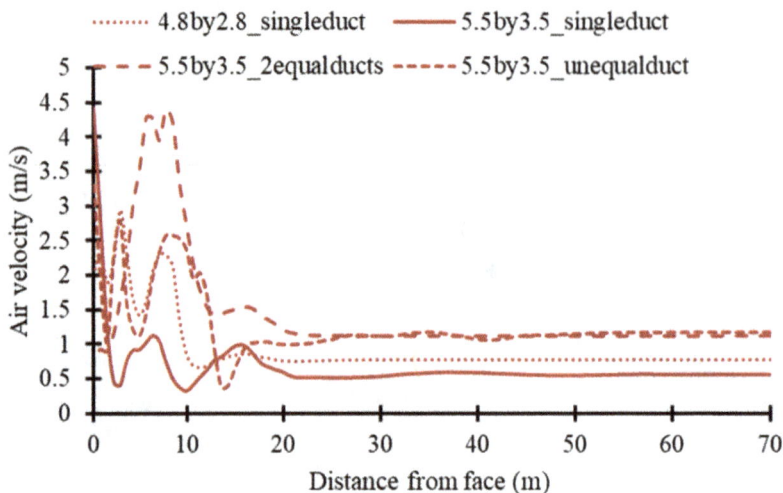

FIG 7 – Variation of air velocity at the centre of tunnel with distance from face for different ventilation configurations.

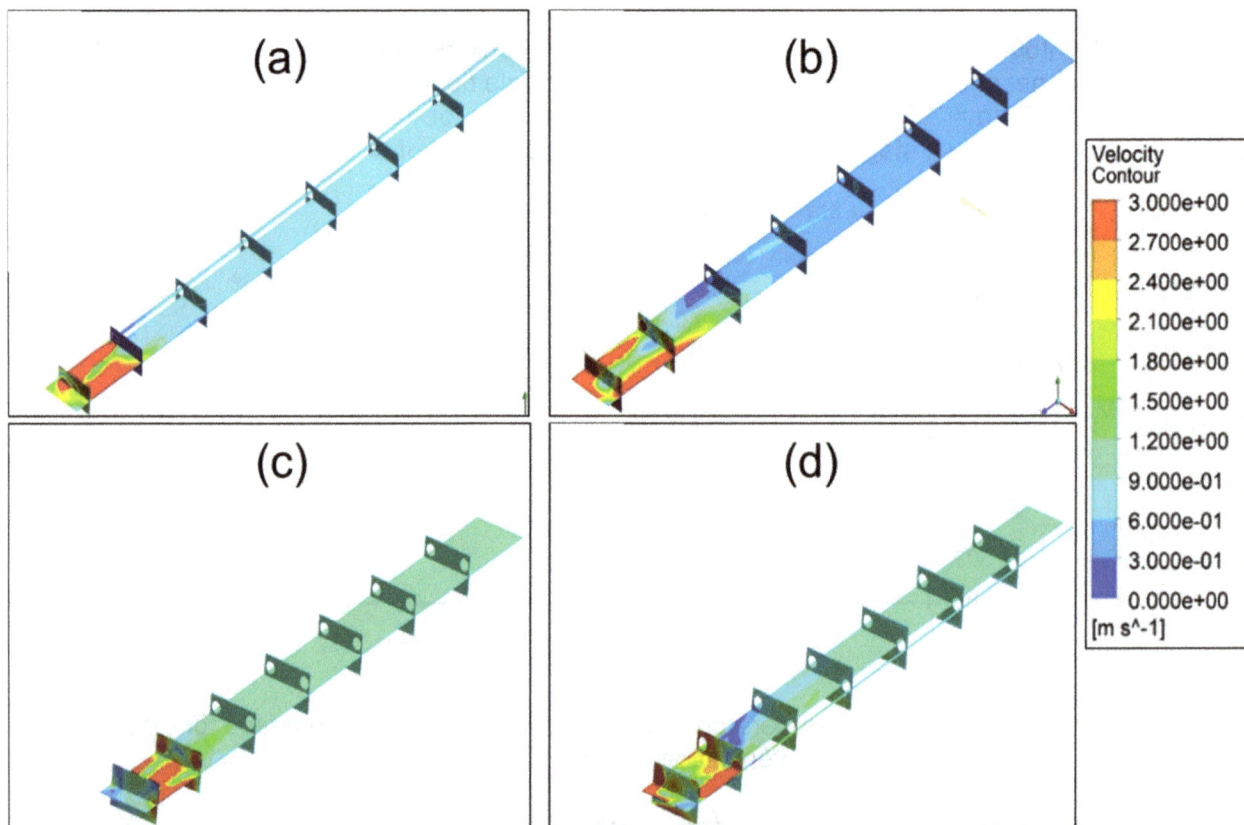

FIG 8 – Contour representation of air velocity for (a) single duct ventilation system for tunnel of cross-section 4.8 m × 2.8 m, (b) single duct ventilation system for tunnel of cross-section 5.5 m × 3.5 m, (c) twin duct ventilation system for tunnel of cross-section 5.5 m × 3.5 m, (d) twin duct staggered ventilation system for tunnel of cross-section 5.5 m × 3.5 m.

Effect of ventilation pattern on methane concentration in tunnel

The methane emission of the active roadway face has an average value of 0.0503 m³/s. Comparisons of methane concentrations for different ventilation systems (refer Figures 9 and 10) show that with increase in distance from the face, the methane concentration reduces. However, the reduction in methane concentration is highest for the twin duct staggered ventilation system. The average methane concentration is also found to be least inside the tunnel for such configuration. When compared to all the other ventilation configurations, the twin duct staggered pattern ventilation system has a blind zone limited only to the corner of the face. This zone can be controlled by reducing the setback distance of the duct from the face.

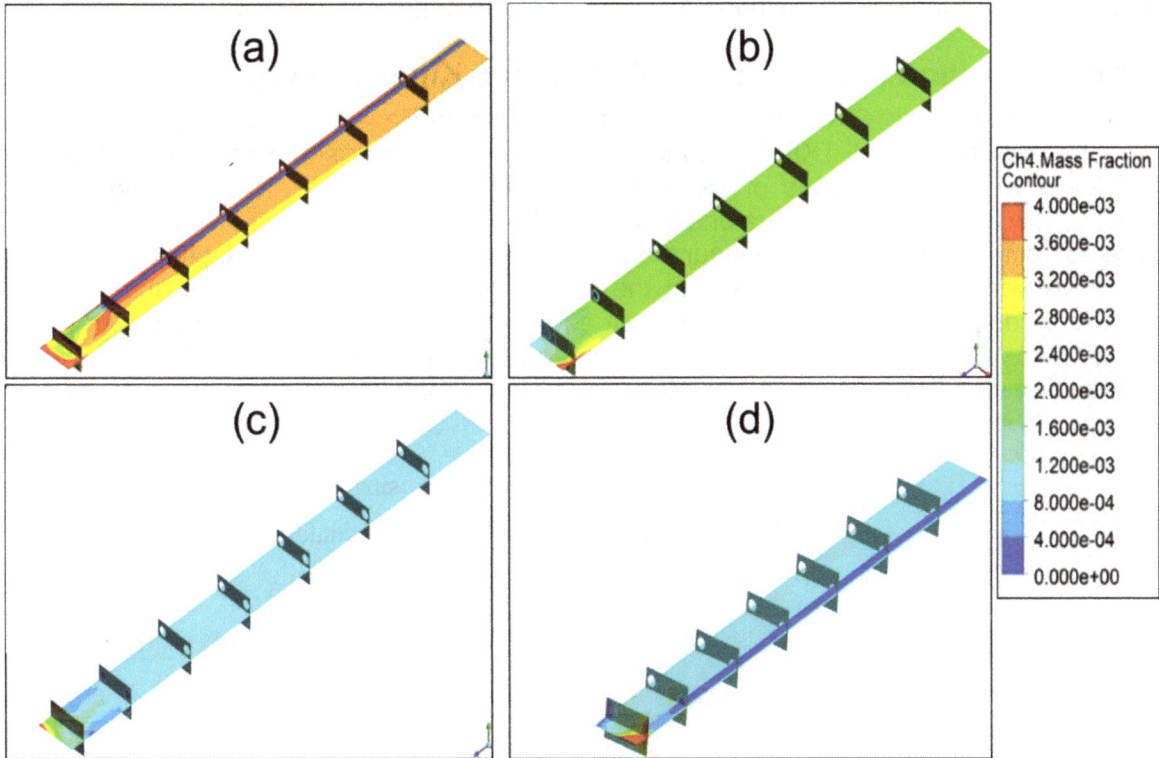

FIG 9 – Contour representation of methane concentration for (a) single duct ventilation system for tunnel of cross-section 4.8 m × 2.8 m, (b) single duct ventilation system for tunnel of cross-section 5.5 m × 3.5 m, (c) twin duct ventilation system for tunnel of cross-section 5.5 m × 3.5 m, (d) twin duct staggered ventilation system for tunnel of cross-section 5.5 m × 3.5 m.

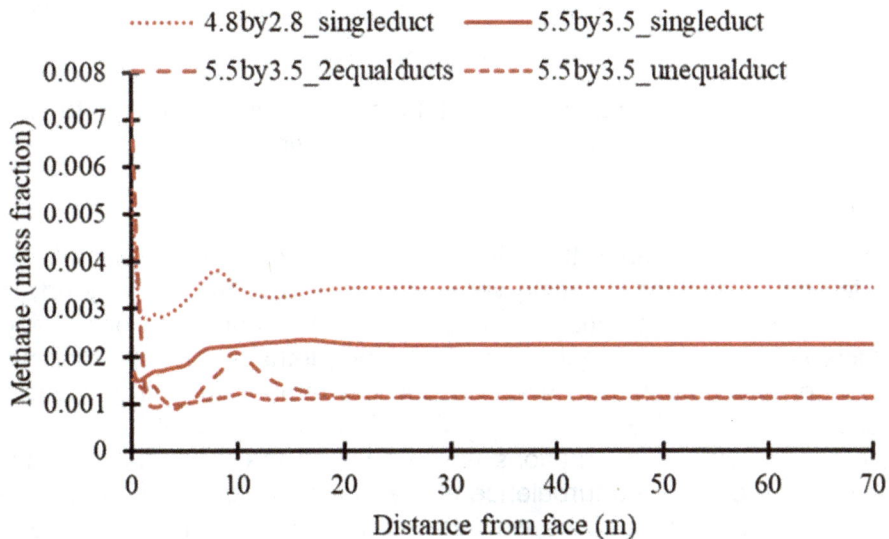

FIG 10 – Variation of methane concentration with distance from face for different ventilation configurations.

Figure 10 graphically shows the variation of methane concentration with distance from face for different ventilation scenarios. The tunnel of dimension 4.8 m × 2.8 m shows higher methane concentration due to its small cross-sectional area. For similar quantity of methane emission from a face of larger cross-section tunnel dilutes the methane when the air quantity increased from 550 m³/min (single duct configuration) to 1100 m³/min (twin duct configuration). But when the ventilation configuration is changed for the larger cross-section tunnel, the methane concentration further reduces due to the modification in position of ducts for ventilation purposes. Thus, clearly, the location of ventilation ducts and quantity of air supplied for ventilation purposes are important design factors to be considered while designing a ventilation system. It can also be deduced that, in this particular case, typical conventional design of ventilation systems in underground mines may not be able to sufficiently dilute methane emission from the larger mining face.

Effect of ventilation pattern on the Turbulence Kinetic Energy

Turbulence kinetic energy (TKE) is an important parameter while studying the methane dispersion and heat transfer in a tunnel. TKE defines the turbulence in the localised environment, and change in TKE leads to the phenomena of momentum transfer across the airflow resulting in species transport in flow domain (Rao, Mishra and Mishra, 2023). Figure 11 depicts the variation in TKE with distance for different ventilation configurations. For twin duct staggered ventilation system, the turbulence is very high. This will result in the higher dilution and dispersion of methane in the computational domain (refer Figure 10) as well as higher heat transfer as evident from Figure 5. Thus, twin duct ventilation system due to its characteristic of larger quantity of air injection along with the greater turbulence would result in better working environment in the working face as well as different points inside the tunnel.

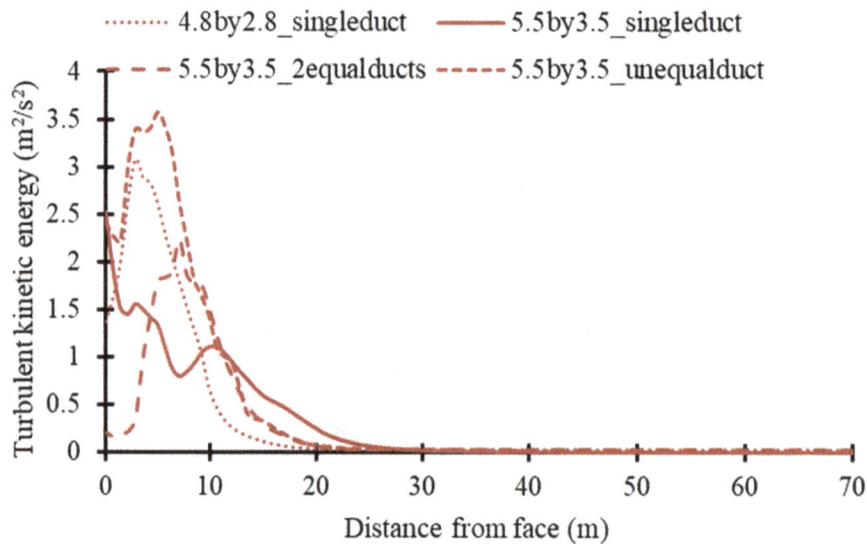

FIG 11 – Variation of turbulence kinetic energy at the centre of tunnel with distance from face for different ventilation configurations.

CONCLUSION

The larger mining machineries require the drivage of larger cross-section tunnel, which cannot be efficiently ventilated by conventional forcing ventilation methods. So, in this study, the novel twin duct staggered ventilation configuration is introduced. This configuration, owing to the duct placement and increased air quantity, not only reduced the methane concentration near the working area inside the tunnel environment, but also reduced the temperature at the face and inside the tunnel which is under the effect of geothermal gradient. Comparative analysis was carried out in this paper with respect to different configurations which showed that twin duct staggered ventilation configuration increases the localised turbulence in the tunnel, which when coupled with increased air quantity reduces the thermal load as well as gas load at the face. However, the wall heat coefficient near the face is also high when using twin duct staggered ventilation system, but the comparative increase in air temperature is low due to high cooling effect of fresh air injected at face by the novel ventilation configuration. Thus, for the well-being of the man and machinery at the larger cross-section working face, twin duct configuration would be a better alternative to the conventional ventilation method.

REFERENCES

ANSYS, Inc., 2013. ANSYS FLUENT Theory Guide, 14.5 edition, SAS IP, Inc., Canonsburg, PA.

Brodny, J and Tutak, M, 2018. Analysis of methane hazard conditions in mine headings, *Tehnicki Vjesnik*, 25:271–276.

Brune, J, 2013. The methane-air explosion hazard within coal mine gobs, *Transactions of the Society for Mining, Metallurgy and Exploration, Inc,* 334:1–16.

Chandra, D, 1992. Jharia Coalfield, Geological Society of India, Bangalore.

Kumar, P, Mishra, D P, Panigrahi, D C and Sahu, P, 2017. Numerical studies of ventilation effect on methane layering behaviour in underground coal mines, *Curr Sci*, 112:1873–1881. https://doi.org/10.18520/cs/v112/i09/1873-1881

Kurnia, J C, Sasmito, A P and Mujumdar, A, 2014. Simulation of a novel intermittent ventilation system for underground mines, *Tunneling and Underground Space Tech*, 42:206–215. https://doi.org/10.1016/j.tust.2014.03.009

Launder, B E and Spalding, D B, 1974. The numerical computation of turbulent flows, *Comput Methods Appl Mech Eng,* 3:269–289. https://doi.org/10.1016/0045-7825(74)90029-2

Mishra, D P, Kumar, P and Panigrahi, D C, 2016. Dispersion of methane in tailgate of a retreating longwall mine: a computational fluid dynamics study, *Environ Earth Sci*, 75. https://doi.org/10.1007/s12665-016-5319-9

Rao, S, Mishra, D P and Mishra, A, 2023. Methane migration and explosive fringe localisation in retreating longwall panel under varied ventilation scenarios: a numerical simulation approach, *Environ Sci Pollut Res*, 30:66705–66729. https://doi.org/10.1007/s11356-023-26959-6

Tutak, M and Brodny, J, 2015. Numerical analysis of airflow and methane emitted from the mine face in a blind dog heading, *Management Systems in Production Engineering*, 4:175–178. https://doi.org/10.12914/MSPE

Versteeg, H K and Malalasekera, W, 2007. *An Introduction to Computational Fluid Dynamics*, second edition (Pearson Education, UK).

Xin, S, Wang, W, Zhang, C, Li, C, Li, H and Yang, W, 2021. Effects of rock-airflow conjugated heat transfer in development headings: A numerical study, *International Journal of Thermal Sciences*, 172:107301. https://doi.org/10.1016/j.ijthermalsci.2021.107301.

A holistic approach to optimising primary fan system efficiency in mines

F Neff[1], M Francoeur[2] and B Spies[3]

1. Aerodynamic and Acoustic Chief Engineer, TLT-Turbo GmbH, Zweibrücken, Rhineland-Palatinate 66482, Germany. Email: f.neff@tlt-turbo.com
2. Mine Ventilation Specialist, Sudbury, Ontario P3C 3K3, Canada. Email: francoeur.myriam@gmail.com
3. Senior Application Engineer – Mining and Wind Tunnel Fan Systems, TLT-Turbo GmbH, Zweibrücken, Rhineland-Palatinate 66482, Germany. Email: b.spies@tlt-turbo.com

ABSTRACT

There is often a mistaken belief that primary fan system efficiencies should align with the fan's aerodynamic efficiency, typically over 80 per cent. Bowling, Schult and Van Diest (2023) demonstrated via field surveys that primary fan installation efficiencies are lower than 65 per cent to 75 per cent, as commonly assumed, especially in hard-rock mines. They argue that these lower-than-expected efficiencies stem from poor installation, inadequate maintenance practices, and fan systems operating outside their peak performance range. They also pointed out that primary fan installation efficiencies depend on the fan itself and various aerodynamic, mechanical, and electrical losses across the system.

However, these reported lower efficiencies may also arise from the balance of factors like project cost, scope and duration of the system design, information availability, expansion of the ventilation network beyond the initial mine planning, and discrepancies between fan system output against the mine total resistance.

In these proceedings, the authors present reference projects and examples demonstrating that strictly selecting capital cost-driven installations might not prove the most cost-effective over the primary fan system's service life and the life-of-mine; this applies especially if energy costs remain elevated or even increase. Achieving and maintaining high overall system efficiency is possible, even as the mine expands and the ventilation requirements evolve, by working with aerodynamically optimised components, smartly selecting fan sizes and speeds, and optimising life-of-mine energy efficiency. The authors also emphasise the need for a good compromise between the primary ventilation system's design, capital costs, and power consumption throughout its operating life: better-optimised fan systems can contribute to underground operations beyond the initially expected life-of-mine with little upgrade.

INTRODUCTION

Ventilation is often said to make up, on average, between 25 per cent and 40 per cent of an underground mine power budget (De Souza, 2018). As a result, mine ventilation fan systems have generally been optimised to reduce power consumption, all to reduce operational expenditure (OPEX) due to increasing electricity rates. Table 1 (Ritchie and Rosado, 2024; Government Offices of Sweden, 2023; Puyo and Zhunussova, 2022; Qu et al, 2023; Government of Northwest Territories, 2023; Mining Association of British Columbia, 2024) provides an overview of the average large-industry electricity rates for underground mining-intensive countries.

TABLE 1

Average industrial electricity rates and equivalent tons of CO_2 emissions for countries with intensive underground mining.

Country	Average large-industry electricity rates (USD/kWh)	Equivalent grams of CO_2 per kWh of electricity	Primary electricity generation method	Secondary electricity generation method	Average carbon tax (USD/t)
Mexico	0.115 (2022)	424	Fossil fuels (74%)	Renewables (23%)	N/A
Sweden	0.120 (2022)	45	Renewables (68%)	Nuclear (30%)	133
Australia (excl. WA)	0.119 (2022)	502	Fossil fuels (67%)	Renewables (33%)	N/A
Australia (WA only)	0.218 (2022)				
Chile	0.115 (2022)	333	Renewables (53%)	Fossil fuels (43%)	5
South Africa	0.065 (2022)	708	Fossil fuels (86%)	Renewables (9%)	10
Brazil	0.154 (2022)	102	Renewables (87%)	Fossil fuels (11%)	N/A
Canada (provinces)	0.093 (2022)	126	Renewables (70%)	Fossil fuels (17%)	25
Canada (territories)	0.330 (2021)	805	Fossil fuels (100%)	N/A	59
United States	0.085 (2022)	368	Fossil fuels (60%)	Renewables (22%)	N/A
World	**N/A**	**438**	**Fossil fuels (61%)**	**Renewables (30%)**	**N/A**

In Table 1, the highest electricity rate is associated with the Canadian territories, located north of the 60th parallel, whose power is predominantly derived from diesel generators.

If electricity rates have been a significant driver in reducing mine ventilation power consumption, eyes are now turning to fossil fuel dependency and greenhouse gas (GHG) emissions for both on- and off-grid operations. Table 1 indicates the weight of CO_2 equivalent emissions per kWh and the two standard electricity generation methods per country. Fossil fuels indicate coal, natural gas, and diesel as power sources; renewables refer to hydro, wind, solar, or alternative means of power production. One notices that about 60 per cent of all electricity for large industrial consumers is still produced worldwide through fossil fuels. While the mining industry is turning towards alternative renewable sources of energy such as solar power fields (Gold Field South Deep, South Africa – Reuters, 2022), wind turbines (Glencore Raglan, Canada – Glencore Canada, 2018), and, potentially, small nuclear modular reactors (Natural Resources Canada, 2021), it still heavily relies on fossil fuel-based power generation, especially for remote, off-grid operations.

However, primary fan installations are often less efficient than anticipated, leading to a gross underestimation of their realistic power consumption and, therefore, their operating cost and impact on a mine's carbon footprint. Bowling, Schult and Van Diest (2023) reported that the average fan system efficiency is about 50 per cent, much less than the assumed 65 per cent to 75 per cent based on the presents of modelling software VUMA-network, version 5.0 (by BBE Group) (65 per cent), VNet, version 1.4 (by SRK Consulting) (65 per cent), or Ventsim®, version 5.4 (by Howden) (75 per cent). Bowling *et al* argue that their findings are likely attributable to poor fan system design, inadequate maintenance practices, and fans operating outside their peak performance range. Yet, these lower-than-expected efficiencies may also arise from other design-based issues, including project cost, scope and allotted system design period, information availability, expansion of the ventilation network system beyond the initial mine planning with disregard to the fan system's output, and discrepancies between the fan installation output against the mine total resistance.

In these proceedings, the authors present examples demonstrating that strictly selecting capital expenditure (CAPEX)-driven installation does not translate into power savings and GHG emission reduction over a primary fan system's service life and/or life-of-mine, specifically during production periods. Achieving and maintaining high overall system efficiencies is possible despite a mine's unforeseen expansion and evolution of its ventilation requirements by adopting aerodynamically optimised components, adequately sizing fans, and maximising life-of-mine system efficiency. The authors also emphasise the need for compromising between the primary fan installation engineering, CAPEX, and power consumption for every stage of its service life to achieve profitable underground operations beyond the initially expected life-of-mine with little upgrades.

RELEVANT DEFINITIONS FOR PRIMARY FAN SYSTEMS

Fan efficiency and system efficiency are often confused and worth specifying clearly. Both definitions represent the ratio between an actual fluid (air hereafter) output power and the electrical (input) power consumed.

The primary fan system, illustrated in Figure 1, is comprised of the following:

- The fans.

- The associated duct sections that connect the fan to the mine. In the case of a surface mine ventilation system, this refers to all ductwork from the shaft connection to the outlet or inlet of the duct section.

- All electrical devices that are required to operate the system.

Direction of airflow

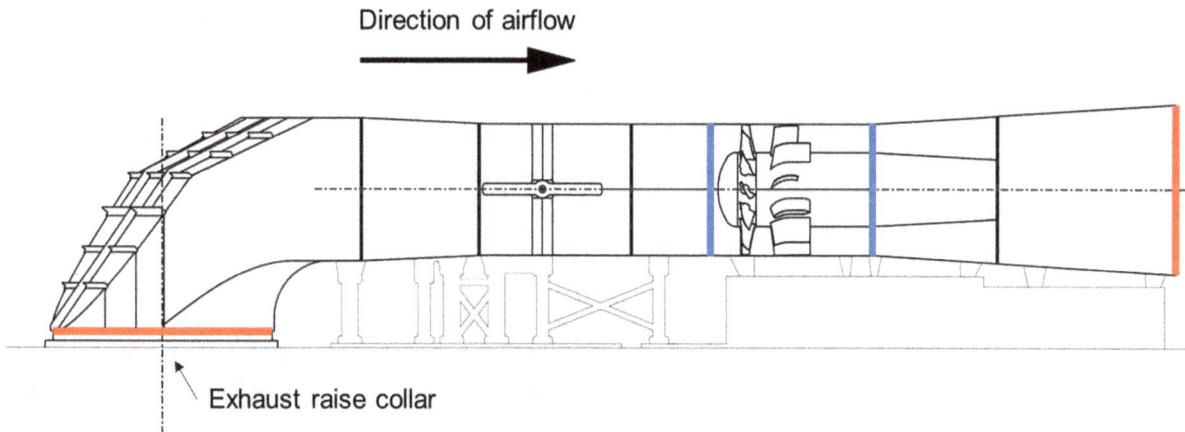

Exhaust raise collar

FIG 1 – Illustration of a surface primary fan system.

Readers should bear in mind that the definition of the fan boundaries (see the blue boundaries in Figure 1) may vary depending on the fan supplier. For example, the 'boundaries' could be the joint planes (flange to flange) of the fan casing. Alternatively, they could simply refer to the core components of the fan–for example, in the case of an axial fan, the fan blades in conjunction with the guide vanes, or in the case of centrifugal fans, the impeller together with the casing volute.

The fan efficiency η_f (Equation 1) is defined as the ratio between the air power supplied at the fan boundaries, $P_{air,f}$, and the power absorbed by the fan, P_{shaft}.

$$\eta_f = \frac{P_{air,f}}{P_{shaft}} \tag{1}$$

The air power (Equation 2) is the product of the mass flow $\dot{m}$ and the total specific work Y_f at the fan boundaries.

$$P_{air,f} = \dot{m} \cdot Y_f \tag{2}$$

The shaft power is calculated as per Equation 3.

$$P_{shaft} = 2 \cdot \pi \cdot T \cdot n, \tag{3}$$

where T is the torque and n is the fan rotational speed.

Alternatively, the shaft power can be given as the product of the electrical power consumed by the fan, P_{el}, to which the efficiency of the motor η_{mo} and, if applicable, the efficiency of the frequency converter η_{fc} are applied (Equation 4).

$$P_{shaft} = P_{el} \cdot \eta_{mo} \cdot \eta_{fc} \tag{4}$$

The definition of fan system efficiency, η_s, aligns with that of fan efficiency, yet it involves distinct boundary definitions. In this case, it is given as the ratio between the air power within the system and the electrical power required to operate it (Equation 5).

$$\eta_s = \frac{P_{air,s}}{P_{el}} \tag{5}$$

$P_{air,s}$ is determined like Equation 2, albeit incorporating values specific to the system's boundaries (Figure 1, red boundaries). The system's total specific work, therefore, includes all internal aerodynamic losses of the individual components composing a system.

As for the electrical power, it now encompasses the aggregate power essential for the operation of the fan system, which accounts for drive motor and frequency converter losses as well as auxiliary systems such as lubrication within the set-up.

Specific work

While specific work is rarely referred to in mine ventilation, it is worth defining it, as it is necessary for the correct calculation of the fan performance (see Equation 1). Especially if the total pressure rise of the fan is rather high ($dp_{tot,f} > 3000 \, Pa$). There are several definitions for specific work. As many fan manufacturers test their fans — or at least models of them — aerodynamically as per ISO 5801:2017, the authors provide it in Equation 6, where p represents the static pressure; ρ, the air density; f_m, the Mach factor, and v, the velocity. Applicable for both fan and fan systems, only the values at the fan inlet (subscript 1) and outlet (subscript 2) or, conversely, as the system's inlet and outlet are to be used.

$$Y = \frac{p_2 - p_1}{\frac{\rho_1 + \rho_2}{2}} + \frac{1}{2} \cdot f_{m2} \cdot v_2{}^2 - \frac{1}{2} \cdot f_{m1} \cdot v_1{}^2 \tag{6}$$

SYSTEM OPTIMISATION

Each component of a fan installation regulates the latter's system efficiency. Typical components of an extracting surface fan system made of two fans include a bifurcated head bend, tapered transitions, isolation doors, straight ducting, and diffusers.

Most of the abovementioned components can be optimised individually. While the authors discuss each item composing an extraction system independently, readers should bear in mind that the effect of an upstream element can influence the performance of all other downstream, ie affect their pressure losses.

Head bend

The head bend, often composed of several outlet branches, redirects the vertical airflow of the raise by 90°. It is available in various configurations, including the miter or lobster-back bend for single-branch systems and the Eschenberg or combo head bend for multiple-branch systems. A plenum arrangement is sometimes used as well.

In all cases, the 90° deflection inevitably leads to pressure losses. Depending on the design (regardless of the type), two further negative effects can occur: flow separation and swirl. In poor designs, this flow deflection is not uniform, so larger flow separations can occur. These naturally cause higher pressure losses and – even worse – propagate into the next component and consequently increase the actual pressure loss of the subsequent components due to local overspeed. The swirl, describing the air velocity components in the circumferential direction, also needs to be addressed. If a significant swirl component is present, this also generally leads to increased pressure losses in the downstream components. At the same time, swirl affects the characteristic curves of the fans. Depending on the direction of the swirl, the ability of the fan to provide the necessary pressure increase is reduced. A reduction in efficiency is to be expected in any case.

Hence, a careful head bend design can prevent flow separation and minimise swirl.

Transitions

Transition pieces that taper the cross-section, ie accelerate the airflow, have a minor aerodynamic effect. However, extreme tapering induces negative effects, such as local overspeed and flow separation. Greater attention is required in the case of diffusers, which are expanding transition components (described in the Diffuser section).

Isolation door

Isolation doors are critical components for maintenance and emergency events as they provide the means to separate and seal off either a fan or ductwork from the primary airflow. Generally, the mining industry uses single-plate, butterfly, or slide gate (guillotine) isolation doors. Their effectiveness and cost depend on their robustness, fast-acting mechanism, and aerodynamic profile in the open state; if isolation door parts sit in the airstream, they must offer the smallest cross-section possible to the airflow to reduce flow separation and swirling.

Fan

At the heart of the system, the fan moves the air and produces a sufficient pressure rise to overcome all losses in the mine network and fan installation. Its performance is primarily determined by the fan affinity laws (Equations 7 to 10). They account for the influences of the fan diameter D and speed n on volume flow Q, pressure rise Δp, and power P. Taking the air density into account as well, the affinity laws are as follows:

$$Q \sim D^3\, n \tag{7}$$

$$\Delta p \sim \rho\, D^2\, n^2 \tag{8}$$

$$Y \sim D^2\, n^2 \tag{9}$$

$$P \sim \rho\, D^5\, n^3 \tag{10}$$

Still, additional factors come into play when determining a fan's performance, including:

1. Blade or impeller design: Every fan manufacturer typically offers a selection of different blade profiles or impeller types. Achieving an efficient fan design involves choosing a blade profile or impeller design tailored to the operating points. The maximum achievable pressure increase, for instance, can be influenced by the blade type or impeller design, which is also affected by the number of installed blades. Furthermore, the blade or impeller type influences the steepness of a fan's characteristic curves, thus, the control of the fan(s). Different blade types can also be adopted.

2. Diameter ratios: Another way to adapt the fan to operating points is to select the appropriate diameter ratio. For an axial fan, the hub-to-tip diameter ratio significantly influences the fan's pressure rise. In the case of a radial fan, the impeller's diameter ratio (outer diameter to inlet diameter) also affects the fan's performance. However, larger hub-to-tip ratios in axial fans result in higher meridional velocities. Under dust-laden conditions, the kinetic energy of the dust particles is therefore higher, which leads to increased wear on the blade tips.

3. Control mechanisms: The type of fan control influences a fan's efficiency and, therefore, the system efficiency at a given duty point. Several control mechanisms can be implemented to regulate speed, vanes, or blade pitch (for axial fans). Such control mechanisms can also be combined in a single fan system. Each control broadens a fan's characteristic map, allowing better efficiencies for a given duty point. Yet, their implementation increases the system cost,

each with dimensional, environmental, or maintenance limitations. Mechanical restrictions may also apply.

4. <u>Adaptable blade configurations:</u> Fans can be tailored to adapt to evolving mine resistance and airflow requirements. Generally designed in the earliest stages of a mine with milestones bound to change, fan duties seldom match development and ramp-up requirements. Retrofitting additional blades or replacing them with another type can solve such an issue.

Nevertheless, fans are subject to limitations and/or boundary conditions that must be accounted for and can influence their performance and, therefore, the system's efficiency. Restrictions that should not be neglected are:

1. <u>Mechanical limitations:</u> A blade or impeller type significantly impacts the rotor's mechanical load. Their designs and materials influence their weight and stress tolerance, translating into maximum permissible stresses. The most suitable blade for a given system may generate large centrifugal forces and mechanical loads on the motor.

2. <u>Wear:</u> The blades of fans operating in dust-laden conditions are subject to abrasion and wear, which is more accurate for axial fans. If nowadays fans can handle high blade tip speeds of up to 160 m/s, greater blade tip speeds are linked to wearing vulnerability. In such conditions, selecting a larger fan can help reduce blade tip speeds and mitigate wear. Selecting blade materials and/or coatings can improve further abrasion resistance, although at the expense of greater blade weight, thus increasing forces on the rotor and bearings.

3. <u>Operational stability:</u> Fan operational stability is paramount during start-up, especially for systems with two or more branches. Fan stall characteristics depend on the blade or impeller type and pitch (for axial fans) and determine if fans can enter a stall regime upon start-up. Anti-stall units or anti-stall rings are countermeasures that can be added, albeit at the expense of efficiency.

Diffuser

Diffusers or cross-section-changing, expanding transition pieces are components with the purpose of converting kinetic energy into static pressure and reducing velocity (pressure recovery). A reduction in the air velocity brings down losses in the downstream ductwork or at the discharge, depending on its location within the fan system. However, the increasing pressure in the diffuser results in another phenomenon. If the area expansion in the diffuser becomes too large, the flow separates. The flow no longer follows the wall boundary. This means that there is no further pressure recovery, ie the diffuser loses its original function. Therefore, flow separation should be avoided to achieve the desired diffuser efficiency.

Numerous fluid mechanical parameters, such as the pressure recovery coefficient, the diffuser efficiency, and the loss coefficient – by far the most important parameter – describe a diffuser's performance. Diffusers are also qualitatively assessed via the diffuser stability diagram.

Other ancillary components

Other components, such as safety screens, silencers, mist eliminators, deflectors, and bends, are ancillary components that cause additional pressure losses in a fan system and reduce its efficiency. Ideally, their integration into a fan system should warrant a reassessment of the design and aerodynamic interactions of all components to reduce the propagation of flow separation, swirl, and turbulence. However, this could prove both time-consuming and cost-intensive.

Interaction of all components

If fan components can be individually tailored to maximise their efficiency, a similar exercise must be carried out with all components interacting within the fan system. For example, as previously mentioned, a head bend in a surface exhaust system can induce flow separation and swirl that propagate throughout the downstream ductwork and up to the fan, producing additional pressure losses and impeding the system's efficiency. Guide vanes and straighteners can be added to given components to improve the homogeneity of the overall flow profile, even if they locally induce pressure losses.

ENERGY OPTIMISATION AND COST ANALYSIS

The repercussions of the aerodynamic optimisation of a primary fan installation can be better understood in the following case study, in which the authors perform a techno-economic analysis of a bifurcated surface exhaust fan system designed for a block cave mine. At its peak, the complete installation must extract 475 m³/s of vitiated air at standard air density and an average temperature of 10°C through a 5.5 m raise. The predetermined static pressure at the shaft's collar is 3200 Pa.

Here, the authors have investigated the performance and total cost of ownership of four different designs, from the most optimal to an inferior aerodynamic configuration, which are described below. Component and pressure losses are detailed in Table 2, where the most significant losses are italicised.

- Optimal system: All fan system components interact ideally with each other, ensuring maximum airflow homogeneity and swirl prevention. The fan diameter has been selected to minimise air velocity and, thus, pressure losses. A bifurcated head bend sits atop the exhaust raise, and a long diffuser helps recover static pressure.

- Optimal system with a smaller fan diameter: Once again, all system components have been selected to ensure airflow uniformity and swirl prevention. The fan diameter has been stepped down to reduce CAPEX, but this has come at the expense of a higher air velocity, thus greater pressure losses.

- Suboptimal system with the same fan diameter as Case 1: This time, system components interact poorly with one another, airflow homogeneity is lost, and swirl most likely occurs. A plenum chamber, rather than a head bend, sits on top of the exhaust raise before the air is channelled into either fan, and the isolation door pressure loss increases due to flow inhomogeneity. The shorter diffuser does not provide as much static pressure recovery as in Case 1. However, this design reduces component and engineering costs for a lower CAPEX.

- Suboptimal system with a smaller fan diameter: Finally, this installation follows the same design strategies as Case 3 and uses a smaller fan diameter to further reduce CAPEX.

TABLE 2

Fan system pressure loss per component (in Pa) per scenario.

Case	Fan diameter (hub diameter) (mm)	Head bend (plenum chamber)	Trans-ition	Isola-tion door	Inlet duct-ing	Anti-stall unit	1st diffu-ser	Carnot loss	2nd diffu-ser	Flow deflec-tor	Exit loss
1	3162 (1778)	*38*	7	*65*	5	35	91	8	5	38	*207*
2	2986 (1778)	*38*	11	*82*	7	49	157	12	7	45	*247*
3	3162 (1778)	*(118)*	3	*196*	5	35	108	13	N/A	67	*369*
4	2986 (1778)	*(119)*	11	*247*	7	49	156	12	N/A	82	*452*

Table 3 provides an overview of the four investigated fan systems, including fan efficiency and system efficiency, as well as the required line power to run at the predetermined duty point. The conversion from shaft to line power is assumed to be 95 per cent efficient and is reflected in the total system efficiency. The CAPEX per fan installation, which does not include any variable-speed drives, is valid as of December 2023. One may notice that while the fan efficiency remains overall the same in all four scenarios (88 per cent), the total system efficiency declines with increasing pressures or line power. Nonetheless, if the system's efficiency decreases, so does the CAPEX. All selected fans run 24/7 at a fixed rotational speed of 890 rev/min on a 60 Hz system; however, it is possible to improve a system's efficiency by a few percent by adjusting the fan speed.

TABLE 3
Investigated fan system performance and cost.

Case	Fan diameter (hub diameter) (mm)	Fan efficiency (%)	Total pressure losses (Pa)	Shaft power (kW)	Total system efficiency (%)	Line power (kW)	CAPEX (M USD)
1	3162 (1778)	88.0	504	918	71.6	966	5.2
2	2986 (1778)	88.0	660	959	68.6	1010	4.9
3	3162 (1778)	88.0	888	1026	65.2	1080	4.9
4	2986 (1778)	88.0	1143	1078	63.9	1135	4.6

The selection of a suboptimal aerodynamic fan system configuration with smaller CAPEX bears long-term financial consequences. To demonstrate the significance of an adequate selection, the authors have compared all scenarios for a period of ten years, which encompasses the average production schedule of an underground mine. Each installation's net present value (NPV) was computed with an 8 per cent discount rate. Figure 2 shows the total cost of ownership (TCO), based on the line powers of Table 3, determined for South Africa (ZA), Western Australia (WA), Australia (AU, excluding WA), and Canadian provinces (CA) and using their respective average energy rate indicated in Table 1.

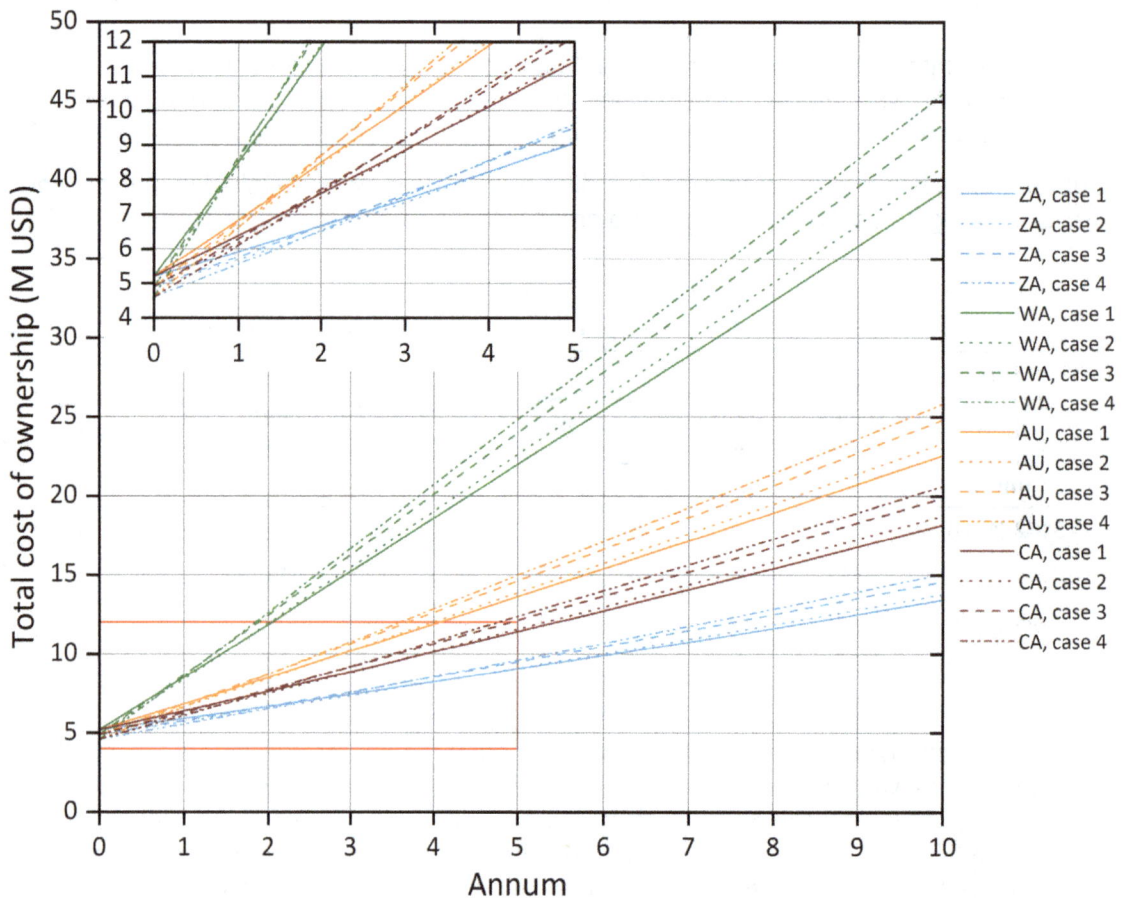

FIG 2 – Scenarios 1 to 4 TCO for various underground mining-intensive countries.

If the first year's TCO favours the selection of lower CAPEX fan systems, readers are quick to realise that the return on investment (ROI) from selecting the most optimal installation (Case 1) always yields the lowest TCO within the mining production window. Additionally, the higher the energy rate, the faster the return on investment is achieved compared to a suboptimal aerodynamic design. For WA, the ROI between Cases 1 and 4 occurs within the first ten months of operations as a result of its significant energy cost, whereas it takes up to 28 months to achieve the ROI between the same scenarios with South African energy rates. Selecting aerodynamically optimised systems also aligns with the mining industry's energy consumption reduction objectives.

Nevertheless, the above analysis assumes that a fan system becomes operational as the underground mine reaches its production period. The authors' case study does not incorporate the early development and production ramp-up stages, each with its specific airflow requirement and corresponding mine resistance, nor does it encompass the installation's TCO throughout the life of the mine. A more accurate analysis should include an entire mine schedule for each milestone and compare the total cost of ownership of different designs for various fan duty points.

Fan system operating carbon footprint

With tighter environmental governance, one may need to examine the CO_2 equivalent emissions linked to operating a fan system. The authors have computed the annual carbon equivalent footprint of all previous scenarios on the CO_2 equivalent emission rates of Table 1 for South Africa, Australia (including WA), and Canada and plotted them in Figure 3. As expected, as fan system efficiency decreases, its carbon footprint increases by as much as 16 per cent if comparing Cases 1 and 4.

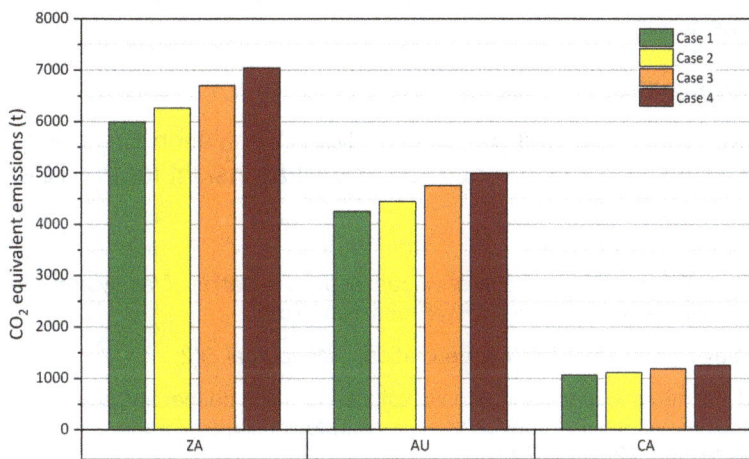

FIG 3 – CO_2 emissions per annum for scenarios 1 to 4 for South Africa (ZA), Australia (AU), and Canada (CA).

Depending on the mining jurisdiction, this increase in carbon emissions can translate into an additional tax penalty. As of May 2023, in South Africa, where a 10 USD/t carbon tax is applied, this would correspond to an added 9540 USD per annum; in Canada, with a provincial average carbon tax of 25 USD/t, the penalty represents about 4250 USD annually. While countries such as Australia or the USA have yet to adopt a carbon tax, the newest policies seem to push it forward. Additionally, carbon taxes are steadily increasing in jurisdictions already applying it: for example, Sweden's industry carbon tax, introduced in 1991, has gone from 7 USD/t to 133 USD/t in 2023 (Government Offices of Sweden, 2023). The Northwest Territories, a remote area in Canada heavily relying on fossil fuel for power generation and with the highest energy rate in Table 1, started applying a carbon tax in 2019; set at 48 USD/t in 2023, it will continue to grow by 11 USD/t per annum until April 2030, where it will reach 126 USD/t (Government of Northwest Territories, 2023).

If current carbon taxes translate into negligible tax penalties at the time of submitting this manuscript (less than 1 per cent of a fan system's annual operating cost), they will undoubtedly become more significant. Additionally, many countries, such as Canada (Deloitte, 2024), offer tax credits or funding to companies that present decarbonisation strategies, including energy reduction.

CONCLUSIONS

To conclude, the authors emphasise the critical importance of selecting fan installation designs that provide the highest system efficiency through the minimisation of pressure losses per component and inhomogeneities in the flow field. It is, therefore, important to consider the interaction between the individual components. The repercussions of aerodynamic optimisation significantly impact the total cost of ownership (TCO) over an underground mine production period, and opting for low-CAPEX fan system designs, albeit initially cost-effective, may result in higher TCO because of increased energy consumption. Considering the global trend of rising electricity prices and increasing carbon taxes, the payback period is likely to become shorter.

The case study in this manuscript presented fan system efficiencies on the higher end of the spectrum; even the worst-case scenario had a nearly 64 per cent total efficiency, which seems to contradict Bowling, Schult and Van Diest's (2023) findings. However, readers must keep in mind that fan vendors present designs at optimal duties; yet an active mine resistance often hardly matches that determined during feasibility or more advanced studies. As a result, fan installation requirements defined years ago on paper may operate poorly once production begins. The authors' previous suggestion of adding vane and/or blade pitch controls or additional blades to complement speed controls could remediate this unwanted scenario.

Finally, aerodynamically optimised fan systems align with the mining industry's objectives of decarbonisation by reducing underground mine energy consumption. Thus, the authors suggest creating and adopting standardised mine ventilation equipment bidding guidelines for ventilation professionals and procurement personnel alike. Such guidelines would streamline procurement procedures and ensure the selection of optimal ventilation solutions with maximum system efficiency and reduced environmental impact.

ACKNOWLEDGEMENTS

The authors are grateful to John Bowling (Principal Mine Ventilation Engineer, SRK Consulting) for providing clarification about his 2023 contribution as well as insight for the TCO analysis.

REFERENCES

Bowling, J, Schult, G and Van Diest, J, 2023. *Practical values for the evaluation of fan system efficiencies in Underground Ventilation* (ed: P Tukkaraja), pp 353–360 (CRC Press: New York).

De Souza, E, 2018. Cost Saving Strategies in Mine Ventilation, *CIM Journal*, 9(2).

Deloitte, 2024. Grants and incentives to reach net-zero emissions – Looking deeper and seizing the opportunities, Available from: <https://www2.deloitte.com/ca/en/pages/strategy/solutions/grants-and-incentives-to-reach-net-zero-emissions.html> [Accessed: 12 March 2024].

Glencore Canada, 2018. Raglan Mine Operates its Second Wind Turbine, Available from: <https://www.glencore.ca/en/media-and-insights/insights/raglan-mine-operates-its-second-wind-turbine> [Accessed: 12 March 2024].

Government of Northwest Territories, 2023. Carbon Tax. Available from: <https://www.fin.gov.nt.ca/en/services/carbon-tax> [Accessed: 12 March 2024].

Government Offices of Sweden, 2023. Sweden's carbon tax. Available from: <https://www.government.se/government-policy/swedens-carbon-tax/swedens-carbon-tax/> [Accessed: 12 March 2024].

Mining Association of British Columbia (MABC), 2024. Carbon pricing. Available from: <https://mining.bc.ca/carbon-pricing/> [Accessed: 12 March 2024].

Natural Resources Canada, 2021. Small Modular Reactors (SMRs) for Mining. Available from: <https://natural-resources.canada.ca/our-natural-resources/energy-sources-distribution/nuclear-energy-uranium/canadas-small-nuclear-reactor-action-plan/small-modular-reactors-smrs-for-mining/22698> [Accessed: 12 March 2024].

Puyo, D M and Zhunussova, K, 2022. Chile: An Evaluation of Improved Green Tax Options, 25 p (International Monetary Fund).

Qu, H, Suphachalasai, S, Thube, S and Walker, S, 2023. South African Carbon Pricing and Climate Mitigation Policy, 5 p (International Monetary Fund).

Reuters, 2022. South Africa's Gold Fields bets on solar to cut costs and carbon. Available from: <https://www.reuters.com/business/sustainable-business/south-africas-gold-fields-bets-solar-cut-costs-carbon-2022-10-13/> [Accessed: 12 March 2024].

Ritchie, H and Rosado, P, 2024. Electricity Mix. Available from: <https://ourworldindata.org/electricity-mix> [Accessed: 12 March 2024].

Jet fan control methodology for overcoming piston effects in underground mining

J Pont[1], F C D Michelin[2] and C M Stewart[3]

1. Mine Ventilation Engineer, Howden Ventsim, Brisbane Qld 4000.
   Email: jason.pont@howden.com
2. Manager, Howden Ventsim, Brisbane Qld 4000. Email: florian.michelin@howden.com
3. Principal Engineer, Minware, Cleveland Qld 4163. Email: craig@minware.com.au

ABSTRACT

Most current mining methods require bulk movement of material to be economically viable. This is often achieved by machines with the largest dimensions possible moving through tunnels with the smallest possible dimensions possible creating a 'piston effect' where air is pushed in front of the machine. This can create undesirable ventilation effects such as forcing contaminants into the fresh air source that are generally mitigated by overcompensating with excess airflow rather than trying to control the piston effect.

Ventilation controls for directing airflow in underground mining such as drop board regulators, walls or louvres all involve a physical barrier that restricts vehicle or personnel movement. In cases where travel is required, single doors can permit vehicle traffic with temporary airflow leakage while double air-lock doors can maintain the ventilation circuit but further slow traffic movement. A solution to this problem may be jet fan control, which is extensively used in road tunnels and some coalmines but rarely used in underground metalliferous mining where large openings and low pressure parallel airways are less common. In road tunnels, jet fans counteract or complement the piston effect of traffic moving through the tunnels to direct suitable airflow without restricting vehicle movement. Where significant piston effect-related changes to airflow can occur, the primary ventilation may be adequate when the machinery is stationary but may not have the ability to compensate for piston effects when it is moving.

This paper will describe a proposed control system algorithm for jet fans in a mining application where the piston effect of mining equipment moving in a tunnel must be overcome. The jet fans are assumed fitted with a variable frequency drive (VFD) operated by a control system. Airflow sensors located in the drive provide feedback to the control system. There is added complexity in this case as an automated Ventilation Control Device (VCD) in each drive controls primary airflow. The algorithm needs to control the jet fans without inefficiently working against the VCD. The response of the jet fan and the piston effect of any vehicle travelling in the extraction drive will add to the feedback within the control loop.

The goal of the research is to simulate a control system algorithm that could be applied to jet fan operation in an underground mining operation with low resistance parallel circuits, such as is common in block caving. The control system is tested using control simulation software to determine the optimum parameters for the algorithm. The simulation considers the piston effect of heavy machinery and primary ventilation controls such as regulators and primary fans. The results of the simulation show that the algorithm can be used to achieve optimum airflow.

INTRODUCTION

Jet fan use underground has historically been more common in civil tunnelling applications and coalmine where large openings and parallel low pressure openings are common (Stewart, 2023) (Stewart, 2023) and have not been generally considered for ventilation control in underground metalliferous mines. In most underground mining methods, jet fans are not required as either primary ventilation or ducted secondary ventilation provides adequate airflow. There are however potential applications for jet fans where ventilation must be controlled and balanced across multiple parallel headings such as in block caving where multiple parallel extraction drives must be provided with flow-through ventilation.

Where multiple parallel drives are present, the airflow in each drive is likely to vary between each drive. Different airway resistances will be caused by unequal distances to intake or exhaust paths, different wall roughness and different airway dimensions. To ensure the minimum required amount of airflow is delivered to parallel drives with unequal resistances, excess ventilation will be needed in some drives in order to meet the minimum requirement in others. The primary airflow required for an extraction level may therefore be much larger than the sum of minimum requirements for each drive.

The 'piston effect' of vehicles moving through drives can also alter airflow, pushing air in the direction of movement. Depending on the location of this effect it can promote or retard local flows. This effect is greater when the vehicle dimensions are closer to the dimensions of the tunnel, such as a loader moving along an extraction drive or a haul truck in a haulage drive. Due to the interaction of multiple loaders moving through parallel airways on a level, it may even be possible for air to be pushed back into the fresh air intake contaminating other drives with dust and diesel exhaust.

Ventilation controls such as drop-board regulators and manual louvres can be used to ensure the resistance in every drive is equalised and similar airflow delivered, however, these ventilation controls do not allow vehicle traffic and while a fast responding set of automatic louvres may be suitable to control piston effects, manually altered ventilation controls would not be suitable to control dynamic effects. Ventilation doors with built-in louvres can allow airflow control with occasional vehicle travel but regular traffic might mean the doors spend much time open, defeating the purpose of the built-in louvres. Jet fans can offer a solution that allows regular vehicle traffic while at the same time still providing ventilation control.

An appropriately designed automated control system can be used to ensure the correct airflow in each tunnel is maintained even with unequal resistance and piston effects. Jet fans with an attached variable frequency drive (VFD) to alter the speed combined with automated louvres to set the initial flow can provide this capability. Stewart (2023) successfully modelled such a system with a theoretical control system that assumed instantaneous response to maintain near perfect airflow control.

APPLICATIONS

Tunnelling

Jet fans are commonly used in road tunnelling applications where ventilation often relies on the piston effect of moving vehicles. When piston effects do not produce the required amount of air through the tunnel, jet fans are used to assist or retard airflow to achieve targets. Airflow sensors are used within a control system to determine the airflow and whether jet fans are needed for ventilation control. Ventilation pressure can be varied with multiple combinations of operating jet fans or by individual fans with VFDs.

Block caving

Block caving represents an opportunity for the application of jet fans where there are many parallel tunnels that require airflow control while still allowing vehicle movement. For example, loader movements along a centrally exhausted extraction drive can restrict or promote airflow within individual airways. Even though some airways would then receive more airflow than necessary, additional ventilation through the exhaust louvres is required to ensure minimum airflow requirements are met in all airways. The additional airflow required across an entire level can be substantial. Jet fans can compensate for vehicle movements, eliminating the need for additional airflow.

Another location is the haulage level where ventilation is required for truck loading chutes across multiple parallel drives. As there is not enough total airflow to ventilate all the chutes at once, each drive has exhaust louvres that only open when a truck is loading. Dust and contaminated airflow may still escape to the rest of the haulage level unless airflow directions are controlled in each drive. Jet fans can help control the ventilation circuit and would normally only need to be operated when a truck was present.

METHODOLOGY

To test this proposal, Ventsim Control was used to model a commonly encountered situation. Ventsim Control is software used to control automated ventilation devices in a ventilation system (Pinedo and Torres Espinoza, 2019). The software controls underground devices connected to a communications network but can also be used with a ventilation model to test and simulate control device performance and ventilation circuit results. A model was set-up in Ventsim with four parallel extraction drives containing operating loaders. Jet fans are placed at each end of the drives with exhaust raises in the middle and intakes at either end. Centrally located automated louvres limit the total airflow in each extraction drive. A primary fan provides airflow from the surface, running at constant speed for this example. Figure 1 shows the typical block caving extraction level layout used as the basis for the extraction model. The workings of one extraction drive are shown.

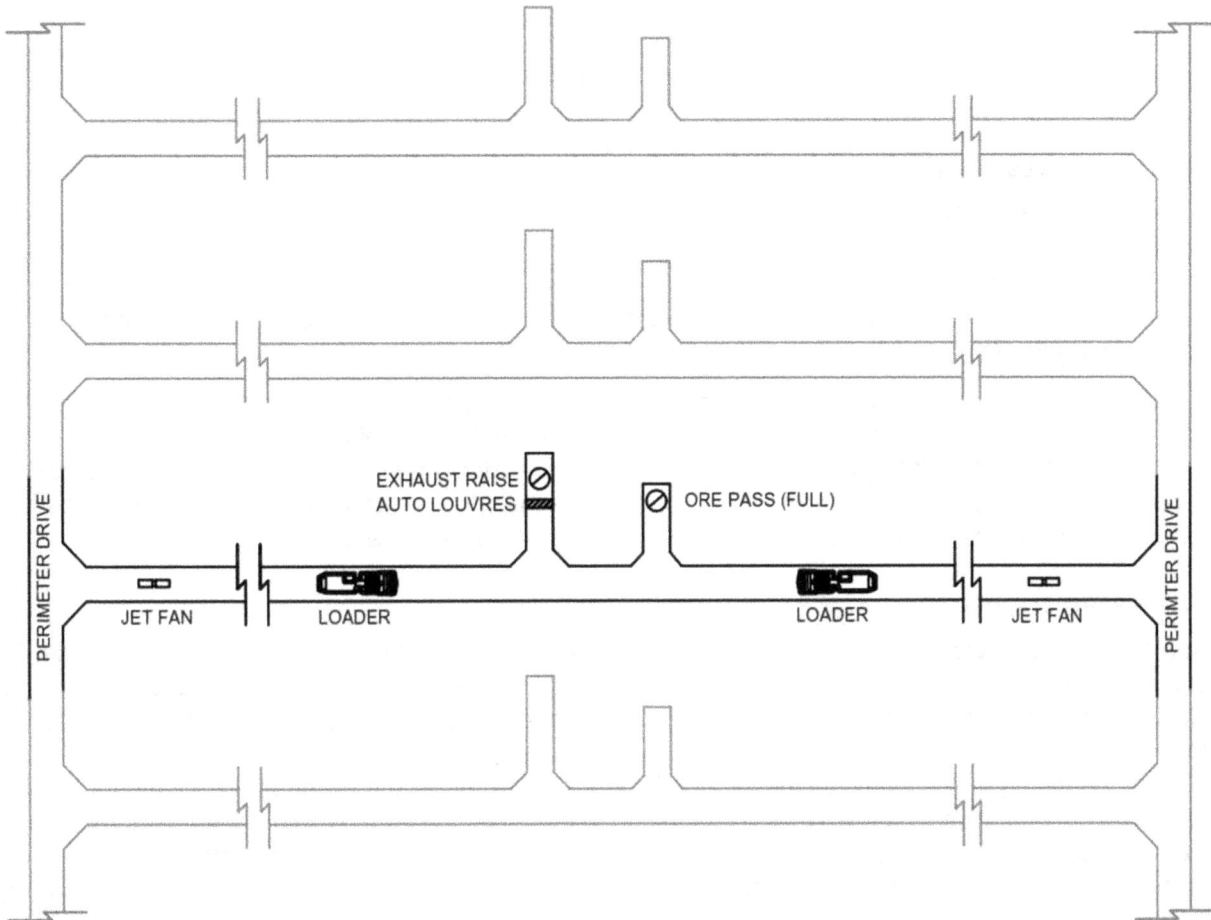

FIG 1 – Typical block caving layout with detail shown for one extraction drive (third from top).

Vehicle movement

Loaders with frontal dimensions of approximately 3.2 m wide and 2.9 m high (8.3 m² area) are assumed, similar to a Cat R2900. For the initial test, a single loader was placed in one drive moving back and forth at 3 m/s. The loader pushes the air in front of it, creating a piston effect induced pressure. This pressure is modelled in Ventsim as a moving pressure differential along an airway.

To calculate the magnitude of this pressure difference, the following equation is first used to calculate the drag force of the loader moving in an airway (Daly, 1992):

$$F_d = \frac{1}{2}\rho u^2 c_d A$$

Where:

F_d = the drag force (N)

ρ = the mass density of air (assume 1.18 kg/m³)

u	= the flow velocity relative to the object (m/s)
c_d	= the drag coefficient (unitless) related to the geometry of the unit and surface friction
A	= the area of the unit normal to the direction of airflow (m²)

As this is the equation applied when in an open area, a correction factor is applied for the situation in a tunnel when there is a high blockage ratio:

$$Correction\ Factor = \frac{1}{1 - \left(\frac{Frontal\ Area}{Tunnel\ Area}\right)^2}$$

Allowing for an airflow of 20 m³/s in a 5 m × 5 m drive with the loader moving against the airflow, the force is approximately 80 N. To calculate the pressure applied the following formula is used:

$$P = \frac{F}{A}$$

Where:

P	= Pressure (Pa)
F	= Force (N)
A	= Area (m²)

Applied across the cross-sectional area of the drive, this results in a pressure of approximately 3 Pa from a moving loader.

This value can change considerably with the air and loader speeds and the effects of other loaders creating piston pressure nearby. A Ventsim Control dynamic simulation makes calculations each iteration based on the changing conditions within the model. In this example, the 3 Pa pressure difference applied by a moving loader was far greater than the estimated less than 1 Pa pressure difference between parallel drives due to airway resistance differences.

Automated louvres

Automated louvres are installed at the centre of each of the extraction drives. Louvres in the centre exhaust raise only control the total airflow in the extraction and cannot control the distribution of airflow at either side of the extraction drive. For this model, 20 m³/s is the target airflow in an airway with a loader and 5 m³/s is the target airflow in an airway without a loader. For example, the total airflow to be set by automated louvres with a loader on one side of the extraction drive but not the other would be 25 m³/s. If there were a loader on each side, the total would be 40 m³/s or if there were no loaders present the total would be 10 m³/s. These values are not determined by the direction or speed of a loader moving in the drive so once the louvres are set they should not change during the simulation.

Jet fans

Jet fans are modelled in Ventsim as a stationary pressure difference induced in an airway. In this typical example of an extraction level, loaders work towards the middle of the extraction drives and the orepass would be located close to the exhaust raise and louvres. As the loader moves along the drive, it pushes the air in front of it towards either end. Jet fans counteract this effect by pushing or pulling air in the opposite direction to the loaders.

There are several possible options for configuration of the jet fans within the extraction drive with either one fan or two fans installed. With a single jet fan installed it would need to operate in both forwards and reverse mode to counteract the operation of the loader in both directions. With two fans installed, they can both be set to run in forwards mode only with one operating at a time. For the purposes of this scenario two fans per extraction drive are used. This has the advantage of locating the jet fans, one at either end of the drive, close to the perimeter drive, more safely away from drawpoint extraction activities. Employing two fans also reduces the wear caused by constantly stopping and changing direction to counteract the effects of a loader moving in two directions along the drive. This is also considered a more responsive system as for the fans to slow down and speed up in the opposite direction would take longer than one fan speeding up while the other is slowing down. Jet fans operating with only on/off functionality are unable to control airflow precisely and would lead to higher or lower flows than the set point. To avoid this undesirable situation, any jet fan used in this application would need to be installed with a variable frequency drive (VFD) to allow any speed depending on the VFD specifications.

Air flow sensors

For the jet fans to operate effectively, airflow sensors are also required. The best place for these is also outside the normal working area of the loaders to limit potential damage while also outside the turbulent airflow created by the fans to allow a more accurate airflow reading. Regardless of whether one or two fans are installed, at least one airflow sensor needs to be installed either side of the exhaust airway in each extraction drive as well as a third sensor in the exhaust as an input to the louvre control system.

The location of airflow sensors may also be determined by the method of airflow measurement. For example, ultrasonic type airflow sensors that directly measure the air velocity are best installed in locations where the airflow is relatively steady and away from excessive dust, that is straight sections of airway with few disturbances. A pressure differential type sensor could be installed across a resistance such as automated louvres. This is particularly useful in the case of limited space for a straight section of airway however it does need detailed data on the resistance, which is a function of the louvre opening percentage, so that the pressure difference can be related to an air velocity.

Control system algorithm

Both the jet fans and the automated louvres are to be controlled with a PID (Proportional, Integral, Derivative) algorithm in this model with the louvres controlling the total flow in the extraction drive and the jet fans controlling the contribution from each side. A PID algorithm is commonly used in control systems (Nise, 2019), to maintain a defined parameter, in this case the airflow by altering a variable, in this case either jet fan speed or louvre opening size. A set point is used as an input to the system, which then tries to match the output, airflow, to the set point. Figure 2 shows an example of a typical PID loop flow chart.

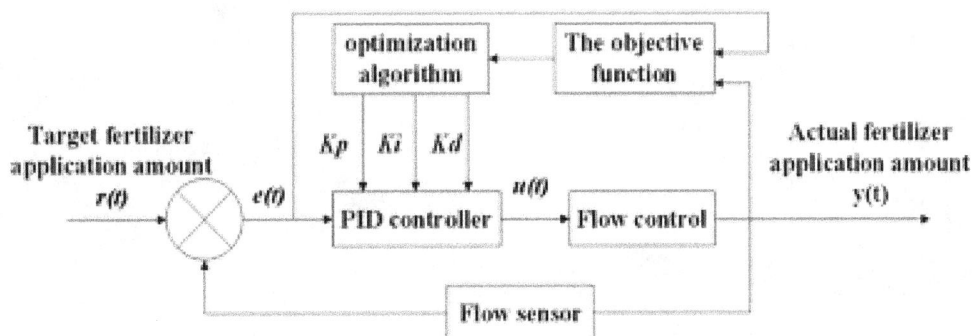

FIG 2 – An example of a PID flow control system.

Within each loop of the PID algorithm, the airflow is measured by sensors positioned within each extraction drive and compared with the set point of 20 m³/s if a loader present and 5 m³/s without a loader present. The difference between the set point and the output is the error, the size of which controls the response of the system.

The PID algorithm uses proportional, integral, and derivative gains to control how the system behaves.

A **Proportional gain** is applied based on the current error. A larger error will result in a larger change to the output, proportional to the error. For example, in the case of the jet fans, a greater error between the measured airflow (output) and the desired output (set point) will result in a greater change in fan speed to reach the set point more quickly.

An **Integral gain** is applied to the past behaviour of the system. If the proportional gain applied to the output results in a constant error, the integral gain builds over time, increasing the fan speed until the output meets the set point and the error is reduced to zero.

A **Derivative gain** is applied based on the future behaviour of the system. If the output appears to be overshooting the set point, the derivative gain will reduce the speed of error correction to reduce the likelihood of overshoot.

Generally, the gains are set based on an estimate and then tuned using trial and error. The behaviour of the system is used to determine how the gains can be altered to achieve a fast system response without overshoot. There are automated methods for calculating the gains, but these require a mathematical model of the system behaviour, which can be difficult to find in complex systems such as an underground ventilation system.

Ventsim model

To test the system, Ventsim is used to model a typical block caving layout with four extraction drives. Two intakes are modelled at either side of the extraction drives and return airways are located beneath the centre of each extraction drive. Jet fans are used at each end of the extraction drive, directing airflow towards the return airway. Each jet fan was given a number from 1–8 and are shown together with the loader operating areas in Figure 3.

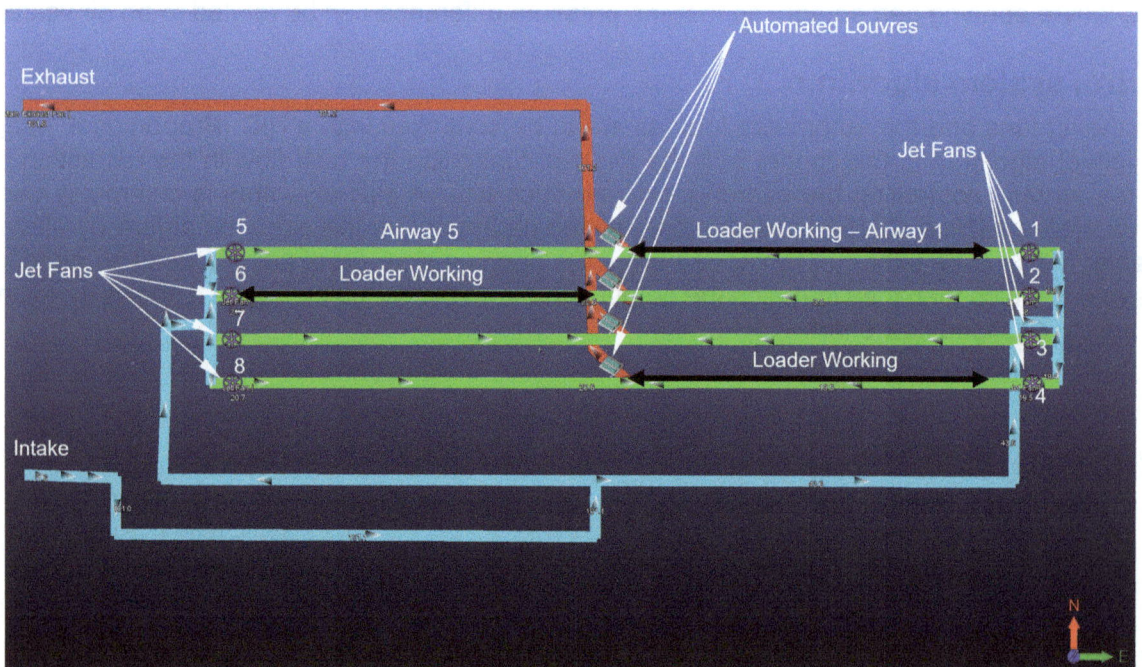

FIG 3 – Ventsim model of the simulation showing jet fan identification numbers and loader working areas.

Ventsim Control model

The Ventsim model was used as a basis for the Ventsim Control system. The model was used to simulate the process and determine the effects of operating the jet fans with a PID algorithm. The activity tracks from the Ventsim model provided the vehicle piston pressure as a disturbance to the steady state model conditions. For louvres, the airflow set point was the sum of the two sides. In this case the set point was applied manually but it is possible to use vehicle tracking to apply the set point automatically if a loader or other equipment is detected (Sanftenberg, 2019).

RESULTS

Two simulations were performed, the first without jet fans and the second with jet fan airflow control using a PID algorithm. Each simulation was run for 30 mins, and the airflow changes recorded at intervals of a few seconds.

The resultant simulated airflows changes for a 500 secs sample period are shown in Figure 4 (loader present) and Figure 5 (no loader present). A negative number indicates the airflow has reversed from the exhaust to the intake. Figure 6 shows the jet fan speed for each airway over the same time span.

In airway 1, with a loader operating and no jet fan present (Figure 4) the flow rate varies from +27 m^3/s to -8 m^3/s with positive values indicating flow from the perimeter drive to the central exhaust. With the jet fan present, the values vary from +5 to +35 m^3/s but these are short spikes as the fan adjusts to vehicle-induced piston effects. In airway 5, with no loader operating and no jet fan present (Figure 5) the flow rate varies from -3 m^3/s to +31 m^3/s. When there is a jet fan present the airflow varies from 0 m^3/s to +21 m^3/s but again the variations from the set point are short spikes as the jet fan adjusts to the vehicle induced piston effects. In both figures there are smaller spikes or step changes with no jet fans caused by the secondary effects of vehicles operating in adjacent airways.

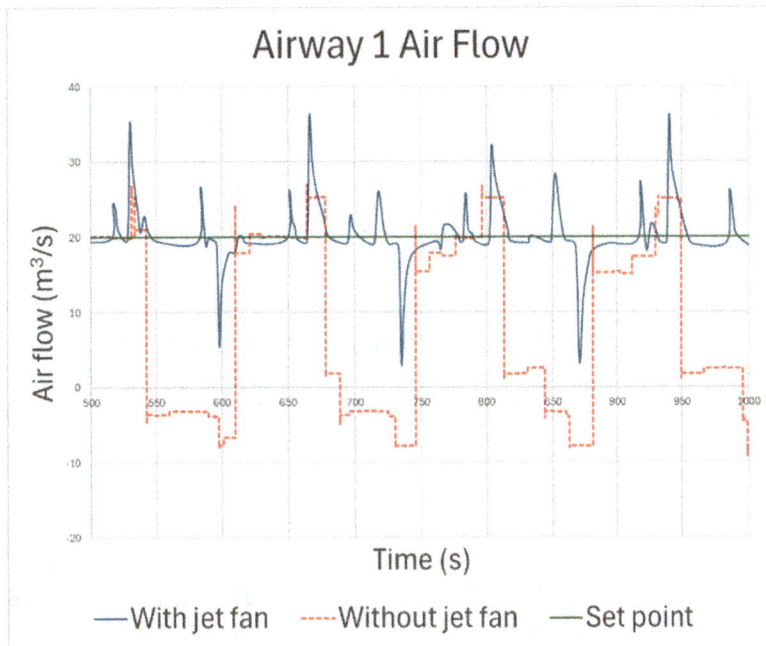

FIG 4 – Airway 1 targeting 20 m^3/s airflow with operating loader – jet fan (Blue), no jet fan (Red).

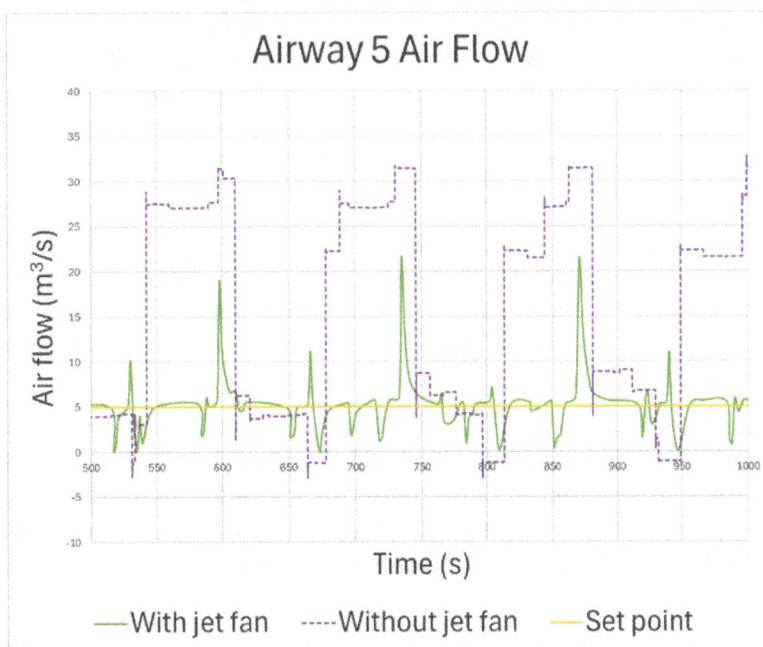

FIG 5 – Airway 5 targeting 5 m^3/s airflow with no loader – jet fan (Green), no jet fan (Purple).

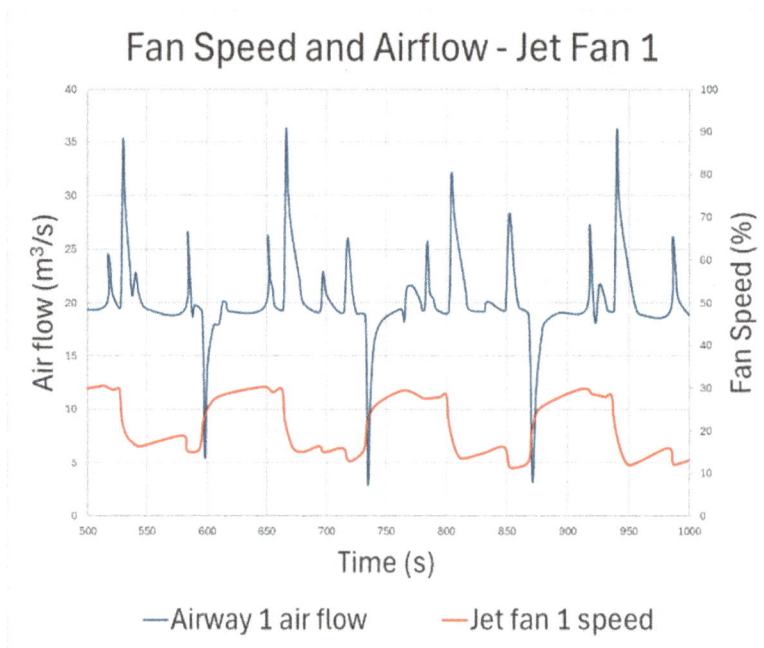

FIG 6 – Jet fan 1 airflow (Blue) compared with fan speed (Red).

DISCUSSION

Figure 3 shows considerable variation in airflow both with a jet fan installed and without a jet fan installed. However, with the jet fan installed, the spikes in airflow corresponding to changes in loader direction are corrected within 10 to 15 secs to restore the airflow to the set point of 20 m³/s. When there is no jet fan installed, the airflow remains lower and there are time periods of over a minute when the airflow is reversed, contaminating the intake with diesel exhaust, heat and dust which enters other extraction drives.

Similar behaviour can be observed in Figure 4, even though there is no loader present. The loader in airway 1 influences the airflow in airway 5 even though the louvres maintain the required total of 25 m³/s from both drives. Even though there is no loader present, the jet fan still has an important role to play in reducing the airflow to the minimum required to allow the correct airflow where loaders are present.

Figure 5 shows the maximum 100 kW jet fan speed required to maintain the required airflow was about 30 per cent of the maximum capacity and the minimum fan speed was about 5 per cent. The results can be used to help specify a reduced more suitable sized fan. Larger spikes in airflow indicate a change in direction of the simulated loader, with smaller spikes indicating airflow changes influenced by loaders operating in other airways. This shows that even though a loader may not be present in the airway, other loaders within the level can still influence the airflow and jet fan control can still be beneficial.

The sudden spikes observed in simulated airflow result from the loader changing direction instantaneously. In a real-world situation there would be a period of deceleration and acceleration, and the control system would be more likely to maintain a steady airflow without the intermittent spikes. At present, Ventsim is unable to model this behaviour which would help create a more accurate model of the system.

The Ventsim model uses a network analysis method of calculating airflows and pressures and while this is appropriate to calculate the far-field (>60 m) effects of the jet fan use on air flow, computational fluid dynamics (CFD) may be required to determine the local effects when the loader is closer to the jet fan. Due to the computational demands of CFD, this method would not be suitable for real time control, and a solution may be to use precalculated effects derived from CFD to guide network analysis behaviour.

From an economic perspective, there is an additional cost for jet fans, airflow sensors and a control system. However, to consistently provide adequate ventilation without this system, it would be necessary to provide greater airflow to ensure the minimum is always met. This is particularly evident when a loader is working on one side only, with the airflow on both sides of the exhaust being roughly the same even though the required airflow is far less on the side with no loader. The alternative would be to accept that there would be times when the airflow provided was not adequate. Apart from the ethical issues in this scenario, there may be additional costs due to downtime while waiting for ventilation to clear the area and associated lost productivity. An economic analysis is not presented here as a case by case approach would be required, however the costs of both implementing and not implementing such a system should be considered.

CONCLUSIONS

Jet fans can help regulate airflow in multiple parallel drives with flow-through ventilation. This is particularly relevant where the piston effect of moving vehicles constantly changes the airflow. Unbalanced airway resistance and piston effects can be mitigated using jet fans controlled by a PID algorithm. The system requires the installation of air flow sensors at a location in the drift controlled by the jet fans in a location safe from damage from moving vehicles. Ideally, the system would be considered in the mine design process to allow space for airflow sensors, jet fans and electrical installations. Any jet fan installed as part of the system would require a VFD in order to be effective.

If airflow needs to be controlled in both directions reversible jet fans or multiple fans pointing in opposite directions will be required. Reversible fans are less power efficient as the blade design needs to accommodate flow in both directions and repeated frequent reversals may shorten the fan or motor life.

Further work in this field could involve more accurate simulation of vehicle piston effects, that consider the acceleration and deceleration of vehicles to help more accurately tune optimum PID parameters. Another area for further research is the use of CFD to assist network analysis where jet fans operate in the proximity of moving vehicles. Although the computational demands of CFD will likely make it unsuitable for real time modelling and control of an entire level, precalculated results or a simplified combination of the two methods may provide a solution. Field test work and tuning is ultimately required to fully prove the effectiveness of the concept.

The work presented in this paper shows using jet fans within a control system including a PID algorithm and VFDs provides a solution for controlling airflow where traditional ventilation controls may not be suitable due to access or control constraints. The method would provide a solution to the problem of providing minimum required airflow to an airway without wastage and preventing airflow reversal that may contaminate other intake airways.

ACKNOWLEDGEMENTS

The author would like to thank Craig Christensen and Martin Griffith of Howden Ventsim for support with software development and advice on the workings of the software used.

REFERENCES

Daly, B B, 1992. *Woods Practical Guide to Fan Engineering* (Woods of Colchester Limited).

Nise, N S, 2019. *Control Systems Engineering*, 8th edition, (Wiley).

Pinedo, J and Torres Espinoza, D, 2019. Implementation of Advanced Control Strategies Using Ventsim Control at San Julian Mine in Chihuahua, Mexico, in *Proceedings of the 17th North American Mine Ventilation Symposium*, pp 435–441 (Canadian Institute of Mining Metallurgy and Petroleum).

Sanftenberg, J, 2019. Mine Ventilation on Demand System at Nickel Rim South Mine; Testing Procedures, System Maintenance and Results, in *Proceedings of the 17th North American Mine Ventilation Symposium*, pp 104–113 (Canadian Institute of Mining Metallurgy and Petroleum).

Stewart, C, 2023. New applications of jet fans in underground mines for haulage ramps and block cave ventilation control, in *Underground Ventilation: Proceedings of the 19th North American Mine Ventilation Symposium*, pp 515–528 (Society for Mining, Metallurgy and Exploration).

Mine fires and emergency response planning

Towards site-wide air mixture monitoring – lessons learned from geostatistics

K Brown Requist[1] and M Momayez[2]

1. Research Assistant, University of Arizona Department of Mining and Geological Engineering, Tucson AZ 85719, USA. Email: katebrown@arizona.edu
2. Professor, University of Arizona Department of Mining and Geological Engineering, Tucson AZ 85719, USA.

ABSTRACT

Real-time air quality monitoring is becoming a standard practice for underground mines worldwide. Many mines have adopted systems for monitoring and control applications. Real-time data is only available at the point-specific level; a site-wide picture of air quality requires the development of mathematical methods to estimate local values. Computational fluid dynamics (CFD) requires substantial computational resources and is time intensive. Mine ventilation network (MVN) methods lack data resolution beyond the branch level. With the use of path-finding algorithms to encode for ventilation conditions, necessary data unfolding is feasible, allowing for the use of geostatistical methods like ordinary kriging. The combination of a novel pathfinding algorithm and ordinary kriging provides a significant reduction in mean absolute error and coefficient of variation of estimates using simulated data in a portion of a room-and-pillar coalmine in Utah, United States.

INTRODUCTION

Real-time monitoring of carbon monoxide (CO) in the belt entries of underground coalmines in the United States has been a prescribed standard since 1995. In the past three decades, real-time monitoring capabilities have expanded, with approximately 14 per cent of underground coalmines in the United States using real-time monitoring as part of an atmospheric monitoring system (AMS) for early-warning fire detection and regulatory compliance (Rowland III, Harteis and Yuan, 2018). Worldwide, AMSs and CO systems have begun to replace monitoring by individuals to reduce the likelihood of exposure to potentially hazardous air mixtures.

Real-time atmospheric monitoring capabilities have expanded in the past 30 years, while capital costs associated with sensing have decreased with the advent of low-cost, typically electro-chemical sensors (Afshar-Mohajer et al, 2018; Zuidema et al, 2021; Ziętek et al, 2020). The implementation of wired and wireless real-time monitoring in the United States has been met with resistance, from a lack of industry consensus on appropriate sensor placement to the time-intensive requirements for inspection, bump testing and calibration (National Archives and Records Administration, 2024; Rowland III, Harteis and Yuan, 2018). While AMS are extensively regulated for underground coalmines in the United States, metal and non-metal mines currently have no regulatory framework for the use of fixed atmospheric monitoring. Alternatively, metal mines can adapt ventilation-on-demand (VOD) systems that include gas monitoring, such as VOD systems seen in Canada's Vale Inc. Coleman Mine and Xstrata Nickel Rim Mine and the United States' Barrick Goldstrike Mine (Shriwas and Pritchard, 2020).

Fixed-location real-time monitoring, while capable of providing reliable representations of current atmospheric conditions within the sampling range of the sensor, fails to provide avenues for the site-wide monitoring of air mixtures. Site-specific implementations of real-time, site-wide monitoring systems have been developed in the past, with mixed levels of success. Notable examples include the use of real-time monitoring in the Waste Isolation Pilot Plant (WIPP) using the WIPPVENT software as developed by the Westinghouse Electric Corporation on behalf of the United States Department of Energy in 1995, and various implementations with Ventsim (Shriwas and Pritchard, 2020; McDaniel and Wallace, 1997; Ruckman and Prosser, 2010; Gillies et al, 2004; Wu and Gillies, 2005). These implementations, however, are based on a ventilation network approach, which critically lacks the data resolution required for a holistic view of airborne contamination.

Computational fluid dynamics (CFD) has been a promising avenue for air quality analysis in underground mines. The time required to construct realistic representations of the area and the computational intensity of CFD poses a barrier to its use as a real-time monitoring method. These models are also incredibly sensitive to the number of cells used in modelling. Too many cells within the CFD model constitutes an increase in processing time and too few cells may cause grid dependency (Mora *et al*, 2002; Xiang, Wei and Haibo, 2017).

With these problems, mine ventilation network (MVN) simulation remains the best current approach. MVN methods are one-dimensional (1D), representing the network as a directional graph, where features within the branches (mine entries) can be determined by considering conservation of mass and energy and applying the Hardy-Cross or similar methods to several governing equations (Sereshki, Saffari and Elahi, 2016). As a 1D method, MVN models cannot estimate behaviours within these mine entries. There is a need for a holistic and high resolution understanding of airborne contamination in underground mines. Shriwas and Pritchard (2020) discuss the importance of the development of mathematical models to estimate local airborne contamination levels from fixed sensing data, especially at the working face.

Airflow impacts the ability to model contamination distributions and must be approached as either a source of locally varying anisotropy or a site-wide trend to address the problem of data stationarity (Boisvert and Deutsch, 2008; Boisvert, Manchuk and Deutsch, 2009). To account for airflow, we have developed a modified approach to the A* pathfinding algorithm that better captures spatial relationships between locations in an underground mine (Brown Requist and Momayez, in prep(a)).

Geostatistics has been used extensively for the determination of contamination distributions in air and soil. Vicedo-Cabrera *et al* (2013) used universal Bayesian kriging for the estimation of children's exposures to nitrogen dioxide and sulfur dioxide in homes in Italy. Sahu and Mardia (2005) developed a Bayesian Kriged-Kalman model to forecast air pollution in New York City. Lin *et al* (2011) used a combination of indicator kriging, logistic regression and regression kriging to assess relationships between heavy metal concentrations in soil and human activity. Von Steiger *et al* (1996) mapped heavy metal concentrations in soil with disjunctive kriging in Switzerland.

Geostatistical methods present a possible means to provide significantly higher spatial resolution than is currently feasible with MVN software. While MVN methods are currently the industry standard for the estimation of contamination distributions, the development of a statistical basis for the estimation of contamination distributions stands to provide mine operators with better understandings of airborne contamination in their mines. Traditionally, geostatistics has used Euclidean distances for the estimation of values, but these Euclidean distances fail to account for air behaviour when attempting to model processes in ventilation systems. Using MVN software as a baseline, we compare two geostatistical methods to assess their potential for use in the monitoring of underground airborne contamination prior to the *in situ* use and validation of geostatistical estimation methods in active mining environments. By using MVN software, we can simulate a wide array of conditions to assess the reliability of one geostatistical estimation method over another.

METHODS

Data unfolding and data spacing

Geostatistical estimation of airborne contaminant concentration requires a strong basis in the underlying dynamic and kinematic processes that determine contaminant transport in an underground mine. The spacing of data and the measurement of these spacings must be approached with caution as to avoid unnecessary clustering of data within the space, especially when atmospheric monitoring sensors can be few and far between. For the measurement of effective distances between points within a ventilation system, we have proposed a modification to the cost function of the A* pathfinding algorithm, which considers airflow direction, mine geometry and drift surface roughness (Brown Requist and Momayez, in prep(a)). These parameters in-form the cost function, generalised in Equation 1.

$$g_f(p_{n+1}) = \begin{cases} s, & \theta_c \leq 45° \\ 2\alpha^{-\frac{1}{2}} * s, & 45° < \theta_c \leq 90° \\ \beta^{-2} * s, & \theta_c > 90° \end{cases} \tag{1}$$

where:

$g_f(p_{n+1})$ is the cost to move from the current position to a neighbouring position

s is the distance between a current position and a neighbouring position

α is the average fraction of right-angle shock loss between two points

β is the surface roughness of the excavation

The theory behind the modified A* pathfinding algorithm is discussed further in Brown Requist and Momayez (in prep(a)).

This provides an avenue to better capture variability within the space by considering ventilation system conditions. Methods like this are well documented for the improvement of kriging outcomes, especially when underlying processes may have an impact on data stationarity (Boisvert and Deutsch 2008; Boisvert, Manchuk and Deutsch, 2009; Beauchamp, de Fouquet and Malherbe, 2017; Van de Kassteele and Stein, 2006; Beauchamp et al, 2018). However, with the use of ventilation conditions in spatial interpolation, the consideration of data clustering requires a different approach than that which is often used in other geostatistical workflows.

The spacing of data should be regular with respect to the distance metric in use. While a common method, cell declustering is not possible with the small number of sensors available and the large spacing between them. As an alternative to the current spacing schema used in United States mines as outlined in Title 30, Code of Federal Regulations 75.351 (National Archives and Records Administration, 2024), careful placement of sensors could provide a more even spatial representation of data using the modified A* algorithm as the source of the distance metric. This placement is achieved using a realisation of a capacity constrained centroidal Voronoi diagram (CCCVD), discussed further in Brown Requist and Momayez (in prep(b)). This distribution generated by the CCCVD approach (Figure 1) is then used to determine monitoring locations for Ventsim 5.4 dynamic simulation of contaminant transport.

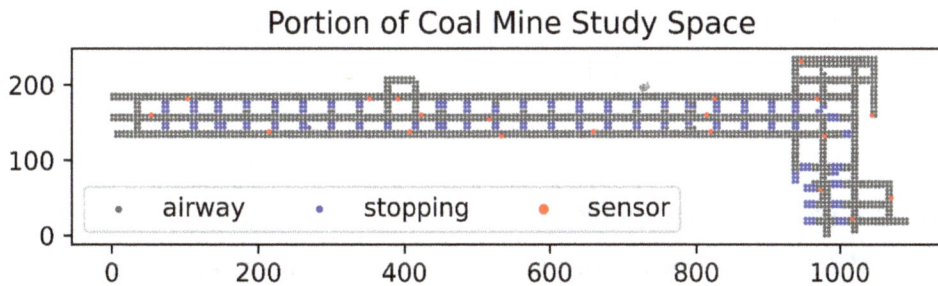

FIG 1 – Optimal placement of 20 sensors (red) in an underground room-and-pillar coalmine in Utah, USA. Both axes are displayed in metres.

Dynamic simulation of input data

Simulation of contaminant spread was performed using Ventsim 5.4 with a ventilation model of an underground room-and-pillar coalmine in Utah, USA to create a set of simulated data at the sensor locations indicated in Figure 1. This simulation spans 180 mins in which a fixed rate release of 25 ppm CO is present in the section's intake air for a period of 90 mins, followed by another 90 min period with no CO present. The contaminated intake air is introduced at a volumetric flow rate of 19 m³/s. Two additional intakes with volumetric flow rates of 19 and 6 m³/s provide a supply of fresh air. Volumetric flow along the section's main beltway ranges from 10 to 20 m³/s to achieve a minimum air velocity of 0.3 m/s. Air is exhausted from two outlets at 22.5 m³/s and 21.5 m³/s for the bottom right and centre right outlets, respectively. Figure 2 displays the path of travel of the contaminated air through the mine.

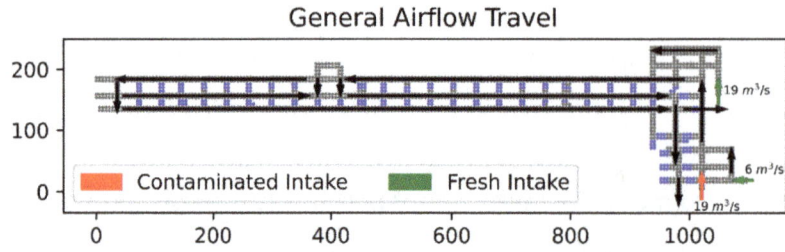

FIG 2 – Path of travel for air in the portion of the mine.

Ordinary kriging of simulated data

Kriging of site-wide air contaminant concentration requires the construction of a semivariogram based on observed data that may be transformed via various methods to approximate a normal distribution. Transform methods for data are numerous, including log transform and square root transform (Webster and Oliver, 2007). To reduce skew in the observed data, lognormal and square root transforms have been applied as necessary for all simulated data. The proper transform is identified by the lowest absolute skewness of observed data for each timestep. A model semivariogram can be fit to experimental semivariance using a spherical model. This model can be used to establish the system of kriging equations as per Equations 2 and 3:

$$\sum_{\beta=1}^{n} \lambda_\beta \bar{C}(v_\alpha, v_\beta) - \mu = \bar{C}(v_\alpha, v_0), \forall \alpha = 1 \ldots n \tag{2}$$

where:

λ $\qquad$ is an unknown set of weights used to estimate local values while kriging

$\bar{C}(v_\alpha, v_\beta)$ $\quad$ is the covariance between two locations, calculated as the sill minus the semivariance

v_0 $\qquad$ is the location where data values are to be estimated with kriging

μ $\qquad$ is the Lagrange parameter

$$\sum_{\beta=1}^{n} \lambda_\beta = 1 \tag{3}$$

To calculate the weights λ, the ordinary kriging system can be represented in matrix form:

$$[\lambda] = \begin{bmatrix} \lambda_1 \\ \vdots \\ \lambda_n \\ -\mu \end{bmatrix} = [K]^{-1}[k] \tag{4}$$

where:

$[K]$ $\qquad$ is the data covariance matrix

$[k]$ $\qquad$ is the data-to-unknown covariance column matrix given by Equations 5 and 6, respectively.

$$[K] = \begin{bmatrix} \bar{C}(v_1, v_1) & \cdots & \bar{C}(v_1, v_n) & 1 \\ \vdots & \ddots & \vdots & \vdots \\ \bar{C}(v_n, v_1) & \cdots & \bar{C}(v_n, v_n) & 1 \\ 1 & \cdots & 1 & 0 \end{bmatrix} \tag{5}$$

$$[k] = \begin{bmatrix} \bar{C}(v_1, v_0) \\ \vdots \\ \bar{C}(v_n, v_0) \\ 1 \end{bmatrix} \tag{6}$$

The covariance matrix $[K]$ should be positive definite, because of the symmetric nature of the semivariogram and covariance: $\bar{C}(v_\alpha, v_\beta) = \bar{C}(v_\beta, v_\alpha)$ (Myers, 1992; Armstrong and Jabin, 1981; Journel 1989; Journel and Huijbregts 1991). This is not the case using the modified A* algorithm and this requirement has been relaxed for the sake of this investigation. In the case where kriging variance (Equation 7) is negative, inverse distance squared weighting (IDW) is used to arrive at a valid estimate:

$$\sigma_{OK}^2 = \bar{C}(v_0, v_0) - \sum_{\alpha=1}^n \lambda_\alpha \bar{C}(v_\alpha, v_0) - \mu \tag{7}$$

Where σ_{OK}^2 is the ordinary kriging variance calculated as the unknown-to-unknown covariance less the linear combination of kriging weights and the Lagrange parameter.

Validation of kriged models

To assess the accuracy of the kriging approach using the modified A* algorithm, the new method can be compared to traditional, Euclidean kriging. By using a contamination simulation conducted in Ventsim 5.4 as a baseline, the two kriging approaches can be directly compared to assess their fitness for this new modelling approach. Two metrics are used to assess model performance: mean absolute error (MAE) and coefficient of variation (CV). These metrics are calculated via leave-one-out cross-validation. MAE (Equation 8) is a measure of the difference between a true value and an estimated value. CV (Equation 9) is a measure of the variation of estimation error. Low CV values are desirable as they indicate that the error of one estimate is close in value to the error of all other estimates:

$$MAE = \frac{1}{n}\sum_{i=1}^n |y_i - x_i| \tag{8}$$

where:

y_i is the estimated value

x_i is the true value

$$CV = \frac{\sqrt{\frac{1}{n}\sum_{i=1}^n (y_i - x_i)^2}}{\bar{x}} \tag{9}$$

Where $\bar{x}$ is the mean of true values.

RESULTS AND DISCUSSION

Construction of semivariograms is relatively straightforward using transformed data. With 180 mins of sensor data simulated in Ventsim 5.4 dynamic simulation, 172 have sufficient data to proceed with ordinary kriging. Because ordinary kriging relies on the spatial relationships of data inputs, a minimum of three non-identical sensor readings (two pairwise semivariance values) is necessary to create a model semivariogram. Only the first 8 mins of simulated data fail to meet this requirement.

The resulting estimated values are low in mean absolute error, with an average MAE of 2.0 ppm using the modified A* distance metric. Compared to an average MAE of 2.3 ppm using traditional ordinary kriging, this yields a significant ($p = 2.11 * 10^{-31}$, SD = 0.93 ppm, n = 172) reduction in MAE using a left-tailed t-test. Furthermore, the average coefficient of variation using the modified A* distance metric is 0.59. Compared with an average CV of 0.68 using traditional ordinary kriging, using the modified A* metric that encodes for airflow direction yields a significant ($p = 2.00 * 10^{-201}$, SD = 0.16, n = 172) reduction in CV.

Resulting estimates using ordinary kriging informed by the modified A* distance metric can be visualised as in Figure 3. The covariance matrices used in estimation are not positive definite because $\bar{C}(v_\alpha, v_\beta) \neq \bar{C}(v_\beta, v_\alpha)$ under the relaxation. The covariance functions remain convex, but the matrices are indefinite and still invertible. As a consequence, the kriging variance of some estimates may be negative, meaning the estimate is unreliable. This requires estimation via IDW. The frequency with which values must be estimated using IDW is relatively low. Further investigation of methods to construct positive definite matrices will further reduce the need to estimate some values using IDW. On average, across 172 mins of input data, 2.4 per cent of values require IDW, with a maximum of 10.0 per cent of values, as displayed in Figure 4.

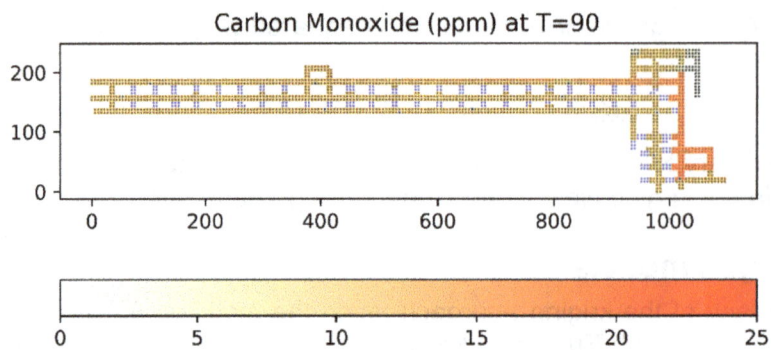

FIG 3 – Estimated data can be visualised with CO values reported in parts per million.

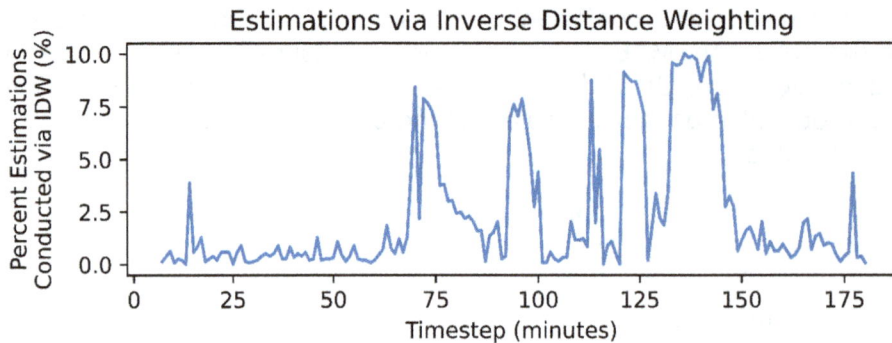

FIG 4 – On average, 2.4 per cent of values must be estimated using inverse distance squared weighting (IDW).

A real-time system with 1 min data resolution is feasible for this mine section. Using an Intel i9 2.50 GHz processor, a space with 1577 estimates and 20 simulated input points requires an average of 50.97 seconds to construct, validate and visualise models using a combination of software written in Python 3.10 and Julia 1.8. Obviously, estimated values generated through simulations in Ventsim 5.4, and the proposed algorithm are no substitute for data collected in an actual underground mining setting. This is especially true without an experimental basis for validation, as verification of the Ventsim model's accuracy is not feasible. Ongoing efforts are being made to validate this modelling technique in the context of subsurface mining through the collection of data from multiple locations and under varying conditions. This validation effort will shed additional light on potential avenues for enhancing the monitoring and modelling of real-time air quality.

Further work is warranted to reduce the MAE and CV for kriged air quality models. The lack of a positive definite covariance matrix poses an issue for the stability of the kriging system. As such, it may be beneficial to investigate the merits of using additional covariates or a nested variogram to ensure matrices remain positive definite. While the example presented is based on a single emission source at a section's intake, real-time emissions are not typically stationary in underground mines, especially in the case of mobile equipment. The use of co-kriging may allow for improved estimation of local air quality, as a model would be able to account for the real-time position of equipment informed by asset tracking systems. This asset tracking could be extended to the personnel level, too, permitting for real-time exposure monitoring for underground workers.

CONCLUSION

Real-time air quality monitoring systems have been piloted in several mines in North America, but these systems are normally based on a 1D MVN approach. These MVN methods are incredibly fast to compute, making them strong candidates for real-time monitoring and control, especially when using volumetric flow and pressure data. However, MVN methods are poorly suited for the site-wide monitoring of airborne contamination, as there is no means to determine local concentration within branches. Instead, these estimates are limited to an average concentration across the network branch.

Alternatively, CFD methods can produce high-quality, high-confidence models that can be used to estimate local contamination levels with low uncertainty. These methods are poorly suited for real-time monitoring. A real-time monitoring method for airborne contamination that can approximate the speed and reliability of MVN methods and the resolution of CFD models is needed. With the use of ordinary kriging and a pathfinding algorithm to encode for relationships between mine geometry and airflow direction, it is possible to create a local air quality monitoring system that can return results for 1577 individual estimates from 20 input points in approximately 51 seconds.

The method presented provides a significant reduction in MAE and CV compared to a traditional ordinary kriged model. With the development of a distributed computing architecture, real-time monitoring using geostatistics is a feasible advance in atmospheric monitoring. Further work is needed to construct positive definite covariance matrices, and the development of additional geostatistical methods for air quality modelling may improve exposure assessment for underground workers. Although the analysis presented here relies on simulation in Ventsim 5.4, the use of a dedicated distance metric as calculated by a modified A* pathfinding algorithm shows a marked improvement over traditional ordinary kriging to estimate airborne contamination distributions. Work is ongoing to validate this estimation method in underground mining environment, and the intermediate results show promise for the use of geostatistics to estimate air quality in underground mines. This method is a step towards better site-wide air quality monitoring and represents a new approach to considering the methods with which ventilation systems are represented and analysed.

ACKNOWLEDGEMENTS

This research was funded by the National Institute of Occupational Safety and Health (NIOSH) under award number U60OH012351.

REFERENCES

Afshar-Mohajer, N, Zuidema, C, Sousan, S, Hallett, L, Tatum, M, Rule, A M, Thomas, G, Peters, T M and Koehler, K, 2018. Evaluation of low-cost electro-chemical sensors for environmental monitoring of ozone, nitrogen dioxide and carbon monoxide, *Journal of Occupational and Environmental Hygiene*, 15(2):87–98. https://doi.org/10.1080/15459624.2017.1388918

Armstrong, M and Jabin, R, 1981. Variogram models must be positive-definite, *Journal of the International Association for Mathematical Geology*, 13:455–459. https://doi.org/10.1007/BF01079648

Beauchamp, M, de Fouquet, C and Malherbe, L, 2017. Dealing with non-stationarity through explanatory variables in kriging-based air quality maps, *Spatial Statistics*, 22:18–46.

Beauchamp, M, Malherbe, L, de Fouquet, C, Létinois, L and Tognet, F, 2018. A polynomial approximation of the traffic contributions for kriging-based interpolation of urban air quality model, *Environmental Modelling and Software*, 105:132–152.

Boisvert, J B and Deutsch, C V, 2008. Kriging in the Presence of LVA Using Dijkstra's Algorithm [online], Center for Computational Geostatistics. Available from: <https://www.ccgalberta.com/ccgresources/report10/2008-110_lva_kriging_dijkstra.pdf>

Boisvert, J B, Manchuk, J G and Deutsch, C V, 2009. Kriging in the Presence of Locally Varying Anisotropy Using Non-Euclidean Distances, *Mathematical Geosciences*, 41:585–601. https://doi.org/10.1007/s11004-009-9229-1

Brown Requist, K and Momayez, M, in prep(a). Minimum Cost Pathfinding Algorithm for the Determination of Optimal Paths under Airflow Constraints, Manuscript available upon request.

Brown Requist, K and Momayez, M, in prep(b). An Algorithm for the Efficient Placement of Air Quality Sensors in Underground Mines, Manuscript available upon request.

Gillies, A D S, Wu, H W, Tuffs, N and Sartor, T, 2004. Development of a real time airflow monitoring and control system, in *Proceedings 10th United States/North American Mine Ventilation Symposium* (eds: S Bandopadhyay and R Ganguli), pp 145–155 (Taylor and Francis: London).

Journel, A G and Huijbregts, C J, 1991. Kriging and the Estimation of in situ Resources, in *Mining Geostatistics*, pp 303–443 (St Edmundsbury Press Ltd: Suffolk).

Journel, A, 1989. Lesson III: Linear Regression under Constraints and Ordinary Kriging, in *Fundamentals of Geostatistics in Five Lessons*, pp 15–21 (American Geophysical Union: Washington, DC).

Lin, Y P, Cheng, B Y, Chu, H J, Chang, T K and Yu, H L, 2011. Assessing how heavy metal pollution and human activity are related by using logistic regression and kriging methods, *Geoderma*, 163(3–4):275–282. https://doi.org/10.1016/j.geoderma.2011.05.004

McDaniel, K H and Wallace, K G, 1997. Realtime mine ventilation simulation, United States Office of Scientific and Technical Information. https://doi.org/10.2172/515492

Mora, L, Gadgil, A J, Wurtz, E and Inard, C, 2002. Comparing Zonal and CFD Model Predictions of Indoor Airflows under Mixed Convection Conditions to Experimental Data, presented at Third European Conference on Energy Performance and Indoor Climate in Buildings (EPIC), 6 p.

Myers, D, 1992. Kriging, Cokriging, Radial Basis Functions and the Role of Positive Definiteness, *Computers Math, Applic*, 24(12):139–148.

National Archives and Records Administration, 2024. Code of Federal Regulations, Title 30, chapter I, subchapter O, part 75, subpart D, § 75.351 – Atmospheric Monitoring Systems [online]. National Archives and Records Administration. Available from: <https://www.ecfr.gov/current/title-30/chapter-I/subchapter-O/part-75/subpart-D/section-75.351>

Rowland III, J H, Harteis, S P and Yuan, L, 2018. A survey of atmospheric monitoring systems in US underground coal mines, *Mining Engineering*, 70(2):37. https://doi.org/10.19150/me.8058

Ruckman, R and Prosser, B, 2010. Integrating ventilation monitoring sensor data with ventilation computer simulation software at the Waste Isolation Pilot Plant facility, in *Proceedings 13th United States/North American Mine Ventilation Symposium* (eds: S Hardcastle and D L McKinnon), pp 237–242.

Sahu, S K and Mardia, K V, 2005. A Bayesian Kriged Kalman Model for Short-Term Forecasting of Air Pollution Levels, *Journal of the Royal Statistical Society*, 54(1):223–244. https://doi.org/10.1111/j.1467-9876.2005.00480.x

Sereshki, F, Saffari, A and Elahi, E, 2016. Comparison of Mathematical Approximation Methods for Mine Ventilation Network Analysis, *International Journal of Mining Science*, 2(1).

Shriwas, M and Pritchard, C, 2020. Ventilation Monitoring and Control in Mines, *Mining, Metallurgy and Exploration*, (37):1015–1021. https://doi.org/10.1007/s42461-020-00231-8

Van de Kassteele, J and Stein, A, 2006. A model for external drift kriging with uncertain covariates applied to air quality measurements and dispersion model output, *Environmetrics*, 17(4):309–322.

Vicedo-Cabrera, A, Biggeri, A, Grisotto, L, Barbone, F and Catelan, D, 2013. A Bayesian kriging model for estimating residential exposure to air pollution of children living in a high-risk area in Italy, *Geospatial Health*, 8(1):87–95. https://doi.org/10.4081/gh.2013.57

Von Steiger, B, Webster, R, Schulin, R and Lehmann, R, 1996. Mapping heavy metals in polluted soil by disjunctive kriging, *Environmental Pollution*, 94(2):205–215. https://doi.org/10.1016/S0269-7491(96)00060-7

Webster, R and Oliver, M A, 2007. Basic Statistics, in *Geostatistics for Environmental Scientists*, 2nd edition, pp 11–35 (Wiley: West Sussex).

Wu, H W and Gillies, A D S, 2005. Real-Time Airflow Monitoring and Control Within the Mine Production System, in *Proceedings of the Eighth International Mine Ventilation Congress*, pp 383–389 (The Australasian Institute of Mining and Metallurgy: Melbourne).

Xiang, Z, Wei, Y and Haibo, H, 2017. Computational grid dependency in CFD simulation for heat transfer, in *Proceedings Eighth International Conference on Mechanical and Aerospace Engineering (ICMAE)*, pp 193–197, https://doi.org/10.1109/ICMAE.2017.8038641

Ziętek, B, Banasiewicz, A, Zimroz, R, Szrek, J and Gola, S, 2020. A Portable Environmental Data-Monitoring System for Air Hazard Evaluation in Deep Underground Mines, *Energies*, 13(23):6331. https://doi.org/10.3390/en13236331

Zuidema, C, Schumacher, C S, Austin, E, Carvlin, G, Larson, T V, Spalt, E W, Zusman, M, Gassett, A J, Seto, E, Kaufman, J D and Sheppard, L, 2021. Deployment, Calibration and Cross-Validation of Low-Cost Electrochemical Sensors for Carbon Monoxide, Nitrogen Oxides and Ozone for an Epidemiological Study, *Sensors*, 21(12):4214. https://doi.org/10.3390/s21124214

Managing the catastrophic risk of underground transformer fires

J J L Du Plessis[1,2], M Biffi[3], F J van Zyl[4], L Kádár[5], J Wu[6] and K Prakash[7]

1. Manager Group HSEC Assurance, Glencore Holdings South Africa (Pty) Ltd, Johannesburg, South Africa. Email: jan.duplessis@glencore.co.za
2. Extra-Ordinary Professor, University of Pretoria, Department of Mining Engineering, South Africa. Email: jan.duplessis@up.ac.za
3. Specialist Engineer, BBE Consulting, Johannesburg, South Africa. Email: mbiffi@bbe.co.za
4. Mechanical Engineer, BBE Consulting, Johannesburg, South Africa. Email: kvanzyl@bbe.co.za
5. Electrical Engineer, Hatch Ltd, Mississauga, Canada. Email: laszlo.kadar@hatch.com
6. Mechanical Engineer, Hatch Ltd, Mississauga, Canada. Email: jenna.wu@hatch.com
7. Analyst, Hatch Ltd, Mississauga, Canada. Email: krishna.prakash@hatch.com

ABSTRACT

Increasing levels of underground mining mechanisation require the transmission and use of greater quantities of energy. This heightens the risk of underground mine fires occurring in confined spaces, eg in transformer bays, on diesel or electric mobile equipment, or in battery charging bays.

This paper describes a comparative assessment tool that was developed for the management of fire risks posed by electrical distribution transformers in underground mining operations. When a transformer experiences power surges due to transformer malfunctions, network overloading, or short-circuiting, the 'fault' energy is released uncontrollably. In worst-case scenarios, the dielectric oil used to cool transformers may be ignited resulting in a significant conflagration dispersing smoke and toxic gases into the mine's ventilation network. The violent reaction could trigger a fire that may result in catastrophic consequences for the mining operation. This risk must be well understood and actively managed.

The fire risk assessment tool described in this paper has been developed for this purpose. It has been used to assess objectively the impact of various protective and mitigating interventions aimed at reducing transformer fire risks. The tool has been applied to evaluate transformer fire risks across a fleet of over 1000 distribution transformers in underground applications globally. The tool uses a high-level screening methodology to provide a semi-quantitative systematic approach to rank the assessed risk values and to assist the selection of appropriate single, or groups of, mitigating and/or preventative measures.

The tool was developed in consultation with underground mining and utility industry experts, relying on industry data, international standards, guidelines, and expert panel judgement. The tool brings rigour and a standardised approach to assess transformer and associated protection measures to yield objective and globally comparable risk rankings across diverse installations.

INTRODUCTION

Electrification in underground mines has been key to support the increase in mechanisation and improvements in mining efficiencies globally. Furthermore, the introduction of battery electric vehicles (BEVs) in the next few decades will have a major impact on mining ore movement, placing more emphasis on the reliability, versatility, and safety of more extensive underground electrical networks.

Increasing levels of underground mining mechanisation and electrification heightens the risk of underground mine fires occurring in confined spaces, eg transformer bays and battery charging bays. Driven by electrification, fast paced expansions in electrical systems further emphasises the importance of managing fire risks associated with the use of electrical energy in underground mining operations.

Underground electrical transformers are an equipment category that may pose a more serious fire hazard which requires suitable management. Electrical transformers are used in underground mining operations to step-up or step-down transmission voltage in the electrical distribution network. Transformers operate under large electrical loads continuously with efficiencies exceeding

90 per cent. However, the inefficiencies generate heat energy which must be transferred to the surroundings. The ability of the transformer to dissipate the heat energy is at the root of the fire hazard posed by electrical transformers. Power surges resulting from sudden increases in line loads and/or load instability, accelerate heat energy generation within the transformer. If the heat loads exceed the operational capacity and design of the transformers and switchgear, overheating, and/or arcing will be triggered which may result in a fire or explosion.

Global commodity producers often operate underground mining operations across a large span of geographic locations where local electrical and fire safety standards differ. Evaluating risks with the intention of providing risk-mitigation strategies supported by reliable decisions across multiple operations can be challenging.

This comparative fire risk assessment tool was developed from a risk framework supporting the allocation of risk reduction resources across a portfolio of underground distribution transformers. The tool has provided an objective assessment of various protective and mitigating measures to reduce transformer fire risks and prioritise the allocation of resources for the implement of identified interventions. The tool has been applied to evaluate transformer fire risks across a fleet of over 1000 distribution transformers in underground applications globally.

ASSESSMENT TOOL DEVELOPMENT

Objectives of the tool

This assessment tool provides a standard and clear framework using a semi-quantitative approach to evaluate and prioritise the fire risks of a large fleet of underground transformers in different underground applications globally.

The tool is intended to compliment engineering studies that implement protection measures or estimate the absolute likelihood of fire to any individual distribution transformer. It is not intended to provide guidance or demonstrate compliance to international guidelines and standards (Factory Mutual Global, 2013; Standards Australia, 2012).

Method overview

The assessment tool was developed with the application of industry empirical data, in conjunction with field experts, and was informed by international standards and guidelines (Factory Mutual Global, 2013; Standards Australia, 2012; National Fire Protection Association (NFPA); ANSI/IEEE C2-2017).

The tool considers several qualitative and quantitative inputs and outputs, includes a risk priority score, interpretation of the numerical outcomes, and provides high-level screening/ranking and guidance that define mitigation actions.

The assessment framework provides a standardised approach to assess and compare various transformer protection measures and allows the user to evaluate the cost of risk reduction measures against the relative level of reduced risk.

The following key steps were applied in the development of the assessment tool:

1. Understanding the mechanisms causing transformer fires, including mitigation strategies.

2. Performing a bow-tie analysis of underground oil-cooled transformer fire risk.

3. Application of a risk assessment framework.

4. Development and field testing of the assessment tool.

Transformer fire mechanism

Types of transformers and associated fire risk

Both oil cooled transformers and dry type transformers are operated in underground mines. Depending on the transformer type, the fire risks differ significantly.

Oil cooled transformers contain dielectric oil in the transformer tank which is the medium for the heat transfer process. Most distribution level transformers use ONAN (natural oil flow, natural air flow) cooling configuration, where heat is dissipated via natural convection to the circulating air in the underground ventilation network. Some underground distribution transformers with higher load capacity may use ONAF (natural oil flow, forced air flow) cooling with a fan or blower forcing air through the external heat exchanger. As a safety precaution against fire, liquid-cooled transformers are hermetically sealed to exclude any air and moisture from the flammable materials present inside the sealed tank to reduce ignition risk. In some transformers, the volume above the dielectric fluid is pressurised with nitrogen to provide the same effect.

Dry type transformers use epoxy resins as the dielectric insulating medium. The windings and core are enclosed in a sealed container filled with pressurised air or gas as the cooling medium. Dry type transformers reduce fire risk by eliminating the flammable dielectric oil. However, have a larger footprint, are more costly than oil cooled transformers for the same capacity and are more sensitive to dusty environments.

Transformers contain flammable materials. Dry type transformers are constructed with resin materials which can combust if exposed to intense heat energy. Oil-cooled transformers are more susceptible to fire incidents due to the use of flammable dielectric fluids. In addition to the transformer contents, ancillary equipment such as electrical cables may also ignite and propagate a fire. As oil cooled transformers contain a large volume of flammable oil that may prolong an initial fire, the consequence of oil cooled transformer ignitions is typically more severe than those involving dry types.

The likelihood of a transformer fire occurring varies significantly between the application and transformer types. However, data from Firetrace International (2023) indicates that the average annual rate of fire events fluctuates between 0.9 per cent and 1.0 per cent. This means that based on a 40-year service life, 2.4 per cent to 4 per cent of all transformers could be exposed to potential fire risk.

Safe operation of distribution transformers is based on proper design and selection for the intended use:

- Designing transformers to operate at the requisite steady-state and transient operations.
- Considering the expected electrical network's static and dynamic loads – which require consideration of varying electrical loads based on the dynamics of mining operations.
- Understanding, determining, and avoiding any conditions or situations that will lead to an uncontrollable power surge through the transformer.
- Selecting sites that are adequately ventilated to meet the anticipated heat energy rejection rates without the transformer overheating.
- Inspections and monitoring of safe operation.
- Ultimately considering the consequences of a transformer fire on the air quality conditions downstream of the site and to the rest of the mine or section.

Causes of transformer fires

A fire requires the presence of oxygen, heat, and fuel to ignite and propagate. Transformer fires are typically caused by an abnormal operating condition, or fault event, resulting in the following conditions that could enable an initial ignition and subsequent fire:

- Oxygen presence – enclosure breach will expose the sealed flammable materials (dielectric oil and gases) in the transformer to oxygen in the atmosphere by the ruptured transformer tank or bushing.
- Heat – intense heat energy caused by an arc, spark or elevated temperature surfaces will enable an ignition.
- Fuel – (hot) flammable material is present in the transformer (insulation, dielectric fluid, combustible vapor), or other surrounding flammable materials will propagate the fire.

Abnormal operating conditions and faults could eventually result in a fire, various studies and papers have explored the causes and protective devices to prevent a transformer fire, such as Hoole *et al* (2017). Figure 1 provides a diagrammatic representation of some plausible fault scenarios and of related protection devices or mechanisms aimed at preventing a fire when the fault occurs. A fault does not always result in a fire. For ignition to occur, either an arc ignition of flammable materials must occur while there is a simultaneous enclosure breach introducing oxygen, or when the flammable dielectric fluid reaches ignition temperature inside the transformer tank, and the enclosure ruptures resulting in an ignition. Various protection mechanisms are available to either prevent enclosure rupture (pressure relief devices, more robust design) control temperature (temperature monitoring interlocks) and prevent arcing (overcurrent protection devices). Using a dielectric oil with a higher flash point will provide more time for protection devices to de-energise the transformer prior to reaching ignition temperature.

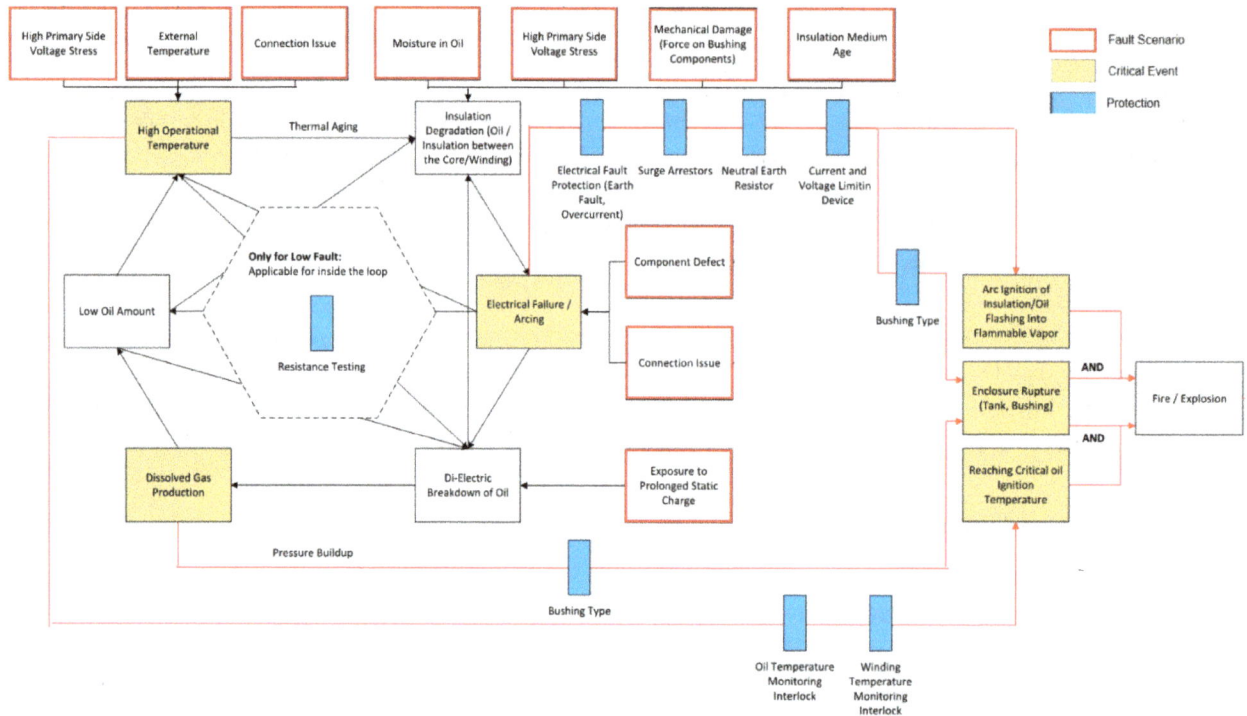

FIG 1 – Transformer tank internal failure loop.

Under normal operating conditions the heat energy dissipated in the transformer's windings will heat-up the dielectric fluid (transformer oil) which circulates between the main enclosure and external heat exchanger (radiator).

In a fault scenario, high energy release may result in one or more failure conditions as shown in Figure 1. These include excessively high operational temperature, low oil level, di-electric breakdown of oil, generation of dissolved gases, electrical failure/arcing, and insulation degradation. Many protection devices or mitigation strategies exist to detect and control the fault events such as temperature monitoring interlocks, pressure relief devices, etc. However, if these mechanisms fail to detect and de-energise the transformer, an internal failure loop can occur, and may give rise to additional failure conditions. Any transformer fault can progressively worsen in this loop unless the fault is cleared. Regardless of what or where the initial fault is, the fault can get progressively worse until it breaks the cycle to produce conditions that will enable a fire as indicated below.

- Electrical faults

  In oil-cooled transformers, electrical faults induce the formation of high-energy electrical arcs or overheating of the dielectric fluid to temperatures beyond its flash point. Due to the substantial energy released in an arcing event, liquid dielectric fluids produce flammable gaseous by-products that could over-pressurise the transformer tank. If the rapid expansion of the gases formed is not prevented by a pressure-relief device, such as rupture disc, pressure relief valve or Buchholz device, the transformer enclosure may be breached bringing the hot flammable gases and liquids into contact with oxygen external to the transformer where it may be ignited by the arc or other energised sources.

- Mechanical faults

Mechanical damage to transformers is usually from external sources that compromise the structural integrity of the transformer's enclosure and/or internal frame. Any loss of enclosure integrity of liquid-cooled transformers may lead to the loss of dielectric fluid. A minimal leak, defined as 'weeping' is generally not seen as a major fire hazard in itself but, the subsequent ingress of air (oxygen and moisture) into the transformer's tank will compromise the inert nature of the contents and the properties of the dielectric fluid, making the transformer more susceptible to any subsequent faults.

In an internal overpressure event, the damaged transformer enclosure may be breached allowing the hot or burning oil to be released into the external atmosphere, enabling ignition. Mechanical faults or defects involving the integrity of the transformer's cable connections (bushings), tap-changers and devices such as rupture diaphragms or Buchholtz devices, may also 'prime' or contribute to the ignition of the gases in the transformer tank.

- Dielectric fluid deterioration

Over the life of a transformer, the condition of the dielectric fluid should be monitored to ensure safe and adequate performance throughout. Deterioration of chemical conditions will affect the dielectric properties of the fluid, reduce transformer efficiency and, more pertinently, lead to the formation of hydrocarbon gas cocktails with corresponding pressure build-up which may result in premature internal arcing.

Regular analyses for the presence of dissolved gases and solid hydrocarbon impurities in the dielectric fluid should be performed by accredited laboratories to provide a record and history of the fluid's condition. This, together with recommendations from the laboratories and a history of low energy faults that may have also occurred, will indicate that the dielectric fluid should be replaced (or filtered) or repaired to reduce the possibility of such occurrences.

For ease of analysis, the transformer fire risk assessment tool further categorises transformer faults qualitatively into low and high energy faults, each requiring appropriate protection mechanisms for fire prevention and detection.

- Low energy fault events

Low energy faults may be either intermittent or continuous. If undetected, these faults could escalate to a more serious fault level due to continual damage and/or degradation of transformer components and insulation properties as described above. The control of low energy faults requires independent detection and protection systems that will de-energise and mitigate the situation when detected timely. These measures include routine maintenance, dissolved gas and particulate analyses, oil and winding temperature monitoring interlocks and employing pressure sensors interlocked with cut-out relays. Inherent in this is assurance that control mechanisms are calibrated, assessed, and kept in good working order as guided by the maintenance schedule.

Low energy faults also include partial energy discharges, or a localised internal overheating – that may be detected through infra-red imaging.

- High energy fault events

High energy electrical fault currents have a significantly greater ability to result in extremely rapid-fire development. Due to the unpredictability of these fault types, there is a possibility that all protective measures may fail to de-energise the transformer promptly, leading to a transformer fire.

These faults are characterised by sudden, high-energy arcing leading to a significant increase in dielectric fluid temperature. This will induce major dielectric fluid decomposition releasing large volumes of gas within the transformer enclosure resulting in a catastrophic mechanical failure of the enclosure. The fire risk is driven by the interaction of the three enabling events: arc occurrence, ignition of the insulating material and transformer enclosure breach. The rapid detection of any of these events and subsequent de-activation of the energy source must be executed promptly to prevent the fire.

Protective tactic

The tactics employed in the prevention of transformer fires may vary in complexity and degree of sophistication depending on the levels of risk assessed or considered to be 'acceptable,' by the user.

The transformer fire risk assessment tool evaluates the effectiveness of response tactics provided for each site. This information is used to rank the operational fire risk levels to which each transformer may be exposed.

In all instances, the promptness with which certain interventions must be activated to prevent an initial ignition source from escalating into a fire, is also a risk-based consideration and an essential aspect in the successful mitigation of any initial fault. Due to this requirement, the tool places significant importance on the level of detection and (automated) response speed in the detection and mitigation of any impending fire event.

The following protective tactics are evaluated by the risk tool:

- Dielectric fluid selection- reduce likelihood of ignition due to higher flashpoint.

  Mineral oil has been used extensively as a dielectric fluid in transformers. The flashpoint of mineral-oil-based fluids varies between 150°C and 170°C. More modern designs use ester-oil based dielectric fluids have flashpoint, exceeding 320°C. As transformer fires involve violent combustion of dielectric liquids or oil-cooled switchgear equipment, the use of lower flammability (higher flashpoint) fluids reduces fire risk (DelFiacco, Luksich and Rapp, 2013).

- Transformer operation monitoring – detect faults, de-energise and set off corrective actions.

  The monitoring of operational parameters such as core and windings temperature, electrical loading, phase load balance, dielectric fluid temperature and levels, provide useful pre-emptive information. This must be complemented by the prompt detection of events such as electrical over-loading, dielectric, core and/or winding overheating, gaseous pressure surges and loss of structural integrity. Real-time condition monitoring and recording of data coupled to routine dielectric fluid sampling and analyses, regular visual inspections and a preventative maintenance program are preferred measures.

- High energy events and quick response intervention.

  High energy events are sudden and involve the generation of flammable gases in a very short period. Rapid pressure release mechanisms include rupture disc, pressure relief valve (PRV) or Buchholtz devices. The implementation of these measures requires adequate design of the protective control and cut-out circuitry to achieve fast and safe response.

Some protective measures work for both low energy fault detection and high energy fault quick response intervention, and some only work for one or the other.

- For low energy faults, the protections are devices or processes that can independently detect the fault and de-energise or clear the fault. The most important protections include:
  - Frequent routine maintenance testing such as dissolved gas analysis.
  - Buchholz relay.
  - Oil temperature monitoring interlock.
  - Winding temperature monitoring interlock.
- For high energy faults, the protections are acting in conjunction and are dependent on each other to quickly de-energise the transformer before enclosure breach and ignition to prevent a fire:

- o De-energization: overcurrent protection, PRV with hardwired trip/Buchholz, neutral earth resistor, temperature interlocks.
- o High flash point oils.
- o Enclosure protection: PRV, Buchholz relay, rupture disc, explosion vent, bushing type.

Bow tie analysis as basis of design

The Transformer Fire Risk Assessment Tool has been developed by considering reviews of Bow Tie Analyses (BTAs) (Wolters Kluwer, 2024; Ruijter and Guldenmund, 2016) applied to transformer fire events across diverse mining operations. These reviews substantiated the need to develop a standard risk assessment tool presented here.

A generic qualitative BTA for the risk of an underground oil-cooled distribution transformer fire was developed. This risk bow-tie 'sample' identifies the failure modes and controls associated with the management transformer fire risks. The BTA only considers an internal failure of the transformer, and excludes external factors such as mechanical impact, sabotage, etc. The BTA outcomes presented in this paper are of a generic nature.

In a BTA, the top event is the point when control over a hazardous situation is lost. For an oil-cooled transformer fire, this was identified as the moment when 'an internal arc in the transformer's oil tank' occurs. An adequately accurate definition of the top event or Materially Unwanted Event (MUE) for any BTA is pivotal in determining relevant associated causes, ensuing impacts or consequences and corresponding preventative and mitigating controls.

Several generic measures can be introduced to prevent the occurrence of an electric arc in the oil-filled transformer tank. Once an arc has occurred, there are additional mitigating controls that can be activated to prevent this from degenerating into a sustained uncontrolled fire or explosion with consequences that, in the case of an underground mines, may result in multiple fatalities.

The BTA was conducted using a BTA facilitator, electrical design engineers and other technical experts responsible for the installation and maintenance of oil-cooled transformers underground. The identified potential transformer fire causes (fuel and ignition sources), and typical transformer failure modes were used to guide the qualitative BTA development. The causes and consequences with identified preventative and mitigating (recovery) controls from the BTA are listed in Table 1.

TABLE 1
Identified oil-cooled transformer fire causes and consequences.

Potential causes	Potential threats
Overvoltage	Electric fault
Di-electric breakdown of oil	Release of toxic gases into the workings resulting in multiple fatalities
Insulation breakdown	Internal arcing
Low oil level	Transformer overheating
Latent manufacturing defects	Unforeseen transformer malfunction
Unintentional interaction with live equipment on the load side	Transformer overloading

Identified generic critical controls

From the generic BTA, critical controls were identified using various criteria, including guidance from the International Council on Mining and Metals (ICMM) guidelines. The critical controls identified are:

Preventative critical controls

1. Overload protection relays.
2. Oil temperature monitoring interlocks.

3. Oil level monitoring interlocks (Buchholz relay trip on low oil level or similar).

4. Insulation resistance testing (Pre-commissioning).

5. Isolate lockout during maintenance.

Mitigating critical controls

1. Electrical fault protection (Sensitive earth fault).

2. Over current/short-circuit protection.

3. Gas over pressure interlock (Buchholz relay trip on low oil level or similar).

4. Selection of low flammability di-electric oil.

5. Automated fire suppression system.

When defining and implementing critical controls, the use of recognised design standards and practices assists in determining the performance requirements of these essential controls.

Risk assessment framework

Overview of framework

This semi-quantitative tool architecture is built around low energy fault detection and response, and high energy fault quick response de-energisation.

The assessment framework considers fire risk in four different components, informed by the fire mechanism and risk bow tie as shown in Figure 2.

1. Initiating fault event and impacts of site-specific operating conditions on the fault event frequency.

2. Risk of protection mechanisms failure on detecting and responding to low energy faults.

3. Risk of protection mechanisms failure in preventing a transformer fire resulting from a high energy fault event.

4. Risk of catastrophic mitigation measure failure due to the fire event (fatality, personnel harm).

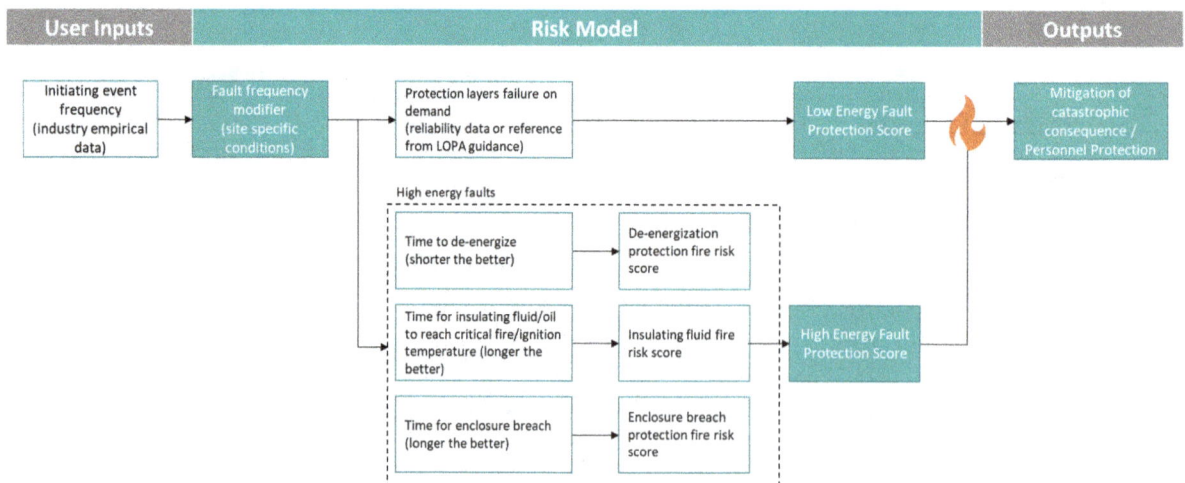

FIG 2 – Risk Assessment Model.

The essential risk analysis model uses quantitative and qualitative transformer metrics from user input, and using expert and industry standard empirical data, calculates a comparative fault frequency. For low energy faults, the layer of protection analysis (LOPA) is used to calculate a low energy fault protection score. Higher energy faults again take into consideration the semi-quantitative industry and expert empirical data to yield a high energy fault protection score. Finally, in the event of a fire/explosion, the qualitative LOPA is again employed to yield a fire mitigation and personnel protection score. Industry research and data such as Kwok-Lung Ng (2007), Petersen (2014) and

Shayan Tariq, Afzal and Khan (2015) were referenced to inform the LOPA to determine the initiating event frequency and the performance of fire protection systems such as fire detection systems.

Quantitative layers of protection analysis

The LOPA method is applied for quantitative risk analysis for its simplicity. It produces order-of-magnitude results which are suitable for high level quantitative risk analysis. The data required to apply LOPA is readily available in the public domain. However, application of LOPA requires the protection layers to be independent of each other, which is not suitable for certain types of risks.

Certain LOPA elements are used as guidance in the development of the underground transformer fire risk assessment tool where applicable. Other qualitative methods are applied where LOPA is not suitable.

The layer of protection analysis (LOPA) is a method of analysis that evaluates the residual risk against a hazardous scenario. LOPA evaluates the several existing, independent, protection layers (IPLs) against the defined hazardous scenario. This is a semi-quantitative approach that operates in order of magnitude approximations. The purpose of LOPA is to determine the efficacy of each independent layer against a specified protection layer, viewed in order of magnitude (Zhu, 2021). LOPA also calculates the frequency of a hazardous situation occurring with the employed layers of protection in place, through a basic mathematical formula. The formula considers two important metrics: an initiating event frequency (IEF) and the probability of failure on demand (PFD) of an IPL (Hlouschko, Cramer, and Pergler, 2020). The initiating event frequency provides the probability of a fault condition, while the PFD estimates the probability that the protection layer will fail to eliminate the fault when it occurs. LOPA can account for conditional modifiers that may affect the protection layer, such as the probability of ignition, probability of personnel in the affected area and probability of fatal injury. Based on guidance from the Center for Chemical Process Safety (2001) the formula to calculate the frequency of the unmitigated consequence is as below:

$$Frequency\ of\ unmitigated\ consequence = IEF * conditional\ modfiers * PFD\ of\ IPL$$

For LOPA to be effective, some basic criteria for the IPLs must be met. The IPL must work effectively and as intended to prevent the consequences. In addition, it must be able to function independently of other protection layers which target the same scenario, and any other initiating event. LOPA experiences limitations when it comes to identify which failure mechanisms can be analysed. Overly complex or poorly understood mechanisms cannot be approached semi-quantitatively, and, in these cases, LOPA does not serve as a substitute for detailed quantitative analysis.

Figure 3 shows schematically the application of LOPA to transformers' low energy faults as used by the tool. Here five IPLs have been identified. The PFD for each IPL has been obtained from available failure occurrence data and the corresponding failure probabilities assigned. These are combined logarithmically to derive an estimated annual failure frequency. By adjusting the failure frequency data for individual transformer types, a relative comparison of the risk across the transformer fleet is thus drawn. Similar analyses for high energy faults and personal protection measures, informed by the respective IPL failure statistics are derived.

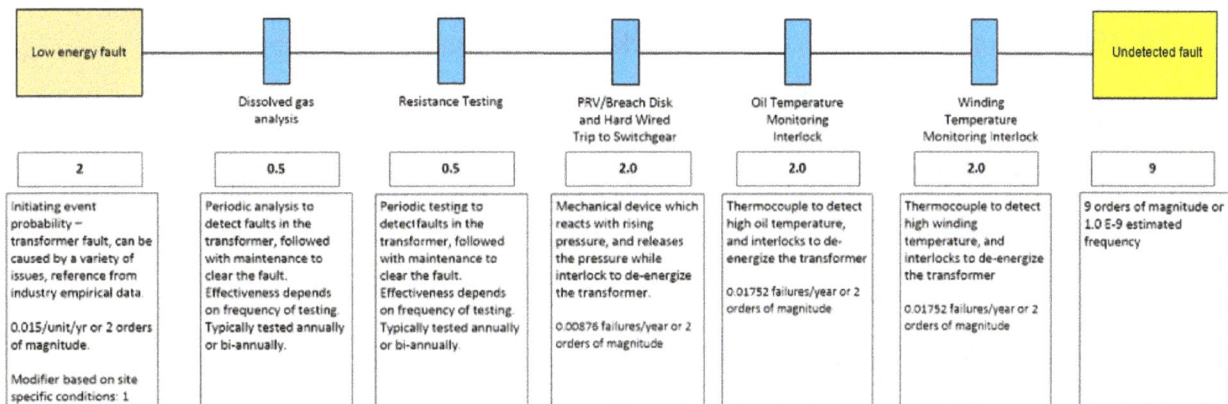

FIG 3 – Example of LOPA applied to low energy fault detection.

Expert panel via Delphi method

Risk of protection mechanisms failure in preventing a transformer fire under high energy fault event, cannot be quantitatively evaluated using the LOPA method, as all protection mechanisms function in conjunction with very short reaction time (milliseconds) to prevent the fire event. For this instance, the assessment tool uses the Delphi method to provide qualitative assessment of the risk.

The Delphi method considers opinions from an expert panel, with careful consideration on limitations of the assumptions and incorporates different perspectives from the expert panel.

A workshop was held to review the case where a high energy fault event occurs in the transformer, and along with the expert panel, each protection mechanism was identified, and its successful operation and dependencies on other protection mechanisms were discussed. Participants contributed their expertise and evidence to support the assumptions applied. Consensus was sought through discussion.

The expert panel provided independent ranking and scoring for the relative effectiveness of each protection mechanism serving similar functions.

FIELD APPLICATION OF THE ASSESSMENT TOOL

Inputs

The input data required by the evaluation tool is subdivided into three broad categories and the input sheet primarily uses a drop-down selection and simple information entry:

- Transformer design parameters and operational conditions:
  - Transformer and site design details including transformer age, ambient versus design elevation, humidity, groundwater ingress, flooding risk, exposure to excessive and repeated ground vibration due to blasting in the vicinity are required.
  - Accessibility of transformer for inspection, maintenance, and repairs.
  - Design and operating conditions such as up-to-date load flow analysis, insulation protection co-ordination studies; typical loading conditions versus design capacity, upstream protective devices.
  - Transformer cooling configuration and dielectric fluid type, including whether it has been retro filled.
  - History of faults, maintenance quality and of dielectric fluid condition from previous dissolved gas analysis, condition of the enclosure.
- Transformer protective measures and devices:
  - Standard fuse protection, earthing/grounding protection, neutral earth resistor, differential phase current relay.
  - Overcurrent protection: mechanical, electronic and fast acting with power interrupt interlock.
  - Winding temperature monitoring interlock, oil temperature monitoring interlock.
  - Overpressure protection with power interrupt interlock (pressure relief valve, Buchholtz relay).
  - Resistance testing, dissolved gas analysis.
  - Cable box pressure relief device or arc-venting mechanism.
- Fire response measures- limit worker exposure to the consequences of the fire:
  - Emergency response procedures that may include worker evacuation to refuge chambers or other places of safety.
  - Forced ventilation, fire detection and suppression systems.

- Provision of fire walls and/or bunds designed to contain the fire and any dielectric fluid spillage from the transformer- thereby also limiting any propagation of the fire.

Outputs and dashboard

To guide operators in the functional use of the tool, five generic transformer installation samples have been incorporated into the tool.

Sample A: a mineral oil-filled transformer with basic protection measures, limited to the use of Buchholz relay, surge arrestors and inert gas sealing of the tank. The transformer has limited maintenance access. This transformer scored above average in transformer failure rate modifier, and low in both the low and high energy fault protection scores. The general guidance output is '_Review and Implement Additional Controls_,' indicating the low fire protection score and the high priority to review the controls.

Sample B: a mineral oil-filled transformer with standard protection measures. In addition to the protection measures of sample A, sample B includes annual dissolved gas analyses, resistance testing, electrical overcurrent protection, and the inclusion of a neutral earth resistor. With these inputs, this sample transformer scores above average in transformer fail rate modifier, and medium in low and high energy protection score. The overall guidance provided is to '_Evaluate additional improvements_,' indicating the need to improve existing fire protection controls by introducing additional protection measures.

Sample C: a mineral oil-filled transformer with high protection standards. This transformer has fast-acting overcurrent protection and winding and oil temperature interlocks. Sample C also contains additional protection including an electrical fault sensitive fault and a current relay. Sample C scores above average for transformer failure rate modifier, a high score for the low energy fault protection, and a medium score for the high energy fault protection. The general guidance given for this transformer is to '_Evaluate Additional Improvements_,' indicating that additional improvements can be implemented.

Sample D: a natural ester oil-filled transformer with standard protection measures, like sample B. With only the difference being the oil type, the sample scored medium in low energy fault protection and high in high energy fault protection. Sample D's guidance is to '_Maintain and Inspect_,' indicating that the transformer fire protection score is comparable to international standards and guidelines and maintenance to ensure the transformer functions normally.

Sample E: a dry-type transformer with standard protection measures where applicable. For this transformer type, there is a low score for the low energy fault protection and a high score for the high energy fault protection. Figure 4 is a screen shot of the tool's dashboard produced by sample analyses of a fleet of transformers. This indicates the site modifiers used, the scoring of the three risk components (Low Energy, High Energy and Personal Protection) and a brief interpretation of these results for the guidance of the user.

No.	Case name	Qty	Transformer fault rate modifier	Low energy fault protection	High energy fault protection score	Personnel fire protection	Overall Interpretation	Condition Details
A	Mineral Oil Basic Protection (SAMPLE)	0	1.1	2.0	4.8	1.0	REVIEW AND IMPLEMENT ADDITIONAL CONTROLS	NOTE - Limited maintenance access////////
B	Mineral Oil Standard Protection (SAMPLE)	0	1.1	3.0	5.5	1.0	EVALUATE ADDITIONAL IMPROVEMENTS	////////
C	Mineral Oil High Protection (SAMPLE)	0	1.1	7.0	6.8	1.0	EVALUATE ADDITIONAL IMPROVEMENTS	////////
D	Natural Ester Oil Standard Protection (SAMPLE)	0	1.1	3.0	8.0	1.0	MAINTAIN AND INSPECT	////////
E	Dry Type (SAMPLE)	0	1.1	0.0	10.0	1.0	MAINTAIN AND INSPECT	////////
F	AS4871 GCAA Input	0	1.0	4.5	5.3	1.0	EVALUATE ADDITIONAL IMPROVEMENTS	////////
G	FM Global Datasheet 4-5Mineral Oil(All recomm	0	1.1	7.0	5.0	6.0	MAINTAIN AND INSPECT	////////
H	FM Global Datasheet 4-5FM Approved (All recom	0	1.1	3.0	7.7	1.0	MAINTAIN AND INSPECT	////////

FIG 4 – Example of assessment tool result dashboard.

All transformers tool inputs are calibrated against two international standards: AS4871 and FM Global. The AS4871 standard provides minimum design, manufacturing, testing and performance requirements for electrical equipment. For transformers specifically, the standard requires that transformer protection must include suitably rated fault-break devices for the primary winding. Oil-filled transformers must also adhere to the risk management clause outlined in the standard. The FM Global data sheet applies to fire protection mechanisms for all transformers and includes loss prevention recommendations in the form of electrical protection, testing, maintenance, and operation. This standard is applicable to distribution, power, and specialty transformer applications. To incorporate consideration and calibration of these standards, the tool uses inputs that comply to these codes.

The assessment tool can provide semi-quantitative comparison ranking for several transformer configurations using different cooling methods, operating conditions, maintenance practices and protective devices. The tool will, indicate visually which transformers in the fleet need to be prioritised for mitigation. Based on the quality of inputs, it can also highlight issues such as lack of maintenance access, lack of load flow study, or poor oil conditions.

LESSONS LEARNED

This risk assessment tool implements an evaluation of the catastrophic risk posed by a major fire in an underground mine involving transformers. When Glencore undertook this development, it was envisioned to investigate the degree of adherence by mining operations across the Globe to a variety of standards using an equitable footing.

In the past, transformer fire risk assessments were performed as part of a broader study on the prevention of catastrophic hazards – those with a low probability of occurring but that could result in multiple fatalities. Initially, the inclusion of transformer fires as risks with such a potential was not well accepted or understood.

The technical intent of this work is justified from a good governance perspective. However, human nature, being what it is, makes individuals feel uncomfortable when their work is placed under scrutiny particularly when one's 'performance' is compared quantitatively with that of a peer in directly measurable terms.

The paradox of such erroneous deductions is that the true intent of this work was lost by alluding to the extremely low probability of such an event ever occurring – particularly at the mine in question. This evidenced that mine management, in this case not being electrical engineers, needed an appreciation that additional controls are well within their capability, that they can effectively reduce such risks to nominal levels and, most importantly, they are also practicable to implement.

Considerable progress was also made by encompassing several additional functions and evaluation steps that identified more complex operational issues and the relationships between them. The implementation of LOPA in two sections of the tool has enabled the comparison of separate independent protective measures aligned with the heuristic study of transformer and network faults.

The need for fire simulation models based on cogent numerical analyses supported by effective visual outputs is also a requirement specified in the tool. This was used to understand the impact of positioning transformers at various locations in the mine, and to visualise the dispersion of toxic atmospheres emitting from the fire. This methodology was also used to assess the safety and effectiveness of worker escape tactics.

Calibration of the tool against several international standards was essential to demonstrate its relevance and value. With time, it became clear that. to remain relevant, the tool must be reviewed regularly to include pertinent technological advancements and changes in operational standards.

The implementation of this tool also resulted in the updating of transformer standards in all regions by prompting the planning and implementation of structured capital projects to meet the target '*Maintain and Inspect*' tool output conditions.

A significant lesson learned from this work is that analyses of complex systems require the use of tools that promote the objective selection of fit-for-purpose corrective interventions. Earlier versions of the tool considered the use of fire detection and suppression on par with interlocking power isolating mechanisms. This effectively contradicted the use of the hierarchy of controls in the decision-making process. For example, in several cases, the allocation of funds for the installation of fast-acting interlocking power cut-outs is preferable to (at times) more costly fire detection and suppression systems.

CONCLUSIONS

The transformer fire risk assessment tool has been developed to provide an objective and consistent evaluation of the fire risk posed by electrical distribution transformers in underground mines.

The principal aim of this development was the provision of a methodology to be used by all of Glencore's operations globally to yield consistent and objective outcomes that would benefit mine management or independent parties in assessing transformer fire risks equitably.

The tool was developed by electrical engineers and fire professionals in consultation with Glencore's technical leadership and experts in the various regions.

At the start of the HSEC Audit's underground assurance campaign, the occurrence of fires in intake airways resulting in an irrespirable atmosphere contaminating the mine's fresh-air network was identified and accepted as a major unwanted event. Oddly, though, the fire risk posed by the operation of liquid-cooled transformers utilising a couple of tons of combustible mineral oil was not considered with the same degree of attention.

Furthermore, the most frequently used intervention mitigating action was the use of fire detection and suppression systems. In reviewing the effectiveness of these and other measures, the expert team realised that the assessment tool should guide through a structured evaluation yielding adequate and effective preventative critical controls.

The roll-out of this tool has contributed to changing these and other similar misconceptions, aligning Glencore's strategy in this area of risk management and generating some common ground across regions to stimulate information sharing. The tool has been used to evaluate the transformer fire risk affecting the full fleet of underground units while assisting fire risk experts in making relevant recommendations.

This tool has combined the expert knowledge from fire engineers, electrical engineers, and practitioners into an uncomplicated application that can be used to evaluate any installation and assess the effectiveness of additional protective controls. Over several years, the tool has been honed and adapted to include any intervening evidence including lessons learned from incidents, recognised technical improvements and changes in international standards that took place during this journey.

This tool is not only an enabler of safe electrical engineering applications but is also a catalyst for fostering a mining safety culture and stimulating its self-improvement.

REFERENCES

Center for Chemical Process Safety, 2001. *Layer of Protection Analysis: Simplified Process Risk Assessment*, Center for Chemical Process Safety of the American Institute of Chemical Engineers (Wiley). Available from: <https://onlinelibrary.wiley.com/doi/pdf/10.1002/9780470935590.fmatter>

DelFiacco, G, Luksich, J and Rapp, K, 2013. Proactive Fire Safety Transformers, Quantifying Risk of Transformer Failures: Proactive versus Reactive Risk Mitigation Approach.

Firetrace International, 2023. Transformer FireProtection. Available from: <https://www.firetrace.com/transformer-fire-protection>

Hlouschko, S, Cramer, M and Pergler, M, 2020. Conference of Metallurgists, in A Semi-Quantitative Catastrophic Risk Likelihood Prioritization Framework for the Metallurgical Industry.

Hoole, P R P, Rufus, S A, Hashim, N I, Saad, M H I, Abdullah, A S, Othman, A-K H, Piralaharan, K, Aravind, C V and Hoole, S R H, 2017. Power Transformer Fire and Explosion: Causes and Control, *International Journal of Control Theory and Applications*.

Kwok-Lung Ng, A, 2007. Risk Assessment of Transformer Fire Protection in a Typical New Zealand High-Rise Building, Master of Engineering thesis, University of Canterbury, Christchurch.

Petersen, A, 2014. Risk Equals Probability Times Consequences Risk Equals Probability Times Consequences, *Transmission and Distribution World,* 66(4):10870849.

Ruijter, A and Guldenmund, F, 2016. The bowtie method: A review, *Journal of Safety Science.*

Shayan Tariq, J, Afzal, R and Khan, A Z, 2015. Transformer Failures, Causes and Impact, International Conference Data Mining, Civil and Mechanical Engineering (ICDMCME'2015).

Wolters Kluwer, 2024. History of the Bow Tie, Wolters Kluwer. Available from: <https://www.wolterskluwer.com/en/solutions/enablon/bowtie/expert-insights/barrier-based-risk-management-knowledge-base/the-historie-of-bowtie>

Zhu, F X X, 2021. Digitalization and Analytics for Smart Plant Performance: Theory and Applications. https://doi.org/10.1002/9781119634140

NORMATIVE STANDARDS AND REFERENCES

Factory Mutual Global, 2013. Transformers, FM Global Property Loss Prevention Data Sheets 5-4, 2013.

Institute of Electrical and Electronics Engineers (IEEE) as ANSI/IEEE C2-2017.

International Council on Large Electric Systems (CIGRE).

International Electro Technical Commission (IEC).

National Fire Protection Association (NFPA). https://www.nfpa.org/Codes-and-Standards/All-Codes-and-Standards/List-of-Codes-and-Standards

Standards Australia, 2012. Australian/New Zealand Standard AS4871.1, Electrical Equipment for Mines and Quarries, 2012.

Underwriters Laboratories (UL).

Synthesis and characteristics of thermosensitive hydrogel for extinguishing coalmines fire

X Huang[1] and L Ma[2]

1. PhD student, College of Safety Science and Engineering, Xi'an University of Science and Technology, Xi'an 710054, PR China. Email: 1079803421@qq.com
2. Professor, College of Safety Science and Engineering, Xi'an University of Science and Technology, Xi'an 710054, PR China. Email: mal@xust.edu.cn

ABSTRACT

Thermosensitive hydrogel shows good application prospect in the prevention of coal spontaneous combustion. In this study, a thermosensitive hydrogel was prepared by adding methylcellulose (MC), sodium polyacrylate (PAAS) and magnesium chloride ($MgCl_2$) as a mixture. Single factor experiments and central composition design (CCD method) were carried out to optimise the gel ratio, the inhibitory effect of the gel on three kinds of bituminous coal was analysed, a test platform was set-up (coal capacity of 200 kg) to test the gel's fire-fighting performance. Results showed that the optimal preparation ratio of gel was MC: PAAS: $MgCl_2$=1.3:2.11:16, resulting in the three response parameters of Y_1 [gelation time, 50 sec], Y_2 [gel strength, 53 KPa], and Y_3 [viscosity, 11.6 Pa·s]. Temperature programmed experiment indicated that the gel can effectively reduce the release of CO and CO_2 emission, reduce the oxygen consumption rate, and delay the initial occurrence temperature of C_2H_4. After injecting the gel in the burning coal, the thermosensitive hydrogel took phase change and absorbed the high temperature as well as thermal energy, reducing the composite speed of the coal oxygen, resulting in the coal fire extinguishing. Research can provide new ideas for the research and application of gel fire extinguishing agents.

INTRODUCTION

Coal has always occupied a leading position in energy production and utilisation in China (Qin *et al*, 2021; Wei *et al*, 2021). Although the incremental substitution effect of the new energy on coal is obvious, the strategic position of coal in China's economic development remains unshakable. However, it is totally observed that the coal spontaneous combustion is one of the most serious disasters among all mine fire disasters, which seriously threaten the sustainable development of mine owning to the occurrence of coal fires in more than half of the coalmines in China (Kong *et al*, 2018; Deng *et al*, 2016). Coal fires will not only burn coal resources and cause huge economic losses (Kong *et al*, 2019), but also produce a large amount of harmful gases such as CO, H_2S, SO_2, it seriously pollutes our living environment (Ajrash, Zanganeh and Moghtaderi, 2017; Zhang *et al*, 2018). In addition, coal spontaneous combustion frequently caused the accidents such as coal fire, gas combustion and explosions, resulting in the serious casualties (Cai *et al*, 2020). most researchers have attempted numerous techniques to prevent coal spontaneous combustion and extend the spontaneous combustion period such as grouting (Ray and Singh, 2007); pumping of inert gases (Zhou *et al*, 2015; Qin *et al*, 2016) or three-phase foams (Qin *et al*, 2015; Lu and Qin, 2015), spraying of inhibitors (Ma *et al*, 2015; Wang *et al*, 2014), chemical fog and claying inorganic foams (Hu, Cheng and Wang, 2014; Lu and Qin, 2015), polymer materials (Hu and Wang, 2013; Guo *et al*, 2019), or other ways. Admittedly, these technologies have an important effect on the coalmine fire preventing and controlling. There are still exist some drawbacks, for instance, the grout material tends to separate from water, resulting in the quantities of slurry loss. the inert gases have a tendency to disperse with leakages. Physical inhibitors such as magnesium chloride and other water-absorbing salts have a short inhibition time, the difference between three-phases foam stability, commonly used chemical inhibitors (such as urea, diammonium borate, ammonium bicarbonate etc) are cost. Therefore, there is a demand for developing some novel materials to effectively prevent coal spontaneous combustion.

When compared with the above-mentioned fire extinction technologies, gel has been continuously developed in the field of fire prevention and extinction in coalmines owning to its integration of many advantages such as the cooling ability, plugging, water retention and air-leakage blocking properties (Ren *et al*, 2019; Tang and Wang, 2018; Zhou and Tang, 2018). Thermosensitive hydrogels, a type of special functional polymer material with wide applications, have been rapidly developed in recent years. It is in the sol state below the lowest critical solution temperature (LCST) and solidifies into a gel above LCST (Ruel-Gariépy and Leroux, 2004; Klouda, 2015). It can be inferred that thermosensitive hydrogel will be promising in the coalmines fire prevention, because it cannot only maintain the advantages of polymer gel such as water retention, cooling and sealing, but also overcome shortcomings including the difficulty of injection and residue processing, due to its high viscosity at the room temperature. A thermosensitive hydrogel based on sodium alginate and N-isopropylacrylamide was synthesised, and the effect of various additives on the swelling rate of hydrogel was discussed (Hu *et al*, 2015). A chitosan-based thermosensitive hydrogel was proposed, in view of its water absorption rate was 200~220 g/g and the LCST was 103°C, which was considered suitable for the coalmine fires prevention (Zhang, Hu and Cheng, 2015). A thermosensitive hydrogel (P (NIPA-co-SA)) was proposed and the inhibition effects with other inhibitors were compared, concluded that P (NIPA-co-SA) produced an endothermic reaction during the liquid-to-gel phase, considerably reduced the total heat released by anthracite and coke coal (Tsai *et al*, 2018)). Chitosan was grafted and polymerised into hydrogel (CTS-g-P (AA-co-MAA)), and proved that it can effectively prevent the initial oxidation of coal (Jiang and Dou, 2020).

In this study, a thermosensitive hydrogel is prepared by using methylcellulose, sodium polyacrylate and magnesium chloride and characterised, the response surface method is utilised to optimise the process of this hydrogel. With the concentration of different additives as the investigating factors, the gelation time, gel strength and viscosity as indicators, the Central Composite Design in Design Expert software is used to obtain the significant relationship among each factor. Furthermore, this research teste the effects of the gel on coal spontaneous combustion characteristics via the temperature-programming experiments, and conducted fire-extinguishing experiments therewith. The research results can provide certain theoretical guiding significance and practical application value for coal spontaneous combustion prevention.

EXPERIMENTS

Gel preparation and performance tests

Samples of MC powder with different concentrations were dispersed in the deionised water at 70°C and dissolved as the temperature gradually dropped to room temperature, required amount of PAAS and $MgCl_2$ powder are added to the MC solution. The mixture is stirred continuously until it becomes transparently. The solution was left to defoam after standing at room temperature for 24 hrs. Figure 1 shows the operation process.

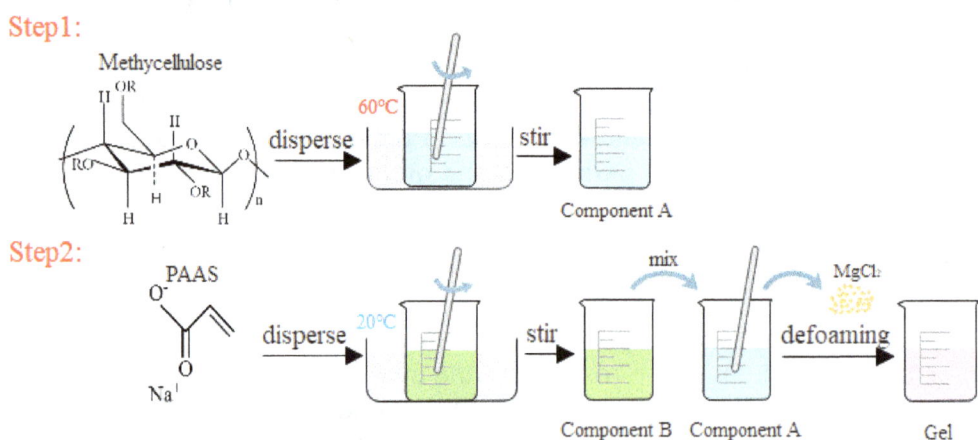

FIG 1 – Schematic diagram of gel preparation.

Measurement of gelation time

The gelation time is measured by the test tube inversion method, 5 mL of liquid hydrogel is put into the colorimetric tube, then put the colorimetric tube into a constant temperature water bath set at 90°C, and start timing from the moment it is placed, taking out the test tube at regular intervals, observing its flowability by inverting 180°, the tine when the gel loses its fluidity is recorded as the gelation time (Nair *et al*, 2007).

Measurement of gel strength

The gel strength of gel is measured by the breakthrough vacuum method. Figure 2 showed the brief diagram of the measuring device. A 25 mL sample solution was put into a colorimetric tube, which is placed in a constant temperature water bath at 90°C for 10 min to form a gel. Then, connecting the colorimetric tube quickly according to Figure 2 and start the 2XZ-2 vacuum pump, recording the maximum value on the vacuum gauge when the air broke through the gel, this is the breakthrough vacuum. Calibrating with water and glycerine each time the device is used, and the breakthrough vacuum of water and glycerine is 0.007 and 0.028 MPa, respectively. Each sample is measured three times, and then the average value is taken as the gel strength of samples (Desbrières, Hirrien and Ross-Murphy, 2000).

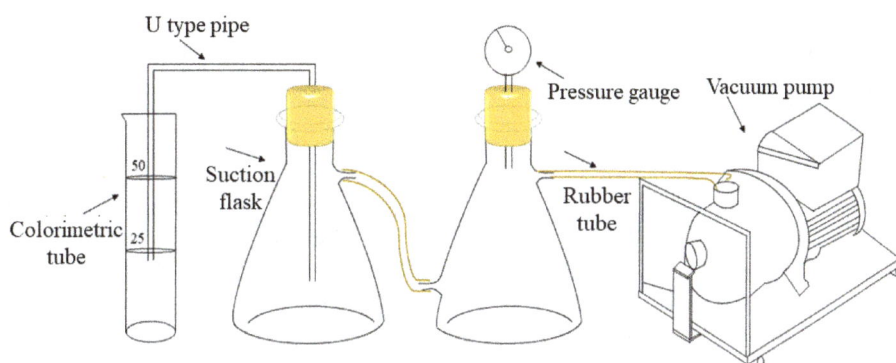

FIG 2 – The breakthrough vacuum device.

Measurement of viscosity

The appropriate rotor and range are selected to test the viscosity of different sample solutions with NDJ-4 viscometer, each sample is tested three times to get the average value.

Measurement of LCST

The critical transition temperature (LCST) of the gel was measured by the test tube inversion method, that is, 5 mL of liquid hydrogel was put into a glass test tube and placed in a constant temperature water bath at a certain temperature. After 15 min, the test tube was taken out, and its flow performance was observed by inverting it at 180°, and the temperature recorded when the gel no longer flowed was LCST. If flowing, continue to heat up 1°C, observe whether the sample flows after 15 mins, and so on until the LCST is determined.

Single-factor experimental design

The single-factor method was used to analyse the effect of different component concentrations on the gel performances. The appropriate concentration range of each factor was selected for gel preparation, and the specific experimental design is shown in Table 1. The effects of the concentration of each component on the gelation time, gel strength, viscosity at room temperature and LCST during the gel transition process were analysed.

TABLE 1

Influencing factors and levels of gel performance.

Level	MC concentration (wt%)	PAAS concentration (wt%)	MgCl$_2$ concentration (wt%)
1	1.1	0.5	4
2	1.2	1	6
3	1.3	1.5	8
4	1.4	2	10
5	1.5	2.5	12
6	1.6	3	14
7	1.7	3.5	16

The response surface method is used to optimise the gel composition ratio. The Central Composite Design (CCD) in Design-Expert 8.0.6 was used for response surface design. Three components were selected as independent variables: MC concentration (X_1), PAAS concentration (X_2) and MgCl$_2$ concentration (X_3), which were set in the range of 1.3–1.6 wt per cent, 1.5–2.5 wt per cent, and 6–16 wt per cent, respectively. In the meanwhile, gelation time (Y_1), gel strength (Y_2), and viscosity (Y_3) were chosen as the response variables. Generally, the CCD includes a factorial runs, axial runs and centre runs six repetitions to estimate the error, as calculated from Equation 1 (Azargohar and Dalai, 2005):

$$N = 2^n + 2n + n_c = 2^3 + 2 \times 3 + 6 = 20 \tag{1}$$

Where N is the total number required for the experiments, n is the total number of factors.

The experimental data is fitted to a quadratic polynomial regression model, and the regression coefficients are obtained. Each response is used to establish an empirical model by using a second-degree polynomial equation as given by Equation 2:

$$Y = \beta_0 + \sum_{i=1}^{k} \beta_i \chi_i + \sum_{i=1}^{k} \beta_{ii} \chi_i^2 + \sum_{i=1}^{k}\sum_{i=1}^{k} \beta_{ij} \chi_i \chi_j + \varepsilon \tag{2}$$

Where Y is the response relating to X; X_i is the experimental variable code β_0 is the coefficient constant; β_i, β_{ii}, β_{ij} represent the liner coefficients, the interaction coefficient and quadratic coefficients, respectively. Table 2 lists the actual values and codes of each factor level.

TABLE 2

Levels of experimental variables.

Independent variables		Levels of coded variables				
		-1.682	-1	0	1	1.682
X_1	MC concentration (wt%)	1.2	1.3	1.45	1.6	1.7
X_2	PAAS concentration (wt%)	1.16	1.5	2	2.5	2.84
X_3	MgCl$_2$ concentration (wt%)	5.27	8	12	16	18.73
Responses		**Type**		**Target***		
Y_1	Gelation time (s)	Numerical		Minimise		
Y_2	Gel strength (KPa)	Numerical		Maximise		
Y_3	Viscosity (Pa·s)	Numerical		Minimise		

Physical performance tests

FTIR analysis

The sample was mixed with KBr and pressed to a pellet, and the structure of gel was confirmed by the Fourier-transform infrared (FTIR) spectra of a Nicolet iN10 spectrometer from the wavenumber range of 400 to 4000 cm^{-1}.

Scanning electron spectroscopy (SEM) analysis

The prepared gel was placed in the refrigerator for 24 hrs, and the liquid nitrogen was quenched and put into a vacuum freeze-drying oven at a temperature of -65°C for freeze-drying. Porous structure of hydrogel was observed by field emission scanning electron microscope.

Thermal stability of the gel

Measuring the rate of weight loss is a standard method to test the thermal stability of hydrogel. The crucible containing the same mass of gel and water are placed in a vacuum drying oven and the temperature is increased from 60°C to 180°C at the same interval. The weight is measured every hour, each sample is measured three times to get the average measurements (Ren *et al*, 2019):

$$n = 1 - \frac{M - m_t}{M}$$ (3)

Where:

n is the water retention rate (%)

M is the original gel mass (g)

m_t is the gel mass (g) at the time instant

The inhibition tests of the gel

Coal samples preparation

Gas-fat coal of Baodian (BD) in Shandong province, non-caking coal of the Changcheng No.3 Coal Mine (CC) in Inner Mongolia Autonomous region and long flame coal of Liu huanggou (LHG) in Xinjiang Uygur Autonomous Region, China, are taken as the test coal samples. The basic data pertaining to the coal samples are shown in Table 1. According to GB/T477-2008, the particle size of coal samples from different regions was screened respectively, the basic data of coal samples are shown in Table 3. The prepared coal sample and gel were mixed in equal proportion, respectively, the samples were dried for 48 hrs in a vacuum oven at 40°C for later use.

TABLE 3
Proximate and ultimate analyses of coal samples used in experiments.

Coal sample	Proximate analyses, wt%				Ultimate analyses, wt%, daf				
	Mad	Aad	Vad	Fcad	Cad	Had	Nad	Sad	Oad
BD	1.91	12.79	13.04	52.76	72.66	4.274	1.276	0.16	21.63
CC	3.03	10.70	11.03	55.09	74.59	4.186	1.402	0.25	19.572
LHG	8.34	4.45	4.85	70.76	74.39	3.497	0.866	0.32	20.927

Programmed temperature experiment

The coal sample is heated by an oil bath temperature program box with Shimadden FP93 temperature control, and the air flow and heating rate are set to 120 mL/min and 0.3°C/min, respectively. During the process of increasing the coal temperature from 30°C to 180°C, gas is collected for analysis every 10°C rise in coal temperature, and gas chromatograph is used to determine the gas components generated by the reaction of coal samples at the different temperatures. A total of six sets of experiments are carried out.

Inhibition rate refers to the reduction rate of index gas phase produced by the suppressed coal sample to original coal sample at the same temperature and time, which is calculated by the Equation 4 (Li *et al*, 2019) as follows:

$$\beta = \frac{S_1 - S_2}{S_1} \times 100\% \tag{4}$$

Where β refers to the inhibition rate of coal, S_1 is the mass fraction of gas released by the raw coal sample (CO, CO_2, C_2H_4), and S_2 is the mass fraction of gas released by the suppressed coal sample (CO, CO_2, C_2H_4).

The oxygen consumption rate is crucial for judging coal spontaneous combustion tendency. The actual oxygen consumption rate of coal can be calculated by Equation 5:

$$V_{O_2}^0(T) = \frac{Q \cdot C_{O_2}^0}{V_m} \cdot In\frac{C_{O_2}}{C_{O_2}^0} \tag{5}$$

Where:

Q	is the air supply volume, cm³/s
s	is the air supply area of the furnace body, cm²
n	is the porosity of the coal sample
z	is the height of the coal sample, cm
C_0	is the oxygen molar concentration at the air inlet, mol/ml, C_0=9.375×10⁻⁶ mol/ml
C_1	is the molar concentration of oxygen at the outlet, mol/ml

Fire-extinguishing experiments

The coal fire extinguishing test system is mainly composed of spontaneous combustion furnace stack, monitoring system, and fire extinguishing system. The schematic diagram of the test system is shown in Figure 3. The test site was selected in the square of the ground grouting station of the Shuiliandong Coal Mine in Shaanxi Province, China. The experimental furnace body is cylindrical, the jaw crusher is used to load the coal collected underground into the coal combustion test furnace when crushing, load 200 kg of coal samples into the furnace body, ignite the coal fire with wood, observe the combustion of coal, and monitor the data. The monitoring system consists of a temperature detection system and a gas product detection system. The thermocouples are arranged on the centre line of the furnace body, which is protected by ceramic tubes. Two sets of multiparameter gas sensors were placed at the S1 and S2 positions respectively; The fire extinguishing system consists of a liquid tank, a high-pressure pump, a high-pressure hose, and related ancillary facilities. The water tank has a volume of 10 litres containing a diluted gel solution. This is shown in the flow chart in Figure 3, stop heating when the temperature of the temperature exceeds 500°C, the six-point steel pipe (burned red) is directly injected into the coal pile, and the fire extinguishing experiment was carried out by pouring gel, the test was finished when there is no large amount of smoke could be observed and no reignition occurred.

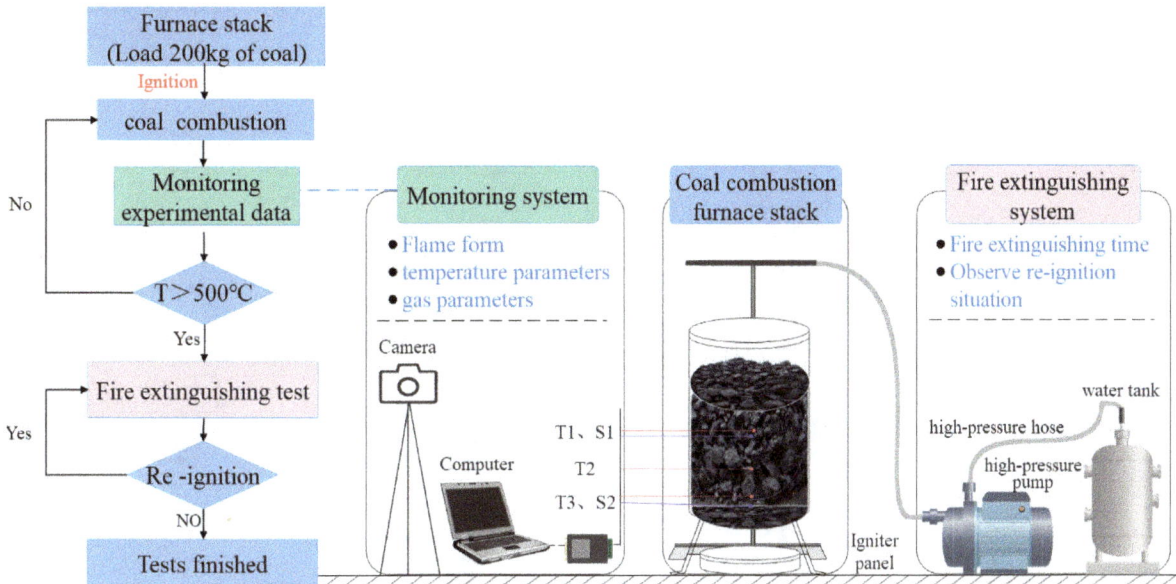

FIG 3 – Schematic diagram of coal fire extinguishing test system.

RESULTS AND DISCUSSION

Signal factor analysis

Figure 4a shows the effect of MC concentration on the gel transition process. With the increase of MC concentration, the intermolecular hydrophobic association in the solution accelerated, and the gelation time was shortened. In this process, the number of physical connections and association strength formed by hydrophobic associations increased, so the gel strength increased (Cho, An and Song, 2006). Furthermore, the higher concentration when MC dissolved in water, the viscosity of the system increases linearly. And the experimental results show that the concentration of MC almost does not affect the LCST of the system.

Figure 4b shows the effect of PAAS concentration on gel time and gel strength. The gelation time decreases with the increase of PAAS concentration, and when the PAAS concentration reaches 3 wt per cent, the gelation time tends to be stable, which is due to the fact that the addition of water-soluble polymer will accelerate the cross-linking of MC. What' more, the gel strength increased first and then decreased, which is similar to the variation of viscosity, and it reached the maximum value when the PAAS concentration reached 1.5 wt per cent, it can be proved that excessive addition of water-soluble polymer would lead to macroscopic phase separation of the heat-curing gel. Moreover, PAAS concentration significantly reduced the LCST of thermosensitive hydrogel solution. It can be seen that the PAAS concentration is in the range of 1.5~2.5 wt per cent, the phase separation of the gel wouldn't occur.

As can be seen from Figure 4c, the gelation time gradually decreases with the increase of $MgCl_2$ concentration, the viscosity gradually increases and tends to stabilise finally, which is due to the fact that $MgCl_2$ promotes gel formation by accelerating hydrophobic association. The gel strength first increased and then decreased in this process, this is due to the gel formation mainly involved hydrophobic association and then formed a three-dimensional network, the number of physical connections and association strength formed by hydrophobic association increased. However, the excess Cl^- would lead to a competition with MC solution for water molecules, the solubility of MC decreases, resulting in gradual precipitation. The LCST showed a decreasing trend with the addition of MgCl$_2$, therefore, it can be concluded that the LCST of the gel could be adjusted by changing the concentration of MgCl$_2$.

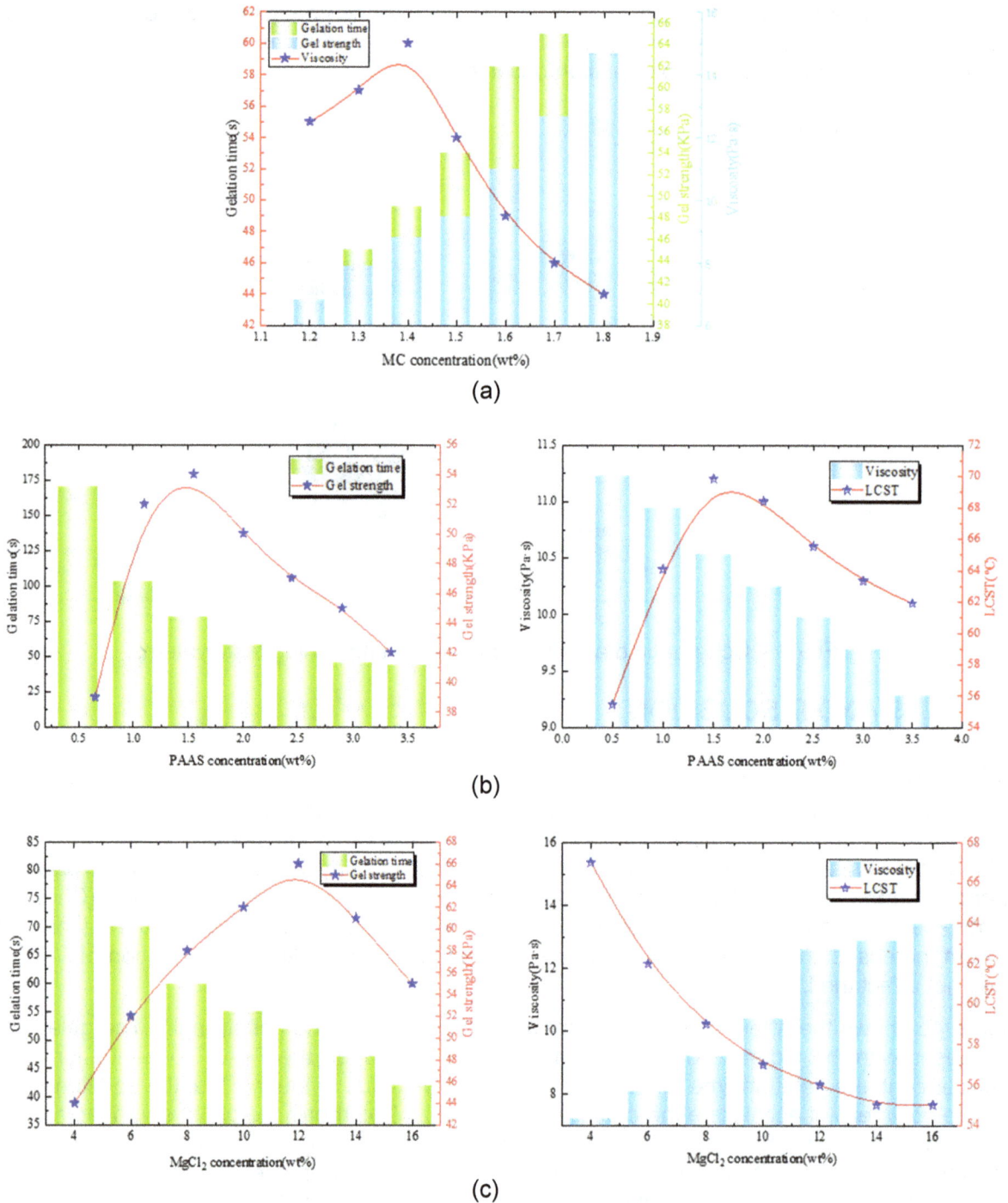

(a)

(b)

(c)

FIG 4 – Effect of different factors on the gel transition: (a) Effect of MC on the gel transition; (b) Effect of PAAS on the gel transition; (c) Effect of MgCl$_2$ on the gel transition.

Gel ratio optimisation

Statistical analyses using the Central composite design

The complete design matrix together the three response values obtained from the experiments are shown in Table 4, the gelation time (s; Y$_1$), gel strength (KPa; Y$_2$) and viscosity (Pa·s; Y$_3$) are found to range from 37 to 70 sec, 37 to 66 KPa, and 9.2 to 18.5 Pa·s, respectively. Differences with p-value <0.05 were regarded significant statistically, a lack of fit referred to the significance between the model-dependent error and the repetitive error, which is not desired (Lack-of-fit >0.05 is desired). In addition, the coefficient of determination (R^2) represents the model's prediction accuracy, the closer the R2 value is to 1, the better the model predicts the response as well. Then, the second-order polynomial regression equations are established, the quadratic model is suggested for further optimising owing to ANOVA results.

TABLE 4

CCD experiment schedule and results.

Run	X_1: MC (wt%)	X_2: PAAS (wt%)	X_3: $MgCl_2$ (wt%)	Y_1: Gelation time (sec)	Y_2: Gel strength (KPa)	Y_3: Viscosity, (Pa·s)
1	1	-1	1	43	61	17.2
2	1.682	0	0	37	66	17.6
3	0	0	0	57	57	14.4
4	0	0	0	56	56	14.3
5	-1.682	0	0	49	48	9.2
6	-1	-1	-1	63	40	11.4
7	0	0	-1.682	64	37	12.8
8	-1	-1	1	57	46	12.3
9	1	1	1	45	62	18.5
10	0	0	0	56	56	14
11	0	0	0	55	58	14.5
12	-1	1	-1	56	42	11.5
13	0	0	0	54	55	14.7
14	-1	1	1	55	50	12.1
15	0	0	1.682	47	60	15.3
16	1	-1	-1	55	50	15.2
17	0	0	0	57	56	14.6
18	0	-1.682	0	70	40	13.8
19	1	1	-1	52	48	17.5
20	0	1.682	0	63	44	16.2

Effects of the independent variables on response variables

Table 5 shows the analysis of varies results for the experimental response parameters, the final quadratic polynomial regression Equations 6, 7 and 8 for Y_1, Y_2, and Y_3 are as follows:

$$Y_1=-4.11X_1-1.59X_2-4.00X_3+1.00X_1X_2-1.50X_1X_3+1.25X_2X_3-4.90X_1^2+3.41X_2^2-0.48X_3^2+55.89 \quad (6)$$

$$Y_2=5.37X_1+0.86X_2+5.69X_3-0.87X_1X_2+1.38X_1X_3+0.63X_2X_3+0.48X_1^2-4.83X_2^2-2.53X_3^2+56.30 \quad (7)$$

$$Y_3=2.58X_1+0.55X_2+0.64X_3+0.46X_1X_2+0.19X_1X_3-0.16X_2X_3-0.29X_1^2+0.28X_2^2-0.06X_3^2+14.41 \quad (8)$$

TABLE 5

Analysis of variance (ANOVA) in the experimental responses for the quadratic model.

Source	DF	Y_1 (Gelation time, s)			Y_2 (Gel strength, KPa)			Y_3 (Viscosity, Pa·s)		
		SS	F	P-value	SS	F	P-value	SS	F	P-value
Model	9	1090.94	37.86	< 0.0001	1285.34	60.89	< 0.0001	105.39	102.68	< 0.0001
X_1	1	231.12	72.20	< 0.0001	393.12	167.60	< 0.0001	90.87	796.77	< 0.0001
X_2	1	34.71	10.84	0.0081	10.07	4.29	0.0651	4.16	36.47	0.0001
X_3	1	218.21	68.17	< 0.0001	441.86	188.38	< 0.0001	5.55	48.65	< 0.0001
$X_1 X_2$	1	8.00	2.50	0.1450	6.12	2.61	0.1372	1.71	15.01	0.0031
$X_1 X_3$	1	18.00	5.62	0.0392	15.13	6.45	0.0294	0.28	2.47	0.1474
$X_2 X_3$	1	12.50	3.90	0.0764	3.13	1.33	0.2752	0.21	1.85	0.2034
X_1^2	1	345.44	107.91	< 0.0001	3.28	1.40	0.2645	1.21	10.63	0.0086
X_2^2	1	167.83	52.43	< 0.0001	335.70	143.12	< 0.0001	1.09	9.60	0.0113
X_3^2	1	3.27	1.02	0.3359	92.12	39.27	< 0.0001	0.052	0.46	0.5104
Residual	10	32.01			23.46			1.14		
Lack of fit	5	25.18	3.68	0.0893	18.12	3.40	0.1028	0.83	2.70	0.1499
Pure error	5	6.83			5.33			0.31		
Total	19	1122.95				1308.80				106.53	
R^2			0.9715			0.9821			0.9893	
R^2_{adj}			0.9458			0.9659			0.9797	

Positive sign in front of the factors has a positive influence on the responses, whereas negative sign indicates an antagonistic effect. The R^2 value for Equations 6, 7 and 8 is 0.9715, 0.9821 and 0.9893, respectively, and the R^2_{adj} value is 0.9458, 0.9659 and 0.9797, respectively. it indicates that 97.15, 98.21 and 98.93 of the total variation in the gelation time, gel strength and viscosity, respectively, was credited with a regression relationship. For the gel strength (Y_1), all X variable showed a negative correlation with the gelation time, which means that an increasement in the concentration of MC, PAAS and $MgCl_2$ shorten the gelation time. In this case, X_1, X_2, X_3, X_1X_3, X_1^2, X_2^2 are significant terms. The results can be obtained from the analysis of variance that, F_{x_1}=72.20, F_{x_2}=10.84, F_{x_3}=68.17, which indicates that the influence order of these three factors on gelation time is X_1, X_3, X_2. This is also consistent with the results of Xu *et al's* (2004) study. A series of 3D response surface and contour plots from the model are used to further analyses the relationships between independent variables and dependent variables. As shown in Figure 5a, the gelation time decreases significantly with the increase of MC and $MgCl_2$ concentration. Therefore, MC and $MgCl_2$ have synergetic effect on gelation time, which is consistent with the AVONA results.

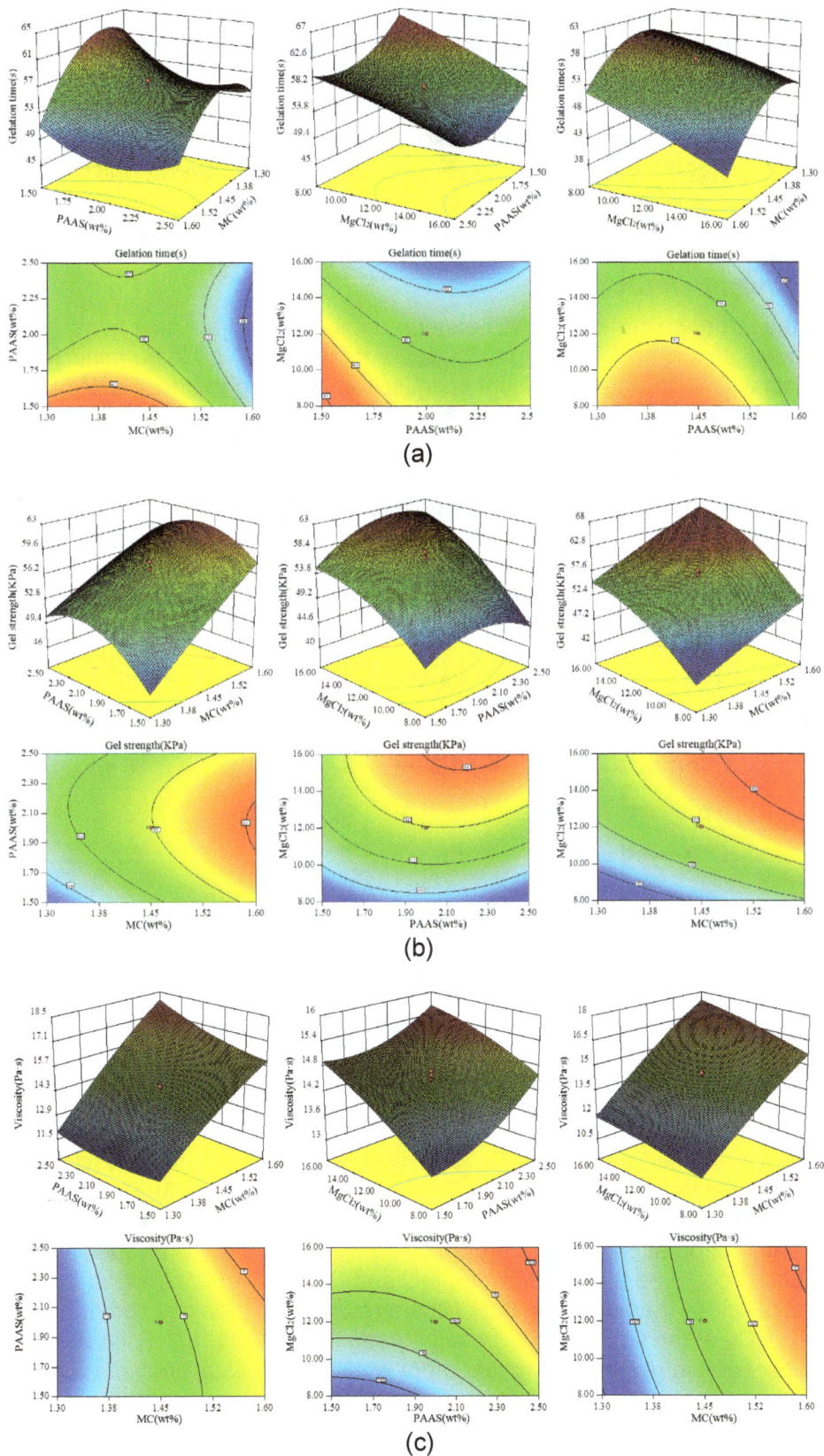

FIG 5 – 3D surface and contour map of the influence of different factors: (a) gelation time; (b) gel strength; (c) viscosity.

For gel strength (Y_2), X_1, X_3, X_2X_3, X_2^2, X_3^2 are significant terms, X_1 and X_3 have the greatest influence on it and the F values are 181.18 and 341.08 (Table 6), respectively, whereas X_2 showed the least influence. Figure 5b displayed the relations between independent variables and gel strength, The gel strength increases significantly with the rise of $MgCl_2$ concentration, the same trend is observed for the MC concentration, and the rise of PAAS concentration plays the least role. The highest gel strength is obtained when all the variables are at the maximum point within the range studied. For viscosity (Y_3), X_1, X_2, X_3, X_1X_2, X_1^2, X_2^2 are significant terms. Figure 5c presented the 3D surface and contour plots of the viscosity versus the X_1, X_2, X_3, it can be found that an increase in the concentration of MC, PAAS and $MgCl_2$ increase the viscosity of hydrogel. The F values of these three independent variables are 796.77, 36, 47 and 48.65, respectively, which means X_1 is more decisive than other X variables owing to MC has the highest molecular weight. In addition, X_1 and X_2 have a synergistic effect on the viscosity of thermosensitive hydrogels.

Numerical optimisation process

In this section, the target desirability for this experimental design is set as follows: gelation time (Y_1) to be minimised, gel strength (Y_2) to be maximised, and viscosity (Y_3) to be minimised. A total of 21 solutions are displayed under the numerical optimisation by using the desirability function, of which solution No.1 is selected owing to its highest desirability values. As shown in Table 6, the optimised composition is 1.3 wt, 2.11 wt and 16 wt for X_1, X_2 and X_3, where the response parameters are predicted as 52.0013 sec (Y_1), 53.4775 KPa (Y_2) and 11.9232 Pa·s (Y_3). In the meanwhile, in order to evaluate the prediction accuracy, the validation experiments are carried out to calculate the error between the experimental and predicted values. The experiment values is measured to be 50 sec (Y_1), 53 KPa (Y_2) and 11.600 Pa·s (Y_3), it is found that the experimental values are in a good agreement with the predicted values from the models, and the errors are all lesser than 5 per cent. Further studies are carried out basing on the optimised composition.

TABLE 6

Observed response values and percentage prediction errors for the optimised value.

Composition	X_1: 1.3 wt	X_2: 2.11 wt	X_3: 16 wt
Predicted value	52.0013	53.4775	11.9232
Observed value	50	53	11.600
Percentage prediction error*	4%	0.9%	2.8%

*Calculated using [(experimental value-predicted value)/experimental value] ×100.

Tests results of the product performance

FT-IR test

The infrared spectra of MC, PAAS, $MgCl_2$ and gel are shown in Figure 6. A wide spectrum can be observed in the wavenumber range of 3600–3100 cm^{-1} (corresponding to the oscillation of O-H). The 2923 cm^{-1} is the expansion vibration peak of the C-H bond, and the 1082 cm^{-1} is the expansion vibration peak of the C-O-C bond. It can be seen that a large number of hydrogen bonds were formed in gel, and the band range at 3411 cm^{-1} narrowed and shifted to the low frequency range, indicating that hydrogen bonds could be formed between MC and PAAS. The band vibration enhancement at 1636 cm^{-1} is the result of the superposition of the stretching vibration absorption peaks of C=C in MC, PAAS and $MgCl_2$ at the same place. It can be seen that the characteristic peaks of MC, PAAS, and $MgCl_2$ were present in gel, indicating that there was only physical cross-linking.

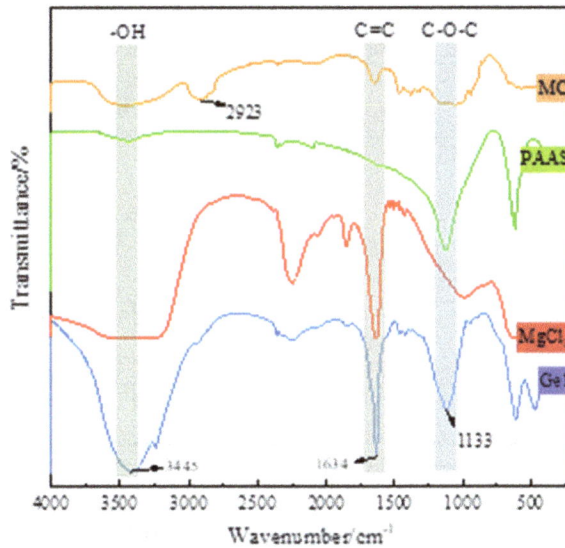

FIG 6 – FT-IR spectra of methylcellulose (MC), sodium polyacrylate (PAAS), magnesium chloride (MgCl₂) and gel.

SEM Test

SEM displayed the micromorphology of the gel, The network structures can be observed in the SEM images (Figure 7), It presents a large number of porous structures, which is conducive to the entry and exit of water molecules, thus enhancing the ability of the gel to absorb and retain water (Xu and Li, 2004; Kim, Park and Park, 2018).

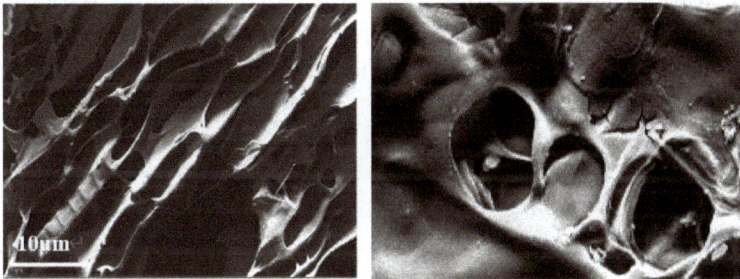

FIG 7 – SEM images of hydrogel.

Thermal stabilities

Since the main component of hydrogel is water, when it is placed in a high-temperature environment for a long time, the internal structure of the gel will be destroyed. As can be seen from Figure 8, the gel quality decreases significantly during the heating process. The higher the temperature, the greater the water loss of the gel. Before 90°C, the gel had a low rate of water loss. When the temperature reached more than 90°C, the strong hydrophobic effect of methoxy groups accelerated the formation of thermosensitive hydrogel, resulting in obvious volume shrinkage of the gel, this process is regarded as endothermic (Deng *et al*, 2011), the water loss of the gel was obvious. When the temperature reached 180°C, the weight loss rate of water reached 77.89 per cent, and the weight loss rate reached 57.87 per cent. The pore structure of the gel plays a water retention effect, and there is a clear binding force between the water inside the gel, which reduces the water loss.

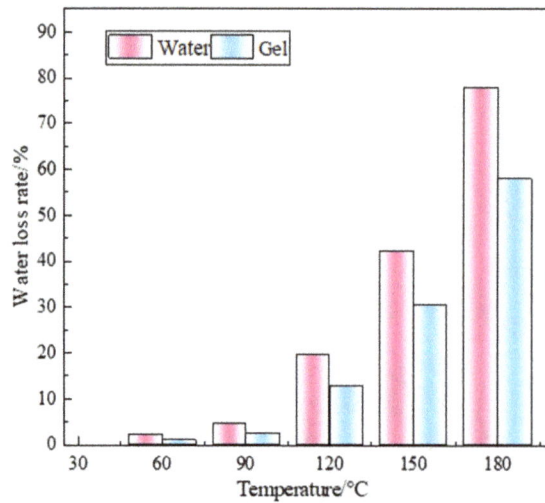

FIG 8 – Water loss rate of gel and water.

Gel inhibitory effect on coal spontaneous combustion

In order to analyse the resistance performance of thermosensitive hydrogels, three coal samples with different degrees of metamorphism were selected to carry out temperature-programming experiments to obtain the indicator gases and the oxygen consumption rate. CO and CO_2 were the decomposition products of carbonyl and carboxyl groups during coal oxidation, respectively, CO can predict coal spontaneous combustion earlier. As shown in Figure 9a, the changes of CO and CO_2 produced by different coal samples in the low-temperature oxidation process were similar. A small amount of CO and CO_2 were generated in a low-temperature stage, and when the temperature exceeded the critical temperature, the gas concentration increased exponentially owing to the accelerated reaction between coal and oxygen. When the degree of coal metamorphism increased, the concentration of CO and CO_2 decreased significantly during the entire oxidation process.

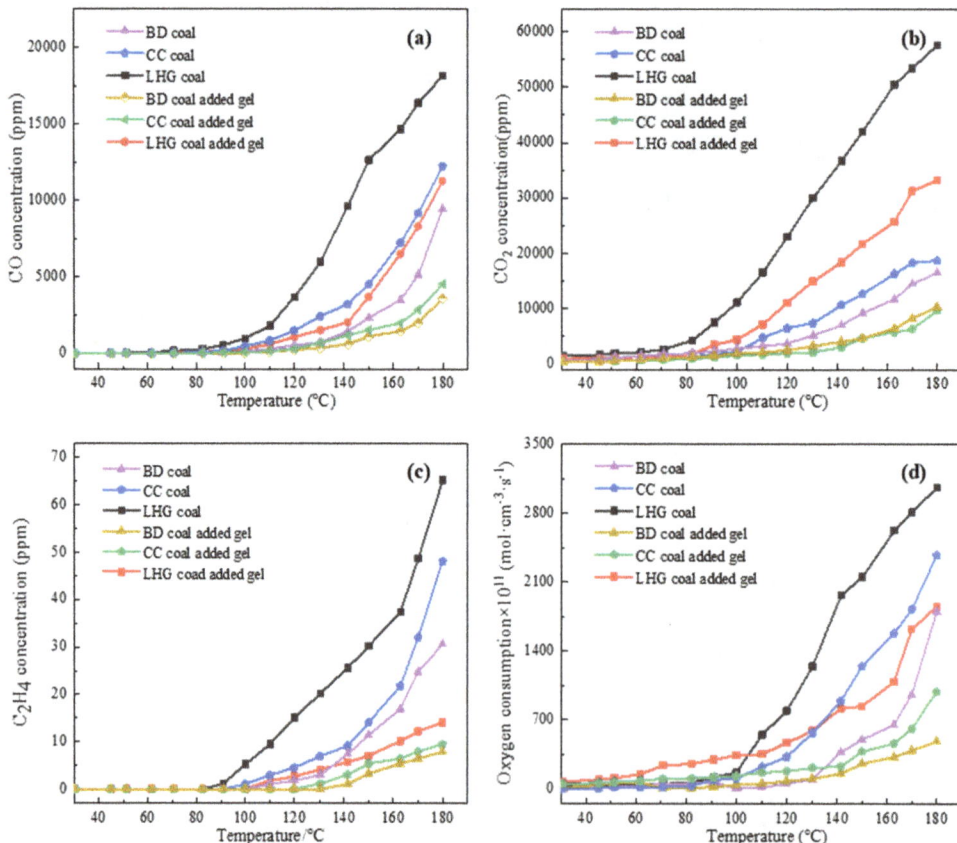

FIG 9 – Inhibitory effect of gel on the spontaneous combustion of different kinds of coal: (a) CO concentration; (b) CO_2 concentration; (c) C_2H_4 concentration; (d) Oxygen consumption.

Therefore, it can be concluded that the lower the coal rank was, the easier it was to react with oxygen. CO and CO_2 concentrations of the three coal samples decreased remarkably after adding the thermosensitive hydrogel, indicating that the coal-oxygen reaction was hindered. In this process, gel covered the surface of the coal and acted as a barrier to isolate oxygen, which led to a slow rate of the oxidation (Deng et al, 2011). In the meanwhile, the gel structure shrank with a certain strength after the temperature rose, and showed a blocking role. So gel illustrated an inhibitory effect during the entire oxidation process of the coal. It can be seen from Table 7 that the CO inhibition rate of gel on the BD, CC and LHG coal were 56.37 per cent, 61.89 per cent and 64.46 per cent, respectively. The gel also showed the effect on increasing the cross point temperature of the coal, which further demonstrated that the thermosensitive hydrogel showed a significant inhibitory effect on coal spontaneous combustion.

C_2H_4 is a marker gas of accelerated oxidation stage in the process of coal spontaneous combustion, and the initial temperature of its formation was higher than that of CO and CO_2. The initial temperature of C_2H_4 was different. As shown in Figure 9c, the inhibitory effect of thermosensitive hydrogel on C_2H_4 was reflected in two aspects. On the one hand, it delayed the appearance of C_2H_4. On the other hand, the addition of gel reduced the concentration of C_2H_4. It can be seen that with the growth of temperature, the concentration of coal gas with gel covering decreased significantly. In the BD, CC and LHG coal, the initial temperature of C_2H_4 climbed from 110°C, 100°C, 90°C to 140°C, 130°C, 110°C respectively, and the gas concentration also decreased by 74.98 per cent, 76.59 per cent and 78.16 per cent, respectively. Therefore, it showed that the thermosensitive hydrogel had the strongest inhibitory effect on the oxidation exothermic process of LHG long flame coal, followed by CC non-stick coal, while the BD gas fat coal had a relatively weaker inhibitory effect.

Figure 9d showed the effect of thermosensitive hydrogel on the oxygen consumption rate during low-temperature oxidation of different coal samples. It can be seen that thermosensitive hydrogel also showed significant inhibition effect on low metamorphic bituminous coal during the heating process. At 180°C, the oxygen consumption of BD coal decreased from 2473.81 to 992.29 ($\times$ 1011 mol·cm^{-3}·s^{-1}), CC coal reduced from 1863.83 to 484.23 ($\times$ 1011 mol·cm^{-3}·s^{-1}), and LHG coal decreased from 2863.64 to 1890.86 ($\times$ 1011 mol·cm^{-3}·s^{-1}).

Fire-extinguishing performance

Temperature and gas product analysis

As shown in Figure 10a, when the coal was burning, a large amount of smoke was generated, and heat was accumulated in this process, the gel was injected when the temperature reached 500°C. The temperature rose briefly, and water vapor formed when the fire was extinguished for 1 min, gel played a cooling role at this moment, and this was an endothermic process. After 5 mins, the smoke almost disappeared, the gel finally formed a covering layer on the surface of the coal and blocked the crack channel, the water vapor gradually disappeared, the smoke decreased significantly, the temperature of coal dropped gradually, and the flame extinguished without re-ignition phenomenon occurrence.

The changes of C_2H_4 and C_2H_2 throughout the test are shown in Figure 10b and 10c. As the coal heated up and the fire sped up, the O_2 concentration witnessed a significant decrease from 20 per cent to less than 5 per cent, which was in correspondence to a rise of the oxygen consumption rate rose. As the gel injected, the temperature of coal saw a sharp decline. In contrast, the O_2 concentration saw a growth. C_2H_4 and C_2H_4 had underwent a similar trend, that is, the concentration began to increase after gel injection. However, some air was brought into the furnace during gel injection process, with a slight fluctuation of gas in much of this period.

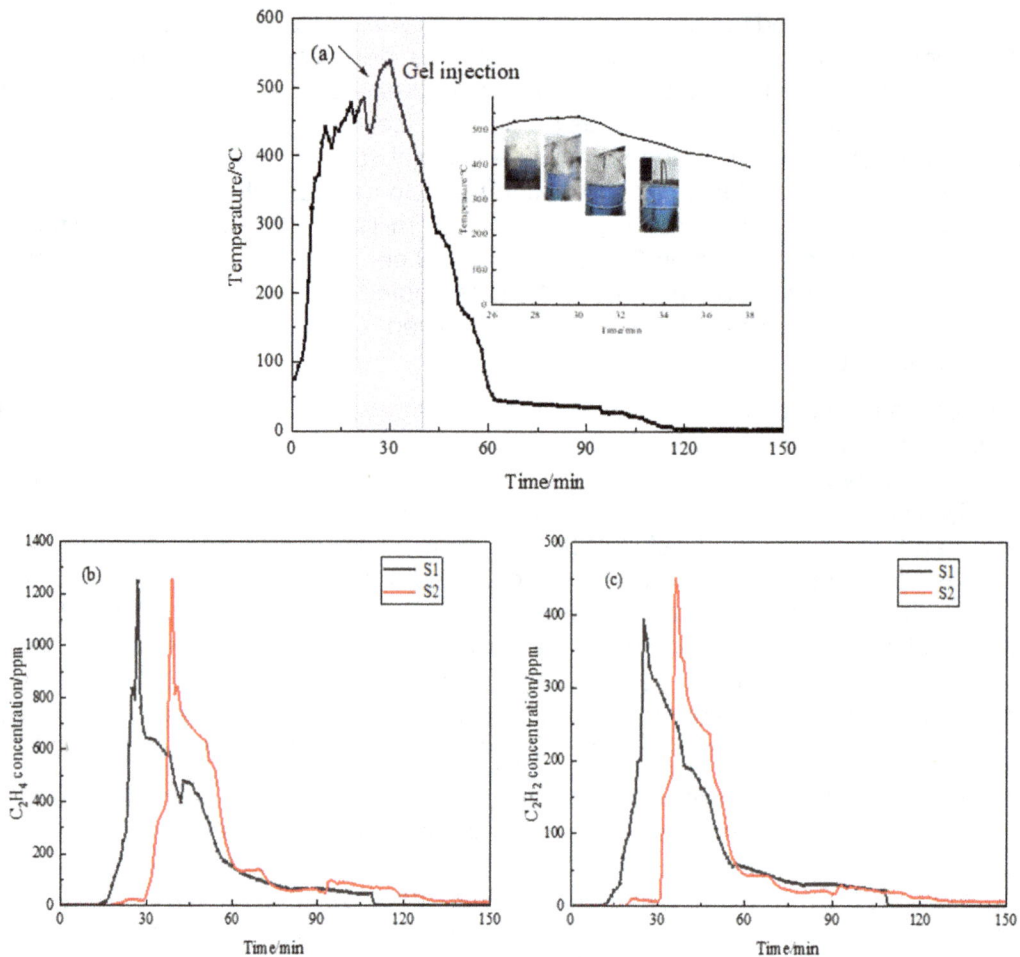

FIG 10 – Variation of monitoring parameter during fire extinguishing test: (a) temperature; (b) C_2H_4; (c) C_2H_2.

Gel fire extinguishing mechanism

The characteristics of thermosensitive gel which underwent the fire extinguishing process are manifested in Figure 11, the phase change took place after the thermosensitive hydrogel was injected into coal of high temperature. At lower temperatures, cellulose ether molecules formed hydrogen bonds with water molecules, which can attribute to the fluidity of gel. The molecule continuously absorbed heat energy as the temperature increased, resulting in the breaking of the hydrogen bond between the MC and the water molecule, with the hydrophobic group exposed and hydrophobic association enhanced, then gel formed. Therefore, the gel undergoing phase change acquired a certain strength, which can fill and plug the fracture and prevent the contact of coal and oxygen. In the meanwhile, gel absorbed a large amount of heat energy of the coal and water vapor formed around here, which accelerated the extinguish of coal fire.

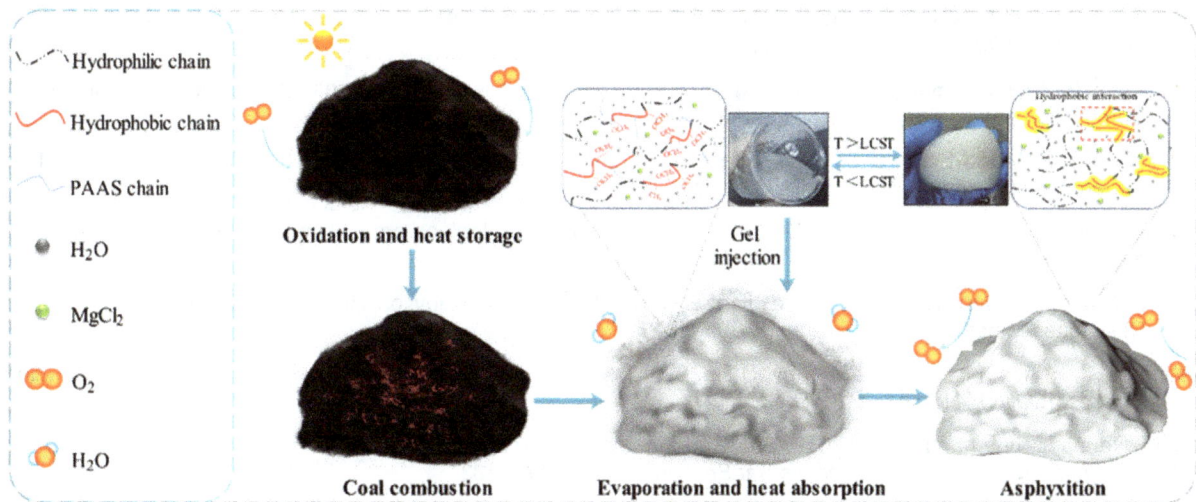

FIG 11 – Fire control and prevention mechanism of thermosensitive hydrogel.

CONCLUSIONS

In this study, a kind of thermosensitive hydrogel based on cellulose was prepared for the fire protection of coal spontaneous combustion, gel properties, inhibition effect and fire extinction effect of the gel were analysed, the main conclusions are summarised as follows:

- Single factor tests showed that the sol-gel phase changed by the interaction of methoxy group, intramolecular and intermolecular hydrogen bonding. At low temperature, macromolecular chains between MC and PAAS intertwined, which enhanced the viscosity of gel. Hydrogen bonds between MC and water molecules were broken when the temperature rose. At this time, Cl⁻ was precipitated and thus weakened the hydrogen bond between water and polymer, which accelerated the sol-gel transition by promoting hydrophobic association;

- Response surface method was used to optimise composition ratio of gel. With a short gel time, a large gel strength and a low viscosity at room temperature as the response values, the optimal ratio was obtained as MC:PAAS:$MgCl_2$=1.3:2.11:16. Under this condition, the gelation time was 50 sec, the gel strength was 53 KPa, and the viscosity was 11.6 Pa·s. SEM test showed that the gel system had a penetrating pore structure, which was further demonstrated in the thermal stability test that it had a good ability of water retaining.

- Programmed temperature experiments were carried out to analyse the inhibition effect of gel on three bituminous coal with different deterioration degrees. The production of CO and CO_2 of coal which was adding with gel saw a notable decline. CO inhibition rate of gel on BD, CC and LHG coal were calculated to be 56.37 per cent, 61.89 per cent and 64.46 per cent, respectively, which demonstrated that the lower the metamorphic degree of bituminous coal, the stronger the inhibition effect of thermosensitive hydrogel.

- Medium-sized test illustrated that thermosensitive hydrogel showed an outstanding fire-fighting performance, resulting in a layer of liquid film forming to wrap to coal of high temperature. In addition, a sol-gel phase transformation took place after reaching the LCST. Gel played a role in absorbing the heat energy of coal, so as to promote the extinguishing of the coal fire

ACKNOWLEDGEMENTS

The authors gratefully acknowledge the financial support from Shaanxi Provincial Science and Technology Innovation Capability Support Program (2023-CX-TD-42), Shaanxi Provincial Department of Education Innovation Team Project (23JP095).

REFERENCES

Ajrash, M J, Zanganeh, J and Moghtaderi, B, 2017. Impact of suspended coal dusts on methane deflagration properties in a large-scale straight duct, *J Hazard Mater*, 338:334–342.

Azargohar, R and Dalai, A K, 2005. Production of activated carbon from Luscar char: Experimental and modeling studies, *Microporous Mesoporous Mater*, 85:219–225.

Cai, J, Yang, S, Zhong, Y, Song, W, and Zheng, W, 2020. A Physical-chemical Synergetic Inhibitor for Coal Spontaneous Combustion and Its Fire Prevention Performance, *Combust Sci Technol*, pp 1–13.

Cho, Y W, An, S W and Song, S C, 2006. Effect of inorganic and organic salts on the thermogelling behavior of poly (organophosphazenes), Macromol Chem Phys, 207:412–418.

Deng, J, Li, B, Wang, K and Wang, C P, 2016. Research status and outlook on prevention and control technology of coal fire disaster in China, *Coal Sci Technol*, 44:1–8.

Deng, J, Wang, N, Wen, H, Chen, X and Jian, Z, 2011. Experiment Study on Retarding Performances of Gel Fire Resistance Material, *Coal Sci Technol*, 39:5–8.

Desbrières, J, Hirrien, M and Ross-Murphy, S B, 2000. Thermogelation of methylcellulose: Rheological considerations, *Polymer (Guildf)*, 41:2451–2461.

Guo, Q, Ren, W, Zhu, J and Shi, J, 2019. Study on the composition and structure of foamed gel for fire prevention and extinguishing in coalmines, *Process Saf Environ Prot*, 128:176–183.

Hu, X M and Wang, D M, 2013. Enhanced fire behavior of rigid polyurethane foam by intumescent flame retardants, *J Appl Polym Sci*, 129:238–246.

Hu, X M, Cheng, W M and Wang, D M, 2014. Properties and applications of novel composite foam for blocking air leakage in coalmine, *Russ J Appl Chem*, 87:1099–1108.

Hu, X M, Cheng, W M, Nie, W and Shao, Z, 2015. Synthesis and characterization of a temperature-sensitive hydrogel based on sodium alginate and N-isopropylacrylamide, Polym Adv Technol, 26:1340–1345.

Jiang, Z and Dou, G, 2020. Preparation and Characterization of Chitosan Grafting Hydrogel for Mine-Fire Fighting, *ACS Omega*, 5:2303–2309.

Kim, M H, Park, H and Park, W H, 2018. Effect of pH and precursor salts on in situ formation of calcium phosphate nanoparticles in methylcellulose hydrogel, *Carbohydr Polym*, 191:176–182.

Klouda, L, 2015. Thermoresponsive hydrogels in biomedical applications A seven-year update, *Eur J Pharm Biopharm*, 97:338–349.

Kong, B, Li, Z, Wang, E, Lu, W, Chen, L and Qi, G, 2018. An experimental study for characterization the process of coal oxidation and spontaneous combustion by electromagnetic radiation technique, *Process Saf Environ Prot*, 119:285–294.

Kong, B, Wang, E, Lu, W and Li, Z, 2019. Application of electromagnetic radiation detection in high-temperature anomalous areas experiencing coalfield fires, *Energy*, 189:116144.

Li, S, Zhou, G, Wang, Y, Jing, B and Qu, Y, 2019. Synthesis and characteristics of fire extinguishing gel with high water absorption for coalmines, *Process Saf Environ Prot*, 125:207–218.

Lu, X, Wang, D, Qin, B, Tian, F and Shi, G, 2015. Novel approach for extinguishing large-scale coal fires using gas–liquid foams in open pit mines, *Environ Sci Pollut Res*, 22:18363–18371.

Lu, Y and Qin, B, 2015. Experimental investigation of closed porosity of inorganic solidified foam designed to prevent coal fires, *Adv Mater Sci Eng*, 2015.

Ma, L, Ren, L F, Ai, S W, Zhang, L R, and Li, B, 2015. Experimental study on the impact of the chloride inhibitor upon the limited parameters of the coal spontaneous combustion, *J Saf Environ*, 15:83–88.

Nair, L S, Starnes, T, Ko, J W K and Laurencin, C T, 2007. Development of injectable thermogelling chitosan-inorganic phosphate solutions for biomedical applications, *Biomacromolecules*, 8:3779–3785.

Qin, B T, Zhong, X X, Wang, D M, Xin, H H and Shi, Q L, 2021. Research progress of coal spontaneous combustion process characteristics and prevention technology, *Coal Sci Technol*, 49:66–99.

Qin, B, Jia, Y, Lu, Y, Li, Y, Wang, D and Chen, C, 2015. Micro fly-ash particles stabilized Pickering foams and its combustion-retardant characteristics, *Fuel*, 154:174–180.

Qin, B, Wang, H, Yang, J and Liu, L, 2016. Large-area goaf fires: a numerical method for locating high-temperature zones and assessing the effect of liquid nitrogen fire control, *Environ Earth Sci*, 75:1–14.

Ray, S K and Singh, R P, 2007. Recent developments and practices to control fire in underground coalmines, *Fire Technol*, 43:285–300.

Ren, X, Xue, D, Li, Y, Hu, X, Shao, Z and Cheng, W, 2019. Novel sodium silicate/polymer composite gels for the prevention of spontaneous combustion of coal, *J Hazard Mater*, 371:643–654.

Ruel-Gariépy, E and Leroux, J C, 2004. In situ-forming hydrogels - Review of temperature-sensitive systems, *Eur J Pharm Biopharm*, 58:409–426.

Tang, Y and Wang, H, 2018. Development of a novel bentonite–acrylamide superabsorbent hydrogel for extinguishing gangue fire hazard, *Powder Technol*, 323:486–494.

Tsai, Y T, Yang, Y, Wang, C, Shu, C M and Deng, J, 2018. Comparison of the inhibition mechanisms of five types of inhibitors on spontaneous coal combustion, *Int J Energy Res*, 42:1158–1171.

Wang, D, Dou, G, Zhong, X, Xin, H and Qin, B, 2014. An experimental approach to selecting chemical inhibitors to retard the spontaneous combustion of coal, *Fuel*, 117:218–223.

Wei, G, Wen, H, Deng, J, Ma, L, Li, Z and Lei, C, 2021. Liquid CO_2 injection to enhance coalbed methane recovery: An experiment and in-situ application test, *Fuel*, 284.

Xu, Y and Li, L, 2005. Thermoreversible and salt-sensitive turbidity of methylcellulose in aqueous solution, Polymer (Guildf), 46:7410–7417.

Xu, Y, Wang, C, Tam, K C and Li, L, 2004. Salt-assisted and salt-suppressed sol-gel transitions of methylcellulose in water, *Langmuir*, 20:646–652.

Zhang, J P, Hu, X M and Cheng, W M, 2015. Preparation and properties of chitosan thermo- sensitive hydrogels for preventing coal spontaneous combustion, *China Saf Sci J*, pp 5–10.

Zhang, Y, Liu, Y, Shi, X, Yang, C, Wang, W and Li, Y, 2018. Risk evaluation of coal spontaneous combustion on the basis of auto-ignition temperature, *Fuel*, 233:68–76.

Zhou, C and Tang, Y, 2018. A novel sodium carboxymethyl ellulose/aluminium citrate gel for extinguishing spontaneous fire in coalmines, *Fire Mater*, 42:760–769.

Zhou, F B, Shi, B B, Cheng, J W and Ma, L J, 2015. A New Approach to Control a Serious Mine Fire with Using Liquid Nitrogen as Extinguishing Media, *Fire Technol*, 51:325–334.

Lab scale investigation of the effectiveness of different fire extinguishing agents on lithium-ion batteries fires

A Iqbal[1] and G Xu[2]

1. PhD student, Missouri University of Science and Technology, Rolla MO 65401, USA. Email: aiqbal@mst.edu
2. Associate Professor, Missouri University of Science and Technology, Rolla MO 65401, USA. Email: guang.xu@mst.edu

ABSTRACT

Despite their advantages, such as high energy density and reliability, the widespread adoption of Lithium-ion Batteries (LIBs) has been impeded due to persistent concerns regarding fire safety. The unique and evolving nature of the fire risks posed by LIBs has rendered traditional classification frameworks inadequate, leaving a crucial gap in the availability of appropriate extinguishing agents. Considering this pressing issue, the primary objective of this comprehensive study is to thoroughly investigate and subsequently enhance the effectiveness of existing fire extinguishers against the complex and dynamic nature of LIB fires, thereby addressing a critical safety concern that has significant implications for various industries. To achieve this goal, the research design encompasses an extensive testing procedure, evaluating the performance of four distinct types of fire extinguishers. These extinguishers will be rigorously assessed on their ability to combat fires originating from five different types of LIB cell chemistries. This precise approach ensures a comprehensive analysis of various extinguisher-LIB chemistry interactions, allowing for a nuanced understanding of their effectiveness in diverse scenarios. The evaluation will take place within a controlled laboratory environment, utilising cutting-edge equipment that accurately replicates real-world fire incidents involving LIBs. First part of the results revealed significant differences in gas emissions between various lithium-ion battery chemistries. Notably, NMC and NCA chemistries exhibited higher emissions, while LFP and LTO generally had lower emissions, making them safer alternatives. The outcomes of this study are expected to contribute significantly to the field of fire safety, providing valuable insights into the optimal strategies for mitigating LIB-related fire risks. The data gathered from these extensive experiments will offer a clear perspective on the performance of different extinguisher types when faced with fires originating from distinct LIB chemistries. Based on the findings, targeted modifications and enhancements will be proposed for the extinguishers, potentially involving the incorporation of innovative additives to maximise their efficacy. By bridging the existing knowledge gap and addressing the intricate challenges posed by LIB fires, this research advances fire safety protocols and underscores its significance in the context of environmental well-being. Ultimately, this study represents a substantial stride towards ensuring the safe and sustainable integration of LIBs across various applications, safeguarding both human lives and the environment from the potential hazards associated with these energy storage technologies.

INTRODUCTION

Lithium-ion batteries (LIBs) are the most commonly used batteries in battery electric vehicles (BEV) due to their high energy density and long lifetime (Larsson, Andersson and Mellander, 2016). However, several BEV accidents involving battery fires have raised concerning alarms about the fire risks of LIBs (Sun et al, 2020). Since 2006, Federal Aviation Administration (FAA) has recorded about 350 fire accidents involving LIBs in aviation (FAA, 2022). In 2012, A Toyota Prius (hybrid) was recorded to have caught fire after being flooded by hurricane sandy (Labovick Law Group, 2021). In 2013, two Tesla Model S were reported to have caught fire after being involved in two separate collisions (Musk, 2013; George, 2013). Further down the timeline, different car makers have experienced BEVs catching fire accidents such as Volkswagen in 2017 (Traugott, 2021), BMW i8 (Bruce, 2019), and two separate events involving Porsche Panamera in 2019 and a Taycan in 2020 happened in Portugal and USA respectively (Labovick Law Group, 2021). Several other events have happened when BEV caught fire while charging or parking idle (Wayland and Kolodny, 2021; Yoney, 2021).

Existing research has identified thermal runaways to be the probable reason for most of the fire accidents in BEVs (Sun *et al*, 2020; Liu *et al*, 2016, 2021). Thermal runaway happens when the battery pack ignites spontaneously due to over-heating, according to literature. It is caused by either electrical abuse, mechanical abuse/collision, or by thermal abuse (Sun *et al*, 2020; Liu *et al*, 2016; Zhong *et al*, 2019; Zhao *et al*, 2021). Once thermal runaway is triggered, chemical reactions start to take place inside the cell/pack, causing the emission of toxic fumes. Due to the presence of fuel and oxidiser inside the cell, these chemical reactions will most likely continue spontaneously, giving rise to heat and toxic gases during the process (Doughty and Roth, 2012). Several studies have analysed the thermal characteristics of LIB fires, for instance, Fu *et al* (2015) concluded from their experimental study that the stored electrical energy has a significant impact on the burning behaviours of LIBs. One contrast, Larsson, Andersson and Mellander (2016) stated that the stored electrical energy has nominal dependence on the thermal properties of the battery.

One of the most under-explored areas of BEV fires is the amount and types of gases emitted during fires. These toxic gases pose an immense threat to human health (Peng *et al*, 2020), as fatalities and injuries are mostly caused by the inhalation of toxic gases in fire accidents (Stec, 2017). Several researchers have performed experiments on the fire behaviour and propagation of the fire in LIB cells (Fu *et al*, 2015; He *et al*, 2020). However, very little information is present on the types and amount of toxic gases as suggested by a review performed by Sun *et al* (2020). The presence of toxic gases such as CO, CO_2, HCl, HF, SO_2, and POF_3 has been confirmed by different studies (Larsson *et al*, 2017; Ribière *et al*, 2012), However, the amount of the gases has still not been well understood. Larsson *et al* (2017) indicated the presence of HF, POF_3, and PF_5, However, the cells used here were of different chemistries and the number of cells in each battery sample varied. In a separate study (Lecocq *et al*, 2016), the fire behaviour of Li-ion batteries with $LiPF_6$ chemistry was analysed and the presence of toxic gases such as HF, and SO_2 was confirmed, however, they were also unable to quantify the amount of the gases and also the experiment was done on only one chemistry of LIB. This research is also endorsed by a different study in which LIB cells were dissembled and the heat value and gases analysis was performed, confirming the presence of PF_5, CO_2, and HF (Kriston *et al*, 2019). However, the quantification and complete analysis of the gases is unavailable.

The main objective of this research paper is to evaluate the toxic fumes emitted during fire incidents involving Li-ion batteries. Similar studies have been performed by other researchers; however, the difference lies in the number of chemistries of LIB used, the types of gases analysed, and the SOC relative to the LIB chemistries tested. The objective of the research will be achieved by first identifying and quantifying the toxic fumes by performing fire abuse tests. Applying fire directly to the battery cells, the cells will be left to burn out completely at different states of charge (SOC). The smoke of the batteries will be analysed for the type and amount of gases, with the help of a multi-gas detector and spectroscopy using FTIR. This analysis will be useful in identifying and quantifying the gases emitted during battery fires, which can further be used in designing specialised protective equipment and fire extinguisher designs, to elevate the health and safety standards of battery-powered vehicles.

METHODOLOGY

Lithium-ion battery incinerator

As shown in Figure 1, A Li-ion battery incinerator was designed and constructed to conduct the experiments. The incinerator measures 20×20×60 inches and is composed of 26-gauge metal sheets and ¾" T-slotted framing rails. It features a 20×20×24" fire chamber, as well as compartments for the data logger and DC power supply. A 4" plastic duct is connected at the top of the chamber, which extends five feet. At the end of the duct, a 4" inline duct fan is installed to pull the smoke out of the chamber. A high-resolution thermal camera (FLIR E4) is used to capture the burning process of the battery through a separate window in the conical section of the chamber.

It is crucial to highlight that the experimental set-up was intentionally designed to reflect real-world conditions as accurately as possible. Accordingly, the incinerator was not made airtight, and the cells underwent testing under ambient environmental conditions. This approach ensures that the findings are applicable to real-world scenarios, providing practical insights into the behaviour of lithium-ion batteries during fires.

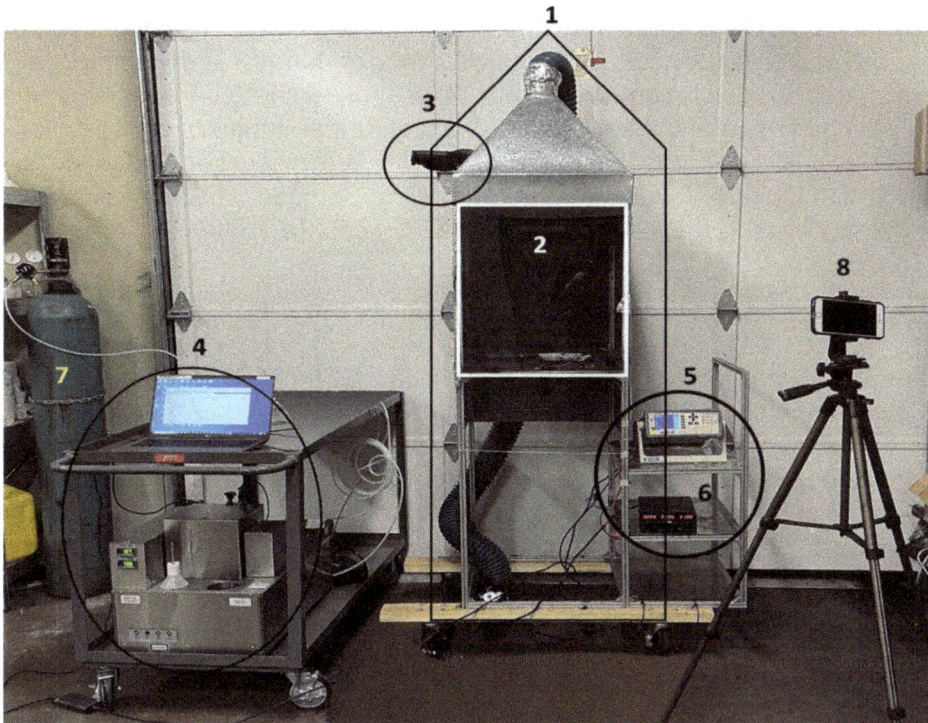

FIG 1 – Lithium-ion battery analysis system with; 1: LIB incinerator, 2: Fire Chamber, 3: Thermal Camera, 4: FTIR including a PC for gas analysis, 5: Thermocouple Data Logger, 6: DC voltage regulator, 7: Nitrogen Tank, 8: A mobile phone for video recording.

CELL SELECTION

LIBs are characterised by medium to high energy density, with a variety of chemistries. These cells have layers of Anode (copper foil coated with a specialty carbon) and Cathode (aluminium foil coated with a lithiated metal oxide or phosphate) separated by a microporous polyolefin separator. The electrode is an organic solvent and a dissolved lithium salt such as lithium-cyclodiflouromethan-1, 1-bis(sulfonyl)imide (LiDMSI) and lithium-cyclohexaflouropropoane-1, 1bis(sulfonyl)imide (LiHPSI) (Younesi *et al*, 2015), the salt provides a medium for lithium-ion transport (Williams and Back, 2014). Existing research have indicated that the emission of toxic gases is primarily affected by the state of charge (SOC) and the cathode chemistry of the LIB. Hence, it was decided to analyse the five chemistries that are the most common among electric vehicle manufacturers ie Lithium Iron Phosphate (LFP), Lithium Nickle Manganese Cobalt (NMC), Lithium Nickle Cobalt Aluminium (NCA), Lithium Titanate (LTO), and Lithium Manganese Oxide (LMO) at five different SOC ie 0 per cent, 25 per cent, 50 per cent, 75 per cent, and 100 per cent. The detailed information of the cells can be seen in Table 1.

TABLE 1

Detailed specifications of the sample cells.

Chemistry	Voltage V	Capacity Ah	Packaging	No of cells	Manufacturer
LFP	3.3	2.5	26650 Cylindrical	10	Lithiumwerks
NMC	3.6	3.0	18650 Cylindrical	20	Samsung
NCA	3.6	3.0	18650 Cylindrical	30	Mohm Tech
LTO	3.2	3.0	26650 Cylindrical	30	Lithium Titanate
LMO	3.7	3.0	18650 Cylindrical	20	Murata

The selection of cell types for this study was focused on those most utilised in the market, emphasizing battery capacity over variations in cell packaging. This approach aligns with industry standards and ensures the practical relevance of our findings. By concentrating on the predominant battery chemistries, this research provides insights that are directly applicable to the majority of battery-operated devices and vehicles in use today. The commercial utilisation of different chemistries is also highlighted by other researchers. Research highlighting the commercial utilisation of various battery chemistries that further supports our selection; Lithium Iron Phosphate (LFP) and Nickel Manganese Cobalt (NMC) are discussed in Goodenough and Kim (2010), Lithium Nickel Cobalt Aluminium Oxide (NCA) in Scrosati, Hassoun and Sun (2011), and Lithium Manganese Oxide (LMO) along with Lithium Titanate (LTO) in Tarascon and Armand (2001).

ABUSE INITIATION

Different studies have used different fire sources including an electric heater (Fu *et al*, 2015), a thermal oven (Larsson and Mellander, 2014), and a propane burner (Larsson, Andersson and Mellander, 2016). In this study, small scale battery fire tests are conducted, using an 18 Gauge Kenthal A-1 Annealed round, Nickle Cobalt resistance wire, which is powered by a DC Power supply to heat up the cells. The wire is wound around the cell as shown in Figure 2b, and gradually heated and powered by the DC Voltage regulator. The voltage of the source is regulated such that it heats the cell at a rate of 6°C per min, until the cell starts smoking off ie the cell surface temperature ≥150°C.

(a)   (b)

FIG 2 – (a) the Nickel Chromium resistance wire, (b) application of heat using Ni Chrome Resistance wire.

HEAT RELEASE RATE

Heat Release rate (HRR) is the rate at which combustion of fuel takes place with the oxygen in air. HRR cannot be measured directly but inferred from other direct measurement such as mass loss measurement (Bryant and Mulholland, 2008). For fuel made of pure substances, the heat of combustion is known, thus the HRR rate can be found out from mass loss measurement. For mass loss measurement, this study make use of sartorius high precision load cell WZA523-N. However, since we are dealing with LIB cells, which is a combination of different materials, thus the HRR measurements are complicated (Bryant and Mulholland, 2008; Ko, Michels and Hadjisophocleous, 2011), hence we will make use of oxygen consumption calorimetry, as introduced by Huggett (1980) and Bryant and Mulholland (2008).

$$q^{\circ} = (\Delta H_c) H C_{vol\,O2} V_e^{\circ} X_{o2}^{\circ} \theta \qquad (1)$$

And:

$$\theta = (X_{o2}^{\circ} - X_{o2})/X_{o2}^2 \qquad (2)$$

where:

X°   is the oxygen volume fraction in the ambient air

X   is the oxygen volume fraction in the duct after the captured smoke

V°   is the measured volume flow rate in the exhaust hood

$(\Delta Hc)HC_{vol\ O2}$ is the average value of heat produced per unit volume of O_2 consumed for generic fuel

For the O_2 measurement the study will make use of the Teledyne R17 sensor with DAQ system.

CELL SURFACE TEMPERATURE

Cell surface temperature data is used for thermal management of the cell, to study the thermal runaway temperature for different cells. The surface temperature of the cell was recorded using K-thermocouples Figure 3e attached to the surface of the sample with the help of heat resistant aluminium adhesive tape. While the data was recorded using the Anabi AT44532 temperature data logger Figure 3a. As illustrated in Figure 3d, two thermocouples are attached on the cell surface at a distance of one-fourth (1/4) from each end. While the cell is heated using a Ni-Chrome resistance coil Figure 3d powered by a DC voltage regulator Figure 3c.

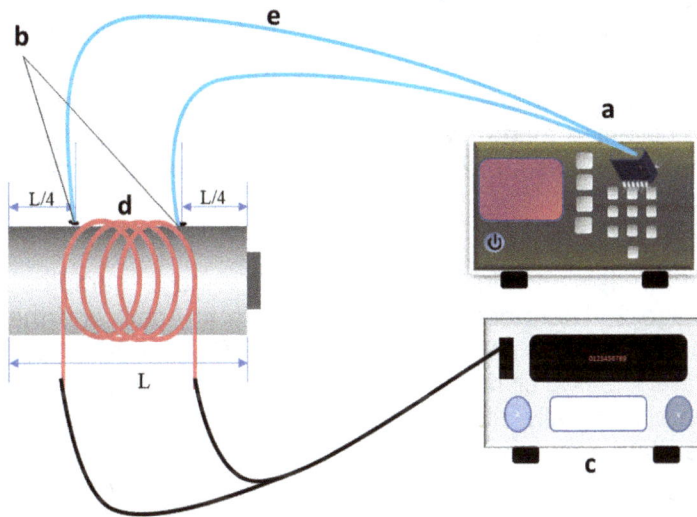

FIG 3 – (a) AT4532 temperature data logger, (b) points at which thermocouples are attached to the cell, (c) DC voltage regulator, (d) Resistance coil.

EMISSIONS ANALYSIS

This study uses Fourier Transform InfraRed (FTIR) spectroscopy (Thermo Scientific Antaris IGS analyser, Figure 4) for the analysis of the smoke. The sample is burned in the fire chamber, and the smoke is pulled out by the inline duct fan at average speed of 40 L/sec. However before the smoke is released to the atmosphere, the sample is sucked into the FTIR gas cell (Figure 4b and c) using an external vacuum pump at 0.05 L/sec. The gas cell was conditioned are set at 180°C and 650 mmHG, and these were the conditions at which the FTIR Library was prepared. The experiment lasted for 45 min and the results are based on real time analysis of the smoke.

Due to the influence of potential gas interference and the sensitivity of the technique to varying gas concentrations, discrepancies in the accuracy of our findings may arise, a common issue with FTIR spectroscopy as noted by Giechaskiel and Clairotte (2021). Nonetheless, this method provides a sufficiently comprehensive understanding of the toxic behaviour of the lithium-ion battery cells under study.

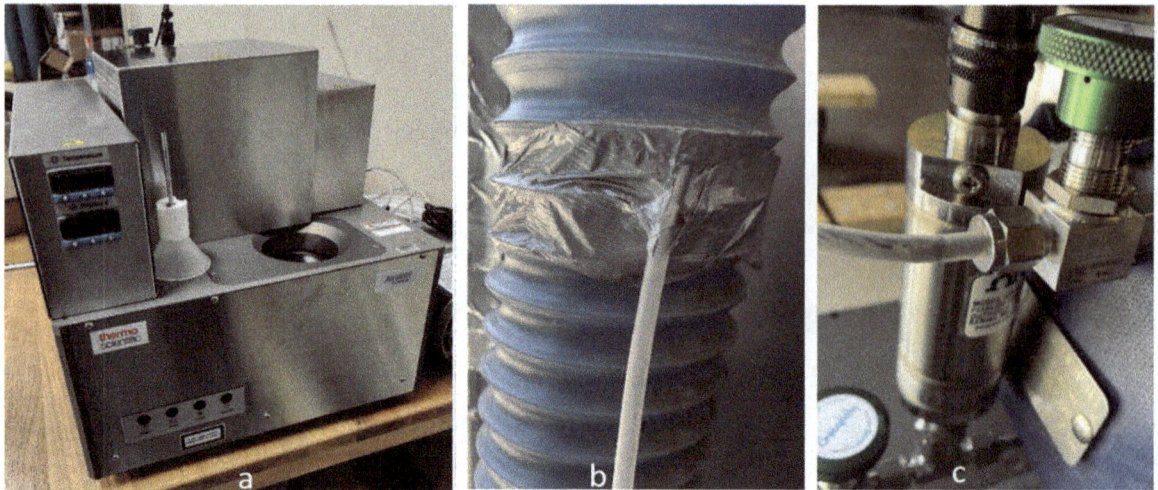

FIG 4 – (a) Thermo Scientific Antaris IGS Analyser, (b) sample point from exhaust, (c) FTIR gas inlet.

The experimental set-up was borrowed and modified from a study performed by Larsson *et al* (2017). once the cell starts to smoke, the smoke sample will be directed toward FTIR for analysis of different gases that includes Carbon Monoxide (CO), Carbon Dioxide (CO_2), Methane (CH_4), Ethene (C_2H_4), Ammonia (NH_3), and Hydrogen Floride (HF). For HRR, Oxygen calorimetry will be utilised oxygen concentration measurement will be done through Teledyne R17-a sensor, data form that sensor will be collected manually. Once the gas sample is analysed, and based on those analyses, PPEs designs, and Fire suppression techniques will be recommended.

EXPERIMENTAL PROCEDURES

The experimental procedure is shown in Figure 5. In the first stage, samples are first charged up to the desired SOC. In stage 2, Thermocouple is attached to the cell, and then the cell is wrapped with the nickel cobalt resistance wire powered with a DC power supply and placed in the incinerator. The cell is then heated till it starts smoking and the smoke is directed towards the FTIR to collect gas concentrations. Finally, the data from the FTIR and the thermocouple are transmitted to a computer for further data analysis.

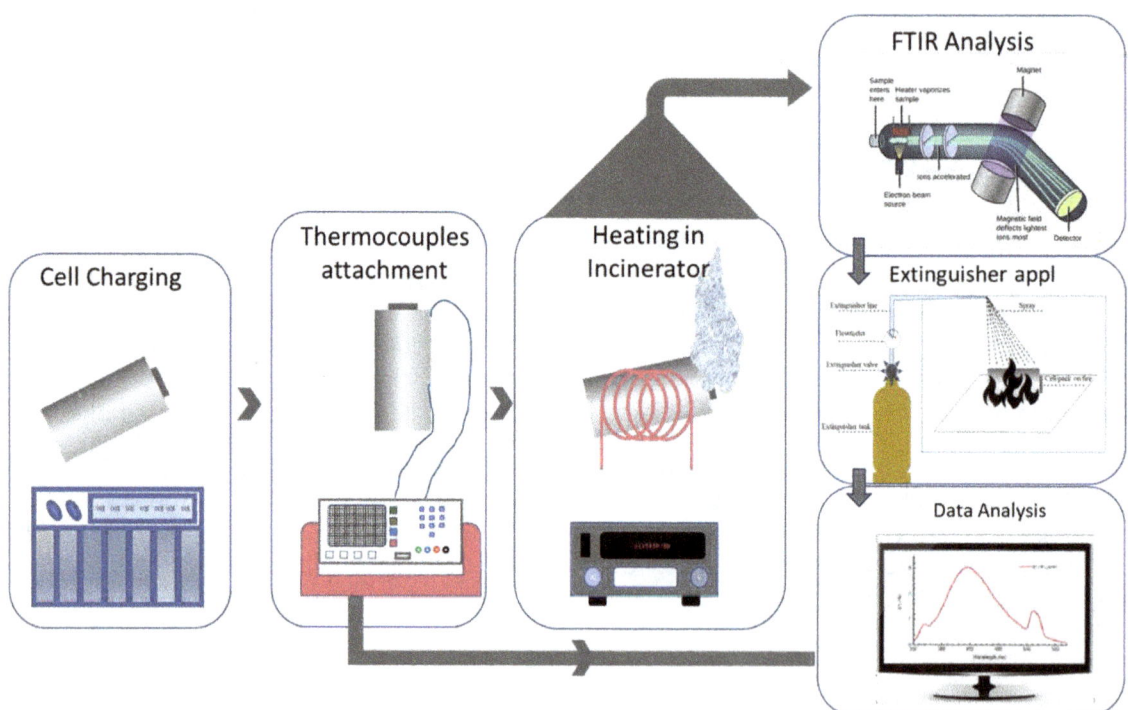

FIG 5 – Experimental set-up used for toxic gas analysis of the LIB emissions.

Statistical analysis

In this section, we'll take a closer look at each gas release and see if there's any connection with two important factors: the State of Charge (SOC%) and the Chemistry of the cell. We're diving into the data to figure out if the amount of gas released has anything to do with how charged the cell is or the specific chemical makeup of the cell. Our goal is to understand whether these factors play a role in determining gas release.

Corbon monoxide (CO)

The ANOVA results indicate that Chemistry has a significant effect on CO emissions in the context of lithium-ion battery (LIB) fires, with an F ratio of 7.3312 and a very small p-value (<.0001). This suggests that the specific chemical composition of the cell has a notable impact on the amount of CO released during LIB fires. However, the State of Charge (SOC%) and the interaction between Chemistry and SOC% do not appear to have a significant effect on CO emissions in this context, as indicated by their higher p-values of 0.6562 and 0.8522 respectively. This suggests that the variability in CO emissions during LIB fires is primarily explained by differences in cell chemistry rather than differences in SOC% or the interaction between SOC% and Chemistry

Carbon dioxide (CO$_2$)

For CO$_2$ gas, neither the State of Charge (SOC%) nor the Chemistry of the cell nor their interaction significantly influence the emissions. The p-values for SOC%, Chemistry, and their interaction are 0.8781, 0.4812, and 0.4429 respectively, indicating that none of these factors have a significant effect on CO$_2$ emissions. Similarly, the overall model does not show statistical significance in predicting CO$_2$ emissions, as indicated by the p-value for the model being greater than the threshold of significance (0.6201). These findings suggest that factors other than SOC% and Chemistry may be more influential in determining CO$_2$ emissions in the context of lithium-ion battery fires.

Methane (CH$_4$)

For Methane gas, the ANOVA results indicate that Chemistry has a significant effect on Methane emissions, with a p-value of 0.0008, indicating statistical significance. This suggests that the specific chemical composition of the cell influences the amount of Methane released. However, the State of Charge (SOC%) and the interaction between Chemistry and SOC% do not appear to have a significant effect on Methane emissions, as their p-values are relatively large (0.9949 and 0.0581 respectively). This implies that the variability in Methane emissions is primarily explained by differences in cell chemistry rather than differences in SOC% or the interaction between SOC% and Chemistry.

Ethene (C$_2$H$_4$)

Chemistry shows a significant effect on Ethene emissions, as indicated by its very small p-value (<0.0001). This suggests that the specific chemical composition of the cell significantly influences the amount of Ethene released. However, the State of Charge (SOC%) does not appear to have a significant effect on Ethene emissions, as its p-value is relatively large (0.6247). The interaction between Chemistry and SOC% shows a marginal significance, with a p-value of 0.0367, suggesting that there may be a modest interaction effect, although it is less pronounced compared to the main effect of Chemistry.

Ammonia (NH$_3$)

Both Chemistry and SOC% show significant effects on Ammonia emissions. Chemistry exhibits a highly significant effect, with a very small p-value (<0.0001), indicating that the specific chemical composition of the cell significantly influences the amount of Ammonia released. Similarly, SOC% also shows a significant effect on Ammonia emissions, with a p-value of 0.0028, suggesting that the State of Charge has a notable impact on the amount of Ammonia released. Additionally, the interaction between Chemistry and SOC% shows a marginal significance, with a p-value of 0.0428, indicating a modest interaction effect between these two factors.

Hydrogen Floride (HF)

Neither SOC% nor Chemistry show significant effects on HF emissions, as indicated by their p-values (0.2448 and 0.1986 respectively). Additionally, the interaction between Chemistry and SOC% does not appear to have a significant effect on HF emissions, with a p-value of 0.2521. This suggests that the variability in HF emissions is not explained by differences in either the State of Charge (SOC%) or the specific chemical composition of the cell. The overall model also does not show statistical significance in predicting HF emissions, as indicated by the p-value for the model being greater than the threshold of significance (0.1873).

RESULTS

Burning process

In this study, controlled heat exposure was applied to lithium-ion battery cells with the help of a DC power supply to investigate their burning characteristics. The heat was incrementally increased from an initial rate of 4W, with increments of 2W every 10 mins, until reaching a maximum of 10W. The heating rate was chosen based on preliminary test results to enable the observation of the cell heating rate and the monitoring of changes in surface temperature during the test and heat application. This careful selection aimed to capture meaningful data on the behaviour of the cells under thermal stress conditions, considering both cell chemistry and state of charge. The heat was systematically applied to the cells, mimicking thermal stress scenarios. Our findings revealed that the burning process varied significantly depending on the chemistry of the cells and their SOC. Different chemistries exhibited distinct burning behaviours, with variations in flame intensity, test duration, smoke duration, and failing temperature, which highlighted insights into the thermal runaway progression of the cells.

Flame intensity

Flame intensity varied among different cell chemistries. LFP and LTO cells did not produce flames. However, at lower state of charges (SOCs) (10 per cent, 25 per cent, and 50 per cent), LFP cells emitted smoke quietly, while at higher SOCs (75 per cent and 100 per cent), the smoke became more intense and vigorous. Similarly, LTO cells did not exhibit flames. Instead, they would suddenly rupture at the moment of failure, expelling their contents outward. The intensity of the rupture was observed to be more intense in higher SOC compared to lower SOCs. Cell swelling was more noticeable in LTO cells as they approached failure.

Fire and explosion were observed in NMC, NCA, and LMO cells, exhibiting varied intensities. At lower state of charges (SOCs), the fire was more subtle and lasted longer, ranging from 50 to 60 secs. However, at higher SOCs, the explosions tended to be more intense and vigorous, although the flame duration was shorter, typically lasting between 10 to 20 secs.

Test duration

Test duration varied among cell chemistries, generally showing an inverse relationship with the state of charge (SOC) except for NMC. Higher SOC resulted in shorter test durations, as illustrated in Figure 6. From Figure 6, the difference in the duration of test is highlighted most prominently in LFP, where the first three test of SOC 10 per cent, took on average 80–85 mins, while the last three tests of SOC 100 per cent, took on average 40–45 mins. Similar trend can be observed in other chemistry cells.

On average, LFP exhibited the longest duration at 70 mins, followed by LTO at 43 mins, while NCA, NMC, and LMO had durations of 22, 25, and 23 mins respectively. However, all chemistries demonstrated an inverse relationship between test duration and SOC, except for NMC. NMC displayed a distorted trend with no specific relationship apparent.

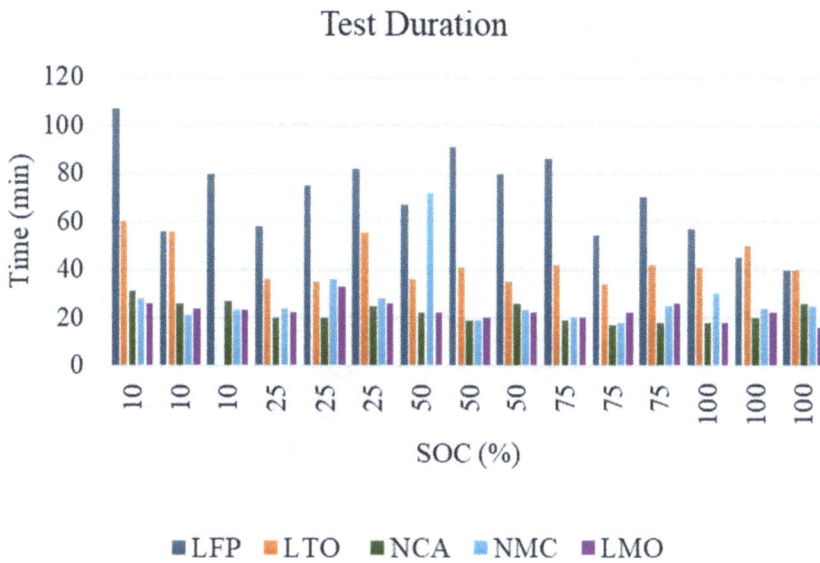

FIG 6 – Test duration for all chemistries and SOCs.

Smoke duration

The duration of smoke production exhibited variability with no clear trend related to SOC, but a relationship was observed with cell chemistry. Figure 7 presents the smoke durations recorded in mins for all the chemistries, the right-side scale is specific to LFP cells, as it recorded the longest smoking duration, approximately 25 mins, which was considerably longer than the others. All other chemistries are related to the left-side scale. As shown in the Figure 7, Following LFP, LTO had the longest average smoking duration of 6 mins. NCA and LMO recorded an average smoking duration of 3 mins each, while NMC had a smoking duration of 2 mins. Additionally, for LTO cells at higher SOCs (75 per cent and 100 per cent), some cells experienced sudden failure without any preceding smoke production, but it had notable swelling.

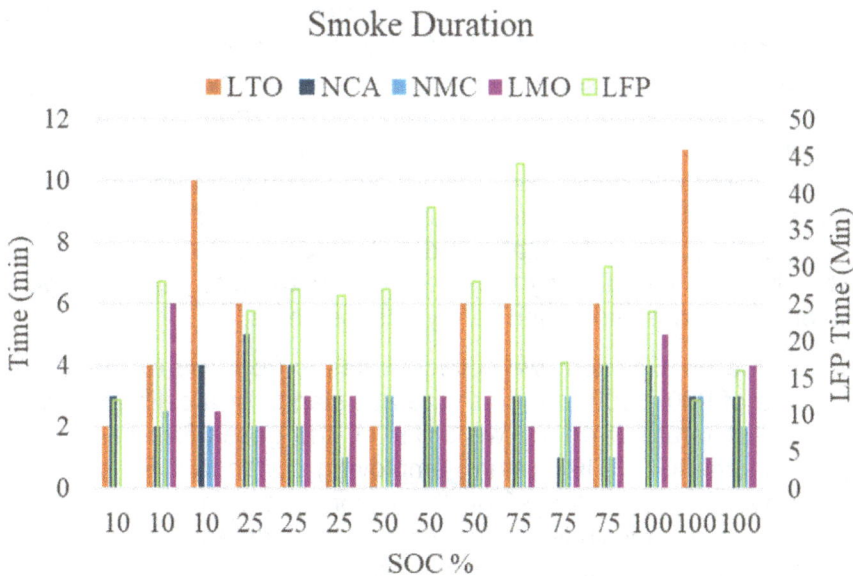

FIG 7 – Smoke duration for all chemistries and SOCs.

Failing temperature

The failing temperature varied among different cell chemistries. Figure 8 represents the failing temperature graphs for all the test combined, as LFP exhibited the highest average failing temperature at 235°C, with some cells reaching up to 300°C. LTO consistently recorded a failing temperature close to 170°C. Similarly, NCA showed an average failing temperature of 190°C, with some exceptions where cells failed at around 165°C. NMC cells recorded a failing temperature of approximately 200°C, except for one exception where the cell failed at 150°C. However, LTO consistently failed at 195°C. No significant difference was observed based on the state of charge (SOC) of the cells, but variations were evident based on cell chemistry.

Failing Temperature

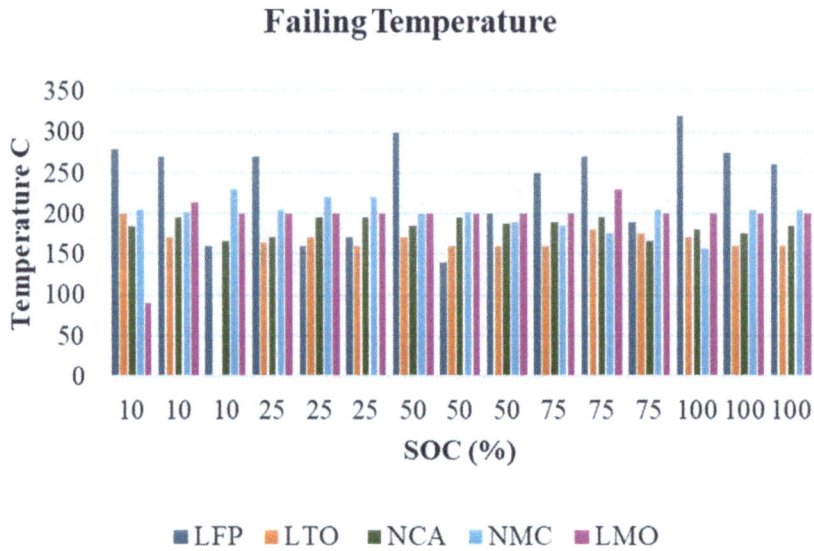

FIG 8 – Failing temperature for all chemistries and SOCs.

Often, when the cell exploded, the thermocouple would detach during the process, making it challenging to precisely determine the maximum surface temperature of the cell explosion. However, for cells that did not explode or in most instances, where we were able to record the maximum temperature of the cell after failure, even if it lasted only for a moment. At that time, the recorded temperature was close to 900°C.

GASEOUS EMISSIONS

In the document, a series of figures are presented that chart the emission patterns of various gases from cells with differing lithium-ion battery chemistries: Lithium Manganese Oxide (LMO), Nickel Manganese Cobalt Oxide (NMC), Nickel Cobalt Aluminium Oxide (NCA), Lithium Titanate Oxide (LTO), and Lithium Iron Phosphate (LFP). Each figure systematically quantifies the gas concentrations over the observation period, capturing the dynamics of emissions. The figures included are representative samples chosen from a broader data set, with only 25 out of 75 tests displayed to maintain clarity and avoid visual overload. The results are organised into categories based on states of charge (SOC) at intervals of 100 per cent, 75 per cent, 50 per cent, 25 per cent, and 10 per cent. This categorisation facilitates a detailed comparative analysis of gas emissions in relation to SOC, highlighting the interplay between battery chemistry and charge state, which has significant implications for performance and safety assessments.

After the cells caught fire or exploded, we collected gases from the exhaust stream using a fan operating at a flow rate of 900–1000 L/m. With the assistance of a calibrated vacuum set to draw gases into the gas cell at a rate of 4 L/m, we collected the gases for analysis. The analysis was conducted using the Antaris IGS FTIR system, employing a standard fire calibration that includes 15 different gases. However, we will focus our discussion on the most prominent results observed in the stream for each gas. That include Carbon Monoxide, Methane, Ethene, Propene, Ammonia, Formaldehyde, Hydrogen Fluoride, Sulfur Dioxide, Carbon Dioxide.

Carbon monoxide (CO)

Figure 9 presents detailed analysis of carbon monoxide (CO) emissions from various lithium-ion battery chemistries revealing significant variations in emission patterns. These emissions are influenced by both the state of charge (SOC) and the specific battery chemistry. For LMO and NMC, CO emissions exhibit pronounced and sharp peaks at higher SOCs, indicating rapid generation and depletion of CO. These peaks become less distinct at lower SOCs, suggesting a decrease in reaction rates. In contrast, NCA shows a single sharp emission peak at high SOCs with reduced emissions as the SOC decreases, which might indicate more stable gas emission behaviour when not fully charged. LTO displays a unique pattern with irregular and frequent peaks across all SOC levels, pointing to multiple mechanisms of CO generation. LFP is notable for significant CO emissions even at lower SOCs, hinting at possible electrochemical instabilities or side reactions at reduced charge states. Across all chemistries, the highest CO concentrations typically occur at 100 per cent SOC, progressively diminishing at lower charge states, highlighting the critical influence of SOC on gas emissions and emphasizing the need for careful management of charging practices to ensure battery safety and longevity.

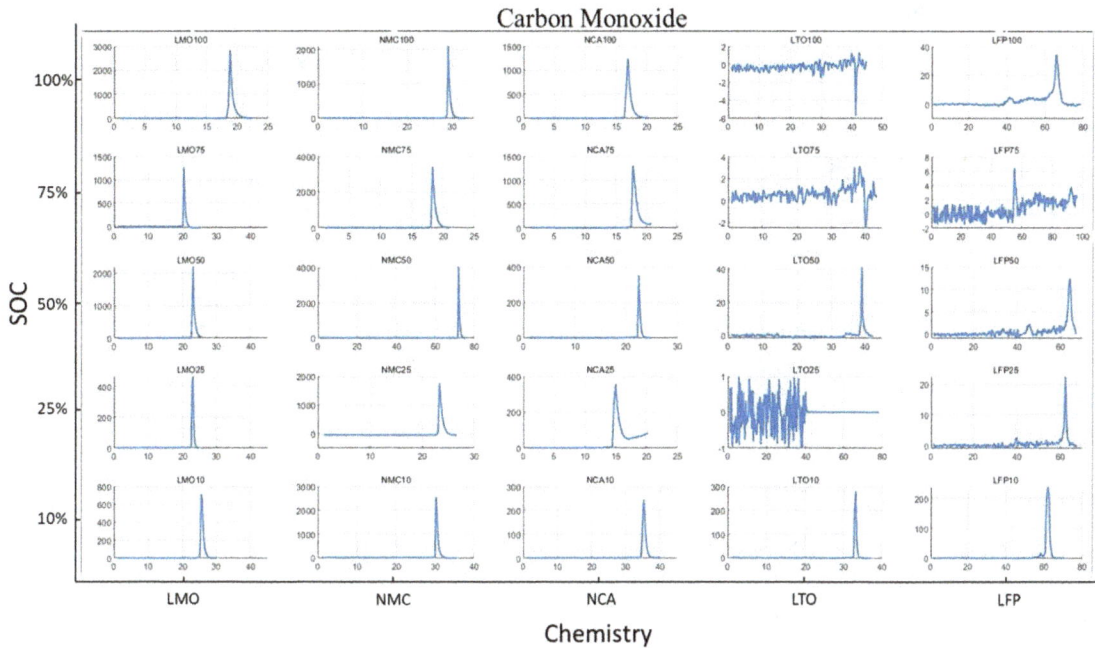

FIG 9 – Carbon monoxide concentration recorded for different tests wrt cell chemistry and SOC.

Methane (CH₄)

Figure 10 illustrates the emission patterns of methane (CH_4) across various states of charge (SOC) in lithium-ion batteries with different chemistries—LMO, NMC, NCA, LTO, and LFP. The general observation indicates that methane emissions are notably chemistry-dependent, displaying distinctive patterns for each battery type.

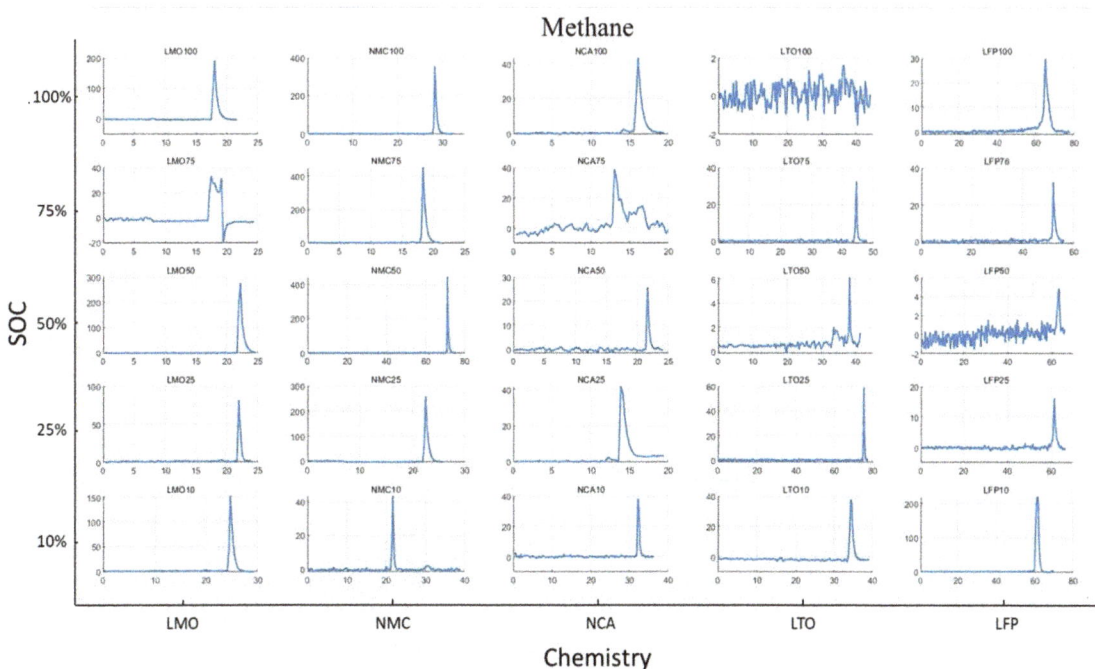

FIG 10 – Methane concentration recorded for different tests wrt cell chemistry and SOC.

LMO and NMC chemistries exhibit notable peaks predominantly at higher SOC levels, with the peaks being more pronounced and sharper for LMO. These pointed peaks suggest a rapid generation of methane, which could be linked to specific reactions at full charge. As the SOC decreases, the peak intensity for both chemistries lessens, indicating a slower rate of methane generation or a possible shift in the reaction dynamics within the battery.

For NCA, there is a clear peak at 100 per cent SOC, which diminishes substantially at lower SOCs. This trend may imply a more controlled emission pattern where methane generation is tightly linked to the battery being at or near full capacity.

The emission pattern for LTO is irregular, showing substantial variability and multiple peaks, especially noticeable at 25 per cent SOC. This irregularity could suggest a more complex set of reactions contributing to methane generation, potentially linked to the unique charge and discharge cycles or the surface chemistry of the LTO.

Lastly, the LFP chemistry shows a significant level of methane emissions at 10 per cent SOC, indicating that methane generation in LFP cells may not be strictly tied to higher charge states and may occur due to other factors or reactions within the cell, even when the charge is low.

Across the board, the figures suggest that methane emissions are impacted by SOC, but the relationship is not linear and varies with the battery chemistry. Understanding these patterns is vital for improving battery safety and performance, as it highlights the need for specific control strategies tailored to each battery type and charge level to mitigate unwanted emissions.

Carbon dioxide (CO$_2$)

Figure 11 presents the carbon dioxide (CO$_2$) emission profiles across different lithium-ion battery chemistries at varying states of charge (SOC), the data reveals distinct emission trends. At 100 per cent SOC, LMO exhibits a sharp peak, indicating a rapid release of CO$_2$, which sharply declines afterward. Similarly, NMC at full charge also shows a significant peak, though less pronounced than LMO, followed by a steep reduction in CO$_2$ emissions as SOC decreases.

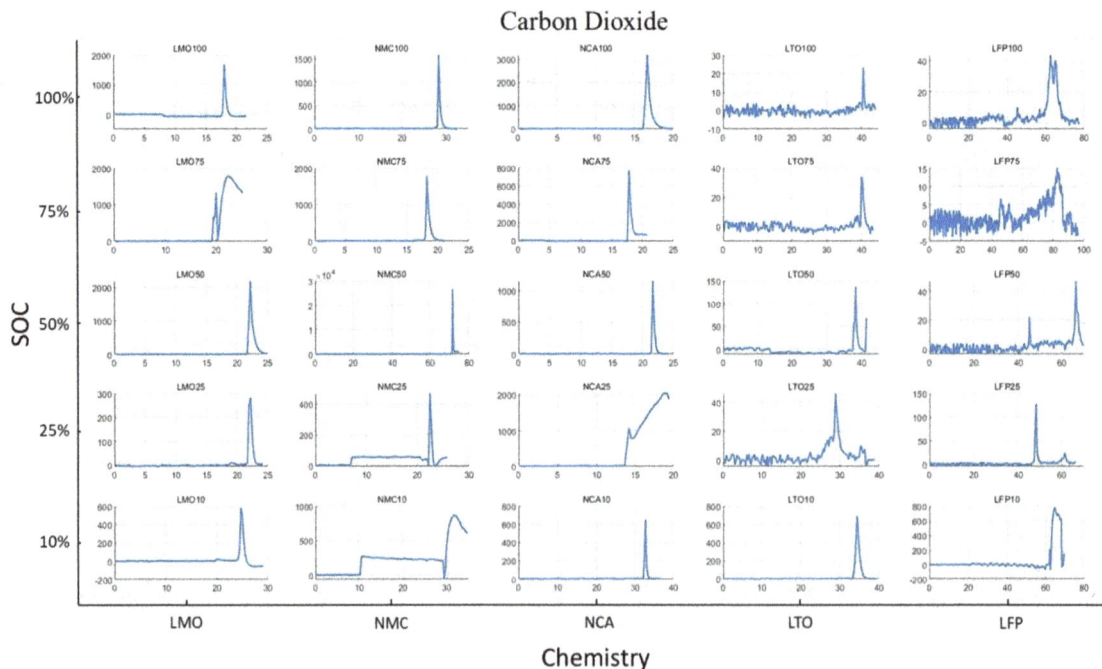

FIG 11 – Carbon dioxide concentration recorded for different tests wrt cell chemistry and SOC.

The NCA curves are notable for a peak at 75 per cent SOC, diverging from the pattern of highest emissions at full charge, seen in LMO and NMC. This indicates a unique CO$_2$ evolution behaviour, possibly due to specific reactions within the NCA chemistry at this SOC level.

For LTO, the CO$_2$ emission patterns are more erratic, with considerable fluctuations at 100 per cent and 75 per cent SOC, and a distinct peak at 25 per cent SOC. This suggests a variable pattern of CO_2 release across the SOC spectrum, potentially attributable to multiple factors influencing CO_2 evolution in LTO cells.

In contrast, LFP displays a relatively stable CO_2 emission profile with lower SOC levels until a significant peak emerges at 10 per cent SOC, indicating a potential increase in CO_2 evolution as the battery discharges to lower levels.

Overall, each battery chemistry exhibits its own characteristic CO_2 emission signature in relation to SOC. LMO and NMC show the highest emissions at full charge, whereas NCA, LTO, and LFP present a more varied relationship with SOC, demonstrating that CO_2 emission is not solely dependent on the level of charge but also on the intrinsic properties of the battery chemistry.

Ammonia (NH₃)

Figure 12, displaying ammonia (NH_3) emissions showcases how different lithium-ion battery chemistries behave at various states of charge (SOC). LMO chemistry shows a distinct peak at 75 per cent SOC, with otherwise minimal emissions. NMC has a marked peak at 75 per cent SOC as well, but also exhibits a smaller peak at 100 per cent SOC and a variable emission pattern at lower SOCs.

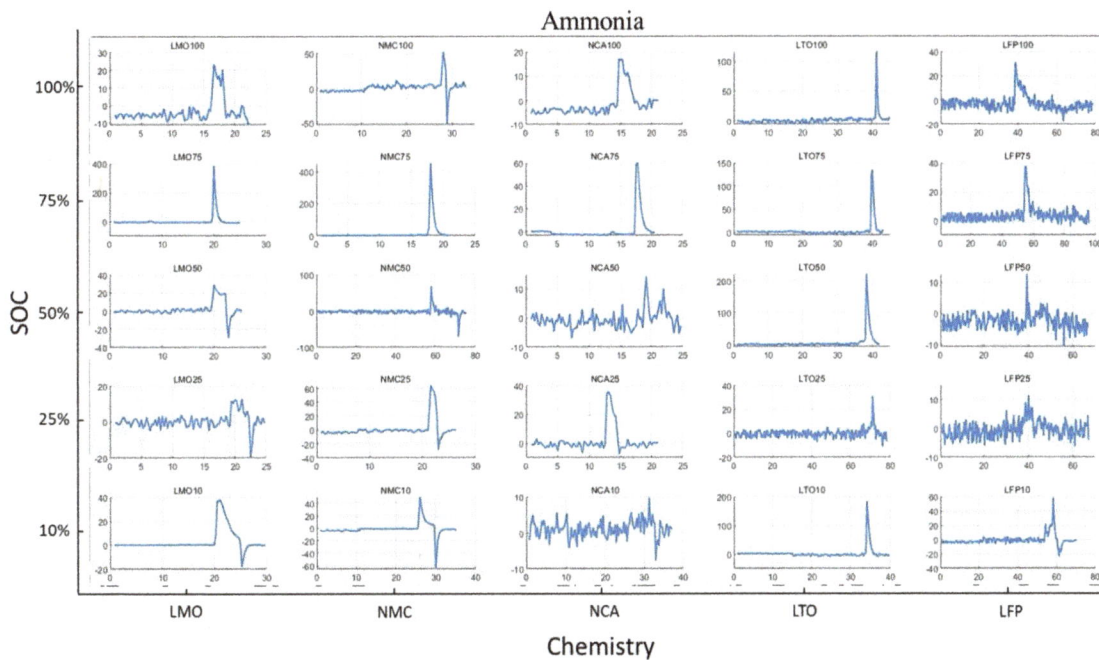

FIG 12 – Ammonia concentration recorded for different tests wrt cell chemistry and SOC.

NCA shows a significant peak at 75 per cent SOC, similar to LMO and NMC, with lower emissions at full charge and more variability as SOC decreases. The pattern suggests that for these chemistries, ammonia emissions are more pronounced at a mid-range SOC rather than at full charge.

LTO presents an entirely different profile, with a notable peak at 10 per cent SOC, indicating increased ammonia emissions as the cell discharges. This is in contrast to the other chemistries where the peak emission does not occur at such low SOC levels.

Lastly, LFP demonstrates irregular emission patterns across all SOCs with no clear peak, suggesting that ammonia emission is less predictable and does not follow a simple trend related to the SOC in LFP batteries.

Overall, the ammonia emission data across the various chemistries indicate that mid-range SOC levels, particularly around 75 per cent, tend to show higher levels of emissions for LMO, NMC, and NCA. In contrast, LTO shows increased emissions at low SOC, and LFP does not exhibit a consistent pattern, highlighting the complex nature of ammonia emissions and their dependence on both the SOC and battery chemistry.

It is important to note that the ammonia emission plots for some battery chemistries, such as LMO at 10 per cent and 50 per cent SOC, exhibit an unconventional feature where the apparent concentration of NH_3 dips into negative values Figure 12. Also present in some cases in NMC, this anomaly is not indicative of a true negative emission but rather a spectroscopic interference in the measurement process. The Fourier Transform Infrared Spectroscopy (FTIR) used to detect gases can sometimes confuse the overlapping spectral signatures of ammonia (NH_3) with those of ethene. Ethene, present in the sampling stream, creates a spectral overlap that the FTIR mistakenly interprets as a reduction in ammonia levels, resulting in what is displayed as negative concentrations. This interference highlights the complexity of accurately measuring gas emissions using spectroscopic techniques, especially when multiple gases with overlapping absorption bands are present.

Ethene (C_2H_4)

Figure 13 represents ethene emissions across different lithium-ion battery chemistries at various states of charge (SOC), the plots display a variety of peak shapes and trends. LMO shows distinct peaks at 75 per cent and 50 per cent SOC, indicating a release of ethene at these specific states. NMC displays a significant peak at 75 per cent SOC and a series of smaller peaks at 50 per cent and 25 per cent SOC, suggesting multiple release events.

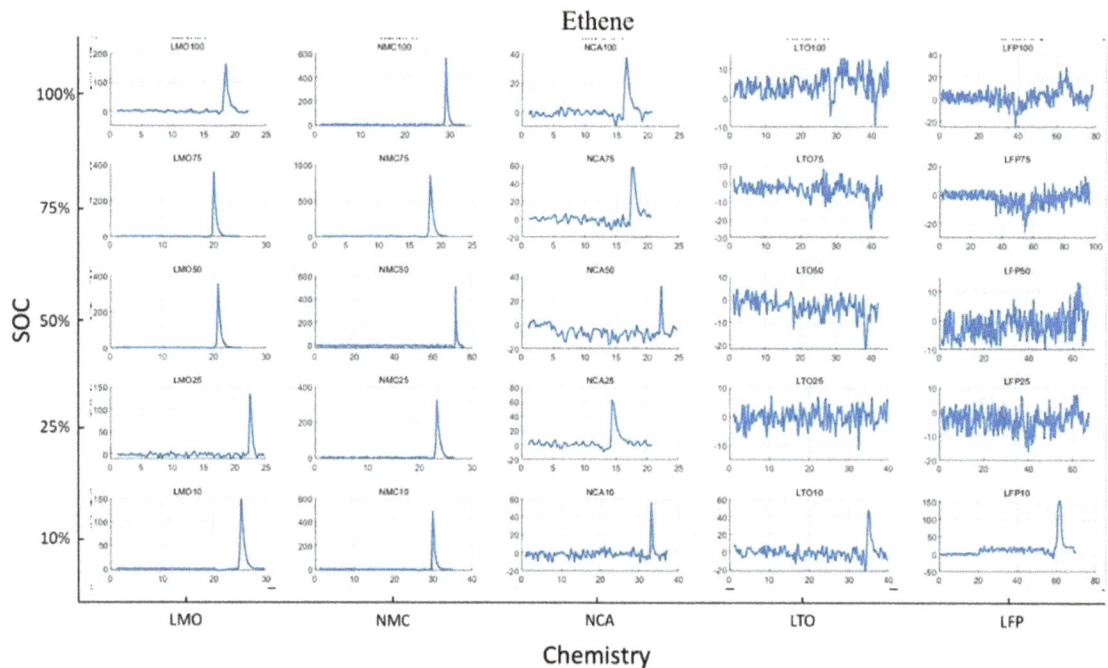

FIG 13 – Ethene concentration recorded for different tests wrt cell chemistry and SOC.

In the case of NCA, there's an interesting peak at 75 per cent SOC with an irregular pattern at lower SOCs, indicating variability in ethene release. Notably, the emissions for NCA are more erratic, with fluctuations seen across the SOC spectrum.

LTO and LFP chemistries demonstrate a different behaviour compared to LMO, NMC, and NCA, with no prominent peaks but rather a more fluctuating emission pattern at all levels of SOC, which might suggest a continuous release of ethene or complex overlapping reactions that contribute to its generation.

Overall, the emission profiles for ethene show that there is no consistent pattern correlating the amount of ethene emitted with SOC across the different battery chemistries, illustrating the complex nature of gas evolution during battery operation.

Hydrogen fluoride (HF)

Figure 14 displays hydrogen fluoride (HF) emissions reveals a wide range of fluctuations across the different lithium-ion battery chemistries at various states of charge (SOC). The plots for each chemistry, LMO, NMC, NCA, LTO, and LFP, do not show pronounced peaks across the higher SOCs, indicating that HF emissions do not exhibit significant spikes at these levels.

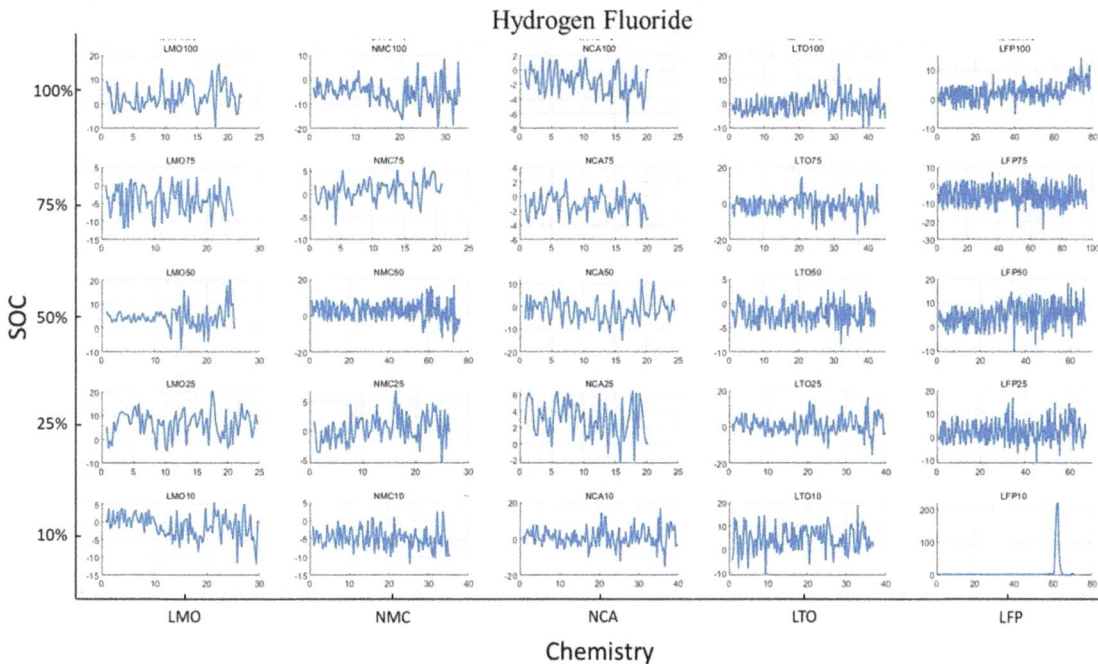

FIG 14 – Hydrogen fluoride concentration recorded for different tests wrt cell chemistry and SOC.

For LMO, the emissions are relatively stable across SOCs with a slight increase at 10 per cent SOC. NMC exhibits a similar trend with more pronounced fluctuations at 25 per cent and 10 per cent SOC. NCA shows relatively low variability in HF emissions until a significant increase at 10 per cent SOC, which is indicative of a change in emission behaviour at lower charge states.

LTO's emission pattern is characterised by consistent fluctuations without any noticeable peaks across the SOC spectrum, suggesting an ongoing release or detection of HF during charge and discharge cycles. In contrast, LFP displays irregular fluctuations across all SOCs, with a substantial peak emerging at 10 per cent SOC. This peak in the LFP plots could point to a more significant HF generation as the battery discharges to lower levels.

Focusing on the Lithium Iron Phosphate (LFP) chemistry, the emission pattern for hydrogen fluoride (HF) across the states of charge (SOC) shows a general trend of fluctuation without distinctive peaks until 10 per cent SOC, where there is a sharp and significant peak. This pronounced increase at lower SOC levels may suggest that HF emissions in LFP chemistry are more likely to occur as the cell discharges and reaches a near-depleted state.

Overall, the emission profiles for HF indicate that while there may be a consistent release of HF across SOC levels, there is a tendency for increased emissions at lower SOCs, particularly for NCA and LFP chemistries. The fluctuating nature of the plots suggests complex underlying mechanisms of HF release that vary by battery type and state of charge. This behaviour underscores the necessity to monitor and manage HF emissions in LFP, particularly at lower SOC levels, to ensure the safety and integrity of LFP-based battery systems. The significant peak at 10 per cent SOC in LFP cells could have implications for end-of-life battery management and recycling processes where cells are often discharged to low SOC levels.

CONCLUSIONS

The research conducted offers a substantial contribution toward the primary objective of evaluating toxic fumes emitted during Li-ion battery fire incidents. By encompassing a broader spectrum of LIB chemistries and analysing a comprehensive range of gases across various SOCs, this study extends beyond the scope of preceding research in this domain. The meticulous identification and quantification of toxic gases through fire abuse tests, conducted at differential states of charge, stand as a testament to the thoroughness of this investigation.

The conclusions drawn from the extensive statistical analysis—particularly highlighting the significant impact of cell chemistry on the emission of certain gases—provides a deeper understanding of the hazardous potential of these chemistries during thermal events. The variations observed in the emissions of carbon monoxide, methane, ethene, propene, ammonia, formaldehyde, hydrogen cyanide, and hydrogen fluoride align with the objectives set out for this study, elucidating the complexities inherent in the release of toxic fumes from Li-ion batteries under fire conditions.

Moreover, the insights gained into the effects of SOC on specific gas emissions—such as the notable impact on hydrogen cyanide and ammonia—enhance our understanding of how charge levels may influence the risks associated with battery fires. This knowledge is crucial in the pursuit of designing specialised protective equipment and effective fire extinguishing methods, as it directly informs the health and safety protocols required for handling battery-powered vehicles.

The practical application of this research is manifold, from informing first responders and safety personnel about the potential risks to aiding manufacturers in improving battery design to mitigate such hazards. The detailed quantification of gases, facilitated by multi-gas detectors and FTIR spectroscopy, paves the way for the development of targeted safety mechanisms and fire suppression solutions tailored to the specific risks posed by different battery chemistries and states of charge.

In essence, the findings of this research not only achieve the stated objectives but also provide a foundational basis for elevating the safety standards in the burgeoning field of battery-powered transportation, ensuring a safer future for such technologies.

RECOMMENDATIONS

Building on the findings from this comprehensive analysis of toxic fumes emitted during Li-ion battery fire incidents, several recommendations can be formulated to guide future research and enhance practical applications:

- Based on the gases identified at different SOCs and battery chemistries, there is a vital need to design specialised protective gear tailored for first responders. Additionally, advancing the development of sophisticated multi-gas detectors could significantly improve real-time hazard assessment during battery fire incidents.

- The study prompts the development of new fire extinguishing methods that are effective against the specific gases released during battery fires. Furthermore, integrating built-in suppression systems within battery designs could pre-emptively address and mitigate the risks of toxic gas emissions.

- It is critical to delve deeper into the specific chemical reactions driving gas emissions across various states of charge. Expanding the scope of study to include more diverse battery chemistries would enrich our understanding and create a broader database of potential hazards during thermal events.

- The establishment of rigorous safety standards and regulations specifically addressing the risks posed by toxic gases from Li-ion batteries is recommended. Industry-wide protocols for the safe handling, storage, and disposal of these batteries are also essential to minimise fire risks.

- Conducting comprehensive health impact assessments for the gases identified can inform medical and safety guidelines for handling exposure. Moreover, assessing the long-term environmental impacts of these toxic gases will be crucial for developing sustainable practices.

- This study suggests developing safety equipment and protocols from our findings but does not delve deeply into how these can be practically implemented. The challenges of adopting new safety technologies and modifying existing battery designs are considerable, involving technical feasibility, cost, and industry acceptance. Future work should involve industry stakeholders to test and improve these safety solutions, ensuring they are effective and feasible for broad implementation.

These recommendations aim to not only extend the research landscape but also to apply the knowledge gained to improve safety standards, emergency responses, and regulatory frameworks surrounding the use of Li-ion batteries. The pursuit of these objectives will substantially contribute to safer deployment of battery technologies in various applications including transportation in underground mining industry.

LIMITATIONS

This study effectively quantifies the emissions from lithium-ion battery cells after ignition, focusing specifically on the types of gases released during such events. While this research does not investigate the chemical reaction mechanisms inside the cells that lead to gas formation, this limitation is intentional. The scope is defined by the primary aim to evaluate emissions post-ignition, which is crucial for developing immediate safety responses. Recognising the importance of understanding chemical kinetics during thermal runaway, it is recommended that future research include both emission characterisation after ignition and an in-depth analysis of the chemical processes to enhance preventative safety measures in battery technology.

Besides, while this study successfully identifies and quantifies the immediate safety and health risks associated with toxic gas emissions from lithium-ion battery fires, it does not address their long-term environmental impacts. The research focuses on acute exposure scenarios crucial for emergency response. However, it overlooks the broader ecological effects, such as contributions to air pollution and long-term ecosystem degradation. Future studies are encouraged to explore these aspects to provide a comprehensive evaluation of the environmental consequences of battery fire emissions

REFERENCES

Bruce, C, 2019. Smoking BMW i8 Dumped In Water By Firefighters, Motor1.com. Available from: <https://uk.motor1.com/news/315605/bmw-i8-fire-dropped-in-water/>

Bryant, R A and Mulholland, G W, 2008. A guide to characterizing heat release rate measurement uncertainty for full-scale fire tests, *Fire and Materials: An International Journal*, 32(3):121–139.

Doughty, D H and Roth, E P, 2012. A general discussion of Li ion battery safety, *The Electrochemical Society Interface*, 21(2):37.

Fu, Y, Lu, S, Li, K, Liu, C, Cheng, X and Zhang, H, 2015. An experimental study on burning behaviors of 18650 lithium ion batteries using a cone calorimeter, *Journal of Power Sources*, 273:216–222.

George, P, 2013. Another Tesla Model S Caught Fire After A Crash In Mexico, Jalopnik. Available from: <https://jalopnik.com/another-tesla-model-s-caught-fire-after-a-crash-in-mexi-1453376349>

Giechaskiel, B and Clairotte, M, 2021. Fourier transform infrared (FTIR) spectroscopy for measurements of vehicle exhaust emissions: A review, *Applied Sciences*, 11(16):7416.

Goodenough, J B and Kim, Y, 2010. Challenges for rechargeable Li batteries, *Chemistry of materials*, 22(3):587–603.

He, X, Restuccia, F, Zhang, Y, Hu, Z, Huang, X, Fang, J and Rein, G, 2020. Experimental study of self-heating ignition of lithium-ion batteries during storage: effect of the number of cells, *Fire technology*, 56(6):2649–2669.

Huggett, C, 1980. Estimation of rate of heat release by means of oxygen consumption measurements, *Fire and materials*, 4(2):61–65.

Ko, Y J, Michels, R and Hadjisophocleous, G, V, 2011. Instrumentation design for HRR measurements in a large-scale fire facility, *Fire technology*, 47:1047–1061.

Kriston, A, Adanouj, I, Ruiz, V and Pfrang, A, 2019. Quantification and simulation of thermal decomposition reactions of Li-ion battery materials by simultaneous thermal analysis coupled with gas analysis, *Journal of Power Sources*, 435:226774.

Labovick Law Group, 2021. Electric Vehicle Fire Incidents And Statistics, Labovick Law Group. Available from: <https://www.labovick.com/blog/electric-vehicle-fire-incidents-and-stats/>

Larsson, F, Andersson, P, Blomqvist, P and Mellander, B-E, 2017. Toxic fluoride gas emissions from lithium-ion battery fires, *Scientific reports*, 7(1):1–13.

Larsson, F, Andersson, P and Mellander, B-E, 2016. Lithium-ion battery aspects on fires in electrified vehicles on the basis of experimental abuse tests, *Batteries*, 2(2):9.

Larsson, F and Mellander, B-E, 2014. Abuse by external heating, overcharge and short circuiting of commercial lithium-ion battery cells, *Journal of The Electrochemical Society*, 161(10), p, A1611.

Lecocq, A, Eshetu, G G, Grugeon, S, Martin, N, Laruelle, S and Marlair, G, 2016. Scenario-based prediction of Li-ion batteries fire-induced toxicity, *Journal of Power Sources*, 316:197–206.

Liu, X, Wu, Z, Stoliarov, S I, Denlinger, M, Masias, A and Snyder, K, 2016. Heat release during thermally-induced failure of a lithium ion battery: Impact of cathode composition, *Fire Safety Journal*, 85:10–22.

Liu, Y, Sun, P, Niu, H, Huang, X and Rein, G, 2021. Propensity to self-heating ignition of open-circuit pouch lithium-ion battery pile on a hot boundary, *Fire Safety Journal*, 120:103081. https://doi.org/10.1016/JF.IRESAF.2020.103081

Musk, E, 2013. Model S Fire, Tesla. Available from: <https://www.tesla.com/blog/model-s-fire>

Peng, Y, Yang, L, Ju, X, Liao, B, Ye, K, Li, L, Cao, B and Ni, Y, 2020. A comprehensive investigation on the thermal and toxic hazards of large format lithium-ion batteries with LiFePO4 cathode, *Journal of hazardous materials*, 381:120916.

Ribière, P, Grugeon, S, Morcrette, M, Boyanov, S, Laruelle, S and Marlair, G, 2012. Investigation on the fire-induced hazards of Li-ion battery cells by fire calorimetry, *Energy and Environmental Science*, 5(1):5271–5280.

Scrosati, B, Hassoun, J and Sun, Y-K, 2011. Lithium-ion batteries, A look into the future, *Energy and Environmental Science*, 4(9):3287–3295.

Stec, A A, 2017. Fire toxicity–The elephant in the room?, *Fire Safety Journal*, 91:79–90.

Sun, P, Bisschop, R, Niu, H and Huang, X, 2020. A review of battery fires in electric vehicles, *Fire technology*, 56(4):1361–1410.

Tarascon, J-M and Armand, M, 2001. Issues and challenges facing rechargeable lithium batteries, *Nature*, 4146861. pp 359–367.

Traugott, J, 2021. Two-Day-Old VW Golf Hybrid Explodes While Driving, CarBuzz. Available from: <https://carbuzz.com/news/two-day-old-vw-golf-hybrid-explodes-while-driving>

Federal Aviation Administration (FAA), 2022. Lithium Battery Incidents, Federal Aviation Administration. Available from: <https://www.faa.gov/hazmat/resources/lithium_batteries/media/Battery_incident_chart.pdf>

Wayland, M and Kolodny, L, 2021. Fires, probes, recalls: The shift to electric vehicles is costing automakers billions, CNBC. Available from: <https://www.cnbc.com/2021/08/19/fires-probes-recalls-automakers-spend-billions-in-shift-to-evs.html>

Williams, F W and Back, G G, 2014. Lithium battery fire tests and mitigation, Naval Research Lab Washington DC Chemistry Div.

Yoney, D, 2021. Hyundai Kona Electric Fires In Norway And Korea Cause Concern, Inside EVs. Available from: <https://insideevs.com/news/515983/kona-electric-fire-norway-korea/>

Younesi, R, Veith, G M, Johansson, P, Edström, K and Vegge, T, 2015. Lithium salts for advanced lithium batteries: Li–metal, Li–O$_2$ and Li–S, *Energy and Environmental Science*, 8(7):1905–1922.

Zhao, J, Xue, F, Fu, Y, Cheng, Y, Yang, H and Lu, S, 2021. A comparative study on the thermal runaway inhibition of 18650 lithium-ion batteries by different fire extinguishing agents, *Iscience*, 24(8):102854.

Zhong, G, Mao, B, Wang, C, Jiang, L, Xu, K, Sun, J and Wang, Q, 2019. Thermal runaway and fire behavior investigation of lithium ion batteries using modified cone calorimeter, *Journal of Thermal Analysis and Calorimetry*, 135(5):2879–2889.

Intelligent fire evacuation routes – leveraging a mine-wide IoT

S Kingman[1], V Androulakis[2] and P Roghanchi[3]

1. Undergraduate Student, New Mexico Institute of Mining and Technology, Socorro NM 87801. Email: shawn.kingman@student.nmt.edu
2. Research Assistant Professor, New Mexico Institute of Mining and Technology, Socorro NM 87801. Email: vasileios.androulakis@nmt.edu
3. Associate Professor, University of Kentucky, Lexington KY 46504. Email: pedram.roghanchi@uky.edu

ABSTRACT

This study proposes a framework and presents proof of concept for a real-time smart evacuation route-planning approach based on graph theory. In the effort to assist mine workers to safely reach the surface or a refuge chamber, a smart system could provide invaluable acquisition of mine-wide situational awareness to the workers. An IoT (Internet of Things) of sensors, such as gas concentration, temperature, smoke, oxygen, and air speed sensors, combined with a real-time path planning algorithm could be a powerful tool to such situations. A mine can be represented by a topological map and every location can be assigned a real-time updated value that quantifies the fire-induced hazard based on data collected by a mine-wide IoT. This combinatory risk considers parameters such as concentrations of toxic gases, oxygen levels, heat, and visibility. Safety and health exposure limits as defined from the various regulatory entities are combined with simulated IoT data to calculate the combined risk. The optimised escape routes could significantly assist mine workers to reach a safe location.

INTRODUCTION

Fires and explosions in mines, which often trigger each other, can lead to worker fatalities and injuries, mine infrastructure destruction, and large economic losses to the industry. Explosions can greatly and rapidly affect workers and infrastructure in the vicinity of their occurrence, while fires can spread through larger areas in a mine with propagation rates more than 0.10 m/s (Conti, 2001). During the period 1900–2008, a number of 35 fire incidents caused 727 fatalities in the USA (Brnich, Kowalski-Trakofler and Brune, 2010). Conti (2001) found that MSHA accident reports for the period 1991–2000 record 137 fire incidents, independently of size or fatalities, in underground coal, metal, and non-metal mines. Although fires and explosions show a significant decline after 2015, they remain a constant concern in the mining industry and require carefully planned preventative or mitigating measures such as appropriate ventilation systems, protective equipment, health and safety training, and designating evacuation routes.

In the event of mine emergencies, optimal miner decision-making, ie identifying and choosing self-evacuation techniques and routes is of vital importance to prevent miners' exposure or entrapment by noxious and toxic gases, to extreme temperatures (Onifade, 2021). However, during emergencies a mine worker's knowledge of the dynamic fire status and fire-induced risks is limited to his immediate vicinity. Therefore, the ability to create, maintain, and utilise mine-wide Internet of Things (IoT) to assist in real-time evacuation path planning would be a beneficial technology for the near-future mines.

Although, several commercially available ventilation simulation software provides insightful evacuation optimisation and assessment tools for mine design, these cannot be used in real-time data analysis, while additional analysis steps are often necessary to acquire the desired output. In a more practical application, Epiroc USA LLC offers the Mobilaris Emergency Support, a software suite for mine evacuation control that allows an emergency leader to efficiently alert and evacuate all personnel from a mine by monitoring and directing evacuees into safe exits or refuge chambers through smart tags, cell phones, and tablets (Epiroc USA LLC, 2024). However, information about the real-time data analysis algorithms and inputs are not available to the public (understandably for trade secrecy reasons). Note: Refuge chambers are movable chambers where the miners can barricade themselves safely in the case of mine emergencies, until the emergency passes or they are extracted by mine rescuers. The chambers provide breathable air, water, food, communication devices, etc. for extended periods of time, and are mandatory in USA coal mines (Margolis *et al*, 2011).

The rest of the paper is structured as follows: Section 2 details the coalmine model and the fire simulations set-up. Section 3 discusses the evacuation path planning algorithms utilised for computing the fire emergency evacuation routes. Section 4 presents the results of the fire simulation and the evacuation path planning, and Section 5 concludes the study.

MINE FIRE SIMULATION – A CASE STUDY

In this study, mine fire emergencies inside a room and pillar mine are simulated with the VentSim™ DESIGN (version 5.4) software package. Subsequently, the fire simulation data are processed using path planning algorithms. The safety constraints, which the algorithms use to eliminate unsafe evacuation routes are defined based on the current MSHA safety and health thresholds. The objective of the data processing is to define safe evacuation paths that minimise the evacuees' exposure to fire-induced risks, such as toxic gas inhalation, high heat, and smoke-obstructed visibility. The model of the room and pillar underground coalmine, the simulated mine wide IoT, and the fire simulation scenarios are described below.

Coalmine ventilation model

The ventilation model of the underground room and pillar coalmine used is one of the built-in models of the VentSim™ DESIGN package. The utilised built-in model simulates a coalmine with 1.8 m wide entries and 3.0 m wide square pillars. Figure 1 shows the coalmine model of the VentSim™ DESIGN. The model can be divided (abstractly) into three coal panels that are accessed by a five-entry system, which expands to seven-entry and nine-entry panels as it moves in by:

- Panel A, accessed directly through the main entry system, has a width of approximately 43 m and runs straight for approximately 213 m (Panel A – Part 1) before it turns, perpendicularly but retaining the same width, and continues for approximately 274 m. A refuge chamber is located at approximately 61 m after the turn (this 61 m long part will be hereafter labelled as Panel A – Part 2). In the fire simulations (explained below), the part of the mine that extends beyond this refuge chamber is not taken into account for the sake of simplicity.

- Panel B, accessed through the bifurcation of the main entry system, has a width of approximately 43 m and runs in parallel to the first part of panel A for approximately 171 m.

- Although the model includes a third panel, which runs in parallel to the second part of panel A, it is sealed off (by design of the built-in model) and does not affect the ventilation of the rest of the mine. Thus, it is not taken into account in the fire simulations.

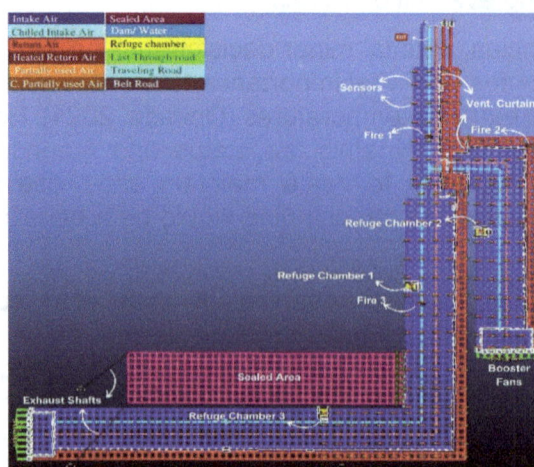

FIG 1 – VentSim™ DESIGN model for an underground coalmine.

There are three refuge chambers throughout the whole mine. The refuge chambers are accessed by exhaust shafts that provide fresh air. Refuge chamber 1 is located in Panel A – Part 1, Refuge chamber 2 is located in Panel B, and Refuge chamber 3 is located in Panel A – Part 2. Additionally, ventilation curtains (also called stoppings) isolate the return air entries throughout the mine. The return air is extracted by booster fans positioned appropriate points (coloured green).

Simulated sensors and fire events

In this study, the parameters of interest include velocity of the airflow (m/s), carbon monoxide concentration (ppm), oxygen concentration (%), wet bulb temperature (°C), and visibility (m). These parameters are a subset of the parameters available in the VentFIRE™ tool. Additionally, the fire events are defined by user-input ignition time, fuel type, and fire characteristics. The duration of the simulations is set to 1 hr which is the approximate time for the fumes to cover the entire mine when only the primary fire event is simulated.

The VentFIRE™ tool was used to create the following on the built-in model:

- **Simulated sensors**: Sensors were added at 216 locations that cover places of interest in the coalmine model. The sensors layout can be seen in both Figures 1 and 2. It should be noted that the option to add sensors on every cross-cut was avoided due to the extremely high number of sensors and nodes this approach would yield. The grids nodes hold a total of 211 sensors, while five additional sensors are added on the locations of the fire events and the refuge chambers.

- **Fire events**: three fire events have been set-up at different locations. Fire 1 is the primary fire event that starts 5 mins after the beginning of the simulation and is located close to the main-entry system within an intake air entry. The primary fire is assumed to start due to coal transportation truck rollover that spills 190 L of diesel and 10 000 kg of average quality coal. Fire 2 starts 20 mins after the beginning of the simulation and is located within the two-entry return air system of Panel B. Lastly, Fire 3, also, starts 20 mins after the beginning of the simulation whereas it is located within the intake air pathways of Panel A - Part 1 and close to Refuge chamber 2. Fires 2 and 3 are assumed to be combustions of 2000 kg average quality coal. After ignition all fires burn until the end of the simulation without reaching the decay stage. Table 1 summarises the inputs for the three fire events.

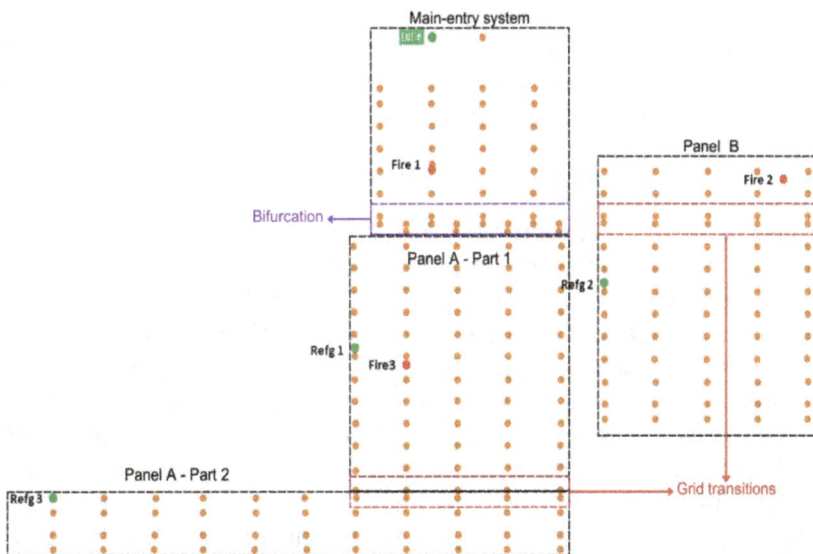

FIG 2 – Layout of sensor grid (orange dots represent sensor positions).

TABLE 1

Fire Event Wizard inputs for VentFIRE™. Growth period is the time for the fire to reach maximum intensity based on fuel and oxygen availability, while decay period is the time after the exhaustion of the fuel for the fire to burn out (Brake, 2013).

ID	Fuel (kg)	Ignition start (min)	Growth period (min)	Decay period (min)
Fire 1	Diesel: 166 and Coal Avg Q: 10 000	5	10	10
Fire 2	Coal Avg Q: 2000	20	5	5
Fire 3	Coal Avg Q: 2000	20	5	5

Fire emergency scenarios

Two simulation scenarios have been analysed in the scope of this study:

- In the first scenario, FE1, three fire events are considered, while only the CO levels are assumed as the only fire-induced risk. In this scenario, the allowable evacuation exits (single-source, multiple-sink case) include all three refuge chambers.

- The second scenario, FE2, has the same set-up as the first scenario, except for the inclusion of temperature, visibility, and oxygen availability to the CO levels as the fire-induced risks.

EVACUATION PATH PLANNING

Pathfinding algorithms

In general, path planning within graph theory is formulated as follows: Let $G = (N, E)$ be a directed graph with the set of nodes N and the set of edges E, where: G models the network of entries and cross-cuts of the coalmine; N is the set of the graph nodes, $i \in N$, the junctions between the entries and the cross-cuts, and E is the set of the graph edges, $(i, j) \in E$, the uninterrupted parts of the entries and cross-cuts of the mine. The size of the node sets is n, and the size of the edges set is m, where $m \geq n - 1$. Additionally, a cost value, $f(i, j)$, is attributed on each edge of the graph. The cost value can be the distance from point i to point j, the travelling time associated with that edge, the energy consumption required to travel that edge, etc. Additionally, some path planning algorithms assign a capacity value, $c(i, j)$, on each edge of the graph that represents the maximum allowable flow or cost for an edge (Goldberg and Tarjan, 1988; Lavrov, 2020).

In this study, the mine fire evacuation problem is formulated through a network flow where: i) the cost function represents the quantified fire-induced hazards, and ii) the capacity value represents the health and safety thresholds as defined by the MSHA regulation. The MFMC problem is solved with the aim to identify viable evacuation route for the miners from workplace points (source nodes) to the surface exit or refuge chambers (sink nodes) while minimising the risk associated with the evacuation route. Three algorithms are selected for this study: a variation of the Ford-Fulkerson algorithm (vFFA), the Network Simplex algorithm (NS), and the NS algorithm enhanced with Capacity Scaling method (NS-CS). The algorithms were tested across the simulated underground coalmine network (described previously) and two fire scenarios. All algorithms are implemented in Python programming language (Python 3.7) by using primarily the 'networkx' library.

The FFA greedily computes all the viable and safe evacuation routes in the network graph between the source (workers workplace) and sink nodes. Each edge of an available evacuation route has a total cost, $f(i, j)$, and a capacity, $c(i, j)$. As long as the cost does not exceed the capacity, $c(i, j) - f(i, j) > 0$, escape through that edge is feasible. Instead, when the residual capacity of the edge becomes zero, $c_r(i, j) = 0$, then the edge is removed from the graph, $(i, j) \notin E$; meaning that all the escape routes that include that edge become invalid and are removed from the set of safe routes (Ford and Fulkerson, 1963). A simple heuristic after the FFA is executed finds the shortest paths. An implementation of this variation of the FFA has been described in Lotero *et al* (2024) for a simpler case study and without considering movement of evacuees over time. In order of highest to lowest priority of optimisation metric, this approach optimises for: a) incremental risk, ie safety of individual edges/pathways, b) overall distance of escape route, and c) overall risk of escape route.

The Network Simplex (NS) algorithm is a MFMC algorithm which is designed to maximise the flow-through a network under certain capacity constraints. The cost function of the network simplex algorithms aims to minimise the cost of the flow within the network. However, the algorithm requires integer cost values. An advantage of the algorithm is that it can handle multiple sources and sinks (Grigoriadis, 1986; Wikipedia contributors, 2021). In this implementation, the flow in the network represents the mine worker's evacuation, the cost is the quantification of the fire-induced health risks, and the capacity constraints are defined based on the MSHA regulations. Moreover, the costs and the capacities are multiplied with a sufficiently big scalar in order to ensure integer costs. This approach optimises for overall risk of escape route.

The extension of NS with capacity scaling (NS-CS) is a successive shortest augmenting path algorithm meant to solve minimum cost flow problems by iteratively scaling the capacity values to integers, where each iteration increases the precision of the scaling by a digit. This variation allows for handling of networks with any types of cost values. Although the additional scaling steps increase the complexity of the algorithm, the precision of the computations is higher (Ahuja and Orlin, 1992; Çalışkan, 2011).

Risk quantification

In this study, the cost function is the fire-induced quantified risk for a mine worker at a given moment. At the same time, the constraints imposed on every path edge, ie the capacity of this edge, are based on the permissible exposure to unhealthy conditions that a miner can sustain without irreversible health damage or death. Feature scaling (ie min-max scaling method) is used for normalisation of the individual risk values (Aminossadati and Hooman, 2014; Nevill and Holder, 1995). Equation 1 represents the normalised value for each parameter:

$$P_n = \begin{cases} \frac{P - P_{min}}{P_{max} - P_{min}}, if\ P_{min}, P_{max} \in \mathbb{R} \\ \begin{cases} 0, if\ P \leq P_{max}\ or\ P \geq P_{min} \\ 1, otherwise \end{cases}, if\ only\ P_{min}\ or\ P_{max} \in \mathbb{R} \end{cases} \quad (1)$$

where:

P_n is the value of the parameter normalised

P is the value of the non-normalised parameter

P_{min} is the minimum allowable value of the parameter

P_{max} is the maximum allowable value of the parameter

The maximum and minimum values are based on the MSHA regulations (Kamon, Doyle and Kovac, 1983; Roy et al, 2022). Table 2 summarises the thresholds used in this study. Subsequently, the linearly combined risk is calculated as:

$$f(i,j) = \sum_{k=0}^{n_p} \lambda_k P_{n_k}(i,j) \quad (2)$$

where:

f is the cost value

k is the number of the individual parameters of interest

λ_k are the weights for the individual parameters

P_{n_k} are the normalised values for each parameter

TABLE 2

Health and safety MSHA thresholds.

Risk parameter	Min	Max
CO (ppm sec)	0	75×900
Temperature (°C)	N/A	27
Oxygen (%)	20	N/A
Visibility (m)	5	N/A

The weight of the individual risks represents the contribution or severeness of the individual risk in comparison to the rest risks. In this study, all the weights are assumed equal, and set to 1. However, it should be noted that the linear combination of the individual risks is a simplifying assumption and may not have a physical meaning to the human health.

RESULTS

The three algorithms are executed for the two scenarios for every node with the output of the 211 sensors from the simulations as input. The results are visualised in graphs where the colour of the node denotes either failure to reach a safe space (red nodes) or the final safe space of the evacuation (blue for surface exit, shades of green for different refuge chambers). Additionally, each successfully evacuated node is labelled with the total distance of the evacuation path (in metres). Table 3 summarises the statistics of the successful evacuations for both fire emergency scenarios. Figures 3a and 3b illustrate the successes/failures for all the nodes of the coal mine with the total length of the evacuation routes for FE1, while Figures 3c and d illustrate the successes/failures for all the nodes of the coal mine with the total length of the evacuation routes for FE2.

TABLE 3

Cumulative statistics of algorithms performance. Notes: '*Avg risk*' averages the total route risk, ie the sum of edge risks as defined by Equations 1 and 2 (0 is low and 1 is high), '*Avg distance*' averages the length of all successful evacuation routes to the closest and/or safest to reach exit or refuge chamber, and *Successes* is the number of starting locations (out of the 211 sensor) which can successfully be evacuated.

Scenario	Fire scenario 1			Fire scenario 2		
Algorithm	Network simplex	Capacity scaling	FFA	Network simplex	Capacity scaling	FFA
Avg risk	0.0570	0.0600	0.0590	0.0051	0.0052	0.0051
Avg distance (m)	68.4	68.4	67.8	62.3	63.1	62.1
Avg execution time for algorithm (s)	0.071	0.790	0.006	0.005	0.385	0.003
Successes	211	211	211	176	178	176

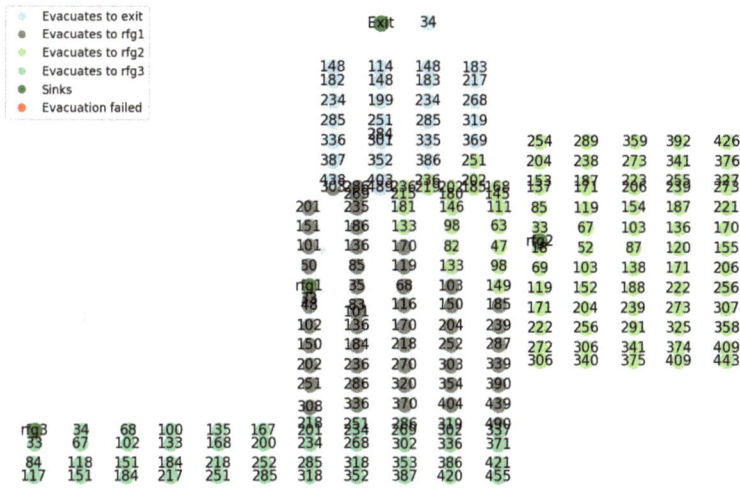

(a)

(b)

(c)

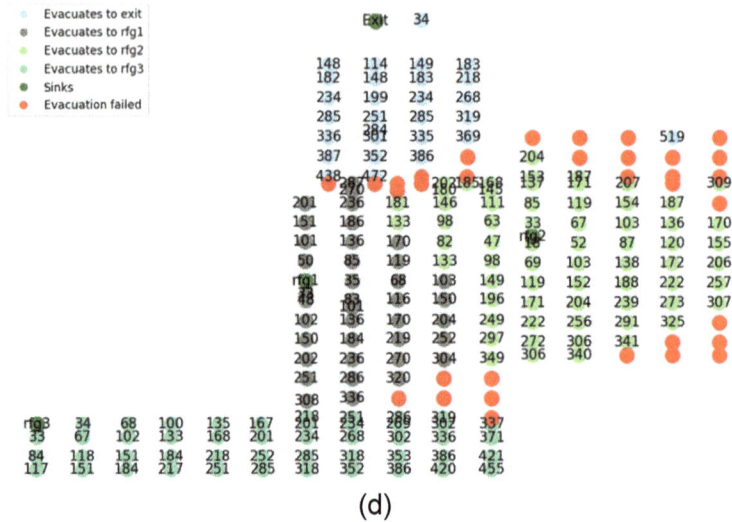

(d)

FIG 3 – Success/failure map of evacuation pathfinding using algorithms: (a) Network Simplex – Scenario 1; (b) Network Simplex with Capacity Scaling – Scenario 1; (c) Network Simplex – Scenario 2; (d) Network Simplex with Capacity Scaling – Scenario 2; (illustrations of FFA variation is omitted due to space limitations but is included in performance statistics tables).

The failed evacuations in FE2 can be attributed to the addition of more risk parameters into the cost function, which as expected yields a graph network of 'harsher' conditions and evacuations. The failures can be attributed to the following reasons: a) The path leads to an unsafe node, which within one time-step will be removed from the residual graph network because all its successive edges exceed the maximum allowable capacity of the cost function. This happens because the algorithms do not have information for future time-steps. A feasible solution would be to penalise potential paths that are directed to spaces of the mine that exhibit higher differential risk increase, and b) The reason some algorithms succeed on the same nodes that other algorithms fail on is due to taking a slightly different pathing at the beginning of the fire simulation. This is attributed to the different prioritisation between incremental risk, total distance, or total risk which the different algorithms try to uphold. The slightly different starting direction can lead to longer paths that will not be able to be followed before the mine reaches a critical point or will lead to the situation of the previous reason of failure. It must be noted, though, the simulations in this study assume that the workers do not have personal protective equipment which otherwise would allow more time and sustain more 'exposure' towards a safe exit or refuge chamber.

CONCLUSIONS

The assumptions considered in this investigation include: a) the vertical dimension in the path planning is not considered, b) the actual effect of the fire-induced risks on the health of a human is assumed to be a linear function of the individual risks and MSHA thresholds, while a conservative capacity is implemented to counteract for the uncertainty, c) the selected grid of sensors is assumed to capture the state of the mine with sufficient accuracy and timeliness, d) the speed of workers is assumed to be constant and equal to 1 m/s, the initiation time of evacuation is assumed as a known factor (2.5 min for both scenarios).

Although, the current work represents a smaller part of an ongoing research plan, the results of the simulations can be insightful to the mining industry in its effort to improve safety and health. The main observation derived is that the execution time of the three algorithms is less than 1 sec. This means that the algorithms could be used in real-time self-evacuation assistance assuming the availability of a real-time mine wide IoT for data collection. On the other side, the differences between the evacuation routes of the different algorithms can be attributed to the nature and the fine-tuning of the algorithms. Different algorithms prioritise differently between optimise for length or risk. Similar reasons apply for the difference between success or fail of the evacuations from slightly different places in the mine.

ACKNOWLEDGEMENTS

This study was funded by the National Institute for Occupational Safety and Health (NIOSH) under the award #U60OH012351. The views and opinions expressed herein are solely those of the authors and do not necessarily reflect the views of NIOSH.

REFERENCES

Ahuja, R K and Orlin, J B, 1992. The scaling network simplex algorithm, *Operations Research*, 40(1):S5–S13.

Aminossadati, S M and Hooman, K, 2014. Numerical Simulation of Ventilation Air Flow in Underground Mine Workings, 12th US/North American Mine Ventilation Symposium.

Brake, D, 2013. Fire modelling in underground mines using Ventsim Visual VentFIRE Software, in *Proceedings the Australian Mine Ventilation Conference*, pp 265–276 (The Australasian Institute of Mining and Metallurgy: Melbourne).

Brnich, M, Kowalski-Trakofler, K M and Brune, J, 2010. Underground coal mine disasters 1900–2010: Events, responses and a look to the future, *Extracting the science: a century of mining research*, pp 363–372.

Çalışkan, C, 2011. A specialized network simplex algorithm for the constrained maximum flow problem, *European Journal of Operational Research*, 210(2):137–147.

Conti, R S, 2001. Responders to underground mine fires, in *Proceedings of the 32nd Annual Conference of the Institute on Mining Health, Safety and Research*, pp 111–121 (University of Utah).

Epiroc USA LLC, 2024. Mobilaris Emergency Support, Epiroc USA LLC. Available from: <https://www.epiroc.com/en-us/products/digital-solutions/safety-solutions/mobilaris-emergency-support> [Accessed: 22 January 2024].

Ford, L and Fulkerson, D, 1963. *Flows in Networks* (Princeton University Press). https://doi.org/10.1515/9781400875184

Goldberg, A V and Tarjan, R E, 1988. A New Approach to the Maximum-Flow Problem, *Journal of the Association for Computing Machinery*, 35(4):921–940.

Grigoriadis, M D, 1986. An efficient implementation of the network simplex method, *Netflow at Pisa*, pp 83–111.

Kamon, E, Doyle, D and Kovac, J, 1983. The oxygen cost of an escape from an underground coal mine, *American Industrial Hygiene Association Journal*, 44(7):552–555.

Lavrov, M, 2020. The Ford–Fulkerson Algorithm, Powerpoint presentation: Math 482, Lecture 26. Available from: <https://misha.fish/archive/docs/482-spring-2020/slides26.pdf>

Lotero, S, Androulakis, V, Khaniani, H, Hassanalian, M, Shao, S and Roghanchi, P, 2024. Optimizing fire emergency evacuation routes in underground coal mines: A lightweight network flow approach, *Tunnelling and Underground Space Technology*, 146:105637. https://doi.org/10.1016/j.tust.2024.105637

Margolis, K A, Westerman, C Y K and Kowalski-Trakofler, K M, 2011. Underground mine refuge chamber expectations training: program development and evaluation, *Safety Science*, 49(3):522–530.

Nevill, A M and Holder, R L, 1995. Scaling, normalizing, and per ratio standards: an allometric modeling approach, *Journal of Applied Physiology*, 79(3):1027–1023. https://doi.org/10.1152/jappl.1995.79.3.1027

Onifade, M, 2021. Towards an emergency preparedness for self-rescue from underground coal mines, *Process Safety and Environmental Protection*, 149:946–957. https://doi.org/10.1016/j.psep.2021.03.049

Roy, S, Mishra, D P, Bhattacharjee, R M and Agrawal, H, 2022. Heat Stress in Underground Mines and its Control Measures: A Systematic Literature Review and Retrospective Analysis, *Mining, Metallurgy and Exploration*, 39:357–383. https://doi.org/10.1007/s42461-021-00532-6

Wikipedia contributors, 2021. Network simplex algorithm, last edited 3 December 2021, Wikipedia. Available from: <https://en.wikipedia.org/wiki/Network_simplex_algorithm> [Accessed: 4 February 2024].

Empirical analysis of fire-induced pressures for mine-wide emergency response planning

G Kolegov[1], K Tom[2], S Bergh[3] and L Botha[4]

1. Mining Solutions Ventilation Engineer, Howden, Renfrew, UK.
   Email: grigorii.kolegov@howden.com
2. Mechanical Engineer, Howden, St-Bruno, Canada. Email: kevin.tom@howden.com
3. Mining Team Leader, Howden, Booysens, South Africa. Email: stephan.bergh@howden.co.za
4. General Manager – Ventsim Engineering, Howden, St-Bruno, Canada.
   Email: leo.botha@howden.com

ABSTRACT

Emergency response planning is an integral part of any modern mine's engineering workload. Potential underground fires and personnel evacuation as well as rescue teams' response algorithms must be considered. The Ventsim®, version 6.0 (by Howden, A Chart Industries Company) offers powerful tools for detailed fire modelling, however when the risk of fire exists everywhere inside the mine, such as in coalmines, planning emergency response for an extensive network could be time-consuming and thus posing a risk of neglecting a potential high-risk issue like a fire-induced recirculation. It is often necessary to assess this recirculation potential since the rescue team must approach the fire source from the fresh air side. In this paper, this issue is illustrated by a model example, where a fire in an incline causes air reversal in the parallel entry, cycling large quantities of toxic gases and decreasing visibility in the loop to nearly zero, making it impossible to approach the source.

To find a solution for quickly highlighting potential recirculation loops caused by fires, a methodology for an underground coalmine fire empirical approximation is proposed. This method uses the airway's area, length, gradient, and initial airflow passing through it to calculate the pressure change caused by the fire at a given moment in time. These values are then applied as fixed pressures, approximating the fires, to an existing mine's model and the results are compared to the VentFire script applications.

INTRODUCTION

Coal remains to this day one of the major sources of electricity generation in the world (IEA report, 2021). Therefore, increasing the safety of the coalminers is still a goal that ventilation engineers should strive to achieve. Routinely designing potential fire scenarios is one of the ways mining operators have been using to approach that task. That involves identifying potential high-risk places inside of a mine where a fire could take place, assessing fuel load, numerically modelling the fire output parameters such as the heat release rate, and drafting adequate response measures.

Extensive work regarding full-scale mine fire experiments has been published (Hansen, 2018, 2019, 2022). However, the author concentrated heavily on metal mine fires, citing the Australian statistics that indicated that fires happened in them more frequently than in coalmines. This, however, might not apply to mines in other parts of the world. Various models and tools for calculating mine fire parameters had been developed as early as the 1980s (Chang and Greuer, 1987). With the onset of mine ventilation networks' modelling and the inclusion of various processes taking place inside the mines, an emphasis was placed on using the ventilation software to predict the hazardous impact of fires on underground mines in terms of transient heat and contaminant distribution (Brake, 2013). The VentFire modelling is also being used to optimise potential escape routes (Sarvestani, 2023).

An underground fire spreading through an entry greatly increases the temperature and volume of the air moving downstream from it, decreasing its density. In a situation when the pressure gradient caused by the fire reaches its critical value and outnumbers the circuit's gauge pressure drop provided by the main fans, the potential for recirculation within that circuit exists. Figure 1 illustrates an example circuit with three inclines, each with a gradient of 15°. The air moves downwards through the rightmost incline, and the ventilation in the other two inclines is upcasting. With the fire source placed in the middle incline, at a certain point during its progression, the fire causes increased pressure, and the air in the leftmost entry starts moving downwards, forming a recirculation loop.

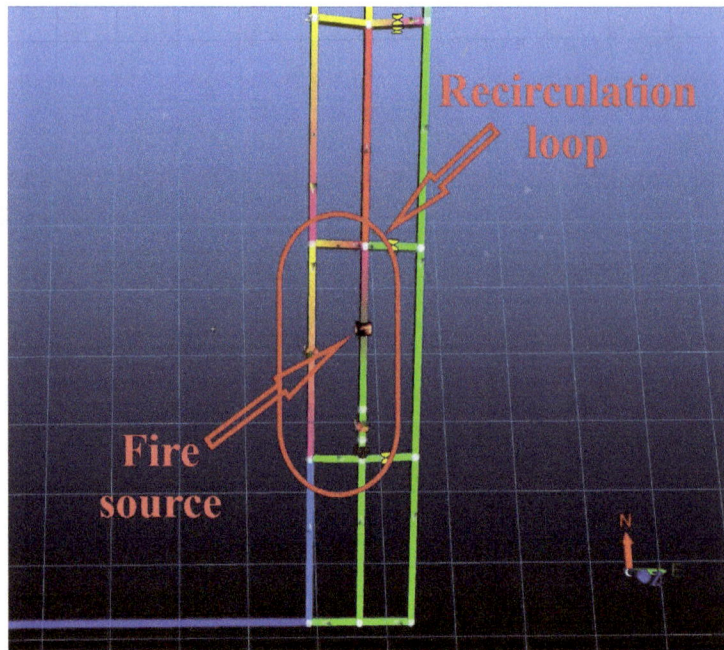

FIG 1 – Recirculation circuits caused by underground fires.

Due to the nature of coal extraction operations, their environment usually has a continuous fuel load, which excludes the option of just letting the fire burn out by itself. Rescue teams' safety and efficiency heavily depend on their ability to approach the fire source from a fresh air entry, provided that the decision to extinguish the fire is made, otherwise they risk being exposed to the temperatures that exceed the upper threshold values of their protection equipment as well as operating in near zero visibility. It is therefore important to be reasonably sure that the approach route the rescue team is going to take does not go through potential recirculation zones caused by the fire.

The dynamic monitors, recording trends of airflow, gas and visibility during the fire simulation, were placed in the airway parallel to the one that caught fire in the example. Figures 2 to 4 illustrate those trends.

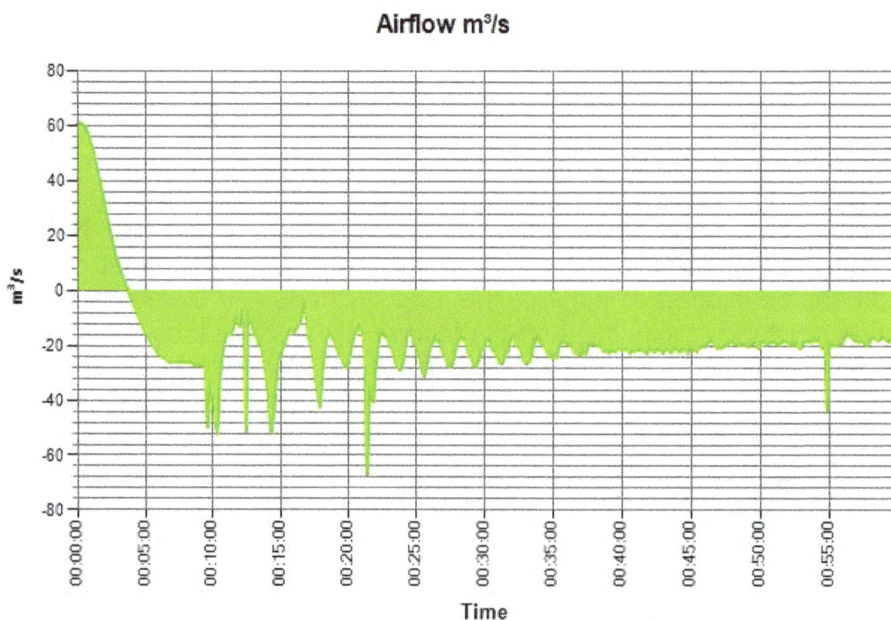

FIG 2 – Airflow in the parallel entry.

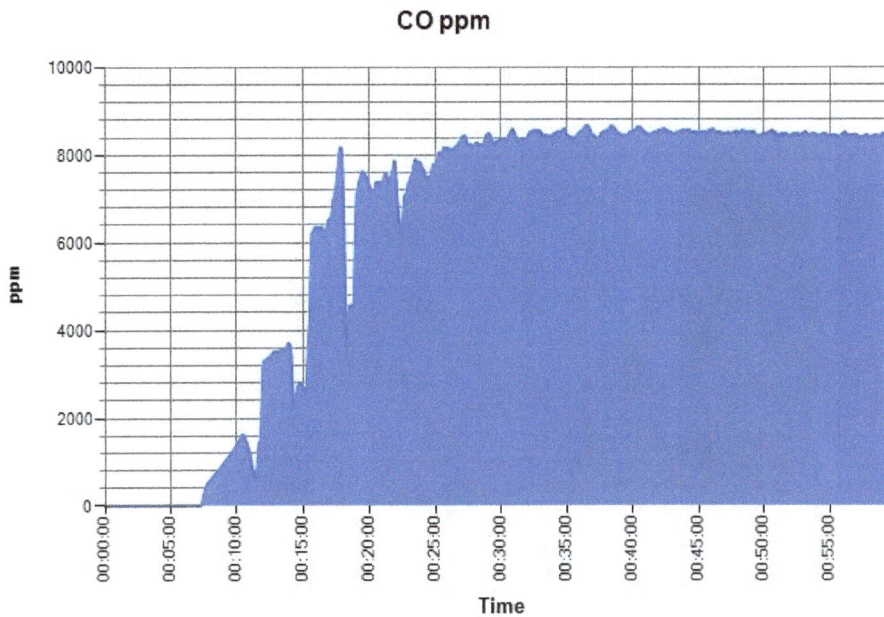

FIG 3 – Carbon monoxide concentration in the parallel entry.

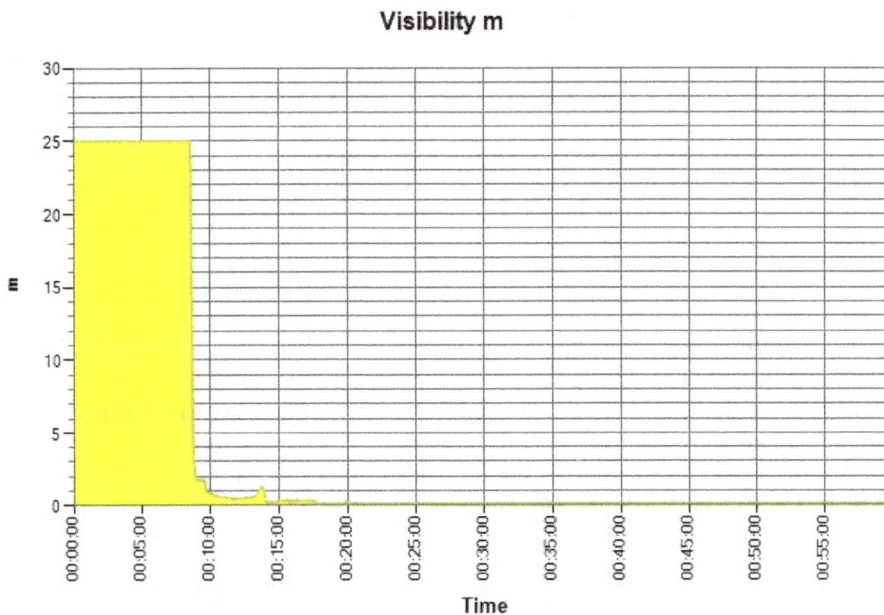

FIG 4 – Visibility in the parallel entry.

It is evident that after 5 mins from the start of the fire, the airflow is reversed, and after 10 mins high concentrations of CO are recirculating through the circuit, and visibility is near zero. Visibility readings from the dynamic monitor placed below the fire, the direction from which the rescue teams are supposed to approach and extinguish it, are shown in Figure 5.

It is evident from the graphs that after 10 mins have passed, and the airflow has reversed, reaching its peak values, it is impossible for the rescue team to safely approach the fire source and extinguish it. In coalmines the recirculation causes the possibility of methane build-up within the circuit, posing an explosion threat.

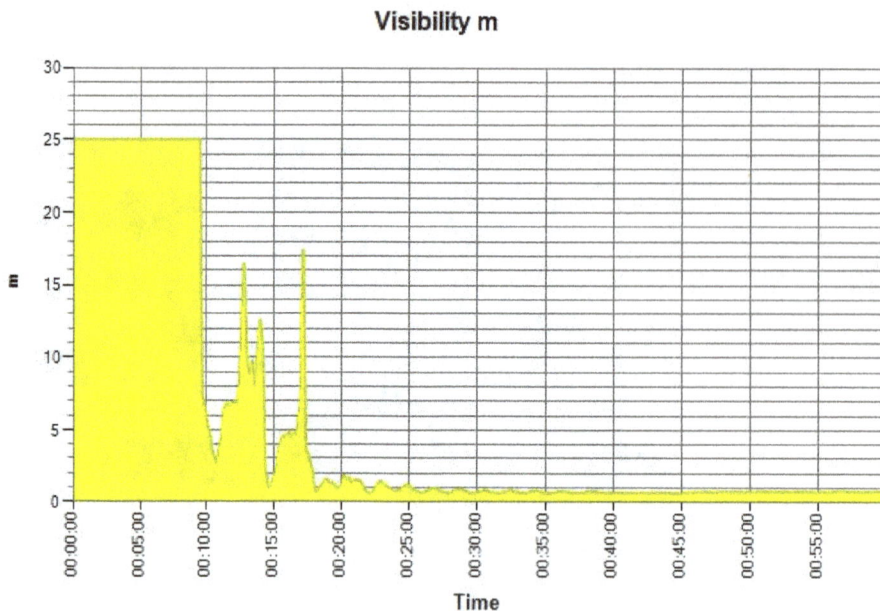

FIG 5 – Visibility below the fire.

Identifying all the potential recirculation caused by underground fires using Ventsim involves implementing the VentFire feature to various circuits where the user suspects it might occur. The length of this identification process is heavily dependent on the user's experience, their familiarity with the mine and the ventilation circuit's complexity. VentFire itself is a complex modelling tool that simultaneously runs dynamic simulations of heat, airflow as well as gas and smoke spread (Brake, 2013). Properly modelling fires in multiple entries could be time-consuming, especially if complex geometry requires smoke rollback pathways to be constructed before simulation (Stewart, Aminossadati and Kizil, 2015).

It is reasonable to assume that when working in conditions of a limited time frame and an extensive ventilation network, a user might neglect their modelling of a potential fire recirculation loop. What can be useful in that case is a method that would apply fixed pressures, approximating peak fire pressure gradient, to quickly test for the potential of recirculation. In the case of the example mine reviewed previously, a fixed pressure of 150 Pa applied to the airway instead of the fire causes the same effect of creating a recirculation loop as demonstrated in Figure 6, but without the time and complexity required to perform a full fire simulation.

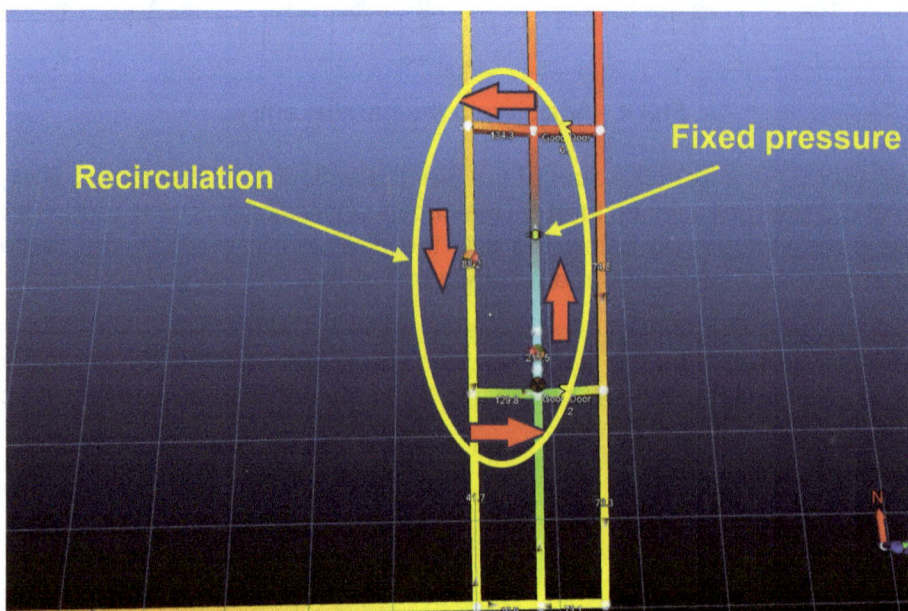

FIG 6 – Substitution of a VentFire preset with a fixed pressure causing recirculation.

International Mine Ventilation Congress 2024 | Sydney, Australia | 12–16 August 2024

The goal of this study then, is to propose a function that could approximate the pressure differentials caused by underground fires, compare it to the VentFire module output with varying initial conditions, and estimate whether its application can be practical to facilitate mines' networks analysis during emergency response planning.

METHODOLOGY

To arrive at a function successfully approximating a worst-case scenario of a fire in an underground coalmine, existing approaches to calculating a fire pressure gradient have been compared to peak pressures modelled in Ventsim with varying 'high Q' coal fuel load. The amount of coal that is going to burn in the actual fire event is assumed to be unknown, and values ranging from 1000 kg to 55 000 kg are estimated.

The continuity of the fire is modelled in the VentFire preset as an event that has growth and sustain periods, but no decay.

For each simulation time step, fire-induced overpressure is calculated as the difference between the pressure loss in the airway at that specific time step and the initial pressure loss:

$$h_{vs} = P_{ti} - P_{init} \text{, Pa} \tag{1}$$

Those resulting values were compared to the ones calculated using an empirical methodology (Bolbat, Levedev and Trofimov, 1992):

Fire zone length:

$$l_f = t\left(0.28 + 0.07\frac{Q}{S}\right) \text{, m} \tag{2}$$

Where

Q   airflow, m³/s

t   time passed from the beginning of the fire, min

Empirical parameter a:

$$a = \frac{\sqrt{S}}{l_f} \tag{3}$$

Relative distance:

$$\bar{x} = \frac{l}{l_f} \tag{4}$$

Fire zone's vertical projection:

$$z = l_f \sin\beta \text{, m} \tag{5}$$

Where β is the airway incline angle, in degrees.

Empirical parameter A:

$$A = \frac{100a}{1.21 + 1.51\frac{S}{Q}} \tag{6}$$

Maximal temperature downstream from the fire:

$$T_M = 1273 - 975e^{-\frac{S}{A}} \text{, K} \tag{7}$$

Temperature at the end of the entry downstream from the fire:

$$T_K = 298 + \left(T_M - 298\right)e^{-\frac{\overline{x}-1}{A}} \text{, K}$$

(8)

Finally, the heat pressure differential is:

$$h_t = 12z\left(0.766 + \ln\left(\frac{T_M}{T_K}\right)\right) \text{, Pa}$$

(9)

To estimate the linear relationship between the overpressure change over time modelled in Ventsim and heat pressure differential over the same period, calculated from Equation 9, the sample Pearson correlation coefficient was calculated for each fuel load mass value:

$$r = \frac{\sum_i h_{VSi}h_{ti} - n\overline{h_{VS}}\,\overline{h_t}}{\sqrt{\sum_i h_{VSi}^2 - n\overline{h_{VS}^2}} \cdot \sqrt{\sum_i h_{ti}^2 - n\overline{h_t^2}}}$$

(10)

Where:

n sample size

h_{vsi}, h_{ti} individual sample points

$\overline{h_{VS}} = \frac{1}{n}\sum_{i=1}^{n} h_{VSi}$ the sample mean

The resulting pressure differentials were applied as fixed pressures to the airways where VentFire scripts had been implemented previously, and occurrences of recirculation loops due to VentFire and approximating empirical function were compared.

RESULTS AND DISCUSSION

The methodology was tested on two examples with different ventilation parameters shown in Figures 7 and 8. The first example represents a real coalmine model, while the second one is an abstract network with considerable differences between surface connections' heights. The purpose of the second example is to investigate the effects of increased natural ventilation pressures on fire stability and its correlation with the proposed approximating function.

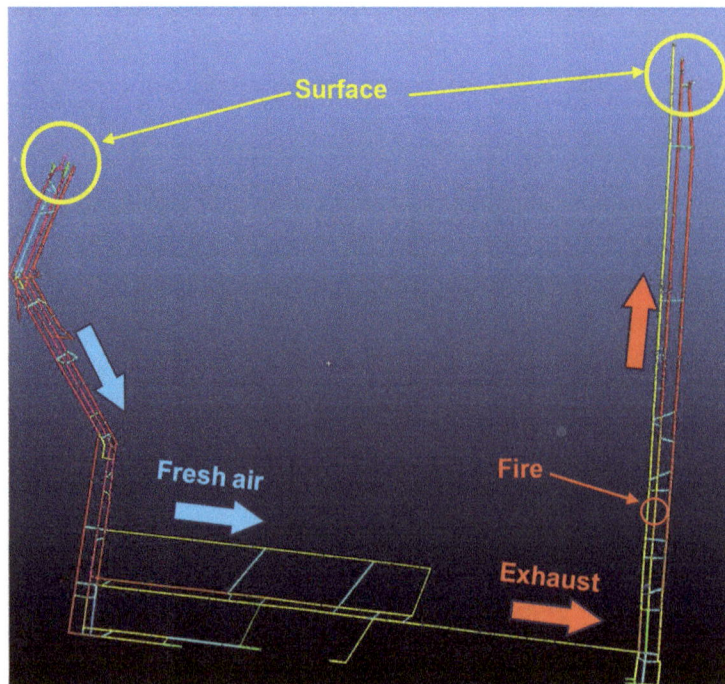

FIG 7 – Mine network example 1.

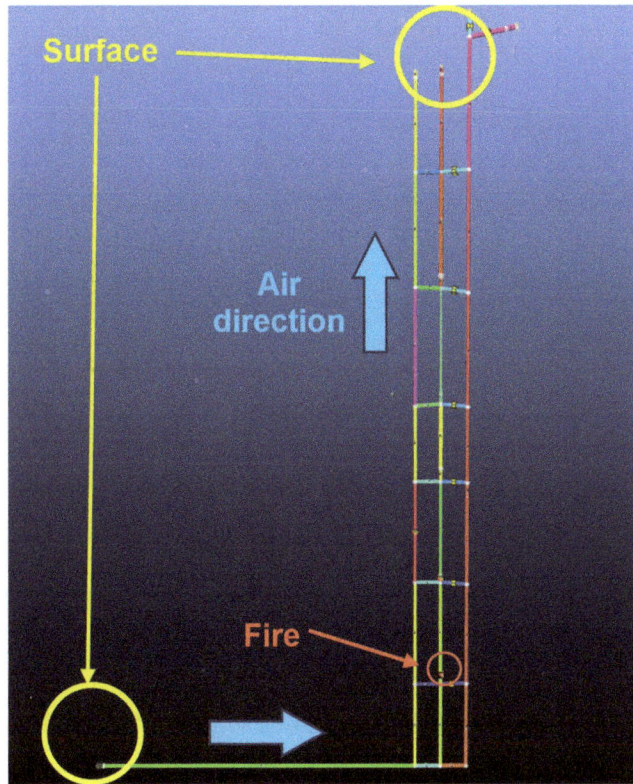

FIG 8 – Mine network example 2.

Figures 9 and 10 illustrate the comparison between fire-induced overpressures modelled in Ventsim and those calculated based on the proposed empirical function for examples 1 and 2 accordingly. In the first example, where the initial natural ventilation pressure effect was less pronounced than in the second one, correlation coefficients for various fuel mass values do not drop below 0.7 which signifies that a considerable linear relationship between the values exists. However, when the network is not stable over time and overpressures tend to fluctuate, as can be seen in Figure 10 for the second example, the correlation coefficient drops below 0.5 with increased fuel load, which means that linear modification of the proposed methodology cannot be applied to better approximate the results obtained by Ventsim modelling.

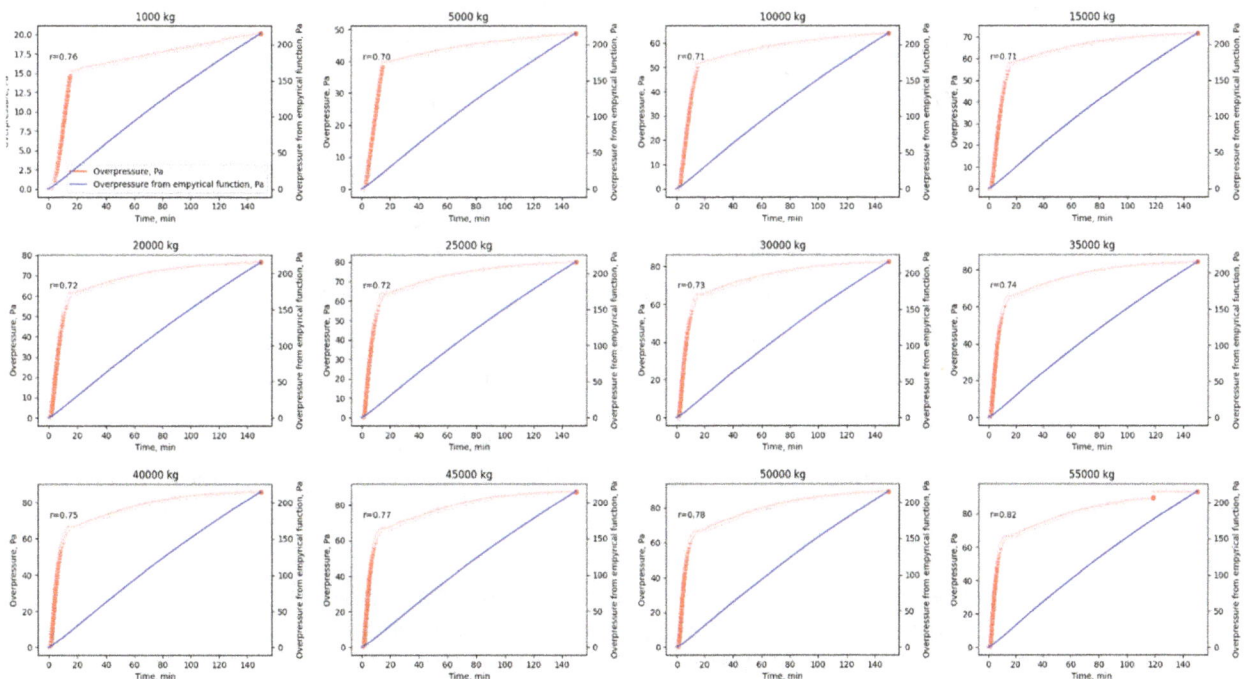

FIG 9 – Pressure change correlation, example 1.

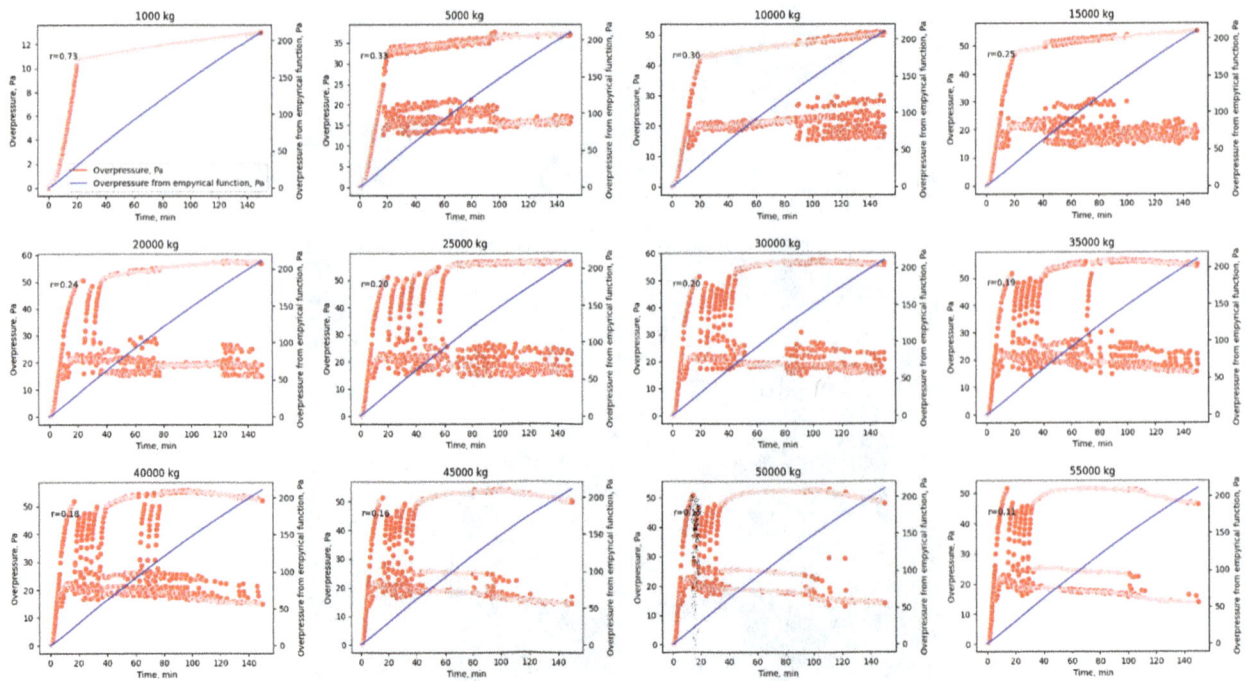

FIG 10 – Pressure change correlation, example 2.

For several incline airways in example 2, the modelling results (H$_{t\ model}$) were compared to the heat pressure differentials H$_{t\ calc}$ calculated from Equations 2–9. After the calculation was complete, pressure differentials from Equation 9 were applied as fixed pressures to the corresponding airways, and the occurrences of the recirculation loops (R$_{l\ calc}$) were compared to the results from VentFire script applications (R$_{l\ model}$). Table 1 contains the comparison between the results of the VentFire modelling and empirical calculations.

TABLE 1

Fire pressure differentials and recirculation comparisons.

Q, m³/s	S, m²	β, deg	l, m	Ht model	Ht calc	D Ht,%	RI model	RI calc
43.9	22	9.6	271.8	220.3	59.8	73	N	N
96.0	21	8.9	84.2	49.9	56.3	13	N	N
105.6	20	9.0	55.1	31.9	57.9	81	N	N
29.6	21	11.5	162.9	175.7	58.3	67	Y	Y
28.3	18	12.6	46.0	47.2	50.6	7	Y	Y
36.4	20	12.2	298.9	278.7	77.7	72	Y	Y
75.1	18	9.2	81.5	38.8	56.4	45	N	N
86.1	22	10.8	207.2	172.5	77.1	55	N	N

The average deviation of the modelled results compared to the empirical fire pressure differentials was 52 per cent, effectively rendering the calculated values unsuitable for any further value-based determinations. However, recirculation loops were found for every case they occurred during the VentFire script applications.

Considerable value deviations can be explained by the fact that natural ventilation pressures during a fire event modelled in Ventsim represent values resulting from calculations applied to every airway in the network as opposed to just the one in which the fire is placed. In the future, the proposed method can potentially be improved by conducting an extensive statistical analysis and obtaining correction factors for the proposed empirical function that would improve the approximation of the values produced by the VentFire script implementations. The function calculating overpressures induced by the fires can be applied to every airway of interest using a script that takes as input a spreadsheet output from Ventsim, containing initial ventilation conditions as well as airways' dimensions, and produces a heatmap-type graph, highlighting the airways with the highest ratio between overpressures produced by the fire and the initial pressure drops in the loops, and thus the highest recirculation risk.

Figure 11 is an example of such a script's output. It might be possible to use the method to quickly highlight potential recirculation loops caused by fires for the user to inspect and apply detailed VentFire modelling to those circuits. It cannot by any means replace the detailed network analysis but could potentially help users during emergency planning find high-risk areas within their mines they possibly have missed.

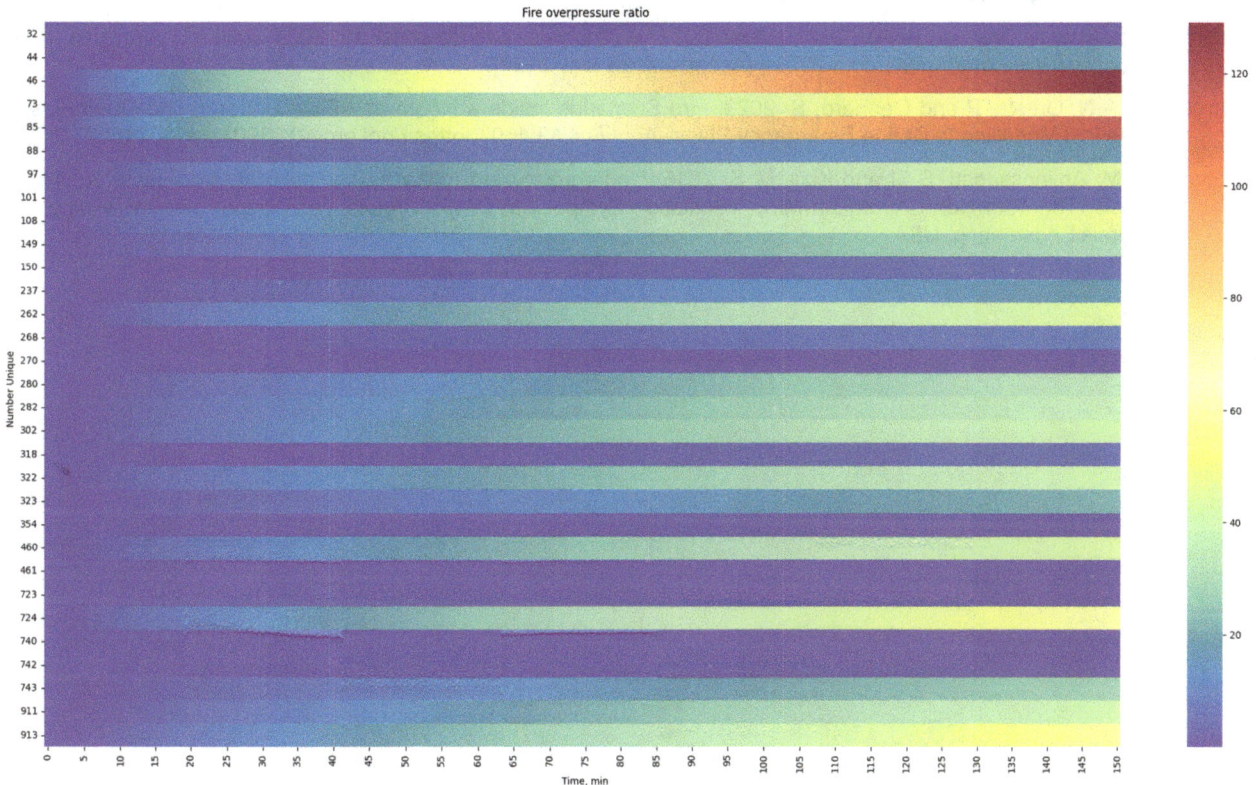

FIG 11 – Fire pressure ratio over time.

CONCLUSION

Approximating fires with fixed pressure differentials can potentially predict recirculation pathways due to fire. The iterative application of the process may guide ventilation professionals into performing more detailed analysis in areas that show a high likelihood of recirculation. To eventually develop such a feature, a function, approximating fire-induced pressures was proposed. Correlation coefficients between the overpressure values produced by the VentFire script application and the proposed empirical function were calculated for various fuel load mass values.

In some cases, the linear relationship between the two data sets existed, in others, when instability caused by the significant impact of the initial natural ventilation pressure values produced overpressure fluctuations, VentFire application results could not be approximated with the proposed empirical function. Overpressure values obtained from the empirical function were compared with the overpressures modelled in Ventsim, the average deviation was 52 per cent, while all resulting recirculation loops matched.

A way to reduce the deviations between the approximating function and VentFire script output by introducing linear coefficients based on further statistical analysis was proposed. A script, iterating overall selected airways and highlighting all branches with high overpressure ratios was created.

REFERENCES

Bolbat, I E, Levedev, V I and Trofimov, V A, 1992. *Emergency ventilation regimes in coal mines*, pp 11-26 (Nedra: Moscow).

Brake, D J, 2013. Fire Modelling in Underground Mines using Ventsim Visual VentFIRE Software, in *Proceedings of the Australian Mine Ventilation Conference*, pp 265–276 (The Australasian Institute of Mining and Metallurgy: Melbourne).

Chang, X and Greuer, R, 1987. A mathematical model for mine fires, in *Proceedings of the 3rd Mine Ventilation Symposium*, pp 453–462 (Society for Mining Metallurgy).

Hansen, R, 2018. Analysis of the average fire gas temperature in a mine drift with multiple fires, *Journal of Sustainable Mining*, 19:226–238. https://doi.org/10.1016/j.jsm.2018.08.001

Hansen, R, 2019. Design of fire scenarios for Australian underground hard rock mines – Applying data from full-scale fire experiments, *Journal of Sustainable Mining*, 18:163–173. https://doi.org/10.1016/j.jsm.2019.07.003

Hansen, R, 2022. Proposed design fire scenarios for underground hard rock mines, *Journal of Sustainable Mining*, 21(4):1. https://doi.org/10.46873/2300-3960.1367

International Energy Agency (IEA), 2021. Coal 2021 – Analysis and forecast to 2024 [online]. Available from: <https://www.iea.org/reports/coal-2021> [Accessed: 25 May 2024].

Sarvestani, A N, Oreste, P and Gennaro, S, 2023. Fire Scenarios Inside a Room-and-Pillar Underground Quarry Using Numerical Modelling to Define Emergency Plans, *Appl Sci*, 13:4607. https://doi.org/10.3390/app13074607

Stewart, C M, Aminossadati, S M and Kizil, M S, 2015. Underground fire rollback simulation in large scale ventilation models, The 15th North American Mine Ventilation Symposium 2015, Virginia Tech Department of Mining and Minerals Engineering, USA.

Control of fire in an opencast coalmine working over developed pillars

D C Panigrahi[1], N Sahay[2] and A Saini[3]

1. Managing Director, PMRC Private Limited, Dhanbad-826004, India.
   Email: pmrcindia5@gmail.com
2. Chief Coordinator, PMRC Private Limited, Dhanbad-826004, India.
   Email: nsahay2k@yahoo.com
3. Principal Research Engineer, PMRC Private Limited, Dhanbad-826004, India.
   Email: abhinavred12@gmail.com

ABSTRACT

In the past, a number of coal seams at shallow depth—around 100 m from the surface—have been developed by bord and pillar method of mining in different Indian coalfields. Most of these developed pillars of different seams are now finally being extracted by opencast method. In more than 90 per cent of such cases, as soon as the pillars are extracted by opencast method, fire breaks out endangering life and property of the mine. The paper discusses one such problem of fire in an opencast coalmine, *viz* Quary AB, West Bokaro Collieries of Tata Steel Limited.

The authors of this paper have studied the above problem with a lot of painstaking efforts and developed a comprehensive methodology for controlling of the fire. It is mainly based on probing method by drilling boreholes from the surface to different contiguously developed seams at strategic locations, periodic monitoring of gas compositions through boreholes and application of water with sodium silicate as retardant. During the application of this method, the gas samples were collected through boreholes in the developed galleries for ascertaining the status of fire. After the fire was controlled, the production from the mine was resumed. Subsequently, liquid nitrogen converted into gaseous form at the site, was injected into the network of galleries to maintain inert atmosphere and it was continued with progressive movement of highwall side of the face. When there was an increase in the CO concentration in samples collected through monitoring boreholes, the injection of sodium silicate mixed with water was resumed till the fire was controlled and the injection of nitrogen gas was brought back after the control of fire. The paper presents details of the investigation undertaken, methodology adopted for controlling of fire and safe extraction of coal from this opencast mine.

INTRODUCTION

Prior to 1972, all the coalmines of India were operated by private owners and subsequently the mining industry was nationalised except a few mines which were operated by Tata Steel Limited. Some of these mines of Tata Steel Limited were operated by underground method and a few mines were operated by opencast method. This problem is pertaining to an opencast coalmine fire of Quarry-AB, West Bokaro Colliery, Tata Steel Limited. This particular mine is located in the central portion of West Bokaro Coalfield, Ramgarh district in the state of Jharkhand. The block is bounded by Latitudes 23° 48'16" to 23° 48'57" and Longitudes 85° 33'07" to 85° 34'34". There are seven seams in the area of the present study, *viz* XI, X, IX, VIII, VII, VI and V seams. The nomenclature of the seams is in descending order towards lower horizon. The top most seam is XI seam and it is not workable. The other seams, *viz* X, IX, VIII and VII, have been developed by bord and pillar method of working through underground mining, whereas VI and V seams are completely virgin. The status of development and de-pillaring of these seams along with the maximum depth of working and other details are given in Table 1.

TABLE 1

Brief geology and development status of seams present in Quarry-AB opencast mine.

Seams	X	IX	VIII	VII	VI
Depth of seam	48.3	52.91	58.11	74.72	102
Thickness of seam	3	2.45	3	7.8	
Gradient of seam	3 to 4 degrees				
Developed by UG/virgin	Developed	Developed and partly virgin	Developed	Developed	Virgin
Ug depillaring status	Caved goaf, standing on pillars	Caved goaf, standing on pillars	Caved goaf, standing on pillars	No	Virgin
Status of entry to UG	Main entry has been excavated out. Drift from X to VII Seam currently connected to existing fire area. Another drift from X to VII seam exists.				

It may be observed from Table 1 that the VI Seam is virgin and X to VII seams were developed by bord and pillar method of mining through underground in 1980s. The depth of all the developed seams lies between 48.30 m to 74.72 m. While de-pillaring, the upper seams, *viz* X, IX and VIII, the fires were observed in underground because of cracks developed on the surface and short circuiting of surface air due to main fan pressure. As a result, mouth of all the openings to the surface were effectively sealed-off due to fire.

The above-mentioned developed seams are now being extracted by opencast method and fires have been observed while working these seams. The study has been carried out to control the fire and extract the coal seams. In the past, a number of such cases have been reported in different coalfields in India, *viz* Chirimiri opencast mine, South Eastern Coalfields Limited (SECL); Block II opencast mine, Bharat Coking Coal Limited (BCCL); Dakra-Bukbuka opencast mine, Central Coalfields Limited (CCL); Ghughus opencast mine, Western Coalfields Limited (WCL); Gautam Khani opencast mine, Singareni Collieries Company Limited (SCCL) etc. A few such cases have also been reported from other countries of the world. It has been observed that while working in these opencast coalmines over developed pillars, different methods, *viz* isolation plug method, flooding of opencast mines from bottom of the quarry to the last bench under extraction, water spraying while lifting the hot material etc were practised (Morris, 1983; Clough and Morris, 1985; Mozumdar, 1995; Panigrahi *et al*, 1996; Vijh and Verma, 1996; Panigrahi, 1997; Sinha, 2014). However, it has been observed that a single method was not applicable in all cases. Therefore, it may be concluded that there is no universal prescription for solving this problem and in each case, the solutions are to be tailor made to suit to the condition of working.

PROBLEMS FACED IN QUARRY-AB OF WEST BOKARO COLLIERIES, TATA STEEL LIMITED

Initially Quarry-AB was started in December 2021 by removing the overburden and X, IX, VIII and VII Seams were exposed in different benches. The fire started with the generation of a low intensity smoke (Stage 1 in Figure 1a), which gradually progressed to high intensity smoke (Stage 2 and Stage 3 in Figure 1b and 1c respectively) and finally converted into blazing fire (Stage 4 in Figure 1d).

a - Fire in stage 1

b - Fire in stage 2

c - Fire in stage 3

d - Fire in stage 4

FIG 1 – Fire in different stages.

Quarry-AB was producing an average production of 2.4 million tonnes per annum (Mtpa) prior to the occurrence of fire. Once the fire activity increased, the production from the mine gradually decreased and the mine management focused on controlling and extinguishing the mine fire.

The genesis of this fire is mainly due to spontaneous heating, which was triggered and accelerated because of slow movement of air through developed galleries exposed from highwall side of the coalfaces. This movement of air supplied oxygen to fallen coal due to spalling from the sides of galleries of developed pillars and roof falls. However, this air movement was so feeble that it could not take away the heat generated in the exothermic reaction of spontaneous heating and as a result, it helped in acceleration of exothermic reaction leading to blazing fire.

FIELD STUDY UNDERTAKEN FOR CONTROLLING OF FIRE

The field study was planned after thorough observation at the site. The salient points of this field study are as mentioned below:

- The fires in X, IX and VIII seams were active close to the highwall side of coal benches. Smoke with the blazing fires have been observed at the mouth of exposed developed galleries on highwall side.

- The fire affected zone from the highwall side is generally up to a distance of one pillar and in no case, it exceeds beyond 50 m from the highwall side of the coal bench.

- If the fire within this 50 m zone from the highwall side is effectively controlled, this will facilitate the production at coalface and the coal producing and hauling machines (shovels, dumpers and drills etc) can be effectively deployed without hampering the safety of men and machines engaged there.

- The control of fire from highwall side can be divided into two separate parts:
  o Controlling the fire in developed galleries with the fallen coal.
  o Controlling the fire in pillars standing by the side of developed galleries, where cracks and fissures might have been developed.

- While controlling the fire within this 50 m zone, it is to be monitored that the fire is not extending beyond that zone which may aggravate in future.

Keeping the above points in view, the field study has been designed for controlling and monitoring of fire. These are as follows:

- The old plans of the workings showing the position of galleries and pillars are required to be studied. The boreholes should be drilled over the galleries to monitor the level of CO and other gases. This baseline data should be compared with the samples to be taken after adoption of control measures. These holes should be drilled between the junction and inbye of the first row of pillars exposed to the highwall side of the coalface.

- The active fire is existing in the galleries of front row of pillars and this is required to be controlled by putting fire treatment holes which will be puncturing the galleries of all three seams. To ensure the stability of holes, perforated casings of MS pipes were put and these perforations in the gallery will allow the water and other fire-retardants to flow out over the fallen coal of different galleries, which may possibly retard and control the fire.

- Since the first row of pillars on the highwall side are also affected, these pillars can also be treated with fire retardant material by putting the boreholes over the pillars from the last overburden bench.

- The water as a fire-retardant material was proposed to be tried initially for treatment of fallen coal in the galleries and also in coal pillars. After studying the efficacy of water as a fire-retardant material in the given circumstances, if required further experimental investigations can be undertaken.

- The coal samples were collected from different seams and the susceptibility to spontaneous heating of coal samples were studied by crossing point and ignition point temperature method before and after addition of retardants like sodium silicate and others. Finally, it was observed that sodium silicate was more effective than other retardants for retarding the spontaneous heating susceptibility of coal samples collected from these seams (Singh, Ghosh and Dhar, 2004; Yadav, 2006).

All the above points were presented to and discussed with the Tata Steel management for its implementation on the mine site. After studying all the pros and cons of the proposed system, the mine management accepted to go ahead with implementation of the same at the mine site. However, they were quite emphatic that PMRC Private Limited should remain associated with the study till its finality.

IMPLEMENTATION OF THE PROPOSED FIELD STUDY AND MONITORING

As per the proposed field study, the scheme was implemented. Before using of fire retardant, the baseline gas samples were collected through the monitoring boreholes. The gas samples were analysed by gas chromatograph and the results for a few boreholes are presented in Table 2.

TABLE 2

Results of baseline gas samples collected through monitoring boreholes on 12 January 2022.

Borehole No.	CO%	CO_2%	CH_4%	C_2H_2%	C_2H_4%	O_2%	H_2S%	H_2%	N_2%	Remarks
801	0.0455	0.4258	NIL	NIL	NIL	20.1278	NIL	0.0196	79.3813	
802	2.2852	2.4813	0.4532	0.0059	0.0175	14.4279	NIL	0.2246	80.1044	
803	2.6838	1.7892	0.3125	0.0099	0.0071	15.4123	NIL	0.6182	79.1669	
804	3.5033	7.1489	1.4382	0.0336	0.0100	8.3129	NIL	1.2932	78.2599	Blazing fire indicates around 804
805	0.6750	2.0175	0.8651	0.0085	0.0177	15.1792	NIL	0.2903	80.9467	

After collection of the above baseline data, fire control measures were taken and water was used as retardant in all the treatment holes. The water injection was carried out for 2 hrs continuously with intermittent break of 2 hrs and this process was continued for 48 hrs. The gas samples from monitoring holes were collected at intermittent interval and analysed to take suitable precaution, if there is a chance of water gas explosion. After 7 days of the treatment, the results of gas sample analysis from the monitoring holes are presented in Table 3.

TABLE 3

Results of gas sample analysis after 7 days of injection of water.

Sl no	Borehole no	CO%	CO_2%	CH_4%	C_2H_2%	C_2H_4%	O_2%	H_2S%	H_2%	N_2%
1	801	0.0015	0.1334	Nil	Nil	Nil	20.8339	Nil	Nil	79.0312
2	802	1.3824	6.8724	0.7895	0.0262	0.042	5.9177	Nil	2.2178	82.752
3	803	0.0420	0.2308	Nil	Nil	Nil	20.8067	Nil	0.0185	78.9020
4	804	1.2289	9.0125	0.9577	0.0589	0.0204	2.7977	Nil	3.4281	82.4958
5	805	0.0750	9.1487	2.5894	0.0055	Nil	19.3127	Nil	0.3875	68.4812

It may be observed from above results that CO% decreased in all the boreholes, the percentage of H_2 and some hydrocarbons increased between the above two days of sampling. Tata Steel management stressed the need for faster reduction of CO% and also bringing down the percentage of H_2 and other hydrocarbons, so that the machines can be deployed for resumption of production.

To achieve the next level of reduction of CO and reduction in combustible gases, the scientists of PMRC Private Limited proposed the injection of water with 1 per cent sodium silicate mixture into the fire area where only water was used as the retardant. It was thought that the silicate may cover the surface of the coal in the form of a layer not to allow the free oxidation of coal. This experiment was carried out on 27 January 2023 and continued till the fire was controlled. The pattern and interval of injection of water with sodium silicate was kept same as earlier as in case of water injection, so that the results can be compared for assessing the efficacy of water with sodium silicate mixture. After injection of this retardant, the air samples were drawn from the monitoring boreholes and analysed by gas chromatograph. The results of gas analysis are presented in Table 4.

TABLE 4

Results of gas sample analysis after injection of water with sodium silicate mixture.

SI no	Borehole no	CO%	CO$_2$%	CH$_4$%	C$_2$H$_2$%	C$_2$H$_4$%	O$_2$%	H$_2$%	N$_2$%
1	801	Nil	2.1809	Nil	Nil	Nil	13.4278	Nil	84.3913
2	811	0.0030	0.1435	Nil	Nil	Nil	19.7352	Nil	80.1183
3	803	Nil	0.1397	Nil	Nil	Nil	20.7163	Nil	79.1440

It may be observed from Table 4 that the CO% has reduced considerably in hole nos 801 and 803. After the results were presented to the Tata Steel management, they deployed the production and hauling equipment for resuming the production from Quarry-AB opencast mine.

The process of injection of water and sodium silicate mixture was continued and the mine was producing the coal between 1.5–2 Mtpa. At this juncture, Tata Steel management wanted to reduce the water consumption, so that coal was less drenched with water for its effective use in their steel plants. Therefore, it was suggested by the scientists of PMRC Private Limited that once the CO level is brought down, liquid nitrogen after vaporisation can be injected to maintain the inert atmosphere and the production can be continued. If the CO% remains under control, the production can be continued without hampering the safety of men and machines deployed. If CO% is slowly increasing after nitrogen injection, water with sodium silicate injection is to be resumed for controlling the fire as earlier.

This was also accepted by Tata Steel management and liquid nitrogen tankers were brought to the site. Liquid nitrogen was converted to the gaseous state by using vaporising coil and injected through treatment holes to the developed galleries to maintain inert atmosphere. However, the water injection continued for the holes drilled into the pillars. This process was continued along with the coal production and the gas samples were collected through the boreholes. The results of a few boreholes are presented below in Table 5.

TABLE 5

Results of gas sample analysis after application of nitrogen gas into the boreholes.

SI no	Borehole no	CO%	CO$_2$%	CH$_4$%	C$_2$H$_2$%	C$_2$H$_4$%	O$_2$%	H$_2$%	N$_2$%
1	801	0.0035	0.1598	Nil	Nil	Nil	20.1053	Nil	79.7314
2	802	Nil	0.1589	Nil	Nil	Nil	20.1722	Nil	79.6689
3	803	Nil	0.1523	Nil	Nil	Nil	20.2125	Nil	79.6352
4	811	0.0030	0.1492	Nil	Nil	Nil	20.2173	Nil	79.6305

However, the process of injection of water with sodium silicate followed by nitrogen continued to maintain the safe atmosphere for the production of coal from Quarry AB. This is ascertained by monitoring of gas concentration in the available boreholes, which were drilled to the subsequent galleries and pillars on the newly exposed highwall side of coal bench.

CONCLUSIONS

On the basis of research work carried out and control measures implemented at the site of Quarry AB, the following conclusions may be drawn:

- Fire is one of the major problems while carrying out opencast mining over developed pillars in coal seams. As a result, it threatens the opencast mining operation in most of the cases and either the mine runs with a low production or sometimes the production is suspended.

- The genesis of fire is the fallen coal in the galleries and availability of oxygen due to slow movement of air through these galleries from the highwall side. This favours the spontaneous combustion of coal and in most of the cases smoke is visible from the gallery mouth on the highwall side. With passage of time, this is converted into blazing fire.

- The active fire has been controlled in the present case to a certain extent by injection of water. However, injection of water mixed with 1 per cent sodium silicate proved very effective in the present case. Once the fire is brought under control, the injection of nitrogen can be carried out to maintain the inert atmosphere inside the galleries for continuance of production in the opencast mine.

- One of the important aspects is to monitor the gas concentrations in the boreholes drilled into the developed galleries.

- If the percentage of CO is showing an increasing trend during the monitoring process, the injection of water with sodium silicate may be resorted to reduce the percentage of increasing CO.

- The method of controlling of fire as described in this paper may not be the universal prescription for all such cases of fires in opencast mines worked over developed pillars by underground method. If sodium silicate mixed with water does not work properly to retard the fire, other retardants may be tried in the laboratory and finally the suitable retardant from laboratory may be used in field condition for controlling fire in all such cases.

ACKNOWLEDGEMENTS

The experimental part of the study has been carried out at the laboratory of PMRC Private Limited and the contributions of the technical and research officers of the laboratory are sincerely acknowledged, which paved the way for initial understanding of the problem. The authors of the paper also sincerely thank the help and support of Tata Steel officials while implementing this study at the mine site.

REFERENCES

Clough, W D and Morris, R, 1985. The opencast mining of previously underground mined coal seams, *Journal of the South African Institute of Mining and Metallurgy*, 85(12):435–439.

Morris, R, 1983. Design of an opencast mine over previously mined areas, in *Proceedings of Second International Surface Mining and Quarrying Symposium*, IMM, pp 289–310.

Mozumdar, B K, 1995. Surface Mining of Developed Coal Seams, Special Number on Innovative Mining Practices, *The Indian Mining and Engineering Journal*, 34(10):57–59.

Panigrahi, D C, 1997. A study of fire problem in an opencast mine and its abatement measures, *The Indian Mining and Engineering Journal*, 36(11):1–13.

Panigrahi, D C, Saxena, N C, Saxena, V K and Jha, F, 1996. Investigation into fire proneness of No. 2 and 3 seams of Chirimiri Opencast and its abatement measures, in *Proceedings of National Seminar on Prevention and control of mine and industrial fires – Trends and challenges*, pp 319–330.

Singh, R V K, Ghosh, S K and Dhar, B B, 2004. Development of coating material for prevention of spontaneous heating in the benches of opencast coal mines, *Mine Tech*, 17(2&3):20–24.

Sinha, S K, 2014. A study of occurrence of fire in opencast coal mines extracting developed pillars, PhD Thesis, IIT(ISM) Dhanbad.

Vijh, K C and Verma, K N P, 1996. Dealing with mine fires in Wardha Vally Coalfield of WCL, in *Proceedings of Seminar on Prevention and control of mine and industrial fires – Trends and challenges*, pp 135–172.

Yadav, M D, 2006. A study of different fire retardants and sealants for preventing and controlling fires in Indian coal mines, PhD Thesis, IIT(ISM) Dhanbad.

Preliminary study on laboratory scale for studying the effect of limestone dust for preventing coal dust explosion in Indonesia

H Prasetya[1], M Ahsan Sulthani Pandu[2], N Priagung Widodo[3] and R Yulianti[4]

1. Researcher, Department of Mining Engineering, Faculty of Mining and Petroleum Engineering, Institut Teknologi Bandung, Bandung 40191, Indonesia. Email: heri.prasetya1511@gmail.com
2. Researcher, Department of Mining Engineering, Faculty of Mining and Petroleum Engineering, Institut Teknologi Bandung, Bandung 40191, Indonesia. Email: mahsansp22@gmail.com
3. Lecturer, Department of Mining Engineering, Faculty of Mining and Petroleum Engineering, Institut Teknologi Bandung, Bandung 40191, Indonesia. Email: npw@itb.ac.id
4. Doctoral student, Department of Mining Engineering, Faculty of Mining and Petroleum Engineering, Institut Teknologi Bandung, Bandung 40191, Indonesia. Email: ririnyuliantii@gmail.com

ABSTRACT

This study aims to analyse the effect of the size and ratio of limestone dust on the pressure value, the rate of change of pressure and the critical ratio of limestone dust immediately before the explosion. This study used a Siwek 20 L Explosion Chamber explosive tube with a coal concentration of 600 g/m^3 and an ignitor energy of 5 kJ. This test uses limestone dust with sizes of 177–74 µm and 74–37 µm and coal dust with sizes of 74–53 µm and 53–44 µm. The results of the first study on coal dust size 53–44 µm showed that the critical ratio between coal dust and limestone dust size 177–74 µm was at a ratio of 44.3:55.7 and for limestone dust 74–37 µm at a ratio of 45.1:54.9. Whereas for coal dust size 74–53 µm, the critical ratio of coal dust to limestone dust size 177–74 µm was obtained at 62.82:37.18 and for limestone dust size 74–37 µm at 81.99:18.01. In terms of the ratio of coal and limestone dust, the result is that the greater the ratio of coal dust compared to the ratio of limestone dust, the greater the pressure and the rate of change of pressure of the coal dust explosion. Based on the size of the limestone, the finer the size of the limestone dust, the lower the pressure and the rate of change of pressure from the coal dust explosion. Meanwhile, based on the size of the coal, the finer the size of the coal, the greater the potential to cause an explosion.

INTRODUCTION

The earliest stage to take advantage of coal is the coal mining stage which can be carried out using open pit and underground mining methods. At this mining stage, there is one aspect that needs to be considered, namely occupational health and safety issues. Occupational health and safety are an effort to create a healthy and safe work environment, to reduce the possibility of work accidents or exposure to disease which can cause deficiencies and demotivation at work.

There are several factors that can cause occupational health and safety problems at the coal mining stage. Especially in underground mines, dust can cause occupational health and safety problems because it can cause dust explosions. A dust explosion is an explosion triggered by combustible dust material (combustible dust) suspended in the air in a closed space and exposed to a heat source (Eckhoff, 2003; US Chemical Safety Board, 2006).

According to analysis results (Yuan et al, 2015), in 2000–2012 there were 193 cases of dust explosions in the world, of which 142 cases occurred in the United States, 92 cases occurred in China, and the rest happens in other countries. The biggest causes of explosions are food dust such as sugar, flour and powdered milk (40 per cent), wood dust (17 per cent), metal (10 per cent), others (10 per cent), coal (9 per cent), rubber (7 per cent), unknown (4 per cent), and inorganic material (3 per cent) according to Yuan et al (2015).

Therefore, coal dust is an element that needs more attention because it can cause explosions in mining activities which can cause destruction and even death of mine workers. There have been several studies carried out to analyse dust explosions, especially coal dust, such as Siwek (1977) who conducted coal dust explosion experiments on a laboratory scale, then experiments on coal dust explosions (Cashdollar, 1996) and ways to prevent or suppress their occurrences. One way to prevent a coal dust explosion is by adding rock dust into a room where there is coal dust.

Several researchers have examined the effect of rock dust on coal dust explosions. As in Mishra and Azam (2017) research which tested the effect of rock dust on coal dust explosions in India. The percentage requirement for limestone rock dust will increase as the size of the rock dust particles decreases. So, from this research it can be concluded that the smaller the particle size, the more effective its use. Rock dust acts as both a heat sink (by increasing the solid heat capacity of the mixture) (Amyotte, Mintz and Pegg, 1992) and a thermal inhibitor (by undergoing endothermic decomposition) (Man and Teacoach, 2009).

METHODOLOGY AND SAMPLE

Method

Coal dust explosion testing will focus on testing on a laboratory scale. There are several steps that need to be carried out to carry out explosion testing of coal dust mixed with limestone dust including, the coal and limestone sample preparation stage, ultimate and proximate testing stage of coal samples, the XRD testing stage to determine the mineral composition of limestone, the dust explosion test equipment preparation stage, and the coal dust explosion testing stage with limestone. The variable that will be varied here is the grain size of coal and limestone. Coal grain size variations are 74–53 µm and 53–44 µm and grain size variations for limestone are 177–74 µm and 74–37 µm.

There are references or standards for laboratory scale coal dust explosiveness tests. The standard used in this explosive test is ASTM E1226 (2010) which explains the standard test method for characterising the explosion of dust in two ways. First by determining whether the dust explodes, which means that dust scattered in the air can spread deflagration, which can cause flashes of fire or explosions. If it can explode, determining the level of explosion or potential explosion hazard from dust can be identified by dust explosion parameters, maximum explosion pressure (P_{max}), maximum pressure increase rate ($(Dp/dt)_{max}$) and deflagration index (K_{st}). In addition, there is another test standard, namely ASTM E1515 (2022) which describes the test standard for determining the minimum explosive concentration (MEC) of a mixture of dust and air that will spread deflagration in a nearly spherical closed vessel with a volume of 20 L or greater.

According to ASTM 1226, coal dust can be said to explode when the Pressure Ratio (PR) ≥ 2. Determination of PR can be done using the following calculations:

$$PR = (P_{ex,a} - \Delta P_{ignitor})/P_{ignition} \tag{1}$$

US Occupational Safety and Health Administration (US OSHA, 2009) classifies dust explosions based on the deflagration index (K_{st}) value, which is the maximum rate of pressure rise obtained from a 1 m³ chamber:

$$K_{st} = (Dp/dt)_{max}.V^{(1/3)} \tag{2}$$

Sample

The coal samples that will be used in the explosion test come from East Jambi District, Jambi Province. Analysis of the characteristics of coal samples begins by preparing a 1.5 kg coal sample. Testing and analysis of HGI, ultimate, proximate, and calorific value of coal samples was carried out at the Laboratory of the TekMIRA Mineral and Coal Testing Center, Bandung. The analysis results are expressed on an air-dried basis (adb) which states the percentage of coal without surface water parameters.

In determining coal ranking, the ASTM or American Society for Testing and Materials D388 standard guideline regarding Standard Classification of Coals by Rank is used. Coal samples according to ASTM D388 (2023b) are classified as **sub-bituminous coal C**.

The limestone samples that will be used in the coal dust explosion prevention test come from Padalarang, West Java, Indonesia. Analysis of the mineral composition of limestone samples begins by preparing a 20 g sample of limestone dust with a size of <177 µm. Testing and analysis of the mineral composition of limestone samples was carried out at the Hydrogeology and Hydrogeochemistry Laboratory, ITB using the XRD method. The X-Ray Diffraction (XRD) method is a fast, non-destructive analytical variable that can be used to identify the phase of crystalline materials and can provide important information regarding the dimensions and composition of unit cells. It is known from the test results that the limestone sample has a composition of 88.3 per cent calcite, 11.3 per cent quartz, and 0.39 per cent wüstite. Table 1 shows the results of the ultimate and proximate analysis of the coal samples used.

TABLE 1

Proximate and ultimate analysis of coal samples.

Analysis parameters	Sample marks	Unit	Basis
HGI	74	-	-
Proximate:			
Moisture in air dried	22.62	%	adb
Ash	4.5	%	adb
Volatile matter	39.88	%	adb
Fixed carbon	33	%	adb
Ultimate:			
Carbon	49.68	%	adb
Hydrogen	6.2	%	adb
Nitrogen	0.6	%	adb
Sulfur	0.24	%	adb
Oxygen	38.78	%	adb
Gross calorific value	4615	Kcal/Kg	adb

The preparation process for coal and limestone samples is carried out in several stages, namely the first is the crushing process using primary and secondary crushers. Coal and limestone with a size >5 cm will be crushed using primary crushing with a jaw crusher then if there is coal and limestone with a size >1 cm it will be crushed using secondary crushing with a roll crusher. Crushing using both crushers is carried out until a finer size is obtained so that it can be continued to the grinding stage using a ball mill.

Then proceed with the drying process. In this process, before the samples are crushed and sifted, the coal and limestone samples have humidity that meets standards and to make it easier to grind and sift, the coal and limestone are dried using an oven for 12–24 hrs. The samples resulting from crushing and drying are continued to be crushed using a ball mill repeatedly until they obtain the desired size. Then finally is the sifting process. From the results of the sieve, the desired size will be obtained, namely 53–44 µm and 74–53 µm for coal in accordance with ASTM 1226, which recommends that coal dust samples be at least 95 per cent < 75 µm. The reason for choosing these two sizes is actually to see the effect of the size of the coal dust grains on the explosive properties of coal dust. Whether the smaller particles make coal dust more explosive or not. Then, Figure 1 shows the sample preparation process from using a jaw crusher, roll crusher to the sieving process.

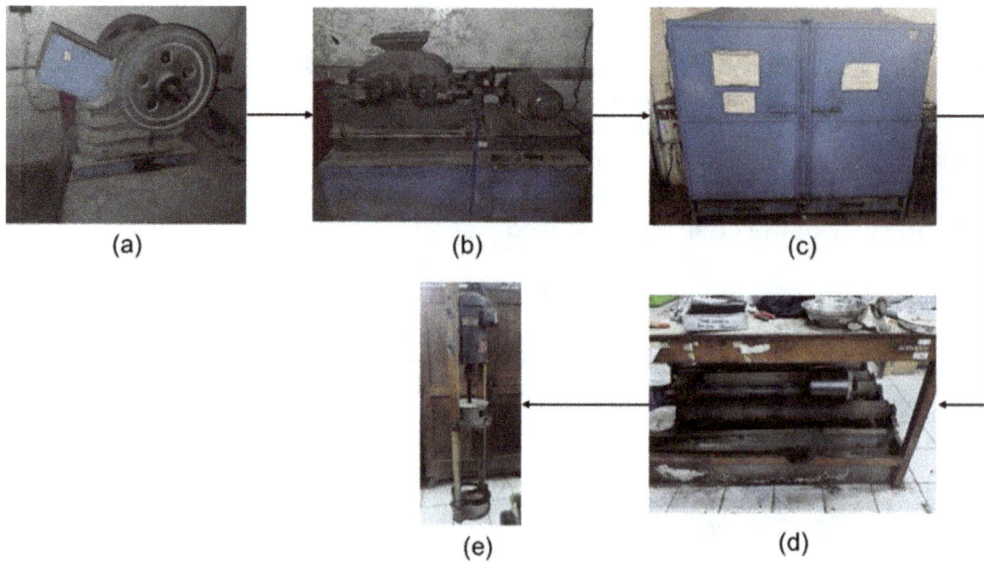

FIG 1 – Sample preparation process with information; (a) jaw crusher, (b) roll crusher, (c) oven, (d) ball mill, (e) sieving.

EXPERIMENTATION AND PROCEDURES

The experiment to test coal dust explosions used a 20 L blast tube which was shaped like a sphere so that the resulting pressure could be damped and could be calculated by the tool in the blast tube and provide homogeneous pressure. Coal dust that is inserted into the chamber will be dispersed, which is expected to be uniform, because it is pushed with compressed air at a pressure of 19 bar. Figure 2 shows a schematic of the explosion chamber.

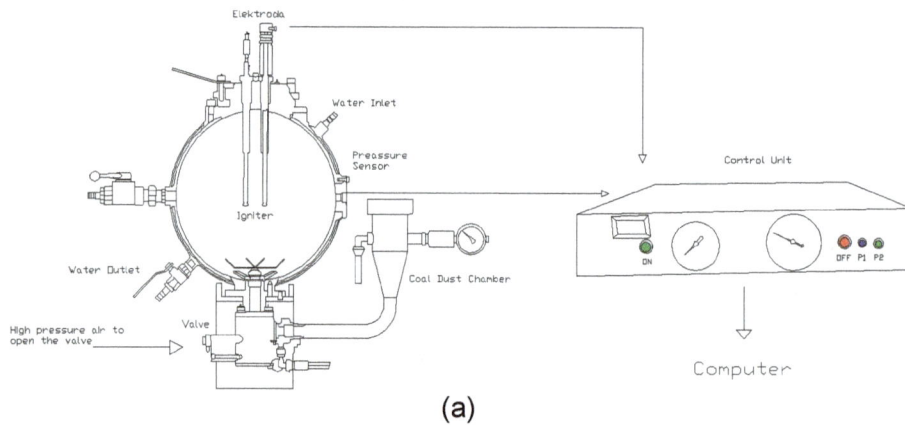

(a)

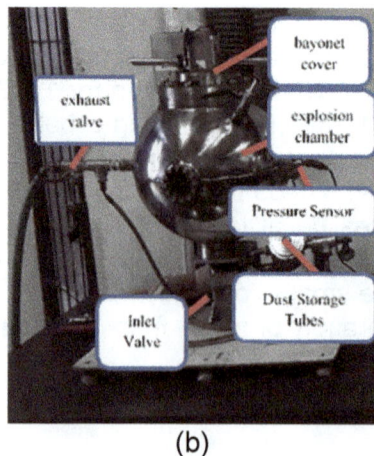

(b)

FIG 2 – The 20 L explosion chamber used for the tests: (a) Schematic of explosion chamber apparatus; (b) Physical appearances of the explosion chamber.

Later this coal dust will burn because it is ignited using pyrotechnics with an energy of 5 kJ (uses two ignitors with 2.5 kJ each). In the explosion tube, a sensor is installed in the form of a flange to read the pressure from the coal dust explosion. The data collection from this explosion test will be processed using the LabView application which acquires data every 0.06 sec or 17 data per second. The LabView application is also used to adjust the delay of the pyrotechnic ignition so that it can ignite simultaneously when the coal dust is dispersed into the explosive tube. Coal and limestone will later be put into the sample vessel according to the concentration and ratio specified. Ignitors used pyrotechnic ignitors type, consisting of zirconium, barium nitrate, and barium peroxide in a ratio of 4:3:3, with a total mass of 1 g (for pyrotechnics energy of 4.2 kJ).

Then the sample will be dispersed into the chamber using compressed air. The explosion test used pyrotechnics with a total energy of 5 kJ. The test was carried out using a ratio of coal dust to limestone dust ratio of 100:0, then decreased until the experimental coal dust explosion did not occur. The reduction is carried out in multiples of a decreasing ratio of ten for coal dust and a gradual increase in the ratio of ten for rock dust. However, if there is a possibility that coal dust will still explode, the ratio will be reduced by twofold.

RESULTS AND ANALYSIS

Pyrotechnic explosion test

The pyrotechnic ignitor explosion test was carried out in conditions without coal dust or limestone dust. The pyrotechnic ignitor explosion test aims to consider and calibrate the 20 L explosion chamber tool, as well as calculating the actual energy of the ignitor used. Bomb Calorimeter data was obtained from the tool's heat capacity of 2440 cal/°C, sample weight of 1 g, temperature rise on the calorimeter of 0.405°C.

$$Pyrotechnic\ energy = \frac{\left(2444\frac{kal}{°C}\right)x(0.405°C)}{1g}$$

$$Pyrotechnic\ energy = 988.2\ \frac{kal}{g}$$

$$Pyrotechnic\ energy = 4125\ J$$

The sample weight used in each pyrotechnic was 0.6 g = 2.481 kJ, two pyrotechnics were used in this test so that a pyrotechnic energy of 4.962 kJ was obtained. Figure 3 shows a graph of the explosion test using two pyrotechnics showing the pressure values of the pyrotechnics.

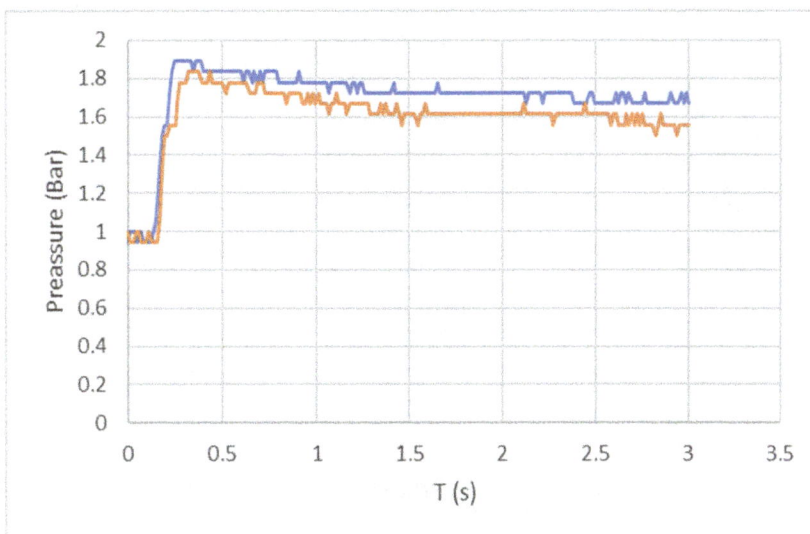

FIG 3 – Graph of pyrotechnic explosion test results.

Data such as $P_{ex,a}$, $P_{ignition}$, and $\Delta P_{ignitor}$ will later be used to correct pressure values using coal dust and/or limestone to obtain PR (Pressure Ratio) values. The $P_{ex,a}$ value (maximum absolute explosion pressure achieved in one explosion test without any consequences of using an ignitor) from pyrotechnics for experiments 1 and 2 was 1.89 bar and 1.83 bar, $P_{ignition}$ value (absolute pressure when the ignitor was activated) was 1.06 bar for experiments 1 and 2, and the $\Delta P_{ignitor}$ value (pressure increase due to activation of the ignitor at atmospheric pressure) is 0.11 bar and 0.06 bar.

Coal explosion test results

Coal dust explosion testing was carried out based on ASTM E1515, ASTM 1226, and research that had been carried out previously with a 20 L laboratory scale explosion tube. Coal dust testing uses coal dust measuring 74–53 µm. According to ASTM E1515, it is recommended to carry out tests with a coal dust concentration of 100 g/m³. If an explosion occurs, the concentration is reduced until no explosion occurs, but if an explosion has not occurred, the concentration must be increased until an explosion occurs. Meanwhile, according to ASTM E1226, it is recommended to carry out tests with a concentration of 250 g/m³.

From Table 2 it can be seen that at concentrations of 250 g/m³ and 400 g/m³ the $P_{ex,a}$ is 1.72 bar, while at the concentration of 600 g/m³ the $P_{ex,a}$ is 10.33 bar. Data processing to obtain PR and K_{st} values with formulations referring to Equations 1 and 2 shows that concentrations of 250 g/m³ and 400 g/m³ have a PR value of less than 2 with both K_{st} values being 0.62 bar.m/s. So that at this concentration there will be no coal explosion. However, at a concentration of 600 g/m³ with a PR value of 7.4 and K_{st} 5.02 bar.m/s, at that concentration an explosion occurred. So, in this study the concentration of 600 g/m³ was used as a reference in determining the influence of limestone dust on coal explosions. Figure 4 shows the value of the coal dust explosion pressure measured at a concentration of 600 g/m³ reaching 10.33 Bar.

TABLE 2

Initial testing of coal dust explosion.

	Initial testing						
Concentration (gr/m³)	$P_{ex,a}$ (Bar)	$P_{Ignition}$ (Bar)	P_{ex} (Bar)	$(dP/dt)_{ex}$ (Bar/s)	K_{st}	PR	Information
250	1.72	1.00	0.72	9.26	0.62	0.63	Did not explode
400	1.72	1.05	0.67	9.26	0.62	0.78	Did not explode
600	10.33	1.06	8.44	111.11	7.41	5.02	explode

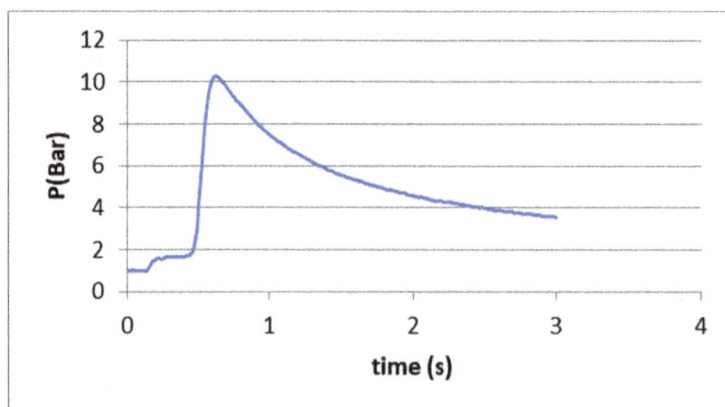

FIG 4 – Graph of coal dust explosion concentration 600 gr/m³

Coal dust and limestone explosion test results

The coal dust explosion ability test was carried out using a 20 L explosion chamber with variations in the mixture ratio of coal dust to limestone dust and two variations of limestone dust sizes 177–74 and 74–37 µm with a coal dust concentration set at 600 g/m³, the size of the coal dust namely 74–

53 µm and 53–44 µm. and pyrotechnic ignitor energy of 5 kJ. In underground mines, the energy of 5 kJ is equivalent to the energy produced by a short circuit for 0.015 sec in a 7LSO type shearer.

The following are the results of coal dust explosion tests with varying mixture ratios and limestone sizes carried out.

Based on Table 3 For coal dust sizes of 74–53 µm, two experiments were carried out for each variation of the ratio of the mixture of coal dust to limestone dust. It was found that in the mixture of coal dust and limestone dust the ratio of coal dust to limestone dust was 100:0 to 70:30 for limestone dust sizes of 177–74 µm an explosion occurs, while for limestone dust sizes of 74–37 µm an explosion occurs at a ratio of coal dust to limestone dust of 100:0 to 90:10.

TABLE 3

Coal dust explosion test data processing results with varying limestone dust mixture ratios (coal dust 74–53 µm).

Limestone dust size	Coal dust: limestone dust ratio	Testing	$P_{ex,a}$ (Bar)	$P_{Ignition}$ (Bar)	$\Delta P_{Ignitor}$ (Bar)	$(dp/dt)_{ex}$ (Bar/s)	K_{st} (Bar.m/s)	PR	Information
177–74 µm	100:0	Test 1	10.00	1.72	0.78	92.59	25.13	5.02	Explode
		Test 2	10.61	1.67	0.72	149.57	40.60	5.93	Explode
	70:30	Test 1	6.44	1.78	0.72	23.15	6.28	3.22	Explode
		Test 2	7.17	1.72	0.78	27.78	7.54	3.71	Explode
	60:40	Test 1	1.67	1.11	0.11	8.55	2.32	1.40	Did not explode
		Test 2	1.67	1.06	0.11	10.42	2.83	1.47	Did not explode
	50:50	Test 1	1.67	1.00	0.06	10.10	2.4	1.61	Did not explode
		Test 2	1.72	1.06	0.06	12.82	3.48	1.58	Did not explode
	40:60	Test 1	1.56	1.06	0.11	12.82	3.48	1.37	Did not explode
		Test 2	1.56	1.00	0.06	8.55	2.32	1.5	Did not explode
	30:70	Test 1	1.72	1.17	0.11	13.07	3.55	1.38	Did not explode
74–37 µm	100:0	Test 1	10.00	1.72	0.78	92.59	25.13	5.35	Explode
		Test 2	10.61	1,67	0.72	149.57	40.60	5.93	Explode
	90:10	Test 1	7.89	1.72	0.56	37.04	10.05	4.26	Explode
		Test 2	6.78	1.72	0.78	25.64	6.96	3.48	Explode
	80:20	Test 1	1.61	0.94	0.06	10.10	2.74	1.65	Did not explode
		Test 2	1.67	1.06	0.11	12.82	3.48	1.47	Did not explode
	70:30	Test 1	1.67	1.06	0.11	12.82	3.48	1.47	Did not explode
	60:40	Test 2	1.72	1.33	0.33	20.83	5.66	1.04	Did not explode

Meanwhile, based on Table 4 for coal dust sizes of 53–44 µm, it is found that in a mixture of coal dust and limestone dust at a ratio of coal dust to limestone dust of 100:0 to 50:50 for limestone dust sizes of 177–74 µm, an explosion occurs, while for limestone dust sizes of 74–37 µm an explosion occurs at a ratio of coal dust to limestone dust of 100:0 to 50:50. The test stopped at a ratio of 30:70 because after reducing the ratio of coal to limestone dust several times, there was no explosion.

TABLE 4

Coal dust explosion test data processing results with varying limestone dust mixture ratios (coal dust 53–44 μm).

Limestone dust size	Coal dust: limestone dust ratio	Testing	$P_{ex,a}$ (Bar)	$P_{Ignition}$ (Bar)	$\Delta P_{Ignitor}$ (Bar)	$(dp/dt)_{ex}$ (Bar/s)	K_{st} (Bar.m/s)	PR	Information
177–74 μm	100:0	Test 1	10.33	1.89	0.83	111.11	7.4	5.02	Explode
		Test 2	10.28	1.78	0.78	111.11	7.4	5.33	Explode
	50:50	Test 1	8	1.83	0.83	45.45	3.03	3.91	Explode
		Test 2	7.5	1.83	0.83	45.45	3.03	3.64	Explode
	40:60	Test 1	1.83	1.78	0.78	45.45	3.03	0.59	Did not explode
		Test 2	1.83	1.61	0.67	17.09	1.14	0.72	Did not explode
	30:70	Test 1	1.89	1.83	0.61	23.15	1.54	0.69	Did not explode
		Test 2	1.94	1.72	0.44	14.81	0.98	0.87	Did not explode
74–37 μm	100:0	Test 1	10.33	1.89	0.83	111.11	7.4	5.02	Explode
		Test 2	10.28	1.78	0.78	111.11	7.4	5.33	Explode
	50:50	Test 1	7.2	1.94	0.94	38.19	2.55	3.22	Explode
		Test 2	7.17	1.83	1	42.74	2.85	3.37	Explode
	40:60	Test 1	1.61	1.51	0.57	12.82	0.85	0.68	Did not explode
		Test 2	1.61	1.5	0.61	13.89	0.93	0.67	Did not explode
	30:70	Test 1	1.61	1.61	0.61	10.42	0.69	0.62	Did not explode
		Test 2	1.67	1.61	0.67	13.89	0.93	0.62	Did not explode

ANALYSIS OF THE EFFECT OF VARYING RATIOS OF COAL DUST MIXTURE WITH LIMESTONE DUST ON THE PREVENTION OF COAL DUST EXPLOSIONS

According to ASTM 1226 and ASTM E1515, coal dust that can still explode when PR ≥ 2 or K_{st} ≥ 1.5 bar.m/s with 0 < K_{st} ≤ 200 according to (US OSHA, 2009) is included in the weak explosion class category.

From the Figure 5 in coal dust measuring 74–53 μm to limestone dust measuring 177–74 μm ratio of 40 per cent (37.18 per cent) no explosion occurred, whereas in coal dust measuring 53–44 μm no explosion occurred when the ratio of limestone dust measuring 177–74 μm is 60 per cent (55.70 per cent). In coal dust measuring 74–53 μm to limestone dust ratio measuring 74–37 μm of 20 per cent (18.01 per cent) no explosion occurred, while coal dust measuring 53–44 μm did not occur when the ratio of limestone dust measuring 74–37 μm is 60 per cent (54.90 per cent).

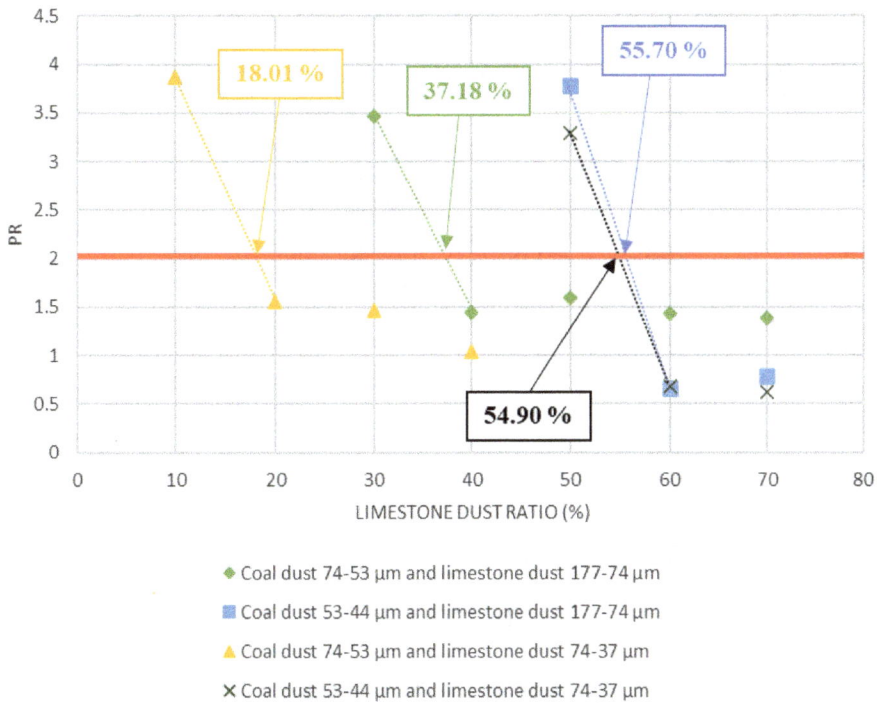

FIG 5 – Graph of coal dust explosion test results, PR versus limestone dust ratio

This shows that the larger the size of the coal dust, the smaller the explosion pressure produced, so that at larger coal dust sizes the need for limestone dust will be lower to reduce or prevent coal dust explosions. Coal dust with a larger size will be more difficult to produce an explosion due to several factors, including a decrease in the reaction surface area of coal dust grains with a larger size and it will be difficult to disperse or form dust clouds in the room because it will easily fall so that it has not yet had time to react to produce an explosion (Coal dust particles cannot be controlled because they are influenced by the mining method and tools used. However, this prevention technique can be done by sampling coal dust in the field (ASTM E1226), then coal dust explosions can be mitigated so that they do not occur).

DISCUSSION

The use of limestone dust is recommended to overcome coal dust explosions. Laws related to the use of rock dust to overcome coal dust explosions also exist in various countries (Luo, Wang and Cheng, 2017). Several countries determine the use of rock dust in the range of 50–80 per cent depending on many factors such as volatile matter, whether there is gas content in the area or not, the channel where the dust will be sprayed and others. Table 5 shows that several countries have set rules for using rock dust in handling coal dust explosions.

TABLE 5

Rules for using rock dust in various countries (Luo, Wang and Cheng, 2017).

Country	Rules
United States of America	US MSHA (Mine Safety and Health Administration) specifies 80% rock dust
Canada	Established 65% on inlet airways (Alberta and British Columbia) 80% on all return airways (Province of Nova Scotia)
Australia	Set 70–85% rock dust (air inlet) 80–85% rock dust (air outlet)
Eastern Europe (Czech, Slovakia, Ukraine, etc.)	Sets 80% rock dust
Poland and Russia	Assign 60–70% air inlet and return (non-gas) Assign 75–80% (gas mine)
South Africa	Sets 80% rock dust
Great Britain	Set 50–75% depending on the Volatile Matter (VM) which varies between 20–35%
Japan	Set 78% rock dust when VM is more than 35%

Mishra and Azam (2017) in Table 6 also conducted research using limestone dust to prevent coal dust explosions with coal type specifications between medium cooking coal – prime cooking coal. Meanwhile, the sizes of limestone dust used range from sizes below 25 μm, 25–38 μm, 38–74 μm, and 74–212 μm. In this construction, the limestone sizes that can be compared are 38–74 μm, and 74–212 μm with the sizes used by painters, namely 37–74 μm, and 74–177 μm.

TABLE 6

Research comparative parameters.

Study	Coal type	Coal class	Volatile matter (%)	Chamber size	Coal size	Limestone size	Concentration
Present research (2024)	Jambi	Subbituminous C	39.88	20 L	74–53 μm and 53–44 μm	37–74 μm and 74–177 μm	600 g/m^3
Mishra and Azam (2017)	Jharia, India	Medium-prime cooking coal (bituminous)	17.73	0.234 L	<212 μm	38–74 μm and 74–212 μm	Varies but reviews are for 600 g/m^3

On Figure 6 There is a comparison between the critical ratio required by limestone dust when an explosion occurs to avoid an explosion. The critical limestone dust ratio for dealing with coal dust explosions in the current study has a smaller value at larger coal dust sizes (74–53 μm) compared to smaller coal dust sizes (53–44 μm).

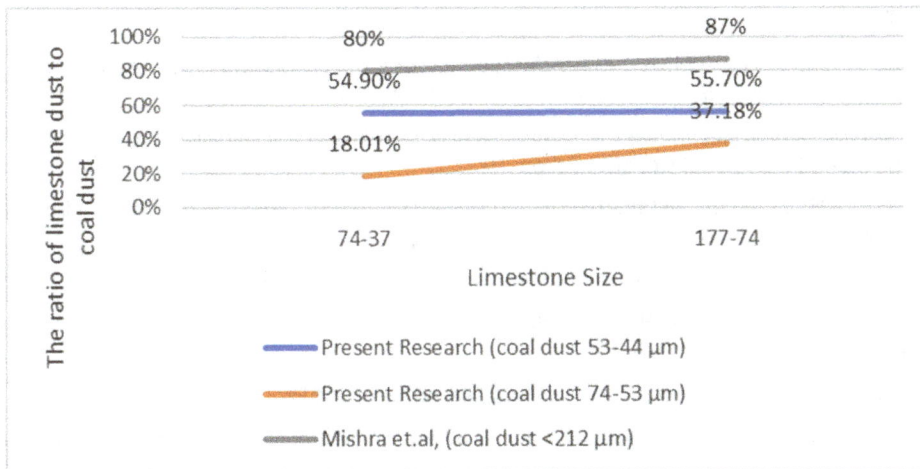

FIG 6 – Comparison chart of critical ratio of limestone dust and limestone size.

The limestone dust ratio required in the current research is smaller than the ratio required in the research of Mishra and Azam (2017). This can occur due to many factors such as differences in the type and origin of coal used, Volatile Matter, as well as the size and type of chamber used.

Indonesia itself has further potential to have more massive underground coalmines, while the regulations regarding the use of rock dust to overcome coal dust explosions are not yet covered, so this research can be initial research to determine the ratio of coal dust to limestone to avoid coal dust explosions. It is necessary to pay attention to the difference between 20 L chamber testing and full-scale testing. In Cashdollar *et al* (1992) it was stated that there were differences such as differences in MEC values where the 20 L room had a lower MEC value than the 1 m³ room test.

CONCLUSION

This research has resulted that limestone is effective in dealing with coal dust explosions depending on the size of the coal dust, the size of the limestone dust, and the concentration of the coal dust. The larger the size of the coal dust, the smaller the explosion pressure produced, so that at larger coal dust sizes the need for limestone dust will be lower to reduce or prevent coal dust explosions. Coal dust with a larger size will be more difficult to produce an explosion due to several factors, including a decrease in the reaction surface area of coal dust grains with a larger size and it will be difficult to disperse or form dust clouds in the room because it will easily fall so that it has not yet had time to react to produce an explosion.

It was found in this study that coal dust measuring 74–53 μm to limestone dust ratio measuring 177–74 μm was 40 per cent (37.18 per cent) without an explosion, whereas coal dust measuring 53–44 μm did not cause an explosion when the ratio of stone dust limestone measuring 177–74 μm is 60 per cent (55.70 per cent). In coal dust measuring 74–53 μm to limestone dust ratio measuring 74–37 μm of 20 per cent (18.01 per cent) no explosion occurred, while coal dust measuring 53–44 μm did not occur when the ratio of limestone dust measuring 74–37 μm is 60 per cent (54.90 per cent). So that it can effectively obtain a ratio range of 20–60 per cent between limestone and coal dust to reduce coal dust explosions depending on the size of the coal dust and its concentration. This research needs to be continued by examining variations in explosion chamber size, coal rank, coal size, rock dust size, rock dust type, etc.

ACKNOWLEDGEMENTS

This research was supported by thank you to lecturers, technicians and colleagues who have helped in Institut Teknologi Bandung, Indonesia. The Authors would like to thank to Equipment and Rock Mechanics Laboratory and Hydrogeology and Hydrochemistry Laboratory of Department of Mining Engineering (DME), Faculty of Mining and Petroleum Engineering, Institut Teknologi Bandung, Indonesia.

REFERENCES

Amyotte, P R, Mintz, K J and Pegg, M J, 1992. Effectiveness of various rock dusts as agents of coal dust inerting, *Journal of Loss Prevention in the Process Industries*, 5(3):196–199.

ASTM, 2022. ASTM E1515-14(2022) – Standard Test Method for Minimum Explosible Concentration of Combustible Dusts, ASTM International (West Conshohocken: PA).

ASTM, 2010. ASTM E1226, Standard test method for pressure and rate of pressure rise for combustible dusts, ASTM International (West Conshohocken: PA).

Cashdollar, K L, 1996. Coal dust explosibility, *Journal of Loss Prevention in the Process Industries*, 9(1):65–76.

Cashdollar, K L, Weiss, E S, Greninger, N B and Chatrathi, K, 1992. Laboratory and large-scale dust explosion research, *Plant/Operations Progress*, 11(4):247–255.

Eckhoff, R, 2003. *Dust Explosion in the Process Industries*, 3rd edition (Gulf Professional Publishing).

Luo, Y, Wang, D and Cheng, J, 2017. Effect of Rock Dusting in Preventing and Reducing Intensity of Coal Mine Explosion, *Int J Coal Sci Technol*, 4(2):102–109.

Man, C-K and Teacoach, K A, 2009. How does limestone rock dust prevent coal dust explosions in coal mines, *Mining Engineering*, 61:69–69.

Mishra, D P and Azam, S, 2017. Rock dust Requirement for Suppression of Coal Dust Explosion in Coal Mines in India-An Investigation, in *Proceedings of the International Conference on NexGen Technologies for Mining and Fuel Industries-NxGnMiFu-2017*, 7 p.

Siwek, 1977. 20-liter laboratory apparatus for determination of explosion characteristics of combustible dusts, Winterthur, Switzerland: Ciba-Geigy AG (Basel) and Winterthur Engineering College.

US Chemical Safety Board, 2006. Combustible Dust Hazard Investigation, US Chemical Safety Board.

US Occupational Safety and Health Administration (US OSHA), 2009. Hazard Communication Guidance for Combustible Dusts, US Department of Labor, OSHA.

Yuan, Z, Khakzad, N, Khan, F and Amyotte, P, 2015. Dust explosions: A threat to the process industries, *Process Safety and Environmental Protection*, 98:57–71.

Numerical study on gas explosion hazard and spontaneous combustion in longwall panel – an Indian perspective

S Rao[1], D P Mishra[2], U S Shukla[3], A Mishra[4] and J S Mahapatra[5]

1. Senior Manager, Moonidih Colliery, Bharat Coking Coal Ltd., 828129, India; Research Scholar, Department of Mining Engineering, IIT(ISM) Dhanbad. Email: sunny.rao@coalindia.in
2. Chair Professor and Head, Dept of Mining Engineering, IIT(ISM) Dhanbad, 826004, India. Email: dpmishra@iitism.ac.in
3. Manager, Dudhichua Project, Northern Coalfield Ltd., India; Research Scholar, Department of Mining Engineering, IIT(ISM) Dhanbad. Email: us.shukla30@gmail.com
4. Assistant Manager, Nigahi Project, Northern Coalfield Ltd., India. Email: aishwaryam.rs.min16@itbhu.ac.in
5. General Manager, WJ Area, Bharat Coking Coal Ltd., 828129, India. Email: gmwj.bccl@coalindia.in

ABSTRACT

The propensity of residual coal towards spontaneous combustion in gas desorption zone of longwall goaf poses a significant safety risk. A numerical model was developed to simulate the flowfield, species transport, reaction kinetics in the porous goaf leading to identification of potential overlap zones of low temperature oxidation, and distinctive explosive gas concentration. The species distribution in the goaf was significantly impacted by the mine ventilation parameters. The model was validated by methane dispersion in the tailgate area of the case study mine and the dispersion was observed to be driven by the dissipation of turbulent kinetic energy in the media. Further, the effect of nitrogen injection in the goaf from the surface holes on the goaf environment and the gas drainage was examined for enhancing the longwall mine safety. The study will, thus, assist in comprehensive evaluation and management of principal hazard in longwall mines.

INTRODUCTION

Underground coal production is integral to the coking coal economy of India. The coking coal reserves in India are seated in Jharia coalfield, which has 8 per cent of the coal reserves and is also a multi-seam coalfield. With the increase in-depth, the hazards associated with the gases are paramount. Air flow behaviour in the working as well as goaf region influences the explosive nature of the environment in a longwall mine (Kumar *et al*, 2017; Ray *et al*, 2022). Longwall mining creates huge disturbances in the strata leading to formation of goaf due to the stress-relief of the overlying strata (Karacan *et al*, 2011). Goaf has been identified as a porous zone and various scholars have postulated an O-shaped permeability of the region. Gases from the overlying as well as underlying coal seams migrate to occupy the void space in goaf, and thus, has been identified as a repository of the gases such as methane, and in case of spontaneous combustion, occasionally CO has also been found from the samples drawn from the goaf environment.

Since, goaf is a porous region, the air generally leaks across it due to the common pressure differential acting across the face and goaf junctions. This leakage of air has often led to the dangerous tail-gate corner methane accumulations and migrations in the working area. The speed of leaked air in goaf is generally below 0.0001 m/s deep inside the goaf (Rao, Mishra and Mishra, 2023). This is due to viscous as well as inertial resistance provided by the goaf. Additionally, due to diffusion, the concentration of constituent gases of air also reduces to zero inside the goaf. The air leakage into goaf induces spatial variation in hazardous gas concentration levels and brings them in explosive zone. Additionally, the air at such a low velocity accelerates the temperature increase due to spontaneous coal combustion and under the effect of heat transfer mechanism between the solid-solid and solid-gas phases driven by the heat of coal oxidation reaction, thereby leading to dual hazard of methane and spontaneous combustion in gassy mines. The identification of such interactive zones of the twin hazards is therefore of great importance for the risk assessment and mitigation program of deep underground coalmines.

The past studies on methane migration in goaf revealed that the air leakage into goaf from the working area is an unavoidable phenomenon leading to the formation of an explosive gas zone (EGZ) in goaf (Gilmore *et al*, 2014; Marts *et al*, 2014a). Therefore, studying the air flow, its diffusion phenomena in goaf, and consequent evolution of the EGZ on the operating mine ventilation parameters holds prime importance for safer working condition. Scholars have also identified the zones in the goaf which may be an ideal seat of spontaneous combustion having 8–18 per cent oxygen and low velocity air movement. Consequently, in this paper, the authors have developed a three-dimensional CFD model to study the oxygen concentration in goaf of a U-type longwall panel which may lead to gas and spontaneous combustion hazards. Additionally, the effect of goaf size on oxygen distribution in goaf is analysed, and consequent evolution of hazard at different ventilation operating conditions has also been examined before advocating for the proactive nitrogen based inertisation program.

STUDY AREA

Moonidih mine of Bharat Coking Coal Ltd (BCCL), India has been considered as the case study mine for this research. The active longwall panel under consideration has been schematically shown in Figure 1.

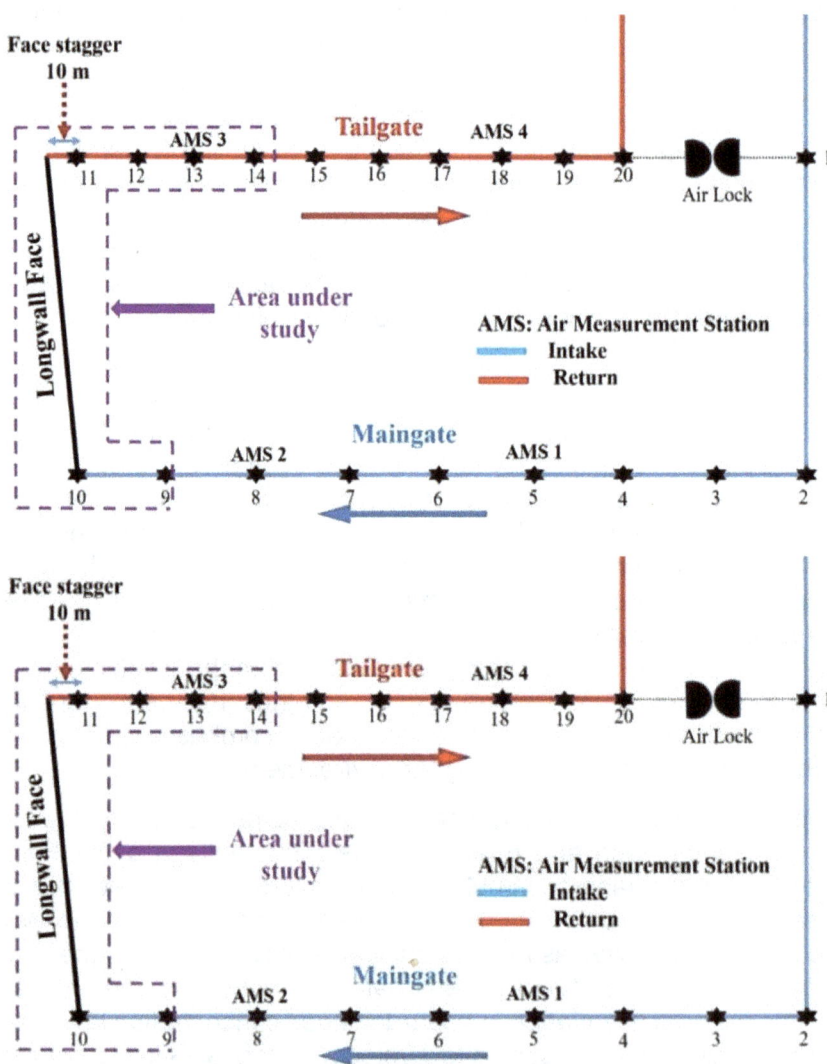

FIG 1 – Schematic diagram of longwall panel in XVI seam of Moonidih mine.

FIELD INVESTIGATIONS

In this study, a ventilation survey was conducted in the longwall panel to calibrate and establish the boundary conditions for computational fluid dynamics (CFD) simulations. The survey included air velocity measurement (see Figure 2a), pressure measurement, and methanometry (see Figure 2b).

This survey covered a total distance of 1650 m along the longwall panel. The net pressure drop across the longwall panel (between stations 1 and 20 in Figure 1) was measured at 400 Pa. It's noteworthy that the high-pressure drop is primarily induced by airflow resistance from various sources such as the maingate auxiliary longwall equipment, longwall face orientation, face conveyor drive head limits, and shearer. Between stations 9 and 14 (indicated by the dotted line in Figure 1), the total pressure drop across the area was measured at 60 Pa, and this value was utilised in developing the computational model. During periods of methane gas-related face stoppage (as shown in Figure 2b), the concentration of methane in the tailgate return ranged from 1.5 to 2.0 per cent.

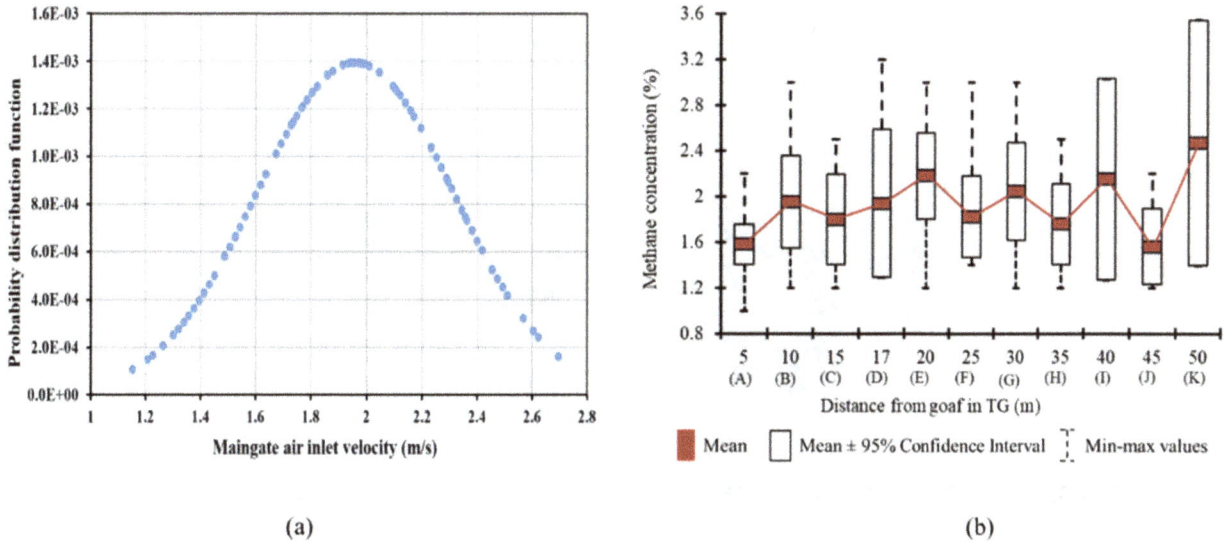

(a)

(b)

FIG 2 – (a) Probability distribution of air velocity measured in the longwall panel of XVI seam, (b) Box plot of methane concentration (%) at different distances from the goaf in tailgate.

DEVELOPMENT OF COMPUTATIONAL DOMAIN

The geometrical model and actual dimensions to study the air flow in longwall panel are shown in Figure 3. The input parameters so used have been obtained from the field investigations. The longwall zone, consisting of the maingate, tailgate, and retreating longwall face, was treated as the fluid zone and incorporated a U-type ventilation system. The goaf zone was regarded as a permeable region, comprising caving rocks stacked in a confined space with a certain level of compaction. The goaf can be viewed as a continuously evolving reservoir with methane present in voids, pores, and cracks. Seepage parameters of the goaf were estimated in FLAC3D using the physico-mechanical properties of the immediate roof, employing the Mohr–Coulomb plasticity model (Esterhuizen and Karacan, 2007). The initial porosity of the goaf was assumed to be 40–50 per cent (Pappas and Mark, 1993), and the final porosity was calculated by subtracting the volumetric strain increment from the initial porosity. The final porosity of the goaf obtained from the model was further utilised to estimate the permeability of the goaf (K_goaf) using the Carman-Kozeny relation (Equation 1) (Behera *et al*, 2019):

$$K_{goaf} = \frac{K_0}{0.241}\left(\frac{n^3}{(1-n)^2}\right)$$

(1)

where:

K_{goaf} is the permeability of goaf

K_0 is the base permeability of rock $1 \times 10^{-3} m^2$ considering it as 'open jointed rock' (Hoek and Bray, 1981)

n is the final porosity of rock

In this study, the permeability of goaf was found to spatially vary across the goaf in the range of 1.7–1.93 × 10⁶ mD. The mean permeability value of 1.85 × 10⁶ mD was used to model the fluid flow in the goaf for better approximation of the seepage flow in the porous medium of the goaf.

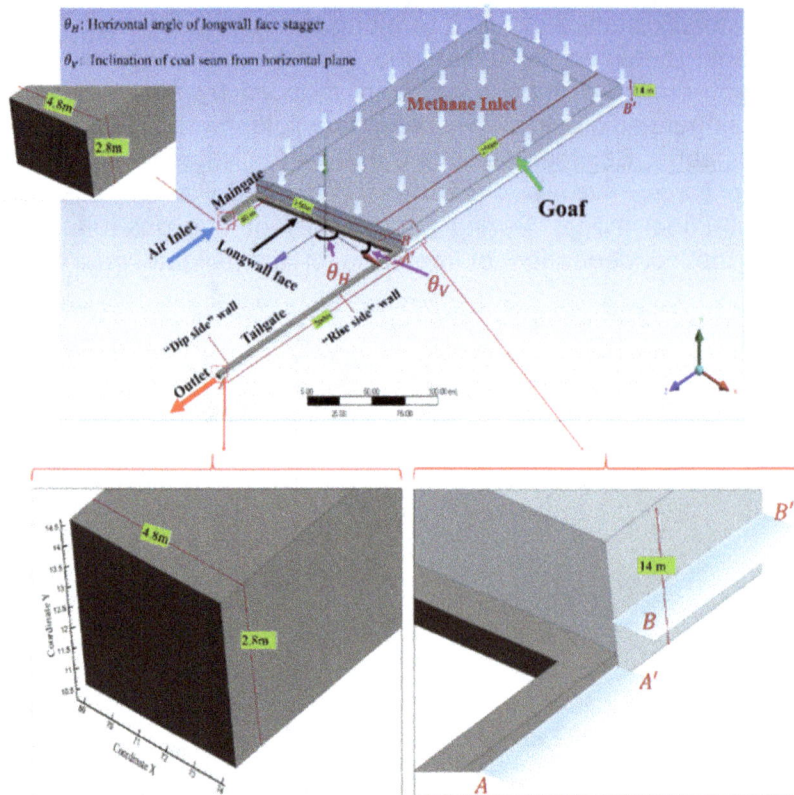

FIG 3 – Computational domain and meshing pattern of longwall panel showing the longwall face, gate roads and goaf.

BOUNDARY CONDITIONS

The boundary condition of velocity inlet was applied at the maingate, while the tailgate was set as pressure outlet of -60 Pa. Therefore, in all the simulations, the pressure outlet condition was set as negative. Further, the pressure drop across the goaf was measured between the maingate and tailgate junction across the face, and its value was estimated as 35 Pa. Given the length of geometry considered in the numerical model, the net pressure difference between the inlet and outlet was set at -60 Pa with maingate inlet of 27 m³/s in the panel so as to incorporate the field conditions of pressure and velocity in the computational model. When the main mine fan underwent shutdown for two hrs at the study site, the methane concentrations in the maingate and tailgate were found nearly identical up to 20 m out bye from the longwall face. Thus, a methane inlet of 0.5 m³/s from the roof of goaf was applied for the simulations. This corresponds to methane inlet at 30 m³/min from the roof of goaf of dimension 250 m × 150 m. The air and methane were assumed to possess low thermo-physical properties, so a constant temperature of 300 K was considered for the simulations. The walls of goaf and longwall were assumed to be smooth, adiabatic, and assigned no-slip boundary conditions. Based on the viscous and inertial loss term, momentum sink was introduced through user defined function and oxygen depletion based on the reaction kinetics and consumption was also included in the source loss term. During the iterations, the convergence was set at 1×10^{-4} as a standard of accuracy mentioned in ANSYS FLUENT Theory Guide (ANSYS, Inc., 2013).

MODEL VALIDATION

The average methane concentration, obtained from Figure 2b as measured in the tailgate between the centre line of the gallery to the rise sidewall, was plotted against the different variants of k-epsilon turbulence model. The simulation results for methane concentration obtained using the standard k-epsilon turbulence model is broadly consistent with the field results. The simulation results as can be seen in Figure 4, varied from the experimental values in the range of 3.2 to 25 per cent.

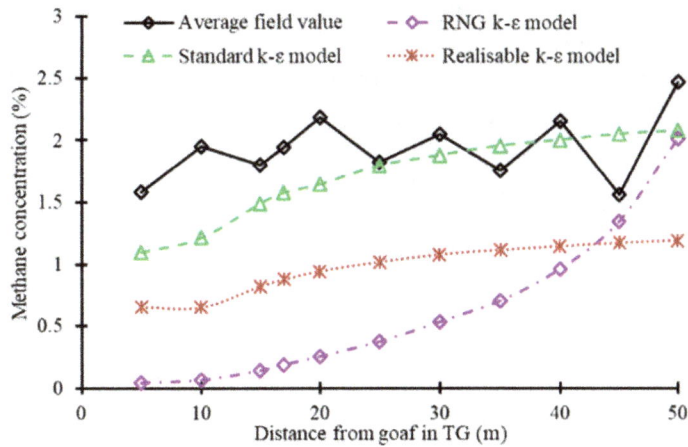

FIG 4 – Validation of the simulation results with the tailgate methane concentration field readings.

RESULTS AND DISCUSSION

The emission of methane from goaf and its dispersion in the tailgate of the longwall panel along with oxygen dispersion in goaf were assessed from the simulations results. Further, computational analysis of auxiliary ventilation was also done so as to assess the reduction in methane concentration.

Distribution of air and methane in goaf zone

In the simulations, setting the tailgate pressure outlet at -60 Pa and the maingate velocity inlet at 2 m/s resulted in a pressure gradient along the maingate, face, and tailgate. This pressure gradient across the face caused air to leak from the maingate into the porous goaf medium and from the goaf into the tailgate of the longwall working. Additionally, the methane velocity inlet from the model roof and pressure gradient across the face in the longwall domain, before reaching a steady state through solution of flow in porous medium, led to a distinct pressure field distribution between the goaf region and tailgate. Figure 5a illustrates the static pressure field in the computational domain. A static pressure gradient is observable in the depth of the goaf, rising from 0 Pa near the longwall face to 145 Pa at the consolidated depth. Moreover, the pressure gradient from the model roof to the floor of the longwall panel suggests that the static pressure distribution across the vertical section decreases from the roof to the floor. This static pressure field distribution resulted in the development of a distinct velocity field in the porous goaf. The velocity vectors inferred from the pressure field distribution depict the movement of air into the goaf from the maingate and methane inlet from the model roof exiting into the tailgate region of the longwall panel (see Figure 5c). As shown in Figure 5b, the air velocity in the goaf decreases with the depth of the goaf. Furthermore, Figure 5d delineates the zones of low air velocity in the goaf, indicating that on average, 60 m behind the longwall face, air in the goaf moves at a low velocity range of 0.001–0.0001 m/s.

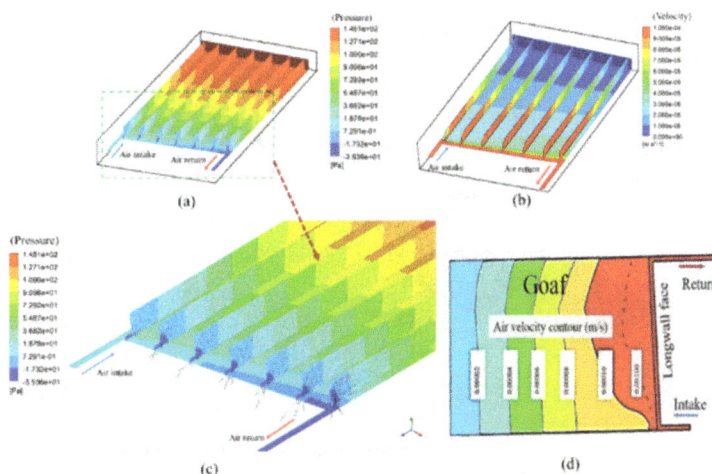

FIG 5 – Air velocity contours in goaf.

The maingate region facilitated airflow into the permeable goaf (refer to Figure 6). This movement caused methane within the goaf to migrate towards the tailgate of the longwall, eventually dispersing into the working area through the tailgate. The pressure gradient between the goaf and tailgate outlet, coupled with air movement in the permeable goaf, influenced the spatial distribution of oxygen and methane within the goaf, as depicted in Figure 6. Li *et al* (2020) also noted a similar distribution pattern of methane and oxygen in the goaf due to differential ventilation pressure and oxygen ingress into the goaf from the maingate (Li *et al*, 2020). Moreover, the oxygen concentration analysis was conducted for a larger dimension goaf, specifically a goaf with a length of 1000 m (Figure 7a). It was found that oxygen diffused up to 250 m inside the goaf from the longwall face. Thus, increasing the length of the goaf, the hazardous zone would be limited within 250 m from the goaf. Consequently, a region up to 250 m inside the goaf needs to be neutralised rather than deep-seated regions inside the goaf.

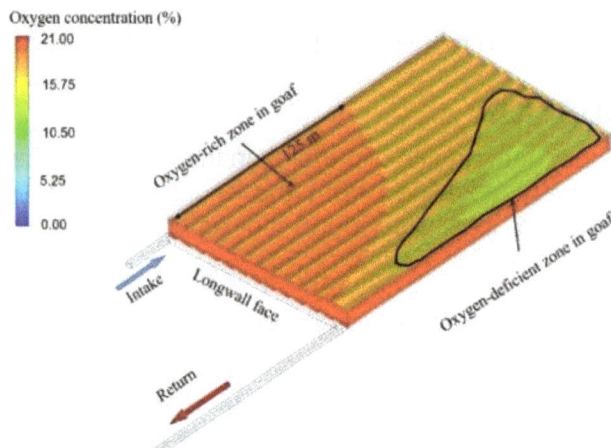

FIG 6 – Oxygen distribution in goaf.

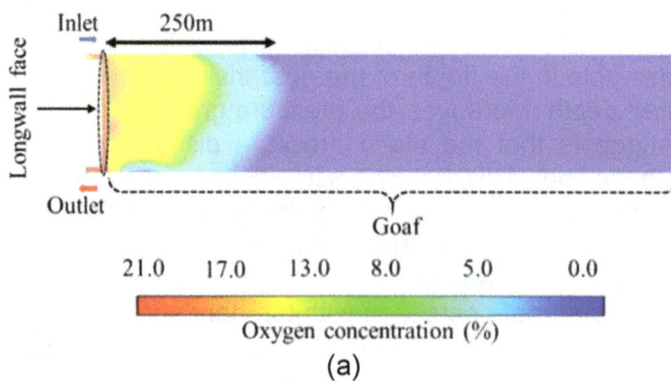

(a)

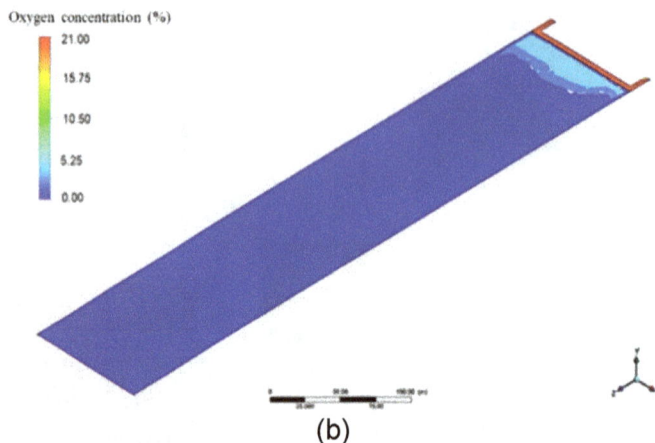

(b)

FIG 7 – Oxygen distribution in the goaf of length 1000 m for (a) 30 m³/min, and (b) 40 m³/min of methane emission rate in goaf.

Further, as make of methane in the goaf was increased up to 40 m³/min by increasing the methane inflow velocity from floor and roof of the goaf, the gas emission in the goaf led to the shrinkage in the oxygen depletion zone. The relevant region was now occupied by the methane and the zone of methane and oxygen diffusion energy exchange was observed in the depth of 50–70 m only (Figure 7b). Thereby, it was concluded that in the presence of high methane emission, the greater risk of fireball driven explosion hazard lies within this depth only.

Localisation of explosive zone in goaf

As illustrated in Figure 8, an increase in inlet air velocity results in the enlargement of the explosive zone. Oxygen ingress into the goaf occurs more prominently from the maingate than the tailgate region, and with an increase in air velocity, the rate of air ingress into the goaf also escalates. It is evident from Figure 8 that the explosive zone expanded up to 100 m in the goaf near the maingate region and 40 m in the tailgate region with the increase in inlet air velocity from 2 to 4 m/s. Thus, the explosive zone existed in the vicinity of the longwall face at various inlet air velocities. Thus, the study identified methane hazards in the longwall panel, with a U-type ventilation system, in the vicinity of the longwall face in the goaf as well as the tailgate.

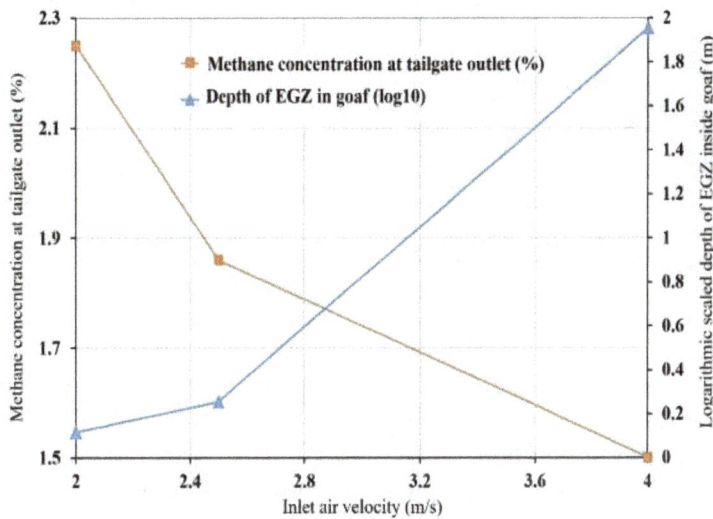

FIG 8 – Variations of explosive zone width and methane concentration at outlet at varying inlet air velocities.

In addition to maingate inlet air velocity, other factors such as the viscous resistance of the goaf (Marts *et al*, 2014b), spatial distribution of goaf permeability (Juganda *et al*, 2020), type of ventilation system adopted (Qin *et al*, 2015), pressure difference across the goaf (Tutak and Brodny, 2015, 2018), and methane inlet rate in the goaf (Queensland Coal Mining Board of Inquiry, 2021) are important factors influencing the formation of the explosive zone in the longwall panel. Furthermore, the discovery of localised methane accumulation in the tailgate corner and the explosive zone inside the goaf necessitates the establishment of real-time mine environmental monitoring stations in the longwall panel as well as auxiliary ventilation systems near the goaf edge for safer longwall mining operations. The methodology used in this study can thus be extended to evaluate the impact of gas drainage and auxiliary ventilation on methane hazards in the longwall panel by considering the aforementioned parameters.

Effect of auxiliary fan in tailgate road on methane dispersion

It can be observed from the Figure 9 that the longwall tailgate corner has a very low air velocity leading to methane enrichment in the region and it's only after a considerable distance into the tailgate that the methane dispersion starts. When analysed with respect to the cascading of the turbulent kinetic energy between the air current, it is observed that the methane dispersion and species transport have a direct correlation. The planes in the figure represent the sections of the tailgate roadway at 5 m intervals and average methane concentration can be observed to be between 1.5–2.0 per cent near the rise side wall, and wherein the velocity currents are abysmally low in the 20 m region out by the longwall face.

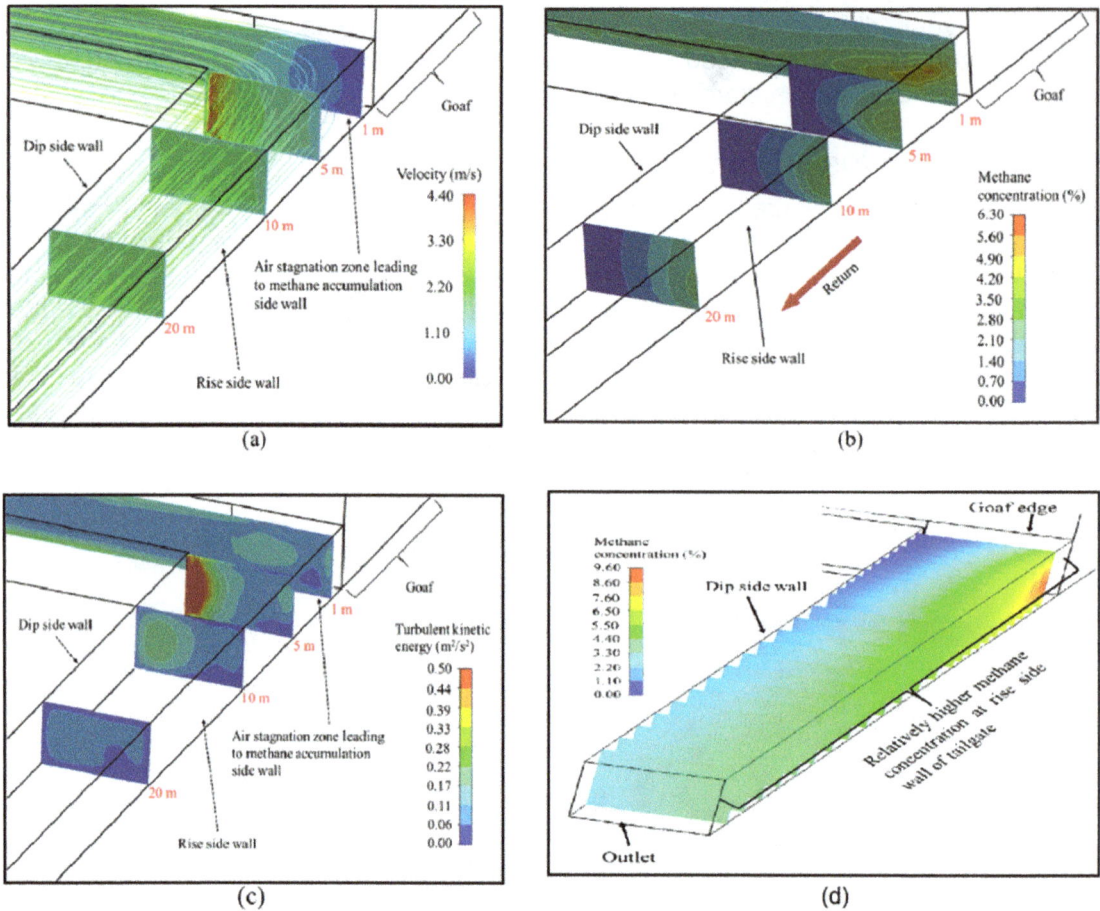

FIG 9 – The variation in different air flow properties and corresponding methane concentration at different positions in the longwall tailgate.

To address the issue of elevated methane concentration near the rise side of the tailgate, an auxiliary ventilation system was introduced in the tailgate (refer Figure 10a). The ventilation arrangement involves usage of L-curtain, auxiliary fan of 74–90 kW and flexible duct of 1000 mm dia. Figure 10b shows that it has substantially reduced the maximum methane concentration in the working zone from 5.68 per cent to 0.78 per cent for air inlet of 2 m/s from the main gate of longwall. Further the maximum methane concentration near the rise side wall has been reduced from 4.8 per cent to 0.5 per cent.

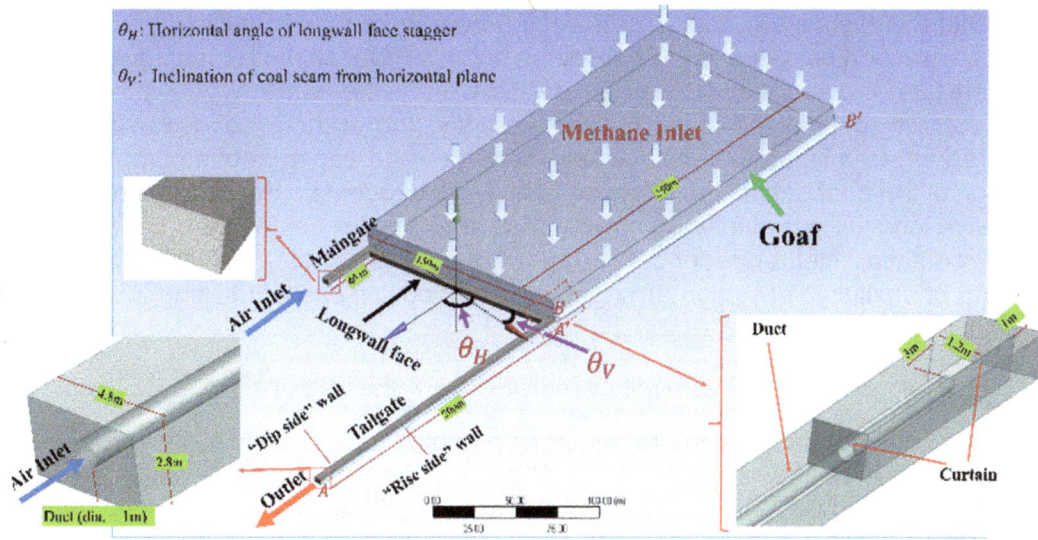

(a)

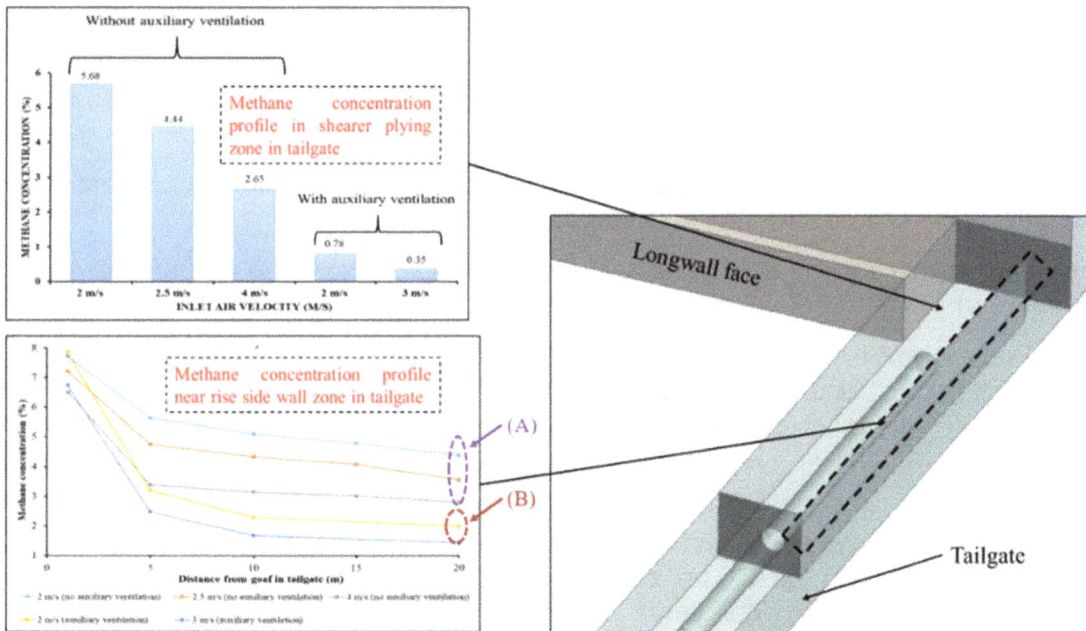

(b)

FIG 10 – (a) The computational model of the longwall panel with auxiliary ventilation arrangement in the tailgate, and (b) its effect under different maingate inlet velocities.

Nitrogen injection in goaf

Yuan and Smith (2014) conducted simulations of nitrogen injection at various locations within the goaf, employing injection rates ranging from 380 cfm to 2000 cfm via seals located at the maingate and boreholes. In a study, nitrogen injection was investigated through a cut-through situated 250 m from the longwall face, as well as through surface boreholes positioned in the tailgate region at depths of 100 m and 600 m, respectively (Qiao *et al*, 2022). Zhou *et al* (2023) explored nitrogen injection from a goaf-side bleeder roadway at a rate of 550 m³/hr, observing the shrinkage of regions prone to spontaneous combustion within the goaf. Similarly, Wang, Qi and Liu (2022) examined nitrogen injection at rates ranging from 550 to 1000 m³/hr within 50 m of the goaf region, noting its impact on the propensity for spontaneous combustion (Wang, Qi and Liu, 2022).

However, this study identifies that elevated methane rates have caused the oxygen depletion zone to shrink from 250 m within the goaf to within 70 m of the longwall face. Furthermore, the longwall panel exhibited roof weighting intervals of 35–40 m, associated with methane emissions in the tailgate that were higher than the average. Previous research has shown that gas drainage wells positioned within this distance in the tailgate region contribute significantly to gas recovery. While the gas drainage suction pressure in the boreholes notably reduces methane concentrations behind the longwall shields, it also leads to high air penetration in the same region, potentially inducing atmospheric volatility behind the shields. Consequently, the authors advocate for prioritising an inertisation program to mitigate hazards arising from gas and spontaneous combustion within the 50–70 m region of the goaf behind the longwall face, while suggesting that the gas recovery program for a progressive longwall panel may be more effective at greater depths than 70 m within the goaf. Further, any potential methane concentration in the tailgate can be mitigated through the implementation of auxiliary ventilation arrangements in longwall panels utilising a U-type ventilation system.

CONCLUSIONS

A computational longwall model incorporating the geo-mining parameter of Moonidih mine, India was developed to study the oxygen dispersion in goaf and consequent methane hazard. The numerical study was validated with field measurements of methane in tailgate of an inactive longwall panel and observed broad agreement despite the limitations of isotropic goaf property and assumption of a model methane inlet roof. The research concluded that:

- The geo-mining parameters distinctly impacted the nature of accumulation of methane in the longwall goaf of the panel.

- The zones of low air velocity have been observed in the goaf. The air moves at a low air velocity range of 0.001–0.0001 m/s at 60 m behind the longwall face in the goaf.

- The methane in goaf settles towards the tailgate of the longwall, and subsequently disperses into the working area through the tailgate.

- The oxygen rich zone in goaf limits itself towards the longwall face with the increase in goaf dimension.

- The formation of explosive zone and its expansion from 5 m to 100 m was observed on maingate side behind the goaf when the inlet air velocity increased from 2 to 4 m/s. The parametric study revealed that at 2.5 m/s maingate inlet velocity, the hazard with respect to explosive gas zone in goaf and methane accumulation in tailgate is minimum of all the variations.

- The usage of auxiliary ventilation in tailgate in the longwall tailgate comprehensively reduced the gas hazard from 1.5–2.0 per cent to less than 0.75 per cent in the longwall tailgate at moderate methane emission rates.

- The normal than average methane emission rates in the longwall goaf appear to shrink the oxygen depletion zone from 250 m to 70 m in the goaf.

- The simulations revealed that inertisation program shall be taken up within 70 m of the face in the goaf and gas recovery through drainage will be taken up at depth of more than 100 m in the goaf, if necessary.

The study advocates to develop a comprehensive gas-drainage and nitrogen inertisation plan for longwall panels designed for greater depths for a safe and productive mining prospect.

REFERENCES

ANSYS, Inc., 2013. ANSYS FLUENT Theory Guide, 14.5 edition, SAS IP, Inc., Canonsburg, PA.

Behera, B, Yadav, A, Singh, G S P and Sharma, S K, 2019. Numerical modeling study of the geo-mechanical response of strata in longwall operations with particular reference to Indian geo-mining conditions, Rock Mech Rock Eng 53:1827–1856. https://doi.org/10.1007/s00603-019-02018-w

Esterhuizen, G and Karacan, C, 2007. A Methodology For Determining Gob Permeability Distributions And Its Application To Reservoir Modeling Of Coal Mine Longwalls, Society for Mining, Metallurgy and Exploration (US), National Institute for Occupational Safety and Health (NIOSH), Denver.

Gilmore, R C, Marts, J A, Brune, J, Bogin, G, Grubb, J W and Saki, S A, 2014. CFD modeling explosion hazards-bleeder vs. progressively sealed gobs, in *10th International Mine Ventilation Congress,* 7 p (The Mine Ventilation Society of South Africa). https://doi.org/10.13140/RG.2.2.28080.56322

Hoek, E and Bray, J D, 1981. *Rock Slope Engineering*, third edition (The Institution of Mining and Metallurgy: London).

Juganda, A, Strebinger, C, Brune, J F and Bogin, G E, 2020. Discrete modeling of a longwall coal mine gob for CFD simulation, *Int J Min Sci Technol*, 30:463–469. https://doi.org/10.1016/j.ijmst.2020.05.004

Karacan, C Ö, Ruiz, F A, Cotè, M and Phipps, S, 2011. Coal mine methane: A review of capture and utilization practices with benefits to mining safety and to greenhouse gas reduction, *Int J Coal Geol*, 86:121–156. https://doi.org/10.1016/j.coal.2011.02.009

Kumar, P, Mishra, D P, Panigrahi, D C and Sahu, P, 2017. Numerical studies of ventilation effect on methane layering behaviour in underground coal mines, *Curr Sci*, 112:1873–1881. https://doi.org/10.18520/cs/v112/i09/1873-1881

Li, Y, Su, H, Ji, H and Cheng, W, 2020. Numerical simulation to determine the gas explosion risk in longwall goaf areas: A case study of Xutuan Colliery, *Int J Min Sci Technol*, 30:875–882. https://doi.org/10.1016/j.ijmst.2020.07.007

Marts, J A, Gilmore, R C, Brune, J and Bogin, G, 2014a. Accumulations of Explosive Gases in Longwall Gobs and Mitigation through Nitrogen Injection and Face Ventilation Method, in 6th Aachen International Mining Symposia, Germany.

Marts, J, Gilmore, R, Brune, J, Jr, G B, Grubb, J and Saki, S, 2014b. Accumulations of Explosive Gases in Longwall Gobs and Mitigation through Nitrogen Injection and Face Ventilation Method, in *6th Aachen International Mining Symposia (AIMS)*, pp 1–13.

Pappas, D and Mark, C, 1993. Behavior of simulated longwall gob material, Report of Investigations/1993.

Qiao, M, Ren, T, Roberts, J, Yang, X, Li, Z and Wu, J, 2022. New insight into proactive goaf inertisation for spontaneous combustion management and control, *Process Safety and Environmental Protection*, 161:739–757. https://doi.org/10.1016/j.psep.2022.03.074

Qin, Z, Yuan, L, Guo, H and Qu, Q, 2015. Investigation of longwall goaf gas flows and borehole drainage performance by CFD simulation, *Int J Coal Geol*, 150–151:51–63. https://doi.org/10.1016/j.coal.2015.08.007

Queensland Coal Mining Board of Inquiry, 2021. Queensland Coal Mining Board: Queensland Coal Mining Board of Inquiry: Report Part II, May 2021.

Rao, S, Mishra, D P and Mishra, A, 2023. Methane migration and explosive fringe localisation in retreating longwall panel under varied ventilation scenarios: a numerical simulation approach, *Environmental Science and Pollution Research*, 30:66705–66729. https://doi.org/10.1007/s11356-023-26959-6

Ray, S K, Khan, A M, Mohalik, N K, Mishra, D, Mandal, S and Pandey, J K, 2022. Review of preventive and constructive measures for coal mine explosions: An Indian perspective, *Int J Min Sci Technol*, 32:471–485. https://doi.org/10.1016/j.ijmst.2022.02.001

Tutak, M and Brodny, J, 2015. Numerical analysis of airflow and methane emitted from the mine face in a blind dog heading, *Management Systems in Production Engineering*, 4:175–178. https://doi.org/10.12914/MSPE

Tutak, M and Brodny, J, 2018. Analysis of the impact of auxiliary ventilation equipment on the distribution and concentration of methane in the tailgate, *Energies (Basel)*, 11. https://doi.org/10.3390/en11113076

Wang, W, Qi, Y and Liu, J, 2022. Study on multi field coupling numerical simulation of nitrogen injection in goaf and fire-fighting technology, *Sci Rep*, 12:17399. https://doi.org/10.1038/s41598-022-22296-9

Yuan, L and Smith, A C, 2014. CFD modelling of nitrogen injection in a longwall gob area, *Int J Min Miner Eng*, 5:164. https://doi.org/10.1504/IJMME.2014.060220

Zhou, X, Jing, Z, Li, Y and Bai, G, 2023. Study on Nitrogen Injection Fire Prevention and Extinguishing Technology in Spontaneous Combustion Gob Based on Gob-Side Entry Retaining, *ACS Omega*, 8:30569–30577. https://doi.org/10.1021/acsomega.3c03879

Suppression of coal dust explosion using limestone and dolomite rock dust inertants

A Sahu[1] and D P Mishra[2]

1. Research Scholar, Department of Mining Engineering, Indian Institute of Technology (Indian School of Mines), Dhanbad, Jharkhand-826 004, India. Email: aashishsahu6@gmail.com
2. Associate Professor and Head, Department of Mining Engineering, Indian Institute of Technology (Indian School of Mines), Dhanbad-826 004, Jharkhand, India. Email: dpmishra@iitism.ac.in

ABSTRACT

Coal dust explosions pose a significant safety concern in underground coalmines. Therefore, prevention and suppression of coal dust explosions in underground coalmines is of paramount significance. This paper experimentally determines the rock dust inertant (limestone and dolomite) requirements for the suppression of coal dust explosion using a Godbert-Greenwald (G-G) furnace. The explosion suppression efficacy of rock dusts is examined at varied rock dust particle sizes of <75, <125, <250, <500, and <850 μm, and coal dust particle sizes of <38, 38–75, 75–125, 125–250, and <850 μm (mine size dust). The detailed characterisation of coal and rock dusts, such as proximate and ultimate analyses of coal, and X-ray diffraction (XRD) analysis, are conducted. A high resolution 4K high-speed camera was used with the G-G furnace to examine the flame suppression behaviour of coal dust explosion. The finer coal dusts (<38 μm) are found to be more explosible due to their greater exposed surface area, which required about 84 per cent of limestone and 85 per cent of dolomite rock dust for explosion suppression. The quantity of rock dust requirement steadily increased with increase in the rock dust particle size. With increase in rock dust particle size from <75 to <850 μm, the average rock dust requirement for suppression of coal dust explosion increased from 64 to 80 per cent and 66 to 81 per cent for limestone and dolomite dust, respectively. The finer rock dusts yielded better explosion suppression efficacy. Overall, the limestone dust is proven to be a better explosion suppressant than the dolomite dust. This study helped in better understanding of the mechanism of coal dust explosion and its suppression by choosing the right type, size, and quantity of rock dust inertants.

INTRODUCTION

Coal is one of the widely used fossil fuels for the generation of electricity. The global coal demand is unceasingly rising. After China, India is the second-largest coal producer and consumer across the globe. Coal mining is a hazardous profession due to the prevalence of various accidents and health hazards. Coal dust generated during coal mining operations is regarded as the most harmful material from the viewpoint of occupational health and safety of the miners. The amount of coal dust generated during mining operations is estimated to be around 3 per cent of the total mass of the excavated coal (Shekarian *et al*, 2021). When fine coal dust is raised in a confined underground coalmine environment, it causes two major hazards, *viz* coal dust explosion and coal workers' pneumoconiosis (CWP). Out of which, coal dust explosions are a major safety concern for underground coalmines and process industries. In the last few decades, the occurrence of mine explosions worldwide has attracted much attention due to the loss of miners' lives and mine properties (Sahu and Mishra, 2023). Hence, considering its significance for the mining industry and society, more impetus should be given on the coal dust explosion studies to curb the explosion hazard.

Numerous studies have been conducted on coal dust explosions using laboratory chambers and in experimental mines. Yet, coal dust explosions frequently occur worldwide and are still a major concern for the coalmine management. For the prevention and suppression of coal dust explosions in coalmines, understanding of explosion characteristics, ie likelihood and consequences of explosions is of paramount importance. Several researchers have investigated the ignition sensitivity and severity of coal dust explosion using different laboratory testing chambers (Cashdollar and Chatrathi, 1993; Addai, Gabel and Krause, 2016; Tan *et al*, 2020; Zhao *et al*, 2021; Sahu and Mishra, 2022; Liu *et al*, 2023a; Liu *et al*, 2023b). Particle size of coal dust is one of the significant parameters which influence the explosion characteristics of coal dust (Mishra and Azam, 2018; Sahu and Mishra, 2022). Finer coal dust particles are more sensitive to ignition, which results in devastating explosions (Addai *et al*, 2017; Cloney *et al*, 2018; Bagaria *et al*, 2019). The ignition sensitivity and explosion severity increase with decrease in the coal dust particle size (Zlochower *et al*, 2018). Moreover, a number of factors, including coal dust concentration, particle size, moisture content, volatile matter content, incombustible content, oxygen concentration, and the existence of firedamp, affect the explosion characteristics of coal dust clouds (Cashdollar, 2000; Going, Chatrathi and Cashdollar, 2000; Sapko, Cashdollar and Green, 2007; Man and Gibbins, 2011; Cao *et al*, 2012; Yuan *et al*, 2012; Mittal, 2013; Ajrash, Zanganeh and Moghtaderi, 2016; Arshad, Taqvi and Buang, 2021).

Suppression of coal dust explosions in underground coalmines is achieved by mixing the coal dust with solid inertant and installing explosion barriers in the mine galleries. The solid inertants used as explosion barriers include rock dust, chemical powders, or composites of rock dust, and chemical powders (Amyotte, 2006; Zhang *et al*, 2021). The limestone and dolomite rock dust inertants are most widely used in underground coalmines for explosion suppression (Amyotte, Mintz and Pegg, 1992; Reddy, Amyotte and Pegg, 1998; Man and Teacoach, 2009; Azam and Mishra, 2019). However, nowadays, chemical inhibitors are most popular among researchers, and several laboratory experiments have been carried out to evaluate their efficacy for explosion suppression (Jiang, Bi and Gao, 2020; Wang *et al*, 2020; Huang *et al*, 2021; Zhang *et al*, 2021; Zhong *et al*, 2022). The finer rock dust inertants are more effective for suppressing coal dust explosions due to their better dispersibility and greater specific surface area, and the same has been demonstrated in previous studies (Amyotte, 2006; Harris *et al*, 2015; Azam and Mishra, 2019; Sahu and Mishra, 2024). The mechanism of explosion suppression using rock dust inertants has been well explained by Sahu and Mishra (2024).

Several studies have been conducted on rock dust inertant requirements for the suppression of coal dust explosions. However, limited studies have been done on the effect of rock dust type, rock dust size, and rock dust proportion on flame suppression behaviour of coal dust explosion. Such a study is particularly sparse for Indian coal. Hence, there is an urgent need for augmenting research on this important topic for developing effective measures for the prevention and suppression of coal dust explosions in underground coalmines. This study investigates the flame propagation and suppression behaviour of coal-rock dust mixtures at varied rock dust proportions and particle sizes using G-G furnace in conjunction with high-speed camera. Moreover, the rock dust inertant requirement for complete suppression of coal dust explosion have been determined by varying the rock dust and coal dust particle sizes. Limestone and dolomite dusts were chosen to assess the coal dust explosion suppression efficacy. The particle size of coal dust and rock dust inertants were varied in the ranges of <75, to <850 µm, and <38 to <850 µm, respectively. This study led to the better understanding of explosion flame suppression behaviour by different proportions and particle sizes of rock dust inertants. The results obtained from this study will be useful for preventing and suppressing coal dust explosions in underground coalmines and related industries.

MATERIALS AND METHODS

Sample collection and preparation

The coal sample for this study was collected from the Karo Seam of Dhory Colliery, where India's most disastrous explosion occurred in 1965, claiming 268 lives and mine properties. Dhory Colliery is a part of the Central Coalfields Limited (CCL), a subsidiary of Coal India Limited (CIL), located in Jharia Coalfield of Jharkhand state, India. The channel sampling method was adopted for the collection of coal samples from the freshly exposed mining face. For the suppression of coal dust explosion, two different types of rock dusts, *viz* limestone ($CaCO_3$) and dolomite [$CaMg(CO_3)_2$], were collected from Arasmeta Limestone Mine of M/s Nuvoco Vistas Corporation Limited, and Hirri Dolomite Mine owned by Bhilai Steel Plant (BSP) of Steel Authority of India Limited (SAIL), respectively. The rock samples were collected from the mines using the grab sampling method.

The particle size of solid inertant is one of the most influencing factors for suppressing coal dust explosions. Therefore, the dolomite and limestone samples were prepared in five different sizes, such as <75, <125, <250, <500, and <850 µm, to evaluate the effect of particle size.

Sample characterisation

The chemical and mineralogical composition of coal and rock dusts plays a significant role in the explosion characteristics and its suppression mechanism (Amyotte, 2006; Azam and Mishra, 2019). The results of the ultimate and proximate analyses of the coal samples are presented in Table 1. The ultimate analysis was conducted to determine the carbon (C), hydrogen (H), nitrogen (N), sulfur (S), and oxygen (O) content of the coal samples as per ASTM:D3176–15, and the proximate analysis was conducted to determine the moisture (M), ash (A), volatile matter (VM), and fixed carbon (FC) contents as per ASTM:D7582–15.

TABLE 1

Results of ultimate and proximate analyses of the coal sample

Ultimate analysis (wt%)					Proximate analysis (wt%)				
C_{ad}	H_{ad}	N_{ad}	S_{ad}	O_{ad}	M_{ad}	A_{ad}	VM_{ad}	FC_{ad}	FR (FC_{ad}/VM_{ad})
74.66	3.97	1.18	0.65	19.74	1.052	19.65	19.33	60.43	3.13

Note: ad – As determined basis.

The X-ray diffraction (XRD) method was used to determine the mineralogy of the coal and rock dust samples. The XRD patterns of coal, limestone, and dolomite dusts are shown in Figure 1. The crystalline phases of coal and rock dust samples were identified by peak positions and intensities using the standard database. The major common minerals found in coal samples are identified as quartz (SiO_2), alumina (Al_2O_3), chromium oxide (Cr_2O_3), and hematite (Fe_2O_3). The most intense peak appeared at $2\theta = 26.65°$ was identified as quartz (SiO_2). Similarly, the major common minerals found in limestone and dolomite rocks are calcite ($CaCO_3$), quartz (SiO_2), alumina (Al_2O_3), and chromium oxide (Cr_2O_3). Moreover, other minerals present in limestone and dolomite dusts are sodium chloride (NaCl), hematite (Fe_2O_3), chromium (Cr), and pyrite (CaF_2). The most intense peaks identified at $2\theta = 29.4°$ and $31.6°$ for limestone and dolomite rock dusts are recognised as calcite and sodium chloride.

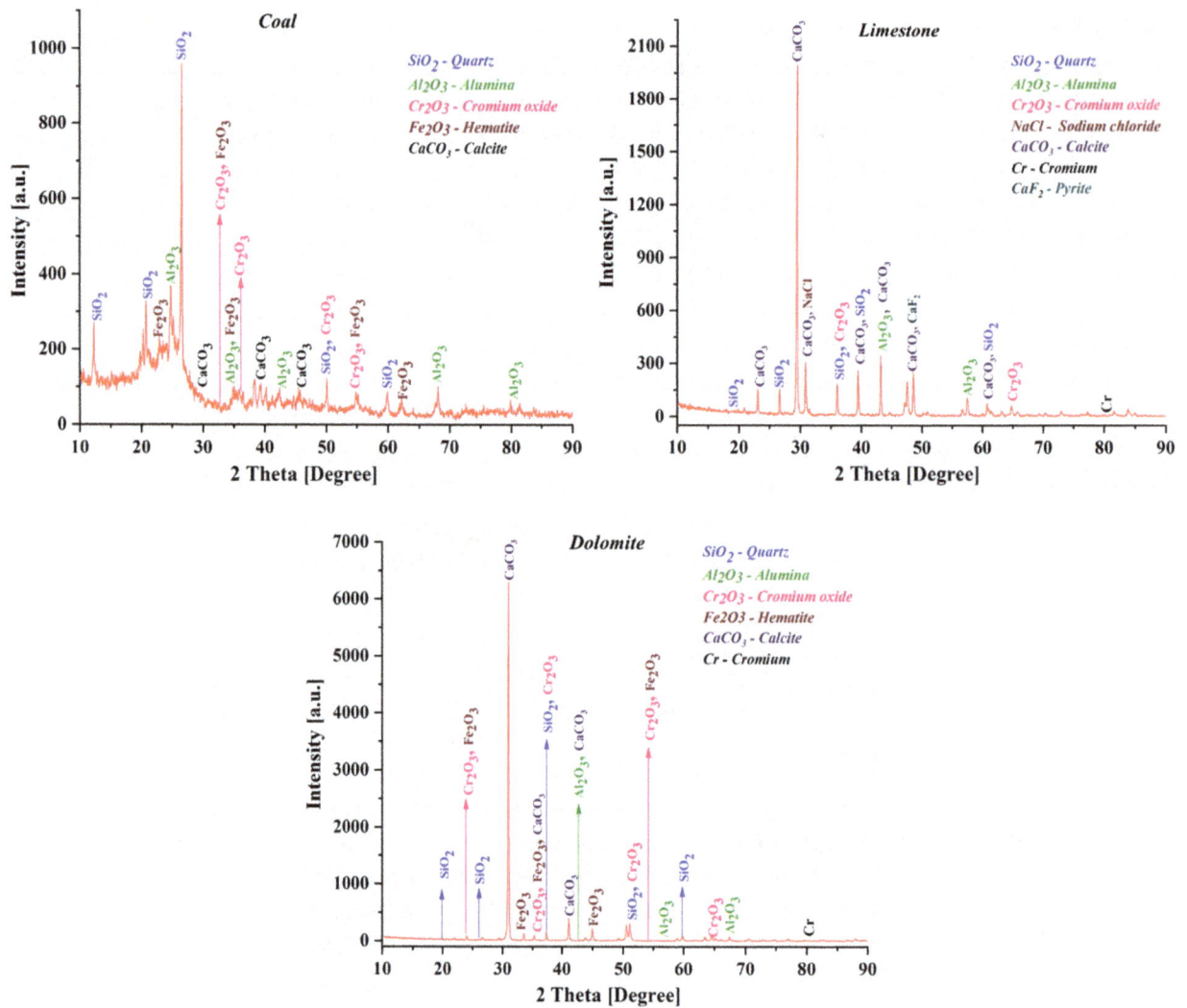

FIG 1 – XRD patterns of different coal and rock samples.

Determination of inertant requirement and explosibility index

Experimental set-up

The G-G furnace was used to determine the ignition sensitivity, solid inertant requirements for the suppression of coal dust explosion, and the explosibility index of coal dust clouds (Azam and Mishra, 2019; Sahu and Mishra, 2024). The experimental set-up of the G-G furnace in conjunction with the high-speed camera is depicted in Figure 2. The high-speed camera was placed in front of the G-G furnace to record the flame propagation and suppression behaviour of coal dust clouds. More information about the experimental set-up has been well described by Sahu and Mishra (2024, 2022).

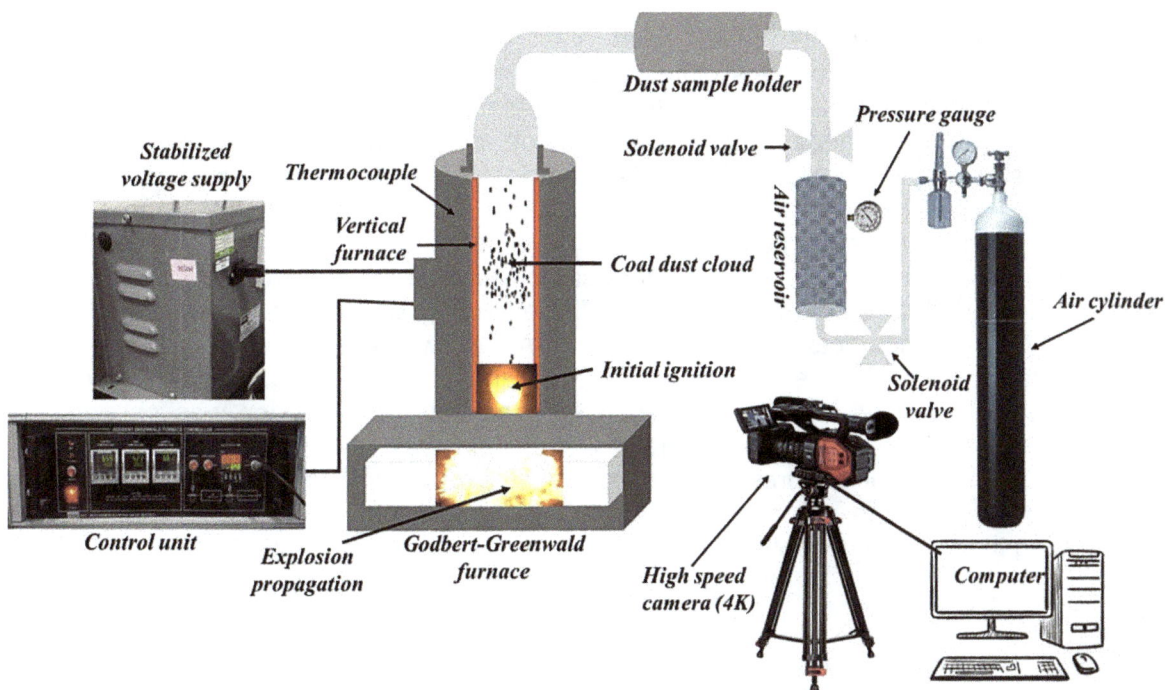

FIG 2 – Experimental set-up of G-G furnace integrated with a high-speed camera.

Experimental procedure

Flame suppression behaviour by rock dust inertant

The proportion of solid inertant in the coal dust significantly affects the flame propagation and suppression behaviour of coal dust explosion. In order to assess the flame suppression behaviour by rock dust inertant, a high-speed camera was positioned in front of the G-G furnace to monitor the intensity of emerging flame in the steel mirror placed at the bottom of the furnace. Firstly, the coal and rock dust of specific particle sizes were thoroughly mixed at a predetermined proportion. The pre-weighed coal-rock dust mixture was kept in the sample holder of G-G furnace. Then, the coal-rock dust mixture was dispersed into the pre-heated furnace at the desired dust dispersion pressure. The emerging flame from the furnace was recorded in the high-speed camera. All these experimental scenarios are repeated for different particle sizes and rock dust proportions, as presented in Table 2. Later, the frames from the recorded videos were extracted using the video-to-JPG converter software. The flame propagation and suppression behaviour were analysed through video frames captured at certain intervals after the first explosion.

TABLE 2

Experimental parameters matrix for flame suppression behaviour of coal dust explosion.

Cases	Parameter variation	Fixed parameters
Case 1: Effect of rock dust particle size (µm)	<75, <125, <250, <500 and <850 µm	T_i: 800°C C_d: 2000 g/m³ P_d: 50 kPa R_{dp}: 20%
Case 2: Effect of rock dust proportion (%)	10–50% at 10°C intervals	D_p: <250 µm C_d: 2000 g/m³ P_d: 50 kPa T_i: 800°C

T_i – Ignition temperature, D_p – Rock particle diameter, C_d – Dust concentration, P_d – Dust dispersion pressure, R_{dp} – Rock dust proportion.

Inertant requirement and explosibility of coal dust

The solid inertant requirement for the suppression of coal dust explosion and the explosibility index of coal dust reflects the degree of severity and intensity of the explosion. The solid inertant requirement and explosibility index of coal dust were determined using the G-G apparatus shown in Figure 2. Firstly, the vertical furnace was preheated at 800°C. The coal and rock dust were thoroughly mixed at a certain proportion. A pre-weighed coal-rock dust mixture was kept in the sample holder of the apparatus, and the air reservoir was charged with air at the desired dust dispersion pressure. Then, the coal dust-inertant sample was dispersed into the preheated vertical furnace with the help of a remote trigger switch. The emerging explosion flame was seen through the steel mirror placed at the bottom of the vertical furnace. Two conditions, either explosion or no explosion, were observed. If an explosion was noticed, the solid inertant proportion in the mixture was progressively increased, and the test was repeated until no explosion was observed. The lastly determined inertant proportion, where no explosion was observed, was considered the inertant requirement (S) for suppressing the coal dust explosion. The procedure was repeated three to five times for each set of experiments. The explosibility index of the coal dusts was calculated using the following equation (Ramlu, 1991):

$$Explosibility\ index = \frac{100}{100-S} \tag{1}$$

where, S is the proportion of rock dust inertant in the mixture that did not yield any explosion. This index also indicates the amount of solid inertant required to prevent the coal dust explosion.

RESULTS AND DISCUSSION

Flame suppression behaviour

Effect of rock dust proportion

The proportion of rock dust inertant mixed with coal dust is vital for suppressing the coal dust explosion. Even a small proportion of rock dust may be effective in lessening the severity of explosion in terms of flame temperature, propagation speed, explosion pressure, and rate of pressure rise. Also, the ignition sensitivity of coal-rock dust mixture is greatly affected by the rock dust proportions (Sahu and Mishra, 2024). The flame suppression behaviour at different proportions of rock dust is depicted in Figure 3. Here, the proportion of rock dust inertant was varied in the range of 10–50 per cent at an interval of 10 per cent, while keeping the rock dust particle size, ignition temperature, dust concentration, and dust dispersion pressure fixed at <250 μm, 800°C, 2000 g/m³, and 50 kPa, respectively. A 4K resolution high-speed camera was used to record the explosion frame at various time intervals to enhance understanding of the explosion propagation and suppression behaviour. The explosion visuals depicted in Figure 3 were captured at different time intervals, beginning with the initial ignition (0 ms) followed by 10 ms, 30 ms, 50 ms, and 70 ms. The figure shows how the intensity of flame propagation is lessened by increasing the rock dust proportion in the coal-rock dust mixture. When the rock dust proportion increased to 30 per cent, the explosion is suppressed to some extent, because the severity and intensity of explosion decreased and its propagation became insignificant after 70 ms. Similarly, when the rock dust proportion was increased to 40 and 50 per cent, the ignition of coal-rock dust mixture occurred, but the flame propagation was completely insignificant because the explosion flame was suppressed by the rock dust inertant due to heat sink and the blanketing effect. The coal-rock dust mixture did not ignite when the proportion of rock dust inertant was 70 per cent or above. However, any proportion of rock dust resulted in the reduction of flame propagation by decreasing the severity and ignition sensitivity of explosion.

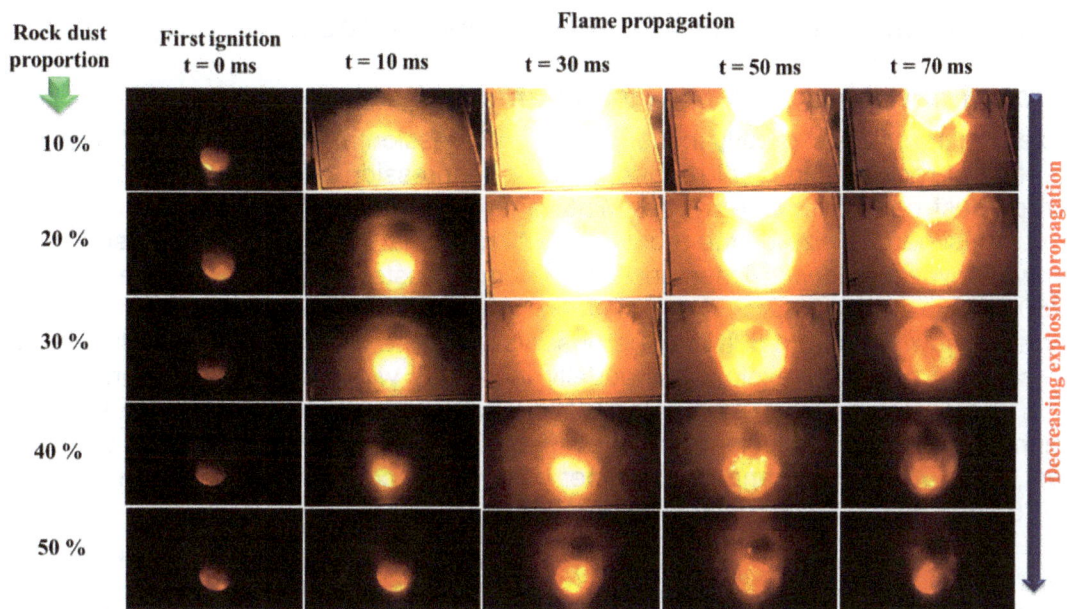

FIG 3 – Explosion flame suppression behaviour at different rock dust proportions.

Effect of rock dust particle size

The suppression mechanism of coal dust explosion is greatly influenced by the particle size of rock dust. Particle size significantly affects the inertant requirement for the suppression of coal dust explosion (Azam and Mishra, 2019). The flame suppression behaviour at different proportions of rock dust inertants is depicted in Figure 4. Here, the particle size of rock dust inertant was varied as <75, <125, <250, <500, and <250 µm, while keeping the rock dust proportion, ignition temperature, dust concentration, and dust dispersion pressure fixed at 20 per cent, 800°C, 2000 g/m^3, and 50 kPa, respectively. The figure demonstrates that the finer rock dust inertant (<75 µm) has better explosion suppression efficacy, even at a smaller rock dust proportion (20 per cent). The explosion flame is significantly diminished after 50 ms from the initiation. This phenomenon occurs because the finer particles are easily dispersed in the air exposing a higher specific surface area towards explosion flame, which thereby acting as a better heat sink and blanket. When the rock dust particle size increases from <125 µm onwards, the explosion suppression is insignificant at a smaller rock dust proportion (20 per cent). However, when the rock dust proportion increased, the explosion intensity is significantly reduced, as depicted in Figure 4, and thereby the explosion flame is suppressed.

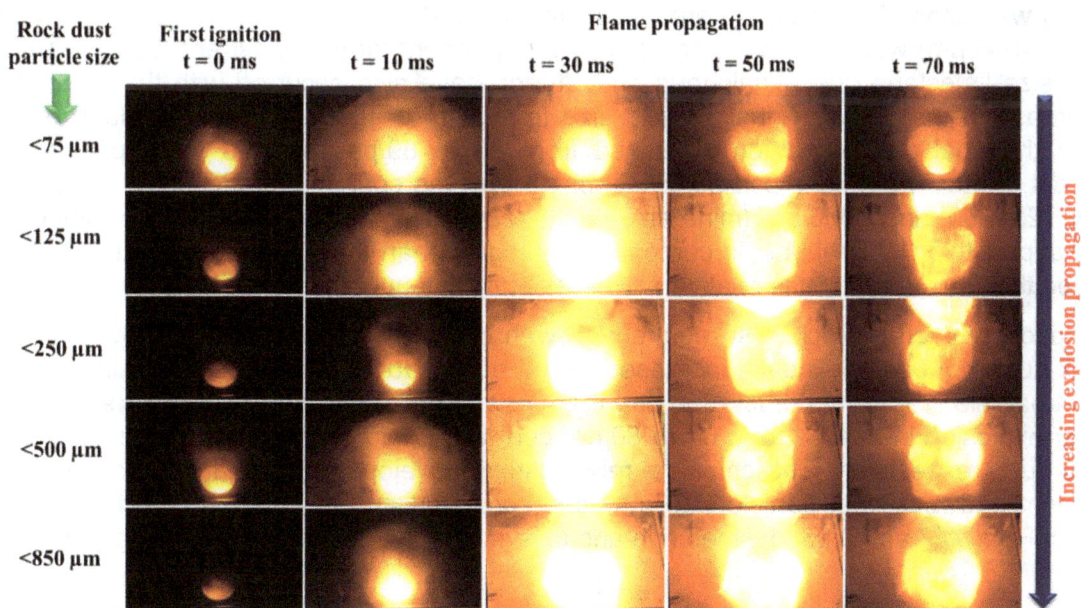

FIG 4 – Explosion flame suppression behaviour at different particle sizes of rock dust inertant.

Rock dust inertant requirement for suppression of coal dust explosion

Inertant requirement based on rock dust particle size

The particle size of rock dust greatly influences the suppression mechanism of coal dust explosion. The proportion of rock dust inertant requirement for suppression of coal dust explosion increases with the rock dust particle size (Amyotte, Mintz and Pegg, 1992; Azam and Mishra, 2019). The effect of rock dust particle size on the requirement of rock dust inertant for suppression of coal dust explosion was investigated with limestone and dolomite dusts of different particle sizes, *viz* <75, <125, <250, <500, and <850 μm. The variation of rock dust inertant requirement with rock dust particle size is shown in Figure 5. It may be observed that the proportion of rock dust requirement for suppression of coal dust explosion gradually increases with the rock dust particle size. With increase in the limestone dust particle size from <75 to <850 μm, the rock dust required for suppression of coal dust explosion increased from 79 to 88 per cent, 77 to 87 per cent, 70 to 87 per cent, 48 to 71 per cent, and 48 to 65 per cent for the coal dust particle size of <38, 38–75, 75–125, 125–250, and <850 μm (mine size dust), respectively.

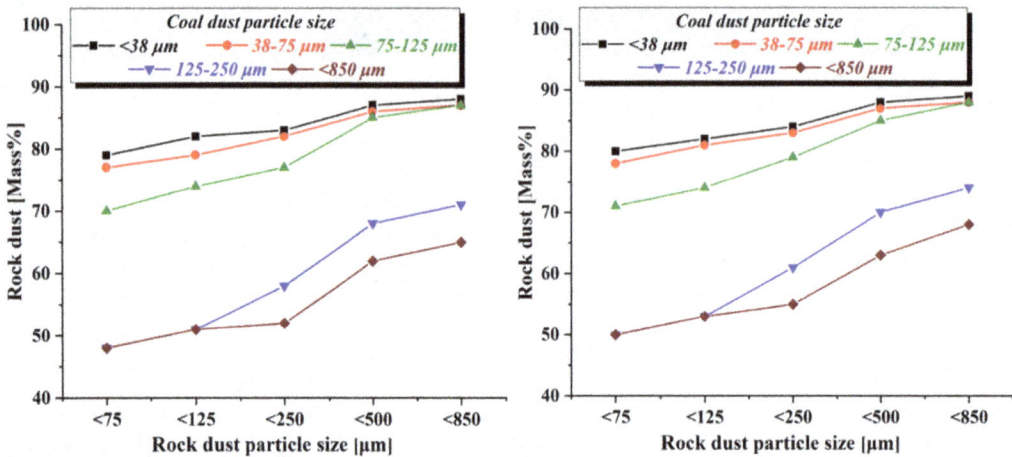

FIG 5 – Variation of rock dust inertant requirement with particle size of (left) limestone dust and (right) dolomite dust.

This analysis revealed that the proportion of limestone dust required to suppress the coal dust explosion is highest for the finest coal dust particle (<38). About 79 per cent finest limestone dust (<75 μm) was required to suppress the coal dust explosion involving the finest coal dust, whereas the limestone dust requirement increased to 88 per cent when the coarsest limestone dust (<850 μm) was used. The minimum amount of limestone dust required to suppress the mine-size coal dust (<850 μm) was determined to be about 48 per cent for the finest and 65 per cent for the coarsest size limestone dust. Analogous phenomena have also occurred with the dolomite dust. With increase in the dolomite dust particle size from <75 to <850 μm, the rock dust required to suppress the coal dust explosion increased from 80 to 89 per cent, 78 to 88 per cent, 71 to 88 per cent, 50 to 74 per cent, and 50 to 68 per cent for the coal dust particle size of <38, 38–75, 75–125, 125–250, and <850 μm (mine size dust), respectively. The requirement of dolomite rock dust for suppression of coal dust is comparatively higher than that of limestone dust. Hence, it is concluded that limestone dust is relatively more effective in suppressing coal dust explosions.

The variation of average rock dust requirement with rock dust particle size is shown in Figure 6. The rock dust requirement linearly increased with the rock dust particle size. With increase in the rock dust particle size from <75 to <850 μm, the average rock dust requirement for suppression of coal dust explosion increased from 64 to 80 per cent and 66 to 81 per cent for the limestone and dolomite dust, respectively. The empirical relationships developed between the rock dust inertant requirement and rock dust particle size through regression analysis, which best fitted the linear curve with $R^2 = 0.97$ and 0.98 for limestone and dolomite rock dust, respectively, are given below:

$$R_L = 4.06D_L + 59.7 \tag{2}$$

$$R_D = 4.12D_D + 61 \tag{3}$$

where, R_L, and R_D are rock dust requirement (mass per cent) of limestone and dolomite dust, and D_L and D_D are rock dust particle size (µm) of limestone and dolomite dust, respectively. These equations can predict the rock dust requirement for the suppression of coal dust explosion with known rock dust particle size up to <850 µm.

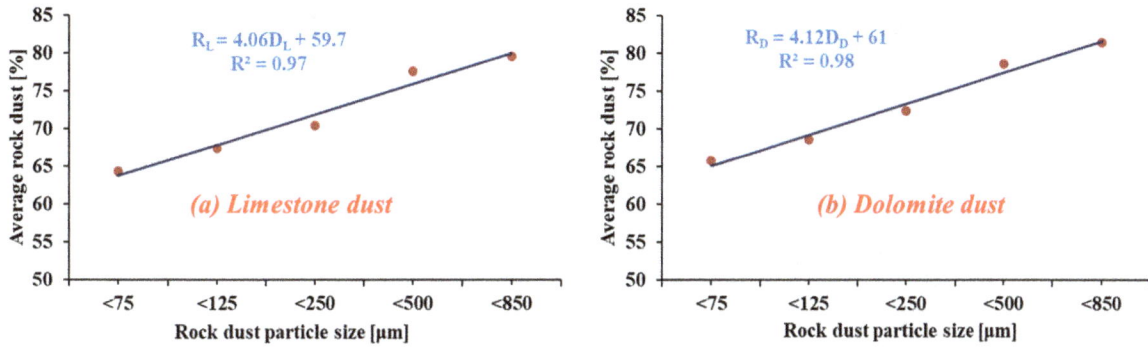

FIG 6 – Variation of average rock dust inertant requirement with particle size of (a) limestone dust and (b) dolomite dust.

Inertant requirement based on coal dust particle size

It is well known that fine coal dust is more explosive than coarser ones due to their greater exposed surface area and lesser energy requirement for devolatilisation (Cao *et al*, 2012; Zlochower *et al*, 2018). The rock dust inertant requirement for suppression of coal dust explosion increases with decrease in the particle size of coal dust (Azam and Mishra, 2019). In order to investigate the effect of coal dust particle size on rock dust inertant requirement, the coal dust particle size was varied as <38, 38–75, 75–125, 125–250, and <850 µm (mine size dust). Figure 7 shows the variation of rock dust inertant requirement with coal dust particle size. The rock dust requirement for suppression of coal dust explosion decreased with increase in the particle size of coal dust. When the coal dust particle size increased from <38 to <850 µm, the rock dust requirement decreased from 79 to 48 per cent, 82 to 51 per cent, 83 to 53 per cent, 87 to 62 per cent, and 88 to 65 per cent for limestone dust particle size of <75, <125, <250, <500, and <850 µm, respectively. Similarly, the dolomite rock dust requirement decreased from 80 to 50 per cent, 82 to 53 per cent, 84 to 55 per cent, 88 to 63 per cent, and 89 to 68 per cent for the dolomite dust particle size of <75, <125, <250, <500, and <850 µm, respectively. The finest coal dust (<38 µm) required highest quantity of rock dust inertant to suppress the coal dust explosion, which varied in the range of 79–88 per cent and 80–89 per cent for different particle sizes of limestone and dolomite dust, respectively. Whereas, the mine size coal dust (<850 µm) required the lowest quantity of rock dust inertant for suppression of coal dust explosion, which varied in the range of 48–65 per cent and 50–68 per cent for different particle sizes of limestone and dolomite dust, respectively.

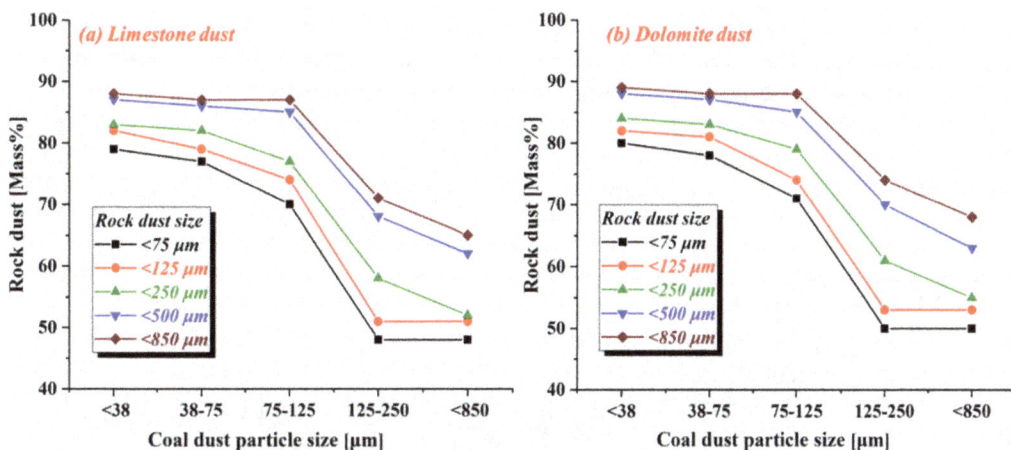

FIG 7 – Variation of rock dust inertant requirement of (a) limestone dust and (b) dolomite dust, with coal dust particle size.

The variation of average rock dust requirement with the particle size of coal dust shown in FIG 8 depicts the decreasing trend of rock dust requirement with coal dust particle size. With increase in the coal dust size from <38 to <850 µm, the average rock dust requirement decreased from 84 to 55 per cent and 85 to 58 per cent for limestone and dolomite rock dust, respectively. The following relationships between the rock dust inertant requirement and particle size of coal dust were obtained, which best fitted the polynomial curve with $R^2 = 0.92$ and 0.92 for dolomite and limestone dust, respectively:

$$R_L = -1.4143(D_C)^2 + 0.5457D_C + 85.8 \tag{4}$$

$$R_D = -1.3571(D_C)^2 + 0.6029D_C + 86.48 \tag{5}$$

where, R_L, and R_D are rock dust requirement (mass per cent) of limestone and dolomite dust, and D_C is coal dust particle size (µm). These empirical equations can be used to predict the rock dust requirement for the suppression of coal dust explosion with known coal dust particle size up to <850 µm.

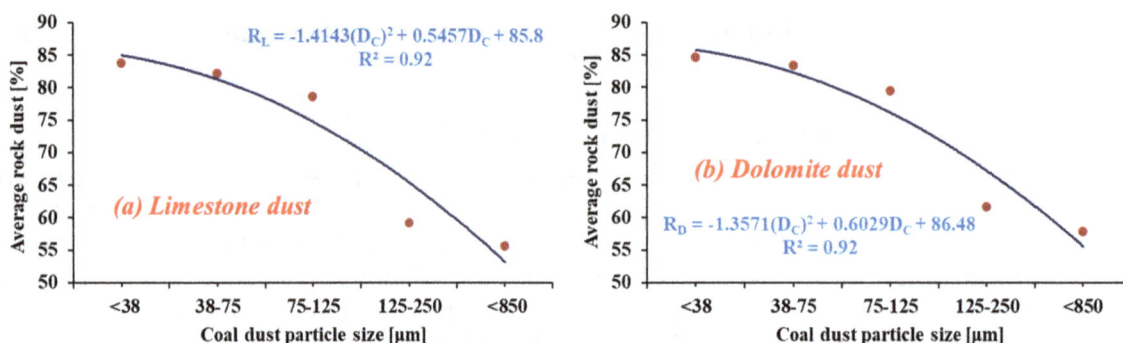

FIG 8 – Variation of average rock dust inertant requirement of (a) limestone dust and (b) dolomite dust, with coal dust particle size.

LIMITATION AND SCOPE OF THE STUDY

This paper investigates the suppression of coal dust explosion at different experimental conditions. Despite its thorough execution, the study has certain limitations. The studies were carried out in a small-scale laboratory chamber (G-G furnace) that simulated explosion suppression in underground coalmines. The coal and solid inertant dispersion inside the laboratory chamber were assumed to be homogeneous. Moreover, the empirical equations developed for average rock dust requirement for suppression of coal dust explosions are applicable for a varied range of experimental conditions performed in this study. The explosion suppression behaviour was analysed based on visualisations of the explosion flame propagation, which were taken by the high-speed camera. The flame temperature and flame propagation velocity were not measured in this study. Despite these limitations, the laboratory tests rather accurately estimated the amount of rock dust needed for the suppression of coal dust explosions.

This study provides a better understanding of the suppression mechanism of coal dust explosions using rock dust inertants. The findings of this research will be useful in deciding the type, quantity, and size of rock dust inertant that is most effective for preventing coal dust explosions in underground coalmines. The outcomes of this study are also beneficial for other industries where large amounts of combustible dusts, such as starch, aluminium, wood, flour, polyethylene etc, are handled.

SUMMARY AND CONCLUSIONS

In this study, pulverised rock dust inertant was used to evaluate the suppression efficacy of coal dust explosions in underground coalmines. The explosion flame suppression behaviour at different rock dust proportions and rock dust particle sizes were examined. Moreover, the proportion of rock dust inertant required for suppression of coal dust explosions was investigated by varying the particle size of coal and rock dusts. The key findings of this study are outlined as follows:

- With an increase in the rock dust proportion from 10 to 50 per cent, the intensity of the explosion flame is significantly reduced. The explosion flame propagation is insignificant after adding 40 per cent rock dust, and the explosion flame is more or less diminished after 50 ms.

- The finer rock dust particle (<75 µm) demonstrated a better explosion suppression efficacy, as the explosion flame is somewhat diminished after 50 ms from the initiation despite its lower proportion (20 per cent).

- The proportion of rock dust requirement for suppression of coal dust explosion is significantly decreased with decrease in rock dust particle size from <850 to <75 µm. The average rock dust requirement for suppression of coal dust explosion decreased from 80 to 64 per cent and 81 to 66 per cent for limestone and dolomite dust, respectively.

- The coal dust explosion initiated by fine particles with a higher exposed surface area required a higher proportion of rock dust inertants for its suppression. The average proportion of rock dust required for suppression of mine size coal dust (<850 µm) is determined 55 to 58 per cent, whereas for the finest particle size coal dust (<38 µm), it is determined about 85 per cent.

Overall, this study revealed that the finer limestone dust is more effective than the dolomite dust for explosion suppression in underground coalmines. Also, it has been observed that any proportion of rock dust is effective in lessening the intensity of explosion flame propagation.

ACKNOWLEDGEMENTS

The authors extend their profound thanks to the mine management of Dhory Colliery, Central Coalfields Limited (CCL), Hirri Dolomite Mine, and Arasmeta Limestone Mine for helping in sample collection for the experiments. The authors are also thankful to IIT (ISM) Dhanbad for providing research facilities for conducting the experiments. This study constitutes a part of the doctoral research of the first author carried out in the Department of Mining Engineering, IIT (ISM) Dhanbad, India.

REFERENCES

Addai, E K, Addo, A, Abbas, Z and Krause, U, 2017. Investigation of the minimum ignition temperature and lower explosion limit of multi-components hybrid mixtures in the Godbert-Greenwald furnace, *Process Safety and Environmental Protection*, 111:785–794. https://doi.org/10.1016/j.psep.2017.09.003

Addai, E K, Gabel, D and Krause, U, 2016. Experimental investigations of the minimum ignition energy and the minimum ignition temperature of inert and combustible dust cloud mixtures, *Journal of Hazardous Materials*, 307:302–311. https://doi.org/10.1016/j.jhazmat.2016.01.018

Ajrash, M J, Zanganeh, J and Moghtaderi, B, 2016. Methane-coal dust hybrid fuel explosion properties in a large scale cylindrical explosion chamber, *Journal of Loss Prevention in the Process Industries*, 40:317–328. https://doi.org/10.1016/j.jlp.2016.01.009

Amyotte, P R, 2006. Solid inertants and their use in dust explosion prevention and mitigation, *Journal of Loss Prevention in the Process Industries*, pp 161–173. https://doi.org/10.1016/j.jlp.2005.05.008

Amyotte, P R, Mintz, K J and Pegg, M J, 1992. Effectiveness of various rock dusts as agents of coal dust inerting, *Journal Loss Prevention in the Process Industries*, 5(3):196–199. https://doi.org/10.1016/0950-4230(92)80024-3

Arshad, U, Taqvi, S A A and Buang, A, 2021. Modelling of the minimum ignition temperature (MIT) of corn dust using statistical analysis and artificial neural networks based on the synergistic effect of concentration and dispersion pressure, *Process Safety and Environmental Protection*, 147(January):742–755. https://doi.org/10.1016/j.psep.2020.12.040

Azam, S and Mishra, D P, 2019. Effects of particle size, dust concentration and dust-dispersion-air pressure on rock dust inertant requirement for coal dust explosion suppression in underground coalmines, *Process Safety and Environmental Protection*, 126:35–43. https://doi.org/10.1016/j.psep.2019.03.030

Bagaria, P, Hall, B, Dastidar, A and Mashuga, C, 2019. Effect of particle size reduction due to dust dispersion on minimum ignition energy (MIE), *Powder Technology*, 356:304–309. https://doi.org/10.1016/j.powtec.2019.08.030

Cao, W, Huang, L, Zhang, J, Xu, S, Qiu, S and Pan, F, 2012. Research on characteristic parameters of coal-dust explosion, *Procedia Engineering*, 45:442–447. https://doi.org/10.1016/j.proeng.2012.08.183

Cashdollar, K L and Chatrathi, K, 1993. Minimum explosible dust concentrations measured in 20-L and 1-m³ chambers, *Combustion Science and Technology*, 87(1–6):157–171. https://doi.org/10.1080/00102209208947213

Cashdollar, K L, 2000. Overview of dust explosibility characteristics, *Journal of Loss Prevention in the Process Industries*, 13(3–5):183–199. https://doi.org/10.1016/S0950-4230(99)00039-X

Cloney, C T, Ripley, R C, Pegg, M J, Khan, F and Amyotte, P R, 2018. Lower flammability limits of hybrid mixtures containing 10 micron coal dust particles and methane gas, *Process Safety and Environmental Protection*, 120:215–226. https://doi.org/10.1016/j.psep.2018.09.004

Going, J E, Chatrathi, K and Cashdollar, K L, 2000. Flammability limit measurements for dusts in 20-L and 1-m^3 vessels, *Journal of Loss Prevention in the Process Industries*, 13(3–5):209–219. https://doi.org/10.1016/S0950-4230(99)00043-1

Harris, M L, Sapko, M J, Zlochower, I A, Perera, I E and Weiss, E S, 2015. Particle size and surface area effects on explosibility using a 20-L chamber, *Journal of Loss Prevention in the Process Industries*, pp 33–38. https://doi.org/10.1016/j.jlp.2015.06.009

Huang, L, Jiang, H, Zhang, T, Shang, S and Gao, W, 2021. Effect of superfine KHCO3 and ABC powder on ignition sensitivity of PMMA dust layer, *Journal of Loss Prevention in the Process Industries*, 72(June):104567. https://doi.org/10.1016/j.jlp.2021.104567

Jiang, H, Bi, M and Gao, W, 2020. Suppression mechanism of Al dust explosion by melamine polyphosphate and melamine cyanurate, *Journal of Hazardous Materials*, 386:121648. https://doi.org/10.1016/j.jhazmat.2019.121648

Liu, T, Gao, Z, Xu, Y, Duan, G and Wang, X, 2023a. Research on Explosion Pressure Characteristics of Long Flame Coal Dust and the Inhibition Effect of Different Explosion Suppressants, *ACS Omega*, 8(39):35919–35928. https://doi.org/10.1021/acsomega.3c03700

Liu, T, Mu, X, Wu, X, Jia, R, Xie, J and Gao, Z, 2023b. Ignition temperature and explosion pressure of suspended coal dust cloud under different conditions and suppression characteristics, *Scientific Reports*, pp 1–10. https://doi.org/10.1038/s41598-023-42117-x

Man, C K and Gibbins, J R, 2011. Factors affecting coal particle ignition under oxyfuel combustion atmospheres, *Fuel*, 90(1):294–304. https://doi.org/10.1016/j.fuel.2010.09.006

Man, C K and Teacoach, K A, 2009. How does limestone rock dust prevent coal dust explosions in coalmines?, *SME Annual Meeting and Exhibit and CMA's 111th National Western Mining Conference 2009*, pp 338–342.

Mishra, D P and Azam, S, 2018. Experimental investigation on effects of particle size, dust concentration and dust-dispersion-air pressure on minimum ignition temperature and combustion process of coal dust clouds in a G-G furnace, *Fuel*, 227:424–433. https://doi.org/10.1016/j.fuel.2018.04.122

Mittal, M, 2013. Limiting oxygen concentration for coal dusts for explosion hazard analysis and safety, *Journal of Loss Prevention in the Process Industries*, 26(6):1106–1112. https://doi.org/10.1016/j.jlp.2013.04.012

Ramlu, M A, 1991. *Mine Disasters and Mine Rescure*, second edition (University Press: Hyderabad).

Reddy, P D, Amyotte, P R and Pegg, M J, 1998. Effect of inerts on layer ignition temperatures of coal dust, *Combustion and Flame*, 114(1–2):41–53. https://doi.org/10.1016/S0010-2180(97)00286-1

Sahu, A and Mishra, D P, 2022. Investigation of lag on ignition of coal dust clouds under varied experimental conditions, *Advanced Powder Technology*, 33(11):103804. https://doi.org/10.1016/j.apt.2022.103804

Sahu, A and Mishra, D P, 2023. Coal mine explosions in India: Management failure, safety lapses and mitigative measures, *Extractive Industries and Society*, 14:101233. https://doi.org/10.1016/j.exis.2023.101233

Sahu, A and Mishra, D P, 2024. Prevention and suppression of coal dust explosion in underground coalmines : Role of rock dust type, particle size, proportion, concentration and thermal properties, *Advanced Powder Technology*, 35(2):104343. https://doi.org/10.1016/j.apt.2024.104343

Sapko, M J, Cashdollar, K L and Green, G M, 2007. Coal dust particle size survey of US mines, *Journal of Loss Prevention in the Process Industries*, 20(4–6):616–620. https://doi.org/10.1016/j.jlp.2007.04.014

Shekarian, Y, Rahimi, E, Rezaee, M, Su, W C and Roghanchi, P, 2021. Respirable coalmine dust: A review of respiratory deposition, regulations an characterization, *Minerals*, 11(7):1–25. https://doi.org/10.3390/min11070696

Tan, X, Schmidt, M, Zhao, P, Wei, A, Huang, W, Qian, X and Wu, D, 2020. Minimum ignition temperature of carbonaceous dust clouds in air with CH4/H2/CO below the gas lower explosion limit, *Fuel*, 264:116811. https://doi.org/10.1016/j.fuel.2019.116811

Wang, Y, Lin, C-D, Qi, Y-Q, Pei, B, Wang, L-Y and Ji, W-T, 2020. Suppression of polyethylene dust explosion by sodium bicarbonate, *Powder Technology*, 367:206–212. https://doi.org/10.1016/j.powtec.2020.03.049

Yuan, J, Huang, W, Ji, H, Kuai, N and Wu, Y, 2012. Experimental investigation of dust MEC measurement, *Powder Technology*, 217:245–251. https://doi.org/10.1016/j.powtec.2011.10.033

Zhang, Y, Wu, G, Cai, L, Zhang, J, Wei, X and Wang, X, 2021. Study on suppression of coal dust explosion by superfine NaHCO3/shell powder composite suppressant, *Powder Technology*, 394:35–43. https://doi.org/10.1016/j.powtec.2021.08.037

Zhao, Q, Liu, J, Huang, C, Zhang, H, Li, Y, Chen, X and Dai, H, 2021. Characteristics of coal dust deflagration under the atmosphere of methane and their inhibition by coal ash, *Fuel*, vol 291. https://doi.org/10.1016/j.fuel.2020.120121

Zhong, Y, Li, X, Jiang, J, Liang, S, Yang, Z and Soar, J, 2022. Inhibition of Four Inert Powders on the Minimum Ignition Energy of Sucrose Dust, *Processes*, 10(2):405. https://doi.org/10.3390/pr10020405

Zlochower, I A, Sapko, M J, Perera, I E, Brown, C B, Harris, M L and Rayyan, N S, 2018. Influence of specific surface area on coal dust explosibility using the 20-L chamber, *Journal of Loss Prevention in the Process Industries*, 54:103–109. https://doi.org/10.1016/j.jlp.2018.03.004

Spontaneous combustion and mine fires in Indian coalmines – characterisation, detection and control measures

P Sahu[1], D C Panigrahi[2] and A Kumar[3]

1. Associate Professor, IIT (ISM) Dhanbad-826004, India. Email: patitapaban@iitism.ac.in
2. Formerly Director, IIT (ISM) Dhanbad-826004, India. Email: dcpanigrahi1961@gmail.com
3. Research Scholar, IIT (ISM) Dhanbad-826004, India. Email: kabhishek2121@gmail.com

ABSTRACT

Mine fires due to spontaneous combustion of coal (SCC) have always been a threat to coalmine industry. Mine fires resulting from SCC not only pose life hazards to miners but also cause loss of coal reserves due to burning, thereby creating serious environmental pollution and blockage of coal reserves in lower seams that adversely affect the economics of mining. Therefore, the determination of susceptibility potential of coals due to spontaneous heating and their classification are essential to plan the production activities. This paper elucidates the factors affecting SCC and early detection methods of spontaneous combustion required for taking control measures in the initial stages. The different experimental techniques, *viz* crossing point temperature, wet oxidation potential difference analysis and adiabatic R70 Analysis of Coal have also been discussed in detail. Multivariate regression models involving intrinsic parameters of the coal samples to fit the results of crossing point temperature (CPT) and wet oxidation potential (WOP) methods have also been discussed in this paper. Further, classifications of coal samples based on the results of CPT and WOP methods have been described. This paper also discusses several mine fire control measures for the prevention of mine fires which can be used by environmentalists and mine planners to take suitable measures, make proper execution plans and select operable parameters accordingly.

INTRODUCTION

Coalmine fires have been a great concern both for the industry and researchers worldwide. Studies carried out by different researchers reveal that, in most of the cases, they are caused by spontaneous heating of coal (Sensogut and Cinar, 2000). Indian coalfields have a historical record of extensive fire activity over the last 100 years. Among all the fire cases in different Indian coalfields, fire in Jharia coalfield (JCF) is very complex as fire moves from one seam to the other. According to the Jharia Master Plan, 70 fires covering an area of 17.32 km^2 spreading over in 41 collieries in the leasehold of Bharat Coking Coal Limited (BCCL) were identified after nationalisation. Due to the fire problem, BCCL is facing severe problems due to coal loss, locking of coal and associated impacts on exploitation of coal resources due to the complications created by fires. About 37 million tonne (Mt) of good quality prime coking coal has been wasted and about 1864 Mt of coal has been locked up due to these fires. In addition, fires in Jharia coalfield emerged as a great cause of concern to the environment, health, safety and well-being of the persons living on or near the fire areas due to evolution of toxic and greenhouse gases. Furthermore, fires posed a threat to the infrastructures like railway lines, roads, buildings, jores/rivers etc by causing surface subsidence. The main cause of fire in JCF is spontaneous combustion due to auto-oxidation of exposed coal. Spontaneous heating of coal depends mainly on two types of factors, such as intrinsic and extrinsic. The intrinsic parameters are mainly associated with the nature of the coal, ie its physico-chemical characteristics, petrographic distribution and mineral make up. On the other hand, the extrinsic parameters are related to atmospheric, geological and mining conditions prevailing during extraction of coal seams and these are mainly site specific.

Jharia coalfield spreads over an area of 450 km^2 and contains 40 identified coal horizons. It has one of the highest coal densities in the world having about 11 000 Mt of coal in proved category up to a depth of 600 m. Jharia has a total of 23 large underground and nine opencast mines and has 70 mine fires of the total 163 such fires identified across the nation. The total coal reserve of JCF is about 19 400 Mt. Bharat Coking Coal Limited, a subsidiary of Coal India Limited (CIL) has suffered a huge loss of coking coal due to the fire. There has been a loss of 4065 crores to BCCL as 37 Mt of coal have been lost and 220 billion tonnes (Bt) are locked up or inaccessible due to fire.

FACTORS AFFECTING SPONTANEOUS HEATING OF COAL

Apart from the natural affinity of coal to self-heating, a number of factors are significant when determining the risk of spontaneous combustion. The important factors that affect spontaneous heating of coal are rank, petrographic composition, methane (CH_4) content, type and amounts of mineral present, moisture, particle size and surface area (Kaymakci and Didari, 2002). A number of physical properties such as porosity, permeability, hardness, thermal conductivity and specific heat can influence the rate of oxidation and thus result in spontaneous combustion. Hardness affects the friability, hence the surface area. The rate of heat transportation from the coal depends on its thermal conductivity. Coals with low thermal conductivity tend more frequently to spontaneous combustion. Mining methods, rate of advance, leakage, and ventilation also play a dominant role in the spontaneous heating of coal (Yuan and Smith, 2012).

EARLY INDICATORS OF SPONTANEOUS HEATING

During initial stage of heating moisture is released from coal and it comes in contact with ventilating air and thereby condenses making haze like formation. In the advanced stage of heating, water droplets form on the roofs, walls and timber supports of the mine. There is a characteristic tarry smell normally found in a coal fire area.

Spontaneous heating of coal is usually accompanied by a rise in temperature before the actual fire occurs. Thus, by measuring the temperature of the area, be it a working face, pillar or sealed off area, the status of heating can be ascertained. Temperature should be measured in places vulnerable to heating like old workings in mines, depillared areas, crushed pillars with cracks and fissures, near gate roads in longwall panels and in return airways. Digital temperature recorders are being extensively used in Indian mines for measuring temperature in sealed off areas. Recently some researchers have utilised infra-red (IR) thermal gun and thermo vision camera for thermal scanning of pillars (Panigrahi and Bhattacharjee, 2004). It has been experienced that IR thermometer provided with scanner can locate hot spots more accurately.

The normal air consists of nitrogen (N_2), O_2 and CO_2 in the proportion of 79.04 per cent, 20.93 per cent and 0.03 per cent, respectively. The deviation of the proportions of these gases in mine air is very useful for assessing the extent of heating (Singh et al, 2007). Combustion also gives rise to pollutants like CO, CO_2, CH_4, oxides of N_2, hydrogen sulfide (H_2S) and other hydrocarbons. Due to the advent of modern monitoring techniques for CO detection, CO has become a very useful indicator of mine fires. Depletion of O_2 in the sealed off fire area is one indicator of an active fire. The rate of O_2 consumption is a guide to distinguish a localised fire from an extensive one. The CO/O_2 deficiency ratio (Graham's ratio) also is an indicator of fire. The CO/O_2 deficiency ratio increases as coal oxidation increases and normally ranges between 0 and 0.40. However, in cases of serious heating, this value can reach 0.50 to 10.00, due to the formation of producer gas and water gas.

EXPERIMENTAL METHODS FOR DETERMINING SPONTANEOUS HEATING OF COAL

Several researchers have used different experimental techniques to predict the self-heating and spontaneous combustion liability of coal. Some of these methods include CPT (crossing point temperature) in India, Russian U-index in Russia, Olpinski index in Poland, adiabatic calorimetry in USA, Wits-Ehac index and Wits-CT index in South Africa and R70 in New Zealand and Australia, average heating rate (AHR), Feng, Chakravorty and Cochrane (FCC) liability index and slope of the time-temperature curve on the CPT (Onifade and Genc, 2020; Sahu and Panigrahi, 2015). These simple indices have been used for decades to predict the spontaneous combustion liability of coal. Some experimental techniques used have been described in the following section.

Crossing point temperature (CPT) of coal

CPT or critical oxidation temperature gives an idea about the proneness of coal to auto-oxidation or self-heating. This is a standard method followed in India for finding out the susceptibility of coal to spontaneous combustion. The coal sample is milled to <250 µm, placed into a 750 mL flask, and then sealed. The flask is then placed into a cold oven and flushed with nitrogen for 1 hr at a rate of 250 mL/min. The oven temperature is then increased to 110°C and the coal sample is dried for

16 hrs. After this period, the oven is turned off, allowing the coal to cool to 40°C under nitrogen. Approximately 70 g of coal from the sealed flask is transferred to a nitrogen-purged spontaneous combustion vessel and placed into a hot storage oven and purged with nitrogen for 50 mins at room temperature. The oven temperature is then raised to 40°C and the coal and oven temperatures are allowed to stabilise for 3–6 hrs under nitrogen before initiating the Crossing Point Test Program.

A thermocouple placed in the centre of the spontaneous combustion vessel is used to monitor the coal temperature while a second thermocouple placed on the vessel monitors the vessel temperature. During the Crossing Point Test, air is flowed through the vessel at approximately 80 mL/min and then heated at 0.6°C/min. The test is complete when the coal temperature has exceeded the oven temperature, or the coal temperature has reached 200°C. Figure 1 shows the experimental set-up for crossing point temperature of coal.

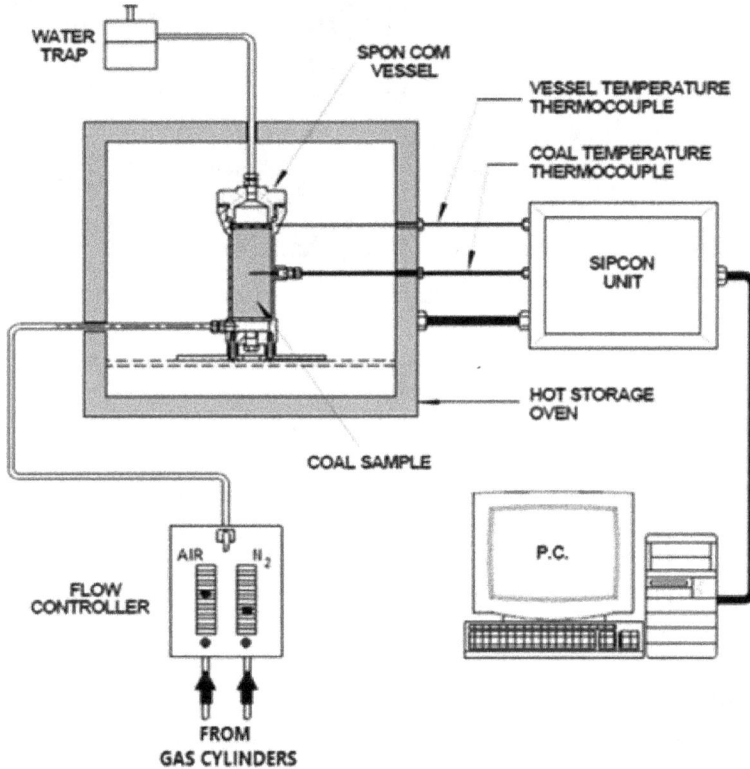

FIG 1 – Experimental set-up for crossing point temperature of coal.

The CPT is defined as the temperature at which the coal temperature exceeds the oven temperature. Where, the lower the crossing point, the greater the propensity of the coal to spontaneously combust.

Wet oxidation potential (WOP) of coal

The experimental set-up comprises of a beaker, one carbon electrode, one calomel electrode, Teflon coated magnetic fish and a millivoltmeter. A photograph of the complete set-up is presented in Figure 2.

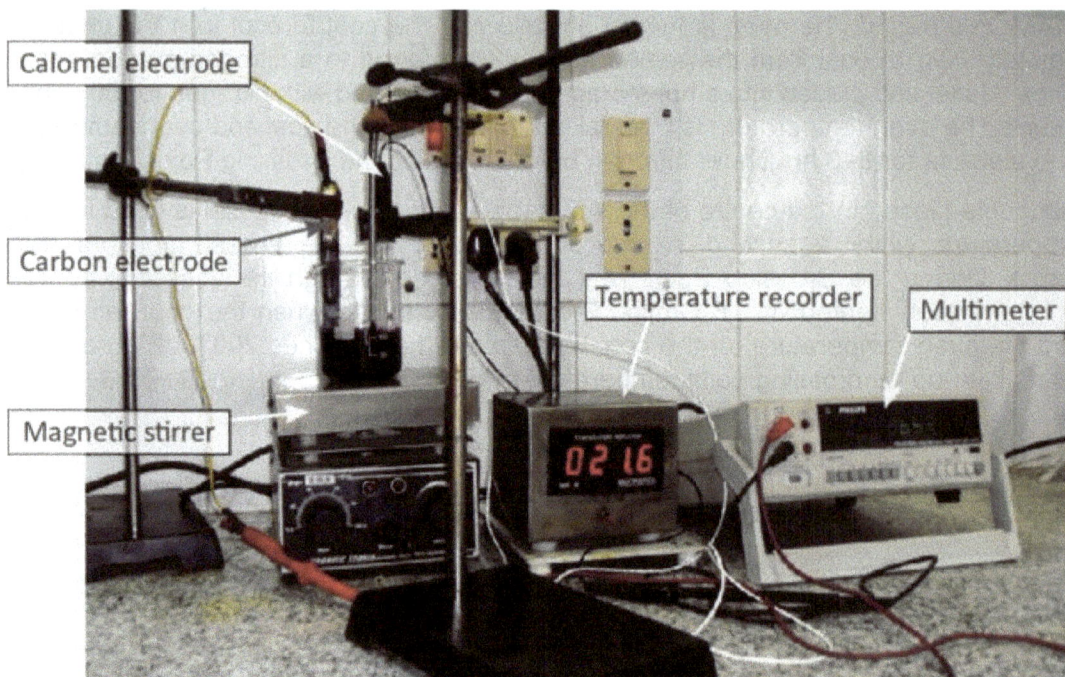

FIG 2 – Photographic view of wet oxidation apparatus (Panigrahi and Ray, 2014).

The beaker along with the electrodes is placed over a magnetic stirrer such that a homogeneous mixture of coal and alkali solution is maintained. The Teflon coated fish of the magnetic stirrer is placed inside the beaker 0.5 g of coal sample of -212 m size is poured onto 100 mL of deci-normal solution of $KMnO_4$ in 1N KOH solution in a beaker and the coal sample is subjected to wet oxidation process. The coal-oxidant suspension is continuously stirred using the magnetic stirrer. The potential difference (EMF: electromotive force) is recorded between the calomel and carbon electrodes over a period of time by using a millivoltmeter till the potential difference attained a nearly constant value. Different samples require different time durations for attaining a nearly constant potential difference (ΔE).

Adiabatic R70 analysis of coal

R70 is defined as a rate of propensity for spontaneous combustion of coal, which is the self-heating rate of coal in a pure oxygen environment where it is measured from 40°C to 70°C or over a period of 72 hrs under the adiabatic environment. This method covers the sample preparation, adiabatic oven operation and data processing for R70 testing. The results obtained provides propensity rating in relation to the spontaneous combustion of coal, where the self-heating rate (by oxidation) is an indicator of the reactivity of the coal sample. Results acquired from this procedure is widely accepted by the coalmine industry as an index of coal's spontaneous combustion property.

An adiabatic process occurs when no heat enters or leaves the system. In this circumstance, the adiabatic oven maintains a specific temperature at a predetermined offset to the measured temperature of the sample vessel. This oven has been specifically designed to simulate the accelerated self-heating process of coal under a pure oxygen environment. With the respective coal sample, it is kept within a thermos flask in the adiabatic oven, thus, any heat loss and/or gain will be compensated by the oven. Allowing any temperature change of the coal to be solely contributed by the oxidisation process.

Once the coal and oven have stabilised at 40.0±0.1°C, the oven is switched to adiabatic mode in which the oxygen flow rate is 50 mL/min. If the coal temperature exceeds 40.1°C during stabilisation, the oven temperature is reduced until it is achieved for the coal temperature to fall before ramping oven back to 40.0°C. The average rate of coal heating from 40 to 70°C is the index R70 (°C/h). The higher the value of such an index, the more susceptibility of the coal to spontaneous combustion (Table 1).

TABLE 1

R70 index (Saffari, Ataei and Sereshki, 2019).

Class	Queensland R70 value (°C/h)	New South Wales R70 value (°C/h)	Propensity rating
I	R70 < 0.5	R70 < 1	Low
II	0.5 ≤ R70 < 1	1 ≤ R70 < 2	Low-medium
III	1 ≤ R70 < 2	2 ≤ R70 < 4	Medium
IV	2 ≤ R70 < 4	4 ≤ R70 < 8	High
V	4 ≤ R70 < 8	8 ≤ R70 < 16	Very high
VI	8 ≤ R70 < 16	16 ≤ R70 < 32	Ultra-high
VII	R70 ≥ 16	R70 ≥ 32	Extremely high

EXPERIMENTAL OBSERVATIONS, ANALYSIS AND RESULTS

Table 2 presents the susceptibility indices of the coal samples collected from 25 different coalmines. Table 3 presents spontaneous combustion risk. The susceptibility indices were determined using CPT and WOP methods. The CPT values of the coal samples collected from different coalmines ranged from 107.13°C to 161.39°C. Table 3 shows CPT risk criteria. The higher the CPT value, the less will be susceptibility of coal to spontaneous heating. The table showed that the CPT values of coal samples were lowest in MINE-1 and highest in MINE-3. Further, it was also observed that the CPT values of the coal samples in the different mines (except MINE-3), was below 140°C. It has been revealed that coal samples in these mines are highly liable to self-heating, and the coal sample in MINE-3 is moderately susceptible to self-heating.

TABLE 2

Susceptibility indices of the coal samples collected from different coalmines.

Mine	CPT (°C)	WOP (mV)
MINE-1	107.13	163.5
MINE-2	125.9	141.4
MINE-3	161.39	---
MINE-4	---	164.9
MINE-5	114.9	168
MINE-6	124.65	186
MINE-7	128.13	189
MINE-8	122.96	174.2
MINE-9	130.46	166.5
MINE-10	----	152.6
MINE-11	124.27	135.5
MINE-12	122.32	154
MINE-13	134.1	165.1
MINE-14	113.19	145
MINE-15	----	150.3
MINE-16	----	165.8
MINE-17	----	152.6

MINE-18	----	161.7
MINE-19	----	97.7
MINE-20	----	133.6
MINE-21	----	136.8
MINE-22	----	151.2
MINE-23	----	125.4
MINE-24	----	148.3
MINE-25	----	119.8

TABLE 3

Risk rating criteria based on CPT values.

CPT (°C)	Spontaneous combustion risk
120–140	Highly liable to self-heating
140–160	Moderately liable to self-heating
160–180	Less liable to self-heating

The wet oxidation potential values of the coal samples collected from different mines ranged from 97.7 mV to 189 mV. The WOP values of coal samples in MINE-7 were highest, and lowest in MINE-19. Plotting potential difference versus time, as shown in Figure 3, is required to compare the susceptibility of different coal samples to spontaneous combustion. Different samples required different time durations to attain a nearly constant potential difference (ΔE). Therefore, the duration of the experiment in each case is different. A constant time frame is chosen to compare the oxidation rate of different samples. The higher the potential difference in the selected time frame, the higher the spontaneous heating tendency of coal.

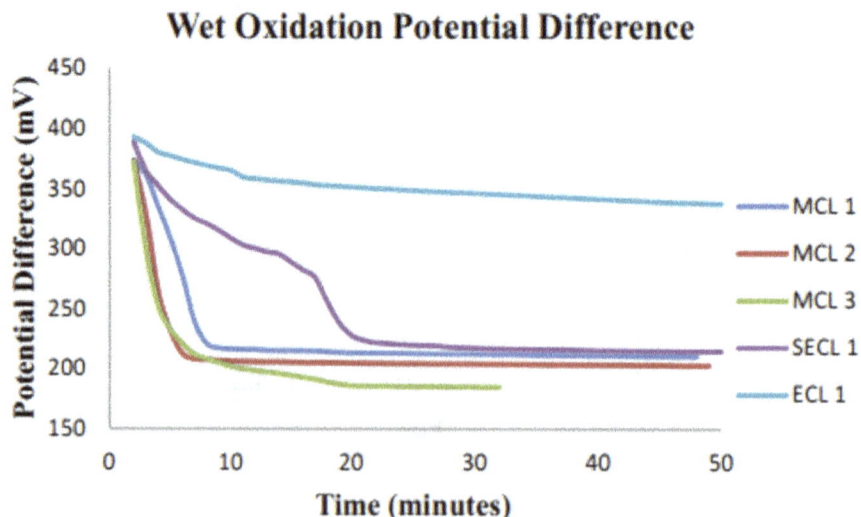

FIG 3 – Plots of EMF versus time for five different coal samples (Rajak, Pradhan and Prince, 2019).

Multiple linear regression

Multivariate regression models involving moisture content, volatile matter, ash content, fixed carbon and gross calorific value (GCV) of the coal samples to fit the results of CPT and WOP methods were performed by Singh (2014). It was reported that the model presented 74.12 per cent of CPT and 73.5 per cent of WOP variations. Ray, Panigrahi and Varma (2014) developed a multiple linear regression analysis involving moisture, volatile matter, carbon, hydrogen, and oxygen to fit the wet oxidation potential of the coal samples. They reported that the model explained 62.89 per cent of the variation of WOP.

PREVENTION AND CONTROL OF SPONTANEOUS HEATING OF COAL

Mine layout is the most important factor in preventing spontaneous heating. Working districts should be designed so that a particular section can be isolated at short notice without affecting production in others. Sites for preparatory seals should be identified and marked during planning. The retreat working method is preferred for coal seams that are highly susceptible to spontaneous combustion. This is because the gob is not subjected to a large difference in ventilation pressure. Coal pillars should be designed to resist excessive crushing. The panel system of working should be preferred and the size of each panel is calculated based on the incubation period and the extraction rate. Ventilation design tools to prevent spontaneous combustion include reducing mine resistance and ventilation pressure, providing a high standard of stoppings and seals and roadway support, using low flow/low-pressure drop roadways alongside goaf areas, using balance chambers to contain sealed areas and injecting inert gases into balance chambers.

N_2 or CO_2 is injected into goaf areas affected by a spontaneous heating to help control the heating by excluding O_2 and to give a cooling effect. The favoured gas is N_2 as it is cheaper and more easily expelled from the mine after the heating has been controlled than is CO_2. Major advances have been made using inert gases to control spontaneous combustion. The objective of inerting is to reduce the concentration of O_2 in the atmosphere. To ensure the prevention of spontaneous combustion, it is necessary to reduce the O_2 content below 3 per cent by volume.

Researchers from different parts of the world have carried out research work on chemical inhibitors to counter spontaneous heating/fire in coalmines with varying degree of success (Smith, Miron and Lazzara, 1988; Panigrahi et al, 2005; Singh, 2013). Some of the common inhibitors that have been used are 10–15 per cent $CaCl_2/MgCl_2.6H_2O$, NH_4Cl, ammonium hydrogen tetraborate, Diapon T (Na-N methyl-N olieltuarte), NaCl, diammonium phosphate etc and are claimed to be surface active agents. Beamish et al (2013) used an antioxidant and found that it significantly reduces the coal self-heating rate and extends the time taken to reach thermal runaway by a factor of three for sub-bituminous coal and by a factor of two for the same application to high volatile C bituminous coal. The search for suitable chemical inhibitors to counter spontaneous heating is in progress in many countries, but a foolproof solution has not yet been achieved.

Sealant combines an inhibitor of coal oxidation, such as calcium chloride ($CaCl_2$), with a binding agent and filler, such as bentonite. Clays such as bentonite flow readily into cracks and swell and fill the cracks. When the bentonite and $CaCl_2$ are homogenised, the resulting mixture has long-term stability. Gypsum-based sealants usually have additives such as perlite, vermiculite, granulated mineral wool, or fibre glass that help increase their strength and elasticity. Such sealants are also of low density and harden rapidly after application. These properties make this type of sealant particularly suitable for providing a thin, air-tight seal to the surface walls and roofs of mine roadways. The low density prevents stressing of the surface material of the mine walls and hence, reduces the possibility of spalling. Urea-formaldehyde foam products have been developed to supersede the polyurethane foams that were toxic and flammable. This foam is suitable for filling between and under chocks or between the shuttering and the side of the pack. The foam sets in 24 hrs and becomes increasingly impermeable under pressure from the strata. Suspected long-term health problems restricted its use.

DISCUSSION AND CONCLUSIONS

The spontaneous heating or auto-oxidation of coal is a result of a number of complex physical and chemical processes, which are accompanied by the absorption of oxygen, formation of coal-oxygen complexes, and their decomposition leading to the liberation of heat. This complexity of the process is enormous because of the great diversity in the coal substance with the associated mineral matter and the conditions of oxidation. During the oxidation of the heterogeneous mass, concurrent and overlapping reactions take place, which are very difficult to separate out. Several researchers have used different experimental techniques to predict the self-heating and spontaneous combustion liability of coal. The techniques are used in different countries with varying experimental standards and therefore, there is no uniformity in application, and sometimes the experimental results are not even comparable.

In wet oxidation potential, different samples require different time durations to attain a nearly constant potential difference for a comparative study. In order to compare the oxidation rate of different samples a constant time frame is chosen. The higher the potential difference in the selected time frame, the higher the spontaneous heating tendency of coal. CPT is a standard method followed in India for finding out the susceptibility of coal to spontaneous combustion. The higher the crossing point temperature value, the less will be the susceptibility of coal to spontaneous heating. The R70 initial self-heating rate parameter normally obtained from the tests is rated using a relative scale. The average rate of coal heating from 40°C to 70°C is considered to be the index R70 (°C/h). The higher the value of such an index, the more prone the coal is to spontaneously combust. Such a method is the most efficient one with respect to the simple characterisation of coals propensity to oxidation and self-heating.

Several investigators assessed the cause of spontaneous combustion in coalmines using various intrinsic and extrinsic properties and different spontaneous combustion tests. However, there is no agreement among researchers for the adoption of a specific technique for the assessment of spontaneous combustion liability of coal. Some studies indicated that some techniques may be attempted to determine accurately the propensity of a certain coal to spontaneous combustion. In view of this, several experimental methods to determine the liability of different coal to experience spontaneous combustion are available but none is superior to the other. This study acknowledges that an evaluation of the effect of spontaneous combustion in coalmines is of significance and in-depth investigation is needed in order to establish a database for the comparisons of the various intrinsic and extrinsic factors affecting the spontaneous combustion liability of coal. It was found that the liability of coal towards spontaneous combustion can be examined using various techniques. These techniques have been proved to be reliable in their application, but the point that no exact test method has become a standard showed that doubt still exists as to the validity of all of them.

Extensive research efforts have been devoted to the prevention of spontaneous combustion in different coalmines. A suitable technique was used to prevent and control spontaneous heating/fire as per site specific situations. Solutions would differ from case to case and mine to mine. a single control measure is not going to solve the fire problem. Hence, the combination of different measures is the best option.

ACKNOWLEDGEMENTS

The authors express their sincere thanks and gratitude to the Indian Institute of Technology (Indian School of Mines) Dhanbad Administration for extending all the help to carry out the laboratory studies.

REFERENCES

Beamish, B, McLellan, P, Endara, H, Turunc, U, Raab, M and Beamish, R, 2013. Delaying spontaneous combustion of reactive coals through inhibition, in *Proceedings of the 13th Coal Operators' Conference*, pp 221–226 (University of Wollongong; The Australasian Institute of Mining and Metallurgy; and Mine Managers Association of Australia).

Kaymakci, E and Didari, V, 2002. Relations between coal properties and spontaneous combustion parameters, *Turkish Journal of Engineering and Environmental Sciences*, 26:59–64.

Onifade, M and Genc, B, 2020. A review of research on spontaneous combustion of coal, *International Journal of Mining Science and Technology*, 30(3):303–311.

Panigrahi, D C and Bhattacharjee, R M, 2004. Development of modified gas indices for early detection of spontaneous heating in coal pillars, *The Southern African Institute of Mining and Metallurgy*, pp 367–379.

Panigrahi, D C and Ray, S K, 2014. Assessment of self-heating susceptibility of Indian coal seams – a neural network approach, *Archives of Mining Sciences*, 59:1061–1076.

Panigrahi, D C, Udaybhanu, G, Yadav, M D and Singh, R S, 2005. Development of inhibitors to reduce the spontaneous heating susceptibility of Indian Coals, in *Proceedings Eighth International Mine Ventilation Congress*, pp 349–353 (The Australasian Institute of Mining and Metallurgy: Melbourne).

Rajak, M K, Pradhan, G K and Prince, M J A, 2019. Assessment of the susceptibility of coals to spontaneous heating using wet oxidation potential difference technique, *International Journal of Engineering and Advanced Technology*, 9:6431–6437.

Ray, S K, Panigrahi, D C and Varma, A K, 2014. An electro-chemical method for determining the susceptibility of Indian coals to spontaneous heating, *International Journal of Coal Geology*, 128–129:68–80.

Saffari, A, Ataei, M and Sereshki, F, 2019. Evaluation of the spontaneous combustion of coal (SCC) by using the R70 test method based on the correlation among intrinsic coal properties (case study: Tabas parvadeh coal mines, Iran), *The Mining-Geology-Petroleum Engineering Bulletin*, 34(3):49–60.

Sahu, H B and Panigrahi, D C, 2015. Spontaneous Heating of Coal — Its Assessment and Control, in *Energy Science and Technology Vol. 2: Coal Energy* (eds: J N Govil, R Prasad, S Sivakumar and U C Sharma), pp 568–613 (Studium Press LLC).

Sensogut, C and Cinar, I, 2000. A research on the tendency of Ermenek district coals to spontaneous combustion, *Mineral Resources Engineering*, 9(4):421–427.

Singh, A K, Singh, R V K, Singh, M P, Chandra, H and Shukla, N K, 2007. Mine fire gas indices and their application to Indian underground coal mine fires, *International Journal of Coal Geology*, 69:192–204.

Singh, P, 2014. An Investigation into Spontaneous Heating Characteristics of Coal and its Correlation with Intrinsic Properties, BTech Thesis, National Institute of Technology, India.

Singh, R V K, 2013. Spontaneous heating and fire in coal mines, in *The Ninth Asia-Oceania Symposium on Fire Science and Technology*, *Procedia Engineering*, 62:78–90.

Smith, A C, Miron, Y and Lazzara, C P, 1988. Inhibition of spontaneous combustion of coal, US Bureau of Mines Report of Investigations RI 9196.

Yuan, L M and Smith, A C, 2012. The effect of ventilation on spontaneous heating of coal, *Journal of Loss Prevention in the Process Industries*, 25:131–137.

Study of the impact of particle size on maximum explosion pressure for subbituminous coal dust explosibility using 20 litre laboratory explosion chamber

F Z Waly[1], N P Widodo[2], S Firdausi[3], M W Putri[4], A Ihsan[5] and Salmawati[6]

1. Master Degree Student, Department of Mining Engineering, Faculty of Mining and Petroleum Engineering, Institut Teknologi Bandung, Bandung 40132, Indonesia. Email: 22122003@mahasiswa.itb.ac.id
2. Associate Professor, Department of Mining Engineering, Faculty of Mining and Petroleum Engineering, Institut Teknologi Bandung, Bandung 40132, Indonesia. Email: npw@itb.ac.id
3. Researcher, Center of Research Excellence of Underground Mining and Mine Safety, Faculty of Mining and Petroleum Engineering, Institut Teknologi Bandung, Bandung 40132, Indonesia. Email: firdabla@gmail.com
4. Researcher, Center of Research Excellence of Underground Mining and Mine Safety, Faculty of Mining and Petroleum Engineering, Institut Teknologi Bandung, Bandung 40132, Indonesia. Email: marlina.intan6@gmail.com
5. Doctorate Degree Student, College of Safety Engineering, China University of Mining and Technology, Xuzhou, Jiangsu, 221116, China. Email: ihsanahmad50@yahoo.com
6. Lecturer, Department of Mining Engineering, Faculty of Mining and Petroleum Engineering, Institut Teknologi Bandung, Bandung 40132, Indonesia. Email: salmawati@itb.ac.id

ABSTRACT

Coal dust explosion is one of high risks that exist in coal mining activities because it can cause mine damage and even fatalities. Research about maximum coal dust explosibility pressure have been conducted by many researchers, however Indonesian subbituminous coal at various coal dust particle sizes has not been studied extensively. This present research was conducted to study maximum explosibility pressure on subbituminous coal sample originating from Jambi Province, Indonesia, using a 20 L Siwek Chamber referring to the ASTM E1226 (2019) procedure. The attesting using 74–44 µm coal dust sample with concentration variations of 100 g/m^3, 200 g/m^3, 250 g/m^3, 500 g/m^3, 750 g/m^3, 1000 g/m^3, 1250 g/m^3. To obtain the maximum pressure of coal dust explosion, the test used a heat source from pyrotechnics of 4.6 kJ. From the tests performed, it is found that the maximum pressure of the explosion is 5.48 bar and the maximum value for the rate of pressure increase is 174.62 bar/s. The deflagration index is 47.40 bar·m/s which categorised as weak explosion class. Compared with previous research from Azwar (2019) at 74–53 µm and Said (2019) at 53–44 µm, this study shows a lower maximum explosion pressure. This is caused by the larger size of coal dust particles and the non-uniform distribution of coal dust particle sizes compared to the two previous research.

INTRODUCTION

Coal dust is one of the elements that can cause an underground coalmine explosion. High levels of coal dust in the air not only affect the health of mine workers, but also can cause a mining explosion, lead infrastructure damage, and fatality on mine workers. According to Roychowdhury (1960), the explosion was caused by several factors, namely the concentration of methane in the air, the concentration of coal dust, and heat sources. In the US there are 281 major accidents regarding combustible dust that occurred between 1980 and 2005, which caused 119 deaths, 718 injuries, and severe damage to industrial facilities (Cao et al, 2017). In Indian coalmines, from 1901 to 2010 there were 40 major accidents caused by explosions in which 1281 people dead and 110 people were seriously injured (Bhattacharjee, Dash and Paul, 2020). In Indonesia, underground coal explosions have occurred several times. One of which occurred at the Sawahlunto coalmine in 2009, where 17 miners died and 23 people were trapped underground (Reuters, 2009).

Several previous studies on coal dust explosions include research on Coal Dust Explosibility (Cashdollar, 1996) which discusses the United States Bureau of Mines (USBM) research on coal dust explosion using the USBM 20 L Chamber to measure some explosion parameters eg maximum explosion pressure (P_{max}), maximum rates of pressure rise (dP/dt) max and examine the relationship between coal particle size and volatile on the explosion parameters. The result is that the higher the volatile and the finer the coal dust are more hazardous.

In Indonesia, especially research for the maximum explosion pressure for coal dust explosion on the explosion chamber was started by Yin *et al* (2011) for bituminous coal from Indonesia. The result shows that maximum explosion pressure increases at first and decreases afterward with the increase of dust concentration. Furthermore, research by Norman, Berghmans and Verplaetsen (2013) conducted on a coal dust explosion test in a 20 L chamber using high volatile bituminous coal from Sebuku, Indonesia. They concluded that the maximum explosion pressure increases almost linearly with increasing oxygen concentration. Recently, coal dust explosion test research in Indonesia was carried out by Widodo *et al* (2019) where testing was carried out in a chamber with a size of 10 L and 20 L using lignite coal samples from Indonesia. Based on research by Widodo *et al* (2019), coal in Indonesia has different explosion characteristics from tests conducted by Cashdollar (1996) and Yin *et al* (2011).

Coal dust explosions are not only influenced by coal rank, the size of coal dust particles is also one of the factors that influence the characteristics of coal dust explosions. Most of the coal dust particle diameter occurring in a coalmine that cause explosion is less than 100 μm, especially particles less than 20 μm will explode at much lower concentrations than particles 20–100 μm (Hertzberg and Cashdollar, 1987). Research related to the influence of coal dust size has been carried out in Indonesia but has not been the focus of discussion in research. The research that has been carried out is on subbituminous coal from Jambi in Centre of Research Underground Mining and Mine Safety Institut Teknologi Bandung by Azwar (2019) at 74–53 μm (with D50 of 63.5 μm) and Said (2019) at 53–44 μm (with D50 of 48.5 μm) that gives P_{max} of 9.58 bar and 12.25 bar, respectively. Based on these two studies, it shows that the finer the particle size can increase the maximum explosion pressure. Further research regarding the influence of dust particle size and distribution is expected to complement the analysis that has been carried out previously for subbituminous coal in Indonesia which can cause explosion hazards in underground coal mining activities.

An investigation into this matter has been carried out in this research which is expected to become a new identification method for characterising coal dust explosions. This research was carried out on subbituminous coal from Jambi with a size range of 74–44 μm at a concentration of 200–1250 g/m^3 using a 4.6 kJ ignitor.

METHODOLOGY

To conduct a coal dust explosion test, the Siwek 20 L Chamber was developed at the Laboratory of Geomechanics and Mine Equipment, Institut Teknologi Bandung as shown in Figure 1. ASTM E1226 (2019) are used as standard guidelines. ASTM E1226 (2019) describes the standard test method for the coal dust explosibility which is used to analyse if a dust dispersion can explode and if it explodes how the severity level will be produced. The general coal explosion test steps, starting with install the 4.6 kJ pyrotechnic on the sensor rod, insert coal dust into the dust chamber, and enter ± 19.5 bar pressurised air. After that, adjust the explosion time, explosion delay time, as well as the amount of data to be obtained, and save the file on LabView software. Then perform a coal dust explosion test. After testing is complete the equipment is cleaned before continuing with further testing.

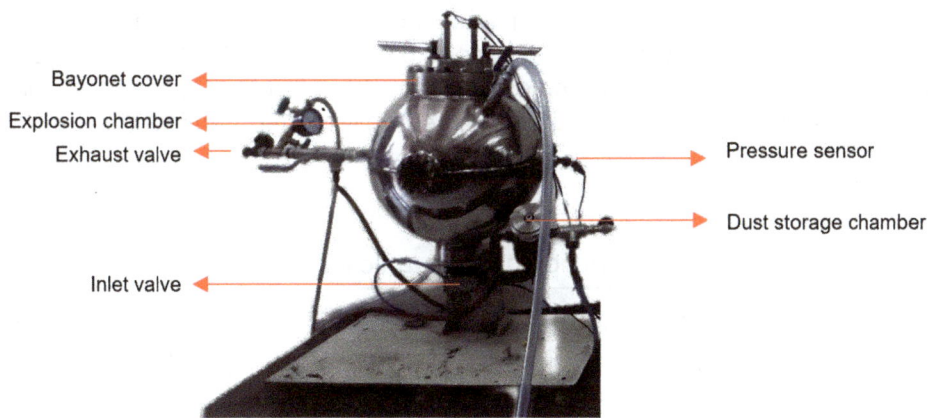

FIG 1 – Siwek 20 L chamber in CORE UMMS, Institut Teknologi Bandung, Indonesia.

Generally, the characteristics of dust explosibility are classified into two categories (Widodo *et al*, 2019). The first group explained the possibility of dust explosion, including: the Minimum Explosible Concentration (MEC), Minimum Ignition Energy (MIE), Oxygen Concentration Limit (LOC), and Minimum Auto Ignition Temperature (MAIT). The second group explains the severity of the dust explosion when it occurs, and this includes the maximum explosion pressure (P_{max}) and the maximum level of pressure increase ($dP/dt)_{max}$, and deflagration index, K_{st}. Occupational Safety and Health Administration (OSHA, 2009) make classification for dust explosions severity based on its deflagration index values (K_{st}) (see Table 1).

TABLE 1

Determination of explosion class based on K_{St} values.

K_{St}	Class
$0 < K_{St} \leq 200$	*St* 1 dust explosion (weak explosion)
$200 < K_{St} \leq 300$	*St* 2 dust explosion (strong explosion)
$K_{St} > 300$	*St* 3 dust explosion (very strong explosion)

This research focused on the second group values, parameters of coal dust explosion obtained are as follows: maximum explosion pressure, P_{max}; maximum value for the rate of increase in pressure, $(dP/dt)_{max}$; and deflagration index, K_{st}. The experimental scenario in these three values is based on ASTM E1226 (2019), at least three series of tests are needed to obtain each of the three maximum pressure rise values (P_{ex}) and the maximum rate of pressure $(dP/dt)_{ex}$ in a single test.

$$P_{max} = \sum_{i=1}^{n} \frac{P_{ex,i}{}^A}{n} \tag{1}$$

$$(dP/dt)_{max} = \sum_{i=1}^{n} \frac{(dP/dt)_{ex,i}{}^A}{n} \tag{2}$$

$$K_{St} = (dP/dt)_{max} \cdot (V)^{\frac{1}{3}} \tag{3}$$

Notes:

A   is maximum score for each test series

n   is number of coal dust testing

V   is volume of explosion chamber (0.02 m³)

EXPERIMENTAL SET-UP

Samples for testing are reduced a particle size of at least 95 per cent (weight) less than 75 µm (ASTM E1515-E14, 2022). In this study, the size variations of particle size are 74–44 µm (see Table 2 for the distribution), the size particle was obtained by using jaw crusher, roll crusher, ball mill, and sieving process. After comminution and sieving, the finer samples are dried until the moisture content of the test sample should not exceed 5 per cent (ASTM E1226, 2019). The process of comminution, sieving, and drying of samples was carried out in the Coal and Mineral Processing Laboratory of Institut Teknologi Bandung. The dried sample is then packed in a zip lock plastic and put in a desiccator.

TABLE 2

Distribution of coal particle size.

Particle size (µm)	Sample Weight	% Mass
74–65	367.84	61.6%
65–53	111.60	18.7%
53–44	117.63	19.7%
Total	**597.07**	**100%**

To obtain the characteristics of coal, the Proximate, Ultimate, and Calorific Test was conducted by the Research Centre for Tekmira Bandung. The characteristics of the coal used can be seen in Table 3. From the characteristics of the coal, a coal rank determination process is carried out based on ASTM D388 (2005) and with a base conversion method refers to ASTM D3180 (1997). The rank of the sample coal used was classified into Subbituminous Coal.

TABLE 3

Coal characteristics.

Parameter analysis	Value	Unit	Basis
Proximate:			
Total moisture	12.58	%	adb
Ash content	5.02	%	adb
Volatile matter	44.31	%	adb
Fixed carbon	38.09	%	adb
Ultimate:			
Total sulfur	0.22	%	adb
Carbon	58.17	%	adb
Hydrogen	4.25	%	adb
Nitrogen	0.88	%	adb
Calorific value	5,257	cal/g	adb

RESULT AND ANALYSIS

Pyrotechnic explosibility test results 4.6 kj

The composition of the ignition source material follows ASTM E1226 (2019) with an energy of 5 kJ for an ignitor (pyrotechnics). However, the actual energy of the pyrotechnics used was determined through tests in a bomb calorimeter. The calorimeter bomb used has a heat capacity of 2.44 kcal/°C. A pyrotechnic sample of 1.2 grams was put into a bomb calorimeter and the resulting temperature increase was seen. A temperature increases of 0.45°C was obtained, so the pyrotechnic energy can be calculated using the method below.

$$\text{Pyrotechnic energy} = 2.44 \text{ kcal/°C} \times 0.45\text{°C}$$

$$= 1.108 \text{ kcal} = 4.6 \text{ kJ}$$

Before the coal dust explosibility test is carried out, the explosibility test without coal dust is first carried out. In this test, only pyrotechnics will explode so that $\Delta P_{ignitor}$ data can be determined. The results and data calculations obtained from the 4.6 kJ pyrotechnic explosibility test are shown in Figure 2.

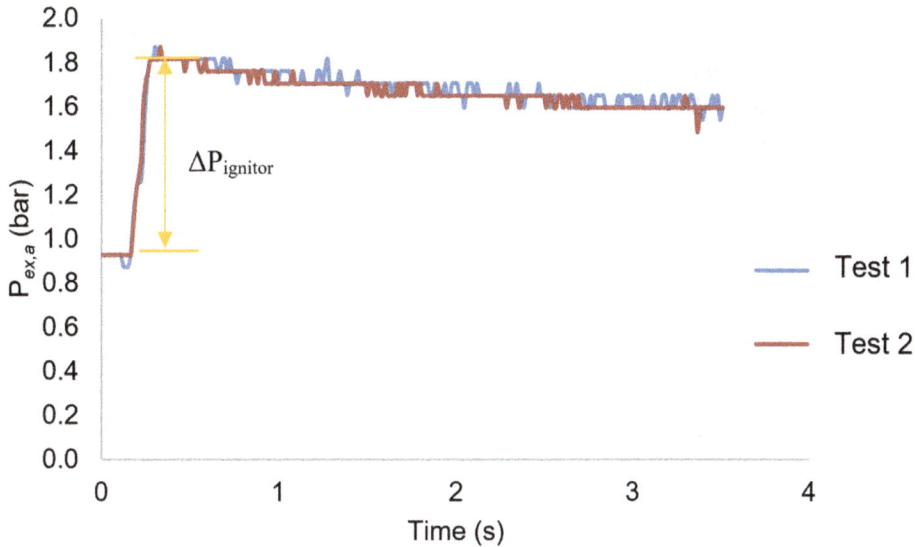

FIG 2 – Pressure–time curve of ignitor (test results for ignitor 4.6 kJ).

It can be seen in Table 4 that the $\Delta P_{ignitor}$ value obtained through Test 1 and Test 2 is 0.94 bar. This value will later be used in calculating the pressure ratio. The pressure ratio value can be calculated using the following equation from ASTM E1226 (2019).

$$PR = \left(P_{ex,a} - \Delta P_{ignitor}\right)\Big/ P_{ignition} \tag{4}$$

Notes:

PR   is a value used to classify a dust sample as explosive or not. The condition for the value is greater than or equal to two at each concentration tested

$P_{ex,a}$ is the maximum absolute pressure in the single test

$P_{ignition}$ is the absolute pressure in the vessel at the time of ignition

Information regarding the $P_{ex,a}$, $P_{ignition}$ and $\Delta P_{ignitor}$ values from the experiments carried out can be seen in the next section.

TABLE 4

Recapitulation of test results for determining P_{max} and $(dP/dt)_{max}$.

Concentration (g/m³)	200		250		500		750		1000		1250	
Explosion data	P_{ex} (bar)	dP/dt (bar/s)	P_{ex} (bar)	dP/dt (bar/s)	P_{ex} (bar)	dP/dt (bar/s)	P_{ex} (bar)	dP/dt (bar/s)	P_{ex} (bar)	dP/dt (bar/s)	P_{ex} (bar)	dP/dt (bar/s)
Test 1	4.11	72.65	4.72	128.21	5.33	_143.52_	5.44	129.63	**5.50**	97.22	4.89	72.65
Test 2	4.72	_179.49_	4.78	111.11	5.28	101.85	**5.39**	89.74	5.17	115.74	4.89	92.59
Test 3	4.39	83.33	5.10	_200.86_	**5.56**	152.78	5.44	101.85	5.33	148.15	5.17	111.11

Notes: The maximum values for each series are highlighted by underlined and bold text.

Coal dust explosibility test results

Data from coal dust explosibility tests will be used to determine the values of P_{max}, $(dP/dt)_{max}$, and K_{St}. Based on the three series of coal dust explosibility test results, values for P_{ex} and dP/dt can be obtained from explosibility tests using various dust concentrations, ranging from 200 g/m³ to 1250 g/m³ (see Table 4).

Based on ASTM E1226 (2019), at least three series of tests are needed to obtain each of the three maximum pressure rise values (P_{ex}) and the maximum rate of pressure $(dP/dt)_{ex}$ in a single test. In Table 4, the results of three series of tests for coal dust concentrations of 200 g/m³ and 250 g/m³ show a $(dP/dt)_{ex}$ value that is two times greater than the other two series of tests. The significant differences in the test results indicate that the coal samples used have heterogeneous properties (Zhu, 2014). Therefore, in coal dust explosion testing to determine the maximum pressure requires three tests to get a pessimistic value (worst case scenario). The test results shown in Figures 3 and 4 are for test results at a concentration of 250 g/m³. The test results at other concentrations show similar curve shapes, the only difference is the maximum P_{ex} obtained.

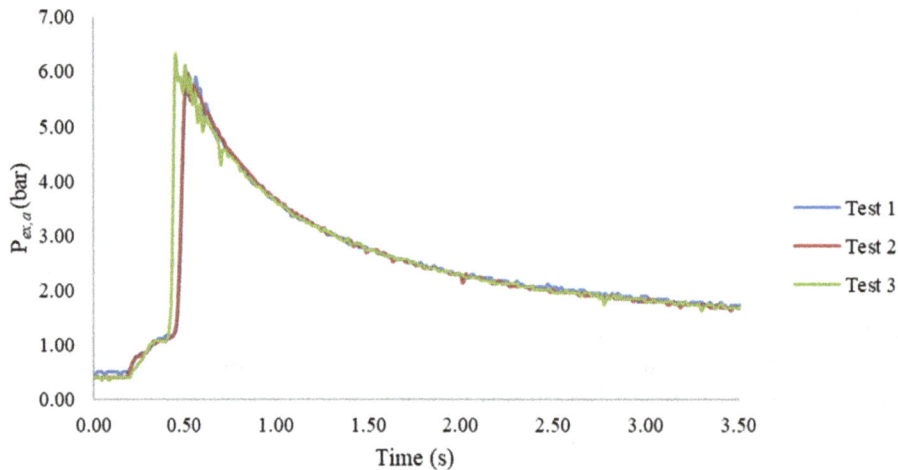

FIG 3 – Pressure – time curve of explosible coal dust (the experiment results at a concentration of 250 g/m³).

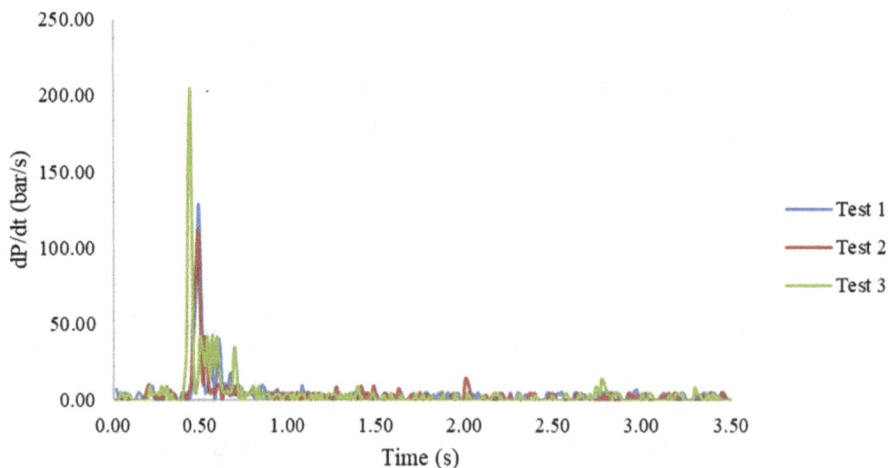

FIG 4 – dP/dt – time curve of explosible coal dust (the experiment results at a concentration of 250 g/m³).

By using data from Table 4 into Equations 1 to 3, the P_{max} and $(dP/dt)_{max}$ values can be obtained as follows:

$$P_{max} = (5.50 + 5.39 + 5.56)/3 = 5.48 \text{ bar}$$

$$(dP/dt)_{max} = (143.52 + 179.49 + 200.86)/3 = 174.62 \text{ bar/s}$$

$$K_{St} = 47.40 \text{ bar·m/s}$$

The three calculated values for P_{ex} and dP/dt_{ex} are estimates of the maximum values when the explosion occurs from each test series based on ASTM E1226 (2019). From the resulting K_{St} value, by referring to Table 1, it can be concluded that the explosion peak belongs to the weak explosion class ($0 < K_{St} \leq 200$).

Comparison with previous experiments

The research that will be used as a comparison is research conducted by Azwar (2019) and Said (2019) who studied the effect of particle size on the explosibility of coal dust for subbituminous coal ranks in the concentration range of 250 g/m³ to 2000 g/m³

Regarding the particle size of the dust samples used, the samples in this study have a larger size range than the sample size ranges in the two previous studies. To obtain the desired particle size, at the preparation stage, Azwar (2019) used a sieve with a size of -200# +270# and Said (2019) used a sieve with a size of -270# +325# hence the coal dust particle size distribution was relatively uniform and the value The D50 is the middle value of the range of sieves used. However, in this study, sample preparation was carried out using sieves -200# +230#, -230# +270#, and -270# +325#. This preparation produces more diverse (not uniform) particle sizes with the largest particle size distribution being in the large size range (74–65 µm) of 61.6 per cent (see Table 2), hence the distribution shows a negative degree of skewness (Figure 5). The large sample non-uniformity in this study affected the resulting P_{max}, $(dP/dt)_{max}$, and K_{St} values.

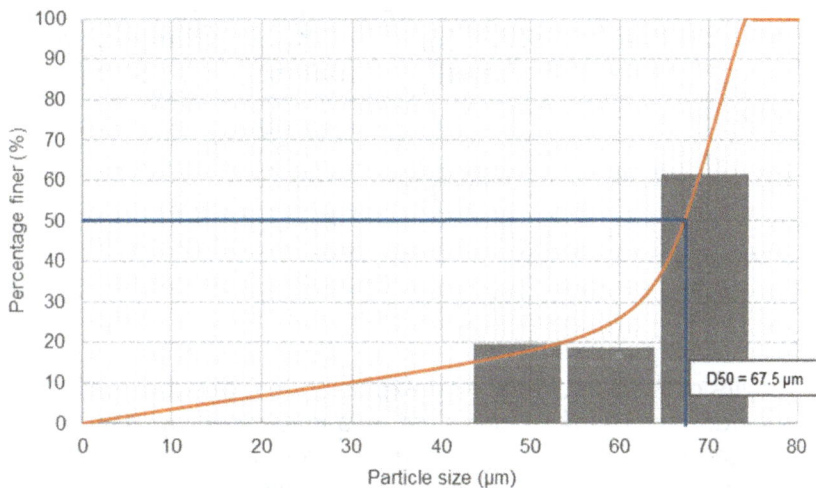

FIG 5 – D50 coal dust particle for presence research.

The smaller particle size (D50 = 48.5 µm) has a larger P_{max} of 12.25 bar in comparison with the larger particle sizes (D50 = 63.5 µm and D50 = 67.5 µm) which produces 9.58 and 5.48 bar, respectively, *vice versa*. Fine particles have a larger surface area and will burn more easily, but their explosive ability will decrease if mixed with larger particles with a smaller surface area (Widodo *et al*, 2023).

Then, based on the P_{max} results obtained, it shows that there is an influence of size distribution on the maximum pressure of coal dust explosion. The explosion starts from the ignitor in the middle, causing an exothermic heat reaction in the nearby coal dust. When it reacts, it causes a chain reaction in other coal dust particles (Khan *et al*, 2022). The more uniform the particle size, the more evenly distributed the heat generated from one particle so that the heat can reach other particles. If it is not uniform, the distance between particles will not be uniform so that the heat of the chain reaction will not be very effective.

In general, non-uniformity with a high percentage mass in large particle size will reduce explosivity of coal dust, resulting in lower P_{max}, $(dP/dt)_{max}$, and K_{St} values. These results are also supported by the results of research conducted by Cashdollar (1996) on high volatile B bituminous coal with a particle size range of 100–20 µm at a concentration of 200–2000 g/m³ (see Figure 6). The trend of increasing P_{ex} values can be used as an indicator of the P_{max} value of coal dust explosions because the values are quite close (see Table 5). Thus, the P_{max} resulted in the Cashdollar (1996) are larger than presence research, while smaller than Azwar (2019) and Said (2019) even though the coal used in the Cashdollar (1996) has a higher ranking (higher carbon content). This can be caused by the size range used by Cashdollar (1996) being quite large in the range of 100–20 μm, so the particle size non-uniformity is bigger than those of Azwar (2019) and Said (2019).

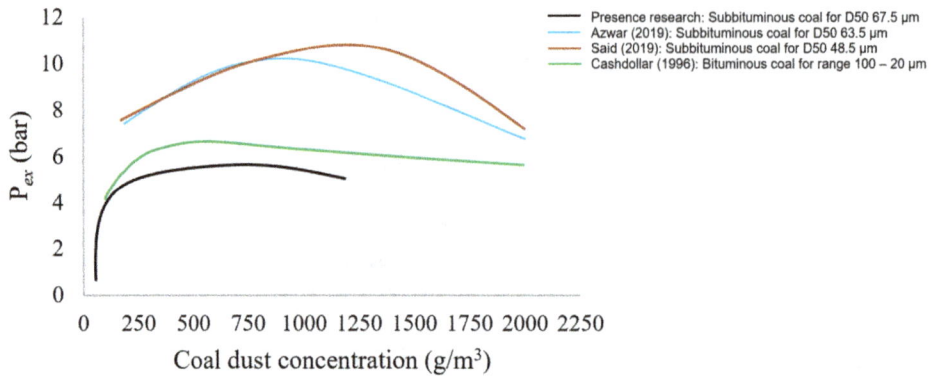

FIG 6 – Explosion pressure – coal dust concentration curve for comparison of the effect of particle size and distribution.

TABLE 5

Comparison of test results and coal sample sizes

Research	P_{max} (bar)	$(dP/dt)_{max}$ (bar/s)	K_{St} (bar·m/s)	Range of particle size (μm)	D50 (μm)
Present research	5.48	174.62	47.40	74-44	67.5
Azwar (2019)	9.58	398.59	108.20	74-53	63.5
Said (2019)	12.25	465.81	126.44	53-44	48.5

CONCLUSIONS

From the research results of the Jambi coal dust explosion test of subbituminous rank with a 4.6 kJ ignitor, the results showed that Pmax of coal dust was influenced by: 1) the size of the coal dust particles, the finer they are, the higher the maximum pressure due to the greater number of reaction surface areas; 2) uniform distribution of coal dust particle size, the more uniform the particle size, the more evenly distributed the heat generated from one particle so that the heat can reach other particles. Therefore, before carrying out an explosion test, it is recommended to carry out a particle size distribution analysis test. Further research is needed to analyse the influence of particle size distribution parameters, including the degree of skewness, coefficient of variance, and its average on the characteristics of coal dust explosions at various rank of coal.

ACKNOWLEDGEMENTS

Authors gratefully acknowledge Institut Teknologi Bandung, Indonesia, supported by the Research, Community Service, and Innovation Programs (2017, 2018 and 2021) of Institut Teknologi Bandung, Indonesia, and China University of Mining and Technology China, for supporting the research facilities. The Authors would like to thank to Center of Research Excellence Underground Mining and Mine Safety (CORE UMMS) and Ventilation Laboratory of Department of Mining Engineering (DME), Faculty of Mining and Petroleum Engineering, Institut Teknologi Bandung, Indonesia.

REFERENCES

ASTM, 1997. D3180–89 – Standard Practice for Calculating Coal and Coke Analyses from As-Determined to Different Bases, ASTM International.

ASTM, 2005. D388–99 – Standard classification of coals by rank, ASTM International.

ASTM, 2019. E1226 – Standard Test Method for Explosibility of Dust Clouds, ASTM International.

ASTM, 2022. E1515 – Standard Test Method for Minimum Explosible Concentration of Combustible Dust, ASTM International.

Said, S H P, 2019. Study of Jambi Coal Dust Concentration 53–44 µm Effect on Explosion Pressure of Coal Using 20 Liter Explosion, Institut Teknologi Bandung, Indonesia. Available from: <https://digilib.itb.ac.id/gdl/view/42760/>

Azwar, U, 2019. Study of Coal Dust Concentration 74–53 µm Effect on Explosion Pressure of Coal Using 20 Liter Explosion Chamber at Laboratory Scale for Coal From Jambi Province, Institut Teknologi Bandung, Indonesia. Available from: <https://digilib.itb.ac.id/gdl/view/42766>

Bhattacharjee, R M, Dash, A K and Paul, P S, 2020. A root cause failure analysis of coal dust explosion disaster–gaps and lessons learnt, *Engineering Failure Analysis*, 111:104229.

Cao, W, Qin, Q, Cao, W, Lan, Y, Chen, T, Xu, S and Cao, X, 2017. Experimental and numerical studies on the explosion severities of coal dust/air mixtures in a 20-L spherical vessel, *Powder Technology*, 310:17–23.

Cashdollar, K L, 1996. Coal Dust Explosibility, *Journal of Loss Prevention in the Process Industries*, Pittsburgh Research Center, Bureau of Mines, US Department of the Interior, Pittsburgh, USA.

Hertzberg, M and Cashdollar, K L, 1987. Introduction to dust explosions, in *Industrial Dust Explosions*, ASTM International.

Khan, A M, Ray, S K, Mohalik, N K, Mishra, D, Mandal, S and Pandey J K, 2022. Experimental and CFD Simulation Techniques for Coal Dust Explosibility: A Review, *Mining, Metallurgy and Exploration*, 39:1445–1463. https://doi.org/10.1007/s42461-022-00631-y

Norman, F, Berghmans, J and Verplaetsen, F, 2013. The minimum ignition energy of coal dust in an oxygen enriched atmosphere, *Chemical Engineering Transactions*, 31:739–744.

Occupational Safety and Health Administration (OSHA), 2009. Hazard communication guidance for combustible dusts, US Department of Labor, Occupational Safety and Health Administration, Washington DC, USA.

Reuters, 2009. Indonesia coal mine blast kills 17, traps 23, official. Available from: <https://www.reuters.com/article/idINIndia-40390720090617> [Accessed: 31 October 2021].

Roychowdhury, S N, 1960. Coal dust explosions and their prevention, Masters Thesis, Missouri School of Mines and Metallurgy.

Widodo, N P, Ihsan, A, Astuti, A W, Karnadiwijaya, R M I, Arisandi, A G, Sulistianto, B and Wahyudi, S, 2019. Study of Critical Concentration on Coal Dust-Air Explosion in 10L and 20L Closed Chambers, *International Symposium on Earth Science and Technology 2019*. Kyushu University, Japan.

Widodo, N P, Putra, S H, Ihsan, A, Gautama, R S, Cheng, J, Waly, F Z and Permadi, D A, 2023. The study of coal dust minimum explosion concentration of subbituminous coal, *Process Safety and Environmental Protection*, 177:1387–1392. http://dx.doi.org/10.1016/j.psep.2023.08.002

Yin, L B, Li, J H, Xu, C H and Wen, Z Y, 2011. Experimental Research on Explosion Characteristics of Indonesian Coal Dust, *J Guangdong Electric Power*, 7.

Zhu, Q, 2014. Coal sampling and analysis standards, IEA Clean Coal Centre, London, United Kingdom.

Mine gases

Application of a new approach to assessing the self-heating status of coal using gas monitoring results

B B Beamish[1], M Brady[2] and J Theiler[3]

1. Managing Director, B3 Mining Services Pty Ltd, Darra Qld 4076.
   Email: basil@b3miningservices.com
2. Director, Joncris Sentinel Services Pty Ltd, Rockhampton Qld 4700.
   Email: michael@joncris.com.au
3. Senior Mining Engineer, B3 Mining Services Pty Ltd, Darra Qld 4076.
   Email: jan@b3miningservices.com

ABSTRACT

TARPs are a very effective frontline tool for spontaneous combustion hazard management. However, to fully understand the self-heating status of coal requires supplementary trending of gas monitoring data to recognise any significant shift from normal background behaviour. It also requires more in-depth analysis to identify stages of further heating development. Various gas indicators have been developed over many years and one of the more common ones in practical use is Graham's Ratio, which is obtained by comparing the carbon monoxide produced to the oxygen consumed by the coal. In most self-heating assessments Graham's Ratio is used as a temperature indicator only, based on results obtained from small-scale gas evolution testing. As such, the trending of the Graham's Ratio value with time is often only thought of from the perspective of its magnitude as an indicator of temperature. However, the nature of the mine environment being sampled with respect to oxygen deficiency can affect the value obtained for the Graham's Ratio. Recent work has shown that the self-heating status can be definitively evaluated by plotting the Graham's Ratio value against the oxygen deficiency value, known as the Beamish Plot. This paper presents gas monitoring results from four heating events as working examples of how to evaluate the self-heating status of coal using this technique. If the Beamish Plot had been available at the time of these events they all could have been detected at an earlier stage.

INTRODUCTION

There are numerous gases and gas ratios recorded from gas monitoring sampling to help provide an indication of the level of oxidation present in a coalmine environment. In Australian coalmines these gas indicators are used to help define the various levels within Trigger Action Response Plans (TARPs). One of the more common gas indicator ratios in practical use is the Graham's Ratio (GR). This ratio is the amount of carbon monoxide produced by the coal for the amount of oxygen consumed and is represented by the equation:

$$GR\ (\%) = (CO \times 100) / (Oxygen\ deficiency)$$

where the oxygen consumed is reported as an oxygen deficiency value, and the ratio is multiplied by 100 to produce a meaningful value since the CO concentration is normally at ppm levels. The nature of Graham's Ratio equation shows that it may asymptotically approach zero when the gas sample has been obtained from a fully inert atmosphere or infinity when the gas sample approaches fresh air.

Since the objective of a sequentially sealed goaf is to create a self-inert atmosphere, it can be expected that over time the GR value of the goaf at each sampling point from the shallow goaf to the deep goaf with respect to the retreating face will asymptotically approach zero or a very small value. In most self-heating status assessments, the GR value is used as a temperature indicator only based on results obtained from small-scale gas evolution testing as shown in Figure 1. As such the trending of the GR value with time is often only thought of from a coal temperature perspective. However, the atmosphere as the hot spot develops tends to move from being in a low oxygen deficiency state to a high oxygen deficiency state as shown in Figure 2. Due to the nature of the Graham's Ratio equation, it can be envisaged that as the heating temperature increases a series of temperature isotherm lines would be developed dependent on the mine environment where the sample is collected as shown in Figure 3. Under normal inerting circumstances the trend should follow along the isotherm line at constant temperature. However, if a trend develops that progresses from T1 to T2 to T3 etc as shown by the direction of the red arrow, this is a clear indication that a heating is developing.

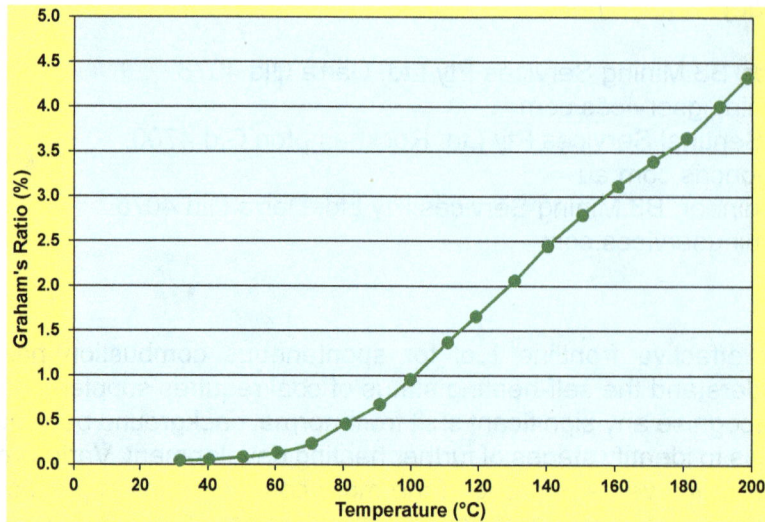

FIG 1 – Graham's Ratio in response to increasing coal temperature for a medium volatile bituminous coal.

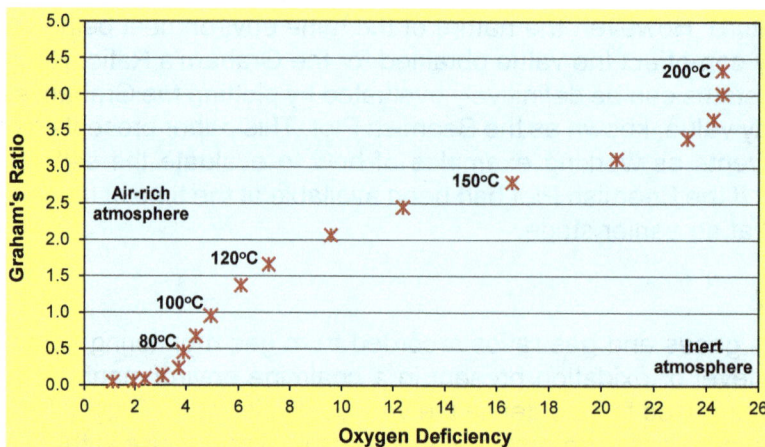

FIG 2 – Relationship between Graham's Ratio and Oxygen Deficiency in response to increasing coal temperature for a medium volatile bituminous coal.

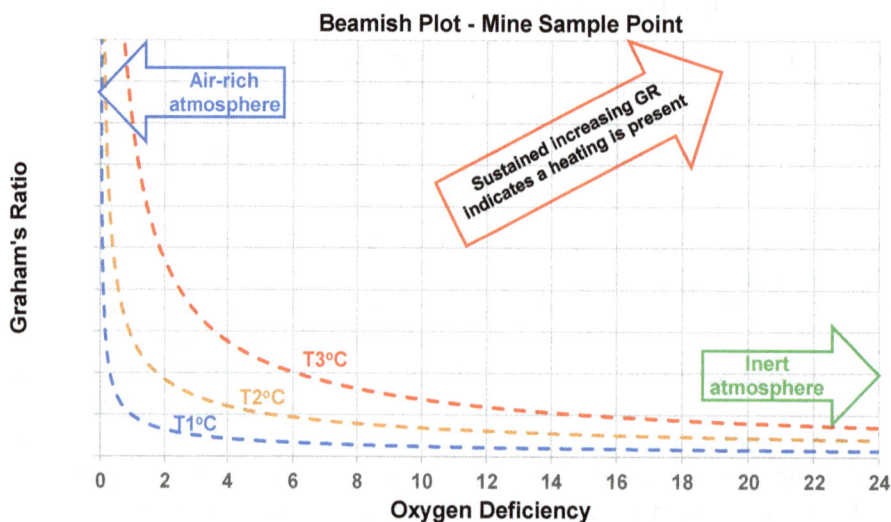

FIG 3 – Relationship between Graham's Ratio and Oxygen Deficiency in response to increasing coal temperature for a medium volatile bituminous coal in different mine environments.

From an operational perspective the data in Figure 3 is best presented on a log-log plot, such that temperature isotherm lines can be constructed for the coal from the gas evolution testing results as shown in Figure 4. The temperature value of the isotherm lines is obtained from the intersection of the gas evolution test results for each coal with the isotherm lines whose slope has been empirically derived from mine experience. Using a log-log scale also increases the sensitivity of any change in trend and it can be seen that a persistent increasing trend on the plot across the temperature isotherm lines indicates a heating is developing. This plot has been routinely used at Narrabri Mine in New South Wales since the beginning of 2023 as a crosscheck of self-heating status and is referred to by the mine operations as the Beamish Plot. It has since been used at several underground coalmines in New South Wales and Queensland on a routine basis. Each mine has its own unique Beamish Plot template since the coals being mined have different reactivities and mine environment boundary conditions.

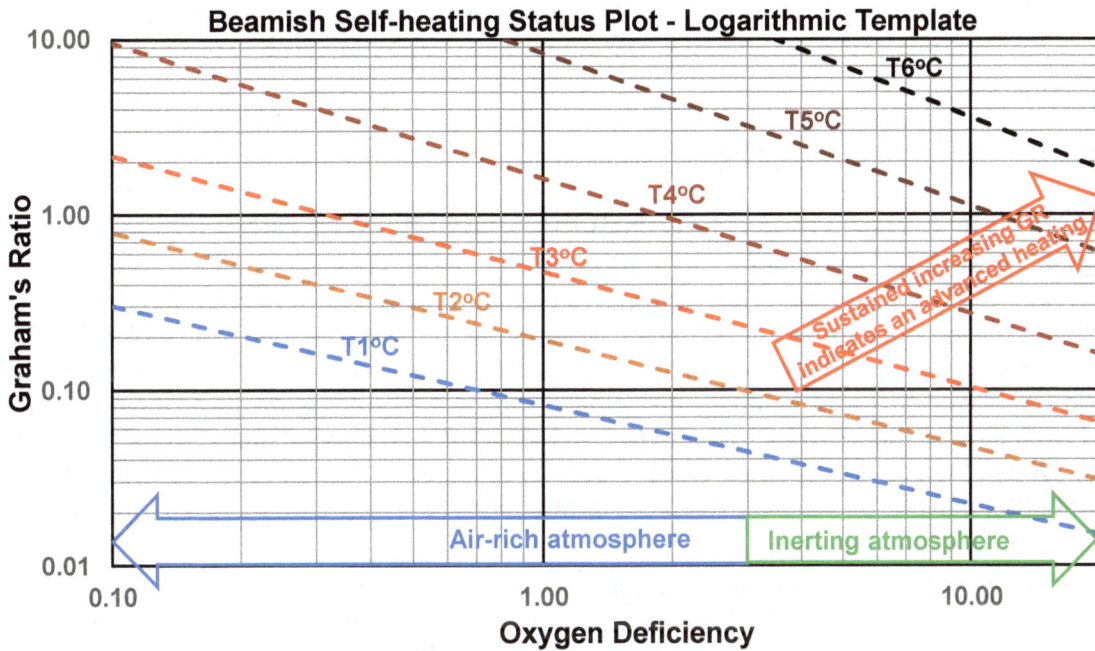

FIG 4 – Beamish Plot (Graham's Ratio versus Oxygen Deficiency) using a logarithmic scale.

Case study examples are presented in this paper that show the anatomy of each heating as recorded by the Beamish Plot and the contrast between different reactivity coals at both low temperature and advanced oxidation levels.

MINING PARAMETERS AND COAL PROPERTIES

The coals presented in the case study examples are from Australian underground coalmines and range in rank from high volatile B bituminous to medium volatile bituminous. The details for each mine are contained in Table 1. All these mines are longwall operations mining in thick seams. The intrinsic reactivity of the coal at three of the mines is rated as low compared to the fourth mine which is rated as high. Additional details for the Dartbrook, North Goonyella and Southland events are contained in NSW Resources Regulator (2021) and Industry and Investment NSW (2011).

TABLE 1

Mine details for heating case studies.

Mine	Carborough Downs	Dartbrook	Southland	North Goonyella
Year of Event	2012	2006	2003	1997
Coal Basin	Bowen	Sydney	Sydney	Bowen
Seam	Leichhardt	Kayuga	Greta	Goonyella Middle
Seam thickness (m)	6	4–4.5	4.8–7.0	6–8
Mining method	Longwall	Longwall	Longwall	Longwall
ASTM coal rank	Volatile bituminous	High volatile B bituminous	High volatile A bituminous	Medium volatile bituminous
R_{70} value	<0.8	3–5	<0.8	<0.5
Intrinsic reactivity	Low	High	Low	Low

ANATOMY OF A HEATING AT CARBOROUGH DOWNS MINE IN 2012

In 2012 the longwall at Carborough Downs Mine was recovering 4.4 m of the coal seam being mined and leaving 1.6 m behind in the goaf. At the beginning of May a fault was encountered with a throw >5 m. The resulting poor mining conditions caused a cavity to form above the shields and in some instances this spanned 20+ shields. This cavity was treated with a variety of support materials including Fenoflex, PUR and FB200, most of which produced exothermic reactions. The constant support work slowed the longwall to the extreme and in some instances up to 6 m of coal was left in the goaf. Due to the coal being gas drained a large quantity of air was used to remove coal dust from the face during cutting. In addition, a lack of a bleeder system resulted in 'basic' monitoring of the goaf.

Gas monitoring of the goaf showed a progressive sequence of the coal self-heating from initial background temperatures of <40°C to 45°C during February as shown by the Beamish Plot in Figure 5a. The coal temperature slowly increased through March as seen in Figure 5b and in April the coal appeared to have reached an equilibrium temperature of approximately 55°C in the goaf environment (Figure 5c). These results are consistent with the incubation self-heating behaviour of the coal as shown in Figure 6.

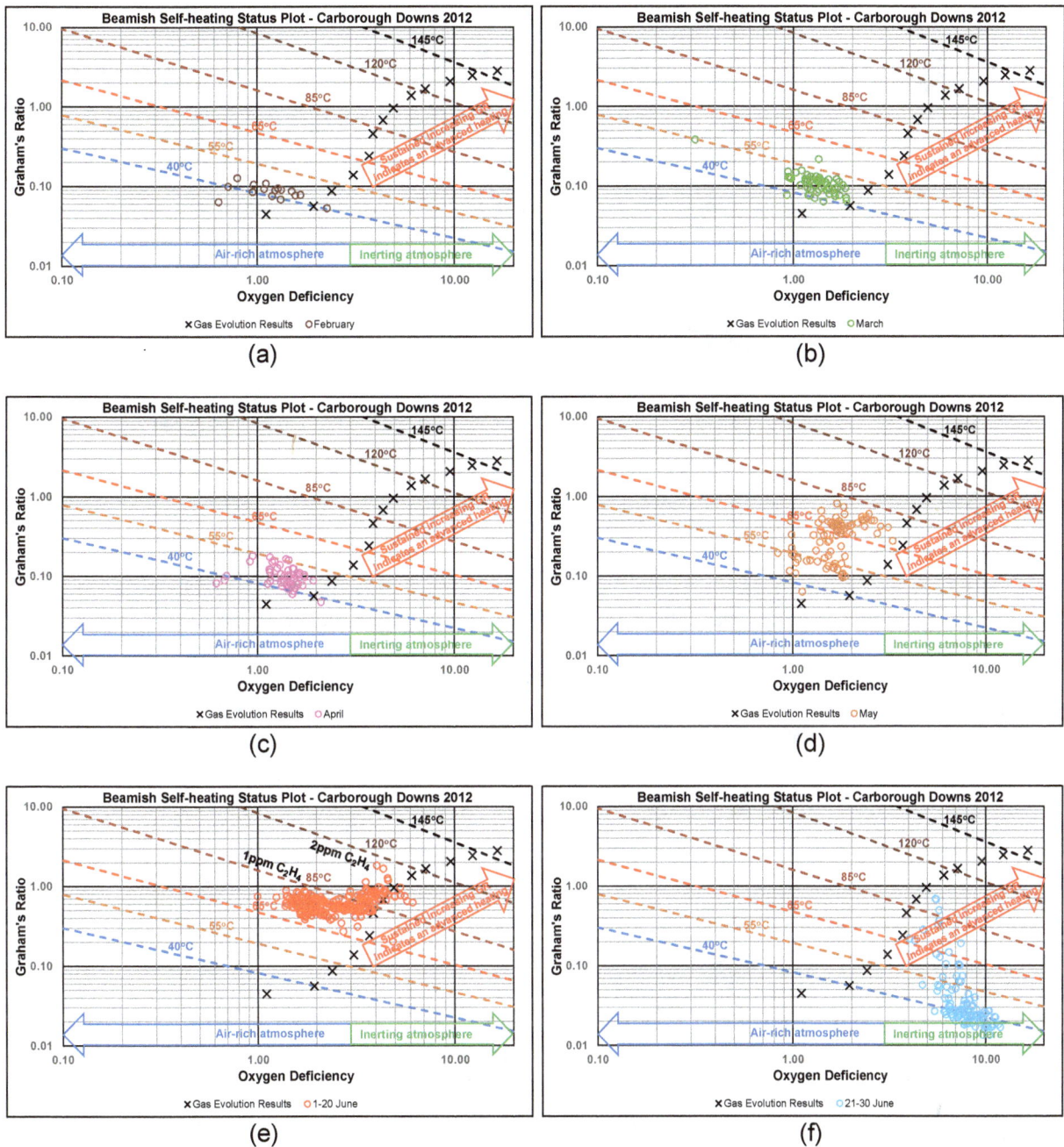

FIG 5 – Carborough Downs 2012 heating event sequence and subsequent inertisation as recorded by the Beamish Plot.

On May 2nd extensive face cavity remedial works were implemented due to the faulted ground encountered by the longwall. This resulted in the carbon monoxide make beginning to gradually trend upward. The Beamish Plot for the month of May (Figure 5d) clearly shows a dramatic increase in the coal temperature, which approached 80°C. This resulted in withdrawal from the mine to the surface on May 28th.

On June 3rd there was a first re-entry of the mine, however on June 5th the IMT again withdrew personnel from the mine. A single re-entry was made into the mine on June 9th to restore power. By June 20th a peak CO gas make of over 240 L/min was recorded reporting to the tailgate tube. The temperature of the coal indicated by the Beamish Plot at this point was approximately 110°C (Figure 5e). Other gas indicators such as ethylene were also recorded during the temperature escalation in June and are shown in Figure 5e.

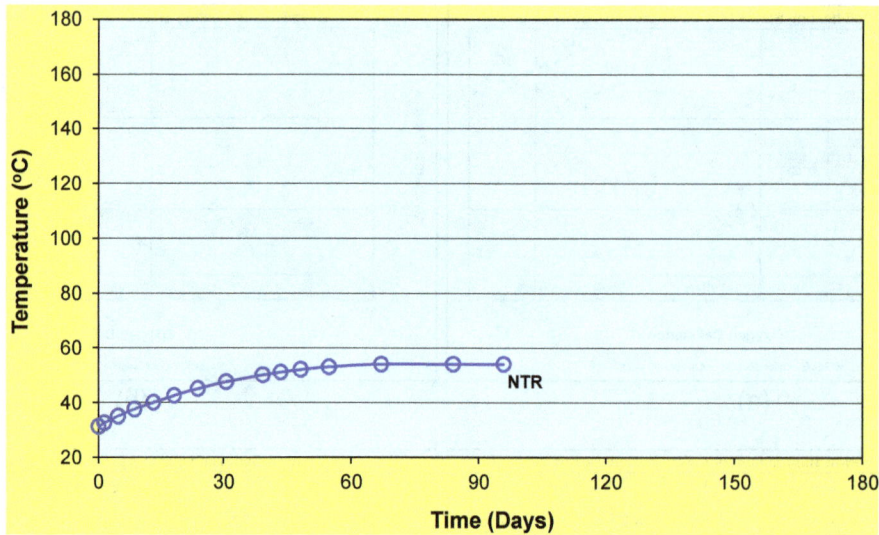

FIG 6 – Self-heating incubation behaviour of Leichhardt seam coal from Carborough Downs Mine from mine ambient temperature.

The incubation self-heating response to the introduction of the cavity fill products is shown in Figure 7. Effectively the additional heat generated by the exothermic reaction during the curing of these products alters the boundary conditions of the coal self-heating and consequently enables the coal to incubate to the point of thermal runaway if no intervention control is applied. Nitrogen inertisation was introduced to the area of the heating on June 21st, which subsequently removed oxygen so that the coal oxidation reaction could not continue. The response is shown on the Beamish Plot in Figure 5f.

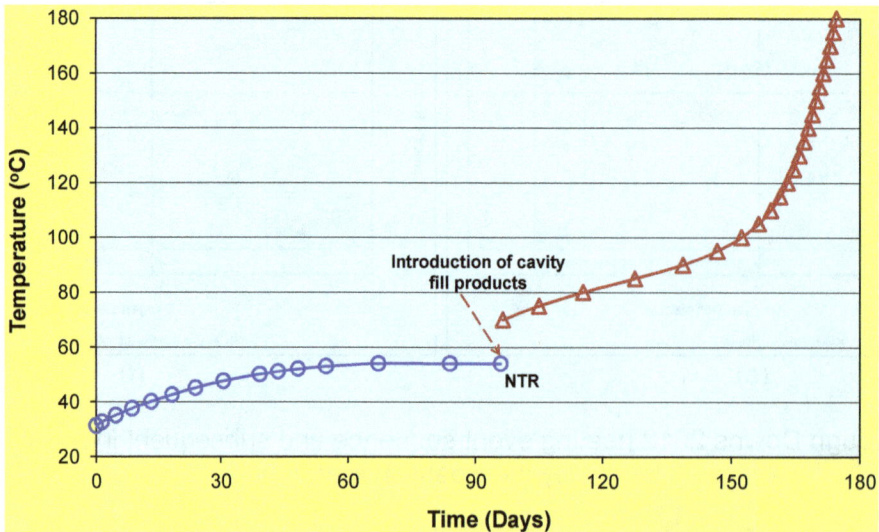

FIG 7 – Self-heating incubation behaviour of Leichhardt seam coal from Carborough Downs Mine from mine ambient temperature and subsequent response to exposure to an additional heat source.

ANATOMY OF A HEATING AT DARTBROOK MINE IN 2006

In mid-January 2006 the LW KA102 encountered deteriorating roof conditions resulting in a tailgate roof fall on January 15th, causing an obstruction and reducing airflow to the longwall face. On January 19th gasbag samples taken from seals were found to contain elevated levels of hydrogen and ethylene. The mine was evacuated and preparations made for inertisation to control the heating suspected to be between 9 and 10CT (Figure 8). The mine was re-entered on January 21st with limited activities. The mine was again evacuated and further inertisation took place with a reduced fan speed. The mine was re-entered on February 18th with limited activities taking place.

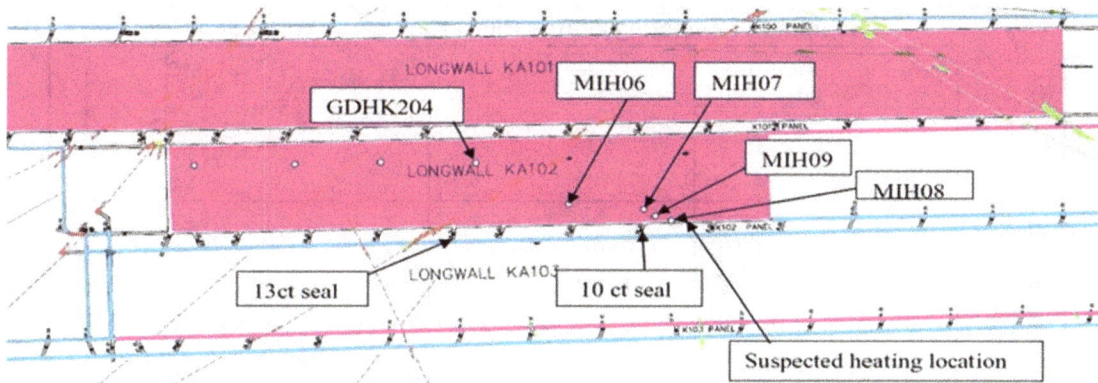

FIG 8 – Dartbrook Mine plan for Kayuga seam workings.

The gasbag data for 10CT, 11CT and 13CT are shown in Figure 9 using the Beamish Plot. The escalation in coal temperature at 10CT and 11CT is clearly seen by December 29th, 2005, with an indicated coal temperature of 60–70°C. An even clearer indication of the heating developing is shown by the results from 13CT (Figure 9c), which showed the coal heating from less than 25°C to 40°C between October 21st, 2005, and November 10th, 2005. By December 5th the temperature indication of the coal at 13CT was 60°C. The interesting thing to note about this response at this location is that it was recorded in a highly oxygen deficient atmosphere. On January 19th the gasbag results from 10CT indicated the coal had reached a temperature of 105°C (Figure 9a). At this temperature due to the high reactivity of Kayuga seam coal, significant hydrogen and ethylene production could be expected and values as high as 308 ppm hydrogen and 11 ppm ethylene were recorded. On February 7th gasbag results at 10CT indicated the coal had reached a temperature of 135°C, with corresponding hydrogen values up to 2250 ppm and ethylene up to 76 ppm.

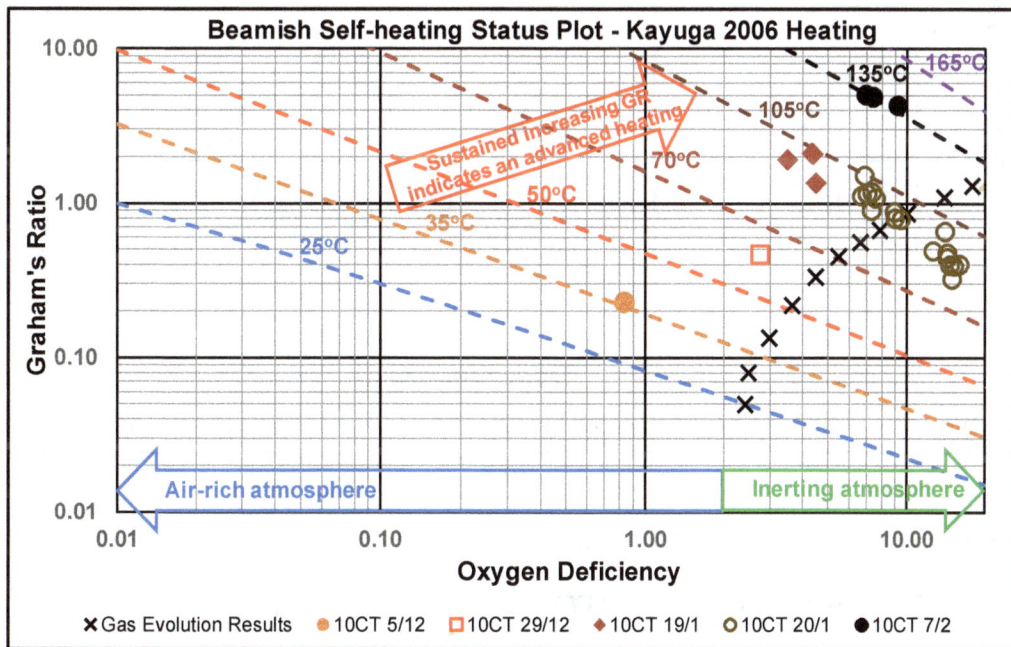

(a)

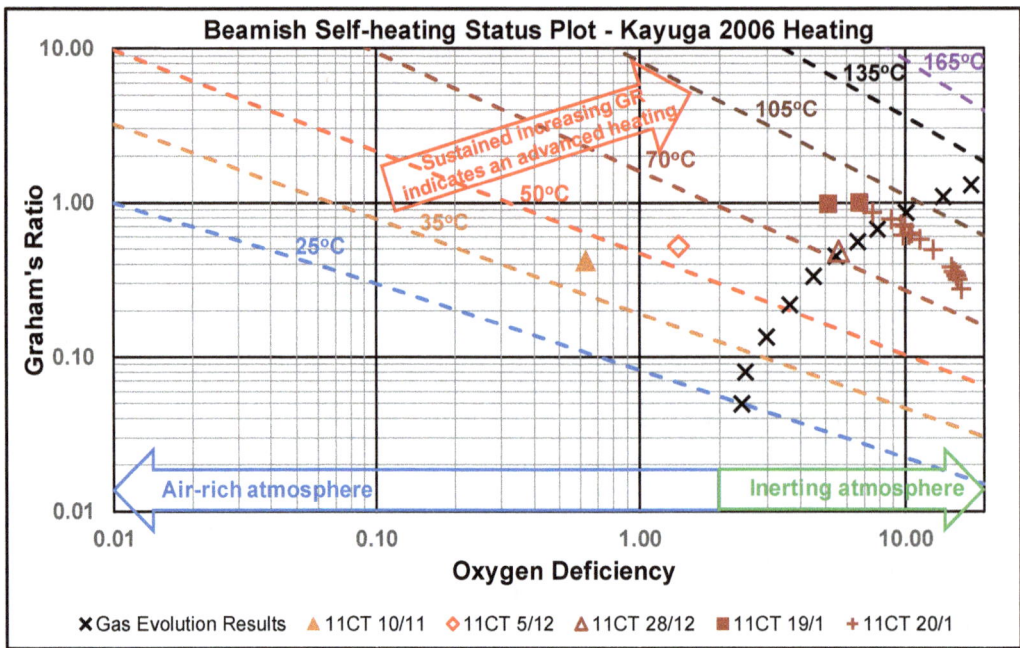

(b)

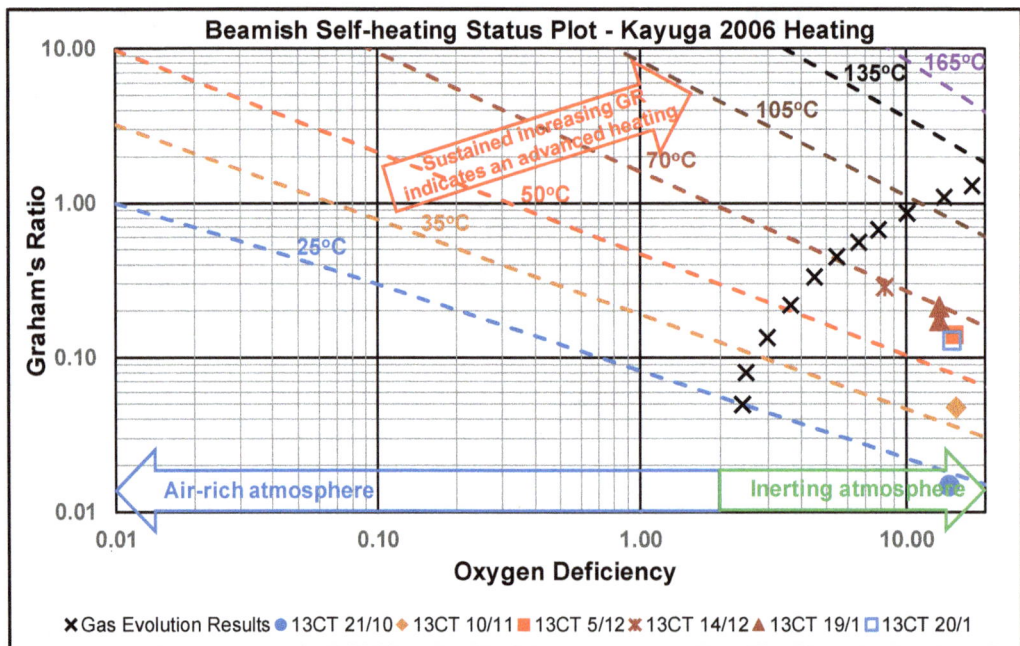

(c)

FIG 9 – Dartbrook 2006 heating event sequence as recorded by the Beamish Plot.

ANATOMY OF A HEATING AT SOUTHLAND MINE IN 2003

The heating took place in the longwall panel adjacent to the active panel as shown in Figure 10. The bottom 3 m of the Greta seam was mined for reasons of coal quality since the upper section of the seam contained high sulfur (pyritic) coal. The seam had a history of heatings despite all the usual spontaneous combustion propensity rating parameters indicating a low propensity (Beamish, Theiler and Ward, 2017). Subsequent testing of the coal using the more recent Incubation test method (Beamish and Theiler, 2019), showed that the pyrite present in the upper part of the Greta seam is reactive and capable of acting as an initiator of the coal self-heating to achieve thermal runaway in a reasonably short time frame under ideal conditions (Beamish, Theiler and Ward, 2017). The self-heating incubation behaviour of the Greta seam is shown in Figure 11, and it can be seen that the amount of moisture present for the pyrite to react with is an important factor, with an optimum moisture content of approximately 7 per cent creating the shortest incubation period in the order of

120 days. In a dry state the coal slowly self-heats as only the carbon is reacting with the oxygen. However, with moisture present it is the pyrite that reacts with the oxygen and water to elevate the coal temperature and initiate more rapid self-heating of the coal. Once the coal dries out the carbon-oxygen reaction takes over to elevate the temperature to the point of ignition. This can occur in a matter of only a couple of days.

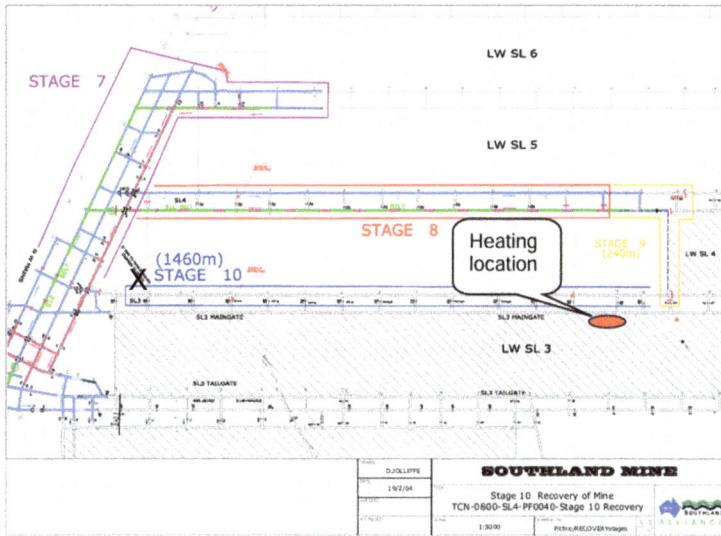

FIG 10 – Southland Mine plan for Greta seam workings.

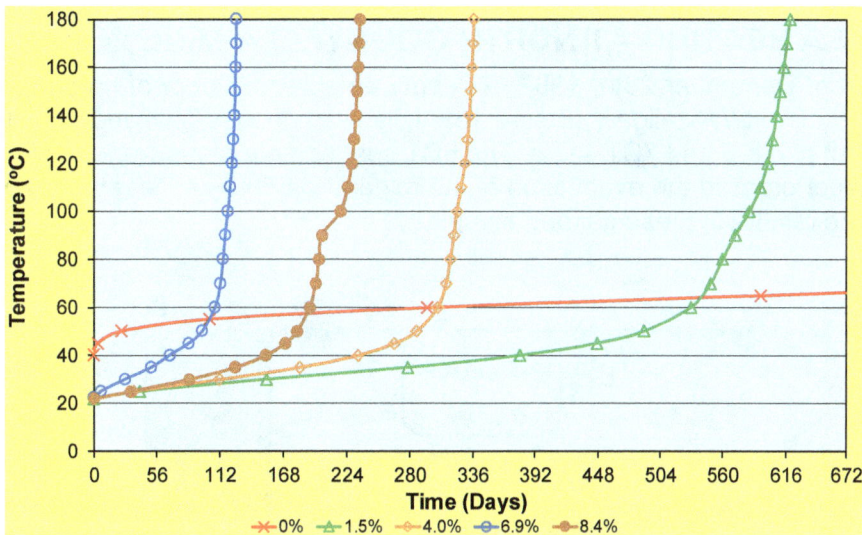

FIG 11 – Self-heating incubation behaviour of Greta seam coal from Southland Mine from mine ambient temperature at different moisture contents.

On December 23rd, a high CO alarm caused the mine to be evacuated. The goaf stopping adjacent to the longwall face crushed and air was entering the adjacent goaf. On December 24th, black smoke issued from the upcast shaft. On December 25th, the colour of the smoke changed to light grey and it was believed that fire had broken out of the goaf into the longwall tailgate. Initial attempts at inertisation were unsuccessful and the mine was then sealed to extinguish the fire. A Tomlinson boiler was used to assist in the re-entry of the mine (Industry and Investment NSW, 2011).

The gas data for SL4 Return AHDG 1–2CT (location X in Figure 10) are shown in Figure 12 using the Beamish Plot. There is a clear indication of heating progression from December 10th to December 17th. By December 23rd the temperature of the heating had escalated considerably and at this stage a well-defined hot spot would have formed and started to rapidly migrate towards the air source feeding it. Within a matter of 24 hrs the coal temperature at the hot spot had reached ignition as indicated by the last two data points obtained on December 24th shown in Figure 12. The gas data trends shown in Figure 12 are consistent with a reactive pyrite-initiated heating. The high oxygen deficiency values at the lower temperatures are a result of the pyrite reaction consuming oxygen. This in turn creates a much lower Graham's Ratio value as the carbon in the coal is not creating the temperature increase. Conversely, at elevated temperatures once the coal becomes dry the carbon reactivity takes over from the pyrite resulting in a rapid increase in the Graham's Ratio.

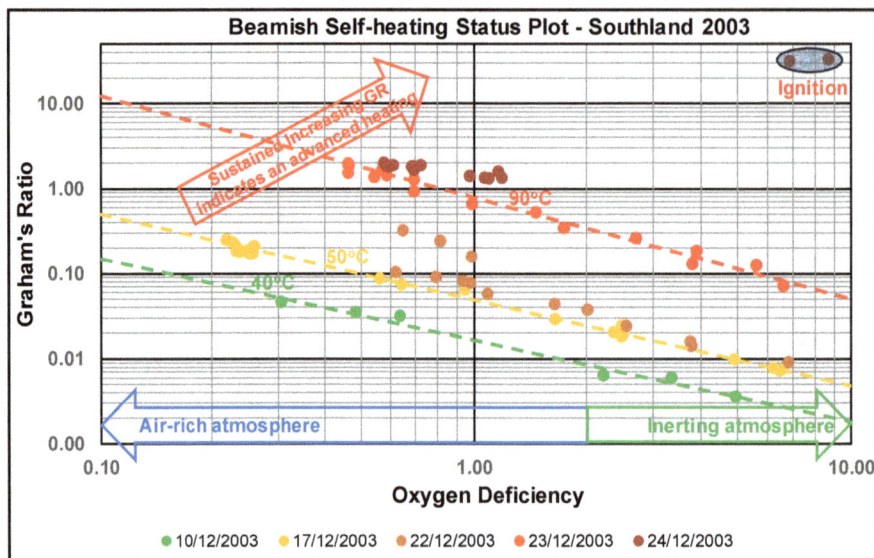

FIG 12 – Southland 2003 heating event sequence as recorded by the Beamish Plot.

ANATOMY OF A HEATING AT NORTH GOONYELLA MINE IN 1997

On the afternoon of December 28th, 1997, a deputy detected 25 ppm of carbon monoxide in the general body of 6CT in LW4 tailgate (Figure 13). This reading was followed up with bag samples from LW3 goaf out of the 5 and 7CT seals. The 6CT seal sample pipes were blocked with mud and water. The manager ordered the evacuation of the mine at 5:55pm on December 29th following the confirmation of the results of these gasbag samples.

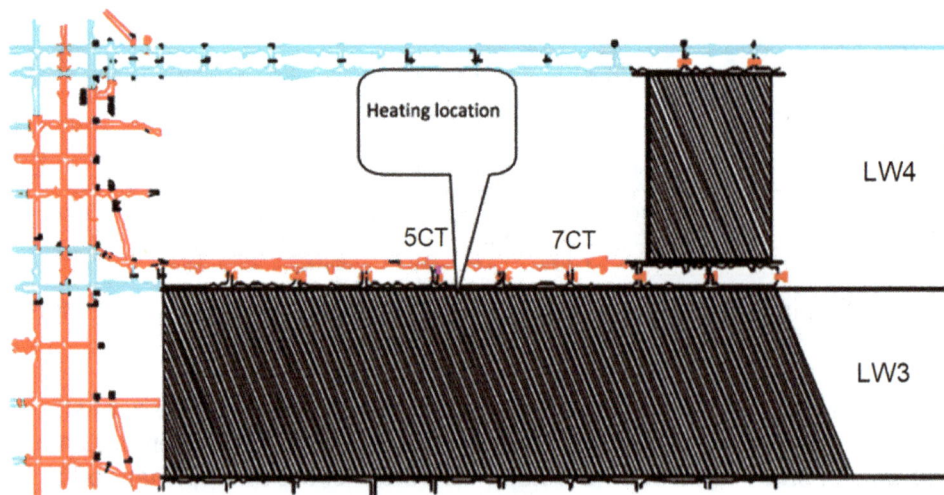

FIG 13 – North Goonyella Mine plan for Goonyella Middle seam workings.

The gasbag data for 5CT and 7CT are shown in Figure 14 using the Beamish Plot. The results from 5CT seal provide a definitive progression of the heating development. In late November to early December the indicated coal temperature had already reached 65°C (Figure 14a), which is well above the mine ambient temperature at the time of 35–40°C. By mid-December the coal temperature had reached 75°C. The heating then accelerated to reach more than 115°C by December 29th. Over the next 24 hrs the coal temperature had reached at least 180°C, although the gasbag results from

7CT were indicating coal temperatures more than 200°C (Figure 14b). Temperatures began to decrease after that following the introduction of inertisation.

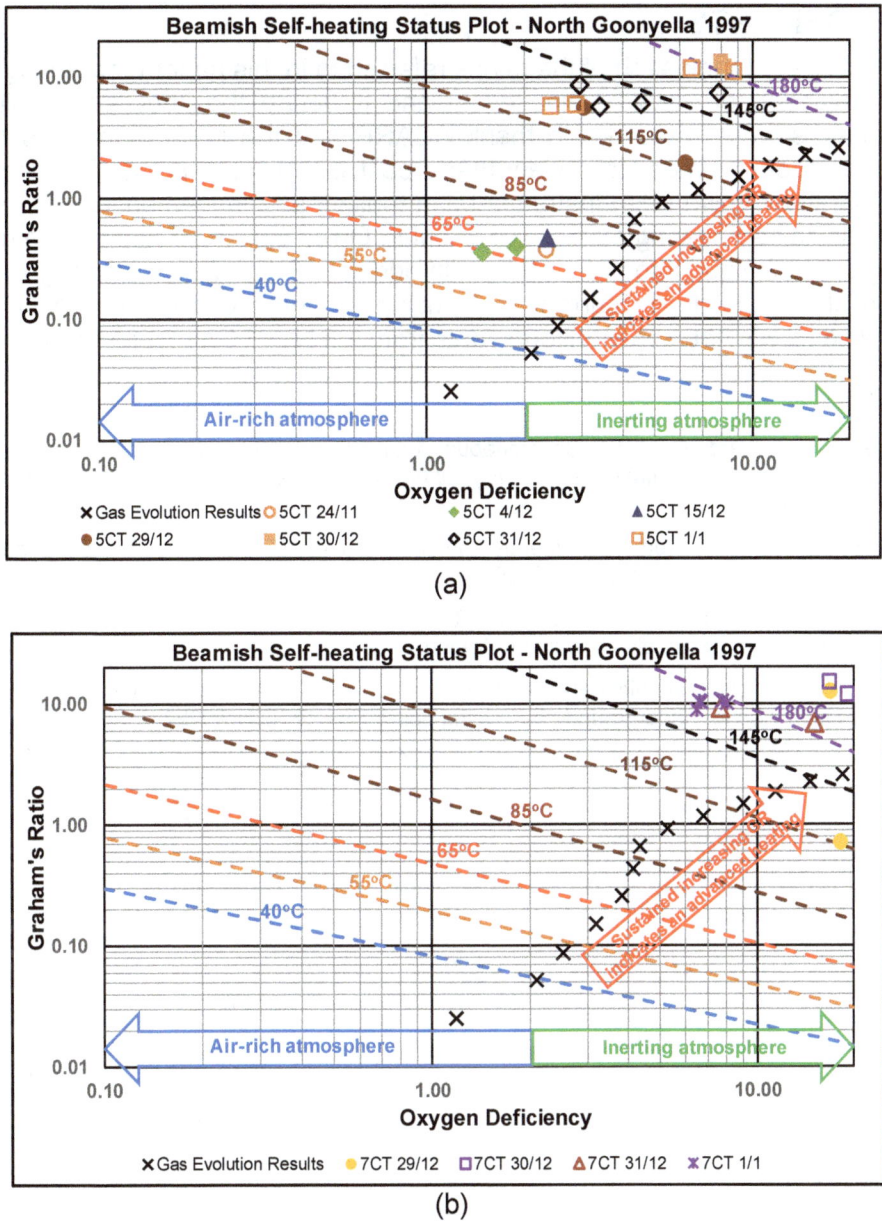

(a)

(b)

FIG 14 – North Goonyella 2006 heating event sequence as recorded by the Beamish Plot.

CONCLUSIONS

A closer examination of previous heating events in underground coalmines has enabled a clearer understanding of the anatomy of the event with respect to the coal self-heating incubation behaviour. The associated gas production from these events provides the opportunity to apply a new approach to assessing the self-heating status of the coal. While the use of gas indicators such as CO Make can track the development of an event, it cannot distinguish between a large mass/low temperature oxidation event and a high temperature hot spot event. In addition, the Graham's Ratio gas indicator parameter is not an absolute temperature indicator without taking into consideration the state of the mine environment, which can vary from air-rich to inert-rich (from seamgas or artificial introduction of nitrogen). Application of the Beamish Plot (Graham's Ratio versus Oxygen Deficiency) on a log-log scale provides a simple, accurate and effective method to monitor the self-heating status of the mine environment under all circumstances. It also overcomes the problem of having to evaluate multiple time series plots of different gases and gas ratio indicators for decision-making. The output of the Beamish Plot enables the early identification of a heating and subsequent progression of hot spot development towards thermal runaway. It can also be used to track the effectiveness of ventilation control measures.

ACKNOWLEDGEMENTS

The authors would like to thank each of the mines referred to in this paper for granting permission to publish the re-evaluation of each heating event and to share the knowledge gained from this exercise. The authors would also like to thank ACARP and the Mine Manager's Association of Australia for their encouragement to review these past events and present findings to industry. Narrabri Coal Operations have been instrumental in trialling and implementing the use of this new approach to assessing self-heating status. In particular, Owen Salisbury has been a great sounding board for putting the technique into practical use. In addition, Sharif Burra challenged us to improve the presentation of the plot to its current state using logarithmic axes. Finally, David Cliff has generously supplied us with the gas monitoring data from the North Goonyella and Dartbrook events.

REFERENCES

Beamish, B, Theiler, J and Ward, C, 2017. Spontaneous combustion behaviour of coal from the Greta seam, in *Proceedings of the 40th Symposium on the Geology of the Sydney Basin*, pp 154–162 (Coalfield Geology Council of New South Wales).

Beamish, B B and Theiler, J, 2019. Coal spontaneous combustion: Examples of the self-heating incubation process, *International Journal of Coal Geology*, 215:103297.

Industry and Investment NSW, 2011. MDG1006 Technical Reference: Technical Reference for Spontaneous Combustion Management Guideline, Mine Safety Operations Branch.

NSW Resources Regulator, 2021. Technical Reference Guide: Development of a spontaneous combustion principal hazard management plan for underground coal mining operations, DOC21/80224.

Improving pre-drainage, reducing gas influx, and lowering emissions from surface and underground coal mining operations using subterranean barrier implementations

R L Johnson Jr[1] and A Mirzaghorbanali[2]

1. General Manager – Technical Services, Novus Fuels, Brisbane Qld 4069.
   Email: ray.johnson@novusfuels.com
2. Associate Professor, The University of Southern Queensland, Brisbane Qld 4300.
   Email: ali.mirzaghorbanali@unisq.edu.au

ABSTRACT

Gas drainage has been performed in advance of coal mining for beneficial use or to reduce levels of gas to safe or regulatory levels required for coal mining. Inadequate pre-drainage can lead to higher levels of gas that may be adequately managed by ventilation systems, potential pressure scenarios, or even outbursts, any of which can affect mine safety and economics. Thus, it may be desirable to more rapidly depressurise the coal seam based on mine schedules, bolster stops and seals, or minimise the effects of carbon costs or emission risk profiles to improve project economics and demonstrate efforts towards emissions reduction.

Non-permeable barriers or 'curtains' to alter subterranean fluid flow patterns in coal have been proposed as a means to enhance pre-drainage in advance of coal mining and reduce mining emissions. This methodology uses hydraulic fracturing with non-permeable materials from horizontal wells to create regions of permeability damage, reducing gas migration. Modelling with representative Bowen Basin data has quantified the possible reduction in emissions from barrier implementation. Finally, previous economics outlined key uncertainties and variables affecting economic outcomes based on the timing of the implementation relative to the start of mining and the economic benefits associated with the emissions trading scheme (ETS) and carbon costing at that time.

This paper will outline the data and efforts required to implement a barrier placement. Reservoir characterisation and a planar, fully 3D fracture model can be used to model the barrier implementation. Reservoir modelling software can investigate a range of potential reservoir parameters (eg porosity, permeability, permeability anisotropy, reservoir dip etc) to assess the impact and sensitivity of parameters and derive the likely effectiveness. Further research into flow modification materials (FMM) and emerging hydraulic fracturing diagnostics can improve barrier placement implementation and verification. Finally, recommendations are made for multi-seam scenarios and operational implementation.

INTRODUCTION

In 2023, the Australian Government introduced the Climate Change Bill with the goals of a 43 per cent reduction in emissions by 2030 and achievement of net zero by 2050. Based on the Green House Gas Protocol of 2001, the world needs to drastically reduce 'Scope 1' and 'Scope 2' emissions. Consequently, emissions-intensive industries, such as metallurgical coal (MC), develop strategies to nearly halve emissions within the next seven years and be firmly established on a credible pathway to net zero and remain viable. Strong demand for Australian MC is forecasted for decades until substitutes are developed and available at scale. As coal is still an essential element for steel production, further emissions reductions via efficiency measures and carbon capture and underground storage (CCUS) technologies are being considered to achieve 'green steel' at some time in the future (McKinsey and Company, 2020; Nurdiawati and Urban, 2021).

Previously, barriers in mining have been applied to reduce water influx into mine works, primarily by selective grouting (cement barriers) around problematic areas. The potential for success of applying artificial barriers to separate areas of drained and undrained coals is considered high, given that natural and artificial barriers have demonstrated success in controlling subterranean flows. For example, naturally occurring barriers (eg faults, igneous intrusions, or dykes) have been shown to effectively separate areas of drained and undrained coals. One instance of this is the Grasstree Mine in Central Queensland, which experienced high initial gas contents despite a decade of effective drainage across a dyke separating Grasstree from the older Central Colliery.

Artificial impermeable barriers have been implemented in civil engineering applications to manage flow or contain contaminant plumes in groundwater and water influx into civil works (Anderson and Mesa, 2006; Hötzl, Eiswirth and Brauns, 2000). In the oil and gas sector, barriers have been proposed to control gas or water leakage or breakthrough in areas of storage, initial production, or secondary recovery processes, mostly involving gas or water egress into outlying reservoirs or unwanted influx into productive intervals (Johnson Jr and Sedaghat, 2023; Jurinak and Summers, 1991). Halliburton Energy Services conducted research and patented some applications of barrier treatments to modify flow and improve recovery in oil and gas reservoirs (Soliman *et al*, 2012).

BACKGROUND

To control open-cut methane emissions, wells can be drilled directly into the face and gas vented or diverted to flare, but the effectiveness can be hampered by continuous mining operations along the face and depth of penetration of at-face drilling. Emissions from underground mining can be from underlying, overlying and down-dip sources. Barrier implementation can reduce down-dip gas influx and aid goaf drainage. Previous modelling has been performed with representative Bowen Basin to quantify the potential reduction in emissions possible from barrier implementation in surface and underground mine operations (Johnson Jr and Sedaghat, 2023; Johnson Jr, 2023). In this section, we will overview the results of that modelling before going into the methodology of implementation.

Implementation of barriers and potential benefits of emissions reduction to surface mining operations

The most readily available solution to reducing emissions and increasing the beneficial use of CMM in surface mining operations is the use of surface-to-inseam (SIS) drainage wells. However, we will show how barrier implementations can further reduce surface methane emissions by up to 57 per cent for a 1 km^2 drainage area.

Example barrier implementation with surface mining operations

An example configuration uses 1000 m-long SIS wells drilled at 90 m spacing along a 1 km^2 drainage area (see Figure 1) parallel to the mining face and in advance of open-cut mining operations with a 'drainage impact zone' advancing across an example 1 km^2 area. To reduce deeper gas influx and improve drawdown, a barrier implementation using a horizontal well can be implemented, down-dip, at the location of the deepest SIS well in the initial drainage pattern, and 'shield' gas influx and improve the efficiency of the drainage pattern. Figure 2 illustrates a barrier implementation 510 m outbye for the same 1 km^2 area with four SIS wells and a barrier well, spaced every 90 m, starting from 150 m outbye, or an area of 50 per cent reduction from initial pressure (P_i) of 1159 kPa. This barrier well could be a single well, as shown in Figure 2, a continuous group of longer drilled barrier wells, strategically placed to manage pre-drainage in advance of mining operations.

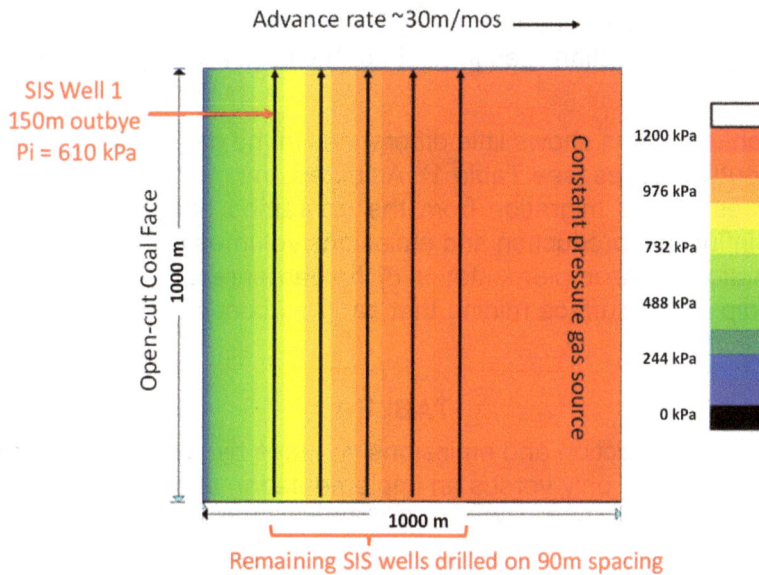

FIG 1 – Depiction of potential SIS well pattern across a 1 km² drainage area ahead of open-cut mining operations without the application of a barrier to lower overall gas emissions (Johnson Jr, 2023).

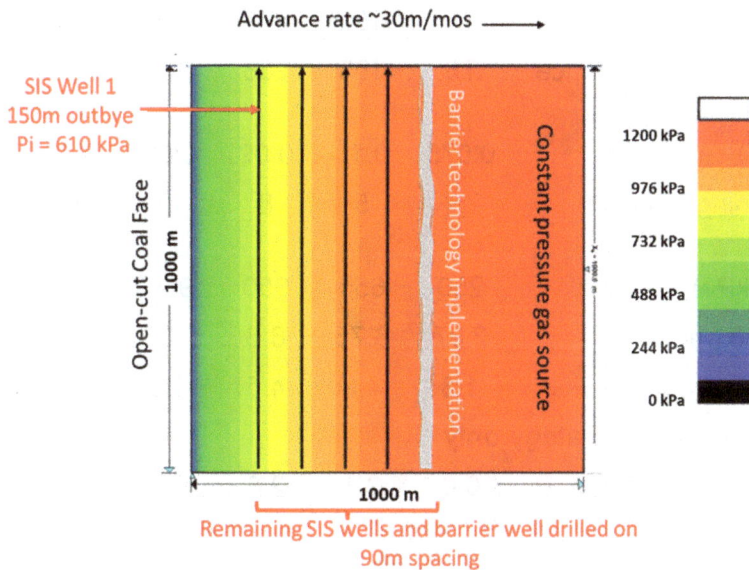

FIG 2 – Depiction of potential SIS well pattern across a 1 km² drainage area ahead of open-cut mining operations with the application of a barrier to lower overall gas emissions (Johnson Jr, 2023).

Modelling of emissions reduction benefits of barrier implementation with surface mining operations

A modelling study has been completed using reservoir properties for a representative Bowen Basin open-cut coal seam. A pressure front at near-initial conditions is assumed to be at ~50 per cent of the 1 km² area based on the initial modelling (Johnson Jr, 2023). This model (based on Figure 1) established an original gas-in-place (GIP) for the area at an estimated 64.6 Mm³ with a remaining GIP volume of 55.3 Mm³ after open-drainage to compare a SIS only versus a barrier with SIS implementation. As only 50 per cent of the area would be within the drainage impact zone, the GIP in that region would bear the reduction of 9.3 Mm³, resulting in a GIP in the targeted zone of 23.6 Mm³.

Two comparative cases could be made to evaluate the potential benefit of barrier implementation. 'Case A' uses three 90 m spaced SIS wells down-dip from the initial 150 m outbye SIS well (see Figure 1). 'Case B' can be estimated for SIS wells with a barrier implementation at 510 m (90 m in from SIS 4) (see Figure 2). These models allow the determination of differences with or without a barrier in volumes of gas production, gas influx into this region with drawdown, and potential total lost or migratory gas.

Modelling of these configurations shows little differences in the gas produced or methane emissions from the initial three outbye cases (see Table 1). As noted, the main difference is the furthest outbye well and the influences of gas migration from the undrained section of the reservoir. The main difference is the gas influx into production and emissions volumes from SIS 4 in Cases A and B (see Table 2). As aforementioned, the implementation of the barrier results in the entrapment of a sizeable volume of gas downdip of the surface mining that can be accessed for beneficial use or third-party sales.

TABLE 1

Resulting beneficial production and emissions volumes by segment for 1 km² study area comparing a plan of SIS wells only versus an implementation of barriers and SIS wells (Johnson Jr, 2023).

| Model region | Well name (case) | | | | | | Subtotals (Mm³) |
	1 Face	2 SIS1	3 SIS2	4 SIS3	5A SIS4 (A)	5B SIS4 (B)	
Well distance from face (m)	0	150	240	330	420	420	
Outer flow region boundary distance (m)	75	195	285	375	1000	510	
Drainage area (km²)	0.075	0.12	0.09	0.09	0.42	0.135	
Advance time to mining front (months)		5	8	11	14	14	
Average pressure (kPa)	250	650	850	960	1159	1075	
Average gas content (m³/t)	2.97	6.75	8.3	9.07	10.3	9.83	
Initial model region GIP (Mm³)	1.3	4.9	4.5	4.9	39.7	8.0	
Case A: SIS well mitigation strategy only							
Gas produced <510 m (Mm³)	0.0	2.1	2.8	3.2	5.2		13.3
Gas influx produced (Mm³)					3.9		3.9
Potential lost/migratory gas (Mm³)	1.3	2.8	1.7	1.7	30.5		38.1
Case B: SIS with barrier well mitigation strategy							
Gas produced (Mm³)	0.0	2.1	2.8	3.2		5.2	13.3
Potential lost/migratory gas (Mm³)	1.3	2.8	1.7	1.7		2.8	10.3

TABLE 2

Summary of production and emissions volumes for 1 km² study area comparing Case A, the implementation of SIS wells only, and Case B, the implementation of barriers and SIS wells (Johnson Jr, 2023).

Case	% Gas produced <510 m	% Gas influx and produced >510 m	% Gas to be lost/capable of migration	% Gas isolated from migration (ie >510 m)
Case A: SIS well mitigation strategy only	24%	7%	69%	0%
Case B: SIS with barrier well mitigation strategy	24%	0	19%	57%

Implementation of barriers and potential benefits of emissions reduction to underground mining operations

Gas influx can become a ventilation issue during mining operations and a management issue after mining operations are completed. Fugitive emissions and influx from abandoned mine works may require continual mitigation (Karacan *et al*, 2007). Fugitive mine emissions are becoming more important as environmental pressures on greenhouse emissions continue to increase (Hayes, 2009; Su *et al*, 2005). The placement of non-permeable barriers around the area of the reservoir planned for pre-drainage and mining has been previously proposed to aid pre-drainage and reduce fluid influx into areas during and post-mining (Johnson Jr, 2014).

Example barrier implementation with underground mining operations

We will now illustrate the process of how barriers or 'curtains' would achieve this. Firstly, assume an area being proposed for longwall mining has an influx of fluids (eg gases, water) into the proposed area for pre-drainage and mining with symmetrical operations on two of the four sides as depicted in Figure 3 to allow no-flow boundary conditions for modelling purposes.

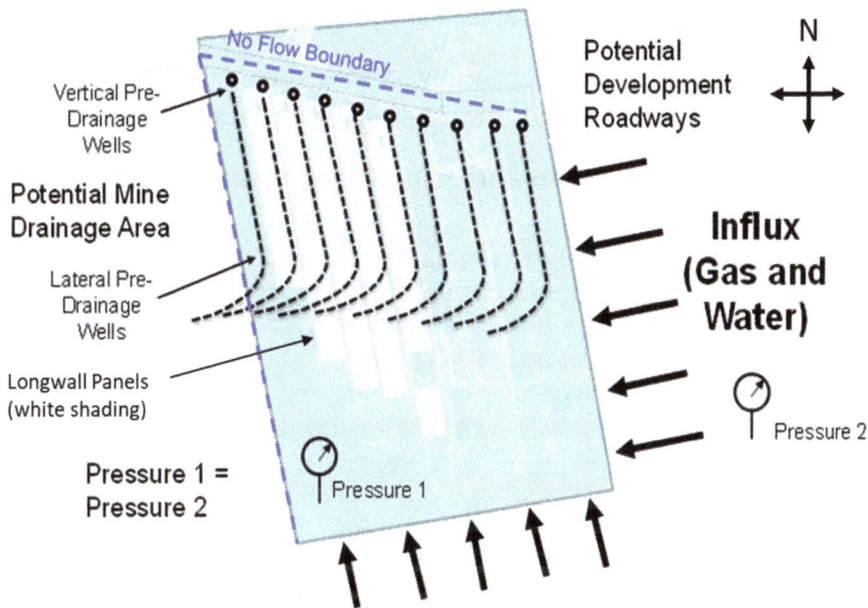

FIG 3 – Depiction of SIS well pattern in conjunction with long-wall mining operations at initial conditions without a barrier to prevent higher pressure, down-dip gas influx and where symmetry is established along no-flow boundaries for modelling purposes (after Johnson Jr, 2014).

After that, barriers could be barriers installed around the area of pre-drainage (see red lines depicted in Figure 4), with the construction and materials for these horizontal barrier implementations discussed in a subsequent section. The effectiveness of the barrier placement on further pre-drainage operations can be quantified by two methods depending on the effectiveness of the barrier implementation, the permeability of the coal, and the effectiveness of the internal drainage pattern (Johnson Jr, 2014, 2023). Firstly, fluids production over time (ie gases and water) would be reduced relative to those volumes modelled or projected and not accounting for the placement of the barriers. Next, if barrier placements are successful, observable, and increasing differential pressures would develop between the interior, or pre-drainage area, and the surrounding reservoir, as noted in Figure 4. Therefore, long-term environmental benefits from fugitive gas emissions could be quantified by measuring the external pressures and gas contents as well as observable, reduced levels of gas influx across the barriers, post-mining operations.

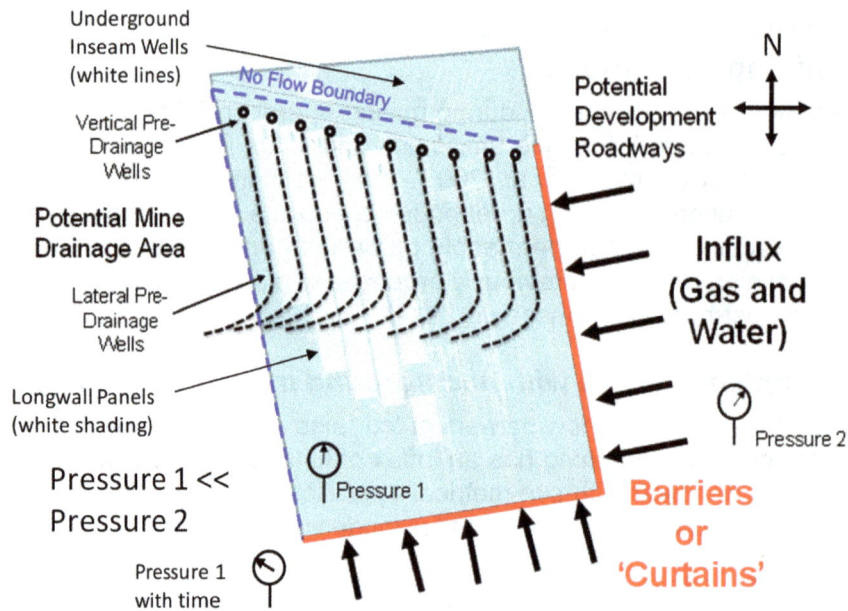

FIG 4 – Pressures in the drainage area decrease with time after installation of barriers or 'curtains' (note Pressure 1 reducing relative to Pressure 2 with time) (after Johnson Jr, 2014).

Modelling of emissions reduction benefits of barrier implementation with underground mining operations

To model emissions reduction in a reservoir simulator, Johnson Jr (2014) developed a model using typical properties for a GM Seam in the Moranbah area, placing SIS wells in the mine area and simulating production with and without a barrier placement over 20 years (see Figure 5). The barrier was modelled based on a blockage efficiency (BE) described by the ratio of the interior permeability of the barrier (k_{block}) relative to its initial permeability ($k_{initial}$) and expressed as a percentage, or BE = 100 per cent - ($k_{block/initial}$) (Johnson Jr, 2014). The general modelling was deterministic, using a single set of reservoir parameters and blockage efficiencies of 0 to 99.9 per cent were investigated. The base permeability value was 18 mD, thus a 99.9 per cent BE resulted in a minimum barrier region permeability of 0.018 mD. Whilst a BE of 99.9 per cent (or $k_{initial} \times 10^{-3}$) would initially appear to be a large reduction for the barrier, the data from colloidal silica testing, applied to unconstrained sand packs indicate that reductions up to 10^{-7} are achievable (McCartney et al, 2011).

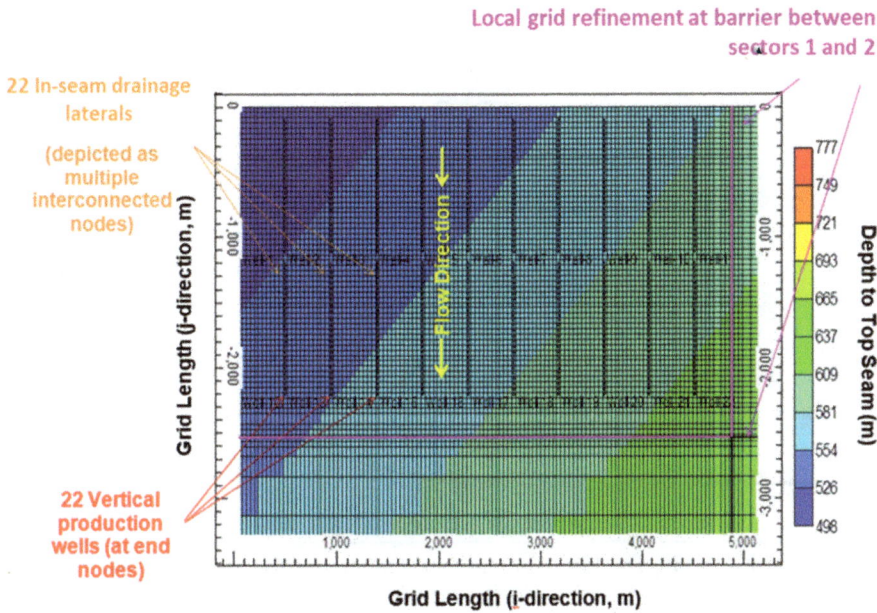

FIG 5 – Expanded drainage area grid, a representation of inner and outer area of Figure 4, as a general model with constant -1° reservoir dip (i- and j-direction) and 500 m pressure datum (Johnson Jr, 2023).

Next, a sensitivity analysis was performed using relationships of drainage efficiency (DE, $P_{time(t)}$/$P_{initial}$, psi/psi) to BE (see Figure 6) and adsorbed gas efficiency (AGE, $Volume_{total}$/$Volume_{initial}$, scf/scf) to describe the interrelationship of parameters versus BE in numerous trials sampling from distributions of reservoir dip in both the i- and j-directions (Johnson Jr, 2014). The sensitivity analyses indicate that permeability is the key parameter for both blocked and unblocked drainage. The dip did not seem to have a major influence on drainage based on these analyses but may influence water production outcomes, an important variable in drainage costs in economic analyses (Johnson Jr, 2014). The key outcome from these sensitivity analyses, comparing a 99.9 per cent BE case relative to 0 per cent BE, is that the barrier placement improves DE and AGE, with the highest impact becoming more apparent in mid- to high-permeability ranges as noted by production rates for gas (see Figure 7) and water (see Figure 8) (Johnson Jr, 2014).

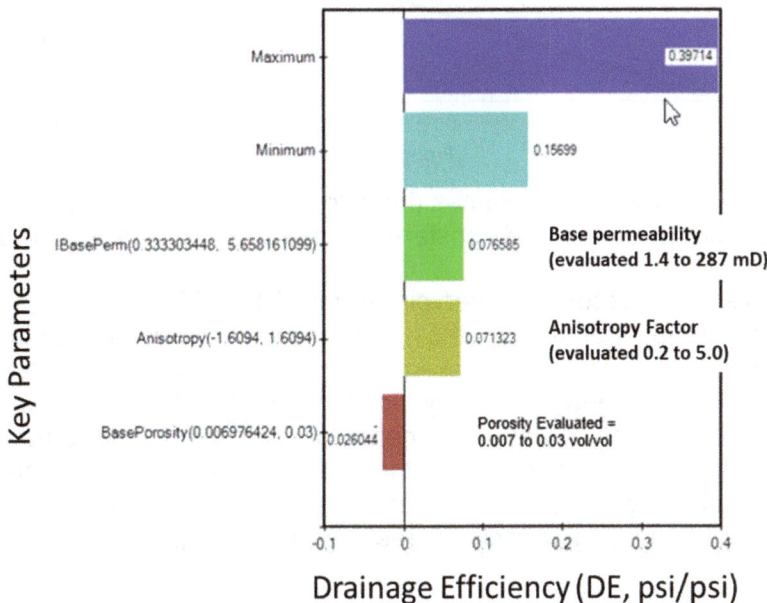

FIG 6 – Sensitivity analysis of ten years of drainage, 0 per cent BE and constant -1° reservoir dip in i- and j-directions (Johnson Jr, 2023).

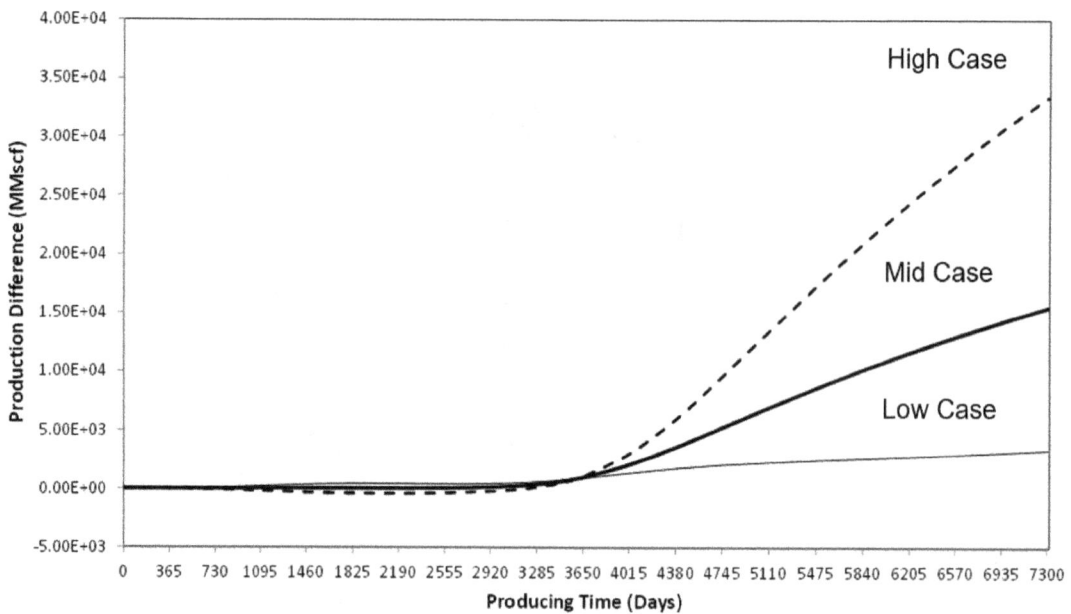

FIG 7 – Differential gas production volumes (high, low, and mid-case) based on 0 per cent and 99.9 per cent BE cases and constant -1° reservoir dip (Johnson Jr, 2014).

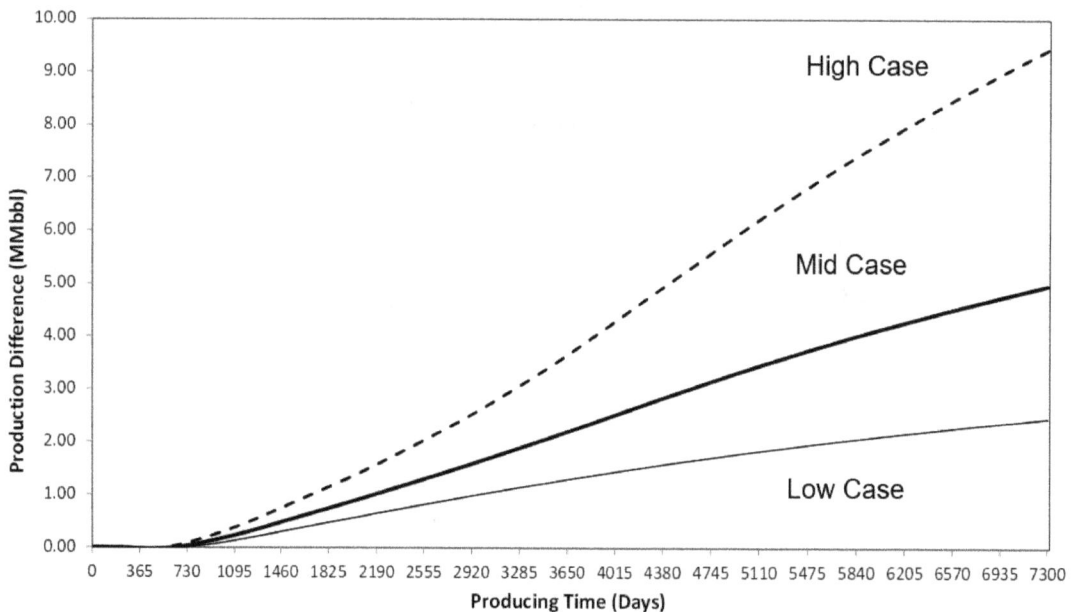

FIG 8 – Differential water production volumes (high, low, and mid cases) based on 0 per cent and 99.9 per cent BE cases and constant -1° reservoir dip (Johnson Jr, 2014).

Johnson Jr (2014) evaluated cases for reservoir dip for -1 to -10°, which did affect water and gas production rates and volumes (not shown); thus, changes in reservoir dip can become a variable to consider when developing drainage costs, gas income, and calculating overall emissions reductions (Johnson Jr, 2014). Finally, it was noted by Johnson Jr (2014) that lower permeability coals experience fewer benefits from barrier implementation versus higher permeability applications. However, the actual application may depend on the overall drainage plan as well as the permeability changes that may be occurring in the outside area beyond the barrier implementation as a result of coal depressurisation and shrinkage effects, increasing permeability.

METHODOLOGY

For both an open-cut and underground barrier implementation, a horizontal well and multi-stage hydraulic fracturing with a damaging and non-permeable material (to be discussed further) would be used similarly to that outlined in greater detail by Johnson Jr (2014). First, an understanding of the permeability anisotropy, *in situ* stress state, and bounding stress conditions would be required similar to any hydraulic fracturing treatment in coal (Johnson Jr *et al*, 2010a). In this case, instead of creating an effective, conductive fracture, the goal of barrier implementation is the creation of a damaged region. Modelling indicates that a three-fold reduction in permeability is effective in altering flow conditions and improving drainage or migration of fluids (Johnson Jr and Sedaghat, 2023; Johnson Jr, 2014).

Determination of the reservoir and geomechanical framework

Absolute permeability and permeability anisotropy are important as they reflect the ability and size of likely materials capable of damaging the coal and creating the barrier. For moderate to high permeability reservoirs, transient testing (eg drawdown build-up, slug, or injection fall-off tests) can determine the bulk permeability. However, another level of detail is required to understand permeability anisotropy and scale effects. Micro-resistivity logging can give a good understanding of cleat frequency and potential anisotropy through detailed processing of the image log data (Johnson Jr *et al*, 2020) and provide scale to transient and DFIT data. Finally, through the use of wells placed strategically along the axes of stress and permeability, multi-well interference testing can provide an understanding of permeability anisotropy (Kabir *et al*, 2011; Koenig and Stubbs, 1986).

A good understanding of the stress regime is essential to model and effectively place the barrier in the fracture geometry created by pumping fluids in the coal and testing should include as much as possible the entire length of the lateral being used for deployment. The integration of the density log can be used to estimate vertical stress (σ_v) and long-spaced sonic data to derive rock mechanical properties (ie Young's modulus and Poisson's ratio). However, diagnostic fracture injection testing (DFIT) is required in the vicinity of the barrier implementation to get a breakdown pressure, pressure-dependent behaviour, and closure stress data. Thereafter, the poroelastic equations of these data can be used to develop estimates of the minimum (σ_{h-min}) and maximum (σ_{H-Max}) horizontal stresses (Thiercelin and Plumb, 1994) and develop an accurate 1D stress profile. Further, the dynamic before-closure and pressure-dependent leak-off data can aid the understanding of potential pressure-dependent permeability effects (Johnson Jr *et al*, 2020). Thus, a scalable understanding of permeability and stress can guide the later selection of barrier materials based on size and required properties at both the ply and seam levels (see Figure 9).

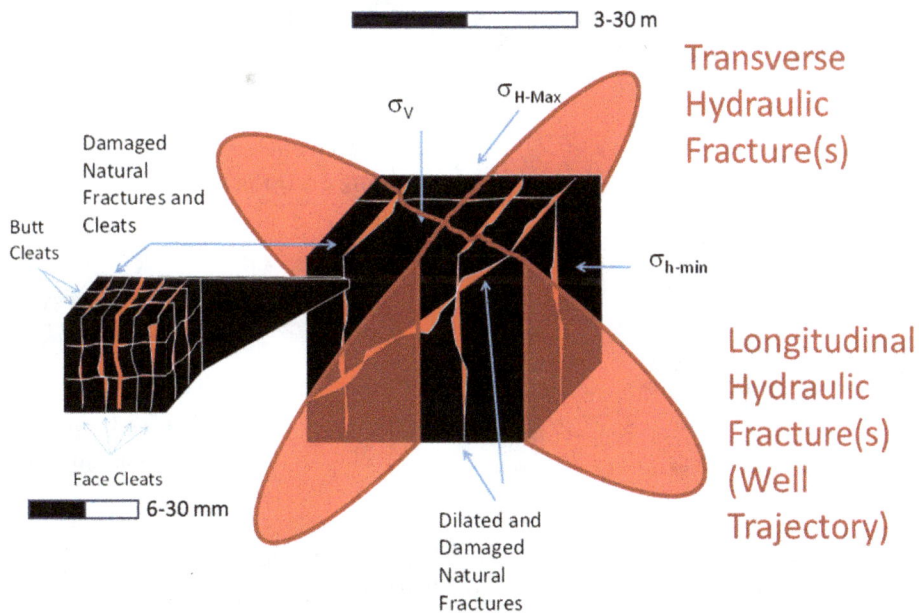

FIG 9 – Representation of created barrier fractures creating permeability blockage in natural fractures, and cleats; face cleats and the majority of natural fractures are assumed aligned with maximum horizontal stress (After Johnson Jr, 2023).

Barrier implementation

For both an open-cut and underground barrier implementation, a horizontal well and multi-stage hydraulic fracturing with a damaging and non-permeable material would be used similarly to that outlined in greater detail by Johnson Jr (2014). With regards to drilling, the well azimuth should be such that damage is orthogonal to the azimuth of maximum permeability. Consideration of the placement within the seam is important if stone bands or vertical variations in the jointing and coal permeability are significant. Performing numerous roof and floor touches whilst drilling may be desirable and could contribute to better placement based on fracture complexity that may result from varying well azimuth and vertical variability. However, the use of branching may be detrimental in containing and concentrating the barrier material to a confined interval and maximising the region of damage. The barrier deployment would require steel casing with swellable or mechanical packer isolation to use ball-actuated frac (staging) sleeves (see Figure 10) being placed in the coal to facilitate the implementation; thus, careful consideration of placement and the mine plan is essential.

Currently, the equipment and ability to deploy require steel components making implementations a special considerations and potential isolation in underground applications unless accommodated in mine plans and recovery considerations in surface mining applications. For widespread implementation, there is no restriction on glass-reinforced epoxy and swell packer implementations in underground implementations; however, the frac sleeves between the swellable packers (see Figure 10c) would need to be constructed from a mineable material.

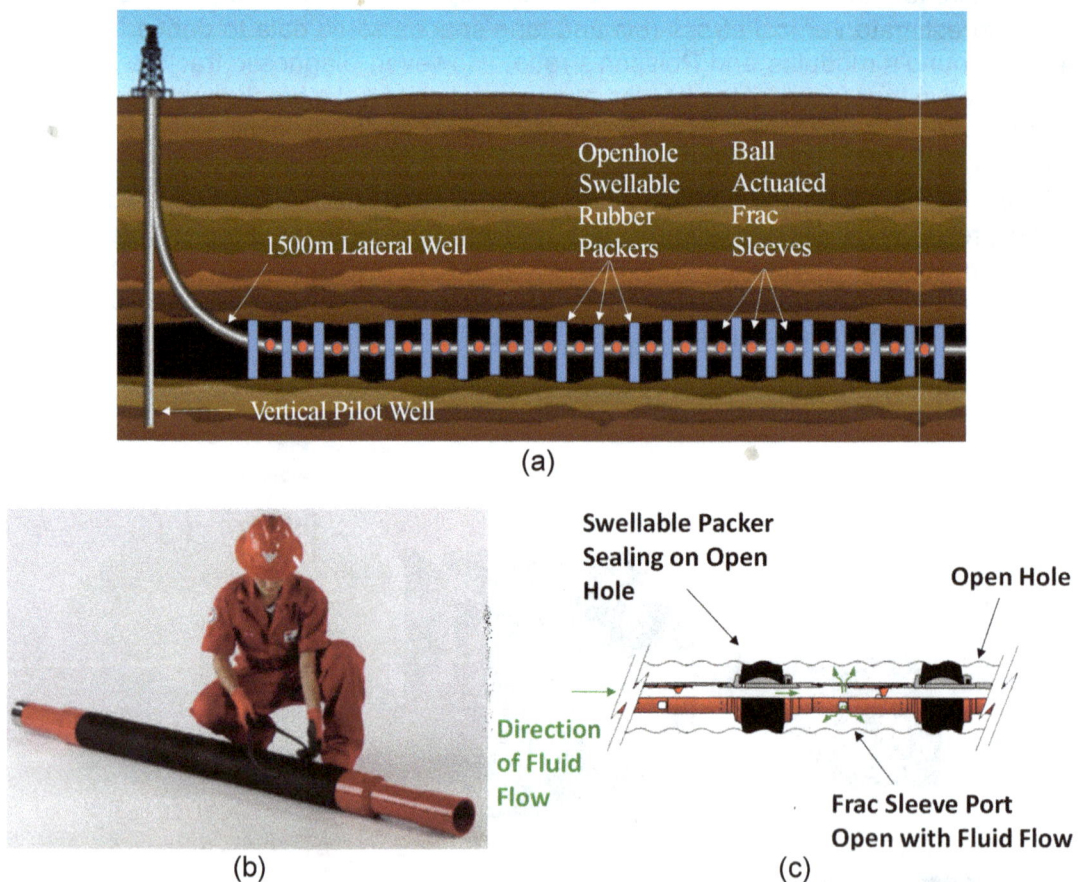

FIG 10 – (a) Side-tracked lateral well with swellable rubber packers and 20 ball-actuated frac sleeves; (b) tool operator callipers a swellable, rubber isolation packer; and (c) ball-actuated frac sleeves in operation during pumping (after Johnson Jr, 2014).

The actual design per stage would be determined using the 1D stress profile and a planar fully 3D fracturing model (Barree, 1983), similar to that used to match results in fully instrumented CSG fracturing experiments in the Bowen and Surat Basins (Gibson *et al*, 2013; Johnson Jr *et al*, 2010b; Pallikathekathil *et al*, 2013; Scott *et al*, 2010). Volumes would be based on achieving a radius of damage and cross-coverage between stages along with viscosity and rate used to generate net pressure and activate pressure-dependent leak-off off thereby increasing the damage.

Laboratory studies have shown that <50 µm particles are capable of entering the cleats and are the basis of microproppant stimulation for coals (Keshavarz et al, 2014). More recent testing indicates that commercially available microproppant materials in the range of 10 to >50 µm can bridge in leak off conditions similar to that being anticipated in barrier treatments (Johnson Jr et al, 2022), thereby increasing fracturing net pressure, pressure-dependent leak off, and overall damage. Thus, the use of a damaging material with 10 to >50 µm particles can achieve damage at a cleat, natural fracture and created fracture scale, making the barrier more effective.

Barrier verification

Hydraulic fracturing diagnostics

Microseismic measurements are typically coarse with low incidence or magnitude in coal and can be less reliable in fracture mapping than in clastics (Warpinski, personal communication, 2010; Zimmer, 2010). Typically, microseismic event magnitudes of -3 to -5 have been observed in published CSG microseismic monitoring studies (Johnson Jr et al, 2010b; Flottman et al 2013). In both documented cases, both studies indicated the best diagnostic for understanding the area and extent of the stimulated reservoir volumes has been surface deformation tiltmeters. Whilst the misinterpretation of 'T-shaped' or horizontal fractures is often suggested for low-magnitude responses in coal if; however, in many cases, a better explanation of low-amplitude horizontal-like responses have been determined to be dilation of fractures and pressure transient effects (Johnson Jr et al, 2010b; Palmer, 1990). Regardless, where horizontal fractures exist in coals, vertical components are almost always observed and can be both longitudinal in the case of a lateral or transverse in both lateral and vertical cases (Jeffrey et al, 1992, 1998; Scott et al, 2010). As the desired damage region is on the radius of 20+ m a possible pattern to image the barrier implementation is shown in Figure 11.

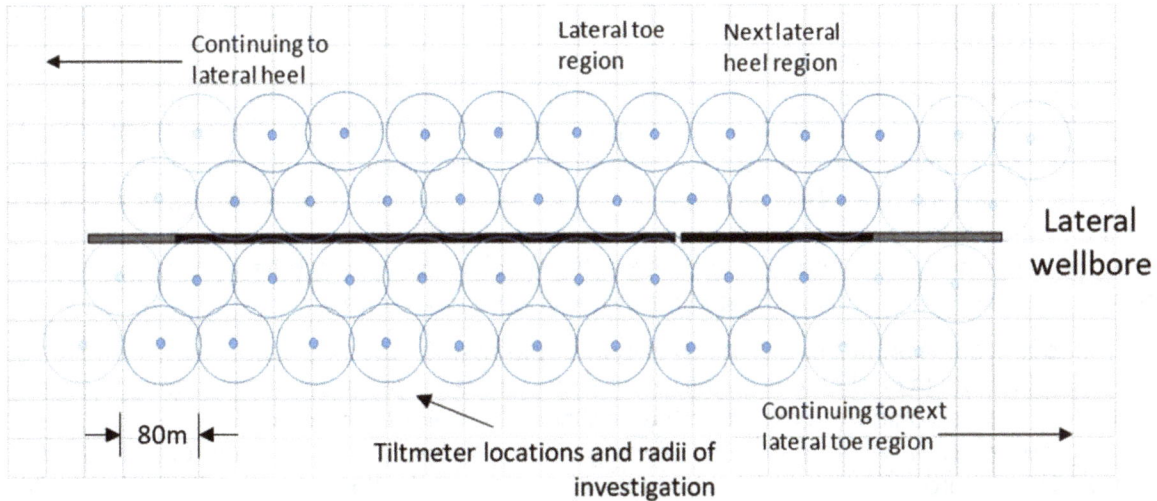

FIG 11 – Potential space out of tiltmeters along the lateral based on 80 m spacing commencing 40 m from the barrier placement (After Johnson Jr, 2014).

Reservoir surveillance to assure barrier effectiveness

Over time, pressure observation wells can be used to characterise the direction and drainage area of the fracture by using interference principles between one or more observation wells and pressure transient analyses solutions (Mavor and Cinco-Ley, 1979; Meehan, Horne and Ramey, 1989; Pierce, Vela and Koonce, 1975; Sawyer, Alam and Rose, 1980; Uraiet, Raghavan and Thomas, 1977). For an underground barrier implementation, observation wells can be placed inside and outside of the proposed barrier placement to monitor pressures and provide estimates of barrier effectiveness (see Figure 12). For a surface barrier implementation, the observation wells would extend on either side of the unidirectional laterals, similar to those depicted in the i-direction in Figure 12.

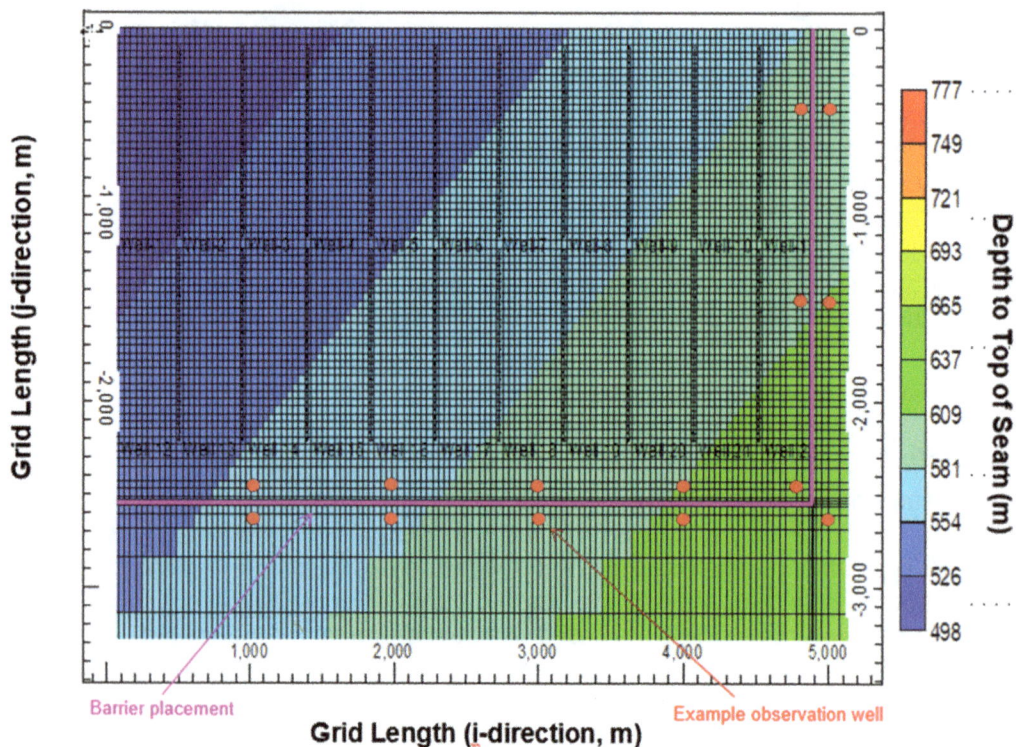

FIG 12 – Location of example observation wells inside and outside of proposed barrier placement for an underground barrier implementation (after Johnson Jr, 2014).

FURTHER RESEARCH

Further research into FMM and emerging hydraulic fracturing diagnostics can improve barrier placement implementation and verification.

Flow modification materials

In the context of reducing methane emissions and enhancing pre-drainage operations, the choice of materials for barrier implementation is paramount. Traditional thixotropic materials, such as cementitious grout, have been considered for their ability to be pumped and form effective barriers. These materials exhibit non-Newtonian behaviours, becoming more viscous under shear stress, allowing for efficient injection into coal seams. Incorporating waste materials, such as glass waste, rubbers, fly ash, and silica fume, into the cementitious grout can contribute to a circular economy concept, promoting sustainability and minimising environmental impact.

Another innovative option is the utilisation of pumpable resins, such as urea silica resin, to create impermeable barriers in coal seams. These resins offer advantages in terms of ease of injection and adaptability to the geological conditions of the coal seam. Similar to the cementitious grout, the resin can be augmented with waste materials, aligning with principles of circular economy and resource efficiency. This approach not only addresses emissions reduction but also contributes to the responsible management of industrial by-products, further supporting the overall environmental sustainability goals.

To systematically evaluate the effectiveness of impermeable barriers in mitigating methane migration within coal seams, a robust testing regime is proposed. The core of this experimental design involves the utilisation of cylindrical coal samples to emulate *in situ* conditions. These samples would be subjected to controlled laboratory conditions, allowing for a comprehensive analysis of the impact of barrier materials on coal permeability in both radial and longitudinal directions.

The first step involves the preparation of cylindrical coal samples representative of the target coal seam. These samples would be carefully selected to capture the heterogeneity and inherent properties of the coal reservoir under investigation. To simulate the pre-drainage and mining scenario, the samples would be subjected to various initial methane pressure conditions within a dedicated chamber.

Subsequently, impermeable barriers would be meticulously applied around select coal samples using materials such as thixotropic cementitious grout and pumpable resin. This step mirrors the intended real-world application of barriers in coal seams for emissions reduction. The choice of materials for the barriers would align with the circular economy principles, potentially incorporating waste materials like glass waste, rubbers, fly ash, and silica fume.

Once the barriers are in place, the coal samples would undergo systematic testing to measure radial and longitudinal permeability values. These permeability tests would be conducted under different methane pressure conditions to replicate diverse scenarios encountered during coal mining operations. By comparing the permeability values of samples with and without barriers, a clear understanding of the barrier's impact on methane migration within the coal seam can be derived. This comprehensive scientific testing regime aims to provide valuable insights into the efficacy of impermeable barriers in controlling methane permeability in coal. Finally, in addition to the technical aspects and effectiveness of barrier materials, the impact on processing barrier-affected coal must be considered to ensure minimal impact on production.

Evaluating barrier effectiveness

One question posed is when mining face advances close to the barrier, how will the mining induced stress redistribution affect the barrier effectiveness, and allow gas flows from the previously blocked area. This can certainly be assessed using a geomechanics model. However, in a practical sense, the likelihood that no drainage is performed on the exterior, undrained side of the barrier for ESG or beneficial use. Several emerging technologies are being implemented that have promise for determining barrier effectiveness. Firstly, fixed fibre-optic installations have been used to measure strains created by the fractures, akin to surface deformation tiltmeter measurements (Bourne *et al*, 2021). For barrier implementation, a fixed fibre optic array in an offsetting monitoring well could give insight into fracture height and the degree of verticality of the barrier implementation based on observed strains. Next, lower-cost, GPS-enabled geophones and sources are coming into use for shallow and mining applications. This means that either real-time or passive 4D seismic may be able to identify changes in amplitude pre-, during and post-barrier implementation (Maxwell *et al*, 2002) as well as changes in coal behaviour over time-based on attribute and curvature analyses (Fisk *et al*, 2010). Both these areas offer promise and potential research areas for a better understanding of barrier implementation and effectiveness.

Economic modelling and benefits

Johnson Jr (2014) performed probabilistic economic modelling based on the costs to install a barrier in a pre-drainage model above otherwise incurred gas pre-drainage costs (ie internal drainage wells, coring, formation evaluation, infrastructure, and operating costs being similar). This probabilistic model considered the base permeability, the BE of the barrier implementation, the gas price at the time, differential gas incomes and water costs, and an emissions trading scheme (ETS) framework in effect at the time, ranging from \$6.20/t CO_{2e} to \$38.00/t CO_{2e}. This model indicated that net present value (NPV) is most affected by the gas solution outcome probability, cash discount rate, and carbon cost as the three key input variables. Although the outcomes of these analyses indicate a log-normal outcome with the results being NPV positive beyond the 91st percentile (see Figure 13). However, the internal rates of return (not shown) indicate that it is cash positive beyond the 3rd percentile outcome (Johnson Jr, 2014). Further work on the model needs to be revisited, and the benefits of barrier implementation using current gas pricing and costs under the Safeguard Mechanisms need to be considered (Australian Government, 2023).

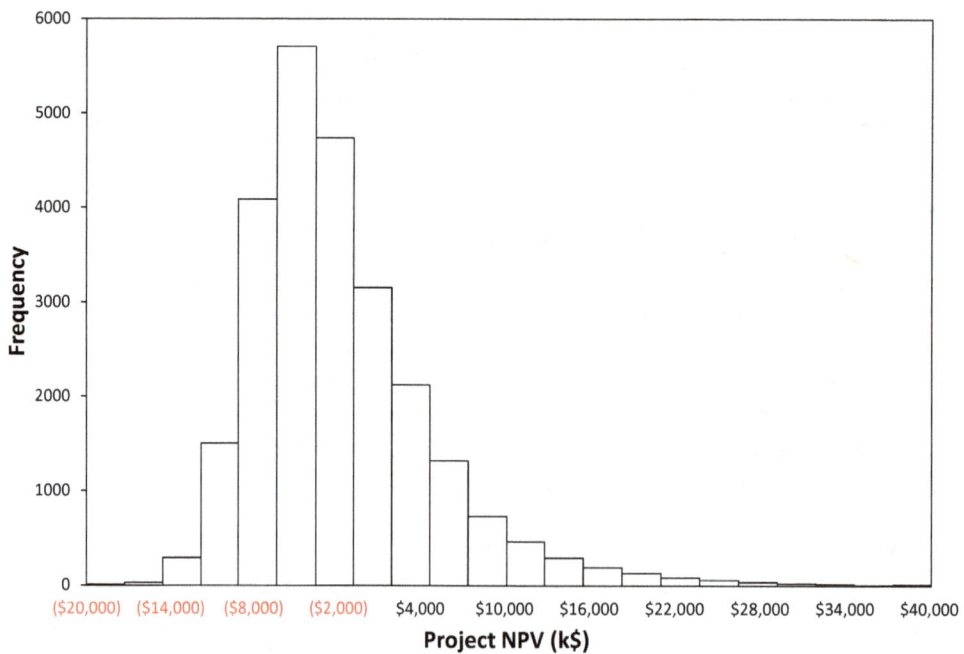

FIG 13 – Histogram of NPV output based on Monte Carlo analysis of economic parameters using -1° constant reservoir dip and current ETS in effect (Johnson Jr, 2014).

CONCLUSIONS

This paper has explored the application of non-permeable barriers, termed 'curtains,' to enhance gas drainage in advance of coal mining and reduce mining emissions. The context is set in Australia's efforts to reduce emissions, particularly in emissions-intensive industries such as metallurgical coal. This paper outlines a detailed methodology for barrier implementation, including reservoir characterisation, geomechanical considerations, and hydraulic fracturing with damaging materials. Whilst the modelling considers various scenarios, primarily focusing on trapping downdip gas for beneficial use, it is recognised that the application of subterranean barriers may not be universally feasible across all coal mining scenarios because of geological variability, directional permeability and stress states, and structural constraints. Further, overlying and underlying seams may be more problematic than inseam influx and overall emissions. Therefore, The study emphasises the importance of fully understanding the reservoir and geomechanical frameworks before implementing a barrier placement.

This study has illustrated the application and methodology of barrier implementation for both surface and underground mining operations. In the context of surface mining, the paper demonstrates that the use of barriers alongside surface-to-inseam (SIS) drainage wells can significantly reduce methane emissions, with a potential reduction of up to 57 per cent for a 1 km² drainage area. For underground mining, the paper has illustrated a realistic gas influx case that can be reduced if non-permeable barriers are implemented around the area planned for pre-drainage and mining. Past modelling indicates that the barrier placement improves drainage efficiency and reduces gas and water production volumes (Johnson Jr, 2014). Detailed sensitivity analyses highlight the key role of permeability in blocked and unblocked drainage (Johnson Jr, 2014). The economic modelling for this case is outlined based on the prior work and the ETS in place at the time (Johnson Jr, 2014). In conclusion, the economic viability of barrier implementation is highly contingent on reservoir and stress conditions and market factors (ie gas prices and emission trading schemes), both of which may vary by area and fluctuate over time, respectively.

This paper has made recommendations for areas of further research including the exploration of FMM and emerging hydraulic fracturing diagnostics to evaluate barrier implementation and effectiveness. The long-term environmental impacts, sustainability of using non-permeable materials for barriers, and impacts on processing barrier-affected coal require further investigation. Further, based on other works, emerging materials (eg microproppant and nano-materials) below 50 μm may improve the damage at the cleat and natural fracture scale, enhancing the overall effectiveness of the barrier. This paper summarised a workable verification framework by Johnson Jr (2014) for barrier effectiveness involving current hydraulic fracturing diagnostics such as surface deformation tiltmeters and reservoir surveillance through pressure observation. In addition, the authors have suggested emerging technologies, including fixed fibre-optic installations and 4D seismic, to improve verification and likely required for substantiating emissions reductions.

ACKNOWLEDGEMENTS

The authors would like to acknowledge James Tauchnitz, Novus Fuels and The University of Southern Queensland for their contributions, support, and permission to publish this work.

REFERENCES

Anderson, E I and Mesa, E, 2006. The effects of vertical barrier walls on the hydraulic control of contaminated groundwater, *Advances in Water Resources,* 29(1):89–98. https://doi.org/10.1016/j.advwatres.2005.05.005

Australian Government, 2023. National Greenhouse and Energy Reporting (Safeguard Mechanism) Rule 2015, Australian Government: Canberra.

Barree, R D, 1983. A Practical Numerical Simulator for Three-Dimensional Fracture Propagation in Heterogeneous Media, SPE 12273, paper presented at the SPE Reservoir Simulation Symposium, California.

Bourne, S, Hindriks, K, Savitski, A A, Ugueto, G A and Wojtaszek, M, 2021. Inference of Induced Fracture Geometries Using Fiber-Optic Distributed Strain Sensing in Hydraulic Fracture Test Site 2, paper URTeC 5472, paper presented at the SPE / AAPG / SEG Unconventional Resources Technology Conference. https://doi.org/10.15530/urtec-2021-5472

Fisk, J C, Marfurt, K J and Cooke, D, 2010. Correlating Heterogeneous Production to Seismic Curvature Attributes In an Australian Coalbed Methane Field, SEG-2010-2323, paper presented at the 2010 SEG Annual Meeting, Colorado.

Flottman, T, Brooke-Barnett, S, Naidu, S K, Paul, P K, Kirk-Burnnand, E, Busetti, S, Hennings, P H and Trubshaw, R L, 2013. Influence of in Situ Stresses on Fracture Stimulation in the Surat Basin, Southeast Queensland, presented at SPE Unconventional Resources Conference and Exhibition-Asia Pacific, paper SPE-167064.

Gibson, M, Mazumder, S, Probst, P and Scott, M, 2013. Application Of Open-hole Diagnostic Fracture Injection Test Results To Regional Stress Interpretation In Bowen Basin Coals, SPE 167074, paper presented at the SPE Unconventional Resources Conference and Exhibition-Asia Pacific, Australia.

Hayes, P, 2009. Capturing fugitive emissions, *Australian Mining.* Available from: <https://www.australianmining.com.au/capturing-fugitive-emissions/>

Hötzl, H, Eiswirth, M and Brauns, J, 2000. Impact of chemical grout injections on urban groundwater, report, Karlsruher Institut für Technologie.

Jeffrey, R G, Byrnes, R P, Lynch, P J and Ling, D J, 1992. An Analysis of Hydraulic Fracture and Mineback Data for a Treatment in the German Creek Coal Seam, presented at SPE Rocky Mountain Regional Meeting, paper SPE-24362.

Jeffrey, R G, Vlahovic, W, Doyle, R P and Wood, J H, 1998. Propped Fracture Geometry of Three Hydraulic Fractures in Sydney Basin Coal Seams, presented at SPE Asia Pacific Oil and Gas Conference and Exhibition, paper SPE-50061.

Johnson Jr, R L, 2014. Modifying subterranean fluid flow patterns using barriers to improve pre-drainage in coal mining, PhD thesis, Mining Engineering, The University of Queensland.

Johnson Jr, R L and Sedaghat, M, 2023. The Implementation of Subterranean Barriers to Reduce Shallow Gas Migration and Coal Mine Methane Emissions from Open-cut Metallurgical-Coal Mines, SPE-217310-MS, paper presented at the Asia Pacific Unconventional Resources Symposium, Australia.

Johnson Jr, R L, You, Z, Ribeiro, A, Mukherjee, S, Salomao de Santiago, V and Leonardi, C R, 2020. Integrating Reservoir Characterisation, Diagnostic Fracture Injection Testing, Hydraulic Fracturing and Post-Frac Well Production Data to Define Pressure Dependent Permeability Behavior in Coal, paper SPE-202330-MS, paper presented at the SPE Asia Pacific Oil and Gas Conference and Exhibition.

Johnson Jr, R L, 2023. The implementation of subterranean barriers with mine pre-drainage to reduce coal mine methane emissions from open-cut and underground metallurgical-coal mines, *International Journal of Coal Geology,* 104402. https://doi.org/10.1016/j.coal.2023.104402

Johnson, R L, Jr, Glassborow, B, Datey, A, Pallikathekathil, Z J and Meyer, J J, 2010a. Utilizing Current Technologies to Understand Permeability, Stress Azimuths and Magnitudes and their Impact on Hydraulic Fracturing Success in a Coal Seam Gas Reservoir, SPE 133066, paper presented at the SPE Asia Pacific Oil and Gas Conference and Exhibition, Australia.

Johnson, R L, Jr, Scott, M, Jeffrey, R G, Chen, Z Y, Bennett, L, Vandenborn, C and Tcherkashnev, S, 2010b. Evaluating Hydraulic Fracture Effectiveness in a Coal Seam Gas Reservoir from Surface Tiltmeter and Microseismic Monitoring, SPE 133063, paper presented at the SPE Annual Technical Conference and Exhibition, Italy.

Johnson, R, Leonardi, C, You, Z, Ribiero, A and Di Vaira, N, 2022. Converting tight contingent CSG resources: Application of graded particle injection in CSG stimulation, The University of Queensland Centre Natural Gas Project Report, The University of Queensland. Available from: <https://natural-gas.centre.uq.edu.au/files/12991/NERA%20 GPI%20Project%20-%20Final%20Report.pdf>

Jurinak, J J and Summers, L E, 1991. Oilfield Applications of Colloidal Silica Gel, *SPE Production Engineering*, 6(4):406–412. https://doi.org/10.2118/18505-PA

Kabir, A H, McCalmont, S, Street, T and Johnson, R L, 2011. Reservoir Characterisation of Surat Basin Coal Seams using Drill Stem Tests, paper SPE-147828-MS, paper presented at the SPE Asia Pacific Oil and Gas Conference and Exhibition, Indonesia. https://doi.org/10.2118/147828-MS

Karacan, C Ö, Esterhuizen, G S, Schatzel, S J and Diamond, W P, 2007. Reservoir simulation-based modeling for characterizing longwall methane emissions and gob gas venthole production, *International Journal of Coal Geology,* 71(2):225–245. https://doi.org/10.1016/j.coal.2006.08.003

Keshavarz, A, Yang, Y, Badalyan, A, Johnson, R L, Jr and Bedrikovetsky, P, 2014. Laboratory-based mathematical modelling of graded proppant injection in CBM reservoirs, *International Journal of Coal Geology*, 136:1–16. https://doi.org/10.1016/j.coal.2014.10.005

Koenig, R A and Stubbs, P B, 1986. Interference Testing of a Coalbed Methane Reservoir, SPE 15225, paper presented at the SPE Unconventional Gas Technology Symposium, Kentucky.

Mavor, M J and Cinco-Ley, H, 1979. Transient Pressure Behavior of Naturally Fractured Reservoirs, SPE California Regional Meeting.

Maxwell, S C, Urbancic, T I, Demerling, C and Prince, M, 2002. Real-Time 4D Passive Seismic Imaging of Hydraulic Fracturing, SPE 78191, paper presented at the SPE/ISRM Rock Mechanics Conference, Texas.

McCartney, J, Nogueira, C L, Homes, D and Zornberg, J G, 2011. Formation of secondary containment systems using permeation of colloidal silica, *J Env Eng*, 137(6):444. https://doi.org/10.1061/(ASCE)EE.1943-7870.0000345

McKinsey and Company, 2020. Decarbonization challenge for steel Hydrogen as a solution in Europe. Available from: <https://www.mckinsey.comDecarbonization-challenge-for-steel.pdf> [Accessed: 14 Oct 2022].

Meehan, D N, Horne, R N and Ramey, H J, 1989. Interference Testing of Finite Conductivity Hydraulically Fractured Wells, presented at SPE Annual Technical Conference and Exhibition, paper SPE-19784.

Nurdiawati, A and Urban, F, 2021. Towards Deep Decarbonisation of Energy-Intensive Industries: A Review of Current Status, Technologies and Policies, *Energies*, 14(9). https://doi.org/10.3390/en14092408

Pallikathekathil, Z J, Puspitasari, R, Altaf, I, Alboub, M, Mazumder, S, Sur, S and Gan, T, 2013. Calibrated Mechanical Earth Models Answer Questions on Hydraulic Fracture Containment and Wellbore Stability in Some of the CSG Wells in the Bowen Basin, presented at the SPE Unconventional Resources Conference and Exhibition-Asia Pacific, paper SPE-167069-MS.

Palmer, I D, 1990. Uplifts and Tilts at Earth's Surface induced by Pressure Transients From Hydraulic Fractures, *SPE Production Engineering*, 5(3):324–332. https://doi.org/10.2118/18538-PA

Pierce, A E, Vela, S and Koonce, K T, 1975. Determination of the Compass Orientation and Length of Hydraulic Fractures by Pulse Testing, *Journal of Petroleum Technology*, 27(12):1433–1438. https://doi.org/10.2118/5132-PA

Sawyer, W K, Alam, J and Rose, W D, 1980. Utilizing Well Interference Data in Unconventional Gas Reservoirs, presented at SPE Unconventional Gas Recovery Symposium.

Scott, M, Johnson Jr, R L, Datey, A, Vandenborn, C and Woodroof, R A, 2010. Evaluating Hydraulic Fracture Geometry from Sonic Anisotropy and Radioactive Tracer Logs, SPE 133059, paper presented at the SPE Asia Pacific Oil and Gas Conference and Exhibition, Australia.

Soliman, M Y, Shelley, R F, East, L E and Cullick, A S, 2012. Methods relating to modifying flow patterns using in-situ barriers, US Patent, AU2011225933, Issue date 06/09/2012.

Su, S, Beath, A, Guo, H and Mallett, C, 2005. An assessment of mine methane mitigation and utilisation technologies, *Prog Ener and Comb Sci*, 31(2):123–170. https://doi.org/http://dx.doi.org/10.1016/j.pecs.2004.11.001

Thiercelin, M J and Plumb, R A, 1994. A Core-Based Prediction of Lithologic Stress Contrasts in East Texas Formations, SPE Formation Evaluation, 9(4):251–258, paper SPE-21847-PA.

Uraiet, A, Raghavan, R and Thomas, G, 1977. Determination of the Orientation of a Vertical Fracture by Interference Tests, *Journal of Petroleum Technology*, 29(1):73–80. https://doi.org/10.2118/5845-PA

Zimmer, U, 2010. Microseismic Mapping of Hydraulic Treatments in Coalbed-Methane (CBM) Formations: Challenges and Solutions, paper, SPE-132958-MS, paper presented at the SPE Asia Pacific Oil and Gas Conference and Exhibition, Australia.

An intelligent design system for gas drainage boreholes in coalmines

S Xue[1] and Y B Li[2]

1. Joint National-Local Engineering Research Centre for Safe and Precise Coal Mining, Anhui University of Science and Technology, Huainan 232001, Anhui, China.
2. School of Safety Science and Engineering, Anhui University of Science and Technology, Huainan 232001, Anhui, China. Email: binyaoli@163.com

ABSTRACT

Gas drainage with boreholes is one of the most effective methods for coalmine gas control. The boreholes need to be properly designed to ensure accurate and efficient gas drainage. The design of the boreholes is often time-consuming and problematic as it requires the considerations of geological conditions, coal seam gas characteristics and mining activities. This paper describes an intelligent design system for gas drainage boreholes in coalmines. The design system has four main subsystems, including the development of an accurate three-dimension mine-scale geological model, division of the geological model into a number of appropriate geological units through the attribute rendering of the model and morphology, division of gas drainage blocks within the geological units based on analyses of mining planning, gas condition and geological structures, and the design of gas boreholes for each of the blocks based on gas emission sources. The four subsystems are interrelated and integrated in the new design system. Taking Liuzhuang Coal Mine in China as an example, the application of the design system is introduced.

INTRODUCTION

China is relatively rich in coal resources, but the occurrence conditions of coal seams vary greatly. For example, coal seams vary greatly in overburden depth, thickness, dip, geological structure and gassiness (Yuan, 2009, 2017; Hu, 2013; Zhang et al, 2009). The continuous increase in coal mining depth leads to the increase of ground stress, gas pressure and gas content, which leads to the increase of the risk of gas-related accidents, which directly affects the safety of coalmine workers and production of coalmines (Yuan, 2016; Wang, Liu and Wang, 2016; Ren et al, 2022; Zhang and Zhang, 2021).

Gas drainage with boreholes is one of the most effective methods in use for coalmine gas control (Zhou, Wang and Xia, 2014; Xin, Ma and Yang, 2015; Wu et al, 2019; Yu, 1992; Wang, Cheng and Wang, 2012). With the increasing mining depth of coal seam, simple gas drainage methods alone can no longer effectively solve the problem of coalmine gas (Wang, Cheng and Wang, 2012; Liu, 2013a; 2013b; Ji, 2015). Gas outburst prone mines have begun to adopt comprehensive gas drainage technologies, which not only solved the gas problem in coalmines, but also achieved coal and gas co-mining (Xiao et al, 2014; Zhang, Han and Cao, 2013; Wang, 2015; Jiang and Li, 2019; Fan, 2019). For example, Yuan (2008) has analysed the mechanical characteristics of the rock in the mining area, determined the development height of the 'O' ring fissure during mining, studied the effective gas drainage method of the surrounding rock of the coal seam, and improved the gas drainage system. Xie and Sun (2013) have proposed to use high-level boreholes to drain gas from both mining seam and adjacent strata to improve the gas drainage rate. Liu et al (2004) have studied the state of the development zone of the ring fissure in the roof, proposed to use long boreholes for gas drainage.

In terms of gas drainage borehole design, current practice in China's coalmines is to design the boreholes manually. The problems associated with the design include low efficiency, poor precision and heavy workload of designers, which is not conducive to the safety, high yield and high efficiency of mines. Many attempts have made in intelligent borehole design (Hao, Fan and Zhao, 2013; Wang, 2023; Shi, Xu and Li, 2015; Chen, 2015). For example, Ma has developed an intelligent design system for directional drilling based on SuperMap Object.Net 6R platform, which has achieved the automatic calculation and mapping of trajectory parameters of directional gas drainage boreholes (Ma, 2017). Chen (2021) has optimised the layout of the boreholes and designed the trajectory of the boreholes in low permeability seams in Yuwu coalmine. Yue (2022) has used the computer design algorithm and process to solve the difficult problem of high-position borehole design through the secondary development of GIS platform. In order to study the gas drainage situation of a fully mechanised caving face. Zhang (2023) has conducted feasibility analysis on gas drainage design in a highly gassy mine in Shaanxi Province. Ahmad and Usman (2020) have analysed the effects of operation parameters in the design of gas drainage boreholes in goaf. This paper proposes a new intelligent design method of coal seam gas drainage borehole, and on this basis a set of intelligent design system of gas drainage boreholes has been developed.

DESIGN PRINCIPLE

This intelligent gas drainage design system includes the construction of three-dimensional geological model, division of gas drainage units based on mining and geological conditions, selection of borehole patterns, and auto generation of borehole layout.

3D geological model construction

The geological model construction includes the collection of relevant gas parameters such as buried depth, thickness, floor elevation, coal seam inclination, gas content and gas pressure in the target mine, the attribute rendering of coal seams, the use of spatial information retrieval to analyse the spatial distribution of coal seams of different conditions, and the use of 3D slicing technology to display the geological conditions of mining coal seams.

The pre-processing of drilling data of the mine is carried out. Through correlation and division of strata, integration, clustering and connection of elements, voxel solid model, fault mathematical model, fold geometry model and well surface model are integrated to show the spatial geometry of complex geological structures and the accurate deployment of gas drainage boreholes.

Based on gas geological data processing, it integrates solid modelling technology and surface model, integrates spatial database, graphic library and knowledge base with 3D dynamic simulation, intuitively, graphically and accurately grasp the local characteristics and overall framework of gas geological data, and establishes 3D geological model from whole to part.

A process of morphological drainage of geological units by attribute rendering on the whole model is proposed. The model can spatially retrieve all kinds of blocks that meet certain geological conditions, and provide data and visual basis for the division of drainage units (Figure 1).

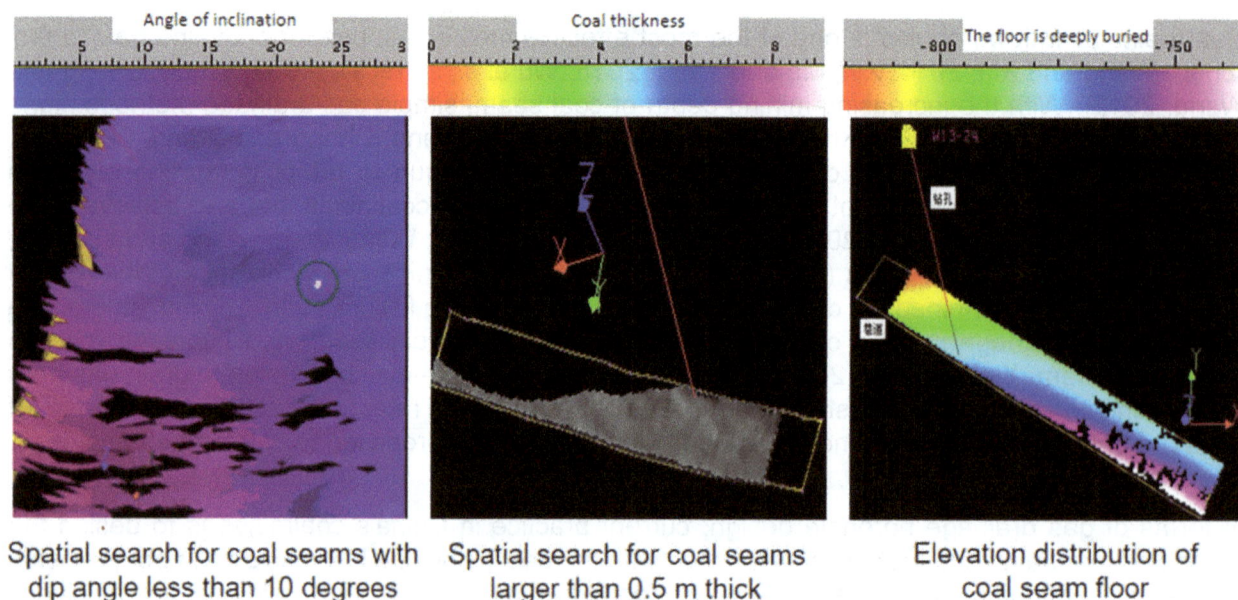

Spatial search for coal seams with dip angle less than 10 degrees   Spatial search for coal seams larger than 0.5 m thick   Elevation distribution of coal seam floor

FIG 1 – Coal seam distribution in mining area 15-21030.

The lithologic combination characteristics of coal seam roof and floor can be observed by using 3D slice, and the degree of influence of surrounding rock combination on coal seam deformation can also be calculated (Figure 2).

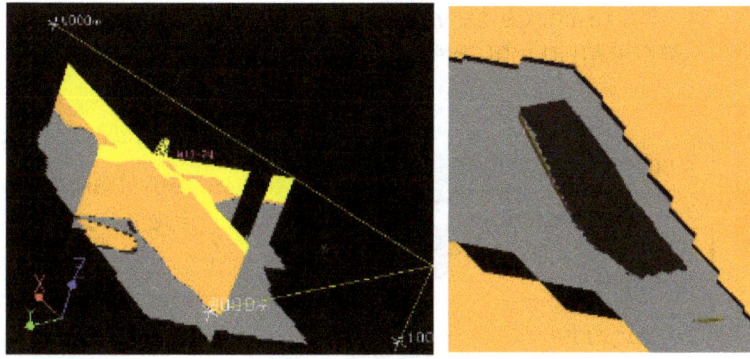

FIG 2 – 3D section of lithology distribution on the top and bottom of coal seam (left: global, right: local).

A method of 3D space distance field transformation and drainage of space elements is proposed and implemented, which provides a fast and accurate means for the calculation of distance separation between design elements of intelligent drilling and the distance between coal seam and design drilling (Figure 3).

FIG 3 – Simulated distance field between borehole and surrounding rock of coal seam.

The generated 3D gas geological model can intuitively show the difference of gas content and pressure in different positions of the working face, and users can refer to the distribution of gas content and pressure on the working face to divide pumping units and intelligent design of pumping boreholes.

Division method of gas drainage unit

The drainage area (unit) is divided according to mine planning gas condition and typical geological structure. The unit division is initially based on mine planning, then the units are subdivided according to gas condition, and finally the subdivided units are further divided by considering geological structures.

Division of drainage unit based on mine planning

The planning area is identified according to the lines presented by the engineering drawings of the 3D geological model. Combined with the average coal cutting rate, the longwall length, the maximum pre-drainage period of the subsequent planning area is calculated automatically. The maximum pre-drainage period determines the reasonable design number or spacing of boreholes, and the succession mine plan determines the pre-drainage method of boreholes in the next working face.

Division of drainage units based on gas condition

Based on the measured gas content or the content distribution predicted by the gas geological model, the mining face is divided into several drainage units according to the gas content, and the borehole design in each unit is carried out based on the gas content in that unit.

In the drainage units with different gas content, the drainage requirement is different, the reasonable pre-drainage period is different, and the borehole parameters are finely designed according to the value of gas content. Pre-drainage requirements for different units are determined according to the value of gas content. The gas drainage serves the needs for outburst prevention, ventilation and other specific standards, as shown in Figure 4.

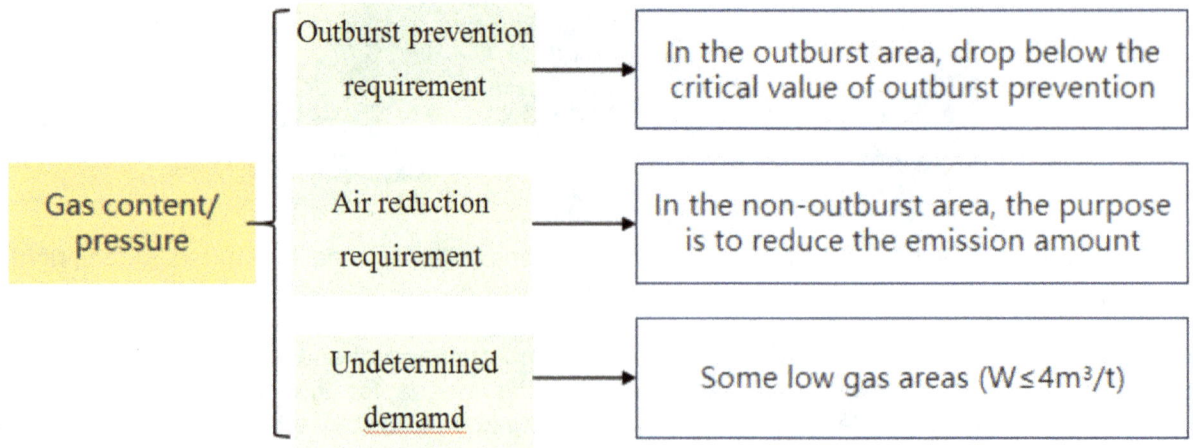

FIG 4 – Standard for drainage of different units.

For outburst-prone areas, the current standard requires that the gas content or pressure be drained below the critical value. For highly gassy areas, the gas pre-drainage rate in each unit η_I is determined according to the gas content value of different drainage units on the mining face or the gas emission amount of the working face, as shown in Table 1. The amount of gas emissions mainly comes from the adjacent seams and surrounding rocks.

TABLE 1

The corresponding table of absolute gas emission and pre-pumping rate of different working faces.

Absolute gas emission of the working face (m³/min)	Pre-drainage rate (%)
$5 \leq Q < 10$	≥ 20
$10 \leq Q < 20$	≥ 30
$20 \leq Q < 40$	≥ 40
$40 \leq Q < 70$	≥ 50
$70 \leq Q < 100$	≥ 60

Measured value Q of adjacent working faces or predicted emission values of working faces can be used:

$$(Q = q_1 + q_2) \tag{1}$$

where:

Q   is the relative gas emission quantity of the working face, the unit is m³/t

q_1 is the relative gas emission quantity of mining layer, the unit is m³/t

q_2 is the relative gas emission of the adjacent layer, the unit is m³/t

$$Q_1 = K_1 \cdot K_2 \cdot K_3 \cdot \frac{m}{M} \cdot (W_0 - W_c) \tag{2}$$

where:

K_1	is the gas emission coefficient of surrounding rock
K_2	is the coefficient of coal gushing in the stoping face, and its value is the reciprocal of the recovery rate
K_3	is the coefficient of pre-discharge of trenching
m	is the thickness of the mining layer, the unit is m
M	is the mining height of the working face, the unit is m
W_0	is the original gas content of coal seam, the unit is m^3/t
W_c	is the residual gas content of coal seam, the unit is m^3/t

$$Q_2 = \sum(W_{0i} - W_{ci}) \cdot \frac{m_i}{M} \cdot \eta_i \tag{3}$$

where:

η_I	is the gas emission rate of the adjacent layer, %
W_{0i}	is the original gas content of each adjacent layer, m^3/t
W_{ci}	is the residual gas content of each adjacent layer, m^3/t
m_i	is the coal thickness of each adjacent layer, m

The residual gas content of coal should be tested, and the desorption index and drainage rate index should be calculated with daily coal production and air flow rate of the mining face, as shown in Table 2.

TABLE 2
Daily output of different working faces, desorption gas quantity corresponding table.

Daily production of working face (t)	Amount of desorption gas (m^3/t)
≤1000	≤8
1001–2500	≤7
2501–4000	≤6
4001–6000	≤5.5
6001–8000	≤5
8000–10 000	≤4.5
>10 000	≤4

The pre-drainage rate under different demands determines the drainage standard conditions in each unit. For enterprises that do not explicitly require the pre-drainage rate, the above calculation method can be used.

Division of drainage units based on geological structure

Borehole length, dip angle and other parameters will be affected by geological structure such as fault, fold and roof collapse column. Borehole design needs to identify these structures and clarify the spatial relationship between the structure and coal seams.

Based on the 3D geological model and the of typical geological structures, the design of gas drainage boreholes is carried out. On the basis of the drainage area divided by gas content, the typical structural influence area is divided into independent drainage units. Under the condition of normal fault, reverse fault and roof collapse column, it should be designed separately. Where the fault affects the roadway, the position and orientation of gas drainage boreholes must be re-determined.

Using the 3D geological model, the area of coal seam thickening or thinning in the mining area can be identified as a single drainage unit. Gas abnormality and stress concentration usually occur in the coal thickness variation area, in which is often associated with the risk of dynamic disasters. The area with large changes in coal thickness is used as an drainage unit and targeted boreholes are designed.

Borehole design process

By constructing the typical target area identification index and method, the target area is locked step by step, and the borehole design is carried out for different gas sources. By constructing the target area and setting the index and method, the target area can be locked step by step, starting from the mine to coal seam or coal seam group, then to mining face, as shown in Figure 5.

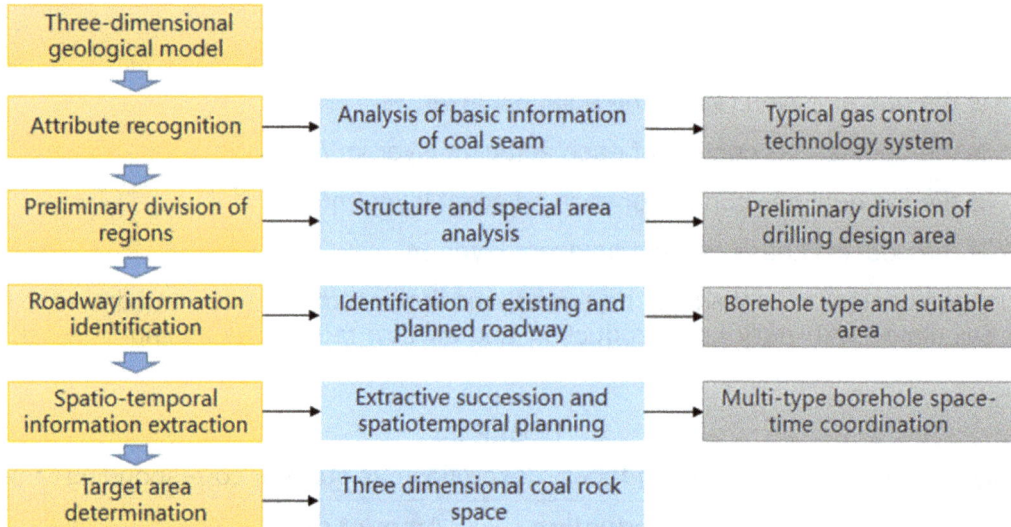

FIG 5 – Basic steps of locking the target region.

The process includes attribute identification, structure and special area identification, access to roadway information and spatio-temporal planning information extraction. For attribute identification, the relevant data is extracted in the 3D geological model to realise the attribute identification of the target area, and then the appropriate gas control technology system is matched, which provides a basis for the subsequent selection and design of relevant parameters. For structure and special area identification, the structure and special area in the target area are automatically identified, and the design area is preliminarily divided, so as to get the area that should be drilled for construction. For access to roadway information. The existing and planned roadway information is identified with the 3D model, so as to obtain the relative spatial position of roadway and coal seam, so as to match the corresponding type and control area. In terms of spatio-temporal planning information extraction, the process includes automatic identification of mining succession and spatio-temporal planning of mining areas, time allocation of different regions, accurate segmentation of target regions. Finally, according to the mining succession planning, different target areas such as mining area and goaf, adjacent seam and local coal seam are designed, as shown in Figure 6.

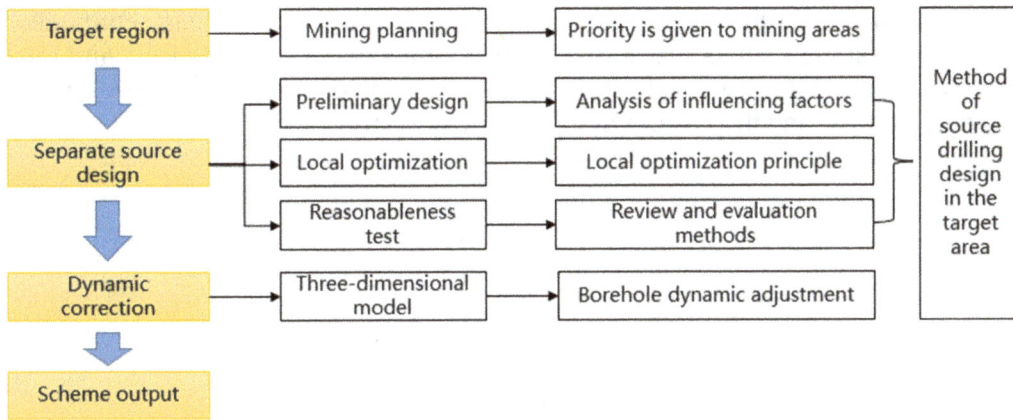

FIG 6 – Points out the basic steps of distinguishing source design.

By studying the constraints of different influencing factors on borehole design, the optimisation level and optimisation method of borehole are proposed, the applicability of differentiated source borehole is studied, and the method of local optimisation and rationality evaluation is proposed.

Drilling intelligent design model and formula

The design rules are written according to the design model of the gas control boreholes. Through the formula, determine the borehole parameters are determined through imbedded formulae. An example is given in Figure 7 where it shows the design model of the example.

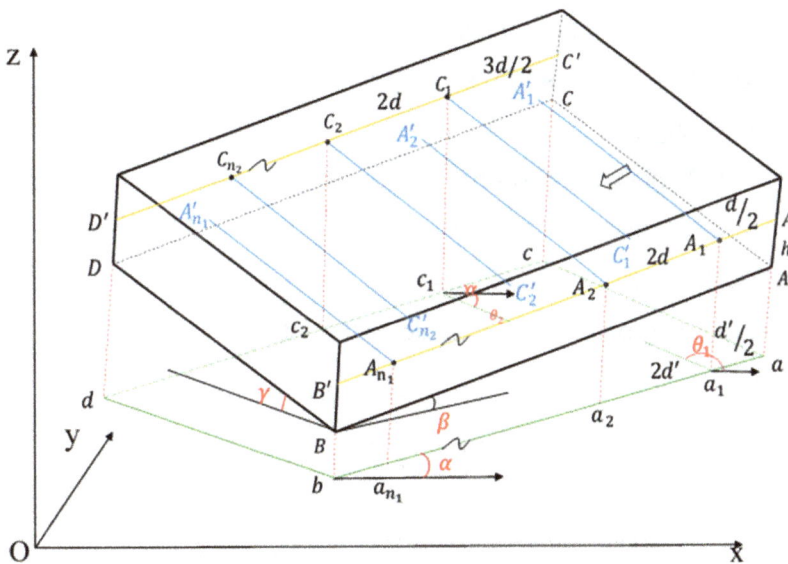

FIG 7 – Design model for horizontal double-sided borehole.

When the maximum drilling capacity of the drilling rig cannot meet the requirements of the single construction drilling, the design and construction of the double-side drilling along the bedding should be considered. And both sides of the drilling holes should be crossed to avoid blank belts.

The formula takes the coordinates of four control points, thickness of coal seam, inclination of coal seam, spacing of boreholes, gas content, gas pressure and length of working face as input parameters. According to the input parameters, the azimuth angle of borehole, length of borehole, coordinates of borehole point (coordinates of inlet air channelling slot, coordinates of return air channelling slot, coordinates of final borehole point (coordinates of inlet air channelling slot) are calculated. Point coordinates of final hole of return air channelling trough) and other parameters.

Figure 7 is the layout of double-sided matching drilling along the strata, where A (x_0, y_0, z_0) is the coordinates of the starting point of the intersection line between the bottom end of the mining face and the bottom of the air inlet channelling, B (x_1, y_1, z_1) is the end point of the intersection line between the stoping line of the mining face and the bottom of the air inlet channelling, and C (x_2, y_2, z_2) is the starting point of the intersection line between the mining face and the bottom of the return channelling. D (x_3, y_3, z_3) is the starting point coordinates of the intersection line between the stoping face and the bottom of the return air channelling.

$$L_{AB} = \sqrt{(x_0 - x_1)^2 + (y_0 - y_1)^2 + (z_0 - z_1)^2}$$
(4)

$$\alpha = \arctan\frac{y_0 - y_1}{x_0 - x_1}$$
(5)

$$\beta = \arcsin\frac{z_0 - z_1}{L_{AB}}$$
(6)

$$\theta_1 = 270° - \alpha$$
(7)

$$\theta_2 = 90° - \alpha$$
(8)

$$n = \operatorname{int}\left(\frac{L_{AB}}{d}\right) + 1$$
(9)

where:

L_{AB}	is the length along the strike of coal seam, the unit is m
α	is the roadway azimuth, the unit is °
β	is the Angle between coal seam strike and horizontal plane, the unit is °
θ_1	is the azimuth Angle of the drilling hole in the air inlet trough, the unit is °
θ_2	is the azimuth Angle of the borehole in the return air trough, the unit is °
d	is the spacing of boreholes, the unit is m
n	indicates the number of holes drilled

The coordinates of the opening point of n_1 hole along the intake roadway $A_{n_1}(x_{1,n_1}, y_{1,n_1}, z_{1,n_1})$:

$$X_{1,n_1} = x_0 - \left(2n_1 - \tfrac{3}{2}\right)d\cos\beta\,\cos\alpha$$
(10)

$$Y_{1,n_1} = y_0 - \left(2n_1 - \tfrac{3}{2}\right)d\cos\beta\,\sin\alpha$$
(11)

$$Z_{1,n_1} = (z_0 + h) - \left[2(n_1 - 1) + \tfrac{1}{2}\right]d\sin\beta$$
(12)

The coordinates of the ending position of n_1 borehole in the intake roadway $A_{n_1}{}'(x_{3,n_1}, y_{3,n_1}, z_{3,n_1})$:

$$X_{3,n_1} = x_0 - \left(2n_1 - \tfrac{3}{2}\right)d\cos\beta\,\cos\alpha - L_z\cos\gamma\,\sin\alpha$$
(13)

$$Y_{3,n_1} = y_0 - \left(2n_1 - \tfrac{3}{2}\right)d\cos\beta\,\sin\alpha + L_z\cos\gamma\,\cos\alpha$$
(14)

$$Z\_(3, n\_1) = (z\_0 + h) - [2\,(n\_1 - 1) + 1/2]d\sin\beta + L\_z$$
(15)

The coordinates of the opening point of n_2 borehole in the return roadway $C_{n_2}(x_{2,n_2}, y_{2,n_2}, z_{2,n_2})$:

$$X_{2,n_2} = x_2 - \left(2n_2 - \tfrac{1}{2}\right)d\cos\beta\,\cos\alpha$$
(16)

$$Y_{2,n_2} = y_2 - \left(2n_2 - \tfrac{1}{2}\right)d\cos\beta\,\sin\alpha$$
(17)

$$Z_{2,n_2} = (z_2 + h) - \left[2(n_2 - 1) + \tfrac{3}{2}\right]d\sin\beta$$
(18)

The coordinates of the ending point of n_2 borehole in the return roadway $C_{n_2}{}'(x_{4,n_2}, y_{4,n_2}, z_{4,n_2})$:

$$X_{4,n_2} = x_2 - \left(2n_2 - \frac{1}{2}\right) d \cos\beta \, \cos\alpha + L_z \cos\gamma \, \sin\alpha \qquad (19)$$

$$Y_{4,n_2} = y_2 - \left(2n_2 - \frac{1}{2}\right) d \cos\beta \, \sin\alpha - L_z \cos\gamma \, \cos\alpha \qquad (20)$$

$$Z_{4,n_2} = (z_2 + h) - \left[2(n_2 - 1) + \frac{3}{2}\right] d \sin\beta - L_z \sin\gamma \qquad (21)$$

Where L_z represents the length of the borehole, the unit is m. The other meanings are the same as above.

INTELLIGENT DESIGN SOFTWARE

Based on the secondary development of MineCAD platform, the software of 'Intelligent design software of gas drainage borehole' is developed with the use of ObjectARX and VisualLISP, and the 3D geological model of the anticipated coal seam is automatically constructed by using the 3D entity discrimination criterion of 'borehole–drainage area–coal body' and the triangle-mesh positioning method of coal seam surface. The control area can be freely selected in the drawing, and the borehole parameter calculation and design drawing can be completed automatically. The software account management function is developed. Users can log in by entering their account and password on the software login interface. The software includes 3D coal seam geological modelling, 3D gas geological modelling, gas drainage unit division, and intelligent design of drainage boreholes.

Introduction of the software functions

3D coal seam geological modelling

3D coal seam geological modelling is set with known parameters, plane parameters, strike profile parameters, dip profile parameters, roadway parameter management, drilling field parameter management and coal seam parameter management as input parameters. The user controls the work surface through the plan, profile and other factors, and adds the corresponding data according to the content of the drawings provided by the mine, so as to generate the 3D model of the work surface. Through the function of 3D coal seam geological modelling, the construction of 3D coal seam geological model is realised. The model can show the geological structure of the coalmine working face, so that it can be understood and used more intuitively and vividly. The 3D coal seam geological model directly reflects the actual situation of underground coal seam, and the accuracy of the model directly affects the design of gas drainage borehole.

3D gas geological modelling

The 3D gas geological model is developed on the basis of the 3D coal seam geological model. Therefore, after all the input parameters of the 3D geological model are set, relevant gas content data or gas pressure data are added to the model according to the specific data measured on the working face to form the corresponding gas content model or gas pressure model. The 3D gas geological model is set by adding gas content data, gas content plane modelling, gas content 3D modelling, gas pressure data, gas pressure plane modelling, and gas pressure 3D modelling. The gas content of the working face shown in the 3D gas geological model is an important reference for the division of the working face drainage unit and the intelligent design of the borehole.

The following division of gas extraction units and intelligent design of extraction drilling holes are mainly determined by the parameter of gas content. Users determine the division of units and the spacing of drilling holes in each unit according to the differences of gas content in different positions of the 3D gas content model of the working face.

Division of gas drainage unit

In order to realise the differentiated source design, the division function of gas drainage unit is developed, which sets the drainage parameter setting and drainage unit division as input parameters. Based on the 3D gas geological model, the gas drainage units is divided according to the relevant regulations of the state and different gas content areas, corresponding to the drainage radius, so as to ensure the best drainage effect. The area with faults is separately divided into one unit for setting.

Intelligent design of gas drainage hole

Due to different construction units, different working faces and other reasons, the corresponding construction conditions are also different. Therefore, considering the universality of the software, rig parameters, opening parameters, opening position, coal piercing mode, and layout mode are taken as the input parameters in the design software. The design combines the input content of 3D coal seam geological modelling, 3D gas geological modelling and gas drainage unit division, then generate the results, and then complete the output of the final borehole design drawing set data.

Software testing

Overview of 151108 working face of Liuzhuang Mine

According to the '11-2 Coal Seam (1511 mining area) Outburst Risk Identification Report' prepared by Chongqing Research Institute of China Coal Science and Industry Group Co., LTD, the 11-2 coal seam in 1511 mining area of Liuzhuang Mine (F8, F18, F20 east to F25 fault west, shallow-762 m level) is identified as non-outburst coal seam. The 151108 working face is in this range, as shown in Figure 8.

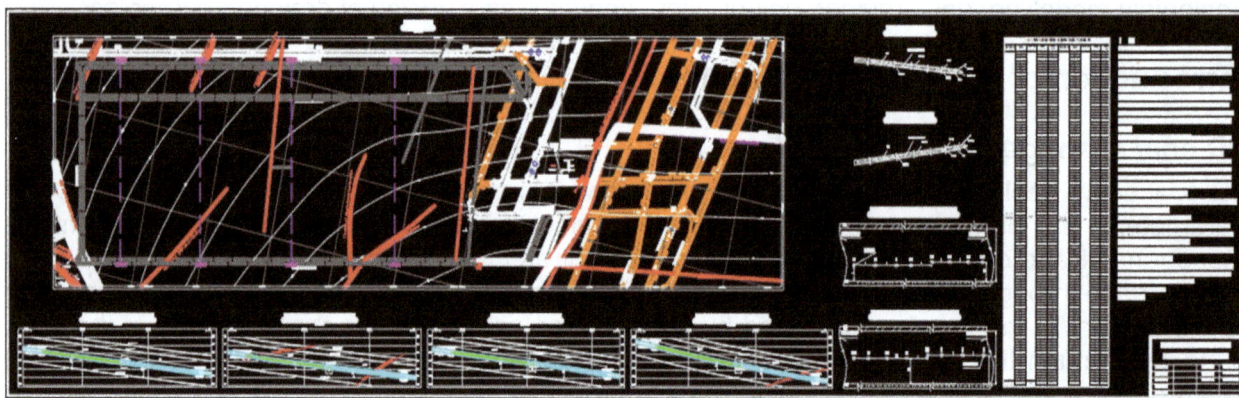

FIG 8 – Drawing of the 151108 working face.

The maximum original gas content during the working face of 151108 is 3.5632 m³/t, the analysable gas content is 2.3362 m³/t, and the atmospheric residual gas content is 1.2270 m³/t. Both the original gas content and the resolvable gas content are lower than the relevant pre-pumping requirements of 'China Coal Xinji [2017] No. 161-- Annex 2' Evaluation System of Coalmine Gas Extraction Standard of China Coal Xinji Company.

The maximum absolute gas emission during the working face of 151108 was 0.78 m³/min, and the maximum gas concentration was 0.06 per cent, all of which were lower than the relevant pre-pumping requirements of 'China Coal Xinji [2017] 161-- Annex 2' China Coal Xinji Company Mine Gas Extraction Standard Evaluation System.

Test method

Part of the 151108 working face is selected to generate a 3D coal seam geological model, a 3D gas geological model and the corresponding drilling design. By comparing the artificial design of relevant workers in Liuzhuang Mine with the intelligent design results automatically generated by the software, a conclusion is drawn to determine whether the software is accurate.

Test content

According to the specific conditions of the 151108 working face drawing, the parameters to be set were respectively set, and the modelling area was selected. The 3D display of the 3D geological model of the 151108 working face initially generated as shown in Figure 9. Input the recorded data of gas content measurement provided by the mine, add the corresponding gas content data to the corresponding position, adjust the gas content interval and corresponding colour according to the gas content data, and the generated 3D gas content model is shown in the figure.

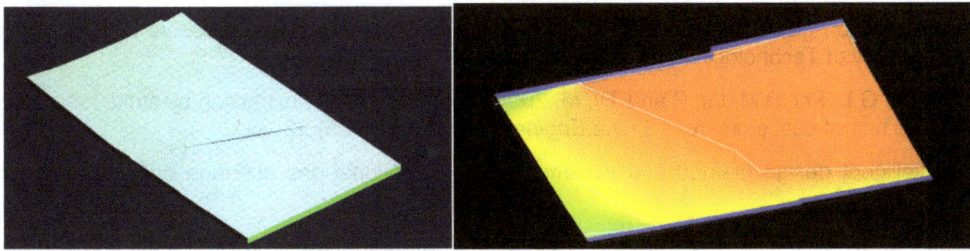

FIG 9 – 3D model of the 151108 working face.

In the duct of the working face, four extraction units are divided between sections 1 and 4, as shown in the figure. In the next step, the borehole intelligent design of the working face is designed according to the area selected at this moment. Because the fault in the extraction area is less than 5 metres, the extraction radius is also designed to be 3 metres.

The drilling design of 151108 working face and its 3D model were generated through the software, as shown in Figure 10.

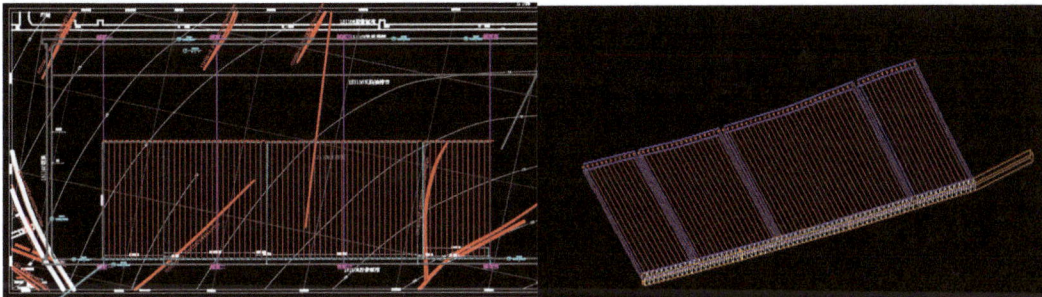

FIG 10 – Design of 151108.

CONCLUSION

In this paper, a complete set of intelligent gas drainage borehole design system is introduced, and based on this, the intelligent design software is developed to realise the automatic drawing of borehole design. Based on the 3D gas geological model, the intelligent design method of coal seam gas drainage borehole is proposed to lock the target area. Based on the secondary development of MineCAD platform, the intelligent design system of gas drainage borehole was developed and achieved the integration of 3D geological modelling, unit division and borehole design. The reliability of the software was verified by field tests in Liuzhuang Mine.

REFERENCES

Ahmad, S S F J B and Usman, M K, 2020. Optimization of gob ventilation boreholes design in longwall mining, *International Journal of Mining Science and Technology*, 30(06):811–817.

Chen, D F, 2015. Research and application of directional drilling technology for thick coal seam roof, *Safety in Coal Mine*, 46(04):122–124; p 127.

Chen, X H, 2021. Layout optimization and trajectory design of long boreholes for gas drainage in yuwu coal mine, *Energy and Energy Conservation*, (10):53–54; p 57.

Fan, Y F, 2019. Study on measurement method of effective radius of gas extraction and optimization of hole distribution parameters in bedding borehole, Xi'an University of Science and Technology.

Hao, T X, Fan, G Y and Zhao, L Z, 2023. Intelligent borehole design system based on AutoCAD, *China Production Safety Science and Technology*, 19(04):71–77.

Hu, W Y, 2013. Study orientation and present status of geological guarantee technologies to deep mine coal mining, *Coal Science and Technology*, 41(08):1–5; p 14.

Ji, C H, 2015. Application and Practice of Coal-gas Co-extraction Technology by Floor Drainage Roadway in Single Low-permeability Outburst Seam, *Mining Safety and Environmental Protection*, 42(3):86–89.

Jiang, X X and Li, C X, 2019. Statistical analysis on coal mine accidents in China from 2013 to 2017 and discussion on the countermeasures, *Coal Engineering*, 51(1):101–105.

Liu, H Z, 2013a. Research on Method and Practice of gas drainage in a coal seam, North University of China.

Liu, Y, 2013b. Research on gas extraction technology of large diameter high level borehole in goaf of high gas mine, Taiyuan University of Technology.

Liu, Z G, Yuan, L, Dai, G L, Shi, B M, Lu, P and Tu, M, 2004. Research on gas extraction by strike long borehole method in ring fracture ring of coal seam roof, *China Engineering Science*, (05):32–38.

Ma, G L, 2017. Intelligent design method and system for long directional gas drainage drilling, *Safety in Coal Mine*, 48(05):103–106.

Ren, S, Wang, F T, Li, S T, Yu, G F and Zhao, D F, 2022. Overburden migration control technology of intelligent filling mining in deep coal seam, *Mining Research and Development*, 42(03):163–167.

Shi, Z J, Xu, C and Li, Q X, 2015. Trajectory design and calculation of nearly horizontal MWD directional borehole in underground coal mine, *Coal Geology and Exploration*, 43(04):112–116.

Wang, H F, Cheng, Y P and Wang, L, 2012. Regional gas drainage techniques in Chinese coalmines, *International Journal of Mining Science and Technology*, 22:873–878.

Wang, J C, Liu, F and Wang, L, 2016. Sustainable coal mining and mining sciences, *Journal of China Coal Society*, 41(11):2651–2660.

Wang, L F, 2015. Research on Gas Control Technology of High Level Borehole in Gob of Wangzhuang Coal Mine, Henan Polytechnic University, Jiaozuo.

Wang, L, 2023. Optimization design of gas drainage boreholes in 3603 working face of horsinghe coal mine, *Modernization of coal mine*, 32(02):26–29.

Wu, Q, Tu, K, Zeng, Y F and Liu, S Q, 2019. Discussion on the main problems and countermeasures for building an upgrade version of main energy (coal) industry in China, *Journal of China Coal Society*, 44(06):1625–1636.

Xiao, L H, Li, Y M, Guo, K M and Zhong, P, 2014. Study on gas drainage technology of screen pipe putting down along full length borehole in soft outburst seam, *Coal Science and Technology*, 42(07):61–64.

Xie, J L and Sun, X Y, 2013. Gas drainage technology of mining fracture developed zone in high gassy and thick seam, *Coal Science and Technology*, 41(05):68–71.

Xin, X P, Ma, G and Yang, C T, 2015. Gas extraction geological unit division and evaluation methods, *Coal Mine Safety*, 46(02):146–150.

Yu, Q X, 1992. *Theory and Technology of Mine Gas Drainage* (China University of Mining and Technology Press: Xuzhou).

Yuan, L, 2008. Key technique of safe mining in low permeability and methane-rich seam group, *Chinese Journal of Rock Mechanics and Engineering*, (07):1370–1379.

Yuan, L, 2009. Theory of pressure-relieved gas extraction and technique system of integrated coal production and gas extraction, *Journal of China Coal Society*, 34(01):1–8.

Yuan, L, 2016. Strategic thinking of simultaneous exploitation of coal and gas in deep mining, *Journal of China Coal Society*, 41(01):1–6.

Yuan, L, 2017. Scientific conception of precision coal mining, *Journal of China Coal Society*, 42(01):1–7.

Yue, J, 2022. Automatic design and software development of high-level drilling for gas drainage, *Coal Mine Machinery*, 43(03):186–188.

Zhang, F Y, Han, Y and Cao, W T, 2013. Study on gas drainage technology by using large aperture high level borehole, *China Mining Industry*, 22(08):120–124.

Zhang, H, Xia, Y J, Zhang, Q, Jin, X L and Jin, D W, 2009. Coal-mining geological conditions and explorations of deep coal deposits: status and problems, *Coal Geology and Exploration*, 37(01):1–11; p 16.

Zhang, J F, 2023. Feasibility analysis of gas extraction in fully mechanized caving face and design of high drilling hole, *Inner Mongolia Coal Economy*, (11):61–63.

Zhang, T G and Zhang, X B, 2021. Study on deformation and instability area of deep coal seam gas drainage boreholes, *Journal of Energy and Environmental Protection, The Lancet*, (04):44–47.

Zhou, F B, Wang, X X and Xia, T Q, 2014. A model of safe drainage of coal seam gas, *Journal of China Coal Society*, 39(08):1659–1666.

Numerical modelling and integration with planning and remote monitoring

Use and limitations of the Atkinson Equation

D W Durieux[1]

1. Principal Advisor – Mine Ventilation and Refrigeration, Rio Tinto, Superior AZ, USA.
   Email: duran.durieux@riotinto.com

ABSTRACT

The Atkinson Equation is almost exclusively used in the mining industry to determine airway (shaft, tunnel, duct etc) pressure loss for a given airflow. Atkinson's pressure loss is a function of the Atkinson Friction Factor (k), airway geometry, velocity, and density. Atkinson's pressure loss is mainly used to determine fan power requirements, size airways, prescribe surface finishes, and forecast electrical power costs. The Atkinson Equation is relatively straightforward to use, and 'k' values are readily available for typical mining airways.

However, Atkinson developed this equation around 1854 before significant advances were made by Reynolds. At the time, Atkinson believed the dimensionless Darcy Friction Factor (f) was constant for a given airway and so Atkinson grouped 'f' and air density to form k, with units of density. Mines were also not very deep at the time and variations in air density were small. Around the 1880s, however, Reynold discovered that 'f' is not constant for a given airway. It is a function of Reynold's Number (Re) and the ratio of the airway absolute surface roughness (e) to diameter. This relationship is well understood in Fluid Mechanics and forms the foundation of pressure loss calculations. Indeed, modern Fluid Mechanics employs the so called Dary-Weisbach Equation to determine pressure losses for any incompressible steady-state fluid flow (gas or liquid). The Moody Chart was also developed to facilitate pipeline calculations and graphically displays the variable nature of the Darcy Friction Factor (f).

This paper aims to outline appropriate use, and limitations, of the Atkinson Equation in the mining industry, particularly on the use of a constant k for a given airway type. It also aims to make the case for replacing the outdated Atkinson Equation for the Darcy-Weisbach Equation. Particular attention is given to the lack of absolute surface roughness (e) data for the mining industry and possible ways to overcome this.

INTRODUCTION

Well respected and accepted mine ventilation literature, such as those of the Mine Ventilation Society of South Africa (MVSSA, 2014) and McPherson's (2007) Subsurface Ventilation and Environmental Engineering, and Hartman's (1997) Mine Ventilation and Air Conditioning, amongst others, use the concept of the Atkinson Friction Factor (k) and the Atkinson Equation to calculate frictional pressure losses in airways. The use of the Atkinson Friction Factor and Atkinson Equation was developed around 1854 and was entrenched in the Mining literature by McElroy's 1935 publication. However, Fundamental Fluid Mechanics literature, for example Cengel and Cimbala (2006) and White (2006), make no mention of the Atkinson Friction Factor nor Atkinson Equation. The seemly different approach to analysing frictional pressure loss for internal fluid flow between the fundamental Fluid Mechanics literature, which employs the so-called Darcy-Weisbach Equation, and the mine ventilation literature has always interested the author. And so, this paper was borne out of both a general curiosity and the desire to reconcile the fundamental Fluid Mechanics versus traditional mine ventilation approach to the analysis of internal flows.

Internal flow, in this paper, refers to any fluid flow that is fully developed inside a solid boundary flowing under steady-state conditions. Any fluid, in either liquid or gaseous state, may be considered but this paper will focus on air, and boundaries will include pipes, ducts and mine airways such as shafts and underground drifts. Furthermore, only incompressible flows will be considered (note the term, incompressible flow and not incompressible fluid). Incompressible flow will be taken as any flow where the Mach Number (Ma, see below equation) is less than 0.3. For air, this can be up to 100 m/s at sea level which is obviously far higher than airflow velocities found in mines.

$$Ma = \frac{\text{instantaneous fluid velocity}}{\text{fluid speed of sound}}$$

In addition to the Fluid Mechanics literature which employs the Darcy-Weisbach Equation, reputable organisations such as ASHRAE and the Sheet Metal and Air Conditioning Contractors' National Association (SMACNA) do not use the Atkinson Equation nor the Atkinson Friction Factor for duct calculations. Instead, these organisations employ the same methods as the Fluid Mechanics literature. This paper aims to develop an understanding of how this came about.

The goal of performing calculations is to predict the behaviour of the physical world. With respect to this paper, we are interested in predicting the behaviour of fluid flow in internal boundaries. To this end, we should be employing the most up-to-date techniques to do so. If we use outdated techniques, which do not fully represent the behaviour of nature, we should not be surprised when our predictions do not come to fruition.

This paper will attempt to show how the Atkinson Friction Factor (k) is not always constant for a given material or surface finish. It is a variable that often needs to be calculated because it is mainly dependant on the airway diameter. This can be shown graphically by means of a simple example, see Figure 1, where a raise borehole of 300 m running at 120 m³/s produces variations in frictional pressure loss for different diameter holes. The dotted line is a plot of the traditional mining approach using a fixed Atkinson Friction Factor (k) and using the Atkinson Equation (Equation 10b) to determine the frictional pressure drop. The solid line represents the Modern Fluid Mechanics approach which uses a fixed absolute surface roughness (e), calculates the Atkinson Friction Factor using the Haaland Equation (Equations 5.2b and 13b), and then uses the Darcy-Weisbach Equation (Equation 4) to determine the frictional pressure drop. At smaller diameters, there is a significant variation in the frictional pressure loss. For example, there is about a 15 per cent variation for a 2.5 m diameter raise bore which will run at about 24.4 m/s in this example. It can however be seen that as the diameter increases, and the velocity decreases, there is not a large variation between using a constant and variable friction factor and so as is typical of engineering problems, the degree to which variations exist depend on the problem at hand.

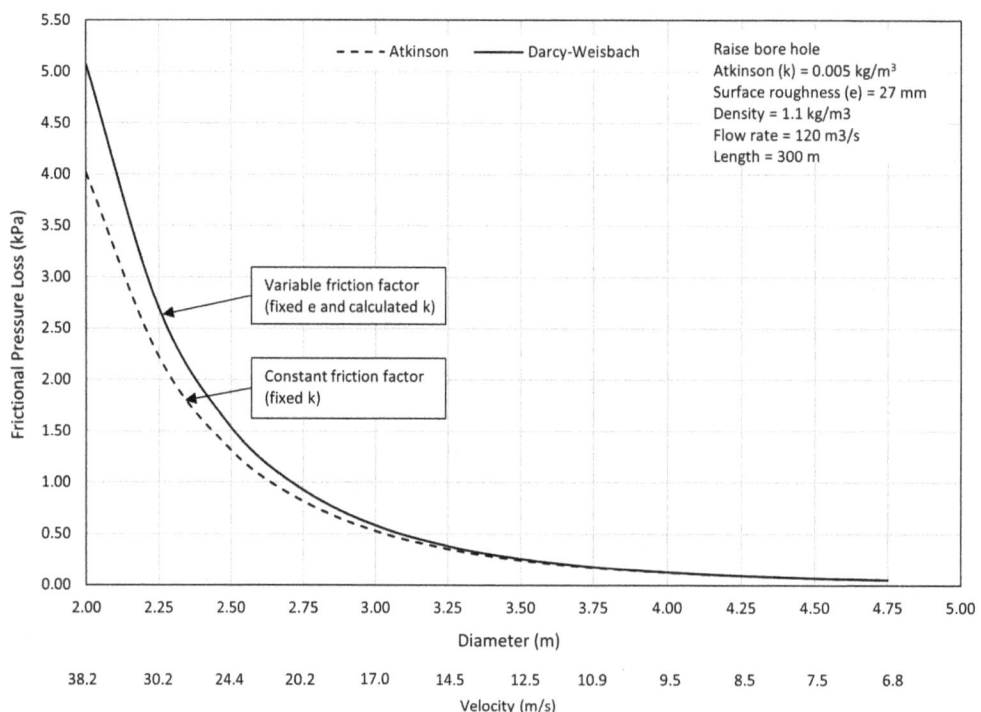

FIG 1 – Frictional pressure loss using constant versus variable friction factor.

FUNDAMENTALS OF INTERNAL FLUID FLOW

Fluid mechanics

Internal fluid flow is well documented and discussed in detail in Fluid Mechanics literature, such as those of White (2006) and Cengel and Cimbala (2006). The aim of this section is to dive straight into the method of analysis with some brief descriptions.

The generalised problem is that of fluid flowing inside a pipe, see Figure 2. The general approach is to modify Bernoulli's equation to include system losses such that the pressure drop can be quantified, and equipment sized. These losses are grouped into either frictional or shock losses such that their sum produces the total pressure loss (ΔP_L).

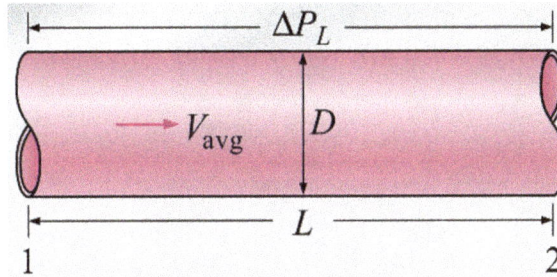

FIG 2 – Generalised internal fluid flow, with flow from point 1 to 2 inside pipe of diameter D and length L.

Reynolds number

The **Reynolds number (Re)** is an important non-dimensional number for internal flows. It relates fluid inertial to viscous forces and takes the form of Equation 1. It also sets limits to various types of flows, see Table 1. In mine ventilation practice, however, laminar flow is seldom experienced and so the theory is omitted in this paper. All flow can be treated as turbulent and some can be classified as fully rough, turbulent flow:

$$\text{Re} = \frac{\rho V D_h}{\mu} \tag{1}$$

where:

ρ = fluid density $\left(\frac{kg}{m^3}\right)$

V = fluid average velocity $\left(\frac{m}{s}\right)$

D_h = boundary hydraulic diamter (m)

μ = fluid dynamic viscosity $\left(\frac{kg}{m.s}\right)$

The **hydraulic diameter (D_h)** is used for non-circular cross-sections such as rectangular ducts and mine airways such as arched or rectangular shapes, see Equation 2.

$$D_h = \frac{4A}{C} \tag{2}$$

where:

A = boundary cross $-$ sectional area (m^2)

C = boundary cross $-$ sectional wetted perimeter (m)

TABLE 1

Reynold number for various internal flow regimes.

Flow regime	Re
Laminar	$Re < 1 \times 10^3$
Transitional	$1 \times 10^3 < Re < 1 \times 10^4$
Turbulent	$Re > 1 \times 10^4$

Friction

As fluid flows inside a boundary, shear stress (τ_w) exist between the fluid and the boundary surface eg duct or drift. Figure 3 shows the velocity profile with no slip at the boundary and τ_w acting in the opposite direction to the fluid flow direction. The magnitude of this shear stress is given by Equation 3.

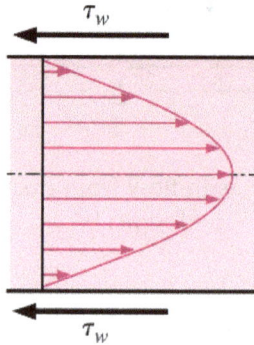

FIG 3 – Shear stress developed at the boundary due to fluid flow with example velocity profile.

$$\tau_w = \left(\frac{f}{4}\right)\left(\frac{1}{2}\rho V^2\right) \tag{3}$$

f = Darcy Friction Factor (−)

$\left(\frac{1}{2}\rho V^2\right)$ = velocity pressure (Pa)

Note the definition of the Darcy Friction Factor (f) in this paper. It relates back to both Equations 3 and 4. This is the definition used by both White (2006) and Cengel and Cimbala (2006) and indeed most American Fluid Mechanics literature. There are however several texts which employ a Friction Factor (λ) that is four times lower ie λ=f/4.

The magnitude of the frictional pressure loss ($\Delta P_{L,f}$) is given by the so-called **Darcy-Weisbach Equation** in Equation 4. Note the use of the velocity pressure and dimensionless friction factor.

$$\Delta P_{L,f} = f\left(\frac{L}{D_h}\right)\left(\frac{1}{2}\rho V^2\right) \tag{4}$$

L = boundary length (m)

The **Darcy Friction Factor (f)** is determined either analytically, using the Colebrook Equation in Equations 5.1 or 5.2b, or by means of the so-called Moody Chart shown in Figure 4. The general problem, before the advent of computers, has been the implicit Colebrook Equation which must be solved iteratively. This can readily be done with an iterative algorithm on modern-day computers, with say a Newton-Raphson iteration. Or the explicit Haaland equation given in Equation 5.2 can be employed.

The implicit **Colebrook Equation** is valid for turbulent flows in smooth and rough pipes.

$$\frac{1}{\sqrt{f}} = -2\log\left(\frac{e/D_h}{3.7} + \frac{2.51}{Re\sqrt{f}}\right) \tag{5.1}$$

e/D_h = relative roughness (−)

e = absolute mean surface roughness (mm)

The explicit **Haaland Equation** is also valid for turbulent flows in smooth and rough pipes. It is accurate to within 2 per cent of the Colebrook Equation stated in Equation 5.1.

$$\frac{1}{\sqrt{f}} = -1.8 \log\left(\left(\frac{e/D_h}{3.7}\right)^{1.11} + \frac{6.9}{Re}\right) \tag{5.2a}$$

And solving for f:

$$f = \left[-1.8 \log\left(\left(\frac{e/D_h}{3.7}\right)^{1.11} + \frac{6.9}{Re}\right)\right]^{-2} \tag{5.2b}$$

The equations above have not been obtained from theoretical analysis, but rather from experimental data – hence are empirical equations. They were obtained by simulating pipe roughness artificially by gluing sand grains of known size on the inside of pipes. The fluid velocity and corresponding pressure drop were measured which allowed the corresponding Darcy friction Factor (f) to be calculated with Equation 4. Furthermore, the equations above, and the Moody Chart below (Figure 4), show that the Darcy Friction Factor (f), in fully developed internal flow, is a function of both the Reynolds No. (Re) and the relative roughness (e/D_h) of the boundary and can be generically expressed with Equation 6.

$$f = fn(Re, e/D_h) \tag{6}$$

Shock losses

So far, the losses associated with friction have been defined. But this is only one type of loss associated with typical mine airflow systems. Determining the magnitude of the total pressure loss is generally of interest. To do this, it is convenient to define a term called shock losses ($\Delta P_{L,s}$) which accounts for other pressure losses such as fittings, dampers, bends, tees, safety screens, fan silencers, non-return doors, etc. For mine airways, these include changes in geometry, cross-cuts, ventilation doors, etc. Typical shock loss coefficients (X) can be found in Appendix C.

Again, the shock loss is defined as a function of velocity pressure and a dimensionless coefficient, Equation 7. The sum of all shock loss coefficients are multiplied by the velocity pressure to obtain the system total shock pressure loss:

$$\Delta P_{L,s} = \sum X \left(\frac{1}{2}\rho V^2\right) \tag{7}$$

X = shock loss coefficient (−)

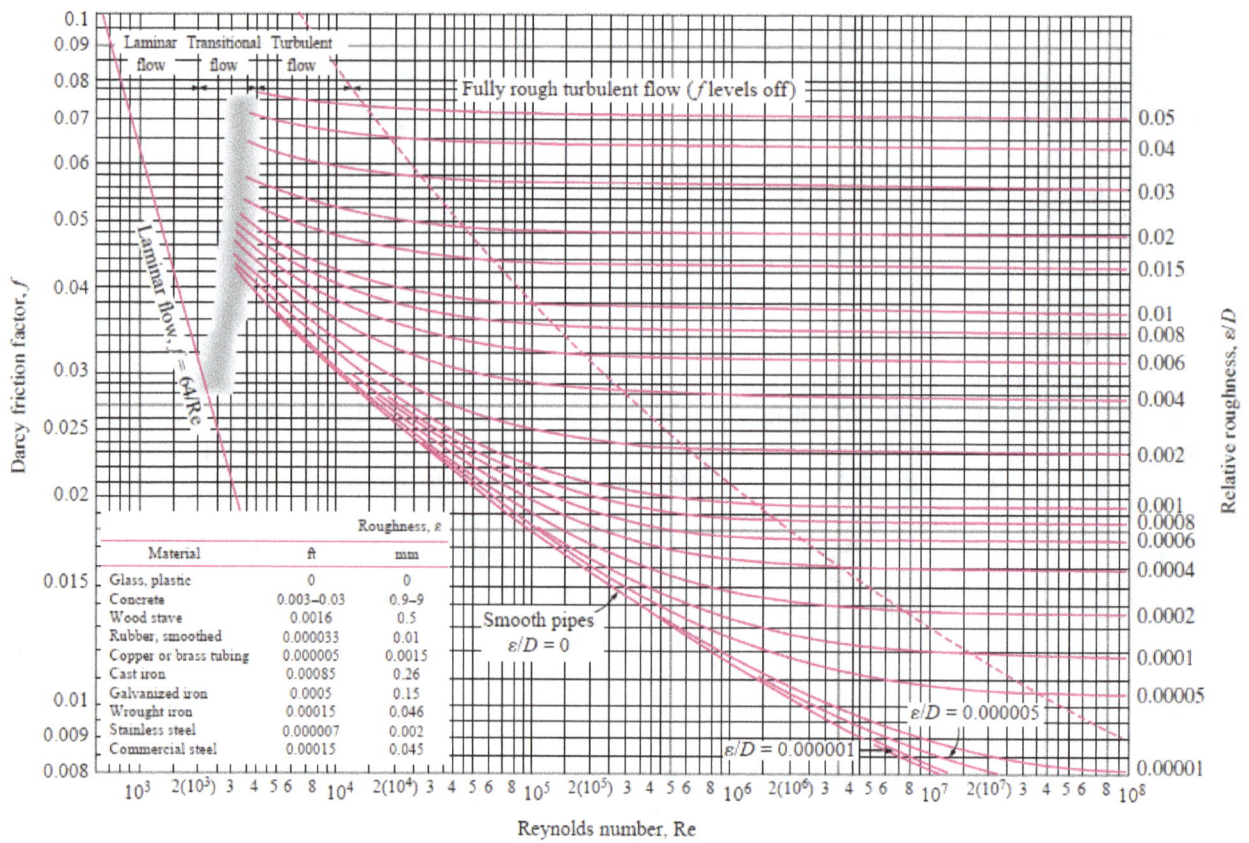

FIG 4 – Standard Moody chart for pipe flow, Cengel and Cimbala (2006).

Modified Bernoulli equation

The total pressure loss, including both frictional (Equation 4) and shock pressure losses (Equation 7), can therefore be summed to produce Equation 8.

$$\Delta P_{L,total} = \Delta P_{L,friction} + \Delta P_{L,shock}$$

$$\Delta P_{L,total} = f\left(\frac{L}{D_h}\right)\left(\frac{1}{2}\rho V^2\right) + \sum X\left(\frac{1}{2}\rho V^2\right)$$

$$\Delta P_{L,total} = \left(f\frac{L}{D_h} + \sum X\right)\left(\frac{1}{2}\rho V^2\right) \tag{8}$$

Although not the focus of this paper, the **modified Bernoulli equation** can then be used to determine the total pressure required ($P_{required}$) to overcome the total pressure loss ($\Delta P_{L,total}$) and also accounting for changes in mechanical and kinetic energy, see Equation 9. This modified Bernoulli equation is valid for incompressible flow (ρ = constant) with no heat transfer and work done. Apply the steady-flow energy equation to the modified Bernoulli equation to account for variations in air density and/or to include heat transfer and/or work done by the system.

$$\text{Energy})1 = \text{Energy})2$$

$$\left[P + \left(\frac{1}{2}\rho V^2\right) + (\rho g Z)\right]_1 + \Delta P_{required} = \left[P + \left(\frac{1}{2}\rho V^2\right) + (\rho g Z)\right]_2 + \Delta P_{L,total}$$

$$\Delta P_{required} = (P_2 - P_1) + \frac{1}{2}\rho(V_2 - V_1)^2 + \rho g(Z_2 - Z_1) + \Delta P_{L,total} \tag{9}$$

$(P_2 - P_1)$ = change in static pressure between point 1 and 2

$\frac{1}{2}\rho(V_2 - V_1)^2$ = change in velocity pressure between point 1 and 2

$\rho g(Z_2 - Z_1)$ = change in hydrostatic pressure between point 1 and 2

Atkinson Equation

The **Atkinson Equation** takes the form outlined in Equation 10a. Note the use of a non-dimensionless friction factor.

$$\Delta P_{L,f} = k \left(\frac{CL}{A^3}\right) Q^2 \tag{10a}$$

k = Atkinson Friction Factor (kg/m^3)

Q = fluid volumetric flow rate (m^3/s)

The Atkinson Friction Factor (k) is generally specified at the sea level air density of 1.2 kg/m^3. This special case k can be written as $k_{1.2}$ and Equation 10a modified so that any air density can be used.

$$\Delta P_{L,f} = \frac{k_{1.2} \times \rho}{1.2} \left(\frac{CL}{A^3}\right) Q^2 \tag{10b}$$

Atkinson Friction Factors (k) can be found in various mining literature, see Appendix A.

Fluid mechanics versus Atkinson

Friction

Both the fluid mechanics and Atkinson Equations have been described and can now be evaluated to determine their relationship with respect to friction only.

Rearranging the Darcy-Weisbach Equation (Equation 4) such that it takes the form of the Atkinson Equation (Equation 10a),

$$\Delta P_{L,f} = f \left(\frac{L}{D_h}\right) \left(\frac{1}{2}\rho V^2\right) \tag{4 repeated}$$

Where the average fluid velocity (V) can be written in the form of the average volumetric flow rate (Q),

$$Q = V \times A$$

Therefore:

$$V = \frac{Q}{A}$$

And:

$$V^2 = \frac{Q^2}{A^2} \tag{11}$$

Substituting Equations 2 and 11 into Equation 4 yields:

$$\Delta P_{L,f} = f \left(\frac{L}{\frac{4A}{C}}\right) \left(\frac{1}{2}\rho\frac{Q^2}{A^2}\right)$$

$$\Delta P_{L,f} = \frac{f\rho}{8} \left(\frac{CL}{A^3}\right) Q^2 \tag{12}$$

Inspection of Equation 10a shows that Atkinson has grouped the Darcy Friction Factor (f) and the air density (ρ) to formulate his Atkinson Friction Factor (k) with units of density, that is:

$$k = \frac{f\rho}{8} \tag{13a}$$

And for standard air density at sea level, Equation 13a can be written as follows:

$$k_{1.2} = \frac{f \times 1.2}{8}$$

$$k_{1.2} = 0.15f \tag{13b}$$

Friction and shock losses

The Atkinson Friction Factor (k) should, as the name suggests, account for friction only. When this is the case, Equation 13b is valid. However, **k is often used to define the total pressure loss** ($\Delta P_{L,total}$) and not just friction. The relationship between k and $\Delta P_{L,total}$ is thus:

$$\Delta P_{L,total} = \left(f\frac{L}{D_h} + \sum X\right)\left(\frac{1}{2}\rho V^2\right) \qquad \text{(8) repeated}$$

Substituting Equations 2 and 11 into Equation 8 yields:

$$\Delta P_{L,total} = \left(f\frac{L}{D_h} + \sum X\right)\left(\frac{8\rho}{C^2 D_h{}^2}\right)Q^2 \qquad \text{(13)}$$

Setting Atkinson equal to the total pressure loss results in the following relationship, Equation 13 = Equation 10b (and using Equation 2):

$$\left(f\frac{L}{D_h} + \sum X\right)\left(\frac{8\rho}{C^2 D_h{}^2}\right)Q^2 = \frac{k_{1.2} \times \rho}{1.2}\left(\frac{CL}{A^3}\right)Q^2$$

$$\left(f\frac{L}{D_h} + \sum X\right)\left(\frac{8}{C^2 D_h{}^2}\right) = \frac{k_{1.2}}{1.2}\left(\frac{CL}{A^3}\right)$$

$$k_{1.2} = \left(f\frac{L}{D_h} + \sum X\right)\left(\frac{9.6 D_h{}^3}{4^3 D_h{}^2 L}\right)$$

$$k_{1.2} = \left(f\frac{L}{D_h} + \sum X\right)\left(\frac{0.15 D_h}{L}\right)$$

$$k_{1.2} = 0.15\left(f + \sum X\frac{L}{D_h}\right) \qquad \text{(14)}$$

Equation 14 shows the relationship between the Atkinson Friction Factor and the total pressure loss as defined by Modern Fluid Mechanics which includes friction and shock losses. Inspection of Equation 14 shows the same relationship already determined for friction (Equation 13b) plus the additional shock loss component. However, this will not be explored further in this paper as the focus is mainly on frictional effects.

THE ABSOLUTE SURFACE ROUGHNESS

It has been shown that both the Darcy Friction Factor (f) and the Atkinson Friction Factor (k) are not independent variables, and thus cannot, and should not, be relied upon for generalised characterisation of friction. That is, k is a function of f (Equation 13b) and f is a function of Re and relative roughness (e/D_h) (Equations 5). This means that if either Re or the relative roughness change, so will the friction factors. Put another way, both f and k will change depending on the fluid velocity, air psychrometric properties, boundary material and fabrication methodology. k can thus be generalised in algebraic form with Equation 15. But the same is not true for the absolute surface roughness (e). It is not dependent on the fluid velocity and psychrometric properties. It is only dependent on the material type and fabrication method ie what is the duct material and how was it manufactured, or what is the drift host rock and what is the surface finish? e can therefore be considered an independent variable.

$$k = fn(f) = fn(Re, e/D_h) \qquad \text{(15)}$$

The current mine ventilation literature lists k values for mine airway types, see Appendix A. This implies that k is constant for a given airway type and is an independent variable. It has been shown that this is fundamentally incorrect and opens the user up to errors in the calculation of real-world frictional pressure drops; particularly at the extremes of typical mine ventilation systems operating conditions. It is true that the Re may not be fundamental in the determination of the friction factor for fully developed turbulent flows, but what about hydraulic diameter changes for a given airway? This will change the relative roughness and thus the friction factor!

Just as k values listed in literature have some uncertainty or error bars, so do e values. But, if empirical data is obtained through experimentation or testing to determine more accurate e values, these can be used to determine friction factors (k or f) which vary depending on air velocity and psychrometric properties. This is how real-world friction behaves in internal fluid flows.

Fluid mechanics literature and reputable organisations such as ASHRAE and SMACNA all publish absolute surface roughness values for a given material and construction type. Either analytical means (eg Colebrook) or graphical (Moody Chart) are then used to determine the darcy friction factor (f). The Darcy-Weisbach equation is then used to evaluate the magnitude of the frictional pressure loss ($\Delta P_{L,f}$). It is the authors opinion that the mine ventilation industry should attempt to use this modern Fluid Mechanics approach.

To enable this and make use of the well-accepted Atkinson Friction Factors ($k_{1,2}$) found in existing mining literature, one can solve for the absolute surface roughness in the explicit Haaland Equation and make use of the frictional relationship between the Darcy (f) and Atkinson Friction Factors ($k_{1,2}$):

$$\frac{1}{\sqrt{f}} = -1.8 \log\left(\left(\frac{e/D_h}{3.7}\right)^{1.11} + \frac{6.9}{Re}\right) \qquad \text{(5.2a) repeated}$$

Solving Equation 5.2a for the absolute surface roughness (e):

$$e = 3.7 D_h \times \left[10^{\left(\frac{-1}{1.8\sqrt{f}}\right)} - \frac{6.9}{Re}\right]^{\frac{1}{1.11}} \qquad \text{(16a)}$$

And substituting Equation 13b into Equation 16a, noting that this is for friction losses only:

$$e = 3.7 D_h \times \left[10^{\left(\frac{-1}{1.8\sqrt{6.67k_{1,2}}}\right)} - \frac{6.9}{Re}\right]^{\frac{1}{1.11}} \qquad \text{(16b)}$$

The challenge now is that the available mining literature does not specify the exact parameters at which published Atkinson Friction Factors ($k_{1,2}$) were determined ie what is the diameter or hydraulic diameter of the airway or duct, and at what flow rate and psychrometric conditions, so that the Reynolds Number (Re) can be determined. Many airway types also seem to account for the total energy loss and not just friction. For airways where friction is the main contributor to the total energy loss, Equation 16b is valid. But for airways where friction is not dominant, such as some shafts where bunton-set drag dominates, care must be taken in the use of Equation 16b.

It will be shown that Re is less important for most mining airways where the flow can be considered fully rough, turbulent. As Re tends to infinity, the friction factor remains constant. The '6.9/Re' term will tend to zero as Re tends to infinity.

Tables 2 and 3 suggest absolute surface roughness (e) values for typical mine airways. Absolute Surface Roughness (e) has been calculated with Equation 16b using estimated air psychrometric properties, velocity, and hydraulic diameter. Changing either of these properties will change the calculated e value and thus Tables 2 and 3 should be used with caution until these values can be validated with field measurements. The validation is mainly for the estimated hydraulic diameter used to calculate the e values. The $k_{1,2}$ values have however been robustly verified with field measurements and accepted in mining literature.

TABLE 2

Absolute surface roughness for typical mine airways (using McPherson $k_{1.2}$) (source: McPherson, 2007).

Source	Estimated						Calculated
Airway	Atkinson Friction Factor	True density	Dynamic viscosity	Average velocity	Hydraulic diameter	Reynolds Number	Absolute surface roughness
	$k_{1.2}$	ρ	μ	V	D_h	Re	e
	[kg/m³]	[kg/m³]	[kg/m.s]	[m/s]	[m]	[-]	[mm]
Shafts							
Smooth lined, unobstructed	0.0030	1.1	1.9E-05	10	6	4E+06	6
Brick lined, unobstructed	0.0040	1.1	1.9E-05	10	6	4E+06	19
Concrete lined, rope guides, pipe fittings	0.0065	1.1	1.9E-05	10	6	4E+06	88
Brick lined, rope guides, pipe fittings	0.0075	1.1	1.9E-05	10	6	4E+06	128
Unlined, well-trimmed surface	0.0100	1.1	1.9E-05	10	6	4E+06	256
Unlined, major irregularities removed	0.0120	1.1	1.9E-05	10	6	4E+06	378
Unlined, mesh bolted	0.0140	1.1	1.9E-05	10	6	4E+06	511
Tubing lined, no fittings	0.0100	1.1	1.9E-05	10	6	4E+06	256
Brick lined, two sides buntons	0.0180	1.1	1.9E-05	10	6	4E+06	798
Two side buntons, each with a tie girder	0.0220	1.1	1.9E-05	10	6	4E+06	1096
Rectangular airways							
Smooth concrete lined	0.0040	1.1	1.9E-05	6	4	1E+06	13
Shotcrete	0.0055	1.1	1.9E-05	6	4	1E+06	36
Unlined with minor irregularities only	0.0090	1.1	1.9E-05	6	4	1E+06	134
Girders on masonry or concrete walls	0.0095	1.1	1.9E-05	6	4	1E+06	152
Unlined, typical conditions no major irregularities	0.0120	1.1	1.9E-05	6	4	1E+06	252
Unlined, irregular sides	0.0140	1.1	1.9E-05	6	4	1E+06	341
Unlined, rough or irregular conditions	0.0160	1.1	1.9E-05	6	4	1E+06	435
Girders on side props	0.0190	1.1	1.9E-05	6	4	1E+06	581
Drift with rough sides, stepped floor, handrails	0.0400	1.1	1.9E-05	6	4	1E+06	1590
Steel arched airways							
Smooth concrete all round	0.0040	1.1	1.9E-05	6	4	1E+06	13
Bricked between arches all round	0.0060	1.1	1.9E-05	6	4	1E+06	46
Concrete slabs or timber lagging between flanges all round	0.0075	1.1	1.9E-05	6	4	1E+06	85
Slabs or timber lagging between flanges to spring	0.0090	1.1	1.9E-05	6	4	1E+06	134
Lagged behind arches	0.0120	1.1	1.9E-05	6	4	1E+06	252
Arches poorly aligned, rough conditions	0.0160	1.1	1.9E-05	6	4	1E+06	435
Coalmines – rectangular entries, roof-bolted							
Intakes, clean conditions	0.0090	1.1	1.9E-05	10	4	2E+06	134
Returns, some irregularities/sloughing	0.0100	1.1	1.9E-05	10	4	2E+06	171
Belt entries	0.0075	1.1	1.9E-05	10	4	2E+06	86
Cribbed entries	0.0100	1.1	1.9E-05	10	4	2E+06	171
Longwall face line with steel conveyor and powered supports							
Good conditions, smooth wall	0.0350	1.1	1.9E-05	10	4	2E+06	1363
Typical conditions, coal on conveyor	0.0500	1.1	1.9E-05	10	4	2E+06	2012
Rough conditions, uneven face line	0.0650	1.1	1.9E-05	10	4	2E+06	2571
Metal mines							
Arch-shaped level drifts, rock bolts and mesh	0.0100	1.1	1.9E-05	10	4	2E+06	171
Arch-shaped ramps, rock bolts and mesh	0.0140	1.1	1.9E-05	10	4	2E+06	341
Rectangular raise, untimbered, rock bolts and mesh	0.0130	1.1	1.9E-05	10	4	2E+06	295
Bored raise	0.0050	1.1	1.9E-05	10	4	2E+06	27
Beltway	0.0140	1.1	1.9E-05	10	4	2E+06	341
TBM drift	0.0045	1.1	1.9E-05	10	4	2E+06	19
Ventilation ducting							
Collapsible fabric ducting (forcing systems only)	0.0037	1.1	1.9E-05	20	0.56	7E+05	1.3
Flexible ducting with fully stretched spiral spring reinforcement	0.0110	1.1	1.9E-05	20	0.56	7E+05	29
Fibreglass	0.0024	1.1	1.9E-05	20	0.56	7E+05	0.18
Spiral wound galvanised steel	0.0021	1.1	1.9E-05	20	0.56	7E+05	0.06

TABLE 3

Absolute surface roughness for typical mine airways (using MVSSA $k_{1.2}$) (Source: MVSSA, 2000).

Source	Estimated						Calculated
Airway	Atkinson Friction Factor	True density	Dynamic viscosity	Average velocity	Hydraulic diameter	Reynolds number	Absolute surface roughness
	$k_{1.2}$	ρ	μ	V	D_h	Re	e
	[kg/m³]	[kg/m³]	[kg/m.s]	[m/s]	[m]	[-]	[mm]
Shafts							
Concrete lined – RSJ buntons	0.0338	1.1	1.9E-05	10	6	4E+06	1956
Brick lined – equipped	0.0074	1.1	1.9E-05	10	6	4E+06	124
Brick lined – two sets buntons	0.0176	1.1	1.9E-05	10	6	4E+06	768
Brick lined – unobstructed	0.0037	1.1	1.9E-05	10	6	4E+06	14
Concrete lined – aerodynamic buntons	0.0148	1.1	1.9E-05	10	6	4E+06	563
Concrete lined – no steel work	0.0040	1.1	1.9E-05	10	6	4E+06	19
Concrete lined – Tubes only	0.0139	1.1	1.9E-05	10	6	4E+06	504
Lined – equipped	0.0275	1.1	1.9E-05	10	6	4E+06	1506
Lined – unequipped	0.0175	1.1	1.9E-05	10	6	4E+06	761
Raise bored	0.0029	1.1	1.9E-05	10	6	4E+06	5
Smooth Lined – unobstructed	0.0037	1.1	1.9E-05	10	6	4E+06	14
Timber lined – single bunton	0.0223	1.1	1.9E-05	10	6	4E+06	1118
Underground Airways							
Brick arches	0.0056	1.1	1.9E-05	6	4	1E+06	38
Coalmine – Bord and Pillar	0.0087	1.1	1.9E-05	6	4	1E+06	124
Concrete slabs with timber	0.0074	1.1	1.9E-05	6	4	1E+06	83
Girders on timber props	0.0186	1.1	1.9E-05	6	4	1E+06	561
Rock – no support	0.0100	1.1	1.9E-05	6	4	1E+06	171
Rock – timbered	0.0175	1.1	1.9E-05	6	4	1E+06	507
Rock -concrete lined	0.0040	1.1	1.9E-05	6	4	1E+06	13
Smooth – concrete lined	0.0037	1.1	1.9E-05	6	4	1E+06	9
Smooth bored	0.0100	1.1	1.9E-05	6	4	1E+06	171
Straight rock tunnels	0.0100	1.1	1.9E-05	6	4	1E+06	171
Unlined – irregular sides	0.0160	1.1	1.9E-05	6	4	1E+06	435
Unlined – uniform sides	0.0120	1.1	1.9E-05	6	4	1E+06	252
Walls with girders	0.0093	1.1	1.9E-05	6	4	1E+06	145
Ventilation ducting							
'Corten' smooth bore ducting	0.00252	1.1	1.9E-05	20	0.57	7E+05	0.24
Fibreglass ducting	0.00237	1.1	1.9E-05	20	0.57	7E+05	0.17
Mild steel hot-dipped galvanised smooth bore ducting	0.00275	1.1	1.9E-05	20	0.57	7E+05	0.38
6 Swage galvanised ducting	0.00364	1.1	1.9E-05	20	0.57	7E+05	1.25
8 Swage galvanised ducting	0.00409	1.1	1.9E-05	20	0.57	7E+05	1.93
Flexible plastic duct (PVC coated)[1]	0.00300	1.1	1.9E-05	20	1.30	2E+06	1.35
Flexible force ducting (longitudinal suspension)	0.00976	1.1	1.9E-05	20	0.57	7E+05	23

[1] Von Glehn and Bluhm (1995).

PROPOSED PROCEDURE FOR MINE VENTILATION CALCULATIONS

Both an analytical and graphical procedure can be used to determine the frictional pressure loss ($\Delta P_{L,f}$). The frictional pressure loss can be used in the evaluation of the total pressure loss ($\Delta P_{L,total}$) which accounts for friction and shock losses. Finally, fans can be sized using the modified Bernoulli Equation such that an energy balance is achieved.

Analytical procedures

Step 1: Calculate the **Hydraulic Diameter (D_h)**:

$$D_h = \frac{4A}{C}$$

Step 2: Calculate the **Reynolds Number (Re)**:

$$Re = \frac{\rho V D_h}{\mu}$$

Step 3: Determine the **absolute surface** roughness **(e)** from Tables 2 and 3, literature in Appendix B, and Vendor datasheets.

Step 4: Calculate the darcy friction factor (f):

$$f = \left[-1.8 \log \left(\left(\frac{e/D_h}{3.7} \right)^{1.11} + \frac{6.9}{Re} \right) \right]^{-2}$$

Step 5: Determine the **overall shock losses ($\sum X$)** from literature (Appendix C) or Vendor datasheets.

Step 6: Calculate the **total pressure loss** due to friction and shock losses:

$$\Delta P_{L,total} = \left(f \frac{L}{D_h} + \sum X \right) \left(\frac{1}{2} \rho V^2 \right)$$

Step 7: Apply Bernoulli Equation between system inlet (1) and outlet (2) to obtain the **total pressure required by the fan**:

$$\Delta P_{required} = (P_2 - P_1) + \frac{1}{2} \rho (V_2 - V_1)^2 + \rho g (Z_2 - Z_1) + \Delta P_{L,total}$$

Graphical procedures

The Graphical procedure is similar to the Analytical procedure, but step 4 is replaced with the proposed Moody Charts to determine the Darcy Friction Factor (f).

Figure 5 shows proposed Moody Charts applicable to the mine ventilation practitioner. An attempt has been made to specify applicable absolute surface roughness (e) values based on Tables 2 and 3. The three graphs offer variations in relative roughness (e/D_h) scale only:

- Figure 5a – For smooth ducts (plotted with low e/D_h)
- Figure 5b – For rough ducts and smooth mine airways (plotted with high e/D_h)
- Figure 5c – For rough mine airways (plotted with very high e/D_h).

Smooth Ducts

Relative Roughness, e/D_h

0.03
0.02
0.01
0.008
0.006
0.004
0.002
0.001
0.0008
0.0006
0.0004
0.0002
0.0001
0.00005
0.00001

Friction Factor [Darcy, f | Atkinson, $k_{1,2}$ (kg/m³)]

$k_{1,2}$	f
0.0098	0.065
0.0083	0.055
0.0068	0.045
0.0053	0.035
0.0038	0.025
0.0023	0.015
0.0008	0.005

1.0E+03 1.0E+04 1.0E+05 1.0E+06 1.0E+07 1.0E+08

Reynold Number, Re

Ducting	e (mm)
Fibreglass	0.17
Hot-dipped galvanised, smooth bore	0.38
6 / 8 Swage galvanised	1.25 / 1.93
Flexible plastic (PVC coated)	1.35
Flexible, longitudinal suspension	23

$$D_h = \frac{4A}{C}$$

$$Re = \frac{\rho V D_h}{\mu}$$

Friction only (chart y-axis):

$$k_{1,2} = 0.15 f$$

Friction + shock (not plotted):

$$k_{1,2} = 0.15 \left(f + \sum x \frac{L}{D_h} \right)$$

Turbulent flow (chart y-axis):

$$f = \left[-1.8 \log \left(\left(\frac{e/D_h}{3.7} \right)^{1.11} + \frac{6.9}{Re} \right) \right]^{-2}$$

(Darcy-Weisbach):

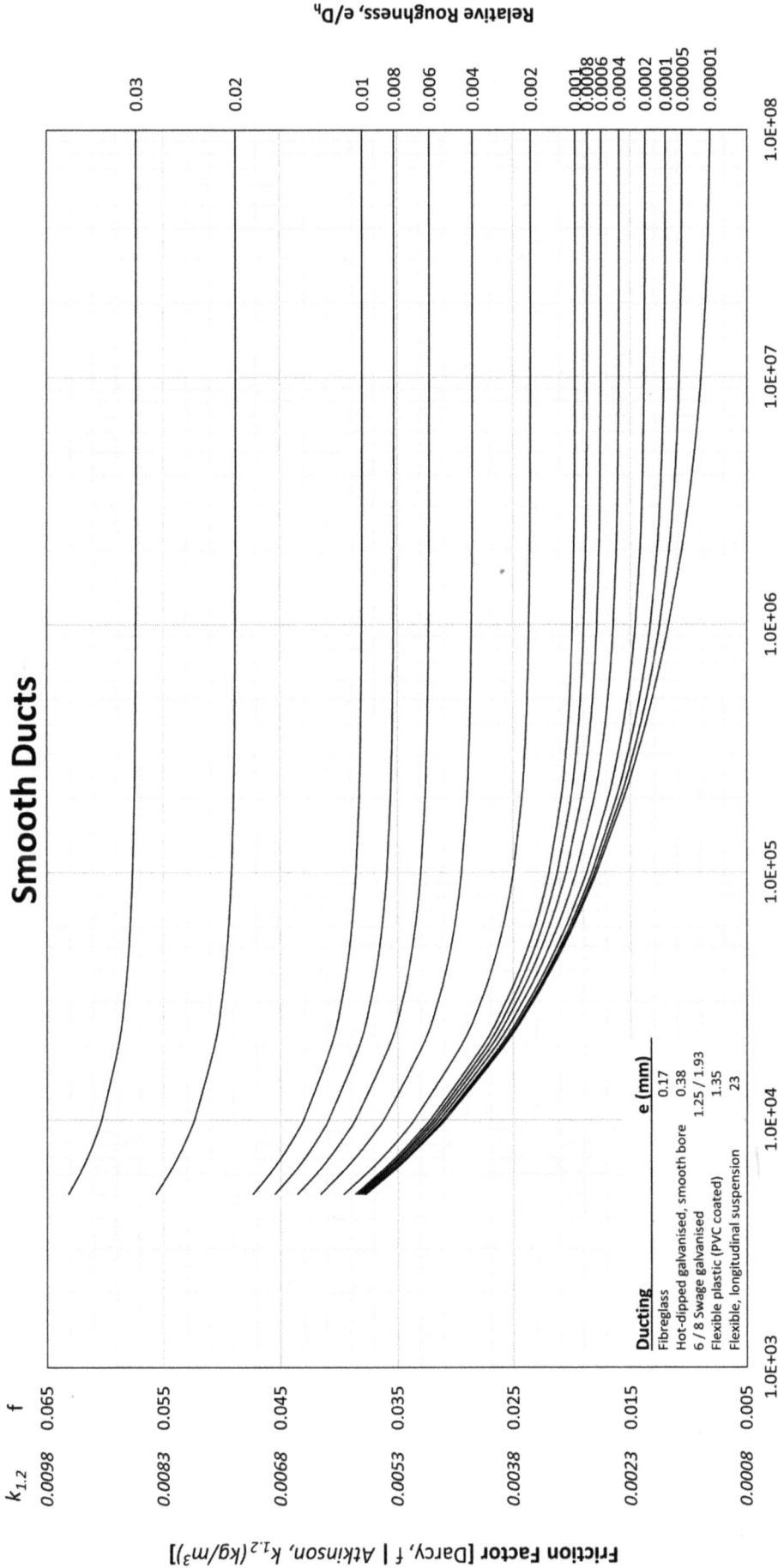

$$\Delta P_{L,f} = f \left(\frac{L}{D_h} \right) \left(\frac{1}{2} \rho v^2 \right) = \frac{f \rho}{8} \left(\frac{CL}{A^3} \right) Q^2$$

(Atkinson):

$$\Delta P_{L,f} = \frac{k_{1,2} \times \rho}{1.2} \left(\frac{CL}{A^3} \right) Q^2$$

FIG 5a – Mine ventilation Moody chart (smooth ducts).

Rough Ducts + Smooth Mine Airways

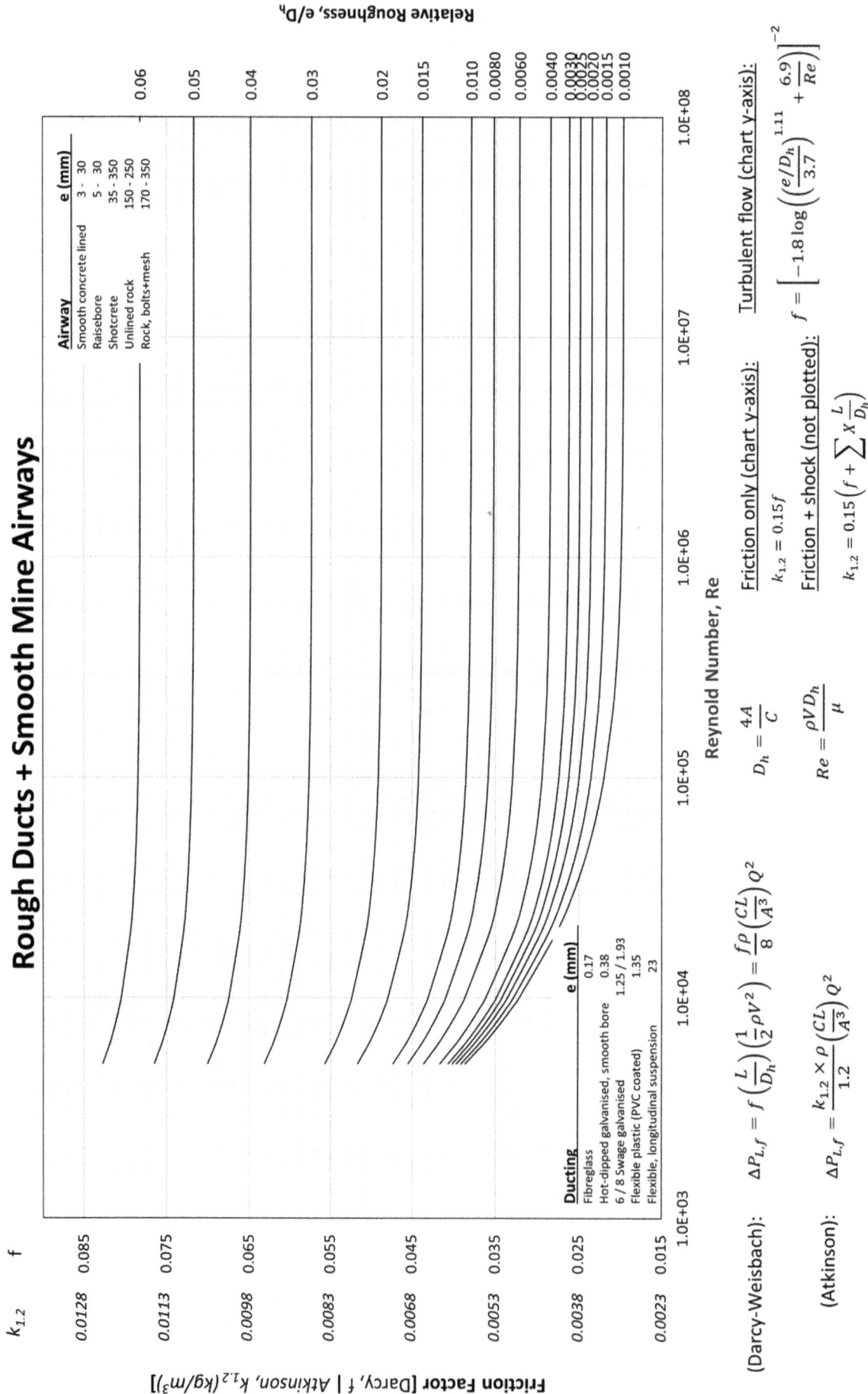

FIG 5b – Mine ventilation Moody chart (rough ducts and smooth mine airways).

Airway	e (mm)
Smooth concrete lined	3 - 30
Raisebore	5 - 30
Shotcrete	35 - 350
Unlined rock	150 - 250
Rock, bolts+mesh	170 - 350

Ducting	e (mm)
Fibreglass	0.17
Hot-dipped galvanised, smooth bore	0.38
6 / 8 Swage galvanised	1.25 / 1.93
Flexible plastic (PVC coated)	1.35
Flexible, longitudinal suspension	23

Friction only (chart y-axis):

$$k_{1.2} = 0.15 f$$

Friction + shock (not plotted):

$$k_{1.2} = 0.15 \left(f + \sum x \frac{L}{D_h} \right)$$

Turbulent flow (chart y-axis):

$$f = \left[-1.8 \log \left(\left(\frac{e/D_h}{3.7} \right)^{1.11} + \frac{6.9}{Re} \right) \right]^{-2}$$

(Darcy-Weisbach):
$$\Delta P_{L,f} = f \left(\frac{L}{D_h} \right) \left(\frac{1}{2} \rho V^2 \right) = \frac{f \rho}{8} \left(\frac{CL}{A^3} \right) Q^2$$

(Atkinson):
$$\Delta P_{L,f} = \frac{k_{1.2} \times \rho}{1.2} \left(\frac{CL}{A^3} \right) Q^2$$

$$D_h = \frac{4A}{C}$$

$$Re = \frac{\rho V D_h}{\mu}$$

Rough Mine Airways

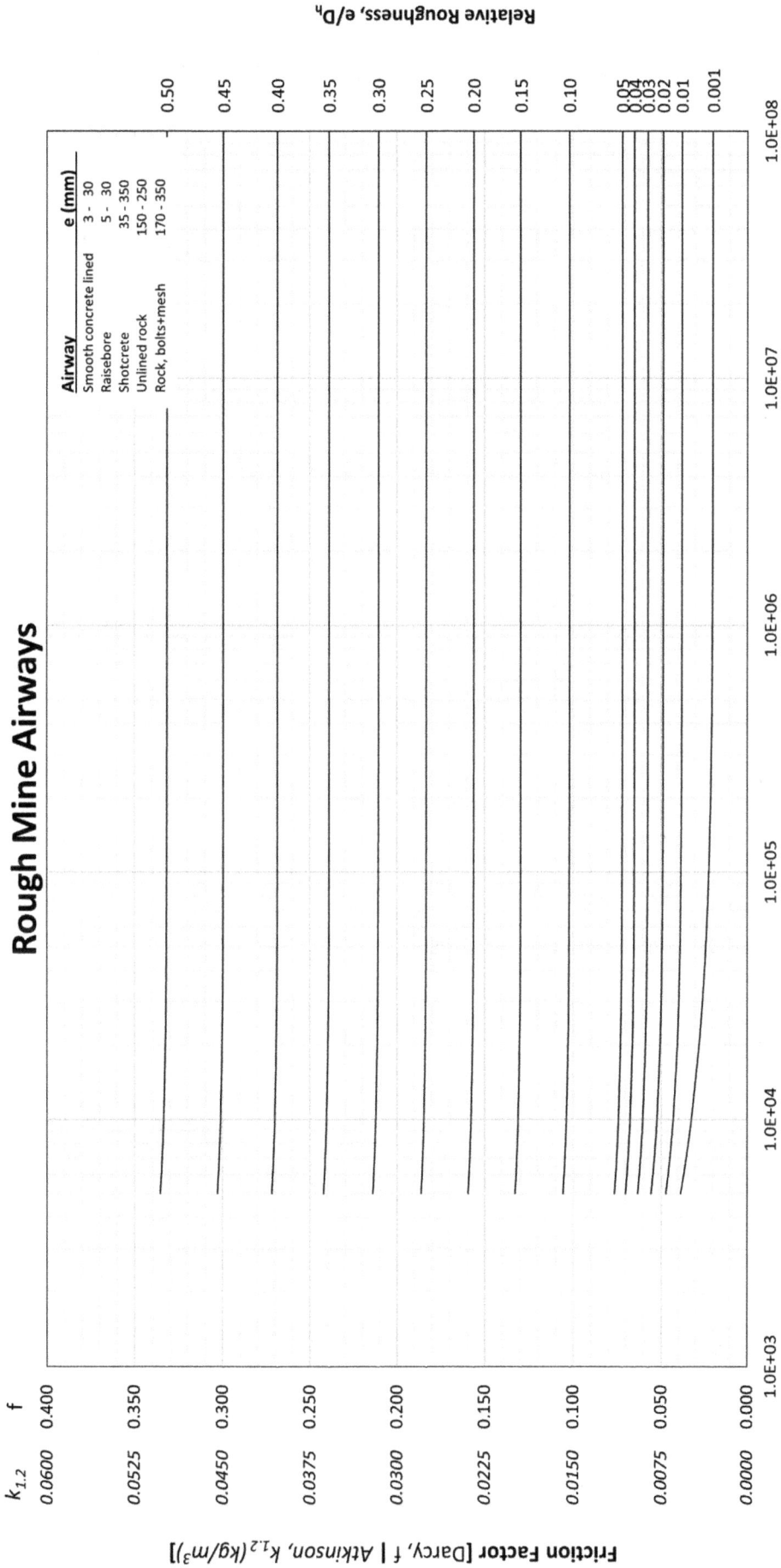

Relative Roughness, e/D_h

Airway	e (mm)
Smooth concrete lined	3 - 30
Raisebore	5 - 30
Shotcrete	35 - 350
Unlined rock	150 - 250
Rock, bolts+mesh	170 - 350

0.50
0.45
0.40
0.35
0.30
0.25
0.20
0.15
0.10
0.05
0.04
0.03
0.02
0.01
0.001

Reynold Number, Re

1.0E+03　1.0E+04　1.0E+05　1.0E+06　1.0E+07　1.0E+08

$k_{1.2}$ | f

0.0600 | 0.400
0.0525 | 0.350
0.0450 | 0.300
0.0375 | 0.250
0.0300 | 0.200
0.0225 | 0.150
0.0150 | 0.100
0.0075 | 0.050
0.0000 | 0.000

Friction Factor [Darcy, f | Atkinson, $k_{1.2}$ (kg/m³)]

Turbulent flow (chart y-axis):

$$f = \left[-1.8 \log \left(\left(\frac{e/D_h}{3.7} \right)^{1.11} + \frac{6.9}{Re} \right) \right]^{-2}$$

Friction only (chart y-axis):

$$k_{1.2} = 0.15 f$$

Friction + shock (not plotted):

$$k_{1.2} = 0.15 \left(f + \sum x \frac{L}{D_h} \right)$$

(Darcy-Weisbach):

$$\Delta P_{L,f} = f \left(\frac{L}{D_h} \right) \left(\frac{1}{2} \rho V^2 \right) = \frac{f \rho}{8} \left(\frac{CL}{A^3} \right) Q^2$$

(Atkinson):

$$\Delta P_{L,f} = \frac{k_{1.2} \times \rho}{1.2} \left(\frac{CL}{A^3} \right) Q^2$$

$$D_h = \frac{4A}{C}$$

$$Re = \frac{\rho V D_h}{\mu}$$

FIG 5c – Mine ventilation Moody chart (rough mine airways).

EXAMPLE PROBLEMS

This section aims to highlight the differences between the Fluid Mechanics (Darcy-Weisbach) and Mining (Atkinson) calculation methods. Mine ventilation systems vary a great deal and so all scenarios cannot be covered, however general principles can be outlined. To this end, four examples are covered; namely, 1) fibreglass duct, 2) collapsible fabric duct, 3) raise borehole, and 4) underground drift.

The aim of this Section is not to debate the merits of the absolute surface roughness (e) or Atkinson Friction Factor (k) chosen to represent the airway, but rather to demonstrate the variable nature of the friction factor and to motivate for the use of the absolute surface roughness (e) as an input parameter to the analysis as opposed to the Darcy Friction Factor (f) or Atkinson Friction Factor (k).

The following holds for the below example problems:

Problem:

Using both the Darcy-Weisbach and Atkinson methods, determine:

1. Friction Factor?
2. Frictional pressure drop?
3. Fan motor size?

Assumptions:

Only the frictional effects are to be considered. Shock losses, changes in velocity pressure, elevation, and static pressure between the inlet and outlet are to be neglected. The effects of heat transfer and/or variations in air density are also to be neglected. Fan-motor efficiency to be taken as 70 per cent. Apply 20 per cent safety margin for motor selection. Neglect leakage.

Properties:

Fluid = moist air; Temperature = 25°C/30°C (wb/db); Pressure = 110 kPa

Therefore, $\rho = 1.25 \text{ kg/m}^3$; $\mu = 18.6 \times 10^{-6} \text{ kg/m.s}$

Example – fibreglass duct

Shape = circular; D = 2 m; Q = 70 m³/s; material = fibreglass; L = 300 m

Analysis:

$Q = V \times A$ therefore, $V = \frac{Q}{A} = \frac{70}{\pi} = 22.3 \text{ m/s}$

$D_h = D$ where, $A = \frac{\pi D^2}{4} = \frac{\pi \times 2^2}{4} = \pi \text{ m}^2$

Modern fluid mechanics (Darcy-Weisbach)	**Traditional mine ventilation (Atkinson)**

e = 0.17 mm (Table 3)

1. Friction Factor?	**1. Friction factor?**

Reynolds No., from Equation 1:

$$Re = \frac{\rho V D_h}{\mu}$$

$$Re = \frac{1.25 \times 22.3 \times 2}{18.6 \times 10^{-6}}$$

$$Re = 3 \times 10^6$$

Relative roughness is:

$$\frac{e}{D_h} = \frac{0.17 \times 10^{-3}\ m}{2\ m} = 0.000085$$

Where, analytically, the darcy friction factor is, Equation 5.2b:

$$f = \left[-1.8 \log \left(\left(\frac{e/D_h}{3.7} \right)^{1.11} + \frac{6.9}{Re} \right) \right]^{-2}$$

Therefore, Darcy friction factor analytically is:

$$f = \left[-1.8 \log \left(\left(\frac{0.000085}{3.7} \right)^{1.11} + \frac{6.9}{3 \times 10^6} \right) \right]^{-2}$$

$$f = 0.012$$

The same result can be obtained from the Moody Chart, Figure 6, by plotting Re versus e/D_h.

On the Atkinson side:

$k_{1.2} = 0.00237\ kg/m^3$ (Table 3)

The duct perimeter is:

$$C = \pi D = \pi \times 2 = 2\pi$$

2. Frictional pressure drop?

Using the Darcy-Weisbach Equation, Equation 4:

$$\Delta P_{L,f} = f \left(\frac{L}{D_h} \right) \left(\frac{1}{2} \rho V^2 \right)$$

$$\Delta P_{L,f} = 0.012 \left(\frac{300}{2} \right) \left(\frac{1}{2} \times 1.25 \times 22.3^2 \right)$$

$$\Delta P_{L,f} = 559\ Pa$$

2. Frictional pressure drop?

Using the Atkinson Equation, Equation 7b:

$$\Delta P_{L,f} = \frac{k_{1.2} \times \rho}{1.2} \left(\frac{CL}{A^3} \right) (Q^2)$$

$$\Delta P_{L,f} = \frac{0.00237 \times 1.25}{1.2} \left(\frac{2\pi \times 300}{\pi^3} \right) (70^2)$$

$$\Delta P_{L,f} = 735\ Pa$$

3. Fan motor size?

Assuming a fan-motor efficiency of 70%:

$$\dot{W}_{absorbed} = \frac{\Delta P_{L,f} \times Q}{\eta}$$

$$\dot{W}_{absorbed} = \frac{559 \times 70}{0.7}$$

$$\dot{W}_{absorbed} = 55.9\ kW$$

Applying a 20% safety margin and selecting the nearest standard motor size:

$$\dot{W}_{rated} = 55.9 \times 1.2$$

$$\dot{W}_{rated} = 67\ kW$$

The nearest standard **motor size is 75 kW.**

3. Fan motor size?

Assuming a fan-motor efficiency of 70%:

$$\dot{W}_{absorbed} = \frac{\Delta P_{L,f} \times Q}{\eta}$$

$$\dot{W}_{absorbed} = \frac{735 \times 70}{0.7}$$

$$\dot{W}_{absorbed} = 73.5\ kW$$

Applying a 20% safety margin and selecting the nearest standard motor size:

$$\dot{W}_{rated} = 73.5 \times 1.2$$

$$\dot{W}_{rated} = 88\ kW$$

The nearest standard **motor size is 90 kW.**

Discussion

The pressure loss due to friction is approximately 30 per cent higher, and the motor size 20 per cent larger, comparing the traditional Mine Ventilation versus modern Fluid Mechanics approach. But, Figure 6 graphically shows how the friction factor, Darcy (f) or Atkinson (k), varies with Re and e/D_h and there is in fact an envelope of possible friction factors depending on the duct diameter and velocity. For the same air properties, the envelope is bound between about 5 m/s and 45 m/s and 0.25 m and 2.00 m duct diameter. The traditional Mine Ventilation approach will not always produce higher frictional pressure losses, especially at very low velocities and/or smaller duct sizes.

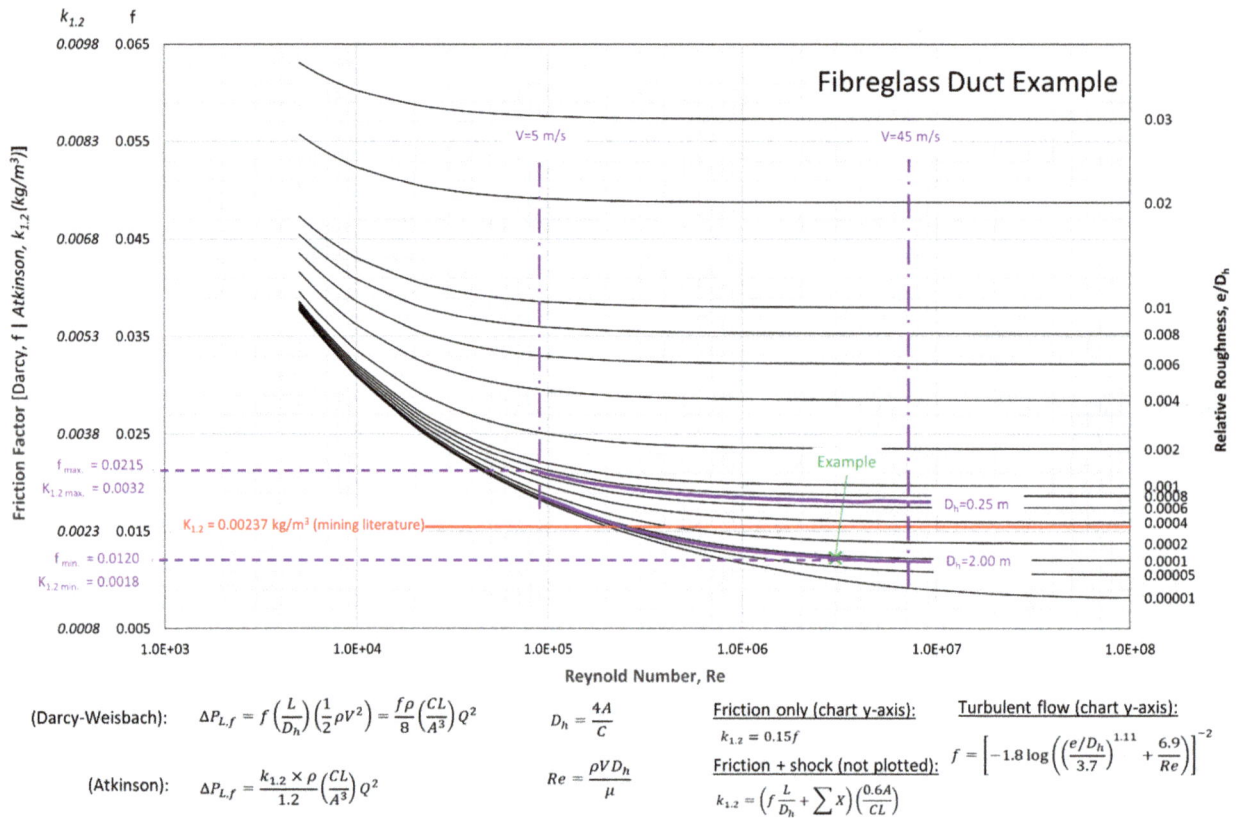

FIG 6 – Moody chart for fibreglass duct example problem.

Example – collapsible plastic duct

Shape = circular; D = 0.5 m; Q = 5 m³/s; material = collapsible fabric; L = 100 m

Analysis:

$$Q = V \times A \qquad \text{therefore, } V = \frac{Q}{A} = \frac{5}{0.196} = 25.5 \text{ m/s}$$

$$D_h = D \qquad \text{where, } A = \frac{\pi D^2}{4} = \frac{\pi \times 0.5^2}{4} = 0.196 \text{ m}^2$$

Modern fluid mechanics **(Darcy-Weisbach)**	**Traditional mine ventilation** **(Atkinson)**

e = 2.00 mm (Tables 2 and 3)

1. Friction Factor?

Reynolds No., from Equation 1:

$$Re = \frac{\rho V D_h}{\mu}$$

$$Re = \frac{1.25 \times 25.5 \times 0.5}{18.6 \times 10^{-6}}$$

$$Re = 7.7 \times 10^5$$

Relative roughness is:

$$\frac{e}{D_h} = \frac{2.00 \times 10^{-3}\ m}{0.5\ m} = 0.004$$

Where, analytically, the darcy friction factor is, Equation 5.2b:

$$f = \left[-1.8 \log\left(\left(\frac{e/D_h}{3.7}\right)^{1.11} + \frac{6.9}{Re}\right)\right]^{-2}$$

Therefore, Darcy friction factor analytically is:

$$f = \left[-1.8 \log\left(\left(\frac{0.004}{3.7}\right)^{1.11} + \frac{6.9}{7.7 \times 10^5}\right)\right]^{-2}$$

$$f = 0.029$$

The same result can be obtained from the Moody Chart, Figure 7, by plotting Re versus e/D_h.

2. Frictional pressure drop?

Using the Darcy-Weisbach Equation, Equation 4:

$$\Delta P_{L,f} = f\left(\frac{L}{D_h}\right)\left(\frac{1}{2}\rho V^2\right)$$

$$\Delta P_{L,f} = 0.029\left(\frac{100}{0.5}\right)\left(\frac{1}{2} \times 1.25 \times 25.5^2\right)$$

$$\Delta P_{L,f} = 2.36\ kPa$$

3. Fan motor size?

Assuming a fan-motor efficiency of 70%:

$$\dot{W}_{absorbed} = \frac{\Delta P_{L,f} \times Q}{\eta}$$

$$\dot{W}_{absorbed} = \frac{2.36 \times 5}{0.7}$$

$$\dot{W}_{absorbed} = 16.8\ kW$$

Applying a 20% safety margin and selecting the nearest standard motor size:

$$\dot{W}_{rated} = 16.8 \times 1.2$$

$$\dot{W}_{rated} = 20\ kW$$

The nearest standard **motor size is 22 kW**.

---

1. Friction Factor?

$$k_{1.2} = 0.0037\ kg/m^3\ (Tables\ 2\ and\ 3)$$

The duct perimeter is

$$C = \pi D = \pi \times 0.5 = \pi/2$$

2. Frictional pressure drop?

Using the Atkinson Equation, Equation 7b:

$$\Delta P_{L,f} = \frac{k_{1.2} \times \rho}{1.2}\left(\frac{CL}{A^3}\right)(Q^2)$$

$$\Delta P_{L,f} = \frac{0.0037 \times 1.25}{1.2}\left(\frac{\pi/2 \times 100}{0.196^3}\right)(5^2)$$

$$\Delta P_{L,f} = 2.01\ kPa$$

3. Fan motor size?

Assuming a fan-motor efficiency of 70%:

$$\dot{W}_{absorbed} = \frac{\Delta P_{L,f} \times Q}{\eta}$$

$$\dot{W}_{absorbed} = \frac{2.01 \times 5}{0.7}$$

$$\dot{W}_{absorbed} = 14.4\ kW$$

Applying a 20% safety margin and selecting the nearest standard motor size:

$$\dot{W}_{rated} = 14.4 \times 1.2$$

$$\dot{W}_{rated} = 17\ kW$$

The nearest standard **motor size is 18.5 kW**.

Discussion

There is a small difference in the two calculated pressure loss due to friction for this example. But again, Figure 7 graphically shows how the friction factor will vary with Re and e/D_h. The friction envelope is bound between about 5 m/s and 45 m/s and 0.25 m and 2.00 m duct diameter. The modern Fluid Mechanics approach will produce higher frictional pressure losses at smaller duct diameters.

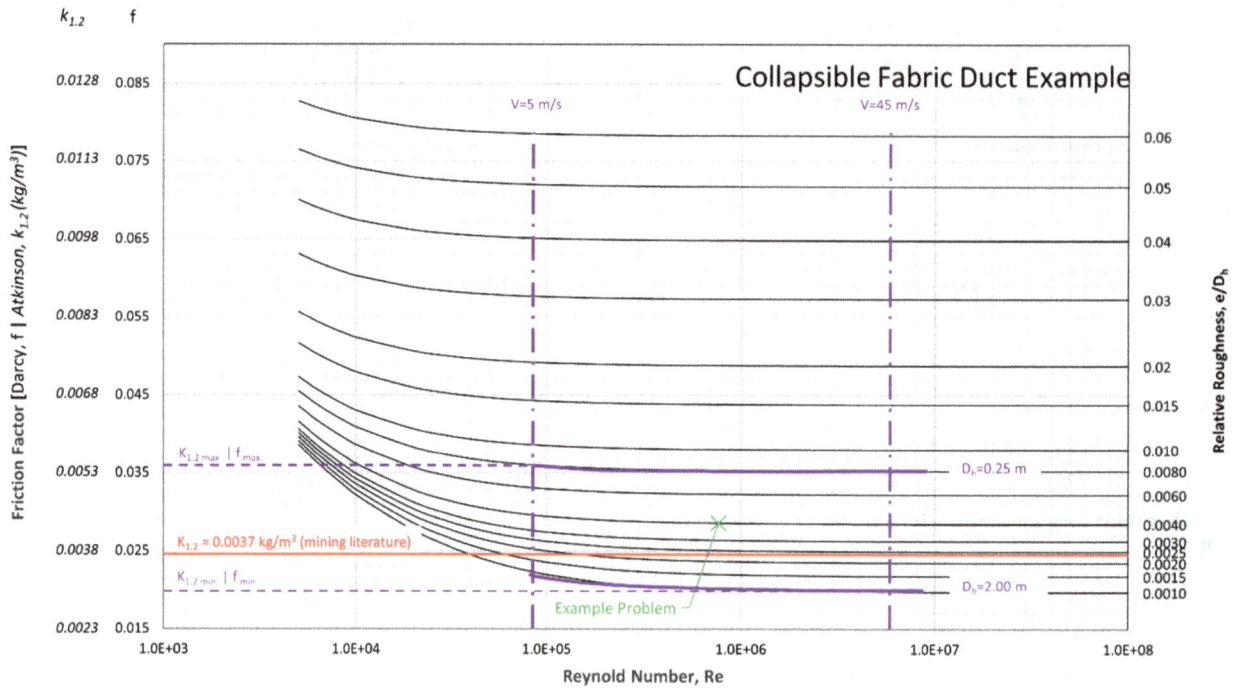

FIG 7 – Moody chart for collapsible fabric duct example problem.

The chart axes and labels:
- Left outer axis: Friction Factor [Darcy, f | Atkinson, $k_{1.2}$ (kg/m³)]
- Right axis: Relative Roughness, e/D_h
- X-axis: Reynold Number, Re
- Title: Collapsible Fabric Duct Example
- Labels: V=5 m/s; V=45 m/s; D_h=0.25 m; D_h=2.00 m
- $K_{1.2\,max}$ | $f_{max.}$
- $K_{1.2}$ = 0.0037 kg/m³ (mining literature)
- $K_{1.2\,min.}$ | $f_{min.}$
- Example Problem

Equations below chart:

(Darcy-Weisbach): $\Delta P_{L,f} = f\left(\frac{L}{D_h}\right)\left(\frac{1}{2}\rho V^2\right) = \frac{f\rho}{8}\left(\frac{CL}{A^3}\right)Q^2$

(Atkinson): $\Delta P_{L,f} = \frac{k_{1.2} \times \rho}{1.2}\left(\frac{CL}{A^3}\right)Q^2$

$D_h = \frac{4A}{C}$

$Re = \frac{\rho V D_h}{\mu}$

Friction only (chart y-axis):
$k_{1.2} = 0.15f$

Friction + shock (not plotted):
$k_{1.2} = \left(f\frac{L}{D_h} + \sum x\right)\left(\frac{0.6A}{CL}\right)$

Turbulent flow (chart y-axis):
$f = \left[-1.8\log\left(\left(\frac{e/D_h}{3.7}\right)^{1.11} + \frac{6.9}{Re}\right)\right]^{-2}$

Example – raise borehole

Shape = circular; D = 2 m; Q = 70 m³/s; material = hard rock; L = 300 m

Analysis:

$Q = V \times A$ therefore, $V = \frac{Q}{A} = \frac{70}{\pi} = 22.3$ m/s

$D_h = D$ where, $A = \frac{\pi D^2}{4} = \frac{\pi \times 2^2}{4} = \pi$ m²

Modern fluid mechanics (Darcy-Weisbach)	Traditional mine ventilation (Atkinson)

e = 27 mm (Tables 2 and 3)

Modern fluid mechanics (Darcy-Weisbach)

1. Friction Factor?

Reynolds No., from Equation 1:

$$Re = \frac{\rho V D_h}{\mu}$$

$$Re = \frac{1.25 \times 22.3 \times 2}{18.6 \times 10^{-6}}$$

$$Re = 2.99 \times 10^6$$

Relative roughness is:

$$\frac{e}{D_h} = \frac{27 \times 10^{-3}\ m}{2\ m} = 0.0135$$

Where, analytically, the darcy friction factor is, Equation 5.2b:

$$f = \left[-1.8 \log \left(\left(\frac{e/D_h}{3.7}\right)^{1.11} + \frac{6.9}{Re} \right) \right]^{-2}$$

Therefore, Darcy friction factor analytically is:

$$f = \left[-1.8 \log \left(\left(\frac{0.0135}{3.7}\right)^{1.11} + \frac{6.9}{2.99 \times 10^6} \right) \right]^{-2}$$

$$f = 0.042$$

The same result can be obtained from the Moody Chart, Figure 8, by plotting Re versus e/D_h.

2. Frictional pressure drop?

Using the Darcy-Weisbach Equation, Equation 4:

$$\Delta P_{L,f} = f \left(\frac{L}{D_h}\right) \left(\frac{1}{2}\rho V^2\right)$$

$$\Delta P_{L,f} = 0.042 \left(\frac{300}{2}\right) \left(\frac{1}{2} \times 1.25 \times 22.3^2\right)$$

$$\Delta P_{L,f} = 1.95\ kPa$$

3. Fan motor size?

Assuming a fan-motor efficiency of 70%:

$$\dot{W}_{absorbed} = \frac{\Delta P_{L,f} \times Q}{\eta}$$

$$\dot{W}_{absorbed} = \frac{1.95 \times 70}{0.7}$$

$$\dot{W}_{absorbed} = 195\ kW$$

Applying a 20% safety margin and selecting the nearest standard motor size:

$$\dot{W}_{rated} = 195 \times 1.2$$

$$\dot{W}_{rated} = 234\ kW$$

The nearest standard **motor size is 250 kW.**

Traditional mine ventilation (Atkinson)

1. Friction Factor?

$$k_{1.2} = 0.005\ kg/m^3\ (Tables\ 2\ and\ 3)$$

The duct perimeter is

$$C = \pi D = \pi \times 2 = 2\pi$$

2. Frictional pressure drop?

Using the Atkinson Equation, Equation 7b:

$$\Delta P_{L,f} = \frac{k_{1.2} \times \rho}{1.2} \left(\frac{CL}{A^3}\right)(Q^2)$$

$$\Delta P_{L,f} = \frac{0.005 \times 1.25}{1.2} \left(\frac{2\pi \times 300}{\pi^3}\right)(70^2)$$

$$\Delta P_{L,f} = 1.55\ kPa$$

3. Fan motor size?

Assuming a fan-motor efficiency of 70%:

$$\dot{W}_{absorbed} = \frac{\Delta P_{L,f} \times Q}{\eta}$$

$$\dot{W}_{absorbed} = \frac{1.55 \times 70}{0.7}$$

$$\dot{W}_{absorbed} = 155\ kW$$

Applying a 20% safety margin and selecting the nearest standard motor size:

$$\dot{W}_{rated} = 155 \times 1.2$$

$$\dot{W}_{rated} = 186\ kW$$

The nearest standard **motor size is 185 kW.**

Discussion

The pressure loss due to friction is approximately 20 per cent lower, and the motor size 25 per cent smaller, comparing the traditional Mine Ventilation versus modern Fluid Mechanics approach. Again, Figure 8 graphically shows how the friction factor varies and the envelope of possible friction factors depends on the bore diameter and velocity. The envelope is bound between about 5 m/s and 45 m/s and 0.5 m and 6.0 m bore diameter. Again, the modern Fluid Mechanics approach will produce higher frictional pressure losses at smaller bore diameters.

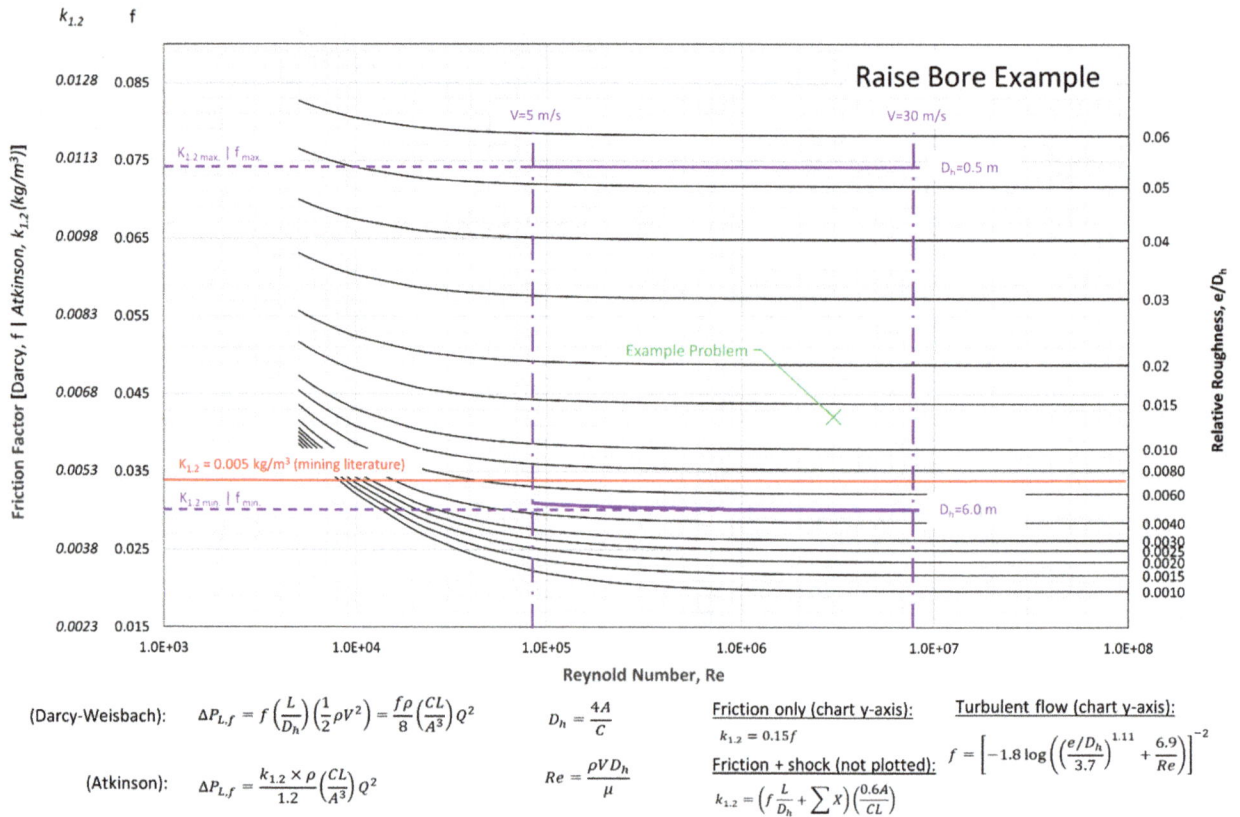

FIG 8 – Moody chart for raise borehole example problem.

Example – underground drift

Shape = arched; H = 4 m (incl. 2 m arch radius); W = 4 m; Q = 100 m³/s; material = Arch-shaped level drifts, rock bolts and mesh; L = 3000 m

Analysis:

$$Q = V \times A \qquad \text{therefore, } V = \frac{Q}{A} = \frac{100}{14.28} = 7.0 \text{ m/s}$$

$$D_h = \frac{4A}{C} = \frac{4 \times 14.28}{18.28} = 3.12 \text{ m} \quad \text{where, } A \approx 14.28 \text{ m}^2 \text{ and } C \approx 18.28 \text{ m}$$

Modern fluid mechanics (Darcy-Weisbach)	Traditional mine ventilation (Atkinson)

e = 170 mm (Tables 2 and 3)

1. Friction Factor?

Reynolds No., from Equation 1:

$$Re = \frac{\rho V D_h}{\mu}$$

$$Re = \frac{1.25 \times 7.0 \times 3.12}{18.6 \times 10^{-6}}$$

$$Re = 1.47 \times 10^6$$

Relative roughness is:

$$\frac{e}{D_h} = \frac{0.17 \text{ m}}{3.12 \text{ m}} = 0.054$$

Where, analytically, the darcy friction factor is, Equation 5.2b:

$$f = \left[-1.8 \log \left(\left(\frac{e/D_h}{3.7} \right)^{1.11} + \frac{6.9}{Re} \right) \right]^{-2}$$

Therefore, Darcy friction factor analytically is:

$$f = \left[-1.8 \log \left(\left(\frac{0.054}{3.7} \right)^{1.11} + \frac{6.9}{31.47 \times 10^6} \right) \right]^{-2}$$

$$f = 0.075$$

The same result can be obtained from the Moody Chart, Figure 9, by plotting Re versus e/D_h.

2. Frictional pressure drop?

Using the Darcy-Weisbach Equation, Equation 4:

$$\Delta P_{L,f} = f \left(\frac{L}{D_h} \right) \left(\frac{1}{2} \rho V^2 \right)$$

$$\Delta P_{L,f} = 0.075 \left(\frac{3\,000}{3.12} \right) \left(\frac{1}{2} \times 1.25 \times 7.0^2 \right)$$

$$\Delta P_{L,f} = 2.21 \text{ kPa}$$

3. Fan motor size?

Assuming a fan-motor efficiency of 70%:

$$\dot{W}_{absorbed} = \frac{\Delta P_{L,f} \times Q}{\eta}$$

$$\dot{W}_{absorbed} = \frac{2.21 \times 100}{0.7}$$

$$\dot{W}_{absorbed} = 315.7 \text{ kW}$$

Applying a 20% safety margin and selecting the nearest standard motor size:

$$\dot{W}_{rated} = 315.7 \times 1.2$$

$$\dot{W}_{rated} = 379 \text{ kW}$$

The nearest standard **motor size is 380 kW**.

---

1. Friction Factor?

$$k_{1.2} = 0.010 \text{ kg/m}^3 \text{ (Tables 2 and 3)}$$

2. Frictional pressure drop?

Using the Atkinson Equation, Equation 7b:

$$\Delta P_{L,f} = \frac{k_{1.2} \times \rho}{1.2} \left(\frac{CL}{A^3} \right) (Q^2)$$

$$\Delta P_{L,f} = \frac{0.010 \times 1.25}{1.2} \left(\frac{18.28 \times 3\,000}{14.28^3} \right) (100^2)$$

$$\Delta P_{L,f} = 1.96 \text{ kPa}$$

3. Fan motor size?

Assuming a fan-motor efficiency of 70%:

$$\dot{W}_{absorbed} = \frac{\Delta P_{L,f} \times Q}{\eta}$$

$$\dot{W}_{absorbed} = \frac{1.96 \times 100}{0.7}$$

$$\dot{W}_{absorbed} = 280 \text{ kW}$$

Applying a 20% safety margin and selecting the nearest standard motor size:

$$\dot{W}_{rated} = 280 \times 1.2$$

$$\dot{W}_{rated} = 336 \text{ kW}$$

The nearest standard **motor size is 355 kW**.

Discussion

There is a small difference in the two calculated pressure loss due to friction for this example. But again, Figure 9 graphically shows how the friction factor will vary with Re and e/D_h. The friction envelope is bound between about 1.0 m and 6.0 m hydraulic drift diameters. Changes in velocity are insignificant. Again, the modern Fluid Mechanics approach produces higher frictional pressure losses at smaller drift hydraulic diameters.

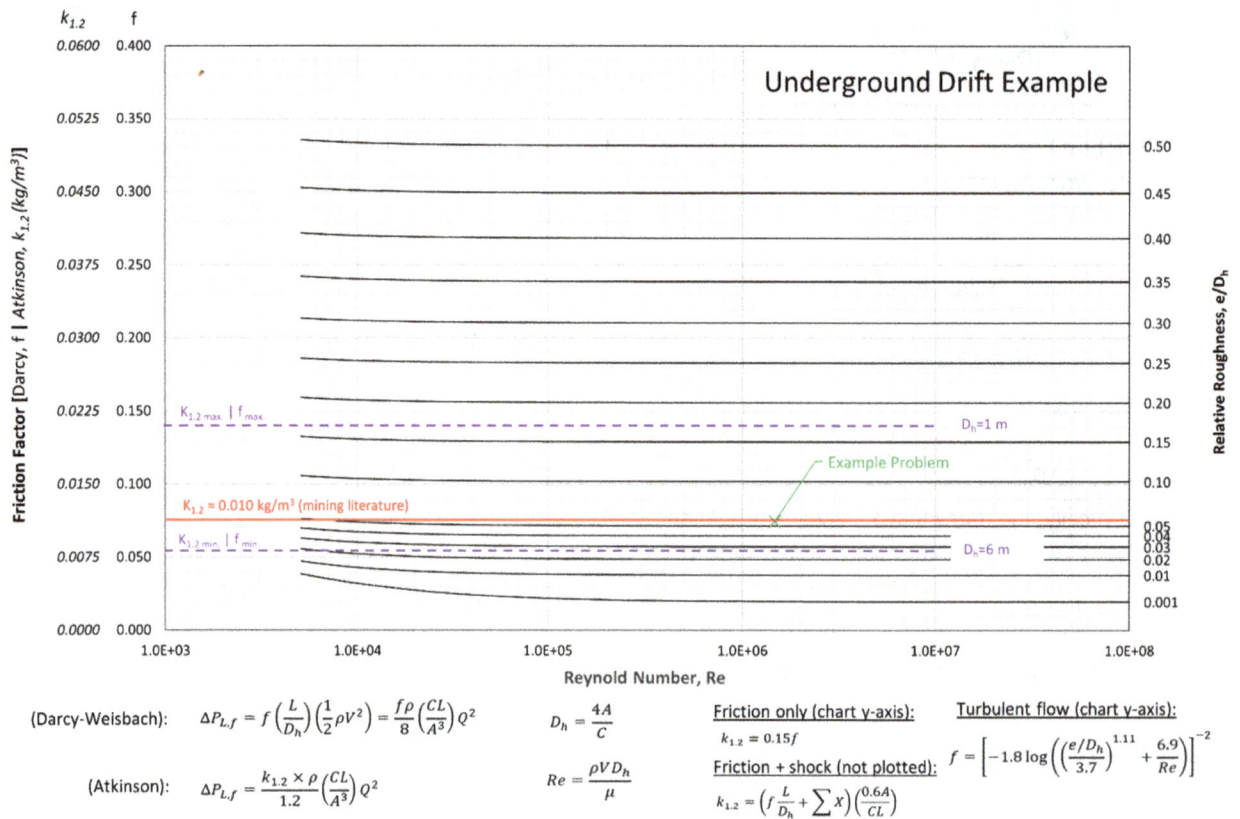

Underground Drift Example

FIG 9 – Moody chart for underground drift example problem.

Equations below chart:

(Darcy-Weisbach): $\Delta P_{L,f} = f\left(\frac{L}{D_h}\right)\left(\frac{1}{2}\rho V^2\right) = \frac{f\rho}{8}\left(\frac{CL}{A^3}\right)Q^2$

(Atkinson): $\Delta P_{L,f} = \frac{k_{1.2} \times \rho}{1.2}\left(\frac{CL}{A^3}\right)Q^2$

$D_h = \frac{4A}{C}$

$Re = \frac{\rho V D_h}{\mu}$

Friction only (chart y-axis):
$k_{1.2} = 0.15f$

Friction + shock (not plotted):
$k_{1.2} = \left(f\frac{L}{D_h} + \sum x\right)\left(\frac{0.6A}{CL}\right)$

Turbulent flow (chart y-axis):
$f = \left[-1.8\log\left(\left(\frac{e/D_h}{3.7}\right)^{1.11} + \frac{6.9}{Re}\right)\right]^{-2}$

NOTE ON EQUIPPED SHAFTS

An equipped shaft example has been intentionally omitted in this paper. The aim of this paper has been to mainly outline the effects of friction, with an understanding that other losses are possible in real-world systems. Other losses have been covered with shock losses and the application of the modified Bernoulli Equation. But equipped shafts include four components which sum to make up the total pressure loss (ΔP_{total}). This is well documented in McPherson (1988), and accounts for the following losses:

- friction
- bunton-set drag
- conveyances
- shock losses.

Applying standard Atkinson Friction Factors ($k_{1.2}$) to shafts must be done with caution. Each shaft layout is different, and the major drivers of total pressure loss must be understood. Furthermore, a change in shaft diameter will change the relative roughness (e/D_h) and hence the friction factor. This must be accounted for.

DISCUSSION

The examples above highlight the variable nature of the friction factor, be it Atkinson or Darcy. The friction factors are dependent variables. They depend on the airway's relative roughness and Reynolds Number. In general, flows within mine airways will be fully rough, turbulent and so the friction factor will not depend on Re. But care must be taken to validate this. But the main takeaway from the above examples is that changes in airway size will change the friction factor. Mining literature lists discrete friction factors for airway types but does not specify the qualifying hydraulic diameter.

Fluid Mechanics empirically discovered the relationship between the Darcy friction factor (f), Reynolds Number (Re) and Relative Roughness (e/D_h). This experimental data was used to produce both the Moody Chart and the Colebrook Equation as a means to determine the Darcy Friction Factor (f). With this in mind, duct Vendors should be encouraged to produce absolute surface roughness (e) values in lieu of Atkinson Friction Factors ($k_{1.2}$) or Darcy Friction Factors (f) for their products. And, where friction dominates, mine ventilation practitioners should work together to establish and validate robust absolute surface roughness values for mine airways. This approach will bring the mine ventilation discipline in line with modern Fluid Mechanics and respected institutions such as ASHRAE and SMACNA.

The modern Fluid Mechanics technique for evaluating internal flow frictional pressure drop, Darcy-Weisbach, is applicable to all incompressible fluid systems encountered in mine ventilation systems, including air flow and water flow. The same method can be used to determine hydraulic system pressure losses and size pumps, just as we have outlined the method for air flow systems.

Although not insurmountable, a drawback of the Atkinson method is the requirement to ensure the Atkinson friction factor measured is corrected to the reference density of 1.2 kg/m³. The Atkinson Friction Factor cannot simply be measure in the field and then used as the value to represent similar airways. It must first be corrected. This is because the Atkinson Friction Factor is not dimensionless. The Darcy friction factor however is dimensionless and requires no correction.

Another area worth highlighting in the mine ventilation literature, is the use of the Atkinson Friction Factor to represent the overall airway losses which includes both friction and shock losses. It has been shown that both the frictional pressure loss and shock losses are functions of velocity pressure. But the frictional pressure loss includes a friction factor and L/D$_h$ term while the shock loss only includes a shock loss coefficient. Using overall Atkinson Friction Factors can readily produce a calculated pressure drop, but significant errors may be encountered in doing so. Particularly at airway sizes that significantly deviate from the originally measure friction factor. Including all the losses into the Atkinson Friction Factor does not reflect the original definition of k, which is for friction only. This is best illustrated with Equation 14.

But we do need to appreciate the simplicity of the Atkinson Equation to rapidly estimate an airway's resistance. Appropriate use would be in conceptual planning type activities where several options are to be readily evaluated and high-level resistances are acceptable.

Shafts have intentionally not been included as an example problem because representing the total energy loss with friction is not true for equipped shafts. Friction makes up one of four components of the total energy associated with the overall energy loss. McPherson outlines this eloquently in his 1988 paper. But the methods presented in this paper can be used for unequipped shafts which are used as dedicated ventilation shafts.

This paper has focused on frictional losses. But shock losses, changes in velocity pressure, elevation, and static pressure between the inlet and outlet can be significant and should be assessed for in real-world systems, see Equation 9. Also, heat transfer and variations in air density have been considered negligible throughout this paper. Again, these must be assessed, and for steady-state systems, the steady-flow energy equation can be used when thermodynamic effects are significant.

Finally, it must be recognised that absolute surface roughness values have an inherent uncertainty of as much as ± 50 per cent according to White (2006). It is pivotal to conduct comprehensive literature reviews and to engage with Vendors for accurate datasheets. Making use of the Darcy-Weisbach Equation will not generate more accurate results if the input variables carry large uncertainties.

CONCLUSIONS

The use of the Atkinson Friction Factor and Atkinson Equation was developed around 1854. It was further entrenched in the Mining literature by McElroy's 1935 publication. Mines of that time were not very deep nor hot, compared to today's mines, and so the grouping of the air density with the darcy friction factor seemed appropriate to form the Atkinson Friction Factor (k). However, there has always been an inherent weakness in making use of the Atkinson Friction Factor, or Darcy Friction Factor for that matter, to characterise a mine airway's resistance. Neither of these friction factors are independent variables. They are both functions of Relative Roughness, and to a lesser degree, the Reynolds Number. And so, an airway's friction factor is not fixed and thus should not be used to define types of airways. The absolute surface roughness is the best input parameter to assess frictional pressure losses.

Either the Atkinson or Darcy Friction Factors can be used to determine an airway frictional pressure loss. But the realisation that neither of these friction factors is fixed for a given airway is fundamental. They must be analytically or graphically determined.

Indeed, modern Fluid Mechanics does not use a friction factor to define an airway resistance. It employs the absolute surface roughness as the independent variable. This is mainly because the absolute surface roughness does not change for variations in Re nor airways size. Modern Fluid Mechanics uses analytical and/or graphical techniques to determine the darcy friction factor which is used in the Darcy-Weisbach Equation to determine an airway's frictional pressure drop. The variable nature of the friction factor is fully captured producing more accurate calculations.

Empirical data is required to verify the proposed absolute surface roughness' calculated in Tables 2 and 3. This can be achieved with input from the mine ventilation fraternity by using techniques outlined in this paper to establish absolute surface roughness values on existing mine airways where frictional resistance dominates. Furthermore, duct Suppliers must be encouraged to state absolute surface roughness values in place of Atkinson Friction Factors for their products. Using the absolute surface roughness within the Darcy-Weisbach Equation to quantify frictional resistance will align the mine ventilation industry with modern Fluid Mechanics.

REFERENCES

ASHRAE, 2021. *ASHRAE Handbook – Fundamentals* (ASHRAE).

Cengel, Y A and Cimbala, J M, 2006. *Fluid Mechanics – Fundamentals and Applications*, first edition (McGraw-Hill).

Hartman, H L, Mutmansky, J M and Wang, Y J, 1982. *Mine Ventilation and Air Conditioning*, second edition (John Wiley and Sons).

Hartman, H L, Mutmansky, J M, Ramani, R V and Wang, Y J, 1997. *Mine Ventilation and Air Conditioning*, third edition (John Wiley and Sons).

McElroy, G E, 1935. Engineering Factors in the Ventilation of Metal Mines, US Department of the Interior, Bureau of Mines, Bulletin 385.

McPherson, M J, 2007. *Subsurface Ventilation and Environmental Engineering*, second edition (Springer).

McPherson, M J, 1988. An Analysis of the Resistance and Airflow Characteristics in Mine Shafts, in Proceedings of the Fourth International Mine Ventilation Congress.

Montecinos, C and Wallace Jr, K, 2010. Equivalent Roughness for Pressure Drop Calculations in Mine Ventilation, in Proceedings of the 13th United States/North American Mine Ventilation Symposium.

Mine Ventilation Society of South Africa (MVSSA), 2000. *Mine Ventilation Practitioner's Data Book*, second edition, Mine Ventilation Society of South Africa.

Mine Ventilation Society of South Africa (MVSSA), 2014. *Ventilation and Occupational Environmental Engineering in Mines*, third edition, Mine Ventilation Society of South Africa.

SMACNA, 1990. HVAC System Duct Design, third edition, Sheet Metal and Air Conditioning Contractors National Association.

Von Glehn, F H and Bluhm, S J, 1995. Ventilation and Cooling of TBM Drives in High Temperature Conditions for the Lesotho Highlands Water Project, *Journal of the Mine Ventilation Society of South Africa*, 48(5).

White, F M, 2006. *Fluid Mechanics*, fourth edition (McGraw-Hill).

APPENDICES

Appendix A – Atkinson Friction Factors

A1: General mining, McPherson (2007)

	Friction factor, k kg/m^3
Rectangular Airways	
Smooth concrete lined	0.004
Shotcrete	0.0055
Unlined with minor irregularities only	0.009
Girders on masonry or concrete walls	0.0095
Unlined, typical conditions no major irregularities	0.012
Unlined, irregular sides	0.014
Unlined, rough or irregular conditions	0.016
Girders on side props	0.019
Drift with rough sides, stepped floor, handrails	0.04
Steel Arched Airways	
Smooth concrete all round	0.004
Bricked between arches all round	0.006
Concrete slabs or timber lagging between flanges all round	0.0075
Slabs or timber lagging between flanges to spring	0.009
Lagged behind arches	0.012
Arches poorly aligned, rough conditions	0.016
Metal Mines	
Arch-shaped level drifts, rock bolts and mesh	0.010
Arch-shaped ramps, rock bolts and mesh	0.014
Rectangular raise, untimbered, rock bolts and mesh	0.013
Bored raise	0.005
Beltway	0.014
TBM drift	0.0045
Coal Mines: Rectangular entries, roof-bolted	
Intakes, clean conditions	0.009
Returns, some irregularities/ sloughing	0.01
Belt entries	0.005 to 0.011
Cribbed entries	0.05 to 0.14
Shafts[1]	
Smooth lined, unobstructed	0.003
Brick lined, unobstructed	0.004
Concrete lined, rope guides, pipe fittings	0.0065
Brick lined, rope guides, pipe fittings	0.0075
Unlined, well trimmed surface	0.01
Unlined, major irregularities removed	0.012
Unlined, mesh bolted	0.0140
Tubbing lined, no fittings	0.007 to 0.014
Brick lined, two sides buntons	0.018
Two side buntons, each with a tie girder	0.022
Longwall faceline with steel conveyor and powered supports[2]	
Good conditions, smooth wall	0.035
Typical conditions, coal on conveyor	0.05
Rough conditions, uneven faceline	0.065
Ventilation ducting[3]	
Collapsible fabric ducting (forcing systems only)	0.0037
Flexible ducting with fully stretched spiral spring reinforcement	0.011
Fibreglass	0.0024
Spiral wound galvanized steel	0.0021

8. Atkinson Friction Factors (Ns²/m⁴) $\left(kg/m^3 \right)$

Tunnels:

Unlined - uniform sides	: 0,012
Unlined - irregular sides	: 0,016
Smooth bored	: 0,01

Shafts:

Raisebored	: 0,008 - 0,01
Lined, unequipped	: 0,01 - 0,025
Lined, equipped	: 0,02 - 0,035

Coal mines:

Bord and pillar	: 0,0087

Appendix B – absolute surface roughness (e)

B1: Duct material, SMACNA (1990)

Duct Material	Roughness Category	Absolute Roughness ϵ, ft	mm
Uncoated carbon steel, clean (Moody 1944) (0.00015 ft) (0.05 mm)	Smooth	0.0001	0.03
PVC plastic pipe (Swim 1982) (0.0003 to 0.00015 ft) (0.01 to 0.05 mm)			
Aluminum (Hutchinson 1953) (0.00015 to 0.0002 ft) (0.04 to 0.06 mm)			
Galvanized steel, longitudinal seams, 4 ft (1200 mm) joints (Griggs 1987) (0.00016 to 0.00032 ft) (0.05 to 0.1 mm)	Medium Smooth	0.0003	0.09
Galvanized steel, spiral seam with 1, 2, and 3 ribs, 12 ft (3600 mm) joints (Jones 1979, Griggs 1987) (0.00018 to 0.00038 ft) (0.05 to 0.12 mm)	(New Duct Friction Loss Chart)		
Hot-dipped galvanized steel, longitudinal seams, 2.5 ft (760 mm) joints (Wright 1945) (0.0005 ft) (0.15 mm)	Old Average	0.0005	0.15
Fibrous glass duct, rigid	Medium rough	0.003	0.9
Fibrous glass duct liner, air side with facing material (Swim 1978) (0.005 ft) (1.5 mm)			
Fibrous glass duct liner, air side spray coated (Swim 1978) (0.015 ft) (4.5 mm)	Rough	0.01	3.0
Flexible duct, metallic, (0.004 to 0.007 ft (1.2 to 2.1 mm) when fully extended)			
Flexible duct, all types of fabric and wire (0.0035 to 0.015 ft (1.0 to 4.6 mm) when fully extended)			
Concrete (Moody 1944) (0.001 to 0.01 ft) (0.3 to 3.0 mm)			

B2: Duct material, ASHRAE (2021)

Duct Type/Material	Absolute Roughness ε, mm	
1	**2**	**3**
	Range	**Roughness Category**
Drawn tubing (Madison and Elliot 1946)	0.00046	Smooth 0.00046
PVC plastic pipe (Swim 1982)	0.009 to 0.046	Medium smooth 0.046
Commercial steel or wrought iron (Moody 1944)	0.046	
Aluminum, round, longitudinal seams, crimped slip joints, 0.91 m spacing (Hutchinson 1953)	0.037 to 0.061	
Friction chart:		
Galvanized steel, round, longitudinal seams, variable joints (Vanstone, drawband, welded. Primarily beaded coupling), 1.22 m joint spacing (Griggs et al. 1987)	0.049 to 0.098	Average 0.09
Galvanized steel, spiral seams, 3.05 m joint spacing (Jones 1979)	0.061 to 0.12	
Galvanized steel, spiral seam with 1, 2, and 3 ribs, beaded couplings, 3.66 m joint spacing (Griggs et al. 1987)	0.088 to 0.116	
Galvanized steel, rectangular, various type joints (Vanstone, drawband, welded. Beaded coupling), 1.22 m spacing[a] (Griggs and Khodabakhsh-Sharifabad 1992)	0.082 to 0.15	
Phenolic duct, aluminum foil on the interior face, sections connected with a four-bolt flange and cleat joint (Idem and Paruchuri 2018)		
1.52 m spacing:	0.149 to 0.391	
3.05 m spacing	0.075 to 0.298	
Wright Friction Chart:		
Galvanized steel, round, longitudinal seams, 0.76 m joint spacing, ε = 0.15 mm	Retained for historical purposes [See Wright (1945) for development of friction chart]	
Flexible duct, nonmetallic and wire, fully extended (Abushakra et al. 2004; Culp 2011)	0.09 to 0.9	Medium rough 0.9
Galvanized steel, spiral, corrugated,[b] Beaded slip couplings, 3.05 m spacing (Kulkarni et al. 2009)	0.54 to 0.91	
Fibrous glass duct, rigid (tentative)[c]	—	
Fibrous glass duct liner, air side with facing material (Swim 1978)	1.52	
Fibrous glass duct liner, air side spray coated (Swim 1978)	4.57	Rough 3.0
Flexible duct, metallic corrugated, fully extended	1.2 to 2.1	
Concrete (Moody 1944)	0.30 to 3.0	

[a]Griggs and Khodabakhsh-Sharifabad (1992) showed that ε values for rectangular duct construction combine effects of surface condition, joint spacing, joint type, and duct construction (cross breaks, etc.), and that the ε-value range listed is representative.
[b]Spiral seam spacing was 119 mm with two corrugations between seams. Corrugations were 19 mm wide by 6 mm high (semicircle).
[c]Subject duct classified "tentatively medium rough" because no data available.

B3: Generic materials and fabrication types, White (2006)

Material	Condition	ε	
		ft	**mm**
Steel	Sheet metal, new	0.00016	0.05
	Stainless, new	0.000007	0.002
	Commercial, new	0.00015	0.046
	Riveted	0.01	3.0
	Rusted	0.007	2.0
Iron	Cast, new	0.00085	0.26
	Wrought, new	0.00015	0.046
	Galvanized, new	0.0005	0.15
	Asphalted cast	0.0004	0.12
Brass	Drawn, new	0.000007	0.002
Plastic	Drawn tubing	0.000005	0.0015
Glass	—	Smooth	Smooth
Concrete	Smoothed	0.00013	0.04
	Rough	0.007	2.0
Rubber	Smoothed	0.000033	0.01
Wood	Stave	0.0016	0.5

B3: Mine airways, Montecinos and Wallace Jr (2010)

Type of excavation and use	Wall finishing	Characteristics of walls and roof	Floor evenness	Area deviation %	Absolute roughness e_r (mm)
Intake Adit	rock surface without bolts	medium roughness	Uneven	1	318
		high roughness	Uneven	5	459
Exhaust Adit	rock surface with bolts	smoothened by dust	Uneven	1	206
Intake Adit	rock surface with bolts & mesh	high roughness	Even	3	554
Exhaust Adits	rock surface with bolts & mesh	smoothened by dust	Even	2	337
		medium roughness	uneven	5	426
		high roughness	even	15	509
Intake Adit	shotcrete	medium roughness	even	-	130
		high roughness	even	6	467
Return Adit	shotcrete	low roughness, smoothened by dust	uneven	-	176
		low roughness, smoothened by dust	even	3	259
		high roughness	very uneven	7	261
Intake Adits	steel frames	spaced at 1 m, protruding 500 mm	uneven	5	305
Return Adit	steel frames	spaced at 1 m, protruding 500 mm	even	-	608
			uneven	1	675
Intake Adits	steel frames with timber lining	lining flushed with flange	even	4	135
Return Adits	steel frames with timber lining	lining flushed with flange	even	7	114
Return Adits	concrete reinforcement, complete	low roughness	even	-	82
			even	6	22
Intake shaft	rock surface with bolts & mesh	round section	-	-	928
Return shaft	rock surface with bolts & mesh	round section with ladder	-	-	976
Intake shaft	Smooth rock (raise borer)	round section	-	-	13

Appendix C – shock loss coefficients (X)

For mining airways, see the following:

- Hartman Appendix A-3 (1982).
- McPherson (2007).
- McElroy (1932).

For ducts, see the following:

- ASHRAE (2021).
- SMACNA.

Study on the impact of mine fire smoke flow on ventilation systems based on Ventsim

L S Huang[1], B Wu[2], C Li[3], J X Wang[4] and B W Lei[5]

1. PhD candidate, China University of Mining and Technology (Beijing), Beijing 100083, China. Email: anquanhls@163.com
2. Professor, China University of Mining and Technology (Beijing), Beijing 100083, China. Email: wbelcy@vip.sina.com
3. PhD candidate, China University of Mining and Technology (Beijing), Beijing 100083, China. Email: cumtb_lic@163.com
4. PhD candidate, China University of Mining and Technology (Beijing), Beijing 100083, China. Email: wjxin0622@126.com
5. Associate Professor, China University of Mining and Technology (Beijing), Beijing 100083, China. Email: leibws@163.com

ABSTRACT

To investigate the flow of air and smoke during mine fires, as well as their impact on ventilation networks, this study employs Ventsim™ software, version 5.1 (by Howden Ventsim). It analyses the airflow and smoke movement in mine roadways during fire incidents and validates the model's effectiveness through physical roadway fire experiments. Finally, it conducts fire simulation research on the Yangchangwan Coal Mine of the Ningxia Coal Group, providing ventilation and disaster relief plans for fire incidents. The research findings indicate that with steeper inclines, occurrences of smoke reversal and airflow reversal are more likely, leading to an increased scope of disaster. However, timely implementation of ventilation control measures effectively prevents smoke from spreading to connected roadways. A comparative analysis between physical roadway fire experiments and simulations reveals some discrepancies in Ventsim's temperature modelling of fire sources and nearby areas, although as distance increases, the simulated temperatures align more closely with reality. This validates the feasibility of Ventsim for large-scale fire simulation studies. In the fire simulation of the Yangchangwan Mine, controlling airflow by short-circuiting the intake side of the working face and reversing airflow on the return side reduces the disaster area to the vicinity of the fire source. Additionally, by directing smoke directly into the return airway through connecting roadways on the intake side of the working face, most areas of the working face and associated roadways downwind of the fire source remain unaffected during fire incidents. The paper provides evacuation times and areas, offering guidance and recommendations for fire prevention and control in the Yangchangwan Coal mine.

INTRODUCTION

Due to its suddenness, great destructiveness, high coupling, unpredictability, and propensity to trigger major compound disasters, the occurrence of underground mine fires greatly increases the difficulty of disaster prevention and emergency rescue. Particularly in cases where fires break out in the main intake airways of mines, improper prevention and control measures can lead to the rapid spread of high-temperature toxic smoke to concentrated areas where personnel are located underground. The unique underground roadway environment, confined spaces, and intricate ventilation networks make it susceptible to the mutual transformation of fires and gas explosions, triggering secondary disasters and resulting in significant casualties and economic losses (Wang *et al*, 2019, 2016). Therefore, understanding the diffusion characteristics of smoke in mine fires, implementing effective ventilation control measures during fire outbreaks, and preventing the uncontrolled spread of fires can rapidly and effectively manage ventilation, smoke evacuation, and personnel evacuation to avoid major disasters (Wu *et al*, 2024).

In recent years, scholars have studied the evolution patterns and destructive effects of fire smoke using a series of disaster smoke flow simulation software such as MFire, Vent-PC, Ventsim, Mine Ventilation Simulation System (MVSS) and Cross Fire (Wu *et al*, 2022; Bracke, Alkan and Müller, 2006; Perera and Litton, 2012; Tilley and Merci, 2013). Simultaneously, extensive research has been conducted on the disaster characteristics and smoke flow migration patterns during fire incidents

(Zhao *et al*, 2019; Cascetta, Musto and Rotondo, 2016; Chow *et al*, 2015; Li *et al*, 2022). The critical airflow velocity calculation model for preventing smoke flow reversal was derived using the Froude number model for inclined mine roadways and it was experimentally validated (Wen *et al*, 2021; Wang *et al*, 2020, 2023). Numerical simulation analysis has been performed to understand the distribution patterns of temperature, smoke concentration, and gas concentration during mine roadway fire disasters, proposing that controlling smoke exhaust velocity reasonably can effectively prevent secondary disasters. However, there has been relatively little research on the evolution patterns of smoke flow in complex ventilation networks during fire incidents. In terms of airflow control theory, Litton and Perera (2012) introduced a ventilation network spatiotemporal control navigation system to simulate evacuation routes during mine disasters. Wang *et al* (2017) developed a remote emergency rescue system for transporting roadway fire smoke and applied it in the Longdong Mine, successfully controlling mine fire smoke during drills. Zhou *et al* (2015) developed a warning and control system for transporting roadway fire disasters and applied it in the Kongzhuang Coal Mine, achieving control of roadway fire and rapid evacuation of personnel during drills. Existing research has mainly focused on the scope of fire smoke fields, lacking studies on the propagation patterns of smoke and airflow in complex ventilation networks and airflow control schemes. This has weakened the safety and applicability of on-site applications of fire smoke emergency linkage control systems.

Therefore, this study analysed the three-dimensional visualisation simulations of smoke spread before and after disaster control schemes based on Ventsim software. This further optimises airflow control strategies, proposing the optimal ventilation and smoke exhaust scheme that simultaneously meets the minimum airflow requirements for smoke exhaust during disasters and the airflow requirements for personnel evacuation and dilution of harmful gases in the mining area. This provides effective references for developing emergency airflow control and smoke exhaust schemes during disasters.

VALIDATION ANALYSIS OF VENTSIM FIRE SIMULATION

Ventilation network roadway fire experiment system

The ignition source is positioned in the middle of the fourth roadway, with combustible materials placed on metal trays supported by metal brackets, raised 40 cm for observation via cameras. Temperature sensors are strategically placed on brackets in areas where smoke may spread, distributing them across the roadway section to comprehensively measure smoke temperature. There are two sets of gas concentration sensors, each comprising a CO_2 sensor with a range of 0–10 per cent, an O_2 sensor with a range of 0–20 per cent and a CO sensor with a range of 0–1000 ppm. The overall layout of the experimental system is shown in Figure 1.

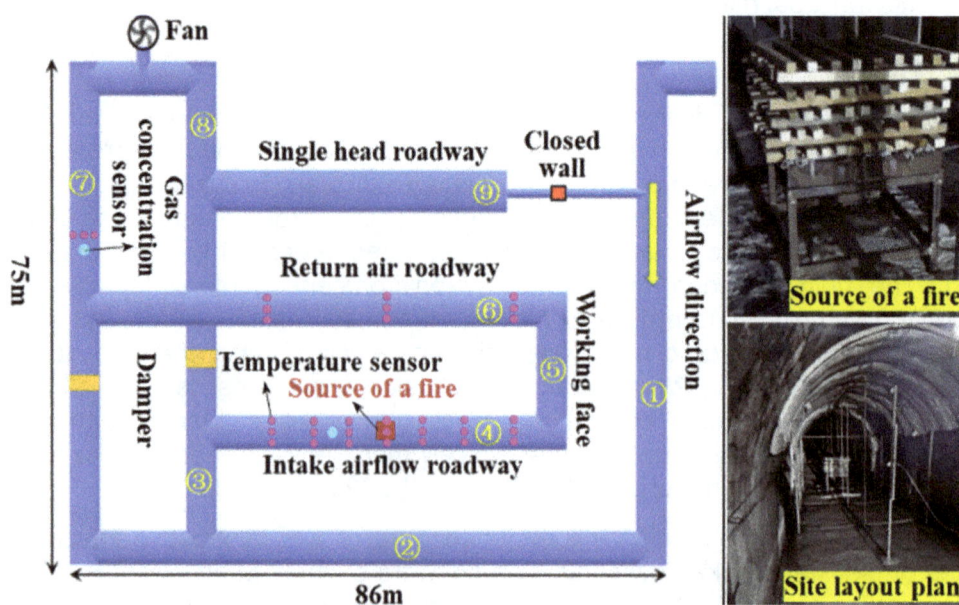

FIG 1 – Overall layout of the experimental system.

Ventilation network roadway fire experiment system

Based on the field-measured data of the Kailuan solid ventilation network roadway, including roadway length and cross-sectional dimensions, a three-dimensional polyline diagram was created using CAD software and imported into Ventsim to generate the roadway model. Ventilation facilities such as air doors and fans were then set-up accordingly. After conducting ventilation simulations, the airflow in each roadway closely matched the measured values. For instance, the simulated negative pressure of the ventilation fan was 28.5 Pa, while the measured value was 28 Pa, resulting in a maximum error of 1.8 per cent between the simulation results and real test data. The established three-dimensional ventilation simulation model meets the accuracy requirements of ventilation simulation. In accordance with the experimental design, the fire source was placed in the middle of the intake airway at the working face and the placement of monitors followed the on-site experimental arrangement. The upwind side of the fire source was set as a stratified roadway to simulate smoke rollback.

Comparison and analysis of simulation and experimental results

CO

There will be a certain degree of smoke backflow in the ignition roadway. Gas analysis devices installed in the experimental roadway can collect CO concentration values from the smoke. Starting from 300 sec, CO concentration values from both the experiment and simulation are selected at intervals of 300 sec to calculate the discrepancies. The comparison of CO concentration simulations and experiments on the upwind side and return airway of the fire source is shown in Figure 2. Combining the simulated results of CO concentration on the upwind side of the fire source and in the return airway, Ventsim demonstrates a trend that closely matches the actual CO concentration in large-scale fire experiments. The errors in simulated CO concentration values are also within a small range. Therefore, Ventsim provides valuable reference for analysing toxic and harmful gases generated by mine fires.

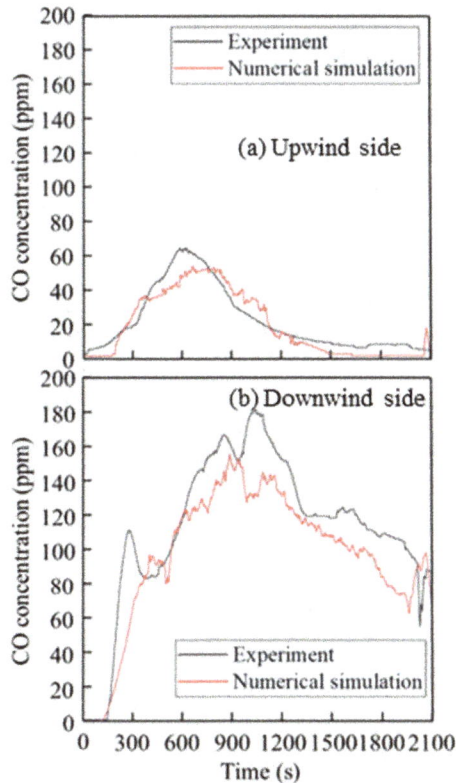

FIG 2 – CO concentration comparison.

Temperature

Since Ventsim fire simulations calculate results by uniformly mixing all components within the cross-section, when processing data from the Kailuan solid roadway experiments, the temperature data measured by temperature sensors at each measuring point within the cross-section are averaged. After obtaining the cross-sectional average data, they are compared with the Ventsim fire simulation data. The comparison between the simulated and experimental temperatures at various measuring points on the ignition roadway is shown in Figure 3.

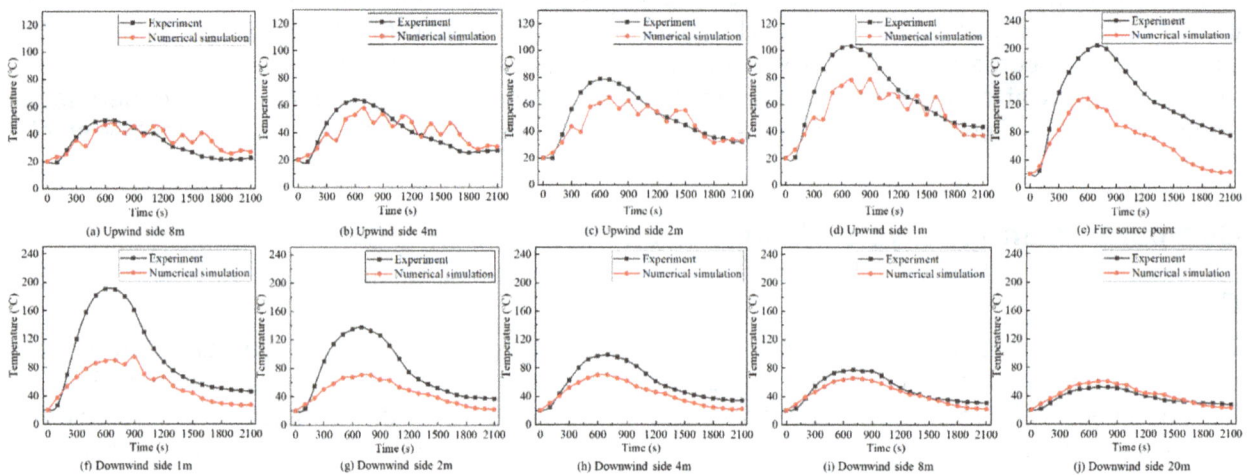

FIG 3 – Comparison of simulated and experimental temperatures.

Analysing the temperature discrepancies at points 1 m to 20 m downwind from the fire source, it can be seen that as the distance increases, the simulation error decreases sharply. This is because, at the downwind side of the fire source, as the distance from the fire source increases, the effect of high temperature on the smoke gradually weakens and the buoyancy generated by heating decreases. As a result, the smoke gradually mixes with the airflow at lower levels of the roadway. As the smoke moves longitudinally along the roadway, the mixing of smoke with airflow becomes more uniform across the cross-section. Consequently, the temperature distribution across the cross-section becomes more uniform, leading to a gradual decrease in the discrepancy between Ventsim's simulated cross-sectional temperatures and the average temperatures measured by all temperature sensors across the roadway section in the actual experiments.

The temperature trend of the upstream retreating smoke flow simulated by Ventsim closely matches the actual trend. However, there are some discrepancies between the simulated and actual temperatures at 1 m and 2 m on the upwind side of the fire source, with maximum deviations of 28.35°C and 18.08°C, respectively. The temperature discrepancies at 4 m and 8 m on the upwind side of the fire source are smaller. This is because as the retreating smoke moves further away from the fire source, the buoyancy effect decreases, causing it to gradually descend and mix with fresh airflow. Consequently, the smoke temperature becomes more uniform across the roadway section where these two measuring points are located, resulting in better agreement between simulated and actual temperatures.

ACTUAL MINE FIRE SIMULATION ANALYSIS

Ventsim model of Yangchangwan Coal Mine

The intersection of the downhill transportation of the sixth coal in Yangchangwan Mine and the transportation heading of the II020613 comprehensive mining face is selected as the ignition point. The airflow at this point flows towards the II020613 comprehensive mining face and the downwind side of the downhill transportation of the sixth coal. Smoke from the fire may spread with the airflow to the working face and most other areas of the mine, with significant smoke dispersion effects. Rubber was chosen as the burning material. To make the fire simulation process more comprehensive, the entire process was divided into growth period, steady period, decay period and recovery period. The maximum power of the fire source is 10 MW, with a maximum combustion rate of 1091 kg/h. Monitoring points were set-up in areas that may be affected by smoke after the fire to observe the impact of fire control measures on the mine fire. The specific parameters for fire simulation are shown in Table 1. The Ventsim model and monitoring point set-up is shown in Figure 4.

TABLE 1

Specific parameters for fire simulation.

Source of a fire	Time (s)	Burning rate (kg·h⁻¹)
	Growth period: 0–900	0–1091
Rubber	stationary period: 900–2700	1091–1091
	Attenuation period: 2700–3600	1091–0
	Recovery period: 3600–5400	0

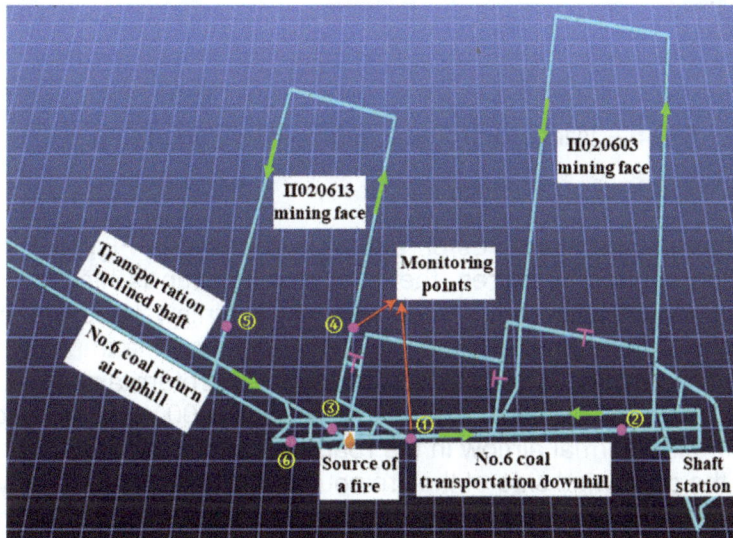

FIG 4 – Ventsim model and monitoring point set-up.

Fire simulation analysis with ventilation control measures

After a fire occurs in the mine, it will have a detrimental impact on the downwind side of the fire source and other related roadways. The ventilation heading of the II020613 comprehensive mining face is connected to the downhill return airway of the sixth coal through a connecting heading. By using a ventilation door as a barrier, at 10 mins into the simulated fire, the ventilation door at this location can be opened, diverting the airflow from the working face's intake side and directing it directly into the downhill return airway of the sixth coal. Simultaneously, the ventilation door at the exit of the return airway of the II020613 comprehensive mining face's return heading can be opened, allowing fresh airflow to enter the II020613 comprehensive mining face. By using the reverse airflow, the smoke entering the working face is forced back into the connecting heading on the intake side of the working face, directly discharging into the main return airway, thus controlling the scope of the fire disaster. The changes in airflow direction before and after opening the ventilation doors and implementing ventilation control measures are illustrated in Figure 5.

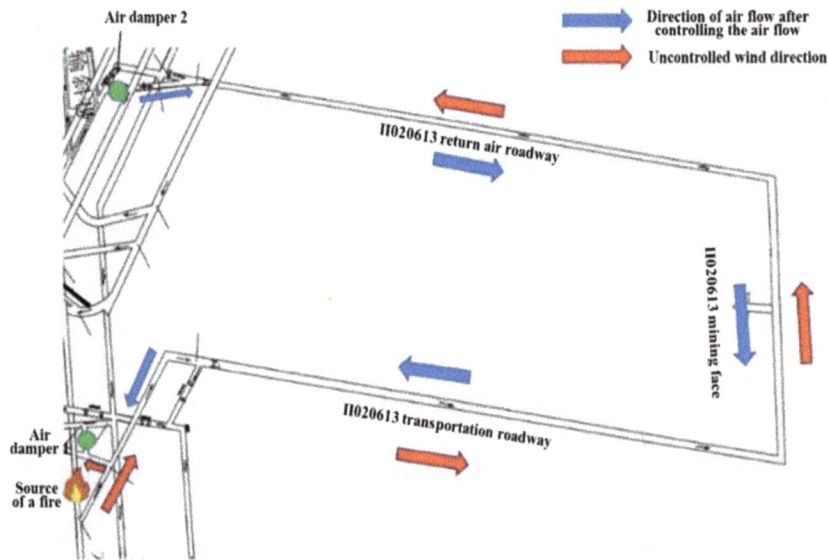

FIG 5 – Wind control and airflow direction changes.

Airflow

The impact of ventilation control measures on airflow is depicted in Figure 6. In Figure 6a, without implementing ventilation control measures, the normal ventilation is hindered by the fire-induced pressure in the downwind ventilation of the fire, causing the airflow at upwind measuring point 6 to decrease to 1250 m^3/min. After 10 mins of the fire occurrence and the implementation of ventilation control measures, the fire heading is directly connected to the return airway, resulting in increased airflow at the upwind side of the fire source. At around 900 sec, it reaches approximately 3000 m^3/min, exceeding the normal airflow in the roadway before the fire occurred. In Figure 6b, at measuring point 2 on the downwind side of the fire source, the airflow is similarly reduced due to the hindrance of normal ventilation by the fire-induced pressure. After 10 mins of implementing ventilation control measures, the direct connection between the downhill roadway of the sixth coal transport and the return airway leads to a significant portion of airflow no longer flowing to the downwind side, directly entering the return airway. This causes a gradual reversal of airflow on the downwind side of the fire source, resulting in a negative airflow, flowing towards the original upwind side. In Figure 6c, at measuring point 4 of the intake heading of the working face, the airflow is increased due to the influence of the fire-induced pressure. After implementing ventilation control measures, the intake airflow of the working face is short-circuited, and the fresh airflow from the return airflow of the working face reverses the airflow direction in the intake heading, resulting in a decrease in airflow and a change to a reverse airflow, with a maximum of approximately 800 m^3/min. In Figure 6d, after opening the ventilation doors to implement ventilation control measures at measuring point 3 of the connecting heading, a large amount of airflow is short-circuited into the return airway. The airflow increases from 140 m^3/min when the doors are closed to 5500 m^3/min.

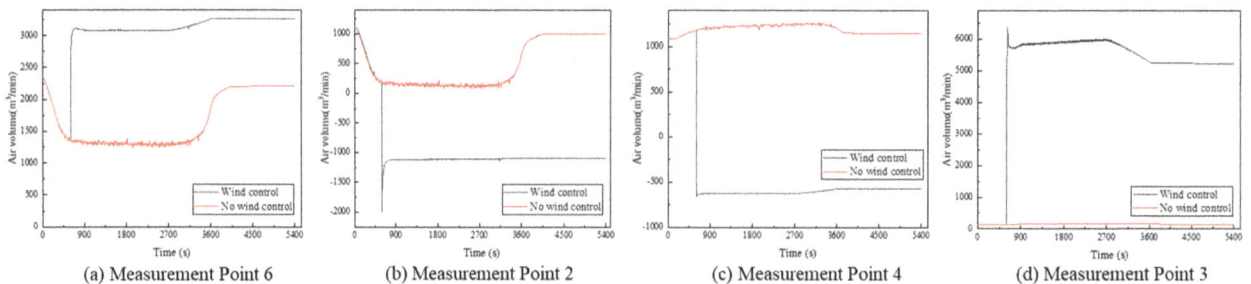

(a) Measurement Point 6   (b) Measurement Point 2   (c) Measurement Point 4   (d) Measurement Point 3

FIG 6 – Effect of ventilation control measures on airflow.

CO

The changes in the spread of CO-containing smoke over time in simulations with and without ventilation control are illustrated in Figure 7. As depicted in Figure 7a, 10 mins after the fire outbreak, smoke containing CO has reached the return air side of the II020613 mining face and the downwind side of the fire source roadway. By 30 mins, the smoke has spread to more associated roadways downstream, and the return side of the working face is filled with smoke, which has also entered the main return airway. From 60 to 70 mins, even after the fire has ended, there is still a significant amount of smoke remaining in the mine, spreading to more downstream associated roadways and entering another working face. At 90 mins, 30 mins after the fire has ended, many roadways still contain CO. In Figure 7b, at 10 mins, the ventilation control measures divert airflow from the intake side of the II020613 mining face, while fresh airflow is introduced on the return side. By 15 mins, smoke from the return side of the working face reverses back to the intake side. At 30 mins, the spread of CO-containing smoke is limited to the vicinity of the ignition point and the main return airway, indicating that the ventilation control measures effectively restrict the spread of smoke. By 61 mins, some of the smoke in the main return airway has been expelled from the mine, leaving only residual smoke near the exit of the return airway, with other areas free of CO-containing smoke. By 63 mins, there are no remaining locations in the entire mine containing CO-containing smoke, demonstrating the significant effectiveness of the ventilation control measures.

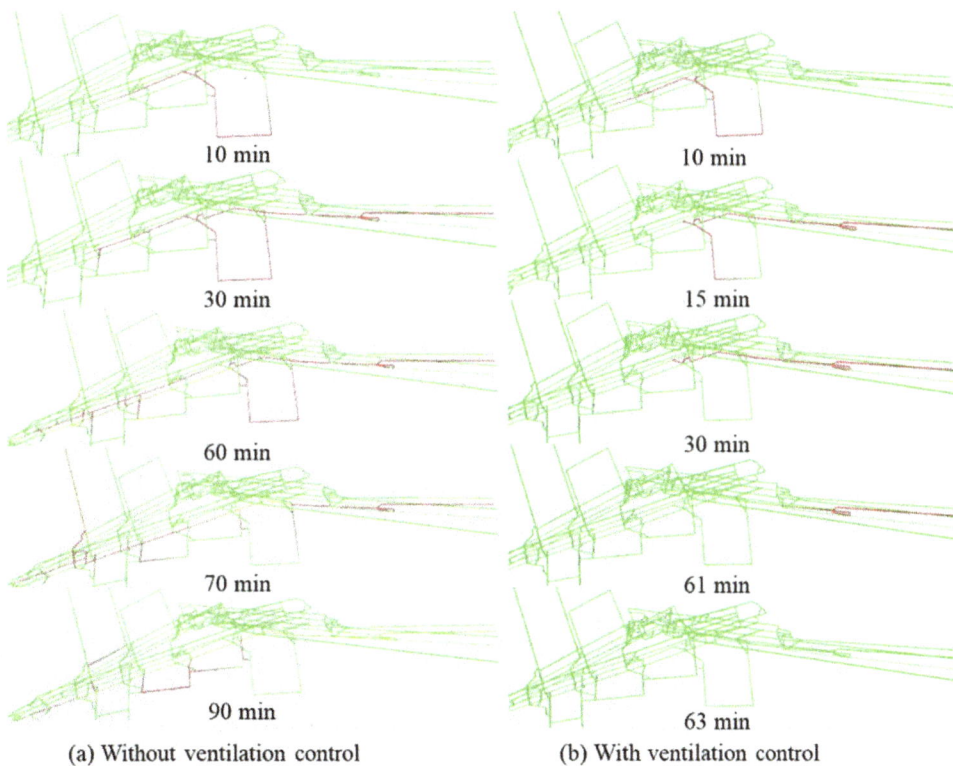

(a) Without ventilation control   (b) With ventilation control

FIG 7 – CO smoke spread range.

The changes in CO concentration at the fire source and the intake side of the working face connection roadway are shown in Figure 8. After implementing ventilation control measures, a short-circuiting of airflow occurs on the downwind side of the fire source, reducing ventilation resistance, increasing airflow in the ignition gallery, promoting more complete combustion at the fire source and subsequently reducing CO concentration. Additionally, the increased airflow quickly carries away smoke, preventing CO accumulation, resulting in lower CO concentrations at the fire source compared to when no ventilation control measures are applied.

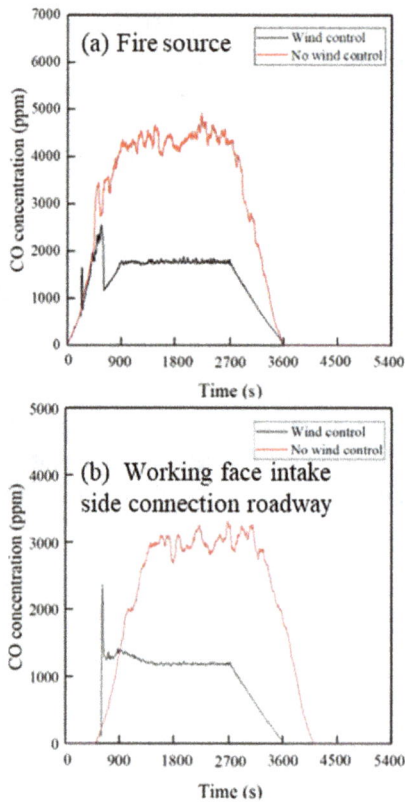

FIG 8 – Changes in CO concentration.

Temperature

Due to the significant impact of implementing ventilation control measures on the II020613 comprehensive mining face, subsequent analysis of temperature and visibility changes mainly focus on monitoring points 4 and 5 of the II020613 face. As shown in Figure 9, it can be observed that the temperature at monitoring point 4 starts to rise shortly after the fire occurs, indicating its proximity to the fire source. Without ventilation control measures, the temperature at monitoring point 4 could rise to 45°C as the fire progresses. However, with ventilation control measures implemented at 10 mins, facilitated by the short-circuiting of the intake air and the counter-flow of the return air, the temperature quickly decreases, returning to normal levels by 900 secs. Monitoring point 5, being farther from the fire source, experiences a reasonable increase in temperature during the fire process even without ventilation control measures. However, with ventilation control measures in place, the temperature decreases, indicating that the fresh airflow brought by the counter-flow can lower the temperature in the roadway, demonstrating the positive impact of ventilation control measures.

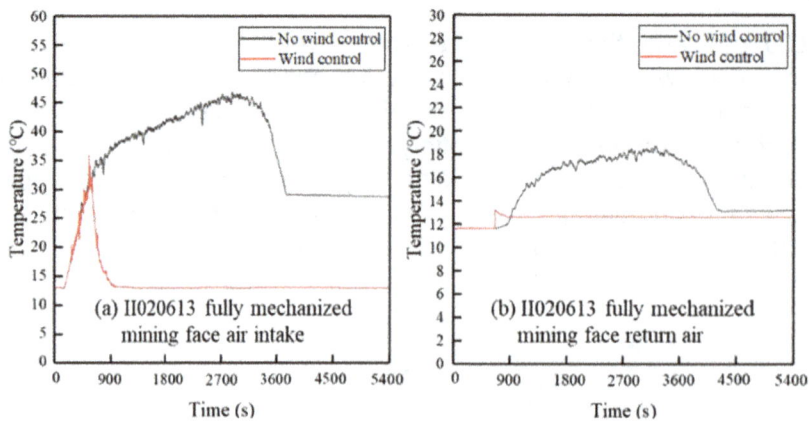

FIG 9 – Variation of temperature at the working face.

Visibility

The changes in visibility at measurement points 4 and 5 on the working face are shown in Figure 10. Without implementing control measures, the visibility across the entire working face would be affected. Visibility begins to decline at measurement point 4 relatively early, with a rapid decrease shortly after the onset of the fire due to smoke influence. Throughout most of the fire duration, visibility at measurement point 4 remains close to 0 m. At 3600 secs, when the fire is extinguished and no new smoke is generated, the smoke at measurement point 4 dissipates and visibility returns to normal levels. Implementing control measures at 10 mins redirects more smoke from the fire source into the return airway and the fresh airflow from the return airflow also helps restore visibility to normal levels around 1300 secs. Measurement point 5 is located on the return airflow side of the working face, farther from measurement point 4. Consequently, visibility begins to decline around 850 secs. After the fire is extinguished, it takes about 450 secs for visibility at measurement point 5 to return to normal levels. Implementing control measures at 600 secs prevents smoke from reaching measurement point 5 throughout the entire fire duration, ensuring that visibility at measurement point 5 remains at normal levels. This ensures that personnel seeking refuge in this area are not affected by decreased visibility.

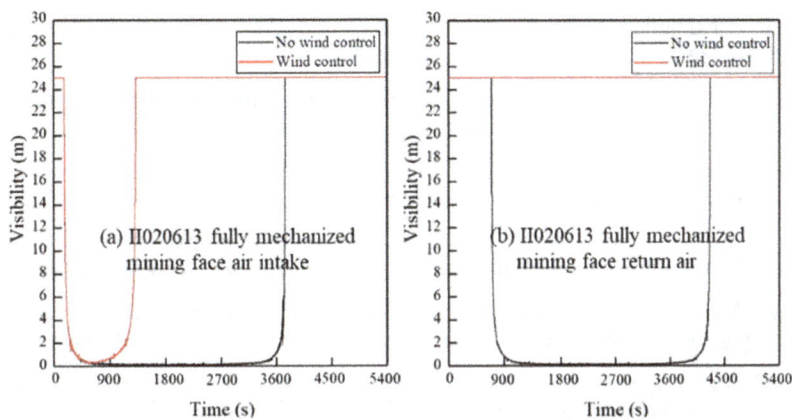

FIG 10 – Changes in visibility at the working face.

CONCLUSIONS

This study utilised the Ventsim software to investigate mine fires. An underground experimental roadway and the Yangchangwan Coal Mine were used to establish a mine fire model, and various factors such as visibility, temperature, carbon monoxide and airflow were analysed. The main conclusions are as follows:

- A comparison between the fire experiments in the Kailuan experimental roadway and the Ventsim fire simulation results revealed that the simulation of fire temperature by Ventsim was not entirely consistent with reality in the areas near the fire source. However, as the distance increased, the mixing of smoke became more uniform, resulting in simulations closer to reality. These comparative results suggest that using Ventsim for fire simulation in large-scale ventilation networks is feasible and can provide valuable insights for fire research.

- In the fire simulation of the Yangchangwan Mine, by adopting the method of short-circuiting the air intake side of the working face and reversing the airflow on the return air side, the disaster area during the fire period is reduced from the working face and downstream side of the fire source associated roadway when no airflow control measures are taken to near the fire source. Smoke is directly discharged into the return airway through connecting roadways, providing personnel with escape time and escape areas. Using Ventsim fire simulation provides guidance for fire prevention, control and rescue as well as disaster avoidance in the Yangchangwan Mine.

- This ventilation control scheme is based on existing ventilation facilities in the mine, making it simple and highly feasible. It provides guidance and suggestions for fire prevention, rescue and disaster avoidance in the Yangchangwan mine.

ACKNOWLEDGEMENTS

This work was supported by the National Natural Science Foundation of China [52374252].

REFERENCES

Bracke, G, Alkan, H and Müller, W, 2006. Modelling underground ventilation networks and radon flow for radiological protection using VUMA, *Uranium in the Environment: Mining Impact and Consequences*, pp 593–599.

Cascetta, F, Musto, M and Rotondo, G, 2016. Innovative experimental reduced scale model of road tunnel equipped with realistic longitudinal ventilation system, *Tunnelling and Underground Space Technology*, 52:85–98.

Chow, W K, Gao, Y, Zhao, J H, Dang, J F, Chow, C L and Miao, L, 2015. Smoke movement in tilted tunnel fires with longitudinal ventilation, *Fire Safety Journal*, 75:14–22.

Li, Q, Kang, J, Wu, Y and Luo, J, 2022. Theoretical and numerical study of smoke back-layering length for an inclined tunnel under longitudinal ventilation, *Fire Technology*, 58 (4):2143–2166.

Litton, C D and Perera, I E, 2012. Evaluation of criteria for the detection of fires in underground conveyor belt haulageways, *Fire Safety Journal*, 51:110–119.

Perera, I E and Litton, C D, 2012. Impact of air velocity on the detection of fires in conveyor belt haulageways, *Fire Technology*, 48:405–418.

Tilley, N and Merci, B, 2013. Numerical study of smoke extraction for adhered spill plumes in atria: Impact of extraction rate and geometrical parameters, *Fire Safety Journal*, 55:106–115.

Wang, K, Hao, H, Jiang, S, Wu, Z, Cai, W and Wang, Z, 2020. Study on fire smoke flow characteristics in the ventilation network and linkage control system in coal mines, *Fire and Materials*, 44 (7):989–1003.

Wang, K, Hao, H, Jiang, S, Wu, Z, Cui, C, Shao, H and Zhang, W, 2019. Escape route optimization by cellular automata based on the multiple factors during the coal mine disasters, *Natural Hazards*, 99:91–115.

Wang, K, He, X, Sun, L, Hu, X, Guo, Y and Zhang, P, 2023. Numerical study on upstream smoke propagation and induced airflow velocity in a tilted channel under natural ventilation, *Thermal Science and Engineering Progress*, 46:102228.

Wang, K, Jiang, S, Ma, X, Wu, Z, Shao, H, Zhang, W and Cui, C, 2016. Numerical simulation and application study on a remote emergency rescue system during a belt fire in coal mines, *Natural Hazards*, 84:1463–1485.

Wang, K, Jiang, S, Wu, Z, Shao, H, Zhang, W, Pei, X and Cui, C, 2017. Intelligent safety adjustment of branch airflow volume during ventilation-on-demand changes in coal mines, *Process Safety and Environmental Protection*, (111):491–506.

Wen, H, Liu, Y, Jin, Y, Zhang, D, Guo, J, Li, R and Zheng, X, 2021. Numerical simulation for mine oblique lane fire based on PDF non-premixed combustion, *Combustion Science and Technology*, 193(1):90–109.

Wu, B, Meng, Y, He, B, Zhao, C and Lei, B, 2022. Study on the migration law of gas explosion disaster products in complex air network of mine, Energy Sources, Part A: Recovery, *Utilization and Environmental Effects*, 44(3):6378–6391.

Wu, B, Meng, Y, Yao, Y, Lei, B, Wang, J and Zhai, J, 2024. Transportation of wind-smoke flow in full-scale laneway fire experiments, *Fire Safety Journal*, 142:104038.

Zhao, X, Chen, C, Shi, C, Chen, J and Zhao, D, 2019. An extended model for predicting the temperature distribution of large area fire ascribed to multiple fuel source in tunnel, *Tunnelling and Underground Space Technology*, 85:252–258.

Zhou, G, Cheng, W, Zhang, R, Shen, B, Nie, W, Zhang, L and Wang, H, 2015. Numerical simulation and disaster prevention for catastrophic fire airflow of main air-intake belt roadway in coal mine—a case study, *Journal of Central South University*, 22:2359–2368.

Efficient and accurate ventilation surveys for model calibration

F C D Michelin[1], S K Ambrosio[2], C A M Jackson[3] and C M Stewart[4]

1. Managing Director, Howden Ventsim Australia, Brisbane Qld 4000.
   Email: florian.michelin@howden.com
2. Mine Ventilation Specialist, Howden Ventsim Australia, Brisbane Qld 4000.
   Email: shane.ambrosio@howden.com
3. Senior Ventilation Engineer, Howden Ventsim Australia, Brisbane Qld 4101.
   Email: christopher.jackson@howden.com
4. Director, Minware Pty Ltd, Brisbane Qld 4101. Email: craig@minware.com.au

ABSTRACT

The increasing use of ventilation simulation software since the 1990s has changed the work of ventilation professionals, decreasing the requirement to do manual calculations and computations. Ventilation simulation software is now used at most mines around the world and is a legal expectation in most Australian mines. Ventilation models are used extensively by ventilation professionals to understand the airflows throughout the mine, troubleshoot potential solutions, and plan for future ventilation design. Ventilation modelling has enabled design improvements in many mines, particularly complex or deep operations with challenging ventilation conditions.

The relative ease of use of ventilation software has enabled less experienced engineers with little training in ventilation to manipulate and design complex circuits. The lack of a core understanding of ventilation fundamentals however may result in inaccurate models that do not align with actual mine workings or designs. The consequences can be dramatic and may result in both unsafe work conditions and costly inadequate design.

This paper describes the core skills required for ventilation professionals who undertake modelling, and the most efficient practices to ensure high model accuracy. Old and new methods to accurately measure underground conditions for model calibration will be discussed such as the correct use of the anemometer (Lambrecht Meteo, 2017) and the measurement of cross-sections to obtain exact airway area and perimeter criteria. From those readings, the steps taken to calibrate a model are described, including the use of appropriate friction factors and how to handle discrepancies in area and perimeter between planned values and actual values. Finally, how to efficiently review a model will also be presented.

INTRODUCTION

Ventilation simulation software is now used at nearly all mine sites, assisting many engineers in their day-to-day activities and long-term planning. Most engineers have a reasonable knowledge of simulation software, yet over time, as many ventilation engineers with different backgrounds modify a model, its quality may decrease as the reasons for certain changes or settings within the model become obscure with time.

When predicting the future requirements of a new mine or extension, an engineer can only use academic references and experience to estimate the resistance of the mine. However, with an existing mine, more information is available. Ensuring the model used is well-calibrated and up to date is an essential part of the role of the site ventilation engineer.

Accurate ventilation models are critical to calculate the operating points for primary fan selection or heating or cooling solution selection. Considering the cost of such equipment it is critical to avoid miscalculations leading to disastrous consequences, both in economic and production delay terms, if the equipment is not selected appropriately.

This paper does not intend to provide a method for an exact calibration with no error but rather to provide a method that relies on the minimum time required for sufficient accuracy.

EXAMPLES

Several examples of the consequences of poorly calibrated models are presented below. The personnel names and mine locations remain anonymous.

Case A – poor fan duty calibration despite correct airflow

A mine claimed their model was accurate and showed good correlation to actual airflow with most readings being within 10 per cent of the model. However, further investigation highlighted that the surface fan operating pressure was double the modelled value.

In this case, the mine resistance was grossly underestimated but the steep fan curve permitted close correlation with measured flow. Using this model to plan for future installation would have likely caused the main fan to stall earlier than expected and the expansion to be stopped.

Case B – poor pressure calibration despite correct fan duty

A mine model's main fan duty (flow and pressure) appeared accurate. During an emergency, it was decided to open a bulkhead to allow additional airflow through. The model suggested a significant pressure against the bulkhead of 400 Pa which should have provided good airflow when opened. When measured underground however, the pressure was found to be only 10 Pa and the opening did not offer the help they hoped.

The emergency team had gone underground with an action plan based on information that wasn't accurate and would not work, potentially endangering personnel.

Case C – the model is calibrated against assumed data

A model was developed remotely using data provided by the site. The model never matched actual measurements and pressure anomalies were present that were never fully accounted for. The off-site engineer used the model to recommend new surface fans in a new mining area of the operation, however once the new fans were commissioned, they immediately stalled.

The engineer visited the site and upon inspection of the existing main fan, he discovered that the blade angle provided in the original data by the site was very different from the one in use. Fortunately, making changes to the operating curve of the main surface fan (ie operating on a lower curve) allowed a system pressure reduction at the new fan installation, allowing these new fans to function within their design curve and without stalling.

By relying on unverified information provided solely by other parties, the study could have had a significant impact on the production if no solutions were found.

CHARACTERISTICS OF A GOOD MODEL

Accurate

A good model needs to represent the mining operations reality accurately. Most would consider a 10 per cent maximum discrepancy in airflow and pressure (difference between measured and modelled) to be reasonable. Any model within 5 per cent would be considered very high accuracy given the limitations and variability of survey measurements. Airflow alone however does not offer enough information to understand the mine resistance and pressures at the main fans and ventilation control devices (VCDs) are critical to understand the mine resistance.

Replicable

A traditional method of building a model is to measure airflow and pressure loss between every junction in a mine (a 'PQ' pressure quantity survey) allowing the engineer to calculate exact resistances (and any inherent shock losses) and derive accurate friction factors if required. A model built correctly this way should be highly accurate, yet it may be difficult for other users to understand, maintain and extrapolate due to previously measured resistances not being suitable for new development. The high turnover of ventilation engineers at many mines exacerbates this problem.

A model with standardised friction factors and resistances (calibrated against a sample of *in situ* measurements) may be slightly less accurate but will be able to be more consistently maintained by different engineers and provide more reliable guidance for future designs and modelling.

Ease of use

Mines are often complex and airway locations can be easily mistaken for others. Modern ventilation modelling software utilises 3D graphics and colours that can assist the user in better understanding the underground ventilation environment. A good model should be easy to comprehend and interpret, with colours, shapes, sizes, layers, animations, and locations representing important information for key parts of the model.

For example, intake and exhaust circuits can be coloured to differentiate air types, and essential data such as quantity, pressure, and temperature can be shown as needed. Dense textual data in large models can be confusing and should be reserved for in-depth analysis only when needed in key parts of a model. Additionally, sources of critical assumptions and factors should be referenced in the model (eg reports, textbooks, measurements).

Useful

George E Box stated that 'All models are wrong, but some are useful.' In that sense it is important to understand the purpose of the initial model. At times a quick model can give you the answers you need but potentially cannot be used for future work. One exception to this is a mine base model which should be kept up to date and fully calibrated. Such a model is often used by the site engineer to troubleshoot future changes, simulate diesel or fire. It will also be sent to consultant to provide advice in the long-term. Such a model needs to be kept accurate in a general term as many will assume it right.

Accuracy can be achieved in many ways, but can sometimes be mimicked by adjusting the wrong parameters. Models are usually used to test what is going to happen before it does, so the true test of a good model is its ability to predict the results of ventilation changes. To do so, a ventilation change such as an open door or a fan turned off would be simulated then measured and both should be within the same accuracy as the initial model. For a mine base model, this should happen naturally, and investigation should take place if the model failed to predict the change.

EFFICIENT MODEL REVIEW

A model review involves a systematic checking procedure and comparison to actual measurements. A few initial checks can often highlight the likelihood a model may require a more extensive review.

Fixed values

Fixing airflow or pressures in a model can assist in estimating unknown data such as the duty requirements of a new fan, however they poorly represent real world items as the flow will not change regardless of any change in resistance or pressure. If a fixed flow is initially used to establish a fan duty it should be removed and replaced with a suitable fan curve where the flow and pressure will change with resistance.

Infinite resistances representing blockages (effectively fixing airflow to zero) allow no leakage regardless of the pressure differential, however true leakage through ventilation controls in large mines may accumulate to 20 per cent or more of the total intake quantity, leading to overestimation of the airflow capable of being delivered to the workplace. Modelling realistic resistances incorporating ventilation control leakage will prevent this risk.

Friction factors

The presence of default (unspecified) friction factors in a model indicates that the user hasn't considered airway friction and the model may need more work. While default factor settings may or may not be reasonable, the lack of applied discipline is a red flag.

Alternatively, models that have been developed over time to use a large number of different friction factors may also be a concern if no comments, explanations, or sources for the factors are provided, as this may indicate unreviewed and potentially inaccurate work has been relied on from past users.

Shock factors

Shock factors represent pressure losses resulting from sudden changes in airflow direction (bends corners and junctions) and velocity (expansions, contractions, duct exits), however they are inconsistently used by many engineers. Applied shock factors should represent significant changes in air direction, with a greater emphasis on airways with higher airflow velocity when shock pressure loss is potentially greatest. Minor shocks losses are usually incorporated into friction values and can normally be ignored in modelling if considered a part of the friction factor.

An example of high shock loss is a series of interconnected sub-vertical shafts or raises ('dog legs') in a high velocity exhaust circuit as shown in Figure 1. Models ignoring shock losses in these arrangements risk significantly underestimating resistance and the pressure required for fans.

FIG 1 – Direction changes in high velocity return systems require shock factors.

MEASURING UNDERGROUND CONDITIONS

Airflow

For airflow measurements to be accurate, both velocity and area need to be measured with accuracy at the same location. The following guidelines will assist in measuring velocity accurately:

- Use instruments that utilise the full cross-section of the airway. Spot measurements can be useful in some circumstances however variation inflow across a section mean they do not provide an accurate enough reading to calibrate a model. Using only centre spot readings for example, will inflate air quantity calculations.

- More than one measurement should be taken and be within 5 per cent of one another (McPherson, 2018). The average of these readings should be taken as the indicated reading.

- Choose a location away from disturbance or junction, at least 15 m past the last junction or major obstruction.

- Anemometers are fragile instruments and should be handled carefully.

- Instruments must be calibrated as per the manufacturer's recommendations (typically once per annum), and the associated calibration chart must be used to apply a correction factor to achieve the true air velocity. If an instrument does not have a calibration certificate or if it cannot be calibrated, it should not be used for ventilation model calibration surveys.

- Use instruments as per the OEM manuals. Some instruments are designed to take measurements using the Full Traverse Method while others make use of the Point Traverse Method (typically an average of 16 points) (Figure 2). The two methods cannot be used interchangeably unless the instrument is designed for that purpose.

- Traversing too fast will increase the measurements, several guidelines exist, the MVSSA (Du Plessis, 2014) recommends 0.2 m/s while McPherson (2018) recommend using 15 per cent of the air velocity (Table 1).

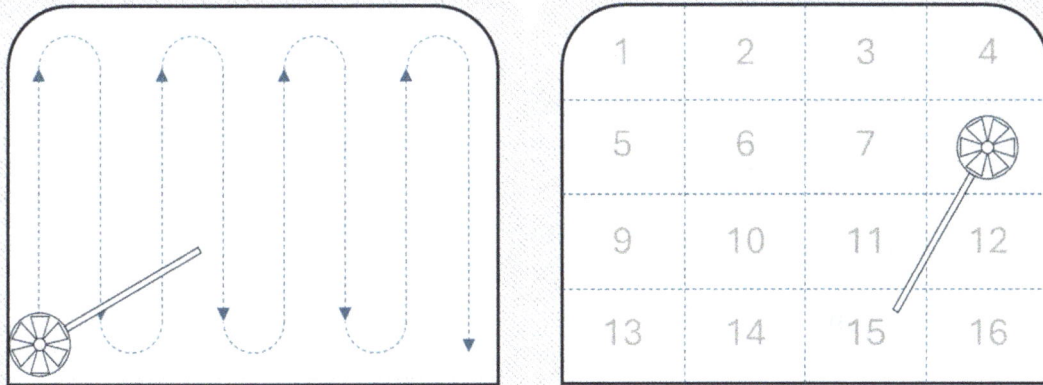

FIG 2 – Full traverse and point traverse method – two different methods for different devices.

TABLE 1

Error induced by traverse speed from MVSSA.

| True velocity (m/s) | Percentage error | | |
| | Traverse speed (m/s) | | |
	0.15	0.3	0.6
1	1.1	4.4	15.6
1.5	0.5	2	8
2	0.3	1.1	4.4
2.5	0.2	0.7	2.8
3	0.1	0.5	1.9

Even with an accurate velocity measurement, any error in the area would impact the airflow. Laser distance metres are now commonly used for measuring airway dimensions however they may not provide sufficient accuracy for irregular shapes or large arched airways. A square approximation of an arched airway may be in error by 10 per cent or more, leading to similar errors in airflow measurement.

If a station is surveyed regularly, an exact cross-section can be obtained from the survey department. If the station is new or the usual measurement point is inaccessible, using a multi-point rotating distance device can provide an accurate estimation of the area and perimeter. Areas can vary significantly over very short distances and it is critical to measure velocity at the same location.

Calculating the airflow while underground allows potential troubleshooting if readings are unusual.

One last source of error is the ongoing changes in a mine, ie fan turning on, truck moving etc. To calibrate a model, measurements with the highest accuracy possible is preferred. As such in addition to the velocity and area measurements considerations, measuring where the confidence in airflow is the highest should be prioritised. For example, the return air drives are usually more stable than the declines.

Pressure

Pressure measurements are an essential part of a model calibration. The most critical location to measure is the pressure of the main fans, whether through measurement of the pressure across the fan or the collar pressure. Similarly, measuring pressures across VCD's is critical as part of routine work to validate the model.

Some areas can be especially challenging to model without pressure reading due to uncertainties in the exact layout, for example a shaft with multiple services included or main returns. For those, measurement of pressures using either a barometer (Ruckman and Bowling, 2012) or gauge and tube method is critical to calculate the correct resistance. The barometric pressure may be easier, add affordable when using a phone (Derrington, 2015), but careful consideration should be taken for elevation and temperature density differences, and the requirement of a second barometer as a reference datum or simultaneous differential point (Rowland, 2012). Once added into a model, the resistance or friction should be well documented to ensure future users do not require to re measure again.

CALIBRATING A MODEL

Getting the base right

Geometry

The base is considered the physical structure of a model, essentially an empty shell that is later filled with fans and control devices. The first step should ensure that all airways are present and have the correct dimensions, which can usually be verified by importing the CAD designed airways and existing survey data. Missing airways or mis-sized airway can have large ramifications for calibration.

It is common for actual airway sizes to differ from design sizes by 10 to 20 per cent (Michelin *et al*, 2014), yet most mine models are calibrated using only the design size and adjusted friction factors are used to compensate for the difference. Using the design size may be logical as it is the base of any future design extensions, however, if friction factors are calculated by measuring resistance and actual airway size, the factors will not be applicable to unadjusted design sizes. The factors may need to be adjusted to the design size only, or else the modelled resistance and pressure may be too high.

Friction factors

A starting basis may be to use researched friction factors, whether from McPherson (2018), or other sources. The number of different factors used should be minimised to keep the model uncomplicated and replicable. Precise naming of friction factors is also important to ensure future engineers can extend the model accurately. For example, friction factors that are named to define declines, level drives, hoisting shafts, open raises etc are all easily identifiable.

In the case of obstruction in an airway such as a belt road, the friction factor may already include the obstruction. If it doesn't, it is essential to take obstructions into consideration as an additional resistance or reduction in airway area.

Ventilation control devices

VCDs often form a disproportionally large part of mine resistance and hence the location and resistance setting of the VCD is crucial for a calibrated model. The resistances of devices may initially come from textbook but should be calculated from actual measurements where possible and entered into the model library for existing calibration and re-use at future installations.

Main fans

Fan curves form a key part of a ventilation model. A common mistake is entering a fan static pressure curve and then applying shock and discharge velocity losses which are already largely excluded by the fan curve choice. A fan total pressure curve is the recommended choice as it includes a potentially usable ventilation velocity pressure component which can be modelled with the application of appropriate installation and discharge losses.

Another common mistake is ignoring the installed fan resistance which plays a crucial role in determining usable system fan performance. Depending on the survey method and location, fan pressure measurements often incorporate these losses, which make calibration to the original fan curve difficult if the losses are not known (the measured pressure is lower than the theoretical fan pressure).

The manufacturer may provide the resistance or fan losses of the installation (including inlet, discharge, and other losses) and this should be incorporated into fan modelling assumptions (Figure 3). If this information is not available, the ventilation engineer must estimate the losses. A method to consider installation losses is to construct a model of the fan installation, applying shock losses to key airway contractions, expansions and bend areas (much like another shock loss sections in the mine). A walled fan should also model calibrated leakage resistance through any access doors.

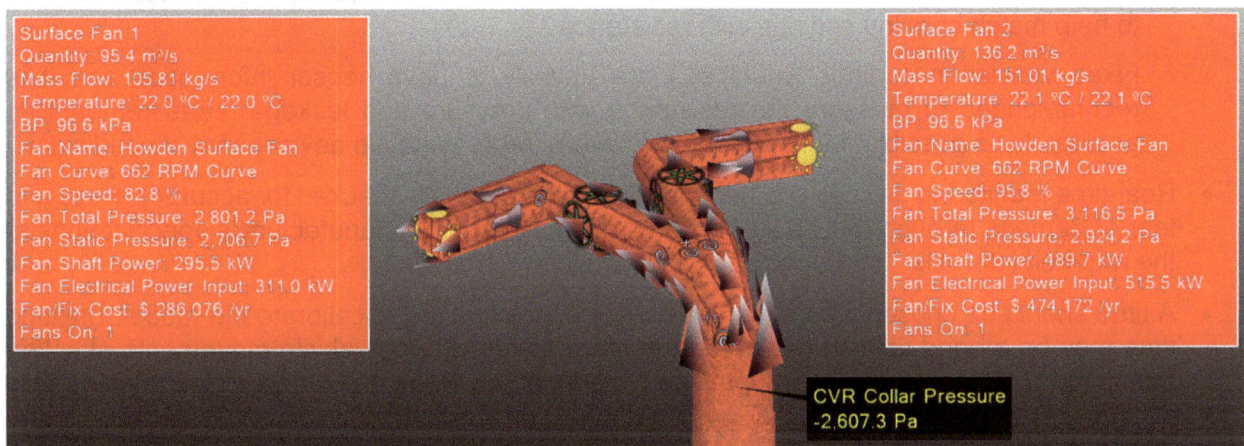

FIG 3 – Fan model including drifts for more accurate fan installation loss.

Adjusting to measurements

One method of calibration is to use fixed flows for open airways and measured fixed pressures for fans in key areas of the system during the calibration. Unknown resistances can then be calibrated until open airway fixed flows no longer add or restrict pressure, and fixed pressure airflows matches measured airflows. Where results are not matching, model information can be reviewed for incorrect or missing data such as incorrect resistance, or incorrect or misplaced ventilation control devices. Where specific measurements have not been made, the preset resistances of common ventilation controls devices (such as access doors) can also be globally adjusted to achieve an expected leakage amount or fan pressure. For example, a higher-than-expected airflow at the bottom of the mine may indicate that additional leakage is present, and by decreasing (or increasing in the opposite case) the preset resistance, the desired airflow may be achieved.

If inaccuracy persists, consider additional measurements such as pressure drop along key drives to improve the friction factors.

NEW TECHNOLOGY AND MODEL CALIBRATION

New technologies in mining may provide a new outlook on model calibration (Brake, 2023).

Calibrating VOD mines

Ventilation devices need to be maintained at the same operating point during a model calibration measurement period. Modern mines are increasingly adopting forms of Ventilation on Demand (Pinedo and Torres Espinoza, 2019) where the ventilation controls continuously adjust to match demand, making consistent measurements impossible unless the controls are 'frozen' for a length period of time which may be impractical for operations.

In this case the individual control devices must be measured and calibrated and a theoretical snapshot in time used to calibrate the model. In more detail:

- Ensures all sensors and ventilation controls operate as expected or adjust the model to reflect the real operation. This will likely take longer than a standard survey, however, the results should be reusable, and all steps would not need to be redone every time. This includes but is not limited to:

  o Main fans: check that the sensors for flow and pressure are working properly and calibrated and that power (which can help verify the fan duty operating point) is recorded accurately.

  o Auxiliary fans ideally should be equipped with flow and pressure sensors. If not, the electrical current can be used to estimate duty from the fan curve if calibrated to flow and pressure. Leakage should also be considered where a forced system draws air through walls to deliver it to the face.

  o Variable ventilation controls require either flow or differential pressure sensors. An accurate resistance curve for the different potential set points of the control devices is recommended to help match opening settings with resistance.

  o Fixed ventilation controls (walls etc) are unlikely to have sensor information and it is recommended to measure the resistance with pressure and leakage measurements as accurately as possible. Measurement of several representative devices may be sufficient.

- Real-time monitoring devices and sensors should be regularly checked for accuracy, wear and tear or damage through routine inspections, and calibrated to manufacturers specifications at the required intervals.

- A time should be chosen where the live data will be extracted to calibrate the model. Blasting time may be a good time as the system is more stable with no vehicle movements affecting the airflow. Once data has been obtained for a time, the steps are similar to those previously described.

Some VOD systems include live simulations based on underground data (Sanftenberg, 2019), which can alert users when discrepancies between simulation and sensor data occur.

Automatic calibration

Recent work has been performed by researchers (Griffith and Stewart, 2022) opening the possibility of using algorithms to solve missing data required for calibration in models. The algorithms focus on adjusting resistances to match measured flow readings, however large data sets are required to calibrate an entire mine and computational requirements may be huge.

Pre-calibration of models to closer accuracy will assist automatic calibration where inconsistent friction factors that do not match expected airway types such as shafts or ramps could be highlighted and corrected. Three dimensional surveys could also be used to get the correct size and location of all airways as well as the roughness (Watson and Marshall, 2018).

Finally, an ideal automatic calibration system would interpret discrepancies in certain areas (a leaking door for example) and adjust with a different resistance (changing a good door for a bad door) to improved model calibration, and also potentially flag further checks required by the engineer.

CONCLUSIONS

The consequences of a poorly calibrated model can be significant, potential contributing to safety hazards such as insufficient airflow or high dust or gas conditions, poor disaster mismanagement for fires or gas, reduced productivity and costly infrastructure mistakes. Initial checks on calibration with key measured values, and the quality of utilised fixed values, friction, and shock factors can help indicate whether a model needs more attention.

A calibrated model starts with getting the basics correct such as geometries and resistance values, with factors and calibrated values used properly described in reference documents. Further calibration focuses on correcting any mistakes by adjusting the ventilation controls and ensuring the corrected pressures are simulated with the correct airflows.

Modern mines with continuous equipment movement or ventilation on demand require special consideration due to variable ventilation conditions yet can still be calibrated using modified methods. As technology evolves, many options will become available to remove human error and improve accuracy. In the age of AI, model calibration is likely to become easier and less time intensive, yet they will still require accurate measurements and critical minds of humans to interpret the results.

ACKNOWLEDGEMENTS

This paper has been written with the input and advice of many ventilation professionals, in particular Shane Ambrosio who articulated and successfully uses the proposed methods of surveying and calibrating and Craig Stewart who has been a keen mentor on how to articulate a technical paper. Others include Chris Jackson, Martin Griffith, Grigorii Kolegov, and many more who contributed to this paper through discussions.

REFERENCES

Brake, D J, 2023. Gauge and tube surveys: What is their future and that of underground measurements generally as mines transition towards greater use of Big Data and Artificail Intelligence systems, North American Mine Ventilation Conference 2023, Rapid City.

Derrington, A S, 2015. Development of a Low-cost Instrument for Barometric Pressure Surveys, in *Proceedings of the Australian Mine Ventilation Conference*, pp 331–336 (The Australasian Institute of Mining and Metallurgy: Melbourne).

Du Plessis, J J L, 2014. *Ventilation and Occupational Environment Engineering in Mines*, third edition, 950 p (Mine Ventilation Society of South Africa: Johannesburg).

Griffith, M D and Stewart, C M, 2022. Automatic Ventilation Model Calibration Using Measured Survey Data, Australian Mine Ventilation Conference 2022, pp 33–42 (The Australasian Institute of Mining and Metallurgy: Melbourne).

Lambrecht Meteo, 2017. Operating Instructions Vane Anemometer (14143). Available from: <https://www.lambrecht.net/upload/manuals/14143_Manual.pdf>

McPherson, M J, 2018. Subsurface Ventilation Engineering Book [online version]. Available from: <https://cornettscorner.com/wp-content/uploads/2019/12/McPherson-Vent.-book.pdf>

Michelin, F, Stewart, C, Griffith, M D and Andreatidis, T, 2019. Calibrating model airway size and resistance with survey asbuilt data, in *Proceedings Australian Mine Ventilation Conference 2019*, pp 363–368 (The Australasian Institute of Mining and Metallurgy: Melbourne).

Pinedo, J and Torres Espinoza, D, 2019. Implementation of Advanced Control Strategies Using Ventsim Control at San Julian Mine in Chihuahua, Mexico, in *Proceedings of the 17th North American Mine Ventilation Symposium*, pp 435–441.

Rowland, J A, 2012. Barometric resistance surveys: A new perspective. in *Proceedings of the 14th North American Mine Ventilation Symposium*, pp 13–22.

Ruckman, R and Bowling, J, 2012. Comparison of barometer pressure surveys with other measuring techniques for determining fictional pressure loss in shafts, in *Proceedings of the 14th North American Mine Ventilation Symposium*, pp 23–29.

Sanftenberg, J, 2019. Mine Ventilation on Demand System at Nickel Rim South Mine; Testing Procedures, System Maintenance and Results. in *Proceedings of the 17th North American Mine Ventilation Symposium*, pp 104–113.

Watson, C and Marshall, J, 2018. Estimating underground mine ventilation friction factors from low, *International Journal of Mining Science and Technology*, 28(4):657–662.

Procedure to enhance temperature calibration analysis in ventilation models developed with Ventsim

D Sepúlveda[1] and R Ugas[2]

1. Senior Ventilation Engineer, Palaris, Wollongong NSW 2500.
   Email: dsepulveda@palaris.com.au
2. Mining Ventilation Engineer, Howden, Santiago 8320000, Chile.
   Email: rodrigo.ugas@howden.com

ABSTRACT

The use of computational ventilation models is crucial for underground mines and tunnels construction and operation, allowing for the simulation of airflow, heat distribution, gas movement, and fire propagation within ventilation systems. Of particular significance is heat simulation, an advanced modelling technique that has become increasingly vital in underground mining scenarios characterised by deepening excavations and escalating production rates. Accurate characterisation of heat loads is essential for ensuring compliance with safety regulations and guiding the design of effective cooling infrastructure and ventilation strategies.

Setting up a heat model in Ventsim requires defining conditions such as surface temperature, the thermodynamic properties of the rock, and the characteristics of the equipment in the mine. The challenge lies in the fact that, except for the rock properties, these parameters are variable and significantly influence the temperature results provided by the software.

The variability of these parameters and their impact on the software output makes it challenging to verify the calibration accuracy of the ventilation model. This discrepancy between the software settings and the actual measurements necessitates adjusting the parameters corresponding to the measurement moment for an accurate comparison.

This study describes three case studies that assess the calibration accuracy of a ventilation model created in Ventsim by comparing it with static setting and dynamic setting parameters corresponding to the measurement time. The modified parameters include dry bulb and wet bulb temperatures on the surface and the operating state of the longwall machine. These parameters correspond to the data collection moment and were modified by using the static script tool in Ventsim.

The results of this methodology improved the calibration accuracy of the heat model by up to 13 per cent, demonstrating its relevance when analysing ventilation and heat models calibration.

BACKGROUND

Ventilation modelling

A mine ventilation model is a mathematical representation of the ventilation system used in an underground mine or tunnel. These models are used to predict and optimise airflow and pressure distribution within the mine to ensure a safe and healthy environment for workers, comply with legal requirements and to improve the efficiency of mineral extraction.

Mine ventilation models typically include variables such as mine geometry, air intake and exhaust locations, ventilation equipment characteristics (such as fans and ductwork), external weather conditions, and sources of heat and contaminants within the mine.

By simulating airflow with these models, engineers can predict how airflow will be distributed and how contaminants and hazardous gases, such as methane, carbon dioxide and DPM, will be diluted. This is critical to ensure worker safety and complying with health and safety regulations in underground mining operations. In addition, ventilation models can help optimise ventilation system design and minimise energy costs associated with the circulation of air.

State-of-the-art

Airflow model calibration

Calibration of an airflow model is a process by which the parameters and characteristics of the model are adjusted to accurately reflect the actual airflow behaviour in the system. This process is critical to ensure that the model predictions are as accurate as possible, and that the ventilation system designed from these predictions will operate effectively.

Calibration involves comparing the model predictions with actual airflow data and/or pressures (obtained from measurements taken at the mine site or tunnel). By comparing model predictions with observed data, discrepancies can be identified, and model parameters can be adjusted to make the model more accurate. Typical used indices are:

- Mean absolute percentage error.
- Mean percentage error.
- Root mean square error.
- Percentage error.
- Weighted percentage error.

Parameters that are typically adjusted during calibration include gallery resistance, ventilation control devices status and main ventilation infrastructure performance.

Calibration is an iterative process that may require multiple adjustments and tests until the model produces results that adequately match the observed data. Once calibrated, the model can be used with greater confidence to predict airflow behaviour under various conditions and to optimise the design and operation of the mine ventilation system.

Heat modelling

Why heat modelling?

Heat modelling in underground mines is critical for several reasons. These simulations are necessary to predict and understand how heat is distributed within the mine, which is critical to ensure worker safety and the efficiency of the ventilation circuit.

By performing heat simulations, it is possible to identify areas of extreme temperature that could pose a risk to the health of workers or the operation of machinery. In addition, these simulations allow the design of adequate ventilation and cooling systems to maintain safe and comfortable working conditions in the mine.

In summary, thermal modelling in underground mines is essential to prevent accidents, optimise productivity, and maintain a safe and healthy working environment for all personnel involved in an underground mining operation.

Heat calibration – Standard procedure

The typical modelling procedure contains at least the following steps:

1. Underground surveys: collection of information specifically on air velocity, wet bulb (WB), dry bulb (DB), relative humidity (RH) and others such as wet bulb globe temperature (WBGT), effective temperature, etc.

2. Set-up the rock parameters:

    o rock density

    o rock specific heat

    o rock thermal conductivity

    o rock age.

3. Set-up surface temperature values.

4. Addition of heat sources to the ventilation/heat model, the main heat sources are:

   ○ fixed heat sources

   ○ mobile heat sources.

5. Data comparison and model validation: As with airflow calibration, several indices are calculated to determine the accuracy of the heat model, the most common being %error.

Figure 1 shows a diagram of the heat model analysis process.

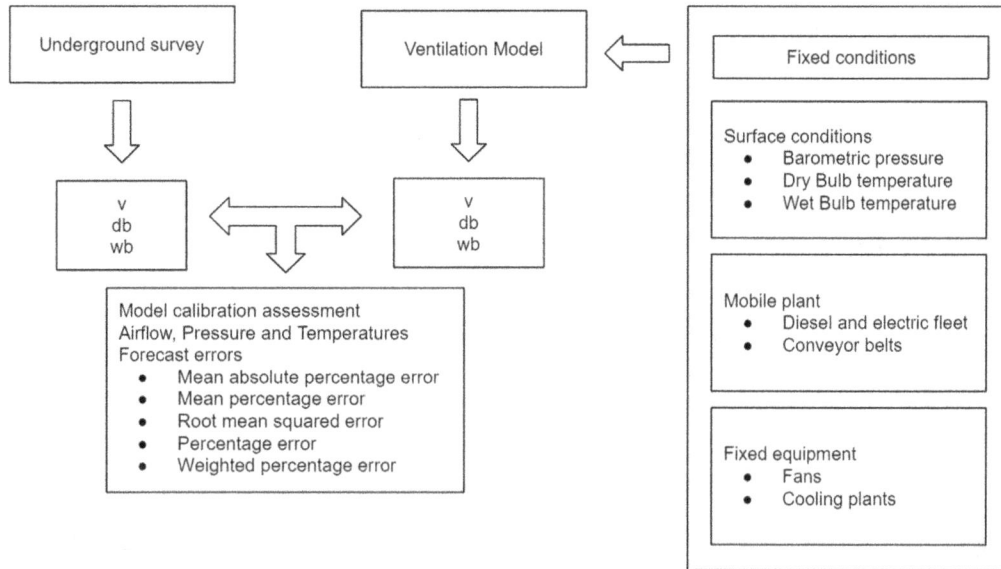

FIG 1 – Heat Model calibration analysis – Standard procedure.

Discussion

The construction of heat models has several points of discussion when comparing measured and modelled data:

- Measurements are usually taken at different times and therefore the environmental conditions (eg surface temperature) vary.

- Heat sources are often added statically and do not account for variations in operation such as number of trucks on the intake roads, conveyor performance, etc.

- Variable surface temperatures affect the temperatures readings underground, especially in mines that are highly influenced by surface conditions.

- Satisfactory statistical indicators may suggest a risk of overfitting, which is critical to the sizing of cooling systems as it can lead to over- or under-estimation of cooling requirements, while unsatisfactory results can often be attributed to modelling with static heat sources.

- There are well defined procedures to validate airflow surveys, considering the variation of barometric pressure (Gyamfi, Halim and Martikainen, 2021; Rowland, 2012). But there are not clear procedures on how to validate temperature surveys with variable heat sources.

- Setting the conditions for each survey is a complex and time-consuming process because it requires configuring as many scenarios as there are surveys in the mine.

PROPOSED PROCEDURE

Based on the points mentioned above, a procedure with the following steps is proposed:

1. Set-up the rock parameters:

   ○ rock density

- o rock specific heat
- o rock thermal conductivity
- o rock age.

2. Surface environmental conditions recording: temperature, humidity, and barometric pressure should be recorded during the mine survey.

3. Mine site measurement: collect information specifically on air speed, WB, DB, and others such as WBGT, effective temperature, etc.

4. Adjustment of ventilation circuit conditions: verify and adjust the conditions of the main ventilation circuit for the periods in which the measurements were taken, the main points of adjustment are:

- o velocity of main fans
- o resistance (regulators, doors etc).

5. Addition of fixed heat sources: add all heat sources whose variation is zero or whose variation minimally affects the environmental conditions, including:

- o water infiltration
- o minor electrical equipment (S/E, pumping stations etc)
- o electrical reticulation.

6. Incorporating variable heat and cooling sources: after identifying the 'variable' sources within the system, it is crucial to analyse these sources in advance to enable their adjustment for each measurement. Table 1 provides some general insights on these variables. At this point, it is highly recommended to explore tools which automatically adjust the conditions of the heat/cooling 'variable' sources. In the case of Ventsim™ DESIGN, the static script feature allows to modify these conditions across multiple roadways simultaneously.

7. Consideration of Natural Ventilation Pressure: this will improve model precision as heat sources are accurately modelled.

8. Airflow Model validation: assess the calibration degree of the airflow model. If the discrepancy is satisfactory, proceed to evaluate the heat model calibration.

9. Heat Model validation: compare the measurements with the modelled data after applying this procedure and calculate the heat model calibration by using the different statistical indicators.

TABLE 1

Variable Heat Sources considerations.

Variables	Comments	Information source
Surface temperature	The surface temperature must be determined/estimated at the time of each measurement inside the mine.	• Environment Department • Weather stations
Diesel equipment	Equipment used at the time, especially hauling equipment.	• Operations • Dispatch
Electric equipment	Power consumed by electrical equipment at the time of measurement, eg conveyor belts in intake circuits.	• Maintenance and or electrical Department
Refrigeration plant	Cooling power or air outlet temperature	• Maintenance • Plant operations

Figure 2 shows a diagram of the proposed procedure.

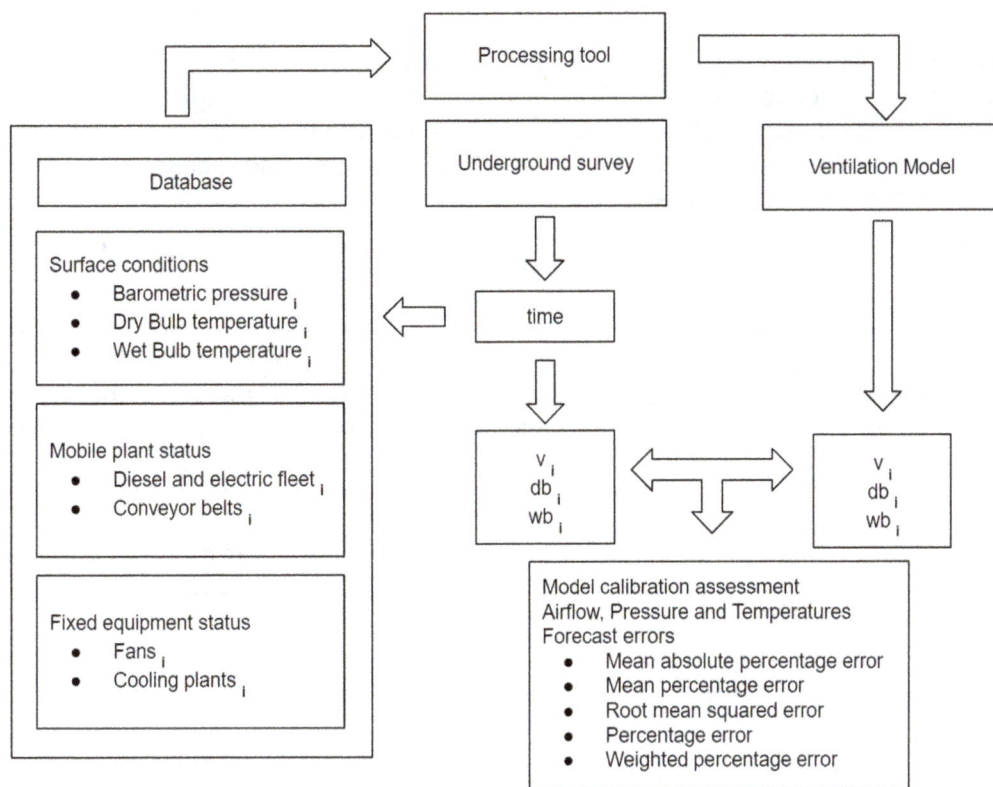

FIG 2 – Proposed procedure.

CASE STUDY 1

Case study 1 – surface temperature adjustment – gold mine

This case corresponds to an analysis of the temperatures measured in an underground gold mine located in Minas Gerais, Brazil during a monthly ventilation survey where 19 measured values are compared with the values provided by the ventilation/heat model in Ventsim. The base case corresponds to the results provided by the Ventsim model with a surface temperature setting equal to the temperatures at the start of the survey.

Information – Surface temperature

The surface temperatures at different times during the ventilation survey are shown in Figure 3. Each point in the graph shows the time at which each measurement was taken, with the first point corresponding to the surface measurement.

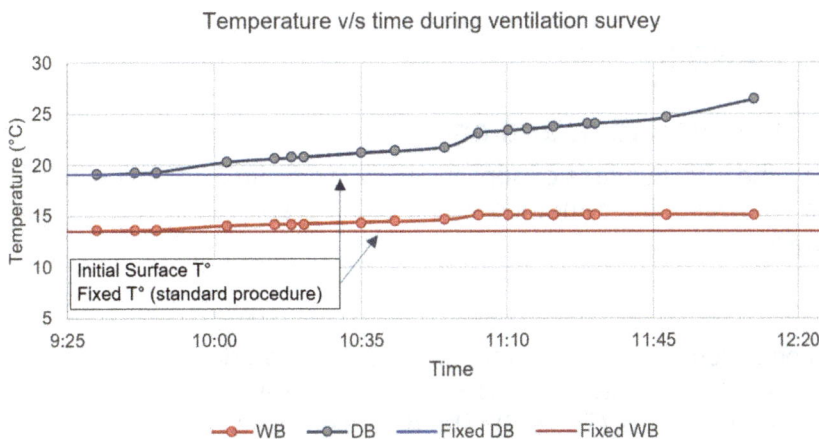

FIG 3 – Surface temperature during surveys. Case 1.

Results

Table 2 shows the calibration results, with the new procedure there is a 5 per cent and 11 per cent reduction in average error for WB and DB respectively. When analysing the behaviour between measured and modelled temperatures (Figure 4), it is important to note that although there are differences in the temperatures, the new procedure improves the accuracy of the model in determining the underground temperatures.

TABLE 2

Statistical indicators.

Procedure	Wet bulb temperature			Dry bulb temperature		
	Average % error	Mean square error	Standard deviation of residuals (°C)	Average % error	Mean square error	Standard deviation of residuals (°C)
Standard procedure	18%	20.9	4.9	20%	44.6	7.1
New procedure	13%	15.8	4.2	9%	10.0	3.4

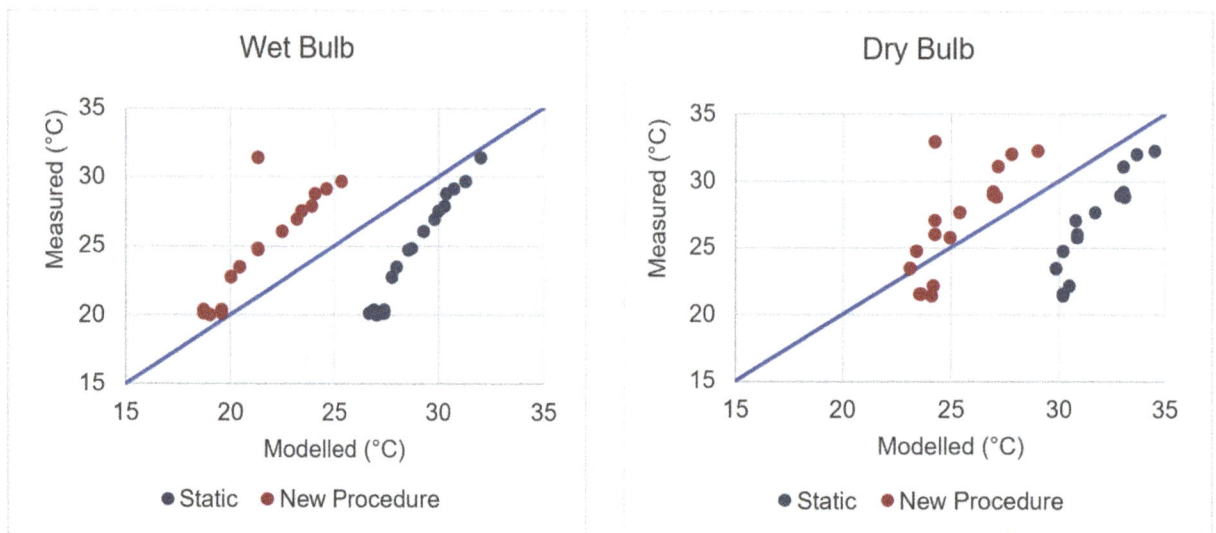

FIG 4 – Scatter Plot. Standard and New Procedure. Case 1.

This procedure reduces result variability by modifying one of the variables influencing the simulation, specifically surface temperature. The remaining discrepancy between measured and modelled data may be attributed to differences in the airflow model and the absence of mobile heat sources, considering that the modelled results are lower than the measured values.

CASE STUDIES 2 AND 3

Surface temperature variation and equipment heat on the LW Face

This case focuses on the effectiveness of applying this process in a coalmine located in Queensland, which employs the longwall mining method and produces six million tonnes per annum (Mt/a) of high-quality hard coking coal.

Case study 2

The present case compares underground measurements taken during a monthly ventilation survey with the values obtained from Ventsim, using surface temperature settings based on both average annual and summer temperatures (Table 3). The data were collected, and the proposed procedure was then applied to adjust the surface temperature according to the timing of these measurements.

TABLE 3

Temperatures for base case – standard procedure.

Variable	Wet bulb °C	Dry bulb °C
Annual average	18.0	23.1
Summer average	22.0	26.2

The ventilation model is first calibrated using airflow and differential pressure readings across the regulators to adjust their resistance. Airflow values are then obtained from the ventilation software and compared to the measured values; any discrepancies are investigated accordingly. It is verified that all adjustments have been made in line with the corresponding ventilation change procedure.

There is not a high traffic of equipment going in and out of the mine, so the heat load of these equipment is negligible, with little impact on the temperature readings.

There is a large heat load focused on the Longwall face. This heat load is composed of Goaf Stream at the VRT temperature, freshly cut rock, and the temperature generated by the shearer friction when cutting coal. This heat source and its impact is analysed on Case Study 3.

Results

The calibration results are presented in Table 4. With the new procedure, there is a 13.3 per cent reduction in WB temperature error and a 12.3 per cent reduction in DB temperature error. In this case, the percentage error in WB temperature closely mirrors the percentage error in airflow.

TABLE 4

Statistical Indicators – Case 2.

Procedure	Airflow		Wet bulb temperature	Dry bulb temperature
	Average % error	Weighted avg % error	Average % error	Average % error
Standard annual average	8.0%	6.0%	21.1%	22.7%
Standard summer average	8.0%	6.0%	10.1%	14.4%
New procedure	9.0%	7.0%	7.8%	10.4%

Analysing the data, airflow measurement/software error is practically the same for the three modelled temperature conditions, therefore, the surface temperature is not the cause of this disparity, but it is due to other variables, such as the modelled resistances, leakage, measurement errors of the device, theoretical/actual fan curve etc.

Case study 3

Case 3 evaluates the effectiveness of the application of the proposed procedure to the temperature in the longwall face, by evaluating the different temperatures results in the chock 168 close to the tail gate. Contrary to the previous case, this temperature is highly dependent on the heat generated on the face, which has a proportional relationship to the production in the shift where the measurements are taken as well as production on previous shifts.

In this case, the values obtained by varying the surface conditions as well as the variation of the heat generated in the longwall face are compared. As in the previous cases, the surface temperature values are correlated with the time at which the measurements are taken.

For the variation of the heat generated on the face, a statistical correlation between production and sensible heat and humidity is used. This relationship is obtained from the statistical analysis of measurements taken by the deputies during the shift. Therefore, in this case study it is verified how the use of the proposed procedure allows to check the degree of calibration of the model for different production levels and to predict the temperatures for the desired case.

The mathematical relationships used are:

- Sensible Heat (kW/m)   = (0.0238 × Production (tpd) + 1974.7) / 354.4 m
- Moisture (g/s/m)   = (0.0077 × Production (tpd) + 632.8) / 354.4 m

For the application of these inputs, the estimated production value per shift was correlated with the measured data and the estimated sensible heat and moisture values were divided by 354.4 m (longwall face length) to make them linear and apply them to each section of the face (Figure 5), while also modifying the surface data.

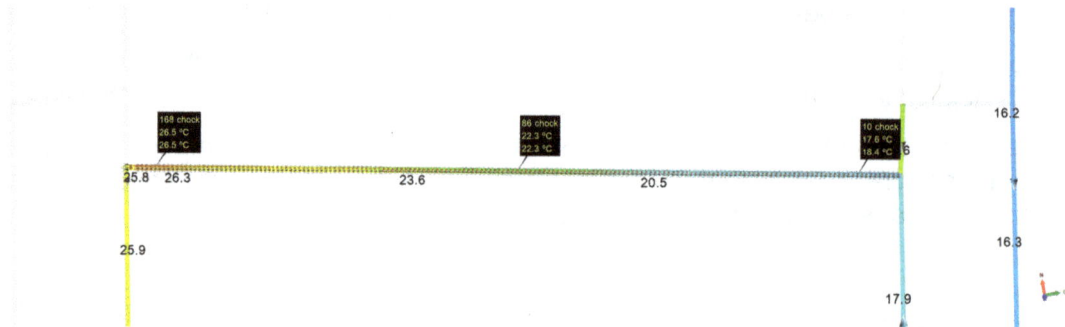

FIG 5 – WB Temperatures on the Longwall Face divided by number of chocks (173); Ventsim 6.0.

Results

The results are shown in Table 5. As in cases 1 and 2, there was a significant reduction in the WB temperature error, while the DB temperature remained stable. The WB results obtained from the heat simulation are more reliable than the DB results due to the nature of the software. Therefore, when assessing the model's calibration, users should focus more on the WB variable.

In this case, it is demonstrated that modifying surface temperatures has a lesser impact than adjusting the heat generated at the longwall face. This is because 35 per cent of the ventilation circuit comprises cooled air with a WB set point of 8°C, which reduces the surface temperatures effect on the air entering the mine.

TABLE 5

Statistical indicators – Case 3.

Procedure	Wet bulb temperature			Dry bulb temperature		
	Average % error	Mean square error	Standard deviation of residuals (°C)	Average % error	Mean square error	Standard deviation of residuals (°C)
Base Case	11.0	10.2	3.3	6.5	4.1	2.1
Proposed – Modified surface	10.0	8.5	3.0	5.6	3.2	1.8
Proposed – Modified longwall	6.6	4.5	2.2	8.2	7.5	2.8
Proposed – Both Modified	5.7	3.9	2.0	7.1	6.2	2.6

The second reason why the sole modification of surface temperature has a lesser impact is due to the distance between the 168 chock and the intake sources, causing temperatures to attenuate as the air travels through the mine to reach the end of the longwall face due to heat exchange with the rock.

On the other hand, the modification of heat generated by the longwall face has a significant impact on the WB temperature error percentages, reaffirming the idea that in certain cases, a specific analysis of point/linear heat sources should be carried out to understand the calibration level of the model.

For example, in the analysis of case 2, the heat generated by the longwall face was irrelevant for comparing measurements since this specific heat source does not affect the measurements being compared.

CONCLUSIONS

- The proposed procedure, which considers the variation of surface temperatures and variable heat sources, improves the value in various statistical indicators, showing enhancements in the accuracy of the heat model.

- It is important to analyse all critical heat sources and their variability to focus the process on these variables (eg diesel truck, longwall machine, conveyor belts). The heat balance is relevant to identify them in first place.

- This type of analysis aims at improving the validity, the statistical indicators, and the robustness of heat models to obtain more accurate projections and sizing of cooling and ventilation infrastructure.

- Ventilation engineers who are unaware of the importance of variability in their model may erroneously conclude that their model is not calibrated, as the difference between modelled and measured temperatures could be significant.

- The programming and automation of scenarios to be evaluated are key, especially in large mines with numerous survey points. In this case Ventsim DESIGN's Static Script tool helps streamline the process.

- The improvement and corroboration of the heat models under this procedure allows improving the behaviour and prediction of natural ventilation pressure.

- It is important to consider that temperature is highly dependent on airflow, therefore, before analysing the degree of calibration of the heat model of a ventilation network, it is relevant to analyse the degree of calibration in airflow. A high degree of temperature calibration, with a low degree of airflow calibration, is not indicative of a calibrated model, since when calibrating the resistances or other parameters that affect the airflow distribution, it is highly probable that the temperature distribution in the ventilation network will change.

- It is crucial to acknowledge the heat absorbed or contributed by the rock, which frequently offsets the variability of other heat sources. With its consistent temperature, the rock serves as a buffer against fluctuations in heat sources. Consequently, in scenarios with significant air-gallery contact, the airflow's temperature and density remain stabilised.

- In metallic mines, when ore and waste rock are transported within the mine's intake circuit, the continuous movement of equipment significantly influences temperature measurements. This alteration necessitates a thorough characterisation of the transportation fleet to enhance the comprehension of the ventilation and heat model.

- Likewise, in massive production methods like panel or block caving, it is essential to accurately characterise the fleet comprising the material handling system. This step is crucial for conducting a precise analysis of the temperatures modelled and measured within the mine, whether involving trucks on a transport level, LHDs on a production level, conveyor belts, railroads, and so forth.

- The user must understand the limitations of Ventsim software for modelling transient phenomena, which should be taken into account when analysing results.

- This method presents an improvement opportunity by adjusting the surface temperature considering the time it takes for the air to reach the measurement point. For example, if measuring 3.6 km from the nearest injection point, with an air velocity of 4 m/s, the air being measured will have entered the mine 15 mins earlier. Thus, the script application should reflect this, which becomes more complex when there is more than one injection source. In such cases, different temperatures should be set at injection points based on the measurement time and air velocity.

- This work only addresses parameter modification for a static simulation, but another tool can provide even more accurate results: applying the same procedure to a dynamic simulation. This type of simulation considers the effects of parameter modification over time in the ventilation model and better represents the buffer effect of the rock, whether absorbing or releasing heat.

REFERENCES

Gyamfi, S, Halim, A and Martikainen, A, 2021. Calibration of LKAB's Konsuln test mine ventilation model using Pressure-Quantity (PQ) survey, in *Proceedings of the 18th North American Mine Ventilation Symposium*, 11 p. https://doi.org/10.1201/9781003188476-3

Rowland, J, 2012. Barometric resistance surveys: 'A new perspective', in 14th United States/North American Mine Ventilation Symposium.

Assessment of using jet fans for ventilation-on-demand in block cave production drives through CFD modelling

J Viljoen[1]

1. Engineering Manager, BBE Consulting Australasia, Joondalup WA 6027.
   Email: joeline.viljoen@bbegroup.com.au

ABSTRACT

Ventilation management is a critical aspect of block cave mining operations, especially when striving to optimise resource usage and maintain efficient airflow in multiple parallel drives on the extraction level. Ventilation-on-demand is a cost-effective strategy, directing full airflow to active production drives while providing reduced airflow to inactive ones. Ensuring a well-balanced airflow to each active production drive is a crucial design consideration in block cave mining operations. Poor airflow distribution can lead to sub-standard conditions, compromising both safety and operational efficiency at the extraction level. The conventional use of ventilation doors with flow regulators or low-pressure fans has limitations, including hindrances to mobile equipment, susceptibility to damage, and temporary flow imbalances during door operation.

This paper introduces a novel approach by suggesting the use of jet fans as an alternative to traditional ventilation doors for controlling, distributing, and balancing airflow in parallel block cave production drives. The study utilises steady-state computational fluid dynamics (CFD) modelling as a powerful tool to simulate and evaluate the proposed jet fan approach to assess whether employing jet fans can effectively achieve the desired airflow balances in parallel production drives.

Preliminary results from the study are promising, indicating that jet fans can successfully overcome moderate pressure differences and improve airflow in individual production drives. Additionally, the findings suggest that the adjustment of jet fan speeds could offer a practical method for controlling and regulating ventilation quantities in specific production drives. This paper delves into the opportunities and challenges associated with the utilisation of jet fans, examining their potential as a dynamic and efficient alternative to traditional ventilation doors in the context of block cave extraction levels. The exploration of jet fans as a ventilation strategy opens possibilities for enhancing airflow control and distribution in block cave mining scenarios.

INTRODUCTION

There are various methods of ventilating the production drives in the extraction level of a block cave mine. The ventilation strategy depends on the mining layout, tip and crusher arrangements, the length of the production drives, and the primary ventilation infrastructure design.

Common mining layouts for the extraction level are the offset Herringbone layout Figure 1a and the El Teniente layout Figure 1b (Gómez *et al*, 2020; Chitombo, 2010). In the Herringbone layout, the load-haul-dump (LHD) unit is not able to back into adjacent drifts and can, therefore, only load ore in one direction. In the El Teniente layout, the LHD can back into the opposite drawpoint and turn around. For this reason, ore can be loaded in both directions. The findings of this paper can be applied to both of these mining layouts.

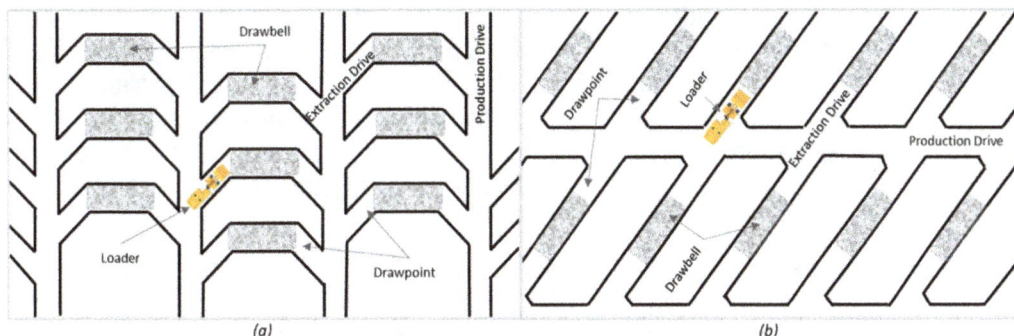

FIG 1 – Extraction level mining layouts.

The movement of the LHDs and the location of the tips and crushers will determine the optimum direction of air movement through the production drives. Two common extraction level ventilation layouts are shown in Figure 2. Due to dust formation at the tips, it is good practice to ventilate the tips to return from the extraction level. Fresh air is supplied from the primary intake infrastructure, through the production drives and out to return through the tips.

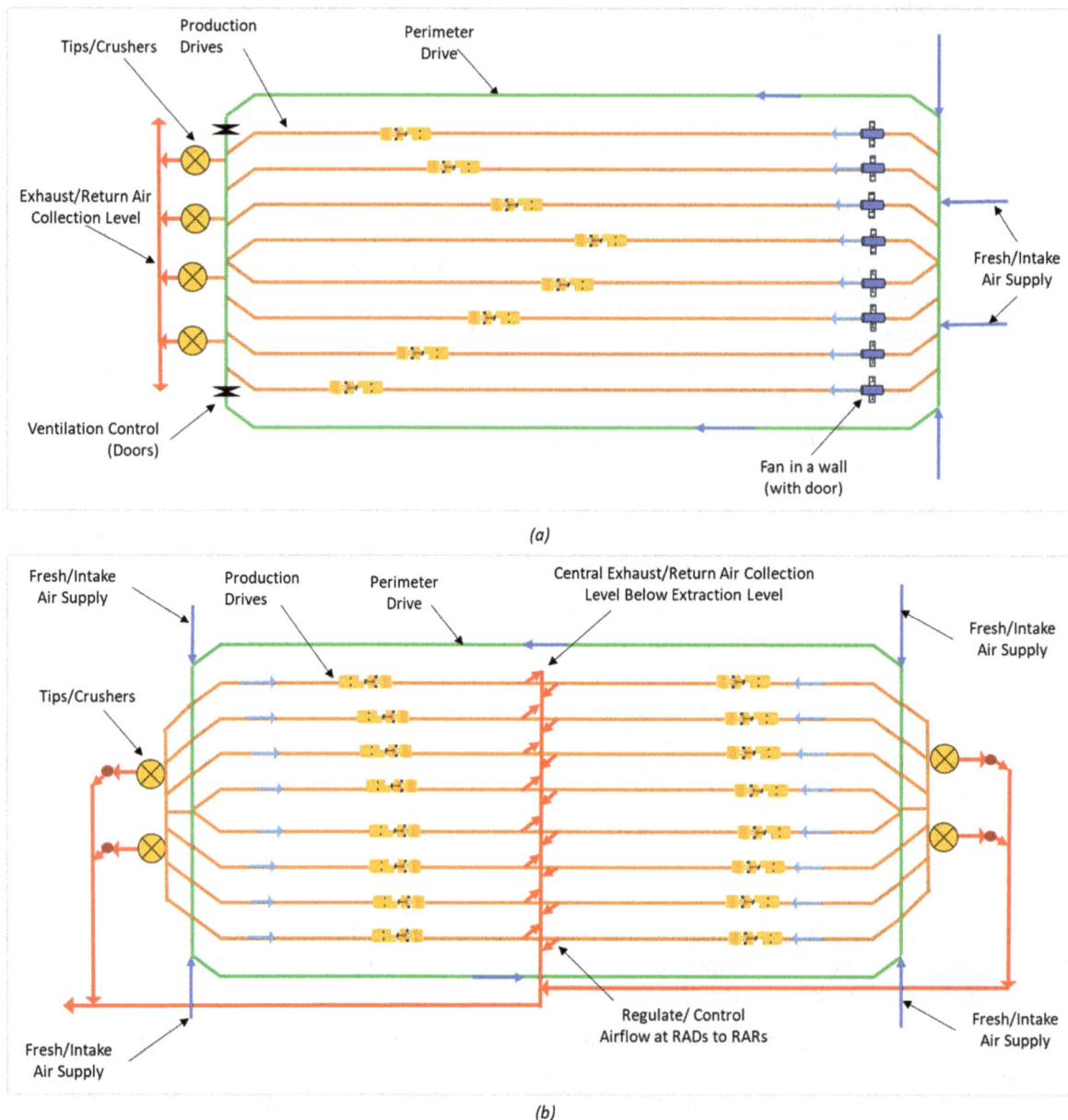

(a)

(b)

FIG 2 – Extraction level ventilation layouts.

The quantity of air required in a production drive will depend on the contaminant removal needs, be it diesel exhaust or heat dilution, radon dilution or minimum air velocity requirements. Active drives will generally require the same quantity of air, but the airflow distribution will vary because of different airway resistances across the production drives. The drives with the lowest resistance will receive higher flow rates, and thus, airflow control is required to achieve an equal distribution across all drives. This is typically achieved by installing a bulkhead wall in the drive with an auxiliary fan and access door, refer to Figure 2a. The auxiliary fans will deliver controlled air quantities to each of the drives.

A second example of an extraction level is shown in Figure 2b. In this case, the production drives might be long, increasing the tramming time and distance for the LHDs. In such a case, the drives can be divided so that they are serviced by two LHDs, one on the east and one on the west, with tips and crushers on both sides of the extraction level. Such a configuration will require fresh air to enter from both ends of each production drive, and a central return/exhaust air system will be required. To ensure that the air distribution through the drives is equal, auxiliary fans or regulators can control the flow in the drive at the return air drive (RAD) connections to the central return air raises (RARs). This method does not require bulkhead walls with doors and fans, allowing free movement of LHDs in production drives. However, this method requires significant ventilation infrastructure to be developed, including a central ventilation collector level.

The conventional methods to control and balance the ventilation flow-through a wall with a door and fan for each production drive can delay the mobile equipment and increase tramming time if the equipment is required to travel through the doors. The doors are also susceptible to damage, and the opening of the ventilation doors results in a temporary flow imbalance.

As an alternative to ventilation doors, this paper explores the potential use of jet fans to control, distribute, and balance the airflow in parallel block cave production drives, utilising computational fluid dynamic (CFD) simulations.

Jet fan operating principles and types

Jet fans are typically freestanding units that use the ejector effect to induce an airflow and pressure change in the surrounding environment (Wolski, 1997). Jet fans are rated in terms of the thrust applied to the surrounding air, with an equal and opposite reaction force borne by the fan supports. The fundamental thrust rating is determined by the momentum flux at the fan outlet, calculated as the product of the mass flow and the average velocity. An ideal jet fan will maximise thrust while minimising the electrical input required.

Jet fans are usually installed on the backs (roof or hanging wall), ensuring they are positioned at an optimal distance not less than one fan diameter from the backs (roof or hanging wall) to avoid a loss of thrust.

Several jet fan types are shown in Figure 3. Skid-mounted jet fans, as shown in Figure 3a, have outlet nozzles that have been optimised to provide a long and focused air stream. These fans are deployed in mines to ventilate development headings, but this is not the subject of the present paper. Jet fans optimised for road and mine tunnels are shown in Figure 3b and 3d and are the subject of this paper. These fans are generally bi-directional, but fan reversibility is not typically exploited in the mining environment. Ductless auxiliary mine fans, as shown in Figure 3c, can be deployed as jet fans, but their efficiency is generally unacceptable.

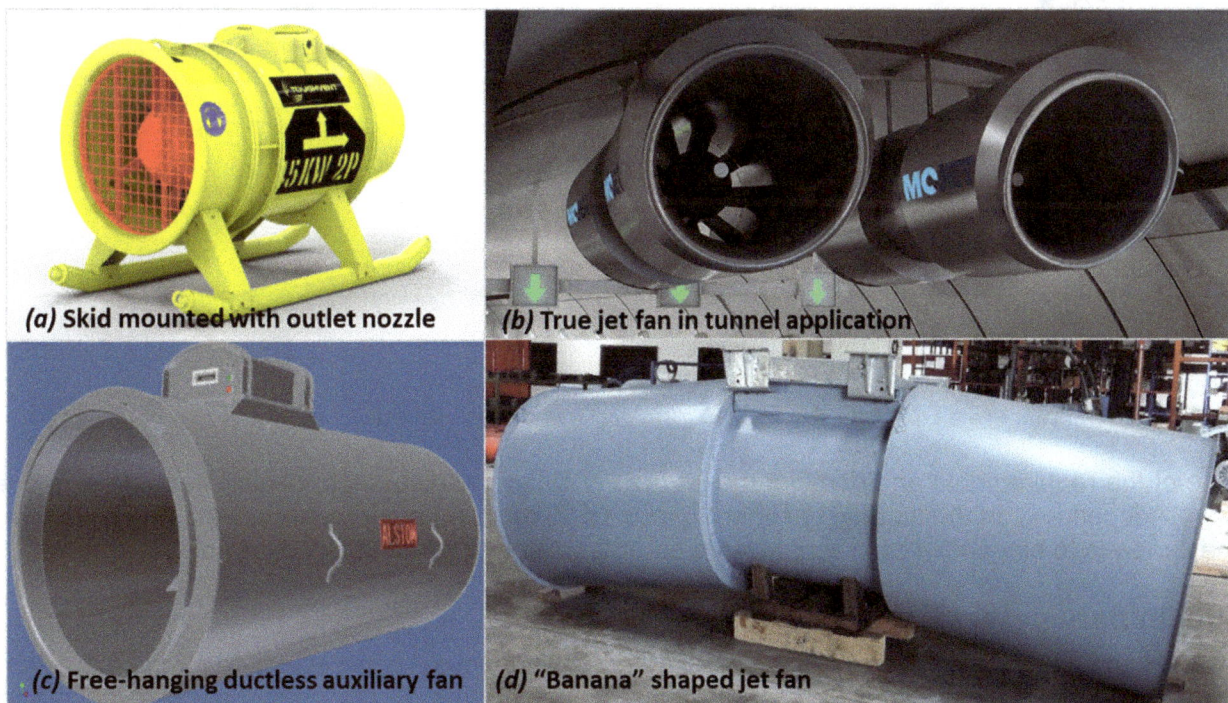

(a) Skid mounted with outlet nozzle   (b) True jet fan in tunnel application

(c) Free-hanging ductless auxiliary fan   (d) "Banana" shaped jet fan

FIG 3 – Various shapes and types of jet fans.

Jet fans applications

Jet fans generally find applications in large, enclosed spaces with minimal obstructions. Typical applications for jet fans include ventilation of road and mine tunnels, ventilation in underground car parks, ventilation of workshops in underground mines and ventilation of mine development headings (Goodman, Taylor and Thimons, 1992).

Jet fans are used to ventilate tunnels because they are easy to install and relocate and because they do not require the construction of stoppings/walls. However, jet fans are known to have low efficiencies and only provide modest pressures. Jet fans are typically used in mine areas with low aerodynamic resistances, such as bord-and-pillar sections.

One challenge associated with jet fans is the potential for significant local air recirculation within the confined air space. This risk must be addressed during design.

A literature search did not find any published case studies for block cave mines or any other mine sites that operate jet fans in the manner proposed in this study. Stewart (2023) has provided a concept for using jet fans in haulage ramps and block cave ventilation supported by Ventsim™ modelling that showed jet fans could be used to correct imbalances within a system. Pont, Michelin and Stewart (2024) expands on this, researching jets fans usage within Ventsim™ Control. Neither investigation used CFD modelling.

VENTILATION MODELLING

Proposed ventilation strategy

The extraction level layout is represented by the schematic given in Figure 4. Fresh air will be supplied from one side of the extraction level via a combination of fresh air raises (regulated or forced), ramps and perimeter drives. Air flows through the production drives and is exhausted to a return air collection level after passing through the tips and crushers on the other side of the level.

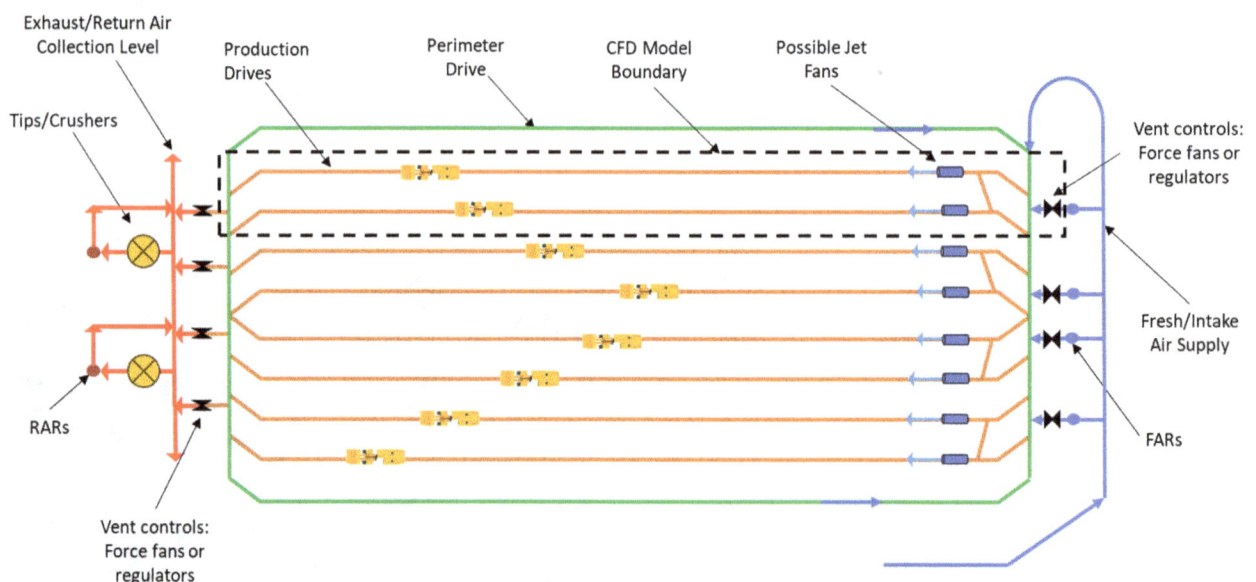

FIG 4 – Extraction level ventilation layout.

One of the requirements and goals of the jet fan ventilation strategy is to control the airflow in the drives, depending on the activity, to either 25 m³/s or 15 m³/s. As equipment cycles between drives, the airflow is to be controlled to supply more air or less air by changing the speed of the jet fans.

Preliminary network modelling

Before conducting the CFD analysis, a Ventsim™ ventilation model was developed for the extraction level to undertake a first-pass assessment of the concept, establish preliminary fan selections and test the ventilation strategy's sensitivity to any changes in the primary ventilation circuit. In addition, the ventilation model was used to determine the boundary conditions for the CFD model.

To understand the principle of jet fans as part of the ventilation philosophy, it is essential to know the 'natural flow distribution' of air to the production drives without any secondary fans. This is done by switching off all the auxiliary fans in the production drives, noting which drives will 'naturally' receive higher and lower flow rates, and how changes in the primary ventilation distribution and controls will affect the drives' natural distribution.

The extraction level is supplied with fresh air from a ventilation level via fresh air raises (FARs), which can either be regulated (given the system has sufficient pressure) or be forced with fans built into walls, as indicated in Figure 4.

On the return air side, the exhaust air can either be pushed to the return collection level or regulated. The fresh and return air distributions and changes in either can influence the 'natural' distribution between the drives.

For the case considered in this paper, it is observed that exhaust fans at the return air side of the extraction level ensure sufficient overall airflow through the production area and are responsible for overcoming a portion of the system's resistance. However, these fans have a limited effect on the distribution between individual drives.

To evaluate what effect the force fans (or regulators) on the intake side will have on the 'natural' distribution through the drives, several scenarios were tested: changes in airflow quantities, locations and distributions. The results show that by adding more frequently located force fans along the intake air side and adjusting the flow delivered by these fans, a better natural distribution can be achieved through the production drives. Thus, the force fans can be used to assist with the pre-distribution of the air.

With a better understanding of the system's behaviour and dependence on the exhaust and fresh air systems, the next component that requires more knowledge is how the jet fans behave when operating in parallel drives when a specific airflow distribution through the production drives is required. The ventilation model was used to evaluate the following:

- The potential use of 30 kW and 37 kW (1070 mm diameter) ductless mine auxiliary fans as 'jet' fans, see Figure **3**c.

- The use of 11 kW and 15 kW (600 mm diameter) true jet fans, see Figure **3**b.

- The effect of LHDs parked in the production drives.

A theoretical production drive airflow distribution was selected to test if the distribution can be achieved by adjusting fan speed only. This was achieved in a drive with the more powerful 37 kW mine auxiliary fan, whereas the 30 kW fans generated insufficient thrust.

For the production area, each individual 'jet' fan had to be modelled at a specific speed to achieve the required air quantity distribution. This level of dynamic auxiliary fan control could prove impractical and unstable.

With true jet fans, the 11 kW option was initially capable of providing the required flow distribution through the drives. However, 15 kW units were required upon including the LHDs. The ventilation simulation confirmed that LHDs parked in the production drives could significantly affect the airflow, with about 6 Pa needed to overcome their influence. Although low in mine pressure loss terms, it indicates that the jet fans are susceptible to minor pressure variations when considering the flows required. The piston effect from the LHDs will further complicate the control of the jet fans.

Network modelling results/CFD inputs

The ventilation modelling has shown what size jet fan are and their behaviour. The flows and system pressure obtained from the ventilation models are the inputs into the CFD models.

Fan speed control (eg VSD) will be required to regulate the jet fan outlet velocity, and thus, jet fans must be installed with VSD-ready motors suitable for speed control. Active control of the airflow in each drive will require a flow monitoring system (eg an ultrasonic flow metre installed in the drive) that provides feedback control of the fan speed using a closed communication loop.

The following sections describe the CFD modelling process, model set-up, results, findings and interpretations, and recommendations.

CFD MODELLING

Computational Fluid Dynamic analysis is a complicated science that must be done by a skilled professional. A detailed general description of the numerical method (including turbulence modelling) is available, but not addressed here.

The extensive analysis presented here focuses on the systematic investigation and findings using steady-state simulations without detailing the computational mesh formation and solver parameters.

The analysis used Siemens' commercial CFD software package STARCCM+. To capture the dynamic behaviour of the airflow, Reynolds–Averaged Navier–Stokes equations for mass, momentum, and energy were solved on each discrete cell of the computational mesh. Turbulence was modelled using the k-omega turbulence model equations, and the boundary layer at the solid interfaces was modelled with logarithmic wall functions.

Model preparation and set-up

The CFD model considers two adjacent production drives, a perimeter drive on the supply/fresh air side with a fresh air drive (FAD) connection where the force fan is installed, and the exhaust side perimeter drive, as shown in Figure 5a.

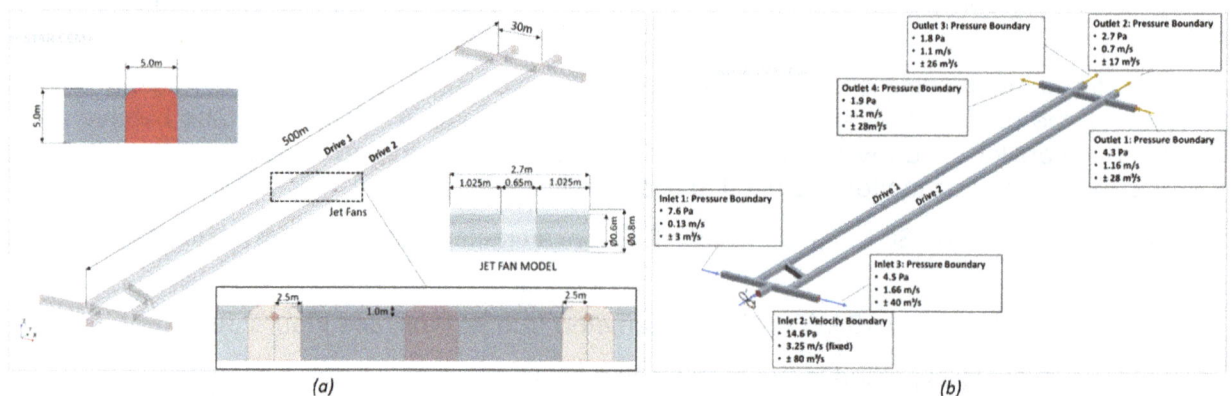

FIG 5 – CFD model set-up and boundary conditions.

Each drive has a jet fan set-up in the middle of the drive, with typical dimensions for a 15 kW, 600 mm jet fan. The jet fans are installed at one fan diameter from the backs (roof/hanging wall), as the literature recommends.

The model boundaries, as well as the conditions at each boundary, are given in Figure 5b. The only fixed boundary condition is the inlet velocity at Inlet 2, which represents the inlet side force fan. The 3.25 m/s at this boundary equates to approximately 80 m³/s. All other relative boundary pressures were set based on the ventilation simulation model's flow quantities/velocities when no jet fans are running (natural distribution).

The natural flow distribution through the two drives, with the jet fans off, is shown in Figure 6. This uneven distribution with Drive 1 at 28 m³/s and Drive 2 at 13 m³/s represents the base case solution.

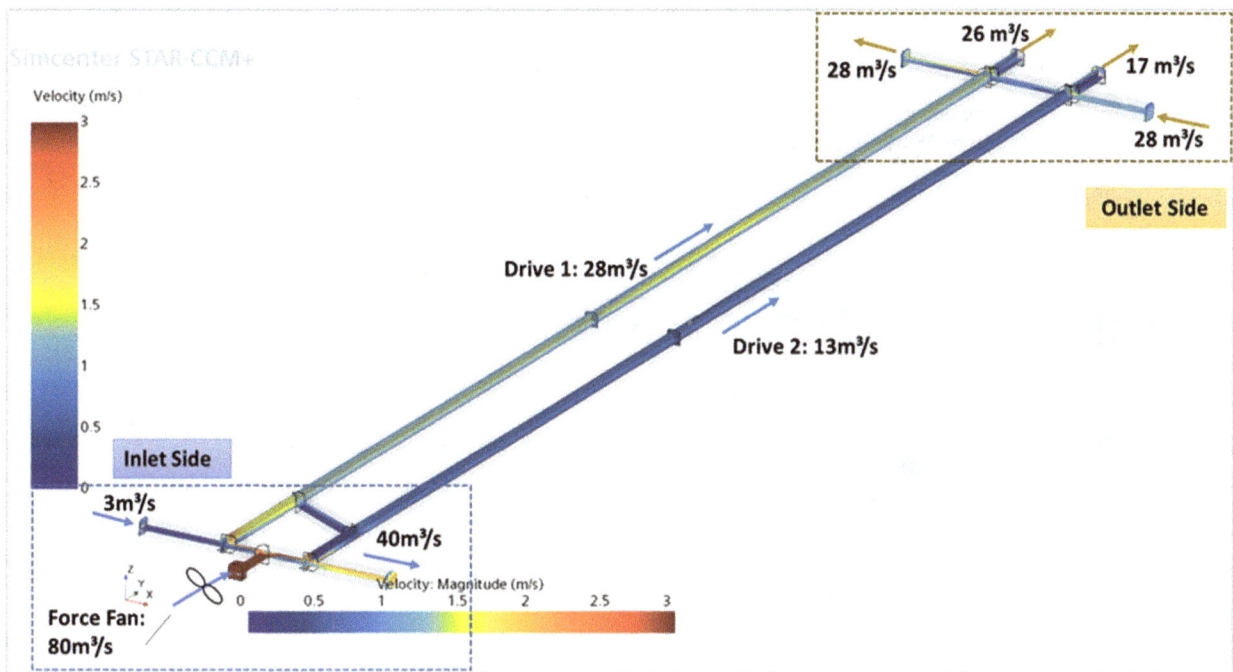

FIG 6 – CFD model with no jet fans operating (base case).

Increasing the fan outlet air velocity

The potential of the jet fans to control the airflow is summarised in TABLE 1 and Figure 7. The fan in Drive 2 is modelled at outlet air velocities of 10, 20 and 30 m/s, respectively, while the jet fan in Drive 1 remains off.

TABLE 1

Effect of increased jet fan outlet velocity.

Scenario	1 (base case)	2	3	4
Drive 1 – jet fan	Off	Off	Off	Off
Drive 2 – jet fan	Off	10 m/s	20 m/s	30 m/s
Drive 1 – air flow	28 m³/s	27 m³/s	26 m³/s	24 m³/s
Drive 2 – air flow	13 m³/s	17 m³/s	30 m³/s	44 m³/s

From the results presented, it is clear that the jet fan outlet air velocity can be used to control the airflow rate in the corresponding drive. However, the jet fan velocity increase in Drive 2 has a smaller disproportional and inverse effect on the airflow rate in adjacent and parallel Drive 1 despite their being connected adjacent to the FAD. As per Figure 7, the increased jet fan velocity in Drive 2 has a more significant effect on flows within the transverse perimeter drives.

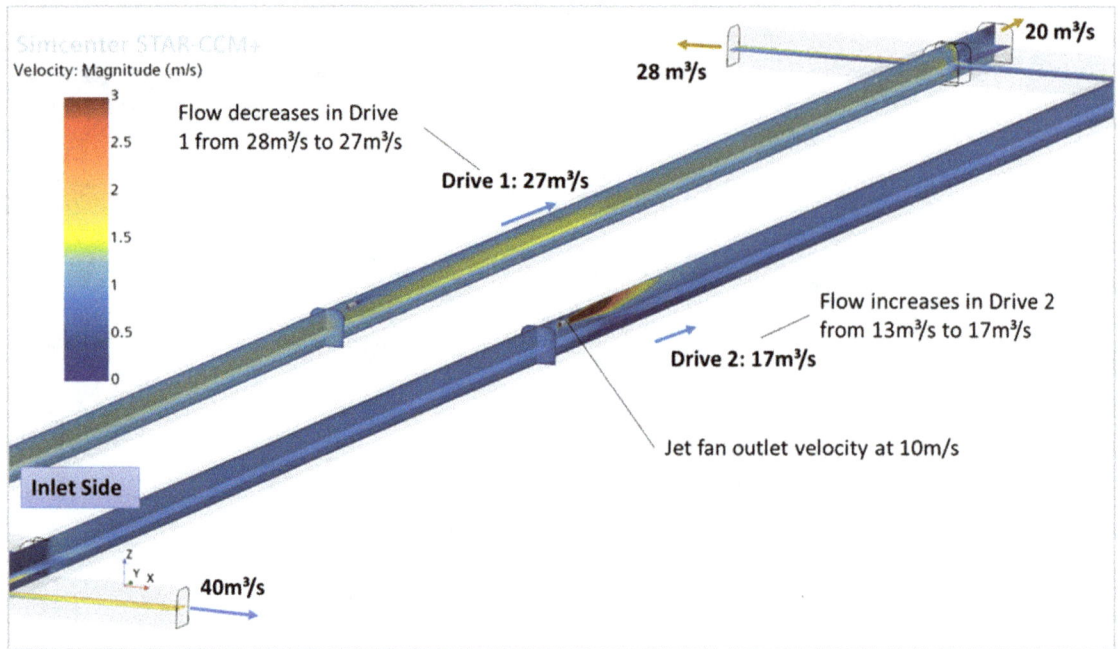

(a)

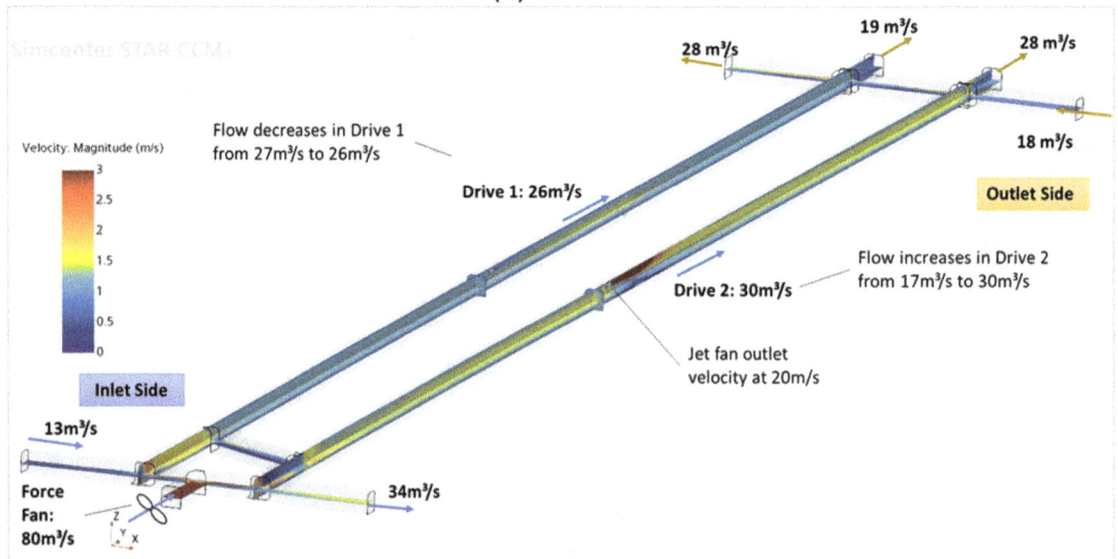

(b)

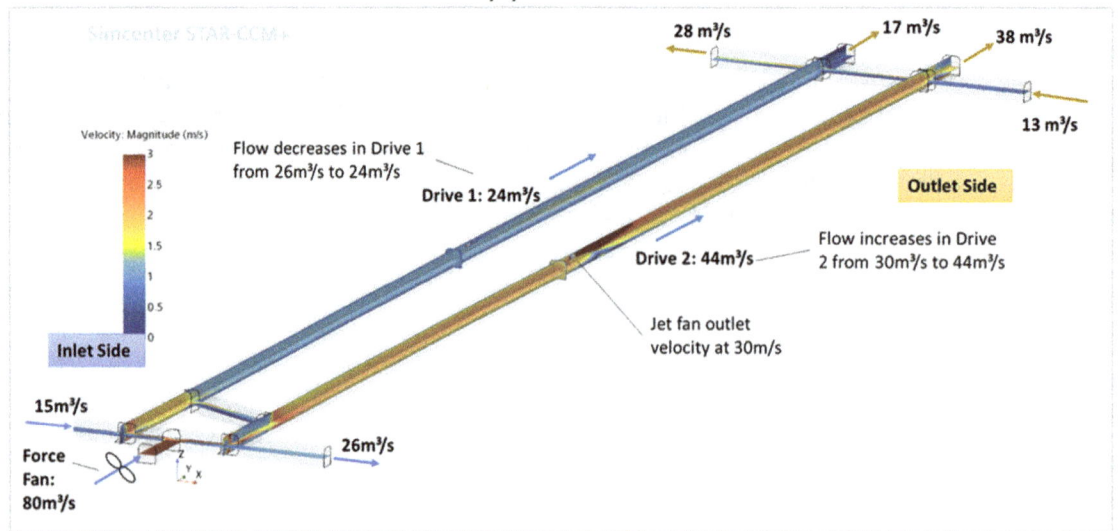

(c)

FIG 7 – CFD modelling results with a jet fan at 10 m/s, 20 m/s and 30 m/s.

Localised jet fan effects

The local effects of jet fan operation at 20 m/s are illustrated with reference to Figure 8a and 8b. The air velocity at the outlet of the jet fan disrupts the overall airflow in the drive and creates local areas of high air velocity, low air velocity and air recirculation. As can be expected, the level of disruption increases with an increase in jet fan outlet air velocity. Regardless, the disruption is local and a uniform air velocity profile is re-established within approximately 55 m.

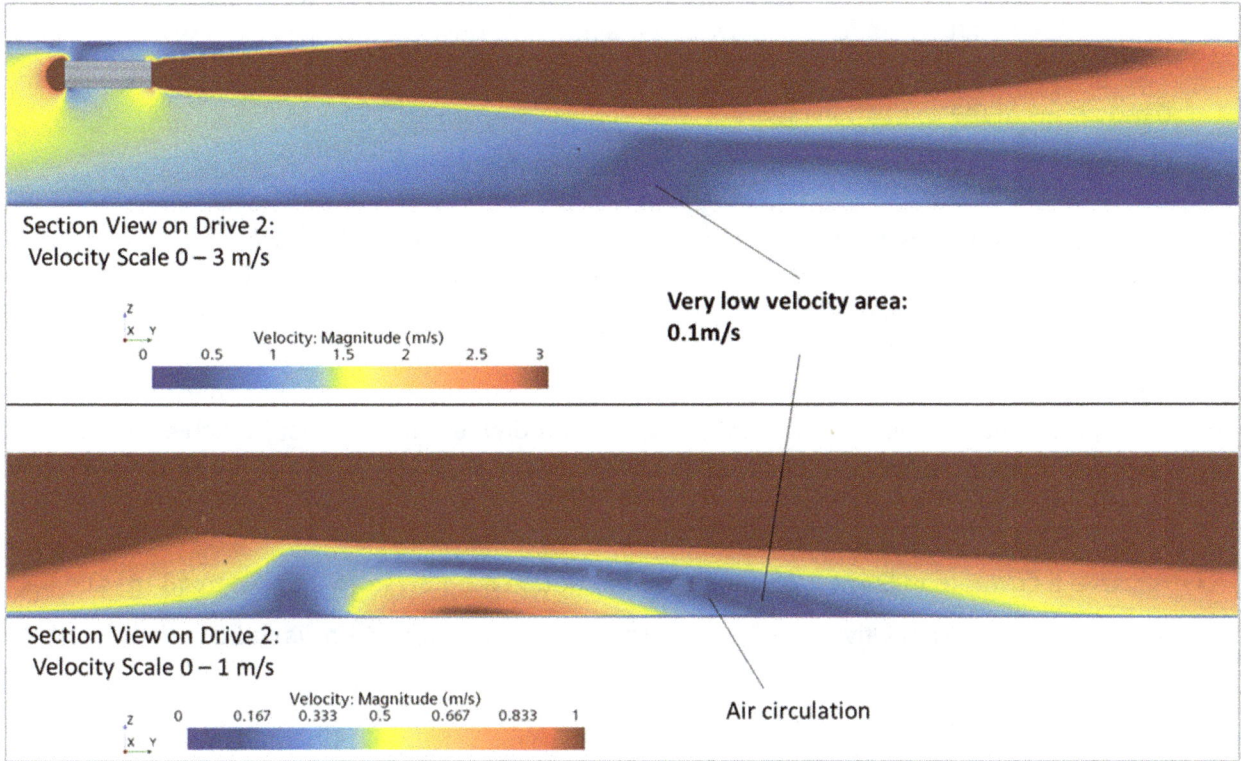

(a)

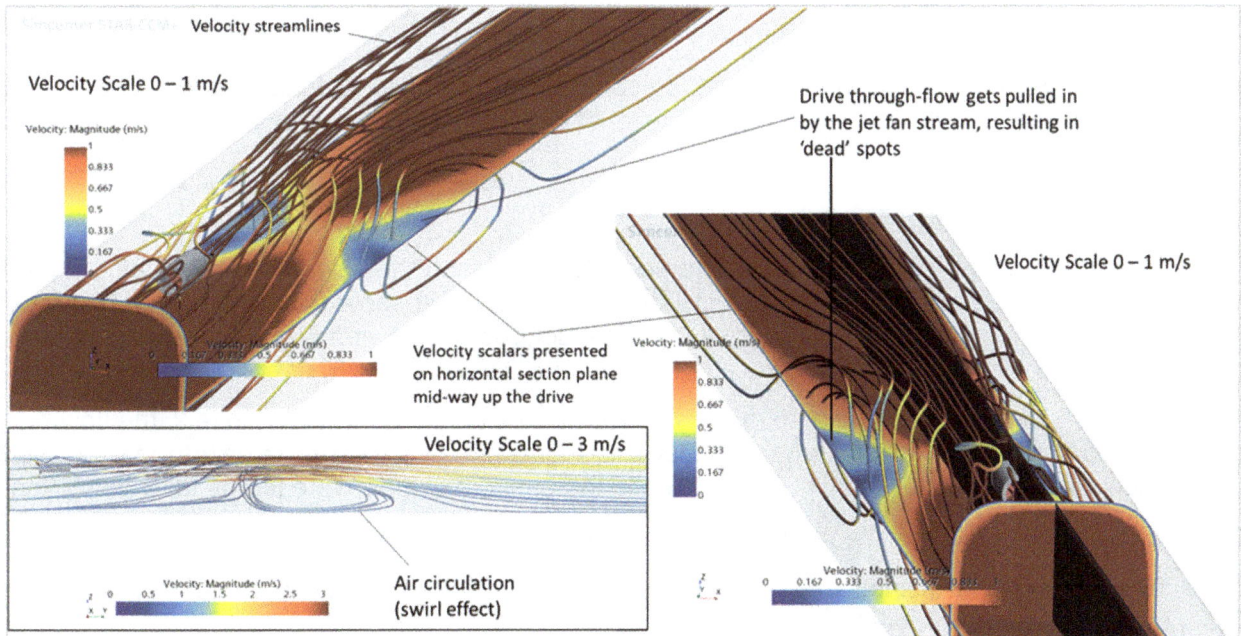

(b)

FIG 8 – CFD modelling results with a jet fan at 20 m/s.

However, there can be concerns. Localised high-velocity zones may promote dust entrainment, and air recirculation zones can also become a problem for dust (and contaminant) build-up, especially if these zones coincide with drawbell drive (extraction drive) locations, which are the main contaminant sources in block cave mines.

Measures to reduce the effect of this the low-velocity zone could include:

- Installing the jet fan at a slight downward orientation to throw the jet stream towards the centreline of the airway (to minimise the wall effect).

- Installing louvres on the fan discharge to achieve the same effect (to deflect the jet stream towards the centreline of the airway).

- Test different installation positions for the jet fans along the length of the drives to minimise the dead zones at the extraction drives.

- Operating a second back-up fan located elsewhere in the drive.

Regulating the airflow to achieve a specific airflow distribution

The CFD results have confirmed that increasing the jet fan outlet air velocity will increase the air flow in the associated drive. However, increasing the flow in one drive will not result in a proportional decrease in air flow in the adjacent parallel drives.

For a block cave operation, the air flow rate in all parallel production drives will need to be controlled. Considering the dynamic nature of the activity within the drives, this may not be achieved using jet fans alone. Consequently, it is suggested that a combination of forced fan control and jet fan control could be used to distribute the air flow in all production drives appropriately. The following scenario was used to investigate this combined control:

- Achieve a flow rate in Drive 1 of 15 m^3/s (thus reduce the flow from 24 m^3/s to 15 m^3/s).

- Achieve a flow rate in Drive 2 of 25 m^3/s (reduce the flow from 44 m^3/s to 25 m^3/s with jet fan operating).

The required air flows were achieved by:

- Reducing the intake forced fan flow from 80 m^3/s to 40 m^3/s. This change could be achieved with fans or regulators.

- Adjusting the jet fan air outlet velocity in Drive 2 to 18 m/s.

Although a combination of forced fan and jet fan control could be used to achieve a required air flow in all parallel production drives. Achieving this level of control in practice across multiple drives may be challenging.

Impact on airflow with LHD in drive

In addition to evaluating the proactive requirements for regulating the flow, the reactive effect of an LHD in the drive is also assessed. The LHD in Drive 2 is represented by a block, with overall dimensions the same as the LHD. The simulations are conducted in a steady state; thus, a stationary LHD is assumed. The set-up of the simulation with the LHD is shown in Figure 9a.

With the jet fan air outlet velocity of 18 m/s and introducing a stationary LHD in Drive 2, the flow rate is decreased from 25 m^3/s to 23 m^3/s, as shown in Figure 9b. To restore the flow to the required distribution, the jet fan outlet velocity must increase to 20 m/s to compensate for the LHD restriction.

From the observed CFD results, the following can be concluded:

- To achieve a flow distribution of 15 m^3/s in Drive 1 and 25 m^3/s in Drive 2, the inlet side force fans and the jet fan outlet velocity (speed) both need to be adjusted.

- The addition of an LHD (or any obstruction) causes the flow in Drive 2 to decrease.

- To re-establish the desired flow in Drive 2, the jet fan outlet velocity must again be adjusted.

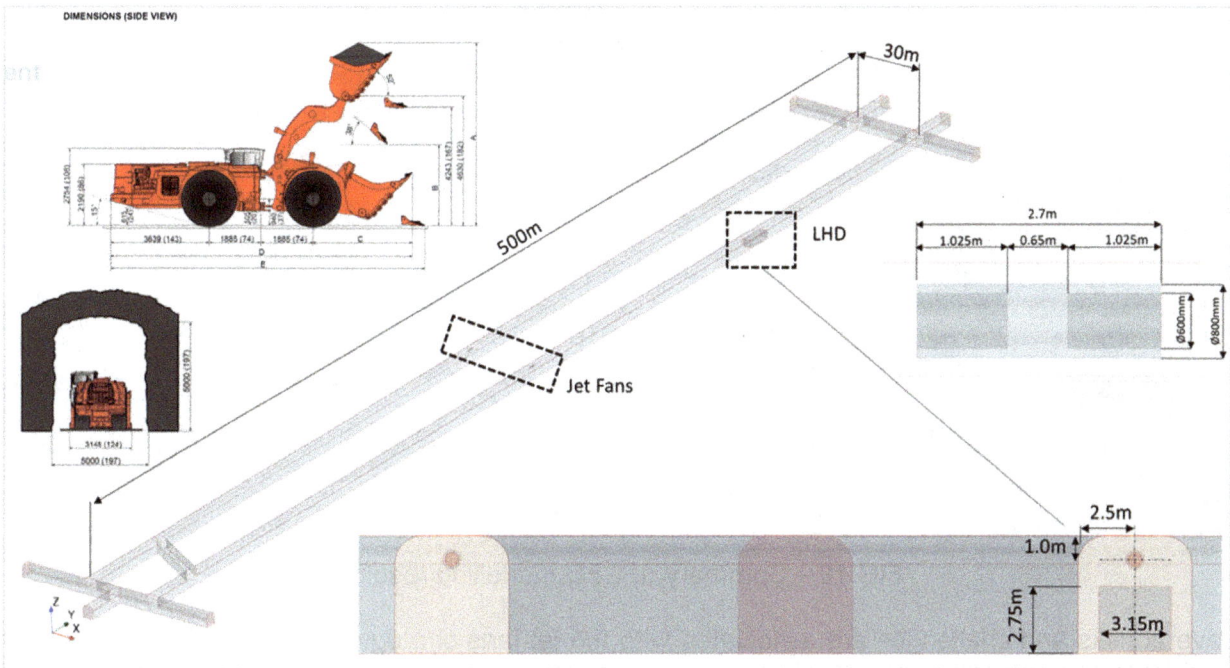

(a)

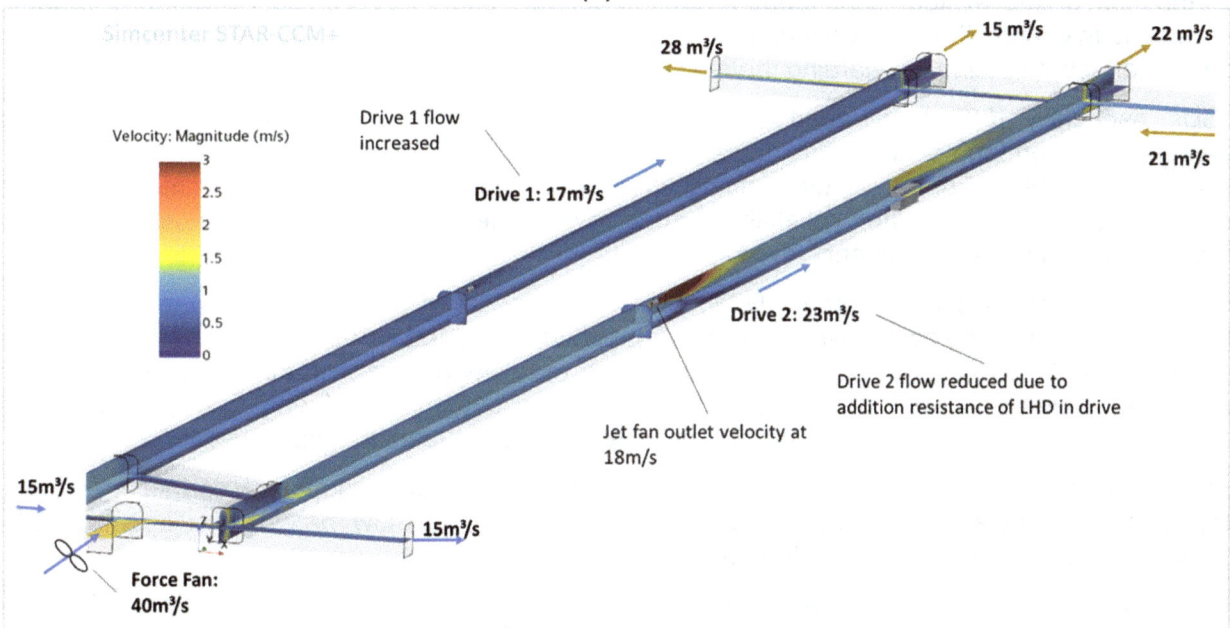

(b)

FIG 9 – CFD modelling results evaluating the impact of an LHD.

DISCUSSION

The results from all the CFD simulations conducted are graphically summarised in Figure 10. It highlights that desired flows can be achieved through fan speed adjustments to adjust the natural distribution of flows and the influence of an LHD. However, the control method needs some consideration.

To regulate the jet fan outlet velocity, fan speed control, eg variable speed drives (VSD), will be required, and the fans must thus be installed with VSD-ready motors suitable for speed control. Active control of the drive's airflow quantity will require a flow monitoring system (eg an ultrasonic flow metre installed in the drive) providing feedback control of the fan speed using a closed communication loop.

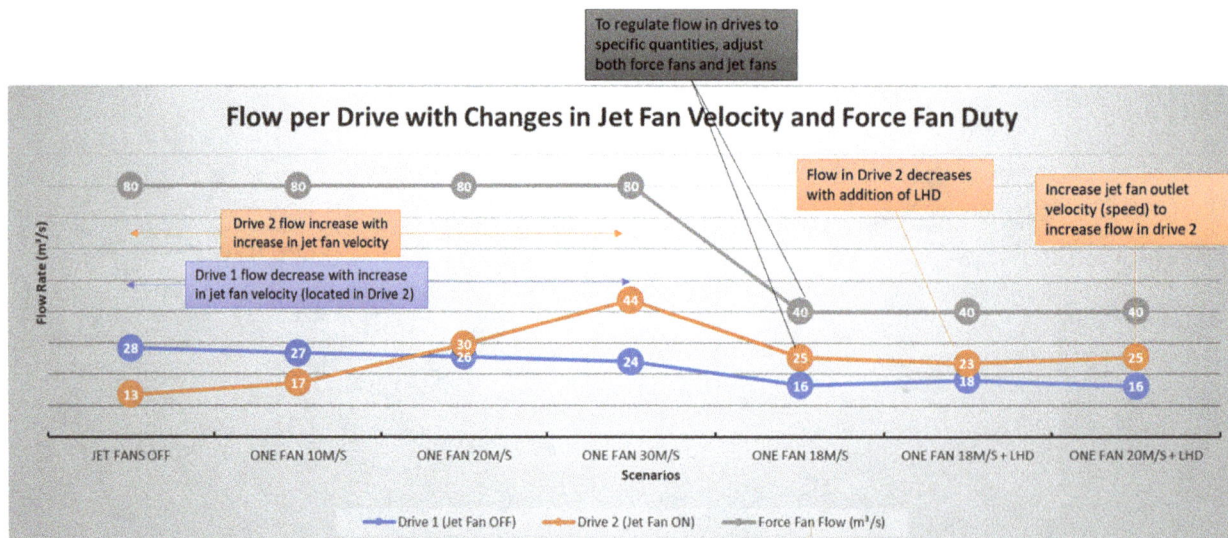

FIG 10 – Summary of CFD modelling results.

Several risks and challenges are associated with the jet fans strategy, such as the overall system complexity, low-pressure differentials, and interdependency of the ventilation control system, as well as the risk of the system continuously hunting for air (shown by ventilation modelling) when attempting to change the airflow dynamically. This could be further complicated by the LHDs creating plunger effects assisting or retarding the set flow.

VSDs, especially 1000 V, still need to be proven reliable underground. With the jet fan strategy relying entirely on fan speed control, lower-voltage fans will be required, raising the need for step-down transformers. Further, if dynamically controlled, the strategy will heavily rely on airflow instruments known to suffer reliability issues over time and are maintenance-intensive in underground mining environments.

The modelling has shown the airflow distribution to be very sensitive to small pressure differentials. Therefore, system wide control will likely be unstable when applied across multiple parallel drives in a real-time VoD-type control system explicitly trying to follow dynamic production activity. The jet fans cannot be relied upon to constantly adjust airflow dynamically during the operating cycle (ie when the number of loaders in the drive changes during the loading loop cycle).

However, if set-up as a 'static' control only occasionally adjusted, or dynamic with certain operational flow bands, jet fans can achieve a slightly more balanced airflow between adjacent parallel production drives to improve the natural airflow split.

Furthermore, control can increase specific airflows to remove heat and contaminants that might otherwise be considered a higher risk for non-production activities, such as work outside air-conditioned cabins/enclosures or performing an inspection.

CONCLUSION

Ventilation management is a critical aspect of block cave mining operations, especially when striving to optimise resource usage and maintain efficient airflow in multiple parallel drives within the extraction level. Ventilation automation could be a cost-saving strategy, directing full airflow to active production drives while reducing airflow to inactive ones.

This paper explored the novel approach of using jet fans as an alternative to traditional ventilation walls, with doors and fans for controlling, distributing, and balancing airflow in parallel block cave production drives. The doors are not popular as they can impede traffic and can be high-maintenance items.

The results from the CFD study are promising, indicating that jet fans can overcome moderate pressure differences and by adjusting the jet fan speed specific production drive flows can be obtained. However, the adjustment in one drive may not be reflected by the opposite effect in an adjacent interconnected drive. Instead, it can influence the flow distribution in within the perimeter airways.

Consequently, some pre-distribution of air to the production drives appears critical to the jet fan strategy. Strategically positioned bulkhead force fans (or regulators) around the extraction level perimeter would be responsible for converting an uncontrolled footprint's natural air distribution into a more balanced arrangement. Jet fans would then be responsible for the fine-tuning of flows combatting low-pressure differentials and LHD effects. Regardless, the pre-distribution force fans (or regulators) will need to be controlled to assist with the overall distribution to the production drives.

Due to the operating principle of jet fans, a low-velocity zone forms downstream of the jet fan. The time/distance it takes for the air in the drive to normalise again to a more fully developed distribution will depend on the fan outlet velocity. However, through the transition, high velocity, low velocity and air recirculation zones develop. The possible build-up (or increased residence time) of contaminants (such as radon, dust, diesel exhaust gases and heat) within these zones has not been assessed or modelled. It might be a potential risk, especially if these zones coincide with an extraction drive (source of contaminants).

This study has demonstrated that jet fans can be used to manage the airflow in the individual drives of a block cave operation. The jet fan concept is suggested to be worth further investigation across other block and panel cave layouts; however, it should not be the base case approach for a project, given that it still relies on an unproven application.

REFERENCES

Chitombo, G P, 2010. Cave mining - 16 years after Laubscher's 1994 paper 'Cave mining – state of the art', in *Caving 2010: Proceedings of the Second International Symposium on Block and Sublevel Caving*, pp 45–61 (Australian Centre for Geomechanics: Perth).

Gómez, R E, Saéz, K, Pino, N, Labbe, E and Marambio, E, 2020. Analysis of extraction level layouts for block caving, in *MassMin 2020: Proceedings of the Eighth International Conference and Exhibition on Mass Mining*, pp 773–786 (Australian Centre for Geomechanics: Perth).

Goodman, G V R, Taylor, C D and Thimons, E D, 1992. Jet fan ventilation in very deep cuts- a preliminary analysis, Report of Investigation 9399, US Bureau of Mines, US Department of the Interior, pp 1–12.

Pont, J, Michelin, F and Stewart, C, 2024. Jet fan control methodology for underground mining, in *Proceedings of the 12th International Mine Ventilation Congress*, pp 92 (The Australasian Institute of Mining and Metallurgy: Melbourne).

Stewart, C, 2023. New applications of jet fans in underground mines for haulage ramps and block cave ventilation control, in *Proceedings of the 19th North American Mine Ventilation Symposium Underground Ventilation,* pp 515–528.

Wolski, J K, 1997. Use of jet fans within mine ventilation networks, *Society for Mining, Metallurgy and Exploration, Inc, Transactions*, 302:166–170.

An assessment of natural versus assisted ventilation of dead-end openings using CFD modelling

J Viljoen[1], S G Hardcastle[2] and L K Falk[3]

1. Engineering Manager, BBE Group Australasia, Joondalup WA 6027.
   Email: joeline.viljoen@bbegroup.com.au
2. Technical Director, BBE Group Canada, Sudbury Ontario P3E 5S1, Canada.
   Email: stephenhardcastle@bbegroup.ca
3. Senior Specialist Mechanical Engineer, Ventilation Systems, Vale Canada Limited, Copper Cliff
   Ontario P0M 1N0, Canada. Email: leif.falk@vale.com

ABSTRACT

Using auxiliary fan-duct systems is a common practice for ventilating dead-end openings to control strata gas emissions, radiation, thermal conditions, and exhaust emissions from diesel vehicles. These dead-ends could be part of development/production activities, service areas, or traffic management. In a large multi-level mine, the number of fan-duct installations needing to be purchased and maintained and their summated power requirements can be significant. Hence, understanding whether these systems can be operated on an on-demand basis, solely using the duct as a flow diverter or the complete fan-duct system eliminated, requires a detailed assessment. Historically, there are some general guidelines based on methane control or blast fume clearance, but little based on diesel dilution, the conditions that could prevail, and the health and safety of personnel.

This paper investigates various options for ventilating dead-end openings adjacent to the primary through ventilation route that, for a short duration, could contain a mobile diesel unit. These were natural circulation, an open duct flow diverter, and the conventional fan-duct system. Consideration was given to the orientation of the opening relative to the main drive at 45° (backward), 90°, and 45° (forward); the length of the opening at 10 m and 20 m; low and high velocities in the primary air route at 0.8 m/s and 4.0 m/s (the minimum and maximum velocity range for the specific mining company); and the presence and orientation of an operating diesel vehicle.

The analysis used steady-state and transient computational fluid dynamics (CFD) modelling. The modelling results are evaluated for their face-flushing ventilation efficiency, dust particle removal and residence times, and the build-up and decay of acute effect gases such as carbon monoxide (CO), nitric oxide (NO) and nitrogen dioxide (NO_2) against exposure limits. The modelling has shown that the adequacy of a ventilation method relies on geometry and main drive velocity and is contaminant dependant. However, specific findings will still require validation through field studies. Overall, the work has indicated the need for a more risk-based analysis, this will be even more critical worldwide where lower exposure limits been accepted or are planned for both NO and NO_2.

INTRODUCTION

A literature review and peer survey has shown there is a limited amount of guidance regarding the depth to which a dead-end opening (heading, cubby etc) can be developed before an auxiliary ventilation system is required. A common premise is that, if less than 10 m in-depth, dead-ends are adequately ventilated by circulation induced by the passing primary flow; and beyond 10 m, additional ventilation controls or mechanical ventilation employing an auxiliary fan-duct system is required to ensure sufficient flow to 'sweep' the face, dilute and remove contaminants.

The review and survey yielded a wide range of findings, from specific values for the length/depth limits of an unventilated dead-end through to an indirect requirement to meet Occupational Exposure Limits. Although regulations, rules of thumb, industry standards, and guidelines exist, in most cases, the depth of unventilated dead-ends is site-specific, the contaminant(s) addressed unidentified and rarely risk-based.

With little or no consideration given to geometry, the velocity of the flow past the dead-end entrance, and the progressive lowering of exposure limits, these historic design criteria may become invalid for global application. Hence the need for them to be reviewed.

Unventilated dead-end guidelines, rules, and regulations

Guidelines and rules by country/state

South Africa

The exact length of dead-ends to be mechanically ventilated is no longer governed by South African legislation. Instead, it is stated that the employer must ensure that the occupational exposure to health hazards of employees is maintained below the limits set out in the Mine Health and Safety Act, 1996, Schedule 22.9(2)(a) and (b) (Mine Health and Safety Act, 1996).

However, a general rule and understanding calls for auxiliary ventilation if the dead-end length is more than twice the width or height (shorter dimension of the two), this is a legacy from the definition of a dead-end from the Managers Guide to Flammable Gas (Chamber of Mines of South Africa, 1989). The understanding at the time was that longer dead-ends might not be cleared through natural draft within the standard 4-hr re-entry time. This was based on a rule made by the Mine Managers Association in the Free State, where flammable gas was a major issue. Dead ends at about 45° were common in these mines, which might have played a role in using the depth-to-width/height ratio over the angular opening. The development of such rules was driven by physical investigations using Drager pumps and safety lamps to assess re-entry.

A SIMRAC project (Meyer, 1993), on coalmine ventilation, includes a recommendation for auxiliary ventilation in headings greater than 10 m in length from the last through road (LTR). This recommendation followed underground trials and early computational fluid dynamics simulations that considered velocity and seam height. From this work, 10 m appears to be the minimum depth of the primary natural ventilation-induced circulation for headings, including a continuous miner. Beyond that, a secondary and opposite circulation could be experienced.

Ghana, Russia

Ghana's Regulations (Minerals and Mining, 2012) and the Russian Regulations (Russian Mining Law, 2014) both stipulate a 10 m depth limit before introducing auxiliary ventilation.

Ontario, Canada

Prior to 2023, Ontario's Mining and Mining Plants Regulation 854, (Government of Ontario, 2023) stated auxiliary ventilation was required for dead-ends in excess of 60 m, this decreases to 10 m where the Control of Exposure to Biological or Chemical Agents Regulation 833 (Government of Ontario, 2020) applies. As of 2023, these depth stipulations have been revoked, and the onus has moved to ensure that a worker's exposure is below occupational exposure limits.

Australia

In Australia, the industry norm is that a dead-end of more than 20 m deep must be mechanically ventilated but still needs to be risk-based.

This is exemplified in Western Australia's regulations (Work Health and Safety, 2020) (and similar for other states). It states that no person in a mine is to be exposed to an airborne concentration exceeding the exposure standard prescribed in the regulations.

Here, the maximum depth of an unventilated dead-end will be risk-based but becomes a function of the currency of the exposure limits reflecting health effect science.

Guidelines and rules by mining group/commodity

Varied guidelines and rules for mining groups, commodities and methods also exist. Unventilated dead-end length specifications, without considering the heading width or height, vary between 9.1 m and 20 m. A practical unventilated dead-end depth of 12 m has also been seen based on the length required to accommodate a load-haul-dump (LHD) diesel unit.

Where dimensions are considered, examples of dead-end's unventilated length limit include:

- 3 × diagonal of the cross-section (for a 5 m × 5 m, the length will be 21 m).

- (3H × 4W)/2(H + W) and thus for a 5 m × 5 m, the length will be 15 m.

Review summary

From the literature review, it is evident that when the maximum unventilated heading depth does not follow a risk-based approach, in almost all cases, the maximum unventilated depth is a specific value or a function of the heading cross-sectional dimensions. There is little supporting evidence of the length limit being a function of the velocity or airflow rate past the heading opening or the angle of the heading relative to the through ventilation route.

Dead-end risk-based analyses

Determining the maximum length of an unventilated heading based on occupational exposure limits will provide more definitive results. However, that distance may not be globally applied due to the significantly different exposure limits that could apply per regulatory jurisdiction.

Table 1 provides an overview of the divergence in the 8-hr time-weighted average (TWA) exposure limit, the short-term exposure limit (STEL) and ceiling exposure limit (C) for carbon monoxide (CO), nitric oxide (NO) and nitrogen dioxide (NO_2) across North America, Europe, and Australia as applicable to mining. Notably, not every jurisdiction has a STEL, which would typically be for a 15-min duration, and even fewer have a ceiling value. In Canada, most of the exposure limits align with those recommended by the American Conference of Governmental Industrial Hygienists (ACGIH), published annually. However, they may not be the latest. The ACGIH has recommended a TWA of 25 ppm for both CO and NO since 1992 and 0.2 ppm NO_2 since 2012; prior to that, from 1996–2011, it was 3 ppm NO_2. As shown in Table 1 with respect to TWA values:

- CO limits range from 50 ppm down to 20 ppm.

- The NO limit, long accepted to be 25 ppm, has been reduced to 2 ppm in the European Union, Sweden, and Germany, and reduces to 2 ppm in Australia as of December 1, 2026. A German advisory group is recommending a further reduction to 0.5 ppm (DFG, 2017).

- NO_2, historically regulated at a 3 ppm limit, is being reduced to 0.5 ppm in the European Union, however it is already at 0.2 ppm in some Canadian jurisdictions.

The above changes respectively represent up to 60 per cent, 92 per cent and 93 per cent reductions in exposure limits. These will directly affect any risk-based analysis of the adequacy of natural ventilation in dead-ends.

TABLE 1

TABLE 1
Summary of occupational exposure limits across.

	Carbon Monoxide, CO			Nitric Oxide, NO			Nitrogen Dioxide NO$_2$		
	TWA	STEL	C	TWA	STEL	C	TWA	STEL	C
	ppm	ppm	ppm	ppm	ppm	ppm	ppm	ppm	ppm
Canada									
Government of Yukon (1986)	50†	400†		25†	35†		5†		
Government of Quebec (2023)	35†	175†		25†			3†	5†	
Government of Alberta (2023)	25†			25†			3†		5†
Government of New Brunswick (2023)	25#			25#			3†	5†	
Government of Northwest Territories (2023); Government of Nunavut (2016)	25†	190†		25#	38†		3†	5†	
Government of Ontario (2020)	25#	75*	125*	25#	75*	125*	3†	5†	15*
Government of Saskatchewan (2019, 2021)	25†			25†			2†		
Government of British Columbia (2022)	25†			25†			1†		
Government of Manitoba (2020); Government of Nova Scotia (2022)	25‡			25‡			0.2‡		
Government of Newfoundland and Labrador (2023)	25‡	75*	125*	25‡	75*	125*	0.2‡	0.6*	1*
USA, MSHA (2013)	50#			25#			5#	5#	
European Union, IFA (2023)	20	100		2			0.5	1	
Sweden, IFA (2023)	20	100		2			0.5	1	
Germany, IFA (2023)	30	60		2	4		0.5	1	
Government of United Kingdom (2020)	20	100		2			0.5	1	
South Africa (Mining), IFA (2023)	30	100		25	35		3	5	
Safe Work **Australia** (2019)	30			25			3	5	
Safe Work Australia (2024)	20			2			3	5	

† – Value specifically stated within regulation; # – Refers to specific ACGIH TLV year; ‡ – Most current ACGIH TLV; * – STEL = 3 × TWA, C = 5 × TWA where not specified by ACGIH.

Dead-end computational fluid dynamics (CFD) modelling

CFD modelling readily permits multi-parameter analysis of complex fluid flows. Prior work was found that focused on diesel particulate matter (DPM) flow patterns in dead-ends with and without auxiliary ventilation (Morla *et al,* 2023, 2022; Morla, Karekal and Godbole, 2020a, 2020b) but without any consideration of gases.

The objective of the modelling reported here was to understand the following:

- Does the flow rate (air velocity) in the main through route/drive, from which the dead-end has been developed, affect the air entrainment into the heading and the face-sweeping potential?

- Does the angle of the dead-end relative to the main drive affect the air entrainment and face-sweeping ability?

- With diesel equipment temporarily located in the dead-end heading, is the ventilation sufficient to control the exhaust emission of acute gases, CO, NO and NO$_2$, to below legislative limits?

The extensive analysis presented here is based upon work originally requested by Vale's Ontario Operations and their input. It also focuses on the systematic investigation and findings using both steady-state and transient simulations without going into details on the computational mesh formation and solver parameters.

The analysis was conducted using the commercial CFD software package from Siemens, STAR-CCM+. To capture the dynamic behaviour of the airflow and gases, time varying Reynolds-Averaged Navier–Stokes equations for mass, momentum and energy were solved on each discrete cell of the computational mesh. Turbulence was modelled using the k-omega turbulence model equations and the boundary layer at the solid interfaces was modelled with logarithmic wall functions.

Model and methodology

The model set-up consists of four drives (D1-D4), each with a different dead-end heading configuration (angle or length), as shown in Figure 1a. These drives combined to a single exhaust; this permitted the simulation of each heading simultaneously.

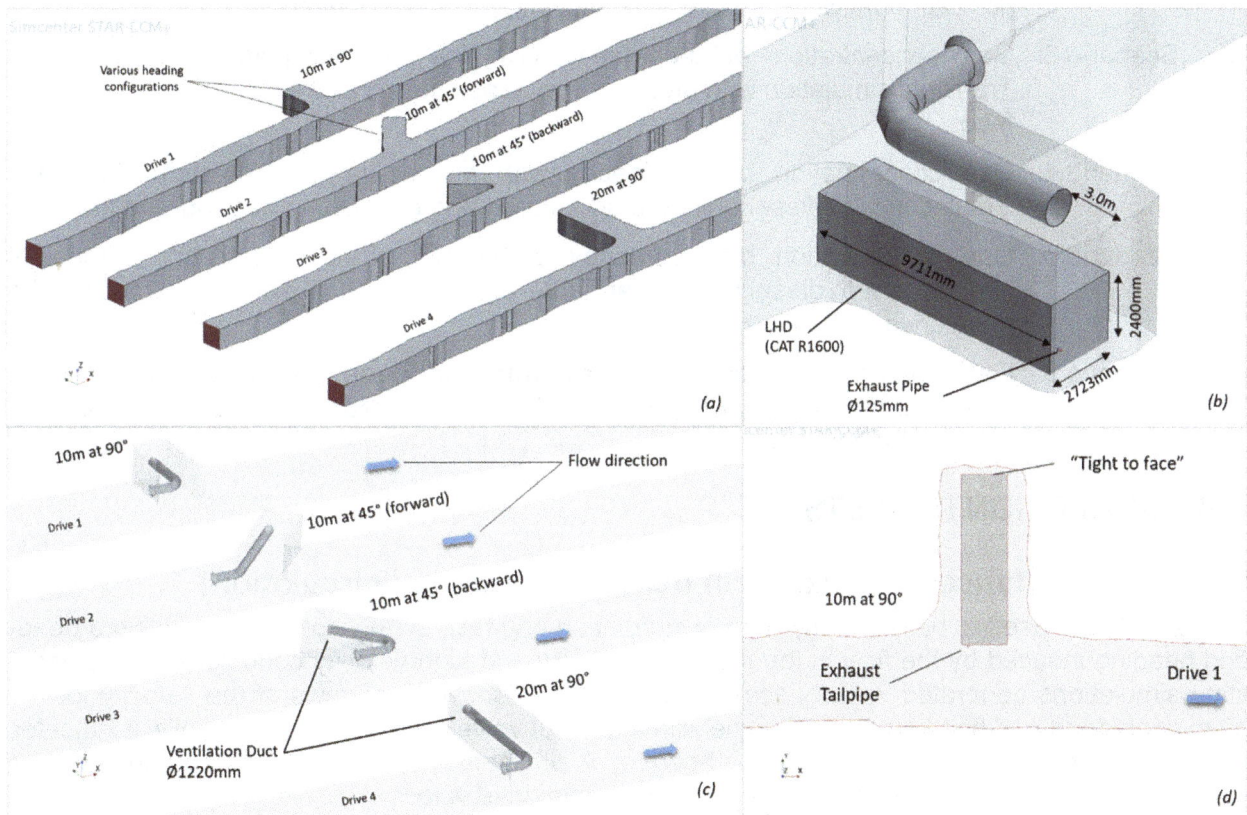

FIG 1 – Model set-up with four drives and headings.

All the drives and headings are approximately 4.9 m [16 ft] (H) × 4.9 m [16 ft] (W). The dead-ends (D1-D3) are all 10 m deep, with angular orientations of 90°, 45° forward and 45° backward, relative to the flow direction. The fourth dead-end (D4) is 20 m deep with an angular orientation of 90°.

Almost exclusively, each simulation was run at two volumetric flow rates, 19 m^3/s (40 000 cfm) and 94 m^3/s (200 000 cfm), producing air velocities of 0.8 m/s (160 fpm) and 4.0 m/s (800 fpm) in each drive. The only exception was the simulation that included an auxiliary fan; it had a drive flow rate and velocity of 11.9 m^3/s and 1.0 m/s, respectively, while the fan delivered 7.0 m^3/s.

Three ventilation scenarios, with and without an operating diesel vehicle present, were considered:

1. Natural eddy circulation, thus no ventilation duct or auxiliary fan.

2. An open duct flow diverter, an open Ø1,220 mm (48") duct with no auxiliary fan FIG 1b and 1c.

3. A conventional fan-duct system.

The diesel unit size details were based upon a Caterpillar R1600 loader at 2.72 m W × 9.71 m L × 2.4 m H but could similarly be a utility vehicle (scissor lift, fuel, or water truck).

The exhaust flow was specified as 1.1 m³/s at a temperature of 149°C from a Ø125 mm tailpipe.

The initial underground air conditions were 110 kPa and 25°C, with zero contaminant concentrations.

The vehicle emissions were modelled to reflect the maximum permissible exhaust concentration for a vehicle returning from maintenance, as 50 ppm NO_2, 700 ppm NO, and 500 ppm CO.

Two diesel unit orientations were also considered: reverse and forward entry into the heading. Figure 1b and 1d show the resultant locations of the vehicle exhaust.

The resultant scenarios simulated are summarised as follows:

Scenario 1: Natural flow circulation in a heading induced by the flow in the through drive. Steady-state only and without diesel equipment.

Scenario 2: Circulation improvements created by an open ventilation diverter/duct (Ø1220 mm). Steady-state only and without diesel equipment.

Scenario 3: Gas concentration build-up/decay with the open ventilation diverter/duct. Transient simulation with diesel unit reverse entered and exhausting at dead-end face.

Scenario 4: Gas concentration build-up/decay with an auxiliary fan system. Transient simulation with diesel unit reverse entered and exhausting at dead-end face.

Scenario 5: Gas concentration build-up/decay under natural flow circulation. Transient simulation with diesel unit forward entered into dead-end exhausting towards the main drive.

Scenario 6: Gas concentration build-up/decay with the open ventilation diverter/duct. Transient simulation with diesel unit forward entered into dead-end exhausting towards the main drive.

CFD SIMULATION RESULTS

Scenario 1 – induced ventilation in dead-ends (natural circulation)

Scenario 1 was used to baseline any natural circulation and face sweeping potential within a dead-end heading induced by the flow in the main drive at low and high-velocity conditions. The steady-state simulations generated velocity scalar plots. Figure 2 shows two views of the velocity for the 10 m deep dead-end heading at a 90° angle with a main drive velocity of 0.8 m/s; visually, it indicates negligible penetration/circulation within the heading. A similar lack of air circulation/entrainment was observed for all dead-end configurations at the low main drive velocity.

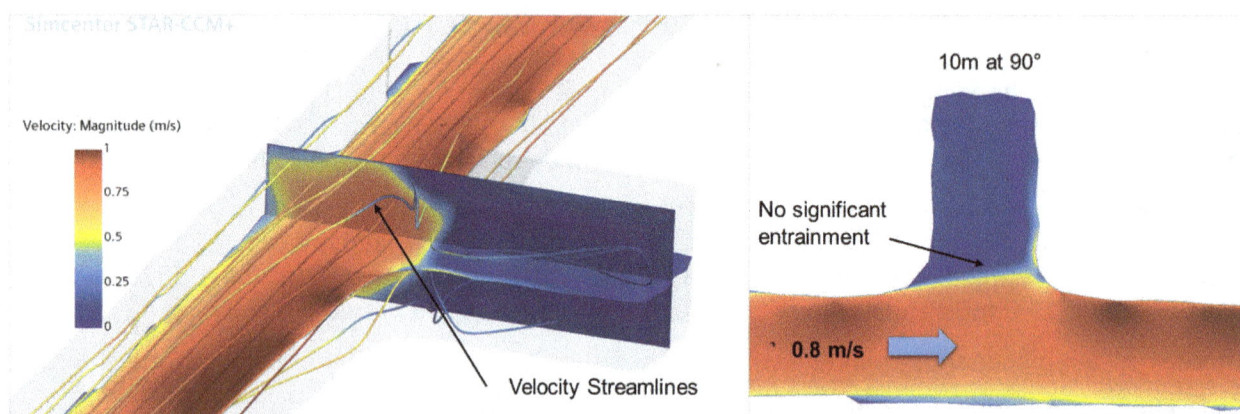

FIG 2 – Example of velocity scalar plots with an average 0.8 m/s velocity in the main drive.

Figure 3 shows the velocity scalars with a main drive velocity of 4.0 m/s for the three 10 m deep headings. Visually, there is an improvement in the air entrainment, but the ability to sweep the face has varied results. The 10 m dead-ends at 90° and 45° (forward) might potentially have sufficient air entrainment to eliminate the use of auxiliary ventilation, whereas the 10 m 45° (backward) and 20 m at 90° dead-ends (not shown) indicate limited penetration and circulation.

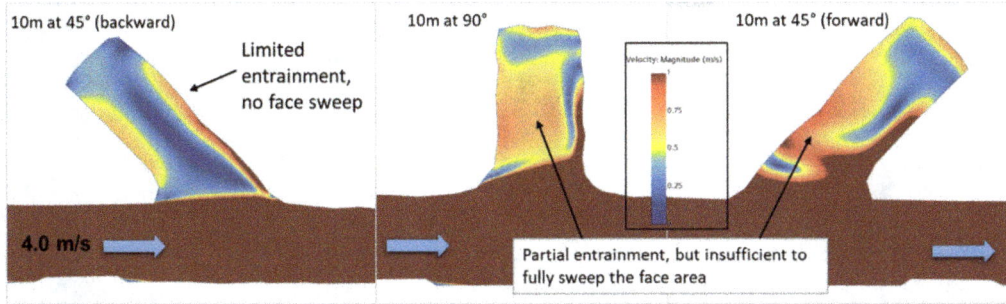

FIG 3 – Velocity scalar plots of 10 m dead-end with an average 4.0 m/s velocity in the main drive.

These results along with velocity vector plots indicated that the flow any plane is multi-directional, included inward and outward flows at the entrance to and within the heading. Consequently, without more detailed analysis, determining the flushing air volume, and distinguishing between what portion of the flow is fresh air and what portion is recirculating air would be problematic.

However, introducing, through simulation, massless particles injected close to the face of the dead-end can indicate face-sweeping potential and help quantify air entrainment. Figure 4 provides a visual comparison of particle tracks at 60 sec, 120 sec and 180 sec for the 10 m at 90° dead-ends for both drive air velocities. Figure 5 summarises the particle residence results for the four heading configurations and the low and high drive velocities. Figure 4, through displaying circulating dust traces, also highlights that at any cross-section of the heading, the flow direction can be both positive and negative compared to the overall flow. These graphs also confirm that both the 10 m dead-ends at 90° and 45° (forward) have significant air entrainment at the face.

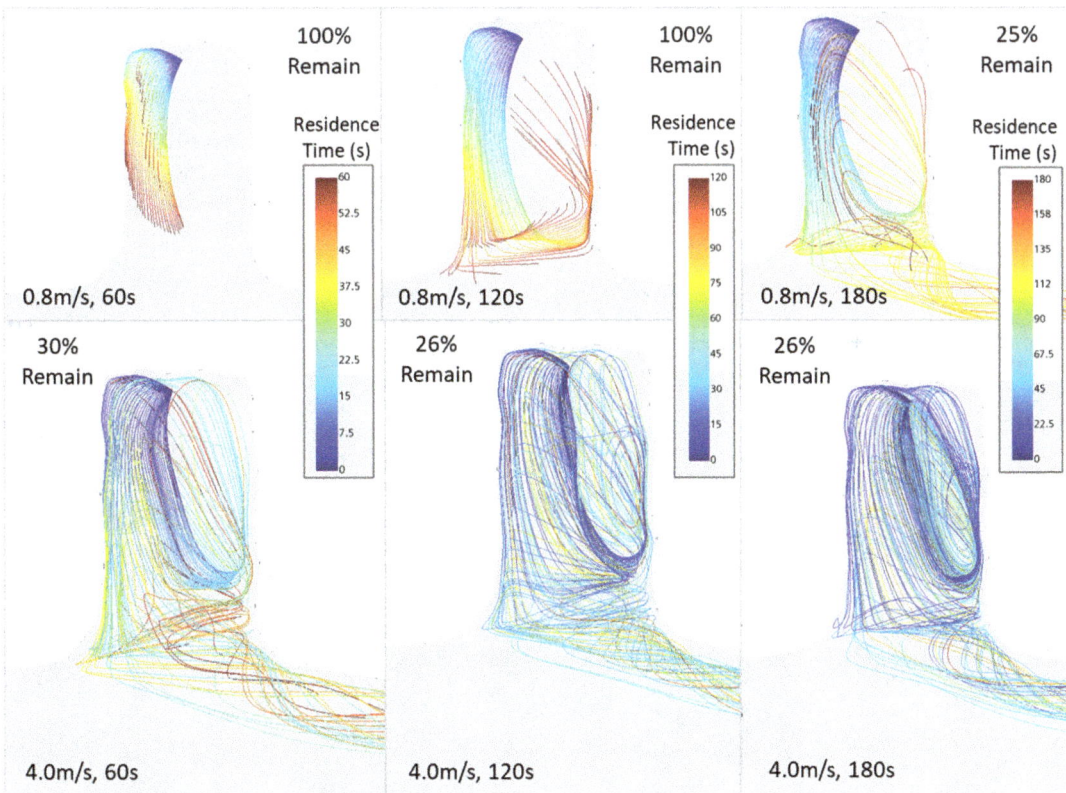

FIG 4 – Particle residence simulations in a 10 m at 90° dead-end as a function of main drive velocity.

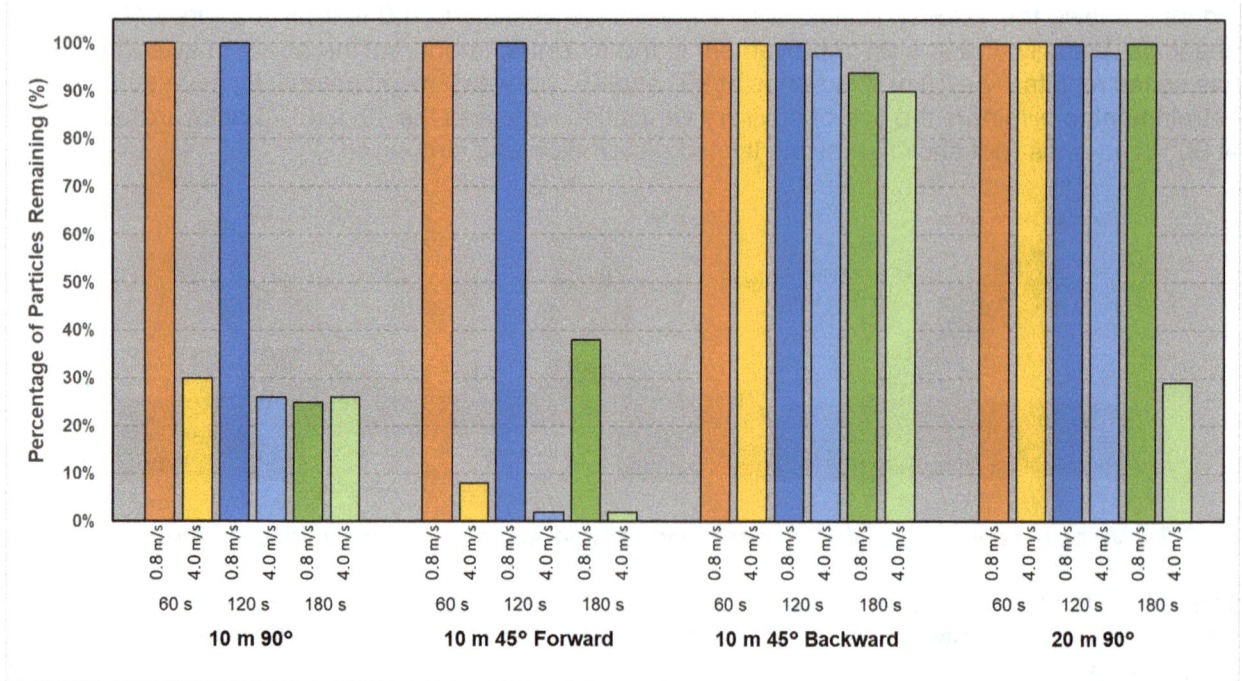

FIG 5 – Qualitative comparison of entrainment/particle residence time in naturally ventilated dead-ends.

Scenario 2 – open duct flow diverter

Scenario 2 evaluated the improvements that a flow diverter may provide across the same models at the high and low drive velocities used in Scenario 1. The flow diverter was an open rigid ventilation duct (Ø1220 mm) with no auxiliary fan. In this scenario, it was possible to determine the air quantity entering the heading by monitoring the duct flow. As shown in

Figure 6, the flow induced into the dead-end headings ranged between 0.8 m³/s at the low main drive velocity and 4.3 m³/s at the high velocity which represents 4 per cent and 5 per cent of the flow in the main drive. It should be noted that this applies to the modelled Ø1220 mm duct size and its length, no other duct sizes have been considered. However,

Figure 6 also shows the variability in total flow (positive and negative) across monitoring planes at a mid-point and near the exit of the dead-end. Due to the circulatory flows, the amount passing through a specific plane can be double that entering the dead-end.

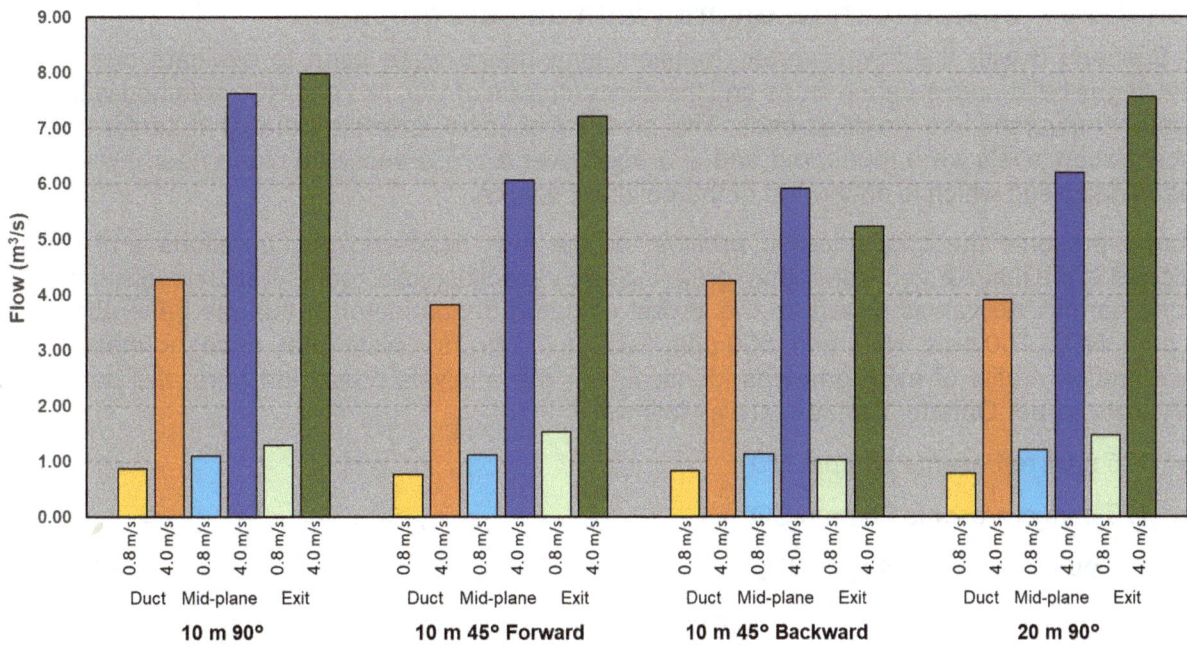

FIG 6 – Comparison of diverter induced flow versus flows within the dead-end heading.

Figure 7 summarises the massless particle injection simulations with the duct diverter. A significant improvement is observed in the 'face-sweeping' potential for the 4.0 m/s main drive velocity across all dead-end heading configurations. However, for the 0.8 m/s main drive velocity, there were minor improvements for the 10 m ends at 90° and 45° (forward) dead-ends, but negligible improvement for the 45° (backward) or the 20 m dead-end configurations.

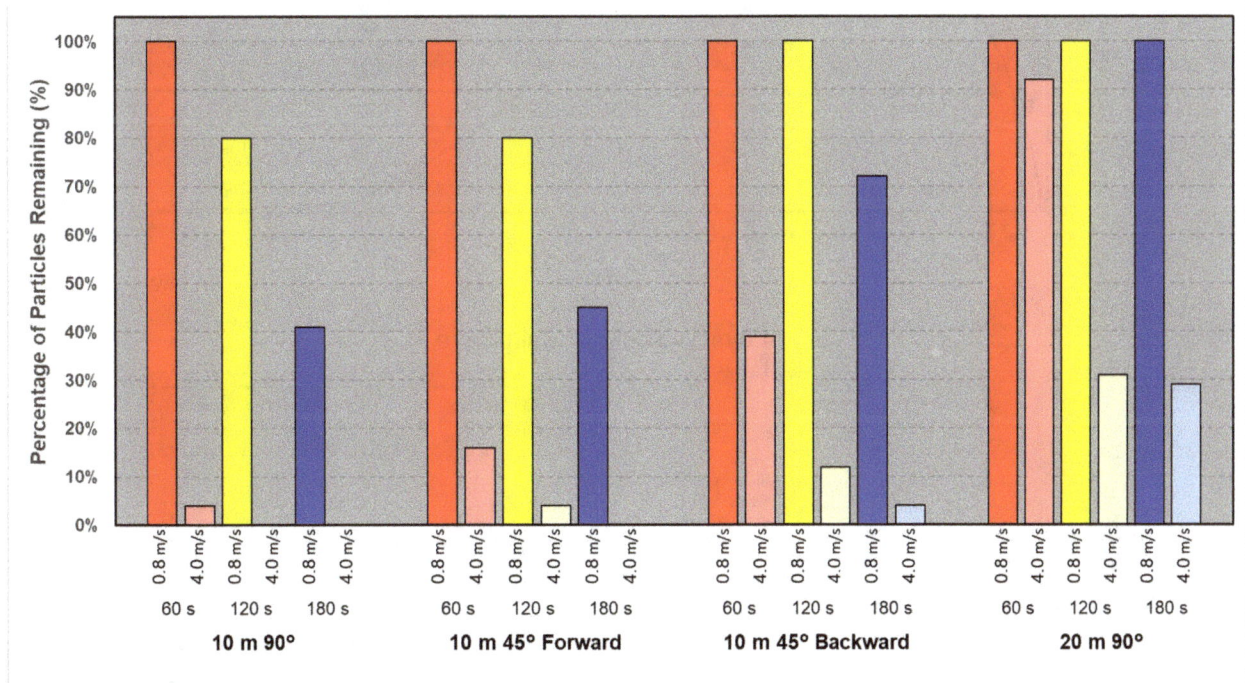

FIG 7 – Qualitative comparison of entrainment/particle residence time dead-ends with a flow diverter.

However, for a risk-based analysis, the performance of a ventilation regime needs to be assessed in terms of pollutant concentrations and occupational exposure limits.

Scenario 3 – open duct flow diverter with diesel equipment

For Scenario 3 with the flow diverter, transient simulations were used to evaluate and monitor exhaust gas build-up/clearance times and the concentration of CO, NO and NO_2 obtained with diesel equipment reversed into the dead-ends. This places the diesel vehicle's exhaust at the far end. The concentration levels were monitored and averaged over a cross-sectional plane near the entrance of each dead-end which is downwind of the pollutant source.

Figure 8 presents the concentration profiles for the four dead-end configurations and the two velocities in the main drive. Rather than graphing each specific gas's concentration levels, this paper reports normalised values based on the lowest common denominator of the gas generation rates (50 ppm NO_2, 700 ppm NO, and 500 ppm CO in 1.1 m^3/s). Note that once normalised, the concentration profile of each gas was identical. For the analysis presented here, the normalised TWA limits as per Ontario, Canada, are as follows:

- 125 ppm represents 25 ppm NO
- 175 ppm represents 25 ppm CO
- 210 ppm represents 3 ppm NO_2.

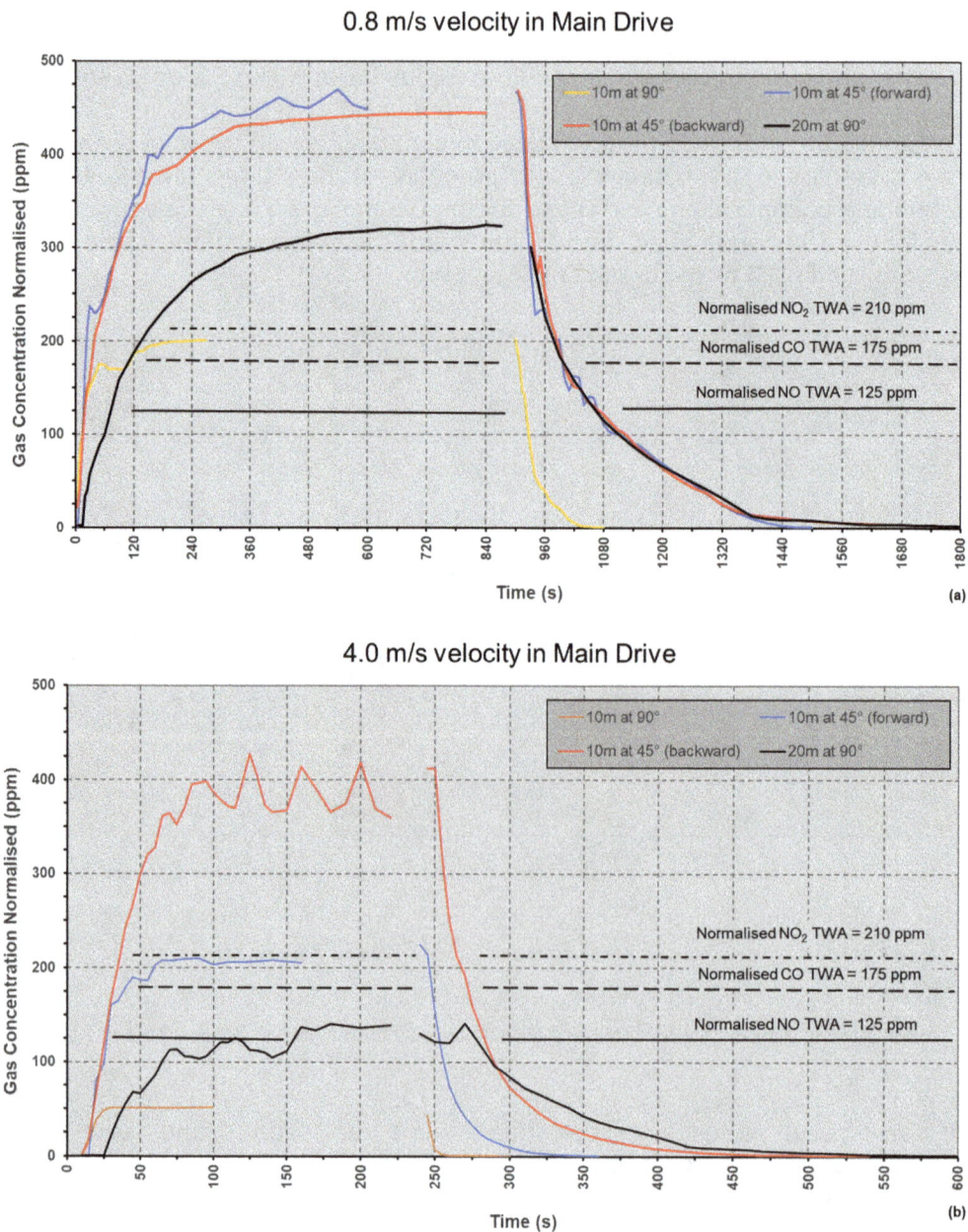

FIG 8 – Gas concentration build-up and clearance with open ventilation duct (diverter).

Consequently, based on the gas generation rates and these exposure limits, the control of NO is the most demanding and NO_2 the least demanding.

Figure 8a and 8b, show that the gas concentration build-up, the steady state concentration, and the concentration decay are all influenced by the dead-end geometry/size and the flow induced into the heading. Except for the 10 m at 45° backward dead-end, all steady state concentrations are notably reduced with the higher drive flow. For all headings, the build-up and clearance times are reduced at the higher main drive flow. The diverter created dead-end flows in the order of 0.8 m^3/s and 4.3 m^3/s respectively for the low and high main drive velocities.

Table 2 summarises the build-up/clearance times, and steady state concentrations. Under the low flow condition virtually all dead-end headings achieve steady state concentrations at or exceeding the TWA limit for all gases, and NO even exceeds the STEL prescribed in Ontario for both 45° dead-ends. However, although it could take up to 14 mins for these final levels to be achieved, certain TWAs are exceeded within 1 min. Under the high flow condition, all build-up/clearance times and steady-state concentrations are reduced, but TWAs are still exceeded for both 45° dead-ends. Note: the concentrations in the 20 m dead-end being lower than the 10 m 45° dead-ends is an attribute of the overall size of 20 m heading and internal circulation.

TABLE 2

Gas concentration, build-up and clearing summary with open ventilation duct (diverter).

Main drive velocity (m/s)	Dead-end configuration	Build-up (min)	Clearance (min)	Steady-state concentration (ppm)		
				NO	CO	NO$_2$
0.8	10 m 90°	≈4	≈3	40[†]	29[†]	3
	10 m 45° forward	≈9	≈9	92[‡]	66[†]	7[†]
	10 m 45° backward	≈14	≈12	91[‡]	65[†]	7[†]
	20 m 90°	≈14	≈14	65[†]	46[†]	5[†]
4.0	10 m 90°	≈1	≈1	10	7	1
	10 m 45° forward	≈2	≈2	41[†]	30[†]	3[†]
	10 m 45° backward	≈2	≈5	82[†]	59[†]	6[†]
	20 m 90°	≈3	≈6	28	20	2

[†] – Exceeds 8-hr TWA; [‡] – Exceeds Ontario STEL (3 × TWA).

Scenario 4 – mechanical ventilation, with auxiliary fan/duct

In this scenario, the velocity in the main drive was increased from 0.8 m/s to 1.0 m/s, which now represents a volumetric flow rate of 24 m^3/s, and the fan/duct system was to deliver 7.0 m^3/s. The vehicle is again reversed into the dead-end. Similar to previous figures, Figure 9 shows the gas concentration profiles in the exhausted air, the results are summarised in Table 3.

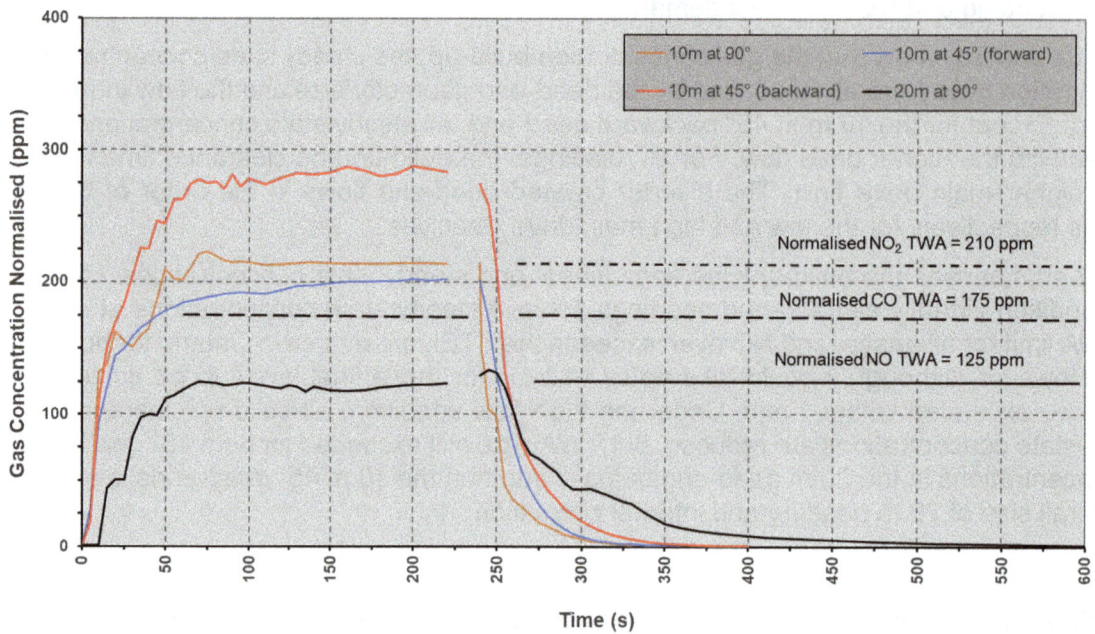

FIG 9 – Gas concentration evaluation at 1.0 m/s main drive velocity, with auxiliary fan and duct.

TABLE 3

Gas concentration, build-up and clearing summary with auxiliary fan system delivering 7 m³/s.

Dead-end configuration	Build-up (min)	Clearance (min)	Steady state concentration (ppm)		
			NO	CO	NO₂
10 m at 90°	≈2	≈2	43[†]	31[†]	3[†]
10 m 45° forward	≈3	≈2	40[†]	29[†]	3
10 m 45° backward	≈2	≈3	57[†]	41[†]	4[†]
20 m 90°	≈2	≈5	25	18	2

[†] – Exceeds 8-hr TWA.

With a velocity of 1.0 m/s in the main drive, some gas will flow upstream in the main drive after leaving the heading. Consequently, some contaminated air might get re-introduced into the dead-ends through the fan/duct system, which can lead to increased clearing times.

When compared to the diverter scenario with a 4.0 m/s velocity in the main drive, in some dead-end configurations, lower concentration levels are observed, and in some configurations, higher concentration levels are observed. This may seem counter intuitive but highlights one of the outcomes of using CFD. For each ventilation and dead-end arrangement, there will be a unique flow/circulation pattern at the discharge. As previously shown in Figure 6, this can include local recirculation with airflows both in to and out of the heading. Consequently, a homogeneous mixture concentration that could simply be defined by the pollutant concentration in the system divided by the airflow is not achieved, and the average concentration is very much a function of the analysis plane location.

Scenarios 5 and 6 – natural and open duct ventilated dead ends with diesel unit exhaust towards main drive

The prior scenarios considered the worst-case situation where the diesel reversed into the dead-end and its exhaust was discharged at the furthest point from the main drive. A series of simulations was also undertaken to evaluate the vehicle driving directly into the dead end with its exhaust directed towards the main drive. As shown in Figure 1d, this resulted in the exhaust tailpipe discharge being into the main drive flow outside of the analysis plane for the 10 m dead-ends; for the 20 m dead-ends, the discharge was 5 m into the heading.

Low velocity in the main drive

In almost all the low-velocity cases, gases are observed to flow upstream within the main drive and then get entrained back into the dead-end, either through the induced circulation (no duct) or the open duct diverter, an example of this is shown in Figure 10. Generally, the averaged concentrations within the dead-end are low, due to dilution from the air in the main drive, and do not exceed TWA limits.

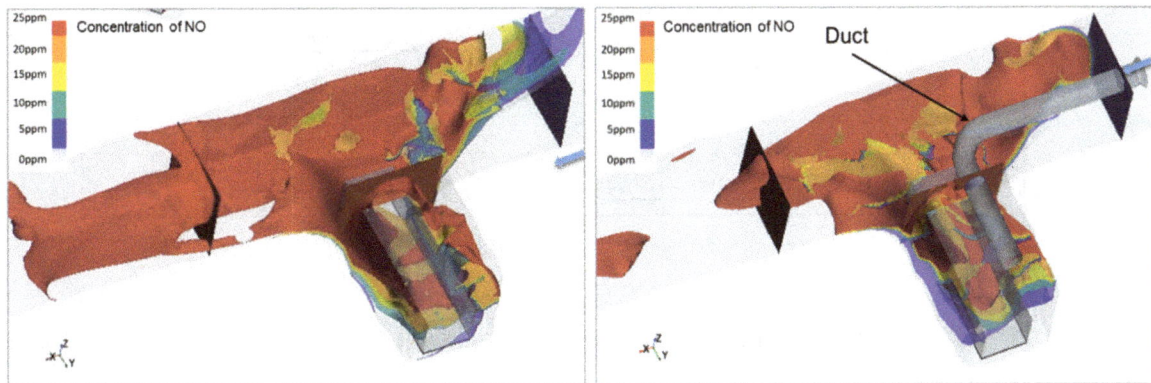

FIG 10 – Upstream flows of NO within a drive at 0.8 m/s, with and without a duct.

High velocity in the main drive

With the high air velocity in the main drive, no exhaust gases migrated upstream to re-enter the dead-end. For the 10 m dead-ends, this was due to the location of the tailpipe and its discharge directly into the main drive flow. Consequently, exhaust gases are only observed downstream of the diesel equipment in the main drive and then only at low concentrations.

TWA VERSUS STEL AND TWA VARIATIONS

In a risk-based approach, using the higher permitted STEL, rather than an 8-hr TWA, may be more relevant to a dead-end analysis where the vehicle and its occupants are only in that location for a short period. However, as highlighted earlier, not every jurisdiction has a STEL. Consequently, it may revert to an individual company's policy regarding whether the TWA can be exceeded at any time.

The CFD analysis interpretation has used the following 8-hr TWA limits: 25 ppm CO, 25 ppm NO and 3 ppm NO_2. These have generally shown either an auxiliary fan/duct system or a flow diverter paired with a high main drive air velocity is required to guarantee circulation and flushing of a dead end. However, as shown in Table 1, lower limits are already in place in certain parts of Canada and apply or are pending in Europe and Australia.

Considering these, the normalised TWA equivalents become:

- 14 ppm would represent 0.2 ppm NO_2 as per Manitoba, Nova Scotia, and Newfoundland and Labrador in Canada.

- 35 ppm would reflect the European 0.5 ppm NO_2.

- 10 ppm would represent the European 2 ppm NO.

- 100 ppm represents the Swedish 20 ppm CO.

Figure 11 is a repeat of Figure 9 but showing these lower normalised TWAs. This figure shows that the gas concentrations in the dead-end will greatly exceed these lower NO_2 and NO limits. For NO_2, assuming a single 5-min exposure at the normalised 200 ppm, this is equivalent to 70 mins at the 0.2 ppm TWA. Similarly, for NO, the normalised 200 ppm is equivalent to 100 mins at the 2 ppm TWA.

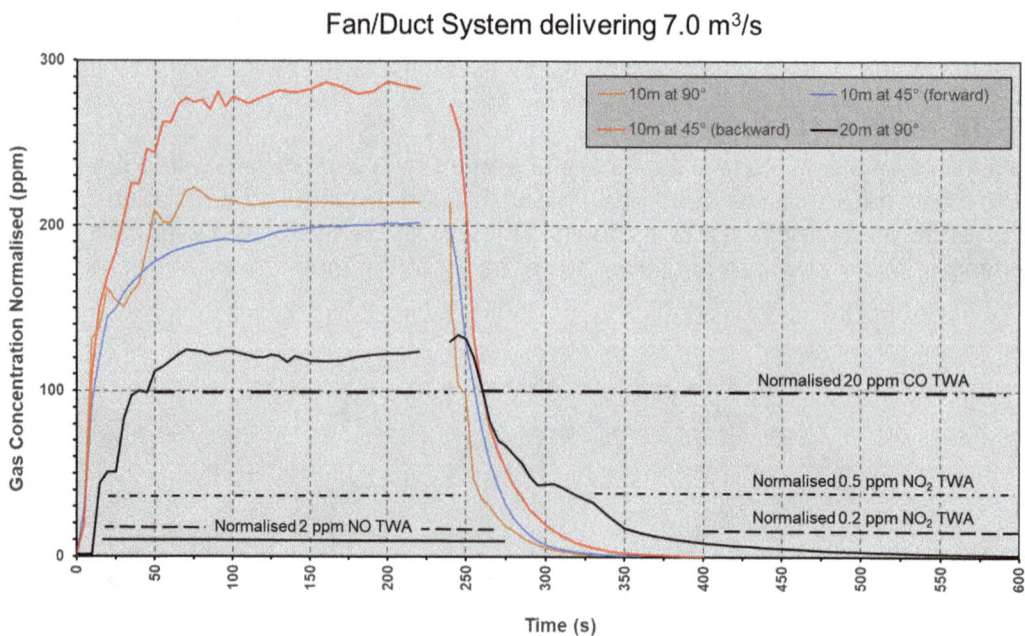

Fan/Duct System delivering 7.0 m³/s

FIG 11 – Gas concentration exceedances against lower CO, NO and NO₂ TWAs.

This indicates that significantly higher ventilation levels may be required in dead-end headings where these lower NO₂ and NO TWA exposures apply.

CONCLUSIONS

The CFD modelling has shown that the orientation of a dead-end, its size, and the air velocity in the main drive passing its entrance influence the flow pattern and volume circulating within the dead-end. This generally agrees with the findings of Meyer (1993) and Morla *et al* (2022, 2023).

Without diverting the air through a duct or forcing it with a fan duct system, the fresh airflow quantity into each end cannot be quantified. The Ø1220 mm open duct/diverter induced a flow equivalent of 4–5 per cent of the main drive volume. The auxiliary fan was located upstream with a prescribed flow. Regardless of the flow generation method, natural eddy, duct diverted, or fan induced, the airflow measured at different planes within dead-end will give higher overall quantities (the combination of flows inward and outward either at the entrance or within the heading), indicating significant local recirculation.

The best clearance of pollutants was achieved with either a high velocity (4.0 m/s) in the main drive or via auxiliary fan-driven systems. Arrangements with a low velocity (0.8–1.0 m/s) in the main drive were generally more problematic as contaminated air could travel upstream and then be recirculated back to the far end of the dead-end through the duct diverter or fan-assisted system. This even occurred when the vehicle tailpipe exhausted towards the main drive under the low flow condition.

With respect to orientation, the 90° and 45° forward configurations had better clearance properties than 45° backward angle across all flow arrangements primarily due to less circulation at the dead-end entry. As for depth, in the absence of an increase in volumetric flow, the shorter 10 m dead-end had a quicker clearance than the 20 m dead-end.

However, most importantly, the modelling, as completed, has failed to indicate that a fixed rule or formula can be applied to determine a maximum dead-end depth. More specifically, the 10 m distance, as stated in several regulations, may not be a valid demarcation point when diesel-related gas concentrations are considered. Furthermore, a more risk-based assessment could be required when NO and NO₂ 8-hr TWA limits are at or below 2 ppm.

Therefore, every mining operation should determine its own guidelines based on the diesel equipment deployed and their contaminant generation rates in association with the locally applicable OELs, plus consider any flammable gas, radiation, or heat management-related ventilation requirements. These should be supported by field investigations because, as shown in this study, CFD modelling may be unable to provide definitive results.

ACKNOWLEDGEMENTS

The authors wish to express their gratitude to Vale Ontario Operations, who requested the initial study.

REFERENCES

Chamber of Mines of South Africa, 1989. Flammable Gas in Metal Mines, A Guide to Managers, Safety and Technical Services, Chamber of Mines of South Africa, October 1989, p 1.

Deutsche Forschungsgemeinschaft (DFG), 2017. List of MAK and BAT Values 2017, Maximum Concentrations and Biological Tolerance Values at the Workplace, Report 53. Available from: <https://onlinelibrary.wiley.com/doi/pdf/10.1002/9783527812127.oth> [Accessed: 15 Feb 2024].

Government of Alberta, 2023. Occupational Health and Safety Code, Schedule 1, Table 2, Alberta Regulation 191/2021. Available from: <https://search-ohs-laws.alberta.ca/legislation/occupational-health-and-safety-code/schedule-1-chemical-substances/#5682> [Accessed: 15 Feb 2024].

Government of Manitoba, 2020. Workplace Safety and Health Regulation, Regulation 217/2006. Available from: <https://web2.gov.mb.ca/laws/regs/current/_pdf-regs.php?reg=217/2006> [Accessed: 5 Feb 2024].

Government of New Brunswick, 2023. New Brunswick Regulation 91–191. Available from: <https://laws.gnb.ca/en/ShowPdf/cr/91-191.pdf> [Accessed: 15 Feb 2024].

Government of Newfoundland and Labrador, 2023. Occupational Health and Safety Regulations, 2012, NLR 5/12. Available from: <https://www.assembly.nl.ca/Legislation/sr/Regulations/rc120005.htm> [Accessed: 15 Feb 2024].

Government of Northwest Territories, 2023. Occupational Health and Safety Regulations, R-039–2015. Available from: <https://www.justice.gov.nt.ca/en/files/legislation/safety/safety.r8.pdf > [Accessed: 15 Feb 2024].

Government of Nova Scotia, 2022. Underground Mining Regulations, NS Reg 296/2008. Available from: <https://novascotia.ca/just/regulations/regs/ohsmine.htm> [Accessed: 15 Feb 2024].

Government of Nunavut, 2016. Occupational Health and Safety Regulations, Nu Reg 003-2016. Available from: <https://www.canlii.org/en/nu/laws/regu/nu-reg-003-2016/124530/nu-reg-003-2016.html> [Accessed: 15 Feb 2024].

Government of Ontario, 2020. Control of Exposure to Biological or Chemical Agents, RRO, 1990, Reg 833, Table 1. Available from: <https://www.ontario.ca/laws/regulation/900833> [Accessed: 15 Feb 2024].

Government of Quebec, 2023. Regulation Respecting Occupational Health and Safety, Schedule 1, Chapter S-2.1, r13. Available from: <https://www.legisquebec.gouv.qc.ca/en/document/cr/s-2.1,%20r.%2013> [Accessed: 15 Feb 2024].

Government of Saskatchewan, 2019. The Mines Regulations, 2018. RRS cS-15.1, Reg 8. Available from: < https://publications.saskatchewan.ca/#/products/100307> [Accessed: 15 Feb 2024].

Government of Saskatchewan, 2021. The Occupational Health and Safety Regulations, 2020, RRS cS-15.1, Reg 10. Available from: <https://publications.saskatchewan.ca/#/products/112399> [Accessed: 15 Feb 2024].

Government of United Kingdom, 2020. EH40/2005 Workplace exposure limits, 4th edition, Health and Safety Executive. Available from: <https://www.hse.gov.uk/pubns/priced/eh40.pdf> [Accessed: 28 May 2024].

Government of Yukon, 1986. Occupational Health Regulations, Table 8, OIC, 1986/164, 17 October 1986. Available from: <https://laws.yukon.ca/cms/images/LEGISLATION/SUBORDINATE/1986/1986-164D/1986-164D_1.pdf?zoom_highlight=Occupational+Health+Regulations> [Accessed: 15 Feb 2024].

Institut für Arbeitsschutz (IFA), 2023. GESTIS International Limits Database. Available from: <https://limitvalue.ifa.dguv.de/> [Accessed: 15 Feb 2024].

Meyer, C F, 1993. Improving underground ventilation conditions in coal mines, SIMRAC (Safety in Mines Research Advisory Committee) Final Project Report COL 29a, November 1993.

Mine Health and Safety Council South Africa, 1996. Mine Health and Safety Act No. 29 of 1996 and Regulations. Available from: <https://www.mhsc.org.za/sites/default/files/public/publications/Mine%20Health%20and%20Safety%20Act%2029%20of%201996%20and%20Regulations%20Final%20Booklet.pdf> [Accessed: 19 Feb 2024].

Mine Safety and Health Administration (MSHA), USA, 2013. Subpart D – Air Quality and Physical Agents: 56.5001: Exposure limits for airborne contaminants, Subchapter K – Metal and Nonmetal Mine Safety and Health, Part 56 – Safety and Health Standards – Surface Metal and Nonmetal Mines. Available from: <https://www.govinfo.gov/content/pkg/CFR-2013-title30-vol1/pdf/CFR-2013-title30-vol1-part56.pdf> [Accessed: 15 Feb 2024].

Minerals and Mining, 2012. Minerals and Mining (Health, Safety and Technical) Regulations, 2012. Section 196(1)(a) and (b), p 129.

Government of British Columbia, 2022. Health, Safety and Reclamation Code for Mines in British Columbia. Table 2.1. Available from: <https://www2.gov.bc.ca/assets/gov/farming-natural-resources-and-industry/mineral-exploration-mining/documents/health-and-safety/code-review/health_safety_and_reclamation_code_nov2022.pdf> [Accessed: 15 Feb 2024].

Morla, R, Chen, J, Karekal, S, Godbole, A, Tukkaraja, P and Chang, P, 2023. Empirical and numerical investigation on the optimal length of eddy airflow in dead-end tunnel, in *Proceedings of the 19th North American Mine Ventilation Symposium 2023* (ed: P Tukkaraja), pp 649–658 (CRC Press).

Morla, R, Karekal, S and Godbole, A, 2020a. CFD simulations of DPM flow patterns generated by vehicles in underground mines for different air flow and exhaust pipe directions, *International Journal of Mining and Mineral Engineering*, 11:51–65.

Morla, R, Karekal, S and Godbole, A, 2020b. Investigation of DPM dispersion in unventilated dead-ends using transient flow modelling, *International Journal of Mining and Mineral Engineering*, 11:121–133.

Morla, R, Karekal, S, Godbole, A, Tukkaraja, P and Chang, P, 2022. Optimum Auxiliary Fan Location to Control Air Recirculation, *Mining, Metallurgy and Exploration*, 39(5):1–9.

Russian Mining Law, 2014. Federal Standards and Rules in the Region of Industrial Safety Regulations during Conduction of Mining Operations and Processing of the Solid Mineral, Regulation 164:2014.

Safe Work Australia, 2019. Workplace Exposure Standards for Airborne Contaminants, December 2019. Available from: <https://www.safeworkaustralia.gov.au/sites/default/files/2024-01/workplace_exposure_standards_for_airborne_contaminants_-_18_january_2024.pdf> [Accessed: 15 Feb 2024].

Safe Work Australia, 2024. Workplace Exposure Standards for Airborne Contaminants, April 2024 Available from: <https://www.safeworkaustralia.gov.au/sites/default/files/2024-04/workplace-exposure-limits-for-airborne-contaminants_april-2024.pdf> [Accessed: 15 Feb 2024].

Work Health and Safety Act, 2020. Work Health and Safety (Mines) Regulations 2022, r.652, p 444.

Occupational health (mine dusts, gases, radon, etc)

What is the value of inhalable coal dust exposure monitoring?

B Belle[1], H Wu[2], Y Jin[3], G Gamato[4], P Wild[4], M Webber[4] and M Kizil[5]

1. School of Minerals and Energy Resources Engineering, The University of New South Wales, Sydney, NSW 2052; The University of Queensland, Brisbane, Qld 4072; University of Pretoria, South Africa; 61DRAWINGS, Australia. Email: bb@61drawings.com
2. Gillies Wu Mining Technology Pty Ltd, Queensland.
3. CSIRO Mineral Resources, Pullenvale Qld 4069.
4. Anglo American, Brisbane Qld 4001.
5. Associate Professor, The University of Queensland, School of Mechanical and Mining, Queensland.

ABSTRACT

The re-emergence of 'Black Lung' or Coal Workers Pneumoconiosis (CWP) in Queensland (Qld) after reporting it being absent for over three decades had cast doubts on the rigour placed in the medical diagnosis and personal exposure assessment data. As of July 2023, Resources Safety Health Queensland (RSHQ) has reported that 68 and 88 cases of CWP and silicosis respectively, across the Qld mining and allied industry since 1984. The Qld Government amended the Qld Coal Mining Safety and Health Regulation 2017 (CMSHR) to reduce the exposure limits for respirable coal dust from 3.0 mg/m^3 to a level of 1.5 mg/m^3, with a respirable silica dust limit of 0.12 mg/m^3 in 2017 to 0.05 mg/m^3. Current exposure limit for inhalable coal dust remained at 10 mg/m^3, with the application of extended shift exposure limit values for compliance determination purposes.

Mining industry worldwide spends a significant amount of technical and financial resources in dust sampling to assess the exposures of health hazards for effectiveness of adequate control measures. Most mining countries carry out personal exposure monitoring for respirable dust. Unlike Australia, very few countries spend their resources in sampling of inhalable coal dust in mining industry. Since its inception in 1920s, the recommended occupational exposure limits of a substance have varied significantly between mining countries worldwide. Over the last two decades, international harmonisation of size-selective sampling curve and instruments which replicate human inhalation of dust particles have changed. This paper discusses the experiences of the inhalable coal dust sampling for exposure and compliance monitoring purposes in the coal industry and shares the shortcomings of current personal exposure monitoring, reporting, assessment and compliance determination challenges using extensive service provider exposure data. Further the paper implores the benefits of inhalable sampling and its value in the long-term personal dose-response curves, non-compliance to 'ideal' inhalable size-characterisation curve and understand potential level of risks. The investigation study indicates sufficient and prior due-diligence of sampler characterisation prior to its industry wide applications. Lastly, it is suggested that any adjustment of inhalable exposure limits using the inhalable samplers that do not meet the size characterisation standards may not benefit all the stakeholders in the industry, let alone compliance monitoring. Until then, what is the value of inhalable coal dust monitoring or how is it being enforced?

INTRODUCTION

Monitoring of coal dust and silica dust in mines is an important task as part of the exposure management journey that requires reliable knowledge of dust sampling devices that intended to collect the harmful dust. There are various means of measuring dust, *viz* personal sampling, area sampling and engineering sampling. Knowledge of routine dust exposure limit values can help workers' and industry focus on protection of workers respiratory health. Against this background, the scrutiny of available sampling devices used for routine sampling and exposure assessment that provides improved accuracy is continuing and appropriate. This paper shares experience of introducing a new instrument for the exposure monitoring that is relevant to similar industries worldwide. In Australia, Inhalable dust is governed by the standard AS 3640. This paper attempts to investigate the 'inhalable' sampler performance, sampling data, proposed limits, issues and its ultimate use in the exposure assessment and medical diagnosis purposes.

Past studies have suggested that the personal sampling method is the most suitable method for assessing, and most representative of, the worker's dust exposure (Leidel, Busch and Lynch, 1977; Kissell and Sacks, 2002). Dust sampling is pursued in mines to understand the level of risk associated with exposure to hazards. Figure 1 provides a typical fraction of dust data in a British colliery taken up by exposed humans during breathing (Gibson, Vincent and Mark, 1987). It was noted that the inspirable dust mass of 38.4 mg contained 6.6 mg of respirable dust, 3.7 mg of tracheobronchial or Inhalable dust, 13.5 mg of thoracic dust.

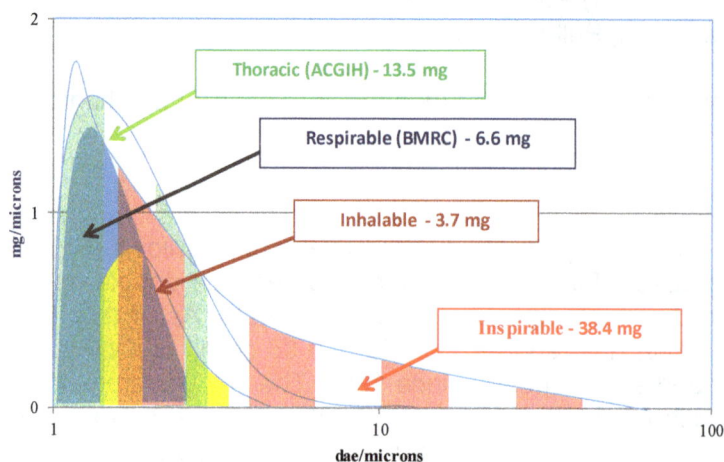

FIG 1 – Illustration of typical respirable fraction of coal dust breathed.

A South African industry study for the introduction of any new dust-monitoring instruments for personal sampling in underground mines to be accepted by the key stakeholders, were required to meet the basic requirements (criteria) as outlined below (Belle, 2002, 2012):

- They must be intrinsically safe for use in underground mines.

- They must sample according to the accepted size-selective criteria at the specified flow rates.

- They must meet the ±25 per cent National Institute for Occupational Safety and Health (NIOSH) accuracy criterion.

- They should preferably use a different 'quick' analysis procedure to the weighting method that is currently used.

- They must be robust enough to withstand the harsh conditions prevailing in mines.

- They must be compact and portable for personal sampling.

- They must offer the possibility of collecting dust samples for further quartz analysis.

The South African extensive multi-commodity mine study noted that the IOM respirable foam sampler failed to meet the NIOSH accuracy criteria and was not pursued further for use in South African mines (Belle, 2012). In Australia, personal respirable dust monitors are to meet the AS2985, definition of respirable dust with specific sampler flow rates mentioned in that standard. As a result of international harmonisation of size-selective dust criteria (ISO, 1995), led to replace the traditional 'total dust' definition by 'inspirable' that later termed as 'inhalable' dust. Therefore, it is expected that the inhalable dust samplers are equally required and independently validated for coal dust to meet the following requirements for any sampling program in the coal mining industry, as inhalable coal dust sampling is practiced widely in the industry:

- ISO 7708:1995 (ISO, 1995) definition of Inhalable dust.

- AS 3640 (Standards Australia, 1989) Method for sampling and gravimetric determination of inhalable dust.

- British Method (Methods for the Determination of Hazardous Substances (MDHS, 2000)) 14/3 method for inhalable dust in air.

History of 'total' dust can be traced back to South African 'sugar tube' that can produce health effects after deposition anywhere in the body, including not only the lung but also other parts of the respiratory tract (eg the nasopharynx), as well as elsewhere in the body if the aerosol material is soluble (Walton and Vincent, 1998). In the USA or rest of the world, traditionally, dust sampling is carried out to measure 'total' and 'respirable' dust as outlined by NIOSH analytical methods. For 'total dust' sampling, a standard 37 mm cassette with a PVC filter membrane, which would collect airborne dust and small enough to fit through the cassette's inlet opening of approximately 4 to 4.5 mm diameter is used. For respirable dust sampling in US mines, currently real-time gravimetric sampler that uses a size-selective HD type cyclone ahead of the filter cassette is used as a compliance device for coal dust exposure assessment.

Unlike harmful metal dust at very low concentrations, coal dust is hydro-phobic and known inhalable dust toxic health risk is less clearly understood. Currently in the USA, there are no personal occupational exposure limits (OEL) or personal exposure limits (PEL) for inhalable dust. Enforcement of those limits was suspended as the Final Rule on Air Contaminants Project (Occupational Safety and Health Administration (OSHA), 1989). While most of the exposure limits were originated in the USA, views submitted to OSHA (1989) when formulating generic 'total' dust or unregulated particulate limit were that there was no evidence of adverse health effects associated with exposure to these particulates. The submissions by the American Iron and Steel Institute at the time noted that effects of such exposures were found to be 'short-term and immaterial'. OSHA has established an 8 hr TWA total dust limit of 10 mg/m³ for all particulates having identified health effects in the toxicological literature but retained a 15 mg/m³ 'total' dust limit for those particulates not specifically linked to health effects other than physical irritation (OSHA, 1989).

In Australia, preliminary industry investigation through engagement with end users (including medical surveillance, (Newbegin, 2020, personal communications)) suggested that the application of the inhalable dust exposure monitoring program implemented at the coal mining operations and 'where' and 'how' these inhalable exposure results in relation to the general health of CMW are used is not clear, well understood, other than for compliance enforcement, where applied. In Australia, Safe Work Australia (SWA) notes that where no specific exposure standard has been assigned and the substance is both of inherently *low toxicity* and free from toxic impurities, exposure to dusts should be maintained below 10 mg/m³, measured as inhalable dust (8 hr TWA) as per AS 3640. This has been the basis for the monitoring of 'inhalable dust' in the coal mining industry. Furthermore, the ambiguities in the differences in measured respirable and inhalable dust has not been scientifically explored. What is clear is that there is inadequate medical evidence for coal dust on the short- and long-term medical health effects associated with exposure to inhalable dust and the reason behind the suspension of 'inhalable' or 'total' coal dust PELs in US or most of the world is not known. It is noted that Australian Coal industry is carrying out relevant inhalable dust research.

INHALABLE DUST, EXPOSURE LIMITS AND SIZE SELECTIVE CURVE

Based on the past epidemiological knowledge (Orenstein, 1960), it has been established that the respirable dust particle size distribution is critical due to its potential health effects and quantifying the risks. Respirable dust refers to particles that settle deep within the lungs that are not ejected by exhaling, coughing, or expulsion by mucus. Since these particles are not collected with 100 per cent efficiency by the lungs, respirable dust is defined in terms of size-selective sampling efficiency curves. This had led to internationally recognised respirable size-selective sampling widely known as the British Medical Research Council (BMRC) definition of the respirable dust fraction or Johannesburg curve with a median aerodynamic diameter of 5 µm collected with a 50 per cent efficiency (D50) (BMRC, 1952). In reality, these size-selective curves represent lung penetration of dust particles that dust sampling instruments attempt to replicate. The International Standards Organisation (ISO) in 1995 recommended that the definition of respirable dust follow the theoretical convention described by Soderholm with a D50 of 4 µm (ISO, 1995; Soderholm, 1989, 1991). An international collaboration for sampling harmonisation has led to the agreement on the definitions of health-related aerosol fractions in the work-place, defined as the inhalable, thoracic and respirable curve (ISO, 1995; ACGIH, 1985, 1999; CEN, 1993).

The American Conference of Governmental Industrial Hygienists (ACGIH) established an Air Sampling Procedures Committee to review available data on regional deposition of inhaled particles and on the collection efficiencies of sampling instruments in 1988. The committee recommended 'Inspirable' Particulate Mass applies to material which is hazardous anywhere in the respiratory tract (Phalen *et al*, 1988). The modern terminology has replaced the term 'inspirable' with 'inhalable' (Kenny, 2003). Inhalable dust refers to the particle size entering the mouth and nose during normal breathing and may be deposited in the respiratory tract. Vincent *et al* (1990) documented that the human respiratory system is an inherent effective size-selective aerosol sampler, and therefore, it is misleading to assume that all airborne particles will enter it. Large particles are excluded from entering the nose and mouth through inertial separation. Personal exposures to this definition of large size dust particles in the workplace may cause physical irritation and respiratory health effects. IARC Monogram (1997) noted that there is no consistent evidence supporting an exposure-response gradient for coalmine dust and stomach cancer.

Vincent *et al* (1990) observed that aspiration or some time referred to as 'inspired' is a function of a number of parameters, including particle size, external air speed, orientation to the prevailing air movement direction, and breathing rate and volume. However, for external wind speeds of a few metre per sec and lower, the probability of a particle entering the mouth or nose (termed inhalable dust particles) may be generalised as being around 100 per cent for dust particles with aerodynamic diameters of a few microns and below, reducing to around 50 per cent at 100 μm aerodynamic diameter. Figure 2 summarises the BMRC and ISO size-selective curves for dust sampling in mines (ISO, 1995; ACGIH, 1985) to demonstrate likely penetration of dust particle sizes to various regions of human respiratory system. It is important to note that it is not only a difference in the D_{50} value but an entire size-selective curve. Interestingly, inhalable sampling for coal dust is rarely practiced worldwide except in Australia and potentially UK. In the early 2000s the ISO standards on respirable and inhalable dust have come to prominence with various sampling devices available for sampling and assessment. Most sampling instruments that purported to measure the 'total' dust were developed without regard to their sampling efficiency characteristics (Ramachandran, 2005). Anecdotally, there were suggestions recently, even to monitor 'thoracic' fraction dust sampling from some dust sampling service providers to the coal mining operations in Australia.

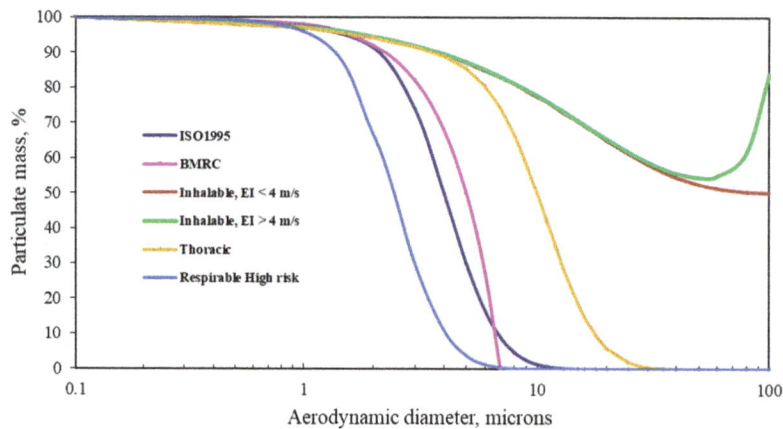

FIG 2 – Respirable and Inhalable dust size selective characteristics (ISO, 1995; ACGIH, 1985).

The inhalable convention is based on particle penetration through the mouth and nose of a breathing mannequin over a range of wind speeds and orientations with respect to the wind, and is defined (Volkwein, Maynard and Harper, 2011; Maynard and Baron, 2004) as:

$$SI(d_{ae}) = 0.5 \times (1 + e^{-0.06 \cdot dae}) \tag{1}$$

for $0 < d_{ae} < 100$ μm. $SI(d_{ae})$ is the inhalable penetration fraction of dust particles entering the system as a function of aerodynamic diameter d_{ae}. It is to be noted that the Figure 2 for wind speeds > 4 m/s demonstrates the limitations of the prescribed ISO (1995) formula, which suggests that it should not be applied to particles with diameter of > 90 μm and win velocities U > 9 m/s. This very same size-characterisation curve implies the current use of inhalable samplers for air velocities under certain mining occupational environments where air velocities are > 4 m/s, as in the longwall face.

In Australia, the term inhalable dust sampling applies to both non-toxic and toxic dusts. Exposure standards for dusts are measured as inhalable dusts unless there is a notation specifying an alternate method, eg silica. In Australia, inhalable dust is defined as same (Table 1) as in ISO 7708:1995 and must be measured according to AS 3640-2009 (Standards Australia, 2009). It is to be noted the Australian health regulations refer to AS 3640-2009 and not ISO 7708:1995. AS 3640-1989 originally recommended either the Casella seven-hole sampler or the IOM sampler for personal sampling of inhalable fraction of airborne dust. The AS 3640-2009 notes that providing the airborne particulate does not contain other hazardous components, compliance with the exposure standard for dusts not otherwise classified should prevent impairment of respiratory function. Where no specific exposure standard has been assigned and the substance is both of inherently low toxicity and free from toxic impurities, exposure to dusts should be maintained below 10 mg/m^3, measured as inhalable dust (8 hr TWA). As expected, the exposure standard for dusts or particles not otherwise classified (PNOC) should not be applied where the particulate material contains other substances which may be toxic or cause physiological impairment at lower concentrations. For example, where a dust contains asbestos or crystalline silica, like quartz, cristobalite or tridymite, exposure to these materials should not exceed the exposure limit values for such substances.

TABLE 1

Inhalable dust definition as per ISO 7708:1995/AS 3640-2009.

Particle equivalent aerodynamic diameter (µm)	Inhalable convention,% for wind speeds < 4.0 m/s	Inhalable convention,% wind speeds > 4 m/s
0	100	100
1	97	97
2	94	95
3	92	92
4	89	90
5	87	87
6	85	85
7	83	83
8	81	81
9	79	79
10	77	78
11	76	76
12	74	75
13	73	73
14	72	72
15	70	71
16	69	69
18	67	67
20	65	65
25	61	62
30	58	59
35	56	57
40	55	56
50	52	55*
60	51	55*
80	50	62*
100	50	84*

*   ISO (1995) inhalable size-selective curve limitations for air velocities > 4 m/s and risks of inhalable samplers where air velocities exceeds 4.0 m/s (see Figure 4).

The origins of inhalable sampler, which is used in Australia, as shown in Figure 3, can be traced back to the aspiration measurements on breathing mannequin research by the Institute of Occupational Medicine (IOM) by Mark and Vincent (1986). There have been various further studies on the shortcomings of IOM inhalable sampler, sample collection and interference, influence of environmental conditions that is expected to follow the ISO (1995) inhalable size convention with particles up to 100 μm aerodynamic diameter (Liden and Kenny, 1994; Kennedy *et al*, 1995; Aitken and Donaldson, 1996; Smith, Bartley and Kennedy, 1998; Roger *et al*, 1998; Li and Lundgren, 1999; Liden and Bergman, 2001; Aizenberg *et al*, 2001). Despite these shortcomings, and in the absence of any other inhalable guidance sampler, it's been accepted in the UK, followed by Australia as a gravimetric method for determining inhalable dust levels (Health and Safety Executive, 1997; Safe Work Australia, 1995).

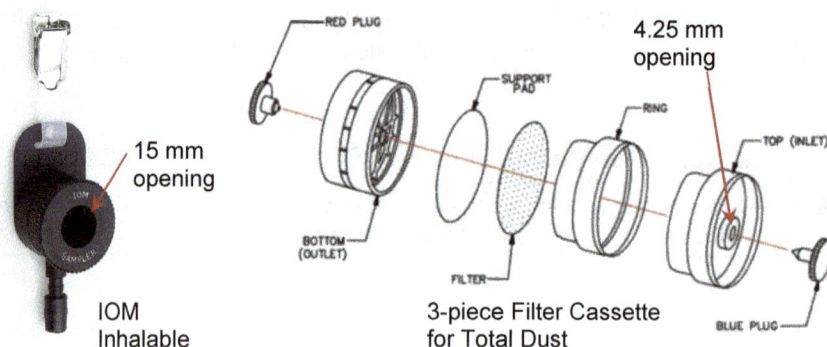

FIG 3 – IOM inhalable sampler and the total dust filter cassette.

The operational issue of 'wall deposits' that is the dust attached to the sampler walls or surfaces was discussed by Harper and Demange (2007) for both IOM sampler and the traditional closed-face 37 mm cassette used in the USA. While no sampler may match the ISO (1995) conventions, understanding the size-ranges collected by the inhalable sampler in the field or manufacturer's size selective curves or sampler bias as a function of test aerosol distributions (Bartley and Breuer, 1982; Liden and Kenny, 1992; Maynard and Kenny, 1995) or dust concentration level is important for practical reasons for workers to understand their risk, operators to improve on controls or to the regulators to adjudge the risk limits.

Both inhalable and total dust samplers operate at a recommended flow rate of 2.0 lpm without any size-selective devices. Inhalable sampler has a larger open circular inlet (15 mm) with a lip that protrudes 1.5 mm outwards, with an aim to minimise the potential for particles deposited on the outer surfaces of the inlet to be carried into the sampler. The 'total' dust closed face three-piece filter cassette has an opening inlet size of 4.25 mm.

Considering the disproportionate attention given in some quarters of the globe to the inhalable sampling and potential limits for inhalable dust, some scrutiny has come. For example, Volkwein, Maynard and Harper (2011), Harper and Muller (2002) and Harper, Akbar and Andrew (2004) have observed that the upper limit of the size range of interest (100 μm) is an *arbitrary* selection, and particles larger than this can be airborne and therefore are available for possible inhalation. Specifically, this single factor of 'upper size' limit alone can be a significant driver in the dust concentration determination values during exposure assessment or compliance determination. They had argued that the inhalable convention does not account for mouth breathing potential and many other physiological variables due to changes in workforce age distribution, fitness, gender, ethnicity, and so on (Liden and Harper, 2006). Adding to the diverse parameters is the coal mining operational environment where the turbulent air velocity conditions of 3 to 5.0 m/s, against the original surface industrial calm air sampler performance evaluation settings of 0.1 to 0.3 m/s (Baldwin and Maynard, 1998; Liden, Juringe and Gudmundsson, 2000; Aitken *et al*, 1999; Kenny *et al*, 1999). What is definitive is that these inhalable samplers and their suitability is not evaluated for underground coal mining conditions where the legislative requirement of minimum air velocity requirements > 0.3 m/s, with normal air velocity ranges of 3 to 5 m/s, that are deployed to dilute and manage safety risks associated with the flammable gases.

Since the design inception of personal gravimetric samplers or cyclones, most to all sampler evaluations were carried out under 'calm air conditions' which is understood to be < 0.1 m/s air velocity to legislated minimum air velocities of 0.3 m/s for underground working environment. Baldwin and Maynard (1998) had noted that typically 80 per cent of the working conditions would have air velocities of up to 0.3 m/s, which influences the efficiencies of personal samplers. However, in modern coal mining conditions, the reality is that air velocities would be an order of magnitude higher or 'turbulent' conditions than these samplers that were designed and evaluated. Figure 4 shows the underground coal mining turbulent air velocity conditions, an operational reality, against laboratory evaluation conditions, which would have impact on sampler performance efficiencies.

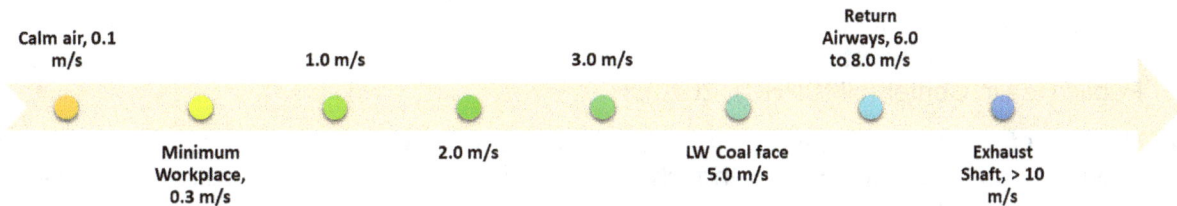

FIG 4 – Laboratory personal sampler evaluation conditions < 0.3 m/s (Baldwin and Maynard, 1998) against underground coal mining ventilation conditions.

In the absence of clear evidence of past medical investigations in relation to coal dust, there are misperceptions and interpretations of health risk definitions used in the literature for respirable and inhalable coal dust. However, what is unquestionable is the critical importance of inhalable monitoring of toxicity of traditional low concentration high risk hygroscopic metal dust (cadmium, led) to asses known health impacts and inhalable exposure data as part of medical diagnosis. The influence of non-conformance to size-selective sampling, flow rates, and measured dust levels and compliance determination has been recently unearthed in Australia that led to the changes to selection of appropriate respirable dust sampler for use (Belle, 2017, 2018). However, this paper attempts to understand if there is a such similar sampler bias in IOM Inhalable coal dust sampler which has a 15 mm diameter inlet orifice, where particles are aspirated into and collected over a 25 mm filter, which is operated at 2.0 Lpm of flow rate.

Following paragraphs provide a glimpse of what could possibly have been the rationale behind inhalable coal dust sampling in Australia, with almost no studies on coalmine inhalable dust monitoring. In the absence of scientific due diligence or apparent significant benefits of inhalable coal dust may potentially lead to unverified confidence in dust controls, when used alongside with respirable dust data.

- Considering the health risks associated with the inhalable wood dust, Hinds (1988) noted a sampling method that accurately measures the amount of inhalable wood dust, including particulate deposited in the nose, is therefore desirable for the evaluation of worker exposures to airborne wood dust. Mark and Vincent (1986) observed that inhalable sampling method is expected to collect more particulate mass than the total dust method.

- A surface lead smelter study (Spear et al, 1997) with side-by-side personal aerosol sampling with 'total' (37 mm sampler) and inhalable (IOM personal sampler) showed the ratio, expressed as IOM (mg/m^3)/37 mm (mg/m^3) of individual paired samples were consistently greater than unity with values ranging from 1.39 to 2.14 aligned with IOM samplers collecting large particles than the 37 mm sampler of airborne dust (Mark et al, 1994; Kenny, 1995).

- A second lead smelter study (Spear et al, 1998) showed the mean mass ratios of inhalable to respirable for different workplaces with likely differing aerosol environment measured as per the ACGIH/ISO/CEN particle size-selective criteria varied from 4 to 10 using the personal inhalable dust spectrometer (PIDS).

- In a carpenter shop exposure study on wood dust, Martin and Zalk (1998) described a comparison of sampling results from air monitoring conducted using total dust and inhalable dust sampling methodologies for the evaluation of wood dust exposures for its association with the health effects referring to the Australian study (Pisaniello, Connell and Muriale, 1991). The Australian health study reference noted that the potential health effects from exposure to wood dust include pulmonary function changes, allergic respiratory responses (asthma), and cancer of the nasal cavity and paranasal sinuses. The Safe Work Australia standards (1995) for inhalable wood dust exposure for hardwood and softwood are 1 mg/m^3 and 5 mg/m^3 respectively and historically 'total' wood dust exposures were measured (Alwis, 1998).

- Martin and Zalk (1998) concluded that the total dust sampling method underestimates the 'true total' inhalable aerosol and suggested that the existing inhalable sampling method needs further research and development before it can be accurately applied for evaluations for wood dust exposure assessment.

- Further two studies related to wood dust (Kim and Lee, 1996; Perrault, Cloutier and Drolet, 1996) reported that Inhalable/total dust ratios of 1.9 to 2.8 and 0.2 to 11.3 respectively and were dependent on dust concentration levels and the type of industry (Navy and Marine Corps Public Health Center (NMCPHC), 2020).

- Vincent *et al* (1997) suggested guidelines for use where it is deemed desirable to adjust exposure data to account for the change in exposure assessment rationale (based on generalisation of results of comparisons between 'total' aerosol as measured using the 37 mm sampler and inhalable aerosol as measured using the IOM sampler) with values 1 to 2.5 and for similar exposure groups, found to take values from close to unity to as large as 4.

- A US defence study (Clinkenbeard *et al*, 2010) that collected breathing-zone air samples for chromium collected for workers engaged in corrosion control maintenance operations on several types of aircraft at several US Air Force bases using pairwise modified 37 mm total dust sampling cassette with an IOM inhalable dust sampler. This approach utilised total chromium as a sensitive surrogate indicator of total aspirated mass. Linear regressions showed that the modified 37 mm cassette over-samples aerosol by 35 per cent compared to the IOM inhalable sampler when a wide range of aerosol concentrations and compositions for multiple field locations are sampled. This is the only study that potentially suggests that total sampling underestimates the dust levels when compared with IOM inhalable sampler.

- Liden *et al* (2000) carried out parallel inhalable personal dust sampling with the open-face filter cassette and the IOM sampler dust for nine types of organic dust. Parallel samples numbering 749 were obtained from 152 plants. The coefficient of regression for each subset ranged between 0.2 and 0.7. Based on the results of this study and the difference in sampling efficiency for large particles between the two samplers, it was concluded that the numerical value of the OEL for inhalable dust may be set at approximately **twice** the numerical value of the corresponding limit value for 'total dust'. If this were to be applied to coal dust, this would suggest that the inhalable limit value would be 20 mg/m^3, considering the current total dust limit value of 10 mg/m^3. This outcome in reality may not be beneficial to coalmine workers, considering the respirable coal dust limit is reduced by half in 2018.

- A field study (Demange *et al*, 2010) on metal exposure results comparing a 37 mm Closed-Face Cassettes and IOM Samplers, noted consistency to those published elsewhere with a ratio IOM/total dust of much higher than 1.

- Verma (1984) studied the measured 40 sets of side-by-side sampling relationship between Inhalable dust using overburden respiratory burden (ORB) sampler developed by Ogden and Birkett (1978), total dust and respirable dust by 10 mm nylon cyclone operated at 1.7 Lpm matching ACCGIH curve and MRE horizontal elutriator (Casella 113A) operated at 2.5 Lpm matching BMRC curve, in an area (static) sampling program at eight selected ferrous and non-ferrous foundries In the foundry environments surveyed, study noted that the total dust correlated highly with the inhalable dust concentration (R^2 = 0.94). The determined relationship from the field evaluations showed that Inhalable/total dust ratio was found to be less than 1.0.

- A Canadian steel industry (including welding) Hexavalent Chromium contaminant exposure assessment study (Shaw *et al*, 2020) showed that inhalable/total dust ratio was found to be 2.2.

- The IARC had classified carbon black as a possible human carcinogen (IARC, 1996) and further literature suggested that inhalation of elemental carbon black may be associated with certain measures of respiratory morbidity (Gardiner *et al*, 1993). Therefore, inhalable carbon black measurement trends were studied using IOM inhalable sampler by Van Tongeren, Kromhout and Gardiner (2000) in the European Carbon Black Manufacturing Industry for overexposure assessment and review of limits.

- Görner *et al* (2010) had carried out an assessment of inhalable dust exposure using five different personal inhalable aerosol samplers in European laboratory wind tunnels under calm air and below 1.0 m/s air velocities using polydisperse glass-beads' test aerosol. Samplers tested were IOM sampler (UK), two versions of CIP 10-Inhalable samplers, 37 mm closed face cassette sampler (USA), 37 mm cassette fitted up with an ACCU-CAP™ insert (USA), and Button sampler (USA). Compared with CEN–ISO–ACGIH sampling criteria for inhalable dust, the experimental results show fairly high sampling efficiency for the IOM sampler. Significant differences between moving air and calm air sampling efficiency were observed for all the studied samplers. What are unknown in this study are the differences in the measured inhalable concentration levels between various inhalable samplers, when exposed to different dust levels, as in the operations. In comparison, for operating coal mining conditions, the air velocities are four times higher than those referred to by Görner *et al* (2010).

- Area sampling performance of six inhalable aerosol samplers was studied using monodisperse, solid particles by Li, Lundrgren and Rowell-Rixx (2000). The study reported that the area sampling performance of the foam sampler is highly dependent on wind orientation, wind speed and particle size. When the measured sampling efficiency was compared with the inhalable convention, the IOM sampler over sampled the large particles (> 20 μm).

- A German study (Wippich *et al*, 2020) to determine conversion functions from inhalable to respirable dust fractions of 15, 120 parallel measurements in German Database with no reference to coal dust concluded that all conversion functions are power functions with exponents between 0.454 and 0.956 and the data do not support the assumption that respirable and inhalable dust are linearly correlated in general.

- IOM report had recommended the Inhalable IOM sampler and Higgins-Dewell Samplers as suitable candidate samplers for measuring personal respirable dust exposure measurements for the Nickel (Ni) industry (Jiménez, Tongeren and Aitken, 2012) based on the historic IOM studies.

AUSTRALIAN OPERATIONAL EVALUATIONS

In Australia, RSHQ data shows the respirable dust limits changed from 3.0 mg/m^3 in 2017 (CWP cases re-identified in 2015) to 2.5 mg/m^3 in 2018 and 1.5 mg/m^3 in 2020 for coal dust. However, there were no changes made to the inhalable dust. In a knowledge share (Figure 5) at the Dust and Respiratory Health forum of 2020, showed an interesting profile of average respirable and inhalable dust levels in a longwall face. While, the respirable dust levels followed the changes in the OEL values, inhalable dust levels did not make a difference at the longwall face. From a worker exposure perspective and engineering control perspective, this is contradictory. Dust suppression efficiency decreases with decreasing particle size (herein respirable dust), and dust control efficiency increases with increasing particle size of the airborne dust. Secondly, higher inhalable average dust values can be attributed to dust sampling program or sampling instrument deficiencies at the anonymous site, although collected D50 of inhalable dust is an order of magnitude higher. While these issues are not readily addressed, it brings into the fore, questioning the value of inhalable sampling and its use or interpretation. It cannot be sure if there is any reasonable worker exposure assessment can be made. Furthermore, there is lack of clarity on the use of divergent worker dust exposure assessment data outcomes on any medical diagnostic significance, let alone the suggestion for pursuing thoracic coal dust sampling.

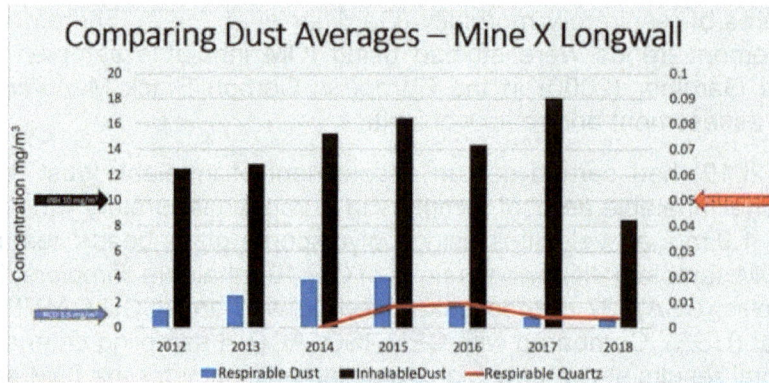

FIG 5 – Relationship between measured respirable and inhalable dust over the years (Source – Qld Dust Forum, 2020).

Another annual survey data of respirable and inhalable dust is shown in Figure 6. The mean respirable coal dust concentration for the period was 0.75 mg/m³, while the average inhalable dust concentration for the period was 10.41 mg/m³, with few higher measured inhalable dust values, exceeding the limit value of 10 mg/m³. What can one make out of the situation based on the inhalable dust data? In the USA, the total dust limit of 10 mg/m³ was equivalent to 2 mg/m³ with less than 5 per cent of silica present in the sample. Figure 7 shows an example of inhalable sample dust collected for high concentration values of 17 mg/m³ and 48 mg/m³. Confidence in these measured inhalable dust levels were questionable, in terms of visibly larger chunks of agglomerated dust on the filters. Situations such as these and absence of past coalmine studies, provoke questions in relation to the value of collecting inhalable dust or the samplers that were used during the collection or how they could be related to compliance respirable dust that is effectively controlled.

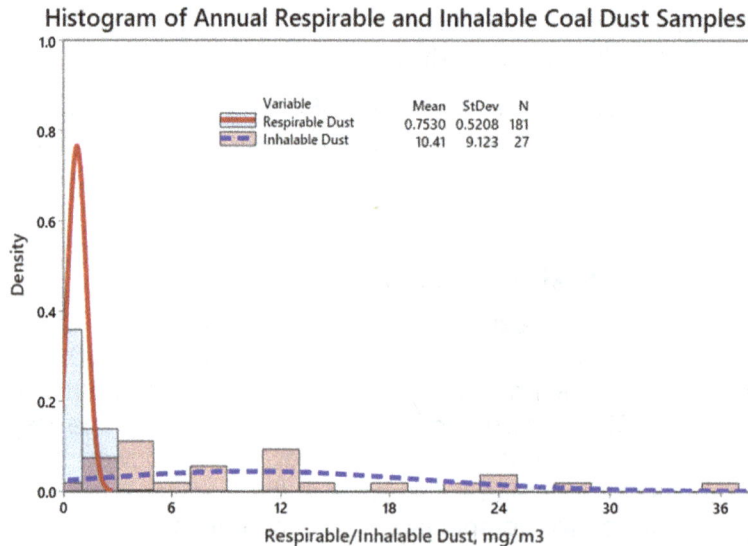

FIG 6 – Example of relationship between measured respirable and inhalable coal dust.

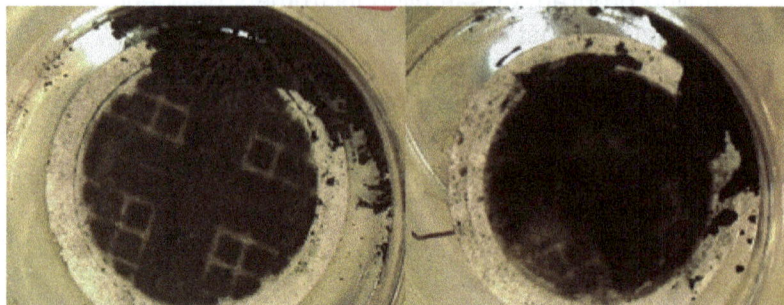

FIG 7 – Examples of inhalable coal dust sample filters with measured dust levels of 48 mg/m³ (left) and 17 mg/m³ (right).

FIELD EVALUATIONS OF SIDE-BY-SIDE INHALABLE, RESPIRABLE AND TOTAL DUST SAMPLERS

Considering the above practical anomalies and differences in measured inhalable dust levels found, a field evaluation comprising of pairwise sampling of inhalable, total and respirable dust samplers was carried out. The samplers were operated as per the sampler operating instructions, which are aligned with the ISO 1995 methodology. This section of the paper discusses the results (Table 2) of the field evaluations on surface of the inhalable, and total dust samplers against the respirable samples collected. A total of 81 filters including blanks were collected and the dust samples were weighed, and concentration levels were determined at an independent accredited Australian laboratory facility with the limit of reporting (LOR) is 0.01 mg.

TABLE 2

Field measurement of three-way sample results of inhalable, respirable and total dust concentrations.

Pair No	Inhalable (I)	Respirable (R)	Total (T)	I/R	T/R	T/I
1	1.239	0.645	1.880	1.92	2.92	1.52
2	1.271	0.713	1.411	1.78	1.98	1.11
3	2.384	0.398	1.832	6.00	4.61	0.77
4	0.747	0.664	1.207	1.12	1.82	1.62
5	1.804	0.368	1.748	4.91	4.75	0.97
6	2.869	0.299	1.659	9.60	5.55	0.58
7	3.030	1.185	3.978	2.56	3.36	1.31
8	1.503	0.800	2.500	1.88	3.12	1.66
9	1.541	0.486	0.708	3.17	1.46	0.46
10	1.667	0.473	2.163	3.52	4.57	1.30
11	0.081*	0.458	1.390	0.18	3.03	17.10
12	0.879	0.377	1.617	2.33	4.29	1.84
13	0.214	0.155	0.026	1.37	0.16	0.12
14	0.733	0.174	0.725	4.21	4.16	0.99
15	0.131	0.037	0.074	3.52	1.98	0.56
16	0.202	0.015	0.073	13.75	4.95	0.36
17	0.105	0.081	0.048	1.30	0.60	0.46
18	5.008	0.717	5.942	6.98	8.28	1.19
19	1.060	0.622	0.385	1.71	0.62	0.36
20	4.058	0.947	1.650	4.29	1.74	0.41
21	1.993	1.062	1.275	1.88	1.20	0.64
22	12.532	0.787	6.141	15.92	7.80	0.49
23	13.293	1.200	5.072	11.07	4.23	0.38
24	10.081	2.913	11.449	3.46	3.93	1.14
Average	**2.85**	**0.65**	**2.29**	**4.52**	**3.38**	**1.56**

* Unusual sample result.

For the gravimetric results, there were issues with loose dust, torn filter and switching filters. Those samples where cases of suspected pump failure or terminated prematurely in some inhalable/total samples were not part of the analyses. A total of 24 pairwise sampling results were available and they were taken over four sampling periods. There were some samples of very high inhalable dust levels with relatively moderate to high total and respirable dust samples (Pairs 22, 23 and 24). These 'high' inhalable dust levels (more than twice of the 'total dust') samples are probably due to large dust particles were deposited into these inhalable dust sample heads as these IOM samplers have much larger sampling inlet (15 mm) than the 'total dust' (the three-piece cassette) with 4.25 mm inlet. Sampling observations noted that large dust particles at higher dust loads were visible near the dust sources where some of the samples were collected. The relationship between the measured values obtained from the side-by-side inhalable-respirable, inhalable-total, total-respirable and inhalable-total dust levels collected under similar test conditions is shown in Figure 8. There was no statistical comparison between these samplers was made, as these samplers are designed and operated to different size-selective performance characteristics. From the plot it is observed that all field measurement values included both compliance and non-compliance levels for the sampling period and that the scatter was wide for both low and high dust concentrations.

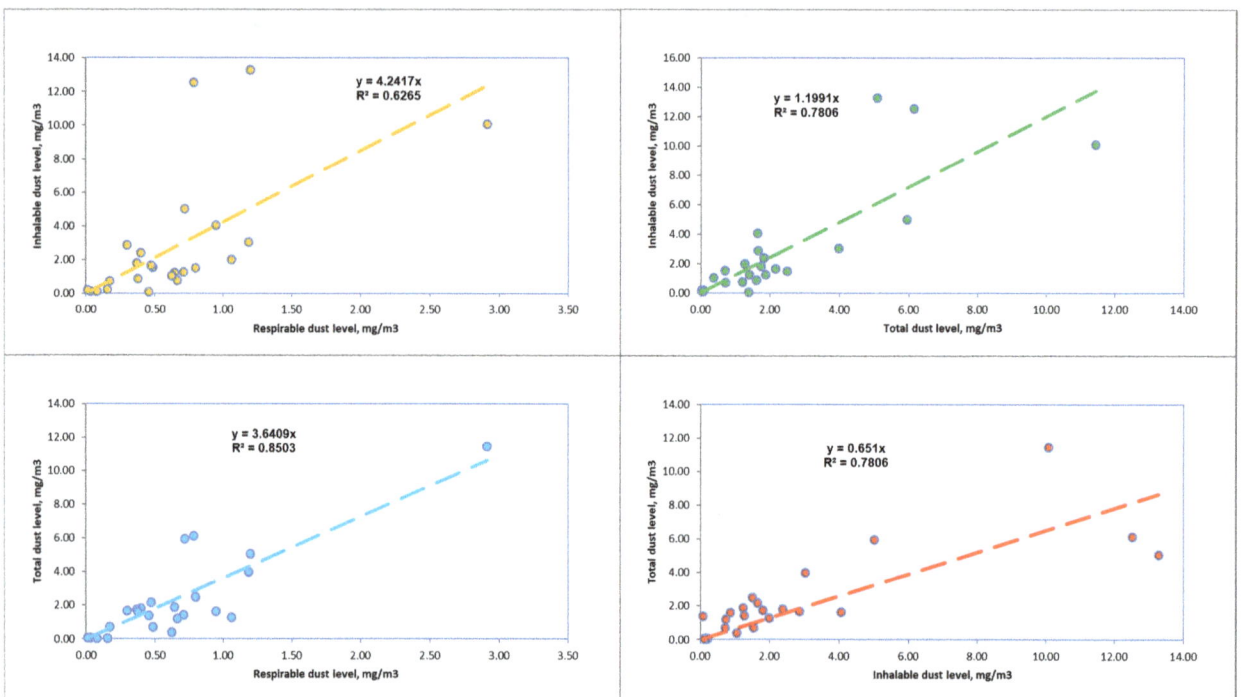

FIG 8 – Relationship between respirable-inhalable-total dust for the three-way samples.

The review of sample data shows the large variation between sampled values using the IOM and total dust samplers depends on the size of large/chunky particles. Inhalable and Total values for sampler 24 are completely different from samplers 22 and 23: inhalable levels are similar but the total dust level for sampler 24 is more than double higher than other two samplers. The existing Inhalable and total dust samplers without size selective device are not able to obtain a confident relationship for any assessment of dust conditions. Even for compliance purpose, different site and operation could generate different varieties of particles with varied large particles. The high sampled concentrations doesn't mean high health hazards because a large amount of collected particles are greatly larger than 100 micron (D50).

The coefficient of determination values (R^2), a quantitative measure of variation attributed, between the inhalable-respirable, total-respirable and inhalable-total dust sample pairs were 0.63, 0.85, and 0.78 respectively. The plot shows a nominal linear relationship and there is a significant difference between the measured dust levels by the inhalable, total and respirable samplers. For the study, the average measured levels (excluding pair #11) of the inhalable, respirable and total dust are 2.97 mg/m³, 0.66 mg/m³ and 2.33 mg/m³ respectively for the field test conditions. The linear relationship model shows relatively poor relationship between Inhalable and respirable sample dust data. When comparing the inhalable and total dust data, excluding large concentration values of 10 mg/m³ for pair # 22, 23 and 24, it is observed that the measured dust levels were approximately 1.5 mg/m³ by the two samplers. This brings to the question, the impact of inhalable samplers at higher dust concentrations with large coal dust clouds and what it means if these samples were to be operated at higher concentration values or dusty conditions, that yielded wide results of inhalable sampler results.

For average inhalable or total concentration below 2 mg/m³, the measured differences between inhalable and total dust sample values are relatively small. If one were to use the respirable dust standard of 1.5 mg/m³, using the relationship between the inhalable and respirable sample data, measured inhalable dust levels would be below 5 mg/m³ for the evaluated conditions, despite the current limit of 10 mg/m³. What is clear from the data set is that at higher dust levels, the confidence in measured inhalable dust is lowered, let alone size characterisation studies associated with it. Therefore, the question, what is the value of inhalable dust sampling and the associated data, if there are no regulatory consequences.

PARTICLE SIZE ANALYSES OF INHALABLE AND TOTAL DUST SAMPLES

In order to understand the particle size distribution (PSD) of collected inhalable and total dust, samples were analysed in an accredited laboratory, where the minimum particle mass required for the PSD analysis is 10 mg. The PSD analyses involved sample preparation to collect enough samples for analyses and coal dust samples on the filters were dispersed in the distilled water/ethanol by sonication. The PSD analyses was performed on Master Sizer 3000 on batch of filter samples that had high filter loading and contains a large amount of dust sample on each filter. In each PSD measurement test, the sample was repeatedly measured five times and D10, D50, D90 and D100 were measured. For example, D10 is the diameter of the particles at which 10 per cent of the sample's volume is comprised of particles with a diameter less than this value. For each type of dust, only four sample filters were processed, which has resulted in an enough amount of dust sample for PSD analysis. The sample preparation ie the dust detachment and dispersion in the solvent was conducted by sonication for a short duration about 5–10 mins. Table 3 and Figure 9 show the sample distribution of the collected inhalable and total dust samples from the field. Based on the analysed size analyses results, following observations are made:

- Maximum particle diameter of the inhalable samples had a mean size value of 450 μm, against the ISO 1995 standard, with larger diameter particles contributing to the increased mass concentrations of dust.

- Similarly, total dust sample data displayed bi-modal particle size distribution with maximum size range of the first mode distribution with an average of 116 μm.

- Average D50 of the collected inhalable dust sample was 16.9 μm, which is greater than the average D50 value of the total dust samples analysed, ie 10 μm, possibly attributed to the inhalable sampler inlet diameter and further contributing to the higher measured dust concentration levels, against the D50 of respirable dust of 4 μm.

- Furthermore, the Relative Standard Deviation (RSD) values of the particle sizes for inhalable sampler were higher than that of total dust.

- One of the key inferences from the Inhalable coal dust sample analyses is that perceived upper limit of 100 μm collected by the inhalable sampler is potentially misleading, as it is observed that the inhalable samples can collect significantly larger than 100 μm that is airborne and therefore are available for possible inhalation. The consequence of this finding is the implication of the existing exposure limit value of 10 mg/m³ as the factor of 'size' alone can be a significant driver in the concentration values in exposure assessment or non-compliance.

TABLE 3

PSD analysis of Inhalable coal dust sample.

Inhalable Dust #	D_{10} (µm)	D_{50} (µm)	D_{90} (µm)	D_{100} (µm)	Total Dust #	D_{10} (µm)	D_{50} (µm)	D_{90} (µm)	D_{100}^* (µm)
1	5.57	18.3	68.6	666	15	3.25	10.1	56.5	127
2	5.22	17.0	55.1	516	16	3.19	10.1	60.0	127
3	4.96	16.4	54.0	454	17	3.12	9.83	53.6	111
4	4.72	15.7	45.9	163	18	3.12	9.98	78.5	98.1
Mean size	5.12	16.9	55.9	450	Mean size	3.17	10.0	62.1	116
RSD (%)	7.11	6.69	16.8	47	RSD (%)	2.03	1.35	18.1	12

* presence of bi-modal size distribution.

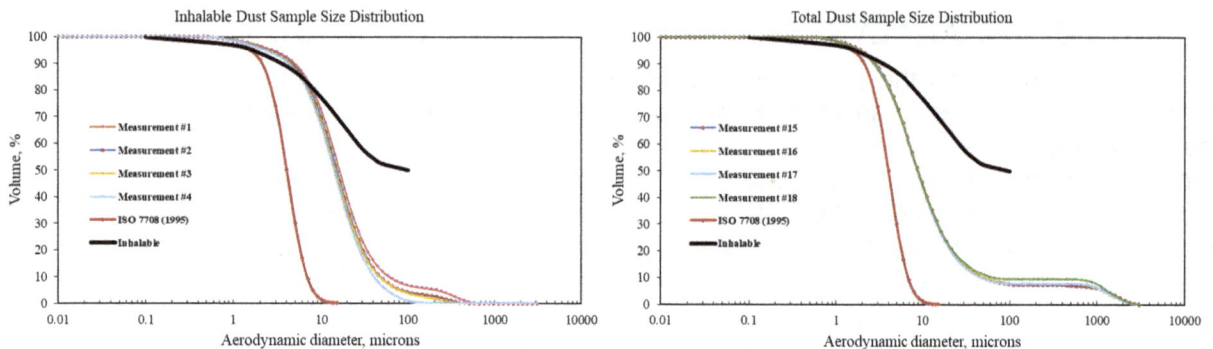

FIG 9 – Particle size distribution of inhalable and total dust samples.

CONCLUSIONS

The re-emergence of 'Black Lung' or CWP in Queensland (Qld) after reporting it being absent for over three decades had cast doubts on the rigour placed in the medical diagnosis and personal exposure assessment data. As of July 2023, RSHQ has reported that 68 and 88 cases of CWP and silicosis respectively, across the Qld mining and allied industry since 1984. The Qld Government amended the Qld CMSHR 2107 to reduce the exposure limits for respirable coal dust from 3.0 mg/m³ to a level of 1.5 mg/m³, with a silica dust limit of 0.12 mg/m³ in 2017 to the current limit of 0.05 mg/m³. Current exposure limit for inhalable coal dust remained at 10 mg/m³, with the application of extended shift exposure limit values for compliance determination purposes.

In the absence of clear evidence of past medical investigations in relation to coal dust, there are misperceptions and interpretations of health risk definitions used in the literature for respirable and inhalable coal dust. However, what is unquestionable is the critical importance of inhalable monitoring of toxicity of traditional low concentration high risk hygroscopic metal dust (cadmium, lead) to assess known health impacts and inhalable exposure data as part of medical diagnosis. The influence of non-conformance to size-selective sampling, flow rates, and measured dust levels and compliance determination has been recently unearthed in Australia that led to the changes to selection of appropriate respirable dust sampler for use. Therefore, this paper attempts to understand if there is a such a similar sampler bias in IOM Inhalable coal dust sampler in terms of its size-selective characteristics when used in coal mining environment.

Mining industry worldwide is spending significant amount of resources in sampling safety and health hazards to ensure adequate control measures are being implemented. Over the years, size-selective sampling curve and instruments which replicate human inhalation have also changed along with dust compliance limits between various mining countries worldwide. Most mining countries sample for respirable dust, however sampling of inhalable dust in mining industry is carried out in very few countries like Australia.

This paper summarises comparative performance through dust concentration results evaluated under field conditions between the Inhalable, total and the 'reference true' Higgins-Dewell UK reference sampler operated in accordance with the to the CEN/ISO/ACGIH size-selective curve at a flow rate of 2.2 L/min (ISO 7708) as side-by-side static samplers. The results of the evaluation are relevant to Australian mines in the context of practices of inhalable personal dust exposure monitoring. The following conclusions can be drawn from these inhalable, total and respirable sampler evaluations:

- The field evaluation was unable to calculate average bias map using the particle size distribution data, due to complex and likely aerosol distributions encountered during the comparative sampling of inhalable dust samples. Based on experiences of monitoring side by side real-time coal dust in the South African coalmines have shown the presence of distinct dust clouds attributed to the dynamic ventilation systems attributed to the significant differences in the measure dust levels.

- Inhalable dust samplers on average measured higher dust levels than 'total dust' samplers. However, there is no consistent relationship between respirable and inhalable sampler measured dust concentrations.

- Using the current coal dust respirable exposure limit of 1.5 mg/m^3, the estimated inhalable coal dust levels, using the inhalable and respirable relationship obtained with this work would be 4.4 mg/m^3.

- As noted by overseas researchers, it is agreed that dust sampler performance can be influenced by airborne dust concentrations, size-characteristics of airborne dust, sampling environment, air velocities and turbulence, sampler orientation that cause degree of uncertainty in measured dust levels. Considering these variable properties, expert judgement is applied to determine compliance with regulatory limits, or the use of data for risk assessment and management of dust control.

- The presence of few coal dust particles of 500 µm to 1000 µm collected by IOM inhalable sampler would have mass value of > 1 mg that definitely skews the measured inhalable dust levels (see Figure 8). While there has been progress in the inhalable dust assessment in the known cancer-causing metal dust types, value of inhalable coal dust is questioned, until a practical relationship is established at the current coal dust exposure limit of 10 mg/m^3. In the interim, comparing the inhalable to respirable dust is unhelpful. On the other hand, it is possible that the current respirable dust collection may not be the reflection of the true dust control at the operations.

It is recognised that the monitoring of inhalable dust for those substance (eg cadmium, lead, manganese) that have immediate human body reaction upon entering the breathing space with known dose-response curve evidence is critical for measuring the likely harm. In this context, there has been no discussions in the Australian industry in relation to the inhalable coal dust sampling (Newbegin et al, 2020) as there is great uncertainty exist and what action must be taken based on the findings presented in the paper. The continued cases of CWP and silicosis in the industry do not provide the confidence in the control effectiveness, which is a lever for predicting future cases of lung diseases and problems at hand. From a medical diagnosis perspective, it is not known, how the inhalable coal dust results are used in the assessment outcome. Notwithstanding the doubts expressed about the inhalable dust measurement, the disparities between inhalable and respirable dust levels indicate that monitoring problems persist in the area of inhalable sampling in the coal mining industry or questions persist on the value of inhalable coal dust sampling data in the mining industry.

Based on the findings of this work, the use of 'inhalable dust' or 'respirable dust' as a criterion or 'key health performance indicator' in evaluating or validating the effectiveness of coal dust control systems or worker's exposure assessment would be problematic, considering contradictory outcome of inhalable and respirable exposure assessment data. This finding is significant in verifying the dust control effectiveness of workers' protection. From an operational perspective, the implications of this findings are significant when compliance and epidemiological determinations are made by using the current approach of using personal inhalable coal dust measurements.

It is hoped that the findings in this paper will assist in navigating with appropriate questions on 'inhalable coal dust sampling' and more importantly, how and where these results are being used by the medical profession or compliance enforcement purposes. In the absence of past studies or references on inhalable coal dust sampler studies in Australia, complicates the approach to the pursuit of reduction in the exposure limits from the existing 10 mg/m^3 to an unknown limit. It is suggested that additional 'controlled' studies replicating underground ventilation conditions be pursued to understand the deficiencies of the inhalable sampling requirements, inhalable size-selective sampler performance curves, compliance determination of worker exposure assessment using inhalable samplers, criterion used for dust control effectiveness using inhalable and respirable data.

The field observations and the exposure data presented herein for the coal industry suggests that the science behind the inhalable sampler may not be well understood yet and require further decomposition and design review of current inhalable sampler may be needed. Despite these findings and insights from this paper, the coalmine worker's expectations for a workplace of health and safety should not be obscured as the reduction of harmful dust is the primary critical control action, while the measurement forms the secondary action. The comparative field evaluation experience shared in this paper suggests sufficient due diligence and prior evaluation of any new instruments for industry wide applications be carried out. Any modifications to sampling methodology or introduction of new instruments must ensure that the exposure data collected is relevant for continued development of long-term dose-response curves and to understand potential level of risks. In the case of inhalable coal dust sampling this is not evident. Finally, it is the consistent approach to inhalable sampling, instruments used, availability of exposure data relationship between respirable and inhalable to develop and understand to correct systematic biases in sampling which in the longer term assists in exposure determination and for continued formulation of dose-response relationships. Until then, what is the value of inhalable coal dust monitoring or how is it being enforced?

ACKNOWLEDGEMENTS

The author hopes that the knowledge sharing of relevant findings presented in this paper will enhance complex issues of dust monitoring, challenges of introducing a new dust monitoring instrument, amendments to the exposure limits and need for data for continued development of dose-response relationships to improve safety and health of workers. Various inputs of all relevant parties are clearly acknowledged. This paper and the work contained herein is an effort to improve the exposure monitoring and improve engineering controls in workplaces.

REFERENCES

Aitken, A J, Baldwin, P E J, Beaumont, G C, Kenny, L C and Maynard, D, 1999. Aerosol inhalability in low air movement environments, *J Aerosol Sci,* 30:613–626.

Aitken, R J and Donaldson, R, 1996. Large Particle and Wall Deposition Effects in Inhalable Samplers, Health and Safety Executive, UK: Report Number 117/1996.

Aizenberg, V K, Choe, S A, Grinshpun, K, Willeke and Baron, P, 2001. Evaluation of personal air samplers challenged with large particles, *J Aerosol Sci,* 32:779–793.

Alwis, K U, 1998. Occupational Exposure to Wood Dust, PhD Thesis, The University of Sydney, NSW, 328 p.

American Conference of Governmental Industrial Hygienists (ACGIH), 1985. Particle Size-Selective Sampling in the Workplace, 1985. ACGIH, USA.

American Conference of Governmental Industrial Hygienists (ACGIH), 1999. Particle Size-Selective Sampling for Particulate Air Contaminants, 1999. J H, Vincent, Ed, ACGIH, USA.

Baldwin, P E J and Maynard, A D, 1998. A survey of wind speeds in indoor workplaces, *Ann Occup Hyg,* 42:303–313.

Bartley, D L and Breuer, G M, 1982. Analysis and optimisation of the performance of the 10mm cyclone, *Am Ind Hyg Assoc J,* 43:520–528.

Belle, B K, 2002. Evaluation of newly developed real-time and gravimetric dust-monitoring instruments for personal dust sampling for South African mines, SIMHEALTH 704, November 2002.

Belle, B, 2012. Experiences of the institute of occupational medicine foam respirable sampler use in mines, in *Proceedings of the 2012 Coal Operators' Conference* (eds: N Aziz and B Kininmonth), pp 202–211.

Belle, B, 2017. Pairwise evaluation of PDM3700 and traditional gravimetric sampler for personal dust exposure assessment, The Australian Mine Ventilation Conference (The Australasian Institute of Mining and Metallurgy: Melbourne).

Belle, B, 2018. Evaluation of gravimetric sampler bias, effect on measured concentration, and proposal for the use of harmonised performance based dust sampler for exposure assessment, *Int J Min Sci Tech*. Available from <https://www.sciencedirect.com/science/article/pii/S2095268618304191>

British Medical Research Council (BMRC), 1952. British Medical Research Council Report, UK.

Clinkenbeard, R E, England, E C, Johnson, D L, Esmen, N A and Hall, T A, 2010. A Field Comparison of the IOM Inhalable Aerosol Sampler and a Modified 37-mm Cassette, pp 622–627.

Comité Européen de Normalisation (CEN), 1993. Workplace atmospheres: size fraction definitions for measurement of airborne particles in the workplace, European Standard EN 481:1993E, European Committee for Standardization.

Demange, M, Görner, P, Elcabache, J and Wrobel, R, 2010. Field Comparison of 37-mm Closed-Face Cassettes and IOM Samplers, pp 200–208.

Gardiner, K, Trethowan, N W, Harrington, J M, Rossiter, C E and Calvert, I, A, 1993. Respiratory health effects of carbon black, A survey of European carbon black workers, *British Journal of Industrial Medicine*, 50:1082–1096.

Gibson, H, Vincent, J H and Mark, D, 1987. A Personal Inspirable Aerosol Spectrometer for Applications in Occupational Hygiene Research, *Ann Occup Hyg*, 31(4A):463–479.

Görner, P, Simon, X, Wrobel, R, Kauffer, E and Witschger, O, 2010. Laboratory Study of Selected Personal Inhalable Aerosol Samplers, *The Annals of Occupational Hygiene*, 54(2):165–187.

Harper, M, Akbar, M Z and Andrew, M E, 2004. Comparison of wood-dust aerosol size-distributions collected by air samplers, *J Environ Monit*, 6:18–22.

Harper, M and Demange, M, 2007. Concerning sampler wall deposits in the chemical analysis of airborne metals, *J Occup Environ Hyg*, 4:D81-D86.

Harper, M and Muller, B S, 2002. An evaluation of total and inhalable samplers for the collection of wood dust in three wood products industries, *J Environ Monit*, 4:648–656.

Health and Safety Executive, 1997. Methods for the Determination of Hazardous Substances 14/3 General Methods for the Gravimetric Determination of Respirable and Total Inhalable Dust (HSE Books: London).

Hinds, W C, 1988. Basis for Particle Size-Selective Sampling for Wood Dust, *Appl Ind Hyg*, 3:67.

International Agency for Research on Cancer (IARC), 1996. Printing Processes and Printing Inks, Carbon Black and Some Nitro Compounds, In *IARC Monographs on the Evaluation of Carcinogenic Risks to Humans*, vol 65, International Agency for Research on Cancer, World Health Organization.

International Agency for Research on Cancer (IARC), 1997. The evaluation of carcinogenic risks to humans-Silica, Some Silicates, Coal Dust and para-Aramid Fibrils, Monograph 68, 521 p.

International Standards Organization (ISO), 1995. ISO 7708:1995 – Air quality: particle size fraction definitions for health-related sampling, International Organization for Standardization.

Jiménez, A S, Tongeren, M and Aitken, R J, 2012. Guidance for Collection of Inhalable and Respirable Dust, Strategic Consulting, p 31.

Kennedy, E R, Fishbach, T J, Song, R, Eller, P M and Shulman, S A, 1995. Guidelines for air sampling and analytical method development and evaluation, DHHS (NIOSH) Publication No. 95–117.

Kenny, L C, 1995. Pilot study of CEN protocols for the performance testing of workplace aerosol sampling instruments, Report of work carried out under EC Contract MAT1-CT92–0047, September 1995, Health and Safety Executive, Sheffield, England.

Kenny, L C, Aitken, R J, Baldwin, P E J, Beaumont, G and Maynard, D, 1999. The sampling efficiency of personal inhalable samplers in low air movement environments, *J Aerosol Sci*, 30:627–638.

Kenny, L, 2003. Scientific Principles and Pragmatic Solutions for the Measurement of Exposure to Inhalable Dust, *Commentary Ann Occup Hyg*, 47(6):437–440.

Kim, H and Lee, D, 1996. A Field Comparison of Total Wood Dust Concentrations by 37 mm Closed-Face Cassette and the Inspirable Particulate Mass Sampler in the Furniture and Sawmill Factories, Poster presentation at the American Industrial Hygiene Conference and Exposition, Washington, DC.

Kissell, F N and Sacks, H K, 2002. Inaccuracy of area sampling for measuring the dust exposure of mining machine operators in coal mines, *Min Eng*, 54(2):17–23.

Leidel, N A, Busch, K A and Lynch, J R, 1977. The inadequacy of general air (area) monitoring for measuring employee exposures, Technical Appendix C in: *Occupational Exposure Sampling Strategy Manual*, NIOSH Publication No. 77–173:75–77.

Li, S, Lundrgren, D A and Rowell-Rixx, D, 2000. Evaluation of Six Inhalable Aerosol Samplers, *AIHAJ*, 61:506–516.

Li, S-N and Lundgren, D A, 1999. Weighing accuracy of samples collected by IOM and CIS inhalable samplers, *Am Ind Hyg Assoc J*, 60:235–236.

Liden, G and Bergman, G, 2001. Weighing imprecision and handleability of the sampling cassettes of the IOM sampler for inhalable dust, *Ann Occup Hyg*, 45(3):241–252.

Liden, G and Harper, M, 2006. The need for an international sampling convention for inhalable dust in calm air, *J Occup Environ Hyg*, 3:D94–D101.

Liden, G and Kenny, L C, 1992. The performance of respirable dust samplers-Sampler bias, precision and inaccuracy, *Ann Occup Hyg*, 36:1–22.

Liden, G and Kenny, L C, 1994. Errors in inhalable dust sampling for particles exceeding 100 µm, *Ann Occup Hyg*, 38:373–384.

Liden, G, Juringe, L and Gudmundsson, A, 2000. Workplace validation of a laboratory evaluation test of samplers for inhalable and total dust, *J Aerosol Sci*, 31:199–219.

Liden, G, Melin, B, Lidblom, A, Lindberg, K and Noren, J O, 2000. Personal sampling in parallel with open-face filter cassettes and IOM samplers for inhalable dust-implications for exposure limits, *Appl Occup Environ Hyg*, 15:263–276.

Mark, D and Vincent, J H, 1986. A new personal sampler for air-borne total dust in workplaces, *Ann Occup Hyg*, 30:89–102.

Mark, D, Lyons, C P, Upton, S L and Kenny, L C, 1994. Wind tunnel testing of the sampling efficiency of personal inhalable aerosol samplers, *J Aerosol Sci*, 25(1):S339–S340.

Martin, J R and Zalk, D M, 1998. Comparison of Total Dust/Inhalable Dust Sampling, Methods for the Evaluation of Airborne Wood Dust, *Applied Occupational and Environmental Hygiene*, 13:3:177–182.

Maynard, A D and Baron, P A, 2004. Aerosols in the industrial environment, in *Aerosols Handbook, Measurement, Dosimetry and Health Effects* (eds: L S Ruzer and N H Harley), pp 225–264 (CRC Press: Boca Raton).

Maynard, A D and Kenny, L C, 1995. Performance assessment of three personal cyclone models, using an aerodynamic particle sizer, *J Aerosol Sci*, 26:671–684.

Methods for the Determination of Hazardous Substances (MDHS), 1997. General Methods for Sampling and Gravimetric Analysis of Respirable and Total Inhalable Dust, HSE Books.

Methods for the Determination of Hazardous Substances (MDHS), 2000. 14/3 General Methods for Sampling and Gravimetric Analysis of Respirable and Inhalable Dust, Health and Safety Executive (HSE), UK.

Navy and Marine Corps Public Health Center (NMCPHC), 2020. Industrial Hygiene Field Operations Manual Technical Manual NMCPHC-TM6290.91–2 12.

Newbegin, K, McBean, R, Kildey, K and Tatkovic, A, 2020. Occupational assessment and centralised repository for Coal Mine Dust Lung Disease (CMDLD), ACARP Australia.

Ogden, T L and Birkett, J L, 1978. An inhalable-dust sampler for measuring the hazard from total airborne particulate, *Annals of Occupational Hygiene*, 21:41–50.

Occupational Safety and Health Administration (OSHA), 1989. Final Rule on Air Contaminants Project, Particulates, OSHA Comments.

Orenstein, A J, 1960. Proceedings of the 1959 Pneumoconiosis Conference, Johannesburg, Churchill, London, UK.

Perrault, G, Cloutier, Y and Drolet, D, 1996. Comparison of Total and Inhalable Samplings of Wood Dust, Poster presentation at the American Industrial Hygiene Conference and Exposition, Washington, DC.

Phalen, R F, Hinds, W C, John, W, Lioy, P J, Lippmann, M, McCawley, M A, Raabe, O G, Soderholm, S C and Stuart, B O, 1988. Particle size selective sampling in the workplace: rationale and Recommended Techniques, *Appl Occup Hyg*, 32:403–411, Supplement 1.

Pisaniello, D L, Connell, K E and Muriale, L, 1991. Wood Dust Exposure During Furniture Manufacture-Results from an Australian Survey and Considerations for Threshold Limit Value Development, *Am Ind Hyg Assoc J*, 52(11):485–492.

Queensland Dust Forum Series, Resources Safety and Health Queensland (RSHQ), 2020.

Ramachandran, G, 2005. *Occupational Exposure Assessment for Air Contaminants*, 181 p (CRC Press Tylor and Francis Group).

Roger, F, Lachapelle, G, Fabries, J F, Gomer, P and Renoux, A, 1998. Behaviour of the IOM aerosol sampler as a function of external wind velocity and orientation, *J Aerosol Sci*, 29:SI133-SI134, Supplement I.

Safe Work Australia, 1995. Guidance on the Interpretation of Workplace Exposure Standards for Airborne Contaminants, Australia.

Shaw, L, Shaw, D, Hardisty, M, Britz-McKibbin, P and Verma, D, 2020. Relationships between inhalable and total hexavalent chromium exposures in steel passivation, welding and electroplating operations of Ontario, *International Journal of Hygiene and Environmental Health*, 230:113601.

Smith, J P, Bartley, D L and Kennedy, E R, 1998. Laboratory investigation of the mass stability of sampling cassettes from inhalable aerosol samplers, *Am Ind Hyg Assoc J*, 59:582–585.

Soderholm, S C, 1989. Proposed International Conventions for Particle Size-Selective Sampling, *Ann Occupational Hygiene*, 33(3):301–320.

Soderholm, S C, 1991. Why Change ACGIH's Definition of Respirable Dust, *Appl Occup Environ Hyg*, 6(4):248–250.

Spear, T M, Werner, M A, Bootland, J, Harbour, A, Murray, E P, Rossi, R and Vincent, J H, 1997. Workers' exposures to inhalable and total lead and cadmium containing aerosols in a primary lead smelter, *Am Ind Hyg Assoc J*, 58:893–899.

Spear, T M, Werner, M A, Bootland, J, Murray, E, Ramachandran, G and Vincent, J H, 1998. Assessment of particle size distributions of health relevant aerosol exposures of primary lead smelter workers, *Ann Occup Hyg*, 42(2):73–80.

Standards Australia, 1989. AS 3640–1989 – Workplace atmospheres – Method for sampling and gravimetric determination of inspirable dust.

Standards Australia, 2009. AS 3640–2009 – Workplace atmospheres – Method for sampling and gravimetric determination of inhalable dust.

Van Tongeren, M J A, Kromhout, H and Gardiner, K, 2000. Trends in Levels of Inhalable Dust Exposure, Exceedance and Overexposure in the European Carbon Black Manufacturing Industry, *Appl Occup Hyg*, 44(4):271–280.

Verma, D K, 1984. Inhalable, total and respirable dust: A Field Study, *The Annals of Occupational Hygiene*, 28(2):163–172.

Vincent, J H, Brosseau, L M, Ramachandran, G, Tsai, P, Spear, T M, Werner, M A and McCullough, N V, 1997. Current Issues in Exposure Assessment for Workplace Aerosols, *Appl Occup Hyg*, 41(Supplement 1):607–614.

Vincent, J H, Mark, D, Miller, B G, Armbruster, L and Ogden, T L, 1990. Aerosol inhalability at higher windspeeds, *J Aerosol Sci*, 21:577–586.

Volkwein, J C, Maynard, A D and Harper, M, 2011. Workplace Measurement, Chapter 25, in *Aerosol Measurement: Principles, Techniques and Applications* (eds: P Kulkarni, P A Baron and K Willeke), third edition (John Wiley and Sons, Inc).

Walton, W and Vincent, J, 1998. Aerosol Instrumentation in Occupational Hygiene: An Historical Perspective, *Aerosol Science and Technology*, 28:5:417–438.

Wippich, C, Rissler, J, Koppisch, D and Breuer, D, 2020. Estimating Respirable Dust Exposure from Inhalable Dust Exposure, *Annals of Work Exposures and Health*, 64(4):430–444.

Key technology and application of dust control by surfactant-magnetised water in underground coalmines

Q Botao[1] and Z Qun[2]

1. Professor, School of Safety Engineering, China University of Mining and Technology, Xuzhou, Jiangsu 221116, China. Email: qbt2003@163.com
2. Associate professor, School of Safety Engineering, China University of Mining and Technology, Xuzhou, Jiangsu 221116, China. Email: qunzhou2016@163.com

ABSTRACT

Aiming at the problems of poor wetting performance and coverage of the available dust removal spray, the novel dust control technology using surfactant-magnetised water (SMW) was proposed. Firstly, surfactant-magnetised water was formed under the synergistic effect between surfactant and magnetisation, of which wetting characteristics were significantly improved than that of the untreated water. For example, the surface tension of surfactant-magnetised water was reduced by 62.9 per cent (to 26.37 mN/m) than that of untreated water. Secondly, the preparation system of surfactant-magnetised water was constructed of a magnetic device, an automatic addition system of surfactant, and a mixer, realising the automatic preparation of surfactant-magnetised water, of which preparation flow rate reached 400 L/min. Additionally, according to the dust production characteristics of mining face, the dust removal spray based on the novel spray devices was constructed to effectively capture dust particles. Field application in an underground coalmine indicated that the respirable dust and total dust suppression efficiency reached more than 87 per cent, which significantly improved the underground working environment.

INTRODUCTION

Recently, coal is still the main energy in China, according to statistics, coal consumption accounted for 55.3 per cent of total energy consumption in 2023 (Wang *et al*, 2024). To meet the demand for coal, the mechanised mining techniques were rapidly developed in underground mines, leading to a substantial increase in coal dust production. It seriously affected the physical and mental health of underground workers. According to statistics, by the end of 2022, more than 915 000 people had been diagnosed with occupational diseases, more than 50 per cent of which were coal workers' pneumoconiosis (Zhou *et al*, 2023). Therefore, the effective prevention method of coal dust had become a major demand for safe and healthy production of coal industry.

At present, water spray was widely applied as an economic method of dust control in the underground coalmines, but the untreated water was poor in wetting and condensing coal dust while water spray could not well enclose dust sources (Peng *et al*, 2022; Wang *et al*, 2020; Zhang *et al*, 2022). It led to the poor dust control efficiency (less than 50 per cent). In order to effectively improve the wetting ability of aqueous solution, surfactant was widely used (Ding *et al*, 2022; Wang, Tang and Tao, 2018). However, the atomisation performance of the surfactant solution was weak due to the high addition usage, not meeting the low-cost and efficient requirements of dust control technologies in underground coalmines. Meanwhile, although magnetisation could improve the wetting performance of the aqueous solution to a certain extent, the dust removal efficiency was only improved by about 10 per cent than that of water spray (Chen, Song and Jiang, 2014; Zhou, Qin and Huang, 2021). It could not effectively reduce the underground dust concentration. In addition, some scholars investigated the influence of spray conditions on the atomising characteristics of the solution. It was found that the wind velocity could affect the atomising characteristics, and the droplet number density was reduced with the increase of wind velocity (Wang, Wang and Wang, 2017). Based on the principle of mechanical wind-assisted negative pressure suction, a novel spray device was proposed to form a well spray field (Nie *et al*, 2023). However, most of the available dust removal spray could not cover the dust source, leading to many dust particles escaping.

Thus, to enhance dust control efficiency of water spray, a dust removal technology based on the synergistic effect between magnetisation and surfactant was proposed to improve the wetting performance of water. Meanwhile, according to the dust production characteristics of mining face, the dust removal spray based on the novel spray devices was constructed, well enclosing dust sources. Field application indicated that the new dust control method of surfactant-magnetised water had a better dust suppression efficiency than that of the traditional spray.

EXPERIMENTAL MATERIAL AND FACILITIES

Coal samples were acquired from Pingmei No.6 coalmine, which is characteristically coking coal. In addition, the main experimental equipment included a surface tensiometer (Sigma 700), contact angle instrument (JGW-360B), high-pressure water pump (HM280), and tablet press machine (FY-24) and dust concentration monitor (DustTrack DRX 8533).

RESULTS AND DISCUSSION

Wetting characteristics of surfactant-magnetised water

In previous research (Zhou *et al*, 2018; Zhou, Qin and Huang, 2021), it was found that surfactant-magnetised water formed by the double effect of the 0.03 wt per cent surfactant solution and a magnetic field could well wet coal dust. To better systematically investigate the dust control characteristics of surfactant-magnetised water, the surface tension, contact angle, dust deposition time of aqueous solution were investigated, shown in Figure 1.

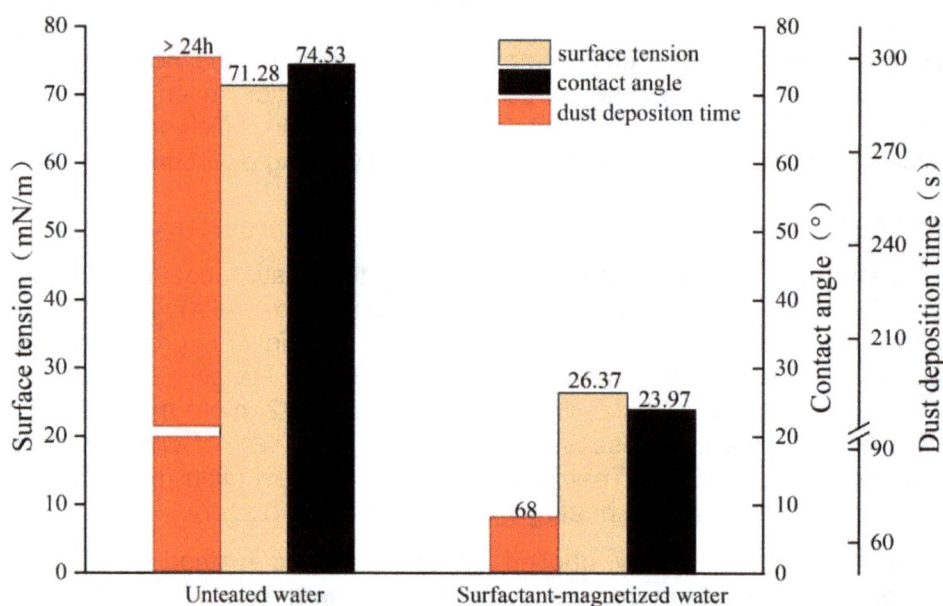

FIG 1 – Wetting characteristics of surfactant-magnetised water.

From Figure 1, it was seen that compared with untreated water, the wetting characteristics of the aqueous solution were significantly improved under the synergistic effect between magnetisation and surfactant. For example, the contact angle of surfactant-magnetised water was only 23.97°, significantly reduced by 67.84 per cent than that of untreated water. This because the magnetisation changed the solution properties by breaking hydrogen bonds between water molecules, making large molecular cluster structures be changed into small molecular groups, which improved the dust removal performance of the solution. Meanwhile, surfactant utilised its own hydrophilic and lipophilic groups to form an isolation layer on the solution surface, significantly improving the wetting characteristics of aqueous solution (Ding *et al*, 2011; ShamsiJazeyi, Verduzco and Hirasaki, 2014). Additionally, under the effect of a magnetic field, the surfactant molecules were more easily distributed on the solution surface, reducing the critical micelle concentration of surfactant, of which usage was only 0.03 per cent.

Preparation system of surfactant-magnetised water

To effectively generate the surfactant-magnetised water and adapt to the limited space in underground coalmine, the preparation system of surfactant-magnetised water was constructed, shown in Figure 2, mainly including a magnetic device, an automatic addition system of surfactant, and a mixer, etc. Because the usage of surfactant used in the preparation process of SMW only accounted for 0.03 wt per cent of the spray water, the paper adopted the dual governing methods between the metering pump and rotameter to achieve the accurate adding of low-concentration surfactant. Thereinto, the surfactant addition system was composed of a metering pump, rotameter, concentrated surfactant solution, and an automatic control system, etc. Through sensing the change of liquid level in the water tank, this system could accurately control the start and stop of the water supply pipe and metering pump, realising the Intelligent supply of the concentrated surfactant solution in the fully mechanised mining face (Figure 2a). Additionally, by sensing the flow change of the water supply pipeline in the fully mechanised excavation face, the addition amount of concentrated surfactant solution could be automatically adjusted to realise the stable supply of small-dose surfactant, which solved the unstable operation problem of the available air adding system (Figure 2b). The mixer was used to mix the surfactant with water. Finally, the surfactant solution was magnetised by a magnetic device, forming the surfactant-magnetised water.

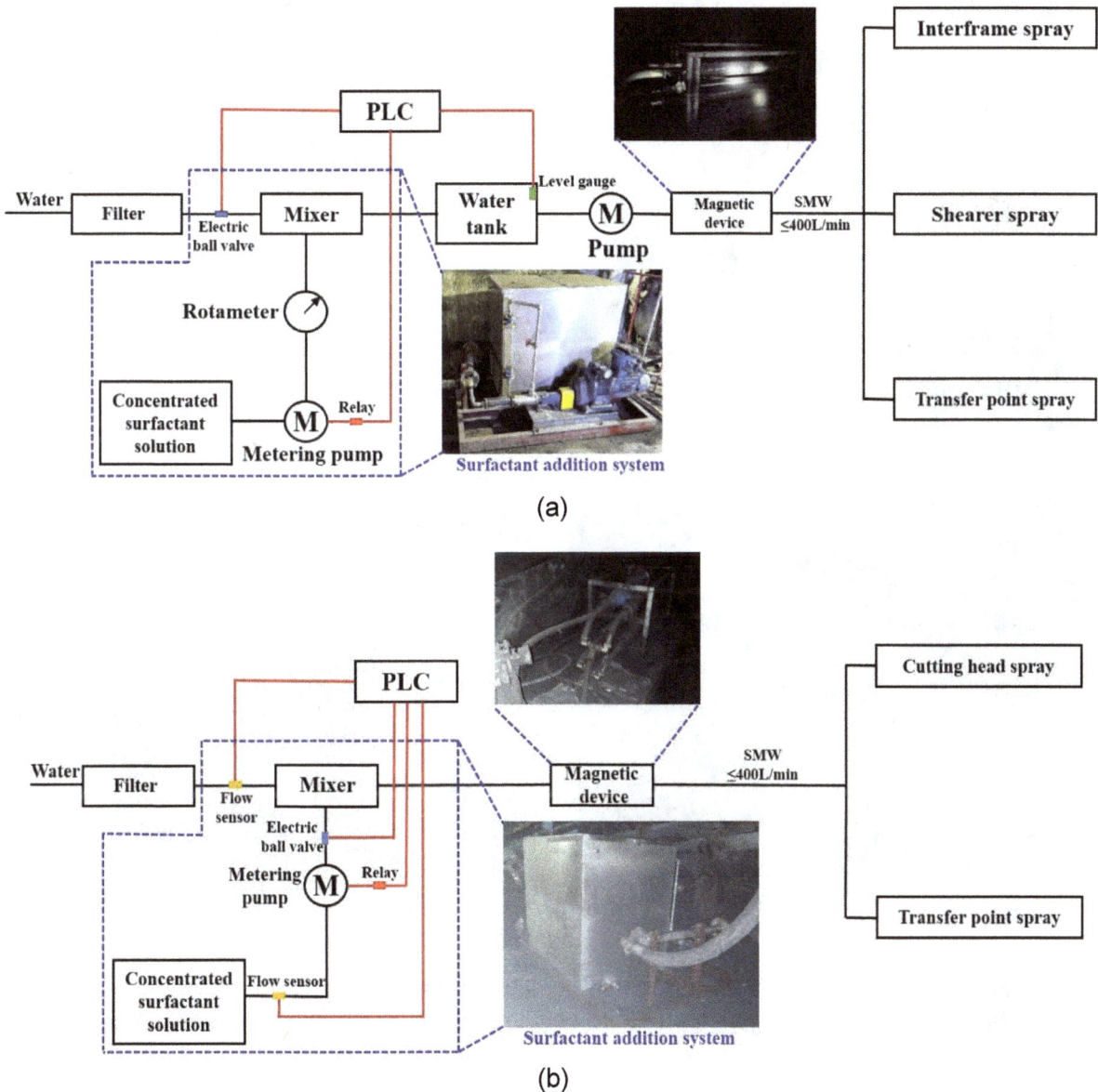

(a)

(b)

FIG 2 – Preparation system of surfactant-magnetised water in underground coalmines for fully mechanised: (a) mining face; (b) excavation face.

Dust removal spray construction of surfactant-magnetised water

Dust control spray used in fully mechanised mining face

To solve the problems of short spray distance, weak anti-wind interference ability of fully mechanised mining face, a hydrodynamic fan spray device was proposed, shown in Figure 3a. It was consisted of three nozzles, fan, rotation axis, and shell. The reaction force formed by high-pressure water ejecting from the nozzle driven the fan to rotate at high speed, forming a large air flow (>110 m^3/min) to improve the atomisation effect and jet distance of a dust removal spray. Compared with that of the original spray, the droplet size decreased from 160.86 μm to 120 μm, which increased the number density of droplets. Meanwhile, the dust removal efficiency of water spray increased significantly from 37.4 per cent to 46.9 per cent. Additionally, to effectively capture the escaping and shifting dust, a negative pressure coiling device was designed based on the Venturi effect. The nozzle could form a negative pressure environment at the back, forming a large suction air volume (>11 m^3/min), which improved the atomisation characteristics. The droplet size was less than 86 μm, while its jet distance was more than 4 m.

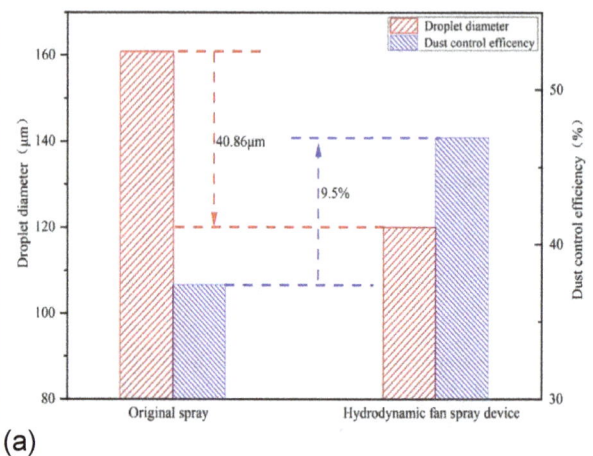

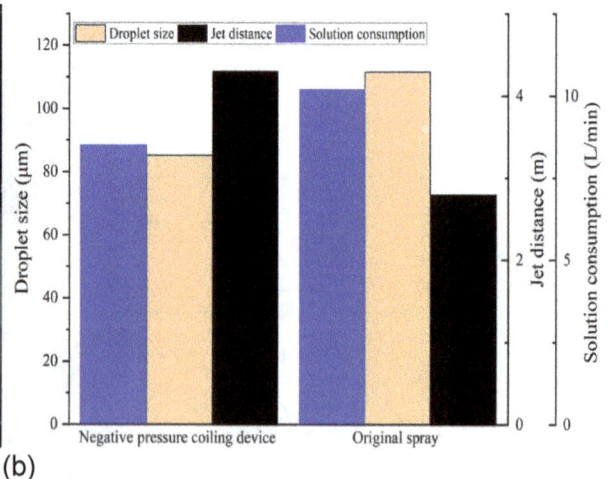

(a)

(b)

FIG 3 – Novel spray device of fully mechanised mining face: (a) Hydrodynamic fan spray device used in the coal cutter; (b) Negative pressure coiling device used in the fully mechanised mining shelf.

Firstly, based on the characteristics of high jet distance and wide atomisation coverage of the hydrodynamic fan spray device, the dust removal spray used to enclose the shearer drum of a coal cutter was constructed, shown in Figure 4. It not only could form a good spray field to capture dust, but also could suction air flow containing dust to achieve dust control again. Secondly, three negative pressure suction nozzles were arranged in one group, arranged in every other support frame along the working face direction. In the normal mining process, three groups of negative pressure suction nozzles were opened before and after the coal cutter to form six water curtains to block and capture dust.

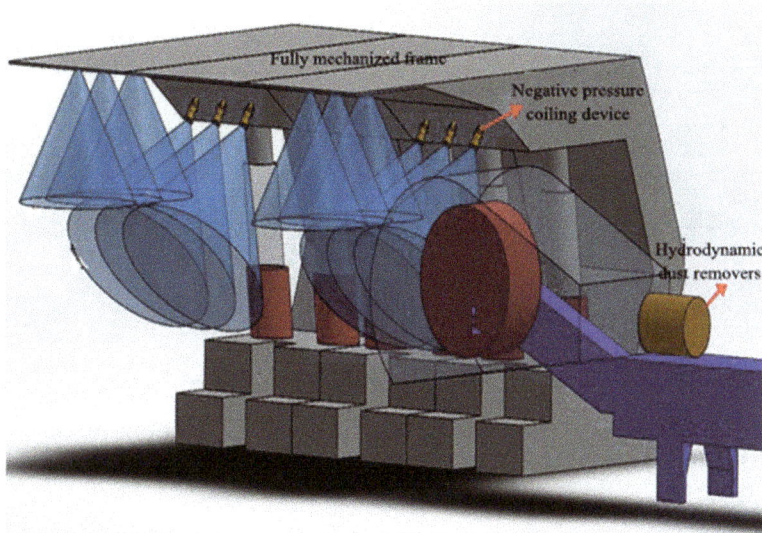

FIG 4 – Spray method used to enclose dust sources of fully mechanised mining face.

Dust control spray used in fully mechanised excavation face

In order to solve the problems of unstable and narrow coverage of the available water spray, an air-water nozzle (Figure 5a) was developed based on the principle of the air-water ultrasonic atomisation. Based on the coupling effect of high-speed air flow, mixing chamber and resonator, the spray solution was highly broken, making the droplet size of the spray field being lower than 100 μm, which was more than 44 per cent lower than that of the traditional spray (Figure 5b). Based on the developed air-water nozzle, the long distance, wide coverage and high atomisation characteristics of a dust removal spray were realised.

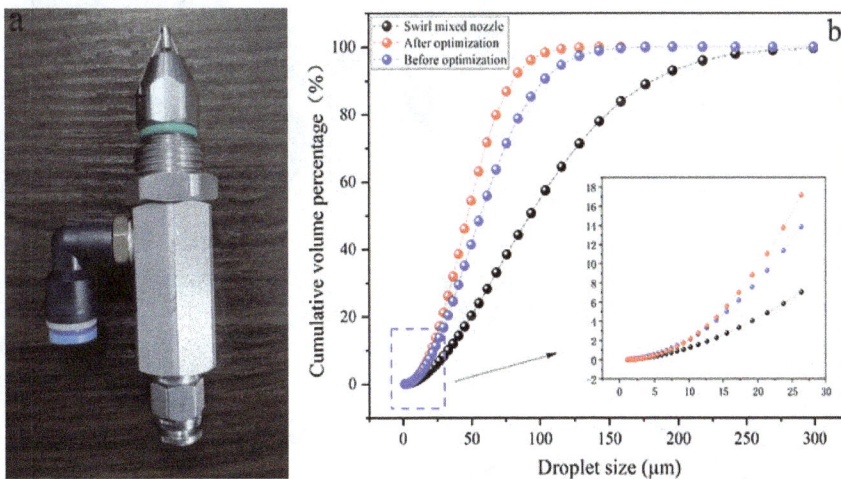

FIG 5 – Spray device used in fully mechanised excavation face.

As shown in Figure 6, to solve the dust control problems of fully mechanised excavation face, such as poor wind resistance ability, the dust source not be covered by spray field, the dust removal technology of surfactant-magnetised water was developed to seal the dust source in the fully mechanised excavation face. Based on the long distance and high atomisation performance of the air-water spray device, the remote arrangement of the spray device from the dust source was realised. Meanwhile, based on the complementary spray arrangement method, the dust removal spray were constructed of three air-water spray devices, which could well cover dust sources, effectively capturing coal dust particle.

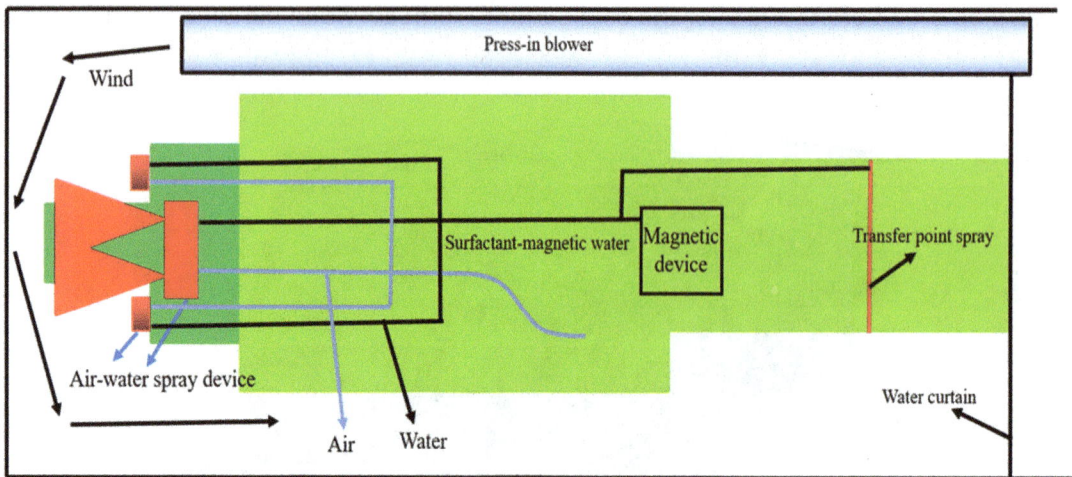

FIG 6 – Spray method used to enclose dust sources of fully mechanised excavation face.

FIELD APPLICATION

Pingmei No.6 coalmine is a modern mine located in Pingdingshan city, Henan province, China. With the continuous increase of the mechanised mining level in underground coalmines, dust hazards had already been a major source of danger in Pingmei No.6 coalmine, which severely threatened the health of workers and the safe production of the mine. To effectively solve dust problems, dust control technology of SMW was successfully applied in coal mining face (Figure 7).

(a)

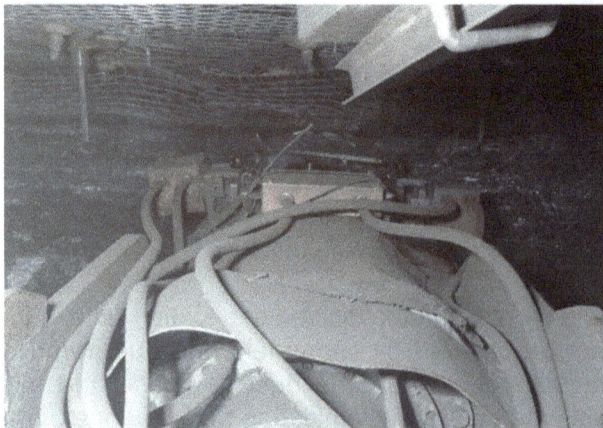

(b)

FIG 7 – Field application of dust control technology using SMW: (a) Hydrodynamic fan spray device Magnetic device; field application of fully mechanised mining face; (b) Arrangement method of air-water spray Atomisation effect of a spray field; Field application of fully mechanised excavation face.

As shown in Figure 8a, under the same conditions of the working environment, the SMW could effectively capture coal dust in fully mechanised mining face. The average efficiencies of respirable dust control and total dust control by surfactant-magnetised water was more than 90 per cent and 93 per cent, respectively, greatly reducing dust concentration in underground coalmine. Additionally, the constructed spray field can better close the dust source of the excavation face. From Figure 8b, it could be seen that compared to that of water, the total dust and respirable dust control efficiency of SMW was increased by 47.1 per cent, 48.9 per cent, respectively, remarkably improving the working environment of the fully mechanised excavation face.

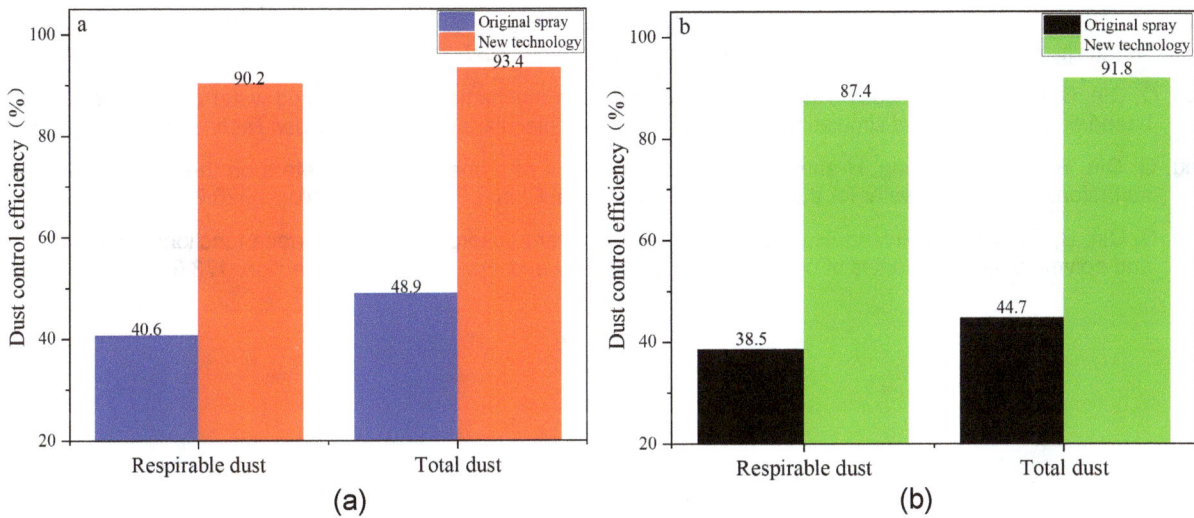

FIG 8 – Dust control efficiency of surfactant-magnetised water in underground coalmines: (a) fully mechanised mining face; (b) fully mechanised excavation face.

CONCLUSIONS

- The wetting characteristics of surfactant-magnetised water was significantly improved than that of untreated water. The surface tension and contact angle of surfactant-magnetised water was reduced by 62.9 per cent, 67.84 per cent than that of untreated water, respectively, reaching 26.37 mN/m, 23.97°, respectively.

- The surfactant-magnetised water generation system were constructed while being successfully applied in coal mining face. Through the constructed surfactant addition system, the low-concentration surfactant was accurately added. Additionally, through the field implication, the respirable dust and total dust control efficiency of SMW reached more than 87 per cent, significantly improving the underground working environment.

REFERENCES

Chen, M L, Song, W C and Jiang, Z A, 2014. Study on Dust Fall Mechanism and Experiment with Magnetized Water Spraying in Coal Mine, *Coal Science and Technology*, 42:65–68.

Ding, X, Wang, D, Luo, Z, Wang, T and Deng, J, 2022. Multi-scale study on the agglomerating and wetting behaviour of binary electrolyte-surfactant coal dust suppression based on multiple light scattering, *Fuel*, 311:122515.

Ding, Z R, Zhao, Y J, Chen, F L, Chen, J Z and Duan, S X, 2011. Magnetisation mechanism of magnetised water, *Acta Physica Sinica*, 60:064701.

Nie, W, Cha, X, Bao, Q, Peng, H, Xu, C, Zhang, S, Zhang, X, Ma, Q, Guo, C, Yi, S and Jiang, C, 2023. Study on dust pollution suppression of mine wind-assisted spray device based on orthogonal test and CFD simulation, *Energy*, 263.

Peng, H, Nie, W, Zhang, X, Xu, C, Meng, X, Cheng, W, Liu, Q and Hua, Y, 2022. Research on the blowing-spraying synergistic dust removal technology for clean environment in large-scale mechanization coal mine, *Fuel*, 324:124508.

ShamsiJazeyi, H, Verduzco, R and Hirasaki, G J, 2014. Reducing adsorption of anionic surfactant for enhanced oil recovery: Part I, Competitive adsorption mechanism, *Colloids and Surfaces A – Physicochemical and Engineering Aspects*, 453:162–167.

Wang, H, Wang, C and Wang, D, 2017. The influence of forced ventilation airflow on water spray for dust suppression on heading face in underground coal mine, *Powder Technology*, 320:498–510.

Wang, L, Li, Z, Xu, Z, Yue, X, Yang, L, Wang, R, Chen, Y and Ma, H, 2024. Carbon emission scenario simulation and policy regulation in resource-based provinces based on system dynamics modeling, *Journal of Cleaner Production*, 460.

Wang, P, Han, H, Tian, C, Liu, R and Jiang, Y, 2020. Experimental study on dust reduction via spraying using surfactant solution, *Atmospheric Pollution Research*, 11:32–42.

Wang, S, Tang, L and Tao, X, 2018. Investigation of effect of surfactants on the hydrophobicity of low rank coal by sliding time measurements, *Fuel*, 212:326–331.

Zhang, T, Zou, Q, Jia, X, Liu, T, Jiang, Z, Tian, S, Jiang, C and Cheng, Y, 2022. Effect of cyclic water injection on the wettability of coal with different SiO2 nanofluid treatment time, *Fuel*, 312:122922.

Zhou, Q, Qin, B and Huang, H, 2021. Research on the formation mechanism of magnetized water used to wet coal dust based on experiment and simulation investigation on its molecular structures, *Powder Technology*, 391:69–76.

Zhou, Q, Qin, B, Wang, J, Wang, H and Wang, F, 2018. Effects of preparation parameters on the wetting features of surfactant-magnetized water for dust control in Luwa mine, China, *Powder Technology*, 326:7–15.

Zhou, Q, Qin, B, Zhou, B and Huang, H, 2023. Effects of surfactant adsorption on the surface functional group contents and polymerization properties of coal dust, *Process Safety and Environmental Protection*, 173:693–701.

Engineering dust controls as enablers for health and safety compliance in underground mining[*]

A De Andrade[1] and P Stowasser[2]

1. Chief Executive Officer, Grydale Solutions Pty Ltd, Brisbane Qld 4500.
   Email: alex.deandrade@grydale.com.au
2. Research Consultant, Brisbane Qld 4500. Email: p.stowasser@hdr.qut.edu.au
* This paper has not been technically reviewed

ABSTRACT

The harmful effects of hazardous airborne contaminants are well documented in the mining industry. Increased knowledge and education, notably around particulates that are respirable in size, has resulted in significant changes to both legislation and practice, each of which aims to protect the health and safety of mine workers and communities. Mandated exposure limits, both nationally and internationally have lowered, resulting in changes to mining processes and practices. These reductions have challenged project managers and supervisors to think more critically about their ventilation plans, and in some instances, how engineered dust controls might support their existing operations. This decision-making process is complex, each mine site is unique, and there are a range of technologies available. Therefore, understanding the contexts in which different dust control technologies can be utilised, becomes crucial to making an informed decision. The aim of this paper is to explore the contexts in which alternate engineered controls (baghouses, mobile dust collectors, wet dust scrubbers) can complement an underground mines main ventilation system to achieve compliance and potentially increase operational, technical, environmental, and economical efficiencies. The paper includes a review of literature related to air quality regulations and interviews with three experienced lead engineers to help explore the key decision-making criteria. The findings are analysed against the criteria and adjusted sub-criteria of an existing model (Analytic Hierarchy Process) known to support complex decision-making in the field. The findings of this paper provide insight to alternate dust control technologies and guidance for anyone looking to maintain or improve compliance with new, more ambitious, clean air regulations.

INTRODUCTION

A concerted effort has been made over the last decade to decrease exposure to hazardous airborne contaminants and improve health and safety conditions for mine workers and communities. Globally, China identified the control of dust in underground mining as a national priority in its Thirteenth Five-Year Plan (Wang *et al*, 2018). In response to the concerning rates of pneumoconiosis for underground coal workers (Nie *et al*, 2022), this attention increased investment in dust control research and initiated positive changes to process and practice. At a similar time, the permissible exposure limit for Respirable Crystalline Silica (RCS) halved in the United States to help combat silicosis and lung disease. This standard lowered from 100 micrograms per cubic metre of air ($\mu g/m^3$) to 50 $\mu g/m^3$ averaged over an 8 hr day (Occupational Safety and Health Administration (OSHA) 2017). This standard has been replicated in Australia, with organisations such as Safe Work Australia and working groups such as NSW Air Quality, increasing awareness of respirable dust through education and recommendations for best practice. In Australia, attention has recently turned to Diesel Particulate Matter (DPM) with Safe Work Australia proposing a workplace exposure standard of 0.015 mg/m^3, which represents a significant reduction from the current standard (0.1 mg/m^3) accepted by industry. Argued in this paper, is that such exposure standards will continue to be lowered in a bid to protect the health and safety of mine workers and communities.

Airborne contaminants include vapours, dusts, particles, fibres, fumes or gases (or combinations of the above), that can be generated by a wide range of tasks or processes in the workplace (WorkSafe, 2020a). Table 1 shows examples of some common airborne contaminants and their properties.

TABLE 1

Airborne Contaminants and their Properties (WorkSafe, 2020a).

Name	Example
Gas	Carbon monoxide and carbon dioxide
Vapour	Acetone, ethanol, chloroform, styrene, petrol
Fibres	Asbestos and glass
Mist	Solution being sprayed (eg paint, steam, electroplating baths)
Fume	Solder and welding fumes
Dust	Flour dust, concrete/cement dust (generated by grinding, cutting, crushing, drilling etc) and silica from stone cutting

Dust is an airborne contaminant common in underground mining and its exposure in excessive amounts can lead to health problems. The harmful effects of dust can vary and are often dependent on size, characteristics, and quantities (WorkSafe, 2020b). For example, large particles of dust can fall to the ground quickly and our body has some natural defence mechanisms (eg cilia hairs, mucus etc) that help to prevent them from entering our respiratory system (SA Health, 2021). In contrast, small dust particles can stay airborne for long periods of time and can penetrate deep into the lungs (WorkSafe, 2020b). For these reasons, the control of inhalable dusts (smaller than 100 μm) and respirable dusts (smaller than 10 μm) remain an industry focus. Table 2 includes examples of Australian workplace exposure standards for dusts that are relevant to underground mining. Important to note is that such 'exposure standards should not be considered as representing an acceptable level of exposure to workers. They establish a statutory maximum upper limit' (Safe Work Australia, 2013).

TABLE 2

Workplace exposure standards for airborne contaminants (Safe Work Australia, 2024).

Airborne contaminant	Time weighted average (mg/m^3)
Coal dust	1.5
Copper dust	1.0
Nickel, metal	1.0
Quartz (respirable dust)	0.05
Respirable crystalline silica	0.05
Silver, metal	0.1

Reduced levels of exposure can present challenges for both existing and new underground operations. Existing operations can encounter changed work environments and equipment malfunctions, which can mean once compliant ventilation systems, are no longer compliant. For new underground operations, lower exposure standards can challenge, but at the same time improve, ventilation planning. De Andrade et al (2024) found reduced exposure standards increased collaboration and planning at a project's commencement, with project managers and supervisors motivated to find solutions to not only protect the health and safety of mine workers but also improve project efficiencies. These authors describe how mobile dust collectors complemented a tunnels main ventilation system by capturing high volumes of dust at the source and then exhausting filtered air into the working area. The improved quality of air allowed for multiple stages of construction to take place simultaneously, which increased the overall production.

In mining, there are a range of methods and technologies to manage airborne contaminants, but the significance of context ensures that some are more suitable than others. For example, wetting methods (eg misting sprays, fogging systems, water cannons) are common to suppress dust, however these methods are reliant on water availability, which can be scarce and tightly policed with short- and long-range modelling projecting changing rainfall patterns, increased evaporation and reduced water availability (AdaptNSW, 2024; Department of Climate Change, Energy the Environment and Water, 2017). Engineered dust controls (eg baghouses, mobile dust collectors, and wet dust scrubbers) offer an alternate solution to complement or potentially substitute such wetting methods. These technologies remove, as opposed to supress, large volumes of dust and are less reliant on water. However, there are important considerations attached to each technology, which dependent on context, can make one more suitable than another. For example, a baghouse may be suited for contexts in which the dust is consistently generated in the one spot, however less suitable if the dust source continually changes. Choosing an engineered control that is closely aligned with the mine's context and operations is therefore crucial.

The complexity attached to 'good' decision-making is well documented in cognitive science research (see Chinn and Rinehart, 2016), and such complexity can be magnified in environments such as mining where there are competing, and often subjective, constraints. In recognition of such complexity, earlier research has engaged decision-making matrices and frameworks to make this process more objective. For example, Kursunoglu and Onder (2015) quantified the weighting attached to different criteria and sub-criteria of the Analytic Hierarchy Process (AHP) to select a main fan for an underground mine in Turkey. They concluded that the method effective for making decisions that are objective and more scientific in nature and suggested that the evaluation criteria could be modified to suit different decision-making contexts. This paper builds on their recommendation by retaining the criteria of the AHP model, but adjusting the sub-criteria and descriptions, to explore the contexts in which alternate engineered (baghouses, mobile dust collectors, wet dust scrubbers) can complement an underground mines main ventilation system-Table 3. In this paper, the criteria and sub-criteria informed the interview questions that the three lead engineers were asked to address. Numeric values to weight each criterion were not assigned as the aim of the paper was to inform future decision-making processes as opposed to selecting one engineered control over another.

TABLE 3

Criteria and sub-criteria for selecting an engineered dust control. Adapted from Kursunoglu and Onder (2015).

Criteria	Sub-criteria	Description
Technical	Air quantity	The volume of air that can be moved per unit of time.
	Filtration	Includes filtration efficiency; filter media; alignment between instream and filter media.
	Pressure	Comprises the pressure affecting on unit area during air transfer by the fan.
	Efficiency	Comprises the amount of air moved per unit of electrical energy input to the fan motor.
	Size and footprint	Speaks to the actual footprint or size of the machine.
Operational	Productivity	The dust control improves productivity by providing better work environments.
	Safety	Some dusts are combustible; made of spark-preventing materials; has a motor that can work with a risk of fire and higher temperatures.
	Flexibility	Adaptive to changes in production conditions; mobile.
Environmental	Noise level	Sound levels are acceptable (dB).
	Waste	Collected dust is disposed of using methods that are both environmentally sustainable and operationally efficient.
	Water	Comprises the responsible use of water; Water outcomes are equitable, environmentally sustainable, and economically beneficial.
Economical	Operating costs	In many instances, the cost of energy consumption over the life of a fan is significantly more than the initial capital cost of the equipment. Also includes the cost and frequency of replacing parts, servicing etc.
	Operating savings	Improved work environments can reduce other maintenance costs; In some circumstances the dust can be recycled or sold for an ROI.

FINDINGS

This paper aimed to explore the contexts in which alternate engineered controls (baghouses, mobile dust collectors, wet dust scrubbers) could complement an underground mines main ventilation system to achieve compliance and potentially increase operational, technical, environmental, and economical efficiencies. Three lead engineers, each with over 15 years of experience in the dust control industry were interviewed, with the criteria and sub-criteria in Table 3 informing the interview questions that were asked. Eight themes emerged from the interviews. These themes represent the decision-making criteria the lead engineers felt important to evaluate at the point of selecting an engineered dust control. The eight themes are now shared along with the direct interview quotations to support their inclusion.

Air volume

Air volume calculations were identified as a logical starting point for choosing a dust control technology. As stated in one of the interviews, 'volume is king in terms of the selection, everything else follows the volume'. The engineered controls discussed in this paper have different capabilities with respect to air volume. Baghouses and mobile dust collectors can move as much as 60 cubic metres of air per second, whereas a wet 'venturi' scrubber has limitations of 42 cubic metres of air per second. However, this does not mean the wet scrubber unable to accommodate large volumes of air. It just means multiple units must work in parallel:

> 'Wet systems will typically use a venturi type scrubber. They are limited to 42 meters cubed per second. To increase the flow, you're going to use two systems in parallel. For flow requirement of 80 cubic meters per second, two wet scrubbers are required. This means, 40 cubic meters per second will flow through one and 40 through the other. Scrubber system efficiency drops quite significantly if 80 cubic meters per second passes through a single wet scrubber'.

It was evident from the interviews that customers can put too much emphasis on air volume alone, and that greater emphasis must be placed on the speed that the dust moves through the system, known also as filtration velocity.

Filtration velocity

Filtration velocity refers to the speed at which the volume of contaminated air passes through the filter media. If the volume of air travels too fast the fibres of the filter cloth can be damaged. To quote one senior engineer, 'A handful of dust at the wrong speed will mess up your system more than a tonne of dust at the right speed'. To put this into context, another engineer described a 10-cube machine with different filter ratios:

> 'If you run a 10-cubic metre per second machine with four filters in it, those four filters are going to last a much smaller period compared to a machine that's got 40 filters in it. The amount of air that needs to be filtered can go through 40 filters a lot slower than what it can be pulled through 4 filters. Therefore, that flow is going to ruin 4 filters very, very quickly'.

Finding the optimal ratio between the surface area of the filter media and the filtration velocity is therefore key for improving the machine's efficiency and increasing the lifespan of the filters. By default, this ratio also influences the overall size of the machine, which can become problematic in an underground environment where the machines operate in confined environments and compete for space with other mining equipment. To give an example, mobile dust collectors are built modularly with cartridge filters, which means the dimensions of the machine (typically length) increase as the number of filters increase.

Spatial requirements

Underground construction takes place in confined environments, so from an operations standpoint, size and footprint are important early considerations. This was reinforced in the interviews with practical considerations around space identified as an obvious starting point for choosing a dust control technology:

> 'In an underground situation, spatial restrictions are key, both in terms of the physical space but also what other equipment and operations are taking place'.

In relation to space, mine operators were challenged to look beyond the needs of today and think more about their operations in 2–3 years' time:

> 'You've got to look at what machinery will want to be used in the future and if the operation is going to be consistent within the next lifespan of the equipment. For a change of operation, for instance going from a construction phase to an operational phase what does that look like? Is the equipment still going to suitable? Have we got enough redundancy within that equipment?'

For situations in which the size of a machine conflicted with the space available, creative, problem-solving solutions for all engineered controls were shared. For example, caverns were cut into walls to store baghouses and/or wet and dry scrubbers; ducting was extended to reach already busy, confined spaces; and in the case of mobile dust collectors, filter configurations were changed to reduce the width of the machine. Spatial requirements, represent one of several highly integrated decision-making criteria already discussed in this paper. However, from these interviews, one got the sense that creative solutions existed for most problems related to spatial requirements.

Temperature

Air expands when heated, which can make a dust control unit more efficient on days with cool air conditions than extremely hot air conditions. Therefore, an understanding of temperature and how it might be controlled in the underground environment becomes another important consideration for choosing an engineered control:

> 'If temperature is high when drilling your unit might work today because it's 30 degrees. But tomorrow you would think, why isn't it extracting as well? Suddenly you're at 50 or 60 degrees and dust collector is not coping because the volume has increased due to expansion'.

One solution offered in the interviews was the use of a venturi wet scrubber to help cool down the air:

> 'The amount of BTU (British Thermal Unit) you need to cool down an air pocket versus cooling down the water is significantly smaller because of the action between the two. It's much more difficult to do it in air. So, what we found is you can do a venturi wet scrubber to cool down the air and then you have a very small cooling tower that cools the water down again and you have that continuous cooling in the in the system and then you cool down the tunnel'.

However, this solution is not without its challenges, as wet scrubbers require access to water, which must either be disposed or recycled. The viability of this solution therefore becomes closely intwined with the environmental considerations discussed later in this paper.

Instream characteristics

The characteristics of the instream were also discussed with conversations centred around the relationship between the characteristics of the dust (eg hard versus soft) and the filter media itself:

> 'When it comes to dry scrubbing, the particulate type forms a very important part of how you select the equipment. Typically, if you do something like Bentonite, which is a very sharp, tough material, you're going to select a certain filter media type. You put something in there that's not going to abrade away quickly'.

Tied closely to filtration velocity, these considerations strengthen the argument that filtration velocity a cornerstone for choosing a dust control technology. The instream moisture levels were also debated, with a hinge-point for choosing a dry or wet dust control technology tentatively suggested:

> 'The moisture level in a gas laden or a dust laden stream. Really, sort of guide you to which technology you're going to use. So typically, between 7 and 8% moisture in a stream will guide you. Under that is a dry system and above that we would go towards more of a wet system because baghouses do not like moisture.... If there is 15% water as liquid droplets or attached to particles in the airstream, your bags will just not last'.

Moisture levels are a useful marker for choosing a dust control technology. However, it was also noted that such a 'hinge point' not as applicable if the cartridge filter already included a hydrophobic membrane.

Mobility

As the name suggests, mobile dust collectors have more mobility than the other technologies presented in this paper. Their mobility is determined by the operations undertaken and how frequently the machines must move:

> 'For the crushing process side of things, the likelihood of having to move the machine often is probably limited. Therefore, a skid machine is more suitable. If it's at the mining side, then a track machine is more suited to mobility requirements.'

The location of a baghouses or wet scrubbers is traditionally fixed, so they do not have this mobility. To achieve mobility these technologies will extend their ducting or add another unit:

> 'They put a baghouse in a tunnel, permanently fixed, and just keep extending the baghouse ducting further down the tunnel or maybe build another permanent baghouse further down the tunnel'.

The decision-making around a fixed or mobile dust control unit is dependent on the operational needs of the mine. If there is certainty and predictability around the source of the dust than a fixed unit may be suitable. A process plant was an example shared in the interviews, 'They put a baghouse up. They duct to it, to all the points where they know it's going to be and sort of that's not a problem anymore'. On the other hand, if the source of the dust is unpredictable and continually changing than a mobile unit could be more suitable. Underground mining, 'where the goal posts are forever moving' through operations such as Drill and Blast, Chasing the Face, and Room and Pillar mining were presented as examples.

Environmental

There were environmental considerations attached to each dust control technology, especially in relation to the disposal of the collected dust or waste. For wet scrubbers, the environmental considerations related largely to water usage and power consumption:

> 'You've also got a treat the outgoing water. It's not just an on-site situation. The water needs to be pumped somewhere else. The wet scrubbers use a lot of water. They are also expensive to run because the fan pressures required are really high. Almost triple the amount of fan pressure is required to be able to pull the same airflow through wet scrubber when compared to a dry scrubber'.

For dry extraction methods, the environmental considerations related to the disposal of the dust:

> 'With the baghouse, disposal of collected dust is always a problem. What do you do with the dust? You've taken the dust out of the airstream and now you are putting it somewhere else. Either into a screw conveyor and that goes to somewhere, or you're putting it into water to make a slurry and it still needs to be disposed of...'

However, regardless of the technology, it was noted that each is exploring ways to beneficiate the waste:

> 'There's a lot of work being done to see how people can reuse the dust and it really does depend on what dust type you have. Cement dust, you can try and sell off as a cheap cement. For sand it may be possible to use it again. There are many processes and ideas that businesses are trying in order to recycle it back into their systems as it becomes expensive to the environment and to themselves to just dispose of it'.

Economic drivers

There emerged one clear economic driver that influenced the decision-making process from beginning to end. This driver spoke to an ongoing tension between Capital Expenditure (CapEx) and Operational Expenditure (OpEx) and the weight given to each during decision-making process. Capital expenditure includes funds used to acquire physical assets, while operational expenditure are funds used in normal business operations. According to those interviewed, an entity would give weight to one at the expense of the other. An entity that placed greater weight on operational expenditure, would decrease upfront capital expenditure and increase operational expenditure (eg servicing and maintenance):

> 'In general, dust extraction is a grudge purchase because it's not providing positive cash flow for the business. It's just a flat loss to the to the bottom line, so they try to keep the CapEx small. Generally, the choice is to live with the inefficiencies of this unit and every three months carry out a service to put in new bags. This pushes the cost to the operational budget and that then increases the OpEx cost to overcome the CapEx small budget'.

In contrast, an entity that places greater weight on capital expenditure, would increase upfront capital expenditure and decrease operational expenditure (eg servicing and maintenance):

> 'Other clients take a different view and request a solution with minimal active monitoring and maintenance requirements. They spend the money up front and do a service every three years instead of every six months or three months. They would then use their OpEx money to do other things than to just look after a baghouse.'

Filter replacement was provided as an example, with cheaper but more frequent baghouse filter replacements compared with more expensive, but less frequent cartridge filter replacements:

> 'The perception is that cartridge replacement costs are much higher than baghouse filters, which they are for a replacement investment. The difference is when you start looking at the life of a cartridge, it can be the same amount as what a baghouse is. It just comes down to the initial engineering size of the machine.'

Arguments of this nature are common and can be problematic to solve. If faced with such a dilemma, those interviewed encourage decision-makers to think more reflectively about the aims and motivations for choosing an engineered control to begin with:

> 'I think the biggest issue…is people miss the concept of what the industry is trying to achieve. We're trying to minimise the amount of respiratory dust that goes into the environment, and what is the cost of that compared to we're just ticking a box'.

DISCUSSION OF FINDINGS

The findings provide helpful considerations for choosing an engineered dust control, yet at the same time demonstrate how a multitude of criteria, often intwined, can complicate the decision. The three steps that follow, synthesize these findings, to simplify the decision-making process. It is argued that a 'good decision' will show an alignment between each of the steps.

Identify aims. The clear identification of aims and values is a recommended starting point for choosing an engineered dust control. For example, if environmentally sustainability a priority, then early consideration should be given to resource availability (eg water, power consumption) and the disposal or recycling of the waste. This is also the time to reflect and challenge any pre-existing values that could become barriers to achieving the identified aims. For example, valuing operational expenditure at the expense of capital expenditure could become a barrier for aims related environmental sustainability as the controls attached to such aims typically require significant upfront capital investment.

Know the context. Choosing a dust control technology that is aligned to the aims is dependent on knowing the context. This includes the consideration of environmental criteria such as, temperature, instream characteristics, and the availability of natural resources (eg water) and operational criteria such as, spatial requirements and mobility. This ongoing evaluation between context and aims will naturally position some engineered controls over the others. To extend the earlier example, if environmental sustainability an aim, yet water in the community scarce, then a baghouse or mobile dust collector seem a more reasoned decision than a wet venturi scrubber. On the other hand, if the underground environment characterised by moist air conditions and water plentiful, then a wet scrubber could be justified.

Technical requirements. Aligning the technical requirements of the chosen engineered control with the context and aims is a final consideration. Key to this decision-making process is an appreciation of the relationship between the different technical criteria. For example, understanding how the lifespan of a filter cartridge is impacted by the speed at which the volume of contaminated air moves through the system. Each of the engineers interviewed stressed that it is not enough to look at the size of the machine alone. The findings also suggest that aligning the technical requirements of the machine with the context will initiate problem-solving opportunities that are both creative and collaborative. For example, mine operators and product design teams working in close collaboration to decide how best to position, move and service the machines. This collaborative problem-solving process is crucial, as it helps to ensure that all decisions made are consistent within the next lifespan of the equipment.

CONCLUSIONS

In this paper an existing framework to support critical decision-making was adapted to explore the contexts in which alternate engineered controls (baghouses, mobile dust collectors, wet dust scrubbers) can complement an underground mines main ventilation system. The findings provide helpful insights for future decision-makers, yet also capture the complexity of the decision-making process, with individuals challenged to evaluate multiple criteria (often intwined) at the same time. Three steps to simplify the decision-making process are proposed and it is argued that a 'good decision' one that shows an alignment between each step. Step one, encourages the decision-maker to clearly identify the aims and values attached to choosing an engineered dust control. Step two focuses on context which included *environmental* criteria such as, temperature, instream characteristics, and the availability of natural resources and *operational* criteria such as, spatial requirements and mobility. It is proposed that the enaction of this step will naturally position one engineered control over another. The final step speaks to the technical requirements of the engineered control. In this step, decision-makers are challenged to think holistically about the relationships between different technical criteria as opposed to thinking about each criterion in isolation.

ACKNOWLEDGEMENTS

The research team would like to acknowledge the lead engineers who agreed to participate in this study and provide such valuable insights.

REFERENCES

AdaptNSW, 2023. Climate change impacts on our water resource, NSW Government. Available from: <https://www.climatechange.environment.nsw.gov.au/impacts-climate-change/water-resources> [Accessed: 28 February 2024].

Chinn, C and Rinehart, R, 2016. Epistemic cognition and philosophy: developing a new framework for epistemic cognition, in *Handbook of epistemic cognition* (eds: J A Greene, W A Sandoval and I Braten), pp 460–478 (Routledge).

De Andrade, A, Fanning, A, Palmer, B and Stowasser P, 2024. Transforming dust control – lessons from tunnelling, in *Proceedings of the International Mine Health and Safety Conference 2024,* pp 33–42 (The Australasian Institute of Mining and Metallurgy: Melbourne).

Department of Climate Change, Energy, the Environment and Water, 2017. Considering climate change and extreme events in water planning and management, Australian Government.

Kursunoglu, N and Onder, M, 2015. Selection of an appropriate fan for an underground coal mine using the analytic hierarchy process, *Tunnelling and Underground Space Technology*, pp 101–109.

Nie, W, Sun, N, Liu, Q, Guo, L, Xue, Q, Liu, C and Niu, W, 2022. Comparative study of dust pollution and air quality of tunnelling anchor integrated machine working face with different ventilation, *Tunnelling and Underground Space Technology incorporating Trenchless Technology Research,* pp 1–15.

Occupational Safety and Health Administration (OSHA), 2017. OSHA's Respirable Crystalline Silica Standard for Construction – OSHA fact sheet. Available from: <https://www.osha.gov/sites/default/files/publications/OSHA3681.pdf> [Accessed: 1 March 2024].

SA Health, 2021. Dust and your health – fact sheet, Government of South Australia. Available from: <https://www.sahealth.sa.gov.au> [Accessed: 1 March 2024].

Safe Work Australia, 2013. Guidance on the interpretation of workplace exposure standards for airborne contaminants, Safe Work Australia.

Safe Work Australia, 2024. Workplace exposure standards for airborne contaminants, Safe Work Australia.

Wang, B, Wu, C, Kang, L, Reniers, G and Huang, L, 2018. Work safety in China's thirteenth five-year plan period (2016–2020): Current status, new challenges and future tasks, *Safety Science*, pp 164–178.

WorkSafe, 2020a. Airborne contaminants. Available from: <https://www.worksafe.qld.gov.au/safety-and-prevention/hazards/hazardous-exposures/airborne-contaminants> [Accessed: 1 March 2024].

WorkSafe, 2020b. Hazardous dusts. Available from: <https://www.worksafe.qld.gov.au/safety-and-prevention/hazards/hazardous-exposures/hazardous-dusts> [Accessed: 1 March 2024].

Characterisation of diesel particulate matter aerosols and control strategies effectiveness in a controlled zone in the underground environment

A Habibi[1], K Homan[2] and A Bugarski[3]

1. Technical Expert – PT Freeport Indonesia, Papua Indonesia. Email: ahabibi@fmi.com
2. Associate Professor, Missouri University of Science and Technology, MO, USA.
3. National Institute for Occupational Safety and Health, Pittsburgh Mining Research Division, Pittsburgh PA, USA.

ABSTRACT

The characterisation of physicochemical properties is important to better understand the health impact of exposure of underground mines to diesel aerosols. A study was conducted in the controlled zone of an underground mine to assess the effects of four diesel emissions control strategies on diesel aerosols. The controlled zone was designed to allow evaluation in the mine atmosphere, good control over ventilation flow rates, and limited interferences from the background aerosols. The objective of this study was to characterise the morphological characteristics of aerosols emitted by diesel-powered vehicles in the underground atmosphere. The sampling strategy was designed to allow for the concurrent collection of grid samples for Scanning Transmission Electron Microscopy (S/TEM) analysis and real-time monitoring of the background particle concentration. The dimensions of agglomerate aerosols were measured, and the results were compared with the number size distribution measured by Fast Mobility Particle Sizer (FMPS) and Electrical Low-Pressure Impactor (ELPI+). All four evaluated control strategies were found to be effective, providing reductions in aerosol number concentration of more than 95 per cent. The microscopic structure analysis of the diesel aerosols in accumulation mode was in reasonable agreement with the results of the FMPS measurements. The FMPS and ELPI+ real-time monitoring data provided valuable data on the number concentration and size distribution. The TEM results provided valuable information on the fractal properties of the agglomerates and primary particles.

METHODOLOGY

A controlled zone was designed to allow for studying the effects of the selected diesel control strategies on submicron aerosols with minimal interference from diesel and other aerosols potentially present in the background air. To minimise the interferences and improve the quality of the samples, the zone was isolated from mining activities and diesel-powered vehicle traffic and ventilated with fresh air supplied from the nearby intake shafts (Figure 1a). In addition, the two-hour long sampling sessions were scheduled between the shift changes to eliminate potential effects of those on the concentrations of background aerosols.

The objective of the study was to characterise the morphological characteristics of diesel aerosols emitted in the underground atmosphere. Control zone sampling made it possible to measure and obtain grid samples simultaneously while the background particle concentration was monitored. The S/TEM images were used to identify the largest agglomerates on a grid. The diameters of agglomerated particles were measured, and the results were compared against the number size distribution obtained concurrently with FMPS.

The zone was set-up at the remote end of the diesel shop area (Figure 1a). The preparation work for the test location consisted of blocking one of the access entries into the shop during evening hours when the mine experienced lower traffic and activities. The mine employees typically travel twice during a single shift: (1) to the assigned working places at the beginning of the shift; and (2) back to the service shaft by the end of their shift. Between those two events, the traffic in the fresh air and the corresponding impact on the concentrations of diesel aerosols in the zone background were relatively modest.

The controlled zone was located in a 300 m (1000 ft) long straight entry with a rectangular cross-section of 6 m × 2.7 m (20 ft × 9 ft) (Figure 1b). The eight cross-cuts toward the active zone of the shop were sealed using temporary curtains. All the curtains were foam sealed to minimise the leakage. The test vehicles were operated over custom-designed cycles within the zone. The zone was closed to the incoming traffic at 1XC, and the traffic was diverted to the other parallel entries outside the controlled zone.

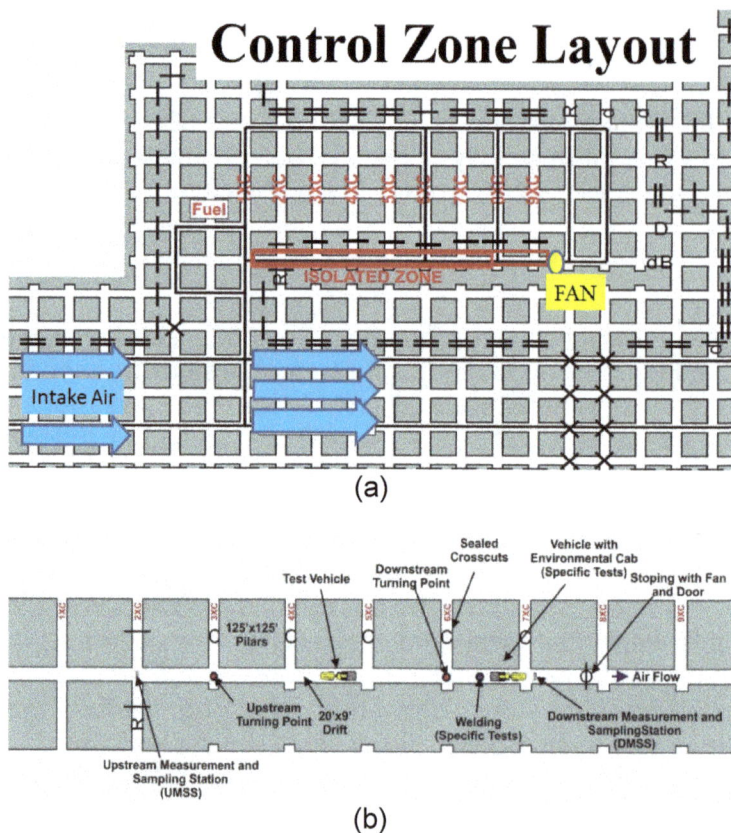

(a)

(b)

FIG 1 – Controlled zone (a) situational map and (b) layout.

Ventilation

The zone was located on a split of intake air from the nearby shaft (Figure 1a). A ventilation system controls were implemented to allow regulation and measurement of the airflow quantities for the controlled zone (Figure 2).

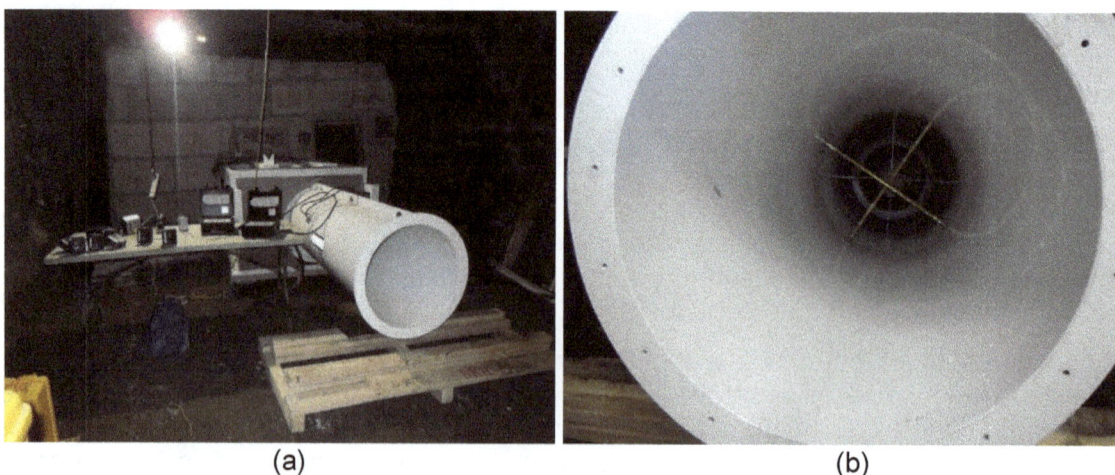

(a)

(b)

FIG 2 – Ventilation system controls (a) a two-stage fan and (b) a VOLU-probe.

A permanent bulkhead was built to house the fan (Figure 2a). The 36 kW (50 HP) two-stage fan (Spendrup Fan Co. Model 60x2-3600 CR FP) provided steady airflow throughout the zone during the testing. The fan was fitted with an 1830 mm long, 610 mm diameter inlet section. Two VOLUprobes were installed inside the inlet section, 1220 mm from the tip (Figure 2b). The VOLU-probe operates on a principle similar to pitot airflow traverse probes. The probe, equipped with multiple sets of total and static sensing ports, had a flow measurement accuracy of 0.15 per cent.

Sampling

Two sampling stations were placed in the zone (Figure 1b). The upstream monitoring sampling station (UMSS) was located at the entrance to the zone (at 2XC). The downstream monitoring sampling station (DMSS) was located at 8XC, 10 m (30 ft) downwind from the fan. The upstream turning point was located 50 m (164 ft) downwind from the upstream monitoring station (3XC). The downstream turning point was located 50 m (165 ft) upwind of the downstream sampling station. This distance allowed for sufficient residence time for the initial formation and transformation of diesel aerosols and the mixing of those in the zone.

Upstream Sampling Station (UMSS)

The results of the UMSS measurements were used to characterise properties and quantify concentrations in the background air. At UMSS, Fast Mobility Particle Sizer (FMPS) (TSI, model 3090) was used to measure aerosol number concentrations and size distributions, and the Vaisala device (Carboncap GM70) was used to continuously record the CO_2 concentrations (Figure 3a). Those results were used to correct the downstream results for the background contribution.

(a)   (b)

FIG 3 – Instrumentation at (a) UMSS and (b) DMSS.

Downstream Sampling Station (DMSS)

The instrumentation at the DMSS was used to measure in real-time the particle size distributions and concentrations (FMPS and electric low-pressure impactor (ELPI), gas concentrations (Fourier Transform Infrared (FTIR) analyser), temperature, and absolute pressure (Figure 3b). The ELPI+ was used to collect the S/TEM samples on the copper grids. The ELPI+ sampling time varied based on the real-time concentration readings obtained from ELPI and FMPS. A spare set of ELPI+ impactor columns was used during each test to: (1) collect TEM samples in multiple stages of the impactor; and (2) record the size, and number and mass concentrations concentration in the stages on which TEM grids were not present. The greased aluminium foils were placed on the stages that did not have the aluminium foils with TEM grids.

The sampling lines for the ELPI and FMPS were connected to the ports mounted on the head of the rotating system. The sampling line for FMPS and ELPI were collected from a fixed location in the centre of the drift, 2.4 m (8 ft) off the ground. All samples were transferred to the instruments using non-heated sampling lines made of conductive rubber tubing. To minimise diffusional losses, the length of the sampling lines was kept to the minimum needed for the operation of the rotating sampling system.

RESULTS

During the study, 12 heavy-duty (HD) and light-duty (LD) vehicles were evaluated in the zone. Several characteristics and properties were analysed.

Fractal dimension and shape factor analysis

The fractal dimension (D_f) is used to characterise the complex structural morphology of agglomerates. The D_f is a function of the radius of gyration (R_g) and the average diameter of primary particles $(\overline{d_p})$ of agglomerates (Megaridis and Dobbins, 1990; Koylu, Xing and Rosner, 1995; Oh and Sorenson, 1997; Lee *et al*, 2003; Park, Kittelson and McMurry, 2004; Song and Lee, 2007). D_f is determined as the slope of the linear fit line of n_p and $ln(2Rg / dp)$. Lee *et al* (2003) measured a fractal dimension of 1.46 to 1.70 for aerosols emitted by LD diesel engines. They also found higher D_f, 1.80 to 1.88 for aerosols emitted by HD diesel engines.

The power law relationship approach is used to identify the fractal dimension of soot agglomerates. In this study, Equation 1 is used as a three-dimensional value of n_p and R_g is used for correlating the agglomerate projected area and primary particle projected area. Koylu, Xing and Rosner (1995) suggested values of α = 1.09 and $k\alpha$ = 1.21 to calculate the fractal dimension for the agglomerates. Oh and Sorenson (1997) considered the primary particles overlapping parameter (d) and suggested the values of α = 1.19 and $k\alpha$ = 1.81 when d = 2. The overlap parameter is the ratio of the primary particle diameter to the distance between the centres to touching primary particles. Where, L_{max} is the maximum length, $k\alpha$ is a correlation perfector, and D_{fl} is the fractal dimension with maximum projected length.

$$n_p = k_\alpha \left(\frac{L_{max}}{d_p}\right)^{D_{fl}} \tag{1}$$

The results show that the D_f for agglomerates collected while LD equipment was operated are lower than those when HD equipment was operated. As shown in Figure 4, the D_f for LD vehicles was 1.59 using Koylu, Xing and Rosner (1995) values and 1.74 using Oh and Sorenson (1997) values. The results from HD analysis show a D_f of 1.74 from Koylu, Xing and Rosner (1995) and 1.90 using Oh and Sorenson (1997) values. The HD D_f results were found to be higher compared to the LD engines. The lower the fractal dimension, the more chain-like the agglomerates.

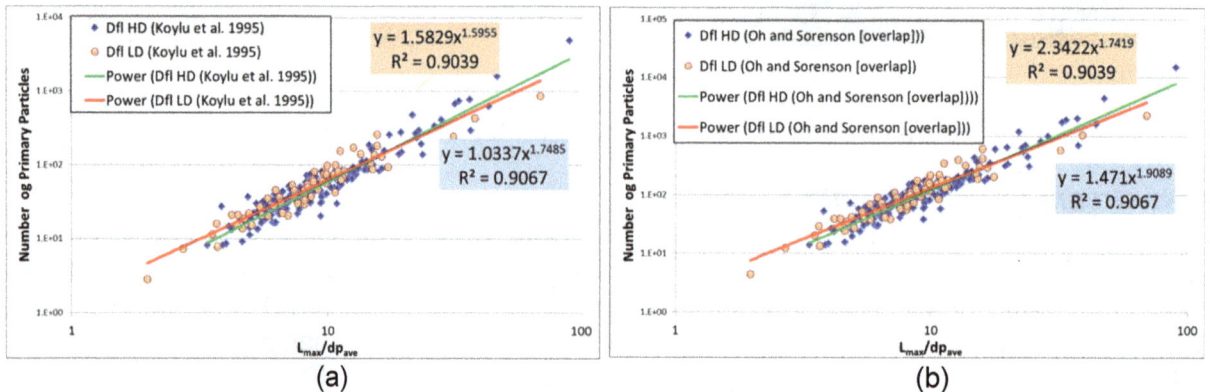

FIG 4 – Fractal dimension analysis for HD and LD equipment: (a) with overlap and (b) without overlap.

The Shape Factor (SF) for HD engines are normally distributed with a mean of 0.63 and a spread of 0.15. The results suggest that the HD equipment agglomerates are somewhat irregular and more spherical rather than chain-like. However, agglomerates with an SF smaller than 0.4 and chain-like shapes were also noticed. The LD SF analysis fitted in a bi-modal distribution with a mean for the first mode at 0.49 with a spread of 0.08. These were the chain-like and elongated agglomerates.

Size distribution and primary particle diameter

The distributions of aerosols at downstream sampling stations measured with FMPS at the selected instance at 6000 sec for three LD vehicles are shown in Figure 5a. Figure 5b shows the FMPS size distribution for HD equipment during controlled zone sampling.

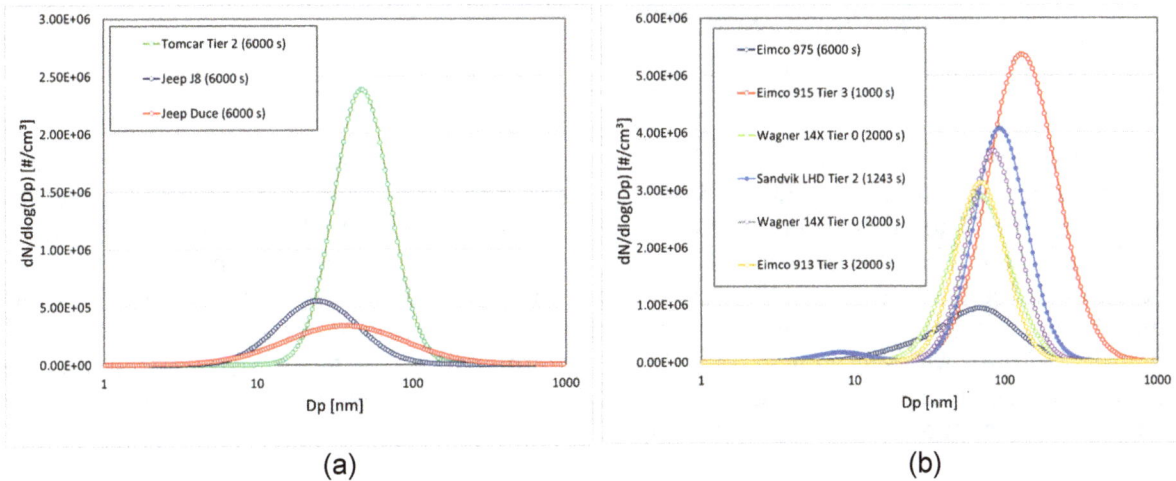

FIG 5 – Size distribution of aerosols for (a) LD vehicles and (b) HD vehicles.

The mean value of primary particle diameter for HD was 23.8 nm with a spread of 6.2, whereas for LD equipment it was 22.6 nm with a spread of 7.2 nm. The results show that the primary particle diameter for HD equipment was slightly higher. Liati and Eggenschwiler (2010) found the size of primary sub-micron soot particles in diesel exhaust is less than 50 nm. Mustafi and Raine (2009) measured the average primary particle diameter of 26 nm.

Control strategies

Several control strategies are used to reduce diesel particulate emissions at the source. The effectiveness of four control strategies was evaluated in this study. This paper discusses the example of retrofit-type sintered metal filters (SMF) in-depth and only summarises the other control strategies.

- Retrofitting EPA Tier 3 engine with SMF system.
- Substituting the LD personnel carrier with a contemporary unit powered by a modern engine.
- Retrofitting EPA Tier 1 engine with the filtration system with a disposable filter element (DFE).
- Utilising vehicle powered by EPA Tier 4 diesel engine.

Retrofitting with SMF system

The effectiveness of retrofit-type SMF systems in curtailment of aerosol emissions from EPA Tier 3 diesel engines in HD vehicles was examined during two tests. The SMF system with Active Regeneration (AR) systems are referred to as SMF-AR. These systems are equipped with heating elements which can initiate the regeneration on demand.

The first test was used to establish the baseline emissions for the Cummins QSB6.7 engine. The second test was used to establish emissions for the same engine retrofitted with the SMF system. The reduction in the elemental carbon (EC) concentrations was estimated to be approximately 97 per cent as shown in Table 1.

TABLE 1

EC concentrations at downstream and upstream sampling stations during SMF-AR.

Vehicle, exhaust configuration	EC at Downstream station (µg/m³)	EC at Upstream station (µg/m³)
Eimco 915, without SMF (baseline)	627.9 ± 60.0	20.4 ± 0.8
Eimco 915, with SMF (main)	25.3 ± 3.7	5.3 ± 2.7

The results of continuous monitoring of concentrations of aerosols at upstream and downstream sampling stations measured with FMPS during two tests conducted to evaluate the SMF system are shown in Figure 6. The contribution of the background (upstream sampling station) to the number concentrations at the downstream sampling station was found to be relatively modest, on average 4.1 per cent for the test with the SMF system and 7.9 per cent for the test without the SMF system. The number of concentrations at the downstream sampling station did not substantially change for the duration of the SMF test, indicating that the performance of the system was not substantially affected by the accumulation of diesel aerosols in the filtration media. The SMF system was found to reduce the concentrations by approximately 93 per cent.

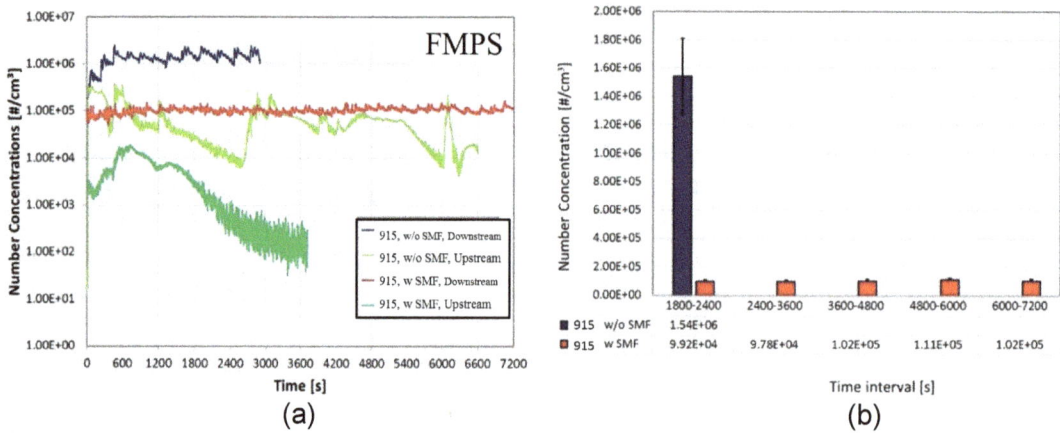

FIG 6 – Number concentrations of aerosols for the SMF tests: (a) FMPS traces, and (b) FMPS averages (ventilation rate of 5.43 m³/s (11 500 ft³/min).

The aerosols at the UMSS were found to be distributed between nucleation and agglomeration modes (Figure 7a). The aerosols at the DMSS were distributed in single agglomeration mode or between nucleation and agglomeration modes, with most of the aerosols in agglomeration mode (Figure 7b). The agglomeration mode aerosols at the downstream sampling station produced by the vehicle operated without SMF had somewhat larger count median diameters (CMDs) than the aerosols produced by the same vehicle retrofitted with the SMF system.

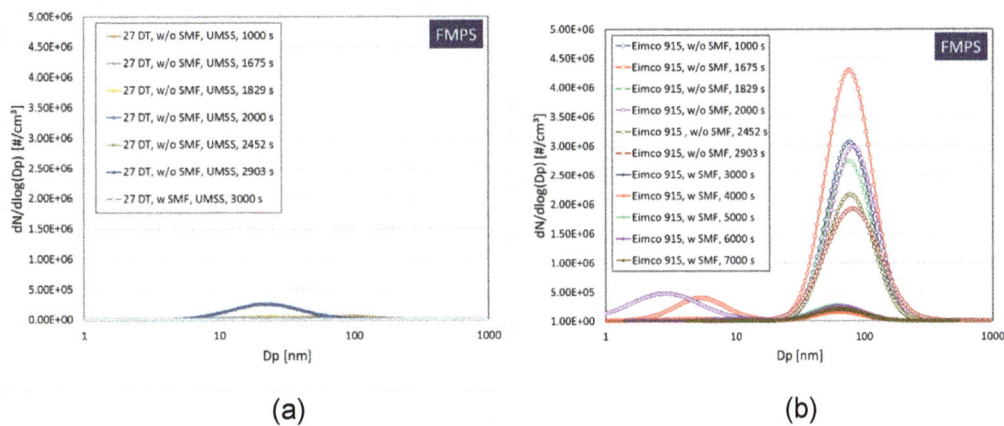

(a)   (b)

FIG 7 – Size distributions of aerosols for SMF tests at (a) UMSS, and (b) DMSS (ventilation rate of 5.43 m³/s (11 500 ft³/min)).

To determine the size distribution using image analysis, the $D_{projected}$ was calculated by measuring multiple attributes of agglomerates. The results show that the size distribution of agglomerates, w/o SMF, follows a normal distribution mostly in the accumulation range. However, for the test with SMF, the aerosols were distributed between nucleation and accumulation mode. It was determined that larger particles were measured by S/TEM analysis in the case where the engine was not fitted with SMF. Tables 2 and 3 show the statistical parameters of the two tests.

TABLE 2

Statistical parameters of $D_{projected}$ size distribution of agglomerates w/o SMF.

Type	Parameter	Estimate	Lower 95%	Upper 95%
Location	μ	143.97	120.50	167.43
Dispersion	σ	59.32	46.72	81.30

TABLE 3

Statistical parameters of $D_{projected}$ size distribution of agglomerates with SMF.

Type	Parameter	Estimate	Lower 95%	Upper 95%
Location	μ	98.95	82.82	115.07
Dispersion	σ	32.42	24.33	48.61

Figure 8 shows the number size distribution measured by FMPS during both tests. As shown in the figure, in the case of the engine w/o SMF, the number size distribution measured by FMPS is higher than for the engine with SMF. Both FMPS size distributions fitted in a normal distribution with the peak CMD of 80 nm for the engine w/o SMF (2452 sec) and 62 nm for the engine with SMF at 6000 sec. The $D_{projected}$ size distribution also fitted normally with the peak of 143 nm w/o SMF and 98 nm with SMF engaged. It is important to note that most of the agglomerate measured during image analyses was in the accumulation range except for some of the ones with SMF engaged. The peak for the $D_{projected}$ for the engine w/o SMF is higher than that measured by FMPS as a result of semi-volatile and volatile agglomerates that were not possible to analyse.

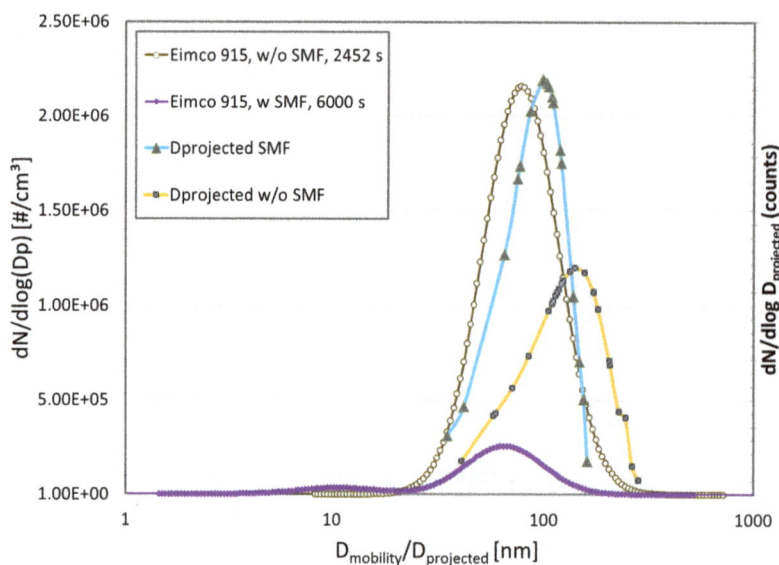

FIG 8 – FMPS number size distribution concentration for the Eimco 915 tests (dN/dlog (Dp)) for various mobility diameters, and projected area equivalent diameter ($D_{projected}$) (nm) versus dN/dlog $D_{projected}$.

DISCUSSION

The diameters of the primary particles were higher for the test when the engine was fitted with SMF than for the test when the engine was operated without SMF-AR. The results show that the shape factors for both tests were similar and that both sets of samples were characterised by a median shape factor of 0.62 to 0.64, suggesting that the agglomerates were rather spherical than elongated. The fractal dimension based on L_{max} analysis showed lower than usual D_{fl} for the samples collected with SMF engaged. The lower D_{fl} suggested more chain-like agglomerates with SMF in operation. The mobility and $D_{projected}$ distribution analysis showed the normally distributed particulates. The size distribution curves showed higher CMD when the engine was operated without the SMF. It was also determined that a minimal number of agglomerates were present in the nucleation mode. This was due to the thermal treatment of SMF, which resulted in the desorption of volatile and semi-volatile components. The other reason could be due to the beam of electrons during S/TEM imaging that resulted in the evaporation of such particles. The 97 per cent reduction in the mass concentration of EC confirms the high effectiveness of the SMF system.

CONCLUSION

The controlled zone study was used to assess the effectiveness of four emissions control strategies. This novel approach was used to assess the physical properties of aerosols emitted by the engines in the LD and HD underground mining vehicles. A replacement of older personal carriers with new ones resulted in a 95 per cent reduction in mass concentrations of elemental carbon. ELPI+ results demonstrated larger aerodynamic particle dimeters (310 nm) were measured for older personal carriers compared to 0.12 µm for new version. The FMPS measurements show the aerosols emitted by the replacement carrier were distributed in the single mode while aerosols emitted by the replaced carrier were distributed single-modally and/or bi-modally. The analysis showed the fractal dimensions between 1.53 and 1.67 for the replaced carrier and between 1.9 and 2.0 for the replacement carrier. The shape factor analysis showed that replaced carrier agglomerates are elongated and long-chained with a peak of 0.58 and spread of 0.18, whereas the shape factor for the agglomerates analysed for a new carrier were more spherical and normally distributed with a peak of 0.68 and spread of 0.14.

The retrofitted SMF with the active regeneration system removed 97 per cent of EC by mass. The diameter of the primary particle was higher for the test where the engine was operated with rather than without SMF. The median shape factor of 0.62 to 0.64 suggests that the agglomerates were spherical rather than elongated. The fractal dimension for the samples collected with SMF suggested the presence of more chain-like agglomerates. The mobility and $D_{projected}$ distribution analysis showed normally distributed particulates with low numbers of agglomerates in the nuclei range.

The DFEs were reduced by at least 95 per cent of EC mass concentrations. The D_{fl} values for the tests conducted with DFEs in place were 1.86, lower than those for the engine operated without DFEs. The shape factor analysis determined the mean values of 0.6 for both tests. The results show that utilisation of DFEs did not impact the values of shape factor and primary particle diameter. The n_p for both tests measured at 24.5 and 23.9 nm for w/o and with DFEs. $D_{mobility}$ and $D_{projected}$ size were normally distributed where CMD for the test w/o DFEs was higher than that with DFEs. It was determined that all four controlled strategies were effective in the removal of EC (over 95 per cent reduction). The results of the statistical microscopic structure analysis of the accumulation size range were in reasonable agreement with the results of FMPS measurements.

ACKNOWLEDGEMENTS

This research was a collaboration of the Missouri University of Science and Technology and the National Institute for Occupational Safety and Health (NIOSH). The assistance of the Diesel Laboratory of NIOSH is very much appreciated.

REFERENCES

Koylu, U O, Xing, Y and Rosner, D E, 1995. Fractal morphology analysis of combustion-generated aggregates using angular light scattering and electron microscope images, *Langmuir*, 11(12):4848–4854.

Lee, O I, Zhu, J, Ciatti, S, Yozgatligil, A and Choi, M Y, 2003. Sizes, graphitic structures and fractal geometry of light-duty diesel engine particulate, SAE Technical Paper 2003, 1-1943.

Liati, A and Eggenschwiler, P, 2010. Characterisation of particulate matter deposited in diesel particulate filters: Visual and analytical approach in macro-, micro- and nano-scales, EMPA (Swiss Federal Laboratories for Materials Science and Technology).

Megaridis, C M and Dobbins, R A, 1990. Morphological description of flame-generated materials, *Combustion Science and Technology*, 71:95–109.

Mustafi, N and Raine, R, 2009. Electron microscopy investigation of particulate matter from dual fuel engine, *Aerosol Science and Technology*, 43:951–960.

Oh, C and Sorensen, C M, 1997. The effect of overlap between monomers on the determination of fractal cluster morphology, *J of Colloid and Interface Science*, 193:17–25.

Park, K, Kittelson, D B and McMurry, P H, 2004. Structural properties of diesel exhaust particles measured by transmission electron microscopy (TEM): Relationships to particle mass and mobility, *Aerosol Science and Technology*, 38:881–889.

Song, J and Lee, K O, 2007. Fuel property impacts on diesel particulate morphology, nanostructures, and NOx emissions. SAE Paper 2007-01-01239.

Enabling crystalline silica analysis of mine respirable dust sampled by real-time gravimetric personal dust monitor (PDM3700)

Y Jin[1], B Belle[2], H Wu[3], R Balusu[4], G Zhao[5], D Tang[6] and Z Qin[7]

1. CSIRO Mineral Resources, Pullenvale Qld 4069. Email: yongang.jin@csiro.au
2. 61Drawings, Australia; University of New South Wales, Sydney NSW 2052; University of Queensland, Brisbane Qld 4000; University of Pretoria, South Africa. Email: bb@61drawings.com
3. Gillies Wu Mining Technology Pty Ltd, Brisbane Qld 4068. Email: hsinwei@minserve.com.au
4. CSIRO Mineral Resources, Pullenvale Qld 4069. Email: rao.balusu@csiro.au
5. CSIRO Mineral Resources, Pullenvale Qld 4069. Email: guangyu.zhao@csiro.au
6. CSIRO Mineral Resources, Pullenvale Qld 4069. Email: dong.tang@csiro.au
7. CSIRO Mineral Resources, Pullenvale Qld 4069. Email: johnny.qin@csiro.au

ABSTRACT

Monitoring of personal exposure levels of respirable coal dust (RCD) and respirable crystalline silica (RCS) is an important step in protecting mine workers from dust related occupational respiratory lung diseases. Timely monitoring and reporting of personal exposure levels is crucial for dust control and exposure management. The personal dust monitor (PDM3700) is a real-time mass-based gravimetric respirable dust sampler in mines. PDM3700 has been successfully used as a compliance monitor in the USA to aid the miners in reducing their exposure to coalmine dust, and also deployed in Australian coalmines since 2016.

The PDM utilises a tapered element oscillating microbalance (TEOM) technology that continuously weighs a filter as the respirable dust is deposited on it, thus monitoring in real time personal exposure levels of RCD. However, as in any currently available compliance gravimetric sampler, it does not provide information specific to RCS without further analytical process. As PDM3700 uses a glass fibre filter for dust sampling, Fourier Transform Infrared (FTIR) and X-ray diffraction (XRD) analytic methods cannot be directly applied to silica analysis due to analytical measurement interference. This paper reports a redeposition methodology to determine the silica content of respirable coalmine dust collected over the PDM glass fibre filter using FTIR method. The newly developed quick CSIRO analytical process has been demonstrated to effectively detach dust particles from the PDM filter without causing sample contamination by glass fibres. In addition, the paper provides an update on the development of a new type of PDM filter with a non-silica filter material and a novel filter assembly structure to enable field-based direct-on-filter silica analysis of PDM-collected dust samples. These new developments greatly advance TEOM based PDM technology to achieve both real-time RCD monitoring throughout the shift and RCS measurement at the end of the shift by using a single PDM sampler. It is envisaged that these would significantly improve RCD and RCS personal exposure monitoring capabilities to identify potential high personal exposure areas and tasks for effective engineering controls, implement on-site quick turnaround compliance assessment by empowering the worker and eliminate the delays in the current gravimetric sampling and off-site silica analysis.

INTRODUCTION

The personal dust monitor (PDM3700) represents the latest advanced technology in legislated personal compliance exposure monitoring of respirable coal dust (RCD) by undertaking real-time continuous gravimetric measurement (Volkwein et al, 2004; Belle, 2017). It was developed and adapted for coalmine use through a two-decade long research and development program in the USA by the National Institute for Occupational Safety and Health (NIOSH) and Mine Safety and Health Administration (MSHA) and is widely regarded as being significantly superior to the traditional gravimetric sampling method for RCD exposure monitoring. It utilises a tapered element oscillating microbalance (TEOM) technology that continuously weighs a filter as dust is deposited on it, and thus monitoring in real time personal exposure levels of RCD. In TEOM monitors, air is drawn through a filter placed on the top of an oscillating glass rod. The air flow rate through the filter is constant and the mass of the particles that attaches onto the filter will influence the oscillation frequency, which in turn makes it possible to calculate the particle mass and express this per volume of air. The key aspect of PDM3700 is that it directly measures the mass of sample dust on a filter according to the principles of physics regardless of dust composition, size and physical characteristics.

The real-time gravimetric PDM has been successfully used to aid miners in reducing their dust exposure by making changes to their work activities based on the continuous reading of the device. Similar to the current manual gravimetric personal dust samplers, it does not provide information specific to personal exposure of respirable crystalline silica (RCS). Miners are exposed to silica-bearing mine dust which can lead to silicosis, a potentially fatal lung disease. The current practice for measuring the exposure to RCS is to submit the filter dust sample collected by the traditional gravimetric sampler to an external laboratory where the mass content of silica in the collected dust sample is measured by the Fourier transform infrared spectroscopy (FTIR) and X-ray diffraction (XRD) analytical methods. The filter media for sampling in the above FTIR and XRD methods are generally PVC membranes which are readily ashed by incineration to redeposit collected dust particles for subsequent silica analysis. Results generated using the above standard laboratory analysis are generally not available for several days to several weeks after sampling, which delays timely and effective intervention for engineering control of dust exposure. For timely RCS measurement, the NOISH has developed a field-based silica analysis method using a compact FTIR spectrometer and the FAST (Field Analysis of Silica Tool) field-based tool to directly measure silica content on-site at the end of the sampling shift. This field-based FTIR method is to directly analyse RCS of loaded dust over a gravimetric PVC filter without ashing and redeposition, which is called the direct-on-filter method (Miller *et al*, 2012; Cauda, Miller and Drake, 2016). It requires a traditional gravimetric sampler so has limited capability of real-time monitoring of RCD. Currently no miniaturised and proven worker-wearable instrument is available to offer real-time silica monitoring. True real-time silica analysis is also hindered by need for sufficient silica mass (ie sample accumulation during sampling). The end-of-shift direct-on-filter silica analysis is considered as the most practical and promising way to provide timely RCS exposure data on-site.

The current commercial PDM filter consists of borosilicate glass fibres with polytetrafluoroethylene (PTFE) polymer binder (Tuchman, Volkwein and Vinson, 2008). The collected dust particles are deposited onto the glass fibre filter media. The FTIR and XRD analytic methods cannot be directly applied to the silica analysis for PDM-collected dust samples due to interference by the glass fibre filter. The major issues with the current PDM filter are the difficulty in removing the collected dust sample from the glass fibre filter by incineration for redeposition and silica analysis, and its inability to accommodate the direct-on-filter analysis due to the blockage of IR beam by the bulky polypropylene filter holder.

Recently we have developed a methodology to enable the silica analysis of coalmine dust sample collected over the PDM glass fibre filter, which involves an invented detachment method to effectively detach the collected dust from the PDM filter into a solvent, followed by a filtration process to redeposit the dust particles onto a PVC membrane filter (Jin *et al*, 2024). The redeposited dust sample onto the PVC membrane will be measured for silica content with FTIR by using an on-filter method without ashing the filter sample. Such an on-filter FTIR analysis method is preferred for field applications at the mine site, due to its simple process and no requirement for complex laboratory sample preparation for polymer filter destruction. Further to methodology development, we have been developing the alternative PDM filter with a non-silica organic membrane material and a novel filter assembly structure to enable the direct-on-filter silica analysis of PDM dust sample with a field-based FTIR method.

The paper provides an update on the progress of both parts of the research with promising results and developments in expanding the capability of PDM to achieve both real-time respirable dust monitoring throughout the shift and silica content measurements at the end of the shift. The overall goal for the research is to develop a rapid reliable field-based technology for optimal monitoring of both RCD and RCS exposure levels by wearing a single dust sampler of PDM, which will significantly enhance the capability of the coal mining industry in monitoring and preventing the health hazards associated with personal exposure to respirable coal and silica dust. The technology has potential to be applied widely in mining and manufacture sectors.

METHODOLOGY FOR SILICA ANALYSIS OF REAL-TIME GRAVIMETRIC PDM DUST SAMPLES

Dust sampling was carried out with a laboratory dust test chamber to obtain a variety of dust samples for methodology development (Figure 1a). In order to validate the measured silica contents of PDM samples, parallel sampling was conducted by using PDM3700 with the commercial glass fibre filter and a traditional gravimetric personal dust sampler (Casella Apex2 Pro) with a PVC filter of 25 mm in diameter. The obtained parallel samples deposited onto the PDM filter and PVC filter will have a same content of respirable silica. The coal contains kaolinite that interferes with the FTIR analysis of quartz (silica dust). As the infrared spectrometry is sensitive to particle size, it is intuitively better to use the same size fractions for quantification of silica and kaolinite (Lee *et al*, 2013). Respirable kaolinite samples (Sigma Aldrich) collected with a personal dust sampler were used for FTIR analysis to establish kaolinite correction following the NIOSH Method 7603 (Schlecht and Key-Schwartz, 2003). The mixture of standard respirable quartz (NIST 1878b) with the coal sample (<125 μm) was fed into the dust chamber for sampling. The mixture ratios and dust feeding parameters were adjusted to vary the dust-loading amount and silica content of collected dust samples.

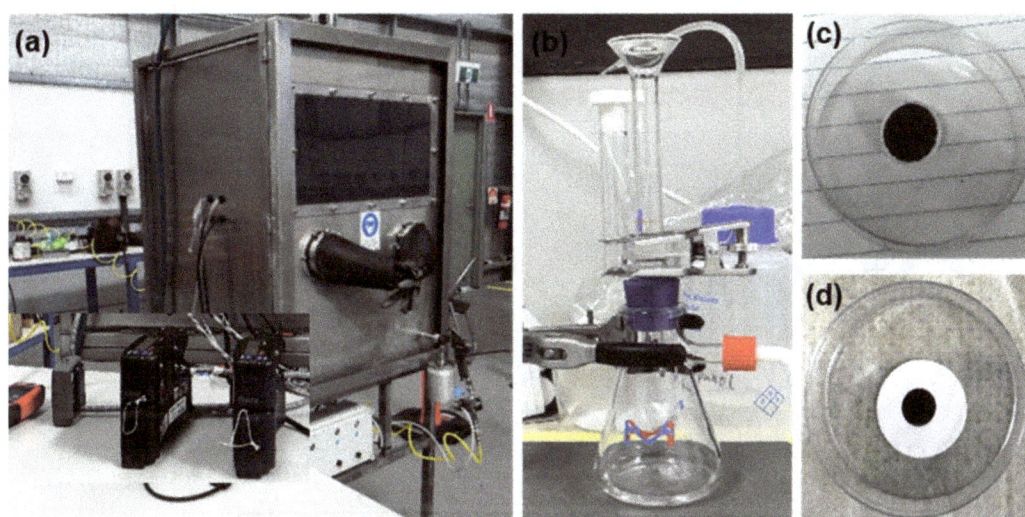

FIG 1 – (a) Laboratory dust test chamber for parallel sampling with PDM3700 and gravimetric samplers, (b) the filtration system for dust redeposition, (c) PDM filter sample and (d) redeposited PDM dust sample onto a PVC membrane.

The key step of the methodology is dust detachment to remove the collected dust from the PDM filter without introducing the contamination of glass fibres from the PDM filter into the detached dust sample. We have devised a novel solvent contact process, in which dust particles are selectively removed from the PDM filter and dispersed in an isopropanol solvent. The redeposition of detached dust in the solvent was carried out by a filtration method similar to the NIOSH 7603 with the filtration apparatus as shown in Figure 1b. Figure 1c and 1d show the PDM sample filter and redeposited dust filter, respectively. FTIR analysis of silica content of redeposited dust filter was conducted with a spectrometer (Thermo Scientific Nicolet iS50) and followed the method as described in the NIOSH 7603 using the IR absorbance peak of 800 cm^{-1} with the baseline from ca 820 to 670 cm^{-1}. The standard respirable quartz (NIST 1878b) was used to establish the calibration curve of silica quantification (silica mass content versus peak height at 800 cm^{-1}). The kaolinite correction was applied in the calculation of the peak height at 800 cm^{-1} for coal dust samples.

For silica analysis of dust samples collected by a gravimetric personal dust sampler, the PVC filter sample was immersed in isopropanol in a beaker. The dust was detached from the PVC filter with an ultrasonic bath and then redeposited onto a PVC filter using the same redeposition process as described above. The redeposited gravimetric sample was analysed for silica content by the FTIR method for comparison with the measured content of the PDM sample collected in the same parallel sampling run. Parallel PDM and gravimetric samples should have a similar silica content.

It has been demonstrated that the developed solvent contact process is efficient to detach the collected dust from the PDM filter with minimum contamination of filter glass fibres. The detailed optical microscope and scanning electronic microscopy (SEM) studies revealed that nearly no glass fibre contamination was observed in the redeposited PDM dust sample. In comparison, significant contamination of glass fibres was observed in the redeposited DPM sample that was detached by the ultrasonic process. The SEM-energy dispersive X-ray (EDX) analysis of the redeposited dust samples was used to identify silica dust particles, glass fibres and kaolinites, and the presence of various silicon-containing components was evident.

Multiple sets of parallel PDM and gravimetric samples with a variety of dust-loading amounts (0.3–2.2 mg) and silica concentrations of (1.0–4.5 wt per cent) were processed using the developed methodology and analysed by FTIR. As shown in Figure 2, the measurement content of silica in the PDM samples are very close to those of gravimetric samples collected in a same run of parallel sampling. The differences of the measured silica contents between PDM and gravimetric samples are less than 10 per cent with the majority of variations below 5 per cent. The FTIR technique is a highly sensitive means for silica quantification, and some of measured samples have a silica content as low as 5–10 microgram.

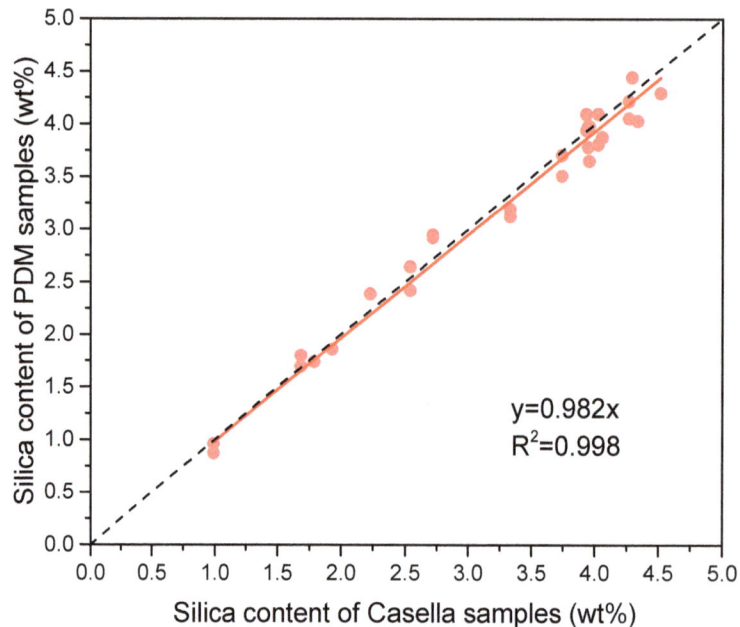

The plot shows silica content of PDM samples (wt%) on the y-axis versus silica content of Casella samples (wt%) on the x-axis, with fitted line $y=0.982x$ and $R^2=0.998$.

FIG 2 – Comparisons of measured silica contents of parallel PDM and gravimetric samples processed by the developed methodology.

The results have clearly demonstrated that the developed methodology is feasible and practical for enabling silica analysis of PDM-collected dust samples. Some of redeposited dust samples were also analysed with a compact FTIR spectrometer (Nicolet Summit), which gives very similar analysis results to those by the Nicolet iS50 FTIR spectrometer. The developed methodology is a simple and rapid process with very simple set-up and ease of operation for end user or the analytical service provider. The set-up is readily installed at mine site, and the entire process for dust detachment, redeposition and FTIR analysis takes about 15 mins. It has significant potential for use in the end-of-shift silica analysis of PDM dust samples at remote mine sites.

NEW PDM FILTERS FOR DIRECT-ON-FILTER SILICA ANALYSIS

The other part of our research is focused to seek the alternative PDM filter with non-silica organic materials and a novel filter assembly structure to enable the direct-on-filter silica analysis of PDM dust sample with a field-based FTIR method. PDM sampling with the developed new PDM filter will completely avoid the issues with the current glass fibre filter. The objective of this work is to enable FTIR silica analysis directly on the dust sample deposited on the new PDM filter at the end of the shift in the field.

To date, the research team has developed a new structured non-silica polymer filter membrane as an alternative to the current glass fibre filter membrane. The new polymer membrane was installed onto the current PDM filter holder for evaluation of sampling performance. The resultant polymer membrane filter was fitted into a PDM3700 for parallel sampling with another PDM3700 that used the current commercial glass fibre membrane filter. Figure 3 shows the comparison of the parallel sampling results of these two PDM3700 samplers. As shown in Figure 3a, the PDM with the new polymer membrane filter exhibits a slightly higher amount of dust collection than that with the commercial filter during a 4 hour dust sampling, with an overall 6.8 wt per cent more at the end of sampling. The measured dust concentrations for the two PDM units are almost same within most of sampling period except those values of at peak dust concentrations inside the chamber (Figure 3b). The dust concentration surge inside the chamber was resulted from the intermittent dust feeding by compressed air. The variation of measured peak dust concentrations between two PDM units is likely due to the inherent errors of the two PDM units used. The detailed sampling tests and evaluation are underway. The new filter holders with a novel design capable of easy removal of filter membrane from the holder for direct-on-filter silica analysis are fabricated and will be evaluated in parallel sampling tests. The key challenge for developing new sample holder is the mass limit of the holder generally requiring below 0.1 mg to meet the PDM3700 operation requirement.

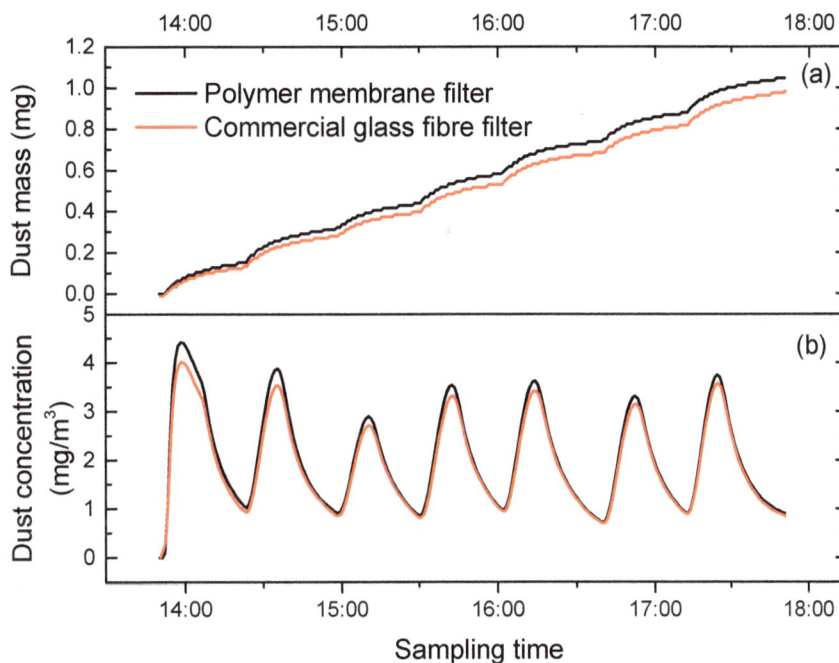

FIG 3 – Parallel sampling results of two PDM3700 units with a commercial glass fibre membrane filter and a newly developed non-silica polymer membrane filter respectively: (a) total mass of collected dust, and (b) dust concentrations throughout the sampling period.

CONCLUSIONS

At the time introduction of real-time gravimetric PDM3700 sampler, there was an incorrect perception created and communicated by various interested parties that the NIOSH (USA) approved and MSHA (USA) legislated PDM3700 sampler is not mass-based gravimetric sampler and do not collect coalmine dust over the filter for silica analyses. These specific misgivings and a result of shortcomings of non-acceptance of US Intrinsic Safety (IS) approval in Australia have resulted in them not being used as an effective compliance tool to empower the workers for personal dust exposure management. This paper updates our research progress on enabling the silica analysis of PDM-collected respirable dust for the deployment of advanced real-time gravimetric TEOM mass-based monitor to achieve both real-time RCD monitoring throughout the shift and RCS measurement at the end of the shift. The methodology developed by the CSIRO, which involves an innovative sample dust detachment process followed by filtration redeposition, has been demonstrated as an efficient technique to measure the silica content of mine dust samples collected by PDM3700 with the current commercial glass fibre filter. The developed methodology combined with a compact field-based FTIR spectrometer has a significant potential to be implemented for end-of-shift silica analysis at the mine site. Moreover, the research team is currently working on the development of alternative

PDM filters with a non-silica organic polymer membrane and a novel filter assembly structure to enable the direct-on-filter silica analysis of PDM dust sample with a field-based FTIR method. The parallel sampling tests with a new polymer membrane filter have showed promising performance with its RCD monitoring result very close to that of the currently used glass fibre dust filter. The success of the research will directly enhance the RCD and RCS monitoring capabilities for the mining industry.

ACKNOWLEDGEMENTS

This research was funded by the Australian Coal Association Research Program (ACARP) and CSIRO. Thanks to CSIRO colleague Xin Yu for his assistance in construction of the dust test chamber.

REFERENCES

Belle, B, 2017. Pairwise evaluation of PDM3700 and traditional gravimetric sampler for personal dust exposure assessment, The Australian Mine Ventilation Conference (The Australasian Institute of Mining and Metallurgy: Melbourne).

Cauda, E, Miller, A and Drake, P, 2016. Promoting early exposure monitoring for respirable crystalline silica: Taking the laboratory to the mine site, *Journal of Occupational and Environmental Hygiene*, 13(3):D39–D45.

Jin, Y, Belle, B, Balusu, R and Guo, H, 2024. Method of processing dust collected on a dust filter of a continuous dust monitoring device for analysis, Patent: WO/2024/011288, World Intellectual Property Organization (WIPO).

Lee, T, Chisholm, W P, Kashon, M, Key-Schwartz, R J and Harper, M, 2013. Consideration of kaolinite interference correction for quartz measurements in coal mine dust, *Journal of Occupational and Environmental Hygiene*, 10(8):425–434.

Miller, A L, Drake, P L, Murphy, N C, Noll, J D and Volkwein, J C, 2012. Evaluating portable infrared spectrometers for measuring the silica content of coal dust, *J Environ Monit*, 14(1):48–55.

Schlecht, P and Key-Schwartz, R, 2003. Quartz in coal mine dust, by IR (redeposition), Method 7603, issue 3, dated on 15 March 2003, National Institute for Occupational Safety and Health (NIOSH) Manual of Analytical Methods (NMAM), Fourth Edition. Available from: <https://www.cdc.gov/niosh/docs/2003–154/pdfs/7603.pd>

Tuchman, D P, Volkwein, J C and Vinson, R P, 2008. Implementing infrared determination of quartz particulates on novel filters for a prototype dust monitor, *J Environ Monit*, 10(5):671–678.

Volkwein, J C, Vinson, R P, McWilliams, L J, Tuchman, D P and Mischler, S E, 2004. Performance of a new personal respirable dust monitor for mine use, US Department of Health and Human Services (DHHS) National Institute for Occupational Safety and Health (NIOSH) Pub. 2004–151, Report of Investigations (RI) 9663. Available from: <https://www.cdc.gov/niosh/mining/UserFiles/works/pdfs/ri9663.pdf>

PVDF-Ti$_3$C$_2$T$_x$ (Mxene) ultrafine nanofiber membrane for effective fine particulate matter $_{(0.1–2.5\ µm)}$ filtration

A Kakoria[1], M M Zaid[2] and G Xu[3]

1. Department of Mining and Explosive Engineering, Missouri University of Science and Technology, Rolla MO 65401, USA. Email: akumar@mst.edu
2. Department of Mining and Explosive Engineering, Missouri University of Science and Technology, Rolla MO 65401, USA. Email: mirzamuhammadzaid@mst.edu
3. Department of Mining and Explosive Engineering, Missouri University of Science and Technology, Rolla MO 65401, USA. Email: guang.xu@mst.edu

ABSTRACT

Coal miners pose serious occupational health issues because of the prolonged exposure to the coal dust ultrafine particulate matters (PM) in coal mines. Fine PM air pollutants (0.1, 0.3 and 2.5 µm in diameter) can quickly enter the human respiratory system and cause health problems. Although an electret filter is effective for fine particulate matter filtration such as 0.1 and 0.3 µm, its use is accompanied by a high-pressure drop, making breathing difficult. Electrostatic interactions in the filter are the dominant filtration mechanism for capturing fine particulate matter. However, these filters are significantly weakened by the high humidity in exhaled breath. This study shows that a filter with an electrostatically rechargeable structure can function with standard breathing air power. A novel supersonically solution blown PVDF-Mxene composite electret nanofiber (PMCEN) membrane is used for the filtration of ultrafine PM. Piezoelectric properties provides improve filtration performance even when air-powered filter bending is used as a regular breathing condition. The air gap between the nanofiber membrane increases air diffusion time. It also preserves electrostatic charges within nanofiber filter membrane under humid air penetration, which is beneficial for coalminers and useful for outdoor filter materials such as face masks and respirators and indoor filtration as well. Hence, this filter can be used in mines and among miners as it has low differential pressure drop (DP) and high filtration efficiency.

INTRODUCTION

Coal is the world's most abundant fossil fuel, accounting for over 30 per cent of worldwide energy demand. Modern mining technology has increased productivity, allowing it to crush hundreds of tons of coal per shift. These activities produce clouds of respirable dust particulate matter (PM) that are harmful for miners. Specifically, airborne coal dust PM of sizes 0.1–2.5 µm in coalmines is the primary source of pollution and has a substantial impact on the health of miners. Despite of giving lot of emphasis, the problem still persists and there is a serious need to abolish prevalence and severity of chronic lung illness caused by coalmine dust exposure. Different PM have different effect on miners' health such as 30 µm diameter dust PM results in bifurcation of trachea, 10 µm dust PM affects respiratory terminal bronchiole, 3–5 µm dust PM affects alveolar duct, 1–2 µm dust PM affects alveolar tract and alveolar sac cavity account for 65 per cent. However, a small portion 0.3 and 0.1 µm dust PM effects on the alveoli. Due to prolonged exposure to these dust PM coalminers have high chances of getting infected and developing the disease spectrum induced by persistent inhalation of mine dust, known as 'coalmine dust lung disease' (CMDLD). CMDLD include not only the traditional interstitial lung disorders (coal worker's pneumoconiosis [CWP], silicosis, and mixed dust pneumoconiosis), but also the more newly described entity of dust-related diffuse fibrosis (DDF). Coalmine dust also causes chronic obstructive lung illness, such as chronic bronchitis and emphysema, which is frequently misdiagnosed because of this exposure.

There had been significant research work performed to understand and reduce the coalmine dust PM such as spray dust fall includes direct jet, rotary and spiral cone, and jet-impinging, etc. On the other hand chemical dust suppression includes wetting dust suppressant, adhesive dust suppressant, cohesive dust suppressant and compound dust suppressant. Coal seam water injection includes hydraulic fracturing, hydraulic slotting, hydraulic blasting and high-pressure, pulse, cyclic etc. Then comes the dust collectors that includes inertial dust removal, gravity dust removal, electrostatic dust removal, bag dust removal and filter cartridge dust removal. However, the dust collecting methods have the limitations of low efficiency and high-power consumption due to high differential pressure, and the expensive cost significantly limits its application. However, a very few works have been done on the air-borne PM particularly less than 0.3 μm removal.

Recent studies confirms that airborne PM removal can be achieved with the use of nanofiber filter membrane. A mask made of these nanofiber membranes is a latest solution and means of personal protection while working in a mine. There are different types of masks, such as fibre-based filter masks, porous carbon masks (such as activated carbon etc), and porous ceramic masks (such as zeolite etc). Fibre-based filters have rapidly developed and are now the most extensively used kind due to the advantages of a simple manufacturing process, good filtration performance, and ease of adjusting the structure and morphology. There are some commercial products available to prevent the miners from coal dust PM that is N95 masks, respirators etc These kinds of product are safety device that covers the nose and mouth and protects the wearer from inhaling harmful substances. It is intended to remove at least 95 per cent of the dust and mold in the air (0.3 μm PM). However, all these masks are incapable of removing the ultrafine dust PM ranging from 0.1–0.3 μm 100 per cent because of the microfibre layer inside instead of a nanofiber layer.

Hence, the objective of our study is to develop a nanofiber-based filter mask to prevent the coalminers from inhaling these ultrafine coal dust PM. In this study a unique surface modified PMCEN membrane was fabricated by a novel method called supersonic solution blowing. The PMCEN membrane demonstrated self-electrostatic charge generation and retention properties in cyclic air blowing and humid air penetration. The PMCEN PM filtration effectiveness is compared based on electrostatic charge generation and retention properties via piezoelectric effect and surface modification. The surface of PMCEN is modified to improve PM filtration rate and provide a lower pressure drop. PMCEN has improved electrostatic charge generation on the filter in a respiration-like environment, which is beneficial for face mask applications. PMCEN facemask is feasible to generate electrostatic charges in the breathing process after wearing. We evaluated the PMCEN membrane for coal dust PM capturing to realise the electret effect, magnetic effect, and surface modification. The filtration efficiency PMCEN membrane is 99.98 per cent under the condition of 0.1–0.3 μm ultrafine coal dust PM, while the resistance is only 141 Pa. This demonstrates that PMCEN membranes can be employed efficiently in coalmine dust PM filtration.

EXPERIMENTAL SECTION

Materials
Sigma Aldrich provided PVDF (Mw = 8×10^5) as well as N, N-dimethylformamide (DMF), Mxene and acetone.

Solution preparation
A 16 wt per cent PVDF solution was prepared by adding N, N-dimethylformamide (DMF) and acetone in (DMF: acetone = 3:1) ratio respectively with PVDF. After adding all together, the solution was kept overnight on a magnetic stirrer for mixing at 60°C and 200 rev/min.

Electrospinning and supersonic solution blowing
The polymer solution was pumped through a 20-gauge needle using a syringe pump (New Era Syringe Pump Inc) during electrospinning (ES). A voltage source (Genvolt) was used to apply a positive high voltage (10–12 kV) to the needle. A 200 rev/min rotating drum was used as both the ground and the collector. The needle-to-collector distance was kept constant at 10–15 cm. The experimental set-up is depicted schematically in FIG 1a. Polymer solution flow rates were varied between 1.0 and 2.5 mL/h. After the process was completed, the nanofiber mats were peeled off and placed in a convection oven at 40°C overnight to evaporate any residual solvent.

The procedure described in (Li and Xia, 2004; Li and Wang, 2013) as followed for supersonic solution blowing (SSB). The process is depicted schematically in FIG 1b. It is made up of an electrospinning scheme coupled with a supersonic converging-diverging de Laval nozzle connected to an air compressor that supplies compressed air at a pressure of 10 bar. The electrospinning needle was placed orthogonal to the nozzle, 2 cm horizontally and 2 cm vertically. Instead of a rotating collector, as in electrospinning, a supersonic air nozzle was used as the ground to attract the electrified polymeric jet exiting the needle to the nozzle. The applied electric voltage is limited to a maximum of 8 kV, but the field intensity is nearly four times that of regular electrospinning. As the electrified jet approaches the core of the supersonic air stream, leaving the nozzle with an exit velocity of 750 m/s, the polymer jet is swept away from the nozzle, as shown by the kink in the polymer jet path in FIG 1. The dried nanofibers were collected on a rotating drum collector 25 cm from the nozzle. As shown in (Liao *et al*, 2014; Kim *et al*, 2016), the final fibre size varied depending on the initial drop size, but the needle gauge size was comparable to electrospinning. The polymer flow rate was kept constant at 0.25–0.3 mL/h, and the process was repeated for 4 hrs to produce mats. This method, like electrospinning, used the same polymer concentrations.

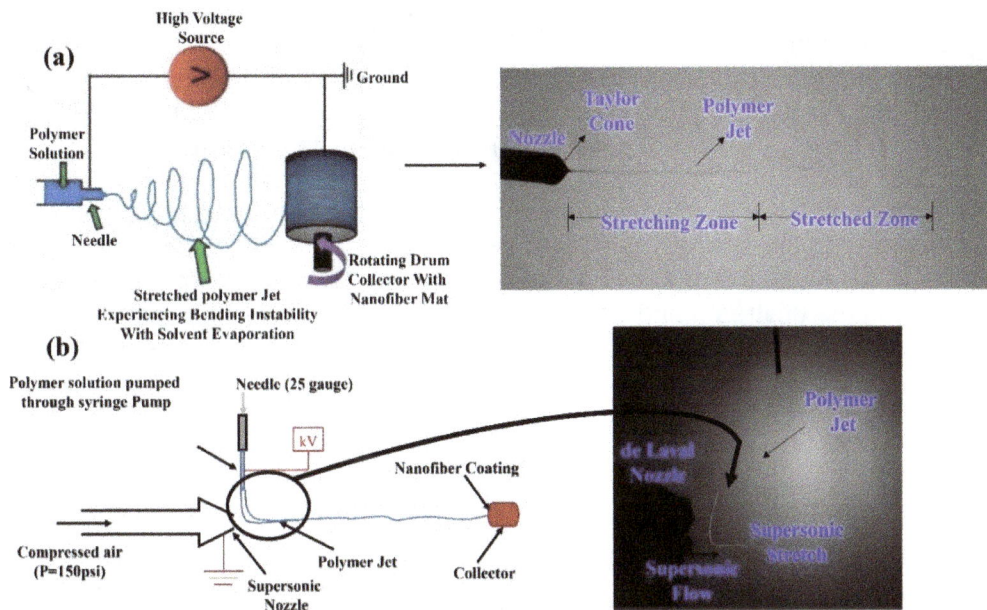

FIG 1 – Shows a schematic diagram of (a) electrospinning and (b) supersonic solution blowing, as well as a high-speed image of the polymer jet path.

Characterisation

All the characterisations in this section are performed to confirm the structural, morphological, and chemical information of PVDF and PMCEN membranes. A field-emission scanning electron microscope (Helios Hydra 5 by Thermo Fisher Scientific) was used to examine the morphology of the nanofibers. A tensile tester was used to test the mechanical properties. Volumetric porosity was calculated by comparing bulk and apparent PVDF densities. Using a powder polycrystalline the X-ray diffraction (XRD) equipment from Philips MPD XRD, XRD pattern of the nanofibers was recorded. An electrostatic tester was used to measure the surface potential of the composite fibre membrane. Polarisation of the nanofiber was obtained by treating them at 150°C and 20 kv DC high voltage. Temperature treatment was done with a temperature-controlled oven with a heating speed of 3°C and a heating temperature range of 25 to 140°C and 20 kv high voltage treatment was done by a high voltage source with 0–3 kv capacity. Attenuated total reflection (ATR) mode Fourier transform infrared spectra (FTIR) were obtained with a Bruker Alpha II equipped with a platinum ATR attachment and a diamond crystal. A differential scanning calorimeter (DSC) and Thermogravimetric analysis (TGA) from NETZSCH was used to determine the melting point, mass loss and total crystallinity of PVDF (DSC 6100, Seiko Instruments, Japan). A heating rate of 10°C per min was used. By comparing the heat of the sample's fusion with 100 per cent crystalline PVDF, the total crystallinity of the polymer was determined. One hundred percent crystalline PVDF has a heat of fusion of 104.7 J/g. Raman spectra were obtained using a He–Ne laser (632.8 nm) as the excitation source and a Horiba–Jobin Yvon LabRam-HR spectrometer (Horiba–Jobin Yvon, Inc, Edison, NJ).

Air filtration set-up and testing protocols

Each mat was sandwiched between two pieces of cotton fabric that were readily available on the market (basis weight: 85 GSM). The mats can be handled easily because of the outer layers, which support the thin nanofiber layers, and the net assembly was used to measure PFE, and pressure drop using the experimental set-up depicted in FIG 2. A condensation particle counter (TSI 3007), an aerosol generator (TSI 3076), and a diffusion dryer are the main components of the experimental set-up. The test section is a 47 mm filter-fitting filter holder section in the middle of a 100 mm PVC pipe housing with two equal parts (net face area: 17.35 cm²). To provide dilution air into the test chamber where neutralised aerosol particles would be released (*cf* FIG 2) after passing through the dryer and charge neutraliser, a compressor (Elgi TS 05LB) with a HEPA filter (TSI Capsule Filter with PFE 99.97 per cent > 0.3 µm) was used. The compressed air flow rate (face velocity of 28–100 cm/s) was adjusted from 30 LPM to 100 LPM. To measuring PFE, the particle concentration (count/cm³) was measured upstream (C_{up}) and downstream (C_{down}) of the filters. The set-up bias was initially measured using two situations—large particles with low velocity (2 µm and 8 cm/s) and small particles with high velocity (0.2 µm and 77 cm/s)—as part of the PFE test protocol, which was modelled after ASTM F2299. Once the particle counts in both the upstream and downstream reached 100±1 per cent, the PFE measurements were started. Four different types of polystyrene spheres with mean diameters of 0.1±0.021, 0.2±0.02, 0.5±0.05, and 2±0.18 m was taken into consideration for PFE measurements. The mean particle count was taken into consideration for PFE estimation along with the upstream and downstream measurements of initial particle concentrations for 2 min each. A digital manometer (Testo 510) was used to measure the pressure drop (DP) across the filter.

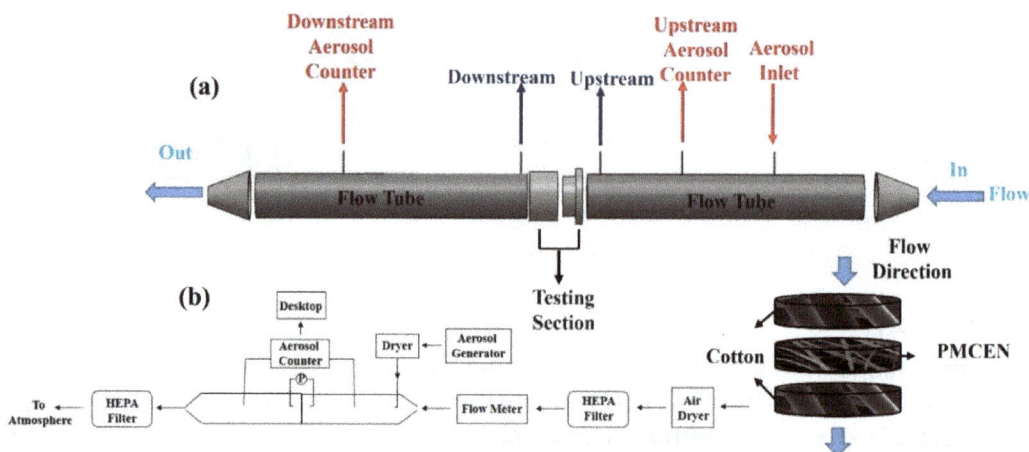

FIG 2 – The line diagram below shows a schematic representation of the air filtration test procedure.

RESULT AND DISCUSSION

The 16 wt per cent ES PVDF nanofibers morphology is shown in FIG 3a with a zoomed-in image in the inset. The fibre size distribution is shown in FIG 3e. The mean fibre size is 0.9 µm with a standard deviation of 0.6 µm. The XRD patterns of PVDF ES nanofiber samples are depicted in FIG 3b. The significant peaks at 19.2° and 20.4° correlate to the existence of g-phase and b-phase, respectively. The α, β and g phases are crystalline domains composed of TGTG (Trans-Gauche-Trans-Gauche conformations) chains and all-TTTT trans planar zigzag chains, respectively. An additional peak was found for PVDF at 37.1° and 41.3°, indicating the existence of g-phase in the sample. All these peaks confirms that the fabricated nanofiber is pure PVDF. The PVDF nanofiber exhibits a developed γ-phase at 2θ = 19.2° (110). The PVDF membrane exhibits significant peaks due to its β-phase crystalline structure at 2θ = 20.4° (110). Improved β-phase ratio in PVDF crystalline structure improves electrostatic charges on nanofiber surfaces through piezoelectricity and mechanical characteristics. The thermal stability of PVDF ES nanofibers was tested to ensure their stability during use in various applications and for future reuse. TGA thermograms of PVDF ES nanofibers are shown in FIG 3c and found to be stable up to 390°C and confirms the existence of PVDF nanofiber. FIG 3d depicts the DSC thermogram of electrospun PVDF ES nanofibers. On the thermogram of pure PVDF ES nanofibers, an endothermic peak at 63.8°C can be clearly seen with a melting peak at 163.4°C. The nature of the melting and annealing peak is explained by the upper glass transition reorganisation within conformationally disordered a-crystals. The melting of a crystalline structure is assigned an endothermic transition at 163.4°C. Both these peaks confirm the existence of PVDF in nanofiber. FIG **3**f represent the Raman shift of pure PVDF ES nanofiber to understand the chemical structure and confirmation of PVDF nanofiber. The band at around 797 cm^{-1} and 835 cm^{-1} is assigned to the rocking motion of CH_2 and is a typical band for a-phase rich PVDF. Bending CH_2 vibrations, which are present in all three crystalline phases of PVDF but primarily in the b and g-phases, cause the band around 1431 cm^{-1} and 1531 cm^{-1}. The band at 2977 cm^{-1} is commonly associated with the b-phase and is attributed to CH_2 symmetric stretching. A total of 16 scans were conducted for every measurement with a 4 cm^{-1} resolution. The stretching bands at the different wavenumbers found α and β phases and are presented in FIG **3**g. The crystalline morphologies of PVDF-α-, β-, and orientated β-phases have been identified, and PVDF in the α phase was identified by vibration band peaks at 611 and 766 cm^{-1}, whereas β-phase PVDF was linked to vibration band peaks at 836 and 508 cm^{-1}. Oriented β-phase PVDF was identified by vibration band peaks at 440 and 475 cm^{-1}.

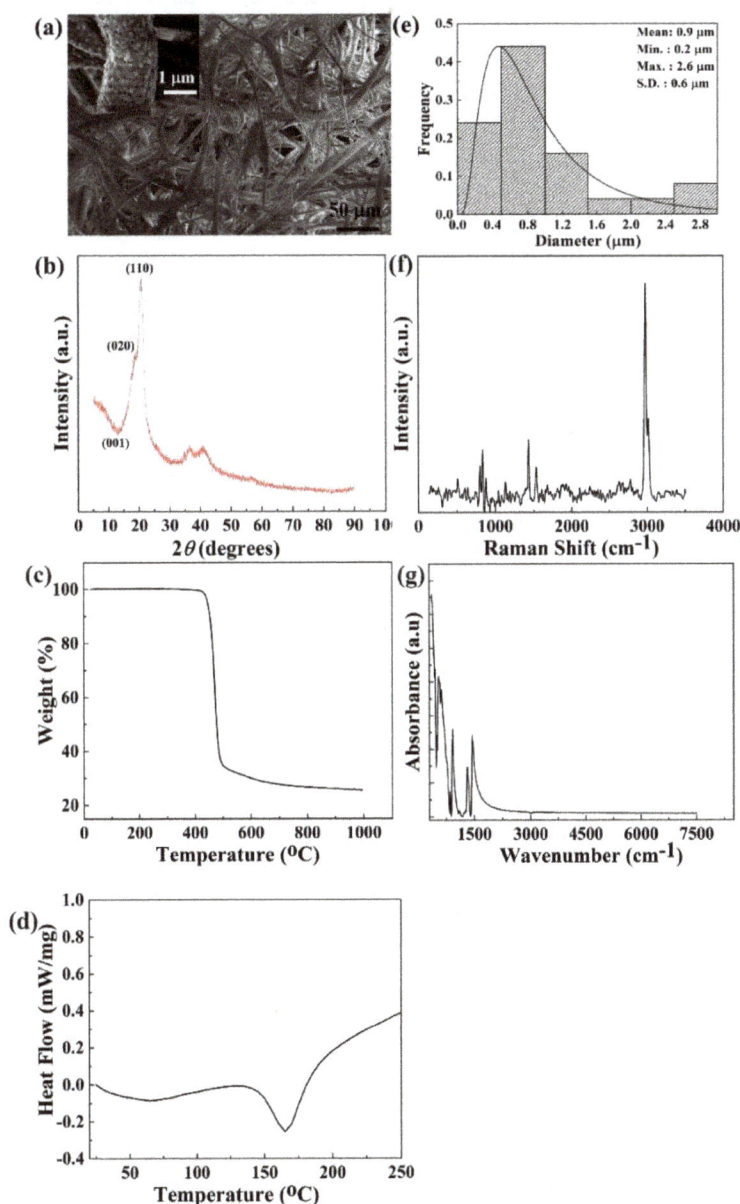

FIG 3 – (a) SEM images of 16 wt per cent PVDF ES nanofibers; (b) X-ray diffraction pattern of 16 wt per cent PVDF ES nanofibers; (c and d) DSC/TGA graphs of PVDF nanofibers; (e) Fibre size distribution of 16 wt per cent of PVDF ES nanofibers; (f) Raman spectra of PVDF nanofibers; (g) FTIR.

CONCLUSION

Nanofibers made of PVDF-Ti$_3$C$_2$T$_x$ (Mxene) show great promise as a material for particulate matter (PM) filtering. By offering several sites for particle collection, the high surface area to volume ratio because of 900 nm nanofiber diameter improves filtration efficiency. PVDF's mechanical strength, hydrophobic qualities, and natural chemical stability all add to its long-term usability and durability in filtration applications. Future, application will be using this PVDF-Ti$_3$C$_2$T$_x$ nanofiber for real field analysis.

REFERENCE

Kim, J F, Jung, J T, Wang, H H, Lee, S Y, Moore, T, Sanguineti, A, Drioli, E and Lee, Y M, 2016. Microporous PVDF membranes via thermally induced phase separation (TIPS) and stretching methods, *Journal of Membrane Science*, 509:94–104. https://doi.org/10.1016/j.memsci.2016.02.050.

Li, D and Xia, Y, 2004. Electrospinning of Nanofibers: Reinventing the Wheel?, *Advanced Materials*, 16(14):1151–1170.

Li, Z and Wang, C, 2013. Effects of Working Parameters on Electrospinning, *One-Dimensional Nanostructures*, pp 15–28.

Liao, Y, Loh, C H, Wang, R and Fane, A G, 2014. Electrospun Superhydrophobic Membranes with Unique Structures for Oil-Water Separation, *ACS Applied Materials and Interfaces*, 6(18):16035–16048.

Investigating the accuracy of the inhalable dust sampling devices

M S Kizil[1], H W Wu[2], G Kizil[3] and B Belle[4]

1. Associate Professor, The University of Queensland, School of Mechanical and Mining, St Lucia Qld 4072. Email: m.kizil@uq.edu.au
2. Mining Technology Pty Ltd, Brisbane Qld 4000. Email: h.wu@minserve.com.au
3. The University of Queensland, School of Mechanical and Mining, St Lucia Qld 4072. Email: g.kizil@uq.edu.au
4. 61Drawings, Belmont Qld 4153. Email: director@61drawings.com

ABSTRACT

Coal mining still has many challenges in terms of health and safety. There are many types of coalmine hazards, for instance methane explosion and spontaneous combustion, and coal dust-related health problems, such as Coal Workers' Pneumoconiosis (CWP). Coal dust in operating mines is mainly produced by mechanical cutting of the coal in both production and development areas. The exposures to coal dust are monitored in terms of inhalable and respirable fractions in Australia. Inhalable dust refers to the particle that can be breathed into the nose or mouth and respirable dust refers to the dust that can be breathed into the gas exchange regions of respiratory system. Inhalable and respirable standards refer to particles with median diameter of 100 µm and 4 µm, respectively.

As recommended in the AS3640 (Standards Australia, 2009), a sampling system for the inhalable dust is essentially, an inhalable dust sampling device with a filter which the dust sample is collected and a pump for drawing the air through the sampling device. The inhalable dust sampling device is to be placed within the worker's breathing zone. A series of tests were carried out to examine the effects of the dust laden airflow orientation on the efficiency of the Institute of Occupational Medicine (IOM) inhalable dust sampler and other most used samplers in Australian coalmines. The experimental parts of the study were executed at Commonwealth Scientific and Industrial Research Organisation (CSIRO) Mineral Resources Pinjarra Hill site utilising a large dust chamber and at operating coalmine site.

The results of the testing show that the coal dust collection efficiency and accuracy of the inhalable dust samplers are significantly affected by the samplers' inlet orientation to the oncoming airflow direction. Based on the particle size distribution analysis results of laboratory and field sampling, it is concluded that the inlet orientation of the IOM sampler to the oncoming airflow also have a significant effect on the coal dust particle size ranges captured by the sampler. Based on these findings, it is concluded that the science behind the inhalable sampler may not be well comprehended yet and further investigation and design evaluation of current inhalable IOM sampler may be required to warrant consistent approach to inhalable dust sampling.

INTRODUCTION

As per the AS3640 – _'Workplace atmospheres – method for sampling and gravimetric determination of inhalable dust'_ (Standards Australia, 2009), the inhalable dust sampling system consist of an inhalable dust sampling head followed by a filter and a pump. The inhalable dust sampling device is to be placed within the worker's breathing zone. The inhalable fraction is collected by using a sampling device should conform with the sampling efficiency curve as specified in the ISO 7708. There are several devices or heads available that generally satisfy the ISO 7708 criteria.

AS3640 provides a few examples of sampling devices available such as modified personal United Kingdom Atomic Energy Authority (UKAEA) sampling head, Institute of Occupational Medicine (IOM) inhalable dusting sampling head and conical inhalable sampling head. It has suggested that UKAEA and IOM sampling heads are more suitable for sampling particles smaller than approximately 30–50 µm equivalent aerodynamic diameter. If a significant proportion of particles are larger than these ranges, IOM sampler should be used in preference to the UKAEA sampling head.

AS3640 also suggests that a sampling bias of less than ± 5 per cent is typical for the IOM inhalable dust sampler, but the conical inhalable and multi-orifice samplers may have larger biases (either positive or negative) under same workplace conditions. Workplaces with high air velocities and large particles are generated by the work process (a common situation in a typical modern Australian longwall production face) often could lead to negative bias. Positive bias usually a resultant from incorrectly handling of the samplers such as the conical and UKAEA after use, as these sampler designs could allow unintended contamination of the filter. This study aimed to identify any potential issues associated with AS3640 and to validate the applicability of the inhalable dust monitoring program currently implemented by Australian coal mining industry.

INHALABLE SAMPLING HEADS

Inhalable sampling heads collect all fractions of particulates (up to 100 μm) which enter the nose and mouth during breathing, ie everything that is available in the air to be inhaled. Table 1 provides a summary of inhalable samplers available currently or in the past. It should be noted that the closed face cassette (CFC) was not designed to be an inhalable sampler but are still used in the US to assess exposures relative to total dust limits.

TABLE 1
Inhalable samplers available.

Sampler	Flow rate (L/min)	Include wall deposits	Region of use	Manufacturer / distributor
Closed face cassette (CFC) - 37 mm	1.0–2.0	No	US ('total dust standards')	SKC Inc., SureSeal Cassette etc.
IOM Inhalable	2.0	Yes	Europe, US, Australia	SKC Inc.
Button Inhalable	4.0	No	US	SKC Inc.
Conical inhalable sampler (CIS)	3.5	No	UK HSE, Germany	Casella CEL, UK
Multi-orifice (7-hole)	2.0	No	UK HSE	Casella CEL, UK
GSP (Gesamtstaub-Probenahmesystem)	3.5	No	Germany	GSMGesellschaft für Schadstoffmesstechnik, GmbH; Neuss-Norf, Germany
Personal Air Sampler (PAS-6)	2.0	No	Netherlands	University of Wageningen, Netherlands
CIP10-I	10.0	No; Ver. 2 reduces loss	France (wood dust)	Arelco ARC, France
PERSPEC	2.0	-	Italy	Lavoro e Ambiente No longer commercially available

With the development of the inhalable sampling criteria, researchers began to develop samplers that collect dust at efficiencies matching this criterion so that occupational exposures could better reflect the true dust concentrations inhaled by workers. The most widely used sampler in the United States and Australia is the IOM inhalable dust sampler. It was developed in the mid-1980s.

The basis for development of the IOM sampler was to design a sampler that represents the amount of aerosol of workers breath into their noses and/or mouths. There are two main components to the IOM sampler, the filter cassette and the sampler housing. The sampler housing is made up of two pieces, which screw together to the filter cassette. The interior filter cassette has a 15 mm opening, and at the opposite side has a 25 mm filter held to the cassette with a mesh screen behind.

Kenny *et al* (1997) investigated the sampling efficiency of the IOM sampler comparing to the inhalable criterion over various air velocities, airflow/sampler orientations, and sampling flow rates. When performing at air velocities of 0.5 m/s and 4.0 m/s, the IOM sampler was found to sample adequately compared to the inhalable convention when facing the dust source. The IOM oversampled compared to the inhalable particulate matter curve, but not enough to cause concern at 0.5 m/s. Sampling efficiency at 4.0 m/s was found to decrease as particle size increased until particle size was larger than 80 μm. Efficiencies at these larger particle sizes were found to be above 60 per cent.

Baldwin and Maynard (1997) found that air velocities may be much lower in some workplaces (around 0.2 m/s) than the testing conditions of the inhalable samplers as tested by Kenny *et al* (1999). Therefore, the IOM sampler was tested at these lower air velocities to determine if the sampler still satisfied the inhalable sampling criteria. Sampling efficiency was found to decrease as particle size increases in low wind speed conditions, which is to be expected. In higher air velocity wind tunnel experiments, sampling efficiency slightly increased. IOM sampler was determined acceptable in low air velocities less than 0.2 m/s (Kenny *et al*, 1999).

For an inhalable sampler to match the inhalable sampling curve, the sampler must follow the inhalable fraction when averaged overall orientations relative to the oncoming airflow. The IOM has been tested under multiple orientations in order to determine whether it does in fact keep to the inhalability standard at all orientations. A study undertaken by Li, Lundgren and Rovell-Rixx (2000) found that the IOM sampler oversampled when facing the oncoming airflow, but under sampled when at orientations of 90° and 180°. When facing the wind, efficiency increased above 100 per cent as sizes of particles increased 60 μm. Zhou and Cheng (2009) also investigated the effect of sampling flow rate on the collection efficiency of the IOM sampler using 10.6 L/min along with the standard 2.0 L/min flow rate and air velocities of 0.6 and 2.2 m/s. The performance of the IOM sampler was reported to be similar when sampling at 10.6 L/min, compared to the standard sampling flow rate, though sampling efficiency was approximately 20 per cent less when particle sizes were larger than 80 μm.

Another possible concern is that AS3640 simply recommends that weighing of filters after sampling is to be undertaken at a suitable time (eg overnight) to allow filters to come to equilibrium with the balance room atmosphere. The IOM sampler could have a large weighing imprecision because of the fact that the entire filter cassette, which can be made of plastic or stainless steel, is weighed as a unit along with the filter. Humidity in this case could cause large weighing imprecisions due to moisture absorbing to the much larger and heavy plastic cassette comparing to the filter alone. Keeping the cassettes in humidity controlled weighing room for seven days before and after use is recommended in order to keep weight fluctuation to ± 0.05 mg. However, a study done by Liden and Bergman (2001) shown that to fully equilibrate the cassette to the weighing environment, 15 to 20 days of equilibration may be required. The results of the same study found that the oils on human hands add a statistically significant amount of weight to the filter cassette.

With the emergence of inhalable samplers and exposure limits based on inhalable criterion, inhalable measurements were compared to historical total dust measurements. It has become a common practice that total dust concentrations to higher inhalable concentrations are compared using a value called the performance ratio (inhalable dust/total dust ratio). A commonly used performance ratio of 2.5 is used for various dusts found in the workplace, after results from multiple studies from 1980 to 1996 were compiled by Werner, Spear and Vincent (1996).

The Button inhalable sampler was developed to minimise the effects of wind direction and velocity on sampling efficiency. This inhalable sampler was developed to allow smooth flow over a front, mesh surface in high air velocities. The Button sampler has a porous curved surface with multiple 381 μm diameter openings, with an overall porosity of 21 per cent. Air velocity and orientation were found to have no effect on the sampling efficiency of the Button sampler in a laboratory setting (Aizenberg *et al*, 2000). While Button inhalable sampler has the advantage of sampling efficiency not affected by air velocity and orientation. it is recommended by the manufacturer that the Button Sampler is better used for low level personal or area inhalable sampling (SKC Ltd, Europe, 2023). Currently, no Button inhalable sampler application was found within the Australian coal mining industry. This could be due to its lower sampling efficiencies and more suitable for low level dust laden condition.

LABORATORY TESTING OF THE IOM INHALABLE DUST SAMPLER

A series of tests was carried out on the IOM inhalable dust sampler commonly used in Australian coalmines at CSIRO Mineral Resources Pinjarra Hill site utilising a large dust chamber designed for another Australian Coal Association Research Program (ACARP) funded project. Tests were set to examine the effects of the oncoming dust laden airflow orientation on the performance efficiency of the IOM inhalable dust samplers.

The laboratory testing undertaken involves placing one IOM inhalable dust with its sampling inlet opening pointing toward the dust laden airflow direction (0°) and the other sampler's opening pointing at 90° that is perpendicular to the oncoming airflow direction within the breathing zone of a mannequin. During the tests, the respirable dust concentrations were measured concurrently using PDM3700. Figure 1 shows the test chamber set-up and details of the two IOM samplers placed in two orientations to the oncoming dust laden airflow within the test chamber.

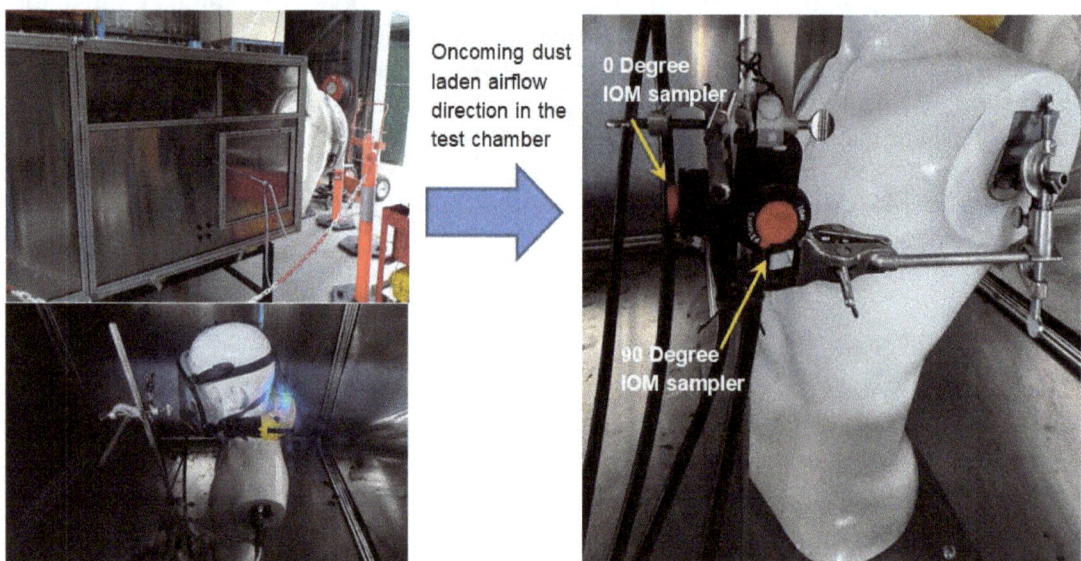

FIG 1 – Set-up of IOM samplers in the test chamber.

Fine coalmine dust sample in the inhalable dust size range for feeding into the test chamber was prepared by the Metallurgical Engineering Laboratory Facilities at The University of Queensland. The coal dust sample sourced from the Coal Handling and Preparation Plant (CHPP) flotation product of a central Queensland coal was screening at 100 per cent passing of 125 µm. This fine coal dust was then fed into the test chamber from the centre top of the chamber about 1.5 m from the mannequin using a precision controlled dust feeder.

Total of eight tests were conducted. Test series were undertaken with sampling period ranging from 3.5 to 5.0 hrs with inhalable dust levels of up to 11.8 mg/m^3 and respirable dust levels of up to 2.7 mg/m^3. Air velocities of oncoming dust laden airstream over the mannequin were ranging from 1.9 to 3.7 m/s which are similar to common air velocity ranges in the face areas of Australian underground coalmine sites. First six test series were done with pairing of two IOM samplers placed in two orientations, 0 and 90° to the oncoming airflow. The last two test series were carried out with pairing of three IOM samples placed in three orientations, 0°, 45° and 90° to the oncoming airflow. However, the sampling pump used for the 90° IOM sampler had failed during the sampling in test series No 7. Table 2 shows summarised results of the test series 1 to 6 with two sampler orientations.

TABLE 2

Summary of test series 1 to 6 results – two orientations.

Test no	Direction to airflow (°)	Flow rate (L/min)	Time (mins)	Dust conc (mg/m³)	90° to 0° ratio
1	0	2.007	242	10.500	31.4%
	90	2.008	242	3.293	
2	0	2.012	265	11.253	34.9%
	90	2.016	265	3.932	
3	0	2.022	235	11.785	34.2%
	90	2.022	270	4.030	
4	0	2.033	272	9.587	30.3%
	90	2.019	273	2.904	
5	0	2.027	308	11.693	31.1%
	90	2.055	308	3.635	
6	0	2.025	272	11.256	29.4%
	90	2.003	272	3.304	

Based on the test results, it is found that the performances (or sampling efficiencies) of IOM inhalable samplers are greatly affected by the orientation of the sampler's inlet to the oncoming airflow. The performance ratio of inhalable coal dust collected between IOM samplers with 0° and 90° orientations is 3.13 to 1. In other words, an IOM sampler with its sampling inlet positioned perpendicular to the oncoming airflow direction only collecting about 32 per cent of inhalable coal dust when compared with that of the IOM sampler with its inlet facing the oncoming airflow. Test series 7 and 8 were done with three sampler inlet orientations namely, 0°, 45° and 90° to the oncoming airflow. As mentioned before, one of the sampling pumps had failed during the test series 8. Table 3 provides a summary of the results from test series 7 and 8.

TABLE 3

Summary of test series 7 and 8 results – three orientations.

Test no	Direction to airflow (°)	Flow rate (L/min)	Time (mins)	Dust conc (mg/m³)	90° or 45° to 0° ratio
7	0	2.028	215	5.964	
	90	sampling pump failure			
	45	1.998	215	3.958	66.4%
8	0	1.989	307	6.387	
	90	2.027	307	2.090	32.7%
	45	2.015	307	4.042	63.3%

Results of test series 8 show that the ratio between 0° and 90° orientations agrees with the ratios observed in test series 1 to 6. The average performance ratio between 0° and 90° orientation of these seven test series is around 3.13 to 1. The average performance ratio between 0° and 45° orientations is around 1.54 to 1. Interestingly, this ratio is somehow proportional to the ratio observed between 0° and 90°. The following figure shows the relationship among ratios of 0°, 45° and 90° orientations with a linear regression.

Figure 2 shows clearly that the coal dust collection efficiency of the IOM inhalable dust sampler is significantly affected by the IOM sampler inlet orientation to the oncoming airflow direction.

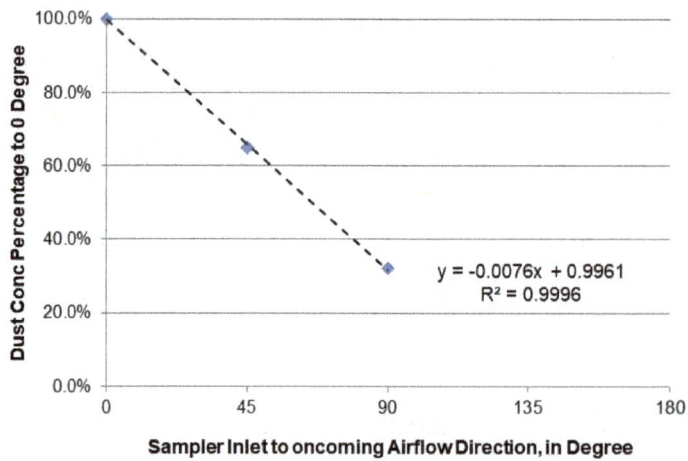

FIG 2 – Inhalable dust collection efficiency and IOM sampler inlet orientation.

Inhalable dust (with orientation variations) and respirable dust concentrations measured in these test series were compared and presented in Table 4. Ratios between inhalable dust (0°), inhalable dust (90°) and respirable dust concentrations are presented in Figure 3. The performance ratio between inhalable dust (0°) and respirable dust concentrations is about 4.3 to 1. A reasonable fit between inhalable dust (0°) and respirable dust concentrations is shown with coefficient of determination, R^2 at an acceptable level of 0.74.

TABLE 4

Comparison of Inhalable and respirable dust concentrations.

Test nso	Respirable dust (PDM3700) (mg/m³)	Inhalable dust (IOM sampler)	
		0° (mg/m³)	90° (mg/m³)
1	2.400	10.500	3.293
2	2.730	11.253	3.932
3	2.113	11.785	4.030
4	2.398	9.587	2.904
5	2.601	11.693	3.635
6	2.686	11.256	3.304
7	1.564	5.964	-
8	1.724	6.387	2.090

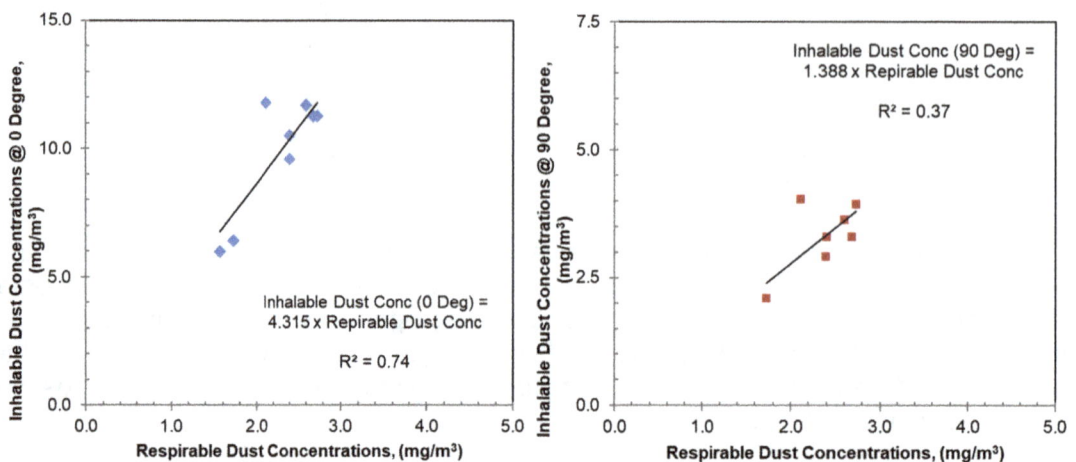

FIG 3 – Ratios of inhalable dust and respirable dust concentrations.

The performance ratio between inhalable dust (90°) and respirable dust concentrations is about 1.4 to 1. However, the goodness-of-fit measure for linear regression ($R^2 = 0.37$) undertaken shows that it is not as reliable as the one observed for the 0° orientation case. Factors could contribute to this are possibly from turbulent flow around samplers, inconsistence in sampler orientation (angle) setting, gravimetric measurement inaccuracy and so on.

MINE SITE TESTING OF AVAILABLE INHALABLE DUST SAMPLING

A field testing plan at a mine site was proposed as part of the project. The planned site testing involved multiple pairwise sampling of IOM inhalable dust sampler, CFC total dust sampler and respirable dust sampler (Casella Higgins Dewell (HD) cyclone with 2.2 L/min flow rate) at available surface and underground locations of the mine site. The current design of the IOM inhalable dust sampler has a larger open circular inlet (15 mm) with a lip that protrudes 1.5 mm outwards as shown in the following figure. The purpose of the sampler inlet lip is to minimise the potential for particles deposited on the outer surfaces of the inlet to be carried into the sampler. Whereas the CFC sampling head used for total dust collection has an opening inlet size of 4.25 mm as shown in Figure 4.

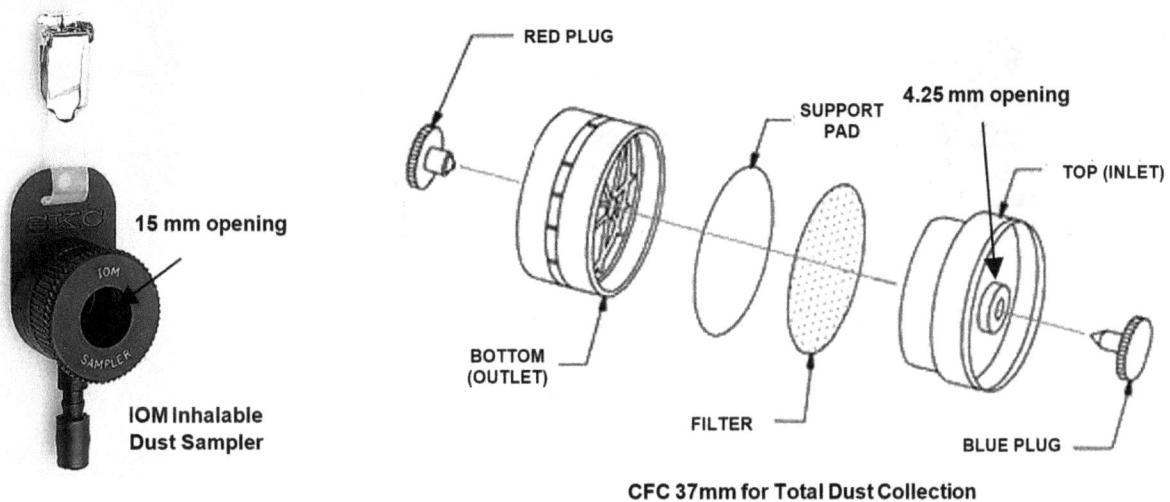

FIG 4 – Sampling inlets of the IOM and CFC samplers.

Thus, there is a high chance of large dust particles 'falling' onto the filter of IOM sampler due to its larger opening if the prevalent airflow direction is in line with the inlet of the IOM sampler. On the other hand, it should be noted that CFC total dust sampler would have a higher average sampling velocity of 2.6 m/s (or higher suction force) via a smaller inlet opening with a sampling flow rate of 1.9 L/min when compared with the average inlet sampling velocity (0.19 m/s) of the IOM inhalable dust sampler (sampling at 2.0 L/min). This could mean that dust particle ranges captured by the total dust sampler would be different from the inhalable dust sampler under the same dustiness condition as the aspiration efficiencies, defined as the free stream air velocity over the sampling velocity, of the two samplers are different in an order of magnitude (Kulkarni, Baron and Willeke, 2011).

Dust samples were analysed gravimetrically as well as for their particle size distributions (PSD). Relevant information such as production rate and ventilation condition around the sampling sites was also to be collected. Due to the availability issue of underground locations at the test mine site, mine site sampling was only undertaken at locations near conveyors in the CHPP plant. Figure 5 shows two set-ups of various samplers at sampling stations.

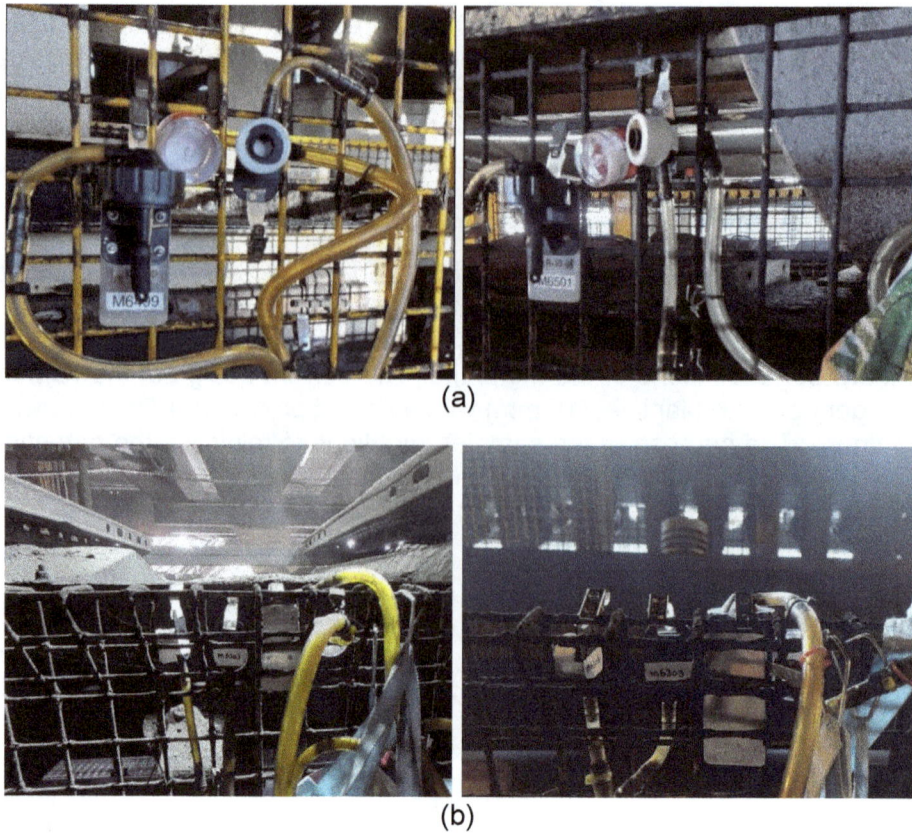

FIG 5 – Photographical views of set-ups at sampling locations: (a) Sampling set-up with IOM, CFC and Respirable dust samplers; (b) Sampling set-up with 2 × IOM, CFC and Respirable dust samplers.

A total of 33 sets of pairwise field samples of inhalable (IOM), total (CFC 37 mm) and respirable dust were taken from the mine site. All these samples were pre and post weighed by an independent third party servicer provider. It should be noted that only 26 sets of pairwise samples are valid as seven sets have incomplete pairwise samplings mainly due to unknown failures of sampling pumps. Table 5 shows a summary of these 26 pairwise sampling results.

Based on the field pairwise sampling of the inhalable, total and respirable dust, the following linear relationships between inhalable and total dust, inhalable and respirable dust, and total and respirable dust are observed.

- Inhalable Dust Conc   = 1.28 × Total Dust Conc; (R^2 = 0.77).
- Inhalable Dust Conc   = 4.60 × Respirable Dust Conc; (R^2 = 0.49).
- Total Dust Conc = 3.49 × Respirable Dust Conc; (R^2 = 0.59).

However, it was found that the relationships between inhalable and total as well as total and respirable pairwise sampling are at acceptable levels with R^2 values at 0.76 and 0.59 respectively. The relationship between inhalable and respirable dust is less reliable and inconclusive with the R^2 value just less than 0.5. It was also found that the performance ratio of 1.28 between inhalable and total dust concentrations is only about half of the performance ratio of 2.5 as compiled and suggested by Werner, Spear and Vincent (1996).

TABLE 5

Summary of pairwise sampling undertaken at an Australian mine site.

Pairwise no	Inhalable dust (mg/m^3)	Total dust (mg/m^3)	Respirable dust (mg/m^3)
1	0.131	0.074	0.037
2	0.202	0.073	0.015
3	0.379	0.573	0.303
4	0.431	0.178	0.118
5	0.733	0.725	0.174
6	0.747	1.207	0.664
7	0.879	1.617	0.377
8	1.060	0.385	0.622
9	1.239	1.880	0.645
10	1.271	1.411	0.713
11	1.503	2.500	0.800
12	1.541	0.708	0.486
13	1.667	2.163	0.473
14	1.780	1.151	0.511
15	1.804	1.748	0.368
16	1.993	1.275	1.062
17	2.384	1.832	0.398
18	2.869	1.659	0.299
19	3.030	3.978	1.185
20	4.058	1.650	0.947
21	4.556	0.889	2.189
22	5.008	5.942	0.717
23	10.081	11.449	2.913
24	12.532	6.141	0.787
25	13.293	5.072	1.200
26	18.536	13.462	2.387
Average	**3.604**	**2.682**	**0.784**

PARTICLE SIZE DISTRIBUTION ANALYSIS

Particle size distribution analyses of selected pairwise dust samples of inhalable and respirable fractions from laboratory testing and field sampling are undertaken. These PSD analyses are done by using a *Mastersizer 3000* (10 nm to 3500 µm, blue light). In each PSD measurement test, the sample was repeatedly measured five times. D_{10} is the diameter of the particles at which 10 per cent of the sample's volume is comprised of particles with a diameter less than this value. D_{50} is the diameter of the particle that 50 per cent of a sample's volume is smaller than this value. D_{90} is the diameter of the particle that 90 per cent of a sample's volume is smaller than this value.

Sample preparation was done with the following procedures:

- the coal dust samples on the filters were dispersed in the distilled water/ethanol by sonication, and/or
- the coal dust samples were dispersed in isopropanol by sonication.

A total of seven coal dust samples from the laboratory testing were selected for the PSD analysis with one sample from the original bulk coal dust sample used for laboratory testing feed and three pairwise sampling sets of inhalable dust at two different sampling inlet orientations, 0° and 90° to the oncoming airflow. It should be noted that the PSD analysis requires a minimum dust particle mass. The respirable portion of coal dust particle collected by respirable dust samplers during the laboratory testing is insufficient. Therefore, they are not analysed for the PSD.

Table 6 shows a summary of the PSD analysis results for these laboratory coal dust samples. It can be seen that inhalable dust sampler with its sampling inlet at 90° to the oncoming dust laden airflow is collecting finer inhalable portion of dust particles when comparing with the sampler's inlet oriented at 0° or directly facing the oncoming airflow. This observation is particularly evident in the larger dust particle fraction (D_{90}) collected by the IOM inhalable dust samplers. The 90° oriented samplers are collected a much smaller particle size in average, 18.8 μm at D_{90} compared with the 0° oriented samplers with the D_{90} particle size at 40.6 μm. However, D_{50} particle sizes (ranging from 3 to 23 μm) of laboratory inhalable coal dust samples captured by the IOM samplers were much lower than the ISO sampling efficiency curve (D_{50} = 100 μm).

TABLE 6

Summary of PSD analysis results for lab inhalable samples.

Sets	No	Orientation (°)	Sample inhalable	Mass on filter (mg)	D_{10} μm	D_{50} μm	D_{90} μm
Bulk coal dust sample feed			-	4.2	23.0	89.5	
A	1	0	A3	6.0	1.6	4.5	30.5
	2	90	A4	2.1	1.3	3.3	10.5
B	3	0	B1	5.6	1.6	4.7	40.8
	4	90	B2	2.2	1.4	3.6	22.7
C	5	0	C1	7.3	1.6	5.2	50.6
	6	90	C2	2.3	1.3	3.8	23.1

As for the dust samples collected in the field sampling program, a total of 12 samples have been selected for the PSD analysis. These samples are from four pairwise sampling sets of inhalable and respirable dust. However, due to much less dust particle mass collected by the respirable dust samplers, PSD analyses of the respirable dust samples are combined into two lots for the PSD analysis. Table 7 shows the PSD analysis results of the field coal dust samples.

TABLE 7

PSD analysis results of the field coal dust samples.

Sets	No	Sample	Filter no	Mass on filter (mg)	D_{10} μm	D_{50} μm	D_{90} μm
A	1	Inhalable 0	6303	14.458	3.8	11.3	30.5
	2	Inhalable 90	6304	14.365	4.1	11.7	32.5
	3	Respirable*	6482	2.756	1.9	5.0	11.4
B	4	Inhalable 0	6309	14.871	3.6	11.9	35.1
	5	Inhalable 90	6310	16.735	3.7	11.7	33.2
	6	Respirable*	6484	4.706	1.9	5.0	11.4
C	7	Inhalable 0	6307	8.689	2.3	6.8	18.5
	8	Inhalable 90	6308	8.834	2.2	6.5	18.2
	9	Respirable*	6483	4.625	1.7	4.4	9.4
D	10	Inhalable 0	6543	6.394	4.8	12.8	36.2
	11	Inhalable 90	6311	6.431	4.0	11.5	30.3
	12	Respirable*	6604	1.088	1.7	4.4	9.4

* Note that respirable dust samples (3 and 6 as a pair and 9 and 12 as another pair) are combined to form enough dust particle mass for the PSD analysis.

Further analysing of the PSD results among inhalable 0°, inhalable 90° and respirable show that in a general trend, the 0° samples have only marginal larger PSD compared with the 90° samples. The effect of sampler's inlet orientation on the PSD is not as obvious in the filed sampling program compared with the laboratory testing. This could be due to the oncoming airflow directions were varying during the field sampling unlike the controlled oncoming airflow setting during the laboratory testing.

Based on the PSD analysis results from laboratory and filed sampling, it is concluded that the inlet orientation of the IOM sampler have a significant effect on the sampling efficiency and resulting variations (D_{50} ranging from 3 to 23 μm) in coal dust particle size captured by the IOM sampler. Even through, this effect observed in the field sampling is not as evident as observed in controlled laboratory testing. The reason for this is probably due to the uncontrolled airflow directions during field setting. However, results from laboratory and filed sampling indicate that coal dust particles captured by the IOM sampler were significantly different from the sampling efficiency curve as specified in the ISO 7708 (D_{50} = 100 μm).

In a typical longwall production face, shearer and chock operators are walking along the longwall face-line between Maingate (MG) and Tailgate (TG) as shearer cutting from MG to TG or TG to MG. The orientation of oncoming airflow to the operator wearing an IOM inhalable dust sampler would vary dramatically. The diagrams in Figure 6 illustrate variations in sampling inlet orientation in common situations in the longwall production face.

a). Shearer Cutting from TG to MG; IOM inlet at 0 degree to oncoming airflow

b). Shearer Cutting from MG to TG; IOM inlet at >0 degree to oncoming airflow

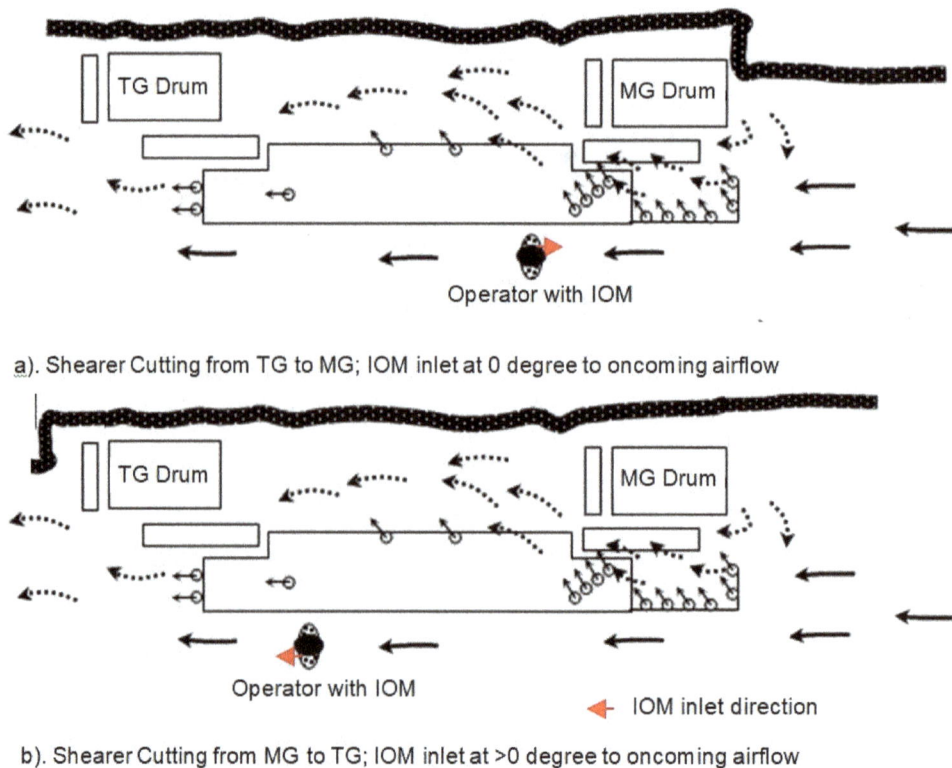

FIG 6 – Examples of operator with IOM samplers at a longwall face.

When shearer is cutting from MG to TG, the operator would have more time with the sampling inlet of the IOM sampler inlet at angles (could be ranging from 15 to 180°) instead of direct facing (at 0°) to the oncoming airflow. When cutting from TG to MG, the operator would face towards MG as the operator walking toward MG direction. In this setting, the sampling inlet of the IOM sampler worn would have more chances facing directly towards oncoming airflow.

In the above situations, the IOM sampler used for sampling inhalable coal dust could either under-sampling or over-sampling the actual inhalable dust level inhaled by the operator. Therefore, it is recommended that a review of the current sampling inlet design the existing IOM inhalable dust sampler should be considered. Button inhalable sampler was developed to reduce the effect of orientation of oncoming airflow to the sampling efficiency of the IOM sampler. However, Button inhalable sampler only has sampling loading limitation issue and only suitable for workplaces with low inhalable dust level applications.

CONCLUSIONS

The laboratory test, delved into a comprehensive exploration of various inhalable dust monitoring devices recommended by AS3640, aiming to provide practical insights into their efficiency and application in Australian coal mining operations. Undertaken at independent testing facilities and coalmine sites, the research extensively evaluates inhalable dust sampling systems, focusing on the AS3640 recommended devices like the modified personal UKAEA sampling head, IOM inhalable dusting sampling head, and conical inhalable sampling head. The lab study examined the performance of sampling systems as per AS3640 criteria. It provided a thorough analysis of inhalable sampling heads, notably the widely used CFC and the IOM sampler. The research highlights the nuances of each sampler, such as the impact of particle size and air velocity on their efficiency. Laboratory tests, conducted at CSIRO Mineral Resources Pinjarra Hill site, scrutinise the effects of airflow orientation on IOM sampler efficiency. The study indicates significant disparities in sampling results based on orientation, emphasising the need for standardised sampling protocols. Field tests, although limited to conveyor areas due to site constraints, provide valuable real-world data on the comparability of inhalable, total, and respirable dust samplers.

In-depth PSD analysis reveals variations in dust particles captured by inhalable samplers oriented differently to airflow and also significantly differences (D_{50} ranging from 3 to 23 µm) to the Inhalable dust size selective curve (D_{50} = 100 µm). The study underscores the importance of orientation in capturing the true inhalable fraction, especially in dynamic workplace conditions like a longwall production face.

The research advocates for a critical review of existing inhalable dust samplers, especially the IOM sampler, concerning airflow orientation. The findings suggest potential under sampling or over sampling scenarios, especially in mobile workspaces. Considering the limitations of existing devices, a re-evaluation of sampler designs is crucial to ensure accurate and representative inhalable dust measurements under various mining scenarios.

This study provided an intricate analysis of inhalable dust sampling devices and highlighted challenges and limitations on their practical applications in coalmines. The results emphasise the need for thoughtful consideration of airflow orientation, urging the industry to re-evaluate current sampling methodologies. As coal mining operations demand precise dust monitoring for worker safety, this research acts as a foundational resource and guiding future advancements in inhalable dust monitoring technologies.

REFERENCES

Aizenberg, V, Grinshpun, S A, Willeke, K, Smith, J and Baron, P A, 2000. Performance characteristics of the button personal inhalable aerosol sampler, *American Industrial Hygiene Association Journal,* 61(3):398–404.

Baldwin, P and Maynard, A, 1997. A survey of air velocities in indoor workplaces, *Annals of Occupational Hygiene,* 42(5):303–313.

Kenny, L C, Aitken, R, Baldwin, P, Beaumont, G and Maynard, A, 1999. The sampling efficiency of personal inhalable aerosol samplers in low air movement environments, *Journal of Aerosol Science,* 30(5):627–638.

Kenny, L C, Aitken, R J, Chalmers, C, Fabriés, J F, Gonzalez-Fernandez, E, Kromhout and Prodi, V, 1997. A collaborative European study of personal inhalable aerosol sampler performance, *Annals of Occupational Hygiene,* 41(2):135–153.

Kulkarni, P, Baron, P and Willeke, K, 2011. *Aerosol Measurement: Principles, Techniques and Applications,* third edition (John Wiley and Sons: Hoboken).

Li, S, Lundgren, D and Rovell-Rixx, D, 2000. Evaluation of six inhalable aerosol samplers, *American Industrial Hygiene Association Journal,* 61(4):506–516.

Liden, G and Bergman, G, 2001. Weighing imprecision and handleability of the sampling cassettes of the IOM sampler for inhalable dust, *Annals of Occupational Hygiene,* 45(3):241–252.

SKC Ltd, Europe, 2023. Button Sampler, SKC Ltd. Available from: <https://www.skcltd.com/products2/sampling-heads/button-sampler.html>

Standards Australia, 2009. AS3640-2009. Workplace atmospheres - method for sampling and gravimetric determination of inhalable dust, Standards Australia.

Werner, M, Spear, T and Vincent, J, 1996. Investigation into the impact of introducing workplace aerosol standards based on the inhalable fraction, *The Analyst,* 121:1207–1214. https://doi.org/10.1039/AN9962101207

Zhou, Y and Cheng, Y, 2009. Evaluation of IOM personal sampler at different flow rates, *Journal of Occupational and Environmental Hygiene,* 7(2):88–93.

An analysis of the inhalable coal dust samples in Queensland

M S Kizil[1], H W Wu[2], G Kizil[3] and B Belle[4]

1. Associate Professor, The University of Queensland, School of Mechanical and Mining, St Lucia Qld 4072. Email: m.kizil@uq.edu.au
2. Mining Technology Pty Ltd, Brisbane Qld 4000. Email: h.wu@minserve.com.au
3. The University of Queensland, School of Mechanical and Mining, St Lucia Qld 4000. Email: g.kizil@uq.edu.au
4. The University of Queensland, School of Mechanical and Mining, St Lucia Qld 4072; The University of New South Wales, Sydney NSW 2000; The University of Pretoria, South Africa; 61Drawings, Belmont Qld 4153. Email: bb@61drawings.com

ABSTRACT

Coal mining faces numerous health and safety challenges, with the generation of dust being a prominent issue throughout the development and mining processes. This paper provides an assessment of the levels of inhalable dust in Queensland underground coalmines to gauge the exposure of coalmine workers. It presents a thorough analysis of over 11 000 inhalable dust samples gathered by Resources Safety and Health Queensland between 2011 and 2021. The samples are analysed based on operation types and similar exposure groups over time, providing insights into exposure and exceedance levels. The findings reveal that the majority of samples were obtained during development production, underground maintenance, and longwall production. Significantly, the study highlights the prevalence of exceedances in specific areas, with longwall face, tailgate, and development production accounting for the highest rates – 36.9 per cent, 33.3 per cent, and 17.0 per cent, respectively. This emphasises the critical need for targeted interventions and safety measures in these specific operational aspects to mitigate risks and ensure the well-being of coalmine workers. The data and analyses presented in this paper contribute valuable insights for developing strategies aimed at minimising dust-related health hazards in the coal mining industry.

INTRODUCTION

Coal is a combustible sedimentary rock in black or brown formed at the strata called the coal seam. The primary constituents of coal are organic carbon compounds. The coal was discovered in the very early period in the history of Australia, dating back to the year 1797, by a survivor of the wreck of a vessel which had landed near NSW. Nowadays, Australia is one of the largest coal producers in the world.

Coal mining still has many challenges in terms of health and safety. There are many coalmine hazards, such as methane explosion and spontaneous combustion, and exposure to coal dust. Exposure to coal dust may result in diseases as Coal Workers' Pneumoconiosis (CWP). Dust is created by various production and development processes such as the cutting and transporting coal. The particle size range of coal dust can vary depending on factors such as the type of coal, the mining or processing methods, and the specific conditions under which the dust is generated. Generally, coal dust particles can range in size from a few microns to several hundred microns in diameter. The air quality within the underground coalmine as well as around the surface infrastructure has the potential to be impacted by the coal dust. The primary dust sources at the underground longwall panels and development sections are mechanical cutting drums on the shearers and continuous miner machines. The primary sources of coal dust at the surface operations are associated with handling, storing, and transporting coal. As before the coal been transported to the port and loaded on trains, coal dust comprises of the small proportion of the total dust that present in the air near the coalmines and the coal export terminals. The size of the coal dust ranges from submicron to 200 microns in diameter. In terms of size, there are currently two size fractions of coal dust sampled for measuring exposure: inhalable dust, and respirable dust. Inhalable dust refers to the particle that can be breathed into the nose or mouth and respirable dust refers to the dust that can be breathed into the respiratory system (Coal Services Pty Limited, 2016). Inhalable dust refers to particles with a median diameter of 100 µm, while respirable dust refers to particles less than 10 µm. Figure 1 shows the inhalable and respirable dust conventions.

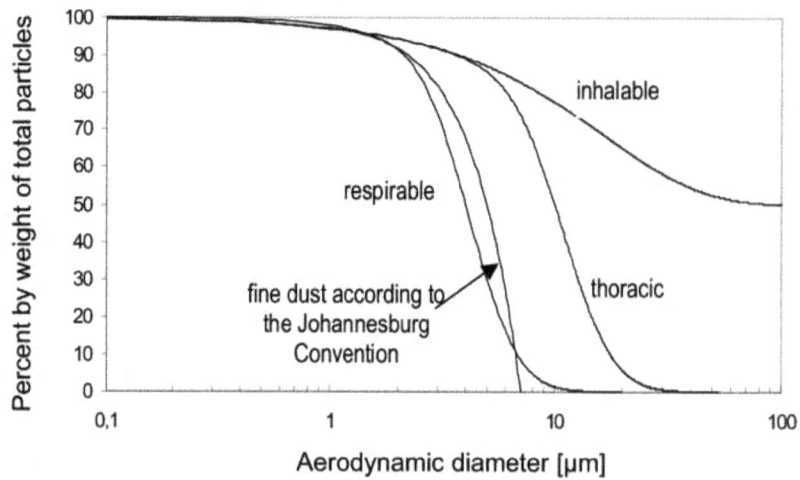

FIG 1 – Inhalable, thoracic and respirable conventions, in percent by weight of total airborne particles according to EN 481, and fine dust fraction according to the Johannesburg Convention (Parlar and Greim, 2005).

The term inhalable dust applies to both non-toxic and toxic dusts. Inhalable dusts that are toxic have an exposure standard based upon the substance of concern. Where the toxic component of the dust is measured, this is satisfactory as long as the exposure standard for dusts not otherwise classified is not exceeded. Exposure standards for dusts are measured as inhalable dusts unless there is a notation specifying an alternate method, eg cotton dust, silica (Safe Work Australia, 2013a).

For dusts without specific exposure standards, it's recommended to maintain exposure below 10 mg/m^3 of inhalable dust (measured over an 8 hr period) if the substance is inherently low in toxicity and free from toxic impurities. However, if the particulate material contains other toxic substances that can cause physiological impairment at lower concentrations, then the exposure standard for the more toxic substance should be applied. For example, if a dust contains asbestos or crystalline silica, exposure should not exceed the appropriate value for those substances (Safe Work Australia, 2013b). Table 1 shows dust exposure standards set by different countries.

TABLE 1

Dust exposure standards by countries.

Country	Organisation	Inhalable dust limit	Comments
US	ACGIH	10 mg/m³	American Conference of Governmental Industrial Hygienists (ACGIH), (1990).
Australia			Adapted from the ACGIH
	Safe Work Australia	10 mg/m³	Safe Work Australia (2013a): Guidance on the Interpretation of Workplace Exposure Standards for Airborne Contaminants
	AIOH	5 mg/m³	Australian Institute of Occupational Hygienists, Inc., (AIOH) (2016); Standards Australia (2009)
	NOHSC	10 mg/m³	National Occupational Health and Safety Commission (NOHSC), 1995
UK	HSE	10 mg/m³	Health and Safety Executive (HSE)
	COSHH	10 mg/m³	Control of Substances Hazardous to Health Regulations (COSHH) (2002) Note that this concentration is not an exposure limit, but it triggers values for application of all the COSHH regulations to the dust.
	IOM	5 mg/m³	IOM (Institute of Occupational Medicine) (2011)
Germany	MAK Commission	4 mg/m³	MAK (maximale Arbeitsplatz-Konzentration) Commission is the German Commission for the Investigation of Health Hazards of Chemical Compounds in the Work Area (MAK Commission, 1983)

INHALABLE COAL DUST DATABASE

The Resources Safety and Health Queensland has provided a database of 11 791 Queensland coal dust monitoring data records collected between 2011 and 2021, with the aim of analysing the data to gain insights into coal dust levels in the region. The coal dust monitoring data is crucial in understanding the potential health risks posed to workers and their exposure levels in the coalmines.

Out of the total 11 791 coal dust monitoring data records, 10 610 (~90 per cent) were found to be usable for analysis. The remaining 1181 (~10 per cent) records were classified as non-usable or deemed to be invalid due to various reasons, which were grouped under the following 13 categories including: damage to filter/sample head, failed post flow, filter overloaded, flow fault, invalidated by lab, not reported, pump damaged, pump failure, pump not collected/returned, short run time, tubing detached, worker removed pump and other reasons.

Sample classification by operation type

The invalid samples have been excluded from the analysis. The remaining valid samples have been classified in to three categories based on what type of operations the samples came from, namely Underground, Surface and Coal Handling and Processing Plant (CHPP) and Run-of-mine (ROM). As seen from Figure 2 more than half of the data have been collected from the underground operations.

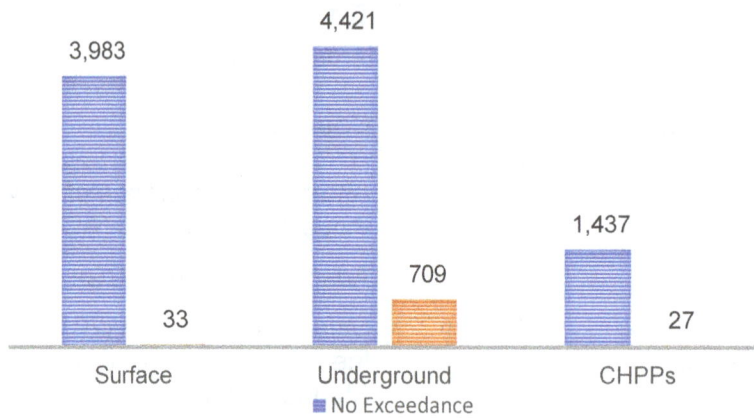

FIG 2 – Number of samples for each operation type.

Exposure to coal dust above threshold limits poses a significant health concern. Therefore, data have been analysed based on exceedances of these threshold limits. The Occupational Exposure Limit (OEL) of 10 mg/m³ is not constant and requires adjustment to account for work shifts that differ from the standard 8 hr workday. Exceedance analyses were performed using these adjusted exposure limits. All exceedance ratios were calculated using Equation 1. If the measured dust concentration is below the corrected exposure limit, the exceedance ratio is considered to be zero.

$$\text{Exceedance Percentage} = \frac{(\text{Measured Dust Concentration-Corrected Exposure Limit})}{\text{Corrected Exposure Limit}} *100 \qquad (1)$$

Number of data with exceeding and non-exceeding the corrected exposure limits based on the mine type has been provided in Table 2.

TABLE 2

Numbers of exceeding and non-exceeding samples.

Mine Type	No Exceedance	Exceedance	Exceedance (%)	Total
Surface	3983	33	0.8	4016
Underground	4421	709	13.8	5130
CHPP and ROM	1437	27	1.8	1464
Overall	9841	769	7.2	10 610

As seen from Table 2, underground mines had the highest exceedance rate, approximately 13.8 per cent while surface mines had a very low number of exceedance cases (0.8 per cent). Overall, the rate of exceedance cases from all operations was 7.2 per cent.

Similar exposure group

Similar exposure group (SEG) codes are used to identify a group of workers who have the same general exposure to risks. This can include:

- similarity and frequency of the tasks performed
- types of materials and processes used to complete tasks or
- similarity of the way tasks are performed.

SEG codes were used to measure the average dust exposure levels and exceedance rate for each SEG. Figure 3 shows the distribution of exceedance percentage based on SEG codes for all underground mines. As seen from the figure, the highest rates of exceedances were recorded for longwall face, tailgate and development production; 36.9 per cent, 33.3 per cent and 17.0 per cent respectively. Figure 4 shows the number of samples collected and the number of exceedances per

SEG activities for underground mines. The highest number of samples were collected development production, underground maintenance and longwall production.

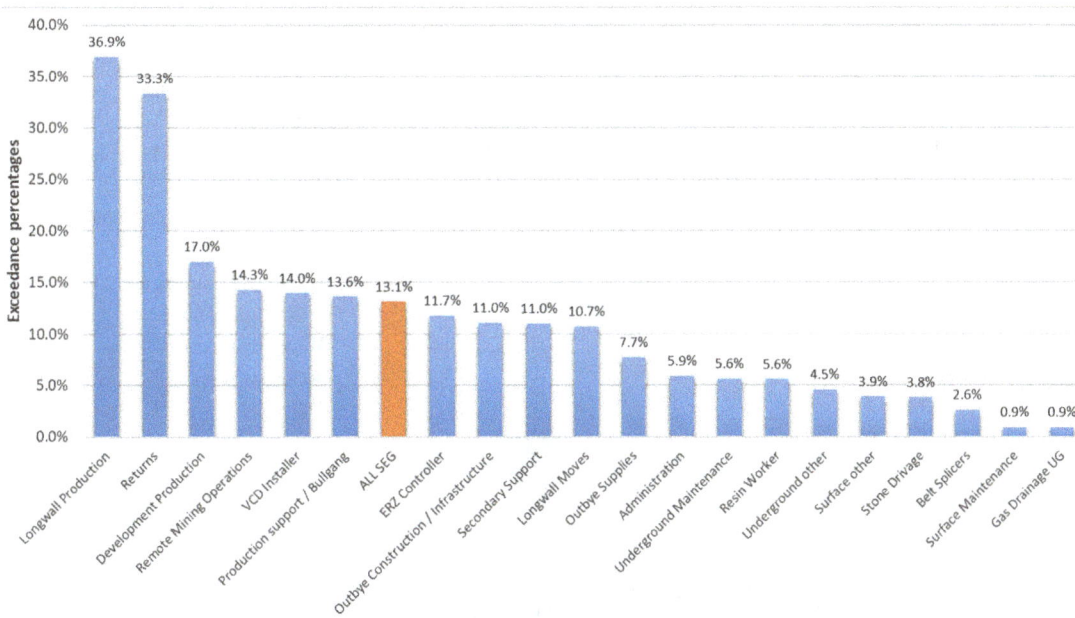

FIG 3 – Percentage of exceedances based on SEG codes in underground mines (overall percentage of all SEGs are shown with orange bar).

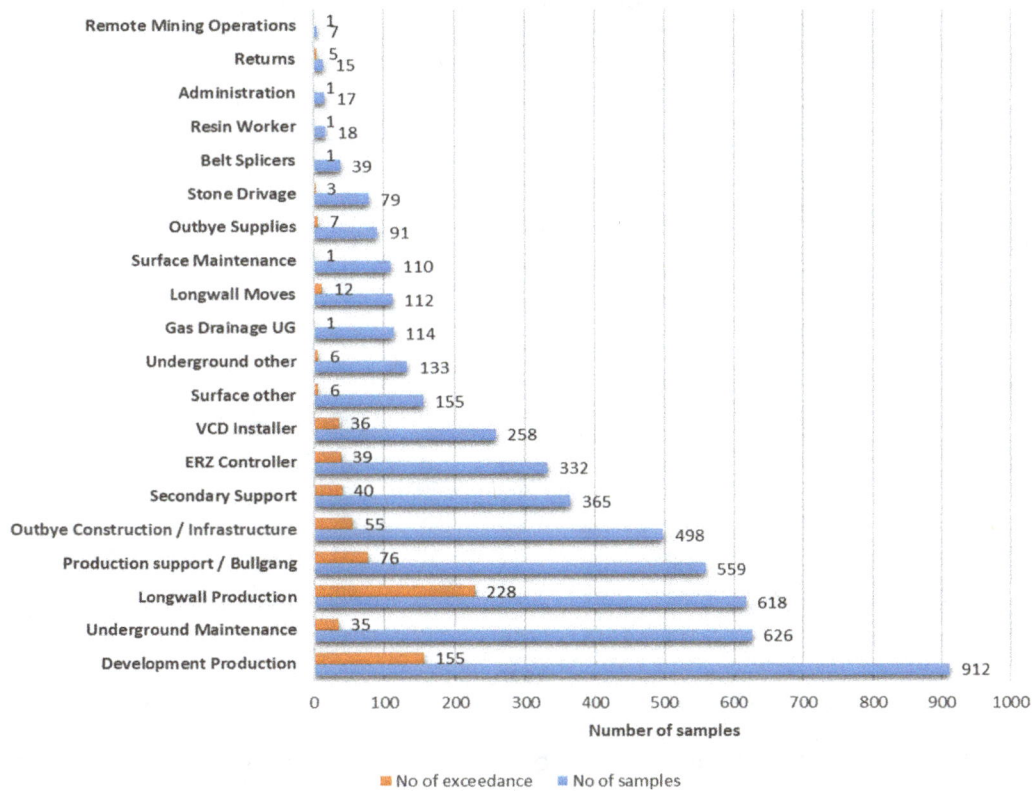

FIG 4 – Number of samples collected and number of exceedances per SEG activities for underground mines.

The Box-Whisker plot of the exceedance ratio for various underground coalmine SEGs is presented in Figure 5. This figure illustrates that the distribution of inhalable dust concentrations across different SEGs is not uniform and varies with different mining activities. The median inhalable dust measurement for the production support/bullgang SEG was the highest among all SEGs. Figure 6 displays the total number of samples, exceedance cases, and exceedance percentages of underground cases on an annual basis. The graph clearly indicates that, although the number of samples collected has been slightly increasing over time, the number of exceedance cases has decreased. After rising from 12 per cent in 2011 to 21 per cent in 2013, the exceedance rate has dropped to 10 per cent by 2021.

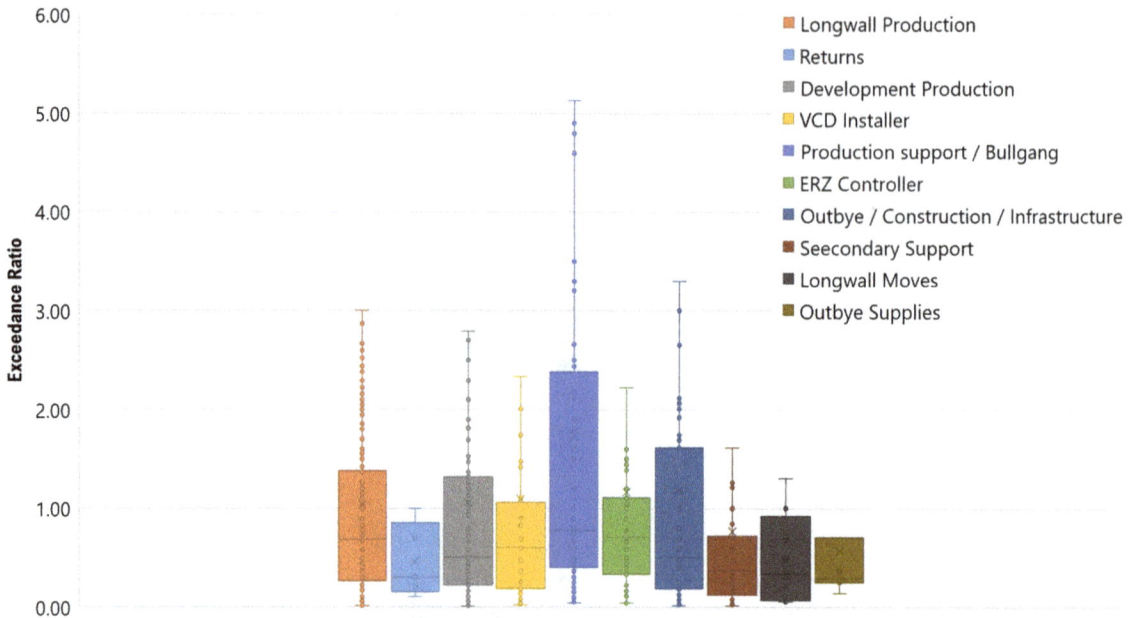

FIG 5 – Box-Whisker plots of the exceedance ratios for some underground coalmines SEGs.

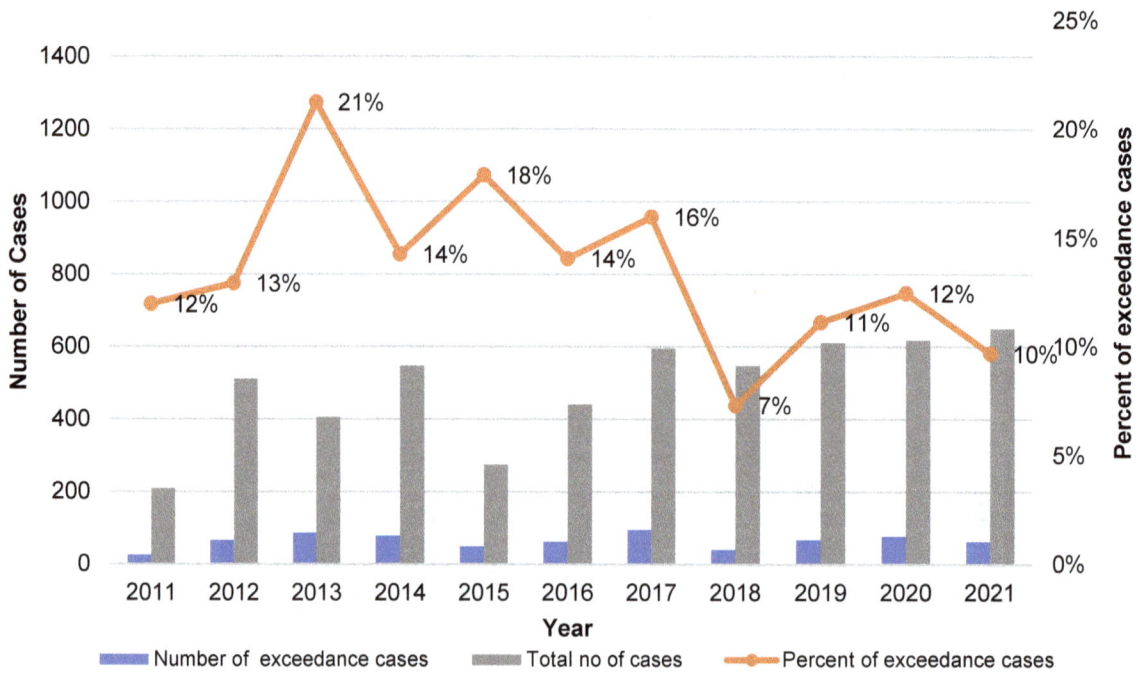

FIG 6 – Total, exceedance, and exceedance percentage of underground cases on annual bases.

Figures 7 and 8 show the average and the maximum measured inhalable dust concentrations for underground coalmines on annual basis, respectively.

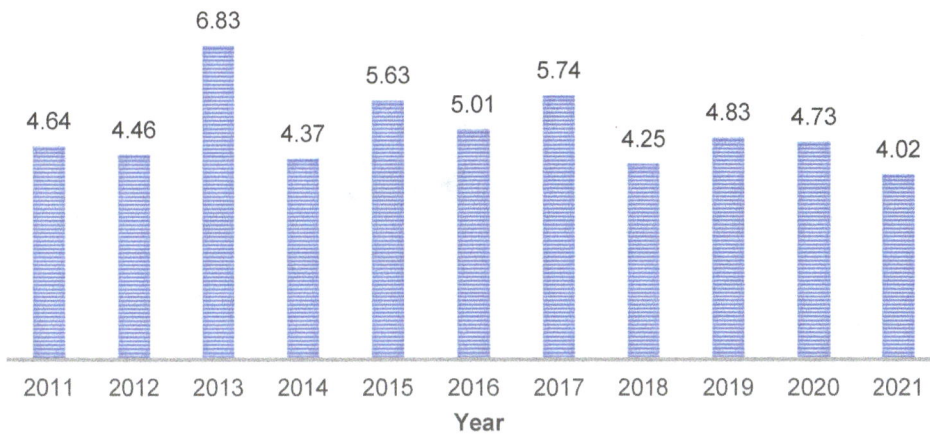

FIG 7 – Average measured inhalable dust concentrations in mg/m³ for underground mines on annual basis.

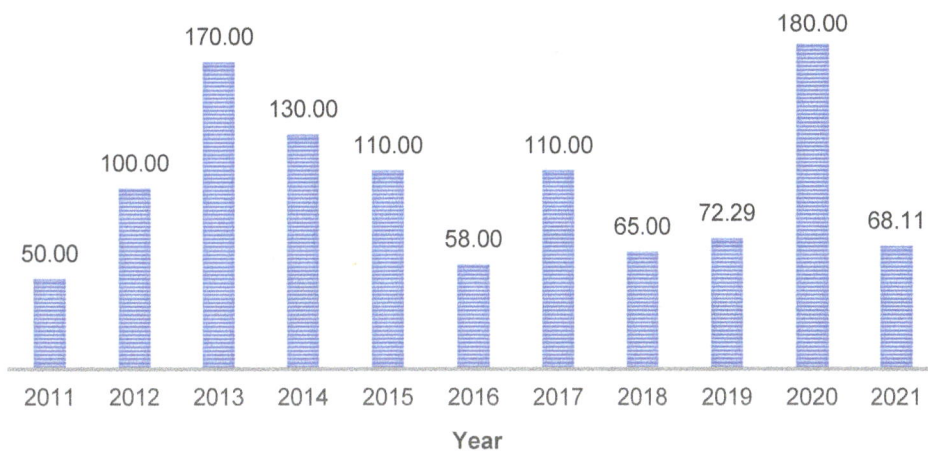

FIG 8 – Maximum measured inhalable dust concentrations in mg/m³ for underground mines on annual basis.

As seen from Figures 7 and 8, there was no specific trend between the maximum and average inhalable dust concentration on year by year bases. The average readings seem to be below the threshold limit while the maximum readings were recorded well above the limit. The underground inhalable dust concentrations show log-normal distribution with positive skewness. The frequency of readings above the threshold limit was less than 0.12, as seen in Figure 9.

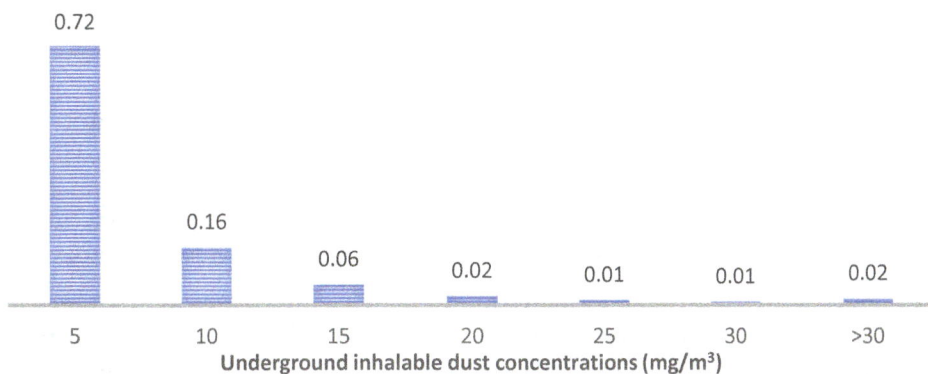

FIG 9 – Relative frequency distribution of inhalable dust concentrations in underground mines.

Personal protection equipment use during dust surveys

The inhalable dust database had a record of weather the coalmine workers were wearing personal protection equipment (PPE) or not during the surveys and whether the PPE was worn for the whole or partial part of the shift. PPE included: P1 half face disposable mask, P2 full face non-disposable mask, P2 half face disposable mask, P2 full face non-disposable mask, P3 full face non-disposable mask and Powered Air Purifying Respirators (PAPR) and half face.

Out of 5408 total cases, no PPE was used in 1003 surveys, and PPE was used for the partial or full duration of the shift in 2614 cases. In 1791 cases, there was no record of whether PPE was used. Among the 709 exceedance cases, there were 299 instances (42 per cent) with no report, 27 instances (4 per cent) where no PPE was worn, and 383 instances (54 per cent) where PPE was worn for the partial or full duration of the shift, as shown in Figures 10 and 11.

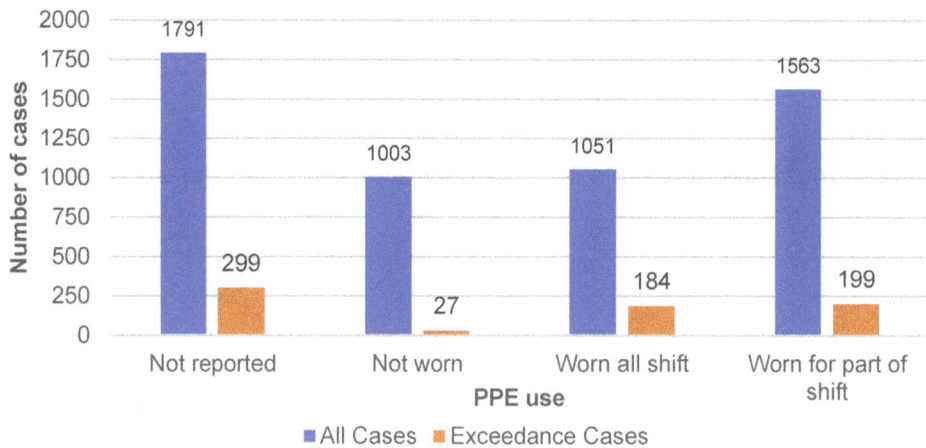

FIG 10 – Number of cases related to PPE use during the survey for all and exceedance cases.

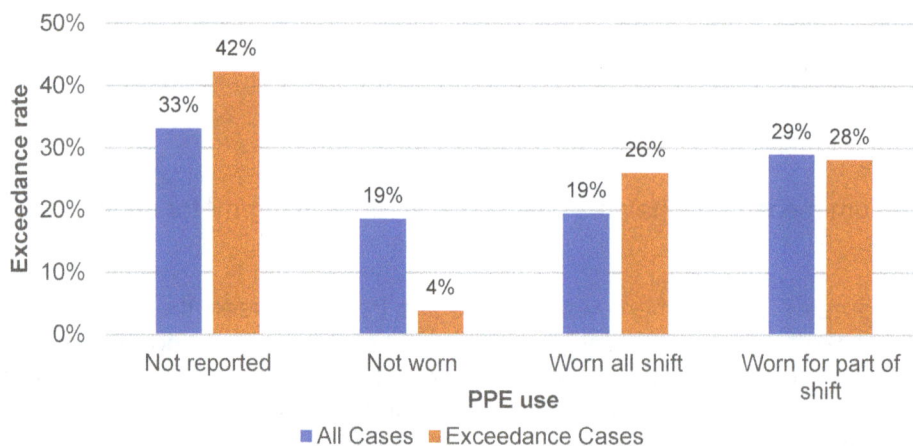

FIG 11 – Percentage of cases related to PPE use during the survey for all and exceedance cases.

CONCLUSIONS

This study analysed a data set comprising 11 791 inhalable dust monitoring records collected by Queensland Resources Safety and Health between 2011 and 2021. The aim was to assess the inhalable coal dust levels to which underground coalmine workers in Queensland were exposed. Of these records, 10 610 (90 per cent) were analysed, while 1181 (10 per cent) were deemed unusable for various reasons. The analysis focused on understanding exposure levels across different mine types (Underground, Surface, CHPP, and ROM) and SEGs.

Exceedance analysis was conducted to assess coal dust exposure levels above threshold limits. Corrected exposure limits were used based on working conditions. Results indicated that underground mines had the highest exceedance rate (13.8 per cent), while surface mines had a minimal rate (0.8 per cent). Overall, the exceedance rate across all operations was 7.2 per cent. SEG-based analysis revealed specific activities, like longwall face and tailgate, had higher exceedance rates (36.9 per cent and 33.3 per cent respectively). Over time, the analysis showed a decrease in exceedance cases, dropping from 21 per cent in 2013 to 10 per cent in 2021, despite a slight increase in the number of samples. However, the distribution of inhalable dust concentration varied across different SEGs, indicating non-uniform exposure levels.

Analysis of PPE use during surveys indicated that in 42 per cent of exceedance cases, there was no report, and in 54 per cent of cases, PPE was worn partially or for the whole shift. This data underscores the importance of monitoring PPE compliance and effectiveness in mitigating exposure risks.

The analysis highlights significant variations in coal dust exposure across different mine types, activities, and specific mines. While there's a positive trend showing a decrease in exceedance cases over the years, it's imperative to continue monitoring and enforcing safety protocols, including proper PPE usage, to ensure the well-being of coalmine workers in Queensland.

REFERENCES

American Conference of Governmental Industrial Hygienists (ACGIH), 1990. 1990–1991 TLVs® and BEIs®; Threshold Limit Values for Chemical Substances and Physical Agents and Biological Exposure Indices. Available from: <https://www.acgih.org/science/tlv-bei-guidelines/>

Australian Institute of Occupational Hygienists, Inc., (AIOH), 2016. Dusts Not Otherwise Specified (Dust Not Otherwise Specified (NOS)) and Occupational Health Issues, position paper, May 2014, AIOH Exposure Standards Committee, p 5. Available from: <https://www.aioh.org.au/product/dust-nos/>

Coal Services Pty Limited, 2016. Protecting against airborne dust exposure in coal mines [online]. Available from: <https://www.coalservices.com.au/wp-content/uploads/2016/12/NEW-CS-Dust-Booklet_Final-artwork.pdf>

Control of Substances Hazardous to Health Regulations (COSHH), 2002. *Control of Substances Hazardous to Health*, fifth edition, Approved Code of Practice and Guidance (HSE Books).

Institute of Occupational Medicine (IOM), 2011. The IOM's position on occupational exposure limits (OEL) for dust. Available from: <https://www.iom-world.org/media/1656/position-paper.pdf>

MAK Commission, 1983. DFG, German Research Foundation – Permanent Senate Commission for the Investigation of Health Hazards of Chemical Compounds in the Work Area, MAK (maximale Arbeitsplatz-Konzentration) Commission. Available from: <https://www.dfg.de/en/dfg-profile/statutory-bodies/senate/health-hazards>

National Occupational Health and Safety Commission (NOHSC), 1995. Proposed National Exposure. Standards for Atmospheric Contaminants in the Occupational Environment (Australian Government Publishing Service: Canberra).

Parlar, H and Greim, H, 2005. Sampling and determining aerosols and their chemical components (chapter), in *The MAK-Collection for Occupational Health and Safety, Part III: Air Monitoring Methods*, vol 9 (Wiley-VCH Verlag). https://doi.org/10.13140/2.1.4181.9202

Safe Work Australia, 2013a. Guidance on the interpretation of workplace exposure standards for airborne contaminant. Available from: <https://www.safeworkaustralia.gov.au/system/files/documents/1705/guidance-interpretation-workplace-exposure-standards-airborne-contaminants-v2.pdf>

Safe Work Australia, 2013b. How to Determine what is Reasonably Practicable to meet a Health and Safety Duty, July 2013. Available from: <http://safeworkaustralia.gov.au/>

Standards Australia, 2009. AS3640-2009. Workplace atmospheres - method for sampling and gravimetric determination of inhalable dust, Standards Australia.

Agnico Eagle Fosterville Gold Mine low frequency noise investigation

J Norris[1]

1. Senior Ventilation Engineer, Fosterville Gold Mine, Vic 3557.
   Email: jeff.norris@agnicoeagle.com

ABSTRACT

The Agnico Eagle Fosterville Gold Mine (FGM) is located approximately 20 km East of Bendigo, Victoria. The mine is located in a semi-rural setting adjacent to native bushland. Residential properties are located to the East, North and South of the mine. The mine commissioned a surface ventilation system in June 2020 and subsequently received complaints of noise disturbances.

The Environmental Protection Agency (EPA) determined that Low Frequency Noise (LFN) in the range of 16–20 Hz, was being emitted from the mine site and beyond the boundaries of the Fosterville Gold Mine. Notices were issued by the EPA to investigate the source and methods to resolve the issue of low frequency noise.

In response to EPA Victoria's findings, operation of the primary ventilation system was modified by reducing the speed of the fans from 920 rev/min to 400 rev/min between the hours of midnight and 06:00 am. The modification of fan speed from 920 rev/min to 400 rev/min resulted in a significant reduction in underground mine air flow/ventilation from 520 m³/s to 220 m³/s.

This paper outlines the program of investigations and modifications undertaken to minimise the risk of LFN emissions from the surface ventilation system fans, so far as reasonably practicable. The low frequency noise control measures implemented include:

- Fan silencer baffle reconfiguration.
- Fan motor housing modifications and additional abatement.
- Fan speed offset modifications.

INTRODUCTION

In June 2020 Agnico Eagle Mines – Fosterville commissioned two Heavy Duty 2-Megawatt mine ventilation fans. The two ventilation fans provide the source of fresh air to the two main production areas supplying approximately 500 to 540 m³/s operating at 920 rev/min. FGM has undertaken an assessment of the acceptable noise contribution from the proposed new ventilation infrastructure to ensure ongoing compliance with the site environmental noise limits. The night-time site limit of 33 dB(A) was chosen as the basis for determining the noise level target for the new ventilation plant. The modelled contribution from the new ventilation infrastructure will need to be at least 10 dB(A) lower than the noise contribution from the other existing mining operations for the overall noise to remain at or below the site noise limits. The noise level target at sensitive receptors located to the south-east of the new ventilation infrastructure, is therefore 23 dB(A).

The noise level target refers to the complete system and must be met by the components of the ventilation infrastructure, including motor noise, VSD induced noise and all other noise sources, whether operating individually or as a whole. Therefore, the noise target for the ventilation system is 23 dB LAeq at the nearest receptors.

Located in a semi-rural setting adjacent to native bushland. Residential properties are located to the East, North and South of the mine, with the nearest receptor (FNB9) approximately 800 m south of the mine and ventilating fans, see Figure 1.

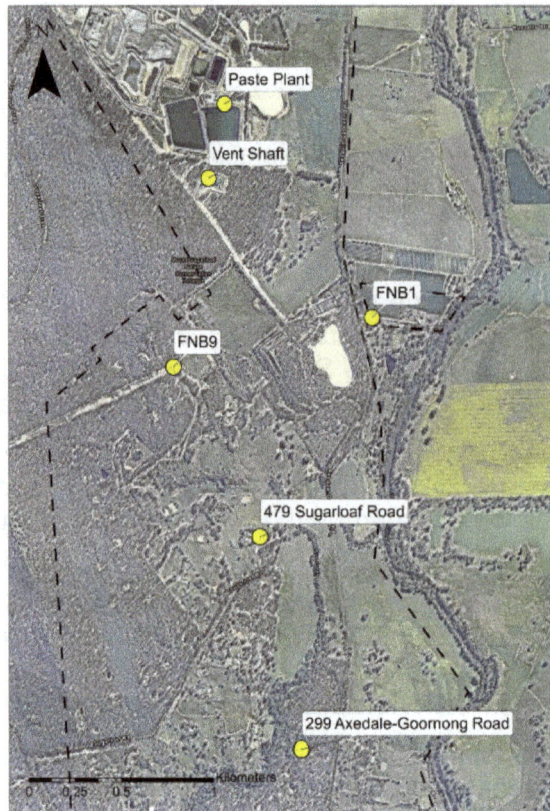

FIG 1 – FGM closest noise receptor at 800 m FNB9.

Due to the low allowable noise levels to be emitted the ventilating fans were fitted with engineered noise abatement; Reactive dissipative type silencers to mitigate noise. Contained within an acoustic enclosure, see Figure 2.

FIG 2 – Surface ventilation fan noise abatement.

Attended noise measurements were undertaken at various locations around the installation sound level metres fitted with a windshield. The microphone was mounted on a tripod at a height of approximately 1.5 m above local ground level under free field conditions. The noise level from the installation were anticipated to be low compared to the background noise at the residential receptors, owing to the considerable distance between them and the installation. Consequently, measurements were taken near the installation and used as a basis for estimating the potential noise level at the receptors.

In addition, a sound level metre fitted with a windshield was installed at a distance of approximately 40 m from the fan outlets for the duration of the measurements to assess the variability in the noise output of the installation. Measurements were undertaken using the 'fast' response time and 'A' frequency network in June 2020. Weather conditions throughout the survey were fine and dry with negligible winds. Monitoring results are shown in Table 1.

TABLE 1
Calculated noise levels at nearby residential receptors.

Receptor location	Measurement distance (m)	Measured level (dBA)	Distance to receptor (m)	Calculated noise level at receptor (dBA)	Criteria (dBA)
FNB1	52	54	1300	<20	23
FNB9	174	40	800	20	23

The results in Table 1 indicate that, given the operating conditions during the measurements, the installation met the night-time noise criteria for all nearby receptors.

ENVIRONMENTAL PROTECTION AGENCY FINDINGS

In response to community reports regarding noise impacts, EPA Victoria undertook a noise investigation within and beyond the boundaries of the Fosterville Gold Mine (FGM). Following this investigation, in December 2021, EPA Victoria issued FGM with Prohibition and Improvement Notices and Notice to investigate the noise source.

The notice observed that:

- The general noise and Low Frequency Noise (LFN) was shown to be emitted from surface drill rigs being operated by FGM, as well as plant equipment that emits more continuous noise, including the surface ventilation system.

- The results found that LFN emanates from the mine in 1/3 octave bands between 16 Hz and 20 Hz. Narrow-band spectral analysis indicated peaks in energy at 18–19 Hz. Measurements taken near the ventilation shaft showed peaks of 15, 18 and 23 Hz.

- This occurrence is particularly evident in noise spectrograms and narrow band analysis between midnight and 06:00 hrs when background levels are at their lowest.

The EPA concluded that the above ground and below ground exploratory and extractive activities being conducted at FGM was generating levels of LFN which could reasonably be expected to have an adverse effect on human health, and that FGM had not minimised the risk of harm to human health so far as reasonably practicable.

FGM was recommended to:

- *Immediately undertake narrow band analysis of the potential sources of low frequency noise at the premises to identify and then inform mitigation measures. Use the following publications to assist with investigations. 'Publication 1826.4 – Noise limit and assessment protocol', and 'Publication 1996-Noise guideline – assessing low frequency noise'.*

- *As narrow band analysis is undertaken, investigate potential noise controls to allow for timely implementation.*

- *Once low frequency noise sources are identified, modify the activities, or install controls at the premises, so that levels of low frequency noise likely to cause harm to human health, is not observed beyond the boundary of your premises.*

and,

by 31/03/22 you must modify the activities or install controls at the premises, so that levels of low frequency noise causing harm beyond the boundary of your premises is minimised as far as reasonably practicable.

RELEVANT LEGISLATION USED TO ASSESS LFN

The key regulations and guidelines relevant to this noise assessment are listed below.

- Environment Protection Act 2017.

- Environment Protection Regulations 2021.

- EPA Publication 1996 – Noise Guideline, Assessing low frequency noise.

- EPA Publication 1826: Noise limit and assessment protocol for the control of noise from commercial, industrial and trade premises and entertainment venues: March 2021.

The Environment Protection Act 2017, updated by the Environment Protection Amendment Act 2018, along with its dependent legislation, became effective on July 1, 2021. This overhaul significantly reformed Victoria's environmental protection laws and the operations of the Environment Protection Authority Victoria (EPA). The Act encompasses environmental obligations and protections for all Victorians, shifting Victoria's focus on environmental protection and human health to a prevention-based approach centred around the General Environmental Duty (GED). It grants the EPA increased powers and tools to prevent and minimise risks to human health and the environment from pollution and waste. Additionally, it enables the EPA to enforce stronger sanctions and penalties against environmental polluters.

The legislation encompasses:

- General Environmental Duty.

- Unreasonable noise.

- Reasonably practicable.

- Environment Protection Regulations.

For thoroughness, the frequency spectrum of LFN emissions at FGM have been compared to the corresponding threshold levels outlined in the EPA's LFN guideline, as shown in Table 2.

TABLE 2
Outdoor one-third octave low frequency noise threshold levels from 10 Hz to 160 Hz.

One-third Octave (Hz)	10	12.5	16	20	25	31.5	40	50	63	80	100	125	160
Leq, dB	92	87	83	74	64	56	49	43	42	40	38	36	34

FOSTERVILLE GOLD MINE LOW FREQUENCY INVESTIGATION

Between August 2021 and December 2021, FGM commissioned acoustic consultant AECOM to complete a comprehensive noise monitoring program to identify the source of low-frequency noise (LFN) reported by residents south of FGM. These monitoring activities were strategically planned around infrastructure outages or modifications to mining operations in a hope to identify and isolate the sources of LFN on-site. These periods of outages and modification provided opportunities to measure changes in on-site LFN and determine if these reductions were mirrored at sensitive receiver locations as various pieces of equipment were turned off and back on again.

Key surface infrastructure was identified, and monitoring conducted at each location. The surface plant and infrastructure considered in this LFN assessment includes:

- Vent shaft and primary fans.

- Paste plant.

- Aster plant.

- Terminal station.

- Drill rigs in operation.

Table 3 summarises the initial narrow-band spectral analysis conducted to investigate the plant and infrastructure potentially contributing to LFN emissions from FGM's premises.

TABLE 3

Summary of initial narrow-band analysis (source: AECOM, 2021).

Plant	Frequencies noted during narrow-band analysis	LFN requiring further investigation
Vent shaft	11–15 Hz, 19 Hz, 21 Hz and 25 Hz	Yes
Drill rigs	40 Hz and 51 Hz	Yes
Paste plant	57 Hz	No
ASTER plant	41 Hz, 49 Hz and 50 Hz	No
Terminal station	50 Hz	No

The noise monitoring conducted during a scheduled complete power outage at the mine, which encompassed the surface ventilation system, in August 2021, indicates that the unweighted sound power levels reach their highest points between 12.5 Hz and 25 Hz under full operating conditions compared to those recorded during the outage, refer to Figure 3. This observation was made at sensitive receiver locations situated 1800 m and 3000 m away.

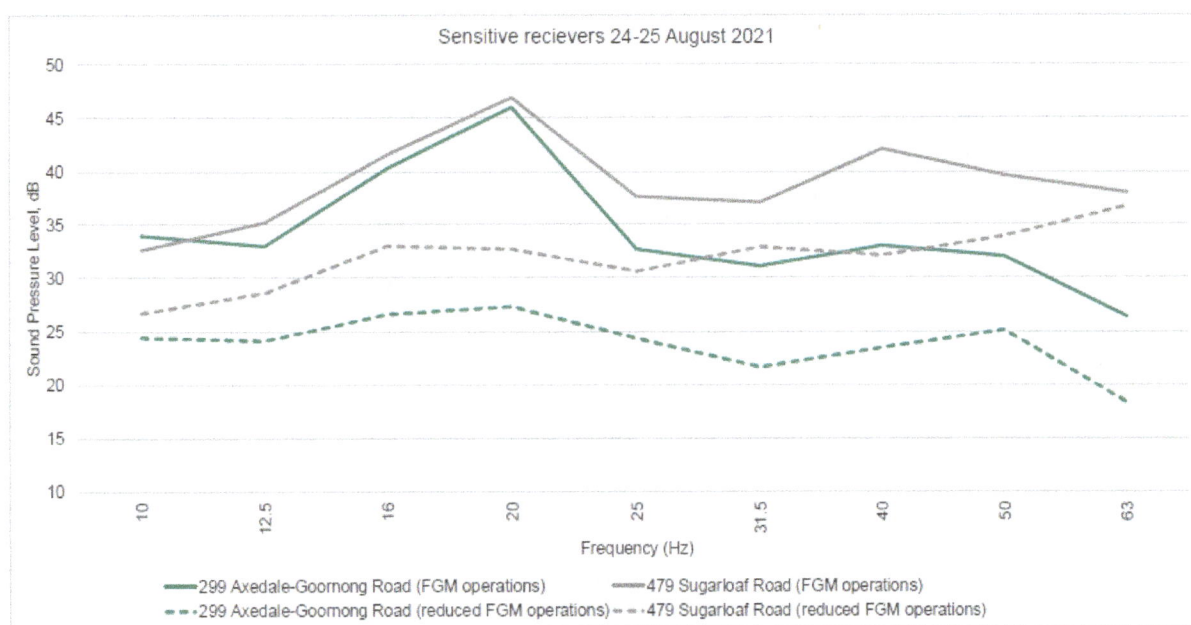

FIG 3 – Comparison of measured levels at low frequencies (source: AECOM, 2021).

Ventilation system low frequency noise investigation

Independent Industry Experts, Noise Consultants and Fan Acoustics and Aerodynamic Engineers were invited to site to assist the working group with FGM staff to assess the factors previously identified as potentially contributing to the LFN.

The working group were able to rule out the following mechanisms as contributors to the LFN issue.

- Ground vibrations traveling along faults and bedding planes.

- Structure born vibrations.

- Internal ducting vibration exciting sound enclosure.

- Underground Shaft vortex shedding induced LFN – ie Similar to the action of a flute.

- Underground mining activity induced frequencies emanating to surface through the shaft.

Noise monitoring by Fan Acoustics and Aerodynamic Engineers at receptors south of the surface fan site, identified the fan noise levels were ~35–40 dB(Z) at a frequency of 15.3 Hz. The frequency, 15.3 Hz, is the fan frequency at 920 rev/min.

MODIFYING ACTIVITIES – FAN SPEEDS

The requirements outlined in the EPA notices required FGM to modify its activities so that levels of low frequency noise likely to cause harm to human health, is not observed beyond the boundary of the premises.

Reducing the fans speed was investigated to determine an operating speed at which activities in the underground workings could continue, however at reduced capacity.

A series of test were completed varying the fan speed between 400 to 920 rev/min. Unattended monitoring data is shown in Figure 4.

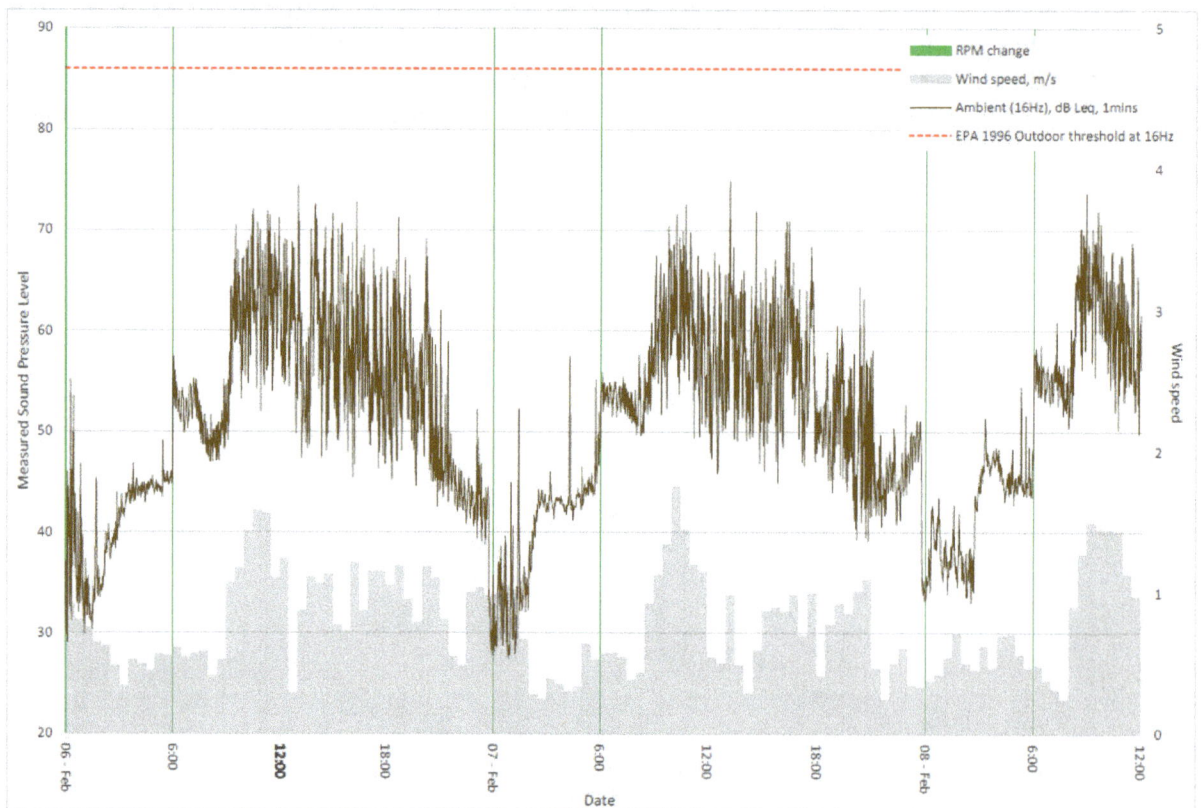

FIG 4 – Measured level at 16 Hz adjacent to the vent fans – December 2021 (source: AECOM, 2021).

The modification to primary ventilation fan speeds has achieved a measurable reduction in LFN intensity at source of up to 19 dB at 16 Hz and 12 dB at 20 Hz. The reduction of these frequencies at receivers in the south is typically between 5–10 dB.

Operation of the primary ventilation system was modified by reducing the speed of the fans from 920 rev/min to 400 rev/min between the hours of midnight and 06:00 am. The modification of fan speed from 920 rev/min to 400 rev/min resulted in a significant reduction in underground mine air flow/ventilation from 500 m³/s to 220 m³/s (approximately 56 per cent reduction). The reduced fan speed required modifications to underground activities, during those times, such that ventilation standards were maintained, and heat levels were managed.

The following operational changes were implemented as a result of the ventilation fan speed modifications:

- No truck haulage in the underground mine areas during periods of reduced fan speed.

- No loader operations in mine areas, apart sporadic short periods.

- Other production and development activities were restricted to match the available airflows.

- Mining activities in some mining areas were restricted such that only one mine level can be operated at a time.

- Modifications and restrictions on daily firing times for development and production.

The modifications outlined above resulted in a change of shift start and finish time for all underground operators, particularly the load and haul crews. Shift start times were brought forward by half an hr to 6:30 am/pm for all operators, and the night shift load and haul crew was changed to 1:00 pm to 1:00 am (previously 7:00 pm to 7:00 am). Approximately 40 truck and loader operators across the four mining crews were affected by this roster change, which significantly and abruptly impacted their lifestyle and families, including changed routines for school drop off/pick up and childcare arrangements. The overlap period between 1:00 pm and 6:30 pm and modified firing times, was required to minimise the impact on productivity during the period with unrestricted mine ventilation.

The impact to productivity brought about by the modification to the primary ventilation system has resulted in a 15–20 per cent reduction in projected underground activities while the modification remains in place.

LFN MANAGEMENT OPTIONS

FGM has progressed a program of investigations and modifications to minimise the risk of low frequency noise (LFN) emissions from the southern surface ventilation system fans, so far as reasonably practicable. The low frequency noise control measures implemented include:

- Fan silencer baffle re-configuration.

- Fan motor housing modifications and additional abatement.

- Fan speed offset modifications.

Fan silencer baffle re-configuration

Each surface fan has been fitted with reactive dissipative type silencers with two pure reactive cells at inlet end of each splitter, refer to Figure 5. Primary airflow passes over a total of eight splitters located within the 5.0 mW × 5.0 mH by 5.5 m long silencers before being discharged to the environment.

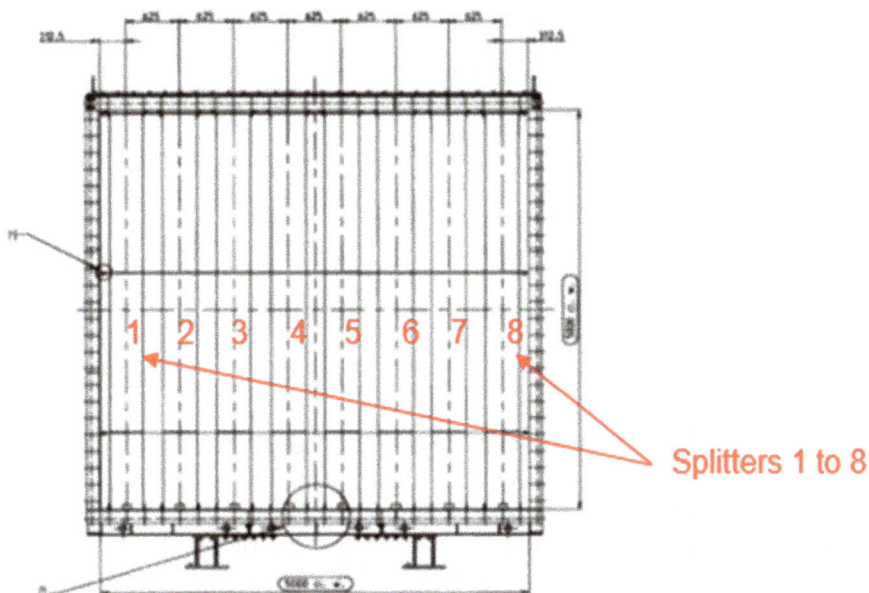

FIG 5 – Silencer splitter box.

In certain circumstances the silencer not only attenuates fan noise ie blade passing frequency (BPF) and its harmonics but can under certain circumstances generate noise. This self-noise is caused through vortex shedding and eddies at the splitter terminations.

The acoustic performance of a silencer is determined, by the splitter width, the width of the air passage and the length to width ratio of the passage. The passage velocity and hence the silencer pressure drop is controlled by the silencer cross-sectional area. Barus Consulting (2022) found that the silencer open area ie (air gap/(air gap + splitter width)) is typically in the range of 39 to 67 per cent.

The existing silencers are fitted with eight 400 mm thick splitters and 225 mm air gaps with a resultant open area of 32 per cent giving a passage velocity at the maximum design airflow of 565 m³/s of 31.4 m/s.

Investigation of the fan system discovered a higher-pressure loss measurement through the silence than assumed in the design phase. Pressure loss is not the only parameter controlled by passage velocity, the silencer regenerated, or self-noise varies as the 6th power of the passage velocity. To reduce the silencer pressure loss closer to the pressure loss estimated during the design phase required the removal of two splitters from eight to six reducing the passage velocity to 27.7 m/s.

The removal of two splitter results is a reduction in the silencer attenuation giving predicted increase of approximately 7 dBA in the sound pressure level at the nearest receptor.

However, this SPL increase is in the audible range at the fan blade passing frequency and its harmonics, at 920 rev/min is 252 Hz. Figure 6 shows predicted noise level comparison with eight and six splitters. This indicates that with 2 splitters removed the regenerated noise is still controlling the total noise below 20 Hz, but it is significantly reduced.

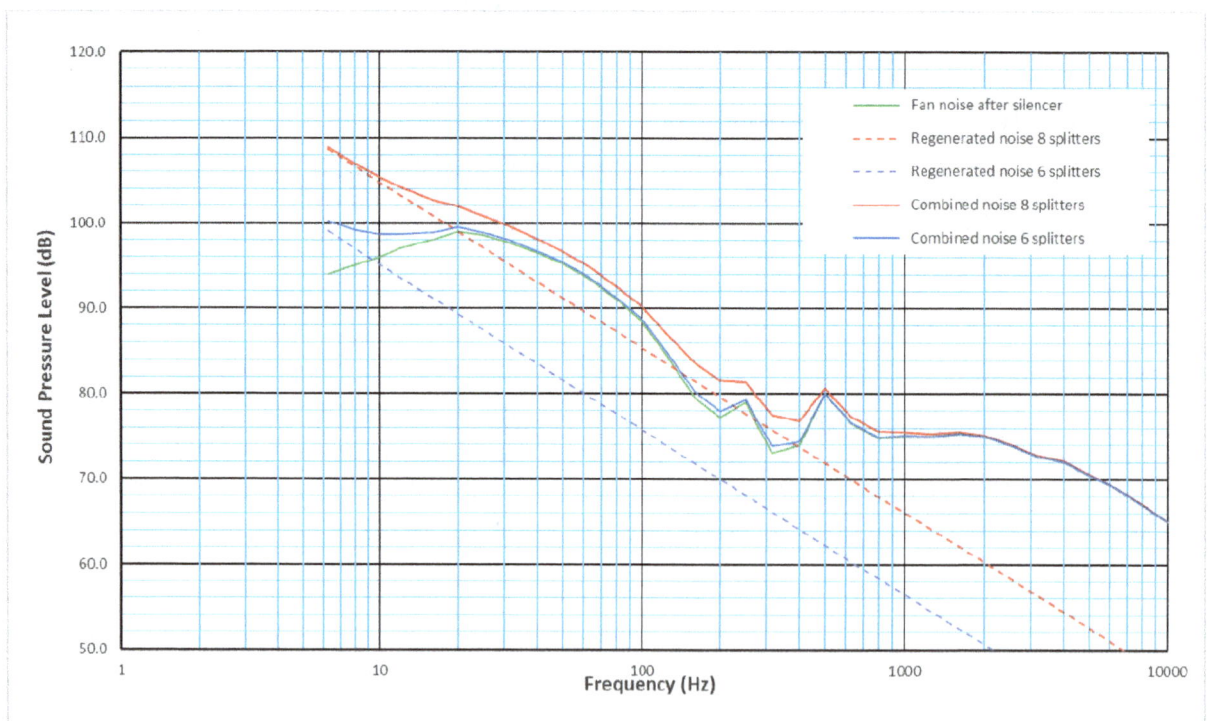

FIG 6 – Predicted splitter reduction noise (source: Barus Consulting, 2022).

Further investigations indicate that to effectively reduce the self-noise from the existing silencers requires the passage velocities to be reduced to below 20 m/s, thus reducing the regenerated noise levels more than 10 dB below the silenced fan noise. To reduce the audible noise and to make up the reduction in attenuation through the existing silencers a second additional pure reactive silencer would be required shown as silencer one Figure 7. To address the low frequency noise a third larger silencer could be added shown as silencer two in Figure 8.

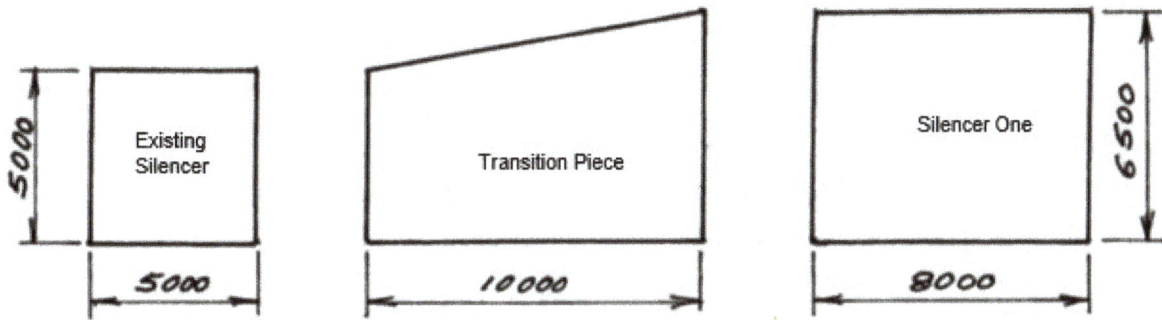

FIG 7 – Transition existing silencer to silencer one.

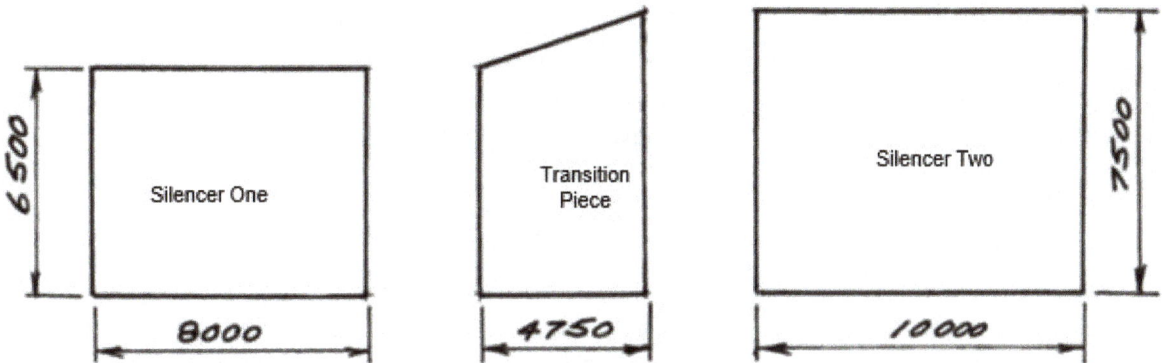

FIG 8 – Transition silencer one to silencer two.

Fan motor housing modifications and additional abatement

Abating low frequency noise emissions from the ventilation fan motors was recommended by the ventilation fan manufacturer. This was highlighted with the use of an acoustic camera during the initial investigations as shown in Figure 9, however the validity of the data was unreliable.

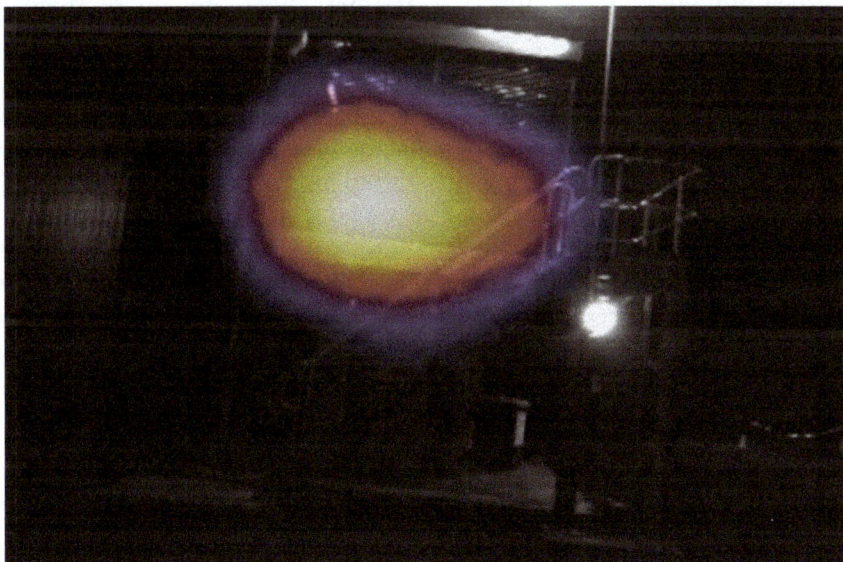

FIG 9 – Acoustic camera imagery at motor enclosure.

January 2023, FGM completed the installation of the additional fan motor enclosure abatement. The original fan motor housing louvers that allowed for air flow to cool the motors did not have any engineered noise abatement properties. These were replaced with bespoke acoustic louvers that were designed by Fan manufacturer and acoustic specialists, as shown in Figure 10. Structural modifications were made to the motor enclosure to allow for the installation. The inside of the enclosure was lined with steel to further attenuate noise produced by the fan motors.

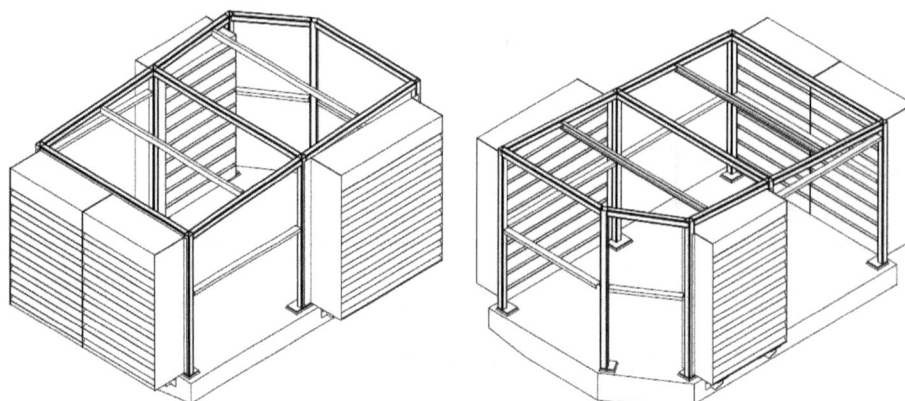

FIG 10 – Bespoke acoustic louvres for fan motor enclosure.

Following noise monitoring, It was found that solid reduction in the A-weighted level (7–9 dB), but no material change in the 16 Hz motor sound power following the completion of the enclosure works, refer to Table 4.

TABLE 4

Summary of Motor SWL pre- and post-measurements.

Item	Pre-louvre L_{WA}	Post-louvre L_{WA} (morning/afternoon)	Pre-louvre 16 Hz L_w	Post-louvre 16 Hz L_w (morning/afternoon)
Fan A motor	98	87 (87/87)	92	93 (93/92)
Fan B motor	96	89 (89/89)	98	99 (100/99)

Fan speed offset modifications

In accordance with EPA Victoria Publication 1996 – Noise Guideline: Assessing low frequency noise (LFN Guideline), consideration should be given to characteristics that can increase the effects of low frequency noise, which may include: its character such as the presence of tones, fluctuations, or pulsing.

During a detailed structural borne noise investigation undertaken in Q4 2022, FGM identified a 16 Hz noise oscillation emitted from the surface ventilation system. The oscillation can be described as fluctuating or pulsing noise characteristic. The identification of the oscillation presented an opportunity to change the character of low frequency noise emitted by the surface ventilation system. In collaboration with the fan manufacturer and acoustic specialists, FGM progressed modifications to the electrical infrastructure and associated software to enable changes to the fan speed control system to permit fan operation at offset speeds. During February 2023, FGM applied a speed offset to the ventilation system, whereby the two fans operated at marginally different speeds, typically offset by 15 or 20 rev/min. This modification balanced the flow velocities, fan power, fan pressures and aligned the fundamental acoustic beat of each fan.

Varying the fans speed resulted in a variation of the oscillation frequency, Figure 11 shows the effect of varying the fan speed. During this period fans were operating at the following speeds.

- 0000–0600 hrs both fans at 920 rev/min.

- 0600–1315 hrs both fans at 950 rev/min.

- 1315–2359 hrs.

  - Fan A 950 rev/min.

  - Fan B 935 rev/min.

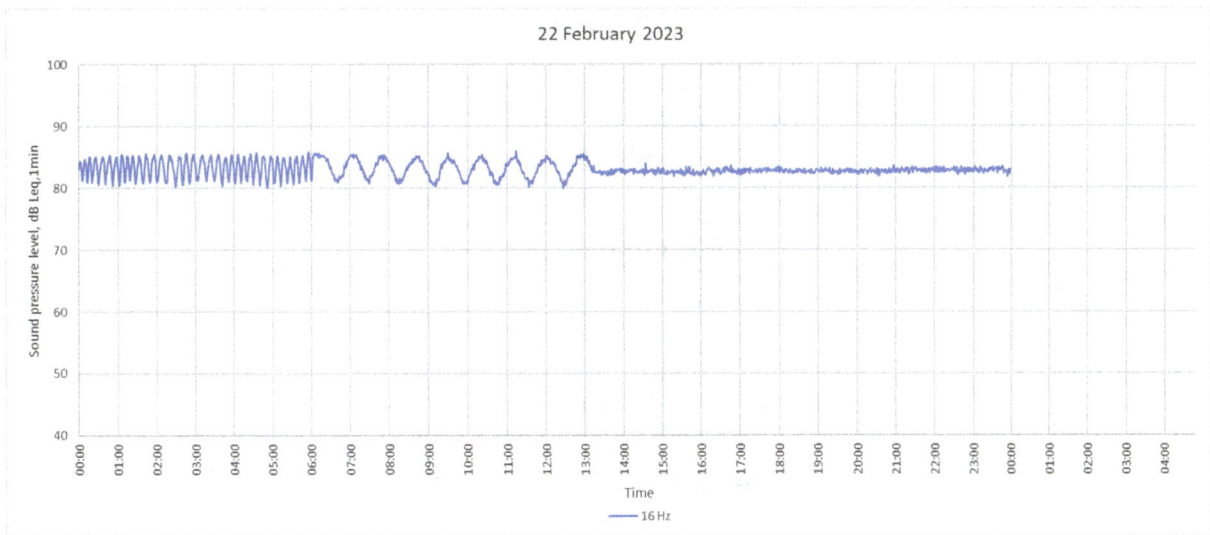

FIG 11 – Varying 16 Hz noise characteristic with offsetting fan speed.

At 1315 hrs, the fan speeds were separated by 15 rev/min, the modulation was removed, and the noise level remained constant at 3 dB below the peak level, which occurred when the fans were at nearly the same speed but with differences sufficient to create beats. Although a 3 dB difference between 83 and 86 dB may not seem significant, it represents a 100 per cent increase in acoustic energy due to the logarithmic nature of the decibel scale.

The decibel scale is used to measure noise because it reflects how the ear perceives sound, which is not linear. A doubling of acoustic energy results in a 3 dB increase in noise level, but for a sound to be perceived as twice as loud, it must increase by 10 dB, a ten-fold increase in energy level. This method is typically used for measuring audible noise. However, for low-frequency noise below 20 Hz, which is not audible, the concern is more about how the body and physical objects, like windows, react to the noise.

In this context, a 3 dB difference is significant and would likely be noticeable in the monitored residences. The variation in measured noise levels over significant periods could also have influenced the analysis of other modifications made to the ventilation system to reduce low-frequency noise, such as reducing the number of splitters in the silencers and removing discharge flow deflectors. These modifications showed little noise reduction, but the expected 3 to 5 dB improvement might have been masked depending on the timing of the measurements.

MODIFICATION ASSESSMENT

Evaluation of the effectiveness of the implemented controls demonstrated the LFN levels at 950 rev/min are similar to the LFN levels pre-modifications at 400 rev/min. Consequently, the risk of harm due to low-frequency noise (LFN) at 950 rev/min (post-modification) is broadly equivalent to the risk of harm due to LFN at 400 rev/min (pre-modification), an operating condition previously accepted by EPA Victoria. Noise monitoring also showed the measured LFN levels at receivers when operating at 950 rev/min post modification, remain significantly below the threshold levels described in EPA Victoria Publication 1996 – Noise Guideline: Assessing low frequency noise.

Noise measurement data taken at sensitive receivers were filtered to select nights with relatively low observed wind speeds (ie ideally less than 5 m/s in accordance with EPA Victoria Publication 1996).

FGM site was working under normal operating conditions during the nights of measurement, with all major plant operating normally and no maintenance or project related activities occurring. The monitoring results are shown in Table 5.

TABLE 5

Post modification noise monitoring results.

	Lowest ambient noise level, LZeq, dB	
	16 Hz	**20 Hz**
Pre-modification		
400 rev/min	40	39
920 rev/min	50	48
Post-modification taken during the 2023 performance evaluation		
400 rev/min	38	34
950 rev/min	44	38

At FNB9, the post-modification noise levels at both 400 rev/min and 950 rev/min are noticeably lower than the pre-modification levels at 400 rev/min and 920 rev/min, respectively. Reductions of up to 10 dB has been observed.

CONCLUSIONS

When considering major ventilation infrastructure such as fans, noise assessment is typically carried out in the A- weighting at frequencies above 63 Hz up to 10 kHz. EPA Publication 1856 Reasonably Practicable states the duty holder's requirement to have and seek out knowledge about the risk activities pose to the environment and human health. The knowledge that a duty holder's actions are assessed against includes:

- What they actually know.

- What someone in their circumstances should reasonably know about the risks.

This information forms the 'state of knowledge' and defines the scope of a duty holder's duties. If there are known, effective controls readily available in industry, it is considered reasonably practicable that these controls are employed. Conversely, there may be little or no guidance or industry precedent for controlling or mitigating a risk, especially an unforeseen risk. In such circumstances the state of knowledge is low, and this is reflected in the determination of reasonably practicable. Nevertheless, the duty holder must continue to investigate and advance the state of knowledge and act on this information as it becomes available.

FGM has contended with multiple technical challenges throughout the surface ventilation noise abatement project, including but not limited to:

- The sound pressure levels of the frequencies of concern are well below the EPA Low Frequency Noise threshold levels, as per EPA Publication 1996, and subsequently were not considered to present a risk of causing adverse impact on human health based on previous LFN investigations.

- There is no industry precedent for abatement of the primary frequencies of concern – No known industry cases were available to draw on for guidance.

- There is no industry precedent for silencers that can abate the primary frequencies of concern. This was reflected in the comments by acoustic engineers at the time *'Infrasonic noise attenuation does not exist in the world at this scale, and there is no literature or research available. All possible measures, have to be done in stages and either tested onsite, or others requiring 1:1 scale tests'*.

The state of knowledge for engineers when considering major ventilation infrastructure installations must consider the levels of low frequency noise below 63 Hz in 1/3 octaves that may potentially be emitted from the installation. Mine sites should also extend their level of knowledge to include current levels of LFN for comparative analysis.

ACKNOWLEDGEMENTS

The author would like to acknowledge Mr Bruce Russell – Barus Consulting for his expertise, and guidance in compiling this paper.

REFERENCES

Barus Consulting, 2022. Investigation into community noise generated by the ventilation system, Unpublished internal company document.

AECOM, 2021. Fosterville Gold Mine Low-Frequency noise investigation, Unpublished internal company document.

Reducing DPM without increasing ventilation

D Rose[1] and F Velge[2]

1. Business Development Manager, Pinssar, Brisbane Qld 4102. Email: drose@pinssar.com.au
2. Managing Director, Pinssar, Brisbane Qld 4102. Email: fvelge@pinssar.com.au

ABSTRACT

As the awareness of the risks associated with diesel particulate matter (DPM) increases, so too does the focus from regulators and government jurisdictions to reduce Occupational Exposure Limits (OEL's).

In mines throughout the world, the risk owners for air quality are the Mine Managers and the Ventilation Engineers. Traditionally, the expectation across the underground mining industry is that the responsibility for DPM mitigation primarily starts with the Ventilation Engineers – with a 'dilution is the solution' mentality.

For various reasons, the Ventilation Engineer(s) may not have the additional ventilation capacity to provide the air required to adequately deal with the emissions in their mine and therefore are looking for additional tools to work in parallel with the ventilation.

This paper explores some real-life examples and case studies of mines on multiple continents using an array of controls to successfully reduce their DPM levels. These examples demonstrate how mitigating DPM is not the sole responsibility of the Ventilation Engineer(s), but is an all-inclusive process, driven by the site based DPM Committee. The mines in the examples have shown world leading methodologies and have not only aimed to be under the industry OEL but have strived for as low as reasonably practicable (ALARP) / as low as reasonably achievable (ALARA) to benefit the health of their workers.

The examples and case studies in this paper include demonstrations of how these mines used Continuous DPM monitoring to manage their controls, by implementing trigger action response plans (TARP). The paper also demonstrates how mines have used the data from Continuous DPM Monitoring to validate their on-site trials of mitigating technologies, including before and after implementations.

The examples will also show how some mines have introduced 'soft' controls to reduce their DPM, which have been more about behavioural actions, rather than introducing technologies.

INTRODUCTION

Primarily, diesel particulate matter (DPM) is diluted by generating more air. However, increasing ventilation is not only capital and operational expensive, it can also substantially increase the carbon footprint of an operation. Bugarski *et al* (2012) states that using ventilation as a single means of controlling DPM exposures would be cost prohibitive in most applications. As companies are driving to reduce their emissions, alternative sources of controls for DPM are being introduced more frequently.

Additionally, a mine may also be at a point in its design life where the ventilation is near capacity, or are operating near capacity until a new shaft is commissioned. Manos (2010) explains that topically by themselves, the improvements in ventilation capacity and distribution to the work areas are neither sufficient nor economically viable methods for reducing the exposure of underground miners to DPM. Hardcastle and Kocis (2004) and Pritchard (2010) highlight that the distribution of the air can be improved based on continuous monitoring of the activities and frequent evaluation and redirection of flow toward working areas with high diesel-powered equipment activity. Bugarski *et al* (2012) also explain for controlling exposures to DPM, dilution by ventilation air is considered to be an essential part of the solution, but it is just one part of the multifaceted approach involving many potential solutions.

Bugarski *et al* (2012) identifies over 40 controls to reduce DPM throughout the publication *Controlling Exposure to Diesel Emissions in Underground Mines*. This paper will be reviewing how underground mining operations were able to reduce their DPM emissions without adding additional air by utilising some of these 40 plus controls in conjunction with monitoring DPM continuously in real time.

BACKGROUND

According to the International Council on Mining and Metals (ICMM; 2022), Diesel-powered mining vehicles can account for anywhere up to 80 per cent of direct emissions at a mine site, depending on the site geography and commodity being mined. Diesel remains the main source of energy in most mines due to a range of reasons varying from efficiency to remoteness of the projects. The same article by ICMM (2022) identifies that there is a genuine interest in moving to newer technologies that will either emit less or eliminate diesel as power source. However, many mining projects will not see a substantial move away from diesel powered vehicles for the foreseeable future.

The issue around diesel particulate matter (DPM) and its associated health risks is nothing new. In 1988 the National Institute for Occupational Health and Safety (NIOSH) in the USA identified that DPM might have a potential to have carcinogenic effects to humans. Subsequent research eventually led to the International Agency for Research on Cancer (IARC; 2012) announcing via press release that the IARC, on behalf of the World Health Organisation (WHO), were declaring DPM as a Group 1 carcinogen.

The WHO 2012 declaration combined with efficiency improvement in the diesel engines has seen substantial downward shift in Occupational Exposure Levels (OEL) for DPM. In most jurisdictions, the mass calculation is recognised as the method of measurement. However, as shown in Table 1, there is no uniform recognition as to what that safe number should be. In the countries that have an OEL for DPM, elemental carbon (EC) particles are used as the reference number. In other countries like the USA the reference is total carbon (TC) which is a compilation of elemental carbon + organic carbon. In both cases, EC or TC are expressed in mg/m^3 or $\mu g/m^3$. The OEL's vary across the globe from as low as 0.05 mg/m^3 / 50 $\mu g/m^3$ in Europe to 1.5 mg/m^3 / 1500 $\mu g/m^3$ in some provinces in Canada.

TABLE 1

DPM OEL's applicable to mining in western countries.

Global DPM OEL's				
Country	Province	mg/m^3	$\mu g/m^3$	TC/EC
Europe (2023)	ALL	0.05	50	EC
UK/Australia	ALL	0.1	100	EC
North America	ON	0.12	120	TC
	USA/SK	0.16	160	TC
	QC/NL	0.4	400	TC
	BC, NB, NS, YK, NWT, NU	1.5	1500	TC

Whilst changes to OEL's do take time to legislate, some of these regions have reduced their OEL's in recent years, including the European Agency for Safety and Health at Work (2004) who in 2019, amended the Directive 2004/37/EC, which came into effect in February 2023 as well as the Ontario Ministry of Labour (2020), who introduced new rules which came into effect September 2023, which is the lowest of any province or state in North America.

Additionally, some specific regions or specific mining companies are taking the proactive steps of mandating OEL's which are lower than the country they operate in. An example of this is the coal mining region of the Bowen Basin in central Queensland, Australia – who have legislated as 0.05 mg/m^3 / 50 $\mu g/m^3$. Another example is mining company BHP, which has an OEL set of as 0.03 mg/m^3 / $\mu g/m^3$ which is detailed in a case study further below in this paper.

Whilst these differences in carbon references of EC and TC can cause considerable discussion at times and operations in various regions are required to focus on EC, Zielinska *et al* (2002) observes many of the organic compounds present in DPM are individually known to have mutagenic and carcinogenic properties. Regardless, it is well understood that DPM is seen as high-risk pollutant and should therefore be considered under the *as low as reasonably practical* (ALARP) protocol. That is to say that whilst we have OEL's for DPM, the aim should always be to reduce the risk to its lowest potential.

Over the years there have been several methods to measure DPM exposure. It is generally accepted that these fell under three categories. Firstly, there is the personal sampling. This measurement will indicate how much EC a person was exposed to during the one shift. The method utilised for this is the NIOSH 5040. The second method is tailpipe monitoring, this will indicate the health of an equipment exhaust and DP filter. The third method is handheld real-time monitors.

Whilst those three methods are either required for compliance and/or quantitatively useful, they are point in time measurements and none of these are continuous. The three methods that are currently used are also prone to human error and can only rarely be duplicated, which can add to the error factor and would make it reasonably difficult to make long-term decisions, as the quality of the data is variable compared to fixed continuous that has no human interference. Fenske (2010) and Lim and D'Souza (2020) describe that since personal samplers were implemented in the 1960s, personal sampling has become a widely accepted practice (or, rather, the reference method) for exposure assessment in occupational hygiene. Fenske (2010) and Pearce and Coffey (2011) also observe that traditionally, personal sampling depends on relatively slow turnarounds between sample collection and subsequent laboratory analysis, which uses standardised methods to generate results, and this can limit the optimal implementation of workplace-risk-mitigation strategies in terms of promptness and efficacy

The environment in a mine is constantly dynamic. Mobile equipment come and go, the conditions of the mobile equipment can change within seconds due to breakdown, extended periods of idling, convoying, etc. Also, the DPM controls that the operation has in place to reduce pollutants might be defective or inefficient due to ware and tare or breakdown. In discussions with mines it was often highlighted that the current sampling methods were good at identifying there was a problem but were of little use in making sure that the operation remained safe 24/7. Some mines consider DPM particles as part of their dust mitigation, however DPM controls usually are quite different to dust controls. Kittelson *et al* (1998) states that most aerosols emitted by diesel engines and other combustion aerosols are submicrometer in size. Cantrell and Volkwein (2001) also explains that diesel aerosols are, in general, one order of magnitude smaller than other respirable aerosols generated in mines from mechanical processes, generally known as dust. Bugarski *et al* (2012) notes that because of their size, DPM aerosols behave in a manner similar to the surrounding gases. They have much longer residence times than larger mechanically generated particles that are removed from the atmosphere quite quickly by gravitational settling.

As mentioned previously, Bugarski *et al* (2012) identified over 40 controls to reduce DPM within an operation. A compiled list of these controls can be found on the Pinssar website (https://pinssar.com.au/wp-content/uploads/2021/11/Hierachy-of-Controls-1-pdf.pdf). However, to identify which of these controls are the most beneficial in reducing DPM at an underground mine, it was generally recognised that continuously monitoring the pollutant was essential first.

Howard *et al* (2022) explains the advances in sensor and communication technology are allowing equipment manufacturers to develop placeable sensors and utilising the new channels of communication to deliver the data to the right person at the right time so to make timely decisions on the quality of the air throughout the operation.

The conventional way of reducing DPM is to increase the ventilation. However, designing and implementing new ventilation systems to match the DPM requirements necessitate long planning periods and capital commitments. The Department of Mines and Petroleum (2013), explains that there are regulatory requirements regarding the air quantity required for underground mines when operating diesel equipment. However, the removal of DPM can involve more complex ventilation design beyond that necessary to meet the legislative requirements for air supply. Adopting good ventilation design standards should be part of the holistic approach to diesel emission management.

Furthermore, running ventilation systems requires a lot of energy. Therefore, more and more ventilation officers require reliable and time relevant information to make the decision they need to take in making sure that the quality of the air is within the operation is at acceptable levels and provide the right air to the people 24/7.

This paper was compiled with information collected from four mines across the globe that used continuous real-time DPM monitoring, this paper will demonstrate how these operations were able to control and reduce DPM emissions by utilising the data from continuous DPM monitoring.

CASE STUDY #1 – RESPONDING TO DPM CHANGES IN REALTIME DURING A LONGWALL MOVE

Background

The Broadmeadow coalmine is located in central Queensland, Australia, 34 km north of the town of Moranbah. The mine was opened in 2005 and is a part of BMA, the 50:50 joint venture between BHP and Mitsubishi. BHP have been a world-leading mining company in relation to the research and governance of DPM.

Both BHP and coalmines in Central Queensland have been at the forefront of management of DPM. Hedges, Djukic and Irving (2007) explained that in 2001, underground coalmines in Central Queensland began to actively monitor personal exposures to DPM. In the Queensland underground coal mining industry, a steering committee was established in February 2002 with representatives from the Inspectorate, industry and the Construction, Forestry, Mining and Energy Union (CFMEU). The committee's objectives were to overview the development and introduction of management plans to minimise the exposure of coalminers to diesel exhaust pollutants including DPM.

Following the WHO declaration in 2012, BHP were quick to be proactive on the topic of DPM. In 2013, BHP commissioned the Driscoll Review, which led BHP to applying the 50 per cent rule throughout their operations globally. This 50 per cent rule was based on the Australian Industry OEL of 100 µg/m³, which resulted in BHP adopting a company-wide OEL of 50 µg/m³.

McDonald (2015) explained further reviews followed in 2014 and in 2015 thru 2016 BHP engaged the Institute of Occupational Medicine to conduct a review of the published literature available. Following this review, BHP mandated a company-wide OEL of 30 µg/m³ in late 2016.

This increased focus on DPM across all BHP sites saw improved declines in DPM exposures. Everson and Batterson (2018) presented at Queensland Mining Industry Health and Safety (QMIHS) conference showing at the Broadmeadow mine, the average DPM recording in 2013 was 98 µg/m³, this improved to be only 12 µg/m³ in 2017. This presentation was recognised with a Highly Commended Award at the conference. The reduction of DPM at Broadmeadow site was largely due to the creation of a site based DPM committee, who was able to get the entire workforce engaged and trialled and implemented initiatives. It is important to acknowledge that the DPM committee could have dissolved quite quickly if they had not campaigned to be world firsts in coal for a lot of new ideas and technologies, as the results provided further momentum and enthusiasm.

Aside from the implementation of the 30 µg/m³ in late 2016, BHP corporate have promoted that the limit for DPM should be *as low as technically feasible* and have encouraged their assets to continue implementing initiatives to further mitigate DPM. In 2019 the BHP group purchased several fixed, continuous DPM monitors, including one by Broadmeadow mine site. These DPM monitors were continuous, real-time monitors suitable for harsh underground environments, which was developed by Pinssar Pty Ltd. These purchases meant BHP was one of the earliest adopters of this technology, as the Pinssar system was commercialised and released to market in 2018.

At Broadmeadow mine, the main focus on DPM was (and still is) during the task of Longwall moves. This focus is based on historical data from personal sampling across all mines in the region published by the local regulator. The mines inspectorate has been collecting and reviewing diesel particulate matter data in underground coalmines since the early 2000s. The Department of Natural Resources, Mines and Energy (2019) reported that despite the improvements observed over time, longwall move activities continue to represent the highest risk similar exposure group (SEG) in underground coalmines.

This case study will explore the journey Broadmeadow mine took with using the continuous DPM monitors to assist in further lowering DPM levels.

Methodology

Considering Broadmeadow mine had purchased DPM monitoring technology which had not been used before, the mine first wanted to test this technology before implementing underground. In 2020 the mine conducted testing and trials on the surface (Figure 1) which concluded '*evidence of linear correlation between Pinssar and NIOSH-5040 when comparing average Pinssar results*'. The testing also resulted in the mine concluding that the Pinssar data should not be assessed on every single sample, but rather a '*mode of interpreting the Pinssar data for control verification purposes that may involve collating & reporting longer term averages (ie 1–2 hourly averages)*'.

FIG 1 – Testing on the surface at Broadmeadow mine.

Considering the main focus on DPM was during the task of Longwall moves, the mine decided to install the one monitor they had underground just prior to their longwall move of September 2021. It was decided to let the monitor run over the entire longwall move to record samples, however these samples were not viewed in real-time. Once the longwall move was completed and the mine was back into production, the mine received the results from their personal NIOSH 5040 monitoring regime. The results from the NIOSH 5040 personal monitors showed that there had been 23 exceedances above the company OEL of 30 µg/m³. It is important to note that none of these 23 samples exceeded the Australian OEL of 100 µg/m³, however the mine was still keen to learn more about these 23 samples.

A retrospective review was conducted comparing the data from the personal monitors with the data from the Pinssar continuous monitor. The comparison is shown in Table 2. The data from the personal monitors showed that 18 of the 23 exceedances recorded occurred within a one week period from 6th October to 12th October. The data from the same period was exported from the Pinssar monitors and compared. The range of differences between the two technologies for these six days was from 0.7 µg/m³ to 12.6 µg/m³.

TABLE 2

Comparison of data.

Date	Shift	Qty gravimetric samples taken	Average value from NIOSH personal gravimetric samples (µg/m³)	12 hr shift average from Pinssar monitor (µg/m³)	Diff (µg/m³)	Did Pinssar 12 hr shift avg exceed BHP OEL 30 µg/m³?
13-Oct	Day	4	57.2	56.4	0.7	Y
6-Oct	Day	2	45.6	42.1	3.5	Y
12-Oct	Night	3	51.7	43.9	7.8	Y
11-Oct	Night	2	52.9	44.5	8.4	Y
6-Oct	Night	1	39.5	28.9	10.6	N
14-Oct	Day	3	48.6	36.0	12.6	Y
12-Oct	Day	3	48.9	17.8	31.0	N

After reviewing the worker logs, there was one day of data which was not deemed comparable, as on this day the site Operators wearing the gravimetric method personal monitors were exposed to a piece of mobile equipment which was found to have a leaking fuel pump. On this day, the difference between the data sets was 31.0 µg/m³, which relates to the fact the Operators were in close proximity to the leaking fuel pump over an extended period of time, whilst the Pinssar monitor was not.

Further observations were made in relation to this comparison of samples:

- The location of the Pinssar reader was in a fixed location @ 10 ct. This fixed reader provided an indication of the DPM levels @ 10 ct on a continuous basis.
- Personal monitor (NIOSH 5040 gravimetric method) samples occurred whilst operators were in the MG15 and MG16 areas.
- MG15 and MG16 areas are inbye of 10 ct. The DPM levels recorded @ 10 ct are an indication of what levels of DPM are heading into MG15 and MG16 areas.
- During the LW move, the MG15 and MG16 DPM levels will be higher than that @ 10 ct, due to activity of diesel equipment in MG15 and MG16 areas.
- The technologies of the personal monitors and the fixed monitors are different.

Upon completion of the review, the conclusion was that if the Pinssar monitor had been viewed in real-time, this would have provided the mine with data to alert the operators that DPM levels were rising and the mine could have responded in a timely manner, potentially eliminating these exceedances.

The report was presented to the Broadmeadow DPM Committee meeting in Feb 2022. During this committee meeting, the following decisions were made in preparation for the 2022 longwall move.

Decisions from Broadmeadow DPM Committee meeting, Feb 2022:

- Increase Pinssar sampling during next LW move, two more monitors purchased.
- Sampling to be in the in Mains and LW take-off (when possible, ERZ v NERZ).
- Site to actively view the Pinssar data in real-time during the longwall move.
- Develop a trial TARP, including assessment of various trigger points.
- Underground District Superintendents are to be notified in real-time when trends start to increase so that they can investigate.

Results

The Pinssar systems were installed underground in September 2022, prior to the longwall move. The longwall move commenced in late September and was completed in October. During this longwall move, the data from the Pinssar monitors were actively viewed in real-time during the longwall move. The 1 hr rolling shift average was the output which was viewed in real-time, which provided the District Superintendents with an warning that was early enough for them to be able to respond to.

The results of this process was that the mine had dramatically reduced their exceedances compared to the previous 2021 longwall move.

2021 Longwall move (not actively using the Pinssar data in real-time):

- Result = 23 exceedances (BHP OEL of 30 µg/m³) recorded by personal samples.

2022 Longwall move (actively using the Pinssar data in real-time):

- Result = five exceedances (BHP OEL of 30 µg/m³) recorded by personal samples.

It was concluded by the mine that the Pinssar system was an important contributor to the significant reduction in exceedances, as the alerts enabled the underground District Superintendents to make changes to the underground environment in real-time before the rise in DPM developed into an exceedance.

Along with the continuous DPM monitoring, other controls and new technologies were also trialled and tested with great success. Modifying its existing fleet of diesel-powered equipment by installing low emission engine upgrades was a major focus. This repowering of machines to low emission standards along with filter technologies provided a significant improvement and the DPM committee having the determination to use the untested methods ie Pinssar and electric vehicles in coal mining has been commended.

It must also be highlighted that the Broadmeadow mine was only able to introduce these new technologies by conducting thorough, detailed and collaborative work risk assessments before introducing, trialling and testing of any new equipment could commence.

Another control implemented by Broadmeadow mine site was to ensure that the contractor's diesel fleet used to do the longwall move was developed to an improved and acceptable standard. Whilst this initiative was an idea which was borne from discussions at the on-site committee meeting, it is also a control which is promoted by Bugarski *et al* (2012) which says the contribution of those vehicles to the overall DPM load should be considered and when possible the contractors fleet should be held to the same emissions standards as the host fleet. This initiative involved detailed collaboration between Broadmeadow and the mine and has resulted in other mines benefitting from a contractors fleet which has a much improved DPM emission output.

Additionally, an administrative control was put in place which included the training of operators to perform their tasks as much as possible in locations that would not be effected by upstream emissions from nearby vehicles. Whilst these locations may seem logical, the reinforcement of training and DPM awareness is a valuable tool for underground operators.

As more and more industry research and monitoring continue, mines will become further educated on how DPM behaves in a confined space and the positioning of operators mentioned above can be validated and/or refined. An example of this research is shown in Morla (2018), the study and modelling of DPM flow patterns using the techniques of computational fluid dynamics (CFD) and validated using field experimental measurements. During this study, DPM field monitoring with a man-riding vehicle (150 kW), the vehicle was stationary and the velocity of ventilation air in the roadway was 1.26 metres per second. The results showed a high concentration of DPM on the exhaust pipe side of the roadway. At 10 m downstream of the vehicle the DPM particles were spread over the entire cross-section of the roadway

Future works

Considering the positive outcomes of using the Pinssar systems in the 2022 longwall move at Broadmeadow, the mine has decided to incorporate this continuous, real-time fixed monitoring as part of their regular longwall move activities. This was replicated in the subsequent 2023 longwall move. Additionally the mine plans to investigate the use of continuous, real-time fixed monitoring for tasks outside of longwall moves, potentially in development area's – which is seen as the second highest DPM risk in a longwall mine, along with continuing to investigate additional mitigating controls.

CASE STUDY #2 – REAL-TIME TRAFFIC MANAGEMENT UNDERGROUND

Background

The Stillwater (including Stillwater east) and East Boulder metalliferous mines are underground mining operations, located near the towns of Nye and McLeod in Montana, USA. The mining assets are owned and operated by the Sibanye Stillwater group and the mines are located on the front range of the Beartooth Mountains with elevations exceeding 2700 m above sea level. These two mines are the lone palladium and platinum producing mines in North America.

Both underground operations were originally mined with some narrow shafts. Over time, as bigger mechanised equipment was introduced, the mine planners were aware that the narrow shafts meant that the emissions would be more confined, compared to what is regarded as a more conventional shaft widths of other mines. Increasing the shaft widths would be an expensive and time-consuming activity and would also require significant engineering. Considering this, the decision was made by

Senior Management to leave the shafts narrow with the understanding that emissions including DPM would need to be closely monitored and managed. This focus ensured that both operations would have the sufficient personnel, equipment and controls to continuously be under the legislated OEL.

For DPM, US legislation stipulates to be below 160 micrograms per cubic metre (0.16 mg/m^3) for total carbon (TC). Sibanye-Stillwater (2022) Integrated Report says to ensure compliance, each mining operation has an industrial hygienist to monitor engineering controls, administrative controls, and employee exposures. Further, the US PGM operations developed a DPM reduction strategy (called the P reduction strategy), which has a three-pronged approach to reducing diesel particulates:

1. Diesel engine maintenance.

2. Provision of adequate dilution ventilation.

3. Operational discipline such as traffic management.

To enable the US PGM Operations to continuously track the progress and success of the P reduction strategy, Pinssar continuous real-time DPM monitors were purchased.

Methodology

As described by Wu and Gillies (2008) and Hardcastle (2014), traditionally the number of vehicles in a section at any given time is regulated using a tag in tag out system. This system is based on using tags placed on a board at the entrance to a section to indicate the number of vehicles that have already entered that area. Bugarski *et al* (2012) then goes on to say additional vehicles are allowed to enter the section only when there is a spare tag position.

At Stillwater Sibanye Montana operations, to monitor and manage operational disciplines, the two mine sites (East Boulder and Nye) have a total of 15 Pinssar units to provide continuous real-time measurement.

Both mines view Pinssar readings in real-time against the USA OEL of 160 µg/m^3.

Mine site rules were put in place around the Pinssar system, the Engineering teams and control rooms both view the Pinssar data on dashboards and use the data for real-time, continuous traffic management.

Rubeli *et al* (2004) shows that although the heavy duty (HD) fleet is typically the primary focus of DPM emissions reduction programs for many mines, the substantial contribution of light duty (LD) vehicles to DPM load in typical underground operations should not be underestimated.

Both underground operations at East Boulder and Nye are large in size and distance and the Engineering team understood that the amount of travel from many light vehicles, such as Kubota's, contributed to the overall emissions load. Considering the majority of the workforce can be transported underground via bulk personnel transportation, the decision was made to limit the number of the light vehicles underground. Rules/mitigating controls were put in place to limit the number of trips by light vehicle operations to five each at any one time. These rules ensured that when the maximum number of five was reached, then only when one light vehicle comes out to the surface, then other can go underground.

Results

The achievements/successes of this control resulted in the DPM levels shown by the Pinssar systems dropped below 160 µg/m^3 – and as the operations further refined the rules, the operations were able to get the levels under 100 µg/m^3 – which was shown on the Pinssar's data.

The Sibanye-Stillwater (2022) Integrated Report says additionally, routine sampling was conducted throughout 2022, and sample results continue to demonstrate improvement in DPM mitigation practices.

Future works

The Sibanye-Stillwater (2022) Integrated Report outlines future plans by stating, the US PGM operations plan to purchase more Pinssar units for Stillwater to continue to get a more granulated picture of their DPM levels throughout the mine. As the units arrive, installation will be determined based on active mining areas and traffic patterns, to determine their optimal positioning in terms of generating leading indicators for mine air quality. The mines will also correlate the Pinssar and gas sensors for better air quality tracking in the future.

Clean fuel initiatives are being implemented at both mines, including filtering closed-loop systems in storage areas. Work continues at both mines to reduce emissions on the small vehicle fleet engines, including traffic management measures. Additionally, the testing of battery-electric LHDs and investing in lower or zero emissions utility vehicles to replace legacy vehicles is planned.

Also, new additional ventilation will be introduces in the West side of the operation.

The mines also intend to continue to progress and enhance understanding from Pinssar DPM monitors and develop the criteria for triggering timely corrective actions to reduce exposure; conduct side by side sampling to further define data relationship.

Future works also include the plan to develop an integrated health management plan, including a sampling schedule and health risk assessment process and further refine and act upon the installed gas and airflow monitoring equipment.

CASE STUDY #3 – REAL-TIME ENVIRONMENTAL MONITORING TO MANAGE FLEET MAINTENANCE

Background

The Khutala operation situated approximately 55 km south-west of Witbank in the Mpumalanga Province of South Africa. Khutala was commissioned in 1984 and has supplied product to the Kendal Power Station since November 1986. The underground portion is a bord and pillar operation and all underground coal is exclusively mined for Eskom's Kendal power station, situated close to the mine shaft. The open cut coal is supplied to both the Kendal power station as well the inland metallurgical market.

Bugarski et al (2012) says generally, the air quality of an underground mine can only be as good as that mine's maintenance of vehicles. A good preventative maintenance program will maintain near original performance of an engine and maximise the equipment productivity and engine life while keeping exhaust emissions at baseline levels. Equipment maintenance is recognised as an indisputable element in reducing the health and safety hazards associated with using diesel powered equipment.

The Khutala mine identified that vehicle maintenance played a major role in their underground emissions. Considering this, the engineers at Khutala embarked on a project to further strengthen the monitoring of vehicles and how they were emitting underground. The Khutala initiative did not only want to monitor DPM underground, they wanted to be able to use the data to improve the emissions underground, by understanding when a particular vehicle may have a change (increase) in emissions and then to be able to rectify this in a timely manner and not wait until the next scheduled maintenance interval. Bugarski et al (2012) advises the philosophy of 'repair on failure' which is still common in some mines, may provide short-term production and cost benefits, but in the long-term, it is best practice to use proactive engine maintenance to ensure cleaner and more efficient engines that run longer and have greater dependability.

The initiative and method developed by Khutala has sometimes been referred to as Emissions assisted maintenance in the mining and other industries. Spears (1997) says emissions assisted maintenance is based on two assumptions:

1. An improperly maintained engine can emit undesirable concentrations of exhaust emissions compared to a well maintained diesel engine.

2. On-site emissions test procedures may be employed to determine whether there is a need for engine maintenance.

When most people think of emissions maintenance, generally the automatic thought is to focus on the vehicle exhaust and filter system, also known as the exhaust aftertreatment system. Whilst the vehicle exhaust aftertreatment system should be a major focus, there are also other areas of the vehicle that should be considered as well, such as the air intake system and the condition of seals and crankcases.

Bugarski *et al* (2012) also states that in a mining application, the intake system is the most critical engine system affecting exhaust emissions. The intake system on a diesel engine must provide an adequate supply of clean air for good combustion at all operating speeds, loads, and operating conditions. This issue will be further compounded if the air quality or even air temperature where the vehicle is operating is compromised. Morla (2018) investigated flow patterns and concentrations of DPM under different ventilation an operating conditions in underground mines. DPM concentration variations were investigated by changing intake air temperature (20°C, 30°C and 40°C) and exhaust fume temperature (50°C, 60°C, 70°C, 80°C). The results showed that the temperature of diesel powered vehicle exhausts had a slight influence on the concentration of DPM on the downstream side of the vehicle. Increasing the temperature of the intake air increased the concentration of DPM at the downstream side of the vehicle.

Bugarski *et al* (2012) says additionally, because of the nature of diesel engine utilisation in underground mines, crank-case emissions can significantly contribute to a miner's exposure to DPM. Hill *et al* (2005) also states that crankcase emissions proved to be an extremely strong contributor to the overall particulate emissions.

Bugarski *et al* (2012) follows up by saying equipment maintenance is recognised as an indisputable element in reducing the health and safety hazards associated with using diesel powered equipment. A good preventative maintenance program will maintain near original performance of an engine and maximise the equipment productivity and engine life while keeping exhaust emissions at baseline levels. However vehicles which have a preventative maintenance regime should still be continuously monitored to quickly detect if a vehicle condition has changed. When considering emissions and emissions assisted maintenance programs, each vehicle will perform differently. No comparison can be made between different engines, even of the same make and model.

Methodology

The Khutala mine purchased the Pinssar continuous DPM monitoring systems as a key part of their emissions assisted maintenance program initiative.

The Pinssar systems were set-up at strategic traffic locations and additionally a video camera was installed on top of the Pinssar monitor. Each vehicle was then driven past the Pinssar systems so that Engineering could establish a baseline of each particular vehicle. Each vehicle was benchmarked, considering the abovementioned statement: no comparison can be made between different engines, even of the same make and model.

Then, during operation, when an outlier reading is detected by the Pinssar systems, Engineering take a look at the footage corresponding to the high reading. The action at this point is for Engineering to radio the operator of the vehicle identified to be emitting highly and the operator is given the directive to take the vehicle to the surface for inspection by the maintenance department.

Results

The achievements/successes of this initiative showed on multiple occasions that once the identified vehicle had been rectified in the workshop and then put back into operation underground, that the vehicle emissions detected by the Pinssar monitors had returned back to what was its baseline reading. This initiative ensured that when a particular vehicle started to emit higher DPM, then the mine was able to rectify this in a timely manner. This initiative contributed to the overall reduction of DPM throughout the mine.

Future works

Possible future works could involve linking the vehicles and the continuous emissions data via tracking software. This would involve the operation incorporating RFID tags on each vehicle.

Aside from tracking vehicles for maintenance, the ventilation team can also benefit from vehicle tracking. The number of vehicles in a section is traditionally established based on the availability of ventilation airflow to dilute pollutants. However, recent developments in instrumentation for continuous and quasi continuous monitoring of gas and DPM concentrations has allowed integration measurement technologies into management systems. Wu and Gillies (2008) and Lethbridge and Good (2010) observe that with the help of continuous monitoring and analysis of the concentration of diesel sourced gases and aerosol the use of diesel powered fleet can be more efficiently controlled.

Additionally, the Khutala mine may choose to use the data from the Pinssar systems to validate vehicle and or fleet upgrades when making capital purchase investments.

CASE STUDY #4 – REAL-TIME TRAFFIC MANAGEMENT UNDERGROUND

Background

The Fruta del Norte deposit is the largest gold deposit in Ecuador. The deposit is part of the Corriente Copper Belt located in the Amazon province of Zamora-Chinchipe, in south-east Ecuador.

In 2006 Aurelian Resources discovered a large deposit at Fruta del Norte, estimated between 6.8 and 10 million ounces of gold and between 9.1 and 14 million ounces of silver.

Lundin Gold acquired the asset in late 2014, began construction in July 2017, poured first gold in November 2019, and declared commercial production in February 2020.

The Fruta del Norte gold mine was the first large scale gold mine in Ecuador and is one of the highest grade, lowest costing gold mines in the world.

Lundin Gold is a Canadian mining company with its head office located in Vancouver, British Columbia, however 88 per cent of Lundin Gold's workforce is from Ecuador.

Considering there is no OEL for DPM in mining in Ecuador, Lundin Gold and have adopted the Ontario/Quebec/Newfoundland/Labrador OEL for the Fruta del Norte mine. The Ontario/Quebec/ Newfoundland/Labrador for DPM at the time of the mine opening, was 400 µg/m^3. Whilst the Quebec/Newfoundland/Labrador OEL remains at 400 µg/m^3, the Ontario Ministry of Labour (2020) legislated through parliament to be reduced to 120 µg/m^3, which came into effect in September 2023 and is now the lowest of any province or state in North America.

The Lundin Gold company OEL currently remains at 400 µg/m^3 which is in line with Canadian provinces of Quebec and Newfoundland and Labrador and is much lower than the province of their headquarters of British Columbia, which is 1500 µg/m^3.

Methodology

Lundin Gold is committed to having a robust health and safety culture at their operations and in local communities. Lundin Gold (2024) explain on their Health and Safety website that they prioritise the resourcing, management, continuous innovation, and monitoring of the effectiveness of their health and safety systems and practices.

Despite there being no OEL legislated for DPM in Ecuador, the company took a proactive approach to DPM.

The Fruta del Norte mine is a relatively new mine which is in a remote location not close to facilities which can provide results from personal gravimetric DPM devices. Considering this, Fruta del Norte investigated the latest DPM monitoring technology and installed three Pinssar systems in June 2022, followed by another three systems in January 2023 and one more in May 2023. The seven systems were installed underground as shown in Figure 2.

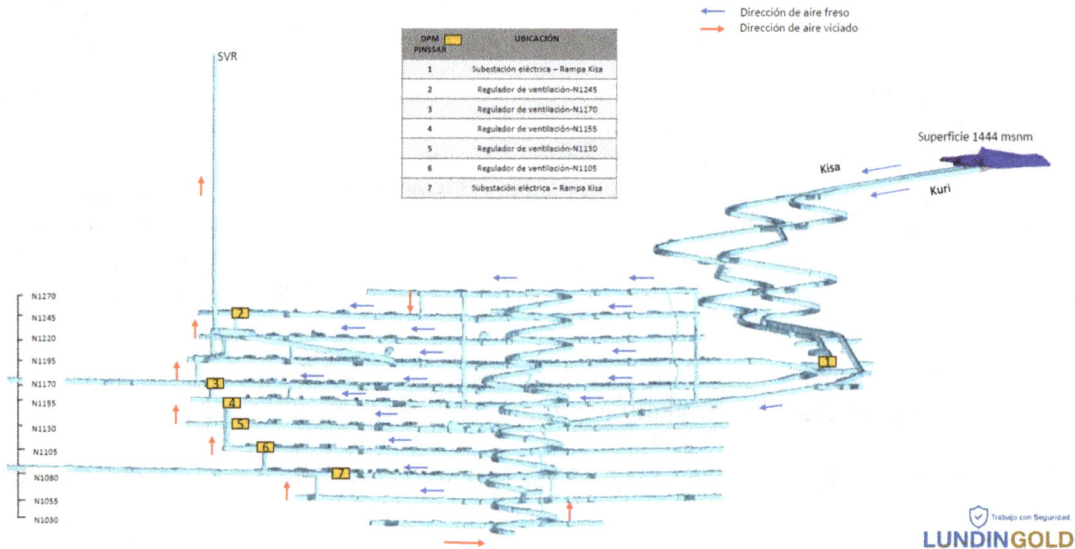

FIG 2 – Locations of seven DPM monitors at Fruta del Norte mine.

The number of vehicles in a section is traditionally established based on the availability of ventilation airflow to dilute pollutants. However, as described by Lethbridge and Good (2010), recent developments in instrumentation for continuous and quasi continuous monitoring of gas and DPM concentrations has allowed integration measurement technologies into management systems. With the help of continuous monitoring and analysis of the concentration of diesel sourced gases and aerosol the use of diesel powered fleet can be more efficiently controlled.

At Fruta del Norte mine, the despatch (control room) at the mine is continuously observing DPM levels from the Pinssar monitors in seven locations underground. If a DPM level goes above a limit, the despatch operator contacts the underground superintendent who then will investigate and put controls in place to reduce back to acceptable levels. Additionally, the mine has mandated PPE in particular areas of the mine.

Some of the initial controls includes removing vehicles from districts and training for changing operator driving and idling behaviours.

This method of using real-time data to respond in a timely manner is gaining uptake amongst the mining industry and has also been promoted across academic publications, such as Fanti *et al* (2022) which states the information on the exposure to the hazard can be available in a more timely way, thus, the implementation of risk mitigation measures may be faster and more efficient (eg workers who receive real-time information can mitigate their own exposure, by changing their behaviour and/or the procedure they are performing).

Results

The results of the abovementioned processes and initial controls were shown to be a success. Four separate reports at four points in time each time showed a decline in the average DPM value.

The reports were presented to the mine in August 2022, February 2023, May and December 2023.

These results in Table 3 have shown that initial behavioural (low cost) soft controls have been effective in reducing DPM levels. The mine plans to further concentrate on these soft controls plus look at additional controls (described below) as part of their continuous improvement initiative.

TABLE 3

Summary of report #4 comparing DPM levels compared to levels at time of install.

Report # 4			
1st June to 30th Nov 2023			
Location	samples taken	sample size (days)	reduction from original install
Kisa Rampa	51,741	180	12%
1245	48,550	169	n/a
1170	51,675	179	47%
1155	43,494	151	49%
1130	50,788	176	22%
1105	24,445	85	0%
1080	50,330	175	33%

Future works

The effort to reduce the exposure of underground miners to diesel pollutants requires the involvement of several key functional groups in the mine, including those responsible for production, health and safety, engine/vehicle/exhaust aftertreatment maintenance, and mine ventilation, as well as the departments responsible for the acquisition of vehicles, engines, exhaust aftertreatment systems, and fuel and lubricating oil. Bugarski *et al* (2012) advises that one of the first steps toward successful execution of such a complex program is the formulation of a comprehensive plan. Organising the layers of command and defining the responsibilities of the functional groups and individuals within the structure (as well as identifying the expected outcomes) is critical to the execution and success of the program.

Fruta del Norte have recently established a site based DPM committee, which includes Management support and input. The committee has been established with the understanding that to reduce DPM levels is not the task of a selected few but rather an all of mine approach. The committee has representation and inputs from various departments, such as engine maintenance, ventilation, OH&S as well as senior management. A key reference for the committee will be the publication *Controlling Exposure to Diesel Emissions in Underground Mines*. Schnakenberg Jr (2006) recommends that a team of experts who have the necessary experience in all critical areas as well as the motivation to properly implement the plan should coordinate this type of program. The team should be coordinated by a team leader, or champion, who should have overarching authority to manage the program. In addition, Mischler and Colinet (2009) recommends the program should be approved and supported by all mine functional groups, including upper corporate and mine management.

The various departments at Fruta del Norte will each bring research and suggestions to the committee, based on the hierarchy of controls. The DPM Committee has appointed a 'champion' who will encourage all departments to pursue continuous improvement.

The DPM Committee at Fruta del Norte have created a plan similar to that described by Bugarski *et al* (2012), the project plan should be based on a broad, multifaceted approach toward the implementation of feasible engineering and administrative controls. The plan should clearly define short-term and long-term strategies and goals. A timeline should be set-up to help implementations of strategies and technologies according to a predefined hierarchy of solutions.

The DPM Committee at Fruta del Norte plan to gradually introduce administrative controls and Engineering controls and will use the Pinssar data to monitor success/failure of mitigation strategies. A major initiative planned is to review the current fleet and make replacements to the high emitting vehicles. It is also planned to install additional monitors to monitor on each level. Smaller initiative are planned also, such as limiting engine idling and road maintenance. Bugarski *et al* (2012) also nominates the emissions from diesel-powered vehicles/equipment could be reduced by limiting unnecessary engine idling. Well maintained roads can help to minimise the frequency of transient events in the duty cycles of mobile diesel powered equipment and therefore reduce Gus and particulate emissions.

The long-term continuous improvement objective will be to reduce the company OEL over time, ie current 400 µg/m³ to 300 µg/m³ then to match USA OEL of 160 µg/m³, then to match Ontario OEL of 120 µg/m³. These future works, along with the works completed to date, is confirming that the Fruta del Norte mine is the most proactive mine in relation to DPM on the South America continent.

SUMMARY

The underground mine sites in the case studies documented within this whitepaper each successfully reduced their DPM levels by using mitigating controls outside of ventilation controls. These controls included:

- On the spot responses by underground superintendents, such as:
  - Removing vehicles.
  - Reducing idling.
  - Increasing DPM awareness of underground operators.
  - Positioning of operators and vehicles.
- Traffic Management underground, monitored by control room/despatch and/or Engineering:
  - Ensuring the correct number of vehicles are in the district.
  - Ensuring a limited number of light vehicles are operating underground at any and all times.
- Fleet maintenance:
  - Ensuring vehicles are not emitting more than their previous baseline.
  - Directing high emitting vehicles to be serviced in real-time.
  - Continuous monitoring of DPF filters.
- Operator management:
  - Reducing or stopping excessive and/or unnecessary idling.
  - Reducing convoying practices.

Whilst these operations use conventional mitigation practices, such as ventilation, diesel particulate filters, low sulfur fuels (where available) etc these operations have proved that human intervention (also known as soft controls of administrative controls) and DPM awareness underground plays an important role in further reducing DPM.

These DPM reductions were achieved by continuously monitoring DPM underground. Publications of successes at other mines around the world are planned for the future.

ACKNOWLEDGEMENTS

The Authors would like to acknowledge the contributions made by the following people. Whether the contributions were with direct inputs to this publication or providing access to information, the knowledge and information sharing is greatly appreciated and should be seen as not only contributing to this publication, but also enabling other mines to learn and improve based on these experiences. The Authors would also like to highlight that the achievements made in the case studies listed where not exclusively due to the work only of those listed below, but was achieved with teamwork and a whole-of mine approach with their respective colleagues:

- BHP/BMA Broadmeadow mine, Australia: Toby Everson, Ian Marshall, Andrew Batterson and DPM committee members (not limited to) Scott Walsh, James Roughan, David Caley, Ken Singer, Kevin Myer.

- Sibanye Stillwater USA Operations: Justus Dean, Domingo Leonis.

- Khutala Colliery, South Africa: Navarre Kruger.

- Lundin Gold Fruta del Norte mine: Geovanni Lucas, Edward Leal, Javier Santiago, Doug Moore, Daniel Teran, Jonathan Jean, David Torres, Carlos Paredes, Thiago Teixeira.

REFERENCES

Bugarski, A D, Samuel, J, Janiskoc, S J, Emanuele, G, Cauda, E G D, Noll, J D, Steven, E and Mischler, S E, 2012. Controlling Exposure to Diesel Emissions in Underground Mines, *Society for Mining, Metallurgy and Exploration (SME)*.

Cantrell, B K and Volkwein, J C, 2001. Mine aerosol measurement, in *Aerosol Measurement: Principles, Techniques, and Applications* (eds: P A Baron and K Willeke), second edition, 10 p.

Department of Mines and Petroleum, 2013. Management of diesel emissions in Western Australian mining operations — guideline, Resources Safety, Department of Mines and Petroleum, Western Australia, 37 p. Available from: <https://www.dmp.wa.gov.au/Documents/Safety/MSH_G_DieselEmissions.pdf> [Accessed: 25 Nov 2023].

Department of Natural Resources, Mines and Energy (DNRME), 2019. Diesel Emissions Management in Underground Coal Mines, Best Practices and Recommendations, Queensland Government, 14 p. Available from: <https://www.resources.qld.gov.au/__data/assets/pdf_file/0009/1438524/diesel-emissions-mgt-underground-coal-mines.pdf> [Accessed: 25 Nov 2023].

European Agency for Safety and Health at Work, 2004. Directive 2004/37/EC - carcinogens, mutagens or reprotoxic substances at work [online], European Agency for Safety and Health at Work. Available from: <https://osha.europa.eu/en/legislation/directive/directive-200437ec-carcinogens-or-mutagens-work>

Everson, T and Batterson, A, 2018. Broadmeadow Mine Diesel Particulate Reduction Program, in *Proceedings of the Queensland Mining Industry Health and Safety Conference (QMIHSC 2018),* 11 p. Available from: <https://www.qmihsconference.org.au/wp-content/uploads/qmihsc-2018-health-bmabroadmeadow.pdf>

Fanti, G, Spinazzè, A, Borghi, F, Rovelli, S, Campagnolo, D, Keller, M, Borghi, A, Cattaneo, A, Cauda, E and Cavallo, D, 2022. Evolution and Applications of Recent Sensing Technology for Occupational Risk Assessment: A Rapid Review of the Literature, *Sensors*, 22(13):4841. https://doi.org/10.3390/s22134841

Fenske, R A, 2010. For good measure: Origins and prospects of exposure science, 2007 Wesolowski Award Lecture, *Journal of Exposure Science and Environmental Epidemiology*, 20:493–502.

Hardcastle, S and Kocis, C, 2004. Mining at Depth, the ventilation challenge, *CIM Bulletin*, 97(1080):51–57.

Hardcastle, S G, 2014. The continuing challenge to provide adequate ventilation and a safe environment in deep mines, in *Deep Mining 2014: Proceedings of the Seventh International Conference on Deep and High Stress Mining* (eds: M Hudyma and Y Potvin), pp 41–54 (Australian Centre for Geomechanics: Perth). https://doi.org/10.36487/ACG_rep/1410_0.3_Hardcastle

Hedges, K, Djukic, F and Irving, G, 2007. Diesel Particulate Matter in Underground Mines – Controlling the Risk (an update), in *Proceedings of the Queensland Mining Industry Health and Safety Conference (QMIHSC 2007),* 15 p. Available from <https://www.qmihsconference.org.au/wp-content/uploads/qmihsc-2007-writtenpaper-hedges_djukic_irving.pdf>

Hill, L B, Force, C A, Zimmerman, N J and Gooch, J, 2005. A multi-city investigation of the effectiveness of retrofit emissions controls in reducing exposure to particulate matter in school buses, Clean Air Task Force report, 68 p.

Howard, J, Murashov, V, Cauda, E and Snawder, J, 2022. Advanced sensor technologies and the future of work, American Journal of Industrial Medicine, 65(1):3–11. https://doi.org/10.1002/ajim.23300

International Agency for Research on Cancer (IARC), 2012. Diesel engine exhaust carcinogenic, IARC press release no. 213, June 12, International Agency for Research on Cancer, World Health Organization. Available from: <https://www.iarc.who.int/wp-content/uploads/2018/07/pr213_E.pdf>

International Council on Mining and Metals (ICMM), 2022. Collaboration for Innovation: Accelerating the Implementation of Zero Emission Vehicles for the Mining and Metals Industry, ICMM. Available from: <https://www.icmm.com/en-gb/stories/2022/accelerating-implementation-of-zero-emission-vehicles> [Accessed: 17 Jan 2024].

Kittelson, D B, 1986. Engines and nanoparticles: a review, *Journal of Aerosol Science,* 29(5–6):575–588.

Lethbridge, T and Good, M, 2010. A multi-focus approach to DPM reduction at the Greens Creek mine, in *Proceedings of the 13th US/North American Mine Ventilation Symposium*, pp 65–71.

Lim, S and D'Souza, C, 2020. A narrative review on contemporary and emerging uses of inertial sensing in occupational ergonomics, *Int J Ind Ergon*, 76:102937.

Lundin Gold, 2024. Health and Safety – Responsible Mining. Available from: <https://lundingold.com/responsible-mining/health-and-safety/>

Manos, E Z, 2010. Continuation of DPM control strategy at the Detroit salt mine using Rypos HDPF/C filters on diesel equipment, in *Proceedings of the 13th US/North American Mine Ventilation Symposium*, pp 79–82.

McDonald, R, 2015. BHP Diesel exhaust management: Our journey of continuous improvement, presentation at Annual Conference of the Australian Institute of Occupational Hygienists, Royal Australasian College of Physicians (RACP) Congress 2016, 8 p. Available from: <https://www.racp.edu.au/docs/default-source/fellows/resources/congress-2016-presentations/racp-16-tuesday-dr-rob-mcdonald.pdf?sfvrsn=586b321a_2> [Accessed: 25 Nov 2023].

Mischler, S E and Colinet, J F, 2009. Controlling and monitoring diesel emissions in underground mines in the United States, in *Mine Ventilation: Proceedings of the Ninth International Mine Ventilation Congress*, 2:879–888. Available from: <https://www.cdc.gov/niosh/mining/userfiles/works/pdfs/camde.pdf>

Morla, R, 2018. Experimental and numerical investigation of distribution patterns of diesel particulate matter and development of control strategies in an underground mine environment, PhD thesis, University of Wollongong.

Ontario Ministry of Labour, 2020. Occupational Health and Safety Act, Regulation 833 – Control of exposure to biological or chemical agents [online]. Government of Ontario Available from: <https://www.ontario.ca/laws/regulation/900833>

Pearce, T and Coffey, C, 2011. Integrating direct-reading exposure assessment methods into industrial hygiene practice, *J Occup Environ Hyg*, 8(5):D31–36.

Pritchard, C J, 2010. Methods to improve efficiency of mine ventilation systems, *CDC – NIOSH Mining*. Available from: <https://www.cdc.gov/niosh/mining/UserFiles/works/pdfs/mtieom.pdf>

Rubeli, B, Gangal, M, Butler, K and Aldred, W, 2004. Evaluation of the Concentration of Light-Duty Vehicles to the Underground Atmosphere Emissions Burden, in *Proceedings of the 10th U.S./ North American Mine Ventilation Symposium* (eds: R Ganguli and S Bandopadhyay) (Balkema).

Schnakenberg, G H Jr, 2006. An integrated approach for managing diesel emissions controls for underground metal mines [online], National Institute for Occupational Safety and Health. Available from: <https://www.cdc.gov/niosh/mining/UserFiles/works/pdfs/aiafm.pdf>

Sibanye-Stillwater, 2022. Integrated Report 2022: Our performance, in Health, Wellbeing and Occupational Hygiene, 11 p. Available from <https://reports.sibanyestillwater.com/2022/download/ssw-IR22-performance-health-wellbeing.pdf>

Spears, M W, 1997. An emissions-assisted maintenance procedure for diesel-powered equipment, Center for Diesel Research. NIOSH contract No. USDI/1432 CO369004. Available from: <http://www.cdc.gov/niosh/mining/eamp/eamp.html>

Wu, H W and Gillies, A D S, 2008. Developments in real time personal diesel particulate in mines, Proceedings of the 12th US/North American Mine Ventilation Symposium 2008.

Zielinska, B, Sagebeil, J, McDonald, J, Rogers C F, Frujita, E, Mousset-Jones, P and Woodrow, J E, 2002. Measuring diesel emissions exposure in underground mines: a feasibility study, in *Proceedings of the research directions to improve estimates of human exposure and risk from diesel exhaust,* Health Effects Institute, pp 181–232.

Mine ventilation acoustics

B A Russell[1]

1. Principal, Barus Consulting, Sydney NSW 2075. Email: baruss@optusnet.com.au

ABSTRACT

The ventilation system is one of the major sources of noise from a mine. The control of noise from the ventilation system has always been important for the well-being of the mine personnel in the near vicinity of the fans and for the community at a distance but has become more important with more restrictive regulations and a more sophisticated community. Historically the primary source of noise has been the fans blade passing frequency and its harmonics, but regulations and community expectations are now placing more attention on the lower frequency portion of the noise spectrum.

The sound levels may be lower at lower frequencies but are much more difficult to attenuate. This is an area in which the mine operators, the fan suppliers, acoustic consultants and the regulators all have limited experience and are entering unknown territory.

This paper will cover the complete acoustic requirements of a modern ventilation system but will address the specific problems which are encountered when dealing with low frequency noise and the realistic responsibilities of the parties involved. The paper discusses regulatory requirements, specification of mine requirements and the required acoustic information from the fan suppliers. It reviews the application of existing technologies and the yet unproven methods which are required to meet the emerging community expectations. Alternative arrangements including placing the primary ventilation system underground are explored.

INTRODUCTION

The control of noise in mining has always been important for the well-being of the mine personnel and for the community at a distance but has become more significant with increasingly restrictive regulations and a more sophisticated community.

The industry is having to deal with a more sophisticated community and the increasing regulation which is being implemented to meet this. This is resulting in the attention being placed in areas which have not previously had to be addressed by the mine operators, the equipment suppliers and the consultants advising the industry.

To meet the increasing requirements the mine and the equipment supplier must work closely, each party has control of part of the solution and must take its responsibility.

The method of predicting the noise or more correctly the sound pressure level at a location remote from an installation is given in ISO 9613-2:1996 the standard defining the method of calculation of attenuation of sound during propagation outdoors as:

$$SPL = SWL + D_C - A$$

(1)

where:

SPL	sound pressure level dB re 2×10^{-5} Pa
SWL	sound power level dB re 10^{-12} W
D_C	The directivity correction.
A	Attenuation from the installation to the receptor. This includes geometrical divergence, atmospheric absorption, ground effect, barriers ectc.

There are three components defining the sound pressure level. The installation sound power level, the directivity of the source and the attenuation between the source and the receptor.

There are three parties to the establishment of the noise at a particular location.

1. The regulatory authority.

2. The mine operator.

3. The supplier of the ventilation installation.

Each can control portions of this relationship and should take responsibility for their component. All parties are having to deal with changing requirements.

The regulatory authority must clearly set the levels to meet the community and political requirements and enforce these in a fair and enlightened manner. They define SPL, the level which must be met.

The ventilation supplier controls SWL and D_C. The ventilation equipment supplier must accurately define the sound power level for the equipment being supplied and by suitably arranging the installation define the directivity.

It is the mine who has the knowledge and control of the factors which make up A. This is perhaps the area which can have the most variability and risk and it is easy for the mine to put the responsibility for it on the equipment supplier. This may appear to a tidy contractual solution but from experience does not always achieve the required outcome. This should be assessed and defined by the mine.

REGULATION AND LOW FREQUENCY NOISE

The most common way of specifying the requirements is as a single number dBA sound pressure level varying with time of day. The lower level is typically 30 or 33 dBA during the night. But a single number assessment does not adequately address low frequency noise which is becoming a more common source of complaint.

Low frequency noise (LFN) is noise is generally considered to be noise with a frequency less than 100 Hz. Noise at frequencies below 20 Hz is also referred to as infra sound.

The audible range for a person with normal hearing is approximately 20 to 18 000 Hz. The sensitivity of the ear varies with the frequency of the sound. It is greatest at mid frequencies between 500 and 3000 Hz and reduces significantly at low frequencies. Various corrections are used to adjust measured spectrums. The most used for environmental and occupational noise is the A weighting which is a good approximation of the ear's response to low level sound. The C weighting is more suitable for higher levels such as aircraft noise. The unweighted or linear level is now referred to as Z. The A and C weighting are shown in Figure 1. At 100 Hz the A scale reduction is 20 dB and at 20 Hz it is 50 dB.

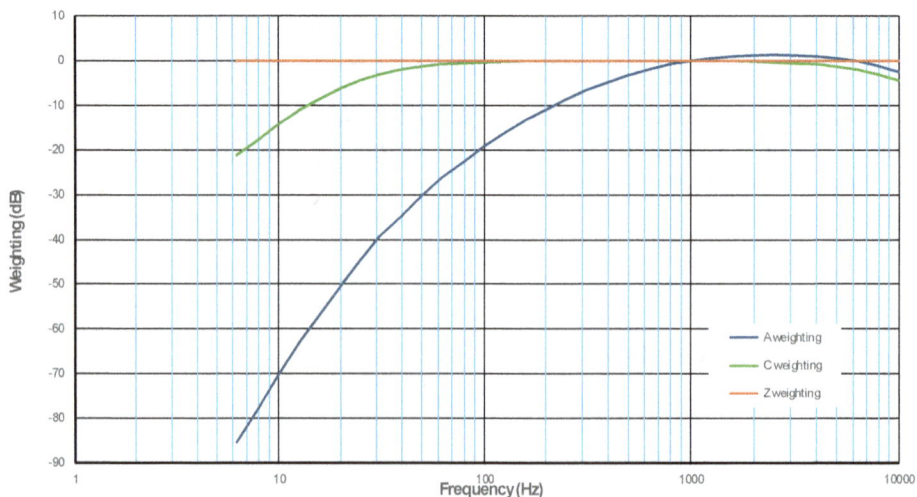

FIG 1 – Sound weighting scales.

The C weighting does not apply as much correction at low frequencies as the A weighting and the difference is one way of defining the presence of low frequency noise.

Though hearing becomes progressively less at low frequency the human body senses low frequency noise in other ways. The German standard DIN 45680:1997 for the measurement and evaluation of low frequency environmental noise states, it

> 'is normally perceived as pulsations or vibration. Subjects report the sensation of pressure in their ears and often express feelings of uneasiness or anxiety. A particular known effect of infrasound is the lowering of breathing frequency. Secondary effects (such as those caused by rattling of windows, doors or glassware, or discernible vibration of objects or parts of the building) often lead to severe discomfort'.

Each country has their specific regulations.

In Australia the environmental regulations are set by each State. The two largest entities do not specify maximum low frequency levels but give threshold levels in 1/3 octaves from 10 to 160 Hz which indicate potential risk of problematic low frequency noise which are given in Figure 2 and compared with the audible limits as defined in ISO 226:2023.

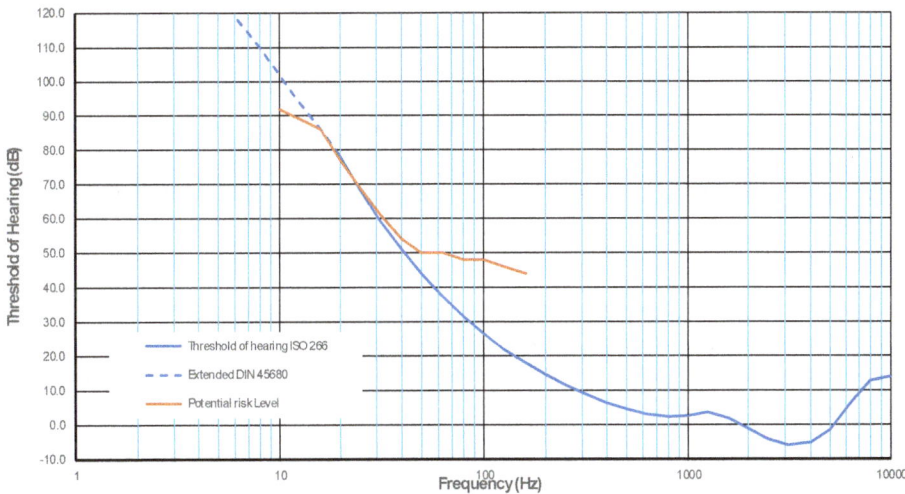

FIG 2 – Potential risk level and threshold of hearing.

Though the levels at 16, 20 and 25 Hz are set at the audible limits as defined in ISO 226 there have been complaints at levels below these and in Victoria a potential harm criterion is now being applied.

The Act requires that the creator of the harm must minimise the risk as far as reasonably possible.

At least three mines in Eastern Australia have had to deal with low frequency noise issues in the last three years. In all cases the problem frequencies were less than 20 Hz.

WHAT IS REASONABLY POSSIBLE

Below 20 Hz there is no auditory perception, noise is perceived in other secondary ways. Between 20 and 60 Hz noise is audible but may be perceived as fluctuations (beats), sensations of booming, vibration or pressure in the head and in ways like those below 20 Hz. Above 60 Hz noise is perceived in a normal manner and is considered as annoying if it has tonal components and secondary effects are not significant.

Because of the use of single number A weighted noise specifications and conventional silencing methods become increasingly ineffective at frequencies below 60 Hz, low frequency noise has not been given the attention it needs.

If the noise in the low frequency audible range, which can be controlled, is allowed to occur and leads to complaints described as low frequency noise this can lead to detailed measurements which show discrete tones at lower frequencies in the inaudible range below 20 Hz. These can become the focus of attention of the receptors and regulatory authority and the true cause overlooked.

This is illustrated in Figure 3 which is a narrow band spectrum of noise from a large ventilation installation. The installation was fitted with silencers of a type which was not effective below 50 Hz.

The focus in finding a solution was the inaudible 12.5 Hz tone while ignoring the high broadband audible levels between 20 and 50 Hz.

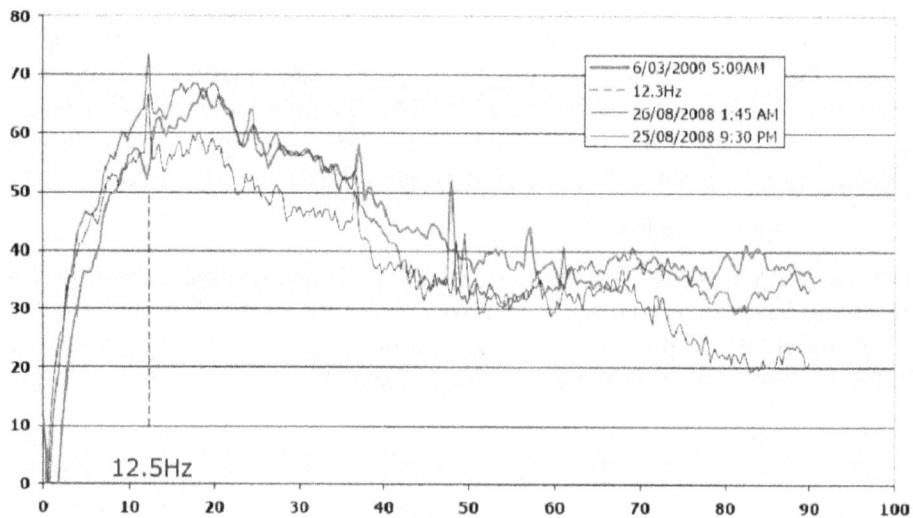

FIG 3 – Large fan noise spectrum.

Measuring noise at low frequencies is difficult and fan manufacturers have avoided supplying information. There are still major suppliers who are only supply sound power data in octave bands and not below 62.5 Hz.

There has not been a large-scale application of technology to reduce noise levels at frequencies below 20 Hz. The methods are understood but the physical limitations are significant. It may be currently impractical but not impossible to remove noise below 20 Hz, but proven methods are available down to 30 Hz.

To show to a regulatory authority that all efforts have been made as far as reasonably possible firstly one should be applying the correct currently available technology to control the noise in the audible range and this may mean working to lower frequencies than suppliers, the mines and consultants are used to and comfortable working to.

NOISE SOURCE

The primary source of the noise is the fan.

Fan noise

The fan should be well designed to minimise the source level. It is the fan supplier's responsibility to provide accurate and reliable sound power levels at all frequencies under review. Fan suppliers have traditionally only supplied noise data in octave bands from 63 Hz to 8 kHz. Currently a limited number will also give data for the 31 Hz octave.

The reluctance reflects the difficulty in accurately measuring sound at lower frequencies. However, with the focus now on noise at lower frequencies fan suppliers must now be able to supply noise data down to the 16 Hz octave and provide it in 1/3 octave bands. Fan suppliers currently do not have this information for all their products.

Recent experience has shown that where data is supplied for lower frequencies it is not always based on reliable testing and appears to be an estimate of what the client would like to see.

This is the area for which the fan supplier is responsible and requires considerably more work.

Traditionally most of the primary ventilation systems have been on the surface though a growing number of large fans are now being installed underground.

For a surface installation the main noise sources from the fan are:

- Fan discharge.

- Fan casing and connecting ducting.
- Drive motors.

Fan discharge noise

The most common method of reducing fan noise from the air stream is a silencer made up of acoustic splitters installed in the fan inlet or discharge ducting. The splitters are of three types:

1. Absorptive.
2. Reactive dissipative.
3. Pure reactive.

Absorptive silencer

The absorptive silencer consists of splitters packed with absorptive material enclosed by perforated plate as shown in Figure 4.

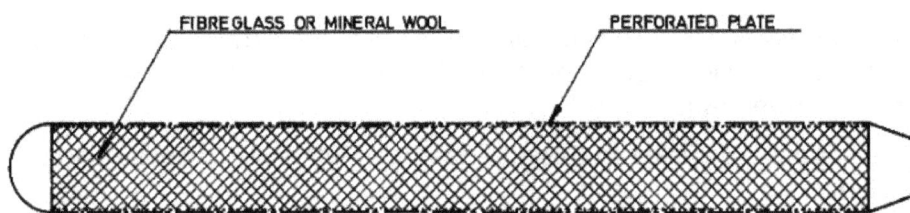

FIG 4 – Absorptive silencer splitter.

The advantage of this form of silencer is its low cost. It has poor low frequency attenuation and is subject to fouling of the perforated plate when used with wet, dirty, dusty air streams including diesel particulate. The performance degrades rapidly with fouling and can fall to less than 25 per cent of initial attenuation within months. Performance can be improved by cleaning but not to the initial level and the degradation is cumulative. After a few cleaning cycles, the splitters must be replaced.

If the silencers are subject to vibration due to direct connection to the fan or from highly turbulent flow the absorptive material breaks down.

The perforated plate is thin and must be maintained. An unmaintained absorptive silencer will eventually result in breakdown of the splitter perforated plates and absorptive material, which can cause a significant increase in mine resistance due to blockage of the silencer passages and possible impeller damage on fans discharging vertically.

This form of silencer is not recommended for primary mine ventilation applications.

Reactive – dissipative silencer

The reactive dissipative silencer consists of splitter with tuned chambers as shown in Figure 5.

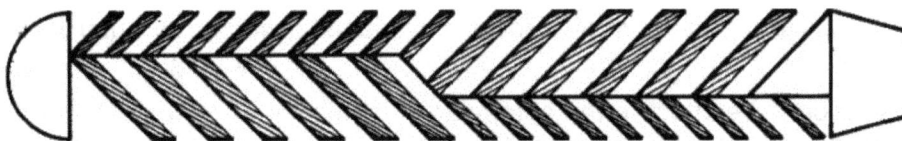

FIG 5 – Reactive dissipative splitter.

The silencer works by sound entering the cavities, being reflected, and when it returns to the chamber entrance cancelling noise entering at a frequency with which it is one half a wavelength out of phase. The chamber depth is 1/4 of the wavelength of its tuned frequency. Typically, the deeper cavities, 2/3 of splitter width, are tuned to the fan blade passing frequency and the shorter cavities, 1/3 of splitter width, to the second harmonic. Full splitter width chambers can be tuned to lower frequencies. If the splitter only had the cavities the attenuation spectrum consists of discrete peaks. Absorptive material is added to the downstream face of the arms to broaden the attenuation spectrum.

This design works well in dirty applications and is ideal when the highest attenuation is required at the fan blade passing frequency, typically in the 250 Hz octave. This is normally where the highest attenuation is required.

It does not deal well with low frequencies where the wavelength becomes large. At 250 Hz a quarter wavelength is 340 mm, at 20 Hz 4.3 m and 10 Hz 8.6 m.

Pure reactive silencers

The pure reactive silencer works on the principle of the Helmholtz resonator. The silencer splitters consist of a series of chambers connected to an air passage by throats of varying width and depth. A typical pure reactive splitter is shown in Figure 6.

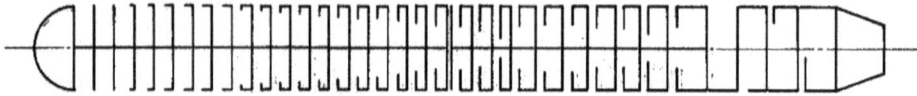

FIG 6 – Pure reactive splitter.

The silencer contains no absorptive material and is ideal for dirty applications. Working on the Helmholtz principle this design does not have the limitations of the 1/4 wavelength reactive dissipative silencer at lower frequencies.

This reactive silencer more complex to design and labour intensive to construct and as a result is the more expensive.

Combined 1/4 wavelength and pure reactive

Where both medium and high frequency attenuation is required in combination with low frequency attenuation an economical solution is to combine both in a single splitter as shown in Figure 7.

FIG 7 – Combined reactive dissipative and pure reactive splitter.

This combines pure reactive chambers tuned to low frequencies with 2/3 and 1/3 splitter width cavities, which would be tuned to blade passing and twice blade passing, and full width cavities tuned to an intermediate frequency.

Noise from fan casing and connecting ductwork

Noise from the fan casing and connecting ductwork can be reduced by:

- Acoustic lagging of the casing and connecting ductwork.
- Enclosing the fan and ducting in an attenuating building.

Acoustic lagging consists of layers of absorptive material enclosed in an outer layer of protective attenuating material.

For acoustic insulation to be effective the outer cladding must be in close contact with the acoustic infill and must be acoustically isolated from the ducting to which it is being applied. If a plate is vibrating due to acoustic excitation and is fitted with a layer of sound absorbing material which is then enclosed with an outer layer of cladding and the cladding is then mechanically connected to the acoustically excited plate the acoustic attenuation through the absorbent material is bypassed or flanked and the outer cladding will vibrate and radiate noise in a similar manner to the original plate.

The alternative is to enclose the entire installation in a correctly designed acoustic enclosure. The decision between the two is normally based on economics. There are arguments on access for maintenance for both.

The key to successes is in the detail. Small openings can have a large effect on the effectiveness of the attenuation as illustrated in Figure 8. When attempting to achieve an attenuation of 40 dB a 1 per cent opening results in a reduction of 50 per cent to 20 dB.

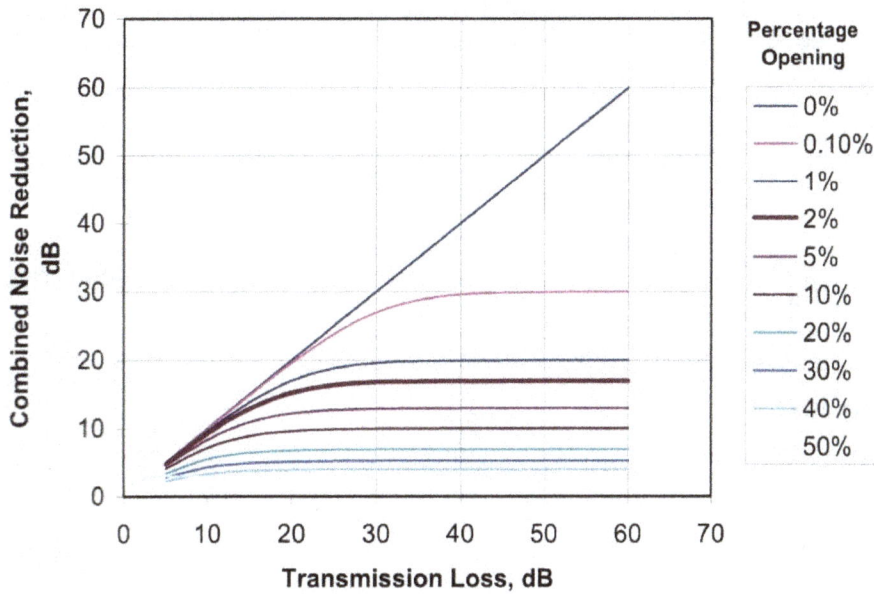

FIG 8 – Effect of openings on attenuation.

Noise from drive motors

The motors can also be enclosed.

The cooling air needs to be handled correctly. Acoustic louvres don't give 40 dB attenuation.

DIRECTIVITY

Sound propagation is directional, the magnitude varying with frequency and the dimensions of the source.

The directivity index based on the work of Day, Hansen and Bennett (2009) is given in Figure 9. The directivity index gives the variation from uniform dispersion of the sound. This is given as a function of direction and Strouhal Number $\frac{fd}{c}$ where f is the frequency, discharge diameter d and c the speed of sound. For a square or rectangular discharge, the hydraulic diameter $D_h = \frac{4A}{P}$ can be used where is A the area and P the perimeter of the discharge.

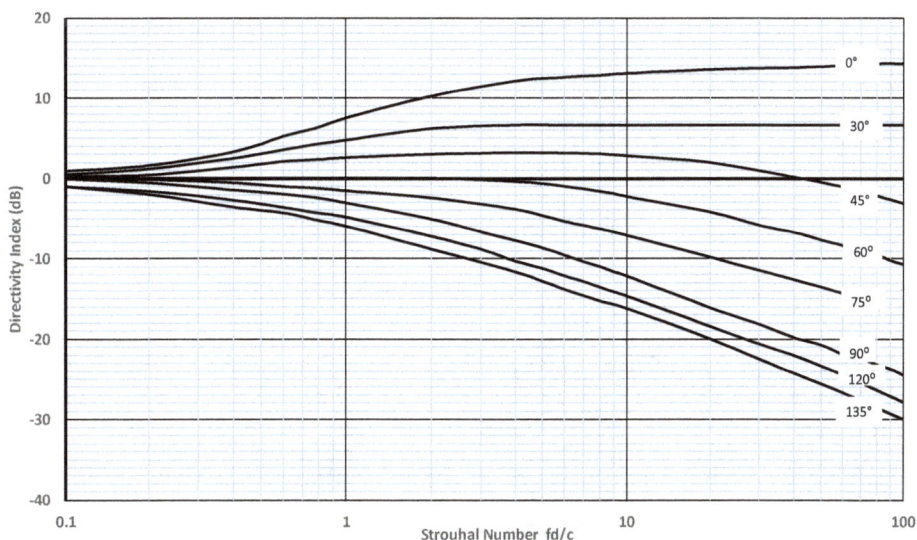

FIG 9 – Directivity index.

ATTENUATION FROM INSTALLATION TO RECEPTOR

The estimation of the attenuation A from the fan installation to a receptor is more complex. This is addressed in ISO 9613 Acoustics – Attenuation of sound propagation outdoors which along with Concawe (1981) is the basis of various computer packages available. For initial studies it can be applied in a simplified spreadsheet form which has been found to give reasonable agreement with actual installation results.

ISO 9613 gives satisfactory results under neutral meteorological conditions but does not handle short-term meteorological effects which are major source of environmental noise complaints.

The code operates in dBA and uses octave bands with a minimum band frequency of 62.5 Hz. This reflects the difficulty in handling lower frequencies. While many, including the International Standards Organisation, handle the situation as if they do not exist, in the mining world frequencies below 62.5 Hz do exist and do create problems.

ISO 9613 divides the Attenuation into five components.

1. Geometrical divergence.
2. Atmospheric absorption.
3. Ground effect.
4. Barriers.
5. Miscellaneous other effects.

ISO 9613 gives guidance for the estimation of the first four but what the fifth covers or more importantly what it does not is why the responsibility for this component should be with the mine as it requires local topographical and long-term detailed meteorological data and local noise experience to access the uncertainties and carry out an appropriate risk analysis.

Placing the responsibility on the equipment supplier will result in an overly conservative offer from a supplier who understands the situation, possibly from experience in the area, and confident low cost offers from sources remote from the locality which will be guaranteed to provide future problems for the operator.

Geometrical divergence

The sound pressure L_P at distance d spreading uniformly from a source of sound power L_W to an area A is:

$$L_P = L_W - 10\,log(A)$$

which for hemispherical divergence at a distance d is:

$$L_P = L_W - 20log(d) - 8$$

Atmospheric absorption

Atmospheric absorption is highly dependent on frequency, temperature and relative humidity. The 1/3 octave attenuation in decibels per kilometre is given in ISO 9613-1:1993.

The attenuation is low at low frequencies but is important at the fan blade passing frequency and its harmonics.

Ground effect

The effect of the ground is due to the sound reflected by the ground surface interfering with the sound propagating directly from the source. It is primarily determined by the ground surfaces near the source and receiver.

At a distance ground effect is not as great as divergence or atmospheric absorption but is significant.

ISO 9613 divides the ground into three regions, source, receiver and middle region. The acoustical properties of each are considered using a ground factor G in relationships given for each region.

It has been found that using G = 0 for the source and receiver and 1 for the middle region gives satisfactory results.

Barriers

ISO 9613 gives detailed methods for handling single and multiple barriers. Conventional height screening barriers do not have a significant influence on levels at distances greater than 500 m.

Miscellaneous other effects

This is a catch all term. The code gives guidance on attenuation through foliage, industrial sites and housing. It does not cover short-term meteorological effects.

ISO 9613 has a meteorological correction C_{met} but this an adjustment of the calculated short-term downwind level to a long-term overall A weighted average. It is a reduction not an increase.

METEOROLOGICAL EFFECTS

ISO 9613 does not adequately allow for all meteorological effects. The predictions are for downwind propagation with wind speeds between 1 and 5 m/s at a height of 3 to 11 m, but it does not address temperature variation or atmospheric turbulence.

Wind

Over open ground wind velocity gradients exist due to friction between the moving air, and he ground. In the absence of turbulence, the velocity typically varies logarithmically up to a height of 30 to 100 m, then remains constant.

As a result of this gradient a sound wave propagating in the direction of the wind will be bent downward and a wave propagating against the wind directed upwards creating an acoustic shadow zone as shown in Figure 10. The radius of the sound path is inversely proportional to the velocity gradient and noise bends to the lower velocity.

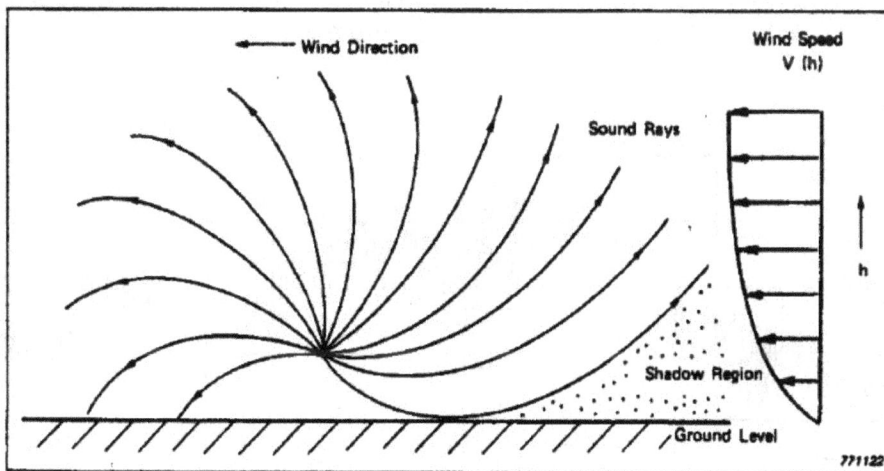

FIG 10 – Effect of wind on sound propagation.

Temperature

A similar effect results from a vertical temperature gradient. The speed of sound is proportional to the square root of the absolute temperature. As with wind this causes the sound wave to be bent in the direction of the lower velocity ie the lower temperature. If the temperature decreases with height the sound rays are bent upwards in all directions and if the temperature increases with height the ray is directed downwards to the ground as shown in Figure 11. As with the wind this creates zones of increased noise and shadow zones. Reductions in noise levels of up to 30 dB are common in shadow zones.

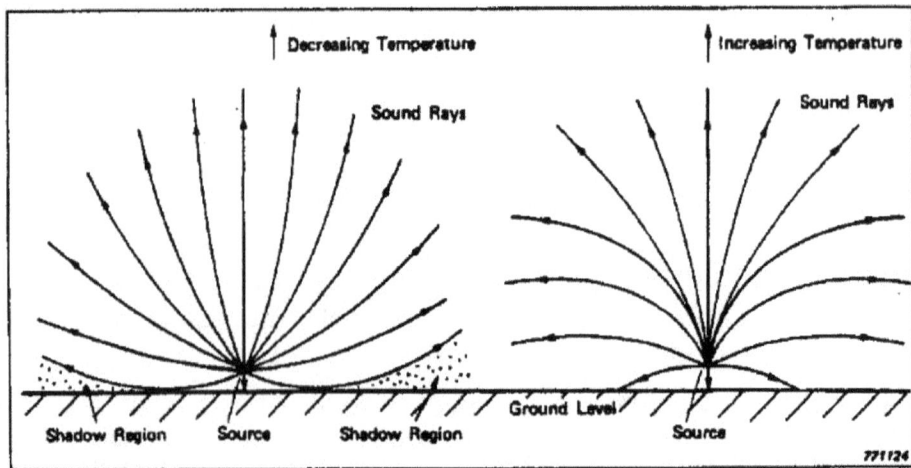

FIG 11 – Effect of temperature on sound propagation.

Under normal conditions there is a negative temperature gradient with increasing height. However, under certain conditions a positive gradient can exist, and these are the conditions when complaints arise with an apparently well-designed ventilation installation. A positive temperature gradient is referred to as a temperature inversion.

In Australia this can typically occur at the end of a hot day as the heat from the air is absorbed by the cooler ground, in colder environments it can occur in winter when the ground is warmer than the air and everywhere on cloudless nights. It is also affected by airflows over the earth's surface and as a result is influenced by local topographical features.

The frequency and severity of these occurrences are location specific as the Australian mining industry on the eastern slopes of the Great Dividing Range have found.

Figure 12 shows the effect of temperature inversion on the noise levels from fans at a residence two kilometres from the installation during the evening of a summer day. The noise level rose 20 dB over a period of 20 mins, remained at this level for approximately 1 hr and then returned to the original level over 30 mins.

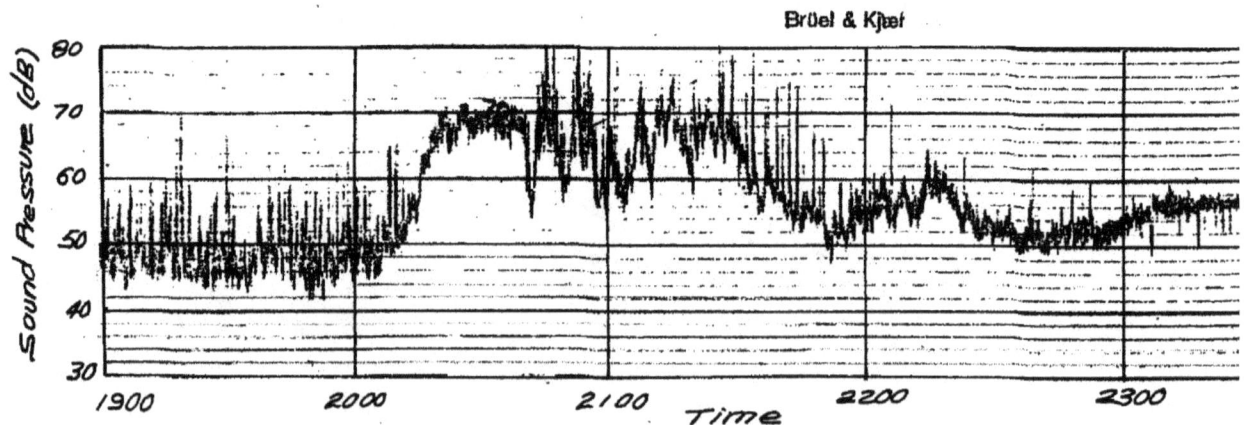

FIG 12 – Effect of temperature inversions on noise levels.

When wind and temperature effects are combined more complex noise patterns of increased noise and shadow zones. It is possible to have low noise close to the installation and much higher levels at a considerable distance

The effect of temperature inversion has been recognised in the past, BHP was indicating an allowance of up to 12 dB should be made in 1990s specifications for mines in the Southern coalfields, but it is still the probably most common cause of environmental noise problems.

The local NSW EPA Noise Policy for Industry (2017) recognises the effect of *noise-enhancing weather conditions including* Pasquill-Gifford categories E, F and G, but these only trigger a modifying factor if the conditions occur more than 30 per cent of the period under consideration and when applied it is only 5 dB. As a result, consultants preparing Environmental Impact Reports do not highlight the issue.

However, it takes much less than this to trigger sufficient complaints for action to be taken against a mine.

UNDERGROUND FAN INSTALLATION

It has always been the practice to use smaller auxiliary and booster fans underground but more recently it has become the practice to install large primary ventilation fans underground. Acoustically this has advantages as the attenuation through the underground airways and shafts reduce the noise entering the environment.

Ventilation and occupational environment engineering in South African mines (Howes, de Koker and Edwards, 2014) gives the following equation for the attenuation in drifts and haulages.

$$Attenuation = 12.6\frac{P}{A}L^{0.8}\alpha^{1.4}$$

Where P and A are the perimeter and area of the airway, L (m) the length, and α the absorption coefficient.

This is a modified form of the original Sabine (1940) relationship for attenuation in lined ducts. The constant is that from the original imperial format and the $\alpha^{1.4}$ term an approximation for the acoustic impedance of a fibrous HVAC duct linings. The $L^{0.8}$ term has been added to predict the reduction in the rate of attenuation with length due the organising of the sound waves with distance.

Using the absorption coefficients for rock surfaces given has not been found to give satisfactory correlation with measured data. To rationalise measurements a simplified form of the relationship has been found to give satisfactory results.

$$Attenuation = \alpha_m\frac{P}{A}L^{0.8}$$

Where α_m is referred to as the mine absorption coefficient. The parameter $\frac{P}{A}$ and L define the geometry of the airway while α_m is defined by the rock conditions, surface irregularities, amount of dust present, porosity of the footwall aggregate amongst other variables.

Currently there is limited measured attenuation data is available. From measurements in a 5.5 m raise bored shaft and in various airways, typically approximately 5 m × 6 m have given the mine absorption coefficients given in Figures 13.

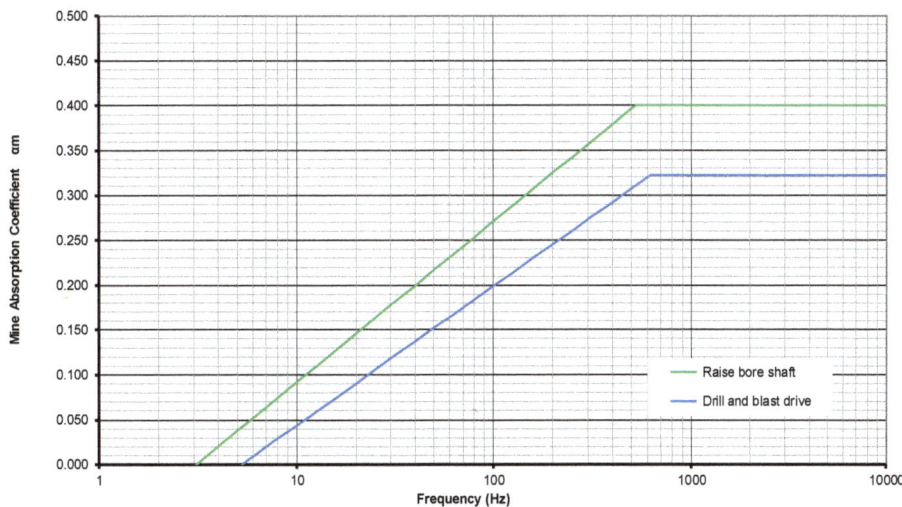

FIG 13 – Mine absorption coefficients in underground airways.

Using this data the attenuation from an underground fan installation and the shaft collar through a 500 m deep 5.5 m shaft and 300 m of 5 m × 6 m airway is shown in Figure 14.

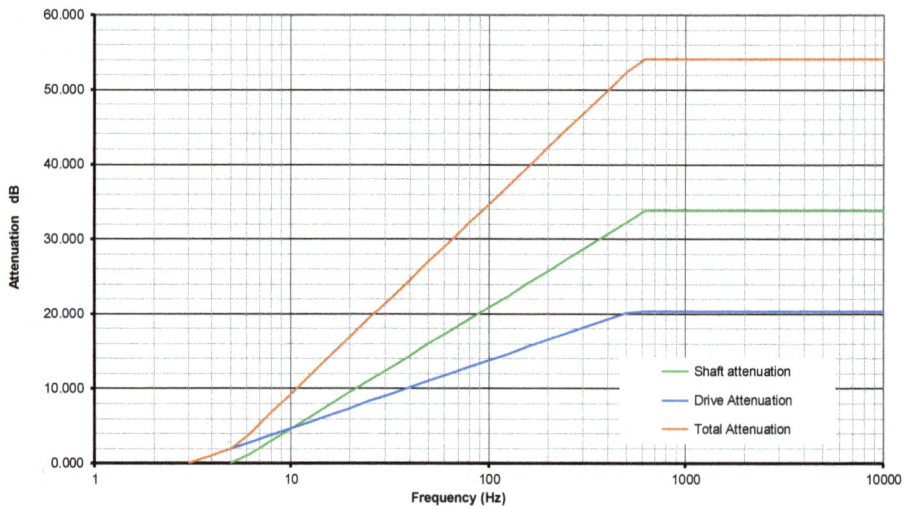

FIG 14 – Noise attenuation in underground airways.

This would be sufficient to meet most environmental noise criteria and gives considerably more attenuation at frequencies below 50 Hz than conventional silencers unless specially designed.

LOW FREQUENCY NOISE SILENCERS

Large scale silencers have yet to be built to reduce noise in the 10 to 30 Hz range.

Splitter type silencers have been built with significant attenuation in the 30 to 50 Hz range using pure reactive principles and this could be comfortably extended to 20 Hz.

Design exercises have been carried out for silencers with significant attenuation between 10 and 20 Hz using multiple Helmholtz resonators arranged in splitter form. These have shown this is feasible but large. This is illustrated in Figure 15 which shows a silencer to provide approximately 10 dB attenuation from 10 to 20 Hz with a flow of 240 m³/s.

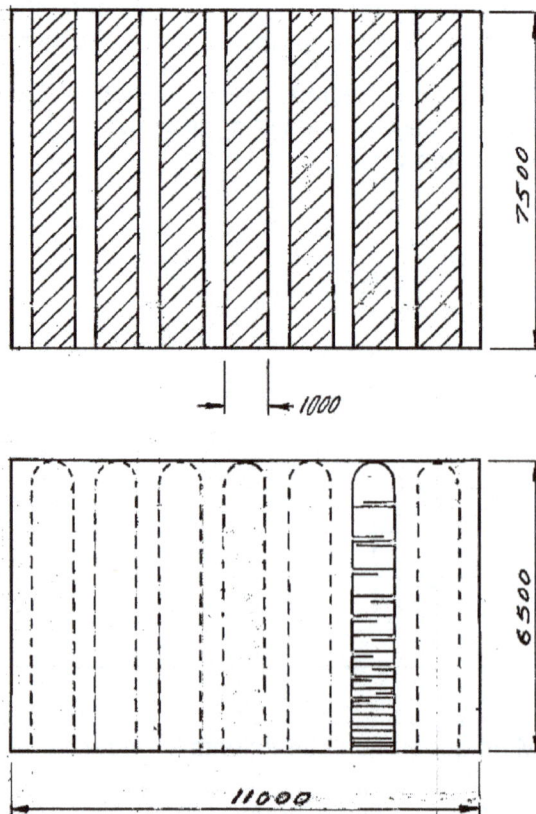

FIG 15 – Silencer to give 10 dB attenuation between 10 and 20 Hz.

Silencers have been proposed using 1/4 wavelength pipes. The pipes are long, 8.6 m at 10 Hz, and a large number are required. These are tuned to one frequency and do not give broadband attenuation. This reflects the attention which is given to discrete frequencies in the low frequency noise spectrum drawing attention from the high-level broadband base. Broadband attenuation can be obtained using 1/4 wavelength principles as shown by Magnani (2021), but this requires a significant number of groups of pipes of differing lengths as shown in Figure 16 to provide significant broadband attenuation.

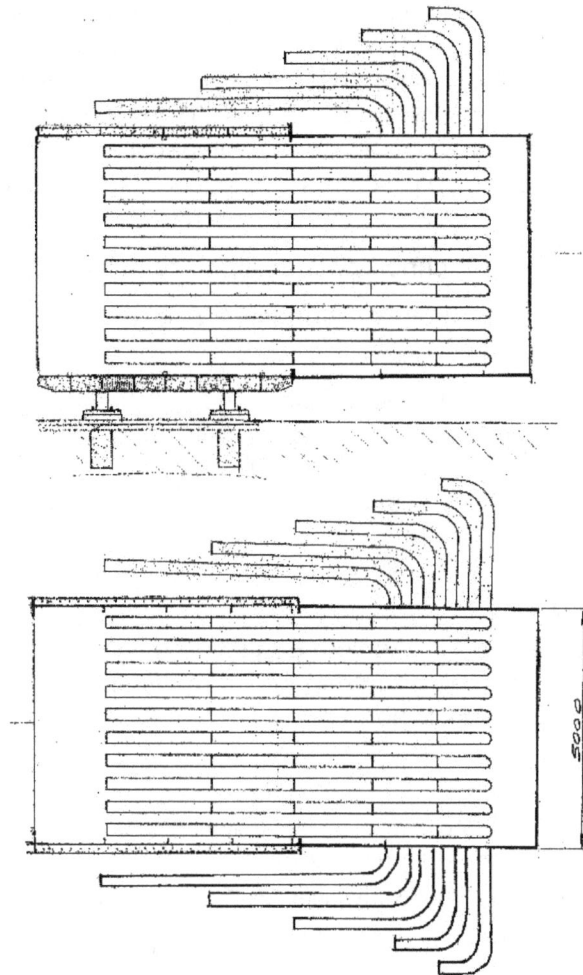

FIG 16 – Multiple 1/4 wavelength pipe silencer with attenuation between 10 and 20 Hz.

CONCLUSIONS

The control of fan noise in the environment has always been important but is becoming increasingly significant and requiring attenuation at lower frequencies an area in which the mines and fan suppliers have limited experience.

In achieving a satisfactory outcome both the mine operator and the fan supplier have responsibilities.

Though it is logical for the mine to specify the sound pressure level which must be achieved in the near vicinity of the ventilation installation it is not logical to specify the required level at a position remote from the fans. This requires modelling which require detailed topographical and meteorological data and experience which is outside the fan suppliers normal field of knowledge. This should be the responsibility of the mine who should specify the required sound power level in 1/3 octave bads from 10 to 10 000 Hz.

The fan supplier must provide accurate sound power data in 1/3 octaves from 10 to 10 000 Hz, arrange the fans in a manner which maximises the effect of directivity in the direction of critical receptors and provide noise control measures to meet the mine operators specified sound power levels.

Low frequency noise is becoming one of the most common sources of complaints and in particular discrete frequencies in the 10 to 20 Hz range. This is below the audible frequency range but the human body senses low frequency noise in other ways.

At low frequencies measurement become more difficult and as a result noise below 62.5 Hz has tendered to be ignored by parties including those preparing measurement and prediction standards, fan suppliers and silencer manufacturers. But because it is difficult does not mean it does not exist and the mining industry must deal with it.

Conventional silencers have little attenuation below 50 Hz. As a result of this there is commonly high levels of noise in the 20 to 50 Hz range. This is in the audible range but is sensed in some of the same forms as noise in the infra sound range below 20 Hz. It is possible that many of the low frequency complaints arise due the high levels in the 20 to 50 Hz area. When the levels are measured discrete frequencies below 20 Hz are found and become the 'source' of the problem. The first step in avoiding low frequency noise problems is to ensure the levels in the audible 20 to 50 Hz band are controlled.

Conventional proven splitter silencer technology is available to achieve this.

No large-scale silencers have been built to attenuate noise in the 10 to 20 Hz range, but the technology exists though they would be large.

Auxiliary and booster fans have always been used underground but it is now becoming more common for the primary ventilation fans to be installed below the surface. This is done for several reasons including reduced noise levels at the collar. Depending on the position in the mine, rock conditions and required community levels this can provide sufficient attenuation such that no additional silencing is required.

ACKNOWLEDGEMENTS

The author would like to acknowledge Agnico Eagle Fosterville Gold for allowing the use of underground attenuation date from their mine and in particular Mr Jeff Norris for his many constructive discussions leading to this paper.

REFERENCES

Concawe, 1981. Report No. 4/81, The Propagation of Noise from Petroleum and Petrochemical Complexes to Neighbouring Communities. Available from: <https://www.concawe.eu/publication/report-no-481/>

Day, A, Hansen, C and Bennett, B, 2009. Duct Directivity Index Applications, *Acoustics Australia*, 37(3):93–97.

Environmental Protection Authority of NSW, 2017. Noise Policy for Industry. Available from: <https://www.epa.nsw.gov.au/your-environment/noise/industrial-noise/noise-policy-for-industry-(2017)>

German Standards, 1997. DIN 45680:1997. Measurement and evaluation of low-frequency environmental noise. Available from: <https://www.normsplash.com/Samples/DIN/143516423/DIN-45680-1997-en.pdf>

Howes, M J, de Koker, M J and Edwards, A L, 2014. Noise Control for Hearing Conservation, chapter 37 in *Ventilation and Occupational Environment Engineering in Mines* (ed: J J L du Plessis), third edition, pp 749–772 (Mine Ventilation Society of South Africa: Johannesburg).

International Organisation for Standardization, 2023. ISO 226:2023, Acoustics – Normal equal-loudness contours. Available from: <https://www.iso.org/obp/ui/#iso:std:iso:226:ed-3:v1:en>

International Organisation for Standardization, 1993. ISO 9613–1:1993 Acoustics – Attenuation of Sound during propagation outdoors, Part 1: Calculation of the absorption of sound by the atmosphere. Available from: <https://www.iso.org/standard/17426.html>

International Organisation for Standardization, 1996. ISO 9613–2:1996 Acoustics – Attenuation of Sound during propagation outdoors, Part 2: General method of calculation. Available from: <https://www.iso.org/standard/20649.html>

Magnami, A, Marescotti, C and Pompoli, F, 2021. Acoustic absorption modeling of single and multiple coiled-up resonators, *Applied Acoustics*, 186(7):108504.

Sabine, H J, 1940. The Absorption of Noise in Ventilation Ducts, *Journal of the American Society of Acoustics*, 11:53–57.

Environmental noise assessment and abatement of surface primary ventilation systems

G R Savage[1]

1. Global Mining Specialist; Howden Australia; Baulkham Hills NSW 2153.
   Email: glenn.savage@chartindustries.com

ABSTRACT

Environmental noise from Primary Ventilation Systems located on surface have and are increasingly being subject to stricter noise attenuation requirements, both for local area personnel safety and far field noise at mine boundaries and critical receiving points. Mine Ventilation System performance and consequently sound power levels are increasing with deeper mines and longer main airways increasing the fan pressure requirements. The resulting impact is that noise attenuation requirements have become a significant element of the Ventilation System capital and operating cost and are increasingly contributing to system reliability issues and problems with meeting requirements. This paper encompasses the specification of noise level requirements, the importance of identifying critical receiving points, applicable standards, applicable tolerances, inversion effect considerations, impact of large duct system surface areas on far field noise, sound power level versus sound pressure level when specifying equipment requirements, inlet silencers versus inlet duct acoustic lagging, silencer type considerations, fan layout and discharge directivity considerations, capital cost impacts of attenuation, assessing risk to operations of noise abatement options, local area safety impacts from noise; assessment of control measures versus attenuation, power consumption increases from silencers, maintenance requirements and operating cost impacts. The author highlights observations in recent years where environmental noise impacts and the consequences of not meeting regulatory requirements and consent conditions has significantly impacted mining operations. The objective of the broad scope of this paper is to make available to the industry the considerations necessary to assess the specification requirements, the equipment options and cost impacts of noise abatement, thus reducing the likelihood of impacts and risk to mining operations.

INTRODUCTION

Mining operations are subject to increasing scrutiny and regulatory pressure regarding noise emissions from Primary Ventilation Systems (PVS). The necessity for stringent noise attenuation measures arises from concerns for the safety and well-being of local personnel, as well as the mitigation of excessive far-field noise levels at mine boundaries and critical receiving points. As mines extend deeper and main airways longer to reach remote extraction sites, this necessitates higher fan pressure requirements, increasing noise emissions. Consequently, noise attenuation has become a pivotal aspect of PVS design, operation, and regulatory compliance.

Environmental noise regulations mandate not only total sound pressure level at critical receiving points but cover noise frequency spectrum limitations to prevent intrusive noise characteristics from tonal and low frequency noise. Not meeting tonal and low frequency noise along with A-weighted total sound pressure level limits represents a risk to mining operations, potentially limiting PVS operation capacity in cases where attenuation design is not adequate. These risks have been highlighted in recent years with several mining operations having to turn down the PVS capacity to meet environmental noise requirements, resulting in unplanned production limitations. Retrofitting noise abatement measures to an existing PVS to achieve design capacity within noise limits and allow the required ventilation necessary to realise the mine production plan can require significant capital expense and PVS downtime.

The purpose of this paper is to provide guidance on steps to ensure noise limits are met and the cost of noise abatement elements is considered through the planning process of a surface PVS project delivery.

SPECIFICATION OF NOISE LEVEL REQUIREMENTS

Effective noise attenuation begins with the specification of noise level requirements tailored to the specific operational context and regulatory mandates. The noise limits need to be clearly communicated to the PVS Contractor. These noise limits should be considered from the earliest concept stages of project planning. Noise abatement can be a high capital and operating cost element of a PVS installation, not including this cost and additional footprint in the early planning stages can result in significant deficiencies in the project plan.

More typically current practice is for the Mine Operator to specify to the PVS Contractor a fan discharge and in most instances a fan casing breakout noise sound pressure level A-weighted limit, commonly in the range of 80 to 85 dB(A). From the perspective of the PVS Contractor this is a definitive measurable limit and is therefore acceptable from a design directive and contractual basis. Whilst specifying in this manner suits the PVS Contractor it is not necessarily in the best interests of the Mine Operator as this specification may only meet local area noise for the safety of personnel and does not necessarily ensure far-field environmental noise requirements are met, particularly in situations where low environmental noise impact is regulated.

In cases where there are no critical receiving points within effected range of the PVS installation the noise level limits may be driven solely by local area noise levels to achieve a safe work environment for personnel. In these cases, a typical specification of $L_{Aeq,\ 15\ minute}$ 83 dB(A) at 1 m from the fan casing is common. Typically, Occupational Health and Safety guidelines state a limit of 85 dB(A) for an 8 hr equivalent continuous A-weighted sound pressure level and a C-weighted peak of 140 dB(C). These limits whilst generally common may vary across different global jurisdictions, local regulations.

Without attenuation the noise from the fan open area is higher than the casing breakout noise. The open discharge zone in the case of exhaust systems is not a working area, personnel are not subject to the open discharge noise in the very near field, therefore the 1 m distance is not applicable. It is recommended that the open discharge noise be evaluated by contribution to the working area noise (outside the exhaust exclusion zone) to assess any requirement for discharge attenuation.

Noise attenuation of ducts open to atmosphere is commonly achieved by incorporating silencers. Silencers increase the capital and operating costs of the PVS, they should only be implemented where there is an explicit requirement to meet safety or regulatory requirements. The PVS specified noise limit should be the lower of regulatory or site policy requirements.

In cases where there are far-field critical receiving points beyond the local working area zone specifying a fan casing and fan discharge noise contribution at 1 m from the fan alone may not meet regulatory/consent criteria at critical receiving points.

Less typically noise specifications of PVS which do have far-field critical receiving points have not specified a sound power level limit of the installation. This paper is recommending a change of practice to specify the PVS noise limit as a single point noise source sound power level limit of the system, in addition to the local area breakout noise limit for local safety.

The sound power level limit specified by the Mine Operator should be maximised within the allowable limits to minimise the capital and operating cost of the PVS installation. If tolerances are taken into account in the specified sound power level limit, then a no negative tolerance should not be specified to ensure the highest project delivery and system operating efficiencies are realised.

Different PVS Contractor system options are likely to have different duct design solutions and fan proportions hence if only specifying breakout noise the far-field noise level will vary for different system proposals. By specifying a PVS sound power level limit the Mine Operator is ensuring the various PVS proposals under consideration meet the criteria.

Specifying a sound power level limit requires the PVS Contractor to consider the noise contribution from all the system elements including the fan casings, openings to atmosphere, number of fans operating, motors, pumps and the duct surface area which can have a higher contribution at far-field than a fan discharge with silencer. Specifying in this way, the PVS Contractor will assess the extent of noise attenuation necessary to achieve compliance.

The following considerations are necessary to establish the allowable PVS Sound Power Level (Lw) limit.

Critical receiving point identification

Typical critical receiving points are mine boundaries and or specified locations in and outside the mine boundary. These critical receiving points need to be considered from a far field noise position relative to the PVS noise source. The noise at each of the critical receiving points is likely to be a combination of the PVS noise source plus the contribution from other mining activities which are assessable by the Mine Operator not the PVS Contractor.

Identifying critical receiving points within and around the mining site is paramount for designing effective noise attenuation strategies. This involves mapping out areas where noise exposure poses the greatest risk to personnel, nearby communities, or sensitive environmental habitats. Critical receiving points may include residential areas, wildlife habitats, or designated quiet zones. Various methodologies, including noise mapping and modelling, can be employed to pinpoint these locations accurately. Regulatory requirements at mine boundaries should be clearly established. Additionally, input from stakeholders, including local residents and regulatory authorities, should be considered when identifying critical receiving points. Background noise levels should be baselined at these locations.

Establishing background noise levels

This involves conducting comprehensive noise surveys to assess existing background noise levels, nighttime and daytime noise levels are required. Typically, any noise 5 dB above the equivalent continuous A-weighted background noise (averaged over 15 mins) at the critical receiving point will be experienced as intrusive. To future proof the installation and reduce the risk to operations that may result from not maintaining the amenity of land use, it is considered prudent to apply this limit even when regulations or consent do not specify this measure. This is particularly relevant to protect Mine Operators from retrospective legislation which has had implications for operations despite originally meeting consent criteria.

Low frequency and tonal noise

Low frequency, tonal, impulsive and intermittent noise may result in annoyance in some cases despite the total sound pressure level at the receiver being within criteria limits. Of these noise characteristics low frequency and tonal noise are potentially most likely applicable to typical PVS noise spectrums. Applying modifying factors (penalties) to account for the impacts of these noise characteristics at the critical receiving points could be required. Regardless of whether or not consent conditions apply this criterion, applying modifying factors where applicable is currently the practiced method to limit impact to future operations.

The quanta of any modifying factors are generally established as part of the environmental approval/consent, if not published guidance is generally available. It is common for different jurisdictions to utilise varying methods for applying modifying factors. The PVS Contractor is generally not expected to be aware of these factors, they must therefore form part of the specification.

There is significant variance in the level of tonality at which penalties apply across various jurisdictions, many of these utilise some variation of the simplified one-third octave band method from ISO 1996-2 (ISO, 2017a), however there is a wide variation between the definition of a tone and the penalty. Europe, New Zealand and other jurisdictions use the narrowband procedures, some for specific applications only. The narrowband assessment procedures defined by ISO 1996-2 or IEC 61400-11 (IEC, 2012) have reportedly been found to have a consistent relationship with the typical subjective response to tonal noise. Where narrowband frequency fan noise is available comparing the modifying factors using the ISO 1996-2 narrowband procedures to the jurisdictions regulatory guidelines is likely to give some measure of the impact of any retrospective regulation change, 'future proofing'.

Tonal noise or tonality; containing a prominent frequency and characterised by a definite pitch, example; fan blade passing frequency. Example, utilising NSW (Australia) Industrial Noise Policy guidelines, a modifying factor of 5 dB should be applied where the level of one-third octave band exceeds the level of the adjacent bands on both sides by; 5 dB or more where the centre frequency of the band containing the tone is above 400 Hz; 8 dB where the tone is 160 to 400 Hz and 15 dB where the tone is below 160 Hz. These guidelines are the least stringent regulation in Australia.

Modifying factors where applicable are applied to the A-weighted sound pressure level at the critical receiving point. Concept stage simplified calculation, assuming no environmental or topographic factors the sound pressure level at the critical receiving point can be nominally estimated as Sound Pressure Level (Lp) = Sound Power Level (Lw)-20.log(r)-8, where; Sound Power Level (Lw) is the PVS Sound Power Level and r is the radius from the PVS to the critical receiving point. Calculation is based on free-field hemispherical propagation.

Low Frequency Noise (LFN); the frequency range defined as LFN varies across different jurisdictions typically within the 10 Hz–250 Hz frequency range. A modifying factor can be applied by assessing the difference between the C and A weighted levels. A 5 dB correction is typically applied if the difference is 15 dB or more, this may vary depending on the jurisdiction.

Where the PVS has tonal and low frequency intrusive noise characteristics in the low frequency range, generally only one modifying factor correction is applied.

The applicable modifying factors and distance to the critical receiving point should be defined in the specification to allow the PVS Contractor to apply the factors where the proposed equipment noise frequency spectrum has the described intrusive characteristics. Stating the 'fans will not have a tonal noise characteristic' or similar in the specification is not enough as all fans will have some tonal characteristic by virtue of the impeller blades passing fixed components.

Meteorological effects

The increase in noise levels that results from atmospheric temperature inversions and wind effects must be considered when assessing the PVS sound power level limit to be specified. Meteorological effects can typically increase noise levels by 3 to 10 dB, in extreme cases by as much as 20 dB.

Inversion occurs where temperatures increase with height above ground level causing sound waves to refract, increasing noise levels at significant distances from the noise source.

Wind drift can also increase the noise level at the receiver where the wind velocities are increasing with height with wind from the source to the receiver. Wind is considered to be a feature where there is a 30 per cent or more occurrence of wind speeds below 3 m/s at 10 m height above ground level.

The first step to assessing meteorological impact is to apply published default meteorological parameters to predict noise levels and thus the required PVS sound power level limit. Where this limit would require attenuation elements for the PVS to meet the criteria then meteorological effect should be considered as potentially significant, further detailed assessment of meteorological effects should be undertaken.

Default meteorological parameters are tabled in most jurisdictions by regulatory authorities as part of industrial noise policies along with methods for determining noise increase. One example of this is the NSW Industrial Noise Policy, published on the NSW Environment Protection Authority website.

Effects of topography/obstructions/barriers

Any permanent shielding provided by natural topography or otherwise between the PVS and the critical receivers can influence the sound pressure level at the receiver location. Whilst a simple assessment of noise levels based on hemispherical free field propagation is likely acceptable for establishing concept level budgets, for later stage planning the effects of these barries and topography require complexed analysis, accepted practice makes use of computer noise models.

Future amenity

Noise levels from planned expansions should be estimated to allow assessment of the cumulative effect so that future amenity is not limited. For example, the planned PVS under consideration may not require attenuation to meet regulatory noise requirements at the mine boundary but the addition of a future ventilation shaft and associated PVS may result in a combined noise level exceeding the criteria. In this case it may be prudent to specify the PVS is designed for the future retrofit of attenuation, for example implementing a fan discharge arrangement and fan capacity to allow for retrofitting discharge silencers.

Specifying PVS sound power level limit at early planning stages

Early in project planning the noise limit criteria may not have been established, it is recommended that EPA/Regulatory Authority policy guidelines be utilised to establish a criterion for planning purposes, ensuring early planning processes capture the cost and footprint of any attenuation requirements. At this point a simplified estimate is likely acceptable.

Example, a greenfield mine development plan at the concept level stage may not have licensed consent noise conditions established however the PVS budget needs to be established for planning purposes. Typically, a rural area nighttime recommended $L_{Aeq, 15\ minute}$ 40 dB(A) sound pressure level residential limit would apply according to amenity criteria published by the jurisdiction NSW EPA (2000) noise policy. Farming residences are located adjacent to the mine boundary, 1500 m distance from the proposed PVS location. The area can be reasonably expected to be impacted by inversion effects increasing the PVS noise level at 1500 m by 5 dB. To ensure the planning budget is conservative given the concept level planning stage it is assumed low frequency noise is significant or the characteristic below 32 Hz is unknown so a modifying factor of 5 dB is applied. Assuming a worst case that the fan blade passing frequency noise has a significant tonal characteristic than a further 5 dB modifying factor is estimated. There are no significant topographic features or constructions between the PVS and the critical receiving point. There are no reasonable expectations that the cumulative noise level from industrial sources could increase in future. Therefore, the concept/pre-feasibility budget for the PVS should be based on a system sound power level limit of 97 dB(A):

> Target max Sound Pressure Level at boundary residence; 40 dB(A)
>
> Inversion effect; 5 dB
>
> Modifying factor; 10 dB
>
> Distance from PVS noise source; 1500 m
>
> PVS Sound Power Level limit = Sound Pressure Level + 20.log(distance) +8
>
> $$= 40-5-10+20.log1500+8$$
>
> $$= 97\ dB(A)$$

Specifying PVS sound power level (Lw) limit at later planning stages

As project planning progresses through stage gates, there is a requirement to increase the proposal accuracy. At these stages consultation with the environmental regulatory authorities and community have progressed and background noise levels have likely been surveyed. Guidance from regulatory authorities and community consultation have allowed a negotiated acceptable noise level limit to be set at the appropriate critical receiving points at and outside the mine boundary, or at least expectations are indicated. It is likely these limits are 5 dB below the surveyed nighttime background noise at least, if not the Mine Operator should consider utilising the background noise survey-5dB as a design condition to preserve amenity, ensure intrusive noise does not impact the environment and future proof the installation.

Regulatory consent levels should be utilised to prescribe the PVS sound power level limit by calculating or modelling the PVS sound power level limit from the regulated $L_{A\ eq, 15\ minute}$ sound pressure level limit at the mine boundary and any other identified critical receiving points, taking into account the applicable environmental factors.

At this stage of planning a simplified estimate based on free-field propagation is likely not appropriate to provide the required planning accuracy in noise sensitive applications. This stage of project planning typically employs noise modelling. Generally, noise Consultant Companies are commissioned to undertake a study which will take into account the site conditions, topography, vegetation, climate, molecular absorption, wind, background noise and the proposed sound pressure level limit at identified critical receiving points in and outside the mine boundary. This study will provide a PVS sound power level limit that can be specified to the PVS Contractor along with the noise limits for local safety of personnel.

PRIMARY VENTILATION SYSTEM SOUND POWER LEVEL LIMIT DESIGN CONSIDERATIONS

In the past it has not been typical for PVS Contractors to design the system to a specified total system sound power level criterion. Following are the considerations and methods to calculate the PVS sound power level.

The PVS sound power level is calculated utilising the number of fans operating at the peak noise design duty.

A surface PVS can include the following equipment, the sound power level will require estimating for each piece of equipment:

- fan/s
- electric motor/s
- ducting
- flexible joints
- lubrication systems
- self-closing doors/dampers
- switch room air-conditioners
- generators
- compressors
- diesel engines
- water washing
- brakes.

Fan sound power

The fan sound power level is the dominant noise source of the PVS. For each fan in the system the fan manufacturer defines a sound power level at each frequency band for the open inlet, open discharge and casing breakout. These values are calculated utilising the 'Fan Laws' as prescribed in ISO 5801 (ISO, 2017b), based on laboratory measurements of scaled models of the fans and take into account the fan casing thickness by applying different attenuation values for material thickness.

Traditionally the sound power level values are published at centre frequency octave bands in the range of 63 Hz to 8000 Hz. As discussed earlier to analyse the tonal noise characteristics it is necessary to have the 1/3 octave band values and potentially the narrow band characteristic. To assess the low frequency characteristics, it is necessary to know the noise characteristics down to 12.5 Hz. If the PVS Contractor cannot provide the 1/3 octave bands and or the low frequency noise data to the 12.5 Hz range, then modifying factors are typically applied by the Mine Operator in line with the regulators guidance to ensure compliance, the contracted sound power level limit of the installation must be specified accordingly.

The fan sound power level spectrum is subject to tolerances according to ISO 13348 (ISO, 2007), the specification of the PVS sound power level limit should detail if the limit includes or excludes this tolerance so that the PVS Contractor can account for the tolerance accordingly.

Electric motor/s sound power

Typically, motor noise levels are provided by the motor manufacturers as sound pressure level at 1 m as a total value in a non-loaded condition. Therefore, unless testing in the loaded condition is undertaken to measure the duty load condition sound pressure level, accepted practice is to add 3 dB to the motor manufacturers quoted noise level.

The estimated operating motor sound power level is equal to sound pressure level + 10.log(surface area @ 1 m from the motor housing m²). Generally, the motor manufacturer will specify a tolerance of 3 dB, this should be added to the sound pressure level value. Example:

Motor manufacture defined unloaded sound pressure level @ 1 m; 76 dB(A)

Motor manufacture quoted tolerance; 3 dB

Projected surface area; 104 m²

Estimated motor noise increase when loaded; 3 dB

Motor sound power level (Lw); 76+3+3+10.log(104) = 102 dB(A)

Diesel engines, compressors, generators, lubrication systems, air conditioners and other proprietary equipment where a total sound pressure level value is provided by the manufacturer should be treated in a similar manner, taking note of whether the quoted noise level is loaded or unloaded and of the tolerance.

Ducting sound power level (Lw)

As PVS capacity has increased, the scale of duct systems has accordingly increased to the point where in some cases the duct breakout sound power level has become of greater significance than the attenuated opening to atmosphere. Commonly in the interest of project schedule fans are being constructed whilst shaft sinking activities are ongoing, to allow this fans are necessarily positioned further from the ventilation shaft or portal, resulting in longer duct lengths hence higher duct breakout sound power level. Assessing duct breakout sound power level is therefore critical to ensure noise criteria limits at critical receivers are achieved.

The calculation of the ductwork sound breakout can be estimated by the formula provided in the ASHRAE (2019) Sound and Vibration Control Chapter 49, issue where:

Sound Power Level $Lw_{(out)}$ = Sound Power Level $Lw_{(in)}$ + 10log(S/A) − TL_{out}

S = the surface area of the outside sound radiating surface, m²

A = the cross-sectional area of the inside of duct, m²

TL_{out} = the normalised duct breakout sound reduction, dB

Note; the ASHRAE guide provides experimentally determined TL_{out} data which have diameters and material thicknesses more typically of HVAC applications. TL_{out} data more appropriate for PVS applications can be found within VDI 3733 July 1996 - Annex B 'Determination of the sound pressure spectrum radiated by a pipe'

Flexible joints considerations

PVS duct systems necessarily utilise flexible joints to allow for expansion and isolate the fans from the duct system. Acoustically, flexible joints act to reduce the propagation of flanking noise and act as acoustic breaks. In noise sensitive applications it is necessary to incorporate acoustic pillows usually from mineral wool fibre and Wave Bar © or similar type flexible element material to ensure the flexible joint is not a leakage path for noise.

Auxiliary equipment sound power level (Lw)

PVS include some intermittent noise sources, including dampers or self-closing doors, brakes, water washing systems and potentially others. They are generally not significant to the time weighted average of noise at far field critical receivers given their use is so sporadic, however they may impact local area safety, in particular self-closing doors on large systems with high back pressure. Typically, Occupational Health and Safety noise limits define a C-weighted peak of 140 dB(C), large self-closing doors have been known to generate a peak noise above this limit when closing, in these cases dampening systems can be employed to reduce the impact noise to allowable limits.

Testing PVS sound power level (Lw)

The PVS Contractor should be expected to undertake a noise test according to ISO 3744 (ISO, 2010) Acoustics Determination of Sound Power Levels. The standard determines sound power via sound pressure measurements on a surface enveloping a sound source located in free field on a reflective surface (ground). The number of measurement positions is determined by the standard and are a function of the size of the envelope.

In addition, the background noise recordings (no fan operations) are required to be recorded immediately after the measurements at each location over the same time interval. It can be problematic to turn the PVS off after testing to measure the background noise due to operational requirements, by agreement the background noise can be determined through measurements carried out at a similar location unaffected by the installation but with effectively the same environmental conditions as the test area.

Flow measurement will need to be taken during the test, the conditions will need to be held stable to ensure air flow rate and fan work remain the same for the duration of the test period. Performance testing should be undertaken with reference to ISO 5802 (ISO, 2001) and ISO 13348 (ISO, 2007) standards.

PRIMARY VENTILATION SYSTEM NOISE ATTENUATION

Should it be determined the PVS will require noise abatement to meet the specified noise criteria the extent of attenuation requirements can vary broadly from simply changing the configuration of the PVS opening to atmosphere to fully lagged and silenced systems.

To allow the PVS Contractor to assess and propose the appropriate noise abatement options they will need guidance on the preferred PVS opening direction to atmosphere, for an exhaust system discharge direction, for an intake system intake direction. This preference can be driven by exhaust plume dispersion, control of exclusion areas, noise directivity, filtration system design, maintenance access, silencer design, silencer length, footprint limits, connection to heating/cooling systems, height limitations, pressure loss, capital and operating cost.

Typically, the first level of attenuation will be incorporating silencers to the duct opening to atmosphere. There are three types of silencers generally employed in PVS: absorptive (dissipative), reactive, reactive-dissipative. Considerations for each of these options are outlined in the following sections.

Intake PVS can require silencing between the fans and shaft or portal connection where personnel are moving through or adjacent to the shaft or portal, example, head frame building with personnel hoist. In these cases, the arrangement must be understood and considered by the PVS Contractor to ensure Occupational noise limits are met.

In higher noise sensitive applications acoustic lagging is employed to attenuate the breakout noise of the fan casing/s and ducts.

Where low noise motors or pumps require attenuation to meet noise criteria than acoustic enclosures are utilised.

Absorptive silencers

Absorptive (dissipative) type silencers utilise a mineral or synthetic fibre wool enclosed within perforated steel sheet, usually in rectangular sections which divide the duct along the flow axis, known as splitters.

This type of silencer provides broadband sound attenuation with relatively little pressure loss by partially converting sound energy to heat through friction as fibres vibrate in the fibrous media. The design is affective in the mid to high frequency range, attenuation is more modest at low frequency.

This type of silencer has the lowest footprint and cost of the three options being discussed. They are suitable for clean environments as contaminants will clog the holes in the perforated sheet and penetrate the fibre continuously degrading the silencers attenuation characteristic. Therefore, absorptive silencers are preferred for intake PVS where clean conditions are generally prevalent.

Commonly however absorptive silencers are being utilised on exhaust PVS applications, depending on the contaminant load, these silencers attenuation will deteriorate at varying rates. The attenuation can typically be expected to degrade rapidly in exhaust PVS applications. If utilising this type of silencer in exhaust PVS applications, consideration should be given to control measures to ensure noise criteria levels are maintained. Cleaning *in situ* to maintain the silencer at design attenuation levels is not considered practical as this would require not only the cleaning of the holes in the perforated sheet but replacement of the fibrous media. In the past where cleaning of the perforated sheets alone has been undertaken, it has been demonstrated that the original design performance was unable to be restored. Depending on the level of fan redundancy of the PVS, quick change out splitters or whole silencer assemblies should be considered. The cost of these control measures will offset the higher cost of the more suitable reactive-dissipative silencer option.

Reactive silencers

Reactive silencer is a general term for reflective and resonator silencers where the majority of the attenuation does not involve sound energy dissipation. Pure reactive silencers are based on the Helmholtz principle, in which the airstream passes through a series of tuned cavities where the sound waves are reflected back on themselves out of phase, in effect cancelling out the noise by mixing it with its mirror image.

This type of silencer is not used for broadband attenuation. The cavities are tuned to cancel out specific frequencies of the fan noise spectrum, commonly combined with absorptive (dissipative) and reactive-dissipative silencers to attenuate low frequency and narrow band noise characteristics such as fan blade passing frequency.

Unlike absorptive silencers reactive silencers are suitable for the typically contaminated air of mine exhaust PVS applications as the air passes over the throat opening of the chamber and does not enter the chamber, contaminants are not deposited but pass through the passages between the splitters. If required they do lend themselves to *in situ* periodic water spray cleaning where contaminants deposit on structural elements.

Reactive-dissipative silencers

A reactive-dissipative silencer combines reactive and dissipative elements, example in Figure 1. The Helmholtz principle is combined with fibrous dissipative sound absorbers. While the reactive process cancels out the specific frequencies generated by the fan, the fibrous material removes a broader spectrum of sound waves.

Like purely reactive chambers the reactive-dissipative chambers are not impacted by contaminants as the fibrous material is inside the chamber and is not impacted by contaminants.

The combination of purely reactive and reactive-dissipative chambers are utilised to provide a broad band spectrum attenuation characteristic tuned to the fan noise characteristic. The footprint and cost of this technology is significantly higher than the absorptive type however the attenuation does not degrade significantly over time in typical mine exhaust applications. This option therefore provides reliable attenuation to meet noise criteria without instigating control measures as with absorptive silencers in contaminated applications.

FIG 1 – Single Reactive-Dissipative splitter example.

Acoustic lagging

Where acoustic lagging of the fan casing and ducting is required the particular lagging system employed needs to balance cost and utility to achieve the required noise criteria. A typical lagging system comprises a defined airgap, fibrous and or porous material thickness (potentially of multiple layers) and an out cladding material.

The typical PVS Contractor will have various options for each of these elements, each system will have different attenuation characteristics. The attenuation characteristic of each system is dependent not only on the material type and thickness but the attachment mechanism as isolation efficiency of the lagging from the structure determines the flanking noise pathways.

It is the responsibility of the PVS Contractor to ensure the lagging system design and application meets the noise criteria.

Exhaust PVS fan inlet duct breakout noise, lagging versus silencer

As discussed, the duct lengths between the fan inlet and the mine shaft or portal connection can be such that the duct breakout noise is significant. This noise can be attenuated by utilising a fan inlet silencer so that only the duct length between the fan and the silencer requires lagging or alternatively lagging the complete inlet duct system.

When assessing these options, the total lifetime cost of the PVS requires analysis. Typically, this analysis has demonstrated lagging to be the preferred option given lagging does not add pressure loss to the system and requires little maintenance, if maintenance of lagging is required it does not generally impact the PVS operation. An inlet silencer would add pressure loss increasing the system power consumption and hence operating cost. Silencer costs should include any operating downtime for silencer maintenance and any additional redundancy costs that have been implemented, if not instigated for other reasons.

If an inlet silencer is found upon analysis to be the selected option, designs must ensure that the inlet silencer does not present a system effect on the fan performance. Transitions to the fan inlet must present acceptable velocity profile distribution to the fan inlet if predicted fan performance is to be achieved, the PVS Contractor should be able to demonstrate impact on inlet flow conditions has been considered in their proposal.

Other attenuation options

Less commonly acoustic barriers can be constructed such as earthen berms, acoustic walls, etc. These options are usually outside of the scope of the PVS Contractor.

Active noise control utilises a secondary source sending out phase shifted noise into a duct to cancel out the fan generated noise. These type of attenuation systems have not typically been utilised in PVS as they are more suited to smaller duct diameters. They do potentially have application to attenuate low frequency noise used in conjunction with the more typical silencers described. Due to length requirements and the typically aggressive environments active noise control has not found common use.

APPLICATION EXAMPLES

The following are actual relevant examples of noise considerations and impacts to mining operations. In each case the site has not been identified. The data provided is necessarily limited due to the potentially sensitive information.

NSW Australia mine – sound power level specified

The Mine Operator in this case specified a sound power level limit of the PVS as per the recommendations of this paper. The installation is located in a highly noise sensitive environment, surrounded by sensitive receivers, inversion effects are relevant. The PVS incorporates a comprehensive attenuation system, utilising acoustic lagging of the ducting and fan casings, reactive-dissipative silencers on the fan discharges and acoustic enclosures for the motors.

The duct surface area of this installation is extensive to allow separation of the PVS construction and shaft sinking activities. This example highlights the relevance of the ducting breakout noise contribution to the total PVS sound power level:

- Estimated total ducting breakout sound power level, L_{WA} = 101 dB(A)
- Estimated silenced discharge sound power level, L_{WA} = 84 dB(A)
- Estimated fans casings breakout sound power level, L_{WA} = 94 dB(A)
- Estimated motors enclosures breakout sound power level, L_{WA} = 91 dB(A)

If the specification had merely specified, the discharge and casing breakout sound pressure noise limits in the way many specifications have traditionally done, then the contribution of the ducting breakout noise would not have been considered in the design:

- Total PVS sound power level without ducting and motors considered, L_{WA} = 94 dB(A)
- Total PVS sound power level including ducting and motors, L_{WA} = 102 dB(A)

The closest sensitive receiver is nominally 550 m from the PVS. For comparison purposes if we assume free-field hemispherical propagation of the noise the difference in sound pressure level at the sensitive receiver is 8 dB(A), demonstrating the significance of including the ducting and motor breakout noise.

VIC Australia mine – EPA restricted use due to low frequency noise

A mine in Victoria Australia has been issued a prohibition notice restricting a surface primary ventilation fan system to 400 rev/min between the hours of midnight and 6 am, due to low frequency noise (LFN) complaints from sensitive receivers.

The noise surveys measured LFN levels significantly below the EPA guidelines (NSW EPA, 2000). LFN levels indoors were found to be below the threshold of audibility for the majority of the population.

The design noise criteria were Lp 23 dB(A) at 800 m, the closest sensitive receiver. Testing after commissioning confirmed the PVS met this design criteria. LFN criteria was not specified to the PVS Contractor however if it had been it would have utilised the applicable EPA guidelines, the PVS noise as measured meets the guidelines.

Despite the noise levels being well within the compliance noise limits and below the EPA guidelines (NSW EPA, 2000), the LFN level at 16 and 20 Hz was considered to cause harm and therefore unreasonable under the act.

Control measures considered relocation of the fan system underground or modification to the attenuation system of the operating surface system. Future ventilation upgrades will position the PVS underground, the surface installation will be decommissioned.

This example demonstrates the importance of understanding the jurisdiction regulations and highlights the impacts of changes to regulations incorporating a health impact reasonability test.

NSW mine – penalty notice issued noise criteria exceeded

Due to tonal low frequency noise characteristics of the mines surface ventilation system noise complaints were received by a NSW mine, reporting an audible low frequency noise with some reports of a non-aural (physical sensation) aspect.

A penalty notice was issued for exceeding operating noise criteria at two noise monitoring stations. A 5 dBA penalty was applied to the noise criteria. Consequently, limiting the operational speed of the installation.

The original absorptive silencer was retrofitted with a reactive-dissipative silencer, the fan discharge to silencer transition was redesigned to improve the flow distribution into the silencer. The reactive-dissipative silencer will provide continuous stable attenuation and provide higher attenuation in the low frequency range.

CONCLUSIONS

This paper is recommending a change from the typical specification of primary ventilation system fans breakout and open inlet or discharge sound pressure level to additionally specifying the total primary ventilation system sound power level. Specifying in this way simplifies the Mine Operators analysis of various supplier system options and noise estimates at critical receiving points. The system supplier is responsible for the total system noise level, ensuring noise criteria are met.

Mine Operators should be aware of the various attenuation options, in particular silencer technology so that a thorough cost benefit analysis including potential impact on operations can be undertaken.

Changing regulations have impacted operations retrospectively, when setting primary ventilation system sound power level limits, current regulatory requirements of the applicable jurisdiction should be compared to the narrowband assessment methodology as defined in ISO 1996-2.

ACKNOWLEDGEMENT

Howden Australia Pty Ltd

REFERENCES

American Society of Heating, Refrigerating and Air-Conditioning Engineers (ASHRAE), 2019. Noise and Vibration Control (TC 2.6, Sound and Vibration), chapter 49, in *ASHRAE Handbook-HVAC Applications*, ASHRAE Technical Committee.

International Electrotechnical Commission (IEC), 2012. IEC 61400-11:2012, Wind turbine – Part 11: Acoustic noise measurement techniques, 3rd edition, 58 p.

International Organization for Standardization (ISO), 2007. ISO 13348, Industrial fans – Tolerances, methods of conversion and technical data presentation.

International Organization for Standardization (ISO), 2017a. ISO 1996-2:2017, Acoustics — Description, measurement and assessment of environmental noise — Part 2:Determination of sound pressure levels.

International Organization for Standardization (ISO), 2010. ISO 3744:2010, Acoustics - Determination of sound power levels and sound energy levels of noise sources using sound pressure – Engineering methods for an essentially free field over a reflecting plane.

International Organization for Standardization (ISO), 2017b. ISO 5801:2017, Fans – Performance testing using standardized airways.

International Organization for Standardization (ISO), 2001. ISO 5802:2001, Industrial Fans – Performance testing in situ.

NSW Environment Protection Authority (EPA), 2000. NSW Industrial Noise Policy, Environment Protection Authority, Sydney, Australia.

Verein Deutscher Ingenieure e.V. (VDI), 1996. VDI 3733, Noise at pipes, Annex B 'Determination of the sound pressure spectrum radiated by a pipe'.

Ultrafine supersonically blown nanofibre membrane for filtration of coalmine dust particles (0.1–0.3 µm)

M M Zaid[1], A Kakoria[2] and G Xu[3]

1. PhD Candidate, Missouri University of Science and Technology, Rolla MO 65401, USA.
   Email: mirzamuhammadzaid@mst.edu
2. Postdoc Fellow, Missouri University of Science and Technology, Rolla MO 65401, USA.
   Email: Akumar@mst.edu
3. Associate Professor, Missouri University of Science and Technology, Rolla MO 65401, USA.
   Email: guang.xu@mst.edu

ABSTRACT

Pneumoconiosis, particularly progressive massive fibrosis (PMF), has rapidly escalated among coalminers, affecting about 15 per cent of individuals. This severe form of the disease has dire consequences, including fatality. With a current estimate of around 60 000 affected miners in the United States and millions more globally, the issue is exacerbated by fine coal dust particles (0.1–0.3 µm). Despite stringent dust regulations, the resurgence of this disease poses significant occupational health hazard. To address this, significant progress has been made in mitigating coal dust through particulate matter filtration during mining operations. Synthetic polymer nanofibrous filter membranes utilised in respirator/mask exhibit substantial potential for coal dust filtration. However, existing technologies such as respirators/masks are typically effective only at the most penetrating particle size (MPPS), around 0.3 µm. This study introduces the utilisation of an Ultrafine Supersonically Blown Nanofibre Membrane with an average nanofibre size less than 100 nm for filtration purposes, targeting particle sizes even smaller than MPPS within coalmines. The fabricated nanofibre membrane enables both transition flow and diffusion-based particle entrapment mechanisms due to its nanofibre size. Consequently, this membrane enhances breathability and particle filtration efficiency beyond existing commercial technologies. This novel Ultrafine Supersonically Blown Nanofibre Membrane holds promise as a potential personal protective equipment (PPE) solution for coalminers, significantly improving safety and health outcomes.

INTRODUCTION

The coal mining industry is crucial in not only providing energy but also essential ingredient in the production of steel. Nevertheless, the mining process exposes workers to various health hazards, including coalmine dust, which can harm their respiratory health. One of the significant hazards faced by coalminers is Pneumoconiosis. Pneumoconiosis is a group of diverse respiratory diseases caused by inhaling coalmine dust, which can lead to varying degrees of lung tissue damage (Su et al, 2023; Zaid, Xu and Amoah, 2023; Zaid et al, 2024). Pneumoconiosis is divided into over a dozen categories in which coal workers' Pneumoconiosis (CWP), silicosis are the most common occupational lung diseases (Wei et al, 2023; Amoah et al, 2024). Overexposure to coal dust is the most common cause of CWP; currently, there is no effective cure for this disease (Shi et al, 2020). The literature has stated that CWP prevalence in the overall United States exceeds 10 per cent and 20.6 per cent in central Appalachia until 2017 (Blackley, Halldin and Laney, 2018). A recent study found that workers exposed to coal dust exhibit changes in insulin and cortisol levels, impairing glucose homeostasis compared to non-exposed workers. Additionally, these workers display increased levels of serum LH and FSH and a decrease in serum testosterone levels. It suggests that exposure to coal dust can compromise the hypothalamo-pituitary-testicular axis, which will negatively impact the metabolic and reproductive health of coalminers (Sulatana et al, 2023).

Various techniques are employed to reduce dust concentration in the working environment. These methods fall under two categories: dust control and personal protection. Currently, dust suppression technologies rely on various methods such as coal seam water injection, dust removal via ventilation, dust collectors, spray dust removal, and chemical dust suppression (Li et al, 2017; Xie et al, 2021; Zhou et al, 2019; Chang et al,; Zhao et al, 2021; Xu et al, 2018). Although the dust control measures mentioned above achieves certain level of success in reducing the dust concentration level in the working environment, they may not always bring the concentration of inhalable particles within the safe range which is 1.5 mg/m^3 required by Mine Safety and Health Administration (MSHA). Additionally, the removal rate for respirable dust remains relatively low due to the low efficiencies of these methods (Zhang, L *et al*, 2021). To address these challenges, enhancing the design and implementation of dust suppression and collection systems, and integrating advanced technologies like nanofibre membranes utilisation in dust collectors and dust scrubbers can improve the air quality in mines and safeguard the health of miners.

Advanced filter technologies are essential for improving workplace safety and miner health in the coal mining industry. Various filtration systems, including dust scrubbers and collectors, are employed to capture dust particles effectively. Specifically, fibrous filters in flooded-bed dust scrubbers have demonstrated substantial efficacy in trapping coal dust in subterranean mining environments, although their cleaning efficiency for 2 μm coal particles remains low at approximately 30 per cent (Gupta, Kumar and Schafrik, 2021). To augment this, non-clogging impingement screen filters have been developed for these scrubbers, increasing the cleaning efficiency for 2 μm particles to 65 per cent (Kumar, Gupta and Schafrik, 2022). Centrifugal wet scrubbers have been studied for their dust capture efficiency in mining applications. In these systems, specific filters like glass fibre filters impregnated with sodium thiosulfate and sorbent modules (such as XAD-2 resin) are used for sample collection and pollutant removal (Cao *et al*, 2018; Pollmann, Ortega and Helmig, 2005). Numerical predictions have been made to assess the dust capture efficiency of centrifugal wet scrubbers, highlighting their potential in reducing dust concentrations in mining and metallurgy processes. The predicted efficiency for particle size between 1–2 μm is 88–90 per cent (Ali, Plaza and Mann, 2018). Other types of filters such as electrostatics precipitators and bag filters have also been used to remove the smaller particle sizes (Zhang, Du and Qi, 2020). Bag filters have been found to have the lowest removal efficiency for particles with a size of 0.3 μm (Zhang *et al*, 2016). Since all above mentioned studies targeting the particle size between 0.3 to 2 μm, therefore, there is a need for innovative filter materials to remove these particles with high efficiency to ensure the safety of miners.

Non-woven fibrous filters are highly regarded for their exceptional filtration efficiency and versatility, making them ideal for use in dust collectors. These filters, created through techniques like electrospinning, mimic the dimensional constituents of extracellular matrix fibres, enhancing their dust particle capturing capabilities (Khorshidi *et al*, 2016). Moreover, nonwoven fibre filters with three-dimensional structures are favoured for their mechanical properties, high surface-to-volume ratio, dust holding capacity, low air pressure-drop, and high filtration efficiency (Luo *et al*, 2021). These filters provide an effective means of capturing particulate matter while maintaining optimal airflow in various filtration applications. Additionally, the use of nanofibres in nonwoven filters has been shown to enhance filtration performance, especially in removing particles larger than 1 mm during decontamination processes (Xu *et al*, 2020). These filters provide an effective means of capturing particulate matter while maintaining optimal airflow in various filtration applications. Additionally, the use of nanofibres in nonwoven filters has been shown to enhance filtration performance, especially in removing particles larger than 1 mm during decontamination processes (Xu *et al*, 2020). Furthermore, nonwoven fibre filters are preferred for their versatility in addressing specific pollutants. For instance, the use of electrospun polymer composite membranes in nonwoven filters has been demonstrated to be effective in capturing PM2.5 particles, showcasing superior thermal stability and chemical resistance for high-efficiency dust removal (Yang *et al*, 2019). Additionally, nonwoven filters made of polytetrafluoroethylene (PTFE) and polyphenylene sulfide (PPS) have been developed for efficient particulate matter removal, highlighting the diverse applications of nonwoven materials in air filtration (Wang *et al*, 2019). The development of technology for producing non-woven filtering materials, particularly through electrospinning, is a crucial scientific pursuit. Electrospinning is recognised as a promising method for creating fibres in the micro- and nanometre range, contributing to the optimisation of water permeability in non-woven filter media for efficient filtration while preserving essential permeability characteristics (Kakoria and Sinha-Ray, 2022; Kakoria, Chandel and Sinha-Ray, 2021). The primary objective of this study is to develop a novel nanofibre filter material. This material aims to achieve superior particle filtration efficiency for sizes ranging from 0.1 to 2.5 μm, especially for particles smaller than 0.3 μm. It seeks to offer significant improvements over current commercial filter technologies in terms of pressure drop and efficiency. The electrospinning and supersonic blowing technique will be utilised to fabricate the nanofibres from waster PET polymers. The resultant nanofibres will be assembled into a filter medium and subjected to rigorous testing. This includes exposure to face velocities varying between 8 and 28 L/min to assess the filter's performance across a spectrum of ultrafine particulate sizes. The filters will be evaluated based on their filtration efficiency for target particle sizes and overall breathability. A key performance metric, the quality factor, will be calculated to quantify the filter's efficacy in a standardised manner. This study introduces a novel application of recycled PET, addressing two pressing issues: occupational health risks in coal mining due to fine particulate exposure and environmental concerns associated with plastic waste. By converting waste plastic bottles into high-efficiency filtration materials, this research not only promotes a synthetic plastic-free environment but also potentially reduces the cost of personal protective equipment (PPE) for miners. This dual benefit underscores the innovative aspect of the research and its contribution to sustainable mining practices.

EXPERIMENTAL

Material

Waste or post-consumer polyethylene terephthalate (PET) bottles of a specific brand, Coca-Cola/Pepsi (500 ml), were procured from diverse, unspecified sources. These bottles are composed of PET material, with a composition of 99.8 wt per cent and devoid of any additional plasticizers (Zander, Gillan and Sweetser, 2016). The commercially available PET bottles of this variant exhibit a molecular weight within the range of 30 to 80 kDa (Strain *et al*, 2015). Prior to incorporation in the research, all recycled PET bottles underwent a meticulous cleaning regimen involving deionised (DI) water and ethanol. Following this cleaning procedure, the bottles were mechanically fragmented into smaller particles as shown in Figure 1. The solvents utilised in the study, namely Trifluoroacetic acid (TFA) and Dichloromethane (DCM), were sourced from Sigma Aldrich.

FIG 1 – PET bottle after cleaning and cutting.

Solution preparation

In the subsequent discourse, the polymer solutions derived from waste PET bottles shall be referred to as PET solutions. To generate these PET solutions, a total of three distinct formulations of PET material were concocted by dissolving Trifluoroacetic acid (TFA) and Dichloromethane (DCM) in a weight ratio of 9:1. This deliberate proportioning resulted in polymer concentrations of 8 wt per cent, 10 wt per cent, and 12 wt per cent. The preparation protocol encompassed the agitation of the mixture at a room temperature for a duration of 8 hrs. This agitation was achieved using a magnetic stirrer integrated with a hot plate.

Electrospinning and supersonic blowing

Nanofibrous sheets were fabricated employing the previously described polymer solutions using both electrospinning and supersonic blowing. During the electrospinning process, the polymer solution was injected through a 20-gauge needle by means of a syringe pump (New Era Syringe Pump Inc). A positive high voltage (ranging from 10 to 12 kV) was applied to the needle, facilitated by a voltage source (GenVolt). A rotating drum, set to rotate at 200 revolutions per min, was utilised as the ground and the collector. The spacing between the needle and the collector was consistently maintained at 10 cm. The schematic configuration of the experimental arrangement is graphically presented in Figure 2a. To achieve a targeted final basis weight of the mats at approximately 25–

27 grams per square metre (GSM) after a 4 hr duration, the flow rates of the polymer solutions were variably adjusted within the range of 1.0 to 1.2 mL/h, with progressive increments in polymer concentration. Upon the culmination of the process, the nanofibre mats were carefully detached and subsequently subjected to an overnight stay within a convection oven at a constant temperature of 40°C. This was executed to effectively eliminate any lingering solvent residues.

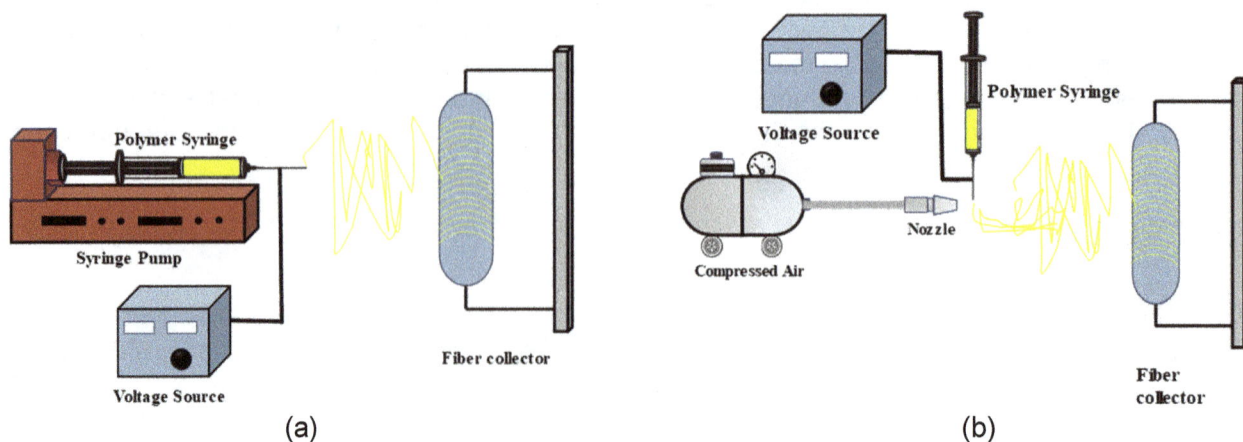

(a)

(b)

FIG 2 – Schematic diagram of: (a) electrospinning; (b) supersonic solution blowing technique employed in this study.

The supersonic blowing process was executed in accordance with the methodologies outlined in prior research endeavours (Yarin, Pourdeyhimi and Ramakrishna, 2014; Thakur *et al*, 2022). The procedural representation of this process is visually presented in Figure 2b. The framework comprises an electrospinning arrangement coupled with a supersonic converging-diverging de Laval nozzle, which is linked to an air compressor responsible for supplying compressed air at a pressure of $P_0 = 10$ bar. In this context, the electrospinning needle was placed orthogonally to the nozzle, with a horizontal distance of 2 cm and a vertical distance of 2 cm. Distinct from conventional electrospinning where a rotating collector is employed, the supersonic air nozzle served as the ground to attract the electrified polymeric jet emanating from the needle towards the nozzle. This arrangement imposes a constraint on the maximum applied electric voltage, restricted to 10 kV; nevertheless, the field intensity experiences an approximate fourfold enhancement compared to regular electrospinning conditions. As the electrified jet approaches the core of the supersonic air stream and exits the nozzle at a velocity of ~1277.59 m/s, the polymer jet encounters a sweeping effect away from the nozzle. This phenomenon is discernible as a curvature in the trajectory of the polymer jet near the needle, as depicted in Figure 2b. The resultant dried nanofibres were amassed on a drum collector positioned 25 cm distanced from the nozzle. The eventual size of the fibres varied based on the initial drop produced at needle to size, as expounded in previous investigations (Sinha-Ray *et al*, 2013); however, the gauge size of the needle remained consistent with that of electrospinning. The polymer flow rate was upheld within the range of 0.25–0.3 mL/h, and the process endured for a duration of 4 hrs to procure mats with a basis weight approximately in the range of 6–7 GSM. Correspondingly, the same polymer concentrations employed in the electrospinning approach were also applied here. The collected mats underwent an analogous treatment of being placed in a convection oven set at 40°C for an overnight period, with the aim of expelling any residual solvent remnants.

Characterisation

Analysis of the fibre size distributions of Polyethylene Terephthalate (PET) nanofibre membranes, as revealed by Scanning Electron Microscopy (SEM) shown in Figure 2, demonstrates a correlation between the PET concentrations and the resulting fibre dimensions. Incremental increases in the PET content from 8 per cent to 12 per cent are reflected in the corresponding histograms. For the PET8 per cent nanofibres, the distribution is tightly clustered around a mean diameter of 0.2 µm, suggesting a relatively uniform fibre structure, suitable for applications requiring high surface area-to-volume ratios. The PET10 per cent fibres show a shift towards a larger mean diameter of 0.4 µm, alongside a doubling in standard deviation, indicative of a broader size range that may influence the membrane's porosity and permeability. The PET12 per cent sample marks a further evolution in fibre morphology, with a substantial increase in mean diameter to 0.8 μm and a maximum diameter extending to 2.7 μm, factors that could significantly affect the mechanical strength and filtration efficiency of the membrane. This trend suggests that higher PET concentrations in the spinning solution led to the formation of larger fibres, potentially due to the increased viscosity and consequent changes in the spinnability and stretching of the fibres during electrospinning. The increased fibre size with higher PET concentrations could have implications for the mechanical properties and the specific application of the nanofibre membranes, such as filtration or as scaffolds in tissue engineering. These results highlight the significant impact of polymer concentration on the morphology of electrospun nanofibres.

The Figure 3a presented illustrate the Raman spectra of polyethylene terephthalate (PET) at different weight percentages (wt per cent). Raman spectroscopy, a non-destructive vibrational spectroscopic technique, exploits the inelastic scattering of monochromatic light to analyse molecular vibrations, phonons, and other low-frequency modes in a system. The incident light interacts with the phonon modes of the sample, which leads to a shift in energy of the scattered light. This energy shift provides information about the vibrational modes associated with the molecular structure of the sample. Each spectrum in the Figure 3a corresponds to PET at varying concentrations: 12 wt per cent, 10 wt per cent, and 8 wt per cent. The vertical axis denotes the transmittance in arbitrary units (a.u.), which represents the intensity of the scattered light, while the horizontal axis indicates the wavenumber in wave numbers (cm^{-1}), a measure of the frequency of the molecular vibrations. Distinct peaks in a Raman spectrum correspond to specific vibrational modes of the molecules within the sample. The figure shows that as the weight percentage of PET changes, there is a variation in the intensity and possibly the position of these peaks. These alterations may be indicative of changes in the molecular environment or the intermolecular interactions as the concentration of PET is altered. For example, the sharp peak observed at approximately 1600 cm^{-1} in the 12 wt per cent PET sample could be attributed to the C=C stretching vibration, which is characteristic of the aromatic ring in the PET molecule. This vibrational mode may result in peaks in the region between 1000 and 1300 cm^{-1}, indicative of the oxygen-containing linkages in PET. In the 10 wt per cent PET spectrum, similar vibrational modes are present but with slightly reduced intensity. In the 8 wt per cent PET spectra, we typically observed aromatic C=C Stretching. The peak may appear even less intense and could potentially shift slightly, reflecting the lowest concentration of PET. Another peak is observed around 1710 cm^{-1}, representing the carbonyl stretch in the ester functional group.

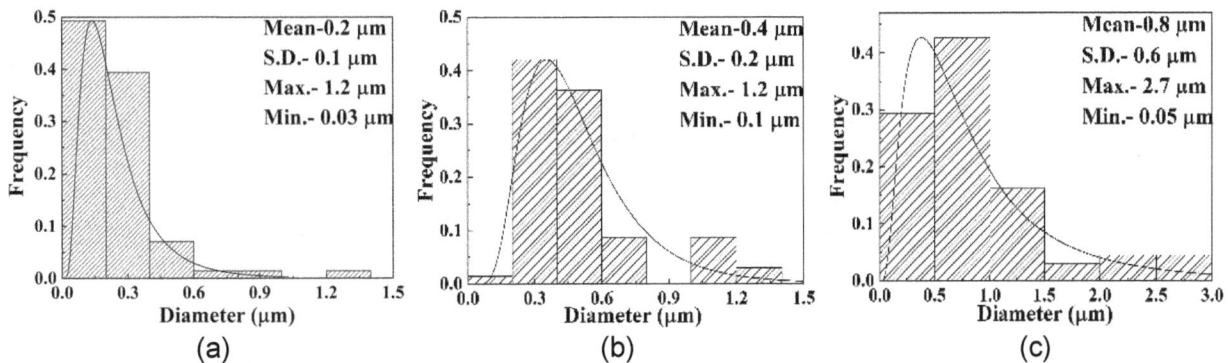

FIG 3 – Mean fibre size from SEM analysis of: (a) PET8 per cent; (b) PET10 per cent; (c) PET12 per cent nanofibre membrane.

Figure 3b displays X-ray diffraction (XRD) patterns for polyethylene terephthalate (PET) samples at different weight percentages: 12 wt per cent, 10 wt per cent, and 8 wt per cent. The XRD technique is utilised to investigate the crystalline structure of materials by measuring the diffraction of X-rays as they interact with the electron clouds within a sample. The horizontal axis, denoted as 2θ, represents the angle between the incident and diffracted X-rays, while the vertical axis measures the intensity of the diffracted rays in arbitrary units (a.u.). The peaks in an XRD pattern correspond to the Bragg's reflections from different sets of lattice planes within the material. For 12 per cent PET fibre, the presence of sharp peaks suggests a higher degree of crystallinity within this concentration of PET. A pronounced peak, which is typically indicative of the (010) or (100) reflection plane, reflects the ordered arrangement of polymer chains. The intensity of these peaks demonstrates that the crystalline domains are well-defined, suggesting a semi-crystalline nature of the PET at this concentration. For 10 per cent PET fibre, the pattern exhibits peaks with reduced intensity compared to the 12 wt per cent PET, implying a decrease in crystallinity. This can be attributed to a lower concentration of PET, leading to less ordered and fewer crystalline regions. The broadening of the peaks also suggests an increase in the amorphous content, which could be due to the polymer chains having more conformational freedom at this concentration. The spectrum for PET8 per cent fibre shows broader and less intense peaks, indicating the lowest degree of crystallinity among the three samples. The amorphous halo, a broad feature centred around 2θ values of 15–25 degrees, is more pronounced here. This is characteristic of the amorphous phase, where the polymer chains are in a disordered state.

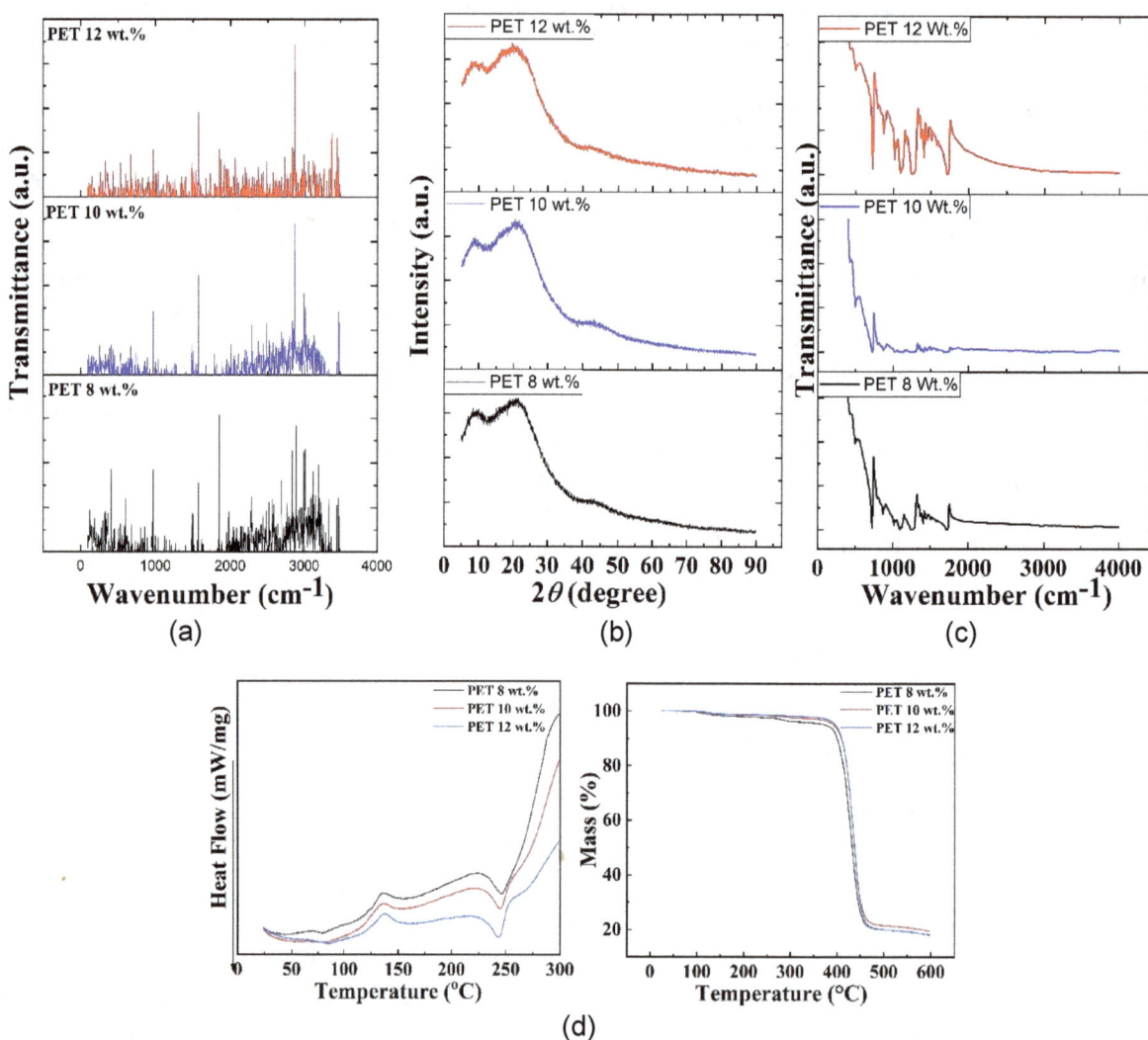

FIG 4 – Raman Spectra (a), X-ray diffraction patterns (b), FTIR (c) and (d) DSC and TGA of ES 8 per cent, 10 per cent and 12 per cent electrospun nanofibre.

The crystallinity of PET is a critical factor that affects its mechanical and barrier properties. The degree of crystallinity can influence the material's stiffness, strength, and transparency. Higher crystallinity (as indicated by sharper and more intense peaks) typically results in a material that is more rigid and less permeable to gases and liquids, which is desirable for certain packaging applications. Conversely, a more amorphous structure (evidenced by broader peaks and the amorphous halo) leads to greater flexibility and transparency. The variation in peak intensity and width across the different concentrations reflects the structural changes in PET as the amount of material changes.

The Figure 3c depicts Fourier-transform infrared (FTIR) spectra for polyethylene terephthalate (PET) at three different concentrations: 12 wt per cent, 10 wt per cent, and 8 wt per cent. FTIR spectroscopy is an analytical technique used to identify organic, polymeric, and, in some cases, inorganic materials by measuring the absorption of infrared radiation at various wavelengths, corresponding to the vibrational frequencies of the molecular bonds within the material. The FTIR spectrum for the 12 wt per cent PET sample, represented in red, manifests a plethora of absorption peaks that are representative of the various functional groups present within the polymer's structure. Notably, a broad absorption band is observed around 3400 cm^{-1}, likely attributable to O-H stretching vibrations, possibly indicative of moisture uptake within the sample. A particularly sharp and pronounced peak emerges around 1710 cm^{-1}, corresponding to the C=O stretching vibrations, a hallmark of the ester groups integral to PET's molecular architecture. Further, in the mid-infrared region, between 1100 and 1300 cm^{-1}, one can discern several bands that are distinctively associated with C-O stretching vibrations within the ester linkages. Additionally, absorption features around 720–730 cm^{-1} are indicative of the out-of-plane bending vibrations of the aromatic C-H bonds. The intensity of these absorption bands is notably high, which is consistent with the increased presence of these functional groups in the polymer matrix, a reflection of the sample's higher concentration of PET.

In the FTIR spectrum of the 10 wt per cent PET sample, depicted in blue, there is a distinct presence of absorption peaks analogous to those found in the 12 wt per cent sample, albeit with some variation in intensity. At approximately 3400 cm^{-1}, a broad absorption peak is apparent, possibly corresponding to O-H stretching vibrations, suggesting the presence of moisture. The sharp and intense peak characteristic of the C=O stretching in the ester group is evident around 1710 cm^{-1}, but it may be less pronounced compared to the higher concentration sample. The region spanning 1100 to 1300 cm^{-1} shows bands that can be attributed to the C-O stretching vibrations of the ester linkages, with a subtle reduction in peak intensity reflecting the lower PET concentration. Additionally, the out-of-plane bending vibrations of the aromatic C-H bonds present around 720–730 cm^{-1} display a relative decrease in intensity. The overall spectral features indicate a slightly reduced concentration of the functional groups due to the lower weight percentage of PET in the sample. Regarding the 8 wt per cent PET spectra, shown in black, the intensity of the absorption peaks is notably lower than in the spectra for the higher concentrations. This decrease in peak intensity across the spectrum is consistent with the lowest concentration of PET among the samples. The broad peak around 3400 cm^{-1}, corresponding to O-H stretching, again suggests moisture within the sample but is less intense. The signature peak of the C=O stretching vibration near 1710 cm^{-1}, while still present, is diminished, paralleling the reduction in PET content. The bands in the range of 1100 to 1300 cm^{-1}, associated with the C-O stretching in the ester linkages, exhibit a drop in intensity and might appear more broadened, indicative of a higher degree of amorphous character at this concentration. The absorption features around 720–730 cm^{-1} related to the aromatic C-H bond bending are also less pronounced. Collectively, these spectral characteristics reveal the effect of PET concentration on the intensity and sharpness of the FTIR spectral bands, providing insight into the material's structural composition at varying levels of PET content. Figure 4d presents the Thermogravimetric Analysis (TGA) and Differential Scanning Calorimetry (DSC) of PET Composites. The TGA thermogram shows As the temperature escalates from ambient to 600°C, all samples exhibit remarkable thermal stability up to approximately 350°C, beyond which a significant decline in mass is observed. This mass loss is indicative of the degradation of PET chains, likely due to the breakdown of the ester linkages within the polymer matrix. The samples manifest a nearly identical pattern of stability followed by a sharp decrease in mass, suggesting a uniform degradation process. It is noteworthy that the onset of degradation slightly shifts to higher temperatures with increasing PET content, hinting at a possible enhancement in thermal stability with higher filler content.

The DSC curves for the PET samples with 8, 10, and 12 wt per cent present two prominent endothermic peaks. The initial peak, common across all samples, occurs at approximately 80°C, which can be ascribed to the glass transition temperature (T_g) of PET, where the polymer transitions from a glassy to a rubbery state. The similarity in T_g for all samples indicates that the varying weight percentages do not significantly impact the glass transition phase. The sec endothermic peak, more pronounced in the 10 and 12 wt per cent samples, corresponds to the melting temperature (T_m) of PET. This peak, situated between 250°C and 260°C, signifies the transition from a semi-crystalline to a completely amorphous state. The 12 wt per cent PET composite demonstrates a slightly higher T_m, suggesting that higher PET content may correlate with increased crystallinity or a more ordered molecular structure within the composite, possibly resulting from a higher degree of polymer chain alignment.

Air filtration set-up and test protocol

The nanofibre mats were enclosed between two commercially available cotton cloth layers (basis weight = 85 GSM). These external cloth layers furnished structural support to the delicate nanofibre sheets and facilitated the handling of the mats. In the context of air filtration testing, a nanofibre mat of 60 mm diameter (fabricated through either the electrospinning (ES) or supersonic blowing (SSB) method) was interposed between two cotton cloth pieces of identical size. This composite assembly was employed for the determination of Particle Filtration Efficiency (PFE, denoted as η) and pressure drop (ΔP), as per the experimental arrangement outlined in Figure 5. The experimental set-up features: (a) An optical particle spectrometer (3330) (b) An aerosol generator (3079A TSI) (c) A charge neutraliser (d) A diffusion dryer. The testing apparatus consists of a 100 mm PVC pipe housing, divided into two equal sections, with a filter holder compartment to accommodate a 47 mm filter (net face area of 17.35 cm^2). For the supply of compressed air, compressor was employed. This compressor channelled air through a HEPA filter (TSI Capsule Filter with PFE 99.97 per cent > 0.3 µm) into the testing chamber. This chamber functioned as the dilution air source, where aerosol particles rendered neutral were released (as depicted in Figure 5). The neutralised particles traversed through the dryer and charge neutraliser. The flow rate of compressed air was subject to variation, spanning from 30 LPM to 80 LPM, consequently translating to face velocities ranging from 28 to 77 cm/s. The concentration of particles (expressed as count/cm^3) was measured both upstream (C_{up}) and downstream (C_{down}) of the filters, serving as a basis for the calculation of Particle Filtration Efficiency (PFE) as per Equation 1.

$$PFE(\eta)= (1 - C_{up}/ C_{down}) \times 100 \tag{1}$$

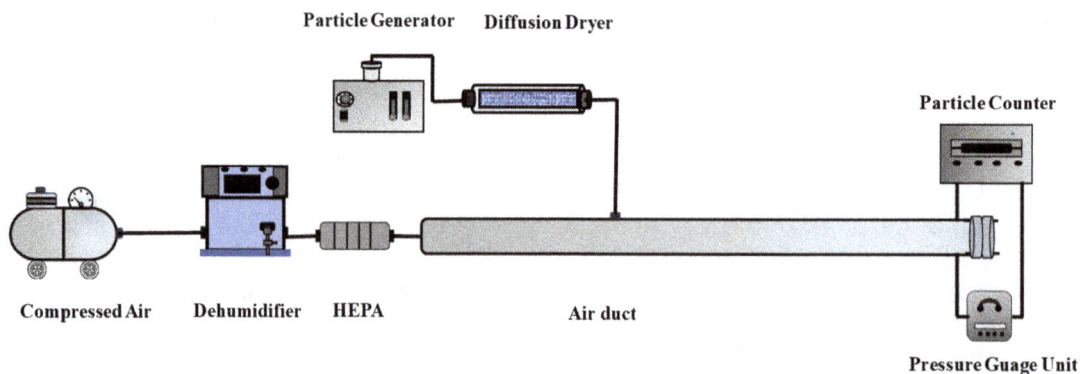

FIG 5 – Schematic representation of air filtration test protocol utilised in this study.

RESULTS

The purpose of this study is to create a novel nanofibre material using PET material to remove most penetrating particle size (MPPS), around 0.3 µm. This study introduces the utilisation of an Ultrafine Supersonically Blown Nanofibre Membrane with an average nanofibre size less than 100 nm for filtration purposes, targeting particle sizes even smaller than MPPS within coalmines. The fabricated nanofibre membrane enables both transition flow and diffusion-based particle entrapment mechanisms due to its nanofibre size. Consequently, this membrane enhances breathability and particle filtration efficiency beyond existing commercial technologies. This novel Ultrafine Supersonically Blown Nanofibre Membrane holds promise as a potential personal protective equipment (PPE) solution for coalminers, significantly improving safety and health outcomes.

Figure 6 depicts the efficiency and pressure drop for PET8 per cent ES fibre across three experimental points. The efficiency shows a general increasing trend, indicating improved particle filtration capability with successive measurements. Specifically, the efficiency starts at approximately 78 per cent and reaches about 86 per cent, reflecting a significant improvement in the fibre's ability to filter particles. This could be attributed to refinements in the fibre structure or operational conditions that enhance the trapping of particles.

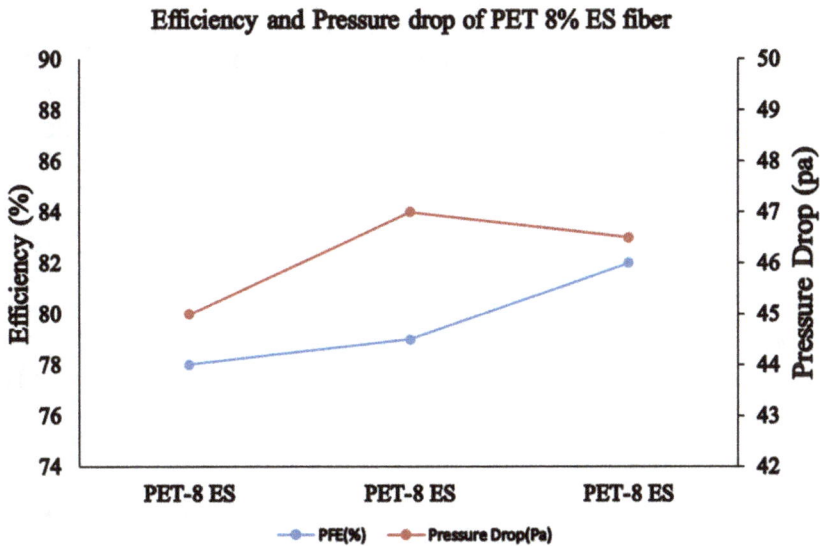

FIG 6 – Performance evaluation of PET8 per cent ES fibres depicted through filtration efficiency and pressure drop measurements.

Conversely, the pressure drop across the fibre also increases, from about 43 Pa to around 49 Pa. This increment in pressure drop could suggest that while the fibre becomes more effective at particle filtration, it also presents a higher resistance to airflow. This trade-off is typical in filter design, where higher efficiency often comes with increased air resistance, potentially impacting the usability in applications requiring minimal airflow obstruction, such as in respiratory masks or air purification systems.

Figure 7 shows results for PET10 per cent ES fibre, where a slightly different trend is observed. The efficiency begins at about 78 per cent and then shows a slight fluctuation before stabilising around 82 per cent. This pattern might indicate variable performance possibly due to inconsistencies in fibre diameter, distribution, or compaction during the electrospinning process. The pressure drop shows a moderate increase from 61 Pa to 67 Pa. This smaller increase compared to the 8 per cent PET fibre suggests that the 10 per cent PET fibre might have a denser or more uniform structure, leading to a steady increase in airflow resistance without abrupt changes. The data suggests that while the filtration efficiency is moderately high, the pressure behaviour is more predictable and controlled, possibly due to a more uniform fibre morphology at this concentration.

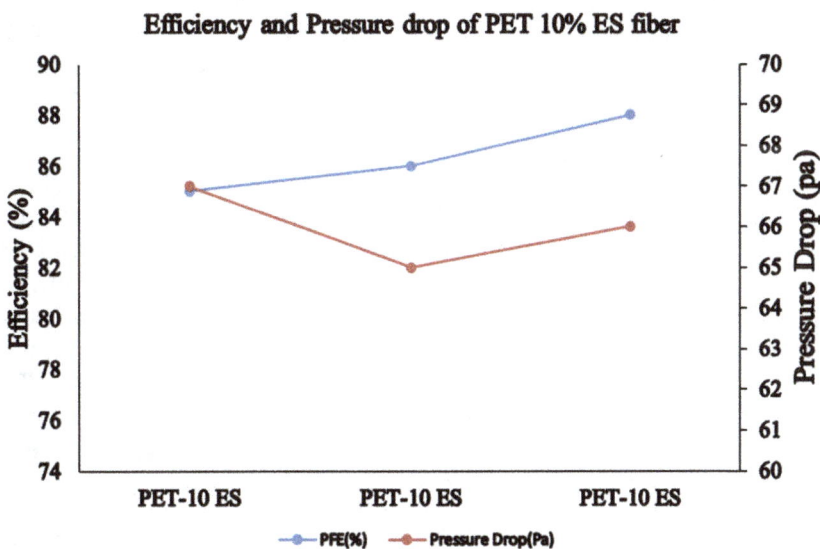

FIG 7 – Performance evaluation of PET10 per cent ES fibres depicted through filtration efficiency and pressure drop measurements.

Figure 8 depicts the PET12 per cent ES fibre displays a notable increase in both efficiency and pressure drop, which is the most pronounced among the three concentrations studied. The efficiency shows a remarkable increase, starting from 91 per cent and reaching up to 95 per cent. This high level of efficiency can be attributed to a potentially tighter and more compact fibre network, which is more effective at trapping smaller particulate matter.

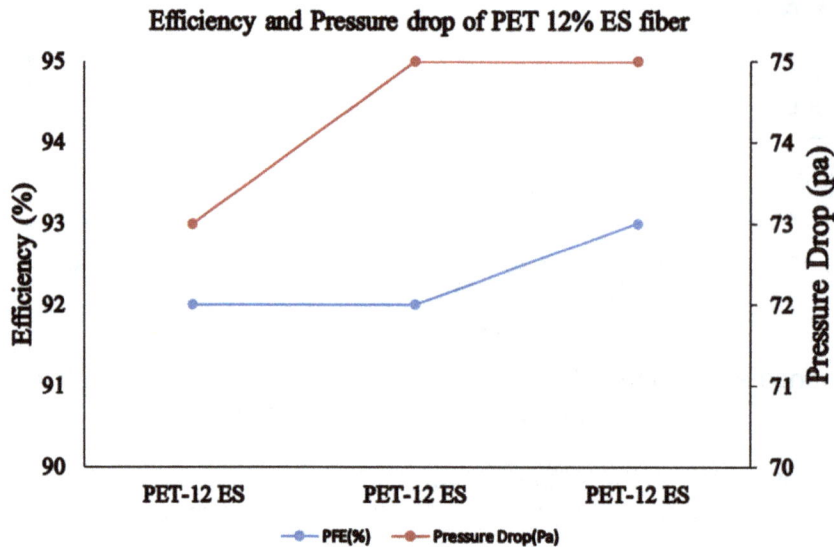

FIG 8 – Analysis of PET12 per cent ES fibres showcasing a significant increase in both filtration efficiency and pressure drop.

The pressure drop follows a similar upward trend, from about 71 Pa to 75 Pa. The substantial rise in pressure drop alongside efficiency suggests that the fibre mat's density or thickness may have increased, enhancing particle capture at the cost of higher resistance to air passage. This result is significant in applications where high filtration efficiency is paramount, even at the expense of higher energy costs for air movement.

CONCLUSION

This study has pioneered the development of an Ultrafine Supersonically Blown Nanofibre Membrane, which represents a significant advancement in personal protective equipment (PPE) for coalminers. The nanofibre membrane, with an average diameter less than 100 nm, is adept at filtering coalmine dust particles, particularly those smaller than the most penetrating particle size (MPPS), which is about 0.3 μm. This capability is crucial as it addresses the critical need for effective protection against pneumoconiosis, specifically progressive massive fibrosis (PMF), a prevalent issue among coalminers.

The results demonstrate that this innovative membrane enhances both breathability and particle filtration efficiency beyond the capabilities of existing commercial technologies. By facilitating both transition flow and diffusion-based particle entrapment mechanisms, the nanofibre membrane offers superior performance in capturing fine particulate matter while maintaining lower pressure drops. The Image analysis from SEM images showed an average diameter of 200 400 and 800 nm for PET8 per cent, 10 per cent and 12 per cent by wt nanofibre membrane. The particles removal efficiency gradually increased from 78 per cent to 94 per cent from PET8 per cent to PET12 per cent. This improvement in filter performance not only enhances the protection of miners against hazardous dust but also promotes better compliance with safety standards due to the increased comfort and reduced breathing resistance.

The use of waste PET polymers in the production of the nanofibre membrane aligns with environmental sustainability goals by reducing plastic waste and promoting the recycling of materials. This approach not only mitigates the environmental impact associated with the disposal of PET bottles but also decreases the manufacturing costs of high-efficiency filtration materials, making this solution both economically and ecologically beneficial.

In conclusion, the development of the Ultrafine Supersonically Blown Nanofibre Membrane from waste PET materials offers a promising solution to the significant challenges of occupational health in the coal mining industry. It provides an effective, sustainable, and cost-efficient approach to improving air quality and protecting miners from the long-term health effects of coal dust exposure. This study sets a precedent for future research and development in the field of protective equipment, aiming to enhance the safety and health outcomes for workers in various industrial environments.

REFERENCES

Ali, H, Plaza, F and Mann, A, 2018. Numerical prediction of dust capture efficiency of a centrifugal wet scrubber, *AIChE J*, 64(3):1001–1012. https://doi.org/10.1002/aic.15979

Amoah, N A, Zaid, M M, Kumar, A R, Chang, P and Xu, G, 2024. Optimized Canopy Air Curtain Dust Protection Using a Two-Level Manifold and Computational Fluid Dynamics, *Min Met and Expl*. https://doi.org/10.1007/s42461-024-01021-2

Blackley, D J, Halldin, C N and Laney, A S, 2018. Continued Increase in Prevalence of Coal Workers 'Pneumoconiosis in the United States, 1970–2017, *American Journal of Public Health*, 108(9):1220–1222. https://doi.org/10.2105/AJPH.2018.304517

Cao, X, Ji, L, Lin, X, Stevens, W R, Tang, M, Shang, F, Tang, S and Lu, S, 2018. Comprehensive diagnosis of PCDD/F emission from three hazardous waste incinerators, *Royal Society Open Science*, 5(7):172056. https://doi.org/10.1098/rsos.172056

Chang, P, Xu, G, Chen, Y, Ghosh, A and Moridi, M A, 2021. Improving coal powder wettability using electrolyte assisted surfactant solution, *Colloids Surfaces A Physicochem Eng Asp*, 613:126042.

Gupta, N, Kumar, A R and Schafrik, S, 2021. Laboratory determination of coal dust cleaning efficacy of a fibrous filter for flooded-bed dust scrubber, *Minerals*, 11(3):1–9. https://doi.org/10.3390/min11030295

Kakoria, A and Sinha-Ray, S 2022. Ultrafine nanofiber-based high efficiency air filter from waste cigarette butts, *Polymer (Guildf)*, 255:125121.

Kakoria, A, Chandel, S S and Sinha-Ray, S, 2021. Novel supersonically solution blown nanofibers from waste PET bottle for PM0, 1–2 filtration: From waste to pollution mitigation, *Polymer (Guildf)*, 234:124260.

Khorshidi, S, Solouk, A, Mirzadeh, H, Mazinani, S, Lagaron, J M, Sharifi, S and Ramakrishna, S, 2016. A review of key challenges of electrospun scaffolds for tissue-engineering applications, *J Tissue Eng Regen Med*, 10(9):715–738. https://doi.org/10.1002/term.1978

Kumar, A R, Gupta, N and Schafrik, S, 2022. CFD modeling and laboratory studies of dust cleaning efficacy of an efficient four stage non-clogging impingement filter for flooded-bed dust scrubbers, *Int J Coal Sci Technol*, 9(1):1–9. https://doi.org/10.1007/s40789-022-00481-5

Li, S, Zhou, F, Wang, F and Xie, B, 2017. Application and research of dry-type filtration dust collection technology in large tunnel construction, *Adv Powder Technol*, 28(12)3213–3221.

Luo, Y, Shen, Z, Ma, Z, Chen, H, Wang, X, Luo, M, Wang, R and Huang, J, 2021. A cleanable self-assembled nano-sio2/(Ptfe/pei)n/pps composite filter medium for high-efficiency fine particulate filtration, *Materials (Basel)*, 14 (24):7853. https://doi.org/10.3390/ma14247853

Pollmann, J, Ortega, J and Helmig, D, 2005. Analysis of atmospheric sesquiterpenes: Sampling losses and mitigation of ozone interferences, *Environ Sci Technol*, 39(24):9620–9629. https://doi.org/10.1021/es050440w

Shi, P, Xing, X, Xi, S, Jing, H, Yuan, J and Fu, Z, 2020. Trends in global, regional and national incidence of pneumoconiosis caused by different aetiologies: an analysis from the Global Burden of Disease Study 2017, *BMJ Open*, pp 1–8. https://doi.org/10.1136/oemed-2019-106321

Sinha-Ray, S, Lee, M W, Sinha-Ray, S, An, S, Pourdeyhimi, B, Yoon, S S and Yarin, A L, 2013. Supersonic nanoblowing: A new ultra-stiff phase of nylon 6 in 20–50 nm confinement, *J Mater Chem C*, 1(21):3491–3498. https://doi.org/10.1039/c3tc30248b

Strain, I N, Wu, Q, Pourrahimi, A M, Hedenqvist, M S, Olsson, R T and Andersson, R L, 2015. Electrospinning of recycled PET to generate tough mesomorphic fibre membranes for smoke filtration, *J Mater Chem A*, 3(4):1632–1640. https://doi.org/10.1039/c4ta06191h

Su, X, Kong, X, Yu, X and Zhang, X, 2023. Incidence and influencing factors of occupational pneumoconiosis: a systematic review and meta-analysis, *BMJ Open*, 13(3). https://doi.org/10.1136/bmjopen-2022-065114

Sulatana, J, Banerjee, O, Singh, S, Mukherjee, S, Ghosh, S and Syamal, A K, 2023. Coal dust affects hypothalamic pituitary testicular axis and impairs glucose homeostasis in coal mine workers, preprint, pp 1–18. https://doi.org/10.21203/rs.3.rs-2556907/v1

Thakur, S S, Chandel, S S, Kakoria, A and Sinha-Ray, S, 2022. Enhancement in pool boiling heat transfer of ethanol and nanofluid on novel supersonic nanoblown nanofiber textured surface, *Exp Heat Transf*, 35(4):516–532. https://doi.org/10.1080/08916152.2021.1919243

Wang, Y, Xu, Y, Wang, D, Zhang, Y, Zhang, X, Liu, J, Zhao, Y, Huang, C and Jin, X, 2019. Polytetrafluoroethylene/Polyphenylene Sulfide Needle-Punched Triboelectric Air Filter for Efficient Particulate Matter Removal, *ACS Appl Mater Interfaces*, 11(51):48437–48449. https://doi.org/10.1021/acsami.9b18341

Wei, F, Xue, P, Zhou, L, Fang, X, Zhang, Y, Hu, Y, Zou, H and Lou, X, 2023. Characteristics of pneumoconiosis in Zhejiang Province, China from 2006 to 2020: a descriptive study, *BMC Public Health*, 23(1):378. https://doi.org/10.1186/s12889-023-15277-8

Xie, Y, Cheng, W, Yu, H and Wang, Y, 2021. Study on spray dust removal law for cleaner production at fully mechanized mining face with large mining height, *Powder Technol*, 389:48–62.

Xu, G, Chen, Y, Eksteen, J and Xu, J, 2018. Surfactant-aided coal dust suppression: A review of evaluation methods and influencing factors, *Sci Total Environ*, 639:1060–1076.

Xu, X, Zhang, G, Wang, S, Lv, S and Zhuang, X, 2020. Fabrication of fibrous microfiltration membrane by pore filling of nanofibers into poly(ethylene terephthalate) nonwoven scaffold, *J Ind Text*, 50(4):566–583. https://doi.org/10.1177/1528083719837733

Yang, X, Pu, Y, Li, S, Liu, X, Wang, Z, Yuan, D and Ning, X, 2019. Electrospun Polymer Composite Membrane with Superior Thermal Stability and Excellent Chemical Resistance for High-Efficiency PM2.5 Capture, *ACS Appl Mater Interfaces*, 11(46):43188–43199. https://doi.org/10.1021/acsami.9b15219

Yarin, A L, Pourdeyhimi, B and Ramakrishna, S, 2014. *Fundamentals and applications of micro-and nanofibers* (Cambridge University Press).

Zaid, M M, Amoah, N, Kakoria, A, Wang, Y and Xu, G, 2024. Advancing occupational health in mining: investigating low-cost sensors suitability for improved coal dust exposure monitoring, *Meas Sci Technol*, 35(2). Available from: <https://iopscience.iop.org/article/10.1088/1361-6501/ad0c2e>

Zaid, M M, Xu, G and Amoah, N A, 2023. Accuracy of low-cost particulate matter sensor in measuring coal mine dust-a wind tunnel evaluation, *Underground Ventilation*, pp 274–284 (CRC Press).

Zander, N E, Gillan, M and Sweetser, D, 2016. Recycled PET nanofibers for water filtration applications, *Materials (Basel)*, 9(4):1–10. https://doi.org/10.3390/ma9040247

Zhang, L, Zhou, G, Ma, Y, Jing, B, Sun, B, Han, F, He, M and Chen, X, 2021. Numerical analysis on spatial distribution for concentration and particle size of particulate pollutants in dust environment at fully mechanized coal mining face, *Powder Technol*, 383:143–158. https://doi.org/10.1016/j.powtec.2021.01.039

Zhang, S, Du, Q and Qi, G, 2020. Experimental study on production and emission characteristics of PM2.5 from industrial fluidized bed boilers, *Therm Sci*, 24(5):2665–2675. https://doi.org/10.2298/TSCI190828001Z

Zhang, S-S, Qi, G, Guan, J, Pan, S, Li, J, Du, Q and Gao, J, 2016. The Removal Efficiency of PM2. 5 with Different Particle Size of Dust Removers, *DEStech Transactions on Engineering and Technology Research*, https://doi.org/10.12783/dtetr/amita2016/3573

Zhao, Z, Chang, P, Xu, G, Ghosh, A, Li, D and Huang, J, 2021. Comparison of the coal dust suppression performance of surfactants using static test and dynamic test, *J Clean Prod*, 328:129633.

Zhou, Q, Qin, B, Wang, F, Wang, H, Hou, J and Wang, Z, 2019. Effects of droplet formation patterns on the atomization characteristics of a dust removal spray in a coal cutter, *Powder Technol*, 344:570–580.

Spontaneous combustion

Broadmeadow Mine solid coal oxidation event

D Caley[1]

1. Ventilation Officer, Broadmeadow Mine, Redhill Rd, Moranbah Qld 4744.
   Email: dave.caley@bhp.com

ABSTRACT

Regardless of the results of any spontaneous combustion propensity testing that may have been completed on the coal at your mine, under ideal conditions, all coal is liable to spontaneous combustion. The migration of oxygen through coal can increase the propensity for spontaneous combustion. An area of exposure for potential oxygen ingress is due to the differential pressures generated by the ventilation network established throughout the mine. Therefore, the management of differential pressures in underground coalmines is critical to ensure that the risk of a spontaneous combustion event is managed to an acceptable level.

Historically in Queensland, and around the world there have been many instances of spontaneous combustion events that have escalated to a point where lives have been lost due to fire and explosion and mines have been sealed.

This paper discusses a recent event which occurred in an intake to return pillar at Broadmeadow Mine and details the various strategies that were put in place to manage the event. The paper will describe the events that led to the oxidation occurring, including mine design and strata management, and will then move on to discuss the discovery of the elevated temperatures in the area. The paper will close with a review of the overall management of the event. This will include diligent inspections, the various inspection methods used in high ventilation pressure areas; ventilation pressure management; strata consolidation processes; inertisation strategies; data collection and interpretation; and the overall management strategy that was implemented to manage the event during and after the event.

INTRODUCTION

Broadmeadow Mine is located in Queensland's Bowen Basin, 34 km north of the township of Moranbah (Figure 1). The Mine is a part of BHP Mitsubishi Alliance (BMA), a 50:50 joint venture between BHP and Mitsubishi Development Pty Ltd. Broadmeadow Mine extracts metallurgical coal utilising the longwall top coal caving system from the Goonyella Middle seam, which is 6.2–7 m thick The Goonyella Middle Seam at Broadmeadow Mine has a low to medium propensity for spontaneous combustion. While there have been no spontaneous combustion events underground at Broadmeadow, in the adjoining lease areas in the Goonyella Middle seam, there have been several spontaneous combustion events.

Spontaneous combustion is recognised as a principal hazard and as such Broadmeadow Mine has developed and implemented a Spontaneous Combustion Management Plan to manage the risk a spontaneous combustion to an acceptable level.

There have been previous pillar events recorded in Australia, including:

- Laleham mine (Blackwater Qld) 1974 – Serious heatings were experienced in pillars between the main intake and return roadways.

- Newlands mine (Glendon Qld) 1998 – A number of small heatings took place near the portal entries with pressure differentials of around 400 Pa.

- Beltana mine (Hunter Valley NSW) 2002 – A heating was detected in the first pillar between the intake and return highwall entries with pressure differentials of around 250 Pa.

- Tasman mine (NSW) 2011 – A heating was detected in a pillar 10 cut throughs away from the main fans with a pressure differential there of around 900 Pa.

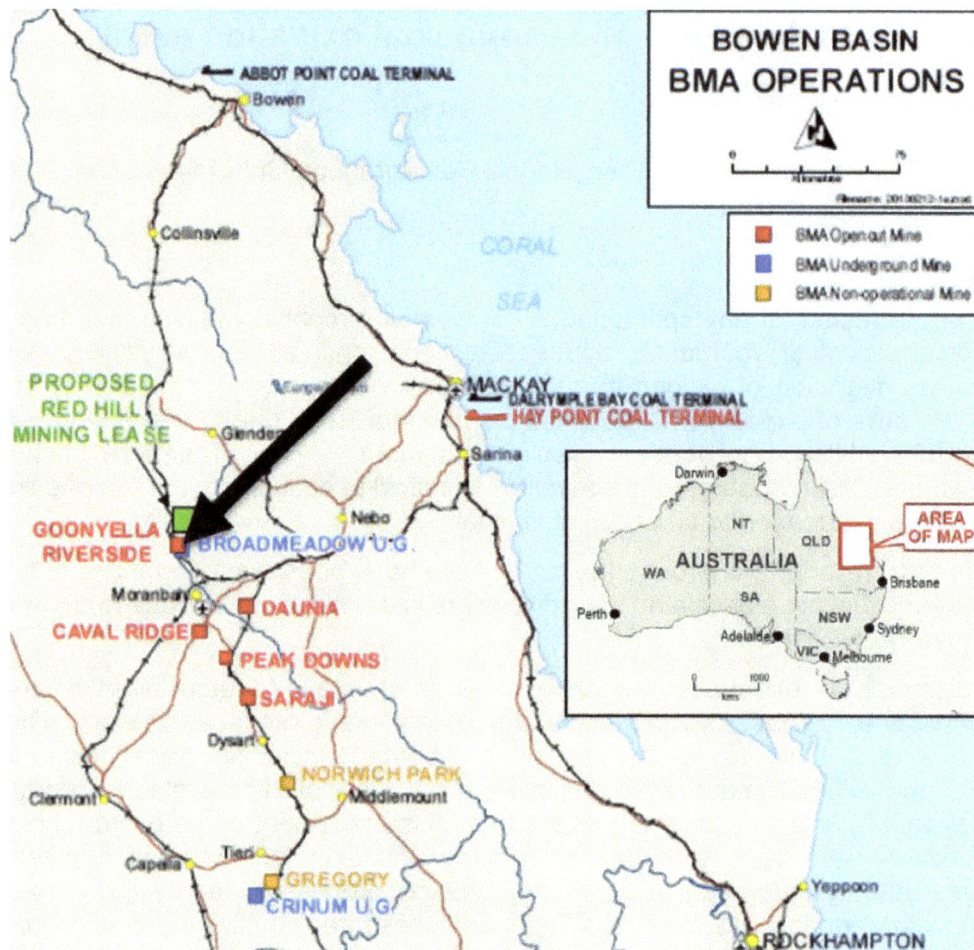

FIG 1 – Location of Broadmeadow Mine.

Broadmeadow Mine transitioned from Punch Highwall Mining to conventional Mains mining after Longwall12. At the time of the oxidation event the mine ventilation system was serviced by two upcast shafts (VS2 and VS3) and one return airway in the box cut (MG11 fans). Main intake airways comprised the two main travel roads and the main conveyor roadway. Mine collar pressure was 3.3 kPa. Previous LW panels north of LW 16 were sealed.

THE OXIDATION EVENT

The site of the oxidation event was in a pillar between intake to return roadways 1.2 km from the mine entry portals. The differential pressure across the pillar was 1360 Pa at the time of discovery.

Initially, only a small hot spot was detected at C-D hdg, 18 ct, but as further remediation works were completed, and the associated monitoring undertaken, it was found that the heating had moved along the length of the pillar to a point where it was detected in the roof at the D19 intersection.

A general layout of the mine is shown in Figure 2. The location of the main fans are shown in blue, with the location of the oxidation event shown in red.

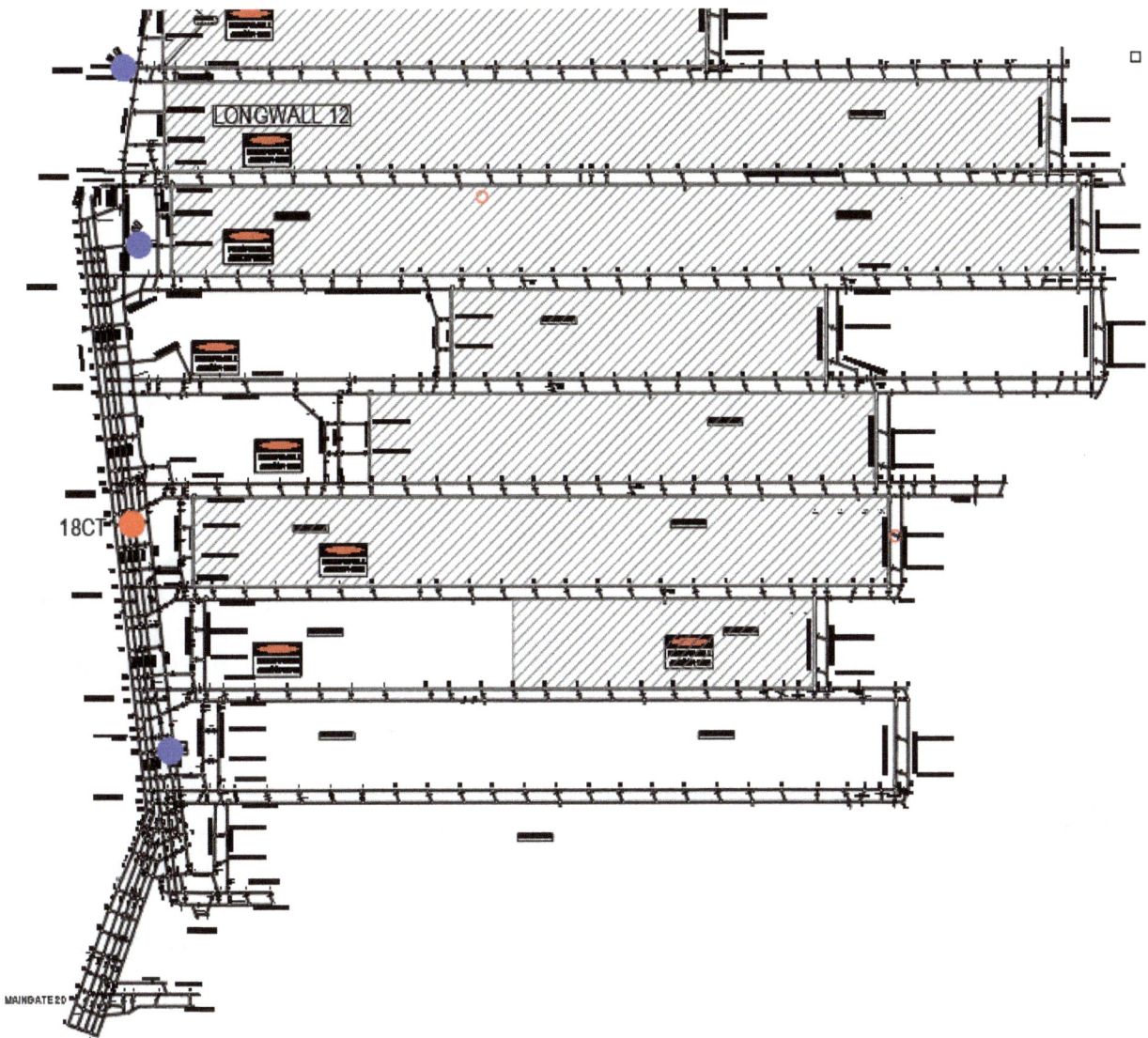

FIG 2 – General layout of Broadmeadow Mine.

Although, by definition, this event was a low-level oxidation, it was reported as a high potential incident (HPI) to the regulator because a Level 2 'Spontaneous Combustion in Solid Coal' Trigger Action Response Plan (TARP) was triggered. This initiated the formation of a Spon Com Management Team to control the management of the event.

CIRCUMSTANCES AROUND THE EVENT

Ventilation arrangement when development mains were first driven

The mains mine design originally had four intakes, (A-D Hdg) and one return, (E Hdg). In late 2017, at the time the area where the event occurred was developed, this ventilation strategy was adequate to manage the ventilation requirements in the mine. However, as the mains expanded and further longwall blocks were driven, it became apparent that the mine ventilation circuit was return constrained.

As part of planned operational requirements for the mine, a niche was driven at 18 ct C-D, which was to have had a borehole drilled into it to use to drop concrete into the mine workings. At the time this was completed, the C-D Hdg pillar at this location was at intake to intake pressure, with no discernible pressure differential between the two adjacent roadways. Figure 3. The location of the niche formed a stook between it and the C Hdg ribline, with approximately 7 m of coal remaining to separate the two areas.

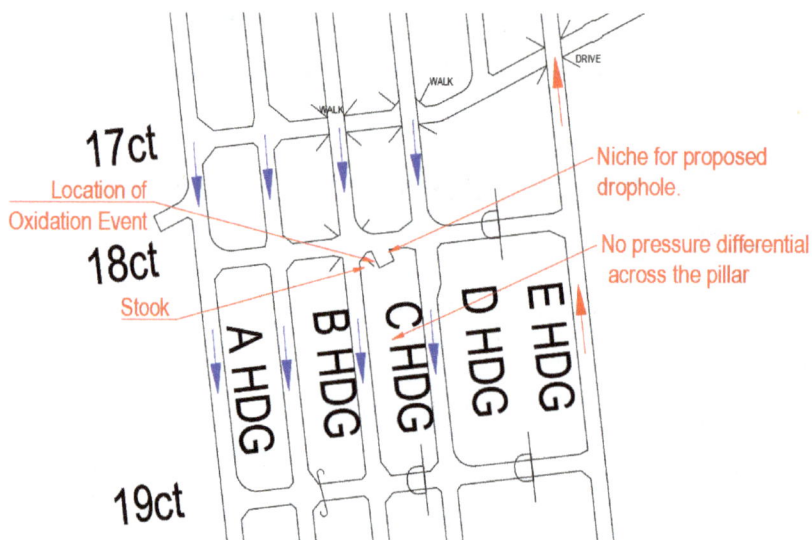

FIG 3 – Four intakes, one return.

Adjacent to the niche is an overcast, (C18 c/t). The strata in this area is broken, therefore the short rib line on top of the overcast was supported with a regime of 8 m cable bolts installed horizontally approximately 1 m above the cut through roof horizon. This VCD was also intake to intake under the original ventilation design, with the overcast providing separation of the belt road from the other intakes but with no meaningful differential pressure across the device.

Ventilation set-up after the ventilation change

Prior to the mine ventilation system becoming unmanageable due to low vent quantities, a major ventilation change was completed to change D Hdg to a return to better balance ventilation circuit losses.

The ventilation change described above was completed in 2018. Due to the large reduction in the resistance in the mine following the addition of a second return, there was a significant increase in the ventilation quantities available to the various districts underground. This change also increased the pressure differential across the pillar where the event occurred from virtually zero to the current intake return pressure which at the time of the event was measured at 1360 Pa. The ventilation design is shown in Figure 4.

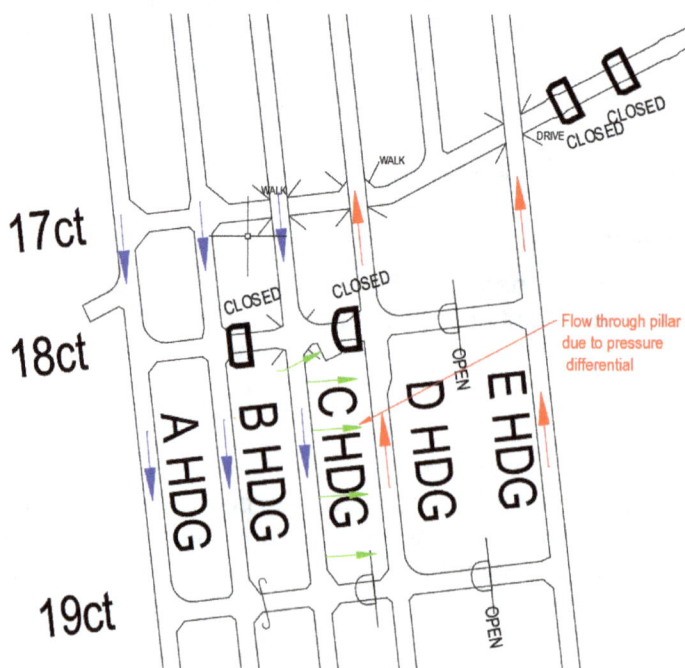

FIG 4 – Ventilation set-up after the vent change.

IDENTIFICATION OF THE OXIDATION EVENT

As part of spontaneous combustion management process at Broadmeadow, it was identified that there should be a system in place to inspect areas of the mine that may be prone to an oxidation event. At Broadmeadow, this process is known as the 'Red Zone Inspection'. By definition, a red zone is any area where there is greater than 1000 Pa pressure differential across a pillar or ventilation control device. Red zone inspections are completed by ERZ Controllers on a weekly basis.

The location of each inspection is identified from routine ERZ Controller inspections where pressure readings are taken from manometers installed at every VCD between intake to return roadways in the mains. Additionally, the mine ventilation model is used to check pressures prior to any major vent change so that new 'Red Zone Inspection' areas locations can be identified and established once the change is complete. Other locations may also be selected by the VO based on risk, eg the location that this paper is based on has continued to be monitored, even though for a long period of time the pressure differential at that point was well below the trigger for inspections to be completed.

The checks require a visual inspection of the area for physical signs of a heating, such as smell or sweating, as well as tests with gas monitoring equipment looking for Carbon Monoxide. Smoke tubes may also be used to look for leaks through the strata, but the key tool used as part of the inspection is a thermal imaging camera (TIC) (Figure 5).

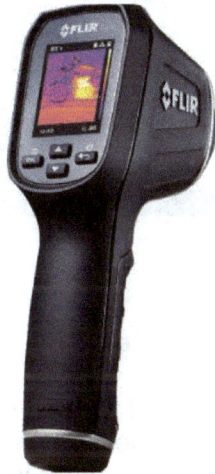

FIG 5 – Thermal imaging camera.

The camera is used to check for hot spots in the coal, as thermal imaging is capable of clearly identifying small changes on the surface of the coal. Once the check is completed, the details of the inspection are recorded on the Red Zone Form shown in Figure 6.

Internal BMA — BHP Mitsubishi Alliance

BRM FRM RED ZONE VCD INSPECTION FORM

Version 3.0 (16 August 2023)   Status: Approved   Business Owner: BMA BRM Officer Mine Control

TO BE COMPLETED BY:	/ /
VCD LOCATION	
DATE AND TIME OF INSPECTION	

CHECKLIST			COMMENTS
General VCD condition	OK ☐	POOR ☐	
Checked with smoke tube	OK ☐	POOR ☐	
Significant leakage paths through coal?	OK ☐	POOR ☐	
Highest CO determined on low pressure side	PPM		
General body CO	PPM		
Maximum temperature differential recorded in the area	Degrees C		
Pressure differential across the area	Pa		

Spontaneous Combustion in Solid Coal TARP

Level	Green Normal	Yellow: Level 1 Response Increase Awareness & Investigate	Orange: Level 2 Response Increase Monitoring & Prepare Contingencies	Red: Level 3 Response Stop Normal Operations and Treat the Problem Directly
TRIGGER CONDITION	Surface of the Coal Coal Temperature measured on the surface of the pillar is uniform with < 2°C change across the measured area. No significant leakage paths	Surface of the Coal Coal Temperature measured on the surface of the pillar is shows an area with a ≤ 2°C < 4°C change across the measured area, OR Air leaking through solid coal	Surface of the Coal Coal Temperature measured on surface of the pillar is shows an area with a ≤ 4°C < 8°C change across the measured area, OR > 2ppm CO above background measured from a crack or fracture, OR smell, sweating or haze.	Surface of the Coal Coal Temperature measured on surface of the pillar is shows an area with a > 8°C change across the measured area, OR Smoke OR as determined by IMT/ SCRT

Sign-Off

	NAME	SIGNATURE	DATE
Inspected By:			
Countersigned by VO:			

FIG 6 – Red Zone form.

The results of the inspections are then reviewed by the Ventilation Officer and any actions are taken based upon the TARP. Should an area of elevated temperature be found, it will be managed according to the TARP which outlines particular actions which will be taken based on predetermined triggers, an example of which is shown in Figure 7.

ALARM LEVEL	Green Normal (In Control) (Please remove this section if not applicable)		Yellow: Level 1 Response Increase Awareness and Investigate (Please remove this section if not applicable)		Orange: Level 2 Response Increase Monitoring and Prepare Contingencies (Please remove this section if not applicable)		Red: Level 3 Response Stop Normal Operations and Treat the Problem Directly (Please remove this section if not applicable)	
	Surface Of the Coal	Measurements From Within the Coal	Surface Of the Coal	Measurements From Within the Coal	Surface Of the Coal	Measurements From Within the Coal	Surface Of the Coal	Measurements From Within the Coal
TRIGGER CONDITION	Coal Temperature measured on the surface of the pillar is uniform with < 2°C change across the measured area. No significant leakage paths	Coal Temperature measured within the pillar is < 45°C.	Coal Temperature measured on the surface of the pillar is shows an area with a ≤ 2°C < 4°C change across the measured area OR Air leaking through solid coal	Coal Temperature measured within the pillar is ≤ 46°C < 60°C.	Coal Temperature measured on surface of the pillar is shows an area with a ≤ 4°C < 8°C change across the measured area. OR > 2ppm CO above background measured from a crack or fracture. OR smell, sweating or haze.	Coal Temperature measured within the pillar is ≤60°C < 100°C. OR >50ppm CO	Coal Temperature measured on surface of the pillar is shows an area with a > 8°C change across the measured area. OR Smoke OR as determined by IMT/ SCRT	Coal Temperature measured within the pillar is > 100°C. OR as determined by IMT/ SCRT

FIG 7 – Triggers for spontaneous combustion in solid coal.

In October 2022, during routine weekly red zone inspections, an elevated temperature was observed at an upper row rib bolt on the return side of the niche area at 18 ct in the mains (Figure 8). The hot spot was identified using a TIC where the surface of the rib was observed to be 33°C in comparison to the surrounding strata of 26°C (Figure 9).

FIG 8 – Location of the hot spot.

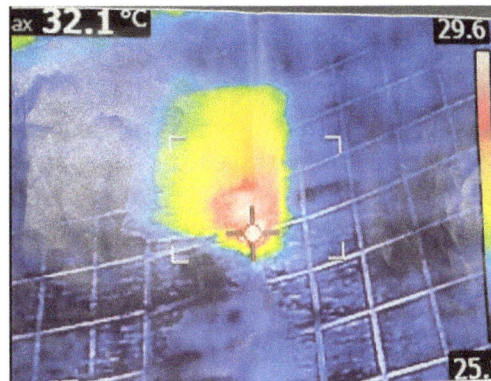

FIG 9 – View of the hotspot using the TIC.

Physical indicators observed were:

- Sweating, very localised to the hot spot at the bottom edge of the rib bolt plate.

- Yellow sulfur stain around the rib bolt plate.

- Small section of jointing directly above the rib bolt and extending through the roof for 1.2 m with a small brown stain directly above the bolt.

- No obvious signs of ventilation leakage through the coal were observed.

- No Carbon Monoxide was detected in the general body atmosphere in the cut through nor immediately adjacent to the hot spot.

- Gas bag sample taken from behind the rib bolt plate returned gas chromatograph analysis of 65 ppm CO, 0.3 ppm C_2H_4, 0.84 per cent CO_2, 16.7 per cent O_2.

MONITORING OF THE AREA

To ensure that the correct action is taken when a change is found, you must first understand how severe the issue is, then as the remediation is being completed, monitoring can indicate if the work is having an impact, either positive or negative. The monitoring from a heating requires an understanding of the temperatures in the area, the gases being liberated from the area, and the pressures across the pillar.

Temperature monitoring – thermistors

Thermistors were sourced that were able to be installed into the strata. These thermistors are 3 m long, and are designed to take a temperature at 3 m, the 2 m and 1 m mark. They are installed in a hole drilled into the seam, and are grouted in. The temperature is calculated by measuring the resistance in the probe. The lower the resistance, the higher the temperature.

Initially, thermistors were installed in the area where the hot spot was detected. As the heating progressed along the pillar, additional thermistors were installed in 18 ct and 19 ct as well as every 20 m along the D Hdg side of the pillar.

We saw an initial measurement of 70°C at the heating site. This was assumed to be from the exothermic reaction of the grout that is used to install the instruments, as the temperature quickly decayed. The thermistors returned a trend over time where we observed temperatures steadily decay by 10°C across all sensors, the highest confident temperature measuring 50°C. It should be noted that the exact location of the heating remains unknown.

The intervals at which the thermistors were read was managed by the Spontaneous Combustion Management Team (SCMT). At the beginning of the event this was twice per shift, but over time, as the event de-escalated, this reduced to once per shift, then daily, and weekly and now to a point where temperatures are no longer taken from the thermistors.

Temperature monitoring – thermal imaging camera

Initially, monitoring of the area once per shift, where the hot spot was located, was completed. Data was recorded and checked for change. As the event expanded, and the issue moved inbye, the frequency and location of the readings was changed in accordance with the requirements of the SCMT. Frequency and locations were reduced as the incident de-escalated. Weekly monitoring of the pillar is still in place to ensure early detection of any change.

Gas monitoring

To ensure that there was full coverage of the area, a comprehensive gas monitoring regime was set-up around the pillar. The monitoring regime was directed by the SCMT and changed as the event unfolded and then de-escalated. The monitoring implemented included:

Installation of gas sampling standpipes into each rib of the stook area 3 m into solid coal and every 20 m into the roof along the D hdg side of the pillar, from 18–19 ct. These sample points were lengths of pipe grouted into the strata with a valve on the end. The pipe allowed samples to be taken from within the pillar using either a personal gas detector with a pump, or a bag sample for the gas chromatograph.

During the event, and prior to the pressure grouting being completed, bag samples taken on the return side of the pillar self-inflated due to the pressure differential across the pillar and the leakage through the coal. The time taken for each bag to fill was used as a measurement on the effectiveness of the remediation. For example, the longer the bags took to inflate, the more effective the remediation was. Initially the bags were filling in seconds, after all works were completed, and the strata was consolidated to minimise leakage, the bags would not self-inflate.

A general body tube bundle monitoring point was installed nearby to the suspected heating area for remote monitoring should a Level 3 TARP be reached whereby all personnel would need to be withdrawn from the mine. All gas monitoring results were reviewed and trended daily by a specialised external consultancy group, an example of which is shown in Figure 10. This figure shows the steady decline in one of the key indicators, Carbon Monoxide. This started at around 550 ppm and decayed to almost NIL after all remediation works were completed.

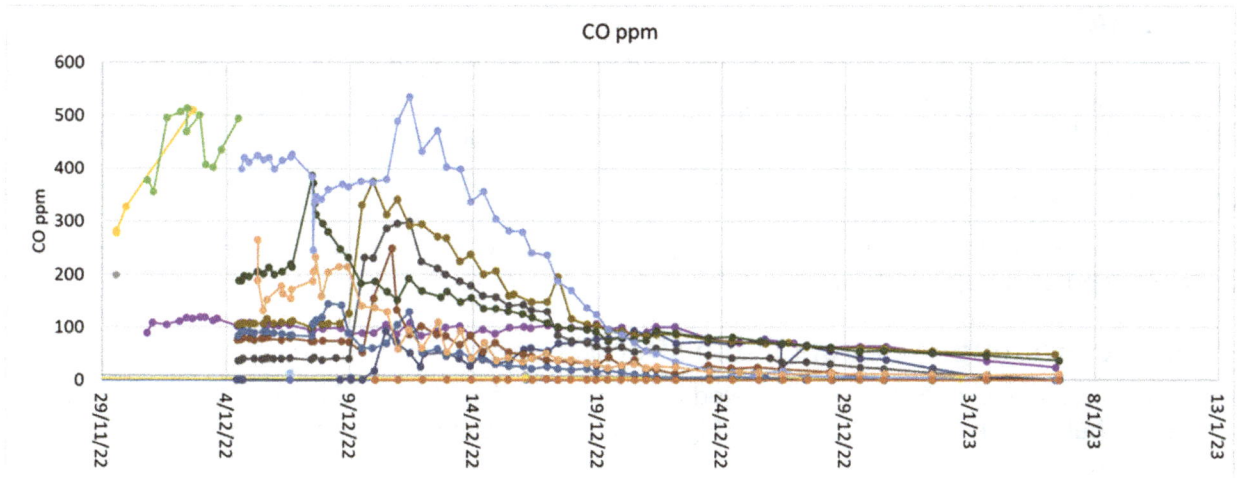

FIG 10 – Carbon monoxide trends from the sample points.

Pressure monitoring

Pressure monitoring was used to understand the extent of the issue. Initial readings allowed us to understand what pressures were influencing the area. This showed a total of 1360 Pa across the pillar where the elevated temperatures were detected as shown in Figure 11. Of note are the 8 m cable bolts above the overcast. Most of these were ungrouted, providing a conduit for ventilation to flow-through the pillar. Following the construction of an isolation stopping along the return side of the pillar, this differential was removed.

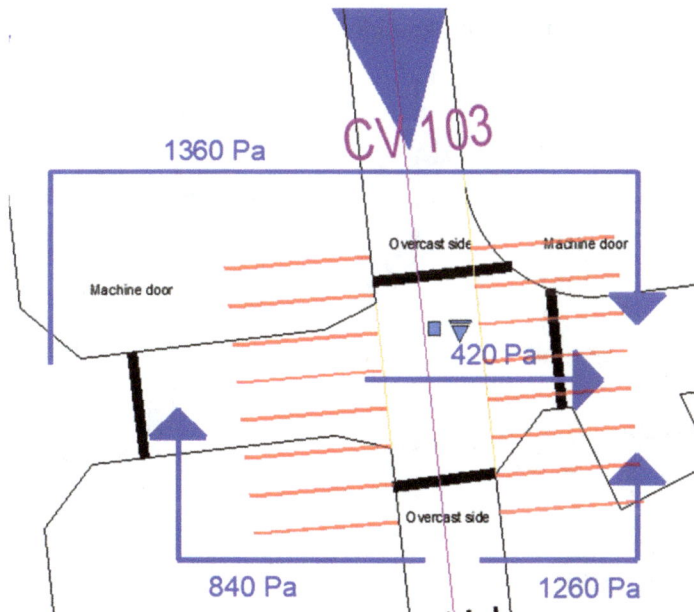

FIG 11 – Pressure sampling location at 18 ct.

Continued monitoring throughout the remediation process allowed changes to be quantified to ensure that anything we did was having a positive impact. These changes included things such as the construction of stoppings, through to main fan speed changes and the hole through of a downcast shaft inbye.

Following the ventilation changes that were made in the mine, the final differential pressure in the area was around 800 Pa.

REMEDIATION PROCESS

In accordance with the TARP, the SCMT ensured that all information about the event was gathered, and the appropriate actions were completed to control the event. To manage the event, it was identified that there were several strategies that could be employed. These included the following.

Reduction of main fan speeds

Fan speeds were reduced at all locations by approximately 10 per cent. This allowed the pit pressure to be reduced by around 600 Pa.

Pressure grouting the strata in the affected area

Extensive pressure grouting was completed around the pillar to tighten the strata and minimise any air paths that may have been present. Figure 12. The pressure grouting was completed at the following locations:

- All cable bolts in the overcast and in the roof of D hdg.

- The ribs in the stook on both the intake and return sides. This was completed in three rows. The lower row at an angle towards the floor, the centre row horizontal and the top row at an angle into the roof.

- The roof on the intake side of 18c t and 19 ct.

- The roof on the return side of the isolation stopping with holes being 1.5–3 m apart, depending on roof conditions.

- The floor on the return side of the isolation stopping.

- Along the C hdg rib 18 ct to19 ct (intake side).

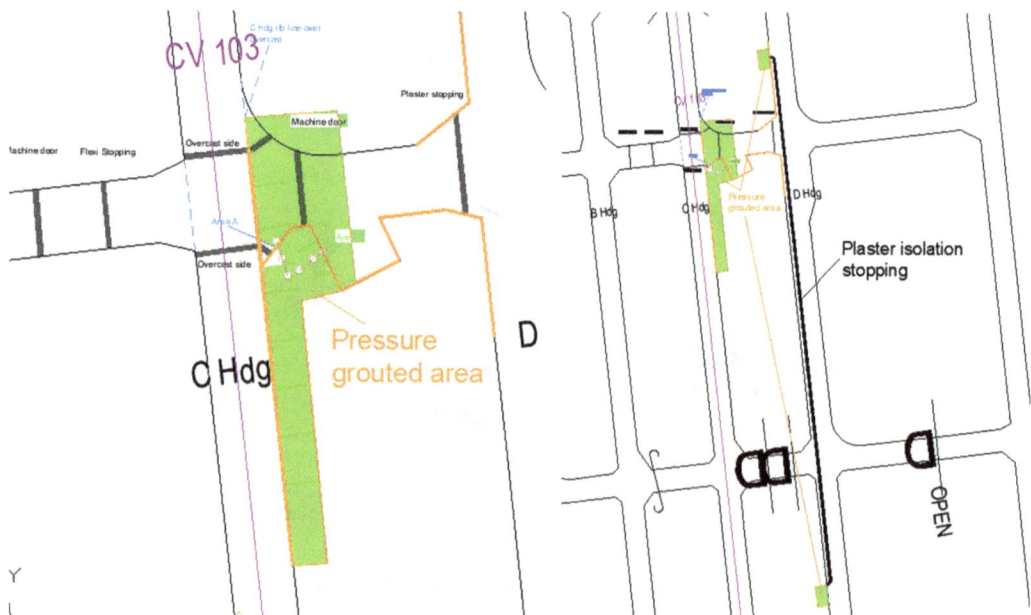

FIG 12 – Pressure grouting of strata.

It was noted during this process that there were areas of roof that took very large amounts of product, and the area still did not pressurise, indicating that there was a large void in that location. There was nothing on the surface of the roof in these areas to indicate any areas of concern at these locations.

Spraying the ribs and roof with plaster

To assist with increasing the resistance across the strata, all surfaces where pressure could influence the heating were sprayed with plaster. These include:

- Around the pillar.

- Through 18 ct and into D hdg.
- On top of the overcast.
- From 18 ct to 19 ct on the C hdg side.
- Inside the Isolation stopping.

Ventilation Pressure Management

A key method for the management of any heating in a pillar is to remove the pressure across the area. Without pressure there can be no flow-through the coal, regardless of the strata conditions. Initially a flexible stopping was constructed adjacent the machine doors at B-C hdg to relieve the pressure on that side of the pillar.

Ultimately, as the event progressed along the pillar into 19 ct, it was decided that full separation of the pillar from intake to return pressure was required. At this time a 145 m long 35 kPa plaster isolation stopping was built from outbye of 18 ct D Hdg to inbye of 19 ct D Hdg. Once the doors in the stoppings at 18 ct and 19 ct were opened the whole pillar became subject to intake pressure removing any differential in this area as shown in Figure 12.

Inert gas injection

Any oxidation event requires oxygen to be present to allow it to propagate. By removing the oxygen, which is part of the fire triangle, there can be no combustion. To assist with this, we injected inert gas into the strata, generated by an inert gas generation plant which is located on-site. This is a Floxal inert gas system which can deliver an oxygen deficient gas mix of around 3 per cent oxygen with the remainder being nitrogen.

To ensure control of not only the pressure and flow, but the location of the injection of the product, standpipes were grouted into the strata at the required locations. These pipes were 3 m long tubes, which were then connected to the Floxal nitrogen system.

The goal with the injection process was not to force the gas into the strata at high pressure, as this could generate more cracks, therefore allowing greater flow of oxygen through the pillar, but rather to present the gas to any areas of negative pressure, allowing the flow to be pulled through the pillar as required. With the use of a flow metre and pressure gauge, the gas was injected at a pressure of 2.5 kPa, with a flow of 10 L/m.

Standpipes were installed approximately 3 m apart at the following locations:

- Into the ribs of C hdg, over the overcast and into the stook.
- Into the rib along C hdg between 18 and 19 ct.
- Into the roof on the intake side at 18 and 19 ct.

Void filling of the isolation stopping

After all other remediation works were completed and the area had stabilised, the void between the D hdg side of the pillar and the isolation stopping was filled with a two part phenolic foam resin which is designed to fill cavities. This void took approximately 1000 m^3 of product which effectively sealed up the whole return side of the pillar.

The void was completely filled from the 18 ct c-D hdg plaster stopping to the flexible stopping in 19 ct C-D. As this product is exothermic, we wanted to monitor the void to ensure we were not making the situation worse by applying additional heat. Thermistors were installed in the area that was being filled to allow temperatures to be monitored during the process. Readings from within the area where this product was applied measured 90°C. Temperatures remained elevated in this area for several weeks. A guidance note on the use of these products is being developed by the regulator using these findings.

Downcast shafts

Coincidentally, during this process there were two downcast shafts being constructed. One, a 1.5 m shaft at the back of the LW18 panel, and the other a 3.5 m shaft in the mains intakes at 31 ct. Once these were brought online, a further reduction in pressure across the 18 ct area was seen, with the pressure across the affected pillar dropping to 760 Pa.

LEARNINGS

- Regular testing of the seam at Broadmeadow is undertaken, and all samples have returned low to medium results for spontaneous combustion propensity, however, as described in the McKenzie Strang mines rescue manual '*all coal is liable to spontaneously combust under the right conditions*'.

- You can design your mine to mitigate the risk, however a mine must have an effective monitoring program and take early intervention measures consistent with good practice. In conjunction with this, it is paramount that diligent inspections are completed.

- Thousands of Red zone inspections have been completed at Broadmeadow since the mine opened many years ago with no heating ever being discovered. It would have very easy for the ERZ Controllers to walk along the roadway and if nothing was obvious, they could have simply walked past, because they have never seen anything previously.

- The use of thermal imaging has proven invaluable in identifying problems well ahead of any gas emissions (fire ladder) that may be detected. For example, low level carbon monoxide evolves at 40–50°C and can be easily diluted in high ventilation quantities, but the TIC gives an instant reading which is not affected by ventilation. It is also simple and does not take any time to complete.

- Ensure the correct risk management processes are used when completing any vent changes. While hindsight is a wonderful thing, we did not identify as part of the original vent change that this could be an issue in the future. While we would still have completed the vent change to ensure that the mine ventilation circuit was able to continue to service the mine requirement, we may have done some additional remedial works to that pillar to minimise the risk of this event occurring.

- Continual reviews of data and updates of associated documents are invaluable. While we had a Spontaneous Combustion in Solid Coal TARP that served us well and allowed us to manage this event, after reviewing all of the data that was collected, we were able to update the document to have more relevant triggers and actions should there be another event in the future.

- The management of an event such as this takes up a huge amount of resources, both time and money. In this case the SCMT met daily, including weekends, for about four months, taking that team away from their normal tasks. There was a significant amount of labour required to complete all the remediation works, preventing day-to-day activities around the mine from being completed, and the cost of remediation was extremely high.

The use and abuse of laboratory-based gas evolution testing in setting triggers for spontaneous combustion Trigger Action Response Plans

D I Cliff[1]

1. MAusIMM, Professor of Occupational Health and Safety in Mining, Minerals Industry Safety and Health Centre, Sustainable Minerals Institute, University of Queensland, Brisbane Qld 4072. Email: d.cliff@mishc.uq.edua.au

ABSTRACT

An integral part of setting triggers for spontaneous combustion based Trigger Action Response Plans (TARPs) is the utilisation of appropriate laboratory-based gas evolution tests. Gas evolution tests have been undertaken in Australia since the early 1980s (Sanders, 1983) and overseas for much longer (Chamberlain, Hall and Thirlway, 1970; Graham, 1914). Routine gas evolution testing has been available in Australia since the early 1990s (Cliff, Bell and O'Beirne, 1992) However, it is important to recognise that these tests have limitations. Not the least of these limitations is that there is no standard test and that there are at least three distinct and different methods on offer, varying in coal mass tested, air flow supplied, and the condition of the coal being tested. This does not mean that the tests should not be used however it does mean that what they do and do not offer needs to be recognised. Gas evolution testing can offer insights into the fundamental mechanisms of coal oxidation, this paper focuses on the potential to apply the results of these tests to the setting of triggers in TARPs.

A typical four level TARP consists of normal operation, abnormal operation, worsening operation and evacuation. The limits to each of these zones is specified by trigger values of a mix of gas concentrations such as carbon monoxide or ethylene and derived indicators such as Graham's ratio or the CO to CO_2 ratio.

The characterisation of the range of the values of these parameters under normal operation is typically derived from analysis of operational data at the mine.

The evacuation point however, is usually determined as the point at which there is a significant potential for an ignition source (coal combustion) and a flammable atmosphere to coexist or to develop rapidly. For most underground coalmines in Australia the potential for a flammable atmosphere to exist somewhere in a goaf is a given and the focus is then on the potential ignition source. Generally, based upon the rapid rise on products of oxidation in the small-scale laboratory tests, it has been assumed that if the temperature of the oxidising coal has exceeded 100°C (the boiling point of water) that the oxidation will proceed to increase in temperature to flame point. This is consistent with the theory that there is no longer any endothermic process due to evaporation and the full surface area of the coal is available to react. Thus, indicators that identify that this temperature has been reached or exceeded are used as the top-level trigger.

Intermediate triggers are then set between the two levels to indicate that the situation has worsened from abnormal but has yet to reach critical – encouraging more strident control actions to be undertaken and preparations for evacuation to occur.

This paper will critique laboratory-based gas evolution testing to indicate how to utilise the test results in assisting in the establishment of indicators for the various TARP levels.

BACKGROUND

Trigger Action Response Plans (TARPs) are used in underground coalmines to trigger appropriate responses to increasing coal oxidation. These plans trigger actions depending upon certain key gas concentrations or derived indicators. The TARPs are usually staged with three levels of increasing severity, culminating in evacuation of the mine due to the imminent risk of a major incident. Evacuate too early and the mine needlessly loses production, evacuate too late and people may be harmed.

Integral to this is setting triggers that relate to the increasing intensity of coal oxidation (spontaneous combustion). Historically laboratory-based gas evolution data has been used to assist in setting these triggers. The challenge in using these small-scale tests is in relating them to the conditions that exist in the underground coalmine. It is not possible to simulate the conditions in an underground mine in a small-scale laboratory test. Indeed, to ensure the test is reproducible, the conditions are idealised – the coal is crushed and dried and a set flow of air through the reactor is used. There are at least three distinct methods offered to undertake these tests, which give differing results. As it is not intended to recommend or criticise the individual methods they will simply be referred to as laboratories 1, 2 and 3.

The basic parameters for the three test methods are outlined in Table 1.

TABLE 1
Basic parameters for the three test methods.

Parameter	Lab 1	Lab 2	Lab 3
Coal mass	150 g	70 g	6.5 kg
dried	yes	yes	Air dried
Crushed diameter	<212 µm	<250 µm	<3 mm
Air flow	10 mL/min	60 mL/min wet air	132 mL/min 20.5% O_2

DISCUSSION

In order to understand the complexities involved in comparing the test methods and distilling from them what may be useful is assisting TARP triggers for spontaneous combustion management it is necessary to establish a basic model of coal oxidation – gas evolution.

The reaction of coal with air at a given temperature can be summarised into three distinct reaction types:

- Pure oxidation – the formation of CO, CO_2 and H_2O.

- Pure pyrolysis – the formation of hydrocarbon fragments and H_2 due to the heat induced breakdown of the coal macromolecule.

- Oxygen catalysed pyrolysis – the formation of H_2 and C_2H_4 at least at lower temperatures (<150°C).

Each of these reaction types displays differing behaviour as the coal temperature increases. Chemical reactions can be strongly or weakly temperature dependent (the Arrhenius factor). It is important to differentiate between if reactions are occurring and our ability to detect whether or not the reactions are occurring. Generally chemical reactions will occur at any temperature, but their reaction rate may be strongly temperature dependent and not be detected until a certain concentration of a gas is detected. Thus, we see products of oxidation now at temperatures as low as 30°C when 30 years ago with higher detection limits this may have been 40°C or even above. Above this the rate of reaction increases strongly with temperature.

The rate of pure pyrolysis seems to be strongly temperature dependent and, in the past, has not been significant until in excess of 150°C.

On the other hand, the oxygen catalysed pyrolysis seems to have little temperature dependence and is really only noticeable from room temperature about 150°C when true pyrolysis begins to dominate.

An additional complexity is that each of these reaction types contain many actual chemical reactions, each potentially with different temperature dependent effects that may change the balance of the output gases produced as the temperature rises.

Some of the parameters that could explain the differences between the results obtained between the different gas evolution test regimes are:

- Air flow – high flows provide more air supply; low flows allow longer reaction times. Hollins (1995) found that increasing the air flow increased the rate of production of CO and CO_2.

- Oxygen concentration – the atmosphere in a mine goaf may contain oxygen concentrations less than fresh air – this will affect the rate of reaction and the balance of products formed. Hollins (1995) found that increasing the oxygen concentration to 34.5 per cent more than doubled the concentrations of CO and CO_2 produced.

- Bed depth/mass of coal – how much coal for a given air flow that is available to react. It is known that at temperatures more than 100°C, the oxygen is completely removed from the air supply by the coal.

- Moisture content – the evaporation of moisture is an endothermic process that can retard the chemical reactions, moisture within the pores of the coal will inhibit oxygen entering the pores. In addition, water is a source of free radical chemicals that can catalyse oxidation and pyrolysis.

- Coal particle size – the potential impact of total surface area. Humphreys (1979) recommended particles <200 μm to remove particle size effect.

- Inherent reactivity of the coal – Is there a variation in output for coals of differing reactivity (as measured by the R70 test, or another test).

- Chemical composition of the coal – the potential impact of ash content, proximate and ultimate analysis and rank of the coal.

Consider the simple chemical model where reactants A and B form compounds C and D.

The rate of production of C or D can then be expressed as:

$$\frac{d[C]}{dt} = k[A][B] \tag{1}$$

Where k is the reaction rate constant, a function of temperature. In the coal oxidation situation, [Coal] would be seen as a constant as it is in excess and [D] is the oxygen concentration.

And the concentration of C as a function of time at constant temperature is then:

$$[C]_t = \int_0^t \frac{d[C]}{dt} \approx \sum_0^t k[A]_n \, [B]_n \tag{2}$$

Where $[A]_n$ is the concentration of A at time n, where the time period t is broken into very small increments.

In all laboratory test methods at temperatures less than 100°C, the oxygen concentration remains constant throughout the reactor and is not significantly reduced in the exhaust in comparison to the inlet. Thus, in the second equation above, both [A] and [B] can be assumed to be constant.

Thus:

- The bigger the t the greater [C] – inverse of air flow – all other things being equal.

- The higher the oxygen concentration – the greater the rate of production of C and conversely the lower the oxygen concentration the slower the rate of production

- If the oxygen concentration is in excess, then [C] is proportional to the time it spends in contact with coal – so [C] increases with increasing bed depth all other things being held constant.

- Increasing the effective surface area of the coal increases the effective reaction rate as there is a greater surface area available to react for a given coal mass. Effective surface area increases as particle size decreases.

- The inherent reactivity of the coal and the chemical composition of the coal affect the rate of reaction.

Coal pyrolysis products are a function of the coal temperature, the length of time that the coal is at that temperature and inversely proportional to the air flow as it is simply a diluent.

Similarly, the products of oxygen catalysed pyrolysis are a function of the coal temperature (though evidence suggests this is not a strong correlation), the length of time that the coal is at that temperature and, provided oxygen is in excess, then inversely proportional to air flow.

It becomes even more complicated when the free oxygen concentration reduces as it passes through the coal in the reactor. Then the second equation above becomes the summation of each slice of coal where the conditions can be assumed to be constant within that slice, but [B] or the oxygen concentration is no longer constant throughout the coal bed. Once the oxygen concentration is exhausted then oxidation can no longer occur but if the coal is at elevated temperature, then pyrolysis can still occur.

It can be expected that these parameters will affect the three reaction types differently and hence cause the variation in results observed between the three commercial gas evolution regimes.

ANALYSIS

Several Australian underground coalmines have provided their laboratory-based gas evolution test results from the three testing bodies. In most cases the mines had only used one or two testing bodies.

From the discussion above it is clear that concentration-based triggers cannot be directly linked to small scale gas evolution tests as the observed concentration of a gas at a particular temperature will depend on the lab test method chosen.

The temperature at which a gas is first detected during the gas evolution studies will depend upon the laboratory method chosen and the sensitivity of the gas chromatograph used. Until 2020 gas evolution test results for example had a lower reporting limit for C_2H_4 of 1 ppm in Australia, whereas overseas labs reported less than 0.1 ppm (Muller, Dhyon and Kelson, 2023; Więckowski, Howaniec and Smoliński, 2020).

Detection/reporting limits are important considerations when the presence of a gas is used as a trigger. A good example of perceived presence versus detection sensitivity can be illustrated by the following example.

During small scale testing of the oxidation of a coal sample gas samples were collected and analysed to US EPA method 14 and 15 for volatile organic compounds (VOC) using a GC Mass Spectrometer system (US EPA, 1999). This system can detect C3 and above hydrocarbons to sub ppb concentrations (approximately 1000 times more sensitive than the standard GC used at mine sites). Figure 1 shows the output from the reactor at 100°C. Note the difference between oxidation – coal in air (blue data) and pyrolysis – coal in nitrogen (blue data). At 400°C there is no perceptible difference between pyrolysis and oxidation indicating that at this temperature the VOC are predominantly generated by pyrolysis (Figure 2). This demonstrates that the balance between the different reaction mechanisms described above shifts with temperature.

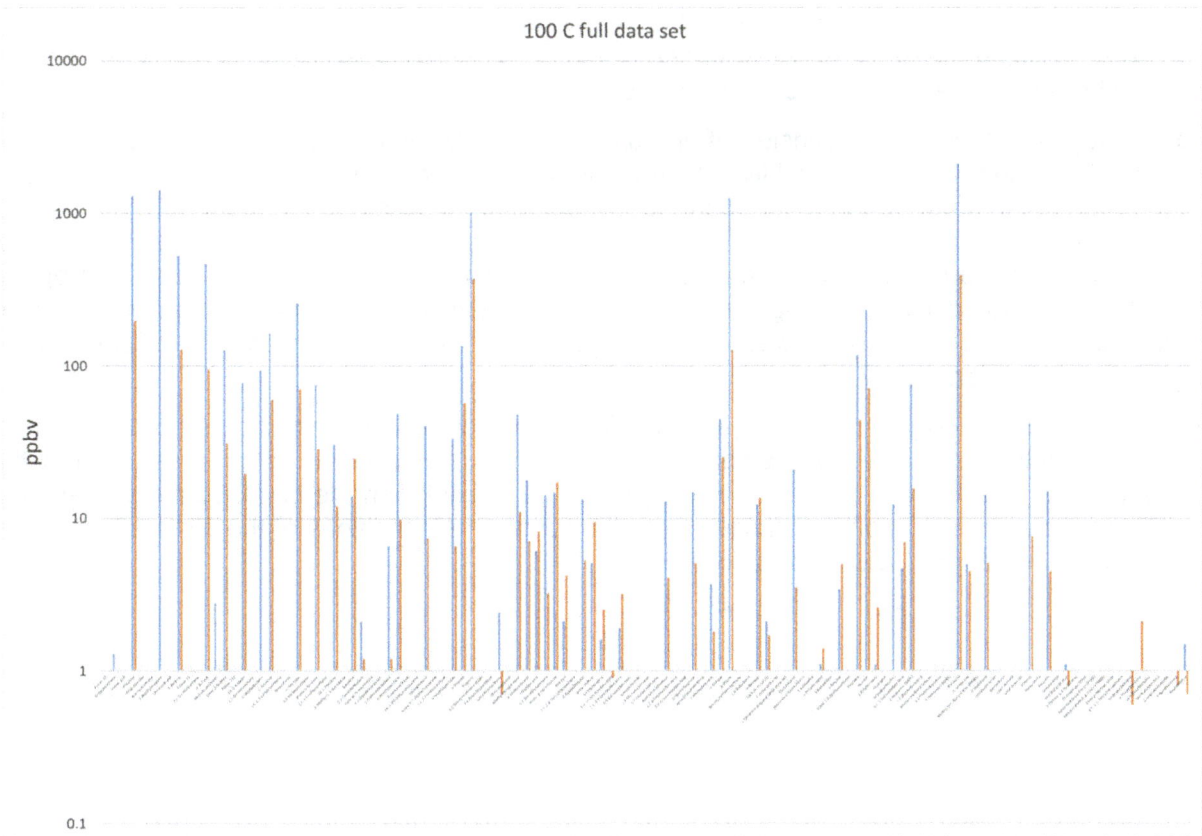

FIG 1 – VOC from small scale gas evolution test at 100°C (Lab 2 conditions). Blue is in air, red is under nitrogen.

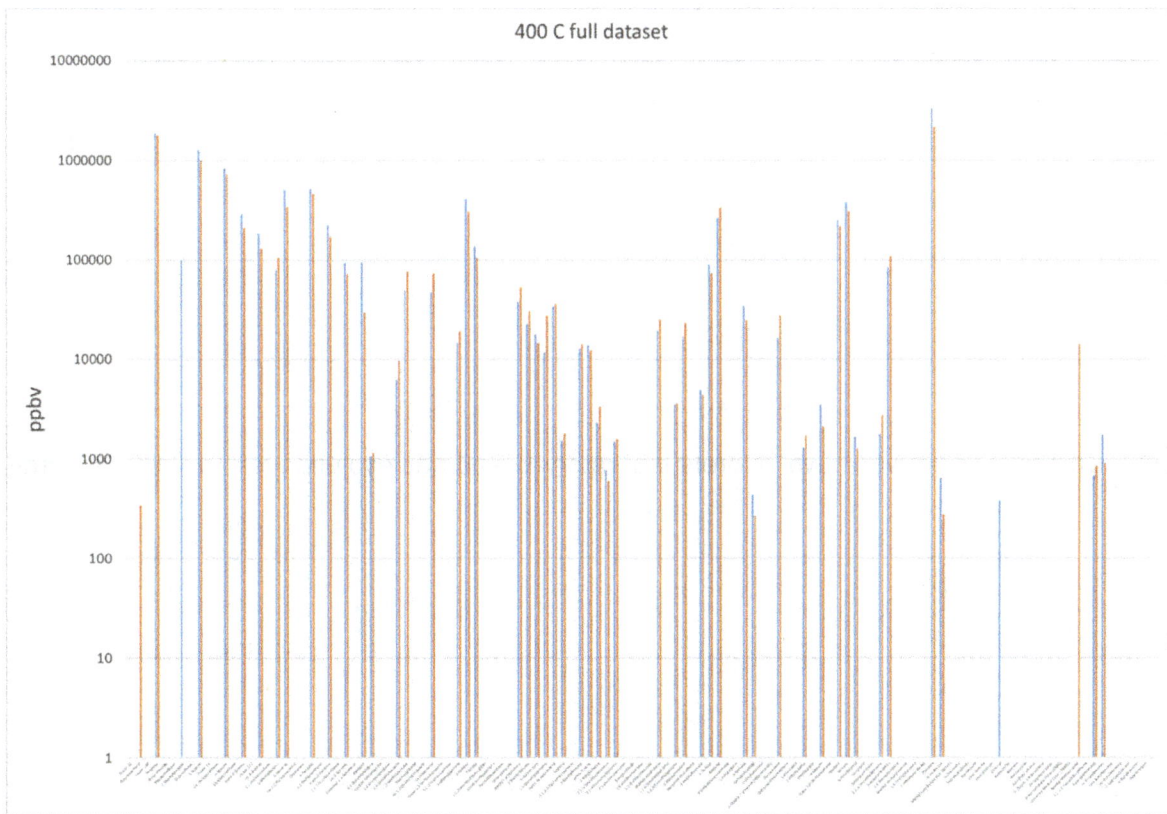

FIG 2 – VOC from small scale gas evolution test at 400°C (Lab 2 conditions). Blue is in air, red is under nitrogen.

Consider what gases would be detected if the threshold was 10 ppb, or 100 ppb or even 1 ppm? At 100°C with a 1 ppm threshold only five gases would be detected whereas for a 0.1 ppm threshold 13 would be detected and for 1 ppb over 50 would be detected.

Another way of looking at is to consider If one small scale gas evolution reactor generates 0.2 ppm C_2H_4 at 60°C, then how much would five in series generate? Reasonably it would be assumed to be 1 ppm. The coal temperature has not changed, oxygen is in excess only the mass of coal exposed to that oxygen has increased. If 1 ppm is assumed to indicate a really serious heating, then when does 60°C become a really serious heating. Equally the reverse can be demonstrated. If 1 ppm of ethylene is generated by a heating and then the post heating gases are blended with another atmosphere to reduce the concentration to 0.2 ppm before it reaches the detection point, is the heating no longer significant?

Figure 3 shows an example of a CO gas evolution curve for two different laboratory based small scale gas evolution tests on the same coal and Figure 4 for H_2. The yellow dots represent a test that was carried out with a lower air flow and a higher mass of coal than the first laboratory used. This would be consistent with a longer reaction time than under the conditions of the first laboratory and hence lead to higher products of oxidation.

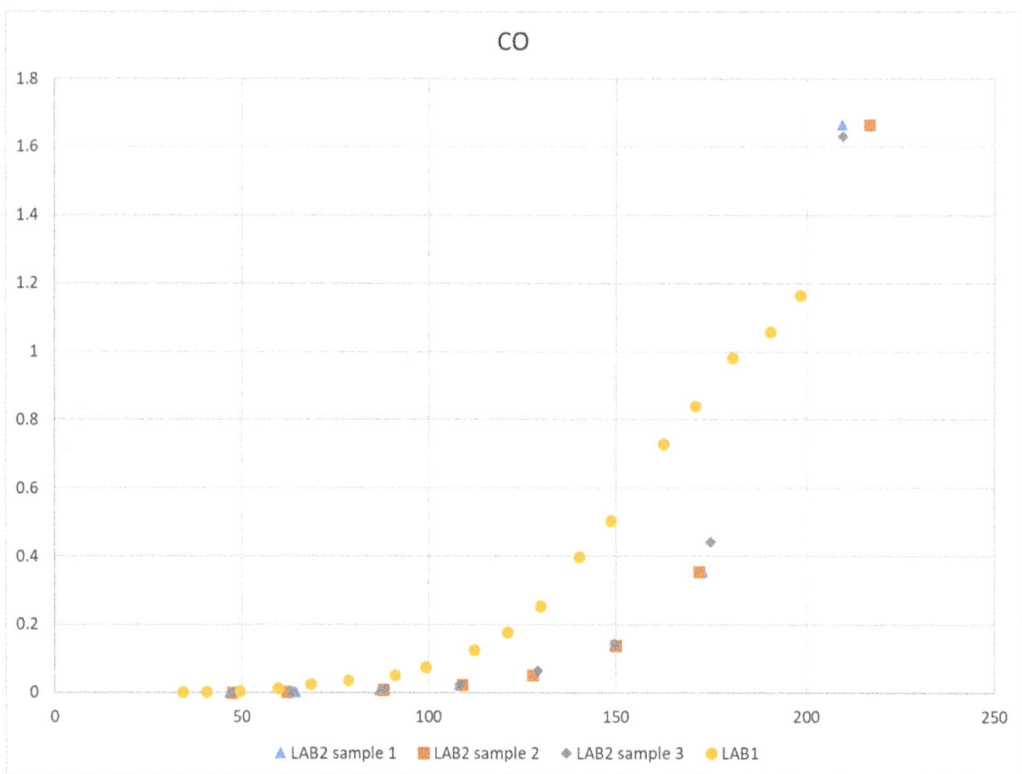

FIG 3 – Comparison between two different laboratory-based gas evolution tests for CO generation.

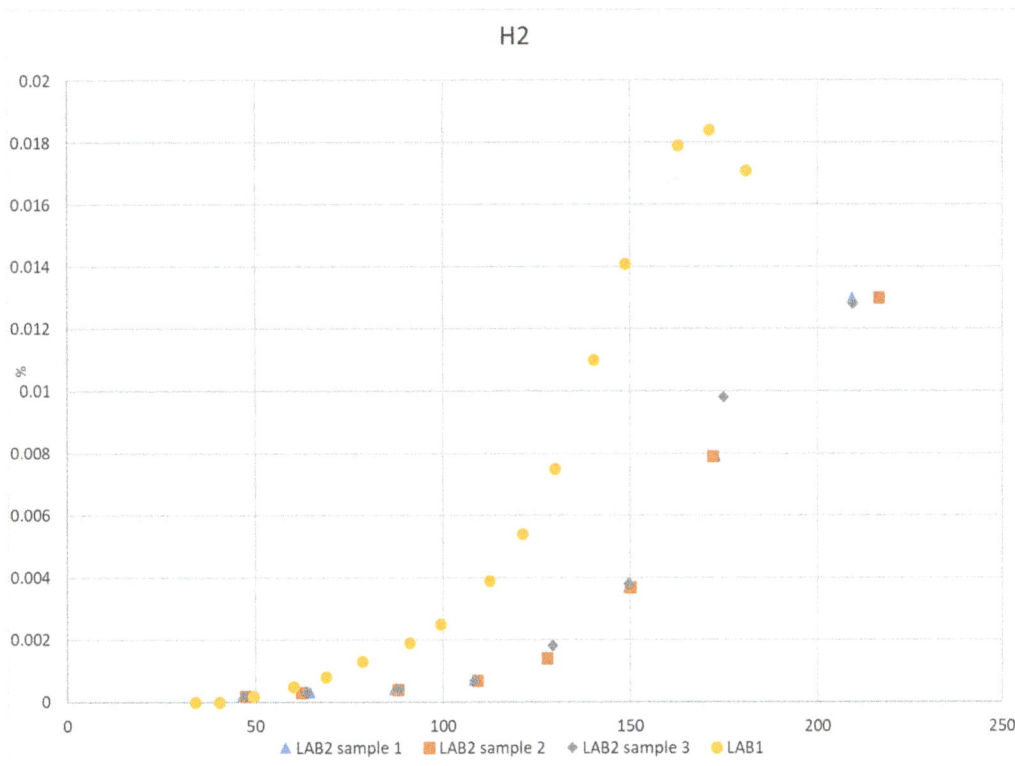

FIG 4 – H_2 evolution for the two laboratory gas evolution tests.

For the gas evolution tests to be representative in any way of the temperature of a coal sample reacting with air then the entire reactor must be a one temperature and exposed to air. If the oxygen concentrations for the above tests are plotted (Figure 5). It is clear that for Lab 2 the oxygen concentration is not uniform throughout the reactor from 50°C. For Lab 1 this decrease is noticeable from 100°C, no doubt due to the higher air flow and lower coal mass. If the concentration of the reactants is not constant, then the chemical processes involved will not be constant. It is clear that at 150°C for Lab 1 for example, that not all the coal in the reactor would be oxidising and thus the source of the CO is only from part of the reactor. Indeed, as the chemical conditions become fuel richer it would be logical to assume that the oxidation reactions would favour CO over CO_2. This means that in Equation 2 above [O_2] would not be constant and the reactor would have to be subdivided into zones of approximately constant concentration and integrated. These concerns were why the original gas evolution tests undertaken in the 1990s at SIMTARS that went to 400°C were subsequently curtailed to 200°C.

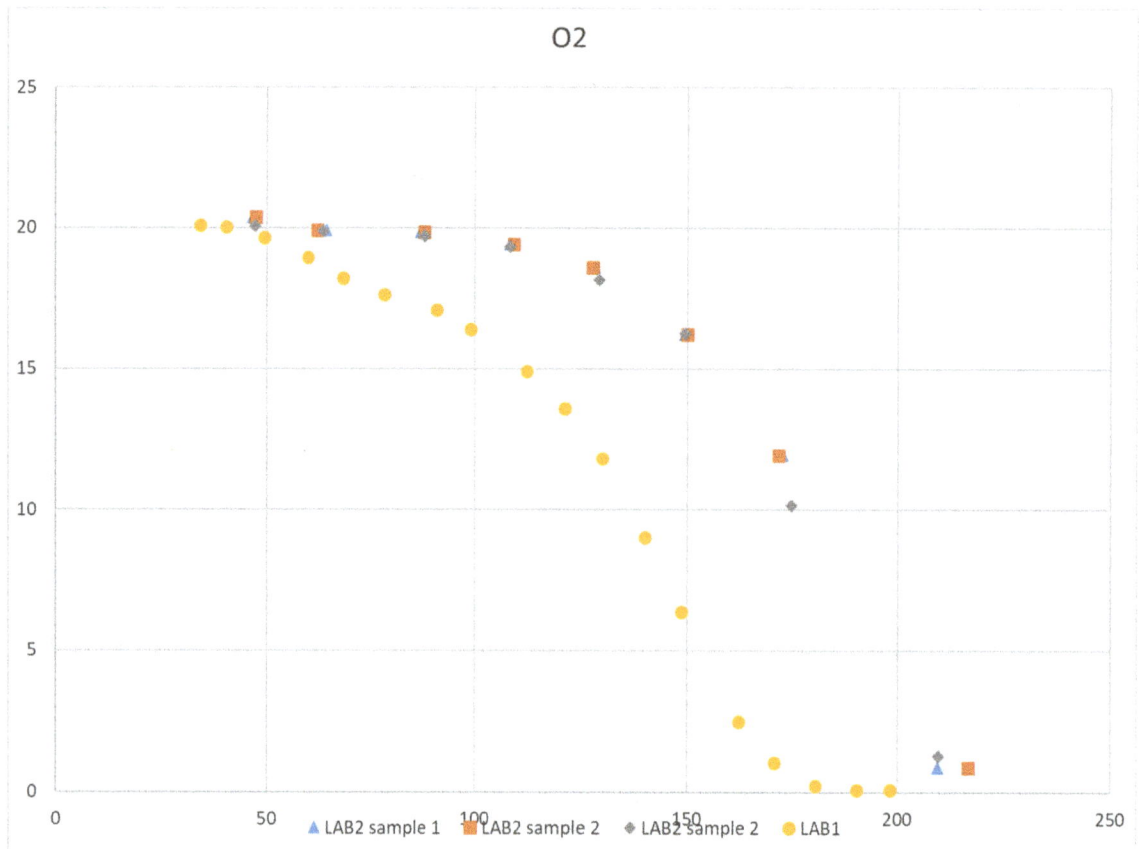

FIG 5 – Oxygen concentration in exhaust of gas evolution tests.

Currently we have limited understanding of the impact of changes in the free oxygen concentration on the chemical reaction processes. This understanding is vital in controlling heatings and using inert gas to reduce the oxygen concentration. There have been some estimates that the rate of oxidation not proportional to $[O_2]$ but $[O_2]^{0.67}$ (Humphreys, 2004). This means that even at $[O_2]$ of 5 per cent the rate of reaction is still 50 per cent of that in fresh air.

As mentioned above there are a number of other factors that can influence the chemistry, for examples: coal in the mine is not finely crushed or air dried. Indeed, the moisture content may vary. Moisture can be a heat sink (evaporation of moisture), and a source of free radicals to participate in the chemical reactions.

MORE COMPLICATED INDICATORS OF SPONTANEOUS COMBUSTION

One stage on from simple concentration-based measurements would be the changes in ratios of either gases such as C_2H_4 to CO or CO to CO_2 or more complicated ones involving oxygen consumed in comparison to products of oxidation – such as Graham's, Jones-Trickett or Morris ratios. The hope would be that they are independent of the gas evolution testing regime used.

- It is important to consider the three reaction systems outlined above: Pure oxidation – the formation of CO, CO_2 and H_2O.

- Pure pyrolysis – the formation of hydrocarbon fragments and H_2 due to the heat induced breakdown of the coal macromolecule.

- Oxygen catalysed pyrolysis – the formation of H_2 and C_2H_4 at least at lower temperatures (<150°C).

The validity of the derived indicator chosen will depend upon whether or not the gases involved are generated or consumed by the same reaction mechanism or not ie ratios that utilise a product of oxidation and a product of pyrolysis will not be equivalent between the test methods.

For components generated by a single reaction mechanism there would be some validity in assuming it should be test method independent. Oxygen deficiency and CO to CO_2 based ratios could be comparable and test method independent. Figures 6 and 7 depict Graham's ratio and CO to CO_2 ratio for the two laboratory testing regimes above, consistent with this assumption. The H_2 to CO ratio shown in Figure 8 does display significant differences between the two labs at temperatures less than 100°C.

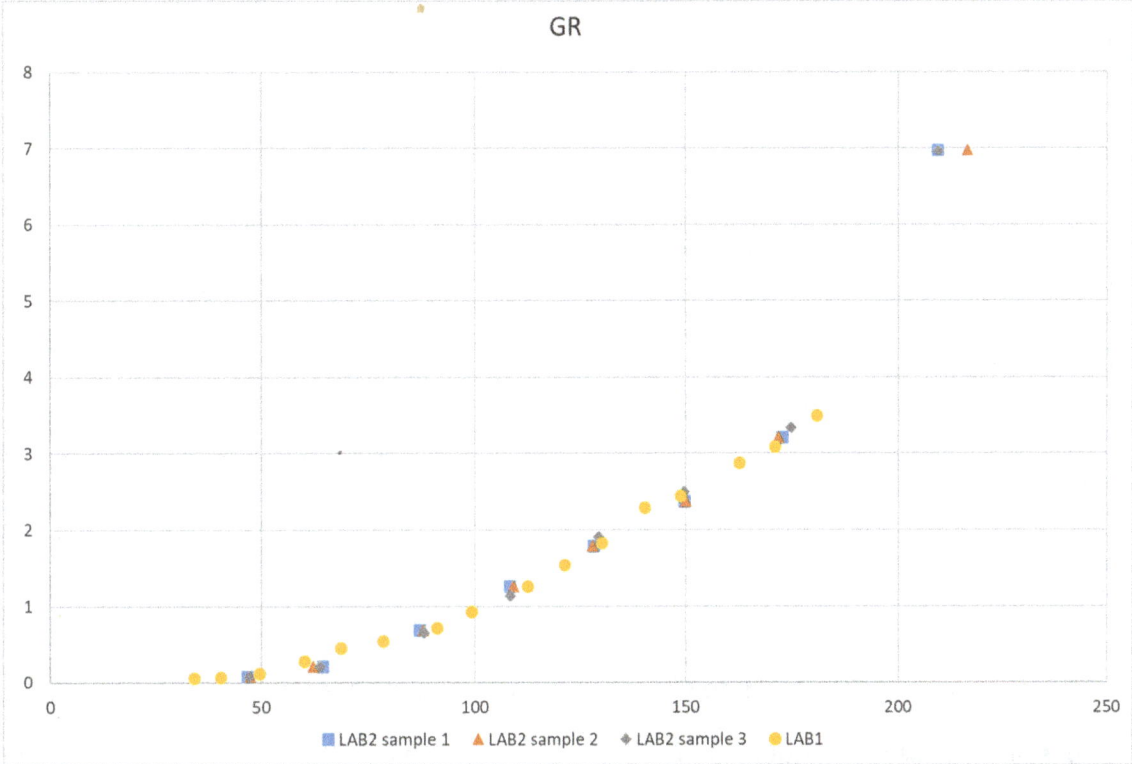

FIG 6 – Graham's ratio in exhaust of gas evolution tests.

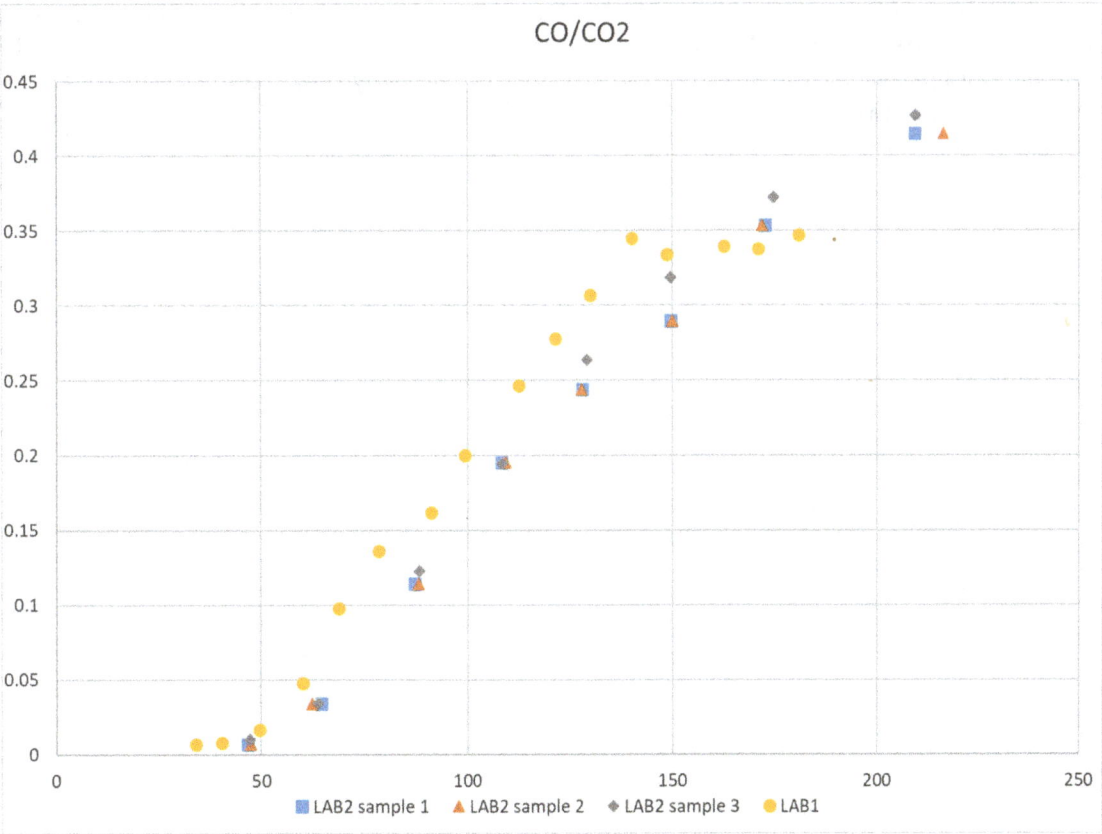

FIG 7 – CO to CO_2 ratio from gas evolution tests.

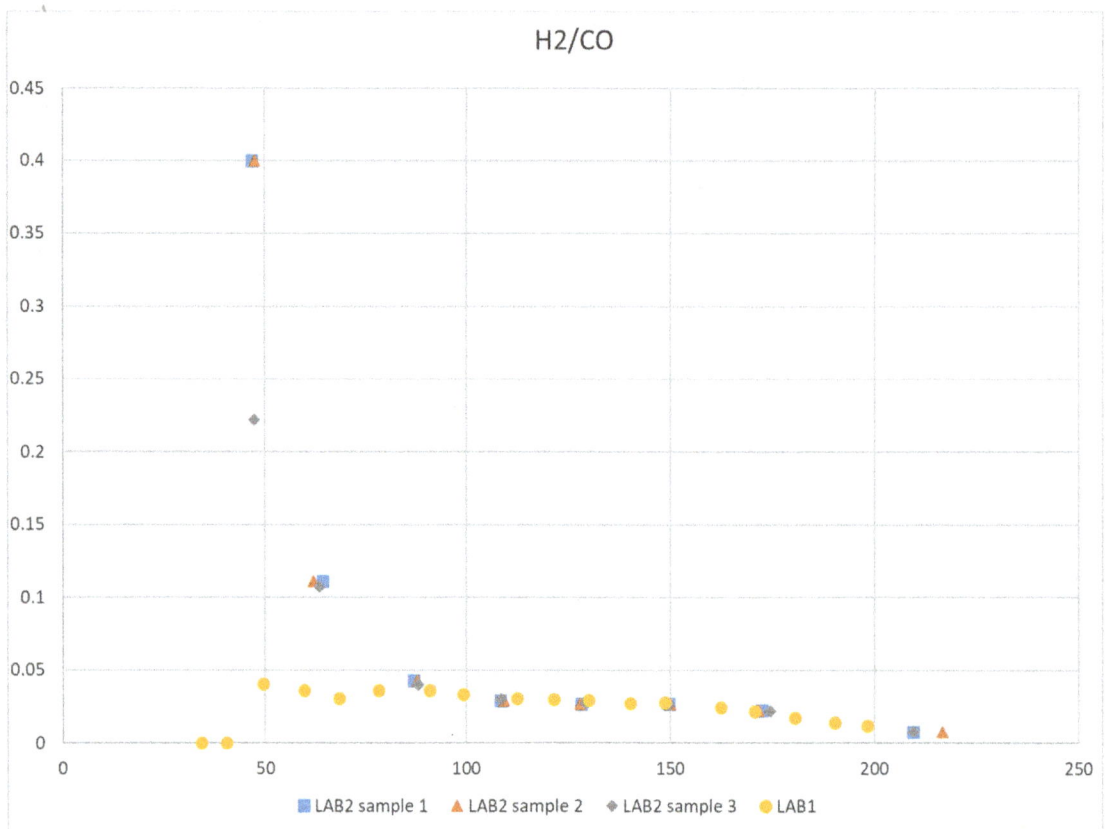

FIG 8 – H$_2$ to CO ratio from gas evolution tests.

Another possibility in setting triggers is to use the ratio of a gas concentration at one temperature in comparison to that at another temperature. Potentially this multiplier could be used in setting triggers in TARPS to indicate that a key coal temperature (say 100°C) has been exceeded. Figure 9 depicts that ratio of [CO] at 100°C to [CO] at 50°C for the three different laboratory techniques across a range of coals. Included in the data are results from a fourth laboratory that no longer does testing. Unfortunately, though within one test method the ratio is relatively coal independent, it can be seen that this ratio varies between the different testing methods.

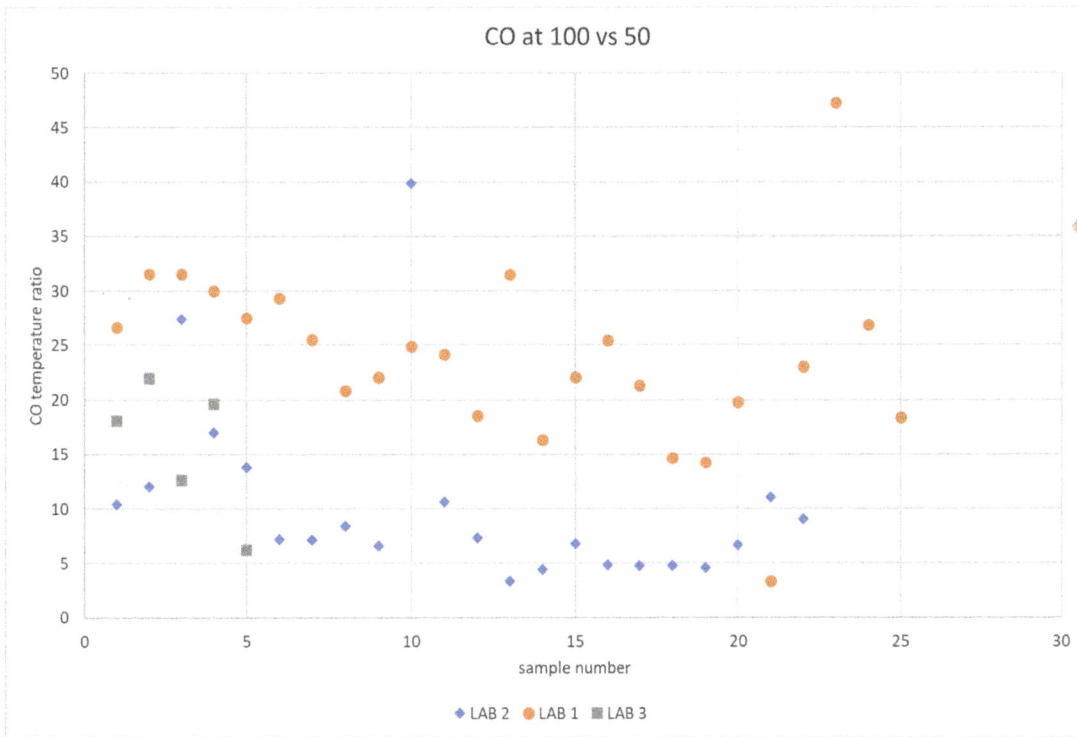

FIG 9 – Comparison of 100°C to 50°C CO concentrations from different laboratories.

CONCLUSIONS

Comparison of the gas evolution data from three different laboratory test methods indicates that there is significant variation in the results between the three laboratories. As such using concentration-based triggers derived from these tests should be used with caution. Oxidation related indicators appear to be independent of the test method and offer greater applicability. Mixed ratios using gas concentrations from different chemical reaction pathways give divergent results between the laboratories.

As the ability to detect gas concentrations improves it is vital that laboratory testing keeps up with this trend and 'presence' of a gas is not used a criterion to indicate spontaneous combustion in TARPS, rather the concentration is specified.

It is important to undertake additional work investigating the potential impact of the parameters outlined above in order to identify robust triggers that can be used in TARPS.

Each test method has been developed in an effort to simulate the oxidation of coal in a coalmine. The diversity of test methods is consistent with the complexity of the mining situation. It is not possible to determine which if any method is the most realistic as all the conditions that exist in mines cannot be determined and are most likely not constant within the area of concern in any case.

It is recommended that any indicator for identifying spontaneous combustion and/or characterising the severity of a heating be shown to be independent of the test method used to derive the key values.

ACKNOWLEDGEMENTS

The data used in this paper has been drawn from laboratory test reports for a number of mines, who wish to remain anonymous. The permission of the mines who have provided the data for this paper is gratefully acknowledged. This paper should not be construed as either endorsing or criticising any particular laboratory-based test method.

REFERENCES

Chamberlain, E A C, Hall, D A and Thirlway, J T, 1970. The ambient temperature oxidation of coal in relation to the early detection of spontaneous heating, Part 1, *Mining Engineer*, 130(121):1–16.

Cliff, D, Bell, S and O'Beirne, T, 1992. Investigation of Bowen Basin Coal Mine fire gas analysis parameters, project no C1463. National Energy Research Demonstration and Development Committee (NERDDC), report 1039, Canberra, Australia.

Graham, J I, 1914. The Absorption of Oxygen by Coal, *Transactions of the Institute of Mining Engineers*, 48:521–534.

Hollins, B, 1995. Fire Gases from Coal Heatings, thesis, University of Queensland, Brisbane.

Humphreys, D R, 1979. A Study of the Propensity of Queensland Coals to Spontaneously Combust, Masters thesis, University of Queensland, Brisbane.

Humphreys, D, 2004. The application of numerical modelling to the assessment of the potential for and the detection of, spontaneous combustion in underground coal mines, PhD thesis, University of Queensland, Brisbane.

Muller, S, Dhyon, S and Kelson, L, 2023. Detection Of Ethylene by Micro Gas Chromatograph and Associated Evolution Temperatures For Gas Evolution Testing, in *Proceedings of 26th World Mining Congress (WMC 2023)*, pp 1765–1775.

Sanders, R, 1983. The analysis and interpretation of the gases resulting from heatings, fires and explosions in certain coal mines, project no C0271, National Energy Research Demonstration and Development Committee (NERDDC), report 189, Canberra, Australia.

US EPA, 1999. Method 14 and 15 Compendium of Methods for the Determination of Toxic Organic Compounds in Ambient Air, Centre for Environmental Research Information Office of Research and Development, US Environmental Protection Agency, Cincinnati, USA.

Więckowski, M, Howaniec, N and Smoliński, A, 2020. Effect of flow rates of gases flowing through a coal bed during coal heating and cooling on concentrations of gases emitted and fire hazard assessment, *Int J Coal Sci Technol*, 7(1):107–121.

Laboratory evaluation of ventilation effects on self-heating incubation behaviour of high reactivity coals

J Theiler[1] and B B Beamish[2]

1. Senior Mining Engineer, B3 Mining Services Pty Ltd, Darra Qld 4076.
   Email: jan@b3miningservices.com
2. Managing Director, B3 Mining Services Pty Ltd, Darra Qld 4076.
   Email: basil@b3miningservices.com

ABSTRACT

The concept of a 'critical velocity zone' in the longwall goaf environment for the development of a spontaneous combustion event has been supported by numerical modelling. However, there is limited experimental data to show ventilation effects on the self-heating incubation behaviour of broken coal piles. Recent incubation testing has been completed on Australian coals as part of ACARP Project C33025 with a wide range of R_{70} self-heating rate values from 0.15°C/h to 11.13°C/h. This paper focuses on incubation test results for two high reactivity coals (R_{70} > 4) and shows the interaction between the moisture/reactivity heat balance under four different flow rate regimes indicative of sluggish ventilation, natural air leakage ventilation, medium ventilation and high ventilation. Coal J is from New South Wales with an R_{70} value of 5.45°C/h and moisture content of 7.4 per cent. Both the sluggish and natural air leakage ventilation conditions result in similar incubation periods, although the shape of each self-heating curve is unique and controlled by the moisture/reactivity heat balance as well as oxygen availability. The practical significance of these results show that this coal has a broader 'critical velocity zone'. Coal U is from Queensland with an R_{70} value of 11.13°C/h and moisture content of 13.8 per cent. The ultra-high reactivity of this coal creates a short incubation period despite the high moisture content. In this case the natural air leakage ventilation conditions results in the shortest incubation period and the medium ventilation conditions is only a couple of days longer. The practical significance of these results shows that for this coal the 'critical velocity zone' would be at a much shallower depth in from the goaf fringe.

INTRODUCTION

The influence of mine ventilation on the possible location for the development of a spontaneous combustion event in a longwall mining environment is illustrated in Figure 1 in terms of the presence of a 'critical velocity zone' (Smith *et al*, 1994). This concept highlights that in the immediate face area adjacent to the goaf fringe the air velocity is too high for heat to accumulate, but as the distance increases into the goaf the air velocity decreases to a critical zone where there is insufficient heat dissipation and a sufficient supply of oxygen to support self-heating. Consequently, in the event of a prolonged face stoppage ideal conditions may be present to allow a caved coal pile to incubate to thermal runaway. Deeper into the goaf the atmosphere becomes too oxygen deficient to support self-heating. While numerical modelling (principally CFD modelling – Yuan and Smith, 2008; Song *et al*, 2017) has often been used in support of this concept, there is no experimental data available for Australian coals to show ventilation effects on the self-heating incubation behaviour of broken coal in the longwall goaf environment.

Laboratory experiments have been conducted on US coal samples using flow rates ranging from 100 to 500 mL/min and a sample mass of 150 g (Yuan and Smith, 2012). The flow to mass ratio used in these experiments is much too high to replicate site conditions and causes evaporation to dominate the heat balance in favour of heat loss. Hence, they partially dried the coal samples prior to testing and heated the coal to a temperature more than 100°C to create a heat balance where heat gain could be achieved to produce thermal runaway. For the three US coals tested a different combination of flow rate and applied temperature increase was necessary for each coal to induce heat gain to thermal runaway. Consequently, practical demonstration and characterisation of ventilation flow rate effects are lacking, particularly with respect to the range of coals being mined in Australia.

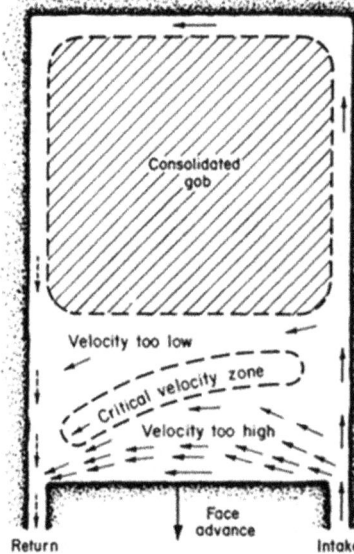

FIG 1 – Schematic diagram of ventilation flow in the vicinity of the goaf fringe on a longwall face (from Smith *et al*, 1994).

Most spontaneous combustion index tests produce relative ratings of spontaneous combustion propensity, or more specifically coal intrinsic reactivity. They do not provide any context of the self-heating behaviour in an actual mine environment and therefore do not assess the likelihood of the coal creating a heating event. They also do not indicate any time frame for an event to occur under mine site conditions. The adiabatic Incubation Test method overcomes these deficiencies (Beamish and Theiler, 2019) and makes it possible to replicate site-specific conditions including different mine ventilation flow rate scenarios. This paper presents the Incubation Test results for two high reactivity coals using four different flow rates indicative of sluggish ventilation 5 mL/min (3.85×10^{-5} m/s), natural air leakage ventilation 10 mL/min (7.7×10^{-5} m/s), medium ventilation 20 mL/min (1.54×10^{-4} m/s) and high ventilation 40 mL/min (3.08×10^{-4} m/s). These coal samples are part of a more extended research program on Australian coals from different basins covering a wide range of reactivities and moisture contents. The full set of results are published under ACARP Project C33025.

DESCRIPTION OF SAMPLES AND ANALYTICAL DATA

The samples used in this laboratory evaluation are from Australian coalmines and are high volatile bituminous in rank. Both mines have recorded minor heating events in the past. Coal analytical and ranking data for the two samples are contained in Table 1. There is a significant variation in coal type as indicated by the location of each sample on a Suggate rank diagram (Figure 2). Coals J plots near the lower boundary of the low-medium vitrinite coal band, whereas Coal U plots near the upper boundary. Both coals have low total sulfur contents and therefore the presence of pyrite is negligible. The as-mined moisture content of coal J is 7.4 per cent, whereas Coals U has a much higher as-mined moisture content of 13.8 per cent.

TABLE 1

Analytical data for high reactivity coal samples.

Proximate analysis (air-dried basis)	Coal J	Coal U
Moisture (%)	5.1	13.2
Ash (%)	23.5	6.2
Volatile matter (%)	23.0	30.2
Fixed carbon (%)	48.4	50.4
Total sulfur (%)	0.23	0.43
Calorific value (MJ/kg)	22.83	26.37
Ultimate analysis (dry ash-free basis)		
Carbon (%)	82.0	80.5
Hydrogen (%)	4.26	5.02
Nitrogen (%)	1.76	2.18
Sulfur (%)	0.32	0.53
Oxygen (%)	11.7	11.8
ASTM rank	hvCb	hvCb

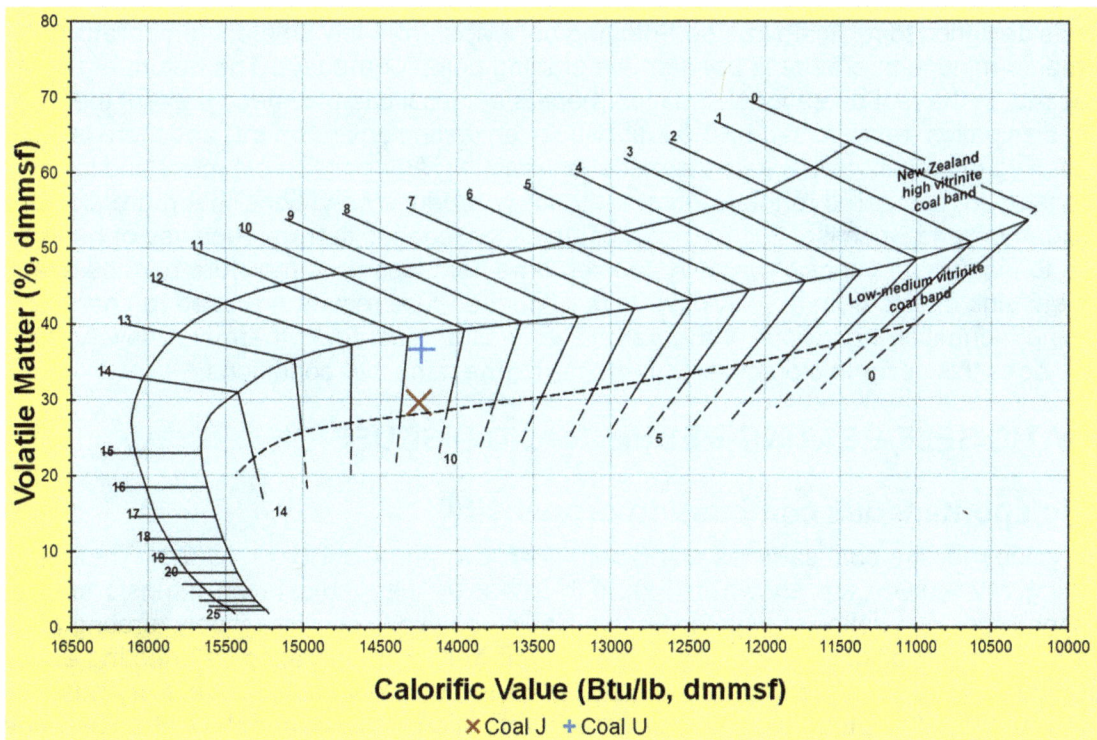

FIG 2 – Suggate rank plot for high reactivity coal samples showing volatile matter and calorific value on a dry mineral matter sulfur free basis.

SELF-HEATING TEST PROCEDURES

Adiabatic oven R_{70} self-heating rate

Full details of the adiabatic oven are given in Beamish, Barakat and St George (2000). The sample to be tested is crushed and sieved to <212 µm in as short a time as possible to minimise the effects of oxidation on fresh surfaces created by the grinding of the coal. A 150 g sample is placed in a 750 mL volumetric flask and a unidirectional flow of nitrogen at 250 mL/min applied to the flask inside a drying oven. Precautions are taken to ensure the exclusion of oxygen from the vessel prior to heating the coal for drying. Hence, the air is flushed from the system at room temperature for a period of one hour. After one hour, the oven is ramped up to 110°C and the coal is dried under nitrogen for at least 16 hr to ensure complete drying of the sample. All R_{70} tests are performed on a dry basis to standardise the test results.

At the completion of drying, the coal is transferred into the reaction vessel and left to stabilise at 40°C in the adiabatic oven with nitrogen passing through it. The reaction vessel is a 450 mL thermos flask inner. When the sample temperature has stabilised, the oven is switched to remote monitoring mode. This enables the oven to track and match the coal temperature rise due to oxidation. The gas selection switch is turned to oxygen with a constant flow rate of 50 mL/min. The temperature change of the coal with time is recorded by a datalogging system for later analysis. The oven limit switch is set at 160°C to cut-off the power to the oven and stop the oxygen flowing when the sample reaches this temperature. When the oven cools down, the sample is removed from the reaction vessel, which is then cleaned in preparation for the next test. The results are used to classify the intrinsic spontaneous combustion propensity of the sample according to the rating scheme published by Beamish and Beamish (2011).

Adiabatic oven self-heating incubation

This test is designed to replicate true self-heating behaviour from low ambient temperature. As such, the normal in-mine temperature is used as the starting point for the test. The nature of the test also assumes that in the real operational situation there is a critical pile thickness present that minimises any heat dissipation (represented by the adiabatic oven testing environment) and there is a sufficient supply of oxygen present to maintain the oxidation reaction. A larger sample mass and lower oxygen flow rate is used, compared to the R_{70} test method, to produce conditions that more closely match reality (Beamish and Beamish, 2011). The sample either reaches thermal runaway or begins to lose heat due to insufficient intrinsic reactivity to overcome heat loss from moisture release/evaporation and/or heat sink effects from non-reactive mineral matter. The results are used to characterise the self-heating incubation behaviour of the sample as well as quantify if thermal runaway is possible and if so, does this occur in a practical time frame for the mine site conditions.

ADIABATIC SELF-HEATING RESULTS AND DISCUSSION

Intrinsic spontaneous combustion propensity

The R_{70} values for the coal samples are Coal J 5.45°C/h and Coal U 11.13°C/h. Their respective self-heating rate curves are shown in Figure 3. These results indicate an intrinsic spontaneous combustion propensity rating of high for Bowen Basin conditions or alternatively a rating of medium for Sydney Basin conditions. This rating does not consider any moderating self-heating effect of the moisture content that is present in the coal since the R_{70} value is obtained on a dry basis with the moisture removed. Also, like most spontaneous combustion index parameters, the R_{70} value does not provide any indication of the time frame for a heating to develop to thermal runaway.

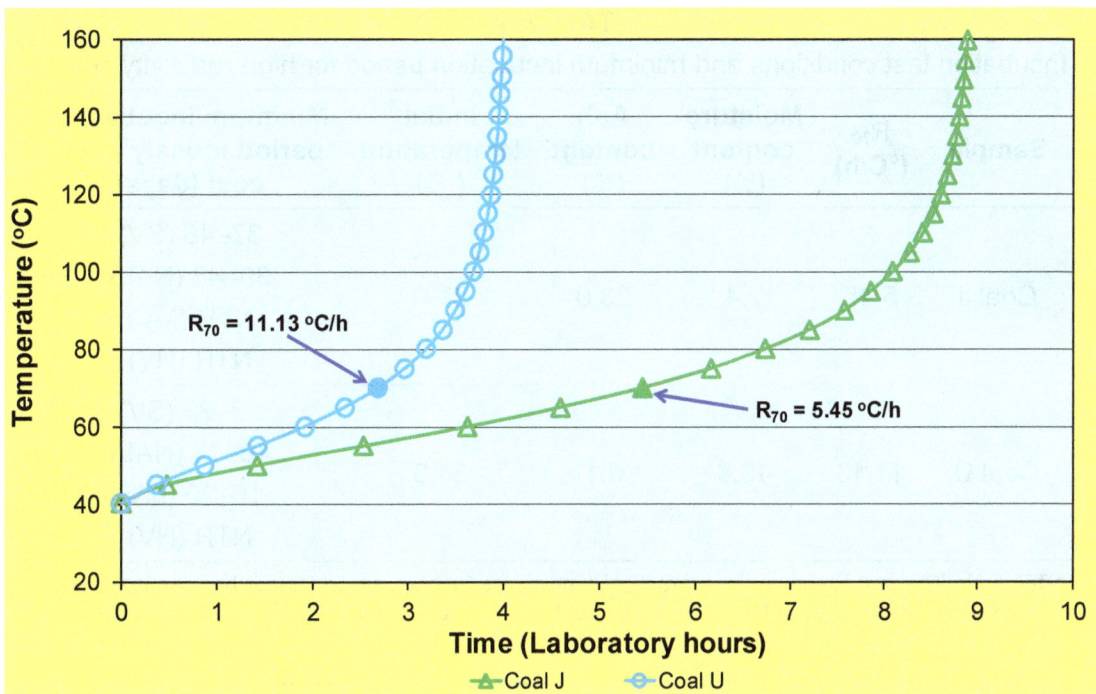

FIG 3 – Adiabatic R_{70} self-heating rate results for high reactivity coal samples.

Self-heating incubation behaviour under different flow rate conditions

The initial conditions for the tests and the inferred minimum incubation periods for the high reactivity coals are contained in Table 2. The corresponding incubation self-heating curves for Coals J and U are shown in Figures 4 and 5 for a loose pile state exposed to the four different ventilation conditions. Both the sluggish and natural air leakage ventilation conditions result in similar incubation periods for Coal J (Figure 4), although the shape of each self-heating curve is unique and controlled by the moisture/reactivity heat balance as well as oxygen availability. The latter is particularly noticeable as the coal proceeds into thermal runaway. With greater oxygen availability under the natural air leakage flow rate the temperature increase is much more rapid. The higher reactivity of Coal U creates a shorter incubation period than Coal J under sluggish and natural air leakage ventilation conditions (Figure 5) despite the high moisture content. It also enables Coal U to incubate to thermal runaway under medium ventilation conditions. The natural air leakage ventilation condition results in the shortest incubation period for Coal U and the medium ventilation condition is only a couple of days longer.

The practical significance of these results show that Coal J has a wide 'critical velocity zone' as shown in Figure 6. Also, the hotspot that would form in a heating event would begin to migrate to the higher flow rate region and therefore as it migrates the hotspot self-heating rate would escalate more rapidly due to the greater oxygen availability. Coal U has a broader 'critical velocity zone' than coal J and can develop at a much shallower depth in from the goaf fringe (Figure 6). The hotspot that would form in a heating event for this coal would escalate in temperature even more rapidly than coal J due to greater oxygen availability.

TABLE 2

Incubation test conditions and minimum incubation period for high reactivity coals.

Sample	R_{70} (°C/h)	Moisture content (%)	Ash content (%)	Initial temperature (°C)	Minimum incubation period loosely piled coal (days)
Coal J	5.45	7.4	23.0	27.0	37–46 (SV) 39–48 (NAL) NTR (MV) NTR (HV)
Coal U	11.13	13.8	6.1	35.0	22–28 (SV) 10–13 (NAL) 15–20 (MV) NTR (HV)

NTR – No Thermal Runaway in a practical time frame without an additional heat source; SV – Sluggish Ventilation; NAL – Natural Air Leakage; MV – Medium Ventilation; HV – High Ventilation.

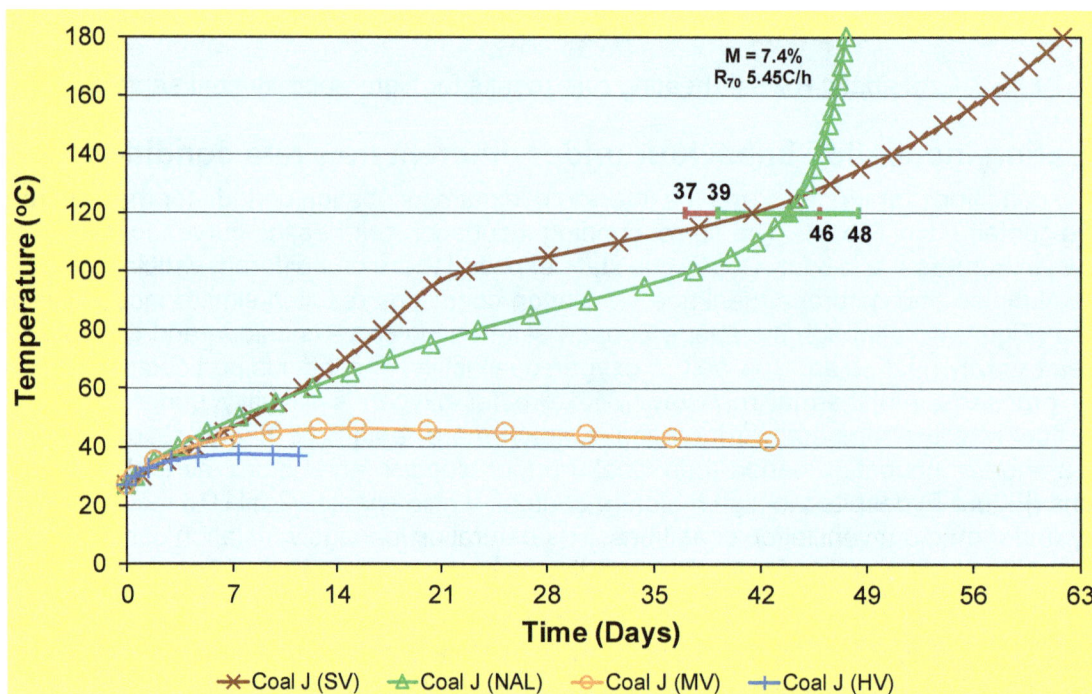

FIG 4 – Adiabatic Incubation Test results for Coal J under a variety of ventilation flow rates with incubation period shown for a loosely piled state under ideal conditions.

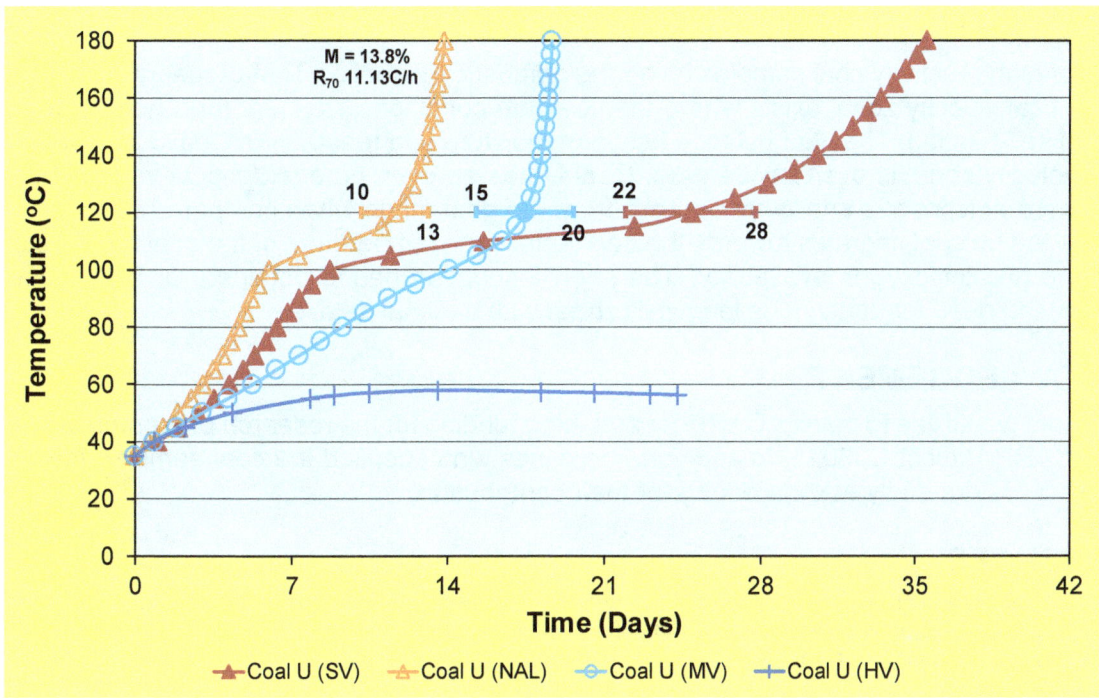

FIG 5 – Adiabatic Incubation Test results for Coal U under a variety of ventilation flow rates with incubation period shown for a loosely piled state under ideal conditions.

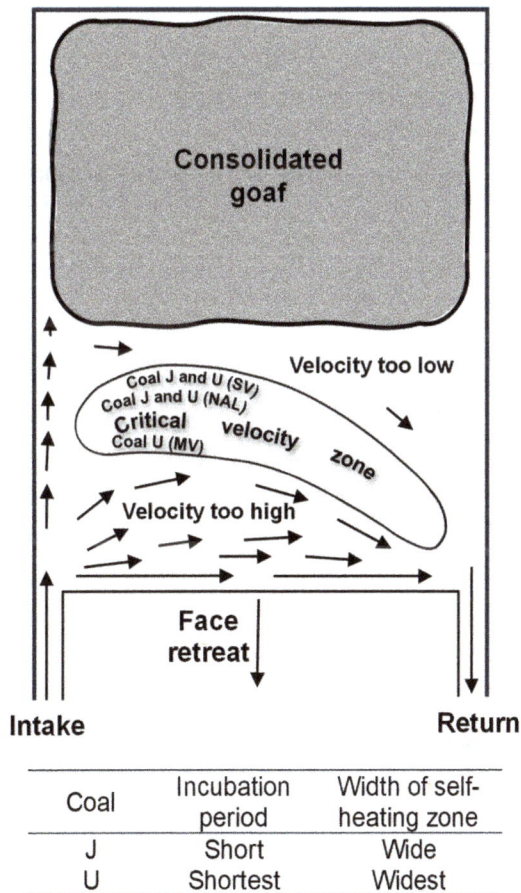

Coal	Incubation period	Width of self-heating zone
J	Short	Wide
U	Shortest	Widest

FIG 6 – Schematic of 'critical velocity zones' for Coals J and U in a goaf behind the face indicating relative incubation period (modified from Smith *et al*, 1994).

CONCLUSIONS

Testing of high reactivity coal samples using the adiabatic Incubation Test procedure has confirmed that a 'critical velocity zone' exists where the flow rate conditions are ideal for a hotspot to develop from coal self-heating. The heat balance between moisture and reactivity of Coal J indicates that the 'critical velocity zone' is over a wide area. Coal U has an even broader 'critical velocity zone' and could form a hotspot at a shallower depth in from the goaf fringe when compared to coal J. In both cases, as the hotspot migrates towards the free surface the temperature of the coal rapidly escalates due to the greater oxygen availability. The short incubation period obtained for both these coals reinforces the need for continuous longwall retreat with minimal delays.

ACKNOWLEDGEMENTS

The authors would like to thank ACARP for providing funding for the research presented in this paper under ACARP Project C33025. In addition, the mines who supplied the coal samples for testing in this project are gratefully acknowledged for their contribution.

REFERENCES

Beamish, B and Beamish, R, 2011. Testing and sampling requirements for input to spontaneous combustion risk assessment, in *Proceedings of the Australian Mine Ventilation Conference*, pp 15–21 (The Australasian Institute of Mining and Metallurgy: Melbourne).

Beamish, B B and Theiler, J, 2019. Coal spontaneous combustion: Examples of the self-heating incubation process, *International Journal of Coal Geology*, 215:103297.

Beamish, B B, Barakat, M A and St George, J D, 2000. Adiabatic testing procedures for determining the self-heating propensity of coal and sample ageing effects, *Thermochimica Acta*, 362(1–2):79–87.

Smith, A C, Diamond, W P, Mucho, T P and Organiscak, J A, 1994. Bleederless ventilation systems as a spontaneous combustion control measure in US coal mines, US Bureau of Mines Information Circular IC9377.

Song, S, Wang, S, Liang, Y, Li, X and Lin, Q, 2017. Influence of air supply velocity on temperature field in the self-heating process of coal, *Sains Malaysiana*, 45(11):2143–2148.

Yuan, L and Smith, A C, 2008. Effects of ventilation and gob characteristics on spontaneous heating in longwall gob areas, in *Proceedings of the 12th US/North American Mine Ventilation Symposium*, pp 141–147 (The Society of Mining, Metallurgy and Exploration Inc., Littleton).

Yuan, L and Smith, A C, 2012. The effect of ventilation on spontaneous heating of coal, *Journal of Loss Prevention in the Process Industries*, 25:131–137.

Pore evolution and reaction mechanism of coal under the combination of chemical activation and autothermal oxidation

H Xin[1], P Zhang[2], J Sun[2], L Lu[2], C Xu[2], B Zhou[2], H Wang[2], Y Yang[2], J Li[2] and D Wang[2]

1. State Key Laboratory of Coal Resources and Safe Mining, China University of Mining and Technology, Xuzhou, Jiangsu 221116, PR China. Email: linyichao2007@126.com; qbt2003@163.com
2. State Key Laboratory of Coal Resources and Safe Mining, China University of Mining and Technology, Xuzhou, Jiangsu 221116, PR China.

ABSTRACT

For the limitations inherent in traditional physical and chemical methods of enhancing coal bed penetration, this study introduces a coal bed methane penetration enhancement method. Techniques, including N_2 and CO_2 adsorption, XPS, and FTIR, were used to investigate the evolution of gas coal matrix pore structure and the changes in active structure under the influence of chemical activation and autothermal oxidation. The findings indicate that upon oxidation at a steady 230°C, there was a substantial augmentation in mesopore and macropore volumes, by 285.8 per cent and 209.1 per cent, respectively, and an overall increase in total pore volume by 63 per cent, accompanied by significant enhancements in the specific surface areas of mesopores, macropores, and total-pores, by 141.11 per cent, 250 per cent, and 53.25 per cent, respectively. Concurrently, the D_2 value decreased. The oxygen-containing functional groups in the activated coal inversely correlate with oxidation temperature, showing a 23.7 per cent increase at 230°C compared to the original coal. Meantime, there was a notable increase of 28.7 per cent in C-O/C-O-C content, alongside a corresponding 23.7 per cent decrease in C-C/C-H content. The presence of highly active radicals enhances the aromatic polymerisation reaction and improves the maturity and aromaticity of the coal. These results underscore the potential of the method for enhancing coal seam gas penetration and extraction.

INTRODUCTION

Coal represents one of the most abundant, widespread, and cost-effective energy resources globally (Qin *et al*, 2018; Tao, Chen and Pan, 2019). Coal bed methane (CBM), a byproduct of coal formation, primarily consists of methane gas. Methane within coal seams predominantly resides within the coal's microporous structure, in both the adsorbed and free states (Zheng *et al*, 2023; Guo *et al*, 2023). As shallow resources diminish and coal mining extends deeper, the complexities of gas disasters intensify, presenting new substantial challenges in prevention and control (Zhang *et al*, 2022; Tang, Yang and Wu, 2017). Concurrently, considerable quantities of methane are directly emitted into the atmosphere, squandering valuable clean energy resources and exacerbating greenhouse gas impacts (Cai *et al*, 2023; Xin *et al*, 2023a). Consequently, efficient extraction of coalbed methane is imperative not only to mitigate gas outburst disasters but also to contribute to energy conservation and emission reduction efforts (Zhu *et al*, 2019; Zhang *et al*, 2023).

To maximise the extraction of gas and enhance the economic benefits of coalbed methane, various coal seam penetration enhancement methods are employed, including CO_2 injection (Zheng *et al*, 2022; Shang *et al*, 2022), microwave radiation (Wang *et al*, 2018; Lu *et al*, 2023a), liquid nitrogen freeze-thawing (Qin *et al*, 2020; Liu *et al*, 2022), and chemical penetration enhancement (Zhao *et al*, 2023; Dang *et al*, 2023; Lu *et al*, 2023b). Among them, chemical penetration enhancement is increasingly favourable due to its cost-effectiveness, environmental friendliness, and high efficiency. Chen *et al* examined the oxidation of coal samples using H_2O_2, NaClO, and ClO_2 oxidisers and found that the treatment dissolved coal's organic matter, thereby augmenting the desorption of coalbed methane (Chen *et al*, 2020). Similarly, Yang Juan *et al* explored the impact of a non-homogeneous cobalt-based activator in ammonium persulfate on coal samples of varying metamorphic degrees, discovering that it could oxidise and dissolve the surfaces, leading to the formation of new fissures and pores (Yang *et al*, 2020). Li *et al* investigated the effects of microwave-assisted Fenton reagent and $Na_2S_2O_8$ on coal oxidation, revealing that while the method expanded pores, it also caused some to collapse and altered the coal's chemical structure by transforming self-associating hydrogen bonds into ether groups, thereby increasing the coal's maturity and aromaticity (Li *et al*, 2022).

This paper introduces a coal seam penetration enhancement method that leverages the synergistic effects of chemical activation and autothermal oxidation (Xin *et al*, 2023b). This technique employs high-energy free radicals generated by persulfate to activate the coal chemically, significantly expediting the coal's autothermal oxidation process. Concurrently, it harnesses the coal's autothermal heating, coupled with chemical activation, to amplify coal seam permeability by fully utilising the coal's *in situ* energy, thus facilitating the rapid expansion of coal pore space. Coal self-heating not only promotes gas desorption but also alters the pore structure, thereby enhancing gas diffusion rates and permeability within the coal seam, offering an effective alternative to heat injection methods. Moreover, chemical activation notably advances nano-pore development within the coal matrix by supplying high-activity free radicals. It not only fosters methane desorption from nano-pores but also expedites the ingress and reaction of oxygen molecules within the coal, significantly improving the rate of coal's autothermal reaction and contributing to the swift enhancement of coal porosity (Li *et al*, 2022).

However, the effect mechanism of this method on the pore and structure of coal is not yet perfect. So the N_2 and CO_2 adsorption techniques were employed to analyse the evolution of the coal matrix pore structure. This analysis revealed a dynamic mechanism of pore expansion and capacity increase during autothermal oxidation prompted by chemical activation. Furthermore, the study utilised XPS and FTIR methods to investigate the changes in the coal matrix's reactive structure and elucidate the evolution of molecular structure during the process. These techniques clarified the coal's pore evolution and reaction mechanisms under the influence of autothermal oxidation and chemical activation. The findings underscore the potential of this autothermal oxidation-driven method in enhancing coal seam gas penetration and extraction.

EXPERIMENTS AND METHODS

Preparation of coal sample

The coal samples used in this experiment were sourced from the Baode coalfield located in Shanxi Province, China. These samples were processed by crushing and sieving to achieve a particle size distribution within the range of 40–80 meshes. The specific coal quality parameters of these samples are detailed in Table 1. To establish a baseline, the raw coal (abbreviated as RC) underwent a drying process at 40°C for 48 hrs to ensure uniform moisture content, serving as a blank test. In this study, $Na_2S_2O_8$ was chosen as the chemical agent, with a concentration of 0.0355 mol/L (Yang *et al*, 2020; Dai *et al*, 2020). The molar ratio of Fe^{2+} to $S_2O_8^{2-}$ was maintained at 1:1 (Wang *et al*, 2020). The coal samples were subjected to chemical activation in a mixed aqueous solution of $Na_2S_2O_8$ and $FeSO_4$ for 72 hrs at a constant temperature of 60°C.

TABLE 1

Proximate and ultimate analyses of raw coal.

Coal type	Proximate analysis, wt%				Ultimate analysis, wt%, daf				
	M_{ad}	A_d	V_{daf}	FC_{ad}	C	H	O	N	S
Gas coal	2.11	41.40	37.46	63.26	51.01	3.34	3.87	1.59	0.42

M_{ad}: moisture; A_d: ash; V_{daf}: volatile matter; FC_{ad}: fixed carbon.

The laboratory's self-designed VDRT-2000 fixed-bed apparatus was employed for programmed heating experiments on activated coal, ensuring high airtightness for complete reactions of coal samples. Research indicates that 80°C marks the transformation onset in the coal's oxidation process, 160°C as the gas coal cross-point temperature, 200°C as the peak of O_2 consumption with notable changes in the coal groups, and 230°C for C_2H_2 gas production. Consequently, coal samples were sequentially heated to 80°C, 160°C, 200°C, and 230°C in an air atmosphere to ensure complete oxidation, each maintained at a constant temperature for one hour. The fully oxidised coal samples were designated such as CA, CA160°C, CA200°C, and CA230°C, respectively.

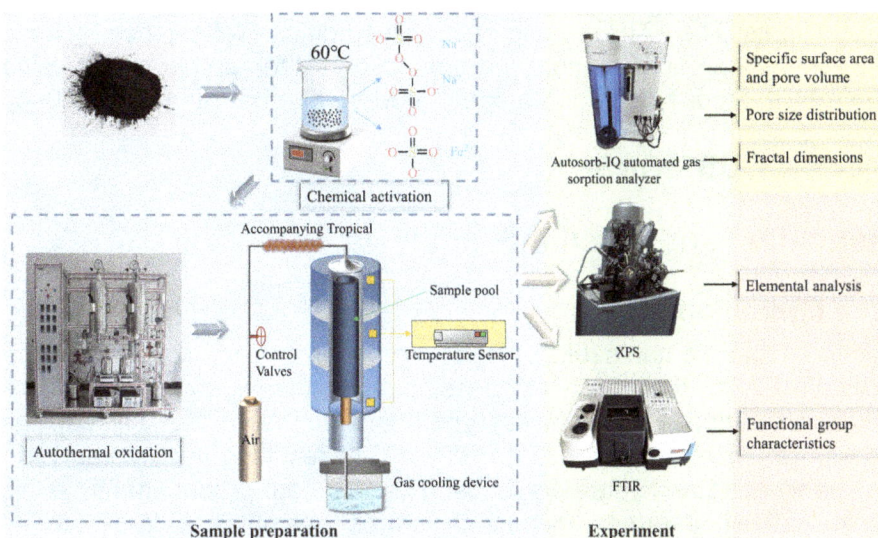

FIG 1 – Experimental procedure.

Coal pore characterisation experiments

This study aimed to analyse the variations in the full range of pore sizes within coal. For this purpose, coal pores were categorised into micropores (<2 nm), mesopores (2–50 nm), and macropores (>50 nm) (Thommes and Cychosz, 2014), adhering to the International Union of Pure and Applied Chemistry (IUPAC) criteria for pore classification. This classification further delineates micropores into ultramicropores (<0.7 nm) and very micropores (0.7–2 nm) (Thommes *et al*, 2015). To precisely assess the pore distribution characteristics of coal driven by autothermal oxidation in the presence of chemical activation, we employed N_2 adsorption at 77 K and CO_2 adsorption at 273 K. These measurements were conducted using the Autosorb-IQ3 Specific Surface Area Analyser (Quantachrome Instruments, USA). This comprehensive approach to pore size analysis allows for a detailed understanding of the pore structure evolution in coal under specified experimental conditions. For each experimental run, a consistent sample mass of 0.5 g of coal was used. Prior to the analysis, all coal samples underwent a degassing process at 110°C for 12 hrs to remove any impurity gases. The N_2 adsorption experiments were performed in a liquid nitrogen bath at 77 K. During these experiments, a total of 41 adsorption and 33 desorption pressure points were recorded, spanning a range of relative pressures from 0.001 to 0.995. Similarly, the CO_2 adsorption experiments were conducted in an ice-water bath at 273 K. Here, 60 adsorption pressure points were collected, covering a relative pressure range from 3×10^{-5} to 0.0278.

X-ray photoelectron spectroscopy (XPS) experiments

To obtain the chemical valence and relative content of major elements in coal under the driving effect of autothermal oxidation by chemical activation, we used an EscalaB 250Xi X-ray photoelectron spectrometer to perform a broad sweep of all the elements in the experimental coal samples, and a narrow sweep of four elements, namely, C, O, S, and Fe. The test conditions of the XPS experiment were Source Gun Type, using an Al k Alpha monochromated aluminium anode target with a beam spot size of 900 μm. All experimental coal samples were broad-scanned in the range of 0–1350 eV, narrow-scanned in the range of 278–297 eV for the element C1s, and narrow-scanned in the range of 524–544 eV for the element O1s.

Fourier-transform infrared spectroscopy (FTIR) experiments

This study aimed to elucidate the changes in species, content, and molecular structure of organic functional groups within coal driven by autothermal oxidation in the presence of chemical activation, particularly under the combined influence of chemical activation and autothermal oxidation. For this purpose, the surface functional groups of coal were analysed using a Nicolet 6700 FTIR spectrometer, manufactured by Nicolet in the USA. To establish a reference baseline, pure ground potassium bromide (KBr) was employed to acquire a reference spectrum. It ensured that all coal sample analyses were conducted against a consistent background, allowing for accurate comparative assessments. The FTIR experiments utilised diffuse reflectance technology to capture the spectral data. The experimental settings were as follows: the number of acquisition scans set to 64, an acquisition resolution of 4 cm^{-1}, and a spectral wave number range of 4000–400 cm^{-1}.

RESULTS AND DISCUSSIONS

Pore characterisation and parameter analysis

Pore adsorption characteristics

Figure 2 presents the experimental results of N_2 adsorption in coal induced by autothermal oxidation and chemical activation. According to IUPAC classifications, the N_2 isothermal adsorption profiles of these coal samples are predominantly a mix of type II and type IV(a). Notably, capillary coalescence occurs in the coal pores at relative pressures $P/P_0 > 0.4$, resulting in hysteresis in the isothermal adsorption lines, manifesting as hysteresis loops. The raw coal predominantly exhibits H2(b)-type hysteresis loops, indicative of intermediary pore structures resembling wide-necked ink bottles; high-temperature oxidation minimally impacts these structures, maintaining their predominance in the RC200°C coal samples. CA coal samples and CA160°C samples display similar H2(b)-type loops, while CA200°C samples transition towards H4-type, suggesting a mix of ink-bottle pores and narrow fissure structures. Significantly, the CA230°C samples demonstrate H4-type hysteresis loops, reflecting an increase in narrow fissure pores as the oxidative temperature rises, signifying a gradual shift in pore structure from H2(b) to H4 type.

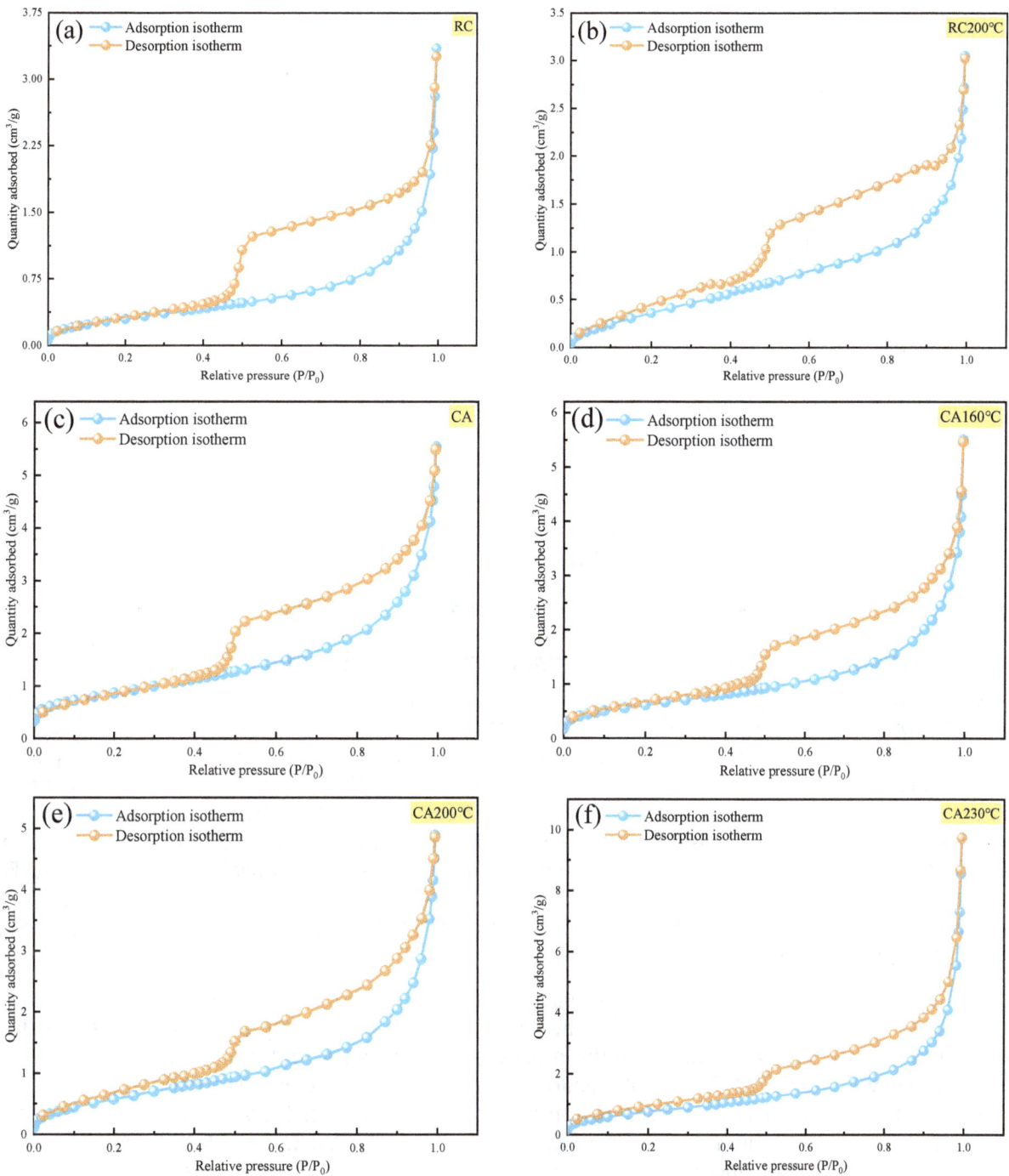

FIG 2 – N_2 adsorption isotherm of coal samples.

Figure 3 illustrates the results of CO_2 adsorption experiments on coal subjected to autothermal oxidation and chemical activation. The CO_2 adsorption isotherms for the six coal sample groups resemble class I adsorption isotherms, suggesting that most micropores in these samples are smaller than 1 nm. The extent of adsorption indirectly reflects the micropore quantity in the coals, with the original coal displaying the lowest adsorption and thus the fewest micropores. As the coal samples undergo warming and oxidation, the adsorption increases variably, with higher oxidation temperatures correlating with greater adsorption and a larger number of micropores. Notably, the CA200°C and CA230°C samples exhibit a substantial increase in adsorption, indicating more developed micropores at temperatures above 200°C. Analysis of adsorption in the low-pressure region reveals that RC200°C, CA200°C, and CA230°C coal samples contain more ultramicropores compared to the RC, CA, and CA160°C samples.

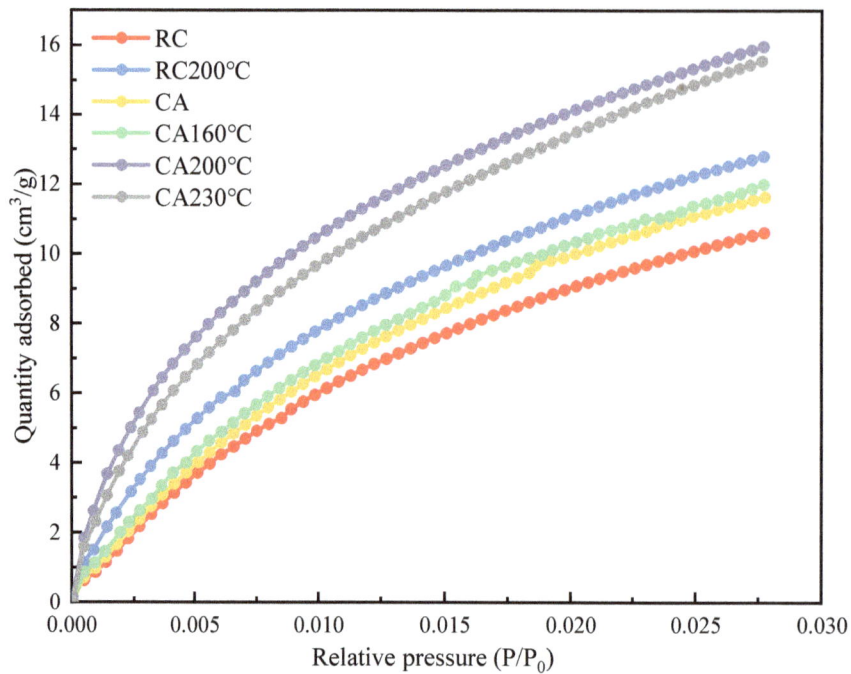

FIG 3 – CO_2 adsorption isotherm of coal samples.

Pore size distribution and specific surface area characteristics

Barrett-Joyner-Halenda (BJH) method, which is predicated on the phenomenon of capillary condensation occurring within pores, is traditionally employed to analyse the distribution of intermediate and large pores in materials. However, it is important to note that the BJH approach tends to underestimate pore sizes below 10 nm, with a significant margin of error — approximately 20 per cent — for pores smaller than 4 nm. On the other hand, the Density Functional Theory (DFT) provides a more accurate determination of pore size distribution, especially for micropores and mesopores. In light of this, our study adopts a combined approach utilising both DFT and BJH methods to characterise the distribution of micropores, mesopores, and macropores in the experimental coal samples. Specifically, we employ the DFT method for accurately characterising pore sizes below 20 nm.

Figure 4 depicts the distribution characteristics of mesopores and macropores in coal induced by autothermal oxidation under chemical activation. The RC coal sample predominantly features mesopores ranging from 4–20 nm, with the most prevalent size being 4.2 nm, akin to the RC200°C sample. However, the RC200°C sample exhibits a significant increase in adsorption at this pore size, suggesting that thermal oxidation disrupts some microscopic pores in the raw coal, leading to a more uniform pore size distribution. Unlike the RC samples, the RC200°C samples demonstrate more frequent 13–14 nm pores, indicating that thermal oxidation ruptures some mesopores, resulting in larger pores. In CA samples, the smallest pores measure 4.2 nm and most mesopores cluster below 20 nm, albeit with uneven distribution. Upon heating to 160°C, the smallest pores in CA samples increase to 4.5 nm and exhibit a multi-peak mesopore distribution, with a decreased adsorption of various-sized mesopores. Further, CA200°C samples show a slight decrease in the size of the most prevalent pores to 4.3 nm, while 5.5 nm pores markedly increase, alongside an overall rise in adsorption. CA230°C samples maintain the most common pore size at 4.2 nm but feature an increased number and adsorption of 6.8 nm pores. The pore size distribution curves reveal that mesopores in coal oxidised at varying temperatures display multi-peak distribution, reflecting non-uniform pore sizes due to elevated temperature oxidation causing greater mesopore development and the merging of smaller pores into larger ones. Generally, as oxidation temperature increases, coal's smaller mesopores evolve into larger ones. However, temperatures below 200°C have a minimal impact on larger pores, with relatively stable adsorption levels.

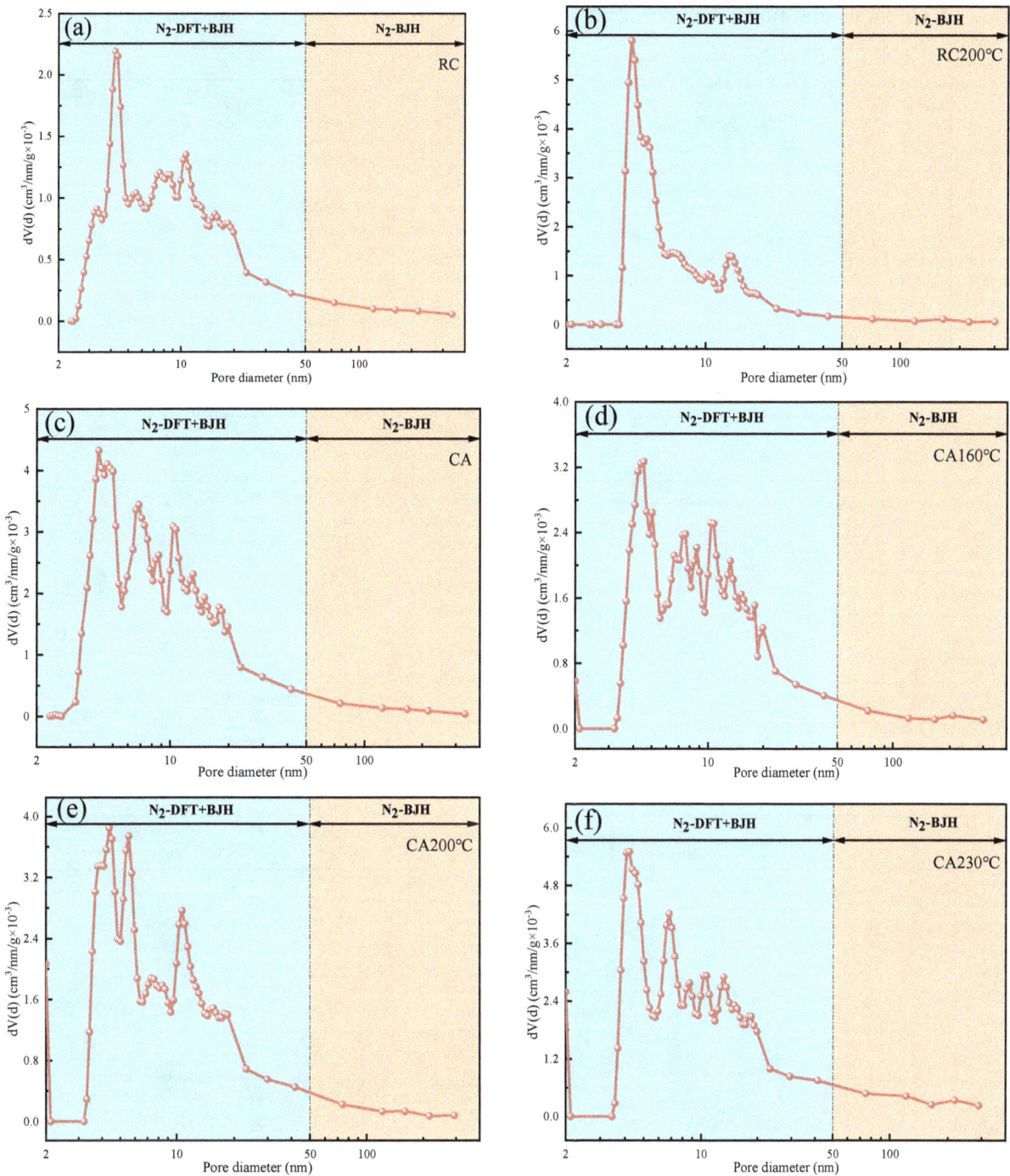

FIG 4 – Characterisation of mesopore and macropore distribution of coal samples based on DFT and BJH methods.

Figure 5 illustrates the microporous distribution characteristics of coals subjected to autothermal oxidation under chemical activation. The predominant pore size in the RC and RC200°C coal samples is 0.48 nm, with an increase in ultramicropores measuring 0.35 nm in the RC200°C sample. Similarly, the CA coal sample features a primary pore size of 0.48 nm, with the second most common size at 0.55 nm, both showing minor differences in adsorption amounts. It suggests that during constant temperature oxidation, some ultramicropores merge to form larger micropores. The CA160°C samples present a slightly larger prevalent pore size of 0.50 nm, indicating that oxidation at 160°C facilitates the merging of micropores. The CA200°C samples maintain this 0.50 nm size, with an enhanced adsorption capacity, particularly noted in 0.79 nm pores. Remarkably, the CA230°C samples exhibit a primary pore size of 0.35 nm, corresponding to a higher adsorption capacity, with larger pores also demonstrating increased adsorption. In summary, as oxidation temperature escalates, new ultramicropores emerge within the coal matrix, and existing micropores coalesce to form larger pores.

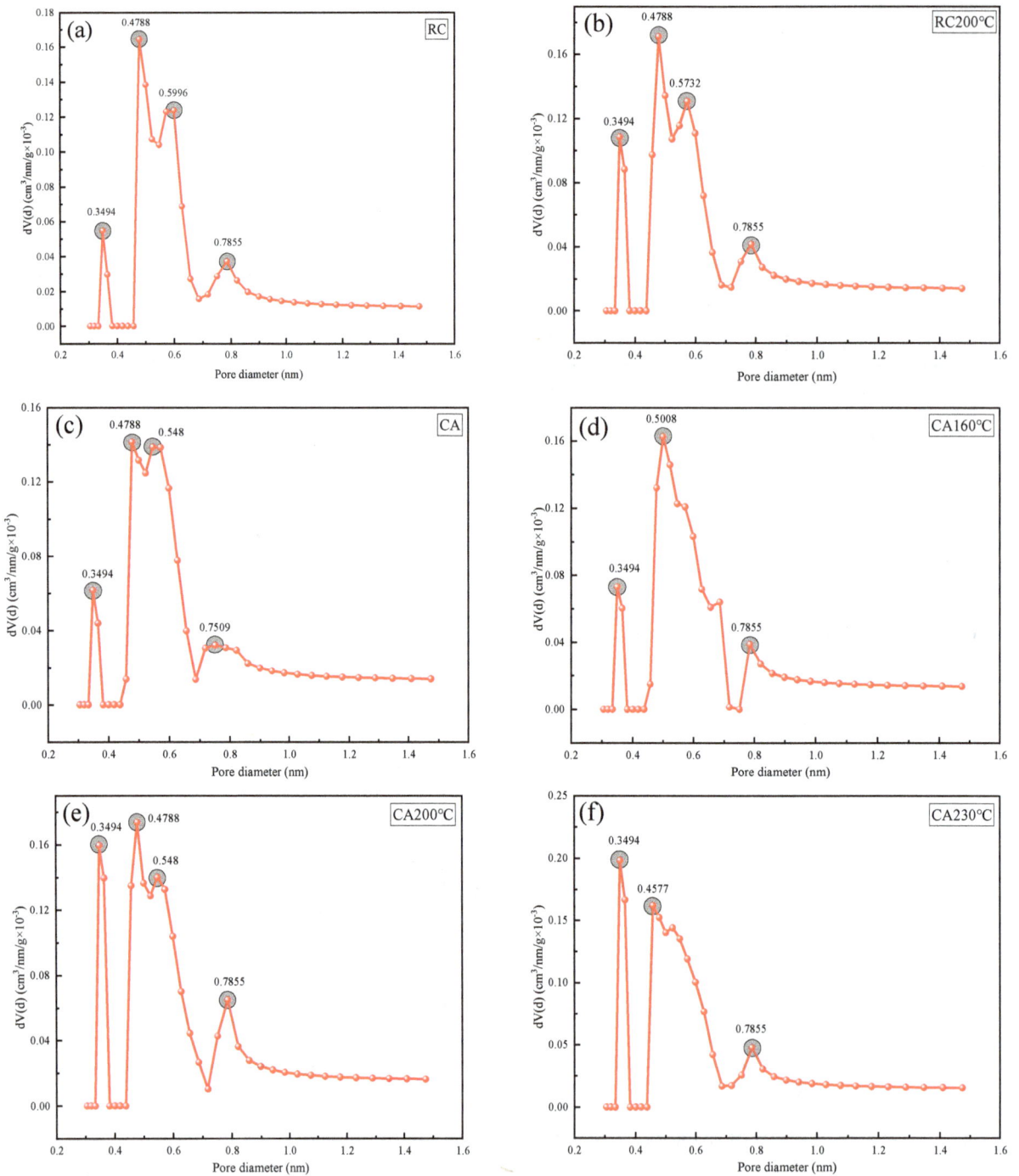

FIG 5 – Characterisation of micropore distribution of coal samples based on DFT methods.

Table 2 and Figure 6 present the changes in pore volumes and specific surface areas of coals induced by autothermal oxidation under chemical activation. Oxidation at 200°C led to a differential increase in pore volumes and specific surface areas in the RC coal samples, with a notable exception of macropores, which decreased. Overall, there was a 15.6 per cent increase in total pore volume and a 21.9 per cent increase in total specific surface area. Before reaching 200°C, higher oxidation temperatures resulted in reduced mesopore and macropore volumes but enlarged micropore volumes. In comparison to CA coal samples, the CA160°C samples exhibited relatively stable microporous volume, a 17.5 per cent decrease in mesopore volume, and a 29.8 per cent increase in macropore volume, likely due to mesopore consolidation into larger pores at elevated temperatures. Beyond 200°C, as temperature escalated, CA230°C samples showed decreased microporous volume and increased meso- and macroporous volumes, with total pore volume rising by 15.7 per cent. This shift is attributed to the release of small molecules from the pores and the unblocking of obstructed pores at higher oxidation temperatures, consequently increasing meso- and macroporous volumes. Notably, the interconnectedness of small pores during high-temperature oxidation contributes to the rise in mesopore volume. After oxidation at a constant temperature of 230°C, there was a dramatic increase in mesopore volume by 285.8 per cent, macropore volume by 209.1 per cent, and total pore volume by 63 per cent, alongside substantial increases in specific surface areas of mesopores, macropores, and total-pores by 141.11 per cent, 250 per cent, and 53.25 per cent respectively. These results underline the significant augmentation in the pore volume and specific surface area of mesopores and macropores of coal samples due to autothermal oxidation combined with chemical activation.

TABLE 2

Pore volume and specific surface area of coal samples.

Sample	Pore volume($cm^3/g^{-1} \times 10^{-3}$)				SSA (m^2/g)			
	Micropore volume	Mesopore volume	Macropore volume	Total volume	Micropore SSA	Mesopore SSA	Macropore SSA	Total SSA
RC	34.83	1.69	2.97	39.66	109.62	0.90	0.08	110.60
RC200°C	40.99	2.65	2.19	45.83	133.51	1.21	0.06	134.78
CA	38.82	5.31	3.36	47.67	121.05	1.83	0.21	123.09
CA160°C	38.92	4.38	4.36	47.74	123.53	1.49	0.12	125.14
CA200°C	48.91	4.60	3.30	55.90	162.72	1.59	0.19	164.50
CA230°C	48.13	6.52	9.18	64.70	167.05	2.17	0.28	169.50

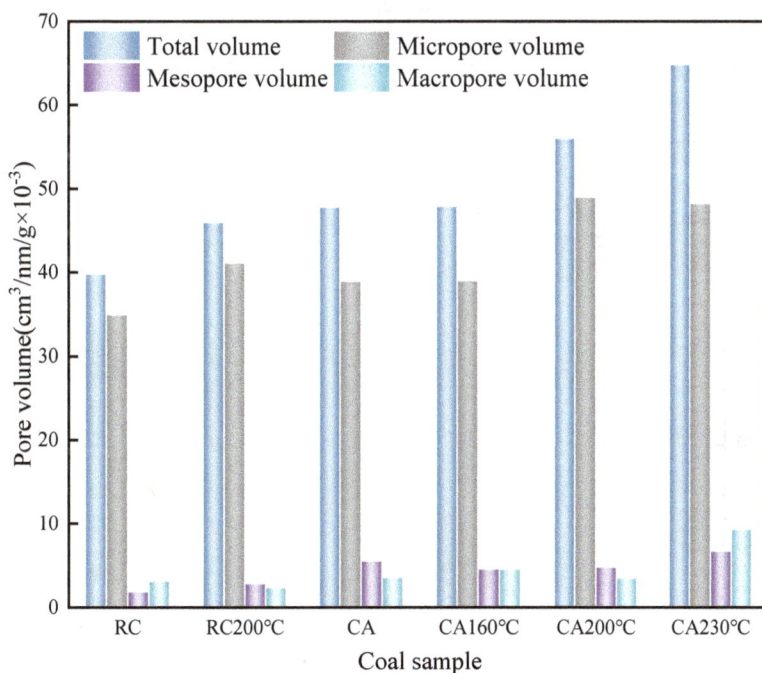

FIG 6 – Change of pore volume in coal samples.

Figures 7 and 8 display the variation in pore volume and specific surface area of coal samples at a comparative oxidation temperature of 200°C. After chemical activation, there was a marked increase in the volume of micropores and mesopores, contributing to the rise in total pore volume. Concurrently, the specific surface area of micropores surged significantly, while that of mesopores rose marginally, collectively enhancing the total specific surface area. The alterations in macropore volume and specific surface area were negligible. Autothermal oxidation at 200°C predominantly expanded microporous structures, accounting for the pronounced elevation in their specific surface area. However, this process might entail some larger pore collapse, slightly diminishing the specific surface area of macropores. The combined impact of chemical activation and autothermal oxidation notably amplified microporous structures, thereby substantially increasing both total pore volume and specific surface area. This underscores the synergistic effect of these methods in oxidative dissolution and expansion of micropores, resulting in modest increments in meso- and macroporous structures as well.

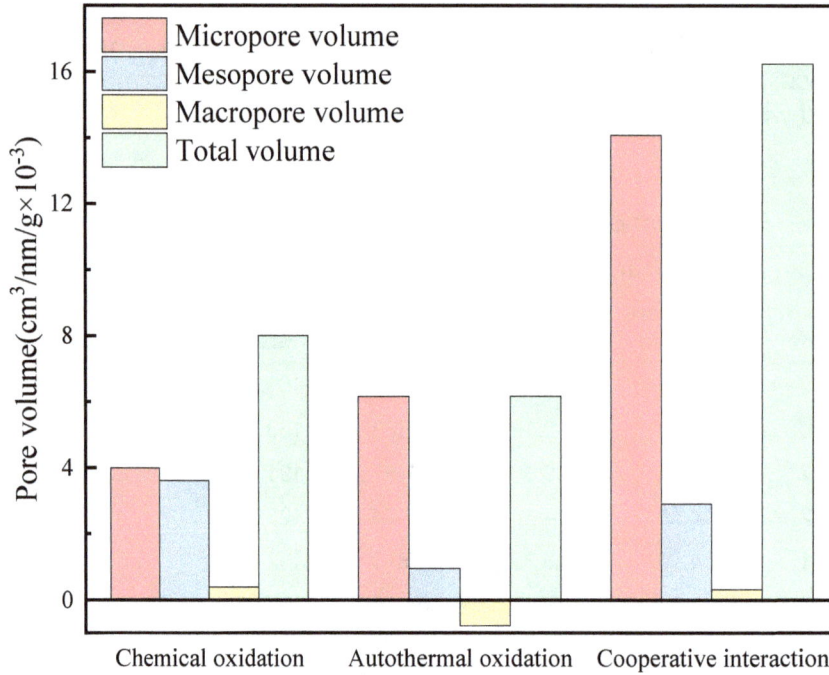

FIG 7 – Changing law of pore volume of coal samples under different roles.

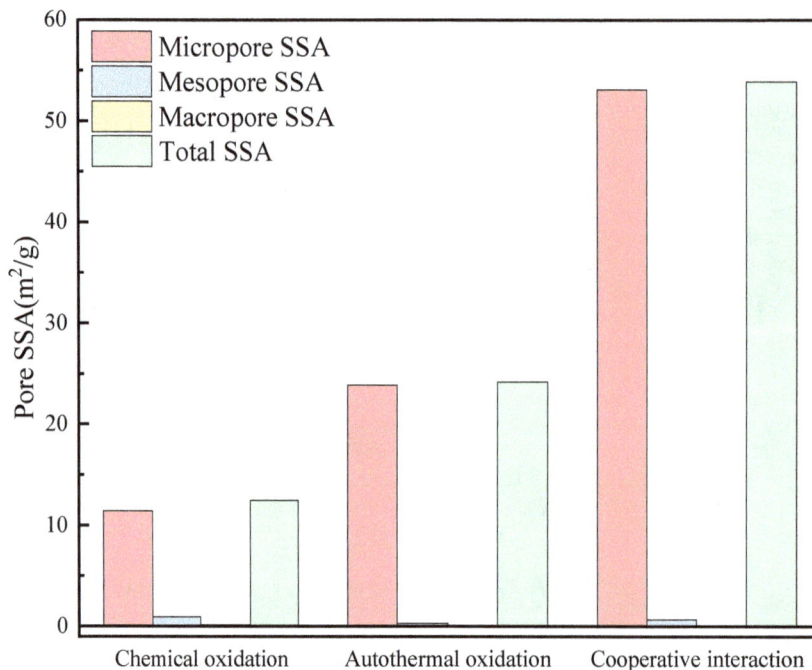

FIG 8 – Changing law of pore specific surface area of coal samples under different roles.

Pore size distribution and specific surface area characteristics

Existing literature highlights that the fractal dimension, denoted as D, is a widely recognised parameter for quantifying the complexity and roughness of pore structures. In this context, the

Frenkel-Halsey-Hill (FHH) model is frequently employed to calculate this fractal dimension. This model is particularly effective when integrated with low-temperature N_2 adsorption experiments. The calculation of the fractal dimension using the FHH model is based on the following equation:

$$ln(V / V_0) = -Aln[ln(P_0 / P)] + C$$

(1)

where:

V   is the N_2 adsorption quantity at an equilibrium pressure in cm^3/g

V_0 is the monolayer adsorption quantity of nitrogen in cm^3/g

A   is the slope of the straight line fitted by the FHH model

slope A   is linearly related to the fractal dimension D

P_0 is the saturated vapor pressure of nitrogen in MPa

P   is equilibrium pressure in MPa, C is a constant

Figure 9 and Table 3 illustrate the fractal dimensions and calculated results of coal samples subjected to autothermal oxidation under chemical activation. For the RC200°C samples, D_1 is less than 2, indicating poorer pore fractals, while an increase in D_2 compared to RC samples signifies a more complex pore structure due to oxidative warming. Between 80–200°C, D_1 in activated coal progressively decreases and D_2 increases, suggesting that rising oxidation temperatures activate more surface functional groups to interact with oxygen, leading to the volatilisation of substances and resulting in smoother pore surfaces and more pronounced pore expansion. From 200–230°C, D_1 increases and D_2 decreases, attributable to an augmented reaction participation of surface groups at higher temperatures, where the rate of volatile substance generation surpasses volatilisation, leaving some residual substances in the pores. In summary, as oxidation temperature increases, more active agents in the coal matrix engage in reactions, leading to a more extensive expansion and deepening of pores. Consequently, D_2 gradually decreases, reflecting a simplification in the coal's pore structure.

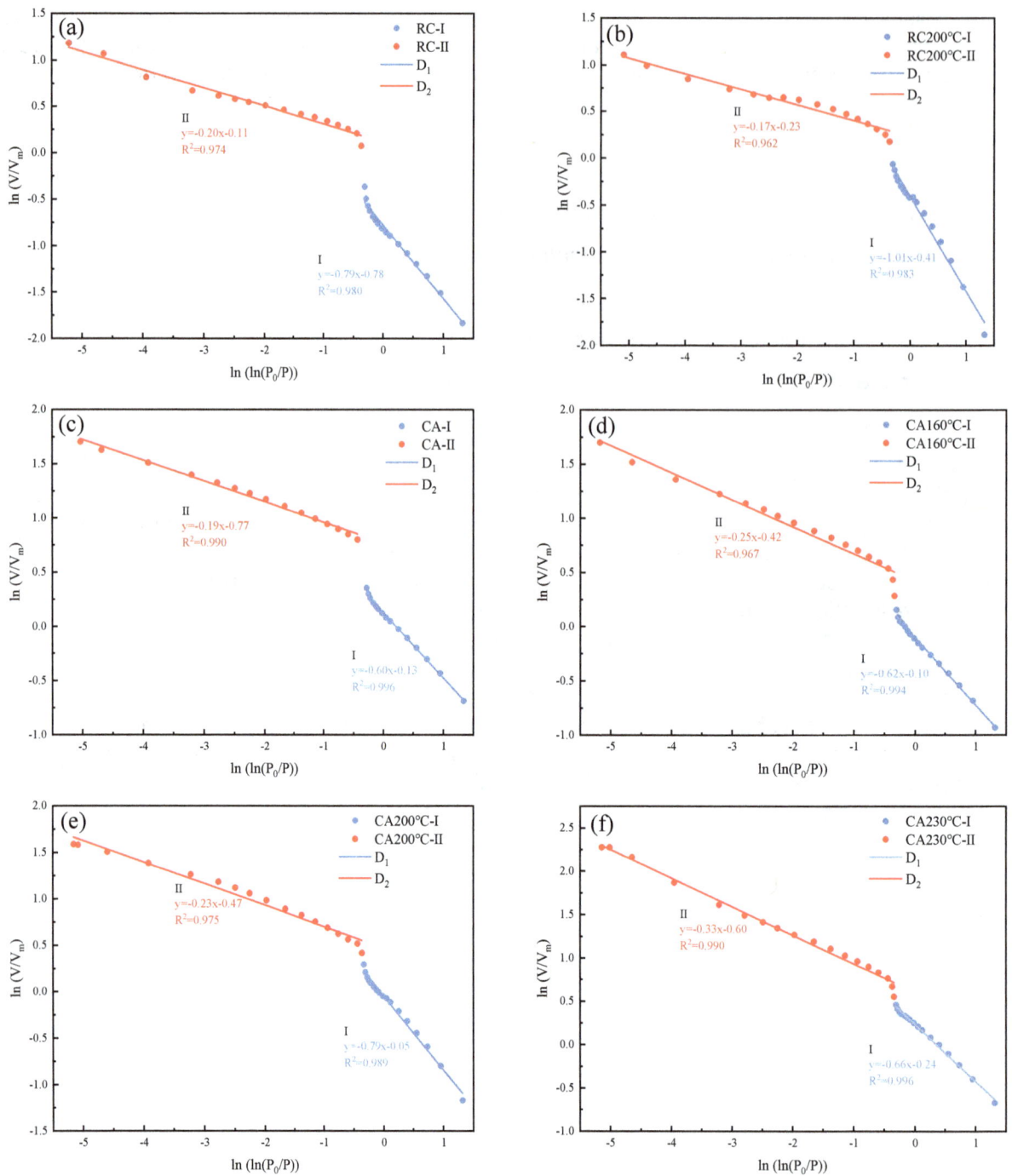

FIG 9 – Fractal dimension of coal samples based on FHH model.

TABLE 3

Results of fractal dimension of coal samples based on FHH model.

Sample	A_1	D_1	R_1^2	A_2	D_2	R_2^2
RC	-0.79	2.21	0.980	-0.20	2.80	0.974
RC200°C	-1.02	1.98	0.983	-0.17	2.83	0.962
CA	-0.60	2.40	0.996	-0.19	2.71	0.990
CA160°C	-0.62	2.38	0.994	-0.25	2.75	0.967
CA200°C	-0.79	2.21	0.989	-0.23	2.76	0.975
CA230°C	-0.66	2.34	0.992	-0.33	2.67	0.996

X-ray photoelectron spectroscopy (XPS)

Elemental Distribution Characteristics

Figure 10 displays the wide-sweep spectra of coals subjected to autothermal oxidation under chemical activation, covering a range of 0–1350 eV. Analysis of the spectra reveals the predominant elements in the coal samples, with their types and relative contents detailed in Table 4. The data indicates that C and O are the most abundant elements, together comprising approximately 90 per cent of the total elemental composition. In contrast, S content remains low across all samples. However, the content of S and Fe increases in activated coal at varying oxidation temperatures compared to RC coal samples, likely due to the immobilisation of some functional groups in conjunction with elements Fe and S in the sodium persulfate solution. The decline in Na content is attributed to the dissolution of soluble sodium salts in the coal during water immersion. In comparing the RC samples with CA230°C samples, there is an observed increase of 1.57 per cent in C and a decrease of 3.08 per cent in O, possibly resulting from an intensified aromatic polymerisation reaction in the coal samples under oxidative warming, leading to the generation of CO and CO_2 gases.

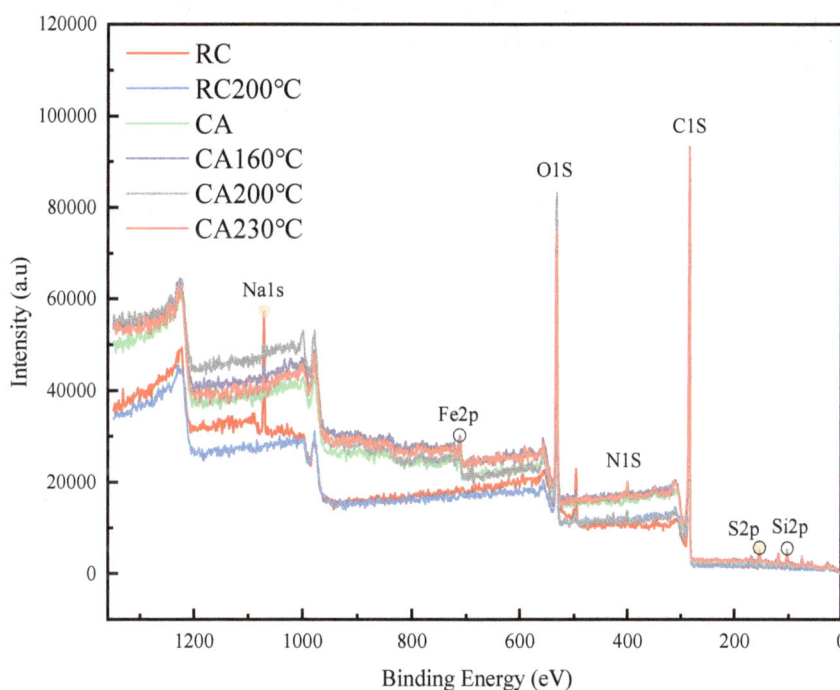

FIG 10 – XPS broad-scan electron spectra of coal samples.

TABLE 4

Categories and contents of elements in nine groups of coal samples.

Sample	Elemental content /%						
	C1s	O1s	S2p	N1s	Si2p	Fe2p	Na1s
RC	62.90	23.97	0.46	3.91	2.39	-	5.57
RC200°C	71.84	21.19	0.68	4.02	3.09	-	-
CA	73.62	19.31	0.79	1.94	2.72	1.49	-69.60
CA160°C	69.60	20.64	0.83	2.19	1.73	1.19	-
CA200°C	68.16	21.35	0.37	2.99	2.39	1.02	-
CA230°C	64.47	20.89	0.70	2.49	3.46	1.24	-

Functional group analysis based on C elements

To investigate the functional group alterations in coal induced by autothermal oxidation and chemical activation, this study conducted narrow scans of the C1s region in six coal samples across a range of 278–297 eV. After calibrating for charge variations, the narrow-scan data were peak-fitted, with the resultant fit for the five coal samples depicted in Figure 11. The congruence between the fitted and experimental curves for all six sample groups indicates a high reliability in the subsequent calculation of relative functional group contents. The details of these calculated functional group contents are in Table 5.

At RC200°C, oxygen-containing functional groups in coal samples increased by 17.5 per cent compared to RC samples. There was a decrease in C-C/C-H content by 12.9 per cent, in COOH by 4.6 per cent, and in C=O by 7.5 per cent in CA coal samples relative to RC ones. This reduction is attributable to the robust oxidising properties of $Fe^{2\pm}$heat-activated sodium polysulfate, which oxidises reducing functional groups like COOH and C=O, leading to an increase in C-O content due to carbon-oxygen double bond breakage. Furthermore, sodium persulfate oxidises C-C bonds in coal, reducing their content. In CA160°C samples, COOH increased by 4.2 per cent, while C=O content change was minimal, reflecting the initial preferential attack of oxygen on C=O at temperatures below 80°C with negligible impact on COOH, and a balance in the generation and consumption of C=O between 80–200°C. Beyond 200°C, the production rate of C=O surpasses its consumption, resulting in increased content. Post 80°C, the production rate of COOH in activated coal exceeds its conversion rate, leading to continued content increase. Overall, as oxidation temperature rises, oxygenated functional groups in activated coal increase, while C-C/C-H content decreases. Specifically, in CA230°C samples, there was a 23.7 per cent decrease in C-C/C-H content, a 28.7 per cent increase in C-O/C-O-C content, and a 23.7 per cent increase in oxygenated functional groups compared to RC samples.

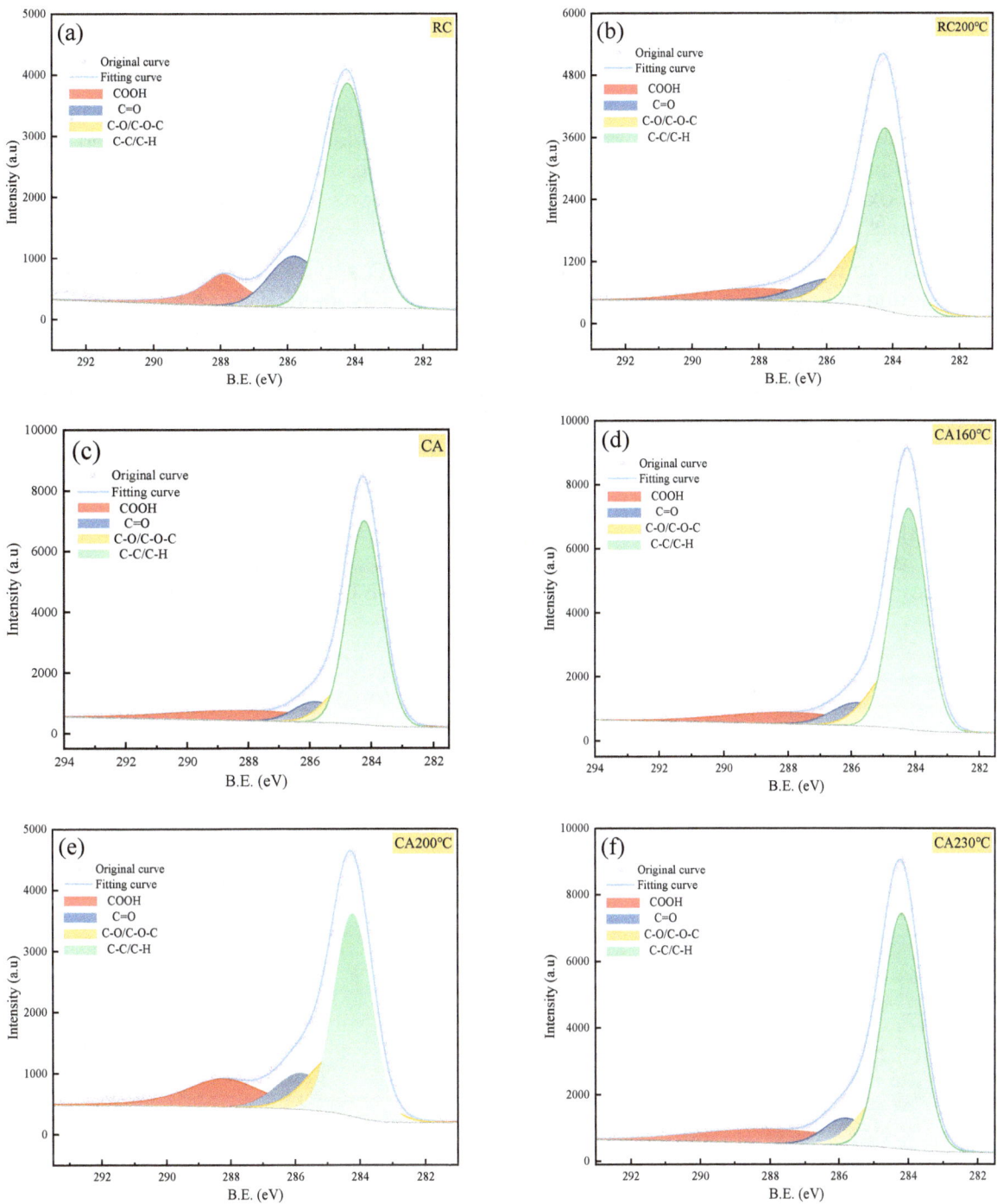

FIG 11 – Peak fitting of C1s of coal samples.

TABLE 5

Changes of functional groups in coal samples.

Sample	C-C/C-H	COOH	C=O	C-O/C-O-C	Oxygenated functional group
RC	65.8%	13.9%	17.1%	3.1%	34.2%
RC200°C	48.3%	10.2%	11.4%	30.1%	51.7%
CA	52.9%	9.3%	9.7%	28.1%	47.1%
CA160°C	49.6%	13.5%	9.2%	26.7%	50.4%
CA200°C	45.5%	15.9%	9.9%	28.6%	54.5%

Functional group analysis based on O elements

To examine the changes in oxygen-containing functional groups in coal due to autothermal oxidation and chemical activation, the O1s regions of six coal sample groups ranging from 524–544 eV were precisely scanned. These scans were then adjusted and peak-fitted using XPSPEAK software, with the O1s peak-splitting results depicted in Figure 12. The relative content of oxygen-containing functional groups in the coal, resulting from the synergistic impact of chemical activation and autothermal oxidation, was deduced from the fitted peak areas for five coal sample groups. The detailed results are in Table 6.

It is important to note that the relative content values of C-O and C=O derived from C1s and O1s fitting differ due to the distinct total peak areas in each fitting. Nonetheless, both fittings consistently show an increase in C-O content and a decrease in C=O content with rising oxidation temperatures, affirming the accuracy of the fitting results. Analysis reveals that inorganic oxygen content in CA coal samples rose by 8.4 per cent compared to RC samples, attributed to the degradation of unsaturated COO bonds after oxidation by activated sodium persulfate solution, and the binding of $SO_4^-\bullet$ radicals with reducing substances in the coal, enhancing inorganic oxygen content. Specifically, in CA230°C samples, inorganic oxygen content increased by 6.9 per cent, likely due to the precipitation of organic oxygen as CO and CO_2 and the detachment of organic volatile substances from the coal matrix, resulting in elevated inorganic oxygen levels. Under the synergistic influence of chemical activation and autothermal oxidation, COO content in CA230°C samples decreased by 11.2 per cent, while inorganic oxygen content increased by 16.6 per cent compared to RC samples.

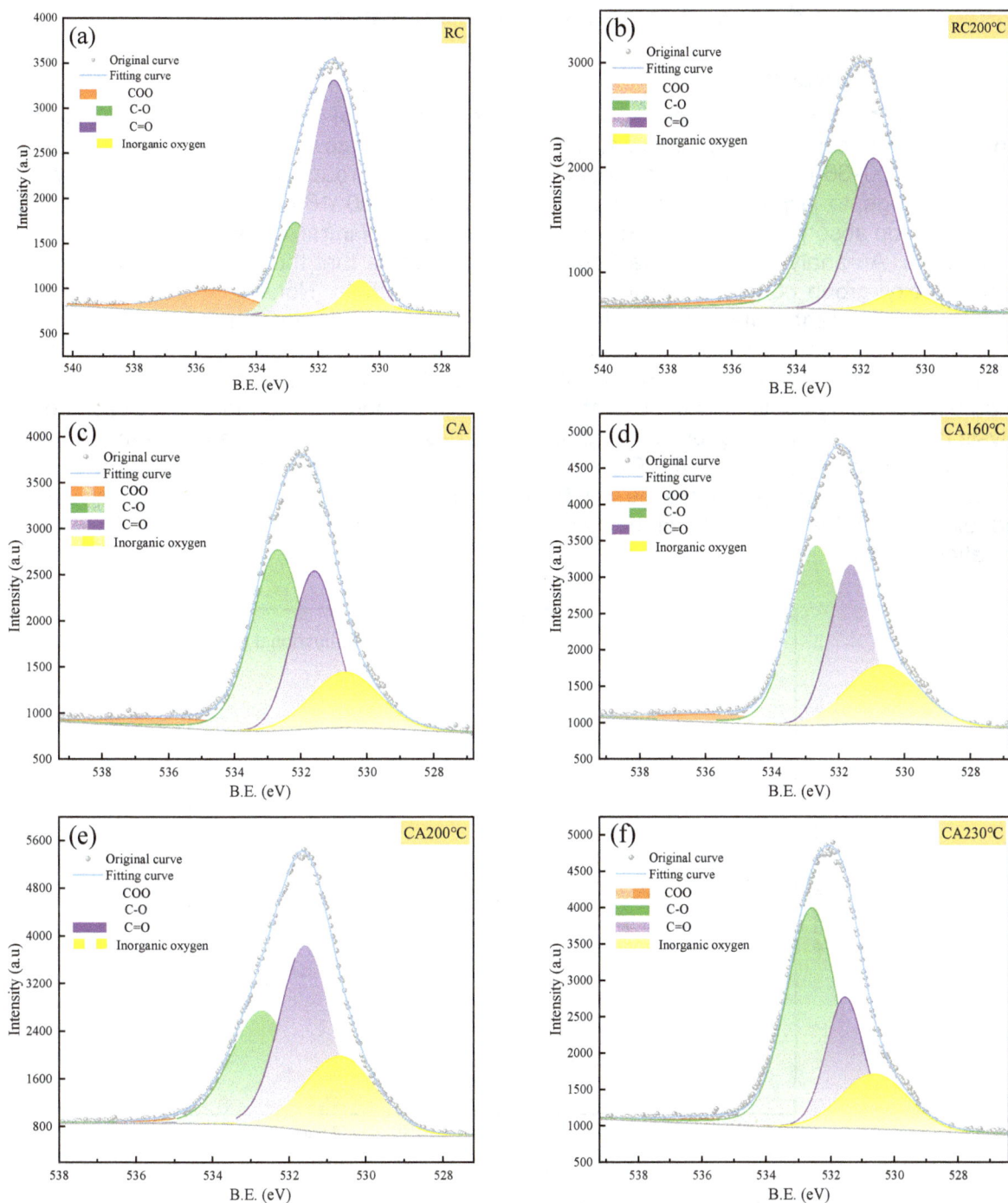

FIG 12 – Peak fitting of O1s of coal samples.

TABLE 6

Changes of oxygen-containing functional groups in coal samples.

Sample	COO	C-O	C=O	Inorganic oxygen
RC	13.0%	18.2%	61.7%	7.2%
RC200°C	7.4%	47.2%	36.5%	8.9%
CA	5.8%	45.7%	31.6%	16.9%
CA160°C	4.7%	44.1%	31.5%	19.8%
CA200°C	2.1%	53.5%	24.3%	20.1%
CA230°C	1.8%	54.0%	20.4%	23.8%

Functional group distribution and structural parameters

Aliphatic hydrocarbon

Prior research has established that coal comprises a variety of chemical functional groups and bonds. Within FTIR spectroscopy spectra, specific regions correlate with different types of absorption peaks: the 3000–3600 cm^{-1} range corresponds to hydroxyl group absorption peaks, the 2800–3000 cm^{-1} range to aliphatic absorption peaks, and the 900–1800 cm^{-1} range to oxygenated functional group absorption peaks. Figure 13 presents the FTIR infrared spectroscopy results for six coal samples. An analysis of the data from Figure 13 reveals that the CA coal sample exhibits a higher hydroxyl group content compared to the RC coal sample. Notably, there is a decrease in hydroxyl group content in the CA sample as the oxidation temperature increases. Regarding the aliphatic hydrocarbons, the content in both CA and RC coal samples shows little difference. However, the oxidation temperature significantly impacts the content of aliphatic hydrocarbons, with a trend of decreasing content as the oxidation temperature rises. Furthermore, the absorption peak analysis for oxygen-containing functional groups indicates that the peak area for these groups in the CA sample is larger than that in the RC sample. This peak area, and thus the content of oxygen-containing functional groups, increases with the rise in oxidation temperature in the CA sample. This trend aligns with the trajectory of oxygen-containing functional groups observed in the XPS calculation results.

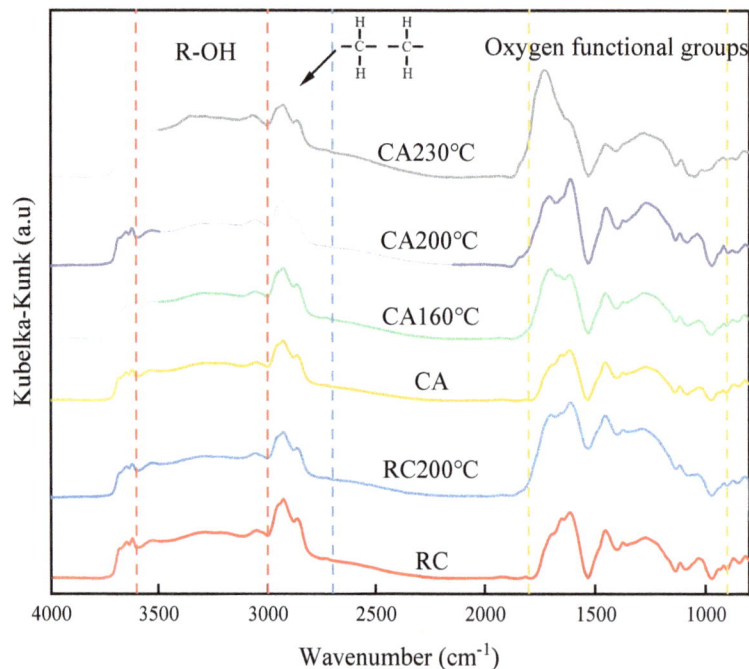

FIG 13 – FTIR infrared spectra of coal samples.

This study aims to quantitatively analyse the variations in aliphatic hydrocarbons within coal, considering the combined effects of chemical activation and autothermal oxidation. To this end, infrared spectra of five coal sample groups were collected in the range of 3000–2800 cm^{-1}. Peak fitting was performed on these spectra to discern specific vibrational modes, and the outcomes of this analysis are depicted in Figure 14. The results in Figure 14 demonstrate a high degree of repeatability between the split-peak fitting results of the aliphatic hydrocarbons and the experimental data for all five coal sample groups. The identified peaks include the antisymmetric stretching vibration of -CH$_3$ at 2960 cm^{-1}, the antisymmetric stretching vibration of -CH$_2$- at 2928 cm^{-1}, the symmetric stretching vibration of -CH$_3$ at 2900 cm^{-1}, and the symmetric stretching vibration of -CH$_2$- at 2858 cm^{-1}. The relative content of each of these functional groups was calculated based on their respective peak areas. These calculated values provide a detailed quantification of the aliphatic hydrocarbon content within the coal samples under study. The results of these calculations are presented in Table 7.

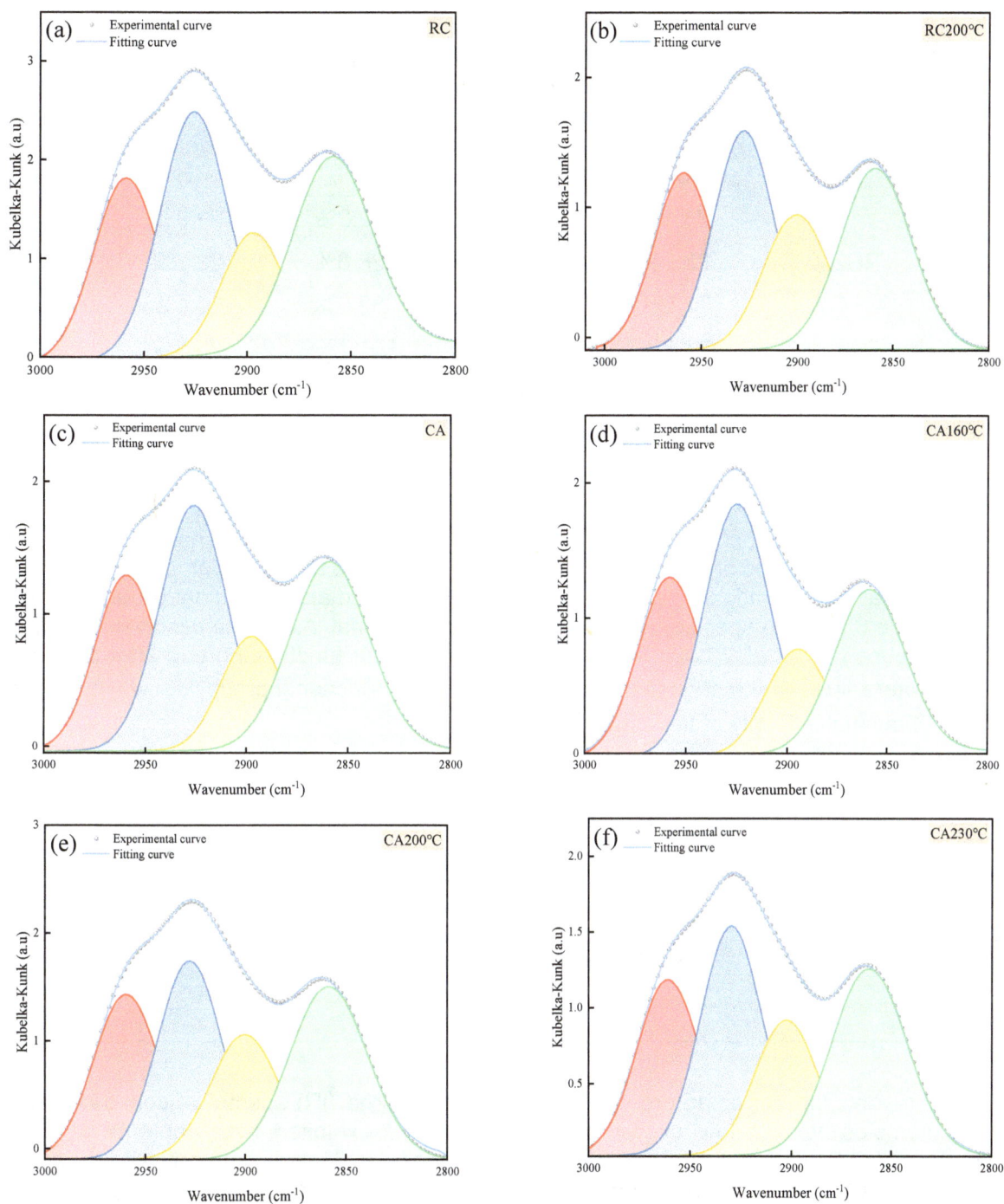

FIG 14 – Peak fitting of aliphatic hydrocarbons in coal samples.

Table 7 illustrates that the highest proportion of antisymmetric stretching vibration -CH_2- is found in the RC coal sample at 32.2 per cent. In the RC200°C sample, there is a decrease in antisymmetric stretching vibration -CH_2- and an increase in symmetric stretching vibration -CH_3, suggesting that the oxidative warming of the coal primarily impacts antisymmetric -CH_2-. Compared to CA samples, both symmetric and antisymmetric stretching vibration -CH_2- contents are reduced in CA160°C samples, likely due to the activation of less reactive -CH_2- at elevated temperatures, resulting in decreased -CH_2- contents and an increased percentage of -CH_3. As oxidation temperature rises, the proportion of -CH_2- in coal diminishes gradually, while antisymmetric stretching vibration -CH_3 content slowly increases when oxidation temperature is below 200°C. Overall, antisymmetric stretching vibration -CH_2- in activated coal is highly sensitive to oxidation temperature, whereas antisymmetric stretching vibration -CH_3 is less influenced.

TABLE 7

Relative content of aliphatic hydrocarbons in coal samples.

Sample	Functional group			
	Asymmetric -CH₃	Symmetric -CH₃	Asymmetric -CH₂-	Symmetric -CH₂-
RC	23.1%	14.8%	32.2%	29.9%
RC200°C	23.2%	19.3%	28.8%	28.7%
CA	22.8%	14.1%	33.0%	30.1%
CA160°C	23.2%	16.6%	31.7%	28.5%
CA200°C	23.5%	19.0%	27.9%	29.6%
CA230°C	24.9%	20.6%	26.1%	28.4%

Characterisation of infrared structural parameters

Due to the significant variance in extinction coefficients among different functional groups, quantum chemical calculations based on theoretical spectra were employed for correction purposes. This step was crucial to accurately adjust the original peak areas of the major functional groups in the coal samples. Table 8 lists the specific wave numbers and vibrational intensities associated with these functional groups (Xin *et al*, 2014). The peak area of the major functional group after correction is $G(v)/f$, where $G(v)$ is the original peak area of the major functional group with wave number v; f is the vibrational intensity of the major functional group.

TABLE 8

Wavenumber and vibration intensity of reactive groups of coal samples.

Functional group	Hydroxyls	Aliphatic hydrocarbons		Carboxyl groups	Carbonyl groups	Aldehyde groups	Benzene
	-OH	-CH3	-CH2-	-COOH	-C=O	-CHO	C=C
Wavenumber	3600–3200	2975–2865	2925–2825	1715–1690	1690–1650	1736–1722	1600
Vibration intensity $f(a.u)$	632.00	77.26	86.88	245.84	85.78	156.94	43.08

Figure 15 presents the corrected group contents of the infrared (IR) spectra of coal samples. The RC coal samples, with a low degree of metamorphism, exhibited high contents of aliphatic hydrocarbons (47.74 per cent) and aromatic hydrocarbons (34.35 per cent), followed by hydroxyl, carbonyl, and carboxyl groups at 8.56 per cent, 4.86 per cent, and 3.59 per cent, respectively, with aldehyde carbonyl groups being the least prevalent at only 0.91 per cent. In RC200°C and CA samples, hydroxyl group content decreased slightly from the original coal but varied with increasing oxidation temperatures, peaking at 10.26 per cent in CA200°C samples. Aliphatic hydrocarbon content generally decreased across activated coal samples except in RC200°C, where it increased, suggesting the consumption of methyl methylene functional groups by the combined effect of chemical activation and autothermal oxidation. Carboxyl content exhibited varying trends, particularly increasing significantly in CA160°C samples, likely due to oxidation of hydroxyl functional groups. In contrast, the carbonyl content decreased in RC200°C, CA, and CA160°C samples compared to RC samples but significantly increased in CA200°C and CA230°C samples, possibly due to dehydration reactions at higher temperatures forming more carbonyl groups. The aldehyde carbonyl content notably increased post-chemical activation and autothermal oxidation, especially in CA200°C samples, indicating potential further oxidation of carbonyl groups or high-temperature cleavage reactions forming aldehyde carbonyl. Benzene ring double bond content rose with increasing temperature up to 200°C, likely due to low-temperature chemical activation promoting aromatisation. However, beyond 200°C, the content decreased, possibly from the accelerated decomposition of aromatic structures exceeding their formation rate.

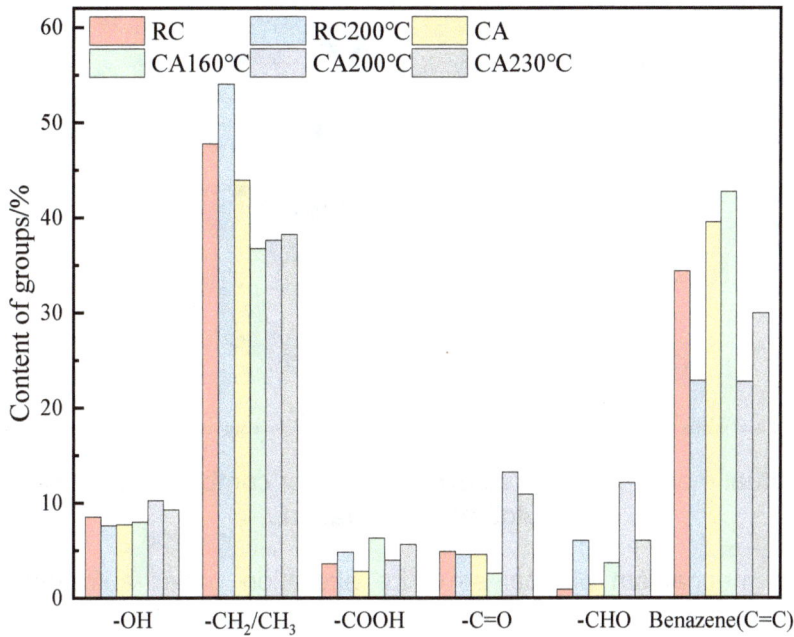

FIG 15 – Group content of coal samples after correction of infrared spectra.

Pore evolution mechanism and reaction mechanism

At 200°C, the active structure of chemically activated coal fully reacts with oxygen, leading to some coal matrix pores breaching their walls and beginning to interconnect, resulting in a more uniform mesopore distribution. Concurrently, small molecules produced during the reaction may remain within the matrix, obstructing pore channels and thus reducing total pore volume. Analysis of the full-size pore volume and specific surface area changes in chemically activated coal post-oxidation at various autothermal temperatures has elucidated the dynamic interchange among micropores, mesopores, and macropores driven by autothermal oxidation under chemical activation. This pore evolution mechanism is conceptualised in Figure 16.

Coal autothermal reaction is a process that further activates the organic structure of coal, enhancing the matrix's pores beyond their original structure. Pore development is influenced by thermal action and structural oxidative reactions. Chemical activation significantly augments the permeability of coal structure's pore channels, facilitating the seepage and reaction of oxygen molecules, accelerating pore evolution, and promoting methane desorption. As oxidation temperature increases, minerals dissolved in the chemically activated coal matrix precipitate, leading to greater pore connectivity. Throughout the oxidation process, pore connectivity in the coal matrix intensifies with temperature. At 230°C, most primary pores rupture and interconnect to form larger pores. Concurrently, the precipitation rate of peroxide small molecules escalates, yet remains lower than their production rate, resulting in fewer residual small molecules within the pore channels.

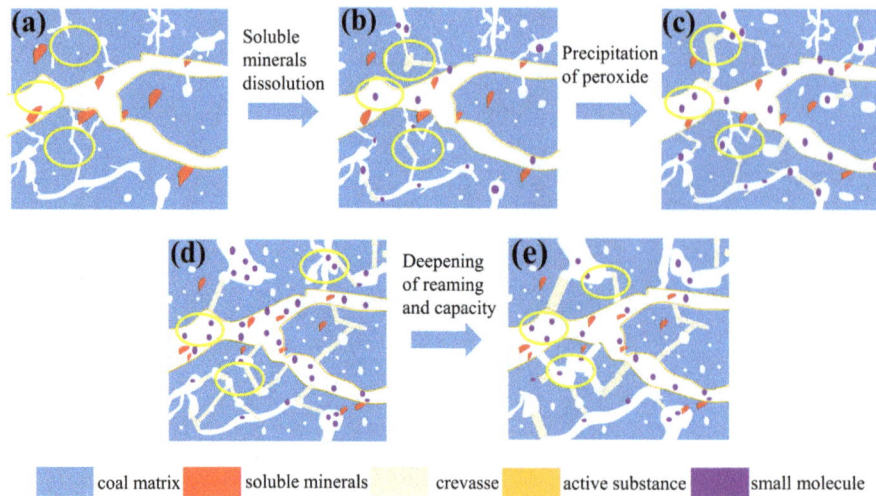

FIG 16 – Conceptual model of coal pore evolution under combined chemical activation and autothermal oxidation.

The overall reaction mechanism of coal structure alteration under autothermal oxidation and chemical activation aligns with the oxidative kinetics of raw coal, with the combined promotional mechanism depicted in Figure 17. Highly active radicals within the coal pores significantly enhance the oxidation of the coal structure, leading to increased depletion of the active structure, an upsurge in carbon-containing free radicals, and a rise in C-O structures due to oxygen incorporation. This depletion serves as the primary impetus for pore evolution during the autothermal process. As temperature rises, pore development transitions from the addition of new micropores to mechanisms involving pore expansion and merging. Consequently, primary micropore ruptures in the coal matrix amplify mesopore and macropore connectivity, thereby increasing the coal structure's permeability.

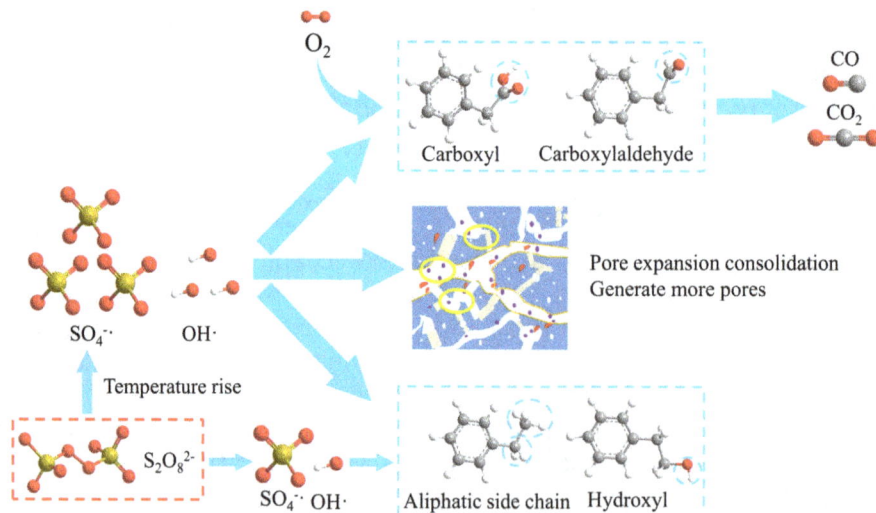

FIG 17 – Promotional mechanisms for the combined action of chemical activation and autothermal oxidation.

This study investigates the evolution of pores and functional groups in coal driven by autothermal oxidation under chemical activation. However, in practical field applications, the effective injection of activated sodium persulfate into the coal body, ensuring its complete reaction with the coal, and controlling the autoxidation temperature of the coal body within a specific range to prevent coal seam ignition, require further investigation.

CONCLUSION

This paper delves into the intrinsic relationship between the molecular structure evolution and pore development in coal due to autothermal oxidation under chemical activation, uncovering both the pore and reaction mechanisms. The main conclusions are as follows:

- As oxidation temperature increases, narrow fissure pores in activated coal expand, along with notable growth in micropore size and capacity. Post-oxidation at 230°C, mesopore volume rises by 285.8 per cent, macropore volume rises by 209.1 per cent, and total pore volume rises by 63 per cent. Specific surface areas also show significant increases, with mesopores at 141.11 per cent, macropores at 250 per cent, and total surface area at 53.25 per cent. This results in more active agents in the matrix, enhanced expansion, and pore deepening, leading to a decrease in D_2 and simpler coal pore structure. Consequently, autothermal oxidation under chemical activation considerably boosts micropore quantity, significantly augmenting total pore volume and surface area.

- Autothermal oxidation under chemical activation triggers a reaction of unstable C-C/C-H bonds in activated coal with oxygen, producing oxygen-containing intermediates. This leads to an increase in C-O/C-O-C and oxygen-containing functional groups, with a decrease in C-C/C-H content. Specifically, in CA 230°C samples compared to RC samples, C-C/C-H content decreases by 28.7 per cent, oxygenated functional groups increase by 23.7 per cent, COO content drops by 11.2 per cent, and inorganic oxygen rises by 16.6 per cent.

- Beyond 80°C, antisymmetric stretching vibration-CH_2- diminishes with rising temperature, while CH_3 stretching shows minimal temperature impact. The reduction in aliphatic hydrocarbons with increasing oxidation temperature indicates the cleavage of longer aliphatic chains. Highly reactive radicals significantly accelerate coal structure oxidation, enhancing the aromatic polymerisation reaction, and improving coal maturity and aromaticity.

CRediT authorship contribution statement

Haihui Xin: Conceptualisation, Investigation, Writing – review and editing, Supervision, Funding acquisition. *Pengcheng Zhang:* Methodology, Writing – original draft, Data curation. *Jianguo Sun:* Validation. *Liang Lu:* Validation. *Chun Xu:* Visualisation. *Banghao Zhou:* Investigation. *Hezi Wang:* Investigation. *Yi Yang:* Investigation. *Junzhe Li:* Investigation. *Deming Wang:* Supervision.

ACKNOWLEDGEMENTS

This work was supported by the National Natural Science Foundation of China (52174220, 52374246), the Key Program of the National Natural Science Foundation of China (52130411), the China Postdoctoral Science Foundation (2022T150553, 2022M710119), and the Jiangsu Science and Technology Association Young Scientist Project (TJ-2022041), and funded by the Postgraduate Research and Practice Innovation Program of Jiangsu Province (SJCX23_1309) and the Graduate Innovation Program of China University of Mining and Technology (2023WLJCRCZL195).

REFERENCES

Cai, Y L, Zhai, C, Yu, X, Sun, Y, Xu, J Z, Zheng, Y F, Cong, Y, Li, Y, Chen, A, Xu, H, Wang, S and Wu, X, 2023. Quantitative characterization of water transport and wetting patterns in coal using LF-NM R and FTIR techniques, *Fuel,* 350:128790–128804.

Chen, M J, Lu, Y, Kang, Y, You, L, Chen, Z, Liu, J and Li, P, 2020. Investigation of enhancing multi-gas transport ability of coalbed methane reservoir by oxidation treatment, *Fuel,* 278:118377–118389.

Dai, J, Yang, J, Xu, S Y, Wei, J P, Li, Y X and Xu, L S, 2020. Experimental study of injecting heat-activated sodium persulfate solution to increase the permeability of soft and low permeability coal seams, *Journal of China Coal Society.* 45:823–832.

Dang, Z, Su, L A, Wang, X M and Hou, S H, 2023. Experimental study of the effect of ClO2 on coal: Implication for coalbed methane recovery with oxidant stimulation, *Energy,* 271:127028–127043.

Guo, H J, Yu, Y J, Wang, K, Yang, Z, Wang, L and Xu, C, 2023. Kinetic characteristics of desorption and diffusion in raw coal and tectonic coal and their influence on coal and gas outburst, *Fuel,* 343:127883–127892.

Li, H, Xu, C, Ni, G, Lu, J, Lu, Y, Shi, S, Li, M and Ye, Q, 2022. Spectroscopic (FTIR, 1H NMR) and SEM investigation of physicochemical structure changes of coal subjected to microwave-assisted oxidant stimulation, *Fuel*, 317:123473–123484.

Liu, S, Yang, K, Sun, H, Wang, D, Zhang, D, Li, X and Chen, D, 2022. Adsorption and deformation characteristics of coal under liquid nitrogen cold soaking, *Fuel*, 316:123026–123037.

Lu, J, Zheng, C, Liu, W, Li, H, Shi, S, Lu, Y, Ye, Q and Zheng, Y, 2023a. Evolution of the pore structure and fractal characteristics of coal under microwave-assisted acidification, *Fuel*, 347:128500–128511.

Lu, Y, Kang, Y L, Ramakrishna, S, You, L J and Hu, Y, 2023b. Enhancement of multi-gas transport process in coalbed methane reservoir by oxidation treatment: Based on the change of the interaction force between coal matrix and gas molecules and knudsen number, *International Journal of Hydrogen Energy*, 48(2):478–494.

Qin, L, Li, S, Zhai, C, Lin, H, Zhao, P, Shi, Y and Bai, Y, 2020. Changes in the pore structure of lignite after repeated cycles of liquid nitrogen freezing as determined by nitrogen adsorption and mercury intrusion, *Fuel*, 267:117214–117225.

Qin, Y, Moore, T A, Shen, J, Yang, Z B, Shen, Y L and Wang, G, 2018. Resources and geology of coalbed methane in China: a review, *International Geology Review*, 60(5–6):777–812.

Shang, Z, Wang, H F, Li, B, Cheng, Y P, Zhang, X H, Wang, Z Y, Geng, S, Wang, Z, Chen, P, Lv, P and Shi, Z, 2022. The effect of leakage characteristics of liquid CO_2 phase transition on fracturing coal seam: Applications for enhancing coalbed methane recovery, *Fuel*, 308:122044–122056.

Tang, Z Q, Yang, S Q and Wu, G Y, 2017. Occurrence Mechanism and Risk Assessment of Dynamic of Coal and Rock Disasters in the Low-Temperature Oxidation Process of a Coal-Bed Methane Reservoir, *Energy and Fuels*, 31(4):3602–3609.

Tao, S, Chen, S D and Pan, Z J, 2019. Current status, challenges and policy suggestions for coalbed methane industry development in China: A review, *Energy Science and Engineering*, 7(4):1059–1074.

Thommes, M and Cychosz, K A, 2014. Physical adsorption characterization of nanoporous materials: progress and challenges, *Adsorption-Journal of the International Adsorption Society*, 20(2–3):233–250.

Thommes, M, Kaneko, K, Neimark, A V, Olivier, J P, Rodriguez-Reinoso, F, Rouquerol, J and Sing, K S W, 2015. Physisorption of gases, with special reference to the evaluation of surface area and pore size distribution (IUPAC Technical Report). *Pure and Applied Chemistry*, 87(9–10):1051–1069.

Wang, H B, Wang, Q, Liu, Y Q, Fu, Y S and Wu, P, 2020. Degradation of diclofenac by ferrous activated persulfate, *Environmental Chemistry*, 39:869–875.

Wang, Z J, Ma, X T, Wei, J P and Li, N, 2018. Microwave irradiation's effect on promoting coalbed methane desorption and analysis of desorption kinetics, *Fuel*, 222:56–63.

Xin, H H, Zhou, B H, Tian, W J, Qi, X Y, Zheng, M, Lu, W, Yang, H, Zhong, X and Wang, D, 2023a. Pyrolytic stage evolution mechanism of Zhundong coal based on reaction consistency analysis of mono/multi molecular models, *Fuel*, 333(2):126371.

Xin, H, Lu, L, Yang, Y, Tang, Z, Wu, J, Xu, Z, Zhang, P and Tian, W, 2023b. Research on coal pore evolution law and permeability enhancement mechanism based on chemical activation by high-energy radicals, *Journal of Mining and Safety Engineering*, 40(6):1335–1346. https://link.oversea.cnki.net/doi/10.13545/j.cnki.jmse.2023.0097

Xin, H, Wang, D, Qi, X, Qi, G and Dou, G, 2014. Structural characteristics of coal functional groups using quantum chemistry for quantification of infrared spectra, *Fuel Processing Technology*, 118:287–295.

Yang, J, Xu, S Y, Dai, J, Wei, J P and Wang, Y G, 2020. Experimental study on the oxidation and permeation of coal samples by activated ammonium persulfate solution, *Journal of China Coal Society*, 45:1488–1498.

Zhang, D, He, H, Ren, Y, Haider, R, Urynowicz, M, Fallgren, P H, Jin, S, Ali, M I, Jamal, A, Sabar, M A, Guo, H, Liu, F-J and Huang, Z, 2022. A mini review on biotransformation of coal to methane by enhancement of chemical pretreatment, *Fuel*, 308:121961–121970.

Zhang, H, Zhang, X B, Zhang, Y G and Wang, Z Z, 2023. The characteristics of methane adsorption capacity and behavior of tectonic coal, *Frontiers in Earth Science*, 10:1034341–1034355.

Zhao, Y, Meng, Y J, Li, K J, Zhao, S J and Ma, H J, 2023. Micropore Structure Changes in Response to H_2O_2 Treatment of Coals with Different Ranks: Implications for Oxidant Stimulation Enhancing CBM Recovery, *Natural Resources Research*, 32:2159–2177.

Zheng, S J, Yao, Y B, Sang, S X, Liu, D M, Wang, M and Liu, S Q, 2022. Dynamic characterization of multiphase methane during CO_2-ECBM: An NMR relaxation method, *Fuel*, 324:124526–124535.

Zheng, Y N, Li, S S, Xue, S, Jiang, B Y, Ren, B and Zhao, Y, 2023. Study on the evolution characteristics of coal spontaneous combustion and gas coupling disaster region in goaf, *Fuel*, 349:128505–128514.

Zhu, C J, Ren, J, Wan, J M, Lin, B Q, Yang, K and Li, Y, 2019. Methane adsorption on coals with different coal rank under elevated temperature and pressure, *Fuel*, 254:115686–115699.

Ventilation air methane (VAM) and GHG management

IOT and AI-based smart centre for particulate matter emissions monitoring for surface coalmines

S Agarwal[1], Y P Chugh[2], V Suresh[3], P Jha[4], A Mukherjee[4] and P Sharma[5]

1. Assistant Professor, IIT-ISM, Dhanbad, India 826004. Email: sagarwal@iitism.ac.in
2. Professor Emeritus, SIUC, Carbondale IL 62901. Email: siu681@siu.edu
3. PhD research scholar, IIT-ISM, Dhanbad, India 826004. Email: 21dr0217@me.iitism.ac.in
4. BTech Student, IIT-ISM, Dhanbad, India 826004. Email: 20je0725@me.iitism.ac.in
5. Research Associate, IIT-ISM, Dhanbad, India 826004. Email: 23pr0049@iitism.ac.in

ABSTRACT

This paper documents the first attempt in India to develop an Integrated Command and Control Centre (ICCC) for real-time data monitoring for particulates and greenhouse gas (GHG) emissions in a surface coalmine. Data collected include particulate matter ($PM_{2.5}$, PM_{10}), humidity, temperature, GHG, including air quality index (AQI). Data are collected through low-cost sensors in a single device, with a 1 min sampling interval from January 2023 to May 2023. The monitoring system has multiple sensor stations, the Internet of Things (IoT), data visualisation and trend tracking through dashboards, web and mobile applications for alerting, and seamless communication for stakeholders. The monitoring network here consists of three locations: the industrial zone (surface Coalmine), the buffer zone (Mining Office and residential colony) 1.5 km from the mine, and the IIT campus located 12 km from the mine. The ICCC has advanced analytics and artificial intelligence (AI) tools to process data, detect patterns, forecast pollutant levels and provide decision support to operators and decision-makers. ANOVA analyses to date show pollution levels around the industrial zone to be significantly higher than the other two zones. Other findings include PM concentrations peaking around 5:00–7:00 am and 4:00–6:00 pm every day since villagers are using coal early morning for cooking and hauling coal for pilferage, and the latter is associated with the blasting schedule. Several AI models are compared for projections but the long, short-term memory model performs the best with 88 per cent accuracy. This research will be used for the strategic location of water sprinklers, finding hot spots of GHG emissions, delineating unit operations contributing to adverse AQI, staying in compliance with statutory air quality standards and mining industry initiatives for tracking particulate and GHG emissions.

INTRODUCTION

Mineral extractive industries are the foundations of our lives and are therefore form an important global industry. The famous quote 'what cannot be grown must be mined' is still true today. The minerals provide resources for infrastructure, energy production, and consumer goods. World Mining Data indicates that the global minerals industry extracted over 17 billion tons of raw materials with a value of about USD2.03 trillion or about 2 per cent of the global GDP (gross domestic product) in 2022 (Zhang and Chugh, 2023).

Minerals-related production activities can disrupt our air, and ecosystem resources (Ghose and Majee, 2000), causing short-term and long-term negative impacts unless disturbed areas are appropriately reclaimed. Although mineral extraction activities provide significant economic, social, and political impacts, the industry must transform slowly under pressure from global competition, environmental regulations, and local communities to ensure sustainability (Patra, Gautam and Kumar, 2016). Surface mining methods dominate the world production of minerals, with almost all non-metallic minerals (over 95 per cent), most metallic minerals (over 90 per cent), and over 60 per cent of coal being mined using this method. Surface mining is also a crucial system in India, contributing about 90 per cent of the annual coal production. Surface mining operations emit air pollutants, such as particulate matter, methane, sulfur dioxide, nitrogen oxides, carbon monoxide, carbon dioxide and volatile organic compounds (VOC), which can negatively impact air quality and human health for workers (Gordon *et al*, 2014) and residents in the surrounding communities, and negatively impact the surrounding ecosystem. The World Health Organization (WHO) documents that about 90 per cent of global population inhales polluted air that results in respiratory health issues and about 7 million deaths per annum.

Most minerals-related extraction and utilisation industrial sites are mandated to deal with air pollution problems through permitting processes and regulatory requirements. Dealing effectively with air quality impacts requires a good understanding of the complex interactions between extractive (mining) and processing activities, associated emissions and their dispersion in the surroundings, and assessment of environmental remediation measures (Hendryx *et al*, 2020). This requires real-time monitoring and characterisation of emissions and transport in and around the vicinity of unit operations and effectiveness of control measures to reduce air pollution. Recent developments and availability of air quality sensors, high-speed data transfer techniques, and rapid processing and analysis of large data allowing feedback loops for control allow us to seek effective air quality management in surface coalmines using the input–output analysis of production system, and IoT-based monitoring approaches. Continuous monitoring can help assess the real-time exposure levels and enable timely interventions to protect the health of the workers (Hendryx *et al*, 2020; Dontala, Reddy and Vadde, 2015; Kahraman and Erkayaoglu, 2021; Wang *et al*, 2016). Almalawi *et al* (2022) and Duarte *et al* (2022) have explored the use of Artificial Intelligence (AI) algorithms (Pandey, Agrawal and Singh, 2014) for developing forecasting models and an air quality index (AQI) model for various pollutants. National Ambient Air Quality (NAAQ) Standards (Beig, Ghude and Deshpande, 2010) and Central pollution control board (CPCB, 2015) are Indian government agencies that do the monitoring for urban and industrial zones in India.

RESEARCH OVERVIEW

The abstract provides a good overview of the overall research. In addition, the author used a Digitalisation Model Structure (DMS) of Chugh and Agarwal (2021) to identify unit operations that may require improvement and to assess alternative solutions. The input-output analysis describes any production system as a combination of inputs processed through one or more processes to produce one or more value-added outputs. It can therefore also be referred to as a 'system of interacting subsystems and/or unit operations'. In order to assess industry operations for digitalisation and analyse data for any selected metric or set of goals, a slightly modified systems analysis-based structure can be used. The DMS was applied in this case study with real-time monitoring of air pollution parameters. The dashboard for data analysis and visualisation sent alerts to mine managers if the air quality parameters exceeded the threshold limit values (TLVs) defined by the National Ambient Air Quality (NAAQ) standards.

The authors attempt to use the proposed DMS for air quality management in this cooperative study with Coal India, Ltd. The surface mine was started in 1990 for one seam mining using a shovel-truck combination. The current mining depth is 100 m with overall stripping ratio of 3.1. Overburden stripping and coal extraction operations require drilling and blasting. There are seven overburden (OB) benches and five coal benches in the mine. The hauled OB is dumped on mined-out seam floor. Digitalisation was planned for air quality sensors (CH_4, CO_2, NO_x, SO_x, $PM_{2.5}$, PM_{10}, VOC). Since available sensors were limited, selected equipment in different unit operations was studied for short periods of time (1–2 weeks). The data from the sensors is entered into an input-output database DMS structure for the production system that allows ease in data management, and data analyses. Analyses should allow for making intelligent process changes based on characteristics of unit operations and air quality emission data. This cooperative project has the potential to significantly improve the profitability and environmental characteristics of the mining operation. To the best of the authors' knowledge such a systematic project has never been conducted in India.

MATERIALS AND METHODS

Data collection and transmission infrastructure through IoT devices

The data was collected from Jan 2023 to April 2023 for a period of four months at 1 minute intervals. The recorded 1440 data points per day were sub-sampled for further analysis. Three monitoring stations were strategically located (Figure 1) in an open space without obstructions. The sensors were installed at a height representative of human exposure. The locations included: an industrial zone at the mine, a buffer zone incorporating the Mine Office and adjoining residential colony, and an environmentally sensitive zone on the university campus (Figure 2). The monitoring stations were located between a humid subtropical climate and a tropical wet and dry climate zone. The monitored data and the analysed graphs can be viewed in real time through a web dashboard and mobile application.

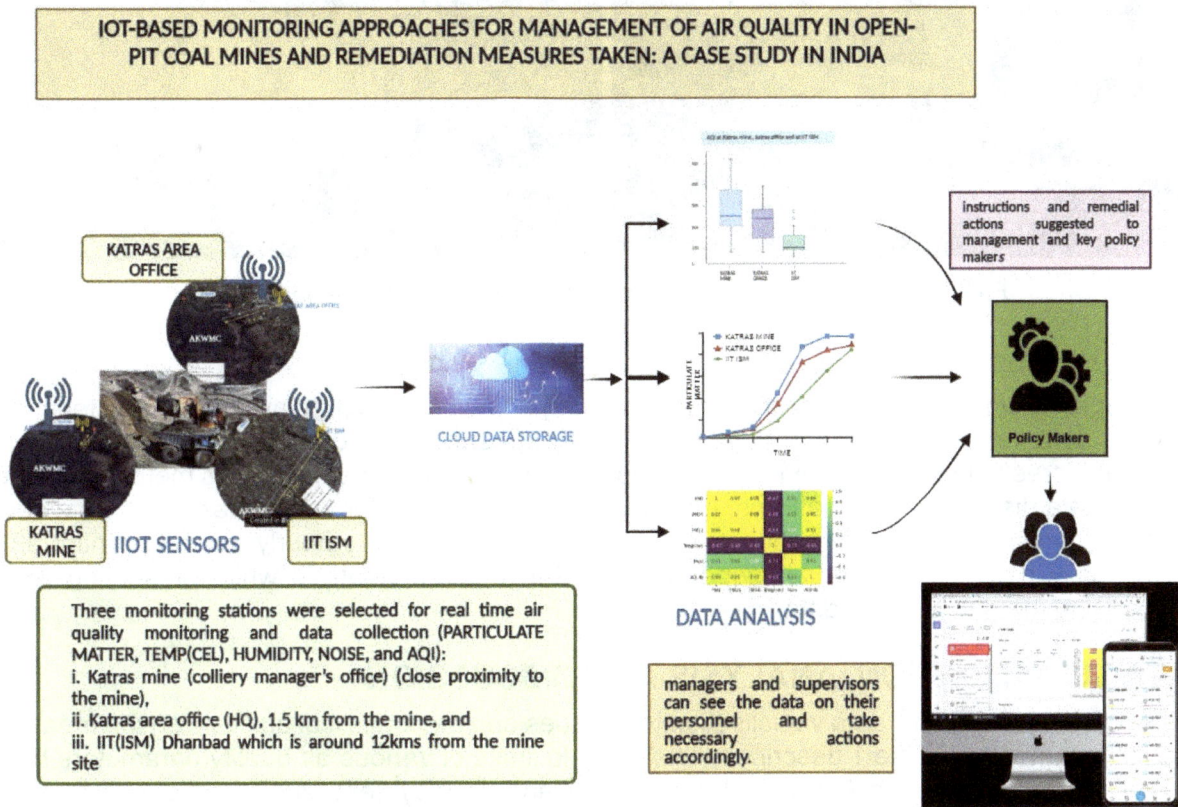

FIG 1 – Overall methodology of IoT-based air pollution monitoring system.

(i) Katras Mine
(23°48'43.3"N, 86°19'29.3"E)

(ii) Katras Area Office
(23°48'46.1"N, 86°20'12.6"E)

(iii) IIT- ISM Dhanbad
(23°48'47.88" N, 86°26' 30.84" E)

FIG 2 – Geographical location and PM sensor measurement areas at: (i) coalmine (ii) mining area office (iii) IIT-premises

The parameters collected by an IOT-based air quality monitor are those required of a Continuous Ambient Air Quality Monitoring System (CAAQMS) depicted in Figure 3. The weather-proof sensor assembly is a low-cost, easy-to-use, portable devices that transmits data via either of these connections GSM, Wi-Fi, and RS-485. Data is stored in both a micro-SD card and AQI cloud storage, which are both accessible through a web dashboard and mobile application.

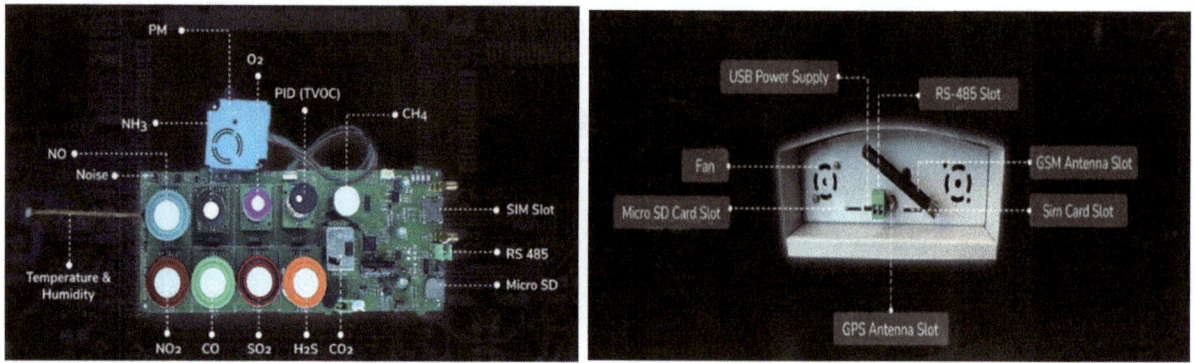

FIG 3 – Air quality low-cost sensors in a housing box and its parts.

Data management

Data preprocessing is required for this study focusing on PM and AQI. It involves three steps: data cleaning, data transformation, and data reduction. The mean of a sliding window with three points ahead and three points behind were used for inputting of any missing values. This method considers the local data distribution and patterns, making it especially useful for time series data. Time series resampling was done for aggregation of data from 1 min intervals to hourly data (1 hr interval) as well as daily data (24 hr interval) depending on the required analysis while preserving and/or summarising the foundational data.

DATA VISUALISATION, ANALYSIS AND DISCUSSION

This section describes the data visualisation techniques and the statistical analysis on the measured data for four months. Table 1 describes the threshold limits of various air quality parameters based on Coal Mines Regulations Act (Ministry of Labour and Employment, 2017), national ambient air quality standards (NAAQS, 2009), and WHO standards as per 2021 (Carvalho, 2021). In case of residential or commercial or industrial place falls within 500 m of any dust generating sources, NAAQS shall be made applicable. Hence, they are applicable in the case of the mines personnel office and IIT-ISM. Air quality standards maintained by coalmines in India, WHO and NAAQS are mentioned in Table 1. It was observed that there $PM_{2.5}$ at coalmine location exceeded the threshold a total of 23 days, all of them occurring in months of January and February. Also five times it happened on two consecutive days. PM_{10} concentration exceeded 21 days in which 20 days are from January and February months. This is a clear evidence high PM concentration in mine area during winter months in India putting mining workers and nearby communities at high risk of health issues.

TABLE 1

Air quality standards of various agencies.

Pollutants	Averaging time	Coal Mines Regulation Act (CMRA) 2017	India (NAAQS)	(WHO, 2021) AQI	No of times threshold limit crossed at coalmine based on CMRA
$PM_{2.5}$ ($\mu g/m^3$)	Annual	215	40	5	23
	24 hr	300	60	15	
PM_{10} ($\mu g/m^3$)	Annual	215	40	5	21
	24 hr	300	60	15	

Particulate matter and AQI analysis

Figures 4 and 5 present box plots that show descriptive statistics for $PM_{2.5}$ and PM_{10} at all three locations. These plots help us understand the behaviour of the data. The orange horizontal line represents the median, and coalmine in Jharkhand area location has a higher median line than mine personnel office and IIT (ISM). The coalmine in Jharkhand area data is more dispersed, as indicated by the wider box, while IIT (ISM) is the least dispersed. This is also evident in the distribution plots on the right. The high concentration of PM in the air is mainly attributed to coal and OB blasting, as well as coal and overburden loading, transportation, and movement of heavy earth moving machinery. These findings highlight the importance of specialised pollution management approaches that consider the individual pollutant types and local circumstances of each location.

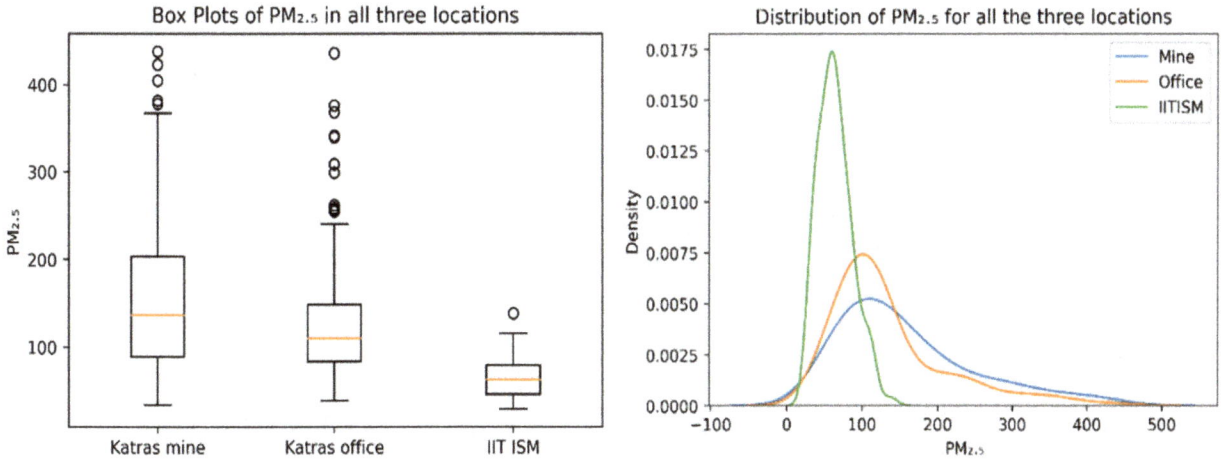

FIG 4 – Box plots and distributions of $PM_{2.5}$ for all three locations.

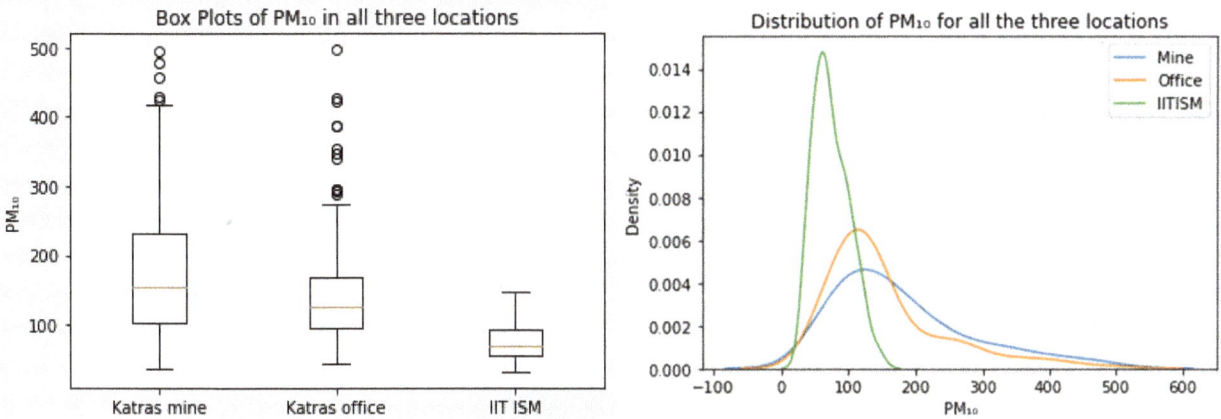

FIG 5 – Box plots and distributions of PM_{10} for all three locations.

Average AQI values for coalmine (Figure 6) fall under the range (201–300) which signifies poor air that may cause breathing discomfort to people on prolonged exposure, and discomfort to people with heart disease. The highest AQI values reach up to 500 which is severe and may cause respiratory impact and serious health impacts related to lung/heart disease.

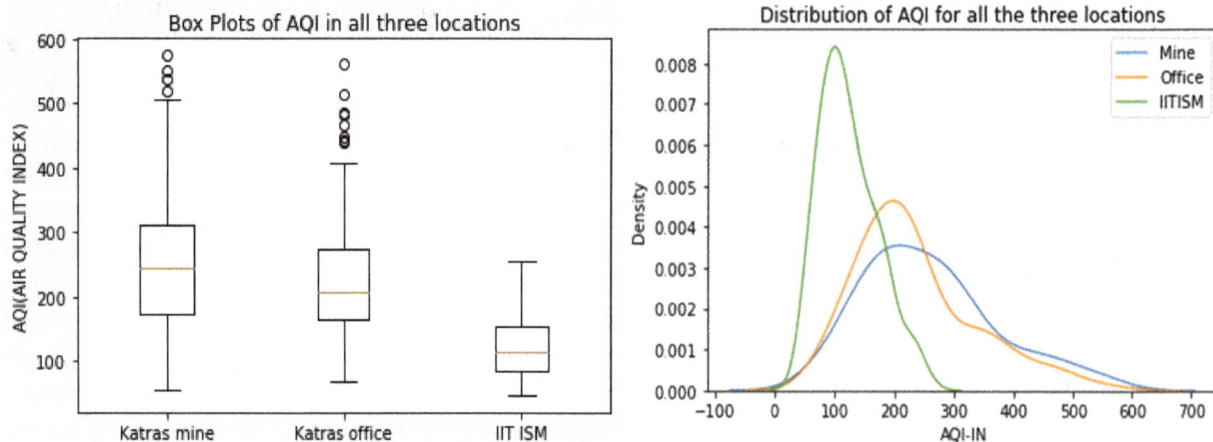

FIG 6 – Box plots and distributions of AQI for all three locations.

During the study period, it was observed that air quality readings over the threshold were reported frequently at coalmine and mine personnel office. On the other hand, IIT (ISM), which is situated farther away from the mining operations, reported AQI readings over the threshold only on nine days in total. This shows that being far from mining sites has a significant positive impact on air quality and highlights the importance of geographic considerations for maintaining it. It is noteworthy that the AQI levels at both coalmine and mine personnel office never fell into the 'good' category throughout the study period, except for just two days at IIT (ISM) (as shown in Figure 6). This clearly indicates the impact of proximity to mining operations on air quality and emphasises the need for targeted air quality control measures in areas with extensive industrial activity.

The correlation heatmap in Figure 7 shows that there is a strong positive correlation between AQI and PM_1, $PM_{2.5}$, and PM_{10}. This is because these particulate matters contribute significantly to air pollution, which in turn is directly related to the decline of air quality. On the other hand, there is a strong negative correlation between particulate matters and temperature. As the temperature increases, gas molecules move faster and spread out, resulting in a more diffuse atmosphere and a decrease in particulate matters. However, humidity has a positive correlation with particulate matter since moisture makes particles heavier, reducing the diffusion of particulate matter.

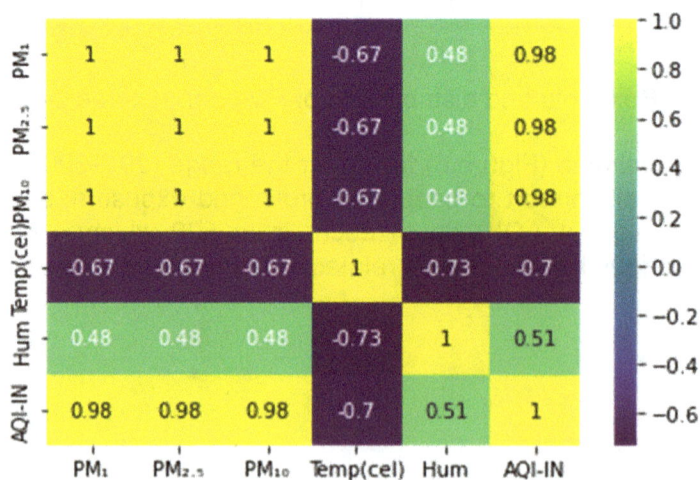

FIG 7 – Heatmap of correlations amongst parameters measured.

Analysis of peaks of PMs in mine area

The hourly plots in Figure 8 provide insight into the analysis of peak PM concentration in the mine area, revealing several key findings. During night-time hours, there is a negative correlation between temperature and PM concentrations (Singh, Singh and Biswal, 2021). This is primarily due to temperature inversion, where cool air forms near the ground, trapping pollutants. Additionally, volatile chemicals condense and wood burning increases, resulting in higher PM levels (Hua *et al*, 2021).

This effect is most noticeable around 6:00 am daily (Luo *et al*, 2021). Secondly, humidity levels are positively correlated with PM concentrations. Increased humidity absorbs a significant amount of solar light, reducing radiation near the Earth's surface. As a result, surface temperatures fall, weakening incoming air currents and increasing PM levels. This association is most prominent during periods of high humidity (Luo *et al*, 2021). Thirdly, $PM_{2.5}$ concentrations at the opencast coalmine fluctuate significantly during peak transportation hours, with a distinct increasing pattern before reaching peak concentrations. This suggests that transportation activities have a significant influence on raising PM levels (Sinha and Banerjee, 1997). Lastly, blasting operations, which occur daily from 3:00 pm to 6:00 pm, are associated with a significant increase in PM concentrations around 6:00 pm. This peak is related to the dust produced during blasting and shift changes, which involve the withdrawal and re-engagement of workers and machinery (Patra, Gautam and Kumar, 2016; Sastry *et al*, 2015). These findings underscore the importance of specialised pollution management approaches that take into account individual pollutant types and site-specific conditions.

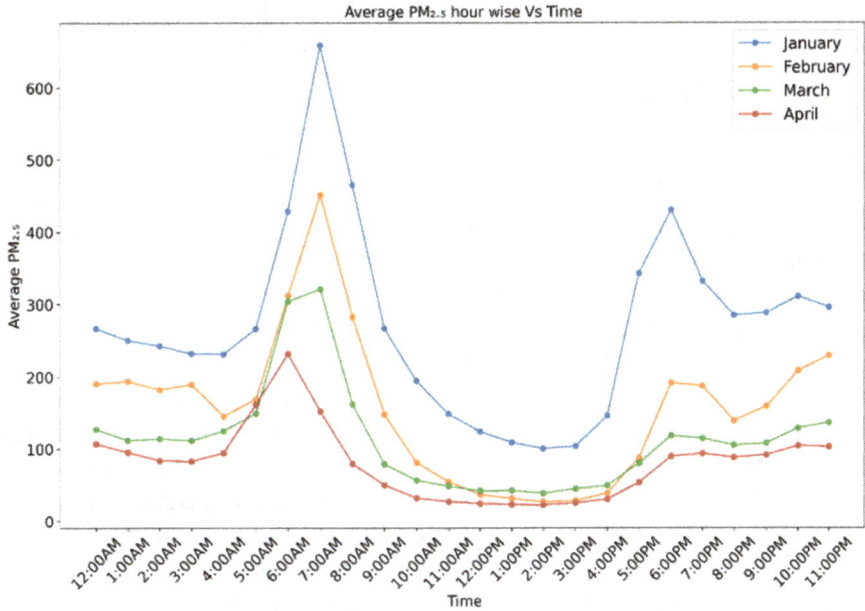

FIG 8 – $PM_{2.5}$ concentration hourly plot averaged for all days in each month from Jan–April.

GHG analysis

Figure 9 depicts the gases emitted in coalmine area and their average concentrations, maximum–minimum values and extremely high observations in a box plot. Clearly all of them have very high mean as compared to WHO and NAAQS standards.

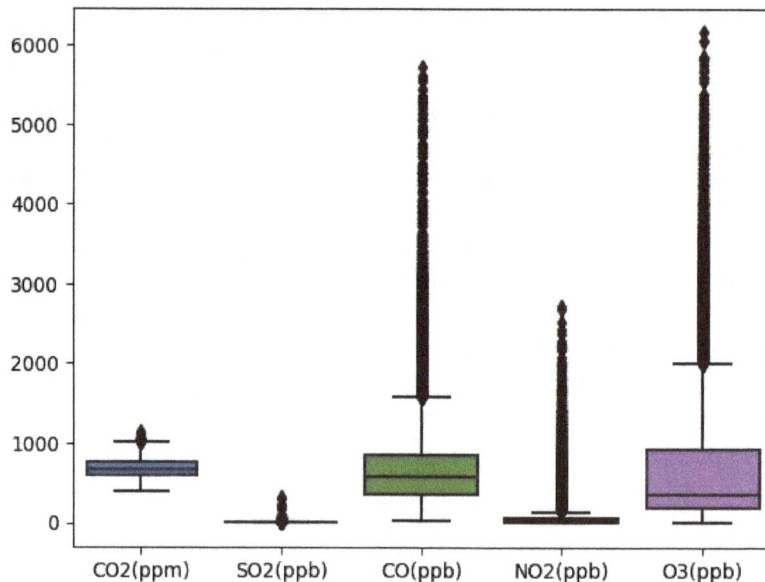

FIG 9 – Box plots of all gases emitted in coalmine located in Jharkhand area.

From Figure 9 it can be observed that the carbon dioxide (CO_2) boxplot shows a narrow range with no visible outliers. The narrow range suggests consistent CO_2 levels, possibly due to regular mining activities. Sulfur dioxide (SO_2) boxplot has a compact interquartile range (IQR) and shows outliers above the upper whiskers. Elevated levels of SO_2 are detected as a result of mine fires triggered by spontaneous combustion of coal with high sulfur content in various sections of the mine. The CO box plot features a larger IQR and the median is closer to the lower quartile. The larger CO levels are due to incomplete combustion caused by SC and use of diesel equipment. In addition, meteorological conditions, such as temperature inversions or stagnant air masses, can exacerbate the buildup of carbon monoxide data in open-cast mines by trapping pollutants close to the ground. The nitrogen dioxide (NO_2) boxplot has an extended upper whisker, indicating variability at higher concentrations and the median is closer to the lower quartile. The extended whisker suggests occasional spikes in NO_2 levels. The high levels of NO_2 can be attributed to several factors such as diesel-powered trucks, loaders, and drills as well as blasting activities.

The ozone (O_3) boxplot has an even longer upper whisker, indicating higher variability at elevated concentrations. O_3 is formed through complex chemical reactions involving precursor pollutants such as nitrogen oxides (NO_x) and volatile organic compounds (VOCs) in the presence of sunlight. Emissions of NOx and VOCs from diesel-powered equipment, blasting activities, and natural sources within the mine can contribute to ozone formation through photochemical reactions.

Different pollutants exhibit varying distributions due to their sources and chemical interactions. Monitoring and controlling of these pollutants are essential for environmental and public health. The following engineering controls/measures that can effectively control the air pollution in opencast mines such as dust suppression techniques; wet spraying and dry fog systems. Wet sprays are used to suppress dust during drilling, blasting, and material handling. Dry fog systems are used to suppress dust from mineral processing plants. Besides planting vegetation around mining areas can serve multiple purposes such as soil stabilisation and trapping dust. Enclosing the conveyor belts, crushers, and other equipment can minimise dust dispersion. Regular maintenance of vehicles and optimising blasting practices can reduce fragmentation and dust production.

CONCLUSIONS

There has been insufficient research and monitoring of real-time particulate matter and air quality in open-pit coalmines in India (Luo *et al*, 2021). This lack of attention could result in various health and environmental hazards for workers, neighbouring communities, and the ecosystem. To tackle this issue, an IoT-based systems, can provide real-time data and insights into pollution sources. IOT based CAAQMS can identify seasonal variations, trends, and the effectiveness of pollution control measures (Jo and Khan, 2017). It can help in managing water sprinkler movements and quantity of water used can be optimised.

ACKNOWLEDGEMENTS

We would like thank the IIT (ISM) for their financial and infrastructure support. The project has allocated funding to the project through faculty research grants. We are thankful to then Director Prof Rajiv Shekhar for his encouragement and support throughout this project.

REFERENCES

Almalawi, A, Alsolami, F, Khan, A I, Alkhathlan, A, Fahad, A, Irshad, K, Qaiyum, S and Alfakeeh, A S, 2022. An IoT based system for magnify air pollution monitoring and prognosis using hybrid artificial intelligence technique, *Environmental Research*, 206:112576.

Beig, G, Ghude, S D and Deshpande, A, 2010. Scientific evaluation of air quality standards and defining air quality index for India, pp 1–27 (Pune: Indian Institute of Tropical Meteorology).

Carvalho, H, 2021. New WHO global air quality guidelines: more pressure on nations to reduce air pollution levels, *The Lancet Planetary Health*, 5(11):e760–e761.

Chugh, Y P and Agarwal, S, 2021. A Digitalization Model Structure for Sustainable and Enhanced Profitability in Industry 4.0, in *2021 International Conference on Simulation, Automation & Smart Manufacturing (SASM)*, pp 1–6 (IEEE).

Duarte, J, Rodrigues, F and Castelo Branco, J, 2022. Sensing technology applications in the mining industry—a systematic review, *International Journal of Environmental Research and Public Health*, 19(4)2334.

Ministry of Labour and Employment, 2017. Coal Mines Regulations, 2017. Available from: <https://www.dgms.net/Coal%20Mines%20Regulation%202017.pdf> [Accessed: 23 Nov 2023].

Central Pollution Control Board (CPBC), 2015. National Air Quality Monitoring Programme, Ministry of Environment and Forests (Government of India). Available from: <http://cpcb.nic.in/air.php>

Dontala, S P, Reddy, T B and Vadde, R, 2015. Environmental aspects and impacts its mitigation measures of corporate coal mining, *Procedia Earth and Planetary Science*, 11:2–7.

Ghose, M K and Majee, S R, 2000. Sources of air pollution due to coal mining and their impacts in Jharia coalfield, *Environment International*, 26(1–2):81–85.

Gordon, S B, Bruce, N G, Grigg, J, Hibberd, P L, Kurmi, O P, Lam, K B H, Mortimer, K, Asante, K P, Balakrishnan, K, Balmes, J and Bar-Zeev, N, 2014. Respiratory risks from household air pollution in low- and middle-income countries, *The Lancet Respiratory Medicine*, 2(10):823–860.

Hendryx, M, Islam, M S, Dong, G H and Paul, G, 2020. Air pollution emissions 2008–2018 from Australian coal mining: Implications for public and occupational health, *International Journal of Environmental Research and Public Health*, 17(5):1570.

Hua, J, Zhang, Y, de Foy, B, Mei, X, Shang, J and Feng, C, 2021. Competing $PM_{2.5}$ and NO2 holiday effects in the Beijing area vary locally due to differences in residential coal burning and traffic patterns, *Science of The Total Environment*, 750:141575.

Jo, B W and Khan, R M A, 2017. An event reporting and early-warning safety system based on the internet of things for underground coal mines: A case study, *Applied Sciences*, 7(9):925.

Kahraman, M M and Erkayaoglu, M, 2021. A data-driven approach to control fugitive dust in mine operations, *Mining, Metallurgy and Exploration*, 38(1):549–558.

Luo, H, Zhou, W, Jiskani, I M and Wang, Z, 2021. Analyzing characteristics of particulate matter pollution in open-pit coal mines: implications for green mining, *Energies*, 14(9):2680.

Pandey, B, Agrawal, M and Singh, S, 2014. Assessment of air pollution around coal mining area: emphasizing on spatial distributions, seasonal variations and heavy metals, using cluster and principal component analysis, *Atmospheric Pollution Research*, 5(1):79–86.

Patra, A K, Gautam, S and Kumar, P, 2016. Emissions and human health impact of particulate matter from surface mining operation—A review, *Environmental Technology and Innovation*, 5:233–249.

Sastry, V, Chandar, K R, Nagesha, K V, Muralidhar, E and Mohiuddin, M S, 2015. Prediction and analysis of dust dispersion from drilling operation in opencast coal mines, *Procedia Earth and Planetary Science*, 11:303–311. https://doi.org/10.1016/j.proeps.2015.06.065

Singh, V, Singh, S and Biswal, A, 2021. Exceedances and trends of particulate matter ($PM_{2.5}$) in five Indian megacities, *Science of the Total Environment*, 750:141461.

Sinha, S and Banerjee, S P, 1997. Characterization of haul road dust in an Indian opencast iron ore mine, *Atmospheric Environment*, 31(17):2809–2814.

Wang, Y, Sun, M, Yang, X and Yuan, X, 2016. Public awareness and willingness to pay for tackling smog pollution in China: a case study, *Journal of Cleaner Production*, 112:1627–1634.

Zhang, R and Chugh, Y P, 2023. Sustainable Development of Underground Coal Resources in Shallow Groundwater Areas for Environment and Socio-Economic Considerations: A Case Study of Zhangji Coal Mine in China, *International Journal of Environmental Research and Public Health,* 20(6):5213.

Effect of catalyst support in the oxidation of fugitive lean methane gas streams

M Bligh[1], M Drewery[2], E M Kennedy[3] and M Stockenhuber[4]

1. PhD Student, University of Newcastle, Newcastle NSW 2308.
   Email: matthew.bligh@uon.edu.au
2. Post Doctorate Research Associate, University of Newcastle, Newcastle NSW 2308.
   Email: matthew.drewery@newcastle.edu.au
3. Professor, University of Newcastle, Newcastle NSW 2308.
   Email: eric.kennedy@newcatsle.du.au
4. Professor, University of Newcastle, Newcastle NSW 2308.
   Email: michael.stockenhuber@newcastle.du.au

ABSTRACT

Natural gas as an energy source in combustion engines is widespread due to its availability, efficient and cleaner burns, and lower carbon dioxide emissions. Natural gas engines are particularly important to supplement intermittent renewable sources because of their fast start-up and shutdown times. However, when operated under the recommended lean burn conditions, methane combustion efficiency is below 100 per cent, and methane is emitted in the natural gas engine exhaust (NGEE). Methane slip is a significant concern as methane has a higher global warming potential than CO_2. The IPCC has indicated a GWP for methane of ca 83 when considering its impact over a 20-year time frame (GWP20) and between ca 30 when considering its impact over a 100-year time frame (GWP100). This has led to increased scrutiny of natural gas engines, prompting alternative methods to reduce methane output. The use of catalyst combustion has proven effective in reducing methane output, with palladium catalyst being the most active. A similar lean methane exhaust gas in ventilation air methane (VAM) was successfully mitigated by combustion over a palladium catalyst. While VAM and NGEE have excess oxygen and lean methane, they differ significantly in water concentration. Water poses issues with the catalyst's Lewis acidity as it acts as a Lewis base. This study compares VAM and NGEE using four catalysts with different supports to highlight how Lewis's acidity affects performance. According to the results, when operated under the VAM conditions, higher activity was observed, with the Pd/TS-1 catalyst exhibiting the highest activity in both scenarios. This increase in performance can be attributed to the neutral framework of the TS-1 catalyst support compared to acidic alumina. Conversely, the ZSM-5-supported catalysts exhibited low activity in VAM and no activity in NGEE, primarily caused by the acidity resulting from surface defects framework tetrahedral alumina. Additional tests indicated that a higher methane concentration led to catalyst deactivation, with the reduction of the catalyst's active site being a significant factor in the deactivation mechanism. Changes observed during on-stream testing between VAM and NGEE resulted in sudden conversion fluctuations, which an equilibrium effect with water could potentially explain.

INTRODUCTION

Waste gas is a significant global concern as a push for a greener energy sector has questioned the extent of emissions. The standard waste gas scrutinised is carbon dioxide due to the quantity emitted daily, but waste methane has caused significant concern. Methane is a much more potent greenhouse gas than carbon dioxide as it has a global warming potential (GWP) of around 86 times that of CO_2 over a 20-year period; this was communicated by workers (Da *et al*, 2020). Waste methane is a consequence of the energy sector, where unburnt trace methane escapes the exhaust. Catalytic oxidation is the primary method proposed to eliminate methane in the exhaust, where combusting methane into carbon dioxide reduces the greenhouse effect significantly. Trace methane can be oxidised at temperatures below 500°C. Workers (Oh, Mitchell and Siewert, 1992) found that palladium is the most active metal in methane combustion.

A major methane emitter is found through the methane slip from natural gas engine exhausts (NGEE). Workers (Cho and He, 2007; Lott and Deutschmann, 2021) communicated that natural gas engines are prevalent worldwide due to their cleaner burn compared to diesel and gasoline-based engines. However, researchers (Lott and Deutschmann, 2021; Cho and He, 2007) have found that incomplete combustion in the engine leads to methane escaping from the exhaust, primarily when run in the more efficient lean burn fashion. Catalytically combusting this methane comes with serious challenges as the exhaust conditions are not entirely favoured for combustion. Methane and other alkanes combusted in the engine lead to a high water vapour concentration. Researchers (Lott and Deutschmann, 2021; Lampert, Kazi and Farrauto, 1997; Gholami, Alyani and Smith, 2015) have found that water concentration can exceed 10 per cent, greater than the oxygen gas concentration. Water is a known catalyst inhibitor, where the Lewis base properties of water interact readily with Lewis acid sites, especially at temperatures below 450°C, as found by workers (Gélin *et al*, 2003; Gholami, Alyani and Smith, 2015). Workers (Dai *et al*, 2018) found that palladium oxide has Lewis acid attributes and shows little interaction with water under low concentrations. Still, under a concentration of 10 per cent, the interaction will be very high. Any Lewis acidity will significantly weaken stability, making the support choice crucial.

Workers (Petrov *et al*, 2018) found that increasing the Si/Al ratio decreases Lewis acidity, but the number of sites the active metal can disperse on the support is reduced. This proposes a balance required to achieve the best stability by altering this property. It was found by workers (Schwartz, Ciuparu and Pfefferle, 2012) that alumina is known to have a very high Lewis acidity, which is suitable for associated dispersion of the active metal but has shown low stability in the presence of water. The workers (Petrov *et al*, 2018) found that zeolites could enhance stability under these water conditions as the silicon-to-alumina ratio can be altered. Research by (Okumura *et al*, 2003) found that ZSM-5-supported catalysts have shown high activity in methane oxidation. Especially so the Lewis acidity can be tailored to a specific use. As communicated by (Hosseiniamoli *et al*, 2020), an altered version of a ZSM-5 catalyst can be found in titanium silicate-1 (TS-1), where the alumina sites are replaced with titanium. It was shown by researchers (Wu *et al*, 2021) that titanium, having a more negligible Lewis acidity than alumina, can still act as a site for the active metal to anchor on and have a decreased interaction with the Lewis base water.

Research performed by workers (Hosseiniamoli *et al*, 2020) showed that palladium on a TS-1 catalyst has a high activity under ventilation air methane or VAM conditions. It was communicated by (Setiawan *et al*, 2014; Hosseiniamoli *et al*, 2020) that the conditions of VAM are similar to NGEE, trace methane, excess oxygen, and water at a varied concentration of 1–3 per cent. It was found by researchers (Schwartz, Ciuparu and Pfefferle, 2012; Hosseiniamoli *et al*, 2020) that under VAM conditions, water inhibits the Pd/Al₂O₃ catalysts, while the Pd/TS-1 catalyst showed high catalytic stability for 2000 hrs on stream. It is hoped that the catalytic developments from VAM can be applied to NGEE. However, the differences could pose serious challenges. The increased water concentration from 1–3 per cent to 10 per cent and the decreased oxygen and methane concentrations will impose challenges.

In this study, catalysts Pd/TS-1, Pd/Al₂O₃, and Pd/ZSM-5 (Si/Al = 23 and 50) were tested under VAM and NGEE conditions to understand how the changes in concentration impact the performance and how the subtle difference in the catalyst handles the conditions. The varied Lewis acidity amongst these catalysts will give insight into how this acidity plays a role under the higher water NGEE conditions.

EXPERIMENTAL

The catalyst was prepared by adding palladium nitrate solution 10 wt per cent in 10 wt per cent nitric acid (Sigma Aldrich) to four dry supports: ZSM-5 (SiO₂/Al₂O₃ = 23 and 50) (Zeolyst), Al₂O₃ and TS-1 (Zeolyst). A mortar and pestle were used to uniformly mix the support and salt-containing solution. Water was added repeatedly between mixing and crushing the catalyst until a toothpaste-like consistency was shown. The almost wet catalyst slurry was then added to a furnace set to 110°C, where thermal decomposition can be achieved. The dried catalyst was sized to 250–425 microns and then loaded into a steel reactor tube. The reactor tube was first prepared by inserting enough quartz wool into the centre of the tube, where the thermocouple is positioned. The catalyst was then calcined under 200 mL/min air at 500°C.

The catalyst was tested under NGEE and VAM conditions to observe overall stability. The GHSV, flow rate, and catalyst mass were kept the same throughout, being 100 000 h⁻¹, 300 mL/min, and

100 mg, respectively. The concentrations used for VAM conditions were water – 1.5 per cent, Oxygen gas – 20 per cent, and methane – 0.7 per cent. The concentrations used for NGEE conditions were water – 10 per cent, Oxygen gas – 8 per cent, and methane – 0.4 per cent. Helium was used as a balance for both conditions, and nitrogen was used as a standard. The water was supplied to the feed via a syringe pump, with a furnace temperature of 450°C.

Activation energy experiments were conducted under the same conditions but with 25 mg of catalyst, resulting in a GHSV of 400 000 h^{-1}. A bubbler supplied the water for a more accurate, well-mixed concentration. The reaction rate for activity experiments was determined via an altered version of the ideal gas law, as shown in Equation 1.

$$rE = \frac{Conv\ (\%) \left(\frac{mol_r}{mol_T}\right) * Flow \left(\frac{m^3}{s}\right) * Pressure\ (Pa)}{8.314\left(\frac{m^3\ Pa}{mol_T\ K}\right) * Temperature\ (K) * Cat.weight\ (g)} \tag{1}$$

Conversion measurements were taken at a series of temperatures to determine how temperature affects the catalyst's activity under VAM and NGEE conditions. The activation energy was determined via the Arrhenius equation shown in Equation 2, substituted into the reaction rate equation, shown in Equation 3, to get Equation 4. A differential reactor is assumed; thus, the gas concentrations are constant throughout the experiments, leading to a simplified Equation 4, treating all the concentrations as 'C'. The natural logarithm was taken for each reaction rate and plotted against the temperature's reciprocal to determine the activation energy, as shown in Equation 5.

$$k = Ae^{-\frac{E_a}{RT}} \tag{2}$$

$$r = k[A]^x[B]^y[C]^z \tag{3}$$

$$r = Ae^{-\frac{E_a}{RT}} * [A]^x[B]^y[C]^z = Ce^{-\frac{E_a}{RT}} \tag{4}$$

$$\ln(r) = \ln(C) - \frac{E_a}{R} * \frac{1}{T} \tag{5}$$

Several nitrogen adsorption experiments were performed on the catalysts to determine a change in surface area and porosity before and after the stability experiments. Due to the microporosity of the three zeolites, the t-plot was used to determine the external and microporous surface area with the Harson and Jura method. Chemisorption experiments were done after catalyst preparation to determine metal dispersion. The catalyst sample was held at 35°C under a low pressure of 1 × 10^{-5} mBar, and a known amount of carbon monoxide was introduced. The dispersion was determined through the pressure loss due to carbon monoxide chemically adsorbing onto the palladium sites. Several scanning electron microscope (SEM) experiments were conducted on the samples prepared for experimental testing under VAM and NGEE. Several snapshots were taken with the SEM at magnifications of 1000 to 60 000 times to understand the morphology of each catalyst sample and observe the brighter backscatter electrons from the heavier palladium atoms. The SEM snapshots were taken with the Carl Zeiss Sigma VP Field-Emission Scanning Electron Microscope, and a coater (LEICA EM ACE200) was used to coat the samples with 6 nm of platinum before testing. Electron dispersive X-ray (EDX) analysis determined the quantity of active metal within it. Powder X-ray diffraction (PXRD) patterns of Pd/TS-1 were collected on a Malvern Panalytical Aeris (Research Edition) geometry using CoKα radiation. The samples were measured at 2θ angles between 5 and 80°.

RESULTS

A stability experiment was conducted to test catalytic stability under VAM conditions, as shown in Figure 1. Catalysts Pd/TS-1, Pd/ZSM-5 (23), Pd/ZSM5 (50) and Pd/Al$_2$O$_3$ were tested for approximately 1 to 2 days to measure the level of deactivation. The gas concentrations were water – 1.5 per cent, oxygen gas – 20 per cent, methane – 0.7 per cent, and an inert gas balance. The reactor was set at 450°C, with 100 mg of catalyst.

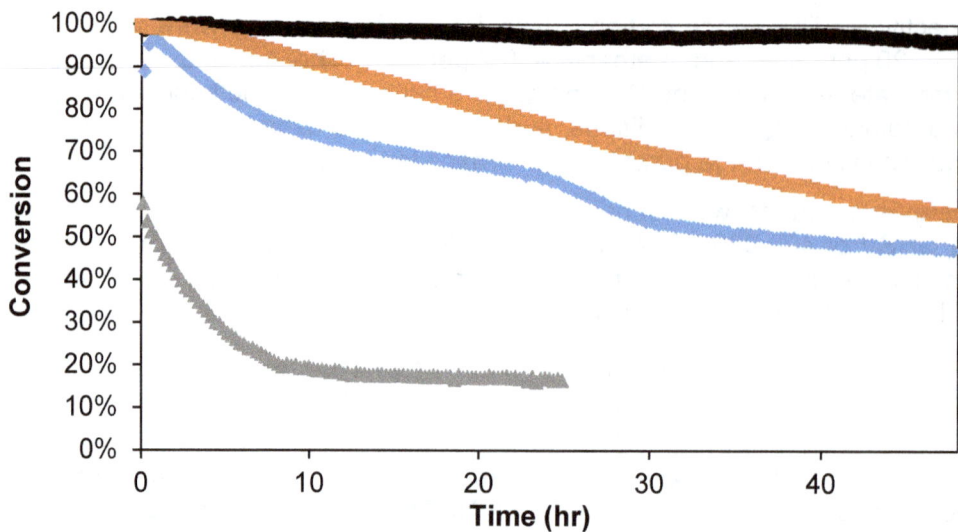

FIG 1 – Catalyst stability under VAM conditions. Feed gas concentration: H_2O – 1.5 per cent, O_2 – 20 per cent, CH_4 – 0.7 per cent and N_2 – Balance. Catalysts: Pd/Al_2O_3 – ■, Pd/TS-1 – ●, Pd/ZSM-5 (50) – ◆ and Pd/ZSM-5 (23) – ▲.

Catalysts with the same concentration of active site have vastly different activities under VAM conditions, highlighting the role of the support. It was communicated by (Schwartz *et al*, 2012) that in methane oxidation, the methane will reduce active sites, meaning oxygen has to be resupplied by the support. The more acidic supports alumina and ZSM-5 (23 and 50) performed much worse than the more neutral support TS-1. So, the acidity of the support affects the interaction with water, altering the overall oxygen transfer. The result of the ZSM-5 catalysts supports this. Increasing the SiO_2/Al_2O_3 and lowering associated Brønsted acid sites greatly improved the stability. In addition, surface defects in the ZSM-5 catalysts facilitate a higher water interaction than alumina. Removing the alumina sites from the ZSM-5 support to create TS-1 has shown a vast improvement where the Lewis acidity is lowered, and titanium allows for an anchoring site for palladium to disperse amongst the support. This great performance is also communicated by (Setiawan *et al*, 2016; Hosseiniamoli *et al*, 2020) exhibiting great stability under VAM conditions.

The same four catalysts were tested under NGEE conditions, shown in Figure 2. Catalysts Pd/TS-1, Pd/ZSM-5 (23), Pd/ZSM-5 (50) and Pd/Al_2O_3 were tested for approximately 1 to 2 days to measure the level of deactivation. The gas concentrations were water – 10 per cent, oxygen gas – 8 per cent, methane – 0.4 per cent, and an inert gas balance. The reactor was set at 450°C, with 100 mg of catalyst loaded into the reactor.

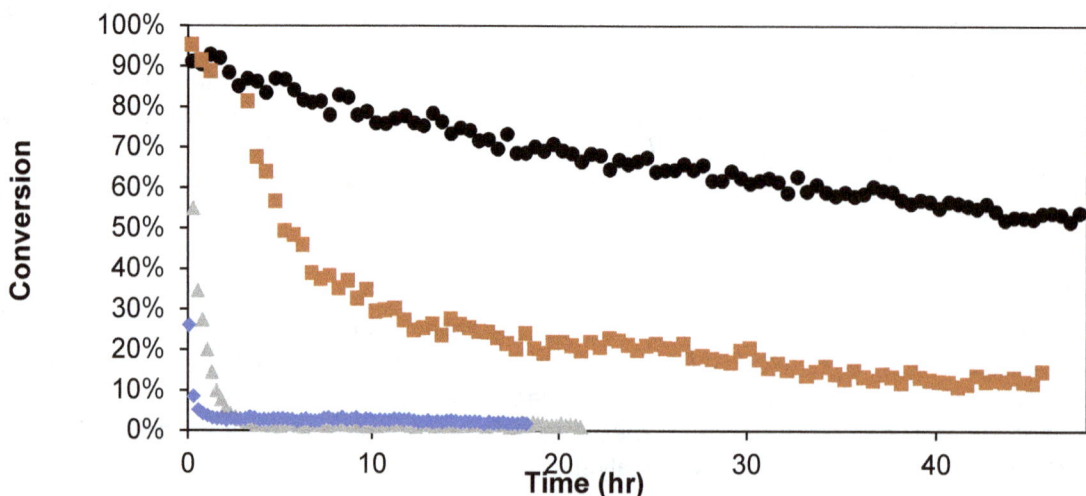

FIG 2 – Stability experiment of varying catalysts under VAM conditions. Concentrations: H_2O – 10 per cent, O_2 – 8 per cent, CH_4 – 0.4 per cent and N_2 – Balance. Catalysts: Pd/Al_2O_3 – ■, Pd/TS-1 – ●, Pd/ZSM-5 (50) – ◆ and Pd/ZSM-5 (23) – ▲.

The catalysts under NGEE conditions showed a much higher deactivation rate than VAM conditions. The low Lewis acidity of the TS-1 support was substantial enough to allow the interaction with the very high-water concentrations. The low oxygen mobility increased the concentration of palladium-reduced sites, reducing the overall conversion. This effect is seen more severely with the other catalysts, with alumina decreasing to a very low conversion of 15 per cent and the ZSM-5 catalysts were completely deactivated after only 2 hrs on stream. The interaction of the catalysts with water either lowers the overall oxygen mobility or sinters the catalysts, reducing active sites. Regardless, removing the acidic elements of the catalysts showed higher activity in methane conversion under high water concentrations. So, to improve the TS-1 catalysts, the Lewis acidity must be decreased further.

The four catalysts were tested under powder X-ray diffraction to understand the changes in their crystalline structure. The results are shown in Figure 3.

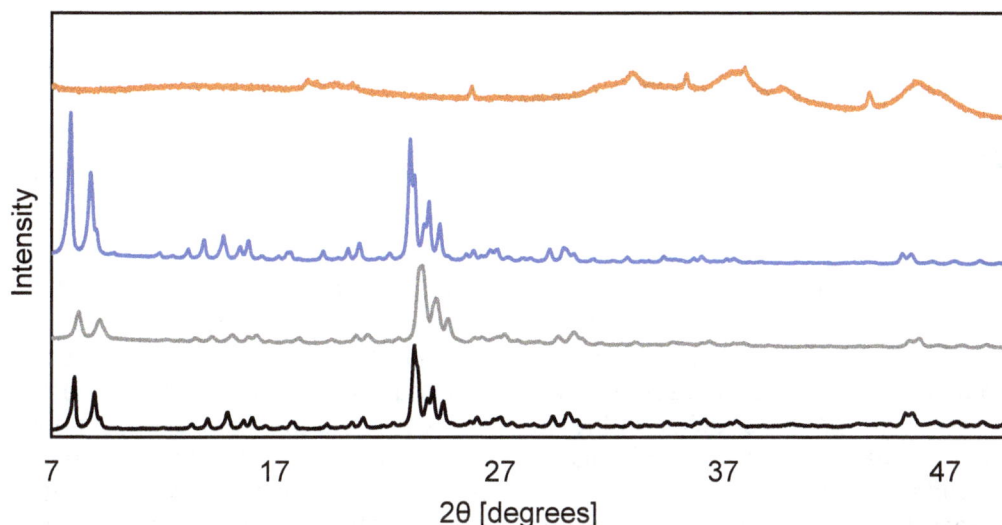

FIG 3 – XRD spectra of the prepared catalysts: Pd/Al$_2$O$_3$; —, Pd/TS-1; —, Pd/ZSM-5 (23); —, Pd/ZSM-5 (50); —.

The XRD spectra of the zeolite catalysts are very similar to each other. This is due to the fact that they are all MFI zeolite, so have the same crystalline structure. This highlights the fact that these three catalysts are very similar, in morphological terms and the only key difference between them being the concentration of alumina and the associated Bronsted acid sites and the existence of titanium which can act as anchor for palladium. The palladium on alumina catalyst has a spectra that is reminiscent of a gamma alumina.

Nitrogen physisorption was utilised to determine the surface area of the prepared catalysts after the experimental testing under NGEE and VAM. The results are shown in Table 1. Chemisorption conducted with CO as the chemisorbed gas was used to determine the palladium dispersion of the prepared catalyst by pressure difference.

TABLE 1

Surface area and pore data for Pd/TS-1, Pd/Al₂O₃, Pd/ZSM-5 (23) and Pd/ZSM-5 (50) catalysts fresh and after VAM and NGEE conditions.

Catalyst	Cond.	Used catalyst			Fresh catalyst		
		SA [m²/g]		Disp. [%] [NGEE]	SA [m²/g]		Disp. [%]
		Ext.	Micro.		Ext.	Micro.	
Pd/Al₂O₃	VAM	203	11	11	205	12	30
	NGEE	177	-				
Pd/TS-1	VAM	100	264	17	132	214	29
	NGEE	105	253				
Pd/ZSM-5 (23)	VAM	82	237	14	45	215	14
	NGEE	78	190				
Pd/ZSM-5 (50)	VAM	142	184	18	82	264	18
	NGEE	118	192				

The combined t-plot surface area of the ZSM-5 catalysts decreased between the NGEE and VAM experiments. Showing levels of sintering or blocking of pores. Since less conversion occurred under NGEE for the ZSM-5 catalysts, it can be suggested that coke formation would be reduced. Therefore, sintering promoted by the higher water concentration must have taken place. The more acidic catalysts showed the highest reduction in the surface area, while the TS-1 catalyst showed a very slight reduction. Therefore, the supports facilitating water adsorption contribute to more sintering and coalescence under high water concentrations. In comparison, the more basic support provides a higher resistance level to potential sintering and coalesce. The high vapour concentrations could lead to the coalescence of palladium sites, which could lead to deactivation. This has been communicated by (Hansen *et al,* 2013) where palladium particles between the 3–10 nm range experience Ostwald ripening and particle migration and coalescence. It was noted by (Hansen *et al,* 2013) that the smaller the particles, the faster the sintering through Ostwald ripening, so the TS-1 catalyst would experience the greatest amount of sintering.

Since the TS-1 catalyst did deactivate significantly under NGEE, the support structure should not play a role in deactivation. Instead, deactivation of the active metal site is the reason for the loss of activity. For example, it has been shown by researchers (Gholami *et al,* 2015; Gélin *et al,* 2003; Persson *et al,* 2007) that palladium oxide, a Lewis acid, could form PdOH under high water concentrations. However it has been communicated by (Schwartz *et al,* 2012) that reduced oxygen mobility could be the primary method for deactivation, as water on the support lowers the overall oxygen transfer between the support structure and the active site. The deactivation of ZSM-5 and alumina catalysts could have occurred through a mixture of sintering, formation of PdOH species and lower oxygen mobility.

Dispersion of Pd strongly affects the methane combustion rate So, catalyst preparation and forming dispersed palladium oxide clusters on the support play a significant role in catalytic activity. It has been suggested by (Hosseiniamoli *et al,* 2020) that he TS-1 support facilitates dispersion because of the pore structure. The higher activity of Pd/TS-1 in methane combustion under both NGEE and VAM conditions is expected to be due to higher Pd dispersion. However, the NGEE experiment shows that the rate of deactivation of Pd/TS-1 is significantly reduced compared to the other catalysts. Therefore, the TS-1 support disperses the active site and minimises water interaction.

A scanning electron microscope was utilised to understand the morphology differences and the palladium particle sizes amongst the four supports; the resulting micrographs are shown in Figure 4. Energy dispersive X-ray spectroscopy was used to understand the composition of the catalysts. Several particles were chosen for this analysis, with the average mass and atomic composition shown in Table 2.

FIG 4 – SEM micrographs of prepared catalysts. A – Pd/ZSM-5 (SiO_2/Al_2O_3 = 50),
B – Pd/ZSM-5 (SiO_2/Al_2O_3 = 23), C – Pd/Al_2O_3, D – Pd/TS-1.
Purple – Zoomed out varied magnifications, Green – Zoomed in to 60 000×.

TABLE 2

EDX analysis of Pd/ZSM-5 (23 and 50), Pd/Al$_2$O$_3$ and Pd/TS-1.

| Catalyst | | Element | | | | | Total | SiO$_2$/ (Al$_2$O$_3$ or Ti) |
		O$_2$	Si	Al	Ti	Pd		
Pd/ZSM-5 (23)	Mass %	57.8	38.1	3.2	0	0.9	100	
	Atom %	70.9	26.6	2.3	0	0.2	100	23
Pd/ZSM-5 (50)	Mass %	56.6	39.9	1.4	0	2.1	100	
	Atom %	70.3	28.2	1	0	0.4	100	56
Pd/Al$_2$O$_3$	Mass %	55.4	0	43.2	0	1.1	99.7	
	Atom %	68.1	0	31.5	0	0.2	99.7	0
Pd/TS-1	Mass %	59	37.8	0	1.8	1.3	100	
	Atom %	72.4	26.6	0	0.8	0.2	100	35

The palladium loadings were reasonably close to the pre-determined loading of 1.2 per cent, with the Pd/ZSM-5 (50) catalyst having a higher loading. Since only three particles were chosen for the EDX analysis, which only goes approximately 1–2 microns deep, this doesn't represent the entire bulk catalyst composition but a surface enrichment., The SiO$_2$/Al$_2$O$_3$ ratios are within error for the ZSM-5 catalysts. The Si/Ti ratio of TS-1 was determined to be approximately 35. Which is lower than the values communicated by (Serrano *et al*, 2007; Moliner and Corma, 2014), reported to be approximately 43–50. The palladium on alumina mass composition had some impurities predicted to be sodium.

The ZSM-5 structures seem to have more of a square platelet-like structure. The ZSM-5 support with a SiO$_2$/Al$_2$O$_3$ = 23 has a larger platelet than the ZSM-5 with a SiO$_2$/Al$_2$O$_3$ = 50, which leads to a smaller external surface area but a larger pore volume, as seen in Table 1. The TS-1 structure has a similar ordered pattern to ZSM-5 but with a more circular shape. The alumina support has a rough morphology, lacking darker microporous regions. The small number of bright spots on the edges of all the support structures means that the palladium particles are too small to be detected by the backscatter electrons, which indicates a highly dispersed catalyst.

Activation energy experiments were conducted on Pd/TS-1, Pd/Al$_2$O$_3$ and Pd/ZSM-5 (23 and 50) under VAM and NGEE conditions. The results are shown in Table 3, and the Arrhenius plots for VAM and NGEE are depicted in Figure 5. The ZSM-5 based catalysts were not analysed through activation energy measurements as the conversion was too low under differential conditions.

TABLE 3

Activation energies and associated temperature range for Pd-based catalysts under VAM and NGEE conditions.

| Catalyst | | VAM | | NGEE | |
		Activation Energy (kJ/mol)	Temperature range (k)	Activation Energy (kJ/mol)	Temperature range (k)
Pd/TS-1		144 ± 1	443–455	215 ± 4	457–481
Pd/Al$_2$O$_3$		114 ± 4	443–467	130 ± 4	443–493
Pd/ZSM-5	(23)	196 ± 3	443–467	-	-
	(50)	164 ± 2	443–467	-	-

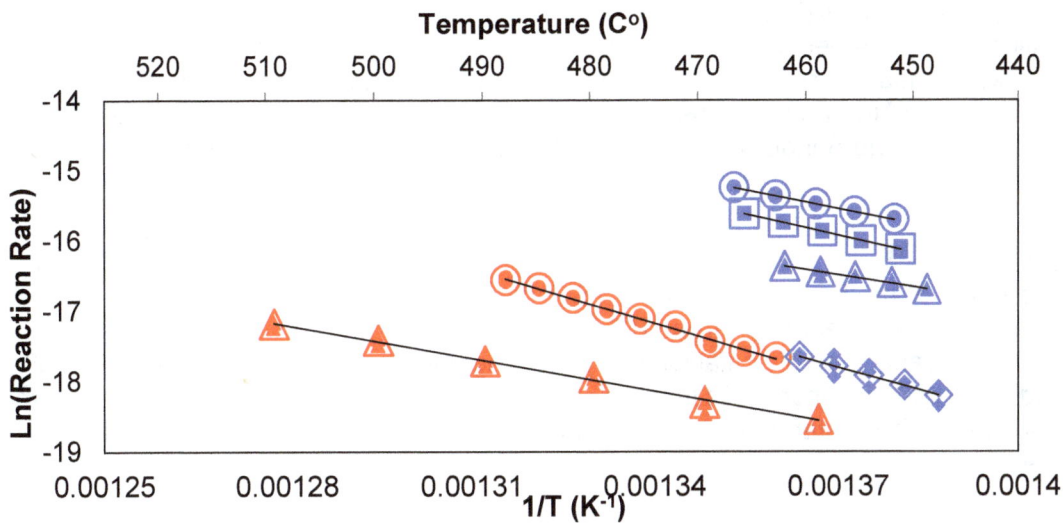

FIG 5 – Logarithm of the reaction rate versus the reciprocal of temperature for palladium-based catalysts under VAM and NGEE conditions. VAM – Blue and NGEE – Red. Average (no fill) and Raw data (fill). Pd/TS-1 – ●, Pd/Al$_2$O$_3$ – ▲, Pd/ZSM-5 (23) – ◆ and Pd/ZSM-5 (50) – ■.

Pd/ZSM-5 (23) had the highest activation energy under VAM conditions, with the next being Pd/ZSM-5 (50). Pd/TS-1 had a higher activation energy for both VAM and NGEE compared to Pd/Al$_2$O$_3$. So, the activation energy did not play a direct role in indicating the catalyst's performance. However, the Pd/TS-1 had a higher dispersion than the palladium on the alumina catalyst, which directly increased the reaction rate of methane under NGEE and VAM at the same temperature. The turnover frequency was also higher at these temperatures, which illustrates the higher catalyst activity. This could be why there is higher catalytic stability under VAM and NGEE conditions, regardless of the activation energy.

Effect of oxygen

In NGEE, the oxygen gas concentration, being 8 per cent, is less than the water concentration and much less than oxygen gas in VAM conditions. Whether oxygen gas concentration affects the catalytic stability between VAM and NGEE is not fully realised. So, palladium on alumina catalyst was put under two NGEE feeds with different oxygen concentrations of 8 per cent and 20 per cent, as shown in Figure 6. The water and methane concentrations were set at 10 per cent and 0.4 per cent, respectively, with a nitrogen balance. The experiment was run at 450°C, with 100 mg of catalyst used for both experiments.

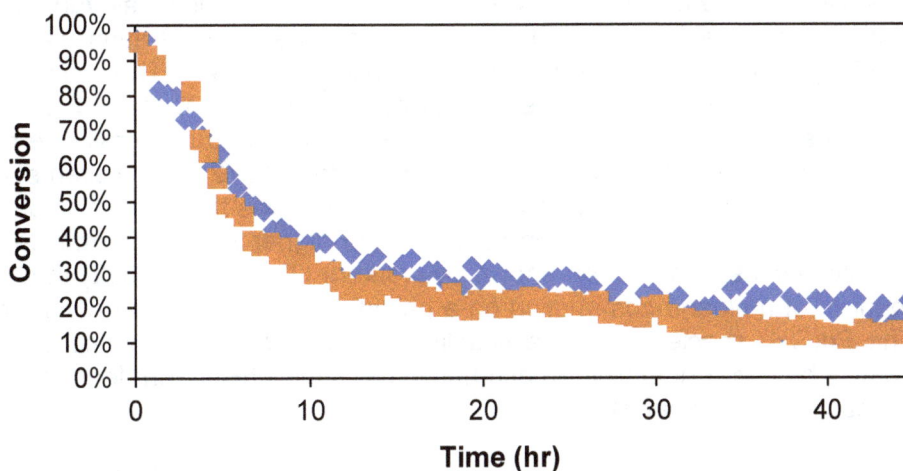

FIG 6 – Methane catalyst combustion of a Pd/Al$_2$O$_3$ catalyst under varied oxygen concentrations in NGEE conditions. Concentrations; H$_2$O – 10 per cent, CH$_4$ – 0.4 per cent, N$_2$-bal. Oxygen concentrations: 20 per cent – ◆, 8 per cent – ■.

The high O_2 gas concentrations do not significantly affect the Pd/Al_2O_3 catalyst. The methane concentration is only 0.4 per cent, and the oxygen gas concentration is well in excess at 8 per cent. Therefore, oxygen gas has reached its maximum effect regarding methane oxidation on a palladium-based catalyst under NGEE conditions. This has been reported by (Monteiro *et al*, 2001; van Giezen *et al*, 1999; Rudham and Sanders, 1972) where the reaction order of oxygen in methane combustion is approximately 0.

Effect of methane

To test the effect of methane, a palladium TS-1 catalyst was put under NGEE conditions with an increasing methane concentration from 1000 to 4000 ppm, with 1000 ppm increases, as shown in Figure 7. The conversion of this catalyst was plotted against time, and the concentration of methane combusted to illustrate the increases in reaction rate. The experiment was run at 450°C, with 100 mg of catalyst used. The water and oxygen gas concentrations were 10 per cent and 8 per cent, respectively.

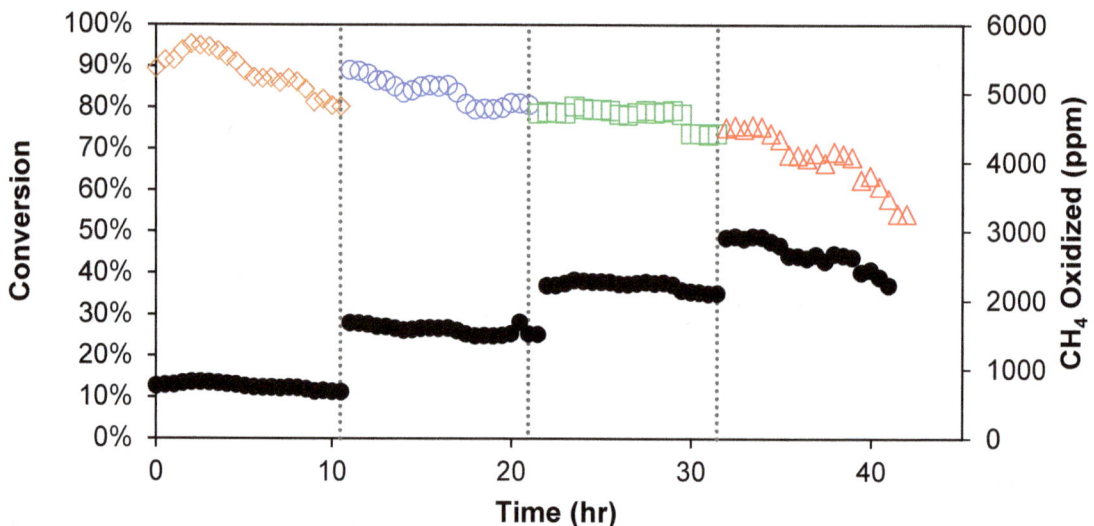

FIG 7 – Catalytic combustion of methane on a Pd/TS-1 catalyst with varied methane concentrations. Methane Oxidised – (Fill), Methane Conversion – (No Fill). Methane concentrations: ◇ – 1000 ppm, ○ – 2000 ppm, □ – 3000 ppm, △ – 4000 ppm.

Under 1000 ppm of methane, the maximum conversion was gradually reached, which could be due to a reduced exotherm. The catalyst is seen to deactivate slightly at this concentration, but the activity reduction of sites was affected only slightly when methane concentration was varied. The methane conversion showed a noticeable but slight increase when the concentration was doubled from 1000 ppm to 2000 ppm. The turnover frequency increased marginally. This was repeated in the following two concentration increases. The conversion remained unchanged, but the turnover frequency showed a significant jump. This indicates an increase in the overall rate and a positive reaction order in methane. Researchers (Rudham and Sanders, 1972; van Giezen *et al*, 1999) also communicated this relationship. At 4000 ppm, the activity loss was the largest among all methane concentrations. Higher methane combustion could lead to higher Pd(OH) formation and a higher reduction of the active sites. It has been shown by workers (Tang *et al*, 2022) that the Mars Van Krevelen mechanism of methane combustion on a palladium catalyst has Pd(OH) formed through combustion. The Pd(OH) species could be stabilised under high water concentrations, and the *catalyst* surface may be deficient in oxygen and rich in methane This in turn leads to carbonaceous deposits which deactivate the catalyst.

NGEE versus VAM

To further test the effect of NGEE compared to VAM, specifically on a Pd/TS-1 catalyst, two cyclic experiments were performed that switched the conditions from NGEE and VAM after a day of each condition, as shown in Figure 8. One experiment started with NGEE, with its respective catalyst A;

the other experiment started with VAM, with its catalyst, catalyst B. Both experiments were run at 450°C, with 100 mg of catalyst used.

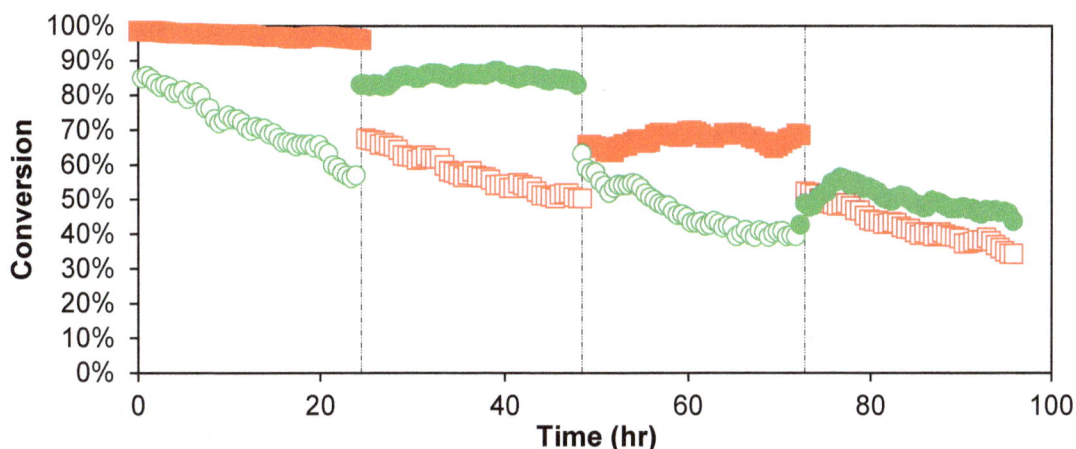

FIG 8 – Pd/TS-1 Catalyst stability under NGEE and VAM conditions. NGEE conditions; H_2O – 10 per cent, O_2 – 8 per cent, CH_4 – 0.4 per cent. VAM conditions; H_2O – 1.5 per cent, O_2 – 20 per cent, CH_4 – 0.7 per cent. Catalyst A; VAM – ●, NGEE – ○. Catalyst B; NGEE – □, VAM – ■.

The catalysts behaved as expected during the first 24 hrs under each condition, behaving like what was shown in Figures 1 and 2. With catalyst B, there is no sign of deactivation under VAM conditions. Still, when conditions were switched, the conversion started noticeably lower than the conversion of catalyst A under NGEE conditions. Therefore, under VAM conditions, the number of active sites was decreasing, but at such a low rate that a conversion of 100 per cent was still achieved. With catalyst A, the first switch from NGEE to VAM showed a rapid increase in conversion. The sudden shifts experienced by both catalysts could indicate an equilibrium change caused by the higher water concentration. Either an increased liberation time of the hydroxide groups or decreased oxygen mobility from the support saturation. After each change in conditions, the catalysts behaved similarly, having a sudden change in conversion, but the respective conversion reduces. This is especially true for the last switch for catalyst A, where the catalyst conversion increases marginally from the NGEE conditions. This could be due to the conversion being so low that an equilibrium shift due to changes in water concentration makes little difference. It should be highlighted that after the conditions switched from NGEE to VAM, the conversion did not return to the original conversions in VAM. This suggests some permanent deactivation, which was shown in Table 1, reinforcing the collapse of palladium sites through the formation of Ostwald ripening. Which was also communicated by (Hansen *et al*, 2013).

CONCLUSIONS

Catalysts showed rapid deactivation and catalytic instability under NGEE conditions, while some deactivation was observed under VAM conditions. The Pd/TS-1 catalysts performed the best under both conditions. The increase in Lewis's acidity was the primary reason for this deactivation between a palladium catalyst of different supports, with the removal of alumina from ZSM-5 increasing stability. Lewis's acidity was found to cause sintering and carbon deposition under high water conditions. A higher oxygen gas concentration has no impact on the activity of a palladium alumina catalyst in oxidation of methane due to oxygen gas being well in excess for both conditions. The activity of a Pd/TS-1 increases with a rise in methane concentration. However, this also results in forming more Pd(OH) and carbonaceous species, which lead to deactivation. The deactivation rate was observed to be the highest at 4000 ppm, which could be due to the water inhibition effect. An equilibrium relationship was shown with water when the catalyst conditions were switched between NGEE and VAM.

REFERENCES

Cho, H M and He, B-Q, 2007. Spark ignition natural gas engines—A review, *Energy Conversion and Management,* 48:608–618.

Da, P, Tao, L, Sun, K, Golston, L M, Miller, D J, Zhu, T, Qin, Y, Zhang, Y, Mauzerall, D L and Zondlo, M A, 2020. Methane emissions from natural gas vehicles in China, *Nature Communications,* 11:4588.

Dai, Q, Zhu, Q, Lou, Y and Wang, X, 2018. Role of Brønsted acid site during catalytic combustion of methane over PdO/ZSM-5: Dominant or negligible?, *Journal of Catalysis,* 357:29–40.

Gélin, P, Urfels, L, Primet, M and Tena, E, 2003. Complete oxidation of methane at low temperature over Pt and Pd catalysts for the abatement of lean-burn natural gas fuelled vehicles emissions: influence of water and sulphur containing compounds, *Catalysis Today,* 83:45–57.

Gholami, R, Alyani, M and Smith, K J, 2015. Deactivation of Pd Catalysts by Water during Low Temperature Methane Oxidation Relevant to Natural Gas Vehicle Converters, *Catalysts,* 5.

Hansen, T W, Delariva, A T, Challa, S R and Datye, A K, 2013. Sintering of Catalytic Nanoparticles: Particle Migration or Ostwald Ripening?, *Accounts of Chemical Research,* 46:1720–1730.

Hosseiniamoli, H, Setiawan, A, Adesina, A A, Kennedy, E M and Stockenhuber, M, 2020. The stability of Pd/TS-1 and Pd/silicalite-1 for catalytic oxidation of methane – understanding the role of titanium, *Catalysis Science and Technology,* 10:1193–1204.

Lampert, J K, Kazi, M S and Farrauto, R J, 1997. Palladium catalyst performance for methane emissions abatement from lean burn natural gas vehicles, *Applied Catalysis B: Environmental,* 14:211–223.

Lott, P and Deutschmann, O, 2021. Lean-Burn Natural Gas Engines: Challenges and Concepts for an Efficient Exhaust Gas Aftertreatment System, *Emission Control Science and Technology,* 7:1–6.

Moliner, M and Corma, A, 2014. Advances in the synthesis of titanosilicates: From the medium pore TS-1 zeolite to highly-accessible ordered materials, *Microporous and Mesoporous Materials,* 189:31–40.

Monteiro, R S, Zemlyanov, D, Storey, J M and Ribeiro, F H, 2001. Turnover Rate and Reaction Orders for the Complete Oxidation of Methane on a Palladium Foil in Excess Dioxygen, *Journal of Catalysis,* 199:291–301.

Oh, S H, Mitchell, P J and Siewert, R M, 1992. Methane Oxidation over Noble Metal Catalysts as Related to Controlling Natural Gas Vehicle Exhaust Emissions, *Catalytic Control of Air Pollution,* American Chemical Society.

Okumura, K, Matsumoto, S, Nishiaki, N and Niwa, M, 2003. Support effect of zeolite on the methane combustion activity of palladium, *Applied Catalysis B: Environmental,* 40:151–159.

Persson, K, Pfefferle, L D, Schwartz, W, Ersson, A and Järås, S G, 2007. Stability of palladium-based catalysts during catalytic combustion of methane: The influence of water, *Applied Catalysis B: Environmental,* 74:242–250.

Petrov, A W, Ferri, D, Krumeich, F, Nachtegaal, M, Van Bokhoven, J A and Kröcher, O, 2018. Stable complete methane oxidation over palladium-based zeolite catalysts, *Nature Communications,* 9:2545.

Rudham, R and Sanders, M K, 1972. The catalytic properties of zeolite X containing transition metal ions: Part 2—Methane oxidation, *Journal of Catalysis,* 27:287–292.

Schwartz, W R, Ciuparu, D and Pfefferle, L D, 2012. Combustion of Methane over Palladium-Based Catalysts: Catalytic Deactivation and Role of the Support, *The Journal of Physical Chemistry C,* 116:8587–8593.

Serrano, D P, Calleja, G, Botas, J A and Gutierrez, F J, 2007. Characterization of adsorptive and hydrophobic properties of silicalite-1, ZSM-5, TS-1 and Beta zeolites by TPD techniques, *Separation and Purification Technology,* 54:1–9.

Setiawan, A J F, Kennedy, E M, Dlugogorski, B Z and Stockenhuber, M, 2014. Catalytic combustion of ventilation air methane (VAM) – long-term catalyst stability in the presence of water vapour and mine dust, *Catalysts Science Technology,* 4:1793–1802.

Setiawan, A, Friggieri, J, Hosseiniamoli, H, Kennedy, E M, Dlugogorski, B Z, Adesina, A A and Stockenhuber, M, 2016. Towards understanding the improved stability of palladium supported on TS-1 for catalytic combustion, *Physical Chemistry Chemical Physics,* 18:10528–10537.

Tang, Z, Zhang, T, Luo, D, Wang, Y, Hu, Z and Yang, R T, 2022. Catalytic Combustion of Methane: From Mechanism and Materials Properties to Catalytic Performance, *ACS Catalysis,* 12:13457–13474.

van Giezen, J C, Van Den Berg, F R, Kleinen, J L, Van Dillen, A J and Geus, J W, 1999. The effect of water on the activity of supported palladium catalysts in the catalytic combustion of methane, *Catalysis Today,* 47:287–293.

Wu, W, Tran, D T, Cheng, S, Zhang, Y, Li, N, Chen, H, Chin, Y-H, Yao, L and Liu, D, 2021. Local environment and catalytic property of External Lewis acid sites in hierarchical lamellar titanium Silicalite-1 zeolites, *Microporous and Mesoporous Materials,* 311:110710.

VAMOX® full scale RTO technology for the abatement of VAM emissions

G Drouin[1], D Kay[2] and J-S D'Amours-Cyr[3]

1. President, Biothermica Technologies Inc., Montreal, QC, Canada.
   Email: guy.drouin@biothermica.com
2. Vice-president, Biothermica Technologies Inc., Montreal, QC, Canada.
   Email: dominique.kay@biothermica.com
3. Mechanical Projects Manager, Biothermica Technologies Inc., Montreal, QC, Canada.
   Email: jean-simon.damours-cyr@biothermica.com

ABSTRACT

Global coalmine methane (CMM) emissions from coalmines amount between 1.2 and 1.5 $GtCO_2e$ each year depending on the reporting agency. Ventilation Air Methane (VAM) represents some 65 per cent of the total CMM emissions. Until now, its dilute methane concentration varying from 0.1 per cent up to more than 1.5 per cent in some very gassy US coalmines has proven to be a barrier to its recovery and use, whereas low volume high concentration gob and drainage gas have been already used for power generation in USA, Australia and China for example. Destroying VAM results in a net reduction of greenhouse gas (GHG) emissions, which activity can generate in USA bankable carbon credits following the adoption by California Air Resource Board (CARB) of the coalmine methane protocol in 2014 and by Quebec province (Canada) in 2016. Biothermica has commissioned in 2009 a first 51 000 m^3/hr (14 m^3/s) VAMOX® system on the VAM bleeder shaft 4–9 at Jim Walter Resources (JWR) underground coalmine in Brookwood, Alabama, which was in operation until 2013. Some 80 766 carbon credits were issued under the Mine Methane Capture (MMC) protocol of the California Air Resources Board (CARB) and sold in the carbon cap-and-trade market. It has been followed by the installation and operation of a first 258 000 m^3/hr (72 m^3/s) and a second full scale 306 000 m^3/hr (85 m^3/s) VAMOX® systems respectively commissioned in July 2022 and in April 2024 on two different vent shafts at a Virginia (USA) met coalmine. The primary goal of these two recent projects was to demonstrate that a large scale VAMOX® system can safely, efficiently, and reliably abate VAM emissions while generating a constant revenue stream from the sale of carbon credits. These projects are achieving combined net emission reductions of approximately 500 000 tCO_2e per annum. The VAMOX® system uses the principle of regenerative thermal oxidation to convert methane into harmless by-products and can safely operate at a methane concentration between 0.2 per cent and 1.2 per cent, dilution air being introduced in the unit to maintain methane concentration at a maximum of 1.2 per cent if VAM concentration is higher than 1.2 per cent.

INTRODUCTION

VAM emissions from underground coalmines could now be thought of as an environmental asset. The developed VAMOX® full scale system can oxidise the dilute methane it contains and, by doing so, generate a revenue stream from the sale of resulting 'carbon credits' and/or energy. Biothermica has recently commissioned two such large scale ventilation air methane (VAM) oxidation projects at an active coalmine in USA (Virginia) which produce carbon credits for the voluntary carbon market. This paper provides information about these first-of-kind large scale projects, the lessons learned from the first 51 000 m^3/hr demo plant, the VAMOX® technology itself and critical considerations for evaluating the feasibility of such projects.

BACKGROUND

Methane released by coal mining operations, or coalmine methane (CMM), is a potent greenhouse gas (GHG) having a global warming potential of 28 over a 100-years period. Only a fraction of worldwide CMM emissions, which represent between 1.2 and 1.5 $GtCO_2e$ each year, is currently being recovered (Olivier, 2022; Tate, 2022; International Energy Agency (IEA), 2023; 2024; Gould, 2024). Some 60 per cent of CMM emissions originate from mine ventilation systems as VAM and is virtually all released to the atmosphere (IEA, 2022). Based on this assumption, VAM emissions amount to about 0.8 to 1.0 $GtCO_2e$ of GHG emissions each year. This represents 2.5 per cent of all man induced GHG emissions. China alone contributes to about 50 per cent of CMM emissions, making this country by far the most important emitter of VAM, followed by Russia, Indonesia, India, USA, Australia and Poland (IEA, 2023).

The concentration of methane in mine ventilation exhaust air in USA is typically less than 1 per cent. Almost all these shafts are bleeder shafts where concentration can legally reach up to 2 per cent methane as per Part 75 of the US mandatory safety standards for underground coalmine sub part 75.323 – Actions for excessive methane, regulated under the authority of the Mine and Safety Health Administration (MSHA). Until now, this extremely dilute content has proven to be a technical barrier to its recovery and use, especially for concentration below 0.25 per cent. At the same time, the world's leading nations now recognise the urgent need to reduce GHG emissions. The Kyoto Protocol set the stage for the exchange and trading of GHG emission offsets, including coalmine methane abatement offsets, followed by the Paris agreement and the Global Methane Pledge approved by some 155 countries (Global Methane Pledge, 2011). Other mandatory emission reduction schemes have since emerged around the world, such as the Safeguard Mechanism Act in Australia and the European Union law of November 15, 2023 to curb methane emissions in their member countries. A voluntary carbon market is also rapidly gaining strength, such in USA and in India, pushed by the deliberate commitment of corporations. The value of this global carbon market was estimated at US$948 billion in 2023 and has been rising sharply over recent years (Global Market Insights, 2023).

The selling of carbon credits now could ensure the profitability of GHG emission reduction projects such as VAM abatement. By securing a revenue stream, carbon credits provide a mean of financing projects that reduce GHG emissions, thereby mitigating climate change. A typical 680 000 m^3/hr coalmine bleeder vent shaft could produce some 500 000 carbon credits at VAM concentration of 0.6 per cent at a typical cost of US$12–15/t$CO_2$e, which is relatively inexpensive compared to other climate mitigation technologies.

VAMOX® TECHNOLOGY

Origin

The BIOTOX® regenerative thermal oxidiser (RTO), commercialised by Biothermica since 1991, has been the base of the development of the VAMOX® RTO system. BIOTOX® technology has been adapted for various non-conventional industrial processes emissions and destroys numerous contaminants such as volatile organic compounds (VOC), condensable organic compounds (COC), total reduced sulfur (TRS) and polycyclic aromatic hydrocarbons (PAH). BIOTOX® technology is recognised as the industry's leading solution for non-traditional RTO applications such as pitch and tar fumes from anode preparation plant, asphalt shingles fumes, dioxin and furane emissions from magnesium smelter, among other. Over the past 20 years, Biothermica has studied the specificities of coal mining ventilation and invested in research and development. As a result, the VAMOX® is specifically designed to maximise the performance and profitability of VAM oxidation projects.

Principle of operation and general description

The VAMOX® converts methane (CH_4) into carbon dioxide (CO_2) and water vapor (H_2O) using the well proven principle of regenerative thermal oxidation (RTO). RTO is based on the cyclic reversal of the airflow-through multiple vessels filled with heat absorbing media to minimise heat losses during the oxidation process. Unlike traditional RTO technologies that process compounds at low concentrations and require the use of an auxiliary heat source to sustain the reaction, the oxidation process on the VAMOX® is flameless and does not use any catalyst. It relies on sufficient CH_4 concentration to guarantee an autothermic oxidation reaction throughout VAM processing. The general arrangement of a VAMOX® system along with its main components is presented in Figure 1. The open end of the inlet duct is directly linked to the mine ventilation shaft's diffuser to capture exhaust air, as illustrated in Figure 2. Physical connection between the duct and the mine ventilation system has been designed to ensure there is no impact on mining ventilation. The VAMOX® does not affect the performance of the mine ventilation whatsoever. Before mine exhaust air is allowed into the system, a gas burner pre-heats the heat recovery vessels filled with specifically designed ceramic media while air is being induced by the centrifugal blower through the fresh air intake. When pre-heat is completed and the vessels are warm enough, the fresh air intake closes, and the cut-off damper opens. Mine exhaust air is then forced into one of the vessels by the blower which is sized to counteract the pressure drop through the system. As it passes through the initially warm vessel, the temperature of the air stream increases to promote exothermic oxidation reactions which release heat into the system. The air stream then passes through the oxidation chamber that allows for the complete oxidation of the methane. Before being evacuated to the atmosphere via the stack, the air stream passes through the other vessel which recovers most of the heat it carries. Over time the 'inlet' vessel cools down as it transfers its heat to the cold incoming air and the 'outlet' vessel warms up from capturing the heat of the processed air. To reverse this tendency, the flow controlling dampers are cycled to invert direction of the air flow-through the system. Incoming air now passes through the warmest vessel first and the heat from the oxidation reaction is captured by the coolest vessel. The dampers are cycled this way every few mins so that the system achieves a dynamic thermal equilibrium. Each vessel is therefore alternatively used to heat the incoming air or to recover heat from it before being released to the atmosphere.

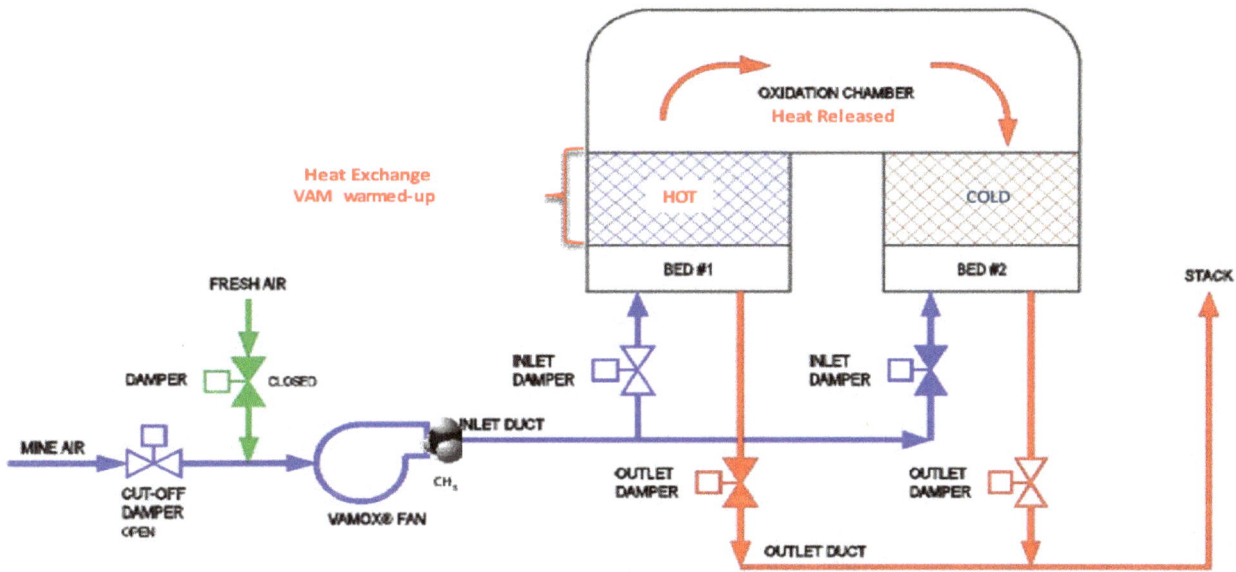

FIG 1 – RTO operating principles.

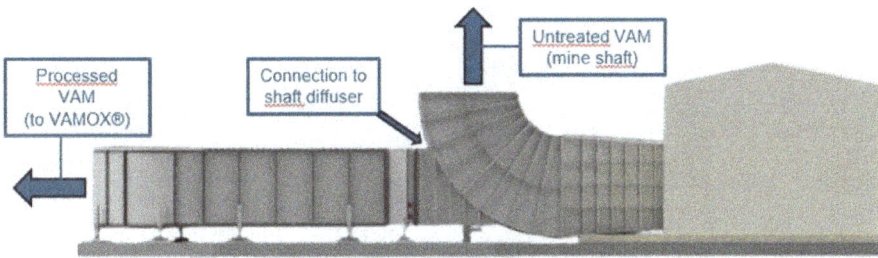

FIG 2 – VAMOX® connection point to typical 600 000 m³/hr US bleeder ventilation shaft.

A specifically designed heat management bypass loop allows the VAMOX® system to operate over a wide range of methane concentrations without overheating. A variable opening damper modulates the flow of air that passes through this loop as required to keep an optimal temperature inside the oxidation chamber. Based on the input of multiple sensors installed throughout the system, a programmable logic controller (PLC) fully automates the operation of the VAMOX®. Methane concentration in incoming mine air is continuously monitored by a gas analyser located upstream of the system. Whenever the methane concentration reaches a pre-determined threshold, the cut-off damper closes and the VAMOX® system shuts down automatically. As a safety precaution, by design no flammable gas mixture can get passed the cut-off damper and into the system. The same shutdown sequence is activated when the PLC detects any abnormal condition. The elaborate heat recovery of the VAMOX® system allows it to reach a very high heat recovery efficiency. This explains why it can oxidise methane at extremely low concentrations without the need for supplemental energy other than the electricity needed to run the centrifugal blower.

The system was designed to exceed applicable safety regulations. All personnel associated with the operation of the VAMOX® systems undergoes training so that they become familiar with the design, operation and safety features of the systems. Safety and control requirements include the detection of CH_4 levels upstream of the cut-off damper. Two CH_4 detectors installed at the vicinity of the tie-in point to the mine ventilation shaft's diffuser ensure, with fast response time and redundancy, that the gas composition stays well below explosive limits. The detector readings are continuously sent to the system safety PLC and displayed in the mine control room while the VAMOX® unit is operating. The detectors are included (in series) in a hardwired connection loop, which also includes the cut-off damper actuator and the fresh air dilution damper actuator. When this hardwired connection loop is opened, the cut-off damper closes, the fresh air damper opens, and an emergency system shutdown sequence is activated.

For control and carbon credits validation purposes, the CH_4 concentration is measured upstream of the oxidation chamber and downstream at the stack. The instruments dedicated for the latter are tuneable diode laser spectrometers (TDLS). Their accuracy complies with CARB's MMC protocol requiring ±5 per cent applied to the reading relative to the project's reference concentration value (CARB, 2014).

VAMOX® FIRST (DEMO) PROJECT – 2009–2016

The main objectives of the first demo VAMOX® project was to: (i) demonstrate that the VAMOX® system can safely, efficiently and reliably destroy VAM at an active coalmine, (ii) demonstrate that this activity can generate carbon credits and ultimately a revenue stream, and (iii) gather technical, operational and financial data for design validation and optimisation as well as scaling up future VAM system mine ventilation flow from a vent shaft typically in the range of 594 000 to 1 700 000 m³/hr.

A 51 000 m³/hr VAMOX® system was financed, built and operated by Biothermica at JWR mine no. 4 near Brookwood, Alabama (USA) from January 2009 to February 2013 (Figure 3) (US EPA, 2019). JWR supervised the operation of the system on a day-to-day basis in close cooperation with Biothermica. This was the first VAM oxidation project ever to be approved by the U.S. Department of Labor's Mine Safety and Health Administration (MSHA) at an active coalmine in North America. The project was also recognised by the U.S. EPA's Coalbed Methane Outreach Program (CMOP) and the Global Methane Initiative (GMI, 2011; US EPA, 2017).

FIG 3 – First VAMOX® demo unit of 51 000 m³/hr capacity (2009).

Based on Biothermica's experience as an integrated developer of GHG emission reduction projects, this project adhered to the strictest applicable carbon standards to maximise revenues from the resulting carbon credits. Some 80 766 carbon credits were generated over 27 000 hrs of operation (CARB, 2016). Initially validated under Climate Action reserve (CAR), the carbon credits were registered under the Mine Methane Capture (MMC) protocol adopted in July 2014 by the California Air Resource Board (CARB) and sold on the Quebec-California cap-and-trade carbon compliance market. Because of the R&D nature of the project, the installed system processes roughly 10 per cent of the total available flow of bleeder shaft no. 4–9. The VAMOX® unit was independent and had no impact on the operation and performance of the mine ventilation system. The average methane concentration in the mine exhaust air was 0.87 per cent during the active mining phase (see Figure 4). Accordingly, the system was designed to operate at concentrations between 0.3 per cent and 1.2 per cent. If the concentration rose above 1.2 per cent, VAM was diluted with fresh air at the system's inlet to keep it at that level. This high concentration limit of 1.2 per cent at the VAMOX's inlet aims to remain within heat resistance limits of standard materials and therefore provide a cost-effective solution.

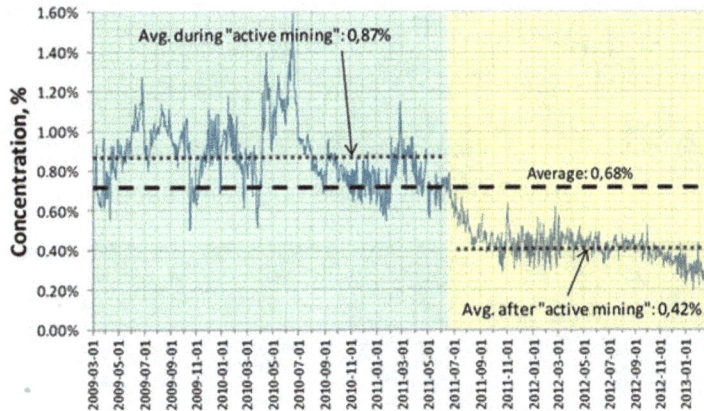

FIG 4 – VAM concentration monitored during VAMOX® Project #1 (2009–2013).

Mine air was routed to the system via an open-ended inlet duct. A 93 kW capacity electrical motor drove a centrifugal blower used to force the air through the system to counteract its pressure drop. The nominal power consumption of the motor was about 100 kW. All dampers were powered by compressed air as supplied by a dedicated on-site compressor driven by an electrical motor. A fast-response gas analyser monitored the methane concentration at both the inlet and the outlet of the system to assess its destruction performance. The inlet concentration reading was also used as the trigger of a safety system to prevent a potentially flammable gas mixture from entering the system. The gas burner required for pre- heating the system ran on propane supplied by an on-site tank and an electrically powered vaporiser. Cold starting the VAMOX® system required that the burner operates during approximately four hrs. As a safety precaution, it was impossible for the gas burner to operate when mine air was being induced into the system. A small mobile office building was located in the immediate vicinity of the system to house and shelter all electrical equipment, switchgear, control instruments and the gas analyser. Since there was no significant need for heat within a reasonable distance from the project site, no heat recovery system was included in the original design. The main demo project outcome was the development of a process simulator which has become a tool used to guide large scales design, develop control strategies and predict system performance.

FIRST AND SECOND FULL-SCALE COMMERCIAL 258 000 M³/HR AND 306 000 M³/HR VAMOX® SYSTEMS (2020–2024)

Following the technical success of the demo VAMOX® system, a first full scale unit of 258 000 m³/hr capacity was commissioned in July 2022 on a 594 000 m³/hr bleeder shaft at Coronado Resources metallurgical underground coalmine in Virginia followed by a second 306 000 m³/hr unit on another shaft in USA. This underground mine uses longwall mining method. Typical longwall characteristics of this underground mine include panel width of approximately 275 m, with lengths ranging from 1675 to 3350 m. The ventilation air that drains out through the bleeder systems, to which the VAMOX® systems are connected, releases 12 to 35 m³ of CH_4 per square metre of mining area (Karacan *et al*, 2024).

Both VAMOX® systems can operate over a wide range of methane concentration from 0.3 per cent up to 1.2 per cent with a destruction efficiency that can reach up to 99 per cent. Figures 5 and 6 are showing photos of both VAMOX® and Table 1 compare the three VAMOX® projects implemented by Biothermica since 2009.

FIG 5 – Third VAMOX® unit of 306 000 m³/hr capacity (2024).

FIG 6 – Second VAMOX® unit of 258 000 m³/hr capacity (2022).

TABLE 1

Overview of the three VAMOX® projects implemented by Biothermica since 2009.

	VAMOX #1 (2009–2013)	VAMOX #2 (2022 – …)	VAMOX #3 (2024 – …)
Concentration range at mine shaft [%CH4]	0.3–1.6	0.3–1.6	0.4–1.0
Flow capacity [am³/hr]*	51 000	258 000	306 000
Carbon credits rates (@0.85%CH4) (see Note 2)	50 000 credits/a	250 000 credits/a	300 000 credits/a
Fan capacity [kW] (see Note 1)	93	671	820
Footprint [m]	10 × 12	16 × 32	16 × 36

*at 20°C, 0.95 atm.

Note 1: Effective power consumption is lower than fan capacity and vary depending on VAM concentration.

Note 2: Carbon credits production rates may vary based on mine shaft methane concentration. Table 1 shows carbon credits rates according to a fixed average methane concentration of 0.85 per centCH4 (assumptions: global warming potential of CH4 of 25 tCO_2e/t CH4 and Emission factor of electricity consumed of 0.47 tCO_2/MWh for Virginia/Carolina subregion (CARB, 2014).

DESIGN AND PROJECT PLANNING CONSIDERATIONS

The main design selection that makes these full scale VAMOX® systems stand out from similar RTO technologies in VAM abatement applications is primarily its size. Bigger-sized units provide a higher capacity of treated VAM per occupied space than opting for multiple smaller-sized units. This is particularly interesting for mine ventilation shafts that exhaust significant quantities of VAM but do not allow for vast equipment footprint due to the limited available terrain at the diffuser's vicinity. However, increasing size means that an interruption of operation brings about greater profit losses.

Furthermore, methane concentration in mine exhaust air is the single most important parameter when evaluating the potential profitability of a VAM oxidation project. It can have a relatively small impact on the initial capital expenditure (CAPEX) but foremost, it dictates the revenues a VAMOX® system generates. For a system with a given air flow capacity and thus, a given capex, revenues double if the concentration doubles. The gas burner, that is only used to pre-heat the system, could run on drained CMM if it is readily available and if its characteristics meet the burner's specifications. By doing so, operational costs could be minimised, and additional carbon credits can potentially be claimed if the CMM source is originally vented or flared. Otherwise, a supply of natural gas or propane is required as the fuel source for the burner.

It can also be beneficial to stabilise the concentration of methane at the inlet of the VAMOX® system by blending in CMM. By doing so, the system's design can be optimised for a narrower or lower range of methane concentration and, as a result, both the initial capital expenditure and the operational costs are lowered. This is especially true when the minimum concentration threshold is decreased under 0.2 per cent. This alternative becomes even more attractive when the CMM induced in the VAMOX® system is not initially valourised since additional carbon credits can be claimed. Of course, this greatly increases the profitability of a VAM oxidation project.

The mine ventilation shaft's service life is critical as well in the evaluation of a project's profitability. Although a VAMOX® system can be moved relatively easily to another site if needed, the associated costs must be accounted in the financial analysis. The heat generated by the system can eventually be recovered as hot water or low-grade steam by means of a heat exchanger or as cold air by a heat pump. For this capability to be attractive, their needs to be a use for the heat or cold within a reasonable distance from the VAMOX® system. Otherwise, the cost of installing the heat/cold carrying pipe(s) becomes prohibitive. These costs are even more significant if a closed-loop system is used since a return pipe also needs to be installed. This is inevitably the case if a water source is unavailable near the VAMOX® system. If an open loop system is used, the capacity and thus the cost of the water treatment facility also becomes a key consideration.

Critical importance of VAM concentration distribution

In typical RTO applications, the level of accumulated energy in the system is regulated by a gas burner to ensure that temperature remains high enough to properly oxidise VOCs within the short residence time of about one second through the system. However, for the VAM abatement application, the use of a gas burner while VAM is being processed has not yet been approved in many jurisdictions including MSHA. As a result, the energy level in the system mainly depends on VAM concentration and flow rate at the inlet. Alternative control strategies must be developed to ensure that the system always remains within the safe and optimal operating range, ie avoiding overheating issues at high VAM concentration, and avoiding the loss of oxidation reactions at low VAM concentration as illustrated in Figure 7.

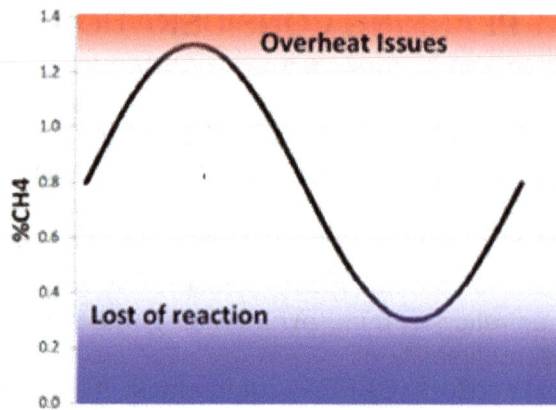

FIG 7 – Challenges of methane concentration variability.

Thus, in applications where the use of supplemental energy sources such as a gas burner is prohibited, the ability of the RTO to sustain oxidation reaction relies on the capacity of its porous ceramic media beds to exchange heat with the gas stream and store the energy released by the exothermic oxidation reactions. At low VAM concentration, less energy is released into the system so media beds of higher heat retention capacity are required to overcome the lack of energy and keep the temperature in the system sufficiently high to sustain oxidation reactions. The size (and cost) of the RTO is thus primarily dictated by the minimum VAM concentration at which the RTO must operate, ie the minimum self-sustain concentration limit. Figure 8 schematically illustrates the impact of this minimum self-sustained concentration limit on the size and cost of an RTO. Cost-effective solutions can easily be implemented for applications that are optimised to treat VAM with concentration above ~0.3 per cent, v/v. However, more complex and expensive solutions must be used to extend operating range below ~0.3 per cent, v/v.

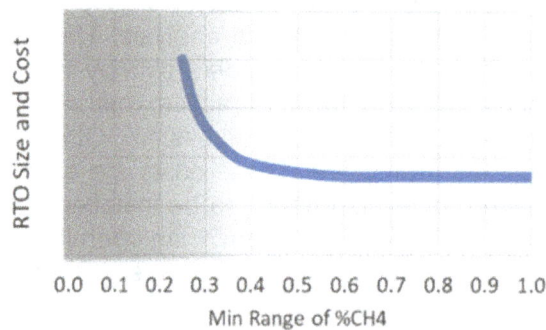

FIG 8 – Impact of minimum self-sustained concentration limit on RTO size and cost.

Increasing the depth of ceramic beds to increase the RTO's heat capacity also results in an increase of flow resistance through the system. Hence, adding ceramic media to a given RTO to extend its operating range to lower concentration will result in either: (1) an increase of electrical power consumption to overcome the additional pressure drop and maintain system's flow capacity, or (2) a loss of flow capacity as illustrated in Figure 9 if the RTO fan has reached full capacity.

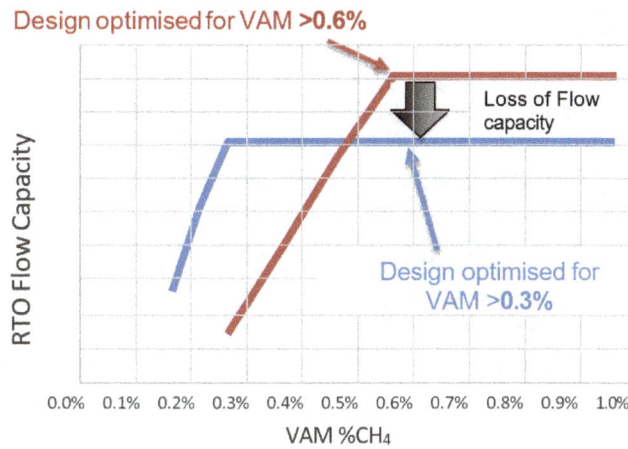

FIG 9 – Loss of flow capacity to extend RTO operating range at lower VAM concentration.

The concentration range of VAM to be treated has thus a huge impact on the design, cost and overall performance of the RTO. Monitoring and defining the expected distribution of methane concentration is therefore a critical project input that is required to design the optimal solution for a given ventilation shaft. Keeping in mind that the objective of VAM abatement projects is to abate as much methane as possible in the most cost-effective way, the RTO optimisation study shall aim to determine the optimum threshold concentration limit at above which the RTO will operate most of the time (eg more than 95 per cent of the time) and maximise its performance above that limit as illustrated. For most applications, it is not necessary to over-design the RTO (and sacrifice performance at nominal concentration) to cover concentration range with very low probability. It should be noted that when concentration drops below this optimum threshold limit, the RTO can still be kept in operation using proprietary control strategies, but at less optimum performance.

CONCLUSION

Biothermica has successfully commissioned two full-scale VAMOX® systems at an active coalmine in the USA. Both systems have the capacity to process VAM from 0.3 per cent up to 1.2 per cent with a destruction efficiency that can reach up to 99 per cent and are producing annual GHG reduction estimated at 500 000 tCO$_2$e. VAM represents about 10 to 15 per cent of all man-induced methane emissions around the world and almost none of it is being captured and utilised. The carbon market is rapidly gaining strength in parallel with regulatory obligation to abate methane. Across the globe, the selling of carbon credits now ensures the profitability of GHG emission reduction projects.

REFERENCE

California Air Resources Board (CARB), 2014. Compliance Offset Protocol Mine Methane Capture Projects [online]. Available from: <https://ww2.arb.ca.gov/sites/default/files/barcu/regact/2013/capandtrade13/ctmmcprotocol.pdf>

California Air Resources Board (CARB), 2016. CARB Offset Issuance [online]. Available from: <https://ww2.arb.ca.gov/sites/default/files/cap-and-trade/offsets/issuance/camm0095-a-d.pdf>

Global Market Insights, 2023. Carbon Credit Market Size – By Type (Voluntary, Compliance), By End Use (Agriculture, Carbon Capture and Storage, Chemical Process, Energy Efficiency, Industrial, Forestry and Landuse, Renewable Energy, Transportation, Waste Management) and Forecast, 2024–2032, GMI7048. Available from: <https://www.gminsights.com/industry-analysis/carbon-credit-market> [Accessed: 22 July 2024].

Global Methane Initiative, 2011. Coal Mine Methane: Reducing Emissions, Advancing Recovery and Use Opportunities [online]. Global Methane Pledge, Highlights from 2023 Global Methane Pledge Ministerial. Available from: <https://www.globalmethanepledge.org/news/highlights-2023-global-methane-pledge- ministerial> [Accessed: 22 July 2024].

Gould, T, 2024. A Global Call to Mobilize Methane Action, Global Methane Forum 2024, Geneva, Switzerland.

International Energy Agency (IEA), 2022. Coal mine methane abatement, proposed modelling approach, 5, table 2.

International Energy Agency (IEA), 2023. Global Methane Tracker 2023 – Analysis [online]. Available from: <https://www.iea.org/reports/global-methane-tracker-2023> [Accessed: 22 July 2024].

International Energy Agency (IEA), 2024. Global Methane Tracker 2024 – Analysis [online]. Available from: <https://www.iea.org/reports/global-methane-tracker-2024> [Accessed: 22 July 2024].

Karacan, C Ö, Irakulis-Loitxate, I, Field, R A and Warwick, P D, 2024. A comparison of methane emission fluxes quantified using PRISMA hyperspectral satellite data with on-site measurements at a US coal mine, Global Methane Forum 2024 Geneva, Switzerland: UN environment programme.

Olivier, J G J, 2022. Trends in Global CO_2 and Total Greenhouse Gas Emissions, PBL Netherlands Environmental Assessment Agency.

Tate, R D, 2022. Bigger than Oil or Gas? Sizing Up Coal Mine Methane, 2022 – Report, Global Energy Monitor.

US EPA, 2017. US Underground Coal Mine Ventilation Air Methane Exhaust Characterization 2011–2015, Colabed Methane Outreach Program, EPA 430-R-21–041.

US EPA, 2019. Ventilation Air Methane (VAM) Utilization Technologies, Methane Coalbed Outreach Program, EPA 430-F-19–023.

The safety of catalytic oxidation of fugitive lean methane emissions[*]

R Noon[1], P Kidd[2], M Drewery[3], E M Kennedy[4] and M Stockenhuber[5]

1. Research Assistant, University of Newcastle, Newcastle NSW 2308.
   Email: ryan.noon@uon.edu.au
2. Research Assistant, University of Newcastle, Newcastle NSW 2308.
   Email: patrick.kidd@uon.edu.au
3. Post Doctorate Research Associate, University of Newcastle, Newcastle NSW 2308.
   Email: matthew.drewery@newcastle.edu.au
4. Professor, University of Newcastle, Newcastle NSW 2308.
   Email: eric.kennedy@newcastle.edu.au
5. Professor, University of Newcastle, Newcastle NSW 2308.
   Email: michael.stockenhuber@newcastle.edu.au
* This paper has not been technically reviewed

ABSTRACT

Catalytic reactors have demonstrated high levels of methane conversion (>90 per cent) at temperatures significantly below the auto-ignition temperature (<500°C). However, challenges arise when dealing with additional impurities in the exhaust streams, notably elevated levels of water vapor. These impurities can lead to catalyst deactivation, necessitating the development of catalysts that are resistant to specific poisoning species.

While it has been shown that palladium-based catalysts are a suitable for methane oxidation, a significant study into the inherent safety of a catalytic system when oxidising ventilation air methane (VAM) has been identified as extremely important by industry.

The current study investigated the safety hazards present during abnormal methane excursions, characterised by short or long duration increases in methane concentration in the stream and/or increase in reactor temperature. Increased reactor temperature could be due to increased exotherm from the reaction under elevated methane concentrations or from the operator controller to increase conversion if gradual deactivation occurs. The most important dangers that present themselves from these abnormal reaction conditions are autoignition of the stream or the formation of carbon monoxide in the event of incomplete oxidation.

It was shown that even under elevated methane concentrations, the catalyst would not produce carbon monoxide unless the stoichiometric limit between methane and oxygen was reached (>10 per cent methane), showing not only that it is very unlikely that carbon monoxide is an intermediate product in the catalyst mechanism, but also that this would be unlikely to be produced in the exhaust gas of a catalytic mitigation system. The catalyst showed excellent ability to oxidise very lean VAM streams, with above 90 per cent conversion achieved for methane concentrations of 0.1–1.0 per cent. The catalyst also exhibited an inherent safety when reactor temperatures exceeded autoignition temperatures with no obvious sign of spontaneous ignition being observed.

INTRODUCTION

An increased need to minimise fugitive greenhouse gases (GHG) emissions has put pressure on industries to reduce their emissions, in some instances required to maintain financial viability. One key sector facing this is the mining industry, where ventilation air methane (VAM) is a significant source of emissions, responsible for 70 per cent of emissions from mining. To reduce these anthropogenic fugitive emissions, a proposed solution is to oxidise the methane to carbon dioxide, as carbon dioxide has a greenhouse potential 21 to 23 times lower than methane as communicated by Solomon (2007) and Baris (2013). This is a difficult process as normal operating conditions see VAM concentrations at less than 0.5 per cent methane, and since the flammability concentration of methane in air at 1 atm and 20°C is 4.6 per cent to 15.8 per cent as shown by Vanderstraeten et al (1997), standard oxidation via flaring is not possible.

Potential solutions to this problem include capturing/concentrating the methane from the stream, thermal decomposition to carbon dioxide, and most promisingly, catalytic oxidation to carbon dioxide.

Catalytic oxidation has many advantages compared to alternative methods such as much lower temperature of operation compared to thermal decomposition, lower complexity than methane capturing/concentration, and the flexibility of being operational at a wide range of concentrations. However, possibly one of the most significant advantages is the possible high degree of safety compared to the high temperatures required for thermal decomposition or the high methane concentrations created by capture/concentration technologies.

Palladium catalysts have shown excellent activity in oxidising methane compared to other noble metal catalysts as shown by the workers (Gélin and Primet, 2002; Grunwaldt, Maciejewski and Baiker, 2003; Setiawan *et al*, 2016), however these catalysts can suffer from poisoning especially from water vapor. The specific catalyst used in this study is a palladium on titanium MFI support (Pd/TS-1). This catalyst provides the excellent oxidation capacity of other palladium catalysts while also providing a great resistance to water vapor poisoning due to the hydrophobicity of the TS-1 support.

VAM Safety Project

While the Pd/TS-1 catalyst has shown excellent performance in both methane oxidation and resistance to water, one unknown factor that remains is the extent of the inherent safety offered by a catalytic system for oxidising methane to carbon dioxide under typical and non-typical VAM conditions. This study focuses on the possible ability of a catalytic system to address safety risks associated with conversion of VAM streams.

The main risks investigated were the potential for safety hazards present during abnormal methane excursions, characterised by short or long duration increases in methane concentration in the stream and/or increase in reactor temperature due to increased exotherm from the reaction or from operator control to increase conversion if gradual deactivation occurs. The most important dangers that present themselves from these abnormal reaction conditions are autoignition of the stream or the potential formation of carbon monoxide. While catalytic systems have an excellent ability for complete oxidation compared to other technologies, the formation of carbon monoxide may occur below stoichiometric methane concentrations as carbon monoxide could be an intermediate product formed on the catalyst surface which may occasionally desorb before complete oxidation to carbon dioxide. Due to this possibility, carbon monoxide concentrations in the product gas were investigated at various methane concentrations.

BACKGROUND

The most prevalent concern for the safety of catalytic oxidation of VAM, is the possibility of propagated ignition or explosion into the feed stream. Research was conducted on what parameters predict if ignition of the VAM stream will occur.

Methane autoignition flammability conditions

Methane ignition in air mixtures has been thoroughly studied. For ignition of methane-air mixtures via ignition source, a concentration of 4.6 per cent to 15.8 per cent methane is required under NTP conditions; this concentration range is assumed to also apply to autoignition requirements. The minimum autoignition temperature (MAIT) of methane has been shown experimentally to be as low as 600°C by Robinson and Smith (1984), and 605°C by Conti and Hertzberg (1988), however conditions where this ignition was noted are not necessarily representative of those which would be seen in an industrial setting (identified in quartz reactor vessels at pressures potentially below atmospheric). Figure 1 shows the MAIT values for varying methane concentrations observed by Robinson and Smith compared to another study performed by Naylor.

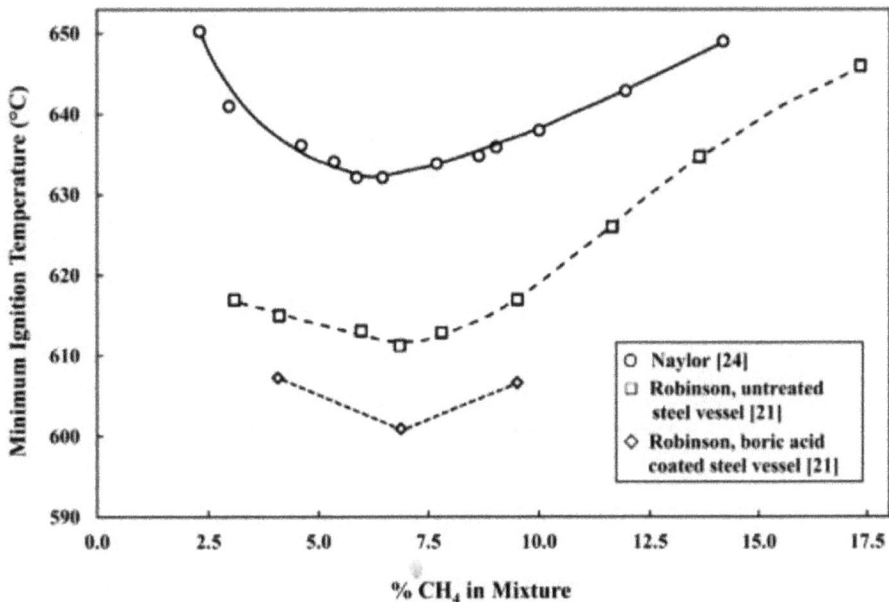

FIG 1 – MAIT for varying methane concentrations (Robinson and Smith, 1984).

These results show that the methane concentration with the lowest MAIT is 7 per cent. It was also noted that the reactor surface has a non-negligible effect on the MAIT, with an untreated steel vessel having a MAIT approximately 10°C higher compared to the same vessel coated with boric acid. Other studies have been conducted into autoignition of methane via a hot surface rather than the whole gas mixture reaching a certain temperature. This essentially has the effect of an autoignition temperature in which the ignition delay is very short since the gas mixture is able to ignite almost instantaneously via contact with a surface. Due to this 'instantaneous' ignition, much higher temperatures are required. Temperatures able to achieve this have been shown to be as low as 1000°C by Coward and Guest (1927), and as low as 936°C by Ungut and James (1999). It is worth noting that both these studies were using an inert surface with Coward and Guest using stainless steel, and Ungut and James using a ceramic plate. Coward and Guest also used natural gas instead of methane, and while natural gas has a very high methane concentration, this could explain the difference in their results along with different surface materials.

A study by Kim et al (1997), also investigated autoignition of methane via a hot surface, investigating comparisons between a catalytic surface and an inert surface. It was found that the surface temperature required for ignition was higher for the platinum catalyst used compared to the inert surface of nickel oxide, with temperatures of 1090°C and 1280°C respectively. This further demonstrates that for hot surface ignitions, temperatures of over 1000°C are required to ignite a methane-air mixture. Also, this shows evidence for catalytic surfaces having higher autoignition temperatures, although since both temperatures are still higher than found by Coward and Guest (1927), this could be simply due to the material and not necessarily the catalytic property.

Critical cell size

It has been shown that outside of concentration and gas temperature, the geometry of the vessel has a significant impact on ignition characteristics of fuel-air mixtures. One property of the vessel or reactor, known as critical cell size, plays a significant role in how a fuel-air mixture will propagate ignition. Detonation of the mixture is possible if the cell size is larger than the critical cell size, otherwise a deflagration will occur. Previous research performed by Knystautas (1984) showed that the critical cell size required for methane was significantly greater than ethane, propane, butane, ethene, acetylene and hydrogen. The minimum cell size for methane was approximately 30 cm. This is similar to research performed by Kundu, Zanganeh and Moghtaderi (2016) who found a minimum critical cell size for methane of 25 cm occurring at 10.5 per cent methane concentration. While critical cell size isn't necessarily a determining variable for ignition to occur, it is still an important variable for understanding how significant the danger of an ignition could be. Figure 2 shows the critical cell size with varying methane concentrations.

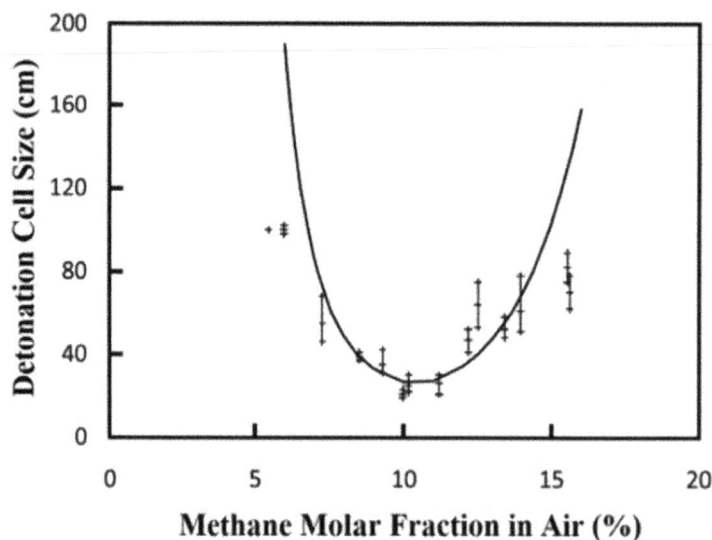

FIG 2 – Detonation critical cell size for methane-air mixture (Kundu, Zanganeh and Moghtaderi, 2016).

EXPERIMENTAL METHODS

Catalyst Preparation

Catalysts with a palladium loading of 1.2 wt per cent, were prepared via wet impregnation, where a palladium solution (Pd(NO$_3$)$_2$) was mixed with TS-1 support using mortar and pestle before being dried at 110°C for 24 hrs. The dried catalyst were sized to 250–425 μm particle diameter. Once catalyst was ready for use, it was packed in the reactor tube and calcined under 300 mL/min dry air at a reactor temperature of 500°C for at least 2 hrs.

Experimental Equipment

All VAM safety experiments were conducted using a tubular stainless steel micro reactor. The reactor temperature was controlled using Eurotherm temperature controllers. The reactor was packed with sized Pd/TS-1 held in place with quartz wool creating a bed within the reactor. Thermocouples were fitted via both ends of the reactor with one thermocouple measuring bed temperature directly at the exit of the bed and one thermocouple measuring the temperature 1–2 cm inside the inlet of the reactor. Baratron's were fitted to the outlet and inlet of the reactor to measure pressure either side of the catalyst bed. Alicat mass flow controllers were used to control the inlet flow of methane and air, with the methane line being fitted with a flashback arrestor upstream for safety. The combined inlet stream has the option to have a saturator attached to humidify the stream. Outlet of the reactor was analysed using gas chromatograph (GC). This reactor configuration is shown in Figure 3.

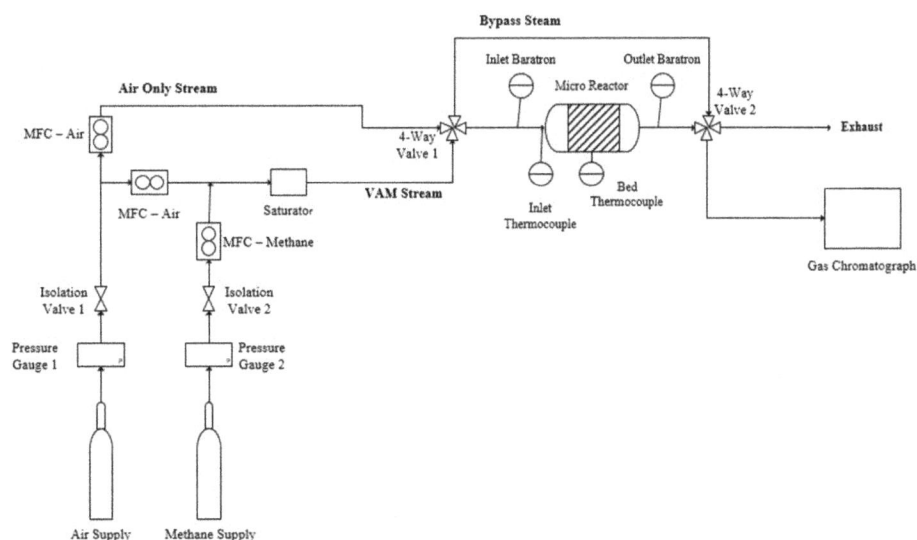

FIG 3 – Schematic of VAM safety system.

Operational Method

To simulate standard VAM conditions, the stream was set to 0.4 per cent methane concentration balanced with air at a total flow rate of 300 mL/min. When a humid stream was required, a saturator was added to the line to humidify the inlet gas. This resulted in approximately 2–3 per cent water in the steam, equivalent to a fully saturated stream at NTP. Reactor was set to 460°C with 100 mg of Pd/TS-1 catalyst forming the bed reactor at approximately 100 000 h^{-1} gas hourly space velocity (GHSV).

Multiple variations were made to the standard operation in order to simulate abnormal VAM conditions. Methane concentrations varied from 0.2 per cent to 15 per cent in air with total flow rates varying from 10 mL/min to 500 mL/min. GHSV did vary in some experiments but was kept at 100 000 h^{-1} for the majority. Reactor temperatures ranged from 200°C to 1100°C in order to investigate reactor behaviour at elevated temperatures and catalyst activity at non-standard operation temperatures.

RESULTS AND DISCUSSION

Step changes in VAM concentration

Pd/TS-1 catalyst has shown excellent activity for the oxidation of methane. However, these experiments have typically involved methane concentrations of 0.4 per cent. To investigate the behaviour of Pd/TS-1 under elevated methane concentrations, methane concentration was 'stepped up' in increasing concentrations with the reactor being allowed to reach steady state before the next concentration increase. Figure 4 shows initial step-up testing with methane concentration starting from standard VAM concentration of 0.4 per cent, increasing to 5 per cent. Due to 5 per cent being within methane flammability concentrations, particular interest was taken in this concentration, although it was understood that ignition shouldn't be possible if the temperature was not within autoignition temperatures.

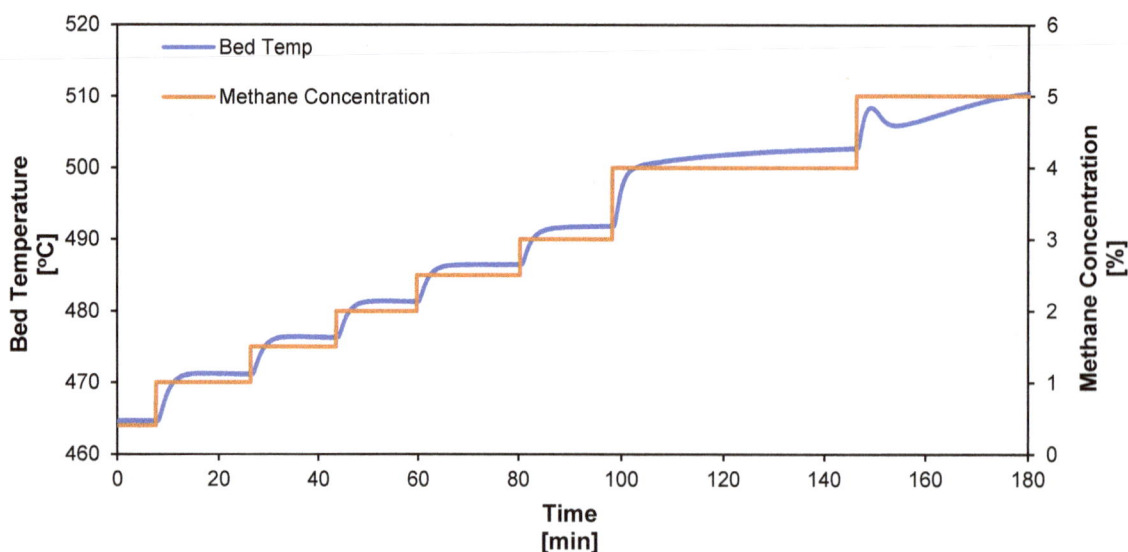

FIG 4 – Temperature data from VAM step-up 0.4 per cent to 5 per cent methane concentration, oxidised by 1.2 wt per cent Pd/TS-1 catalyst. Total flow rate: 300 mL/min, GHSV: 100 000 h^{-1}, 100 mg catalyst, dry conditions.

As expected, bed temperature closely correlates with methane concentration as higher concentration means more exotherm produced by the reaction. It was observed that bed temperature would increase quickly following an increase in methane concentration, followed by a gradual plateau as the reactor approached steady state. The only abnormal temperature trend was observed at 5 per cent methane concentration. The temperature initially increased but shortly after decrease before steadily increasing until steady state condition. This temperature inflection is most likely due to a small channel suddenly forming in the bed, causing the conversion to decrease and therefore decrease the exotherm from the reaction.

To extend the experimental findings shown in Figure 4, fresh Pd/TS-1 catalyst was tested again via step-up concentration. Two changes were made with the methane concentration range being 0.4 per cent to 10 per cent and also 'stepping down' the methane concentration back to 0.4 per cent. The reason for the step-down was due to interest in how the catalyst might perform when methane concentration returns to normal compared to initially. Bed behaviour was similar during step-up to the results shown in Figure 4. However, during step-down the bed temperature was drastically lower. The disparity is most likely due to temporary physical water interference as a significant amount of water is being formed at these high methane concentrations. This is shown to be further plausible by conversion data shown in Figure 5.

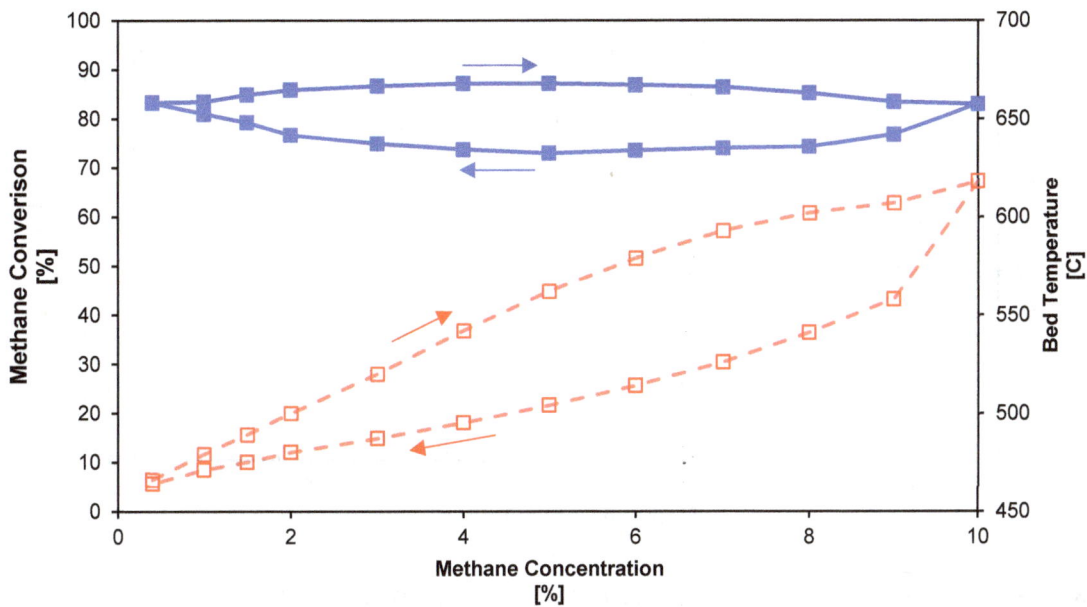

FIG 5 – Conversion data from VAM step-up, step-down of 0.4 per cent to 10 per cent to 0.4 per cent methane concentration, oxidised by 1.2 wt per cent Pd/TS-1 catalyst. Total flow rate: 300 mL/min, GHSV: 100 000 h⁻¹, 100 mg catalyst, dry conditions; ■ = methane conversion, □ = bed temperature.

It is clear that conversion was notably lower during step-down, going from 87.1 per cent at the highest during step-up, to 72.9 per cent during step-down. This further confirms an effect of the significant amount of water vapour formed during high methane concentrations. This also supports the assumption that the water vapour is not causing a permanent deactivation of the as the conversion and bed temperature returned to the same values measured initially.

To further confirm the phenomenon observed during the step-up, step-down experiments, identical experiments were performed with step-down occurring first. Most notable effect during step-down is the bed temperature decreasing without reaching steady state in a reasonable duration of time. This is most noticeable at high methane concentrations with the bed temperature able to reach steady state quickly at 4 per cent methane concentration and lower. This is once again most likely due to the significant amount of water vapour, formed during high methane concentrations, having a temporary effect on conversion. It is also observed that again the reactor behaves as expected during the step-up segment. Figure 6 shows the conversion and steady state bed temperatures for this experiment.

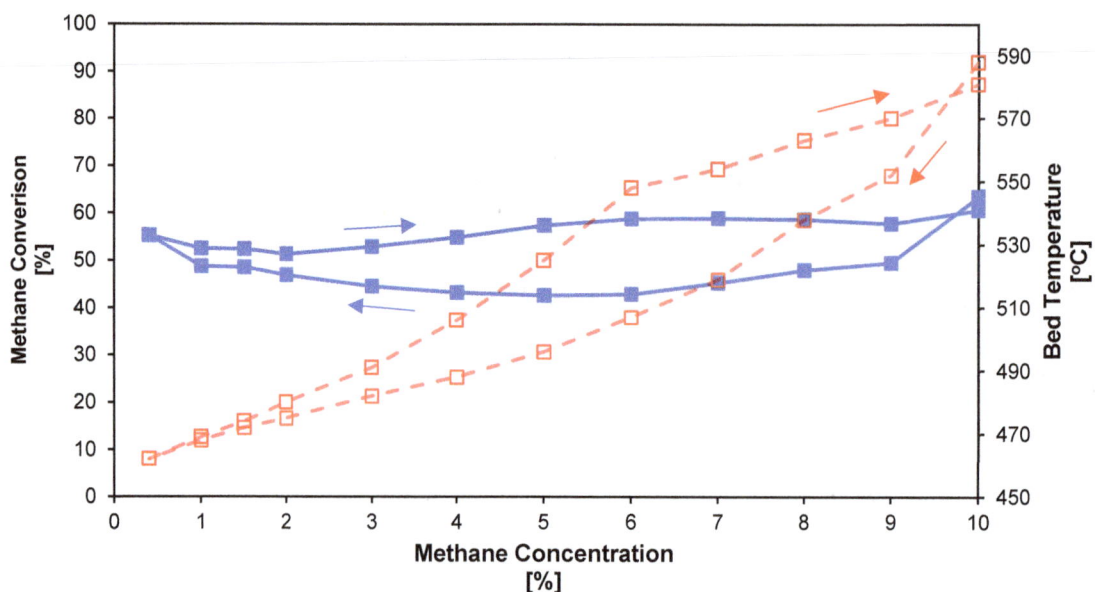

FIG 6 – Conversion data from VAM step-down, step-up of 10 per cent to 0.4 per cent to 10 per cent methane concentration, oxidised by 1.2 wt per cent Pd/TS-1 catalyst. Total flow rate: 300 mL/min, GHSV: 100 000 h⁻¹, 100 mg catalyst, dry conditions; ■ = methane conversion, □ = bed temperature.

It is shown that conversion is lower during step-down and conversion improves during step-up. The fact that conversion improves during step-up shows once again that the effect of high levels of water production reducing the conversion, is seemingly only temporary. This is most likely due to water vapour accumulating faster than it can leave the reactor which over time produces the constantly decreasing bed temperatures show in Figure 6. Once the methane concentration drops, and in turn the water vapour production, the reactor is able to clear the water from the catalyst and conversion increases. It can also be noted that the overall conversion values are lower than during the step-up, step-down run. This is likely due to a channel forming immediately from the sudden large amount of carbon dioxide formed when the methane concentration is set from 0 per cent to 10 per cent. This rapid change could disrupt the bed enough to form significant channelling and therefore reduce the overall conversion.

To investigate the effects of very high methane excursions, a step-up experiment was performed from a methane concentration of 1 per cent to 15 per cent. The main interest of this was to understand how the catalyst behaves once the methane concentration is well past the stoichiometric limit. Despite the significant methane concentration of 15 per cent, the bed temperature steadily increased before approaching steady state during the whole run. Also of note, the pressure in the reactor closely followed the bed temperature with no spikes shown to indicate any ignition. In Figure 7, it was shown that conversion was steady between 3 per cent and 12 per cent methane concentration. A dip in conversion at per cent methane concentration is observed which is attributed to the lack of oxygen for further methane oxidation. This also is shown by the bed temperature dropping by nearly 10°C despite more methane being available for oxidation.

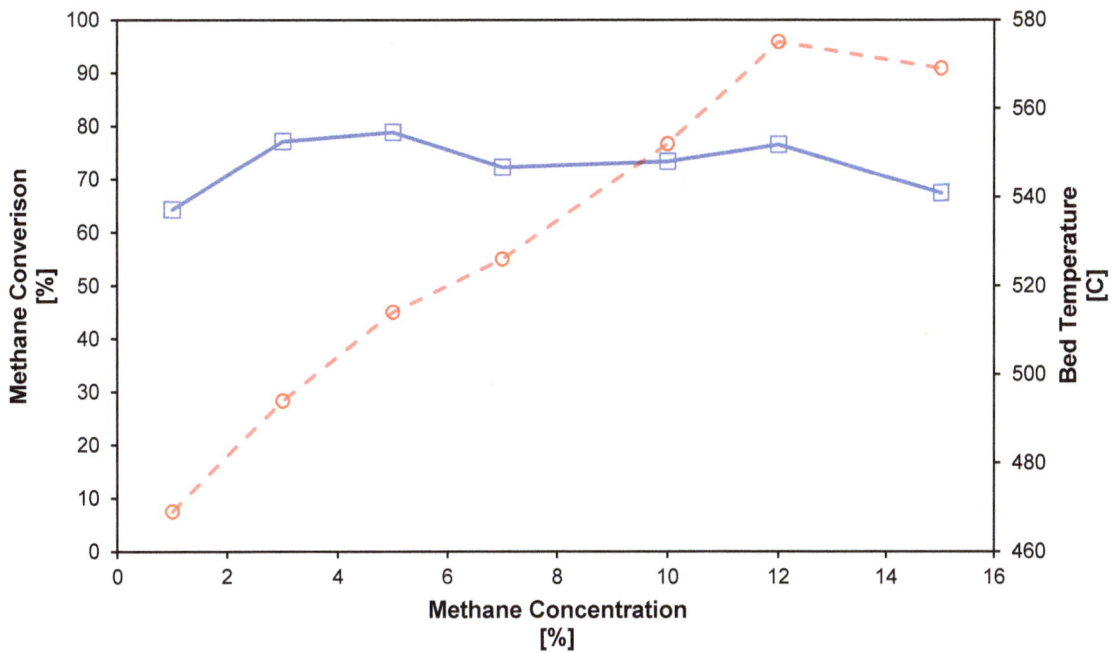

FIG 7 – Conversion data from VAM step-up of 1 per cent to 15 per cent methane concentration, oxidised by 1.2 wt per cent Pd/TS-1 catalyst. Total flow rate: 300 mL/min, GHSV: 100 000 h⁻¹, 100 mg catalyst, dry conditions; □ = methane conversion, ○ = bed temperature.

Carbon monoxide formation

A significant safety consideration of methane oxidation is the possibility of carbon monoxide formation due to partial or incomplete oxidation. This can be especially apparent in catalytic oxidation as carbon monoxide could be an intermediate compound formed on the catalyst surface. Carbon monoxide formation was tested for Pd/TS-1 at elevated methane concentration. Shown in Figure 8, carbon monoxide was not present in the product stream until the feed stream reached 11 per cent methane, with CO concentration measured as 0.01 per cent. Carbon monoxide level did not significantly increase until 14 per cent and 15 per cent methane concentration with CO concentration reaching 2.4 per cent and 3.3 per cent respectively. While the stoichiometric limit for methane in air is approximately 10 per cent, since the bed was not at 100 per cent conversion, the true stoichiometric limit was most likely reached at 14 per cent methane concentration, with the small trace of CO formed at 11 per cent methane concentration likely being formed to small region in the bed being starved of oxygen due to imperfect mixing. As carbon monoxide wasn't formed until methane concentration was well past the stoichiometric limit of oxygen available in the stream, it is unlikely that carbon monoxide is an intermediate product formed on the surface of Pd/TS-1. This is an ideal result as it is very unlikely that on-site VAM concentrations will ever approach stoichiometric limit and therefore it is extremely unlikely for any carbon monoxide to form during on-site catalytic VAM oxidation.

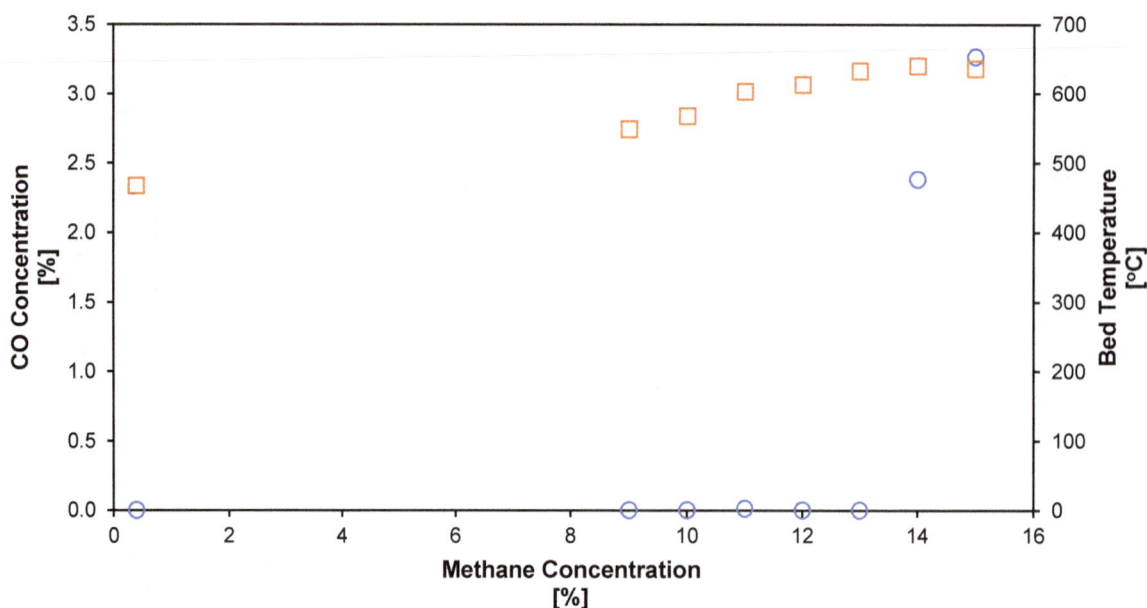

FIG 8 – carbon monoxide concentration in product gas from VAM stream oxidised by 1.2 wt per cent Pd/TS-1 catalyst. Total flow rate: 300 mL/min, GHSV: 100 000 h⁻¹, 100 mg catalyst, dry conditions; □ = bed temperature, ○ = carbon monoxide concentration.

Catalyst performance on lean VAM concentrations

As mentioned previously, one significant challenge of mitigating VAM streams is the low methane concentration of 0.4 per cent. This however is only the average VAM concentration with even lower concentrations of 0.1 per cent being possible for extended periods. This further complicates the problem for other VAM mitigation technologies, while this is a significant advantage of catalytic systems. To investigate if this was the case for the Pd/TS-1 catalyst, methane concentrations of 0.1–0.5 per cent were flowed through the catalyst bed at standard reactor operation temperature of 460°C. Conversion for dry conditions remained above 90 per cent for all five methane concentrations while humid conditions produced marginally lower conversion with values from 85–88 per cent. This leads to temperature changes in the bed for dry conditions being higher than humid. This difference between dry and humid conditions could be due to water vapour in the stream acting as a weak heat sink as water has a higher heat capacity than air and methane. This could cause the marginally varied bed temperature increases and in turn decrease conversion. It is also possible that this variation could be due to a small amount of channelling present in the bed as this is always a possible source of error with sized catalyst beds. Regardless, Pd/TS-1 was shown to have excellent performance despite the lean methane concentration of the stream.

VAM at elevated reactor temperatures

A significant safety risk involved in oxidising methane is the ignition and flashback of the feed stream. As noted in previous literature, when methane concentrations are at 7 per cent in air, the MAIT is lowest at 600°C. To investigate the behaviour of the micro reactor and the catalyst, at or above MAIT, a reactor ramp experiment was conducted where the reactor temperature was steadily increase at a constant rate. This was first performed with a blank reactor to investigate under what conditions stream ignition would occur. Bed temperature remained stable with no temperature spikes noted. Pressure also remained stable until close to 1000°C where small pressure spikes are observed, although the magnitude is low at only approximately 0.5 mbar. While the reactor size is below the critical cell size for detonation, deflagration is still possible which is most likely the cause of these small pressure spikes.

GC data, presented in Figure 9, confirmed that methane did oxidise to carbon dioxide. No conversion was observed until 850°C and complete conversion occurred at 925°C. These temperatures are higher than expected. This is most likely due to the stream flowing through the reactor quick enough that the gas stream does not achieve MAIT until the reactor has reached significantly higher temperatures. This is supported by hot surface ignition being well over 1000°C and therefore seems logical that the flowing system used would produce ignition at temperatures between the MAIT and hot surface ignition temperatures. It is also noted that between 850°C and 925°C carbon monoxide was formed, before total ignition of the stream at 925°C. This is most likely due to the stream 'puffing', where the stream briefly ignites but doesn't propagate and then extinguishes in a repetitive cycle, causing incomplete combustion.

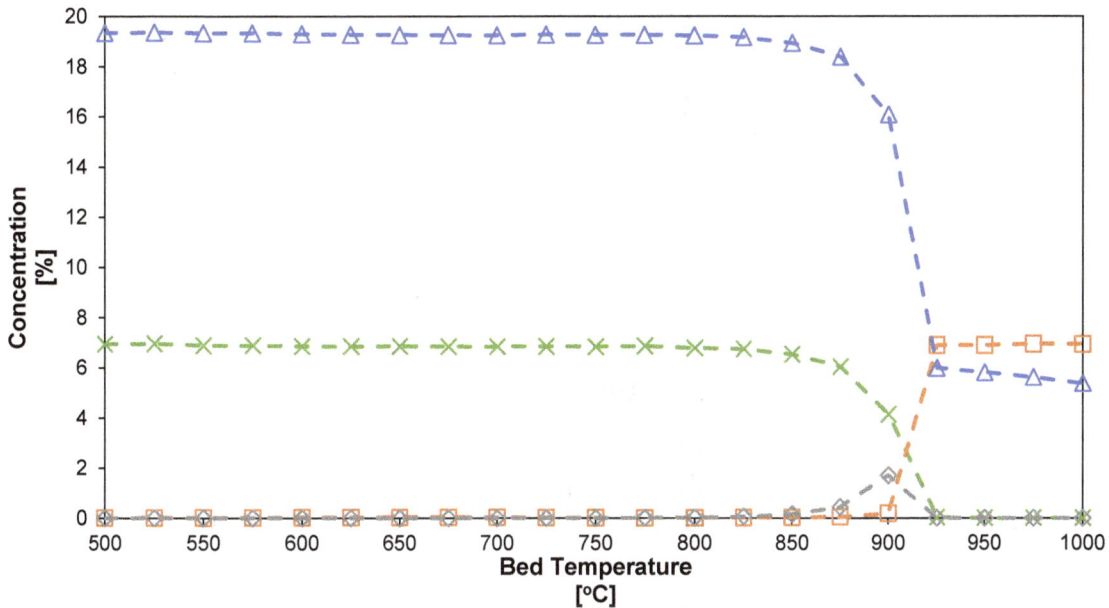

FIG 9 – Product stream composition of VAM stream through empty stainless-steel reactor at temperatures of 500°C to 1000°C and at a methane concentration of 7 per cent. Total flow rate: 300 mL/min; ◇ = carbon monoxide, □ = carbon dioxide, × = methane, △ = oxygen.

To validate the assumption that the ignition temperature is affected significantly by the residence time, the same reactor ramp experiment was performed at 100 mL/min flowrate. Methane concentration was increased to 10 per cent to increase the likelihood of observing evidence of sudden ignition as this concentration equates to the smallest cell size and possibly a stronger pressure response to be detected.

Once again, no obvious signs of ignition in the form or temperature or pressure spikes were observed. The reduction in flowrate did have the expected effect of reducing the temperature required to achieve oxidation of the stream. First signs of methane oxidation occurred at approximately 675°C, which is a decrease of 175°C compared to 300 mL/min flowrate. Total ignition also occurred 25°C cooler at 900°C. Carbon monoxide was observed to be present even after complete ignition due to the lack of oxygen available as this run was at stoichiometric limit, unlike in Figure 9.

To compare to a system using a catalyst, a reactor ramp experiment was performed with Pd/TS-1 in the reactor. This was performed with a methane concentration of 7 per cent and a flow rate of 100 mL/min. The ramp was performed at a temperature range of 250°C to 1000°C. As observed in the previous two experiments, no obvious signs of sudden ignition were observed. A gentle increase in temperature occurred at approximately 275°C due to the catalyst beginning to oxidise the methane in the stream. The product stream analysis is shown in Figure 10.

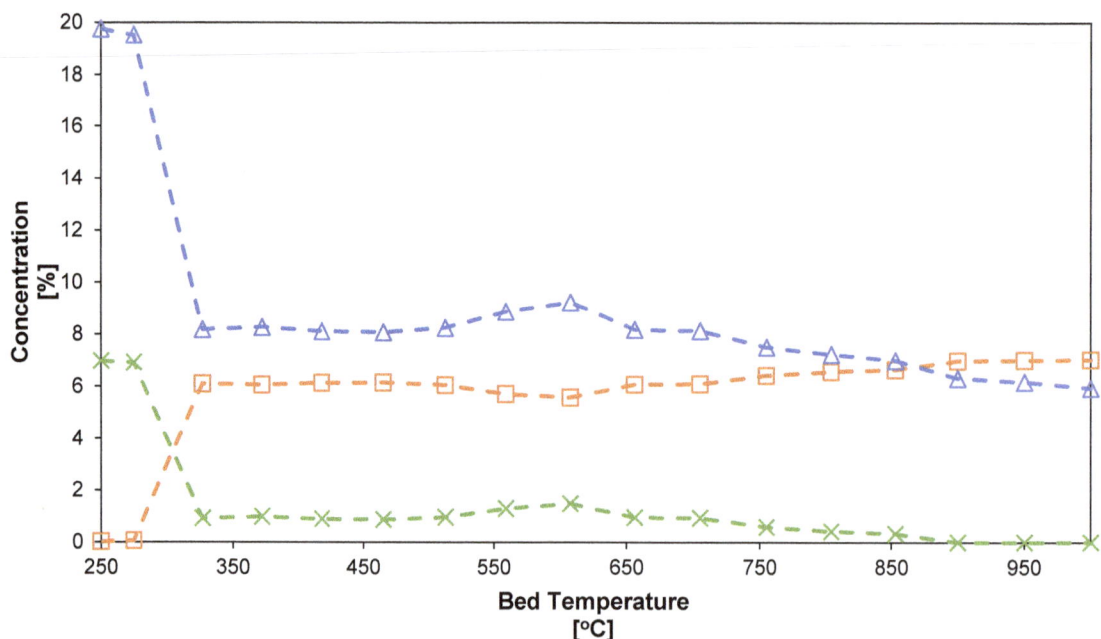

FIG 10 – Product stream composition of VAM stream oxidised by 1.2 wt per cent Pd/TS-1 at temperatures of 250°C to 1000°C and at a methane concentration of 7 per cent. Total flow rate: 100 mL/min, GHSV: 100 000 h^{-1}, 33 mg catalyst; □ = carbon dioxide, × = methane, △ = oxygen.

It was found that little to no methane conversion occurred while the bed temperature was below 275°C. Once bed temperature reached approximately 300°C, conversion rapidly increased to 86 per cent where it remained reasonably stable until complete ignition at approximately 900°C. Small fluctuations in conversion where noted, for example at temperatures between 550°C and 600°C, conversion decreased to 77 per cent. This is most likely due to a small channel being formed in the bed causing some of the stream to pass through without contact with the catalyst. Conversion steadily increases after 600°C due to increased reaction rate with temperature. One important observation was no production of carbon monoxide compared to an empty tube. This is an excellent example of the advantages of a catalyst system with Pd/TS-1.

CONCLUSIONS

The safe operation of catalytic reactors for the oxidation of VAM operating under abnormal reaction conditions was investigated. Pd/TS-1 catalyst was used due to its high activity for oxidising methane and significant resistance to water poisoning, both in the feed stream or formed by the reaction.

Under elevated VAM concentrations, the catalyst showed excellent stability and activity. When concentrations were increased, bed temperature increases were gentle before reaching steady state. No significant pressure changes were observed, further demonstrating the stable environment in the reactor while the catalyst is oxidising methane. It was observed that during a step-down, the conversion is lower in comparison to the same methane concentration during step-up. This was assumed to be due to a significant amount of water vapour being produced when methane concentration was high, and temporarily reducing catalyst activity. Once concentration returned to lower levels, the reactor was able to clear the water vapour and the catalyst activity returned to initial values.

It was shown that the catalyst would not produce carbon monoxide unless the stoichiometric limit between methane and oxygen was reached. This limit would be extremely unlikely to occur during on-site operation and therefore the catalyst is deemed safe in terms of not producing carbon monoxide.

The catalyst showed excellent ability to oxidise very lean VAM streams, with above 90 per cent conversion achieved for methane concentrations of 0.1–0.5 per cent.

The catalyst also exhibited an inherent safety when reactor temperatures exceeded autoignition temperatures.

Further VAM safety research will be performed using a larger reactor and using a Pd/TS-1 coated structured support instead of a packed bed of sized Pd/TS-1 particles. This will allow the investigation into the changes in catalyst or reactor behaviour when reactor volume increases. One significant advantage will be the complete mitigation of channelling causing errors, as the support structure will remain constant under VAM operating conditions.

REFERENCES

Baris, K, 2013. Assessing ventilation air methane (VAM) mitigation and utilization opportunities: A case study at Kozlu Mine, Turkey, *Energy for sustainable development,* 17:13–23.

Conti, R S and Hertzberg, M, 1988. Thermal autoignition temperatures for hydrogen-air and methane-air mixtures, *Journal of fire sciences,* 6:348–355.

Coward, H and Guest, P, 1927. Ignition Of Natural Gas-Air Mixtures By Heated Metal Bars1, *Journal of the American Chemical Society,* 49:2479–2486.

Gélin, P and Primet, M, 2002. Complete oxidation of methane at low temperature over noble metal based catalysts: a review, *Applied Catalysis B: Environmental,* 39:1–37.

Grunwaldt, J-D, Maciejewski, M and Baiker, A, 2003. In situ X-ray absorption study during methane combustion over Pd/ZrO 2 catalysts, *Physical Chemistry Chemical Physics,* 5:1481–1488.

Kim, H, Enomoto, H, Kato, H, Tsue, M and Kono, M, 1997. A study of the ignition mechanism of methane-air mixtures by inert and catalytic hot surfaces, *Combustion science and technology,* 128:197–213.

Knystautas, R, 1984. Measurement of cell size in hydrocarbon-air mixtures and predictions of critical tube diameter, critical initiation energy and detonability limits, *Progress in Astronautics and Aeronautics,* 94:23–37.

Kundu, S, Zanganeh, J and Moghtaderi, B, 2016. A review on understanding explosions from methane–air mixture, *Journal of Loss Prevention in the Process Industries,* 40:507–523.

Robinson, C and Smith, D, 1984. The auto-ignition temperature of methane, *Journal of hazardous materials,* 8:199–203.

Setiawan, A, Friggieri, J, Hosseiniamoli, H, Kennedy, E M, Dlugogorski, B Z, Adesina, A A and Stockenhuber, M, 2016. Towards understanding the improved stability of palladium supported on Ts-1 for catalytic combustion, *Physical Chemistry Chemical Physics,* 18:10528–10537.

Solomon, S, 2007. Climate change 2007-the physical science basis: Working group I contribution to the fourth assessment report of the IPCC (Cambridge University Press).

Ungut, A and James, H, 1999. Autoignition of gaseous fuel-air mixtures near a hot surface, *Institution Of Chemical Engineers Symposium Series 2000,* Institution of Chemical Engineers, pp 487–502.

Vanderstraeten, B, Tuerlinckx, D, Berghmans, J, Vliegen, S, Van'T Oost, E and Smit, B, 1997. Experimental study of the pressure and temperature dependence on the upper flammability limit of methane/air mixtures, *Journal of Hazardous Materials,* 56:237–246.

Bending the curve on global coalmine methane emissions

F Ruiz[1], C Talkington[2] and N Butler[3]

1. Director, International Methane Partnerships and Outreach, Clean Air Task Force, Boston MA 02019, USA. Email: fruiz@cleanairtaskforce.org
2. Vice President, Advanced Resources International, Inc., Arlington VA 22203, USA. Email: ctalkington@adv-res.com
3. Chief Engineer, Peak Carbon Ltd, London EC2V 7BG, UK. Email: neil.butler@peakcarboncapital.com

ABSTRACT

The launch of the Global Methane Pledge has garnered support from more than 150 countries to date, with leaders worldwide committing to reduce global warming and meet the 1.5°C target pledged under the Paris Climate Agreement. However, as shown by the Intergovernmental Panel on Climate Change (IPCC) Working Group III report, we are rapidly running out of time to meet this commitment (IPCC, 2022). The US National Oceanic and Atmospheric Administration (NOAA) has found in new analysis that atmospheric levels of methane spiked by a record amount in 2021 and are increasing at the fastest-ever recorded rate. This is combined with the fact that carbon dioxide emissions hit their highest level in recorded history. While reducing methane is our best tool for rapidly addressing the impacts of climate change, much is still needed to highlight its importance worldwide. This is especially apparent in the coal sector. According to a recent report from Global Energy Monitor, global methane emissions from operating coalmines equal 52.3 million tons, which is larger than annual emissions from oil or gas. This does not include abandoned mines, nor new mines slated for opening. The report also highlights one province alone – Shanxi Province in China – emits roughly the same amount of coalmine methane (CMM) as the rest of the world combined.

While the dire need to reduce methane emissions remains active in the public eye, CMM projects have stalled and emissions continue to rise, even though proven technologies exist to capture and use the gas. Financing exists but is not always prioritised for the coal sector. Policies designed to reduce emissions are not always enforced or leave large gaps in how they might be implemented. A well-integrated effort from policymakers, industry, civil society, and the financial community will be essential for bending the curve on CMM emissions.

INTRODUCTION

Odourless, invisible, and with a global warming potential more than 80 times that of carbon dioxide (CO_2) in the short-term, methane has contributed approximately 0.5°C of global warming we are experiencing today. Quickly and significantly reducing methane emissions is among one of the most important opportunities we can take to slow the rate of global warming and avoid near term, irreversible climate tipping points. In 2022, momentum on addressing methane pollution led to unprecedented action, with more than 100 countries signing on to the Global Methane Pledge (GMP) at COP26, collectively committing to voluntarily reducing global methane emissions by at least 30 per cent from 2020 levels by 2030 (US DOS, 2022). Signatories now exceed 150 nations.

Despite this historic pledge, we are rapidly running out of time to meet our commitment. The US National Oceanic and Atmospheric Administration (NOAA) has found that atmospheric levels of methane spiked by a record amount in 2021 and are increasing at the fastest-ever recorded rate (NOAA, 2022). This is combined with the fact that CO_2 emissions hit their highest level in recorded history. While reducing methane is our best tool for rapidly addressing the impacts of climate change, much is still needed to highlight its importance to policy and decision-makers, industry, the financial community, and the public worldwide. This is especially apparent in the coal sector.

The coal mining industry has traditionally vented methane from mine ventilation shafts to ensure safe working conditions, as methane is a hazardous gas and highly flammable; however, this release of methane results in significant emissions of greenhouse gases. According to a recent report from the Global Energy Monitor, methane emissions from the world's operating coalmines equal 52.3 million tons (Mt), which is larger than annual emissions from oil or gas (Global Energy Monitor,

2022). This estimate does not include closed or abandoned mines, nor does it account for new mines slated for opening. The report also highlights the fact that one province alone – Shanxi Province in China – emits roughly the same amount of coalmine methane (CMM) as the rest of the world combined.

While the dire need to reduce methane emissions is clear and well documented, CMM projects have stalled and emissions continue to rise, even though proven technologies exist to capture and use or, at a minimum, destroy the gas. Financing exists but is not always prioritised for the coal sector. Policies designed to reduce emissions are not always enforced or leave large gaps in how they might be implemented.

Methane is emitted from numerous sources and operations, including ventilation air and degasification systems from underground mines, abandoned or closed mines; surface or open pit mines, as well as fugitive emissions from post-mining operations, such as coal transport and storage.

Globally, ventilation air methane, or VAM, is the largest source of emissions from underground mining, responsible for roughly 70 per cent of total methane emissions. However, dilute concentrations of VAM have prevented its wide scale use and destruction, even though technologies are available and proven in the field to abate it in large quantities. This paper explores CMM emissions sources, including VAM, options for mitigating methane emissions, barriers to project development, and policy considerations to spur additional research, financing, and project development in the coal sector.

COALMINE METHANE EMISSIONS

Emissions from underground mines

In gassy mines, methane is contained within the coal matrix but can also be present in the free space, for example fractures within the coal or adjacent strata. Massive quantities of methane are 'liberated' into the mine workings during mine development and coal production. In an underground mine these gases are removed from the mine for health and safety reasons. This can occur before, during or after production. Without this necessary intervention, an explosion may occur with potentially catastrophic human, infrastructure, and financial consequences.

In addition to the methane stored in the mined seam, gas bearing strata above and below the mined seam, if present, may also contribute to the total gas volume released into the mine workings. This occurs when the roof collapses and the floor heaves after the longwall passes through forming the goaf (or 'gob'). As noted in the United Nations Economic Commission for Europe (UNECE's) Best Practice Guidance for Effective Methane Drainage and Use in Coal Mines:

> European studies (Creedy et al, 1997; April) have shown that a de-stressed arch or zone of disturbance, within which gas is released, forms above a longwall typically extending 160 m to 200 m into the roof and below the longwall to about 40 m to 70 m into the floor. (UNECE, 2016)

If the strata hold considerable volumes of gas, these adjacent strata can contribute considerable volumes of gas into the workings.

The mine's ventilation system provides the first 'line of defence' with large fans drawing massive quantities of fresh air into the mine while also capturing and transporting dilute methane, other gases, and dust to the surface where it is released to the atmosphere. At many mines around the world, mine ventilation alone is not sufficient to maintain methane concentrations at safe levels in the underground environment. In such situations, mine operators often employ gas drainage to supplement the mine's ventilation systems. Gas drainage systems range from one or more vertical wells drilled from the surface to complex systems of boreholes drilled in-mine and from the surface that are connected through an extensive gas gathering system of pipes and pumps. In some countries, gas drainage galleries connected to the gas collection system are also popular in-mine gas drainage techniques.

The balance of ventilation versus gas drainage depends on operating practices, regulatory requirements and historical practice. In some countries operators are given the flexibility to determine the optimal balance of ventilation and gas drainage. In other countries, however, prescriptive regulations may require gas drainage systems to be installed when the *in situ* gas content of the coal being mined exceeds certain thresholds, for example 10 cubic metres per tonne (m^3/t).

The mine ventilation exhaust fans, gas drainage wells vented to the atmosphere or connected to a flare or other control device, and the gas drainage collection stations serve as methane emission point sources at an underground mine. The distinct location of each and the fact that all point sources are operating under pressure (usually negative pressure), makes it easy to identify emission sources while also making it manageable to establish emission baselines, measure and monitor emissions from all point sources, develop mitigation strategies and implement mitigation projects, and ultimately quantify emission reductions.

Not all underground coalmines are 'gassy' mines, nor do they emit methane in sufficient quantities for emissions to be a significant safety or environmental threat. For example, in 2022 there were 185 underground coalmines operating the United States of which 105 had detectable methane emissions (US EPA, 2023a), but only 61 of those reported to the US's Greenhouse Gas Reporting Program (GHGRP). The GHGRP tracks mine methane emissions from mines that liberate 36.5 million cubic feet of methane per annum, equivalent to 1.03 million cubic metres or 701 tonnes of methane annually (Talkington, 2023). However, care must always be taken to effectively manage methane liberated into the mine workings to ensure mine worker safety and the integrity of the mine itself and surrounding areas, and every reasonable effort should be made to minimise or eliminate methane emissions to achieve near zero emissions, mitigating the environmental impact.

Mitigating emissions

Technologies to deliver near-zero methane emissions during coal mining are available and in operation in projects worldwide. Mitigation options range from directing gas drainage to on-site boilers to use in combined heat and power units and even use as vehicle fuel. The methane contained in mine ventilation air, known as ventilation air methane (VAM), is very dilute, normally less than one per cent (1 per cent) of air by volume, although higher concentrations up to two per cent can be allowed in certain instances, for example in bleeder shafts in the United States. VAM can be captured and destroyed through use of regenerative thermal oxidisers or catalytic oxidisers.

CMM end-use technologies can operate on a range of gas concentrations from 25 per cent to near 100 per cent, or pipeline quality, methane. For safety reasons methane concentrations in gas drainage should not fall below this threshold (that is, into the explosive range) and combustion of CMM at or near the mining facility or any way connected to the mining facility should never occur at concentrations below 25 per cent. While many CMM technologies can take gas drainage without any significant modifications, good practice is to maximise gas capture through the drainage system. Thus, the first step in reducing emissions for many operators and project developers is through installation and operation of an effective gas drainage system, if needed, and implementation of procedures and practices to improve the quantity and quality of gas produced by the system. This may entail practices such as improvements to borehole sealing, better regulation of vacuum pressure, greater attention to the orifice cross-sections at the wellhead and ensuring the integrity of the gas gathering system to reduce air ingress. It may also involve changing drainage methods or deployment of a portfolio of drainage techniques. Undertaking these efforts delivers important co-benefits of improving mine safety while also increasing gas available for use or destruction which, in turn, creates greater opportunities for GHG emission reductions.

Examples of technologies that are in use at coalmines today to capture, control, or use methane include:

- On-site and off-site boilers to provide heating and hot water for the mine and local community.
- Power production and combined heat and power using gas engines.
- Flaring.
- Industrial burners for mine heating, fertiliser production, coal drying and other uses.
- Desiccant cooling.

- Delivery into high pressure natural gas transmission and distribution pipelines.

- Delivery into low pressure local distribution systems (also known as 'Town Gas').

- Compressed Natural Gas (CNG) for vehicle fuel.

- Gas drainage vacuum pumps.

The low concentration and large quantities of VAM historically prevented its use; however, regenerative thermal oxidation (RTO) and regenerative catalytic oxidation (RCO) technologies have been proven at commercial scale to be effective and safe in destroying VAM emissions.

Facility level emissions

Although some investors, lenders and carbon offset buyers have been hesitant to support CMM and VAM mitigation projects due to their association with the coal industry, many others find such projects to be attractive investment opportunities compared to other types of GHG mitigation projects. This is especially true for methane emissions from underground mines and is primarily due to the large volume of emissions available for mitigation at the facility-level and the limited number of point sources, mine ventilation shafts, gas drainage collection systems, and gas wellheads. Combined, these factors allow for project developers and other stakeholders to quantify methane emissions, establish emission baselines, implement mitigation options, and quantify emission reductions more easily and confidently. The large scale of emissions on a facility-specific basis makes coalmines especially attractive when formulating policy to encourage GHG emission reductions. Table 1 compares GHG emissions by sector for major methane emitting industries in the United States that report to the US Greenhouse Gas Reporting Program (GHGRP). The data are from 2022 which are the most recently published data by the GHGRP at the time this paper was finalised.

TABLE 1

US Greenhouse Gas Reporting Program reporting facilities and fugitive methane emissions in million metric tonnes CO_2 equivalent ($MtCO_2e$) (2022).

Sector	Total CH_4 ($MtCO_2e$)	Facilities	Facility Avg CH_4 ($MtCO_2e$)
Underground coalmines	29.6	61	0.49
Oil and gas onshore production	33.5	459	0.07
Municipal landfills	101.1	1452	0.07

*US Environmental Protection Agency (US EPA) uses a global warming potential of CH_4 = 28.

Statistics from the US's GHGRP demonstrate how the magnitude of methane emissions at a single underground coalmine can far exceed those at typical oil and gas and municipal waste facilities. (It is important to note, however, that measured emissions at oil and gas facilities and landfills can greatly exceed reported levels).

There is also growing recognition that despite the transition to a decarbonised world, demand for coal will continue in the near and mid-terms and potentially longer as energy systems and economies incur the infrastructure and costs of transitions. The International Energy Agency (IEA) forecasts that coal demand will decrease by 40 per cent between 2022 and 2050 based on current stated policies among major coal producing and consuming countries. However, this still means global annual coal production will be around 3.5 billion tonnes per annum in 2050 (IEA, 2023a). Project developers, investors, carbon credit buyers and policymakers see the benefit of reducing GHG emissions in the coal sector where mitigation options are well understood and proven in the field.

The advantages of monitoring from a limited number of discrete emission point sources and designing and operating methane abatement projects at those point sources is also favourable for the coal industry when compared to certain other industries. In addition to general facility data, reporters to the GHGRP also provide counts of emission pathways such as ventilation shafts and degasification wells for coalmines, equipment counts for the oil and gas industry, and measurement points for landfills. Underground coalmines in 2022 had less than 600 emission point sources. By comparison, the number of possible sources at a single oil and gas production facility can be in the hundreds with more diffuse emissions among sources. As Table 2 shows, there are more than one million emission point sources in the US onshore oil and gas production segment. Moreover, while coalmines and municipal landfills are limited in the physical extent of the facility, an oil and gas production facility in the GHGRP includes all production equipment at the basin level. The Permian Basin in Texas and New Mexico measures approximately 400 km × 480 km. A company reporting for its Permian operations would, therefore, consider all wells and other emitting equipment to be part of the facility over this broad expanse.

TABLE 2

GHG emission source points for US GHGRP reporting sectors (2022) (US EPA, 2023b).

Sector	Emission points	Type
Underground coalmines	582	Shafts (390)
		Wells and gas drainage collection (192)
Oil and gas onshore production	3 976 200	Pneumatic devices (795 000)
		Pneumatic pumps (45 600)
		Acid gas removal units (24)
		Dehydrators (6200)
		Liquid unloading events (2 220 000)
		Well completions/workovers (26 000)
		Storage tanks (281 000)
		Associated gas wells (38 800)
		Flare stacks (63 000)
		Compressors (28 800)
		Wells for estimating leaks (471 800)
Municipal landfills	842	Gas collection systems

VENTILATION AIR METHANE (VAM)

Global VAM emissions are believed to account for approximately 70 per cent of total methane emissions from coal mining, with the actual per centage varying by country. In the United States, VAM emissions comprise 50 per cent of total methane liberated from underground mines and 89 per cent net emissions from underground mining after accounting for methane destruction and utilisation (US EPA, 2024).

Technologies to abate VAM emissions

Technologies are available to destroy and use VAM emissions, and in some cases have been demonstrated successfully at commercial scale. In addition, other VAM technologies are under development as policymakers support continued research and development to address this large source of methane emissions.

Regenerative Thermal Oxidation (RTOs) and Regenerative Catalytic Oxidation (RCOs)

Foremost among VAM abatement options is the use of RTOs and RCOs. The technologies have been used in thousands of industrial applications and have been adapted from their traditional uses of destroying low concentration volatile organic compounds (VOCs) and odours to oxidise low concentration VAM. RTO technology has been demonstrated at commercial scale at live mining operations in several countries. RCOs have been demonstrated at bench scale and commercial scale in controlled laboratory conditions; however, the cost of catalysts, the additional residence time required to ensure the catalyst is effective and the potential for catalyst replacement have deterred its use in commercial operations. Still there remains great interest in using catalysts. Ongoing research and development projects, including three funded by the US Department of Energy, are investigating the use of catalytic oxidation to lower oxidation temperature, reduce costs, and address potential safety concerns.

The first commercial RTO was placed into service in 2007 in Australia. MEGTEC and BHP Billiton also successfully demonstrated the first commercial-scale VAM destruction plant at the West Cliff Colliery from 2007 to 2017. Since then, commercial projects have continued to operate in the United States and China, and many others are currently in the development or construction phase. The scale of these projects and their abatement potential is significant. Commencing operation in September 2018, the world's largest operating VAM project is located at the Gaohe Mine of the LuAn Mining Group in Shanxi Province, China. The project utilises 12 RTOs with a total throughput capacity of 300 Nm^3/s. The RTOs were manufactured by Dürr, and the project was developed by Fortman (Beijing) Clean Energy Technology Ltd. of China (Wan, 2018). In addition to destroying VAM emissions, the project produces sufficient superheated steam to supply a 30-MW turbine for electricity generation. Following completion and successful start-up and operation, Fortman developed a second project with six RTO's that can manage ventilation air throughput of 150 Nm^3/s and incorporates a 15-MW turbine at Yangquan Mining Company's Mine Number 2 in Shanxi Province using RTOs manufactured by Anguil Environmental (Wan, 2018). Four additional projects in China consist of 9 RTOs in total with a combined throughput capacity of 225 Nm^3/s (Grzanka, 2024). Together these projects have the capacity to reduce methane emissions by 1.9 million tonnes CO$_2$ equivalent per annum (1.9 MtCO$_2$e) at 0.50 per cent methane concentration in the mine ventilation air, 2.8 MtCO$_2$e at 0.75 per cent methane and 3.8 MtCO$_2$e at 1.0 per cent methane. In addition to these projects, the China Coal Information Institute (CCII) identified other Chinese VAM projects at the Global Methane Forum held in March 2024 in Geneva, Switzerland (Han, 2024).

The United States has hosted three commercial VAM projects. The first project, developed by Biothermica, operated from 2009 through 2012 at the Warrior Met Coal Mine Number 4 in Alabama abating 80 766 tonnes CO$_2$e (Biothermica, 2024). The second VAM project was located at the Marshall County Mine in West Virgina. The project ceased operation in January 2024 after almost 12 years of successful and safe operation. In total, the project reduced methane emissions from mine by 1 660 307 tonnes CO$_2$e (ACR, 2024). The Coronado Coal Company together with NextERA, a project developer, and Biothermica, developed and operates a large-scale VAM destruction project at Coronado's Buchanan underground coalmine. The project commenced operation in 2022, and is expected to reduce emissions from the mine by 300 000 metric tonnes CO$_2$e per annum (Mining Technology, 2022). A second project is currently under development at the Buchanan mine and is expected to begin operation in 2024. Additional projects are reported to be in the planning and development stages, underwritten by projected revenues from the California Cap-and-Trade program allows emission reductions from US coalmine methane projects to be traded as offsets in the California market.

Use as Combustion Air

VAM has been used as combustion air in internal combustion engines at the Appin Colliery in New South Wales, Australia. The plant consisted of 54 VAM/CMM-driven internal combustion engines to power generators that produced 55.6 MW of electricity for the mine. VAM used as combustion air in large utility or industrial boilers has also reportedly been demonstrated at pilot scale at the Vales Point Power Station in Australia (US EPA, 2019). Critical considerations are location near a mine ventilation fan, corrosion, and dust contamination.

Direct Power Generation

Direct use of VAM in lean-burn gas turbine systems is especially attractive as it would use all the methane to produce power directly without large-scale oxidation. Several companies and research organisations have targeted this market. CSIRO in Australia has led the development of VAM use in catalytic combustion gas turbine systems through the design of the VAMCAT. According to CSIRO, the VAMCAT 'uses a novel catalytic combustion gas turbine system to oxidise methane to carbon dioxide and water, generating electricity from an otherwise explosive waste product.' A 25 KWe prototype was successfully demonstrated at an underground coalmine operated by the Huainan Coal Mining Group in China (CSIRO, 2024).

Barriers to VAM project development

Even though the first industrial-scale demonstration projects were undertaken in the 1990s and early 2000s, wide-scale uptake of VAM destruction and use projects has been slow due to a range of barriers, notably regulatory incentive, safety, cost, and equipment limits.

Safety

The primary concern within the mining community and among safety regulators has been the potential for an oxidiser or other VAM abatement or use technology to ignite a methane-rich VAM stream. Process safety of these installations is of paramount concern to coalmines. If the mine has a sudden emission of methane underground, this may lead to a hazard at the surface where the methane emission abatement equipment is installed. It is essential that the team designing and developing the project, whether under the umbrella of the mine owner/operator or a third-party developer, must include a specialist engineering design team with knowledge and competence in coal mining, mine ventilation, mine gas management, abatement system technology and process safety of mine to abatement equipment interface. The risk assessment starts with an understanding of mine safety, and the likelihood of a flammable atmosphere being created from the mine and transported to the surface. Permitted designs incorporate sensors, dampers, explosion prevention devices and other equipment to minimise explosion risk. It is noteworthy that to date, there are no known or reported explosions or other fire or explosion-related incidents with respect to VAM oxidation technologies or other VAM utilisation technologies in development. Existing and previously operated projects have demonstrated that VAM abatement projects can be carried out safely.

Cost

Equipment and project costs have been another prohibitive factor in deploying VAM technologies. Capital (Capex) costs can be high for large abatement projects and are influenced by fluctuations in materials costs and availability, mobilisation costs and labour costs. Operating costs (Opex) can account for a large share of total life cycle costs, especially in mining regions where electricity prices are high. Mattus (2024) estimates that total Capex for VAM processing plant capable of handling a ventilation airflow of 140 Nm^3/s is USD14 million ±15 per cent assuming uncomplicated conditions. This equates to an average cost in real (uninflated) terms of USD18 per tCO_2e per annum. Additional soft costs and other unexpected costs are likely to increase total average project CAPEX to USD23 per tCO_2e per annum. Although there are some economies of scale, Mattus reports that many costs are linear. Therefore, doubling the plant capacity would cost around USD25 million. Mattus (2024) estimates OPEX costs just below USD9 per tCO_2e. Capital costs are expected to decline as additional projects are implemented, equipment designs are refined, and project developers become more efficient.

Revenue

Revenues are generated almost exclusively from environmental markets where there is a value for the carbon emission reduction. In the US, the California Cap-and-Trade Program allows emission reductions from CMM projects, including VAM, to be traded in the markets as offsets. VAM-to-power projects are unlikely to be common, as they can only be carried out where VAM methane concentrations are high enough, but in some instances will provide additional revenue or cost savings to the project if the cost of power is high. In regions with low power prices, the cost of the power project may exceed any revenues produced. An alternative to voluntary or compliance carbon markets is legislatively mandated caps on emissions, such as in Australia's Safeguard Mechanism and the European Union's recently finalised emissions requirements for thermal coalmines.

Technical limitations

Although most VAM emissions typically emanate from a limited number of the gassiest mines, there are many mines with lower concentration shafts that together produce large volumes of VAM. For these situations, use or destruction has been inhibited by the technical lower concentration limits of equipment available. Historically, RTOs were limited to oxidising methane at concentrations greater than 0.20 per cent to 0.30 per cent. Below those concentrations, the oxidisers could not self-sustain the autoignition process that allows oxidisers to run continuously after start-up. Refinements in the design of RTOs and the addition of catalysts allows oxidation of lower concentrations, reportedly down to below 0.10 per cent. Similarly, use of VAM in catalytic lean-burn gas turbines has also been limited to consistent concentrations at or above 0.80 per cent methane (US EPA, 2019).

Familiarity

Mine gas management is a critical aspect of any mine operator's portfolio. However, abating methane emissions is generally outside the core competencies of most mining operators, and operators are not familiar with the abatement process technology. In addition, as new suppliers enter the market, many are not familiar with coal mining and applicable mine safety regulatory requirements. This means that several projects from mining organisations have stopped or stagnated because they have proven unable or unwilling to broach the perceived risks involved in being the first to push this type of project to full scale development. In CMM power generation or pipeline projects, it is common to outsource these activities to third party specialist, Build Own Operate businesses. Given that mining companies' core focus is on mining, it seems likely that future VAM projects may be developed by third parties working very closely with the mine owner/operator in a business model like CMM.

Mobilising equipment

Mobilising equipment can be difficult in areas with challenging terrain, such as mountainous locations where roads, tunnels and other features may not be able to accommodate transfer of oxidisers and other related equipment. Location near the shaft fan may also be difficult and cannot impair access to the fan. The weight of the equipment can also be challenging and may require reinforcing the foundation, which will impact costs, although generally the weight of the plant is spread over a large area generating a relatively low, uniformly distributed load.

Risks

Aside from ignition risk, mine operators and regulators have expressed concern that a VAM processing plant will place back pressure on the shaft fan impacting safety and operations but also costs as additional fan power will be required to offset the back pressure. This is being successfully addressed by various designs that use fans to draw the ventilation air into the oxidisers or allowing a gap between the shaft fan and oxidisers to indirectly connect to the fan housing, or only allowing the VAM plant to take a slipstream of the total shaft flow, for example 60 per cent.

Legislative and regulatory barriers

Use of gas drainage has been commonplace for many years in many different countries. In contrast, reducing VAM emissions is a relatively new concept, leaving mine safety regulators to develop the appropriate regulatory framework for VAM from the ground up. Thus, regulations are being developed as the technologies are being developed, tested, and commercially applied. Gas ownership has also acted as a hurdle for some projects where rights to the gas in mine ventilation air at the point of emission are not clear. Mines generally have the right to vent methane for safety reasons, but without the permission of the gas rights holder they may not have the right to use or destroy the methane and secure carbon credits or other environmental commodities. This is especially true where rights may be privately held.

POTENTIAL FOR VAM IN AUSTRALIA

Australia is the world's fifth largest coal producer after China, India, Indonesia, and the United States. Annual coal production is roughly 439 million tonnes (Mt), about the same as Russia (IEA, 2023b). The country's mining industry depends heavily on exports which are expected to grow to 351 Mt in 2024–2025 as several new mines open. Metallurgical coal accounts for almost half of all exports (Qu, 2024).

Annual GHG emissions from coalmines total 28 $MtCO_2e$ per annum or six per cent of Australia's total GHG emissions and 57 per cent of fugitive emissions from the energy sector. Underground mines emitted 17.4 $MtCO_2e$ in 2021 for a four per cent share (DCCEEW, 2023). VAM is assumed to account for 70–80 per cent of Scope 1 emissions for underground coalmines.

There is substantial experience with methane capture and use in Australia. Of the 99 operational mines, 30 are underground operations and approximately 16 of those have implemented CMM recovery and use projects, all gas drainage projects (Qu, 2024).

As previously noted, mining companies and regulators have been exposed to VAM abatement technologies. MEGTEC conducted a successful trial of RTO technology at the Appin Colliery and BHP and MEGTEC completed a full field demonstration project at the West Cliff Colliery. Moreover, CSIRO is a world leader in research and development of innovative VAM technologies. Based on mine ventilation data that Peak Carbon Ltd. compiled and analysed, we estimate that implementation of 10 to 12 VAM abatement projects at Australia's gassiest underground coalmines would deliver 8 million tonnes CO_2e per annum of on-site abatement, equivalent to 25 per cent of the coal sector's reported Scope 1 emissions and two per cent of Australia's national emissions.

Policy drivers are also in place to deliver VAM reductions. National legislation introducing the Safeguard Mechanism aims to reduce the emissions of Australia's largest industrial facilities over time. In addition to emissions caps, the Mechanism also provides for emissions trading, placing a market value on emission reductions. At the state level the Queensland Government has introduced an AU$520 million Low Emissions Investment Partnerships (LEIP) Program to accelerate investment in projects that will drive down emissions in Queensland's highest emitting facilities, with an initial focus on the metallurgical coal sector. Since 2009, the New South Wales State Government has administered the $100 million Coal Innovation (CINSW) fund to assist emerging technologies (such as VAM use and destruction technology to reduce GHG emissions associated with coal mining and coal use) (Qu, 2024). However, it is critical to ensure that mine owner/operators and safety regulators understand and accept the safe design of the equipment, the process engineering and risk management associated with VAM project implementation.

VAM MITIGATION IN THE US

Coalmines in the United States produced 539 008 metric tonnes of coal in 2022 with underground mines producing 37 per cent at 185 operating underground mines (US EPA, 2024). Opencast mines produced the remaining 63 per cent (EIA, 2023).

Fifty per cent of methane liberated into the workings during mining operations, or 28 $MtCO_2e$ at US underground mines is captured by mine ventilation systems and emitted to the atmosphere. When considering that a large share of mine gas drainage is recovered for use, 87 per cent, ventilations systems account for 89 per cent of net methane emissions from underground mines (US EPA, 2024). There has been a significant decrease in the population of underground mines from a decade earlier as many unprofitable and smaller mines have closed; however, the mines that remain open and many new mines are high production longwall mines with significant emissions.

The mining industry has a long history of capturing and using drained methane beginning in the late 1970s and early 1980s, pioneering the modern coalbed methane industry. Initially most drained gas was sold into natural gas transmission pipelines. Although gas pipeline sales still account for the largest share of end-use for gas drainage projects, due primarily to one mine comprising 80 per cent of all CMM pipeline sales (US EPA, 2024), the number of gas sales projects has declined as flaring has become popular (ACR, 2024; Coté, 2022). This is due in part to the lower cost of flaring projects but also the eligibility of such projects in the California Cap-and-Trade program whereas offsets generated by natural gas pipelines sales are not eligible. Coté (2022) found that by 2020, the US coal industry had the fewest number of CMM pipeline sales projects in the previous 30 years. Moreover, he found that six coalmines with historically significant pipeline sales volumes, a combined 141.6 million cubic metres as recently as 2011, were selling virtually no gas into pipelines by 2020 (Coté, 2022). On the other hand, the California market has also spurred increased interest in VAM. As noted earlier in the paper, a new project started in 2022 at the Buchanan mine and a second project will begin operation at the Buchanan mine in 2024. There are also several other projects reportedly in the planning or development stage.

Market prices in the California market are sufficient to sustain VAM projects at shafts with consistently high methane concentrations, primarily bleeder shafts. The market has also demonstrated sustainability, providing investors with confidence that the market will endure. A critical challenge facing VAM projects, though, are limits on the number of offsets allowed into the market and additional limits on the number of offsets that can originate outside of California. Thus, there is a significant incentive to move quickly in placing projects at the mines with the most potential.

Regulatory issues have largely been resolved following substantial engagement with the US Mine Safety and Health Administration beginning in the early 2000s.

VAM MITIGATION IN CHINA

China is the world's largest coal producer and the world's largest CMM emitting country. In 2022, IEA reported that China's coal production grew by almost 9 per cent to 4374 Mt which was an all-time high (IEA, 2023b). It is very difficult to find statistics on China's CMM emissions; however, the IEA estimates emissions to be 9685 kt methane or 271 $MtCO_2 3e$ in 2023 (IEA, 2024). VAM emissions are believed to comprise 70 to 80 per cent of China's total CMM emissions but this cannot be verified.

Like Australia and the US, China has a long history of recovering and using CMM in a range of applications from combined heat and power to vehicle fuel. China has also taken a very active role in leading development of VAM projects with most of the early projects supported through carbon financing provided by the Clean Development Mechanism (CDM). Newer projects are believed to be financed through a portfolio of financial instruments including self-financing from mining companies to offset emissions, voluntary carbon markets, and cost savings where the VAM project includes power production. As with other countries, a great deal of effort was put forward to work with the State Administration of Worker Safety to agree to a safe design criteria and safe operational practices. Today China hosts at least six projects, and Han (2024) reports that four more projects are known to be in development.

Although there is no specific legislation targeting VAM abatement, the central government of China has prioritised mine methane capture to reduce emissions while also producing energy. Thus, project developers are encouraged to include heat recovery at a minimum and, if possible, power production.

The abatement potential in China is very large given the continued commitment to coal mining, the large scale of mining, and presence of large gas reserves in the mining regions. There is also substantial demand for energy generated from use of VAM to deliver power to the electricity grid.

POLICY AND FUNDING CONSIDERATIONS

Achieving near-zero emissions mining will require a portfolio of policy tools that must address VAM emissions. The priority should be to increase gas drainage where feasible and cost-effective. Higher concentration gas produced from gas drainage systems can be more easily and cost-effectively managed while driving down the quantities of methane captured by the mine's ventilation system. However, it will be impossible to capture all gas through drainage. Even in situations where gas drainage is very effective, gassy underground mines will continue to emit very large quantities of methane through ventilation. Therefore, consideration should be given to implement policies that encourage and/or mandate mitigation of VAM emissions.

Place regulatory limits on VAM emissions

Although market-based incentives may be the initial preference in many cases, it may be advisable to adopt regulatory limits on VAM emissions. This is especially true recognising the critical necessity of reducing GHG emissions on a large scale to meet global targets. Regulatory mandates such as Australia's Safeguard Mechanism and the recently adopted European Regulations on methane emissions which limit venting of methane from thermal coalmines are at the forefront of regulatory action to limit CMM emissions, providing frameworks that should result in lower emissions. Regulatory caps can also be combined with emissions trading to provide flexibility in meeting the emissions targets.

Improve monitoring, reporting and verification of CMM and VAM

Accurate monitoring, reporting and verification (MRV) of VAM emissions and emission reductions is paramount to fully understand the industry's emissions profile while also necessary to have confidence in any claimed emission reductions. Detailed greenhouse gas reporting schemes such as those already implemented by Australia and the US and that will be implemented with the new EU regulations provide the policymakers, the mining industry, investors and the public with reliable data on which to make informed decisions. However, improvements to data collection methods, frequency of measurements and other refinements will improve gas emission inventories further enhancing decision-making. It should be noted that there are practical difficulties in accurately measuring air flow in huge quantities in mines, due to variations in air quality, and due to turbulence at point of measurement.

Develop carbon markets

There is currently no financial incentive to undertake a VAM project absent a regulatory driver or carbon market. Without a value for the emission reduction, it is unlikely that the first VAM projects in the US would have been constructed as it is the driver for projects under development now. However, an adequate price signal is crucial to justify the investment decision to move forward with a project. Outside of regulatory mandates, the additionality of VAM abatement is without question. The projects will not proceed without an economic or regulatory incentive. It should be noted however that concerns arise about subsidies from carbon markets to coalmines (assuming some profit from executed VAM projects), and the fact that the investments made in voluntary projects, based on revenues from carbon markets, have been used to argue *against* regulatory standards in other sectors, such as oil and gas.

Promote research and development

The Australian, Chinese and US governments have provided funding for continued research and development to improve existing technical options while designing and testing new and innovative technologies. This type of support is essential to build a suite of technologies that can manage a wide range of VAM concentrations and air flows at manageable cost while also ensuring the safe operation of the equipment.

Support pilot studies at operating mines

Successful VAM field demonstrations using RTOs have already proven the validity of the technology under 'real world' conditions while improving the operational practices to prepare for full industry acceptance. Such studies have been critical for wide-scale adoption of VAM mitigation. New pilot projects are essential to commercially test catalyst technologies.

CONCLUSIONS

Reducing methane emissions has become a cornerstone for global efforts to achieve net zero emissions. Much attention is being paid to the emissions from the coal industry as several studies, including those by the Global Energy Monitor and the International Energy Agency, are indicating that emissions from mining have been under-reported. Mines have limited geographic scope and emission sources at those mines are easily identifiable points sources, thus making the coal industry an attractive and cost-effective priority for emission reductions.

VAM is the largest source of methane emissions accounting for around 70 per cent of emissions from underground mining. Although destruction of VAM has been technically challenging since the concept was first explored in the 1990s, RTO technology has derisked most of the technical barriers. Projects in Australia, the US and China underscore the technical viability of RTO technology. The barriers remaining are largely cost and mine safety related. The safe operation of the VAM projects deployed to date addresses many safety concerns; however, each country or jurisdiction must determine whether the approaches employed to mobilise and operate a VAM processing plant meet the legal and cultural criteria in that country.

Going forward it will be essential to continue supporting new projects in Australia and R&D to further improve on existing technologies including RTOs and RCOs while at the same time underwriting development of new technologies to give stakeholders a suite of options from which to choose. Fostering carbon markets or other incentive-based markets is also vital, and this might extend to promulgating regulatory limits on VAM emissions if voluntary progress is slow to deliver results.

REFERENCES

American Carbon Registry (ACR), 2024. Green House Gas Registry. Available from: <https://acrcarbon.org/acr-registry/> [Accessed: 18 June 2024].

Biothermica, 2024. Ventilation Air Methane in the USA. Available from: <https://www.biothermica.com/content/ventilation-air-methane-usa> [Accessed: 17 July 2024].

Coté, M, 2022. Recent Developments with Coal Mine Methane and Abandoned Mine Methane Projects in the US, Presentation to 2022 Global Methane Forum. Available from: <https://www.youtube.com/watch?v=IVNdd1t7xps>

CSIRO (Commonwealth Scientific and Industrial Research Organisation), 2024. VAMCAT: Ventilation Air Methane Catalytic Turbine. Available from: <https://www.csiro.au/en/work-with-us/industries/mining-resources/mining/fugitive-emissions-abatement/vamcat> [Accessed: 19 June 2024].

DCCEEW (Department of Climate Change, Energy, the Environment and Water), 2023. Paris Agreement Inventory, 2021. Available from: <https://www.greenhouseaccounts.climatechange.gov.au/>

Energy Information Administration (EIA), 2023. Annual Coal Report 2022. US Department of Energy. Available from: <https://www.eia.gov/coal/annual/pdf/acr.pdf>

Global Energy Monitor, 2022. Sizing Up Coal Mine Methane, March 2022. Available from: <https://globalenergymonitor.org/wp-content/uploads/2022/03/GEM_CCM2022_final.pdf>

Grzanka, R, 2024. Ventilation Air Methane and Coal Mine Methane Abatement and Utilization, Global Methane Forum. Available from: <https://www.globalmethane.org/documents/2024Forum/Coal%20Sessions%20-%20Wednesday/Session%204%20-%20Updates%20on%20Capabilities%20to%20Mitigate%20VAM%20(Part%201)/Grzanka_Coal_3.pdf>

Han, J, 2024. China's Progress in VAM Utilization and Emission Reduction, Global Methane Forum. Available from: <https://www.globalmethane.org/documents/2024Forum/Coal%20Sessions%20-%20Wednesday/Session%204%20-%20Updates%20on%20Capabilities%20to%20Mitigate%20VAM%20(Part%201)/Jiaye_Coal_2.pdf>

Intergovernmental Panel on Climate Change (IPCC), 2022. Climate Change 2022: Mitigation of Climate Change, Contribution of Working Group III to the Sixth Assessment Report of the Intergovernmental Panel on Climate Change (eds: P R Shukla, J Skea, R Slade, A Al Khourdajie, R van Diemen, D McCollum, M Pathak, S Some, P Vyas, R Fradera, M Belkacemi, A Hasija, G Lisboa, S Luz and J Malley) (Cambridge University Press: Cambridge). https://doi.org/10.1017/9781009157926

International Energy Agency (IEA), 2023a. World Energy Outlook 2023, October.

International Energy Agency (IEA), 2023b. Coal 2023. Available from: <https://iea.blob.core.windows.net/assets/a72a7ffa-c5f2-4ed8-a2bf-eb035931d95c/Coal_2023.pdf>

International Energy Agency (IEA), 2024. Global Methane Tracker 2024. Available from: <https://www.iea.org/data-and-statistics/data-tools/methane-tracker> [Accessed: 19 June 2024].

Mattus, R, 2024. Best Practice Guidance on Ventilation Air Methane, ECE/ENERGY/GE, 4/2024/3. 4 January 2024. Available from: <https://unece.org/sed/documents/2024/01/working-documents/best-practice-guidance-ventilation-air-methane-vam>

Mining Technology, 2022. Biothermica commissions mine methane abatement facility in US, 11 November 2022. Available from: <https://www.mining-technology.com/news/biothermica-methane-abatement-us/>

National Oceanic and Atmospheric Administration (NOAA), 2022. Increase in atmospheric methane set another record during 2021. Available from: <https://www.noaa.gov/news-release/increase-in-atmospheric-methane-set-another-record-during-2021#:~:text=NOAA's%20preliminary%20analysis%20showed%20the,during%202020%20was%2015.3%20ppb>

Qu, Q, 2024. Australia CMM Update, Presented to the Global Methane Initiative Coal Subcommittee, 21 March 2024. Available from: <https://www.globalmethane.org/documents/2024Forum/Coal%20Sessions%20-%20Monday/Session%202%20-%20GMI%20Coal%20Mines%20Subcommittee%20and%20Sector%20Updates/Australia%20Update_Qu.pdf>

Talkington, C, 2023. United States Greenhouse Gas Reporting Program for Coal Mine Methane, 2nd Meeting of the Task Force on Methane Emissions Reductions. Available from: <https://unece.org/info/events/event/382721>

United Nations Economic Commission for Europe (UNECE), 2016. Best Practice Guidance for Effective Methane Drainage and Use in Coal Mines, ECE Energy Series No 47, 2nd edition. Available from: <https://unece.org/DAM/energy/cmm/docs/BPG_2017.pdf>

US Department of State (US DOS), 2022. Global Methane Pledge: From Moment to Momentum – United States Department of State, United States Department of State. Available from: <https://www.state.gov/global-methane-pledge-from-moment-to-momentum/#:~:text=Achieving%20the%20Global%20Methane%20Pledge>

US Environmental Protection Agency (US EPA), 2019. Ventilation Air Methane (VAM) Utilization Technologies. Available from: <https://www.epa.gov/sites/default/files/2019-11/documents/vam_technologies.pdf>

US Environmental Protection Agency (US EPA), 2023a. Inventory of US greenhouse gas emissions and sinks: 1990–2021. Available from: <https://www.epa.gov/ghgemissions/inventory-us-greenhouse-gas-emissions-and-sinks-1990-2021>

US Environmental Protection Agency (US EPA), 2023b. Greenhouse Gas Reporting Program data. Available from: <https://www.epa.gov/ghgreporting/find-and-use-ghgrp-data>

US Environmental Protection Agency (US EPA), 2024. Inventory of US Greenhouse Gas Emissions and Sinks: 1990–2022, EPA 430-R-24-004. Available from: <https://www.epa.gov/ghgemissions/inventory-us-greenhouse-gas-emissions-andsinks-1990-2022>

Wan, S, 2018. Destruction and Utilization of Ventilation Air Methane in China, World Mining Congress 2018, UN Economic Commission for Europe and Global Methane Initiative Workshop 'Turning Coal Mine Methane into an Asset: Implementing Best Practices'. Available from: <https://unece.org/fileadmin/DAM/energy/images/CMM/CMM_CE/18._Wan.pdf>

Considerations for oxidation of ventilation air methane using regenerative thermal oxidisers in the Australian coal industry

A Zander[1], W Harris[2], B Wright[3], S Nelson[4], D Middleton[5] and S Arsenault[6]

1. Process Engineer, Hatch, Brisbane Qld 4000. Email: amy.zander@hatch.com
2. Australasia Regional Director, Climate Change, Hatch, Brisbane Qld 4000. Email: wendy.harris@hatch.com
3. Process Engineer, Hatch, Perth WA 6000. Email: beau.wright@hatch.com/
4. Lead Instrumentation and Controls Engineer, Hatch, Brisbane Qld 4000. Email: scott.nelson@hatch.com
5. Coal Handling and Preparation Plants (CHPP) Infrastructure Lead- Coal, Hatch, Brisbane Qld 4000. Email: dave.middleton@hatch.com
6. Senior Engineer – Climate Change Practice, Hatch, Trail BC V1R 4A7, Canada. Email: simon.arsenault@hatch.com

ABSTRACT

Fugitive ventilation air methane (VAM) emissions contribute to an underground coalmine's direct CO_2e emissions. To maintain a safe working environment underground, methane is diluted with ventilation air to maintain concentrations below the lower explosive limit. This results in large ventilation flow rates with low methane concentrations (typically <1 per cent) that are challenging to abate. Regenerative thermal oxidisers (RTOs) have been applied to combust methane in the mine exhaust air for coalmines in China and the United States. This paper provides an overview of the technology selection process for achieving efficient methane destruction at low concentrations and discusses critical decisions that need to be made early in the design process to ensure that a safe, efficient, and reliable system is implemented in an Australian context.

INTRODUCTION

Australian commitments for fugitive emissions reduction

Methane (CH_4) is the second most important greenhouse gas (GHG) after carbon dioxide (CO_2) as a contributor to climate change. Methane has a higher global warming potential (GWP) than CO_2: 28 times that of CO_2 over 100 years (IPCC, 2014). This impact is even larger in the short-term with a GWP of 84 tCH_4/tCO_2 over 20 years. Although methane has a shorter atmospheric half-life compared to CO_2 (10–15 years compared to 1000 years), it absorbs more energy while it is in the atmosphere. Therefore, a reduction in methane emissions has the potential for near-term improvements in the rate of increase of atmospheric GHG concentrations compared against a business-as-usual trajectory.

Australian Federal Government policies and commitments that motivate fugitive methane emission reductions include the Safeguard Mechanism and the Global Methane Pledge. In 2023, the Safeguard Mechanism was reformed such that Australia's largest emitters are held accountable to emissions above a declining target referred to as a 'baseline'. A facility's baseline (ie their target) will be reduced in line with Australia's emission reduction targets of 43 per cent below 2005 levels by 2030 and net zero by 2050, with other intermediate steps) (Clean Energy Regulator, 2023a, 2023b).

Whilst the Safeguard Mechanism relates to a facility's overall CO_2 equivalent emissions, Australia also joined the Global Methane Pledge in 2022, which aims to specifically reduce methane emissions by at least 30 per cent below 2020 levels by 2030 (Commonwealth of Australia, 2022). The Pledge also includes initiatives across the energy and waste sectors, including capturing and/or avoiding fugitive methane emissions from coalmines and gas infrastructure.

State Governments have also released emission reduction targets that highlight the need for the reduction of emissions, particularly for potent GHGs such as methane.

Coalmine fugitive emissions

GHGs including methane and CO_2 are naturally produced in the formation of coal seams and escape during both open-cut and underground mining operations (CSIRO, 2021). Methane is a safety concern for underground mining operations, as methane at concentrations between 5–15 vol per cent in air mixtures is considered an explosion risk.

Underground coal operations use ventilation systems to move fresh air into the mine while removing methane and other unwanted gases from the mine. This ventilation strategy ensures the safe working conditions required for the mine's operation. The resulting exhaust air is traditionally vented to atmosphere. Methane that is exhausted as part of this stream is known as ventilation air methane (VAM). Figure 1 presents a basic schematic of an underground mining set-up, highlighting VAM compared to other sources of methane from underground mines (pre-drainage and goaf drainage) that are sold, utilised for power generation or flared (ie combusted to form CO_2) and therefore emitted at a lower GWP. Pre-drainage involves removing methane from the coal seams before mining activity commences (ahead of the longwall). Goaf drainage removes methane from the area just mined, as mining activity releases more methane from the seams.

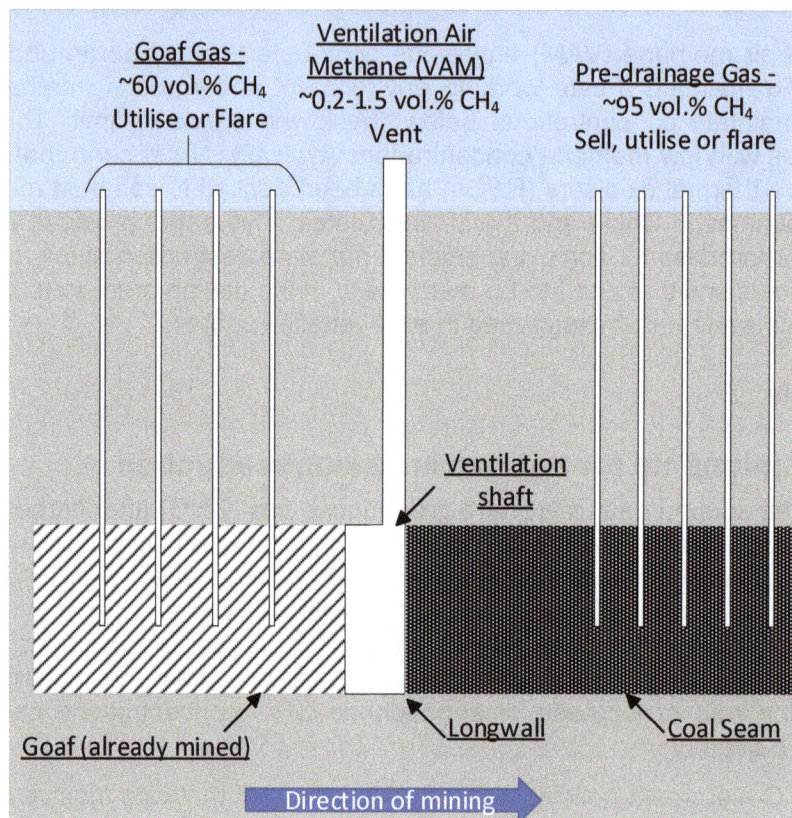

FIG 1 – Underground mine fugitive methane emissions.

The impact of coalmine fugitive methane on the total emissions in Australia is significant. Based on 2019 data shown in Figure 2, coalmine methane is approximately 25 $MtCO_2e$ per annum or 4 per cent of Australia's total emissions (Assan, 2022). Breaking down the coalmine methane emissions further, 70 per cent is produced from underground mines while the rest comes from open-cut mines. The biggest contributor to Australia's underground coal mining fugitive methane emissions is VAM, making up 82 per cent of the emissions or ~14 $MtCO_2e/a$. If all methane emitted through VAM were fully oxidised to CO_2 this would reduce to 0.5 $MtCO_2e/a$ and represent approximately a 55 per cent reduction from Australia's current coal mining methane emissions (25 to 11.1 $MtCO_2e/a$). This makes VAM an important emissions source for coal producers to address as they plan pathways to achieve their GHG targets.

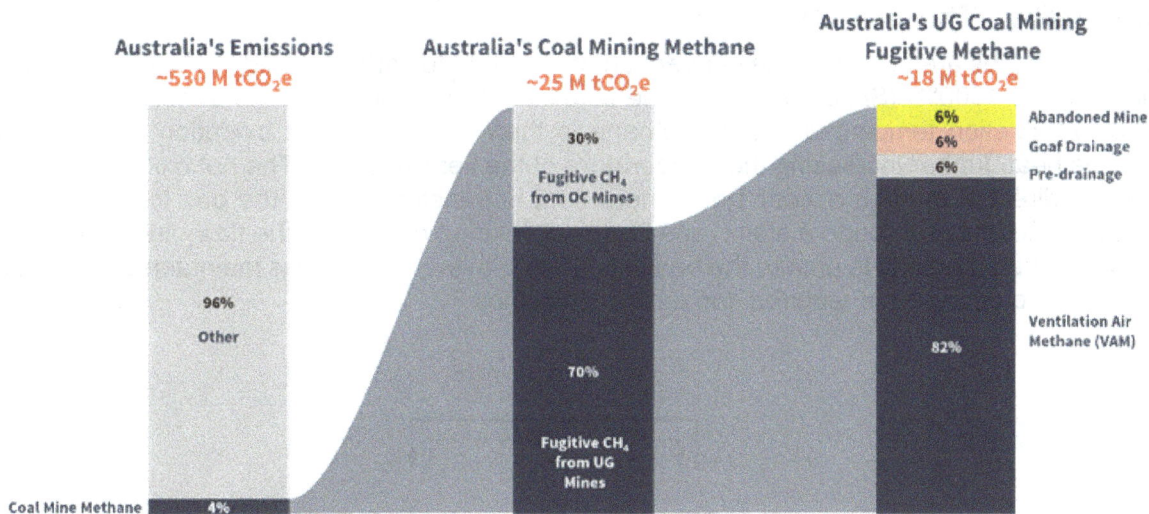

FIG 2 – Typical proportion of emissions – 2019 data (Assan, 2022).

Australian coal industry context

As reported in 2022, Australia has 27 underground coalmines within Queensland and New South Wales (NSW) (Assan, 2022) and none are currently operating VAM abatement systems. In Australia, West Cliff VAM project (WestVAMP) operated an regenerative thermal oxidiser (RTO) for abatement of 20 per cent of the available ventilation air at BHP's Appin Colliery from 2007 to 2017 (EPA, 2019) (Ashton Coal, 2011). South32 Illawarra Metallurgical Coal and CSIRO have collaborated in development and site trials of a VAM abatement technology from 2019 to 2023 (CSIRO, 2024). In 2023, BHP also announced that a ventilation air methane destruction project has begun for BMA's Broadmeadow coalmine (BHP, 2023).

VAM abatement systems must consider local state mining and safety regulations in Australia, as well as specific site standards. Currently, underground coal mining legislation does not specifically consider VAM abatement. As such, VAM abatement projects need to evaluate how to meet applicable legislation that will satisfy key internal and external stakeholders (regulators, safety inspectors, union/operations representatives, and corporate/management).

For example, in Queensland, the *Coal Mining Safety and Health Regulation 2017* states that unless otherwise undertaking authorised hot work, 'a person must not use equipment underground if its external temperature is more than 150°C'. Whilst a VAM abatement technology would be on the surface, it is connected to underground, therefore a sufficient control at the boundary between surface and underground for the VAM abatement system may be required to satisfy stakeholders.

Given that Australian mining legislation does not specifically address VAM abatement, the safety risks of connecting to an underground mine ventilation system, and the limited precedents in Australia, this paper aims to fill a gap by presenting some considerations when designing a VAM abatement plant in Australia.

VAM TECHNOLOGIES LANDSCAPE

VAM is a low methane concentration stream (typically around 0.2–1.5 vol per cent) with a high overall gas volumetric flow rate (in the range of 400 Am^3/s per shaft) (UNECE, 2016). This combination means methods for VAM destruction or use are difficult to implement due to high costs and technical complexity. As the concentration can vary and has large impacts on the system design, it is important to understand site-specific VAM flow rates and concentrations over an extended period and during different operating conditions to evaluate and design abatement opportunities.

VAM abatement technologies that have been commercially operated at coalmines in China and the USA employ the regenerative thermal oxidiser (RTO) technology. Details of RTOs and a brief overview of some other technologies that could be considered and developed for the coal mining industry are provided below.

Regenerative Thermal Oxidisers (RTO)

RTO technologies are based on the conversion of methane to carbon dioxide at high temperatures, in the range of 1000°C (Shi *et al*, 2019). The incoming ventilation air enters a 'hot' bed where it is heated to oxidation temperature. The methane is then oxidised in the oxidation chamber which generates heat, further increasing the temperature of the ventilation air. The hot oxidised ventilation air then is directed through a 'cold' bed, where heat is transferred from the gas to the bed before exiting to atmosphere through a stack (see Figure 3). During operation, the flow direction within the RTO is switched regularly to ensure the bed in contact with the inlet VAM is maintained at a sufficient temperature to achieve the required conversion efficiency.

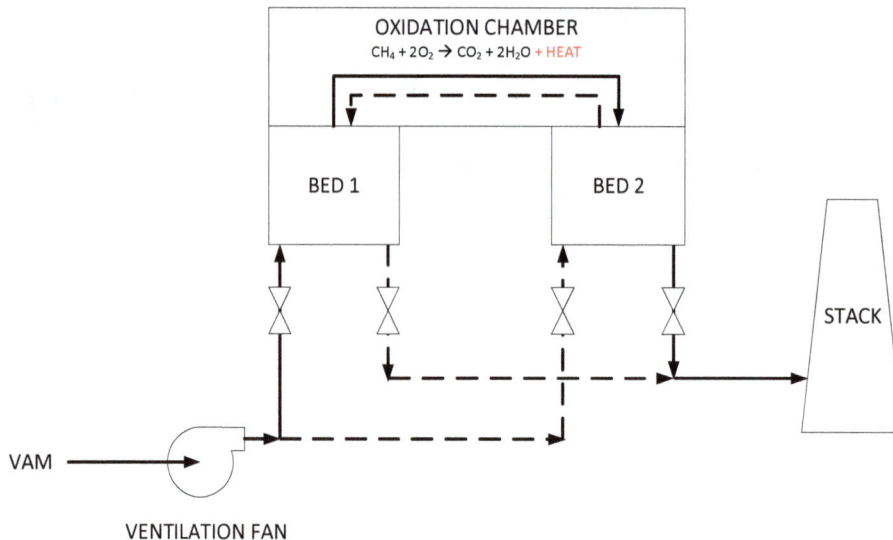

FIG 3 – Basic RTO diagram.

There are various parameters impacting the design of an RTO system, including VAM concentrations, dust levels and humidity of the ventilation air, and emissions from the RTOs (noise, NO_X and CO) (Shi *et al*, 2019). Furthermore, VAM RTO plants need to consider a safety system to address the additional risk and consequence of a methane-air mixture within the explosive range (methane at 5 to 15 vol per cent concentrations in air) given the presence of an ignition source. Safety considerations to address this risk are discussed later in this paper.

Other VAM abatement technologies

Table 1 describes VAM abatement technology alternatives to RTOs. The table includes a ranking for each with respect to Technology Readiness Level (TRL) and Commercial Readiness Index (CRI). TRL is a scale from 1 to 9 with TRL1 as blue-sky research to TRL9 as demonstrated at all expected operating conditions. CRI is a scale from 1 to 6 with CRI1 as a hypothetical commercial proposition and CRI6 as a bankable asset class. As there are operating RTOs in VAM abatement applications in USA and China, this indicates TRL9 and CRI5. On a global scale, RTOs for VAM abatement are experiencing market demand driving development, however in Australia there has not been a full-scale implementation of an RTO for VAM abatement.

TABLE 1

Comparison of alternative VAM abatement technologies.

Technology	Description	TRL[1]	CRI[2]	Comparison to RTOs

Regenerative Catalytic Oxidisers (RCOs)	Utilises a catalyst bed as part of the oxidation process that allows oxidation to occur at a lower temperature compared to RTOs.
	
	Similar to RTOs, RCOs use a flow reversal process and heating cycles within multiple beds in order to keep the reactor core temperature constant, ensuring maximum thermal efficiency.
		9	2	Advantages:
				• Operating temperature is lower than RTO. Ongoing research and development has targeted temperatures below the autoignition temperature for VAM.
				• Potential use for low concentration VAM.
				• Trialled in coalmines:
				   ○ CSIRO catalytic VAM abatement piloted at South32 Illawarra Metallurgical Coal (hot trials in August and September 2023) (CSIRO, 2024).
				   ○ RCO commissioned in coalmine in Duerping, China (CH4MIN™ developed by CSIRO CANMET and licensed by Sindicatum).
				• Electricity generation incorporated if VAM is maintained >0.8 vol per cent CH_4 (CSIRO, 2019).
				
				Disadvantages:
				• Dust can reduce the effectiveness of the catalyst.
				• Expensive and sensitive catalysts (typically platinum or palladium).
				• Lower adoption rate for VAM projects than RTOs (due to higher costs).

Photocatalytic oxidation	Photocatalysts absorb UV light, producing electrons and 'holes' along the surface which then become sites for methane oxidation, forming oxygenates and eventually carbon dioxide (Johannisson and Hiete, 2022).	3	1	Advantages: • No ignition source therefore it allows a safe connection to the ventilation system. • Potential application for open cut methane abatement (large surface area can be covered). Disadvantages: • High catalyst cost, catalyst decays over time due to photo-corrosion. • Dust treatment/removal is required. • Low TRL. • Only works when the sun is shining, so there is less abatement (VAM needs to be vented at night). Otherwise, cost trade-off of using artificial UV light for improved abatement.

Technology	Description	TRL[1]	CRI[2]	Comparison to RTOs
Chemical looping	Methane is oxidised by an oxygen carrier (eg metal oxides) – oxygen molecule in the metal oxides reacts with methane to form carbon dioxide and water. The de-oxygenated carrier metal is then regenerated (oxidised) (Daneshmand-Jahromi, Sedghkerdar and Mahinpey, 2023). Requires a reaction temperature of approximately 500–600°C (Yongxing, 2014).	5	1	**Advantages:** • Operating temperature is lower than RTOs but further investigation/development is needed to ensure max temperatures are sufficiently below auto-ignition temperature of VAM. • Demonstrations have shown good potential for abatement with very low inlet VAM concentrations (Doroodchi and Moghtaderi, 2012). **Disadvantages:** • Dust impacts the effectiveness of metal oxides (increases purging). • Low TRL – demonstrated at bench and pilot scales at University of Newcastle. Project ongoing between UoN and ASCON Energy to demonstrate the VAMCO (Versatile Advanced Methods of Cleaning Off-Gases) chemical looping technology to increase its TRL (currently TRL5) to a commercial product (TRL9) (CSIRO, 2023).
Methane absorption/ adsorption	Methane is selectively adsorbed/absorbed into a material to produce an exit stream with higher methane concentration. The exit concentration of ~30 vol per cent is viable to blend and sell to an energy producer (eg mixed with pre-drainage gas) or flare or use in a gas turbine (Shi et al, 2019).	5	1	**Advantages:** • Safe connection to ventilation system – produces methane above the upper explosive limit and is low in temperature. • Opportunity to enrich methane to usable concentration, ie blend with drainage gas to sell or use in processes that locks carbon such as pyrolysis (produces solid carbon and hydrogen). **Disadvantages:** • Low TRL (demonstrated in batch experiments but not through continuous operation). • High electrical power required to regenerate the sorbent. • Design, number of columns and operating conditions must ensure that the enriched methane is not an explosive mixture with air (5–15 per cent) (Shi et al, 2019).

1 – TRL: Technology Readiness Level (scale 1 to 9); 2 – CRI: Commercial Readiness Index (scale 1 to 6).

While the VAM abatement technologies in Table 1 have some potential advantages compared with RTOs, they generally have a combination of lower technology readiness levels, expected higher costs, fewer precedents in coal mining applications and/or reduced abatement performance compared to RTOs.

For coal mining companies that aim to develop industrial scale VAM abatement projects within the next five years to meet 2030 GHG emission targets, RTO is the option with the highest CRI and TRL.

Advancement of alternative technologies through further research and piloting may result in technologies other than RTOs emerging as preferred for VAM. This could include technologies that use/convert methane to other carbon products instead of emitting it as CO_2. Examples include pyrolysis (produces solid carbon and hydrogen), and methane conversion to methanol or single cell protein through reaction with enzymes such as methane monooxygenase (Blanchette *et al*, 2016). Single cell protein can be used as an input to animal feed production (fisheries/pets).

VAM RTO DESIGN FOR THE AUSTRALIAN COAL INDUSTRY

For the purposes of this paper, a VAM abatement plant includes:

- A capture system that takes ventilation air from the discharge of existing mine ventilation fan(s) connected to the mine ventilation shaft.

- Surface ductwork (duct and dampers) that connects from the capture system to the RTO(s).

- The RTO unit(s) to oxidise the methane in ventilation air.

- Service fan(s) that will draw ventilation air from the capture system, through the RTO(s) to the discharge of the stack(s).

The following discusses safety, technical, and project considerations for implementing RTOs in Australian coalmines.

Safety

Safety is an important design consideration for the use of RTOs in coalmines. Each RTO project needs to apply an appropriate safety assessment to address the risks and consequences associated with a potential methane in air mixture, above the lower explosive limit (LEL) of 5 per cent, reaching an RTO operating above the auto-ignition temperature of methane.

Various factors make it unlikely that methane concentrations above the LEL will be experienced in the VAM stream in normal operations:

- Pre-drainage of the mining area is applied to remove as much methane from the seam as practicable before mining. However, during the act of mining coal, seams are fractured, and more methane is released. Draining the area already mined (goaf drainage) is another important control for VAM concentrations.

- Underground gas exceedance control measures applied to coalmines are required to measure and monitor methane concentration using gas detectors.

- Ventilation systems in underground coalmines are controlled to keep the methane concentration in the ventilation system well below the LEL.

With respect to this last point, in Queensland for example, a methane concentration of equal to or greater than 2.5 per cent in air is considered dangerous under the Coal Mining Safety and Health (Methane Monitoring and Ventilation Systems) Amendment Regulation 2019 (Qld). If this occurs the regulation stipulates that underground workers should be withdrawn from the mine and an inspector from Coalmines Inspectorate must be notified. In New South Wales, the reportable concentration is 2 per cent methane under Work Health and Safety (Mines and Petroleum Sites) Regulation 2022 (NSW). Likewise, in Western Australia, under the Work Health and Safety (Mines) Regulations 2022 (WA), the atmosphere must not rise above 1.25 per cent by volume of methane in a production area, 0.25 per cent in the intake air, and 2 per cent in the return air.

There is a risk that pockets of methane in concentrations above the typical 0.2 to 1.5 per cent can occur for under various unplanned or atypical process and operational reasons, including:

- Improper drainage (failure or reduced capacity of in-seam drainage pumps).
- Strata failure causing a surge in methane.
- Goaf collapse that is bigger than expected.
- Improper ventilation changes made underground.
- Start-up of the ventilation system after a mine shutdown.

The hazard scenarios applicable will depend on the specific underground mine and ventilation system being studied. Assessment and confirmation of the required risk reduction using established safety assessment methodologies is vital in RTO projects. Many options are available for technical risk analysis and it is highly recommended that projects utilise a structured approach to the identification and treatment of technical risk, such as that defined in Australian Standards 'AS 61508 Functional safety of electrical/electronic/programmable electronic safety-related systems' and/or 'AS IEC 61511 Functional safety – Safety instrumented systems for the process industry'.

This systematic approach requires the development of a plan, including identification of risk analysis methodology and key stakeholders, prior to the risk analysis being undertaken. All projects require a specific risk analysis to establish a target risk reduction for the new facility that takes into account the existing controls in place underground and the risk tolerance of the organisation. As a project moves through the phases of the life cycle, verification steps ensure the outputs of each phase are appropriate based on the activities undertaken.

Preventative measures

Safety systems should be considered to detect high concentrations of VAM approaching the LEL. This will allow actions to be taken to prevent the mixture from entering the RTO in combustible form. Actions could be in the form of isolation and diversion of the mixture away from the RTO and/or dilution of the VAM to methane concentrations below LEL.

Several items that impact the effective detection, dilution, and diversion. The following items will be further discussed in this paper with respect to being an effective preventative measure:

1. Methane detection sensors (including type and location).
2. Means of diversion of ventilation air from RTO (passive or actuated).
3. Impact of connection configuration (including damper location) on safety.

Methane detection sensors

The methane sensors used in a VAM RTO context are a key component of the safety system and should be fast acting and sensitive to the specific range of VAM concentrations. The preferred gas sensor for the VAM RTO application may be different to gas sensors existing on-site. The sensors must be configured at a trip point that allows the RTOs to operate within the expected range of methane concentration, but low enough to allow for the safety system to respond given the response curve for gas detectors.

The location of methane detection is an important consideration to ensure there is sufficient time between the detection of a high VAM concentration triggering an action and the dampers completing an action (either closing for isolation or opening for dilution and diversion). Sensor location can broadly be categorised as underground (eg within the mine or at the base of the ventilation shaft in the mine) or on the surface (downstream of the duct connection to the ventilation shaft). Figure 4 shows example configurations for VAM safety systems with underground and aboveground sensors.

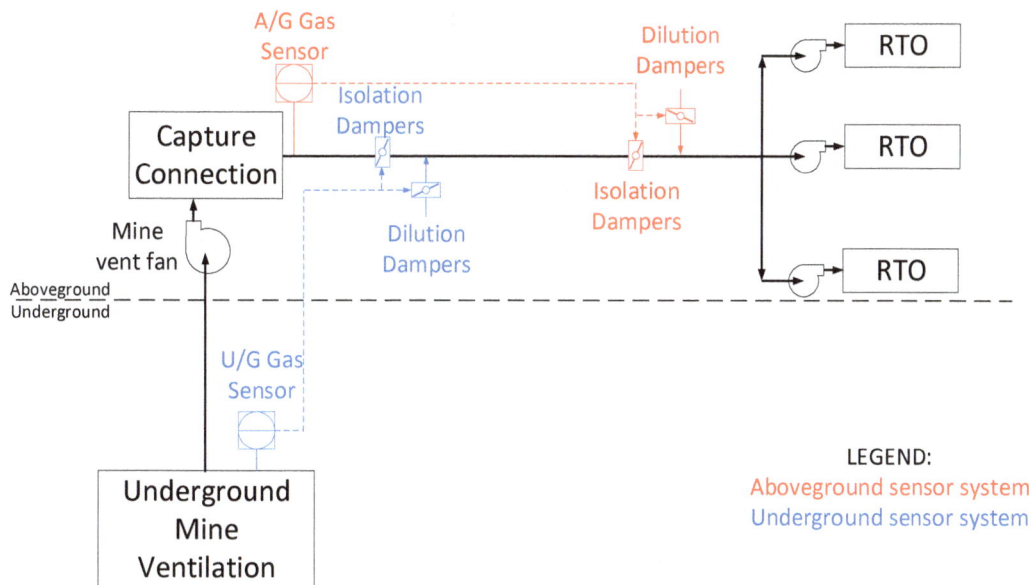

FIG 4 – Example configuration for VAM RTO safety systems with underground and aboveground sensors.

For underground sensors, the maximum allowable time between detection and the dampers completing an action to avoid methane above the LEL entering the RTO is determined by the ventilation air velocity and distance through the shaft and surface duct to the dampers. For sensors at the surface, the allowable time is a function of the velocity in the duct and the duct length to the RTO. Therefore, there are potential cost and footprint savings associated with minimising the surface ductwork by using an underground sensor. Faster acting dampers also help to mitigate the duct length required.

There are a number of considerations associated with underground sensor selection for application in VAM abatement projects:

- Sensors used underground are restricted to those explicitly approved by the testing organisations identified by the Regulator or other approver for use in the environment.

  o Stand-alone methane detectors which carry an appropriate IEC 61508 safety certificate in this application underground are not widely available, since SIL (Safety Integrity Level) rated gas detection in underground coal mining applications are part of an overall 'system' certification. This will impact the achieved functional safety of a gas sensing arrangement.

  o Two sensor types are approved for underground use- catalytic bead and infrared. It is common that both catalytic bead and infrared sensors are deployed underground for methane sensing. Catalytic bead type sensors are vulnerable to 'poisoning' from excessive methane and/or low oxygen concentrations. This is a concern for a safety system in RTO VAM installations because poisoning can compromise its ability to measure a methane exceedance. Thus, once an exceedance is detected, the sensor must be refurbished/replaced and tested; maintenance intervals need to be set with this in mind. The hazardous area classification is different for surface installations, allowing for a wider variety of measurement types which can be less affected by poisoning.

  o Technologies involving fibre optic methane sensors, capable of achieving intrinsically safe certification, will provide a high-speed alternative to typical catalytic bead and infrared detectors currently used underground. However, these products are not yet commercially available and projects should monitor their development.

- Sensors used in underground applications follow the requirements of 'AS/NZS 60079.29.1 Explosive atmospheres Gas detectors – Performance requirements of detectors for flammable gases' which may not provide sufficient confidence in the efficacy of the detector. Options to address this include:

- o Testing of the sensors in return airway flow conditions.
- o Implementing frequent maintenance of the sensors.
- o Triggering the underground sensor at set points lower than the T90 measuring range (time taken to reach 90 per cent of the final reading) to improve the response time (as accuracy is not the primary means of achieving functional safety).

Aboveground sensing, installed in the surface ducting, allows for the use of detection technologies not explicitly approved for use in underground coalmines, assuming appropriate controls at the boundary between underground and surface facilities. These devices have extremely fast response times, are highly accurate and can achieve SIL 3 in redundant architectures. There are a number of manufacturers with product offerings, and common applications include gas turbine enclosure ventilation and process off-gas monitoring. Installations of RTOs in VAM applications in other jurisdictions also use similar sensors at the inlet to individual RTO units to monitor for the presence of methane and react to close the isolation damper and open the dilution damper. Careful sensor selection, specification, and offline/online maintenance considerations are required to cater for the potential high flow/gas velocity, dust concentration and condensation expected in the surface duct application.

Diversion of ventilation air

Diverting ventilation air away from the RTOs (to atmosphere) in the event of a trip or isolation of the VAM RTO plant is required to prevent a trip of the main mine ventilation fans or compromising the performance of the underground mine ventilation system.

The connection of the VAM RTO plant ductwork to the mine shaft can be designed to have passive diversion such that ventilation air will be directed to the atmosphere when the RTOs are isolated. Examples of connections that will achieve passive diversion include a decoupled connection (ie a hood extracts the ventilation air from the outlet of the ventilation shaft) or a partially coupled connection which has both a connection to the VAM plant and a vent to atmosphere. A fully coupled connection would not be able to provide passive diversion. Figure 5 shows simplified schematics of a fully coupled, a partially coupled, and a decoupled connection. Any capture connection will need to be designed to minimise ingress air during normal VAM RTO operation and maximise capture efficiency of VAM for abatement. Design considerations on the capture connection is discussed further in the paper.

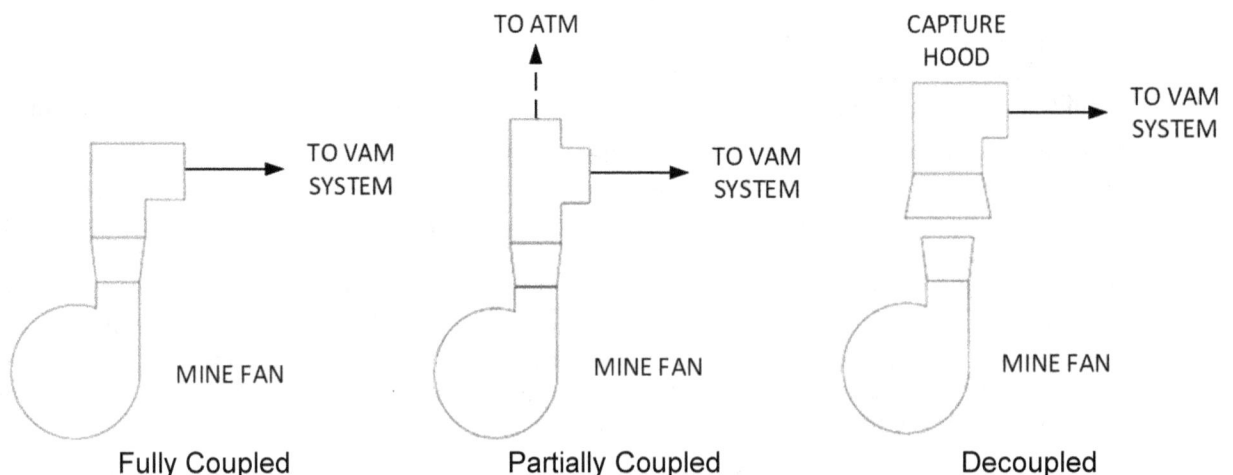

FIG 5 – Example capture systems.

Diversion can also be achieved through use of actuated dampers installed upstream of the isolation dampers. These dampers would simultaneously open when the isolation dampers are closed to allow venting of the ventilation air to atmosphere, maintaining mine ventilation in fully coupled configurations. Figure 6 shows an example configuration with actuated diversion dampers.

Compared to passive diversion, consideration must be given to the speed and reliability of opening the damper when required.

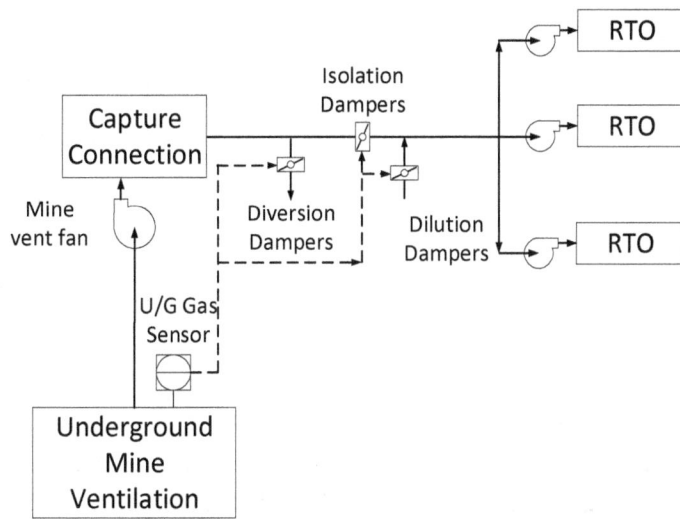

FIG 6 – Typical configuration of actuated diversion dampers for an underground gas sensor system.

Connection configuration

The air flow rates from a typical, single mine vent shaft exceed the capacity of commercially available RTO units. Because of this multiple RTOs units may be required to maximise the abatement from a single mine vent shaft. Several configurations can be used for connection of multiple RTO units in parallel to the mine vent shaft. Each has different implications for the safety system.

Table 2 and Figure 7 outline three possible configurations indicating isolation and dilution actuated dampers. In each, a passive diversion system is assumed and not shown on the figures for simplicity. For the purposes of this configuration discussion, a typical mine ventilation shaft is assumed, with three mine vent fans (2 duty/1 standby) having a capacity of 150 Am^3/s per fan.

TABLE 2

Comparison of example configurations for VAM RTO safety system.

Parameter	Configuration 1	Configuration 2	Configuration 3
Description	All vent fans discharge to a common manifold which feeds a bank of RTOs (isolation and dilution occurring on the common manifold).	All vent fans discharge to a common manifold which feeds two RTO banks (isolation and dilution are at the common feed to each bank).	Each mine vent fan discharge duct has isolation and dilution and is connected to a dedicated RTO bank. This requires more installed RTO capacity than the other configurations with common manifolds.
Dampers	Isolation and dilution dampers on large diameter common manifold would require longer actuation time for close/open action.	Isolation and dilution dampers on the individual banks require less actuation time than on a single common manifold.	Isolation and dilution dampers on the discharge duct for each individual fan are of moderate size, requiring less actuation time than on configurations with common manifolds.
Functional safety implications	Only one set of dampers feeding RTOs. Failure modes can be addressed through standard engineering practices.	Isolation and dilution of two separate lines. Failure modes are more complex and additional engineering required (multiple, independent safety systems).	Isolation and dilution of three independent separate lines. Failure modes are more complex and additional engineering required (multiple, independent safety systems).

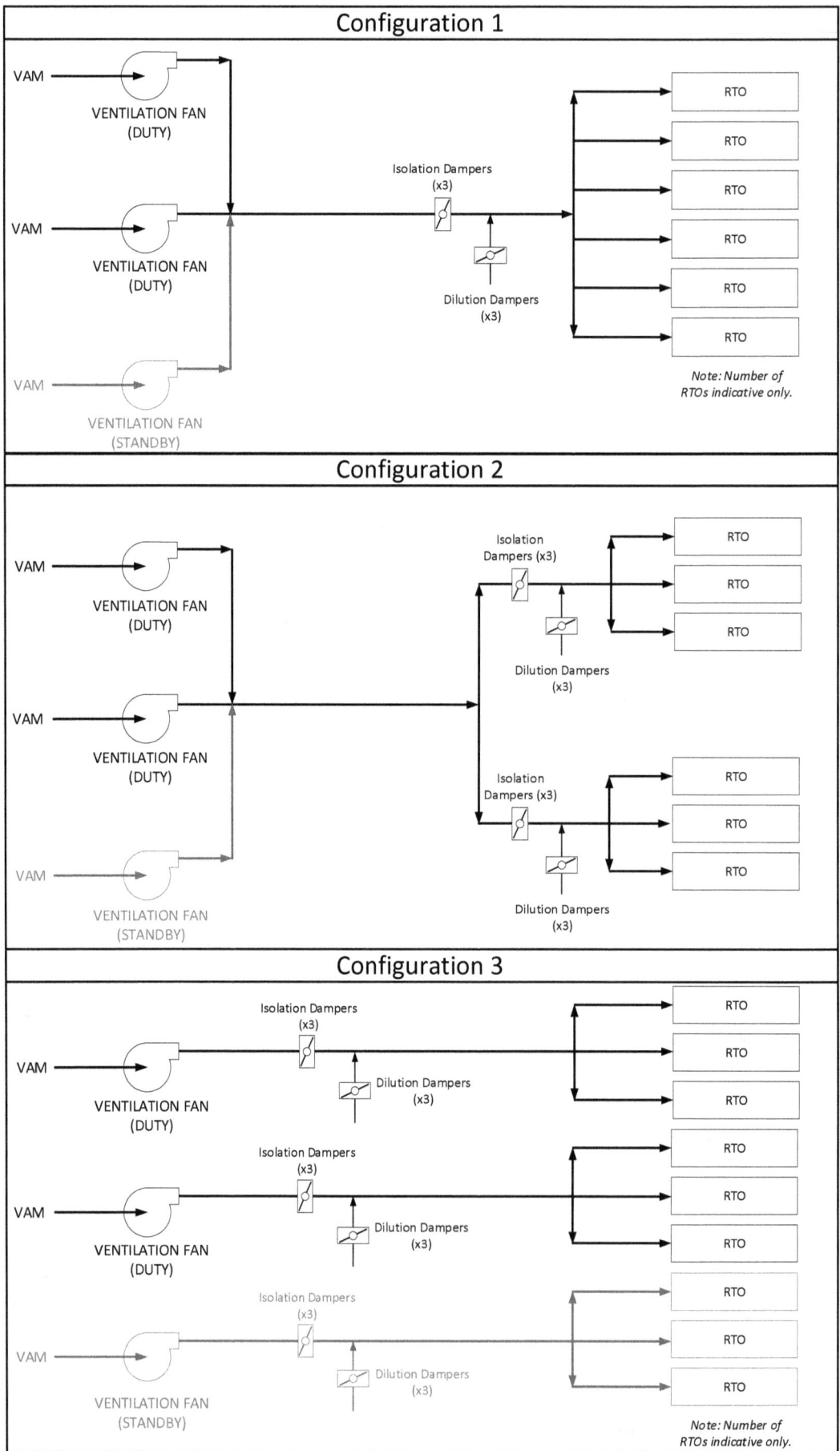

FIG 7 – Example configurations for VAM RTO safety system.

Mitigation measures

Effective mitigation measures can prevent propagation back to the mine and lessen the impacts of an explosive methane-air mixture that has entered an RTO, including secondary impacts to surrounding areas with personnel or equipment. Several deflagration and detonation barrier/arrestor systems can be considered for the VAM RTO duty, drawing on experience from other flammable gas handling industries. Many of these are commercially available in those industries. Examples include:

- Pressure relief or venting. This is a passive mitigation measure to reduce the explosion pressure but will not alone prevent propagation.

- Suppression of the flame front through passive (flame arrestors) or active (fast acting detection and chemical suppression) measures.

- Containment of the flame by isolating the expected explosion/flame through separation or rapid acting dampers (physical barriers).

The key challenges with applying these technologies to a VAM application are:

- Multiple layers of safety systems are likely required to both mitigate and arrest the flame front and minimise the blast impact on equipment and surrounding plant/buildings.

- Passive dry flame arrestor systems, while less complex than active suppression systems, are commercially available in other applications, but are still being developed for the large diameter VAM ducts and are prone to dust build-up that can impact performance. Regular cleaning would need to be considered.

- Water and stone dust flame suppressions systems currently used in underground coalmine workings are not preferred in ductwork applications and would need to be engineered for the different conditions expected in the VAM ductwork (eg higher gas velocities).

In addition, as another form of mitigation, blast impact assessments are recommended to: 1) help locate the RTO to avoid potential impacts on equipment or buildings within the blast radius, and 2) determine the level of protection required for critical equipment or occupied buildings within the vicinity of the VAM equipment, if locating everything outside the blast radius is not practical.

Technical considerations

Important technical considerations for VAM abatement plants in Australian coalmines include VAM concentrations exceeding the capability of the RTO equipment and high dust in the ventilation air. Other general aspects of design including capture design optimisation, mobility of the VAM abatement plant, humidity of the ventilation air, and preheating after shutdowns are not unique to Australian mines.

VAM concentrations exceeding RTO capabilities

As described in the *Safety* section, ventilation practices for Australian coalmines follow the state regulations for methane concentrations, which result in typical VAM concentrations of 0.2 to 1.5 vol per cent. However, VAM concentrations should be characterised (ie measured) on a continuous basis over an extended period of time and during different mine operations (including for different ventilation flow rates and pressures as well as during longwall operation). This will allow equipment operating requirements to be assessed against realistic conditions that are specific to the site. Characterisation of VAM concentration will determine requirements of the RTO design such that the lower specification limit will influence bed sizing and media type and the upper limit will determine the need for heat rejection, both of which ultimately influence cost.

The allowable range of operating methane concentration for RTOs is based on fundamental technology limitations. High VAM concentrations can lead to overheating beyond the equipment design temperature. Minimum VAM concentration for RTO can go as low as 0.2 vol per cent, however 0.4 vol per cent and above is preferred for sustained operation. When the methane concentration is too low, the RTO units cannot achieve a self-sustaining temperature due to reduced heat generated from oxidation.

If VAM concentration is too high and causes high temperature in the RTO, a hot side bypass mode can be triggered whereby a portion of oxidised ventilation air from the oxidation chamber is sent to stack instead of to the ceramic bed.

Where the VAM concentration, on average, is at or below the RTO operating range, further engineering is required to ensure cost-effective RTO design, stable operation and adequate efficiency. This could be achieved in several ways and may be a unique solution for each mine. Options include:

- Optimising ventilation practices underground to target higher concentrations at the mine vent shaft. This could potentially be achieved by:

  o Introducing a dedicated upcast shaft for the purpose of having the VAM concentration optimal for RTOs. There are multiple places where the upcast shaft could be located, and it is dependent on what contributes most to VAM emissions. For example, if the longwall of a mine contributes the most to VAM, an option could be to construct a shaft or raise located in the longwall tailgate to service longwall panels. An advantage of this option is that the dedicated VAM stream is targeted for optimal concentration and flow rates for RTOs with minimal impact to underground operations. The disadvantages include high CAPEX for a new dedicated vent route and losing the opportunity of 100 per cent VAM capture since the other vent shafts will not be at the required RTO concentrations. It will however reduce the overall volume and improve the economics of the RTO unit per kg of VAM abated.

  o Optimising ventilation flow rates and shaft positions to ensure certain existing shafts have VAM at required concentrations. The amount captured will depend on whether the system can be optimised for all shafts to be at concentrations ideal for RTOs. It may be that only some vents will be targeted for VAM and others will be too low for efficient capture. An advantage of this option is low CAPEX whilst a disadvantage is that it requires modification of the ventilation system underground.

- Increasing the methane concentration of VAM by enriching it with high concentration methane such as pre-drainage or goaf drainage gas. Considerations for this option include:

  o If coalmine gas, used for enrichment, could otherwise be sold or used to produce power, the net operating cost will be impacted. If coalmine gas is not available for this option, other gaseous fuels can be used but this is also less attractive due to higher operating cost and lower GHG emissions reduction.

  o A safety critical system will be required to control and monitor enrichment gas injection to prevent a hazardous event of an ignition resulting from build-up of methane within the enrichment gas duct. Supplemental fuel injection (typically natural gas) when volatile organic compound (VOC) loads are low is used in existing RTO plant installations in other industrial processes controlling VOC emissions (Anguil, 2024). Controls include physically limiting the flow-through orifices or pipe sizes, and employing starting sequences for the units to ensure fresh air is flushed through the enrichment system prior to start-up of RTOs.

Dust in ventilation air

Dust in ventilation air is generated from various activities underground and consists of coal dust and inert stone from any stone dusting activities. RTO vendors typically require the inlet dust loading to be below ~10 mg/m³ (vendor specific). Organic dust can be oxidised periodically in the RTO in a 'bake-out' mode where each bed is exposed to elevated temperatures. However, inert dust cannot be oxidised and leads to increased risk of RTO bed fouling which causes increased pressure drops, blockages, and will require replacement/cleaning of media. BHP's WestVAMP identified issues with dust in ventilation air which caused premature replacement of the ceramic medium in the RTO beds (Ashton Coal, 2011).

Dusty ventilation air can also lead to build-up of dust in the VAM plant ductwork, especially at low velocities. An option to mitigate this is to design the ductwork to maintain velocities reasonably high to avoid solids dropout but low enough to avoid erosion issues. Additionally, duct design should consider ease of access for cleaning.

Measurement of dust levels and characterisation of the dust is recommended to support the RTO design. If levels are higher than recommended, then the following options are to be considered:

- Install dust collection system, such as a baghouse, upstream of the RTO units. This option will have significant CAPEX impact and will have additional operating considerations/risks associated with handling coal dust.

- If high dust levels only occur during certain infrequent and/or short-term activities, it may be possible to divert VAM away from the RTOs during these periods.

- Include an operational plan and budget for regular maintenance/replacement of the media bed. It is recommended to work with the RTO vendor to ensure the equipment design caters for easy and safe change out of ceramic media.

Other considerations

Other important aspects of VAM RTO plant design that are not unique to the Australian context include:

- Factors influencing the number of RTOs includes the profile of the VAM flow rate, the desired capture rate of VAM, and the required turndown rate of the RTOs (ratio of minimum operating flow rate and design flow rate), as well as standard vendor sizes and modularisation requirements. Equipment specification and selection should take the required turndown into account.

- The design of the capture and connection from the ventilation shaft to the RTO duct should be optimised considering the following:

  o The VAM abatement system must not impact the availability or compromise the performance of the mine ventilation system, particularly during a trip of the VAM abatement system.

  o Design to capture and abate as much VAM as possible, ie minimise VAM discharge to atmosphere.

  o Minimise VAM dilution to avoid oversizing the abatement plant and operating outside the RTOs' range for inlet methane content.

- As the mine development follows the coal seams, the location of the active ventilation shaft will move throughout a mine's lifetime. Design of the RTO unit and upstream common ductwork components should consider mobility including sizing to fit on a standard truck, and modular designs to assist with ease of relocation.

- If ventilation air humidity is too high, water vapour may condense as the air decompresses to surface conditions and contacts ductwork or equipment surfaces at lower temperatures. Corrosion and buildup is a concern for the ductwork, dampers and various components on the inlet side of the RTO units. Corrosion and buildup could result in reduced availability of the system through increased downtime for repairs. Corrosion could also result in significant ingress air over time which would dilute the VAM and reduce the capacity of the RTO abatement plant to capture VAM from the vent shaft. Options to mitigate corrosion and buildup due to humidity include:

  o Specifying appropriate materials of construction for corrosion protection such as insulation of main common ductwork, or using stainless steel for inlet components to the RTO. The cost of this will need to be weighed against the corrosion risk considering the likelihood and cost of replacement of components.

  o Reducing relative humidity of the ventilation air by, for example, increasing the temperature.

- RTOs require preheating of the ceramic beds after shutdown periods. This preheat can be achieved either through electric heaters or burners fired on natural gas, diesel, or coalmine drainage gas. The preheat method will be dependent on the vendor selected as well as the quantity and quality of drainage gas available on-site.

TECHNICAL RISK MITIGATION

A common method to mitigate the risk associated with implementing a new facility or system is through piloting. Depending on the nature of the facility and the commercial landscape, piloting could consist of operating a vendor-supplied modular pilot or construction of a purpose-built pilot.

For VAM RTOs, risks outlined in the section *Technical Considerations* can be addressed and key areas of uncertainty reduced, by staging the project implementation starting with a single RTO module and gaining operating experience prior to implementing the remaining units. The single RTO module essentially acts as a pilot facility and can then transition to one of the commercial scale modules. This will assist in building operator and stakeholder confidence that the VAM system will not compromise safety or impact operations.

While some components can be investigated through stand-alone tests off-site, the following items are recommended to be investigated as part of the single module operation, before implementing full-scale operation.

- Methane detectors: test response time, accuracy, and repeatability of sensors in actual operating conditions. This is important for both underground and aboveground mounted sensors, to validate the ability of sensors to activate the necessary safety control response (this could even be done in advance of the single module).

- Dampers (isolation, dilution and diversion): test the reliability and response time of the dampers before RTOs are connected to the VAM duct or brought online. Response time of dampers can be tested before installation in system, ie initially in cooperation with vendors at their facilities and then during cold commissioning of the single module RTO.

- Safety related control system: test response time and reliability of PLC and communications with sensor and actuator, again during commissioning and before the RTOs are connected or brought online.

- Connection to ventilation system: confirm and optimise design to maximise percentage of VAM captured and minimise air infiltration from the connection. Validate that mine ventilation system is unaffected by VAM abatement plant operation/trips.

- RTO: test the following:
  - Ability to handle different methane concentrations (check destruction efficiency and minimum CH_4 for self-sustaining oxidation).
  - Tolerance of RTO units to stone dust, coal dust and water vapour.
  - Start-up, shutdown, and operability testing.
  - Frequency of ceramic media replacement due to stone dust and other contaminants.

- All components: validate required maintenance requirements.

- Enrichment system (if applicable): validate location and control strategy for the enrichment gas injection.

CONCLUSIONS

Recent Australian Government policies have incentivised the Australian coal mining industry to take action on addressing fugitive methane emissions. VAM is the largest contributor to fugitive coalmine methane and whilst there are various solutions to abate VAM, RTOs offer the lowest technical and commercial risk solution. This is because they are already being used in coalmines in both the USA and China, and they have the highest TRL and CRI values of the technologies reviewed.

The paper outlines the importance for VAM RTO project implementation in Australia to include rigorous design of safety systems with sufficient reliability and redundancy. With respect to safety, upset conditions have the potential to exceed the allowable and safe inlet VAM concentration for an RTO. Assessment and confirmation of the required risk reduction using established safety assessment methodologies is important to comply with Australia's mining regulations and standards.

Several technical challenges have engineered solutions as outlined in this paper, however all solutions start with characterisation of the ventilation air (determine flow rate, VAM composition, dust and humidity levels over a variety of operating scenarios). The impact of some of these parameters present a risk to the operability and success of the RTO system. As such, Operators may want to consider a program of test work and piloting trials before full scale implementation to validate that the proposed engineering solutions are appropriate for a safe and optimised VAM abatement system. Given that the RTO system will likely be modular in nature for a VAM application, a practical way to pilot an RTO is proposed. The suggested approach is a sequential implementation, starting with installation and operation of the first module as a pilot unit. Once the required information has been gathered, and the system safety has been demonstrated, a pathway to full scale commercial can be developed.

REFERENCES

Anguil, 2024. Overview of Emission Control Technologies [online]. Available from: <https://anguil.com/resources/overview-of-emission-control-technologies/> [Accessed: March 2024].

Ashton Coal, 2011. Ashton Coal Mine Greenhouse Gas Abatement Investigation Report [online]. Available from: <https://www.ashtoncoal.com.au/icms_docs/254828_modification-10-environmental-assessment-appendix-9.pdf> [Accessed: March 2024].

Assan, S, 2022. Tackling Australia's Coal Mine Methane Problem [online], Ember. Available from: <https://ember-climate.org/insights/research/tackling-australias-coal-mine-methane-problem/>

BHP, 2023. Annual Report 2023 [online]. Available from: <https://www.bhp.com/investors/annual-reporting/annual-report-2023>

Blanchette, C, Knipe, J, Stolaroff, J, DeOtte, J, Oakdale, J, Maiti, A, Lenhardt, J, Sirajuddin, S, Rosenzweig, A and Baker, S, 2016. Printable enzyme-embedded materials for methane to methanol conversion, *Nature Communications*, 7(11900).

Clean Energy Regulator, 2023a. Quarterly Carbon Market Report September Quarter 2023, Australian carbon credit units (ACCUs) [online]. Available from: <https://www.cleanenergyregulator.gov.au/Infohub/Markets/Pages/qcmr/september-quarter-2023/Australian-carbon-credit-units-(ACCUs).aspx> [Accessed: February 2024].

Clean Energy Regulator, 2023b. The Safeguard Mechanism [online]. Available from: <https://www.cleanenergyregulator.gov.au/NGER/The-Safeguard-Mechanism> [Accessed: December 2023].

Commonwealth of Australia, 2022. Australia joins Global Methane Pledge [online]. Available from: <https://minister.dcceew.gov.au/bowen/media-releases/australia-joins-global-methane-pledge> [Accessed: October 2023].

CSIRO, 2019. Site Trials of a Suite of Novel VAM Technologies [online]. Available from: <https://publications.csiro.au/rpr/download?pid=csiro:EP20728&dsid=DS2> [Accessed):1 September 2022].

CSIRO, 2021. Mine ventilation air methane abatement [online]. Available from: <https://www.csiro.au/en/work-with-us/industries/mining-resources/Mining/Fugitive-emissions-abatement/Mine-ventilation-air-methane-abatement> [Accessed: Feb 2024].

CSIRO, 2023. Achieving Negative Emissions in Production of Green Steel and Green Chemicals Using the VAMCO Family of Gas Separation Technologies [online]. Available from: <https://research.csiro.au/hyresearch/achieving-negative-emissions-in-production-of-green-steel-and-green-chemicals-using-the-vamco-family-of-gas-separation-technologies/> [Accessed: February 2024].

CSIRO, 2024. Development and site trials of a novel pilot ventilation air methane catalytic mitigator [online]. Available from: <https://www.resourcesregulator.nsw.gov.au/sites/default/files/2024–05/csiro-s32-vam-project-final-report.PDF> [Accessed: May 2024].

Daneshmand-Jahromi, S, Sedghkerdar, M H and Mahinpey, N, 2023. A review of chemical looping combustion technology: Fundamentals and development of natural, industrial waste and synthetic oxygen carriers, *Fuel*, 341.

Doroodchi, E and Moghtaderi, B, 2012. Chemical looping removal of ventilation air methane, Australia, Patent No, AU2012315483A1.

EPA, 2019. Ventilation Air Methane (VAM) Utilisation Technologies [online]. Available from: <https://www.epa.gov/sites/default/files/2019-11/documents/vam_technologies.pdf> [Accessed: September 2023].

IPCC, 2014. Climate Change 2014: Synthesis Report, Geneva, Switzerland: Intergovernmental Panel on Climate Change.

Johannisson, J and Hiete, M, 2022. Exploring the photocatalytic total oxidation of methane through the lens of a prospective LCA, *Atmospheric Environment: X*, 16(100190).

Su, S, Yin, J, Yu, X, Bae, J, Jin, Y, Cunnington, M, Villella, A, Loney, M, Ashby, M, Hipsley, R and Jara, M, 2019. Site Trials of a Suite of Novel VAM Technologies (CSIRO).

UNECE, 2016. Best Practice Guidance for Effective Methane Drainage and Use in Coal Mines (United Nations).

Zhang, Y, 2014. Utilisation of ventilation air methane in chemical looping systems, PhD thesis, University of Newcastle.

LEGISLATION

Queensland Government, 2020. *Coal Mining Safety and Health (Methane Monitoring and Ventilation Systems) Amendment Regulation 2020* (Qld).

NSW Government, 2022. *Work Health and Safety (Mines and Petroleum Sites) Regulation 2022* (NSW).

Department of Energy, Mines, Industry Regulation and Safety, 2022. *Work Health and Safety (Mines) Regulations 2022* (WA).

Development path of coal and gas co-extraction technology in dual-carbon strategy

J Zhang[1], S G Li[2], H F Lin[3], M Yan[4] and Y Bai[5]

1. Doctor, College of Safety Science and Engineering, Xi'an University of Science and Technology, Xi'an, Shaanxi 710054, China. Email: 782356383@qq.com
2. Professor, College of Safety Science and Engineering, Xi'an University of Science and Technology, Xi'an, Shaanxi 710054, China. Email: lisg@xust.edu.cn
3. Professor, College of Safety Science and Engineering, Xi'an University of Science and Technology, Xian, Shaanxi 710054, China. Email: lhaifei@xust.edu.cn
4. Professor, College of Safety Science and Engineering, Xi'an University of Science and Technology, Xian, Shaanxi 710054, China. Email: minyan1230@xust.edu.cn
5. Professor, College of Safety Science and Engineering, Xi'an University of Science and Technology, Xian, Shaanxi 710054, China. Email: 978486384@qq.com

ABSTRACT

Since the proposal of the 'dual-carbon' target, China's ecological civilisation development has advanced to a crucial point, with carbon reduction serving as the key strategic direction. The position of coal as the primary energy source won't change in the near future. The development route of coal and gas co-extraction technology in dual carbon strategy is formulated, and the main technical issues are examined, based on the accurate analysis of the challenges faced by coalmine CH_4-CO_2 dual carbon emission reduction in carbon peak and carbon neutral stage. In order to achieve zero emissions from CH_4 extraction and low emissions of CO_2, the development characteristics of 'low-carbonisation of the extraction system (L)-accurate drilling construction (A)-intellectualisation of equipment development (I)-dynamic regulation of whole process (D)' are proposed for AGE (accurate gas extraction) at the stage of carbon peak. China's 'dual-carbon' strategy target will eventually be accelerated by building a technically sound, economically sound, long-term safe carbon emission reduction technology system at the carbon neutrality stage. The use of 'CCUS + ecological carbon sink', full-concentration gas gradient utilisation, and other carbon-negative technologies will be reinforced. The goal is to offer creative development ideas for the accomplishment of carbon neutrality in China's coal sector through the progressive adoption of this concept.

INTRODUCTION

Global warming is one of the biggest threats to human life and growth in this century (Rogelj *et al*, 2016). The usage of fossil fuels like coal is directly related to carbon emissions, the main source of the greenhouse effect. The greenhouse gases that contribute most to the global increase in radiation stress are CO_2 and CH_4. In response to climate change, 191 countries have accepted the Paris Agreement and committed to transitioning to a low-carbon economy (Meinshausen *et al*, 2022). In light of the critical time that China's ecological civilisation construction has entered a critical strategic direction of carbon reduction, the Central Committee for Comprehensively Deepening Reform held its second meeting in July 2023. The committee emphasised the need to strengthen the regulation of total energy consumption and intensity and gradually transition to the dual control of total carbon emissions and intensity (Udeagha and Muchapondwa, 2023). Many firms are directed by the primary idea of 'carbon peak and carbon neutrality' (dual-carbon) to implement precise regulations, modify policies in response to local circumstances, and improve their own emission reduction technology system from several angles (Zhang *et al*, 2022). China has a resource endowment that is heavily dependent on coal, which places significant restrictions on the amount of space available for carbon emissions and the timing of emission reduction. Furthermore, coal and gas co-extraction must now adhere to stricter environmental and low-carbon requirements (Saint-Vincent and Pakney, 2020).

To achieve the green and low-carbon development of the coal and gas co-extraction technology system and advance the idea of CH_4-CO_2 near-zero carbon emissions, the position of coal as the basic security energy is unshakable (Elik and Zelik, 2023; Su *et al*, 2024). To reassess the paradigm for new collaborative emission reduction technology and the production and use of gas. In light of 'dual-carbon', deep integration of low-carbon integration technology and negative carbon technology is the only way for the coal sector to grow in a high-quality and sustainable way (Law *et al*, 2024). Building a sustainable development model is necessary in the process of gas drainage. The gas utilisation link also needs to further improve the high-efficiency and low-carbon operation mechanism through the cascade utilisation of total gas concentration (Hasanzadeh, Azin and Fatehi, 2023).

The reduction of CO_2 emissions from gas use must focus on negative carbon technologies like CCUS (carbon capture, utilisation, and storage) in order to reach nearly zero carbon emissions of CH_4-CO_2 in coalmines (Eyitayo *et al*, 2024). The primary target geological body for CO_2 storage in coalmines is the deep unmineable coal seam, but it is also important to consider the possibility of sealing mine goaf as an unconventional possible storage geological body. The collection and storage of CO_2 in the closed mine goaf, along with its absorption and storage of CO_2, has a host of applications in the secondary use of goaf resources that have been abandoned, and it is a critical negative carbon technical reserve to address the issue of carbon emission in the coal industry.

Numerous academics and industry professionals have produced insightful study findings in low-carbon development and effective resource use in coalmines. Intelligent low-carbon and high-efficiency technologies will undoubtedly advance gas simultaneous mining into a new stage when the 'dual-carbon' action is further developed. The author and his team presented the idea of a coal and gas co-extraction system with low carbon fusion technology and negative carbon technology as the core in response to the new situational demands of the 'dual-carbon' background. They also developed the coal and gas co-extraction system's development path. The discussion and prospecting of the major technologies provides a theoretical framework for encouraging the achievement of CH_4-CO_2 emission reduction targets in China's coal sector.

DEVELOPMENT PATH OF COAL AND GAS CO-EXTRACTION

With the transformation of the world's energy consumption structure, 'developed countries to reduce coal production, developing countries to control coal production' has become an unavoidable trend. As the key point of the coal and gas co-extraction system in the dual-carbon strategy, how to realise the near-zero carbon emission or even zero emission vision of CH_4-CO_2 should be combined with the whole industrial layout of the dual-carbon strategy.

Based on the current state and objective of reducing carbon emissions in the coal industry, the author and his colleagues have developed the development route of coal and gas co-extraction technology in the dual-carbon strategy, as illustrated in Figure 1. The AGE (accurate gas extraction) of coal seam, cascade exploitation of full concentration of gas, and 'CCUS + ecological carbon sink' are the core technologies in this technology system. Accurate extraction is the cornerstone of high-quality growth of coalmine gas control and use. The cascade utilisation of full gas concentration is the core of reducing methane carbon emissions. 'CCUS + ecological carbon sink' finely regulates and reduces the emission of CO_2 produced in the stage of gas extraction and comprehensive utilisation.

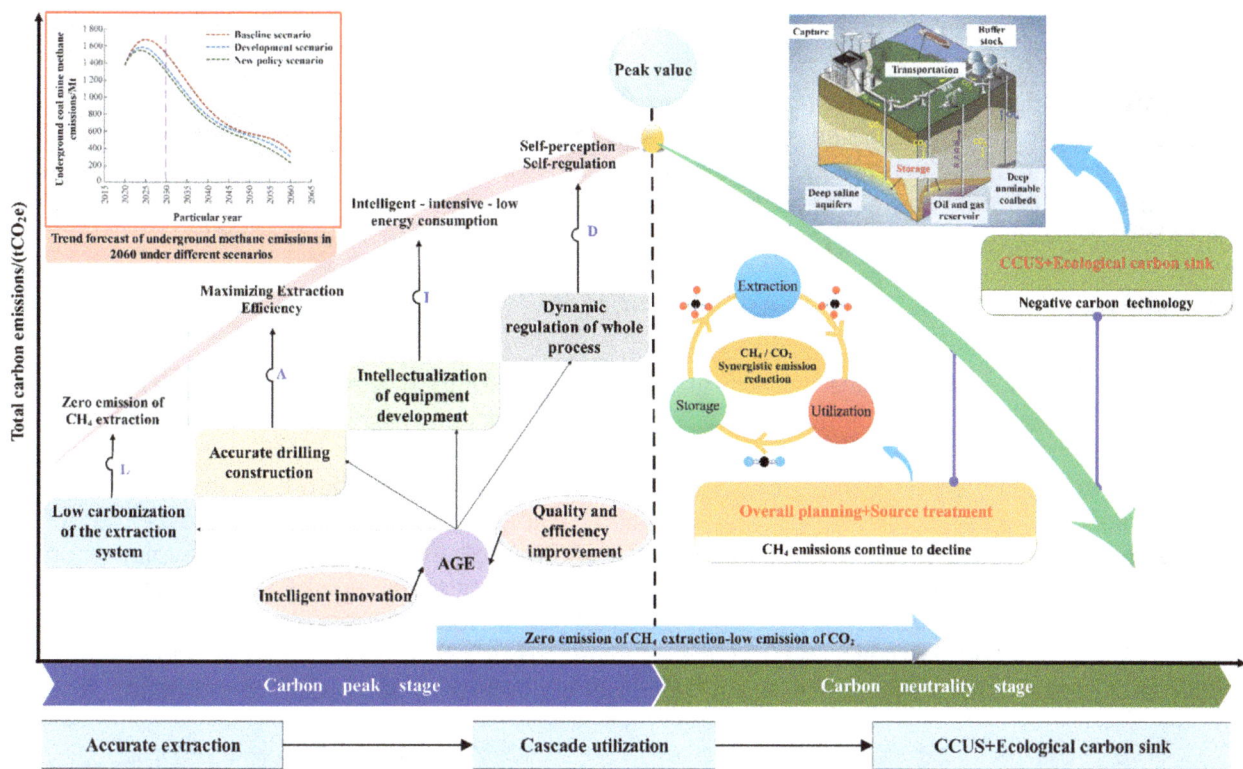

FIG 1 – Development path of coal and gas co-extraction technology system under double carbon strategy.

The 'L-A-I-D' development features (L: low-carbonisation of the extraction system; A: precise drilling construction; I: intellectualisation of equipment development; D: dynamic regulation of the entire process) ought to be adhered to during the carbon peak period. During the carbon neutrality stage, it is important to emphasise the joint emission reduction across the entire industrial chain and the ongoing implementation of source governance measures. lower CO_2 emissions, quicken the deployment of the carbon peak, and lower the peak value even more. The application of trustworthy, affordable, and secure negative carbon technologies, such as 'CCUS + ecological carbon sink', which are rewarded, speeds up the growth of my country's 'dual-carbon' aim.

KEY TECHNOLOGY OF COAL AND GAS CO-EXTRACTION SYSTEM

Green low-carbon technology for AGE

As the basis of low-carbon development of coalmine gas, AGE is not only related to safe production, but also directly affects the fine control and efficient utilisation of CH_4. How to realise the precision of gas extraction under complex reservoir conditions and achieve the purpose of improving quality and efficiency, the applicability of key technologies and long-term effectiveness are the key points. To precisely locate the high efficiency gas extraction area, fracture evolution, fracture seepage, and sub-domain determination are used. Whole-domain precision drilling of coal seams is combined with the 'replacing roadways with holes' and intelligent sealing technology to improve the precise construction of gas extraction.

Transparent geological guarantee technology

AGE (accurate gas extraction) is dependent on the transparent geological system being constructed. Several geological information elements must be integrated at the outset of gas drainage design in order to focus on the gas drainage process as a whole. Transparent mine must also gradually become more intelligent, digital, and informatic, with technologies such as transparent geological big data interpretation and transparent geological dynamic model reconstruction being realised. The enhancement of the application and precision of different exploration technologies, such as dynamic high-precision exploration and the progressive transparent geological exploration technology of 'advance drilling-excavation roadway-mining face', should be the main focus of the reconstruction of the transparent geological dynamic model.

Transparent geological big data interpretation technology should focus on the integration and visualisation of massive geological data collection, collation, analysis and feedback, and actively develop advanced digital mine technologies such as big data analysis and decision-making system, dynamic fusion of multi-source heterogeneous geological information, and '3S' (GPS, GIS, RS) information collection to realise dynamic real-time interpretation (Horrocks *et al*, 2024).

Key technologies for extraction and identification of gas target area for pressure relief

The identification of gas enrichment target area is the basis for improving the reliability of extraction design and enhancing the efficiency of coal seam gas extraction, which can effectively reduce the methane carbon emission in coal mining. The main methods used to identify the target area for gas enrichment are numerical simulation, physical simulation in the lab, and theoretical models. In recent years, based on the theory of mining-induced fracture elliptic paraboloid zone (Figure 2), the author and his team have constructed a comprehensive effect model of compaction zone evolution under the influence of multiple factors, and proposed a precise and efficient extraction technology of pressure relief gas in goaf, which is 'accurate identification of annular pressure relief gas enrichment zone in goaf + efficient extraction design in enrichment zone'. Combined with the above results, it laid a theoretical foundation for maximising the effect of coal seam gas precision extraction.

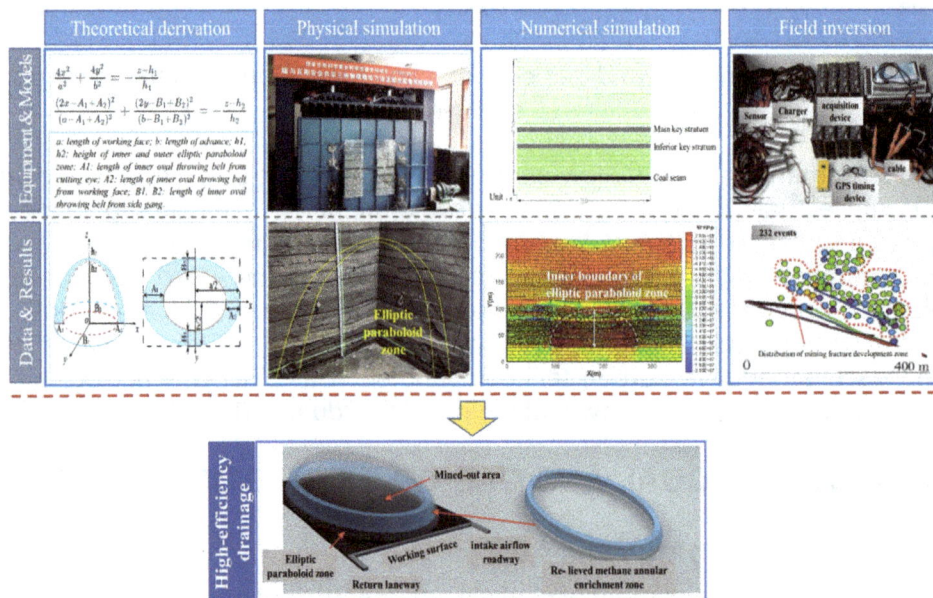

FIG 2 – Accurate identification of pressure relief gas efficient extraction area (Li *et al*, 2021).

Accurate design of gas extraction and drilling technology in various formations and mining times

It is crucial to categorise and make use of coal seam gas extraction if coalmine gas is to be developed in a cost-effective and low-carbon manner. In order to optimise the design of crucial design units such a reasonable construction horizon, drilling parameters, construction parameters, sealing/sealing parameters etc, drilling design primarily uses the method of integrating theory with field. The major research techniques include theoretical calculations, physical simulation test platforms (such as real triaxial 'stress-seepage-energy' integration test benches, two-dimensional and three-dimensional physical similarity simulation test benches), and numerical simulation.

The directional drilling of gas extraction in coalmine is the core technology to realise the precise extraction of gas. It includes the global gas extraction mode of time and space scale and the high-power directional drilling rig, which effectively guarantees the accuracy and effectiveness of gas extraction construction. In recent years, the directional drilling equipment with higher adaptability has greatly improved the construction efficiency. For example, the underground rotary geo-steering directional drilling technology equipment (gas extraction in the large plate area of medium-hard coal seam), the double-power head double-pipe directional drilling technology equipment (progressive extraction in the gas area of broken soft coal seam) etc, can further improve the intelligent level of directional drilling equipment in coalmines, build a digital platform for directional drilling, and support adaptive autonomous navigation drilling decision-making.

These components are concentrated examples of the strength of 'hard and core technology'. Drilling efficiency is increased and drilling engineering is streamlined by the consistency of extraction design and drilling effect. The gas extraction technology of ultra-longhole directional drilling in medium-hard coal seam large area, the gas extraction mode of 'hole instead of roadway' for mining pressure relief gas, and the regional gas extraction of broken soft coal bed methane dynamic directional drilling are formed with the aim of comprehensive and accurate gas extraction in underground coalmines (Figure 3).

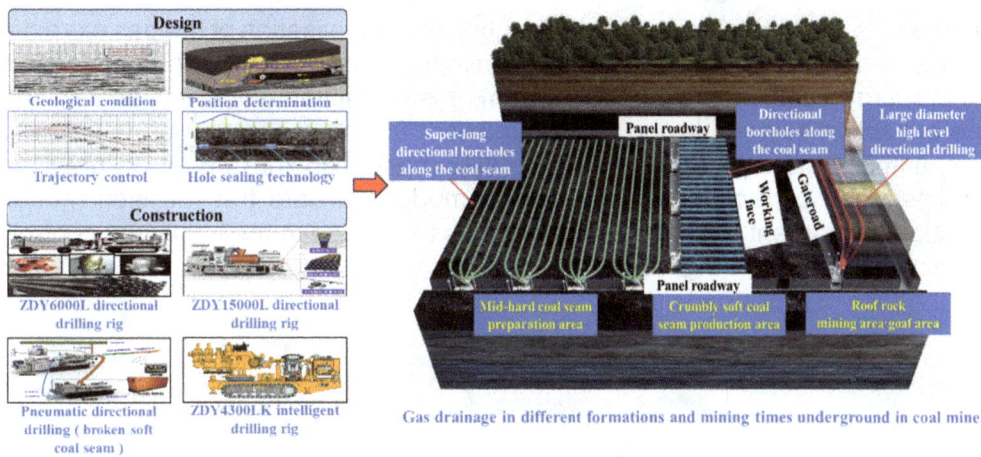

FIG 3 – Accurate design of gas extraction and global drilling technology (Hao and Peng, 2019).

Multi-scene accurate permeability-increasing technology

The principle of hydraulic permeability-increasing technology is to use water as a medium to transfer energy to expand the fracture network of coal and rock mass, generate gas seepage advantage channels to improve the extraction effect, including hydraulic fracturing (range fracture expansion antireflection) and hydraulic slotting, high pressure water jet (point ring permeability-increasing) and so on. Many scholars at home and abroad have developed sand fracturing, activated magnetic hydraulic fracturing, variable frequency pulse fracturing, self-excited oscillation pulse water jet slotting etc, which further expand the adaptability of anti-reflection technology to complex formations.

Such as pressure relief fracture zone seepage diffusion, segmented hydraulic (sand) fracturing, and high-pressure water displacement of methane double impact to enable effective extraction; The deep hole pre-splitting blasting technology causes radial and tangential fracture damage in the coal seam as a result of long-distance blasting, relieving pressure and improving coal seam permeability. Hydration-free fracturing permeability-increasing is primarily based on liquid nitrogen and carbon dioxide fracturing gas injection displacement technology. Recent years have seen the development of several innovative permeability-increasing technologies, including controlled shock wave permeability-increasing, calcium oxide flameless blasting, and others. An innovative approach to resource development in deep low-permeability coal seams is offered by the author and his team's synergistic hydraulic fracturing-ultrasonic excitation technique (Zhang *et al*, 2022). The full permeability-increasing system of 'point-surface-body' is gradually constructed based on the precise identification and construction design of gas-efficient extraction region, which contributes technical support for the advancement of accurate gas extraction in coalmines (Figure 4).

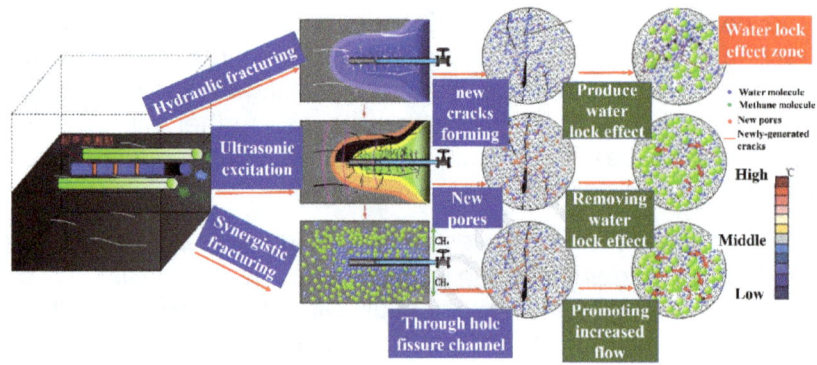

FIG 4 – Hydraulic fracturing-ultrasonic excitation synergistic flow increasing technology.

Intelligent control technology of AGE

In addition to ensuring the high efficiency, intelligence, and visualisation of the gas extraction system, the intelligent cloud platform for gas extraction monitoring is being built with the intention of further embedding the energy consumption control system, gas extraction efficiency, and emission control system of the extraction system, laying the groundwork for the realisation of global zero methane emission. The author built a cloud platform for AGE based on the aforementioned variables. It primarily consists of an extraction process control module, a data transmission module, and an analytical decision adjustment module. The platform is then used in the intelligent control of the gas extraction pipeline (as shown in Figure 5).

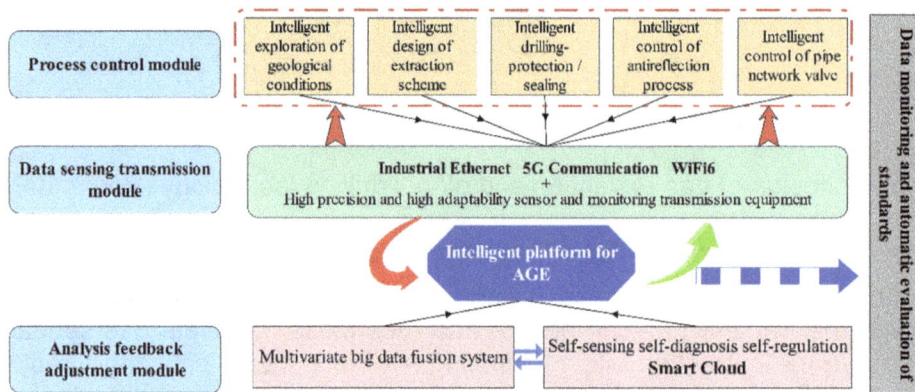

FIG 5 – Intelligent control platform for AGE.

Cascade utilisation of total gas concentration in coalmine

Cascade utilisation of total gas concentration, which is a major component of coalmine gas low carbon emission technology, has a direct impact on the effectiveness of the double carbon target implementation in the coal and electricity industries. The technical approach to full concentration cascade coalmine gas utilisation is primarily based on the direct utilisation of high concentration gas, the concentration and efficiency improvement technology of ventilation air and low concentration gas, and gradually develops a stepped comprehensive utilisation system (Figure 6). In order to achieve the near-zero carbon emission target of CH_4-CO_2 in coalmines, the CO_2 created by the efficient use of gas is captured and used in the field of CCUS technology.

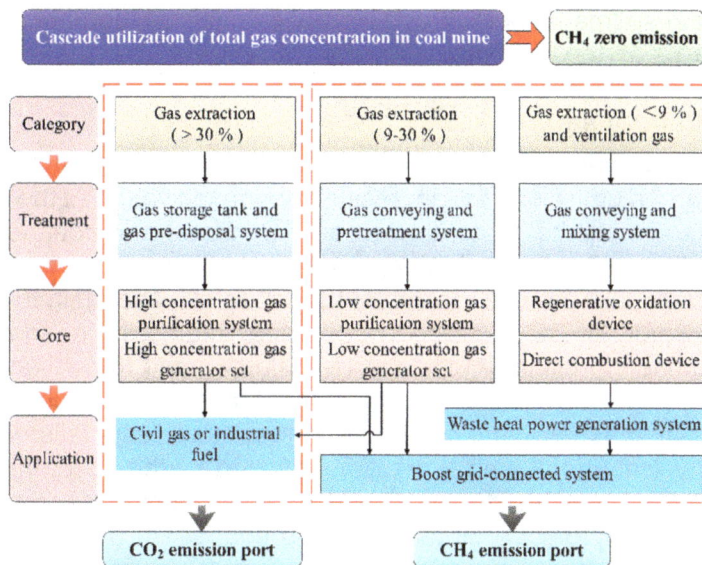

FIG 6 – Cascade utilisation of total gas concentration in coalmine.

The primary sources of CO_2 emissions during the gas utilisation stage are the direct combustion power generation of high concentration gases with volume fractions larger than 30 per cent and civil/industrial fuels; at this stage, the focus is on low carbon operation processes and combustion efficiency. The green low-carbon mode, which combines high-power and high-efficiency generator sets with carbon capture technology, has been steadily gaining traction thanks to intelligent mines' advanced methods. Meanwhile, the regional coal base extraction, concentration, utilisation, and capture technology system has advanced quickly.

The primary sources of CH_4 emissions from coalmines are ventilation air (<1 per cent) and low concentration gas (<30 per cent). One of the main responsibilities is to use technology to increase efficiency and focus. Currently, the predominant technical framework is produced by combining catalytic oxidation with thermal storage oxidation. Many advanced technologies have been promoted, including lean gas turbines (LBGT), rotary thermal storage oxidation, thermal counter-current reactors (TFRR), monolithic catalytic reactors (CMR), catalytic counter-current reactors (FFRR), and regenerative burners (VAMRAB). A low-concentration gas (<8 per cent), ventilation gas (0.2 per cent), blending power generation combined with an organic Rankine cycle (ORC) flue gas waste heat cascade utilisation technology system is formed by Dafosi Coalmine in Binchang Mining Area, which uses the concept of 'resource utilisation and waste heat and pressure cascade utilisation' as a guiding principle. This effectively realises the green and low-carbon development mode of 'zero emission' of gas in the entire mine (Zhao *et al*, 2023).

The security status of coal is crucial given the strong push for clean energy, and gas's significance as a resource used in conjunction with coal cannot be understated. Technical research should be done in the areas of effective coalmine gas extraction, gas power generation, utilisation of ventilation air gas, heat storage, and oxidation of low-concentration gas in light of the development of full-concentration cascade utilisation technology of coalmine gas. This will help to promote the improvement of gas extraction and utilisation as a whole and create a positive feedback loop where 'pumping to ensure use and use to promote pumping'. This is an important way to realise the deep emission reduction of coalmine gas and help China achieve the goal of 'dual carbon'.

Key technologies of 'CCUS + ecological carbon sink'

Low-carbon integration and negative carbon technology research are not only necessary for industrial transformation, but also a hotbed of global rivalry, since they are the focus of green and low-carbon development in the coal sector. The achievement of the carbon neutrality objective can be effectively aided by mature, reliable, and sustainable carbon emission reduction technologies. The extraction-utilisation link in the coal and gas co-extraction technology system generates carbon emissions (CO_2), which CCUS, a typical negative carbon technology, is crucial in absorbing and offsetting. Coal seams and goafs are the principal geological bodies in the coalfield that store carbon.

By displacing CO_2 in coal seams, a method that increases coalbed methane production is relatively mature and has reached a certain commercial size.

CO_2 sequestration in coal seam

By displacing coalbed methane (gas), CO_2 sequestration in deep coal seams serves the dual purposes of reducing carbon emissions and increasing economic value. At the moment, fundamental research and pilot tests are still being conducted on CO_2 storage in coal seams. The use of gas injection and displacement technologies in deep coal seams, as well as the safety assessment of CO_2 storage in coal seams, are the main topics of the research. The main theoretical foundation is the assessment of the safety of CO_2 storage in deep coal seams. It undertakes research on CO_2 leakage pathways, engineering disturbances, meso-scale damage to coal and rock reservoirs and caprocks, and macro-geological disasters, with a primary focus on the structural instability of sealed geological entities. The main factors that contribute to the meso-scale failure of coal-rock reservoir and caprock are the physical and chemical reactions caused by the supercritical CO_2 injection into deep coal seam, CO_2-CH_4 gas adsorption replacement, diffusion seepage and displacement mechanism, and the efficiency of CO_2 storage in deep coal seam (Faizan, Bana and Gugulothu, 2024). China United Coal and other units undertook the first domestic research project of deep coal seam injection/burial CO_2 mining coalbed methane technology with the help of the cooperation between China and the Canadian government in terms of equipment technology and application. Liulin and Shizhuang in the Qinshui-Ordos Basin undertook a number of test injection and monitoring studies, focusing primarily on a number of key issues like improved development of coalbed methane and CO_2 burial, evaluation of CO_2 injection to improve coalbed methane recovery, numerical simulation and economic evaluation of pilot tests, research and equipment development of deep coal seam gas injection displacement technology, which provided a crucial basis for future research (Du et al, 2018).

Mechanism of adsorption storage, transportation and carbon sequestration of vegetation in goaf of CO_2

A technical issue that must be resolved quickly in the field of low-carbon development and utilisation in the coal industry is the exploration of large-scale, affordable CO_2 underground storage technology, which has significant theoretical relevance and application prospects. The number of subterranean areas, such as shuttered mines and goafs, is growing as domestic green coal resources are gradually depleted. Statistics show that during **the 13th Five-Year Plan period** (outline of the 13th five-year plan for national economic and social development of the people's republic of China), 5500 coalmines were shutdown nationwide, and during **the 14th Five-Year Plan period** (outline of the 14th five-year plan for national economic and social development of the people's republic of China), the number of coalmines would be further reduced. The author and his team proposed the key technology of 'CCUS + ecological carbon sink' for CO_2 coalmine goaf absorption and vegetation carbon sequestration (Figure 7). This technology was developed in order to address the issue of secondary utilisation of abandoned/closed mines while focusing on the concept of near-zero carbon emission of CH_4-CO_2 in the entire life cycle of coalmines (Li et al, 2022).

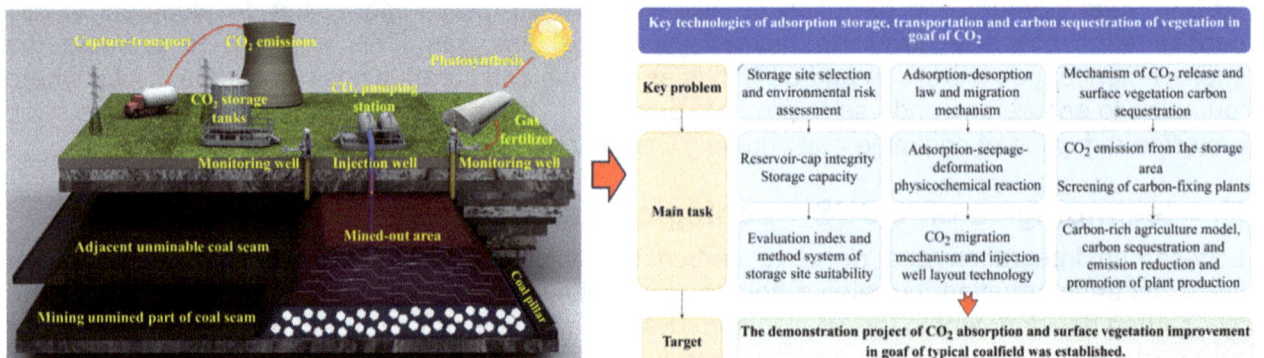

FIG 7 – Key technologies of adsorption storage, transportation and carbon sequestration of vegetation in goaf of CO_2.

The coalmine goaf is a potentially useful geological body for CO_2 absorption since it is an unusual type of geological trap structure. Following are some financial benefits of using coalmine waste for CO_2 absorption:

- The geological conditions of the coalmine goaf are complete, and the early exploration investment cost is low.

- The absorption and storage of goaf is injected by gaseous CO_2, and the operation cost is low.

- Under the stable conditions of the reservoir and caprock, goaf has a high permeability and a wide range of CO_2 diffusion.

- It is possible to carry out the secondary extraction of methane in goaf, which has some financial advantages.

- Coalmines are located near large thermal powerplants and other centralised CO_2 emission sources, which allows for local CO_2 storage and reduces long-distance CO_2 transportation.

Understanding the dual function of CO_2 geological storage and surface biomass compensation and absorption will help us better understand the key technologies of CO_2 storage and vegetation carbon sequestration in coalmine goaf. This will be important for China's secondary utilisation of coalmine goaf and the implementation of the 'carbon neutrality' goal in the coal industry.

The general concept is that the injection of CO_2 will compete with the remaining CH_4 in the goaf (the adsorption advantage of CO_2 is stronger than that of CH_4), and the gas in the goaf will be promoted and the fire control will be carried out simultaneously. Lastly, by controlling the ground monitoring wells (or other control units) to release CO_2 within a reasonable range as gas fertiliser, it is used for vegetation carbon sequestration in agricultural greenhouses. Site selection, environmental risk assessment, the law of adsorption and desorption and movement mechanisms, CO_2 release, and the method for surface vegetation carbon sequestration are among the major concerns. The technological safety and economic viability of CO_2 storage are the crucial aspects to implement this concept since a unique geological body created by coal mining disruption. The advancement of essential technology theory and the investigation of low-cost operation are generally necessary for the safe and economically viable development of CO_2 coalmine gas storage and vegetation carbon sequestration technology, which also identifies the direction for future study.

CONCLUSIONS

- As the core point of the coalmine gas simultaneous mining technology system under the 'dual-carbon' target, the low-carbon fusion and negative carbon technology mainly covers the gas extraction stage, the gas utilisation stage and the capture-storage-utilisation stage of CO_2 produced in the utilisation process. The general idea is to create a closed cycle in the whole life cycle of coal seam gas extraction and utilisation, extract carbon from coal seam in the form of CH_4, capture the used CO_2, and finally seal it underground in the form of CO_2 through CCUS and other negative carbon technology means to displace coal seam gas or carry out other carbon fixation means, so as to complete the 'internal closed circle digestion', aiming at realising the vision of near zero carbon emission of CH_4-CO_2 in coalmine and promoting the green cycle development of coal industry in China.

- At the carbon peak stage, AGE should follow the development path of 'L-A-I-D', ultimately achieving zero emission from CH_4 extraction-low emissions of CO_2. In the stage of carbon neutrality, we should actively build a carbon emission reduction technology system with reliable technology, low cost and long-term safety, and accelerate the realisation of 'dual-carbon' strategic goal.

ACKNOWLEDGEMENTS

This study was financially supported by the National Natural Science Foundation of China (Nos. 5237-4227, 5207-4217), and Shaanxi Provincial Natural Science Basic Research Program Project (2021JLM-26).

REFERENCES

Du, Y, Sang, S, Wang, W, Liu, S, Wang, T and Fang, H, 2018. Experimental study of the reactions of supercritical CO_2 and minerals in high-rank coal under formation conditions, *Energy Fuels*, 32(2):1115–1125.

Elik, A and Zelik, Y, 2023. Investigation of the effect of caving height on the efficiency of the longwall top coal caving production method applied in inclined and thick coal seams by physical modeling, *International Journal of Rock Mechanics and Mining Sciences*, 162:105304.

Eyitayo, S I, Okere, C J, Hussain, A, Gamadi, T and Watson, M C, 2024. Synergistic sustainability: Future potential of integrating produced water and CO_2 for enhanced carbon capture, utilization and storage (CCUS), *Journal of Environmental Management*, 351.

Faizan, M, Bana, D and Gugulothu, H K, 2024. Comparative Investigation on the catalytic potential of borylated triazine-based FLPs for CO_2 sequestration, *Chemistry Select*, 11:9.

Hao, S J and Peng, X, 2019. Research on direction drilling in accurate connecting roadway technology with long-distance and large-elevation in underground mine, *International Journal of Coal Science and Technology*, 47:47–52.

Hasanzadeh, M, Azin, R and Fatehi, R, 2023. Investigation of optimum conditions in gas-assisted gravity drainage using Taguchi experimental design, *Geoenergy Science and Engineering*, 211630:2–10.

Horrocks, T, Wedge, D, Hackman, N, Green, T and Holden, E J, 2024. Automated geological logging from FTIR reflectance spectra utilising geochemical and historical logging constraints, *Ore Geology Reviews*, 168.

Law, L C, Mastorakos, E, Othman, M R and Trakakis, A, 2024. A thermodynamics model for the assessment and optimisation of onboard natural gas reforming and carbon capture, *Emission Control Science and Technology*, 10:52–69.

Li, S, Yang, E, Lin, H and Zhao, P, 2021. Construction and practice of accurate gas drainage system for pressure relief gas in deep mining, International *Journal of Coal Science and Technology*, 49:1–10.

Li, S, Zhang, J, Shang, J, Lin, H, Wang, S, Ding, Y, Hou, E and Zhao, H, 2022. Conception and connotation of coal and gas co-extraction technology system under the goal of carbon peak and carbon neutrality, *Journal of China Coal Society*, 47(4):1416–1429.

Meinshausen, M, Lewis, J, McGlade, C, Gütschow, J, Nicholls, Z, Burdon, R, Cozzi, L and Hackmann, B, 2022. Realization of Paris Agreement pledges may limit warming just below 2 °C, *Nature*, 604:304–309.

Rogelj, J, den Elzen, M, Höhne, N, Fransen, T, Fekete, H, Winkler, H, Schaeffer, R, Sha, F, Riahi, K and Meinshausen, M, 2016. Paris Agreement climate proposals need a boost to keep warming well below 2 °C, *Nature*, 534:631–639.

Saint-Vincent, P and Pakney, N, 2020. Beyond-the-meter: unaccounted sources of methane emissions in the natural gas distribution sector, *Environmental science and technology*, 54(1):39–49.

Su, K, Ouyang, Z, Wang, H, Zhang, J, Ding, H and Wang, W, 2024. Experimental study on municipal sludge/coal co-combustion preheated by self-preheating burner: Self-preheating two-stage combustion and No_x emission characteristics, *Energy*, 290:130222.

Udeagha, M C and Muchapondwa, E, 2023. Achieving regional sustainability and carbon neutrality target in Brazil, Russia, India, China and South Africa economies: Understanding the importance of fiscal decentralization, export diversification and environmental innovation, *Sustainable Development*, 31(4):2620–2635.

Zhang, J, Lin, H, Li, S, Yang, E, Ding, Y, Bai, Y and Zhou, Y, 2022. Accurate gas extraction (AGE) under the dual-carbon background: Green low-carbon development pathway and prospect, *Journal of Cleaner Production*, 13:134372.

Zhao, W, Cao, H, Ma, T, Yang, F, Cui, J and Chen, X, 2023. Research on the coupled hazardous zones of coal self-ignition and CH_4 blast in goaf during upward mining, *Case Studies in Thermal Engineering*, 52:103733.

Ventilation economics and optimisation

Case study – ventilation on demand at Fosterville Mine

C A M Jackson[1], J Norris[2] and F C D Michelin[3]

1. Senior Ventilation Engineer, Howden Ventsim Australia, Brisbane Qld 4101.
   Email: christopher.jackson@howden.com
2. Senior Ventilation Engineer, Fosterville Gold Mine, Vic 3559.
   Email: jeff.norris@agnicoeagle.com
3. Managing Director, Howden Ventsim Australia, Brisbane Qld 4101.
   Email: florian.michelin@howden.com

ABSTRACT

Fully automated Ventilation On Demand (VOD) systems have been slow to gather momentum in Australia, despite success and advances in this technology abroad. Fosterville in Victoria is the first mine to have successfully implemented the use of heavy equipment tracking based ventilation automation (Level 4 VOD) in Australia.

Level 4 VOD utilises software and hardware automation to control secondary fans and airflow based on tagged equipment location. Plans are now underway for the introduction of level 5 (highest level) VOD, which features automated control of the mine's surface fans based on underground activity to further reduce power consumption while ensuring airflow compliance throughout the mine.

This paper discusses the difficulties and challenges which needed to be overcome when installing this system, while providing insights on how to avoid problems with future installations. For example, Fosterville's incorporation of dual speed fans into the VOD system to reduce power consumption is a significant innovation which required changes to site hardware and third-party software to use of this type of fan.

The paper describes the use of Ventsim Control Software when connected to dual-speed fans, as well as the additional benefits experienced with equipment tagging and tracking. Data will be presented on the reduction in auxiliary fan power consumption, improved environmental conditions, and reduction in re-entry times experienced after VOD was installed while showing there has been no negative impact upon production. Finally, the completion strategy to introduce level five VOD (incorporating surface fans) into the mine will be discussed and presented.

INTRODUCTION

Fosterville Mine is a high-grade, low-cost underground gold mine, located 20 km east of the city of Bendigo in the State of Victoria, Australia. Fosterville gold mine is the largest gold producer in Victoria.

The mine is accessed via a decline and is approximately 1.35 km deep. Predominately sub-level stoping is used with Cemented Rock Fill (CRF) and Paste backfill to backfill stopes. The current projected Life-of-mine (LOM) for Fosterville mine is approximately ten years, however this is subject to change depending on exploration results, production rates and economic conditions.

The site senior ventilation engineer, is responsible for:

- Maintaining the Primary Ventilation Circuit.
- Life-of-mine network design and modelling.
- Life-of-mine heat and refrigeration design and modelling.
- Project management of short- and long-term projects.

As the manager of these responsibilities the site senior ventilation engineer was aware of high-power cost experienced in 2017 ($0.17 kWh) which resulted in auxiliary fan power cost approaching $3.5M per annum (pa). In subsequent years, mine power cost has reduced substantially through negotiation with power providers reducing to $0.08 kWh, however there was still room for improvement as annual auxiliary power cost for the 17 fans in this review is close to $1.6M pa.

As the need for increased energy efficiency and environmental sustainability continues to grow, so too has interest in implementation of Ventilation On Demand (VOD) software, which, when coupled with suitable hardware has helped to achieve these goals. Power prices, which are projected to increase before the end of 2024 by 20 per cent, place new focus onto the VOD system and further gains which can be achieved with new levels of control.

VOD systems have a variety of schemes or levels of control which to date have not been firmly defined across different manufactures. This study references the Howden Ventsim definition of VOD levels as shown below in Figure 1 (Pinedo and Torres Espinoza, 2019).

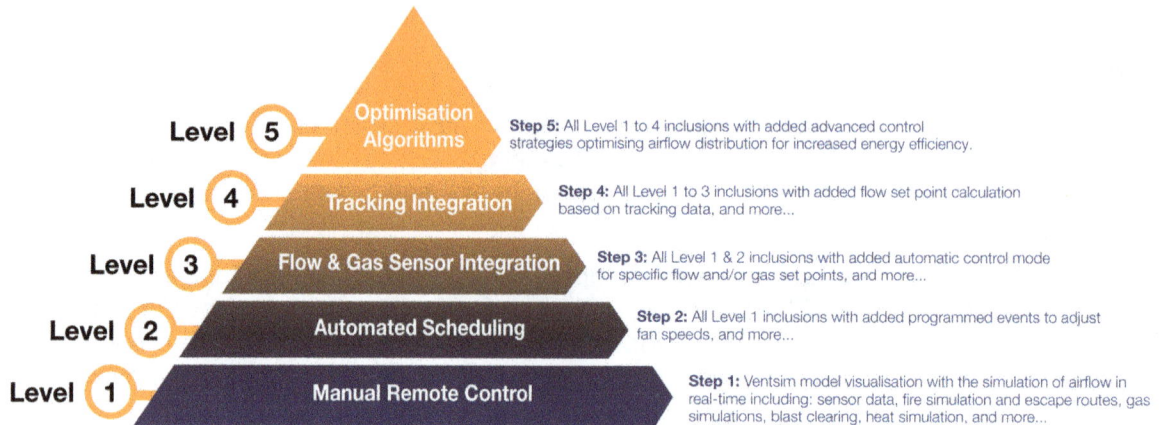

FIG 1 – VOD Levels of Control (Howden Ventsim).

PLANNING – HARDWARE CHOICES FOR VOD

During the planning process, it was decided to use a staged approach when installing VOD. This approach would enable the success of each stage to be proved in achieving its goals, while ensuring the impact of setbacks or roadblocks which may be experienced during the installation process are reduced for future work.

The following is a list of items which Fosterville considered for implementation when installing the VOD system on-site.

Air quality stations

Six Maestro Vigilante Air Quality Stations (AQS) were chosen for installation at key Return Air Raise (RAR) sites in the mine to measure air flowing into the exhaust and airborne contaminants in the system. Measurements at each station included the following.

- airflow velocity
- temperature – wet bulb and dry bulb
- carbon monoxide
- nitrogen dioxide.

Measurements taken by these units are relayed to the surface and displayed in the Ventsim Control tables and the Ventsim model in real time from each AQS. Conditions in blasted areas, which were often inaccessible for long periods after firing are now available in real time via the AQS monitors and displayed in the control room on the surface. The software has the ability to model and predict air flow and gas levels downstream from each AQS based on real time gas readings throughout the mine. This provides additional information to enable educated decisions to be made in relation to equipment management, temperature and gases levels present in the mine and can be used to reduce re-entry times and increase productivity.

Auxiliary fans

Seventeen Auxiliary fan sites were chosen for introduction into the VOD system. Auxiliary fan options considered are summarised in Table 1.

TABLE 1

Comparison of different fan types for VOD.

Option	Pros	Cons
• Variable Inlet Vane Fans (VIV)	• Allows to achieve exact airflow required • Robust	• High cost (fan and starter) • Low efficiency • Heavier and harder to install
• Variable Speed Drives (VSD)	• Allows to achieve exact airflow required • Highest efficiency and potential for cost reduction	• High starter cost • At the time of selection, lack of references for 1000V VSD for auxiliary fans
• Dual Wound/Dual Stage Fans (DW) • Lo – 1 stage, 6 Pole • LoLo – 2 stage, 6 Pole • Hi – 1 stage, 4 Pole • HiHi – 2 stage, 4 Pole	• High efficiency • Only minimally more expensive than the On/Off fan • Allows to choose from five different states (Off, Lo, Hi, LoLo, HiHi)	• Doesn't allow precise selection of flow • Cost of motor replacement
• On/Off Fans	• Industry standard • Saving on Level 3 would remain	• Would not allow savings on level 4 when minimal airflow needs to be maintained

The Dual Wound/Dual Stage option was selected over full variable speed control due to lower cost, reliability and sufficient flexibility to achieve savings. The existing fans were sent away and retrofitted, with the existing motors rewound as dual speed motors to enable the fans to run at different speeds (4 pole – 1500 rev/min and 6 pole – 1000 rev/min). The cast impellors were not altered and remained in the fans.

Regulators

Five automatic Clemcorp regulators were installed at key exhaust locations in the mine. The introduction of automated regulators into the ventilation system was primarily to improve air quality by ensuring the correct volume of air is removed off active levels, preventing heat and blast fumes re-entering the active work zones downstream through the primary circuit. Control over this infrastructure was considered a crucial component to improving work conditions and safety of personnel downstream.

When tagging and tracking is introduced into the system, automated control over exhaust regulators will enable introduction of level 4 and 5 control over the primary circuit to continue to improve air quality in worked zones while reducing power consumption.

Previous control system

The fans, sensors and regulators when first installed were connected to the network and the Supervisory Control and Data Acquisition (SCADA) system. This allowed the mine to monitor and control the equipment remotely. However, drawbacks with of the SCADA system arose early on with delays in updating the system as it progressed and expanded due to the availability of qualified engineers to update the SCADA system as the mine changed and progressed.

Adding complex capabilities and flexibility was not practical or economically viable using a SCADA system. The mine decided to use a more flexible solution and Ventsim CONTROL was chosen as it provided greater ventilation control functionality at a lower cost. Importantly, the ventilation engineer could modify or update the system when required without the need to contact external specialised software developers.

LEVEL 3 – SCHEDULING

Overview

Level 3 savings have been experienced using the scheduling tool in Ventsim Control. Each auxiliary fan, when connected to the software was configured to turn off for an hour during firing time at the start and end of each shift (twice a day).

Level 3 also allows supervisors to set selected exhaust regulators at specified flows that are maintained through automation. This would ensure that if changes are made in the system on other levels, it would not impact upon the active automated levels. It also ensures that if operators change louvre setting, this can be seen and rectified from the control room if needed.

Set-up

The Ventsim CONTROL set-up occurred over four months. The two most common bottle necks were the server set-up and the PLC hardware automation programming.

Server set-up and software installation

When beginning the VOD installation, Fosterville Gold Mine was required to set-up two virtual servers. Once prepared, Howden Ventsim specialists were given remote access to install Ventsim CONTROL onto these servers. Completing this installation at the beginning enabled uninterrupted communication between the servers and underground installations throughout the entire process.

One server is used as a control server, sending commands to the underground equipment, reading sensors and all other background information. The control server needed to be able to communicate with the underground equipment. The other server is the Human Machine Interface (HMI), allowing users to access, view and adjust the software. The HMI communicates with user's personal computers, allowing them to log in using remote desktop software.

Until the start of the commissioning, Ventsim CONTROL is set to read only, meaning that data can be read but equipment cannot be controlled. This prevents premature or unexpected control of underground fans, louvres and other ventilation controls.

PLC programming

The automation operated by Programmable Logic Controllers (PLCs) must be controllable by three methods: communication with Ventsim CONTROL, the SCADA system and via local manual control. To ensure that the commands from each device do not clash with each other, some PLC programming was required. Additionally, any fail-safe in case of loss of communication with Ventsim CONTROL needed to be programmed into the PLC. The most common option in case of loss of communication is the 'bump less' option where the equipment is kept at the last recorded automated setting, however for safety or operational reasons some applications may require the fan to ramp up to full speed if a loss of communication occurs.

Equipment configuration

To be able to communicate with the underground equipment, the following information at a minimum was required to be known to allow Ventsim CONTROL to read and write to the PLC:

- IP address: which indicates the address of the device on the network: example: Regulator 2050RL: 10.165.155.122

- Tags: which indicate where on the PLC the information is stored, example: Airflow: 40005

- Data type: which indicate the number format the information comes in, for example 'Integer'.

The different devices were added into Ventsim CONTROL with the appropriate tags. As the tags were added, the information received is checked to ensure it is accurate. Before commissioning, the system was tested to ensure the program could read all underground information.

Software modifications

Introduction of Dual Wound (DW) motors into underground auxiliary fans for underground mining was a new concept, however the Ventsim Control software was only designed to control Variable Speed Drives (VSD) and Variable Inlet Vane (VIV) fans. As a result, the software was reconfigured to enable separate control of each winding by recognising these fans as dual stage/dual wound fans with effectively four motors in each dual stage fan.

The safety of operators working on a fan while the fan is connected to the software was also considered. A maintenance button was added to the Programmable Logic Controller (PLC) on the fan starter,, which, when activated prevented control of the fan so that operators installing duct or completing maintenance and repairs could do so safely.

Commissioning

The commissioning process included ensuring that commands from Ventsim CONTROL are actioned as expected and that all unexpected behaviours were handled properly. This process was planned to be performed on-site by a Howden employee, however due to COVID-19 travel constraints restricting site visits, commissioning of the system was conducted by local mine personnel via Microsoft (MS) Teams meetings with great success. This format was continued by a core team consisting of a Howden engineer, the site ventilation engineer and a PLC programmer through to completion of the project with no further issues

To streamline the process, the new PLC code was initially loaded on a test PLC not connected to underground fans. This allows for easier changes of the PLC code and did not impact personnel safety or production. Some challenges arose around ensuring that the emergency stop (E-Stop) was properly handled and reported into Ventsim CONTROL. A few additional statuses were also added to the PLC and Ventsim CONTROL during the commissioning. Once the test PLC was commissioned, the code was loaded onto an underground fan which was then commissioned. Following the successful commissioning of the first underground fan, the site handled the commissioning of the remaining fans using the same procedure. The process of commissioning the test fan followed and underground fans took three days and commissioning of regulators was completed in a similar fashion.

Training

Due to COVID-19 constraints, training was also conducted over MS Teams with the ventilation engineer and the supervisors.

Supervisor

FGM has a mine control room with mine control operators. The mine control operators, while very good at their role, are not necessarily familiar with daily mining operations. For this reason, once the new Ventsim Control system came online, daily control of the system was handed over from mine control operators to shift supervisors.

Shift supervisors are more familiar with firing procedures, production locations, positioning of the fans, the level which they ventilate and the circuit/s which each regulator affects. It was felt that this knowledge placed shift supervisors in a better position to make VC related decisions.

Supervisors were taught how to turn on the ventilation on specific levels based on predefined script in the software created by the ventilation engineer. Their login access is set by the senior ventilation engineer and only allows them to do the tasks they have been trained to perform.

Senior Ventilation Engineer

The senior ventilation engineer was given full administrator rights to the software, allowing access to complete all required modifications to the system and add users as required. The training included not only how to operate the system but also how to change and configure it. Following the training, the senior ventilation engineer was able to add and change the location of the fans efficiently and without additional assistance, ensuring the system was kept up to date.

Results

The benefits of using level 3 VOD scheduling were measured over the first 13 months (before the system transitioned to level 4 VOD), and returned an average 15.7 per cent cost reduction for the 17 DW auxiliary fans. This saving effectively paid for the software purchase and installation in less than six months.

While harder to quantify, implementation of Ventsim CONTROL level 3 VOD provided additional benefits, by reducing re-entry times and improving air quality on levels through use of automated regulators.

Fosterville used level 3 VOD control for almost two years before recently moving to level 4 VOD control. All data has been included in this graph shown in Figure 2.

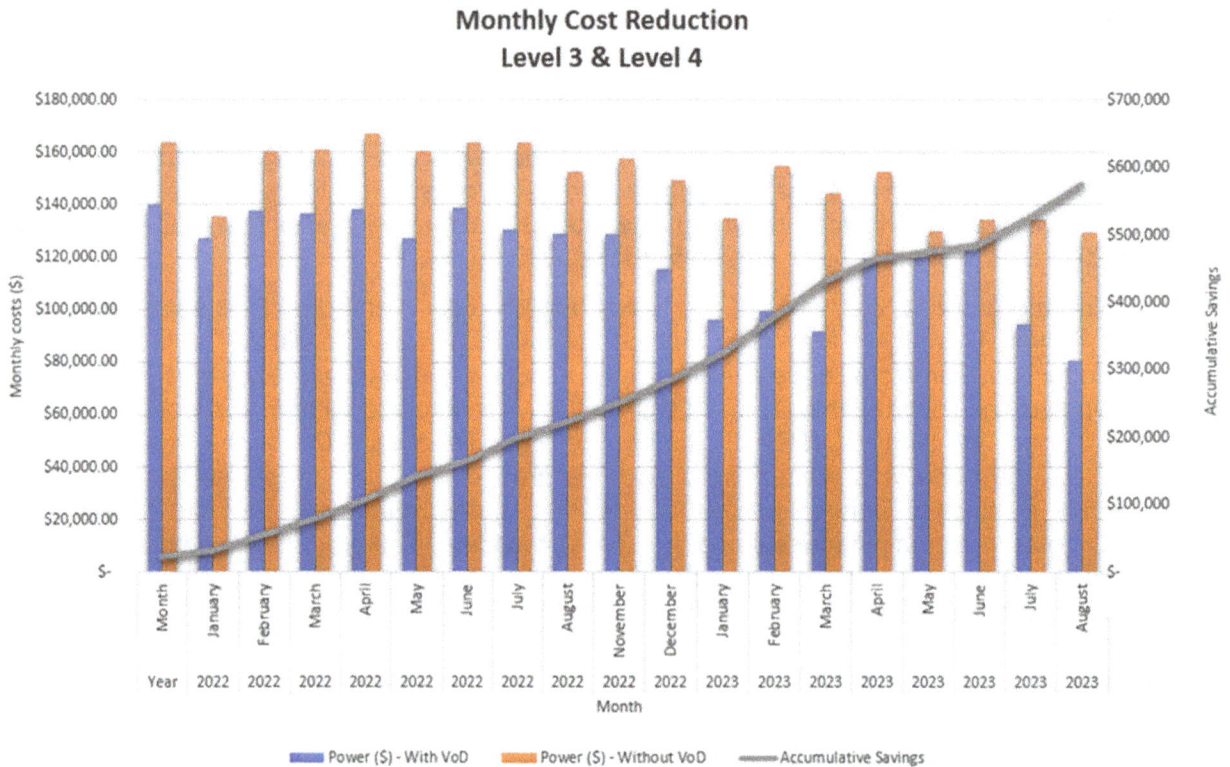

FIG 2 – Level 3 and Level 4 – savings by month.

LEVEL 4 – INTEGRATION WITH TRACKING SYSTEM

How it works

To control ventilation based on tracking, the mine is divided into zones, each zone has a dedicated ventilation requirement setting, for example a level. The tracking system will indicate which zone each vehicle is located. In this system, if no personnel or vehicles are present, the fan will be running at the lowest setting (Lo). If personnel or vehicles are detected, the fan will run at the highest setting (HiHi). Due to the long length of ducts being used, it was decided not to use lower levels of ventilation based on engine kW rating at present.

Integration with tracking system

There are several ways to integrate a VOD control system with a tracking system. The easiest way to integrate is for the database where the tracking system stores information to create a view of where the vehicles are. The view can then be shared with Ventsim CONTROL allowing this information to be read and used.

In this case, this was not recommended by Mobilaris (tracking provider) who preferred to use their Application Programming Interface (API) to access their data programmatically. Howden Ventsim then programmed an application to read the latest vehicle location data using the API and entered it in a database for Ventsim CONTROL to access.

Once the program was checked it was installed onto a test server which could read data from site without controlling devices. The tracking locations read by the Ventsim CONTROL were then compared to the tracking provider information and validated. Once validated, the system was installed on the main server.

An initial challenge was some inaccuracies in the tracking system, leading to vehicles not necessarily being recorded at the correct location. Due to its impact on the ventilation system, most areas remained set on manual, flow or scheduling. Once more confidence was obtained in the tracking information, more devices were switched to VOD mode. Other mine events also forced the system to stop the tracking and/or scheduling, however the system can change modes easily based on what information is possible at the time.

Results

Savings experienced using Level 4 VOD increased from the baseline level 15.7 per cent to 36.6 per cent reducing power cost to approximately $630 kpa as shown in Figure 3.

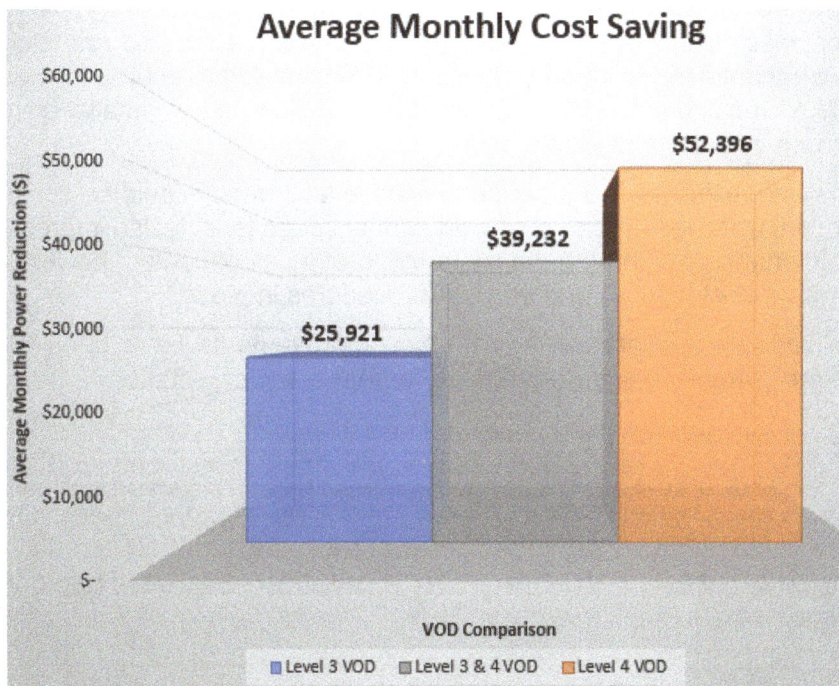

FIG 3 – Average monthly saving since project conception.

FUTURE IMPROVEMENT

Dual speed full range usage

So far, the fan speed is always set to 'HiHi' when vehicles or personnel are present. However, a loader and a person inspecting the area do not require the same amount of ventilation. This could be improved by adding the power rating of each vehicle. A flow factor representing the air requirement per kW of diesel can then be used (for example 0.05 m^3/s/Kw). This would provide the airflow demand set point for the fan and the airflow is then passed on as a set point for the relevant devices.

As there is no flow or pressure sensors at the fan, and no VSD's, only fixed points can be chosen for the DW fans. To find the appropriate set points, the resistance of the duct is calculated based on the length, area, perimeter and friction factor which are configured in Ventsim CONTROL. It is then compared to the fan curve for each of the four settings (Lo, Hi, LoLo, HiHi). The lowest fan curve which matches the requirement could then be chosen. In this arrangement, the duct length would need to be updated in the Ventsim model consistently to ensure appropriate flow reaches the heading.

It is also possible to use the Ventsim DESIGN model simulation to estimate the flow. This method has an advantage, with the ability to choose the start or end of the duct as the location of the modelled flow value. However, it also requires frequent and accurate Ventsim DESIGN model updates to ensure airflow requirements are not under-predicted.

Main fan automation

Level 5 implementation is currently being planned where main fan speeds will be adjusted based on the underground requirements. A specific control philosophy has not been decided yet, but some are already in use at other mines (Sanftenberg, 2019).

To trial different control options without affecting the operation, a trial server which reads data from the mine will be used. By controlling the Ventsim DESIGN model, different control strategies can then be applied and compared to the normal operation.

CONCLUSION

The VOD system at Fosterville Mine has been running since 2020 with very few issues. Similarly, the set-up and commissioning were successful despite some unexpected roadblocks (such as the COVID19 lockdown) thanks to having a good team: a motivated project leader to see it happen, a knowledgeable electrician and responsive IT department. Following the installation, the site was able to run the system with very limited outside help.

Significant savings were achieved, 15.7 per cent using level 3, which could be achieved using on/off fans and 36.6 per cent using level 4 and dual wound/dual speed fans. It also improved the operability of the system. Maintenance is critical to the system operating as expected, however, at present this is well managed with well-structured and efficient procedures in place.

While already delivering significant savings, further improvements have been identified and the changes required are currently being investigated for future implementation.

REFERENCES

Pinedo, J and Torres Espinoza, D, 2019. Implementation of Advanced Control Strategies Using Ventsim Control at San Julian Mine in Chihuahua, Mexico, in *Proceedings of the 17th North American Mine Ventilation Symposium*, pp 435–441.

Sanftenberg, J, 2019. Mine Ventilation on Demand System at Nickel Rim South Mine; Testing Procedures, System Maintenance and Results, in *Proceedings of the 17th North American Mine Ventilation Symposium*, pp 104–113.

Prospects for mine intelligent ventilation technology in the future

S G Li[1], J Gao[2], F Wu[3], Z Yan[4] and X Chang[5]

1. Professor, College of Safety Science and Engineering, Xi'an University of Science and Technology, Xi'an 710054, China. Email: lisg@xust.edu.cn
2. Lecturer, College of Safety Science and Engineering, Xi'an University of Science and Technology, Xi'an 710054, China. Email: 943157956@qq.com
3. Professor, College of Safety Science and Engineering, Xi'an University of Science and Technology, Xi'an 710054, China. Email: 15038537@qq.com
4. Associate Professor, College of Safety Science and Engineering, Xi'an University of Science and Technology, Xi'an 710054, China. Email: 393826629@qq.com
5. Professor, College of Safety Science and Engineering, Xi'an University of Science and Technology, Xi'an 710054, China. Email: changxt@xust.edu.cn

ABSTRACT

Mine intelligent ventilation is an important safety technology guarantee for the construction of intelligent mines. The new generation of information technology represented by cloud computing, big data, artificial intelligence, and industrial internet of things is deeply integrated with mine ventilation safety technology, which has promoted the high-quality development of intelligent ventilation technology in China's mining industry. This paper summarises the research progress of theories and overall architecture of intelligent ventilation in mines. From the aspects of intelligent perception, intelligent decision-making, and intelligent regulation of mine ventilation system, this paper elaborates on the research progress of core technologies such as the development of high-precision monitoring sensors, optimal sensor layout, accurate and rapid acquisition of wind resistance, real-time calculation of ventilation network, intelligent identification and diagnosis of ventilation anomaly, intelligent analysis of air leakage flow field in gob and intelligent linkage control of main ventilation equipment/facilities is introduced. At the same time, this paper introduces the on-site applications of precise wind speed measurement, real-time calculation of ventilation network, and linkage control of ventilation facilities and equipment. Finally, this paper points out the problems and development directions in the construction process of mine intelligent ventilation are pointed out, in order to provide ideas and references for future research and development of mine intelligent ventilation.

INTRODUCTION

Intelligent ventilation in mines is a key technical means to ensure intelligent mining. With the deep integration of new generation information technology represented by cloud computing, big data, artificial intelligence, and industrial Internet of Things with mine ventilation safety technology, it has promoted the high-quality development of intelligent ventilation technology in China's mining industry. As of the end of 2022, China has initially built over 500 mines with intelligent ventilation system (Zhou *et al*, 2023). The ventilation system is the 'breathing system' of the mine and the first line of defence to protect the safety of personnel. During the production process of the mine, it is necessary to ensure the quality of air supply in real time according to the temperature, humidity, harmful gases, and dust concentration at various locations underground, in order to meet the needs of timely and on-demand air supply at various locations during normal and disaster periods (Lu and Yin, 2020). Therefore, the intelligent ventilation system should be an intelligent safety and health assurance system, which meets the intelligent perception of ventilation parameters, operation status of ventilation facilities and equipment, and disaster information, as well as the intelligent decision-making of daily ventilation management and disaster emergency control plans, and the intelligent control of the implementation of ventilation regulation decision plans (Zhou *et al*, 2023; Liu, 2020).

In recent years, many scholars and enterprises at home and abroad have extensively explored the construction of intelligent ventilation in mines. In terms of theoretical research, the calculation of natural air distribution, and the on-demand optimisation and control of air flow models and algorithms as the theoretical core of mine ventilation, have been gradually improved, which has laid a theoretical foundation for the realisation of mine ventilation intelligence. In terms of key technologies, the main focus is on accurate and rapid determination of ventilation parameters, optimal arrangement of sensors, intelligent diagnosis of ventilation faults, real-time calculation of ventilation network and other technologies. In terms of equipment system research and development, it is mainly reflected in ventilation parameter measurement and monitoring equipment, intelligent ventilation control platform, intelligent variable frequency ventilator, and remote intelligent control air door and window, local intelligent ventilation system, disaster emergency wind control equipment, etc. At the same time, China has also successively carried out research on the goals, connotations, logical structure, key technology systems and application scenarios of mine ventilation intelligent construction, aiming to strengthen top-level design, unify construction standards, clarify implementation paths, and improve mine ventilation intelligence level. These research results have promoted the development of intelligent mine ventilation. The author reviews and summarises the theory of intelligent ventilation in mines, the overall architecture of intelligent ventilation system in mines, real-time online accurate monitoring of mine ventilation status parameters, intelligent analysis and decision-making in mines, intelligent regulation and emergency response of mine ventilation system, and introduces the current development status of intelligent ventilation technology in mines in China. The author analyses the problems and development directions in the current process of intelligent ventilation construction in mines, in order to provide ideas and references for future research and development of intelligent ventilation technology in mines.

DEVELOPMENT OF INTELLIGENT VENTILATION THEORY

With the continuous promotion of intelligent construction of mine ventilation in China, multiple scholars have conducted in-depth discussions on the definition, connotation, key technologies, and system composition of intelligent ventilation in mines. Based on the Internet of Things technology, automation equipment, and intelligent system in mines, Lu and Yin (2020) proposed the theory and technology of intelligent ventilation in mines. Liu (2020) discussed the key scientific and technological issues that urgently need to be solved in intelligent ventilation. Shao, Yu and Chen (2020) believes that intelligent ventilation system should be based on underground Internet of Things, personnel positioning and other systems, and adopt theories such as system engineering technology to achieve real-time calculation of ventilation networks, as well as dynamic real-time adjustment of air doors, air windows, and fans. Zhou et al (2020) systematically studied the principle, key technologies, and system composition of intelligent ventilation in mines, and clearly explained the definition and connotation of intelligent ventilation in mines. Wang et al (2023) studied the principle of intelligent ventilation supply-demand matching and linkage control theory in mines. Zhang, Wang and He (2023) elaborated on the basic theory and core ideas of the definition of intelligent ventilation in mines. Zhang and Liu (2024) proposed the process of intelligent ventilation in mines, clarifying the input and output elements of each link of intelligent ventilation in mines and the logical relationship between each link.

DEVELOPMENT OF OVERALL ARCHITECTURE OF INTELLIGENT VENTILATION SYSTEM IN MINES

In order to make the construction of intelligent ventilation system more orderly and controllable, Chinese mining industry experts have conducted in-depth research on the architecture of Intelligent ventilation system in mines. Xie and Han (2012) proposed an intelligent system architecture design scheme from the perspectives of informatisation, visualisation, integration, and intelligence, and preliminarily explored various comprehensive integration methods of mine ventilation design and decision-making technology. This study combines data management and disaster simulation functions to form the framework of an intelligent mine ventilation system. Zhang, Yao and Zhao (2020) proposed that intelligent perception, intelligent decision-making, and intelligent control are the key research and development directions for achieving intelligent ventilation. He pointed out in detail that the key core technologies for achieving unmanned and intelligent mine ventilation are automatic mapping of the ventilation system, rapid and accurate acquisition of ventilation parameters, data-driven network model construction method, linkage analysis and decision-making technology of ventilation network and ventilation regulation, automatic identification and alarm technology of ventilation hazards, accurate identification method of ventilation disasters, and disaster control equipment. Lu and Yin (2020) has developed a four-layer intelligent ventilation system architecture consisting of a mining 4DGIS platform, a mining Internet of Things platform, intelligent ventilation software, and a ventilation intelligent control system, providing reference for the formation of an intelligent mine ventilation system. Wang (2022) proposed to build an integrated intelligent ventilation system that integrates 'accurate online monitoring of air volume, decision-making of air flow control plans, and response to air flow isolation/regulation' around intelligent monitoring perception, intelligent decision-making, and intelligent regulation. The intelligent ventilation system consists of four subsystems: an accurate online monitoring system for mine ventilation parameters, a three-dimensional intelligent decision-making software platform for mine ventilation control, intelligent adjustment facilities for mine ventilation, and an intelligent control system for ventilation power. Based on the three major fields of state perception, control algorithms, and control strategies of intelligent ventilation control system, Zhang *et al* (2023) proposed a design principle for an intelligent ventilation precision control system that is 'safe and reliable, integrated sensing and control, and collaborative linkage', as well as the logical framework with the edge's integration of perception and control as the core. Zhang and Liu (2024) studied the architecture of an intelligent analysis and decision-making platform for mine ventilation using a logical layering approach, which was decomposed into an intelligent ventilation perception system, a data centre, an intelligent ventilation control platform, and intelligent ventilation control equipment.

According to the current research status of intelligent ventilation architecture, the mine intelligent ventilation system is a deep integration of geographic information, communication, mobile internet, Internet of Things, big data, artificial intelligence, cloud computing, automatic control, surveying and monitoring, intelligent equipment etc with mine ventilation technology and equipment, integrating a comprehensive and autonomous perception of mine ventilation and climate environment parameters, the operation status of ventilation facilities and equipment, the real-time and efficient interconnection of ventilation perception system with ventilation networks, ventilation facilities and equipment, disaster prevention system, and major production system, ventilation air volume adjustment plans, emergency air flow control strategies for disasters, and precise coordinated control of ventilation facilities and equipment to achieve intelligent linkage control of the entire process of mine ventilation. Accordingly, the mine intelligent ventilation technology system mainly includes intelligent sensing technology, intelligent analysis and decision-making technology, and intelligent control technology. Based on this, the overall architecture of intelligent ventilation in mines is summarised as shown in Figure 1.

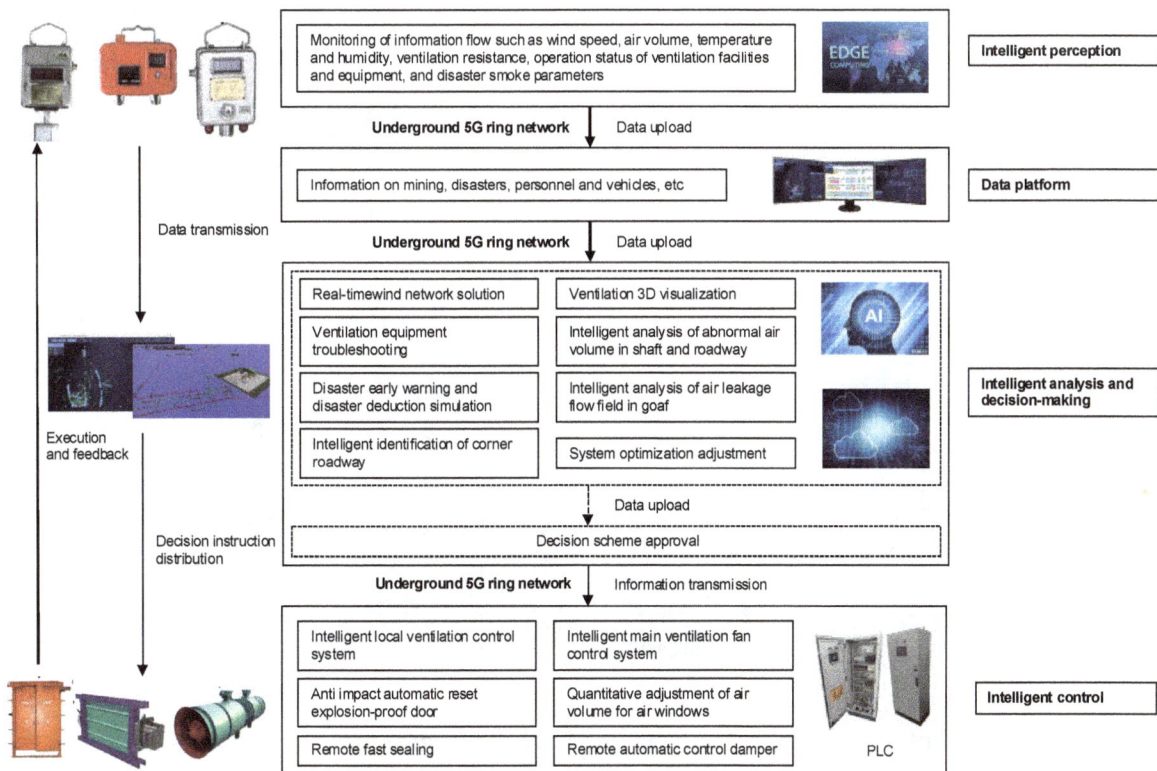

FIG 1 – Overall architecture of intelligent ventilation system.

RESEARCH PROGRESS IN INTELLIGENT PERCEPTION TECHNOLOGY OF MINE VENTILATION PARAMETERS

Whether it is online accurate perception of the operating status of the wind network, real-time accurate calculation and intelligent accurate control, accurate ventilation parameters are required as the basis. Therefore, the development of accurate and rapid testing technology and equipment for ventilation parameters is of great significance for mine intelligent ventilation.

At present, the mine wind speed sensors are mostly thermal, differential pressure and vortex street sensors, with low measurement accuracy and insensitive response, which cannot meet the requirements of intelligent ventilation for accurate and rapid wind speed testing. Based on the advantages of wide testing range and high accuracy of ultrasonic speed measurement technology (Figure 2a shows the principle of ultrasonic time difference method for wind measurement), China has developed a high-precision ultrasonic wind measurement device (as shown in Figure 2b) (Song *et al*, 2022), with a wind measurement error of ±0.1 m/sec and a lower wind measurement limit of 0.1 m/sec. We have developed a full section multi-point mobile fully automatic wind measurement system based on the 'nine point' collection method for the average wind speed area of the roadway section (Sun *et al*, 2022), as shown in Figure 2c, achieving the upgrade of monitoring from 'point wind speed' to 'full-section wind speed' in the roadway. Based on the diagonal ultrasonic wind speed sensor and the principle of integrating multiple line wind speeds into full-section wind speeds in roadway, a full-section scanning fully automatic wind measurement system has been developed (as shown in Figure 2d) (Zhang *et al*, 2022), the full-section wind speed field scanning monitoring of roadway was achieved. The Wangpo Coalmine in Shanxi, China has put into use a multi-point mobile wind measurement device, and the wind measurement accuracy has reached more than 95 per cent. The Zhangjiamao coalmine in Shaanxi, China has equipped with an automatic mine wind measurement system, which takes only two minutes to complete the entire mine wind measurement.

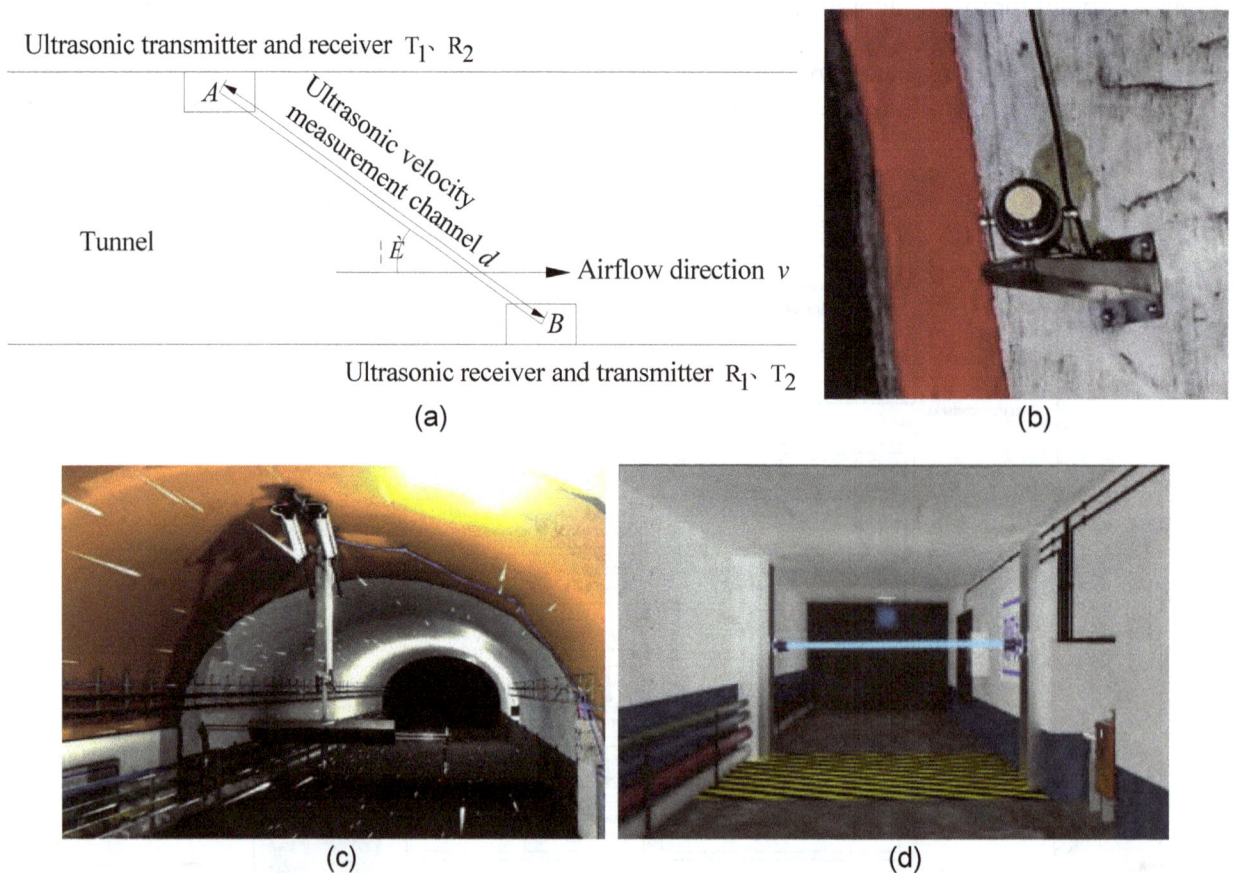

FIG 2 – Measurement of wind speed in roadway based on ultrasonic technology: (a) Principle of ultrasonic accurate wind measurement; (b) Ultrasonic high-precision wind measurement device;

(c) Full-section multi-point mobile automatic wind measurement system; (d) Full-section scanning automatic wind measurement system (Song *et al*, 2022; Sun *et al*, 2022; Zhang *et al*, 2022).

The inhomogeneity of the wind speed distribution in the roadways indicates that the wind speed at the roadway point tested by the sensor cannot represent the average wind speed. The works (Ji, 1997; Lu, Zhang and Fan, 2015; Zhao *et al*, 2014; Zhang, 2019; Wang *et al*, 2022) studied the roadway wind speed field and the relationship between point wind speed and average wind speed by analysis, laboratory similar model test, numerical simulation and other methods.

The real-time and accurate measurement of wind pressure is an important means to monitor the distribution of ventilation resistance in mines, and to identify abnormal wind flow conditions and ventilation structure failures. It is also of great significance to the real-time calculation of mine ventilation network and the intelligent regulation of wind flow. Commonly used mine ventilation parameter detection instruments have low measurement accuracy, resulting in low reliability of resistance measurement results for large-section and small-resistance roadways. Wind pressure sensors mostly use capacitance or semiconductor pressure-sensing principles, which have low measurement accuracy, insensitive response, and cannot meet the requirements of mine intelligent ventilation for real-time and accurate monitoring of wind pressure. In recent years, China adopted the American MENSOR silicon resonant principle, developed the CPD120 smart wind pressure gauge, the absolute pressure resolution is accurate to 1 pascal, and has functions such as automatic storage of monitoring data and wireless transmission, and has developed a high-precision pressure sensor with automatic compensation with a measurement accuracy of 1 pascal, supplemented by ventilation resistance measurement and verification, which realises real-time online accurate perception of mine ventilation resistance, the technical principle is shown in Figure 3 (Zhou *et al*, 2020), which solves the time-consuming and labour-intensive problem of manual resistance measurement.

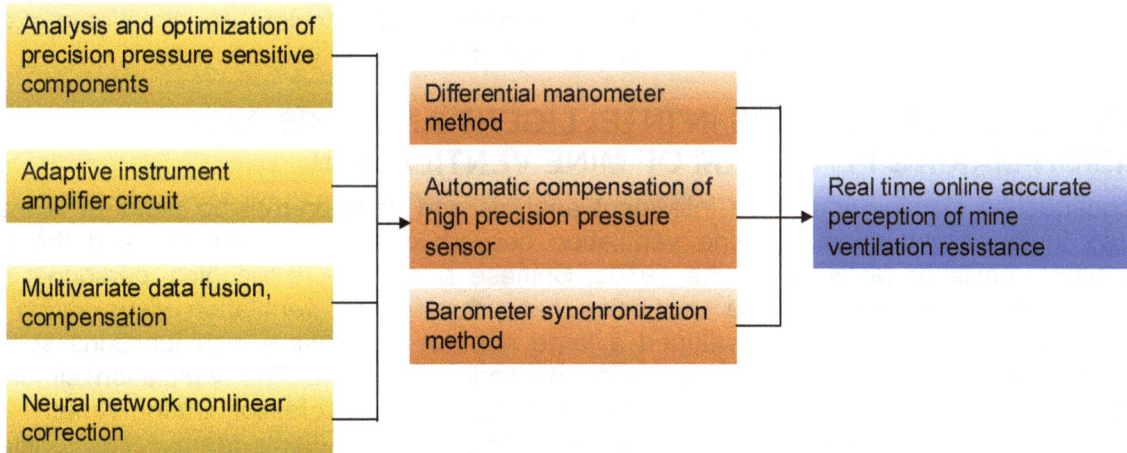

FIG 3 – Real time online accurate perception of mine ventilation resistance.

Whether it is optimisation of ventilation system, system transformation, real-time calculation of wind network or intelligent control of wind flow, the wind resistance of each branch in the wind network needs to be known. At present, the wind resistance obtained by the calculation of ventilation resistance and the Atkinson formula has a large error, which cannot meet the requirements of mine intelligent ventilation. Chen and Zhao (1994) studied the measurement adjustment method of mine ventilation network, and improved the resistance measurement accuracy. Wu, Zhao and Lei (2021) given a friction and wind resistance calculation method that can characterise the three-dimensional surface roughness of the roadway based on fractal theory, and combines the three-dimensional laser scanning technology to obtain roadway friction and wind resistance quickly and accurately. Compared with the time-consuming and laborious work of measuring ventilation resistance, the wind measurement operation is relatively easy. The works (Zhou, Lyu and Liu, 2004) studied the wind resistance measurement technology, its technical principle: take the air volume of the roadway and the working point of the main ventilator as the known quantity, changing the wind resistance of one or several branches in the wind network, the air volume distribution of the wind network under different working conditions is obtained, and then the number of equations for solving the wind resistance is increased, and then the roadway wind resistance is inversely calculated. Wu, Zhao and Wang (2024) established a double constraint optimisation model for correcting air resistance, and developed a corresponding software system, which achieved intelligent correction of shaft and roadway wind resistance using measured air volume with quickly and accurately obtained wind resistance parameters.

Artificial neural networks, support vector machines and other artificial intelligence technologies have superior nonlinear processing capabilities and are suitable for the prediction of ventilation resistance coefficients. Wei *et al* (2018) used artificial neural networks, support vector machines and other methods to establish a roadway friction resistance coefficient prediction model. Deng (2014) established a mathematical model of ventilation resistance coefficient inversion based on the principle of least squares, and applied intelligent optimisation algorithms such as genetic algorithm and particle swarm algorithm to the ventilation resistance coefficient inversion problem.

The optimisation problem of sensor layout to meet the needs of accurate perception of ventilation parameters of the whole wind network and to optimise the economy is the focus of mine intelligent ventilation research. Liu, Ma and Yang (2017) used variable fuzzy set theory to construct a variable fuzzy optimal model for wind speed sensors, accordingly, calculates relative wind speed sensors, which obtain the reasonable weight of the wind speed sensor installed on each branch, use the breadth-first search algorithm to obtain the minimum spanning tree of the ventilation network, and determine the basic loop of the ventilation network combined with the root-finding method, so as to quickly and accurately determine the specific loop branch where the wind speed sensor is installed, the optimal location scheme of the wind speed sensor is obtained, and the real-time dynamic monitoring of the air volume of the ventilation network without blind spots is realised. Li *et al* (2021) used the minimum spanning tree algorithm to solve the minimum spanning tree of the ventilation network graph with weights assigned to the branches, and obtains the network residual branch combination with the largest weight in the ventilation network graph, determine the monitoring distribution point as the largest residual branch combination, and calculate the air volume of each branch of the ventilation system by monitoring the residual branch air volume metre.

CONSTRUCTION PROGRESS IN INTELLIGENT ANALYSIS AND DECISION-MAKING PLATFORM OF MINE VENTILATION

The construction of mine ventilation intelligent analysis and decision-making platform is mainly reflected in the development of mine ventilation network calculation software, and the rapid diagnosis of ventilation faults. Since the 1970s, Chinese scientific and technical personnel have invested a lot of research in mine ventilation network analysis, ventilation visualisation and ventilation simulation, and have developed a large number of software with functions such as ventilation network diagram drawing and wind network calculation. The ventilation simulation system for mining construction projects (CFIRE) developed by Professor Chang Xintan; Mine Ventilation Simulation System (MVSS) developed by Professor Liu Jian's team; Mine Ventilation Management System (MVMS) developed by Professor Wang Deming, etc. These software have been widely used in the actual mine ventilation design and daily management. After nearly 50 years of development, China's software for static analysis of ventilation system has become more mature, and the main functions of the software have also become more complete.

In recent years, based on mine ventilation theory, geographic information, communication, cloud computing, internet, sensors, Internet of Things and other technologies, Chinese researchers have used ventilation parameter monitoring sensors with full wind network coverage to optimise site selection, monitoring data noise reduction and pre-processing, and multiple data information fusion and redundancy analysis and other methods to analyse in real-time and in-depth mining for mine ventilation monitoring data, have established a real-time calculation model of mine ventilation network, the mine ventilation network calculation software and ventilation monitoring system are integrated. Developed a series of mine ventilation intelligent auxiliary decision-making system that integrate the functions of real-time calculation of mine ventilation network, formulation of wind flow control schemes in daily and disaster periods, and online ventilation monitoring and early warning, and realise the show of the mine ventilation system diagram, ventilation parameter changes, real-time calculation results of ventilation network, ventilation abnormal diagnosis and early warning through the form of two-dimensional and three-dimensional dynamic visualisation. Fang and Ma

(2016) proposed an integrated algorithm of ventilation monitoring and ventilation simulation, through the method of sharing the mine monitoring system database, realises the dynamic solution of the wind network and the online real-time simulation of the ventilation system based on the internet technology. Luo *et al* (2019) established a data-driven dynamic calculation model of the ventilation network, optimised the sensor arrangement process for monitoring ventilation parameters, and developed an online monitoring system for mine ventilation, realised the real-time monitoring of mine ventilation simulation, intelligent online solution and three-dimensional (3D) dynamic online. Xi'an University of Science and Technology has developed a visual mine ventilation network solution model based on the mine ventilation network parallel solution model, the interface is shown in Figure 4a, which realises fast and accurate calculation of large-scale ventilation networks. An intelligent ventilation software system has been developed, and the interface is shown in the Figure 4b, the system realises the functions of safety tracking, monitoring and early warning, daily ventilation management, medium and long-term analysis and design, disaster deduction and emergency ventilation disposal. Based on cloud computing technology, a three-dimensional intelligent real-time dynamic solution system for mine ventilation network is developed, it has realised the functions of whole wind network monitoring, real-time calculation of wind network, three-dimensional ventilation simulation, abnormal air volume analysis, demonstration of disaster avoidance route, and formulation of ventilation plan. The three-dimensional display of the ventilation system is shown in Figure 4c.

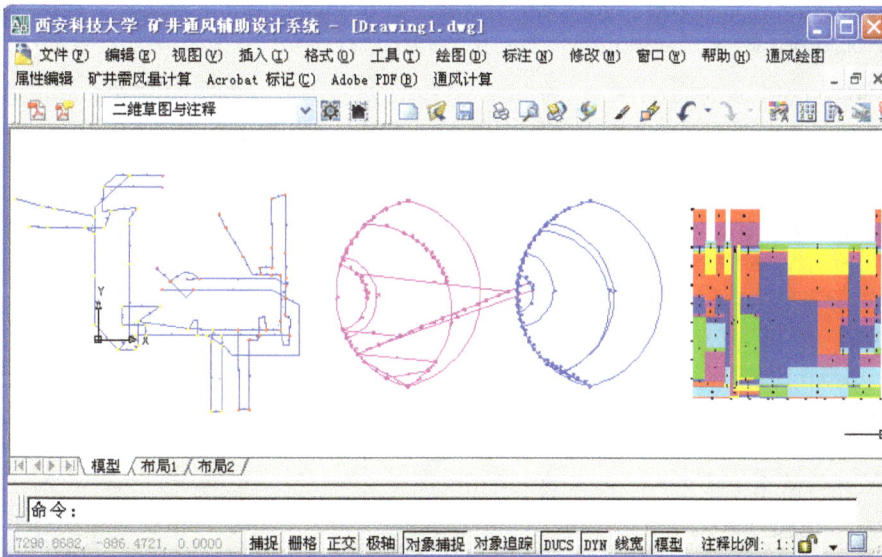

(a)

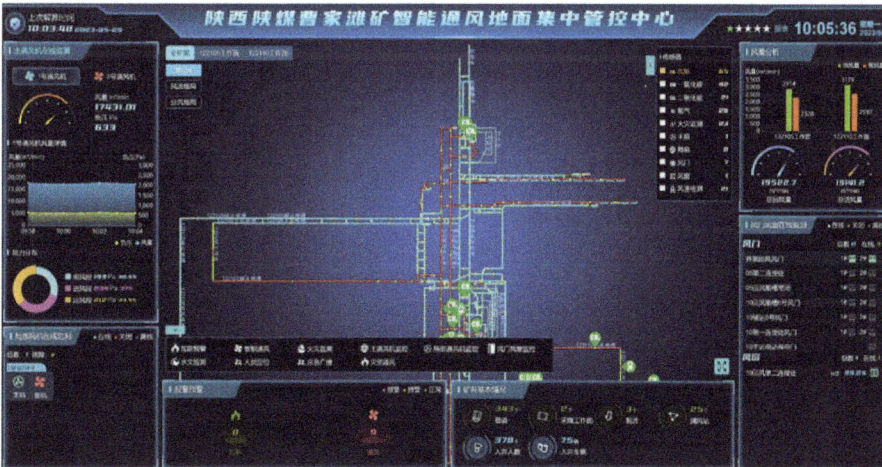

(b)

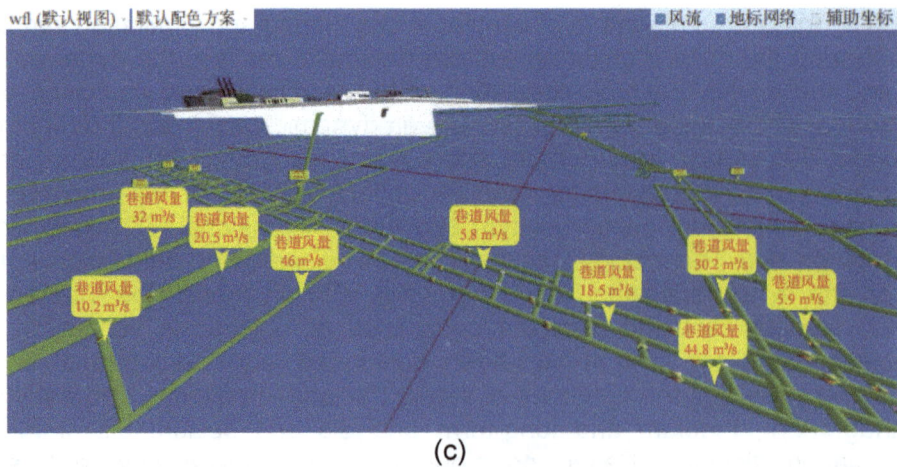

(c)

FIG 4 – Intelligent ventilation comprehensive control platform: (a) Mine ventilation simulation software; (b) Mine intelligent ventilation platform; (c) Three-dimensional display of mine ventilation system.

The intelligent diagnosis of mine ventilation faults mainly includes abnormal analysis of the air volume of the roadway, optimised layout of monitoring points, identification of the angle-connected roadway, and fault diagnosis of ventilation equipment.

From the perspective of the entire mine ventilation network, the reasons for the abnormal air volume in the roadway (the collapse of the roadway, the unclosed or damaged air door, the extension and scrapping of the roadway, and the emptying of the coal bunker) can all be attributed to the change of the branch wind resistance. The mine wind network is an overall structure, and the change of the wind resistance of any branch will cause the change of the air volume of the relevant branch or even all branches in the whole network. The sensitivity of the branch air volume to the change of other branch wind resistance is defined as sensitivity.

The works (Wang, Wu and Wang, 2008; Wu, 2011a, 2011b; Jiang, 2011) applied the sensitivity theory to the research of air volume regulation, outlier analysis of air volume, identification of corner-connected roadways, and analysis of wind network stability. Xi'an University of Science and Technology developed a sensitivity calculation program and integrated it into the ventilation network calculation software, and realised the calculation and view of the sensitivity of complex networks under Auto CAD.

Research on automatic and accurate identification of corner-joined roadways. Zhao, Liu and Yang (2001) proposed a seven-tuple algorithm, the automatic identification program of the angle-connected roadway is used to realise the accurate and rapid identification of the angle-connected roadway, and the air leakage and seepage area in the gob is analysed as an air channel in the wind network. Zhao *et al* (2009) studied the mathematical model of corner-connected structure recognition based on parallel computing, and realised automatic identification of corner-connected structures in super-large wind networks. Yan *et al* (2018) proposed an equivalent attribute graph model, and established a high-risk area identification algorithm based on the subgraph isomorphism method. While analysing and comparing the topology of the wind network, the attributes of the roadway are included in the judgment. Through the relationship between roadway attributes and risks, structures that do not have production risks are eliminated, and finally the high-risk areas of corner connection in the complex wind network are automatically identified.

In the research of intelligent analysis technology of air leakage flow field in gob. The works (Wu *et al*, 2016, 2017) studied the Jacobian matrix characteristics of mine ventilation network and its parallel solution technology, realised the rapid solution of large-scale ventilation network, and provided technical support for the use of network solution method to study gob flow field. The works (Wu, Luo and Chang, 2019; Wu, Gao and Chang, 2020; Wu and Luo, 2020) studied the ventilation problem of the coupling of mine ventilation network and gob flow field, proposed a numerical simulation method for coupling the one-dimensional mine ventilation network and the two/three-dimensional (2D/3D) gob flow field-the finite flow tube method (as shown in Figure 5a), developed a visualised mine ventilation network and gob air leakage integrated simulation software (i-MVS), achieved generation of gob flow pipes, a coupling model of mine ventilation network and gob flow field based on the finite-flow pipe method is established (as shown in Figure 5b), and developed the software can realise the partition of the finite element mesh of the gob, ventilation network modelling, and dynamic determine the areas with gas explosion and coal spontaneous combustion hazards in gob and mine ventilation network (as shown in Figure 5c), and can also analyse the risk analysis in various mines ventilation scheme.

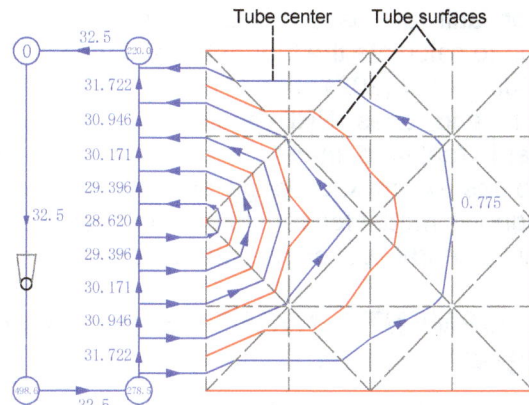

(a)

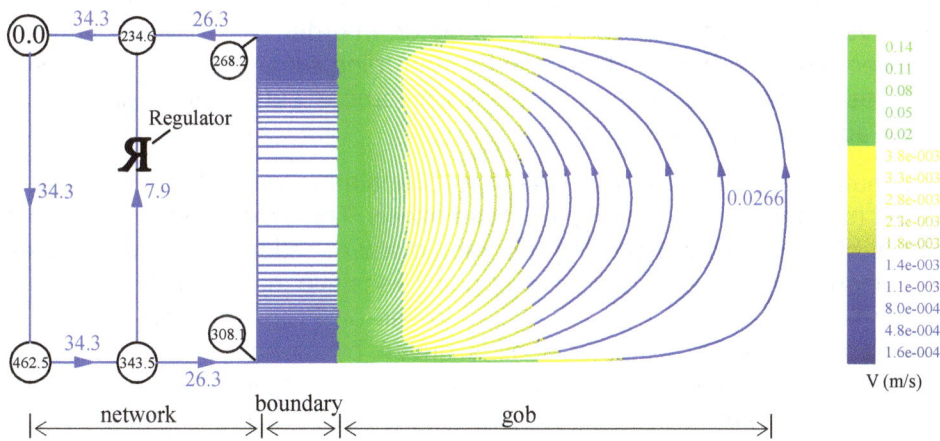

(b)

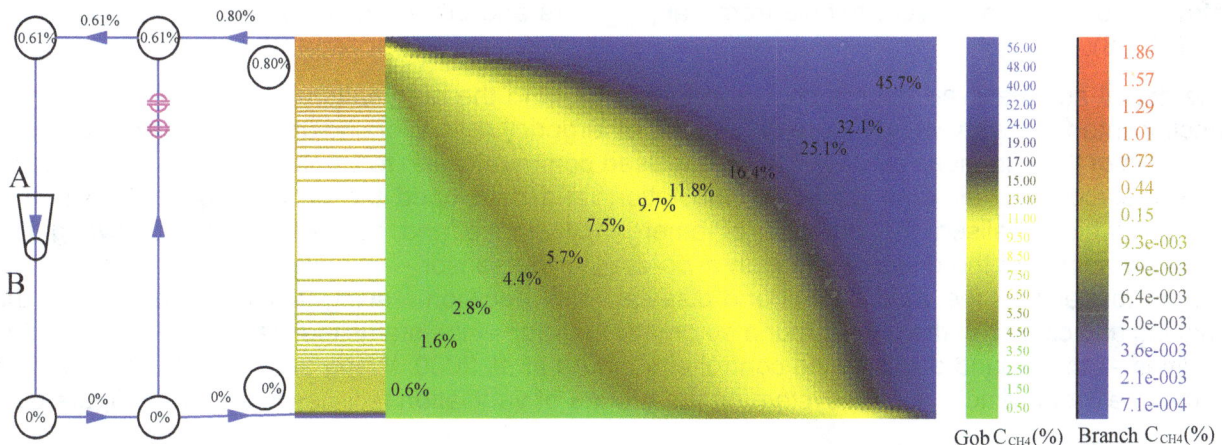

(c)

FIG 5 – Intelligent analysis technology of air leakage flow field in gob: (a) The coupling network (blue colour) of the one-dimensional mine ventilation network and the 2D/3D gob flow field; (b) The coupling simulation results of mine ventilation network and gob flow field; (c) Methane concentration in the ventilation system.

With the development of modern digital signal processing, computer, virtual instrument, artificial intelligence and other technologies, the fault diagnosis technology of mine ventilator has entered the field of intelligence. China has applied artificial intelligence technology and related theories to develop intelligent diagnosis technology for mine ventilator faults, and has achieved a series of fruitful results. Li and Li (2010) applied the holographic spectrum technology of multi-sensor information fusion theory to the fault diagnosis of mine ventilators, and combined with virtual instrument technology, developed a virtual instrument for ventilator fault diagnosis, and realised the accurate diagnosis of ventilator faults. Cui and Cheng (2021) used fuzzy reasoning theory to convert the constructed main ventilator fault rule base into a fuzzy matrix of fault relations, and uses the maximum membership function to calculate the highest confidence level, based on which to find the cause of the ventilator failure. Jing *et al* (2021) used Unity3D, 3dsMax, SciFEA etc to build a digital twin model of the ventilator, and uses PREspective to communicate with the ventilator's Programmable Logic Controller in real time, maps the ventilator's operating status to the digital twin model in real time, and combines expert knowledge and machine learning, historical data etc to establish a predictive fault diagnosis model for ventilators, and use the improved whale optimisation algorithm to optimise probabilistic neural network to conduct predictive fault diagnosis for ventilators.

RESEARCH AND DEVELOPMENT PROGRESS IN INTELLIGENT REGULATION EQUIPMENT OF MINE VENTILATION

The main function of mine ventilation facilities is to control the air flow, the air door and window are distributed throughout the underground, it takes a lot of manpower to manually adjust the air volume. More importantly, when a disaster occurs in the underground, timely control of the ventilation system can effectively prevent the spread of the disaster. In order to realise the intelligent regulation of ventilation system and improve the technical level of mine emergency and disaster relief, China has carried out research and development work on remote automatic control of air doors and remote automatic adjustment of air windows.

The remote automatic control air door is an intelligent ventilation facility for mine air flow. At present, automatic air door has been widely used in Chinese mines. This type of air door sends and receives signals by means of mechanical triggering, ultrasonic wave, radar microwave, infrared induction, electromagnetic induction, pulsed light source photosensitive etc, and controls the mechanical transmission mechanism through relay, microcontroller or Programmable Logic Controller, the automatic opening and closing of the air door is driven by means of compressed air, hydraulic pressure or motor, which solves various problems of manual opening and closing, the pressure structure around the door frame solves the problem of the opening and closing failure of the air door due to the deformation of the roadway. At the same time, it has voice sound and light alarm prompt automatic air door lock, intelligent video surveillance technology to identify vehicles and pedestrians, infrared detection to prevent people from trapping cars and other functions (Wu *et al*, 2010; Shao *et al*, 2008).

The actual automatic air door is shown in Figure 6a, and the automatic air door video monitoring function interface is shown in Figure 6b. For the emergency rescue of underground fire accidents, on the basis of the automatic air door, China has used communication technology to design the ground central station and air door controller substation, and developed the remote automatic air door system, which realised the purpose of emergency smoke and gas evacuation, following this workflow: monitors the disaster situation according to the sensor → the signal is transmitted to the ground to alarm → the wind control and disaster relief decision-making → the ground control the opening and closing of the underground automatic air door → the status feedback of the wind control facilities → the ground display wind control effect (Zhu *et al*, 2000; Wang *et al*, 2021). The interface of the intelligent remote control system on the ground of ventilation facilities is shown in Figure 6c.

(a)

(b)

(c)

FIG 6 – Remote automatic control air door: (a) Automatic air door of Caojiatan Mining Company; (b) Remote monitoring system of automatic air door in Xiaobaodang Mining Company; (c) Ground intelligent remote control system for ventilation facilities in Xiaobaodang Mining Company.

The remote automatic adjustment of air windows is an intelligent ventilation facility for quantitative adjustment of mine air volume. By integrating automation, sensors and Internet of Things technologies, China has developed high-precision, high-sensitivity intelligent air doors and windows, the quantitative relationship model between wind resistance increment and wind window adjustment parameters is constructed, forming the intelligent remote control technology of air volume, so as to realise the intelligent and precise control of the mine ventilation system. Shandong Lion King Software Co., Ltd. has designed a multi-layer push-pull type, louver type, shutter type and other structures of intelligent control air door and window, as shown in Figure 7. The design principle is to use a stepper motor to accurately control the movement of the windshield through the synchronous belt, the material level opening sensor feeds back the opening of the wind window in real time, the controller accepts the remote control command to drive the actuator, and feeds back the collected sensor and wind window status data to the remote system in real time, realising the remote automatic closed-loop precision control of the wind window. This type of intelligent control air door and window has several control modes such as remote automatic adjustment, local remote control adjustment, automatic and manual switching adjustment etc, which can meet the emergency response under different ventilation conditions (Lu and Yin, 2020; Zhou *et al*, 2020) based on the adjustment of the characteristic curve of the wind and window, integrating the automatic control technology, the structural characteristics of the adjusting wind window and the network structure characteristics of the branch, a proportional-integral-derivative adjustment model of the branch air volume is constructed, and the continuous and precise control of the adjusting wind window is realised.

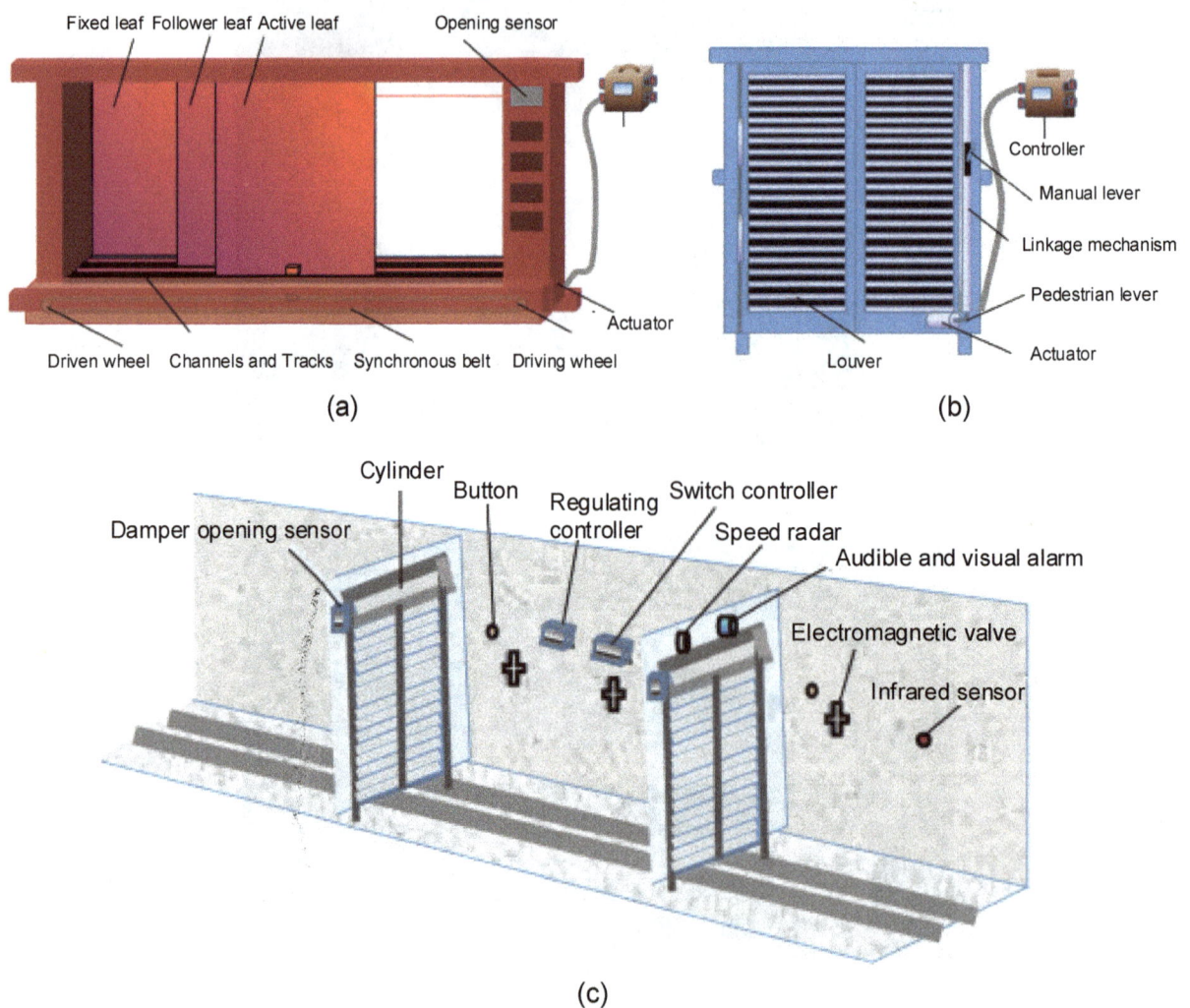

FIG 7 – Mine ventilation intelligent regulation facility: (a) Multi-layer push-pull adjustable window; (b) Louvered regulating window; (c) Intelligent control air door.

The ventilation rate of the traditional tunnelling face is determined by quantitative calculation in advance according to factors such as temperature, gas emission, carbon dioxide emission, and the maximum number of workers, and after takes its maximum value, According to this, the quantitative air supply of the local ventilator belongs to the ventilation mode of one wind blowing, It is impossible to intelligently control the air flow according to the changes of environmental parameters, resulting in waste of energy consumption caused by excessive air volume or gas exceeding the limit caused by insufficient air volume. In recent years, integrating communication, sensors, Internet of Things, automation, frequency conversion control and other technologies, China has developed an intelligent control system for local ventilators as shown in Figure 8. The system can intelligently sense environmental parameters through sensors and feed back to the intelligent control system in real time to adjust the frequency and speed of the local ventilators, thereby changing the output air volume and air pressure of the ventilators, so as to realise on-demand air supply, automatic gas discharge, automatic switching of the main and standby ventilators, and remote unmanned start-up, thereby ensuring the safety, efficiency and energy saving of local ventilation (Cui, 2016; Rui, 2021; Liu, 2017; Cheng, 2020). Zhang (2008) combined the pulsating air supply technology with the frequency conversion speed regulation technology, and developed an intelligent ventilation control system for the excavation face, the system analyses and determines the gas concentration collected by the gas sensor, changes the ventilator speed to adjust the air volume, air pressure, and the switching cycle and switching ratio of the local ventilator, thereby realising intelligent control of the ventilation of the excavation face. Ping An Electrical Group has developed an intelligent local ventilation system, which consists of local ventilator, mine flow channel inverter, intelligent control switch etc systematically adopts a control system composed of Programmable Logic Controller and inverter, used the change of gas concentration as the main reference to adjust the ventilator speed, so as to realise the automatic adjustment of air volume and control of gas discharge (Chen, Xiao and Hu, 2009; Wang et al, 2012).

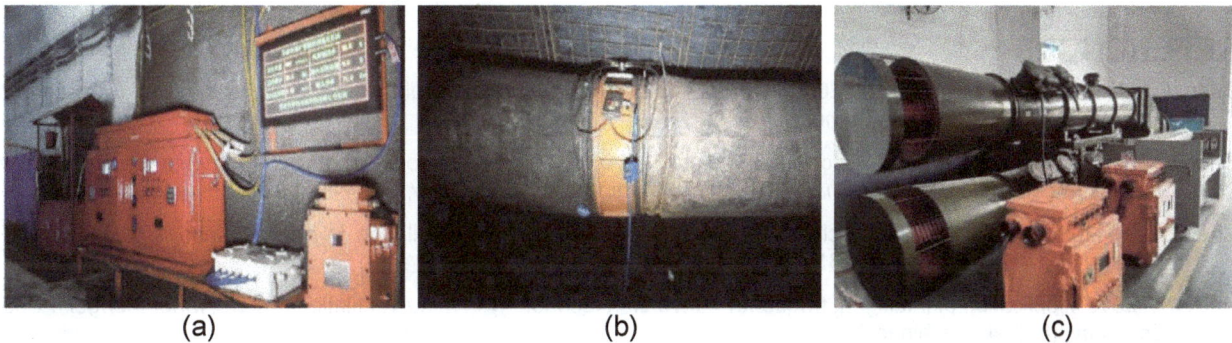

FIG 8 – Mine intelligent local ventilation system: (a) Frequency converter; (b) Wind speed sensor; (c) Mine intelligent local ventilator.

THE PROBLEMS AND DEVELOPMENT PROSPECTS OF ON-SITE IMPLEMENTATION OF INTELLIGENT VENTILATION SYSTEM IN MINES

Based on the above research progress in the theory and technology of intelligent ventilation in mines, it can be seen that intelligent ventilation in mines has initially improved the timeliness of obtaining basic ventilation parameters, calculating ventilation networks, and regulating ventilation facilities and equipment, and reduced the intensity of ventilation management work. However, its on-site implementation still faces many challenges. At present, the development of intelligent ventilation in mines is still in its early stages, while the theory and technology of artificial intelligence have reached a very mature stage. It deeply integrates mine ventilation technology with big data analysis, data mining and other technologies, further strengthens the application of intelligent technology in ventilation system fault intelligent diagnosis, disaster intelligent warning and other aspects, verifies the applicability of intelligent warning analysis technology on-site, promotes scientific and reliable intelligent ventilation technology, allowing intelligent ventilation technology to truly play a role in mine safety production and the occupational health of miners. Due to harsh conditions such as high humidity and heavy dust in mines, new high-precision sensors may have operational failures, resulting in unreliable ventilation basic data and low credibility of intelligent perception, intelligent analysis and decision-making, and intelligent regulation, which may increase the operational risks of ventilation system. To address this issue, it is necessary to conduct in-depth research on the mechanisms and laws that affect sensor accuracy, and establish an operation and maintenance guarantee system for intelligent equipment. This lays the foundation for truly building an automated and unmanned overall mine intelligent ventilation system.

CONCLUSIONS

With the in-depth integration of big data, artificial intelligence, Internet of Things, cloud computing and other high-tech technologies and mine ventilation technology and equipment, Intelligent integrated management and control of mine ventilation characterised by 'comprehensive and autonomous perception of ventilation parameters, deep and efficient integration of monitoring data, intelligent analysis and decision-making of ventilation status, and precise coordinated control of ventilation facilities/equipment' has become an important mode of mine ventilation intelligent upgrade in China. In recent years, mines in China have firmly grasped the development opportunities brought by the new generation of information technology, and have achieved a series of important achievements in intelligent ventilation technology, software and equipment. However, judging from the current development status of mine intelligent ventilation technology in China, the intelligent construction of mine ventilation is still in its infancy, and it is necessary to continue to explore and tackle key technologies, so as to promote the development of mine ventilation engineering from experience to high-tech, and truly realise the role of intelligent ventilation in increasing safety, reducing personnel, and improving efficiency.

ACKNOWLEDGEMENTS

The authors sincerely thank the scholars who provided references, whose papers provided important references for this paper. This work is supported by the National Natural Science Foundation of China [grant numbers 51974232].

REFERENCES

Chen, C H X, Xiao, W L and Hu, X M, 2009. Introduction of intelligent local ventilation system, *Jiangxi Coal Science and Technology*, (3):109–112 (in Chinese).

Chen, K Y and Zhao, Y H, 1994. The correlated adjustment of condition equations for resistance measurement of mine ventilation network, *Journal of China University of Mining and Technology*, 23(1):80–89 (in Chinese).

Cheng, L H, 2020. Application research on intelligent control system of ventilation in heading face, *Shandong Coal Science and Technology*, (11):131–133 (in Chinese).

Cui, B W, 2016. Application of intelligent frequency conversion technology in mine ventilation system, *Inner Mongolia Coal Economy*, (8):3–4 (in Chinese).

Cui, H F and Cheng, F L, 2021. Fault diagnosis of mine main ventilator based on improved fuzzy reasoning theory, *Coal Technology*, 40(10):112–115 (in Chinese).

Deng, L J, 2014. *Study on mine ventilation resistance coefficient inversion*, PhD thesis, Liaoning University of Engineering and Technology, Fuxin (in Chinese).

Fang, B and Ma, H, 2016. Mine ventilation network application monitoring database and application dynamic solver, *Journal of Liaoning Technical University (Natural Science)*, 35(12):1439–1442 (in Chinese).

Ji, C H S, 1997. Analysis of Wolonin's basic theory of mine ventilation, *Nonferrous Metals Engineering*, (2):1–5 (in Chinese).

Jiang, S H, 2011. Theory and application of mine ventilation network sensitivity, *Safety in Coal Mines*, 42(1):96–99 (in Chinese).

Jing, H X, Huang, Y R, Xu, S H Y and Tang, C H L, 2021. Research on the predictive fault diagnosis of mine ventilator based on digital twin and probabilistic neural network, *Industry and Mine Automation*, 47(11):53–60 (in Chinese).

Li, M and Li, Y, 2010. Virtual instrument of fault diagnosis for mine ventilator based on holographic spectrum, *Coal Science and Technology*, 38(3):93–96, p 108 (in Chinese).

Li, Y J, Wu, J K, Li, Y H and Yao, P Y, 2021. Optimization on monitoring layout of mine ventilation network based on the principle of minimum spanning tree, *Mining Research and Development*, 41(7):172–175 (in Chinese).

Liu, J, 2020. Overview on key scientific and technical issues of mine intelligent ventilation, *Safety in Coal Mines*, 51(10):108–111, p 117 (in Chinese).

Liu, P, 2017. The research and application of intelligent mine local ventilation equipment, *Coal Mine Modernization*, (5):87–88 (in Chinese).

Liu, Y X, Ma, H and Yang, H R, 2017. Mine wind sensor variable fuzzy optimization model, *Journal of Liaoning Technical University (Natural Science)*, 36(10):1031–1035 (in Chinese).

Lu, G L, Zhang, M H and Fan, C Q, 2015. Analysis of fixed-point wind speed measurement method based on CFD, *Coal Technology*, 34(5):156–157 (in Chinese).

Lu, X M and Yin, H, 2020. The intelligent theory and technology of mine ventilation, *Journal of China Coal Society*, 45(6):2236–2247 (in Chinese).

Luo, G, Zou, Y H, Ning, X L and Jin, N Q J, 2019. Research and application of online monitoring technology for mine ventilation network, *Mining Safety and Environmental Protection*, 46(5):47–50, p 55 (in Chinese).

Rui, G X, 2021. Study on intelligent reconstruction scheme of ventilation system in Chahasu coal mine, *Journal of North China Institute of Science and Technology*, 18(8):40–44 (in Chinese).

Shao, H, Jiang, S H G, Qin, J H, Wu, Z H Y and Wang, L Y, 2008. Program design on auto air-door base on PLC control, *Coal Science and Technology*, 36(4):66 (in Chinese).

Shao, L B, Yu, B C and Chen, X Y, 2020. Key technology of mine intelligent ventilation, *Safety in Coal Mines*, 51(11):121–124 (in Chinese).

Song, T, Wang, J W, Wu, F L, Zhang, G Q, Chen, F, Feng, X and Li, L Q, 2022. Real-time calculation method of mine ventilation network based on ultrasonic full-section wind measurement, *Journal of Mine Automation*, 48(4):114–120, p 141 (in Chinese).

Sun, Y X, Zhang, L, Yang, X, Liu, Y Q, Ma, Q, Li, W, Zhang, H J, Zhao, K and Duan, S G, 2022. Development and application of automatic on-line measuring device for roadway air volume, *Safety in Coal Mines*, 53(9):251–256 (in Chinese).

Wang, G F, 2022. New technological progress of coal mine intelligence and its problems, *Coal Science and Technology*, 50(1):1–27 (in Chinese).

Wang, H G, Wu, F L and Wang, Y, 2008. The airflow abnormal value analysis of mine ventilation on network based on the sensitivity, *Safety in Coal Mines*, (9):86–88 (in Chinese).

Wang, H, Qiu, L M, He X Q, Song, D Z and Zhao, Y J, 2022. Study on the wind speed distribution law of coal mine roadway section under different factors, *Mining Research and Development*, 42(7):125–132 (in Chinese).

Wang, K, Cai, W Y, Gao, S W, Chen, X Y and Zhang, Y C H, 2021. Research on the linkage reliability of mine fire wind and smoke flow emergency control system, *Journal of China University of Mining and Technology*, 50(4):744–754 (in Chinese).

Wang, K, Pei, X D, Yang, T, Chen, R D, Hao, H Q, Jiang, S G and Sun, Y, 2023. Study on intelligent ventilation linkage control theory and supply-demand matching experiment in mines, *Chinese Journal of Engineering*, 45(7):1214–1224 (in Chinese).

Wang, W C, Qiao, W, Li, G and Tian, C H, 2012. Application of mine intelligent partial ventilation system in Huhewusu coal mine, *Safety in Coal Mines*, 43(6):114–116 (in Chinese).

Wei, N, Sun, Y S H N, Deng, L J, Huang, D and Guo, X, 2018. Influence factors analysis and prediction on mine ventilation resistance coefficient based on SVM, *Journal of Safety Science and Technology*, 14(4):39–44 (in Chinese).

Wu, B, Zhao, C H G and Lei, B W, 2021. Characterization and calculation method of friction resistance based on fractal theory of roadway rough surface, *Journal of China University of Mining and Technology*, 50(4):633–640 (in Chinese).

Wu, F L and Luo, Y, 2020. An innovative finite tube method for coupling of mine ventilation network and gob flow field: Methodology and Application in Risk Analysis, *Mining, Metallurgy and Exploration*, 37(5):1517–1530.

Wu, F L, 2011a. Sensibility calculation and stability analysis on a complicated mine ventilation network, *Science and Technology Review*, 29(19):62–65 (in Chinese).

Wu, F L, 2011b. Using sensitivity to optimize mine air regulation point and parameter, *Mining Safety and Environmental Protection*, 38(5):1–3, p 7 (in Chinese).

Wu, F L, Gao, J N, Chang, X T and Li, L Q, 2016. Symmetry property of Jacobian matrix of mine ventilation network and its parallel calculation model, *Journal of China Coal Society*, 41(6):1454–1459 (in Chinese).

Wu, F L, Gao, Y C H and Chang, X T, 2020. Boundary coupling model and solution technology of mine ventilation network and gob flow field, *Journal of Central South University (Science and Technology)*, 51(8):2333–2342 (in Chinese).

Wu, F L, He, X C H, Chang, X T, Ma, L and Li, C H, 2017. Research on simulation technology of surface air leakage of shallow-buried goaf based on network calculation, *Industry and Mine Automation*, 43(12):64–69 (in Chinese).

Wu, F L, Luo, Y and Chang, X T, 2019. Coupling simulation model between mine ventilation network and gob flow field, *Journal of the Southern African Institute of Mining and Metallurgy*, 119(10):783–792.

Wu, F L, Zhao, H and Wang, T, 2024. Development and implementation of mine ventilation network calibration using a two-step method, *Mining, Metallurgy and Exploration*, 41(1):193–205.

Wu, H W, Zhang, Y M, Wu, Z H Y, Jiang, S H G, Wang, L Y and Wang, J, 2010. Development, analysis and comparison of technologies of mine-used automatic air doors, *Industry and Mine Automation*, 36(1):61–65 (in Chinese).

Xie, X P and Han, M W, 2012. Research on informationization and intellectualization of mine ventilation, *Yunnan Metallurgy*, 41(5):1–7 (in Chinese).

Yan, Z H G, Chang, X T, Fan, J D, Wang, Y P and Zhao, P X, 2018. Recognization on high risk region of mine ventilation system in coal mine based on subgraph isomorphism, *Coal Science and Technology,* 46(11):63–68 (in Chinese).

Zhang, G J, 2008. Design and application of intelligent ventilation control system to mine roadway heading face, *Coal Science and Technology,* 36(5):76–79 (in Chinese).

Zhang, J G, Wang, Q Y and He, X, 2023. Research status and system design of intelligent mine ventilation, *Mining Safety and Environmental Protection,* 50(5):37–42 (in Chinese).

Zhang, L and Liu, Y Q, 2024. Research on technology of key steps of intelligent ventilation in mines, *Coal Science and Technology*, 52(1):178–195 (in Chinese).

Zhang, L, Yao, H F, Li, W, Yang, Y, Sun, X G, Liu, Y Q and Zhen, Y, 2022. Research and application of complete set of technical equipment for mine intelligent ventilation, *Journal of Intelligent Mine*, 3(3):71–79 (in Chinese).

Zhang, Q H, Yao, Y H and Zhao, J Y, 2020. Status of mine ventilation technology in China and prospects for intelligent development, *Coal Science and Technology*, 48(2):97–103 (in Chinese).

Zhang, S L, 2019. Study on measurement and change law of wind speed in cross-section of coal mine ventilation roadway, *Mining Safety and Environmental Protection,* 46(4):17–20 (in Chinese).

Zhang, Z T, Li, Y C, Li, J Q, Zhang, J and Li B L, 2023. Architecture and implementation of intelligent ventilation precise control system, Journal of China Coal Society, 48(4):1596–1605 (in Chinese).

Zhao, D, Huang, F J, Chen, S, Wang, D and Wang, D W, 2014. Relationship between point air velocity and average air velocity in circular roadway, *Journal of Liaoning Technical University (Natural Science)*, 33(12):1654–1659 (in Chinese).

Zhao, D, Liu, J, Pan, J T and Ma, H, 2009. Analysis of identifying diagonal structure of ventilation network based on parallel computing, *Journal of China coal society*, 34(9):1208–1211 (in Chinese).

Zhao, Q L, Liu, J and Yang, C H X, 2001. A new approach to automatic identification of diagonal branches in a complicated mine ventilation network, *Journal of Safety and Environment,* 1(6):19–21 (in Chinese).

Zhou, F B, Wei, L J, Xia, T Q, Wang, K, Wu, X Z H and Wang, Y M, 2020. Principle, key technology and preliminary realization of mine intelligent ventilation, *Journal of China Coal Society,* 45(6):2225–2235 (in Chinese).

Zhou, F B, Xin, H H, Wei, L J, Shi, G Q and Xia, T Q, 2023. Research progress of mine intelligent ventilation theory and technology, *Coal Science and Technology*, 51(6):313–328 (in Chinese).

Zhou, L H, Lyu J and Liu, X J, 2004. The principle and implementation of calculating resistance through surveying airflow quantity, *Journal of Xi'an University of Science and Technology,* (2):148–150; 165 (in Chinese).

Zhu, H Q, Liu, X S H, Zhang, W K and Zhang, F Y, 2000. Research and application of auto air flow control system at initial mine fire accident, *Coal Science and Technology,* (2):26–28 (in Chinese).

Ventilation monitoring and control

Unusual mine ventilation appliance failures

E De Souza[1]

1. President, AirFinders Inc., Kingston ON, Canada. Email: info@airfinders.ca

ABSTRACT

The uninterrupted operation of a ventilation system is critical to ensure the mine meets its scheduled production while securing the health and safety of all personnel working underground.

This paper shares real-life experiences pertinent to the occurrence of upset conditions as a result of the improper operation and management of mine ventilation system appliances.

A series of case studies are presented to describe disruptions to mine ventilation systems due to system catastrophic failures. Special focus is given to mistakes made and on the resulting serious consequences. The case studies also demonstrate how detailed on-site engineering assessments were employed, how each upset condition was evaluated, and solutions were found permitting the mines to safely resume production activities.

INTRODUCTION

Operational issues associated with the proper functioning of a mine ventilation system will prevent the mine from operating safely and within its target production rates. When upset conditions occur, they should be promptly resolved in order for the system to function properly, in accordance with design and in compliance with regulations.

Nine case studies are presented to demonstrate how fundamental design or operational mistakes can lead to adverse conditions or to catastrophic consequences, how detailed on-site engineering assessments can be employed, and how simple, low- cost, engineering solutions can be implemented to successfully resolve the issues and return the ventilation system to compliance, permitting the mines to safely resume production activities.

CASE STUDY 1 – INADEQUATE MAINTENANCE OF PRIMARY FANS

Two vertically mounted axial flow fans, in parallel configuration, are used to exhaust 103.8 m³/s from a mine via a 2.74 m diameter raise. The fan diameter is 1.524 m and the hub diameter is 0.66 m, fitted with 149 kW motors running at 1770 rev/min. The fans were installed with arrangement 'D', at which the motor is located on the outlet side of the fan. Figure 1 shows the characteristics for a single fan operating at standard air density. Each fan operates at 51.91 m³/s and 2.34 kPa total pressure.

The mine did not have an established management and maintenance program; lack of regular maintenance can cause ventilation assets to operate below their design performance levels or lead to system breakdown.

Failure of one fan occurred prompting an immediate and thorough inspection of the installation. The motor twisted in the casing and destroyed the rotor and fan blades. The motor stabiliser rods and the vanes were broken. The fan was replaced with a new fan.

Inspection of the second fan that was in poor condition. The blades were very pitted and corroded. The metal vanes were in poor condition and six of them were broken away from the flange. There were no stabiliser rods attaching the motor to the fan casing. A decision was made not to reinstall the fan since there was a good possibility that the motor/impeller assemblage could have fallen out and down the raise. This fan was replaced with a spare fan. The fan was rebuilt and kept as a spare.

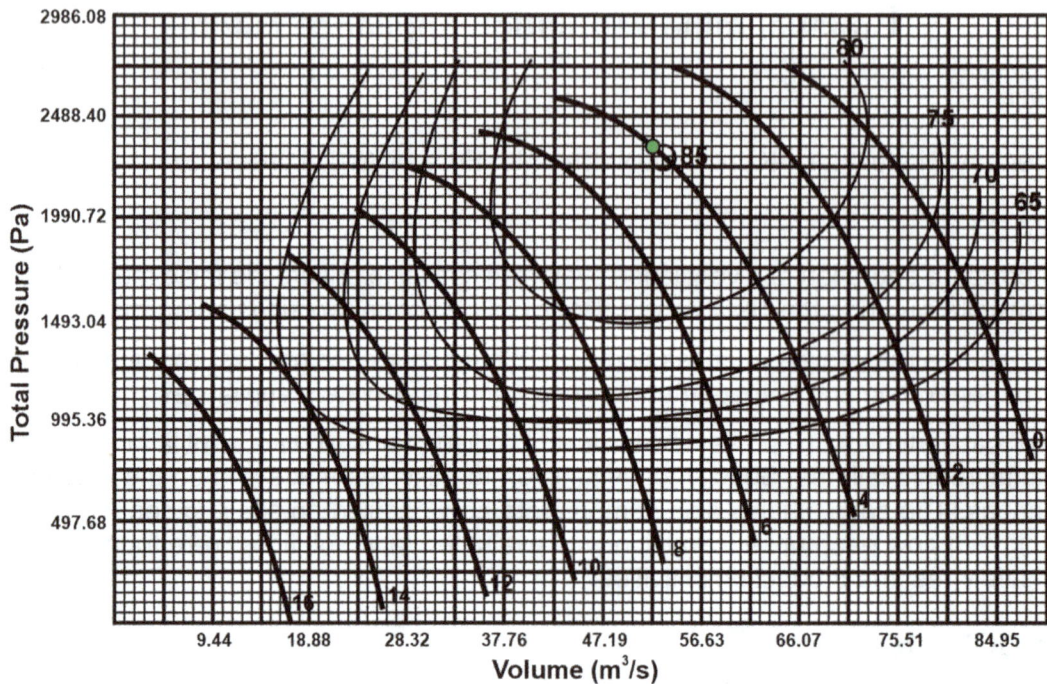

FIG 1 – Case study 1 – Main surface exhaust fan characteristics and operating point.

CASE STUDY 2 – UNDERGROUND BOOSTER FAN OUTLET ISSUE

A large underground booster fan station, illustrated in Figure 2, was designed to supply 70.8 m^3/s to a new mining block. The fan diameter was 1.943 m and the hub diameter was 1.27 m. A 224 kW motor running at 800 rev/min was specified. Fan assemblage components included inlet bell, screen, podless silencers and an outlet diffuser. The designed diffuser was 2.58 m in outlet diameter and 5.18 m long.

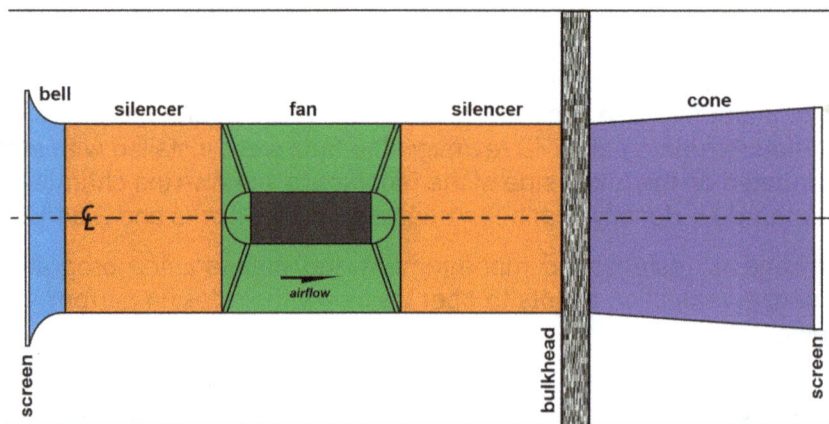

FIG 2 – Case study 2 – Schematic of booster fan installation.

When installed, the booster fan was properly fitted with the inlet bell, screen and silencers, but it did not have an outlet diffuser installed in it, as prescribed by the design. The installation of booster fans without diffusers was common practice in the mine.

Operation of the fan resulted in eventual failure of the system. The fan operating points as per design and without a diffuser are shown in Figure 3. The design operating point was 2.39 kPa total pressure at 70.8 m^3/s and 197 kW brake power. Without a diffuser, the operating point (3.13 kPa total pressure at 70.8 m^3/s) falls outside the fan total pressure range and exceeds the motor capacity.

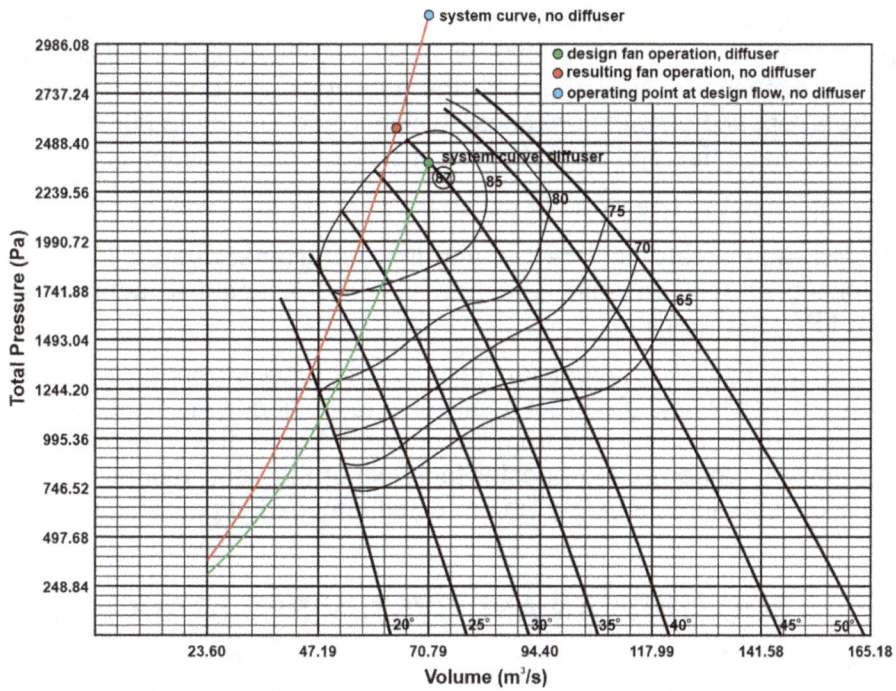

FIG 3 – Case study 2 – Booster fan characteristics and operating point.

The resulting unavailability of this primary exhaust raise was unacceptable from an operational point of view, preventing the mine from operating safely and from meeting its production schedule.

A new fan was installed, including the designed diffuser. Following the corrective action, diffusers were installed in all other booster fan installations in the mine.

CASE STUDY 3 – BOOSTER FAN AND SHOTCRETE OPERATIONS

An underground booster fan, mounted on a bulkhead, supplies 68.9 m³/s of fresh air to a section of a hard rock mine. The fan diameter is 1.676 m and the hub diameter is 0.66 m, fitted with a 112 kW motor running at 1770 rev/min. Figure 4 shows the characteristics for the fan operating at standard air density and the fan operating point of 68.9 m³/s and 0.995 kPa total pressure. The fan operating point is located on a stable region of the fan curve.

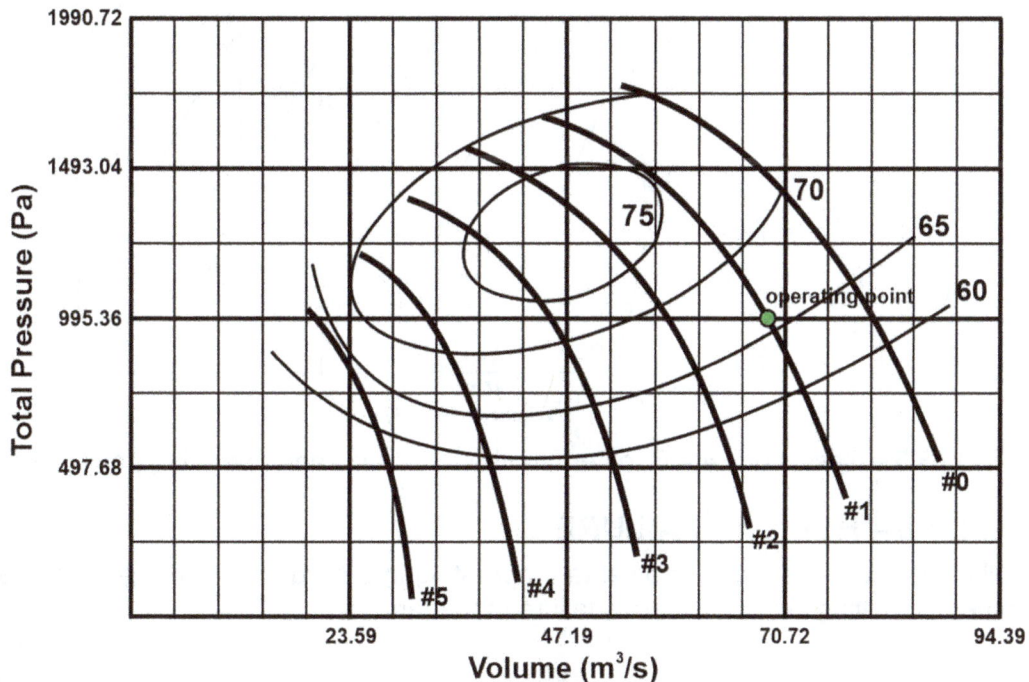

FIG 4 – Case study 3 – Booster fan characteristics and operating point.

A ventilation bulkhead was being constructed upstream of the fan and, during shotcreting operations, the booster fan was not protected from shotcrete spray overshooting carried by the ventilation air. The drift air velocity exceeded 6 m/s and, with the relatively high drift air velocity and excessive turbulence, shotcrete rebound entrainment became significant. Shotcrete build-up on the fan blades created an imbalance on the fan and an instantaneous catastrophic failure occurred.

The fan had to be replaced by a new one because the cost of repair was prohibitive.

The new half-bladed fan diameter is 1.829 m and the hub diameter is 0.686 m, and fitted with a 112 kW motor running at 1780 rev/min. The new fan was selected to achieve a higher flow of 77.9 m³/s at 0.896 kPa total pressure.

Two control factors are critical to avoid fan failure occurrences during shotcreting: control the shotcrete spraying process by using correct application techniques to reduce the rebound rate, and ventilation control by preventing rebound from becoming airborne. It was recommended that, when shotcreting in proximity to underground booster fans, either the ventilation flow should be reduced to some 1 m/s to avoid entrainment or the fan protected (stopped, if possible) to prevent damage to the fan.

CASE STUDY 4 – BOOSTER FAN AND SYSTEM RESISTANCE PRESSURES

A booster fan station, designed to ventilate a main ore handling level with rockbreaker, crusher and ore bin, was installed downstream from a storage facility. The fan was designed to supply 40.6 m³/s to the level and the level static resistance pressure was estimated at 1.306 kPa. The fan diameter was 1.676 m and the hub diameter was 0.66 m. A 93.2 kW motor running at 1200 rev/min was specified. Fan assemblage components included inlet bell, inlet and outlet screen, podless silencers and an outlet diffuser. The designed diffuser was 2.03 m in outlet diameter and 1.02 m long. The fan curve and design operating point are shown in Figure 5. The design operating point was determined at 1.49 kPa total pressure at 40.6 m³/s and 82 kW brake power.

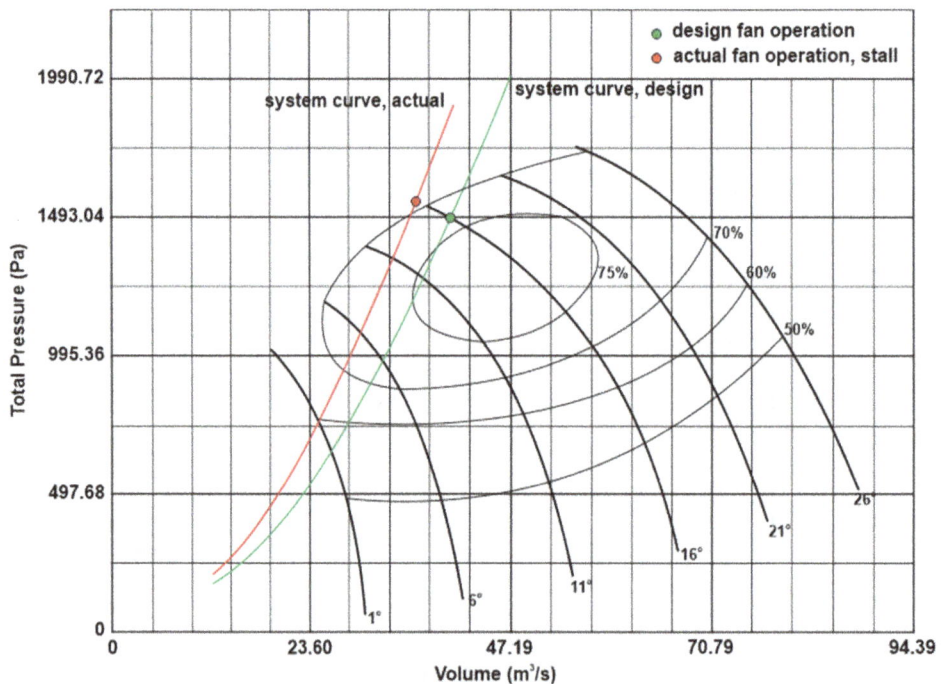

FIG 5 – Case study 4 – Booster fan characteristics and operating point.

CASE STUDY 5 – HUB CAP FAILURE

A single bulkheaded exhaust booster axial fan, operating in a hard rock mine, experienced a hub cap detachment, resulting in instantaneous failure of the fan.

The hub cap, upon detachment from the rotor hub, impacted a rotating blade, resulting in warping of the blade and consequential dislodgement from the hub. The failed blades were thrown against the screen and it prevented the blades from being thrown clear of the fan.

Inspection indicated fracture of the blade at the point of impact, with breakage occurring close to the blade fastener element to the rotor hub. The fastener element remained attached to the hub. It was also observed that the bolts securing the nosecap to the hub element were rusted and came unsecured. The bolts installation torque could not be verified.

Causes of loose bolts include under-tightening, vibration, embedding, gasket creep, differential thermal expansion, and shock.

This is one fan assemblage component that is often ignored during inspection and maintenance activities. Recommended steps to prevent loose bolts include – during installation of the hub cap the torque applied to secure the nuts should follow manufacturer specifications, – use of washers and adhesives and – periodic inspection. The bolts should also be primed with corrosion protection products to prevent rusting. In addition, use of corrosion resistant bolts, such as stainless steel was recommended.

CASE STUDY 6 – PRIMARY SURFACE FAN FOUNDATION ISSUES

A new raise was driven to exhaust 273.7 m³/s of air from a section of a mine and two 2.58 m axivane full-bladed fans, fitted with 298 kW HP motors running at 710 rev/min, were installed in parallel configuration, each fan passing approximately 136.9 m³/s at 1.294 kPa total pressure. The fan curve and operating point are shown in Figure 6. The motors were installed external to the fan, illustrated in Figure 7.

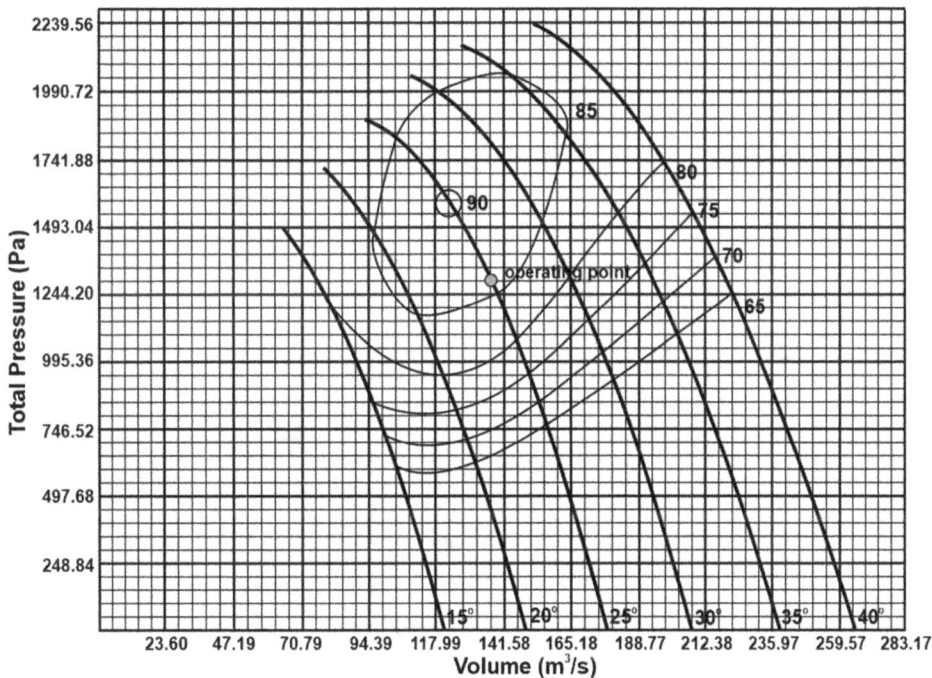

FIG 6 – Case study 6 – Booster fan characteristics and operating point.

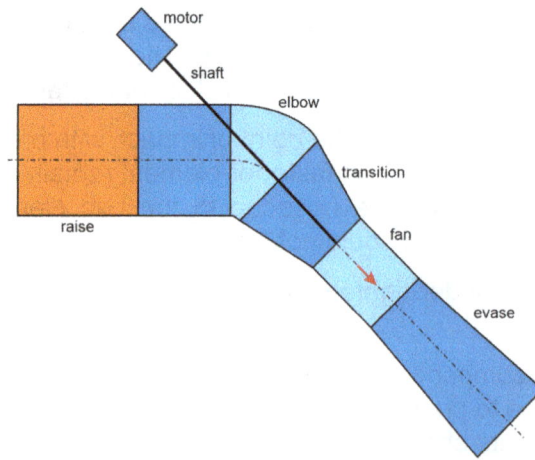

FIG 7 – Case study 6 – Schematic of external motor installation.

The fan foundation, comprising poured concrete of normal strength, was designed to be capable of absorbing expected fan vibrations. Anchoring to bedrock was effected to ensure rigidity, since foundation stiffness is critical to ensure proper balancing of the fan to the required operating vibration level. A rigid level foundation support was formed, and designed to be capable of absorbing normal expected fan assemblage vibration levels. Fan anchor bolts were embedded in the foundation, set during concrete pour. The top surface of the foundation was level to permit proper anchoring at appropriate centreline alignment. Shimming and grouting allowances were also provided in design and construction to allow for proper angular alignment. Fan centreline assemblage alignment and levelling was carefully performed with laser alignment equipment.

During operation, motor misalignment occurred, with the drive shaft not lining up with the load, causing harmful vibrations that wore down the motor, including premature failure of the motor bearings.

A site inspection indicated that the motor foundation was formed in soil of low bearing capacity, permeable ground, subject to freeze/thaw cycles, and that, unlike the fan foundation, was not anchored to bedrock. Over time, misalignment of the motor occurred, followed by shaft looseness, creating excessive clearance between mechanical components, leading to premature failure of the motor bearings, and precipitating motor failure.

CASE STUDY 7 – DUCT PUSH-PULL SYSTEM FAILURE

A push and a pull system were installed in parallel side-by-side on a development heading, illustrated in Figure 8. Layflat and polyethylene ducts of 1.07 m diameter were used, with fans of same diameter. The fans, fitted with inlet bells and screens and silencers, were properly secured with chains. Proper ratchet straps were used to attach the ducting to the fans. Both duct columns were straight and level, hung from messenger cables, and in good condition, with good duct section joint connections. The ventilation tubing was maintained in a condition such that air leakage was minimised.

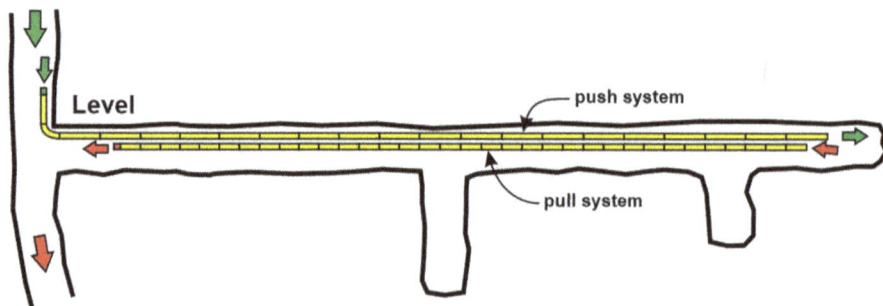

FIG 8 – Case study 7 – Schematic of duct push-pull system installation.

During operation, the end of the force duct column got dislodged and, when knocked loose, blocked the entry to the rigid duct. This resulted in the collapse of the entire pull duct column, however the entire column staying hung in place. Undoubtedly, proper hanging of the rigid column ensured a safe working area for the workers.

This unusual occurrence underscores the need for properly securing the end of a force column thus preventing it from swaying and knocking itself loose.

CASE STUDY 8 – MAIN SURFACE FAN OVERLOAD

A primary surface fan system, consisting of two fans in parallel, is used to supply 245.4 m³/s of heated air to a hard rock mine. The fan diameter is 1.943 m and the hub diameter is 1.27 m, fitted with 485 kW motors running at 1180 rev/min. Both fans are fitted with variable frequency drives. The fan assemblage consists of a heater building, inlet bell and screen, discharge cone, damper, discharge transition and junction, and discharge elbow. The fan heater building has a 12.2 m wide × 6.7 m high inlet bird screen.

The fan system operating point, shown in Figure 9, is 122.7 m³/s per fan at 2.49 kPa total pressure, located on a stable region of the fan curve.

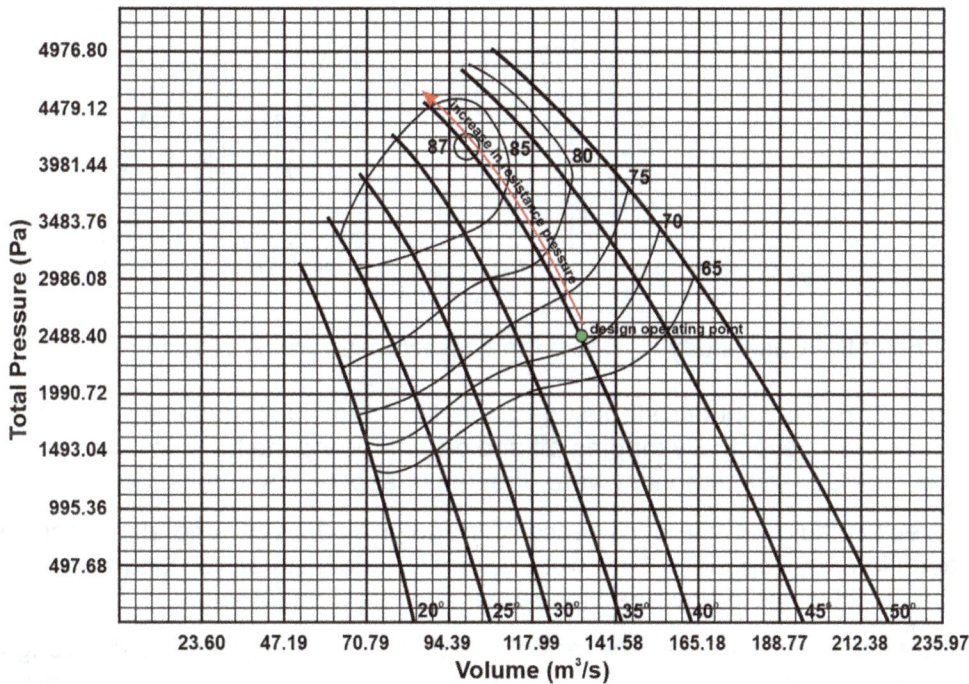

FIG 9 – Case study 8 – Main fan characteristics and operating point.

On a colder, drier day, a massive dump of fresh snow occurred, producing powder snow. The snow quickly accumulated on the heater inlet bird screen, completely blocking the air intake to the fan, resulting in a rapid surge in fan pressure. All mining activities were halted until the situation was investigated.

This distinct incident resulted in increases in fan noise and vibration and damage to the heater building. Had the heater building not been designed and built with steel columns and beams and structural panels to resist very high live non-static negative loads, the building would have collapsed.

Based on observations it is believed that the fans reached stall. Immediate action was taken to reduce the excessive pressures and unstable conditions by reducing the fans rev/min. The building service doors could then be opened to further relieve the resistance pressure. The accumulated snow was cleared from the screen and the fan operation was safely restored.

CASE STUDY 9 – MAIN FAN BLADE WEAR

Spent air is exhausted from a hard rock mine via a production shaft with two surface axivane fans installed in parallel configuration. From the shaft, the exhaust air travels through a plenum and enters the fans at surface collar elevation The fan diameter is 2.578 m and the hub diameter is 1.397 m, fitted with 224 kW motors running at 710 rev/min. The fans are rated to move 132.1 m³/s of air each at 1.394 kPa total pressure, shown in Figure 10, operating on a stable region of the fan curve.

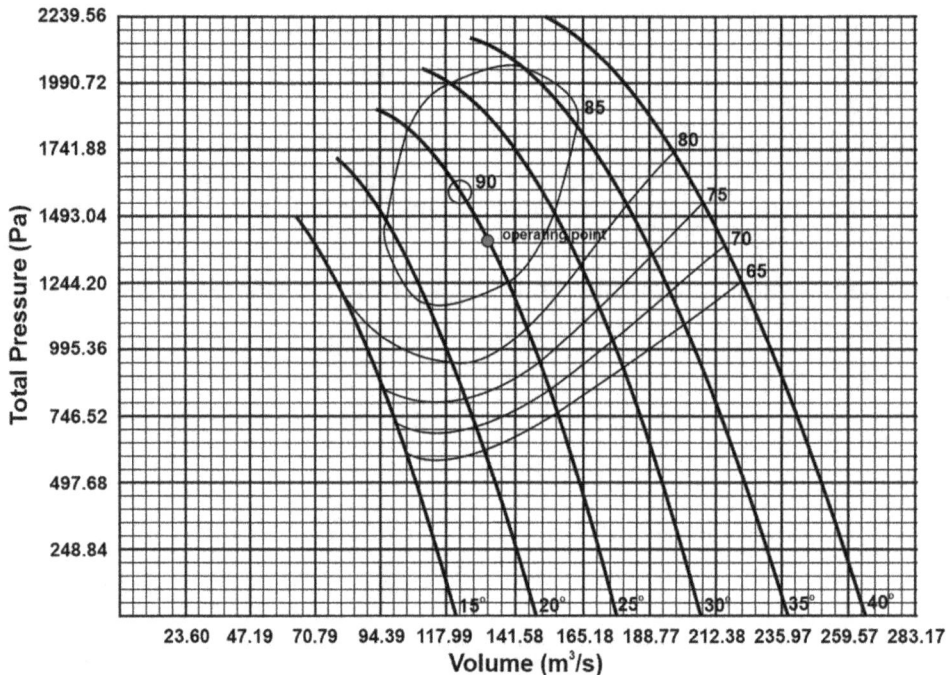

FIG 10 – Case study 9 – Main fan characteristics and operating point.

After a long time span of continuous stable operation, fan vibration started increasing gradually, with the fan vibration sensors eventually tripping off and de-energizing the fans. In response a mine-wide evacuation was ordered, and personnel were directed to escape via egress facilities.

A detailed inspection indicated the occurrence of irregular blade wear and also significant wear all along the blades leading edge. The fans were operating under abrasive conditions in exhaust air, leading to debris buildup and premature blade wear and blade degradation.

The blades were replaced, and the impeller balanced with the fan expeditiously brought back into operation. It was recommended that specialty coatings be applied to the blades to control blade degradation.

Exceptionally, during fan start-up it was observed that a worker, while cleaning the shaft station, was sweeping collected dirt down the shaft. The material was being drawn by the fans, travelling through the plenum, further exacerbating the fan imbalance issue. Immediate action was taken to eliminate this unacceptable practice.

CONCLUSIONS

When mistakes are made during the design, management or operation of a mine ventilation system, adverse conditions or catastrophic consequences can occur leading to disruptions in production and compromising the health and safety of operators. Nine case studies demonstrate the consequences of fundamental mistakes in design or operation of mine ventilation systems. This paper also demonstrates how detailed on-site engineering assessments, supported by extensive scientific and technical experience, were effectively used to expeditiously and successfully restore the mine ventilation operations to conformity.

Construction of an intelligent sensing system for mine ventilation parameters based on fluid network theory

T Jia[1] and H Ma[2]

1. PhD candidate, Liaoning Technical University, Huludao Liaoning 125100, China. Email: fsblcq1119@163.com
2. Professor, Liaoning Technical University, Huludao Liaoning 125100, China. Email: mahenglgd@163.com

ABSTRACT

The basis for realising intelligent ventilation in mines is to construct an air velocity sensor network that can accurately monitor the state of the ventilation system for a mine-wide intelligent sensing of the ventilation parameters. To solve the sensor position domain and complete the digital siting of the sensor network, this study applied a step-by-step solution method based on the fluid network theory. First, *The Coal Mine Safety Rules* was used as the benchmark for the initial localisation of the position-domain elements. Second, to avoid the situation in which joint branches with disordered air velocity attributes belong to the position domain, the minimum number of sensors required to achieve mine-wide branch-air-velocity monitoring was taken as the objective, and an optimisation model for the number of position-domain elements was constructed. The number of position-domain elements was then solved using the improved cut-set matrix algorithm. Moreover, to avoid the overlapping of sensor monitoring ranges, a multifactor indicator decision-making model was established to complete the state selection of the position-domain elements by calculating the comprehensive fitness scores of the branch-installed sensors. The accurate positioning of the sensor position domain was completed, and the digital siting of the sensor network was realised. To verify the feasibility of the model in mine production, taking the Gucheng coalmine as an engineering research subject, the sensor position domain required for a mine-wide ventilation parameter monitoring was realised by constructing an optimisation model for the number of position-domain elements and a multifactor indicator decision-making model for this coalmine. This study can serve as a basis for the intelligent monitoring of mine ventilation systems and thus help ensure mine safety production.

INTRODUCTION

An intelligent ventilation system is vital to improving the reliability of mine safety production (Liu *et al*, 2020; Zhou et al, 2020). The construction of such a system is based on an intelligent sensing system for the ventilation parameters that can accurately monitor the state of the ventilation system (Liu, 2020; Zhang *et al*, 2023). With the ventilation parameters obtained using the intelligent sensing system as the regulatory information source and by linking quantitative decision-making with intelligent control, the core functions of the intelligent ventilation system can be realised: air supply optimisation during production and emergency air control in the event of a disaster (Huang and Liu, 2023).

The construction of a parameter sensing system depends on the linkage of the underground multiparameter sensor network, and the core aspect behind the construction of this network is the air velocity sensor (hereinafter referred to as the sensor) network (Ali *et al*, 2022; Cui *et al*, 2023). The sensor can not only monitor the ventilation parameters (air velocity, volume, and pressure) of the branches in a mine network but can also act as a weight influencing factor in the construction of a methane sensor network and help monitor the air volume at the main gas outflow point in real time (Baris and Aydin, 2020; Gong *et al*, 2022). In other words, the basis of the construction of the parameter sensing system is the siting of the sensor network.

If the site selection of the sensor network is specified only in accordance with the provisions of *The Coal Mine Safety Rules*, the monitoring of only the main airflow site can be achieved; the system sensing provided cannot cover the entire mine. If each branch is included in the sensor network, the same branch is monitored by multiple sensors, leading to a wastage of resources. Hence, it is necessary to select the sensor network based on a comprehensive consideration of the number of sensors and mine monitoring coverage.

Extensive research has been conducted on the siting of sensor networks, mainly along two directions: Inversion solution of sensor network positioning in combination with fault diagnosis theory (Cohen *et al*, 1987; Li *et al*, 2022; Liu *et al*, 2021; Ni *et al*, 2021; Sharma *et al*, 2022; Zhang *et al*, 2021; Zhao and Pan, 2011) and mathematical modelling of site selection based on graph theory and using an algorithm to solve the model (He, 2013; Jia, Ma and Gao, 2024; Liu *et al*, 2022; Rodriguez-Diaz *et al*, 2020; Yan *et al*, 2023; Yang *et al*, 2023). Both the types of research are based on a qualitative analysis of the branches in a mine network. However, for large mines built in the early stage, because a joint branch is the simplest way to connect the mining areas and roadways, it accounts for a large proportion of the weight in the ventilation network topology, and it is easily selected in the siting process of the sensor network. In actual production, due to the characteristics of short length and large friction, the airflow in a joint branch is disordered. The accuracy of the ventilation parameters measured using the sensors installed in joint branches is low, that is, the proportion of joint branches should be reduced in the siting process. However, if joint branches are directly deleted when building the ventilation network topology, there will be monitoring omission, and the overall operation state of the ventilation network will not be accurately reflected. Hence, the site selection of a large-mine sensor network requires performing a qualitative–quantitative analysis of the mine branches.

With this background, to realise a scientific siting of large-scale mine sensor network, in this study, the siting of the sensor network was transformed into the solution of the sensor position domain based on the fluid network theory. With the goal of solving the sensor position domain under multiple constraints, the step-by-step solution method was used to solve the target. First, the initial localisation of the position-domain elements was based on *The Coal Mine Safety Rules* and the original sensor network of the mine. The second step was to build an optimisation model for the number of position-domain elements, taking the branch air volume as the global search weight and using the cut-set matrix algorithm to solve the model, which effectively reduced the probability of joint branches being selected in the sensor position domain. The third step was to build a multifactor indicator decision-making model for the state selection of the position-domain elements to avoid any overlap of the sensor monitoring ranges. Finally, the site optimisation of the sensor network was realised based on a comprehensive consideration of the number of sensors and the coverage of mine monitoring.

INITIAL LOCALISATION OF POSITION-DOMAIN ELEMENTS

The Coal Mine Safety Rules (hereinafter referred to as *The Safety Rules*) (Ministry of Emergency Management of the People's Republic of China, 2022, pp 110–115) stipulates that in mines equipped with mine safety monitoring systems, air velocity sensors should be installed in each air-return roadway in the mining area, each wing air-return roadway, and main air-return roadway.

Through a general survey of the mine ventilation system, the attribute classification of the roadway should be summarised, and the corresponding roadway according to the requirements of *The Safety Rules* can be classified into the sensor network as initial position-domain elements. In the subsequent calculations, the air volumes of the initial localisation branch elements are considered to be known.

OPTIMISATION OF THE NUMBER OF POSITION-DOMAIN ELEMENTS

Let a mine ventilation network diagram represented by $G_r = (V, E)$ be a (*m*, *n*) directed graph with node set $V = \{v_1, v_2, v_3, \ldots, v_m\}, m = 1, 2, 3 \cdots$ and edge set $E = \{e_1, e_2, e_3, \ldots, e_n\}, n = 1, 2, 3 \cdots$.

> **Lemma 1**: *The air volume at the edge of each tree is equal to the algebraic sum of the air volumes of the cotree edges in the independent circuit to which the tree edge belongs or is equal to the algebraic sum of the air volumes of the cotree edges in the basic cut set determined by the tree edge (Chen, 2018; Liu, Jia and Zheng, 2002).*

Model principle

Because the number of tree edges in G_r is *m*−1 and the number of cotree edges is *n*−*m*+1, G_r has *m*−1 basic cut sets, and each basic cut set can be used as a row vector to generate a (*m*−1)×*n*-order basic cut-set matrix **S**. The law of conservation of the air volume can be generalised to an equal flow inflow and outflow of the cut set closed surface. This can be mathematically expressed as in Equations 1–3:

$$T_r = E(T_r) = \{e_{1a}, e_{2a}, \ldots e_{(m-1)a}\}, \quad \overline{T_r} = E(G_r) - E(T_r) = \{e_{1b}, e_{2b}, \ldots e_{(n-m+1)b}\} \tag{1}$$

$$\mathbf{S} = \{S_i\}^{\mathrm{T}} = \{s_{ij}\}_{(m-1)\times n}, \quad S_i = \{e_{1b}, e_{2b}, \ldots e_{ia}, \ldots e_{n-m+1}\}, e_{ia} \in T_r, e_{jb} \in \overline{T_r} \tag{2}$$

$$\mathbf{BG} = \mathbf{SG} = 0 \tag{3}$$

where:

T_r	is the spanning tree of the graph G_r
$\overline{T_r}$	is the cotree of G_r
S	is the basic cut-set matrix of G_r
S_i	is a basic cut set of G_r
e_{ia}	is a tree edge of G_r, i = 1, 2,…, $m-1$
e_{jb}	is a cotree edge of G_r, j = 1, 2,…, $n-m+1$
B	is the basic incidence matrix of G_r
G	is the air volume matrix of G_r

The branch air volume was considered an unknown number. According to the fluid network theory, it can be seen that only the basic cut set needs to be calculated. The $n-m+1$ cotree edges in the basic cut set are selected as the position-domain elements, and then, according to the law of conservation of air volume and **Lemma 1**, a linearly independent branch-air-volume equation set, shown in Equation 4, is constructed to solve the air volumes in the tree edges. Thus, the monitoring and calculation of the ventilation parameters of the entire mine can be realised.

$$[\mathbf{B}_{11}, \mathbf{B}_{12}][\mathbf{G}_y, \mathbf{G}_S]^{\mathrm{T}} = [\mathbf{S}_{11}, \mathbf{I}_S][\mathbf{G}_y, \mathbf{G}_S]^{\mathrm{T}} = \mathbf{0} \Rightarrow \mathbf{G}_S = -\mathbf{B}_{12}^{-1}\mathbf{B}_{11}\mathbf{G}_y = -\mathbf{S}_{11}\mathbf{G}_y \tag{4}$$

In the formula,

$\mathbf{G_S}$	is the matrix of the air volume in the tree
$\mathbf{G_y}$	is the cotree air volume matrix
$\mathbf{B_{11}}, \mathbf{B_{12}}$	are the sub-matrices of **B**
$\mathbf{S_{11}}$	is the sub-matrix of **S**
$\mathbf{I_S}$	is the unit matrix

Optimisation model for the number of position-domain elements

The minimum number of branches Z with sensors required for ventilation parameter monitoring in all the branches of the mine was set as the objective function. Based on the above principles, an optimisation model for the number of position-domain elements can be expressed as follows:

$$\begin{cases} \min F = Z, Z \geq j \\ s.t. \\ \mathbf{SG} = 0 \end{cases} \tag{5}$$

Model calculation

Solving the basic cut-set matrix of the model

The principle of the cut-set matrix algorithm is to use the network structure of the ventilation system to solve the spanning tree T_r and cotree $\overline{T}_r$, basic incidence matrix **B**, and basic circuit matrix **C**. Based on the conversion relationship between the matrices, the basic cut-set matrix **S** is solved.

To avoid joint branches with disordered airflow properties from being in the sensor position domain, an improvement was made to the cut-set matrix algorithm. The improvement principle was to reduce the probability of joint branches being selected as elements in the cotree set $\overline{T}$ based on the low-flux airflow characteristics of the joint branches. The detailed improvement measures are as follows.

After building the network topology of the ventilation system, the branch air volume was added as the global search weight, and an in-depth search of the global network was performed on the basis of the relationship between the nodes and edges to solve the minimum spanning tree T and cotree $\overline{T}$ of the network ($T \in T_r$, $\overline{T} \in \overline{T}_r$). At the same time, the incidence matrix element b_{ij} was assigned using the following formula to generate $\mathbf{B} = \{b_{ij}\}_{(m-1) \times n}$.

$$b_{ij} = \begin{cases} 1, \text{Node } v_i \text{ is connected with edge } e_i \text{ and } v_i \text{ is the start node} \\ 0, \text{Node } v_i \text{ is connected with edge } e_i \text{ and } v_i \text{ is the end node} \\ -1, \text{Node } v_i \text{ is not connected with edge } e_i \end{cases}$$

(6)

T is taken as the matrix column vector sorting standard, and the formula $\mathbf{BC}^{\mathrm{T}} = \mathbf{0}$ is used to calculate the basic circuit matrix **C**:

$$\mathbf{BC}^T = [\mathbf{B}_{11}, \mathbf{B}_{12}][\mathbf{I}_C, \mathbf{C}_{12}^{\mathrm{T}}]^{\mathrm{T}} = \mathbf{B}_{11} + \mathbf{B}_{12}\mathbf{C}_{12}^{\mathrm{T}} = \mathbf{0} \Rightarrow \mathbf{C}_{12} = -\mathbf{B}_{11}^{\mathrm{T}}(\mathbf{B}_{12}^{-1})^{\mathrm{T}}$$

(7)

Similarly, $\overline{T}$ is taken as the matrix column vector sorting standard, and the formula $\mathbf{CS}^{\mathrm{T}} = \mathbf{0}$ is used to calculate the basic cut-set matrix **S**:

$$\mathbf{CS}^{\mathrm{T}} = [\mathbf{I}_C, \mathbf{C}_{12}^{\mathrm{T}}][\mathbf{S}_{11}^{\mathrm{T}}, \mathbf{I}_S]^{\mathrm{T}} = \mathbf{S}_{11}^{\mathrm{T}} + \mathbf{C}_{12} = \mathbf{0} \Rightarrow \mathbf{S}_{11} = -\mathbf{C}_{12}^{\mathrm{T}} = \mathbf{B}_{12}^{-1}\mathbf{B}_{11}$$

(8)

Calculating the number of position-domain elements

Based on **S** calculated by the model, $m-1$ basic cut-set groups were determined, from which the number of elements in the position domain is identified as $n-m+1$.

However, according to *The Safety Rules*, at the initial localisation of the position-domain elements, the branch in which the sensor needs to be installed has been considered the known branch. The initially localised branches are not necessarily the cotree edges solved using the improved cut-set matrix algorithm, which will lead to the phenomenon of multi-sensor monitoring of the same branch. Therefore, it is necessary to select the state of the multiple position-domain elements that monitor the same branch. After filtering, the number of position-domain elements is still $n-m+1$.

STATE SELECTION OF POSITION-DOMAIN ELEMENTS

Building a multifactor indicator decision-making evaluation model

The fitness of the sensor installed in the branch is defined as the state of the branch. Using the relative average multi-index and the quantitative selection method, a multifactor indicator decision-making comprehensive evaluation model for comprehensive measurement and evaluation of the state selection scheme of the position-domain elements was built. The defined domain element of the evaluation model is the branch that requires state selection. The procedural framework of the comprehensive evaluation model is as follows:

- Determining the evaluation goal of the comprehensive evaluation object:
  - Evaluation object: Branches that need state selection.

- o Evaluation goal: Determine the optimal scheme for the state selection of the position-domain elements according to the fitness of the branch-installed sensors.
- Establishing a comprehensive evaluation indicator system model.

The Application and Management Standard for Coal Mine Safety Monitoring System and Detector (AQ1029–2007) (hereinafter referred to as *The Standard*) (Standards China, 2019, p 16) stipulates that the air velocity sensor should be set within 10 m before and after the roadway where there is no turning, barrier-free, no change in the cross-section, and where the air volume can be accurately calculated.

According to *The Standard*, the main factors influencing the state selection of the position-domain elements are the length, height, air volume, friction air resistance, and humidity of the branch. The significance of each factor in the state selection is explained below.

- Branch length: When a short branch is chosen for sensor installation, it is difficult for the distribution of the airflow in the branch to reach an equilibrium state, which is prone to vortex and turbulence, leading to an increase in the monitoring error of the sensor.

- Branch height: The air velocity in the branches is unevenly distributed. Theoretically, there is a certain point at the centre of the cross-section of the branch. If the air velocity monitoring value at a point is equal to the average air velocity, this point is the sensor installation point. However, when the branch height is low, the sensor installation point affects pedestrians and transportation.

- Branch air volume: The sensor has a certain reasonable monitoring range. When the branch in which the sensor is installed is in a breeze state or the air volume is too low, it will increase the monitoring error of the sensor.

- Friction air resistance: The greater the friction air resistance value of the installed sensor branch, the higher the unevenness of the rock wall in the underground environment, and the more unstable the airflow in the branch, leading to an increase in the monitoring error of the sensor.

- Branch humidity: The greater the humidity in the sensor-installed branch, the wetter the underground. High humidity coupled with a high dust concentration can easily block the sensor test port, which reduces the accuracy of monitoring signal acquisition.

Based on the feedback direction of the influencing factors, a multifactor decision-making index was formed, and a comprehensive evaluation model for multifactor indicator decision-making with the goal of achieving the state selection of the position-domain elements was established, as shown in Figure 1.

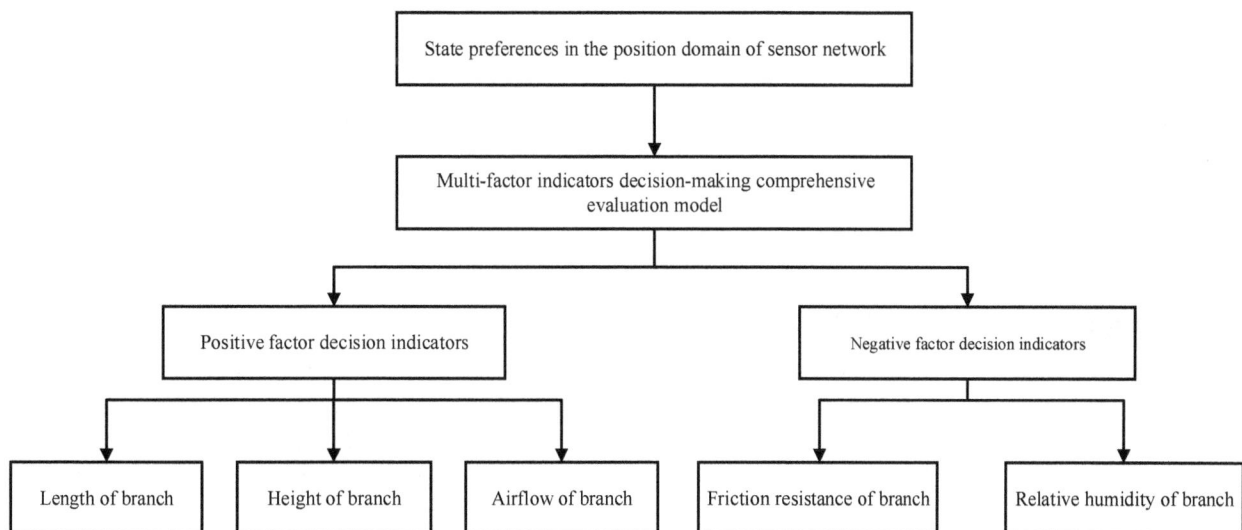

FIG 1 – Comprehensive evaluation model for multifactor indicator decision-making.

Model processing

It is assumed that the system has p branches, and each branch has q indicator eigenvalues to evaluate its advantages and disadvantages.

Step 1 – Standardised processing of indicator data

Due to the different dimensions between multifactor indicators, the indicators need to be standardised first. The basic processing principle is to standardise the value of the positive indicators using Equation 9 and reverse standardise the value of the negative indicators using Equation 10.

$$x'_{ij} = \left(x'_{ij} - \min\{x'_{1j}, x'_{2j}, \cdots, x'_{pj}\}\right) \Big/ \left(\max\{x'_{1j}, x'_{2j}, \cdots, x'_{pj}\} - \min\{x'_{1j}, x'_{2j}, \cdots, x'_{pj}\}\right)$$
(9)

$$x'_{ij} = \left(\max\{x'_{1j}, x'_{2j}, \cdots, x'_{pj}\} - x'_{ij}\right) \Big/ \left(\max\{x'_{1j}, x'_{2j}, \cdots, x'_{pj}\} - \min\{x'_{1j}, x'_{2j}, \cdots, x'_{pj}\}\right)$$
(10)

where:

x_{ij} is the eigenvalue of the evaluation factor for branch i corresponding to indicator j

x'_{ij} is the value of indicator j for branch i, $i \in \{1, 2, \cdots, p\}, j \in \{1, 2, \cdots, q\}$

Step 2 – Indicator weight calculation

$$w_j = \left[1 + k \sum_{i=1}^{p}\left(\left(x'_{ij}\Big/\sum_{i=1}^{p}x'_{ij}\right) \times \ln\left(x'_{ij}\Big/\sum_{i=1}^{p}x'_{ij}\right)\right)\right]\Bigg/\left[q + k \sum_{j=1}^{q}\left(\left(x'_{ij}\Big/\sum_{i=1}^{p}x'_{ij}\right) \times \ln\left(x'_{ij}\Big/\sum_{i=1}^{p}x'_{ij}\right)\right)\right]$$
(11)

where:

k is a constant, $k = 1/\ln(p)$

w_j is the weight value of indicator j

Step 3 – Comprehensive score calculation of branch

$$\tau_i = \sum_{j=1}^{q} w_j \left(x'_{ij}\Big/\sum_{i=1}^{p}x'_{ij}\right)$$
(12)

where:

τ_i is the score of branch i under the comprehensive indicators

Step 4 – Sorting based on the comprehensive evaluation results

Based on a comprehensive evaluation model for multifactor indicator decision-making, the comprehensive scores of the branches were calculated as the evaluation result to sort the state of the branches. The higher the comprehensive score, the higher the fitness of the branch-installed sensor, and the higher the probability of the branch being selected in the sensor position domain. Therefore, the optimal scheme for the state selection of the position-domain elements can be determined for an accurate positioning of the sensor position domain.

MODEL FEASIBILITY VERIFICATION

Theoretical feasibility of model solving

The feasibility of the proposed theory and algorithm was verified by randomly establishing a mine ventilation network model (referred to as the network model) with multiple joint branches, as shown in Figure 2. Table 1 presents the initial parameters of the model, where e_{18} is the air intake shaft, and e_{19} is the return shaft. To increase the connectivity of the network topology, e_{18} and e_{19} are considered the same virtual branch.

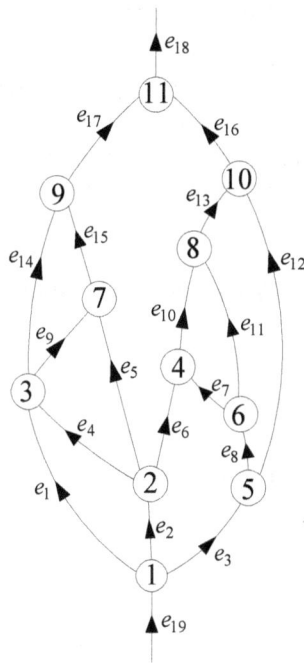

FIG 2 – Structure of the stochastically built mine ventilation network model.

TABLE 1

Factor indicators for each branch in the network model.

Branch number	Branch property	Branch length L/m	Branch height H/m	Branch air volume Q/m^3·s^{-1}	Friction air resistance R/N·s^2·m^{-8}	Branch humidity φ/%
e_1	Main intake airway	1200	4.20	32.85	0.042	92.87
e_2	Track main way	800	4.20	27.70	0.055	93.11
e_3	Main haulage way	1800	4.20	39.45	0.028	92.23
e_4	Crossheading	400	4.00	7.37	0.057	93.62
e_5	Track main way	600	3.60	10.82	0.053	93.65
e_6	Crossheading	200	3.60	9.51	0.063	93.64
e_7	Crossheading	200	3.80	4.05	0.027	92.75
e_8	Main haulage way	500	3.70	15.65	0.016	93.45
e_9	Crossheading	500	4.00	10.55	0.028	93.14
e_{10}	Haulage way	400	3.80	13.56	0.021	92.98
e_{11}	Main return airway	500	4.00	11.60	0.032	92.69
e_{12}	Shaft bottom	1000	3.70	23.80	0.057	92.55
e_{13}	Return incline airway	400	4.00	25.16	0.038	93.42
e_{14}	Crossheading	1000	3.90	29.67	0.016	92.17
e_{15}	Main return airway	400	4.00	21.37	0.024	93.21
e_{16}	Main return airway	1200	4.00	48.96	0.029	93.52
e_{17}	Main return airway	600	4.50	51.04	0.033	93.47
e_{18}	Return shaft	650	4.20	100.00	0.049	92.02
e_{19}	Intake shaft	700	4.50	100.00	0.071	92.74

Initial localisation of position-domain elements

From Table 1, it can be seen that the branch properties of e_{11}, e_{16}, and e_{17} are main return airways (main air-return roadway), and e_{18} is the air return shaft. Based on the requirements of *The Safety Rules*, the branches e_{11}, e_{16}, e_{17}, and e_{18} require sensors. Because e_{18} is a vertical shaft, the conditions cannot meet sensor installation requirements. The total return air condition can be monitored simply by installing sensors in the shaft bottom roadways e_{16} and e_{17}. That is, the initial localisation position domain is $\{e_{11}, e_{16}, e_{17}\}$.

Optimisation of the number of position-domain elements

Solving the basic cut-set matrix of the model

By inputting node number $m = 11$ and edge number $n = 18$, a model ventilation network topology was built. Using the air volume as the global search weight, the Prim algorithm was used to solve the minimum spanning tree T and cotree $\bar{T}$ as follows:

$$T = \{e_2, e_4, e_6, e_7, e_8, e_9, e_{11}, e_{12}, e_{15}, e_{16}\}$$

$$\bar{T} = \{e_1, e_3, e_5, e_{10}, e_{13}, e_{14}, e_{17}, e_{18}\}$$

Based on the model minimum spanning tree T, the model adjacency matrix $\mathbf{A}$, basic incidence matrix $\mathbf{B}$, and basic circuit matrix $\mathbf{C}$ were calculated. According to the formula $\mathbf{S}_{11} = -\mathbf{C}_{12}^{T} = \mathbf{B}_{12}^{-1}\mathbf{B}_{11}$, the data inversion of the basic cut-set matrix $\mathbf{S}$ was performed, and the basic cut-set matrix $\mathbf{S}$ of 10×18 order was obtained. A cut-set matrix element distribution graph that reflects the relationship between the matrix element values, cut set numbers, and branch numbers was generated, as shown in Figure 3.

$$\mathbf{S} = \begin{bmatrix}
1 & 1 & 0 & 0 & 0 & 0 & 0 & -1 & 1 & 0 & 0 & 0 & 0 & 0 & 0 & 0 & 0 & 0 \\
1 & 0 & 1 & 0 & 0 & 0 & -1 & 0 & 0 & 1 & 0 & 0 & 0 & 0 & 0 & 0 & 0 & 0 \\
0 & 1 & 0 & 0 & 0 & 0 & 1 & -1 & 0 & 0 & 1 & 0 & 0 & 0 & 0 & 0 & 0 & 0 \\
0 & -1 & 0 & -1 & 0 & 0 & -1 & 1 & 0 & 0 & 0 & 1 & 0 & 0 & 0 & 0 & 0 & 0 \\
0 & -1 & 0 & 0 & -1 & 0 & -1 & 1 & 0 & 0 & 0 & 0 & 1 & 0 & 0 & 0 & 0 & 0 \\
0 & 0 & 1 & 0 & 0 & 1 & -1 & 0 & 0 & 0 & 0 & 0 & 0 & 1 & 0 & 0 & 0 & 0 \\
0 & 0 & 0 & 1 & -1 & 0 & 0 & 0 & 0 & 0 & 0 & 0 & 0 & 0 & 1 & 0 & 0 & 0 \\
0 & 0 & 0 & 0 & 1 & 0 & 1 & -1 & 0 & 0 & 0 & 0 & 0 & 0 & 0 & 1 & 0 & 0 \\
0 & 0 & 0 & 0 & 0 & 1 & -1 & 0 & 0 & 0 & 0 & 0 & 0 & 0 & 0 & 0 & 1 & 0 \\
0 & 0 & 0 & 0 & 0 & 0 & 1 & -1 & 0 & 0 & 0 & 0 & 0 & 0 & 0 & 0 & 0 & 1
\end{bmatrix}$$

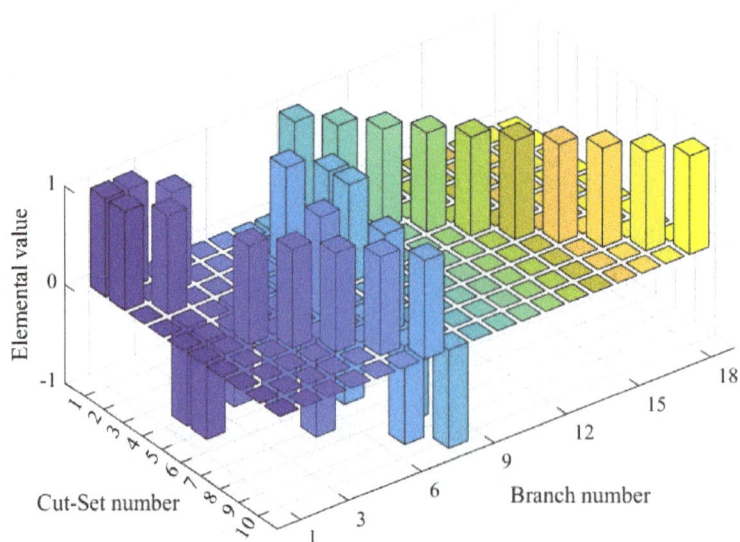

FIG 3 – Distribution of the elements in the cut-set matrix of the network model.

Calculating the number of position-domain elements

According to **S**, ten basic cut sets were listed, and a calculation formula for the branch air volume was determined based on the mathematical relationship between $\mathbf{G_s}$ and $\mathbf{G_y}$, as listed in Table 2.

TABLE 2
Branch air volume calculation based on network topology.

Basic cut sets	Cotree edges in the basic cut set	Air volume calculation formula
$S_1=\{e_1, e_2, e_3, e_{18}\}$	e_1, e_3, e_{18}	$Q_{18}=Q_1+Q_2+Q_3$
$S_2=\{e_1, e_4, e_5, e_{17}\}$	e_1, e_5, e_{17}	$Q_4=Q_{17}-Q_1-Q_5$
$S_3=\{e_3, e_6, e_{17}, e_{18}\}$	e_3, e_{17}, e_{18}	$Q_6=Q_{18}-Q_{17}-Q_3$
$S_4=\{e_3, e_7, e_{10}, e_{17}, e_{18}\}$	$e_3, e_{10}, e_{17}, e_{18}$	$Q_7=Q_3+Q_{10}+Q_{17}-Q_{18}$
$S_5=\{e_3, e_8, e_{13}, e_{17}, e_{18}\}$	$e_3, e_{13}, e_{17}, e_{18}$	$Q_8=Q_3+Q_{13}+Q_{17}-Q_{18}$
$S_6=\{e_5, e_9, e_{14}, e_{17}\}$	e_5, e_{14}, e_{17}	$Q_9=Q_{17}-Q_{14}-Q_5$
$S_7=\{e_{10}, e_{11}, e_{13}\}$	e_{10}, e_{13}	$Q_{11}=Q_{13}-Q_{10}$
$S_8=\{e_{12}, e_{13}, e_{17}, e_{18}\}$	e_{13}, e_{17}, e_{18}	$Q_{12}=Q_{18}-Q_{17}-Q_{13}$
$S_9=\{e_{14}, e_{15}, e_{17}\}$	e_{14}, e_{17}	$Q_{15}=Q_{17}-Q_{14}$
$S_{10}=\{e_{16}, e_{17}, e_{18}\}$	e_{17}, e_{18}	$Q_{16}=Q_{18}-Q_{17}$

The initial localisation position domain is $\{e_{11}, e_{16}, e_{17}\}$, and Q_{11}, Q_{16}, and Q_{17} are considered to be known in subsequent calculations. Q_{18} can be calculated according to the S_{10} air volume calculation formula, and Q_{18} is considered to be known in subsequent calculations. According to the S_9 air volume calculation formula, Q_{15} can be calculated simply by installing sensors in e_{14}. In the subsequent calculation, Q_{14} and Q_{15} are considered to be known. According to the S_7 and S_8 air volume calculation formulae, Q_{12} and $Q_{10}(Q_{13})$ can be calculated by selecting only one branch in e_{10} and e_{13} for sensor installation. According to the S_6 air volume calculation formula, Q_9 can be calculated simply by installing the sensor in e_5. According to the S_3, S_4, and S_5 air volume calculation formulae, Q_6, Q_7, and Q_8 can be calculated simply by installing sensors in e_3. According to the S_1 and S_2 air volume calculation formulae, Q_2 and Q_4 can be calculated simply by installing sensors in e_1.

In summary, sensors need to be installed in e_{11}, e_{16}, and e_{17} due to the branch property, which belong to the position domain. According to the cut-set matrix algorithm and the branch air volume calculation formula, it is determined that sensors need to be installed in e_1, e_3, e_5, and e_{14}, which belong to the position domain. Moreover, it is necessary to select a branch in e_{10} and e_{13} to install sensors. That is, the number of position-domain elements is 8.

State selection of position-domain elements

A zero matrix is built and assigned according to the multifactor indicators values of the network model shown in Table 1, and the main function of data reading is established. The number of samples and indicators are determined according to the data reading, and the positive and negative pointing permissions of each indicator are assigned values, respectively. The standardised interval end point is determined, the data are standardised, and the indicators weight w_j are calculated, and the branch comprehensive fitness score r_j, which requires state selection, is calculated. Table 3 presents the fitness score r_j of all the branches of the network model. Figure 4 presents the fitness score curves for each branch.

TABLE 3

Branch fitness scores for network model.

Branch number j	Fitness score r_j	Branch number j	Fitness score r_j
e_1	49.36	e_{10}	28.23
e_2	36.70	e_{11}	34.68
e_3	70.98	e_{12}	36.60
e_4	14.03	e_{13}	25.86
e_5	10.65	e_{14}	55.27
e_6	3.31	e_{15}	30.38
e_7	24.92	e_{16}	43.69
e_8	22.54	e_{17}	45.65
e_9	28.67	e_{18}	73.29

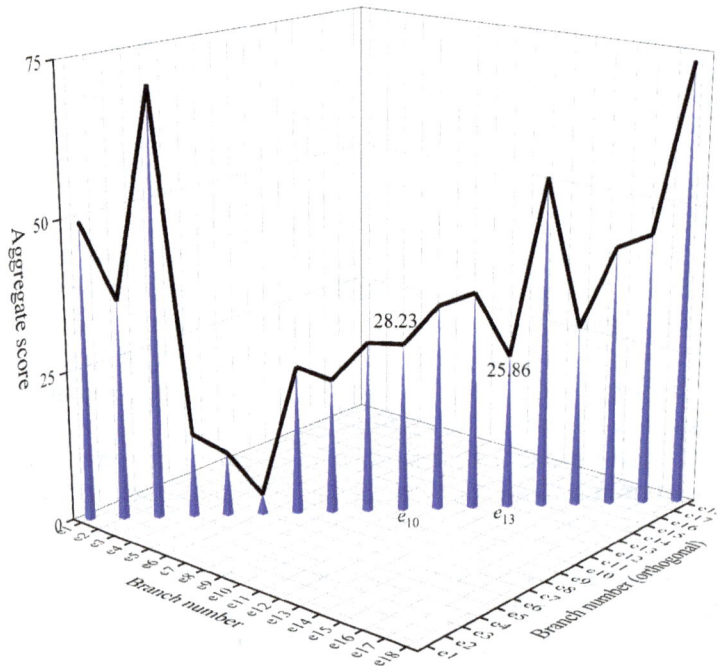

FIG 4 – Comprehensive fitness score curves for each branch.

Since $r_{10} = 28.23 > r_{13} = 25.86$, e_{10} is preferred as the position-domain element. The sensor position domain of the network model is $\{e_1, e_3, e_5, e_{10}, e_{11}, e_{14}, e_{16}, e_{17}\}$. The ventilation parameters of the entire mine can be monitored simply by installing sensors in the position-domain elements.

Production feasibility of model solving

To verify the feasibility of the proposed model and algorithm in actual production, the Gucheng coalmine in China was used as the engineering research subject for a case study.

Analysis of the ventilation system of Gucheng coalmine

Analysis of ventilation network characteristics

A 'three-in and two-return' area ventilation scheme was adopted for the Gucheng coalmine. The main inlet shaft and the secondary inlet shaft were set-up to undertake the task of air intake and the central return vertical shaft to undertake the task of air return. The Taoyuan inlet shaft was set-up to undertake the task of air intake and the Taoyuan return shaft to undertake the task of air return. Three main intake airways and two main return airways were set-up in each wing of the mine. To facilitate pedestrian and traffic movement, multiple crossheadings were set-up between the airways. At the same time, a layout of gob-side entry retaining was adopted for the working face, and multiple crossheadings were set-up between adjacent haulage gateway and return gateway. Reflected in the characteristics of the ventilation network is a complex ventilation network with multiple return shafts and multiple joint branches, with a total of 804 branches.

The equivalent ventilation resistance theory was used to simplify the ventilation system. The simplified system has 256 branches. Figure 5 shows the system structure. The general survey and ventilation resistance measurement of the Gucheng coalmine were performed to obtain the initial value of the air volume and multifactor indicators, providing a data basis for solving the sensor position domain.

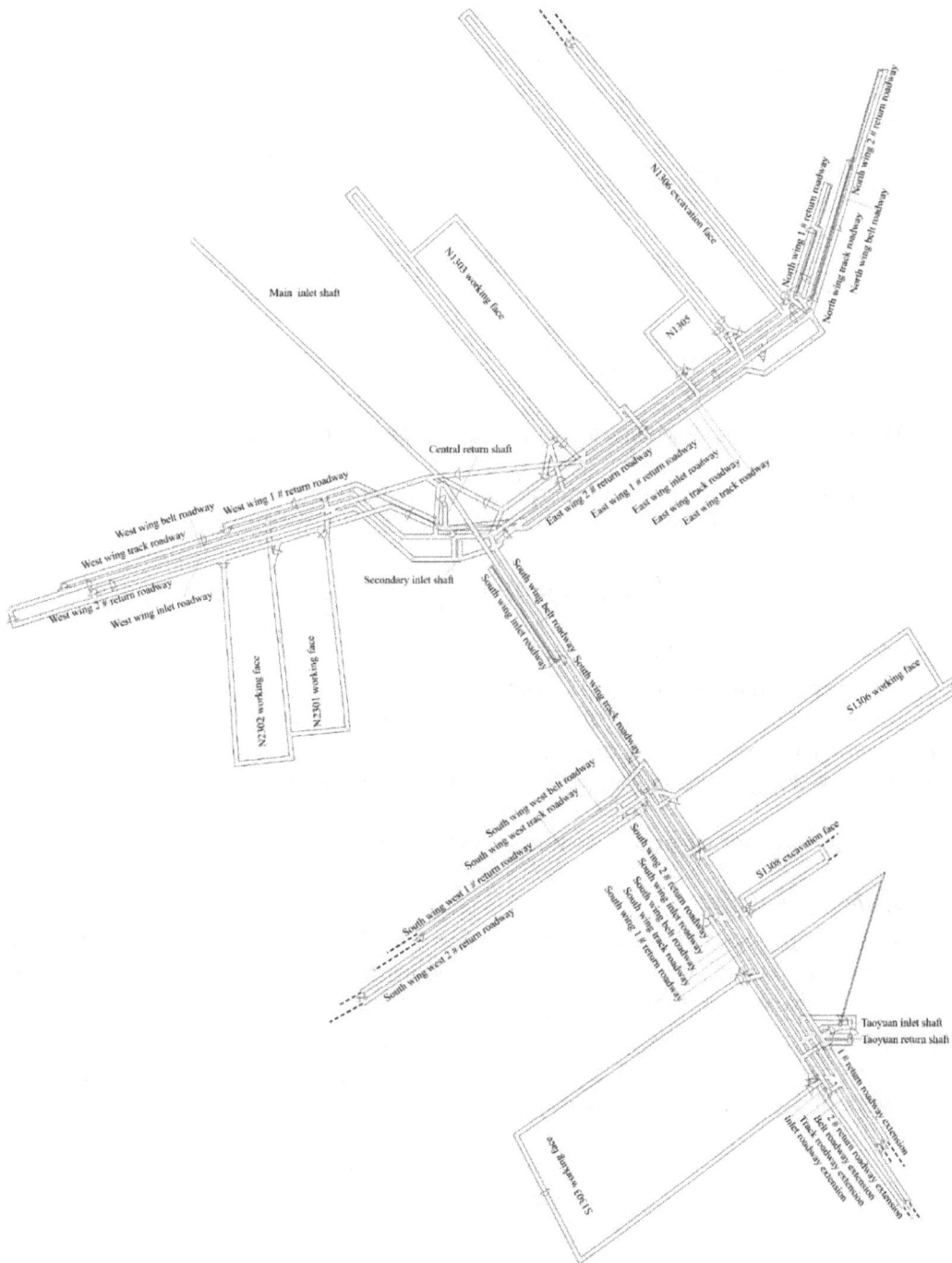

FIG 5 – Simplified ventilation system of the Gucheng coalmine.

Analysis of the characteristics of the sensing system

Currently, Gucheng coalmine adopts the single-sensor local point arrangement, and 27 sensors are installed at the locations stipulated in *The Safety Rules* and the main ventilation route. The sensors are not interactively connected through the monitoring sub-station. This means that the single-sensor local points that exist in this mine are not coordinated to form a sensor network; this can only realise the monitoring of the ventilation parameters of the production mining area and cannot accurately monitor the global state of the ventilation system in real time. The existing single sensor local points are marked in the ventilation network of the Gucheng coalmine, as shown in Figure 6.

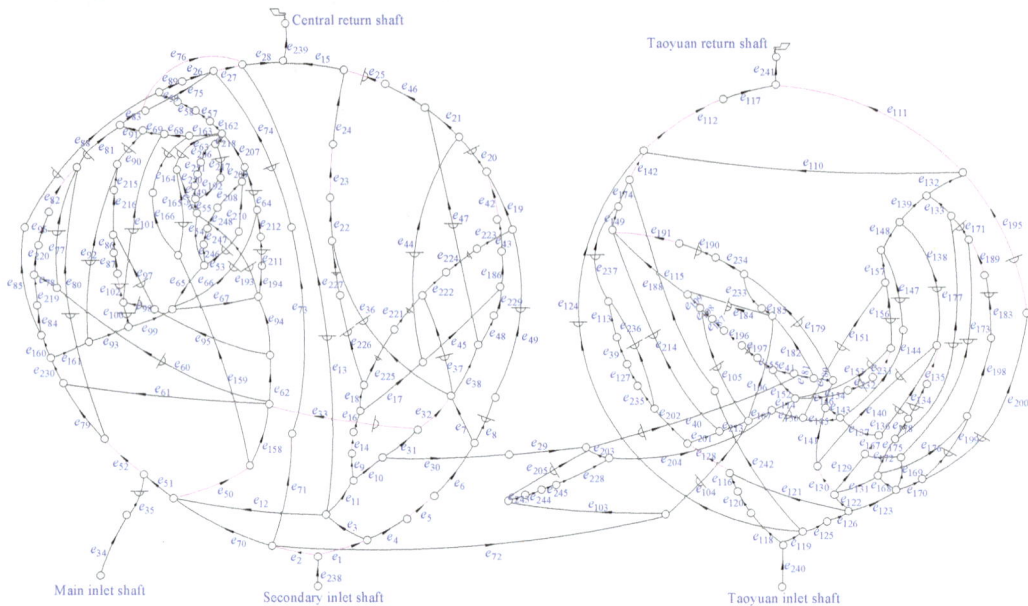

FIG 6 – The single-sensor local point arrangement in Gucheng coalmine.

As shown in Figure 6, the existing set of single-sensor local points is Y_1 = { e_1, e_2, e_5, e_{14}, e_{23}, e_{25}, e_{27}, e_{32}, e_{33}, e_{42}, e_{50}, e_{52}, e_{76}, e_{82}, e_{102}, e_{111}, e_{112}, e_{128}, e_{130}, e_{135}, e_{142}, e_{167}, e_{189}, e_{191}, e_{195}, e_{196}, e_{231} }.

Sensor network siting in Gucheng coalmine

Initial localisation of position-domain elements

Based on the census results, a total of 21 branches, namely e_1, e_2, e_{23}, e_{25}, e_{27}, e_{42}, e_{76}, e_{82}, e_{102}, e_{111}, e_{112}, e_{128}, e_{130}, e_{135}, e_{142}, e_{167}, e_{189}, e_{191}, e_{195}, e_{196}, and e_{231}, are all main return airways, mining area return roadways, or excavation roadways. Therefore, sensors need to be installed in the above 21 branches, which belong to the initial localisation position domain.

Optimisation of the number of position-domain elements

Solving the basic cut-set matrix

By inputting node number m = 182 and edge number n = 262 (to improve the accuracy of network topology construction, six virtual branches e_{256}–e_{262} were added from the return shaft to the inlet shaft), a ventilation network topology structure of the model was constructed. The model was solved to generate the basic cut-set matrix **S**, minimum spanning tree T, and cotree $\bar{T}$ (the virtual branch search weight was set to 0). Figure 7 shows the element distribution of the cut-set matrix of the ventilation network of the Gucheng coalmine.

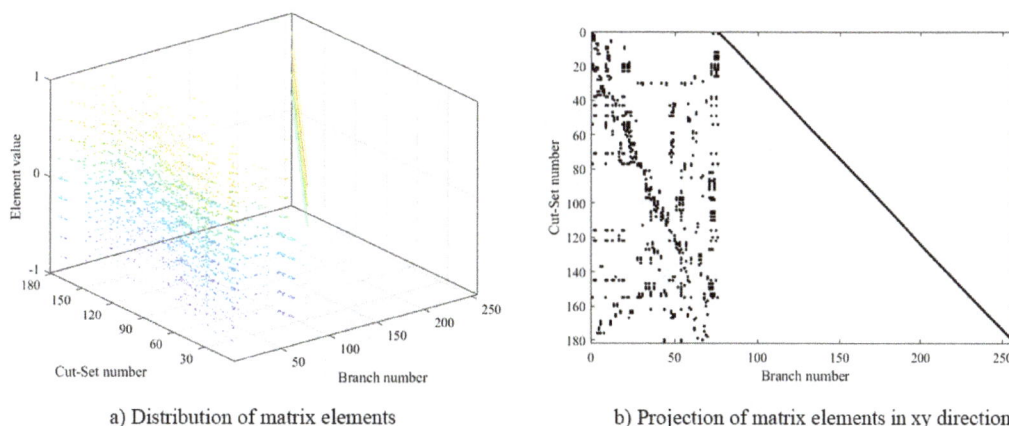

a) Distribution of matrix elements

b) Projection of matrix elements in xy direction

FIG 7 – Distribution of the elements of the cut-set matrix of the ventilation network of the Gucheng coalmine.

Calculating the number of position-domain elements

As there are 181 cut sets in the mine, only the branches where a choice needs to be made are listed.

The branches e_1, e_2, e_{23}, e_{25}, e_{27}, e_{42}, e_{76}, e_{82}, e_{102}, e_{111}, e_{112}, e_{128}, e_{130}, e_{135}, e_{142}, e_{167}, e_{189}, e_{191}, e_{195}, e_{196}, and e_{231} require sensors due to branch properties and belong to the position domain.

According to the cut-set matrix algorithm and branch air flow calculation formulae, it is determined that sensors need to be installed in the following 50 branches: e_5, e_{14}, e_{15}, e_{19}, e_{28}, e_{32}, e_{33}, e_{34}, e_{48}, e_{50}, e_{52}, e_{62}, e_{74}, e_{89}, e_{91}, e_{94}, e_{99}, e_{113}, e_{116}, e_{123}, e_{125}, e_{131}, e_{139}, e_{145}, e_{148}, e_{149}, e_{154}, e_{155}, e_{157}, e_{159}, e_{161}, e_{162}, e_{164}, e_{165}, e_{166}, e_{172}, e_{174}, e_{180}, e_{188}, e_{209}, e_{212}, e_{213}, e_{218}, e_{221}, e_{223}, e_{228}, e_{252}, e_{253}, e_{254}, and e_{255}. The above branches belong to the position domain.

Four position-domain elements need to be identified by selecting one branch for sensor installation in each of the four sets of the branch state selection sets: $\{e_{18}, e_{65}\}$, $\{e_{54}, e_{225}\}$, $\{e_{63}, e_{119}\}$, and $\{e_{93}, e_{207}\}$.

In summary, the number of sensor position-domain elements in the Gucheng coalmine is 75.

State selection of position-domain elements

A comprehensive evaluation model for multifactor indicator decision-making was built, and the comprehensive fitness scores (τ_j) of the four branches that need to be selected should be calculated. Table 4 presents the scores. Figure 8 presents the fitness score curves.

TABLE 4

Branch comprehensive fitness scores.

Branch number j	Fitness score τ_j	Branch number j	Fitness score τ_j
e_{18}	58.72	e_{65}	13.36
e_{54}	21.92	e_{225}	42.29
e_{63}	70.93	e_{119}	18.32
e_{93}	34.27	e_{207}	14.26

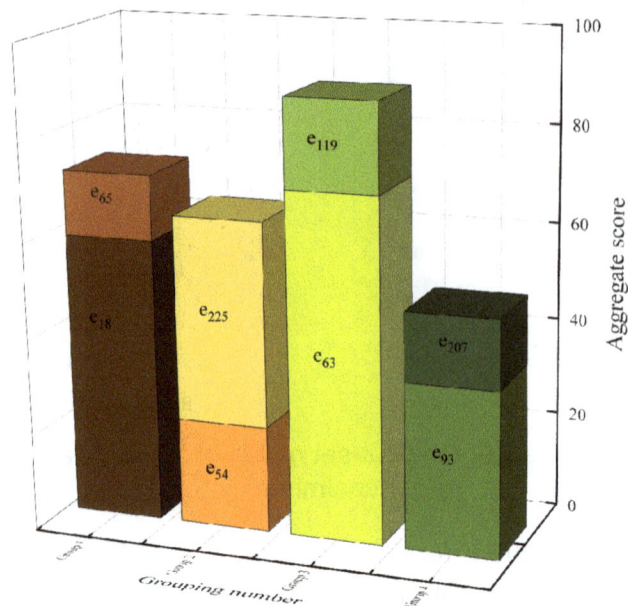

FIG 8 – Comparison of the scores for state selection branches.

For group $\{e_{18}, e_{65}\}$, $\tau_{18} = 58.72 > \tau_{65} = 13.36$, e_{18} is preferred as the position-domain element.

For group $\{e_{54}, e_{225}\}$, $\tau_{225} = 42.29 > \tau_{54} = 21.92$, e_{225} is preferred as the position-domain element.

For group $\{e_{63}, e_{119}\}$, $\tau_{63} = 70.93 > \tau_{119} = 18.32$, e_{63} is preferred as the position domain-element.

For group $\{e_{93}, e_{207}\}$, $\tau_{93} = 34.27 > \tau_{207} = 14.26$, e_{93} is preferred as the position domain-element.

In summary, the sensor position domain of the Gucheng coalmine is $\{ e_1, e_2, e_5, e_{14}, e_{15}, e_{18}, e_{19}, e_{23}, e_{25}, e_{27}, e_{28}, e_{32}, e_{33}, e_{34}, e_{42}, e_{48}, e_{50}, e_{52}, e_{62}, e_{63}, e_{74}, e_{76}, e_{82}, e_{89}, e_{91}, e_{93}, e_{94}, e_{99}, e_{102}, e_{111}, e_{112}, e_{113}, e_{116}, e_{123}, e_{125}, e_{128}, e_{130}, e_{131}, e_{135}, e_{139}, e_{142}, e_{145}, e_{148}, e_{149}, e_{154}, e_{155}, e_{157}, e_{159}, e_{161}, e_{162}, e_{164}, e_{165}, e_{166}, e_{167}, e_{172}, e_{174}, e_{180}, e_{188}, e_{189}, e_{191}, e_{195}, e_{196}, e_{209}, e_{212}, e_{213}, e_{218}, e_{221}, e_{223}, e_{225}, e_{228}, e_{231}, e_{252}, e_{253}, e_{254}, e_{255}\}$. The monitoring of the ventilation parameters for the entire mine can be realised simply by installing sensors in the position-domain elements.

Comparative analysis of single-sensor local points and sensor network

According to the original single-sensor local point arrangement diagram and combined with the characteristics of the sensing system of the Gucheng coalmine, it can be seen that, in addition to the installation of sensors on the 21 branches as stipulated in *The Safety Rules*, sensors were chosen to be installed on six more branches; these were used to monitor the ventilation parameters of the main mining area to improve the safety and reliability of the production, and realise the monitoring of the ventilation parameters of the key air-using areas of the system (working face, main intake airway, bottom of the central return shaft, bottom of the Taoyuan return shaft, and bottom of the secondary inlet shaft). Thus, the ventilation safety of the air-using areas could be ensured; however, it is not possible to accurately grasp the operation state of the entire ventilation network.

In the sensor network localisation, the six sensor branches selected as the original single-sensor local points have a large weight because they are in the key air-using areas, and the roadway conditions are good, so they are also selected as the sensor position-domain elements in the global search. Hence, to realise the ventilation monitoring of the entire mine while also avoiding the wastage of resources, 48 sensor installation points are additionally added on the basis of the existing 27 single-sensor local points to form the sensor network.

Compared with the original single-sensor local point arrangement, the optimised sensor network is more inclined to use the ventilation network structure analysis to realise the function of monitoring the ventilation parameters of the entire mine. Therefore, it focuses on dividing the roadway into segments and then monitoring these segments to grasp the global ventilation information.

Figure 9 shows a schematic comparison of the original single-sensor local point arrangement with the optimised sensor network arrangement in the ventilation system of the Gucheng coalmine. Figure 10 shows a partial enlargement of the comparison results.

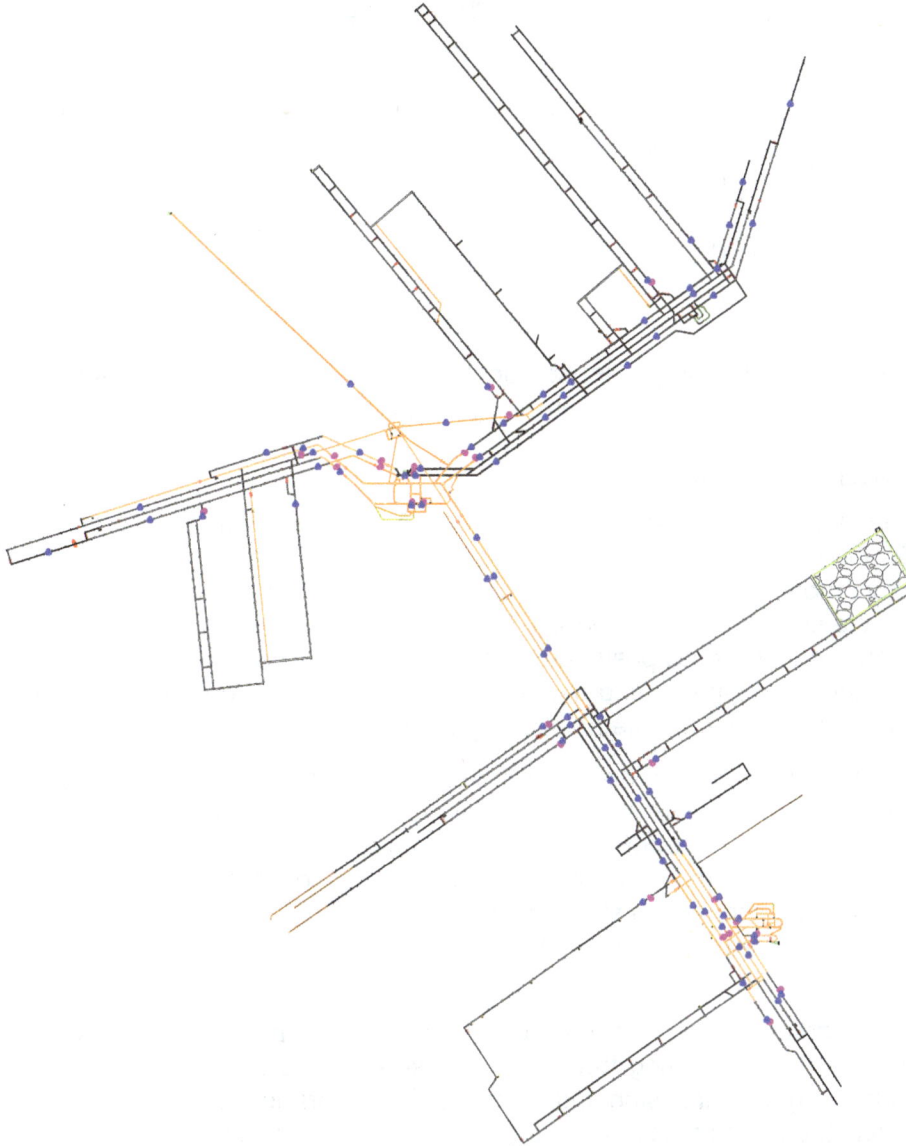

FIG 9 – Comparison of single-sensor local point and optimised sensor network.

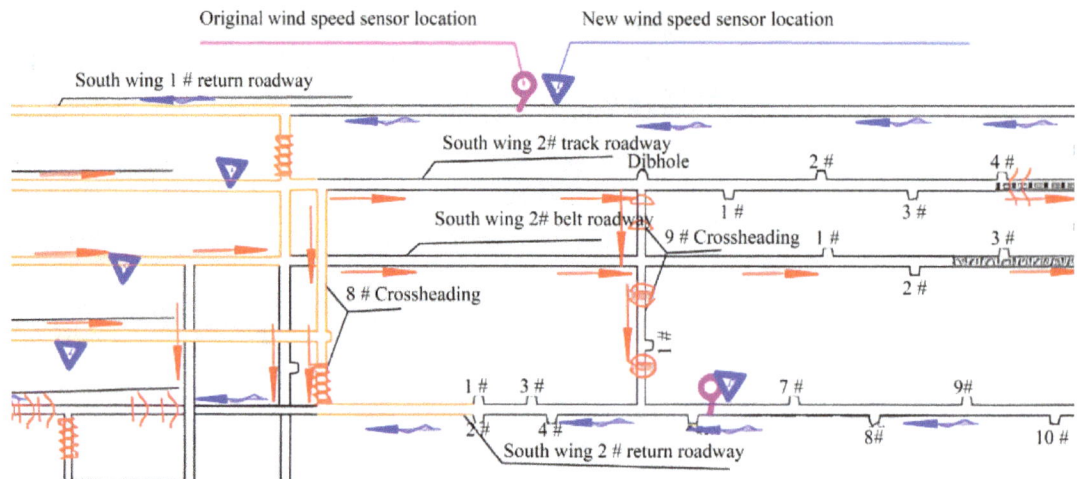

FIG 10 – Comparison of the single-sensor local point and the sensor network somewhere in the south wing of the mine.

CONCLUSIONS

- The construction of an underground air velocity sensor network to form an intelligent sensing system for the ventilation parameters is the basis for intelligent mine ventilation. Based on the fluid network theory, the optimisation problem of the sensor network was transformed into the solution of the sensor position domain. Using the step-by-step solution method and based on the initial localisation of the position-domain elements, an optimisation model for the number of position-domain elements and a multifactor indicator decision-making model were built. By solving the models, an accurate positioning of the sensor position domain was realised.

- Typically, joint branches with disordered airflow properties in actual production are searched as position-domain elements when global search is performed on a ventilation network during the process of solving the above optimisation model. To avoid this situation, an improved cut-set matrix algorithm with the branch air volume as the search weight was proposed. The ventilation network connection structure was used to directly reduce the probability of the joint branch being searched, and the number of branch elements in the sensor position domain was confirmed.

- A multifactor indicator decision-making model was built to avoid any overlap of the sensor monitoring ranges. By solving the comprehensive fitness scores of the branches under the positive and negative decision indicators, suitable branches for sensor installation were selected, and the state selection of the position-domain elements was completed.

- The feasibility of the proposed model and algorithm in the actual production of mines was verified by taking the Gucheng coalmine as the engineering research subject. More specifically, through the construction and calculation of an optimisation model for the number of position-domain elements and a multifactor indicator decision-making model, the sensor position domain for this mine was solved. Finally, a sensor network optimisation of the Gucheng coalmine was performed.

ACKNOWLEDGEMENTS

This project was funded by the Natural Science Foundation of China (No. 52074148).

REFERENCES

Ali, M H, Al-Azzawi, W K, Jaber, M, Abd, S K, Alkhayyat, A and Rasool, Z I, 2022. Improving coal mine safety with internet of things (IoT) based Dynamic Sensor Information Control System, *Phys Chem Earth*, 128:103225.

Baris, K and Aydin, Y, 2020. Atmospheric monitoring systems in underground coal mines revisited: a study on sensor accuracy and location, *Int J Oil Gas Coal Technol*, 23(3):325–350.

Chen, K Y, 2018. *Ventilation Network Analysis*, pp 10–18 (China University of Ming and Technology Press: Beijing).

Cohen, A F, Fisher, T J, Waston, R A and Kohler, J L, 1987. Location Strategy for Methane, Air Velocity and Carbon Monoxide Fixed-Point Mine-Monitoring Transducers, *IEEE Trans Ind Appl*, IA-23(2):375–381.

Cui, Y M, Liu, S Y, Li, H S, Gu, C C, Jiang, H X and Meng, D Y, 2023. Accurate integrated position measurement system for mobile applications in GPS-denied coal mine, *ISA Trans*, 139:621–634.

Gong, W H, Hu, J, Wang, Z W, Wei, Y B, Li, Y F, Zhang, T T, Zhang, Q D, Liu, T Y, Ning, Y N, Zhang, W and Grattan, K T V, 2022. Recent advances in laser gas sensors for applications to safety monitoring in intelligent coal mines, *Front Phys*, 10:1058475.

He, S X, 2013. A graphical approach to identify sensor locations for link flow inference, *TRANSPORT RES B-METH*, 51:65–76.

Huang, X and Liu, Y L, 2023. Research and design of intelligent mine ventilation construction architecture, *Int J Low-Carbon Technol*, 17:1232–1238.

Jia, T, Ma, H and Gao, K, 2024. Intelligent optimization of wind speed monitoring based on cut-set matrix algorithm, *J Lanchow Univ, Nat Sci*, 60(1):8–15.

Li, J Q, Li, Y C, Zhang, J, Li, B L, Zhang, Z T, Dong, J Y and Cui, Y N, 2022. Accurate and real-time network calculation for mine ventilation without wind resistance measurement, *J Wind Eng Ind Aerod*, 230:105183.

Liu, H, Mao, S J, Li, M and Lyu, P Y, 2020. A GIS Based Unsteady Network Model and System Applications for Intelligent Mine Ventilation, *Discrete Dyn Nat Soc*, 1041927.

Liu, J, 2020. Overview on Key Scientific and Technical Issues of Mine Intelligent Ventilation, *Saf Coal Mines*, 51(10):108–111;117.

Liu, J, Jia, J Z and Zheng, D, 2002. *Fluid Network Theory*, pp 161–170 (Emergency Management Press: Beijing).

Liu, J, Jiang, Q H, Liu, L, Wang, D, Huang, D, Deng, L J and Zhou, Q C, 2021. Resistance variant fault diagnosis of mine ventilation system and position optimization of wind speed sensor, *J China Coal Soc*, 46(6):1907–1914.

Liu, Y J, Liu, Z Y, Gao, K, Huang, Y H and Zhu, C Y, 2022. Efficient Graphical Algorithm of Sensor Distribution and Air Volume Reconstruction for a Smart Mine Ventilation Network, *Sensors*, 22(6):2–17.

Ministry of Emergency Management of the People's Republic of China, 2022. *The Coal Mine Safety Rules*, April 2022.

Ni, J F, Le, X R, Chang, L F and Deng, L J, 2021. Resistance variant fault diagnosis and optimized layout of sensors for mine ventilation based on decision tree, *J Saf Sci Technol*, 17(2):34–39.

Rodriguez-Diaz, O, Novella-Rodriguez, D F, Witrant, E and Franco-Mejia, E, 2020. Control strategies for ventilation networks in small-scale mines using an experimental benchmark, *Asian J Control*, 23(1):72–81.

Sharma, A N, Dongre, S R, Gupta, R, Pandey, P and Bokde, N D, 2022. Partitioning of Water Distribution Network into District Metered Areas Using Existing Valves, *Comput Model Eng Sci*, 131(3):1515–1537.

Standards China, 2019. AQ1029–2019-The Application and Management Standard for Coal Mine Safety Monitoring System and Detector, August 2019.

Yan, Z G, Wang, Y P, Fan, J D, Huang, Y X and Zhong, Y H, 2023. An Efficient Method for Optimizing Sensors' Layout for Accurate Measurement of Underground Ventilation Networks, *IEEE Access*, 11:72630–72640.

Yang, S G, Zhang, X F, Liang, J and Xu, N, 2023. Research on Optimization of Monitoring Nodes Based on the Entropy Weight Method for Underground Mining Ventilation, *Sustainability*, 15(20):14749.

Zhang, T Q, Yao, H Q, Chu, S P, Yu, T C and Shao, Y, 2021. Optimized DMA Partition to Reduce Background Leakage Rate in Water Distribution Networks, *J Water Resour Plann Manage*, 147(10):04021071.

Zhang, Z T, Li, Y C, Li, J Q, Zhang, J and Li, B L, 2023. Architecture and implementation of intelligent ventilation precise control system, *J China Coal Soc*, 48(4):1596–1605.

Zhao, D and Pan, J T, 2011. Fault Source Diagnosis for Mine Ventilation Based on Improved Sensitivity Matrix and Its Wind Speed Sensor Setting, *China Saf Sci J*, 21(2):78–84.

Zhou, F B, Wei, L J, Xia, T Q, Wang, K, Wu, X Z and Wang, Y M, 2020. Principle, key technology and preliminary realization of mine intelligent ventilation, *J China Coal Soc*, 45(6):2225–2235.

Integrated leak detection for tube bundle systems – ACARP Project C27035

S Muller[1], Z Taylori[2], R Irwin[3] and A Chong[4]

1. Principal Scientist, Simtars – Resources Safety and Health Queensland, Australia.
   Email: sean.muller@simtars.com.au
2. Analytical Chemist, Simtars – Resources Safety and Health Queensland, Australia.
   Email: zachary.taylor@simtars.com.au
3. Analytical Chemist, Simtars – Resources Safety and Health Queensland, Australia.
   Email: rebecca.irwin@simtars.com.au
4. Computer Systems Engineer, Simtars – Resources Safety and Health Queensland, Australia.
   Email: aaron.chong@simtars.com.au

ABSTRACT

Tube bundle systems (TBSs) are a critical component gas monitoring system utilised in underground coalmines in Australia. Kilometres of tubing are installed from locations underground to the analyser room located on the surface of the mine. Samples are drawn to the surface under vacuum to allow for automated sampling and analysis by infrared/paramagnetic analysis or by Gas Chromatograph (GC) (Cliff, Brady and Watkinson, 1999). The ability to bring samples to the surface to be analysed on a common instrument is a major advantage of this technique. This is crucial in an emergency where the underground environment is inaccessible to workers and power has been lost, thus rendering real time monitoring and discrete bag sampling obsolete (Brady, Watkinson and Harrison, 2015). Tube bundle analysis is particularly effective for the monitoring of sealed locations for indications of coal self-heating and assessment of flammability of atmospheres underground.

Tubes must be correctly installed, joined, and maintained to ensure that a representative sample is collected on the surface. The maintenance and calibration of electrical equipment in underground coalmines (AS 2290.3 2018), prescribes the requirements for leak testing of tubes for gas monitoring, however the process allows the operator to perform this task in a manner fit for purpose (Standards Australia, 2018). As there is no standardised method prescribed for leak testing, this opportunity has resulted in the development of several different methods for leak testing (Forrester, 2017).

This paper details preliminary results generated for ACARP Project C27035 for the demonstrated application of an integrated leak testing prototype developed in collaboration with Deltamation. The prototype employs a process which purges the tube with inert gas before drawing it back through an analyser under vacuum. This process allows for routine leak testing, as well as the determination of tube length and location of potential damaged tubes. This method will enable the assessment of tube status during emergency response, where a fire, explosion or strata failure has occurred underground. This prototype has been tested extensively at Simtars with a variety of tube infrastructure and simulated leaks to provide proof of concept and determine the advantages as well as the limitations of this method of leak testing.

INTRODUCTION

Tube bundle systems are used extensively in the Australian underground coal mining industry as a means for monitoring goafs and return roadways from working panels (Simtars, 2016). They provide an early indication of spontaneous combustion and monitor the atmosphere for explosive gases such as methane, particularly during the seal up of a panel. Leaking tubes lead to dilution of the underground sample and can therefore lead to missed opportunities to identify spontaneous combustion, and incorrect assessment of the atmosphere post explosion. With current practices it is often difficult to determine the sampling location of a tube post explosion when tubes are often damaged either partially or completely (Simtars, 2016). Several weeks were required to analyse the information from the tube bundle system at Moura No. 2, to determine if and where tubes were broken.

One drawback of Tube Bundle Systems is the reliance on the physical integrity of the tubes. Tubes must be installed and joined correctly and maintained to ensure that a representative sample is collected on the surface. To ensure this, the maintenance and calibration of electrical equipment in underground coalmines (AS2290.3 2018), prescribes the requirements for leak testing of tubes for gas monitoring, however the process allows the operator to perform this task in a manner fit for purpose. This has resulted in several different methods for leak testing used across the coal mining industry.

The current manual integrity testing methodology is time consuming; it involves manual application of nitrogen at the sampling point and requires significant human resources. This is not possible when access to underground is restricted due to an evacuation and/or mine fire/explosion.

An update to AS2290.3 in 2018 (Electrical equipment for coalmines – Maintenance and overhaul Part 3: Maintenance of gas detecting and monitoring equipment) allows the mine to choose the way that leak testing is performed (Forrester, 2017), therefore allowing the use of automated nitrogen purging on the tube, before drawing the sample back through an analyser. Previously, a full integrity test was required every month by the application of a known gas to the sampling point. The addition for monthly leak testing does not remove the requirement for integrity testing every six months or at the initial deployment of a tube at a sampling location.

Delta Automation has developed a leak testing method that involves pushing nitrogen down the tube and drawing it back up again. Although this method is performed entirely from the surface, the methodology is a discreet service, requiring Delta Automation equipment and personnel to be transported to mine site to perform the testing. As part of ACARP project C27035, Simtars in collaboration with Deltamation have developed a prototype based on this method of leak detection to be retrofitted to an existing tube bundle systems to perform leak testing on a permanent basis from the surface.

Further to the requirements for leak testing, the development and testing of this process is advantageous for the determination of the status of the existing tubes underground in an emergency where the sampling location and integrity of the tubes may have been compromised. The ability for comparison of the data generated from the system operating on a normal basis and post explosion is critical in determining the status of the tubes underground.

SCOPING VISITS AT UNDERGROUND COALMINES

Scoping visits were undertaken at three different underground coalmines in Australia to determine any constraints or considerations which may impact the viability or functionality of an integrated tube bundle leak testing system. A visit to Deltamation was undertaken to assess the current hardware and software used in manual leak testing system and to identify any possible constraints with the engineers and technicians who operate the equipment.

A survey on tube bundle hardware and maintenance was requested from the at several underground coalmines in Queensland and New South Wales. The purpose of the survey was to determine an industry wide and individual mine site viability for running the leak testing system and to identify any tube bundle system modifications to be implemented for the successful operation of an integrated leak testing system.

From the scoping visits and surveys, the following requirements for the requirements for a leak testing prototype were identified:

- The ability to force a gas through the tube in reverse (from the surface down underground), without over pressuring or damaging the tube or tube infrastructure.

- The tube infrastructure must not introduce oxygen to the tube.

- Ability to drawback nitrogen from full length of tube in a reasonable time frame.

- Appropriate measurement of oxygen.

- Appropriate space for required hardware on the surface, located inside/next to tube bundle facilities.

DEVELOPMENT OF PROTOTYPE

A prototype of the leak detection system was designed in collaboration with Deltamation engineers, taking into consideration the requirements of the system and the information gained during the mine surveys and site visits.

Hardware specifications for automatic leak detection system

The system was designed to operate with any tube bundle system installed in underground coalmines in Australia. As per the requirements for a permanent leak detection system installation, a reliable, easy to maintain system was developed.

The system was designed to be retrofitted to any tube bundle systems manufactured by various tube bundle system manufacturers. To satisfy this requirement, the leak detection system was designed to connect to the tubes as they enter the tube shed, so that the system can operate independently of various hardware variations which would be encountered on systems from different suppliers. The prototype was designed to be used open to atmosphere to comply with hazardous area requirements.

As the automatic leak detection system is intended to be deployed alongside an existing tube bundle infrastructure, the prototype size and transportability of the prototype was designed accordingly for the ease of logistics with a limited footprint. Considering the harsh mine site environment and the remote location of tube bundle system, hardware on this system such as the gas cylinder cage, pumps, variable speed drive and sensors must be weather protected from rain and the direct heat of the sun.

The system uses a paramagnetic sensor configured specifically for the accurate monitoring of the full range of oxygen (O_2) concentrations encountered in underground coalmines (0–21 per cent) with an accuracy of ±1 per cent of full scale. Together with the readings from the systems flow sensors and pressure sensors, the calculation of the leakage factor and the locations of leak are possible.

Sensors are utilised to measure the following parameters:

- injection flow (L/min) purging
- sample flow (L/min) drawing
- O_2 concentration (%)
- injection pressure (kPa)
- sample pressure (kPa)
- pump pressure (kPa).

Purging of the tubes is facilitated through the pressure of a G size (approximately 9.3 m^3) nitrogen cylinder regulated to a maximum of 30 kPa delivery pressure. A flow metre is used to further limit and measure the purging flow of the (high purity) nitrogen during this phase of the sequence. A variable speed drive pump and mass flow controller was utilised for the drawing phase of the test and to supply the appropriate sample to the analyser. The system utilises a Programmable Logic Controller (PLC) to control the solenoids and valves. Selected PLC and relevant inputs and outputs are specified as Allen Bradley Compact Logix processor with ethernet/IP communication interfaces. Analogue and digital inputs and outputs are also selected to retrieve and send data accordingly. A software program for the control of solenoids and monitoring of sensor data was developed to assess the outcomes of the testing. Figure 1 shows the schematic of the leak tester prototype.

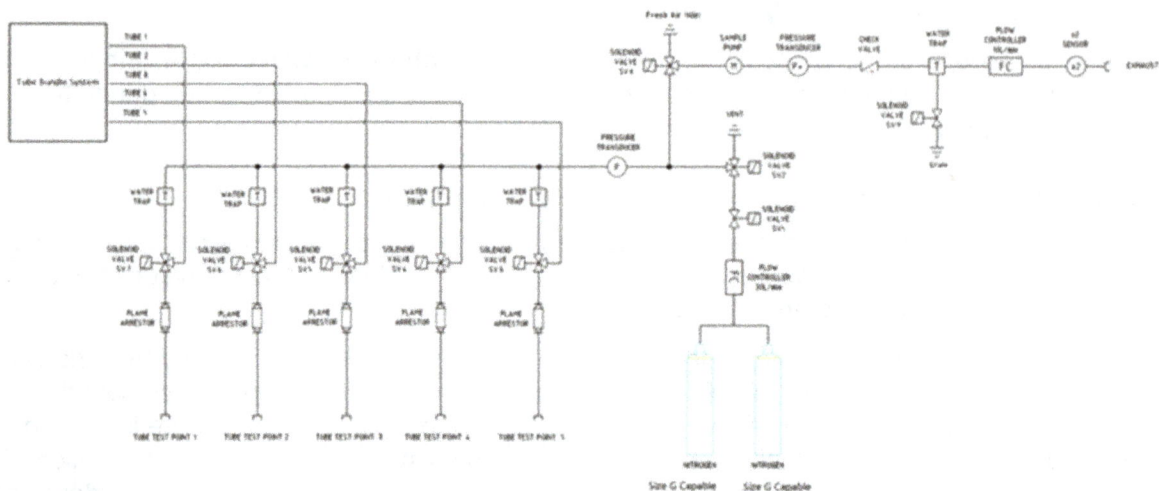

FIG 1 – Schematic of leak tester prototype.

The leak testing prototype is designed to test one tube at a time, however the capacity to connect five tubes simultaneously to the system was implemented in the prototype. Each connection point is installed with a flame arrestor and water trap. The prototype was mounted on a manifold transportable by forklift. Figure 2 shows the commissioned prototype during testing.

FIG 2 – Commissioned leak tester prototype.

TESTING AND RESULTS

Comprehensive testing of the integrated leak tester prototype was undertaken. The objective of the testing was to:

- Determine the viability of the prototype to conduct leak testing for a length of tube resulting in a pass//failure based on the measured oxygen percentage (less than 0.5 per cent).

- Determine the length of the tube using the dimensions of the tubes and the flow rate.

- Test the ability to determine the location of leaks present in the tube.

- Test the effects of tube infrastructure on the ability to conduct leak testing using the prototype.

- Measure the effect of partial or fully blocking of the tube prior to conducting leak testing.

The testing was undertaken at Simtars using lengths of tube representative of what was used in underground operations from initial scoping visits. This enabled the simulations of different sizes and lengths of tube, testing with a variety of common tube infrastructure, and the ability to test consistent leaks at different points in the tubing.

Test method

Testing of the prototype and tubing was undertaken as follows:

1. Tube bundle tube was connected to one of the testing points on the leak testing system using stainless steel or brass Swagelok fittings. Testing was conducted one tube at a time.

2. Extra rolls of tubing were connected using Swagelok fittings depending on the length required for the experiment as demonstrated in Figure 3. Tubes of 5/8 outer diameter (OD) inch (5/8") tubing was supplied in rolls of 200 m. Tubes of 1/2 inch OD (1/2") tubing was supplied in rolls of 500 m. Lengths of tubing were left on the rolls during testing.

3. Tube infrastructure and or small lengths of leaking tube was installed in the tube using Swagelok fittings for required testing depending on the experiment.

4. The leak testing software was initiated, which begins recording of all sensors (flow, pressure, oxygen) for the duration of the test.

5. The tube was purged with nitrogen under pressure from the ultra-high purity nitrogen cylinder. This gas was delivered to the system at 30 kPa before being further downregulated to a purging flow rate of approximately 2–3 L/min. A portable gas monitor was used to determine the gas exiting the tube during the purge. When the oxygen concentration was equal to 0.5 per cent oxygen or less, the tube purging was considered complete, and the purge time recorded.

6. The 'drawing phase' was initiated at the completion of the purging phase and the nitrogen in the tube was drawn back though the leak testing manifold by vacuum using the variable speed drive pump.

7. The test was considered complete when the oxygen sensor reached approximately 20.8 per cent oxygen and the recording of sensor information was ceased.

8. The length of the tubing when no leaks were present was calculated by determining the time at which the oxygen reached 0.5 per cent.

9. In the case of determining leakage, the length of tubing before the leak was calculated using the time at which the oxygen reached 0.5 per cent. The remaining length of tubing was calculated by determining the time at which the oxygen level reached more than 0.5 per cent above the maximum fraction of oxygen measured from the length of the leaking tube.

FIG 3 – Tube bundle tubing used during testing.

Testing results for 5/8 inch tubing

Four kilometres of 5/8 inch tubing was acquired in lengths of 200 m. The dimensions of the tube are listed in Table 1.

TABLE 1

Dimensions for 5/8" tubing.

Tube inner diameter	1/2"
Tube outer diameter	5/8"
Length of tube roll	200 m

Determination of purging times

Purging of the tube at lengths of 200 m, 800 m, 1000 m, 1600 m, 2400 m and 3600 m was undertaken and the purge times calculated using the oxygen readings from the portable gas monitor. The results of the purge testing are shown in Figure 4. This testing was undertaken without tube infrastructure connected.

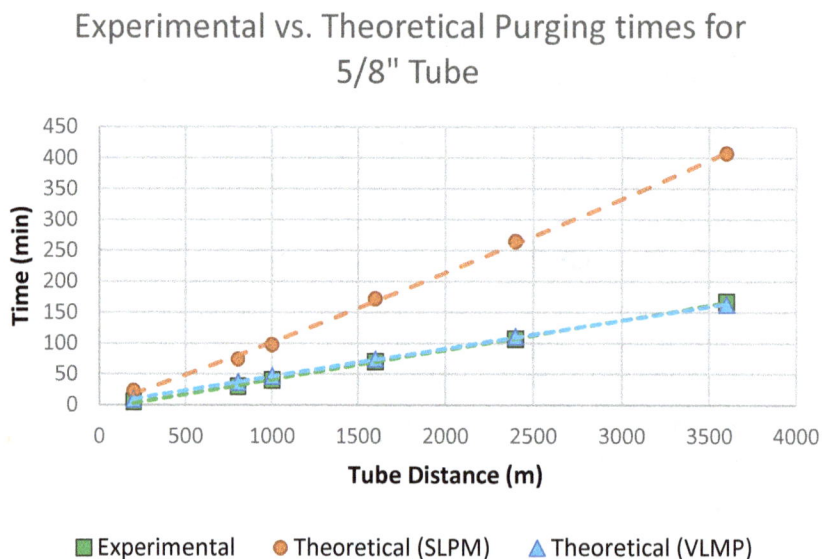

FIG 4 – Purge time for various lengths of 5/8" tubing.

As demonstrated by the purge times in Figure 4, the theoretical and experimental purging time for each tube was very closely aligned based on the flow rate measured by the flow metre VLMP (volumetric litres per minute). The longest length of tube at 3600 m required 2 hrs and 40 mins for complete purging. This step was the most time-consuming step in the process and represents a drawback when using this technique for 5/8" tubing. Despite this, it may still be reasonable to remove a tube from sampling sequence for this amount of time using a remote and automated system to complete the testing while temporarily removing the tube from sampling sequence. The purging for a tube length of 1000 m took less than 50 mins.

Determination of draw times

After the competition of the purging phase of the tubes demonstrated in Figure 4, the pump was used to reverse the flow in the tubes and draw the nitrogen back through to the analyser until fresh air was detected. The time from beginning of the drawing until the time at which the oxygen reading increased to 0.5 per cent was determined to be the travel time used to calculate the length of the tubing. The results of the tube drawing for 5/8 inch tube are shown in Figure 5.

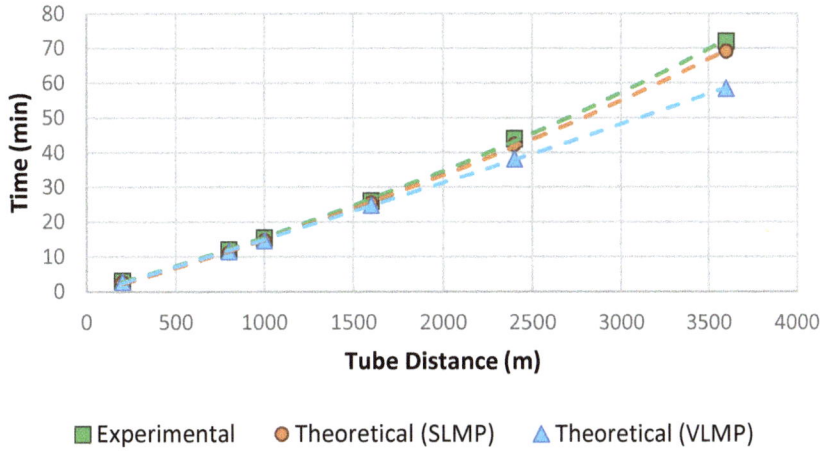

Experimental vs. Theoretical Drawing times for 5/8" tube

FIG 5 – Draw time for various lengths of 5/8" tubing.

As demonstrated by the draw times in Figure 5, the theoretical and experimental drawing time for each tube was very closely aligned based on the flow rate measured by the flow metre SLPM (standard litres per minute). The longest length of tube at 3600 m required approximately 70 mins for drawing. The combined purge and draw time for the longest tube tested at 3600 m was completed in less than 4 hrs. The theoretical draw time calculated using the flow rate for the 3600 m tube was approximately 4.3 per cent lower than the experimental length, resulting in an error of 4.3 per cent or 150 m discrepancy in estimation for the entire length of tube.

Testing of 1/2 inch tube

Two kilometres of 1/2 inch tubing was acquired in lengths of 500 m. The dimensions of the tube are listed in Table 2.

TABLE 2

Dimensions for 1/2" tubing.

Tube inner diameter	3/8"
Tube outer diameter	1/2"
Length of tube roll	500 m

Determination of purging times

Purging of the tube at lengths of 500 m, 1000 m, 1500 m and 2000 m was undertaken and the purge times calculated using the oxygen readings from the portable gas monitor. The results of the purge testing are shown in Figure 6. This testing was undertaken without tube infrastructure connected.

Experimental vs. Theoretical Purging times for 1/2" Tube

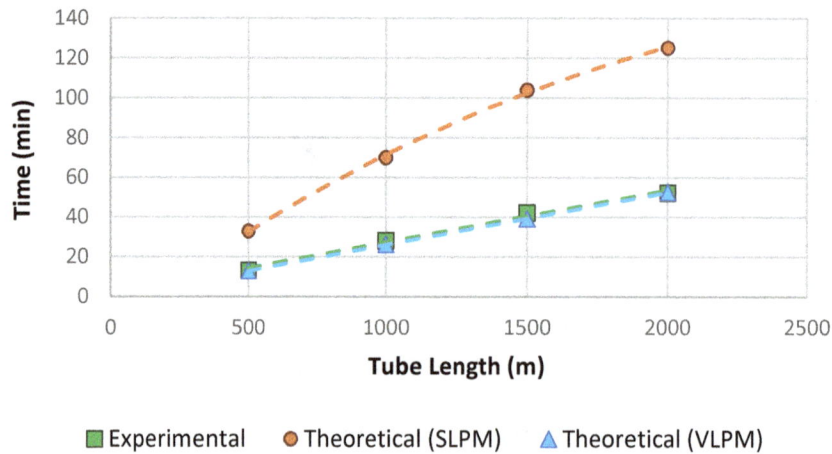

FIG 6 – Purging times for various lengths of 1/2" tube.

As demonstrated by the purge times in Figure 6, the theoretical and experimental purging time for each tube was very closely aligned based on the flow rate measured by the flow metre VLMP which is consistent with the purging of the 5/8" tube shown in Figure 4. The longest length of tube at 2000 m required 50 mins for complete purging. By extrapolation a length of 4000 m of 1/2" tubing would take approximately 1 hr and 40 mins, which is less than the 5/8" tubing of 3600 m length (approximately 2 hrs 40mins). The reduction in draw times for 1/2" tubing is consistent with the lower volume of gas required due to the reduced internal diameter of the 1/2" tubing. The draw time for 1000 m of 1/2" tubing was approximately 30 mins compared to approximately 50 mins with the 5/8" tubing.

Determination of draw times

After the competition of the purging phase of the tubes demonstrated in Figure 6, the pump was used to reverse the flow in the tubes and draw the nitrogen back through to the analyser until fresh air was detected. The time from beginning of the drawing until the time at which the oxygen reading increased to 0.5 per cent was determined to be the travel time used to calculate the length of the tubing. The results of the tube drawing for ½ inch tube are shown in Figure 7.

Experimental vs. Theoretical Drawing times for 1/2" Tube

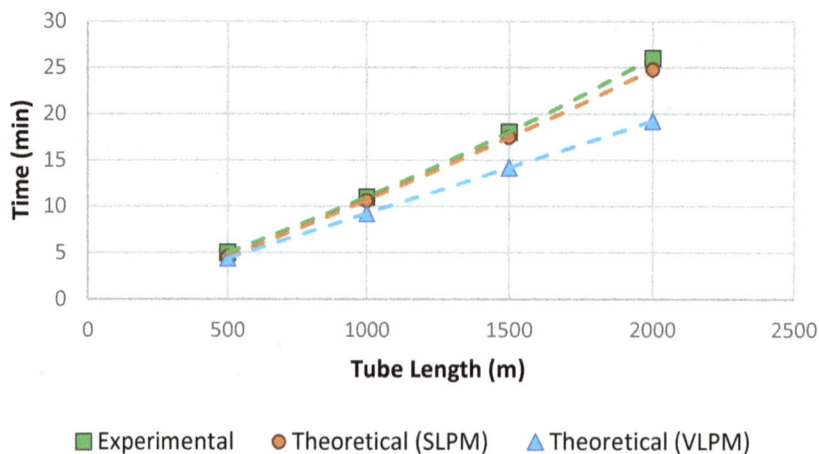

FIG 7 – Draw times for various lengths of 1/2" tube.

As demonstrated by the draw times in Figure 7, the theoretical and experimental drawing time for each tube was very closely aligned based on the flow rate measured by the flow metre SLPM, which is consistent with the drawing of the 5/8" tube shown in Figure 5. The drawing of the 1/2" tube took a maximum of approximately 26 mins with 2000 m of tubing. By extrapolation a length of 4000 m of 1/2" tubing would take approximately 1 hr, which is less than the 5/8" tubing of 3600 m length (approximately 70 mins). The reduction in draw times for 1/2" tubing is consistent with the lower volume of gas required due to the reduced internal diameter of the 1/2" tubing. The draw time for 1000 m of 1/2" tubing was approximately 11 mins compared to approximately 15 mins with the 5/8" tubing.

Detection of tube leakage

Testing for tube leakage was simulated by installing a short length of damaged tube between existing rolls of tube used for testing. This allowed for testing of a consistent leak at different points of the tube and for various lengths and sizes of tubing. The leak in the length of damaged tubing was approximately 2.5 mm long and 1 mm wide, representing a major perforation to the tubing. A leak of this nature may occur underground in the case of acute physical damage to the tube, or repeated friction on the tube – for example the tube rubbing on a hook used to support the tube, over a duration of months or years. Figure 8 demonstrates testing of 2000 m of 5/8" tube with the leak located at the half-way point (1000 m). This test was undertaken using an end of line sample filter and flame arrestor at the end of the tube.

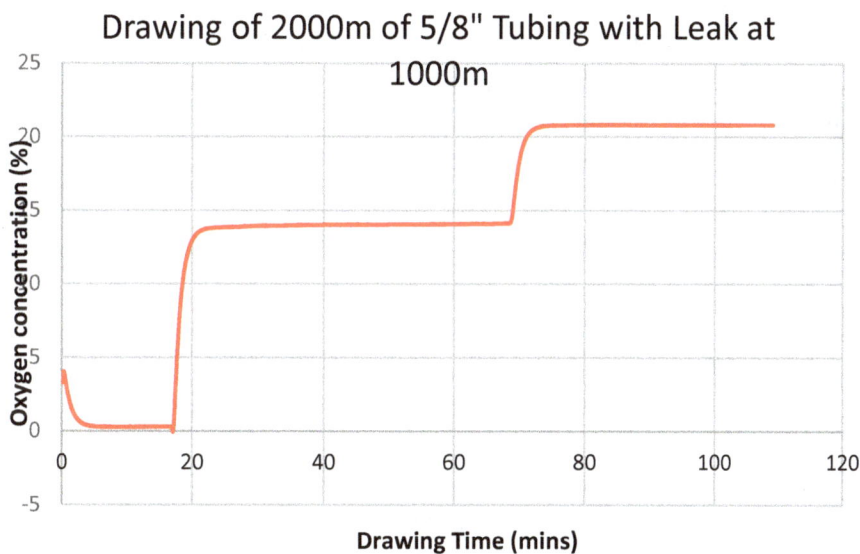

FIG 8 – Identification of tube leakage during drawing phase.

The first increase in oxygen at approximately 17 mins on Figure 8 represents the draw time for the length of tubing prior to the leak (1000 m). The second increase at approximately 69 mins represents the draw time for the entire length of tube (2000 m), including the leak. The oxygen concentration (approximately 14 per cent) between the leak and the end of the tube was used to calculate the leakage factor, and thus the relative dilution of the sample from the end of the tube and corresponding flow rate. The drawing phase for this tube took approximately 75 mins to complete.

Table 3 shows the calculations for total tube length and the location of tube leakage compared to the true or experimental lengths and locations. The calculated leakage point was within 63 m of the true leakage point, and the calculated total length of tubing was within 118 m of the true length. These differences between true and calculated represent approximately +6 per cent error margin between experimental and calculated tube lengths.

TABLE 3
Leakage calculations for 2000 m of 5/8" tubing with a leak at 1000 m.

Total tube length	2000 m
Calculated total tube length	2118 m
Location of leak	1000 m
Calculated location of leak	1063 m

Water blocked tubing

A common maintenance problem for tube bundle monitoring systems is the accumulation of water in the tubes due to condensation. This can be particularly prevalent in locations where a sample from a particularly warm or humid environment (such as a goaf) is being drawn through tubing located in an environment of lower temperature, (such as an intake roadway). Water traps used at seal panels and additional strategic locations (marshalling panels) can reduce or eliminate the accumulation of water in tubes. In the case of ineffective/absent water traps and excessive condensation, water may accumulate in tubes which may substantially increase the resistance or completely block tubes from being sampled. Figure 9 shows the result of purging a 600 m length of 5/8" tubing containing 1 L of water in a 1 m dip in the tubing close to the end of the tube. This test was undertaken using an end of line sample filter and flame arrestor at the end of the tube.

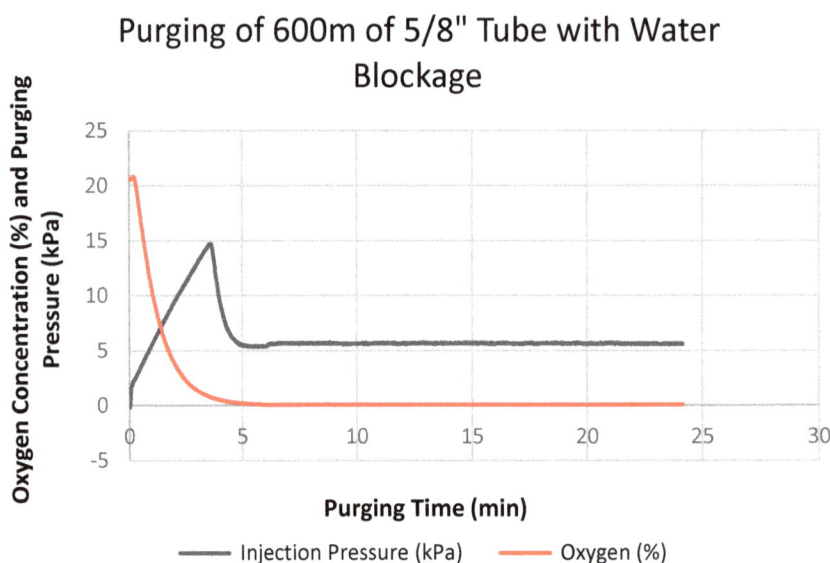

FIG 9 – Purging pressure for a tube blocked with water.

The measured pressure in Figure 9 demonstrates a linear increase in purging pressure to 15 kPa before the water was forced from the line and out of the end of line filter at approximately 3.7 mins. This indicates that under the right circumstances the pressure of the nitrogen purging will be enough to effectively blow the water out of the tube as would normally be done with compressed air for maintenance.

Blocked or kinked tubes

The potential for tubes to become completely blocked or kinked is a possibility during underground events involving roof falls, spontaneous combustion events or explosions. The mechanism for this to occur may include physical crushing of the tube, kinking of the tube from explosion, or melting of the tube due to heat or combustion. Under normal circumstances, tubes may become completely blocked due to excessive water or ice during extreme weather.

Figure 10 demonstrates the purging of a physically blocked tube and a kinked tube. The fully blocked tube was purged until the pressure reached 14 kPa and the blockage was released to avoid damage to the testing apparatus. The gradient of the pressure increase is roughly linear for the blocked tube. The kinked (partially blocked) tube exhibited a logarithmic increase in pressure in contrast to the fully blocked tube.

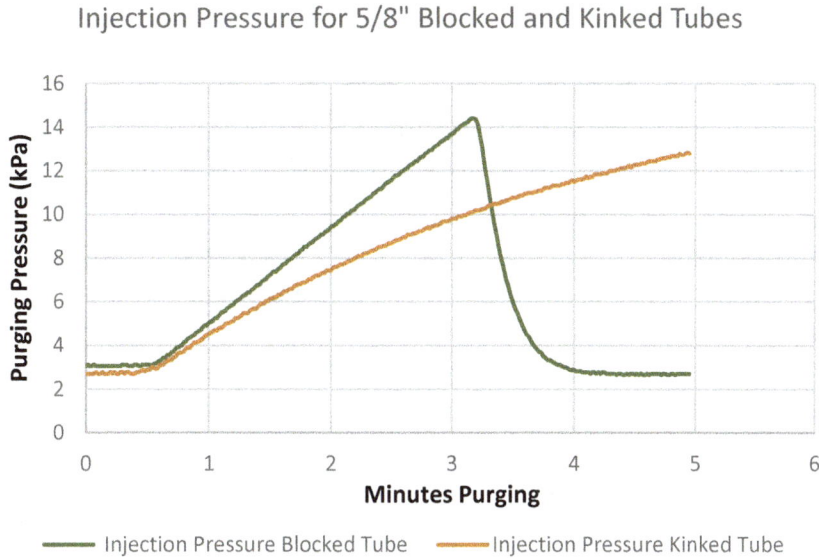

FIG 10 – Purging pressure gradient increase for physically blocked or kinked tubes.

DISCUSSION OF RESULTS

Effectiveness of the integrated leak detection system for routine leak testing

The testing of the prototype demonstrated the ability for the system to perform leak testing to an accuracy fulfilling the requirements for monthly leak testing by AS2290.3;2018.

Due to the practicalities of finding and repairing tube leaks underground (walking and inspecting the tube line and remaking fittings), this technique represents a substantial advantage in being able to narrow down the location of tube leaks and may represent a labour saving of hours or days in searching for leaks in the tube.

Application of the integrated leak detection system for emergency response

During an event underground involving a fire, explosion, or roof fall, the possibly exists for the tubing to be kinked, sealed shut, or damaged resulting in a major leak in the tube from a point in the tubing which is not the same as the original intended sampling location (Simtars, 2016). Under these circumstances it can be challenging or impossible to determine the status of the tubing and where it is now sampling from. Completely blocked or kinked tubes can be identified using the tube vacuum from the tube bundle system, however due to the time taken for a full vacuum to be pulled using the sample pump, the system may have to be taken offline for a substantial time, thus compromising the acquisition of data from effective sampling locations.

By using the integrated leak detection system, compromised tubes may be taken offline and examined without disturbing the sampling sequence of the tube bundle system. Kinked or sealed tubes may be identified relatively quickly by an increase in pressure while trying to purge the lines. Tube lines which are not blocked may be tested using the system to determine the effective length of tubing. By comparing this to the expected length of the tube, and location of the tube underground, the effective sampling point and integrity of the tubing may be determined. This process will be most effective when comparison of tubing is available from the last successful leak test under normal circumstances.

CONCLUSIONS

Preliminary testing for the integrated leak detection prototype was completed using 5/8" and 1/2" tube configured to simulate conditions for tube installation in underground coalmines. Tube infrastructure from various suppliers were tested including end of line filters, seal panel filters, flame arrestors and water traps. Leaks at various distances in the tubing were tested to establish the effectiveness of the prototype in the determination of the approximate location of the leakage in the tube. The testing demonstrated the following outcomes:

- The leak detection prototype allows leak testing to be performed on functioning tubes to an accuracy which fulfils the requirements stated by AS2290.3:2018.

- The time for purging and drawing of the tubes is proportional to the volume of the tube. This allows for an estimation of tube length based on the draw time of the tube and a prediction of purge time based on the length of the tube. 5/8" tubing requires longer purge and draw times than 1/2" tubing.

- The approximate location of tube leakage can be estimated using the analysers oxygen reading and draw time. Accuracy of estimation of leakage points in the tube was within ±6 per cent relative distance of the actual leakage point.

- In the case of a minor leak, the entire tube length may still be estimated based on the oxygen readings and the draw time, however excessive tube purging may be required in this case. Under these circumstances the estimated purge time for a non-leaking tube will be sufficient to reveal the leaking tube and the estimated location of the leak.

- The integrated leak testing system will not function on blocked or kinked tubes. A kinked tube or major blockage can be detected by the system using the pressure reading during the purging phase. It was demonstrated during the testing that a pressure exceeding 15 kPa was indicative of a blockage or kinked tube.

- Testing demonstrated a 15.5 kPa pressure from nitrogen purging was sufficient to overcome the resistance in the tube provided by at least 1 L of water in a length of tube more than 1 m below horizontal at the sampling location. Self-draining seal panel water traps from two different suppliers were not an obstacle to the purging of water from the lines. It was determined during the testing that a pressure of 15 kPa was indicative of water accumulation in the tubing, or a physical tube blockage.

- From the scoping visits and testing, it was determined that the presence of tube infrastructure was not detrimental to the application of the leak detection prototype apart from non-return valves. Tube end of line filters, seal panel filters, flame arrestors, and water traps provided minimal extra resistance to the overall tube. The most significant factors for resistance of the tube were the length of the tubing and the accumulation of water.

- Overall, it was determined that the integrated leak detection system was an effective method for the monthly leak testing of tubes and provides substantial advantages for detecting the location of leaks when compared to traditional leak detection methods. This method of leak testing grants unique advantages for the determination of tube status during a mine emergency where the integrity of the tubes may be compromised and inaccessible.

- The design of a tube bundle conducive to the application of the integrated leak detection system, requires the absence of non-return valves and strategic installation of tubing and water traps to prevent accumulation of water condensation in the tubes.

REFERENCES

Brady, D, Watkinson, M and Harrison, P, 2015. The application of Tube Bundle Systems in the prevention of mine fires and explosions and post event response, prepared for the National Institute for Occupational Safety and Health (NIOSH), contract number 200-2013-56949, Simtars.

Cliff, D, Brady, D and Watkinson, M, 1999. *Spontaneous Combustion in Australian Coal Mines, The Green Book*, Simtars, University of Queensland, Mine Safety Institute of Australia.

Forrester, L, 2017. Tube bundle integrity testing methodologies, in *Proceedings Australian Mine Vent Conference 2017*, pp 129–136 (The Australasian Institute of Mining and Metallurgy: Melbourne).

Standards Australia, 2018. AS/NZS, 2290.3:2018 – Electrical equipment for coal mines — Introduction, inspection and maintenance, Part 3: Gas detecting and monitoring equipment.

Intelligent modelling of ventilation fan performance curves using machine learning

B J Viviers[1] and J C F Martins[2]

1. Computer Engineer, Air Blow Fans, Pretoria Gauteng 0133, South Africa.
   Email: bernard@airblowfans.co.za
2. Engineering Manager, Air Blow Fans, Pretoria Gauteng 0133, South Africa.
   Email: jose@airblowfans.co.za

ABSTRACT

Designing a new range of ventilation fans typically requires extensive CFD (Computational Fluid Dynamics) simulations to cover various operating points. Key parameters such as fan diameter, hub-to-tip ratio, number of blades, and blade angle need optimisation to achieve maximum efficiency. This study leverages existing CFD simulation data to predict fan performance for unseen configurations, thereby reducing the need for additional costly simulations. A two-stage machine learning approach is employed: the first model optimises fan parameters for maximum efficiency at specified duty points, while the second model predicts performance curves (volume flow versus pressure and volume flow versus efficiency). The data set includes 54 axial ventilation fans with consistent blade design and performance data for 21 blade angles. Parameters and performance metrics were normalised to improve model performance. Various regression models, including linear regression and random forest with polynomial features, were trained and evaluated using Mean Squared Error (MSE), Mean Absolute Error (MAE), and R-squared (R^2).

Results show that efficiency can be accurately predicted by both models with an approximate error of 0.3 per cent. Performance data sequences for volume flow, pressure and efficiency are predicted with high accuracy using the second model, including duty point information enhances this model's performance. R^2 values very close to 1 are achieved, indicating that the model explains nearly all the variance in the test data. Future work will explore advanced learning techniques, additional parameters, and real-world data integration to further refine predictions. This approach demonstrates the feasibility of using machine learning to optimise fan designs and predict performance, potentially reducing reliance on expensive CFD simulations.

INTRODUCTION

When designing a new range of ventilation fans, the basic design may be achieved using analytical techniques, however it is typically necessary to produce a large amount of CFD (Computational Fluid Dynamics) simulation data to cover various fan operating ranges or predict performance at given duty points. Key parameters that can be adjusted to generate fans optimised for different duty points include fan diameter, fan hub-to-tip ratio, number of blades, blade shape, aspect ratio, taper ratio, twist, camber and blade angle. While all the above parameters affect the fan efficiency this study concentrates on developing models to predict fan performance based on four parameters only, these being fan diameter, hub-to-tip ratio, blade count and blade angle. This allows the model to predict the performance of geometrically similar rotor and stator blades for variations of the modelled parameters mentioned above.

The simulation data includes performance metrics for multiple fan diameters, hub-to-tip ratios, number of blades and different blade angles. Traditionally, generating new performance data for different fan geometries requires expensive and time consuming CFD simulations to obtain accurate characteristic performance curves. The models developed for this study seeks to minimise the CFD simulations to determine an optimal fan design for a given duty point and to then produce the performance data for that design. Given a substantial amount of simulation data, it can be beneficial to leverage this data to predict performance curves for configurations not included in the original data set. Machine learning models can be trained on this data to predict fan performance for new, unseen fan configurations. Optimising fan efficiency often requires configurations not present in the initial simulation data. Traditional fan laws, which scale performance based on fan diameter (assuming geometric similarity), air density, and speed, do not account for changes in the hub-to-tip ratio and blade count, necessitating additional CFD simulations.

The model's performance hinges on the quality and accuracy of the provided simulation data. If the simulation data does not accurately reflect real-world performance, the model's predictions will be unreliable. Therefore, it is assumed that CFD simulation data used for training correlates well with the performance of actual production fans.

The primary goal is to determine whether existing simulation data can be used to predict new fan performance data. This involves two main tasks: optimising fan parameters for a specified duty point to maximise fan efficiency and predicting the fan curve for these optimised parameters (ie volume flow versus pressure and volume flow versus efficiency curves). Matching the predictions to real-world fan performance is beyond the scope of this paper at this stage but remains an interesting avenue for future research. This work also does not aim to verify the CFD simulation data; it assumes that the generated data accurately represents real-world fan performance.

DATA PREPARATION AND PROCESSING

A data set containing 54 primary axial ventilation fans with similar blade design was used. All fans in the data set operate at the same speed, with performance curves for 21 different blade angles. The data was organised into seven parameters: fan outer diameter, fan hub-to-tip ratio, number of blades, blade angle, volume flow, pressure, and efficiency. For each blade angle, there are 21 corresponding curve points for volume flow, pressure, and efficiency.

Machine learning models generally perform better when input and output data are scaled to a similar range, typically between 0 and 1. Consequently, all fan performance data (volume flow, pressure, and efficiency) was made dimensionless. Dimensionless parameters allow for the direct comparison of fans of different sizes and operating conditions, facilitating the evaluation and optimisation of various fan configurations on a consistent and universal basis(Zumsteeg and Karadzhi, 2019). Henceforth, dimensionless performance parameters are referred to as being normalised.

The following equations were used to convert the performance data into dimensionless units (Zumsteeg and Karadzhi, 2019):

$Efficiency(\eta)$

$$\eta = \frac{\Delta p_t \cdot Q}{P_w}$$

η: $\quad$ *Dimensionless efficiency of the fan*

Δp_t: *Total pressure difference(Pa)*

Q: $\quad$ *Airflow(m³/s)*

P_w: *Power on the shaft(W)*

$Pressure\ Coefficient(\psi)$

$$\psi = \frac{\Delta p_t}{\left(\frac{\rho}{2} \cdot u_2^2\right)}$$

ψ: $\quad$ *Dimensionless pressure coefficient*

Δp_t: *Total pressure difference(Pa)*

ρ: $\quad$ *Density of the medium(kg/m³)*

u_2: $\quad$ *Rotational Blade tip speed(m/s)*

$Airflow\ Coefficient(\phi)$

$$\phi = \frac{Q}{u_2 \cdot \left(\frac{\pi \cdot d_2^2}{4}\right)}$$

ϕ: *Dimensionless airflow coefficient*

Q: *Airflow* (m^3/s)

u_2:   Rotational *Blade tip speed* (m/s)

d_2:   *Impeller diameter* (m)

Other fan configuration parameters (fan outer diameter, number of blades, and blade angle) were also scaled to between 0 and 1. The selection and structuring of input and output data are important considerations in this process, as they have a large impact on the model's performance and success.

It is also important to consider the potential use of model combinations, where the output from one model can serve as the input for another. Input representation will strongly influence model performance, as the model needs to learn the interactions and relationships between input features and target variables effectively. Insufficient information in the input data will prevent the model from accurately learning these underlying relationships. Therefore, careful preprocessing and thoughtful data structuring are essential to achieving accurate and reliable model predictions (Burkov, 2020).

MODEL IMPLEMENTATION

The model implementation follows a two-stage approach, where each model can operate independently, but they are typically used sequentially in the process of fan selection. The workflow begins by selecting a fan configuration that provides the highest efficiency for a user-specified duty point (volume flow and pressure value). Subsequently, performance curves (volume flow versus pressure and volume flow versus efficiency) can be generated for this duty point using the optimised fan configuration to achieve maximum efficiency.

- **First Stage:** The first model predicts the maximum efficiency by optimising key fan parameters: fan outer diameter, fan hub-to-tip ratio, number of blades, and blade angle. For training it is assumed that optimal fan parameters will be selected by the optimisation algorithm.

- **Second Stage:** The second model takes the optimised fan configuration and optionally the duty point of maximum efficiency to generate detailed performance curves.

By following this approach, users can determine the most efficient fan configuration for their specific requirements and visualise the associated performance curves.

Optimal efficiency prediction based on fan configuration

To predict the maximum efficiency at a given duty point, the solution targets the points on each performance curve corresponding to maximum efficiency. The data is processed to include only these optimal efficiency points along with the associated fan configuration for each blade angle. Despite not utilising most points on the curve, the model accurately predicts efficiency using only the points of maximum efficiency.

Model inputs and outputs

Inputs: Fan diameter, fan hub-to-tip ratio, number of blades, blade angle, normalised volume flow, and pressure—each as single values corresponding to the points of maximum efficiency. All inputs are scalars.

Outputs: Maximum normalised efficiency (scalar).

Given the data's structure, the model naturally predicts an optimal efficiency value for the presented fan configuration as it is trained only on configurations that provide the maximum efficiency. Two models are explored:

- **Random Forest regressor:** An ensemble method known for handling complex relationships and interactions within the data.

- **Random Forest regressor with polynomial features:** This model enhances the Random Forest by incorporating polynomial features of the input variables, capturing more complex relationships.

Both models are evaluated using five-fold cross-validation, with metrics such as R-squared (R^2), Mean Absolute Error (MAE), and Mean Squared Error (MSE). Fan diameter and the number of blades are treated as categorical features, as these parameters typically fall into a discrete set of values.

The machine learning procedure includes loading and preparing the data. A randomly selected fan is excluded from the data set which cross validation is performed on. The remaining data set (53 fans) is split into different variations of training and test sets during five-fold cross validation. During training and testing, performance data is shuffled. A single fan is excluded to verify model accuracy and allow plotting of efficiency for all blade angles associated with this fan. Random Forest is a versatile and widely used ensemble learning method that constructs multiple decision trees during training and outputs the mean prediction of the individual trees (Burkov, 2019). It is known for its robustness and ability to handle many input variables without overfitting.

Fan performance curve prediction

This second model aims to predict fan performance curve points—normalised volume flow, normalised pressure, and normalised efficiency—based on given fan configuration inputs. The fan configuration parameters include fan outer diameter, fan hub-to-tip ratio, number of blades, and blade angle. Optionally, the model can also use as input the duty point (volume flow and pressure) and corresponding maximum efficiency value predicted by the first model.

Model inputs and outputs

Inputs: Fan outer diameter, fan hub-to-tip ratio, number of blades, blade angle.

Optionally: normalised volume flow and pressure and corresponding efficiency. All inputs are scalars.

Outputs: Three arrays of 21 values each representing performance points for normalised volume flow, normalised pressure, and normalised efficiency.

Initially, the data preprocessing involves aggregating sequences of the target variables and padding them to ensure uniform length. Feature engineering includes encoding of categorical variables and polynomial transformation to capture non-linear relationships. Polynomial features are generated by raising the input features to different powers and creating interaction terms, which allows the model to capture more complex, non-linear relationships between the input features and the target variables (Burkov, 2019). The inclusion of polynomial features allows the models to better capture the underlying patterns and interactions in the data, thus improving the predictive accuracy of the fan curve points.

The predictive modelling involves training and evaluating various regression models. The models tested include linear regression and random forest regressor with polynomial features of varying degrees (second and third degree), both with and without the inclusion of the duty point features. The data is split into training and testing sets, and models are trained separately for each target variable. Performance metrics—MSE, MAE, and R^2 scores—are computed to evaluate model performance. A separate model is trained for each performance parameter sequence (volume flow, pressure and efficiency).

This model can be used independently to predict performance curve data given a fan configuration. This is useful for exploring performance curves related to unknown fan configurations without explicitly stating the duty point and efficiency. However, providing the model with the duty point and corresponding efficiency improves the accuracy and performance of the curve prediction.

RESULTS

A two-stage modelling approach was used on a data set of 54 axial ventilation fans. The first model predicts the maximum efficiency given a fan configuration, and the second model generates performance curves and can optionally utilise the duty point as additional input. Various regression models, including linear regression and random forest with polynomial features, were trained and evaluated using MSE, MAE, and R^2 scores. The results for these models are presented below.

Evaluation metrics

To assess model performance, three key metrics were used (Rajawat, Mohammed and Nath Shaw, 2022):

- **Mean Squared Error (MSE):** Measures the average squared difference between predicted and actual values, with higher values indicating larger errors. MSE is sensitive to outliers.

- **Mean Absolute Error (MAE):** Measures the average absolute difference between predicted and actual values, offering a straightforward interpretation of prediction errors. MAE is less sensitive to outliers than MSE.

- **R-squared (R^2):** Indicates the proportion of variance in the dependent variable explained by the independent variables, ranging from 0 to 1. Higher values suggest better model performance.

These metrics provide a comprehensive view of model performance, highlighting error magnitude and explanatory power.

Optimal efficiency prediction results

Multiple regression models were trained to predict fan efficiency, including Linear Regression and Gradient Boosting. However, the Random Forest Regressor consistently outperformed these models. Both the standard Random Forest and the Random Forest with Polynomial Features were evaluated using five-fold cross-validation to ensure accurate performance results.

Table 1 presents the MAE, MSE and R^2 scores from cross-validation for both models. Both models exhibit high R^2 values, indicating strong predictive power and a good fit to the training data. The slight improvement in R^2 for the model with polynomial features suggests that including these features enhances the model's ability to explain variance in the data. MAE values are very low for both models, there is only an approximate error of 0.3 per cent when predicting the efficiency.

TABLE 1

Comparison of cross-validation scores.

Evaluation metric	Random Forest	Random Forest with polynomial features
MAE	0.003212	0.003139
MSE	0.000044	0.000041
R^2	0.919211	0.924981

Table 2 provides performance metrics for the excluded fan. The Random Forest with Polynomial Features model outperforms the standard Random Forest across all metrics. The MAE and MSE are lower for the polynomial model, indicating higher accuracy and smaller prediction errors. Specifically, the MAE shows that the standard Random Forest model makes an average error of 0.33 per cent in efficiency prediction, while the polynomial model has an average error of 0.13 per cent. Both errors are small, indicating good prediction accuracy. The higher R^2 value for the polynomial model suggests it explains nearly all the variance in the test data, indicating superior predictive capability.

TABLE 2

Model result comparison (for excluded fan).

Evaluation metric	Random Forest	Random Forest with polynomial features
MAE test	0.003304	0.001325
MSE test	0.000016	0.000003
R^2 test	0.938602	0.989528

Figures 1 and 2 show the actual and predicted efficiency values at different blade angles for the two models on the excluded fan. Visual inspection confirms that the actual and predicted efficiency values are very close. The slight accuracy improvement from the polynomial features model (Figure 2) over the standard model (Figure 1) are not very noticeable in these visualisations.

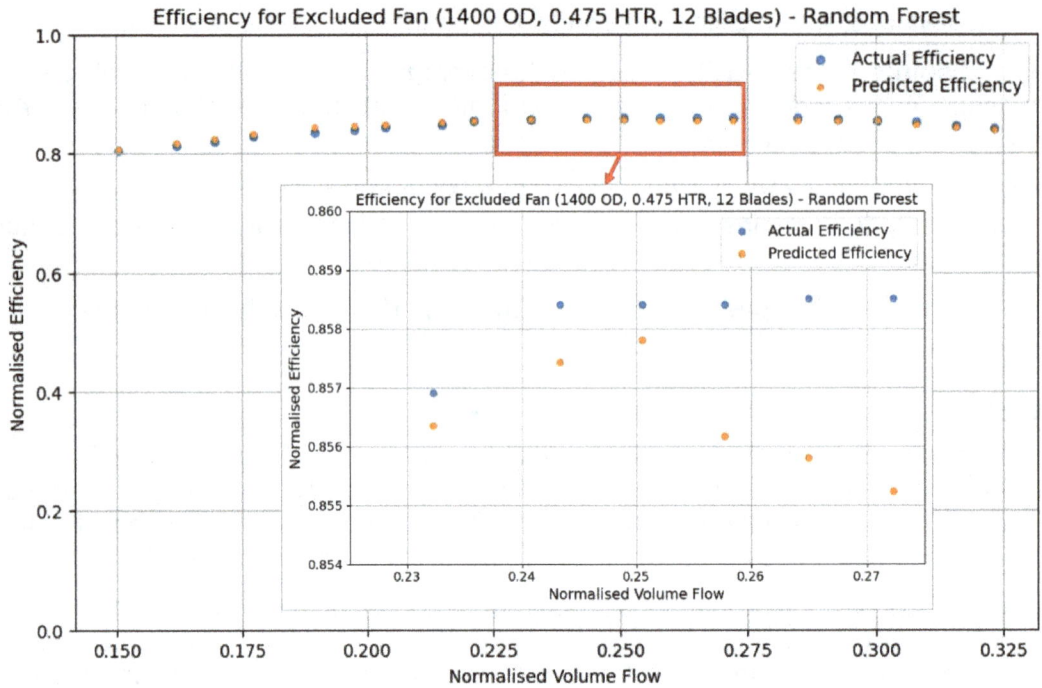

FIG 1 – Efficiency values for Random Forest model.

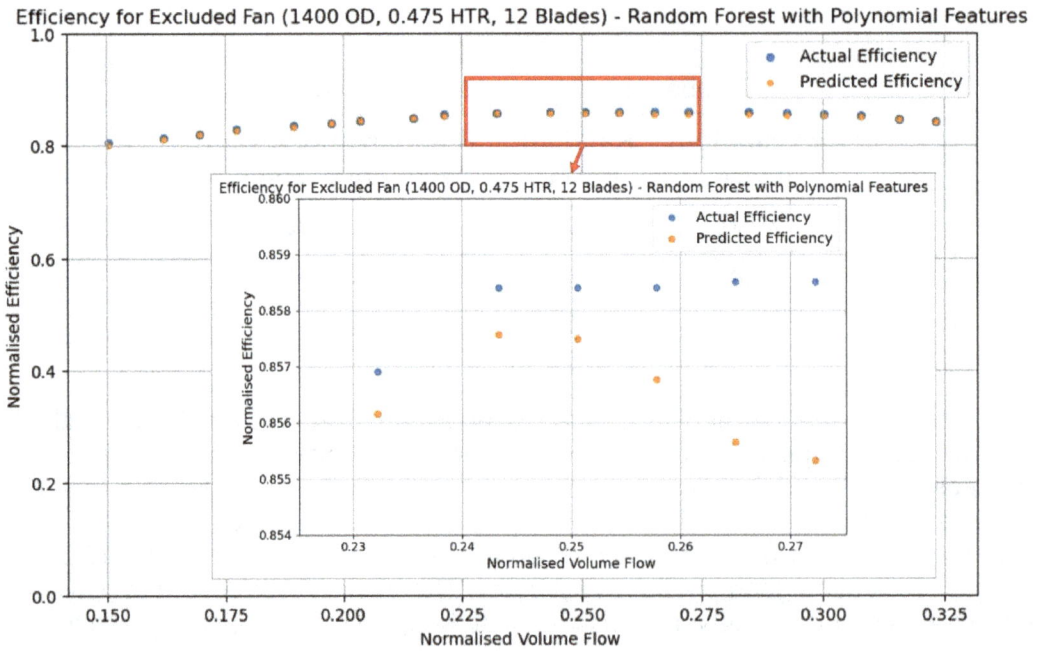

FIG 2 – Efficiency values for Random Forest with polynomial features model.

Performance curve prediction results

The performance curve prediction models predict sequences for volume flow, pressure, and efficiency. Each sequence prediction is evaluated separately. Six model variations were tested, incorporating different polynomial degrees (second and third) and using either linear regression or random forest as regressors. Models were also tested with and without duty point information in the inputs. The data set was split into training and testing sets using an 80 per cent/20 per cent ratio, and results were derived from the test set evaluation.

Figure 3 presents evaluation metrics for the different models on each performance curve sequence. The models struggle more with efficiency sequence prediction, as indicated by higher MAE and lower R^2 scores. Including duty point information generally improves model performance by a noticeable margin (see blue versus orange bars and green versus purple bars). Even without duty point information, models perform well, with low MAE and MSE across the board, particularly the third-degree Polynomial with Random Forest (red bar). The third-degree polynomial with duty point and linear regression model (purple bar) is the best performing overall, showing very low MSE and MAE values and R^2 values close to 1.

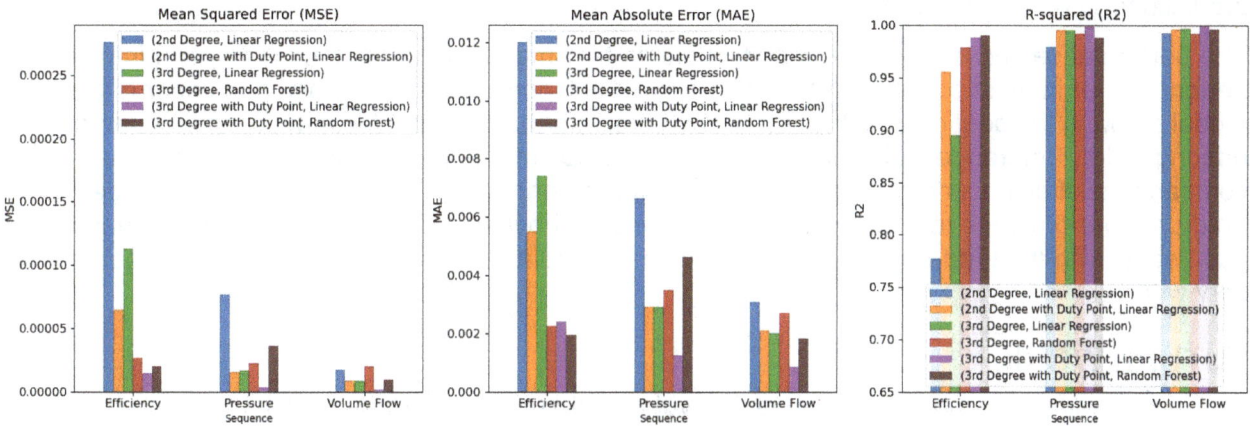

FIG 3 – Performance curve prediction evaluation metrics.

Figure 4 illustrates prediction performance for two randomly selected fan configurations from the test set. For simplicity, only four models are shown. The third-degree polynomial with duty point and linear regression model (green line) follows the curves of the first fan the closest, matching the evaluation metrics results. There is greater prediction divergence from the actual values on the first fan for all models. All models closely follow the actual curves for the second fan, demonstrating good accuracy.

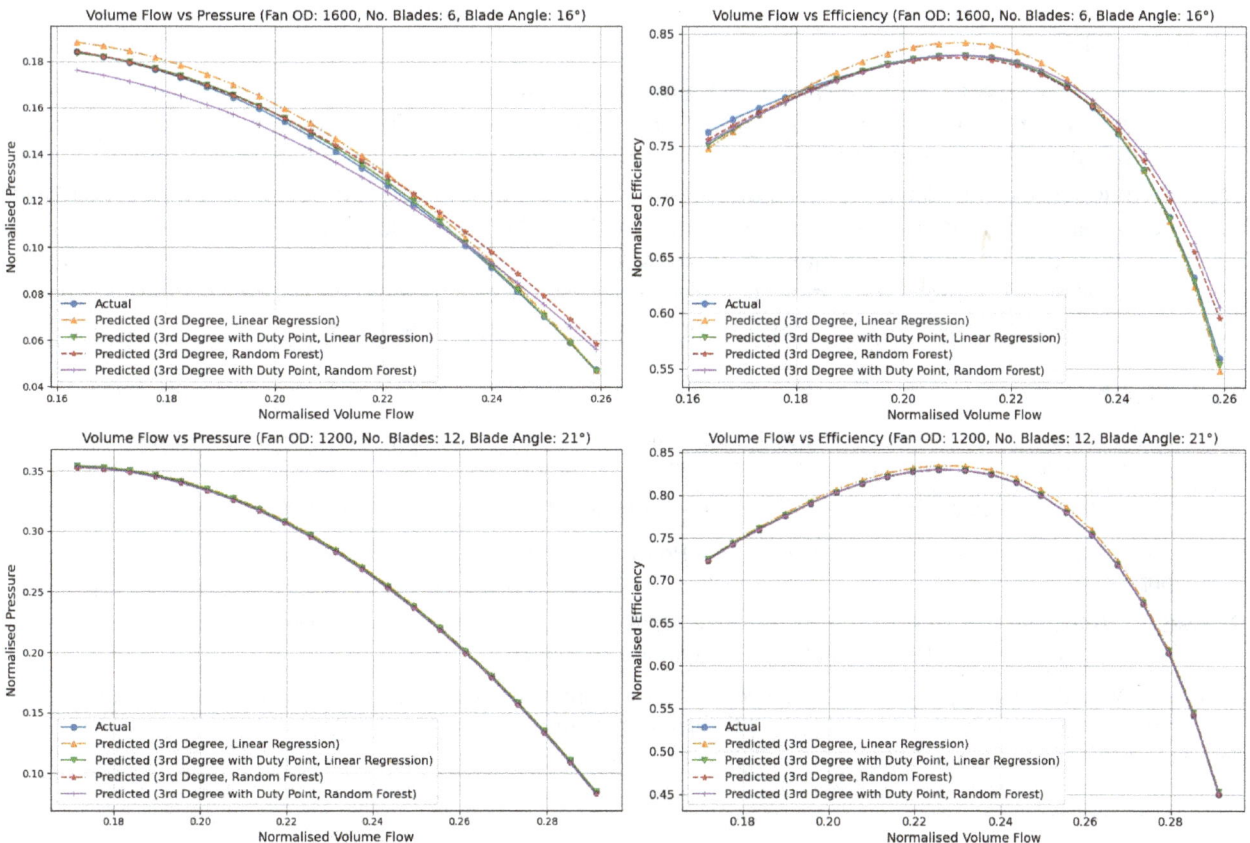

FIG 4 – Performance curve prediction examples.

DISCUSSION OF RESULTS

The evaluation of regression models for predicting fan efficiency highlights the effectiveness of the Random Forest Regressor, especially when enhanced with polynomial features. The high R^2 values observed during cross-validation indicate strong predictive capabilities for both the standard Random Forest and the Random Forest with Polynomial Features models. The slight improvement with polynomial features suggests better capture of underlying data relationships. When evaluating the models on the excluded fan, the Random Forest with Polynomial Features model outperforms the standard Random Forest. The lower MAE and MSE for the polynomial model indicate more precise predictions with smaller errors.

The performance curve prediction results indicate that models incorporating duty point information perform better in most cases, with the third-degree polynomial with duty point and linear regression model showing the best results. Including duty point information provides a noticeable advantage by offering a key reference point, improving input-output mapping and prediction accuracy. Visual comparisons of actual and predicted efficiency values confirm the high accuracy of the models. The prediction performance for different fan configurations shows that all models follow the actual curves accurately, with the third-degree polynomial with duty point and linear regression model performing very well.

FUTURE WORK

Advanced learning techniques such as Convolutional Neural Networks (CNNs), Recurrent Neural Networks (RNNs), and Long Short-Term Memory networks (LSTMs) can be explored to better capture complex feature interactions and sequential data patterns. Additionally, hyperparameter tuning using methods like grid search or random search can optimise the current models for improved performance. Expanding the model to include additional blade design parameters such as platform, aspect ratio, twist, camber, aerofoil shape etc, and incorporating real-world fan test data will further enhance accuracy (Angelini, Corsini and Delibra, 2019). Prioritising real-world data in model training will ensure predictions closely reflect actual performance. These steps will make the models more robust and applicable to a broader range of fan configurations and performance conditions.

CONCLUSION

The study demonstrates that existing CFD simulation data can be effectively leveraged to predict fan performance parameters and curves for new configurations. The Random Forest Regressor, particularly when enhanced with polynomial features and duty point information, shows strong predictive accuracy and reliability. This approach allows for optimising fan designs based on existing data, potentially reducing the need for additional costly and time consuming CFD simulations. Future work should focus on validating these models with real-world data to further confirm their efficacy and applicability in practical scenarios. Furthermore, the models should be expanded to account for fan design parameters which affect performance but are not yet modelled.

REFERENCES

Angelini, G, Corsini, A and Delibra, G, 2019. Exploration of Axial Fan Design Space with Data-Driven Approach, *Internation Journal of Turbomachinery Propulsion and Power (MDPI)*.

Burkov, A, 2019. *The Hundred-Page Machine Learning Book.* Available from: <https://themlbook.com/wiki/doku.php>

Burkov, A, 2020. *Machine Learning Engineering,* True Positive Inc. Available from: <https://www.mlebook.com/wiki/doku.php>

Rajawat, A, Mohammed, O and Nath Shaw, R, 2022. Renewable energy system for industrial internet of things model using fusion-AI, *Applications of AI and IOT in Renewable Energy.*

Zumsteeg, M and Karadzhi, S, 2019. *Systemair Technical Handbook: Ventilation,* Systemair GmbH. Available from: <https://www.systemair.com/en/contact-and-support/good-to-know/technical-handbook>

Ventilation planning for coal and metalliferous mines (case studies)

Longwall tailgate proactive sponcom and gas management strategy in hot underground mines – Australian safety share

B Belle[1], R Balusu[2] and K Tanguturi[2]

1. University of New South Wales, Sydney NSW 2052; University of Queensland, Brisbane Qld 4000; University of Pretoria, South Africa; 61Drawings, Australia. Email: bb@61drawings.com
2. CSIRO Mineral Resources, Pullenvale Qld 4069.

ABSTRACT

The reference to the subject of spontaneous combustion (sponcom) was made in the 17th century by Plot (1866) and later in various UK coal mining references (Lord, 1986). The evolution of major coal oxidation and resulting sponcom incidents are sudden, unlikely to predict the time of the event and may result in catastrophic negative safety outcome or result in the withdrawal of persons and closure of longwall panels/mines. Historically, gassy longwall workings in Australian Goonyella Middle (GM) seam (late-1990s to mid-2010s), experienced increasing trend in carbon monoxide (CO) levels associated with different stages of coal oxidation indicator gases and major safety incidents were due to oxygen (O_2) ingress on the longwall maingate side. The original Australian active longwall goaf gas drainage system designs are based on the past work of the CSIRO, verified by the operational experiences. The CSIRO based studies in gassy and hot coalmines had carried out numerical and field data investigations on goaf hole gas flow mechanisms and proactive inertisation strategies for preventative spontaneous combustion (sponcom) and goaf gas management. This critical foundational knowledge work contributed to the original goaf gas drainage and sponcom management strategies in other Australian longwall mines and potentially extended to rest of the world. Considering the risks associated with sponcom, GM seam operations were the first operations in Australia to introduce proactive N_2 injection along the MG in mid-to-late-2000s, to manage sponcom linked fire and explosion risks in an active goaf.

Over two decades ago, active goaf gas drainage flow rates were moderate (2000 L/s to 3000 L/s) and the O_2 ingress on TG side was not a major concern. With increasing goaf gas drainages rates up to 10 000 L/s and manual or automated mode operation of goaf wells to extremes to address higher goaf gas emissions, TG O_2 ingress and air wash zones deeper in the goaf has become a major issue recent years, necessitating the introduction of both MG and TG inertisation strategies now to address this emerging issue. Introduction of MG proactive inertisation strategy had reduced the number of high CO or intensive oxidation incidents over two decades. This paper provides historical context of sponcom management using N_2 gas and use of oxidation indicator gas CO and its trend in delineating various levels of heating. The practical benefits of longwall MG and TG inertisation using proactive N_2 injection supported by the original computational fluid dynamics (CFD) modelling studies in collaboration with the industry, during various phases of longwall production and stoppages as part of essential gas and sponcom management strategy for worker's safety is shared.

INTRODUCTION

The original and modified longwall seam gas drainage system design evolutions in the Australian coalmines are essentially based on the joint industry led historic initiatives and work of the CSIRO, supported by the GM seam operations (Balusu *et al*, 2001, 2002, 2004, 2011; Balusu, Ren and Humphries, 2005; Balusu, Belle and Tanguturi, 2017, 2019). The historic review of goaf drainage introduction in Australia and the rest of the world, drainage design and evolution of operational practices are summarised elsewhere (Belle, 2015, 2017). In addition, the impact of longwall width and TG hole positioning study in GM seam operations by the CSIRO for the GM seam operations is summarised elsewhere (Khanal *et al*, 2021). The science-based CSIRO studies in gassy, sponcom prone hot coalmines had been carried out through numerical and field data investigations on goaf hole gas flow mechanisms and proactive inertisation strategies for preventative spontaneous combustion management. This critical operational knowledge contributed significantly to the original Australian underground longwall mine goaf gas drainage and sponcom management strategies and potentially extended to rest of the coal mining world. The contradictory nature of the GM seam sponcom led fire and highly gassy mines necessitated the maximised goaf drainage capacity systems and the need to reduce oxygen ingress into the MG and TG active goaf required careful operational strategies. The term 'sponcom' is used in this paper to discuss the various stages of coal oxidation to the development of fully uncontrollable combustion when large quantities of coal left in the goaf due to geotechnical and mining safety considerations.

Similarly, 'air wash zone' a term commonly used is typically referred to as the concentration of relative O_2 in the goaf atmosphere aiding the leftover coal oxidation, and typically referred to O_2 concentration values of 5 per cent to 21 per cent. Fresh air is the concentration that is representative of longwall fresh air intake. It is to be noted that O_2 at high concentrations can be present even in situations where there is no or sluggish airflow movement in the goaf. Considering the risks associated with sponcom, GM seam mines were the first operations in Australia to introduce proactive N_2 injection to manage sponcom related fire and explosion risks in active or sealed goaf areas. In this context, it is important to note that frequent recommendations and inappropriate use of ventilation driven controls for longwall tailgate gas management that are practiced elsewhere in Australia or in the world, are not necessarily apt for gassy, steep geothermal gradient and known sponcom prone seam coalmines.

Sponcom risk facts

Over the last ten years, Australian coal industry in both NSW and Qld have suffered safety consequences of coal heating related sponcom events. While the root cause analyses or learnings from the events are not readily available or events purporting to be not attributed to the risk of sponcom in coalmines, following coal mining safety risk facts globally provide the reader to make an independent assessment of sponcom risk perceptions and resulting fires and explosion led safety consequences. The statistics on the incidents of heating and methane ignitions are fraught with difficulty, unless its legislated and readily available during risk assessment. Sometimes lack of clarity compounds the classification, whereby the operator may report only when a heating event is progressed to a stage where its visible to workers or stopping the production. The following facts provide historic evidence in relation to the sponcom risk in global coal mining.

- In German hard coal mining, approximately one or two cases of sponcom are recorded per 10 million tonnes (Mt) of saleable output. In approximately every 20th sponcom event, methane is ignited (Hermulheim and Beck, 1997).

- In Polish collieries, sponcom fires (or called as 'concealed' fires) caused 705 of the underground fires (Wactawik, Branny and Cygankiewicz, 1997) with 675 of the fires occurred below the mining depth of 500 m with 56 per cent sponcom fires took place in low sponcom propensity coalmines against 44 per cent fires in high propensity coalmines. Typical mining retreat rates from these mines ranged from 65 to 130 m/month with airflow of 15 m³/s across the longwall face with VRT of 25°C. In comparison, Qld operations currently operate with VRT of 40°C and expected to operate at 50°C at 500 m depth (Belle and Biffi, 2018).

- In Poland, between 1945 to 1970s, the number of fires in hard coalmines reached the value of several thousands of cases per annum. About 80 per cent of coalmine fires were attributed to sponcom. In addition, a total 54 fire incidents have occurred in Polish hard coalmines between 1997 and 2006 (Wachowicz, 2008).

- Ren and Edwards (1995) noted that despite the 'low propensity' of UK coal for sponcom, 255 of the regulatory notifications had contributed to 10 per cent of the UK underground fires during the period of 1987 to 1990.

- In South Africa, during 1980s, an average of more than five incidents per annum occurred in underground thermal coalmines (Wade, Phillips and Gouws, 1987). Despite the very low gassy coalmines, sponcom was found to be the major cause of fires, being responsible for more than 30 per cent of the 254 fires reported in the 1980s, ie an average of 4.3 incidents per annum (Gouws and Knoetze, 1995). Similarly, study suggested that between 1970s and 1990s, sponcom was responsible for 34 per cent of the underground fires (Gouws and Knoetze, 1995).

- Several studies in the USA, have highlighted the seriousness of sponcom led fires. Of the 16 reported fires attributed to sponcom between 1978 and 1986, 14 occurred in gob areas (Stephan, 1986). In the USA, approximately 15 per cent of the 164 total reported fires in underground coalmines for the period between 1978 to 1990 were caused by sponcom (Yuan and Smith, 2009). For the period 1990–2006, 25 reported fires in underground coalmines were caused by sponcom (DeRosa, 2004).

- In French coalmines, a survey of more than 100 reported cases of sponcom between 1960 and 1972 showed that 64 per cent occurred in gob areas (Jeger and Froger, 1975).

- In Australia, since1972, sponcom has caused three underground mine explosions with a total loss of 41 lives in Queensland and several extended and permanent pit closures in New South Wales (Ham, 2019), culminating in the closure of Southland Colliery in December 2003. Since 2010, there were known sponcom events in Australia, that had resulted in injuries and loss of assets both in Qld and NSW, although no statistical analysis of sponcom fires or incidents that are readily available.

- Indian coalmines had a historical record of over 70 per cent fire activity due to sponcom over 140 years (Mohalik et al, 2016).

- In Turkey, there were six mine fires in Karadon Colliery due to spontaneous combustion between 1990 and 2000 (Cakir and Barris, 2009).

Sponcom detection, monitoring and gas data indicators

The reference to the subject of sponcom was made in the 17th century by Plot (1866) and later in various UK coal mining references (Lord, 1986). The quest for early detection and preventative controls for sponcom risk has been addressed over a century ago. It is learned that whenever coal is exposed to O_2, absorption occurs and, even at normal temperatures, the coal and O_2 combine resulting in the production of heat or increased temperatures. The temperature will then rise at a rate dependent upon how quickly the heat is dissipated. Any rise in temperature will then cause the rate of oxidation to increase; this in turn produces more heat and the process is self accelerating and eventually to spontaneous combustion or ignition.

Historically, two methods of detecting underground heatings was documented and appropriate herein for the reader to reflect on their current usage (Storrow and Graham, 1925), namely, (a) that of the underground coal miner, by the utilisation of various senses, that of smell is the one which probably gives the miner most help and the application of powers of observation in noticing the slightest signs of abnormality when going his rounds of inspections; and (b) by the careful interpretation of accurate gas analysis. It was recognised since the early days that neither method is infallible, and in many cases believed that 'firemen' may discover a heating before it can be found by chemical analysis; in other cases, chemical analyses of gas samples indicated abnormality before a trained observer can detect trouble. Storrow and Graham (1925), opined about the advantages of thorough collaboration between observant workers and the trends of composition of gas samples analysed.

Sponcom heating signals of firemen

Storrow and Graham (1925) recorded from the field observations and exchanges with the UK Firemen, who have had great experience in the detection and aftertreatment of heatings, and readily discern any unusual change in the underground condition for heating. Storrow and Graham (1925) had noted that these heating signals varied in different pits through observations and gas composition data. Various sponcom related training materials in Australia refers to coalminer's observations and human responses to changing atmosphere using various terminologies such as 'sweating, smell, odour, hearing etc'. Storrow and Graham (1925) noted various stages of heating or signs of firemen in UK coalmines with the first sign of anything abnormal is frequently termed as 'sweating' a deposition of moisture, with rise in humidity of the air, and sometimes rise of air temperature, as a consequence of the warm gases leaving the heated material being saturated with moisture and depositing the latter on the cooler sides of the roadway. Sweating is the terminology used to describe the condensation effect of hot air coming into contact with the cold surface with the formation of water droplets. In the Australian coalmines with steep thermal gradient with extensive mine cooling, where the hot goaf stream of 35°C wet bulb temperature (WBT) meets the longwall ventilation air at 25°C WBT, the 'sweating' may possibly prone to misinterpretation. With further increase in heating, it was noted that a peculiar musty odour (due to small traces of volatile oxidation products) and by a petrol-like smell. Graham (1921) noted that with experiments on the oxidation of coal indicate that no smell would be produced at an average oxidation temperature of 70°C, unless oxidation were taking place on a large scale, or the ventilation were poor (normal values were 6 to 9 m^3/s), the air of a main return is hardly likely to be affected. The composition of oxidised air coming from such a heating shows a considerable proportion of oxides of carbon combined with 'combustible gases' such as methane, ethane, and the vapors of higher hydrocarbons, eg pentane (C_5H_{12}) and hexane (C_6H_{14}) (Storrow and Graham, 1925). If the heating develops into a fire, a strong tarry odour is usually evident (and hydrogen is as a rule found in the gases which are being drawn from the seat of the trouble). Storrow and Graham (1925) cautioned that *these stages of heating is not universal in occurrence*, whereby some pits little sweating is observed, and, again, on perhaps rare occasions, the paraffin or petrol-like smell may be absent. Various indications (evidence) upon which UK firemen relied for information concerning abnormal occurrences in the old workings or goaf is summarised in Table 1.

TABLE 1

Signs of firemen (Storrow and Graham, 1925).

Goaf leakage	Slight heating	Advanced heating	Heating to fire
Stronger smell than normal	Fusty, peculiar, sweet smell with slight suspicion of oil	Distinct oily smell and characteristic	Firestink smell beyond oily stink stage
Hearing	Sweating more pronounced and saturated	Temperature rising depends on the distance from the seat of fire and may be constant	Considerably higher temperature
Sight by dust test	Temperature rise in thermometers		A strong haze prior to seeing fire
Sweating may be seen			After fire is out distinct 'dead smell'

In the USA, there were examples of detection of heating by inspections and thermocouple sensors in the goaf or gob (Koenning and Boulton, 1997). Signs included the appearance of a haze, sweating of the strata, a characteristic smell, and smoke with specially trained personnel. Thermocouple sensors were found to be ineffective, let alone if they could be approved for use in Australian coalmines. The appearance of H_2 in gas samples is treated with urgency because of the heating reaching advanced stage, although in Australia, H_2 as a gas indicator is not given importance in workers safety training, other than it being seen to be compromised, although this is most unfortunate. Two sponcom events at Cyprus mine longwall (USA) retreated to 600 m and 2100 m respectively from start-up face, noted that the O_2 content at the location had little or no O_2 depletion (ie 20 per cent) for a prolonged period. Lessons learned from the event was to have a primary goal of the prevention and detection efforts is to maintain an O_2 deficient atmosphere in the goaf as a proactive and reactive approach. The likelihood of a heating developing in a severely O_2 deficient atmosphere is remote (Koenning and Boulton, 1997), although specific O_2 value was not discussed.

Heating and sponcom indicator gases

It was recognised that the constituents of coal oxidising atmosphere is by the evolution of gases from the coal-seam or neighbouring strata or as a result of various oxidation processes as a result of the absorption of oxygen by coal (Graham, 1915). Graham (1921) showed that a convenient method of expressing the impairing air quality (termed as *vitiation*) is to represent the production of carbon dioxide (CO_2) and carbon monoxide (CO) as a percentage of the oxygen (O_2) which has disappeared. When expressed in this way, Graham (1921) suggested the rise of temperature on the production of the oxides of carbon relative to the amount of O_2 absorbed. Historically, the potential sources of return air of workings are not only confined to the oxidation of coal or other carbonaceous material but can also attributed to the breathing of men and animals, the decay of timber, the oxidation of pyrites and other inorganic material or the gas emissions from the seams. In the majority of cases in UK, the amount of CO_2 and CO in the gases given off from the coal seam or neighbouring strata has been found to be negligible.

The laboratory studies of coal oxidation by Graham (1921) showed that the higher the temperature, the greater is the production of *carbon dioxide (CO₂) and carbon monoxide (CO), with certain amount of water-vapour, a fact that was previously showed by Dr. Haldane using powdered coal.* In addition to the proportion being affected, the actual production of these two gases is considerably more, since rise of temperature produces a very considerable increase in the rate of oxidation (Graham, 1915). Graham (1921) concluded that the higher CO found in some pits usually coexist with old working of considerable coal left in the goaf. *Graham (1921) clearly noted that the estimation of small quantities of CO in the mine air will prove of considerable help in the detection and investigation of underground heatings.* During those early years and today, despite the plethora of development of complex gas composition ratios, continued development of new gas ratios and recent plots linking to CO or CO_2 gases as helpful indicators, increased production of CO continue to afford a reliable '*fingerprint*' indication of heating. These generated CO and CO_2 gases as a result of absorbed oxygen to coal devised by Graham (1921) are commonly termed as 'Graham's Ratios', not referred to by the originator, Graham (1922), but these indices are primarily linked to CO and CO_2 levels.

Another term, 'oxygen deficiency' that is likely to be misunderstood or misinterpreted, was not used by Graham (1922) for the current longwall conditions. Unlike, the historic coal mining methods, for the current Australian longwall mining conditions, it is difficult to estimate its magnitude in the old workings or active goaf areas, for the operating longwall ventilation flow dynamics, and goaf gas drainage systems. Traditionally, in the USA, the OSHA's 29 CFR 1910.146 confined space standard defines 'oxygen deficient' atmosphere as any atmosphere containing less than 19.5 per cent O_2 by volume (OSHA, 2024). In most to all historic papers, early detection of coal heating and its intensity were associated with CO and CO_2 trends of the gas samples collected and analysed. For comparison purposes, in the historic UK collieries, where Graham (1921) made the sponcom heating observations, ventilation flow rate per person provided in Australia today exceeds 30 times more than those UK collieries and observations made during the 1920s, implying the importance of monitoring of longwall goaf and continued data analyses CO and CO_2 levels.

Graham (1921) had noted that it is difficult from UK coal mining experiences, to give *definite* values for the CO and CO_2 production levels as a percentage to the O_2 absorbed, above which one may suggest coal heating or fire is present; rather conditions governing *each case* must be considered. From the survey of UK coalmines, Graham (1921) noted that in South Yorkshire pits, a value of 0.5 per cent, in the CO production (as a percentage of O_2 absorbed) demands attention, whilst a value of 1 per cent was viewed with grave suspicion along with CO_2 production. Graham (1921) summarised results of the composition of return air in various UK pits and pointed out that CO produced as a percentage of O_2 absorbed may be found normally in return air in quantity varying from 0.3 to 1.5 per cent. *Very importantly, that there were references by Graham (1921) whereby slight heatings were observed where the CO production as a percentage to O_2 absorbed was at 0.2 and 0.3.* Interestingly, it was noted by Penman (Graham, 1921) that the proposed values of CO and CO_2 generated as a percentage of the absorbed oxygen values by Graham (1921) may not be universally applicable and to Indian Coalmines that had different mining method with thick seams and the galleries high and wide, and one can get quite close to the seat of the fire, an observation shared by this paper co-author, Balusu (2020). Therefore, continued suggestions in Australia of values between 0.0 to 0.4, a ratio of CO production as a percentage of O_2 absorbed, *as normal* is misleading and lacks verifiable technical references. Recent coal mining events over the last few years, are suggesting to the industry that the current 'normal conditions below 0.4' must be revised. In addition, for each coalmine continued analyses of CO and CO_2 trends will be able to aid in the detection of early coal heating.

Historically, the CO and CO_2 generation values as a result of O_2 absorbed by the coal differed from mine to mine depending on the temperature and length of exposure, with differing level of coal oxidation to fires. What is certainly established is that the increase in CO and CO_2 generation suggests that coal oxidation intensity increases. It was observed that during a fire one usually gets a much greater proportion of CO_2 and CO, especially the former. Actual *fire* implies that a fair supply of air is getting to the highly heated combustible material, and under these conditions a certain amount of the CO may be burnt, but the CO_2 production will always be high. Based on the historical observations in UK, it is noted (Graham, 1921) that the efficient prevention of serious heating depends upon the detection and rapid sealing of leakages and slight heatings, which very seldom give definite early indications. What is a definitive and effective known strategy is ensuring reduced oxygen levels through minimised air leakage or induced air into the goaf. Slight heating will cause the O_2 to be absorbed more rapidly and as the heating increases CO_2 is formed more rapidly in proportion to the amount of oxygen absorbed.

One of the useful considerations from the above historic knowledge share for a ventilation engineer is the less than adequate confidence in the magnitude of O_2 absorbed by the coal and its role in the oxidation of an active dynamic goaf. Since there are serious shortcomings in the understanding of the *air wash zone*, airflow dynamics in the goaf, every responsible person, statutory or enforcement official must use gas trends, in particular CO and CO_2 gas levels that has been the ally in determining the level of coal oxidation. In the absence of experienced workers with underground exposure to document various physical and visual signals of coal oxidation, the reliance of gas monitoring systems and analyses of *gas trends* to evaluate the status of current sponcom management practices. In the current context of longwall mining in Australia, sponcom indicators and outsourced trigger value determinations may become a 'detriment' or serious 'hurdles' without critical thinking of gas trends. For a successful sponcom monitoring, there should be rigorous analyses of the gas composition data trends on a continuous basis by the ventilation engineer, supported by the mine worker, mine manager with added assistance, to identify the 'seat of the sponcom trouble.'

History of N_2 inertisation and longwall goaf monitoring

Inertisation in this paper is referred to as a preventative proactive safety control method to avoid the potential coal heating by creating an atmosphere in the longwall goaf area is such that the environment cannot sustain coal heating, including potential ignitions leading to explosions of methane, can extinguish the combustion process (fire), and is therefore 'inert' and devoid of O_2 rich atmosphere for heating or ignitions. The O_2 concentration can be lowered to levels below that of normal air through consumption by slow or fast coal oxidation processes or by the addition of inert gas such as N_2 which do not participate in the oxidation or combustion processes (Mucho *et al*, 2005). The specific objective of inert gas injection operations is to reduce the goaf O_2 levels to 5 per cent to 8 per cent (with a factor of safety of 1.5 on the explosive nose limit of 12 per cent) before methane concentration reaches the lower explosive limit of 5 per cent or potential for coal heating at O_2 levels > 5 per cent.

While it may seem to be surprising to some at the time of a major heating, or fire and explosion event, use of an inert gas to combat the fire safety event in Belgium was recorded, when gases from a coke oven were actually ducted into the downcast ventilation shaft (Plot, 1886). While the CO_2 were attempted, due to its relatively high freezing point is a considerable disadvantage along with decomposition when contact with heat. Other observed disadvantages from flue gases were that they either contain CO or formation of CO on contact with the heating source. Furthermore, it is noted that the monitoring of heating activity very much uncertain. On the other hand, N_2 has the advantage of being inert when exposed to heat and nontoxic. It has low boiling point and does not contain CO and has very low O_2 content and cost being the disadvantage. Despite the recent misgivings in Australia due to its interference in calculating certain ratios, globally, use of N_2 in coal mining countries such as UK, Germany, France, Russia, India, Bulgaria, Poland, Czech Republic has been extensive (Adamus, 2002). Following facts provide the extent of its use for sponcom management and its appropriateness for Australian Coalmines.

- Nitrogen was first used in underground coalmines in UK in Fernhill Colliery in South Wales in 1962, as a result of shot firing heating event (Vaughan-Thomas, 1964). The recovery operation had demonstrated that premature cutting off the N_2 to the sealed area with 1.8 per cent O_2 resulted in increase of O_2 levels to 7.5 per cent, the CO levels to 300 ppm and H_2 reappearing at a concentration of 30 ppm resulting in unsafe working conditions for rescue teams. The experience at Fernhill Colliery suggested that '…consideration should be given to some form of N_2 generator which could be set-up on the site of a colliery…'; Experiences of Fernhill

Colliery by Vaughan-Thomas further suggested that maintaining O_2 levels below 2 per cent for a period would result in fire being extinguished.

- In the 1980s, at Daw Mill (UK) longwall goaf heating was managed for the first time by N_2 inertisation after unsuccessful attempts of managing the event for many weeks of combat the coal heating (Harris, 1981). Harris (1981) suggested that presence of N_2 in the goaf atmosphere results in the increased safety factor for likely ignitions, heating and resulting explosions.

- In UK, one of the most pertinent and valuable observations (Harris, 1981) made at the time, that is being recently ignored in Australia in pursuit of complicated unhelpful ratios, that the ratio of CO generated to O_2 absorbed (Graham, 1921) cannot be calculated and unhelpful. Harris (1981) unambiguously noted that the sponcom propensity or coal heating and control effectiveness has to be exercised by monitoring CO, CO_2, O_2, CH_4, and N_2 injected. Under the high CH_4 and CO_2 emitting mines, the heating indicator gas ratios of CO and CO_2 generated to O_2 absorbed would not be appropriate for assessment purposes. However, trends of sponcom heating indicator gases would benefit the ventilation engineer or risk assessor.

- The flow N_2 inertisation in UK mines were ranging from 166 L/s to 666 L/s, with longwall airflows were up to 10 m^3/s, resulting in reduced oxidation levels with CO levels coming down from 350 ppm CO to 15 ppm CO levels (Thomas, 1986).

- The UK longwall experiences of 1960s and 1980s have suggested that N_2 injection is an effective method of achieving a temporary, and virtually total, reduction in goaf heating activity. UK experiences have suggested that the injected N_2 replaces the resident methane, as well as O_2, resulting in safer atmosphere in the vicinity of heating (Thomas, 1986).

- Polish reference (Kukuczka, 1982; Thomas, 1986) discussions had noted the use of low temperature liquid N_2 in the goaf heating environment to remove the localised heat.

- Use of N_2 in Polish coalmines and as part of the Polish Central Mines Rescue Station has been recorded as early as 1980s and deployed to suppress sponcom coalmine fires in sealed areas resulting in the concentration of O_2 at 3 per cent (Bradecki, Matuszewski and Nowak, 1987; Kajdasz, Golstein and Buchwald, 1989; Kajdasz and Stefanowicz, 2002).

- In Bulgarian mines, the use of N_2 due to its solubility in water 55 times less than of CO_2 was attempted through foam slurry using fly ash from electricity generating powerplants (Michaylov and Vlasseva, 1995; Michaylov, 1996).

- In an another Bugarian lignite mine with a high propensity to sponcom at depths of 380 m to 410 m below surface at the Babino Colliery had used N_2 at 830 L/s (Adamus, 2002).

- In the Czech republic, N_2 was first used for gob sponcom fire at the Doubrava mine in the Upper Silesian Coalfield in 1949 (Adamus, Hajek and Posta, 1995) using N_2 central pipeline. The Czech coal mining studies had noted that at 7 per cent to 10 per cent O_2, coal are susceptibility for heating. The Czech safety regulations recommend N_2 flow rates of 166 L/s to 250 L/s for retreating longwall faces with longwall ventilation flow rates of 10 m^3/s for a face width of 100 m (Adamus and Vicek, 1997).

- German coalmines had a CO alarm threshold limit of 10 L/min, with gas emissions of 600 L/sec producing 3 Mt/a at 900 m depth, that were sometimes referred to in the Australin coalmines, that had disproportionate gas emissions. Ethylene was a well proven indicator for temperature > 150°C.

- In Germany (1974), N_2 inertisation system was used at Osterfeld colliery at a flow rate of 1000 L/s to guard against the danger of an explosion during salvage operations in a section of the mine in which a heating had developed and applied in 41 coal mining operations (Both, 1981).

- German coalmines of Ruhr had used the N_2 inertisation and during one event in 1995, after the methane deflagration in low gassy mines at a rate of 3300 L/s to 833 L/s for recovery and reduce the O_2 to below 5 per cent. Authors had advocated that for safety reasons, independent

- of economic reason, N_2 was made available for sponcom management (Hermulheim and Beck, 1997).

- In India, the use of N_2 to control preventative sponcom heatings can be traced to the Salma seam in the Eastern Coalfield of India. The first trials at Laikdih Colliery (1981), had used one inert gas generator of 138 L/s to combating a blazing underground waste fire at Singareni Colliery using N_2 systems and continued to be a reliable control method (Ramaswamy and Katiyar, 1988; Varma, Mehta and Mondal, 2000; Zutshi, Ray and Bowmick, 2001).

- In Australia, nitrogen injection (700 m^3 of liquid N_2) was used in number of mine fire incidents, with varying degrees of success. For instance, nitrogen was pumped into Moura No 4 mine after the explosion to render the mine atmosphere inert for rescue teams to enter and control an active fire created by the explosion (Lynn, 1987).

- At Ulan Colliery (1991) of NSW, Australia, major spontaneous combustion incident was successfully controlled using Nitrogen (Healey, 1995).

- The TBM operation in Grosvenor (Qld, Australia), when encountered with the heating related gases and methane from surface to the working seams, Nitrogen was injected into the cutterhead successfully, that later was deployed worldwide (Belle and Foulstone, 2014).

- Dartbrook Colliery of NSW (Australia) goaf heating (2006) was managed by N_2 inertisation systems injections up to 1200 L/s (Gillies and Wu, 2007).

- In South Africa, five cases of the fighting of underground fires using N_2 have been attempted, of which Springfield colliery had used 1200 t of liquid N_2 to control sponcom fire underground. Inert gas generating units developed in Poland were deployed in the Gold mines and no recorded analyses of their success or otherwise were available. In 2008, South African Mines Rescue services fire management system included the Floxal Nitrogen inertisation system for sponcom management of the Witbank Collieries (Belle, Thomson and De Klerk, 2009).

- In Romania N_2 was used for the first time at the Dalja and Vulcan mines in the Petrosani coal basin in 1979–1980 at a flow rate of 200 L/s to manage the sponcom fires (Adamus, 2002).

- In Russian and Siberian coalmines, the theory of the smothering of mine fires by unreactive gas was explained by Sucharevskij (1952) who documented the use of nitrogen, as early as 1950s.

- In France, first trials of N_2 injection in the French coalfield was done in 1976 primarily for the control of sponcom when 30 L/min of CO is encountered (Benech, 1977; Froger, 1985), in comparison to current 'normal' TARP of 45 L/min at some of the Australian operations. The Air Liquide floxal system with N_2 pipeline (1983) had the capacity of 2800 L/s at 99.8 per cent N_2 with the pipeline diameters measuring 250, 200 or 150 mm delivering at 1 MPa (Amartin, 2001). Recent numerical flow modelling studies on the use of N_2 also proved to be beneficial (Pokryszka et al, 1997).

- In the USA, Greuer (1975) noted that Cyprus mine did not use the bleeder ventilation system to minimise the flow of air through the goaf, had coal heating events.

- In the USA, bleederless ventilation (Smith et al, 1994) is common where the prevention of sponcom is a key parameter for the ventilation design of an active panel as per the MSHA regulations. Bleederless ventilation or U-ventilation system is an attempt to render the gob inert in that it permits the accumulation of CH_4 and other non-flammable gases, while limiting the introduction of O_2. This ventilation design creates an inert atmosphere that will not sustain the self-combustion of coal and, therefore, limits the potential for these types of hazardous fire events or as a potential ignition source for an explosion.

- A well-known US example of employing N_2 inertisation was at BHP's San Juan Mine near Farmington, NM utilising bleederless ventilation system, similar to Australian mines. The system injected N_2 from the surface to then be horizontally injected through pipes into the Gob at around 315 L/s (Bessinger et al, 2005). The San Juan Mine has a target O_2 level of less than 2 per cent for inside the isolated Gob areas to prevent sponcom, with a scientific basis of low O_2 and methane levels prevent explosions as well as sponcom.

- High explosive detonation investigations in the UK in normal air and inert N_2 atmosphere, showed that the luminosity of the fireball (and, therefore, the temperature and presumably the presence of ongoing chemical reactions) on the surface of the fireball are far less pronounced in the N_2 atmosphere with 2.1 per cent O_2 (Tyas, 2019) than in the normal air of 21 per cent O_2.

In Summary, the use of N_2 to manage sponcom was known for over centuries and its known influence on gas composition analyses and recommendations on its continued use of CO and CO_2 trending for analyses of coal heating. One of the most valuable observations (Harris, 1981) made at the time, that is being recently ignored in Australia in pursuit of complicated unhelpful ratios, that the ratio of CO generated to oxygen absorbed (Graham, 1921) cannot be calculated and unhelpful. Harris (1981) **unambiguously** noted that the sponcom propensity or coal heating and control effectiveness has to be exercised by monitoring CO, CO_2, O_2, CH_4, and N_2 injected. Under the high CH_4 and CO_2 emitting mines, the heating indicator gas ratios of CO and CO_2 generated to O_2 absorbed would not be appropriate for assessment purposes. However, trends of sponcom heating indicator gases would benefit the ventilation engineer or expert fire risk assessor. More recently, Chamberlain, Hall and Thirlaway (1970) showed that CO is the most sensitive indicator of the early stages of self heating and recommended that continuous monitoring of this gas would provide the earliest detection of spontaneous combustion. It is impressive to note that well-known experts of the day like Graham or Stowell or Lord had ever suggested the 'masking' of the goaf atmosphere as a result of N_2 injection. One of the salient observations of those successful sponcom management or elevated heating were the use of oxygen, CO levels, and CO and CO_2 generated to the O_2 absorbed estimations that can equally confidently provide the sponcom status. The discussions did not suggest or merely use the ratio of CO or CO_2 generated to O_2 absorbed by the coal (Graham, 1921) based on the sample results due to deficiencies in ascertaining the 'true' O_2 levels absorbed by the coal in the current LW operations.

Longwall goaf monitoring

The use of gas analysis and various ratios of gases in the detection of coal oxidation was recognised and evaluated early in this century. Early detection is critical for the prevention and control of sponcom (Smith *et al*, 1994). In the modern longwall mines, the analysis of gaseous products of combustion is the primary method against underground manual observations, for early fire detection. However, only in cases of advanced heatings were gas analysis data able to identify heatings (Storrow and Graham, 1925). Chamberlain, Hall and Thirlaway (1970) showed that CO is the most sensitive indicator of the early stages of self heating and recommended that continuous monitoring of this gas would provide the earliest detection of sponcom and thus is the preferred method of sponcom detection worldwide using real-time sensor based or tube bundle gas monitoring systems.

In Australia, the continuous monitoring of longwall goaf management strategies was developed for gas and sponcom management purposes. Active goaf continuous monitoring is paramount to understanding of goaf gas distribution, sponcom management and the effectiveness of the proactive inertisation. Where an opportunity is possible whereby goaf gas monitoring is possible on both sides of the LW panel, it is recommended to install monitoring points on both sides of the longwall. This would enable, detailed goaf gas distribution on both sides of these initial longwall panel or extended LW panel from the previous panel. It is very rare to have access to such valuable data for multi-hazard longwall ventilation, gas and sponcom management to understand the goaf dynamics.

In remaining longwall panels, one can only monitor on the maingate side of the panel due to access issues and potentially consider the manual goaf hole samples. Figures 1 and 2 provide a general monitoring locations as a guidance. Mining operations may amend the monitoring point locations with qualified reasons in order to obtain sufficient gas composition data for reviewing the spontaneous combustion indicator gas levels from the active goaf, inertisation effectiveness and for developing appropriate control responses.

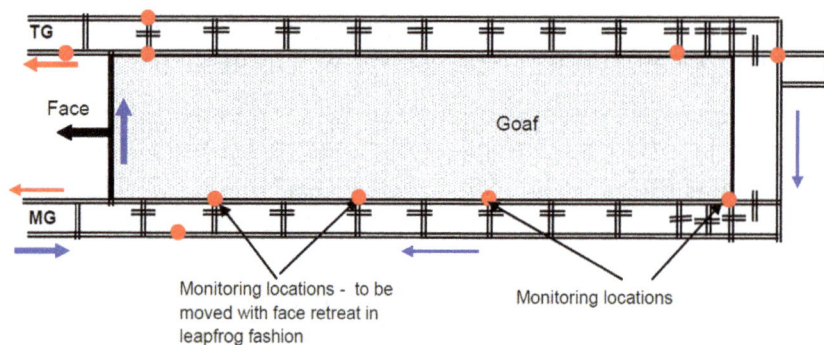

FIG 1 – General layout of monitoring locations with TG seals are accessible.

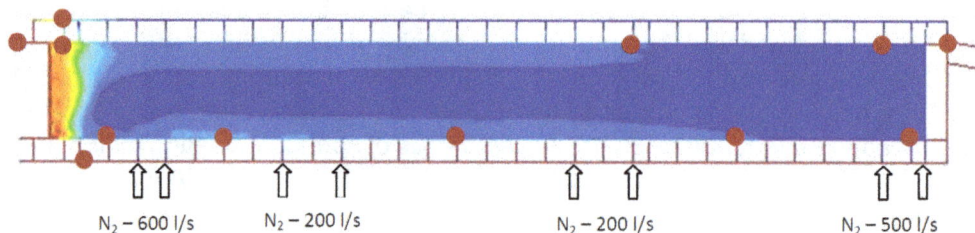

FIG 2 – Generalised layout of monitoring and MG inert gas injection locations.

BACKGROUND TO GAS AND SPONCOM MANAGEMENT STRATEGIES IN MODERN AUSTRALIAN HIGH PRODUCTION GASSY AND HOT LONGWALLS

Gassy, deep, hot and known sponcom prone GM seam longwall operations require a greater understanding the goaf gas behaviour, post gas drainage control strategies for TG gas management, and well balanced gas and sponcom management strategies of high magnitude gas reservoir mines. Following paragraphs below highlight the spectrum of scientific applied research based engineering controls and monitoring systems developed and improved over the last two decades in the Australian longwall operations.

Longwall goaf gas distribution patterns

In order to provide a visual understanding of goaf gas flow patterns for coalmine workers, operators and ventilation engineers, CFD models of operating longwalls were developed with operating panel geometries of longwall panel (Balusu *et al*, 2001) covering 1.0 km length of longwall goaf using actual floor contours for 2 gate and 3 gate road development scenarios. A typical longwall schematic with U ventilation system with the total ventilation quantity of 50 m³/s flows across the longwall face is used in CFD modelling studies are shown in Figure 3.

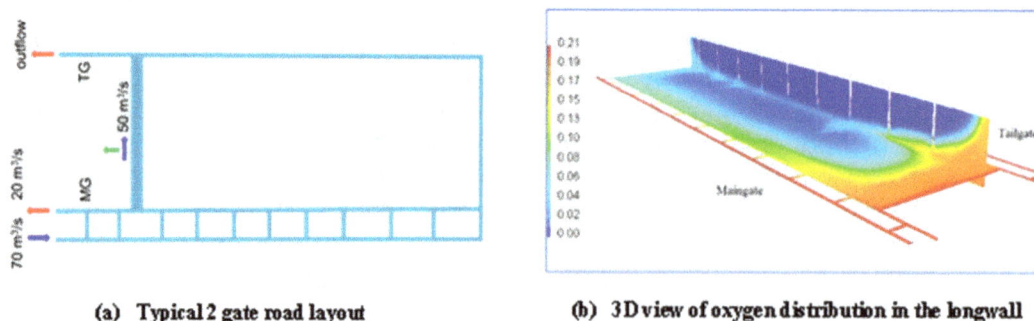

(a) Typical 2 gate road layout

(b) 3D view of oxygen distribution in the longwall

FIG 3 – Typical 2 gate road Longwall ventilation system (Balusu *et al*, 2002).

At the tailgate (TG) return, an outflow boundary condition was specified in the modelling simulations. The longwall panel width is 300 m and the roadway width on both maingate (MG) and tailgate (TG) sides of the face is 5.4 m. The goaf height up to 80 m above the working seam and the floor strata down to 10 m below the working seam is included in all the CFD models. The CFD models incorporated MG and TG cut-throughs of 5 m in width and cut-throughs spaced at 100 m intervals along the panel and goaf drainage holes replicated the drainage conditions of the operating site. The total number of finite volume cells used for meshing are around 2.0 million, for obtaining the grid independent solutions in simulations.

For visual understanding purposes, methane and oxygen gas distributions patterns in longwall goafs under two different conditions using operational longwall panel gas emissions and goaf gas drainage conditions with total gas emissions into the longwall goaf of around 9000 L/s with 98 per cent methane (CH_4) is shown Figure 4. The total goaf gas drainage rate was around 8000 L/s, with gas concentration in different vertical goaf holes varying between 80 per cent and 95 per cent with adjacent sealed panel goaf drainage of 800 L/s for a typical U ventilation system in a 2 gate road panel. In the CH_4 and O_2 gas distribution colour contours below, the red colour indicates higher gas concentration and the blue colour indicates lower methane or oxygen gas concentration. As noted herein, the presence of O_2 and continued CH_4 emissions are contradictory controls requiring finer balance and continued vigilance in goaf drainage operations and highly reliable gas trend monitoring.

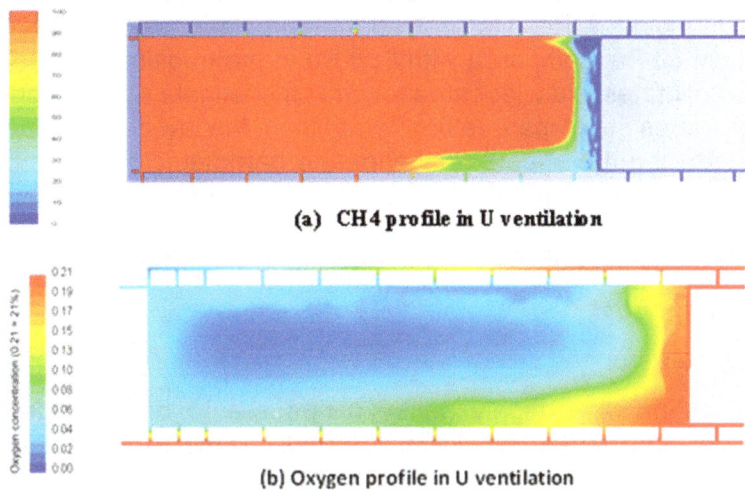

(a) CH4 profile in U ventilation

(b) Oxygen profile in U ventilation

FIG 4 – Visualisation of methane and oxygen gas distributions patterns in longwall goafs under two different conditions.

Various parametric studies by the CSIRO verified by the operational data from Australian mines had indicated that the gradient of the seam, both across the face and along the panel, had a major effect on the distribution of gases within the goaf. Similarly, results showed that gas emission rate and face airflow have a substantial effect on oxygen ingress into the goaf, particularly into the deep goaf (> 1 to 2 km behind the longwall face). Results show that intake airflow influenced airwash zone in the goaf with over 10 per cent O_2 concentration levels has extended further into the goaf with increase in intake airflow. In base case simulations the high oxygen level zone extended up to 150–200 m behind the face, whereas in the case of high intake airflow the high O_2 zone extended up to 300 m into the goaf (Figure 4).

In the recent years, there is often a preferential emphasis put on the 'pressure differentials' or loosely termed 'pressure' in an active goaf for ventilation and gas management. Historic work in relation to the static pressure distribution in the goaf with traditional practice of gas drainage from two goaf holes near the longwall face is presented visually in Figure 5 for surface goaf holes closest to the face operating at the total flow rate of about 1500 L/s. Results show that with this type of goaf hole drainage, gas static pressure in the goaf measured at the MG seal builds up to 180 Pa, that is sensitive various other mining and natural factors. The time based static pressure distribution model indicates that all the goaf gas migrates towards the tailgate corner of the goaf and a major proportion of the goaf gas may escape into the tailgate return airway, particularly during low barometric pressure periods. Therefore, both the goaf gas drainage strategy and its operation are paramount in addition to the goaf hole design.

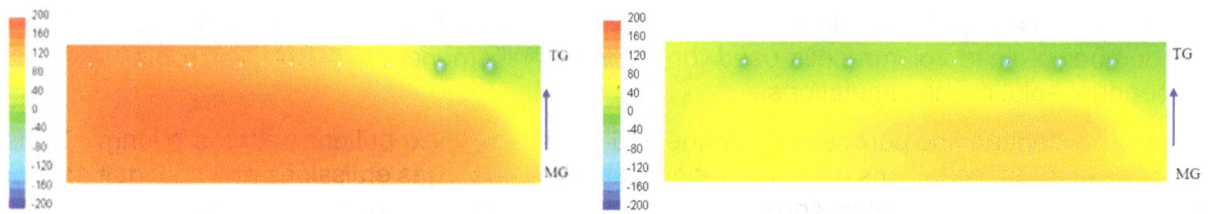

FIG 5 – Goaf gas pressure distribution in the goaf with few operating wells behind the face (Left) to maximised goaf drainage system (right).

In order to improve the gas drainage system efficiency, deep goaf holes gas drainage strategy was introduced into the modelling simulations. Deep goaf hole drainage in this paper refer to those vertical goaf drainage holes from surface that drain the goaf gas located greater than 1 to 2 km behind the operating longwall face and yet times even near the longwall start-up face area for very long panels. The static goaf gas pressure distribution in the goaf with maximised gas drainage strategy is presented in Figure 5. Results showed that the static pressure development in the goaf with this strategy is only 80 Pa, compared with 180 Pa in the traditional strategy scenario. Results indicated that goaf gas migrates to a wider area towards the tailgate side and only a minor proportion of goaf gas escapes towards the tailgate return. In addition, the low goaf gas pressure development in the goaf helps in reducing the effects of changes in barometric pressure on return gas levels. These results indicate that the optimum gas drainage strategy should incorporate goaf holes near the face as well as deep goaf holes with optimised increased numbers in the panel in order to improve the gas drainage system efficiency with large longwall block well retreated with large goaf gas reservoir size.

Goaf gas management

As a leading traditional practice, highly gassy mines are to be managed through extensive predrainage techniques long before the actual longwall mining to take place. During the active longwall mining, goaf drainage systems are used as the *primary* control for gas management with adequate drainage capacity for maximised drainage along with ventilation as the *secondary* control for gas management as a dilution control (Belle, 2015). Longwall gas emissions have increased significantly in recent years in some Australian longwall mines due to increased seam gas reservoir size with multiple upper and lower seams, higher production rates and increase in mining depths. There have been mines, that had previously deployed 3 gate road systems in their longwall panels for continued access to diesel vehicles during maintenance periods and for gas management. With the greater understanding with extensive and flexible goaf drainage systems and capacity, Australian coalmines use 2 gate road U ventilation system for longwall panel development and extraction. Extensive scientific and field work has been carried out to develop optimum gas and spontaneous combustion control strategies for 2 gate road longwall panels in Australia (Balusu *et al*, 2001, 2002, 2004, 2011; Balusu, Ren and Humphries, 2005; Balusu, 2020, 2021; Belle, 2014, 2015, 2017; Balusu, Belle and Tanguturi, 2017, 2019; Balusu and Tanguturi, 2019; Balusu, Tanguturi and Belle, 2021).

Although 3 gate road system provides more ventilation capacity during gate road development and assists in providing more ventilation dilution capacity in tailgate during longwall extraction, its effect on goaf gas distribution and explosive fringe gas profiles in the longwall goaf areas was historically unknown. There is a continued perception that as the 3 gate road system provides more ventilation capacity for gas dilution in the longwall tailgate return, it would also reduce the explosive fringe gas distribution profile near the tailgate area in the longwall goaf to manage the explosion risk. The results of the CSIRO CFD modelling simulations calibrated with field conditions indicated that there is a significant difference in the spread of explosive fringe gas distribution profiles in the longwall goaf under 2 gate road and 3 gate road conditions, ie *a significant increase* in the spread of explosive fringe (or close to explosive range) zone in the goaf under 3 gate road conditions. Based on the results of these investigations, appropriate strategies have been developed for gas control and minimisation of the spread of explosive fringe gas distribution in the longwall goaf (Balusu, Tanguturi and Belle, 2021). Therefore, based on the extensive operational gas management experience, it is the predrainage and longwall goaf gas drainage management is the primary gas management control rather than ventilation engineering controls to manage the major gas hazard.

An example comparison of the methane and oxygen gas concentration distribution patterns in the longwall goaf near the tailgate area in a 3 gate longwall retreat scenario are presented in Figure 6. Results of this simulations indicate that the methane gas distribution inbye of the longwall face is close to the explosive range in both cases which reflects the goaf drainage hole design and their operational effectiveness. The contour scale provides the methane and oxygen distribution. For example, 0.207 per cent equal to 20.7 per cent oxygen and 0.40 per cent equal to 40 per cent methane. In the Figure 5, 'closed at location 1' signifies the temporary roadway seal. The results of the CFD modelling simulations indicate that there is a significant difference in the spread of explosive fringe gas distribution profiles in the longwall goaf under 2 gate road and 3 gate road conditions, ie a significant increase in the spread of explosive fringe (or close to explosive range) zone in the goaf under 3 gate road conditions or partial 3 gate road LW operations.

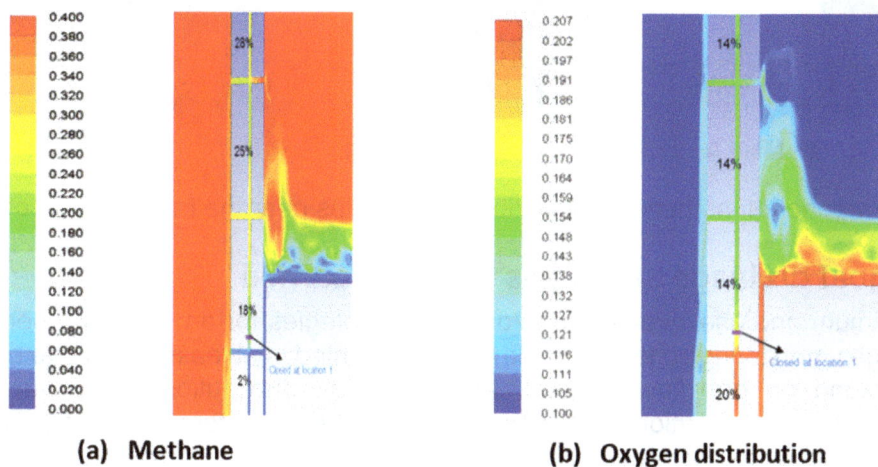

(a) Methane   (b) Oxygen distribution

FIG 6 – Close up views of methane and oxygen distribution inbye of the longwall face (Balusu, Tanguturi and Belle, 2021).

One of the major difficulties in the ventilation and gas flow dynamics is our inability to visualise the complex likely gas concentration profiles, ie methane or oxygen in the active goaf with time and nonconstant retreating longwall. The advances in the CFD numerical calculations have enabled the industry by providing an understanding of gas management or the extent of 'air wash zone' that is often used colloquially during risk assessment or emergency situations. In this paper, air wash zone is typically referred to as the concentration of relative oxygen in the goaf atmosphere aiding the leftover coal oxidation, and typically referred to with oxygen concentration values of 3 per cent to 21 per cent. Fresh air is the concentration that is representative of longwall fresh air intake. It is to be noted that due to almost no relative airflow movement that can be measured in the goaf may also mean the presence of oxygen even at levels of 2 per cent to 3 per cent. The CSIRO studies (Balusu *et al*, 2002) provided a visual scenario of potential oxygen distribution in an active LW goaf and goaf well nearer to the LW face. The field studies have noted that (Balusu, Ren and Humphries, 2005) the O_2 concentration was above 19 per cent for up to 100 m behind the longwall face and reduced to 6 per cent at 250 m behind the face in the absence of any inertisation control. This air penetration distances of 250 m to 350 m may be mainly attributed due to poor MG brattice control practices along with the increased longwall airflow rates for gas dilution purposes and inadequate gas drainage. The tracer gas studies have revealed that the goaf at 300 m behind the face is highly consolidated and does not allow direct travel of air from the intake side to return side of the TG.

Similarly, methane gas concentration distribution profiles at the tailgate region of different mines vary significantly depending on the geological, gas, mining and operational conditions. Gas concentration distribution profiles (Balusu, 2020) at the tailgate area under three mining conditions are presented in Figure 7. The white box in the plots show the TG motor area and white line replicating the Bretby across the face. In addition, the gas distribution profiles even at the same mine can vary significantly depending on number of conditions, including changes in barometric pressures, goaf falls, gas emission rates, goaf gas drainage efficiency, face location with respect to goaf holes and cut-throughs, face creep, floor contours, caving conditions behind the face and in gate roads, coal production rates, face ventilation, face cutting and chock advance sequences. Gas concentration distribution profiles and potential flammable gas mixture zones near the tailgate roadway area of the longwall face are dynamic and complex in nature and varies widely depending on the changes in above parameters during mining operations. Thus the behaviour of goaf gas composition is complex and influenced by mining engineering and fluid dynamics and are to be assessed by suitably skilled, qualified and experienced expert to provide appropriate guidance to safe operation of coalmines.

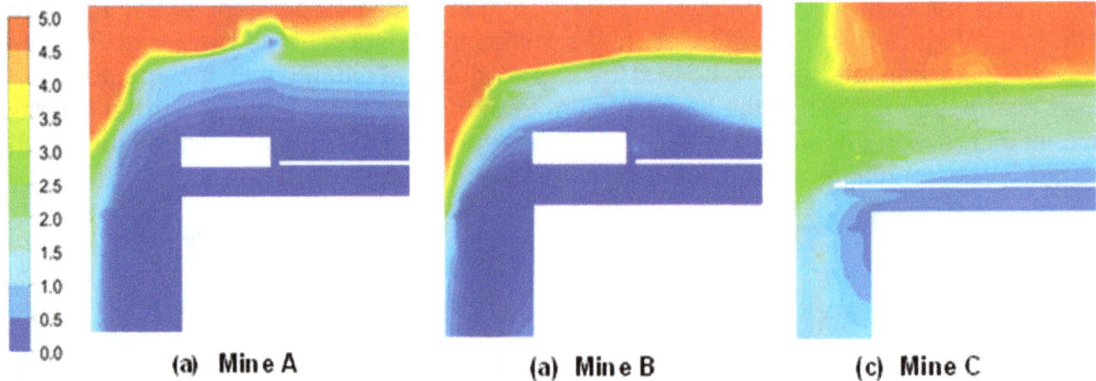

FIG 7 – A snapshot of methane gas distributions profiles near the tailgate area of longwalls.

Maximised goaf drainage strategies lessons learned

To develop optimum and effective goaf gas drainage strategies for any new or operating mine, an extensive goaf gas monitoring scheme should be implemented in at least one or two panels to obtain detailed information on gas flow patterns and goaf gas distribution under various operating circumstances for the site conditions and geometry. A number of factors including goaf gas emission flow rates and composition, panel ventilation, coal seam gradients, overlying and underlying coal seams, face retreat rates, caving characteristics, and goaf gas flow patterns need to be considered during development of goaf gas drainage strategy and goaf hole operations. In many cases, the standard practice of draining gas from two to five goaf holes near the face operating its peak capacity would not solve the tailgate gas problems but exacerbate the O_2 ingress into the deeper portion of tailgate area of an active LW goaf. Based on the results of various CSIRO studies and investigations supported by the coal operators over the last two decades, the following practical guidelines are recommended for optimum maximised goaf gas drainage strategies at highly gassy mines:

- Surface goaf holes for gas drainage provide the highest capacity, flexibility and lowest cost option for goaf gas drainage under most circumstances.

- Goaf holes should be drilled on the return side of the goaf, preferably at 20 to 70 m from gate road depending on the longwall caving conditions.

- Goaf holes are to be drilled 80 m to 100 m away from faults/dyke areas or geological structures.

- Uniform, stable and continuous operation of goaf holes (sudden peaks and lows in goaf drainage flow rate increases the coal oxidation potential resulting in sponcom risk).

- Goaf gas drainage hole diameter should be in the range of 250 to 400 mm for optimum flow rates and the goaf holes may be drilled at 50 m to 300 m spacing depending on the goaf gas emissions and other mining conditions.

- The total capacity of the goaf gas drainage plants should be around two to three times the expected goaf gas emissions to cater for deep goaf holes gas drainage, shifting of goaf plants or goaf hole connection changes and reduced plant efficiencies due to high pressure losses. Provision of a high-capacity and flexible gas drainage system allows optimisation of goaf gas drainage strategies, maximised goaf drainage with retreating longwall, flexibility, improves the overall efficiency and provides better gas control on the longwall face.

- The goaf gas drainage system should include a combination of goaf holes near the face and deep goaf holes in the panel in order to improve the overall gas drainage efficiency and to reduce the effects of barometric pressure changes on tailgate gas levels.

- The strategy of continuous operation of deep goaf holes at moderate capacity should be implemented ie intermittent operation of deep goaf holes at high capacity may not improve the overall efficiency and may lead to problems such as oxygen ingress into deep goaf.

- Goaf gas drainage should be carried out from around maximised number of goaf holes with the retreating longwall goaf in the panel (including deep holes), instead of the practice of gas drainage from just few goaf holes closest to the face.

- Application of increased suction pressure to drain more gas from goaf holes closest to the face might result in increased air dilution, without any net increase in gas drainage flow rates.

- The ventilation system in the panel should be designed to minimise oxygen ingress into the goaf, including immediate sealing-off all the cut-throughs behind the face, MG tight brattice control in order to improve overall gas drainage efficiency.

- Oxygen concentration level in the goaf hole flow should be less than 5 per cent for extended periods of time in goaf holes beyond 100 m from the LW face line to reduce sponcom risk in the longwall goafs.

- Gas drainage from adjacent old goafs should also be carried out where possible, depending on the goaf gas emission flow rates and adjacent seal strengths.

Sponcom management

Major events in Queensland demonstrates the critical importance of proactive sponcom management for underground coalmines extracting/working in known sponcom prone Moranbah region GM seams. Widely referred low sponcom propensity (R_{70}) of coal risk ratings in Principal Hazard Management Plan (PHMP) documents and frequency of their testing may be misleading the likely initiation or risk frequency estimations. For example, both German Creek and Goonyella Middle Seam R_{70} values are similar in magnitude but the leftover roof coal in GM seam goaf increase the oxidation risk with increasing depth due to steep geothermal gradient.

The evolution of major coal oxidation and sponcom incidents are sudden and may result in catastrophic negative safety outcome or result in the withdrawal of persons and closure of panels/mines. In view of the recent incidents in a number of mines working in GM seam (irrespective of the cause of the incidents), elevated oxidation reaction may potentially become an ignition source and inadequate control may result in the undesirable safety outcome.

Based on the two decades of close collaboration between the coal operators, ACARP and CSIRO, resulted in the current sponcom and gas management knowledge. The learning from these studies, have resulted in the following summary and context behind the proactive preventative sponcom and gas management goaf inertisation strategies:

- Historically, gassy longwall workings in Australian Goonyella Middle (GM) seam (late-1990s to mid-2010s), experienced increasing trend in CO levels associated with coal oxidation and sponcom indicator gases and major safety incidents were due to oxygen ingress on the maingate side. To address this issue, MG proactive inertisation strategy was introduced at GM seam operations mid-to-late-2000s, which ultimately reduced the number of high CO incidents over the next decade.

- When MG proactive inertisation strategy developed by the CSIRO and the operators was first introduced, there was no precedence in Australia and there was no field data to validate its effectiveness (prior to its implementation). However, it is to be noted that during longwall operations of GM seams, it's the additional proactive N_2 inertisation strategy that was essential to successfully manage the sponcom and resulting major fire risks during long periods of production stoppages due to geotechnical and mining related matters.

LW GAS AND PROACTIVE MG AND TG INERTISATION AND MONITORING STRATEGY

In order to manage the elevated oxidation levels with increased goaf gas drainage in high production gassy mines to effectively manage the active longwall TG gas levels, following proactive inertisation strategy was implemented in an operating mine. The optimum locations of the TG and MG holes for both gas and sponcom management with proactive inertisation for high gassy mines were based on the original and fundamental goaf gas and sponcom management work by the CSIRO, calibrated with the operational experiences over the last decade (Balusu, 2021).

Data collation and limitations

The study had analysed extensive longwall panel data, that had implemented for the first time, a strategy that incorporated MG and TG nitrogen (97 per cent N_2) inertisation and goaf gas management. The limitations and characteristics of the data used for analyses are summarised briefly below:

- The goaf drainage hole data which recorded CO, CO_2, CH_4, flow were averaged daily for individual holes collected for the entire longwall panel.

- Active goaf gas composition was based on the daily individual average goaf hole data comprising, flow, CH_4, O_2, CO, CO_2 composition.

- Goaf gas flow data with flow rates of >200 L/s are used in the final analyses, as some of the new goaf holes were usually checked for validating the goaf connections once the longwall had retreated 20 m to 30 m behind the longwall face line.

- It is to be noted that the data flow rate <200 L/s is usually associated with the start-up of the goaf hole prior to LW intersections with higher levels of O_2 (>14 per cent) for short periods.

- Negative O_2 readings were removed from the data set considering the likely data flaws or sensor measurement errors.

- It was noted that the data set yet times showed that during the normal operation of goaf wells probably based on TARP trigger levels, resulted in undesirable gas composition mixtures.

- The data contained over 90 vertical surface goaf drainage holes with a combination of deep goaf, adjacent goaf, MG and TG goaf week operations of the entire longwall panel for effective longwall TG gas management.

LW gas management and proactive MG and TG inertisation assessment

An extensive data analyses was carried out for the entire longwall panel for the gas and sponcom management effectiveness and are discussed hereafter. Figure 8 shows the profiles of active longwall goaf gas composition with retreating longwall with likely less than effective inertisation and the absence of TG inertisation.

Key observations from the analyses are as follows:

- The effectiveness of the N_2 injection and the operation of goaf gas flow rate management is reflected in the gas composition of the active longwall goaf gases, *viz*, CO, O_2, CH_4.

- The composition of higher levels of oxygen behind the longwall face in the TG region (up to 300 m behind the face) may be attributed to the very higher flow rates and goaf holes coming online at the time of goaf formation.

- Similarly, higher daily average CO levels, demonstrate the presence of the excessive and conducive oxygen rich environment as a result of deeper air wash zone in the TG area of the goaf contributing towards early oxidation of left over coal in the goaf.

- It is equally noted that the methane concentration behind the active longwall in overall terms increases in the purity as a result of continued desorption of overlaying undrained coal seams with greater gas reservoir size. The failure to drain the deeper portion of the active longwall goaf will eventually travel towards the tail gate region as a result of increasing goaf gas reservoir size with the longwall retreat and definitively contributes in higher gas levels during the steep drop in barometric pressures.

- It is equally important to note that the uncontrolled deep goaf hole operation at very high flow rates, may further exacerbate the oxygen rich environment deeper in the TG goaf area in the absence of effective TG inertisation.

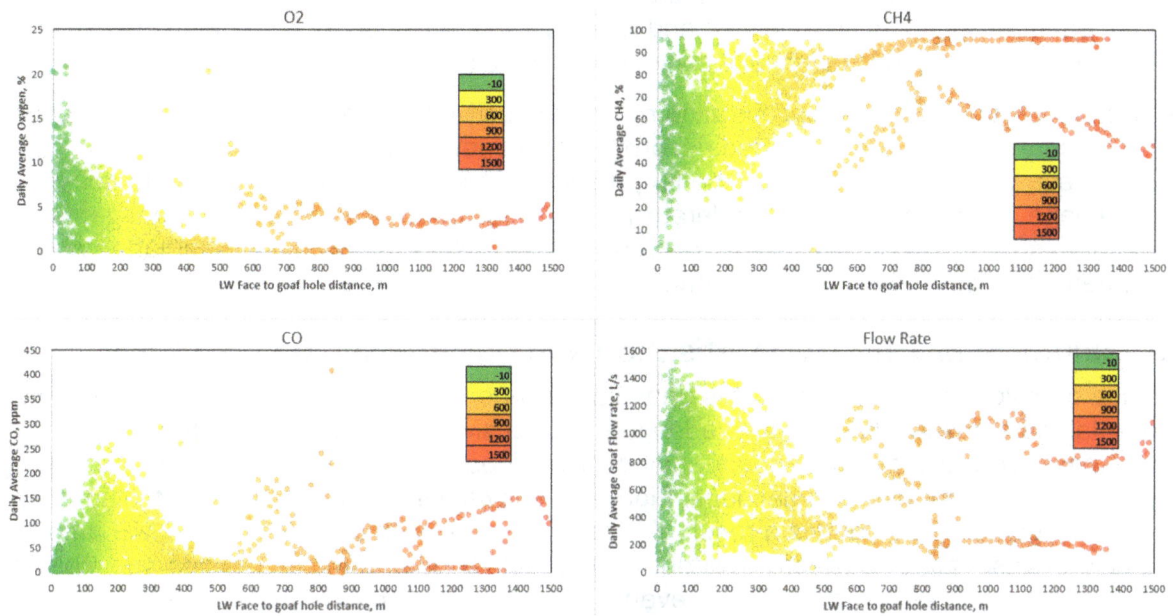

FIG 8 – Goaf hole gas composition without the TG inertisation with retreating LW face oxygen (top left), CH_4 (top right), CO (bottom left); flow rate (bottom right).

Similarly, Figure 9 shows the profiles of active longwall goaf gas composition with retreating longwall with effective MG inertisation and with the introduction of TG inertisation for the first time.

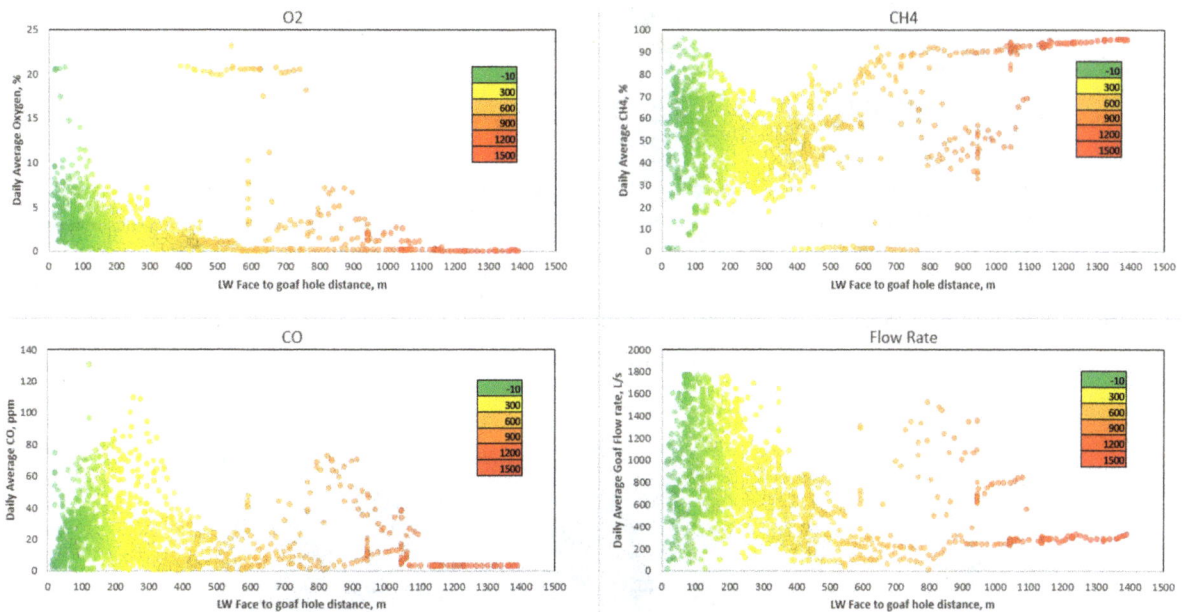

FIG 9 – Goaf hole gas composition with MG and TG inertisation with retreating LW face oxygen (top left), CH_4 (top right), CO (bottom left); flow rate (bottom right).

Key observations from the analyses are as follows:

- Combined MG (1500 L/s) and TG inertisation rate of 750 L/s with greater goaf gas management is clearly reflected in the favourable gas composition of the active longwall goaf gases, *viz*, CO, O_2, CH_4.

- The composition of reduced levels of oxygen behind the longwall face in the TG region (up to 100 m behind the LW face) may be attributed to the N_2 injection in the tailgate minimising the oxygen ingress into the goaf. There is a significant difference in the oxygen levels at less than 5 per cent beyond 100 m of the inbye goaf assisting in minimising the risk of coal oxidation and oxygen presence in the deeper portion of the goaf.

- Similarly, daily average CO levels in the active goaf have significantly reduced, demonstrating the effectiveness of the N_2 injection and less than adequate conducive oxygen presence in the TG airwash zone, thus potentially minimising early oxidation.

- As before, the methane concentration behind the active longwall in overall terms increases in the purity as a result of continued desorption of overlaying undrained coal seams with greater gas reservoir size and aided by the N_2 presence. This environment enables the deeper goaf drainage practices at low to moderate flow rate for maximised goaf gas management and drainage efficiency, to reduce the gas reservoir as well as ventilation air methane (VAM) and greenhouse gas (GHG) management.

Discussions of MG and TG inertisation and gas management strategy

Presence of oxygen in the active goaf is unavoidable when carrying out goaf drainage activities to manage tail gate gas levels of an active longwall. Extensive goaf seal monitoring activities by the industry have provided greater understanding of the goaf dynamics and input to the goaf gas composition studies for over two decades. Figure 10 shows one such historic work that informs the impact of oxygen profile extent as a result of proactive N_2 injection on both TG and MG areas. The reduced air wash zone is clearly evident in the active goaf thus minimising the conducive environment for any elevated oxidation events (Balusu, Ren and Humphries, 2005). Figure 11 provides the latest CSIRO study using calibrated CFD model of MG and TG inertisation strategy for an operating high gassy longwall mine.

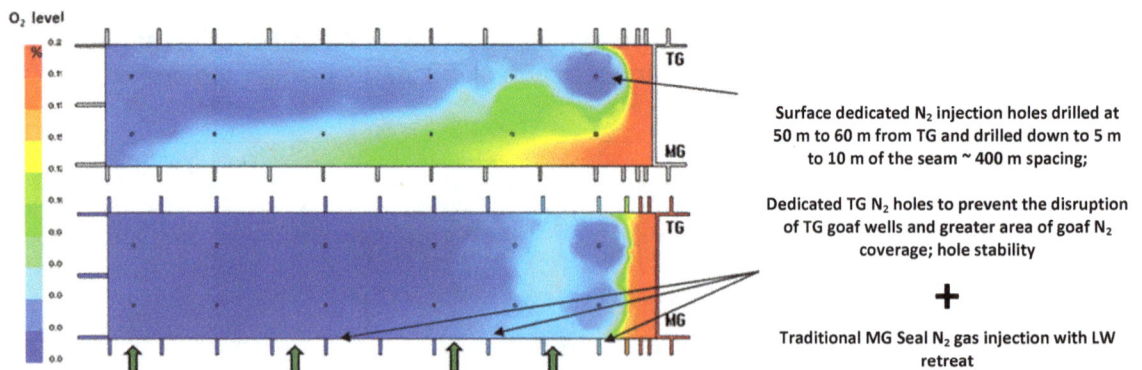

FIG 10 – Historic MG and TG inertisation conceptual strategy for longwalls (Balusu, Ren and Humphries, 2005).

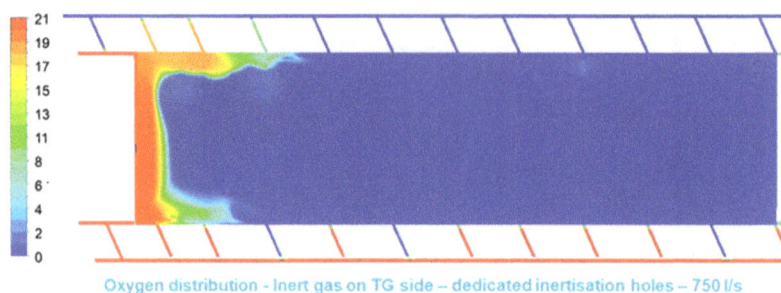

Oxygen distribution - Inert gas on TG side – dedicated inertisation holes – 750 l/s

FIG 11 – Latest MG and TG inertisation strategy for an operating gassy longwall mine (Balusu, 2021).

Based on the aforementioned gas composition analysis of field goaf hole production data, the goaf gas distribution under intensive gas drainage is summarised as a conceptual model is shown in

Figure 12 (Xiang *et al*, 2021). This drawing may be very subjective—specifically airwash zone contours can shift depending on the complex goaf hole operations for U ventilation system with varying goaf hole operational controls and designs without proactive inertisation. The goaf drainage operation suggests that greater oxidation may be possible in the TG, while there is minimal occurrence of it at MG area with greater ventilation controls such as tight brattice controls and timely build-up of seals.

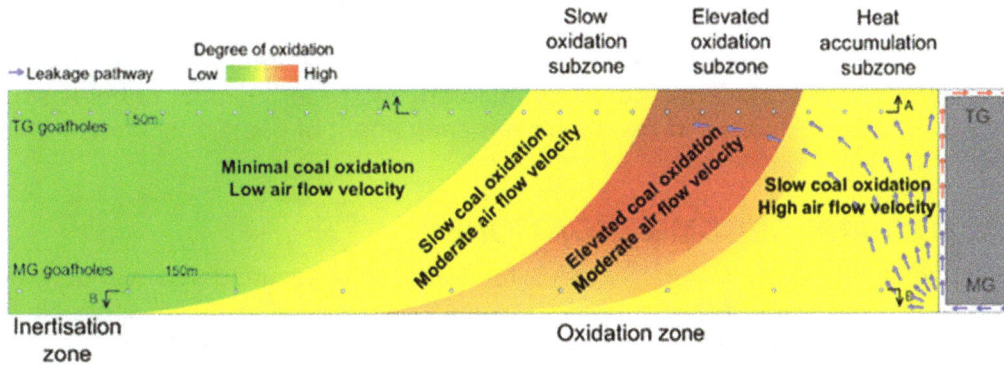

FIG 12 – A conceptual model of goaf gas environment under intensive goaf gas drainage impact.

The historic field studies (including tracer gas and goaf hole shut off studies) of CSIRO have shown that reducing and increasing the flow capacity of goaf holes, including complete shutoff of the holes and measuring corresponding changes in return gas levels and response times have shown that 80 per cent to 90 per cent of goaf gas migrated to longwall return within 1 to 3.0 hrs when goaf holes were 100 to 400 m from the longwall face. Response times varied between two mins (150 m behind the face) to an hour (1000 m behind the face) along the MG and on the TG goaf holes with few minutes to several hours depending on the goaf hole designs and caving characteristics. Field studies have shown that even when goaf holes located more than 1000 m from the face, had a substantial effect on gas flow dynamics and on longwall return gas levels. The recommended strategy was that the deep goaf holes are to be operated at low to moderate flow rates continuously as long as the oxygen levels below 3 per cent and no increased trend in the CO levels.

One of the additional challenges often faced during an unfortunate gas event leads to plethora of suggestions without having appropriate science and data based studies or specialist expertise in the field. One such view is that the goaf drainage flow rate has no impact on oxygen in goaf well or active goaf. Figure 13 shows the relationship between goaf hole flow rates and goaf hole oxygen levels at various distances from the active longwall face.

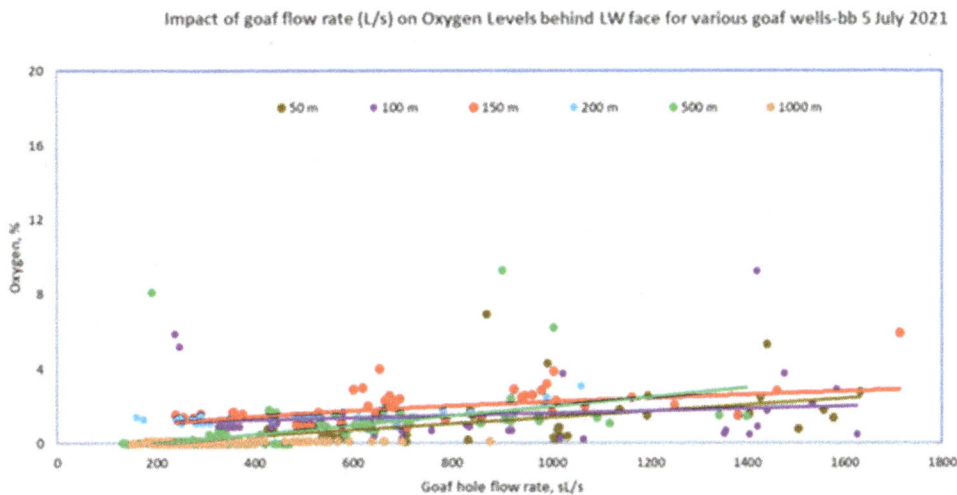

FIG 13 – A conceptual model of goaf gas environment under intensive goaf gas drainage impact.

It is observed from the above plot that when the goaf holes are deep, the O_2 levels are typically very low and lower flow rates help to minimise the goaf reservoir size with the retreating longwall with major rate of barometric pressure drop. The general observation is that the increase in goaf flow rate to the extreme regions causes significant increase in the O_2 levels in the goaf, which would require inertisation to minimise the potential oxidation. It's the moderate flow rate with increased number of operating wells against few number of wells operating at extreme flow rates assists in managing the TG gas levels. While the sole intent is to minimise the TG gas levels, the risk of operating few wells at maximum flow rates behind the longwall face would certainly bring the fresh air or oxygen into the active goaf of a retreating longwall. It must be noted that the goaf hole design including its distance from the working seam (historic design of ideally 10 m or 40 ft as in the USA) will have the impact on TG gas levels as well as oxygen levels in the active goaf. Therefore, it's the maximised goaf drainage of active goaf with moderate goaf flow rates and increased number of holes supported by the active inertisation would assist in the effective management of gas and sponcom management risks.

CONCLUSIONS AND RECOMMENDATIONS

Conjointly managing the gas and sponcom risk is fundamental to securing a safe underground place of work at GM seam longwall operations. With the increasing gassy and known sponcom prone coal seams and a working depth with steep geothermal gradient is contributing towards the step changes in controls required for gas and sponcom management in order to be compliant with the safe TG gas limits as well as reduced ventilation air methane (VAM) emissions. This strategy reinforces the fundamental importance of the predrainage systems with long lead drainage time prior to longwall mining. Over two decades ago in Australia, goaf gas drainage rates in Qld and NSW were low to moderate (1000 L/s to 3000 L/s) and the ingress of airwash zone on longwall TG side was not a major concern. However, in the recent years, with increasing goaf gas drainages rates up to 6000 L/s, 10 000 L/s and manual or automatic mode operation of goaf wells to the extreme flow rates to address higher goaf gas emissions, TG oxygen ingressing deeper in the goaf has become a major issue, necessitating the introduction of both TG and MG inertisation strategies now to address this emerging coal oxidation risk. Furthermore, contrary to the views in relation to elevated coal oxidation and sponcom events related to goaf hole spacing and maximised goaf drainage practices, it is prudent to note that there have been historic cases of sponcom events with 400 m to 200 m goaf hole spacing, and even with no goaf drainage practices.

Following gas and sponcom management strategies aided with both MG and TG inertisation using proactive N_2 injection during various phases of longwall production and stoppages provides appropriate guidance:

- In the absence of continuous proactive N_2 inertisation, maintaining 5 per cent to 8 per cent O_2 levels in the active goaf would be very difficult and may exacerbate the oxidation in an active goaf when the longwall retreat rates slows down or stops for weeks and months.

- Any reduction in goaf drainage to manage the O_2 ingress in deep goaf area of TG region will significantly increase the longwall TG gas levels. This approach would put operations in a dangerous position from gas management perspective.

- It is the operational management of goaf wells (not sudden or automatic operation of the goaf wells to the extreme flow rates, rather stepwise increase) with proactive LW active goaf inertisation for sponcom management will enable the appropriate maximised goaf drainage gas management to manage the longwall return gas levels and oxidation risks.

- The introduction of TG inertisation assists in reducing the airwash zone along the TG side of the active goaf. Every longwall operation need to continue with the leapfrogging of well-established MG seal inertisation strategy (including quicker and timely MG seal build, and tight MG brattice control), and dedicated tube bundle monitoring points for goaf gas monitoring.

- As a general long-term strategy, known sponcom prone longwall operations to ensure flexible and contingency inertisation infrastructure is readily available and to remain in a state that it is able to be recommissioned within a single shift at any time during the monitoring period.

- The association of slow retreat due to known or unknown geological structures in the longwall hazard plan and known historic oxidation related incidents, have reinforced the benefits of proactive inertisation system with around 2200 L/s of inert flow into the active panel (around

1200 L/s on the MG side and 1000 L/s on the TG side using dedicated vertical N_2 holes drilled 5 to 10 m from the working seam height).

- Inert gas injection into the active longwall must be through multiple injection points distributed appropriately, depending on the oxygen ingress profiles. One of the inert gas injection points on the MG side should be at around 200 to 300 m behind the longwall face. Inert gas locations on the TG side should be designed based on the goaf gas drainages strategies behind the operating longwall face.

- Considering the various uncertainties associated with the LW operations in sponcom prone seams, recommendation of inertisation holes at 200 m spacing on TG side is appropriate for long-term risk management planning and design purposes. If the evidence suggests otherwise, ie constant increased retreat or no stoppages, then TG proactive inertisation may be carried out through inertisation holes at increased spacing (ie alternate inertisation holes) in those areas. This would mean that the intermediate dedicated holes and other old/deep inertisation holes can be equally used for oxygen ingress or airwash zone monitoring and sponcom monitoring on the TG side of the goaf as per the oxygen limit recommendations of an active goaf.

- It is to be noted that these gas and sponcom management strategies are not merely based on the LW retreat rate, but also includes the inherent nature of seam propensity for sponcom despite it being equally rated as 'low' risk sometimes, geothermal gradient, existence of faults and structures, amount of coal left behind, changes to the goaf hole designs, delay in operational related building of MG seals, inadequate MG brattice leakage control for long periods. other engineering related uncertainties associated with the LW equipment, strata control uncertainties associated with moisture/water in TG roadways, and uncertainties associated with cavity control measures.

- Strengthening the proactive inertisation strategy on both MG and TG with flexible inertisation capacity and responding to the up to date trigger response values of the oxidation scenario rapidly developing into an advanced stage using appropriate early monitoring strategy is essential for the future proofing of underground sponcom risk.

Finally, sponcom has been an ever present problem in coalmines resulting in fires and explosion. Managing the sponcom risk through N_2 has been evolved and as a preventative as well as reactive safety control has been well established worldwide. Various risk indicators or warning signs have been developed over the years starting from Graham (1925) using early gas indicators and underground observations. The historic studies and operational experiences note that, from as an early warning indicator for sponcom under proactive inertisation controls, amongst the wide array of existing indicator gases, it's the monitoring of rate of changes of CO as a most reliable and historically proven indicator gas on a continued basis, even when inertisation is used as a sponcom control action. With appropriate site based ownership of sponcom monitoring gases of different underground locations provide confident 'signals' of deterioration or safe control of the underground environment to workers.

ACKNOWLEDGEMENTS

The authors would like to thank various local and overseas researchers, operators, workers and regulators for their constructive comments, inputs support and encouraging remarks towards this safety work. Also acknowledge Australian mining industry for continued support in advancing the safety management through collaborations and support. The views expressed herein are of the authors and do not necessarily represent the views of any organisation.

REFERENCES

Adamus, A and Vicek, J, 1997. The optimisation of the Nitrogen Infusion Technology, Proceedings of the 6th International Mine Ventilation Congress, 566 p.

Adamus, A, Hajek, L and Posta, V A, 1995. A review of experience on the use of Nitrogen in Czech Coal Mines, in *Proceedings of the 7th US Mine Ventilation Symposium*, pp 237–241.

Adamus, A, 2002. Review of the use of nitrogen in mine fires, *Trans Instn Min Metall, Sect A: Min Technol*, 307:A89–A98.

Amartin, J P, 2001. Optimization of nitrogen injection for inertization of longwall faces goaf in CdF coal mines, in *Proceedings of the 7th Int Mine Ventilation Congress*, pp 849–53.23.

Balusu, R, Mallett, C, Xue, S, Wendt, M and Worrall R, 2002. Mine Gas Control – JCOAL, CSIRO and Dartbrook Collaborative project Final report, CSIRO Exploration and Mining Report No. 1012C, March/October 2002, 183 p.

Balusu, R, Tanguturi, K and Belle, B, 2021. Goaf gas distribution profiles near the longwall tailgate area, Proceedings of the 18th North American Mine Ventilation Symposium, (ed: P Tukkaraja).

Balusu, R, 2020. Longwall Tailgate Gas Compositions, Industry Note, Australia.

Balusu, R, 2021. Unpublished CFD Model Study on MG and TG inertisation Strategies, CSIRO.

Balusu, R, Belle, B and Tanguturi, K, 2017. Development of gas and spontaneous combustion control strategies for 6.0 km long longwall panels, in *Proceedings of the 16th US Mine Ventilation Symposium*, pp 18-11–18-18.

Balusu, R, Belle, B and Tanguturi, K, 2019. Development of Goaf Gas Drainage and Inertisation Strategies in 1.0-km- and 3.0km-Long Panels, *Mining, Metallurgy and Exploration*, 36(6):1127–1136. https://doi.org/10.1007/s42461-019-0071-9.

Balusu, R, Deguchi, G, Holland, R, Moreby, R, Xue, S, Wendt, M and Mallett, C, 2001. Goaf gas flow mechanics and development of gas and sponcom control strategies at a highly gassy coal mine, Australia-Japan Technology Exchange Workshop, Australia, 18 p.

Balusu, R, Ren, T X and Humphries, P, 2005, Proactive Inertisation Strategies and Technology Development, ACARP Project C12020, CSIRO Exploration and Mining Report P2006/26, 114 p.

Balusu, R, Schiefelbein, K, Ren, T, O'Grady, P and Harvey, T, 2011. Prevention and control of fires and explosions in underground coal mines, in *Proceedings of the Australian Mine Ventilation Conference*, pp 69–77.

Balusu, R and Tanguturi, K, 2019. Gas management and risk mitigation strategies for longwalls, ACARP Project C25066, CSIRO Energy Report EP191657, March 2019, 210 p.

Balusu, R, Tuffs, N, Peace, R, Harvey, T, Xue, S and Ishikawa, H, 2004. Optimisation of Goaf Gas Drainage and Control Strategies, ACARP Project C10017, CSIRO Exploration and Mining Report 1186F, 149 p.

Belle, B, 2014. Evaluation of barometric pressure and cage effect on longwall tailgate gas levels for prevention of explosion in coalmines, International Mine Ventilation Congress, Johannesburg.

Belle, B, 2015. Innovative tailgate mobile goaf gas management in two gateroad longwall panels – concept to implementation, The Australian Mine Ventilation Conference (The Australasian Institute of Mining and Metallurgy: Melbourne).

Belle, B, 2017. Optimal goaf hole spacing in high production gassy Australian longwall mines – operational experiences, The Australian Mine Ventilation Conference (The Australasian Institute of Mining and Metallurgy: Melbourne).

Belle, B and Foulstone, A, 2014. Explosion Prevention in Coal Mine TBM Drifts – An Operational Safety Knowledge Share, *Procedia Earth and Planetary Science*, 11:15–28.

Belle, B and Biffi, M, 2018. Cooling pathways for deep Australian longwall coal mines of the future, *International Journal of Mining Science and Technology*, 28(6):865–875.

Belle, B, Thomson, C A S and De Klerk, C, 2009. Inertisation of Mine fires for South African Collieries-A Business Case, International Mine Ventilation Congress.

Benech, M, 1977. Expériences d'injection d'azote dans les arrièretaillesà soutirage, *Industrie Minérale*, pp 363–371.

Bessinger, S L, Abrahamse, J F, Bahe, K A, McCluskey, G E and Palm, T A, 2005. Nitrogen Inertization at San Juan Coal Company's Longwall Operation, SME Annual Meeting, Preprint 05–32.

Both, W, 1981. Fighting mine fires with nitrogen in the German coal industry, *The Mining Engineer*, pp 797–804.

Bradecki, W, Matuszewski, K and Nowak, H, 1987. The use of nitrogen gas in suppression of spontaneous heating in wastes at the mine Sosnica, Preglond Gorniczy, (7/8):3–10.40.

Cakir, A and Baris, K, 2009. Assessment of an Underground Coal Mine Fire: A Case Study From Zonguldak, Turkey, in Proceedings of the 2009 Coal Operators' Conference (eds: N Aziz and B Kininmonth).

Chamberlain, E A C, Hall, A D and Thirlaway, J T, 1970. The Ambient Temperature Oxidation of Coal in Relation to the Early Detection of Spontaneous Heating, *Min Eng*, 130(121):1–16.

DeRosa, M, 2004. Analysis of mine fires for all US underground and surface coal mining categories, 1990–1999. NIOSH Information Circular 9470.

Froger, C E, 1985. Fire fighting expertise in French underground mines, in *Proceedings of the 2nd US Mine Ventilation Symposium*, pp 3–10.

Gillies, A D S and Wu, H W, 2007. Inertisation and mine fire simulation using computer software, ACARP Project C14025, p 215.

Gouws, T P and Knoetze, M J, 1995. Coal self-heating and explosibility, *Journal of the Southern African Institute of Mining and Metallurgy*, 95(1):37–43.

Graham, J I, 1922. The Composition of Mine Air and its Relation to the Spontaneous Combustion of Coal Underground, *Fuel in Science and Practice*, pp 51–58.

Graham, J I, 1921. The Normal Production of Carbon Monoxide in Coal Mines, *Transactions-Institution of Mining Engineers, 1920–1921*. 6:222–234.

Graham, 1915. The absorption of oxygen by Coal, *Trans Inst Min Eng, 1914–1915*, XLVIII:503–549.

Greuer, E R, 1975. Study of mine fire fighting using inert gases, Research report of the commissioned from Department of Mining Engineering, Michigan Technological University, Department of the Interior, Bureau of Mines, 1975.

Ham, B, 2005. A Review of Spontaneous Combustion Incidents, in *Proceedings of the 2005 Coal Operators' Conference* (eds: N Aziz and B Kininmonth).

Harris, L, 1981. The Use of Nitrogen to Control Spontaneous Combustion heatings, *Min Engr* 140, 237:883–892.

Healey, P, 1995. 1991 Ulan heating, paper presented at Dept Mineral Resources spontaneous combustion seminar.

Hermulheim, W and Beck, K D, 1997. Inertization as Means for Reducing Down Time and the Explosion Risk in Cases of Spontaneous Combustion, in *Proceedings of the 6th International Mine Ventilation Congress*, pp 299–303.

Jeger, C and Froger, C, 1975. Conditions for the Initiation of Spontaneous Combustion-Application to Prevention and Monitoring, *Preprints in XVI International Conference on Coal Mine Safety Research*, BuMines OFR 83(1)-78, pp 3.1–3.14.

Kajdasz, Z and Stefanowicz, T, 2002. Emergency rescue operation at Zasjadzko mine in Donetsk basin, Email communication from CMRS Bytom, Poland, 9 January 2002.

Kajdasz, Z, Golstein, Z and Buchwald, P, 1989. Modern methods ofinertization of mine atmospheres for improving fire safety in the Polish coal industry, *Przeglond Gorniczy*, (10):21–5.41.

Khanal, M, Poulsen, B, Adhikary, D, Balusu, R, Wilkins, A and Belle, B, 2021. Numerical Study of Stability and Connectivity of Vertical Goaf Drainage Holes, *Geotechnical and Geological Engineering*, 39(3):2669–2679. https://doi.org/10.1007/s10706-020-01650-6

Koenning, T and Boulton, J, 1997. Spontaneous Combustion Experience at Cyprus Shoshone Coal Mine, in *Proceedings of the 6th International Mine Ventilation Congress*, pp 295–298.

Kukuczka, M, 1982. A new method for determining explosibility of complex gas mixtures, *Mechanizacja I Automatuzacja Gornictwa*, 164(11):36–39.

Lord, S B, 1986. Some Aspects of Spontaneous Combustion Control, *The Mining Engineer*, 296:479–488.

Lynn, K P, 1987. Warden's inquiry–Report on an accident at Moura No 4 underground mine on Wednesday, 16th July 1986 (Queensland Government Press).

Michaylov, M, 1996. Analiz na prilaganeto na metogite za namaljavanena pozarnite riskove i na tengenciine za izpolzvane na azot vrudnik'Babino, University of Mining and Geology Sofia, Research Report 123, October 1996.

Michaylov, M and Vlasseva, E, 1997. Three Phase Foam Production for Sponcom Fighting in Underground Mines, in *Proceedings of the 6th International Mine Ventilation Congress*, pp 305–312.

Mohalik, N K, Lester, E, Lowndes, I S and Singh, V K, 2016. Estimation of greenhouse gas emissions from spontaneous combustion/fire of coal in opencast mines – Indian context, *Carbon Management*, 7(5–6):317–332. https://doi.org/10.1080/17583004.2016.1249216

Mucho, T P, Houlison, I R, Smith, A C and Trevits, M A, 2005. Coal Mine Inertisation by Remote Application, *Proceedings of the National Coal Show*, pp 1–14 (Mining Media, Inc).

Occupational Safety and Health Administration (OSHA), 2024. Title 29 of the CFR, USA. Available from: <https://www.ecfr.gov/current/title-29>

Plot, R, 1866. Spontaneous Combustion in Heaps, History of Staffordshire, UK.

Pokryszka, Z, Tauzière, C, Carrau, A, Bouet, R and Saraux, E, 1997. Application of numerical gas flows modelling to optimization of nitrogen injection in the goaf, in *Proceedings of the 27th International Conference of Safety in Mines Research Institutes*, pp 411–20.22.

Ramaswamy, A and Katiyar, P S, 1988. Experiences with liquid nitrogen in combating coal fires underground, *J Minerals, Metals and Fuels*, pp 415–424.

Ren, T X and Edwards, J S, 1997. Research into the Problem of Spontaneous Combustion of Coal, in *Proceedings of the 6th International Mine Ventilation Congress*, pp 317–322.

Smith, A, Diamond, W P, Mucho, T P and Organiscak, A, 1994. Bleederless Ventilation Systems as a Spontaneous Combustion Control Measure, in U S Coal Mines By US Department of the Interior, Bureau of Mines IC 9377.

Stephan, C R, 1986. Summary of Underground Coal Mine Fires, January 1 1978 to July 31 1986, MSHA Rep. 06–357–86, Dec 1986, 18 p.

Storrow, J T and Graham, J I, 1925. The Application of Gas Analysis to the Detection of Gob-Fires, *Transactions Institution of Mining Engineers, 1924–1925.* LXVIII:408–430.

Sucharevskij, V M, 1952. The sealing of underground fires with application of inert gases 1952, 192 p (Moscow: Ugletechnika).

Thomas, G H, 1986. Discussions on the paper, Some Aspects of Spontaneous Combustion Control, South Staffs and South Midlands Branch of IMinE, held at Birch Coppice Miners Welfare Centre, *The Mining Engineer 1986,* p 488.

Tyas, A, 2019. Blast Loading from High Explosive Detonation: What we Know and Don't know, in Proceedings of the 13th International Conference on Shock and Impact Loads on Structures.

Varma, S K, Mehta, S R and Mondal, P K, 2000. Spontaneous heating in a longwall face—a case study of Jhanjra project, International Mine Environment and Ventilation Symposium, Dhanbad (India).

Vaughan-Thomas, T, 1964. The use of Nitrogen in controlling an Underground Fire at Fernhill Colliery, *Min Engr,* 123(42):311–336.

Wachowicz, J, 2008. Analysis of underground fires in Polish hard coal mines, *Journal of China University of Mining and Technology,* 18(3):332–336.

Wactawik, J, Branny, M and Cygankiewicz, 1997. A Numerical Simulation of Spontaneous combustion of coal in Goaf, in *Proceedings of the 6th International Mine Ventilation Congress,* pp 313–316.

Wade, L, Phillips, H R and Gouws, M J, 1987. Spontaneous Combustion of South African Coals, *Advances in Mining Science and Technology,* 1:65–73.

Xiang, Z, Si, G, Wang, Y, Belle, B and Webb, D, 2021. Goaf gas drainage and its impact on coal oxidation behaviour: A conceptual model, *International Journal of Coal Geology,* 248:103878.

Yuan, L and Smith, A C, 2009. Numerical Study On Spontaneous Combustion Of Coal in U S Longwall Gob Areas, *Mine Ventilation: Proceedings of the Ninth International Mine Ventilation Congress* (ed: D C Panigrahi), pp 263–271 (Oxford and IBH Publishing Co).

Zutshi, A, Ray, S K and Bowmick, B C, 2001. Indian coals vis-a-vis spontaneous heating problems, *Journal of Mines, Metals and Fuels,* pp 123–135.

Evaluation of the benefits of controlling the longwall shearer's cutting speed under methane inflow conditions based on simulations

W Dziurzyński[1], J Krawczyk[2] and P Skotniczny[3]

1. Professor, Strata Mechanics Research Institute of Polish Academy of Sciences, 30-059 Kraków, Poland. Email: wdziurzyn@gmail.com
2. Professor, Strata Mechanics Research Institute of Polish Academy of Sciences, 30-059 Kraków, Poland. Email: krawczyk@imgpan.pl
3. Director, Strata Mechanics Research Institute of Polish Academy of Sciences, 30-059 Kraków, Poland. Email: skotnicz@imgpan.pl

ABSTRACT

Ventilation air methane causes hazards and production limitations. The effects of currently applied preventive measures may be augmented by controlling the longwall shearer's cutting speed in response to the methane sensors indications.

Prior to a field test, the relevance of such a solution has been verified by computer simulations. This paper presents results of such a verification. It used a calibrated Virtual Longwall Model bound with an automatic system, based on a proportional–integral–derivative (PID) controller. The model applies a derivative of the in-house developed Mine Ventilation Network Simulator Ventgraph™. The simulator is capable of showing the flow and gas concentration transients caused by the shearer operation and related variability of the methane emission. Properties of the simulator and the verification of its credibility have been outlined and referenced. Two hard coalmines and one brown coalmine longwalls, all applying U-type ventilation systems have been considered. First, the computer models have been developed and calibrated upon the actual data obtained from mines and field measurements. The models have been tuned so as to obtain the best fit to actual readings of the flow and methane sensors monitoring mine environment during several cycles of mining. Then the effects of the control system has been compared with a standard pre-set constant speed of mining. The longwalls have operated at moderate hazard levels therefore no exceedances were recorded, so in the prediction of the effectiveness of the control system, the level of methane emissions was increased, preserving the other parameters of the models. Simulation results indicate that although the control system occasionally reduces the cutting speed considerably, in response to unfavourable trends in methane concentration, these limitations do pay off, as the flexible cutting velocity settings allow to accelerate and offset the temporary reduction of output and protect from the losses generated by the methane related emergency downtimes. The obtained results constitute the basis for further steps toward implementation of the system in real conditions.

INTRODUCTION

Previous experience in operating underground coalmines under conditions of methane hazards has resulted in a number of preventive measures, implemented both at the stage of design and operation. In Poland, like in many other countries methane concentrations are continuously monitored by pellistor or infrared sensors at specified locations. During the production planning, methane emissions are predicted taking into account the properties of adjacent strata (Janoszek and Krawczyk, 2022) and the technical conditions of operation, and are selected accordingly so as to minimise shutdowns (see Krawczyk and Wasilewski, 2009).

Despite those measures, during operation, for a number of reasons that are difficult to predict, local and temporary increases in methane emissions occur, which by regulations in many countries enforce an immediate shutdown of electric devices and withdrawal of personnel from the endangered zone. For example, in Poland the shutdown threshold is 2 per cent vol CH_4 for longwalls, while for return airways it is 1 per cent, or 1.5 per cent if monitoring is applied. For exhaust shafts the limit is 0.75 per cent. Mining may be restarted after checking that the concentration is below this value. Such non-productive period is termed as a forced break. In addition, in connection with the conduct of successive mining cycles, we have to deal with technological breaks resulting from changes in the direction of mining and the reconstruction of equipment (shields, conveyors).

In response to the variability of the level of hazard in time, the idea was to apply an automatic control system, which would set the mining speed of the shearer, according to the current methane concentration monitored by a dedicated set of sensors.

This idea inspired a number of research works, in particular the EU-funded international PICTO project (Krawczyk *et al*, 2023). As part of the project, virtual models of longwalls, representative of typical mining conditions of Polish single-layer longwalls and multilayer brown coal underground mining in Slovenia, were built through a series of *in situ* experiments in the longwall areas, analysis of data obtained during mining by the mine's methane hazard monitoring systems and numerical simulations. In particular, the numerical models of the longwalls for the Bielszowice, Jankowice and Velenje mines have been developed and calibrated.

For the aforementioned longwalls, the analyses have been conducted in a two-step procedure. First, the numerical models were calibrated so that the simulation results were as close as possible to the actual indications of the monitoring systems. The studied facilities operated at moderate hazard levels and no exceedances were recorded. Therefore, in the prediction of the effectiveness of the control system, the increased level of methane emissions was assumed, preserving the other parameters of the models. It should be noted that the parameters characterising mining and geological conditions for the three longwalls are identical to those during their operation, and the increased methane inflow is due to the assumption of an increased value of methane release during coal mining, which corresponds to a higher absolute coal methane content. Such cases are often encountered in mining practice are due to the uneven distribution of methane content in the exploited coal seam. For each longwall, several mining cycles were simulated both at a presumed constant speed and with the control system in operation. Then the results were compared, in particular the level of extraction and total methane emissions into the atmosphere. The results presented in section 3 show the benefits of using the control system.

TOOLS AND METHODOLOGY FOR NUMERICAL SIMULATION OF THE CONTROL OF THE MINING SPEED OF THE LONGWALL SHEARER

Virtual models of longwalls have been implemented in the version 17.0 of the Ventgraph™ mine ventilation networks simulator named VentgraphPlus™ (by Strata Mechanics Research Institute of Polish Academy of Sciences), customised to meet these task requirements. This in-house developed program is based on a one-dimensional flow approximation, a classic implementation of the Hardy-Cross method used in many simulators of this type (Hinsley and Scott, 1951). The functionality of Ventgraph™ has been significantly extended by adding a number of mathematical and numerical models (Dziurzyński, Pałka and Krach, 2021). They allow simulations of the flow transients, including inflow and transport of gases such as methane and filtration flow in goafs adjacent to workings. It is also possible to take into account the influence the longwall shearer operation, represented as moving local resistance and a distributed, time-varying source of methane (Dziurzyński, Krach and Pałka, 2018; Dziurzyński *et al*, 2007, 2020). These functionalities have been verified in a number of research projects and expert opinions (Pritchard, 2010, Dziurzyński *et al*, 2010).

The basic prerequisite for the correctness of the results of computer simulation of the ventilation of a longwall region with a shearer in operation is the proper selection of input values for the mathematical models used in the simulation program. These models describe the dependence of such parameters of the air ventilating the longwall as the volume flows and the volume share of methane in these streams. The Ventilation Engineer program VentgraphPlus™ allows to calculate the distribution of air and methane in a mine's ventilation network using a database that includes the topology of the ventilation network and the parameters of its elements, such as the aerodynamic resistances and geometric dimensions of airways and goafs, the elevation of junctions being the network nodes and air densities in the branches, and the distribution of methane inflow sources.

Sources of methane inflow in a longwall with a working shearer are the cut seam, the coal transported on conveyors, inflows from the longwall goaf and adjacent goafs, as well as from the seams below and above the mined panel (Drzęźla and Badura, 1980; Dziurzyński *et al*, 2019). The VentgraphPlus™ program includes mathematical models for calculating the distribution and intensity of methane inflow to the air flow in the longwall from the mined seam and the output on the longwall conveyor, taking into account the variability resulting from operation of the shearer and the conveyor. It is also possible to calculate the distribution and evolution of methane inflow from the coal on the conveyor in the haulage roadway. The calculated fluxes of inflowing methane allow to determine the time-varying distributions of methane concentration in the longwall and adjacent galleries. For these calculations, it is necessary to enter input data into the program. These are:

- Methane content of the mined seam.

- The trajectory of movement of the shearer in the longwall taking into account all technological phases like turn around, actual cutting and brakes after cut or for a shift change.

- The cut width and height of the mined seam.

- Speeds of conveyors.

In order to know the values of the mentioned quantities, it is necessary to collect data from documentation and mine monitoring systems and, if possible, to make in field measurements of the quantities that are feasible under the conditions of the working longwall.

The process of mining of coal with a longwall shearer is accompanied by a time-varying distribution of methane emissions from the coal seam being mined and from the coal transported by the longwall armored face conveyor (AFC) and further by the system of belt conveyors (Dziurzyński *et al*, 2021). In practice, the shearer moves at a variable speed, depending (among other things) on the current methane inflow conditions. In addition, the shearer may have stoppages, may mine in both directions of movement along the longwall, and conveyors may also stop when the shearer's power is turned off. Hence, the need arose to extend the model, taking into account such movement of the shearer and conveyors. The paper by Dziurzyński, Krach and Pałka (2018) presents an extended mathematical model considering those phenomena. Based on the modified models, new algorithms were developed and implemented in the VentgraphPlus™, describing methane emissions as a function of time, reflecting the effect of current state and history of the longwall operation.

The second important addition to the VentgraphPlus™ computer program is the introduction of a control system model, which automatically predicts the maximum speed of the shearer's advance taking into account the value of the measured methane concentration and the rate of its change. Previous work on validation of the model and calculation algorithms has assumed the recorded course of the shearer's operation in the form of its schedule (Dziurzyński *et al*, 2019, 2021).

In the control system tests, it was assumed that the shearer starts work with a known mining speed, and its speed is modified depending on the measurement by a virtual set of methane concentration sensors located at the longwall and methane meters in the air current flowing in and out of the longwall region. It is assumed that the optimal determination of the position of this sensor or several sensors will be possible on the basis of measurement experiments in the longwall region and multivariate simulations of air and methane flow during the simulation of the operation of the virtual mining set. The control algorithm, based on a proportional–integral–derivative (PID)-type controller adjusts the speed of coal extraction by the shearer depending on the increase or decrease of the measured values of methane concentrations during the operation of the shearer.

ESTIMATING BENEFITS BASED ON SIMULATIONS USING VIRTUAL LONGWALL MODELS

Air and methane flow analysis processes are too complex to allow easy prediction of the effectiveness of individual solutions. Empirically searching for a solution at a site operated in a potentially explosive atmosphere without previous extended verification is too risky. The proposed simulation tests provide an opportunity to study the problem.

The purpose of the conducted research was to determine the possibility of controlling the operation of a longwall shearer, in particular its speed by means of an automatic PID-type controller (Figure 1).

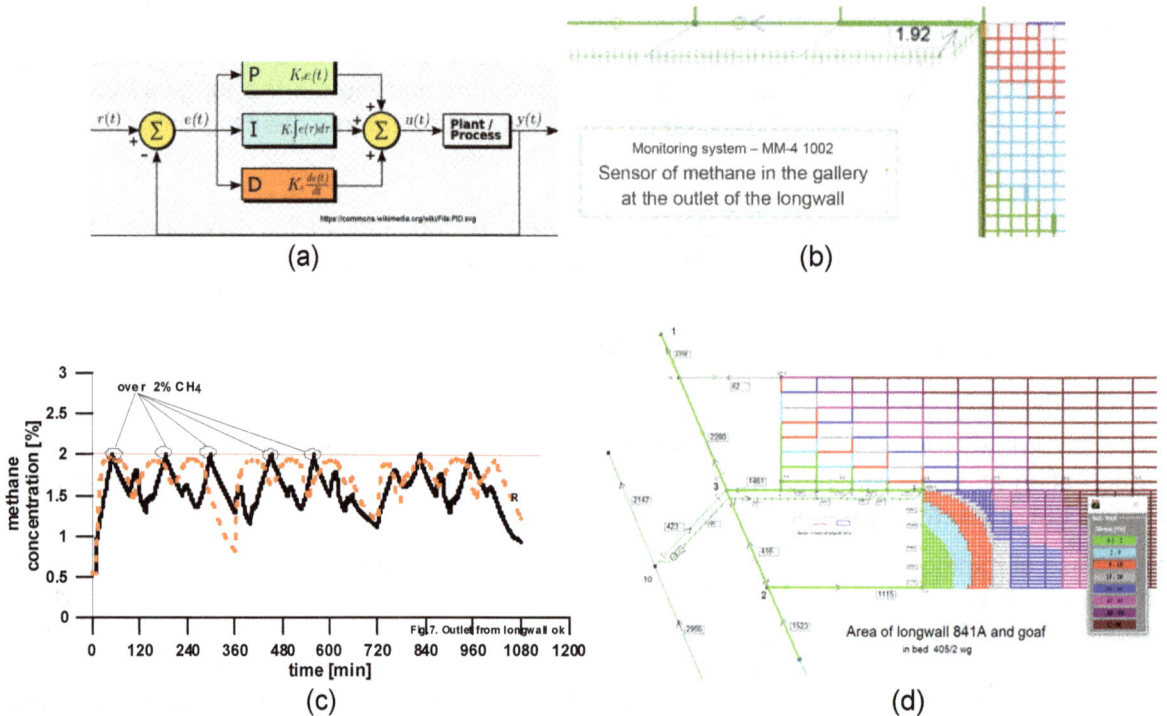

(a)

(b)

(c)

(d)

FIG 1 – The idea of control system modelling and verification: (a) PID controller, (b) virtual methane sensor position, (c) comparison of transients of methane concentration, (d) longwall region model coloured by methane concentration distribution.

The analyses considered three examples of simulations of mining speed control for numerical models of three longwalls (properties listed in Table 1). Prior to testing they have been calibrated on the basis of the actual object's data (ie for Example 1 reusing Dziurzyński *et al* (2010) data), with the examples of operation of a longwall shearer and haulage of excavated material under conditions of increased methane inflow considered:

- Example 1 – longwall 841A in the Bielszowice mine (Figure 2), PGG SA Poland.

- Example 2 – longwall BK-95, PMV Velenje, Slovenia applying LTCC system (Si *et al*, 2015).

- Example 3 – longwall Z-11A of the Jankowice mine, PGG SA Poland.

All three longwalls were ventilated by the 'U' system but had different lengths, seam heights and the level of methane inflow resulting from the mining and geological conditions of a particular longwall. No pre-drainage has been applied and only the Polish longwalls had gas drainage system operated during production. In all cases the mines had a long history and a complex multi-level structure and at least two longwalls operating simultaneously. For example, Bielszowice mine had three ventilation shafts and total exhaust of three fans was 47 000 m³/min. Jankowice mine also had same number of exhaust shafts and fans of total exhaust 37 800 m³/min, and the longwall seam 408/1 methane content was 4137 m³ CH₄/Mg. For examples 1 and 3 a full model of ventilation system has been developed. For the Slovenian mine a simplified network has been designed to provide adequate flow conditions for the longwall region. Those models have been calibrated by comparison with the mine monitoring data and available results of field measurements.

FIG 2 – Example 1 mine ventilation network scheme.

For each longwall the monitoring data did not contain records of shutdowns. During the simulations testing the control system the release of methane has been slightly increased without alteration of all other model properties.

TABLE 1

Parameters of longwalls.

Item	Longwall extent and exploitation parameter	Unit	Example 1	Example 2	Example 3
1	Longwall length	(m)	130	160	140
2	Longwall height	(m)	2.9	4	2.5
3	Panel length	(m)	550	400	220
4	Shearer cut width	(m)	0.8	1.0	0.7
5	AFC speed	(m/s)	1.33	1.08	1.08
6	Belt conveyer speed	(m/s)	2.0	1.33	1.33
	Ventilation and methane inflow parameters	unit			
7	Methane inflow stream provided with fresh air current	(m³/min)	3.00	1.09	0.68
8	Estimated absolute methane release (sum no.= 9+10+11)	(m³/min)	39.66	12.05	18.78
9	Average methane intake by methane drainage	(m³/min)	11.16	0	**30.83**
10	Methane inflow to goaf	(m³/min)	15.40	5.4	4.2
11	Methane inflow from the face of the longwall and the mined coal seam	(m³/min)	13.10	6.65	14.58
12	Airflow at longwall	(m³/min)	1085	1818	1350
13	Air duct flow	(m³/min)	259	237	0
14	Ventilation methane emission	(m³/min)	20.9	10.27	**5.2**

The scheme of the Ventgraph's ventilation network model for longwall 841A is shown in Figure 2. The longwall uses U type ventilation with dilution of methane by auxiliary ventilation duct (denoted with line) discharging air at the longwall outlet. The shearer is represented by increase in branch resistance and a specific distribution of methane inflow in its vicinity. As the shearer moves, the resistance position and methane inflow is modified accordingly. Rectangular grids represent the filtration flow in goafs adjacent to previous and currently operated longwall. The lines representing the galleries and flow paths in the adjacent goaf are coloured according to methane concentration. For calculations of the amount of methane released during the simulation of mining, the conditions at the end of a gallery discharging used air from the longwall (Figure 1b), namely the volumetric flow rate and methane concentration were recorded.

The calculations were carried out for the case of using the mining speed control system and, for comparison of the results, for the case of a pre-set constant mining speed. Transients of methane concentration for both modes of operation presented in Figure 1c shows that for constant cutting speed the maximum allowed concentration has been exceeded several times, which caused emergency power shutdowns and production breaks. Properly tuned controller timely reduces the cutting speed marinating the concentration below the threshold. Obtained results indicate that controlling the shearer's mining speed is possible and can result in higher output. They are summarised in Table 2 for longwall 841A of the Bielszowice Mine, in Table 3 for the longwall Bk-95 of Velenje Mine, and in Table 4 for the longwall Z-11A of Jankowice Mine. They contain a comparison total amount of methane flowing out of a region, total tonnage of production and a volume of methane per ton of mined coal. In all cases this ratio has been smaller for the controlled operation and the reduction level termed as Ratio difference varies from 5.1 to 22.6 per cent.

TABLE 2

Example 1 controlled versus constant cutting speed – comparison of output and CH_4 emission.

Bielszowice longwall 841			
Total methane inflow in 300 min (m^3)	Mining (t)	Ratio (m^3/t)	Ratio difference (%)
with speed control	7689	1013	7.6	5.1
without speed control	7079	885	8.0	

TABLE 3

Example 2 controlled versus constant cutting speed – comparison of output and CH_4 emission.

Velenje longwall BK-95			
Total methane inflow in 300 min (m^3)	Mining (t)	Ratio (m^3/t)	Ratio difference (%)
with speed control	9359	3325.6	2.8	20.7
without speed control	8351	2353.2	3.5	

TABLE 4

Example 3 controlled versus constant cutting speed – comparison of output and CH_4 emission.

Jankowice longwall Z11a				
	Total methane inflow in 300 min (m^3)	Mining (t)	Ratio (m^3/t)	Ratio difference (%)
with speed control	6470	1357	4.8	22.6
without speed control	5615	913	6.2	

Figures 3, 4 and 5 show the amount of coal extraction obtained when mining the shearer with the control system on (blue curve), and in the second case with the control system off (green curve).

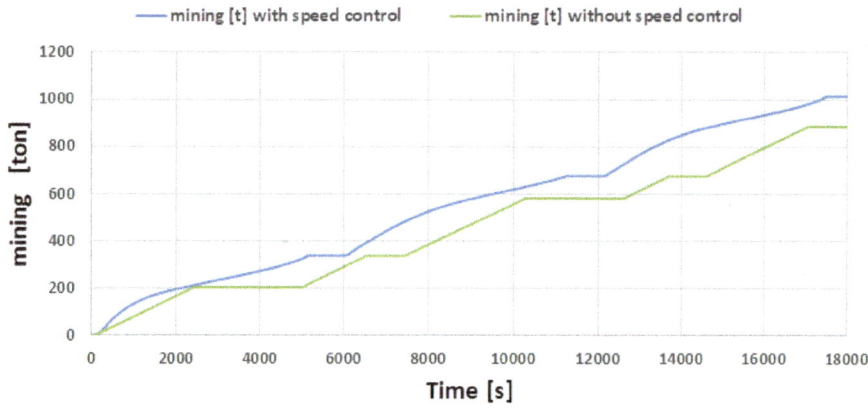

FIG 3 – Mining output (tons), Bielszowice longwall 841A.

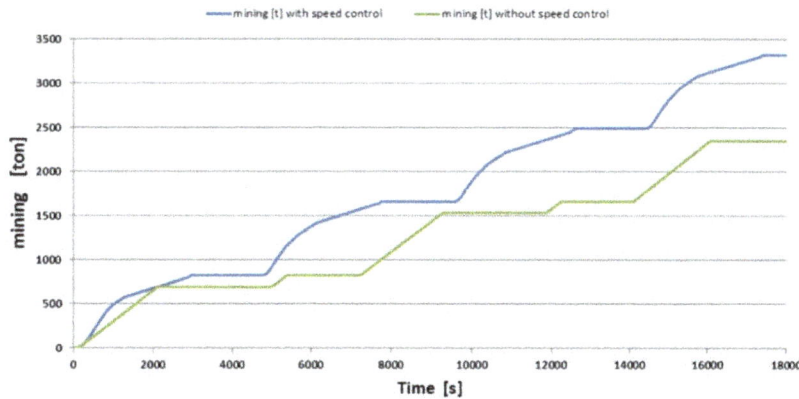

FIG 4 – Mining output (tons), Velenje longwall BK-95.

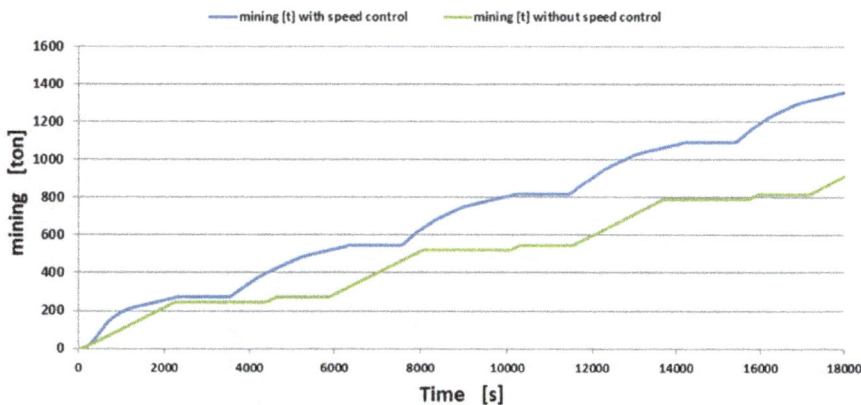

FIG 5 – Mining output (tons), Jankowice longwall Z11A.

Figures 6, 7 and 8 show the amount of methane per unit of coal mining (the amount of methane per unit of coal mining) during the operation of the shearer with the control system on (brown curve) and in the other case with the control system off (purple curve).

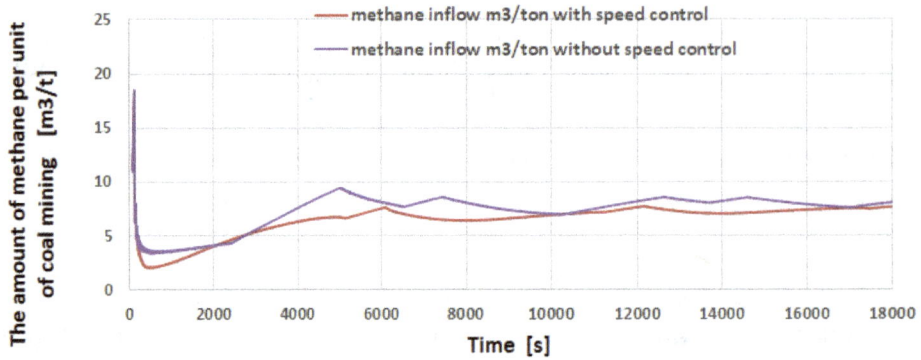

FIG 6 – The amount of methane per unit of coal mining Bielszowice longwall 841A.

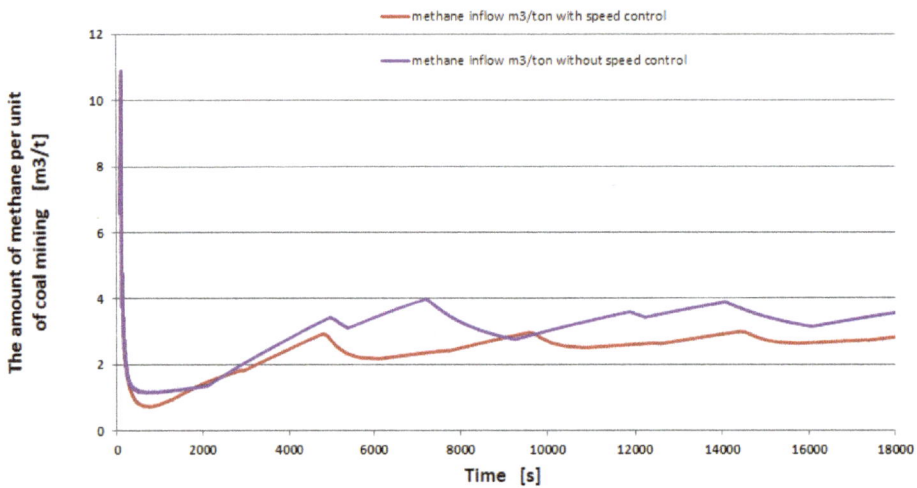

FIG 7 – The amount of methane per unit of coal mining Velenje longwall BK-95.

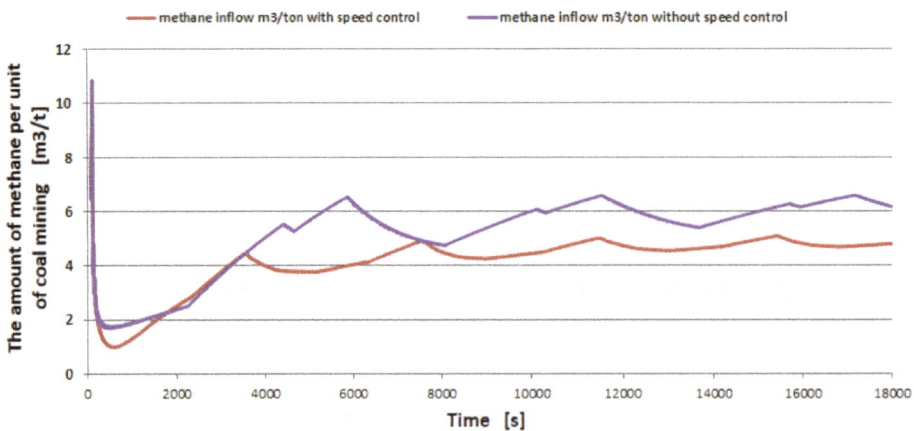

FIG 8 – The amount of methane per unit of coal mining Jankowice longwall Z11A.

Simulations and calculations show that with the use of a shearer controller, the amount of methane released is lower, assuming the same mining volume. This is an interesting conclusion, showing the additional beneficial effect of using an automatic control system for the mining speed of the shearer. The percentage of reduction in the Ventilation Air Methane emission resulting from the cutting speed control, is listed in the last column of Tables 2, 3 and 4. Shearer controls have been used in Australia for over a decade and these should not be seen as a primary control over gas pre-drainage prior to longwall mining, unless the pre-drainage is not effective due to the geological conditions, like the ones specific for Poland.

CONCLUSION

As shown in section 3, the simulations performed on calibrated models of longwalls indicate, that the use of a shearer mining speed controller allows a higher extraction while reducing the release of methane into the atmosphere. The benefits obtained are summarised in Table 2 for longwall 841A of the Bielszowice Mine, in Table 3 for longwall Bk-95 of the Velenje Mine, and in Table 4 for longwall Z-11A of the Jankowice Mine. Computer simulations have shown that although the control system sometimes significantly slows the shearer when trends in methane density are unfavourable, these limitations pay off, as flexible settings of the mining speed allow acceleration where possible and protect against losses generated by methane-related emergency stoppages (Figure 1c). Simulations and calculations also show that the amount of methane released is lower with the shearer controller, assuming the same mining volume which could be an additional benefit of the automatic control system.

Such a system will be a fully innovative extension of the automation of longwall operation under methane hazard conditions, aimed at:

- increase productivity as well as safety by reducing the number of emergency stoppages
- reducing the presence of crew in the longwall area
- reducing the negative environmental impact of uncontrolled methane release from coalmines.

Simulations carried out on calibrated models for three mines showed the wide possibilities of the benefits obtained, which depend mainly on the conditions of methane release. The developed simulation tool provides a lot of cognitive value of the longwall ventilation process, in which mining is carried out with control of the mining speed of the shearer. The estimated benefits using the virtual longwall model depend on a number of adopted parameters and of particular importance is the choice of parameters of the control controller itself.

The results obtained provide the basis for the planned implementation of the system in real conditions.

By now the automatic control of longwalls consider mining and geological conditions as well as technical parameters of applied devices and systems (Fiscor, 2017, 2024; Peng *et al,* 2020). However, so far there is no information on the practical implementation of the automation with regard to the gas hazards; however this subject has attracted interest of researchers (Trenczek *et al,* 2020).

ACKNOWLEDGEMENTS

The study was carried out as part of the PICTO research project titled 'Production Face Environmental Risk 403 Minimization in Coal and Lignite Mines', No. 800711, financed by the European Community Research Programme of the Research Fund for Coal and Steel (RFCS) and Polish MNiSW W93/FBWiS/2018. Authors also acknowledge the contribution of the many researchers and experts from the PICTO project consortium.

REFERENCES

Drzęźla, B and Badura, H, 1980. Przybliżony rozkład stężenia metanu emitowanego z urobku (Approximate distribution of methane concentrations emitted from mined coal), *Arch Min Sci,* 25(2).

Dziurzyński, W, Krach, A and Pałka, T, 2018. Shearer control algorithm and identification of control parameters, *Arch Min Sci,* 63(3):537–552.

Dziurzyński, W, Krach, A, Krawczyk, J and Pałka, T, 2020. Numerical Simulation of Shearer Operation in a Longwall District, *Energies.* https://doi.org/10.3390/en13215559

Dziurzyński, W, Krach, A, Pałka, T and Wasilewski, S, 2007. Validation of the computer ventilation simulation program VentMet for the longwall area, taking into account the time-varying methane sources associated with the shearer's cyclic operation (in Polish), *Trans Strata Mech Res Inst,* 9:3–26.

Dziurzyński, W, Krach, A, Pałka, T and Wasilewski, S, 2010. Digital simulation of the gas dynamic phenomena caused by bounce, experiment and validation, *Arch Min Sci,* 55:403–425.

Dziurzyński, W, Krach, A, Pałka, T and Wasilewski, S, 2019. Methodology for determining methane distribution in a longwall district, *Arch Min Sci,* 64:467–485. https://doi.org/10.24425/ams.2019.129363

Dziurzyński, W, Krach, A, Pałka, T and Wasilewski, S, 2021. The Impact of Cutting with a Shearer on the Conditions of Longwall Ventilation, *Energies,* 14(21):6907. https://doi.org/10.3390/en14216907

Dziurzyński, W, Pałka, T and Krach, A, 2021. Modele matematyczne programu VentGraph Plus (Mathematical models of VentGraph Plus program), monograph ed, Instytut Mechaniki Górotworu PAN 2021.

Fiscor, S, 2017. Improved safety and productivity through advanced shearer automation, *Coal Age,* 122:16–23. https://www.coalage.com/flipbooks/july-august-2017/

Fiscor, S, 2024. Longwall operators add capacity for 2024. *Coal Age,* pp 20–27, https://www.coalage.com/flipbooks/january-february-2024/

Hinsley, F B and Scott, R, 1951. *Ventilation Network Theory,* Coll Eng, 28:29.

Janoszek, T and Krawczyk, J, 2022. Methodology development and initial results of CFD simulations of methane distribution in the working of a longwall ventilated in a short Y manner, *Arch Min Sci,* 67(1):3–24. https://doi.org/10.24425/ams.2022.140699

Krawczyk, J and Wasilewski, A, 2009. Migration of Methane into Longwall and Tailgate Crossing, 9th Int Mine Vent Congress, India, vol 1, pp 483–494.

Krawczyk, J, Dziurzyński, W, Ostrogórski, P, Skoczylas, N, Kudasik, M, Durucan, S and Pierburg, L, 2023. Controlled Longwall Shearer Operation Aimed at the Methane Hazard Reduction, in *Proceedings of the 26th World Mining Congress,* pp 3555–3567 (Commonwealth Scientific and Industrial Research Organisation (CSIRO)).

Krawczyk, J, Dziurzyński, W, Wasilewski, S, Skotniczny, P W, Krach, A, Pałka, T, Jamróz, P, Janus, J, Ostrogórski, P, Skoczylas, N, Kudasik, M, Pajdak, A, Skiba, M, Krause, E, Krzemień, A, Skiba, J, Janoszek, T, Jura, B, Hildebrandt, G, Durucan, S, Korre, A, Yuan, X, Agrawal, H, Cao, W, Pierburg, L, Deimel, T, Haarmann, D, Bezak, B, Potoczek, H, Jędrzejek, K, Słowik, A, Rošer, J, Jamnikar, S, Uranjek, G and Sedlar, J, 2022. 'Production Face Environmental Risk Minimisation in Coal and Lignite Mines' (PICTO), report of the RFCS 800711 PICTO project, European Commission Research Programme of the Research Fund for Coal and Steel Technical Group, Instytut Mechaniki Górotworu Polskiej Akademii Nauk, 63 p. Available from: <https://imgpan.pl/projekty/picto/>

Peng, S S, Du, F, Cheng, J and Li, Y, 2020. Automation in US longwall coal mining: A state-of-the-art review, *International Journal of Mining Science and Technology.* https://doi.org/10.1016/j.ijmst.2019.01.005

Pritchard, C J, 2010. Validation of the Ventgraph program for use in metal/non-metal mines, 13th United States/North American Mine Ventilation Symposium (eds: S Hardcastle and D L McKinnon), (MIRARCO: Sudbury).

Si, G, Shi, J Q, Durucan, S, Korre, A, Lazar, J, Jamnikar, S and Zavšek, S, 2015. Monitoring and modelling of gas dynamics in multi-level longwall top coal caving of ultra-thick coal seams, Part II: Numerical modelling, *Int J Coal Geology,* 144:58–70.

Trenczek, S, Lutyński, A, Dylong, A and Dobrzaniecki, P, 2020. Controlling the longwall coal mining process at variable methane hazard, *Acta Montanistica Slovaca,* 25(2):159–169. https://doi.org/10.46544/AMS.v25i2.3

Research on the process of cloud formation and the induced explosion from smouldering deposited coal dust with varying degrees

J Fan[1], L Ma[2] and X L Fan[3]

1. PhD student, College of Safety Science and Engineering, Xi'an University of Science and Technology, Shaanxi Xi'an 710054, China. Email: fanjing@stu.xust.edu.cn
2. Professor, College of Safety Science and Engineering, Xi'an University of Science and Technology, Shaanxi Xi'an 710054, China. Email: mal@ xust.edu.cn
3. PhD student, College of Safety Science and Engineering, Xi'an University of Science and Technology, Shaanxi Xi'an 710054, China. Email: fanxinli1997@163.com

ABSTRACT

In thermal environments, the accidental uplift of self-igniting coal dust can lead to the formation of a dust cloud and trigger combustion, which is a key factor in dust explosion incidents within pulverising systems. To reveal the explosion process and dynamic mechanisms induced by self-ignition of coal dust in thermal environments, this paper establishes an experimental system for coal dust combustion explosions and investigates the ejection and explosion processes of deposited coal dust under different smouldering degrees. The study also discusses the self-ignition characteristics of deposited coal dust, dust explosion behaviour, and the resulting product gases. Using the Fluent software, the smouldering coal dust-induced explosion process was simulated, analysing the spatiotemporal evolution of particle trajectories, velocity, and temperature during the diffusion process. The results indicate that when the temperature of smouldering coal dust is between 275°C and 395°C, the resulting dust cloud can induce an explosion, and the intensity of the explosion first increases and then decreases with the degree of smouldering. The formation of the coal dust cloud includes three stages: rapid ejection of particles, decelerated diffusion, and free diffusion, with the dust cloud explosion primarily occurring during the latter two stages. Combustible gases such as CO accumulated during the smouldering stage are ignited by high-temperature particles, leading to gas-phase combustion, which along with the high-temperature particles igniting the coal dust, is the main cause of the smouldering coal dust-induced explosions. Before implementing coal dust blowing procedures in pulverising systems, detecting and reducing CO concentration is an important measure to reduce the risk of coal dust explosions.

INTRODUCTION

Coal is a key energy source in industries like thermal power, coal chemical production, and cement manufacturing (Chang *et al*, 2016). Commonly, coal is milled into a fine powder for secondary processing applications (Yan *et al*, 2022; Yadav and Mondal, 2019). However, the production, transport, and storage of pulverised coal generate substantial dust. When this dust settles on hot surfaces and remains there, it can self-ignite. Furthermore, the airborne spread of such combusted coal dust heightens explosion risks. Consequently, investigating the formation of coal dust clouds and how explosions initiate is vital to mitigate explosion hazards.

Currently, numerous scholars have conducted studies on the mechanism of coal dust explosions. Coal dust, like other combustible dusts, requires specific conditions to explode, including oxygen, an ignition source, a certain concentration, and degree of dispersion (Eckhoff and Li, 2021). Khan *et al* (2022) identifies four ignition sources: electric sparks, open flames, static electricity, human-induced ignition, and friction, the presence of a dispersed coal dust cloud is a prerequisite for such explosions. Moroń and Ferens (2024) found through experiments that there is no interdependence between the parameters of dust fire and explosion. The critical factor for dust ignition is the temperature at which volatile release rate peaks, rather than the dust's physicochemical properties. Shi *et al* (2021) reports that thermal runaway of pulverised coal particles takes place through the ignition of volatiles at temperature exceeds 1100K. Nematollahi *et al* (2020) studied the preheating, drying, and pyrolysis processes of coal dust particles, as well as heterogeneous and homogeneous reactions. Scholars have also used CFD software to investigate the mechanisms of dust explosions. Glushkov, Strizhak and Vershinina (2014) and Glushkov, Kuznetsov and Strizhak (2015) examined the heat transfer, thermal decomposition, volatile diffusion, and oxidation reactions in coal dust particle explosions.

The study found that in addition to non-homogeneous phase reactions, coal particle explosions also have a volatile fraction gas phase and partial ignition of coke. Zhi-ming and Xiao-qiang (2010) believes that the amount of volatiles is critical for the occurrence of homogeneous combustion during coal dust explosions.

Scholars have also conducted a large amount of research on the explosion process of deposited coal dust. Jia and Tian (2022) analysed how gas explosions trigger coal dust detonations, detailing the sequence and strength of carbon particle blasts within combustible gas-rich dust clouds. According to Lin *et al* (2020), deposited coal dust can intensify gas explosions, resulting in longer and brighter flames. Guo *et al* (2020) suggests that higher volatile and gas content in coal dust can increase explosion intensity, the gaseous products of coal dust explosions are mainly CO, H_2, and CH_4 (Qian *et al*, 2023). Song and Li (2020) determined that external factors cause suspended coal dust to ignite with increased violence, reaching peak flame temperatures of approximately 4500°C. Liu *et al* (2021a) found that as the weight of deposited coal dust increased, the explosion flame propagation distance and duration initially increased and then decreased. Scholars have also used numerical simulations to study the effect of particle size on explosion (Song and Zhang, 2020; Liu *et al*, 2021c). Ray *et al* (2020) simulated the dispersion process of coal dust particles in a 20 L sphere using the Discrete Particle Model (DPM). Rao *et al* (2023) combined experimental comparisons on the basis of simulated and found that heterogeneous processes mainly control the explosion. Wang *et al* (2020) also described the particle tracking and devolatilisation process under high-temperature conditions. The fluent is well suited to simulate the flame propagation, CO/CO_2 generation, and thermochemical conversion process characteristics of coal dust explosion (Liu *et al*, 2021d, Islas *et al*, 2023).

Most of the current studies are in the context of coalmine roadways, focusing on the external ignition energy, optimal ignition time, and explosion intensity during coal dust explosions. However, In industrial settings, smouldering coal dust may also transition to and induce explosions. These explosions, initiated by high-temperature coal dust as an ignition source, involve two different scales of combustion processes. The mechanisms of such occurrences have been less frequently studied by scholars. This paper investigates the self-ignition and subsequent explosion of deposited coal dust and reveals the mechanisms behind the formation of self-igniting coal dust clouds and the evolution of induced explosions.

THEORY AND EXPERIMENT

Experimental set-up

Based on the hot plate in ASTM E2021-2015 and the BAM oven from ASTM E1491-06 (2019), a coal dust combustion-explosion experimental apparatus system was constructed, as shown in Figures 1 and 2. The self-ignition chamber is a rectangular body with a cross-sectional size of 15 cm × 15 cm, and the explosion chamber is cylindrical with a radius of 15 cm. The former is used to generate self-igniting dust, which is then injected into the latter by high-pressure gas.

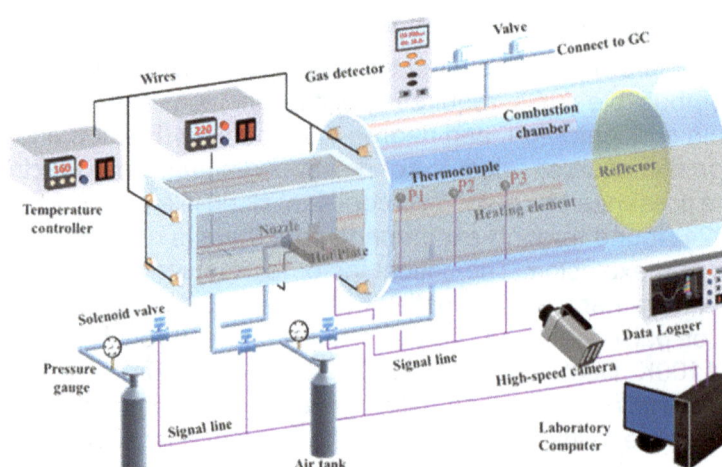

FIG 1 – Schematic diagram of the device.

FIG 2 – Coal dust combustion-explosion device.

The experimental furnace maintains an internal temperature of 120°C, while the hot plate is set to 230°C. A sieving machine is used to obtain lignite coal dust samples (d=58~75) µm. These samples (the results of industrial analysis are shown in Table 1) are then placed in a drying oven at 60°C for 8 hrs. A 6.0 g sample of coal dust is placed on the hot plate, and four thermocouples are arranged at 2 mm intervals within the coal dust layer to monitor the internal temperature and smouldering degree. Self-ignition occurs within the coal dust when the temperature measured by the central thermocouple exceeds the hot plate temperature by 50°C. Coal dust is transported into the explosion chamber using high-pressure gas at 0.4 Mpa and the injection time is 0.05 sec. An explosion is deemed to have occurred when a flame longer than 60 mm is formed inside the explosion chamber (Wu *et al*, 2016).

TABLE 1

Industrial analysis results

Parameter	Mad/%	Aad/%	Vad/%	FCad/%
Value	11.97	4.55	35.98	46.12

Simulation theory

This paper models and simulates the development of self-igniting coal dust clouds and their explosion mechanisms, applying conservation of mass and energy and chemical equilibrium.

1. Continuous phase and discrete phase model: The N-S equation and k-ε model were used to describe the gas-phase flow and turbulent flow processes, respectively. Particle motion and coal particle combustion are traced using the Discrete Phase Model (DPM) (Li *et al*, 2020).

2. The droplet evaporation model: When coal dust particles move close to a high-temperature ignition area, the moisture within the coal dust particles will be removed.

3. Combustion Model: During the evaporation and pyrolysis processes of coal particles, a single kinetic rate model is employed to describe the volatile release from coal particles. The composition of the combustible volatiles is very complex, comprising CO, H_2, $CxHy$, and others (Si *et al*, 2021; Shi *et al*, 2021; Glushkov, Kuznetsov and Strizhak, 2018), which are typical of homogeneous combustion (Liu *et al*, 2021b). The overall reaction rate is controlled by turbulent mixing, the chemical reaction rate is controlled by the large eddy mixing time scale, k/ε. The homogeneous gas reaction rate is determined by the minimum value between the chemical reaction rate and the turbulent mixing rate (Cao *et al*, 2017; Chen *et al*, 2017). The combustion of solid char is assumed to be a two-step or multi-step parallel reaction (Yuan, Restuccia and Rein, 2020).

4. Radiation Model: The P-1 radiation model is used to accurately depict thermal radiation interactions between gas and particle phases in a coal dust explosion (Wang *et al*, 2014).

Simulation parameter settings

The model is constructed using fluent meshing and is the same size as the actual model. The computational domain is the interior region of the explosion chamber, with the exterior being an open boundary. The grid is divided into hexahedra with mesh refinement at the boundary layer, achieving a mesh quality of 0.75. The model is shown in Figure 3. Assuming the coal dust particles are spherical, and considering the collisions and thermal radiation between coal particles, the chemical reactions and intermediate species of combustion can be neglected. The temperature of the self-igniting coal dust layer is simplified to discrete particles with the same temperature. Other parameters remain consistent with the experimental set-up. The injection velocity is calculated based on the injection pressure, using Bernoulli's equation and the continuity equation.

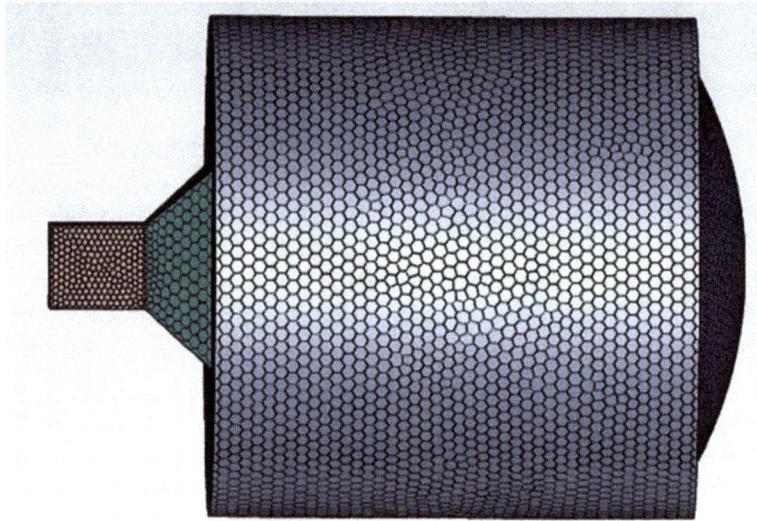

FIG 3 – Diagram of the simulation model.

EXPERIMENTAL RESULTS ANALYSIS

Coal dust self-ignition process

The degree of spontaneous combustion of coal dust is an important parameter in determining the injection time. Figure 4 shows the self-ignition process of stacked coal dust at an ambient temperature of 130°C and a hot plate temperature of 235°C.

- In 0~920 secs, thermal conductivity dominates, and the internal temperature of the coal dust layer rises gradually, and the closer to the hot plate the higher the temperature. The emitted moisture forms white smoke on the surface of the pulverised coal.

- After the internal temperature is the same as that of the hot plate, it enters the negative combustion stage, and the chemical reaction exotherm causes the internal temperature to rise sharply, and cracks appear on the surface of the coal dust. According to the definition of thermal runaway, the ignition delay time of this experiment is 1050 secs, and the thermal runaway temperatures of the positions with heights of 6.0, 4.0, and 2.0 mm are 339, 297, and 271°C, respectively.

- 1300~1600 secs is the combustion stage, the surface cracks gradually expand, more oxygen and coal dust react, the combustion rate and temperature are relatively stable.

- The coal dust enters the decay stage after the combustion is completed, the upper surface of the coal dust layer is exposed to more oxygen, the decay occurs earlier, and the temperature of the coal dust decreases.

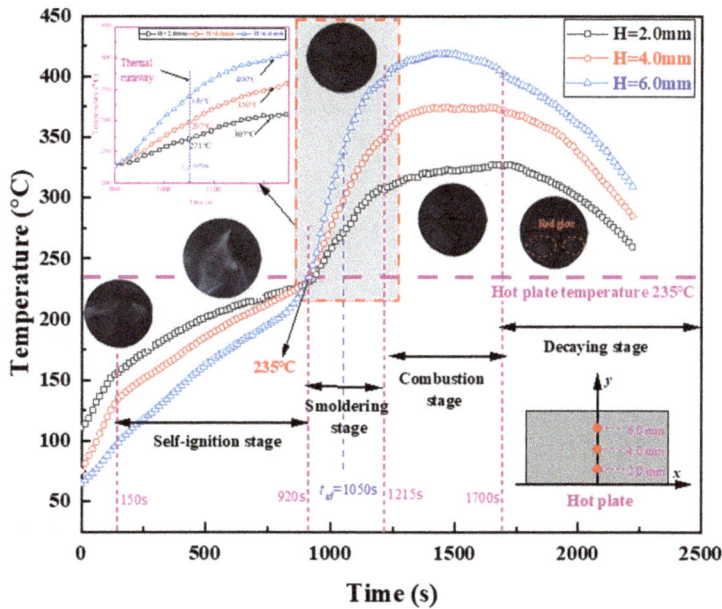

FIG 4 – Temperature distribution of coal dust.

The TG/DTG results depicted in Figure 5 exhibit commendable consistency with the hot plate experiment outcomes. The ignition point temperature is identified to be 340°C, which is close to the maximum temperature at the moment of thermal runaway. The combustion stage temperature ranges between 397°C and 512°C, similar to the initial temperature of the combustion phase in Figure 4, indicating a vigorous reaction between coal dust and oxygen during this stage.

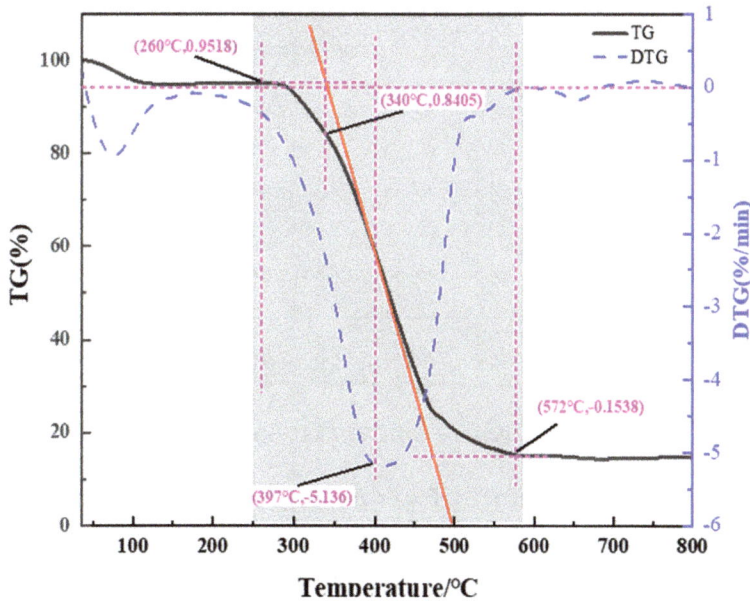

FIG 5 – TG/DTG curves under air at 10°C/min.

Figure 6 shows the minimum ignition temperatures for coal dust layers of varying thicknesses. A 10 mm layer ignites around 230°C. With increased thickness, enhanced internal exothermic oxidation and reduced surface convective heat loss lower the required hot plate temperature for thermal runaway. Based on the fitting results of Frank-Kamenetskii theory (Frank-Kamenetskii, 2015) as shown in Figure 7, the activation energy E of the coal dust is obtained as 108 kJ/mol, and the pre-exponential factor qA value is 1.39×10^{11} J/kg sec.

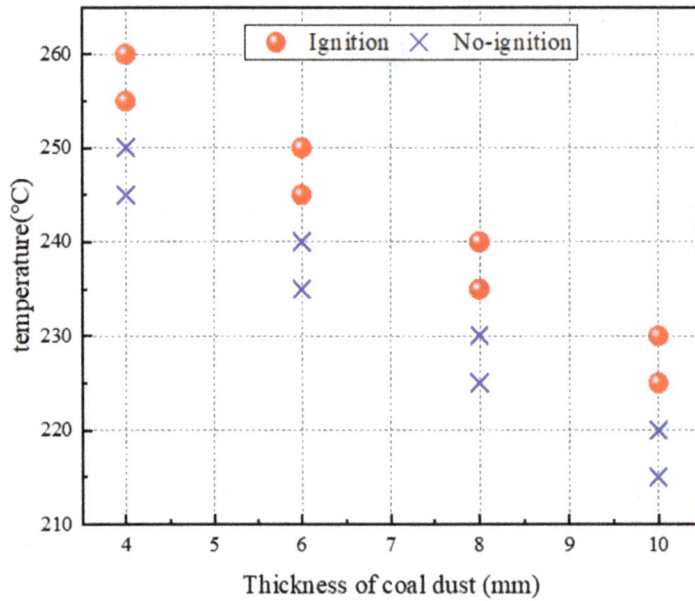

FIG 6 – Minimum ignition temperature at different thicknesses.

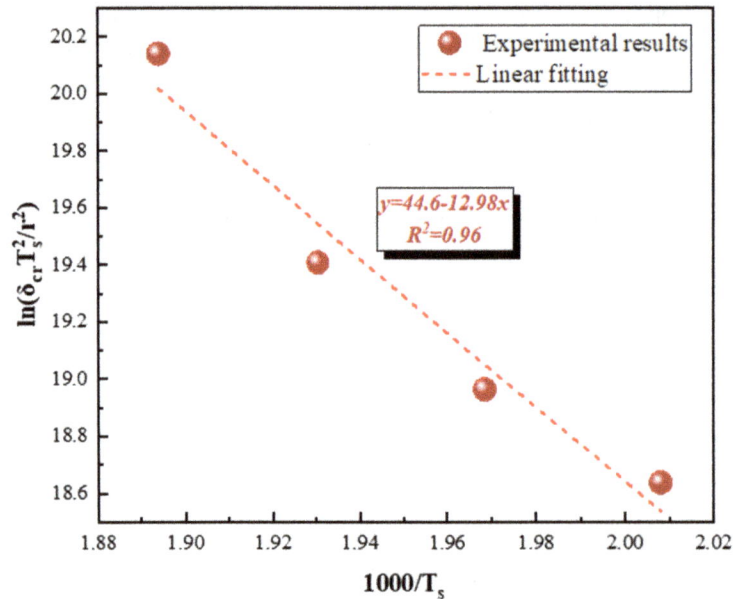

FIG 7 – Linear fit results.

Coal dust induced explosions at different smouldering intensities

After smouldering coal dust is dispersed into the air, the temperature of the particles and the content of volatiles are key to inducing an explosion. Figure 8 shows the results of the internal temperature changes in coal dust before ejection (an explosion occurred after injecting). The temperature of the dust layer reached 234°C after 912 secs. The temperatures at thermal runaway were 340°C, 291°C, and 264°C, respectively, and the temperatures at the time of ejection were 400°C, 350°C, and 300°C, respectively. These results are consistent with the self-ignition results of the dust layer, indicating that the self-ignition process experiments of the coal dust layer are repeatable.

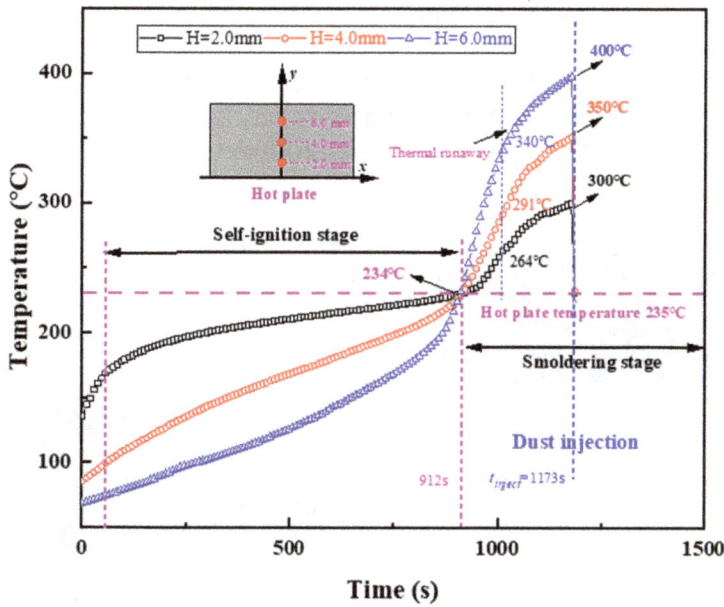

FIG 8 – Temperature in self-igniting dust layer.

Figure 9 illustrates the gas composition within the combustion chamber before coal dust dispersal. CO and H_2, products of the coal dust layer's self-ignition, are the main volatile gases. Pre-explosion CO levels were measured at 3940 ppm, dropping to 842 ppm post-explosion, while H_2 concentrations decreased from 762 ppm to 98.6 ppm after the blast. Oxygen content was recorded at 17.7 per cent before the explosion and fell to 15.25 per cent afterwards. Alkane levels remained low throughout, with ethylene (C_2H_4) being the most significant at 27 ppm. The consumption of these volatile gases during the explosion significantly contributes to their reduced concentrations.

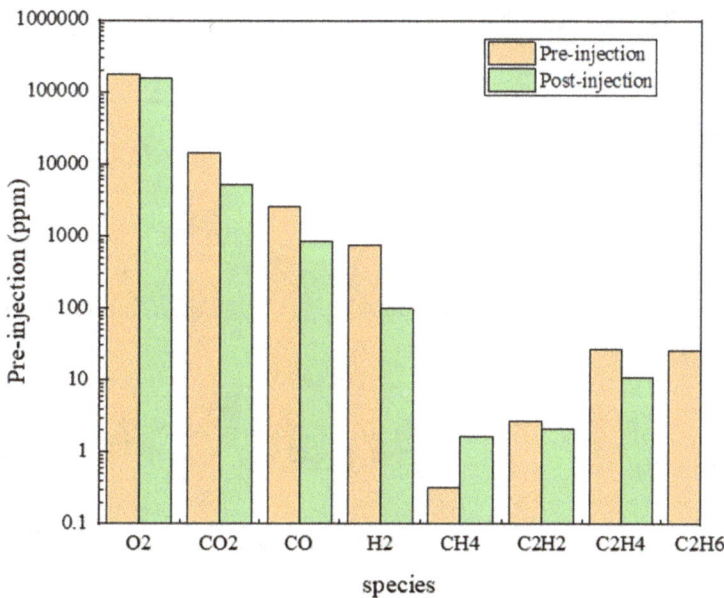

FIG 9 – Gas composition before dust injection.

To examine the correlation between the intensity of self-ignition and the propensity to trigger an explosion, the temperature at the 4.0 mm mark served as a benchmark. Injections at varying temperatures post thermal runaway were conducted to determine the likelihood of an explosion. The outcomes of these tests are depicted in Figure 10. When the smouldering temperature is 280°C or 360°C, the sparks from the dust injection do not ignite; between temperatures of 300°C to 340°C, the degree of explosion does not linearly correlate with the degree of smouldering. The flame is brightest at 320°C, with deep yellow unburned substances still present at 300°C, and partial combustion occurs at 340°C, with the flames not filling the combustion chamber. At low levels of smouldering, more heat is required for an explosion, and the heat generated from the combustion of particles or volatiles is insufficient to maintain flame propagation; at high smouldering levels, the content of incompletely burned particles is low, and the solid residues from combustion suppress the explosion. Consequently, The temperature range at which smouldering coal dust can explode is shown in Figure 11, that is, when the internal temperature is between 270°C and 392°C, there is a risk of explosion after the coal dust is dispersed.

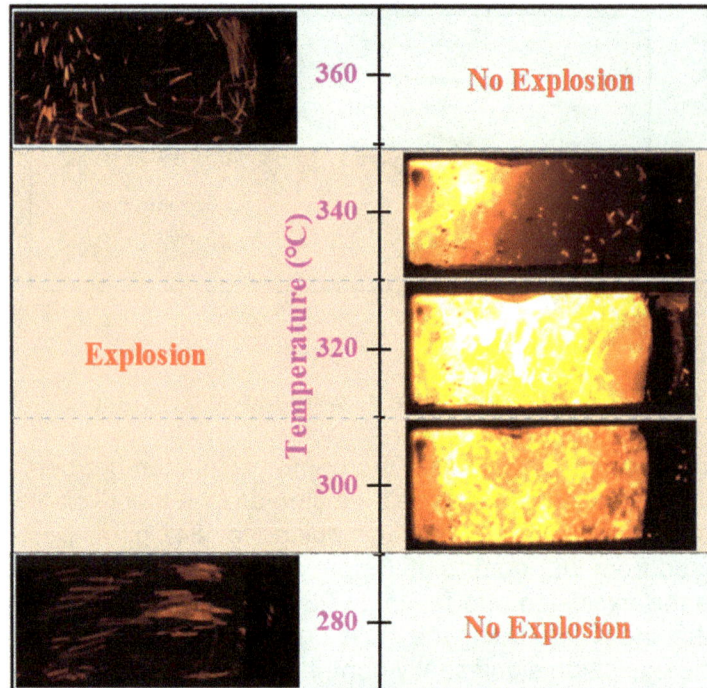

FIG 10 – Results of self-ignition induced explosions.

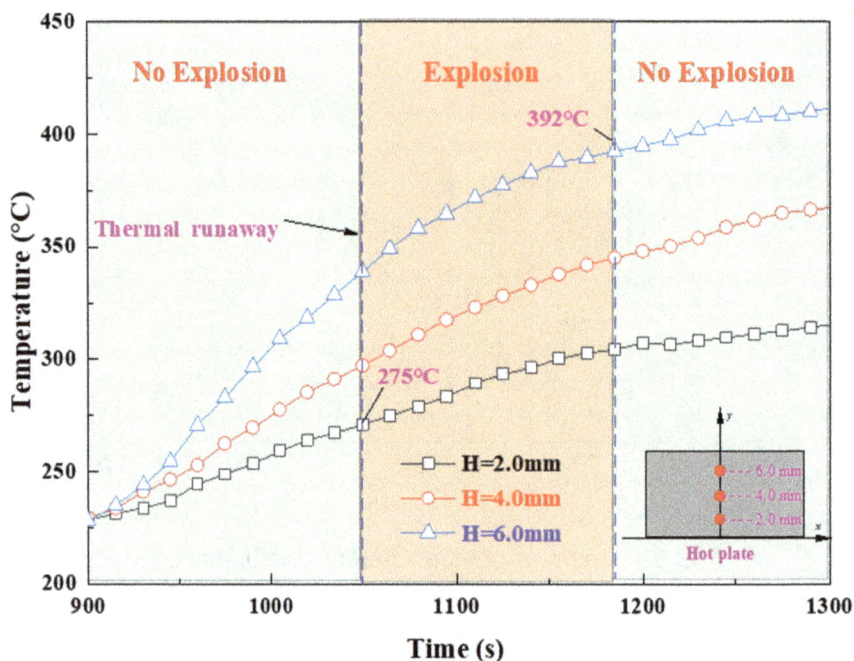

FIG 11 – Range of explosion.

Figure 12 depicts the explosion process following the ejection of smouldering dust. Within 90 ms after the dust is ejected, the brightness gradually increases as particle oxidation releases heat and further raises the temperature. At 120 ms, ignited particles ignite the gas above, forming a small flame; after an additional 30 ms, the flame rapidly expands to form a bright explosion flame. This indicates that the ignition of gas-phase combustion by the high-temperature particles from the dispersed dust is a significant cause of explosions.

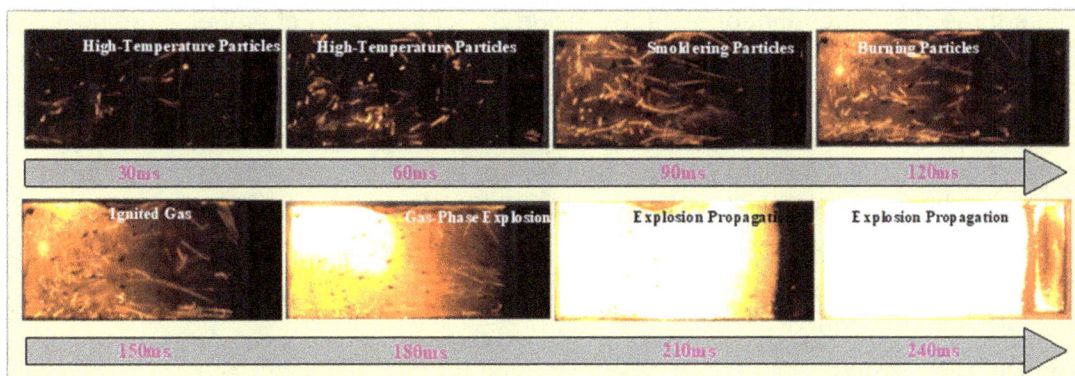

FIG 12 – Flame during the explosion of coal dust (350°C).

Formation process of coal dust clouds

The DPM model simulation in Fluent for cathodic combustion coal dust injection simplifies the coal dust layer into discrete particles at varying temperatures. Based on TG/DTG curves, the volatile content and temperature of these particles at different heights are set, as illustrated in Figure 13. The primary volatile components are CO and H_2. To conserve computational resources, the reaction mechanism for simulations has been simplified, as shown in Figure 14.

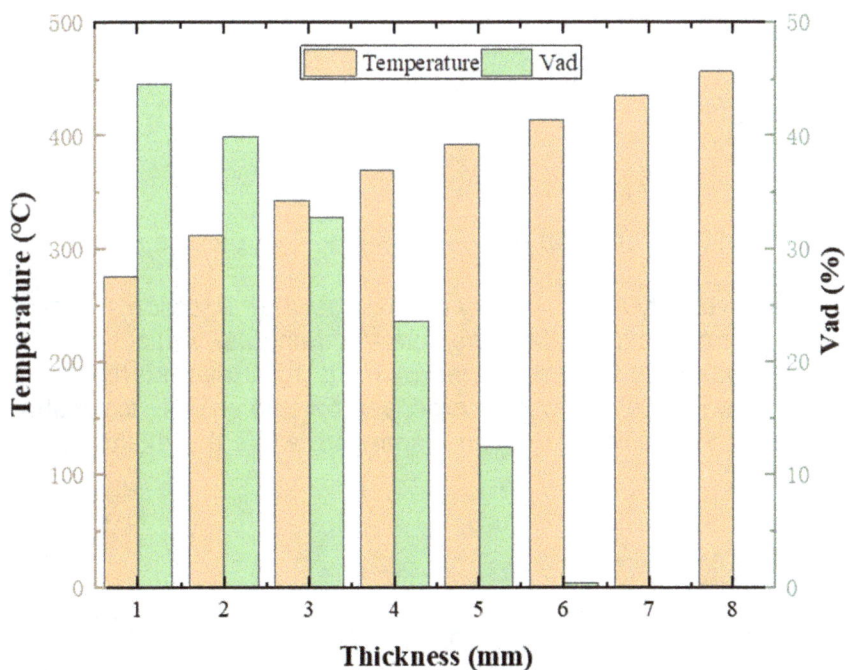

FIG 13 – Coal dust temperature and volatile content.

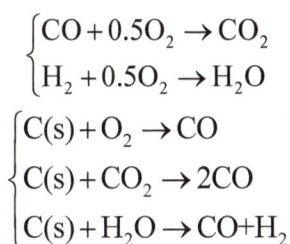

$$\begin{cases} CO + 0.5O_2 \rightarrow CO_2 \\ H_2 + 0.5O_2 \rightarrow H_2O \end{cases}$$

$$\begin{cases} C(s) + O_2 \rightarrow CO \\ C(s) + CO_2 \rightarrow 2CO \\ C(s) + H_2O \rightarrow CO + H_2 \end{cases}$$

FIG 14 – Reaction mechanism for smouldering coal dust explosion.

Figure 15 illustrates the temporal distribution of particles within the explosion chamber. The particle colours denote their residence time, with red representing those that have remained the longest, and gradually lighter shades indicating more recently injected particles. According to Figure 15, coal dust experiences three distinct continuous stages inside the combustion chamber, forming a coal dust cloud:

1. Rapid Injection Phase (0~50 ms): Under the action of high-pressure gas, the particles move into the explosion chamber in a conical shape. Due to the Saffman lift force, they diverge, with particles on the sides moving upward at a slower speed. As the number of entering particles increases, the flow resistance grows, prompting the start of dispersion

2. Deceleration and Dispersion Phase (50~90 ms): After colliding with the deflector plate, the particles diverge and slow down. Affected by thermophoresis, the particles move closer to the wall, forming a cylindrical stream of particles

3. Free Diffusion Phase (90 ms and beyond): Even as their velocity decreases, small-diameter particles can remain suspended; the unburned coal dust particles are consumed by the explosion, with those closer to the injection point reacting sooner.

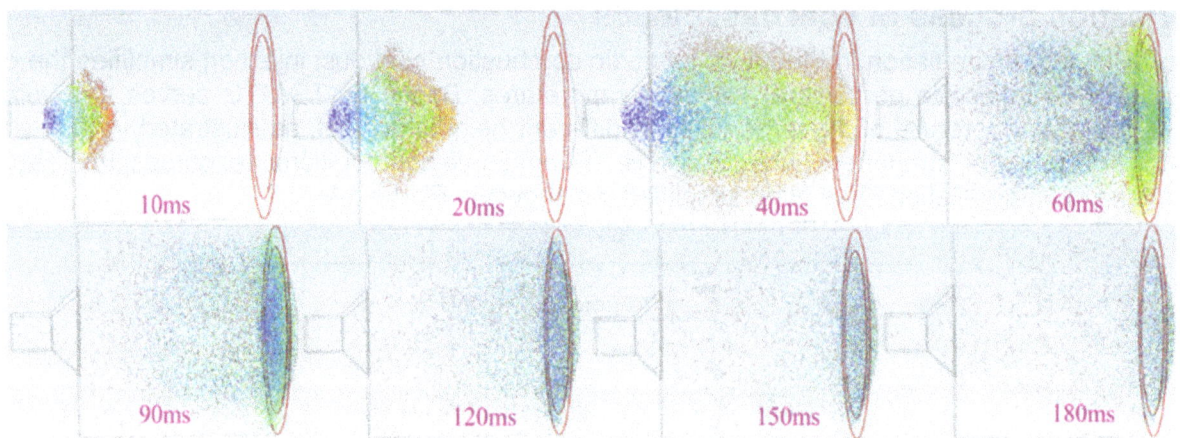

FIG 15 – Formation process of a coal dust particle cloud.

Figure 16 shows the flow velocity and trajectory of coal dust after injection at different time intervals, where a swift horizontal particle stream and two inverted triangular vortices form due to turbulence. By 40 ms, the vortices fade, and the particle jet, peaking at 60 m/s, starts to exit the combustion chamber. Post-50 ms, new vertical vortices emerge, and particle velocity diminishes. By 180 ms, the deflector plate hinders particle flow, promoting dispersion, while the explosion unburned particles towards the plate at 3.5 m/s.

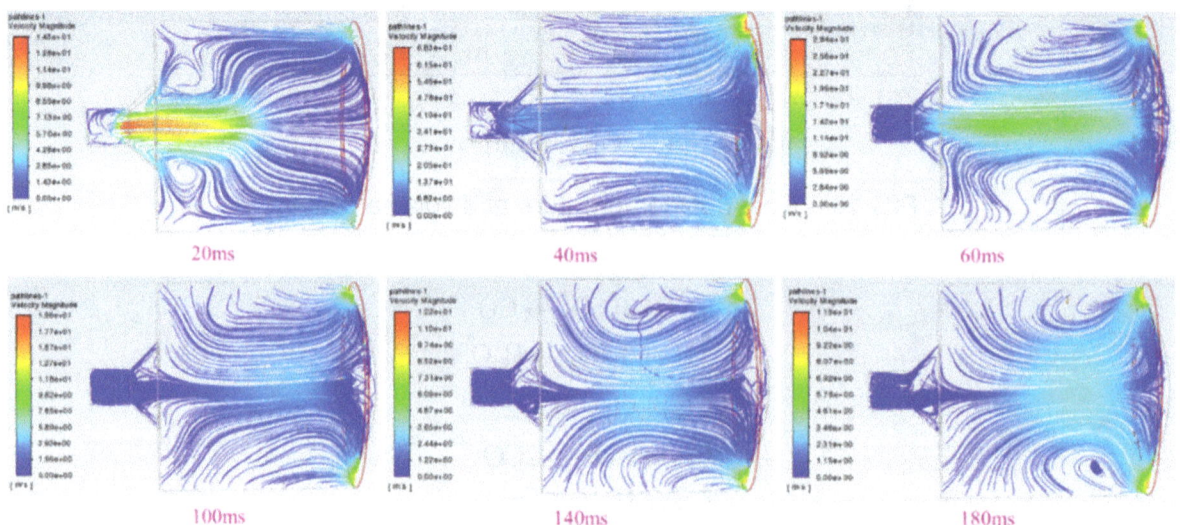

FIG 16 – Dispersion paths and velocity of dust cloud.

Explosion temperature of the coal dust cloud

The shape sequence of smouldering coal dust explosion are obtained as shown in Figure 17.

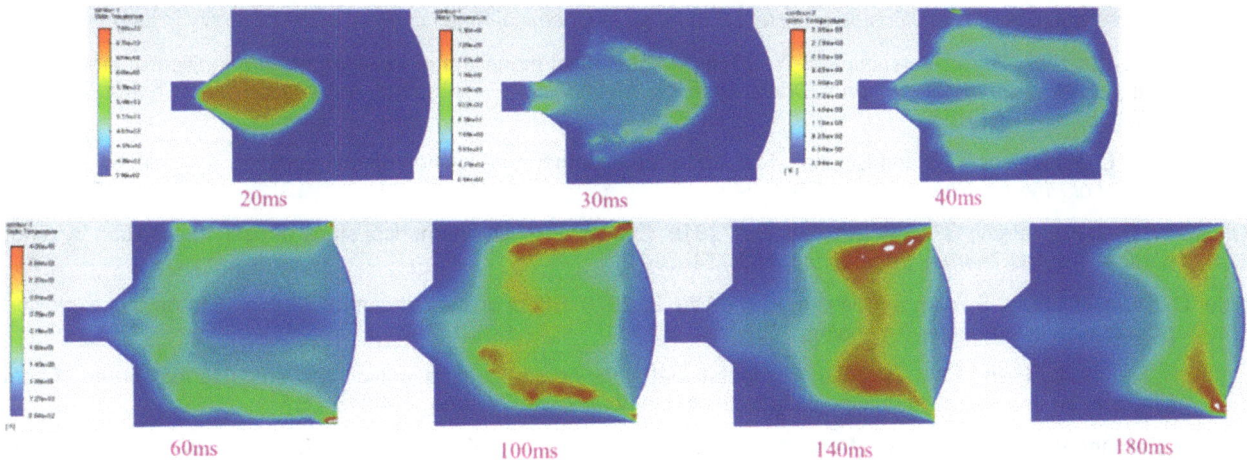

FIG 17 – Temperature and shape sequence of smouldering coal dust explosion.

Within 20 ms after particle injection, the highest temperature reaches 675°C, which is below the maximum initial temperature, indicating that no combustion reaction has occurred; at 30 ms, the highest temperature of the particles is 878°C, and temperatures are higher away from the injection point, mainly due to the more thorough contact between the ejected particles and oxygen from the accumulated coal dust, facilitating further oxidation of the high-temperature particles. During the 40~60 ms, volatile gases combust, and the flame temperature stabilises at around 1500°C. Between 100~180 ms, the flame centre temperature is high, approximately 3000°C, while the leading edge temperature is around 2700°C. The volatiles' thermal decomposition followed by homogeneous combustion outside and heterogeneous combustion of the char in the middle are significant reasons for the higher temperatures at the flame centre.

CONCLUSION AND DISCUSSION

The study of the self-ignition and subsequent explosion processes of accumulated coal dust in high-temperature environments led to the following key conclusions:

- The thicker the coal dust accumulation, the lower the temperature of the hotplate required for self-ignition, with heat conduction dominating the ignition phase and exothermic oxidation reactions prevailing during the self-ignition phase.

- The temperature of smouldering dust is between 275°C and 395°C, which possesses explosive potential after being disturbed, increasing with the degree of smouldering before decreasing.

- The formation of a coal dust cloud includes three stages: rapid injection, deceleration and dispersion, and free diffusion, with explosions primarily occurring during the deceleration and free diffusion stages.

- The explosion induced by deposited coal dust involves both homogeneous and heterogeneous combustion processes, where smouldering particles oxidise upon ejection, increasing their temperature, then ignite combustible gases like CO produced during the smouldering stage, and finally, gas combustion ignites unburned coal dust particles, which is the key mechanism triggering an explosion.

REFERENCES

Cao, W, Qin, Q, Cao, W, Lan, Y, Chen, T, Xu, S and Cao, X, 2017. Experimental and numerical studies on the explosion severities of coal dust/air mixtures in a 20-L spherical vessel, *Powder Technology*, 310:17–23.

Chang, S, Zhuo, J, Meng, S and Yao, Q, 2016. Clean coal technologies in China: current status and future perspectives, *Engineering*, 2:447–459.

Chen, T, Zhang, Q, Wang, J, Liu, L and Zhang, S, 2017. Flame propagation and dust transient movement in a dust cloud explosion process, *Journal of Loss Prevention in the Process Industries*, 49:572–581.

Eckhoff, R K and Li, G, 2021. Industrial Dust Explosions, A Brief Review, *Applied Sciences*, 11:1669.

Frank-Kamenetskii, D A, 2015. *Diffusion and Heat Exchange in Chemical Kinetics* (Princeton University Press).

Glushkov, D, Strizhak, P and Vershinina, K Y, 2014. Mathematical modelling of low-temperature ignition of small-sized coal particles, *2014 International Conference on Mechanical Engineering, Automation and Control Systems (MEACS)*, pp 1–4 (IEEE).

Glushkov, D O, Kuznetsov, G V and Strizhak, P A, 2015. Low-temperature ignition of coal particles in an airflow, *Russian Journal of Physical Chemistry*, B, 9:242–249.

Glushkov, D O, Kuznetsov, G V and Strizhak, P A, 2018. Experimental and numerical study of coal dust ignition by a hot particle, *Applied Thermal Engineering*, 133:774–784.

Guo, C, Shao, H, Jiang, S, Wang, Y, Wang, K and Wu, Z, 2020. Effect of low-concentration coal dust on gas explosion propagation law, *Powder Technology*, 367.

Yuan, H, Restuccia, F and Rein, G, 2020. Computational study on self-heating ignition and smouldering spread of coal layers in flat and wedge hot plate configurations, *Combustion and Flame*, 214:346–357.

Islas, A, Fernández, A R, Betegón, C, Martínez-Pañeda, E and Pandal, A, 2023. Biomass dust explosions: CFD simulations and venting experiments in a 1 m3 silo, *Process Safety and Environmental Protection*, 176:1048–1062.

Jia, J and Tian, X, 2022. Propagation and attenuation characteristics of shock waves in a gas–coal dust explosion in a diagonal pipeline network, *Scientific Reports*, 12.

Khan, A M, Ray, S K, Mohalik, N K, Mishra, D, Mandal, S and Pandey, J K, 2022. Experimental and CFD Simulation Techniques for Coal Dust Explosibility: A Review, *Mining, Metallurgy and Exploration*, 39:1445–1463.

Li, H, Deng, J, Chen, X, Shu, C M, Kuo, C H, Zhai, X, Wang, Q and Hu, X, 2020. Transient temperature evolution of pulverized coal cloud deflagration in a methane–oxygen atmosphere, *Powder Technology*, 366:294–304.

Lin, S, Liu, Z, Wang, J, Qian, Z and Gu, 2020. Flame Characteristics in a Coal Dust Explosion Induced by a Methane Explosion in a Horizontal Pipeline, *Combustion Science and Technology*, pp 1–14.

Liu, T, Cai, Z, Sun, R, Wang, N, Jia, R and Tian, W, 2021a. Flame Propagation Characteristics of Deposited Coal Dust Explosion Driven by Airflow Carrying Coal Dust, *Journal of Chemical Engineering of Japan*, 54(12):631–637.

Liu, T, Jia, R, Sun, R, Tian, W, Wang, N and Cai, Z, 2021b. Research on Ignition Energy Characteristics and Explosion Propagation Law of Coal Dust Cloud under Different Conditions, *Mathematical Problems in Engineering*, 2021. 1–8.

Liu, T, Tian, W, Sun, R, Jia, R, Cai, Z and Wang, N, 2021c. Experimental and numerical study on coal dust ignition temperature characteristics and explosion propagation characteristics in confined space, *Combustion Science and Technology*, 195:2150–2164.

Liu, T, Wang, N, Sun, R, Cai, Z, Tian, W and Jia, R, 2021d. Flame propagation and CO/CO$_2$ generation characteristics of lignite dust explosion in horizontal pipeline, *International Journal of Low-Carbon Technologies*, 16:1384–1390.

Moroń, W and Ferens, W, 2024. Analysis of fire and explosion hazards caused by industrial dusts with a high content of volatile matter, *Fuel*, 355:129363.

Nematollahi, M, Sadeghi, S, Rasam, H and Bidabadi, M, 2020. Analytical modelling of counter-flow non-premixed combustion of coal particles under non-adiabatic conditions taking into account trajectory of particles, *Energy*, 192:116650.

Qian, J, Liu, Z, Lin, S, Liu, H, Ali, M and Kim, W, 2023. Re-explosion hazard potential of solid residues and gaseous products of coal dust explosion, *Advanced Powder Technology*, 34(9):104129.

Rao, Y, Huang, T S, Tian, C, Xiang, C and Su, G, 2023. Investigation on the Explosion Process and Chemical Activity of Pyrite Dust, *Russian Journal of Physical Chemistry A*, 97:894–901.

Ray, S K, Mohalik, N K, Khan, A M, Mishra, D, Varma, N K, Pandey, J K and Singh, P K, 2020. CFD modeling to study the effect of particle size on dispersion in 20l explosion chamber: An overview, *International Journal of Mining Science and Technology*, 30:321–327.

Shi, X, Zhang, Y, Chen, X, Zhang, Y and Ma, Q, 2021. Characteristics of coal dust ignited by a hot particle, *Process Safety and Environmental Protection*, 153:225–238.

Si, M, Cheng, Q, Yuan, L, Luo, Z, Yan, W and Zhou, H, 2021. Study on the combustion behavior and soot formation of single coal particle using hyperspectral imaging technique, *Combustion and Flame*, 233.

Song, B and Li, Y, 2020. Study on propagation characteristics of the secondary explosion of coal dust, *International Journal of Low-carbon Technologies*, 15:89–96.

Song, Y and Zhang, Q, 2020. Criterion and propagation process of spark-induced dust layer explosion, *Fuel*, 267:117205.

Wang, C, Dong, X, Ding, J and Nie, B, 2014. Numerical investigation on the spraying and explosibility characteristics of coal dust, *International Journal of Mining, Reclamation and Environment*, 28:287–296.

Wang, D, Qian, X, Wu, D, Ji, T, Zhang, Q and Huang, P, 2020. Numerical study on hydrodynamics and explosion hazards of corn starch at high-temperature environments, *Powder Technology*, 360:1067–1078.

Wu, D, Norman, F, Verplaetsen, F and Van den Bulck, E, 2016. Experimental study on the minimum ignition temperature of coal dust clouds in oxy-fuel combustion atmospheres, *Journal of Hazardous Materials*, 307:274–280.

Yadav, S and Mondal, S S, 2019. A complete review based on various aspects of pulverized coal combustion, *International Journal of Energy Research*, 43:3134–3165.

Yan, D, Li, M, Zou, L, Gu, M, Li, M and Wang, F, 2022. A study on fragmentation and emissions characteristics during combustion of injected pulverized coal, *Fuel*, 309:122152.

Zhi-Ming, X and Xiao-Qiang, W, 2010. A Support Vector Machine model on correlation between the heterogeneous ignition temperature of coal char particles and coal proximate analysis, Asia-Pacific Power and Energy Engineering Conference.

Considering climate change for potential long-life mine cooling projects located in the subarctics

F K R Klose[1], C McGuire[2], D L Cluff[3] and A L Martikainen[4]

1. Research Engineer, LKAB, Malmberget 98381, Sweden. Email: frederic.klose@lkab.com
2. Ventilation and Refrigeration Engineer, Hatch, Mississauga Ontario L5K 2R7, Canada. Email: chris.mcguire@hatch.com
3. CEO, CanMIND Associates, Sudbury Ontario P3C 2C5, Canada. Email: daniel.cluff@deepmining.ca
4. Ventilation Specialist, LKAB, Malmberget 98381, Sweden. Email: anu.martikainen@lkab.com

ABSTRACT

Luossavaara-Kiirunavaara Aktiebolag (LKAB) owns and operates two sublevel-caving iron ore mines in Malmberget and Kiruna. The mines are located above the arctic circle in Norrbotten county and are currently being evaluated for expansions that could potentially increase their mine lives to beyond 2070. Heating of intake air during the cold winter months is necessary to prevent shaft freezing. On the other hand, large-scale mine cooling is not currently practiced.

As future depth, production rates and mining method are being evaluated, this may change. The necessity for cooling would be a significant departure from current practices. Potentially large capital investments and operational expenditures could prevent or limit the expansion projects. Additionally, the area above the arctic circle is considered vulnerable to climate change. Improving the understanding of how climate change could impact long-life projects in this region in terms of cooling requirements is the principal motivation for this paper.

A hypothetical long-life, deep mining project in the region is presented and future climate data is constructed to allow assessment of any potential impacts over the duration of the project. The constructed data is based on actual historical climate data and projections made by the representative concentration pathway (RCP) scenarios. Critical depths and corresponding cooling requirements are then calculated for the entire project duration using an accepted workplace temperature limit and the constructed climate data.

The impact of climate change on long-life projects is then assessed by using the annual peak cooling requirements to identify upgrade/replacement times for the corresponding model plants, which are sized using the 95th and 98th temperature percentiles of the five previous years.

INTRODUCTION

Luossavaara-Kiirunavaara Aktiebolag (LKAB) is a state-owned producer of iron ore products, operating both open pit and underground mines in northern Sweden above the arctic circle. The operations at Malmberget and Kiruna are considered to be the largest underground iron ore mines in the world and, based on resources, are set to operate for decades to come.

Climate change is becoming an important topic when evaluating long-term projects. It refers to an observable long-term departure from established weather patterns, which manifests itself in shifting in mean temperatures and increasing frequency of extreme weather events (after IPCC, 2021). The arctic and subarctic climate zones, the latter of which LKAB's mines are located in, are thought to experience higher than average temperature increases due to climate change (IPCC, 2021). This means that current heating and especially cooling practices are likely to change.

Ventilation and air conditioning by means of heating and cooling are some of the most energy-, and indeed, cost-intensive aspects of an underground mining operation. Any change may therefore affect the viability of an underground mining operation in the long-term. The purpose of this paper is therefore to assess how climate change could affect a long-life operation, using available climate data.

BACKGROUND AND APPROACH

Mine

A fictional mining scenario based on one of LKAB's operations is presented to help answer questions. This operation is located close to the Swedish town of Gällivare, Norrbotten county, and has a hypothetical mine life until 2100. The location coincides with that of the weather station so that no correction is required. The starting depth is 1000 m beneath datum in the year 2020, with a sink rate between 10 m/a and 20 m/a to account for different mining practices. Peak air mass flow requirements in kg/s are not considered to make the results more generally applicable. During the winter season, the intake air is heated to +1°C dry bulb temperature (DB). No cooling solutions are currently implemented.

Cooling plant

The critical depth concept will be used to assess whether cooling will likely be required. Brake (2001) defines critical depth as 'the depth [...] at which air will exceed the underground target WB [wet bulb] temperature solely through auto-compression [...]'. The calculation of the adiabatic lapse rate is thus required. An arbitrary temperature limit value (TLV) of 26°C WB will be referred to throughout the paper to calculate the minimum cooling requirements. It is assumed that this value will remain the same until the year 2100.

The approach to defining the need for installation of cooling capacity is that a first model refrigeration plant, based on the identified minimum cooling requirements, is installed once the TLV at depth is exceed for 18 hrs annually. The 18 hrs translate to three or two consecutive cooling days when assuming a cooling requirement for six or nine consecutive hours during the warmest times of a day, separated by a night-time cooling gap. This approach addresses the lack of historical cooling demand data when critical depth is first reached. The plant is then sized using the 98th and 95th percentiles (P_{98} and P_{95}) to compare a conservative to standard approach.

CONSTRUCTING A FUTURE TYPICAL METEOROLOGICAL YEAR

Climate change effects and assumptions

Climate models utilise representative concentration pathways (RCP). These describe the development of the atmospheric concentration of anthropogenic greenhouse gases for the 21st century, following the adoption of different mitigation policies. The data provided by the Swedish Meteorological and Hydrological Institute (Sveriges Meteorologiska och Hydrologiska Institut) is the aggregated result of multiple climate models (SMHI nd(a)). Either absolute or anomaly values are available for each of the climate indicators. The 10th, 50th and 90th percentiles of the modelling outcomes are provided for three scenarios: RCP2.6 representing a 'low-emissions' scenario, an 'intermediate' RCP4.5 scenario with 'emissions peaking around 2040', and a 'high-emissions scenario' RCP8.5 (IPCC, 2021). Indicator changes are tracked relative to the standardised 30-year reference period of 1971–2000.

SMHI projects DB to increase in the coming decades. Greater increases are expected for the winter than for the summer seasons, and daily temperature minima are projected to increase faster than maxima, especially during winter. This makes DB anomaly values the primary climate change indicator of interest.

Relative humidity (RH) is also of interest. IPCC (2021) show that global average surface RH levels are declining. Douville and Willett (2023) write that a decline of global surface RH is 'consistent' with increasing temperatures and give an example of a 5 per cent decline is given in the year 2100 for a high-emission scenario in the 'northern midlatitudes'. However, with projected data on RH unavailable from SMHI, RH is assumed constant. This can be considered as conservative with regards to eventual WB calculation.

Station pressure (P), which is also of interest, is linked to weather patterns and hence does not feature prominently as a climate indicator. Regardless, the increased likelihood of extreme weather events implies a higher fluctuation in P. However, changing the value of P has only minor effect on the calculation on WB, which is why P is assumed constant in this paper.

Constructing a future typical meteorological year

A general approach in constructing design data is the compilation of multi-year weather data into a typical meteorological year (TMY). By averaging temperatures over several years, the natural variability is reduced and data gaps are removed. For the analysis to be done in this paper, a TMY was compiled using hourly data for the years 2015–2019 to be both representative of the most recent years. The TMY was then considered to start in 2020. Weather data is sourced from the meteorological station 'Gällivare A', which is operated by the Swedish national weather service SMHI (nd(b)). DB and RH data were used as recorded, whereas the pressure was converted from mean sea level pressure to P based on Equations 1 through 3c as provided by SMHI (2022). Equation 1 represent a version of the well-known barometric formula. February 29 was ignored as it occurs only during leap years.

$$P = \frac{P_{msl}}{exp\left(\frac{HB}{T_1}\right)} \tag{1}$$

$$B = 0.034163(1 - 0.0026373\cos(2L)) \tag{2}$$

$$T_1 = 1.07T + 274.5 \qquad \text{for DB} \geq 2°\text{C} \tag{3a}$$

$$T_1 = 0.535T + 275.6 \qquad \text{for } -7°\text{C} \leq \text{DB} < 2°\text{C} \tag{3b}$$

$$T_1 = 0.500T + 275.0 \qquad \text{for } -7°\text{C} < \text{DB} \tag{3c}$$

Where:

P	is station pressure (Pa)
P_{msl}	is pressure at mean sea level (Pa)
H	is station elevation (m)
T	is DB (°C)
L	is station latitude (°)
T_1	represents a corrected temperature based on DB at the recording time
B	represents the quotient of the standard gravitational acceleration, the molar mass of air and the universal gas constant, corrected for the location of the weather station

The 50th percentiles of the projected temperature anomalies provided by SMHI for Norrbotten were then used to create the future TMY (FTMY). Seasons, separated into three-month periods, were taken into account by using the corresponding data. It is assumed for ease of calculation that the data is equally valid for all of the county. It is also assumed that the duration of each season remains unchanged into the future. This is following the structure of the data. However, it is noted that a change in temperature also causes changes to the duration of seasons. Finally, to change the reference period from 1971–2000 to the TMY, the starting year anomaly was subtracted from subsequent ones while considering seasonal differences. Adding the results to the hourly TMY data then produces a simple FTMY. As an example outcome, the RCP8.5 summer DB forecasts for the year 2100 are 3.1°C higher when based on the TMY compared to 4.7°C higher when based on the original reference period.

Calculating wet-bulb temperature

Constructing the FTMY yields data for 80 years, each year containing 8760 hours. The hourly DB, RH and P values can then be used to derive surface WB as the key parameter to identify the critical depth and cooling requirements.

Various approaches exist to estimate WB. These can be iterative and computationally intensive (eg ASHRAE 2001) or based on statistical curve-fitting to limit their reliance on few input parameters at the cost of a narrower validity range (eg Stull, 2011; Sadeghi et al, 2013). Common limitations are a standard pressure of 1013.25 hPa and an insufficient coverage to low temperature ranges, which are expected in the (sub)arctic climate zone.

Given the extent of the data, a curve-fitted solution is preferred. The simplest solution found, relying only on RH and DB as input parameters, is that by Stull (2011). Limitations to standard pressure and temperatures above -20°C were regarded as acceptable. The average annual station pressure is 966 hPa, while heating of the air to +1°C DB during the colder seasons, reducing the error due to low temperatures.

$$WB = DB\tan^{-1}\left(0.151977\sqrt{RH + 8.313659}\right) + \tan^{-1}(DB + RH) - \tan^{-1}(RH - 1.676331) +$$
$$\sqrt[3]{RH}\tan^{-1}(0.023101RH) - 4.686035 \tag{4}$$

Heating of the air is regarded as sensible, meaning that moisture content remains unchanged. Assuming pressure remaining unchanged as well, it follows that vapor pressure is identical before and after the heating process, thus allowing RH to be calculated using Equation 5. Huang's (2018) expressions for saturated vapor pressure were inserted into this equation in the form of Equation 6a or 6b, depending on the DB. McPherson's (2009) approach was not used, as it is valid for positive temperature ranges only. A new WB can then be calculated according to Equation 4.

$$RH_2 = RH_1\frac{e_{sd1}}{e_{sd2}} \tag{5}$$

$$e_{sd} = \frac{\exp\left(34.494 - \frac{4924.99}{DB+237.1}\right)}{(DB+105)^{1.57}} \qquad \text{for DB > 0°C} \tag{6a}$$

$$e_{sd} = \frac{\exp\left(43.494 - \frac{6545.8}{DB+278}\right)}{(DB+868)^2} \qquad \text{for DB ≤ 0°C} \tag{6b}$$

Where e_{sd} is saturation vapor pressure at DB (Pa).

Critical depth and sigma heat

With the inlet conditions defined, the WB at the shaft bottom can be found by rearranging Equation 7 for the adiabatic lapse rate (after McPherson, 2009) after solving for its components using Equations 8 through 13. Critical depth can be found by solving for Z_2 and substituting the shaft bottom WB_2 for the TLV:

$$WB_2 = WB_1 + (Z_2 - Z_1)\left(\frac{-g}{V}\right)\frac{0.286\left[(1+6078X_s)\frac{WB}{P}+\frac{L_wX_s}{287.04(P-e_{sw})}\right]}{\left[1+6078X_s+\frac{L_w{}^2PX_s}{463810(P-e_{sw})WB^2}\right]} \tag{7}$$

$$V = \frac{287.04(DB+273.15)}{P-e} \tag{8}$$

$$e = \frac{PX}{0.622+X} \tag{9}$$

$$X = \frac{L_wX_s-1005(DB-WB)}{L_w+1884(DB-WB)} \tag{10}$$

$$L_w = (2502.5 - 2.386WB)1000 \tag{11}$$

$$X_s = 0.622\frac{e_{sw}}{P-e_{sw}} \tag{12}$$

$$e_{sw} = 610.6\exp\left(\frac{17.27WB}{237.3+WB}\right) \tag{13}$$

Where L_w is the latent heat of evaporation at WB (J/kg moisture evaporated), X is the moisture content (kg/kg dry air), X_s is the moisture content at saturation (kg/kg dry air), e is vapor pressure (Pa), e_{sw} is saturation vapor pressure at WB (Pa), V is the apparent specific volume of dry air (kg/m³) and is Z is elevation (m).

Finally, the cooling deficit is expressed as the change in sigma heat (Equation 14) to reduce shaft bottom WB to the TLV by applying Equation 15.

$$S = L_wX_s + 1005WB \tag{14}$$

$$(S_{TLV} - S_1) - (S_2 - S_1) = S_{TLV} - S_2 \tag{15}$$

Where S is sigma heat (J/kg dry air).

RESULTS

Effects of climate and sink rate

Six different cases can be defined based on the combination of cooling scenario and sink rate, which are summarised in Table 1. For all cases, cooling requirements peak during the warmest days of the FTMY around the end of July. With cooling hours and days spreading around this time, the cooling period can be regarded as lasting from mid-June to mid-September. Despite increasing temperatures and depths, no overlap of this 'cooling season' with the heating season occurs within the data, shown in Figure 1.

Based on critical depth data shown in Figure 2, a cooling deficit appears to develop between the late 1950s and early 1970s. It is obvious from this figure that sink rates affect the starting year and ultimate cooling deficit much more than the actual climate change scenarios themselves. In the most extreme comparison of cases 5 and 6, cooling deficits develop 19 years apart. Besides cooling deficits developing farther into the future, the magnitude of the deficits is also substantially lower when the sink rates differ. The peak deficit for case 6 is shown in Table 1 to be almost three times greater than for case 5, while the total cooling hours and days increase by factors of about 16 and 8, respectively. On the other hand, the peak deficit of case 6 shows that the maximum cooling deficit is only about 1.6 and 1.8 times larger than those of cases 4 and 2. The difference in the duration of the deficit are also apparent. On the other hand, no substantial cooling deficit develops for cases 1 and 3, which are omitted from further analysis.

It is important to reiterate that the cooling deficit under discussion is on the basis of the critical depth based on auto-compression alone and does not account for underground heat loads. It is therefore expected that the actual refrigeration duty required to achieve the TLV is greater than the values presented here and beyond the scope of this analysis.

TABLE 1

Summary of start and peak deficit values for different scenarios as indicated in Figure 2.

Scenario	Sink rate	Case	Initial cooling deficit				Peak cooling deficit			
			Year	Total hours	Total days	Max. deficit	Year	Total hours	Total days	Max. deficit
	m/a	-	-	-	-	kJ/kg	-	-	-	kJ/kg
RCP2.6	10	1	-	0	0	0	-	0	0	0
	20	2	2067	1	1	-1.1	2096	254	28	-16.9
RCP4.5	10	3	2091	1	1	-1.7	2091	1	1	-1.7
	20	4	2062	1	1	-0.4	2096	337	36	-18.8
RCP8.5	10	5	2072	1	1	-1.1	2098	66	9	-10.8
	20	6	2059	1	1	-0.9	2098	1051	71	-29.5

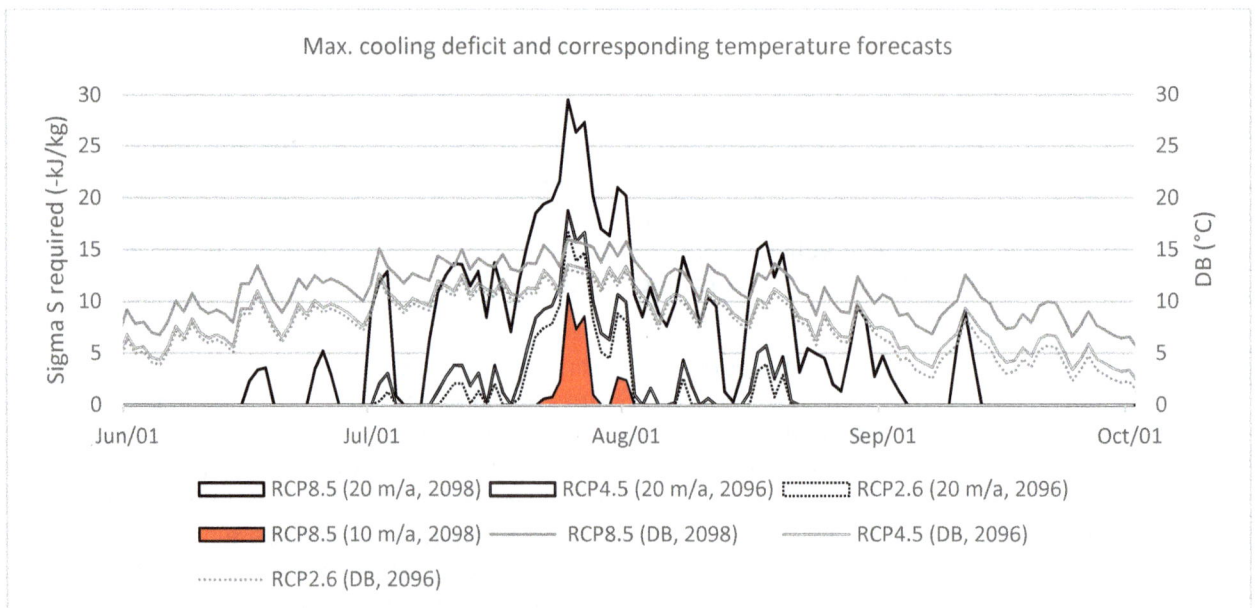

FIG 1 – Maximum cooling deficits for different climate change scenarios and sink rates (areas), with minimum daily DB overlap (lines).

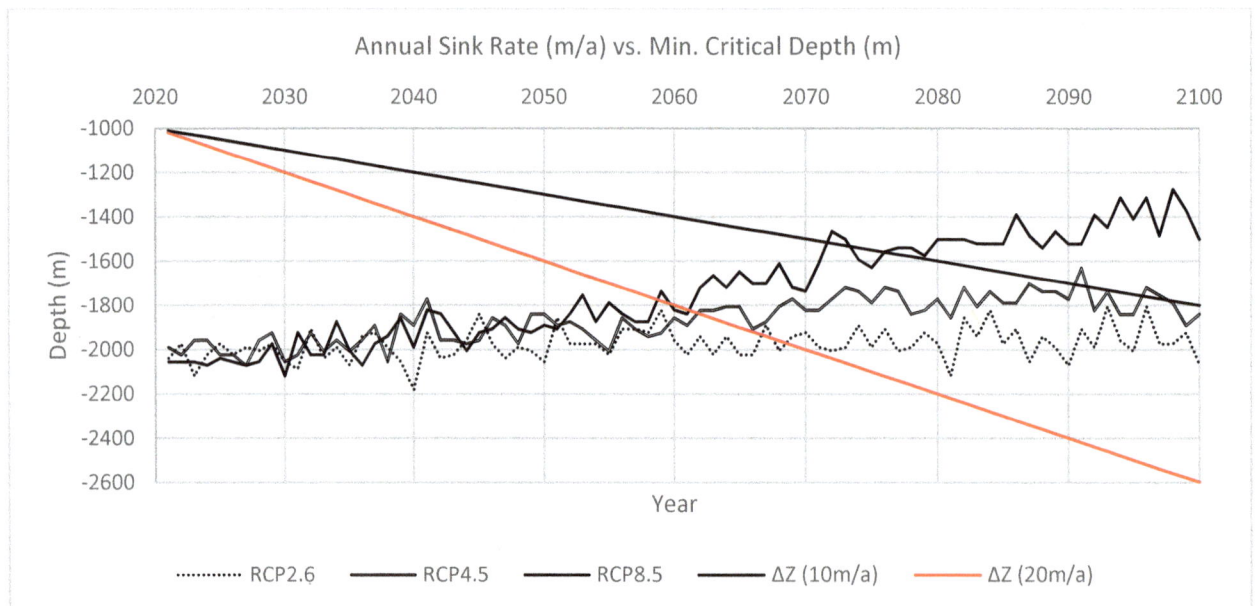

FIG 2 – Critical depth as calculated from chosen TLV value versus different sink rates.

Cooling deficits and model plant sizing

More optimistic climate scenarios lead to cooling plants required farther into the future, as already shown in Figure 2. For example, based on the procedure outlined earlier, cases 2 and 4 require initial cooling plant operation 5 and 12 years later than case 6 due to the shallower depth of mining. The number of upgrades required throughout the lifetime is also decreased for these cases as the increase in auto-compression heat load is more gradual at the reduced sink rate.

The effect of sink rates is again shown by comparing cases 5 and 6, which in Figure 3 represent the opposite extremes with regards to cooling plant establishment date, number of required upgrades and the lower peak cooling requirement.

As for sizing of the cooling plants, the differences between the P_{98} and P_{95} approach are around 4 kJ/kg range for case 6, reducing as climate change scenarios become more optimistic to about 1.7 kJ/kg and 1.3 kJ/kg for cases 4 and 2, respectively.

It should be noted that the values presented in Table 2 should be interpreted as an illustration of the effects of depth and climate change on the cooling deficit to reach TLV at the shaft bottom and not as sufficient plant cooling capacities.

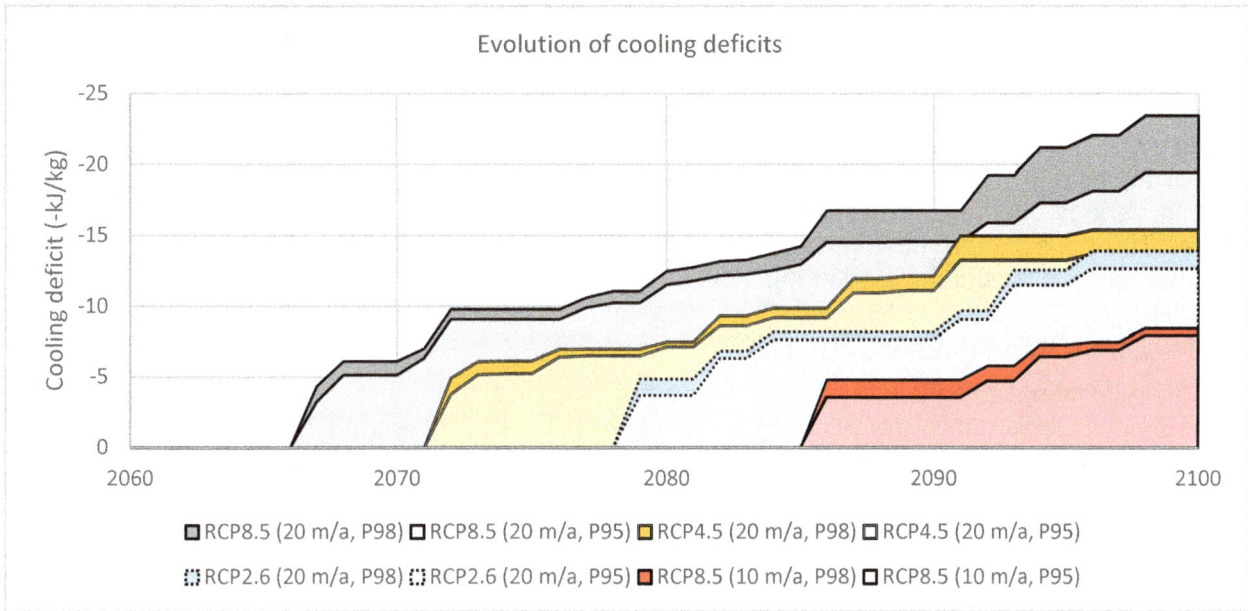

FIG 3 – Evolution of cooling deficits following a mock plant design approach.

TABLE 2

Summary of start and end values for four cases (see also Figure 3 and Table 1).

Case	Design percentile	Initial establishment		Final establishment		
		Year first established	Initial capacity	Year of final capacity	Final Capacity	Number of increases
		-	kJ/kg	-	kJ/kg	-
2	P_{98}	2079	-4.9	2096	-13.9	5
	P_{95}	2079	-3.7	2096	-12.6	5
4	P_{98}	2072	-4.9	2096	-15.4	11
	P_{95}	2072	-3.8	2096	-13.7	11
5	P_{98}	2086	-4.8	2098	-8.5	4
	P_{95}	2086	-3.5	2098	-7.9	4
6	P_{98}	2067	-4.4	2098	-23.4	16
	P_{95}	2067	-3.3	2098	-19.4	17

DISCUSSION

The analysis shows that cooling is required starting from the second half of the century, even when excluding underground heat loads at the model mine. Cooling is required for only about three months when a conservative climate change scenario is combined with an aggressive sink rate. The critical depths calculated for the pessimistic RCP8.5 scenario approach the 1000–1500 m level range found in several historical and operating mines located in subarctic Canada and Europe, including those of LKAB. It can be expected that many of these will continue to operate well into the future.

The need to address cooling deficits in future mine planning has the potential to impact the project economics. Frequent upgrade requirements suggest that a stable cooling plant size is not possible without oversizing. Even a 1 kJ/kg difference translates to 1 MW of cooling for a dry air mass flow of 1000 kg/s, which is less than the mass flow capacity of each of LKAB's existing underground mines.

Appropriate and flexible strategies that reduce or mitigate future cooling deficits, and do so efficiently when considering future upgrades or interim turndown operation, are in the interest of any long-life mining project.

No overlap between the cooling and heating period has been identified in this analysis. However, this is based on constructed weather data up to 80 years into the future and a chosen TLV, in this case 26°C WB. Lower workplace temperature limits can generally be expected to receive higher acceptance from the workforce, particularly in a region that is already characterised by long, cold winters and short, cool summers. More restrictive WB values are expected to lead to shallower critical depths and *vice versa*. Moreover, there would also be an increased probability of heating and cooling periods overlapping, as lower WB values at depth require lower intake WB as well. This makes TLV possibly the most important variable influencing the outcome of the analysis, although no such analysis was completed at this time.

CONCLUSIONS AND OUTLOOK

The overall findings are that cooling requirements, as based on the critical depth approach, appear to be limited unless great target depths are considered in the second half of the century. The calculation of critical depth as presented here is a simplistic but straightforward approach that provides the basis for discussing the effects of climate change, and the variability of the projections, on cooling requirements. Strategies to tackle the potential impacts can then be devised in due time.

It is important to reiterate the limitations of the critical depth approach. The analysis conducted relies first and foremost on trustworthy data for climate change scenarios, which dictates the shaft inlet conditions and thus the outcome of the analysis through the calculation of the adiabatic lapse rate. The second important variable is depth, which is represented here by an annual sink rate. Higher sink rates lead to critical depths being exceeded sooner. Finally, the target WB likely has a considerable impact and is identified as a topic of further investigation.

Ultimately, it is clear that the analyses conducted in this paper can only partially illustrate the climate change situation that sub-arctic mining projects could be exposed. There are large uncertainties with regards to technological developments and future mining projects in the region, so that the situation will ultimately be somewhere in-between.

As far as LKAB's mines are concerned, the next step of the analysis will include fine-tuning of the assumptions presented in this paper. Underground heat loads will also be included. Climate simulations of different life-of-mine stages will then be conducted using adjusted surface conditions and representative thermal parameter data as model inputs. The effect of extreme events such as heat waves will also be evaluated. It is expected that this approach will help in assessing the implications that climate change may have and devise and evaluate cooling strategies, either conventional or unconventional.

ACKNOWLEDGEMENTS

LKAB is gratefully acknowledged for permission to publish this paper. The authors also wish to thank the two anonymous reviewers for their contributions and valuable suggestions that improved this paper.

REFERENCES

American Society of Heating, Refrigeration and Air-Conditioning Engineers (ASHRAE), 2001. Psychrometrics, in ASHRAE Fundamentals Handbook (SI), chapter 6, 17 pp (ASHRAE: Atlanta, GA).

Brake, D J, 2001. The application of refrigeration in mechanised mines, The AusIMM Proceedings 306(1):1–10.

Douville, H and Willett, K M, 2023. A drier than expected future, supported by near-surface relative humidity observations, *Sci Adv*, 30(9):9. https://doi.org/10.1126/sciadv.ade6253

Huang, J, 2018. A simple accurate formula for calculating saturation vapor pressure of water and ice, Journal of Applied Meteorology and Climatology 67(6): 1265–1272.

Intergovernmental Panel on Climate Change (IPCC), 2021. Climate change 2021: The physical basis. Contribution of Working Group I to the Sixth Assessment Report of the Intergovernmental Panel on Climate Change (eds: Masson-Delmotte, V, Zhai, P, Pirani, A, Connors, S L, Péan, C, Berger, S, Caud, Chen, Y, Goldfarb, L, Gomis, M I, Huang,

M, Leitzell, K, Lonnoy, E, Matthews, J B R, Maycok, T K, Waterfield, T, Yelekci, O, Yu, R, Zhou, B), pp 2391 (Cambridge University Press: Cambridge, New York).

McPherson, M J, 2009. Subsurface ventilation engineering [online]. Available from: <https://www.srk.com/download/file/594> [Accessed: 12 January 2024].

Sadeghi, S-H, Peters, T, Cobos, D R, Loescher, H W, Campbell, C S, 2013. Direct calculation of thermodynamic wet-bulb temperature as a function of pressure and elevation, Journal of Atmospheric and Oceanic Technology 30(8): 1757–1765.

Stull, R, 2011. Wet-bulb temperature from relative humidity and air temperature, Journal of Applied Meteorology and Climatology 50(11): 2267–2269.

Sveriges Meteorologiska och Hydrologiska Institut (SMHI), 2022. Hur mäts lufttryck? (in Swedish) (How is air pressure measured?) [online]. Available from: <https://www.smhi.se/kunskapsbanken/meteorologi/lufttryck/hur-mats-lufttryck-1.23830> [Accessed: 19 January 2024].

Sveriges Meteorologiska och Hydrologiska Institut (SMHI), nd(a). Fördjupad klimatscenariotjänst (in Swedish) (In-depth climate scenario service) [online]. Available from: <https://www.smhi.se/klimat/framtidens-klimat/fordjupade-klimatscenarier/met/norrbottens_lan/medeltemperatur/rcp45/2071-2100/year/anom> [Accessed: 19 January 2024].

Sveriges Meteorologiska och Hydrologiska Institut (SMHI), nd(b). Ladda ner meteorologiska observationer (in Swedish) (Download meteorological observations) [online]. Available from: <https://www.smhi.se/data/meteorologi/ladda-ner-meteorologiska-observationer#param=airtemperatureInstant,stations=core,stationid=180760> [Accessed: 19 January 2024].

Mitigating saline water discharge challenges in Gwalia Mine exhaust shafts

R Morla[1], P Tukkaraja[2], S Jayaraman Sridharan[3], M Arnold[4], P Chang[5], R Chhabra[6] and B Nguyen[7]

1. Senior Ventilation Engineer, Genesis Minerals Limited, Perth WA 6000. Email: ramakrishna.morla@genesisminerals.com.au; Adjunct Research Fellow, Western Australian School of Mines, Minerals, Energy and Chemical Engineering, Curtin University, Bentley WA 6102. Email: ram.morla@curtin.edu.au
2. Associate Professor, South Dakota School of Mines and Technology, Rapid City SD 57701, USA. Email: pt@sdsmt.edu
3. Research Scientist, South Dakota School of Mines and Technology, Rapid City SD 57701, USA. Email: srivatsan.jayaramansridharan@sdsmt.edu
4. Manager Underground, Genesis Minerals Limited, Perth WA 6000. Email: matthew.arnold@genesisminerals.com.au
5. Lecturer, Western Australian School of Mines, Minerals, Energy and Chemical Engineering, Curtin University, Bentley WA 6102. Email: ping.chang@curtin.edu.au
6. Alternate Underground Manager, Genesis Minerals Limited, Perth WA 6000. Email: raghav.chhabra@genesisminerals.com.au
7. Ventilation Engineer, Genesis Minerals Limited, Perth WA 6000. Email: brian.nguyen@genesisminerals.com.au

ABSTRACT

Saline water discharge from underground mine exhaust shafts poses significant challenges to both the environment and operational efficiency. This paper presents a case study conducted at an underground mine, examining the complexities of mitigating saline water discharge issues. The study investigates the behaviour of saline water droplets within the exhaust shaft and explores methods to minimise water carry-over and its detrimental effects on ventilation systems. Numerical simulations using CFD were performed to analyse airflow and water droplet dynamics within the shaft under varying conditions. Results indicate that water droplet entrainment and build-up, termed 'water blanketing,' occur within specific velocity ranges, significantly increasing airflow resistance. Field data from the Gwalia Mine ventilation system validate these findings, highlighting the trade-off between reducing water discharge and maintaining ventilation efficiency. The study underscores the importance of considering both water carry-over and water blanketing effects in developing effective mitigation strategies for saline water discharge in underground mine exhaust shafts.

INTRODUCTION

Underground mines situated in areas with substantial groundwater aquifers or experiencing high precipitation may require dewatering pump capacities exceeding 100 000 L/min to prevent the inundation of operational areas (Lottermoser, 2003).

Saline mine water discharge presents a serious environmental and economic problem. High salinity levels can harm nearby vegetation and infrastructure threat (Derrington, 2002; Gregory, Ward and John, 2009; Zgórska et al, 2016). Studies have shown that the total cost of these issues can be significant, with discharge fees and mitigation strategies reaching millions of dollars annually. Researchers have looked into the utilisation of saline water from a coalmine by using a combination of nanofiltration, electrodialysis, evaporation and crystallisation (Turek, Dydo and Surma, 2005). This process would not only eliminate environmental impact but could also be profitable due to salt production and reduced discharge fees.

Saline water discharge from exhaust shafts in mines poses a significant challenge for the mining industry. This water can have a detrimental impact on the environment, causing damage to vegetation and infrastructure near the mine. Additionally, the salt spray can lead to corrosion, erosion, and reliability problems in the ventilation fans themselves, increasing maintenance costs and downtime (Derrington, 2002).

Water transport in vertical airways like mine shafts is a balancing act between gravity and airflow. At low speeds, water simply falls down. As air velocity increases, smaller droplets get carried upwards, potentially reaching an equilibrium state where they suspend mid-air (Kolesov *et al*, 2023; Semin and Zaitsev, 2020). This water accumulation in upcast shafts, also known as water blanketing, can significantly increase airflow resistance. This phenomenon can potentially lead to ventilation issues and reduced airflow through the mine (Kolesov *et al*; Semin and Zaitsev). However, higher airflows can also re-entrain water films from the shaft walls back into the airstream. Previous studies have identified a critical velocity of approximately 8 m/s as the most common speed at which water build-up occurs in such mine shafts. This critical velocity can vary between 7 m/s and 12 m/s depending on various influencing factors (Viljoen and von Glehn, 2019). There are no studies available for water droplet size in the exhaust shaft. A typical drop size distribution in a nozzle spray (Chen and Trezek, 1977) is used by Viljoen and von Glehn in the numerical study of the upcast shaft. The study used volume of fraction (VOF) to model the water in the airflow. VOF's inability to capture air-water interfaces and droplet coalescence/breakup limits its accuracy for simulating water droplet behaviour in turbulent airflow.

Several methods can be employed to prevent water from entering the return air shaft in the first place. Installing a lining along the shaft acts as a physical barrier against groundwater. Alternatively, dewatering involves drilling holes around the shaft to lower the surrounding water table, reducing the pressure that pushes water into the shaft. These preventative measures can be highly effective but come with their own drawbacks, such as potentially higher costs and longer implementation times.

If preventing all water ingress isn't feasible, various techniques can enhance water capture within the shaft. A simple solution is directing airflow downwards into a sump at the base to collect water droplets. However, this doesn't address entrained water already in the air. Water drop-out boxes and cyclones can be installed within the ductwork to separate water from the air stream using gravity or centrifugal force. While cyclones are more efficient, they create significant pressure drops that can hinder ventilation. The selection of the most suitable mist mitigation method depends on factors such as the desired capture rate, cost-effectiveness, and potential impacts on airflow.

GWALIA MINE – CASE STUDY

The Gwalia mine is situated 3 km from the town of Leonora in Western Australia. Presently, the mine reaches a depth of 1.9 km, with plans to extend its Life-of-mine (LOM) to 2.3 km. The mine has high-grade gold mineralisation. All ore and waste are transported out of the mine on the decline using a fleet of 65 ton diesel haul trucks, making it recognised as the world's deepest trucking mine.

The mine comprises three intake airways (VR2, VR7 and Decline) and two return airways (VR3 and VR6), as shown in Figure 1. The VR3 shaft is equipped with two centrifugal fans, while the VR6 shaft is equipped with three centrifugal fans. The total intake air quantity for the mine amounts to 830 m^3/s, and the mine has 16.6 MW of NH$_3$ and LiBr cooling systems.

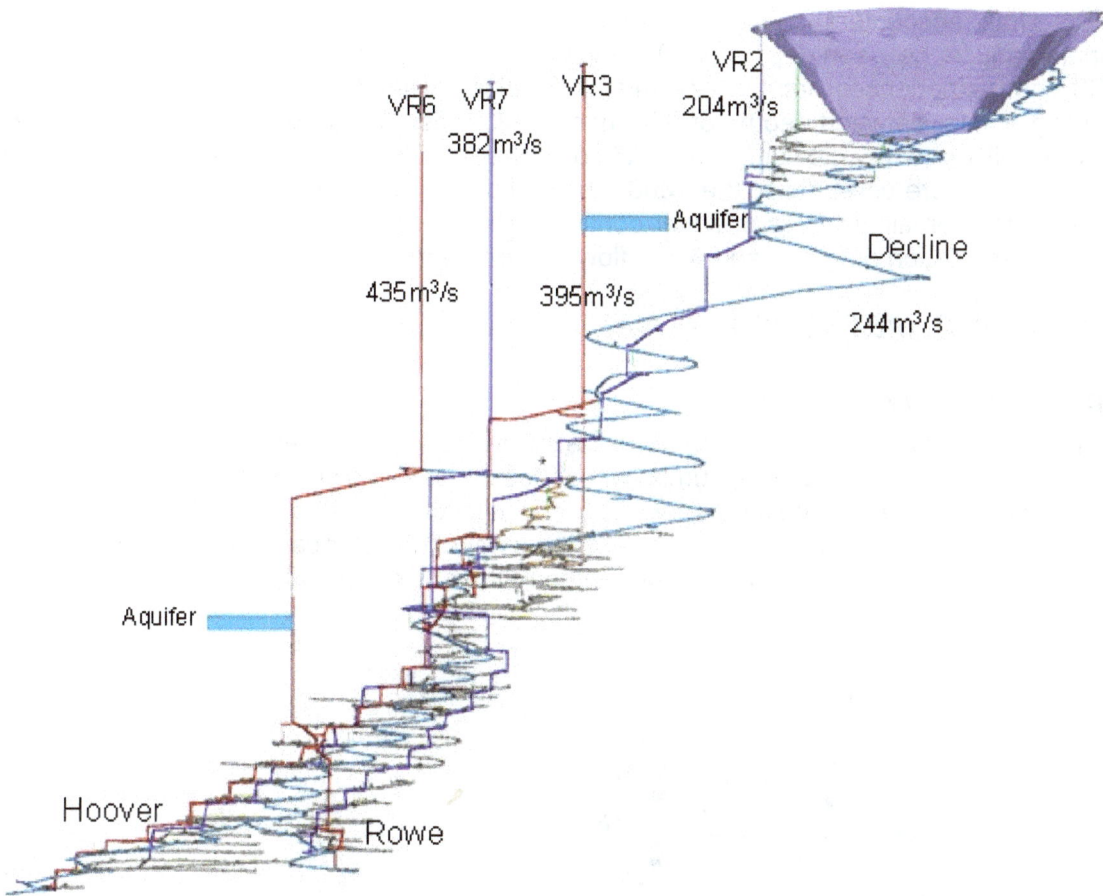

FIG 1 – Details of Gwalia mine ventilation network system.

The return air shaft, VR6, is a dedicated exhaust air ventilation rise with a diameter of 5.0 m, extending from the surface to a depth of 1000 metres below surface (mbs). Beyond this depth, it is connected to another Ø5 m shaft from the 1000 mbs level to the 1520 mbs. On the surface, the VR6 rise is equipped with three 1.7 MW AirEng centrifugal fans. Currently, all fans are running at 82 per cent capacity, exhausting 435 m³/s of airflow at a pressure of 5.6 kPa. From Figure 2a, it can be seen that the air quantity at the aquifer is 362 m³/s, whereas at the fans, it is 435 m³/s. The difference in quantities is due to density variations, and some of the air is joining in at 1000 mbs.

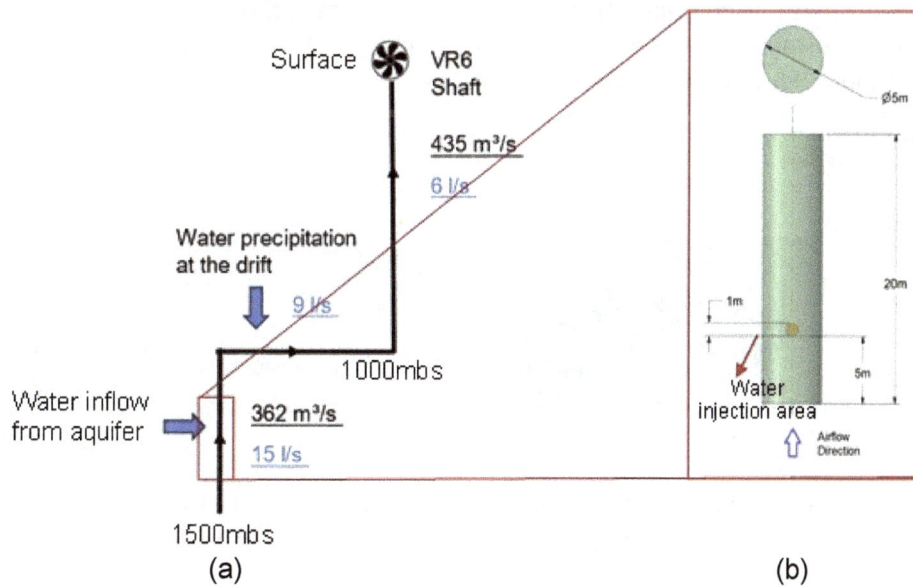

FIG 2 – Schematic of the VR6 air and water flow rate: (a) Scimatic of VR6 shaft; (b) Dimensions of the shaft section used in the numerical analysis.

The VR6 shaft encounters substantial saltwater infiltration, approximately 54 m³/hr, from the surrounding aquifer between depths of 1000 mbs to 1520 mbs, as illustrated in Figure 2a. Similarly, the VR3 shaft experiences notable saltwater infiltration, estimated at around 22 m³/hr, from the surrounding aquifer between depths of 600 mbs to 800 mbs. The air within the shafts is lifting and transporting all the water upward. For VR6, out of the 15 L/s of water, approximately 9 L/s of precipitated water are collected at the 1000 mbs sump. Pumps located there then lift this water to the surface. The remaining 6 L/s of water continues to move with the air to the surface. Figure 2a shows a representation of the air and water flow rate in the VR6 shaft.

MIST CONTROL METHODS IN GWALIA MINE

Sump enclosed by a fence

The mine uses a sump with a fence to control the saline water mist entering the environment at VR3 shaft. Figure 3 shows the VR3 fans, sumps with fences. The sump was designed to discharge mine water at 45° angle into a rock lined embankment. The diffuser at the end of the duct was incorporated to slow the discharge velocity and hence reduce the amount of water carry-over due to rebound. The 3 m high bund effectively prevents mist from entering the environment.

FIG 3 – Sumps with closed bund.

Mist eliminator

The mine employs a low-pressure loss mist eliminator system, featuring vertical corrugated parallel plates within a horizontal duct (Figure 4). As mist-laden air traverses these plates, it undergoes multiple directional changes, causing denser mist droplets to follow a more direct path due to their higher inertia (Figure 4a). With each rotation of the vane blades, these droplets collide with the plate surfaces, gradually forming a cohesive film. This film is then directed into a vertical lip channel and drains downward into a sump at the eliminator's base.

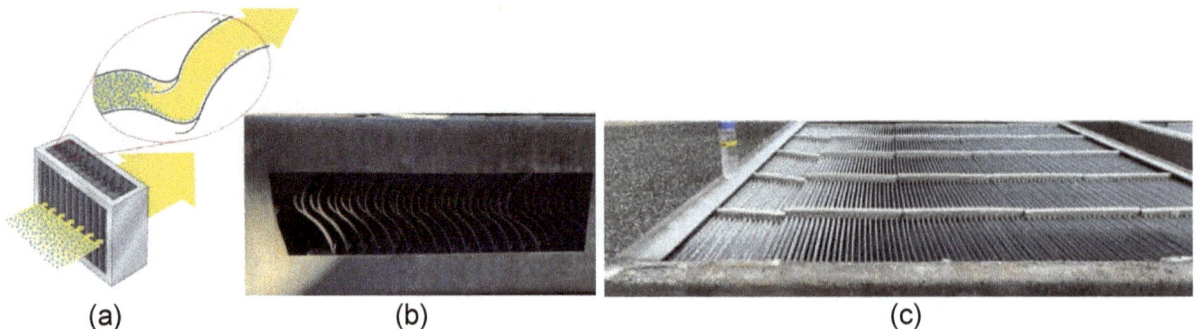

(a)   (b)   (c)

FIG 4 – Principle and photographic views of the mist eliminator panel: (a) Principle; (b) Mist eliminator side view; (c) Mist Eliminator isometric view.

Moreover, mist eliminator panels are seamlessly integrated into the fan inlet ducting to capture water entrained in the airflow and direct it to the water catchment dam. This process involves deflecting water-laden air around a separator profile within the inlet ducting, where droplets collide and merge, forming a cohesive liquid film. Guided by gravity, this film flows downward to the bottom of the profiles, facilitated by a sump frame beneath the mist eliminator profile, which efficiently channels water into outlet pipes for further management.

In the absence of mist eliminators, the airflow laden with water would be pulled into the fan impeller. This would lead to a combination of impeller velocity and increased air friction, resulting in partial vaporisation of water at the air inlet and deposition of solids on both the fan impeller and casing. Eventually, this solid buildup would impair fan performance, necessitating operational shutdowns for high-pressure cleaning to remove the accumulated solids. However, mist eliminator panels effectively remove a significant portion of the water and solids, thereby reducing the frequency of fan impeller and duct cleanings. Figure 5a shows the location of the mist eliminators and fans.

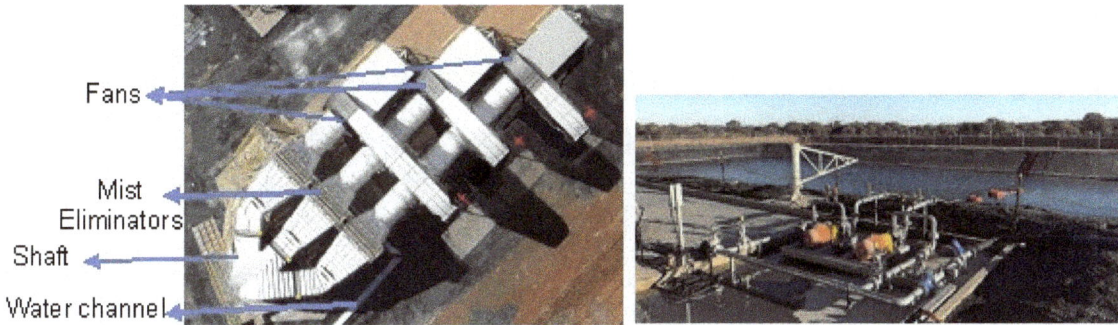

FIG 5 – VR6 mist eliminator and water sump: (a) VR6 fans and mist eliminator; (b) VR6 water dam.

The cleaning intervals for the VR6 mist eliminators and fan impellers have typically between 10 and 20 weeks, respectively, since their operation began. The water collected by the mist eliminators is directed through piping into the VR6 Dam. In comparison, VR3 Primary Vent Fans, which lack mist eliminators, require cleaning every 6–8 weeks.

VR6 water dam and discharge

The VR6 water catchment dam holds a capacity of approximately 530 m³ when reaching overflow levels (Figure 5b). Its water, saline in nature, exhibits a total dissolved solids (TDS) level of around 10 000 mg/L as per testing. The salinity of the water in the dam ranges from 75 g/L–150 g/L ie two to four times of the sea water salinity. To prevent wildlife access and potential entrapment, the entire fan compound and dam area are securely fenced. The VR6 discharge pumps operate with self-priming high head units, configured in a duty/standby arrangement. Moreover, the water flow from the mist eliminator measures about 9 L/s.

NUMERICAL SIMULATION SET-UP

In this study, a numerical analysis of airflow at the VR6 shaft along with the water mass is performed. To optimise computational efficiency while capturing the essential physics, a representative section of the exhaust shaft was chosen for numerical investigation as shown in Figure 2b. The 3D Realisable $k - \varepsilon$ turbulence model was used to describe the turbulent airflow within the shaft. This model is based on transport equations for k (turbulent kinetic energy) and ε (its rate of dissipation). The transport equation for k was derived from the exact equation, whereas the transport equation for ε was formulated based on physical reasoning, resulting in a form that differs significantly from its exact mathematical counterpart. The $k - \varepsilon$ model assumes fully turbulent flow and negligible molecular viscosity, making it suitable for modelling mine air, which is considered fully turbulent. The equations governing k and ε are as follows:

$$\frac{\partial}{\partial t}(\rho k) + \frac{\partial}{\partial x_i}(\rho k u_i) = \frac{\partial}{\partial x_j}\left[\left(\mu + \frac{\mu_t}{\sigma_k}\right)\frac{\partial k}{\partial x_j}\right] + G_k + G_b - \rho\varepsilon - Y_M + S_k \tag{1}$$

$$\frac{\partial}{\partial t}(\rho\varepsilon) + \frac{\partial}{\partial x_i}(\rho\varepsilon u_i) = \frac{\partial}{\partial x_j}\left[\left(\mu + \frac{\mu_t}{\sigma_\varepsilon}\right)\frac{\partial\varepsilon}{\partial x_j}\right] + C_{1\varepsilon}\frac{\varepsilon}{K}(G_k + C_{3\varepsilon}G_b) - C_{2\varepsilon}\rho\frac{\varepsilon^2}{K} + S_\varepsilon \tag{2.}$$

where G_b is the generation of turbulent kinetic energy due to buoyancy, and G_k is the production of turbulent kinetic energy due to the mean velocity gradient.

For tracking the motion of individual particles, such as water droplets, within a gas flow (like air in an exhaust shaft), the Discrete Phase Model (DPM) offers a compelling solution. The DPM model incorporates a breakup, collision, and coalescence model to account for the complex behaviour of the water droplets as they interact with the air stream. Water particles are tracked using the Lagrangian method in the discrete phase, and the particle or droplet trajectories are computed individually at specified intervals during the fluid phase calculation. The force balance equation relates the particle inertia with the forces acting on the particle and can be written as:

$$\frac{d\overrightarrow{u_p}}{dt} = F_D(\vec{u} - \overrightarrow{u_p}) + \frac{\vec{g}(\rho_p - \rho)}{\rho_p} + \vec{F} \tag{3}$$

Where $\vec{F}$ is an additional acceleration (force/unit particle mass), $F_D(\vec{u} - \overrightarrow{u_p})$ is the drag force per unit particle mass, and:

$$F_D = \frac{18\mu}{\rho_p d_p^2} \frac{C_D Re}{24} \tag{4}$$

Here $\vec{u}$ is the fluid phase velocity, $\overrightarrow{u_p}$ is the particle velocity, μ is the dynamic viscosity of the fluid, ρ is the fluid density, ρ_p is the density of the particle material, and d_p is the particle diameter. Re is the relative Reynolds number, which is defined as:

$$Re = \frac{\rho d_p |\overrightarrow{u_p} - \vec{u}|}{\mu} \tag{5}$$

The additional forces induced on the particle due to the fluid surrounding the particle due to growth in the boundary layer is called the virtual mass force and is given by:

$$\vec{F} = C_{vm} \frac{\rho}{\rho_p} \left(\overrightarrow{u_p} \nabla \vec{u} - \frac{d\overrightarrow{u_p}}{dt} \right) \tag{6}$$

Where C_{vm} is the virtual mass factor with a default value of 0.5, the fluid and the particle are coupled together mathematically in the form of slip velocity.

This approach reflects the concept that water flowing down the shaft from the aquifer breaks up into droplets due to interaction with the air. Initially the water is injected as water drops from designated locations on the shaft walls using a typical water drop size distribution (Viljoen and von Glehn, 2019). Figure 2b shows the dimensions of the shaft geometry, airflow direction, and the water injecting surface. The computational domain was discretised using a structured mesh consisting of approximately 350 000 elements. This mesh resolution ensures a balance between capturing the flow details and computational efficiency. A uniform velocity profile is used as the inlet boundary condition, and the outlet boundary is at atmospheric pressure. The inlet air is assumed to be saturated, and the shaft walls are at uniform temperature.

Water droplet and airflow behaviour were investigated at various air velocities (6 m/s to 20 m/s) and noted the corresponding inlet pressure changes. The simulation was configured to run for a total simulated flow time of 30 secs. To mimic the water inflow scenario, water droplet injection from the shaft walls was initiated after 5 secs of simulated airflow and sustained for a duration of 10 secs at a flow rate of 20 L/s. This set-up allows the simulation to capture the initial airflow behaviour before introducing the water droplets and examine the subsequent water build-up within the shaft. The numerical analysis was performed using ANSYS Fluent 2023 R2 software.

RESULTS

This section describes the findings of the numerical simulation studies on water droplet and airflow behaviour in an exhaust shaft. Water droplet behaviour in the shaft, water flow rate distribution, and airflow resistance are analysed at different velocities. The findings are then compared to field data on ventilation parameters in the exhaust shaft.

Water droplet behaviour in the shaft

The simulated water droplet behaviour in the exhaust shaft during injection period (5–15 secs) is visually represented in Figure 6. The size of each ball represents the diameter of a water droplet, and the associated lines show the path of that water droplet. The colour of the arrows indicates the air velocity magnitude. These numerical results shown in Figure 6 suggests that at 6 m/s, most of the water droplets fall or drain down the shaft, there is minimal breakup of droplet, and very small sized water drop is carried up the shaft. As the air velocity increases, the water droplet size and the amount of water carried up the shaft increases. From the figure, it can be seen that at velocities higher than 16 m/s, all the large water droplets (>5 mm) breakup into smaller size droplets.

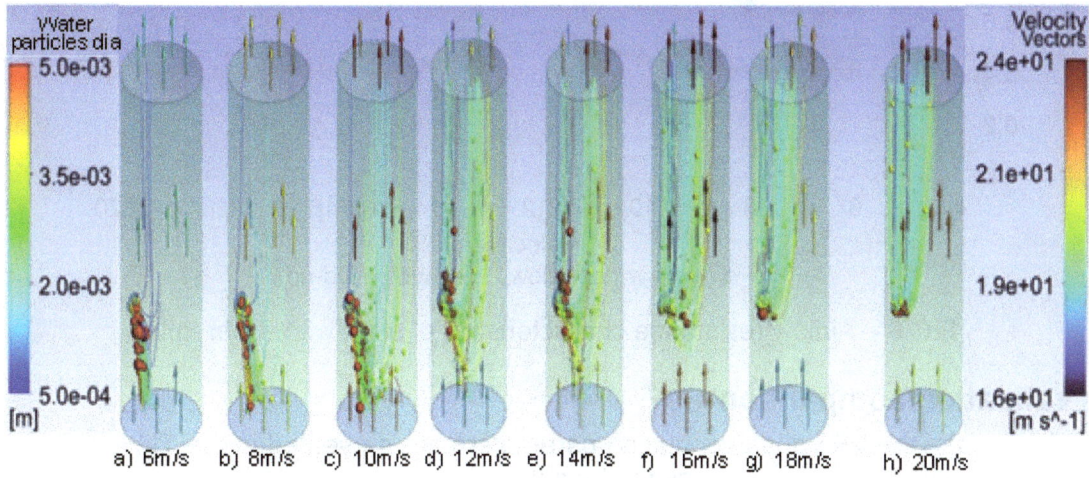

FIG 6 – Water droplet behaviour in exhaust shaft at different air velocities.

Water flow rate distribution

Figure 6 shows that the water droplets in the shaft can drain down the shaft, get suspended and build-up in the shaft, or escape out of the shaft with exhaust air. A quantitative analysis of these water flow rate at different velocities is performed. The stacked plot in Figure 7 illustrates how the distribution of water flow rate categories changes with velocity. Build-up in the shaft (orange) is the rate of water droplets in suspension. Significant water build-up or blanketing effect is observed in the 12–14 m/s velocity range.

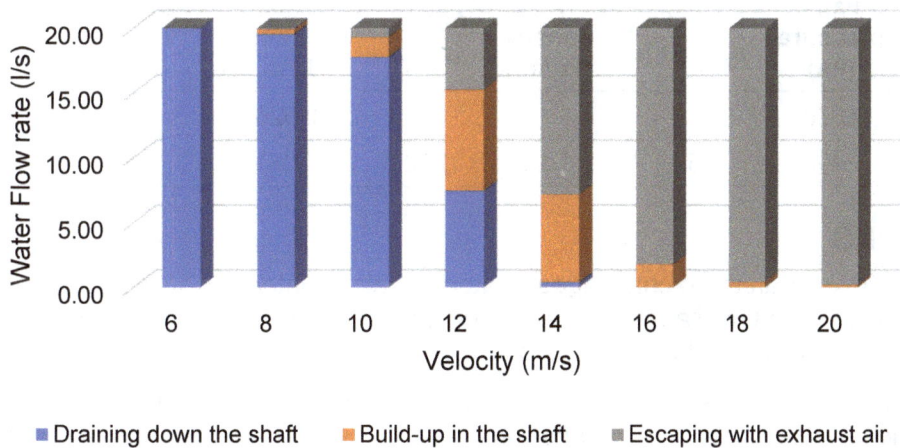

FIG 7 – Water flow rate distribution in the shaft at different velocities.

Shaft resistance characteristics

Pressure values at the inlet boundary are monitored during the simulation and it was observed that the inlet pressure increased during the water inflow/injection. Figure 8 shows the airflow resistance of the shaft at different velocities during water inflow and shortly after the water inflow event (with build-up). The increased resistance to airflow due to water blanketing can be clearly seen in the Figure 8. Even though less water is carried out with exhaust air at 12 m/s, the shaft resistance is maximum.

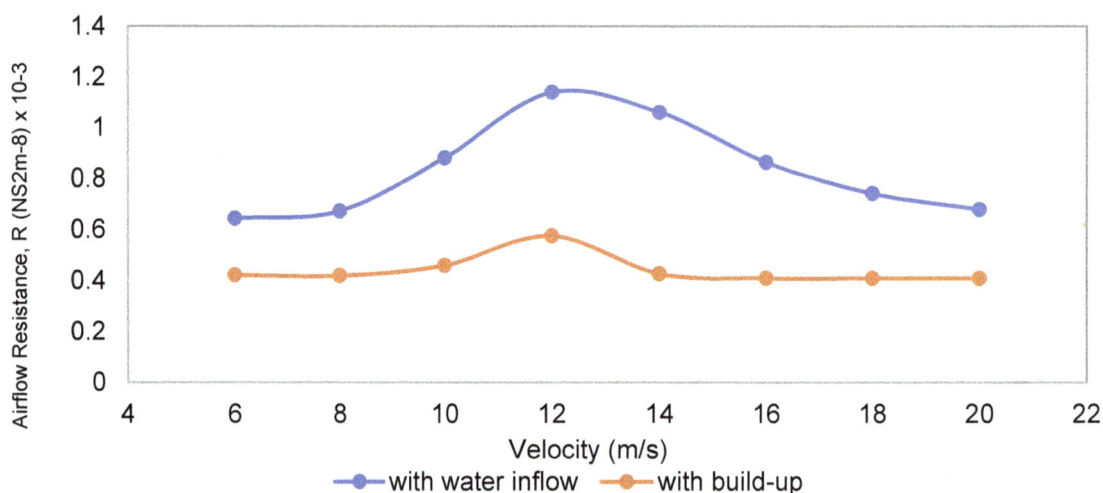

FIG 8 – Airflow resistance characteristics of shaft with water inflow.

Field validation/comparison

The field values of the VR6 shaft ventilation parameters are presented in the Table 1. In the table, airflow and velocity at aquifer are values for the shaft section between 1520 mbs and 1000 mbs. The scenario 1 corresponds to the ventilation scenario presented in Figure 2. To minimise the saline water discharge from the shaft, the fan operating speed is reduced to 57 per cent when the operation allows for it. In this second scenario, most of the water is precipitated and collected at the 1500 mbs sump, and no water is carried to the surface. But the total mine resistance for scenario 2 is higher than scenario 1. This is in accordance with the shaft airflow resistance characteristics shown in Figure 8. The numerical simulation results suggest that the increased airflow resistance is due to the water blanketing effect in the aquifer shaft section (1500–1000 mbs).

TABLE 1

VR6 ventilation parameters.

Scenario	Fans pressure (Pa)	Operating speed (%)	Air density (kg/m³)	Airflow (m³/s)	Total resistance (Ns²m⁻⁸)	Airflow at aquifer (m³/s)	Velocity at aquifer (m/s)
1	4878	82	1.12	435	0.026	362	18.4
2	2492	57	1.13	282	0.031	239	12.2

CONCLUSIONS

This case study investigated the challenges related to saline water discharge from exhaust shafts in underground mines. The Gwalia mine was utilised as a real-world example, highlighting the issues of saline water carry-over in exhaust shafts and the implementation of effective control systems.

The study employed numerical simulations to investigate water droplet behaviour within the exhaust shaft at varying air velocities. The results indicated that water droplet entrainment and build-up ('water blanketing') occur within a specific velocity range (12–14 m/s). This phenomenon significantly increases the resistance to airflow in the shaft, as confirmed by pressure measurements in the real-world scenario.

The trade-off between water carry-over and ventilation efficiency was highlighted. Reducing fan speed to minimise water discharge at the surface resulted in higher total mine airflow resistance due to water blanketing within the shaft. Our findings suggest that effective ventilation strategies for saline exhaust shafts should consider not only water carry-over but also the impact of water blanketing on airflow resistance. Future studies could explore mitigation strategies for water saline removal techniques to address water build-up in critical zones.

International Mine Ventilation Congress 2024 | Sydney, Australia | 12–16 August 2024

REFERENCES

Chen, K H and Trezek, G J, 1977. Effect of heat transfer coefficient, local wet bulb temperature and droplet size distribution function on the thermal performance of sprays, *J Heat Transfer*, 99(3):381–385. https://doi.org/10.1115/1.3450706

Derrington, A S, 2002. Control of Water Discharge From Mine Ventilation Shafts, in Proceedings of the 8th Underground Operators' Conference 2002 (The Australasian Institute of Mining and Metallurgy: Melbourne).

Gregory, S, Ward, M and John, J, 2009. Changes in the chemistry and biota of Lake Carey: a large salt lake impacted by hypersaline discharge from mining operations in WA, *Hydrobiologia*. https://doi.org/10.1007/s10750-009-9744-6

Kolesov, E, Kazakov, B, Shalimov, A and Zaitsev, A, 2023. Study of the Water Build-Up Effect Formation in Upcast Shafts, *Mathematics*, 11(6):1288. https://www.mdpi.com/2227-7390/11/6/1288

Lottermoser, B, 2003. Mine Water, in *Mine Wastes* (Springer: Berlin). https://doi.org/10.1007/978-3-662-05133–7_3

Semin, M and Zaitsev, A, 2020. On a possible mechanism for the water build-up formation in mine ventilation shafts, *Thermal Science and Engineering Progress*, 20:100760. https://doi.org/10.1016/j.tsep.2020.100760

Turek, M, Dydo, P and Surma, A, 2005. Zero Discharge Utilization of Saline Waters From 'Wesola' Coal-Mine, *Desalination*. https://doi.org/10.1016/j.desal.2005.03.082

Viljoen, J and von Glehn, F H, 2019. Investigation of water build-up in vertical upcast shafts through CFD analysis, in *Proceedings of the 11th International Mine Ventilation Congress*, pp 1003–1014 (Springer: Singapore). https://doi.org/10.1007/978-981-13-1420-9_86

Zgórska, A, Trząski, L and Wiesner, M, 2016. Environmental risk caused by high salinity mine water discharges from active and closed mines located in the Upper Silesian Coal Basin (Poland), in *Proceedings IMWA*, 85:85.

Development of strategies for improving air quantity in Indian underground metal mine – a case study

G D N Raju[1], M K Shriwas[2] and R Morla[3]

1. Research Scholar, Department of Mining Engineering, National Institute of Technology, Rourkela Odisha 769008, India. Email: gdnraju@yahoo.com
2. Assistant Professor, Department of Mining Engineering, National Institute of Technology, Rourkela Odisha 769008, India. Email: shriwasmk@nitrkl.ac.in
3. Senior Ventilation Engineer, Genesis Minerals Limited, Perth WA 6000. Email: ramakrishna.morla@genesisminerals.com.au; Adjunct Research Fellow, Western Australian School of Mines, Minerals, Energy and Chemical Engineering, Curtin University, Bentley WA 6102. Email: ram.morla@curtin.edu.au

ABSTRACT

The mine ventilation system is one of the most integral part of any underground mine, ensuring adequate quantity and quality of airflow in the subsurface mine workings. A minimum air quantity requirement is necessary for the safety and health of miners and improving productivity. The ventilation survey was conducted in an underground metal mine located in India as a case study to determine the quantity of air flowing and pressure drop at different mine branches. The ventilation survey data was used to develop a ventilation network model using Ventsim software for simulations. The eight cases were simulated using Venstim to supply adequate quantity of air for the next extracting level. The case 4 is best suited in supplying sufficient air quantity to the upcoming new level out of eight cases. The simulation results help in developing strategies for improving air quantity in the mine.

INTRODUCTION

Mine ventilation is one of the essential and integral parts of any underground mine, providing fresh airflow to the underground workings to dilute and remove dust and toxic gases and regulate temperature within acceptable limit. The ventilation system of every mine must be designed to deliver sufficient fresh air to all mine workings (Wang *et al*, 2022). In a mine, the airways that carry fresh air along the main airways and exhaust contaminated air from the working areas to the surface are collectively referred to as the primary ventilation system. These airways might be vertical, near vertical, or horizontal. On the other hand, a ventilation system that supplies fresh air to the production areas by use of several auxiliary fans and auxiliary ducting systems is referred to as an auxiliary ventilation system; this kind of arrangement is made for supplying fresh air to the dead-end headings where there is no availability of sufficient of airflow. Auxiliary ventilation system is to be planned and designed separately from the mine's primary ventilation system to not affect airflow distribution within the mine's primary ventilation system (Brune, 2019).

The ventilation network and distribution of airflow changes frequently due to the regular development of the underground workplace. This will cause various complications and ultimately affect the quality and efficiency of the underground ventilation system (Yang, Yao and Wang, 2022). An emphasis has always been placed on measuring airflow as it is required to fulfil most governmental ventilation regulations. As per Indian mining statutory norms, there is need to provide adequate ventilation to dilute the gases that are inflammable and noxious so as to render them harmless and also to contain not less than 19 per cent oxygen.

Production, productivity, worker health, and safety in underground mines depend on the working environment. The effectiveness of workers and equipment is decreased in a polluted environment, which eventually affects the mine's profitability (Mishra, Sugla and Singha, 2013; Rudakov, 2020). Proper ventilation of mine workings helps in safe and efficient operation in mines; therefore, it is crucial to regularly do a ventilation study to track changes in the surrounding environment and make it more conducive to functioning (Semin and Levin, 2019). A ventilation survey essentially consists of pressure-quantity measurement to determine airflow, pressure, and temperatures. The assessment is often performed to determine the ventilation system's efficiency, leakages, and the necessary actions for corrections as necessary. The types of ventilation surveys generally involve quantity and pressure surveys. In the quantity survey, the velocity and the quantity of air flowing at various places are determined, whereas the pressure drop in each ventilation branch can be determined by carrying the pressure survey. The quantification of differential pressures and airflows is essential for developing an accurate ventilation model (McPherson, 2012).

In this study, an extensive field investigation has been carried out in an underground metal mine to determine the quantity of air flowing at different working levels; strategies are developed to improve air quantity in order to meet the statutory norms and to ensure the safety and productivity.

Background and ventilation system of the selected mine

The selected underground mine has been accessed by a decline from the surface. Figure 1 shows the ventilation circuit of the study mine. The surface elevation of the decline is 167mRL and developed up to -147mRL with a gradient of 1 in 8. From the decline, different levels at 145mRL, 132mRL, 85mRL, 72mRL, 25mRL, 12mRL, and -35mRL, -48mRL, -95mRL, and -108mRL have been driven. To establish the second entry to the underground workings and to facilitate better ventilation of the development headings, one vertical shaft has been sunk from 200mRL and has been connected with the decline at all the levels. The vertical shaft is circular with a finished diameter of 4.5 m. Currently development work is going on for opening next working level ie -155mRL.

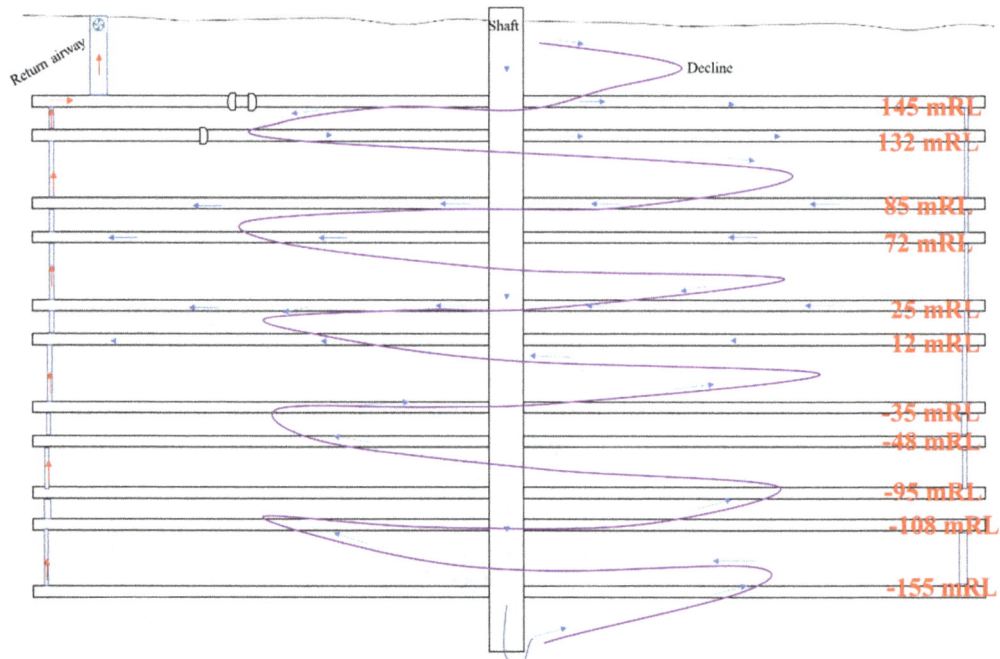

FIG 1 – Ventilation circuit of selected mine.

The selected mine has two ventilation fans in parallel located on return airway shaft at surface, each having a power rating of 250 kW. The capacity of the unit fan is 7500 m³/min. Both fans can be operated with variable frequency drive through a PLC (Programmable Logic Controller) system as well as manually. The existing ventilation system has two intakes and one return airway. The intakes are a shaft and a decline.

METHODOLOGY

A Pressure and Quantity (P-Q) survey has been carried out to determine the quantity of air flowing throughout the ventilation circuit of the mines, and to determine the pressure drop. Further, simulation models were developed with eight cases for supplying the adequate air quantity to the proposed working level (-155mRL) for future operations. The methodology adopted for this study is depicted in Figure 2.

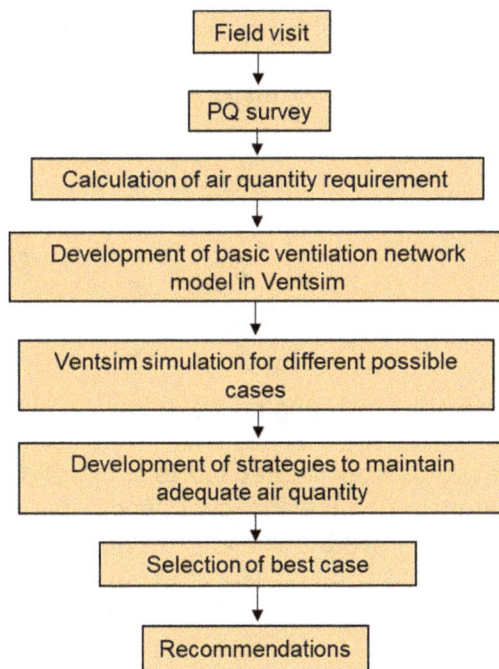

FIG 2 – Methodology.

Pressure and Quantity (P-Q) survey

The systematic process of gathering information that quantifies the distributions of airflow, pressure, and air temperatures throughout the main airflow channels of a ventilation circuit in a mine is called a ventilation survey.

Pressure survey

In order to find out the airflow (Q) recorded in each branch of a survey route and the corresponding frictional pressure drop (P), a pressure survey was conducted throughout the mine, including all the working levels and entries. The gauge and tube survey method was selected for this study as it gives better results than other methods (Prosser and Loomis, 2004).

During the pressure survey, the entire airway is split into various branches. Two pitot tubes were held against the airflow direction at the starting and end of each branch; one end of the nylon tube was connected to the pitot tube, whereas the other ends of both nylon tubes were connected to the Magnehelic pressure gauge, as shown in Figure 3, the difference shown in the gauge was noted.

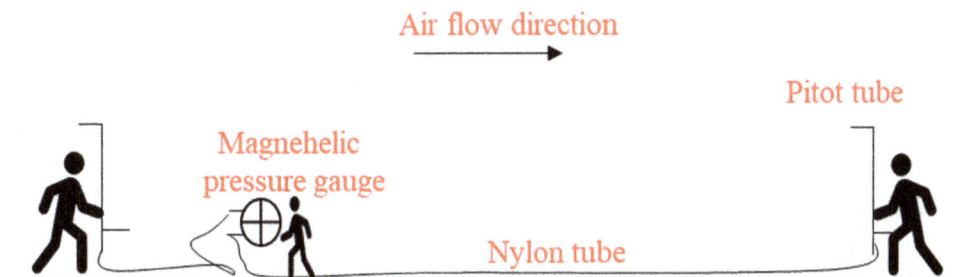

FIG 3 – Pressure survey by gauge and tube method.

Quantity survey

The majority of manually operated underground air velocity measurements are obtained using Vane-anemometer. The angled vanes rotate at an angular velocity nearly proportional to the air velocity when they are kept in a flowing airstream due to the force of the air passing through the instrument. The gearing mechanism and clutch arrangement connect the vanes to a digital counter or analogue pointer.

The velocity of the air was measured at different points throughout the cross-section, as shown in Figure 4. The average value was taken as the velocity of air at that point. The area of the roadway was measured with the help of a laser distance metre.

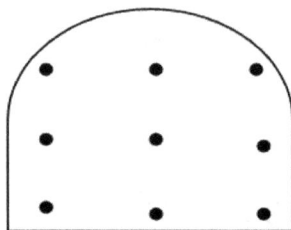

FIG 4 – Velocity measuring points in the roadway gallery.

In the traversing method, the anemometer is attached to a rod of 1.5 m in length; the attachment mechanism was made such that it permits the options of allowing the anemometer to hang vertically or to be fixed at a constant angle to the rod. Now, the observer keeps the rod against the airflow by maintaining some gap from his body. To commence the traverse, the instrument is held in the lower side corner of the airway by activating the clutch of the anemometer, and it is traversed, as shown in Figure 5. The time taken to complete the one traverse is noted with the help of a stopwatch. The distance measured by anemometer is divided by the time in order to calculate the air velocity (V). The cross-sectional area (A) is calculated from dimensions measured by laser distance metre.

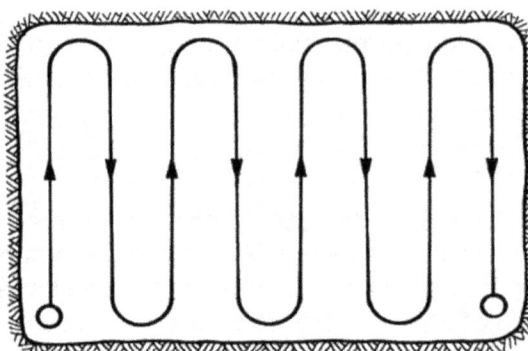

FIG 5 – Traversing path of anemometer.

The quantity of air can be calculated with the help of following equation

$$Q = V \times A \tag{1}$$

where:

Q   is quantity of air m^3/s

V   is velocity of air in m/s

A   is area of the cross-section in m^2

Development of strategies and simulation models using Ventsim

The underground metal mine ventilation is mostly governed by the maximum rated diesel engine deployed. As per DGMS (2018), there is need to supply minimum 0.06 m^3/s/kW of the maximum rated engine, where the diesel operated equipment are being used. Table 1 shows the air requirement for the level -155 mRL as per proposed total capacity of maximum rated diesel engine.

TABLE 1

Air requirement for the -155 mRL level.

Capacity of maximum rated diesel engine (kW)	Quantity of air required as per diesel vehicle (m³/min)	Actual quantity of air requirement considering leakages @10%
260	936	1030

Ventilation simulation software (VentSim) enables ventilation engineers to simulate heat, fire, airflow pressure, and several other factors in a specific mine ventilation model. Additionally, engineers may use this tool to simulate a mine ventilation circuit and gain a comprehensive knowledge of the behaviour of the airflow, fan pressures, and consequences during specific mining operations (Maleki, Sotoudeh and Sereshki, 2018).

The basic network model of the selected mine was developed in VentSim software with the help of survey data collected during the field investigation, which is depicted in Figure 6, and the same model was used to simulate the airflow under eight different cases as mentioned below:

Case 1: One main exhaust fan with no booster fan.

Case 2: Combination of one main exhaust fan and booster fan of 100 per cent rev/min.

Case 3: Combination of one main exhaust fan and booster fan of 75 per cent rev/min.

Case 4: Combination of one main exhaust fan and booster fan of 50 per cent rev/min.

Case 5: Two main exhaust fans with no booster fan.

Case 6: Combination of two main exhaust fans and booster fan of 100 per cent rev/min.

Case 7: Combination of two main exhaust fans and booster fan of 75 per cent rev/min.

Case 8: Combination of two main exhaust fans and booster fan of 50 per cent rev/min.

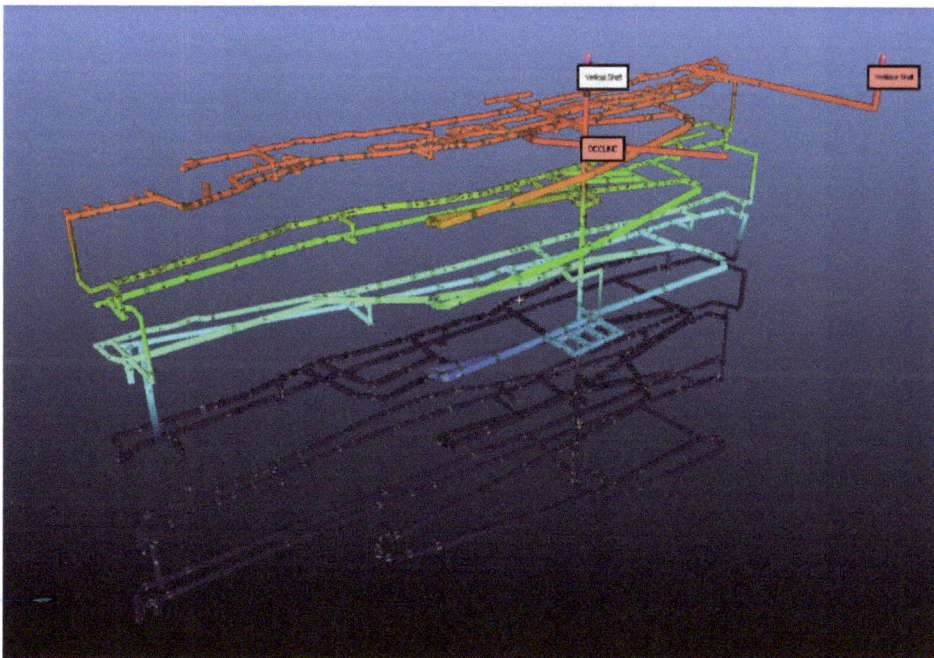

FIG 6 – Ventilation network model of the selected mine.

OBSERVATIONS AND DISCUSSIONS

The pressure quantity survey was conducted to find the airflow and pressure drop at each ventilation circuit. Significant pressure drops were observed, where there was a sudden increase or decrease in the cross-sectional area. Eight different cases are considered for the simulations to select the best case for providing the required airflow to the level -155mRL mine workings; the simulation results of eight cases are presented in Table 2, it can be seen that there is no change in the main fan performance after installing the booster fan in the -155mRL level as booster fan is located far from the main fan. From Figure 7, it is noticed that the maximum quantity of airflow is simulated in case 2. However, case 4 (combination of one main exhaust fan and booster fan of 50 per cent rev/min) can be adopted to meet the minimum air requirement at the -155mRL level (ie 1030 m³/min). The power requirement for the main and booster fans are 159 and 2.6 kW respectively.

TABLE 2
Simulation results under different cases.

Case	Main fan				Booster fan			
	Quantity (m³/min)	Static pressure (Pa)	Total pressure (Pa)	Power (kW)	Quantity (m³/min)	Static pressure (Pa)	Total pressure (Pa)	Power (kW)
Case 1	5942	1078	1176	159	408	-	-	-
Case 2	5942	1078	1176	159	2048	284.2	343	20.7
Case 3	5942	1078	1176	159	1525	98	205.8	9
Case 4	5942	1078	1176	159	1030	68.6	88.2	2.6
Case 5	7641	1715	1813	337.7	522	-	-	-
Case 6	7641	1715	1813	337.7	2224	313.6	372.4	21.2
Case 7	7641	1715	1813	337.7	1667	176.4	205.8	9
Case 8	7641	1715	1813	337.7	1122	68.6	78.4	2.5

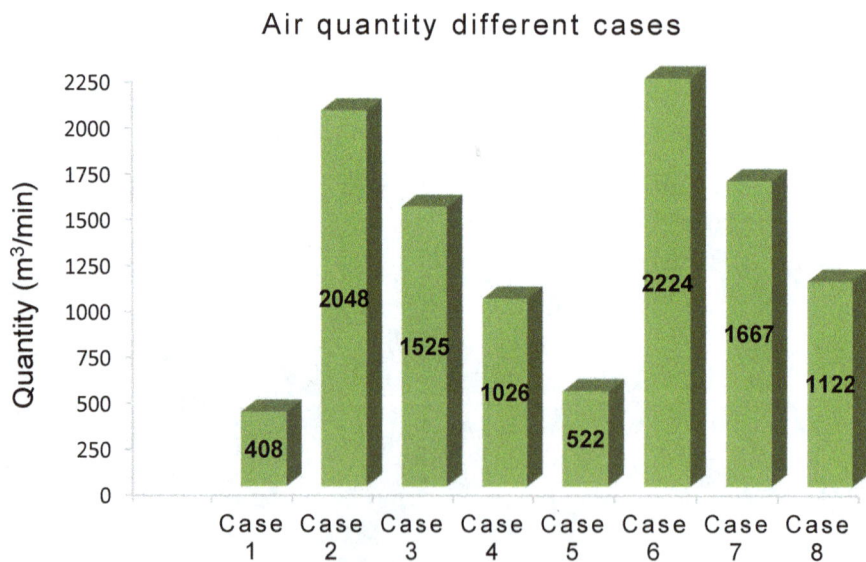

FIG 7 – Quantity of air flowing under different cases.

CONCLUSIONS

Mine ventilation is necessary to ensure a safe and healthy environment and productivity in the mine. Hence, proper attention is to be given to it. This study has been carried out to supply the adequate ventilation at the proposed level (-155mRL) to determine the best case suited to mine operation. Therefore, simulation models were developed using Ventsim software by incorporating the data collected from the PQ survey. Eight different cases were considered, and case 4 was suited the best for ensuring the required quantity of airflow. However, this may change as the mine advances further. Hence, it is recommended to do the ventilation and pressure survey whenever any major changes occur so that further necessary modifications can be implemented accordingly. Additionally, the performance of the main exhaust fan located at the surface was found to be unaffected by the booster installed at the -155mRL level. This is likely because of their long distance from one another and less capacity of booster fan.

REFERENCES

Brune, J F, 2019. Mine ventilation networks optimized for safety and productivity, *Advances in Productive, Safe and Responsible Coal Mining*, pp 83–99 (Woodhead Publishing).

DGMS Tech, Circular No. 1 of 2018 Standards and Safety Provisions of Diesel Equipment for using in belowground coal and metalliferous mines.

Maleki, S, Sotoudeh, F and Sereshki, F, 2018. Application of VENTSIM 3D and mathematical programming to optimize underground mine ventilation network: A case study, *Journal of Mining and Environment*, 9(3):741–752.

McPherson, M J, 2012. *Subsurface Ventilation and Environmental Engineering* (Springer Science and Business Media).

Mishra, D P, Sugla, M and Singha, P, 2013. Productivity improvement in underground coal mines-a case study, *Journal of Sustainable Mining*, 12(3):48–53.

Prosser, B S and Loomis, I M, 2004. May, Measurement of frictional pressure differentials during a ventilation survey, in *Proceedings of the 10th US Mine Ventilation Symposium,* pp 59–66.

Rudakov, M L, 2020. Assessment of environmental and occupational safety in mining industry during underground coal mining, *Journal of Environmental Management and Tourism (JEMT)*, 11(03):579–588.

Semin, M A and Levin, L Y, 2019. Stability of air flows in mine ventilation networks, *Process Safety and Environmental Protection*, 124:167–171.

Wang, J, Jia, M, Bin, L, Wang, L and Zhong, D, 2022. Regulation and optimization of air quantity in a mine ventilation network with multiple fans, *Archives of Mining Sciences*, 67(1).

Yang, B, Yao, H and Wang, F, 2022. A review of ventilation and environmental control of underground spaces, *Energies*, 15(2):409.

Case study – ventilation design solutions for post-excavation phases in tunnelling projects

M J Shearer[1] and C M Stewart[2]

1. Ventilation and Compliance Manager, CPB Contractors, Tunnelling, Sydney NSW 2060. Email: michael.shearer@cpbcon.com.au
2. Principal Engineer, Minware, Cleveland Qld 4163. Email: craig@minware.com.au

ABSTRACT

The Rozelle Interchange (RIC) currently serves as the largest and most complex road tunnelling project in Australia and is located underneath the city of Sydney, New South Wales. This paper will discuss some critical ventilation control solutions used in tunnelling, and the processes used to design and maintain a high level of safety for project personnel.

Ventilation Control Devices (VCDs) are not a new concept or unique in mining and tunnelling ventilation and control air movement from one location to another within an operation. A variation was developed for the RIC which uses an integrated dual airlock brattice wall that contains a roller-door system with or without booster fans acting as a VCD and network control.

The modular airlock system was identified by the tunnelling project to be a viable and cost-effective solution for the creation of a hybrid ventilation network. This temporary solution provided the opportunity to identify and create independent (modular) ventilation zones, referred to as the Primary and Secondary Ventilation Districts during the post-excavation phase of the project. The creation of the Primary and Secondary modular districts provided the opportunity to remove >95 per cent of rigid and flexible ducting from the overall tunnel network.

The design of the proposed modular network was scrutinised by using the Ventsim™ modelling software to demonstrate the practical effectiveness of the control system. The project used the Activity Tracking and Velocity Heat Mapping functions in the Ventsim™ software to stress test and validate the concept for overall compliance of the design. The results from the Ventsim™ software modelling also provided the project with relevant feedback for assessing the impact of maintenance tasks and emergency preparedness.

INTRODUCTION

The Rozelle Interchange (RIC) tunnelling project is located underneath the city of Sydney, New South Wales, Australia. The project commenced tunnelling in December 2019 and was completed and fully operational for traffic use in November 2023.

The RIC is essentially an underground labyrinth or turnpike type road system that is approximately 24 km in total length. The bottom of the mainline tunnel is located approximately 80 metres below the existing topography as indicated in Figure 1.

The project forms part of the WestConnex underground motorway scheme and provides greater flexibility for road users via its connection to the M4 and M4–M8 tunnels and existing above-ground road infrastructure. During the excavation phase, the project used 16 × 355 kW axial flow fans with ducting up to 2800 mm in diameter to deliver approximately 1290 m³/s of fresh air throughout the underground workings to maintain a healthy underground atmosphere.

The ventilation network in the RIC is designed differently from most Australian tunnelling projects, resembling a metalliferous mine rather than a typical road tunnel. One key difference is that tunnelling projects generally do not incorporate dedicated return air raises during production or development phases due to their unique locations.

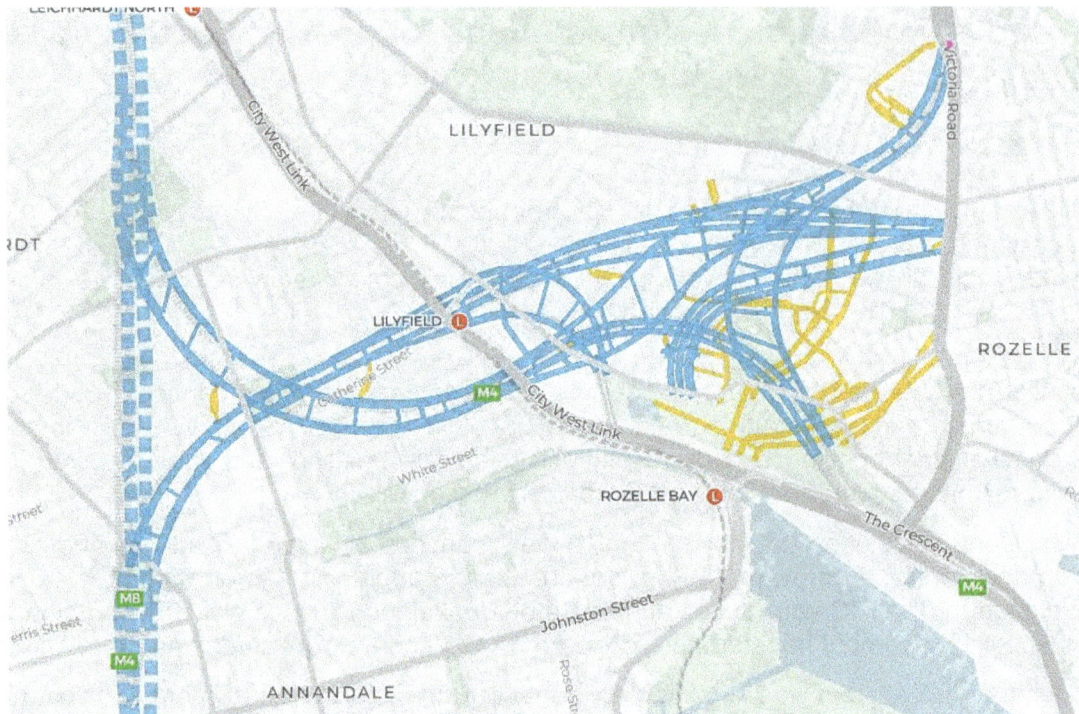

FIG 1 – Location of Rozelle Interchange Tunnel project (WestConnex RIC, 2024).

The legislative requirements for tunnelling in Australia for diesel dilution is more conservative than most mining regulatory standards with tunnelling requiring a minimum of 0.067 m^3/s per kW of rated power, which is between 10 per cent (0.06 m^3/s) – 25 per cent (0.05 m^3/s) more air required than underground mines to manage contaminates.

This paper will discuss some critical solutions such as the Ventilation Control Devices (VCD) employed during the transition from the excavation phase and migration into the post-excavation phase of the project. The paper will also discuss further value-add engineering processes employed to conceptualise, design, and install ventilation network controls to maintain a high level of safety for personnel over the lifespan of the project.

VENTILATION DESIGN AT ROZELLE INTERCHANGE

The overall tunnel layout of RIC is mined through the Sydney Sandstone Basin and is extraordinarily complex with up to three separate tunnels constructed over the top of one another. The sizes of the excavation profiles within the project range from Cross Passages (XP) @ approximately 10 m^2 up to extremely large caverns with a roof height of 23 m @ approximately 450 m^2 used to install the Permanent Vent Facility fans and infrastructure.

The ventilation design during the excavation phase delivered approximately 1290 m^3/s and consisted of the following control methods to manage and maintain optimal conditions:

- Forced systems delivered via flexible and rigid ducts.

- Forced overlap systems using mobile dust scrubber units via flexible and rigid ducts.

- Negative pressure systems using mobile dust scrubber units via a rigid duct.

- Push/Pull systems utilising brattice/inflated walls to isolate work areas.

Prior to the excavation sequence being completed, the removal of the vent duct from the 24 kms of mined tunnels was required while maintaining adequate air volumes and velocities to meet regulatory requirements. As the project was still isolated from the WestConnex tunnel network, parallel carriageways were used as pathways for air to migrate from one area to another.

Tunnels directly connected to the surface with an available inlet and outlet were identified as the Primary Ventilation Districts. Opposing carriageway tunnels on each of the connections were used in the same ventilation circuit with one carriageway used for the fresh air intake whilst the opposite is used for the return (RIC, 2020). Tunnels not directly connected to the surface and located between the Primary Districts were identified as the Secondary Ventilation Districts. In total, there were four Primary Ventilation Districts and four Secondary Ventilation Districts, with the management of the underground zones controlled by the relationship between the primary and secondary ventilation systems (Shearer and Mutton, 2015).

Network analysis of the individual Primary and Secondary circuits was critical to quantifying the minimum amount of fresh air volume required for delivery into and out of the tunnel network. This analysis also provided the opportunity to identify parallel airflow interactions and minimise the residency time of the ventilating air in and out of the secondary districts.

The total fresh air volume required was designed at 1050 m^3/s based on an average air velocity of 0.7 m/s within the mainline tunnels of the Primary and Secondary Districts and required 6.9 MW of installed fan power to operate. The use of Ventilation Control Devices (VCD) was tabled as part of the overall strategy to enable the project to migrate ventilating air throughout the underground workings between the Primary and Secondary Districts. The newly designed ventilation layout incorporated the following components as indicated in Table 1.

TABLE 1

Ventilation network control system components (RIC, 2020).

Ventilation control type	Number of units
Primary Fans (200 kW–355 kW)	16
Auxiliary Fan (160 kW–55 kW)	5
Twin Interlock Doors (brattice walls)	9
Automatic Roller Doors	25
Brattice Walls (including prior to XP door installation)	66
Access doors for 72 XP's (progressively installed)	144
Vent Bulkhead	1

The introduction of the Primary and Secondary districts allowed the vent circuit to be separated into eight ventilation zones across the project footprint. A check of fan operation and a total of 69 underground atmospheric readings were undertaken daily to verify the consistent performance of the ventilation network.

The estimated combined peak of kilowatts of rated diesel power for the project during the post-excavation phase was approximately 12 500 kW. The designed volume of fresh air of 1050 m^3/s provided a theoretical dilution value of 125 per cent for the 12 500 kW rated power.

CHALLENGES

Establishing air pathways

One of the major challenges was finding suitable locations and pathways to create a Hybrid or Artificial ventilation network system from the existing footprint of the RIC post-excavation. A detailed map of the RIC excavated tunnel footprint is indicated in Figure 2.

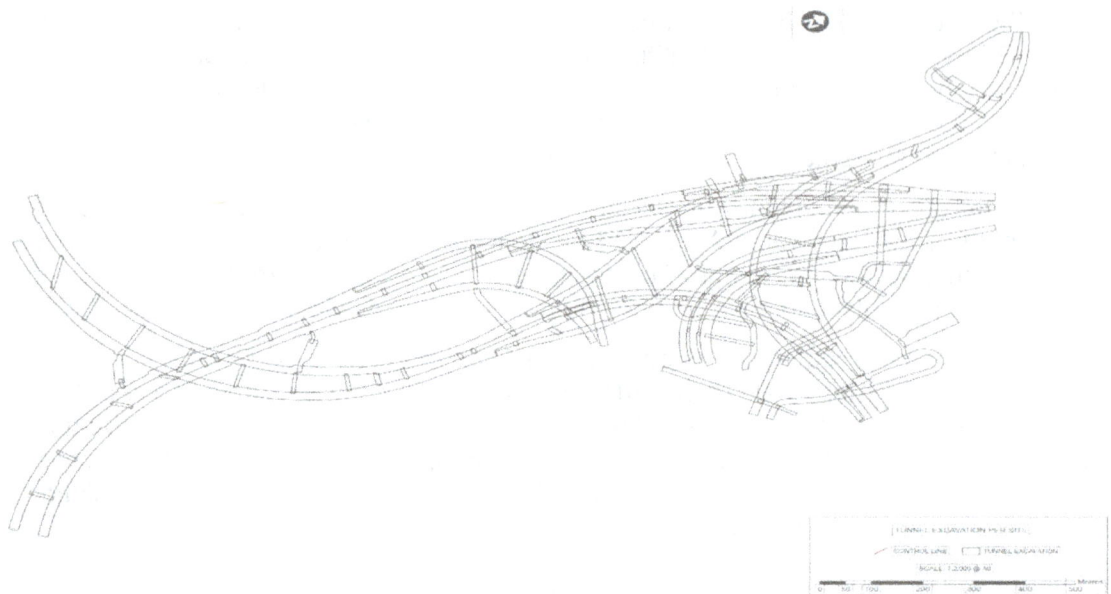

FIG 2 – Rozelle Interchange mined tunnel excavation plan (RIC, 2020).

The challenges of the post-excavation project sequence revolved around the need to establish a functional ventilation network within the RIC. The initial analysis identified that the RIC would remain isolated from the WestConnex interconnecting tunnels and that civil construction interactions at the Portals and Cut and Cover structures were still being completed. Another complexity the project team faced was the completion of permanent infrastructure works being conducted simultaneously. This required careful planning to sufficiently ventilate the workings and meet legislative compliance.

The team identified available locations and pathways throughout the excavated tunnel network to allow fresh air into the network and provide pathways for the migration or exchange of ventilating air between parallel or merging airways to reduce residency time. It was identified that some pathways would be Primary Districts, while other interconnected airways would become Secondary Districts. A combination of approaches was used to evaluate required airflows, and directions using ventilation software models to validate findings as indicated below in Figure 3. This initial phase of the project set the foundation for the subsequent steps in creating an efficient and effective ventilation network.

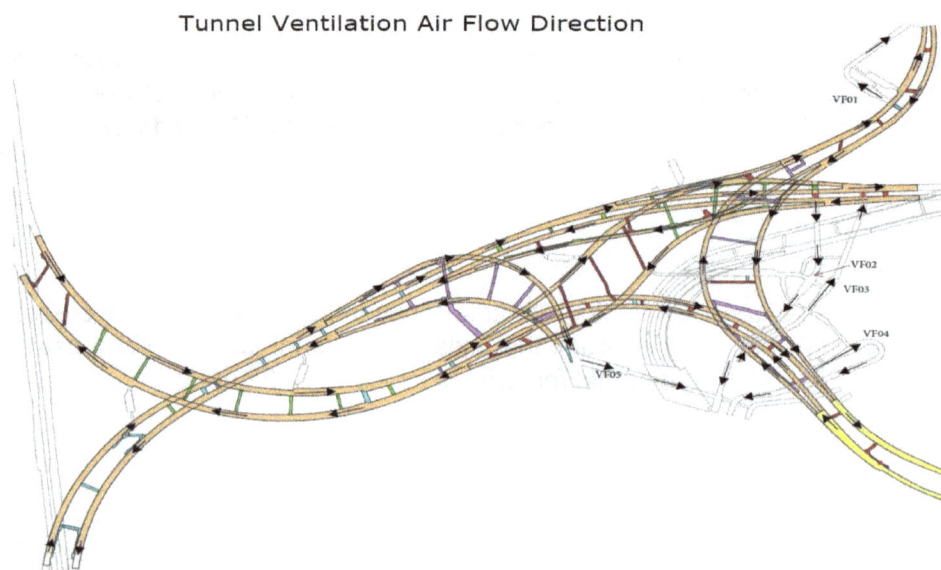

FIG 3 – Proposed alternative airflow direction (RIC, 2020).

Meeting airflow requirements

Once the potential network pathways were identified for the post-excavation phase, the next challenge was assigning sufficient airflows to support project completion and meet legislative compliance. The schedule and distribution of works for each quarter across the project footprint were quantified using the diesel fleet loading as shown below in Figure 4 to create a base case to justify predicted contaminate dilution rates.

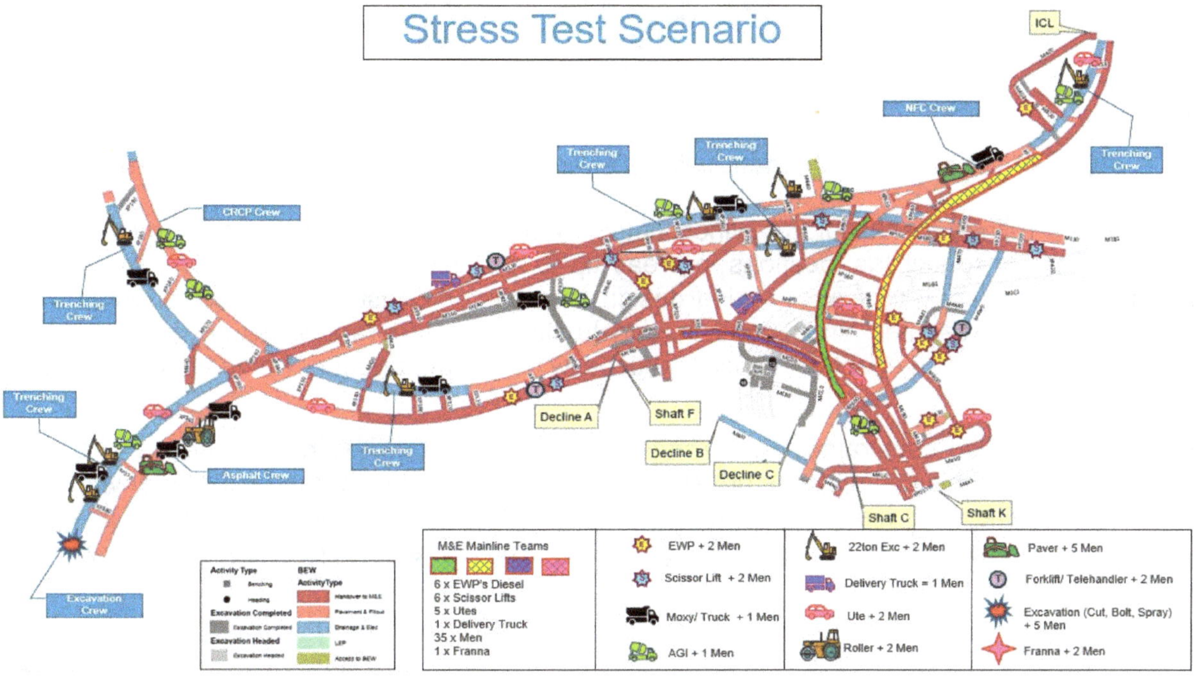

FIG 4 – Operational stress test (RIC, 2020).

It became apparent that not all ventilating air could be exchanged via a direct pathway to and from the surface. Approximately >95 per cent of the installed flexible and rigid ducting from the 24 kms of tunnel workings was removed to clear work zones and help create the complex ventilation system.

Ventilation modelling software (Ventsim™) was used to evaluate air movement, velocity heat mapping, and to mitigate potential recirculation pathways throughout the network. This was necessary to identify compliant airflows and demonstrate optimal underground conditions could be met with the proposed system.

The analysis also scrutinised the upcoming works schedule to identify the peak diesel loadings of approximately 12 500 kW of rated power for the project would occur during the first quarter of the following year and would require approximately 840 m³/s of fresh air at a dilution rate of 0.067 m³/s. The minimum air velocity of 0.5 m/s within the mainline tunnel profile was increased to 0.7 m/s to increase air volume to 1050 m³/s, providing a further 25 per cent in dilution capacity. The daily results from the ventilation monitoring positions throughout the underground workings further validated the positive outcome from reducing the overall residency time for ventilating air travelling through the Secondary Districts.

Ventilation control

The project required an easily manageable and robust set of Ventilation Control Devices (VCDs) that could be designed and implemented within the existing tunnel profile. In total, more than 200 VCDs were required, which included ventilation fans installed with podded silencers to reduce noise, brattice-lined timber walls with pressure loadings, stoppings for XP's, roller doors for access, power allocation and cameras. The double airlock walls with interlocking roller doors were the most challenging VCD for the project to integrate the design and implement. The Interlocking doors needed to provide continued functionality, operate with and without fans, mitigate recirculation and maintain safe access for pedestrian walkways and traffic flow as shown in Figure 5.

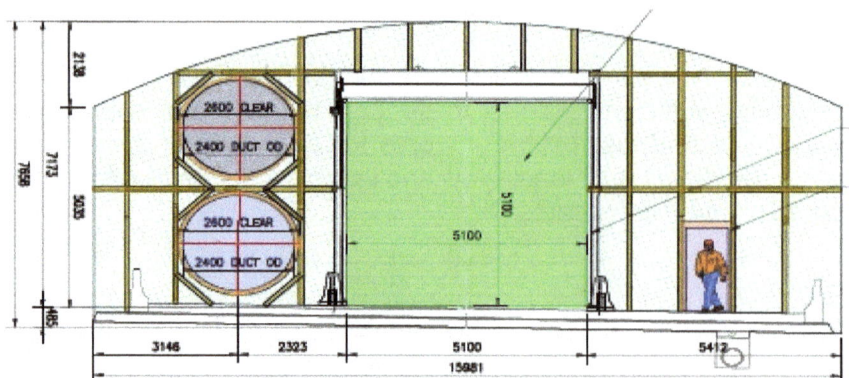

FIG 5 – VCD concept schematic with roller door and pedestrian access (RIC, 2020).

The pressure ratings for the brattice-lined timber walls used for the VCDs required a designed loading of 300 Pa to suit the locations in areas up to 130 m² in size. The brattice walls required specialised temporary works engineering design to allow the construction of the structures to be used. This also presented challenges in constructing a temporary structure atop permanent pavement to maintain an effective airlock functionality due to the ongoing works being completed in most districts.

The project was also constrained by a range of factors, such as the need to reuse and recycle materials due to supply chain issues driven by the global pandemic and the existing environmental constraints. Additionally, the project faced challenges in fitting the airlocks within the available footprints, whilst maintaining airflow throughout the Primary and Secondary districts as shown below in Figure 6.

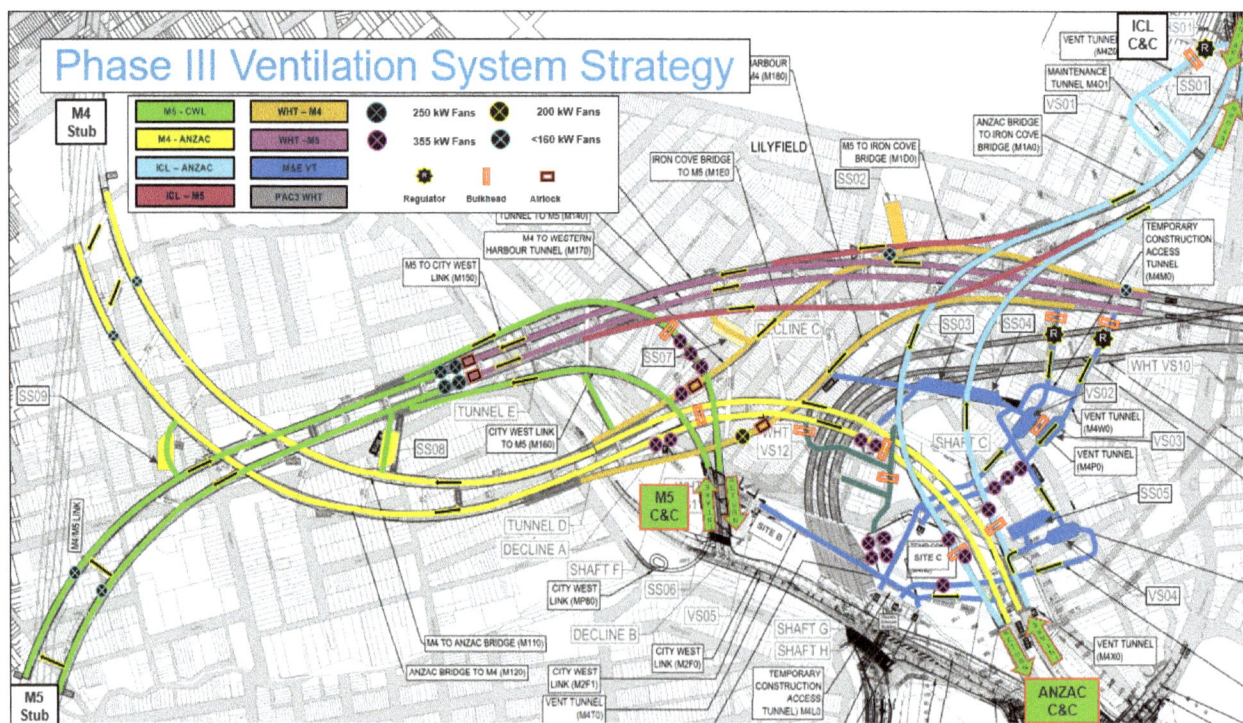

FIG 6 – Overall Primary and Secondary air pathway districts (RIC, 2020).

The power supply availability to the project also impacted the locations of electrical infrastructure, resulting in a low-pressure high-volume ventilation network that lowered peak demand and utilised existing fans and infrastructure. The design of a high-volume ventilation network utilising existing fans and infrastructure was a requirement to also meet environmental and noise requirements across the project in areas closely connected to the open community. This resulted in identifying that a suitable fan and airlock system would need to operate at a low pressure of ≤2000 Pa and suitable locations chosen within the tunnel could meet and maintain noise compliance with the aid of silencers fitted to the intake and outlet of fans.

The temporary post-excavation ventilation network did not have any real-time monitoring equipment to collect underground atmospheric conditions, fan data, or operational status of VCD's across the project footprint. This posed a challenge as the ventilation network required ongoing daily monitoring and analysis to ensure it was stable and operational . To validate the network, and to identify any potential issues or malfunctions, a total of 69 underground monitoring stations were introduced. The field data and mapping were manually checked for any changes and trends, and this was digitally tracked using the Ventsim™ tool Ventlog.

System optimisation

During the project, there was an opportunity to recycle and reuse some of the Primary and Secondary fans and ducting utilised during the excavation phase. The modular airlock system with roller doors became the primary VCD and were evaluated for vehicle/equipment movements prior to being finally chosen. The airlock system required a temporary works engineering design to manage a pressure loading of 300 Pa on the walls. The airlock system needed to be easy to install, relocate, and be non-intrusive as shown below in Figure 7.

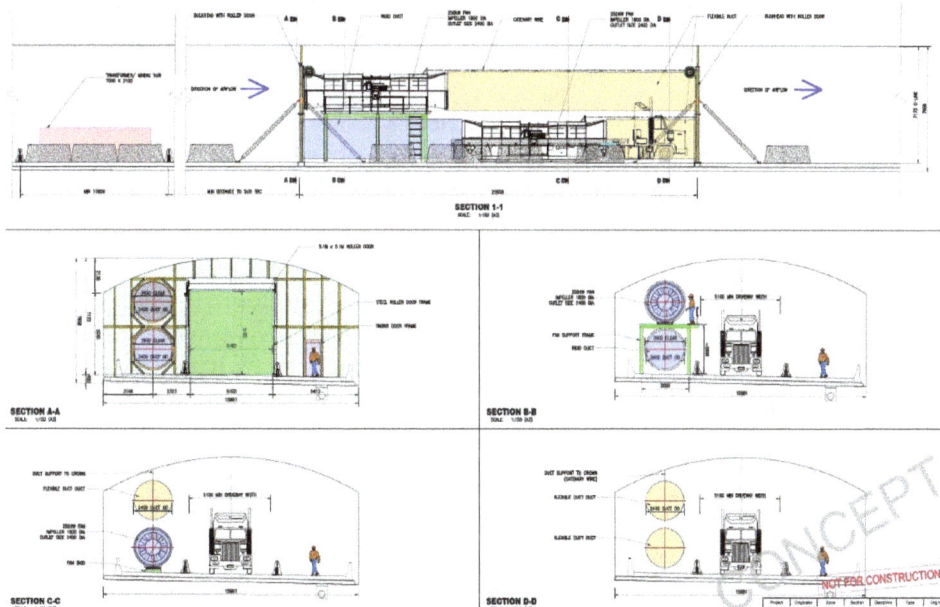

FIG 7 – Interlock door system and schematics (RIC, 2020).

The installed power loading of the 21 fans used for the post-excavation phase required to manage the 1050 m³/s of air volume was 6.9 MW. A reduction of the fan operating points was designed to below 2000 Pa to create a high volume-low pressure push-pull network that reduced the electrical loading to approximately 5.3 MW. This further reduced the environmental footprint and functioned as a reverse-load shedding system. The operating points of the utilised fans and the locations of VCDs helped the project to improve compliance through the reduction in noise constraints and community impacts.

The operational capabilities of the automatic airlock doors were faced with the demand during shift changes, the number of cycles per hour, potential damage, or the need to shutdown direct access through interconnected districts that could create disruptions. A robust control solution was required to be identified to mitigate ventilation network impacts and provide smooth operation of the airlock system.

A total of 69 vent monitoring stations, broken into eight zones, were installed and this allowed project staff to visually check the VCD operations during the daily inspections. The sequencing of operating doors and providing access through the airlock system was mapped out with a

Programmed Logic Control (PLC) to manage the ongoing interaction between personnel and traffic movements and further mitigate the potential for impacts to the ventilation network.

SUMMARY

The Rozelle Interchange (RIC) faced several ventilation challenges which were solved with careful analysis and design. Finding suitable locations and pathways to create a hybrid ventilation system within the existing tunnel footprint was solved by the project team utilising mud maps, airflow calculations, and ventilation software models to identify available and suitable pathways that would migrate fresh air throughout the network.

Providing sufficient airflows to support project completion and meet legislative compliance was ensured by using ventilation modelling software to evaluate air movement and velocity, ensuring compliant air quality was developed to maintain optimal underground conditions. Additionally, the project required a range of over 200 Ventilation Control Devices (VCDs) to be designed and modelled to control and manage air movement including fans fitted with silencers, brattice lined timber walls, roller doors, and access doors over the course of approximately 18 months.

Temporary works engineering design and integration faced challenges, such as constructing temporary structures atop of permanent pavement and fitting airlocks within available footprints of the tunnel system. The vertical movement and speed of the roller door units was something that required further investigation to increase serviceability caused by impacts and the PLC connections.

The project also considered its environmental and noise footprint, resulting in a low-pressure high-volume design that lowered power consumption by approximately 1.6 MW and further reduced its environmental footprint by using some existing infrastructure from the excavation phase.

CONCLUSION

Clear communication, coordination, and stakeholder engagement were crucial in overcoming these challenges and ensuring the success of the hybrid ventilation network. By addressing each challenge with a systematic and innovative approach, the RIC project was able to create an efficient and effective ventilation system that supported the project's lifespan and contributed to the overall success of the WestConnex underground motorway scheme.

The project team successfully identified suitable locations and pathways to create a hybrid ventilation system within the existing tunnel footprint while maintaining compliance with legislative requirements for the lifespan of the post excavation portion of the project whilst supporting the commissioning stage. The design and utilisation a robust VCDs such as the airlock roller doors, both with and without fans as a primary control to coordinate the sufficient distribution of airflow throughout the site footprint also contributed to the project success.

Overall, the solutions implemented for the ventilation challenges in the RIC project demonstrate the importance of thorough planning, analysis, and innovation in large-scale infrastructure projects.

ACKNOWLEDGEMENTS

The author would like to thank the Rozelle Interchange management team for the collaborative support and permission to present this paper, and gratefully acknowledge the assistance and support provided by various tunnel managers, engineers, and Adrienne Frame.

REFERENCES

RIC, 2020. RIC Design Manual, Internal document, 2020–2024.

Shearer, M and Mutton, V, 2015. Advanced Ventilation Control: An Analysis of Innovation and Environmental Sustainability, in Proceedings of the Australian Mine Ventilation Conference 2015, pp 55–61 (The Australasian Institute of Mining and Metallurgy: Melbourne).

WestConnex RIC, 2024. Rozelle Interchange tunnel location. Available from: <https://caportal.com.au/tfnsw/wcxri/wa-surface?mapView=notifCard_6b05dfea-5d41-4897-9aec-3f06558ea146#?mapView=notifCard_6b05dfea-5d41-4897-9aec-3f06558ea146>

Optimised ventilation system for hard rock metalliferous mine implementing a drill-and-blast bord-and-pillar mining method

M Van Der Bank[1] and W Marx[2]

1. Specialist Ventilation Consultant, BBE Consulting, Bryanston 2191, South Africa. Email: mvanderbank@bbe.co.za
2. Principal Engineer, BBE Consulting, Bryanston 2191, South Africa. Email: wmarx@bbe.co.za

ABSTRACT

Hard rock metalliferous bord-and-pillar mines are distinctive for their underground infrastructure developed within the reef horizon, imposing unique constraints on ventilation distribution. The design criteria for these mines differ significantly from the well-established coalmine bord-and-pillar ventilation systems due to factors such as drill-and-blast rock breaking, distinct mining cycles, and the need for heat and blast fume clearance.

Traditionally, in most bord-and-pillar operations, the production sections are ventilated in series. Fresh air intake occurs through a main decline system and is distributed along the strikes to the production section faces. The air then returns through the series of worked-out areas, reaching the top-most collector level before being exhausted from the mine, typically via a centralised upcast shaft equipped with the main exhaust fans. While a 'series' ventilation approach facilitates centralised main fan infrastructure and relatively easy ventilation control, it comes with significant drawbacks. These include extended blast-fume clearance times, the accumulation of diesel emissions and other contaminants, and significantly increased consequences in the event of an underground fire, as all workings downstream of the fire are completely contaminated.

This paper examines an alternative, cost-effective, and practical ventilation strategy that utilises the concept of ventilation districts to mitigate the shortcomings of traditional 'series' ventilation. The proposed strategy employs smaller ventilation districts, each comprised of a fixed number of production sections, with a dedicated intake-return system. To further mitigate the inherent fire risk, the strategy also involves ventilating the main dip conveyor belt directly to return within its own ventilation district.

INTRODUCTION

The adoption of new mining methods, procedures, and technological innovations in the global mining industry has traditionally been slow due to entrenched practices. However, recent years have seen a notable increase in innovation and technological advances in areas such as rock breaking, ore transport, energy efficiency, software, and mineral processing. Similarly, there have been significant advances in mine ventilation, including monitoring and control, ventilation-on-demand, software tools, and ventilation management and control equipment. Together with enhancements in fan and refrigeration and cooling system technology, these advancements have led to ventilation systems characterised by improved business risk analysis, enhanced safety, healthier working environments, and optimised capital and operating costs.

This study was based on feasibility and pre-feasibility work that identified the opportunities and benefits of dedicated ventilation districts as an optimised alternative to the traditional once-through series ventilation system implemented in hard rock metalliferous drill-and-blast, bord-and-pillar operations. However, the pre-feasibility ventilation designs revealed several constraints, such as cumbersome return airway requirements, extensive crossovers, and high ventilation leakage potential as described by Deglon and Hemp (1992).

The mining engineering group provided detailed production schedules and Life-of-Mine (LOM) layouts, including mining depth and strike distances. The ventilation design involved trade-offs to propose the optimal primary ventilation system for various ventilation district layout options. These options typically included determining the number of sections in each ventilation district, whether strike belts should be ventilated to return or form part of the intake/district, and whether the return airways should consist of an off-reef decline corridor or dedicated Return Air Raises (RARs).

Study results are from a feasibility study recently completed for a South African platinum mine using drill-and-blast bord-and-pillar mining methods. The final primary ventilation system design included independent ventilation districts, each consisting of several production sections ventilated in series to dedicated RARs. For the specified mine, strike conveyor belt drives deliver fresh air to the face, with the main dip conveyor belt situated in a dedicated ventilation district ventilated directly to return. Strike belts will pass through conveyor belt seals to transfer ore to the main dip belt.

This paper emphasises the ongoing potential for innovation and optimisation in well-established primary mine ventilation systems. These ventilation improvements do not always necessitate the adoption of new technology; sometimes, they merely require a fresh approach that combines first principles and experience.

VENTILATION DESIGN CRITERIA AND CONSTRAINTS

The mine is designed for mechanised on-reef, drill-and-blast, bord-and-pillar operations. Geological features of the specific mine include a narrow reef platinum orebody dipping at 9°, accessed from a surface portal and decline system, reaching a depth below surface of 900 m, and extending on strike to 3000 m on both sides of the decline. Mine production ultimately reached 200 ktons per month.

Access to the mine is through a four-barrel on-reef decline cluster developed from a surface portal, consisting of two roadway declines, a chairlift decline, and a conveyor decline. Production sections will consist of 15 production panels, with the blasted ore cleaned and transported to the strike conveyors by diesel-powered low-profile Load Haul Dumpers (LHDs). Each production section will be served by two strike conveyors that in turn feed onto the main decline dip conveyor. A typical overall mine layout is shown in Figure 1.

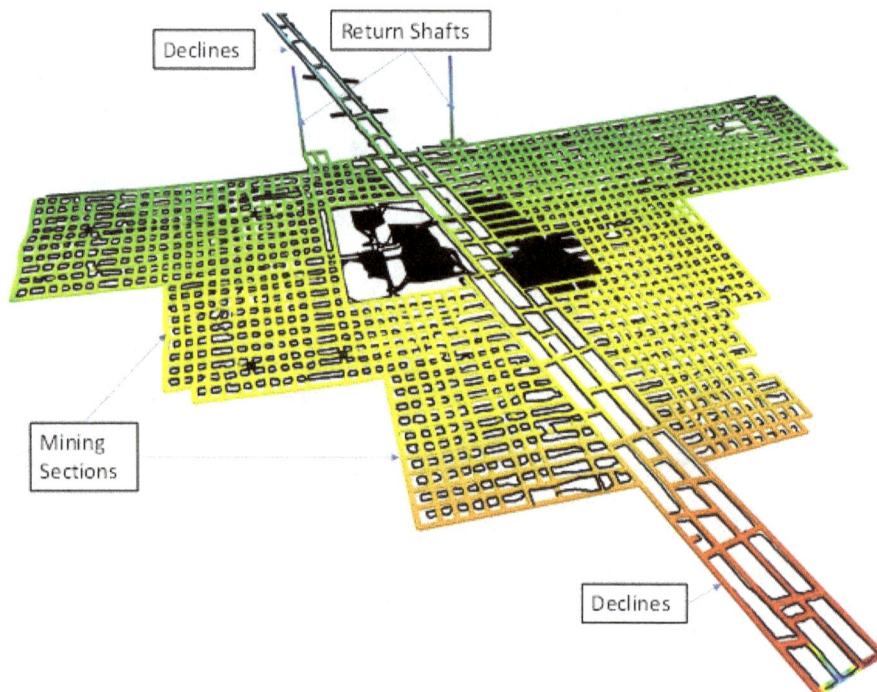

FIG 1 – Typical overall mine layout for hard rock, bord-and-pillar mine.

The project plan provided dimensions for the main declines, lateral connections, and mining sections. Specific geothermal rock properties and temperature gradients (Jones, 2015), along with ambient weather data, were used (Marx *et al*, 2001) in the air mass and heat energy balance calculations and simulations.

Airway velocity criteria set by the mine were used to determine the required capacity of shafts and the decline cluster, as well as to identify additional Fresh Air Raises (FARs) to increase air intake capacity down the decline. The velocity criteria for the last through roads, strike conveyor excavation, and other travel ways were used to determine ventilation quantities for the production sections. This, in turn, helped establish the optimal number of sections in each ventilation district. Other factors and criteria taken into account included primary and secondary leakage, thermal criteria, re-entry periods for blast fume clearance, Diesel Particulate Matter (DPM) buildup, and diesel exhaust gas criteria prescribed in Du Plessis (2014).

CONVENTIONAL DECLINE BORD-AND-PILLAR VENTILATION SYSTEMS

Ventilation of mines accessed via decline systems from the surface with shallow, wide dipping orebodies typically includes an on-reef airway network comprising parallel fresh air and return air roadways configured in a bord-and-pillar layout as described in Tonderai, Phillips and Cawod (2022). These systems are primarily designed to provide sufficient air for the dilution of gases and dust, and for cooling when the mine heat load exceeds the cooling capacity of the ventilation air. When the fresh air introduced through the decline system from the surface reaches its dilution limit, additional vertical airway connections to the surface augment the fresh air capacity. As mining progresses down dip over the life of the mine, these Fresh Air Raises (FARs) convert to Return Air Raises (RARs) equipped with surface fans.

On a macro scale, mining advances from the main access infrastructure towards the extremities of the mine. Bord-and-pillar panels are developed in an advancing arrangement, with developing sections opening new panels ahead of the operating sections and main access infrastructure utilising (Jordaan, 2003) conveyors or load-haul-dump trucks to convey the broken rock.

The primary ventilation system for bord-and-pillar mines, in its simplest configuration, involves regulating adequate airflow from the fresh air intake mains into respective strike sections via access roads and strike conveyor belts. The majority of the fresh air is drawn to the lowest production sections, from where it is sequentially circulated and reused in series with adjacent up-dip sections before returning to the exhaust shafts. This is generally referred to as a conventional once-through system.

The risks and constraints associated with a once-through system include long re-entry periods after blasting, the need for all blasting to occur within one time window, the potential for fires to affect the entire mine, and the buildup of pollutants such as heat, dust, and gas throughout the series ventilation circuit. However, there are some benefits, such as requiring fewer and simpler ventilation controls and a limited number of large central main ventilation fans (Figure 2).

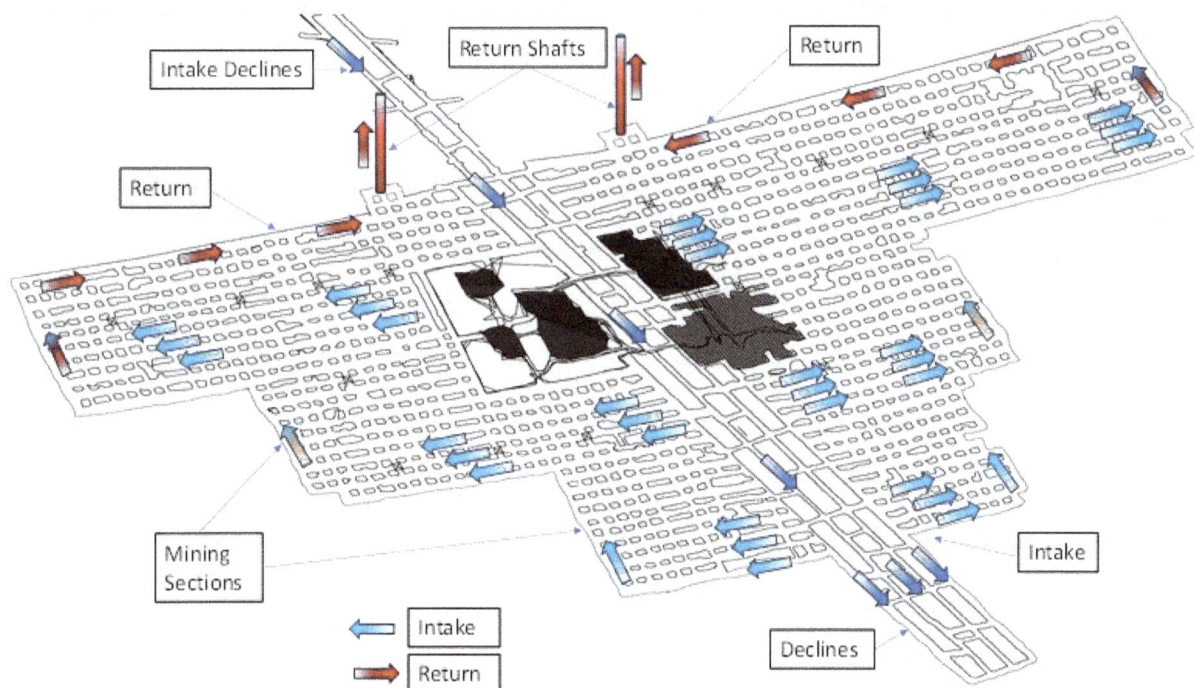

FIG 2 – Conventional once-through, series ventilation system.

BASE VENTILATION SYSTEM TO BE OPTIMISED

The pre-feasibility primary ventilation design allowed for each production section to be ventilated separately, relying on the establishment of secure ventilation crossovers to separate return air from the production section and the intake declines. Although the intention is to minimise worker exposure to contaminants, reduce re-entry periods, and facilitate unconstrained mining, this approach presents several challenges.

The pre-feasibility design necessitates the installation of competent airtight seals (Schophaus, Bluhm and Funnell, 2005) to separate the intake declines from the Return Airway (RAW) system that runs parallel to each other for the entire decline length. To satisfy airflow velocity criteria, in the upper portion of the mine, the RAW comprises up to five airways. Secure crossovers are also necessary in the RAW for routing the section strike conveyor belts to the main dip conveyor. Each section requires two sets of crossovers: the first set traverses the two redundant intake bords and the bottom conveyor, while the second set traverses the traveling way, roadway, and top conveyor. In this design, all conveyor belt installations are ventilated directly to return to mitigate fire risk. Figure 3 illustrates the complexity of these crossovers and the potential for leakage between the intake and RAWs.

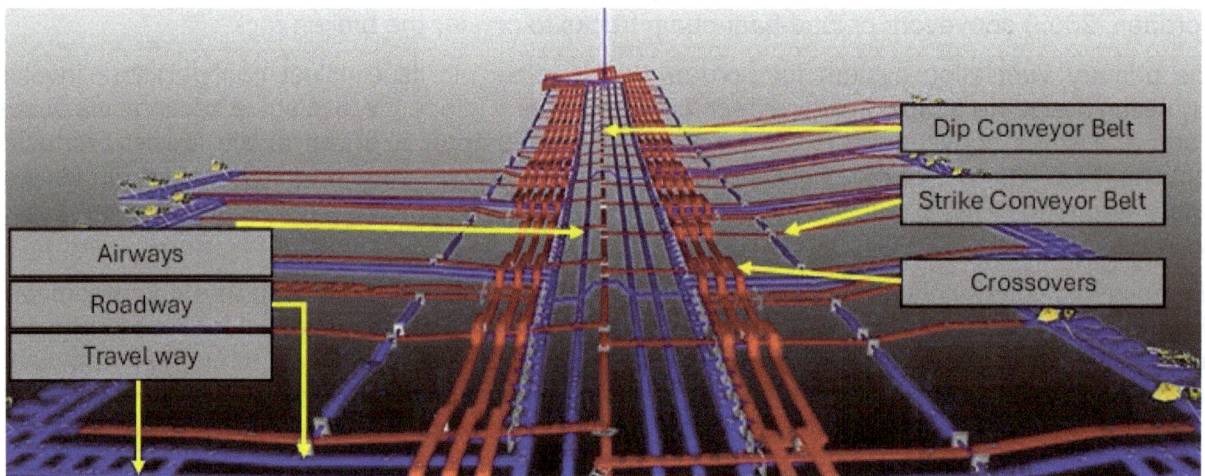

FIG 3 – On-reef, decline return airway system with multiple crossovers.

The secondary ventilation system for the production sections draws fresh air from the decline cluster through the bottom two roadways. Additional fresh air is introduced by regulating airflow into the roadway and travel way associated with each strike conveyor belt. The strike belts exhaust air from the production section into the main dip conveyor, resulting in all conveyor belts being ventilated directly to return. The two top-most roadways return the remaining air in the section to the dip RAW. Figure 4 demonstrates the section layout.

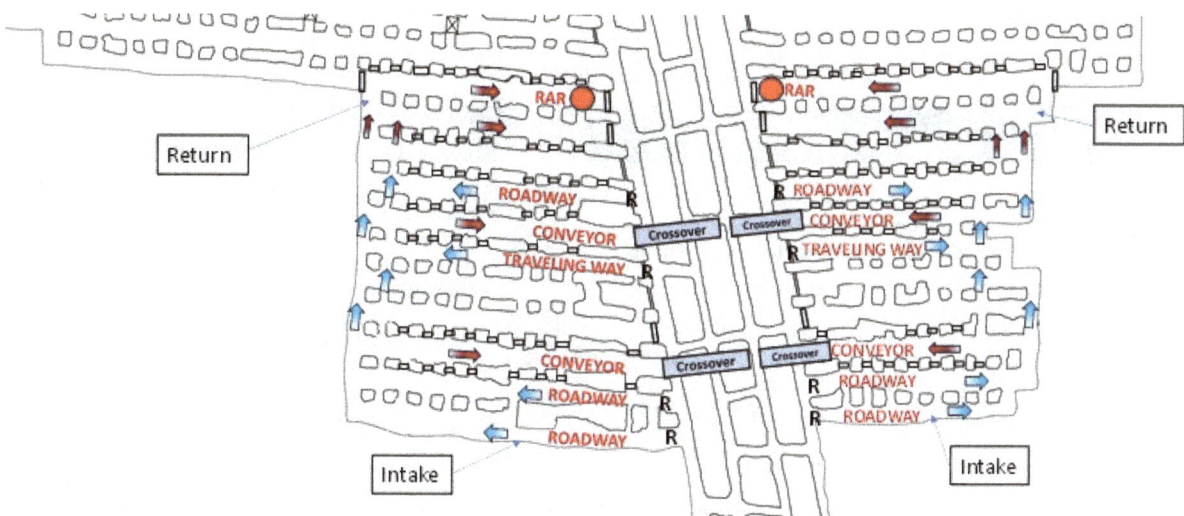

FIG 4 – Typical pre-feasibility section ventilation layout.

OPTIMISED VENTILATION SYSTEM DESIGN

The geology of the specific hard rock mine posed challenges in the development and installation of multiple crossovers, a situation common to most hard rock bord-and-pillar mines. Practically, this number of crossovers would significantly impede mine development and introduce inherent risks associated with mining into the hanging wall. The pre-feasibility work highlighted the excessive costs associated with the complex ventilation control system, affirming the necessity for an optimal design between the two extremes of a once-through series ventilation system and ventilating each production section within its own ventilation district.

Creating ventilation districts with dedicated intake and Return Airways (RAWs) supplying sections of a mine is a proven ventilation method for hard rock mines confirmed by Wallace *et al* (2008) using ventilation zones for Red Lake Mine. Applying the same approach to the study mine preserved many benefits of a one-section-per-ventilation-district system while mitigating the risks and constraints associated with a once-through system.

The ventilation district strategy must meet ventilation design criteria such as re-entry periods and limits on contaminant build-up while also being safe, practical, and cost-effective. To determine the optimal number of production sections in each ventilation district, multiples of sections, starting with two, were compared based on overall air quantity requirements, re-entry period, and cumulative contamination of the air stream as it moved from section to section.

Additionally, the exhaust RAW from each ventilation district could be one of two options: either a RAR to a surface fan station, or a footwall RAW system excavated at an appropriate depth below the reef horizon under the decline complex. A footwall return system would necessitate raise boreholes to connect the overlying reef to the RAW, with the raises located at the same positions as the RARs in the previous option. The second option would also require large centrifugal fans and an airlock system located near the surface decline portal.

Ventilation district Return Airway configuration

First, trading-off the two RAW options produced comparative Table 1.

TABLE 1

Comparison of return airway configurations.

Parameter	Footwall Return Corridor		RAR station per ventilation district	
Construction	Footwall return airway corridor may be developed concurrently with decline cluster from the portal area	✗ ○ ✓	Surface RAR infrastructure requires establishment of surface access road in mountainous terrain. Potential difficulty in transporting and fitting (the larger) reamer heads underground in a low-profile environment	✗ ○ ✓
Waste Rock Handling and Disposal	Waste rock from the footwall development operations must be trucked to surface for life of mine – over increasingly longer routes. A surface waste rock dump facility has to be constructed	✗ ○ ✓	During the establishment of each RAR, LHDs will be mucking raise bore chips which can be stowed underground or used for roadway construction / maintenance directly	✗ ○ ✓
Surface Fan Stations	This arrangement will require the establishment of two large main fan stations near the portal position. Failure to a fan station will result in a larger production loss impact.	✗ ○ ✓	In a configuration where three production sections are allocated to one district, a total of sixteen satellite fan stations will be required along the 7km surface access road. "Leapfrogging" of RAR stations will be implemented. Failure of a fan station will have a reduced impact.	✗ ○ ✓
Power supply	Power supply is required for the two surface main fan stations and therefore it will be centralized near the portal infrastructure.	✗ ○ ✓	Power must be distributed on surface along the access road to the respective satellite fan stations	✗ ○ ✓
Surface Infrastructure	Limited to the two surface fan stations. Easy to maintain and operate. The underground return airway corridor will be separated from all other infrastructure.	✗ ○ ✓	Access road and fan pads need to be established and maintained along with the 16 surface satellite fan stations, associated servitudes and raise boreholes	✗ ○ ✓
Operational Risk	Damage to the integrity of the return airway corridor could compromise production. Return airway capacity is fixed by the configuration of the underground excavations	✗ ○ ✓	Higher risk of physical damage to return airway infrastructure due to vandalism, theft, sabotage, surface fires and severe weather. The use of RAR station may be modified or extended to provide further capacity if so required.	✗ ○ ✓
Estimated Capital Cost	US$ 94,765,600	✗ ○ ✓	US$ 71,857,000	✗ ○ ✓

Although the footwall return corridor option has advantages over the RAR-to-surface option in certain areas, the burden of waste rock handling, increased operational risk, and 50 per cent higher cost disqualified this option.

Ventilation of strike conveyor belt drives

There are two available options: routing intake air across the strike belts or returning air across the belt, which affects the ventilation district air volume requirements. Within the airway velocity criteria, ventilating strike conveyors as intake to the section increases the overall air quantity. There is a significant difference in system performance between the strike conveyor belt ventilation intake and return modes. For example, in the strike belt exhaust mode, re-entry periods increase with the increasing number of sections, confirming the cumulative nature of the contamination. However, although still noticeable, the effect is considerably reduced by allowing the strike conveyor belts to introduce fresh air (intake mode) into the production areas.

Table 2 summarises the results of the trade-off and shows that ventilating the strike belts to return presents significant challenges compared to the alternative configuration.

TABLE 2
Strike conveyor ventilation trade-off.

Parameter	Strike Conveyor Belt Roadway – Intake Air		Strike Conveyor Belt Roadway – to Return	
Fire Risk	A conveyor belt fire in the strike drive will diffuse smoke and gases to the section. Only the district concerned downstream will be affected. There is a risk of the fire extending to the dip belt.	✗ 🟡 ✓	Smoke from a conveyor belt fire in the strike drive will be exhausted to the main RAW. Dip belt infrastructure will be affected until the fire is extinguished. There is a risk of the fire extending to the dip belt.	✗ 🟡 ✓
Crossovers	No crossovers required.	✗ ⚪ ✓	Numerous crossovers are required to isolate the return airway system from the fresh air infrastructure.	✗ ⚪ ✓
Ventilation Control	Conveyor belt roadways are segregated by an extensive strike wall network, airflow regulators and ventilation doors.	✗ 🟡 ✓	Conveyor belt roadways are segregated by an extensive strike walls, airflow regulators and ventilation doors.	✗ 🟡 ✓
Blast Fume Exposure	Conveyor belts operate in fresh air without the risk of blast fume exposure. Dust released in the conveyor belt road and heat energy from conveyor belt operation are transferred to the production section.	✗ ⚪ ✓	Blast fumes will flow from the production area over the strike conveyor belt potentially affecting the conveyor belt operation, maintenance and integrity.	✗ ⚪ ✓
Primary Air Requirement	The system requires 30% more air quantity to compensate for the additional heat energy loading.	✗ 🟡 ✓	Lesser airflow requirement for the two options	✗ ⚪ ✓
Conveyor Belt Protection	Continuous belt monitoring for smoke and gas emission, electrical overloading and excessive operational controls are required. An effective fire suppression or fire-fighting system is required. The use of fire-retarded conveyor belt lining should be mandatory.	✗ 🟡 ✓	Continuous belt monitoring for smoke and gas emission, electrical overloading and excessive operational controls are required. An effective fire suppression or fire-fighting system is required. The use of fire-retarded conveyor belt lining should be mandatory.	✗ 🟡 ✓

An example of the benefits of using the strike belts for intake is evident in Figure 5 that shows the buildup of DPM for both options. The buildup for strike belts to intake shows a much lower buildup than the option with strike belts to return.

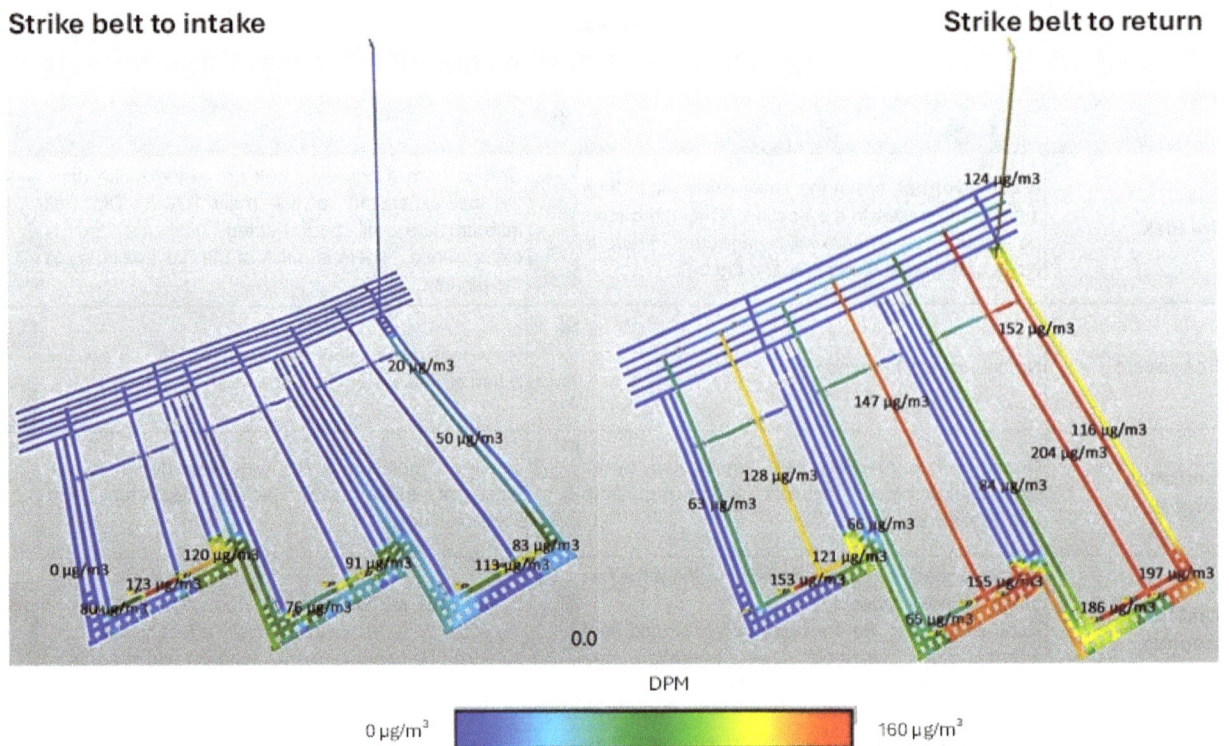

Strike belt to intake

Strike belt to return

FIG 5 – DPM buildup for strike conveyor ventilation options.

Ventilation district configuration

Ventilation districts consisting of a minimum of two and up to a maximum of five production sections were compared based on overall air quantity requirements, re-entry periods, and cumulative contamination of the air stream as it progressed between sections. Factors influencing the optimum district ventilation strategy include the physical layout of primary airways and the strategy for ventilation of strike and dip conveyor belts.

Advantages of increasing the number of production sections in each ventilation district include a reduction in the number of RARs required, a decrease in the total air quantity in circulation, simplification and reduction of ventilation control infrastructure such as regulators and airlocks, and a reduction in the potential mixing of intake and Return Airways (RAWs) and leakage. Disadvantages include increased RAR diameter, higher reject air temperature, elevated cumulative air contamination (from Diesel Particulate Matter (DPM), dust, and blast fumes), longer post-blast re-entry periods, and a reduced potential for unconstrained mining.

Table 3 compares the qualitative (simplicity, complexity, risk etc) and quantitative simulated (contaminant buildup, blast fume clearance, airflow quantity and leakage etc) criteria for the number of production sections. Figure 6 compares re-entry period, RAR diameter, and DPM buildup in a ventilation district for the number of production sections in the district and overall ventilation quantity. Although the comparative trends were anticipated, the graph demonstrated the extent to which the number of production sections in each district impacted the results.

TABLE 3
Ventilation district size trade-off comparison.

	2 Sections per Vent District	3 Sections per Vent District	4 Sections per Vent District	5 Sections per Vent District
Simplicity of ventilation control	(✗) Complex control required to balance between districts, and extensive sealing	(✓) Much easier control with only two-thirds of districts compared to 2-Sections option and aligning with mine schedule	(○) Complexity of control increases because fractions of districts will have to be controlled to meet mine schedule	(✓) Easier to control with only two districts required to align with mine schedule
Complexity of constructing ventilation districts	(✗) Complex construction with extensive vent sealing required	(✓) Easier construction with acceptable vent sealing required	(✓) Easy construction, with limited sealing required	(✓) Very easy construction
Degree of unconstrained mining	(✓) High Degree of unconstrained mining, very flexible blasting, airflow per district easily adjustable	(✓) Although less than 2-Sections option, still good degree of unconstrained mining	(○) Mining becomes constrained with only a few vent districts and fractions of districts	(✗) 5-Sections approaches a once-through, series vent system and constrains mining flexibility
Safety risk level	(✓) Very safe option as fire or other emergency are confined to a small portion of the mine	(✓) A safe option with fire and emergency confined to a number of vent districts	(✓) Better than a once-through system, however fire and emergencies will affect a significant portion of the mine	(○) This option approaches a once-through, series vent system and fires and emergencies will affect a large portion of the mine
Contaminant build-up (heat, gas, DPM)	(✓) Low contaminant build-up due to limited backlength and air residence time, exceeds design criteria	(✓) Acceptable contaminant build-up, meeting design criteria	(○) Significant contaminant build-up, exceeding design criteria	(✗) Unacceptable contaminant build-up, exceeding design criteria
Blast fume clearance	(✓) Beat design criteria	(✓) Meet design criteria	(✓) Exceed design criteria	(○) Exceed design criteria
Overall airflow quantity, leakage risk	(✗) Highest overall airflow requirement and leakage	(✓) Reasonable overall airflow requirement and leakage	(○) Low overall airflow requirement and leakage	(✓) Least overall airflow requirement and leakage

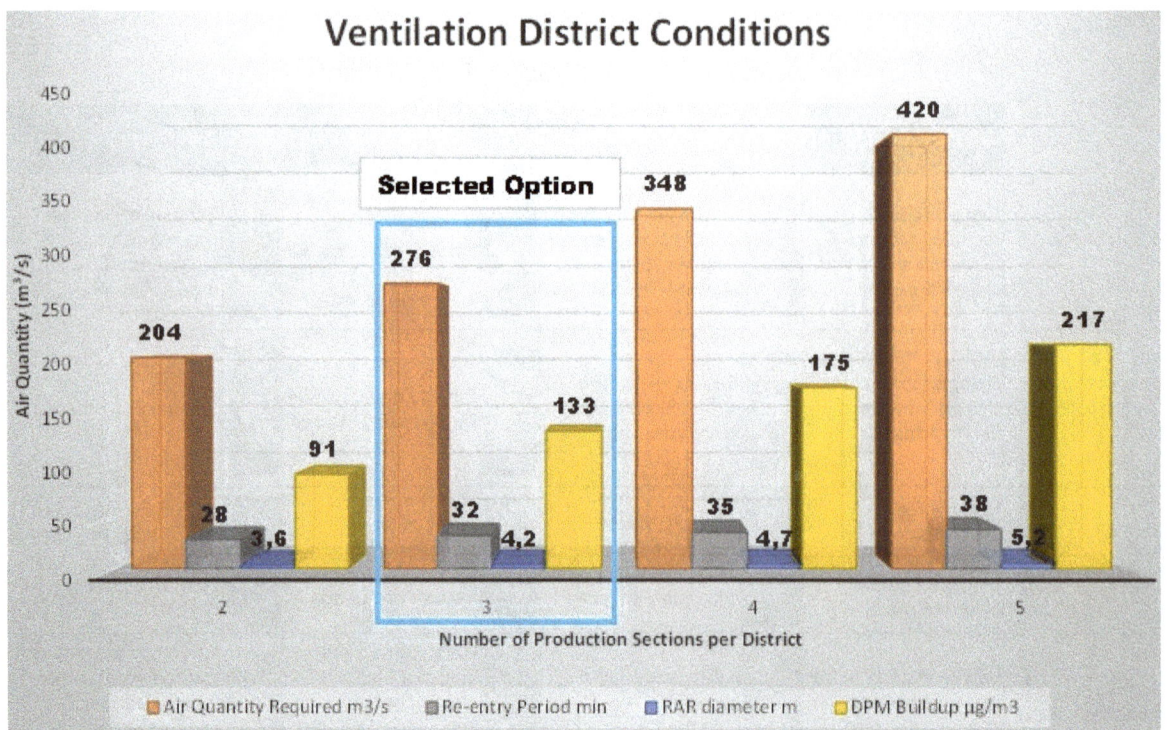

FIG 6 – Ventilation district condition comparison.

The comparative approach provided an optimised solution between the once-through system and the independent ventilation of production sections.

Eliminating the need for a significant number of crossovers had significant benefits, including capital cost, shorter construction schedule, simplification of ventilation system layout and controls, and reduction in section preparation lead times and physical risks. Based on both a quantitative and qualitative comparison the analysis indicated that for the specific case having three sections in each ventilation district was the optimal configuration to meet all design criteria; specifically in terms of the DPM build-up and re-entry as shown in Marais (2022).

CONCLUSIONS

The study highlights the importance of innovative and optimised primary ventilation systems in the mining industry, particularly for hard rock, bord-and-pillar operations. By evaluating and addressing the constraints of conventional once-through series ventilation systems, the study has demonstrated significant potential for improvements in safety, efficiency, and cost-effectiveness through the adoption of ventilation districts.

Although ventilation district designs have been used in most underground mines, commodities and mining methods, the design described in this paper demonstrated a new and innovative application of ventilation districts. Initially the concept was doubted because the idea of 20 RARs in a mountainous area versus two to four larger RARs near the portal area of the mine did not seem financially or practically viable.

However, the proposed approach of creating independent ventilation districts with dedicated intake and return airways (RAWs) for each production section offers a balanced solution. This strategy preserves many benefits of a one-section-per-ventilation-district system while mitigating the risks associated with a conventional once-through system. The study's quantitative and qualitative analysis identified that having three sections in each ventilation district is optimal. This configuration effectively balances the need for reduced cumulative air contamination, shorter post-blast re-entry periods, and lower diesel particulate matter (DPM) build-up against the cost and complexity of the ventilation infrastructure.

Moreover, the study underscores the advantages of ventilating strike conveyor belts as intakes rather than returns. This method significantly reduces re-entry periods and contamination levels, thus enhancing overall system performance. Additionally, the elimination of numerous crossovers simplifies the ventilation system layout, reduces capital costs, shortens construction schedules, and minimises physical risks associated with mining into the hanging wall.

In conclusion, the study results presents a compelling case for the adoption of optimised ventilation district configurations in hard rock mining operations. By integrating first principles with innovative approaches, the study offers practical and scalable solutions that enhance the safety, efficiency, and economic viability of mining ventilation systems. This optimisation not only addresses the immediate operational challenges but also paves the way for future advancements in mine ventilation technology.

ACKNOWLEDGEMENTS

The authors acknowledge the Mining Group that commissioned and funded this work, the entire project team's inputs during the study, and contributions from BBE colleagues, specifically Mark Butterworth, Hendrik Botma, Louis Matthysen, Jennifer de Beer and Janine Bester.

REFERENCES

Deglon, P and Hemp, R, 1992. An evaluation of parameters to be used in colliery ventilation, Proceedings Fifth International Mine Ventilation Congress (Mine Ventilation Society of South Africa: Johannesburg).

Du Plessis, J, 2014. *Ventilation and occupational Environment Engineering in Mines*, 3rd edition, Johannesburg: Mine Ventilation Society of South Africa.

Jones, M Q W, 2015. Thermophysical properties of rocks from the Bushveld Complex, *Journal of the Southern African Institute of Mining and Metallurgy*, 115(2):153–160.

Jordaan, J, 2003. Bord-and-pillar mining in inclined orebodies, *Journal of the Southern African Institute of Mining and Metallurgy*, 103(02)101–110.

Marais, B, 2022. Analysis of alternative bord-and-pillar ventilation strategies, *Journal of the Mine Ventilation Society*, Mine Ventilation Society of South Africa: Johannesburg.

Marx, W M, von Glehn, F H, Bluhm, S and Biffi, M, 2001. VUMA (ventilation of underground mine atmospheres) a mine ventilation and cooling network simulation tool, Proceedings Seventh International Mine Ventilation Congress, Cracow, Poland.

Schophaus, N, Bluhm, S and Funnell, R, 2005. Effects of Ventilation leakages in deep, Hot Room + Pillar operations. Brisbane, Proceedings 8th International Mine Ventilation Congress.

Tonderai, C, Phillips, H and Cawod, F, 2022. Ventilation optimization through digital transformation, *Journal of the Southern African Institute of Mining and Metallurgy*, 122(12):687–696.

Wallace, K G, Tessier, M, Pahkala, M and Sletmoen, L, 2008. Ventilation planning at the Red Lake Mine, in *Proceedings Eighth International Mine Ventilation Congress* (ed: A Gillies), pp 447–455 (The Australasian Institute of Mining and Metallurgy: Melbourne).

Ventilation and cooling considerations with the conversion of low-profile diesel LHDs to battery electric in a room-and-pillar mine

R van der Westhuizen[1], I K Taunyane[2] and K Reynders[3]

1. Senior Engineer, BBE Consulting, Bryanston, South Africa.
   Email: rvanderwesthuizen@bbe.co.za
2. Ventilation and Occupational Hygiene Manager, Impala Platinum, Illovo, South Africa.
   Email: ishmael.taunyane@implats.co.za
3. Managing Director, Rham Equipment, Olifantsfontein, South Africa.
   Email: kreynders@rham.co.za

ABSTRACT

This paper considers a mechanised/trackless room-and-pillar platinum mining operation in South Africa using low-profile diesel LHD vehicles (load-haul-dump). The mine is serviced from a central decline system. Mine expansion will see the room-and-pillar operation move further from this central decline system which will place a strain on the existing primary ventilation infrastructure. To satisfy the design criteria for diluting heat and diesel exhaust pollutants, an additional intake and return shaft from surface would be required. This would require a significant investment and will have a negative impact on the schedule due to long implementation time required for planning, permits and execution.

An alternative, to mitigate the additional shaft infrastructure, is the conversion of the diesel LHDs to battery electric drives; the conversion retains the original LHD chassis and hydraulic drives. A site-specific trade-off study, with emphasis on ventilation and cooling and, with production and mine layouts unchanged, was undertaken to quantify the potential benefits and identify the infrastructure requirements for the conversion.

The study involved close interaction with the mine team and Rham, the LHD OEM (original equipment manufacturer). The work involved underground on-site assessments, ventilation modelling, specification of the battery electric vehicle and operational characteristics, and high-level risk assessment. The study excluded the detailed impact on mining, logistics and engineering.

This paper discusses the methodology used in the trade-off, compares ventilation and cooling requirements for diesel and battery electric LHDs and highlights the practical ventilation and cooling issues that influence the decision to convert to battery electric LHDs.

BACKGROUND

BEV benefits

There is a general belief that the introduction of battery electric vehicles (BEVs), to replace diesel powered units, in underground mines will reduce a mine's need for ventilation and cooling, and thereby reduce operating power costs. Obviously BEVs do not generate any harmful exhaust products needing dilution, and due to their better efficiency, they generate less heat.

One Canadian in-mine study (Armburger and McGuire, 2018) specifically assessed a single load-haul-dump (LHD) vehicle using a simulated tramming cycle within a heading. It indicated that at a comparable duty, the energy input of a diesel unit was ten times that of a BEV, and the air temperature increase up to seven and a half times more. However, the energy input of the BEVs was reduced through regeneration when travelling down slope during the cycle; the authors also recommend further testing. Additionally, in a later real-world comparison (McGuire *et al*, 2022) between a diesel LHD and a BEV LHD, for similar duty, the heat output was approximately five times less for the BEV than the diesel LHD in that application.

Another in-mine study performed in Finland (Halim *et al*, 2022), looked at the combined LHD and trucking activity within a level *en masse*. This has shown significantly lower dry and wet bulb temperatures with BEVs, t_{db} 9–10°C/t_{wb} 7.5–8°C, compared to diesels, t_{db} 10–16°C/t_{wb} 9–13°C. As would be expected, this study also showed the absence of diesel particulate matter (DPM), carbon monoxide (CO) and nitrogen dioxide (NO_2) when only battery equipment was operating. Respirable mineral dust concentrations were also less with the BEVs, however beyond the suggestion the diesel unit's tail-pipe exhaust was raising dust from the walls, a cause was not presented.

Halim and Kerai (2013) recommended a design requirement of 0.04 m^3/s/kW, based on an equivalent diesel power, for a deep mine with the assumption of some air cooling, decreasing to 0.025 m^3/s/kW for a cool shallow mine. This deep mine design value was used as the departure point for this study.

Merensky Mine

The study was site specific with the objective to determine if, by replacing the existing LHD vehicle fleet with battery powered equivalents, the ventilation requirements would reduce to the point where additional intake and return raise-bore holes (RBH) identified in early project work, could be eliminated. The RBHs negatively impacted capital cost and were a risk to the mine plan and production schedule due to the long implementation time required for planning, permits and execution.

The mine is located on the western limb of the Bushveld Igneous Complex. Parallel Upper Group 2 (UG2) and Merensky reefs, separated by approximately 70 m of non-pay ground (middling) are mined independently, with the Merensky reef located above the UG2. Mining activity is effectively divided in two; the upper, early mine (above 24 Level) and the lower mine to 28 Level, 1270 m below collar with a virgin rock temperature of approximately 51°C. The upper mine is accessed through a vertical shaft where both orebodies are mined using conventional stoping methods with hand-held drills and winches scraping into off-reef footwall haulages. The lower mine features a mix of conventional and bord-and-pillar (mechanised/trackless) mining; UG2 is mined conventionally and Merensky utilises on-reef, mechanised, bord-and-pillar. Initially, the Merensky areas below 24 Level were mined conventionally from footwall infrastructure, but after approximately 800 m strike development mining changed to the current on-reef mechanised bord-and-pillar method; this mining is best suited for shallow dipping (typically less than 10°) orebodies.

At present the ventilation system downcast approximately 1000 kg/s through the Main Shaft into the upper mine. A decline system developed between the reefs feeds the lower mine; in addition, a dedicated RBH from surface feeds into the lower decline section between 26 and 28 Levels. To maintain average reject air temperatures below 28.5°C bulk air cooling (BAC) is installed on surface, nominally 14 MW (measured duty) at the Main Shaft that mainly serves the upper mine and a 12.2 MW BAC at the intake RBH, serving the lower mine. Return air utilises exhaust fans installed on surface at No.1 and No.2 ventilation shafts. Figure 1 shows the mine cross-section.

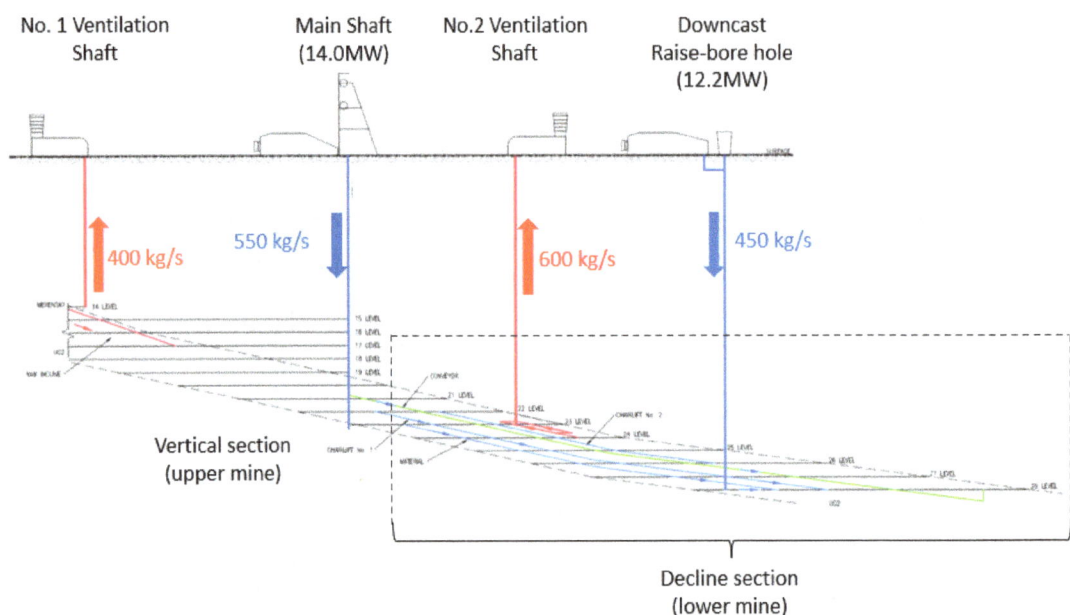

FIG 1 – Mine layout and existing shaft infrastructure.

Air-tonnage ratios can be influenced by many factors, however, from experience, for conventional platinum mining at depths of 1200 m the ratio is approximately 4.5 kg/s/ktpm. In comparison, mechanised bord-and-pillar operations require up to 50 per cent more air due to the diesel dilution requirement. As the primary ventilation system was originally designed for conventional mining only, the introduction of mechanised on-reef mining has placed a strain on airway capacity, particularly servicing the lower Merensky. The deficit in capacity is further amplified by a short-term imbalance in production between the lower north and south Merensky sections. The mining schedule requires a transition from a relatively even spilt to a point in 2027/2028 (FY28, the critical year) when almost 90 per cent of mining will be in the north. The imbalance has resulted in insufficient capacity to adequately ventilate and deliver cooling to the mechanised Merensky mining area north of the decline. Rescheduling production to balance it was not an option.

OVERVIEW OF CURRENT BORD-AND-PILLAR VENTILATION

The bord-and-pillar sections operate low profile diesel LHDs that move ore from the face to strike conveyors that typically connect to a main decline trunk conveyor. The sections are arranged in a grid pattern of 7 m × 7 m bords (headings) separated by 7 m pillars with a general footwall to hanging wall height of 2.0 m. The headings are ventilated in series using air from the preceding heading to ventilate the next; jet fans are used to scavenge air in the bord. Sufficient ventilation must be supplied to each section to satisfy the greater of either diesel exhaust gas dilution or minimum velocity in the 1st and 2nd roads. For planning bord section ventilation volumes, the mine standards prescribe a minimum diesel dilution factor of 0.11 m^3/s/kW of the rated power per bord section and a road air minimum velocity of 0.8 m/s.

Figure 2 shows a simplified ventilation arrangement where a large portion of the ventilation is introduced from the bottom of the decline and is supplemented with additional air introduced to the successive mining sections. Leakage from the lower strike to the back-areas mixes with the fresh air introduced to the successive sections. Air returns to the top sections from where it enters the mine exhaust system.

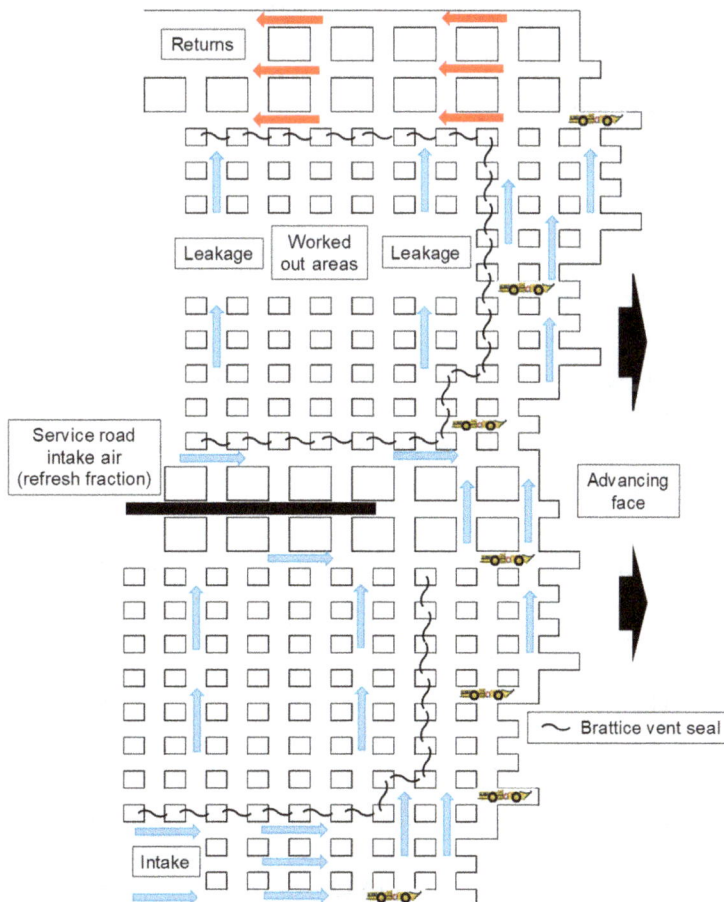

FIG 2 – Simplified overview of typical bord and pillar ventilation.

Low profile diesel LHD

The mine operates Rham model 20HD low profile LHDs, shown in Figure 3. The vehicles operate mostly on a level plane in rough terrain. The vehicle's hydraulic pump maximum input power is limited to 120 kW but, including a nominal allowance for overshoot and transfer efficiency, the engine must provide 140 kW shaft power output to operate. This value was used to calculate the ventilation rate required for diesel exhaust gas dilution.

FIG 3 – Rham model 20HD low profile LHD (https://miningtara.co.za).

Ventilation shortfall in critical year 2027/2028 (FY28)

In FY28, the mine plans to operate 38 active LHDs, of which 34 will be in the north. Using the adjusted shaft output power of 140 kW, 62 m^3/s and 524 m^3/s will be required, at the underground reference density, to ventilate the south and north respectively for diesel exhaust gas dilution.

For the decline trackless section air intakes from the decline on 25, 26, 27 and 28 Levels north, and 27 and 28 Levels south (footwall drives), with some air entering through a ramp below 28 Level. The footwall drives each have a nominal area of 9 m^2 and convey up to 72 m^3/s (each) at the design velocity. The maximum quantity available to the north, based on intake airway carrying capacity, is restricted to approximately 360 m^3/s, this includes intake air from the ramp.

In the critical year, in addition to the Merensky trackless mining, two conventional UG2 areas are mined from the north decline with the result that the Merensky north intake infrastructure is shared between conventional UG2 and Merensky trackless mining. After allocating ventilation to the UG2 areas, the shortfall to the north trackless section was estimated to be 220 m^3/s. To mitigate the shortfall several options were considered including:

- Divert intake air from Merensky south to the north through old workings.

- Footwall development of Merensky north twin drives (currently single drives) to increase carrying capacity to north.

- The inclusion of new downcast and upcast ventilation RBHs from surface, targeting the north Merensky.

The first two options proved inadequate and practically inexecutable due to existing shaft limitations, the third option, while providing sufficient airflow, would delay production buildup and was not financially viable. To proceed airflow demands for the north block needed to be reduced.

Reduce airflow

Airflow required in the north block was dictated by exhaust gas dilution for the diesel fleet, options to reduce ventilation were:

- reduce the number of diesel LHDs

- replace the diesel LHDs with an alternative requiring less air, ie electric powered.

A smaller number of diesel units would have an impact on production and was not a viable option. Electric production units could be tethered, or battery powered. Tethered vehicles were not considered due to mine layout and hard rock environment. The LHD original equipment manufacturer (OEM) was approached by the mine operator with a view to supplying BEV LHD equivalents. The response was that the diesel engines fitted in the LHDs could be exchanged for an electric motor connected to the existing hydraulic drive. The modifications would keep the same vehicle chassis and maintain the same capabilities with no disruption to planned equipment availability. The BEV LHDs would accommodate battery replacement during the shift and battery charge and change-out bays were specified for the project.

As a departure point for conceptual planning, an airflow requirement of 0.04 m^3/s/kW was applied to the total equivalent diesel power to estimate the ventilation required for BEVs (Halim and Kerai, 2013). After allowing 20 per cent for primary ventilation leakage and using 140 kW as the shaft output power, it was estimated that approximately 30 m^3/s and 230 m^3/s would be required at the underground reference density to ventilate the south and north trackless sections respectively (compared to 62 m^3/s and 524 m^3/s with diesel LHDs). The reduced airflows brought the ventilation requirement to within the capacity of existing infrastructure without the need for additional raises. Although encouraging, the conversion of existing diesel to BEV LHDs required validation through more detailed analysis and ventilation modelling. To quantify the benefits of the conversion from the existing diesel fleet to BEVs a ventilation and cooling trade-off study and risk assessment were undertaken with the objective to determine if the replacement could practically reduce ventilation sufficiently to eliminate the additional upcast and downcast raises.

TRADE-OFF

The trade-study required the mine owner and LHD OEM to confirm the operational requirements for the converted BEV equivalent to the Rham model 20HD low profile LHD, with the significant advantage that the ability of the proven LHD to operate in a narrow reef environment was not compromised.

The trade-off focused on ventilation and cooling and used a calibrated ventilation model with the mine production schedule and visualiser included to show the production and active face positions through to the critical year when the additional raises were required. Two scenarios were developed within the modelling:

1. Diesel option: this included the new downcast and upcast RBHs into the north Merensky block with surface cooling at the new downcast and a new fan station at the upcast RBH.

2. BEV replacement: no new raises but an additional, 3rd fan was added at No.2 Ventilation Shaft with supplemental cooling added at the existing downcast RBH.

With tonnage matched to the production profile the models were used to estimate air distribution and identify practical shortcomings such as high airway velocity and ambient temperatures. The models also proved useful in identifying locations for the additional infrastructure required for BEV LHDs, such as battery charging bays and their impact on ventilation. Analysis and results are summarised below for both cases.

Ventilation for diesel LHDs

To fully satisfy the design criteria, the total primary airflow required in FY28 was 1250 kg/s, of this 680 kg/s was utilised in the trackless section. Main fan absorbed power was approximately 7.6 MW. For the diesel trackless section, the overall ventilation rate was approximately 6.4 kg/s/ktpm. An additional nominal 1.6 MW bulk air cooling was added to the new downcast raise and total absorbed power for cooling was equivalent to 8.0 MW to achieve 27.8 MW of total surface air cooling.

A (rounded) breakdown of the heat added in the FY28 model is shown in Figure 4, the dominant contributors are auto-compression and heat flow from the surrounding rock. The diesel vehicle heat contribution amounts to approximately 7.4 MW. Other sources include auxiliary fans and electrical reticulation losses. Vehicle heat contributed 15 per cent of the overall mine heat load.

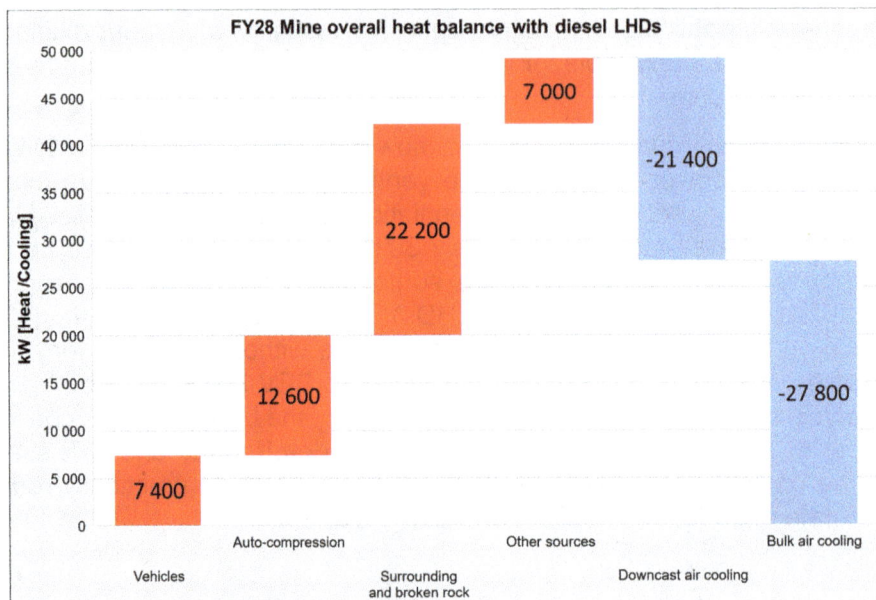

FIG 4 – 2028 Mine overall heat balance with diesel LHDs (rounded).

The CAPEX estimate for the new ventilation and cooling infrastructure, including the shafts, is USD29 M.

Ventilation for BEV LHDs

The ventilation design criteria were fully satisfied without the need for additional intake and return RBHs. The total primary airflow in FY28 was 1050 kg/s, of this 360 kg/s was utilised in the trackless section. This represents a 16 per cent reduction in overall primary ventilation and a 47 per cent reduction for the section. Predicted main fan absorbed power reduced slightly to approximately 7.5 MW. However, an additional fan (240 m^3/s @ 5.5 kPa) at Ventilation Shaft No.2, and a further 1.7 MW nominal air cooling at the existing decline section downcast RBH would be required. For the BEV trackless section, the overall ventilation rate was approximately 3.5 kg/s/ktpm. With the planned increase in primary ventilation for the decline section, while retaining existing shafts, there is an expected increase in both shaft velocities and overall system resistance. The existing surface main fans on the decline section ventilation shaft were checked for operating near the estimated stall point. This is not a concern, and the fans should operate comfortably under the estimated stall point. Currently, high intake air velocities that exceed the design criteria of 6 m/s are experienced at the main intake decline. With the proposed third (additional) fan at the decline section, ventilation shaft intake air velocities would increase, consequently an additional level intake from the downcast shaft is planned (10 m^2 × 410 m long). This intake will be on-reef (Merensky) and connect the downcast shaft to north trackless worked-out areas (WOAs), spanning over the decline. With the addition of this main intake, velocities will return to within design criteria limit.

Total absorbed power for cooling will remain unchanged at approximately 8.0 MW to achieve 27.9 MW of total surface air cooling. The slight increase in cooling duty required for the BEV option is due to a lower positional efficiency of the current downcast shaft compared to new downcast raise in the diesel option. The significant reduction in natural air cooling compared to the diesel case can be attributed to less primary ventilation for the BEVs, higher primary leakage fraction with poor overall positional efficiency due the absence of the new downcast raise, in the diesel option, feeding cold air directly into the north Merensky production area.

Heat from the major sources is shown in Figure 5, heat added for the LHDs included charger inefficiency of 5 per cent (Global Mining Guidelines Group, 2022) and reticulation losses to the charger bays. Other heat includes conveyors, auxiliary fans, and electrical reticulation losses (common to both options). As with the diesel option heat from rock and auto compression accounted for the bulk of the heat. BEVs contributed 8.2 per cent of the overall mine heat load, whereas diesels contributed 15.0 per cent.

The CAPEX estimate for the new ventilation and cooling infrastructure, including the additional intake airway, is USD9.1 M.

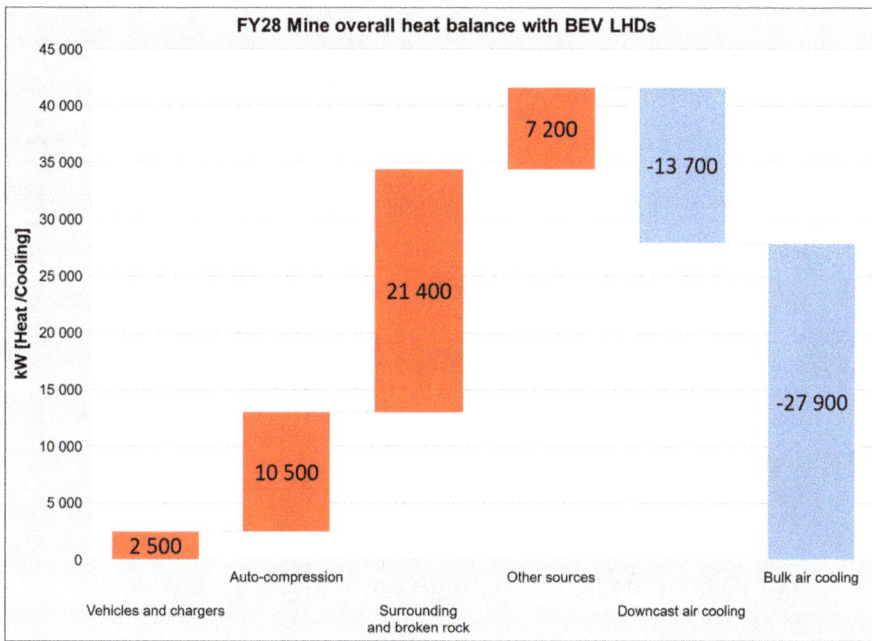

FIG 5 – 2028 Mine overall heat balance with BEV LHDs (rounded).

Battery charging and exchange

At this conceptual stage, the charging stations are not ventilated to the return airway system but are ventilated in series with the production sections. Charging bays will utilise existing redundant excavations between pillars, requiring minimal additional mining and will be re-established as the mining advances. Although located in the worked-out areas, the charging bays will be approximately 200 m from the last through roads. The location of the charging stations will be subject to further risk review as the project design progresses. The battery charging process will not produce gases, however, the battery charging bay equipment will each add approximately 30 kW of heat to the air stream. As the air is to be re-used down-stream this heat is included in the primary cooling requirement.

ISSUE BASED RISK ASSESSMENT

A Workplace Risk Assessment and Control (WRAC) analysis was carried out with stakeholders. The WRAC was followed-up with a barrier-based risk management session (bow-tie risk assessment, BTA) to visualise risks using a simple diagram to clearly differentiate between proactive and reactive risk management.

Battery bay design

Due to the ventilation layout of the bord-and-pillar sections (Figure 2), it was considered impractical to ventilate all battery charge bays to return. Under normal operating conditions the charging station concept provides for slipstream ventilation and cooling, and for containment during an emergency condition. The conceptual layout, used for costing, is shown in Figure 6.

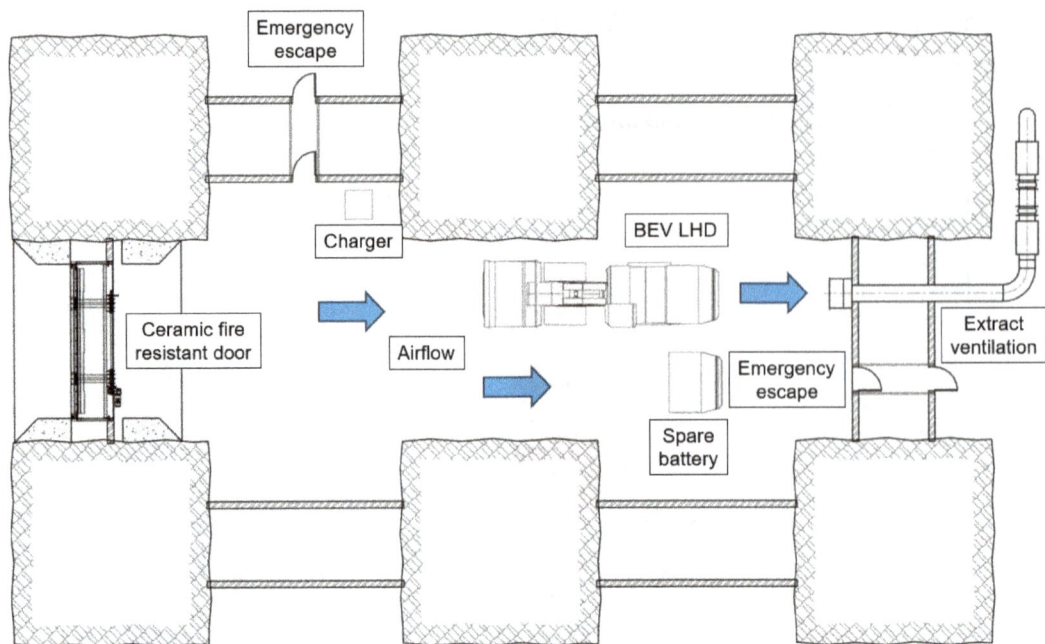

FIG 6 – Electric LHD charge bay conceptual layout.

Conceptually, a battery charging station consists of a number of individual charging bays, one in each dedicated split; orientated along strike and each separately earmarked for the charging of one battery pack offloaded from an LHD.

Each bay is ventilated with air drawn from the section's intake air stream by a ducted fan system mounted at the back wall of each bay. The bays are isolated by double containment brick walls on three sides, and a fire-resistant roll-up ceramic fibre door located at the entrance. Second/emergency egress is provided by two separate, interlocked, escape airlock doors.

Under normal operation, the air used in each bay is drawn in by the respective ducted fans and discharged (extraction ventilation) where it flows into the section's main intake or to the next charging station intake directly downstream.

When LiFePO4 batteries are affected by thermal runaway, combustible products are liberated, and inhalation of these combustion products is hazardous and should be avoided. A series of fire simulations at selected locations in the sections, including the charging station are planned to confirm the consequences of a fire event.

The charging station layout is intended to contain any fire within the confines of the bay affected by such an event while a suppression system will be activated.

The concept layout of the charging station currently includes the following main elements:

- Extraction ventilation with fusible link fire dampers.
- Ceramic cloth fire resistant vehicle door and two emergency escape doors built into double-skin fire walls.
- Smoke and heat detection with fire suppression.
- Battery charge bays to have at least one empty road (nominally 7.0 m wide) between them to prevent fire spread.

The battery charging station layout described is still conceptual (since the focus is on ventilation and fire risks) and it requires further Rational Fire Design and specific Risk Assessments.

CONCLUSION AND OBSERVATIONS

Table 1 lists primary results of the trade-off between the diesel and BEV options. The study objective was to confirm whether the replacement of the existing diesel LHD fleet with battery powered equivalents could eliminate the need for new intake and return raise bored shafts from surface to the north Merensky without impacting production. The study was able to show that by replacing the existing LHD diesel drives with battery electric motors the ventilation in the north Merensky could be reduced from 524 m³/s to 293 m³/s which is within the capacity of the current primary ventilation shaft infrastructure. For the BEV option an additional horizontal intake airway was required to keep velocities within the design criteria.

TABLE 1

FY2028 diesel versus BEV LHD trade-off primary results.

Description	Diesel LHD option	BEV LHD option	Reduction
Additional infrastructure	• New downcast shaft and surface BAC • Upcast shaft and fan station • Each ø5.5 m, 1200 m deep	• Third fan at existing upcast shaft • New surface BAC at existing downcast shaft • Underground intake airway (10 m² × 410 m long)	• No new downcast and upcast shafts
Main fan power (MW)	7.6	7.5	1.3%
Total mine heat (MW)	49.2	41.6	15.4%
Total mine cooling (MW)	27.8	27.9	-0.3%
Primary ventilation to north Merensky trackless (kg/s)	680	360	47%
Total vehicle heat (MW)	7.4	3.4 *	54%
Vehicles as percentage of overall heat (%)	15	8.2	45%
CAPEX for additional ventilation and cooling infrastructure (USD)	29.5 M **	9.1 M	69%

* Includes battery charger heat.

** Includes USD18.8 M for the new intake and exhaust raise-bore shafts.

Capital costs comparisons for conversion to BEV LHDs was not part of the study but was part of the mine's overall financial model to consider it as an option.

Site-specific observations

The mine specific case study has shown:

- The BEV option eliminated the need for the additional raises and reduced the ventilation and cooling infrastructure capital expenditure estimate by approximately USD18.8 M compared to the diesel solution that required the new raises.

- In both options the main fan electrical power was essentially the same despite less ventilation being required for the BEVs. This can be attributed to the higher ventilation system resistance without the new North Merensky downcast and upcast shafts required by the diesel option.

- Despite slightly less mine heat with BEVs, the cooling was similar for both options. This can be explained by the new intake shaft provided in the diesel option greatly improving the positional efficiency of refrigerated air supplied to the North Merensky from surface and a reduction in downcast cooling capacity of the reduced airflow.

The last two points highlight the importance of avoiding blanket conclusions across all types of mining methods with respect to the ventilation and cooling benefits that may be achieved by the introduction of BEVs. The general expectation that BEVs will reduce both ventilation and cooling needs and produce commensurate reductions in power requirements/operating costs has been shown as partly invalid in this instance. It still may be valid for other mines. BEVs met the primary objective of reducing overall volume requirements; however, the main fan power and cooling duty remained relatively unchanged from the diesel case.

This outcome highlights the importance of tempering anticipated benefits for deeper complex mines in line with site specific conditions and variables.

Finally, the issue-based risk assessment indicated that battery charge bays could be ventilated in series within the production areas with isolation controls. However, the finding will be assessed further before implementation and a recommendation is for pilot projects on surface and underground for proof of concept.

REFERENCES

Armburger, J and McGuire, C, 2018. Battery vs. diesel underground LHDs: Direct comparison of heat generation [presentation], Mining Diesel Emissions Council Conference. Available from: <https://www.mdec.ca/2018/S7P3_Jay_Armburger.pdf> [Accessed: 21 Feb 2024].

Global Mining Guidelines Group, 2022. Recommended practices for battery electric vehicles in underground mining, version 3. The Electric Mine Working Group, Section 4.7.4.2, p 32. Available from: <https://gmggroup.org/wp-content/uploads/2022/06/2022–06–23_Recommended-Practices-for-Battery-Electric-Vehicles-in-Underground-Mining.pdf> [Accessed: 21 Feb 2024].

Halim, A and Kerai, M, 2013. Ventilation Requirement for 'Electric' Underground Hard Rock Mines - A Conceptual Study, in *Proceedings The Australian Mine Ventilation Conference*, pp 215–220 (The Australasian Institute of Mining and Metallurgy: Melbourne).

Halim, A, Lööw, J, Johansson, J, Gustafsson, J A, van Wageningen, A and Kocsis, K, 2022. Improvement of Working Conditions and Opinions of Mine Workers When Battery Electric Vehicles (BEVs) Are Used Instead of Diesel Machines — Results of Field Trial at the Kittilä Mine, Finland, *Mining, Metallurgy and Exploration*, 39:203–219.

McGuire, C, Witow, D, Mayhew, M and Bowess, K, 2022. Comparison of heat, noise and ore handling capacity of battery-electric versus diesel LHD, in *Proceedings of the Australian Mine Ventilation Conference 2022*, pp 348–359 (The Australasian Institute of Mining and Metallurgy: Melbourne).

Experimental study on friction resistance of roadway based on the surface roughness characteristics

B Wu[1], C G Zhao[2], B W Lei[3] and C Li[4]

1. Professor, School of Emergency Management and Safety Engineering, China University of Mining and Technology (Beijing), Beijing 100083. Email: wbelcy@vip.sina.com
2. Lecturer, School of Environmental and Safety Engineering, Nanjing Polytechnic Institute, Jiangsu 210044, China. Email: zcgzgkydx12@163.com.
3. Associate professor, School of Emergency Management and Safety Engineering, China University of Mining and Technology (Beijing), Beijing 100083. Email: leibws@163.com
4. PhD candidate, School of Emergency Management and Safety Engineering, China University of Mining and Technology (Beijing), Beijing 100083. Email: cumtb_lic@163.com

ABSTRACT

In order to accurately evaluate the friction resistance of the underground roadway by the roughness characteristics, a rough pipe experimental platform was designed and constructed to study the effect of the distance between rough elements on the friction resistance. Through the airflow resistance experiments of five pipes with different roughness, a new friction resistance calculation model based on fractal parameters to characterise surface roughness is established. Finally, the traditional Nikuradse experimental results were used to verify the new model, and the comparison results show that the maximum error is 8.12 per cent, which verifies the reliability of the new model. The research results show that the new model can effectively reflect the influence of three-dimensional rough surface on airflow resistance, and provide a basis for the theoretical development and technical application of friction resistance.

INTRODUCTION

Mine intelligent ventilation system is the cornerstone of mine intelligent construction and one of the core technologies to ensure the transformation and upgrading of China's coal safety. The establishment of an accurate ventilation network model is pivotal for the regulation and optimisation of ventilation systems. In the construction of network models, the resistance determination method is commonly employed to obtain the friction resistance of each roadway, and the ventilation network model is derived through network calculation. However, resistance determination methods currently by manual or semi-manual processes, thereby failing to meet the requirements for the intelligent development of mine operations. There is an urgent need for a novel method capable of rapidly and accurately determining the friction resistance.

In recent years, three-dimensional laser scanning technology has been rapidly developed in various fields because it can quickly obtain the three-dimensional shape of the object surface (Yao et al, 2020). The rough surface of roadway can be quickly obtained by three-dimensional laser scanning technology, and then the friction resistance of roadway can be calculated. However, a significant challenge of this method is the characterisation of wall surface roughness. Several scholars have undertaken relevant studies in this regard. Bråtveit, Lia and Olsen (2012) used a 3D laser scanner to obtain the point cloud of waterway tunnel and calculated the wall roughness based on the point cloud data. Watson and Marshall (2018) used point cloud data to obtain the mine roadway wall roughness and calculate its resistance. However, the roughness still has only one-dimensional roughness, failing to adequately capture the variation in the three-dimensional characteristics of roadway surfaces.

Fractal geometry, which was developed in the late 1970s, is a new subject with irregular geometry as the research object. Because the fractal characterisation of rough surface can well avoid the problem of too many roughness parameters in statistical models (Ge and Chen, 1999a), fractal theory has been widely applied in the characterisation of rough surface (Gagnepain and Roques-Carmes, 1986; Berry and Lewis, 1980). Fractal geometry is applied to geotechnical mechanics, and the fractal characterisation of rock roughness is discussed in detail (Xie, 1992). Sun et al (2019) divided the rock joint roughness into first-order large protrusion and second-order small protrusion, and used the double-order fractal dimension to refine the characterisation of roughness. Based on the fractal approach, Ban *et al* (2019) proposed a roughness evaluation system which can reflect the anisotropy of joint surface while not affected by the measurement scale. Ge and Chen (1999b) combined fractal dimension with scale coefficient, and proposed a characteristic roughness parameter to characterise the surface roughness.

As demonstrated above, fractal theory has found widespread application in characterising three-dimensional rough surfaces, which is also used to characterise wall surface roughness, and a new friction resistance model is established through experiments, which provides a reference for the theoretical development of resistance.

THEORY

In fractal theory, there are two basic fractal parameters used to characterise rough surfaces, fractal dimension D and scale parameter C. The fractal dimension D mainly influences the complexity of the rough surface, while the scale parameter C primarily affects the height of the rough surface. The calculation method of the two fractal parameters is as follows.

Fractal dimension

The fractal dimension D is computed by the three-dimensional box-counting method, where small cubic boxes with a side length r are employed to cover the rough surface, and the number of boxes covering the rough surface is tallied. By varying the side length r of the boxes, different counts of boxes Nr(A) covering the surface are obtained. As the box side length r approaches zero, there exists a value D satisfying the following formula.

$$D = \lim_{r \to 0} \frac{\log N(r)}{\log(1/r)}$$

(1)

Scaling parameter

The scale parameter C represents a comprehensive characterisation of the roughness element height on the rough surface. It is calculated using the method of average height calculation, which calculated as the following formula.

$$C = \frac{\sum h_r}{m_r}$$

(2)

where:

h_r is the height of the roughness element

m_r is the number of the roughness element

EXPERIMENT

Experimental devices and condition

The experimental platform consists of three components: the rough pipe, the fan, and the measurement instruments, as shown in Figure 1. The pipeline section is 0.2 m × 0.2 m square, which is formed by splicing four 1 m long pipelines. The upper sides of the pipe are made of transparent acrylic material, and the bottom is made of iron plate. They are assembled together during the test. The fan is a variable frequency fan and with an outlet diameter of 0.2 m, capable of delivering a maximum airflow of 1250 m³/min. To ensure a seamless connection with the pipe, a cross-section conversion is installed at the fan outlet and sealed with aluminium foil tape. The wind velocity sensor has a measurement range of 0–20 m/s, with an accuracy of FS 0.02 per cent. It can provide measurements accurate to 0.01 m/s. And the differential pressure sensor has a measurement range of 0–50 Pa and an accuracy of FS 0.02 per cent. It can provide measurements accurate to 0.1 Pa.

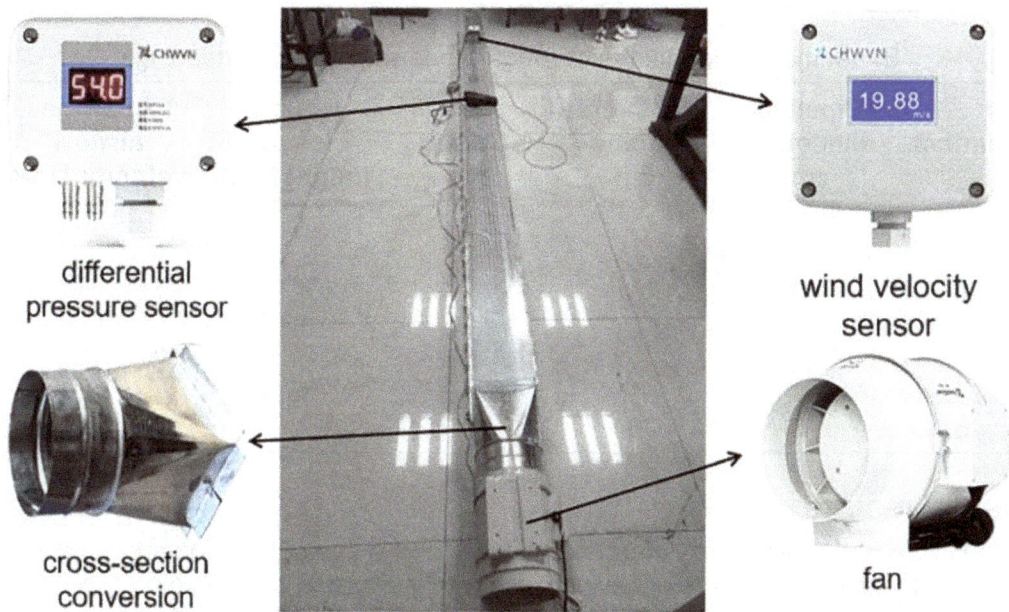

FIG 1 – Test devices diagram.

To simulate surface roughness, glass hemispheres were uniformly adhered to the wall surface as rough elements in the test, as shown in Figure 2. The diameter of the hemisphere is 10 mm. On the pipe wall, the axial direction parallel to the air flow direction and the transverse direction perpendicular to the air flow direction are defined. Variations in the spacing of the rough elements along the axial and transverse directions at the same height will affect the overall resistance of the pipe. In order to study the influence of the variation of the roughness element spacing on the friction resistance, five sets of experiments were conducted in this study. The axial and transverse spacings between rough elements were as follows: 1 cm × 1 cm, 2 cm × 1 cm, 3 cm × 1 cm, 1 cm × 2 cm and 1 cm × 3 cm.

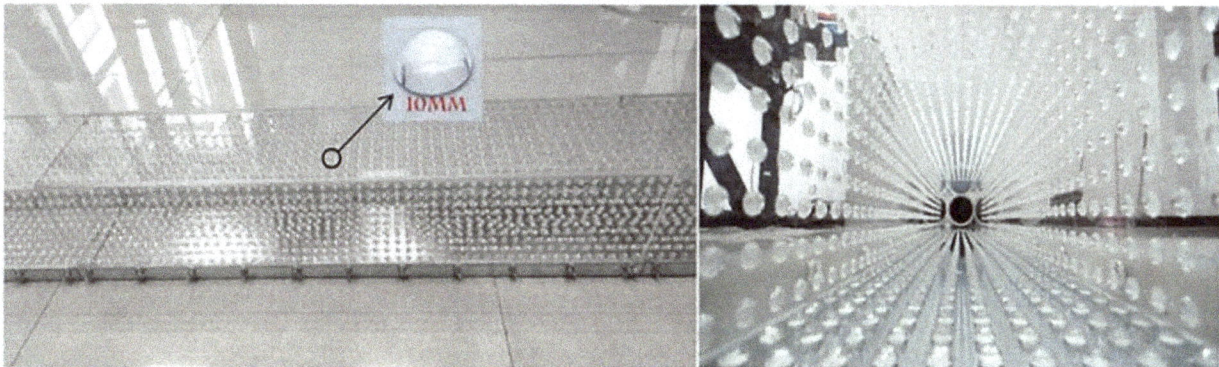

FIG 2 – Roughness test diagram.

The experimental conditions are shown in Table 1. Static pressure differential values were recorded at each airflow velocity. With an average airflow velocity of 1.6 m/s in the duct, the Reynolds number was calculated to be 21573.04, indicating turbulent flow. Therefore, within the specified range of airflow velocities, the flow state in the experimental pipe is consistent with that in the mine roadway, both being turbulent flow.

TABLE 1

Experimental conditions of rough pipes with different spacing.

Number	Pipeline diameter / cm	Transverse spacing / cm	Axial spacing / cm	Rough element height / cm	Average airflow velocity / m/s
1			1		
2		1	2		1.6, 2, 2.4,
3	20		3	0.5	2.8, 3.2, 3.6, 4
4		2			
5		3	1		

Experimental results

As shown in Figure 3, the pressure differential values for the five sets of experiments with different roughness spacings exhibit an exponential increase with increasing airflow velocity. Furthermore, as the spacing between rough elements varies, the pressure differential values follow the trend: 1 cm × 1 cm > 2 cm × 1 cm > 1 cm × 2 cm > 3 cm × 1 cm > 1 cm × 3 cm. This indicates that smaller roughness spacings result in greater surface roughness and higher airflow resistance, while larger roughness spacings lead to reduced surface roughness and lower airflow resistance. Notably, the arrangement of 2 cm × 1 cm and 1 cm × 2 cm, as well as 3 cm × 1 cm and 1 cm × 3 cm, which follows the same arrangement. However, the airflow resistance measurement results demonstrate that 2 cm × 1 cm > 1 cm × 2 cm and 3 cm × 1 cm > 1 cm × 3 cm, indicating that the roughness elements arranged along the axial direction have a greater impact on the airflow resistance than those arranged along the transverse direction.

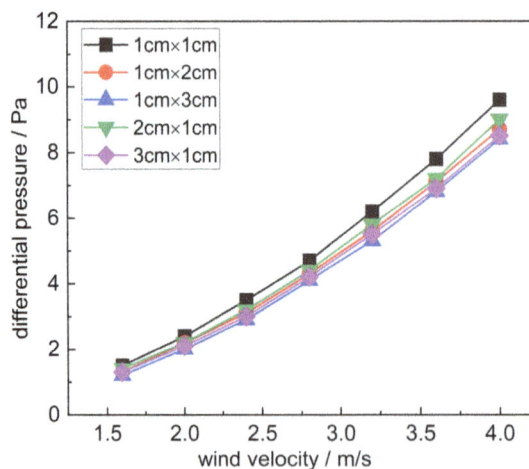

FIG 3 – Experimental results.

Resistance model based on fractal parameters

As shown in Figure 4 the fractal dimensions obtained using the box-counting method for different rough pipes are plotted on a double logarithmic coordinate system, where the relationship between box size r and count Nr appears as a straight line. This indicates that the different rough pipes have different fractal characteristics. It can be observed that as the spacing between rough elements increases, the fractal dimension gradually decreases. This is because variations in the spacing between rough elements affect the overall complexity of the rough surface, thereby influencing its fractal dimension. Although the rough surfaces with transverse spacing of 1 cm and axial spacing of 2 cm have the same arrangement of rough elements as those with transverse spacing of 2 cm and axial spacing of 1 cm, the total number of rough elements is different due to boundary constraints. As a result, the fractal dimensions calculated using the box-counting method are also different.

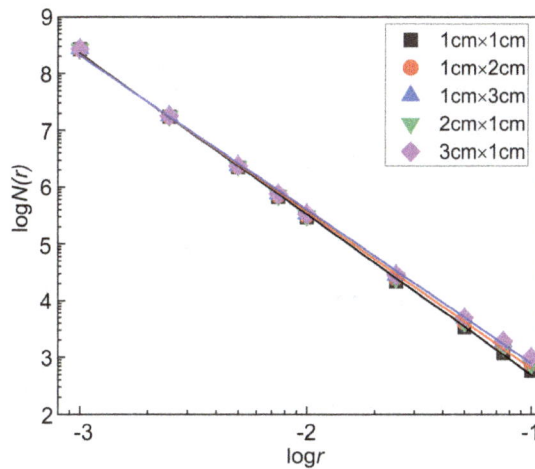

FIG 4 – Calculation results of fractal dimension.

According to the theory, the fractal parameters influencing surface roughness include fractal dimension D and scale parameter C. The scale parameter C is related to the height of the rough elements, and according to Equation 2, the scale parameter C is calculated to be 5 mm. Table 2 presents the fractal dimensions and scale parameters of five different arrangements of rough elements.

TABLE 2

Calculation results of fractal dimension of rough pipes with different spacing.

Number	Rough element spacing	Slope	Fractal dimension d	Scale parameter c
1	1 cm × 1 cm	-2.828	2.828	
2	1 cm × 2 cm	-2.762	2.762	
3	1 cm × 3 cm	-2.688	2.688	5 mm
4	2 cm × 1 cm	-2.745	2.745	
5	3 cm × 1 cm	-2.688	2.688	

According to the Atkinson formula, the frictional resistance coefficients f for five types of rough surfaces are calculated. Considering that the frictional resistance coefficient f is solely dependent on roughness, the equivalent roughness height he, for different operating conditions in the pipes is computed using the Colebrook-White formula (Equation 3), as shown in Table 3.

$$f = \left[4 \log_{10}(\frac{e/d}{3.7}) \right]^{-2}$$

(3)

Where:

e   is the absolute roughness, m

d   is the pipe diameter, m

e/d   is the relative roughness, a dimensionless parameter

f   is the frictional resistance coefficient

TABLE 3

Parameters of rough pipes with different spacing

Number	Rough element spacing	R / kg/m^7	f	He/mm
1	1 cm × 1 cm	375.75	0.0167	8.603
2	1 cm × 2 cm	342	0.0152	6.835
3	1 cm × 3 cm	351	0.0156	7.373
4	2 cm × 1 cm	326.25	0.0145	6.210
5	3 cm × 1 cm	333	0.0148	6.520

The equivalent roughness height he has a functional relationship with the fractal dimension D and the scale parameter C:

$$h_e = F(D,C) \tag{4}$$

In order to obtain the relationship expression of the equivalent roughness height he, five groups of experimental data were fitted. Since all rough element heights in the experiments are 5 mm, the influence of the scale parameter C is neglected in Equation 9, resulting in the relationship between the equivalent height he and the fractal dimension D at the same roughness height. As the fractal dimension D approaches 2, the rough surface tends toward a smooth plane, hence the equivalent roughness height he approaches 0 mm. Setting D=2 as a boundary condition, a power function curve fitting is performed in the graph, as shown in Figure 5.

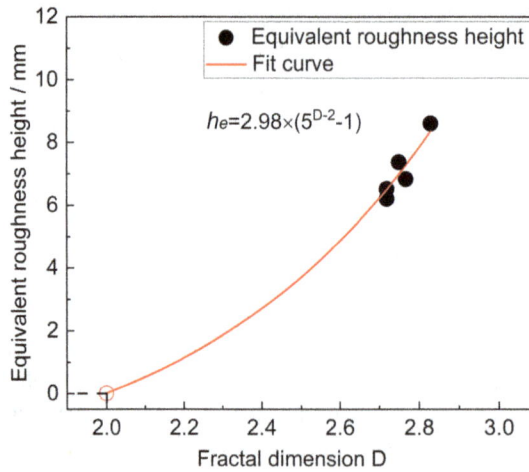

FIG 5 – Curve fitting of equivalent roughness height.

The equivalent roughness height obtained by fitting is shown in Equation 5, which shows that the equivalent roughness height h_e is proportional to the D-2 power of the scale parameter C.

$$h_e = 2.98 \times (5^{D-2} - 1) \tag{5}$$

Extend it to roughness of different heights:

$$h_e = 2.98 \times (C^{D-2} - 1) \tag{6}$$

Substituting Equation 6 into Equation 3, the frictional resistance coefficient is:

$$f = \left[4\log_{10}\left(\frac{2.98 \times (C^{D-2} - 1)/d}{3.7}\right) \right]^{-2} \tag{7}$$

Finally, the friction resistance model based on fractal parameters is established as follows:

$$R = \frac{\rho}{2} \cdot \left[4\log_{10}(\frac{2.98 \times (C^{D-2} - 1)/d}{3.7}) \right]^{-2} \cdot \frac{LU}{S^3}$$

(8)

MODEL VALIDATION

In order to verify the accuracy of the new friction resistance model based on fractal representation, Nikuradse experimental data was used for verification (Moody, 1944). In the experiment, Nikuradse tightly covered the pipe wall with sand particles, as shown in Figure 6, which is a schematic diagram of the tight arrangement of sand particles in the Nikuradse experiment.

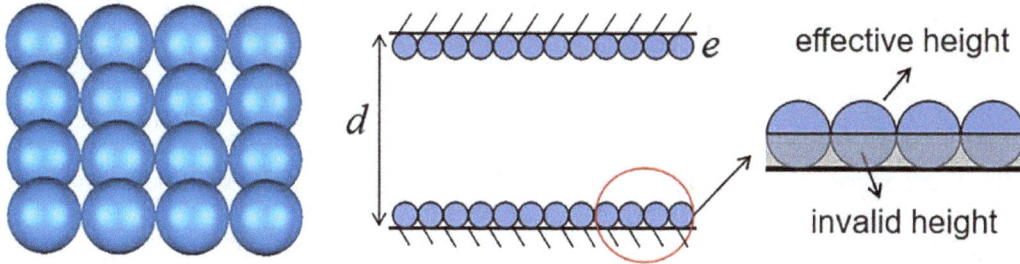

FIG 6 – Rough element arrangement diagram of Nikuradse experiment.

The sand particles are spherical particles with a diameter of e. Due to the close arrangement of sand particles, the lower half of the sand particles is an invalid height, and the effective diameter of the whole pipeline is d-e, so the effective diameter dt is:

$$d_t = d - h$$

(9)

Due to the small height of the rough element compared to the diameter of the pipeline, the effective diameter dt of the pipeline and the real diameter d of the pipeline can be regarded as equal, dt ≈ d. The scale parameter C in the Nikuradse experiment is equal to the radius of the sand particles. When the rough sand particles in the Nikuradse experiment are arranged in the way shown in Figure 6, the fractal dimension calculation result of the rough surface is shown in Figure 7. In the case of tight arrangement, the fractal dimension D of the rough surface in the Nikuradse experiment is 2.983.

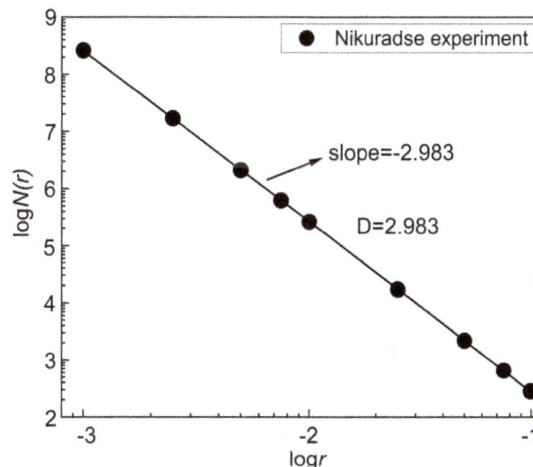

FIG 7 – Calculation results of fractal dimension of rough pipe wall in Nikuradse experiment.

The Nikuradse experiment carried out six groups of experiments with different sand particle diameters, with relative roughness values of 1/30, 1/61.2, 1/120, 1/252, 1/507, and 1/1014, respectively. Assuming that the length and cross-sectional size of the Nikuradse pipeline are the same as the experimental pipe, Equation 8 is used to calculate the friction resistance. Compared with the results of the Nikuradse experiment, as shown in Figure 8, it can be seen that the overall error is small, with a maximum error of 8.12 per cent, proving the accuracy of using the formula to calculate friction resistance for rough pipelines at different heights.

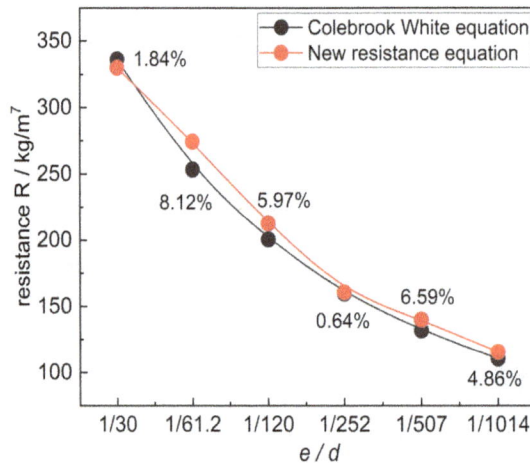

FIG 8 – Verification of friction resistance coefficient of rough pipes with different heights.

CONCLUSIONS

This study established a pipeline roughness experimental platform, and conducted five sets of roughness elements spacing experiments by pasting roughness elements on the pipe wall. The effect of roughness elements with the same roughness height on pipeline resistance at different spacing was studied. And a friction resistance model based on fractal characterisation was constructed, the main conclusions are as follows:

- The experimental results show that the smaller the spacing between rough elements, the rougher the surface of the pipeline, and the greater the friction resistance of the pipeline. The arrangement of rough elements in the axial direction has a greater impact on airflow resistance compared to the arrangement in the transverse direction.

- Based on fractal theory, a friction resistance model based on fractal parameters was constructed, which can effectively reflect the influence of three-dimensional rough surface on air flow resistance.

- To verify the accuracy of the new model, the validation study of Nikuradse experiment was conducted. The maximum error of the Nikuradse experiment was 8.12 per cent, indicating the reliability of the new model.

ACKNOWLEDGEMENTS

This work was supported by the National Natural Science Foundation of China [52374252].

REFERENCES

Ban, L R, Qi, C Z, Yan, F Y, Liu, Y, Zhu, C and Tao, Z G, 2019. A new method for determining the JRC with new roughness parameters, *Journal of China Coal Society*, 44(4):1059–1065.

Berry, M V and Lewis, Z V, 1980. On the Weierstrass-Mandelbrot fractal function, in *Proceedings of the Royal Society of London A Mathematical and Physical Sciences*, 370(1743):459–484.

Bråtveit, K, Lia, L and Olsen, N R B, 2012. An efficient method to describe the geometry and the roughness of an existing unlined hydro power tunnel, *Energy Procedia*, 20:200–206.

Gagnepain, J J and Roques-Carmes, C, 1986. Fractal approach to two-dimensional and three-dimensional surface roughness, *Wear*, 109(1/2/3/4):119–126.

Ge, S R and Chen, G A, 1999a. Fractal prediction models of sliding wear during the running-in process, *Wear*, 231(2):249–255.

Ge, S R and Chen, G A, 1999b. Characterization of surface topography changes during running-in process with characteristic roughness parameter, *Journal of China University of Mining and Technology*, 28(3):4–7.

Moody, L F, 1944. Friction factors for pipe flow, *Transactions of the American Society of Mechanical Engineers*, 66(8):671–678.

Sun, S Y, Li, Y C, Tang, C A and Li, B, 2019. Dual fractal features of the surface roughness of natural rock joints, *Chinese Journal of Rock Mechanics and Engineering*, 38(12):2502–2511.

Watson, C and Marshall, J, 2018. Estimating underground mine ventilation friction factors from low density 3D data acquired by a moving LiDAR, *International Journal of Mining Science and Technology*, 28(4):657–662.

Xie, H p, 1992. Fractal geometry and its application to rock and soil materials, *Chinese Journal of Geotechnical Engineering*, 14(1):14–24.

Yao, X H, Zhou, Y M, Jia, S and Zhao, W, 2020. Realization method and application of building reverse modeling based on 3d laser scanning, *Industrial Construction*, 50(3):178–181.

Vale Coleman Mine ventilation system design strategies when considering heat

H Zhang[1] and C Allen[2]

1. Principal Engineer, Vale Base Metals, Copper Cliff, Ontario P0M 1N0, Canada.
   Email: hongbin.zhang@vale.com
2. Technical Leader, Vale Base Metals, Copper Cliff, Ontario P0M 1N0, Canada.
   Email: cheryl.allen@vale.com

ABSTRACT

Underground mine ventilation system design is critical for maintaining a safe environment and ensuring the well-being and productivity of mine personnel. As mines deepen, heat introduced from air auto compression, rock strata, mechanised equipment, and fans becomes vital to consider as part of the ventilation system design strategies. The results of excessive heat in the underground working environment include risks to workers' health and safety, productivity due to frequent work-rest regimes (or stop work) and equipment malfunction. Hence, heat constraints are converted to production and development constraints in the life-of-mine schedule to quantify the impact. Common heat-control strategies include reducing sources of heat (eg diesel equipment, fans), cooling plants (either on the surface or underground) and increasing fresh air volume. This paper presents a comprehensive investigation into the design strategies employed by Vale Coleman Mine to address the challenges associated with heat as the mine expands. The ventilation system design strategies include conducting a trade-off study between battery-electric and diesel-powered mobile equipment, converting the ventilation system from a forcing to an exhausting system to reduce heat generated from fans entering the working areas, optimising the size of fresh air and return air raises, and selecting high-efficiency fans best fit the life-of-mine plan. Modelling of the different options was conducted, and results were analysed. The preferred design option was to use a hybrid diesel-electric equipment fleet within an existing forcing ventilation system. This option is energy efficient, operationally friendly, lower capital investment, less overall heat and can achieve production and development targets in the life-of-mine plan.

INTRODUCTION

Coleman Mine is located in Levack, Ontario. The mine began operation in 1970 and of important value to Vale Base Metals North Atlantic Operations (Vale, 2021). Coleman Mine is an underground hard rock mine currently developing the Lower 170 OB at 1850 m depth. The mine has a 10.3 MW$_R$ surface cooling plant (Hardcastle, Shaw and Allen, 2023), which was commissioned in 2021. The cooling plant ran at half capacity in 2021, full capacity in 2022, and upgrades (ie side shields) were done in 2023. The design air temperature at the collar of McCreedy fresh air raise (FAR) is 15.4°C wet-bulb (WB) and 18.2°C dry-bulb (DB) (Hardcastle, Shaw and Allen, 2023). The surface ambient temperature basis of 26°C DB and 19°C WB represents a 98th percentile. The surface temperatures in Sudbury climates can range from 5°C to 15°C WB in the spring and fall and 15°C to 21°C WB in the summer, averaged by Environment Canada from 2005 to 2010 (Government of Canada, 2024). As the mine deepens, heat continues to be an issue mainly due to auto compression and diesel equipment. The Lower 170 OB consists of three main production levels, which are 5850L, 5950L, and 6050L. Mining methods used in L170 OB are sublevel stoping and cut-and-fill. Currently, only two (5950L and 6050L) of the three levels are active because of ventilation constraints (ie air and potential heat issues). To meet production targets, the L170 OB requires all three levels to be in production. This paper summarises the ventilation system design strategies recommended to address the heat issues. The modelling work leading to the design strategies was completed in 2021 while at the same time going through a redesign (including new FARs and return air raises (RARs)) as well as the life-of-mine (LOM) changes. Hence, the ventilation design strategy work was conducted with assumptions and estimations in a high-level study to support the ongoing L170 OB redesign (eg LOM design, mining sequence).

The focus of this paper is a trade-off study conducted between battery-electric vehicles (BEVs) and diesel equipment for the L170 OB considering capacity of available air to remove heat generated by diesel equipment and auxiliary fans. The study also sought potential opportunities for increasing the L170 production while adhering to design and working in heat guidelines. A network modelling software, VentSim DESIGN™, version 5.4 (by Howden), was used to predict airflow distributions in the mine and climatic simulation program, CLIMSIM™, version 1.3 (by Mine Ventilation Services Inc.), was used to perform heat simulation with the existing ventilation network. Results from the heat simulations were analysed, and recommendations provided. Because of the limited operating time of the new cooling plant, insufficient temperature data has been collected with the cooling plant actively running. Therefore, this trade-off study should be treated as a qualitative study with less precision placed on exact temperature results.

Six scenarios (or heat simulations) were evaluated in CLIMSIM as follows:

1. Base case – current diesel fleet.

2. Current fleet with the equivalent BEV (based on the current diesel fleet).

3. Current fleet with a mix of BEV and diesel (diesel trucks and battery-electric scoops).

4. Future diesel fleet (larger scoops and additional equipment).

5. Future diesel fleet (larger scoops and additional equipment) with an additional 63.5 m³/s (or 135 kcfm) of air via the new FAR on 5800L in L170.

6. Potential BEV fleet for increased production opportunities (unconstrained mine plan).

Other ventilation system design strategies considered to reduce heat to L170 OB include:

- Switching the existing forcing ventilation system (meaning fresh air is sent to the L170 OB using booster fans, which brings fan heat into the OB) to an exhausting ventilation system so that the fan heat does not enter the working levels

- Using an underground spot cooler only for L170 OB.

The exhausting ventilation system would require installing an airlock system (at least two doors) on the main haulage ramp in L170 OB, removing existing fresh air booster fans, and installing new return air booster fans. After discussing with operators and managers in the mine, this option was not accepted due to the production impact during the changeover period (eg ordering fans, building new booster fan stations, installing doors, removing existing booster fan stations) and potentially increasing maintenance on the doors in the haulage ramp. For the option of using a spot cooler, the procurement and construction lead time is long careful maintenance is needed, thus this option was not selected as well.

Optimising the size of FARs and RARs is a routine practice at Vale North Atlantic mines completed by the central ventilation team which is passed to the mid-range and long-range mine design teams for scheduling and procurement. The optimal size is determined based on the lowest net present value of the raise through its life. Hence, both capital cost (ie construction cost) and operational cost (ie power cost for sending required air volume via the raise) are considered in the calculation sheet. Other considerations include various raise sizes, shapes (eg round, rectangle), ground support condition (ie with or without ground support), construction methodology (eg pilot-slashing, raise bored, Alimak), number of raises (eg one single raise versus two rises in parallel), and constructability (eg certain raise bore machines have a maximum drillable size at a specific length range). After the raise size is determined, the next step is selecting high-efficiency fans based on operating points over the Life-of-mine. For L170 OB, the size of the new raises were optimised based on the current LoMP.

Another feasibility study (FEL2 level) was carried out as part of the L170 OB redesign in 2023. The conclusions and recommendations from this study were distributed to the Vale mine design teams and project teams as a reference to support the L170 OB redesign.

HEAT MODELLING

The climatic simulations include the geometry from the McCreedy FAR to the L170 OB and do not include the other OBs or airways which do not contribute heat to the L170 OB. The branch table for scenario 1 in CLIMSIM is shown in Table 1. Each branch represents an airway in the ventilation network which includes the airway's physical properties (eg DB and WB temperatures, wetness factor, depth, length) as the model's input.

TABLE 1

Example of branch table in the CLIMSIM model with current diesel fleet.

Branch	Name	Dry bulb (°C)	Wet bulb (°C)	Pressure (kPa)	Quantity (m³/s)	Length (m)	Depth in (m)
1	Ambient air	26	19	97.32	88	2	0
2	Air cooler outlet	13.2	13	97.32	478	2	0
3	#1 McCreedy East Fresh Air Shaft (Surf to 3770L)	15.11	13.9	97.32	566	1169	0
4	3770 L-FAR Transfer #1 (south of #1 shaft to intersection)	26.84	20.24	110.81	190	22	1169

Branch	Depth out (m)	Friction (kg/m³)	Wetness	VRT in (°C)	Geo. step (m/°C)	Conductivity (W/m°C)	Diffusivity (m²/s×10⁻⁶)
1	0	0.002	0.05	3.1	1000	3.6	1.6
2	0	0.002	0.05	3.1	1000	3.6	1.6
3	1169	0.0076	0.17	3.1	63	3.6	1.6
4	1172	0.0129	0.09	21.7	63	3.6	1.6

Assumptions

The following statements on the air quantities, friction factors, wetness factors, virgin rock temperatures, and geothermal steps apply to all six scenarios.

In Table 1, the air quantity on each branch comes from the current calibrated VentSim model. The DB and WB temperatures on branches 1 and 2 are the surface cooling plant design input temperatures. The McCreedy fresh air fans (2×1500 HP) were included in branch 3, 10 m from the surface. The fan heat increased air temperature from 15.1°C DB and 13.9°C WB to 18.2°C DB and 15.3°C DB. The values match with the above-mentioned design air temperature at the collar of the FAR. This shows that the cooling plant was correctly modelled in the CLIMSIM model. A part of the results table for branch 3 is shown in Table 2.

TABLE 2

A part of the results table for branch 3.

Row	Distance (m)	Dry Bulb (°C)	Wet Bulb (°C)
1	0	15.11	13.90
2	10	15.20	13.95
3	20	18.17	15.27

The friction factors in Table 1 were entered based on the Vale ventilation design guidelines.

From Table 1, the wetness factor varies for branches. The wetness factors for all the branches except branch 3 (McCreedy fresh air shaft) came from a previous internal study, which had a calibrated CLIMSIM model for the L170 OB. The wetness factor for the McCreedy fresh air shaft (branch 3 in Table 1) was calibrated to 0.17 with the raw temperature data collected in 2016.

The virgin rock temperatures (VRT) for branches 1 and 2 in Table 1 were calculated to 3.1°C based on the historical borehole data. 3.1°C represents the VRT 15.2 m (50 ft) below the surface. Similarly, the geothermal step was calculated to be 0.87°F/100 ft, equivalent to 63 m/°C in the CLIMSIM model.

The geothermal step for branches 1 and 2, shown in Table 1, was set to 1000 m/°C because the two branches are on the surface and do not have any surrounding rock.

Model set-up

For visualisation purposes, the airways were labelled with the branch numbers in the VentSim model shown in Figures 1 and 2. An enlarged view of the L170 OB levels (5850L, 5950L and 6050L) with clearer views of branch numbers are shown in Figures 3, 4 and 5, respectively. Green airways represent fresh air, and red airways represent return air.

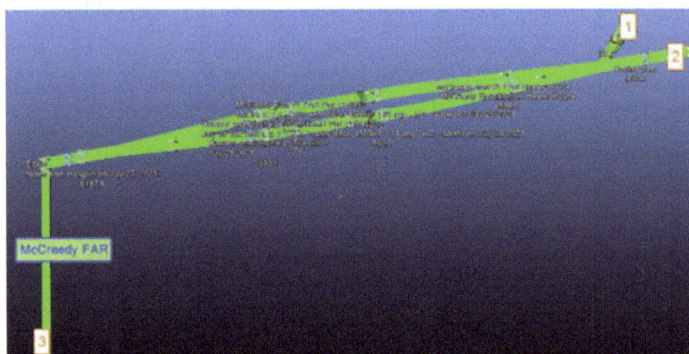

FIG 1 – Branches 1, 2, and 3 in the VentSim model.

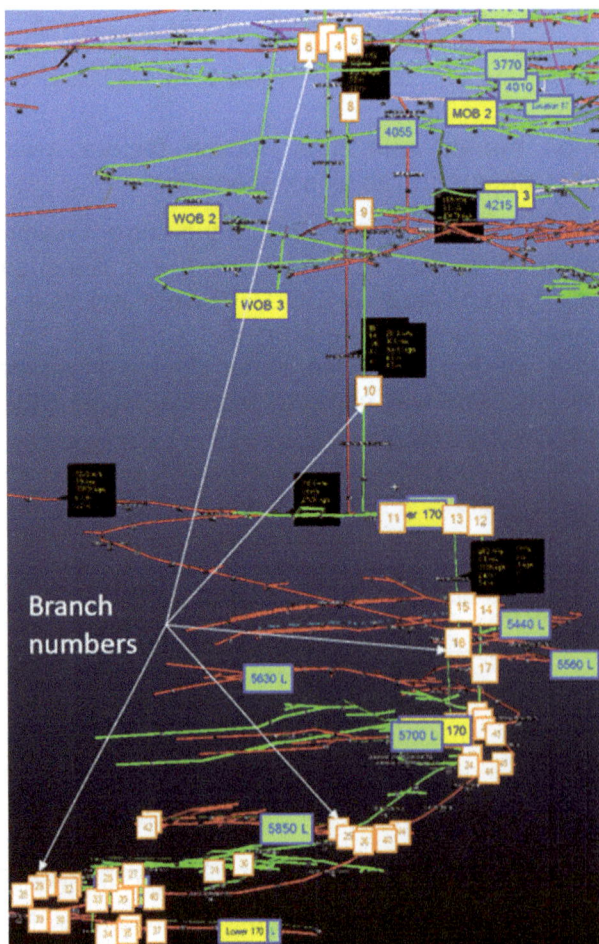

FIG 2 – Branches (from branch 4 to branch 45) in the VentSim model.

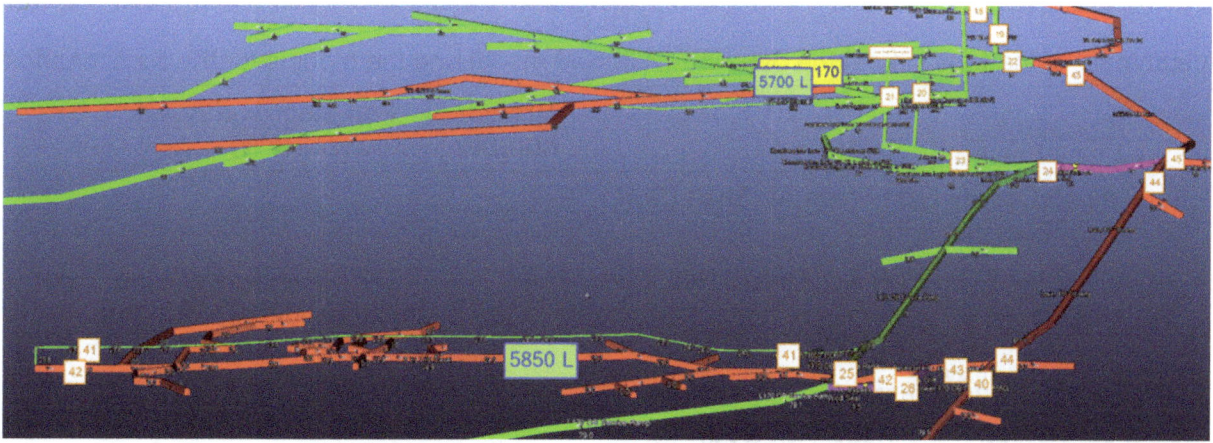

FIG 3 – Branch numbers on the 5850L.

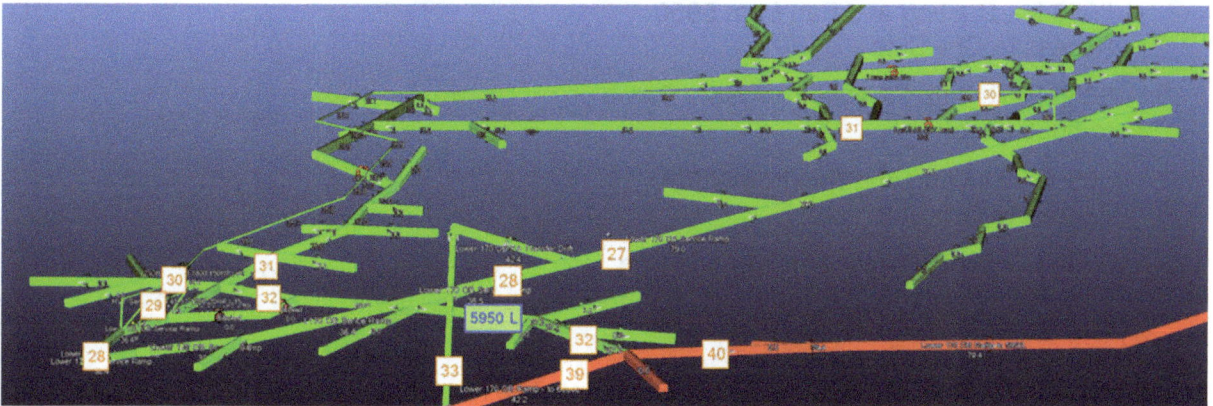

FIG 4 – Branch numbers on the 5950L.

FIG 5 – Branch numbers on the 6050L.

Model calibration

A baseline model was calibrated using the raw DB temperature and relative humidity data collected in 2016 from surface and two locations underground (3770L and 5700L). The raw data were the same data mentioned above.

The baseline model contains heat input from the fans (both on the surface and underground) but does not include the equipment underground or the cooling plant on surface. A summary of the fans included in the climatic simulation is shown in Table 3. Note that the power references that actually consumed by the fan (electrical demand) instead of the fan shaft power as the consumed electrical power is all converted to heat in the air stream.

TABLE 3

A summary of the fans in both the baseline CLIMSIM model and the models for the six scenarios.

Location	Branch number	Fan electrical power (kW)
#1 McCreedy East Fresh Air Shaft (Surf to 3770L)	3	1927
5160 L – 5700 FAR Acc Dr to 5160L Fan station	11	277 kW in the baseline model and 589 kW in the six scenarios
5950L – 170 Serv Ramp – 27 HP (or 20 kW) fan (fans in the bulkhead)	28	20
5950L – 170 Serv Ramp – 20 HP fan	29	15
5950 L – Ducting and Fan from 5950 Access to Ore Cuts	30	100
6050 FAR – 12' FA Rse from Serv Ramp to 6050 Level	33	40
6050L – Ducting and Fan from 6050 Access to Ore Cuts	36	100
5850 L – 5850 Access Ducting and Fan from Haul Ramp to Ore Cuts	41	100

The baseline model does not include the cooling plant. The reason for calibrating the baseline model is that the input properties (eg wetness factor, geothermal step, VRT, thermal conductivity, thermal diffusivity) can be validated with the raw data collected in 2016. These input properties were constant in the six scenarios (with the cooling plant) because the rock properties did not change significantly. Table 4 shows the result comparisons between the data from the baseline model and the raw temperature data collected in 2016.

TABLE 4

Comparisons between the raw temperature data collected in 2016 and the data from the baseline model in CLIMSIM.

Temperature	Surface		3770 FAR		5700 FAR	
	Average raw data	CLIMSIM model	Average raw data	CLIMSIM model	Average raw data	CLIMSIM model
DB (°C)	26.0	26.0	35.4	35.5	39.9	39.6
WB (°C)	19.0	19.0	23.0	24.7	24.1	27.0

Table 4 shows that the WB temperatures in the 3770 FAR and 5700 FAR from the CLIMSIM model are greater than the field data. The DB temperatures from the model are close to the field data. It is recommended that more temperature data should be collected with the cooling plant actively running on the surface to provide better model calibration. Hence, this trade-off study should be treated as a qualitative study with less precision placed on exact temperature results.

Six scenarios (or heat simulations) in CLIMSIM

As per alignment with Coleman Mine, the equipment used in the L170 OB are:

- One 8 yard scoop
- Two 30 ton trucks

- 1 Maclean bolter
- 1 Boltec
- 1 two boom jumbo
- 2 one boom jumbos
- Two 2.5 yard scoops (for the veins)
- SCR dozer (roadway maintenance and or cleaning floors)
- Two Kubota forklifts (services or moving supplies)
- One bulk loader (will be used at one point during the shift to load production rounds)
- Two jeeps.

The trade-off study considered the scoops, trucks, one Maclean bolter and one scissor lift. This trade-off study did not consider the heat from the BEV charging stations. The six scenarios mentioned above were carried out to understand the heat distributions with different types and numbers of equipment. The summarised equipment input in the CLIMSIM model can be found in Table 5. The utilisation factors and water to fuel ratios were entered based on the Vale CLIMSIM user guideline. The water to fuel ratio only applies to diesel equipment.

TABLE 5

Equipment input in the CLIMSIM model.

Scenario	Level	Branch number	Equipment	Number of equipment	Power per unit (kW)	Utilisation (%)	Water/Fuel (L/L)
1. Current diesel fleet	5950	31	Diesel - 2.5 yd LHD	1	81	75	4
			Diesel - 8 yd LHD	1	263	75	5
	6050	37	Scissor Lift	1	78	40	4
			MacLean Bolter	1	147	40	4
	5950 upramp	40	Diesel - 30T Truck	1	305	75	5
	5850	42	Diesel - 2.5 yd LHD	1	81	75	4
	5850 upramp	44	Diesel - 30T Truck	1	305	75	5
2. Current BEV fleet	5950	31	BEV - 2.5 yd LHD	1	65	50	NA
			BEV - 8 yd LHD	1	200	50	NA
	6050	37	Scissor Lift	1	78	40	4
			MacLean Bolter	1	147	40	4
	5950 upramp	40	BEV - 30T Truck	1	331	50	NA
	5850	42	BEV - 2.5 yd LHD	1	65	50	NA
	5850 upramp	44	BEV - 30T Truck	1	331	50	NA
3. Hybrid BEV and Diesel fleet	5950	31	BEV - 2.5 yd LHD	1	65	50	NA
			BEV - 8 yd LHD	1	200	50	NA
	6050	37	Scissor Lift	1	78	40	4
			MacLean Bolter	1	147	40	4
	5950 upramp	40	Diesel - 30T Truck	1	305	75	5
	5850	42	BEV - 2.5 yd LHD	1	65	50	NA
	5850 upramp	44	Diesel - 30T Truck	1	305	75	5

Table 5 continued ...

	5950	31	Diesel - 8 yd LHD	2	263	75	5
			Scissor Lift	1	78	40	4
4. Future diesel fleet	6050	37	MacLean Bolter	1	147	40	4
			Diesel - 8 yd LHD	1	263	75	5
	5950 upramp	40	Diesel - 30T Truck	1	305	75	5
	5850	42	Diesel - 8 yd LHD	1	263	75	5
	5850 upramp	44	Diesel - 30T Truck	1	305	75	5
	5950	31	Diesel - 8 yd LHD	2	263	75	5
5. Future diesel fleet with an additional 63.5 m³/s (or 135 kcfm) of air			Scissor Lift	1	78	40	4
	6050	37	MacLean Bolter	1	147	40	4
			Diesel - 8 yd LHD	1	263	75	5
	5950 upramp	40	Diesel - 30T Truck	1	305	75	5
	5850	42	Diesel - 8 yd LHD	1	263	75	5
	5850 upramp	44	Diesel - 30T Truck	1	305	75	5
	5950	31	BEV - 8 yd LHD	2	200	50	NA
	6050	37	Scissor Lift	1	78	40	4
6. Potential BEV fleet for opportunities			BEV - 8 yd LHD	2	200	50	NA
	5950 upramp	40	BEV - 30T Truck	2	331	50	NA
	5850	42	BEV - 8 yd LHD	2	200	50	NA
	5850 upramp	44	BEV - 30T Truck	1	331	50	NA

From Table 5, scenario 1 represents the existing diesel fleet provided by Coleman Mine. Scenario 2 represents the current fleet with the equivalent BEV (based on the current diesel fleet).

Regarding the hybrid BEV and diesel fleet in scenario 3, the diesel-powered scoops in the current diesel fleet were replaced with battery-electric scoops based on easier transition than haulage trucks with significant travel distance.

As discussed with the Vale central mine design team, the drift size in the three levels would be increased from 2.7 m (or 9 ft) to 4.6 m (or 15 ft), and the 8 yard scoops would be increasingly used instead of the small 2.5 yard scoops. The equipment list in Table 5 is updated to represent this change in scenario 4 (with the future diesel fleet).

Scenario 5 has the same equipment fleet as in scenario 4. Scenario 5 represents a solution (introducing an extra 63.5 m³/s (or 135 kcfm) of fresh air on 5800L via a new FAR and booster fans) to accommodate the 8 yard scoops. The airflow differences in the key areas in the two scenarios are shown in Table 6.

TABLE 6
Differences in air quantities between scenarios 4 and 5.

Location	Branch number	Air quantity (m³/s) in Scenario 4	Air quantity (m³/s) in Scenario 5
5950L	30	36	60
6050L	36	28	40
5850L	41	22	42

Scenario 6 (heat simulation with the potential BEV fleet for opportunities) represents the opportunities to increase the number of battery-electric scoops and trucks while keeping the level reject air temperature under the Vale design limit, 26.5°C WB. The results show the potential to have more scoops and trucks on the three levels, which can increase production while speeding up the cycle time of the equipment.

Because the scoops and trucks are the main heat sources in the equipment fleet, only the engine power rating and heat output for the diesel and equivalent BEV fleets in the CLIMSIM model are presented in Tables 7 and 8, respectively. In general, the efficiency of a diesel engine is one-third of that of a battery-electric motor (Allen and Stachulak, 2019). Hence, with identical work output, the heat generated from a diesel engine is about three times the heat generated from a battery-electric motor (Allen and Stachulak, 2019). In CLIMSIM, the heat generated from a diesel engine is 2.83 times of the heat generated from a battery-electric motor.

TABLE 7

The diesel engine power rating and heat output.

	Engine rating (HP)	Engine rating (kW)	Utilisation factor (%)	Total power (kW)	Heat output (kW)
2.5 yard Scoop	109	81	75[1]	61	173
8 yard Scoop	353	263	75	198	559
30T Truck	409	305	75	229	648

[1]: The utilisation factor represents the motor capacity utilisation. For example, a scoop is not always loading a truck and there are times when the scoop is level tramming or idling.

TABLE 8

The equivalent BEV engine power rating and heat output.

	Engine rating (kW)	Utilisation factor (%)	Total power (kW)	Heat output (kW)
2.5 yard Scoop	65[1]	50[4]	32	32
8 yard Scoop	200[2]	50[4]	100	100
30T Truck	331[3]	50[4]	166	166

[1] Rationing based on Epiroc ST14 8 yard (diesel 250 kW (Epiroc, 2022)) scoop versus Epiroc ST14 8 yard (BEV 200 kW (Epiroc, 2022)) is 1.25. This ratio may vary, depending on the manufacturer and specific equipment. The 2.5 yard BEV scoop will have an estimated engine power of 65 kW (=81/1.25).

[2] According to the technical specifications of the Epiroc ST14 (8 yard) BEV scoop (Epiroc, 2022)

[3] Ratioing based on Sandvik 40T (diesel 405 kW) (Sandvik, 2017) versus Artisan (Z40) 40T (BEV 440 kW) (Artisan Vehicles, 2021) is 0.92. This ratio may vary, depending on the manufacturer and specific equipment. The 30T BEV truck will have an estimated engine power of 331 kW (=305/0.92).

[4] The utilisation factor for a BEV is lower than that for diesel equipment. The reason being the BEV motor stops working when idle while the diesel equipment's engine still works when idle.

The detailed location information (associated with the branch numbers in Table 5) is presented in Table 9. These branches can also be found in Figures 3, 4 or 5.

TABLE 9

Association between the location and branch number in the CLIMSIM model.

Location	Branch number
5950 L – 1st Cut to 5950 Access (return from Duct)	31
6050 L – 1st Cut to Ore Acc (return from duct)	37
170 Haul Ramp – 5950 L Access to 5850 L Access	40
5850 L – Ore Cuts to 5850 Bridge Access (return from duct)	42
170 Haul Ramp – 5850 L Access to Remuck #1	44

A summary of the number of trucks and scoops in the six scenarios is presented in Table 10.

TABLE 10

Summary of the number of trucks and scoops in the six scenarios.

Scenario	Total Number of 2.5 yard Scoop	Total Number of 8 yard Scoop	Total Number of 30T Truck	Total Number of Equipment (including scissor lift and MacLean bolter)
1. Base Case – Current diesel fleet	2	1	2	7
2. Current fleet with the equivalent BEV	2	1	2	7
3. Hybrid BEV and Diesel fleet	2	1	2	7
4. Future diesel fleet	0	4	2	8
5. Future diesel fleet with an additional 63.5 m³/s (or 135 kcfm) of air	0	4	2	8
6. Potential BEV fleet for opportunities	0	6	3	10

From Table 10, scenario 6 (potential BEV fleet for opportunities) has the most equipment among the six scenarios. Scenarios 1, 2, and 3 have the same equipment as the only difference among the three scenarios is whether the trucks and scoops are battery-electric or diesel-powered. Scenario 4 (future diesel fleet) and 5 (future diesel fleet with additional 63.5 m³/s (or 135 kcfm) of air) has more 8 yard scoops than that in scenario 1.

RESULTS

In this trade-off study, the reject air temperatures from branches 32, 38, and 43 in the CLIMSIM model should not exceed 26.5°C WB to prevent temperatures from triggering a work/rest or shutdown requirement based on the Vale ventilation design guidelines. The results from the six scenarios are shown in Table 11.

TABLE 11

Reject air temperatures at the three levels (5850L, 5950L and 6050L).

Location	Branch number	Reject air WB Temperature (°C)					
		Scenario 1: Current diesel fleet	Scenario 2: Current fleet with the equivalent BEV	Scenario 3: Hybrid BEV and diesel fleet	Scenario 4: Future diesel fleet	Scenario 5: Future diesel fleet with additional 63.5 m³/s (or 135 kcfm) of air	Scenario 6: Potential BEV fleet for opportunities
5950L access	32	27.4	24.7	24.7	29.0	26.3	25.1
6050L access	38	25.5	25.5	25.5	27.9	26.2	25.6
5850L access	43	26.1	24.8	24.8	29.4	26.2	26.3

From Table 11, scenario 1, the reject air temperature on 5950L is above the limit while the reject air temperature on 5850L is 0.4°C below the limit. Hence, 5950L is heat constrained with the equipment usage listed in Table 5 and a work-rest schedule could be required. The number of diesel equipment on the 5850L would be closely monitored to avoid potential overheated environment.

Both scenarios 2 (equivalent BEV fleet) and 3 (hybrid BEV and diesel fleet) show no heat constraint with the equipment usage listed in Table 5.

The three levels in scenario 4 (future diesel fleet) are all exposed to an overheated working environment with the equipment listed in Table 5. Hence, managing the heat from replacing the 2.5 yard scoops with 8 yard scoops will be a challenge without the additional fresh air.

The three levels in scenario 5 (future diesel fleet with additional 63.5 m³/s (or 135 kcfm) of air) are all free of heat constraints allowing the change to 8 yard scoops but it will be challenging to increase the air quantity from the current 36 m³/s (or 76 kcfm) to 60 m³/s (or 127 kcfm) on 5950L (shown in Table 6) through the existing ventilation system. The current system on 5950L relies on a single duct with two fans in series used to deliver 36 m³/s (or 76 kcfm) of air after the FAR on the 6050L is completed.

Scenario 6 (potential BEV fleet for opportunities) has the most equipment, with six 8 yard scoops and three 30T trucks in the scenario. From Table 11, two levels (5950L and 6050L) are free of heat constraint, and the 5850L is 0.2°C below the limit.

CONCLUSIONS AND RECOMMENDATIONS

Based on the analysis and results shown above, conclusions from this study are as follows:

- With the existing diesel fleet and summer month surface temperatures (26°C DB and 19°C WB), 5950L is heat constrained, and 5850L can be heat constrained if more equipment is active on the level. Heat management at Vale will be followed if working temperatures exceed the limit. Hence, the number of diesel equipment (eg scoops, trucks) on a level will be restricted impacting the associated mining activities and production.

- Without extra fresh air, all three levels in the L170 OB will be heat constrained if the 2.5 yard diesel-powered scoops are replaced with the 8 yard diesel-powered scoops.

- With an additional 63.5 m³/s (or 135 kcfm) of fresh air on 5800L in the L170 OB, all three levels will be free of heat constraint if the 2.5 yard diesel-powered scoops are replaced with the 8 yard diesel-powered scoops.

- The study includes a financial analysis to evaluate combinations of ventilation changes, equipment fleet, and/or underground cooling options.

The recommendations from a ventilation perspective are:

- Additional real-time temperature data will be collected to confirm the surface fan intake temperature variations throughout the year post-commissioning the cooling plant.

- Either replace diesel scoops with BEV scoops (ie hybrid fleet) or introduce additional fresh air to eliminate the heat and air volume constraint that currently exists with the current diesel fleet.

- A BEV fleet and/or a new FAR should be considered and planned to address the predictable heat issues and avoid summer shutdowns due to the overheated working environment.

- 5950L will remain heat constrained with the future diesel fleet even if an additional 63.5 m³/s (or 135 kcfm) of air is introduced on 5800L. Therefore, BEV scoops are being evaluated on 5950L to remove heat constraints.

- Replacing diesel equipment with BEV equipment, in the Lower 170 OB, will allow additional equipment and additional levels to be active in the orebody to increase production and reduce the equipment cycle time facilitating mining of the L170 OB earlier in the LOMP.

ACKNOWLEDGEMENTS

The authors want to thank Vale Base Metals for the support of this work and approval to publish.

REFERENCES

Allen, C and Stachulak, J, 2019. Mobile Equipment Power Source—Impact on Ventilation Design, in *Proceedings of the 11th International Mine Ventilation Congress*, pp 3–16 (Springer: Singapore).

Artisan Vehicles, 2021. Z40 Haul Truck Specifications. Available by request from: <https://www.rocktechnology.sandvik/en/products/technology/electrification/> [Accessed: 19 July 2021].

Epiroc, 2022. Technical specifications - Epiroc Scooptram ST14 S. Available from: <https://www.epiroc.com/en-ss/products/loaders-and-trucks/diesel-loaders/scooptram-st14> [Accessed: 3 August 2023].

Government of Canada, 2024. Historical weather and climate data. Available from: <https://climate.weather.gc.ca> [Accessed: 2 March 2024].

Hardcastle, S G, Shaw, J K and Allen, C, 2023. Selection, design challenges and construction of Vale's Coleman Mine 10 MW surface refrigeration plant, in *Proceedings of the 19th North American Mine Ventilation Symposium* (ed: P Tukkaraja), pp 163–173 (CRC Press/Balkema).

Sandvik, 2017. Technical specifications - Sandvik TH540 Underground Truck. Available from: <https://www.rocktechnology.sandvik/en/products/equipment/trucks/> [Accessed: 23 July 2021].

Vale, 2021. Sudbury Operations Update 2020–2021 [online]. Available from: <https://vale.com/documents/d/guest/2044_sudbury_final> [Accessed: 2 March 2024].

AUTHOR INDEX

Roghanchi, P	411	van Vuuren, A R J	175
Rose, D	733	van Zyl, F J	359
Ruiz, F	893	Van Zyl, K	161
Russell, B A	749	Velge, F	733
Sabanov, S	13	Viljoen, J	603, 617
Sahay, N	431	Viviers, B J	1003
Sahu, A	463	Waly, F Z	487
Sahu, P	477	Wang, C	97
Saini, A	431	Wang, D	821
Sakinala, V	195	Wang, H	821
Salmawati	487	Wang, J X	227, 573
Sani, R	303	Wang, L	39
Savage, G R	763	Wang, Y	263
Sepúlveda, D	593	Wang, Y B	55
Sharma, P	847	Webber, M	635
Shearer, M J	1089	Widodo, N P	487
Shriwas, M K	1081	Wild, P	635
Shukla, U S	317, 451	Wright, B	907
Si, G	49, 263	Wu, B	227, 573, 1119
Skotniczny, P	1037	Wu, F	945
Song, Z	39	Wu, H	635, 683
Spies, B	329	Wu, H W	695, 709
Stewart, C M	205, 339, 583, 1089	Wu, J	359
Stockenhuber, M	857, 879	Wu, W	3
Stowasser, P	663	Wu, X	49
Sun, J	821	Xin, H	821
Suresh, V	847	Xu, C	821
Talkington, C	893	Xu, G	393, 689, 775
Tang, D	683	Xue, S	527
Tanguturi, K	1013	Yan, M	55, 925
Taunyane, I K	1109	Yan, Z	945
Taylori, Z	989	Yang, D	247
Theiler, J	499, 813	Yang, W	39
Tian, C	39	Yang, Y	821
Tom, K	215, 421	Yulianti, R	439
Tukkaraja, P	111, 1071	Zaid, M M	689, 775
Ugas, R	593	Zander, A	907
Van Der Bank, M	1097	Zhang, H	1129
van der Westhuizen, R	1109	Zhang, H X	55